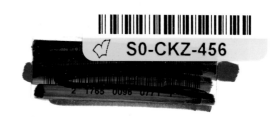

COLD SPRING HARBOR SYMPOSIA ON QUANTITATIVE BIOLOGY

VOLUME LIX

COLD SPRING HARBOR SYMPOSIA ON QUANTITATIVE BIOLOGY

VOLUME LIX

The Molecular Genetics of Cancer

COLD SPRING HARBOR LABORATORY PRESS
1994

COLD SPRING HARBOR SYMPOSIA ON QUANTITATIVE BIOLOGY
VOLUME LIX

© 1994 by The Cold Spring Harbor Laboratory Press
International Standard Book Number 0-87969-067-4 (cloth)
International Standard Book Number 0-87969-068-2 (paper)
International Standard Serial Number 0091-7451
Library of Congress Catalog Card Number 34-8174

Printed in the United States of America
All rights reserved

COLD SPRING HARBOR SYMPOSIA ON QUANTITATIVE BIOLOGY

Founded in 1933 by
REGINALD G. HARRIS
Director of the Biological Laboratory 1924 to 1936

Previous Symposia Volumes

I (1933) Surface Phenomena
II (1934) Aspects of Growth
III (1935) Photochemical Reactions
IV (1936) Excitation Phenomena
V (1937) Internal Secretions
VI (1938) Protein Chemistry
VII (1939) Biological Oxidations
VIII (1940) Permeability and the Nature of Cell Membranes
IX (1941) Genes and Chromosomes: Structure and Organization
X (1942) The Relation of Hormones to Development
XI (1946) Heredity and Variation in Microorganisms
XII (1947) Nucleic Acids and Nucleoproteins
XIII (1948) Biological Applications of Tracer Elements
XIV (1949) Amino Acids and Proteins
XV (1950) Origin and Evolution of Man
XVI (1951) Genes and Mutations
XVII (1952) The Neuron
XVIII (1953) Viruses
XIX (1954) The Mammalian Fetus: Physiological Aspects of Development
XX (1955) Population Genetics: The Nature and Causes of Genetic Variability in Population
XXI (1956) Genetic Mechanisms: Structure and Function
XXII (1957) Population Studies: Animal Ecology and Demography
XXIII (1958) Exchange of Genetic Material: Mechanism and Consequences
XXIV (1959) Genetics and Twentieth Century Darwinism
XXV (1960) Biological Clocks
XXVI (1961) Cellular Regulatory Mechanisms
XXVII (1962) Basic Mechanisms in Animal Virus Biology
XXVIII (1963) Synthesis and Structure of Macromolecules
XXIX (1964) Human Genetics
XXX (1965) Sensory Receptors
XXXI (1966) The Genetic Code
XXXII (1967) Antibodies
XXXIII (1968) Replication of DNA in Microorganisms
XXXIV (1969) The Mechanism of Protein Synthesis
XXXV (1970) Transcription of Genetic Material
XXXVI (1971) Structure and Function of Proteins at the Three-dimensional Level
XXXVII (1972) The Mechanism of Muscle Contraction
XXXVIII (1973) Chromosome Structure and Function
XXXIX (1974) Tumor Viruses
XL (1975) The Synapse
XLI (1976) Origins of Lymphocyte Diversity
XLII (1977) Chromatin
XLIII (1978) DNA: Replication and Recombination
XLIV (1979) Viral Oncogenes
XLV (1980) Movable Genetic Elements
XLVI (1981) Organization of the Cytoplasm
XLVII (1982) Structures of DNA
XLVIII (1983) Molecular Neurobiology
XLIX (1984) Recombination at the DNA Level
L (1985) Molecular Biology of Development
LI (1986) Molecular Biology of *Homo sapiens*
LII (1987) Evolution of Catalytic Function
LIII (1988) Molecular Biology of Signal Transduction
LIV (1989) Immunological Recognition
LV (1990) The Brain
LVI (1991) The Cell Cycle
LVII (1992) The Cell Surface
LVIII (1993) DNA and Chromosomes

Front Cover (Paperback): Retinoblastoma protein forms "corrals" in the nucleus. (For details, see W.-H. Lee et al., p. 103, this volume.)

Back Cover (Paperback): H & E staining and immunohistochemical staining with antibodies against von Willebrand factor of lungs of mice after removal of the primary tumor. (For details, see O'Reilly et al., p. 479, this volume.)

Authorization to photocopy items for internal or personal use, or the internal or personal use of specific clients, is granted by Cold Spring Harbor Laboratory for libraries and other users registered with the Copyright Clearance Center (CCC) Transactional Reporting Service, provided that the base fee of $5.00 per article is paid directly to CCC, 222 Rosewood Dr., Danvers, MA 01923. [0-87969-067-4/94 $5.00 + .00]. This consent does not extend to other kinds of copying, such as copying for general distribution, for advertising or promotional purposes, for creating new collective works, or for resale.

All Cold Spring Harbor Laboratory Press publications may be ordered directly from Cold Spring Harbor Laboratory Press, 10 Skyline Drive, Plainview, NY 11803. Phone: 1-800-843-4388 in Continental U.S. and Canada. All other locations (516)349-1930. FAX: (516)349-1946.

Symposium Participants

ABEL, KENNETH, Dept. of Internal Medicine, University of Michigan, Ann Arbor
ADYA, NEERAJ, Dept. of Infectious Diseases, Case Western Reserve University School of Medicine, Cleveland, Ohio
AGARWAL, MUNNA, Dept. of Molecular Biology, Cleveland Clinic Foundation, Cleveland, Ohio
ALBERTS, BRUCE, Dept. of Biochemistry and Biophysics, University of California, San Francisco
AMARIGLIO, NINETTE, Institute of Hematology, Chaim Sheba Medical Center, Tel Hashomer, Israel
AMFO, KWASI HENRY, Dept. of Medical Genetics, Free University of Brussels, Belgium
ANDERS, ROBERT, Dept. of Biochemistry, Mayo Clinic, Rochester, Minnesota
ANDO, KIYOSHI, Tokai University School of Medicine, Kanagawa, Japan
ANDRULIS, IRENE, Dept. of Molecular and Medical Genetics, Samuel Lunenfeld Research Institute, Mount Sinai Hospital, Toronto, Ontario, Canada
APLAN, PETER, Dept. of Pediatrics and Molecular Medicine, Roswell Park Cancer Institute, Buffalo, New York
ARASON, ADALGEIR, Dept. of Pathology and Cell Biology, University Hospital of Iceland, Reykjavik
ARIGA, HIROYOSHI, Dept. of Molecular Biology, Hokkaido University, Sapporo, Japan
AURIGEMMA, ROSEMARIE, Dept. of Molecular Oncology, National Cancer Institute, Frederick Cancer Research Facility, Frederick, Maryland
AXELROD, AMY, *Cell*, Cambridge, Massachusetts
AYTAY, SAIKA, Applied Genetics Laboratories, Melbourne, Florida
BADE, ERNESTO, Dept. of Biology, University of Konstanz, Konstanz, Germany
BALMAIN, ALLAN, Dept. of Medical Oncology, Beatson Institute for Cancer Research, Glasgow, Scotland, United Kingdom
BAND, VIMLA, Dept. of Radiation Oncology, New England Medical Center, Boston, Massachusetts
BARKARDÓTTIR, ROSA, Dept. of Pathology and Cell Biology, University Hospital of Iceland, Reykjavik
BARNES, DEBORAH, *Journal of NIH Research*, Washington, D.C.
BARRETT, J. CARL, Lab. of Molecular Carcinogenesis, NIH-National Institute of Environmental Health Sciences, Research Triangle Park, North Carolina
BARTEK, JIRI, Danish Cancer Society, Copenhagen, Denmark
BASILE, THOMAS, Chadbourne and Parke, New York, New York
BEACH, DAVID, Cold Spring Harbor Laboratory, Cold Spring Harbor, New York
BENFIELD, PAM, Dept. of Molecular Biology, Du Pont Merck Pharmaceutical Co., Wilmington, Delaware
BENZOW, KELLIE, Virginia Piper Cancer Institute, Abbott Northwestern Hospital, Minneapolis, Minnesota
BEPLER, GEROLD, Dept. of Medicine and Molecular Cancer Biology, Duke University Medical Center, Durham, North Carolina
BERNDT, NORBERT, Dept. of Pediatrics, Childrens Hospital of Los Angeles, California
BERNS, ANTON, Dept. of Molecular Genetics, The Netherlands Cancer Institute, Amsterdam
BI, WANLI, Dept. of Anatomy and Cell Biology, University of Cincinnati, Ohio
BIRAN, HAIM, Familial Cancer Program, Sheibe Institute of Oncology, Kaplan Hospital, Rehovot, Israel
BISHOP, J. MICHAEL, Dept. of Microbiology, University of California, San Francisco
BLACH, JAROMIR, Swiss Institute for Experimental Cancer Research, Epalinges, Lausanne, Switzerland
BLAESE, MICHAEL, Dept. of Cell Immunology, National Institutes of Health, Bethesda, Maryland
BLOOM, THEODORA, *Current Biology*, London, United Kingdom
BODRUG, SHARON, Div. of Oncogene and Tumor

Suppressor Gene Research, La Jolla Research Foundation, La Jolla, California
BOISE, LAWRENCE, Howard Hughes Medical Institute, University of Chicago, Chicago, Illinois
BOON, THIERRY, Ludwig Institute for Cancer Research, Brussels, Belgium
BORG, ÅKE, Dept. of Oncology, University Hospital, Lund, Sweden
BOSSELUT, REMY, Lab. of Viral and Cellular Oncology, Institut Curie-Biology, Centre Universitaire, Orsay, France
BOUCK, NOEL, Dept. of Microbiology and Immunology, Northwestern University, Chicago, Illinois
BOYLAN, JOHN, Du Pont Merck Pharmaceutical Co., Wilmington, Delaware
BRASH, DOUGLAS, Lab. of Radiobiology, Yale University School of Medicine, New Haven, Connecticut
BRENNER, MALCOLM, Dept. of Bone Marrow Transplantation, St. Jude Children's Research Hospital, Memphis, Tennessee
BRISKEN, CATHERINE, Dept. of Biology, Massachusetts Institute of Technology, Whitehead Institute for Biomedical Research, Cambridge, Massachusetts
BROK-SIMONI, FRIDA, Institute of Hematology, Chaim Sheba Medical Center, Tal Hashomer, Israel
BROWN, ANTHONY, Dept. of Cell Biology and Anatomy, Cornell University Medical College, New York, New York
BROWNE, SARA, Dept. of Pathology and Microbiology, University of Bristol, Bristol, United Kingdom
BRUSKIN, ARTHUR, Oncogene Science, Inc., Uniondale, New York
BUSTELLO, XOSE, Dept. of Molecular Biology, Bristol-Myers Squibb Pharmaceutical Research Institute, Princeton, New Jersey
BUTLER, ALISON, Lab. of Molecular Endocrinology, Imperial Cancer Research Fund, London, United Kingdom
CALIGO, MARIA ADELAIDE, Instituto Di Anatomia Patologica, Universitá Di Pisa, Pisa, Italy
CAMPBELL, BRUCE, Dept. of Molecular Biology, Diagnostic Products Corp., Los Angeles, California
CAMPISI, JUDITH, Dept. of Cell and Molecular Biology, Lawrence Berkeley Laboratory, University of California, Berkeley
CANTOR, GLENN, Dept. of Veterinary Microbiology, Washington State University, Pullman
CASEY, GRAHAM, Dept. of Cancer Biology, Cleveland Clinic Research Institute, Cleveland, Ohio
CASTILLA, LUCIO, Dept. of Biology, University of Michigan, Ann Arbor
CERNI, CHRISTA, Dept. of Tumor Virology, Institute of Tumorbiology and Cancer Research, University of Vienna, Austria
CHAGNOVICH, DANIEL, Robert H. Lurie Cancer Center, Northwestern University, Chicago, Illinois
CHANG, HENRY, Dept. of Molecular and Cell Biology, University of California, Berkeley
CHANG, KAI-HSUAN, Institute of Biotechnology, University of Texas Health Sciences Center, San Antonio
CHANG, SHENG-YUNG, Roche Molecular Systems, Alameda, California
CHANG, ZEE-FEN, Dept. of Biochemistry, Chang Gung Medical College, Tao-Yuan, Taiwan
CHAPMAN, CAROLINE, Beatson Institute for Cancer Research, Glasgow, Scotland, United Kingdom
CHAY, CHRISTOPHER, Lab. of Human Carcinogenesis, National Cancer Institute, National Institutes of Health, Bethesda, Maryland
CHEN, CHANG-YAN, Cancer Research Center, Boston University School of Medicine, Boston, Massachusetts
CHEN, GANG, Institute of Biotechnology, University of Texas Health Sciences Center, San Antonio
CHEN, YUMAY, Institute of Biotechnology, University of Texas Health Sciences Center, San Antonio
COBRINIK, DAVID, Whitehead Institute for Biomedical Research, Cambridge, Massachusetts
COHEN, DALIA, Dept. of Oncology, Sandoz Research Institute, East Hanover, New Jersey
COHEN, ROBERT, Research Discovery, Genentech, Inc., South San Francisco, California
COLLINS, COLIN, Lawrence Berkeley National Laboratory, University of California, San Francisco
COLLINS, FRANCIS, Natl. Center for Human Genome Research, National Institutes of Health, Bethesda, Maryland
CORY, SUZANNE, Walter and Eliza Hall Institute of Medical Research, Royal Melbourne Hospital, Victoria, Australia
COUCH, FERGUS, Dept. of Internal Medicine, University of Michigan, Ann Arbor
COULOMBE, JOSEE, Dept. of Biochemistry and Medicine, University of Ottawa, Ontario, Canada

CRISPIN, DAVID, Dept. of Gastroenterology, University of Washington, Seattle
CRIST, KEITH, Medical College of Ohio, Toledo
CROSBY, RENAE, Dept. of Cell Biology and Medicinal Chemistry, Glaxo Research Institute, Research Triangle Park, North Carolina
CUI, WEI, Dept. of Medical Genetics, University of Glasgow, Scotland, United Kingdom
CULVER, KENNETH, Iowa Methodist Medical Center, Human Gene Therapy Research Institute, Des Moines, Iowa
D'ANDREA, ALAN, Div. of Pediatric Oncology, Dana-Farber Cancer Institute, Boston, Massachusetts
DAI, KANG, Institute of Biotechnology, University of Texas Health Sciences Center, San Antonio
DALBAGNI, GUIDO, Dept. of Urology, Memorial Sloan-Kettering Cancer Center, New York, New York
DALLA-FAVERA, RICCARDO, Dept. of Pathology, Columbia University College of Physicians & Surgeons, New York, New York
DAS, GOKUL, Dept. of Molecular Biology, University of Texas Health Center, Tyler
DEBANT, ANNE, Dept. of Tumor Immunology, Dana-Farber Cancer Institute, Boston, Massachusetts
DE BRUIJN, DIEDERIK, Dept. of Human Genetics, University of Nijmegen, The Netherlands
DEICHMAN, ALEXANDER MAZKOVICH, Lab. for Experiments in Diagnosis and Therapy of Tumors, Oncology Center, Moscow, Russia
DE LEEUW, BERTIE, Dept. of Human Genetics, University Hospital, University of Nijmegen, The Netherlands
DEVEREUX, THEODORA, Lab. of Molecular Carcinogenesis, NIH-National Institute of Environmental Health Sciences, Research Triangle Park, North Carolina
DIAMANDIS, ELEFTHERIOS, Dept. of Clinical Biochemistry, Toronto Hospital, Toronto, Ontario, Canada
DIERKS, MEGHAN, Dept. of Surgery, Washington University School of Medicine, St. Louis, Missouri
DONEHOWER, LAWRENCE, Dept. of Molecular Virology, Baylor College of Medicine, Houston, Texas
DOSHI, ASHITA, Dept. of Molecular Biology and Microbiology, Tufts University School of Medicine, Boston, Massachusetts
DOVE, WILLIAM, McArdle Laboratory for Cancer Research, University of Wisconsin, Madison
DOWDY, STEVEN, Whitehead Institute for Biomedical Research, Cambridge, Massachusetts
DOWNES, STEPHEN, Dept. of Zoology, University of Cambridge, Cambridge, United Kingdom
DRABKIN, HARRY, Dept. of Medical Oncology, University of Colorado Health Sciences Center, Denver
DRAETTA, GIULIO, Dept. of Biological Research, Mitotix, Inc., Cambridge, Massachusetts
DUROCHER, FRANCINE, Dept. of Molecular Endocrinology, CHUL Research Center, Quebec, Canada
EBLEN, SCOTT, Dept. of Thoracic Research, Mayo Clinic, Rochester, Minnesota
EISENMAN, ROBERT, Div. of Basic Sciences, Fred Hutchinson Cancer Research Center, Seattle, Washington
EVAN, GERARD, Lab. of Biochemistry of the Cell Nucleus, Imperial Cancer Research Fund, London, United Kingdom
FEARON, ERIC, Boyer Center for Molecular Medicine, Yale University School of Medicine, New Haven, Connecticut
FEINBERG, ANDREW, Dept. of Medicine and Molecular Biology and Genetics, Johns Hopkins University School of Medicine, Baltimore, Maryland
FILATOV, LEONID, Dept. of Pathology, St Jude Children's Research Hospital, Memphis, Tennessee
FLAVELL, RICHARD, Sect. of Immunobiology, Yale University, New Haven, Connecticut
FLICK, PARKE, United States Biochemical Corporation, Cleveland, Ohio
FOLKMAN, JUDAH, Dept. of Surgery and Surgical Research, Children's Hospital, Harvard University Medical School, Boston, Massachusetts
FOULKES, J. GORDON, Oncogene Science, Inc., Uniondale, New York
FREDERICKS, WILLIAM, The Wistar Institute, Philadelphia, Pennsylvania
FREIJE, DIHA, Dept. of Genetics, Boyer Center for Molecular Medicine, Yale University, New Haven, Connecticut
FRIEDMAN, LORI, Dept. of Molecular and Cell Biology, University of California, Berkeley
FRIEND, STEPHEN, Massachusetts General Hospital Cancer Center, Charlestown, Massachusetts
FRISCHAUF, ANNA-MARIA, Imperial Cancer Research Fund, London, United Kingdom
FUJITA, DONALD, Dept. of Medical Biochemistry,

University of Calgary Medical Center, Calgary, Alberta, Canada
FURNEAUX, HENRY, Dept. of Pharmacology and Therapeutics, Memorial Sloan-Kettering Cancer Center, New York, New York
FURTH, MARK, Dept. of Molecular Sciences, Glaxo Research Institute, Research Triangle Park, North Carolina
FUTREAL, ANDY, Lab. of Molecular Carcinogenesis, NIH-National Institute of Environmental Health Sciences, Research Triangle Park, North Carolina
GABRIELSEN, ODD, Dept. of Biochemistry, University of Oslo, Norway
GALLOWAY, DENISE, Fred Hutchinson Cancer Research Center, Seattle, Washington
GAO, CHUAN, Lab. of Urology Research, University of Texas M.D. Anderson Cancer Center, Houston
GEBHARDT, MARK, Dept. of Orthopedic Surgery, Massachusetts General Hospital, Boston, Massachusetts
GEMMILL, ROBERT, Eleanor Roosevelt Institute, Denver, Colorado
GERMAN, JAMES, New York Blood Center, New York, New York
GERONDAKIS, STEVE, Walter and Eliza Hall Institute of Medical Research, Royal Melbourne Hospital, Victoria, Australia
GESSLER, MANFRED, Dept. Physiologische Chemie, Biozentrum der Universität Würtzburg, Würtzburg, Germany
GIAM, CHOU-ZEN, Dept. of Infectious Diseases, Case Western Reserve University School of Medicine, Cleveland, Ohio
GIBSON, NEIL, Dept. of Neuroscience and Cancer, Pfizer, Groton, Connecticut
GIDEON, RECHAVI, Institute of Hematology, Chaim Sheba Medical Center, Tel Hashomer, Israel
GIL-GOMEZ, GABRIEL, Dept. of Biochemistry, University of Barcelona, Barcelona, Spain
GILLE, HENDRIK, Spemann Lab., Max-Planck-Institut für Immunbiologie, Freiburg, Germany
GILLILAND, GARY, Dept. of Medicine, Brigham and Women's Hospital, Harvard Medical School, Boston, Massachusetts
GODLEY, LUCY, National Cancer Institute, National Institutes of Health, Bethesda, Maryland
GOLUB, TODD, Dept. of Medicine, Brigham and Women's Hospital, Harvard Medical School, Boston, Massachusetts
GOODFELLOW, HUGH, Institute of Molecular Medicine, John Radcliffe Hospital, Headington, Oxford, United Kingdom
GOONEWARDENA, PONMANI, Dept. of Genetics, Kaiser Medical Center, San Jose, California
GORBOVITSKAYA, MARY, Dept. of Genetics and Selection, St. Petersburg University, St. Petersburg, Russia
GOROSPE, MYRIAM, Gerontology Research Center, National Institute on Aging, National Institutes of Health, Baltimore, Maryland
GOTTESMAN, MICHAEL, Dept. of Molecular Biology, National Cancer Institute, National Institutes of Health, Bethesda, Maryland
GRAY, JOSEPH, Dept. of Cytometry, University of California, San Francisco
GREEN, MAURICE, Dept. of Molecular Virology, St. Louis University Medical Center, St. Louis, Missouri
GRETARSDOTTIR, SOLVEIG, Lab. of Molecular and Cell Biology Research, Icelandic Cancer Society, Reykjavik
GREULICH, HEIDI, Lab. of Molecular Oncology, Rockefeller University, New York, New York
GRONER, BERND, Tumor Biology Center, Institute for Experimental Cancer Research, Freiburg, Germany
GUALBERTO, ANTONIO, Lineberger Comprehensive Cancer Center, University of North Carolina, Chapel Hill
GUPTA, SOHAN, Hipple Cancer Research Center, Dayton, Ohio
HABER, DANIEL, Massachusetts General Hospital Cancer Center, Charlestown, Massachusetts
HAKIMI, JANETTE, Dept. of Urology, Johns Hopkins University School of Medicine, Baltimore, Maryland
HALL, FREDERICK, Dept. of Orthopedic Surgery, Childrens Hospital of Los Angeles, University of Southern California, Los Angeles
HALL, KEARA, National Cell and Tissue Culture Centre, Dublin City University, Glasnevin, Dublin, Ireland
HALL, STEPHEN, Brooklyn, New York
HANAHAN, DOUGLAS, Dept. of Biochemistry, Hormone Research Institute, University of California, San Francisco
HANLEY-HYDE, JOAN, Lab. of Genetics, National Cancer Institute, National Institutes of Health, Bethesda, Maryland
HANNINK, MARK, Dept. of Biochemistry, University of Missouri, Columbia
HANSEN, LISE LOTTE, Dept. of Human Genetics, Aarhus University, Aarhus, Denmark
HARADA, HISASHI, Institute for Molecular and

Cell Biology, Osaka University, Osaka, Japan
HARLEY, CALVIN, Geron Corp., Menlo Park, California
HARLOW, ED, Massachusetts General Hospital Cancer Center, Charlestown, Massachusetts
HARON, JAY, Dept. of Biotech Research and Development, R.W. Johnson Pharmacology Research Institute, San Diego, California
HARPER, WADE, Dept. of Biochemistry, Baylor College of Medicine, Houston, Texas
HARRIS, CURTIS, Lab. of Human Carcinogenesis, National Cancer Institute, Bethesda, Maryland
HARTER, NIKKI, Dept. of Molecular Biology, Cleveland Clinic Foundation, Cleveland, Ohio
HARTWELL, LELAND, Dept. of Genetics, University of Washington, Seattle
HARVEY, RICHARD, Dept. of Research Development, Gen-Probe, San Diego, California
HATA, AKIKO, Dept. of Molecular Oncology, Rockefeller University, New York, New York
HEGI, MONIKA, Institute of Neuropathology, University Hospital, Zurich, Switzerland
HENGST, LUDGER, Dept. of Molecular Biology, Scripps Research Institute, La Jolla, California
HEO, DAE, MSRBI, University of Michigan, Ann Arbor
HERZOG, CHRISTOPHER, Dept. of Pathology, Medical College of Ohio, Toledo
HILBERG, FRANK, Bender and Co., Ernst-Boehringer-Institut, Vienna, Austria
HOEIJMAKERS, JAN, Dept. of Cell Biology and Genetics, Erasmus University, Rotterdam, The Netherlands
HOGGARD, NIGEL, Dept. of Cancer Genetics, Paterson Institute for Cancer Research, Manchester, United Kingdom
HORVITZ, H. ROBERT, Dept. of Biology, Howard Hughes Medical Institute, Massachusetts Institute of Technology, Cambridge, Massachusetts
HOSTOMSKA, ZUZANA, Dept. of Biochemistry, Agouron Pharmaceuticals, Inc., San Diego, California
HOUSMAN, DAVID, Center for Cancer Research, Massachusetts Institute of Technology, Cambridge, Massachusetts
HOWLEY, PETER, Dept. of Pathology, Harvard Medical School, Boston, Massachusetts
HSIEH, LING-LING, Dept. of Public Health, Chang-Gung Medical College, Tao-Yuan, Taiwan
HUANG, DOLLY, Dept. of Pathology, Chinese University of Hong Kong, Shatin
HUANG, MARY, Dept. of Pathology and Immunology, Tufts University, Dana-Farber Cancer Institute-Children's Hospital, Boston, Massachusetts
HUDSON, KEVIN, Dept. of Cancer Research, Zeneca Pharmaceuticals, Macclesfield, Cheshire, United Kingdom
HUFF, ANNE, Sterling Winthrop, Inc., Collegeville, Pennsylvania
HUNG, PAUL, Dept. of Papillomavirus Research, Wyeth-Ayerst Research, Radnor, Pennsylvania
IGGO, RICHARD, Swiss Institute for Experimental Cancer Research, Epalinges, Switzerland
IRIE, SHINJI, La Jolla Cancer Research Foundation, La Jolla, California
ISAACS, WILLIAM, Dept. of Urology, Johns Hopkins University School of Medicine, Baltimore, Maryland
ISHIKAWA, TOSHIHISA, Dept. of Experimental Pediatrics, University of Texas M.D. Anderson Cancer Center, Houston
ISHIZAKA, YUKIHITO, Dept. of Molecular Biology, The Cleveland Clinic Foundation, Cleveland, Ohio
JACKS, TYLER, Center for Cancer Research, Massachusetts Institute of Technology, Cambridge, Massachusetts
JACOB, WILLIAM, Genetic Tharapy, Inc., Gaithersburg, Maryland
JAKOBS, RAINER, German Cancer Research Center (DKFZ), Heidelberg, Germany
JAMES, GUY, Dept. of Molecular Genetics, University of Texas Southwest Medical Center, Dallas
JARAMILLO, MARIA, Dept. of Medicine, University of Ottawa, Ontario, Canada
JOHANNSSON, OSKAR THOR, Dept. of Oncology, University Hospital, Lund, Sweden
JOHNSON, EDWARD, Dept. of Pathology, Mount Sinai School of Medicine, New York, New York
JOHNSON, LEISA, Center for Cancer Research, Massachusetts Institute of Technology, Cambridge, Massachusetts
KAHN, SCOTT, Institute for Cancer Research, Columbia University, New York, New York
KAMB, ALEXANDER, Dept. of Genomics, Myriad Genetics, Inc., Salt Lake City, Utah
KASTAN, MICHAEL, Dept. of Oncology, Johns Hopkins University School of Medicine, Baltimore, Maryland
KELLY, THOMAS, Dept. of Molecular Biology and

Genetics, Johns Hopkins University School of Medicine, Baltimore, Maryland
KELMAN, ZVI, Dept. of Microbiology, Cornell University Medical Center, New York, New York
KERBEL, ROBERT, Sunnybrook Health Science Center, University of Toronto, Ontario, Canada
KIBERSTIS, PAULA, Science, Washington, D.C.
KING, MARY-CLAIRE, Epidemiology Program, School of Public Health, University of California, Berkeley
KINSELLA, ANNE, Dept. of Surgery, Royal Liverpool University Hospital, University of Liverpool, United Kingdom
KIPREOS, EDWARD, Dept. of Biology, Johns Hopkins University, Baltimore, Maryland
KIRSCH, ILAN, Navy Medical Oncology Branch, National Cancer Institute, National Institutes of Health, Bethesda, Maryland
KITAJEWSKI, JAN, Dept. of Pathology, Columbia University College of Physicians & Surgeons, New York, New York
KO, TIEN, Dept. of Surgery, University of Texas Medical Branch, Galveston
KOIKE, KATSURO, Cancer Institute of Japan, Tokyo
KOLODNER, RICHARD, Dana-Farber Cancer Institute, Harvard Medical School, Boston, Massachusetts
KORSMEYER, STANLEY, Howard Hughes Medical Institute, Washington University School of Medicine, St. Louis, Missouri
KOSKI, RAYMOND, Old Lyme, Connecticut
KRASIKOV, NATALIE, Integrated Genetics, Santa Fe, New Mexico
KRONER, GABRIELE, Institute of Molecular Biology, University of Essen Medical School, Essen, Germany
KRTOLICA, ANA, Dept. of Biochemistry, University of Rochester Medical Center, Rochester, New York
KRUSE, ULRICH, Dept. of Molecular and Experimental Medicine, Scripps Research Institute, La Jolla, California
KRUZELOCK, RUSSELL, Dept. of Molecular Genetics, University of Texas M.D. Anderson Cancer Center, Houston
KUNG, HSING-JIEN, Dept. of Molecular Biology and Microbiology, Case Western Reserve University, Cleveland, Ohio
KURTZ, DAVID, Dept. of Pharmacology, Medical University of South Carolina, Charleston
KWON, HO JEONG, Dept. of Agricultural Chemistry, University of Tokyo, Japan
LADANYI, MARC, Dept. of Pathology, Memorial Sloan-Kettering Cancer Center, New York, New York
LAND, HARTMUT, Imperial Cancer Research Fund, London, United Kingdom
LANDERS, JOHN, Dept. of Genetics, University of Pennsylvania, Philadelphia
LANE, DAVID, Dept. of Biochemistry, Cancer Research Campaign Laboratories, University of Dundee, Scotland, United Kingdom
LEE, AMANDA, Dept. of Cell Biology, Glaxo Research Institute, Durham, North Carolina
LEE, EVA Y.-H.P., Dept. of Biotechnology, University of Texas Health Sciences Center, San Antonio
LEE, STEPHEN, Dept. of Pathology, Washington University School of Medicine, St. Louis, Missouri
LEE, WEN-HWA, Institute of Biotechnology, University of Texas Health Sciences Center, San Antonio
LEMOINE, NICHOLAS, Oncology Unit, Hammersmith Hospital, Imperial Cancer Research Fund, London, United Kingdom
LEUTZ, ACHIM, Zentrum für Molekulare Biologie, University of Heidelberg, Germany
LEVINE, ARNOLD, Dept. of Molecular Biology, Princeton University, Princeton, New Jersey
LI, JIAN, Dept. of Molecular and Experimental Medicine, Scripps Research Institute, La Jolla, California
LI, WAN-CHENG, Dept. of Ophthalmology, Columbia University, New York, New York
LIN, RONG TUAN, Lady Davis Institute for Medical Research, Montreal, Quebec, Canada
LIN, SHENCAI, Dept. of Medicine, Howard Hughes Medical Institute, University of California at San Diego, La Jolla
LING, VICTOR, The Ontario Cancer Institute, Toronto, Ontario, Canada
LISITSYN, NIKOLAI, Cold Spring Harbor Laboratory, Cold Spring Harbor, New York
LIU, WEI, Dept. of Cell and Molecular Biology, Dana-Farber Cancer Institute, Boston, Massachusetts
LIU, XIAO-LONG, Dept. of Radiation Oncology, New England Medical Center, Boston, Massachusetts
LIU, YUSEN, Gerontology Research Center, National Institute on Aging, Baltimore, Maryland
LIVINGSTON, DAVID, Dept. of Neoplastic Disease Mechanisms, Dana-Farber Cancer Institute, Boston, Massachusetts

LOEWENSTEIN, PAUL, Dept. of Molecular Virology, Health Science Center, St. Louis University, St. Louis, Missouri
LUDLOW, JOHN, Dept. of Tumor Biology, University of Rochester, Rochester, New York
LUNDGREN, KAREN, Dept. of Pharmacology, Glaxo Research Institute, Research Triangle Park, North Carolina
LUTCHMAN, MOHINI, Dept. of Neuroscience, McGill University, Montreal, Quebec, Canada
LUTZGER, STUART, Dept. of Molecular Biology, Princeton University, Princeton, New Jersey
MA, CHI, Metabolism Branch, National Cancer Institute, National Institutes of Health, Bethesda, Maryland
MACKENZIE, KAREN, Dept. of Clinical Pharmacology, St. Vincent's Hospital, Rushcutters Bay, New South Wales, Australia
MACLEOD, KAY, Center for Cancer Research, Massachusetts Institute of Technology, Cambridge, Massachusetts
MAI, SABINE, Basel Institute for Immunology, Basel, Switzerland
MAIHLE, NITA, Dept. of Biochemistry and Molecular Biology, Mayo Clinic, Rochester, Minnesota
MAK, YIN, Dept. of Cancer Studies, CRC Laboratories, University of Birmingham Medical School, Birmingham, United Kingdom
MANOHAR, CHITRA, Robert H. Lurie Cancer Center, Northwestern University, Chicago, Illinois
MANSOUR, SAM, Dept. of Chemistry and Biochemistry, University of Colorado, Boulder
MARTIN, GERALD, Dept. of Radiation Oncology, Glasgow University, Glasgow, Scotland, United Kingdom
MARX, JEAN, *Science*, Washington, D.C.
MARX, MARIA, Dept. of Retroviral and Molecular Oncology, Institut Curie-Biology, Centre Universitaire, Orsay, France
MARX, STEPHEN, Sect. of Genetics and Endocrinology, National Institutes of Health, Bethesda, Maryland
MAVROTHALASSITIS, GEORGIOS, Lab. of Molecular Oncology, National Cancer Institute, Frederick, Maryland
MCALOON, ELIZABETH, Chadbourne and Parke, New York, New York
MCCORMICK, FRANK, ONYX Pharmaceutical Corp., Richmond, California
MCMAHON, JERRY, Sugen, Inc., Redwood City, California

MERAJVER, SOFIA, Dept. of Internal Medicine, University of Michigan, Ann Arbor
MERLO, GIORGIO, Friedrich Miescher Institute, Basel, Switzerland
MILLER, JULIA, Dept. of Research and Development, Gen-Probe, Inc., San Diego, California
MINNA, JOHN, Simmons Cancer Center, University of Texas Southwestern Medical School, Dallas
MITCHELL, LLOYD, Dept. of Pathology, George Washington University, Washington, D.C.
MIZUKAMI, TAMIO, Tokyo Research Laboratories, Kyowa Hakko Kogyo Co., Ltd., Tokyo, Japan
MOELLING, KARIN, Institute of Medical Virology, University of Zurich, Switzerland
MOLL, UTE, Dept. of Pathology, State University of New York, Stony Brook
MONDELLO, CHIARA, Istituto Di Genetica Biochimica ed Evoluzionistica, Consiglio Nazionale Delle Ricerche, Padua, Italy
MOREIRA, ANTONIO, Dept. of Genetics, Universidade Nova de Lisboa, Lisbon, Portugal
MÖRÖY, TARIK, Molekularbiologie und Tumorforschung, Phillips Universität Marburg, Marburg, Germany
MOSHINSKY, DEBORAH, Div. of Toxicology, Massachusetts Institute of Technology, Cambridge, Massachusetts
MOUSSES, SPYRO, Dept. of Molecular and Medical Genetics, Samuel Lunenfeld Institute, Mount Sinai Hospital, Toronto, Ontario, Canada
MOYNAHAN, MARYELLEN, Dept. of Medicine, Cell Biology and Genetics, Memorial Sloan-Kettering Cancer Center, New York, New York
MÜLLER, ROLF, Institut für Molekularbiologie und Tumorforschung, Phillips Universität Marburg, Marburg, Germany
MUTHUSWAMY, SENTHIL, Dept. of Molecular Biology and Biotechnology, McMaster University, Hamilton, Ontario, Canada
NABEL, GARY, Dept. of Internal Medicine, Howard Hughes Medical Institute, University of Michigan, Ann Arbor
NASI, SERGIO, Centro Acidi Nucleici, Università La Sapienza, Rome, Italy
NATHANS, DANIEL, Dept. of Molecular Biology and Genetics, Johns Hopkins University, Baltimore, Maryland
NAUMANN, ULRIKE, Institut für Strahlen und Zellforschung, Universität Wurzburg, Germany
NEIMAN, PAUL, Div. of Basic Science, Fred Hutch-

inson Cancer Research Center, Seattle, Washington
NGAN, BO, Dept. of Pathology, Sunnybrook Health Science Centre, University of Toronto, Ontario, Canada
NOMURA, HITOSHI, Dept. of Molecular Science, Chugai Pharmaceutical Co., Ltd., Shizuoke, Japan
NORDHEIM, ALFRED, Institute for Molecular Biology, Hannover Medical School, Hannover, Germany
NUCIFORA, GIUSEPPINA, Dept. of Medicine, University of Chicago, Chicago, Illinois
O'CONNOR, PATRICK, Dept. of Molecular Pharmacology, National Cancer Institute, Bethesda, Maryland
O'NEILL, EDWARD, Dept. of Molecular Immunology, Merck, Sharp, & Dohme Research Laboratories, Rahway, New Jersey
ONEL, KENAN, Dept. of Microbiology, Cornell University Medical College, New York, New York
OREN, MOSHE, Dept. of Chemical Immunology, Weizmann Institute of Science, Rehovot, Israel
ORTH-TAUSSIG, KIM, Dept. of Biochemistry, University of Texas Southwestern Medical Center, Dallas, Texas
OSTERMEYER, ELIZABETH, Dept. of Molecular and Cell Biology, University of California, Berkeley
ÖSTMAN, ARNE, Biomedical Center, Ludwig Institute for Cancer Research, Uppsala, Sweden
OZÇELIK, HILMI, Dept. of Molecular and Medical Genetics, Samuel Lunenfeld Institute, Mount Sinai Hospital, Toronto, Ontario, Canada
OZER, HARVEY, Dept. of Microbiology and Molecular Genetics, UMDNJ-New Jersey Medical School, Newark, New Jersey
PADILLA, DAVID, Amgen, Inc., Thousand Oaks, California
PANDITA, TEJ, Center for Radiological Research, Columbia University, New York, New York
PATY, PHILLIP, Dept. of Surgery, Memorial Sloan-Kettering Cancer Center, New York, New York
PAVLETICH, NIKOLA, Dept. of Cellular Biochemistry and Biophysics, Memorial Sloan-Kettering Cancer Center, New York, New York
PAWELEK, JOHN, Dept. of Dermatology, Yale University School of Medicine, New Haven, Connecticut
PELLS, STEPHEN, Dept. of Pathology, The University of Edinburgh Medical School, Scotland, United Kingdom
PELTOKETO, HELLEVI, Biocenter and Dept. of Clinical Chemistry, University of Oulu, Finland
PEREIRA-SMITH, OLIVIA, Dept. of Molecular Virology, Huffington Center on Aging, Baylor College of Medicine, Houston, Texas
PERLMAN, RIKI, Whitehead Institute for Biomedical Research, Cambridge, Massachusetts
PERUCHO, MANUEL, California Institute of Biological Research, La Jolla, California
PETERS, GORDON, St. Bartholomew's Hospital, Imperial Cancer Research Fund, London, United Kingdom
PETRITSCH, CLAUDIA, Dept. of Pathology, Institute of Molecular Pathology, Vienna, Austria
PHILIPP, ANGELIKA, Zentrum für Molekulare Biologie (ZMBH), Heidelberg, Germany
PINTZAS, ALEXANDER, Institut de Chimie Biologique, Faculte Medicin, CNRS-LGME, INSERM, Strasbourg, France
PIPAS, JAMES, Dept. of Biological Sciences, University of Pittsburgh, Pittsburgh, Pennsylvania
PRENDERGAST, GEORGE, The Wistar Institute, Philadelphia, Pennsylvania
PRESTON, GLORIA, Lab. of Molecular Carcinogenesis, NIH-National Institute of Environmental Health Sciences, Research Triangle Park, North Carolina
PRIVES, CAROL, Dept. of Biological Sciences, Columbia University, New York, New York
QIN, XIAOFENG, Lab. of Molecular Immunology, Rockefeller University, New York, New York
RADFORD, DIANE, Washington University School of Medicine, St. Louis, Missouri
RANZANI, G.N., Dept. of Genetics and Microbiology, University of Padua, Italy
RAO, NAGARAJA, Dept. of Cancer Research, Eli Lilly & Company, Indianapolis, Indiana
RAULF, FRIEDRICH, Dept. of Preclinical Research, Sandoz Pharma, Ltd., Basel, Switzerland
RAUSCHER, FRANK, The Wistar Institute, Philadelphia, Pennsylvania
REICH, NANCY, Dept. of Pathology, State University of New York, Stony Brook
REID, BRIAN, Div. of Gastroenterology, University of Washington, Seattle
RINKER-SCHAEFFER, CARRIE, Oncology Center, Johns Hopkins University School of Medicine, Baltimore, Maryland
ROBERTS, JAMES, Dept. of Basic Sciences, Fred

Hutchinson Cancer Center, Seattle, Washington

ROGLER, CHARLES, Dept. of Medicine, Liver Research Center, Albert Einstein College of Medicine, Bronx, New York

ROHN, JENNIFER, Dept. of Microbiology, University of Washington, Seattle

RON, DAVID, Dept. of Medical and Cell Biology, New York University Medical Center, New York, New York

ROWLEY, JANET, Dept. of Medicine, University of Chicago, Chicago, Illinois

ROYER-POKORA, BRIGITTE, Institute of Human Genetics, University of Heidelberg, Germany

SABE, HISATAKA, Dept. of Biological Responses, Institute for Viral Research, Kyoto University, Kyoto, Japan

SABERS, CANDACE, Dept. of Pharmacology, Mayo Clinic, Rochester, Minnesota

SAGER, RUTH, Dept. of Cancer Genetics, Dana-Farber Cancer Institute, Boston, Massachusetts

SAMBROOK, JOSEPH, Dept. of Biochemistry, University of Texas Southwestern Medical Center, Dallas

SANDBERG NORDQVIST, ANN-CHRISTIN, Dept. of Oncology-Pathology, Karolinska Hospital, Stockholm, Sweden

SANDS, ARTHUR, Dept. of Molecular Genetics, Baylor College of Medicine, Houston, Texas

SARDET, CLAUDE, Whitehead Institute for Biomedical Research, Cambridge, Massachusetts

SCHERNECK, SIEGFRIED, Dept. of Tumor Genetics, Max Delbrück Center for Molecular Medicine, Berlin

SCHIMKE, ROBERT, Dept. of Biological Sciences, Stanford University School of Medicine, Stanford, California

SCHLEGEL, RICHARD, Dept. of Pathology, Georgetown University Medical Center, Washington, D.C.

SCHLEGEL, ROBERT, Lab. of Toxicology, Harvard School of Public Health, Boston, Massachusetts

SCHLESSINGER, JOSEPH, Dept. of Pharmacology, New York University Medical Center, New York, New York

SCHOENBERG, MARK, Dept. of Urology, Johns Hopkins Hospital, Baltimore, Maryland

SCHOTT, DAVID, Geraldine Brush Cancer Research Institute, San Francisco, California

SCUDERI, RICHARD, Lab. of Hematology, Karolinska Hospital, Stockholm, Sweden

SEIZINGER, BERND, Dept. of Oncology Drug Research, Bristol-Myers Squibb Pharmaceutical Research Institute, Princeton, New Jersey

SENGSTAG, CHRISTIAN, Institute of Toxicology, Swiss Federal Institute of Technology and University of Zurich, Schwerzenbach, Switzerland

SETH, ARUN, Lab. of Molecular Oncology, National Cancer Institute, Frederick Cancer Research and Development Center, Frederick, Maryland

SETO, MASAO, Lab. of Chemotherapy, Aichi Cancer Center Research Institute, Nagoya, Japan

SETTLEMAN, JEFFREY, Dept. of Medicine, Harvard Medical School, Massachusetts General Hospital Cancer Center, Charlestown, Massachusetts

SGOURAS, DIONYSSIOS, National Cancer Institute, National Institutes of Health, Frederick Cancer Research and Development Center, Frederick, Maryland

SHAN, BEI, Institute of Biotechnology, University of Texas Health Sciences Center, San Antonio

SHARP, STEPHEN, Dept. of Biological Sciences, R.W. Johnson Pharmaceutical Research Institute, San Diego, California

SHEA, MARY JOAN, Oyster Bay Cove, New York

SHELLING, ANDREW, Institute of Molecular Medicine, John Radcliffe Hospital, Imperial Cancer Research Fund, Oxford, United Kingdom

SHERR, CHARLES, Dept of Tumor Biology, Howard Hughes Medical Institute, St. Jude Children's Research Hospital, Memphis, Tennessee

SHORT, NICHOLAS, *Nature,* London, United Kingdom

SIMARD, JACQUES, Dept. of Molecular Endocrinology, CHUL Research Center, Laval University, Quebec, Canada

SINGH, BALRAJ, Dept. of Molecular Pathology, University of Texas M.D. Anderson Cancer Center, Houston

SKLAR, JEFFREY, Dept. of Pathology, Brigham and Women's Hospital, Harvard Medical School, Boston, Massachusetts

SKOLNICK, MARK, Dept. of Genetic Epidemiology, University of Utah, Salt Lake City

SKOULTCHI, ARTHUR, Dept. of Cell Biology, Albert Einstein College of Medicine, Bronx, New York

SLAMON, DENNIS, Dept. of Hematology/Oncology,

University of California School of Medicine, Los Angeles
SMALL, MICHAEL, Dept. of Microbiology, New Jersey Medical School, Newark, New Jersey
SPAARGAREN, MARCEL, ONYX Pharmaceutical Corp., Richmond, California
STACEY, DENNIS, Dept. of Molecular Biology, The Cleveland Clinic Foundation, Cleveland, Ohio
STACK, MARIA, Dept. of Cancer Genetics, Paterson Institute for Cancer Research, Manchester, United Kingdom
STAMPFER, MARTHA, Dept. of Cell and Molecular Biology, University of California, Berkeley
STANBRIDGE, ERIC, Dept. of Microbiology and Molecular Genetics, University of California, Irvine
STAUDT, LOUIS, Metabolism Branch, National Cancer Institute, National Institutes of Health, Bethesda, Maryland
STAWICKI, MARY, Sterling Winthrop, Inc., Collegeville, Pennsylvania
STEPHENSON, JOHN, Santa Cruz Biotechnology, Inc., Santa Cruz, California
STERNBERG, PAUL, Dept. of Biology, California Institute of Technology, Pasadena, California
STILLMAN, BRUCE, Cold Spring Harbor Laboratory, Cold Spring Harbor, New York
STUBDAL, HILDE, Dana-Farber Cancer Institute, Boston, Massachusetts
SU, WEN, Dept. of Cell and Molecular Biology, Dana-Farber Cancer Institute, Boston, Massachusetts
SU, YAN, Lab. for Cancer Genetics, Natl. Center for Human Genome Research, National Institutes of Health, Bethesda, Maryland
SUPER, HEIDI GILL, Dept. of Molecular Genetics and Cell Biology, University of Chicago, Chicago, Illinois
SWIRNOFF, ALEXANDER, Dept. of Pathology, Washington University School of Medicine, St. Louis, Missouri
SZEKELY, LASZLO, Microbiology and Tumor Biology Center, Karolinska Institute, Stockholm, Sweden
TAKENAKA, IVONE, Dept. of Molecular Genetics and Cell Biology, Bristol-Myers Squibb Pharmaceutical Research Institute, Princeton, New Jersey
TAKIMOTO, MASATO, Dept. of Molecular and Cellular Biology, Osaka University, Osaka, Japan
TAM, WAYNE, Dept. of Molecular Biology, Memorial Sloan-Kettering Cancer Institute, New York, New York

TARUSCIO, DOMENICA, Ultrastructures, Istituto Superiore de Sanita, Rome, Italy
TAVASSOLI, MAHVASH, Dept. of Genetics and Development, University of Sussex School of Biological Sciences, Brighton, West Sussex, United Kingdom
TAYA, YOICHI, National Cancer Center Research Institute, Tokyo, Japan
TEPPER, ROBERT, Dept. of Tumor Biology, Massachusetts General Hospital, Charlestown, Massachusetts
THEILLET, CHARLES, Dept. of Molecular Biology, CNRS, University of Montpellier, Montpellier, France
THOMAS, GILLES, Lab. of Tumor Genetics, Institut Curie, Paris, France
THOME, KELLY, Dept. of Pathology, Tufts University School of Medicine, Boston, Massachusetts
THORLACIUS, STEINUNN, Lab. of Molecular and Cell Biology Research, Icelandic Cancer Society, Reykjavik
THORNTON, GEORGE, Dept. of Biological Sciences, R.W. Johnson Pharmaceutical Research Institute, San Diego, California
TIKHONENKO, ANDREI, Dept. of Basic Sciences, Fred Hutchinson Cancer Research Center, Seattle, Washington
TLSTY, THEA, Dept. of Pathology, Lineberger Cancer Research Center, University of North Carolina, Chapel Hill
TONIN, PATRICIA, Dept. of Medicine, McGill Center for Human Genetics, Montreal, Quebec, Canada
TOPOL, LILIA, Dept. of Molecular Oncology, Frederick Cancer Research Facility, Frederick, Maryland
TOUMBIS, CONSTANTINE, Dept. of Cell Biology, Albert Einstein College of Medicine, Bronx, New York
TSAI, MIAW-SHEUE, Dept. of Biological Chemistry, University of California School of Medicine, Davis
TUREK, LUBOMIR, Dept. of Pathology, University of Iowa College of Medicine, Iowa City
VALANCIUS-MANGEL, VICKY, Dept. of Molecular Genetics, University of Cincinnati, Cincinnati, Ohio
VANDE WOUDE, GEORGE, Lab. of Molecular Oncology, ABL-Basic Research Program, NCI-Frederick Cancer Research and Development Center, Frederick, Maryland
VAN DYKE, TERRY, Dept. of Biochemistry and

Biophysics, University of North Carolina, Chapel Hill
VAN MEIR, ERWIN, Ludwig Institute for Cancer Research, La Jolla, California
VAN ZEIJL, MARJA, Dept. of Molecular Biology, American Cyanamid Company, Pearl River, New York
VARMA, VIJAY, Dept. of Pathology, Atlanta VA Medical Center, Decatur, Georgia
VARMUS, HAROLD, Director's Office, National Institutes of Health, Bethesda, Maryland
VOGELSTEIN, BERT, Dept. of Oncology, Johns Hopkins University Oncology Center, Baltimore, Maryland
VOGT, PETER, Dept. of Molecular and Experimental Medicine, Scripps Research Institute, La Jolla, California
WADGAONKAR, RAJ, Dept. of Oncology, Hoffmann-La Roche, Inc., Nutley, New Jersey
WATSON, ROGER, St. Mary's Hospital Medical School, Ludwig Institute for Cancer Research, London, United Kingdom
WEBER, BARBARA, Dept. of Internal Medicine and Genetics, University of Michigan Medical Center, Ann Arbor
WEINBERG, ROBERT, Dept. of Biology, Whitehead Institute for Biomedical Research, Cambridge, Massachusetts
WEINMANN, ROBERTO, Bristol-Myers Squibb Pharmaceutical Research Institute, Princeton, New Jersey
WESTPHAL, EVA-MARIA, Lab. of Molecular Carcinogenesis, NIH-National Institute of Environmental Health Sciences, Research Triangle Park, North Carolina
WHITE, EILEEN, Dept. of Biological Sciences, Center for Advanced Biotechnology and Medicine, Rutgers University, Piscataway, New Jersey
WIGLER, MICHAEL, Cold Spring Harbor Laboratory, Cold Spring Harbor, New York
WILLEMS, LUC, Dept. of Molecular Biology, Sciences Agronomiques de Gembloux, Gembloux, Belgium
WILLIAMS, BART, Dept. of Biology, Center for Cancer Research, Massachusetts Institute of Technology, Cambridge, Massachusetts
WILLIAMSON, KATHLEEN, MRC Human Genetics Unit, Western General Hospital, Edinburgh, Scotland, United Kingdom
WIRNITZER, UTA, Pharmaforschungszentrum, Institut für Toxikologie IC, Wuppertal, Germany
WISEMAN, ROGER, Lab. of Molecular Carcinogenesis, NIH-National Institute of Environmental Health Sciences, Research Triangle Park, North Carolina
WITTE, OWEN, Howard Hughes Medical Institute, University of California, Los Angeles
WOOD, ALEXANDER, Dept. of Oncology and Virology, Hoffmann-LaRoche, Inc., Nutley, New Jersey
WYLLIE, ANDREW, Dept. of Pathology, CRC Laboratories, University of Edinburgh Medical School, Edinburgh, Scotland, United Kingdom
XIANG, JIALING, Dept. of Tumor Cell Biology, St. Jude Children's Research Hospital, Memphis, Tennessee
XIONG, YUE, Dept. of Biochemistry and Biophysics, University of North Carolina, Chapel Hill
XU, GANGFENG, NDM Dept., Dana-Farber Cancer Institute, Boston, Massachusetts
XU, JUNZHE, Dept. of Internal Medicine, University of Michigan, Ann Arbor
YANG, YILI, Dept. of Immune Cell Biology, National Institutes of Health, Bethesda, Maryland
YANG-YEN, HSIN-FANG, Institute of Molecular Biology, Academia Sinica, Nang-Kang, Taipei, Taiwan
YEH, HEIDI, Lab. of Human Carcinogenesis, National Cancer Institute, National Institutes of Health, Bethesda, Maryland
YEN, JEFFREY J.Y., Institute of Biomedical Services, Academia Sinica, Nang-Kang, Taipei, Taiwan
YIN, XIAO-MING, Howard Hughes Medical Institute, Washington University Medical Center, St Louis, Missouri
YOSHIDA, MINORU, Dept. of Agricultural Chemistry, University of Tokyo, Japan
YOU, MING, Dept. of Pathology, Medical College of Ohio, Toledo
YOUNG, PAUL, Dept. of Biology, Queen's University, Kingston, Ontario, Canada
ZHANG, HAILAN, Center for Advanced Biotechnology and Medicine, Piscataway, New Jersey
ZHANG, JIAO JIAO, Mt. Sinai School of Medicine, New York, New York
ZHANG, QIANG, Dept. of Pathology, University of Chicago, Chicago, Illinois
ZILLIAN, OLAV, Institut für Molekularbiologie, Universität Zürich, Switzerland
ZUR HAUSEN, HARALD, Deutsches Krebsforschungszentrum, Heidelberg, Germany

First row: P. Neiman; H. Varmus; F. Collins, C. Rogler
Second row: P. Young, R. Schimke; R. Sager, P. Vogt
Third row: K. Culver; J. Sklar; B. Vogelstein; A. Kamb
Fourth row: C. Greider, D. Ron; M.C. King; A. Nordheim

First row: D. Stewart, F. McCormick; M. Oren; N. Short, T. Jacks
Second row: G. Peters, J. Witkowski, M. Bishop, C. Greider; E. Harlow, C. Harris, D. Beach
Third row: D. Galloway, N. Harter; T. Kelly, B. Stillman
Fourth row: Picnic

Foreword

Investigation of the mechanistic aspects of cancer has its roots in studies on tumor viruses and their effects on cell proliferation, function, and growth. This outstanding progress is well documented in previous *Cold Spring Harbor Symposia on Quantitative Biology*. The classic 1974 Symposium on Tumor Viruses, followed by the 1979 Symposium on Viral Oncogenes, chronicled the valuable insights that the study of viruses provided regarding the cancer problem. Studies on the genetics of human cancer arose from cytological observations of the abnormalities that are present in the chromosomes of tumor cells. Quite a number of the genes affected by either a translocation breakpoint, amplification, deletion, or point mutation had been identified by the time the Laboratory hosted a comprehensive meeting on the Origins of Human Cancer in our centennial year of 1990.

In the early to mid 1980s, progress on the development of chromosome mapping strategies and the accumulation of DNA probes that identified polymorphisms, encouraged by the International Human Genome Project, enabled the identification of other genes that contributed to familial inheritance of high susceptibility to specific cancers. This approach was very successful and led to a degree of optimism that one aspect of cancer, the multistep genetic process from early neoplasia to metastatic tumors, was beginning to be understood. It therefore seemed appropriate that the 59th Symposium on Quantitative Biology focus attention on the Molecular Genetics of Cancer. The concept was to combine the exciting progress on the identification of new genetic alterations in human tumor cells with studies on the function of the cancer gene products and how they go awry in tumor cells. Such an undertaking would have been most difficult without valuable advice from many colleagues. Principal among these were Ed Harlow, Eric Fearon, Bert Vogelstein, Thea Tlsty, and Michael Gottesman. I also thank David Stewart, Director of the meetings program at the Laboratory, who made this organizational task more efficient.

The meeting was held from June 1 to June 8, 1994 and was attended by a record 421 participants, with 86 oral presentations in 14 sessions and 185 poster presentations on three separate afternoons. Outstanding first night introductory lectures by Leland Hartwell, Robert Weinberg, Bert Vogelstein, and Suzanne Cory set a high standard that was maintained throughout the next 7 days. Harold Varmus, whom we are fortunate to have as the current Director of the National Institutes of Health, presented a masterful Dorcas Cummings Lecture to members of our local community. I am grateful to Ed Harlow who generously agreed to summarize the meeting. He provided the audience with sound advice, good humor, and great insight.

Essential funds that supported this meeting were obtained from the National Cancer Institute (a division of the National Institutes of Health) and the U.S. Department of Energy, Office of Health and Environmental Research. Additional support came from our ever more important and needed Corporate Sponsors: Alafi Capital Company; American Cyanamid Company; Amgen Inc.; Becton Dickinson and Company; Biogen; Bristol-Myers Squibb Company; Chugai Pharmaceutical Co., Ltd.; Ciba-Geigy Corporation; Diagnostic Products Corporation; The Du Pont Merck Pharmaceutical Company; Forest Laboratories, Inc.; Genentech, Inc.; Glaxo; Hoffmann-La Roche Inc.; Johnson & Johnson; Kyowa Hakko Kogyo Co., Ltd.; Life Technologies, Inc.; Mitsubishi Kasei Institute of Life Sciences; Monsanto Company; New England BioLabs, Inc.; Oncogene Science, Inc.; Pall Corporation; The Perkin-Elmer Corporation; Pfizer Inc.; Research Genetics; Sandoz Research Institute; Schering-Plough Corporation; SmithKline Beecham Pharmaceuticals; Sterling Winthrop Inc.; Sumitomo Pharmaceuticals Co., Ltd.; Takeda Chemical Industries, Ltd.; Toyobo Co., Ltd.; Wyeth-Ayerst Research; and Zeneca Group PLC.

Once again, I thank the efficient staff in our meetings office, now under the direction of David Stewart, particularly Diane Tighe, Mikki McBride, and Andrea Stephenson for handling all the administrative aspects. Herb Parsons and his staff provided excellent audiovisual assistance. The organization of this meeting greatly relied on the marvelous work of my assistant Delia Costello who made this time-consuming task a pleasure. Finally, it was again a pleasure to work with the efficient Publications Office, under the direction of John Inglis, particularly Nancy Ford, Patricia Barker, and Joan Ebert. Their efficiency made this volume possible.

Bruce Stillman
January 17, 1995

Contents

Symposium Participants	v
Foreword	xix

Control of Cell Cycle and Cell Growth

The Cancer Cell and the Cell Cycle Clock *M. Hatakeyama, R.A. Herrera, T. Makela, S.F. Dowdy, T. Jacks, and R.A. Weinberg* — 1

D-Type Cyclins and Their Cyclin-dependent Kinases: G_1 Phase Integrators of the Mitogenic Response *C.J. Sherr, J. Kato, D.E. Quelle, M. Matsuoka, and M.F. Roussel* — 11

p21 Is a Component of Active Cell Cycle Kinases *H. Zhang, G.J. Hannon, D. Casso, and D. Beach* — 21

Cyclins, Cdks, and Cyclin Kinase Inhibitors *J.M. Roberts, A. Koff, K. Polyak, E. Firpo, S. Collins, M. Ohtsubo, and J. Massagué* — 31

Role of a Cell Cycle Regulator in Hereditary and Sporadic Cancer *A. Kamb* — 39

$p16^{INK4}$ Mutations and Altered Expression in Human Tumors and Cell Lines *A. Okamoto, D.J. Demetrick, E.A. Spillare, K. Hagiwara, S.P. Hussain, W.P. Bennett, K. Forrester, B. Gerwin, M.S. Greenblatt, M. Serrano, M. Shiseki, J. Yokota, D.H. Beach, and C.C. Harris* — 49

Molecular Genetic Approaches to the Study of Cellular Senescence *T.J. Goletz, J.R. Smith, and O.M. Pereira-Smith* — 59

Molecular and Cell Biology of Replicative Senescence *G.P. Dimri and J. Campisi* — 67

Growth Suppression by Members of the Retinoblastoma Protein Family *L. Zhu, G.H. Enders, C.-L. Wu, M.A. Starz, K.H. Moberg, J.A. Lees, N. Dyson, and E. Harlow* — 75

The Adenovirus E1A-associated 300-kD Protein Exhibits Properties of a Transcriptional Coactivator and Belongs to an Evolutionarily Conserved Family *R. Eckner, Z. Arany, M. Ewen, W. Sellers, and D.M. Livingston* — 85

The Corral Hypothesis: A Novel Regulatory Mode for Retinoblastoma Protein Function *W.-H. Lee, Y. Xu, F. Hong, T. Durfee, M.A. Mancini, Y.-C. Ueng, P.-L. Chen, and D. Riley* — 97

The Max Transcription Factor Network: Involvement of Mad in Differentiation and an Approach to Identification of Target Genes *P.J. Hurlin, D.E. Ayer, C. Grandori, and R.N. Eisenman* — 109

BCL-6 and the Molecular Pathogenesis of B-Cell Lymphoma *R. Dalla-Favera, B.H. Ye, F. Lo Coco, C.-C. Chang, K. Cechova, J. Zhang, A. Migliazza, W. Mellado, H. Niu, S. Chaganti, W. Chen, P.H. Rao, N.Z. Parsa, D.C. Louie, K. Offit, and R.S.K. Chaganti* — 117

Functional Analysis of the *TAN-1* Gene, a Human Homolog of *Drosophila Notch* *J. Aster, W. Pear, R. Hasserjian, H. Erba, F. Davi, B. Luo, M. Scott, D. Baltimore, and J. Sklar* — 125

Novel Oncogenic Mutations in the *WT1* Wilms' Tumor Suppressor Gene: A t(11;22) Fuses the Ewing's Sarcoma Gene, *EWS1*, to *WT1* in Desmoplastic Small Round Cell Tumor *F.J. Rauscher III, L.E. Benjamin, W.J. Fredericks, and J.F. Morris* — 137

Ras Signal Transduction Pathway in *Drosophila* Eye Development *H.C. Chang, F.D. Karim, E.M. O'Neill, I. Rebay, N.M. Solomon, M. Therrien, D.A. Wassarman, T. Wolff, and G.M. Rubin* — 147

Molecular Genetics of Proto-oncogenes and Candidate Tumor Suppressors in *Caenorhabditis elegans* *P.W. Sternberg, C.H. Yoon, J. Lee, G.D. Jongeward, P.S. Kayne, W.S. Katz, G. Lesa, J. Liu, A. Golden, L.S. Huang, and H.M. Chamberlin* — 155

Proto-oncogenes and Plasticity in Cell Signaling *J.M. Bishop, A.J. Capobianco, H.J. Doyle, R.E. Finney, M. McMahon, S.M. Robbins, M.L. Samuels, and M. Vetter*	165
Activation of Ras and Other Signaling Pathways by Receptor Tyrosine Kinases *J. Schlessinger and D. Bar-Sagi*	173
Ras Partners *L. Van Aelst, M.A. White, and M.H. Wigler*	181
Proteins of the 14-3-3 Family Associate with Raf and Contribute to Its Activation *E. Freed, F. McCormick, and R. Ruggieri*	187
Regulation of the Cryptic Sequence-specific DNA-binding Function of p53 by Protein Kinases *T.R. Hupp and D.P. Lane*	195
DNA-binding Properties of the p53 Tumor Suppressor Protein *C. Prives, J. Bargonetti, G. Farmer, E. Ferrari, P. Friedlander, Y. Wang, L. Jayaraman, N. Pavletich, and U. Hubscher*	207
Functions of the p53 Protein in Growth Regulation and Tumor Suppression *J. Lin, X. Wu, J. Chen, A. Chang, and A.J. Levine*	215
Targets for Transcriptional Activation by Wild-type p53: Endogenous Retroviral LTR, Immunoglobulin-like Promoter, and an Internal Promoter of the *mdm2* Gene *Y. Barak, A. Lupo, A. Zauberman, T. Juven, R. Aloni-Grinstein, E. Gottlieb, V. Rotter, and M. Oren*	225
E6-AP Directs the HPV E6-dependent Inactivation of p53 and Is Representative of a Family of Structurally and Functionally Related Proteins *J.M. Huibregtse, M. Scheffner, and P.M. Howley*	237
p53-Dependent Apoptosis In Vivo: Impact of *p53* Inactivation on Tumorigenesis *H. Symonds, L. Krall, L. Remington, M. Sáenz-Robles, T. Jacks, and T. Van Dyke*	247

Checkpoints and Genome Stability

Cell Cycle Checkpoints, Genomic Integrity, and Cancer *L. Hartwell, T. Weinert, L. Kadyk, and B. Garvik*	259
Genomic Integrity and the Genetics of Cancer *T.D. Tlsty, A. White, E. Livanos, M. Sage, H. Roelofs, A. Briot, and B. Poulose*	265
DNA Damage Responses: p53 Induction, Cell Cycle Perturbations, and Apoptosis *C.E. Canman, C.-Y. Chen, M.-H. Lee, and M.B. Kastan*	277
Lymphocyte-specific Genetic Instability and Cancer *I.R. Kirsch, J.M. Abdallah, V.L. Bertness, M. Hale, S. Lipkowitz, F. Lista, and D.P. Lombardi*	287
Cell Cycle Checkpoint Control Is Bypassed by Human Papillomavirus Oncogenes *D.A. Galloway, G.W. Demers, S.A. Foster, C.L. Halbert, and K. Russell*	297
Telomerase, Cell Immortality, and Cancer *C.B. Harley, N.W. Kim, K.R. Prowse, S.L. Weinrich, K.S. Hirsch, M.D. West, S. Bacchetti, H.W. Hirte, C.M. Counter, C.W. Greider, M.A. Piatyszek, W.E. Wright, and J.W. Shay*	307
Three Unusual Repair Deficiencies Associated with Transcription Factor BTF2(TFIIH): Evidence for the Existence of a Transcription Syndrome *W. Vermeulen, A.J. van Vuuren, M. Chipoulet, L. Schaeffer, E. Appeldoorn, G. Weeda, N.G.J. Jaspers, A. Priestley, C.F. Arlett, A.R. Lehmann, M. Stefanini, M. Mezzina, A. Sarasin, D. Bootsma, J.-M. Egly, and J.H.J. Hoeijmakers*	317
Human Mismatch Repair Genes and Their Association with Hereditary Non-Polyposis Colon Cancer *R.D. Kolodner, N.R. Hall, J. Lipford, M.F. Kane, M.R.S. Rao, P. Morrison, L. Wirth, P.J. Finan, J. Burn, P. Chapman, C. Earabino, E. Merchant, D.T. Bishop, J. Garber, C.E. Bronner, S.M. Baker, G. Warren, L.G. Smith, A. Lindblom, P. Tannergard, R.J. Bollag, A.R. Godwin, D.C. Ward, M. Nordenskjold, R.M. Liskay, N. Copeland, N. Jenkins, M.K. Lescoe, A. Ewel, S. Lee, J. Griffith, and R. Fishel*	331
Defects in Replication Fidelity of Simple Repeated Sequences Reveal a New Mutator Mechanism for Oncogenesis *M. Perucho, M.A. Peinado, Y. Ionov, S. Casares, S. Malkhosyan, and E. Stanbridge*	339

CONTENTS

Ovarian Tumors Display Persistent Microsatellite Instability Caused by Mutation in the Mismatch Repair Gene *hMSH-2*   K. Orth, J. Hung, A. Gazdar, M. Mathis, A. Bowcock, and J. Sambrook ... 349

Loss of Imprinting in Human Cancer   A.P. Feinberg, L.M. Kalikin, L.A. Johnson, and J.S. Thompson ... 357

Enhanced Cell Survival and Tumorigenesis   S. Cory, A. Strasser, T. Jacks, L.M. Corcoran, T. Metz, A.W. Harris, and J.M. Adams ... 365

Apoptosis

The Genetics of Programmed Cell Death in the Nematode *Caenorhabditis elegans*   H.R. Horvitz, S. Shaham, and M.O. Hengartner ... 377

Bcl-2 Gene Family and the Regulation of Programmed Cell Death   X.-M. Yin, Z.N. Oltvai, D.J. Veis-Novack, G.P. Linette, and S.J. Korsmeyer ... 387

Control of p53-dependent Apoptosis by E1B, Bcl-2, and Ha-*ras* Proteins   E. White, S.-K. Chiou, L. Rao, P. Sabbatini, and H.-J. Lin ... 395

Apoptosis in Carcinogenesis: The Role of p53   A.H. Wyllie, P.J. Carder, A.R. Clarke, K.J. Cripps, S. Gledhill, M.F. Greaves, S. Griffiths, D.J. Harrison, M.L. Hooper, R.G. Morris, C.A. Purdie, and C.C. Bird ... 403

Cellular Senescence and Cancer   J.C. Barrett, L.A. Annab, D. Alcorta, G. Preston, P. Vojta, and Y. Yin ... 411

Apoptosis and the Prognostic Significance of p53 Mutation   S.W. Lowe, S. Bodis, N. Bardeesy, A. McClatchey, L. Remington, H.E. Ruley, D.E. Fisher, T. Jacks, J. Pelletier, and D.E. Housman ... 419

Genetic Models

Transgenic Approaches to the Analysis of *ras* and *p53* Function in Multistage Carcinogenesis   C.J. Kemp, P.A. Burns, K. Brown, H. Nagase, and A. Balmain ... 427

Mouse Model Systems to Study Multistep Tumorigenesis   A. Berns, N. van der Lugt, M. Alkema, M. van Lohuizen, J. Domen, D. Acton, J. Allen, P.W. Laird, and J. Jonkers ... 435

Tumorigenic and Developmental Effects of Combined Germ-line Mutations in *Rb* and *p53*   B.O. Williams, S.D. Morgenbesser, R.A. DePinho, and T. Jacks ... 449

Insulin-like Growth Factor II Is Focally Up-regulated and Functionally Involved as a Second Signal for Oncogene-induced Tumorigenesis   P. Naik, G. Christofori, and D. Hanahan ... 459

Angiostatin: A Circulating Endothelial Cell Inhibitor That Suppresses Angiogenesis and Tumor Growth   M.S. O'Reilly, L. Holmgren, Y. Shing, C. Chen, R.A. Rosenthal, Y. Cao, M. Moses, W.S. Lane, E.H. Sage, and J. Folkman ... 471

The *p53* Tumor Suppressor Gene Inhibits Angiogenesis by Stimulating the Production of Thrombospondin   K.M. Dameron, O.V. Volpert, M.A. Tainsky, and N. Bouck ... 483

Defining the Steps in a Multistep Mouse Model for Mammary Carcinogenesis   H.E. Varmus, L.A. Godley, S. Roy, I.C.A. Taylor, L. Yuschenkoff, Y.-P. Shi, D. Pinkel, J. Gray, R. Pyle, C.M. Aldaz, A. Bradley, D. Medina, and L.A. Donehower ... 491

The Adenomatous Polyposis Coli Gene of the Mouse in Development and Neoplasia   W.F. Dove, C. Luongo, C.S. Connelly, K.A. Gould, A.R. Shoemaker, A.R. Moser, and R.L. Gardner ... 501

Genetic Instability and Apoptotic Cell Death during Neoplastic Progression of v-*myc*-initiated B-cell Lymphomas in the Bursa of Fabricius   P.E. Neiman, J. Summers, S.J. Thomas, S. Xuereb, and G. Loring ... 509

Colorectal Cancer and the Intersection between Basic and Clinical Research   B. Vogelstein and K.W. Kinzler ... 517

Human Cancer Genes and Their Products

Green Pigs, Red Herrings, and a Golden Hoe: A Retrospective on the Identification of *BRCA1* and the Beginning of Its Characterization *E.A. Ostermeyer, L.S. Friedman, E.D. Lynch, C.I. Szabo, P. Dowd, M.K. Lee, S.E. Rowell, and M.-C. King*	523
Progress toward Isolation of a Breast Cancer Susceptibility Gene, *BRCA1* *B.L. Weber, K.J. Abel, F.J. Couch, S.D. Merajver, S.C. Chandrasekharappa, L. Castilla, D. McKinley, P.P. Ho, K. Calzone, T.S. Frank, J. Xu, L.C. Brody, and F.S. Collins*	531
RNA Genetics of Breast Cancer: Maspin as Paradigm *R. Sager, S. Sheng, A. Anisowicz, G. Sotiropoulou, Z. Zou, G. Stenman, K. Swisshelm, Z. Chen, M.J.C. Hendrix, P. Pemberton, K. Rafidi, and K. Ryan*	537
Acute Myeloid Leukemia with Inv(16) Produces a Chimeric Transcription Factor with a Myosin Heavy Chain Tail *P. Liu, N. Seidel, D. Bodine, N. Speck, S. Tarlé, and F.S. Collins*	547
Genetic Alterations in the Chromosome 22q12 Region Associated with Development of Neuroectodermal Tumors *G. Thomas, O. Delattre, J. Zucman, P. Merel, C. Desmaze, T. Melot, M. Sanson, K. Hoang-Xuan, B. Plougastel, P. Dejong, G. Rouleau, and A. Aurias*	555
Molecular Genetic Changes Found in Human Lung Cancer and Its Precursor Lesions *A.F. Gazdar, S. Bader, J. Hung, Y. Kishimoto, Y. Sekido, K. Sugio, A. Virmani, J. Fleming, D.P. Carbone, and J.D. Minna*	565
Molecular Characterization of *QM*, a Novel Gene with Properties Consistent with Tumor Suppressor Function *E. Stanbridge, A. Farmer, A. Mills, T. Loftus, D. Kongkasuriyachai, S. Dowdy, and B. Weissman*	573
Barrett's Esophagus: A Model of Human Neoplastic Progression *K. Neshat, C.A. Sanchez, P.C. Galipeau, D.S. Cowan, S. Ramel, D.S. Levine, and B.J. Reid*	577
Detection of Genetic Loss in Tumors by Representational Difference Analysis *N.A. Lisitsyn, F.S. Leach, B. Vogelstein, and M.H. Wigler*	585
Genetic Approaches to Defining Signaling by the CML-associated Tyrosine Kinase BCR-ABL *D.E.H. Afar, A. Goga, L. Cohen, C.L. Sawyers, J. McLaughlin, R.N. Mohr, and O.N. Witte*	589
The *AML1* Gene in the 8;21 and 3;21 Translocations in Chronic and Acute Myeloid Leukemia *G. Nucifora and J.D. Rowley*	595
Expression of P-Glycoprotein in Normal and Malignant Rat Liver Cells *C.H. Lee, G. Bradley, and V. Ling*	607
Genes Coding for Tumor-specific Rejection Antigens *T. Boon, B. Van den Eynde, H. Hirsch, C. Moroni, E. De Plaen, P. van der Bruggen, C. De Smet, C. Lurquin, J.-P. Szikora, and O. De Backer*	617
Pathogenesis of Cancer of the Cervix *H. zur Hausen and F. Rösl*	623
Autocrine Mechanism for *met* Proto-oncogene Tumorigenicity *S. Rong and G.F. Vande Woude*	629
Studies of the Deleted in Colorectal Cancer Gene in Normal and Neoplastic Tissues *E.R. Fearon, B.C. Ekstrand, G. Hu, W.E. Pierceall, M.A. Reale, and S.H. Bigner*	637
Molecular Cytogenetics of Human Breast Cancer *J.W. Gray, C. Collins, I.C. Henderson, J. Isola, A. Kallioniemi, O.-P. Kallioniemi, H. Nakamura, D. Pinkel, T. Stokke, M. Tanner, and F. Waldman*	645
Genetic Alterations in Prostate Cancer *W.B. Isaacs, G.S. Bova, R.A. Morton, M.J.G. Bussemakers, J.D. Brooks, and C.M. Ewing*	653
Multicellular Resistance: A New Paradigm to Explain Aspects of Acquired Drug Resistance of Solid Tumors *R.S. Kerbel, J. Rak, H. Kobayashi, M.S. Man, B. St. Croix, and C.H. Graham*	661

Genetic Methods for Diagnosis and Cancer Therapy

Overcoming Complexities in Genetic Screening for Cancer Susceptibility *S.H. Friend, R. Iggo, C. Ishioka, M. Fitzgerald, I. Hoover, E. O'Neill, and T. Frebourg*	673
Exploiting Multidrug Resistance to Treat Cancer *M.M. Gottesman, S.V. Ambudkar, B. Ni, J.M. Aran, Y. Sugimoto, C.O. Cardarelli, and I. Pastan*	677
Toxicity and Immunologic Effects of In Vivo Retrovirus-mediated Gene Transfer of the Herpes Simplex-Thymidine Kinase Gene into Solid Tumors *K.W. Culver, D.W. Moorman, R.R. Muldoon, R.M. Paulsen, Jr., J.L. Lamsam, H.W. Walling, and C.J. Link, Jr.*	685
Gene Transfer and Bone Marrow Transplantation *M.K. Brenner, H.E. Heslop, D. Rill, C. Li, T. Nilson, M. Roberts, C. Smith, R. Krance, and C. Rooney*	691
Molecular Genetic Interventions for Cancer *G.J. Nabel, E.G. Nabel, Z. Yang, B.A. Fox, G.E. Plautz, X. Gao, L. Huang, S. Shu, D. Gordon, and A.E. Chang*	699
An Introduction to the Puzzle *E. Harlow*	709

Author Index   725

Subject Index   729

COLD SPRING HARBOR SYMPOSIA
ON QUANTITATIVE BIOLOGY

VOLUME LIX

The Cancer Cell and the Cell Cycle Clock

M. HATAKEYAMA,* R.A. HERRERA,* T. MAKELA,* S.F. DOWDY,*
T. JACKS,† AND R.A. WEINBERG*

*Whitehead Institute for Biomedical Research, Department of Biology, Massachusetts Institute of Technology;
†Howard Hughes Medical Institute, MIT Center for Cancer Research, Cambridge, Massachusetts 02142

Cancer is increasingly viewed as a cellular disease. We now perceive the macroscopic tumor as being little more than a large population of similar cells sharing a common set of genetic aberrations. Such thinking focuses attention on the single cancer cell and mechanisms responsible for its transformation from a normal to a malignant growth state. Indeed, in the eyes of some, the disease of cancer will be understood by describing the processes associated with cell transformation.

This reduction would seem to provide the means for unifying the field of cancer research within a single conceptual framework. The reality, however, is that our field remains highly fragmented. Contemporary research on the cancer cell operates using three distinct and quite unrelated conceptual paradigms. Each paradigm creates its own logical space, its own distinct way of thinking about the cancer cell.

Until the mid 1970s, cancer cell research focused largely on the biology of the cancer cell. Cell transformation was described in terms of cell physiology. Then came oncogenes and, later on, tumor suppressor genes, and cell transformation was studied in terms of individual genes and their gene products. Molecular genetics promised to explain the complex physiology of cancer cells in terms of genes and proteins. The promise remains largely unfullfilled: The biology of the cancer cell has not been resolved simply by cloning genes and studying their encoded proteins. The worlds of genes and cell physiology remain essentially unconnected.

This has necessitated a third approach, a new way of thinking about the biology of cancer cells and the molecules within them. Within the past several years, a new paradigm has emerged that promises to offer a bridge between cancer cell physiology and molecular biology. Many in our field have been engaged in assembling the molecules involved in regulating cell proliferation into complex signaling cascades and tying these in turn together into elaborate signal-processing systems that very much resemble electronic circuitry. These signaling cascades all converge on a control apparatus that lies at the heart of this circuitry. This apparatus is the cell cycle clock, the central controller of cell proliferation. This clock holds the promise of unifying and rationalizing the complex molecular phenomenology of oncogenes and tumor suppressor genes. Further down the road, it may enable us to tie oncogenes and suppressor genes directly to the complex physiology of the cancer cell.

MITOGENIC SIGNALING AND THE CYTOPLASM

Even before the debut of the cell cycle clock, limited connections had already been forged between cancer genes and cancer cell biology. These enabled us to understand how certain distinct aspects of cell biology can be related to the mutation of specific target genes in the genome of a developing tumor cell. The clearest connection here was made between the biological trait of mitogen independence and the mutant genes that specify a variety of cytoplasmic and nuclear oncoproteins.

We can rationalize the reduced mitogen dependence of cancer cells quite simply. Many oncogenes encode proteins normally involved in the acquisition and processing of mitogenic signals (Bishop 1991). Some of these proteins, once altered through mutations, flood the cytoplasm with growth-stimulating signals mimicking those experienced by the normal cell that has been exposed to an exogenous mitogen. The cytoplasmic oncoproteins encode either growth factor receptors or the signal-transducing proteins that are activated by them within the cytoplasm. Within the past several years, these proteins have been interconnected in a single cytoplasmic circuitry (Ahn et al. 1992; Egan and Weinberg 1993). Included in this pathway are cytoplasmic proteins encoded by oncogenes such as *erbB*, *erbB2*, *ras*, *src*, and *raf*.

Oncoproteins expressed in the nuclei of cancer cells are found in the high levels usually seen only in normal, mitogen-stimulated cells. Examples here are the products of the *myc* and *fos* oncogenes. The constitutive expression of these two proteins by their respective genes obviates the growth factors normally required for their expression. Thus, the problem of the reduced serum dependence shown by tumor cells, at least in its outlines, appears to have been solved. At least one cancer cell phenotype has been explained in terms of genotype.

ESCAPE FROM NEGATIVE SIGNALS

Cancer cells show a second profound biological abnormality when interacting with their surroundings. They exhibit nonresponsiveness to extracellular growth-inhibitory signals. This type of change requires a different set of physiological and biochemical explanations from those used to rationalize mitogen independence.

This nonresponsive phenotype has been especially difficult to study. The difficulty stems in large part from the nature of the inhibitory signals exchanged between cells (Weiss 1970). Some of these are conveyed by soluble factors such as transforming growth factor β (TGF-β). The biochemical mechanisms responsible for transducing these negative signals have begun to yield to the same experimental approaches previously used for mitogenic factors. This is exemplified by the recent successful isolation of genes encoding TGF-β receptors and interferon signal transducers (Lopez-Casillas et al. 1991; Wang et al. 1991; Lin et al. 1992; Franzen et al. 1993; Ihle et al. 1994; Massagué et al. 1994; Shuai 1994). Other equally important inhibitory signals appear to be conveyed by cell-to-cell contacts and adhesion to extracellular matrix components (Wieser and Oesch 1986; Yamasaki 1990). These signals are not readily addressed by the standard experimental strategies of protein biochemistry.

At least two paradigms have been developed to explain the changes inside cancer cells that allow them to transcend or ignore growth-inhibitory signals received from their surroundings. Both derive from molecular genetics of well-studied genes. The first of these is stimulated by the study of the *myc* gene. Its product is a transcription factor that appears to be a central controller that drives the cell into and through its growth cycle. In normal cells, this gene is up-regulated by mitogens and down-regulated by negative factors such as interferon and TGF-β (Resnitzky et al. 1986; Pietenpol et al. 1990). The oncogenic alleles of *myc* are constitutively expressed; this enables their ongoing expression in the presence of negative signals that would normally succeed in shutting them down.

The second strategy for escaping growth inhibition depends on the loss of components responsible for transducing growth-inhibitory signals from the cell exterior into the nucleus. This scheme invokes receptors for anti-mitogens as well as tumor suppressor proteins. Loss of these transducers from negative signaling cascades interrupts the flow of inhibitory signals into the cell. Such loss in turn renders the cancer cell oblivious to such signals. One example here is provided by certain tumor cells that have lost TGF-β receptors and hence acquired resistance to growth inhibition by TGF-β (Kimchi et al. 1987).

Our information in this area is still fragmentary. As mentioned above, we already have a well-defined diagram of the transducers of the positive, mitogenic signals and how these transducers connect with one another in the cellular signaling circuitry, but the layout of the corresponding growth-inhibitory circuitry is still quite obscure. Do tumor suppressor proteins assemble to form their own negative cascades which operate in parallel to the mitogen cascades, converging with the latter on some critical target in the nucleus? Or are tumor suppressor proteins salted here and there within the mitogen-driven signaling cascades, sitting astride critical choke points? Here the mapping between genes and cell physiology remains elusive.

THE CELL CYCLE AND ITS CLOCK

The traits of mitogen independence and refractoriness to growth inhibition represent only a small part of the complex biology of cancer cells. Moreover, even these traits are explained only superficially by descriptions of the molecules that transduce positive and negative growth signals. The ultimate answers lie at a deeper level. The ultimate answers will come only from answering a different type of question: How does the cell—normal and malignant—process and integrate the complex array of signals flowing through it and decide when to grow, to become quiescent, to differentiate, or to die?

The discovery of the cell cycle clock now offers us the prospect of integrating the diverse array of molecules involved in cell signaling into a single, centrally operating, signal-processing circuit. Once fleshed out in detail, this circuit should inform us how the normal cell makes its important proliferative decisions and how these decisions are derailed as the cell evolves toward the malignant growth state.

Until very recently, the cell cycle clock was studied in splendid isolation. It was the purview of yeast geneticists who dissected it into its component parts through study of controlling genes and embryologists who focused on its roles in frog oocyte maturation and early development (Hartwell and Weinert 1989; Hunt 1989; Nurse 1990; Pines and Hunter 1990; Hunter and Pines 1991; Murray 1992). Only during the past several years has the cell cycle clock made its way into the thinking of those interested in cancer (Hunt 1991; Motokura et al. 1991; Hartwell 1992). Only now do we begin to realize a simple and profound truth about this clock: It represents the ultimate recipient of all the positive and negative signals that influence cell proliferation. All of the circuits that control cell growth must eventually feed into and affect the cell cycle clock. The actions of oncogene and tumor suppressor proteins must ultimately be understood in terms of how they influence the workings of the cell cycle clock.

The connections forged to date between the oncogenes, tumor suppressor genes, and the clock are still tenuous or nonexistent. The interface between these genes and the clock is formed almost entirely from the study of two suppressor genes, the retinoblastoma (*RB*) gene and the *p53* gene and their respective products, pRB and p53. pRB operates in the midst of the clock machinery; p53 stands back from the machinery, controlling it from a short distance (El-Deiry et al. 1993). In contrast, the connections between cellular oncogene proteins and the clock machinery remain totally obscure.

We focus here largely on pRB and on its role in a discrete phase of the cell cycle in mid/late G_1. As argued below, this mid/late G_1 transition represents a critical time in the life of the cell during which a cell decides between continued proliferation and escape from the cell cycle. The deregulation of this transition would appear to be central to the creation of the malignant growth state. It remains to be seen whether deregulation of the cell cycle clock affects other, equally important transitions in other phases of the cell cycle.

The seminal observation connecting pRB with the cell cycle clock and this period of G_1 was a simple one. pRB undergoes a conversion from a hypo- to a hyperphosphorylated form several hours before the end of the G_1 phase of the cell cycle (Buchkovich et al. 1989; Chen et al. 1989). This finding provided the first hint of its involvement in controlling cell cycle progression and took on additional significance when the precise timing of this phosphorylation was put into the context of older, cell biological studies.

As Pardee demonstrated almost two decades ago, the mammalian cell passes by an important milestone at a point several hours before the transition from the G_1 into the S phase of its growth cycle (Pardee 1974, 1989). He termed this milestone the restriction (R) point and defined it operationally in two ways (Fig.1). First, cell cycle progression depends on mitogen stimulation from the beginning of G_1 up to the R point; thereafter, further cell cycle progression is essentially mitogen-independent. Second, during most of G_1 prior to the R point, cell cycle progression can be inhibited by low levels of cycloheximide; thereafter, the cell seems oblivious to this drug. More recently, a third operational definition of the R point has been put forward by others: Prior to this point, TGF-β, acting as an anti-mitogen, succeeds in shutting down the advance of the cell through its growth cycle; after this point, TGF-β loses its anti-mitogenic influence (Laiho et al. 1990; Howe et al. 1991; Geng and Weinberg 1993; Koff et al. 1993).

The model inspired by these observations is a simple one. Through much of G_1 up to the R point, the cell receives and integrates a variety of positive and negative growth-regulating signals. Arriving at the R point, the cell decides whether or not the signals accumulated since the beginning of G_1 are propitious for growth. If they are, it passes through the R point and commits itself to traverse the remainder of the cell cycle through mitosis. If they are not, the cell may either tarry in front of the R gate until conditions are more favorable or take more drastic action such as reverting to its G_0, quiescent state or entering irreversibly into a postmitotic state.

In either event, it is clear that the R point transition represents a centrally important decision in cell cycle control. Indeed, it is so important that it causes one to divide the active cell cycle into five periods rather than the customary four: early G_1 before R, late G_1 after R, S, G_2, and M. The fact that pRB undergoes hyperphosphorylation contemporaneously with the R point transition suggests a more than casual association between the two processes.

CONTROL OF pRB PHOSPHORYLATION

The hyperphosphorylation of pRB appears to be tightly linked to its functioning as a growth-suppressing protein. The evidence for this is severalfold. DNA tumor virus oncoproteins that compromise pRB function associate preferentially with the underphosphorylated form of pRB (Ludlow et al. 1989; Imai et al. 1991; Templeton and Weinberg 1991). This suggests that they are able to eliminate the pool of functional pRB within the cell by focusing their attentions on this form of the protein. Second, stimuli that cause pRB hyperphosphorylation also favor its functional neutralization (Hinds et al. 1992). Finally, forms of pRB that are refractory to this phosphorylation remain in a constantly growth-suppressing configuration (Hamel et al. 1992; Hinds et al. 1992).

This pRB hyperphosphorylation is the handiwork of the cyclin:cdk complexes that operate as the central constituents of the cell cycle clock. In the case of pRB hyperphosphorylation, the G_1 cyclins (Sherr 1993) and their cdk partners are strongly implicated. Evidence presented to date has implicated both cyclin E and the D cyclins—D1, D2, and D3.

The precise contributions of each of these cyclins to pRB hyperphosphorylation are not yet resolved. Human osteosarcoma cells transfected with vectors forcing the ectopic expression of cyclins A and E cause pRB hyperphosphorylation whereas vectors causing cyclin D1 overexpression do not (Hinds et al. 1992). This suggests that cyclin E, which becomes active in mid/late G_1, may play a central role in driving pRB hyperphosphorylation and attendant functional inactivation.

A conflicting series of observations comes from in vitro biochemistry, in which cyclin D has been found to cause pRB hyperphosphorylation in the presence of its partner kinase, cdk4 (Matsushime et al. 1992; Ewen et al. 1993; Kato et al. 1993). This would suggest that cyclin D plays a central role. However, the pRB substrate used in many experiments has been a truncated version of this protein and it remains unclear whether

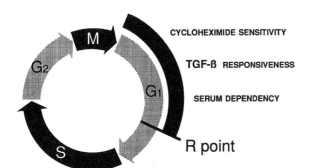

Figure 1. Phases of the cell cycle and the restriction (R) point defined by Pardee (1974). Before the R point, cell cycle advance is dependent on serum mitogens and sensitive to inhibition by low levels of cycloheximide and TGF-β. Thereafter, its progression into S and the remainder of the cell cycle is unresponsive to these agents.

or not a fully intact pRB can serve as a good cyclin D:cdk4 substrate in vitro.

A resolution of the conflicting models arising out of these results may come from our own recent work using a yeast model of pRB phosphorylation (Hatakeyama et al. 1994). This yeast model suggests that both interpretations may be correct. More specifically, cyclin E and cyclin D (1, 2, or 3) may collaborate with one another to effect pRB hyperphosphorylation. This conclusion comes from the introduction of a human pRB expression vector into cells of Saccharomyces cerevisiae. Unlike the case with mammalian cells, ectopically expressed pRB appears to have no observable effect on yeast cell proliferation. However, pRB undergoes a phosphorylation that is almost indistinguishable from that seen in mammalian cells. As a consequence, the modification of pRB by phosphorylation can be studied in the absence of the confounding effects arising from any effects it might have on cell proliferation.

The similarities between the observed phosphorylation events in mammalian and yeast cells are striking and manifold. The pattern of phosphopeptide spots is almost indistinguishable: 23 out of 25 are the same. Moreover, this phosphorylation occurs in late G_1 well before the G_1/S transition. Yeast mating pheromone, which blocks yeast cells at START, a landmark that may correspond to the mammalian R point (Reed 1980, 1991; Marsh et al. 1991), prevents pRB hyperphosphorylation, just as TGF-β does when applied to mammalian cells. Mutants of the cdc28 kinase that block cell cycle progression in G_1 prevent pRB hyperphosphorylation, whereas those that block in G_2 permit phosphorylation to occur.

Most informative are the results flowing from use of yeast cells that lack one or more of their G_1 cyclin genes. Any one of these—CLN1, CLN2, or CLN3—is sufficient to propel the yeast cell through G_1. Significantly, yeast cells lacking CLN3 are able to grow quite well yet do not hyperphosphorylate their pRB. Those lacking CLN1 and CLN2 are similarly affected. From these experiments, we have concluded that the phosphorylation of pRB in yeast cells depends on the activity of CLN3 plus either CLN1 or CLN2.

This suggests that a similar collaboration may operate in the G_1 phase of mammalian cells. To begin, we noted certain similarities between the mammalian and yeast G_1 cyclins. Thus, cyclins A and E are expressed in specific time windows in G_1, as are Clns 1 and 2. The D cyclins and Cln3 do not appear to cycle periodically as the cell passes through its growth cycle.

More compelling are experiments in which defects in certain CLN genes of S. cerevisiae are complemented by introduced mammalian G_1 cyclin genes. Mammalian cyclin E is able to rescue much of the pRB hyperphosphorylating activity that is lost upon inactivation of Cln1 plus Cln2. Conversely, cyclin D1 is able to stand in place of Cln3 in restoring pRB hyperphosphorylation. Such data provide the beginnings of a proof that pRB hyperphosphorylation requires the intervention of two G_1 cyclins operating in the mammalian G_1 phase and

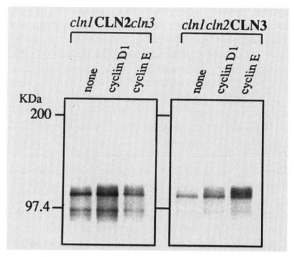

Figure 2. The *cln1 CLN2 cln3* or *cln1 cln2 CLN3* yeast strain containing an expression vector for human pRB was further transformed with an expression vector for human cyclin D1 or E under the control of *GAL1, 10* promoter. Their transformants co-expressing human pRB and cyclin proteins were grown to mid-log phase in synthetic minimal medium containing 2% galactose and cellular proteins were labeled with [^{35}S]methionine. Cell lysates were then prepared and pRB was immunoprecipitated from the lysates with anti-human pRB. The immune complexes were analyzed on 7.5% SDS-polyacrylamide gel electrophoresis.

suggest that pRB serves to integrate the converging signals conveyed by at least two types of G_1 cyclins (Fig. 2) (Hatakeyama et al. 1994).

pRB AS CONTROLLER OF CYCLINS

The above results depict pRB as the object of control by G_1 cyclins. In fact, pRB may also serve to control G_1 cyclins, specifically those of the D class. The evidence here stems from observations that pRB is able to bind D cyclins in vitro and in vivo (Dowdy et al. 1993; Ewen et al. 1993). This binding depends on a sequence motif, LXCXE, that is present at the amino termini of these three cyclins and is seen as well within the structures of the three DNA tumor virus oncoproteins, E1A, large T, and E7, known to bind pRB (Figge et al. 1988). Use of site-directed mutagenesis to alter residues in this motif results in the loss of binding ability in the case of both the D cyclins and the viral oncoproteins (Dowdy et al. 1993).

The suggestion here is that both classes of proteins depend on similar means to bind to pRB. Moreover, they would seem to target the same region of pRB: the domain defined as its "pocket." Mutant, functionally defective forms of pRB isolated from human tumors lose their ability to bind both the viral oncoproteins and the D cyclins. Moreover, an oligopeptide derived from the LXCXE region of adenovirus E1A protein is able to compete effectively with cyclin D1 for pRB binding (Dowdy et al. 1993).

This similarity in binding might suggest a similar mode of action. Thus, like the DNA tumor virus oncoproteins, the D cyclins may control pRB by occupying its pocket domain, effectively sequestering pRB in the process. In fact, a quite different mechanism seems to be operating. A mutant of cyclin D that has lost its pRB-binding ability actually shows an increased growth-promoting ability, both in human osteosarcoma cells and in Rat-1 cells (Hinds et al. 1993; S. Dowdy and R.A. Weinberg, unpubl.).

Even more persuasive are the results of recent experiments performed in a strain of *S. cerevisiae* that has been deprived of all three of its *CLN* genes. The growth of cells of this strain is dependent on an introduced cyclin D1 gene (M. Hatakeyama and R.A. Weinberg, unpubl.). When a plasmid expressing wild-type pRB is introduced into these cells, their growth is slowed dramatically; this is in contrast to the situation observed in wild-type yeast cells where pRB has no effect on cell proliferation. Mutant pRBs encoded by RB alleles isolated from human tumors have also been introduced into these yeast cells. These pRBs have lost their ability to bind D cyclins and, concomitantly, their ability to suppress the cyclin D-dependent proliferation of these yeast cells. We conclude that pRB and cyclin D are able to associate and interact with one another functionally within the living cell and that this association leads to the sequestration and functional inactivation of cyclin D1. We predict that pRB has similar interactions with the related cyclins D2 and D3.

As is the case with other ligands captured by pRB, notably the E2F transcription factors, cyclin D sequestration is reversed upon pRB hyperphosphorylation (Dowdy et al. 1993). This leads in turn to a model in which hypophosphorylated pRB captures D cyclins in early G_1 and releases them upon becoming hyperphosphorylated several hours before S phase.

Taken with the earlier results, these appear to be confounding. How can cyclin D be both a sequestered captive of pRB and, at the same time, an agent that contributes to pRB phosphorylation? We suggest the following model. In early G_1, D cyclins bind pRB and contribute to its partial phosphorylation. This phosphorylation is necessary for and anticipates the definitive, inactivating hyperphosphorylation that occurs at the mid/late G_1 transition. Mutant forms of pRB that retain their cdk phosphorylation sites but fail to bind cyclin D do not undergo hyperphosphorylation. This is consistent with an essential role played by the cyclin D:pRB complex in the process of pRB phosphorylation.

While bound by pRB, cyclin D and its partner kinase, cdk4, are prevented from attacking other, non-pRB substrates. This only occurs when cyclin E appears on the scene. It acts with its partner kinase, cdk2, to force pRB hyperphosphorylation and, as a consequence, causes the liberation of cyclin D from pRB control. Cyclin D and cdk4 can then proceed to modify other substrates. Such modification, in turn, is essential to the late G_1 processes that lead on into S phase (Fig. 3).

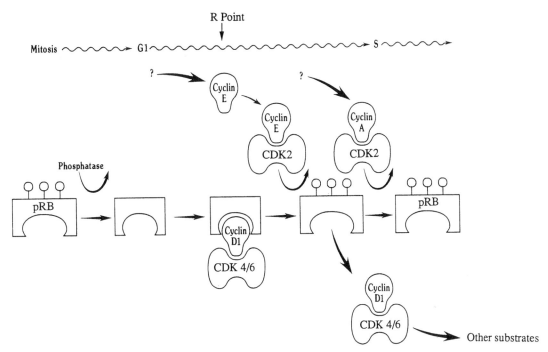

Figure 3. Scheme describing the interaction of G_1 cyclins and the retinoblastoma protein (pRB). pRB emerges from mitosis in a hypophosphorylated state. This enables it to bind cyclin D1 and, via cyclin D1, CDK4/6. The latter then promote partial phosphorylation of pRB which is completed by cyclin E:cdk2 complexes which become active in mid/late G_1. Once hyperphosphorylated, pRB releases cyclin D1 (as well as other sequestered proteins such as the transcription factor E2F). The released cyclin D1, together with its associated CDK4/6 kinase, is then able to proceed to phosphorylate other substrates, enabling entrance into S phase. Following the G_1/S transition, cyclin A:cdk2 complexes assume the task of maintaining pRB phosphorylation.

CONTROL OF CYCLIN EXPRESSION

If pRB acts, as proposed, to integrate the afferent signals transduced by cyclin E and the D cyclins, this suggests an additional puzzle: What controls the expression of these cyclins and their respective activities? These cyclins must sit at the ends of complex signaling cascades. What lies at the top of these cascades?

Few experimental data are available to address these questions at present. One answer will eventually derive from cell biology. The ultimate controllers of the G_1 cyclins must be extracellular mitogens and antimitogens impinging on the cell during the first portion of G_1. But here, as mentioned before, the connections between the signaling cascades activated by these factors and the cyclins are tenuous, indeed nonexistent. At present, no one knows how these extracellular factors control G_1 cyclins, and through these cyclins, the forward motion of the clock.

We have recently come across one mechanism that also appears to play an important role in controlling cyclin expression, doing so without referring to signals originating from the cell's extracellular environment. This mechanism of cyclin control depends on the cell cycle clock itself. Such internal control suggests the operation of a closed feedback loop, in that the clock is controlling one of its own regulators. The operation of this feedback loop, in turn, may provide an explanation of how the R point transition is effected and, once effected, how the decision to proceed through the remainder of the cell cycle becomes quasi-irreversible.

These recent insights depend on use of early passage mouse embryo fibroblasts that lack functional *RB* gene copies (Jacks et al. 1992). Such homozygous mutant cells were prepared from mouse embryos whose chromosomal *RB* gene copies were altered by the techniques of homologous recombination. These mutant embryo fibroblasts are 40% smaller than their wild-type counterparts. Their cell cycle is also distorted, in that they show a truncated G_1 phase and a lengthened S phase (R.E. Herrera et al., in prep.).

Analysis of synchronized cells indicates that gene expression in early G_1 and late G_1 is unaffected by the absence of pRB in these cells. However, a mid/late G_1 event, the induction of cyclin E expression, is profoundly altered. It occurs 5–6 hours prematurely in these mutant cells. This indicates that pRB normally operates to suppress cyclin E expression, allowing it to appear only 10–12 hours after entrance into G_1. Such control would appear to fly in the face of earlier observations that indicated an opposite relationship—control of pRB by cyclin E and its partner kinase.

A reconciliation comes here from a scheme in which pRB and cyclin E each controls the other. pRB acts to suppress cyclin E expression; cyclin E functions to inactivate pRB. Together they form a positive, self-reinforcing feedback loop.

In fact, this feedback loop requires the involvement of a third component, a transcription factor that links pRB to control of cyclin E expression. Thus, pRB

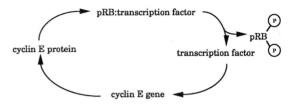

Figure 4. A simple, self-amplifying positive feedback loop involving pRB and cyclin E. Hyperphosphorylation of a small amount of pRB by cyclin E (and an associated kinase) in mid/late G_1 causes it to release a transcription factor that in turn activates expression of the cyclin E gene. This yields more cyclin E protein and in turn more hyperphosphorylated pRB.

would appear to sequester a factor whose activity is essential for cyclin E gene expression. Initially, hyperphosphorylation of a small amount of pRB in mid/late G_1 will cause liberation of a small amount of this transcription factor that in turn will contribute to induction of cyclin E gene expression. The latter, through its product, will hyperphosphorylate and thus inactivate a bit more pRB, resulting in the liberation of yet more transcription factor and amplification of this signal. Once ignited, this feedback loop is self-sustaining until extraneous signals inactivate or destroy one or more of its components. This loop is illustrated in its simplest form in Figure 4.

We note that control of transcription factors by pRB is well documented; hypophosphorylated pRB sequesters factors of the E2F class and releases the factors upon hyperphosphorylation (Bagchi et al. 1991; Chellappan et al. 1991; Chittenden et al. 1991). Moreover, other investigators studying yeast have also proposed a positive feedback loop in late G_1 that, although differing in its details, also serves to stabilize the commitment by the cell to proceed into S phase and the remainder of the cell cycle (Dirick and Nasmyth 1991).

THE R POINT SWITCH

Passage through the R point gate and associated commitment to traverse the remainder of the cell cycle represents an important decision for a cell and must be regulated in a complex fashion. This decision must be executed only after weighing a series of countervailing signals including those released by mitogen and antimitogen receptors, surface receptors monitoring contact with extracellular matrix and with other cells, and intracellular monitors of metabolism.

The feedback loop, described above in its simplest form, would seem to present opportunities for just such multilevel control, in that each of the steps in the loop (Fig. 4) may be modulated by extrinsic signals. The result is a centrally acting circuit that we term the R point switch which serves as the common target of a diverse array of afferent regulatory signals. The opportunities for regulation by a variety of incoming signals are illustrated in Figure 5.

Among the control modes made possible by these incoming signals are (1) control of cyclin D levels by

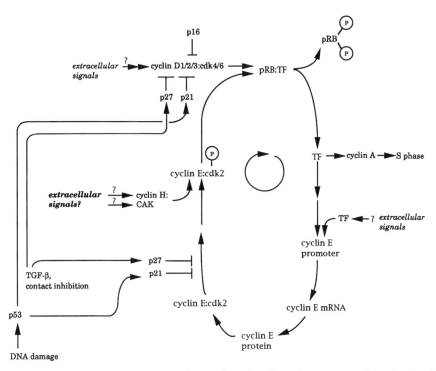

Figure 5. A positive feedback loop involving pRB and cyclin E and regulated by a diverse array of incoming signals. This scheme is an elaboration of that presented in Fig. 4. As indicated, the firing of this positive feedback loop can be suppressed by negatively acting agents such as p16, p21, and p27. Conversely, it may be promoted by CAK:cyclin H complexes, cyclin D:cdk4/6 complexes, and possibly a series of transcription factors that collaborate with the pRB-associated transcription factor to induce cyclin E gene expression; each of the latter in turn may be controlled ultimately by extracellular growth-regulating signals.

mitogens (Matsushime et al. 1991; Ajchenbaum et al. 1993) since, as mentioned above, cyclin D plays an essential role in pRB phosphorylation; (2) control of cdk2 and cdk4 activities by cdk-activating kinases (CAKs) that activate cdks through phosphorylation of critical threonine residues (Desai et al. 1992; Poon et al. 1992; Solomon et al. 1992, 1993; Fresquet et al. 1993; Kato et al. 1994); (3) control of cdk:cyclin complexes by ancillary inhibitory subunits, namely, the recently described p16, p21, and p27 proteins (Gu et al. 1992; El-Deiry et al. 1993; Harper et al. 1993; Serrano et al. 1993; Xiong et al. 1993; Dulic et al. 1994; Polyak et al. 1994a,b; Toyoshima and Hunter 1994); and (4) control of the cyclin E promoter by additional, non-pRB-dependent transcription factors. Other signaling pathways that converge on this feedback loop may well be uncovered in coming years.

The p16, p21, and p27 proteins seem to provide connections with a variety of physiological signals that act to inhibit transit through the R point gate. p27 is reported to be controlled by TGF-β and contact inhibition (Polyak et al. 1994a). p21 is up-regulated by p53; the latter in turn is activated by DNA damage and by the generational clock that orchestrates cell senescence and crisis (El-Deiry et al. 1993). Both cdk2 and cdk4 complexes are essential for firing of this feedback circuit (see Fig. 5) and are also inhibitable by one or more of these p16/p21/p27 regulators. This presents multiple opportunities for intervention by physiological signals that antagonize cell proliferation.

One additional attraction of this circuit stems from older observations of oncogene collaboration. These earlier experiments ascribed similar mechanisms of action to the DNA tumor virus oncoproteins, to mutant p53 alleles, and to the *myc* oncogene, thereby placing all of these genes within a common functional group. Members of this group, like its archetype the *myc* oncogene, were found to be able to collaborate with a *ras* oncogene in the transformation of primary embryo cells (Weinberg 1989). The various viral oncoproteins can act to short-circuit the R point feedback loop by removing pRB, thereby liberating cyclin E from its control; some can also block p53 action, uncoupling the feedback loop from this important conduit of negative signals. Chromosomal mutations in the p53 gene can serve a similar end (Hinds et al. 1989). Unexplained by all this is how the archetype of these genes—*myc*—interacts with this feedback circuit. We suspect that it may act as a positive regulator of cyclin gene expression. Indeed, evidence has already been presented supporting just such a view (Jansen-Durr et al. 1993).

Importantly, the firing of this loop can be deregulated by a variety of molecular mechanisms known to be operating in cancer cells. Included here are the aforementioned DNA tumor virus oncoproteins, mutant forms of the p53 gene (Hinds et al. 1989), deletions or inactivations of the p16 gene (Cairns et al. 1994; Kamb 1994), and overexpression of the cyclin D1 gene (Lammie et al. 1991; Motokura et al. 1991; Withers et al. 1991; Jiang et al. 1992).

CAKs

As mentioned above, the CAKs may offer an important conduit for funneling regulatory signals into the R point feedback loop. In fact, little is known about how they are regulated. The CAK enzyme described to date—the mammalian homolog of *Xenopus* MO15 kinase—phosphorylates cdk2 on its Thr-160 residue. Both the resulting phosphate group and an association with a cyclin subunit appear to be essential for cdk activity toward substrates. The phosphorylation of cdk2 is regulated in a cell-cycle-specific manner: Phosphorylation of Thr-160 of cdk2 first appears in late G_1 concomitant with the activation of cdk2:cyclin E (Desai et al. 1992; Poon et al. 1992; Solomon et al. 1992, 1993; Fresquet et al. 1993; Kato et al. 1994). This raises an interesting question: Is CAK activity modulated, and if so, does this present an additional opportunity for regulation by physiological signals?

To address this possibility, we began recently to search for CAK-associated proteins, doing so by using the yeast two-hybrid screening system. In the course of this work, we isolated a gene encoding a novel, CAK-associated protein that has all the earmarks of being a cyclin of ancient origins: It shows homology with the yeast proteins mcs2 and Ccl1 and with mammalian cyclin C. We term this protein cyclin H. Cyclin H binds efficiently to CAK in mammalian cells, and its association with CAK appears to be important for the phosphorylation of cdk2 by CAK.

Importantly, the CAK:cyclin complex operates in quite a different way from the other cdk complexes characterized to date. The latter form part of the core machinery of the cell cycle clock and operate to send out signals to more peripheral proteins that then execute cell cycle transitions. The CAK:cyclin complex, in contrast, has at its substrate a component of the clock itself and thus stands upstream of the entire clock machinery. It remains to be seen whether elaborate signaling cascades are assembled by hierarchies of cyclin:kinase complexes that activate each other in succession. Also unanswered is the question of how CAK and cyclin H are regulated. We hope to find their physiological regulators.

COALESCENCE

Our research field is now in a critical phase. A decade-long accumulation of experimental results has presented us with a wealth of information on a large variety of cellular proteins involved in cell growth control. Now for the first time we can begin to arrange this large and diverse array of proteins into larger schemes—signal-transducing cascades and feedback loops and other depictions of the circuitry that enable the cell to process complex growth-controlling signals by perturbing the cell cycle clock. All this work is beginning to coalesce into a single, unifying picture.

We can hope that the stunning successes in assembling the cytoplasmic signaling cascades will be echoed in the nucleus. The R point loop proposed here will undoubtedly be superseded by other more precise descriptions of nuclear signaling circuits. This will happen quickly, as new connections between hitherto-unconnected components are being forged almost monthly. By the end of the decade, the long-sought central circuitry that processes growth-regulating signals will be fleshed out in detail. For the first time, we will indeed be able to understand cell growth physiology in terms of the molecules within the cell.

Even with these successes, many larger problems will still remain. The largest and most subtle of these is closely connected to these issues of cell growth control: How does a normal cell, using its signal-processing machinery, relate to the complex array of cells around, integrate their diverse signals, and converge on the decision to grow, differentiate, or die? We leave that to another Symposium a decade or more down the road.

ACKNOWLEDGMENTS

This work is supported by grants from the American Cancer Society and the National Cancer Institute to R.A.W. and T.J. R.A.W. is an American Cancer Society Research Professor. M.H. was supported by a grant from the Leukemia Research Foundation, R.A.H. was supported by a grant from the Anna Fuller Fund (AFF-737), and T.M. was supported by a grant from the Human Frontier Science Program (LT-259/93). S.F.D. was supported by grants from the Cancer Research Fund of the Damon Runyon-Walter Winchell Foundation and the Margaret and Herman Sokol Biomedical Postdoctoral Award in Cancer Research.

REFERENCES

Ahn, N.G., R. Seger, and E.G. Krebs. 1992. The mitogen-activated protein kinase activator. *Curr. Opin. Cell Biol.* **4:** 992.

Ajchenbaum, F., K. Ando, J.A. DeCaprio, and J.D. Griffin. 1993. Independent regulation of human D-type cyclin gene expression during G_1 phase in primary human T lymphocytes. *J. Biol. Chem.* **268:** 4113.

Bagchi, S., R. Weinmann, and P. Raychaudhuri. 1991. The retinoblastoma protein copurifies with E2F-1, an E1A-regulated inhibitor of the transcription factor E2F. *Cell* **65:** 1063.

Bishop, J.M. 1991. Molecular themes in oncogenesis. *Cell* **64:** 235.

Buchkovich, K., L.A. Duffy, and E. Harlow. 1989. The retinoblastoma protein is phosphorylated during specific phases of the cell cycle. *Cell* **58:** 1097.

Cairns, P., K. Tokino, Y. Eby, and D. Sidransky. 1994. Homozygous deletions of 9p21 in primary human bladder tumors detected by comparative multiplex polymerase chain reaction. *Cancer Res.* **54:** 1422.

Chellappan, S.P., S. Hiebert, M. Mudryj, J.M. Horowitz, and J.R. Nevins. 1991. The E2F transcription factor is a cellular target for the RB protein. *Cell* **65:** 1053.

Chen, P.-L., P. Scully, J.-Y. Shew, J.Y.J. Wang, and W.-H. Lee. 1989. Phosphorylation of the retinoblastoma gene product is modulated during the cell cycle and cellular differentiation. *Cell* **58:** 1193.

Chittenden, T., D.M. Livingston, and W.G. Kaelin. 1991. The T/E1A-binding domain of the retinoblastoma product can interact selectively with a sequence-specific DNA-binding protein. *Cell* **65:** 1073.

Desai, D., Y. Gu, and D.O. Morgan. 1992. Activation of human cyclin-dependent kinases in vitro. *Mol. Biol. Cell* **3:** 571.

Dirick, L. and K. Nasmyth. 1991. Positive feedback in the activation of G1 cyclins in yeast. *Nature* **351:** 754

Dowdy, S.F., P.W. Hinds, K. Louie, S.I. Reed, A. Arnold, and R.A. Weinberg. 1993. Physical interaction of the retinoblastoma protein with human cyclins. *Cell* **73:** 499.

Dulic, V., W.K. Kaufmann, S.J. Wilson, T.D. Tlsty, E. Lees, J.W. Harper, S.J. Elledge, and S.I. Reed. 1994. p53-dependent inhibition of cyclin-dependent kinase activities in human fibroblasts during radiation-induced G1 arrest. *Cell* **76:** 1013.

Egan, S.E. and R.A. Weinberg. 1993. The pathway to signal achievement. *Nature* **365:** 781.

El-Deiry, W.S., T. Tokino, V.E. Velculescu, D.B. Levy, R. Parsons, J.M. Trent, W.E. Mercer, K.W. Kinzler, and B. Vogelstein. 1993. *WAF1*, a potential mediator of p53 tumor suppression. *Cell* **75:** 817.

Ewen, M.E., H.K. Sluss, C.J. Sherr, D.M. Livingston, and H. Matsushime. 1993. Functional interactions of the retinoblastoma protein with mammalian D-type cyclins. *Cell* **73:** 487.

Figge, J., T. Webster, T.F. Smith, and E. Paucha. 1988. Prediction of similar transforming regions in simian virus 40 large T, adenovirus E1A, and myc oncoproteins. *J. Virol.* **62:** 1814.

Franzen, P., P. ten Dijke, H. Ichijo, H. Yamashita, P. Schultz, C.H. Heldin, and K. Miyazono. 1993. Cloning of a TGF beta type I receptor that forms a heteromeric complex with the TGB beta type II receptor. *Cell.* **75:** 681.

Fresquet, D., J.-C. Labbé, J. Derancourt, J.-P. Capony, S. Galas, F. Girard, T. Lorca, J. Shuttleworth, M. Doree, and J.-C. Cavadore. 1993. The *MO15* gene encodes the catalytic subunit of a protein kinase that activates cdc2 and other cyclin-dependent kinases (CDKs) through phosphorylation of Thr 161 and its homologues. *EMBO J.* **12:** 311.

Geng, Y. and R.A. Weinberg. 1993. TGF-β effects on expression of G1 cyclins and cdks. *Proc. Natl. Acad. Sci.* **90:** 10315.

Gu, Y., C.W. Turck, and D.O. Morgan. 1992. Inhibition of CDK2 activity *in vivo* by an associated 10K regulatory subunit. *Nature* **366:** 707.

Hamel, P.A., R.M. Gill, R.A. Phillips, and B.L. Gallie. 1992. Regions controlling hyperphosphorylation and conformation of the retinoblastoma gene product are independent of domains required for transcriptional repression. *Oncogene* **7:** 69.

Harper, J.W., G.R. Adami, N. Wei, K. Keyomarsi, and S.J. Elledge. 1993. The p21 CDK-interacting protein Cip1 is a potent inhibitor of G1 cyclin-dependent kinases. *Cell* **75:** 805.

Hartwell, L. 1992. Defects in a cell cycle checkpoint may be responsible for the genomic instability of cancer cells. *Cell* **71:** 543.

Hartwell, L.H. and T.A. Weinert. 1989. Checkpoints: Controls that ensure the order of cell cycle events. *Science* **246:** 629.

Hatakeyama, M., J.A. Brill, G.A. Fink, and R.A. Weinberg. 1994. Collaboration of G1 cyclins in the functional inactivation of the retinoblastoma protein. *Genes Dev.* **8:** 1759.

Hinds, P., C. Finlay, and A.J. Levine. 1989. Mutation is required to activate the *p53* gene for cooperation with the *ras* oncogene and transformation. *J. Virol.* **63:** 739.

Hinds, P.W., S.F. Dowdy, E. Ng-Eaton, and R.A. Weinberg. 1993. Function of a human cyclin gene as an oncogene. *Proc. Natl. Acad. Sci.* **91:** 709.

Hinds, P.W., S. Mittnacht, V. Dulic, A. Andrew, S.I. Reed, and R.A. Weinberg. 1992. Regulation of retinoblastoma protein functions by ectopic expression of human cyclins. *Cell* **70:** 993.

Howe, P.H., G. Draetta, and E.B. Leof. 1991. Transforming growth factor β1 inhibition of p34^{cdc2} phosphorylation and histone H1 kinase activity is associated with G1/S-phase growth arrest. *Mol. Cell. Biol.* **11:** 1185.

Hunt, T. 1989. Maturation promoting factor, cyclin and the control of M-phase. *Curr. Opin. Cell Biol.* **1:** 268.

———. 1991. Cell cycle gets more cyclins. *Nature* **350:** 462.

Hunter, T. and J. Pines. 1991. Cyclins and cancer. *Cell* **66:** 1071.

Ihle, J.N., B.A. Witthuhn, F.W. Quelle, K. Yamamoto, W.E. Thierfelder, B. Kreider, and O. Silvennoinen. 1994. Signaling by the cytokine receptor superfamily: JAKs and STATs. *Trends Biochem. Sci.* **19:** 222.

Imai, Y., Y. Matsushima, T. Sugimura, and M. Terada. 1991. Purification and characterization of human papillomavirus type 16 E7 protein with preferential binding capacity to the underphosphorylated form of retinoblastoma gene product. *J. Virol.* **65:** 4966.

Jacks, T., A. Fazeli, E. Schmitt, R.T. Bronson, M. Goodell, and R.A. Weinberg. 1992. Effects of an RB mutation in the mouse. *Nature* **359:** 295.

Jansen-Durr, P., A. Meichle, P. Steiner, M. Pagano, K. Finke, J. Botz, J. Wessbecher, G. Draetta, and M. Eilers. 1993. Differential modulation of cyclin gene expression by MYC. *Proc. Natl. Acad. Sci.* **90:** 3685.

Jiang, W., S.M. Kahn, N. Tomita, Y.J. Zhang, S.H. Lu, and I.B. Weinstein. 1992. Amplification and expression of human cyclin D gene in esophageal cancer. *Cancer Res.* **52:** 2980.

Kamb, A. 1994. A cell cycle regulator potentially involved in genesis of many tumor types. *Science* **264:** 436.

Kato, J.-Y., M. Matsuoka, D.K. Strom, and C.J. Sherr. 1994. Regulation of cyclin D-dependent kinase 4 (cdk4) by cdk4-activating kinase. *Mol. Cell. Biol.* **14:** 2713.

Kato, J.-Y., H. Matsushime, S.W. Hiebert, M.E. Ewen, and C.J. Sherr. 1993. Direct binding of cyclin D to the retinoblastoma gene product (pRb) and pRb phosphorylation by the cyclin D-dependent kinase CDK4. *Genes Dev.* **7:** 331.

Kimchi, A., X.-F. Wang, R.A. Weinberg, S. Cheifetz, and J. Massagué. 1987. Absence of transforming growth factor-β receptors and growth inhibitory responses in retinoblastoma cells. *Science* **240:** 196.

Koff, A., M. Ohtsuki, K. Polyak, J.M. Roberts, and J. Massagué. 1993. Negative regulation of G_1 in mammalian cells: Inhibition of cyclin E-dependent kinase by TGF-β. *Science* **260:** 536.

Laiho, M., J.A. DeCaprio, J.W. Ludlow, D.M. Livingston, and J. Massagué. 1990. Growth inhibition of TGF-beta linked to suppression of retinoblastoma protein. *Cell* **62:** 175.

Lammie, G.A., V. Fantl, R. Smith, E. Schuuring, S. Brookes, R. Michalides, C. Dickson, A. Arnold, and G. Peters. 1991. D11S287, a putative oncogene on chromosome 11q13, is amplified and expressed in squamous cell and mammary carcinomas and linked to BCL-1. *Oncogene* **6:** 439.

Lin, H.Y., X.-F. Wang, E. Ng-Eaton, R.A. Weinberg, and H.F. Lodish. 1992. Expression cloning of the TGF-β type II receptor, a functional transmembrane serine/threonine kinase. *Cell* **68:** 775.

Lopez-Casillas, F.S., S. Cheifetz, J. Doody, J.L. Andres, W.S. Lane, and J. Massagué. 1991. Structure and expression of the membrane proteoglycan betaglycan, a component of the TGF-β receptor system. *Cell* **67:** 785.

Ludlow, J.W., J.A. DeCaprio, C.M. Huang, W.H. Lee, E. Paucha, and D.M. Livingston. 1989. SV40 large T antigen binds preferentially to an underphosphorylated member of the retinoblastoma susceptibility gene product family. *Cell* **56:** 57.

Marsh, L., A.M. Neiman, and I. Herskowitz. 1991. Signal transduction during pheromone response in yeast. *Annu. Rev. Cell Biol.* **7:** 699.

Massagué, J., L. Attisano, and J.L. Wrana. 1994. The TGF-β family and its composite receptors. *Trends Cell Biol.* **4:** 172.

Matsushime, H., M.F. Roussel, R.A. Ashmun, and C.J. Sherr. 1991. Colony-stimulating factor 1 regulates novel cyclins during the G1 phase of the cell cycle. *Cell* **65:** 701.

Matsushime, H., M.E. Ewen, D.K. Strom, J.Y. Kato, S.K. Hanks, M.F. Roussel, and C.J. Sherr. 1992. Identification and properties of an atypical catalytic subunit (p34^{PSK-J3}/cdk4) for mammalian D type G1 cyclins. *Cell* **71:** 323.

Motokura, T., T. Bloom, H.G. Kim, H. Juppner, J.V. Ruderman, H.M. Kronenberg, and A. Arnold. 1991. A novel cyclin encoded by a *bcl*1-linked candidate oncogene. *Nature* **350:** 512.

Murray, A.W. 1992. Creative blocks: Cell-cycle checkpoints and feedback controls. *Nature* **359:** 599.

Nurse, P. 1990. Universal control mechanism regulating onset of M-phase. *Nature* **344:** 503.

Pardee, A.B. 1974. A restriction point for control of normal animal cell proliferation. *Proc. Natl. Acad. Sci.* **71:** 1286.

———. 1989. G1 events and regulation of cell proliferation. *Science* **246:** 961.

Pietenpol, J.A., R.W. Stein, E. Moran, P. Yaciuk, R. Schlegel, R.M. Lyons, M.R. Pittelkow, K. Munger, P.M. Howley, and H.L. Moses. 1990. TGF-β1 inhibition of c-*myc* transcription and growth in keratinocytes is abrogated by viral transforming proteins with pRB binding domains. *Cell* **61:** 777.

Pines, J. and T. Hunter. 1990. p34^{cdc2}: The S and M kinase? *New Biol.* **2:** 389.

Polyak, K., J.-Y. Kato, M.J. Solomon, C.J. Sherr, J. Massagué, J.M. Roberts, and A. Koff. 1994a. p27^{Kip1}, a cyclin-cdk inhibitor, links TGF-β and contact inhibition to cell cycle arrest. *Genes Dev.* **8:** 9.

Polyak, K., M.-H. Lee, H. Erdjument-Bromage, A. Koff, J.M. Roberts, P. Tempst, and J. Massagué. 1994b. Cloning of p27^{Kip1}, a cyclin-dependent kinase inhibitor and a potential mediator of extracellular antimitogenic signals. *Cell* **78:** 59.

Poon, R.Y.C., K. Yamashita, J.P. Adamczewski, T. Hunt, and J. Shuttleworth. 1992. The cdc2-related protein p40^{MO15} is the catalytic subunit of a protein kinase that can activate p33^{cdk2} and p34^{cdc2}. *EMBO J.* **12:** 3123.

Reed, S.I. 1980. The selection of *S. cerevisiae* mutants defective in the START event of cell division. *Genetics* **95:** 561.

———. 1991. Pheromone signaling pathways in yeast. *Curr. Opin. Genet. Dev.* **1:** 391.

Resnitzky, D., A. Yarden, D. Zipori, and A. Kimchi. 1986. Autocrine β-related interferon controls c-*myc* suppression and growth arrest during hematopoietic cell differentiation. *Cell* **46:** 31.

Serrano, M., G.J. Hannon, and D. Beach. 1993. A new regulatory motif in cell-cycle control causing specific inhibition of cyclin D/CDK4. *Nature* **366:** 704.

Sherr, C.J. 1993. Mammalian G1 cyclins. *Cell* **73:** 1059.

Shuai, K. 1994. Interferon-activated signal transduction to the nucleus. *Curr. Opin. Cell Biol.* **6:** 253.

Solomon, M.J., J.W. Harper, and J. Shuttleworth. 1993. CAK, the p34^{cdc2} activating kinase, contains a protein identical or closely related to p40^{MO15}. *EMBO J.* **12:** 3133.

Solomon, M.J., T. Lee, and M.W. Kirschner. 1992. Role of phosphorylation in p34^{cdc2} activation: Identification of an activating kinase. *Mol. Biol. Cell* **3:** 13.

Templeton, D. and R.A. Weinberg. 1991. Nonfunctional mutants of the retinoblastoma protein are characterized by defects in phosphorylation, viral oncoprotein association, and nuclear tethering. *Proc. Natl. Acad. Sci.* **88:** 3033.

Toyoshima, H. and T. Hunter. 1994. P27, a novel inhibitor of G1 cyclin-cdk protein kinase activity, is related to p21. *Cell* **78:** 67.

Wang, X.-F., H.Y. Lin, E. Ng-Eaton, J. Downward, H.F. Lodish, and R.A. Weinberg. 1991. Expression cloning and characterization of the TGF-β type III receptor. *Cell* **67:** 797.

Weinberg, R.A. 1989. Oncogenes, antioncogenes and the molecular bases of multi-step carcinogenesis. *Cancer Res.* **49:** 3713.

Weiss, R.A. 1970. The influence of normal cells on the proliferation of tumor cells in culture. *Exp. Cell Res.* **63:** 1.

Wieser, R.J. and F. Oesch. 1986. Contact inhibition of growth of human diploid fibroblasts by immobilized plasma membrane glycoproteins. *J. Cell Biol.* **103:** 361.

Withers, D.A., R.C. Harvey, J.B. Faust, O. Melnyk, K. Carey, and T.C. Meeker. 1991. Characterization of a candidate *bcl*-1 gene. *Mol. Cell. Biol.* **11:** 4846.

Xiong, Y., G.J. Hannon, H. Zhang, D. Casso, R. Kobayashi, and D. Beach. 1993. p21 is a universal inhibitor of cyclin kinases. *Nature* **366:** 701.

Yamasaki, H. 1990. Role of cell-cell communication in tumor suppression. *Immunol. Ser.* **51:** 245.

D-Type Cyclins and Their Cyclin-dependent Kinases: G_1 Phase Integrators of the Mitogenic Response

C.J. SHERR,*† J. KATO,† D.E. QUELLE,*† M. MATSUOKA,*† AND M.F. ROUSSEL†
*Howard Hughes Medical Institute, †Department of Tumor Cell Biology,
St. Jude Children's Research Hospital, Memphis, Tennessee 38105

The processes of DNA replication (S phase) and mitosis (M phase) must be properly coordinated to ensure that the complement of genetic material in the proliferating somatic cells of an organism remains constant. Genes that control the orderly progression of S phase to M phase are vital for cell division, and their encoded products likely represent the core building blocks of the cell cycle machinery. Examples of proteins that play roles in these processes include the $p34^{cdc2/CDC28}$ protein kinase, its regulatory subunit cyclin B, and the kinases (e.g., wee1/mik1) and phosphatases (e.g., cdc25) that posttranslationally regulate $p34^{cdc2}$/cyclin B kinase activity to control mitotic entry and exit in all eukaryotic cells (Norbury and Nurse 1992). Superimposed on the basic replicative and mitotic apparati are checkpoint controllers, which guarantee that an essential process is completed before another begins by resetting the cell cycle clock when certain conditions are improperly met (Hartwell and Weinert 1989; Murray 1992). Checkpoint controls are exerted at various points in the cycle and involve processes such as those that ensure the completion of DNA synthesis before the onset of mitosis and those that monitor the correct assembly of the mitotic spindle before actual cell division occurs.

In metazoans, early embryonic cell divisions are represented by repeated S and M phases without intervening gaps. These divisions can occur with great rapidity, because the necessary building blocks of the cell cycle machine are maternally derived and not initially rate-limiting. However, subsequent cell growth and organismal development depend on zygotic transcription, and the somatic cell cycles which ensue include two gap phases, G_1 and G_2, which separate M from S phase and S from M phase, respectively. During development, progression through G_1 comes under the control of a complex array of extracellular inducers, mitogens, and growth inhibitory molecules which act in concert to determine whether cells divide, differentiate, or undergo apoptosis. Hence, G_1 progression in higher eukaryotes requires an elaborate sensing mechanism that couples extracellular signals with the cell cycle clock, and here, the rate-limiting biochemical integrators are represented, at least in part, by multiple G_1 cyclins and a host of cyclin-dependent kinases (cdks) related to, but distinct from, $p34^{cdc2/CDC28}$ itself (Fig. 1).

The D- and E-type G_1 cyclins and their cdks are now recognized to be key regulators of G_1 to S phase progression in mammalian cells (Reed 1992; Sherr 1993). Overexpression of cyclins D_1 or E in fibroblasts shortens their G_1 interval, decreases cell size, and renders the cells less dependent on mitogens for S-phase entry (Jiang et al. 1993; Ohtsubo and Roberts 1993; Quelle et al. 1993; Resnitzky et al. 1994), indicating that both classes of G_1 cyclin can be rate-limiting for the cell's entry into S phase. Recent data indicate that these effects on the cell cycle are additive, implying that D- and E-type cyclins function in parallel pathways (D. Resnitzky and S. Reed, pers. comm.). Microinjection of antisense plasmids or monoclonal antibodies to cyclin D_1 into serum-stimulated fibroblasts during early to mid G_1 phase prevents cellular DNA synthesis, but these manipulations are without effect once the cells have neared the G_1/S phase transition (Baldin et al. 1993; Quelle et al. 1993), implying that cyclin D_1 executes its necessary rate-limiting function in mid to late G_1 phase and is not required thereafter. The assembly of the catalytic (cdk) and regulatory (G_1 cyclin) subunits into functional enzymes depends on mitogenic

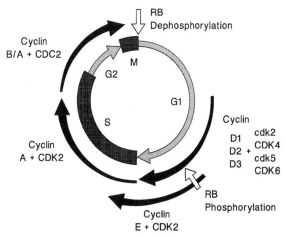

Figure 1. Combinatorial interactions of cyclins and cdks during the cell cycle. The D-type cyclins have been detected in complexes with cdk2, cdk4, cdk5, and cdk6 in various cell types, but cdk4 and cdk6 appear to be their major catalytic partners. Cyclins E and A interact with cdk2, and cyclins A and B form complexes with cdc2 (cdk1) later in the cycle. The timing of initial pRb hyperphosphorylation occurs late in G_1 but prior to the G_1/S transition. (Adapted, with permission, from Sherr 1993 [copyright Cell Press].)

signals required throughout G_1 phase, and conversely, many inhibitors of cell growth operate to prevent the activation of these enzymes and the cell's entry into S phase. Thus, G_1 cyclins and their associated catalytic subunits are integrators of both positive and negative growth regulatory signals and translate these inputs into critical phosphorylation events that ultimately trigger the onset of DNA replication.

By determining the readiness of cells to initiate DNA replication, the D-type cyclins function, at least in part, in G_1 checkpoint control. Although some such actions might well prove to be dispensable to the workings of the core cell cycle machinery, their disruption would be expected to predispose cells to malignancy. Like proto-oncogenes that encode proteins strategically positioned along mitogen-activated signal transduction pathways, G_1 cyclins and their cdk partners are indeed now recognized to be targets of genetic perturbations which contribute to human cancer (Motokura and Arnold 1993; Marx 1994). Conversely, their functions in the cell cycle are intimately tied to the actions of known tumor suppressor genes, such as the retinoblastoma gene (*RB*) and *p53*, whose functional elimination disrupts G_1 checkpoint controls.

RESULTS AND DISCUSSION

D-Type Cyclin-dependent Kinases Are Growth Factor Sensors

Three D-type cyclins, first discovered via three independent investigational routes (Matsushime et al. 1991; Motokura et al. 1991; Xiong et al. 1991), are combinatorially expressed in a cell lineage-specific manner in proliferating mammalian cells. Although D-type cyclins can be immunoprecipitated from cell lysates in complexes with cdks 2, 4, 5, and 6 (Matsushime et al. 1992; Xiong et al. 1992; Meyerson and Harlow 1994; Bates et al. 1994b), cdk4 and cdk6 are their major catalytic partners and, in combination with D-type cyclin, have a distinct substrate preference in vitro for the retinoblastoma protein (pRb) versus histone H1 (Matsushime et al. 1992, 1994; Meyerson and Harlow 1994). Each of the D-type cyclins can assemble into active holoenzymes with cdk4 or cdk6 when reconstituted in insect Sf9 cells coinfected with appropriate baculovirus vectors, but neither of these two catalytic subunits is activated by cyclins A, B, C, or E (Kato et al. 1993). In the same assay system, cdk2 also productively interacts with cyclins D_2 and D_3, but not D_1 (Ewen et al. 1993b), and cdk5 fails to form active enzyme complexes with any of the three D-type cyclins (J.-Y. Kato and C.J. Sherr, unpubl.). Among the known cdks, the predicted amino acid sequences of cdk4 and cdk6 are more nearly identical to one another than either is to cdc2/CDC28 (cdk1), cdk2, and cdk3 (Meyerson et al. 1992). The latter enzymes can complement conditionally defective CDC28 mutants in yeast cells, whereas cdk4 and cdk6 cannot; in the same vein, cyclins E, A, and even mitotic cyclin B complement yeast G_1 cyclins (Clns) far more efficiently than the D-type regulatory subunits (Koff et al. 1991; Lew et al. 1991; Xiong et al. 1991). Together, these results imply that cyclin D/cdk complexes play specialized roles in regulating the cell cycles of higher eukaryotes.

In mouse macrophages and fibroblasts, complexes between cyclin D_1 and cdk4 represent the major D-type cyclin-dependent kinase assembled during G_1 phase (Matsushime et al. 1994). Indeed, cyclin D_1 was first cloned from mouse bone-marrow-derived macrophages stimulated to enter the cell cycle with colony-stimulating factor-1 (CSF-1), by use of a subtractive hybridization screen designed to isolate delayed early response genes (Matsushime et al. 1991). In these cells, CSF-1 deprivation results in the arrest of the cell cycle in early G_1. When starved macrophages are restimulated with the growth factor, they enter the cycle synchronously, and cyclin D_1 mRNA synthesis is induced within the first 2 hours after stimulation. The cdk4 protein is detected at relatively low levels in cells deprived of CSF-1, and the "free" form of the D_1 protein accumulates in excess in the first few hours after growth factor stimulation (Matsushime et al. 1992). cdk4 is induced later in mid G_1 phase, so that functional holoenzyme complexes between the two subunits assemble more rapidly and accumulate as cells move into late G_1 phase (Matsushime et al. 1991, 1992, 1994). The half-life of D-type cyclins, whether free or bound to cdk4, is less than 20 minutes (Matsushime et al. 1992), so that removal of CSF-1 before late G_1 leads to the rapid disappearance of cyclin D/cdk4 complexes and correlates with the failure of the cells to enter S phase; however, once cells have passed the G_1/S boundary, removal of CSF-1 and cyclin D degradation no longer affect the ability of these cells to complete S phase and divide (Matsushime et al. 1991). Again, these results are consistent with the notion that cyclin D/cdk4 function is required prior to, but not after, the G_1/S transition.

Similar kinetics of activation and assembly of the cyclin D_1/cdk4 complex occur in serum-stimulated fibroblasts (Won et al. 1992; Winston and Pledger 1993; Matsushime et al. 1994). Likewise, in peripheral blood T lymphocytes, signals through the antigen receptor trigger the synthesis of cyclin D_2, and cyclin D_3 is induced somewhat later in G_1 phase in response to interleukin-2 (Ajchenbaum et al. 1993). In T cells, both subunits assemble preferentially with cdk6, whose overall kinetics of activation mirror those observed for cdk4 in macrophages and fibroblasts (Meyerson and Harlow 1994). In each of these systems, D-type cyclin-dependent kinase activity is first detected in mid G_1 and rises progressively until soon after cells enter S phase (Matsushime et al. 1994; Meyerson and Harlow 1994). The appearance of the active kinases precedes that of cyclin E/cdk2, which is expressed periodically (Koff et al. 1991; Lew et al. 1991) and is maximal at the G_1/S transition (Fig. 1) (Dulic et al. 1992; Koff et al. 1992).

A clear functional link between growth factor signaling and cyclin D function has been recently established using partially defective CSF-1 receptor (CSF-1R) mutants. Introduction of the human CSF-1 receptor into mouse NIH-3T3 fibroblasts enables them to proliferate at the expense of serum-derived growth factors in response to human recombinant CSF-1 (Roussel et al. 1987; Roussel and Sherr 1989). A single substitution of Tyr-809 (a site of ligand-dependent autophosphorylation) by phenylalanine in the human CSF-1 receptor abolishes its ligand-induced mitogenic activity in this system, and cells expressing the mutant receptor remain arrested early in G_1 phase after CSF-1 treatment. Ligand stimulation of CSF-1R 809F activates the receptor tyrosine kinase and induces the expression of immediate early response genes of the *fos* and *jun* families, but c-*myc* is poorly induced, and its mRNA levels are not properly sustained after CSF-1 stimulation (Roussel et al. 1990). Ectopic expression of c-*myc* in these cells resensitizes them to the mitogenic effects of CSF-1, indicating that receptor signals must minimally bifurcate to stimulate *fos/jun* on the one hand and c-*myc* on the other and that activation of the latter is necessary for CSF-1-induced mitogenicity (Roussel et al. 1991). We recently found that any of the three D-type cyclins can reconstitute CSF-1 responsiveness in cells expressing CSF-1R 809F, even though c-*myc* mRNA levels in response to CSF-1 remain low and are poorly sustained in the proliferating cells (M.F. Roussel et al., unpubl.). Therefore, D-type cyclins can either bypass c-*myc* in a signaling pathway required for G_1 progression or, alternatively, their overproduction can potentiate the biologic effects of subthreshold c-*myc* levels. Whatever the interpretation, these data functionally link receptor-mediated signaling and c-*myc* expression with cyclin D activity during G_1 progression. Interbreeding of mice that express cyclin D_1 and c-*myc* transgenes under the control of an immunoglobulin heavy-chain enhancer significantly accelerates the age-dependent onset of c-*myc*-induced B-cell lymphoma, revealing synergy between the two genes in an in vivo setting (Bodrug et al. 1994).

Posttranslational Regulation of Cyclin D/cdk Activity

Although transcription of the D-type cyclin and cdk genes is regulated by mitogens, formation of the active holoenzymes is subject to additional posttranslational controls, at least some of which are growth factor dependent. In fibroblasts engineered to ectopically express both a D-type cyclin and cdk4, constitutively synthesized subunits did not coassemble efficiently in serum-starved cells, and no active D-type cyclin-dependent kinase activity was detected. Serum stimulation facilitated the assembly of the ectopically produced subunits, and because the transcriptional requirement for their synthesis was bypassed, assembly occurred relatively rapidly and led to a greatly increased level of cyclin-D-dependent pRb kinase activity as compared to that detected in normal cells (Matsushime et al. 1994).

We presume that the ability to more rapidly generate threshold levels of cyclin-D-dependent kinase is in turn responsible for the shortening of G_1 phase and the relaxed serum requirement observed for cells over-producing D-type cyclin (Jiang et al. 1993; Quelle et al. 1993; Resnitzky et al. 1994). However, we do not as yet understand why cyclin D/cdk4 assembly does not occur in the absence of a mitogenic signal and have suggested that some "assembly factor" must be necessary to mediate the process (Matsushime et al. 1994). We use this term in its most general sense: The factor could be a third protein necessary for complex formation per se or a regulator that posttranslationally modifies either the cyclin or cdk to alter their affinities for one another or their locations within the cell.

Another "upstream" activity necessary for the formation of functional cyclin D/cdk4 complexes is a cdk-activating kinase (CAK). When cdk4 is bound to a D-type cyclin, it undergoes phosphorylation on a single threonine residue (Thr-172), and this modification is required for activation of the holoenzyme (Kato et al. 1994a). Other cdks, including cdc2 and cdk2, do not require CAK to assemble with cyclins, but they must undergo phosphorylation by CAK to become active (Desai et al. 1992; Solomon et al. 1992). It remains unclear whether the phosphorylation of cdk4 on Thr-172 is mediated by cdc2/cdk2 CAK or by a related but distinct enzyme. CAK for cdc2/cdk2 is itself a multisubunit enzyme composed of a catalytic subunit (designated MO15) (Fesquet et al. 1993; Poon et al. 1993; Solomon et al. 1993) and a regulatory subunit now designated cyclin H (Fisher and Morgan 1994). We have cloned mouse MO15, and specific antisera raised to the recombinant protein can both deplete cdk4 CAK activity from mammalian cell lysates and precipitate an active enzyme (Matsuoka et al. 1994). This suggests that cdk4 CAK contains either MO15 itself or an antigenically related catalytic subunit.

MO15 protein levels remain constant and cdk4 CAK activity is readily detected in lysates of murine macrophages or fibroblasts throughout the cell cycle (Matsuoka et al. 1994), but its access to cyclin D/cdk4 complexes may be subject to negative regulation. For example, addition of dibutyryl-cAMP or cAMP inducers, such as prostaglandin E_2, to mouse macrophages stimulated with CSF-1 to enter the cell cycle from quiescence blocks cell cycle progression in mid G_1 (Jackowski et al. 1990; Vairo et al. 1990). Cyclin D_1/cdk4 complexes assemble in cAMP-treated cells, but the catalytic subunit lacks detectable Thr-172 phosphate, and the complex is inactive (Kato et al. 1994b). We were unable to activate recombinant cyclin D/cdk4 or cyclin A/cdk2 complexes in vitro using lysates from cAMP-treated G_1-arrested cells, but surprisingly, active CAK could be precipitated from such extracts using antiserum to MO15. Therefore, certain inhibitor(s) of cyclin D/cdk4 complexes (see below) which accumulate in cAMP-arrested cells can prevent CAK from phosphorylating cyclin-bound cdk4 and cdk2. Growth arrest of mink lung epithelial cells by transforming growth factor-β (TGF-β) inhibits the activity of cyclin E/cdk2 at least in part by interfering with CAK-induced phosphorylation of the cyclin-bound catalytic subunit (Koff et al. 1993; Slingerland et al. 1994).

Polypeptide Inhibitors of G_1 Progression

Whereas CAK and a putative assembly factor represent positive controllers of cyclin-D-dependent kinase activity, a group of recently discovered polypeptide inhibitors function to extinguish holoenzyme activity under particular circumstances. A 21-kD protein (p21, alias Cip1, WAF1, Sdi1) acts as a p53-inducible inhibitor of cyclin A/cdk2 or cyclin E/cdk2 and cyclin D/cdk4 complexes to invoke cell cycle arrest in response to ionizing radiation (El-Deiry et al. 1993; Gu et al. 1993; Harper et al. 1993; Xiong et al. 1993b; Dulic et al. 1994; Noda et al. 1994). The p21 protein was first observed in quaternary complexes containing cyclin/cdks and proliferating cell nuclear antigen (PCNA, a subunit of DNA polymerase-δ), and these are formed in normal cells but are disrupted in tumor cells (Xiong et al. 1992, 1993a). Interestingly, p21 copurifies with active cyclin/cdk complexes and only inhibits their kinase activity when bound to them at high stoichiometry (Zhang et al. 1994), suggesting that it is a multifunctional regulator. One speculation is that p21 or other proteins in this family (see below) might act as "assembly factors" for binary cyclin/cdk complexes but, in higher proportions, might inhibit their enzyme activities. The *p21* gene is positively regulated through the transcriptional activity of the *p53* tumor suppressor gene, and the protein likely contributes to the G_1 arrest induced by high p53 levels (El-Deiry et al. 1993, 1994; Xiong et al. 1993a; Dulic et al. 1994), which accumulate in response to ionizing radiation (Kastan et al. 1991; Kuerbitz et al. 1992). In agreement, ectopic overexpression of the gene encoding p21 in human brain, lung, and colon cancer cell lines that lack functional *p53* caused substantial growth suppression, similar to that observed after introduction of the *p53* gene itself (El-Deiry et al. 1993).

Another cdk inhibitory protein, p27^{Kip1}, accumulates in an active form in cells arrested by contact inhibition or after treatment of susceptible epithelial cells with TGF-β (Polyak et al. 1994a; Slingerland et al. 1994). Kip1 inhibitory activity is detected at a threshold level in lysates of arrested cells, but not in those from proliferating cells. However, boiling extracts from proliferating cells releases heat-stable Kip1 from latent stores, implying that the protein may be present throughout the cell cycle, either in a sequestered (bound) or active (free) form. Kip1 inhibits the histone H1 kinase activity of preformed cyclin E/cdk2 complexes, but addition of cyclin D/cdk4 complexes to inhibitory lysates from TGF-β-treated mammalian cells stoichiometrically titers out Kip1, enabling subsequently added cyclin E/cdk2 complexes to retain H1 kinase activity (Polyak et al. 1994a). Because preformed cyclin D/cdk4 complexes containing mutationally inactivated cdk4 subunits were equally active in this in vitro competition assay, it was reasoned that the accumulation of cyclin D/cdk4 complexes during G_1 phase might similarly titrate threshold levels of Kip1 in normal cells, thereby enabling cyclin E/cdk2 to be activated later in the cycle (Fig. 2). Consistent with this hypothesis were data indicating that TGF-β inhibits both the synthesis of cdk4 (Ewen et al. 1993a) and the activity of the assembled cyclin E/cdk2 holoenzyme in mammalian cells (Koff et al. 1993); enforced ectopic expression of cdk4, but not cdk2, and reformation of cyclin D/cdk4 complexes renders such cells resistant to TGF-β and leads to reactivation of cyclin E/cdk2 enzyme activity (Ewen et al. 1993a).

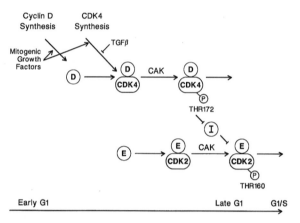

Figure 2. Posttranslational regulation of cyclin/cdk complexes during G_1 phase. Cyclin D and cdk4 synthesis are transcriptionally induced by mitogenic growth factors, but their assembly into functional holoenzymes also requires mitogenic stimulation. Cdk-activating kinase activity (CAK) is required to activate both the cyclin D- and E-dependent kinases by phosphorylating the bound cdks on a single threonyl residue (Desai and Morgan 1992; Solomon et al. 1992; Kato et al. 1994). TGF-β inhibits cdk4 protein synthesis, resulting in the collapse of cyclin D/cdk4 complexes (Ewen et al. 1993a) and inhibition of the activity of the preformed cyclin E/cdk2 complex (Koff et al. 1993). The TGF-β block can be overridden by ectopic expression of cdk4, but not cdk2 (Ewen et al. 1993a). Threshold levels of two potential cdk4 and cdk2 inhibitors (p21^{Cip1} and p27^{Kip1}, denoted by the capital I in the figure) are found in higher-order complexes with the G_1 cyclins and their cdk partners in normal cycling cells, and their inhibitory effects may be determined by their stoichiometry in the complexes (Xiong et al. 1993a; Polyak et al. 1994a,b; Toyoshima and Hunter 1994; Zhang et al. 1994). TGF-β treatment is proposed to shift p27^{Kip1}, which may normally be "titrated" by cyclin D/cdk4, into complexes with cyclin E/cdk2, thereby inhibiting its kinase activity. During the normal cycle, titration of p27^{Kip1} by cyclin D/cdk4 might help set the order of "firing" of cyclin D- and cyclin E-dependent kinases. p21^{Cip1} levels increase in response to p53-mediated induction (El-Deiry et al. 1993; Xiong et al. 1993a; Dulic et al. 1994). Inducers of cAMP induce mid G_1 arrest, at least in part by inhibiting the activity of CAK. This growth inhibitory effect of cAMP is not directly mediated through CAK itself, but through inducible activities (most likely, the cdk inhibitors) which, apart from inhibiting active cyclin/cdk complexes, can block the access of CAK to the cyclin-bound cdks (Kato et al. 1994b). (Adapted, with permission, from Sherr 1994 [Elsevier Trends Journals].)

The amino acid sequences of portions of purified Kip1 were recently determined and used to clone the gene (Polyak et al. 1994b). Independently, other workers obtained the same cDNA in a two-hybrid screen which employed cyclin D_1 as the bait (Toyoshima and Hunter 1994). $p27^{Kip1}$ and $p21^{Cip1}$ share related amino acid sequences in their amino-terminal moieties, and the conserved region alone can inhibit cyclin D/cdk4, E/cdk2, and A/cdk2 complexes assembled in Sf9 extracts (Polyak et al. 1994a,b; Toyoshima and Hunter 1994). As for *p21*, *p27* overexpression in mammalian cells induces G_1 phase arrest. Immunoblotting analyses have suggested that total *Kip1* levels are invariant during the cell cycle, and that a major proportion of the protein forms complexes with cyclin D/cdk4 in proliferating cells (Toyoshima and Hunter 1994). Therefore, cyclin D/cdk4 complexes appear to serve at least two functions. Stoichiometrically, they titer Kip1 activity and alleviate cyclin E/cdk2 from a potential inhibitory constraint, establishing an order of activation of the two kinases during G_1. When the Kip1 threshold is exceeded by the level of cyclin D/cdk4 complexes, they can now function catalytically to phosphorylate substrates whose modifications are necessary for G_1 exit. Unlike results obtained with wild-type cdk4, the ectopic overexpression of a catalytically inactivated cdk4 subunit does not override a TGF-β block (M. Ewen, pers. comm.), implying that both the stoichiometric and catalytic functions of cyclin D/cdk4 complexes are important for G_1 progression in normal cells.

An entirely different class of inhibitors is represented by "p16" (ink4/MTS1), a polypeptide composed of four loosely conserved ankyrin motifs (Serrano et al. 1993). This 16-kD protein is not normally detected in cdk complexes, but it is found in association with cdk4, at the expense of cyclin D, in *RB*-negative tumor cells, including those transformed by pRb-binding oncoproteins (Serrano et al. 1993; Xiong et al. 1993a). $p16^{ink4}$ is a specific *in*hibitor of the cyclin D/cd*k4* holoenzyme, and unlike $p21^{Cip1}$ and $p27^{Kip1}$, it has no apparent activity on cdk2 in complexes with cyclins E or A. Because cyclin D/cdk4 complexes are disrupted in *RB*-negative cells (Lukas et al. 1994; Tam et al. 1994), and cdk4 accumulates in an inactive form with $p16^{ink4}$, such cells appear to no longer require cdk4 kinase activity for G_1 progression. In agreement, microinjection of monoclonal antibodies to cyclin D_1 into *RB*-negative tumor cells does not prevent their entry into S phase, whereas *RB*-positive cells (whether *p53*-positive or -negative) remain sensitive (Tam et al. 1994). Recent reports suggest that *ink4/MTS1* may act as a tumor suppressor gene, because the locus on human chromosome 9p21 is deleted or mutated in a very high percentage of tumor cell lines (Kamb et al. 1994; Noburi et al. 1994). Given the interplay between pRb and $p16^{ink4}$, it would be interesting to know whether pRb plays any role in regulating the *ink4/MTS1* gene (see below) or whether genetic inactivation of the two tumor suppressors are usually mutually exclusive events in cell transformation.

Functional Linkage between Cyclin D/cdk4 and pRb

The product of the *RB* gene has properties of a cell cycle oscillator in the sense that it is initially phosphorylated in mid to late G_1 phase, accumulates phosphate later in the cycle, and is dephosphorylated as cells exit mitosis (Ludlow et al. 1990). DNA oncoproteins, such as adenovirus E1A, SV40 T antigen, and human papillomavirus E7, bind and inactivate pRb's growth-suppressive function (DeCaprio et al. 1988; Whyte et al. 1988; Dyson et al. 1989), and because they associate only with the G_1-specific hypophosphorylated forms of pRb, it was predicted that only the latter species were active in growth suppression (Ludlow et al. 1990). In normal cells, phosphorylation of pRb during G_1 would analogously serve to inactivate this growth-suppressive function, thereby facilitating G_1 exit. As predicted, pRb can block G_1 exit when introduced into certain *RB*-negative human tumor cell lines, but it is inactive as a growth suppressor later in the cycle (Goodrich et al. 1991). pRb phosphorylation in G_1 phase occurs at multiple cdk sites (Lees et al. 1991; Lin et al. 1991; Hu et al. 1992), implying that the process is mediated by one or more G_1 cyclin-dependent kinases expressed during this interval (Fig. 3). The introduction of *RB* into *RB*-negative human Saos-2 cells induces G_1 arrest, but cotransfection of *RB* with cyclins D, E, or A (but not B) can induce pRb phosphorylation and cancel its inhibitory effects on G_1 exit (Hinds et al. 1992; Ewen et al. 1993b). Hypophosphorylated pRb binds to transcription factors, such as E2F, and prevents their activities on certain target genes. pRb phosphorylation releases E2F from this inhibitory constraint converting it to a transcriptional activator, and this may be critical

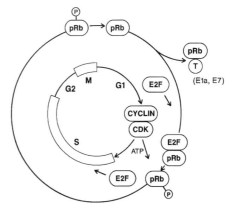

Figure 3. G_1 cyclin/cdk complexes can regulate G_1 exit by phosphorylating pRb. pRb undergoes phosphorylation in mid to late G_1 by cyclin-dependent kinases. Both cyclin D- and E-dependent enzymes may contribute to this process, possibly in defined sequence, as cells approach the G_1/S boundary. pRb can also be inactivated by binding of oncoproteins such as T antigen, E1A, and E7. pRb inactivation by oncoprotein binding or cdk-mediated phosphorylation releases pRb-tethered transcription factors, such as E2F, which are thought to change patterns of cellular transcription from a G_1 to S phase mode. (Adapted, with permission, from Sherr 1994 [Elsevier Trends Journals].)

in switching cells from a G_1-specific to an S-phase transcriptional state (Fig. 3) (Nevins 1992). Intriguingly, the introduction of E2F into quiescent, serum-deprived cells can induce a significant proportion of them to replicate DNA (Johnson et al. 1993).

Several lines of evidence suggest that the initial pRb phosphorylations observed during mid to late G_1 phase might be catalyzed by the cyclin-D-dependent kinases (Sherr 1994). First, unlike the results previously obtained with other cyclin/cdk complexes, pRb, and not histone H1, is a highly preferred substrate for the D-type cyclin-dependent kinases in vitro (Matsushime et al. 1992, 1994; Meyerson and Harlow 1994). Second, the timing of activation of cyclin-D-dependent kinase activity during G_1 corresponds to the time at which pRb is initially phosphorylated and occurs before cyclin E/cdk2 activity is manifest (DeCaprio et al. 1989; Ludlow et al. 1990). In fibroblasts engineered to express inducible D_1 and E cyclins, the induction of cyclin D_1 during G_1 phase accelerates pRb phosphorylation, whereas cyclin E induction has no immediate effect (D. Resnitzky and S. Reed, pers. comm.). Third, the D-type cyclins can bind to hypophosphorylated forms of pRb directly, and they contain an amino-terminal pentapeptide sequence (Leu-X-Cys-X-Glu) not found in other cyclins but shared with the pRb-binding oncoproteins of DNA tumor viruses, which helps to mediate their physical association with pRb (Dowdy et al. 1993; Ewen et al. 1993b; Kato et al. 1993). When baculoviruses encoding D-type cyclin and pRb are coexpressed in insect Sf9 cells, stable complexes between the two proteins are formed in a near 1:1 molar ratio (Kato et al. 1993). Coinfection of these cells with a third virus encoding cdk4 led to pRb phosphorylation at physiologically relevant cdk sites and to the destabilization of binary complexes between pRb and the cyclins. However, coinfection with a virus encoding catalytically inactive cdk4 enabled stable ternary complexes between the three proteins to be formed. We therefore reasoned that one property of D-type cyclins was to physically target pRb (and possibly other pRb-related proteins) for phosphorylation by cdk4 (or cdk6). The ability of pRb to undergo phosphorylation by D-type cyclin-dependent kinases does not preclude the possibility that pRb is progressively phosphorylated by other cdks as cells traverse the G_1/S transition and, indeed, it would be interesting to determine if such phosphorylations occur in a defined sequence and are interdependent.

Although the introduction of D-type cyclin expression vectors together with RB into RB-negative human Saos-2 osteosarcoma cells was able to override pRb-induced growth arrest (Dowdy et al. 1993; Ewen et al. 1993b), cyclin D_1 mutants disrupted in their LXCXE motif were reported to be more effective in cancelling pRb's ability to prevent S-phase entry (Dowdy et al. 1993). This suggested an alternative model; namely, that D-type cyclins might be subject to pRb sequestration and that pRb phosphorylation by another cdk might release D-type cyclin from its pRb tether, enabling it to subsequently combine with cdks to phosphorylate "downstream" substrates. Although this idea is intriguing, it rests on the ability of mutant D-type cyclin to function better than the wild-type protein in such assays. In contrast, we recently found that D-type cyclins disrupted in their LXCXE pentapeptides and thus disabled in pRb binding are less active than their wild-type counterparts in rescuing CSF-1 responsiveness in cells bearing a receptor mutant (CSF-1R 809F) that is partially defective in signal transduction (see above). Moreover, overexpression of an analogous mutant D-type cyclin in a fibroblast cotransformation assay performed with ras and mutant E1A has not so far revealed potentiating activity (Hinds et al. 1994). If pRb acts to sequester D-type cyclins, one would expect that stable cyclin D/pRb complexes would accumulate in cells during G_1 phase, but these have at best been difficult to detect in vivo (Dowdy et al. 1993; Ewen et al. 1993b). In contrast, if the role of D-type cyclins is to target pRb phosphorylation, the putative complexes would be unstable unless cdk4 were functionally inactivated (Kato et al. 1993). The latter conditions are realized in vivo in murine erythroleukemia cells in which cdk4 synthesis is inhibited by inducers of cell differentiation and where cyclin D moves into stable complexes with pRb (Kiyokawa et al. 1994).

Disruption of pRb function in tumor cells does not result in increased complex formation between D-type cyclins and their cdk partners, but instead paradoxically leads to a loss of cyclin D binding to either cdk4 or cdk6 (Serrano et al. 1993; Xiong et al. 1993a; Bates et al. 1994a; Lukas et al. 1994; Tam et al. 1994). The turnover of "free" cyclin D is faster than that of its cdk-bound forms, and cyclin D levels fall in RB-negative tumor cells (Bates et al. 1994a; Lukas et al. 1994; Tam et al. 1994). Under such circumstances, cdk4 and cdk6 move into complexes with $p16^{ink4}$ (Xiong et al. 1993a; Serrano et al. 1993), and microinjection of antibodies to cyclin D_1 into such cells cannot prevent S-phase entry (Tam et al. 1994). These findings led Serrano and coworkers (1993) to speculate that pRb (and/or proteins in the pRb pathway) might be the only crucial substrates of D-type cyclin kinases during G_1. We would like to know whether similar data would be obtained using cells from RB "knock-out" mice, or whether other genetic perturbations characteristic of tumor cells contribute to the above phenotype. An idea consistent with the currently available data might be that the $p16^{ink4}$ gene is itself regulated by pRb-tethered transcription factors, so that its expression is increased as pRb is phosphorylated and cells exit G_1 phase (Fig. 4). Hypothetically, $p16^{ink4}$ might function as a feedback controller that acts to extinguish cdk4 and cdk6 kinase activity after pRb's growth-suppressive function is cancelled. pRb-negative tumor cells would therefore synthesize $p16^{ink4}$ constitutively, maintain cdk4/$p16^{ink4}$ complexes throughout the cycle, exhibit uncharacteristically low levels of D-cyclin expression, and be refractory to G_1 arrest induced by microinjection of anti-D-cyclin antibodies. At face value, such a hypoth-

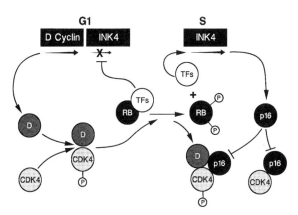

Figure 4. What is the role of p16^{ink4}? It seems counterintuitive that cdk4, a kinase seemingly required for G$_1$ progression, is sequestered by an inhibitor in tumor cells lacking functional pRb. One possibility is that *ink4* transcription is normally up-regulated in S phase in response to pRb phosphorylation and the subsequent release of pRb-bound transcription factors, such as E2F. p16 would then act to reduce cyclin D/cdk4 kinase activity as cells enter S phase. In *RB*-negative cells, *ink4* expression might then be constitutive, resulting in the formation of inactive cdk4/p16 complexes. Conversely, in cultured cells that have sustained *ink4* deletions, the activity of the cyclin D/cdk kinase would not be expected to oscillate during the cell cycle. At this writing, the model is purely hypothetical, but it makes some interesting and readily testable predictions, the most important being that cyclin D/cdk4 might be dispensable for G$_1$ progression in *RB*-negative cells.

esis contradicts direct measurements of cyclin D/cdk4 kinase activity, which remains elevated during S phase in rodent fibroblast and macrophage cell lines (Matsushime et al. 1994). However, *Ink4/MTS1* is deleted or mutated in many such cell lines (Kamb et al. 1994; Noburi et al. 1994), and measurements of cdk4 kinase activity during the cell cycle need to be carried out with synchronized cell strains that retain p16^{ink4} function. It will therefore be important to determine whether p16^{ink4} is induced as a G$_1$ checkpoint controller or is expressed periodically during the normal cell cycle.

Whatever the mechanistic role of p16^{ink4}, the behavior of cyclin D/cdk4 complexes in pRb-negative tumor cells clearly differs from that observed in cells that maintain pRb-mediated G$_1$ phase controls. Experiments which were designed to test whether cointroduced D-type cyclins could override the growth-suppressive effects of transiently expressed *RB* alleles in human *RB*-negative Saos-2 osteosarcoma cells may therefore require reinterpretation. Ectopically expressed cyclins A and E were significantly more effective than cyclin D$_1$ in overriding pRb-induced G$_1$ arrest (Hinds et al. 1992). However, in Saos-2 cells, cdk4 would presumably be inactivated through its association with p16^{ink4} and would be unavailable to interact with D cyclins. The ability of a cyclin to cancel the growth-suppressive effects of a transfected wild-type *RB* gene in such cells should then depend on its capacity to associate with another cdk partner, such as cdk2. Cyclin D$_1$ is unable to form functional complexes with this catalytic subunit when the two proteins are coexpressed at high levels in insect cells. In contrast, cyclin D$_2$ was more effective than D$_1$ in overriding pRb-induced G$_1$ arrest in Saos-2 cells, and it retains the ability to functionally interact with cdk2 (Ewen et al. 1993b). In a complementary series of experiments, cyclins D$_2$ and D$_3$ were able to inhibit granulocyte differentiation when introduced into myeloid blasts by forming complexes with cdk2, whereas cyclin D$_1$ or cyclin D$_2$ mutants disrupted in their LXCXE motifs were inactive (Kato and Sherr 1993). In retrospect, because of the frequency of p16^{ink4} inactivation in cultured cells and the ability of ectopically overexpressed D-type cyclins to be driven into complexes with cdk2, such experiments may be of marginal value in pinpointing the cyclin/cdk complexes that physiologically phosphorylate pRb. Nonetheless, it seems reasonable to propose a role of D-type cyclins in regulating G$_1$ checkpoint control in response to environmental signals, which is mediated, at least in part, through cancellation of pRb's growth-suppressive function.

ACKNOWLEDGMENTS

We thank R. Fisher, D. Morgan, D. Resnitzky, S. Reed, H. Kiyokawa, M. Ewen, J. Massagué, H. Toyoshima, and T. Hunter for communicating unpublished data. This work was supported in part by National Institutes of Health grant CA-47064 to C.J.S., by Cancer Center Core Grant CA-21765, and by the American Lebanese Syrian Associated Charities (ALSAC) of St. Jude Children's Research Hospital.

REFERENCES

Ajchenbaum, F., K. Ando, J.A. DeCaprio, and J.D. Griffin. 1993. Independent regulation of human D-type cyclin gene expression during G1 phase in primary human T lymphocytes. *J. Biol. Chem.* **268:** 4113.

Baldin, V., J. Lukas, M.J. Marcote, M. Pagano, and G. Draetta. 1993. Cyclin D1 is a nuclear protein required for cell cycle progression in G1. *Genes Dev.* **7:** 812.

Bates, S., D. Parry, L. Bonetta, K. Vousden, C. Dickson, and G. Peters. 1994a. Absence of cyclin D/cdk complexes in cells lacking functional retinoblastoma protein. *Oncogene* **9:** 1633.

Bates, S., L. Bonetta, D. Macallan, D. Parry, A. Holder, C. Dickson, and G. Peters. 1994b. CDK6 (PLSTIRE) and CDK4 (PSK-J3) are a distinct subset of the cyclin-dependent kinases that associate with cyclin D1. *Oncogene* **9:** 71.

Bodrug, S.E., B.J. Warner, M.L. Bath, G.J. Lindeman, A.W. Harris, and J.M. Adams. 1994. Cyclin D1 transgene impedes lymphocyte maturation and collaborates in lymphomagenesis with the *myc* gene. *EMBO J.* **13:** 2124.

DeCaprio, J.A., J.W. Ludlow, D. Lynch, Y. Furukawa, J. Griffin, H. Piwnica-Worms, C. Huang, and D.M. Livingston. 1989. The product of the retinoblastoma susceptibility gene has properties of a cell cycle regulatory element. *Cell* **58:** 1085.

DeCaprio, J.A., J.W. Ludlow, J. Figge, J. Shew, C. Huang, W. Lee, E. Marsilio, E. Paucha, and D.M. Livingston. 1988. SV40 large tumor antigen forms a specific complex with the product of the retinoblastoma susceptibility gene. *Cell* **54:** 275.

Desai, D., Y. Gu, and D.O. Morgan. 1992. Activation of human cyclin-dependent kinases *in vitro*. *Mol. Biol. Cell* **3:** 571.

Dowdy, S.F., P.W. Hinds, K. Louis, S. I. Reed, A. Arnold, and R.A. Weinberg. 1993. Physical interactions of the retinoblastoma protein with human cyclins. *Cell* **73:** 499.

Dulic, V., E. Lees, and S.I. Reed. 1992. Association of human cyclin E with a periodic G1-S phase protein kinase. *Science* **257:** 1958.

Dulic, V., W.K. Kaufmann, S.J. Wilson, T.D. Tlsty, E. Lees, J.W. Harper, S.J. Elledge, and S.I. Reed. 1994. p53-Dependent inhibition of cyclin-dependent kinase activities in human fibroblasts during radiation-induced G1 arrest. *Cell* **76:** 1013.

Dyson, N., P.M. Howley, K. Munger, and E. Harlow. 1989. The human papilloma virus-16 E7 oncoprotein is able to bind to the retinoblastoma gene product. *Science* **243:** 934.

El-Deiry, W.S., T. Tokino, V.E. Velculescu, D.B. Levy, R. Parsons, J.M. Trent, D. Lin, E. Mercer, K.W. Kinzler, and B. Vogelstein. 1993. WAF1, a potential mediator of p53 tumor suppression. *Cell* **75:** 817.

El-Deiry, W.S., J.W. Harper, P.M. O'Connor, V.E. Velculescu, C.E. Canman, J. Jackman, J.A. Pietenpol, M. Burrell, D.E. Hill, Y. Wang, K.G. Wiman, W.E. Mercer, M.B. Kastan, K.W. Kohn, S.J. Elledge, K.W. Kinzler, and B. Vogelstein. 1994. WAF1/Cip1 is induced in p53-mediated G1 arrest and apoptosis. *Cancer Res.* **54:** 1169.

Ewen, M.E., H.K. Sluss, L.L. Whitehouse, and D.M. Livingston. 1993a. TGF-β inhibition of cdk4 synthesis is linked to cell cycle arrest. *Cell* **74:** 1009.

Ewen, M.E., H.K. Sluss, C.J. Sherr, H. Matsushime, J. Kato, and D.M. Livingston. 1993b. Functional interactions of the retinoblastoma protein with mammalian D-type cyclins. *Cell* **73:** 487.

Fesquet, D., J. Labbé, J. Derancourt, J. Capony, S. Galas, F. Girard, T. Lorca, J. Shuttleworth, M. Dorée, and J. Cavadore. 1993. The MO15 gene encodes the catalytic subunit of a protein kinase that activates cdc2 and other cyclin-dependent kinases (CDKs) through phosphorylation of Thr161 and its homologues. *EMBO J.* **12:** 3111.

Fisher, R.P. and J.O. Morgan. 1994. A novel cyclin associates with MO15/CDK7 to form the CDK-activating kinase. *Cell* **78:** 713.

Goodrich, D.W., N.P. Wang, Y. Qian, Y.P. Lee, and W. Lee. 1991. The retinoblastoma gene product regulates progression through the G1 phase of the cell cycle. *Cell* **67:** 293.

Gu, Y., C.W. Turek, and D.O. Morgan. 1993. Inhibition of CDK2 activity *in vivo* by an associated 20K regulatory subunit. *Nature* **366:** 707.

Harper, J.W., G.R. Adami, N. Wei, K. Keyomarsi, and S.J. Elledge. 1993. The p21 cdk-interacting protein Cip1 is a potent inhibitor of G1 cyclin-dependent kinases. *Cell* **75:** 805.

Hartwell, L.H. and T.A. Weinert. 1989. Checkpoints: Controls that ensure the order of cell cycle events. *Science* **246:** 629.

Hinds, P.W., S.F. Dowdy, E.N. Eaton, A. Arnold, and R.A. Weinberg. 1994. Function of a human cyclin gene as an oncogene. *Proc. Natl. Acad. Sci.* **91:** 709.

Hinds, P.W., S. Mittnacht, V. Dulic, A. Arnold, S.I. Reed, and R.A. Weinberg. 1992. Regulation of retinoblastoma protein functions by ectopic expression of human cyclins. *Cell* **70:** 993.

Hu, Q., J.A. Lees, K.J. Buchkovich, and E. Harlow. 1992. The retinoblastoma protein physically associates with the human cdc2 kinase. *Mol. Cell. Biol.* **12:** 971.

Jackowski, S., C.W. Rettenmier, and C.O. Rock. 1990. Prostaglandin E_2 inhibition of growth in a colony-stimulating factor 1-dependent macrophage cell line. *J. Biol. Chem.* **265:** 6611.

Jiang, W., S.M. Kahn, P. Zhou, Y. Zhang, A.M. Cacace, A.S. Infante, S. Doi, R.M. Santella, and I.B. Weinstein. 1993. Overexpression of cyclin D1 in rat fibroblasts causes abnormalities in growth control, cell cycle progression, and gene expression. *Oncogene* **8:** 3447.

Johnson, D.G., J.K. Schwarz, W.D. Cress, and J.R. Nevins. 1993. Expression of transcription factor E2F1 induces quiescent cells to enter S phase. *Nature* **365:** 349.

Kamb, A., N.A. Gruis, J. Weaver-Feldhaus, Q. Liu, K. Harshman, S.V. Tavtigian, E. Stockert, R.S. Day III, B.E. Johnson, and M.H. Skolnick. 1994. A cell cycle regulator involved in genesis of many tumor types. *Science* **264:** 436.

Kastan, M.B., O. Onyekwere, D. Sidransky, B. Vogelstein, and R.W. Craig. 1991. Participation of p53 protein in the cellular response to DNA damage. *Cancer Res.* **51:** 6304.

Kato, J. and C.J. Sherr. 1993. Inhibition of granulocyte differentiation by G1 cyclins D2 and D3 but not D1. *Proc. Natl. Acad. Sci.* **90:** 11513.

Kato, J., M. Matsuoka, D.K. Strom, and C.J. Sherr. 1994a. Regulation of cyclin D-dependent kinase 4 (cdk4) by cdk4-activating kinase. *Mol. Cell. Biol.* **14:** 2713.

Kato, J., M. Matsuoka, K. Polyak, J. Massagué, and C.J. Sherr. 1994b. Cyclic-AMP-induced G1 phase arrest mediated by an inhibitor ($p27^{Kip1}$) of cyclin-dependent kinase-4 activation. *Cell* (in press).

Kato, J., H. Matsushime, S.W. Hiebert, M.E. Ewen, and C.J. Sherr. 1993. Direct binding of cyclin D to the retinoblastoma gene product (pRb) and pRb phosphorylation by the cyclin D-dependent kinase, CDK4. *Genes Dev.* **7:** 331.

Kiyokawa, H., V.M. Richon, R.A. Rifkind, and P.A. Menks. 1994. Suppression of cyclin-dependent kinase-4 during induced differentiation of erythroleukemia cells. *Mol. Cell. Biol.* **14:** 1195.

Koff, A., M. Ohtsuki, K. Polyak, J.M. Roberts, and J. Massagué. 1993. Negative regulation of G1 progression in mammalian cells: Inhibition of cyclin E-dependent kinase by TGF-β. *Science* **260:** 536.

Koff, A., F. Cross, A. Fisher, J. Schumacher, K. Leguellec, M. Philippe, and J.M. Roberts. 1991. Human cyclin E, a new cyclin that interacts with two members of the *CDC2* gene family. *Cell* **66:** 1217.

Koff, A., A. Giordano, D. Desai, K. Yamashita, J.W. Harper, S. Elledge, T. Nishimoto, D.O. Morgan, B.R. Franza, and J.M. Roberts. 1992. Formation and activation of a cyclin E-cdk2 complex during the G1 phase of the human cell cycle. *Science* **257:** 1689.

Kuerbitz, S.J., B.S. Plunkett, W.B. Walsh, and M.B. Kastan. 1992. Wild-type p53 is a cell cycle checkpoint determinant following irradiation. *Proc. Natl. Acad. Sci.* **89:** 7491.

Lees, J.A., K.J. Buchkovich, D.R. Marshak, C.W. Anderson, and E. Harlow. 1991. The retinoblastoma protein is phosphorylated on multiple sites by human cdc2. *EMBO J.* **10:** 4279.

Lew, D.J., V. Dulic, and S.I. Reed. 1991. Isolation of three novel human cyclins by rescue of G1 cyclin (Cln) function in yeast. *Cell* **66:** 1197.

Lin, B.T-Y., S. Gruenwald, A. Morla, W. Lee, and J.Y.J. Wang. 1991. Retinoblastoma cancer suppressor gene product is a substrate of the cell cycle regulator cdc2 kinase. *EMBO J.* **10:** 857.

Ludlow, J.W., J. Shon, J.M. Pipas, D.M. Livingston, and J.A. DeCaprio. 1990. The retinoblastoma susceptibility gene product undergoes cell cycle-dependent dephosphorylation and binding to and release from SV40 large T. *Cell* **60:** 387.

Lukas, J., M. Pagano, Z. Staskova, G. Draetta, and J. Bartek. 1994. Cyclin D1 protein oscillates and is essential for cell cycle progression in human tumor lines. *Oncogene* **9:** 707.

Marx, J. 1994. How cells cycle toward cancer. *Science* **263:** 319.

Matsuoka, M., J. Kato, R.P. Fisher, D.O. Morgan, and C.J. Sherr. 1994. Activation of cyclin-dependent kinase-4 (CDK4) by mouse MO15-associated kinase. *Mol. Cell. Biol.* **14:** 7265.

Matsushime, H., M.F. Roussel, R.A. Ashmun, and C.J. Sherr. 1991. Colony-stimulating factor 1 regulates novel cyclins during the G1 phase of the cell cycle. *Cell* **65:** 701.

Matsushime, H., D.E. Quelle, S.A. Shurtleff, M. Shibuya, C.J. Sherr, and J. Kato. 1994. D-type cyclin-dependent kinase activity in mammalian cells. *Mol. Cell. Biol.* **14:** 2066.

Matsushime, H., M.E. Ewen, D.K. Strom, J. Kato, S.K. Hanks, M.F. Roussel, and C.J. Sherr. 1992. Identification and properties of an atypical catalytic subunit (p34^{PSKJ3}/CDK4) for mammalian D-type G1 cyclins. *Cell* **71:** 323.

Meyerson, M. and E. Harlow. 1994. Identification of a G1 kinase activity for cdk6, a novel cyclin D partner. *Mol. Cell. Biol.* **14:** 2077.

Meyerson, M., G.H. Enders, C. Wu, L. Su, C. Gorka, C. Nelson, E. Harlow, and L. Tsai. 1992. The human cdc2 kinase family. *EMBO J.* **11:** 2909.

Motokura, T. and A. Arnold. 1993. Cyclins and oncogenesis. *Biochim. Biophys. Acta.* **1155:** 63.

Motokura, T., T. Bloom, H.G. Kim, H. Juppner, J.V. Ruderman, H.M. Kronenberg, and A. Arnold. 1991. A novel cyclin encoded by a *bcl*1-linked candidate oncogene. *Nature* **350:** 512.

Murray, A.W. 1992. Creative blocks: Cell cycle checkpoints and feedback controls. *Nature* **359:** 509.

Nevins, J.R. 1992. E2F: A link between the Rb tumor suppressor protein and viral oncoproteins. *Science* **258:** 424.

Noburi, T., K. Miura, D.J. Wu, A. Lois, K. Takabayashi, and D.A. Carson. 1994. Deletions of the cyclin-dependent kinase-4 inhibitor gene in multiple human cancers. *Nature* **368:** 753.

Noda, A., Y. Ning, S.F. Venable, O.M. Pereira-Smith, and J.R. Smith. 1994. Cloning of senescent cell-derived inhibitors of DNA synthesis using an expression screen. *Exp. Cell Res.* **211:** 90.

Norbury, C. and P. Nurse. 1992. Animal cell cycles and their control. *Annu. Rev. Biochem.* **61:** 441.

Ohtsubo, M. and J.M. Roberts. 1993. Cyclin-dependent regulation of G1 in mammalian fibroblasts. *Science* **259:** 1908.

Polyak, K., J. Kato, M.J. Solomon, C.J. Sherr, J. Massagué, J.M. Roberts, and A. Koff. 1994a. p27Kip1, a cyclin-cdk inhibitor, links transforming growth factor-β and contact inhibition to cell cycle arrest. *Genes Dev.* **8:** 9.

Polyak, K., M. Lee, H. Erdjument-Bromage, A. Koff, J.M. Roberts, P. Tempst, and J. Massagué. 1994b. Cloning of p27^{Kip1}, a cyclin-dependent kinase inhibitor and a potential mediator of extracellular antimitogenic signals. *Cell* **78:** 59.

Poon, R.Y.C., K. Yamashita, J.P. Adamczewski, T. Hunt, and J. Shuttleworth. 1993. The cdc2-related protein p40^{MO15} is the catalytic subunit of a protein kinase that can activate p33^{cdk2} and p34^{cdc2}. *EMBO J.* **12:** 3123.

Quelle, D.E., R.A. Ashmun, S.A. Shurtleff, J. Kato, D. Bar-Sagi, M.F. Roussel, and C.J. Sherr. 1993. Overexpression of mouse D-type cyclins accelerates G1 phase in rodent fibroblasts. *Genes Dev.* **7:** 1559.

Reed, S.I. 1992. The role of p34 kinases in the G1 to S-phase transition. *Annu. Rev. Cell. Biol.* **8:** 529.

Resnitzky, D., M. Gossen, H. Bujard, and S.I. Reed. 1994. Acceleration of the G1/S phase transition by expression of cyclins D1 and E with an inducible system. *Mol. Cell. Biol.* **14:** 1669.

Roussel, M.F. and C.J. Sherr. 1989. Mouse NIH/3T3 cells expressing human CSF-1 receptors overgrow in serum-free medium containing human CSF-1 as their only growth factor. *Proc. Natl. Acad. Sci.* **86:** 7924.

Roussel, M.F., J.L. Cleveland, S.A. Shurtleff, and C.J. Sherr. 1991. *Myc* rescue of a mutant CSF-1 receptor impaired in mitogenic signalling. *Nature* **353:** 361.

Roussel, M.F., S.A. Shurtleff, J.R. Downing, and C.J. Sherr. 1990. A point mutation at tyrosine 809 in the human colony-stimulating factor 1 receptor impairs mitogenesis without abrogating tyrosine kinase activity, association with phosphatidylinositol 3-kinase, or induction of *fos* and *jun*B genes. *Proc. Natl. Acad. Sci.* **87:** 6738.

Roussel, M.F., T.J. Dull, C.W. Rettenmier, P. Ralph, A. Ullrich, and C.J. Sherr. 1987. Transforming potential of the c-*fms* proto-oncogene (CSF-1 receptor). *Nature* **325:** 549.

Serrano, M., G.J. Hannon, and D. Beach. 1993. A new regulatory motif in cell cycle control causing specific inhibition of cyclin D/cdk4. *Nature* **366:** 704.

Sherr, C.J. 1993. Mammalian G1 cyclins. *Cell* **73:** 1059.

———. 1994. The ins and outs of RB: Coupling gene expression to the cell cycle clock. *Trends Cell Biol.* **4:** 15.

Slingerland, J.M., L. Hengst, C.H. Pan, D. Alexander, M.F. Stampfer, and S.I. Reed. 1994. A novel inhibitor of cyclin-cdk activity detected in transforming growth factor β-arrested epithelial cells. *Mol. Cell. Biol.* **14:** 3683.

Solomon, M.J., J.W. Harper, and J. Shuttleworth. 1993. CAK, the p34^{cdc2} activating kinase, contains a protein identical or closely related to p40^{MO15}. *EMBO J.* **12:** 3133.

Solomon, M.J., T. Lee, and M.W. Kirschner. 1992. Role of phosphorylation in p34^{cdc2} activation: Identification of an activating kinase. *Mol. Biol. Cell.* **3:** 13.

Tam, S.W., A.M. Theodoras, J.W. Shay, G.F. Draetta, and M. Pagano. 1994. Differential expression and regulation of cyclin D1 protein in normal and tumor human cells: Association with cdk4 is required for cyclin D function in G1 progression. *Oncogene* **9:** 2663.

Toyoshima, H. and T. Hunter. 1994. p27^{Pic2}, a novel inhibitor of G1 cyclin/cdk protein kinase activity, is related to p21^{Pic1}. *Cell* **78:** 67.

Vairo, G., S. Argyriou, A.M. Bordun, G. Whitty, and J.A. Hamilton. 1990. Inhibition of the signaling pathways for macrophage proliferation by cyclic AMP: Lack of effect on early responses to colony stimulating factor-1. *J. Biol. Chem.* **265:** 2692.

Whyte, P., K.J. Buchkovich, J.M. Horowitz, S.H. Friend, M. Raybuck, R.A. Weinberg, and E. Harlow. 1988. Association between an oncogene and an anti-oncogene: The adenovirus E1A proteins bind to the retinoblastoma gene product. *Nature* **334:** 124.

Winston, J. and W.J. Pledger. 1993. Growth factor regulation of cyclin D1 expression through protein synthesis-dependent and -independent mechanisms. *Mol. Biol. Cell* **4:** 1133.

Won, K., Y. Xiong, D. Beach, and M. Gilman. 1992. Growth-regulated expression of D-type cyclin genes in human diploid fibroblasts. *Proc. Natl. Acad. Sci.* **89:** 9910.

Xiong, Y., H. Zhang, and D. Beach. 1992. D-type cyclins associate with multiple protein kinases and the DNA replication and repair factor PCNA. *Cell* **71:** 505.

———. 1993a. Subunit rearrangement of the cyclin-dependent kinases is associated with cellular transformation. *Genes Dev.* **7:** 1572.

Xiong, Y., T. Connolly, B. Futcher, and D. Beach. 1991. Human D-type cyclin. *Cell* **65:** 691.

Xiong, Y., G.J. Hannon, H. Zhang, D. Casso, R. Kobayashi, and D. Beach. 1993b. p21 is a universal inhibitor of cyclin kinases. *Nature* **366:** 701.

Zhang, H., G. Hannon, and D. Beach. 1994. p21-Containing cyclin kinases exist in both active and inactive states. *Genes Dev.* **8:** 1750.

p21 Is a Component of Active Cell Cycle Kinases

H. ZHANG, G.J. HANNON, D. CASSO, AND D. BEACH
Howard Hughes Medical Institute, Cold Spring Harbor Laboratory, Cold Spring Harbor, New York 11724

Much of our current understanding of the regulation of the cell division cycle has emerged from studies of a family of protein kinases (cdc2, CDC28, and generically CDK) and their inhibitors and activators (for review, see Sherr 1993). A critical step in understanding cell cycle control was the discovery that CDKs interact with cyclins, proteins that serve as essential activating subunits and specificity determinants of the kinases (Draetta 1990; Sherr 1993). In human cells, multiple cyclins and CDKs interact in a relatively promiscuous fashion to form a large family of related cyclin kinases, each of which is presumed to play a specific role in cell cycle progression.

The view that mammalian cyclin kinases exist predominantly in a binary (cyclin/CDK) state prevailed until these enzymes were examined in normal human fibroblasts, rather than in the many oncogenically transformed cell types that had been investigated previously (Xiong et al. 1992). In normal fibroblasts, the major population of multiple cyclin kinases exists in quaternary complexes consisting of cyclin, CDK, proliferating cell nuclear antigen (PCNA), and a protein of apparent M_r 21,000, p21 (Zhang et al. 1993). However, in fibroblasts that are transformed with a variety of tumor viral oncoproteins, the quaternary complexes essentially reduce to a binary state. Deregulation of cell proliferation is a hallmark of neoplastic transformation. Difference in the composition of cell cycle kinases between normal and transformed cells suggests that fundamental changes in cell cycle regulation contribute to neoplastic transformation.

Recently, it has become apparent that p21 is a universal inhibitor of cyclin/CDK catalytic activity (Gu et al. 1993; Harper et al. 1993; Xiong et al. 1993b). Each member of the cyclin kinase family is inhibited by p21, but their relative affinities vary with each enzyme (Gu et al. 1993; Harper et al. 1993; Xiong et al. 1993b). The concept that p21 is a universal cell cycle kinase inhibitor creates a paradox, since in normal proliferating fibroblast cells, the majority of multiple kinases exist in quaternary complexes (Zhang et al. 1993). Here, we present the unexpected finding that p21 is a component of catalytically active cyclin kinases. It appears that p21-containing enzymes can change between active and inactive states, probably through changes in the stoichiometry of the p21 subunit. These findings have broad implications for our understanding of the normal cell cycle and the subversion of control pathways in tumor cells.

MATERIALS AND METHODS

Cell culture. The culture of human normal fibroblasts (WI38), the Li-Fraumeni cell line (LSC041), the SV40 virus-transformed human fibroblast line (VA13), and HeLa cells was described previously (Xiong et al. 1993a). The colorectal carcinoma cell line (RKO) was kindly provided by Dr. Michael Kastan (The Johns Hopkins University, Baltimore, Maryland) and the myeloid leukemia line (HL60) was obtained from American Type Culture Collection (Rockville, Maryland). These cells were cultured similarly. Insect Sf9 cells were cultured in Grace's medium supplemented with 10% heat-inactivated fetal bovine serum, lactalbumin hydrolysate, and yeastolate ultrafiltrate at 27°C as described previously (Xiong et al. 1993b). The expression of baculoviruses encoding p21, PCNA, various CDKs, and cyclins in Sf9 cells was carried out as described previously (Xiong et al. 1993b).

Plasmids and DNA recombination techniques. For the production of MBP-p21 fusion protein, a DNA fragment encoding the human p21 gene was cloned in frame to the carboxyl terminus of maltose-binding protein (MBP) in the pMAL-c2 vector between SalI and HindIII sites. The fusion protein was produced in *Escherichia coli* (BL21) and purified on amylose resin according to the instructions provided by the manufacturer (New England Biolabs). The MBP protein was similarly purified. The purified proteins were dialyzed into kinase buffer (50 mM Tris, pH 7.5, 10 mM $MgCl_2$, 0.5 mM DTT) and stored at −80°C. MBP and MBP-p21 proteins were more than 95% pure.

Assembly of active and inactive quaternary complexes. Typically, insect Sf9 cells were individually infected with baculoviruses encoding cyclin, CDK, or p21. At 48–72 hours postinfection, cells were labeled with 100 µCi/ml of [^{35}S]methionine for 3 hours. Cell lysates were prepared in kinase buffer as described previously (Xiong et al. 1993b). The cell lysates containing cyclin or CDK2 were mixed in the presence of 1 mM ATP, incubated for 30 minutes at 30°C, and then cooled on ice for 5 minutes to form cyclin/CDK kinase complexes. Increasing amounts of p21-containing lysates were then added to each 10 µl of kinase lysate. After incubation on ice for 15 minutes to allow complex formation, the reaction was further incubated at 30°C for 15 minutes. The reaction was stopped by addition of 20 mM EDTA. Complexes were recovered by immunoprecipitation. The kinase assay using histone H1 as the substrate was conducted as described previously (Xiong et al. 1993b). The relative kinase activities were determined from the band intensities on a Fuji BAS2000 bioimager.

RESULTS

p21 Is Regulated by p53

We have previously demonstrated the association of p21 and PCNA with multiple cyclin kinases in normal fibroblast cells (Zhang et al. 1993). In certain transformed cells, p21 and PCNA disappear from the cell cycle kinases and the kinases exist mostly in a cyclin/CDK binary state. The appearance of p21/PCNA/cyclin/CDK quaternary complexes appears to correlate with the p53 status in the cell (Xiong et al. 1993a). We have found that the p21 genomic clone contains the p53 transcriptional regulatory motif (data not shown). In fibroblast cells derived from mice that have both p53 alleles deleted from their genome, p21 expression is depressed (Fig.1A, top panel). p21 expression is further stimulated by DNA damage agents only in cells that contain wild-type p53 (Fig.1A, bottom panel). Overexpression of p21, similar to the overexpression of p53, in Saos-2 osteosarcoma cells leads to growth suppression (Fig. 1B). We have also raised a specific p21 antiserum to examine the full spectrum of p21-associated proteins in vivo. This serum immunoprecipitated from lysates of ^{35}S-labeled normal human diploid fibroblasts (WI38) a prominent 21-kD protein and a number of additional polypeptides (Fig. 1C, lane labeled anti-p21). Partial V8 protease digestion revealed identity between the 21-kD polypeptide band and p21 protein derived from in vitro translation of the p21 cDNA (data not shown). To aid in identification of the proteins that coimmunoprecipitate with p21, immunoprecipitates of CDC2, CDK2, CDK4, cyclin A, cyclin B_1, and cyclin D_1 were electrophoresed in parallel. By comparison, virtually all of the p21-associated proteins could be identified (see Fig. 1C). These results suggest that the cyclin kinases and PCNA are the major p21-associated proteins in normal fibroblasts. Since p21 is a universal inhibitor of cell cycle kinases (Gu et al. 1993; Harper et al. 1993; Xiong et. al. 1993b), these results suggest that p21 may serve as an effector of cell cycle arrest in response to activation of the p53 checkpoint pathway. A similar suggestion has also been made by another group (El-Deiry et al. 1993).

Active and Inactive Forms of the p21-associated Cyclin Kinases In Vitro

p21 is a universal inhibitor of cyclin kinases in vitro (Gu et al. 1993; Harper et al. 1993; Xiong et al. 1993b). Thus, it was predicted that all p21-containing complexes would lack catalytic activity. However, in previous in vitro experiments, progressive addition of p21 to cyclin kinases did not result in a precise, corresponding loss of kinase activity but instead caused an abrupt transition from full activity to essentially complete inhibition (Xiong et al. 1993b). Here we have examined the seeming cooperativity of p21 inhibition of cyclin kinases in greater detail.

Lysates were prepared from metabolically ^{35}S-labeled insect cells infected with baculoviruses directing the expression of human cyclin A, CDK2, and p21. Progressive addition of the p21-containing lysate to preformed cyclin A/CDK2 complexes resulted in the accumulation of p21/cyclin A/CDK2 ternary complexes (Fig. 2A). At a specific point in the titration, the histone H1 kinase activity in cyclin A immunoprecipitates was abruptly lost (Fig. 2A). As noted previously, this transition between active and inactive enzyme was not directly reflected in the binding of p21 to cyclin A/CDK2, which was approximately proportional to the amount of p21 added (Fig. 2A). The discrepancy between p21 binding and inhibition suggested that p21 might fail to act as a cyclin kinase inhibitor at subsaturating levels. To test this possibility, we specifically recovered the p21-containing complexes using the p21 antiserum (Fig. 2B). Contrary to previous predictions, complexes formed at subsaturating p21 concentrations possessed substantial histone H1 kinase activity. With addition of p21, p21-associated kinase activity progressively increased, mirroring the accumulation of ternary complexes. This was followed by abrupt loss of p21-associated kinase at the same p21 concentration that abolished activity in parallel cyclin A immunoprecipitates. At its maximum, p21-associated kinase was approximately 70% of that detected in CDK2 immunoprecipitates from the same mixtures (Fig. 2). These results demonstrate that p21-containing cyclin kinase complexes can exist in both active and inactive states. Inhibition is achieved only as p21 reaches a saturating concentration.

p21 is a universal inhibitor of cyclin kinases, and nonlinear p21 inhibition profiles have been observed with cyclin A/CDK2, cyclin E/CDK2, and cyclin D1/CDK4 enzymes (Fig. 3) (Xiong et al. 1993b), suggesting that the association of p21 with active cyclin kinases might be a general phenomenon. Consistent with this prediction, we detected substantial kinase activity in p21/cyclin E/CDK2 and p21/cyclin B/CDC2 complexes formed at subsaturating p21 concentrations (Fig. 3A,B). In the case of cyclin E/CDK2, p21-associated activity reached approximately 80% of that detected in parallel cyclin E immunoprecipitates.

We have previously shown that the inclusion of PCNA had no effect on the inhibition of cyclin kinases by p21. Similarly, PCNA had no effect on the association of p21 with active enzyme (data not shown).

p21 Is a Component of Active Cyclin Kinases In Vivo

We have previously shown that in normal human diploid fibroblasts (e.g., WI38), the majority of cyclin

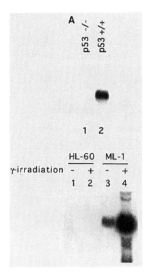
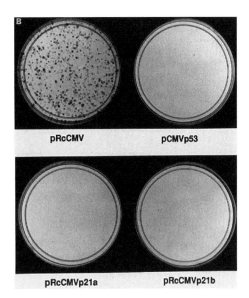
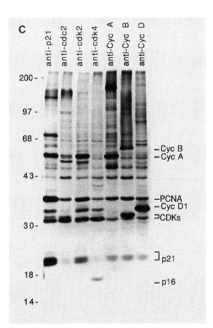

Figure 1. Regulation of p21 expression and p21-associated proteins. (*A*) p21 expression is regulated by p53. (*Top* panel) Total RNA was isolated from mouse embryonic fibroblasts that either contain wild-type p53 (p53 +/+) or have both p53 alleles deleted (p53 -/-). (*Bottom* panel) RNA isolated from human HL60 (p53 deficient) or ML-1 (wild-type p53) cells that received either no irradiation (-) or 5 Gy γ irradiation (+). The level of p21 transcript was determined by ^{32}P-labeled p21 DNA probe. (*B*) Saos-2 cells were transfected with 20 μg of pRcCMV vector; pCMVp53 or pRcCMVp21 (a and b were same construct but different DNA preparation) and grown in the presence of G418 for 2–3 weeks. G418-resistant colonies were stained with crystal violet. (*C*) p21-associated proteins were immunoprecipitated from cell lysates of ^{35}S-labeled normal human fibroblasts (WI38) and compared with the proteins immunoprecipitated by anti-CDC2, CDK2, CDK4, cyclin A, cyclin B, or cyclin D1 antisera as indicated. The positions of protein molecular-weight markers, electrophoresed in parallel, are indicated.

kinases exist in complexes containing p21 (Zhang et al. 1993). Thus, the finding that p21 is an inhibitor of cyclin kinases raised questions concerning the nature of the active kinases in normal cells. The association of p21 with active cyclin kinase in vitro suggested that p21-containing complexes might also possess activity in vivo. Therefore, we immunoprecipitated p21 from several cell lines and assayed for coprecipitation of histone H1 kinase activity. In all cell lines that contain detectable levels of p21 protein, we found catalytically active p21 complexes (Fig. 4, panels WI38, RKO, HeLa, VA13). In WI38 cells that contain functional p53, p21 is relatively abundant and is associated with substantial histone H1 kinase (Fig. 4, panel WI38). Comparable activity is recovered in CDC2 and CDK2 immunoprecipitates from the same cell lysates. RKO cells also possess functional p53 and active p21-associated cyclin kinases. However, p21-associated activity is exceeded severalfold in these cells by the activity found in CDK2 immunoprecipitates. HeLa and VA13

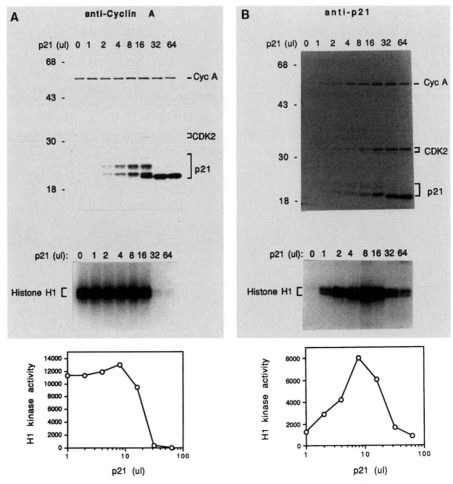

Figure 2. Active and inactive p21/cyclin A/CDK2 kinase complexes. Increasing amounts of ^{35}S-labeled p21 containing Sf9 cell lysates were added to preformed cyclin A/CDK2 kinase complexes. Proteins were recovered from the mixtures by immunoprecipitation with either anti-cyclin A (A) or anti-p21 (B) antibodies. Composition of the immunoprecipitates was monitored by direct visualization of the labeled proteins (*top* panels). Activity of the complexes was assayed using histone H1 (*middle* and *bottom* panels) as the substrate. The relative kinase activities were determined from the band intensities on a Fuji BAS2000 bioimager. The exposure time for anti-p21 is twice that for anti-cyclin A.

cells have approximately fivefold reduced levels of p21 due to the presence of p53-inactivating viral oncoproteins (see below), but these cells still contain noticeable p21-associated kinase. In HL60 cells and in the Li-Fraumeni line, LCS041, functional p53 is absent, and p21 is virtually undetectable. CDC2 and CDK2 immunoprecipitates from these cells contain high levels of histone H1 kinase but, as expected, activity is absent from anti-p21 immunoprecipitates.

Active and Inactive States of the p21-associated Enzyme

Our results indicate that p21-associated kinases exist in active and inactive states. Transition between these states could be explained by two broadly different classes of models. In the first, modification of one constituent of the kinase complex could regulate its activity. For example, phosphorylation of the cyclin, CDK, or p21 subunit might immunize the complex against inhibition by bound p21. In this regard, kinase-bound p21 displayed an altered electrophoretic mobility under conditions in which cyclin A/CDK2/p21 was active as a histone H1 kinase (Fig. 2A,B, lanes 1–16), and these mobility shifts reflect p21 phosphorylation (data not shown). In an alternative model, changes in subunit stoichiometry might cause the transition from active to inactive p21-associated kinase. In particular, the nonlinear p21 inhibition profiles (see Figs. 2,3) suggest that multiple p21 subunits are necessary for kinase inhibition.

To distinguish between these alternative mechanisms, we took the approach outlined in Figure 5A. p21/cyclin A/CDK2 complexes were formed at a variety of p21 concentrations, yielding both active and inactive p21-associated enzymes. These were immuno-affinity purified using the p21 antiserum (see Fig. 5A, step b). Complexes were then challenged with a purified fusion protein consisting of p21 and the maltose-binding protein (MBP-p21, step c). According

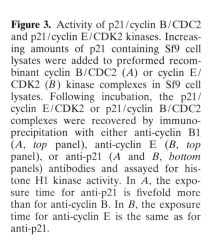
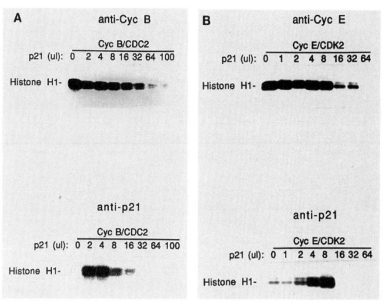

Figure 3. Activity of p21/cyclin B/CDC2 and p21/cyclin E/CDK2 kinases. Increasing amounts of p21 containing Sf9 cell lysates were added to preformed recombinant cyclin B/CDC2 (A) or cyclin E/CDK2 (B) kinase complexes in Sf9 cell lysates. Following incubation, the p21/cyclin E/CDK2 or p21/cyclin B/CDC2 complexes were recovered by immunoprecipitation with either anti-cyclin B1 (A, top panel), anti-cyclin E (B, top panel), or anti-p21 (A and B, bottom panels) antibodies and assayed for histone H1 kinase activity. In A, the exposure time for anti-p21 is fivefold more than for anti-cyclin B. In B, the exposure time for anti-cyclin E is the same as for anti-p21.

to the first model (Fig. 5A, top) changes in the phosphorylation state of complex constituents are necessary for changes in activity. Thus, assuming no p21 subunit exchange and assuming that potential modifying enzymes have been removed from immunoprecipitated complexes, incubation with additional MBP-p21 should have no effect (Fig. 5A, top, step d).

If, on the other hand, changes in p21 stoichiometry account for changes in activity (Fig. 5A, bottom), MBP-p21 should be incorporated into the complex and inhibit kinase. The experimental results are consistent with the latter scheme (Fig. 5B,C). Addition of saturating amounts of MBP-p21 (see Fig. 5B), but not MBP alone, converted purified, active, p21-containing

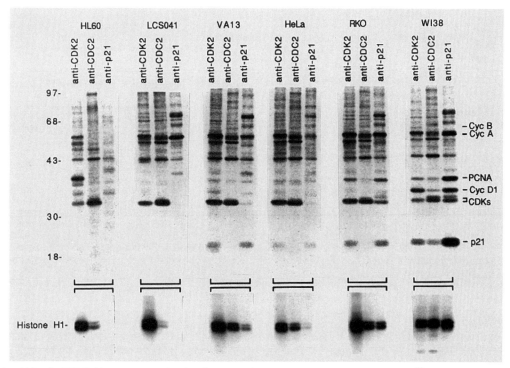

Figure 4. p21/cyclin/CDK kinases are active in vivo. Proteins were immunoprecipitated from ^{35}S-labeled cell lysates from HL60, LCS041, VA13, HeLa, RKO, or WI38 cells (as indicated) with either anti-CDK2, anti-CDC2, or anti-p21 antibodies as indicated. The immunoprecipitates were split and used for direct visualization of immunoprecipitated proteins (upper panel) or for histone H1 kinase assays (lower panel). The positions of protein size markers are indicated.

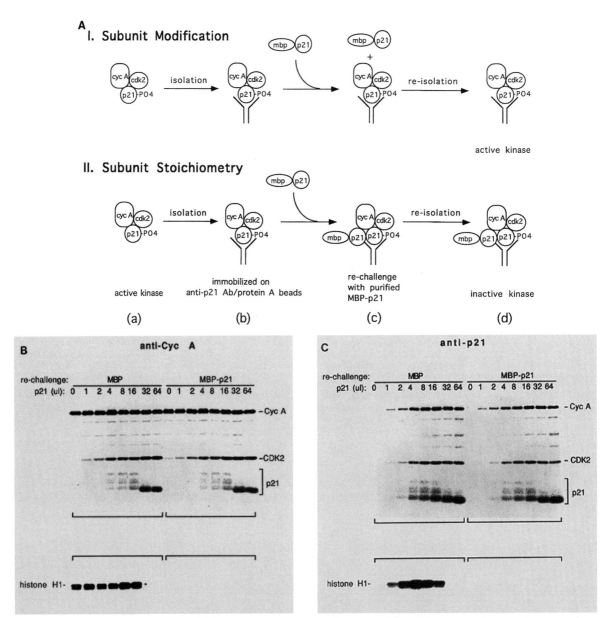

Figure 5. Active p21-associated enzymes are inhibited by additional p21. (*A*) Experimental design. (*a*) Formation of active p21/cyclin/CDK kinase at a subsaturating concentration of p21 in insect cell lysates. p21 is phosphorylated during this reaction. (*b*) Isolation of the active p21-containing kinase by immunoprecipitation with anti-p21 antibody immobilized on protein A beads. (*c*) Challenge of the isolated active p21/cyclin/CDK kinase complex with purified MBP-p21. Outcomes predicted from two possible models are presented. (*Top*) If protein modification is responsible for the maintenance of active p21/cyclin/CDK kinase, challenge of the kinase with excess MBP-p21 will not affect the activity of the kinase. (*Bottom*) If increased p21 stoichiometry accounts for kinase inhibition, incorporation of MBP-p21 into the active p21/cyclin/CDK kinase complex should result in formation of an inactive kinase complex. (*B*) Formation of ^{35}S-labeled p21/cyclin A/CDK2 kinase complex at various p21 concentrations in the insect cell lysates. The kinase complexes were immunoprecipitated with anti-cyclin A antibody. The purified immunoprecipitates were then challenged with either 2 μg of purified MBP (*left* panel) or MBP-p21 fusion protein (*right* panel). The protein complexes were reisolated and assayed for labeled protein composition (*top*) and histone H1 kinase activity (*bottom*). (*C*) Mixtures described in *B* were precipitated with the p21 antiserum prior to the rechallenge analysis.

complexes to an inactive state (Fig 5C). Examination of labeled proteins present in the immunoprecipitates revealed that inhibition occurred without changes in the phosphorylation state of p21, exchange of p21 subunits, or disruption of the cyclin and CDK subunits of the previously active enzyme. Similar results have been obtained using a glutathione-*S*-transferase-p21 fusion protein (GST-p21; data not shown).

In parallel experiments, we have identified the phosphorylation sites that cause p21 mobility shifts (see Figs. 2,3) as serines 98 and 130. Conversion of either serine to alanine had no obvious effects on either the association of p21 with active kinase or the ability of p21 to inhibit cyclin kinases (data not shown), suggesting that phosphorylation of these sites is irrelevant to the activity of p21-associated enzymes.

Multiple p21 Subunits in Inactive Complexes

The previous experiments suggested that active and inactive p21-containing enzymes might differ in stoichiometry of their p21 subunits and might therefore be physically distinguishable. To test this possibility, complexes containing p21, cyclin A, and CDK2 were formed at subsaturating p21 concentrations, and lysates were fractionated by gel filtration chromatography. p21-containing complexes were recovered from column fractions using p21 antiserum, and immunoprecipitates were tested for both protein composition and histone H1 kinase activity (Fig. 6A). p21-associated complexes distributed into two resolvable peaks. One migrated near the position predicted for a p21/cyclin A/CDK2 ternary complex (fractions 30–36; see Fig. 6) whereas a second migrated at a much higher apparent molecular weight (fractions 16–20). p21-associated histone H1 kinase activity resided almost exclusively in the low-molecular-weight fractions; these complexes possess 20-fold greater activity than the high-molecular-weight fractions, although they contain only about 3-fold more CDK subunit (see Fig. 6). Identical results were obtained when complexes were formed in the presence or absence of an unfused maltose-binding protein.

To test the p21 subunit composition of complexes in the low- and high-molecular-weight fractions, we formed p21/cyclin A/CDK2 complexes using a mixture of physically distinct p21 proteins. In this experiment, cyclin A, CDK2, and one of the two types of p21 subunits, native p21, were derived from baculoviral lysates and were ^{35}S-labeled. The second type of p21 was the unlabeled MBP-p21 protein purified from bacteria (see above). If any complexes contain multiple p21 subunits, then native, ^{35}S-labeled p21 should coprecipitate with tagged p21 protein on an MBP-specific amylose resin.

Complexes were formed by incubating the four components at subsaturating p21 concentrations, and low- and high-molecular-weight complexes were separated by gel filtration chromatography. Only cyclin A and CDK2 were recovered in association with MBP-

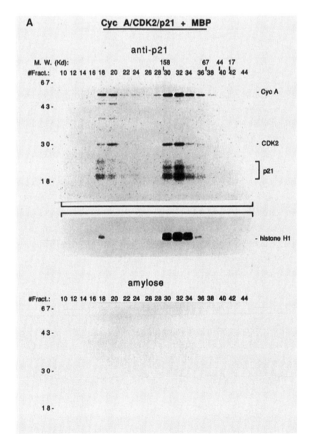
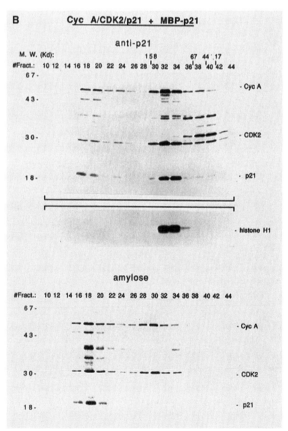

Figure 6. Inactive p21-containing kinases contain multiple p21 subunits. ^{35}S-labeled p21/cyclin A/CDK2 kinase complexes were formed at subsaturating concentrations of p21 in insect cell lysates. The lysates were then mixed with subsaturating amounts of either MBP (*A*) or MBP-p21 (*B*). The reaction mixtures were loaded onto a superose 6 gel filtration column and fractionated at 0.5 ml/fraction. Proteins were recovered using anti-cyclin A or anti-p21 antibodies or amylose agarose. Protein complexes were assayed for labeled protein composition (*top* and *bottom* panels) or histone H1 kinase activity (*middle* panels). Migration of protein molecular-weight markers in the superose 6 column are indicated above the fraction numbers in each panel.

p21 from the low-molecular-weight fractions, implying that they contained MBP-p21/cyclin A/CDK2 ternary complexes. However, from the high-molecular-weight fractions, native p21, cyclin A, and CDK2 were recovered with MBP-p21 by binding to the amylose resin. This strongly suggests that inactive complexes present in the high-molecular-weight fractions contain multiple p21 subunits and that those in the low-molecular-weight fractions contain a single p21 molecule.

To confirm that partition into low-molecular-weight and high-molecular-weight complexes is due to differences in p21 stoichiometry, we isolated low-molecular-weight, p21-containing complexes (fractions 30–36, Fig. 6A) and challenged them by incubation in the presence of either MBP-p21 or MPB alone. Subsequent refractionation by gel filtration showed that challenge with MPB-p21 shifted a significant portion of the low-molecular-weight complexes into the high-molecular-weight fraction, and this was accompanied by incorporation of the tagged p21 protein into the complexes (data not shown). Challenge with MBP alone had no effect on the chromatographic profile of the active complexes. Taken together, our results indicate that inhibition of cyclin kinases requires binding of more than one p21 subunit.

DISCUSSION

The most striking finding of this study is that p21, which was thought to function as an inhibitor of cyclin kinases (Gu et al. 1993; Harper et al. 1993; Xiong et al. 1993b), can be a component of catalytically active enzymes both in vitro and in vivo. We have previously shown that in normal human diploid fibroblasts, the major fraction of cyclin kinases exists in quaternary complexes containing a cyclin, a CDK, PCNA, and p21 (Zhang et al. 1993). Thus, the discovery that p21 functions as a kinase inhibitor was paradoxical, since it predicted that normal cells should contain virtually no active cyclin kinase. By demonstrating that p21-containing cyclin kinases exist in both active and inactive states, we have rationalized the inhibitory role of p21 with the ability of normal cells to progress through the cell cycle. It is apparent that some fraction of quaternary complexes are catalytically active in vivo. In fact, our results are consistent with the notion that the quaternary complex may represent the normal state of active cyclin kinases and that this form of the complex controls cell cycle progression in normal cells.

The expression of the p21 gene is regulated by the tumor suppressor protein, p53 (El-Deiry et al. 1993; Xiong et al. 1993b). Thus, it has been speculated that p21 might be the key effector of cell cycle arrest induced by activation of the p53 checkpoint pathway (Xiong et al. 1993b). Although our results do not affect the basic premise of this model, they do suggest that p53 may also play a more subtle role in the regulation of normal cell growth. For example, the presence of p21 in normal cells may be sufficient to inactivate low levels of cyclin/CDK complexes. Thus,

normal cells might need to overcome this p21 threshold before active cyclin kinases promote cell cycle progression. The level of p21 is greatly decreased (>50-fold) in cells lacking functional p53 (see Fig. 4; LCS041, HL60), and therefore, p53-deficient cells would not be subject to this type of control. Cyclin kinases exist in different conformations in normal and p53-deficient cells. In normal cells, active quaternary complexes contain PCNA and p21 subunits which are absent from the binary cyclin/CDK enzymes present in many transformed cells. Although the presence of PCNA and of a single p21 subunit appear neutral in our in vitro assays, it is likely that the quaternary and binary cyclin kinases have as-yet-uncharacterized functional differences in vivo.

It is clear that p21-containing complexes exist in physically distinct, active and inactive states. Our results are consistent with a model in which these states differ in the relative stoichiometry of the p21 subunit. Active complexes appear to contain a single p21 molecule, whereas inactive complexes possess multiple p21 subunits. Changes in p21 stoichiometry are sufficient to account for the conversion of active to inactive complexes in our in vitro experiments. However, association of cyclin kinases with p21 must be intertwined with other modes of regulation in vivo. Cyclin kinases are subject to a number of activating and inhibitory phosphorylations. For example, most members of the cyclin kinase family require an activating phosphorylation at a threonine (a.a. 161 or its equivalent) residue by CDK-activating kinase (CAK) and are further regulated by inhibitory phosphorylation of threonine and tyrosine residues near their amino termini (for review, see Draetta 1990; Sherr 1993). Although binding of a single p21 subunit appears neutral in our in vitro experiments, it is unknown what effect association with p21 might have on the function of CDK-modifying enzymes in vivo. It is also possible that p21 binding might alter the subcellular localization of active kinase complexes. Further study will be required to fully understand how all of these regulatory pathways integrate to affect growth control in normal cells.

ACKNOWLEDGMENTS

We thank Drs. David Morgan (University of California at San Francisco) and Stephen Gruenwald (Pharmingen) for providing recombinant baculoviruses. We are also grateful to Gretchen Hannon and Sanae Matsumoto for technical assistance and to Drs. Scott Davey and Brad Nefsky for discussions. We thank Mike Ockler, Jim Duffy, and Phil Renna for preparation of the figures. Finally, we thank Judi Smith for help in manuscript preparation. H.Z. is a postdoctoral fellow of the Howard Hughes Medical Institute, and G.J.H. is a postdoctoral fellow of the Damon Runyon-Walter Winchell Cancer Research Fund. D.B. is an investigator of the Howard Hughes

Medical Institute. This work was supported in part by a grant from the National Institutes of Health.

REFERENCES

Draetta, G. 1990. Cell cycle control in eukaryotes: Molecular mechanisms of cdc2 activation. *Trends Biol. Sci.* **15:** 378.

El-Deiry, W.S., T. Tokino, V.E. Velculescu, D.B. Levy, R. Parsons, J.M. Trent, D. Lin, W.E. Mercer, K.W. Kinzler, and B. Vogelstein. 1993. *WAF1*, a potential mediator of p53 tumor suppression. *Cell* **75:** 817.

Gu, Y., C.W. Turck, and D.O. Morgan. 1993. Inhibition of CDK2 activity in vivo by an associated 20K regulatory subunit. *Nature* **366:** 707.

Harper, J.W., G.R. Adami, N. Wei, J. Keyomarsi, and S.J. Elledge. 1993. The p21 cdk-interacting protein Cip1 is a potent inhibitor of G1 cyclin-dependent kinases. *Cell* **75:** 805.

Sherr, C. 1993. Mammalian G1 cyclins. *Cell* **73:** 1059.

Xiong, Y., H. Zhang, and D. Beach. 1992. D type cyclins associate with multiple protein kinases and the DNA replication and repair factor PCNA. *Cell* **71:** 505.

———. 1993a. Subunit rearrangement of the cyclin-dependent kinases is associated with cellular transformation. *Genes Dev.* **7:** 1572.

Xiong, Y., G.J. Hannon, H. Zhang, D. Casso, R. Kobayashi, and D. Beach. 1993b. p21 is a universal inhibitor of cyclin kinases. *Nature* **366:** 701.

Zhang, H., Y. Xiong, and D. Beach. 1993. Proliferating cell nuclear antigen and p21 are components of multiple cell cycle kinase complexes. *Mol. Biol. Cell* **4:** 897.

Cyclins, Cdks, and Cyclin Kinase Inhibitors

J.M. ROBERTS,* A. KOFF,* K. POLYAK,† E. FIRPO,*
S. COLLINS,‡ M. OHTSUBO,* AND J. MASSAGUɆ

*Department of Basic Sciences and ‡Department of Molecular Medicine, Fred Hutchinson Cancer Research Center,
Seattle, Washington 98104; †Cell Biology and Genetics Program, Howard Hughes Medical Institute,
Memorial Sloan Kettering Cancer Center, New York, New York 10021

After each mitotic cycle, a cell either remains in the cell cycle and divides once again, or withdraws from the cell cycle and adopts an alternative, nonproliferating cell fate (e.g., quiescence or terminal differentiation). This "decision," first called Start by Lee Hartwell and coworkers, was discovered and initially characterized by genetic analyses of the cell cycle of the yeast, *Saccharomyces cerevisiae* (Hartwell et al. 1974). Upon completion of Start, yeast become committed to complete the mitotic cycle and simultaneously initiate the parallel pathways leading to spindle pole body duplication, bud emergence, and DNA replication. An analogous set of events is thought to take place during the G_1 phase of the mammalian cell cycle (Zetterberg and Larson 1985), and this is often called the restriction point (Pardee 1974). Prior to the restriction point, but not after, the cell exits the cell cycle if specific mitogenic and growth signals are absent.

Cells respond to both extracellular and intracellular cues when deciding to begin a new mitotic cycle (Fig. 1). Extracellular signals that promote cell proliferation include growth factors, mitogens, nutrients, and cell–substratum interactions. Examples of signals that cause withdrawal from the cell cycle are antimitogens (such as TGF-β) and cell-cell contact. Pathways intrinsic to the cell itself also promote or inhibit cell proliferation by signaling the completion or failure of particular cell cycle events. These homeostatic pathways are known as "checkpoints" (Weinert and Hartwell 1988). They ensure that cell cycle events occur in a specific sequence and that the cell cycle arrests if these events are not executed correctly. For example, entry into S phase depends on successful completion of the previous mitosis, adequate cell growth, and the repair of any DNA damage that might be present (Hartwell and Weinert 1989; Li and Murray 1991).

What are the molecular processes that comprise the decision called Start (or the restriction point), and how are they affected by the various stimulatory or inhibitory signals that control cell proliferation? During the past 5–10 years, a paradigm to explain cell cycle control has emerged from genetic experiments with yeast, together with molecular and biochemical observations made on marine invertebrates, amphibians, and mammalian cells (Nurse 1990; Hartwell 1991). The basic idea is that the key transitions in the cell cycle are controlled by the activation of a special family of protein kinases called the cyclin-dependent kinases (Cdks). A simple picture is that at Start, or the restriction point, the Cdks phosphorylate and activate specific proteins necessary to transit from the uncommitted to the committed part of G_1, and at the conclusion of G_2, Cdks phosphorylate and activate a different set of proteins that promote mitosis and

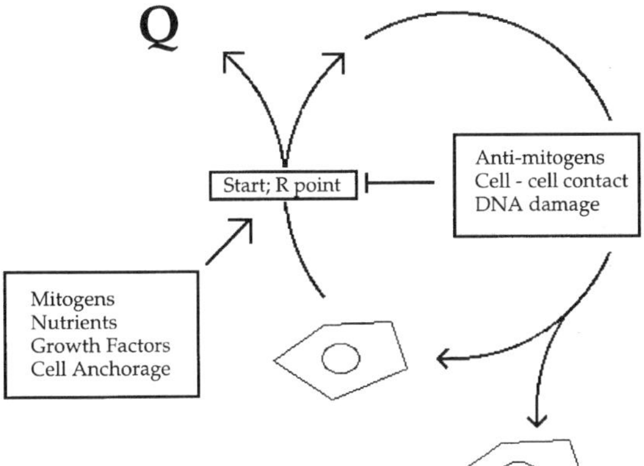

Figure 1. Cell cycle commitment occurs in G_1. The physiological event that results in commitment to a new mitotic cycle is called Start (in yeast) or the Restriction point (in mammalian cells). Most signals that control cell proliferation act at this commitment point to promote either a new mitotic cycle or entry into a quiescent (Q) state.

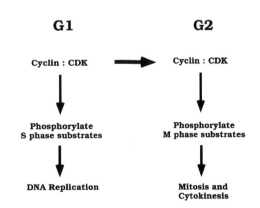

Figure 2. Cell cycle transitions are controlled by cyclin-dependent kinases.

cytokinesis (Fig. 2). However, this is clearly an oversimplification because other transitions in the cell cycle may also be regulated similarly (e.g., entry into S phase at the end of G_1) (Girard et al. 1991; Epstein and Cross 1992; Pagano et al. 1992; Schwob and Nasmyth 1993).

Much progress has been made in understanding the biochemical steps and molecular interactions that control activation of the Cdks during the cell cycle. One essential step in Cdk activation is the assembly of a cyclin/Cdk complex. Cyclins are a family of proteins that are periodically synthesized (or activated) during the cell cycle (Rosenthal et al. 1980), and specific cyclin/Cdk complexes are assembled and activated at different stages in the cell cycle (Hunt 1989; Sherr 1993). Presumably, each different cyclin/Cdk complex executes a unique cell cycle function, and the sequential assembly and activation of these complexes will, at least in part, explain the order in which cell cycle events occur. An important unanswered question is how the activation of a given cyclin/Cdk complex is linked to the activation of cyclin/Cdk complexes that either precede or follow it in this sequence. In addition, the molecular mechanism by which cyclin binding allows Cdk activation is not understood. In vivo, only the Cdks that are bound to cyclins are phosphorylated on an essential threonine residue (Thr-160 for Cdk2) (Solomon et al. 1990; Gu et al. 1992). This phosphorylation is performed by Cdk activating kinase (CAK) (Solomon et al. 1993) and is thought to induce a structural change in the Cdk protein that moves a polypeptide loop out of the substrate binding site (DeBondt et al. 1993). Perhaps the association of a Cdk with cyclin facilitates Cdk recognition by CAK or inhibits dephosphorylation of Cdk by phosphatases.

CYCLIN E

By 1991, a number of independent experimental systems had shown that Cdk function was necessary for the G_1/S phase transition, not only in yeast (Hartwell et al. 1974; Nurse and Bisset 1981), but in higher eukaryotes as well (Blow and Nurse 1990; D'Urso et al. 1990; Furakawa et al. 1990; Fang and Newport 1991; see also Pagano et al. 1993; Tsai et al. 1993; Van den Heuval and Harlow 1994). Studies on Cdc2 regulation at the G_2/M transition had shown that Cdk activation required assembly into a cyclin/Cdk complex, and in yeast it appeared that different classes of cyclins were important at Start and at G_2/M (Cross 1988; Surana et al. 1991). Therefore, it seemed reasonable to expect that higher eukaryotes might express cyclins that were necessary for completion of G_1 and entry into S phase, and that those cyclins would be different from the B-type cyclins necessary for mitosis. A number of different experimental approaches confirmed this prediction and led to the discovery of at least two types of mammalian G_1 cyclins, cyclins D and E (Koff et al. 1991; Lew et al. 1991; Matsushime et al. 1991; Motokura et al. 1991; Xiong et al. 1991). Both of these cyclins bind to and activate Cdks during G_1, and both are necessary for S phase to begin. The discovery and characterization of the D-type cyclins is discussed by Sherr et al. (this volume).

Cyclin E was discovered by genetic complementation of an *S. cerevisiae* strain that was conditionally mutant for endogenous G_1-cyclin expression (Koff et al. 1991; Lew et al. 1991). Molecular and physiological experiments have since confirmed that cyclin E is a G_1 cyclin in mammalian cells. The cyclin E protein is periodically synthesized during the mammalian cell cycle, reaching a peak in abundance in late G_1 and early S phase (Fig. 3) (Dulic et al. 1992; Koff et al. 1992). Overall, its level changes approximately 5-fold between early G_1 and the peak at late G_1, and this change in the cyclin E protein correlates with a similar change in the level of the cyclin E mRNA (Lew et al. 1991). Cyclin E protein is also present in some quiescent cells, and its abundance approximates that seen in early G_1 cells. Cyclin E is a nuclear protein, and significant amounts of cyclin E can first be detected in the nucleus in late G_1 just prior to expression of cyclin A and the onset of DNA synthesis (Fig. 4). In mammalian cells, cyclin E associates primarily with

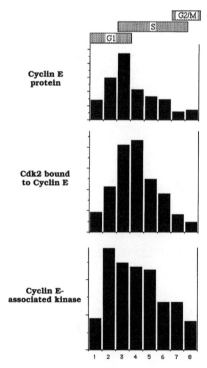

Figure 3. Cyclin E expression and activity during the cell cycle. A human lymphoid cell line (MANCA) was separated by centrifugal elutriation into eight sequential fractions representing progression from G_1 to mitosis. The cell cycle stage of each fraction was determined by flow cytometry to measure nuclear DNA content. The levels of the cyclin E protein, the amount of Cdk2 bound to cyclin E, and cyclin E-associated histone H1 kinase activity are indicated for each fraction.

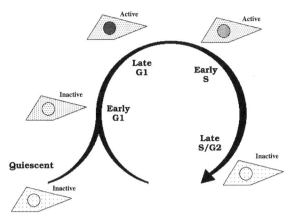

Figure 4. Schematic representation of cyclin E expression and activity in normal human fibroblasts during the cell cycle, as studied by immunofluorescence. Primary human diploid fibroblasts were synchronized by serum starvation. After restimulation, cells were fixed and stained with affinity-purified anti-cyclin E antibodies. Cyclin E accumulates in the nucleus, reaching a peak in late G_1 cells, just prior to DNA synthesis and just prior to expression of cyclin A. Also indicated for each cell cycle position is whether cyclin E-associated kinase is detected upon immunoprecipitation of cyclin E from cell extracts.

Cdk2, and in actively proliferating cells the abundance of the cyclin E/Cdk2 complex and the kinase activity of this complex directly correlate with the abundance of cyclin E itself (Fig. 3). Genetic experiments have shown that cyclin E is necessary for S phase entry in *Drosophila* embryos (Knoblich et al. 1994), and antibody microinjection experiments show that this is true in mammalian cells as well (M. Pagano and J. Roberts, unpubl.).

In addition to being necessary for G_1 transit, the cyclin D1/Cdk4 and the cyclin E/Cdk2 kinases are both rate-limiting for completion of G_1 (Ohtsubo and Roberts 1993; Quelle et al. 1993). Thus, overexpression of either cyclin D1 or cyclin E shortens G_1. It is not possible to tell from these experiments whether cyclins D1 and E are performing overlapping functions, or whether they govern independent rate-limiting steps in G_1 progression. The rate-limiting roles of these cyclin kinases in G_1 suggested that they were likely to be the targets of signals that controlled G_1 progression. In support of this hypothesis, it was found that cyclin accumulation remained rate-limiting even when G_1 was lengthened by growing cells in suboptimal amounts of serum growth factors (Ohtsubo and Roberts 1993). In light of these observations, one approach toward understanding how mitogenic signals promote cell proliferation has been to study the pathways that link these signals to the molecular events necessary for Cdk activation. Our experiments have focused on cyclin E and the cyclin E/Cdk2 complex, primarily because this complex becomes active at the time when the cell becomes committed to a new mitotic cycle, at the restriction point.

MITOGENIC AND ANTIMITOGENIC SIGNALS CONTROL CYCLIN E ACTIVITY

We approached the problem of how proliferative signals control cell cycle proteins by looking both at mitogens that promote cell proliferation and at antimitogens that arrest the cell cycle. The experimental system we developed for studying the effects of positively acting mitogens on cell cycle proteins was stimulation of primary, peripheral blood T lymphocytes by antigen and interleukin-2 (IL-2) (Firpo et al. 1994). At the same time, we explored how an antimitogen, transforming growth factor-β (TGF-β), stopped the cell cycle in G_1 (Koff et al. 1993; Polyak et al. 1994a,b). Remarkably, these two experimental approaches uncovered the same molecular mechanism of cell cycle control. We found the cyclin E/Cdk2 kinase is controlled by p27Kip1, a protein that binds directly to this complex and blocks its activity. p27 is now known to be one member of a family of Cdk inhibitory proteins that are induced by signals which negatively control cell proliferation and down-regulated by signals which promote cell proliferation. Independent work on the mechanism of cell cycle arrest caused by TGF-β has come to very similar conclusions (Slingerland et al. 1994).

TGF-β is a naturally occurring polypeptide that arrests the cell cycle in G_1. It was known that cells arrested by TGF-β contain only unphosphorylated Rb protein (Laiho et al. 1990), implying that they had failed to activate G_1 cyclin/Cdk complexes, the putative Rb kinases. Initial experiments confirmed that cells arrested in G_1 by TGF-β did not contain active cyclin E/Cdk2 kinase but, surprisingly, also revealed that the arrested cells contained just as much cyclin E and Cdk2 protein as did actively proliferating cells (Koff et al. 1993). To characterize the molecular mechanism blocking cyclin E/Cdk2 activation, cell extracts were tested for their ability to assemble active cyclin E/Cdk2 complexes. Although assembly and activation of cyclin E/Cdk2 complexes occurred readily in extracts from proliferating cells, only inactive cyclin E/Cdk2 complexes assembled in extracts from TGF-β-arrested cells (Koff et al. 1993).

Earlier work had isolated inactive cyclin B/Cdc2 complexes as intermediates in the activation of this kinase at the G_2/M transition (Solomon et al. 1990). These inactive Cdc2 complexes were either missing the activating phosphorylation of Thr-161 and/or contained an inhibitory phosphorylation of Tyr-15. Moreover, in some organisms, checkpoint pathways that arrest the cell cycle in G_2 in response to unreplicated or damaged DNA cause accumulation of tyrosine-phosphorylated, inactive Cdc2 (Gould and Nurse 1989; Smythe and Newport 1992). Initially, we pursued the idea that cell cycle arrest in response to antimitogenic signals might be an example of a checkpoint that coordinates the G_1/S transition with the extracellular environment and that the pathway for regulation of cyclin B/Cdc2 at G_2/M would provide a paradigm for the biochemical mechanisms that could regulate cyclin E/Cdk2 at the G_1/S checkpoint. It quickly became clear that there was no increase in Cdk2 tyrosine phosphorylation at the TGF-β cell cycle block. However, the question of whether their inactivity could be explained by the absence of Thr-160 phosphorylation was more complicated. Although extracts from TGF-β-arrested cells contained normal amounts of active CAK (the kinase that phosphorylates Cdk2 on Thr-160), the inactive cyclin E/Cdk2 complexes assembled in those extracts were not phosphorylated by CAK (Koff et al. 1993; Polyak et al. 1994b). In addition, Cdk2 activation was not rescued when extracts from TGF-β-blocked cells were combined with extracts from control cells. Instead, arrested cell extracts contained a factor that prevented Cdk2 activation in control cell extracts (Polyak et al. 1994a). Together these observations were most consistent with the hypothesis that TGF-β induced a factor that blocked Cdk2 activation by CAK. Other experiments showed that this factor did not inhibit CAK directly, but rendered cyclin E/Cdk2 inaccessible to that kinase.

A key experiment that led to the identification of the Cdk2 inhibitor in TGF-β-treated cells was the demonstration that it could be depleted from cell extracts using cyclin E/Cdk2 affinity chromatography (Fig. 5) (Polyak et al. 1994a). This showed that the inhibitor bound directly to cyclin E/Cdk2 and was consistent with an earlier result that the inhibitory activity in cell extracts was titratable with excess cyclin E/Cdk2 complexes. It was then possible to search for proteins that bound specifically to cyclin E/Cdk2 in extracts from TGF-β-treated cells, but not in extracts from control cells. A protein with an apparent molecular size of 27 kD was found to exhibit these properties, and further purification and functional assays showed that this protein was able to block Cdk2 activation (Polyak et al. 1994a). This protein was named p27Kip1 (Cdk inhibitory protein).

Peptides derived from purified p27 were microsequenced, and this information was used to clone a cDNA encoding p27Kip1 (Polyak et al. 1994b). The sequence of p27 revealed that it was structurally related to another Cdk inhibitory protein, p21 (Fig. 6) (El-Deiry et al. 1993; Gu et al. 1993; Harper et al. 1993; Xiong et al. 1993). Together these proteins define a family of cyclin kinase inhibitors that share a number of functional and biochemical properties. Both p21 and p27 are heat-stable proteins that tightly bind to and inhibit all known cyclin A/Cdk, cyclin B/Cdk, cyclin D/Cdk, and cyclin E/Cdk complexes, although the cyclin B/Cdc2 complex is inhibited relatively poorly. The Cdk inhibitors can both block activation of newly assembled cyclin/Cdk complexes and turn off active complexes. In the latter case, the inactivation occurs without altering the phosphorylation state of the active Cdk (i.e., there is no dephosphorylation of T160), indicating that the block to CAK phosphorylation is not sufficient to explain p27's mechanism of inhibition. The Cdk inhibitors are found in cyclin/Cdk complexes in vivo, and overexpression of these proteins in transient expression assays causes cell cycle arrest. It is likely that cell cycle arrest is caused by inhibition of Cdk activity, but it is also possible that these inhibitors have additional targets. p21, for example, binds to and inhibits the activity of proliferating cell nuclear antigen (PCNA), a polymerase δ processivity factor (Waga et al. 1994).

Our studies with human T lymphocytes were designed to find out how mitogenic signals activate cell cycle proteins, and complemented the work on the antimitogenic effects of TGF-β. Most cell types require more than one mitogenic stimulus to complete G_1 and enter S phase (Pardee 1989). This idea was first explored in detail with fibroblasts, which have well-characterized requirements for platelet-derived growth factor in sequence with insulin-like growth factor or epidermal growth factor. T lymphocytes display an analogous requirement for two sequential mitogenic signals. They must first be stimulated through the antigen receptor, which induces expression of the high-affinity IL-2 receptor and competence to respond to IL-2. Exposure to IL-2 then allows completion of G_1 and entry into S phase (Ullman et al. 1990).

We found that the first mitogenic signal, stimulation of the T-cell-antigen receptor, induced expression of

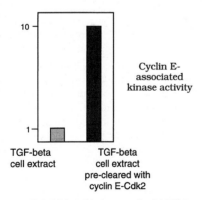

Figure 5. A Cdk inhibitor binds to cyclin E/Cdk2 complexes. Extracts were prepared from TGF-β-arrested mink lung epithelial cells. Recombinant cyclin E was added to the extracts. Assembly and activation of cyclin E/Cdk complexes was then assayed by immunoprecipitation with anti-cyclin E antibodies and testing the immunoprecipitates for histone H1 kinase. Cyclin E was unable to activate Cdk2 in arrested cell extracts. However, cyclin E activation of Cdk2 was restored after pre-clearing the extract by adsorption to a cyclin E/Cdk2 affinity matrix.

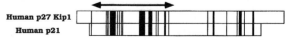

Figure 6. A family of Cdk inhibitory proteins. p27Kip1 is structurally related to p21. Identical amino acids in human p27 and human p21 are indicated by black bars. The primary protein sequence of p27 revealed a 60-amino-acid domain that was 44% identical to p21. This domain, indicated by the double-headed arrow, is sufficient for Cdk inhibitory activity.

cyclin E, cyclin A, and Cdk2, but was not sufficient to promote the transition from G_1 to S (Firpo et al. 1994). Moreover, cyclin E/Cdk2 and cyclin A/Cdk2 complexes assembled but were not active kinases. The second mitogenic signal, IL-2, both allowed Cdk2 activation and caused the cells to enter S phase. The accumulation of inactive G_1 cyclin/Cdk complexes in G_1 lymphocytes clearly paralleled the observations we were making on the cell cycle block caused by TGF-β. In fact, we were able to show that G_0 and G_1 lymphocytes contained a heat-stable 27-kD protein that bound to and inactivated cyclin E/Cdk2 and cyclin A/Cdk2 complexes and that this inhibitor was down-regulated by IL-2. After p27Kip1 had been cloned, it was possible to show that the lymphocyte Cdk inhibitor was human p27Kip1. These results lead to the idea that different mitogenic signals can have independent but complementary effects on cell cycle proteins. In the specific case of T lymphocytes, antigen stimulation promotes the synthesis and assembly of cyclin/Cdk complexes, but the activity of those complexes is kept in check until the lymphocytes receive the second of the two required mitogenic signals, IL-2. This second signal reverses the Cdk block and permits S phase to begin.

REGULATION OF CDK INHIBITORS

Our current picture of the Cdk inhibitors suggests that they set an adjustable threshold for cyclin-dependent activation of the Cdks. According to this model, antimitogens raise the amount of Cdk inhibitor so that it exceeds the amount of cyclin/Cdk complexes in the cell. Conversely, mitogenic signals lower the inhibitor threshold below the level of cyclin/Cdk complexes, permitting Cdk activation and cell cycle progression (Figs. 7 and 8A,B). There is good evidence that p27 levels are regulated in the manner predicted by this model. For example, the p27 protein level decreases more than tenfold when G_1 lymphocytes are mitogenically activated by IL-2 and increases more than tenfold when proliferating lymphocytes stop dividing after IL-2 is withdrawn (Fig. 7A). p27 mRNA levels were also found to increase in response to specific inducers of differentiation, such as retinoic acid (Fig. 7B). HL60 and K562 are immortalized cell lines that can differentiate into specific hematopoietic cell types. HL60 cells, for example, stop dividing and differentiate into macrophages when exposed to retinoic acid, and this is associated with a five- to tenfold increase in the level of p27 mRNA. In contrast, when TPA is used to cause HL60 differentiation into granulocytes, there is no increase in p27, suggesting that p27 may have lineage-specific or inducer-specific effects on cell proliferation. K562 cells ordinarily are not responsive to retinoic acid. Introduction of the retinoic acid receptor renders them responsive to the antiproliferative effect of retinoic acid, and this is associated with induction of p27 (Fig. 7B). We have also found that p27 protein levels decrease five- to tenfold

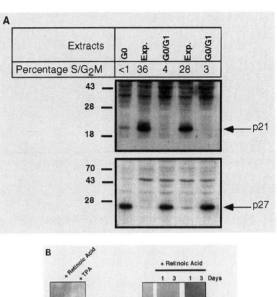

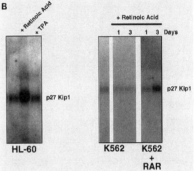

Figure 7. Regulation of p27 by mitogens and antimitogens. (*A*) Peripheral blood human T lymphocytes were stimulated to proliferate using a combination of PHA, IL-2, and anti-CD28 antibodies. After several days of proliferation, the cells were washed and allowed to become quiescent. This cycle was repeated again, and cell extracts from each round of proliferation and quiescence were assayed for the presence of p27 and p21 proteins by immunoblotting. At each round of stimulation and quiescence, cellular DNA content was measured by flow cytometry, and the percentage of cells in S/G_2M is indicated. p27 is expressed in quiescent cells and p21 is expressed in proliferating cells. (*B*) Expression of the p27 mRNA was measured by Northern blot. HL60 cells were induced to differentiate into granulocytes by addition of retinoic acid and into macrophages by TPA. Total cellular RNA was prepared 3 days after induction of differentiation, and p27 expression was compared to proliferating HL60 cells. K562 cells, or K562 cells infected with a retrovirus expressing the retinoic acid receptor (RAR), were treated with retinoic acid for 1 and 3 days, and total RNA was prepared. K562 cells expressing the RAR stop dividing after exposure to retinoic acid, but control cells continue to proliferate.

when quiescent fibroblast cells are stimulated by growth factors to re-enter the cell cycle. Surprisingly, however, p27 protein levels are not affected by TGF-β in mink cells. Instead, TGF-β turns off Cdk4 expression (Ewen et al. 1993), and this releases the p27 that would ordinarily be bound to cyclin D/Cdk4 complexes. The elevated amount of free p27 now exceeds the level of cyclin E, and this prevents Cdk2 activation. These results imply that the interaction of p27 with both cyclin D/Cdk4 and cyclin E/Cdk2 renders cyclin E/Cdk2 activation dependent on prior assembly of a sufficient number of cyclin D/Cdk4 complexes to

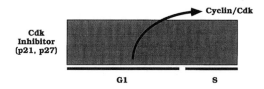
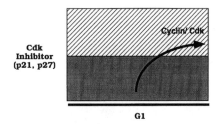
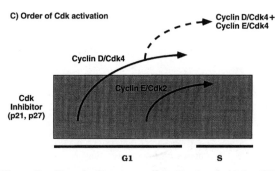

Figure 8. p27 and p21 set an adjustable threshold for Cdk activation. (*A*) During a normal mitotic cycle, the level of p27 or p21 set a cyclin threshold for Cdk activation. When cyclin/Cdk levels are greater than inhibitor levels, the Cdks are activated and the cell cycle proceeds. (*B*) Antimitogenic signals increase the level of Cdk inhibitor so that it exceeds the level of cyclin/Cdk complexes. Since cyclin/Cdk levels never are greater than inhibitor levels, the kinases remain inactive and the cell cycle stops. (*C*) Cyclin D/Cdk4 titrates the Cdk inhibitors allowing activation of cyclin E/Cdk2. In the absence of cyclin D/Cdk4, the levels of cyclin E/Cdk2 would not be sufficient to overcome the inhibitor threshold.

titrate p27 (Fig. 8C). This helps to ensure that these G_1 cyclins are active in a specific sequence during recovery from cell cycle arrest and may also be important in enforcing their order of action during the mitotic cycle.

The results described above implicate p27 as a determinant of cell cycle activation in response to extracellular mitogenic stimuli. On the other hand, a related Cdk inhibitor, p21, may play a more important role in controlling cell cycle progression in response to intracellular checkpoint pathways. p53, a transcription factor for the p21 gene, is required for cell cycle arrest at the G_1/S checkpoint in response to DNA damage. p21 RNA and protein levels increase after DNA damage, and this increase is dependent on p53 (El-Deiry et al. 1993). The increase in p21 inactivates cyclin/Cdk complexes and presumably prevents cell cycle progression (Dulic et al. 1994). When we compared p27 and p21 expression in lymphocytes, we found that p21 increases and p27 decreases as quiescent lymphocytes enter the cell cycle (Fig. 7A). Thus, at least in lymphocytes, the predominant Cdk inhibitor during the transition from G_0 to G_1 is p27, whereas the predominant Cdk inhibitor in cycling cells is p21. This is consistent with the idea that p27 controls entry into and exit from the cell cycle, whereas p21 controls transitions within the cell cycle.

An unresolved issue is whether Cdk inhibitors establish a cyclin threshold for Cdk activation during a normal, unperturbed mitotic cycle. If cells have a constitutive, basal level of Cdk inhibitory activity, then one possibility is that Cdk activation will only occur when the number of cyclin/Cdk complexes is greater than the number of inhibitory molecules (Fig. 8A). The presence of a threshold for Cdk activation in G_1 could play at least two important roles in cell cycle control. First, it has been shown that G_1 cyclin levels in yeast are controlled by a positive feedback loop in which active G_1 cyclin/Cdk complexes promote transcription of the G_1 cyclin genes (Cross and Tinkelenberg 1991; Dirick and Nasmyth 1991). Thus, the cell rapidly and irreversibly progresses from a state in which G_1 cyclins are scarce to one in which they are abundant. In higher eukaryotes there may also be transcriptional feedback loops for cyclin expression, perhaps involving the Rb and E2F protein families. The presence of a threshold for Cdk activation would suppress inappropriate activation of a cyclin-positive feedback loop by small, unscheduled changes in cyclin protein levels. Second, the requirement for a threshold level of cyclin is one way to fix the duration of G_1. Thus, the length of G_1 will be determined by the time it takes to synthesize sufficient cyclin molecules to exceed the inhibitor threshold. There are problems with this model because, in its simplest form, it is not clear how it could account for the effect of cell size on G_1 length (see Cross et al. 1989). However, if we postulate that the inhibitor is present in a constant amount per cell, then it might be possible to explain size control and the timing of G_1 progression by such a mechanism.

Both p21 and p27 can be detected in cyclin/Cdk complexes in proliferating cells, and this is consistent with the idea that these inhibitors might set a threshold for Cdk activation during a normal mitotic cycle. However, it is not known whether the level of p21 or p27 determines the length of G_1. A further complication is that the stoichiometry of the cyclin/Cdk/inhibitor complex has not been determined, and it is unclear whether complexes containing just one inhibitor molecule are active or inactive (i.e., binding of two or more inhibitor molecules may be needed to inactivate a cyclin/Cdk complex). The majority of cyclin/Cdk complexes in proliferating cells may contain just one inhibitor molecule and therefore be active. Thus, the presence of Cdk-associated p21 or p27 in proliferating cells may not reflect the presence of an inhibitor threshold that needs to be overcome for G_1 progression. Instead, p21 and p27 might have other functions, such as promoting the assembly of cyclin/Cdk complexes and directing nuclear uptake of cyclin/Cdk complexes. Cdk inactivation would occur after additional inhibitor molecules bound to the complex.

THE CELL CYCLE AND CANCER

Although tumor cells proliferate under conditions where normal cells do not, there is little evidence that the defects in tumor cells are, or should be, in the cell cycle machinery itself. The rate-limiting steps for cell proliferation in vivo are poorly understood and may bear little relationship to the limiting events studied in cell culture. One extreme example is angiogenesis, a major component of tumor growth in vivo that has no obvious counterpart for cells grown in vitro. On the other hand, cyclin expression is rate-limiting for G_1 progression in cultured cells, and cyclin overexpression has been observed in some tumors (Bartek et al. 1993; Keyomarsi et al. 1994; for review, see Hunter and Pines 1991; Morgan 1992). It would be premature, however, to assume that cyclin expression can be a driving force for tumor progression, rather than just providing a proliferative advantage to a cell that has already acquired other changes that render it tumorigenic. Nevertheless, tumor cells clearly evade normal controls on cell proliferation, and the idea that this is due to a breakdown in the coupling of mitogenic (or antimitogenic) signals to cell cycle proteins is intriguing. The cyclin kinase inhibitors, p21 and p27, link mitogenic signals to the cell cycle. Therefore, mutations in these proteins, or in other proteins that regulate their expression or activity, are likely to give rise to at least some of the properties we associate with oncogenic transformation.

ACKNOWLEDGMENTS

A.K. was supported by a postdoctoral fellowship from the National Institutes of Health. M.O. was supported by a postdoctoral fellowship from the Leukemia Society of America. J.M. is a Howard Hughes Medical Institute Investigator, and support for this work was provided in part by the Howard Hughes Medical Institute. J.M.R. was a Lucille P. Markey Scholar in the Biomedical Sciences, and this work was supported in part by grants to J.M.R. from the Lucille P. Markey Charitable Trust and from the National Institutes of Health.

REFERENCES

Bartek, J., Z. Staskova, G. Draetta, and J. Lukas. 1993. Molecular pathology of the cell cycle in human cancer cells. *Stem Cells* **11:** 51.

Blow, J.J. and P. Nurse. 1990. A Cdc2-like protein is involved in the initiation of DNA replication in *Xenopus* egg extracts. *Cell* **62:** 855.

Cross, F. 1988. *DAF1*, a mutant gene affecting size control, pheromone arrest, and cell cycle kinetics of *Saccharomyces cerevisiae*. *Mol. Cell. Biol.* **8:** 4675.

Cross, F.R. and A.H. Tinkelenberg. 1991. A potential positive feedback loop controlling *CLN1* and *CLN2* gene expression at the start of the yeast cell cycle. *Cell* **65:** 875.

Cross, F., J. Roberts, and H. Weintraub. 1989. Simple and complex cell cycles. *Annu. Rev. Cell Biol.* **5:** 341.

De Bondt, H., J. Rosenblatt, J. Jancarik, H. Jones, D. Morgan, and S.-H. Kim. 1993. Crystal structure of cyclin-dependent kinase 2. *Nature* **363:** 595.

Dirick, L. and K. Nasmyth. 1991. Positive feedback in the activation of G_1 cyclins in yeast. *Nature* **351:** 754.

Dulic, V., E. Lees, and S. Reed. 1992. Association of human cyclin E with a periodic G1-S phase protein kinase. *Science* **257:** 1958.

Dulic, V., W. Kaufman, S. Wilson, T. Tlsty, E. Lees, J.W. Harper, S. Elledge, and S. Reed. 1994. p53-dependent inhibition of cyclin-dependent kinase activities in human fibroblasts during radiation-induced G1 arrest. *Cell* **76:** 1013.

D'Urso, G., R.L. Marraccino, D.R. Marshak, and J.M. Roberts. 1990. Cell cycle control of DNA replication by a homologue from human cells of the p34-Cdc2 protein kinase. *Science* **250:** 786.

El-Deiry, W., T. Tokino, V. Velculescu, D. Levy, R. Parsons, J. Trent, D. Lin, E. Mercer, K. Kinzler, and B. Vogelstein. 1993. *WAF1*, a potential mediator of p53 tumor suppression. *Cell* **75:** 817.

Epstein, C. and F. Cross. 1992. CLB5: A novel B cyclin from budding yeast with a role in S phase. *Genes Dev.* **6:** 1695.

Ewen, M., H. Sluss, L. Whitehouse, and D. Livingston. 1993. cdk4 modulation by TGF-β leads to cell cycle arrest. *Cell* **74:** 1009.

Fang, F. and J. Newport. 1991. Evidence that the G1-S and G2-M transitions are controlled by different Cdc2 proteins in higher eukaryotes. *Cell* **66:** 731.

Firpo, E., A. Koff, M. Solomon, and J. Roberts. 1994. Inactivation of a Cdk2 inhibitor during IL-2 induced proliferation of human T-lymphocytes. *Mol. Cell. Biol.* **14:** 4889.

Furakawa, Y., H. Piwnica-Worms, T.J. Ernst, Y. Kanakura, and J.D. Griffin. 1990. *Cdc2* gene expression at the G_1 to S transition in human T lymphocytes. *Science* **250:** 805.

Girard, F., U. Strausfeld, A. Fernandez, and N. Lamb. 1991. Cyclin A is required for the onset of DNA replication in mammalian fibroblasts. *Cell* **67:** 1169.

Gould, K.L and P. Nurse. 1989. Tyrosine phosphorylation of the fission yeast Cdc2$^+$ protein kinase regulates entry into mitosis. *Nature* **342:** 39.

Gu, Y., J. Rosenblatt, and D.O. Morgan. 1992. Cell cycle regulation of CDK2 activity by phosphorylation of Thr160 and Tyr15. *EMBO J.* **11:** 3995.

Gu, Y., C. Turck, and D.O. Morgan. 1993. Inhibition of Cdk2 activity in vivo by an associated 20K regulatory subunit. *Nature* **366:** 707.

Harper, J., G. Adami, N. Wei, K. Keyomarsi, and S. Elledge. 1993. The p21 Cdk-interacting protein Cip1 is a potent inhibitor of G1 cyclin-dependent kinases. *Cell* **75:** 805.

Hartwell, L. 1991. Twenty-five years of cell cycle genetics. *Genetics* **129:** 975.

Hartwell, L. and T. Weinert. 1989. Checkpoints: Controls that ensure the order of the cell cycle events. *Science* **246:** 629.

Hartwell, L., J. Culotti, J. Pringle, and B. Reid. 1974. Genetic control of the cell division cycle in yeast. *Science* **183:** 46.

Hunt, T. 1989. Maturation promoting factor, cyclin and the control of M-phase. *Curr. Opin. Cell Biol.* **1:** 268.

Hunter, T. and J. Pines. 1991. Cyclins and cancer. *Cell* **66:** 1071.

Keyomarsi, K., N. O'Leary, G. Molnar, E. Lees, H. Fingert, and A. Pardee. 1994. Cyclin E, a potential prognostic marker for breast cancer. *Cancer Res.* **54:** 380.

Knoblich, J., K. Sauer, L. Jones, H. Richardson, R. Saint, and C. Lehner. 1994. Cyclin E controls S phase progression and its down-regulation during *Drosophila* embryogenesis is required for the arrest of cell proliferation. *Cell* **77:** 107.

Koff, A., M. Ohtsuki, K. Polyak, J. Roberts, and J. Massagué. 1993. Negative regulation of G1 in mammalian cells: Inhibition of cyclin E-dependent kinase by TGF-β. *Science* **260:** 536.

Koff, A., F. Cross, A. Fisher, J. Schumacher, K. Leguelle, M. Philippe, and J.M. Roberts. 1991. Human cyclin E, a new cyclin that interacts with two members of the *CDC2* gene family. *Cell* **66:** 1217.

Koff, A., A. Giordano, D. Desai, K. Yamashita, W. Harper, S. Elledge, T. Nishimoto, D. Morgan, R. Franza, and J. Roberts. 1992. Formation and activation of a cyclin E/CDK2 complex during the G1 phase of the human cell cycle. *Science* **257:** 1689.

Laiho, M., J. DeCaprio, J. Ludlow, D. Livingston, and J. Massagué. 1990. Growth inhibition by TGF-β linked to suppression of retinoblastoma protein phosphorylation. *Cell* **62:** 175.

Lew, D.J., V. Dulic, and S.I. Reed. 1991. Isolation of three novel human cyclins by rescue of G_1 cyclin (Cln) function in yeast. *Cell* **66:** 1197.

Li, R. and A. Murray. 1991. Feedback control of mitosis in budding yeast. *Cell* **66:** 519.

Matsushime, H., M. Roussel, R. Ashmun, and C.J. Sherr. 1991. Colony stimulating factor 1 regulates novel cyclins during the G1 phase of the cell cycle. *Cell* **65:** 701.

Morgan, D. 1992. Cell cycle control in neoplastic cells. *Curr. Opin. Genet. Dev.* **2:** 33.

Motokura, T., T. Bloom, H.G. Kim, H. Juppner, J.V. Ruderman, H.M. Kronenberg, and A. Arnold. 1991. A BCL1-linked candidate oncogene which is rearranged in parathyroid tumors encodes a novel cyclin. *Nature* **350:** 512.

Nurse, P. 1990. Universal control mechanism regulating onset of M-phase. *Nature* **344:** 503.

Nurse, P. and Y. Bisset. 1981. Gene required in G1 for commitment to cell cycle and in G2 for control of mitosis in fission yeast. *Nature* **292:** 558.

Ohtsubo, M. and J. Roberts. 1993. Cyclin dependent regulation of G1 in mammalian fibroblasts. *Science* **259:** 1908.

Pagano, M., P. Pepperkok, F. Verde, W. Ansorge, and G. Draetta. 1992. Cyclin A is required at two points in the human cell cycle. *EMBO J.* **11:** 961.

Pagano, M., R. Pepperkok, J. Lukas, B. Baldin, W. Ansorge, J. Bartek, and G. Draetta. 1993. Regulation of the cell cycle by the cdk2 protein kinase in cultured human fibroblasts. *J. Cell Biol.* **121:** 101.

Pardee, A.B. 1974. A restriction point for control of normal animal cell proliferation. *Proc. Natl. Acad. Sci.* **71:** 1286.

———. 1989. G_1 events and regulation of cell proliferation. *Science* **246:** 603.

Polyak, K., J.-Y. Kato, M.J. Solomon, C.J. Sherr, J. Massagué, J.M. Roberts, and A. Koff. 1994a. p27Kip1, a cyclin-Cdk inhibitor, links transforming growth factor-β and contact inhibition to cell cycle arrest. *Genes Dev.* **8:** 9.

Polyak, K., M.H. Lee, H. Erdjument-Bromage, A. Koff, J. Roberts, P. Tempst, and J. Massagué. 1994b. Cloning of p27Kip1, a cyclin-dependent kinase inhibitor and a potential mediator of extracellular antimitogenic signals. *Cell* **78:** 59.

Quelle, D., R. Ashmun, S. Shurtloff, J.Y. Kato, D. Bar-Sagi, M. Roussel, and C.J. Sherr. 1993. Overexpression of mouse D-type cyclins accelerates G1 phase in rodent fibroblasts. *Genes Dev.* **7:** 1559.

Rosenthal, E.T., T. Hunt, and J.V. Ruderman. 1980. Selective translation of mRNA controls the pattern of protein synthesis during early development of the surf clam, *Spisula solidissima*. *Cell* **20:** 487.

Schwob, E. and K. Nasmyth. 1993. CLB5 and CLB6, a new pair of B cyclins involved in S phase and mitotic spindle formation in *S. cerevisiae*. *Genes Dev.* **7:** 1160.

Sherr, C. 1993. Mammalian G1 cyclins. *Cell* **73:** 1059.

Slingerland, J., L. Hengst, C.-H. Pan, D. Alexander, M. Stampfer, and S. Reed. 1994. A novel inhibitor of cyclin-Cdk activity detected in transforming growth factor β-arrested epithelial cells. *Mol. Cell. Biol.* **14:** 3683.

Smythe, C. and J. Newport. 1992. Coupling of mitosis to the completion of S phase in *Xenopus* occurs via modulation of the tyrosine kinase that phosphorylates p34 CDC2. *Cell* **68:** 787.

Solomon, M.J., J. Harper, and J. Shuttleworth. 1993. CAK, the p34 CDC2 activating kinase, contains a protein identical or closely related to p40 MO15. *EMBO J.* **12:** 3133.

Solomon, M.J., M. Glotzer, T.H. Lee, M. Phillippe, and M.W. Kirschner. 1990. Cyclin activation of p34-Cdc2. *Cell* **63:** 1013.

Surana, U., H. Robitsch, C. Price, T. Schuster, I. Fitch, A.B. Futcher, and K. Nasmyth. 1991. The role of CDC28 and cyclins during mitosis in the budding yeast *S. cerevisiae*. *Cell* **65:** 145.

Tsai, L., E. Lees, B. Haha, E. Harlow, and K. Riabowol. 1993. The cdk2 kinase is required for the G_1-to-S phase transition in mammalian cells. *Oncogene* **8:** 1593.

Ullman, K.S., J.P. Northrop, C.L. Verweij, and G.R. Crabtree. 1990. Transmission of signals from the T lymphocyte antigen receptor to the genes responsible for cell proliferation and immune function: The missing link. *Annu. Rev. Immunol.* **8:** 421.

Van den Heuval, S. and E. Harlow. 1994. Distinct roles for cyclin-dependent kinases in cell cycle control. *Science* **262:** 2050.

Waga, S., G. Hannon, D. Beach, and B. Stillman. 1994. The p21 inhibitor of cyclin-dependent kinases controls DNA replication by interaction with PCNA. *Nature* **369:** 574.

Weinert, T. and L. Hartwell. 1988. The *RAD9* gene controls the cell cycle response to DNA damage in *Saccharomyces cerevisiae*. *Science* **241:** 317.

Xiong, Y., T. Connolly, B. Futcher, and D. Beach. 1991. Human D-type cyclin. *Cell* **65:** 691.

Xiong, Y., G. Hannon, H. Zhang, D. Casso, R. Kobayashi, and D. Beach. 1993. p21 is a universal inhibitor of cyclin kinases. *Nature* **366:** 701.

Zetterberg, A. and O. Larson. 1985. Kinetic analysis of regulatory events in G1 leading to proliferation of quiescence of Swiss 3T3 cells. *Proc. Natl. Acad. Sci.* **82:** 5365.

Role of a Cell Cycle Regulator in Hereditary and Sporadic Cancer

A. KAMB

Myriad Genetics, Inc., Salt Lake City, Utah 84108

An assault on the mechanisms that underlie carcinogenesis has come during the last two decades from the directions of biochemistry and genetics. Fundamental contributions have been made by studies in organisms as diverse as yeast and mammals. A host of molecules has been identified which play various roles in tumor formation, most falling into two general categories: oncogene products and tumor suppressors. Not surprisingly, these two basic types include sets of homologs or analogs that act in similar ways and often share structural features.

Biochemical analysis of the apparatus that controls the cell cycle is beginning to identify oncogenes and tumor suppressor genes that participate directly in growth control decisions. At the center of the cell division choice is a family of molecules known as the cyclin-dependent kinases (CDKs), which phosphorylate a number of key substrates to drive the cell through DNA replication and mitosis (for review, see Sherr 1993). The behavior of CDKs is controlled by several classes of regulatory molecules, including a set of associated activating factors, the cyclins, and an emerging family of negative regulators such as p21 (Waf-1, Cip-1, CDKN1), p16 (cdk^{ink4}, MTS1, CDKN2), and p27 (Kip-1) (El-Deiry et al. 1993; Gu et al. 1993; Harper et al. 1993; Serrano et al. 1993; Xiong et al. 1993; Polyak et al. 1994). In addition, the behavior of CDKs is modulated by regulatory kinases and phosphatases that together determine the phosphorylation state of specific CDKs and, hence, their activity. Some of the CDK regulators, including cyclin D1 and p21, have been shown to function as oncogene products or tumor suppressors (Lammie et al. 1991; Motokura et al. 1991; Rosenberg et al. 1991; Withers et al. 1991; Jiang et al. 1992; El-Deiry et al. 1993).

Genetic analysis of cancer-prone families and of somatic tumor cells is also uncovering tumor suppressor genes and oncogenes. As recently as 20 years ago, cancer was not suspected to be a genetic disease but, rather, the result of cumulative environmental insults and stochastic processes. During the last 8 years, however, study of rare familial cancers has revealed numerous genes that play a role in susceptibility to cancer (for review, see Knudson 1993). Genetic studies of more common cancers such as colon cancer, breast cancer, and melanoma have demonstrated that these cancers also have significant genetic components (Hall et al. 1990; Cannon-Albright et al. 1992; Peltomaki et al. 1993). Indeed, estimates of the hereditary component of cancer range from 5% to 10% for most organ sites (Lynch 1967). Positional cloning of several familial cancer genes has led to molecular analysis of new tumor suppressors (WT1, Rb, VHL, NF1, NF2, APC), DNA repair enzymes (hMSh2, hMLh1), and oncogenes (*MEN2A*) (Knudson 1993; Bodmer et al. 1994). In each case where the familial gene has been studied extensively, mutations have been shown to arise spontaneously in sporadic, nonhereditary tumors as well. Thus, the genetic approach to human cancer identifies genes that contribute not only to hereditary cancer, but also to sporadic tumorigenesis.

In this paper, I describe the discovery of a tumor suppressor gene that may play a role in a wide variety of human cancers. The discovery illustrates the interplay between genetics and biochemistry that is helping to illuminate the causes of cancer. The tumor suppressor gene in question has been designated *CDKN2*, but goes by other names as well (e.g., *MTS1*, cdk^{ink4}), and the protein it encodes is referred to as p16, in accordance with its apparent molecular weight assessed by SDS gels (Serrano et al. 1993). I discuss the role of *CDKN2* in sporadic and hereditary cancer, present some arguments for its place in tumor suppression pathways, and also touch on the potential of *CDKN2* as a diagnostic screening target and as a route toward development of anticancer therapeutics.

MLM: A MELANOMA SUSCEPTIBILITY LOCUS

In 1992, Cannon-Albright et al. mapped a principal determinant for melanoma predisposition to a region (9p21) on the short arm of human chromosome 9. This gene, called *MLM*, is linked to a set of polymorphic markers that give a combined LOD score of 12.4 for a set of seven Utah and one Texas melanoma-prone families. Other groups studying kindreds in America, Holland, and Australia have confirmed this result (Gruis et al. 1993; Nancarrow et al. 1993). In only one study has evidence been found for genetic heterogeneity indicative of a second predisposition locus on chromosome 1 (Goldstein et al. 1994). However, three of the families that show evidence of linkage to 1p also show evidence of linkage to 9p21. Thus, a single locus, *MLM*, is the predominant genetic basis for differences in melanoma susceptibility. Certain predisposing al-

Figure 1. Genetic map of a portion of the 9p21 region showing the relative positions of genetic markers and sequence tagged sites referred to in the text. IFNA-l is the polymorphic microsatellite marker used in genetic studies. IFNA-s is a nonpolymorphic fragment that amplifies along with the polymorphic one. D9S736 is a microsatellite marker developed during the course of the cloning of *CDKN2* (Cannon-Albright et al. 1994).

leles of *MLM* increase melanoma risk 20-fold (Cannon-Albright et al. 1993, 1994). Such *MLM* alleles confer dominant susceptibility to melanoma, yet the gene is believed to act somatically as a recessive according to the classic model of tumor suppressor genes proposed by Knudson (1971). A single defective copy of *MLM* predisposes to cancer, and a second mutational event on the other homolog completes one of the steps toward malignancy.

Subsequent to mapping *MLM*, Cannon-Albright et al. (1994) identified several recombinant chromosomes among the individual carriers in the melanoma-prone kindreds. These recombinants were analyzed in detail using genetically linked markers and other markers developed during the efforts to clone *MLM*. In the large Texas kindred, two informative recombinant chromosomes were identified allowing localization of *MLM* to a region flanked by markers D9S736 on the distal (telomeric) side and D9S171 on the proximal (centromeric) side (Fig. 1). This region of 9p21 may encompass more than 1 Mb of DNA.

Previous studies pinpoint 9p21 as the site of frequent chromosomal aberration in melanoma cell lines, gliomas, and leukemias (Diaz et al. 1988, 1990; Lukeis et al. 1990; Middleton et al. 1991; Fountain et al. 1992; Nobori et al. 1992; Cheng et al. 1993; James et al. 1993; Olopade et al. 1993; Cairns et al. 1994; Merlo et al. 1994; van der Riet et al. 1994; Weaver-Feldhaus et al. 1994). In some cases, the genetic lesion involves homozygous deletion of 9p21 sequences, stretching several megabases in certain tumor cell lines (Fountain et al. 1992). These large deletions include markers known to be linked genetically to *MLM*, such as D9S126 and IFNA-l. Assuming that the deletion events are causal in the formation of the tumors, this finding independently indicates the presence of a tumor suppressor gene in 9p21. Ample precedent exists for the observation that familial cancer genes are mutated in sporadic cancer. Thus, a simple hypothesis that explains the presence of chromosomal aberrations in the immediate vicinity of a melanoma susceptibility gene is that *MLM* and the 9p21 tumor suppressor gene are the same.

LOCALIZATION OF A 9p21 TUMOR SUPPRESSOR GENE (*CDKN2*)

The assumed correspondence between *MLM* and the 9p21 tumor suppressor gene suggests a strategy for localizing the gene based on analysis of deletions in tumor DNA. Such approaches have been successful previously in mapping other tumor suppressor genes, including the Wilms' tumor gene (*WT1*), and the *d*eleted in *c*olon *c*ancer gene (*DCC*) (Fearon et al. 1990; Ton et al. 1991). Success depends on identification of homozygous deletions in tumor cell lines or primary tumors using molecular probes within the region of interest. The approximate breakpoints of the deletions are then located and the minimum region of deletion is identified. This approach narrows the region that must be screened for candidate genes. Broken down into steps, the strategy for cloning *MLM* consists of (1) identification of the approximate region of homozygous deletion in tumor cell lines using existing 9p21 markers; (2) construction of a physical map and associated new markers from the region; (3) examination of a large panel of tumor cell lines for homozygous deletions using these markers; (4) identification of the smallest region of deletion; (5) isolation of candidate genes within the region; (6) analysis of the candidate genes for somatic mutations in tumor DNA; and (7) analysis of the candidate genes in kindreds that segregate 9p21-linked melanoma susceptibility.

The initial phase of the experiments identified the area between IFNA-s and D9S171 as the site of a tumor suppressor gene based on loss of this region in several tumor cell lines (Weaver-Feldhaus et al. 1994). Using IFNA-s and D9S171, a contiguous physical map was assembled that stretches between the two markers with one gap. The map consists of yeast artificial chromosome (YAC) clones, P1 clones, and cosmid subclones that were used to generate 60 sequence-tagged sites (STSs) in the region. The STSs were used in turn to screen a total of 290 tumor cell lines for homozygous deletions. This analysis identified a cosmid subclone (c5) of one of the YACs that lies within the smallest area of deletion. Indeed, certain STSs within c5 are deleted homozygously from tumor cell lines, whereas others in c5 are retained. This suggests that a tumor suppressor gene lies at least partly within cosmid c5.

DNA sequence analysis of cosmid c5 reveals two sequences of interest. One, named *CDKN2*, is the gene for a previously defined cDNA that encodes a 16-kD inhibitor of CDKs named p16 by its discoverers (Serrano et al. 1993). The other sequence, *MTS2*, is very similar to part of *CDKN2*. Comparison of the genomic *CDKN2* sequence with the p16 cDNA defines 3 coding exons: E1 (150 bp), E2 (307 bp), and E3 (11 bp) (Fig. 2) (Kamb et al. 1994a). *MTS2* contains sequences 93% identical to E2 of *CDKN2*, but lacks about 50 nucleotides from the 3′ end of E2. Recently, cDNA clones corresponding to *MTS2* have been isolated, indicating that *MTS2* is a functional, expressed gene (S. Stone et al., unpubl.). The sequence similarity between *CDKN2* and *MTS2* suggests that p16 and the protein encoded by *MTS2* may be functional homologs, although this possibility remains to be proved. Southern blot experiments using E2 as a

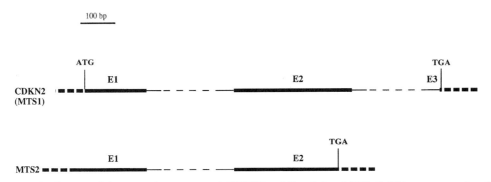

Figure 2. Schematic diagram of the genomic structure of *CDKN2* and its relative *MTS2*. Solid bars represent coding exons; dashed bars represent untranslated regions. Dashed lines represent introns of undefined length. The initiation codon of *MTS2* is not known.

probe fail to find evidence for sequences closely related to *CDKN2* other than *MTS2*. However, it is possible that other more distantly related genes exist that are not detected easily by genomic Southern blots.

INVOLVEMENT OF *CDKN2* IN CANCER

Analysis of the region including *CDKN2* and *MTS2* reveals homozygous deletions in nearly half of 290 tumor cell lines examined (Kamb et al. 1994a). A second group working independently has also observed homozygous deletions at the *CDKN2* locus in tumor cell lines (Nobori et al. 1994). The frequency of homozygous deletions ranges from 82% (14/17) in astrocytoma cell lines to 0 (0/20) in colon cancer lines (Table 1). Of the 14 types of tumor cell line tested, 12 have at least one representative that contains homozygous deletion of *CDKN2* sequences. Considering all cell lines together, the maximum number of deletions is detected with a marker immediately 3′ to E2 in *CDKN2*. When the different tumor cell types are considered separately, all have homozygous deletion frequencies that peak in or very near *CDKN2* (Fig. 3).

No evidence has been found for deletions specifically affecting *MTS2*.

To help exclude the possibility that homozygous deletions might inactivate a tightly linked alternative gene, sequences from *CDKN2* (E1, E2) and *MTS2* (E2) were screened for smaller genetic lesions such as point or frameshift mutations. Of the subset of melanoma lines that do not contain homozygous *CDKN2* deletions, 16/42 contain missense, nonsense, or frameshift mutations (Kamb et al. 1994a; N.A. Gruis et al., in prep.). One line contains 2 heterozygous mutations yielding a total of 17 mutations. Of these 17, 11 result in truncation of the p16 molecule and are, therefore, likely to be disruptive to *CDKN2* function. Similar studies of other types of tumor cell lines reveal a rather low frequency of such genetic lesions; nonetheless, 5/13 tumor cell line types analyzed to date contain at least one example of a point or frameshift mutation (Table 1). DNA sequence analysis of *MTS2* has not revealed any mutations likely to affect gene function. All *MTS2* sequence variants are located in presumptive introns, dozens of nucleotides from intron-exon junctions, and are relatively common in the normal population (Kamb et al. 1994a and in prep.).

Table 1. Genetic Lesions in Tumor Cell Lines

Tumor type	Deletions	% Deletions	Cell line mutations	% Cell line mutations	Tumor mutations	% Tumor mutations
Astrocytoma	3/17	18	0/3	0		
Bladder	6/16	38	3/10	30	3/33	33
Breast	6/10	60	0/6	0		
Colon	0/20	0	0/15	0		
Esophagus					14/27	52
Glioma	32/43	74	0/9	0		
Leukemia	9/18	50	0/4	0		
Lung	19/70	27	2/29	7		
NSC lung			1/7	14		
Melanoma	65/112	58	17/42	40	5.34	15
Neuroblastoma	0/10	0	0/8	0		
Sarcoma	3/5	60	4/10	40		
Ovary	2/7	29	0/5	0		
Renal	5/9	56	0/6	0		
Total	160/336	48	27/144	19	22/94	23

The ratios represent the number of lines having *CDKN2* lesions of a particular type over the number screened. For mutations (defined as nonsense, missense, and frameshift mutations), only cell lines without homozygous deletions were included. Thus, the denominator does not include those cell lines known to have homozygous deletions. The numbers include data from Nobori et al. (1994) and Mori et al. (1994).

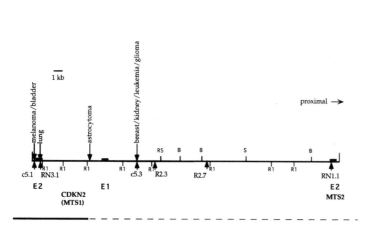

Figure 3. Position of homozygous deletion frequency maxima for various tumor cell lines shown above the restriction map of cosmid c5 with coding exons E1 and E2 from *CDKN2* and E2 from *MTS2* shown. (B) *Bam*HI, (S) *Sal*I, (R1) *Eco*RI, (R5) *Eco*RV. The direction of transcription of *CDKN2* and *MTS2* is from right to left. Nine markers that cover a region including *CDKN2* and *MTS2* were tested against cell lines derived from various types of tumors (Kamb et al. 1994a). The position of the most frequently deleted marker in each cell line type is shown by an arrow. When two markers were deleted at equal frequency, the midpoint is shown as the site of most frequent deletion. Stubby arrows show the positions of the six markers that lie within c5. Only cell line types with several examples where both deletion breakpoints are roughly mapped are shown. Below are shown the approximate extent of the most proximal deletion breakpoint observed in the analysis of 290 cell lines (*upper* line) and the most distal breakpoint (*lower* line). Dashed lines show lost genomic DNA.

Analysis of *CDKN2* mutations in primary melanoma and bladder tumors demonstrates the presence of point mutations at frequencies similar to those observed in melanoma and bladder cell lines, respectively (N.A. Gruis et al., in prep.). Moreover, the types of mutations found in cell lines are similar to those found in the corresponding primary tumor. For example, in melanoma cell lines, point mutations occur primarily in pyrimidine-rich areas of sequence. In two cases, the point mutations involve substitution of two adjacent bases. This type of double mutation is also seen in primary melanomas and is expected from mutagenesis induced by UV irradiation (Brash et al. 1991; Hutchinson 1994). Although the tumor samples were not tested for homozygous deletions, these results suggest that mutations occur in bladder cancer and melanoma primarily in vivo and may not occur to a significant extent in culture. However, analysis of homozygous deletions in these cancer types, as well as other types, is needed to quantify the frequency of inactivation of *CDKN2* in different primary tumors. The relatively low frequencies of point mutations detected in most cell lines contrast with the high frequency of homozygous deletions. In all types of tumor cell lines examined, the frequency of homozygous deletions is higher (Q. Liu et al., in prep.). These results suggest that deletion may be the predominant means of inactivation of *CDKN2* (Table 1). However, a recent report demonstrates that frameshift and missense mutations are observed frequently (14/27) in esophageal carcinomas (Mori et al. 1994). Thus, the spectrum of disruptive *CDKN2* mutations likely depends on the tumor type in question.

If *CDKN2* mutations originate primarily in vivo, it is important to know when in the course of tumor development they occur. All cancer types, regardless of the tissue of origin, progress through different stages of cell proliferation leading to malignancy. As illustrated by the elegant analysis of colon cancer progression, certain genotypic changes correlate with the growth and invasiveness of specific tumor stages (for review, see Fearon et al. 1992; Bodmer et al. 1994). In colon cancer, no evidence of *CDKN2* involvement has been found (Kamb et al. 1994a; Q. Liu et al., in prep.). However, melanomas and bladder carcinomas are known to progress through stages that involve loss of heterozygosity (LOH) of 9p. Indeed, 9p LOH is one of the early events detected in progression of both melanoma and bladder carcinoma (Fountain et al. 1990; Dalbagni et al. 1993). This finding implies that, if *CDKN2* is one of the genes affected by LOH, mutation at *CDKN2* occurs early in tumor formation.

ROLE OF *CDKN2* IN GENETIC SUSCEPTIBILITY TO MELANOMA

Several features confer on *CDKN2* the appearance of an ideal candidate for the melanoma susceptibility gene *MLM*: (1) *CDKN2* lies within the 2-cM interval where *MLM* maps genetically; (2) *CDKN2* is deleted or mutated at high frequency in melanoma cell lines; (3) *CDKN2* is mutated in primary melanomas; and (4) *CDKN2* encodes a protein that appears to function as an inhibitor of the cell cycle. For these reasons, we have undertaken a thorough search for germ-line mutations in the coding sequence and associated splice junctions of *CDKN2*.

Analysis of *CDKN2* in 13 kindreds that segregate 9p-linked melanoma susceptibility, as well as in 38 additional melanoma-prone families, uncovered 2 candidate predisposing mutations, resulting in a Gly→Trp substitution in one case, and a Val→Asp substitution in the other (Kamb et al. 1994b). Both are rare in the normal population and are linked to the predisposition-carrying chromosome. Neither involves a conservative substitution. Thus, these two missense substitutions are candidates for predisposing mutations. However, no clearly disruptive mutations such as nonsense or frameshift mutations have been found in germ-line DNA. A single lymphoblastoid cell line derived from a patient with dysplastic nevus syndrome, a condition characterized by abnormal skin moles, contains an apparently heterozygous nonsense mutation in *CDKN2* (Nobori et al. 1994; Q. Liu, unpubl.). However, lymphoblastoid lines characteristically exhibit variable aneuploidy and, therefore, may not contain a faithful representation of germ-line DNA. In the absence of further confirmatory evidence, it is difficult to rule out the possibility that this mutation occurred during culture of the cells.

Because of the prevalence of *CDKN2* homozygous deletions in tumor cell lines, we also tested the 9p21-linked families and other melanoma-prone families for heterozygous deletions or DNA rearrangements at this locus. No evidence for such deletions was found, a result that fits with analysis of point mutations where, apart from the lymphoblastoid mutation mentioned above, no unequivocally disruptive point mutations were detected (Kamb et al. 1994b). These findings contrast sharply with the presence of a substantial percentage of disruptive mutations in melanoma cell lines: 58% of cell lines contain homozygous deletions of *CDKN2*, and 11% contain either nonsense or frameshift mutations. If these changes were distributed in the same way in the germ line as in somatic cells, roughly the same percentage of kindreds should display disruptive mutations (Table 2). Because the screening analysis detected only two candidate missense mutations, the possibilities must be considered that either *CDKN2* is not *MLM*, or that the majority of germ-line predisposing mutations fall outside the *CDKN2* coding sequence. This latter possibility might be expected if, for example, *CDKN2* mutations were haploinsufficient; that is, if heterozygous, germ-line, loss-of-function mutations were lethal. Under such circumstances, reduced function mutations that affect mRNA level might persist, whereas more severe mutations would be selectively lost from the gene pool.

The possibility of a second gene involved in melanoma formation has been raised by Coleman et al. (1994). These workers have found evidence for a homozygous deletion in a melanoma cell (SK-Mel-133) line that does not overlap the region that includes *CDKN2*. However, this result involves a single cell line that contains a highly fragmented chromosome 9 (Coleman et al. 1994). Moreover, the homozygous deletion defined by Coleman et al. falls outside the region within which *MLM* maps genetically (Cannon-Albright et al. 1994). Finally, we have examined SK-Mel-133 for homozygous deletions and found that *CDKN2* is indeed homozygously deleted. This result suggests that either there is confusion surrounding the cell lines used in the analysis, or SK-Mel-133 contains two adjacent homozygous deletions, one removing *CDKN2* and another removing sequences proximal to D9S126 (Fig. 1). With these considerations, it seems unlikely that SK-Mel-133 will lead to another gene that proves to be *MLM*.

BIOCHEMICAL FUNCTION OF *CDKN2*

A potential downstream target for p16 has been proposed by Serrano et al. based on biochemical studies of the *CDKN2* gene product, p16. In these experiments, p16 behaves as a specific inhibitor of CDK4, with relatively poor inhibitory potency against CDK2 (Serrano et al. 1993). This result, coupled with the involvement of *CDKN2* at high frequency in tumor cell lines, suggests a major role for CDK4 in control of the cell cycle and in tumor formation. CDK4 phosphorylates the retinoblastoma protein Rb, causing Rb to release a group of transcription factors, including E2F, which in turn are believed to facilitate the expression of a set of genes involved in the transition from G_1 to S phase (Matsushime et al. 1992; Serrano et al. 1993). In addition, the primary cyclin-binding partner of CDK4 is cyclin D1, believed to be one of the critical factors in driving the cell from G_1 into S phase (Baldwin et al. 1993; Bates et al. 1994). Together, these results support a model in which p16 blocks the transition from G_1 to S phase in the cell cycle (Fig. 4A). However, it is not clear that the biochemical experiments reflect the exact role of p16 in vivo. Indeed, no experiments have yet demonstrated convincingly that CDK4 participates in the cell cycle (van den Heuvel 1993). Moreover, the puzzling observation that p16 is found complexed with CDK4 in transformed cells seems to contradict its presumptive role as an inhibitor of the cell cycle that acts through CDK4 (Serrano et al. 1993). Thus, it is unclear if CDK4 is the only target for p16 in vivo, or if another molecule, perhaps a different CDK, is the relevant target. In the absence of definitive data, it is also possible that p16 may act at other points of the cell cycle, for example, by trapping cells in G_0 arrest. At least in the mouse, Rb appears to be responsible for preventing reentry of differentiated myotubes into the

Table 2. Genetic Lesions in the Germ Lines of Individuals from Melanoma-prone Families, in Melanoma Cell Lines, and in Melanoma Primary Tumors

	No.	Deletion	Missense	Nonsense	Frameshift	Splice site
Germ lines	51	0	2	0	0	0
Cell lines	99	57	6	8	3	0
Tumors	34	n.d.	3	2	0	0

The column at the far left shows the number of each type screened. Other columns show the number of examples detected in each class of lesion. n.d. indicates not done.

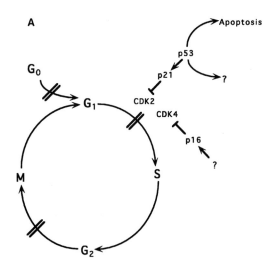

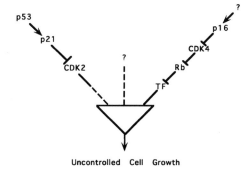

Figure 5. Diagram of three parallel pathways of tumor suppression converging on a central regulatory barrier. Mutation of elements in all three pathways is required to progress through the barrier. The number of pathways, as well as the identity of many of the elements, is unknown. (TF) Transcription factor.

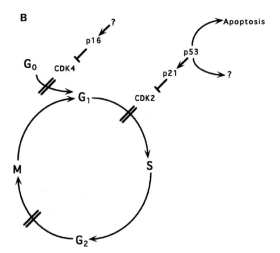

Figure 4. Diagrams showing possible roles for p16 in the cell cycle. Various checkpoints are represented by double bars. Arrows represent activating signals, flat-head lines represent inhibitory signals. Regulators of p16 activity are unknown. (A) Model for p16 function at the G_1 to S transition; (B) model for p16 function at the G_0 to G_1 boundary.

cell cycle (Schneider et al. 1994). As a potential regulator of Rb activity, this finding implies a role for p16 in the process of G_0 arrest (Fig. 4B). However, further experiments are needed to distinguish among these and other possible models.

Although the upstream regulators of *CDKN2* expression are unknown, one reasonable candidate for controlling *CDKN2* expression is p53. *TP53* is mutated in a wide range of tumors and is known to activate transcription of p21, like p16 a presumptive negative regulator of CDKs. However, genetic experiments indicate that p53 does not regulate *CDKN2* expression. The evidence for this conclusion comes from analysis of *TP53* mutations and *CDKN2* mutations in bladder primary carcinomas and cell lines, and in melanoma primary and metastatic tumors and cell lines (N.A. Gruis et al., in prep.). In all these cell types, *TP53* and *CDKN2* mutations occur independently of each other. Tumors and cell lines are found that are mutant for both *CDKN2* and *TP53*. If, for example, *CDKN2* and *TP53* functioned in the same pathway of tumorigenesis, the set of tumors and lines containing *CDKN2* mutations should not overlap the set containing *TP53* mutations. Instead, the sets overlap at a frequency that corresponds to the product of the individual mutation frequencies. This result implies that *TP53* and *CDKN2* function in separate pathways, both of which must be bypassed or inactivated to progress toward malignancy.

What might these pathways be? *TP53* participates in several general processes such as cell cycle checkpoint control, apoptosis, and maintenance of genome stability. The primary biochemical function of *TP53* appears to be as a transcriptional activator, and it regulates the expression of several proteins, including p21. If p21 is one of the important negative regulators of the cell cycle whose function must be removed to manifest malignancy, the mutational analysis suggests that tumor cells often require elimination of both p16 and p21 activities. Thus, p16 and p21 (and p53) may function in parallel pathways that represent largely independent systems for regulating cell division. Progression to frank malignancy may require inactivation of these parallel pathways via mutation of one of the elements in each pathway (Fig. 5). On the basis of this model, tumors should not contain loss-of-function mutations in multiple elements of one pathway; rather, tumors at a particular stage of progression should contain mutations in one of the elements that comprise a specific tumor suppression pathway.

DIAGNOSTIC AND THERAPEUTIC POTENTIAL OF *CDKN2*

As yet, there is no clear basis for developing a diagnostic screening test for genetic predisposition using *CDKN2* mutations. However, in each instance studied, tumor suppressor genes contribute to at least some hereditary cancers. Further studies are needed to determine the role of *CDKN2* in familial melanoma, as well as possible roles in other cancer types. With the gene in hand, this endeavor should be straightforward.

A role for *CDKN2* in somatic diagnosis of cancer seems likely to emerge. As mutations at the *TP53* locus correlate with bladder carcinomas that respond poorly to treatment, *CDKN2* mutational analysis may also be a useful analytical tool for pathologists and oncologists (Elledge et al. 1993; Mitsudomi et al. 1993; Hamelin et al. 1994). Such genetic diagnoses may eventually supersede cruder forms of histological analysis in current practice. Genotypic characterization of tumors may facilitate prediction of phenotypic behavior, potentially very useful in tailoring different types of anticancer therapy. This utility will be tested in the coming months as the role of *CDKN2* in tumor formation is clarified, opening the way for clinical correlation studies.

Therapeutic possibilities for *CDKN2* are exciting although far from certain. In general, tumor suppressor genes offer two avenues toward therapeutic development: gene therapy and functional mimicry. If mutation of *CDKN2* plays an important role in tumor cell growth, replacement of a wild-type *CDKN2* gene or functional mimicry of the p16 protein may provide new routes to therapy. The potential involvement of *CDKN2* in a wide variety of human tumors suggests that it acts as a tumor suppressor in many cell types. Thus, therapeutic compounds based on *CDKN2* might be effective in a substantial percentage of human cancers. On the other hand, if p16 acts in many normal cell types as well, it may be difficult to target tumor cells selectively and avoid the awful side effects that accompany standard chemotherapy. The usefulness of *CDKN2* as a therapeutic agent thus depends on the response of tumor cells and normal cells to a constitutive or pulsed supply of p16 activity. These responses remain to be explored thoroughly.

CONCLUSION

A molecular understanding of cancer is emerging that involves genetic pathways in which various elements are activated to become oncogenes, or inactivated to become nonfunctional tumor suppressor genes. One of the most promising areas of cancer research at present involves study of the basic apparatus that controls cell division. This approach is exciting because many of the molecules that regulate cell division are present in all cells in the body. Thus, it is likely that these central molecules are involved in tumor suppression in a wide range of tissues. Molecules that regulate cell division at sites more removed from the cell cycle apparatus are more likely to be involved in specific cancer types (Fig. 6). The discovery of *CDKN2* as a potential tumor suppressor gene active in a wide range of tissues supports the view that molecules that regulate the cell cycle are prime candidates for important genes in tumorigenesis.

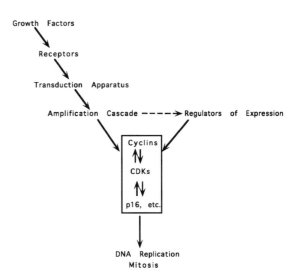

Figure 6. Diagram showing the different classes of molecules that participate in cell growth decisions. The various signals converge on the cell cycle control apparatus itself, shown inside the box.

Note Added in Proof

After this manuscript was submitted, two relevant articles appeared. One reports the sequence and expression of a full-length *MTS2* cDNA. This clone directs the synthesis of a 15-kD (p15) protein with similar biochemical properties to p16 (Hannon and Beach 1994). A second paper describes candidate predisposing mutations in *CDKN2* in several melanoma-prone kindreds, including the nonsense mutation reported by Nobori et al. (1994) (Hussussian et al. 1994).

ACKNOWLEDGMENTS

I thank M. Skolnick, L. Cannon-Albright, S. Tavitigian, K. Harshman, and S. Stone for comments on the manuscript, and F. Bartholomew for expert assistance in manuscript preparation.

REFERENCES

Baldwin, V., J. Likas, M.J. Marcote, M. Pagano, J. Bartek, and G. Draetta. 1993. Cyclin D1 is a nuclear protein required for cell cycle progression in G1. *Genes Dev.* **5:** 812.

Bates, S., L. Bonetta, D. MacAllan, D. Parry, A. Holder, C. Dickson, and G. Peters. 1994. CDK6 (PLSTIRE) and CDK4 (PSK-J3) are a distinct subset of the cyclin-dependent kinases that associate with cyclin D1. *Oncogene* **9:** 71.

Bodmer, W., T. Bishop, and P. Karran. 1994. Genetic steps in colorectal cancer. *Nat. Genet.* **6:** 217.

Brash, D.E., J.A. Rudolph, J.A. Simon, A. Lin, G.J. McKenna, H.P. Baden, J. Halperin, and J. Ponten. 1991. A role for sunlight in skin cancer: UV-induced p53 mutations in squamous cell carcinoma. *Proc. Natl. Acad. Sci.* **88:** 10124.

Cairns, P., K. Tokino, Y. Eby, and D. Sidransky. 1994. Homozygous deletions of 9p21 in primary human bladder tumors detected by comparative multiplex polymerase chain reaction. *Cancer Res.* **54:** 1422.

Cannon-Albright, L.A., W.P. McWhorter, L.J. Meyer, D.E. Goldgar, C.M. Lewis, J.J. Zone, and M.H. Skolnick. 1993. Penetrance and expressivity of the chromosome 9p melanoma susceptibility gene. *Am. J. Hum. Genet. (suppl.)* **53:** 171. (Abstr.)

Cannon-Albright, L.A., D.E. Goldgar, S. Neuhausen, N.A. Gruis, D.E. Anderson, C.M. Lewis, M. Jost, T.D. Tran, K. Nyguen, A. Kamb, J. Weaver-Feldhaus, L.J. Meyer, J.J. Zone, and M.H. Skolnick. 1994. Localization of the 9p melanoma susceptibility locus to a 2cM region between D9S736 and D9S171. *Genomics* (in press).

Cannon-Albright, L.A., D.E. Goldgar, L.J. Meyer, C.M. Lewis, D.E. Anderson, J.W. Fountain, M.E. Hegi, R.W. Wiseman, E.M. Petty, A.E. Bale, O.I. Olopade, M.O. Diaz, D.J. Kwiatkowski, M.W. Piepkorn, J.J. Zone, and M.H. Skolnick. 1992. Assignment of a locus for familial melanoma MLM, to chromosome 9p13-22. *Science* **258:** 1148.

Cheng, J.Q., S.C. Jhanwar, Y.Y. Lu, and J.R. Testa. 1993. Homozygous deletions within 9p21-22 identify a small critical region of chromosomal loss in human malignant mesotheliomas. *Cancer Res.* **53:** 4761.

Coleman, A., J.W. Fountain, T. Nobori, O.I. Olopade, G. Robertson, D.E. Housman, and T.G. Lugo. 1994. Distinct deletions of chromosome 9p associated with melanoma versus glioma, lung cancer, and leukemia. *Cancer Res.* **54:** 344.

Dalbagni, G., J. Presti, V. Reuter, W.R. Fair, and C. Cardon-Cardo. 1993. Genetic alterations in bladder cancer. *Lancet* **342:** 469.

Diaz, M.O., C.M. Rubin, A. Harden, S. Ziemen, R.A. Larson, M.M. Le Beau, and J.D. Rowley. 1990. Deletions of interferon genes in acute lymphoblastic leukemia. *N. Engl. J. Med.* **322:** 77.

Diaz, M.O., S. Ziemin, M.M. Le Beau, P. Pitha, S.D. Smith, R.R. Chilcote, and J.D. Rowley. 1988. Homozygous deletion of the α- and β_1-interferon genes in human leukemia and derived cell lines. *Proc. Natl. Acad. Sci.* **85:** 5259.

El-Deiry, W.S., T. Tokino, V.E. Velculescu, D.B. Levy, R. Parsons, J.M. Trent, D. Lin, W.E. Mercer, K.W. Kinzler, and B. Vogelstein. 1993. WAF1, a potential mediator of p53 tumor suppression. *Cell* **75:** 817.

Elledge, R.M., S.A. Fuqua, G.M. Clark, P. Pujol, D.C. Allred, and W.L. McGuire. 1993. Prognostic significance of p53 gene alterations in node-negative breast cancer. *Breast Cancer Res. Treat.* **26:** 225.

Fearon, E.R., K.R. Cho, J.M. Nigro, S.E. Dern, J.W. Simons, J.M. Ruppert, S.R. Hamilton, A.C. Preisinger, G. Thomas, K.W. Kinzler, and B. Vogelstein. 1990. Identification of a chromosome 18q gene that is altered in colorectal cancers. *Science* **247:** 49.

Fountain, J.W., S.H. Bale, D.E. Housman, and N.C. Dracopoli. 1990. Genetics of melanoma. *Cancer Surv.* **9:** 645.

Fountain, J.W., M. Karayiogou, M.S. Ernstoff, J.M. Kirkwood, D.R. Vlock, L. Titus-Ernstoff, B. Bouchard, S. Vijayasaradhi, A.N. Houghton, J. Lahti, V.J. Kidd, D.E. Housman, and N.C. Dracopoli. 1992. Homozygous deletions within human chromosome band 9p21 in melanoma. *Proc. Natl. Acad. Sci.* **89:** 10557.

Goldstein, A.M., N.C. Dracopoli, M. Engelstein, M.C. Fraser, W.H. Clark, and M.A. Tucker. 1994. Linkage of cutaneous malignant melanoma/dysplastic nevi to chromosome 9p, and evidence for genetic heterogeneity. *Am. J. Hum. Genet.* **54:** 489.

Gruis, N.A., L.A. Sandkuijl, J.L. Weber, A. van der Zee, A.-M. Borgstein, W. Bergman, and R.R. Frants. 1993. Linkage analysis in Dutch familial atypical multiple mole-melanoma (FAMMM) syndrome families. Effect on naevus count. *Melanoma Res.* **3:** 271.

Gu, Y., C.W. Turck, and D.O. Morgan. 1993. Inhibition of CDK2 activity in vivo by an associated 20K regulatory subunit. *Nature* **366:** 707.

Hall, J.M., M.K. Lee, B. Newman, J.E. Morrow, L.A. Anderson, B. Huey, and M.-C. King. 1990. Linkage of early-onset familial breast cancer to chromosome 17q21. *Science* **250:** 1684.

Hamelin, R., P. Laurent-Puig, S. Oschwang, N. Jego, B. Asselain, Y. Remvikos, J. Girodet, R.J. Salmon, and G. Thomas. 1994. Association of p53 mutations with short survival in colorectal cancer. *Gastroenterology* **106:** 42.

Hannon, G.J. and D. Beach. 1994. p15^{INK4B} is a potential effector of TGF-β-induced cell cycle arrest. *Nature* **371:** 257.

Harper, J.W., G.R. Adami, N.Wei, K. Keyomarsi, and S.J. Elledge. 1993. The p21 cdk-interacting protein Cip1 is a potent inhibitor of G1 cyclin-dependent kinases. *Cell* **75:** 805.

Hussussian, C.J., J.P. Struewing, A.M. Goldstein, P.A.T. Higgins, D.S. Ally, M.D. Sheahan, W.H. Clark, Jr., M.A. Tacker, and N.C. Dracopoli. 1994. Germline p16 mutations in familial melanoma. *Nat. Genet.* **8:** 15.

Hutchinson, F. 1994. Induction of tandem-base change mutations. *Mutat. Res.* (in press).

James, C.D., J. He, V.P. Collins, M.J. Allalunis-Turner, and R.S. Day. 1993. Localization of chromosome 9p homozygous deletions in glioma cell lines with markers constituting a continuous linkage group. *Cancer Res.* **53:** 3674.

Jiang, W., S.M. Kahn, N. Tomita, Y.J. Zhang, S.H. Lu, and I.B. Weinstein. 1992. Amplification and expression of the human cyclin D gene in esophageal cancer. *Cancer Res.* **52:** 2980.

Kamb, A., N.A. Gruis, J. Weaver-Feldhaus, Q. Liu, K. Harshman, S.V. Tavtigian, and M.H. Skolnick. 1994a. A cell cycle regulator potentially involved in genesis of many tumor types. *Science* **264:** 436.

Kamb, A., D. Shattuck-Eidens, R. Eeles, Q. Liu, N.A. Gruis, W. Ding, C. Hussey, T. Tran, Y. Miki, J. Weaver-Feldhaus, M. McClure, J.F. Aitkin, D.E. Anderson, W. Bergman, R. Frants, D.E. Goldgar, A. Green, R. MacLennan, N.G. Martin, L.J. Meyer, P. Youl, J.J. Zone, M.H. Skolnick, and L.A. Cannon-Albright. 1994b. Analysis of the p16 gene (CDKN2) as a candidate for the chromosome 9p melanoma susceptibility locus. *Nat. Genet.* **8:** 22.

Knudson, A.G. 1971. Mutation and cancer: Statistical study of retinoblastoma. *Proc. Natl. Acad. Sci.* **68:** 820.

———. 1993. All in the (cancer) family. *Nat. Genet.* **5:** 103.

Lammie, G.A., V. Fantl, R. Smith, E. Schuuring, S. Brookes, R. Michalides, C. Dickson, A. Arnold, and G. Peters. 1991. D11S287, a putative oncogene on chromosome 11q 13, is amplified and expressed in squamous cell and mammary carcinomas and linked to BCL-1. *Oncogene* **6:** 439.

Lukeis, R., L. Irving, M. Gason, and S. Hasthorpe. 1990. Cytogenetics of non-small-cell lung cancer: Analysis of consistent non-random abnormalities. *Genes Chromosomes Cancer* **2:** 116.

Lynch, H.T. 1967. *Hereditary factors in carcinoma.* Springer-Verlag, New York.

Matsushime, H., M.E. Ewen, D.K. Strom, J.Y. Kato, S.K. Hanks, M.F. Hanks, M.F. Roussel, and C.J. Sherr. 1992. Identification and properties of an atypical catalytic subunit for mammalian D type cyclins. *Cell* **71:** 323.

Merlo, A., E. Gabrielson, F. Askin, and D. Sidransky. 1994. Frequent loss of chromosome 9 in human primary non-small cell lung cancer. *Cancer Res.* **54:** 640.

Middleton, P.G., R.A. Prince, I.K. Williamson, P.R.A. Taylor, M.M. Reid, G.H. Jackson, F. Katz, J.M. Chessells, and S.J. Proctor. 1991. Alpha interferon gene deletions in adults, children and infants with acute lymphoblastic leukemia. *Leukemia* **5:** 680.

Mitsudomi, T., T. Oyama, T. Kusano, R. Nakanishi, and T. Shirakusa. 1993. Mutations of the p53 gene as a predictor of poor prognosis in patients with non-small-cell lung cancer. *J. Natl. Cancer Inst.* **85:** 2018.

Mori, T., K. Mire, T. Ai, T. Nishihira, S. Mori, and Y.

Nakamura. 1994. Frequent somatic mutation of the MTS1/CDK41 (multiple tumor suppressor/cyclin-dependent kinase 4 inhibitor) gene in esophageal squamous cell carcinoma. *Cancer Res.* **54:** 3396.

Motokura, T., T. Bloom, H.G. Kim, H. Juppner, J.V. Ruderman, H.M. Kronenberg, and A. Arnold. 1991. A novel cyclin encoded by a BCL1-linked candidate oncogene. *Nature* **350:** 512.

Nancarrow, D.J., J.M. Palmer, N.K. Walters, M.B. Kerr, G.J. Hafner, L. Garske, R. McLeod, and N.K. Hayward. 1993. Confirmation of chromosome 9p linkage in familial melanoma. *Am. J. Hum. Genet.* **53:** 936.

Nobori, T., J.M. Cowan, J.D. Rowley, and M.O. Diaz. 1992. Molecular analysis of deletions of the short arm of chromosome 9 in human gliomas. *Cancer Res.* **52:** 2523.

Nobori, T., K. Mire, D.J. Wu, A. Lois, K. Takabayashi, and D. Carson. 1994. Deletions of the cyclin-dependent kinase-4 inhibitor gene in multiple human cancers. *Nature* **368:** 753.

Olopade, O.I., D.L. Buchhagen, K. Malik, J. Sherman, T. Nobori, S. Bader, M.M. Nau, A.F. Gazdar, J.D. Minna, and M.O. Diaz. 1993. Homozygous loss of the interferon genes defines the critical region on 9p that is deleted in lung cancers. *Cancer Res.* **53:** 2410.

Peltomaki, P., L.A. Aaltonen, P. Sistonen, L. Pylkkanen, J.P. Mecklin, H. Jarvinen, J.S. Green, J.R. Jass, J.L. Weber, F.S. Leach, G.M. Petersen, S.R. Hamilton, A. de la Chapelle, and B. Vogelstein. 1993. Genetic mapping of a locus predisposing to human colorectal cancer. *Science* **260:** 810.

Polyak, K., J. Kato, M.J. Solomon, M.J. Sherr, J. Massagué, J.M. Roberts, and A. Koff. 1994. P27^{kip1}, a cyclin-Cdk inhibitor, links transforming growth factor-β and contact inhibition to cell cycle arrest. *Genes Dev.* **8:** 9.

Rosenberg, C.L., E. Wong, E.M. Petty, A.E. Bale, Y. Tsujimoto, N.L. Harris, and A. Arnold. 1991. PRAD1, a candidate BCL1 oncogene: Mapping and expression in centrocytic lymphoma. *Proc. Natl. Acad. Sci.* **88:** 9638.

Schneider, J.W., W. Gu, L. Shu, V. Mahdavi, and B. Nadal-Ginard. 1994. Reversal of terminal differentiation mediated by p107 in Rb-/- muscle cells. *Science* **264:** 1467.

Serrano, M., G.J. Hannon, and D. Beach. 1993. A new regulatory motif in cell-cycle control causing specific inhibition of cyclin D/CDK4. *Nature* **366:** 704.

Sherr, C.J. 1993. Mammalian G1 cyclins. *Cell* **73:** 1059.

Ton, C.C., V. Huff, K.M. Call, S. Cohn, L.C. Strong, D.E. Housman, and G.F. Saunder. 1991. Smallest region of overlap in Wilms tumor deletions uniquely implicates an 11p13 zinc finger gene as the disease locus. *Genomics* **10:** 293.

van den Heuvel, S. and E. Harlow. 1993. Distinct roles for cyclin-dependent kinases in cell cycle control. *Science* **262:** 2050.

van der Riet, P., H. Nawroz, R.H. Hruban, R. Corio, K. Tokino, W. Koch, and D. Sidransky. 1994. Frequent loss of chromosome 9p21-22 early in head and neck cancer progression. *Cancer Res.* **54:** 1156.

Weaver-Feldhaus, J., N.A. Gruis, S. Neuhausen, D. Le Paslier, E. Stockert, M. Skolnick, and A. Kamb. 1994. Localization of a putative tumor suppressor gene using homozygous deletions in melanomas. *Proc. Natl. Acad. Sci.* **91:** 7563.

Withers, D.A., R.C. Harvey, J.B. Faust, O. Melnyk, K. Carey, and T.C. Meeker. 1991. Characterization of a candidate bcl-1 gene. *Mol. Cell. Biol.* **11:** 4846.

Xiong, Y., G.J. Hannon, D. Casso, R. Kobayashi, and D. Beach. 1993. p21 is a universal inhibitor of cyclin kinases. *Nature* **366:** 701.

p16^{INK4} Mutations and Altered Expression in Human Tumors and Cell Lines

A. OKAMOTO,* D.J. DEMETRICK,† E.A. SPILLARE,* K. HAGIWARA,* S.P. HUSSAIN,*
W.P. BENNETT,* K. FORRESTER,* B. GERWIN,* M.S. GREENBLATT,* M. SERRANO,†
M. SHISEKI,‡ J. YOKOTA,‡ D.H. BEACH,† AND C.C. HARRIS*

*Laboratory of Human Carcinogenesis, National Cancer Institute, National Institutes of Health, Bethesda, Maryland 20892; †Cold Spring Harbor Laboratory, Cold Spring Harbor, New York 11724; ‡National Cancer Center Research Institute, 1-1, Tsukiji 5-chome, Chuo-ku, Tokyo 104, Japan

The orderly progression of cells through the cell cycle is governed by genes encoding proteins transmitting positive (e.g., activated cyclin and cyclin-dependent kinases [Cdk]) and negative (e.g., inhibitors of Cdk) signals (Norbury and Nurse 1992; Reed 1992; Hunter 1993; Sherr 1993; Ron 1994). Dysregulation of these genes can lead to premature entry into the next phase of the cell cycle prior to completion of critical macromolecular events, including repair of DNA damage, and generate genomic instability and neoplastic transformation (Hartwell 1992). Negative regulation of the cell cycle occurs at G_1 and G_2 checkpoints (Hartwell and Weinert 1989; Pardee 1989; Hartwell 1992). Phosphorylation of the Rb protein by Cdk and the release of Rb-associated proteins, e.g., the transcription factor E2F, are correlated with the transition across the G_1 checkpoint (Buchkovich et al. 1989; Chen et al. 1989; DeCaprio et al. 1989; Mihara et al. 1989; Helin et al. 1992; Kaelin et al. 1992; Cress et al. 1993; Ludlow et al. 1993; Dulic et al. 1994). The free E2F is then available to transcriptionally activate genes encoding proteins critical for S-phase function, including deoxynucleotide biosynthesis (Kim and Lee 1991; Pearson et al. 1991; Slansky et al. 1993).

Three inhibitors of activated cyclin/Cdk complexes controlling the G_1 checkpoint of mammalian cells have recently been identified. A gene (*p21*, *WAF1*, *Cip1*, *Sdi1*) encoding a 21-kD inhibitor of multiple cyclin/Cdk complexes is one of the downstream effectors of p53-mediated G_1 arrest and apoptosis in response to DNA damage (Xiong et al. 1992; El-Deiry et al. 1993, 1994; Harper et al. 1993; Jiang et al. 1993; Dulic et al. 1994; Noda et al. 1994). p27^{Kip1}, an inhibitor of cyclin D2/Cdk4, has been linked to G_1 arrest of cells either exposed to transforming growth factor-β_1 (TGF-β_1) or undergoing contact inhibition in vitro (Polyak et al. 1994). A third inhibitor, p16^{INK4}, complexes with Cdk4/cyclin D1 or 2 and may act in a regulatory feedback circuit with Cdk4, D-type cyclins, and Rb and Rb-related proteins (Serrano et al. 1993). Since genes encoding these and other inhibitors of activated cyclin/Cdk complexes are candidate tumor suppressor genes, our investigation of this class of putative tumor suppressor genes was initiated by examining the genomic structure and expression of p16^{INK4} in human cell lines and primary tumors (Okamoto et al. 1994).

MATERIALS AND METHODS

Immunoprecipitation and Western blot analysis. Protein lysates were prepared as described previously (Xiong et al. 1992). 300 µg of either ^{35}S-labeled or unlabeled protein was incubated with the following antibodies: anti-p16^{INK4} (Serrano et al. 1993), anti-cyclin D1 (UBI), or anti-Rb (Santa Cruz) for 4–12 hours and electrophoresed on 15%, 10%, or 7.5% SDS-polyacrylamide gels. Gels were fixed, enhanced with 1.5 M sodium salicylate, and dried. For Westerns, proteins were electrophoretically transferred to PVDF membranes (Millipore) for 3–18 hours at 150 mA. Membranes were blocked with 2% bovine serum albumin, incubated with one of the above antibodies for 4 hours, and incubated with donkey anti-rabbit immunoglobin (Amersham). Detection of proteins was measured by chemiluminescence (DuPont). The Western analysis of p16^{INK4} was corroborated by immunoprecipitation of [^{35}S]methionine-labeled cellular proteins using anti-p16^{INK4} antibodies.

Northern blotting. Total RNA (10 µg), prepared by acid guanidium thiocyanate-phenol chloroform extraction, was electrophoresed in a 1% denaturing agarose gel and electrophoretically transferred to nylon membranes. Membranes were hybridized with human p16^{INK4} cDNA and GAPDH [^{32}P]dCTP-labeled probes.

Southern blotting. HindIII-digested DNA, prepared by proteinase K digestion and phenol-chloroform extraction, was electrophoresed on a 1% agarose gel and transferred to nylon membranes. Membranes were hybridized with human p16^{INK4} cDNA and ERCC3 (XPB) [^{32}P]dCTP-labeled probes.

Polymerase chain reaction analysis. PCR intronic primers were identified from the p16^{INK4} genomic sequence. Primers for exon 1 were (5′–3′):
1 CGGAGAGGGGGAGAACAG (sense)
2 TCCCCTTTTTCCGGAGAATCG (antisense)

PCR conditions for amplification of exon 1 consisted of a 5-minute denaturation at 94°C, followed by 40 cycles of 40 seconds at 94°C, 40 seconds at 55°C, and 90 seconds at 72°C. Primers for exon 2 were (5′–3′):
1 CTCTACACAAGCTTCCTTTCC (sense)
2 GGGCTGAACTTTCTGTGCTGG (antisense)
PCR conditions for amplification of exon 2 consisted of a 5-minute denaturation at 95°C, followed by 40 cycles of 1 minute at 95°C, 1 minute at 60°C, and 1 minute at 72°C.

PCR-single-strand conformation polymorphism (SSCP) analysis. Genomic DNA (100 ng) was amplified in a total volume of 40 μl in a buffer containing 2.0 μl of [^{32}P]dCTP (3000 Ci/mmole; 10 Ci/ml). The PCR product (2 μl) was diluted 10-fold with a buffer containing 96% formamide. The diluted sample (4 μl) was heat-denatured and applied to a 6% neutral polyacrylamide gel. In addition, 2%, 5%, or 10% glycerol was added to the gels for analysis of exons 1 and 2. For exon 2, the PCR product was digested with *Sma*I before loading.

Sequencing. The PCR product was purified and sequenced by the dideoxy chain termination method using a Sequenase kit (USB).

Plasmids. The 960-bp *Eco*RI fragment of the p16^{INK4} cDNA (Serrano et al. 1993) was removed from Bluescript (Stratagene) vector and ligated into the *Xba*I site of the vector pRc/CMV (Invitrogen). Both orientations were used to create sense and antisense constructs.

Transfection. Cells were transfected with Lipofectin reagent (GIBCO/BRL) according to the provided protocol using 10 μg of either pCMV-p16 sense or pCMV-p16 antisense. Transfections were terminated after 5 hours. For Calu 6 and SK-OV-3 cell lines, cells were replated at a 1:4 dilution 48 hours after transfection and selected with 225 μg/ml and 525 μg/ml of G418 (Geneticin, GIBCO/BRL), respectively, for 5 days. Colonies were isolated using cloning cylinders after 21 days, then expanded, and protein was extracted for Western blot analysis.

Colony-forming efficiency experiments. 48 hours after transfection, HCE-4 cells were replated at a density of 3.3 × 10^5 cells per dish. Cells were selected for 9 days in 199 media (Biofluids) supplemented with 10% fetal bovine serum (Biofluids) containing 275 μg/ml G418. M9K cells were replated at a density of 1.0 × 10^5 cells per dish and were selected for 8 days in LHC-MM media (Biofluids) supplemented with 3% fetal bovine serum (Biofluids) containing 600 μg/ml G418. Colonies were counted using an Autocount image analyzer (Dynatech).

RESULTS

Alteration of the p16^{INK4}, Rb, and Cyclin D1 Expression in Cell Lines

We examined the genomic status and steady-state levels of p16^{INK4} mRNA and protein in cell lines (Fig. 1; Table 1). Expression of the p16^{INK4} gene was analyzed in 29 tumor cell lines (Fig. 1B; Table 1). p16^{INK4} mRNA and protein were detected in normal WI38 human fibroblasts and the nontumorigenic SV40 T-antigen-immortalized human cell lines HET1A, BEAS2B, and THLE5B. In contrast, p16^{INK4} mRNA and protein were not detected in 12 of 28 (43%) and 23

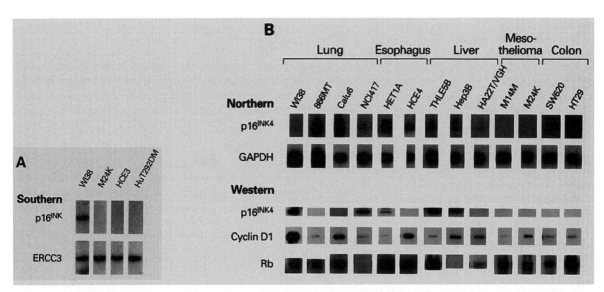

Figure 1. (*A*) Homozygous deletions were detected in cancer cell lines (mesothelioma, M24K; esophageal carcinoma, HCE3; lung carcinoma, HUT 292DM), when compared to normal human fibroblasts (WI38). (*B*) Steady-state levels of p16^{INK4} and GAPDH mRNAs, and p16^{INK4}, Rb, and cyclin D1 proteins in representative cell lines.

Table 1. Expression of the $p16^{INK4}$, Cyclin D1, and Rb Genes and Genomic Status of $p16^{INK4}$ and p53 Genes in Normal and Neoplastic Human Cells

Cell type	$p16^{INK4}$ DNA (SSCP) exon 1	$p16^{INK4}$ DNA (SSCP) exon 2	$p16^{INK4}$ mRNA	$p16^{INK4}$ protein	p53[a] protein	Rb	cyclin D1
Fibroblast							
WI38[b]	P(W)	P(W)	+	+	WT	+	+
Esophagus							
HET1A[c]	P(W)	P(W)	+	+	WT	+	+
HCE4	P(W)	P(W)	+	−	M	+	+
Lung							
BEAS2B[c]	P(W)	P(W)	+	+	WT	+	+
866MT	A(M)	P(W)	+	−	M	+	+
Calu 1	P(W)	P(W)	+	−	M	+	+
Calu 6	P(W)	P(W)	+	−	M	+	+
NCIN417	P(W)	P(W)	+	+	M	−	+
NCIH1155	P(W)	P(W)	+	+	M	−	−
NCIH358	P(W)	P(W)	+	−	M	+	+
NCIH157	P(W)	P(M)	+	−	M	+	+
NCIH322	D	D	−	−	M	+	+
NCIH596	P(W)	P(n.d.)	n.d.	+	M	−	+
Liver							
THLE5B[c]	P(W)	P(W)	+	+	WT	+	+
Hep3B	P(W)	P(W)	+	+	M	−	+
HepG2	P(W)	P(W)	+	+	WT	+	+
HuH4	P(W)	P(W)	+	−	M	+	+
Ha22T/VGH	A(M)	P(W)	+	−	M	+	+
HB611	P(W)	P(W)	+	+	M	+	+
Pancreas							
ASPC1	P(W)	n.d.	+	−	M	+	+
Mesothelioma							
M9K	D	D	−	−	WT	+	+
M24K	D	D	−	−	WT	+	+
M14M	D	D	−	−	WT	+	+
M19	D	D	−	−	WT	+	+
M33K	D	D	−	−	WT	+	+
Colon							
DLD1	P(M)	P(W)	−	−	M	+	+
SW620	P(W)	P(W)	+	−	M	+	+
HT29	P(W)	P(n.d.)	+	−	M	+	+
LS174T	P(W)	P(W)	−	−	WT	+	+
HCT116	P(M)	n.d.	−	−	WT	+	+
SW948	P(W)	P(n.d.)	−	−	n.d.	+	−
SW403	P(W)	P(W)	−	−	n.d.	+	−
HCT15	P(M)	P(W)	−	−	n.d.	+	+

(+) Expressed; (−) not detected; (n.d.) not done; (A) abnormal; (P) present; (D) deleted; (WT) wild type; (M) mutant. All cell lines were obtained from the American Type Culture Collection or established in the authors' and coworkers' laboratories. GAPDH mRNA was positive in all cell lines.

[a] Greenblatt et al. (1994).
[b] Human diploid cell line from normal embryonic lung tissue.
[c] Human epithelial cell line immortalized by SV40 T antigen.

of 29 (79%) tumor cell lines, respectively (Table 1). Rb and cyclin D1 proteins were detected in 26 of 29 lines (90%).

Homozygous Deletions and Intragenic Mutations of the $p16^{INK4}$ Gene in Cell Lines

Homozygous deletions of $p16^{INK4}$ were found in tumor cell lines by amplifying DNA using the PCR with specific intronic primers, and were confirmed by Southern blot analysis in 5 of 5 (100%) pleural mesothelioma, 7 of 18 (39%) lung carcinoma, 1 of 3 (33%) esophageal carcinoma, 1 of 8 (13%) liver carcinoma, 2 of 3 (67%) acute lymphocytic leukemia, 2 of 4 (50%) breast carcinoma, 1 of 1 glioma, and none of 13 colon carcinoma cell lines (Fig. 1A; Table 1) (Okamoto et al. 1994). Since tumor cell lines expressing $p16^{INK4}$ mRNA, but not its translated protein, may harbor nonsense and frameshift mutations in the $p16^{INK4}$ gene, DNA from such cell lines was amplified by PCR using specific primers and then analyzed by SSCP and DNA sequencing. A lung (866MT) and a liver (Ha22T/VGH) carcinoma cell line had abnormal SSCP patterns, and DNA sequencing revealed deletions in exon 1 of the $p16^{INK4}$ gene (Fig. 2; Table 2). Several other tumor cell lines had abnormal SSCP patterns in exon 1 (DLD1, HCT116, HCT15, and CAPAN2) or exon 2

A

Exon 1

cagcgggcggcggggagcagc ATG GAG CCG GCG GCG GGG AGC AGC ATG GAG CCT TCG GCT GAC TGG CTG
GCC ACG GCC GCG GCC CGG GGT CGG GTA GAG GAG GTG CGG GCG CTG CTG GAG GCG GGG GCG CTG
CCC AAC GCA CCG AAT AGT T<u>AC GGT CGG AGG CCG ATC CAG</u>gtgggtagagggtctgcagcgggagcagggg

Exon 2

tctctggcag GTC ATG ATG ATG GGC AGC GCC CGA GTG GCG GAG CTG CTG CTG CTC CAC GGC GCG GAG
CCC AAC TGC GCC GAC CCC GCC ACT CTC ACC CGA CCC GTG CAC GAC GCT GCC CGG GAG GGC TTC
CTG GAC ACG CTG GTG GTG CTG CAC CGG GCC GGG GCG CGG CTG GAC GTG CGC GAT GCC TGG GGC
CGT CTG CCC GTG GAC CTG GCT GAG GAG CTG GGC CAT CGC GAT GTC GCA CGG TAC CTG CGC GCG
GCT GCG GGG GGC ACC AGA GGC AGT AAC CAT GCC CGC ATA GAT GCC GCG GAA GGT CCC TCA
Ggtgaggactg

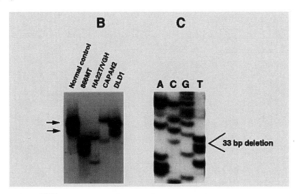

Figure 2. (*A*) Genomic sequence of p16^INK4 exons 1 and 2 including corrections from previously published sequence (Serrano et al. 1993). (*B*) Examples of SSCP analysis of cell lines (866MT, Ha22T/VGH, CAPAN2, DLD1) and (*C*) DNA sequence analysis of 866MT. Arrows indicate mobilities of DNA fragments amplified from normal esophageal tissue sample. 866MT has a 33-bp deletion which is underlined in *A*.

(NCIH157), and some of them were confirmed by DNA sequencing (Table 2).

Alteration of the p16^{INK4} Gene in Primary Tumors

No tumor-specific SSCP patterns were observed in 12 esophageal carcinomas, 38 lung carcinomas, 6 liver carcinomas, 43 colon carcinomas, and 18 ovarian carcinomas. Mobility shifts were found in the germ line and tumors of three patients with lung or ovarian carcinomas (Table 3). Sequence alterations in three of these patients were confirmed at codon 119, GCA^{ala} to TCA^{ser}; at codon 127, GGG^{gly} to GGA^{gly} (lung primary); and at codon 58, CAC^{his} to CGC^{arg} (ovary primary) (Table 2). Figure 3 depicts the mutation pattern of the p16^{INK4} gene, summarizing data from our laboratory and others (Cairns et al. 1994; Kamb et al. 1994; Nobori et al. 1994; Okamoto et al. 1994; S. Kern, pers. comm.). p16^{INK4} mutations usually (74%) produce major protein alterations (nonsense 29%, frameshift 39%, splicing 6%); only 20% were missense, 2% inframe deletion/insertion, and 4% silent.

Functional Assay of the p16^{INK4} Gene

The functional significance of the loss of p16^{INK4} was investigated by transfection of a vector containing the

Table 2. Mutations of p16^{INK4} in Cell Lines and Primary Tumors

Samples	Mutation	Coding effect	Location[a]
866MT (lung)	33-base deletion	splice alteration	2nd base of codon 44– nucleotide 13 in intron 1
HA22T (liver)	23-base deletion	stop codon in exon 1	1st base of codon 21– 2nd base of codon 28
Capan 2 (pancreas)	6-base insertion	(ACG + GCC) Thr + Ala	between codons 19 and 20
HCT 116 (colon)	1-base insertion	stop codon in exon 1	codon 23
Lung primary[b] (N, T)	GCA → TCA	Ala → Ser	codon 127
Lung primary[b] (N, T)	GGG → GGA	none	codon 135
Ovary primary[b] (N, T)	CAC → CGC	His → Arg	codon 66
	GGG → GGA	none	codon 135

[a] These codon numbers are based on numbering 1 at the start codon ATG (Fig. 3A).
[b] This mutation was detected in both tumor and the corresponding normal tissue.

Table 3. p16^{INK4} Gene Alterations in Primary Tumors and Metastases by PCR-SSCP and Southern Analysis

	Esophagus	Lung	Liver	Colon	Ovary	Total
PCR-SSCP	0/12	1/38	0/6	1/43	1/18	3/117
Southern	n.d.	0/28	n.d.	0/43	0/18	0/89
Total	0/12	1/38	0/6	1/43	1/18	3/117

p16^{INK4} cDNA driven by a cytomegalovirus promoter into a human esophageal carcinoma cell line (HCE-4). The colony-forming efficiency of the p16^{INK4}-transfected cells was 112 ± 8 (mean ±S.D.) compared to 192 ± 3 in the cells transfected with antisense p16^{INK4} (68%). A similar degree of inhibition was found using a human mesothelioma cell line (M9K) that had homozygous deletion of p16^{INK4} (53%), whereas Rb-deficient cell line Hep3B was inhibited to a lesser extent (83%). We transfected the p16^{INK4} expression vector into Calu 6 and SK-OV-3, which lack p16^{INK4} expression, and isolated clones were expanded. SK-OV-3 showed rearrangement of the p16^{INK4} gene. Although p16^{INK4} mRNA expression appeared transiently at 48 hours, p16^{INK4} protein was not detected by Western analysis in the expanded cell populations.

DISCUSSION

Previous cytogenetic and allelic deletion analyses have demonstrated frequent deletions and rearrangements of chromosome 9p in carcinoma of the human bladder, lung, head and neck, and esophagus (Vanni et al. 1988; Lukeis et al. 1990; Whang-Peng et al. 1991; Cairns et al. 1993a,b, 1994a; Olopade et al. 1993; Ruppert et al. 1993; Knowles et al. 1994), brain glioma (Miyakoshi et al. 1990; James et al. 1991; Olopade et al. 1992; James et al. 1993), leukemia (Diaz et al. 1988, 1990), mesothelioma (Hagemeijer et al. 1990; Pelin-Enlund et al. 1990; Taguchi et al. 1993), and melanoma (Pedersen et al. 1986; Cowan et al. 1988; Fountain et al. 1992). The p16^{INK4} gene was localized by fluorescence in situ hybridization (FISH) to chromosome 9p21-22 (Okamoto et al. 1994). Our data are consistent with the cytogenetic and allelic analyses of human cancers and tumor cell lines showing homozygous deletions and the recently described mutations of the p16^{INK4} gene, which has been termed the multiple tumor suppressor 1 (*MTS1*) (Kamb et al. 1994) or cyclin-dependent kinase 4 inhibitor gene (Nobori et al. 1994) in human cancers.

Tumor cell lines expressing p16^{INK4} mRNA, but not its translated protein, may harbor nonsense and frameshift mutations in the p16^{INK4} gene. These deletions and insertions are consistent with the DNA polymerase slippage model of endogenous mutagenesis (for review, see Kunkel 1990; Greenblatt et al. 1994) and may have occurred either in vitro or in vivo. The remaining tumor cell lines have normal SSCP patterns, suggesting either defects of p16^{INK4} mRNA at the level of processing and translation or enhanced degradation of the p16^{INK4} protein.

We also initiated a survey of primary tumors analyzed by SSCP prescreening and DNA sequencing. No somatic mutation was detected in 117 primary tumors. Two germ-line mutations were confirmed by sequencing. Further studies are needed to determine if this alteration is related to cancer-proneness or is a polymorphism unrelated to cancer risk. These results and others (Cairns et al. 1994b; Okamoto et al. 1994) indicate that the frequency of intragenic p16^{INK4} gene mutations in some types of primary tumors (i.e., lung, bladder, kidney, head and neck, brain) is quite low.

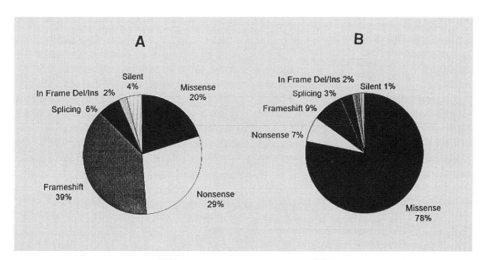

Figure 3. Classes of mutations of the (*A*) p16^{INK4} and (*B*) p53 genes. Most p16^{INK4} mutations (*n* = 49) result in truncation of the protein by base substitutions, which create nonsense codons or alter splicing or frameshift deletion or insertions. Most p53 mutations (*n* = 2837) are missense substitutions, which result in the production of a full-length but mutant protein.

Nevertheless, Mori et al. (1994) recently reported that intragenic mutations of the p16^{INK4} gene were detected in 52% of Japanese esophageal carcinomas. Although we examined microdissected Chinese esophageal carcinoma samples, we did not find any mobility shifts by PCR-SSCP in exons 1 and 2. One possibility is that there are some false negative cases by PCR-SSCP. However, we used several conditions for PCR-SSCP analysis, and PCR products of less than 300 base pairs were analyzed by SSCP, which generally has a false negative rate of less than 10% (Hayashi and Yandell 1993). Another possibility is that there are different genetic and environmental factors for esophageal carcinogenesis between Japanese and Chinese.

Our functional results suggest that expression of transfected p16^{INK4} inhibits growth and is selected against in these cells. The finding of homozygous deletions and nonsense point mutations in cancer cell lines and inhibition of their growth by expression of a transfected p16^{INK4} gene is genetic and functional evidence consistent with the hypothesis that p16^{INK4} is a recessive tumor suppressor gene (Fig. 1A; Tables 1, 2) (Cairns et al. 1994b; Kamb et al. 1994, Nobori et al. 1994; Okamoto et al. 1994; S. Kern, pers. comm.). However, additional studies are required to search for germ-line mutations that segregate with cancer in families and to determine the frequency of p16^{INK4} mutations in a wide spectrum of primary tumors.

The spectrum of p16^{INK4} mutations differs from that of the p53 gene. p16^{INK4} mutations are primarily frameshift, deletions, and nonsense substitutions (74%) (Tables 2, 3) (Cairns et al. 1994b; Kamb et al. 1994; Nobori et al. 1994; Okamoto et al. 1994; S. Kern, pers. comm.). This indicates a complete loss or loss of function of the p16^{INK4} gene product with inhibition of activated cyclin D/Cdk4 phosphorylation of cell-cycle-related substrates including the Rb protein. However, 25% of p16^{INK4} mutations were missense. This may indicate that the presence of some mutant p16^{INK4} proteins can provide cells with a selection advantage, similar to p53 mutations. However, a portion of the missense p16^{INK4} mutations may be genetic polymorphisms. Further investigation is required to determine the functional significance of missense p16^{INK4} mutations. p53 mutations are primarily missense mutations (78%) (for review, see Hollstein et al. 1991; Harris and Hollstein 1993). Missense mutations of p53 can produce both loss of suppressor function and gain of oncogenic activity (for review, see Greenblatt et al. 1994).

The G$_1$ checkpoint can be abrogated by genetic and epigenetic mechanisms. Mutations in the Rb gene can inhibit Rb protein neutralization of the transcriptional activity of E2F (Fig. 4) (for review, see Nevins 1992). Therefore, we examined the status of Rb protein in the cell lines (Fig. 1B; Table 1), since loss of Rb protein is positively correlated with Rb mutations (for review, see Weinberg 1992). The tumor cell lines containing Rb protein rarely expressed p16^{INK4} protein, and conversely, p16^{INK4}-expressing tumor cell lines rarely had detectable Rb protein. This inverse relationship between

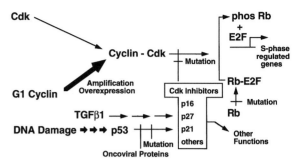

Figure 4. Examples of dysregulated control of the cell cycle G$_1$ checkpoint. The G$_1$ checkpoint can be overridden by enhanced expression of G$_1$ cyclins, e.g., D1, or mutational inactivation of Cdk inhibitors, e.g., p16^{INK4}, p21, or p27^{Kip1}, which would lead to increased activity of Cdk and increased phosphorylation of Rb and Rb-related proteins. G$_1$ arrest in response to DNA damage would be negated either by p53 mutations and reduced *trans*-activation of p21 by mutant p53 protein or by the mutational inactivation of p21. Oncoviral proteins, e.g., SV40T, adenovirus E1B, human papillomavirus E6, or hepatitis B virus X protein, also would inactivate p53 *trans*-activation of p21. Interruption of the TGF-β1 pathway of G$_1$ arrest mediated by *trans*-activation of p27^{Kip1} also would cause cyclin/Cdk activity to go unchecked. Mutational inactivation of Rb also would release E2F to transcriptionally activate critical S-phase genes. The schematic diagram is an oversimplification; e.g., E2F is a family of proteins, and Rb-related proteins, such as the p107 and p130, are involved in cell cycle regulation.

expression of p16^{INK4} and Rb protein is statistically significant ($p < 0.0006$, Fisher exact) and indicates that these proteins participate in the same G$_1$ checkpoint pathway and that a mutation in either Rb or p16^{INK4} is sufficient to disrupt the pathway. p53 mutations do not correlate with either p16^{INK4} or Rb expression in these tumor cell lines ($p = 0.38$ and 0.13, respectively), indicating that p53 participates in an independent pathway such as regulation of the G$_1$ checkpoint in response to DNA damage.

Since cyclin D1 overexpression accelerates entry of cells into S phase (Ando et al. 1993; Ohtsubo et al. 1993; Quelle et al. 1993), negates the G$_1$ arrest induced by Rb introduced into Rb-null cell lines (Dowdy et al. 1993; Ewen et al. 1993), and neoplastically transforms rat embryo fibroblasts alone or in cooperation with the E1A or Ha-*ras* oncogene (Jiang et al. 1993; Hinds et al. 1994; Lovec et al. 1994), the steady-state levels of cyclin D1 were analyzed (Fig. 1B; Table 1). Although the expression of p16^{INK4} and cyclin D1 is not correlated, the positive correlation between Rb and cyclin D1 proteins ($p = 0.05$) is consistent with the recent observation of Rb protein regulating the expression of cyclin D1 (Müller et al. 1994).

Continuous, long-term culture may select for cells with loss of senescence genes. p16^{INK4} is a candidate for the cellular senescence gene mapped to chromosome 9 by cell-cell hybrid analysis of immortal cells (Barrett and Preston 1994). The G$_1$ arrest state of senescent cells is associated with persistent hypophosphorylated Rb protein and inactive cyclin/Cdk complexes, and thus, has much in common with quiescent mortal cells (Stein et al. 1990). Functional defects of senescence genes, including p21 (Noda et al. 1994) and perhaps p16^{INK4}, would allow cells to escape programmed senescence and enhance the probability of their neoplastic transformation.

The overexpression and gene amplification of G_1 cyclins, the mutations in the p16^{INK4}, p53, and Rb genes, and the inactivation of p53 and Rb by oncoproteins of certain DNA viruses such as human papillomavirus, implicate dysregulated control of the G_1 checkpoint in carcinogenesis and tumor progression. The low frequency of p16^{INK4} mutations in primary tumors when compared to the high frequency in tumor cell lines indicates that one should be cautious in extrapolating data from cell lines to primary tumors. Since p16^{INK4} may be inactivated in the late stages of tumor progression, a comparison of primary tumors and their metastases is warranted.

ACKNOWLEDGMENTS

The cell lines we obtained from the American Type Culture Collection are WI38 (human diploid cell line from normal embryonic lung tissue), A549, A427, A2182, 866MT, HUT292DM, DM592, SW1271, SK-Lu-1, Calu 1, Calu 6, NCI-H526, NCI-N417, NCI-H1155, NCI-H358, NCI-H157, NCI-H322, NCI-H596, NCI-H446 (lung cancer cell lines); Hep3B, HepG2, SK-Hep-1, HuH-4, HuH-7, HA22T/VGH, HB611, 2.2.15 (liver cancer cell lines); CAPAN-2, ASPC-1 (pancreatic cancer cell lines); SK-OV-3 (ovarian cancer cell line); DLD-1, COLO320, SW480, SW620, HT29, WIDR, LS174T, HCT116, SK-CO1, SW-948, SW-48, SW-403, HCT-15 (colon cancer cell lines); H9, CCL119, CCL120 (leukemia cell lines); MCF7, HLB100, HTB126, ZRB75 (breast cancer cell lines); and U118 (glioma cell line). HET-1A, HCE-3, HCE-4, and HCE-7 were provided by Dr. G. Stoner. M9K, M24K, M14M, and M19 were provided by Dr. K. Linnainmaa. tc4N was provided by Dr. M. Noguchi. Beas2B and THLE5B were established in our laboratory. We appreciate the aid of our colleagues Stephan Ambs, Peter Shields, Susan Ungar, Tsung-Tang Sun, Judith Welsh, Mohammed Khan; the editorial assistance of Dorothea Dudek; and the expert photomicrography of Ricardo V. Dreyfuss. We also thank Dr. Y. Terashima, Dr. S. Yokoyama, and Dr. H. Takano for providing ovarian cancer samples. This work was supported in part by a grant from the Japanese Overseas Cancer Fellowship of the Foundation for Promotion of Cancer Research (to A.O.).

REFERENCES

Ando, K., F. Ajchenbaum-Cymbalista, and J.D. Griffin. 1993. Regulation of G1/S transition by cyclins D2 and D3 in hematopoietic cells. *Proc. Natl. Acad. Sci.* **90:** 9571.

Barrett, J.C. and G. Preston. 1994. Apoptosis and cellular senescence: Forms of irreversible growth arrest. In *Apoptosis II: The molecular basis of apoptosis in disease* (ed. L.D. Tomei and F.O. Cope), p. 253. Cold Spring Harbor Laboratory Press, Cold Spring Harbor, New York.

Buchkovich, K., L.A. Duffy, and E. Harlow. 1989. The retinoblastoma protein is phosphorylated during specific phases of the cell cycle. *Cell* **58:** 1097.

Cairns, P., M.E. Shaw, and M.A. Knowles. 1993a. Preliminary mapping of the deleted region of chromosome 9 in bladder cancer. *Cancer Res.* **53:** 1230.

———. 1993b. Initiation of bladder cancer may involve deletion of a tumour-suppressor gene on chromosome 9. *Oncogene* **8:** 1083.

Cairns, P., K. Tokino, Y. Eby, and D. Sidransky. 1994a. Homozygous deletion of 9p21 in primary human bladder tumors deleted by comparative multiplex polymerase chain reaction. *Cancer Res.* **54:** 1422.

Cairns, P., L. Mao, A. Merlo, D.J. Lee, D. Schwab, Y. Eby, K. Tokino, P. van der Riet, and D. Sidransky. 1994b. Low rate of p16 (MTS-1) mutations in primary tumors with 9p loss. *Science* **265:** 415.

Chen, P.L., P. Scully, J.Y. Shew, J.Y. Wang, and W.H. Lee. 1989. Phosphorylation of the retinoblastoma gene product is modulated during the cell cycle and cellular differentiation. *Cell* **58:** 1193.

Cowan, J.M., R. Halaban, and U. Francke. 1988. Cytogenetic analysis of melanocytes from premalignant nevi and melanomas. *J. Natl. Cancer Inst.* **80:** 1159.

Cress, W.D., D.G. Johnson, and J.R. Nevins. 1993. A genetic analysis of the E2F1 gene distinguishes regulation by Rb, p107, and adenovirus E4. *Mol. Cell. Biol.* **13:** 6314.

DeCaprio, J.A., J.W. Ludlow, D. Lynch, Y. Furukawa, J. Griffin, H. Piwnica-Worms, C.M. Huang, and D.M. Livingston. 1989. The product of the retinoblastoma susceptibility gene has properties of a cell cycle regulatory element. *Cell* **58:** 1085.

Diaz, M.O., C.M. Rubin, A. Harden, S. Ziemin, R.A. Larson, M.M. Le Beau, and J.D. Rowley. 1990. Deletions of interferon genes in acute lymphoblastic leukemia. *N. Engl. J. Med.* **322:** 77.

Diaz, M.O., S. Ziemin, M.M. Le Beau, P. Pitha, S.D. Smith, R.R. Chilcote, and J.D. Rowley. 1988. Homozygous deletion of the α- and β-1-interferon genes in human leukemia and derived cell lines. *Proc. Natl. Acad. Sci.* **85:** 5259.

Dowdy, S.F., P.W. Hinds, K. Louie, S.I. Reed, A. Arnold, and R.A. Weinberg. 1993. Physical interaction of the retinoblastoma protein with human D cyclins. *Cell* **73:** 499.

Dulic, V., W.K. Kaufmann, S.J. Wilson, T.D. Tlsty, E. Lees, J.W. Harper, S.J. Elledge, and S.I. Reed. 1994. p53-dependent inhibition of cyclin-dependent kinase activities in human fibroblasts during radiation-induced G1 arrest. *Cell* **76:** 1013.

El-Deiry, W.S., T. Tokino., V.E. Velculescu, D.B. Levy, R. Parsons, J.M. Trent., D. Lin, W.E. Mercer, K.W. Kinzler, and B. Vogelstein. 1993. WAF1, a potential mediator of p53 tumor suppression. *Cell* **75:** 817.

El-Deiry, W.S., J.W. Harper, P.M. O'Connor, V.E. Velculescu, C.E. Canman, J. Jackman, J.A. Pietenpol, M. Burrell, D.E. Hill, Y. Wang, K.G. Wiman, W.E. Mercer, M.B. Kastan, K.W. Kohn, S.J. Elledge, K.W. Kinzler, and B. Vogelstein. 1994. WAF1/CIP1 is induced in p53-mediated G1 arrest and apoptosis. *Cancer Res.* **54:** 1169.

Ewen, M.E., H.K. Sluss, C.J. Sherr, H. Matsushime, J. Kato, and D.M. Livingston. 1993. Functional interactions of the retinoblastoma protein with mammalian D-type cyclins. *Cell* **73:** 487.

Fountain, J.W., M. Karayiorgou, M.S. Ernstoff, J.M. Kirkwood, D.R. Vlock, L. Titus-Ernstoff, B. Bouchard, S. Vijayasaradhi, A.N. Houghton, J. Lahti, V.J. Kidd, D.E. Housman, and N.C. Dracopoli. 1992. Homozygous deletions within human chromosome band 9p21 in melanoma. *Proc. Natl. Acad. Sci.* **89:** 10557.

Greenblatt, M.S., W.P. Bennett, M. Hollstein, and C.C. Harris. 1994. Mutations in the p53 tumor suppressor gene: Clues to cancer etiology and molecular pathogenesis. *Cancer Res.* **54:** 4855.

Hagemeijer, A., M.A. Versnel, E. Van Drunen, M. Moret, M.J.

Bouts, T.H. van der Kwast, and H.C. Hoogsteden. 1990. Cytogenetic analysis of malignant mesothelioma. *Cancer Genet. Cytogenet.* **47:** 1.

Harper, J.W., G.R. Adami, N. Wei, K. Keyomarsi, and S.J. Elledge. 1993. The p21 cdk-interacting protein Cip1 is a potent inhibitor of G1 cyclin-dependent kinases. *Cell* **75:** 805.

Harris, C.C. and M. Hollstein. 1993. Clinical implications of the p53 tumor-suppressor gene. *N. Engl. J. Med.* **329:** 1318.

Hartwell, L. 1992. Defects in a cell cycle checkpoint may be responsible for the genomic instability of cancer cells. *Cell* **71:** 543.

Hartwell, L.H. and T.A. Weinert. 1989. Checkpoints: Controls that ensure the order of cell cycle events. *Science* **246:** 629.

Hayashi, K. and D.W. Yandell. 1993. How sensitive is PCR-SSCP? *Hum. Mutat.* **2:** 338.

Helin, K., J.A. Lees, M. Vidal, N. Dyson, E. Harlow, and A. Fattaey. 1992. A cDNA encoding a pRB-binding protein with properties of the transcription factor E2F. *Cell* **70:** 337.

Hinds, P.W., S.F. Dowdy, E.N. Eaton, A. Arnold, and R.A. Weinberg. 1994. Function of a human cyclin gene as an oncogene. *Proc. Natl. Acad. Sci.* **91:** 709.

Hollstein, M., D. Sidransky, B. Vogelstein, and C.C. Harris. 1991. p53 mutations in human cancers. *Science* **253:** 49.

Hunter, T. 1993. Braking the cycle. *Cell* **75:** 839.

James, C.D., J. He, V.P. Collins, M.J. Allalunis-Turner, and R.S. Day. 1993. Localization of chromosome 9p homozygous deletions in glioma cell lines with markers constituting a continuous linkage group. *Cancer Res.* **53:** 3674.

James, C.D., J. He, E. Carlbom, M. Nordenskjold, W.K. Cavenee, and V.P. Collins. 1991. Chromosome 9 deletion mapping reveals interferon alpha and interferon beta-1 gene deletions in human glial tumors. *Cancer Res.* **51:** 1684.

Jiang, W., S.M. Kahn, P. Zhou, Y.J. Zhang, A.M. Cacace, A.S. Infante, S. Doi, R.M. Santella, and I.B. Weinstein. 1993. Overexpression of cyclin D1 in rat fibroblasts causes abnormalities in growth control, cell cycle progression and gene expression. *Oncogene* **8:** 3447.

Kaelin, W.G., Jr., W. Krek, W.R. Sellers, J.A. DeCaprio, F. Ajchenbaum, C.S. Fuchs, T. Chittenden, Y. Li, P.J. Farnham, M.A. Blanar, D.M. Livingston, and E.K. Flemington. 1992. Expression cloning of a cDNA encoding a retinoblastoma-binding protein with E2F-like properties. *Cell* **70:** 351.

Kamb, A., N.A. Gruis, J. Weaver-Feldhaus, Q. Liu, K. Harshman, S.V. Tavtigian, E. Stockert, R.S. Day, B.E. Johnson, and M.H. Skolnick. 1994. A cell cycle regulator potentially involved in genesis of many tumor types. *Science* **264:** 436.

Kim, Y.K. and A.S. Lee. 1991. Identification of a 70-base-pair cell cycle regulatory unit within the promoter of the human thymidine kinase gene and its interaction with cellular factors. *Mol. Cell. Biol.* **11:** 2296.

Knowles, M.A., P.A. Elder, M. Williamson, J.P. Cairns, M.E. Shaw, and M.G. Law. 1994. Allelotype of human bladder cancer. *Cancer Res.* **54:** 531.

Kunkel, T.A. 1990. Misalignment-mediated DNA synthesis errors. *Biochemistry* **29:** 8003.

Lovec, H., A. Sewing, F.C. Lucibello, R. Muller, and T. Moroy. 1994. Oncogenic activity of cyclin D1 revealed through cooperation with Ha-*ras*: Link between cell cycle control and malignant transformation. *Oncogene* **9:** 323.

Ludlow, J.W., C.L. Glendening, D.M. Livingston, and J.A. DeCaprio. 1993. Specific enzymatic dephosphorylation of the retinoblastoma protein. *Mol. Cell. Biol.* **13:** 367.

Lukeis, R., L. Irving, M. Garson, and S. Hasthorpe. 1990. Cytogenetics of non-small cell lung cancer: Analysis of consistent non-random abnormalities. *Genes Chromosomes Cancer* **2:** 116.

Mihara, K., X.R. Cao, A. Yen, S. Chandler, B. Driscoll, A.L. Murphree, A. T'Ang, and Y.K. Fung. 1989. Cell cycle-dependent regulation of phosphorylation of the human retinoblastoma gene product. *Science* **246:** 1300.

Miyakoshi, J., K.D. Dobler, J. Allalunis-Turner, J.D. McKean, K. Petruk, P.B. Allen, K.N. Aronyk, B. Weir, D. Huyser-Wierenga, D. Fulton, R.C. Utrasun, and R.S. Day III. 1990. Absence of IFNA and IFNB genes from human malignant glioma cell lines and lack of correlation with cellular sensitivity to interferons. *Cancer Res.* **50:** 278.

Mori, T., K. Miura, T. Aoki, T. Nishiura, S. Mori, and Y. Nakamura. 1994. Frequent somatic mutation of the MTS1/CDK4I(multiple tumor suppressor/cyclin-dependent kinase 4 inhibitor) gene in esophageal squamous cell carcinoma. *Cancer Res.* **54:** 3396.

Müller, H., J. Lukas, A. Schneider, P. Warthoe, J. Bartek, M. Eilers, and M. Strauss. 1994. Cyclin D1 expression is regulated by the retinoblastoma protein. *Proc. Natl. Acad. Sci.* **91:** 2945.

Nevins, J.R. 1992. E2F: A link between the Rb tumor suppressor protein and viral oncoproteins. *Science* **258:** 424.

Nobori, T., K. Miura, D.J. Wu, A. Lois, K. Takabayashi, and D.A. Carson. 1994. Deletions of the cyclin-dependent kinase-4 inhibitor gene in multiple human cancers. *Nature* **386:** 753.

Noda, A., Y. Ning, S.F. Venable, O.M. Pereira-Smith, and J.R. Smith. 1994. Cloning of senescent cell-derived inhibitors of DNA synthesis using an expression screen. *Exp. Cell Res.* **211:** 90.

Norbury, C. and P. Nurse. 1992. Animal cell cycles and their control. *Annu. Rev. Biochem.* **61:** 441.

Ohtsubo, M. and J.M. Roberts. 1993. Cyclin-dependent regulation of G1 in mammalian fibroblasts. *Science* **259:** 1908.

Okamoto, A., D.J. Demetrick, E.A. Spillare, K. Hagiwara, S.P. Hussain, W.P. Bennett, K. Forrester, B. Gerwin, M. Serrano, D. H. Beach, and C.C. Harris. 1994. Mutations and altered expression of p16^{INK4} in human cancer. *Proc. Natl. Acad. Sci.* (in press).

Olopade, O.I., R.B. Jenkins, D.T. Ransom, K. Malik, H. Pomykala, T. Nobori, J.M. Cowan, J.D. Rowley, and M.O. Diaz. 1992. Molecular analysis of deletions of the short arm of chromosome 9 in human gliomas. *Cancer Res.* **52:** 2523.

Olopade, O.I., D.L. Buchhagen, K. Malik, J. Sherman, T. Nobori, S. Bader, M.M. Nau, A.F. Gazdar, J.D. Minna, and M.O. Diaz. 1993. Homozygous loss of the interferon genes defines the critical region on 9p that is deleted in lung cancers. *Cancer Res.* **53:** 2410.

Pardee, A.B. 1989. G1 events and regulation of cell proliferation. *Science* **246:** 603.

Pearson, B.E., H.P. Nasheuer, and T.S. Wang. 1991. Human DNA polymerase α gene: Sequences controlling expression in cycling and serum-stimulated cells. *Mol. Cell. Biol.* **11:** 2081.

Pedersen, M.I., J.W. Bennett, and N. Wang. 1986. Nonrandom chromosome structural aberrations and oncogene loci in human malignant melanoma. *Cancer Genet. Cytogenet.* **20:** 11.

Pelin-Enlund, K., K. Husgafvel-Pursiainen, L. Tammilehto, M. Klockars, K. Jantunen, B.I. Gerwin, C.C. Harris, T. Tuomi, E. Vanhala, K. Mattson, and K. Linnainmaa. 1990. Asbestos-related malignant mesothelioma: Growth, cytology, tumourigenicity and consistent chromosome findings in cell lines from five patients. *Carcinogenesis* **11:** 673.

Polyak, K., J.Y. Kato, M.J. Solomon, C.J. Sherr, J. Massagué, J.M. Roberts, and A. Koff. 1994. p27^{Kip1}: A cyclin-Cdk inhibitor, links transforming growth factor-β and contact inhibition to cell cycle arrest. *Genes Dev.* **8:** 9.

Quelle, D.E., R.A. Ashmun, S.A. Shurtleff, J.-Y. Kato, D. Bar-Sagi, M.F. Roussel, and C.J. Sherr. 1993. Overexpression of mouse D-type cyclins accelerates G1 phase in rodent fibroblasts. *Genes Dev.* **7:** 1559.

Reed, S.I. 1992. The role of p34 kinases in the G1 to S-phase transition. *Annu. Rev. Cell Biol.* **8:** 529.

Ron, D. 1994. Commentary: Inducible growth arrest: New mechanistic insights. *Proc. Natl. Acad. Sci.* **91:** 1985.

Ruppert, J.M., K. Tokino, and D. Sidransky. 1993. Evidence for two bladder cancer suppressor loci on human chromosome 9. *Cancer Res.* **53:** 5093.

Serrano, M., G.J. Hannon, and D. Beach. 1993. A new regulatory motif in cell-cycle control causing specific inhibition of cyclin D/CDK4. *Nature* **366:** 704.

Sherr, C.J. 1993. Mammalian G1 cyclins. *Cell* **73:** 1059.

Slansky, J.E., Y. Li, W.G. Kaelin, and P.J. Farnham. 1993. A protein synthesis-dependent increase in E2F1 mRNA correlates with growth regulation of the dihydrofolate reductase promoter. *Mol. Cell. Biol.* **13:** 1610.

Stein, G.H., M. Beeson, and L. Gordon. 1990. Failure to phosphorylate the retinoblastoma gene product in senescent human fibroblasts. *Science* **249:** 666.

Taguchi, T., S.C. Jhanwar, J.M. Siegfried, S.M. Keller, and J.R. Testa. 1993. Recurrent deletions of specific chromosomal sites in 1p, 3p, 6q, and 9p in human malignant mesothelioma. *Cancer Res.* **53:** 4349.

Vanni, R., R.M. Scarpa, M. Nieddu, and E. Usai. 1988. Cytogenetic investigation on 30 bladder carcinomas. *Cancer Genet. Cytogenet.* **30:** 35.

Weinberg, R.A. 1992. Tumor suppressor genes. *Science* **254:** 1138.

Whang-Peng, J., T. Knutsen, A. Gazdar, S.M. Steinberg, H. Oie, I. Linnoila, J. Mulshine, M. Nau, and J.D. Minna. 1991. Nonrandom structural and numerical chromosome changes in non-small-cell lung cancer. *Genes Chromosomes Cancer* **3:** 168.

Xiong, Y., H. Zhang, and D. Beach. 1992. D type cyclins associate with multiple protein kinases and the DNA replication and repair factor PCNA. *Cell* **71:** 505.

Molecular Genetic Approaches to the Study of Cellular Senescence

T.J. Goletz,[*][†][‡] J.R. Smith,[*][†][‡][§] and O.M. Pereira-Smith[*][†][‡][§]
[*]Roy M. and Phyllis Gough Huffington Center on Aging, [†]The Cell and Molecular Biology Program,
[‡]Division of Molecular Virology, [§]Departments of Cell Biology and Medicine, Baylor College of Medicine, Houston,
Texas 77030

Numerous positive and negative regulatory mechanisms function to maintain homeostasis; any alterations of these processes could have deleterious consequences at both the cellular and organismal levels. Changes in cellular proliferation can manifest as either loss of proliferation, as observed in age-related changes in the immune system, or as uncontrolled proliferation, as demonstrated by tumor formation. These are two of many ways in which alterations in normal regulatory processes participate in the development and progression of disease, as well as the normal aging process.

PHENOMENON OF CELLULAR SENESCENCE

It has been well established that normal human cells have a finite proliferative potential in vitro (Hayflick and Moorhead 1961). Cells that have reached the end of their in vitro life span and are unable to undergo further rounds of division are referred to as senescent. Cellular senescence is not a process of programmed cell death, since these cells remain viable in culture for long periods, ranging from several months up to 3 years (Matsumura et al. 1979b; J. Smith, unpubl.). However, the distinguishing feature of senescent cells is an irreversible block in the cell cycle. These cells are unable to synthesize DNA (Matsumura et al. 1979b), yet they remain metabolically active (Matsumura et al. 1979a). As a population of cells approaches senescence, the time required to transit the cell cycle is longer, primarily due to increases in the G_1 phase of the cell cycle (Grove et al. 1976); at senescence, cells arrest in the cell cycle with a G_1 DNA content (Yanishevsky et al. 1974; Yanishevsky and Carrano 1975; Schneider and Fowlkes 1976). Actively proliferating, early-passage fibroblasts can be induced to enter a quiescent (G_0) state by the removal of mitogenic stimuli; however, this state is reversible when appropriate growth factors are reintroduced (Pardee et al. 1978). In contrast, no known combination of mitogens can stimulate DNA synthesis and/or cell division in senescent cells (Cristofalo et al. 1989). The inability of mitogens to induce DNA synthesis in senescent cells does not appear to be due to a lack of DNA synthesis machinery, because these cells can be induced to synthesize DNA by infection with SV40 (Gorman and Cristofalo 1985) or cytomegalovirus (Ide et al. 1984). These data suggest that senescence is the result of an active, genetically controlled process(es).

The observation of the phenomenon of cellular senescence led to the proposal that the limited division potential of these cells in vitro is "an expression of aging or senescence at the cellular level" (Hayflick 1965). A variety of human cell types have been found to undergo senescence, including epidermal keratinocytes (Rheinwald and Green 1975), smooth muscle cells (Bierman 1978), lens epithelial cells (Tassin et al. 1979), glial cells (Blomquist et al. 1980), endothelial cells (Thornton et al. 1983), T lymphocytes (Effros and Walford 1984; Pereira-Smith et al. 1990), and adrenocortical cells (McAllister and Hornsby 1987). Several other lines of evidence support the study of senescence as an experimental model for the study of aging at the cellular level: (1) Cells derived from shorter-lived animal species have fewer population doublings in vitro than cells derived from longer-lived species (LeGuilly et al. 1973; Goldstein 1974; Rohme 1981). (2) In vitro replicative potential is inversely related to the age of the tissue donor (Martin et al. 1970; Schneider and Mitsui 1976; Goldstein et al. 1978). (3) Fibroblasts obtained from patients with genetic disorders that mimic premature aging, such as Hutchinson-Gilford and Werner's syndromes, undergo fewer population doublings than cells derived from normal, age-matched controls (Martin et al. 1965; Goldstein 1969, 1979; Norwood et al. 1979). Although these observations do not prove the validity of in vitro cellular senescence as a model system for the study of in vivo cellular aging, they do suggest that a strong relationship exists between the intrinsic rate of aging of an organism and the proliferative capacity of its individual cells.

APPROACHES TO THE STUDY OF CELLULAR SENESCENCE

Comparison of Young and Senescent Cells

To gain a better understanding of the mechanism(s) responsible for replicative senescence, many studies have made comparisons between young and senescent cells. This has resulted in a large amount of data regarding morphological, biochemical, and molecular changes that occur during the course of the in vitro life span of the cell (Finch 1990). However, the significance of these changes as a cause or a result of senescence is not fully understood. What is clear is that the program of cellular senescence results in the inability of metabolically active cells to initiate DNA synthesis. Therefore, it has been hypothesized that the end point of the senescence program is the production of a protein that inhibits initiation of DNA synthesis. Inhibitors of DNA synthesis are thought to be involved in normal tissue homeostasis (Bullough 1964), and several inhibitors have been isolated from primary tissues and established cell lines (Nakamura et al. 1983; Iype and McMahan 1984). Potent inhibitors such as TGF-β (Moses et al. 1981; Roberts et al. 1981) and growth inhibitory protein (GIP) (Wittenberger et al. 1978; Lieberman et al. 1981; Raben et al. 1981) demonstrate that cell cultures can produce agents capable of inhibiting DNA synthesis. The identification of a senescence-specific inhibitor of DNA synthesis has been a major goal of investigators characterizing the in vitro model of cellular senescence.

The hypothesis that senescent human diploid fibroblasts express a DNA synthesis inhibitor is supported by several lines of evidence. Norwood et al. (1974) and Yanishevsky and Stein (1980) demonstrated that heterokaryons formed when senescent human fibroblasts were fused with young, proliferation-competent cells resulted in an inability of either nucleus to initiate DNA synthesis, as measured by tritiated thymidine incorporation. This phenomenon was not observed in fusions between young cells, but heterokaryons created by the fusion of young quiescent cells and proliferating young cells also exhibited a decrease in DNA synthesis initiation (Rabinovitch and Norwood 1980; Stein and Yanishevsky 1981). Rabinovitch and Norwood (1980) further demonstrated that the inhibitor(s) acted early in the G_1 phase of the cell cycle but was unable to block cells that had already entered S phase. Collectively, these data suggested that senescent cells produce an inhibitor of DNA synthesis that can diffuse through the common cytoplasm of a heterodikaryon to affect a young nucleus. Further work by Stein and Yanishevsky (1979) demonstrated that senescent cells could inhibit initiation of DNA synthesis when fused with various immortal cell lines. However, the inhibition did not occur in fusions with cell lines that were immortalized by DNA tumor viruses (Norwood et al. 1975; Stein et al. 1982) because of the counteraction by specific DNA tumor viral genes (i.e., SV40 T antigen and papillomavirus E6, E7). Carcinogen-derived immortal cell lines were unable to overcome the senescent cell inhibitor, suggesting that elements encoded by DNA tumor viruses might interact with the senescent cell-inhibitory mechanism.

Drescher-Lincoln and Smith (1983) and Burmer et al. (1983) performed experiments in which senescent cytoplasts were fused with young proliferating cells and found that the senescent cell cytoplast retained the inhibitory activity. If RNA (Burmer et al. 1983) or protein synthesis (Drescher-Lincoln and Smith 1984) was blocked in the cytoplasts prior to fusion, the inhibitory activity was lost. These experiments indicated that the senescent cytoplasm contains an inhibitor which is actively produced, and that the inhibition is not the result of merely diluting positive factors present in young cells that are missing in senescent cells (Pendergrass and Norwood 1985). Similar results were obtained when quiescent, young cytoplasts were fused to young diploid cells that had not been deprived of growth factors. However, the inhibitory activity in quiescent cells was insensitive to treatment with protein synthesis inhibitors, indicating subtle differences between the inhibitors (quiescent vs. senescent).

Additional studies localized the inhibitor to surface membranes. Conditioned medium from senescent cells was found not to contain the activity, and proteins loosely associated with the membrane did not inhibit DNA synthesis in young cells. Mild trypsin treatment destroyed the inhibitory activity in quiescent and senescent cells (Pereira-Smith et al. 1985), indicating an association with the cell membrane. Furthermore, surface-membrane-enriched preparations from senescent and quiescent cells contained the inhibitory activity when added to cultures of young cells (Pereira-Smith et al. 1985; Stein and Atkins 1986). Treatment with periodate inactivated the inhibitory activity, suggesting that the inhibitor was a glycoprotein (Stein and Atkins 1986).

Additional evidence for the DNA synthesis inhibitor was obtained from studies with senescent cell mRNA. Microinjection of mRNA isolated from senescent cells into young cells resulted in inhibition of initiation of DNA synthesis (Lumpkin et al. 1986); however, this activity was not present following RNase treatment or in the non-poly(A)$^+$ fraction of the senescent RNA. These results supported the hypothesis that senescent cells produced RNA transcripts and proteins that were specific for the inhibitory activity. The presence of an inhibitory mRNA expressed by senescent cells was used to devise an expression screen of a senescent cell cDNA library. This led to the identification and isolation of several *s*enescent cell-*d*erived *i*nhibitors of DNA synthesis (SDI1, 2, 3). These inhibitory transcripts have been partially characterized for their ability to inhibit DNA synthesis in young cells following transfection and microinjection (Noda et al. 1994). mRNA levels of SDI2 and 3 remained unchanged throughout the in vitro life span of fibroblasts. However, SDI1 mRNA levels were up-regulated 10- to 20-fold in both senescent cells and proliferation-competent cells made quiescent by serum deprivation. Although functional characterization of this protein is ongoing, results demonstrate that SDI1 has several functional characteristics that are similar to the proposed inhibitors described by whole-cell fusion studies. In particular, cotransfection of SDI1 and SV40 T antigen into young cells revealed that SV40 T antigen antagonizes the ability of SDI1 to inhibit initiation of DNA synthesis (Harper et al. 1993). Other groups have also identified the SDI1 gene product as a Cdk-interacting protein (CIP1/p21) and a p53-*trans*-activated gene (*WAF1*) (El-Deiry et al. 1993; Harper et al. 1993). The inhibitory activity is thought to be the result of interaction with and inhibition of Cdk-cyclin kinase activity, which may interfere with phosphorylation of Rb (Harper et al. 1993; Dulic et al. 1994). In addition, the gene is regulated by a p53-dependent pathway(s) (El-Deiry et al. 1993). However, p53 levels in senescent and quiescent human cells are low, suggesting an alternative mechanism for SDI1 up-regulation. Johnson et al. (1994) have provided evidence for the same by the treatment of normal human cells with a variety of growth-arresting agents. Treatment with hydroxyurea and prostaglandins, agents that do not work via p53, resulted in an up-regulation of SDI1 and growth inhibition. More direct evidence was obtained by treating immortal human cells that did not express wild-type p53 with some of these agents. Up-regulation of SDI1 did not occur following γ irradiation, since this is known to act by the p53 pathway to cause growth arrest. However, other agents such as hydrogen peroxide and hydroxyurea caused an elevation in RNA levels of SDI1 and growth arrest.

The importance of SDI1 in cell survival is indicated by the fact that we have been unable to detect mutations or deletions in this gene in a large number of immortal human cell lines. Putative mutations were found to be polymorphisms when normal human cell lines were examined. This is in contrast to the p53 gene, which acts to inhibit growth through SDI1 and in which many mutations are observed in immortalized cells. Collectively, these data suggest that SDI1 is required for growth arrest to allow for DNA repair, and loss of function would result in genetic catastrophe.

Somatic Cell Hybrid Studies

An alternative approach to the study of cellular senescence is to examine the mechanisms by which cells escape senescence to become immortal (immortalization). This approach should enable the identification of important growth-regulatory genes/pathways, alterations that could play a role in tumor formation. The frequency with which cells escape senescence to become immortal varies between species. For example, both spontaneous and induced immortalization occurs much more frequently in rodent than in human cells. Since human cells are refractory to spontaneous immortalization, and induced immortalization using various viral or chemical carcinogens occurs rarely (approximately $<10^{-8}$), human cells are preferred for studies of cellular aging.

Fusion of Normal Cells with Normal Cells

Normal human fibroblasts have a limited in vitro life span of approximately 40–60 population doublings (PD). This greatly limits long-term study of hybrids generated by fusing normal cells, since generation of mutants enabling the biochemical selection of hybrids would expend a significant number of PD (i.e., 1×10^6 cells = ~20 PD). In early experiments, attempts were made to fuse young proliferating cells (HPRT$^-$) to nondividing senescent cells (Littlefield 1973). Since senescent cells are unable to divide and HAT would select against the young cells, only hybrids should have proliferated in these conditions. Although hybrids were generated, they did not proliferate sufficiently to allow for karyotypic analysis but did indicate that senescence was dominant. Additional fusion studies were conducted using populations of young (many PD remaining) and old (few PD remaining) cells. The parental cells expressed different isoenzymes of glucose-6-phosphate dehydrogenase (G6PD), enabling identification of hybrid cells as those expressing a G6PD heteropolymer band. The resulting hybrids demonstrated an intermediate life span (Hoehn et al. 1978).

Duthu et al. (1982) reported culture of a human fetal lung fibroblast cell line with an unusually long in vitro life span, one which was capable of attaining nearly 100 PD. This enabled the generation of a double mutant derivative strain which was resistant to both ouabain (O^r) and 6-thioguanine (TG^r). This double mutant continued to proliferate for approximately 45 PD and was therefore useful as a universal hybridizer. Pereira-Smith and Smith (1982) reported the isolation of an O^r mutant clone from a Lesch-Nyhan (HPRT$^-$) cell strain with approximately 28 PD remaining. These clonal cell strains enabled the determination of the proliferative potential of hybrids generated by fusing normal young and old cells. Such studies demonstrated that the proliferative potential of the hybrids resembled that of the old parental cells, suggesting that the senescent phenotype is dominant. In addition, fusion of cells at the end of their in vitro life span with each other did not produce hybrids with a life span greater than either of the parental cell lines. This indicates a lack of complementation of positive-acting factors to result in increased life span and provides further evidence for a dominant negative regulatory senescence phenotype.

Fusion of Normal Cells with Immortal Cells

Fusions of normal cells with immortal cell lines has provided further evidence for a genetic dominant basis of senescence. For example, if the accumulation of random damage is the underlying cause of senescence, one would assume that immortal cells have acquired mechanisms with which they cope or repair such damage. These "new" mechanisms should be reflected in the phenotype of somatic cell hybrid clones generated by fusing immortal cells with normal cells such that the resulting hybrids would have an immortal phenotype. However, fusions utilizing clonal populations of various immortal and normal (mortal) human cells yielded hybrid cells with a limited doubling potential (Bunn and Tarrant 1980; Muggleton-Harris and De Simone 1980; Pereira-Smith and Smith 1981, 1983). These studies demonstrated that the senescence phenotype is dominant and that the immortal cells arise from recessive mutations and strongly supported a genetic basis for cellular senescence.

In early studies of tumorigenicity, normal and immortal human cells were fused, and the resulting hybrids were analyzed for growth properties (Croce and Koprowski 1974; Stanbridge 1976; Klinger 1980). On the basis of these studies, the phenotype of unlimited growth potential (immortal) was thought to be dominant. However, later careful studies demonstrated that fusion of human fibroblasts with human cervical-carcinoma-derived cells, HeLa, resulted in some hybrids with a limited division potential (Bunn and Tarrant 1980). These results were questioned, as the genetic stability of the parental HeLa cell line was suspect. The limited life span of the hybrids might have resulted from the loss of "immortalizing" HeLa DNA; however, this seemed unlikely since Bunn and Tarrant (1980) observed that nondividing hybrids could be maintained in culture and that, over time, some of the nondividing cultures gave rise to foci of dividing cells at a low frequency (1×10^5 to 2×10^5 cells). These dividing cells had regained the immortal phenotype and could be cultured indefinitely. It was most likely such immortal variants that were analyzed in the earlier tumorigenicity studies.

Muggleton-Harris and De Simone (1980) fused normal human cells with SV40-transformed (immortal) cells. They reported limited division in 98% of the hybrids (<6 PD). The remaining 2% were large clones that were thought to be immortal and were not further analyzed. Pereira-Smith and Smith (1981) extended these observations by fusing normal human fibroblasts with SV40-transformed immortal cells. They reported limited proliferation (<7 PD) in 70% of the hybrids. However, approximately 30% of the hybrids continued proliferating and eventually achieved between 16 and 52 PD, after which division ceased. Expression of SV40 large T antigen was assessed and found to be present in all hybrids that had stopped dividing. As reported by Bunn and Tarrant (1980), immortal variants did arise from nondividing populations at a low frequency.

Fusion of different immortal cell lines with highly differentiated cell types, such as normal human T cells and endothelial cells, also yielded hybrids with limited division potential (Pereira-Smith et al. 1990). Collectively, these data support a genetic basis for aging and clearly indicate that cells escape senescence to become immortal due to recessive changes such as loss or inactivation of gene(s). They also suggest that the dominance of senescence is governed by a common mechanism(s) in very different cell types.

Fusion of Immortal Cells with Immortal Cells

To determine the number of genes involved in cellular senescence, a series of experiments were performed in which different immortal human cell lines were fused to each other. These experiments were based on the idea that if the immortal parental lines have the same recessive changes, the resulting hybrids would be immortal and the cell lines would assign to the same complementation group. However, fusion of immortal cells with different recessive changes should result in complementation of these defects. The resulting hybrids would have a limited in vitro life span, and these cell lines would assign to different complementation groups. Using this approach, more than 30 different cell lines have assigned to four complementation groups for indefinite division (Pereira-Smith and Smith 1983, 1988; Whitaker et al. 1992; Duncan et al. 1993; Goletz et al. 1994). In earlier studies (Pereira-Smith and Smith 1988), group assignment did not show a correlation with cell type, tumor type, embryonal layer of origin, expression of an activated oncogene, or constitutive expression of high levels of proto-oncogenes c-*myc* or c-Ha-*ras*. However, a recent study revealed that 7 independently derived immortal T- and B-cell lines all assigned to the same complementation group for indefinite division (Goletz et al. 1994). These data argue strongly for a common mechanism of immortalization in human lymphocytes.

The identification of only four complementation groups for indefinite division indicate that a limited number of genes or gene pathways are altered to produce immortal cells and that a small number of genes are involved in the process of senescence. These results are consistent with the fact that spontaneous immortalization of human cells from normal individuals has not been reported and that the frequency of induced immortalization of human cells is low. The complementation group studies greatly facilitate a focused approach to the identification of group-specific changes responsible for immortalization and lend further support to the hypothesis that cellular senescence is the result of genetically programmed processes.

Microcell-mediated Chromosome Transfer Studies

The assignment of many immortal human cell lines to four complementation groups for indefinite division permitted a focused approach to identify the chromosomes or genes involved in cellular senescence. One approach was to utilize the technique of microcell-mediated chromosome transfer to introduce single human chromosomes into immortal human cell lines assigned to the complementation groups for indefinite division (Ning et al. 1991a,b; Ogata et al. 1993; Hensler et al. 1994). The basis of these experiments was to identify chromosomes encoding senescence-related genes specifically altered in cell lines assigning to a given complementation group and not altered in cell lines in the other groups.

Since cellular senescence has been proposed to be an antitumor mechanism (Cairns and Logan 1983; O'Brien et al. 1986; Sager 1989), chromosomes implicated in tumor suppression have been examined for the ability to restore finite proliferation to immortal cell lines. Human chromosome 11 was reported to suppress tumorigenicity in several tumor cell lines, and a region of chromosome 11 was implicated in Wilms' tumor (Saxon et al. 1986; Weissman et al. 1987; Koi et al. 1989; Misra and Srivatsan 1989). This chromosome was assessed for the ability to restore finite proliferation in cell lines assigning to the various complementation groups but was found to have no effect on cell lines representative of all four complementation groups (Ning and Pereira-Smith 1991; Ning et al. 1991a). Therefore, there is no cell-senescence-related gene on human chromosome 11.

Whole-cell fusion experiments had demonstrated that the tumorigenic phenotype of HT-1080, a human fibrosarcoma cell line, was suppressed when fused with normal human diploid fibroblasts. Concordance/discordance analysis of nontumorigenic hybrids and hybrids that regained the tumorigenic phenotype (tumorigenic segregates) showed a correlation between loss of chromosomes 1 and 4 and acquisition of tumorigenic potential (Benedict et al. 1984). Therefore, chromosome 4 was introduced into cell lines assigning to the complementation groups. These studies demonstrated that chromosome 4 could reverse the immortal phenotype and induce cellular senescence in three tumor lines assigned to complementation group B, but had no effect on the proliferation potential of cell lines representative of the other complementation groups. This result suggests that a gene, altered in immortal cell lines assigned to complementation group B, resides on human chromosome 4, and the normal counterpart of this gene is required for expression of the senescent phenotype. Work is ongoing to identify the gene(s) involved.

Human chromosome 1 has also been implicated in growth control. Sugawara et al. (1990) reported human chromosome 1 was able to induce senescence in an immortal hamster cell line. In these experiments, interspecies hamster/human fusions between normal diploid human fibroblasts and an immortal Syrian hamster cell line produced hybrids that exhibited limited life spans similar to what has been observed in human cell fusions. In hybrids that failed to undergo senescence, karyotypic analysis revealed that all of the clones had lost both copies of human chromosome 1, whereas all other human chromosomes were observed in at least some of the immortal hybrids. With the hypothesis that human chromosome 1 is important in the control of cellular senescence, microcell-mediated chromosome transfer was used to introduce a normal human chromosome 1 into various immortal human cell lines representing the complementation groups for indefinite division. Hensler et al. (1994) reported that human chromosome 1 is able to reverse the immortal phenotype of three cell lines which assign to complementation group C. Therefore, human chromosome 1 contains a gene or set of genes important in the control of cellular senescence, which are defective in group C cell lines. Again, this effect was group specific in that introduction of human chromosome 1 into cell lines assigning to the remaining complementation groups (A, B, or D) had no effect on proliferation. This activity maps to the long arm of chromosome 1 (1q), and efforts are in progress to identify and clone this senescence-related gene(s).

Work by several other laboratories has implicated other chromosomes in normal growth control or senescence. For example, Klein et al. (1991) reported a loss of cellular proliferation following the introduction of a normal human chromosome X into a nickel-transformed rodent cell line, and Ogata et al. (1993) reported that human chromosome 7 suppresses proliferation of the cell lines, SUSM-1 and KMST1, 4NQO-transformed human fibroblast cell lines. It is possible that these chromosomes do carry genes involved in cell senescence. However, since the chromosomes were transferred into cell lines chosen somewhat randomly, it is difficult to judge the specificity of the effect and to distinguish the result as due to aberrations in the particular cell line used. The strength of using cell lines already assigned to complementation groups is that it allows one to test that multiple cell lines within a group respond by losing proliferation following the introduction of a particular chromosome. More importantly, cell lines assigned to the other groups are not affected by the transfer of this chromosome.

SUMMARY

Cellular senescence is an inability of cells to synthesize DNA and divide, which results in a terminal loss of proliferation despite the maintenance of basic metabolic processes. Senescence has been proposed as a model for the study of aging at the cellular level, and the basis for this model system and its features have been summarized. Although strong experimental evidence exists to support the hypothesis that cellular senescence is a dominant active process, the mechanisms responsible for this phenomenon remain a mystery.

Investigators have taken several approaches to gain a better understanding of senescence. Several groups have documented the differences between young and senescent cells, and others have identified changes that occur during the course of a cell's in vitro life span. Using molecular and biochemical approaches, important changes in gene expression and function of cell-cycle-associated products have been identified. The active production of an inhibitor of DNA synthesis has been demonstrated. This may represent the final step in a cascade of events governing senescence.

The study of immortal cells which have escaped senescence has also provided useful information, particularly with regard to the genes governing the senescence program. These studies have identified four complementation groups for indefinite division, which suggests that there are at least four genes or gene pathways in the senescence program. Through the use of microcell-mediated chromosome transfer, chromosomes encoding senescence genes have been identified; efforts to clone these genes are ongoing. The isolation and characterization of these genes and their protein products will be an important step in understanding cellular senescence.

Collectively, these data demonstrate that disruption of specific gene products can alter the proliferative capacity of a population of cells and may result in immortalization. Although conclusive evidence has not been accumulated to distinguish between the role of these gene products as an immediate cause of cellular senescence and their role as products which are expressed in response to the changes that result from senescence, the identification, isolation, and characterization of the genes governing the senescence program will provide a greater understanding of the changes that occur during normal aging. Since these genes clearly play regulatory roles, understanding their normal function is also likely to provide insight into the mechanisms underlying a number of diseases. This information could prove useful in the design of therapies for the treatment of disorders that result from altered growth regulation. For example, further characterization of the inhibitor of DNA synthesis may make this protein a target for manipulation in diseases such that the inhibitor could be introduced or up-regulated in diseases of uncontrolled proliferation (e.g., tumors) or down-regulated when proliferation is needed (e.g., augmentation of wound healing). Clearly, an understanding of the mechanisms responsible for the phenomenon of cellular senescence will contribute to our understanding not only of aging, but also of growth-regulatory processes in general.

ACKNOWLEDGMENTS

This work was supported by National Institutes of Health grants R37 AG-05333, PO1 AG-07123, and RO1 AG-11066.

REFERENCES

Benedict, W.F., B.E. Weissman, C. Mark, and E.J. Stanbridge. 1984. Tumorigenicity of human HT1080 fibrosarcoma × normal fibroblast hybrids: Chromosome dosage dependency. *Cancer Res.* **44:** 3471.

Bierman, E.L. 1978. The effect of donor age on the *in vitro* life span of cultured human arterial smooth muscle cells. *In Vitro* **14:** 951.

Blomquist, E., B. Westermark, and J. Ponten. 1980. Aging of human glial cells in culture: Increase in the fraction of nondividers as demonstrated by a mini cloning technique. *Mech. Ageing Dev.* **12:** 173.

Bullough, W.S. 1964. Growth regulation by tissue-specific factors, or chalones. In *Cellular control mechanisms and cancer* (ed. P. Emmelot and O. Muhlbock) p. 124. Elsevier, Amsterdam.

Bunn, C.L. and G.M. Tarrant. 1980. Limited lifespan in somatic cell hybrids and cybrids. *Exp. Cell Res.* **37:** 385.

Burmer, G.C., H. Motulsky, C.J. Zeigler, and T.H. Norwood. 1983. Inhibition of DNA synthesis in young cycling human diploid fibroblast cells upon fusion to enucleate cytoplasts from senescent cells. *Exp. Cell Res.* **15:** 79.

Cairns, J. and J. Logan. 1983. Step by step into carcinogenesis. *Nature* **304:** 582.

Cristofalo, V.J., P.D. Phillips, T. Sorger, and G. Gerhard. 1989. Alterations in the responsiveness of senescent cells to growth factors. *J. Gerontol.* **44:** 55.

Croce, C.M. and H. Koprowski. 1974. Positive control of transformed phenotype between SV-40 transformed and normal human cells. *Science* **184:** 1288.

Drescher-Lincoln, C.K. and J.R. Smith. 1983. Inhibition of DNA synthesis in proliferating human diploid fibroblasts by fusion with senescent cytoplasts. *Exp. Cell Res.* **144:** 455.

———. 1984. Inhibition of DNA synthesis in senescent-proliferating human cybrids is mediated by endogenous proteins. *Exp. Cell Res.* **153:** 208.

Dulic, V., W.K. Kaufmann, S.J. Wilson, T.D. Tlsty, E. Lees, J.W. Harper, S.J. Elledge, and S.I. Reed. 1994. p53-dependent inhibition of cyclin-dependent kinase activities in human fibroblasts during radiation-induced G1 arrest. *Cell* **76:** 1013.

Duncan, E.L., N.J. Whitaker, E.L. Moy, and R.R. Reddel. 1993. Assignment of SV40-immortalized cells to more than one complementation group for immortalization. *Exp. Cell Res.* **205:** 337.

Duthu, G.S., K.I. Braunschweiger, O.M. Pereira-Smith, T.H. Norwood, and J.R. Smith. 1982. A long-lived human diploid fibroblast line for cellular aging studies: Applications in cell hybridization. *Mech. Ageing Dev.* **20:** 243.

Effros, R.B. and R.L. Walford. 1984. T cell cultures and the Hayflick limit. *Hum. Immunol.* **9:** 49.

El-Deiry, W.S., T. Tokino, V.E. Velculescu, D.B. Levy, R. Parsons, J.M. Trent, D. Lin, W.E. Mercer, K.W. Kinzler, and B. Vogelstein. 1993. WAF1, a potential mediator of p53 tumor suppression. *Cell* **75:** 817.

Finch, E. 1990. *Longevity, senescence and the genome.* The University of Chicago Press, Chicago, Illinois.

Goldstein, S. 1969. Life span of cultured cells in progeria. *Lancet* **I:** 424.

———. 1974. Aging in vitro: Growth of cultured cells from the Galapagos tortoise. *Exp. Cell Res.* **83:** 297.

———. 1979. Studies on age-related diseases in cultured skin fibroblasts. *J. Invest. Dermatol.* **73:** 19.

Goldstein, S., E.L. Moerman, J.S. Soeldner, R.E. Gleason, and D.M. Barnett. 1978. Chronologic and physiologic age affect replicative life span of fibroblasts from diabetic, prediabetic, and normal donors. *Science* **199:** 781.

Goletz, T.J., S. Robetorye, and O.M. Pereira-Smith. 1994. Genetic analysis of indefinite division in human cells: Evidence for a common immortalizing mechanism in T and B lymphoid cell lines. *Exp. Cell Res.* **215:** 82.

Gorman, S.D. and V.J. Cristofalo. 1985. Reinitiation of cellular DNA synthesis in BrdU-selected nondividing senescent WI-38 cells by simian virus 40 infection. *J. Cell. Physiol.* **125:** 122.

Grove, G.L., E.D. Kress, and V.J. Cristofalo. 1976. The cell cycle and thymidine incorporation during aging *in vitro*. *J. Cell Biol.* **70:** 133a. (Abstr.)

Harper, J.W., G.R. Adami, N. Wei, K. Keyomarsi, and S.J. Elledge. 1993. The p21Cdk-interacting protein Cip1 is a potent inhibitor of G1 cyclin-dependent kinases. *Cell* **75:** 805.

Hayflick, L. 1965. The limited *in vitro* lifetime of human diploid cell strains. *Exp. Cell Res.* **37:** 614.

Hayflick, L. and P.S. Moorhead. 1961. The serial cultivation of human diploid cell strains. *Exp. Cell Res.* **25:** 585.

Hensler, P.J., L.A. Annab, J.C. Barrett, and O.M. Pereira-Smith. 1994. A gene involved in control of human cellular senescence on human chromosome 1q. *Mol. Cell. Biol.* **14:** 2291.

Hoehn, H., E.M. Bryant, and G.M. Martin. 1978. The replicative life span of euploid hybrids derived from short lived and long lived human skin fibroblast cultures. *Cytogenet. Cell Genet.* **21:** 282.

Ide, T., Y. Tsuji, S. Ishibashi, Y. Mitsui, and M. Toba. 1984. Induction of host DNA synthesis in senescent human diploid fibroblasts by infection with human cytomegalovirus. *Mech. Ageing Dev.* **25:** 227.

Iype, P.T. and J.B. McMahan. 1984. Hepatic proliferation inhibitor. *Mol. Cell. Biochem.* **59:** 57.

Johnson, M.J., D. Dimitrov, P.J. Vojta, J.C. Barrett, A. Noda, O.M. Pereira-Smith, and J.R. Smith. 1994. Evidence for a p53 independent pathway for upregulation of SDI1/CIP1/WAF1/p21 RNA in human cells. *Mol. Carcinog.* **11:** 59.

Klein, C.B., K. Conway, X.W. Wang, R.K. Bhamra, X.H. Lin, M.D. Cohen, L. Annab, J.C. Barrett, and M. Costa. 1991. Senescence of "nickle-transformed" cells by an X chromosome possible epigenetic control. *Science* **32:** 796.

Klinger, H.P. 1980. Suppression of tumorigenicity in somatic cell hybrids. I. Suppression and re-expression of tumorigenicity in diploid human × D98AH2 hybrids and independent segregation of tumorigenicity from other cell phenotypes. *Cytogenet. Cell Genet.* **27:** 254.

Koi, M., H. Morita, H. Yamada, S. Satoh, J.C. Barrett, and M. Oshimura. 1989. Normal human chromosome 11 suppresses tumorigenicity of human cervical tumor cell line SiHa. *Mol. Carcinog.* **2:** 12.

LeGuilly, Y., M. Simon, P. Lenoir, and M. Bourel. 1973. Long term culture of human adult liver cells: Morphological changes related to in vitro senescence and effect of donors age on growth potential. *Gerontologia* **19:** 303.

Lieberman, M.A., D. Raben, and L. Glaser. 1981. Cell surface associated growth inhibitory proteins. *Exp. Cell Res.* **133:** 413.

Littefield, J.W. 1973. Attempted hybridization with senescent human fibroblasts. *J. Cell. Physiol.* **82:** 129.

Lumpkin, C.K., J.K. McClung, O.M. Pereira-Smith, and J.R. Smith. 1986. Existence of high abundance antiproliferative messenger RNA in senescent human diploid fibroblasts. *Science* **232:** 393.

Martin, G.M., C.A. Sprague, and C.J. Epstein. 1970. Replicative life span of cultivated human cells: Effects of donor's age, tissue, and genotype. *Lab. Invest.* **23:** 86.

Martin, G.M., S.M. Gartler, C.J. Epstein, and A.G. Motulsky. 1965. Diminished lifespan of cultured cells in Werner syndrome. *Fed. Proc.* **24:** 678.

Matsumura, T., E.A. Pfendt, and L. Hayflick. 1979a. DNA synthesis in the human diploid cell strain WI-38 during in vitro aging: An autoradiographic study. *J. Gerontol.* **34:** 328.

Matsumura, T., Z. Zerrudo, and L. Hayflick. 1979b. Senescent human diploid cells in culture: Survival, DNA synthesis and morphology. *J. Gerontol.* **34:** 328.

McAllister, J.M. and P.J. Hornsby. 1987. Improved clonal and nonclonal growth of human, rat, and bovine adrenocortical cells in culture. *In Vitro Cell. Dev. Biol.* **23:** 677.

Misra, B.C. and E.S. Srivatsan. 1989. Localization of HeLa cell tumor suppressor gene to the long arm of chromosome 11. *Am. J. Hum. Genet.* **45:** 565.

Moses, H.L., E.L. Branum, J.A. Proper, and R.A. Robinson. 1981. Transforming growth factor production by chemically transformed cells. *Cancer Res.* **41:** 2842.

Muggleton-Harris, A. and D. De Simone. 1980. Replicative potentials of various fusion products between WI-38 and SV40 transformed WI-38 cells and their components. *Somatic Cell Genet.* **6:** 689.

Nakamura, T., K. Yoshimoto, Y. Nakayama, Y. Tomita, and A. Ichihara. 1983. Reciprocal modulation of growth and differentiated functions of mature rat hepatocytes in primary culture by cell-cell contact and cell membranes. *Proc. Natl. Acad. Sci.* **80:** 729.

Ning, Y. and O.M. Pereira-Smith. 1991. Molecular genetic approaches to the study of cellular senescence. *Mutat. Res.* **256:** 303.

Ning, Y., J.W. Shay, M. Lovell, L. Taylor, D.H. Ledbetter, and O.M. Pereira-Smith. 1991a. Tumor suppression by chromosome 11 is not due to cellular senescence. *Exp. Cell Res.* **192:** 220.

Ning, Y., J.L. Weber, A.M. Killary, D.H. Ledbetter, J.R. Smith, and O.M. Pereira-Smith. 1991b. Genetic analysis of indefinite division in human cells: Evidence for a cell senescence related gene(s) on human chromosome 4. *Proc. Natl. Acad. Sci.* **88:** 5635.

Noda, A., S.F. Venable, O.M. Pereira-Smith, and J.R. Smith. 1994. Cloning of senescent cell-derived inhibitors of DNA synthesis using an expression screen. *Exp. Cell Res.* **211:** 90.

Norwood, T.H., W.R. Pendergrass, and G.M. Martin. 1975. Reinitiation of DNA synthesis in senescent human fibroblasts upon fusion with cells of unlimited growth potential. *J. Cell Biol.* **64:** 551.

Norwood, T.H., H. Hoehn, D. Salk, and G.M. Martin. 1979. Cellular aging in Werner syndrome: A unique phenotype? *J. Invest. Dermatol.* **73:** 92.

Norwood, T.H., W.R. Pendergrass, C. A. Sprague, and G.M. Martin. 1974. Dominance of the senescent phenotype in heterokaryons between replicative and post-replicative human fibroblast-like cells. *Proc. Natl. Acad. Sci.* **71:** 2231.

O'Brien, W., G. Stenman, and R. Sager. 1986 Suppression of tumor growth by senescence in virally transformed human fibroblasts. *Proc. Natl. Acad. Sci.* **83:** 8659.

Ogata, T., D. Ayusawa, M. Namba, E. Takahashi, M. Oshimura, and M. Oishi. 1993. Chromosome 7 suppresses indefinite division of nontumorigenic immortalized human fibroblast cell lines KMST-6 and SUSM-1. *Mol. Cell. Biol.* **13:** 6036.

Pardee, A.B., R. Dubrow, J.L. Hamlin, and R.F. Kletzien. 1978. Animal cell cycle. *Annu. Rev. Biochem.* **47:** 715.

Pendergrass, W.R. and T.H. Norwood. 1985. Cell size and the regulation of DNA polymerase alpha in senescent human diploid fibroblast-like cultures. In *Molecular biology of aging: Gene stability and gene expression* (ed. R.S. Sohal et al.), p. 195. Raven Press, New York.

Pereira-Smith, O.M. and J.R. Smith. 1981. Expression of SV40 T antigen in finite life span hybrids of normal-SV40 transformed fibroblasts. *Somatic Cell Genet.* **7:** 411.

———. 1982. Phenotype of low proliferative potential is dominant in hybrids of normal human fibroblasts. *Somatic Cell Genet.* **8:** 731.

———. 1983. Evidence for the recessive nature of cellular immortality. *Science* **221:** 964.

———. 1988. Genetic analysis of indefinite division in human cell: Identification of four complementation groups. *Proc. Natl. Acad. Sci.* **85:** 6042.

Pereira-Smith, O.M., S.F. Fisher, and J.R. Smith. 1985. Senescent and quiescent cell inhibitors of DNA synthesis: Membrane associated proteins. *Exp. Cell. Res.* **160:** 297.

Pereira-Smith, O.M., S. Robetorye, Y. Ning, and F.M. Orson. 1990. Hybrids form fusion of normal human T lymphocytes with immortal human cells exhibit limited life span. *J. Cell. Physiol.* **144:** 546.

Raben, D., M.A. Liberman, and L. Glaser. 1981. Growth inhibitory protein (GIP) in the 3T3 cell plasma membrane. Partial purification and dissociation of growth inhibitory events from inhibition of amino acid transport. *J. Cell. Physiol.* **108:** 35.

Rabinovich, P.S. and T.H. Norwood. 1980. Comparative heterokaryon study of cellular senescence and the serum deprived state. *Exp. Cell Res.* **130:** 101.

Rheinwald, J.G. and M. Green. 1975. Serial cultivation of strains of human epidermal keratinocytes: The formation of keratinizing colonies from single cells. *Cell* **6:** 331.

Roberts, A.B., M.A. Anzano, L.C. Lamb, J.M. Smith, and M.B. Sporn. 1981. New class of transforming growth factors potentiated by epidermal growth factor: Isolation from non-neoplastic tissues. *Proc. Natl. Acad. Sci.* **78:** 5339.

Rohme, D. 1981. Evidence for a relationship between longevity of mammalian species and life spans of normal fibroblasts in vitro and erythrocytes in vivo. *Proc. Natl. Acad. Sci.* **78:** 5009.

Sager, R. 1989. Tumor suppressor genes: The puzzle and the promise. *Science* **246:** 1406.

Saxon, P.J., E.S. Srivatsan, and E.J. Stanbridge. 1986. Introduction of human chromosome 11 via microcell transfer controls tumorigenic expression of HeLa cells. *EMBO J.* **5:** 3461.

Schneider, E.L. and B.J. Fowlkes. 1976. Measurement of DNA content and cell volume in senescent fibroblasts utilizing flow mutiparameter single cell analysis. *Exp. Cell Res.* **98:** 298.

Schneider, E.L. and Y. Mitsui. 1976. The relationship between *in vitro* cellular aging and *in vivo* human age. *Proc. Natl. Acad. Sci.* **73:** 3584.

Stanbridge, E.J. 1976. Suppression of malignancy in human cells. *Nature* **260:** 17.

Stein, G.H. and L. Atkins. 1986. Membrane associated inhibition of DNA synthesis in senescent human diploid fibroblasts: Characterization and comparison to quiescent cell inhibition. *Proc. Natl. Acad. Sci.* **83:** 9030.

Stein, G.H. and R.M. Yanishevsky. 1979. Entry into S phase is inhibited in two immortal cell lines fused to senescent human diploid cells. *Exp. Cell. Res.* **120:** 155.

———. 1981. Quiescent human diploid cells can inhibit entry into S phase in replicative nuclei in heterodikaryons. *Proc. Natl. Acad. Sci.* **78:** 3025.

Stein, G.H., R.M. Yanishevsky, L. Gordon, and M. Beeson. 1982. Carcinogen-transformed human cells are inhibited from entry into S phase by fusion to senescent cells but cells transformed by DNA tumor viruses overcome the inhibition. *Proc. Natl. Acad. Sci.* **79:** 5287.

Sugawara, O., M. Oshimura, M. Koi, L.A. Annab, and J.C. Barrett. 1990. Induction of cellular senescence in immortalized cells by human chromosome 1. *Science* **247:** 707.

Tassin, J., E. Malaise, and Y. Courtois. 1979. Human lens cells have an *in vitro* proliferative capacity inversely proportional to the age. *Exp. Cell Res.* **123:** 388.

Thornton, S.C., S.N. Mueller, and E.M. Levine. 1983. Human endothelial cells: Use of heparin in cloning and long-term serial cultivation. *Science* **222:** 623.

Weissman, B.E., P.J. Saxon, S.R. Pasquale, G.R. Jones, A.G. Geiser, and E.J. Stanbridge. 1987. Introduction of a normal human chromosome 11 into Wilms' tumor cell line controls its tumorigenic expression. *Science* **236:** 175.

Whitaker, N.J., E.L. Kidston, and R.R. Reddel. 1992. Finite life span of hybrids formed by fusion of different simian virus 40-immortalized human cell lines. *J. Virol.* **66:** 1202.

Wittenberger, B., D. Raben, M.A. Lieberman, and L. Glaser. 1978. Inhibition of growth of 3T3 cells by extract of surface membranes. *Proc. Natl. Acad. Sci.* **75:** 5457.

Yanishevsky, R. and A.V. Carrano. 1975. Prematurely condensed chromosomes of dividing and non-dividing cells in aging human cell cultures. *Exp. Cell Res.* **90:** 169.

Yanishevsky, R.M. and G.H. Stein. 1980. Ongoing DNA synthesis continues in young human diploid cells (HDC) fused to senescent HDC, but entry into S phase is inhibited. *Exp. Cell Res.* **126:** 469.

Yanishevsky, R., M.L. Mendelsohn, B.H. Mayall, and V.J. Cristofalo. 1974. Proliferative capacity and DNA content of aging human diploid cells in culture: A cytophotometric and autoradiographic analysis. *J. Cell. Physiol.* **84:** 165.

Molecular and Cell Biology of Replicative Senescence

G.P. DIMRI AND J. CAMPISI
*Department of Cell and Molecular Biology, Lawrence Berkeley Laboratory,
University of California, Berkeley, California 94720*

Normal eukaryotic cells have only a limited capacity for cell division. This property has been termed the finite replicative life span of cells, and the process that limits cell division has been termed cellular or replicative senescence.

The finite replicative life span of cells was formally described about 30 years ago (Hayflick 1965). Since then, the phenomenon has been studied by a number of investigators, primarily by studying the proliferation of cells in culture. Several important features of replicative senescence have emerged (for review, see Stanulis-Praeger 1987; Goldstein 1990; McCormick and Campisi 1991; Peacocke and Campisi 1991).

First, replicative senescence is particularly stringent in human cells, which virtually never spontaneously immortalize. This is in sharp contrast to cells derived from several rodent species, which spontaneously immortalize at low but detectable frequencies. Nonetheless, when cell cultures are monitored carefully, it is clear that cells derived from very many, if not all, species and organs divide only a limited number of times.

Second, senescence is a stable state and is not apoptotic or programmed cell death. Senescent cells arrest growth with a G_1 DNA content but remain viable and metabolically active for long periods of time. Moreover, many genes (but not all genes) remain mitogen-inducible in senescent cells, despite the fact that the cells have lost the ability to respond to mitogens by initiating DNA replication.

Third, both the finite replicative life span phenotype and the final senescent state are dominant phenotypes in somatic cell fusion experiments. Moreover, at least four genetic complementation groups for immortality have been identified. Thus, replicative senescence appears to be a genetically dominant trait and to be controlled by multiple genetic loci.

Fourth, in addition to the cessation of cell proliferation, replicative senescence entails selected changes in differentiated functions. Thus, by the criteria of a stable and irreversible growth arrest and altered cell function, senescent cells resemble terminally differentiated cells. In contrast to terminal differentiation, however, replicative senescence occurs solely as a consequence of completing a finite number of cell divisions.

Despite substantial progress in understanding the prevalence, genetics, and molecular correlates of cell senescence, very little is known about the primary regulatory mechanisms that establish a finite replicative life span and maintain the senescent state. In addition, the biological significance of replicative senescence is incompletely understood. In this paper, we review our results, and those of other workers, which support the concepts that replicative senescence reflects processes which occur during the aging of organisms and that it constitutes a tumor suppressive mechanism. We also summarize our recent findings which suggest that senescent cells express one or more dominant transcription factors that may act to repress the proliferation of senescent cells.

REPLICATIVE SENESCENCE AND AGING

The biological significance of replicative senescence has been viewed from two, not mutually exclusive, perspectives. One of these holds that replicative senescence occurs throughout the life of multicellular organisms and that aged organisms accumulate senescent cells. Implicit in this view is the more controversial hypothesis that senescent cells cause or contribute to age-related pathology. These ideas have remained highly speculative and controversial primarily because it is not possible to identify senescent cells in vivo. Replicative senescence is generally studied in culture, where cells can be stimulated to proliferate in a controlled fashion for the many population doublings required for most or all cells in a culture to senesce, as judged by their failure to synthesize DNA in response to physiological stimuli. However, measurements of DNA synthesis do not distinguish senescent cells from other nonreplicating cells, such as quiescent or terminally differentiated cells, in complex tissues.

Support for the idea that replicative senescence in culture reflects processes that occur during the aging of organisms comes from three lines of evidence. First, cell cultures established from short-lived species senesce more rapidly than cultures established from long-lived species (Goldstein 1974; Hayflick 1976; Röhme 1981). In addition, cells from human donors afflicted with hereditary premature aging syndromes senesce much more quickly in culture than cells from age-matched controls (Goldstein 1969; Martin et al. 1970). These findings suggest that the replicative life span of cells and the chronological life span of organisms are related and perhaps controlled by a common set of genes. Finally, cell cultures established from old individuals tend to senesce more rapidly than cultures from younger individuals (Martin et al. 1970; Schneider and Mitsui 1976). This suggests that cells in renewable tissues undergo replicative senescence in vivo and accumulate in aged tissues. Taken together, these findings—although providing strong correlative evidence for the relationship between replicative senescence and aging—suffer from a lack of direct evidence for such a link.

A BIOMARKER FOR REPLICATIVE SENESCENCE

We recently found that senescent human cells of diverse tissue origins express an unusual β-galactosidase which has a pH optimum of about 6 (G.P. Dimri et al., in prep.). The pH optimum of this activity, which we refer to as the senescence-associated β-galactosidase (SA-β-gal), differed from that of the nearly ubiquitous lysosomal β-galactosidase (pH optimum of 4) and that of the bacterial β-galactosidase that is commonly used as a reporter enzyme (pH optimum ~7.5). The origin and function of this activity are unknown. However, it provides an ideal marker for testing the idea that senescent cells exist and accumulate with age in vivo: It lends itself to simple histochemical staining, and its detection does not rely on measurements of DNA synthesis.

First, we found that this activity was expressed by senescent human fibroblasts and keratinocytes in culture (Fig. 1), as well as by senescent human umbilical vein endothelial cells, adult mammary and surface ovarian epithelial cells, and neonatal melanocytes. However, SA-β-gal was not expressed by presenescent, quiescent, or terminally differentiated cells in culture (Fig. 1). It also was not expressed by a wide variety of cultured immortal or tumor cells (Fig. 1 and data not shown) but could be induced in two immortal human cell lines by normal human chromosomes 1 and 4, which have been shown to reverse the immortal phenotype in these cells (Fig. 1) (Ning et al. 1991; Hensler et al. 1994). Thus, this marker was tightly linked to the senescent phenotype of a variety of human cells in culture.

A pertinent feature of this marker was its expression in human skin samples obtained from different-aged donors. Although we found age-independent staining in the hair follicle and eccrine glands of the skin, there was a striking age-dependent staining pattern in the dermis and epidermis. Skin sections from relatively young donors (<39 years of age) showed minimal to no SA-β-gal staining in the dermis and epidermis (Fig. 2). In contrast, skin sections from relatively old donors (>69 years of age) showed moderate to strong SA-β-gal staining in either the dermis, epidermis, or both (Fig. 2). The cell type giving rise to this staining in the dermis was identified as fibroblasts, and the positive-staining cells in the epidermis were identified as basal (undifferentiated) keratinocytes. Thus, SA-β-gal activity serves as a biomarker for replicative senescence in culture and in vivo and provides the first in situ

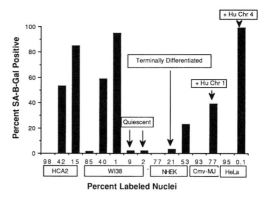

Figure 1. SA-β-gal and proliferative capacity in cultured human cells. HCA2 and WI38 are normal neonatal foreskin and fetal lung human fibroblasts, respectively. Cells at early, mid-, and late passage were labeled with [^{3}H]thymidine for 72 hr (Seshadri and Campisi 1990), fixed for 3–5 min at room temperature in 2% formaldehyde-0.2% glutaraldehyde, and stained for SA-β-gal using X-gal as a substrate and citric acid/sodium phosphate buffer, pH 6.0 (G.P. Dimri et al., in prep.). After autoradiography, 100–500 cells were counted at 200× magnification and the percentages of radiolabeled nuclei (% LN) and SA-β-gal-positive cells were determined. WI38 cells were made quiescent by depriving them of serum for 3 (9% LN) or 7 (2% LN) days, radiolabeled and stained for SA-β-gal. Neonatal human epidermal keratinocytes (NHEK) were obtained from Clonetics (San Diego, California), passaged in serum-free medium according to the supplier's instructions, and radiolabeled and stained for SA-β-gal at early (77% LN) and mid- (53% LN) passage. Early-passage NHEK were induced to terminally differentiate by addition of 2 mM CaCl$_2$ to the medium; 4 days later, they were radiolabeled and stained for SA-β-gal. CMV-MJ are immortal human fibroblasts that are induced to senesce within several doublings after normal human chromosome 1 is introduced by microcell fusion (Hensler et al. 1994). The parent and fusion cultures were radiolabeled and stained for SA-β-gal 8–10 doublings after fusion. HeLa are human cervical carcinoma cells that senesce upon introduction of human chromosome 4 (Ning et al. 1991). The parent culture and hybrid clones containing 60 cells or less were radiolabeled and stained for SA-β-gal.

evidence that senescent cells exist, and accumulate with age, in vivo.

REPLICATIVE SENESCENCE AND TUMOR SUPPRESSION

A second view of replicative senescence holds that it constitutes a tumor suppressive mechanism. The evidence for this view is more solid than the evidence that senescence is related to aging, because it derives from both correlations as well as mechanistic studies on etiology and progression of malignant tumors.

First, both naturally occurring and experimentally induced tumors often contain cells that have escaped the limits imposed by replicative senescence. That is, tumors often contain immortal cells or cells that have an extended replicative life span (Ponten 1976; Sager 1989). Moreover, immortality greatly increases a cell's susceptibility to progression toward a more malignant phenotype because it permits the extensive cell divi-

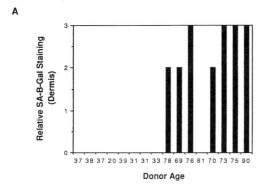

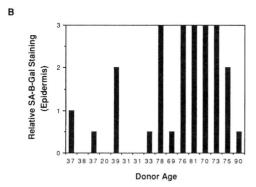

Figure 2. SA-β-gal staining in human skin samples from young and old donors. Human skin specimens were frozen in liquid nitrogen, thin (4 μm) sections were cut, and the sections were fixed and stained for SA-β-gal as described for cultured cells in Fig. 1. The sections were evaluated in a blinded fashion for the prevalence and intensity of SA-β-gal staining in the dermis and epidermis and were then assigned an overall staining value between 0 and 3.

sion that is often necessary for cells to acquire successive mutations.

Second, it is evident that some oncogenes, including those carried by certain human viruses, act at least in part by immortalizing or extending the replicative life span of cells (Weinberg 1985; Galloway and McDougall 1989; Shay and Wright 1991). Thus, genetic lesions, or the strategies of certain oncogenic viruses, that lead to tumorigenesis may and do involve mechanisms that permit cells to escape replicative senescence.

Finally, among the cellular genes that are known to be critical for establishing and/or maintaining the senescent phenotype are two well-recognized tumor suppressor genes: the retinoblastoma susceptibility gene (Rb) and the p53 gene.

ROLE OF p53 AND Rb IN REPLICATIVE SENESCENCE

Thus far, the only genes that are known to reliably immortalize or extend the replicative life span of human cells are the oncogenes of certain DNA tumor viruses. These genes include the large T antigen of the SV40 virus, and the combination of the E1A and E1B genes of adenoviruses and E6 and E7 genes of the papillomaviruses (Shay and Wright 1991). T antigen is known to bind and inactivate both the Rb and p53 proteins, whereas E1A and E7 are known to inactivate Rb, and E1B and E6 are known to inactivate p53. Taken together, these findings suggest that the actions of both Rb and p53 are critical for replicative senescence. They further suggest that inactivation of both Rb and p53 is essential in order for cells to circumvent the normal proliferative constraints imposed by replicative senescence.

Support for the importance of Rb and p53 in replicative senescence derives primarily from studies of life span extension or immortalization by SV40 T antigen. T antigen extends the replicative life span of human cell cultures; at the end of their replicative life span, T-antigen-expressing cultures enter a state termed crisis, in which cell proliferation and cell death occur, and from which rare immortal variants may emerge (Shay and Wright 1989).

Human fibroblasts grown in the presence of antisense oligonucleotides (oligos) that inhibited the expression of Rb and p53 behave very similarly to cells expressing SV40 T antigen (Fig. 3) (Hara et al. 1991). That is, the Rb and p53 antisense oligos extended replicative life span to the same extent as T-antigen expression. Of interest were the effects of the Rb and p53 antisense oligos alone. The Rb antisense oligo, when used alone, extended proliferation about half as well as T antigen. In contrast, the p53 antisense oligo had little effect on replicative life span when used alone, but potentiated the effect of the Rb antisense oligo (Fig. 3). Thus, inactivation of Rb and p53 could account for all the life span extension caused by T antigen.

In addition, we found that p53 and Rb were important for maintaining the growth-arrested senescent state. We constructed expression vectors bearing either a wild-type SV40 T antigen or T antigens that were mutated in either the p53 or Rb binding domains. These expression vectors were introduced into senes-

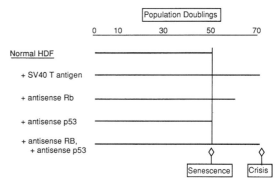

Figure 3. Rb and p53 inactivation accounts for the life span extension caused by SV40 T antigen. The replicative life span of TIG-1 normal human fibroblasts, grown in the absence or presence of T antigen or antisense oligonucleotides that inhibited expression of Rb or p53, was determined as described by Hara et al. (1991).

Table 1. Stimulation of DNA Synthesis by Wild-type, Rb, and p53 Binding-Deficient SV40 T Antigens

Condition	Plasmid	% Labeled nuclei
Presenescent, quiescent	CMV-T	80 (70–100)
	CMV-T[K1]	59 (50–75)
	CMV-T[p]	54 (52–56)
	CMV-βgal or none	6 (1–10)
Senescent	CMV-T	78 (65–78)
	CMV-T[K1]	9 (4–15)
	CMV-T[p]	24 (21–29)
	CMV-βgal or none	2 (0–5)
	CMV-T[K1] + CMV-T[p]	65 (102–152)

Expression vectors (CMV, human cytomegalovirus early region promoter) for wild-type SV40 T antigen (CMV-T) or T antigens mutated in the p53 (CMV-T[p]) or Rb (CMV-T[K1]) binding domains were microinjected into the nucleus of quiescent (serum-deprived) or senescent human diploid fibroblasts. An expression vector for the *E. coli* β-galactosidase was coinjected and served to identify injected cells by histochemical staining at pH 7.5 (Dimri et al. 1994). The injected cells were given [^{3}H]thymidine (10 μCi/ml) for 48 hr, processed for autoradiography, and 100–500 cells were counted under bright field microscopy to determine the percentage of injected cells that synthesized DNA (% labeled nuclei). The numbers in parentheses show the range of values over 2–5 independent experiments, except the last entry (CMV-T[K1] + CMV-T[p]) shows the number of injected cells with labeled nuclei/total number of injected cells in a single experiment.

cent cells by microinjection, and their ability to induce DNA synthesis was monitored by the incorporation of [^{3}H]thymidine and autoradiography (Table 1). Only the wild-type T was capable of inducing a large fraction of senescent cells to initiate DNA synthesis. The Rb-binding-defective T was a very poor inducer, showing only 10–15% of wild-type activity. The p53-binding-defective T was a weak inducer, showing about 30% of wild-type activity. Both mutants showed 65–70% of wild-type activity in quiescent presenescent cells. The results suggest that Rb and p53 are important for maintaining the growth-arrested state of senescent cells.

Taken together, the antisense and T-antigen mutant experiments support the idea that the Rb and p53 tumor suppressor genes are critical for establishing and maintaining replicative senescence. Rb and p53 are, of course, among the genes that frequently suffer a loss-of-function mutation in tumors. Thus, cell senescence and tumorigenesis may be viewed as alternate processes, each strongly influenced by certain tumor suppressor genes (Fig. 4).

SELECTIVE REPRESSION OF GROWTH REGULATORY GENES

The growth arrest associated with replicative senescence is exceedingly stringent. It is refractory to all known physiological mitogens, and of course ensures essentially complete resistance to tumorigenic transformation. Even SV40 T antigen, which induces senescent cells to initiate DNA synthesis (Table 1), fails to induce a complete cell cycle: SV40-infected senescent cells initiate and complete DNA replication but fail to undergo mitosis (Gorman and Cristofalo 1985).

The immediate cause for the failure of senescent cells to proliferate may be the selective repression of a subset of genes whose expression is essential for cell proliferation (use here interchangeably with growth). These genes include two early response genes and several genes whose expression is induced just prior to the start of S phase (Fig. 5).

The first growth-regulatory gene that was shown to be repressed in senescent cells was the c-*fos* proto-oncogene (Seshadri and Campisi 1990). This gene is an immediate-early growth response gene, and it encodes one component of the AP-1 transcription factor. The expression of c-*fos* is selectively repressed in senescent human fibroblasts at the level of transcription. Other immediate-early response genes, such as the c-*myc* proto-oncogene, ornithine decarboxylase, JE, and KC, are expressed similarly in presenescent and senescent cells (Rittling et al. 1986; Seshadri and

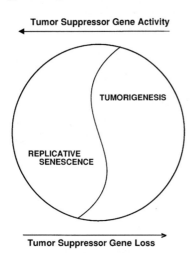

Figure 4. Replicative senescence and tumorigenesis may be alternative processes, each dependent (albeit in opposing ways) on the function of tumor suppressor genes such as *Rb* and *p53*.

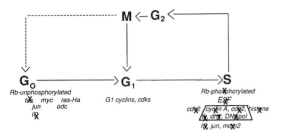

Figure 5. Growth-related gene expression in senescent human fibroblasts. Shown are a subset of the genes that are induced by mitogen in early-passage cells, and the genes whose induction fails in senescent cells (marked by an X). Of several immediate-early genes, only two (*fos* and *Id*) are repressed in senescent cells. In contrast, several late G_1 genes are repressed in senescent cells, but many of these may be the result of the repression of E2F activity.

Campisi 1990). Thus, senescent cells may fail to proliferate due, at least in part, to repression of the c-*fos* proto-oncogene. However, attempts to reactivate DNA synthesis by providing senescent cells with an expressible c-*fos* gene have not met with success (Table 2).

Recently, we found a second immediate-early gene whose expression is necessary for growth and whose expression was repressed in senescent human fibroblasts. This was the Id gene, which encodes a helix-loop-helix (HLH) protein that inhibits the activity of basic helix-loop-helix (bHLH) transcription factors (Sun et al. 1991). Human fibroblasts expressed three Id genes: *Id-1H* and *Id-1H'*, which are derived by alternate splicing of a single gene, and *Id-2H*. All three Id genes were induced as immediate-early genes upon growth factor stimulation of quiescent fibroblasts and showed a second peak of expression in mid- to late G_1. All three Id genes were repressed in senescent human fibroblasts (Hara et al. 1994). Thus, a second reason why senescent cells fail to proliferate may be the repression of Id genes. However, Id expression vectors, whether alone or in combination with a c-*fos* expression vector, were incapable of stimulating senescent cells to initiate DNA replication (Table 2).

As noted earlier, several late G_1 genes are also repressed in senescent human fibroblasts. These include the replication-dependent histones, cdc2, cyclin A, dihydrofolate reductase, thymidine kinase, and DNA polymerase α (see Dimri et al. 1994). Many of these genes are positively regulated by a common transcription factor, E2F. E2F acts as a heterodimer, and two E2F components, E2F-1 and DP-1, have been shown to be particularly synergistic for *trans*-activating activity. We found that DP-1 was expressed similarly in presenescent and senescent cells. In contrast, E2F-1 was repressed in senescent cells, and senescent cells were markedly deficient in E2F-binding activity (Dimri et al. 1994). Thus, senescent cells may fail to express several late G_1 genes due to repression of E2F-1, which leads to a deficiency in E2F activity (Fig. 5). However, an E2F-1 expression vector did not stimulate senescent cells to initiate DNA synthesis (Dimri et al. 1994), nor did it synergize with c-*fos* or Id expression vectors to do so (not shown).

Taken together, these findings suggest that the immediate cause for the growth arrest associated with senescence is the selective repression of growth regulatory genes. These genes include the c-*fos* and Id immediate-early genes, and the E2F-1 gene, whose repression in turn may be responsible for the failure of senescent cells to express several late G_1 genes. Although any one of these gene repressions can explain the growth arrest, it is clear that the expression of none of these genes, alone or in combination, is sufficient to reverse the growth arrest. It is entirely possible that other, yet-to-be discovered, growth regulatory genes are also repressed in senescent cells. On the other hand, as noted earlier, the senescent phenotype is dominant in somatic cell hybrids. This finding predicts that senescence should not be reversed by restoring deficiencies in gene expression. Rather, the dominance predicts that senescent cells express one or more inhibitors of cell proliferation whose inhibitory activity is dominant over the growth stimulatory activities of c-*fos*, Id, and E2F.

DOMINANT GROWTH SUPPRESSORS

What is the nature of dominant growth suppressors expressed by senescent cells? First, senescent cells may express very high levels of growth inhibitors present in quiescent cells. This is clearly the case for p21/sdi-1, the Cdk2 inhibitor, which was first identified as a transcript that is present at higher levels in senescent cells relative to quiescent cells (Noda et al. 1994). In addition, our preliminary evidence suggests that senescent cells express at least two novel transcription factors that may act to suppress growth.

We suggest that one of these factors is a bHLH protein. This hypothesis derives from our search for cellular genes that are capable of complementing the T-antigen mutants in Rb and p53 binding. In these studies, we found that the cellular *mdm2* gene complemented the p53-binding-deficient T antigen. This was not a surprising result because it is known that *mdm2* encodes a negative regulator of p53 activity

Table 2. Restoration of Early-response Gene Expression Does Not Reactivate DNA Synthesis in Senescent Human Fibroblasts

Plasmid injected	% Labeled nuclei
None	8
CMV-T	86
CMV-fos	8
CMV-Id-1H	19
CMV-Id-2H	9
CMV-Id-1H, c-fos	20

Expression vectors for wild-type SV40 T antigen, c-*fos*, or human *Id* genes were microinjected into senescent human fibroblasts, and the cells were monitored for ability to initiate DNA replication, as described in Table 1.

(Momand et al. 1992). What was surprising was our finding that the *Id-1H* gene complemented the Rb-binding-deficient T antigen. Thus, the Rb-binding-deficient T was unable to stimulate senescent cells to initiate DNA synthesis (~10% of wild-type T activity), although it was capable of stimulating quiescent cells (~70% of wild-type activity). This finding alone suggests that the activity of Rb, or an Rb-related protein, differs between senescent and presenescent cells. When the Rb-binding-deficient T was introduced into senescent cells together with a variety of cellular genes, only the *Id-1H* gene restored this mutant to near wild-type activity. Our working hypothesis is that senescent cells express a bHLH protein that cooperates with Rb, or an Rb family member, to suppress growth. We propose that this bHLH protein may act in a dominant fashion and that Id acts to inactivate it. Thus, senescent cells may differ from presenescent cells both in the expression of this putative growth suppressor and in the repression of its potential inactivator.

We found a second potential senescence-associated negative growth regulator in the course of our studies on E2F activity in presenescent and senescent cells. As noted earlier, senescent human fibroblasts were markedly deficient in E2F-binding activity, but they expressed relatively high levels of a novel protein/DNA complex. This complex was not an E2F-specific complex (Dimri et al. 1994) and its binding site was outside the E2F-binding site in the oligonucleotide probe. Our preliminary data suggest that this binding site, when cloned upstream of a heterologous promoter, acts as a transcriptional silencer sequence in senescent cells, but not in presenescent cells. Thus, this complex, which we term senescence factor or SNF, may act to suppress transcription in senescent cells.

Taken together, our results suggest that it is not surprising that restoration of AP-1, E2F, or Id expression to senescent cells is not sufficient to induce DNA replication because senescent cells express dominant growth inhibitors, of which a bHLH factor and SNF are candidates (Fig. 6).

ACKNOWLEDGMENTS

We thank our many colleagues, particularly M. Acosta, G. Basile, E. Hara, M. Linskens, J. Nehlin, M. Peacocke, O. Pereira-Smith, E. Medrano, J. Oshima, C. Roskelley, G. Scott, T. Seshadri, A. Uzman, and K. Yoshimoto, who contributed experiments and/or discussions to the work presented here. This work was supported by grants from the National Institutes of Health (AG-09909 and AG-11658).

REFERENCES

Dimri, G.P., E. Hara, and J. Campisi. 1994. Regulation of two E2F-related genes in presenescent and senescent human fibroblasts. *J. Biol. Chem.* **269:** 16180.

Galloway, D.A. and J.K. McDougall. 1989. Human papillomaviruses and carcinomas. *Adv. Virus Res.* **37:** 125.

Goldstein, S. 1969. Lifespan of cultured cells in progeria. *Lancet* **I:** 424.

———. 1974. Aging in vitro: Growth of cultured cells from the Galapagos turtle. *Exp. Cell Res.* **83:** 297.

———. 1990. Replicative senescence: The human fibroblast comes of age. *Science* **249:** 1129.

Gorman, S.D. and V.J. Cristofalo. 1985. Reinitiation of cellular DNA synthesis in BrdU-selected nondividing senescent WI38 cells by simian virus 40 infection. *J. Cell. Physiol.* **125:** 122.

Hayflick, L. 1965. The limited in vitro lifetime of human diploid cell strains. *Exp. Cell Res.* **37:** 614.

———. 1976. The cell biology of human aging, *N. Engl. J. Med.* **295:** 1302.

Hensler, P.J., L.A. Annab, J.C. Barrett, and O. M. Pereira-Smith. 1994. A gene involved in control of human cellular senescence on human chromosome 1q. *Mol. Cell. Biol.* **14:** 2291.

Hara, E., H. Tsuri, S. Shinozaki, and K. Oda. 1991. Co-operative effect of antisense-Rb and antisense-p53 oligomers on the extension of lifespan in human diploid fibroblasts, TIG-1. *Biochem. Biophys. Res. Commun.* **179:** 528.

Hara, E., T. Yamaguchi, H. Nojima, T. Ide, J. Campisi, H. Okayama, and H. Oda. 1994. Id-related genes encoding helix-loop-helix proteins are required for G1 progression and are repressed in senescent human fibroblasts. *J. Biol. Chem.* **269:** 2139.

Martin, G.M., C. A. Sprague, and C.J. Epstein. 1970. Replicative lifespan of cultivated human cells. Effect of donor's age, tissue and genotype. *Lab. Invest.* **23:** 86.

McCormick, A. and J. Campisi. 1991. Cellular aging and senescence. *Curr. Opin. Cell Biol.* **3:** 230.

Momand, J., G.P. Zambetti, D.C. Olson, D. George, and A.J. Levine. 1992. The mdm2 oncogene product forms a complex with the p53 protein and inhibits p53-mediated transactivation. *Cell* **69:** 1237.

Ning, Y., J.L. Weber, A.M. Killary, D.H. Ledbetter, J.R. Smith, and O.M. Pereira-Smith. 1991. Genetic analysis of indefinite division in human cells: Evidence for a cell senescence-related gene(s) on human chromosome 4. *Proc. Natl. Acad. Sci.* **88:** 5635.

Noda, A., Y. Ning, S. F. Venable, O. M. Pereira-Smith, and

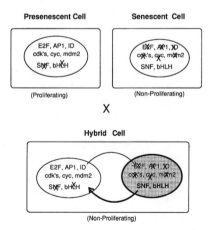

Figure 6. Dominance of the senescent state. When proliferating presenescent and senescent cells are fused, the resulting hybrid fails to initiate DNA replication. This is the case despite the fact that the presenescent cell can provide the senescent cell with activities that it lacks. Senescent cells are believed to express dominant acting growth suppressors. We propose that two of these suppressors may be a bHLH protein and SNF.

J.R. Smith. 1994. Cloning of senescent cell-derived inhibitors of DNA synthesis using an expression screen. *Exp. Cell Res.* **211:** 90.

Peacocke, M. and J. Campisi. 1991. Cellular senescence: A reflection of normal growth control, differentiation, or aging? *J. Cell. Biochem.* **45:** 147.

Ponten, J. 1976. The relationship between in vitro transformation and tumor formation in vivo. *Biochim. Biophys. Acta* **458:** 397.

Röhme, D. 1981. Evidence for a relationship between longevity of mammalian species and lifespans of normal fibroblasts in vitro and erythrocytes in vivo. *Proc. Natl. Acad. Sci.* **78:** 5009.

Rittling, S.R., K.M. Brooks, V.J. Cristofalo, and R. Baserga. 1986. Expression of cell cycle-dependent genes in young and senescent WI38 fibroblasts. *Proc. Natl. Acad. Sci.* **86:** 3316.

Sager, R. 1989. Tumor suppressor genes: The puzzle and the promise. *Science* **246:** 1406.

Schneider, E.L. and Y. Mitsui. 1976. The relationship between in vitro cellular aging and in vivo human aging. *Proc. Natl. Acad. Sci.* **73:** 3584.

Seshadri, T. and J. Campisi. 1990. Repression of c-*fos* transcription and an altered genetic program in senescent human fibroblasts. *Science* **247:** 205.

Shay, J.W. and W.E. Wright. 1989. Quantitation of the frequency of immortalization of normal human diploid fibroblasts by SV40 large T antigen. *Exp. Cell Res.* **184:** 109.

———. 1991. Defining the molecular mechanisms of human cell immortalization. *Biochim. Biophys. Acta* **1072:** 1.

Stanulis-Praeger, B.M. 1987. Cellular senescence revisited: A review. *Mech. Ageing Dev.* **38:** 1.

Sun, X.H., N.G. Copeland, N.A. Jenkins, and D. Baltimore. 1991. Id proteins Id1 and Id2 selectively inhibit DNA binding by one class of helix-loop-helix proteins. *Mol. Cell Biol.* **11:** 5603.

Weinberg, R.A. 1985. The action of oncogenes in the cytoplasm and in the nucleus. *Science* **230:** 770.

Growth Suppression by Members of the Retinoblastoma Protein Family

L. Zhu,* G.H. Enders,*† C.-L. Wu,* M.A. Starz,‡ K.H. Moberg,‡
J.A. Lees,‡ N. Dyson,* and E. Harlow*

*Massachusetts General Hospital Cancer Center, Charlestown, Massachusetts 02129;
†Gastrointestinal Unit, Massachusetts General Hospital, Boston, Massachusetts 02114;
‡Center for Cancer Research, Department of Biology, Massachusetts Institute of Technology,
Cambridge, Massachusetts 02139

Cellular changes induced by oncogenes of small DNA tumor viruses have provided unique insights into the mechanisms of cell transformation and growth regulation. Studies of adenovirus E1A, SV40 large T antigen, and E7 proteins of human papillomaviruses suggest that these viral oncoproteins use similar mechanisms to transform cells. An important feature of these oncoproteins is that they physically interact with key regulators of cellular proliferation. One such cellular target is the retinoblastoma protein (pRB). pRB is the product of the prototype tumor suppressor gene, which is mutated or deleted in all retinoblastomas and a wide variety of other tumors (for review, see Weinberg 1991, 1992). Overexpression of wild-type pRB in tumor cell lines that lack functional pRB causes a growth arrest in the G_1 phase of the cell cycle in some cells and often reduces their tumorigenic potential (Huang et al. 1988; Bookstein et al. 1990; Goodrich et al. 1991; Hinds et al. 1992). From these observations, it appears that the normal role of pRB is as a negative regulator of cell proliferation.

In cells infected with the DNA tumor viruses, it is thought that physical association with the viral oncoproteins functionally mimics the mutation of the *RB-1* gene in tumor cells (DeCaprio et al. 1988; Whyte et al. 1988; Dyson et al. 1989a,b). This hypothesis has been supported by several lines of experimental evidence. Detailed mutagenesis studies of E1A, T antigen, and E7 proteins have demonstrated that the ability to bind to pRB is required for transformation (Egan et al. 1989; Ewen et al. 1989; Munger et al. 1989; Whyte et al. 1989). Although pRB is phosphorylated in a cell-cycle-dependent manner, viral oncoproteins bind preferentially to the hypophosphorylated forms of pRB, which represent the active forms of pRB predominant in the G_1 phase of the cell cycle (Buchkovich et al. 1989; Chen et al. 1989; DeCaprio et al. 1989; Ludlow et al. 1989; Mihara et al. 1989; Dyson et al. 1992). Growth suppression mediated by pRB leads to an accumulation of unphosphorylated protein and is reversed by coexpression of the viral oncoproteins (Hinds et al. 1992).

The regions of E1A, T antigen, and E7 proteins that bind to pRB also interact with other cellular proteins (Dyson et al. 1989a,b, 1992; Ewen et al. 1989; Giordano et al. 1989, 1991; Whyte et al. 1989). By analogy with pRB, these cellular proteins may also be involved in cell cycle control. The isolation and cloning of p107 and p130, two such E1A-binding proteins, have revealed that they are structurally related to pRB and may therefore have similar functional properties (Ewen et al. 1991; Hannon et al. 1993; Li et al. 1993; Mayol et al. 1993). The greatest similarity is in the so-called pocket domain. This region of pRB, comprising two noncontiguous segments (A and B) separated by a spacer, is both necessary and sufficient for stable interaction with E1A (Hu et al. 1990; Huang et al. 1990; Kaelin et al. 1990). This region also harbors most of the naturally occurring mutations detected in human tumors and is essential for the growth suppression activity of pRB (Hu et al. 1990; Qin et al. 1992). The spacer regions of p107 and p130 are homologous to one another but are unrelated to the spacer region of pRB. Interestingly, the spacer regions of p107 and p130 help confer stable association with cyclin-dependent kinases containing cyclin A and cyclin E (Ewen et al. 1992; Faha et al. 1992; Li et al. 1993).

A common biochemical property has been established for pRB, p107, and p130. All three proteins form complexes with the E2F transcription factor (Bandara et al. 1991; Chellappan et al. 1991; Cao et al. 1992; Devoto et al. 1992; Shirodkar et al. 1992; Cobrinik et al. 1993). To date, the interactions of E2F with pRB and of p107 with E2F have been studied in most detail. pRB and p107 repress the activity of E2F through physical interaction that requires an intact pocket domain and at least a portion of the carboxyl terminus (Hiebert et al. 1992; Qin et al. 1992; Helin et al. 1993a; Schwarz et al. 1993; Zamanian and La Thangue 1993; Zhu et al. 1993). These interactions are disrupted by E1A. There are multiple lines of evidence suggesting that the release of E2F may lead to inappropriate cell proliferation. E2F-binding sites are found in promoter regions of many genes whose products are required for DNA synthesis (for review, see Nevins 1992; Helin and Harlow 1993). The overexpression of E2F-1 is sufficient to drive quiescent cells into S phase (Johnson et al. 1993), can prevent cycling cells from exiting the cell cycle (Johnson et al. 1993), and can transform immortalized cells

(Singh et al. 1994). Thus, it is proposed that the repression of E2F activity by pRB and p107 is an important component of the mechanisms by which these proteins negatively regulate cell proliferation.

To investigate the biological functions of p107, we have isolated a full-length human p107 cDNA and expressed it in vivo under the control of the CMV promoter. In these experiments, p107 is a cell-type-specific growth suppressor that arrests cell cycle progression in G_1 phase. To investigate this arrest, we have performed structure-function analysis of p107 and attempted to rescue the arrest by the coexpression of additional genes. These experiments show that p107 can arrest the cell cycle by two distinct mechanisms. One of these appears to be similar to the growth suppression domain of pRB and involves repression of E2F activity. The second mechanism for growth suppression is distinct from pRB and is mediated by physical association with cyclin A/cdk or cyclin E/cdk complexes.

RESULTS

p107 Is a Repressor of E2F-mediated *trans*-Activation

pRB and p107 associate independently with the transcription factor E2F. A reporter construct that allows expression of the CAT gene under the control of the adenovirus E2 promoter (containing two E2F-binding sites) was used to determine the functional consequences of the p107/E2F interaction. As shown in Figure 1, the E2F CAT reporter has a measurable activity that is due to the endogenous E2F of the cell. Cotransfection of p107 or pRB expression plasmids with the reporter construct repressed the activity of the E2 promoter. This repression was apparently mediated through E2F, since an E2 promoter containing mutations in the ATF-binding site was repressed by pRB and p107, but a mutant promoter in which both E2F-binding sites were mutated was not significantly affected. Thus p107, like pRB, is a potent repressor of E2F activity.

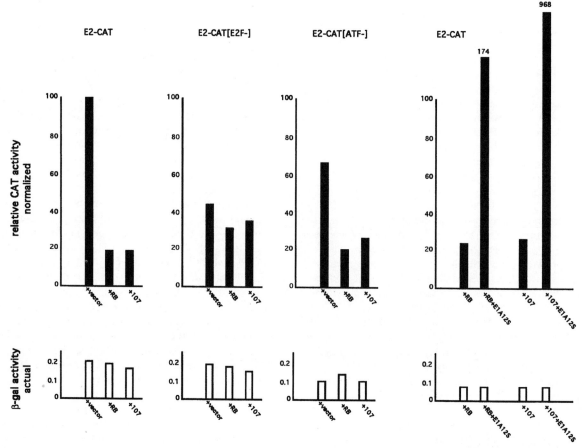

Figure 1. pRB and p107 are repressors of E2F activity. Saos-2 cells were transiently transfected using the calcium phosphate precipitation method with the following plasmids: 5 μg CAT reporter plasmid (E2-CAT, E2-CAT[E2F-], or E2-CAT[ATF-] as indicated); 5 μg pRSVβ-gal; 5 μg of the CMV expression constructs (vector, pCMVpRB, or pCMVp107 as indicated) and 5 μg of pCMVE1A12S as indicated. The total amounts of DNA in transfection were adjusted to 25 μg with pBSK DNA. Cell extracts were prepared 40 hr after transfection. β-gal activity was determined with the O-nitrophenol-β-D-galactoside method and presented as the actual OD_{420} reading. CAT activity was normalized against the β-gal activity to be presented as arbitrary units.

p107 Is a Cell Type and Cell Cycle Stage-specific Growth Suppressor

The growth suppression activity of p107 has been observed in three different assay systems. Growth suppression was first detected in colony formation assays (Zhu et al. 1993). In these assays, a p107 expression construct carrying a neomycin resistance gene was transfected into Saos-2 cells. Transfected cells were selected to grow into colonies in the presence of the drug G418. The transfection of p107 or pRB consistently resulted in about tenfold decreases in colony numbers as compared with the vector control (data not shown), indicating that overexpression of p107 or pRB was not compatible with cell growth.

A second assay system was used to determine the position in the cell cycle of the p107-induced growth arrest. p107 or pRB expression constructs were cotransfected with an expression construct for the cell surface protein CD20 that is normally expressed only in B-lineage lymphocytes. Transfected cells were identified by staining with an anti-CD20 antibody. The DNA content and the cell cycle profile of the transfected cells were measured by flow cytometry analysis (van den Heuvel and Harlow 1993). In this assay system, transfection of Saos-2 cells with p107 or pRB caused a significant increase in the population of cells in G_1 with corresponding decreases in S and G_2/M phase populations (Fig. 2, column A). These changes were consistent with cell cycle arrest in G_1 phase but could also be caused by faster progression of cells through S, G_2, and M phases. To distinguish between these possibilities, p107- or pRB-transfected cells were incubated with nocodazole. Nocodazole disrupts microtubule formation and blocks cells at G_2/M (Zieve et al. 1980). Although the addition of nocodazole to cycling cells led to an accumulation of cells in G_2/M, the p107- or pRB-transfected cells remained arrested in G_1 (Fig. 2, column B).

The growth suppression activities of p107 and pRB are not only seen in transfected cells but are also apparent in stable Saos-2 cell lines that express p107 or pRB under an inducible promoter. These cell lines were established using the tetracycline repressible system described by Gossen and Bujard (1992). Induction of p107 or pRB expression resulted in about tenfold increases in protein levels and was accompanied by efficient growth suppression. In Figure 3, [^{3}H]thymidine incorporation was used to measure cell proliferation before and after induction of p107 or pRB. Significant inhibition of [^{3}H]thymidine incorporation could be seen as early as 12 hours after induction and was complete by 24 hours.

Osteosarcoma cell line U2OS was included in the above three types of growth suppression assays. Although both the pRB and p107 proteins were expressed at high levels, the growth characteristics of U2OS cells were not affected (data not shown). This result indicates that the growth suppression activities of pRB and p107 are cell-type-specific (also see below).

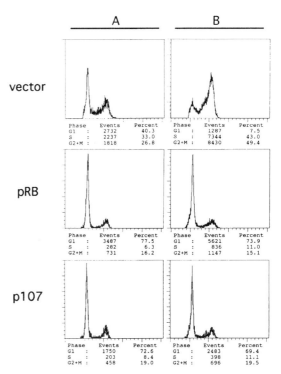

Figure 2. Cell cycle arrest in G_1 induced by pRB or p107. Saos-2 cells were transfected with 10 μg of vector, pCMVpRB, or pCMVp107 as indicated, together with 2 μg of pCMVCD20 and 10 μg of pBSK DNA. Cells were harvested for flow cytometry analysis 48 hr later (A) or treated with 40 ng/ml nocodazole from 24 to 48 hr after transfection (B). Events are numbers of cells (CD20 positive) counted for each cell cycle phase, and percentages of cells in each phase are listed under Percent.

p107 Contains Two Growth Suppression Domains

Using the flow cytometry growth suppression assay described above, we carried out systematic structure-function studies of the p107 molecule in an attempt to identify the mechanisms through which p107 functions to block cell cycle progression. Figure 4A shows a set of representative p107 mutants. L29 is the pocket region (schematically represented by segment A, spacer, and segment B) of p107 that is both necessary and sufficient for E1A binding. N385 includes the pocket domain of p107 together with the carboxy-terminal fragment. L19 contains the amino terminus of p107 plus the A segment of pocket and the spacer sequences. L30 differs from L19 in that it contains only the amino-terminal half of the spacer. Amino acid residues at the carboxyl terminus of p107 (1035–1068) were added to all the mutants that lack a complete carboxyl terminus to facilitate efficient nuclear localization (L. Zhu, unpubl.). The mutant proteins were tagged with the hemagglutinin epitope (HA) to enable them to be distinguished from the endogenous protein and expressed under the control of the CMV promoter. The levels of proteins synthesized from these constructs

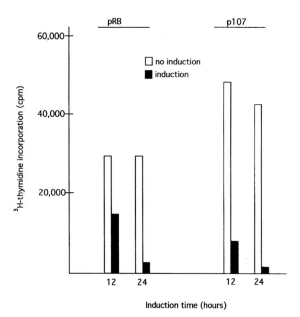

Figure 3. Inhibition of DNA synthesis in pRB- or p107-inducible cell lines. pRB- or p107-inducible cell lines were generated using the tetracycline repressible system (Gossen and Bujard 1992; our unpublished results). Cells were cultured in media containing 1.0 μg/ml tetracycline, and induction of protein synthesis was achieved by complete withdrawal of tetracycline from the medium. At 12 and 24 hr after induction, cells were incubated in media containing 2 μCi [^{3}H]thymidine/ml for 1 hr, and TCA precipitable [^{3}H]thymidine counts were measured by liquid scintillation counting.

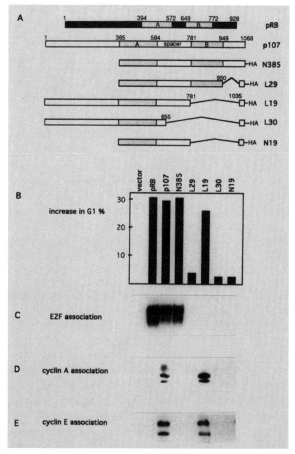

Figure 4. Growth suppression by p107 through two different mechanisms. (*A*) Schematic drawing of pRB, p107, and p107 mutants. The *A* and *B* segments contain sequences homologous between pRB and p107; sequences outside these two segments are different between the two proteins, as illustrated by the black and white colors. Amino acid residue numbers are noted above deletion or other transition boundaries. HA represents the hemagglutinin epitope: YPYDVPDYASL (single amino acid code). (*B*) Growth suppression activities. Saos-2 cells were transfected with the indicated expression constructs and analyzed by flow cytometry as in Fig. 2A. The ability to cause G_1 arrest was presented as net increases in G_1 phase population (%) compared with vector transfected cells. (*C*) E2F association. Saos-2 cells were transfected with the indicated plasmids, and cell extracts were prepared for immunoprecipitation with anti-pRB MAb XZ77 (Hu et al. 1991) (vector, and pRB lanes), anti-p107 MAb SD4 (Dyson et al. 1993) (p107 lane), anti-HA MAb 12CA5 (N385, L29, L19, L30, and N19 lanes). The immunoprecipitates were treated with 0.8% sodium deoxycholate followed by band shift with ^{32}P-labeled E2F-binding-site-containing oligonucleotides. The free, unbound oligonucleotides had run off the gel. (*D,E*) Same set of experiments was performed as above, except the antibodies were covalently coupled to protein A–Sepharose beads and the immunoprecipitates were separated on an 8% SDS-PAGE and blotted. Two identical blots were probed with anti-cyclin A MAb BF683 (Faha et al. 1992) (*D*) or anti-cyclin E MAb HE12 (Dulic et al. 1993) (*E*). The blots were developed with enhanced chemiluminescence (ECL).

after transient transfection were similar (data not shown).

Figure 4B shows the cell cycle profile of Saos-2 cells transfected with the indicated expression constructs. The extent of cell cycle arrest is presented as a percentage increase in the G_1 phase population compared to the vector control. The N385 mutant was as active as the full-length p107 protein, a feature similar to that observed previously with the analogous domain of pRB. As has been reported for pRB, the deletion of sequences carboxy-terminal to the pocket of p107 to generate the so-called "small pocket" of p107 (L29) inactivated the growth suppression properties of p107.

Unexpectedly, however, mutant L19 also arrests cells with an efficiency only slightly lower than the wild-type p107 protein. This was surprising, since L19 lacks segment B of the p107 pocket domain, and previous studies have shown that the analogous region of pRB is essential for its growth suppression activities. Comparison of the structures of N385 and L19 suggested that the sequences in common to these two mutants might be sufficient for growth suppression. This, however, is not the case. Mutant N19 containing segment A plus the spacer was unable to cause a cell cycle arrest. The sequences in the amino-terminal half of the spacer appear to be important for the growth suppression activity of L19, since further deletion of the spacer up to amino acid residues 655 (mutant L30) eliminated this activity. These results suggest that p107 contains two different growth suppression domains. For the remainder of this paper we use mutants N385 and L19 to represent these activities. Further work defining the boundaries of the domains in more detail is described by L. Zhu et al. 1994 (in prep.).

The Two Growth Suppression Domains Function through Different Mechanisms

To investigate the mechanisms involved in growth suppression by the two domains of p107, we have studied the biochemical properties of the p107 mutants. The interaction of these mutants with E2F and cyclins was examined after transient transfection of Saos-2 cells. p107 proteins were immunoprecipitated with a monoclonal antibody to the HA epitope tag. E2F proteins associated with p107 were released from the immune complexes by treatment with 0.8% sodium deoxycholate and detected by band shift assays. As shown in Figure 4C, E2F activity was readily detected associated with p107 in Saos-2 cells. Immune complexes prepared from pRB-deficient Saos-2 cells with a monoclonal antibody to pRB had no associated E2F activity unless the cells were transfected with a pRB expression construct. When the p107 mutant constructs were assayed in this system, E2F activity was only found associated with the N385 growth suppression domain. Further analysis of the mutants that delineate the carboxyl terminus of the growth suppression domain contained in N385 has revealed a tight correlation between E2F binding and ability to cause cell cycle arrest (L. Zhu et al., in prep.). Importantly, mutant L19, which contains sequences from the amino terminus of p107 and also possesses growth suppression function, was unable to associate with E2F.

The same panel of mutants were assayed for association with cyclin A and cyclin E in the transfected cells. Cell extracts were immunoprecipitated with anti-pRB, anti-p107, or anti-HA antibodies. The immunoprecipitates were separated on SDS-PAGE, and the gels were blotted with specific antibodies against cyclin A (Fig. 4D) or cyclin E (Fig. 4E). Consistent with previous observations, cyclin A and cyclin E were detected in p107 immunoprecipitates but not in pRB immunoprecipitates. L19, which lacked association with E2F, bound to cyclins A and E as effectively as the full-length p107 protein. In contrast, N385, which bound to E2F, lacked the cyclin association. L30, which was inactive in growth suppression assays, also lost the cyclin-binding ability. Further analysis of the spacer sequences deleted in L30 but present in L19 has revealed a tight correlation between cyclin binding and the growth suppression activities of these mutants (L. Zhu et al., in prep.). Thus, the two growth suppression domains of p107 have different biochemical functions that correlate with their growth suppression activities. The activity of the domain represented by N385 correlates with E2F binding; the activity of the domain represented by L19 correlates with cyclin binding. In other studies, we have shown that growth arrest by the E2F-binding domain of p107 can be rescued by overexpression of an E2F that binds to p107. Analogously, the arrest caused by the cyclin-binding domain can be rescued by overexpression of cyclin A or cyclin E (L. Zhu et al., in prep.).

Cervical Carcinoma Cell Line C33A Is Sensitive to Growth Suppression by p107 but Not pRB

Using the flow cytometry analysis of transfected cells, a number of human tumor cell lines have been tested for their ability to be arrested by pRB or p107. Most cells lines tested were not clearly sensitive to overexpression of either pRB or p107 by this approach (Zhu et al. 1993). The cervical carcinoma cell line C33A provided the most interesting observation; C33A cells were arrested in G_1 by overexpression of p107 but not pRB (Fig. 5) (Zhu et al. 1993). Experiments using the N385 and L19 mutants of p107 to examine the two growth suppression domains of p107 separately indicate that this difference between p107 and pRB was due to the cyclin-binding domain of p107 (Fig. 5). C33A cells were arrested in G_1 by the L19 mutant but were unaffected by the N385 mutant. We interpret this result as a further indication that p107 uses two different mechanisms to regulate cell growth. C33A cells share with Saos-2 cells the sensitivity to functional inhibition of cyclins A and E, but differ from Saos-2 cells in responsiveness to repression of E2F activity.

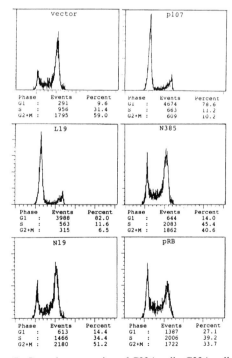

Figure 5. Growth suppression of C33A cells. C33A cells were used in the growth suppression assay as in Fig. 2 except that the cells were treated with 40 ng/ml nocodazole from 24 to 36 hr after transfection and then harvested. Transfection plasmids are indicated within the histograms.

Sensitivity to p107-mediated Growth Suppression Is Not Due to a Nonfunctional Endogenous p107

To date, only cells that lack functional pRB have been found to be readily growth arrested by the transfection of pRB. However, the fact that C33A cells have mutant pRB but are not growth inhibited by overexpression of wild-type pRB appears to have broken this limited correlation between a mutant pRB and the sensitivity to pRB growth suppression. Other factors must also be involved in determining the sensitivity to growth suppression. In this regard, one possible explanation for the observation that Saos-2 cells are sensitive to expression of N385 was that this domain of p107 is mutated in Saos-2 cells. We have examined the abilities of the Saos-2 cell endogenous p107 to associate with E1A, E2F, and cyclin A. On the basis of these criteria, this protein appears to be wild type. We next cloned the large pocket of p107 from Saos-2 cells through reverse transcription (RT)-PCR and tested it for growth suppression activity. As shown in Table 1, two independent PCR clones containing the Saos-2 p107 large pocket sequences were able to arrest cells in G_1. Thus, the sensitivity of these cells to p107 appears to be determined by a factor other than the status of p107.

p107 and pRB Are Also Growth Suppressors in Untransformed Fibroblasts

The above observations from tumor cell lines prompted us to examine the growth suppression effects of pRB and p107 in untransformed NIH-3T3 fibroblasts. NIH-3T3 cells were arrested in G_0/G_1 by incubation in media containing 0.1% serum for 40 hours. Cells were trypsinized and replated in 10% serum media, and progression of cells into S phase was monitored by incorporation of bromodeoxyuridine (BrdU). Protein expression constructs, or empty vector as control, were microinjected into the cells at about 1–4 hours after serum stimulation, and effects on DNA synthesis were monitored using a pulse label of BrdU. The expression plasmids were coinjected with β-galactosidase expression construct to identify successfully injected cells, and protein expression was confirmed by immunofluorescence staining. Results from a representative experiment are shown in Figure 6. 50% of cells injected with a control plasmid scored as BrdU-positive in this assay. Injection of plasmids expressing wild-type pRB or p107 reduced the level of BrdU-positive cells to 25%, suggesting that p107 and pRB provide a block to entry into S phase. Similar effects were seen with the N385 and L19 mutants of p107, indicating that both suppression domains of p107 are active in this assay. Importantly, these results indicate that the cell cycle arrest caused by the overexpression of pRB and p107 is not a feature that is restricted to tumor cell lines.

p107 and pRB Associate with Different E2Fs

We and other investigators have now isolated an extensive family of E2F genes from human cells (Helin et al. 1992; Kaelin et al. 1992; Shan et al. 1992; Girling et al. 1993; Ivey-Hoyle et al. 1993; Lees et al. 1993; Beijersbergen et al. 1994; C.-L. Wu et al., in prep.). One possible explanation for the large number of genes is the possibility that specific E2F heterodimers are regulated by each pocket protein. This issue has been investigated using antibodies that are specific for individual proteins, and an example, examining E2F-1 protein complexes, is shown in Figure 7. In this experiment, ML-1 cell extracts were immunoprecipitated with monoclonal antibodies specific for p107, pRB, E2F-1,

Table 1. Growth Suppression by the Saos-2 Cell Endogenous p107

Expression plasmid	G_1	S	$G_2 + M$
Vector	34.2	25.4	40.4
p107	75.7	8.8	15.5
N385	82.2	7.5	10.3
SN-1	80.3	7.7	12.0
SN-2	78.5	9.0	12.5

Saos-2 cells were transfected with 23 µg of the indicated expression plasmids plus 2 µg of pCMVCD20. Two days later, cells were harvested for flow cytometry analysis to determine the cell cycle profiles (percentages of cells in each phase) of the transfected cells as in Fig. 1. SN-1 and SN-2 were two independent clones obtained from RT-PCR of Saos-2 cell RNA.

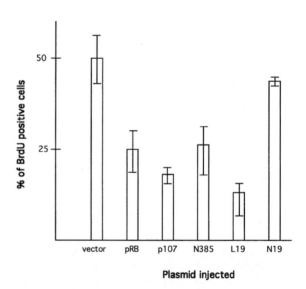

Figure 6. Inhibition of DNA synthesis in NIH-3T3 cells by microinjection of plasmids expressing pRB, p107, and related proteins. Quiescent NIH-3T3 cells were restimulated and microinjected with plasmids expressing each of the noted proteins together with a marker plasmid expressing β-galactosidase. The mean percentage of β-galactosidase-positive cells that incorporated detectable BrdU is shown ± the ranges from two (expressing plasmids) or three (vector) independent experiments. The cell numbers from which these means are derived are as follows: vector 23/54, 26/49, 15/27; pRB 5/28, 14/49; p107 11/54, 3/19; N385 11/60, 10/32; L19 7/43, 1/15; N19 16/37, 15/34.

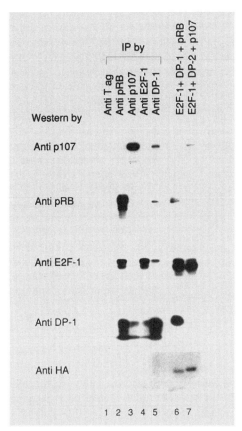

Figure 7. E2F-1 associates with pRB whereas DP-1 associates with both pRB and p107 in vivo. ML-1 extracts (600 μg per reaction) were immunoprecipitated with monoclonal antibodies coupled with protein G Sepharose. (Lane *1*) Anti-SV40 T antigen (PAb419; Harlow et al. 1981); (lane *2*) anti-pRB (XZ77; Hu et al. 1991); (lane *3*) anti-p107 (SD2, 4, 6, 9, and 15; Dyson et al. 1993); (lane *4*) anti-E2F-1 (KH95; Helin et al. 1993b); and (lane *5*) anti-DP-1 (WTH16 and WTH24; C.-L. Wu et al., in prep.). Immunoprecipitates were separated on an 8% SDS-PAGE and blotted onto nitrocellulose paper. Proteins were detected by immunoblotting sequentially with anti-p107 (SD9), anti-pRB (XZ77), anti-E2F-1 (KH22), anti-DP-1 (WTH16), and anti-HA tag antibody 12CA5. C33A cell extracts (15 μg) transfected with E2F-1/HA-DP-1/pRB (lane *6*) or E2F-1/HA-DP-2/p107 (lane *7*) were used as controls for the antibodies.

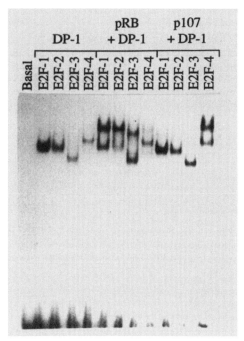

Figure 8. E2F-1, -2, -3, and -4 bind differentially to pRB and p107. C33A cells were transiently transfected with various combinations of E2F and DP expression vectors in the absence or presence of either pCMV, pCMVpRB, or pCMVp107, as indicated. Cell extracts were prepared 24 hr after transfection and then tested for their ability to bind to the consensus E2F site in gel retardation assays.

and DP-1 or with control antibodies (lanes 1–5). The immunoprecipitates were separated through SDS-PAGE, blotted, and then probed with the same panel of antibodies. As seen in lanes 2 and 3, E2F-1 was coimmunoprecipitated with pRB but not p107, whereas DP-1 is present in both pRB and p107 immunoprecipitates.

Experiments with E2F-1, E2F-2, E2F-3, and E2F-4 proteins show that they can all form functional heterodimers with either DP-1 or DP-2 (Bandara et al. 1993; Helin et al. 1993b; Krek et al. 1993; Beijersbergen et al. 1994; C.-L. Wu et al., in prep.). Figure 8 demonstrates specific interactions between pRB or p107 and these functional dimers. Combinations of genes were transfected into C33A cells and the E2F complexes generated were detected on a band shift assay using a probe containing an E2F-binding site. The coexpression of pRB resulted in a higher-molecular-weight complex with E2F-1, E2F-2, and E2F-3, but not E2F-4. Conversely, coexpression of p107 generated a similar higher-molecular-weight complex only with E2F-4, not E2F-1, E2F-2, or E2F-3. These observations indicate that pRB and p107 may regulate different E2F complexes. Because of these observations and previous results showing that pocket proteins associated with E2F at overlapping but distinct phases of the cell cycle (Mudryj et al. 1991; Lees et al. 1992; Shirodkar et al. 1992; Schwarz et al. 1993), we believe it likely that the p107/E2F and pRB/E2F complexes have distinct functions in the regulation of E2F-dependent transcription.

DISCUSSION

The pRB-binding sequences of adenovirus E1A, SV40 large T antigen, and HPV E7 proteins are essential for the transforming properties of these viral oncoproteins. These same regions also provide stable interaction with p107 and p130, cellular proteins that are structurally related to pRB. We have investigated the biochemical and functional properties of p107 to determine whether p107, like pRB, acts as a negative regulator of cell proliferation.

Several different experimental approaches show that the overexpression of p107 can arrest cells in G_1 phase of the cell cycle. Analysis of p107 mutants reveals that p107 contains two separable growth suppression domains that appear to act in independent ways. One of these domains resides in the large pocket domain of p107. This portion of p107 contains the sequences with greatest homology to pRB, and like pRB, the activity of this domain correlates with the ability to bind to E2F. These observations are supported by experiments showing rescue of pRB or p107 growth arrest by the coexpression of specific genes. E1A 12S protein has been shown to release E2F from pRB- and p107-mediated repression, and the expression of E1A 12S rescues the growth suppression by either pRB or p107. Further analysis of the pRB/E2F and p107/E2F complexes has suggested that these contain different forms of E2F. Consistent with this, cell cycle arrest caused by pRB can be rescued by the coexpression of E2F-1, whereas G_1 arrest caused by the large pocket of p107 is overcome by the coexpression of E2F-4/DP-1. Thus, the inhibition of E2F activity appears to be at least part of the mechanism through which pRB and the large pocket domain of p107 can arrest cells in G_1 phase of the cell cycle.

In addition to the large pocket domain, p107 contains a second growth suppression domain that has no equivalent in pRB. In structure-function analyses, the activity of this domain correlates with p107 binding to cyclins A and E. This suggests that p107 may arrest cells in G_1 by sequestering essential components of the cell cycle machinery. This mechanism is distinct from the growth arrest caused by pRB and the large pocket domain of p107. A survey of tumor cell lines revealed one line, C33A cells, that was arrested in G_1 by p107 but not pRB. By independently testing the activities of the two growth suppression domains of p107, it was found that C33A cells were arrested by the cyclin-binding mutants but not by the E2F-binding mutants. Thus, the differential sensitivity of C33A cells to pRB and p107 appears to be due to the presence of an additional growth suppression domain in p107.

Several important questions remain to be answered. First, in normal cells does p107 act to cause growth arrest in G_1? Results obtained from microinjection experiments with NIH-3T3 cells demonstrate that pRB and p107 (including both growth suppression domains) can inhibit DNA synthesis in these untransformed fibroblasts when cells are stimulated to progress from G_0 into S phase. This shows that the cell cycle arrest activities of pRB and p107 are not limited to tumor cell lines. However, to date, p107-induced growth arrest has only been observed when p107 is overexpressed. It is unclear whether the arrest is solely a consequence of overexpression or whether it is part of the normal function of p107. It should be pointed out that the same issue remains unresolved for pRB. It is interesting to note that the accumulation of p107 to relatively high levels has been observed during myogenic differentiation of $pRB^{-/-}$ cells (Schneider et al. 1994).

The observation that many tumor cell lines are apparently insensitive to pRB- and p107-induced growth arrest raises a second issue. Do pRB and p107 fail to repress E2F in these cells, or do these cells not need E2F activity to proliferate? Experiments looking at E2F protein complexes and E2F-responsive promoters in these cells indicate that p107/E2F and pRB/E2F complexes are increased and E2F activity is repressed. Further work is needed to determine the long-term effects of pRB and p107 on the expression of putative E2F-regulated genes in these cells.

Finally, if p107 can act like pRB to suppress cell growth, then can the inactivation of p107 lead to inappropriate cell proliferation? E2F-4, the regulatory target of p107, has been shown to have potent tumorigenic activity (Beijersbergen et al. 1994), and one might expect to see the inactivation of p107 in some tumor cells. Surprisingly, no p107 mutations have been detected in human tumors. One potential explanation for this paradox is that the functions of pRB, p107, and p130 may overlap and be, to some extent, functionally redundant. Thus, although p107 and p130 cannot compensate for the loss of pRB in tumor cells, the presence of pRB and/or p130 may compensate for mutation of the p107 gene.

ACKNOWLEDGMENTS

We thank other members of our laboratories for helpful discussions. L.Z. is a fellow of the Leukemia Society of America, and G.H.E. is a recipient of a Clinical Investigator Award from the National Cancer Institute. J.A.L. was a special fellow of the Leukemia Society of America during a portion of this work, and C.-L.W. was supported by a postdoctoral fellowship from the National Institutes of Health. E.H. is an American Cancer Society Research Professor. This work was supported by grants from the National Institutes of Health to E.H and N.D.

REFERENCES

Bandara, L.R., J.P. Adamczewski, T. Hunt, and N. La Thangue. 1991. Cyclin A and the retinoblastoma gene product complex with a common transcription factor. *Nature* **352:** 249.

Bandara, L.R., V.M. Buck, M. Zamanian, L.H. Johnston, and N.B. La Thangue. 1993. Functional synergy between DP-1 and E2F-1 in the cell cycle-regulating transcription factor DRTF1/E2F. *EMBO J.* **12:** 4317.

Beijersbergen, R.L., R.M. Kerkhoven, L. Zhu, L. Carlée, P.M. Voorhoeve, and R. Bernards. 1994. E2F-4, a new member of the E2F gene family, has oncogenic activity and associates with p107 in vivo. *Genes Dev.* **8:** 2680.

Bookstein, R., J.Y. Shew, P.L. Chen, P. Scully, and W.H. Lee. 1990. Suppression of tumorigenicity of human prostate carcinoma cells by replacing a mutated RB gene. *Science* **247:** 712.

Buchkovich, K., L.A. Duffy, and E. Harlow. 1989. The retinoblastoma protein is phosphorylated during specific phases of the cell cycle. *Cell* **58:** 1097.

Cao, L., B. Faha, M. Dembski, L.-H. Tsai, E. Harlow, and N. Dyson. 1992. Independent binding of the retinoblastoma protein and p107 to the transcription factor E2F. *Nature* **355:** 176.

Chellappan, S., S. Hiebert, M. Mudryj, J. Horowitz, and J. Nevins. 1991. The E2F transcription factor is a cellular target for the RB protein. *Cell* **65:** 1053.

Chen, P.-L., P. Scully, J.-Y. Shew, J. Wang, and W.-H. Lee. 1989. Phosphorylation of the retinoblastoma gene product is modulated during the cell cycle and cellular differentiation. *Cell* **58:** 1193.

Cobrinik, D., P. Whyte, D.S. Peeper, T. Jacks, and R.A. Weinberg. 1993. Cell cycle-specific association of E2F with the p130 E1A-binding domain. *Genes Dev.* **7:** 2392.

DeCaprio, J.A., J.W. Ludlow, D. Lynch, Y. Furukawa, J. Griffin, H. Piwnica-Worms, C.M. Huang, and D.M. Livingston. 1989. The product of the retinoblastoma susceptibility gene has properties of a cell cycle regulatory element. *Cell* **58:** 1085.

DeCaprio, J.A., J.W. Ludlow, J. Figge, J.Y. Shew, C.M. Huang, W.H. Lee, E. Marsilio, E. Paucha, and D.M. Livingston. 1988. SV40 large tumor antigen forms a specific complex with the product of the retinoblastoma susceptibility gene. *Cell* **54:** 275.

Devoto, S.H., M. Mudryj, J. Pines, T. Hunter, and J.R. Nevins. 1992. A cyclin A-protein kinase complex possesses sequence-specific DNA binding activity: p33cdk2 is a component of the E2F-cyclin A complex. *Cell* **68:** 167.

Dulic, V., L.F. Drullinger, E. Lees, S.I. Reed, and G.H. Stein. 1993. Senescent human fibroblasts have an overabundance of cyclin D1 and cyclin E, but lack cyclin E-associated kinase activity. *Proc. Natl. Acad. Sci.* **90:** 11034.

Dyson, N., K. Buchkovich, P. Whyte, and E. Harlow. 1989a. The cellular 107K protein that binds to adenovirus E1A also associates with the large T antigens of SV40 and JC virus. *Cell* **58:** 249.

Dyson, N., P. Guida, K. Munger, and E. Harlow. 1992. Homologous sequences in adenovirus E1A and human papillomavirus E7 proteins mediate interaction with the same set of cellular proteins. *J. Virol.* **66:** 6893.

Dyson, N., P.M. Howley, K. Munger, and E. Harlow. 1989b. The human papilloma virus-16 E7 oncoprotein is able to bind to the retinoblastoma gene product. *Science* **243:** 934.

Dyson, N., M. Dembski, A. Fattaey, C. Nguwu, M. Ewen, and K. Helin. 1993. Analysis of p107-associated proteins; p107 associates with a form of E2F that differs from pRB associated E2F-1. *J. Virol.* **67:** 7641.

Egan, C., S.T. Bayley, and P.E. Branton. 1989. Binding of the Rb1 protein to E1A products is required for adenovirus transformation. *Oncogene* **4:** 383.

Ewen, M., B. Faha, E. Harlow, and D. Livingston. 1992. Interaction of p107 with cyclin A independent of complex formation with viral oncoproteins. *Science* **255:** 85.

Ewen, M.E., Y. Xing, J.B. Lawrence, and D.M. Livingston. 1991. Molecular cloning, chromosomal mapping, and expression of the cDNA for p107, a retinoblastoma gene product-related protein. *Cell* **66:** 1155.

Ewen, M.E., J.W. Ludlow, E. Marsilio, J.A. DeCaprio, R.C. Millikan, S.H. Cheng, E. Paucha, and D.M. Livingston. 1989. An N-terminal transformation-governing sequence of SV40 large T antigen contributes to the binding of both p110Rb and a second cellular protein, p120. *Cell* **58:** 257.

Faha, B., M. Ewen, L.-H. Tsai, D. Livingston, and E. Harlow. 1992. Interaction between human cyclin A and adenovirus E1A-associated p107 protein. *Science* **255:** 87.

Giordano, A., C. McCall, P. Whyte, and B.R. Franza. 1991. Human cyclin A and the retinoblastoma protein interact with similar but distinguishable sequences in the adenovirus E1A gene product. *Oncogene* **6:** 481.

Giordano, A., P. Whyte, E. Harlow, B.J. Franza, D. Beach, and G. Draetta. 1989. A 60 kd cdc2-associated polypeptide complexes with the E1A proteins in adenovirus-infected cells. *Cell* **58:** 981.

Girling, R., J.F. Partridge, L.R. Bandara, N. Burden, N.F. Totty, J.J. Hsuan, and N.B. La Thangue. 1993. A new component of the transcription factor DRTF1/E2F. *Nature* **362:** 83.

Goodrich, D.W., N.P. Wang, Y.-W. Qian, E.Y.-H.P. Lee, and W.-H. Lee. 1991. The retinoblastoma gene product regulates progression through the G1 phase of the cell cycle. *Cell* **67:** 293.

Gossen, M. and H. Bujard. 1992. Tight control of gene expression in mammalian cells by tetracycline-responsive promoters. *Proc. Natl. Acad. Sci.* **89:** 5547.

Hannon, G.J., D. Demetrick, and D. Beach. 1993. Isolation of the Rb-related p130 through its interaction with CDK2 and cyclins. *Genes Dev.* **7:** 2378.

Harlow, E., L.V. Crawford, D.C. Pim, and N.M. Williamson. 1981. Monoclonal antibodies specific for simian virus 40 tumor antigens. *J. Virol.* **39:** 861.

Helin, K. and E. Harlow. 1993. The retinoblastoma protein as a transcriptional repressor. *Trends Cell Biol.* **3:** 43.

Helin, K., E. Harlow, and A.R. Fattaey. 1993a. Inhibition of E2F-1 transactivation by direct binding of the retinoblastoma protein. *Mol. Cell. Biol.* **13:** 6501.

Helin, K., J.A. Lees, M. Vidal, N. Dyson, E. Harlow, and A. Fattaey. 1992. A cDNA encoding a pRB-binding protein with properties of the transcription factor E2F. *Cell* **70:** 337.

Helin, K., C. Wu, A. Fattaey, J. Lees, B. Dynlacht, C. Ngwu, and E. Harlow. 1993b. Heterodimerization of the transcription factors E2F-1 and DP-1 leads to cooperative transactivation. *Genes Dev.* **7:** 1850.

Hiebert, S.W., S.P. Chellappan, J.M. Horowitz, and J.R. Nevins. 1992. The interaction of pRb with E2F inhibits the transcriptional activity of E2F. *Genes Dev.* **6:** 177.

Hinds, P.W., S. Mittnacht, V. Dulic, A. Arnold, S.I. Reed, and R.A. Weinberg. 1992. Regulation of retinoblastoma protein functions by ectopic expression of human cyclins. *Cell* **70:** 993.

Hu, Q.J., N. Dyson, and E. Harlow. 1990. The regions of the retinoblastoma protein needed for binding to adenovirus E1A or SV40 large T antigen are common sites for mutations. *EMBO J.* **9:** 1147.

Hu, Q., C. Bautista, G. Edwards, D. Defeo-Jones, R. Jones, and E. Harlow. 1991. Antibodies specific for the human retinoblastoma protein identify a family of related polypeptides. *Mol. Cell. Biol.* **11:** 5792.

Huang, H.J., J.K. Yee, J.Y. Shew, P.L. Chen, R. Bookstein, T. Friedmann, E.Y.-H.P. Lee, and W.H. Lee. 1988. Suppression of the neoplastic phenotype by replacement of the RB gene in human cancer cells. *Science* **242:** 1563.

Huang, S., N.P. Wang, B.Y. Tseng, W.-H. Lee, and E.Y.-H.P. Lee. 1990. Two distinct and frequently mutated regions of retinoblastoma protein are required for binding to SV40 T antigen. *EMBO J.* **9:** 1815.

Ivey-Hoyle, M., R. Conroy, H. Huber, P. Goodhart, A. Oliff, and D.C. Heinbrook. 1993. Cloning and characterization of E2F-2, a novel protein with the biochemical properties of transcription factor E2F. *Mol. Cell Biol.* **13:** 7802.

Johnson, D.G., J.K. Schwartz, W.D. Cress, and J.R. Nevins. 1993. Expression of transcription factor E2F1 induces quiescent cells to enter S phase. *Nature* **365:** 349.

Kaelin, W.G.J., M.E. Ewen, and D.M. Livingston. 1990. Definition of the minimal simian virus 40 large T antigen- and adenovirus E1A-binding domain in the retinoblastoma gene product. *Mol. Cell. Biol.* **10:** 3761.

Kaelin, W.G., W. Krek, W.R. Sellers, J.A. DeCaprio, F. Ajchanbaum, C.S. Fuchs, T. Chittenden, Y. Li, P.J. Farnham, M.A. Blanar, D.M. Livingston, and E.K. Flemington. 1992. Expression cloning of a cDNA encoding a retinoblastoma-binding protein with E2F-like properties. *Cell* **70:** 351.

Krek, W., D.M. Livingston, and S. Shirodkar. 1993. Binding to DNA and the retinoblastoma gene product promoted by complex formation of different E2F family members. *Science* **262:** 1557.

Lees, E., B. Faha, V. Dulic, S.I. Reed, and E. Harlow. 1992. Cyclin E/cdk2 and cyclin A/cdk2 kinases associate with p107 and E2F in a temporally distinct manner. *Genes Dev.* **6:** 1874.

Lees, J.A., M. Saito, M. Vidal, M. Valentine, T. Look, E. Harlow, N. Dyson, and K. Helin. 1993. The retinoblastoma protein binds to a family of E2F transcription factors. *Mol. Cell. Biol.* **13:** 7813.

Li, Y., C. Graham, S. Lacy, A.M.V. Duncan, and P. Whyte. 1993. The adenovirus E1A-associated 130-kD protein is encoded by a member of the retinoblastoma gene family and physically interacts with cyclins A and E. *Genes Dev.* **7:** 2366.

Ludlow, J.W., J.A. DeCaprio, C.M. Huang, W.H. Lee, E. Paucha, and D.M. Livingston. 1989. SV40 large T antigen binds preferentially to an underphosphorylated member of the retinoblastoma susceptibility gene product family. *Cell* **56:** 57.

Mayol, X., X. Grana, A. Baldi, N. Sang, Q. Hu, and A. Giordano. 1993. Cloning of a new member of the retinoblastoma gene family (pRb2) which binds to the E1A transforming domain. *Oncogene* **8:** 2561.

Mihara, K., X.R. Cao, A. Yen, S. Chandler, B. Driscoll, A.L. Murphree, A. T'Ang, and Y.K. Fung. 1989. Cell cycle-dependent regulation of phosphorylation of the human retinoblastoma gene product. *Science* **246:** 1300.

Mudryj, M., S.H. Devoto, S.W. Hiebert, T. Hunter, J. Pines, and J.R. Nevins. 1991. Cell cycle regulation of the E2F transcription factor involves an interaction with cyclin A. *Cell* **65:** 1243.

Munger, K., B.A. Werness, N. Dyson, W.C. Phelps, E. Harlow, and P.M. Howley. 1989. Complex formation of human papillomavirus E7 proteins with the retinoblastoma tumor suppressor gene product. *EMBO J.* **8:** 4099.

Nevins, J.R. 1992. E2F; A link betwen the Rb tumor suppressor protein and viral oncoproteins. *Science* **258:** 424.

Qin, X.-Q., T. Chittenden, D. Livingston, and W.G. Kaelin. 1992. Identification of a growth suppression domain within the retinoblastoma gene product. *Genes Dev.* **6:** 953.

Schneider, J.W., W. Gu, L. Zhu, V. Mahdavi, and B. Nadal-Ginard. 1994. Reversal of terminal differentiation mediated by p107 in $Rb^{-/-}$ muscle cells. *Science* **264:** 1467.

Schwarz, J.K., S.H. Devoto, E.J. Smith, S.P. Chellappan, L. Jakoi, and J.R. Nevins. 1993. Interactions of the p107 and Rb proteins with E2F during the cell proliferation response. *EMBO J.* **12:** 1013.

Shan, B., X. Zhu, P.-L. Chen, T. Durfee, Y. Yang, D. Sharp, and W.-H. Lee. 1992. Molecular cloning of cellular genes encoding retinoblastoma-associated proteins: Identification of a gene with properties of the transcription factor E2F. *Mol. Cell. Biol.* **12:** 5620.

Shirodkar, S., M. Ewen, J.A. DeCaprio, D. Morgan, D. Livingston, and T. Chittenden. 1992. The transcription factor E2F interacts with the retinoblastoma product and a p107-cyclin A complex in a cell cycle-regulated manner. *Cell* **68:** 157.

Singh, P., S.H. Wong, and W. Hong. 1994. Overexpression of E2F-1 in rat embryo fibroblasts leads to neoplastic transformation. *EMBO J.* **13:** 3329.

van den Heuvel, S. and E. Harlow. 1993. Distinct roles for cyclin-dependent kinases in cell cycle control. *Science* **262:** 2050.

Weinberg, R.A. 1991. Tumor suppressor genes. *Science* **254:** 1138.

———. 1992. The retinoblastoma gene and gene product. *Cancer Surv.* **12:** 43.

Whyte, P., N.M. Williamson, and E. Harlow. 1989. Cellular targets for transformation by the adenovirus E1A proteins. *Cell* **56:** 67.

Whyte, P., K.J. Buchkovich, J.M. Horowitz, S.H. Friend, M. Raybuck, R.A. Weinberg, and E. Harlow. 1988. Association between an oncogene and an anti-oncogene: The adenovirus E1A proteins bind to the retinoblastoma gene product. *Nature* **334:** 124.

Zamanian, M. and N.B.L. Thangue. 1993. Transcriptional repression by the Rb-related protein p107. *Mol. Biol. Cell* **4:** 389.

Zhu, L., S. van den Heuvel, K. Helin, A. Fattaey, M. Ewen, D. Livingston, N. Dyson, and E. Harlow. 1993. Inhibition of cell proliferation by p107, a relative of the retinoblastoma protein. *Genes Dev.* **7:** 1111.

Zieve, G.W., D. Turnbull, J.M. Mullins, and J.R. McIntosh. 1980. Production of large number of mitotic mammalian cells by use of the reversible microtubule inhibitor nocodazole. *Exp. Cell. Res.* **126:** 397.

The Adenovirus E1A-associated 300-kD Protein Exhibits Properties of a Transcriptional Coactivator and Belongs to an Evolutionarily Conserved Family

R. Eckner, Z. Arany, M. Ewen, W. Sellers, and D.M. Livingston
Dana-Farber Cancer Institute and Harvard Medical School, Boston, Massachusetts 02115

The development of a tumorigenic phenotype in a previously untransformed cell is a multistep process requiring both the activation of oncogenes and the inactivation of tumor suppressor genes (Fearon and Vogelstein 1990; Scrable et al. 1990). The E1A and E1B proteins encoded by the early region of adenovirus can elicit a tumorigenic phenotype and have provided insights into the interplay between the products of DNA tumor viral oncogenes and certain tumor suppressor genes. The study of these two viral genes has also helped to elucidate the phenomenon of cooperativity among certain oncogenes (Ruley 1983).

The two major E1A polypeptides are synthesized from alternatively spliced transcripts and are designated 12S and 13S E1A mRNA products. Both proteins display transforming activity in the presence of E1B. The oncogenic activity of E1A is tightly linked to its ability to bind to a set of cellular proteins. These proteins recognize discrete E1A regions that have been conserved among various adenovirus serotypes. Conserved regions 1 and 2 (termed CR1 and CR2) serve as binding sites for members of the retinoblastoma protein family (also known as nuclear pocket proteins). This family includes pRB, the retinoblastoma tumor suppressor protein itself; p107; and p130. In contrast, a carboxy-terminal segment of CR1 and the nonconserved amino terminus of E1A are interaction sites for a cellular 300-kD protein (p300). Binding of all of the above proteins appears to be essential for the full expression of E1A oncogenic activity (for reviews, see Dyson and Harlow 1992; Moran 1993).

In recent years, p300 has attracted considerable interest because the transforming potential of E1A depends, in part, on its interaction with p300 (Subramanian et al. 1988; Whyte et al. 1989). In addition, a number of other discrete biological effects are linked to the formation of E1A/p300 complexes. For example, association of E1A with p300 is sufficient to stimulate entry of quiescent cells into S phase (Howe et al. 1990; Stein et al. 1990), suggesting that p300 may participate in the control of cell growth. Furthermore, complexes between E1A and p300 are associated with inhibition of terminal differentiation in multiple cell types, repression of several tissue-specific enhancers/promoters (Webster et al. 1988), and silencing of certain viral enhancers, e.g., the SV40 enhancer (Borelli et al. 1984; Velcich and Ziff 1985). It seems likely that the repressive action of E1A on transcription constitutes an important component of the ability of E1A to block cellular differentiation.

In an effort to elucidate the cellular functions of p300, we have cloned the cDNA for p300. Its imputed sequence contains certain motifs characteristic of transcriptional coactivator proteins. In keeping with such a function, p300 molecules lacking an intact E1A-binding site can bypass E1A repression of the SV40 enhancer and restore, to a significant extent, the transcriptional activity of this element (Eckner et al. 1994). Interestingly, p300 is highly related, although not identical, to the recently cloned CREB binding protein, CBP (Chrivia et al. 1993). CBP appears to function as a coactivator for certain cAMP-responsive transcription factors. Moreover, since a protein which is highly related in primary sequence to p300/CBP has been found in *Caenorhabditis elegans* (Wilson et al. 1994), p300 and CBP seem to be members of an evolutionarily conserved protein family (Arany et al. 1994).

EXPERIMENTAL PROCEDURES

Isolation of p300 cDNAs. Immunoscreening of a λZap cDNA library (Ewen et al. 1991) with a polyclonal α-p300 mouse antiserum was carried out as described previously (Harlow and Lane 1988). The insert from one of the two positive clones obtained was used for isolating longer p300 cDNA fragments by hybridization (Church and Gilbert 1984). To isolate cDNA clones covering the complete p300 coding unit, additional libraries, including a HeLa λZap library (Xiao et al. 1991), were screened. The entire p300 cDNA was sequenced on both strands using the dideoxy-chain-termination method and a series of p300-specific primers.

Cell culture, transfections, and adenovirus infections. U-2 OS (human osteosarcoma cells) were cultured in Dulbecco's modified essential medium (DMEM) supplemented with 10% fetal calf serum (FCS, Hyclone). Cotransfections of luciferase reporter and/or other expression plasmids were performed using the HEPES-CaPO$_4$ method (Ausubel et al. 1987). 12 hours after addition of the precipitate, cells were washed twice with phosphate-buffered saline (PBS) and incubated another 36 hours in DMEM + 10% FCS. Cells were then labeled with [^{35}S]methionine or processed for luciferase assays. Adenovirus infections were conducted in 3 ml of serum-free DMEM at a moi of 10 for 1 hour. DMEM containing 10% FCS was then added, and cells were labeled 10 hours later with [^{35}S]methionine.

Immunoprecipitations. Generally, cells were labeled with 0.5 mCi [^{35}S]methionine per 90-mm dish for 4 hours in DMEM containing 5% dialyzed FCS. Lysis of labeled cultures was performed for 20 minutes at 4°C in 1 ml of EBC (50 mM Tris-HCl [pH 8.0], 170 mM NaCl, 0.5% NP-40, and 50 mM NaF) containing 10 μg/ml of aprotinin, leupeptin, and phenylmethylsulfonyl fluoride (PMSF). The lysate was cleared of debris by centrifugation for 10 minutes in an Eppendorf centrifuge at full speed. 50 μl of a 1:1 slurry of protein A–Sepharose in 4% bovine serum albumin (BSA) was then added to the supernatant in order to preclear the lysate (30 min rocking). Subsequently, 100 μl of a relevant monoclonal antibody was added to the precleared lysate, and the tube was rocked for 1 hour. Immunocomplexes were harvested by adding 25 μl of protein A–Sepharose and rocking for another 30 minutes. Beads were washed five times with 1 ml of NETN (10 mM Tris-HCl [pH 8.0], 250 mM NaCl, 5 mM EDTA, and 0.5% NP-40), resuspended, and boiled in SDS-sample buffer (Ausubel et al. 1987), and the eluates were analyzed by SDS-PAGE.

Construction of p300 expression plasmids. An HA-tag was attached to p300 either on the carboxyl or amino terminus. The carboxy-terminal tag was created by ligating the HA epitope-encoding oligonucleotide 5'-CTAGCCCCGGGATGGCCTACCCATACGAC GTGCCTGACTACGCCTCCCTCGGATA-3' and its complementary strand to the *Nhe*I site near the carboxyl terminus of p300. The amino-terminal tag was added by polymerase chain reaction (PCR)-assisted mutagenesis. In brief, a 420-bp fragment from the 5' end of the p300 coding region was amplified with the upstream primer: 5'-CCTGGATCCACCATGGCAT ACCCATACGACGTGCCTGACTACGCCTCCGC CGAGAATGTGGTG-3' and downstream primer 5'-GTAGGACCCTGATTTGGTC-3'. The PCR product was digested with *Bam*HI and *Spe*I and ligated to the *Spe*I site near the amino terminus of p300. A mammalian expression vector, CMVβ, was employed to express full-length p300 and its derivatives in mammalian cells. cDNA inserts in this vector are transcribed by the CMV enhancer/promoter, and transcripts are processed by splicing and polyadenylation signals from the rabbit β-globin gene.

The two internal deletion mutants in the E1A-binding domain of p300 (del30 and del33) were obtained by resecting *Eco*47III-cut p300 DNA with *Bal*31 exonuclease. Unidirectional deletions toward the carboxyl terminus of p300 were generated by first cleaving the resected DNA with *Hin*dIII to release the 3' end of the cDNA. The released fragment was ligated to the remainder of the p300 cDNA digested with *Eco*47III and *Hin*dIII.

Templates for in vitro transcription/translation were generated by ligating oligonucleotides encoding a eukaryotic consensus translation initiation signal to the restriction sites noted in Figure 5D.

Peptide maps. Partial proteolysis by *Staphylococcus aureus* V8 protease was carried out as described previously (Cleveland et al. 1977). Cleavage by *N*-chlorosuccinimide was performed according to the method of Draetta et al. (1987).

Luciferase assays. To assay the activity of the SV40 enhancer, the pGL2-control plasmid (Promega) was utilized as a reporter. 5 μg of pGL2-control and 1 μg of CMV-lacZ (as an internal control) were transfected together with varying amounts of CMV12SE1A and 8 μg of the indicated p300 expression plasmid or 8 μg of pBluescript. The total amount of transfected DNA was between 15 μg and 20 μg and was dependent on the amount of E1A expression vector present. The freeze/thaw method was applied to lyse the cells for β-galactosidase and luciferase assays (Ausubel et al. 1987). For monitoring E1A levels by Western blotting, one-sixth of each transfected dish was directly lysed in SDS-sample buffer and loaded on an SDS protein gel. The anti-E1A MAb, M73, served as primary antibody. The secondary antibody was a goat anti-mouse IgG coupled to alkaline phosphatase.

RESULTS

Cloning of p300

p300 was purified from 293 cells using an anti-E1A immunopurification procedure (see Experimental Procedures). Three mice were immunized with gel-purified p300, resulting in three high-titer antisera. One was employed to screen a 293 cell cDNA library, leading to the isolation of two phage clones. Since both reacted with all three antisera, they were considered candidate p300 cDNA clones.

One of the cDNA inserts was used to screen other libraries by hybridization. After several screening rounds, cDNA fragments spanning 9 kb were obtained. Northern blot analysis indicated that this length matches the size of the putative p300 messenger (data not shown). The gene for p300 was mapped to human chromosome 22q13.2-q13.3 (Eckner et al. 1994), a region that, thus far, has not been found to be consistently deleted in a specific type of human cancer.

To determine whether the protein encoded by our cDNA could interact with various E1A mutants, we generated a segment of this polypeptide (encompassing the carboxy-terminal ~200 kD of the open reading frame) in U-2 OS cells (Fig. 1, lane 14). These cells were then infected with adenoviruses encoding wild-type or a mutant E1A gene, and the ability of the 200-kD protein (termed 5.3 ATG) to bind E1A was assayed by immunoprecipitating ^{35}S-labeled extracts from these cells with the E1A MAb, M73 (Harlow et al. 1985). Figure 1 shows that the coprecipitation patterns of the 200-kD fragment paralleled that of the endogen-

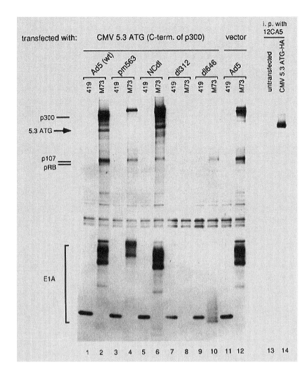

Figure 1. A carboxy-terminal 200-kD fragment of cloned p300 interacts with E1A in a manner indistinguishable from endogenous 300-kD protein. Lanes *1–10* show the ^{35}S-labeled immunoprecipitation products from lysates generated from U-2 OS cells transiently transfected with the CMV 5.3 ATG vector (encoding the carboxy-terminal 200 kD of p300) and subsequently infected with the adenovirus species indicated on top of the autoradiogram. Lysates from each dish were split in half and incubated with either pAb 419 (control) or the anti-E1A MAb M73 (Harlow et al. 1985). Cells used for lanes *11* and *12* were transfected with the vector backbone alone and infected with wt Ad5. In lanes *13* and *14*, lysates from untransfected U-2 OS cells or from cells transfected with a plasmid expressing an HA-tagged 200-kD carboxy-terminal fragment of p300, respectively, were immunoprecipitated with the anti-HA MAb 12CA5. On the left side, the migration positions of E1A and its associated proteins are indicated. (Reprinted, with permission, from Eckner et al. 1994.)

ous 300-kD protein. Wild-type E1A (lane 2) or the E1A mutant NCdl (lane 6), which is missing the region between CR1 and CR2, bound both the exogenous 200-kD and the endogenous wild-type 300-kD proteins. In contrast, neither protein bound to the two E1A mutants, pm563 (lane 4), which carries a substitution at amino acid 2, and dl646 (lane 10), which lacks an intact CR1 sequence. The high-molecular-weight protein in lane 4 is likely p400, another E1A-associated protein, which appears to be structurally related to p300 (Barbeau et al. 1994). None of the E1A-associated proteins was detected when lysates from U-2 OS cells infected with the dl312 virus, which lacks the E1A gene, were incubated with M73 (Fig.1, lane 8). Likewise, no protein of 200 kD was coprecipitated by M73 from cells which were infected with wild-type adenovirus but were not transfected with the expression vector encoding the 200-kD fragment (lane 12). We conclude that the 200-kD protein encoded by a fragment from our putative cDNA binds E1A with the same specificity as the endogenous p300 protein.

Features of the p300 Sequence

Inspection of the p300 sequence revealed that this protein contains three regions rich in cysteines and histidines (see Fig. 2A for a schematic representation). Each of these three regions exhibits a unique arrangement of these two amino acids. The spacing of the cysteine and histidine residues within these discrete regions does not conform to that encountered in other Cys/His-rich motifs, e.g., the various types of zinc fingers (Harrison 1991), the LIM domain (Li et al. 1991), or the RING motif (Freemont et al. 1991). Therefore, it appears that the three Cys/His-rich regions of p300 may form structurally novel motifs.

A computer analysis identified a bromodomain in the center of p300. This motif includes approximately 65 amino acids and is conserved from yeast to man (Haynes et al. 1992; Tamkun et al. 1992). Figure 2B compares the sequences of the bromodomains from several human proteins with that of p300. Although the function of the bromodomain is not known, its occurrence in several putative transcriptional coactivator proteins suggests a role in transcription. Among these coactivators are the human (Sekiguchi et al. 1991; Hisatake et al. 1993; Ruppert et al. 1993) and *Drosophila* (Kokubo et al. 1993; Weinzierl et al. 1993) 250-kD TATA-binding protein-associated factors (TAF$_{II}$250/CCG1) and the yeast protein, GCN5 (Georgakopoulos and Thireos 1992).

Another motif with similarity to a known protein is present within the third Cys/His-rich region. A stretch of about 30 amino acids displays significant homology with the yeast ADA2 protein (Fig. 2C), another putative coactivator protein (Berger et al. 1992). ADA2 likely contributes to the activation potential of certain transcription factors containing an acidic activation domain.

A Family of p300-like Proteins

After acceptance of the manuscript describing the cloning of p300 (Eckner et al. 1994), further database searching revealed striking homologies of human p300 with two recently identified proteins, the CREB-binding protein CBP (Chrivia et al. 1993) and a hypothetical *Caenorhabditis elegans* protein discovered during the nematode chromosome III sequencing project (Wilson et al. 1994). Figure 3A illustrates the regions of homology among these three proteins. Remarkably, all three polypeptides are large (>2000 residues) and contain, in their central portions, a segment of approximately 800 amino acids with largely uninterrupted, high-level homology. This central region includes the bromodo-

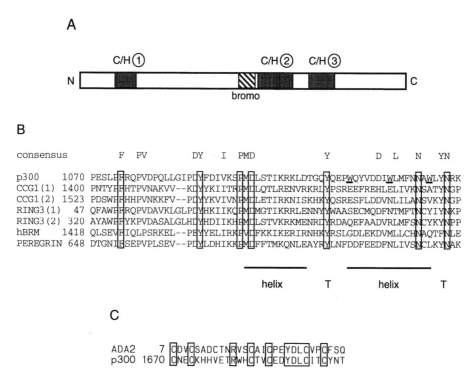

Figure 2. Sequence motifs present in p300. (*A*) Schematic representation of the main motifs of p300. The relative locations of the bromodomain and the three cysteine and histidine (C/H)-rich motifs are indicated. The p300 cDNA sequence has been deposited in GenBank and is available under the accession number U01877. (*B*) Comparison of the bromodomain sequences of multiple human bromodomain-containing proteins. Residues that are conserved in all seven bromodomains are boxed. Residues that are present in at least six of the seven bromodomains are noted in the top line (*consensus*). The region of the bromodomain predicted to form two α-helices followed by reverse turns (Haynes et al. 1992) is marked below the alignment. (Reprinted, with permission, from Eckner et al. 1994.) (*C*) Alignment of the homology region of yeast ADA2 and human p300. The region of homology of p300 to the ADA2 protein is located within the E1A-binding domain (Cys/His-rich region 3).

main and the second and third Cys/His-rich region. The sequence alignment for the third Cys/His-rich region is given in Figure 3B. Two other segments with a high percentage of identity among these three proteins exist in the amino-terminal half of the three proteins. These regions are the first Cys/His-rich domain and a part of the region of CBP which has been defined as the site of interaction with CREB (Chrivia et al. 1993). Interestingly, the spacing of the aforementioned homology regions is also well conserved among the three proteins.

From an analysis of a partial sequence of murine p300, we asked whether p300 and CBP are products of a common gene (Arany et al. 1994). Although the two proteins share certain regions of profound identity, the sequence comparison analysis clearly indicated that p300 and CBP are products of different genes.

CBP is suspected of acting as a coactivator for CREB. This transcription factor binds to regulatory elements in the promoters of cAMP-inducible genes and stimulates transcription in response to increased intracellular cAMP levels. CBP might, therefore, be an integral part of the cAMP signal transduction pathway. Given the strong homology between p300 and CBP in the CREB-binding domain of CBP, it is possible that p300 also interacts with CREB or a related protein and is involved in the cAMP signaling pathway (Arany et al. 1994; see also Discussion).

Comparison of the Peptide Maps of Cloned p300 and E1A-associated 300-kD Protein

To distinguish exogenous from endogenous p300, we produced, after transient transfection, a hemagglutinin (HA)-tagged version of p300 in U-2 OS cells. ^{35}S-labeled extracts from these cells were immunoprecipitated with an α-HA antibody and analyzed in parallel with an α-E1A immunoprecipitate from 293 cells. Figure 4A shows that exogenous and E1A-associated 300-kD proteins comigrated (cf. lanes 2 and 3). In contrast, in-vitro-translated p300 exhibited a somewhat faster mobility than either exogenous or endogenous p300 (lane 1), probably because it lacks certain posttranslational modifications.

Partial proteolysis represents a useful method for testing the structural similarity of proteins (Cleveland et al. 1977). To this end, *S. aureus* partial V8 protease digestion was employed to partially digest HA-tagged, cloned p300 and E1A-bound p300. As shown in Figure 4B, the proteolytic digestion patterns of the two proteins were very similar, although not identical. Partial

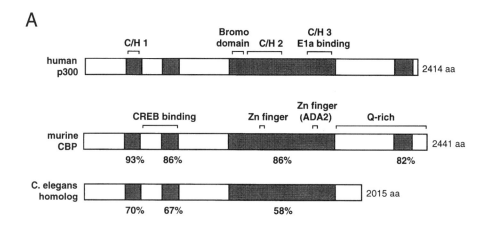

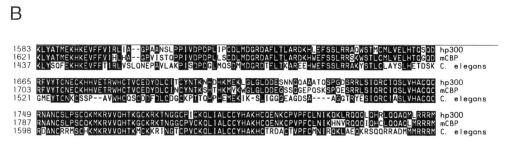

Figure 3. A family of p300 proteins conserved in evolution. (*A*) Schematic representation of human p300, murine CBP, and the *C. elegans* homolog. Shaded boxes mark the regions of highest homology, which is expressed as percentage of identical residues relative to p300 below the boxes. Domain structures of CBP and p300 are represented above the shaded boxes and are given according to the respective publications (Chrivia et al. 1993; Eckner et al. 1994). Abbreviations: (C/H) Cys/His-rich region; (Q-rich) glutamine-rich domain; (aa) amino acids. (*B*) Sequence comparison of the third Cys/His-rich region of human p300, murine CBP, and the *C. elegans* homolog. In human p300, this region represents the binding site of E1A. Residues that are identical between at least two of the three proteins are shaded. (Reprinted, with permission, from Arany et al. 1994 [copyright held by Cell Press].)

cleavage by *N*-chlorosuccinimide (NCS) was used to test the similarity of the two proteins by a second method (Fig. 4C). The obtained patterns were again highly similar, although not identical. Taken together, the results of the peptide maps support the conclusion that the isolated cDNA codes for p300 or a very closely related protein. The small differences between cloned p300 and E1A-associated p300 may be related to the fact that cloned p300 was overproduced after transfection. Conceivably, the relatively high expression levels of transfected p300 could titrate out certain modifying enzymes, leading to incomplete modification. Alternatively, the small differences between exogenous and endogenous 300-kD proteins could be caused by the HA tag attached to the cloned p300. A third possibility is that "endogenous" p300, represented by the p300 protein which associates with E1A, is really a collection of similar, but not identical, proteins. In that case, any one member of the group might be characterized by a partial proteolytic banding pattern which is similar, but not identical, to that of the mixture.

In search of additional evidence for the identity of the protein encoded by the p300 cDNA, we raised a series of monoclonal antibodies against a carboxy-terminal fragment of the p300 product, synthesized in bacteria. In all cases, these antibodies precipitated a 300-kD protein whose partial V8 map was indistinguishable from E1A-bound p300 (Eckner et al. 1994; and data not shown).

E1A Binds to the Third Cys/His-rich Region of p300

To determine the E1A-binding site on p300, we utilized a set of in-vitro-translated p300 polypeptides which are characterized by the presence of progressive deletions from either the carboxyl or amino terminus. The ^{35}S-labeled translation products were incubated with unlabeled 293 extracts (as a source of E1A), followed by M73 immunoprecipitation. As illustrated in Figure 5A for the amino-terminal truncation series, removal of p300 sequences up to the *Sma*I site (amino acid 1572) did not affect the ability of the translation product to bind E1A (lanes 6–8). A p300 protein initiated at the *Mun*I site (amino acid 1752) was partially defective in E1A binding, whereas a protein initiated at the *Aha*II site (amino acid 1869) was completely unable to bind to E1A. Thus, the amino-terminal border of the E1A-binding domain of p300 is located between the *Sma*I and *Mun*I sites (see Fig. 5D for a

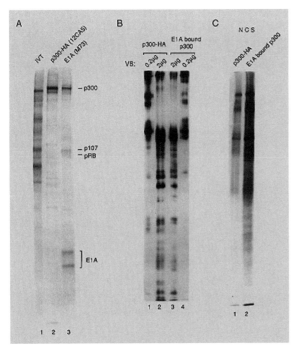

Figure 4. Comparison of cloned p300 with E1A-associated p300. (*A*) Electrophoretic mobility of in-vitro-translated p300 (lane *1*) is compared with that of HA-tagged, cloned p300 synthesized in U-2 OS cells (lane *2*) and with E1A-associated 300-kD protein (lane *3*). (*B*) Cloned, HA-tagged p300 and E1A-bound p300 were partially digested with two different concentrations of *S. aureus* V8 protease. The ^{35}S-labeled digestion products were electrophoresed in a 15% protein gel. (*C*) Peptide maps of HA-tagged p300 and the E1A-associated 300-kD protein, employing *N*-chlorosuccinimide (NCS) as cleaving agent. (Reprinted, with permission, from Eckner et al. 1994.)

schematic diagram of the truncations). The carboxy-terminal border of the E1A-binding domain is situated near the *Pvu*II site (amino acid 1818), since the relevant truncated product readily bound E1A (Fig. 5B, lane 11), whereas further truncation to the *Mun*I site completely abrogated it (lane 12). This in vitro approach suggests that the third Cys/His-rich region of p300 represents the E1A interaction site (see summary of data in Fig. 5D).

To test the validity of this assignment by an in vivo assay, we introduced two internal deletions into the third Cys/His-rich domain (named del30 and del33) and produced these mutants as 200-kD carboxy-terminal fragments with attached HA tags in U-2 OS cells. These truncations allowed us to separate transfected from endogenous p300, on the basis of size. A 12S E1A expression vector was included in all transfections. ^{35}S-labeled cell lysates from each dish were split in half: One half was immunoprecipitated with HA antibody and the other half with M73. Wild-type, del30 and del33 200-kD fragments were efficiently synthesized, as illustrated in Figure 5C, lanes 1, 3, and 5, respectively. However, only the wild-type fragment was coprecipitated by M73 (cf. lane 2 with lanes 4 and 6 in Fig. 5C).

Importantly, endogenous, full-length 300-kD protein was readily visible in all three cases (lanes 2, 4, 6). This experiment supports the above-noted assignment of the third Cys/His-rich region as the E1A-binding domain.

Relief of E1A-mediated Repression of the SV40 Enhancer

The presence of the bromodomain and of the ADA2 homology region suggested that p300 may exert part of its biological role as a transcriptional coactivator. Historically, the SV40 enhancer was the first transcriptional control element shown to be repressed by E1A (Borelli et al. 1984; Velcich and Ziff 1985). Repression of this enhancer is dependent on the presence of an intact p300-binding site in E1A (Jelsma et al. 1989). To examine the involvement of p300 in SV40 enhancer function, we first tried to overcome E1A-imposed repression by overproducing p300. For this purpose, we employed, as a reporter, the luciferase gene driven by the SV40 enhancer/promoter. This plasmid was transfected into U-2 OS cells together with a p300 expression vector and variable amounts of a 12S E1A plasmid. The activity of the reporter in the absence of E1A was set to a value of 100. At low concentrations of the E1A expression vector, repression of the SV40 enhancer was partial and cotransfected p300 increased, to some extent, its transcriptional activity (Fig. 6A, columns with 0.1 and 0.25 μg E1A). Western blot analysis, carried out in parallel, indicated that E1A levels were about threefold higher in cells that were transfected with p300 and E1A as compared to cells which only received E1A (Fig. 6B). Hence, p300 did not stimulate enhancer activity by reducing the levels of E1A. However, immunofluorescence experiments indicated that overproduced p300 tended to form intracellular aggregates (Eckner et al. 1994; and data not shown). Therefore, we reasoned that p300 might sequester E1A in these aggregates, thereby lowering the intracellular concentration of active E1A molecules. This could result in the observed reactivation of the enhancer in the presence of p300. Consistent with this possibility was the fact that the repressive effect of high levels of E1A could not be overcome by p300 (Fig. 6A, columns with 2.5 and 5 μg of E1A), implying that, under these conditions, E1A was in clear excess over p300.

To avoid the possible sequestration and inactivation of E1A by cotransfected p300, we substituted full-length p300 with either the del30 or del33 internal deletion in the E1A-binding region for wild-type p300 in these experiments. As shown in Figure 5C, these p300 versions are unable to bind E1A and, therefore, cannot sequester E1A. Figure 6C shows that both p300 mutants restored SV40 enhancer activity to about half-maximal levels, even at very high E1A concentrations. Western blots indicated that the E1A concentrations were comparable, whether or not a p300 plasmid was cotransfected (Fig. 6D). Collectively, these results demonstrate that p300 can bypass E1A-mediated silencing

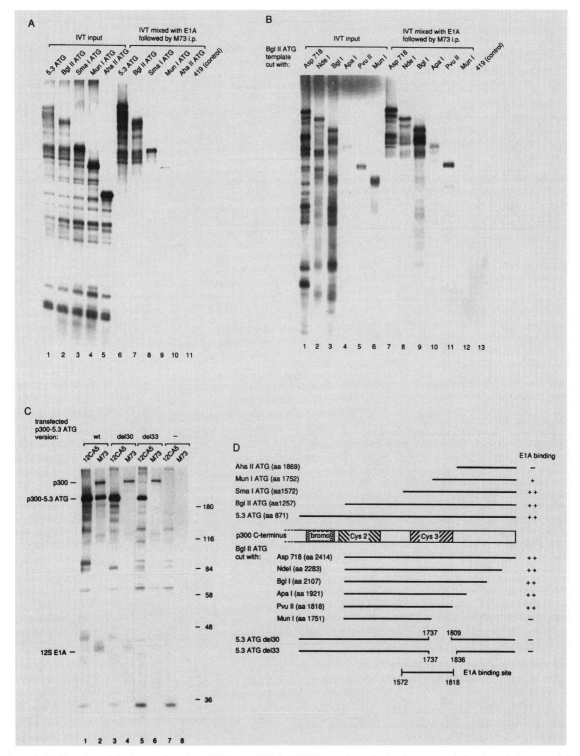

Figure 5. Delineation of the interaction site of E1A on p300. (*A*) DNA templates for in vitro transcription/translation containing progressive 5' deletions were generated by ligating a translation initiation codon to the restriction sites indicated on top. (Lanes *1–5*) Input of the translates. (Lanes *6–10*) Translation products recovered after mixing them with cold E1A protein, followed by immunoprecipitation with the α-E1A MAb, M73. For lane *11*, translation products from the 5.3 ATG template were incubated with the control antibody, pAb 419. (*B*) A series of templates with progressive 3' terminal deletions prepared by cutting the *Bgl*II ATG plasmid with the restriction enzymes noted above each lane. (Lanes *1–6*) Untreated translation products; (lanes *7–12*) translation products binding to E1A. In lane *13*, the products of the *Bgl*II ATG template cut with Asp-718 were mixed with E1A and precipitated with pAb 419 (control). (*C*) Two internal deletions in the E1A-binding region of p300 (del30 and del33) were constructed in the context of the 200-kD carboxy-terminal fragment of p300. HA-tagged versions of these two deletions and the wild-type 200-kD fragment were expressed in U-2 OS cells together with 12S E1A. Half of the lysate from each transfected dish was immunoprecipitated with the anti-HA MAb, 12CA5 (odd-numbered lanes), or with M73 (even-numbered lanes). In lanes *7* and *8*, the lysate from untransfected cells was incubated with the two antibodies. (*D*) Schematic drawing of deletion mutants used in the experiments described in *A–C*. The E1A-binding properties of each are indicated on the right. (Reprinted, with permission, from Eckner et al. 1994.)

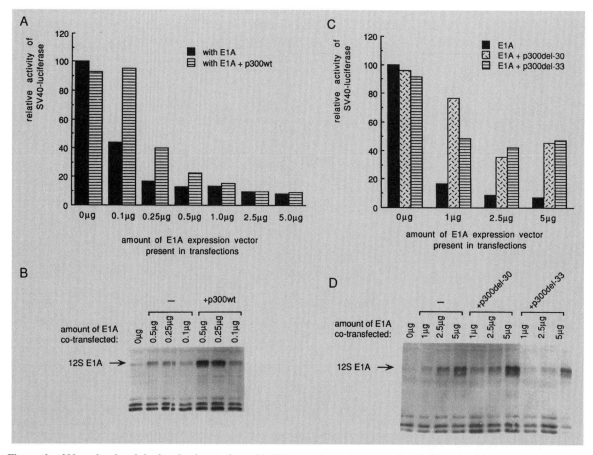

Figure 6. p300 molecules defective for interaction with E1A can bypass E1A-mediated SV40 enhancer repression. (*A*) A luciferase reporter gene driven by the SV40 enhancer/promoter (5 μg) was transfected into U-2 OS cells along with a plasmid encoding β-galactosidase (1 μg, internal reference) and increasing amounts (see numbers below each pair of bars) of a vector encoding 12S E1A. In addition, one half of the dishes were also transfected with 8 μg of pBluescript (black bars), and the other half with 8 μg of a vector encoding full-length, wild-type p300 (bars with horizontal lines). The luciferase activity obtained with the reporter alone was arbitrarily set to 100. The luciferase activity of all other transfections is relative to this value. Data represent the average of four independent experiments. (*B*) Western blot monitoring the expression levels of E1A in the transfection experiments described in *A*. Proteins extracted from one sixth of each transfected culture were subjected to Western blotting, employing M73 as the E1A antibody. The amount of cotransfected E1A plasmid is noted above each lane. (*C*) U-2 OS cells were transfected as outlined in *A*, except that, instead of the expression vector encoding wild-type p300, a vector encoding either deletion mutant p300del30 (stippled bars) or mutant p300del33 (bars with horizontal lines) was present in the cotransfection mixture. Values represent the average of five independent experiments. (*D*) Western blot showing the E1A levels present in cells transfected as described in *C*. (Reprinted, with permission, from Eckner et al. 1994.)

of the SV40 enhancer and suggest that p300 is directly involved in modulating the activity of this enhancer.

DISCUSSION

In this work, we described the isolation and initial characterization of a cDNA encoding full-length human p300. Several lines of evidence indicate that the isolated cDNA codes for p300. First, cloned p300 interacted with wild-type and mutant E1A in a manner indistinguishable from endogenous 300-kD protein. Second, peptide maps comparing cDNA-encoded p300 with E1A-bound p300 were largely similar. Third, monoclonal antibodies raised against cloned p300 recognized endogenous p300 (data not shown; see Eckner et al. 1994).

The presence of two motifs frequently found in transcriptional adapter or coactivator proteins, namely the bromodomain and the ADA2 motif, strongly suggested a coactivating function for p300. Further evidence for a coactivator role is provided by our experiment with the SV40 enhancer. p300 molecules bearing a deletion in their E1A-binding domain bypassed E1A-imposed enhancer repression and restored to about half-maximal levels the activity of this element. The results of these experiments are consistent with, although not proof of, a role for p300 as a coactivator. Finally, the well-documented ability of E1A to repress via its amino terminus and CR1 so many seemingly unrelated enhancers or promoters (Hen et al. 1985; Webster et al. 1988; Rochette-Egly et al. 1990; Stein et al. 1990; Braun et al. 1992) also argues in favor of the coactivator model. In contrast to classical transcription factors, which interact in a sequence-specific manner with DNA, coactivators, in general, do not bind DNA by themselves but rather are recruited to the promoter by protein-protein interactions.

Why was the activity of the SV40 enhancer not fully restored? There are several possible explanations for the incomplete relief of E1A-imposed repression. On one hand, it is possible that the removal of part of the E1A-binding region impairs the ability of p300 to coactivate normally. On the other hand, p300 may not be the only protein targeted (and presumably inactivated) by E1A. For full restoration, the putative "other" protein would have to be produced in reasonable quantities, and perhaps in a form selectively unable to bind E1A, together with p300. Alternatively, as shown previously (Eckner et al. 1994), p300 has a tendency to form aggregates when overexpressed, presumably leading to the inactivation of p300. It is conceivable that we were unable to produce sufficiently high levels of active p300 because of this aggregation phenomenon.

An unanticipated finding was the discovery that p300, at least in mammals, is a member of a family of proteins. The CREB-binding protein, CBP, is both structurally and functionally (as a coactivator) closely related to p300. Moreover, at least one protein of the size and structure of p300/CBP exists also in *C. elegans*. Remarkably, of the major E1A-binding proteins, p300 is the first one with a homolog in invertebrates; so far, no Rb-like protein has been described in either *C. elegans* or *Drosophila*. This strong evolutionary conservation of p300 suggests that it (or its family members) exerts very basic functions.

The CREB protein belongs, together with ATF1 and CREM, to a family of transcription factors which are rapidly activated by increases in intracellular concentrations of the second messenger cAMP. This increase leads to the activation of protein kinase A (PKA), which phosphorylates CREB on Ser-133, thereby rendering it transcriptionally active (for review, see Lalli and Sassone-Corsi 1994). CBP was isolated as a protein that specifically interacts with CREB phosphorylated on Ser-133. On the basis of this property, and its ability to activate transcription as a Gal4 fusion protein (Chrivia et al. 1993), CBP has been postulated to function as a coactivator for the CREB family of proteins.

The striking similarity of p300 and CBP in all three Cys/His-rich segments, the bromodomain, and the CREB-binding region raises a number of important questions. First, does CBP also bind E1A? If it does, what has been called "p300" would be a collection of at least two different protein species. Second, does p300 also interact with CREB? If so, its function would be linked to the cAMP signaling pathway.

Previous studies have demonstrated that E1A can interfere with cAMP-stimulated transcription in a CR1-dependent manner (Janaswami et al. 1992; Kalvakolanu et al. 1992). CR2 is dispensable for the repressive effect of E1A, suggesting that members of the Rb family are not involved. Rather, targeting of p300 appears to be key to E1A's ability to interfere with cAMP-dependent transcription. A possible molecular mechanism for this negative action of E1A is provided by the observation that CBP contains a consensus PKA site in the middle of the third Cys/His-rich region, the putative E1A-binding domain. This PKA site is precisely conserved in p300. Conceivably, binding of E1A to p300 or CBP blocks phosphorylation of this site by PKA and prevents a PKA-induced change in p300/CBP function. Consistent with this hypothesis, Chrivia et al. (1993) have reported that the *trans*-activation potential of CBP itself (when assayed as a Gal4-CBP chimeric protein) can be modulated by cAMP.

In certain cell types, cAMP has a profound effect on cell growth. In fibroblasts and macrophages, for example, elevation of the intracellular cAMP concentration leads to G_1 arrest (Vairo et al. 1990). Furthermore, cAMP can induce cellular differentiation and concomitant exit from the cell cycle. Considering the well-established growth-stimulating capacity of E1A and its ability to interfere with cAMP-induced transcription, it would not be surprising to find that E1A can counteract this antiproliferative effect of cAMP.

In light of the antagonistic effects of cAMP and E1A, some interesting predictions about the mechanistic basis of cooperativity among oncogenes can be made. E1A cooperates efficiently with activated *ras* to transform primary rodent cells. In contrast, cAMP treatment leads to down-regulation of *ras*-pathway signaling (Burgering et al. 1993; Cook and McCormick 1993), and it can reverse the transformed phenotype of certain *ras*-transformed cells (Pastan and Willingham 1978). Thus, one reason for the cooperativity of E1A and *ras* may lie in the ability of E1A to inactivate cellular control pathways that act negatively on the growth-stimulatory effects of *ras*. If this is true, one would predict that certain cells which produce sufficient E1A or are transformed by E1A and *ras* would be resistant to some of the negative effects of cAMP on growth. In that case, a fundamental question will be whether E1A perturbation of p300, CBP, and/or other as-yet-unknown members of this family of presumed coactivators is a key step in the mechanism of immunity to cAMP.

REFERENCES

Arany, Z., W.R. Sellers, D.M. Livingston, and R. Eckner. 1994. E1A-associated p300 and CREB-associated CBP belong to a conserved family of coactivators. *Cell* **55:** 799.

Ausubel, F.M., B. Roger, R.E. Kingston, D.D. Moore, J.A. Smith, J.G. Seidman, and K. Struhl. 1987. *Current protocols in molecular biology*. Wiley, New York.

Barbeau, D., R. Charbonneau, S.G. Whalen, S.T. Bayley, and P.E. Branton. 1994. Functional interactions within adenovirus E1A protein complexes. *Oncogene* **9:** 359.

Berger, S.L., B. Pina, N. Silverman, G.A. Marcus, J. Agapite, J.L. Regier, S.J. Triezenberg, and L. Guarente. 1992. Genetic isolation of ADA2: A potential transcriptional adaptor required for function of certain acidic activation domains. *Cell* **70:** 251.

Borelli, E., R. Hen, and P. Chambon. 1984. Adenovirus-2 E1A

products repress enhancer-induced stimulation of transcription. *Nature* **312:** 608.
Braun, T., E. Bober, and H.H. Arnold. 1992. Inhibition of muscle differentiation by the adenovirus E1a protein: Repression of the transcriptional activating function of the HLH protein myf-5. *Genes Dev.* **6:** 888.
Burgering, B.M.T., G.J. Pronk, P.C.V. Weeren, P. Chardin, and J.L. Bos. 1993. cAMP antagonizes $p21^{ras}$-directed activation of extracellular signal-regulated kinase 2 and phosphorylation of mSos nucleotide exchange factor. *EMBO J.* **12:** 4211.
Chrivia, J.C., R.P.S. Kwok, N. Lamb, M. Hagiwara, M.R. Montminy, and R.H. Goodman. 1993. Phosphorylated CREB binds specifically to the nuclear protein CBP. *Nature* **365:** 855.
Church, G. and W. Gilbert. 1984. Genomic sequencing. *Proc. Natl. Acad. Sci.* **81:** 1991.
Cleveland, D.W., S.G. Fischer, M.W. Kirschner, and U.K. Laemmli. 1977. Peptide mapping by limited proteolysis in sodium dodecyl-sulfate and analysis by gel electrofocusing. *J. Biol. Chem.* **252:** 1102.
Cook, S.J. and F. McCormick. 1993. Inhibition by cAMP of ras-dependent activation of raf. *Science* **262:** 1069.
Draetta, G., L. Brizuela, J. Potashkin, and D. Beach. 1987. Identification of p34 and p13, human homologs of the cell cycle regulators of fission yeast encoded by $cdc2^+$ and $suc1^+$. *Cell* **50:** 319.
Dyson, N. and E. Harlow. 1992. Adenovirus E1A targets key regulators of cell proliferation. *Cancer Surv.* **12:** 161.
Eckner, R., M.E. Ewen, D. Newsome, M. Gerdes, J.A. DeCaprio, J.B. Lawrence, and D.M. Livingston. 1994. Molecular cloning and functional analysis of the adenovirus E1A-associated 300-kD protein (p300) reveals a protein with properties of a transcriptional adaptor. *Genes Dev.* **8:** 869.
Ewen, M., Y. Xing, J.B. Lawrence, and D.M. Livingston. 1991. Molecular cloning, chromosomal mapping, and expression of the cDNA for p107, a retinoblastoma gene product-related protein. *Cell* **66:** 1155.
Fearon, E.R. and B. Vogelstein. 1990. A genetic model for colorectal tumorigenesis. *Cell* **61:** 759.
Freemont, P.S., I.M. Hanson, and J. Trowsdale. 1991. A novel cysteine rich sequence motif. *Cell* **64:** 483.
Georgakopoulos, T. and G. Thireos. 1992. Two distinct yeast transcriptional activators require the function of the GCN5 protein to promote normal levels of transcription. *EMBO J.* **11:** 4145.
Harlow, E. and D. Lane. 1988. *Antibodies: A laboratory manual.* Cold Spring Harbor Laboratory, Cold Spring Harbor, New York.
Harlow, E., B.R. Franza, and C. Schley. 1985. Monoclonal antibodies specific for adenovirus early region 1A proteins: Extensive heterogeneity in early region 1A products. *J. Virol.* **55:** 533.
Harrison, S.C. 1991. A structural taxonomy of DNA-binding domains. *Nature* **353:** 715.
Haynes, S.R., C. Dollard, F. Winston, S. Beck, J. Trowsdale, and I.B. Dawid. 1992. The bromodomain: A conserved sequence found in human, *Drosophila* and yeast proteins. *Nucleic Acids Res.* **20:** 2603.
Hen, R., E. Borelli, and P. Chambon. 1985. Repression of the immunoglobulin heavy chain enhancer by the adenovirus-2 E1A products. *Science* **230:** 1391.
Hisatake, K., S. Hasegawa, R. Takada, Y. Nakatani, M. Horikoshi, and R.G. Roeder. 1993. The p250 subunit of native TATA box-binding factor TFIID is the cell-cycle regulatory protein CCG1. *Nature* **362:** 179.
Howe, J.A., J.S. Mymryk, C. Egan, P.E. Branton, and S.T. Bayley. 1990. Retinoblastoma growth suppressor and a 300kDa protein appear to regulate cellular DNA synthesis. *Proc. Natl. Acad. Sci.* **87:** 5883.
Janaswami, P.M., D.V.R. Kalvakolanu, Y. Zhang, and G.C. Sen. 1992. Transcriptional repression of interleukin-6 gene by adenovirus E1A proteins. *J. Biol. Chem.* **267:** 24886.
Jelsma, T.N., J.S. Howe, J.S. Mymryk, C.M. Evelegh, N.F.A. Cunniff, and S.T. Bayley. 1989. Sequences in E1A proteins of human adenovirus 5 required for cell transformation, repression of a transcriptional enhancer, and induction of proliferating cell nuclear antigen. *Virology* **170:** 120.
Kalvakolanu, D.V.R., J. Liu, R.W. Hanson, M.L. Harter, and G.C. Sen. 1992. Adenovirus E1A represses the cyclic AMP-induced transcription of the gene for phosphoenolpyruvate carboxykinase (GTP) in hepatoma cells. *J. Biol. Chem.* **267:** 2530.
Kokubo, T., D.-W. Gong, S. Yamashita, M. Horikoshi, R.G. Roeder, and Y. Nakatani. 1993. *Drosophila* 230kD TFIID subunit, a functional homolog of the human cell cycle gene product, negatively regulates DNA binding of the TATA box-binding subunit of TFIID. *Genes Dev.* **7:** 1033.
Lalli, E. and P. Sassone-Corsi. 1994. Signal transduction and gene regulation: The nuclear response to cAMP. *J. Biol. Chem.* **269:** 17359.
Li, P.M., J. Reichert, G. Freyd, H.R. Horvitz, and C.T. Walsh. 1991. The LIM region of a presumptive *Caenorhabditis elegans* transcription factor is an iron-sulfur- and zinc-containing metallodomain. *Proc. Natl. Acad. Sci.* **88:** 9210.
Moran, E. 1993. DNA tumor virus transforming proteins and the cell cycle. *Curr. Opin. Genet. Dev.* **3:** 63.
Pastan, I. and M. Willingham. 1978. Cellular transformation and the "morphological phenotype" of transformed cells. *Nature* **274:** 645.
Rochette-Egly, C., C. Fromental, and P. Chambon. 1990. General repression of enhanson activity by the adenovirus-2 E1A proteins. *Genes Dev.* **5:** 1200.
Ruley, H.E. 1983. Adenovirus early region 1A enables viral and cellular transforming genes to transform primary cells in culture. *Nature* **304:** 602.
Ruppert, S., E.H. Wang, and R. Tjian. 1993. Cloning and expression of human $TAF_{II}250$: A TBP-associated factor implicated in cell-cycle regulation. *Nature* **362:** 175.
Scrable, H.J., C. Sapienza, and W.K. Cavenee. 1990. Genetic and epigenetic losses of heterozygosity in cancer predisposition and progression. *Adv. Cancer Res.* **54:** 25.
Sekiguchi, T., Y. Nohiro, Y. Nakamura, N. Hisamoto, and T. Nishimoto. 1991. The human *CCG1* gene, essential for progression of the G_1 phase, encodes a 210-kilodalton nuclear DNA-binding protein. *Mol. Cell. Biol.* **11:** 3317.
Stein, R.W., M. Corrigan, P. Yaciuk, J. Whelan, and E. Moran. 1990. Analysis of E1A-mediated growth regulation functions: Binding of the 300-kilodalton cellular product correlates with E1A enhancer repression function and DNA synthesis-inducing activity. *J. Virol.* **64:** 4421.
Subramanian, T., M. Kuppaswamy, R.J. Nasser, and G. Chinnadurai. 1988. An N-terminal region of adenovirus-E1A essential for cell-transformation and induction of an epithelial-cell growth factor. *Oncogene* **2:** 105.
Tamkun, J.W., R. Deuring, M.P. Scott, M. Kissinger, A.M. Pattatucci, T.C. Kaufman, and J.A. Kennison. 1992. *brahma*: A regulator of *Drosophila* homeotic genes structurally related to the yeast transcriptional activator SNF2/SWI2. *Cell* **68:** 561.
Vairo, G., S. Argyriou, A.-M. Bordun, G. Whitty, and J.A. Hamilton. 1990. Inhibition of the signaling pathways for macrophage proliferation by cAMP. *J. Biol. Chem.* **265:** 2692.
Velcich, A. and E. Ziff. 1985. Adenovirus E1a proteins repress transcription from the SV40 early promoter. *Cell* **40:** 705.
Webster, K.A., G.E.O. Muscat, and L. Kedes. 1988. Adenovirus E1A products suppress myogenic differentiation and inhibit transcription from muscle-specific promoters. *Nature* **332:** 553.
Weinzierl, R.O.J., B.D. Dynlacht, and R. Tjian. 1993. Largest subunit of Drosophila transcription factor IID directs assembly of a complex containing TBP and a coactivator. *Nature* **362:** 511.
Wilson, R., R. Ainscough, K. Anderson, C. Baynes, M. Berks, J. Bonfield, J. Burton, M. Connell, T. Copsey, and J.

Cooper. 1994. 2.2 Mb of contiguous nucleotide sequence from chromosome III of *C. elegans*. *Nature* **368:** 32.

Whyte, P., N.M. Williamson, and E. Harlow. 1989. Cellular targets for transformation by the adenovirus E1A proteins. *Cell* **56:** 67.

Xiao, J.H., I. Davidson, H. Matthes, J.-M. Garnier, and P. Chambon. 1991. Cloning, expression, and transcriptional properties of the human enhancer factor TEF-1. *Cell* **65:** 551.

The Corral Hypothesis: A Novel Regulatory Mode for Retinoblastoma Protein Function

W.-H. Lee, Y. Xu, F. Hong, T. Durfee, M.A. Mancini,
Y.-C. Ueng, P.-L. Chen, and D. Riley

Institute of Biotechnology / Center for Molecular Medicine, University of Texas Health Science Center at San Antonio, San Antonio, Texas 78245

The retinoblastoma susceptibility gene product functions to prevent neoplastic transformation in a variety of cell types. Inactivation of the retinoblastoma gene (*RB*) is genetically linked to the formation of a pediatric cancer, retinoblastoma (Knudson 1971; Cavenee et al. 1983; Friend et al. 1986; Fung et al. 1987; Lee et al. 1987a), and has also been found in many different adult cancers including osteosarcoma (Toguchida et al. 1988), bladder cancer (Horowitz et al. 1989), small cell lung carcinoma (Harbour et al. 1988; Yokota et al. 1988), and breast carcinoma (Lee et al. 1988). In addition, supplementing human cancer cells containing mutated *RB* genes with wild-type genes leads to suppression of the cells' malignant properties and supports the notion that *RB* is indeed a tumor suppressor gene (Huang et al. 1988; Bookstein et al. 1990; Sumegi et al. 1990; Takahashi et al. 1991; Chen et al. 1992; Goodrich et al. 1992).

Evidence for the critical role of Rb protein in the regulation of cell growth and differentiation also stems from studies that alter the expression of Rb protein in intact animals. Overexpression of the human *RB* gene under control of its own promoter in transgenic mice affects prenatal and postnatal development and results in dose-dependent growth retardation (Bignon et al. 1993; Chang et al. 1993). Underexpression of Rb protein was accomplished using gene "knockout" experiments, in which *RB* was inactivated in the germ line of mice. Embryos nullizygous for *RB* die prenatally and exhibit massive neuronal cell death and abnormal erythropoiesis (Clarke et al. 1992; Jacks et al. 1992; Lee et al. 1992). Mice carrying one mutant and one wild-type *RB* allele live to adulthood but develop characteristic tumors of the pituitary intermediate lobe, with nearly 100% penetrance, by the time they are 10 months old (Hu et al. 1994). Comparison of the pituitary tumor DNA with surrounding normal tissue DNA indicated that the remaining wild-type *RB* allele was lost in all the tumors examined. These observations in mice are thus consistent with the recessive property and "two-hit" kinetics of tumor suppressor genes known since their existence was first proposed (Knudson 1971). The Rb-deficient mouse is therefore an appropriate and useful model for studying the effects of Rb in development and in tumorigenesis.

The *RB* gene encodes a nuclear phosphoprotein containing 928 amino acids (Lee et al. 1987b). Analysis by partial proteolytic digestion and genetic deletion mapping (Hensey et al. 1994) shows that Rb protein can be divided into three regions. The amino-terminal region, from amino acids 1 to 380, comprises two globular domains, N and R (Hensey et al. 1994). This region has no clearly defined biological function according to our present knowledge but is required for the oligomerization of purified Rb protein in vitro. Several reports have also shown that amino-terminal domain sequences are necessary for complete phosphorylation (Hinds et al. 1992; Qian et al. 1992; Qin et al. 1992). Moreover, one nuclear matrix protein, p84, which localizes to sites of RNA processing (Durfee et al. 1994), specifically binds to this region and may link the amino terminus of Rb to a specific nuclear subcompartment (Mittnacht and Weinberg 1991; Templeton et al. 1991). The central portion of the Rb molecule, which spans about 390 amino acids, contains two noncontiguous regions required for simian virus 40 T-antigen binding, connected by a spacer sequence. This region, often called the T-binding domain, is used to interact with several DNA tumor virus oncoproteins such as SV40 large T antigen (DeCaprio et al. 1988), adenovirus E1A (Whyte et al. 1988), and papillomavirus-16 E7 (Dyson et al. 1989), as well as with an expanding number of cellular proteins including E2F-1 (Helin et al. 1992; Kaelin et al. 1992; Shan et al. 1992), RBP-1, RBP-2 (Defeo-Jones et al. 1991), protein phosphatase type 1 catalytic subunit (Durfee et al. 1993), p48 (Qian et al. 1993), and a human homolog of yeast nuc2 (P.-L. Chen et al., in prep.). The T-binding domain of Rb is also the one most often mutated in human tumors (Bookstein and Lee 1991). Finally, the third region of Rb consists of the carboxy-terminal 150 amino acids, which contain sequences for both nuclear localization and nonspecific DNA binding (Lee et al. 1987b; Wang et al. 1990b). To date there is no evidence to suggest that Rb protein recognizes sequence-specific DNA.

The phosphorylation state of Rb protein fluctuates in concert with the cell cycle. The protein is hypophosphorylated in the G_1 phase, and distinct phosphorylation events occur in late G_1 phase, early S phase, and at the G_2-to-M transition point (Buchkovich et al. 1989; Chen et al. 1989; DeCaprio et al. 1989). The parallel between this cyclic phosphorylation pattern and the activity of the cyclin-dependent protein kinase complexes has engendered the hypothesis that cyclin-dependent kinases (Cdks) are responsible, at least in part, for Rb phosphorylation (Hollingsworth et al. 1993). Although Rb is a reasonably good substrate for many different kinds of Cdks in vitro, the precise enzyme responsible for Rb phosphorylation in vivo remains undetermined. Nevertheless, Rb is believed to have a role in regulating cell cycle progression. Microinjection of hypophosphorylated Rb protein, for example, inhibits G_1 progression of the cell cycle (Goodrich et al. 1991). This result is consistent with the observations of the tumor suppressive activity of Rb when a normal *RB* gene is replaced in rapidly proliferating cancer cells. It also fits nicely with the role of Rb in regulating transcription factors such as E2F-1 that are necessary for progression to committed phases of the cell cycle (Nevins 1992). Interestingly, the hypophosphorylated form of the Rb protein, but not the hyperphosphorylated form, is responsible for binding to DNA and to all of the Rb-associated proteins, as well as for tethering to the nuclear matrix (Mittnacht and Weinberg 1991; Mancini et al. 1994). Phosphorylation may inactivate Rb and thereby remove constraints on proliferation in many cells. There are at least ten potential Cdk phosphorylation sites distributed throughout the Rb molecule (Lees et al. 1991). However, the precise mechanism by which phosphorylation of specific amino acid residues influences Rb function remains to be determined.

Recently, we have characterized the purified, recombinant human retinoblastoma protein, $p110^{RB}$, expressed from *Escherichia coli*. Purified $p110^{RB}$ was found to form oligomeric structures when analyzed by nondenaturing polyacrylamide gel electrophoresis (Hensey et al. 1994). Electron microscopy of the purified protein revealed linearly extended, macromolecular structures. These properties require the amino terminus of the retinoblastoma protein, since an amino-terminal truncated form, $p56^{RB}$, could not oligomerize. Using the yeast two-hybrid system, we confirmed that Rb could associate with itself through interactions between its amino and carboxyl termini.

Given this background, we have proposed the "corral" hypothesis as a mechanism for Rb to execute its function in regulating cell proliferation and suppressing tumor growth. The hypothesis states that hypophosphorylated Rb protein in vivo may form macromolecular complexes that corral and compartmentalize certain nuclear proteins necessary for cell cycle progression (Lee et al. 1991). Correlates of the hypothesis predict the following points: (1) Rb complexes should bind to several cellular proteins and regulate their function; (2) they should bind to different motifs in individual proteins with different affinities; (3) many Rb-associated proteins, with different specific functions, should coordinate their functions in a temporal manner; and (4) the corral or nuclear subcompartment created by the Rb complexes should dissociate and release proliferation-promoting proteins when Rb is phosphorylated.

In this paper, we describe several pieces of evidence to support and strengthen the corral hypothesis. First, we demonstrate the ability of the Rb protein to bind specifically but with different affinities to a group of proteins identified by the yeast two-hybrid system and the "Rb-sandwich" method, as described previously (Shan et al. 1992; Durfee et al. 1993). These proteins may function either during the G_1/S transition or during M-phase progression. Second, we show that the oligomerized form of Rb protein binds to an associated protein and that phosphorylation of Rb protein by Cdks leads to a significant attenuation of the oligomerizing property, suggesting that phosphorylation may serve to control the oligomerization. These results, together with those described previously, provide convincing evidence that Rb protein regulates other nuclear proteins in a unique, specific, and coordinated manner. Subcompartments created by the complexes of Rb protein with other nuclear proteins are postulated to account for phenomena observed in human cells and animals.

MATERIALS AND METHODS

Purification of human recombinant retinoblastoma protein. $p110^{RB}$ and $p56^{RB}$ were purified by conventional chromatography from an *E. coli* expression system (Hensey et al. 1994). Full-length Rb protein was also obtained from a baculovirus-based system, as described previously (Wang et al. 1990a). p48, expressed and purified from *E. coli*, is described elsewhere (Qian et al. 1993). Protein concentrations were determined according to the method of Bradford (1976).

Nondenaturing PAGE and immunoblot analysis. For nondenaturing gel electrophoresis, a gel containing 6% acrylamide, 0.16% BIS, and 123 mM Trisglycine (pH 8.7) was set at 4°C for 24 hours. Typically, Tris-HCl (pH 8.67) and glycerol were added to each protein sample to final concentrations of 100 mM and 10%, respectively, incubated for 20 minutes on ice, loaded, and electrophoresed for 16 hours at 4°C using a constant 100 V setting. We have subsequently found that the addition of 0.002% SDS to the protein sample, followed by incubation for 1 hour on ice, facilitates the reproducibility of the multiple band pattern of the full-length Rb protein. To perform immunoblot analysis, the electrophoresed gel was incubated in a 1% SDS solution for 30 minutes, after which samples were transferred onto nitrocellulose paper by electroblotting for 20 hours at 4°C. A constant 200 mA current was used in a buffer composed of 49 mM Tris-HCl glycine buffer (pH 8.9). The SDS-PAGE and immunoblot analysis were performed as described previously (Hensey et al. 1994).

Electron microscopy. To perform the immunogold labeling experiment, purified Rb protein (0.1 mg/ml) in (20 mM $NaPO_4$ [pH 7.0], 1 mM β-mercaptoethanol, 1 mM EDTA) was placed on a carbon/formvar coated grid. The grid was incubated for 10 minutes with BSA

(0.1 mg/ml) in HN buffer (10 mM HEPES [pH 7.4], 100 mM NaCl), then incubated with the first antibody (1:20 dilution) in HN containing BSA (0.1 mg/ml) for 10 minutes, washed 5 times in HN, incubated with the second antibody (1:20 dilution) in HN with BSA (0.1 mg/ml) for 10 minutes, washed 5 times in HN, incubated in a staining solution (1 part saturated uranyl acetate solution, 9 parts dd H_2O), and finally viewed with an electron microscope (Hitachi).

Yeast two-hybrid system. The yeast strain Y153 (*MATa leu2-3,112, ura3-52, trp1-901, his3-D200, ade2-101, gal4D, gal80D, URA3::GAL-lacZ, LYS2::GAL-HIS3*) was used for all experiments. A cDNA encoding the Rb carboxyl terminus (301–928) and the corresponding regions of p107 were cloned in frame with sequences for the Gal4 DNA-binding domain present on the expression vector pAS1, to create pASRB-1 and pASp107-1. A second expression plasmid, pSE11, containing the Gal4 *trans*-activation domain, was originally used to create a cellular cDNA library, and positive clones were initially isolated from this library. The yeast strain Y153 was cotransformed with the above two plasmids. Cotransformants were assayed for their ability to activate transcription of the *lacZ* gene, and the resulting β-galactosidase activity was measured by the colony lift method and quantitated using a chlorophenol red-β-D-galactopyranoside (CPRG) assay, as described previously (Durfee et al. 1993).

Immunofluorescence and image analysis. CV1 cells were fixed in cold methanol for 10 minutes at −20°C. All antibody and washing solutions were made in 0.1 M Tris-HCl (pH 7.4), 0.15 M NaCl, 0.1% Tween-20, 5% (w/v) dry milk, and 0.1% sodium azide. MAb 11D7 was used to label Rb protein and secondarily detected by goat anti-mouse IgG conjugated to Texas Red. Confocal optical sections were obtained with a Zeiss 310 LSM, equipped with argon and helium-neon lasers. A Z-series (0.25-μm steps) of digital images was processed for 3-D reconstruction first by kernel filtering ($N = 3$) the raw data with VoxelMath 2.0 (VitalImages, Inc., Fairfield, Iowa), using an average smoothing operation. A pseudo color table (red = intense fluorescence; blue = low fluorescence), opacity table (opacity made inverse to intensity), and reflective lighting were used to accentuate differences in the amount of Rb protein present throughout the nucleus. A seed function algorithim was used from various X, Y, Z points to identify subvolumes of Rb concentration based on thresholding and adjacency, which were pseudo-colored in red. Voxels below an 80% threshold and not adjacent were subtracted from the volume.

RESULTS

A major biochemical activity of the hypophosphorylated forms of the Rb protein is the formation of stable complexes with a group of cellular proteins and the transforming proteins of several related DNA tumor viruses. The T antigen of SV40, the E1A protein of adenovirus type 5, and the E7 protein of human papillomavirus type 16 all bind to two domains of Rb formed in its carboxy-terminal half (DeCaprio et al. 1988; Whyte et al. 1988; Dyson et al. 1989; Chellappan et al. 1992). These interactions between Rb and tumor virus oncoproteins may be important to the cellular function of Rb, since many naturally occurring Rb mutants are clustered at this region, and the transforming functions of the viral oncoproteins bind specifically to this region. We have employed three different approaches, namely, Rb affinity chromatography (Qian et al. 1993), the Rb-sandwich method (Shan et al. 1992), and the yeast two-hybrid system (Durfee et al. 1993), to identify a set of cellular proteins that also bind to this carboxy-terminal half of the Rb protein.

Rb Binds to Many Cellular Proteins with Different Affinities

Using Rb-affinity chromatography, we have identified about 10 distinct spots on a two-dimensional gel (Lee et al. 1991; Qian et al. 1993). Among them, p48 has been molecularly cloned and shown by immunoprecipitation to have Rb-binding activity (Qian et al. 1993). Identification of Rb-associating proteins by direct screening using the Rb-sandwich method has isolated 10 different genes. Among them, E2F-1 and mitosin have been extensively characterized (Shan et al. 1992; X. Zhu et al., in prep.). By the yeast two-hybrid system method, an additional 25 distinct genes have been isolated. Among them, pp1α, *a*cidic *R*b-associated *p*rotein (ARP) (P.-L. Chen et al., in prep.), a human homolog of yeast $nuc2^+$ (H-nuc) (P.-L. Chen et al., in prep.), and many other clones have been characterized. This set of genes, together with clones previously identified by clever guessing, makes at least 50 of the cellular proteins that can interact with Rb either in vitro or in vivo. However, because methods to determine association are based on a single biochemical property, namely the ability of two proteins to bind qualitatively, the biological relevance of such association has not yet been established. The significance of the interaction is not easy to resolve, since there are limited biological assays for the functional activity of Rb protein. To help determine the biological relevance of Rb-associated proteins, quantitative assessment of a potential interaction, together with knowledge of the relative in vivo concentrations of the proteins, should be useful.

We have used both in vitro biochemical methods and the two-hybrid system to determine the binding affinities between the Rb protein and its associated proteins. Six Rb-associated proteins, including a truncated SV40 T antigen (amino acids 1–273), and five cellular proteins were quantitatively tested for their ability to bind in the two-hybrid system and in vitro. In both assays, a consistent order in the strength of binding to the Rb protein emerged. In fact, this relationship is linear when plotted logarithmically (T. Durfee et al., in prep.). Thus, for the proteins tested, the results obtained in vitro reinforce those obtained in yeast and indicate that the strength of binding measured in the two-hybrid assay reflects the intrinsic ability of these proteins to interact in a consistent manner.

Different Proteins Bind to Rb with Distinct Motifs

We have used the above methods to measure the β-galactosidase activity as a convenient indirect means of assessing the relative affinity of Rb for its associated proteins. As shown in Figure 1A, after taking experimental deviation into account, at least three different categories of binding strength can be identified and defined. First there are the strong binders, with β-gal activity greater than 400 units. These include T antigen, ARP, E2F-1, and seven other novel gene products. The second group of intermediate binders has β-gal activity in the range of 50–400 units. This group includes H-nuc and 8 other proteins. The third group consists of relatively weak binders, with β-gal activity between 5 and 50 units. It includes pp1α and 14 other gene products. Interestingly, all these cellular proteins, regardless of

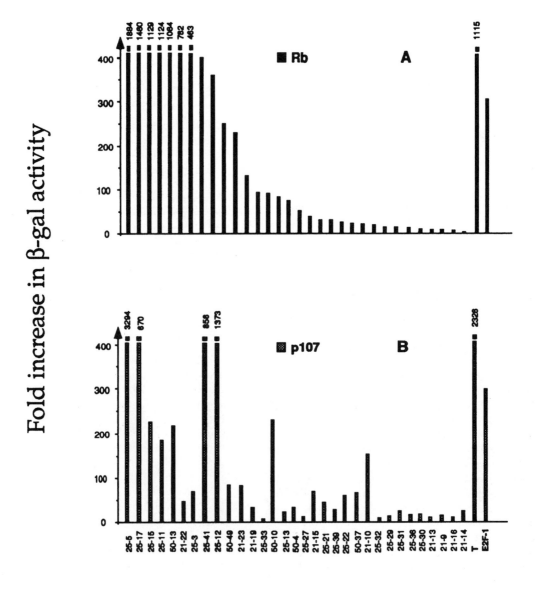

Figure 1. Rb and a member of the Rb family of proteins interact with many cellular proteins, and the interactions occur with different affinities. Rb-binding proteins were found using the yeast two-hybrid system, as explained in the text and in Durfee et al. (1993). The increase in β-gal activity (vertical axis) above baseline represents a quantitative measure of the relative binding affinities of various associated protein clones (horizontal axis) for Rb protein (A) and p107 (B). Binding affinities can be divided into strong (>400 units of β-gal activity), intermediate (50–400 units), and weak (5–50 units) categories. A unit of β-gal activity is defined as the fold increase in β-galactosidase activity as measured by OD_{600} in the CRPG assay (Durfee et al. 1993). The β-gal axis is discontinuous in order to illustrate the large differences in affinity of Rb-associated proteins for p107.

their differing affinities, bind to Rb in the region similar to that to which the T antigen binds. Nonetheless, these associated proteins use a variety of motifs to bind. For example, ARP binds via its carboxy-terminal L-x-C-x-E motif, which is identical to that of T antigen or E1A. E2F-1, in contrast, uses a unique 18-amino-acid motif to interact with Rb; H-nuc binds to Rb with its entire TPR domain; and pp1α uses a different, large, carboxy-terminal region. The diversity of protein motifs used partially explains the different affinities with which the cellular proteins interact with Rb. In fact, however, even if proteins use the same L-x-C-x-E motif to bind to Rb, their binding activities can be different. We have used T-antigen mutants with single amino acid substitutions adjacent to the L-x-C-x-E motif to test the importance of the amino acids surrounding the motif in contributing to the binding affinity. A gradient of β-galactosidase activity of between 3.5× and 17.5× was observed with these T mutants (T. Durfee et al., in prep.). This indicates the wide variation in β-gal activity determined in the yeast two-hybrid system can be the result of the altered affinities of identical individual motifs for the Rb protein, conferred by mutation of sequences flanking the L-x-C-x-E core.

Many Rb-associated Proteins Function at Discrete Time Points during Cell Cycle Progression

We have further characterized the potential biological function for several Rb-associated proteins. One of these clones has been identified as E2F-1, a cellular transcriptional factor that regulates many cellular genes functioning during the G_1/S transition. Ongoing characterization of eight clones among the other three dozen isolated so far has revealed an interesting picture. At least three gene products, pp1α, mitosin, and H-nuc, have roles in M-phase progression, whereas three other genes (E2F-1, *h*ighly *e*xpressed in *c*ervical carcinoma or Hec [Y. Chen and W.-H. Lee, unpubl.], and ARP) may function during the G_1/S transition.

To test the specificity of the interactions of these proteins with Rb, we have used an Rb-related protein, p107 (Ewen et al. 1991), which shares some sequences with Rb in T-binding domains, to perform similar binding assays in the yeast two-hybrid system. As shown in Figure 1B, many of the proteins isolated by association with Rb fail to interact strongly with p107. The T antigen appears to bind both proteins with reasonably high affinity. These results suggest that Rb interacts with a different set of proteins than p107 and may perform cellular activities quite distinct from those of p107.

Purified p110^{RB} Forms Microscopic Filaments In Vitro

When retinoblastoma protein expressed in baculovirus purified was analyzed by electron microscopy, linearly extended, macromolecular structures were observed (Fig. 2), as previously reported (Lee et al. 1991;

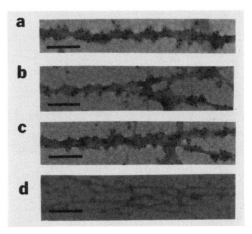

Figure 2. Linearly extended structures formed by purified Rb protein. Immunogold labeling was performed on structures observed with purified full-length Rb protein, expressed via a baculovirus-based system. A monoclonal anti-Rb antibody (Ab245, 0.10 mg/ml, panels *a–c*) or an anti-p53 antibody (Ab122, 0.11 mg/ml, panel *d*) was used as the primary antibody. For the second antibody, anti-mouse IgG goat antibody conjugated to 5 nm gold particles (Amersham) was used. Bars: (*a–c*) 133 nm; (*d*) 175 nm. Linearly extended structures are labeled specifically by anti-Rb antibodies but not by anti-p53 antibodies. (Adapted from Hensey et al. 1994.)

Hensey et al. 1994). The protein was placed directly on carbon/formvar-coated copper grids, negatively stained with uranyl acetate, and viewed with an electron microscope. Immunogold labeling using an anti-Rb antibody, Ab245, shows numerous gold particles bound, often in an undulating or spiral pattern along the structures, indicating that the macromolecular structures contain the Rb protein (Fig. 2, a–c). When the experiment was repeated without adding Rb protein, the labeled structures did not appear, indicating that the structures do not originate from any of the reagents used in the immunogold-labeling experiment. The linearly extended structures were not detected either when purified p56^{RB} was examined, which suggests that the amino-terminal portion of Rb is required for the formation of macromolecular filaments. Neither an anti-p53 antibody (Fig. 2d), nor an antibody directed against SV40 large T antigen (data not shown) labeled the oligomeric structures. These data suggest that the observed structures consisted mainly of the Rb protein and were not artifacts.

Rb Oligomers Retain the Ability to Bind an Associated Protein

To test whether the oligomeric forms of Rb retained the ability to bind to an associated protein, a purified Rb-associated protein, p48 (Qian et al. 1993), was incubated with the full-length Rb protein in vitro and the mixture was separated by nondenaturing PAGE. Multiple bands migrating at positions different from the Rb oligomers appeared along with a diminished p48 band (Fig. 3, lane 3). To demonstrate that these differ-

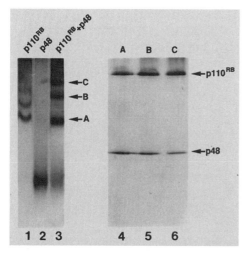

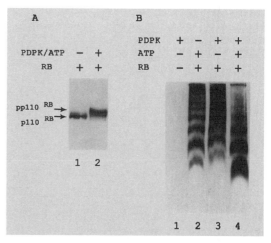

Figure 3. Multimeric complexes formed between Rb and its associated protein, p48. Three samples, including purified full-length Rb protein (0.3 mg) expressed using a baculovirus-based system (lane *1*), purified p48 protein (0.3 mg) (Qian et al. 1993) expressed in *E. coli* (lane *2*), and a mixture containing full-length Rb and p48 (0.3 mg each) proteins (lane *3*), were separated by nondenaturing PAGE and stained with silver. Bands marked A, B, and C, shown in lane *3*, were excised from duplicate gels, incubated in SDS-sample buffer, electrophoresed through a 7.5% SDS-polyacrylamide gel, and stained again with silver (lanes *4–6*). Letters above lanes *4–6* correspond to the bands in lane *3* that were excised.

Figure 4. Phosphorylation of full-length Rb protein by PDPK diminishes Rb's oligomerizing property. (*A*) PDPK-kinased, full-length Rb protein exhibits slower mobility on an SDS/polyacrylamide gel, and is characteristic of phosphorylated Rb species. The kinase reaction was performed by incubating purified, full-length Rb protein, expressed in a baculovirus-based system, with purified PDPK (provided by F. Hall), in a buffer composed of 100 mM Tris-acetate (pH 7.5), 20 mM $MgCl_2$, and 0.2 mM ATP, at 30°C for 1.5 hr. After the kinase reaction, the samples were separated by 7.5% SDS-PAGE and analyzed by Western blotting using Ab245 as the primary antibody. Untreated Rb protein (0.2 mg) (lane *1*); PDPK-kinased Rb protein (0.2 mg) (lane *2*). (*B*) Analysis of PDPK-kinased Rb protein by nondenaturing PAGE. The following components were added to each reaction sample: PDPK alone (lane *1*); Rb protein and ATP (lane *2*); Rb protein and PDPK (lane *3*); Rb protein, PDPK, and ATP (lane *4*). Where stated, 0.5 µg of purified, full-length Rb protein, expressed in a baculovirus-based system, or 0.2 mM of ATP was added. The kinase reaction was performed as described in *A*. After the reaction, the pH of each sample was adjusted to 8.5 prior to the electrophoresis. The electrophoresed gel was analyzed by Western blotting using Ab245.

ent bands were due to complexes formed by p48 and Rb proteins, the bands marked A–C in lane 3 were excised from duplicate gels and analyzed by SDS/PAGE. The results shown in lanes 4–6 clearly show the presence of both p48 and Rb proteins in each band excised. This finding suggests that the oligomeric forms of Rb retain their ability to associate with p48, an indication that oligomers are still capable in this state of interacting with other proteins.

In Vitro Phosphorylation of Rb Diminishes Its Oligomerization

The effect of phosphorylation on oligomerization of the Rb protein was determined by using a purified proline-directed protein kinase (PDPK), which consists of cdc2 and cyclin A, to convert the Rb protein into its phosphorylated forms. Phosphorylated Rb migrates more slowly when analyzed by SDS-PAGE (Fig. 4A, lane 2). Rb protein incubated with ATP alone (Fig. 4B, lane 2), or with PDPK alone in the absence of ATP (Fig. 4B, lane 3), showed the typical oligomerization pattern, with at least six higher order bands. When the phosphorylated Rb protein was subjected to analysis by nondenaturing PAGE, significantly reduced amounts of Rb oligomer were obtained, with three faster-migrating bands of greatly increased intensity being observed. Although phosphorylation of Rb protein by PDPK did not completely abolish the oligomerization, this may have resulted from incomplete phosphorylation. Alternatively, phosphorylation by additional kinases may be required to disrupt the oligomeric structure completely. The increased mobilities of certain Rb protein forms (compare lane 4 with lanes 2 and 3 in Fig. 4B) are probably due to additional negative charges gained by the Rb protein after phosphorylation. These results demonstrate that phosphorylation of the full-length Rb protein may regulate the oligomerization process. In addition to specific primary sequence requirements for the formation of oligomers, conformational properties of the protein are probably essential. Thus, phosphorylation of Rb protein, which likely causes a conformational change, seems to be a plausible mechanism for the attenuation of oligomerization. Regulation via phosphorylation has been reported for various other proteins which form oligomeric structures, including nuclear lamins (Heald and McKeon 1990).

This result is consistent with what we have observed using sucrose gradient centrifugation to determine whether Rb forms macromolecular structures in the nucleus. We have previously shown that the differentially phosphorylated isoforms of human Rb protein migrate between the 105-kD and 114-kD positions in SDS-polyacrylamide gels, with the form lacking phosphate groups migrating the fastest (Lee et al. 1991). Hypophosphorylated Rb sedimented at the highest molecular weight in sucrose gradients, between 180 kD and 250 kD. This analysis, however, was limited to the soluble fraction of Rb protein. Insoluble fractions containing the Rb protein may represent larger molecular complexes. The apparent molecular weight of native Rb protein appears to be inversely correlated with its degree of phosphorylation. We interpret these findings as evidence that hypophosphorylated Rb protein, and possibly partially phosphorylated forms, participate in high-molecular-weight complexes in the cell, whereas Rb phosphorylated maximally exists mainly as monomers.

Compartmentation of Rb Protein in the Nucleus as Analyzed by Confocal Immunofluorescence Microscopy

We have used a specific monoclonal antibody to localize Rb protein in the nuclei of cultured CV1 cells. Digitization of confocal optical sections, obtained by laser scanning fluorescent microscopy, followed by three-dimensional reconstruction (i.e., volume rendering), reveals that Rb is unevenly distributed in the nucleus (Fig. 5A). In an effort to determine if there is a higher-order Rb concentration gradient, we used a "seed volume" algorithm (originating at a Rb hot spot within the nucleus) to extract a subvolume (Fig. 5B). This substructure, representing pixels with values of at least 80% of the threshold intensity of the original seed, and immediately adjacent in the X, Y, or Z directions, shows that an anastamosing network of relatively high concentrations of Rb protein exists within the nucleus. Similar images were obtained from hot spots elsewhere in the nucleus. This provides a heretofore unseen view of how Rb protein is organized in vivo and suggests that a mechanism is in place to provide geometric partitioning. The best candidate that would provide this organization is the nuclear matrix, the insoluble framework that remains after detergent and high salt extraction of most cellular proteins (Berezney and Coffey 1975; Nickerson and Penman 1992).

Using immunogold microscopy, Rb protein has been shown to be differentially clustered on the dense, fibrogranular assemblies of the nuclear matrix (Mancini et al. 1994). These assemblies are considered to be the structural location of RNA processing and of DNA replication. Indeed, the nuclear compartmentation of Rb, in terms of both spatial distribution and solubility, is highly complex. The association of Rb with the nuclear matrix supports the hypothesis that Rb can corral other proteins into nuclear subcompartments. We are now only beginning to identify the complexities of such compartments.

DISCUSSION

As suggested by some sequence similarity to the intermediate filament family of proteins, of which the nuclear lamins are members, Rb protein can polymerize into microscopic filaments in vitro. We have also shown that phosphorylation of the retinoblastoma protein by a proline-directed protein kinase, a member of the cdc2 kinase family, diminishes the capacity of Rb protein to oligomerize. Oligomeric forms of the retino-

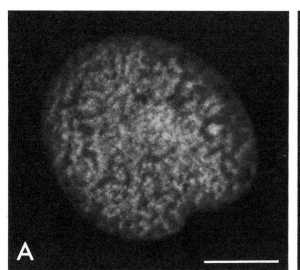
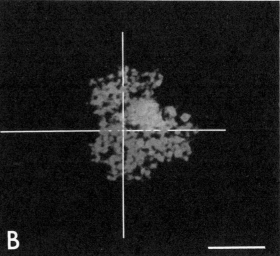

Figure 5. Rb distributes heterogeneously throughout the nucleus. Laser scanning confocal immunofluorescence was used to identify the localization of Rb in the nucleus. Monoclonal antibody 11D7 was used to immunolabel Rb in methanol-fixed, normal monkey kidney CV1 cells. (A) A Z-series of optical sections was reconstructed (volume rendered) using psuedo-color and opacity tables, as well as reflective lighting, in order to highlight high concentrations of fluorescent intensity (Rb concentration). Note the numerous "hot spots" (red) throughout the nucleus. Bar, 5 μm. (B) To visualize the organization of Rb concentration, a seed function alogrithm was used to extract an interconnected subvolume of Rb fluorescence that is based on an 80% threshold and direct adjacency to an X,Y, and Z seed point (cross hairs). The subvolume image illustrates that a high degree of order exists for the spatial distribution of Rb in vivo. Bar, 5 μm.

blastoma protein retain the ability to bind to a Rb-associated protein, p48. In addition, the Rb protein forms complexes with many different cellular proteins that may function in cell cycle progression. These proteins bind to Rb via distinct motifs and with differing affinities. Moreover, Rb is organized into "connected" clusters in the nucleus and is linked to a distinct portion of the nuclear matrix. On the basis of these findings and previous discoveries, we propose a model in which the Rb protein acts as a negative regulator of cell proliferation by sequestering or "corralling" into inactive subnuclear structures or compartments a variety of proteins that function to promote proliferation. This corral may wholly or partially disassemble upon phosphorylation of the Rb subunits. In the intact cell, there may be a dynamic equilibrium between Rb multimers, monomers, and all states in between. Regulation by phosphorylation may drive the equilibrium to one extreme or the other and thereby alter the contents and structure of the corral.

Several structural proteins are known to be intimately involved in the control of cell growth. For instance, vinculin, a protein which may tether cytoskeletal filaments to anchoring proteins in the cell membrane, appears to be a primary target of the pp60^{v-src} protein kinase (Jove and Hanafusa 1987). The activity of this oncogenic kinase is known to contribute to neoplastic transformation of susceptible cells and may do so by altering normal cytoskeletal structure. Another example is histone H1, the linker histone in eukaryotic chromatin; it can repress the transcription and, upon being phosphorylated, affect the chromatin condensation required for progression through the mitotic phase of the cell cycle (Bradbury et al. 1974). In addition, the nuclear lamins, to which a portion of Rb is related, form much of the peripheral nuclear matrix. The matrix is the subnuclear structure which organizes transcription, RNA trafficking, and many other nuclear transactions (Berezney and Coffey 1975; Ciejek et al. 1983; Zeitlin et al. 1989). It is worthwhile to point out that the ability of the nuclear lamins to perform their structural functions, in a manner similar to those of vinculin and histone H1, is regulated by phosphorylation (Heald and McKeon 1990). The oligomerization of the Rb protein appears likewise to be regulated by phosphorylation, as described in this paper.

Recently, a novel nuclear matrix protein, p84, was identified by its binding to the amino-terminal 300-amino-acid region of Rb. p84 is thus far the only protein found to interact specifically with Rb at this region. Unlike Rb, it associates with the nuclear matrix throughout interphase (Durfee et al. 1994). In fact, p84 colocalizes with many Rb hot spots to intranuclear speckles associated with RNA processing. It is possible that Rb oligomers can bind to p84, as well as to other proteins, and anchor them directly at the nuclear matrix. Indeed, several Rb-binding viral oncoproteins have an association with the nuclear matrix (Chatterjee and Flint 1986; Deppert and von der Weth 1990; Greenfield et al. 1991). Another possibility is that the Rb

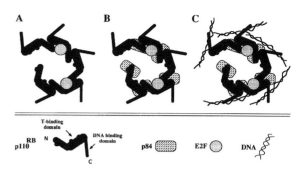

Figure 6. Schematic diagram illustrating the corral hypothesis by which Rb may regulate other proteins that promote proliferation. (*A*) A part of the nuclear pool of hypophosphorylated p110^{RB} forms multimers, apparently via an interaction between an amino-terminal domain and the T-antigen-binding domains. Rb multimers bind to and "corral" into an inactive compartment factors such as E2F that promote cell proliferation. (*B*) Complexes consisting of Rb multimers and corralled proteins may also bind to nuclear matrix proteins such as p84. Rb thereby may serve to bring associated proteins to important subnuclear compartments (e.g., RNA processing and splicing centers) and anchor them there. (*C*) The DNA-binding property of Rb protein may also bring corralled proteins in proximity to DNA, where transcription factors and other proliferation-promoting proteins eventually function if external signals and phosphorylation events release them from the Rb corral. The Rb corral may also serve to compartmentalize and *activate* other proteins important for *restricting* progression of the cell cycle.

multimer will bind to DNA with its carboxy-terminal 150-amino-acid region. As illustrated in Figure 6, Rb, by itself or by interacting with cellular proteins such as p84, or with DNA, may then form specific compartments for sequestering transcription factors and other cellular proteins. Cdk-induced phosphorylation of Rb could then lead to disassembly of Rb complexes and release factors that promote cell proliferation.

To date, many results have accumulated to suggest that a function of Rb is to restrict entry into committed phases of the cell cycle. Rb may also function to promote cells to differentiate. An important question remains: How does Rb normally achieve its function to restrict passage through G_1 phase? As demonstrated in the results described above, it may do so by regulating an array of cellular proteins that function in cell cycle progression. It must regulate this set of proteins in a coordinate manner, precisely at a limited time and in a limited space, since the diverse group of Rb-associated, growth-promoting proteins will subsequently branch out to control different pathways. Cells that lose functional Rb protein, and for which no suitable redundant protein is available, will cycle constantly and fail to exit from the cycle appropriately.

According to the corral hypothesis, Rb regulates many associated proteins by a novel mechanism, i.e., by sequestering them in specific subnuclear compartments. Phosphorylation of a limited target substrate, Rb protein, can disrupt the macromolecular "corral" and allow the released proteins to perform their collective functions at a narrow, critical window of opportunity created by appropriate external signals. This mode of regulation can respond efficiently to the external signals without de novo protein synthesis and the need for new transcription; thus, it can be altered nearly instantaneously if necessary. Activation of proteins that promote proliferation is effected in many instances by specific kinases which recognize and phosphorylate several different substrates with similar motifs. We suggest that kinases might also recognize certain exposed residues on critical coordinating substrates, such as Rb macromolecular complexes, to release and effectively activate corralled Rb-associated proteins. Such regulation would be similar in lag time and tight control to phosphorylation and dephosphorylation of the corralled proteins themselves, but might be more energy efficient, since fewer phosphorylations would be required to achieve similar results. In addition, since the phosphorylation of Rb protein in vivo is known to occur sequentially (Chen et al. 1989; DeCaprio et al. 1989; Furukawa et al. 1990), a portion of the corral that Rb sets up in G_1 phase may not be completely disrupted by the initial G_1 phosphorylation or phosphorylations. Such partially opened parts of the corral may still serve to bind certain molecules so that they can be coordinately regulated at other times or other places within the nucleus.

Another example of similar regulation can be seen in transcriptional control by specific transcription factors. E2F-1, for example, will turn on several genes during the G_1/S transition because their promoters contain functional E2F-1 recognition sites. This type of regulation, however, unlike those produced by phosphorylation/dephosphorylation and the Rb corral, requires time for de novo protein synthesis and therefore cannot adapt as quickly as the regulation detailed above. Nonetheless, the different modes by which protein function is regulated together serve to achieve similar specific missions in a coordinate fashion. By these modes of regulation, cells can efficiently respond by proliferation or differentiation to changes in the external environment, if all their parts are intact and expressed in a timely fashion. When crucial regulatory molecules like the retinoblastoma protein are missing or altered in certain cells, coordination of programs and signals to divide or differentiate can go awry, as they do in many cancer and embryonic cells containing only mutated *RB* genes.

ACKNOWLEDGMENTS

This work was supported by grants from the National Institutes of Health and the Council for Tobacco Research. The technical assistance of Y.-W. Qian is also appreciated.

REFERENCES

Berezney, R. and D.S. Coffey. 1975. Nuclear protein matrix: Association with newly synthesized DNA. *Science* **189:** 291.

Bignon, Y.-J., Y. Chen, C.-Y. Chang, D.J. Riley, J.J. Windle, P.L. Mellon, and W.-H. Lee. 1993. Expression of a retinoblastoma transgene results in dwarf mice. *Genes Dev.* **7:** 1654.

Bookstein, R. and W.H. Lee. 1991. Molecular genetics of the retinoblastoma suppressor gene. *Crit .Rev. Oncog.* **2:** 211.

Bookstein, R., J.Y. Shew, P.-L. Chen, P. Scully, and W.-H. Lee. 1990. Suppression of tumorigenicity of human prostate carcinoma cells by replacing a mutated *RB* gene. *Science* **247:** 712.

Bradbury, E.M., R.J. Inglis, and H.R. Matthews. 1974. Control of cell division by very lysine rich histone (f1) phosphorylation. *Nature* **247:** 257.

Bradford, M.M. 1976. A rapid and sensitive method for the quantitation of microgram quantities of protein utilizing the principle of protein-dye binding. *Anal. Biochem.* **72:** 248.

Buchkovich, K., L.A. Duffy, and E. Harlow. 1989. The retinoblastoma protein is phosphorylated during specific phases of the cell cycle. *Cell* **58:** 1097.

Cavenee, W.K., T.P. Dryja, R.A. Phillips, W.F. Benedict, and R. Godbout. 1983. Expression of recessive alleles by chromosomal mechanisms in retinoblastoma. *Nature* **305:** 779.

Chang, C.-Y., D.J. Riley, E.Y.-H. Lee, and W.-H. Lee. 1993. Quantitative effects of retinoblastoma gene on mouse development and tissue-specific tumorigenesis. *Cell Growth Differ.* **4:** 1057.

Chatterjee, P.K. and S.J. Flint. 1986. Partition of E1A proteins between soluble and structural fractions of adenovirus-infected and -transformed cells. *J. Virol.* **60:** 1018.

Chellappan, S., V.B. Kraus, B. Kroger, K. Munger, P.M. Howley, W.C. Phelps, and J.R. Nevins. 1992. Adenovirus E1A, simian virus 40 tumor antigen, and human papillomavirus E7 protein share the capacity to disrupt the interaction between transcription factor E2F and the retinoblastoma gene product. *Proc. Natl. Acad. Sci.* **89:** 4549.

Chen, P.-L., Y. Chen, B. Shan, R. Bookstein, and W.-H. Lee. 1992. Stability of retinoblastoma gene expression determines the tumorigenicity of reconstituted retinoblastoma cells. *Cell Growth Differ.* **3:** 119.

Chen, P.-L., P. Scully, J.-Y. Shew, J.Y.-J. Wang, and W.-H. Lee. 1989. Phosphorylation of the retinoblastoma gene product is modulated during the cell cycle and cellular differentiation. *Cell* **58:** 1193.

Ciejek, E.M., M.J. Tsai, and B.W. O'Malley. 1983. Actively transcribed genes are associated with the nuclear matrix. *Nature* **306:** 607.

Clarke, A.R., A.R. Maandag, M. van Roon, N.M.T. van der Lugt, M. van der Valk, M.L. Hooper, A. Berns, and H. te Riele. 1992. Requirement of a functional *Rb-1* gene in murine development. *Nature* **359:** 328.

DeCaprio, J.A., J.W. Ludlow, D. Lynch, Y. Furukawa, J. Griffin, W.H. Piwnica, C.M. Huang, and D.M. Livingston. 1989. The product of the retinoblastoma susceptibility gene has properties of a cell cycle regulatory element. *Cell* **58:** 1085.

DeCaprio, J.A., J.W. Ludlow, J. Figge, J.-Y. Shew, C.-M. Huang, W.-H. Lee, E. Marsillo, E. Paucha, and D.M. Livingston. 1988. SV40 large tumor antigen forms a specific complex with the product of the retinoblastoma susceptibility gene. *Cell* **54:** 275.

Defeo-Jones, D., P.S. Huang, R.E. Jones, K.M. Haskell, G.A. Vuocolo, M.G. Hanobik, H.E. Huber, and A. Oliff. 1991. Cloning of cDNAs for cellular proteins that bind to the retinoblastoma gene product. *Nature* **352:** 251.

Deppert, W. and A. von der Weth. 1990. Functional interaction of nuclear transport-defective simian virus 40 large T antigen with chromatin and nuclear matrix. *J. Virol.* **64:** 838.

Durfee, T., M.A. Mancini, D. Jones, S.J. Elledge, and W.-H. Lee. 1994. The amino-terminal region of the retinoblastoma gene product binds a novel nuclear matrix protein that localizes to RNA processing domains. *J. Cell Biol.* **127:** 609.

Durfee, T., K. Becherer, P.-L. Chen, S.-H. Yeh, Y. Yang, A.E. Kilburn, W.-H. Lee, and S.J. Elledge. 1993. The retinoblastoma protein associates with the type 1 protein phosphatase catalytic subunit. *Genes Dev.* **7:** 555.

Dyson, N., P.M. Howley, K. Munger, and E. Harlow. 1989. The human papilloma virus-16 E7 oncoprotein is able to bind to the retinoblastoma gene product. *Science* **243:** 934.

Ewen, M.E., Y.G. Xing, J.B. Lawrence, and D.M. Livingston. 1991. Molecular cloning, chromosomal mapping, and expression of the cDNA for p107, a retinoblastoma gene product-related protein. *Cell* **66:** 1155.

Friend, S.H., R. Bernards, S. Rogelj, R.A. Weinberg, J.M. Rapaport, D.M. Albert, and T.P. Dryja. 1986. A human DNA segment with properties of the gene that predisposes to retinoblastoma and osteosarcoma. *Nature* **323:** 643.

Fung, Y.K.T., A.L. Murphree, A. T'Ang, J. Qian, S.H. Hinrichs, and W.F. Benedict. 1987. Structural evidence for the authenticity of the human retinoblastoma gene. *Science* **236:** 1657.

Furukawa, Y., J.A. DeCaprio, A. Freedman, Y. Kanakura, M. Nakamura, T.J. Ernst, D.M. Livingston, and J.D. Griffin. 1990. Expression and state of phosphorylation of the retinoblastoma susceptibility gene product in cycling and noncycling human hematopoietic cells. *Proc. Natl. Acad. Sci.* **87:** 2770.

Goodrich, D.W., Y. Chen, P. Scully, and W.-H. Lee. 1992. Expression of the retinoblastoma gene product in bladder carcinoma cells associates with a low frequency of tumor formation. *Cancer Res.* **52:** 1968.

Goodrich, D.W., N.P. Wang, Y.-W. Qian, E.Y.-H.P. Lee, and W.-H. Lee. 1991. The retinoblastoma gene product regulates progression through the G1 phase of the cell cycle. *Cell* **67:** 293.

Greenfield, I., J. Nickerson, S. Penman, and M. Stanley. 1991. Human papillomavirus 16 E7 protein is associated with the nuclear matrix. *Proc. Natl. Acad. Sci.* **88:** 11217.

Harbour, J.W., S.-H. Lai, J. Whang-Peng, A.F. Gazdar, J.D. Minna, and F.J. Kaye. 1988. Abnormalities in structure and expression of the human retinoblastoma gene in SCLC. *Science* **241:** 353.

Heald, R. and F. McKeon. 1990. Mutations of phosphorylation sites in lamin A that prevent nuclear lamina disassembly in mitosis. *Cell* **61:** 579.

Helin, K., J.A. Lees, M. Vidal, N. Dyson, E. Harlow, and A. Fattaey. 1992. A cDNA encoding a pRB-binding protein with properties of the transcription factor E2F. *Cell* **70:** 337.

Hensey, C.E., F. Hong, T. Durfee, Y.-W. Qian, E.Y.-H.P. Lee, and W.-H. Lee. 1994. Identification of discrete structural domains in the retinoblastoma protein: Amino-terminal domain is required for its oligomerization. *J. Biol. Chem.* **269:** 1380.

Hinds, P.W., S. Mittnacht, V. Dulic, A. Arnold, S.I. Reed, and R.A. Weinberg. 1992. Regulation of retinoblastoma protein functions by ectopic expression of human cyclins. *Cell* **70:** 993.

Hollingsworth, R.E., C.E. Hensey, and W.-H. Lee. 1993. Retinoblastoma protein and the cell cycle. *Curr. Opin. Genet. Dev.* **3:** 55.

Horowitz, J.M., D.W. Yandell, S.H. Park, S. Canning, P. Whyte, K. Buchkovich, E. Harlow, R. Weinberg, and T. Dryja. 1989. Point mutational inactivation of the retinoblastoma antioncogene. *Science* **243:** 937.

Hu, N., A. Gutsmann, D.C. Herbert, A. Bradley, W.-H. Lee, and E.Y.-H.P. Lee. 1994. Heterozygous $Rb-1^{\Delta 20}/+$ mice are predisposed to tumors of the pituitary gland with nearly complete penetrance. *Oncogene* **9:** 1021.

Huang, H.J., J.K. Yee, J.Y. Shew, P.-L. Chen, R. Bookstein, T. Friedmann, E.Y.-H.P. Lee, and W.-H. Lee. 1988. Suppression of the neoplastic phenotype by replacement of the RB gene in human cancer cells. *Science* **242:** 1563.

Jacks, T., A. Fazeli, E.M. Schmitt, R.T. Bronson, M.A. Goodell, and R.A. Weinberg. 1992. Effects of an *Rb* mutation in the mouse. *Nature* **359:** 295.

Jove, R. and H. Hanafusa. 1987. Cell transformation by the viral *src* oncogene. *Annu. Rev. Biochem.* **3:** 31.

Kaelin, W.G., W. Krek, W.R. Sellers, J.A. DeCaprio, F. Ajchenbaum, C.S. Fuchs, T. Chittenden, Y. Li, P.J. Farnham, M.A. Blanar, D.M. Livingston, and E.K. Flemington. 1992. Expression cloning of a cDNA encoding a retinoblastoma-binding protein with E2F-like properties. *Cell* **70:** 351.

Knudson, A.G. 1971. Mutation and cancer: Statistical study of retinoblastoma. *Proc. Natl. Acad. Sci.* **68:** 820.

Lee, E.Y.-H.P., H. To, J.-Y. Shew, R. Bookstein, P. Scully, and W.-H. Lee. 1988. Inactivation of the retinoblastoma susceptibility gene in human breast cancers. *Science* **241:** 218.

Lee, E.Y.-H.P., C.-Y. Chang, N. Hu, Y.-C.J. Wang, C.-C. Lai, K. Herrup, W.-H. Lee, and A. Bradley. 1992. Mice deficient for RB are nonviable and show defects in neurogenesis and hematopoiesis. *Nature* **359:** 288.

Lee, W.-H., R. Bookstein, F. Hong, L.J. Young, J.Y. Shew, and E.Y.-H.P. Lee. 1987a. Human retinoblastoma susceptibility gene: Cloning, identification, and sequence. *Science* **235:** 1394.

Lee, W.-H., R.E. Hollingsworth, Jr., Y.-W. Qian, P.-L. Chen, F. Hong, and E.Y.-H.P. Lee. 1991. RB protein as a cellular "corral" for growth-promoting proteins. *Cold Spring Harbor Symp. Quant. Biol.* **56:** 211.

Lee, W.-H., J.Y. Shew, F.D. Hong, T.W. Sery, L.A. Donoso, L.J. Young, R. Bookstein, and E.Y.-H.P. Lee. 1987b. The retinoblastoma susceptibility gene encodes a nuclear phosphoprotein associated with DNA binding activity. *Nature* **329:** 642.

Lees, J.A., K.J. Buchkovich, D.R. Marshak, C.W. Anderson, and E. Harlow. 1991. The retinoblastoma protein is phosphorylated on multiple sites by human cdc2. *EMBO J.* **10:** 4279.

Mancini, M., B. Shan, J. Nickerson, S. Penman, and W.-H. Lee. 1994. The retinoblastoma gene product is a cell-cycle dependent, nuclear-matrix associated protein. *Proc. Natl. Acad. Sci.* **91:** 418.

Mittnacht, S. and R.A. Weinberg. 1991. G1/S phosphorylation of the retinoblastoma protein is associated with an altered affinity for the nuclear compartment. *Cell* **65:** 381.

Nevins, J.R. 1992. E2F: A link between the Rb tumor suppressor protein and viral oncoproteins. *Science* **258:** 424.

Nickerson, J.A. and S. Penman. 1992. The nuclear matrix: Structure and involvement in gene expression. In *Molecular and cellular approaches to the control of proliferation and differentiation* (ed. G. Stein and J. Lian), p. 340. Academic Press, New York.

Qian, Y., C. Luckey, L. Horton, M. Esser, and D.J. Templeton. 1992. Biological function of the retinoblastoma protein requires domains for hyperphosphorylation and transcription factor binding. *Mol. Cell. Biol.* **12:** 5363.

Qian, Y.-W., Y.-C.J. Wang, R.E. Hollingsworth, Jr., D. Jones, N. Ling, and E.Y.-H.P. Lee. 1993. A cDNA encoding a retinoblastoma-binding protein, RbAp48, has properties related to msi1, a negative regulator of *ras* in yeast. *Nature* **364:** 648.

Qin, X.-Q., T. Chittenden, D.M. Livingston, and W.G. Kaelin. 1992. Identification of a growth suppression domain within the retinoblastoma gene product. *Genes Dev.* **6:** 953.

Shan, B., X. Zhu, P.-L. Chen, T. Durfee, Y. Yang, D. Sharp, and W.-H. Lee. 1992. Molecular cloning of cellular genes encoding retinoblastoma-associated proteins: Identification of a gene with properties of the transcription factor E2F. *Mol. Cell. Biol.* **12:** 5620.

Sumegi, J., E. Uzvolgyi, and G. Klein. 1990. Expression of the RB gene under the control of MuLV-LTR suppresses tumorigenicity of WERI-Rb-27 retinoblastoma cells in immunodefective mice. *Cell Growth Differ.* **1:** 247.

Takahashi, P.R., T. Hashimoto, X. Hong-Ji, S.-X. Hu, T. Matsui, T. Miki, H. Bigo-Marshall, S.A. Aaronson, and W.F. Benedict. 1991. The retinoblastoma gene functions as a growth and tumor suppressor in human bladder carcinoma cells. *Proc. Natl. Acad. Sci.* **88:** 5257.

Templeton, D.J., S.H. Park, L. Lanier, and R.T. Weinberg. 1991. Nonfunctional mutants of the retinoblastoma protein are characterized by defects in phosphorylation, viral oncoprotein association, and nuclear tethering. *Proc. Natl. Acad. Sci.* **88:** 3033.

Toguchida, J., K. Ishizaki, M.S. Sasaki, M. Ikenaga, M. Sugimoto, Y. Kotoura, and T. Yamamuro. 1988. Chromosomal reorganization for the expression of recessive mutation of retinoblastoma susceptibility gene in the development of osteosarcoma. *Cancer Res.* **48:** 3939.

Wang, N.-P., Y. Qian, A.E. Chung, W.-H. Lee, and E.Y.-H.P. Lee. 1990a. Expression of the human retinoblastoma gene product pp110^{RB} in insect cells using the baculovirus system. *Cell Growth Differ.* **1:** 429.

Wang, N.-P., P.-L. Chen, S. Huang, L.A. Donoso, W.-H. Lee, and E.Y.-H.P. Lee. 1990b. DNA-binding activity of retinoblastoma protein is intrinsic to its carboxyl-terminal region. *Cell Growth Differ.* **1:** 233.

Whyte, P., K.J. Buchkovich, J.M. Horowitz, S.H. Friend, M. Raybuck, R.A. Weinberg, and E. Harlow. 1988. Association between an oncogene and an anti-oncogene: The adenovirus E1A proteins bind to the retinoblastoma gene product. *Nature* **334:** 124.

Yokota, J., T. Akiyama, Y.-K.T. Fung, W.F. Benedict, Y. Namba, M. Hanaoka, M. Wada, T. Terasaki, Y. Shimosato, T. Sugimura, and M. Terada. 1988. Altered expression of the retinoblastoma (RB) gene in small-cell carcinoma of the lung. *Oncogene* **3:** 471.

Zeitlin, S., R.C. Wilson, and A. Efstradiatis. 1989. Autonomous splicing and complementation of in vivo-assembled spliceosomes. *J. Cell Biol.* **108:** 765.

The Max Transcription Factor Network: Involvement of Mad in Differentiation and an Approach to Identification of Target Genes

P.J. HURLIN, D.E. AYER, C. GRANDORI, AND R.N. EISENMAN
Division of Basic Sciences, Fred Hutchinson Cancer Research Center, Seattle, Washington 98104

The Myc protein family (including c-, N-, and L-Myc) belongs to the larger class of basic-helix-loop-helix-zipper (bHLHZip) transcription factors. Members of this protein class are highly conserved throughout evolution and function in multiple aspects of cell behavior. The involvement of Myc in cell proliferation, apoptosis, differentiation, and oncogenesis has been particularly well documented (for recent reviews, see Dang 1991; DePinho et al. 1991: Marcu et al. 1992). Although we still lack a detailed understanding of how Myc specifically controls these cellular functions, we have over the last several years been able to gain some important insights into the molecular properties of Myc. One of these is the highly specific association of Myc family proteins with the bHLHZip protein Max (for reviews, see Blackwood et al. 1992a; Marcu et al. 1992).

Max is also a member of the bHLHZip protein class but, outside of this region, lacks any significant homology with other known proteins. Max was originally identified by screening a λgt11 protein expression library with the Myc carboxy-terminal bHLHZip domain as a probe (Blackwood and Eisenman 1991). The Max protein forms homodimers and specifically associates with c-, N-, and L-Myc; all these interactions are mediated through the HLHZip regions of both partners in the dimer (Blackwood and Eisenman 1991; Prendergast et al. 1991). An important consequence of dimerization is the ability of Myc:Max heterodimers and Max:Max homodimers to interact with the E-box sequence, CACGTG, as well as other related DNA sequences (Blackwell et al. 1990, 1993). Binding to DNA requires the basic regions of both partners. The recent solution of the X-ray co-crystallographic structure of Max homodimers with their CACGTG-binding sites at 2.9 Å resolution has provided an elegant structural basis for these interactions (Ferre-D'Amare et al. 1993). In the co-crystal, Max homodimers form a parallel four-helix bundle composed of two long helices contributed by each partner. One of these consists of the basic region, which becomes helical upon contact with DNA and is continuous with helix 1. The second extended helix consists of helix 2 and the zipper region. The loop regions, which separate the two long helical segments, also contact the DNA backbone. The overall stability of the structure is derived from interhelical contacts, a hydrophobic core, and the coiled-coil interactions of the two Zip regions. There is every reason to believe that the Myc:Max heterodimeric structures are substantially the same.

Because Myc does not homodimerize at physiological concentrations, it requires heterodimerization with Max in order to interact with DNA. Furthermore, Max, although it readily forms homodimers, appears to preferentially form heterocomplexes with Myc. Both Myc:Max and Max:Max bind the CACGTG sequence, but the heterodimers bind with higher apparent affinity than Max homodimers, as suggested by gel-mobility shift analyses and off-rate measurements (Bousset et al. 1993; Kretzner et al. 1993). Therefore, the picture that has emerged, largely from in vitro studies, is that Max can dimerize and bind DNA alone. However, the addition of Myc, itself incapable of homodimerization, leads to formation of Myc:Max heterodimers with higher DNA-binding affinity.

The Myc:Max heterocomplex functions as a transcriptional activator in transient transfection assays employing reporter genes with promoter-proximal CACGTG-binding sites (Amati et al. 1992; Kretzner et al. 1992). In contrast, Max homodimers appear to be transcriptionally inert, and overexpression of Max leads to repression of transcription from the reporter (Kretzner et al. 1992). These activities are consistent with the findings that Myc contains amino-terminal sequences which can function as transcriptional activators, whereas Max lacks any activation regions (Kato et al. 1990, 1992; Amati et al. 1992). Taken together, these results have led to a model in which transcriptionally inactive Max homodimers are converted into transcriptionally active Myc:Max heterodimers upon synthesis of Myc protein (for review, see Blackwood et al. 1992a).

The idea that Myc acts as the rate-limiting component of the complex is consistent with studies on the biosynthesis of Myc and Max in cells. In vivo, Myc protein has a short half-life ($t_{1/2}$ = 15–30 min), and its synthesis is tightly coupled to cell proliferation (for review, see Lüscher and Eisenman 1990). On the other hand, Max is highly stable and appears to be in excess of Myc. Max levels are constant in both proliferating and nonproliferating cells (Blackwood et al. 1992b). Therefore, newly synthesized, rapidly degraded Myc is the regulated and rate-limiting component of the complex. A plausible hypothesis is that the induction of Myc during mitogenesis leads to conversion of preexisting Max homodimers into Myc:Max heterodimers, capable of activating expression of target genes. In contrast, during differentiation, the shutoff of Myc results in the loss of active complexes.

The models for Myc and Max regulation and transcriptional activity can explain why the deregulation of *myc* expression is frequently associated with oncogenesis. The overexpression and loss of normal control over *myc*, frequently encountered in diverse neoplasms, would be expected to lead to augmented formation of Myc:Max heterocomplexes and an inability to down-regulate the transcriptional activity of the complexes in response to appropriate signals. Domain-swapping experiments in which the dimerization specificities of Myc and Max have been exchanged demonstrate that Max is required for the cooperative transformation of primary cells by deregulated *myc* and *ras*, as well as for *myc*-induced apoptosis (Amati et al. 1993a,b). In the following sections, we examine two interesting questions raised by earlier work on Myc:Max complex formation and transcriptional activity: (1) Can Max confer similar activities on other proteins and (2) what are the targets regulated by Myc:Max heterodimers?

IDENTIFICATION OF NOVEL MAX-BINDING PROTEINS: MAD AND MXI1

Max appears to be present in excess of Myc during proliferation and is also detected during periods when Myc is not synthesized (see below). This suggests that Max might be available for interaction with other proteins. Two such novel Max-binding proteins, Mad (Ayer et al. 1993) and Mxi1 (Zervos et al. 1993), have recently been identified by employing interaction cloning strategies with Max as a probe. Mad and Mxi1 are both members of the bHLHZip family and display approximately 40% overall identity to each other. Outside of their bHLHZip domains, these proteins do not show significant homology with other proteins in the database. In terms of their in vitro binding activities, both Mad and Mxi1 appear similar to Myc: They do not homodimerize or bind DNA alone but form stable complexes with Max capable of binding the CACGTG sequence. No interaction with Myc or other bHLHZip proteins has been demonstrated. Furthermore, as determined by heterodimer formation in gel-shift assays using the CACGTG sequence, Mad can compete with Myc for interaction with Max.

To determine the transcriptional activity of Mad, a *mad* expression vector was transfected into mammalian cells together with a reporter gene containing four promoter-proximal CACGTG-binding sites. This resulted in a relative decrease in the basal level of expression from the reporter (Ayer et al. 1993). Repression by Mad is reversed by transfection of increasing amounts of Myc and Max. Conversely, transcriptional activation due to Myc is repressed by expression of Mad and Max. Both Mad-induced repression and Myc-induced activation are abrogated using a dominant interfering form of Max, demonstrating that Max is critical for the activities of both proteins. The similarity of Mxi1 to Mad predicts that it will also act as a repressor; however, the transcriptional activity of Mxi1 has not yet been reported.

The properties of Mad (and to a certain extent of Mxi1) suggest that it may act to antagonize the function of Myc. Such antagonism could occur through competition between Myc and Mad for available Max and by competition between Myc:Max and Mad:Max for target binding sites. In either case, the formation of Mad:Max complexes would be expected to repress transcription of genes activated by Myc:Max complexes.

MAD IS INDUCED DURING MYELOID DIFFERENTIATION

To explore the possibility that Mad opposes the function of Myc, we attempted to identify the Mad protein in vivo and determine its pattern of expression. A Northern blot analysis of different human tissues is shown in Figure 1A. Two Mad polyadenylated mRNAs are detected in all tissues, with lung and placenta having high levels relative to other organs. The 3.5-kb mRNA is the size expected based on the *mad* cDNA. We do not know in what regions the 3.5-kb and 7.0-kb mRNAs differ. In other experiments, we found low or undetectable levels of Mad mRNA or protein in several proliferating cell types. However, upon testing the human myeloid leukemia cell lines ML-1 and U937, we found that treatment with 12-O-tetra-decanoyl-phorbol 13-acetate (TPA), an agent that induces differentiation of these cell lines along the monocyte/macrophage lineage, leads to the rapid induction of *mad* expression (Ayer and Eisenman 1993). The 3.5-kb and 7-kb *mad* mRNAs that can be detected at low levels in proliferating, undifferentiated ML-1 and U937 cells both increase after addition of TPA (Fig. 1B) (Ayer and Eisenman 1993). In U937 cells this increase also occurs in the presence of cycloheximide, suggesting that *mad* induction does not require new protein synthesis and is therefore likely to be a direct response to TPA (Ayer and Eisenman 1993; Larsson et al. 1994). Importantly, treatment of murine fibroblasts or HeLa cells with TPA does not induce differentiation and also does not result in increased *mad* expression (data not shown). However, other agents that produce U937 cell differentiation, such as vitamin D_3, dimethyl sulfoxide, retinoic acid, and interferon-γ, also result in the induction of *mad* (Ayer and Eisenman 1993; Larsson et al. 1994). In addition, transforming growth factor β (TGF-β) treatment of ML-1 cells causes differentiation as well as induction of *mad* (Fig. 1B). Thus, the increase in *mad* expression in these cells appears to be closely tied to differentiation.

In further experiments employing the monoblastic U937 cells, we showed that Mad protein is present at low levels in undifferentiated cells and increases continuously between 2 hours and 48 hours after TPA treatment. In contrast, Myc is easily detected in prolif-

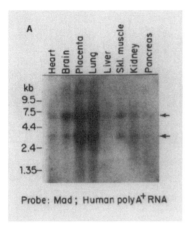

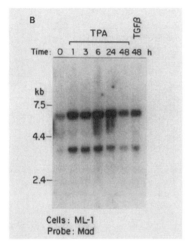

Figure 1. *mad* RNA expression in human tissues and during differentiation of ML-1 cells. (*A*) Northern blot analysis of poly(A)$^+$ RNA isolated from the indicated adult human tissues. The ^{32}P-labeled *mad* cDNA was used as probe. (*B*) The human myeloblastic cell line ML-1 was treated with 5 nM TPA or 10 ng/ml TGF-β and poly(A)$^+$ RNA was prepared at different times. Northern blot analysis of the RNAs was carried out using the ^{32}P-labeled *mad* cDNA as probe.

erating cells, increases by 6 hours following TPA, and then slowly declines. Max levels remain constant throughout differentiation (Ayer and Eisenman 1993). Examination of low-stringency anti-Max immunoprecipitates (Blackwood et al. 1992b) reveals a switch from Myc:Max to Mad:Max complexes between 24 hours and 48 hours after TPA addition (Ayer and Eisenman 1993). A somewhat different picture emerges from studies on the expression of these proteins in the myeloblastic ML-1 cell line. Despite the fact that *mad* mRNA is induced rapidly following TPA treatment (see Fig. 1B), Mad protein synthesis only occurs between 6 hours and 24 hours (Fig. 2A). Figure 2B shows that Mad protein is synthesized up to 96 hours after TPA addition. Myc is also detected at this time, while Max synthesis appears to decline. Immunoprecipitation of Max complexes demonstrates that Mad and Max are associated following overnight treatment with TPA (Fig. 2C; association with Myc was not examined in these cells).

The comparison between U937 and ML-1 therefore reveals differences in the temporal patterns of expression of Myc, Max, and Mad proteins. The different patterns of expression may be related to the fact that these cell lines represent unique stages in hematopoietic differentiation. The differences between *mad* mRNA and protein levels in the ML-1 cells may reflect translational or posttranslational controls over Mad that do not operate in the U937 cells. Our results show that Mad:Max complexes are formed in both cell lines even in the presence of Myc protein. This suggests that Myc may in some manner be inhibited from interacting with Max, thereby allowing Mad to associate. The induction of Mad in a series of myeloid and monocyte cell lines, and the association between Mad and Max, suggest that these proteins may play roles in differentiation of hematopoietic cells (Figs. 1, 2) (Ayer and Eisenman 1993; Larsson et al. 1994).

INDUCTION OF MAD DURING DIFFERENTIATION OF PRIMARY HUMAN KERATINOCYTES

We decided to investigate *mad* expression during keratinocyte differentiation for two reasons. First, keratinocytes represent a well-characterized differentiation system distinct from hematopoietic cells. Second, keratinocytes can be readily obtained from human foreskin, making it possible to study differentiation in primary cells, as opposed to the established cell lines used previously. When primary human keratinocytes are grown in culture and treated with TPA, they undergo cell cycle arrest (Wille et al. 1985) and commitment to terminal differentiation, as evidenced by expression of involucrin (Jones 1994) and development of a cornified envelope (Wille et al. 1985). We used anti-Myc, anti-Max, and anti-Mad antisera in immunoprecipitation assays of human primary keratinocytes (HPK) labeled with [^{35}S]methionine at relatively early times following treatment with TPA. The results, shown in Figure 3, indicate that the Mad protein is present in the untreated and 2-hour posttreatment HPK. By 9 hours, there is a significant increase in Mad protein synthesis. Both Myc and Max are also easily detected in the untreated HPK. However, Myc shows a sharp decline by 2 hours and is undetectable by 9 hours. Max synthesis decreases more slowly and is still clearly present by 9 hours. Therefore, by 9 hours after treatment with the differentiation inducer, Myc appears to no longer be synthesized, whereas both Mad and Max are produced.

To determine whether the changes in Myc, Max, and Mad synthesis in HPK affected complex formation with Max, we carried out low-stringency immunoprecipitations with anti-Max at different times following TPA addition. The precipitates were then dissociated and reprecipitated with a mixture of anti-Mad and anti-Myc. Figure 4 shows that Myc is clearly present in Max

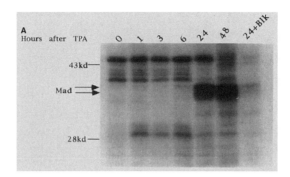
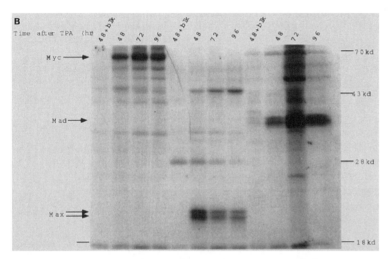
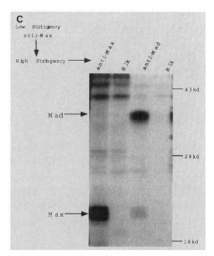

Figure 2. Immunoprecipitation analyses of proteins synthesized during ML-1 differentiation. (*A*) Cells were labeled with [^{35}S]methionine for 45 min at the indicated times following treatment with 5 nM TPA. The cells were lysed under high-stringency conditions (see Blackwood et al. 1992b) and immunoprecipitated with anti-Mad antibodies (Ayer and Eisenman 1993). The 35-kD Mad protein can be seen as a doublet at 24 and 48 hr. The bands between 35 kD and 44 kD are cross-reacting proteins that are not Mad-related. (*B*) Cells were labeled with [^{35}S]methionine for 60 min at the indicated times following TPA treatment. Following lysis under high-stringency conditions, the lysate was divided into three aliquots and immunoprecipitated with anti-Myc, anti-Max, and anti-Mad antibodies. The positions of the proteins are indicated by arrows at the left. (*C*) Association between Mad and Max. At 24 hr after TPA treatment, cells were lysed under low-stringency conditions (Blackwood et al. 1992b) and immunoprecipitated with anti-Max and then with anti-Mad and anti-Myc at high stringency as for *B*. Blk indicates a competitive inhibition of the antibodies with the cognate immunogen.

immunocomplexes in the proliferating cells but is absent by 2 hours following treatment with TPA. Low levels of Mad can be detected in a complex with Max even in proliferating cells, but peak complex formation appears at 24 hours after treatment. The widely spaced time points in this experiment preclude a detailed time course of heterocomplex switching, but other experiments indicate that the highest levels of Mad can be detected in the complex at 9 hours, followed by decline by 24 hours (P. Hurlin, in prep.). In other experiments, we have shown that Mad expression occurs in human papillomavirus (HPV)-immortalized HPK cells that can be induced to differentiate, but not in tumorigenic variants which do not differentiate. This provides another link between differentiation and Mad induction (P. Hurlin, in prep.).

MAX HETEROCOMPLEX SWITCHING AND DIFFERENTIATION

The increase in Mad expression and the switch from Myc:Max to Mad:Max complexes that occur upon induction of differentiation in monoblasts, myeloblasts, and keratinocytes open the possibility that Mad may play a role in the differentiation of these rather diverse cell types. We have previously suggested that the Max heterocomplex switch observed in U937 cells reflects a change in the transcriptional regulation of target genes for these complexes (Ayer and Eisenman 1993). As mentioned above, reporter genes with promoter-proximal binding sites for Max heterocomplexes can be activated by Myc:Max and repressed by Mad:Max (Ayer et al. 1993). Therefore, we predict that the switch to Mad:Max complexes results in decreased expression of genes previously activated through Myc:Max binding.

The notion that Max complex switching is involved in differentiation is compatible with evidence that the down-regulation of c-Myc expression, which occurs upon differentiation of numerous cell types, is important in terminal differentiation. Previous studies have demonstrated that enforced expression of Myc blocks terminal differentiation in erythroid (Coppola and Cole 1986), PC12 (Maruyama et al. 1987), and myeloid cells

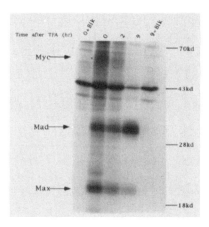

Figure 3. Myc, Max, and Mad protein synthesis during differentiation of HPK. HPK were treated with 10 nM TPA and labeled with [^{35}S]methionine at the times indicated. Anti-Myc, anti-Max, and anti-Mad antibodies were added together to identify all three proteins. The positions of Myc, Max, and Mad are indicated by arrows. Note that the Myc band is diffuse and its position is also indicated by an asterisk. Blk indicates a competitive inhibition of the antibodies with the cognate immunogen.

(including U937, Larsson et al. 1988), as well as in adipocytes (Freytag 1988) and myoblasts (Miner and Wold 1991). This suggests that *myc* expression needs to be turned off in order for differentiation to proceed. Nonetheless, there have been reports that *myc* may not be turned off during differentiation of some cell types, including keratinocytes (Dotto et al. 1986), neuronal cells (Bernard et al. 1992), and the ML-1 myeloid line (Craig et al. 1993). Our results with TPA-induced differentiation of keratinocytes clearly demonstrates rapid down-regulation of *myc* expression (Fig. 3). Consistent with the results of Dotto et al. (1986), who used calcium as the differentiation-inducing agent, treatment of human keratinocytes with calcium does not result in down-regulation of *myc* expression (B. Lüscher and R.N. Eisenman, unpubl.). Therefore, the response of keratinocytes is likely to vary depending on the inducing agent (Yuspa et al. 1989) and possibly on the species of origin (Fuchs 1990). We are currently determining whether Mad is induced by calcium in HPK.

Our results with ML-1 cells are in agreement with those of Craig et al. (1993) in that we detect Myc protein even at late times (Fig. 2B). These authors argue that Myc may be localized to the cytoplasm during ML-1 differentiation and, therefore, sequestered from Max. This may account for the formation of Mad:Max complexes even in the presence of Myc (Fig. 2B,C) and may be an example of how *myc* expression could be uncoupled from its activity. These differing patterns of expression could be related to the particular properties of the leukemic cell lines used. On the other hand, the variation may represent lineage- and stage-specific mechanisms of differentiation.

The model that Mad can antagonize Myc function has received some support from experiments showing that the introduction of *mad* into primary rat embryo fibroblasts can block their transformation by *myc* and *ras* (Lahoz et al. 1994; P. Koskinen, unpubl.). However, more information on the biological effects of *mad* overexpression, as well as of disruption of *mad* function during differentiation, will be required in order to understand the role that Mad plays in differentiation and development. Our experiments indicate that *mad* is not expressed during the differentiation of all cell types. The recent identification of four new Max interacting proteins opens the possibility that other factors may carry out functions similar to Mad in specific lineages (P. Hurlin, unpubl.).

IDENTIFICATION OF MYC:MAX TARGET GENES

Implicit in the models for Myc:Max activation and Mad:Max repression is the assumption that these complexes regulate transcription of a specific set of genes whose functions are important in cell behavior. However, the identification of genes directly regulated by Myc:Max heterocomplexes has been fitful and slow. Progress has been impeded by the fact that it is difficult to generate cells in which normal *myc* expression is truly conditional. In addition, variations in Myc levels are usually accompanied by significant changes in the growth state of cells. Thus, experiments that rely on subtractive hybridization or differential display to identify Myc target genes may result in isolation of genes that are regulated due to indirect effects of Myc. Nonetheless, several candidate genes have been identified using these approaches. These include α-prothymosin (Eilers et al. 1991) and ECA39, an embryonic gene of unknown function (Benvenisty et al. 1992). Another method has been to examine genes that are simply thought to be likely targets of Myc. Ornithine decarboxylase (ODC) (Bello-Fernandez et al. 1993), EIF2α (Rosenwald et al. 1993), and p53 (Reisman et al. 1993) have now been reported as candidates for direct Myc regulation.

What criteria can be used to determine whether putative target genes are regulated by Myc:Max? Cer-

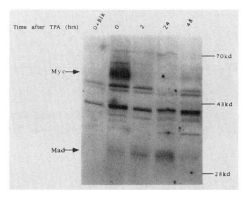

Figure 4. Max heterocomplex switching following differentiation of HPK. HPK were treated with 10 nM TPA and lysed under low-stringency conditions at the indicated times. Anti-Max immunoprecipitates were disrupted and re-immunoprecipitated with a mixture of anti-Myc and anti-Mad antibodies. The Myc and Mad proteins detected were in a heterocomplex with Max in the initial immunoprecipitate.

tainly the presence of Myc:Max-binding sites in the sequences of putative target genes; this would include the E-box sequence CACGTG and other sequences that are bound with somewhat lower affinity (Blackwell et al. 1993). However, such sequences are also binding sites for other more abundant transcription factors (see Blackwell et al. 1993), and the hypothetical frequency of any given hexamer, if it were to occur randomly, is about once every four kilobases. This high frequency and the fact that multiple proteins can bind such sites makes the simple presence of a Myc-Max-binding site insufficient evidence for its being an in vivo target. Therefore, it would be important to demonstrate that the binding sites detected are required for Myc-regulated expression of the putative target. ECA39, α-prothymosin, ODC, and p53 all have CACGTG sites. In general, these are located downstream from the transcription start site. In all four cases, the CACGTG sites appear to be important for transcriptional activity in transient expression assays using overexpressed Myc. For ECA39 and ODC, the expression pattern is consistent with their being Myc-regulated. However, other transcription factors, such as USF, TFE3, and TFEB, also recognize the CACGTG consensus, and it is uncertain to what extent these might contribute to expression.

IMMUNOISOLATION OF CLONE MJ143

We have used another approach to identify direct Myc:Max targets. It is based on an immunoisolation procedure previously employed to identify targets for *Drosophila* transcription factors (Gould et al. 1990) and recently used in our laboratory to isolate promoters regulated by the thyroid hormone receptor (Bigler and Eisenman 1994). The method, which is outlined in Table 1, involves immunoprecipitation of chromatin fragments with anti-Myc and anti-Max antibodies. The DNA associated with Myc and Max is then cloned, and pools of independent clones are tested for their ability to bind Myc:Max protein in electrophoretic mobility shift assays. In principle, this approach should enrich in Myc:Max-bound DNA segments and should avoid the complication of identification of secondary or later downstream effects of Myc:Max. Some of these genomic binding sites might be proximal to transcribed regions whose expression is dependent on Myc:Max.

In the following, we describe the properties of one clone, MJ143, identified using the immunoisolation procedure. The chromatin fragments were obtained from a lymphocyte cell line overexpressing both Myc and Max (CBMM) (Gu et al. 1993) and were immuno-

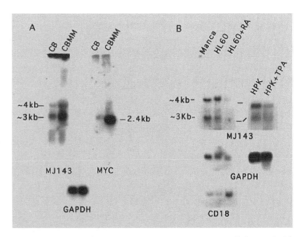

Figure 5. Northern blot analysis of clone MJ143. (*A*) poly(A)⁺ mRNA from human cord blood lymphocytes (CB) and CB cells overexpressing Myc and Max proteins (CBMM) was probed with human c-*myc*, GAPDH, and clone MJ143. (*B*) RNAs were prepared from the Manca Burkitt's lymphoma cell lines, from proliferating and RA-treated HL60 cells, and from proliferating and TPA-treated HPK. The blots were probed with MJ143, GAPDH, and CD18.

precipitated with anti-Max followed by anti-Myc antibodies. Clones that were capable of binding Myc:Max in a gel shift assay were sequenced and found, with only two exceptions, to contain at least one, and frequently several, binding sites, as defined by previous in vitro experiments (Blackwell et al. 1993). A number of candidates were used to probe Northern blots to determine whether they might contain any transcribed regions. Several were found to be positive. One of these, clone MJ143, was selected for further study.

Figure 5A shows a Northern blot of poly(A)⁺ RNA prepared from CBMM cells and from the parental cell line expressing only endogenous Myc and Max. The right-hand portion of Figure 5A shows that the 2.4-kb c-*myc* mRNA is overexpressed sevenfold in the CBMM cells relative to the CB controls. Using the MJ143 clone as probe, two RNAs, sized approximately 4 kb and 3 kb, are detected. These two RNAs are overexpressed in the CBMM cells relative to the CB cells. When normalized to GAPDH levels, we calculate approximately threefold overexpression. Note that the RNA detected with the GAPDH probe is the same in both cells. The Manca Burkitt's lymphoma cell line, possessing a rearranged c-*myc* allele, also expresses the MJ143 RNAs (Fig. 5B).

The results of the Northern blot analysis in Figure 5A indicate that Myc overexpression correlates with an increase in MJ143 expression in the CB cells. We next

Table 1. Immunoisolation of Myc-Max Target Genes

1. Isolate nuclei from Myc-Max-transformed lymphocytes
2. Digest with DNase-I or micrococcal nuclease
3. Immunoprecipitate soluble chromatin fragments with anti-Max antibody and elute Myc-Max with antibody-specific peptide
4. Repeat step 3 with anti-Myc antibody
5. Clone Myc-Max-complexed DNA and assay for in vitro Myc-Max binding
6. Northern analysis with the clones containing Myc-Max-binding sites

sought to determine whether decreased Myc expression would lead to a down-regulation of MJ143. We therefore examined two cell systems which are known to shut off c-*myc* expression upon differentiation. Figure 5B shows that the HL60 promyelocytic leukemia cell line contains the 3-kb and 4-kb MJ143 RNAs, but that neither of these RNAs is detected following retinoic acid (RA)-induced differentiation. We note that this is not a general effect on all RNAs, since GAPDH mRNA decreases only slightly, and an mRNA encoding CD18, a differentiation marker for HL60 cells, is markedly increased following RA treatment. Furthermore, the MJ143 RNAs decrease upon TPA-induced differentiation of human primary keratinocytes.

Our findings with MJ143 show that its expression levels correlate with c-*myc* expression and with differentiation in several different cell types. Furthermore, this clone contains a CACGTG-binding site in a context that can activate transcription when placed upstream of a heterologous promoter (C. Grandori, unpubl.). These results are consistent with the idea that clone MJ143 represents a direct Myc:Max target gene. Further studies of its expression during the cell cycle and as a function of ectopic c-*myc* expression will be needed to support this prediction. Such studies are under way for MJ143 and several other immunoselected clones that display similar behavior.

SUMMARY

The small bHLHZip protein, Max, was originally identified through its interaction with Myc family proteins and appears to be an obligate partner for Myc function. Max has now been found to interact with at least two other proteins, Mad and Mxi1. These also belong to the bHLHZip class but are otherwise unrelated to Myc. Mad has been shown to abrogate the positive transcriptional activity of Myc and to inhibit Myc in co-transformation assays. This suggests that Mad may antagonize Myc function. Mad is rapidly induced upon differentiation, a time when Myc is frequently down-regulated. We show here evidence for Mad expression upon differentiation of myeloblasts, monoblasts, and keratinocytes. Mad:Max complexes are detected during differentiation and appear to replace the Myc:Max complexes present in proliferating cell populations. Since these complexes appear to form even in the presence of Myc, there may exist mechanisms that act to inhibit Myc:Max, or to promote Mad:Max, complex formation. We speculate that Max complex switching causes a change in the transcriptional activity of groups of target genes. Mad is not induced in all differentiating cell types, suggesting that other, possibly tissue-restricted, proteins might act in similar switch mechanisms to effect changes in transcriptional programs. We have also developed an approach to identification of the gene targets for Myc:Max complexes. By employing an immunoisolation procedure, we have begun characterization of several clones whose expression levels correlate with those of c-*myc*. Further identification of Myc-regulated genes may allow us to determine the molecular mechanism by which Myc governs cell proliferation and differentiation.

ACKNOWLEDGMENTS

We thank Quentin Lawrence and Jaclynn Mac for expert technical assistance and Katherine Tietje for her help with the ML-1 experiments. We are grateful to Riccardo Dalla-Favera for a gift of the CB and CBMM cell lines. This research was supported by a National Institutes of Health Postdoctoral Fellowship to P.H., a Virology Training Grant to D.E.A., and by National Institutes of Health/National Cancer Institute grants RO1CA-20525 and RO1CA-57138 to R.N.E.

REFERENCES

Amati, B., T.D. Littlewood, G.I. Evan, and H. Land. 1993a. The c-Myc protein induces cell cycle progression and apoptosis through dimerization with Max. *EMBO. J.* **12:** 5083.

Amati, B., M.W. Brooks, N. Levy, T.D. Littlewood, G.I. Evan, and H. Land. 1993b. Oncogenic activity of the c-Myc protein requires dimerization with Max. *Cell* **72:** 233.

Amati, B., S. Dalton, M.W. Brooks, T.D. Littlewood, G.I. Evan, and H. Land. 1992. Transcriptional activation by the human c-Myc oncoprotein in yeast requires interaction with Max. *Nature* **359:** 423.

Ayer, D.E. and R.N. Eisenman. 1993. A switch from Myc:Max to Mad:Max heterocomplexes accompanies monocyte/macrophage differentiation. *Genes Dev.* **7:** 2110.

Ayer, D.E., L. Kretzner, and R.N. Eisenman. 1993. Mad: A heterodimeric partner for Max that antagonizes Myc transcriptional activity. *Cell* **72:** 211.

Bello-Fernandez, C., G. Packham, and J.L. Cleveland. 1993. The ornithine decarboxylase gene is a transcriptional target of c-*myc*. *Proc. Natl. Acad. Sci.* **90:** 7804.

Benvenisty, N., A. Leder, A. Kuo, and P. Leder. 1992. An embryonically expressed gene is a target for c-Myc regulation via the c-Myc-binding sequence. *Genes Dev.* **6:** 2513.

Bernard, O., J. Drago, and H. Sheng. 1992. L-*myc* and N-*myc* influence lineage determination in the central nervous system. *Neuron* **9:** 1217.

Bigler, J. and R.N. Eisenman. 1994. Isolation of a T3-responsive gene by immunoprecipitation of thyroid hormone receptor-DNA complexes. *Mol. Cell. Biol.* **14:** 7621.

Blackwell, T.K., L. Kretzner, E.M. Blackwood, R.N. Eisenman, and H. Weintraub. 1990. Sequence-specific DNA-binding by the c-Myc protein. *Science* **250:** 1149.

Blackwell, T.K., J. Huang, A. Ma, L. Kretzner, F.W. Alt, R.N. Eisenman, and H. Weintraub. 1993. Binding of Myc proteins to canonical and noncanonical DNA sequences. *Mol. Cell. Biol.* **13:** 5216.

Blackwood, E.M. and R.N. Eisenman. 1991. Max: A helix-loop-helix zipper protein that forms a sequence-specific DNA binding complex with Myc. *Science* **251:** 1211.

Blackwood, E.M., L. Kretzner, and R.N. Eisenman. 1992a. Myc and Max function as a nucleoprotein complex. *Curr. Opin. Genet. Dev.* **2:** 227.

Blackwood, E.M., B. Luscher, and R.N. Eisenman. 1992b. Myc and Max associate in vivo. *Genes Dev.* **6:** 71.

Bousset, K., M. Henriksson, J.M. Lüscher-Firzlaff, D.W. Litchfield, and B. Lüscher. 1993. Identification of casein kinase II phosphorylation sites in Max: Effects on DNA-binding kinetics of Max homo- and Myc/Max heterodimers. *Oncogene* **8:** 3211.

Coppola, J.A. and M.D. Cole. 1986. Constitutive c-*myc* oncogene expression blocks mouse erythroleukemia cell differentiation but not commitment. *Nature* **320:** 760.

Craig, R.W., H.L. Buchan, C.I. Civin, and M.B. Kastan. 1993. Altered cytoplasmic/nuclear distribution of the c-myc protein in differentiating ML-1 human myeloid leukemia cells. *Cell Growth Differ.* **4:** 349.

Dang, C.V. 1991. c-Myc oncoprotein function. *Biochim. Biophys. Acta* **1072:** 103.

DePinho, R.A., N. Schreiber-Agus, and F.W. Alt. 1991. *myc* family oncogenes in the development of normal and neoplastic cells. *Adv. Cancer Res.* **57:** 1.

Dotto, G.P., M., Gilman, M. Maruyama, and R.A. Weinberg. 1986. c-*myc* and c-*fos* expression in differentiating mouse primary keratinocytes. *EMBO J.* **5:** 2853.

Eilers, M., S. Schirm, and J.M. Bishop. 1991. The MYC protein activates transcription of the α-prothymosin gene. *EMBO J.* **10:** 133.

Ferre-D'Amare, A.R., G.C. Prendergast, E.B. Ziff, and S.K. Burley. 1993. Recognition of Max by its cognate DNA through a dimeric b/HLH/Z domain. *Nature* **363:** 38.

Freytag, S. 1988. Enforced expression of the c-*myc* oncogene inhibits cell differentiation by precluding entry into a distinct predifferentiation state in G_0/G_1. *Mol. Cell. Biol.* **8:** 1614.

Fuchs, E. 1990. Epidermal differentiation. *Curr. Opin. Cell Biol.* **2:** 1028.

Gould, A.P., J.J. Brookman, D.I. Strutt, and R.A. White. 1990. Targets of homeotic gene control in *Drosophila*. *Nature* **348:** 308.

Gu, W., K. Cechova, V. Tassi, and R. Dalla-Favera. 1993. Opposite regulation of gene transcription and cell proliferation by c-Myc and Max. *Proc. Natl. Acad. Sci.* **90:** 2935.

Jones, K.T. 1994. Staurosporine, a non-specific PKC inhibitor, induces keratinocyte differentiation and raises intracellular calcium, but RO31-8220, a specific inhibitor does not. *J. Cell. Physiol.* **159:** 324.

Kato, G.J., J. Barrett, M. Villa-Garcia, and C.V. Dang. 1990. An amino-terminal c-Myc domain required for neoplastic transformation activates transcription. *Mol. Cell. Biol.* **10:** 5914.

Kato, G.J., W.M.F. Lee, L. Chen, and C.V. Dang. 1992. Max: Functional domains and interaction with c-Myc. *Genes Dev.* **6:** 81.

Kretzner, L., E.M. Blackwood, and R.N. Eisenman. 1992. The Myc and Max proteins possess distinct transcriptional activities. *Nature* **359:** 426.

Kreztner, L., E.M. Blackwood, J. Mac, and R.N. Eisenman. 1993. Transcriptional repression by Max proteins p21 and p22. *Colloq. INSERM* **229:** 75.

Lahoz, E.G., L. Xu, N. Schreiber-Agus, and R.A. DePinho. 1994. Suppression of Myc, but not E1a, transformation activity by Max-associated proteins, Mad and Mxi1. *Proc. Natl. Acad. Sci.* **91:** 5503.

Larsson, L.-G., M. Pettersson, F. Öberg, K. Nilsson, and B. Lüscher. 1994. Expression of *mad*, *mxi1*, *max*, and c-*myc* during induced differentiation of hematopoietic cells: Opposite regulation of *mad* and c-*myc*. *Oncogene* **9:** 1247.

Lüscher, B. and R.N. Eisenman. 1990. New light on Myc and Myb. I. Myc. *Genes Dev.* **4:** 2025.

Marcu, K.B., S.A. Bossone, and A.J. Patel. 1992. *myc* function and regulation. *Annu. Rev. Biochem.* **61:** 809.

Maruyama, K., S. Schiavi, W. Huse, G.L. Johnson, and E. Ruley. 1987. *myc* and E1A oncogenes alter the responses of PC12 cells to nerve growth factor and block differentiation. *Oncogene* **1:** 361.

Miner, J.H. and B.J. Wold. 1991. c-*myc* inhibition of MyoD and myogenin-initiated myogenic differentiation. *Mol. Cell. Biol.* **11:** 2842.

Prendergast, G.C., D. Lawe, and E.B. Ziff. 1991. Association of Myn, the murine homolog of Max, with c-Myc stimulates methylation-sensitive DNA binding and *ras* co-transformation. *Cell* **65:** 395.

Reisman, D., N.B. Elkind, R. Baishali, J. Beamon, and V. Rotter. 1993. c-Myc *trans*-activates the p53 promoter through a required downstream CACGTG motif. *Cell Growth Differ.* **4:** 57.

Rosenwald, I.B., D.B. Rhoads, L.D. Callanan, K.J. Isselbacher, and E.V. Schmidt. 1993. Increased expression of eukaryotic translation initiation factors eIF-4E and eIF-2α in response to growth induction by c-*myc*. *Proc. Natl. Acad. Sci.* **90:** 6175.

Wille, J.J., Jr., M.R. Pittelkow, and R.E. Scott. 1985. Normal and transformed prokeratinocytes express divergent effects of a tumor promoter on cell cycle-mediated control of proliferation and differentiation. *Carcinogenesis* **6:** 1181.

Yuspa, S.H., A.E. Kilkenny, P.M. Steinert, and D.R. Roop. 1989. Expression of murine epidermal differentiation markers is tightly regulated by restricted extracellular calicum concentrations in vitro. *J. Cell Biol.* **109:** 1207.

Zervos, A.S., J. Gyuris, and R. Brent. 1993. Mxi1, a protein that specifically interacts with Max to bind Myc-Max recognition sites. *Cell* **72:** 223.

BCL-6 and the Molecular Pathogenesis of B-Cell Lymphoma

R. Dalla-Favera,* B.H. Ye,* F. Lo Coco,* C.-C. Chang,* K. Cechova,* J. Zhang,*
A. Migliazza,* W. Mellado,* H. Niu,* S. Chaganti,† W. Chen,† P.H. Rao,† N.Z. Parsa,†
D.C. Louie,† K. Offit,† and R.S.K. Chaganti†

*Division of Oncology, Department of Pathology, and Department of Genetics and Development, College of Physicians & Surgeons, Columbia University, New York, New York 10032; †Cell Biology and Genetics Program, and the Departments of Human Genetics and Pathology and Medicine, Memorial Sloan-Kettering Cancer Center, New York, New York 10021

Non-Hodgkin's lymphomas (NHL) include a group of neoplasms which share a common target tissue, lymphoid cells, yet are characterized by a high degree of biological and clinical heterogeneity (for review, see Magrath 1990). Most NHL derive from the B-cell lineage and, in particular, from mature B cells, which are characterized by rearranged immunoglobulin (Ig) heavy- (IgH) and light-chain genes and by the expression of cell-surface Ig and B-cell-associated markers. NHL are usually classified according to their degree of clinical aggressiveness correlated with their stage of differentiation and pattern of growth: Low-grade NHL include small lymphocytic lymphoma (SLL) and the majority of follicular lymphomas (FL); intermediate-grade NHL include a subset of FL and diffuse large cell lymphoma (DLCL); high-grade NHL include immunoblastic and lymphoblastic NHL as well as Burkitt's lymphomas (BL).

The wide clinico-pathological heterogeneity of NHL is reflected at the cellular level by the variety of molecular pathways underlying NHL pathogenesis (Table 1) (Gaidano and Dalla-Favera 1993). Very little is known about the pathogenesis of non-FL low-grade NHL, except for "mantle zone" lymphomas, which are associated in 50% of cases with the t(11;14) translocation involving the juxtaposition of the IgH locus to the BCL-1/PRAD-1/cyclin D1 gene coding for a protein involved in the control of cell cycle progression (Tsujimoto et al. 1984; Motokura et al. 1991; Raffeld and Jaffe 1991). In follicular NHL, the t(14;18) translocation juxtaposes the IgH locus to BCL-2 (Tsujimoto et al. 1984; Bakhshi et al. 1985; Cleary and Sklar 1985), a gene coding for a protein that prevents programmed cell death or apoptosis (Korsmeyer 1992). After years of indolent course, a significant fraction of FL undergoes histological transformation and clinical progression to DLCL, an event which is associated with loss or mutations of p53 tumor suppressor gene (Lo Coco et al. 1993). In BL, the t(8;14), t(8;22), and t(2;8) chromosomal translocations lead to the deregulated expression of the c-myc proto-oncogene by juxtaposition to one of the Ig loci (Dalla-Favera et al. 1982, 1983; Taub et al. 1982; Dalla-Favera 1991). A sizable fraction (35%) of BL are also associated with loss or mutations of p53 (Gaidano et al. 1991).

These well-characterized genetic lesions are associated, however, with only a fraction of NHL cases. In fact, a number of additional recurrent NHL-associated

Table 1. Distribution of Molecular Lesions in B-NHL

NHL type	Molecular lesions				
	BCL-1	BCL-2	c-MYC	p53	BCL-6
Low grade					
small lymphocytic	50%[a]	—	—	—	—
follicular, small cell	—	90%	—	—	6%
follicular, mixed	—	90%	—	—	—
Intermediate grade					
follicular, large cell	—	90%	—	—	—
diffuse, small cell	—	20%	10%	—	—
diffuse, mixed	—	20%	10%	—	20%
diffuse, large cell	—	20%	10%	—[d]	40%
High grade					
immunoblastic[b]	—	—	20%	—	20%
lymphoblastic	—	—	NA[c]	—	—
Burkitt's	—	—	100%	30%	—

[a] Restricted to cases of mantle-zone lymphomas.
[b] In immunoblastic lymphomas of HIV-infected individuals.
[c] NA, not assessed.
[d] In a previous study (Lo Coco et al. 1993), 4 out of 6 DLCL with previous FL history had p53 mutations.

chromosomal translocations exist which have not been characterized in their genetic components and, in addition, no genetic lesion has been identified for some relevant NHL subtypes (Chaganti et al. 1989). This paper focuses on the review of recent progress obtained on the identification of a molecular lesion associated with DLCL, the most frequent and most lethal human lymphoma, which accounts for ~40% of initial NHL diagnoses and is often the final stage of progression of FL (Magrath 1990).

Figure 1. Schematic representation of the BCL-6 protein based on homology with other zinc-finger proteins (Ye et al. 1993b; and see text). The approximate positions of the zinc-finger motifs (Zn + +) and the NH$_2$-POZ domain are indicated.

CHROMOSOMAL TRANSLOCATIONS AFFECTING 3q27 IN DLCL: CLONING OF CHROMOSOMAL BREAKPOINTS

Cytogenetic analysis of large panels of DLCL cases has revealed relatively frequent (10–12%) chromosomal alterations affecting band 3q27 in this NHL subtype (Offit et al. 1989; Bastard et al. 1992). These alterations involve reciprocal translocations between the 3q27 region and various partner chromosomal sites including, but not limited to, those carrying the Ig heavy- (14q32) or light- (2p12, 22q11) chain loci. These observations suggest that 3q27 may be the site of a proto-oncogene whose structural lesion may be critical for DLCL pathogenesis.

As a first step toward the molecular characterization of the putative 3q27 proto-oncogene, we have cloned the chromosomal breakpoints of several cases of (3;14)(q27;q32) translocations in which the involvement of the immunoglobulin locus on 14q32 provided a probe for the cloning of the translocation junctions. The same genomic region within 3q27 was cloned from all the t(3;14)(q27;q32) cases analyzed (Baron et al. 1993; Deweindt et al. 1993; Kerckaert et al. 1993; Ye et al. 1993a; Miki et al. 1994). Furthermore, DNA probes from this region identified rearrangements in this locus in 13 of 17 cases carrying 3q27 alterations, irrespective of the partner chromosomes involved in the translocations (Ye et al. 1993b). These results indicated that the chromosomal breakpoints clustered within a restricted genomic region in various DLCL cases (see Fig. 4), further suggesting that the affected genetic locus may be important for lymphomagenesis.

THE *BCL-6* GENE AND ITS PROTEIN PRODUCT

A search for transcribed sequences led to the identification of a gene, called *BCL-6*, whose sequences span the translocation breakpoints (Ye et al. 1993b). The corresponding cDNA was cloned, sequenced, and demonstrated to code for a novel protein with essential features illustrated in Figure 1.

The carboxy-terminal region of *BCL-6* contains six C_2H_2 zinc-finger motifs and a conserved stretch of seven amino acids (the H/C link) connecting the successive zinc-finger repeats, as is typical of various members of the *Krüppel*-like subfamily of zinc-finger proteins (Rosenberg et al. 1989). Zinc-finger-encoding genes are plausible candidate oncogenes, as they have been shown to participate in the control of cell proliferation, differentiation, and organ pattern formation (for review, see El-Baradi and Pieler 1991). In addition, alterations of zinc-finger genes have been detected in a variety of tumor types including PLZF and PML in acute promyelocytic leukemia (de Thé et al. 1991; Kakizuka et al. 1991; Pandolfi et al. 1991; Chen et al. 1993), EVI-1 in mouse and human myeloid leukemia (Morishita et al. 1988; Fichelson et al. 1992), TTG-1 in T-cell acute lymphoblastic leukemia (ALL) (McGuire et al. 1989), HTRX in acute mixed-lineage leukemia (Djabali et al. 1992; Gu et al. 1992; Tkachuk et al. 1992), and WT-1 in Wilm's tumor (Haber et al. 1990).

The amino-terminal region of *BCL-6* is devoid of the FAX (Knochel et al. 1989) and KRAB (Dellefroid et al. 1991) domains sometimes seen in *Krüppel*-related zinc-finger proteins, but it does have homologies (Fig. 1) with other zinc-finger transcription factors including the human ZFPJS protein, a putative human transcription factor that regulates the major histocompatibility complex II promoter (M.M. Sugawara, unpubl.), the *Tramtrack* (*ttk*) and *Broad-complex* (*Br-c*) proteins in *Drosophila* that regulate transcription during development (Harrison and Travers 1990; Dibello et al. 1991), the human KUP protein (Chardin et al. 1991), and the human PLZF protein. Such homology is also found in proteins (e.g., VA55R) of the poxvirus family (Koonin et al. 1992) as well as in the *Drosophila kelch* protein involved in nurse cell/oocyte interaction (Xue and Cooley 1993). This homology domain, recently named POZ (for *POX*/Zinc finger), has been found in an additional zinc-finger protein, ZID, and shown to act as a specific protein-protein interaction domain capable of regulating DNA binding by the zinc-finger domain (Bardwell and Treisman 1994).

These structural features of *BCL-6* suggest that it may function as a DNA-binding transcription factor possibly involved in the control of cell differentiation and tissue development. This hypothesis is confirmed in part by our recent results showing that *BCL-6* is localized in the nucleus (C.-C. Chang et al., unpubl.), is able to bind to a specific DNA sequence (B.H. Ye et al., unpubl.), and contains a transcriptional repressor domain (C.-C. Chang et al., unpubl.).

Table 2. Expression of the *BCL-6* Gene in Cell Lines

Cells	Phenotype	BCL-6 RNA
B-lineage		
697	pre-B (acute lymphoblastic leukemia)	±
Daudi	B cell (EBV⁺ Burkitt's lymphoma)	++
Ramos	B cell (EBV⁻ Burkitt's lymphoma)	++
Bjab	B cell (B cell lymphoma)	++
LCL[a] (6 tested)	B cell (EBV-immortalized)	−
U266	plasma cell (multiple myeloma)	−
T-lineage		
CEM	pre-T (acute lymphoblastic leukemia)	−
Hut-102	T cell (adult T-cell leukemia)	±
Other		
K562	erythroid (chronic myelogenous leukemia)	−
U937	monocyte (acute monocytic leukemia)	−
HL-60	myeloid (acute promyelocytic leukemia)	−

[a] LCL: Lymphoblastoid cell line.

THE *BCL-6* GENE IS EXPRESSED IN MATURE B CELLS

The pattern of *BCL-6* mRNA expression was investigated in a variety of human and mouse cell lines by Northern blot analysis. A single 3.8-kb RNA species was readily detectable in all cell lines derived from mature B cells except for Epstein-Barr virus (EBV)-immortalized lymphoblastoid cell lines (LCL) (Table 2; see Fig. 2 for representative results). Low or undetectable amounts of *BCL-6* mRNA were found in cells preceding (pre-B) or following (plasma cells) mature B cells in the B-cell differentiation pathway. Cell lines derived from T cells, from other hematopoietic cell lineages (see Table 1), or from other normal tissues also lacked notable levels of *BCL-6* mRNA, except for skeletal muscle, in which the *BCL-6* message was clearly detectable (B.H. Ye et al., unpubl.). Conversely, normal B cells derived from human peripheral blood also contained detectable levels of *BCL-6* RNA under both resting and mitogen-stimulated conditions (J.

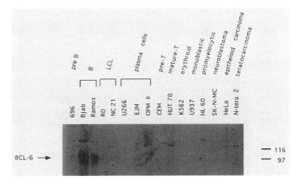

Figure 3. Western blot analysis of BCL-6 protein expression in various tumor cell lines. See Table 2 for description of the cell lines. A rabbit antiserum (N-70-6) raised against a synthetic KLH-conjugated oligopeptide corresponding to the amino-terminal region of *BCL-6* was used.

Zhang et al., unpubl.). These results suggest that *BCL-6* is selectively expressed in mature B cells under normal physiological conditions.

The tissue distribution of the BCL-6 protein has also been investigated by Western blot analysis using a rabbit antiserum raised against a synthetic peptide corresponding to the amino terminus of *BCL-6*. Figure 3 shows that a protein is detectable in the cell types where the *BCL-6* mRNA was found, namely, cell lines displaying a mature B-cell phenotype (Ramos and Bjab) and normal B cells from human tonsils (H. Niu et al., unpubl.). The selective expression of *BCL-6* in differentiated B cells suggests that *BCL-6* may play a role in the control of normal B-cell differentiation and lymphoid organ development.

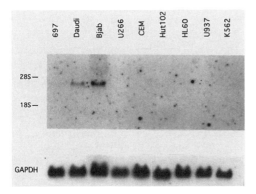

Figure 2. Northern blot analysis of *BCL-6* mRNA expression in various tumor cell lines (*top* panel). See Table 2 for description of the cell lines. The *bottom* panel shows the result of rehybridization of the same filter to a GAPDH probe as controls for RNA amounts.

REARRANGEMENTS OF THE *BCL-6* GENE IN DLCL

To define the incidence of *BCL-6* rearrangements in various lymphoproliferative diseases, the entire *BCL-6*

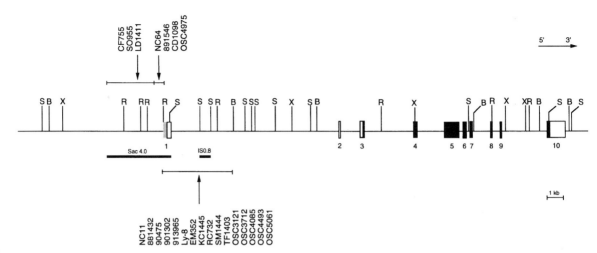

Figure 4. Exon-intron organization of the *BCL-6* gene and mapping of breakpoints detected in DLCL. Coding and non-coding exons are represented by filled and empty boxes, respectively. Patient codes (NC11, 891546, etc.) are grouped according to the rearranged patterns displayed by tumor samples. Arrows indicate the breakpoint position for each sample as determined by restriction digestion/hybridization analysis. Restriction enzyme symbols are: S, *Sac*I; B, *Bam*HI; X, *Xba*I; H, *Hin*dIII, R, *Eco*RI.

genomic locus was cloned. Figure 4 shows that this gene contains 10 exons spanning approximately 26 kb of genomic DNA. Sequence analysis indicated that the first exon is non-coding and that the transcription initiation codon is located within the third exon (Ye et al. 1993b). Using various probes from this locus, we analyzed a panel of cases not previously selected on the basis of 3q27 alterations but representative of the major subtypes of NHL as well as of other lymphoproliferative diseases, including ALL, chronic lymphocytic leukemia (CLL), and multiple myeloma (MM) (Lo Coco et al. 1994).

The results of this analysis are summarized in Table 3 and representative data are shown in Figure 5. All cases of ALL, CLL, and MM showed a normal *BCL-6* gene. Among distinct NHL histological subtypes, rearrangements were detected in 35% of DLCL, but significantly less frequently in FL (6%). All except two of the DLCL cases displaying *BCL-6* alterations lacked *BCL-2* rearrangements. Although cytogenetic data were not available for the panel of tumors studied, the frequency

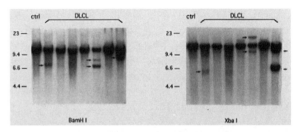

Figure 5. Analysis of *BCL-6* rearrangements in DLCL. DNAs were digested with *Bam*HI or *Xba*I as indicated and hybridized to probe Sac4.0 (Lo Coco et al. 1994). Human placenta DNA was used as *BCL-6* germ-line control (ctrl). Bands corresponding to rearranged *BCL-6* alleles are indicated by arrows.

of *BCL-6* rearrangements detected by Southern blot analysis far exceeded that expected for 3q27 aberrations (10–12% in DLCL) at the cytogenetic level, suggesting that *BCL-6* rearrangements can occur as a consequence of submicroscopic chromosomal aberrations.

REARRANGEMENTS AFFECT THE 5′ NON-CODING REGION OF *BCL-6*

The positions of the breakpoints within the *BCL-6* locus have been mapped, allowing for a preliminary understanding of their effect on the structure/function of the *BCL-6* gene (Ye et al. 1993b). All the observed breakpoints could be mapped within the 5′-flanking region, within the first exon, or within the first intron of *BCL-6* (Fig. 4). As a result, the coding domain of *BCL-6* is left intact, whereas the 5′ regulatory region containing the promoter sequences is either completely removed (in case of truncation within the first exon or intron) or truncated.

Table 3. Rearrangements of the *BCL-6* Gene in Lymphoid Tumors

Tumor[a]	Rearranged/tested	%
NHL		
SLL	0/8	0
FL	2/31	6
DLCL	16/45	35
SNCL	0/22	0
ALL	0/45	0
CLL	0/51	0
MM	0/23	0

[a] (NHL) Non-Hodgkin's lymphoma; (ALL) acute lymphoblastic leukemia; (CLL) chronic lymphocytic leukemia; (MM) multiple myeloma; (SLL) small lymphocytic lymphoma; (FL) follicular lymphoma; (DLCL) diffuse large cell lymphoma; (SNCL) small non-cleaved cell lymphoma.

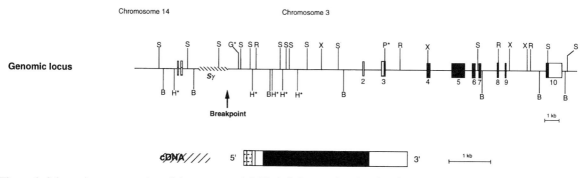

Figure 6. Schematic representation of the rearranged *BCL-6* allele spanning the t(3;14) translocation breakpoint cloned from a phage library constructed from Ly-8 genomic DNA (*top* panel) and of its corresponding cDNA cloned from a Ly-8 cDNA library. Ig region exons (stippled boxes), Sγ sequences, non-coding (open boxes) and coding (filled boxes) *BCL-6* exons are also indicated. Restriction enzyme symbols are: S, *Sac*I; B, *Bam*HI; X, *Xba*I; H, *Hin*dIII; R, *Eco*RI; P, *Pst*I; G, *Bgl*II. Restriction sites marked by an asterisk have been only partially mapped.

To precisely determine the consequences of these rearrangements on *BCL-6* mRNA and protein structure, the rearranged genomic locus and the corresponding cDNA were cloned and characterized from a number of cases. Figure 6 shows a schematic representation of these clones derived from a representative DLCL cell line, called Ly-8. This cell line carries a t(3;14)(q27;q32) translocation and, as a result, the *BCL-6* gene is truncated within its first intron and linked to the IgH locus within the Switch-γ (Sγ) recombination region. The corresponding cDNA shows that the *BCL-6* first exon sequences are replaced by sequences derived from the I region of the γ3 locus (Kuze et al. 1991). The I region is located upstream of Sγ, and its promoter is normally responsible for expression of the sterile transcripts in germ-line IgH genes. These sterile transcripts contain I region sequences spliced to the downstream sequences from the Ig constant region. In Ly-8 cells, I sequences are spliced to *BCL-6* exon 2 sequences, resulting in a fusion transcript that retains the entire *BCL-6* coding domain. Analysis of this cDNA predicts that the functional consequence of this *BCL-6* rearrangement is the expression of a normal BCL-6 protein under the control of a heterologous promoter sequence leading to the loss of its normal pattern of regulation. This model is valid for various translocations involving 3q27 and suggests that the *BCL-6* coding domain (exons 3–10) may be linked downstream to heterologous sequences which, based on cytogenetic analysis, can originate from different chromosomes in different cases.

BCL-6 REARRANGEMENTS ARE ASSOCIATED WITH A DISTINCT CLINICO-PATHOLOGICAL SUBSET OF DLCL

DLCL are highly heterogeneous in terms of pathological manifestations and prognosis. In order to determine whether *BCL-6* rearrangements could identify any distinct subset of tumors, 102 DLCL cases were analyzed for *BCL-6* rearrangements and the results were correlated with histology, *BCL-2* rearrangement status, age, disease stage, clinical status, and treatment outcome (Offit et al. 1994). *BCL-6* rearrangements were noted in a statistically significant subset of cases characterized by involvement of extranodal tissues, lack of bone marrow involvement, and favorable prognosis (Table 4). Overall, *BCL-6* status was found to be an independent prognostic marker of survival and freedom from disease progression in the multivariate model and added predictive value to established clinical prognostic models (Offit et al. 1994). These findings suggest that *BCL-6* rearrangements may identify a biologically distinct subset of DLCL and may serve as a diagnostic and prognostic marker in the management of patients with this disease.

SUMMARY AND CONCLUSIONS

The results presented identify the first genetic lesion associated with DLCL, the most clinically relevant form of NHL. Although no proof yet exists of a role for

Table 4. Clinico-pathological Features of DLCL Cases Displaying *BCL-6* Rearrangements

BCL-6	No. of cases	*BCL-2*[a]	Extranodal[b]	Survival[c]
Rearranged	23	0 (0%)	16 (70%)	19 (83%)
Germ-line	79	21 (26%)	27 (34%)	40 (50%)

[a] Number of cases displaying *BCL-2* rearrangements.
[b] Number of cases in which the tumor appeared outside lymph nodes as extranodal disease.
[c] Number of cases free of disease progression 36 months after diagnosis.

these lesions in DLCL pathogenesis, the feature of the BCL-6 gene product, its specific pattern of expression in B cells, and the clustering of lesions disrupting its regulatory domain strongly suggest that deregulation of *BCL-6* expression may contribute to DLCL development. A more precise definition of the role of *BCL-6* in normal and neoplastic B-cell development is the goal of ongoing study of transgenic mice engineered either to express *BCL-6* under heterologous promoters or lacking BCL-6 function due to targeted deletions.

In addition to contributing to the understanding of DLCL pathogenesis, the identification of *BCL-6* lesions may have relevant clinical implications. DLCL represent a heterogeneous group of neoplasms which are treated homogeneously despite the fact that only 50% of patients experience long-term disease-free survival (Schneider et al. 1990). The fact that *BCL-6* rearrangements identify biologically and clinically distinct subsets of DLCL suggests that these lesions may be useful as markers in selection of differential therapeutic strategies based on different risk groups. Furthermore, the *BCL-6* rearrangements can be used to identify and monitor the malignant clone with sensitive PCR-based techniques. Since clinical remission has been observed in a significant fraction of DLCL cases, these markers may serve as critical tools for sensitive monitoring of minimal residual disease and early diagnosis of relapse (Gribben et al. 1993).

ACKNOWLEDGMENTS

This work has been supported by National Institutes of Health grants CA-44029 (to R.D.F.), and CA-34775 and CA-08748 (to R.S.K.C.). F.L.C. is presently at Sezione di Ematologia, Dipartimento di Biopatologia Umana of the University "La Sapienza" of Rome, and was partially supported by Associazione Italiana contro la Leucemia (AIL). K.O. is a recipient of a Clinical Oncology Cancer Development award from the American Cancer Society. B.H.Y. is a Leukemia Society of America Fellow.

REFERENCES

Bakhshi, A., J.P. Jensen, P. Goldman, J.J. Wright, O.W. McBride, A.L. Epstein, and S.J. Korsmeyer. 1985. Cloning the chromosomal breakpoint of t(14;18) human lymphomas: Clustering around J_H on chromosome 14 and near a transcriptional unit on 18. *Cell* **41:** 889.

Bardwell, V.J. and R. Treisman. 1994. The POZ domain: A conserved protein-protein interaction motif. *Genes Dev.* **8:** 1664.

Baron, B.W., G. Nucifora, N. McCabe, R. Espinosa III, M.M. Le Beau, and T.W. McKeithan. 1993. Identification of the gene associated with the recurring chromosomal translocations t(3;14)(q27;q32) and t(3;22)(q27;q11) in B-cell lymphomas. *Proc. Natl. Acad. Sci.* **90:** 5262.

Bastard, C., H. Tilly, B. Lenormand, C. Bigorgne, D. Boulet, A. Kunlin, M. Monconduit, and H. Piguet. 1992. Translocations involving band 3q27 and Ig gene regions in non-Hodgkin's lymphoma. *Blood* **79:** 2527.

Chaganti, R.S.K., L.A. Doucette, K. Offit, D.A. Filippa, G.J. Allen, M.R. Condon, S.C. Jhanwar, B.D. Clarkson, and P.H. Lieberman. 1989. Specific translocations in non-Hodgkin's lymphoma: Incidence, molecular detection, and histological and clinical correlations. *Cancer Cells* **7:** 33.

Chardin, P., G. Courtois, M.-G. Mattei, and S. Gisselbrecht. 1991. The *KUP* gene, located on human chromosome 14, encodes a protein with two distant zinc fingers. *Nucleic Acids Res.* **19:** 1431.

Chen, Z., N.J. Brand, A. Chen, S. Chen, J.-H. Tong, Z.-Y. Wang, S. Waxman, and A. Zelent. 1993. Fusion between a novel *Krüppel*-like zinc finger gene and the retinoic acid receptor-α locus due to a variant t(11;17) translocation associated with acute promyelocytic leukaemia. *EMBO J.* **12:** 1161.

Cleary, M.L. and J. Sklar. 1985. Nucleotide sequence of a t(14;18) chromosomal breakpoint in follicular lymphoma and demonstration of a breakpoint-cluster region near a transcriptionally active locus on chromosome 18. *Proc. Natl. Acad. Sci.* **82:** 7439.

Dalla-Favera, R. 1991. Chromosomal translocations involving the c-myc oncogene and their role in the pathogenesis of B cell neoplasia. In *Origins of human cancer: A comprehensive review* (ed J. Brugge et al.), p. 543. Cold Spring Harbor Laboratory Press, Cold Spring Harbor, New York.

Dalla-Favera, R., S. Martinotti, R.C. Gallo, J. Erikson, and C.M. Croce. 1983. Translocation and rearrangements of the c-*myc* oncogene locus in human undifferentiated B-cell lymphomas. *Science* **219:** 963.

Dalla-Favera, R., M. Bregni, J. Erickson, D. Patterson, R.C. Gallo, and C.M. Croce. 1982. Human c-*myc* oncogene is located on the region of chromosome 8 that is translocated in Burkitt lymphoma cells. *Proc. Natl. Acad. Sci.* **79:** 7824.

de Thé, H., C. Lavau, A. Marchio, C. Chomienne, L. Degos, and A. Dejean. 1991. The PML-RARα fusion mRNA generated by the t(15;17) translocation in acute promyelocytic leukemia encodes a functionally altered RAR. *Cell* **66:** 675.

Dellefroid, E.J., D.A. Poncelet, P.J. Lecocq, O. Revelant, and J.A. Martial. 1991. The evolutionarily conserved Krüppel-associated box domain defines a subfamily of eukaryotic multifingered proteins. *Proc. Natl. Acad. Sci.* **88:** 3608.

Deweindt, C., J.-P. Kerckaert, H. Tilly, S. Quief, V.C. Nguyen, and C. Bastard. 1993. Cloning of a breakpoint cluster region at band 3q27 involved in human non-Hodgkin's lymphoma. *Genes Chromosomes Cancer* **8:** 149.

Dibello, P.R., D.A. Withers, C.A. Bayer, J.W. Fristrom, and G.M. Guild. 1991. The *Drosophila Broad-Complex* encodes a family of related proteins containing zinc fingers. *Genetics* **129:** 385.

Djabali, M., L. Selleri, P. Parry, M. Bower, B.D. Young, and G.A. Evans. 1992. A trithorax-like gene is interrupted by chromosome 11q23 translocations in acute leukemias. *Nat. Genet.* **2:** 113.

El-Baradi, T. and T. Pieler. 1991. Zinc finger proteins: What we know and what we would like to know. *Mech. Dev.* **35:** 155.

Fichelson, S., F. Dreyfus, R. Berger, J. Melle, C. Bastard, J.M. Miclea, and S. Gisselbrecht. 1992. Evi-1 expression in leukemic patients with rearrangements of the 3q25-q28 chromosomal region. *Leukemia* **6:** 93.

Gaidano, G. and R. Dalla-Favera. 1993. Biologic and molecular characterization of non-Hodgkin's lymphoma. *Curr. Opin. Oncol.* **5:** 776.

Gaidano, G., P. Ballerini, J. Gong, A. Neri, E.W. Newcomb, I.T. Magrath, D.K. Knowles, and R. Dalla-Favera. 1991. p53 mutations in human lymphoid malignancies: Association with Burkitt lymphoma and chronic lymphocytic leukemia. *Proc. Natl. Acad. Sci.* **88:** 5413.

Gribben, J.G., D. Neuberg, A.S. Freedman, C.D. Gimmi, K.W. Pesek, M. Barber, L. Saporito, S.D. Woo, F. Coral, N. Spector, S.N. Rabinowe, M.L. Grossbard, J. Ritz, and L. Nadler. 1993. Detection by polymerase chain reaction of residual cells with the BCL-2 translocation is associated with increased risk of relapse after autologous bone marrow transplantation for B-cell lymphoma. *Blood* **81:** 3449.

Gu, Y., T. Nakamura, H. Alder, R. Prasad, O. Canaani, G. Cimino, C.M. Croce, and E. Canaani. 1992. The t(4;11) chromosome translocation of human acute leukemias fuses the *ALL*-1 gene, related to *Drosophila trithorax*, to the *AF*-4 gene. *Cell* **71:** 701.

Haber, D.A., A.J. Buckler, T. Glaser, K.M. Call, J. Pelletier, R.L. Sohn, E.C. Douglass, and D.E. Housman. 1990. An internal deletion within an 11p13 zinc finger gene contributes to the development of Wilms' tumor. *Cell* **61:** 1257.

Harrison, S.D. and A.A. Travers. 1990. The *tramtrack* gene encodes a *Drosophila* finger protein that interacts with the *ftz* transcriptional regulatory region and shows a novel embryonic expression pattern. *EMBO J.* **9:** 207.

Kakizuka, A., W.H. Miller, Jr., K. Umesono, R.P. Warrell, Jr., S.R. Frankel, V.V. Murty, E. Dmitrovsky, and R.M. Evans. 1991. Chromosomal translocation t(15;17) in human acute promyelocytic leukemia fuses RARα with a novel putative transcription factor, PML. *Cell* **66:** 663.

Kerckaert, J.-P., C. Deweindt, H. Tilly, S. Quief, G. Lecocq, and C. Bastard. 1993. *LAZ3*, a novel zinc-finger encoding gene, is disrupted by recurring chromosome 3q27 translocations in human lymphoma. *Nat. Genet.* **5:** 66.

Knochel, W., A. Poting, M. Koster, T. El Baradi, W. Nietfeld, T. Bouwmeester, and T. Pieler. 1989. Evolutionary conserved modules associated with zinc fingers in *Xenopus laevis*. *Proc. Natl. Acad. Sci.* **86:** 6097.

Koonin, E.V., T.G. Senkevich, and V.I. Chernos. 1992. A family of DNA virus genes that consists of fused portions of unrelated cellular genes. *Trends Biochem. Sci.* **17:** 213.

Korsmeyer, S.J. 1992. Bcl-2 initiates a new category of oncogenes: Regulators of cell death. *Blood* **80:** 879.

Kuze, K., A. Shimizu, and T. Honju. 1991. Characterization of the enhancer region for germline transcription of the gamma 3 constant region gene of human immunoglobulin. *Int. Immunol.* **3:** 647.

Lo Coco, F., G. Gaidano, D.C. Louie, K. Offit, R.S.K. Chaganti, and R. Dalla-Favera. 1993. p53 mutations are associated with histologic transformation of follicular lymphoma. *Blood* **82:** 2289.

Lo Coco, F., B.H. Ye, F. Lista, P. Corradini, K. Offit, D.M. Knowles, R.S.K. Chaganti, and R. Dalla-Favera. 1994. Rearrangements of the BCL-6 gene in diffuse large-cell non-Hodgkin lymphoma. *Blood* **83:** 1757.

Magrath, I. 1990. Lymphocyte ontogeny: A conceptual basis for understanding neoplasia of the immune system. In *The non-Hodgkin's lymphoma* (ed. I. Magrath), p. 29-48. Williams and Wilkins, Baltimore, Maryland.

McGuire, E.A., R.D. Hockett, K.M. Pollock, M.F. Bartholdi, S.J. O'Brien, and S.J. Korsmeyer. 1989. The t(11;14)(p15;q11) in a T-cell acute lymphoblastic leukemia cell line activates multiple transcripts, including Ttg-1, a gene encoding a potential zinc finger protein. *Mol. Cell Biol.* **9:** 2124.

Miki, T., N. Kawamata, S. Hirosawa, and N. Aoki. 1994. Gene involved in the 3q27 translocation associated with B-cell lymphoma, BCL-5, encodes a *Krüppel*-like zinc-finger protein. *Blood* **83:** 26.

Morishita, K., D.S. Parker, M.L. Mucenski, N.A. Jenkins, N.G. Copeland, and J.N. Ihle. 1988. Retroviral activation of a novel gene encoding a zinc finger protein in IL3-dependent myeloid leukemia cell lines. *Cell* **54:** 831.

Motokura, T., T. Bloom, K.H. Goo, H. Juppner, J.V. Ruderman, H.M. Kronenberg, and A. Arnold. 1991. A novel cyclin encoded by a bcl-1 linked candidate oncogene. *Nature* **350:** 512.

Offit, K., S. Jhanwar, S.A. Ebrahim, D. Filippa, B.D. Clarkson, and R.S.K. Chaganti. 1989. t(3;22)(q27;q11): A novel translocation associated with diffuse non-Hodgkin's lymphoma. *Blood* **74:** 1876.

Offit, K., F. Lo Coco, D.C. Louie, N.Z. Parsa, D. Leong, C. Portlock, B.H. Ye, F. Lista, D.A. Filippa, A. Rosenbaum, M. Ladanyi, R. Dalla-Favera, and R.S.K. Chaganti. 1994. Rearrangement of the *BCL6* gene as a prognostic marker in diffuse large cell lymphoma. *N. Engl. J. Med.* **331:** 74.

Pandolfi, P.P., F. Grignani, M. Alcalay, A. Mencarelli, A. Biondi, F. Lo Coco, F. Grignani, and P.G. Pelicci. 1991. Structure and origin of the acute promyelocytic leukemia myl/RARα cDNA and characterization of its retinoid-binding and *trans*activation properties. *Oncogene* **6:** 1285.

Raffeld, M. and E.S. Jaffe. 1991. Bcl-1, t(11;14), and mantle zone lymphomas. *Blood* **78:** 259.

Rosenberg U.B., C. Schroder, A. Preiss, A. Kienlin, S. Côté, I. Riede, and H. Jäckle. 1989. Structural homology of the product of the *Drosophila Krüppel* gene with *Xenopus* transcription factor IIIA. *Nature* **319:** 336.

Schneider, A.M., D.J. Straus, A.E. Schluger, D.A. Lowenthal, B. Koziner, B.J. Lee, G. Wong, and B.D. Clarkson. 1990. Treatment results with an aggressive chemotherapeutic regimen (MACOP-B) for intermediate and some high grade non-Hodgkin's lymphomas. *J. Clin. Oncol.* **8:** 94.

Taub, R., I. Kirsch, C. Morton, G. Lenoir, D. Swan, S. Tronick, S. Aaronson, and P. Leder. 1982. Translocation of c-*myc* gene into the immunoglobulin heavy chain locus in human Burkitt lymphoma and murine plasmacytoma cells. *Proc. Natl. Acad. Sci.* **79:** 7837.

Tkachuk, D.C., S. Kohler, and M.L. Cleary. 1992. Involvement of a homolog of *Drosophila Trithorax* by 11q23 chromosomal translocations in acute leukemias. *Cell* **71:** 691.

Tsujimoto, Y., J. Yunis, L. Onorato-Showe, J. Erikson, P.C. Nowell, and C.M. Croce. 1984. Molecular cloning of the chromosomal breakpoint on chromosome 11 in human B-cell neoplasms with the t(11;14) chromosome translocation. *Science* **224:** 1403.

Xue, F. and L. Cooley. 1993. *Kelch* encodes a component of intercellular bridges in *Drosophila* egg chambers. *Cell* **72:** 681.

Ye, B.H., P.H. Rao, R.S.K. Chaganti, and R. Dalla-Favera. 1993a. Cloning of bcl-6, the locus involved in chromosomal translocations affecting band 3q27 in B-cell lymphoma. *Cancer Res.* **53:** 2732.

Ye, B.H., F. Lista, F. Lo Coco, D.M. Knowles, K. Offit, R.S.K. Chaganti, and R. Dalla-Favera. 1993b. Alterations of a zinc finger-encoding gene, BCL-6, in diffuse large cell-lymphoma. *Science* **262:** 747.

Functional Analysis of the *TAN-1* Gene, a Human Homolog of *Drosophila Notch*

J. Aster,* W. Pear,‡ R. Hasserjian,* H. Erba,*† F. Davi,*
B. Luo,* M. Scott,‡ D. Baltimore,‡ and J. Sklar*

*Division of Molecular Oncology, Department of Pathology, and †Division of Hematology-Oncology,
Brigham and Women's Hospital, Harvard Medical School, Boston, Massachusetts 02115;
‡Department of Biology, Massachusetts Institute of Technology, Cambridge, Massachusetts 02142*

The *TAN-1* gene was originally discovered at the breakpoint of a recurrent (7;9)(q34;q34.3) chromosomal translocation found in a subset of human T-lymphoblastic leukemias (Reynolds et al. 1987; Smith et al. 1988; Ellisen et al. 1991). This translocation joins roughly the 3' half of *TAN-1* head-to-head with the 3' portion of the β T-cell-receptor gene (*TCRB*) beginning at the 5' boundary of one or the other J segment. Intact *TAN-1* is normally transcribed into an 8.2-kb transcript that is present in many tissues, most abundantly in developing thymus and spleen (Ellisen et al. 1991). This tissue distribution and the apparent involvement of an altered version of the gene in T-cell cancers have suggested that *TAN-1* normally has some special function in lymphocytes or their precursors.

Nucleotide sequence analysis of the *TAN-1* transcript has revealed a single open reading frame of 2555 codons. The predicted amino acid sequence is highly homologous to the product of the *Drosophila* gene *Notch*. This gene encodes a transmembrane protein that has been implicated in cell fate decisions between alternative differentiative pathways in a variety of tissues during both embryonic and adult fly development. Most data suggest that the protein product of *Notch* functions as a receptor in a signaling pathway (Fortini and Artavanis-Tsakonis 1993; Ghysen et al. 1993). The similarity of the *TAN-1* and *Notch* proteins (referred to here as tan-1 and notch) includes a series of shared sequence motifs within the primary structures of the two proteins (summarized in Fig. 1). Other *Notch* homologs have been identified recently in mice (Franco del Amo et al. 1992; Reaume et al. 1992; Lardelli and Lendahl 1993), rats (Weinmaster et al. 1991, 1992), zebrafish (Bierkamp and Campos-Ortega 1993), and *Xenopus* (Coffman et al. 1990). In mice, there are at least three closely related homologs, *Notch1–3*, in addition to one more distantly related homolog, *Int-3*, which shares most structural motifs with the other genes (Robbins et al. 1992). A partial sequence of human *Notch2*, termed *hN* (Stifani et al. 1992), has also been reported, indicating the existence of a *Notch* gene family in humans as well as mice.

We decided to investigate the *TAN-1* gene and its role in oncogenesis for a number of reasons. First, homology between genes at chromosomal breakpoints in human cancers and developmental genes from *Drosophila* has become a common theme in molecular oncology. *TAN-1* was among the first such genes identified, and we were interested in the interrelationship between tumor formation and normal differentiation. We were also curious about the significance of alterations in *TAN-1* with respect to malignant transformation. The karyotypes of neoplasms with the t(7;9)(q34;q34.3) contain few other cytogenetic abnormalities. Nevertheless, it is always possible that certain obscure mutations play a more fundamental role in transformation and that translocations involving *TAN-1* merely provide some marginal proliferative or surviv-

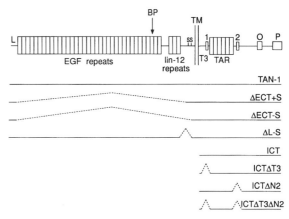

Figure 1. Schematic representation of the tan-1 protein and polypeptides expected to be expressed from various cDNA constructs. The positions of structural motifs within the full-length protein based on the cDNA sequence are shown at the top. (L) Leader sequence; (EGF repeats) epidermal growth factor-like repeated sequences; (BP) position at which the amino acid sequence of tan-1 would be disrupted by the breakpoints in the t(7;9)(q34;q34.3) in all such translocations analyzed to date; (lin-12 repeats) repeated sequences resembling those found in the *Lin-12* gene of *C. elegans*; (SS) highly conserved cysteines found in all *Notch*-related genes and located 49 and 42 amino acids external to the transmembrane domain in tan-1; (TM) transmembrane domain; (1 and 2) two putative nuclear localization signals; (T3) region between the transmembrane domain and the ankyrin repeats used as an immunogen; (TAR) ankyrin-like repeats; (O) glutamine and proline-rich region resembling an opa sequence; (P) region containing proline, glutamate, serine, and threonine, which resembles a PEST sequence. Portions of the full-length tan-1 protein expected to be expressed by the cDNA constructs are shown as solid lines at the bottom; interstitial deleted segments are indicated by dashed lines.

al advantage to cells that had already undergone transformation due to the effects of these other mutations. Additionally, the fact that only a fraction (perhaps 3–5%) of T-lymphoblastic neoplasms contain the t(7;9)(q34;q34.3) and that these neoplasms are not readily distinguishable from other T-lymphoblastic tumors by any outstanding morphological, histochemical, or immunologic feature also suggests that alterations in *TAN-1* may constitute only a secondary phenomenon which follows the primary transformation event. Finally, we believed that insights into the function of any of the several structural motifs within tan-1 might provide clues to the role of similar regions in the relatively large number of proteins that share these motifs. With these considerations in mind, we have proceeded in analysis of the *TAN-1* gene along five different lines of investigation: (1) structural characterization of tan-1 in cells with and without the t(7;9)(q34;q34.3); (2) analysis of the subcellular location of tan-1; (3) in vivo transformation of murine bone marrow stem cells by *TAN-1* cDNA; (4) identification of an intracellular ligand of tan-1; and (5) detection of a possible role for tan-1 in transcriptional activation.

EXPERIMENTAL PROCEDURES

Materials. Enzymes used in cloning procedures were obtained from New England Biolabs. Unless otherwise indicated, cell culture reagents were obtained from Gibco-BRL.

Cells and cell lines. The cell lines SUP-T1 and Jurkat were grown in RPMI 1640 supplemented with 10% heat-inactivated fetal calf serum. 293T cells were grown in Dulbecco's modified Eagle's medium supplemented with 10% heat-inactivated calf serum; NIH-3T3 cells were grown in the same medium supplemented with 10% donor calf serum.

Synthesis of cDNA expression constructs. *TAN-1* constructs were assembled from previously described cDNAs (Ellisen et al. 1991) using standard cloning methods. DNA sequences encoding various portions of the protein (Fig. 1), including the *TAN-1* translational start site and 5' signal peptide, the T3 region, and the TAR region, were amplified from cDNA templates by PCR. The cDNA expression constructs assembled from these PCR products encode the following polypeptides: TAN-1, amino acids 19–2555; ΔECT + S(L), amino acids 1673–2555; ΔECT-S(L), amino acids 1704–2555; ICT, amino acids 1768–2555; T3, amino acids 1762–1879; TAR, amino acids 1872–2150. ΔL-S was derived from the full-length cDNA by creating an internal deletion that removed base pairs 4684–5109 (codons 1562–1703).

Construction and in vitro transcription/translation of a full-length BCL-3 cDNA. An overlapping 3' fragment of the *BCL-3* cDNA encoding the carboxyl terminus of the protein was PCR amplified from a cDNA library prepared from Jurkat cell poly(A)$^+$ RNA. This fragment was ligated to the 5' end of the *BCL-3* cDNA that had been isolated from the human endothelial cell cDNA library. The *BCL-3* construct was ligated into pBluescript and the structure confirmed by both restriction endonuclease digestion and DNA sequence analysis. Deletion constructs were prepared by PCR of the full-length *BCL-3* cDNA using appropriate oligonucleotide primer pairs. In vitro transcription with T3 RNA polymerase was performed according to the recommendations of the supplier (Stratagene). In vitro translation of the *BCL-3* transcripts was performed in a rabbit reticulocyte lysate in the presence of [^{35}S]methionine according to the recommendations of the manufacturer (Stratagene).

Polypeptide synthesis and antibody preparation. cDNAs encoding the T3 and TAR regions of tan-1 were cloned into the vector pGEX-4T (Pharmacia) and expressed in *Escherichia coli* as glutathione-S-transferase (GST) fusion proteins. GST, GST-T3, and GST-TAR were purified from bacterial extracts by affinity chromatography on glutathione-Sepharose columns (Pharmacia). Chickens and rabbits were immunized with purified antigen emulsified in Freund's complete adjuvant and boosted periodically with antigen in Freund's incomplete adjuvant. Serum from immunized chickens and rabbits was cleared of antibody against GST by passage over a GST-AffiGel column (BioRad), and the flowthrough was then applied to AffiGel-GST-T3 or AffiGel-GST-TAR columns. Bound antibody was eluted with 0.2 M glycine (pH 2.7), neutralized with 1 M Tris (pH 8.0), and dialyzed against PBS.

Transient and stable expression of cDNA. For transient expression studies, cDNAs were cloned into the *Bam*HI site of the vector pcDNAI (Invitrogen). 293T cells were transfected with CsCl-banded DNA using a modified calcium phosphate precipitation method (Pear et al. 1993). For stable expression studies, cDNAs were cloned into the *Bcl*I site of the retroviral shuttle vector pGD (Daley et al. 1990). Helper-free retroviral stocks produced by transfection of the packaging cell line Bosc23 were used to infect NIH-3T3 cells (Pear et al. 1993), and cells containing stably integrated provirus were selected with G418 (1 mg/ml).

Bone marrow transplantation. Transfection of Bosc23 cells with retroviral vectors, cocultivation of Bosc23 cells and mononuclear cells isolated from 5-FU-treated bone marrow, and injection of infected mononuclear cells into lethally irradiated BALB/cByJ recipients were performed as described previously (Pear et al. 1993).

Preparation and Western blot analysis of protein extracts. Protein extracts were prepared from cell lines and peripheral blood mononuclear cells using RIPA (50 mM Tris [pH 7.4] containing 150 mM NaCl, 1 mM EDTA, 1% NP-40, 0.5% sodium deoxycholate, 0.1% SDS, 10 mg/ml aprotinin, 10 mg/ml leupeptin, 1 mM PMSF) by standard methods (Harlow and Lane 1988). For Western blot analysis, proteins were subjected to discontinuous polyacrylamide gel electrophoresis

(SDS-PAGE) (Laemmli 1970) and transferred to nitrocellulose membranes (Towbin et al. 1979). Blots were blocked and incubated with primary and secondary antibodies as described previously (Dutta et al. 1993). Blots were developed using the ECL method (Amersham).

Western blots using radiolabeled GST-TAR as probe (Far Western blots) were performed on bacterial cell extracts prepared by sonication in the presence of 0.5% NP-40. Extracts were fractionated by SDS-PAGE and proteins were transferred to nitrocellulose (in the absence of methanol) by electroelution. The filters were blocked with 5% milk. Proteins were renatured by washing in a decreasing guanidine-HCl gradient. GST-TAR fusion protein was radiolabeled with ^{32}P and bovine heart muscle kinase according to the method of Kaelin et al. (1992). Binding of radiolabeled GST-TAR was performed overnight at 4°C in HYB-75 (20 mM HEPES [pH 7.7], 75 mM KCl, 0.05% NP-40, 0.1 mM EDTA, 2.5 mM $MgCl_2$) and 1% milk. Filters were washed in HYB-75 at 4°C. Identical binding and wash conditions were used to screen the human endothelial cell cDNA library in λgt11 (Blackwood and Eisenman 1991; Kaelin et al. 1992).

Immunoprecipitation of metabolically labeled polypeptides. For standard immunoprecipitations, about 5×10^6 Jurkat or SUP-T1 cells were washed with Hank's buffered saline and incubated twice for 1 hour at 37°C in methionine-free RPMI 1640 containing 10% dialyzed fetal calf serum (Sigma). Cells were then incubated for 3 hours in fresh methionine-deficient media supplemented with 100 mCi to 1 mCi of [^{35}S]methionine (New England Nuclear). For pulse-chase analysis, 5×10^7 cells depleted of methionine as described above were resuspended in 8 ml of warm (37°C) methionine-deficient medium containing 5 mCi of [^{35}S]methionine. After 10 minutes at 37°C, cells were spun at 500g for 5 minutes, washed once with complete medium, and resuspended in 40 ml of complete medium containing 2 mM unlabeled methionine. At various time points thereafter, 5-ml aliquots were removed and diluted into 15 ml of ice-cold Hank's buffered saline, and then washed once with ice-cold Hank's buffered saline. RIPA extracts prepared from washed cell pellets were "cleared" by mixing for 30 minutes at 4°C with 20 μl of protein A–Sepharose beads (Pharmacia). After centrifugation, supernatants were incubated with 1 μl of rabbit anti-T3 serum for 1 hour at 0°C, then mixed with 20 μl of protein A–Sepharose beads for 30 minutes at 4°C. Beads were washed 3 times with RIPA and boiled for 10 minutes in 0.1 ml of 0.5% SDS. Supernatants were adjusted to RIPA conditions (final volume, 0.5 ml), and cross-reactive polypeptides were reprecipitated as above using 1 μl of rabbit anti-TAR serum. Control immunoprecipitates were prepared by sequential incubation with pre-immune sera.

Peptide mapping. Proteins were metabolically labeled with [^{35}S]methionine, immunoprecipitated, and subjected to discontinuous SDS-PAGE. After visualization of the position of bands by fluorography, one-dimensional peptide maps were prepared from excised bands using V8 protease (Cleveland et al. 1977).

Immunolocalization. For immunoperoxidase staining, 1×10^4 SUP-T1 or Jurkat cells were deposited on slides by cytocentrifugation. For immunofluorescent staining, cells were grown on 8-chamber slides (Lab-Tek, Permanox). In each case, cells were fixed in 2% paraformaldehyde, permeabilized in $-20°C$ methanol, and stained with affinity-purified rabbit anti-tan-1 antibodies using previously described incubation and washing conditions (Pinkus et al. 1985).

RESULTS

Structural Analysis of tan-1 in Cells with and without the t(7;9)

To investigate the structure of the tan-1 proteins produced in cells containing or lacking the t(7;9)(q34;q34.3), antibodies raised against portions of the cytoplasmic domain of tan-1 were used in Western blot analysis of whole-cell extracts prepared from SUP-T1 and Jurkat cells (Fig. 2). SUP-T1 is a cell line established from a T-lineage lymphoblastic lymphoma that contained two identical copies of the t(7;9) (q34;q34.3), no normal chromosome 9, and therefore no normal copy of *TAN-1*. Jurkat, a control cell line also derived from a T-lymphoblastic tumor, lacks any

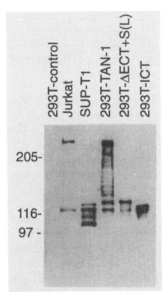

Figure 2. Western blot of cross-reacting polypeptides detected with anti-tan-1 antibodies. Whole-cell protein extracts from Jurkat cells (50 μg of protein), SUP-T1 cells (50 μg), or 293T cells (500 ng) transiently transfected with empty vector or vector carrying *TAN-1* cDNAs were fractionated by SDS-PAGE in 6% gels prior to electrophoretic transfer to nitrocellulose. After incubation with affinity-purified chicken anti-T3 antibody (1 μg/ml) and secondary rabbit anti-chicken antibody linked to horseradish peroxidase (1:2500), cross-reactive polypeptides were detected using chemiluminescence. The positions of size markers are shown in kilodaltons at the left.

known abnormality of chromosome 9 or *TAN-1*. Initial studies performed with antibodies raised in chickens against a portion of the cytoplasmic domain close to the transmembrane domain (termed T3; see Fig. 1) detected a series of bands in SUP-T1 extracts corresponding to polypeptides of ~125–100 kD. This antibody also detected two major bands in Jurkat extracts—one band corresponding to a large polypeptide of ~350 kD (hereafter referred to as p350) and a second band corresponding to a smaller polypeptide of ~120 kD (referred to as p120). The size of the p350 polypeptide is compatible with that predicted for the full-length tan-1 protein, and the size of the p120 polypeptide matches that observed for one of the major truncated products present in SUP-T1 cells.

Other antisera raised in chickens and rabbits against T3 and the region (termed TAR) containing a series of six so-called ankyrin-like repeats also recognized p350 and p120 in extracts from Jurkat cells and polypeptides of 100–125 kD in extracts from SUP-T1 cells (not shown), suggesting that these polypeptides are encoded by the wild-type and t(7;9)(q34;q34.3) *TAN-1* alleles, respectively. The specificity of various antibodies was confirmed in several ways. The polypeptides detected with affinity-purified antibodies against tan-1 were not stained by pre-immune sera, and staining with specific antibody was inhibited by pre-incubation with purified antigen. Furthermore, Western blots of extracts from 293T cells transiently transfected with expression vectors containing full-length *TAN-1* cDNAs showed greatly increased levels of cross-reacting polypeptides.

To characterize the structure of the various cross-reactive polypeptides found in Jurkat and SUP-T1 cells, full-length *TAN-1* cDNAs and cDNAs carrying deletions from the 5′ end of the gene were transiently transfected into 293T cells (Figs. 1 and 2). Transfection with the full-length cDNA resulted in expression of a series of polypeptides, the most prominent of which was close to the size of p350. In contrast, the sizes of the largest of the SUP-T1 polypeptides and p120 were similar to sizes of polypeptides encoded by cDNAs with deletions removing most of the coding sequence for the extracellular domain of tan-1. To further study the structural similarities between polypeptides, a method for specific immunoprecipitation of cross-reactive polypeptides was developed (Fig. 3A). Partial digestion of ^{35}S-labeled immunoprecipitated proteins with V8 protease generated a set of comparable bands in one-dimensional gels from p120 and several of the truncated polypeptides of SUP-T1 (Fig. 3B). Therefore, the compositions of p120 and the truncated polypeptides associated with chromosomal translocation appear similar. Partial digestion of a polypeptide precipitated from 293T cells transfected with a truncated cDNA construct, ΔECT-S(L), yielded a one-dimensional map closely corresponding to that of p120 and the largest truncated polypeptides of SUP-T1. This expression construct encodes a mature polypeptide with an amino terminus at codon 1703, a position 30 amino acids external to the transmembrane domain and just inter-

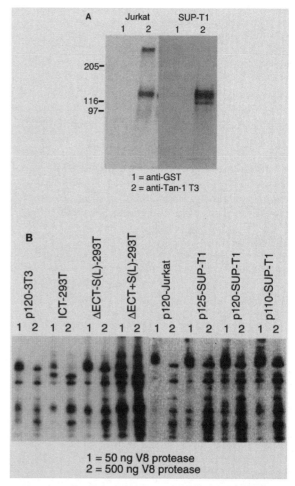

Figure 3. Peptide mapping of tan-1-related polypeptides. (*A*) Immunoprecipitations of tan-1 polypeptides from extracts of Jurkat and SUP-T1 cells. Cells were metabolically labeled with [^{35}S]methionine and proteins were immunoprecipitated with rabbit anti-T3 antiserum, followed by resolubilization of complexes by boiling and reprecipitation with rabbit anti-TAR antiserum. Control immunoprecipitations were performed with rabbit anti-GST antiserum during the first stage. Results were analyzed by SDS-PAGE in a 6% gel and fluorography of the gel. Comparison to the results in Fig. 2 demonstrates the specificity of sequential immunoprecipitation using the two antisera. (*B*) Peptide mapping of immunoprecipitated protein performed as in *A*. Immunoprecipitation was carried out on the NIH-3T3 cells infected with pGD-TAN-1, transiently transfected 293T cells, and Jurkat and SUP-T1 cells metabolically labeled with [^{35}S]methionine. Precipitated polypeptides were separated by SDS-PAGE, located by fluorography, excised from the gel, and subjected to partial proteolysis with 50 ng or 500 ng of V8 protease during electrophoresis in 12% polyacrylamide gels.

nal to two conserved cysteine residues encoded by codons 1683 and 1690. Maps of polypeptides from cells transfected with cDNA constructs including more or less coding sequence near this site (e.g., ΔECT + S[L] and ICT) differed from maps of p120 and the truncated polypeptides. Therefore, the amino terminus of p120 and the largest truncated polypeptides in SUP-T1 are very close and possibly just carboxy-terminal to the two conserved cysteines.

These data show that the size and structure of the tan-1 polypeptides found in extracts of SUP-T1 are consistent with translation of polypeptides from a truncated *TAN-1* transcript. Northern blot analyses of SUP-T1 cells have shown the presence of abundant novel mRNA species that are derived from the 3′ portion of *TAN-1* and vary from 4 to 6 kb (Ellisen et al. 1991), suggesting activation of cryptic promoters in DNA near the breakpoint. The exact positions of the transcriptional start sites are not known and could be either 5′ or 3′ of the breakpoint. The precise translational start site or sites are also uncertain. The translocation breakpoint in SUP-T1 cells and two other tumors containing the t(7;9)(q34;q34.3) lie in an intron within DNA encoding EGF repeat 34, such that the first in-frame codon 3′ of the breakpoint in all tumors begins at bp 4015 of the cDNA (Ellisen et al. 1991). Transfection of 293T cells with a cDNA beginning at bp 4015 and continuing through the 3′ stop codon results in synthesis of several polypeptides of lower abundance, which are larger than those seen in SUP-T1 cells (not shown). This suggests that sequences 3′ of the breakpoint that normally are intronic may contribute one or more translational start sites, and raises the possibility that some of the cross-reactive polypeptides in SUP-T1 cells have sequences at their amino termini found nowhere in the full-length protein. If such sequences exist, however, they are not required for oncogenesis (see below).

The results of Western blots and peptide mapping suggested that p120 is generated by proteolytic cleavage of p350. This interpretation is consistent with the presence of a band at the 120 kD position in blots of extracts from 293T cells and NIH-3T3 cells transiently or stably overexpressing full-length *TAN-1* cDNA, respectively. Northern blots of Jurkat RNA also support this interpretation, since these analyses fail to show *TAN-1* transcripts other than the 8.2-kb species. Therefore, translation of p120 from either an alternatively spliced mRNA or a transcript originating from a second internal promoter is unlikely.

To confirm that p350 is the precursor of p120, a pulse-chase analysis was performed with Jurkat cells. This demonstrated a clear inverse relationship between the amount of p350 and p120, with progressively more p120 accumulating over time (Fig. 4A). These results are compatible with the derivation of p120 from p350 by proteolytic cleavage. On the basis of peptide map data, it was predicted that this cleavage site should be just external to the transmembrane domain. This conclusion is supported by inhibition of proteolytic processing by a deletion removing the amino acid sequence between the lin-12 repeats and the transmembrane domain (Fig. 4B).

Analysis of the Subcellular Location of tan-1

The location of tan-1 proteins within cells was studied by immunofluorescence and immunohistochemistry using the anti-T3 and anti-TAR antibodies.

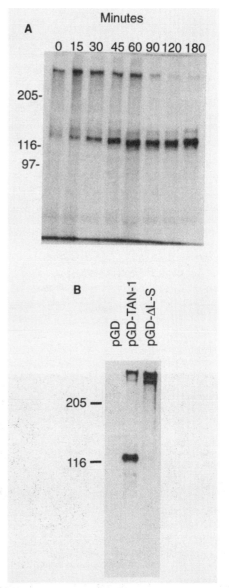

Figure 4. Processing of full-length tan-1 protein. (*A*) Pulse-chase analysis. Jurkat cells were metabolically labeled with [^{35}S]methionine for 10 min, resuspended in medium containing unlabeled methionine, and harvested thereafter at intervals shown at the top of the figure. Polypeptides were immunoprecipitated and analyzed as in Fig. 3A. (*B*) Western blot analysis of NIH-3T3 cells stably expressing tan-1 from retroviral vectors. Whole-cell extracts were prepared from cells infected by the pGD vector containing no cDNA insert, a full-length *TAN-1* cDNA, or a *TAN-1* cDNA deleted for a region lying between the lin-12 repeats and the transmembrane domain (ΔL-S). From each extract, 10 μg was analyzed, using chicken anti-T3 antibody as in Fig. 2.

Because cells lacking the t(7;9)(q34;q34.3) showed only weak staining with these reagents, attempts were made to determine the normal location of tan-1 by infecting NIH-3T3 cells with a recombinant retrovirus that produces relatively large amounts of p350 and processed p120. This procedure resulted in staining of the cytoplasm in a speckled, perinuclear pattern, suggesting localization to the endoplasmic reticulum (Fig. 5A). The N-linked glycosyl residues of p350 were also found to be endoH-sensitive (J. Aster and J. Sklar, unpubl.), further supporting retention of tan-1 in the endoplasmic reticulum. Similar cytoplasmic retention of notch has been demonstrated in overexpressing insect cells, although both notch and tan-1 are predicted to function as transmembrane receptors and therefore to be expressed on plasma membrane. These findings indicate that surface expression of tan-1 and notch may be tightly regulated and dependent on cofactors which have yet to be discovered.

In contrast to the cytoplasmic location of tan-1 in cells expressing full-length tan-1, cross-reacting polypeptides were found predominantly in the nuclei of SUP-T1 cells (Fig. 5B). This staining pattern was also observed in 293T cells transiently expressing the cytoplasmic domain (ICT) of tan-1 (Fig. 5C). Since the polypeptides expressed in SUP-T1 lack signal peptides, insertion of these polypeptides into membranes may occur inefficiently or not at all, thereby explaining the difference in subcellular localization of these polypeptides and polypeptides encoded by the ΔECT(L) expression constructs. Taken together, these findings imply that signals permitting efficient nuclear transport exist within the intracellular portion of tan-1. Four short stretches of intracellular sequence relatively rich in basic amino acids represent potential nuclear localization signals (NLSs). The first two of these lie within the T3 region of the protein, and the other two lie just carboxy-terminal to the ankyrin repeats. Deletion of both of these regions led to retention of most of the tan-1 polypeptide in the cytoplasm (Fig. 5D). Deletion of either region separately was much less effective in causing the redistribution of protein from nucleus to cytoplasm, indicating an additive effect of NLSs in these two regions.

In Vivo Transformation of Murine Bone Marrow Stem Cells by *TAN-1* cDNA

The role of tan-1 protein in malignant transformation was directly tested by transduction of truncated *TAN-1* cDNAs into murine bone marrow cells and infusion of these cells back into lethally irradiated syngeneic mice. Two cDNA constructs, ΔECT + S(L) and ΔECT-S(L),

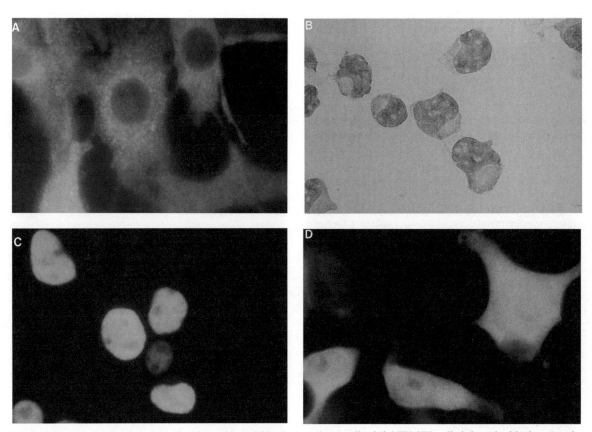

Figure 5. Immunolocalization of tan-1 polypeptides within tissue culture cells. (*A*) NIH-3T3 cells infected with the retrovirus pGD-TAN-1, overexpressing full-length tan-1; (*B*) SUP-T1 cells; (*C*) 293T cells transiently transfected with expression vectors carrying a cDNA construct for the cytoplasmic domain of *TAN-1* (ICT); and (*D*) 293T cells transiently transfected with an expression vector identical to that in *C*, except for the deletion from the *TAN-1* coding sequence of two putative nuclear localization signals (ICTΔT3ΔN2). Cells in each panel were stained with affinity-purified rabbit anti-T3 or anti-TAR antibodies, followed by sheep anti-rabbit antibody linked to fluorescein thiocyanate (*A*, *C*, and *D*) or goat anti-rabbit antibody linked to horseradish peroxidase (*B*). Cells in *A*, *C*, and *D* were photographed through a fluorescence microscope. Cells in *B* were photographed through a standard visible light microscope after developing the image by incubation with diaminobenzidine and counterstaining with methyl green. Magnification, 650×.

were inserted into the retroviral shuttle vector pGD, which was transfected into the packaging cell line Bosc23. Bone marrow mononuclear cells were infected with the resultant helper-free retrovirus by co-cultivation with the Bosc23 cells. Following transplantation, mice were monitored by biweekly bleeding and by observation for cachexia. Mice with significantly elevated leukocyte counts or grossly apparent disease were sacrificed and autopsied.

To date, 7 of 20 mice have developed leukemia with a latency of between 11 and 40 weeks posttransplantation. Mice that developed leukemia were observed with both cDNA constructs (Table 1). Flow cytometry indicated that the leukemic cells in each case had an antigenic profile compatible with an immature T-cell phenotype. All tumors have proven to be readily transplantable into syngeneic recipients, and one stable cell line has been derived (T6E).

Immunostaining of various tissues with tan-1 antibodies demonstrated high levels of cross-reactive protein in infiltrating tumor cells relative to surrounding normal tissue (not shown). To determine the intracellular distribution of staining, additional immunohistochemistry was performed on cytospin preparations of the T6E cell line and disaggregated tumors. This analysis revealed most staining to be cytoplasmic in a vesicular, perinuclear distribution, with less intense nuclear membrane staining evident in a subset of cells (Fig. 6A). The precise identity of the vesicular structures is not yet known. Western blot analysis performed on extracts of these tissues with anti-T3 and anti-TAR antibodies detected one to several prominent bands at the position of polypeptides of about 120 kD, the approximate size of the products expected to be synthesized from the ΔECT + S(L) and ΔECT-S(L) cDNA constructs (Fig. 6B). Additionally, particularly in extracts prepared from the T6E cell line, a band of about 350 kD was also present, corresponding to the size of full-length notch-related proteins. Further work has shown that this band also cross-reacts with an antibody directed against a region just amino-terminal of the opa sequence (J. Aster et al., unpubl.). This region is found in murine tan-1 (notch1), but is absent from notch2, notch3, and int-3, suggesting that the cross-reactive band is the product of endogenous murine *Notch1*. It is unclear

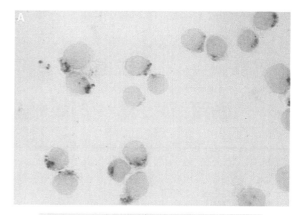

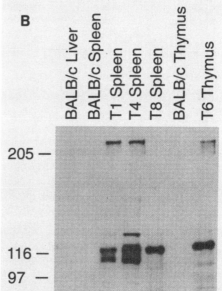

Figure 6. Analysis of tan-1 polypeptide in murine leukemias induced by infection of bone marrow cells with retroviral vectors carrying deleted forms of the *TAN-1* cDNA. (*A*) Immunolocalization of tan-1 polypeptide in the T6E in vitro cell line. The cells were stained with rabbit anti-T3 antibody and the image was developed and photographed as in Fig. 5B. Magnification, 650×. (*B*) Western blot analysis of extracts prepared from organs of leukemic mice which showed heavy visceral involvement by tumor and from organs of normal BALB/c control animals. T1, 4, 6, and 8 correspond to the designations in Table 1. Ten μg from each extract were analyzed in Western blots using rabbit anti-T3 antibody.

Table 1. Characteristics of Lymphoblastic Neoplasms Arising in Mice Which Underwent Transplantation with Bone Marrow Infected with Retroviral Vectors Carrying ΔECT + S(L) and ΔECT-S(L) cDNAs.

Animal #	Cys repeats	latency (weeks)	Terminal WBC (k/μl)	Sites of gross involvement				FACS					2° mice
				node	spleen	liver	thymus	CD3	CD4	CD8	β	
T1	−	11	70	+	+	−	−	+	+	+	+	+
T3	−	16	4	−	−	−	−	+	−	+	+	+
T4	−	12	68	+	+	+	−	+	+	+	+	+
T6	+	15	42	+	+	−	+	+	−	+	+	+
T7	+	40	65	+	+	+	+	n.d.	−	−	n.d.	n.d.
T8	+	15	41	+	+	−	−	+	+	+	+	+
T14	+	16	55	+	+	−	−	+	+	+	+	+

The first three mice received ΔECT-S(L) (indicated by − in the second column) and the next four ΔECT + S(L) (indicated by + in the second column). FACS = results of flow cytometry. n.d. = not done. WBC = white blood cell count. β = T-cell receptor β chain.

whether expression of notch1 is due to induction by the truncated tan-1 protein or to the selective transformation of T-cell precursors in which the gene is normally expressed.

Southern blot analysis with probes specific for the retroviral vector sequences has shown that six of seven tumors analyzed contain a single integrated provirus, with the seventh having two proviruses (not shown). Southern blot analyses showed that all sites of insertion into the genome were unique. Additionally, analyses with Jβ1 and Jβ2 probes indicated that all tumors except for that in animal T6 had clonal rearrangements of the β T-cell-receptor gene.

These studies demonstrate that tumors arising in recipients of marrow infected with retrovirus carrying truncated forms of *TAN-1* develop clonal T-lymphoblastic neoplasms at high frequency. An additional seven tumors have now been observed in other transplant cohorts, all of which have also demonstrated an immature T-cell phenotype (W.S. Pear et al., unpubl.), further emphasizing the T-cell oncotropism of *TAN-1*. In contrast, of approximately 200 animals receiving marrow transduced with pGD carrying other cDNA constructs, none has developed a T-cell tumor (W.S. Pear et al., unpubl.). Although recombination of defective retrovirus with an endogenous provirus to produce competent Moloney leukemia virus is a theoretical concern, transcripts encoding the Moloney envelope have not been detected in leukemic animals (W.S. Pear et al., unpubl.), making this possibility unlikely.

Identification of an Intracellular Ligand of tan-1 Protein

The presence of both large extracellular and intracellular domains within the tan-1 protein suggests that this molecule may mediate the transduction of some signal across the plasma membrane by binding or interacting with ligands on either side of the membrane. Indeed, in *Drosophila*, direct physical interaction has been demonstrated between the extracellular portion of notch and the protein products of two genes, *Delta* and *Serrate* (Rebay et al. 1991). Our initial efforts to identify ligands of tan-1 have concentrated on the intracellular region of the molecule. Since ankyrin repeats have been implicated previously in interactions between other proteins (Lux et al. 1990; Nolan and Baltimore 1992), we have searched for ligands that might bind to the region of tan-1 containing these repeats.

A 750-bp fragment encoding the ankyrin repeats (TAR) was amplified by PCR from the cloned *TAN-1* cDNA, ligated into the bacterial glutathione-*S*-transferase (GST) expression plasmid pGEX2TK, and expressed in *E. coli* as a fusion polypeptide with GST at the amino terminus. The GST-TAR fusion polypeptide labeled with ^{32}P was then used to screen plaques for TAR-binding proteins from a human endothelial cell cDNA library cloned in λgt11 (Ginsburg et al. 1985). This randomly primed cDNA library was chosen for screening because it is known to be highly representative of a vast array of RNAs and because it is the same library from which cDNA on the 5' end of the *TAN-1* mRNA was originally isolated (Ellisen et al. 1991). Several plaques showed binding of the GST-TAR fusion polypeptide in this assay. Bacteriophage from the single plaque that continued to show binding after three rounds of screening were purified, and the nucleotide sequence of the cDNA inserted within the vector DNA was determined. Comparison of sequences in the GenBank database demonstrated a match between the isolated cDNA and a large 5' fragment of the coding sequence for the gene *BCL-3*.

To confirm the binding of the GST-tan-1 fusion polypeptide to bcl-3, portions of the *BCL-3* cDNA were inserted into the pGEX2TK plasmid, and GST-bcl-3 fusion polypeptides were expressed in *E. coli*. Protein extracts from these bacteria were separated in SDS-polyacrylamide gels and transferred to nitrocellulose membranes. The membranes were then incubated with radiolabeled GST-TAR fusion polypeptide. Autoradiograms of the membranes revealed that the GST-TAR polypeptide had specifically bound the amino-terminal fragment of bcl-3 protein containing the first six of the seven ankyrin repeats found in bcl-3 (Fig. 7A). As few as three bcl-3 ankyrin repeats were sufficient for some residual binding by GST-TAR. To demonstrate that the interaction between TAR and bcl-3 polypeptides was not due to an artifact arising from truncations of the bcl-3 polypeptide, full-length bcl-3 radiolabeled with [^{35}S]methionine was synthesized in vitro. Bcl-3 protein synthesized in this fashion bound to GST-TAR glutathione-Sepharose beads, but not GST glutathione-Sepharose beads (Fig. 7B).

Direct co-immunoprecipitation of tan-1/bcl-3 complexes with anti-tan-1 and/or anti-bcl-3 antibodies from tissue culture cells was complicated by the low amounts of bcl-3 produced in most cells and by the low affinity of the anti-bcl-3 antibodies which we have either produced ourselves or obtained from other sources. To overcome the problem of low antigen concentration within cells, we have conducted co-immunoprecipitation studies using transfected 293T cells co-expressing epitope-tagged bcl-3 and various forms of tan-1. For these studies, the *BCL-3* cDNA was inserted into the vector pCGN, which produces a fusion polypeptide containing an epitope from the influenza virus hemagglutinin (HA) at the amino terminus of the fusion protein. Immunoprecipitates of extracts from the doubly transfected cells using the mouse monoclonal antibody 12CA5, which recognizes the HA epitope, brought down p350 together with p120 tan-1 polypeptides presumably bound to the HA-bcl-3 fusion protein (Fig. 7C). These data demonstrate that interaction of tan-1 and bcl-3 can occur in intact cells.

Evidence of a Possible Role for tan-1 in Transcriptional Activation

We unexpectedly uncovered evidence suggesting a possible role for tan-1 in transcriptional activation

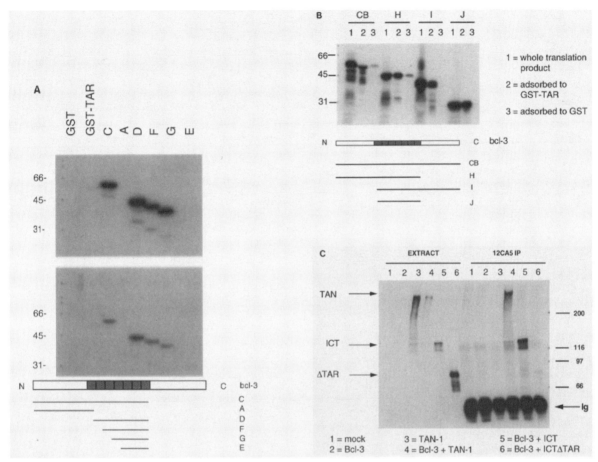

Figure 7. Analysis of protein-protein interactions between tan-1 and bcl-3. (*A*) Far Western blot of GST-bcl-3 polypeptides expressed in *E. coli*, using radiolabeled GST-TAR as a probe. Protein extracts from 400 μl saturated bacterial cultures were loaded in each lane; Coomassie blue staining of a duplicate gel showed equivalent amounts of protein in the lanes. To demonstrate the specificity of binding, the membrane in the top panel was incubated with probe in the presence of unlabeled GST and in the lower panel in the presence of unlabeled GST-TAR. At the bottom are shown fragments of bcl-3 joined to GST in various fusion polypeptides analyzed. The seven tandem ankyrin repeats in bcl-3 are indicated by shaded boxes. (*B*) Adsorption of ^{35}S-labeled bcl-3 polypeptide translated in vivo to GST or GST-TAR bound to glutathione-Sepharose beads. In vitro translation products were separated by SDS-PAGE in a 6% gel immediately following translation (lane *1*) or after adsorption to glutathione-Sepharose beads preloaded with either GST-TAR (lane *2*) or GST alone (lane *3*). For lanes 2 and 3, fivefold more translation product was incubated with the beads as compared to the amount of product loaded directly in lane *1*. The gel was dried and fluorographed. (*C*) Co-immunoprecipitation of tan-1 and epitope-tagged bcl-3 polypeptides from transfected cells. 293T cells were infected with pCGN-HA-BCL-3 and/or pCGN-TAN-1 expression vectors, some of which contained deletions in the *TAN-1* cDNA. NP-40 extracts were prepared after two days of culture, and HA-bcl-3 fusion polypeptide was precipitated with the 12CA5 antibody directed against the HA epitope. Whole extracts (*left*) and immunoprecipitates (*right*) were analyzed by Western blotting as in Fig. 2. 25-fold more extract was used for immunoprecipitation compared to the amounts of whole extract loaded directly on the gel.

when we attempted to utilize the so-called yeast two-hybrid system (Zervos et al. 1993) to identify intracellular ligands of tan-1. In preparation for using the two-hybrid system, we found that a "bait" plasmid expressing a portion of the cytoplasmic domain of tan-1 (including the ankyrin repeats) fused to the DNA-binding protein LexA showed significant transcriptional activation of a β-galactosidase reporter gene. In fact, the degree of activation was in the range produced by a LexA-gal4 fusion protein, a natural strong transcriptional activator in *Saccharomyces cerevisiae* (Table 2). On the basis of these results, the intracellular domain of tan-1 apparently has the capacity, by itself, to activate transcription in yeast when it is supplied with the proper DNA-binding site.

DISCUSSION

Our studies have established several major points about *TAN-1* and have also yielded a number of paradoxical observations. As demonstrated by experiments in which cDNA fragments of the *TAN-1* gene were transduced into murine bone marrow cells that were then infused into irradiated hosts, truncated forms of the gene have potent oncogenic activity and exhibit a striking oncotropism for T-cell precursors. This ob-

Table 2. Transcription Activation by *TAN-1* cDNA Fragments in the Yeast Two-hybrid System

	β-Galactosidase
LEXA	5,1
LEXA-GAL4	1196,1744
LEXA-TAN-1 1760-2124	1596,759

On the left is the DNA insert ligated into the bait plasmid of the yeast two-hybrid system. *LEXA* is the coding sequence for the *E. coli* DNA-binding protein. *GAL4* is the full-length cDNA of the *S. cerevisiae* transcription factor. *TAN-1* 1760-2124 is the portion of the *TAN-1* cDNA containing codons 1760–2124 of the tan-1 protein (a region beginning on the intracellular side of the transmembrane domain and ending just carboxy-terminal to the ankyrin repeats. Each of these constructs was expressed in yeast and scored for the ability to increase the expression of a β-galacosidase reporter gene (Zervos et al. 1993). β-Galactosidase activity was measured in arbitrary optical units. The results of two separate cultures for each bait plasmid are shown.

servation confirms the inference that the 3′ portion of this gene is related to the primary transforming event in the leukemia of those patients whose malignant T cells carry the t(7;9)(q34;q34.3). The frequency with which the recipient mice developed leukemia suggests that the oncogenic potential of this gene is high despite the fact the t(7;9)(q34;q34.3) occurs only in a minority of T-cell lymphoblastic neoplasms. However, the precise efficiency of truncated *TAN-1* in inducing transformation cannot be determined, since we do not yet know the number of murine bone marrow cells which actually acquire the *TAN-1* cDNA in the transplant experiments or the identity of the target cell that is susceptible to transformation. The frequency of leukemias in the transplanted mice is probably also affected to some degree by the stochastic acquisition of other mutations, which, in addition to transduction of altered *TAN-1*, are probably required for malignant transformation of T cells. However, the high overall incidence of leukemia in the transplanted mice and the relatively short latency period suggest that the number of additional genetic events required for transformation is small.

Given the potency of truncated *TAN-1* in transforming T cells, the relative rarity of human leukemias with the t(7;9)(q34;q34.3) may reflect a narrow interval during T-cell development during which the translocation can exert its oncogenic effect. Alternatively, some structural feature of DNA within the *TAN-1* gene may make translocation in the critical region of the gene unlikely. These experiments also show that production of a chimeric gene product, for instance, between the β T-cell receptor and tan-1, is not essential for transformation. In light of this fact, perhaps the proper positioning or activation of a promoter that can function in T cells is the most important feature of translocations which will transform T cells through truncation of *TAN-1*.

The structure and subcellular localization of the polypeptides encoded by *TAN-1* cDNAs that exhibit transforming activity are difficult to reconcile in several respects with the tan-1 polypeptides detected in cells either containing or lacking the t(7;9)(q34;q34.3). Analyses of the tan-1 polypeptides in cells with intact *TAN-1* genes show extensive proteolytic cleavage of the full-length protein, p350, within the extracellular part of the protein near the transmembrane domain. The proteolytic product containing the intracellular portion of tan-1, p120, closely resembles the major truncated tan-1 polypeptide found in the cell line SUP-T1, containing the t(7;9)(q34;q34.3). Furthermore, both of these polypeptides are similar to the transforming polypeptides expressed from the ΔECT + S(L) and ΔECT + S(L) cDNAs. These observations suggest that p120 ought to be functionally similar to polypeptides encoded by transforming cDNAs. However, all evidence at present indicates that p120 has no transforming potential in T-lineage cells, even when generated presumably at high levels from viruses carrying a full-length *TAN-1* cDNA in additional murine bone marrow transplant experiments (W.S. Pear et al., unpubl.).

One obvious difference between the normal p120 and the truncated product in SUP-T1 is the subcellular location of the proteins. At least in transfected cells, p120 seems to be located predominantly in the cytoplasm, whereas truncated tan-1 in SUP-T1 is found primarily in the nucleus. These locations are consistent with the divergent mechanisms by which the proteins are produced. p120 is presumably derived from p350 after insertion into the membrane of the endoplasmic reticulum, but the truncated product in SUP-T1 may be inefficiently inserted into membranes due to the lack of a signal peptide. However, like p120, both the ΔECT + S(L) and ΔECT + S(L) polypeptides in transformed murine T cells accumulate preferentially in the cytoplasm, as expected on the basis of the structure of the encoded proteins. Therefore, neither the cytoplasmic location nor the membrane association appears to account for the inability of p120 to transform T cells, despite the resemblance of this polypeptide to oncogenic versions of tan-1. This paradox suggests that p120 may differ from transforming polypeptides in tertiary structure and/or in its ability to associate with unknown regulatory factors involved in tan-1 signaling.

The teleologic reason for the intense nuclear localization of the truncated tan-1 polypeptides in SUP-T1 is not clear. As discussed above, abundant intranuclear tan-1 polypeptide does not seem to be required for transformation. Therefore, if transformation indicates some exaggerated physiological effect of tan-1, nuclear localization is not a prominent feature of tan-1 function. Moreover, no cross-reacting tan-1 polypeptide could be detected by immunofluorescence in the nuclei of transfected cells expressing large amounts of p350 and p120.

Nevertheless, the presence of some nuclear tan-1 polypeptide cannot be ruled out. It may be that small amounts of tan-1 polypeptide do move to the nucleus and that very little protein within the nucleus is necessary for transformation. This scenario is reminiscent of the situation recently described for the sterol-regulatory element binding protein (SREBP-1) (Wang et al. 1994). The precursor form of this protein is inserted into the endoplasmic reticulum and nuclear envelope of cells, where it is proteolytically processed in response to low sterol levels. A fragment of the processed protein migrates to the nucleus and up-regulates expression of genes for enzymes involved in sterol synthesis. SREBP-1 is very unstable within the nucleus, being undetectable in the absence of a specific inhibitor that blocks intranuclear degradation of the protein.

The analogy of tan-1 to SREBP-1 would be even more complete if tan-1 served some role in transcription activation, as suggested by our data on the ability of tan-1 to activate transcription in yeast. The relevance of transcriptional activity detected in this heterologous system remains to be determined, although the magnitude of activity displayed by the tan-1 sequences is unusual for a purely artifactual result. If tan-1 possesses physiological intranuclear activity of any kind, such activity would depend on an additional cleavage of the full-length or previously processed protein at some position on the cytoplasmic side of the transmembrane domain in order to free the polypeptide from the membrane. To date, we have detected no evidence of such a cleavage. On the other hand, the presence of the highly efficient nuclear localization signals identified within the intracellular portion of the tan-1 protein would certainly be consistent with intranuclear function. The alternative explanation for these signals is that they are merely fortuitous.

Studies of notch in *Drosophila* suggest that this protein transmits developmental signals across the plasma membrane of cells in a number of tissues. This transduction is likely to involve the binding ligands on both the extra- and intracellular sides of the membrane. In humans, the nature of the extracellular ligand is completely unknown; however, our work has demonstrated that bcl-3 is a strong candidate for an intracellular ligand. The interaction between bcl-3 and tan-1 was revealed through a screen for possible ligands expressed by a library of human endothelial cell cDNAs. This library was utilized for this purpose solely because it was known to contain *TAN-1* cDNAs as well as cDNAs for a wide variety of other genes. Despite the random and unbiased nature of this screen, the protein detected to bind to the intracellular portion of tan-1 is one associated with lymphocytes, having originally been discovered as the product of a gene lying at the breakpoint of a translocation, t(14;19)(q32;q13), found in the neoplastic B cells of certain cases of chronic lymphocytic leukemia (McKeithan et al. 1987, 1990).

The binding of bcl-3 to tan-1 has at least two important implications. First, interaction between bcl-3 and tan-1 occurs through ankyrin repeats present in both proteins, a type of interaction that has not been described previously. Second, since bcl-3 has been shown to be a regulator of NF-κB (Franzoso et al. 1992; Kerr et al. 1992; Fujita et al. 1993; Nolan et al. 1993), interaction of tan-1 with bcl-3 suggests that tan-1 may be involved indirectly in the transcriptional control of genes regulated by NF-κB. The role of tan-1 within the NF-κB pathway of gene regulation is not evident at this time. Perhaps, under certain circumstances, tan-1 serves to sequester bcl-3 in a perinuclear location and thereby alters transcription of genes under NF-κB control. This model by no means eliminates the possibility of other actions by tan-1 in lymphoid or other tissues. However, involvement of tan-1 in the NF-κB pathway does highlight the expanding number of positions at which this pathway is vulnerable to genetic modifications which can contribute to malignant transformation. Along with *BCL-3* and the rel family of oncogenes (Nolan and Baltimore 1992), including *LYT-10*, another translocation-associated human oncogene (Neri et al. 1991), *TAN-1* brings to three the number of genes with transforming potential related to the NF-κB pathway. Very likely, additional genes producing products that impinge on NF-κB will be found to have similar properties in the near future.

REFERENCES

Bierkamp, C. and J.A. Campos-Ortega. 1993. A zebrafish homologue of the *Drosophila* neurogenic gene *Notch* and its pattern of transcription during early embryogenesis. *Mech. Dev.* **43:** 87.

Blackwood, E.M. and R.N. Eisenman. 1991. Max: A helix-loop-helix zipper protein that forms a sequence-specific DNA-binding complex with Myc. *Science* **251:** 1211.

Cleveland, D.W., S.G. Fischer, M.W. Kirschner, and U.K. Laemmli. 1977. Peptide mapping by limited proteolysis in sodium dodecyl sulfate and analysis by gel electrophoresis. *J. Biol. Chem.* **252:** 1102.

Coffman, C., W. Harris, and C. Kintner. 1990. Xotch, the *Xenopus* homologue of *Drosophila Notch*. *Science* **249:** 1438.

Daley, G.Q., R. van Etten, and D. Baltimore. 1990. Induction of chronic myelogenous leukemia in mice by the P210$^{bcr/abl}$ gene of the Philadelphia chromosome. *Science* **247:** 824.

Dutta, A., J.M. Rupert, J.C. Aster, and E. Winchester. 1993. Inhibition of DNA replication factor RPA by *p53*. *Nature* **365:** 79.

Ellisen, L.W., J. Bird, D.C. West, A.L. Soreng, T.C. Reynolds, S.D. Smith, and J. Sklar. 1991. *TAN-1*, the human homolog of the *Drosophila Notch* gene, is broken by chromosomal translocations in T lymphoblastic neoplasms. *Cell* **66:** 649.

Fortini, M.E. and S. Artavanis-Tsakonis. 1993. *Notch*: Neurogenesis is only part of the story. *Cell* **75:** 1245.

Franco del Amo, F., D.E. Smith, P.J. Swiatek, M. Gendron-Maguire, R.J. Greenspan, A.P. McMahon, and T. Gridley. 1992. Expression pattern of *Motch*, a mouse homolog of *Drosophila Notch*, suggests an important role in post-implantation development. *Development* **115:** 737.

Franzoso, G., V. Bours, S. Park, M. Tomita-Yamaguchi, K. Kelly, and U. Siebenlist. 1992. The candidate oncoprotein Bcl-3 is an antagonist of p50/NF-κB-mediated inhibition. *Nature* **359:** 339.

Fujita, T., G. Nolan, H. Liou, M. Scott, and D. Baltimore. 1993. The candidate proto-oncogene *bcl-3* encodes a transcriptional coactivator that activates through NF-κB p50 homodimers. *Genes Dev.* **7:** 1354.

Ghysen, A., C. Dambly-Chaudiere, L.Y. Jan, and Y.N. Jan. 1993. Cell interactions and gene interactions in peripheral neurogenesis. *Genes Dev.* **7:** 723.

Ginsburg, D., R.I. Handin, D.T. Bonthron, T.A. Donlon, G.A.

Bruns, S.A. Latt, and S.H. Orkin. 1985. Human von Willebrand factor (vWF): Isolation of complementary DNA (cDNA) clones and chromosomal localization. *Science* **228:** 1401.

Harlow, E. and D. Lane. 1988. *Antibodies: A laboratory manual*. Cold Spring Harbor Laboratory, Cold Spring Harbor, New York.

Kaelin, J., W.G., W. Krek, W.R. Sellers, J.A. DeCaprio, F. Ajchenbaum, C.S. Fuchs, T. Chittenden, Y. Li, P.J. Farnham, M.A. Blanar, D.M. Livingston, and E.K. Flemington. 1992. Expression cloning of a cDNA encoding a retinoblastoma-binding protein with E2F-like properties. *Cell* **70:** 351.

Kerr, L.D., C.S. Duckett, P. Wamsley, A. Zhan, P. Chiao, G. Nabel, T.W. McKeithan, P.A. Baeuerle, and I.M. Verma. 1992. The proto-oncogene *BCL-3* encodes an IκB protein. *Genes Dev.* **6:** 2352.

Laemmli, U.K. 1970. Cleavage of structural proteins during the assembly of the head of bacteriophage T4. *Nature* **227:** 680.

Lardelli, M. and U. Lendahl. 1993. *MotchA* and *MotchB*: Two mouse *Notch* homologs co-expressed in a wide variety of tissues. *Exp. Cell Res.* **204:** 364.

Lux, S.E., K.M. John, and V. Bennett. 1990. Analysis of cDNA for human erythrocyte ankyrin indicates a repeated structure with homology to tissue-differentiation and cell-cycle control proteins. *Nature* **334:** 36.

McKeithan, T.W., H. Ohno, and M.O. Diaz. 1990. A transcriptional unit lies adjacent to the breakpoint on chromosome 19 in the t(14;19) in chronic lymphocytic leukemia. *Genes Chromosomes Cancer* **1:** 247.

McKeithan, T.W., J.D. Rowley, T.B. Shows, and M.O. Diaz. 1987. Cloning of the chromosome translocation breakpoint junction of the t(14;19) in chronic lymphocytic leukemia. *Proc. Natl. Acad. Sci.* **84:** 9257.

Neri, A., C.-C. Chang, L. Lombardi, M. Salina, P. Corradini, A.T. Maiolo, R.S.K. Chaganti, and R. Dalla-Favera. 1991. B cell lymphoma-associated chromosomal translocation involves candidate oncogene *lyt-10*, homologous to NF-κB p50. *Cell* **67:** 1075.

Nolan, G.P. and D. Baltimore. 1992. The inhibitory ankyrin and activator Rel proteins. *Curr. Opin. Genet. Dev.* **2:** 211.

Nolan, G., T. Fujita, K. Bhatia, C. Huppi, H. Liou, M. Scott, and D. Baltimore. 1993. The *bcl-3* proto-oncogene encodes a nuclear IκB-like molecule that preferentially interacts with NF-κB p50 and p52 in a phosphorylation-dependent manner. *Mol. Cell. Biol.* **13:** 3557.

Pear, W.S., G.P. Nolan, M.L. Scott, and D. Baltimore. 1993. Production of high-titer helper-free retroviruses by transient transfection. *Proc. Natl. Acad. Sci.* **90:** 8392.

Pinkus, G.S., E.M. O'Connor, C.L. Etheridge, and J.M. Corson. 1985. Optimal immunoreactivity of keratin proteins in formalin-fixed paraffin-embedded tissues requires the use of preliminary trypsinization. An immunoperoxidase study of various tumours using polyclonal and monoclonal antibodies. *J. Histochem. Cytochem.* **33:** 465.

Reaume, A.G., R.A. Conlon, R. Zirngibl, T.P. Yamaguchi, and J. Rossant. 1992. Expression analysis of a *Notch* homologue in the mouse embryo. *Dev. Biol.* **154:** 377.

Rebay, I., R.J. Fleming, R.G. Fehon, L. Cherbas, P. Cherbas, and S. Artavanis-Tsakonas. 1991. Specific EGF repeats of *Notch* mediate interactions with *delta* and *serrate*: Implications for *Notch* as a multifunctional receptor. *Cell* **67:** 687.

Reynolds, T.C., S.D. Smith, and J. Sklar. 1987. Analysis of DNA surrounding the breakpoints of chromosomal translocations involving the β T cell receptor gene in human lymphoblastic neoplasms. *Cell* **50:** 107.

Robbins, J., B.J. Blondel, D. Gallahan, and R. Callahan. 1992. Mouse mammary tumor gene *int-3*: A member of the *notch* gene family transforms mammary epithelial cells. *J. Virol.* **66:** 2594.

Smith, S.D., R. Morgan, R. Gemmill, M.D. Amylon, M.P. Lind, C. Linker, B.K. Hecht, R. Warnke, B.E. Glader, and R. Hecht. 1988. Clinical and biological characterization of T cell neoplasias with rearrangements of chromosome 7 band q34. *Blood* **71:** 395.

Stifani, S., C.M. Blaumueller, N.J. Redhead, R.E. Hill, and S. Artavanis-Tsakonas. 1992. Human homologs of a *Drosophila Enhancer of Split* gene product define a novel family of nuclear proteins. *Nat. Genet.* **2:** 119.

Towbin, H., T. Staehelin, and J. Gordon. 1979. Electrophoretic transfer of proteins from polyacrylamide gels to nitrocellulose sheets: Procedure and some applications. *Proc. Natl. Acad. Sci.* **76:** 4350.

Wang, X., R. Sato, M.S. Brown, X. Hua, and J.L. Goldstein. 1994. SREBP-1, a membrane-bound transcription factor released by sterol regulated proteolysis. *Cell* **77:** 53.

Weinmaster, G., V.J. Roberts, and G. Lemke. 1991. A homolog of *Drosophila Notch* expressed during mammalian development. *Development* **113:** 199.

———. 1992. *Notch2*: A second mammalian *Notch* gene. *Development* **116:** 931.

Zervos, A.S., J. Gyuris, and R. Brent. 1993. Msi1, a protein that specifically interacts with Max to bind Myc-Max recognition sites. *Cell* **72:** 223.

Novel Oncogenic Mutations in the *WT1* Wilms' Tumor Suppressor Gene: A t(11;22) Fuses the Ewing's Sarcoma Gene, *EWS1*, to *WT1* in Desmoplastic Small Round Cell Tumor

F.J. RAUSCHER III, L.E. BENJAMIN, W.J. FREDERICKS, AND J.F. MORRIS

The Wistar Institute, Philadelphia, Pennsylvania 19104

The differentiated phenotype of a cell is largely determined by the set of genes that are actively being transcribed at a given moment. Hence, the set of nuclear transcription factors present and active that control expression of those target genes specify the output from this system and therefore the cellular phenotype. Since a single transcription factor may regulate hundreds of "downstream" target genes via binding to their regulatory promoter/enhancer elements, inactivation or alteration of DNA binding or transcriptional regulatory functions (via mutation or deletion) could result in catastrophic effects on cellular phenotype or proliferative potential. Thus, it is not surprising that alteration of transcription factor expression/function has emerged as a major theme among molecular mechanisms of oncogenesis (Lewin 1991). Many examples of sequence-specific DNA-binding proteins with oncogenic potential have been encountered in the guise of viral/cellular oncogenes, tumor suppressor genes, and developmental control genes (Bishop 1991). Many of these proto-oncogene transcription factors normally function to control cell differentiation/proliferation and tissue development during embryogenesis. Their normal roles may be subverted via a number of mechanisms, including (1) mutation or overt deletion, (2) translocation and disruption of the transcription-factor-encoding gene, or (3) inappropriate tissue-specific expression via fusion of the gene to new regulatory elements (for review, see Cleary 1991; Nichols and Nimer 1992; Cline 1994). Genetically, these oncogenic alterations essentially fall into two main categories: loss or gain of function.

A very interesting theme which has recently emerged from the study of human disease-related mutations is that different types of mutations in a single transcription-factor-encoding gene can result in dramatically different disease phenotypes. These findings can be partly reconciled by the facts that most transcription factors comprise independent functional domains (i.e., DNA binding, transcriptional regulation, nuclear localization, etc.), each of which is subject to alteration, and that each mutation will have different consequences on cell or tissue function (Pabo and Sauer 1992). There are also examples of spatially independent mutations in a single gene that inactivate a common biochemical property of the protein (DNA-binding activity) yet result in dramatically different tumor phenotypes (Levine et al. 1994).

A recent example of the struggle to unite genotype, alteration of the biochemical function of a transcription factor, and phenotype has come from the study of the pediatric soft-tissue sarcoma, *a*lveolar *r*habdo*m*yo*s*arcoma (ARMS). In ARMS, a t(2;13) translocation results in fusion of the paired homeobox transcription factor gene *PAX3* to a novel forkhead-domain-containing transcription factor *FKHR* (Galili et al. 1993; Shapiro et al. 1993). We have shown that the resulting fusion protein (PAX-FKHR) retains the DNA-binding domain and activity of PAX3 and gains a highly potent transcriptional activation function from the FKHR domain (Fredericks et al. 1995). Remarkably, the *PAX3* gene is also a target for mutations in Waardenburg syndrome type 1, an autosomal dominant disorder characterized by sensorineural hearing loss, pigmentary disturbances, and other developmental defects (Baldwin et al. 1992; Tassabehji et al. 1992). However, in these cases, the mutations sustained by *PAX3* all occur in the DNA-binding domain and functionally inactivate DNA-binding activity. Thus, two independent genetic and biochemical alterations of *PAX3* result in dramatically different disease phenotypes.

We have been studying the *WT1* Wilms' tumor suppressor gene product in an effort to correlate structure/function aspects of transcriptional regulation with disease phenotype with particular reference to Wilms' tumor itself. In the course of these studies, it has become clear that different types of alterations in *WT1* also manifest themselves as different disease processes. In this paper, we describe a novel translocation involving the *WT1* gene in a pediatric sarcoma, *d*esmoplastic *s*mall *r*ound *c*ell *t*umor (DSRCT), which results in fusion of the amino-terminal domain of the Ewing's sarcoma gene (*EWS1*) to the carboxy-terminal zinc finger region of *WT1*, thereby creating a potent oncogene.

WT1 AND WILMS' TUMOR

Wilms' tumor is a pediatric nephroblastoma that occurs in both heritable and sporadic forms and comprises 6% of all pediatric neoplasms (for review, see Rauscher 1993). A long series of cytogenetic and chromosomal mapping studies (for review, see Coppes et al. 1993) culminated in the positional cloning of the chromosome 11p13 *WT1* Wilms' tumor gene (Bonetta et al. 1990; Call et al. 1990; Gessler et al. 1990). The biologic and genetic properties of *WT1* and its encoded protein are summarized in Figure 1. As expected for a tumor suppressor gene, *WT1* is mutated or deleted in subsets of sporadic and familial Wilms' tumor, in 100% of Wilms' tumors associated with the Denys-Drash syndrome (Pelletier et al. 1991), and in mesotheliomas of the lung (Park et al. 1993). The *WT1* gene encodes a 52- to 54-kD nuclear protein that contains four zinc fingers of the C_2H_2 class in the carboxyl terminus (Morris et al. 1991). The *WT1* protein binds to oligonucleotides containing the EGR core consensus sequence (5'-GCGGGGGCG-3'), and mutations in the zinc finger region of WT1 identified in Wilms' tumor specimens inactivate its DNA-binding activity (Rauscher et al. 1990). In transfection assays, WT1 functions as a potent repressor of transcription (Madden et al. 1991) and DNA replication (Anant et al. 1994), a property dependent on integrity of the proline, glutamine-rich amino terminus. The genes encoding insulin-like growth factor-II, its receptor, and platelet-derived growth factor-A chain are probable physiologically relevant target genes for WT1-mediated repression during kidney development (Drummond et al. 1992; Gashler et al. 1992; Wang et al. 1992; Werner et al. 1994). The *WT1* gene is expressed primarily in the kidney during early stages of development characterized by mesenchymal to epithelial cell differentiation events leading to the formation of the primitive renal vesicle and glomerulus (Pritchard-Jones et al. 1990). Homozygous loss of *WT1* function results in embryonic lethality and almost total gonad and kidney agenesis (Kreidburg et al. 1993). The *WT1* mRNA is also detectable in the adult ovaries and testes and in the mesothelial cell linings of all organs in the abdomen, suggesting a role for WT1 in development of the genitourinary system and a broad role in mesenchymal cell function (Sharma et al. 1992). An important question is whether tumors that arise from other non-kidney organs which normally express *WT1* will contain alterations in the gene.

EWS1 AND EWING'S SARCOMA

The group of pediatric soft-tissue sarcomas designated "small round cell tumors" include Ewing's sarcoma, embryonal/alveolar rhabdomyosarcomas, Askins' tumor, and peripheral neuroectodermal tumor (for review, see Dellatre et al. 1994). Many of these *s*mall *r*ound *c*ell *t*umors (SRCT) are characterized by recurrent chromosomal translocations involving the *EWS1* gene (which encodes a putative RNA-binding transcription factor) on chromosome 22q12 and a number of other loci depending on tumor type (Fig. 2). The wild-type *EWS1* gene encodes a novel protein with transcription-factor-like characteristics (Fig. 2): an RNA-binding domain homologous to hnRNP proteins; amino acid segments rich in proline, arginine, or glycine; and an amino-*t*erminal *d*omain (NTD) comprising approximately 31 repeats of the hexapeptide SYSQQS, reminiscent of the carboxy-terminal domain of RNA polymerase II (Dellatre et al. 1992). The NTD displays about 50% amino acid sequence homology with the *TLS/FUS* gene, which is translocated to *CHOP*, a bZIP domain transcription factor in myxoid liposarcoma (Crozat et al. 1993; Rabbits et al. 1993). When involved in a translocation, the NTD of *EWS1* is fused in-frame to (1) the ETS class DNA-binding domains of *Fli-1* (t[11;22]) (Dellatre et al. 1992) or *Erg-1* (t[21;22]) (Sorenson et al. 1994) in Ewing's sarcoma and peripheral neuroectodermal tumors, or (2) the bZIP DNA-binding domain of *ATF-1* (t[12;22]) (Zucman et al. 1993) in soft-tissue clear cell sarcoma and malignant melanoma of soft parts (Fig. 2). Each translocation involves fusion of a highly potent transcriptional activation NTD from EWS to a novel DNA-binding domain and creates a potent tumor-type-specific chimeric transcription factor with oncogenic potential. With this paradigm in mind, we have molecularly characterized a newly recognized clinico-pathologic subtype of SRCT, namely *d*esmoplastic

WT1

Location:	Chromosome 11p13
Size:	~50 kilobase genomic locus
Structure:	10 exons, 2 alternative splices
mRNA:	~3.5 kb
Expression Patterns:	Embryonic: Kidney (condensing metanephric blastema and podocytes), Mesothelial Lining (all organs), Gonadal Ridge Mesothelium, Spleen, Brain (area postrema), Spinal cord (ventral horn motor neurons) Adult: Kidney (Glomerular epithelium), Ovary (Granulosa cells), Testis (Sertoli cells), Uterus (Decidual cells)
Protein Product:	52-54 kDa, nuclear protein
Structural Motifs:	4 Cys_2-His_2 Zinc-Fingers, Glutamine-Proline-Glycine-rich Transcriptional Regulation Domain
Interacting Proteins:	p53
DNA Binding Site:	EGR consensus sequence: 5'-GGAGCGGGGGCG-3'
Target Genes:	IGF-II, IGF-II-Receptor, Egr-1, Pax-2, PDGF-A, CSF-1, TGF-beta1
Diseases associated with:	Wilms' Tumor, Denys-Drash Syndrome, Mesothelioma, Desmoplastic Small Round Cell Tumor

Figure 1. Summary of structure-function-expression characteristics of the WT1 Wilms' tumor suppressor gene and protein. The adult and embryonic expression patterns are compiled from human, mouse, and rat data. (Modified from Rauscher 1993.)

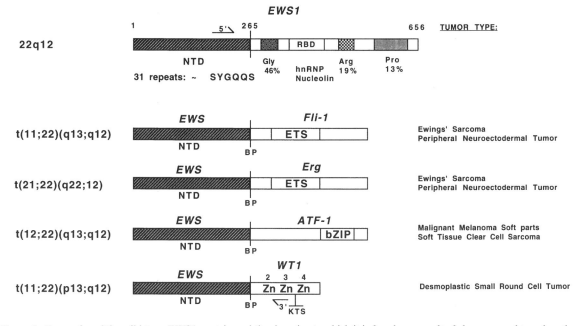

Figure 2. Properties of the wild-type EWS1 protein and the domains to which it is fused as a result of chromosomal translocation. The EWS1 protein contains a putative RNA-binding domain (RBD) and an amino-terminal domain (NTD) composed of a hexapeptide repeat sequence. As a result of the indicated chromosomal translocations, the NTD is fused in-frame to the DNA-binding domains of the indicated genes. (BP) Breakpoint. The arrows indicate approximate positions of the PCR primers in *EWS1* and *WT1* used for RT-PCR analysis. References for the data summarized here are found in the text.

*s*mall *r*ound *c*ell *t*umor (DSRCT), which also contains a recurrent chromosomal translocation involving the 22q12 *EWS1* locus. We show that the *EWS1* gene is fused to the zinc finger region of the *WT1* Wilms' tumor suppressor gene as a result of the t(11;22) (p13;q12) in DSRCT (Fig. 2). This in-frame fusion of EWS to WT1 both alters the DNA-binding activity of WT1 and converts WT1 from a repressor into a potent activator of transcription.

METHODS

RNA and DNA were isolated from frozen specimens of DSRCT and matched normal tissue (a gift from Dr. Kathleen R. Cho, Johns Hopkins School of Medicine) using standard techniques (MacDonald et al. 1987). Reverse-transcriptase (RT)-PCR was performed as described previously (Galili et al. 1993) using a 5′ EWS primer 5′-GGATCCTACAGCCAAGCTCCA-3′ and a 3′ WT1 zinc finger region primer, 5′-CTTCGT TCACAGTCCTTGAAGTC-3′. The PCR products were cloned and the DNA was sequenced on both strands. Southern blot analysis was performed using 16 µg of genomic DNA and approximately 1.8 kb human WT33 cDNA probe (Call et al. 1990). Full-length coding region chimeric *EWS-WT1* genes were constructed via PCR-mediated overlapping gene fusion techniques (Rauscher et al. 1990), and the resulting genes were subcloned into the pCMV-CB6+ expression vector (Madden et al. 1991). The procedures for COS-1 cell transfections, immunoprecipitation analysis, transient transfections in NIH-3T3 fibroblasts, CAT assays, nickel-chelate purification of recombinant zinc finger proteins, and gel shift analysis have been described previously (Rauscher et al. 1990; Madden et al. 1991; Morris et al. 1991).

RESULTS

The salient clinicopathologic features (compiled from Ordonez et al. 1989, 1993; Gerald et al. 1991) of DSRCT (also known as intra-abdominal DSRCT) are summarized in Figure 3A. This highly aggressive, rare tumor occurs most frequently in adolescent males and is located almost exclusively in the abdomen. Patients at presentation often have tumor involvement of many abdominal organs and also in the serosal lining of the gut, a feature of the tumor which has hampered identification both of the primary site of tumor development and of the target cell for oncogenic transformation. Histopathology (Fig. 3B) shows a "nested" pattern of tumor cell growth containing islands of densely packed small round cells among the characteristic (and diagnostic) desmoplastic stroma. Remarkably, the immunohistochemical analysis demonstrates that DSRCTs often coexpress epithelial, mesenchymal, and neuronal markers. Together, these findings have led pathologists to consider other designations for this tumor based on this apparently primitive "blastomatous" cellular phenotype, most notably mesothelioblastoma, peritoneal blastoma, or extrarenal Wilms' tumor. This last designation stems apparently in part from similarities among DSRCTs and "classic" triphasic Wilms' tumors,

A
Desmoplastic Small Round Cell Tumor

- Most Frequent in Adolescent Males
- Almost Exclusive Intra-Abdominal Location
- Nested pattern of cell growth
- Intense Desmoplastic reaction
- Express Epithelial, Mesenchymal and Neural Cell Markers
- Recurrent t(11;22)(p13;q12)

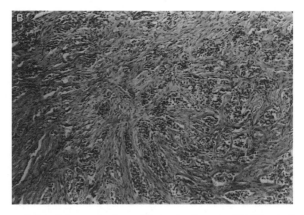

Figure 3. (*A*) Summary of clinical and histological characteristics of desmoplastic small round cell tumor. (*B*) Histologic features typical of DSRCT, which include nests of small round cells among a desmoplastic stroma. H&E stain.

which also contain epithelial, mesenchymal, and stromal cell elements (Beckwith et al. 1990). Cytogenetics have been reported in a number of DSRCTs by independent investigators and have identified a translocation t(11;22) (p13;q12) (Sawyer et al. 1992; Shen et al. 1992; Biegel et al. 1993; Rodriguez et al. 1993) possibly involving the 22q12 *EWS1* gene and the *WT1* gene at 11p13. On the basis of (1) these published cytogenetic findings, (2) the similar histopathologic profiles of triphasic Wilms' and DSRCT, (3) the knowledge that the mesothelial cell lining of the abdomen is a site of normal *WT1* expression, and (4) the fact that *EWS1* is almost exclusively found fused to a DNA-binding domain when involved in a chromosomal translocation in SRCTs (Fig. 2), we characterized *WT1* and *EWS1* genes in DSRCT.

A Chimeric EWS1-WT1 Transcript in DSRCT

The tumor depicted in Figure 3B was diagnosed as a DSRCT and occurred in a 16-year-old male who presented with infiltrating tumor in omentum, liver, and bladder. Characteristic aggregates of small round blue cells surrounded by a desmoplastic fibrous stroma are seen, and immunohistochemical staining showed strong positives for cytokeratin, desmin, and neuron-specific enolase (K. Cho, pers. comm.). RNA was extracted from the tumor and subjected to RT-PCR analysis using a 5' EWS1 primer located 5' to the most common breakpoint in the *EWS1* gene (Plougastel et al. 1993) and a 3' WT1 primer positioned in the second zinc finger of WT1. The WT1 primer could detect a fusion of the EWS-NTD to either exon 8 (zinc finger 2) or exon 7 (zinc finger 1) of WT1. RT-PCR analysis of tumor RNA revealed an approximately 100-base pair product present in the DSRCT sample, but not present in RNA extracted from adjacent normal tissue (omentum). The PCR product was dependent on the presence of the PCR primers in the reaction. The length of the PCR product obtained was consistent with fusion of the EWS-NTD at exon 7 (the most common *EWS1* breakpoint in other SRCTs) to exon 8 (zinc finger 2) of *WT1*

(Haber et al. 1991; Gessler et al. 1992). To identify the exact fusion junction, the PCR product shown in Figure 4A was cloned and sequenced (Fig. 4B), revealing an in-frame fusion of the EWS-NTD to zinc finger 2 (exon 8) of the WT1 zinc finger region. A similar RT-PCR analysis in 11 of 14 other independent DSRCT specimens showed the exact same PCR product using *EWS1*- and *WT1*-specific PCR primers (F.J. Rauscher and F.G. Barr, unpubl.). We next analyzed the long-range genomic structure of the *WT1* gene via Southern blot hybridization (Fig. 5). This analysis revealed a rearrangement of the *WT1* gene in tumor, but not normal, DNA as evidenced by novel restriction fragments at about 5.6 and 6.7 kb. Additional Southern blot analysis of three independent DSRCTs has revealed the presence of a new restriction fragment in each tumor which hybridized strongly with both *WT1* and *EWS1* cDNA probes (F.G. Barr, pers. comm.). These data strongly suggest that fusion of *WT1* and *EWS1* genes and production of a chimeric transcript are common events in DSRCTs.

DNA-binding Activity of EWS-WT1 Proteins

The junction (Fig. 4B) between *EWS1* and *WT1* genes detected by RT-PCR suggests that a fusion protein containing the amino-terminal NTD of EWS1 and the carboxy-terminal three zinc fingers of WT1 is produced in DSRCTs. Thus, the amino-terminal transcriptional repression domain of WT1 is lost as well as zinc finger 1. To determine the consequences of these alterations on the DNA binding and transcriptional regulatory properties of WT1, we constructed full-length EWS-WT1 expression vectors and generated recombinant, purified DNA-binding domain proteins in *Escherichia coli*. The following proteins were produced as histidine-fusion proteins (Fig. 6A,B) and utilized in gel shift assays with the EGR-binding site oligonucleotide: WTZF; wild-type WT1 zinc finger region, EWS-WT1 (+/− KTS); zinc finger region of WT1 which contains amino-terminal EWS sequences but lacks finger 1 and

FUSION OF *EWS1* AND *WT1* GENES 141

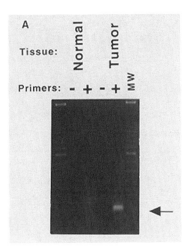

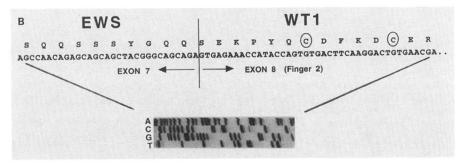

Figure 4. Detection of a fusion transcript between EWS1 and WT1 in DSRCT RNA. (*A*) RNA extracted from the tumor shown in Fig. 3B or adjacent normal tissue was reverse-transcribed and utilized in PCR analyses with 5′ EWS1- and 3′ WT1-specific oligonucleotide primers. The arrow indicates the ~100-bp PCR product obtained in reactions that contained tumor RNA but not normal tissue RNA. This PCR product was cloned and sequenced (*B*), revealing in-frame fusion of EWS1 exon 7 to WT1 exon 8 (zinc finger 2).

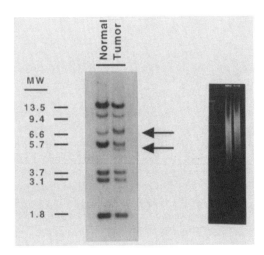

Figure 5. Rearrangement of the *WT1* locus in DSRCT. DNAs extracted from tumor or normal tissue were digested with *Eco*RI and subjected to Southern blot analysis using a human WT1 cDNA probe. The panel on the right is an ethidium bromide-stained agarose gel demonstrating integrity of the genomic DNA. Molecular-weight markers indicate the expected seven restriction fragments for *Eco*RI-digested WT1 (Call et al. 1990). Arrows indicate new restriction fragments present in tumor DNA.

EGRZF; the zinc finger region of EGR-1 which also recognizes the EGR-binding site oligonucleotide.

Gel shift analysis (Fig. 6C) revealed that the three-finger EGRZF protein bound with much higher affinity than the four-finger WTZF protein, as we have previously reported (Madden and Rauscher 1993). However, the EWS-WTZF (−KTS) protein (which lacks finger 1) binds with much higher affinity than WTZF and more closely resembles the EGRZF profile. The binding activity of the EWS-WTZF (+KTS) is completely abolished by the three-amino-acid splice insertion, similar to what has been observed with the WTZF (+KTS) (Rauscher et al. 1990). These results suggest that truncation of the WT1 DNA-binding domain (e.g., loss of finger 1) via fusion to the EWS-NTD creates a more active DNA-binding protein at the EGR binding site.

Activation of Transcription by the EWS-WT1 Fusion Protein

To determine the transcriptional regulatory potential of the EWS-WT1 protein in comparison to wild-type WT1, cotransfection experiments were performed with CMV-promoter-based expression vectors (Fig. 7A). Proper expression of WT1 and EWS-WT1 proteins was

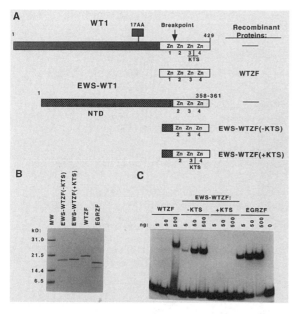

Figure 6. The EWS-WT1 zinc finger region displays increased DNA-binding affinity compared to wild-type WT1 zinc finger region. (A) The indicated recombinant zinc finger region proteins were purified from E. coli as histidine fusion proteins. WTZF has been described previously (Rauscher et al. 1990). The EWS-WTZF fusion proteins contained 21 EWS-encoded amino acids amino-terminal of the fusion point (see Fig. 3B) and were produced with (+KTS), or without (−KTS) the three-amino-acid insertion (lysine-threonine-serine) which occurs in wild-type WT1 as a result of alternative mRNA splicing. (B) The purified proteins (0.5 μg) were resolved by 15% SDS-PAGE and stained with Coomassie blue. (C) The indicated amount of each protein was utilized in a gel-shift assay using the ^{32}P-labeled EGR-binding site oligonucleotide as a probe (Rauscher et al. 1990).

verified in COS-1 cells using an antiserum (anti-WTZF) that recognizes the zinc finger region of WT1 (W. Fredericks, unpubl.). We used the promoter sequence of EGR-1 fused to the CAT gene as a target for regulation by either WT1 or EWS-WT1. In cotransfection assays, increasing amounts of CMV-WT1 expression vector resulted in potent repression of EGR-CAT transcription, as we have reported previously (Fig. 7A) (Madden et al. 1991). The ability of WT1 to repress transcription is dependent on integrity of the amino-terminal repression domain and presence of a functional zinc finger DNA-binding domain. In contrast to the repression mediated by WT1, titration of EWS-WT1 into the cotransfection assay with EGR-CAT resulted in activation of transcription. These results suggest that the EWS-NTD, when fused to the WT1 zinc finger region, functions as a potent activation domain. To verify that both the WT1 amino terminus and the EWS-NTD contain intrinsic repression and activation functions, respectively, we tested each domain as a fusion to the heterologous DNA-binding domain GAL4-1-147. Each GAL4 fusion construct was then cotransfected with a CAT reporter plasmid containing binding sites for GAL4. Under these conditions, GAL4-EWS-NTD was a very potent activator of transcription, as has been shown by other investigators (May et al. 1993; Ohno et al. 1993; Bailly et al. 1994), whereas GAL4-WT1 functioned as a repressor of transcription (Madden et al. 1993; W. Fredericks, unpubl.). Thus, the fusion of EWS1 and WT1 creates a chimeric transcription factor containing a high-affinity, zinc finger DNA-binding domain derived from WT1, which is fused to a potent transcriptional activation domain contributed by the EWS-NTD.

DISCUSSION

The data presented here show that the recurrent chromosomal translocation t(11;22) (p13;q12) in DSRCT results in the fusion of the Ewing's sarcoma gene (*EWS1*) to the Wilms' tumor suppressor gene *WT1*. The chromosomal breakpoints appear to be intronic, the gene is transcribed from the endogenous EWS promoter, and a hybrid RNA transcript is pro-

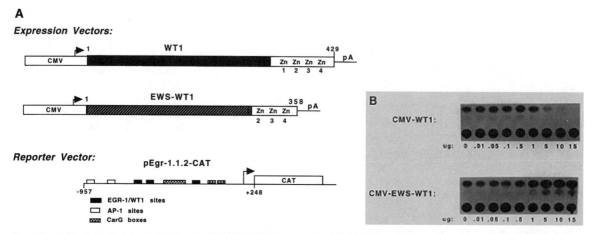

Figure 7. Activation of transcription by the EWS-WT1 fusion protein. (A) CMV-based expression vectors containing full-length WT1 or EWS-WT1 coding regions were individually cotransfected into NIH-3T3 fibroblasts with the EGR-CAT reporter plasmid (2 μg) and an internal control plasmid (CMV-β-galactosidase). (B) The indicated amount (μg) of each expression plasmid was used in transfection. After 48 hr, cell extracts were prepared and normalized amounts were used for CAT assays.

duced in which exon 7 of EWS is fused to exon 8 (zinc finger 2) of WT1. The nucleotide sequence of the fusion junction shows that the reading frame of the WT1 zinc finger region is preserved and that production of an EWS-WT1 fusion protein is possible. We have provisionally detected this fusion protein via Western blot analysis of DSRCT lysates using EWS- and WT1-specific antisera (J.F. Morris and L.E. Benjamin, unpubl.). An identical EWS-WT1 hybrid transcript has been detected in 11 of 14 independent DSRCT samples (F.G. Barr and F.J. Rauscher, unpubl.), many of which showed no cytogenetic evidence of a t(11;22) (p13;q12) translocation. Other investigators have recently shown evidence of a similar EWS-WT1 transcript in 3 of 5 DSRCTs examined (Ladanyi and Gerald 1994). Fusion of *EWS* and *WT1* genes appears to be a common feature of this subclassification of small round cell tumors, and the RT-PCR-based assay of fusion transcripts may serve as a molecular diagnostic tool for distinguishing DSRCT from other sarcomas. Thus, almost all known histopathologic variants of small round cell tumors contain different tumor-specific translocations involving EWS that are detectable by molecular methods (Delattre et al. 1994). From a practical standpoint, it will be interesting to determine whether this information ultimately improves diagnosis, prognosis, or approach to therapy for these patients.

It is important to consider how in-frame fusion of *EWS* and *WT1* alters the function of each protein and, alternatively, if the hybrid protein gains a new, potentially oncogenic function. From a formal genetic standpoint, the wild-type function(s) of the proteins encoded at each locus affected by the t(11;22) (p13;q12) is inactivated. For *WT1* this interpretation is consistent with its role as a tumor suppressor/recessive oncogene, in that loss of WT1 function has been associated with Wilms' tumorigenesis and other disease processes. However, our analysis of the biochemical functions associated with the EWS-WT1 fusion protein suggest that WT1 has sustained a gain-of-function mutation.

First, and counterintuitively, loss of zinc finger 1 appears to activate the DNA-binding potential of the remaining three zinc fingers when assayed with the EGR-consensus oligonucleotide sequence. We have previously shown that finger 1 provides additional specificity for DNA recognition and destabilizes DNA-binding activity of the wild-type, four-finger WT1 protein at some DNA-binding sites (Madden and Rauscher 1993; Drummond et al. 1994). Furthermore, loss of finger 1 function (via site-directed mutagenesis or overt deletion) enhances DNA-binding affinity. Thus, the results of our in vitro structure/function analysis of requirements for WT1 DNA binding parallel the effects of naturally occurring mutations in DSRCT. The loss of sequence specificity and/or gain in DNA-binding affinity displayed by the EWS-WT1 fusion protein may expand the repertoire of downstream target genes normally subject to WT1-mediated transcriptional control. In addition, since fingers 2, 3, and 4 of WT1 most closely resemble the three-finger EGR proteins, the normal target genes for EGRs may be deregulated by EWS-WT1.

A second and dramatic alteration of WT1 is replacement of its normal amino terminus with the NTD of EWS. The glutamine-proline-rich amino terminus of WT1 is capable of mediating both repression and activation of transcription depending on cell-type and binding-site context within the target promoter (Madden et al. 1991). Presumably, these functions are normally highly regulated via posttranslational modification and/or presence of associated proteins (Maheswaran et al. 1993). However, in DSRCTs, the amino terminus of WT1 is replaced by a potent transcriptional activation domain derived from EWS. When assayed under conditions where wild-type WT1 functions as a transcriptional repressor, EWS-WT1 functions as a potent activator of transcription. Thus, the oncogenic potential of the EWS-WT1 fusion (which has yet to be experimentally demonstrated) may stem from both its expanded DNA-binding capabilities (due to loss of finger 1) and to activation of physiologically relevant target genes normally repressed by wild-type WT1. Since a number of target genes for regulation by WT1, relevant to Wilms' tumorigenesis, have been identified (Fig. 1), it will be interesting to determine the transcriptional status of these genes in DSRCT-derived cells. However, this analysis has been hampered by the lack of cell lines or in vivo experimental tumor models for DSRCT.

With regard to the potential biologic and molecular connections between known functions of WT1 and DSRCT, it is interesting to note the following: (1) Both Wilms' tumors and DSRCTs appear to arise from primitive blastematous transformed cells capable of multilineage differentiation; (2) the peritoneal mesothelium normally expresses high levels of *WT1* and is the presumptive target cell population for the somatic transformation event leading to DSRCT; and (3) mesotheliomas of the lung have been shown to sustain mutations in *WT1* that convert it into an activator of transcription (Park et al. 1993), thus functionally analogous to the situation where *EWS* is fused to *WT1* in DSRCT. These observations further underscore the relevance of deregulated *WT1* function and DSRCT tumorigenesis.

It is remarkable that a common domain (the NTD) from a single gene (*EWS*) can be translocated and fused in-frame to a number of different types of DNA-binding domains (Fig. 2), ultimately resulting in the creation of oncogenic transcription factors with different disease specificities. A number of major questions arise from these observations which require speculation: First and foremost is whether the tumor type is specified by the type of DNA-binding domain to which EWS is fused. This is an attractive concept, since it can simply be argued that each type of DNA-binding domain may target an independent set of promoters and that deregulation of these different sets of target genes is reflected in the various tumor cell phenotypes. However, this mechanism as the only one for generating disease specificity is not consistent with what we know of in vitro DNA-binding specificities. For instance, the DNA sequence recognized by Fli-1 and Erg (both ETS type) domains is quite short, rather degenerate, and is found in many different promoter/enhancer regions, suggesting that similar sets of target genes are regulated by each factor. It will be interesting to determine whether production of transgenic animals expressing each type of EWS-DNA-binding domain fusion will recapitulate disease specificity.

A second major issue for speculation is the actual mechanism of transformation. A number of plausible mechanisms must be considered. First, the EWS-NTD may simply provide a highly potent, unregulated activation domain and, as a consequence, overexpression of normal target genes results in cell transformation. Second, the possibility exists that the EWS-NTD (which is usually tethered to RNA via the RNA-binding domain of EWS) is mislocalized to chromatin in the nucleus as a consequence of fusion to a DNA-binding domain and that this mediates cell transformation. Finally, it is formally possible that the sequence-specific activity of each DNA-binding domain is not required but simply serves as a nuclear localization function. In this scenario, the constant nuclear presence of the potent EWS activation domain could function as a titrator or squelcher of transcription. All of these are experimentally testable hypotheses.

SUMMARY

These studies suggest that the *WT1* tumor suppressor gene, originally identified as a recessive oncogene in Wilms' tumors, is capable of sustaining a gain-of-function mutation which results in its contribution to a completely different disease entity: desmoplastic small round cell tumor. Two independent biochemical functions of WT1, DNA-binding activity and mode of transcriptional regulation, are altered as a consequence of the chromosomal translocation and fusion with EWS. The fusion of *EWS* and *WT1* genes in DSRCT thus provides a unique paradigm for a means by which different alterations of transcription factor function can lead to diverse oncogenic processes.

ACKNOWLEDGMENTS

We thank Dr. K.R. Cho, Johns Hopkins Medical School, for tissue samples and for the histologic section shown in Figure 3B; Dr. F.G. Barr, University of Pennsylvania, for sharing unpublished data and helpful discussions; and Dr. C. Denny, UCLA, for the GAL4-EWS plasmid. These studies were supported in part by grants CA-52009, CA-47983, core grant CA-10815, and training grant CA-09171 (W.J.F.) from the National Institutes of Health; grant BIR-9318027 from the National Science Foundation; and grants from the Hansen Memorial Foundation and the Mary A.H. Rumsey Foundation. J.F.M. is a Cancer Research Institute Fellow. F.J.R. is a Pew Scholar in the Biomedical Sciences.

This paper is dedicated to the memory of Dr. Frank J. Rauscher, Jr.

REFERENCES

Anant, S., S.A. Axenovich, S.L. Madden, F.J. Rauscher III, and K.N. Subramanian. 1994. Novel replication inhibitory function of the developmental regulator/transcription repressor protein WT1 encoded by the Wilms' tumor gene. *Oncogene* **9:** 3113.

Bailly, R.-A., R. Bosselut, J. Zucman, F. Cormier, O. Delattre, M. Roussel, G. Thomas, and J. Ghysdael. 1994. DNA binding and transcriptional activation properties of the EWS-FLI-1 fusion protein resulting from the t(11;22) translocation in Ewing's sarcoma. *Mol. Cell. Biol.* **14:** 3230.

Baldwin, C.T., C.F. Hoth, J.A. Amos, E.O. daSilva, and A. Milunsky. 1992. An exonic mutation in the HuP2 paired domain gene causes Waardenburg's syndrome. *Nature* **355:** 637.

Beckwith, J.B., N.B. Kiviat, and J.F. Bonadio. 1990. Nephrogenic rests, nephroblastomatosis and the pathogenesis of Wilms' tumor. *Pediatr. Pathol.* **10:** 1.

Biegel, J.A., K. Conard, and J.J. Brooks. 1993. Translocation (11;22) (p13;q12): Primary change in intra-abdominal desmoplastic small round cell tumor. *Genes Chromosomes Cancer* **7:** 119.

Bishop, J.M. 1991. Molecular themes in oncogenesis. *Cell* **64:** 235.

Bonetta, L., S.E. Kuehn, A. Huang, D.J. Law, L.M. Kalikin, M. Koi, A.E. Reeve, B.H. Brownstein, H. Yeger, B.R.G. Williams, and A.P. Feinberg. 1990. Wilms' tumor locus on 11p13 defined by multiple CpG island-associated transcripts. *Science* **250:** 994.

Call, K.M., T. Glaser, C.Y. Ito, A.J. Buckler, J. Pelletier, D.A. Haber, E.A. Rose, A. Kral, H. Yeger, W.H. Lewis, C. Jones, and D.E. Housman. 1990. Isolation and characterization of a zinc finger polypeptide gene at the human chromosome 11 Wilms' tumor locus. *Cell* **60:** 509.

Cleary, M.L. 1991. Oncogenic conversion of transcription factors by chromosomal translocations. *Cell* **66:** 619.

Cline, M.J. 1994. The molecular basis of leukemia. *N. Engl. J. Med.* **330:** 28.

Coppes, M.J., C.E. Campbell, and B.R.G. Williams. 1993. The role of *WT1* in Wilms' tumorigenesis. *FASEB J.* **7:** 886.

Crozat, A., P. Aman, N. Mandahl, and D. Ron. 1993. Fusion of CHOP to a novel RNA-binding protein in human myxoid liposarcoma. *Nature* **363:** 640.

Delattre, O., J. Zucman, T. Melot, X.S. Garau, J.-M. Zucker, G.M. Lenoir, P.F. Ambros, D. Sheer, C. Turc-Carel, T.J. Triche, A. Aurias, and G. Thomas. 1994. The Ewing family of tumors—A subgroup of small-round-cell tumors defined by specific chimeric transcripts. *N. Engl. J. Med.* **331:** 294.

Delattre, O., J. Zucman, B. Plougastel, C. Desmaze, T. Melot, M. Peter, H. Kovar, I. Joubert, P. de Jong, G. Rouleau, A. Aurias, and G. Thomas. 1992. Gene fusion with an ETS DNA-binding domain caused by chromosome translocation in human tumours. *Nature* **359:** 162.

Drummond, I.A., S.L. Madden, P. Rowher-Nutter, G.I. Bell, V.P. Sukhatme, and F.J. Rauscher III. 1992. Repression of the insulin-like growth factor-II gene by the Wilms' tumor suppressor *WT1*. *Science* **257:** 674.

Drummond, I.A., H.D. Rupprecht, P. Rohwer-Nutter, J.M. Lopez-Guisa, S.L. Madden, F.J. Rauscher III, and V.P. Sukhatme. 1994. DNA recognition by variants of the Wilms' tumor suppressor, *WT1*. *Mol. Cell. Biol.* **14:** 3800.

Fredericks, W.J., N. Galili, S. Mukhopadhyay, G. Rovera, J. Bennicelli, F.G. Barr, and F.J. Rauscher III. 1995. The PAX3-FKHR fusion protein created by the t(2;13) translocation in alveolar rhabdomyosarcoma is a more potent transcriptional activator than PAX3. *Mol. Cell. Biol.* (in press).

Galili, N., R.J. Davis, W.J. Fredericks, S. Mukhopadhyay, F.J.

Rauscher III, B.S. Emanuel, G. Rovera, and F.G. Barr. 1993. Fusion of a *fork head* domain gene to PAX3 in the solid tumor alveolar rhabdomyosarcoma. *Nat. Genet.* **5:** 230.

Gashler, A.L., D.T. Bonthron, S.L. Madden, F.J. Rauscher III, T. Collins, and V.P. Sukhatme. 1992. Human platelet-derived growth factor A chain is transcriptionally repressed by the Wilms' tumor suppressor *WT1*. *Proc. Natl. Acad. Sci.* **89:** 10984.

Gerald, W.L., H.K. Miller, H. Battifora, M. Miettinen, E.G. Silva, and J. Rosai. 1991. Intra-abdominal desmoplastic small round-cell tumor. Report of 19 cases of a distinctive type of high-grade polyphenotypic malignancy affecting young individuals. *Am. J. Surg. Pathol.* **15:** 499.

Gessler, M., A. Konig, and G.A.P. Bruns. 1992. The genomic organization and expression of the *WT1* gene. *Genomics* **12:** 807.

Gessler, M., A. Poustka, W. Cavenee, R.L. Neve, S.H. Orkin, and G.A.P. Bruns. 1990. Homozygous deletion in Wilms' tumours of a zinc finger gene identified by chromosome jumping. *Nature* **343:** 774.

Haber, D.A., R.L. Sohn, A.J. Buckler, J. Pelletier, K.M. Call, and D.E. Housman. 1991. Alternative splicing and genomic structure of the Wilms' tumor gene *WT1*. *Proc. Natl. Acad. Sci.* **88:** 9618.

Kreidburg, J.A., H. Sariola, J.M. Loring, M. Maeda, J. Pelletier, D. Housman, and R. Jaenisch. 1993. *WT1* is required for early kidney development. *Cell* **74:** 679.

Ladanyi, M. and W. Gerald. 1994. Fusion of the *EWS* and *WT1* genes in the desmoplastic small round cell tumor. *Cancer Res.* **54:** 2837.

Levine, A.J., M.E. Perry, A. Chang, A. Silver, D. Dittmer, M. Wu, and D. Welsh. 1994. The 1993 Walter Hubert Lecture: The role of the p53 tumour-suppressor gene in tumorigenesis. *Br. J. Cancer* **69:** 409.

Lewin, B. 1991. Oncogenic conversion by regulatory changes in transcription factors. *Cell* **64:** 303.

MacDonald, R.J., G.H. Swift, A.E. Przybala, and J.M. Chirgwin. 1987. Isolation of RNA using guanidinium salts. *Methods Enzymol.* **152:** 219.

Madden, S.L. and F.J. Rauscher III. 1993. Positive and negative regulation of transcription and cell growth mediated by the EGR family of zinc finger gene products. *Ann. N.Y. Acad. Sci.* **684:** 75.

Madden, S.L., D.M. Cook, and F.J. Rauscher III. 1993. A structure-function analysis of transcriptional repression mediated by the *WT1*, Wilms' tumor suppressor protein. *Oncogene* **8:** 1713.

Madden, S.L., D.M. Cook, J.F. Morris, A. Gashler, V.P. Sukhatme, and F.J. Rauscher III. 1991. Transcriptional repression mediated by the *WT1* Wilms' tumor gene product. *Science* **253:** 1550.

Maheswaran, S., S. Park, A. Bernard, J.F. Morris, F.J. Rauscher III, D.E. Hill, and D.A. Haber. 1993. Interaction between the p53 and Wilms' tumor (*WT1*) gene products: Physical association and functional cooperation. *Proc. Natl. Acad. Sci.* **90:** 5100.

May, W.A., S.L. Lessnick, B.S. Braun, M. Klemsz, B.C. Lewis, L.B. Lunsford, R. Hromas, and C.T. Denny. 1993. The Ewing's sarcoma *EWS/FLI-1* fusion gene encodes a more potent transcriptional activator and is a more powerful transforming gene than FLI-1. *Mol. Cell. Biol.* **13:** 7393.

Morris, J.F., S.L. Madden, O.E. Tournay, D.M. Cook, V.P. Sukhatme, and F.J. Rauscher III. 1991. Characterization of the zinc finger protein encoded by the *WT1* Wilms' tumor locus. *Oncogene* **6:** 2339.

Nichols, J. and S.D. Nimer. 1992. Transcription factors, translocations, and leukemia. *Blood* **80:** 2953.

Ohno, T., V.N. Rao, and S.P. Reddy. 1993. EWS/Fli-1 chimeric protein is a transcriptional activator. *Cancer Res.* **53:** 5859.

Ordonez, N.G., R. Zirkin, and R.E. Bloom. 1989. Malignant small-cell epithelial tumor of the peritoneum co-expressing mesenchymal-type intermediate filaments. *Am. J. Surg. Pathol.* **13:** 413.

Ordonez, N.G., A.K. El-Naggar, J.Y. Ro, E.G. Silva, and B. Mackay. 1993. Intra-abdominal desmoplastic small cell tumor: A light microscopic, immunocytochemical, ultrastructural, and flow cytometric study. *Hum. Pathol.* **24:** 850.

Pabo, C.O. and R.T. Sauer. 1992. Transcription factors: Structural families and principles of DNA recognition. *Annu. Rev. Biochem.* **61:** 1053.

Park, S., M. Schalling, A. Bernard, S. Maheswaran, G.C. Shipley, D. Roberts, J. Fletcher, R. Shipman, J. Rheinwald, G. Demetri, J. Griffin, M. Minden, D.E. Housman, and D.A. Haber. 1993. The Wilms' tumor gene *WT1* is expressed in murine mesoderm-derived tissues and mutated in a human mesothelioma. *Nat. Genet.* **4:** 415.

Pelletier, J., W. Bruening, C.E. Kashtan, S.M. Mauer, J.C. Manivel, J.E. Striegel, D.C. Houghton, C. Junien, R. Habib, L. Fouser, R.N. Fine, B.L. Silverman, D.A. Haber, and D. Housman. 1991. Germline mutations in the Wilms' tumor suppressor gene are associated with abnormal urogenital development in Denys-Drash syndrome. *Cell* **67:** 437.

Plougastel, B., J. Zucman, M. Peter, G. Thomas, and O. Delattre. 1993. Genomic structure of the *EWS* gene and its relationship to EWSR1, a site of tumor-associated chromosome translocation. *Genomics* **18:** 609.

Pritchard-Jones, K., S. Fleming, D. Davidson, W. Bickmore, D. Porteous, C. Gosden, J. Bard, A. Buckler, J. Pelletier, D. Housman, V. van Heyningen, and N. Hastie. 1990. The candidate Wilms' tumour gene is involved in genitourinary development. *Nature* **346:** 194.

Rabbits, T.H., A. Forster, R. Larson, and P. Nathan. 1993. Fusion of the dominant negative transcription regulator CHOP with a novel gene *FUS* by translocation t(12;16) in malignant liposarcoma. *Nat. Genet.* **4:** 175.

Rauscher, F.J., III. 1993. The *WT1* Wilms' tumor gene product: A developmentally regulated transcription factor in the kidney that functions as a tumor suppressor. *FASEB J.* **7:** 896.

Rauscher, F.J., III, J.F. Morris, O.E. Tournay, D.M. Cook, and T. Curran. 1990. Binding of the Wilms' tumor locus zinc finger protein to the EGR-1 consensus sequence. *Science* **250:** 1259.

Rodriguez, E., C. Sreekantaiah, W. Gerald, V.E. Reuter, R.J. Motzer, and R.S.K. Chaganti. 1993. A recurring translocation t(11;22) (p12;q11.2) characterizes intra-abdominal desmoplastic small round-cell tumors. *Cancer Genet. Cytogenet.* **69:** 17.

Sawyer, J.R., A.F. Tryka, and J.M. Lewis. 1992. A novel reciprocal chromosome translocation t(11;22) (p13;q12) in an intraabdominal desmoplastic small round-cell tumor. *Am. J. Surg. Pathol.* **16:** 411.

Shapiro, D., J.E. Sublett, B. Li, J.R. Downing, and C.W. Naeve. 1993. Fusion of PAX3 to a member of the *forkhead* family of transcription factors in human alveolar rhabdomyosarcoma. *Cancer Res.* **53:** 5108.

Sharma, P.M., X. Yang, M. Bowman, V. Roberts, and S. Sukumar. 1992. Molecular cloning of rat Wilms' tumor complementary DNA and a study of messenger RNA expression in the urogenital system and the brain. *Cancer Res.* **52:** 6407.

Shen, W.P., B. Towne, and T.M. Zadeh. 1992. Cytogenetic abnormalities in an intra-abdominal desmoplastic small cell tumor. *Cancer Genet. Cytogenet.* **64:** 189.

Sorenson, P.H.B., S.L. Lessnick, D. Lopez-Terrada, X.F. Liu, T.J. Triche, and C.T. Denny. 1994. A second Ewing's sarcoma translocation t(21;22) fuses the *EWS* gene to another ETS-family transcription factor, ERG. *Nat. Genet.* **6:** 146.

Tassabehji, M., A.P. Read, V.E. Newton, R. Harris, R. Balling, P. Gruss, and T. Strachan. 1992. Waardenburg's syndrome patients have mutations in the human homologue of the Pax-3 paired box gene. *Nature* **355:** 635.

Wang, Z.Y., S.L. Madden, T.F. Deuel, and F.J. Rauscher III. 1992. The human platelet-derived growth factor A-chain

(PDGF-A) gene is a target for repression by the *WT1* Wilms' tumor protein. *J. Biol. Chem.* **267:** 21999.

Werner, H., F.J. Rauscher III, V.P. Sukhatme, I.A. Drummond, C.T. Roberts, Jr., and D. LeRoith. 1994. Transcriptional repression of the insulin-like growth factor I receptor (IGF-I-R) gene by the tumor suppressor *WT1* involves binding to sequences upstream and downstream of the IGF-I-R transcription start site. *J. Biol. Chem.* **269:** 12577.

Zucman, J., O. Delattre, C. Desmaze, A. Epstein, G. Stenman, F. Speleman, C.D.M. Fletchers, A. Aurias, and G. Thomas. 1993. *EWS* and *ATF-1* gene fusion induced by t(12;22) translocation in malignant melanoma of soft parts. *Nat. Genet.* **4:** 341.

Ras Signal Transduction Pathway in *Drosophila* Eye Development

H.C. Chang, F.D. Karim, E.M. O'Neill, I. Rebay, N.M. Solomon,
M. Therrien, D.A. Wassarman, T. Wolff, and G.M. Rubin

Department of Molecular and Cell Biology, Howard Hughes Medical Institute, University of California, Berkeley, California 94720-3200

Ras plays a central role downstream from many receptor tyrosine kinases (RTKs) which function in diverse cellular processes (for review, see Schlessinger 1993). In the *Drosophila* eye, Ras1 is required for the fate determination of all photoreceptors (Simon et al. 1991). The *Drosophila* compound eye is composed of an orderly array of 800 unit eyes, or ommatidia (Wolff and Ready 1993). Each ommatidium contains 8 photoreceptor neurons, R1–R8, as well as 4 lens-secreting cone cells, and 7 other accessory cells. Although little is known about the role of Ras1 in the development of R1–6 and R8, its role in the determination of photoreceptor R7 has been extensively studied (for review, see Dickson and Hafen 1994; Krämer and Cagan 1994; Zipursky and Rubin 1994). Ras1 transduces the signal initiated by Sevenless (Sev), a RTK expressed in a subset of cells including the presumptive R7 precursor (Tomlinson et al. 1987; Bowtell et al. 1989). In the absence of Sev signaling, the R7 precursor cell fails to initiate neuronal differentiation and instead is thought to adopt the fate of a nonneuronal cone cell (Tomlinson and Ready 1986).

Proteins that link Sev and Ras1 were identified in screens for mutations that alter the signaling effectiveness of the Sev RTK. These include *Son of sevenless* (*Sos*; Rogge et al. 1991; Simon et al. 1991), *downstream of receptor kinase* (*drk*; Olivier et al. 1993; Simon et al. 1993), and *gap1* (Gaul et al. 1992), the *Drosophila* homologs of guanidine nucleotide exchange factor (GNEF; Boguski and McCormick 1993), the GRB2/sem-5 adapter protein (Clark et al. 1992; Lowenstein et al. 1992), and GTPase activating protein (GAP; Boguski and McCormick 1993), respectively (Fig. 1). It is thought that Sos forms a complex with activated Sev via Drk and elevates the level of the active (GTP-bound) form of Ras1. Gap1, a negative regulator of Ras1, completes the cycle by stimulating the intrinsic GTPase activity of Ras1, returning Ras1 to the inactive (GDP-bound) state. Constitutive activation of Ras1, by either expression of a deregulated Ras1 (Fortini et al. 1992) or removal of Gap1 (Gaul et al. 1992), can bypass the requirement for Sev in R7 differentiation, suggesting that activation of Ras1 might be the sole function of Sev signaling.

Mutations in *Drosophila* homologs of Raf (*Draf*; Dickson et al. 1992), MEK/MAP kinase kinase (*Dsor1*; Tsuda et al. 1993; F. Karim et al., in prep.), and MAP kinase (*rolled* [*rl*]; Biggs et al. 1994; Brunner et al. 1994b) alter the strength of the *sev/Ras1* signaling pathway, suggesting that they have a role in the determination of R7 fate. Consistent with data from other systems, genetic analysis places *Draf*, *Dsor1*, and *rl* downstream from *Ras1*, indicating that the role of this MAPK cassette (Raf, MEK, and MAPK) in Ras signaling is evolutionarily conserved. Sina, a nuclear factor whose mechanism of function remains unknown, also affects R7 determination (Carthew and Rubin 1990).

Here we present a summary of our recent progress in understanding how the signal initiated from the Sev RTK at the cell membrane results in the commitment to a neuronal fate by R7 precursor cells. Using a genetic screen for modifiers (suppressors and enhancers) of a gain-of-function *Ras1* allele, we have identified several new components in the *sev/Ras1* signaling cascade. Our result suggests that a type I geranylgeranyl transferase (*ggt1*) is required for membrane anchorage of Ras1 (M. Therrien et al., in prep.), and protein phosphatase 2A (*PP2A*) acts as an antagonist downstream from Ras1 (D. Wassarman et al., in prep.). Two Ets-related transcription factors, products of the *pointed* (*pnt*) and

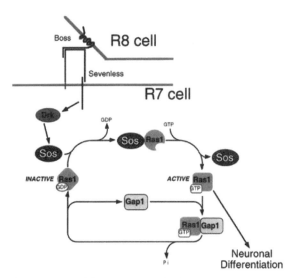

Figure 1. A model for genes functioning upstream of Ras1 in the Sev cascade. Boss, a membrane protein expressed on the surface of R8 cells during the critical period for R7 determination, is the ligand for the Sev RTK (Zipursky and Rubin 1994).

yan genes, have been implicated as direct targets of MAPK (O'Neill et al. 1994). Finally, *phyllopod* (*phyl*) encodes a novel protein whose expression appears to be regulated by the Ras1/MAPK cascade (Chang et al. 1995).

A GENETIC SCREEN FOR DOMINANT MODIFIERS OF ACTIVATED RAS1

Many genes encoding components of the Sev/Ras1 signaling cascade, such as *Ras1*, *Draf*, and *rl*, are also required for earlier stages of development, suggesting that genetic screens aimed at identifying components of this pathway may miss genes required for viability. To circumvent this problem, a genetic screen for dominant modifiers of an activated *Ras1* allele, $Ras1^{V12}$, was used to identify new genes acting downstream from *Ras1*. Expression of $Ras1^{V12}$ under control of the *sev* enhancer (sev-$Ras1^{V12}$) results in the transformation of some of the four cone cell precursors into supernumerary R7 cells and a variety of other defects, which disrupt the regular array of ommatidia and cause a rough eye (Fortini et al. 1992). Flies bearing two copies of the sev-$Ras1^{V12}$ construct have rougher eyes than those which contain only one copy, suggesting that the degree of eye roughness is dependent on the dose of $Ras1^{V12}$ protein. Since there is a correlation between the degree of eye roughness and the strength of the Ras1 signaling pathway, a twofold reduction in the dose of a gene (i.e., by mutating one copy) acting downstream from Ras1 might alter the signaling strength of Ras1 sufficiently to visibly modify the rough eye phenotype of flies carrying the sev-$Ras1^{V12}$ transgene (Fig. 2).

On the basis of this premise, we screened about 200,000 EMS-mutagenized and about 650,000 X-ray-mutagenized flies and isolated mutations in a number of genes (F. Karim et al., in prep.). The results are summarized in Table 1. More than 500 suppressors and enhancers were isolated, many of which were assigned to 31 multimember lethal complementation groups. Most of the suppressors are recessive lethal, confirming our suspicion that they have essential roles outside of eye development. Another advantage of using the modifier screen approach was shown by the observation that many enhancers (not included in Table 1) are homozygous viable with no apparent phenotype. Mutants of this class would not have been recovered by most eye-phenotype-based screens.

Alleles of *Draf*, *Dsor1*, and *rl* were obtained from the sev-$Ras1^{V12}$ screen (Table 1), indicating that the screen is effective in identifying genes in the pathway. Mutations in *Dsor1* and *rl* also suppress the rough eye phenotype of sE-raf^{Torso}, a construct expressing an activated Draf kinase under the control of *sev* enhancer (Dickson et al. 1992), suggesting that they function downstream from *Draf* (Fig. 3). It is interesting to note that although we isolated 108 *rl* alleles, we only recovered one *Draf* and two *Dsor1* alleles. The dominant modifier screen is designed to isolate genes when a twofold reduction in dose can alter the pathway enough to visibly affect the sev-$Ras1^{V12}$ phenotype. For genes like *rl*, where we isolated many independent mutations, this is likely to be the case. However, if a protein is in great excess, a twofold reduction in dose may not have an effect on the rough eye phenotype of sev-$Ras1^{V12}$ and we would only recover mutations in such genes if the mutations were dominant-negative; that is, the mutant when heterozygous with wild-type results in less than 50% of wild-type activity. It is likely that the *Draf* and *Dsor1* alleles we isolated are of this type. Other factors, such as gene size and differential susceptibility to mutagens, could also contribute to the observed distribution of mutants.

RAS1 IS BROUGHT TO THE CELL MEMBRANE BY GERANYLGERANYLATION

One of the suppressor groups, *SR2-2*, exhibits no interaction with sE-raf^{Torso}, suggesting that *SR2-2* functions either between *Ras1* and *Draf*, downstream from *Ras1* in a pathway parallel to *Draf*, or as an enzyme required for the posttranslational modification and membrane anchorage of Ras1. The Ras-like family of small GTPases are targeted to the cell membrane by prenylation, a posttranslational attachment of farnesyl (15-carbon) or geranylgeranyl (20-carbon) groups to a cysteine residue near the carboxyl terminus (Cys-186 for *Drosophila* Ras1). This membrane anchorage is essential for Ras function (Boguski and McCormick 1993), suggesting that mutations in genes required for prenylation may alter (i.e., suppress) the sev-$Ras1^{V12}$ phenotype. To investigate the nature of the suppression of sev-$Ras1^{V12}$ by *SR2-2*, we constructed the $Dsrc90$-$Ras1^{V12\Delta CAAX}$ gene, which encodes a $Ras1^{V12}$ without the prenylation signal (CAAX) fused with the amino-terminal 90 amino acids of *Drosophila* Src (Simon et al. 1985), which contains a signal for myristylation, an alternative membrane targeting modification. Flies bearing this myristylated $Ras1^{V12}$ construct expressed under the control of *sev* enhancer (Bowtell et al. 1991) have the same rough eye phenotype as our regular sev-$Ras1^{V12}$, whereas expression of a $Ras1^{V12}$ without any membrane targeting signal produces no phenotype. When we tested *SR2-2* with sev-$Dsrc90$-$Ras1^{V12\Delta CAAX}$, it did not suppress the rough eye phenotype, suggesting that *SR2-2* is required for the membrane localization of $Ras1^{V12}$ protein. Consistent with the prediction of our genetic analysis, we found that *SR2-2* encodes a *Drosophila* homolog of the β subunit of type I geranylgeranyl transferase (M. Therrien et al., in prep.). Interestingly, unlike most other Ras proteins which are farnesylated, *Drosophila* Ras1 is predicted to be geranylgeranylated, since the carboxy-terminal CAAX box of Ras1 contains a leucine residue at the "X" position (Boguski and McCormick 1993).

PP2A APPEARS TO ACT AS A NEGATIVE REGULATOR IN THE RAS1 PATHWAY

ER2-6 corresponds to the *Drosophila* homolog of protein phosphatase 2A (PP2A; Orgad et al. 1990; D. Wassarman et al., in prep.), indicating that PP2A acts as

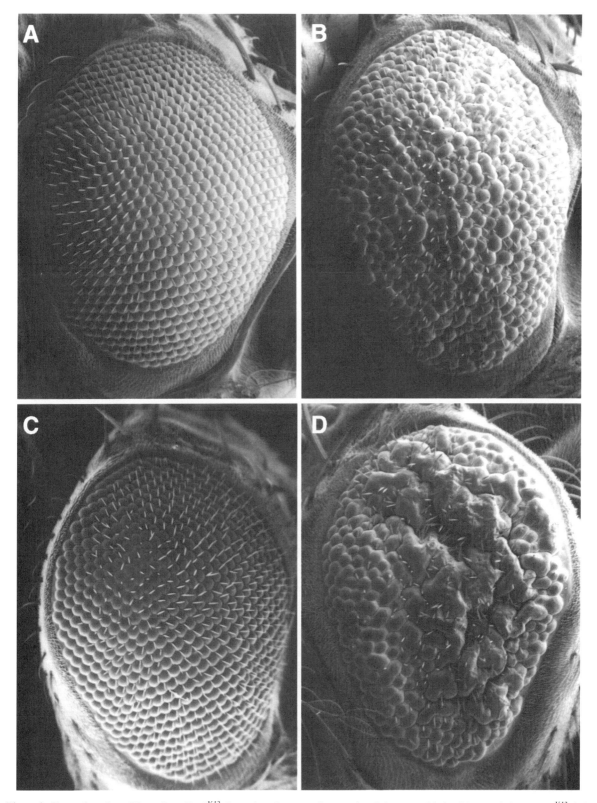

Figure 2. Examples of modifiers of *sev-Ras1^{V12}*. Scanning electron micrographs of the eyes of (*A*) wild-type; (*B*) *sev-Ras1^{V12}*; (*C*) *rl/+ , sev-Ras1^{V12}*; and (*D*) *yan/+ , sev-Ras1^{V12}*. Anterior is to the right.

Table 1. Screen for Modifiers of Activated *Ras1*

No. flies screened	EMS 250,000	X-ray 600,000	Total 850,000
Suppressor groups		Enhancer groups	
	No. alleles		No. alleles
First chromosome		*Second chromosome*	
Raf (l(1)polehole) [MapKKK]	1	ER2-1	7
Dsor [Map KK]	2	ER2-2 (yan)	148
Notch	8	ER2-5	22
		ER2-6 (PP2A)	19
Second chromosome		ER2-7	4
SR2-1 (rolled) [MapK]	108	ER2-8	15
SR2-2 (ggt1)	7	ER2-9	2
SR2-3 (phyllopod)	46	ER2-10	5
SR2-4 (mastermind)	7	ER2-11	2
"single hits"	4		
		Third chromosome	
		ER3-1A (hedgehog)	21
Third chromosome		ER3-1B	5
SR3-1	19	ER3-2	11
SR3-2	3	ER3-3	10
SR3-3	6	ER3-4	3
SR3-4	11	ER3-5	2
SR3-5	3	ER3-6	5
SR3-6	2	ER3-7	2
SR3-7	5		
SR3-9	5		
"single hits"	3	Total	283
			17 groups
Total	242		
	15 groups		

Results of the screen for modifiers of *sev-Ras1^{V12}*. Suppressors are designated as SR (*S*uppressor of *Ras1^{V12}*), and enhancers are designated as ER (*E*nhancer of *Ras1^{V12}*). Several mutants are alleles of known genes. These include *Draf, Dsor1, Notch, rolled, mastermind, yan,* and *hedgehog*. Many enhancers, some of which are homozygous viable, have not been assigned to any of the groups and are not included in the table.

a negative regulator downstream from Ras1 or in a parallel pathway. Since a MAP kinase cassette appears to relay the signal from Ras1, this phosphatase may counteract a critical phosphorylation event and help to return the cascade to the inactive state.

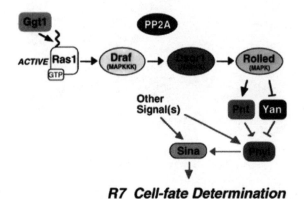

Figure 3. A model for genes functioning downstream from Ras1 in the Sev cascade. Arrows shown in gray are based solely on inferences from genetic analysis (see text for details).

ETS-RELATED TRANSCRIPTION FACTORS FUNCTION DOWNSTREAM FROM MAP KINASE

Differentiation of R7 photoreceptors is accompanied by many changes in gene expression, some of which must result from the modification of the activity of one or more preexisting transcription factors by the Ras1/MAPK cascade in R7 precursor cells. Our analysis of Pnt and Yan suggests that these two Ets-related proteins mediate transcriptional regulation by the Ras1/MAPK cascade (Fig. 3). The Ets family of transcription factors share a DNA-binding domain called the ETS domain (Karim et al. 1990; Wasylyk et al. 1993). In several mammalian cell culture systems, elements in Ras-responsive promoters have been identified that correspond to Ets-binding sites (EBS), suggesting that members of this family of transcription factors might act downstream from Ras (Wasylyk et al. 1991; Conrad et al. 1994).

Mutations in *pnt* suppress the rough eye phenotypes of *gap1*, *sev-Ras1^{V12}*, and *sE-rl^{sem}* (a construct expressing an activated rl/MAPK under the control of *sev* enhancer [Brunner et al. 1994b]), suggesting that it functions downstream from *rl* in the Sev cascade

(Brunner et al. 1994a; O'Neill et al. 1994). In agreement with its role in determination of photoreceptor fate, Pnt expression is detected in cells several hours prior to the time when they begin to differentiate as neurons. Pnt transcription is initiated from two promoters, called P1 and P2, which drive the expression of two alternately spliced transcripts that code for two related proteins, referred to here as Pnt^{P1} and Pnt^{P2} (Klämbt 1993). The two proteins contain a common carboxy-terminal sequence that includes the ETS domain, but have different amino-terminal sequences. Pnt^{P2} contains a single MAPK consensus phosphorylation site (PXS/TP; Clark-Lewis et al. 1991; Gonzales et al. 1991), suggesting that it might be a target of Rl/MAPK.

To demonstrate that Pnt^{P2} activity is modulated by Rl/MAPK, a reporter construct (E$_6$BCAT) was generated which consists of six tandem copies of the high-affinity Ets-1-binding site (Nye et al. 1992) upstream of the E1B basal promoter followed by the bacterial chloramphenicol acetyltransferase (CAT) gene (O'Neill et al. 1994). *Drosophila* tissue culture cells were cotransfected with E$_6$BCAT and a second plasmid expressing either the P1 or P2 form of Pnt under the control of *Drosophila* actin 5C promoter. Lysates from these transiently cotransfected cells were assayed for CAT activity. We found that both Pnt^{P1} and Pnt^{P2} activate transcription of the E$_6$BCAT reporter, with Pnt^{P1} being a significantly stronger activator than Pnt^{P2}. Cotransfection of plasmid expressing either Ras1^{V12} or Rl/MAPK^{sem} stimulates the ability of Pnt^{P2} to activate transcription, suggesting that the Ras1/MAPK cascade can modulate the activity of Pnt^{P2}. In contrast, no stimulation by Ras1^{V12} or Rl/MAPK^{sem} was observed with Pnt^{P1}. Mutation of the putative phosphoacceptor, Thr-151, in the Pnt^{P2} consensus site to an alanine residue abolishes its response to activation of Ras1, suggesting that the ability of Pnt^{P2} to activate transcription is regulated through direct phosphorylation by Rl/MAPK.

Yan is a another Ets-related protein and contains eight MAPK consensus phosphorylation sites (Lai and Rubin 1992). Flies that are homozygous for hypomorphic *yan* mutations contain extra R7 cells, indicating that *yan* acts as a negative regulator of the pathway (Lai and Rubin 1992). Mutations in *yan* enhance the rough eye phenotypes of *gap1*, *sev-Ras1^{V12}*, and *sE-raf^{Torso}*, suggesting that *yan* acts downstream from *Draf* (*yan* corresponds to *ER2-2*, Table 1; Lai and Rubin 1992; F. Karim et al., in prep.). Consistent with this negative regulatory role, cotransfection of plasmid expressing Yan in the tissue culture assay described above represses the transcriptional activation of the E$_6$BCAT reporter by Pnt^{P1}. The repression by Yan is alleviated by cotransfection of plasmid expressing either Ras1^{V12} or Rl/MAPK^{sem}, suggesting that the ability of Yan to repress transcription can be modulated by the Ras1/MAPK cascade.

The presence of eight consensus MAPK phosphorylation sites suggests that Yan might be directly phosphorylated in response to activation of the Ras1/MAPK cascade. Yan protein isolated from cells co-transfected with plasmid expressing either Ras1^{V12} or Rl/MAPK^{sem} had a slower electrophoretic mobility than did Yan protein isolated from cells transfected with only the Yan-expressing plasmid. This altered mobility appears to be due to phosphorylation of Yan, since treatment of the extracts with phosphatase causes the more slowly migrating forms to comigrate with Yan protein derived from cells transfected with only the Yan-expressing plasmid. Thus, there is an inverse correlation between the phosphorylation state and repressor activity of Yan, suggesting that Yan is negatively regulated by phosphorylation. Down-regulation of Yan may occur by reducing either its DNA binding or transcriptional repression activity. Alternatively, phosphorylation by MAPK could alter the stability or subcellular localization of the Yan protein.

PHYL, A COMPONENT OF THE SEV/RAS1 CASCADE REQUIRED IN ONLY A SUBSET OF PHOTORECEPTORS

Mutations in *SR2-3*, which we have named *phyllopod* (*phyl*), dominantly suppress *sev-Ras1^{V12}* and *sE-raf^{Torso}*, placing *phyl* downstream from *Draf* in the *sev* signaling cascade (H. Chang et al., in prep.). In homozygous *phyl$^-$ yan$^-$* double mutant ommatidia, the extra R7 cell phenotype of *yan* is completely suppressed, suggesting that *phyl* acts downstream from *yan* (Fig. 3). Ectopic expression of *phyl* under the control of *sev* enhancer (*sev-phyl*) results in the formation of supernumerary R7 cells and a rough eye, phenocopying *sev-Ras1^{V12}*. This extra R7 phenotype of *sev-phyl* can bypass the requirement for *sev* but still requires *sina*, suggesting that *phyl* acts downstream from *sev* and upstream of *sina*.

We have cloned *phyl* and the cDNA contains a single open reading frame of 1200 bp, which predicts a protein of 400 amino acids (H. Chang et al., in prep.). The conceptually translated Phyl protein exhibits no significant sequence similarity to known proteins and has no identifiable functional motifs. The primary sequence of the Phyl protein has two notable features: a basic stretch in the central region (amino acids 147–217), and an acidic and glutamate-rich region located at the carboxyl terminus (amino acids 323–375), where 20 of the 53 residues are glutamate.

Unlike *Ras1*, which is required for the development of all photoreceptors, *phyl* is required for only three of the eight photoreceptors. In the absence of *phyl*, ommatidia contain R2, R3, R4, R5, and R8, the first five photoreceptors to be recruited during normal development, and extra cone cells. Our phenotypic analysis indicates that the precursors of R1, R6, and R7, the last three photoreceptors to be recruited, fail to initiate neuronal differentiation and adopt an alternative non-neuronal cone cell fate. In situ RNA hybridization showed that *phyl* RNA is specifically localized in the precursors of R1, R6, and R7, consistent with its role in the fate determination of these cells. Even though *phyl* is required for the determination of only a subset of the photoreceptors, *phyl* appears to have a more general role in neuronal fate specification. *phyl* is also required for embryonic peripheral nervous system (PNS) development and bristle development. Expression of the *phyl* cDNA in the nonneuronal cone cells can induce ectopic R7 differentiation, and in pigment cells can induce the expression of neuronal antigens, suggesting that *phyl* plays an instructive role in neuronal fate specification.

Since *phyl* expression is not detected in the cone cell precursors during normal development, the requirement for *phyl* function in the transformation of the cone cell precursors to supernumerary R7 cells by *sev-Ras1^{V12}* suggests that one role of Ras1 in photoreceptor determination is to induce *phyl* expression. This is consistent with the observations that *phyl* acts downstream from *yan*, a putative transcription repressor (Lai and Rubin 1992; O'Neill et al. 1994), and that *sev-phyl* and *sev-Ras1^{V12}* transgenes produce a similar phenotype (Fortini et al. 1992). If *phyl* expression does depend on Ras1 activation, Ras1 is not likely to be the sole regulator of *phyl* expression. Ras1 is required for the development of all photoreceptors (Simon et al. 1991); however, *phyl* transcript is only detected in the precursors of R1, R6, and R7, suggesting that cell-specific signals must be integrated to generate the observed expression pattern of *phyl*.

Similar to R7 differentiation, activation of the Ras/MAP kinase cascade in rat pheochromocytoma (PC12) cells induces neuronal differentiation (Thomas et al. 1992; Wood et al. 1992), suggesting that the role of Ras in neuronal determination is conserved in diverse organisms. The function of Ras activation in these neuronal precursor cells may be to switch on genes that are required for initiating neuronal differentiation. The identification of such genes will be important in understanding how the Ras/MAP kinase pathway confers neuronal identity on certain cells.

The Phyl protein may act at the crossroads where the general Ras1 pathway diverges into specific developmental pathways. For example, the regulation of *phyl* expression might be a point where multiple signals are integrated to mediate the divergence of the general Ras1/MAP kinase cascade into the specific developmental pathways of R1, R6, and R7. An understanding of the regulation and the downstream effectors of Phyl should further our understanding of the mechanisms by which Ras serves as the common central switch in multiple cellular processes.

REFERENCES

Biggs, W.H., K.H. Zavitz, B. Dickson, A. van der Straten, D. Brunner, E. Hafen, and S.L. Zipursky. 1994. The *Drosophila rolled* locus encodes a MAP kinase required in the Sevenless signal transduction pathway. *EMBO J.* **13:** 1628.

Boguski, M.S. and F. McCormick. 1993. Proteins regulating Ras and its relatives. *Nature* **366:** 643.

Bowtell, D.D.L., B.E. Kimmel, M.A. Simon, and G.M. Rubin. 1989. Regulation of the complex pattern of *sevenless* expression in the developing *Drosophila* eye. *Proc. Natl. Acad. Sci.* **86:** 6245.

Bowtell, D.D.L., T. Lila, W.M. Michael, D. Hackett, and G.M. Rubin. 1991. Analysis of the enhancer element that controls expression of *sevenless* in the developing *Drosophila* eye. *Proc. Natl. Acad. Sci.* **88:** 6853.

Brunner, D., K. Dücker, N. Oellers, E. Hafen, H. Scholz, and C. Klämbt. 1994a. The ETS domain protein pointed-P2 is a target of MAP kinase in the Sevenless signal transduction pathway. *Nature* **370:** 386.

Brunner, D., N. Oellers, J. Szabad, W.H. Biggs III, S.L. Zipursky, and E. Hafen. 1994b. A gain of function mutation in *Drosophila* MAP kinase activates multiple receptor tyrosine kinase signalling pathways. *Cell* **76:** 875.

Carthew, R.W. and G.M. Rubin. 1990. *seven in absentia*, a gene required for specification of R7 cell fate in the *Drosophila* eye. *Cell* **63:** 561.

Chang, H.C., N.M. Solomon, D.A. Wassarman, F.D. Karim, M. Therrien, and T. Wolff. 1995. *phyllopod* functions in the fate determination of a subset of photoreceptors in *Drosophila*. *Cell* (in press).

Clark, S.G., M.J. Stern, and H.R. Horvitz. 1992. *C. elegans* cell signalling gene *sem-5* encodes a protein with SH2 and SH3 domains. *Nature* **356:** 340.

Clark-Lewis, I., J.S. Sanghera, and S.L. Pelech. 1991. Definition of a consensus sequence for peptide substrate recognition by p44mpk, the meiosis-activated myelin basic protein kinase. *J. Biol. Chem.* **266:** 15180.

Conrad, K.E., J.M. Oberwetter, R. Vaillancourt, G.L. Johnson, and A. Gutierrez-Hartmann. 1994. Identification of the functional components of the Ras signaling pathway regulating pituitary cell-specific gene expression. *Mol. Cell. Biol.* **14:** 1553.

Dickson, B. and E. Hafen. 1994. Genetics of signal transduction in invertebrates. *Curr. Opin. Genet. Dev.* **4:** 64.

Dickson, B., F. Sprenger, D. Morrison, and E. Hafen. 1992. Raf functions downstream of Ras1 in the Sevenless signal transduction pathway. *Nature* **360:** 600.

Fortini, M.E., M.A. Simon, and G.M. Rubin. 1992. Signalling by the *sevenless* protein tyrosine kinase is mimicked by *Ras1* activation. *Nature* **355:** 559.

Gaul, U., G. Mardon, and G.M. Rubin. 1992. A putative Ras GTPase activating protein acts as a negative regulator of signaling by the *sevenless* receptor tyrosine kinase. *Cell* **68:** 1007.

Gonzalez, F.A., D.L. Raden, and R.J. Davis. 1991. Identification of substrate recognition determinants for human ERK1 and ERK2 protein kinases. *J. Biol. Chem.* **266:** 22159.

Karim, F.D., L.D. Urness, C.S. Thummel, M.J. Klemsz, S.R. McKercher, A. Celada, C. Van Beveren, R.A. Maki, C.V. Gunther, J.A. Nye, and B.J. Graves. 1990. The ETS-domain: A new DNA-binding motif that recognizes a purine-rich core DNA sequence. *Genes Dev.* **4:** 1451.

Klämbt, C. 1993. The *Drosophila* gene pointed encodes two ETS-like proteins which are involved in the development of the midline glial cells. *Development* **117:** 163.

Krämer, H. and R.L. Cagan. 1994. Determination of photoreceptor cell fate in the *Drosophila* retina. *Curr. Opin. Neurobiol.* **4:** 14.

Lai, Z.-C. and G.M. Rubin. 1992. Negative control of photoreceptor development in *Drosophila* by the product of the *yan* gene, an ETS domain protein. *Cell* **70:** 609.

Lowenstein, E.J., R.J. Daly, A.G. Batzer, W. Li, B. Margolis, R. Lammers, A. Ullrich, E.Y. Skolnik, D. Bar-Sagi, and J. Schlessinger. 1992. The SH2 and SH3 domain-containing protein GRB2 links receptor tyrosine kinases to ras signaling. *Cell* **70:** 431.

Nye, J.A., J.M. Petersen, C.V. Gunther, M.D. Jonsen, and B.J. Graves. 1992. Interaction of murine ets-1 with GGA-binding sites establishes the ETS domain as a new DNA-binding motif. *Genes Dev.* **6:** 975.

Olivier, J.P., T. Raabe, M. Henkemeyer, B. Dickson, G.

Mbamalu, B. Margolis, J. Schlessinger, E. Hafen, and T. Pawson. 1993. A *Drosophila* SH2-SH3 adaptor protein implicated in coupling the sevenless tyrosine kinase to an activator of Ras guanine nucleotide exchange, Sos. *Cell* **73:** 179.

O'Neill, E.M., I. Rebay, R. Tjian, and G.M. Rubin. 1994. The activities of two Ets-related transcription factors required for *Drosophila* eye development and modulated by the Ras/MAPK pathway. *Cell* **78:** 137.

Orgad, S., N.D. Brewis, L. Alphey, J.M. Axton, Y. Dudai, and P.T. Cohen. 1990. The structure of protein phosphatase 2A is as highly conserved as that of protein phosphatase 1. *FEBS Lett.* **275:** 44.

Rogge, R.D., C.A. Karlovich, and U. Banerjee. 1991. Genetic dissection of a neurodevelopmental pathway: *Son of sevenless* functions downstream of the *sevenless* and EGF receptor tyrosine kinases. *Cell* **64:** 39.

Schlessinger, J. 1993. How receptor tyrosine kinases activate Ras. *Trends Biochem. Sci.* **18:** 273.

Simon, M.A., G.S. Dodson, and G.M. Rubin. 1993. An SH3-SH2-SH3 protein is required for p21Ras1 activation and binds to *sevenless* and Sos proteins in vitro. *Cell* **73:** 169.

Simon, M.A., B. Drees, T. Kornberg, and J.M. Bishop. 1985. The nucleotide sequence and the tissue-specific expression of *Drosophila* c-src. *Cell* **42:** 831.

Simon, M.A., D.D.L. Bowtell, G.S. Dodson, T.R. Laverty, and G.M. Rubin. 1991. Ras1 and a putative guanine nucleotide exchange factor perform crucial steps in signaling by the *sevenless* protein tyrosine kinase. *Cell* **67:** 101.

Thomas, S.M., M. DeMarco, G. D'Arcangelo, S. Halegoua, and J.S. Brugge. 1992. Ras is essential for nerve growth factor- and phorbol ester-induced tyrosine phosphorylation of MAP kinases. *Cell* **68:** 1031.

Tomlinson, A. and D.F. Ready. 1986. *Sevenless*: A cell specific homeotic mutation of the *Drosophila* eye. *Science* **231:** 400.

Tomlinson, A., D.D. Bowtell, E. Hafen, and G.M. Rubin. 1987. Localization of the *sevenless* protein, a putative receptor for positional information, in the eye imaginal disc of *Drosophila*. *Cell* **51:** 143.

Tsuda, L., Y.H. Inoue, M.-A. Yoo, M. Mizuno, M. Hata, Y.-M. Lim, T. Adachi-Yamada, H. Ryo, Y. Masamune, and Y. Nishida. 1993. A protein kinase similar to MAP kinase activator acts downstream of the raf kinase in *Drosophila*. *Cell* **72:** 407.

Wasylyk, B., S.L. Hahn, and A. Giovane. 1993. The Ets family of transcription factors. *Eur. J. Biochem.* **211:** 7.

Wasylyk, C., A. Gutman, R. Nicholson, and B. Wasylyk. 1991. The c-Ets oncoprotein activates the stromelysin promoter through the same elements as several non-nuclear oncoproteins. *EMBO J.* **10:** 1127.

Wolff, T. and D.F. Ready. 1993. Pattern formation in the *Drosophila* retina. In *The development of* Drosophila melanogaster (ed. M. Bate and A. Martinez-Arias), p. 1277. Cold Spring Harbor Laboratory Press, Cold Spring Harbor, New York.

Wood, K.W., C. Sarnecki, T.M. Roberts, and J. Blenis. 1992. *ras* mediates nerve growth factor receptor modulation of three signal-transducing protein kinases: MAP kinase, Raf-1, and RSK. *Cell* **68:** 1041.

Zipursky, S.L. and G.M. Rubin. 1994. Determination of neuronal cell fate: Lessons from the R7 neuron of *Drosophila*. *Annu. Rev. Neurosci.* **17:** 373.

Molecular Genetics of Proto-oncogenes and Candidate Tumor Suppressors in *Caenorhabditis elegans*

P.W. STERNBERG, C.H. YOON, J. LEE,* G.D. JONGEWARD,†
P.S. KAYNE, W.S. KATZ, G. LESA, J. LIU,
A. GOLDEN, L.S. HUANG, AND H.M. CHAMBERLIN‡

Howard Hughes Medical Institute and Division of Biology, California Institute of Technology, Pasadena, California 91125

Analysis of *C. elegans* vulval differentiation has led to the definition of a signaling pathway involving nematode homologs of mammalian proto-oncogenes (for review, see Sternberg et al. 1992; Clark et al. 1992b). Results of this analysis led to the placement of ras downstream from a receptor tyrosine kinase (Aroian et al. 1990; Han and Sternberg 1990), placement of SEM-5/GRB-2 upstream of ras (Clark et al. 1992a), and raf downstream from ras (Fig. 1) (Han et al. 1993). These results helped set the stage for the biochemical definition of a "universal" tyrosine kinase/ ras pathway (for review, see Dickson and Hafen 1994; J. Schlessinger and D. Bar-Sagi, this volume). Our current efforts use vulval differentiation to identify new genes involved in intercellular signaling and to elucidate their roles with respect to this universal signaling pathway.

The current molecular genetics of *C. elegans* utilizes the powerful self-fertilizing hermaphrodite genetics (Brenner 1974), facile description of phenotypes at a cellular level (see, e.g., Sulston and Horvitz 1981), easy construction of transgenic animals (Fire 1986; Mello et al. 1991), and an advanced genomic map with almost complete ordering of cosmid and YAC clones and more than 6% of genomic sequence as well as partial random cDNA sequence (Coulson et al. 1988; Sulston et al. 1992; Waterston et al. 1992; Wilson et al. 1994). Here we describe our uses of *C. elegans* molecular genetics to identify new genes that act in, or regulate, this universal signaling pathway. This pathway mediates several intercellular signaling events during nematode development, including vulval induction in the hermaphrodite (essentially a female that makes sperm as well as oocytes) and spicule development in the male copulatory tail.

THE LET-23 SIGNALING PATHWAY AND VULVAL DIFFERENTIATION

Activation of the LET-23 signaling pathway results in vulval differentiation. In the presence of the anchor cell of the somatic gonad, a source of a localized ligand, three of six potential vulval precursor cells (VPCs) undergo three rounds of mitosis and generate vulval cells (Fig. 2). The other three VPCs divide once and generate nonspecialized epidermis. In the absence of the ligand, all six VPCs generate nonspecialized epidermis, and the resulting hermaphrodites are vulvaless.

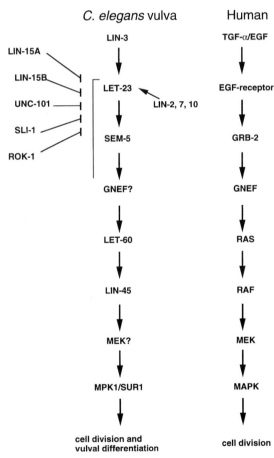

Figure 1. Genetic pathway of vulval induction. A simplified genetic pathway for the induction of vulval differentiation by the anchor cell in *C. elegans* hermaphrodites. Arrows indicate positive regulation of protein activity. Bars represent negative regulation. For the main "universal" pathway, corresponding human proteins are shown. The role of a guanine nucleotide exchange factor (GNEF) for ras in vulval induction is inferred from studies in mammalian and *Drosophila* cells. (EGF) Epidermal growth factor; (TGF) transforming growth factor-α; (MEK) MAP/ERK kinase; (MAPK) MAP kinase.

Present addresses: *University of California, Berkeley; †University of California, San Francisco; ‡Simon Fraser University, Burnaby, British Columbia, Canada.

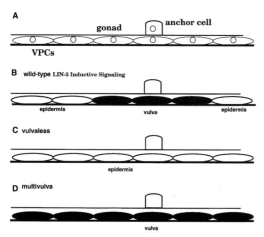

Figure 2. Schematic of vulval induction. (*A*) Relative positions of the vulval precursor cells (VPCs) and the anchor cell of the somatic gonad. (*B*) In wild-type hermaphrodites, the anchor cell induces the nearest three VPCs with the LIN-3 signal. White ovals represent uninduced VPCs, which generate nonspecialized epidermis. Black ovals represent induced VPCs, which generate vulval cells (the two types of induced VPCs are lumped together for the purposes of this discussion). (*C*) In vulvaless mutants, either the signal is not present, or the VPCs fail to respond to the signal, as depicted. (*D*) In multivulva mutants, more than three VPCs generate vulval cells.

Such hermaphrodites can neither lay eggs nor copulate with males. If this signaling pathway is hyperactivated, all six VPCs generate vulval cells. Such hermaphrodites often have a functional vulva as well as additional vulva-like structures ("multivulva"). The genes *lin-3*, *let-23*, *sem-5*, and *let-60* are required for vulval differentiation; *lin-45* and *sur1/mpk1* appear to be required as well (Ferguson and Horvitz 1985; Ferguson et al. 1987; Beitel et al. 1990; Han et al. 1990, 1993; Aroian and Sternberg 1991; Clark et al. 1992a; Lackner et al. 1994; Wu and Han 1994). Genetic tests of epistasis indicate that these genes act in a common pathway and suggest the order of their action as *lin-3* via *let-23* via *sem-5* via *let-60* via *lin-45* and *sur1/mpk1* (Ferguson et al. 1987; Beitel et al. 1990; Han and Sternberg 1990; Han et al. 1990, 1993; Hill and Sternberg 1992; Stern et al. 1993; W.S. Katz et al., in prep.; J. Liu and P. Sternberg, unpubl.). For example, the excessive vulval differentiation caused by hyperactivity of LET-23 (see below) requires wild-type activities of SEM-5, LET-60, and LIN-45, but not LIN-3 (W.S. Katz et al., in prep.). The genes *lin-2*, *lin-7*, and *lin-10* are required for full level of vulval differentiation but are not absolutely required as

is the case for *lin-3*, *let-23*, and *let-60* (Ferguson and Horvitz 1985; Ferguson et al. 1987; Kim and Horvitz 1990). *lin-2*, *lin-7*, and *lin-10* most likely do not act in the main signaling pathway, but rather act to facilitate signaling.

The LET-23 pathway is activated by an inductive signal from one cell in the somatic gonad, the anchor cell (Kimble 1981). In the absence of the anchor cell no vulval differentiation occurs. The inductive signal from the anchor cell is encoded by the *lin-3* gene (Hill and Sternberg 1992; R. Hill et al., unpubl.). LIN-3 proteins are predicted to have an extracellular domain with a single epidermal growth factor (EGF) motif, a transmembrane domain, and a cytoplasmic domain. This architecture matches that of a family of ligands for the EGF-receptor subfamily of tyrosine kinases (Carpenter and Wahl 1990; Holmes et al. 1992; Wen et al. 1992). The mammalian versions of this family are transforming growth factor α (TGF-α), amphiregulin, heparin-binding EGF, EGF, and the heregulins. Overexpression of LIN-3 from multiple copies of a wild-type genomic clone in transgenic worms also results in excessive vulval differentiation (Hill and Sternberg 1992; R. Hill et al., unpubl.). Thus, as with overexpression of TGF-α (Massagué 1990), growth factor deregulation can result in hyperactivation of this conserved signaling pathway.

let-23 encodes a nematode homolog of all four members of the human EGF-receptor subfamily (HER, HER2/c-neu, HER3, and HER4 in humans; Xmrk in fish; and DER/flb/top/Elp in *Drosophila*) (Ullrich and Schlessinger 1990). Like *lin-3*, *let-23* is necessary for vulval differentiation as well as other aspects of development (see below). We have recently found that a dominant mutation of *let-23* causes excessive vulval differentiation. This autosomal dominant multivulva mutation was extracted from a mutant strain with complex genotype kindly sent to us by Jim Thomas (University of Washington, Seattle). This mutation, *sa62*, causes excessive vulval differentiation as a homozygote, an incompletely penetrant phenotype as a heterozygote (W.S. Katz et al., in prep.), and partial vulval differentiation in the absence of the inductive signal. We mapped *sa62* to the *let-23* region of chromosome II, and demonstrated its allelism with *let-23* by a *cis-trans* test. Sequence of the *let-23* protein coding region from the *sa62* mutant revealed a single tyrosine substitution for Cys-359 in the first cysteine-rich domain of the extracellular domain of LET-23 (Fig. 3). To confirm that this mutation is sufficient to cause the

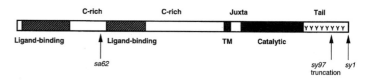

Figure 3. Schematic of the LET-23 protein. The approximate positions of *sa62*, *sy97*, and *sy1* mutations are shown. The sequence is reported by Aroian et al. (1990), the mutations by Aroian et al. (1994) and W.S. Katz et al. (in prep.). *sa62* is a point mutation in the first cysteine (C-rich) domain. *sy97* truncates the last three of eight tyrosine sites in the tail domain. *sy1* truncates six amino acids from the carboxyl terminus. (TM) Transmembrane domain; (Juxta) juxtamembrane domain; (Catalytic) protein tyrosine kinase domain; (Tail) carboxyl regulatory tail (see Ullrich and Schlessinger 1990).

multivulva phenotype, we engineered it into a 15-kb *let-23* genomic clone. Transgenic *C. elegans* hermaphrodites carrying this plasmid have a multivulva phenotype. In contrast, the 15-kb wild-type genomic clone does not cause excessive vulval differentiation (Aroian et al. 1990). Because *sa62* results in at least partial signal-independent activity of LET-23, we speculate that this region is involved in the mechanism by which ligand binding promotes dimerization and hence activation of the receptor, e.g., by a conformational change or release of an inhibitor of dimerization.

In addition to the multivulva phenotypes caused by hyperactive LIN-3 or LET-23, hyperactivity of *let-60 ras*, either by a point mutation at codon 13 or by overexpression (Beitel et al. 1990; Han and Sternberg 1990), results in a ligand-independent excessive vulval differentiation. All five independent activating mutations of *let-60* affect the same codon. We believe that only the weakly activating mutant of ras (Gly13Glu) was recovered in the genetic screens that demanded viability, because high copy of the wild-type gene results in lethality, although lower copy number results in excessive vulval differentiation (Han and Sternberg 1990).

In summary, activation of the LET-23 pathway by point mutation (*let-23*, *let-60*) or overexpression (*lin-3*, *let-60*) results in a multivulva phenotype. The analogous mutations of their mammalian counterparts result in cell transformation.

NEW POSITIVE FUNCTIONS IN THE PATHWAY

We have taken several approaches to find additional positively acting components in the LET-23 pathway. One approach is to go from clone to function. We have characterized cDNA and genomic clones of a MAP kinase kinase homolog (*mek-1*) first identified by random cDNA sequencing (Waterston et al. 1992). MEK-1 is approximately 29% identical to human MEK (P.S. Kayne et al., unpubl.). *mek-1* maps to the *C. elegans* X chromosome near *dpy-7*. We are testing mutations in this region with effects on vulval induction to determine whether they are defective in *mek-1* (P.S. Kayne et al., unpubl.).

Our second approach is to identify extragenic suppressors of overexpressed LIN-3 (J. Liu and P. Sternberg, unpubl.). Most of the genes in the LET-23 signaling pathway were identified by vulvaless phenotypes (*lin-3*, *let-23*), by suppression of the multivulva phenotype of *lin-15* (*sem-5*, *lin-45*, *let-60*), or by suppression of the multivulva phenotype of activated *let-60* (*sur-1/ mpk-1*); screens for extragenic suppressors of multivulva mutants have thus proven to be a powerful method to specifically obtain additional positive acting genes in this pathway. We screened for suppressors of overexpressed LIN-3 by starting with a strain, *syIs1*, carrying a transgene of multiple copies of *lin-3* coding sequences integrated in the *C. elegans* genome that results in a completely penetrant multivulva phenotype.

This multivulva phenotype depends on activity of LET-23 and thus behaves as expected. We mutagenized strains carrying *syIs1* and screened the F_2 grand-progeny of these animals for phenotypic reversion (non-multivulva) (Fig. 4). In principle, such revertants might result from deletion of the transgene. However, of over 50,000 gametes screened, we have not detected such events. Two classes of suppressor mutations recovered suggest that this approach will yield informative mutations. We obtained one new allele of *let-23*, the presumed receptor for LIN-3, and two mutations that appear to be dominant alleles of *lin-12*. *lin-12* gain-of-function mutations are expected because animals bearing such dominant *lin-12* alleles lack an anchor cell, the cell that expresses *lin-3* during vulval induction, and hence would prevent production of LIN-3. We found at least one new locus; mutations in this locus appear to act upstream of LET-23 (J. Liu and P. Sternberg, unpubl.). A gene defined by such a mutation might be necessary for LIN-3 expression, processing, or transport, or a cofactor of LET-23.

MULTIPLE NEGATIVE REGULATORS

As described above, gain-of-function mutations that result in excessive vulval differentiation are analogous to dominant oncogenic mutations in the human homologs of the nematode genes: overexpression of LIN-3 (TGF-α), activation or overexpression of LET-60 (ras), activation of LET-23 (EGF-receptor, *neu*).

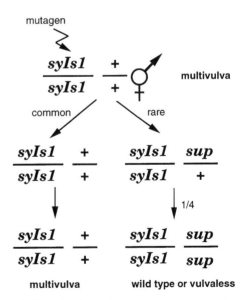

Figure 4. Genetic scheme to recover suppressors of overexpressed LIN-3. Hermaphrodites homozygous for the *syIs1* integrated transgene, which contains multiple copies of wild-type *lin-3* genomic DNA, are mutagenized and allowed to self-fertilize. If a suppressor (*sup*) mutation is induced in either a sperm or ovum of the mother, one F_1 progeny will be heterozygous for that suppressor mutation (*sup*/+). That heterozygote will segregate one-quarter homozygous suppressor progeny, assuming the *sup* mutation is recessive and homozygous viable.

Recessive mutations that cause excessive vulval differentiation define negative regulators of LET-23-mediated signaling. We hypothesize that such mutations might define the nematode homologs of tumor suppressor genes.

We have found what we believe are five distinct negative regulatory activities, defined by LIN-15A, LIN-15B, UNC-101, SLI-1, and ROK-1. These are defined by recessive mutations that confer, in general, no defect on their own, but result in excessive vulval differentiation of animals defective in two regulators. Thus, there is considerable redundancy in their action. These mutations were identified in three ways: by recovering double mutants, as extragenic suppressors of vulvaless phenotype of *let-23*, or in screens starting with a mutation in one of the negative regulatory pathways.

LIN-15 A AND B PATHWAYS

Ferguson and Horvitz (1989) defined two redundant sets of genes involved in negative regulation of vulval development. Class A genes were defined by *lin-8*; class B genes were defined by *lin-9*. In addition to genes that only provide A function, and others that provide B function, they discovered that *lin-15* provides both functions. Specifically, an animal carrying only an A mutation is wild type; an animal carrying only a B mutation is also wild type; however, an animal homozygous for both an A and a B mutation is multivulva. These observations are consistent with two pathways (or protein complexes), A and B. We focused on *lin-15* because it appeared to act in both A and B pathways (Ferguson and Horvitz 1989) and because it acts independently of two known signals controlling the fates of the VPCs, the anchor cell inductive signal and a lateral signal among the VPCs (Sternberg 1988; Herman and Hedgecock 1990). Molecular cloning of *lin-15* revealed that it encodes two transcripts, both required for full activity of the locus (Clark et al. 1994; Huang et al. 1994). These two transcripts encode two proteins, LIN-15A and LIN-15B, which are hydrophilic predicted proteins with no obvious similarity to each other or other known proteins. *lin-15* apparently acts in cells other than the VPCs and the anchor cell (Herman and Hedgecock 1990). The novelty of LIN-15 proteins is perhaps not surprising, as negative regulators of growth factor receptor-mediated signaling that act in surrounding tissues might not have been discovered in assays based on tissue culture in which mixed cell populations are not present. Further analysis of *lin-15* and interacting class A and class B genes should reveal the nature of this novel cell regulatory pathway.

UNC-101

By screening for extragenic suppressors of the vulvaless phenotype caused by a severely compromised but non-null *let-23* mutation, we isolated mutations in a number of loci. These loci include *unc-101*, *sli-1* (see below), and an activating mutation of *let-60* ras (Jongeward et al. 1995). *unc-101* was first identified by D. Riddle (pers. comm.) due to its uncoordinated movement. *unc-101* encodes a homolog of mammalian AP47, medium chain of the so-called *trans*-Golgi adaptin complex (Fig. 5) (Lee et al. 1994). Adaptins link transmembrane proteins such as EGF receptor to clathrin and can drive clathrin coat assembly onto membranes in vitro (Keen 1990; Pearse and Robinson 1990; Schmid 1992). One adaptin heavy chain (α adaptin) has been shown to associate with activated EGF receptor (Sorkin and Carpenter 1993). UNC-101 is approximately 70% identical over 422 amino acids to mouse AP47, and approximately 40% identical to mammalian and nematode AP50, a subunit of a distinct adaptin complex (Lee et al. 1994). Mouse AP47 can functionally replace UNC-101 in transgenic nematodes (Lee et al. 1994).

In addition to a variety of phenotypes, apparent null mutations of *unc-101* are semi-lethal; 50% of homozygous animals from homozygous mothers die (Lee et al. 1994). One reason that 50% of the *unc-101* mutant animals survive might be that *C. elegans* contains a second AP47 homolog (J. Lee and P. Sternberg, in prep.). The predicted product of this gene, APM-1, is identical at 72% of residues with UNC-101 and AP47. Expression of *apm-1* cDNA under the presumed transcriptional and translational control of *unc-101* provides UNC-101 function in transgenic animals. We do not yet know whether *apm-1* and *unc-101* are expressed in the same cells.

UNC-101 most likely complexes with nematode homologs of the other three polypeptides of the *trans*-Golgi-associated adaptin complex (β', γ, and AP19; Fig. 5); we have also identified a γ adaptin homolog but do not yet have mutations that disrupt its function (J. Lee et al., unpubl.). Myriam Peyrard and Jan Dumanski

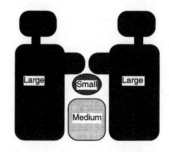

Subunit	"Golgi-associated" AP-1	"plasma membrane-associated" AP-2	C. elegans
Large	γ β'	α β	γ
Medium	AP47	AP50	UNC-101 APM-1 APM-2
Small	AP19	AP17	

Figure 5. Adaptin complexes. Schematic of adaptin complex, consisting of two large chains, one medium chain, and one small chain. The *C. elegans* proteins homologous to the mammalian proteins are listed. APM-1 and UNC-101 are AP47-like; APM-2 is AP50-like.

(pers. comm.) have recently found that a human meningioma tumor suppressor locus encodes a β adaptin. This important observation strongly suggests that adaptins can act as tumor suppressors in some cell types. Our results suggest one mechanism whereby this might occur: namely, negative regulation of a tyrosine kinase pathway.

Adaptins may thus belong to a somewhat special but potentially large class of potential tumor suppressor genes: genes that function in a wide variety of cell types but act as tumor suppressors only in some. *unc-101* is a case in point. Although adaptins are broadly used components of membrane trafficking, elimination of adaptin function in *Saccharomyces cerevisiae* is not lethal (Phan et al. 1994). It is possible that in a given cell type, a general function might have a specific effect on a cell regulatory pathway. If such a function is not essential for the survival of that cell type, then it can be mutated and thereby reveal its potential to promote cancer initiation or progression in susceptible cell types.

SLI-1

A second locus, *sli-1* (*s*uppressor of cell *li*neage defect), is defined by extragenic suppressors of severely compromised but non-null *let-23* vulvaless mutations (Jongeward et al. 1995). Our genetic analysis indicates that *sli-1* is a negative regulator of LET-23-mediated signaling. *sli-1* mutations are reduction-of-function mutations; a deletion of the *sli-1* locus has a semidominant effect on vulval differentiation. We cloned *sli-1* based on its genetic map position and its suppressor phenotype (C. Yoon et al., in prep.). *sli-1* encodes a protein with greater than 30% identity over its approximately 600 amino acids with the CBL protein, identified by Blake et al. (1991) as the transforming gene of the murine CAS-1 virus (Fig. 6). Both include a "RING finger" (Lovering et al. 1993) as well as a proline-rich region that includes potential SH3-binding sites. Oncogenic forms of CBL truncate the carboxyl terminus including the RING finger and the proline-rich regions. One way of rationalizing our genetic results with the oncogenic potential of CBL is that it normally acts as a tumor suppressor and truncation results in a dominant negative form. Since we can assay increase or decrease of *sli-1* activity, we are testing whether a similar truncation results in a dominant phenotype in transgenic *C. elegans*, and if so, whether it is dominant positive or dominant negative.

sli-1 mutations suppress severe reduction-of-function alleles of *let-23* and *sem-5*, but not similarly defective alleles of *lin-45 raf* (Jongeward et al. 1995). A partially defective allele of *let-60* is slightly suppressed. However, loss of *sli-1* activity does not compensate for complete loss of *let-23* activity. Thus, in the absence of *sli-1* activity, vulval differentiation may be driven by an alternative pathway from LET-23, leading to activation of LIN-45 raf. Although *let-23* is clearly necessary, we cannot rule out an involvement of ras in this alternative pathway. Since a *sli-1* mutation in combination with an allele of *let-23, sy97* (see Fig. 3), results in animals that are essentially wild type, we are screening for mutations that block signaling in this strain with the expectation of identifying genes responsible for LIN-45 raf activation in the presence of decreased SEM-5 activity (A. Golden and P. Sternberg, unpubl.).

Mutation of either *lin-2*, *lin-7*, or *lin-10* results in a partial vulvaless phenotype similar to the *let-23(sy1)* mutation (Sulston and Horvitz 1981; Ferguson et al. 1987; Kim and Horvitz 1990; Aroian and Sternberg 1991). Mutation of *sli-1* suppresses the vulvaless phenotype of either *let-23(sy1)*, *lin-2*, *lin-7*, or *lin-10* to a phenotype of excessive vulval differentiation that is signal-dependent. Previous analysis suggested that *let-23*, *lin-2*, *lin-7*, and *lin-10* are necessary for induction of a negative regulatory activity, as well as stimulation of vulval differentiation (Ferguson and Horvitz 1985; Aroian and Sternberg 1991; J.D. Jongeward and P.W. Sternberg, in prep.). Thus, removal of one negative regulatory activity due to these mutations and a second negative regulatory activity (SLI-1) results in excessive vulval differentiation (see below).

ROK-1

Double mutants defective in both *sli-1* and *unc-101* are multivulva even though single *sli-1* or *unc-101* mutants are not multivulva. To identify additional genes that negatively regulate LET-23-mediated signaling, we screened for mutations that cause a multivulva phenotype in a *sli-1* mutant background. Starting with a strain homozgyous for a reduction-of-function *sli-1* mutation (Fig. 7), we screened for multivulva animals. We isolated an additional allele of *unc-101*, multiple alleles of *lin-2* and *lin-10* (all expected because *sli-1 lin-2* or *lin-10; sli1* double mutants are multvulva; see above), and at least one new locus, *rok-1* (*r*egulator *o*f *k*inase-mediated signaling; J. Lee and P. Sternberg, in prep.). The *rok-1* mutation results in a temperature-sensitive multivulva phenotype. *rok-1* in combination with *unc-101* results in a multivulva phenotype (Table 1), and suppresses the vulvaless phenotype of *let-23(sy1)*, although not as strongly as do *sli-1* mutations. A triple mutant defective in *unc-101*, *rok-1*, and *sli-1* has a more excessive vulval differentiation than do any of the three double mutant combinations. With the caveat that *rok-1* and, conceivably, *sli-1* mutations used in this study might not be null mutations, these observations imply that UNC-101, ROK-1, and SLI-1 define three separate negative regulatory activities.

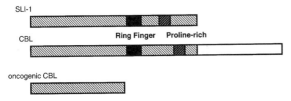

Figure 6. SLI-1 and CBL comparison. See text for details.

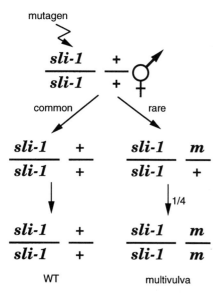

Figure 7. Genetic scheme to obtain mutations synergistic with *sli-1*. The grandprogeny of mutagenized *sli-1/sli-1* homozygotes are screened for multivulva animals. See Fig. 4 and text for rationale. (*m*) Interacting mutation.

RELATIVE ROLES OF THE FIVE NEGATIVE REGULATORS

We believe that existing mutations in these genes severely reduce or eliminate activity of each gene, and thus the synergistic phenotypes observed imply that each gene defines a distinct activity. By this criterion, LIN-15A is distinct from LIN-15B, and UNC-101 is distinct from SLI-1. In contrast, there are other class A and B genes (see above); combination of two class A mutations or of two class B mutations does not result in a multivulva phenotype (Ferguson and Horvitz 1989). Since UNC-101 and SLI-1 do not strongly synergize with LIN-15A and LIN-15B, we argue they represent distinct activities. However, the partitioning is not complete, as a few combinations of mutations result in a multivulva phenotype at 25°C (high temperature for *C. elegans*). The *rok-1* mutation synergizes with all but *lin-15*B mutations, and causes a multivulva phenotype on its own at 25°C. We tentatively assign it to a unique class because class B mutations display no multivulva phenotype in the absence of additional mutations.

The observed apparent redundancy might be due to a threshold effect; a certain degree of negative regulation might be unnecessary, and consequently, release of only one pathway from negative regulation would not result in excessive vulval differentiation. We imagine that there is a quantitative effect such that removal of one negative regulatory activity does not cause increased vulval differentiation, but removal of two activities lowers the negative regulatory capacity past a threshold level and results in excessive vulval differentiation. However, since not all combinations of mutations are equivalent, there must be different modes of negative regulation.

We can partition the functions of these negative regulatory activities in two ways: how they interact with the LET-23 pathway, and whether they act on basal or ligand-stimulated activity of the LET-23 pathway. LIN-15A and LIN-15B apparently act upstream of LET-23, whereas UNC-101, SLI-1, and ROK-1 are more likely to act at the LET-23 step or downstream, based on the following arguments.

lin-15 might act by indirectly negatively regulating LET-23 signaling. Reduction-of-function mutations in *let-23* but not in *lin-3* are epistatic to *lin-15* null mutations, suggesting that the role of *lin-15* is to prevent action of *let-23* (Ferguson et al. 1987; Huang et al. 1994; R. Hill and P. Sternberg, unpubl.). Extragenic suppressors of *lin-15* include *let-23*, *sem-5*, *let-60*, and *lin-45*, but not *lin-3* (Han et al. 1990, 1993; Aroian and Sternberg 1991; Clark et al. 1992a, b), supporting the hypothesis that *lin-15* acts upstream of LET-23.

We believe that SLI-1 regulation is exerted near the LET-23/SEM-5 step in the pathway (Jongeward et al. 1995). *sli-1* loss-of-function mutations strongly suppress a mutation that deletes the carboxyl half of the cytoplasmic tail of LET-23, including those tyrosine-containing peptides that are candidate SEM-5-binding sites (*sy97*; Aroian et al. 1994). In addition, *sli-1* mutations suppress a *sem-5* mutation. Since *sli-1* mutations do not suppress null mutations of *let-23*, they do not allow bypass of the LET-23 pathway. *sli-1* mutations also partially suppress leaky mutations in *lin-3* and *let-60* but not *lin-45* (Jongeward et al. 1995). We infer from this spectrum of suppression that SLI-1 regulation acts at or near transduction from LET-23 to LET-60. For example, SLI-1 might inhibit an alternative pathway of ras activation: In the absence of SLI-1 this alternative pathway is active or more active, allowing bypass of the LET-23 carboxyl terminus and of SEM-5.

The spectrum of *unc-101* suppression similarly suggests that UNC-101 regulation more likely acts at the LET-23 step in the pathway than upstream or downstream (Lee et al. 1994). Apparent *unc-101* null mutations strongly suppress certain mutations in *let-23*, but only weakly mutations in *lin-3* and *lin-45*.

One difference between these negative regulators is their effect on basal and ligand-stimulated activity of

Table 1. Negative Regulators

Genotype			Extent of vulval differentiation[a]	
			+signal	−signal
Wild type			3	0
lin-15			6	6
let-23(sy97)			0	0
let-23(sy97)		*lin-15*	0.4	0
unc-101	+	+	3	0
+	*rok-1*	+	3	0
+	+	*sli-1*	3	0
unc-101	*rok-1*	+	3.2	0
+	*rok-1*	*sli-1*	3.8	0
unc-101	+	*sli-1*	3.8	1.9
unc-101	*rok-1*	*sli-1*	4.9	2.0

[a] Number of vulval precursor cells generating vulval cells.

LET-23 signaling. A simple test of whether a negative regulator affects basal or ligand-stimulated activity of the LET-23 pathway is to compare the extent of vulval differentiation in animals that either have or lack endogenous ligand, due to the presence of the gonad and, hence, the anchor cell. Since the effects of the identified negative regulators can only be examined in the background of another defective negative regulator, comparison of the ligand-independent vulval differentiation among pairwise combinations of *lin-15A*, *lin-15B*, *unc-101*, *sli-1*, and *rok-1* are necessary. We compared vulval differentiation among many double mutants in the presence and absence of the gonad (Table 1). These results suggest that LIN-15A and LIN-15B act on basal activity, that ROK-1 acts on stimulated activity, and that UNC-101 and SLI-1 act on stimulated and possibly basal activity of the LET-23 signaling pathway (Huang et al. 1994; G.D. Jongeward and P.W. Sternberg, in prep.; J. Lee and P. Sternberg, unpubl.). For example, deletion of *lin-15*A and *lin-15*B results in all VPCs forming vulval tissue in the absence of signal or LIN-3. We have not formally ruled out a role of *lin-15* in stimulated activity. ROK-1 contributes only to negative regulation in the presence of inductive signal. One complication is that a severe defect in negative regulation of stimulated activity could result in cells hypersensitive to ligand. The LIN-3 signal is expressed in the anchor cell, but also likely in other cells earlier in development; thus, it is conceivable that there is ligand present even in the absence of the anchor cell. Since we are not assaying in the absence of LIN-3, SLI-1 and UNC-101 might, according to this hypothesis, only regulate stimulated activity of the LET-23 pathway.

CELL TYPE DIFFERENCES IN ACTION OF THE LET-23 PATHWAY

The LET-23 pathway is required not only for vulval differentiation, but also for viability past the first larval stage, male spicule development, hermaphrodite fertility, and posterior ventral ectoderm development (Aroian and Sternberg 1991; Chamberlin and Sternberg 1994). Analysis of the genes required for these other functions of LET-23 signaling might help reveal the basis for the tissue-specific action of a "universal" signaling pathway. Particular *let-23* mutations can result in a defect in a subset of these functions. For example, a truncation of half of the carboxyl tail by the *sy97* mutation results in defects in all but hermaphrodite fertility, whereas another mutation (not yet sequenced) results specifically in a defect in male spicule development (Aroian and Sternberg 1991; Aroian et al. 1994; H.M. Chamberlin and P.W. Sternberg, unpubl.).

Analysis of male spicule development has revealed that the LET-23 pathway is required for induction of three precursor cell fates by the F and U cells. *lin-3*, *let-23*, *sem-5*, *let-60*, and *lin-45* are necessary for the induced fates (Chamberlin and Sternberg 1994). Ectopic expression of LIN-3 results in ectopic induced fates. Comparison of the three induced fates, α, γ, and ϵ, indicates that this signaling pathway controls the choice between alternative cell fates; each fate has a characteristic extent of proliferation (Chamberlin and Sternberg 1993, 1994). The α cell generates fewer progeny than the uninduced β cell; the γ cell generates more progeny than the uninduced δ cell (Sulston et al. 1980). During spicule development, *lin-15* appears to play a minor role in negative regulation of the LET-23 pathway. Fewer than half of *lin-15* males have abnormal lineages, in contrast to the completely penetrant effect on excessive vulval differentiation. Although *lin-15* may play a smaller role in spicule development than in vulval development, it nonetheless has the same role: Mutations in *let-23* and *lin-45* suppress the defect of a *lin-15* null mutation. We suggest that this partial requirement for *lin-15* during spicule development may be due to the presence of other regulatory pathways, which have been identified by cell ablation experiments but not yet genetically (Chamberlin and Sternberg 1993, 1994).

PROSPECTS

C. elegans development provides an opportunity to use intensive molecular genetics to study a tyrosine kinase/ras-mediated signaling pathway and its regulation. We have developed powerful genetic screens for negative regulators of LET-23-mediated signaling. Such negative regulators are formally analogous to tumor suppressor genes because the pathway they regulate involves nematode homologs of human proto-oncogenes. There may be many tumor suppressors that normally act to negatively regulate tyrosine kinase/ras signaling. By performing this analysis in whole animals, we can identify negative regulators that act between cells as well as intracellularly, even those signals coming from cells not previously identifed as playing a role in the signaling process being studied. One of our goals is to identify all such negative regulators and their human homologs to allow a test of this hypothesis.

ACKNOWLEDGMENTS

This research was supported by grants to P.W.S from the U.S. Public Health Service (HD-23690) and the March of Dimes Birth Defects Foundation, and by the Howard Hughes Medical Institute, of which P.W.S. is an investigator and P.S.K. and W.S.K are associates. A.G. is an American Cancer Society California Division Senior postdoctoral fellow. G.L. is a Merck graduate fellow.

REFERENCES

Aroian, R.V. and P.W. Sternberg. 1991. Multiple functions of *let-23*, a *C. elegans* receptor tyrosine kinase gene required for vulval induction. *Genetics* **128:** 251.

Aroian, R.V., G.M. Lesa, and P.W. Sternberg. 1994. Mutations in the *Caenorhabditis elegans let-23* EGFR-like gene define elements important for cell-type specificity and function. *EMBO J.* **13:** 360.

Aroian, R.V., M. Koga, J.E. Mendel, Y. Ohshima, and P.W. Sternberg. 1990. The *let-23* gene necessary for *Caenorhabditis elegans* vulval induction encodes a tyrosine kinase of the EGF receptor subfamily. *Nature* **348:** 693.

Beitel, G., S. Clark, and H.R. Horvitz. 1990. The *Caenorhabditis elegans ras* gene *let-60* acts as a switch in the pathway of vulval induction. *Nature* **348:** 503.

Blake, T.J., M. Shapiro, H.C. Morse III, and W.Y. Langdon. 1991. The sequences of the human and mouse c-*cbl* proto-oncogenes show v-*cbl* was generated by a large truncation encompassing a proline-rich domain and a leucine zipper-like motif. *Oncogene* **6:** 653.

Brenner, S. 1974. The genetics of *Caenorhabditis elegans*. *Genetics* **77:** 71.

Carpenter, G. and M.I. Wahl. 1990. The epidermal growth factor family. In *Handbook of experimental pharmacology* (ed. A. Sporn), p. 69. Springer-Verlag, Heidelberg.

Chamberlin, H.M. and P.W. Sternberg. 1993. Multiple cell interactions are required for fate specification during male spicule development in *Caenorhabditis elegans*. *Development* **118:** 297.

———. 1994. The *lin-3/let-23* signaling pathway mediates induction during male spicule development in *Caenorhabditis elegans*. *Development* **120:** (in press).

Clark, S.G., X. Lu, and H.R. Horvitz. 1994. The *C. elegans* locus *lin-15*, a negative regulator of a tyrosine kinase signaling pathway, encodes two different proteins. *Genetics* **137:** 987.

Clark, S.G., M.J. Stern, and H.R. Horvitz. 1992a. *C. elegans* cell-signalling gene *sem-5* encodes a protein with SH2 and SH3 domains. *Nature* **356:** 340.

———. 1992b. Genes involved in two *Caenorhabditis elegans* cell-signaling pathways. *Cold Spring Harbor Symp. Quant. Biol.* **57:** 363.

Coulson, A., R. Waterston, J. Kiff, J. Sulston, and Y. Kohara. 1988. Genome linking with yeast artificial chromosomes. *Nature* **335:** 184.

Dickson, B. and E. Hafen. 1994. Genetics of signal transduction in invertebrates. *Curr. Opin. Genet. Dev.* **4:** 64.

Ferguson, E. and H.R. Horvitz. 1985. Identification and characterization of 22 genes that affect the vulval cell lineages of *Caenorhabditis elegans*. *Genetics* **110:** 17.

———. 1989. The multivulva phenotype of certain *C. elegans* mutants results from defects in two functionally-redundant pathways. *Genetics* **123:** 109.

Ferguson, E.L., P.W. Sternberg, and H.R. Horvitz. 1987. A genetic pathway for the specification of the vulval cell lineages of *Caenorhabditis elegans*. *Nature* **326:** 259.

Fire, A. 1986. Integrative transformation of *Caenorhabditis elegans*. *EMBO J.* **5:** 2675.

Han, M. and P.W. Sternberg. 1990. *let-60*, a gene that specifies cell fates during *C. elegans* vulval induction, encodes a ras protein. *Cell* **63:** 921.

Han, M., R. Aroian, and P.W. Sternberg. 1990. The *let-60* locus controls the switch between vulval and non-vulval cell types in *C. elegans*. *Genetics* **126:** 899.

Han, M., A. Golden, Y. Han, and P.W. Sternberg. 1993. *C. elegans lin-45 raf* gene participates in *let-60 ras* stimulated vulval differentiation. *Nature* **363:** 133.

Herman, R.K. and E.M. Hedgecock. 1990. The size of the *C. elegans* vulval primordium is limited by *lin-15* expression in surrounding hypodermis. *Nature* **348:** 169.

Hill, R.J. and P.W. Sternberg. 1992. The *lin-3* gene encodes an inductive signal for vulval development in *C. elegans*. *Nature* **358:** 470.

Holmes, W.E., M.X. Sliwkowski, R.W. Akita, W.J. Henzel, J. Lee, J.W. Park, D. Yansura, N. Abadi, H. Raab, G.D. Lewis, H.M. Shepard, W. Kuang, W.I. Wood, D.V. Goeddel, and R.L. Vandlen. 1992. Identification of heregulin, a specific activator of p185^{erbB2}. *Science* **256:** 1205.

Huang, L.S., P. Tzou, and P.W. Sternberg. 1994. The *lin-15* locus encodes two negative regulators of *C. elegans* vulval development. *Mol. Biol. Cell* **5:** 395.

Jongeward, G.D., T.R. Clandinin, and P.W. Sternberg. 1995. *sli-1*, a negative regulator of *let-23*-mediated vulval differentiation in *C. elegans*. *Genetics* (in press).

Keen, J.H. 1990. Clathrin and associated assembly and disassembly proteins. *Annu. Rev. Biochem.* **59:** 415.

Kim, S.K. and H.R. Horvitz. 1990. The *Caenorhabditis elegans* gene *lin-10* is broadly expressed while required specifically for the determination of vulval cell fates. *Genes Dev.* **4:** 357.

Kimble, J. 1981. Lineage alterations after ablation of cells in the somatic gonad of *Caenorhabditis elegans*. *Dev. Biol.* **87:** 286.

Lackner, M.R., K. Kornfeld, L.M. Miller, H.R. Horvitz, and S. Kim. 1994. A MAP kinase homolog, *mpk-1*, is involved in *ras*-mediated induction of vulval cell fates in *Caenorhabditis elegans*. *Genes Dev.* **8:** 160.

Lee, J., G.D. Jongeward, and P.W. Sternberg. 1994. *unc-101*, a gene required for many aspects of *C. elegans* development and behavior, encodes a clathrin-associated protein. *Genes Dev.* **8:** 60.

Lovering, R., I.M. Hanson, K. Borden, S. Martin, N.J. O'Reilly, G.I. Evan, D. Rahman, D.J.C. Pappin, J. Trowsdale, and P.S. Freemont. 1993. Identification and preliminary characterization of a protein motif related to the zinc finger. *Proc. Natl. Acad. Sci.* **90:** 2112.

Massagué, J. 1990. Transforming growth factor-α. *J. Biol. Chem.* **265:** 21393.

Mello, C.C., J.M. Kramer, D. Stinchcomb, and V. Ambros. 1991. Efficient gene transfer in *C. elegans* after microinjection of DNA into germline cytoplasm: Extrachromosomal maintenance and integration of transforming sequences. *EMBO J.* **10:** 3959.

Pearse, B.M.F. and M.S. Robinson. 1990. Clathrin, adaptors, and sorting. *Annu. Rev. Cell Biol.* **6:** 151.

Phan, H.L., J.A. Finlay, D.S. Chu, P.K. Tan, T. Kirchhausen, and G.S. Payne. 1994. The *Saccharomyces cerevisiae APS1* gene encodes a homolog of the small subunit of the mammalian clathrin AP-1 complex: Evidence for a functional interaction with clathrin at the Golgi complex. *EMBO J.* **13:** 1706.

Schmid, S.L. 1992. The mechanism of receptor-mediated endocytosis. *BioEssays* **14:** 589.

Sorkin, A. and G. Carpenter. 1993. Interaction of activated EGF receptors with coated pit adaptins. *Science* **261:** 612.

Stern, M.J., L.E.M. Marengere, R.J. Daly, E.J. Lowenstein, M. Kokel, A. Batzer, P. Olivier, T. Pawson, and J. Schlessinger. 1993. The human *GRB2* and *Drosophila Drk* genes can functionally replace the *Caenorhabditis elegans* cell signaling gene *sem-5*. *Mol. Biol. Cell* **4:** 1175.

Sternberg, P.W. 1988. Lateral inhibition during vulval induction in *Caenorhabditis elegans*. *Nature* **335:** 551.

Sternberg, P.W., R.J. Hill, G. Jongeward, L.S. Huang, and L. Carta. 1992. Intercellular signaling during *C. elegans* vulval induction. *Cold Spring Harbor Symp. Quant. Biol.* **57:** 353.

Sulston, J. and H.R. Horvitz. 1981. Abnormal cell lineages in mutants of the nematode *Caenorhabditis elegans*. *Dev. Biol.* **82:** 41.

Sulston, J.E., D.G. Albertson, and J.N. Thomson. 1980. The *Caenorhabditis elegans* male: Postembryonic development of nongonadal structures. *Dev. Biol.* **78:** 542.

Sulston, J., Z. Du, K. Thomas, R. Wilson, L. Hillier, R. Staden, N. Halloran, P. Green, J. Thierry-Mieg, L. Qiu, S. Dear, A. Coulson, M. Craxton, R. Durbin, M. Berks, M. Metzstein, T. Hawkins, R. Ainscough, and R. Waterston. 1992. The *C. elegans* genome sequencing project: A beginning. *Nature* **356:** 37.

Ullrich, A. and J. Schlessinger. 1990. Signal transduction by receptors with tyrosine kinase activity. *Cell* **61:** 203.

Waterston, R., C. Martin, M. Craxton, C. Huynh, A. Coulson, L. Hillier, R. Durbin, P. Green, R. Shownkeen, N. Halloran, M. Metzstein, T. Hawkins, R. Wilson, M. Berks, Z. Du, K. Thomas, J. Thierry-Mieg, and J. Sulston. 1992. A survey of expressed genes in *Caenorhabditis elegans*. *Nat. Genet.* **1:** 114.

Wen, D., E. Peles, R. Cupples, S.V. Suggs, S.S. Bacus, Y. Luo, G. Trail, S. Hu, S.M. Silbiger, R.B. Levy, R.A. Koski, H.S. Lu, and Y. Yarden. 1992. Neu differentiation factor: A transmembrane glycoprotein containing an EGF domain and an immunoglobulin homology unit. *Cell* **69:** 559.

Wilson, R., R. Ainscough, K. Anderson, C. Baynes, M. Berks, J. Bonfeld, J. Barton, M. Connell, T. Copsey, J. Cooper, A. Coulson, M. Craxton, S. Dear, Z. Du, R. Durbin, A. Favello, A. Fraser, L. Fulton, A. Gardner, P. Green, T. Hawkins, L. Hillier, M. Jier, L. Johnston, M. Jones, and 30 others. 1994. 2.2 Mb of contiguous nucleotide sequence from chromosome III of *C. elegans*. *Nature* **368:** 32.

Wu, Y. and M. Han. 1994. Suppression of activated Let-60 Ras protein defines a role of *C. elegans* Sur-1 MAP kinase in vulval differentiation. *Genes Dev.* **8:** 147.

Proto-oncogenes and Plasticity in Cell Signaling

J.M. Bishop,* A.J. Capobianco,* H.J. Doyle,* R.E. Finney,† M. McMahon,‡
S.M. Robbins,* M.L. Samuels,‡ and M. Vetter*

*The G.W. Hooper Research Foundation, University of California, San Francisco, California 94143-0552;
†Cell Therapeutics Inc., Seattle, Washington 98119; ‡DNAX Research Institute of
Molecular and Cellular Biology, Palo Alto, California 94304

Cancer cells contain genetic damage of two sorts (Bishop 1991): dominant, with targets known as proto-oncogenes; and recessive, with targets known most commonly as tumor suppressor genes. The lesions in proto-oncogenes cause an unwanted gain of function, whereas the damage to tumor suppressor genes typically causes loss of function. Proto-oncogenes and tumor suppressor genes are generally thought of as governors of cellular proliferation, with the former serving as accelerators, the latter as brakes. Jam an accelerator or remove a brake, and the cell is unleashed to relentless division. Useful as this view may be, it is erroneously simplistic. In reality, many proto-oncogenes and at least some tumor suppressor genes have diverse, even paradoxical functions that vary with cellular and physiological context: There is plasticity in the machine. Until we can understand the nature and genesis of that plasticity, it is unlikely that we will be able to fully understand the biochemical maladies of cancer cells.

The plasticity displayed by proto-oncogenes is better understood than that of tumor suppressor genes and is based on the physiological role of these genes in the lives of metazoan organisms. The proteins encoded by proto-oncogenes serve as relays or junction boxes in the elaborate biochemical circuitry that governs the phenotype of eukaryotic cells (Cantley et al. 1991). That circuitry is designed to permit extravagant pleiotropism in the response to extracellular signals. Here we summarize our efforts to elucidate several examples of such pleiotropism, using both genetic and biochemical strategies.

MATERIALS AND METHODS

All procedures were carried out as described in the cited references.

RESULTS

Pleiotropic Outcomes from a Single Signaling Pathway

The *torso* gene of *Drosophila melanogaster* commands a signaling pathway that is required for proper development at the ends of the embryo (Perrimon 1993). *torso* encodes a transmembrane receptor with protein-tyrosine kinase activity and a topography resembling in some ways that of the vertebrate receptor for platelet-derived growth factor (PDGF) (Sprenger et al. 1989). The *torso* receptor is expressed over the entire surface of the *Drosophila* egg. Signaling occurs when maternal cells of the follicular epithelium begin to produce the ligand for the *torso* receptor, at the outset of embryogenesis. The signaling is localized by virtue of where the ligand is produced. Thus, the *torso* protein is a vital link in a chain of events that transmits a localized signal from the follicular epithelium of the mother to the transcriptional apparatus of the embryo.

The function of *torso* provides an accessible model for the dissection of a complete signaling pathway, aimed at the details of signal transduction and the manner in which it plays on the transcriptional apparatus. We have performed the dissection by using a mutation of *torso* that causes a gain of function: a single amino acid substitution in the extracellular domain of the receptor protein that constitutively activates the kinase and leads to a grotesque distortion of embryonic phenotype (Klingler et al. 1988). We screened for dominant mutations that would blunt the impact of the *torso* mutation and, thus, suppress the mutant phenotype (Doyle and Bishop 1993). We anticipated that the suppressor mutations might unveil two sorts of functions: those that regulate the action of *torso* and those that transmit the signal from it. To date, we have found only the latter.

Our screen produced more than 70 dominant suppressor mutants for *torso*, representing at least seven complementation groups. By the use of these and other mutants, we demonstrated that signaling from *torso* is transduced through a *RAS* protein (pRas) and its attendant apparatus (Fig. 1) (Doyle and Bishop 1993). The principal event in transduction is activation of guanine nucleotide exchange, but inhibition of the GTPase of pRas (by means of inhibiting the GTPase activating protein, or GAP; see Fig. 1) could also augment transduction, and GAP itself can be used to suppress the *torso* mutant (Doyle and Bishop 1993).

Further analysis of *torso* has uncovered a complete cytoplasmic signaling pathway (Fig. 2). Activation of pRas in turn triggers a cascade of protein kinases, which begins with the product of the *RAF* proto-oncogene and continues through a mitogen-activated protein-kinase kinase (MAPKK) and a mitogen-activated protein kinase (MAPK). This cascade eventually

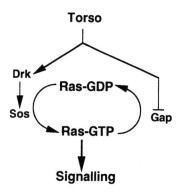

Figure 1. *torso* activates signaling through pRas. The diagram illustrates how activation of the *torso* receptor in turn activates signaling through pRas. Arrowheads indicate activation, perpendicular bars indicate inhibition. The individual functions are designated according to the names of the corresponding *Drosophila* genes: (Torso) receptor protein-tyrosine kinase; (Drk) adapter protein with SH2 and SH3 domains; (Sos) guanine nucleotide exchange factor; (Ras) a *Drosophila* counterpart of a vertebrate pRaf; and (Gap) a GTPase activating protein.

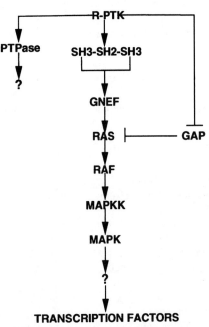

Figure 2. A cytoplasmic signaling pathway with diverse purposes. The diagram outlines one means by which a signal is tranduced from a receptor protein-tyrosine kinase to the transcriptional apparatus of the nucleus. The same pathway serves diverse purposes in metazoan cells, including either differentiation or proliferation in various cellular lineages (see text for more details). Abbreviations: (R-PTK) receptor protein-tyrosine kinase; (SH2-SH3-SH2) protein-protein binding domains within an adapter protein; (PTPase) protein-tyrosine phosphatase; (GNEF) guanine nucleotide exchange factor; (RAS) a small GTPase of the *RAS* family; (GAP) GTPase activating protein; (RAF) a protein-serine/threonine kinase; (MAPKK) mitogen-activated protein-kinase kinase; (MAPK) mitogen-activated protein kinase.

plays on the transcriptional program. In parallel, *torso* also signals through a protein-tyrosine phosphatase (see Fig. 2) (Perkins et al. 1992); the details of this signaling are not yet known, but they are likely to involve an independent pathway to the nucleus (M.A. Simon, pers. comm.). The striking point is that genetic analysis has shown signaling through both pRas and the phosphatase to be essential for the wild-type function of *torso*. Thus, it is possible that both pathways are necessary, but neither sufficient.

As the signaling machinery for *torso* unfolded, plasticity came into view. An identical pathway also serves the *sevenless* receptor of *Drosophila* (Simon et al. 1993), which is required for development of a photoreceptor cell in the retina, and the *vulvaless* pathway of the nematode *Caenorhabditis elegans* (Han et al. 1993), required for the formation of the vulva. In addition, work in mammalian cells has implicated the same pathway in activation of the cell cycle (for review, see Pelech 1993). How can a single signaling pathway serve such diverse phenotypic outcomes? The apparent explanation invokes cellular context: The response to signaling from the cytoplasmic cascade may be determined by which set of transcription factors is deployed to receive the incoming signal. The variation in deployment becomes a molecular exemplification of cellular determination and is now a major puzzle to be solved.

Examining Nodal Points in the Cascade

Genetic analysis provides a clear ordering of events within a signaling cascade, but it can occasionally overlook intermediaries and it can be a blunt instrument for examining the biochemical events at individual steps. The signaling pathway commanded by *torso* is also represented in mammalian cells. In that setting, we have examined the biochemical nature of signal transduction at two of the steps defined by genetics: from pRas to pRaf, and from pRaf to MAPKK and MAPK. The issues were the same for both instances: Has the sequence of events been portrayed correctly by genetics, and do the functions portrayed in tandem actually communicate directly or through as yet unidentified intermediaries?

To examine transduction from pRas to pRaf, we used immunoprecipitation to demonstrate physical complexes involving the two proteins (Fig. 3A). Antibodies directed against pRas also precipitated pRaf from extracts of cells that had been transformed by a mutant allele of *RAS* and, thus, contained a constitutively activated version of pRas (Fig. 3B) (Finney et al. 1993). We also asked whether the interaction between pRas and pRaf could be induced acutely as part of a signaling cascade. To do this, we exploited the previous finding that stimulation of the T-cell receptor leads to immediate activation of pRas (Fig. 4A) (Downward et al. 1990). In this setting, we found that the interaction between pRas and pRaf could be induced in concert with the activation of pRas, reaching a maximum within 2 minutes and declining to baseline within 30 minutes

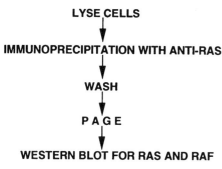

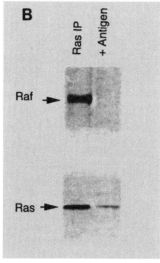

Figure 3. pRas and pRaf form a physical complex in mammalian cells. All procedures were as described by Finney et al. (1993). (*A*) Protocol for coprecipitation of pRas and pRaf. (PAGE) Polyacrylamide gel electrophoresis. (*B*) Representative data, showing precipitation of both pRas and pRaf by antisera directed against pRas.

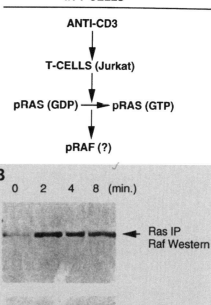

Figure 4. The complex between pRas and pRaf forms in concert with signal transduction in T cells. All procedures were carried out as described by Finney et al. (1993). (*A*) Outline of the experiment. The Jurkat line of T cells was activated with antibody directed against the CD3 component of the T-cell receptor. This leads to activation of pRas by exchange of GTP for GDP. (*B*) Representative data, showing coprecipitation of pRaf by antisera against pRas following stimulation of the T-cell receptor.

(Fig. 4B) (Finney et al. 1993). Thus, pRaf appears to be one of the elusive and long-sought downstream effectors of pRas. Other investigators have reached the same conclusion from a variety of data (for review, see Moodie and Wolfman 1994).

The mechanics of signal transduction from pRas to pRaf have now been explored in some detail. pRas must first be activated by the binding of GTP. Transduction of the signal to pRaf then occurs, requiring the effector domain of pRas and the regulatory domain of the pRaf protein kinase. During the course of transduction, pRaf is phosphorylated on serine and, by some accounts, tyrosine (Fabian et al. 1993), but the role of these phosphorylations in the regulation of pRaf remains uncertain. Phosphorylation of tyrosine is claimed as necessary for activation of the pRaf kinase (Fabian et al. 1993), whereas phosphorylation of serine may represent, in part, a feedback to terminate signaling from the protein (Nishida and Gotoh 1993; and see below). However, the mechanism by which pRas activates pRaf remains unclear. Indeed, no one has yet reported enzymatic activation of pRaf as a consequence of binding to pRas in vitro.

What of signaling downstream from pRaf? To examine these events, we created an ectopic and conditional version of pRaf in which the activity of the protein is under the direct control of estrogen (Samuels et al. 1993). Using this device, we were able to show that activation of pRaf led within seconds to activation of MAPKK and, thence, to activation of MAPK itself (Fig. 5) (Samuels et al. 1993), and that withdrawal of pRaf activity led to a prompt interruption of the MAPK cascade (Samuels et al. 1993). These findings provide in vivo evidence that pRaf signals directly through MAPKK, although it may have other outputs, as well (Porras et al. 1994; and see below). We also noted that activation of the ectopic pRaf led to phosphorylation of endogenous pRaf (Samuels et al. 1993), even though no signaling event upstream of pRaf had been deployed. This phosphorylation was not accompanied by activation of the kinase of endogenous pRaf. Others have

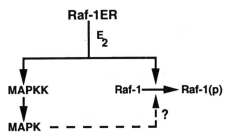

Figure 5. Signal transduction from pRaf to MAPKK and MAPK. The activity of pRaf was placed under the direct control of estrogen (E_2) as described by Samuels et al. (1993). Activation with estrogen in Rat-1A cells led to the prompt phosphorylation of MAPKK, followed by phosphorylation of MAPK and the endogenous pRaf. The dashed line represents hypothetical feedback from MAPK to pRaf, as proposed by others (Nishida and Gotoh 1993) and sustained by the data outlined here.

proposed that MAPK provides a feedback inhibition of pRaf by phosphorylating the latter protein (Fig. 5) (Nishida and Gotoh 1993), thus delimiting the signal from pRaf. Our results sustain this view.

Multiple Outputs from pRaf

We have described signaling through pRas and pRaf as a simple linear progression. In reality, of course, matters are more complex. As an explicit example, consider the effects of pRaf on different cells (Table 1). Both mouse 3T3 and Rat-1 cells are fibroblastic in nature, and both can be flagrantly transformed by a mutant allele of RAF-1. In both cellular contexts, pRaf activates MAPKK, but this leads to the anticipated activation of MAPK only in the 3T3 cells (Table 1) (Samuels et al. 1993). Our evidence indicates that the failure of MAPK to respond in Rat-1 cells is apparently due to a constitutively active phosphatase (Samuels et al. 1993).

We can draw two conclusions from these results. First, there must be more than one output from pRaf/MAPK, else the transformation of Rat-1 cells could not be explained. Other workers have also reported evidence for multiple outputs from pRaf (Porras et al. 1994). Second, transformation by RAF genes may be executed by means other than MAPK.

Table 1. Multiple Outputs from pRaf

Gene	Cells	TF	MAPKK	MAPK
Raf-1	3T3	+	+	+
Raf-1	Rat-1	+	+	−

Estrogen-controlled versions of pRaf were expressed in Rat-1 and mouse 3T3 cells by means of a retroviral vector. The kinase of pRaf was activated at will by the application of estrogen to the cells. Responses to the activation are given as + or −, for the following parameters: (TF) transformation; (MAPKK) mitogen-activated protein-kinase kinase; (MAPK) mitogen-activated protein kinase.

Pleiotropism Served by Multiple Outputs from a Single Receptor

The cellular receptor for PDGF is served by diverse outputs, each of which apparently originates with a physical interaction between a phosphotyrosine within the receptor and an SH2 domain of another signaling protein (Pazin and Williams 1992). Figure 6 summarizes the currently identified outputs, although the details remain poorly explained. To explore how the various outputs determine the phenotypic response to activation of the PDGF receptor, we have examined the function of the receptor in two disparate cellular contexts.

The PDGF receptor is a natural component of most fibroblasts, where its activation leads to cellular proliferation. In contrast, the PDGF receptor is not normally produced in PC12 cells (rat pheochromocytoma cells that differentiate to a partially neuronal phenotype in response to several polypeptide factors). We used a retroviral vector to express the PDGF receptor in PC12 cells and discovered that stimulation of the receptor by PDGF induced not proliferation, but differentiation—of the sort elicited by nerve growth factor (NGF) and its receptor (M. Vetter, unpubl.). Similar findings have been reported by other workers (Heasley and Johnson 1992). Biochemical analysis revealed that, in PC12 cells, stimulation of the PDGF receptor activated several signaling outputs, including phosphatidyl-inositol-3' kinase (PI-3K), phospholipase-C-γ (PLC), and the kinase cascade controlled by pRas. The intensity of activation was roughly equivalent to that seen in response to other ligands that elicit differentiation of PC12 cells through their corresponding receptors.

These results again exemplify sharing among signaling pathways, as illustrated before by the work with *Drosophila*. Without sharing, how could PC12 cells

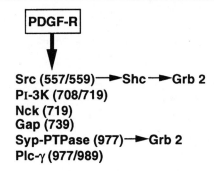

Figure 6. Multiple outputs from the PDGF receptor. Activation of the PDGF receptor leads to autophosphorylation of tyrosine residues within the receptor, which in turn bind to SH2 domains in the various signaling proteins indicated in the figure. Numbers in parentheses represent residues of phosphorylated tyrosine within the receptor. Abbreviations for the signaling proteins are defined in the text, with the exceptions of Nck, Shc, and Grb-2, which are all SH2/SH3 adapter proteins akin to the one shown in Figs. 1 and 2.

respond to a protein-tyrosine kinase they have never before seen? More importantly, the phenotypic response to the PDGF receptor in PC12 cells is the opposite of what we might have expected. How is this influence of cellular context achieved?

In search of an answer, we used mutations in the PDGF receptor that eliminate the interaction between the receptor and one or another component of the signaling circuitry (all kindly provided by L.W. Williams). We found that at least four of the outputs from the receptor appeared to be dispensable for differentiation in response to PDGF: GAP, PI-3K, PLC, and SYP (Table 2). Even combined deficiencies failed to cripple differentiation.

The easiest way to explain these findings is to suggest that there may be redundancy among several of the signaling pathways. For example, work by other investigators has found redundancy in the signaling for mitogenesis by the PDGF receptor in fibroblasts (Valius and Kazlauskas 1993). A more provocative finding also emerged: a striking contrast between the results obtained for differentiation of PC12 cells and those obtained for mitogenesis of fibroblasts (Table 2). Mutations that created a variety of either single or combined defects in the outputs through PI-3K, GAP, PLC, and SYP all crippled or eradicated mitogenesis in fibroblasts but had no effect on differentiation of PC12 cells. We see here a hint that the response to a single signaling device, a single protein-tyrosine kinase, can be diversified downstream—presumably as a function of cellular context, perhaps as a function of which substrates are available. Here again is a form of cellular determination, of plasticity in the machine.

A Pleiotropic Response to a Single Receptor in a Single Cellular Setting

Our use of the PDGF receptor in PC12 cells was artificial, in a sense, since we manufactured a comparative response between fibroblasts and PC12 cells. However, diversification of phenotypic response to seemingly similar signaling devices can also be observed within the same cell. Consider the responses of PC12 cells to two different transmembrane protein-tyrosine kinases,

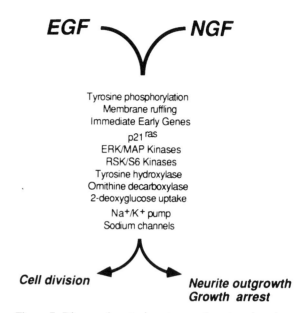

Figure 7. Diverse phenotypic outcomes from tyrosine phosphorylation in PC12 cells. Stimulation of PC12 cells with EGF elicits proliferation, whereas stimulation with NGF elicits differentiation; yet the intracellular responses to the ligands have many features in common, as shown.

the receptors for NGF and EGF. Activation of these receptors elicits a large array of intracellular events that are held in common (Fig. 7). Despite this extensive commonality, the ultimate phenotypic responses are diametric opposites: proliferation in response to EGF, differentiation in response to NGF. We have attempted to explore the genesis of this seeming paradox.

At first glance, the function of the EGF receptor offers no help, since it appears to utilize outputs similar to those employed by the PDGF receptor (Carpenter and Cohen 1990). We did obtain a clue from reports of a difference between the enzymatic responses of PC12 cells to EGF and NGF (Qiu and Green 1992; Traverse et al. 1992): Activation of the MAPK cascade in PC12 cells by NGF is sustained for several hours, whereas activation of the same enzymes by EGF is relatively transient. This finding prompted the hypothesis that a difference merely in duration of signal might diversify phenotypic response.

To test this hypothesis, we utilized a mutant of the EGF receptor that abolishes down-regulation of the receptor (Chen et al. 1989; Wells et al. 1990). As a result, a single exposure to EGF elicits sustained signaling rather than the sharply delimited signaling characteristic of the wild-type receptor. PC12 cells expressing this mutant displayed prolonged activation of the EGF receptor-kinase and differentiated in response to EGF, in a manner indistinguishable from the response to NGF (Fig. 8). In contrast, the response of fibroblasts to the wild type and mutant receptors did not differ in kind (A.J. Capobianco and M. Vetter, unpubl.), although they may differ in degree (see Chen et al. 1989; Wells et al. 1990). Moreover, ectopic expres-

Table 2. Signaling Outputs Required for Phenotypic Responses to the PDGF Receptor

	Mitogenesis of fibroblasts	Differentiation of PC12 cells
ΔPI-3K	−	+
ΔPI-3K/ΔGAP	−	+
ΔPLCγ/ΔSyp	↓	+
ΔPI-3K/ΔPLCγ/ΔSyp	−	+

Mutants of the PDGF receptor were expressed in PC12 cells by means of a retroviral vector. The mutations eliminated one or more of the signaling outputs from the receptor, as indicated, and were kindly provided by L.T. Williams. Results with fibroblasts were as reported previously by various sources. Responses are designated as follows: (+) wild type; (−) eliminated; (↓) diminished.

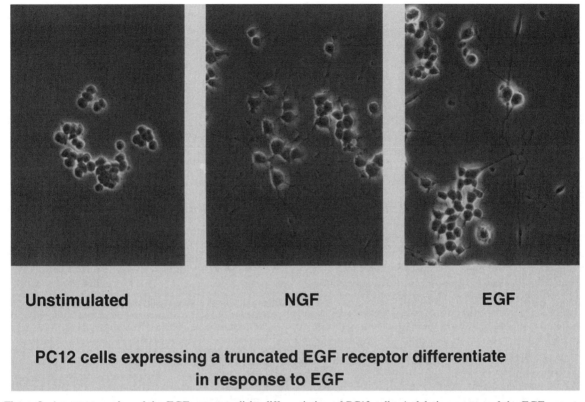

Figure 8. A mutant version of the EGF receptor elicits differentiation of PC12 cells. A deletion mutant of the EGF receptor described previously (Chen et al. 1989) was expressed in PC12 cells with a retroviral vector. The mutant is not internalized after stimulation by ligand and, thus, gives sustained signaling from the surface of the cell. The figure illustrates cells expressing the mutant in the absence of any stimulus, in the presence of NGF, and in the presence of EGF. Normal PC12 cells do not differentiate in response to EGF, but instead, proliferate (data not shown).

sion of the wild-type EGF receptor, in quantities equivalent to those of the mutant, did not produce differentiation of PC12 cells in response to EGF (data not shown). Only when the wild-type receptor was expressed in extreme abundance could it mediate differentiation. Thus, it appears that prolonging the signal from the EGF receptor may have no qualitative effect on the phenotypic response in fibroblasts, yet may switch the response in PC12 cells from mitogenesis to differentiation.

DISCUSSION

Plasticity in Cell Signaling

We have illustrated three apparent means by which plasticity in signaling can be achieved. Cells can deploy different targets for the same input; combinatorial variations can be manipulated, both within and among signaling pathways; and a variation in the duration of a signal can be translated into different phenotypic responses. There may well be others.

These are descriptions, not explanations. Attributing specificity in outcome to the deployment of targets begs the question of how the deployment is governed. Attribution of phenotypic specificity to particular combinations of signaling pathways tells us nothing of how each combination gives a different outcome. There is, as yet, no explanation for how variation in the duration of a signal might change the phenotypic outcome.

The evolutionary purpose of plasticity seems obvious: genetic parsimony. The whole is much greater than the sum of its parts because the parts are combined not in one way, but in many ways. Plasticity of signaling may eventually explain why major elements in signaling appear to be involved in the genesis of some types of cancer, but not of others; the ubiquitously expressed *RAS* genes provide an accessible example.

Redundancy in Cell Signaling

We have made only passing mention of a second major issue in signaling: redundancy. In reality, redundancy is a prominent feature of signaling pathways in vertebrate cells; the machinery seems to have become deliberately complex. What is the purpose of this complexity? Just as plasticity may be the evolutionary product of genetic parsimony, redundancy may arise from the evolutionary search for security—a biochemical failsafe that could blunt the impact of single defects. In this scheme, tumorigenesis necessarily becomes a multistep event: Multiple genetic lesions would be required to foil the failsafe.

Cell Signaling and Tumorigenesis

There is now a reasonably clear view of how malfunctioning proto-oncogenes contribute to the pathogenesis of cancer. Consider the pathway through pRas, which can conduct a proliferative signal from the cell surface to the transcriptional apparatus (Fig. 2). Close to half the components of the pathway have surfaced as proto-oncogenes; in each instance, gain-of-function mutations short-circuit the pathway, leading to sustained signaling and consequent cellular proliferation. For good measure, the archetypical proto-oncogene, *SRC*, plays on this pathway by phosphorylating the adapter protein that mediates between the transmembrane receptor and the guanine nucleotide exchange factor. This scheme makes clear why there is hope that pharmacological agents directed at products of proto-oncogenes might prove useful in the treatment of cancer.

Most experimental strategies until now have treated signaling pathways as simple linear progressions or cascades, and it is fortunate that signaling seems to be amenable to analysis in those terms. The truth of the matter is far more complex, with diverse pathways intersecting and counterbalancing one another in ways that we have glimpsed only dimly (Pelech 1993). It will be necessary to understand the dynamic equilibrium of the signaling circuitry before we can safely claim to have understood the circuitry itself.

REFERENCES

Bishop, J.M. 1991. Molecular themes in oncogenesis. *Cell* **64:** 235.

Cantley, L.C., K.R. Auger, C. Carpenter, B. Duckworth, A. Graziani, R. Kapeller, and S. Soltoff. 1991. Oncogenes and signal transduction. *Cell* **64:** 281.

Carpenter, G. and S. Cohen. 1990. Epidermal growth factor. *J. Biol. Chem.* **265:** 7709.

Chen, W.S., C.S. Lazar, K.A. Lund, J.B. Welsh, C.-P. Chang, G.M. Walton, C.J. Der, H.S. Wiley, G.N. Gill, and M.G. Rosenfeld. 1989. Functional independence of the epidermal growth factor receptor from a domain required for ligand-induced internalization and calcium regulation. *Cell* **59:** 33.

Downward, J., J.D. Graves, R.H. Warne, S. Rayter, and D.A. Cantrell. 1990. Stimulation of p21^{ras} upon T-cell activation. *Nature* **346:** 719.

Doyle, H.J. and J.M. Bishop. 1993. Torso, a receptor tyrosine kinase required for embryonic pattern formation, shares substrates with the Sevenless and EGF-R pathways in *Drosophila*. *Genes Dev.* **7:** 633.

Fabian, J.R., I.O. Daar, and D.K. Morrison. 1993. Critical tyrosine residues regulate the enzymatic and biological activity of Raf-1 kinase. *Mol. Cell. Biol.* **13:** 7170.

Finney, R.E., S.M. Robbins, and J.M. Bishop. 1993. Association of pRas and pRaf-1 in a complex correlates with activation of a signal transduction pathway. *Curr. Biol.* **3:** 805.

Han, M., A. Golden, Y. Han, and P.W. Sternberg. 1993. *C. elegans lin-45 raf* gene participates in *let-60 ras*-stimulated vulval differentiation. *Nature* **363:** 133.

Heasley, L.E. and G.L. Johnson. 1992. The β-PDGF receptor induces neuronal differentiation of PC12 cells. *Mol. Biol. Cell* **3:** 545.

Klingler, M., M. Erdelyi, J. Szabad, and C. Nüsslein-Volhard. 1988. Function of *torso* in determining the terminal anlagen of the *Drosophila* embryo. *Nature* **335:** 275.

Moodie, S.A. and A. Wolfman. 1994. The 3Rs of life: Ras, Raf and growth regulation. *Trends Genet.* **10:** 44.

Nishida, E. and Y. Gotoh. 1993. The MAP kinase cascade is essential for diverse signal transduction pathways. *Trends Biol. Sci.* **18:** 128.

Pazin, M.J. and L.T. Williams. 1992. Triggering signaling cascades by receptor tyrosine kinases. *Trends Biol. Sci.* **17:** 374.

Pelech, S.L. 1993. Networking with protein kinases. *Curr. Biol.* **3:** 513.

Perkins, L.A., I. Larsen, and N. Perrimon. 1992. Corkscrew encodes a putative protein tyrosine phosphatase that functions to transduce the terminal signal from the receptor tyrosine kinase torso. *Cell* **70:** 225.

Perrimon, N. 1993. The Torso receptor protein-tyrosine kinase signaling pathway: An endless story. *Cell* **74:** 219.

Porras, A., K. Muszynski, U.R. Rapp, and E. Santos. 1994. Dissociation between activation of Raf-1 kinase and the 42-kDa mitogen-activated protein kinase/90-kDa S6 kinase (MAPK/RSK) cascade in the insulin/ras pathway of adipocytic differentiation of 3T3 L1 cells. *J. Biol. Chem.* **269:** 12741.

Qiu, M.-S. and S.H. Green. 1992. PC12 cell neuronal differentiation is associated with prolonged p21^{ras} activity and consequent prolonged ERK activity. *Neuron* **9:** 705.

Samuels, M.L., M.J. Weber, J.M. Bishop, and M. McMahon. 1993. Conditional transformation of cells and rapid activation of the mitogen-activated protein kinase cascade by an estradiol-dependent human raf-1 protein kinase. *Mol. Cell. Biol.* **13:** 6241.

Simon, M.A., G.S. Dodson, and G.M. Rubin. 1993. An SH3-SH2-SH3 protein is required for p21^{ras1} activation and binds to Sevenless and Sos proteins in vitro. *Cell* **73:** 169.

Sprenger, F., L.M. Stevens, and C. Nüsslein-Volhard. 1989. The *Drosophila* gene *torso* encodes a putative receptor tyrosine kinase. *Nature* **338:** 478.

Traverse, S., N. Gomez, H. Paterson, C. Marshall, and P. Cohen. 1992. Sustained activation of the mitogen-activated protein (MAP) kinase cascade may be required for differentiation of PC12 cells. *Biochem. J.* **288:** 351.

Valius, M. and A. Kazlauskas. 1993. Phospholipase C-gamma 1 and phosphatidylinositol 3 kinase are the downstream mediators of the PDGF receptor's mitogenic signal. *Cell* **73:** 321.

Wells, A., J.B. Welsh, C.S. Lazar, H.S. Wiley, G.N. Gill, and M.G. Rosenfeld. 1990. Ligand-induced transformation by a noninternalizing epidermal growth factor receptor. *Science* **247:** 962.

Activation of Ras and Other Signaling Pathways by Receptor Tyrosine Kinases

J. Schlessinger* AND D. Bar-Sagi†

*Department of Pharmacology, New York University Medical Center, New York, New York 10016;
†Cold Spring Harbor Laboratory, Cold Spring Harbor, New York 11724

Polypeptide growth factors such as epidermal growth factor (EGF), fibroblast growth factor (FGF), or platelet-derived growth factor (PDGF) play an important role in the control of cell growth and differentiation. Peptide growth factors mediate their diverse biological responses by binding to and activating cell-surface receptors with intrinsic protein tyrosine kinase activity (Schlessinger and Ullrich 1992). Growth factor binding to the extracellular domains of their surface receptors induces receptor dimerization. Receptor dimerization is responsible for activation of the intrinsic protein tyrosine kinase activity and for autophosphorylation: Both processes are mediated by an intermolecular process (Schlessinger 1988). Receptor autophosphorylation functions as a molecular switch. Autophosphorylation of receptors such as insulin, insulin-like growth factor 1 (IGF1), or FGF receptors maintains their intrinsic protein tyrosine kinases in an active state. Autophosphorylation sites of other receptors such as EGF receptor compete with exogenous substrates for the catalytic domain. In the case of the EGF receptor, tyrosine autophosphorylation decreases the K_m for substrate phosphorylation, resulting in a release of an autoinhibition (Ullrich and Schlessinger 1990).

Most autophosphorylation sites on growth factor receptors function as binding sites for *src homology* 2 (SH2) domains of cytoplasmic signaling molecules. SH2 domains are independent protein modules composed of approximately 100 amino acids. SH2 domains bind specifically to a phosphotyrosine residue within the context of short, immediate amino acid sequence (Mayer and Baltimore 1992; Pawson and Schlessinger 1993). Most known binding sites for SH2 domains comprise a phosphotyrosine residue together with 3–5 amino acids carboxy-terminal to the phosphotyrosine moiety. Residues amino-terminal to the phosphotyrosine moiety may also play a role in determining the binding specificity of certain SH2 domains (Pawson and Schlessinger 1993; Schlessinger 1994). The binding of various SH2 domains to autophosphorylation sites in growth factor receptors plays an important role in determining the specificity of signal transduction pathways which are activated by growth factor receptors (Pawson and Schlessinger 1993).

THE RAS SIGNALING PATHWAY

Ras functions as a molecular switch for various signals which are initiated at the cell surface and transmitted to the nucleus to control cell growth and differentiation (Satoh et al. 1992). Ras is inactive in the GDP-bound form and active in the GTP-bound form. The interconversion between these two forms is regulated by the balanced action of guanine nucleotide releasing factors (GNRF) and GTPase activating proteins (GAP). GNRF release Ras-bound GDP molecules, which are rapidly replaced by cellular GTP molecules. GTPase activating proteins such as GAP or NF1 accelerate the intrinsic GTPase activity of Ras. Using neutralizing anti-Ras antibodies microinjected into living cells and dominant interfering mutants of Ras, it was demonstrated that Ras has an important role in signaling by growth factor and lymphokine receptors (Nakafuka et al. 1992; Marshall 1994). The predominant mechanism by which receptor tyrosine kinases control Ras activation is by stimulation of guanine nucleotide exchange activity (McCormick 1994).

Various approaches have been used to analyze the signal transduction pathway which is initiated at the cell surface by receptor tyrosine kinases, leading to Ras activation and culminating in cell proliferation. Biochemical studies in animal cells and genetic screening in *Drosophila* and *Caenorhabditis elegans* provided a consistent picture concerning the various components of a highly conserved signal transduction pathway to relay signals from the cell surface to the nucleus (Schlessinger 1993).

EGF stimulation leads to the activation of Ras. One of the tyrosine autophosphorylation sites of EGF receptor serves as a binding site for the adapter protein Grb2. Grb2 is a small protein composed of one SH2 domain flanked by two SH3 domains (Clark et al. 1992; Lowenstein et al. 1992). Grb2 is bound via its two SH3 domains to a proline-rich region in the carboxy-terminal tail of the GNRF Sos (Schlessinger 1994). The binding of Grb2/Sos to Tyr-1068 on the EGF receptor brings Sos to the plasma membrane in the vicinity of Ras. This increases the local concentration of Sos at the plasma membrane, leading to the exchange of GDP for

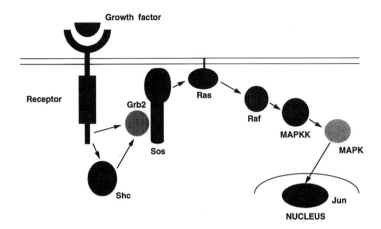

Figure 1. The Ras signaling pathway relays signals from the cell surface to the nucleus. Key components of this pathway include growth factors that bind to and activate receptor tyrosine kinases. Activated receptor tyrosine kinase recruits Grb2/Sos complex directly and via tyrosine-phosphorylated Shc. Membrane-associated Sos activates Ras by exchange of GDP for GTP. GTP-bound Ras recruits Raf, which is activated by an unknown mechanism. Raf activates MAPKK, which phosphorylates and activates MAPK. The targets of MAPK are both cytoplasmic and nuclear, such as Jun. Constitutive activation or overexpression of key elements of this pathway such as growth factors (v-Sis), receptors (v-erbB), Ras, Raf, and Jun leads to oncogenesis.

GTP and activation of Ras (Fig. 1). The GTP-bound active form of Ras binds the Ser/Thr kinase Raf, which is activated by an as yet unknown mechanism. Raf phosphorylates and activates mitogen-activated protein-kinase kinase (MAPKK), which in turn phosphorylates and activates mitogen-activated protein kinase (MAPK). The targets for MAPK are both nuclear and cytoplasmic (Marshall 1994). MAPK is thought to phosphorylate and hence regulate the activity of transcription factors (Fig. 1).

In response to EGF activation, the adapter protein Shc is tyrosine-phosphorylated and becomes associated with the autophosphorylated EGF receptor at Tyr-1173. Both Shc and EGF receptor contain a consensus binding site for Grb2 with the canonical sequence pY V/LN (Rozakis-Adcock et al. 1992; Batzer et al. 1994). These tyrosine phosphorylation sites serve as binding sites for the SH2 domain of Grb2, leading to the recruitment of Grb2/Sos. EGF receptor is therefore able to recruit Grb2/Sos complex directly at Tyr-1068 and indirectly via Shc at Tyr-1173 (Batzer et al. 1994). The functional significance of the multiplicity of Grb2/Sos recruitment by EGF receptor is not known.

It has been shown that Sos is phosphorylated on serine and threonine residues in response to growth factor stimulation (Li et al. 1993; Rozakis-Adcock et al. 1993). Expression of dominant interfering mutants of Ras prevents EGF, FGF, or insulin from inducing Sos phosphorylation, indicating that the kinase(s) responsible for the phosphorylation of Sos is dependent on Ras activation (K. Degenhardt et al., in prep.). MAP kinase can phosphorylate, in vitro, specific serine and threonine sites in the carboxy-terminal tail of Sos. These sites and other Ser/Thr residues are also phosphorylated in vivo, indicating that MAPK probably phosphorylates Sos in response to growth factor stimulation of intact cells. The function of Sos phosphorylation is not known. Phosphorylation is probably not required for activation of Sos, since changes in enzymatic activity of Sos could not be detected in response to growth factor stimulation (Gale et al. 1993). Moreover, MAPK activation precedes Sos phosphorylation, indicating that signaling via Ras activation occurs prior to Sos phosphorylation. Phosphorylated Sos molecules are found preferentially in the cytosol. It is possible that Sos phosphorylation may function as a negative feedback signal to terminate growth factor activation of Ras.

Two mechanisms were proposed for activation of Ras by Grb2/Sos complex. According to an allosteric mechanism, the binding of tyrosine-phosphorylated proteins to the SH2 domain of Grb2 will cause a conformational change in the molecule that will be transmitted to Sos via the two SH3 domains that are bound to the carboxy-terminal tail of the molecule. According to the translocation model, Grb2 is bound to an already active Sos molecule that is translocated to the plasma membrane, where Ras is located by virtue of Grb2 binding to the tyrosine-phosphorylated EGF receptor. The translocation model was tested by exploring the biological activity of a membrane-targeted farnesylated mutant of Sos in which the CAAX farnesylation signal from Ras was inserted at the carboxy-terminal tail of Sos. Similarly, Sos mutants with src myristilation sites inserted at the amino-terminal region of Sos were also generated. The expression of either membrane-targeted Sos mutant caused activation of the Ras signaling pathway, leading to cell transformation (Aronheim et al. 1994). However, point mutations in Sos constructs that eliminate membrane targeting failed to activate the Ras signaling pathway, indicating that membrane localization of Sos is sufficient for Ras activation (Aronheim et al. 1994).

Quantitative binding experiments using isothermal titration calorimetry (ITC) and BIAcore measurements of phosphopeptides to the SH2 domain and a proline-rich peptide from Sos to the SH3 domains of Grb2 indicate that ligand occupation of the SH2 domain does not affect ligand binding to the SH3 domains (M.A. Lemmon et al., in prep.). These experiments suggest that ligands binding to the SH2 and SH3 domains of Grb2 are independent. Moreover, the inability to detect changes in the activity of Sos following growth factor stimulation is also consistent with the translocation mechanism for the activation of Ras by membrane-associated Grb2/Sos complex.

The Ras signaling pathway is highly conserved in evolution. All the known components of this pathway are found in *C. elegans* and *Drosophila* (Pawson and Schlessinger 1993; Marshall 1994). Five of the ten components of this pathway were independently identified as retroviral or cellular oncogenes, including growth factors (v-Sis), growth factor receptors (v-erbB), Ras, Raf, and Jun (Fig. 1). Moreover, membrane-localized forms of Sos (Aronheim et al. 1994), activated mutants of MAPKK (Cowley et al. 1994), and overexpression of Shc (Rozakis-Adcock et al. 1992) can also lead to cell transformation. These results indicate that the Ras signaling pathway has an important role in the control of cell proliferation and that mutations that activate key elements of this pathway can lead to oncogenesis. Moreover, oncogenic proteins such as Bcr-Abl are able to recruit Grb2/Sos and thus feed into the Ras signaling pathway. Elimination of Grb2-binding sites at Tyr-177 of Bcr-Abl prevents Bcr-Abl-induced transformation (Pendergast et al. 1993). These experiments provide an additional example for the central role of the Ras signaling pathway in human cancers. Development of specific inhibitors for key components of this pathway will open many opportunities for the development of novel approaches for pharmaceutical intervention of many types of human cancers.

SIGNALING MOLECULES CONTAINING SH2 AND SH3 DOMAINS

We have used the phosphorylated carboxy-terminal tail of EGF receptor as a specific probe to clone from cDNA expression libraries various signaling molecules that bind to tyrosine-phosphorylated growth factor receptors through their SH2 domains (Skolnik et al. 1991; Margolis et al. 1993). We call this expression/cloning method CORT for *c*loning *o*f *r*eceptor *t*argets, and every protein we clone with CORT we call Grbs for *g*rowth factor *r*eceptor *b*ound. Twelve Grb proteins containing SH2 domains were cloned using the CORT expression/cloning method. It is possible to divide the known signaling proteins containing SH2/SH3 domains into two main groups (Fig. 2). One group contains within the same gene product distinct enzymatic activities together with SH2 and usually also SH3 domains. Several examples are presented in Figure 2A, including the protein tyrosine kinase fyn (cloned as Grb5), phospholipase Cγ (PLCγ) (cloned as Grb6), ras-GAP, a protein tyrosine phosphatase PTP1D/Syp (cloned as Grb9), among many others. An additional small protein module containing approximately 120 amino acids termed pleckstrin homology (PH) domain was found in many SH2- and SH3-containing signaling molecules including ras-GAP and Vav (Fig. 2). It was proposed that, like SH2/SH3 domains, PH domains mediate protein-protein interactions in signal transduction (Mayer et al. 1993; Musacchio et al. 1993). The solution structures of PH domains of pleckstrin and spectrin were recently determined by nuclear magnetic resonance (NMR) spectroscopy (Macias et al. 1994; Yoon et al. 1994). Interestingly, the structure of the PH domain of pleckstrin is similar to the structure of retinol-binding proteins (Yoon et al. 1994). It is noteworthy that the targets for many PH-containing proteins are localized in cell membranes. For example, enzymes involved in the regulation of Ras activity, such as ras-GAP, Sos, and GNRF, all contain PH domains. Because of the structural similarity to retinol-binding protein, it is possible that the targets for PH domains are fatty acids in membranes or fatty acids bound to proteins. It is possible that PH domains function as membrane localization signals for signaling proteins whose targets are localized in cell membranes.

The second class of signaling molecules are called adapters, some of which are depicted in Figure 2B. Examples include Grb2, Crk (cloned as Grb3), Nck (cloned as Grb4), Grb7, and Shc (cloned as Grb12).

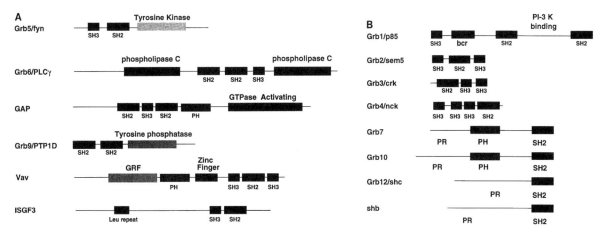

Figure 2. Signaling molecules containing SH2 and SH3 domains. (*A*) Signaling proteins containing various catalytic activities. Fyn is a protein tyrosine kinase (cloned as Grb5). PLCγ is phospholipase Cγ. GAP is *ras* GTPase activating protein. PTP1D/Syp is a protein tyrosine phosphatase. Vav is a putative guanine nucleotide releasing factor for an unknown G protein. ISGF3 is a transcription factor. (*B*) Adapter proteins. p85 (Grb1) is a regulatory subunit of PI-3 kinase. Grb10 is a homolog of Grb7. Both Grb7 and Grb10 contain a proline-rich amino-terminal region and a single pleckstrin homology (PH) domain.

Adapter proteins such as Grb2, Nck, or Crk contain only SH2 and SH3 domains (Mayer and Baltimore 1992; Pawson and Schlessinger 1993). These proteins utilize their SH2 domains to interact with tyrosine-autophosphorylated receptors or other tyrosine-phosphorylated proteins and their SH3 domains to interact with proline-rich sequences in downstream effector proteins. Although Grb2 is not phosphorylated in response to growth factor stimulation, Nck and Crk are subject to growth-factor-induced tyrosine, serine, and threonine phosphorylation. The function of growth-factor-induced phosphorylation of Crk and Nck is not known. Shc is a ubiquitously expressed adapter protein that is tyrosine-phosphorylated in response to stimulation by many growth factors and lymphokines. Moreover, Shc is also tyrosine-phosphorylated in v-*src*-transformed cells. It was shown that tyrosine-phosphorylated Shc binds the SH2 domain of Grb2, providing an alternate link between protein tyrosine kinases and the Ras signaling pathway (Rozakis-Adcock et al. 1992; Pawson and Schlessinger 1993).

The cytoplasmic domain of the PDGF receptor contains nine tyrosine autophosphorylation sites (Fig. 3A) that function as binding sites for the SH2 domains of signaling proteins (Fantl et al. 1993; Kazlauskas 1994; A.K. Arvidson et al., in prep.). In the juxtamembrane region, Tyr-576 and Tyr-581 function as binding sites for pp60^{c-src}. In the kinase insert region, Tyr-716 serves as a binding site for Grb2, and Tyr-740 and Tyr-751 serve as binding sites for the SH2 domains of phosphatidyl inositol (PI)-3 kinase (pYMXM). Tyr-751 serves also as a binding site for Nck, whereas Tyr-770 serves as a binding site for ras-GAP. In the carboxy-terminal tail, Tyr-1009 serves as a binding site for PTP1D/SYP. PTP1D is tyrosine-phosphorylated in response to PDGF receptor activation, creating a binding site for Grb2. Finally, Tyr-1021 functions as a binding site for PLCγ. Synthetic phosphopeptides which correspond to individual autophosphorylation sites block the binding of specific SH2-containing proteins, whereas nonphosphorylated peptide counterparts do not block binding. For example, a phosphopeptide containing tyrosine residues 579 and 581 of PDGF receptor blocks the binding of pp60^{c-src} and does not affect the binding of other signaling molecules to the PDGF receptor. Similarly, elimination of individual autophosphorylation sites by site-directed mutagenesis prevents the binding of specific signaling molecules and does not affect the binding of signaling molecules that bind to other autophosphorylation sites.

The EGF receptor contains five tyrosine autophosphorylation sites in the carboxy-terminal tail of the receptor (Fig. 3B) in a region composed of approximately 200 amino acid residues (Ullrich and Schlessinger 1990). The major binding sites for PLCγ, Grb2, and Shc are Tyr-992, Tyr-1068, and Tyr-1173, respectively (Fig. 3B). However, in the EGF receptor, other tyrosine autophosphorylation sites can serve as secondary binding sites for these signaling molecules (Rotin et al. 1992; Batzer et al. 1994). The binding of ras-GAP and PI-3 kinase is absolutely dependent on EGF receptor autophosphorylation (Margolis et al. 1990). Yet, specific binding sites for these signaling molecules could not be assigned. In all, the specificity of binding of signaling molecules to the autophosphorylation sites of the EGF receptor is less stringent as compared to the binding of the same signaling molecules to the PDGF receptor. This may be because the local concentration of the tyrosine autophosphorylation sites in the carboxy-terminal tail of the EGF receptor is much higher than the concentration of the tyrosine phosphorylation sites in the PDGF receptor, which are distributed over the entire cytoplasmic domain (Fig. 3).

OTHER SIGNALING PATHWAYS AND MECHANISMS FOR RECEPTOR SPECIFICITY

Several other signaling pathways were discovered in recent years. The interaction between the SH2 domains of PLCγ and tyrosine autophosphorylation sites in EGF, PDGF, or FGF receptors links these receptors to PI hydrolysis. The binding of PLCγ to growth factor

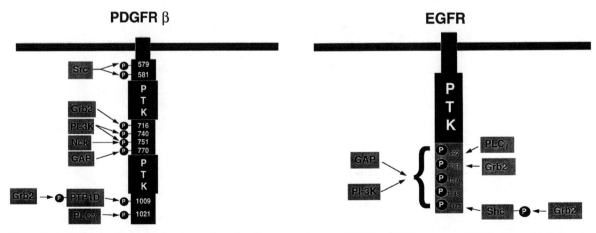

Figure 3. Assignment of binding sites for various signaling molecules on a PDGF or EGF receptor. Grb2 is able to bind directly and indirectly to both EGF and PDGF receptor.

receptor is followed by tyrosine phosphorylation of PLCγ, which is essential for activation of enzymatic activity (Schlessinger and Ullrich 1992). The activation of PLCγ leads to the generation of diacylglycerol and inositol triphosphate (IP_3), which in turn activate protein kinase C and release Ca^{++} from internal stores, respectively (Ullrich and Schlessinger 1990). There is good evidence that this signaling pathway can feed into the Ras/MAPK pathway, probably by phosphorylation and activation of Raf (J. Huang et al., in prep.). Elimination of PLCγ binding sites does not have a major effect on MAPK activity, and various cell types expressing a mutant FGF receptor that does not stimulate PI hydrolysis (Y766F) are fully mitogenic in response to FGF stimulation (Schlessinger and Ullrich 1992). Similarly, neuronal differentiation of PC12 in response to FGF is also not affected by elimination of FGF-induced PI hydrolysis. Nerve growth factor (NGF)-induced neuronal differentiation of PC12 cells is reduced by elimination of Shc-binding sites. However, complete inhibition of neuronal differentiation of PC12 cells is accomplished only after elimination of (Obermeier et al. 1994; Stephens et al. 1994) both Shc and PLCγ binding sites. These experiments suggest that processes that are dependent on phosphatidylinositol hydrolysis are able to feed into and probably participate in the fine tuning of the Ras/MAPK signaling pathway. Moreover, PKC activation and release of intracellular Ca^{++} may lead to the activation of unknown target proteins crucial for differentiation of PC12 cells.

The Jak-Stat pathway is another signaling pathway that relays signals from the cell surface to the nucleus (Darnell et al. 1994). It was initially thought that this pathway is stimulated only by interferons. However, various studies indicate that many growth factors and lymphokines are able to activate this pathway. The known components of this pathway are presented in Figure 4. Interferon stimulation of cell-surface receptors leads to activation of members of the Jak family of non-receptor protein tyrosine kinases. It is thought that these kinases tyrosine-phosphorylate members of the ISFG3 or Stat family of transcription factors that are localized in the cytoplasm. The tyrosine-phosphorylated Stat proteins (p91 in Fig. 4) translocate to the nucleus, bind specific DNA sequences, and promote transcription. Several laboratories demonstrated that EGF, PDGF, and other growth factors are able to activate Jak kinases, to induce tyrosine phosphorylation of Stat proteins, and to direct transcription. Moreover, it was shown that the Jak-Stat pathway is independent of the Ras signaling pathway, since a dominant interfering mutant of Ras did not affect growth-factor-induced Stat transcriptional activity (Silvennoinen et al. 1993). EGF is clearly able to activate Jak kinases and to induce tyrosine phosphorylation of p91. However, it is not clear yet whether Jak kinases are responsible for EGF-induced tyrosine phosphorylation of p91 or whether p91 is directly phosphorylated by the EGF receptor kinase. Nevertheless, the Jak-Stat path-

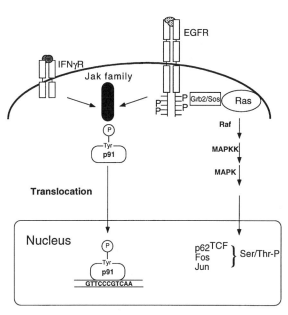

Figure 4. Two signaling pathways that relay signals from the cell surface to the nucleus. (*Right*) The Ras signaling pathway. (*Left*) The Jak-Stat signaling pathway.

way is not essential for EGF- or PDGF-induced cell proliferation, since expression of a dominant interfering mutant of Ras blocks growth-factor-induced mitogenesis but does not affect Stat transcription activity (Silvennoinen et al. 1993).

The fact that most growth factors and lymphokines stimulate similar signaling pathways raises an important question concerning the factors which determine the specificity of signaling pathways generated by various receptor tyrosine kinases. Moreover, it is now well established that the same receptor tyrosine kinase can inhibit or stimulate cell proliferation when expressed in the context of different cell types (Schlessinger and Ullrich 1992). For example, activation of NGF receptors in PC12 cells or sympathetic neurons leads to neuronal differentiation and cell survival, whereas in the context of fibroblasts, NGF receptors induce cell proliferation and oncogenesis. The biological response is determined both by the properties of the receptor tyrosine kinase and by the cellular environment in which the receptor is expressed.

Stimulation of PC12 cells by NGF leads to growth arrest and neurite outgrowth, whereas EGF or insulin stimulates the proliferation of PC12 cells. Stimulation with NGF leads to sustained activation and nuclear translocation of MAPK, whereas both insulin and EGF induce transient activation of MAPK without pronounced nuclear translocation of the enzyme (Qiu and Green 1992). The addition of insulin to PC12 cells overexpressing the receptors induces neuronal differentiation of the receptor-overexpressing cells. Interestingly, MAPK activation in response to insulin stimulation of cells overexpressing the insulin receptor is similar to MAPK activation in response to NGF stimulation of overexpressing or parental PC12 cells: sustained activation and nuclear translocation of MAPK (Dikic et al. 1994). Similar results were obtained in PC12 cells overexpressing EGF receptor that were stimulated with EGF (Traverse et al. 1994).

It is now well established that the Ras signaling pathway is both necessary and sufficient for differentiation of PC12 cells (Noda et al. 1985; Howe et al. 1992; Wood et al. 1992; Cowley et al. 1994). The differential subcellular localization and duration of MAPK induced by insulin, EGF, or NGF may explain the difference in the biological activities of these factors in PC12 cells. It is possible that the strength of the signal generated by a receptor tyrosine kinase may influence the circuitry of the Ras signaling pathway, leading to cell differentiation instead of cell proliferation.

Most known signaling molecules such as Grb2, Sos, PI-3 K, and PLCγ are ubiquitously expressed. We have previously proposed that specificity of receptor tyrosine kinases may be determined in part by a combinatorial mechanism in which different receptors recruit different combinations of various SH2/SH3-containing signaling molecules from a common pool (Schlessinger and Ullrich 1992). However, certain SH2/SH3-containing proteins have restricted tissue expression patterns. For example, the SH3/SH2-containing Vav protein is expressed only in hematopoietic cells (Katzav et al. 1989), Grb7 is expressed only in kidney and liver (Margolis et al. 1993), and one form of *src* (neuronal *src*) is expressed only in neuronal tissue. These cell-type-specific SH2/SH3 proteins may participate in cell-type-specific cellular responses initiated by receptor tyrosine kinases. Finally, signal specificity may be also determined in part by tissue-specific protein tyrosine phosphatases, which regulate tyrosine phosphorylation in concert with the action of protein tyrosine kinases (Mournay and Dixon 1994). The delicate balance between these activities is crucial for the control of normal cell growth and differentiation. Specific aberrations in key elements of these pathways are likely to play an important role in human cancers.

ACKNOWLEDGMENTS

This work was supported by a grant from SUGEN (J.S.). We acknowledge the excellent secretarial help of J. Small.

REFERENCES

Aronheim, A., D. Engleberg, N. Li, J. Schlessinger, and M. Karin. 1994. Membrane targeting of the nucleotide exchange factor Sos is sufficient for activating the Ras signaling pathway. *Cell* **78:** 1949.

Batzer, A.G., D. Rotin, J.M. Ureña, E.Y. Skolnik, and J. Schlessinger. 1994. Hierachy of binding sites for Grb2 and Shc on the EGF-receptor. *Mol. Cell. Biol.* **14:** 5192.

Clark, S.G., M.J. Stern, and H.R. Horvitz. 1992. C. elegans cell signaling gene Sem-5 encodes a protein with SH2. *Nature* **356:** 340.

Cowley, S., H. Paterson, P. Kemp, and C.J. Marshall. 1994. Activation of MAP kinase is necessary and sufficient for PC12 differentiation and for transformation of NIH3T3 cells. *Cell* **77:** 841.

Darnell, J.E., I. Kerr, and G.R. Stark. 1994. Jak-Stat pathway and transcriptional activation in response to IFN's and other extracellular signaling proteins. *Science* **264:** 1415.

Dikic, I., J. Schlessinger, and I. Lax. 1994. PC12 cells overexpressing the insulin receptor undergo insulin dependent neuronal differentiation. *Curr. Biol.* **4:** 702.

Fantl, W.J., D.E. Johnson, and L.T. Williams. 1993. Signaling by receptor tyrosine kinases. *Annu. Rev. Biochem.* **62:** 453.

Gale, N.W., S. Kaplan, E. Lowenstein, J. Schlessinger, and D. Bar Sagi. 1993. Grb2 mediates the EGF dependent activation of guanine nucleotide exchange on Ras. *Nature* **363:** 88.

Howe, L.R., S.J. Leevers, N. Gomez, S. Nakienly, P. Cohen, and C.J. Marshall. 1992. Activation of the MAP kinase pathway by the protein kinase raf. *Cell* **71:** 335.

Katzav, S., D. Martin-Zanca, and M. Barbacit. 1989. Vav, a novel human oncogene derived from a locus ubiquitously expressed in hematopoietic cells. *EMBO J.* **8:** 2283.

Kazlauskas, A. 1994. Receptor tyrosine kinases and their targets. *Curr. Opin. Genet. Dev.* **4:** 5.

Li, N., A. Batzer, R. Daly, V. Yajnik, E. Skolnik, P. Chardin, D. Bar-Sagi, B. Margolis, and J. Schlessinger. 1993. Guanine nucleotide releasing factor hSos1 binds to Grb2 and links receptor tyrosine kinase to Ras signaling. *Nature* **363:** 85.

Lowenstein, E.J., R.J. Daly, A.G. Batzer, W. Li, B. Margolis, R. Lammers, A. Ullrich, E.Y. Skolnik, D. Bar-Sagi, and J. Schlessinger. 1992. The SH2 and SH3-containing protein Grb2 links receptor tyrosine kinase to Ras signaling. *Cell* **70:** 431.

Macias, M.J., A. Musacchio, H. Ponsting, M. Nilges, M. Saraste, and H. Oshkinat. 1994. Structure of the Pleckstrin homology domain from β spectrin. *Nature* **369:** 675.

Margolis, B., O. Silvennoinen, F. Comoglio, C. Roonpraput, E. Skolnik, A. Ullrich, and J. Schlessinger. 1993. High-efficiency expression/cloning of epidermal growth factor-receptor-binding proteins with Src homology 2 domains. *Proc. Natl. Acad. Sci.* **89:** 8894.

Margolis, B., N. Li, A. Koch, M. Mohammadi, D. Hurwitz, A. Zilberstein, A. Ullrich, and J. Schlessinger. 1990. The tyrosine phosphorylated carboxy terminal tail of the EGF receptor is the binding site for GAP and PLCγ. *EMBO J.* **9:** 4375.

Marshall, C.J. 1994. MAP kinase kinase kinase, MAP kinase kinase and MAP kinase. *Curr. Opin. Genet. Dev.* **4:** 82.

Mayer, B.J. and D. Baltimore. 1992. Signaling through SH2 and SH3 domains. *Trends Cell Biol.* **3:** 8.

Mayer, B.J., R. Ren, C.K. Clark, and D. Baltimore. 1993. A putative modular domain present in diverse signaling proteins. *Cell* **73:** 629.

McCormick, F. 1994. Activators and effector of ras P21 proteins. *Curr. Opin. Genet. Dev.* **4:** 71.

Mourney, R.J. and J.E. Dixon. 1994. Protein tyrosine phosphatases: Characterization of extracellular and intracellular domains. *Curr. Opin. Genet. Dev.* **4:** 31.

Musacchio, A., T. Gibson, P. Rice, J. Thompson, and M. Saraste. 1993. The PH domain: A common piece in the structural patchwork of signaling proteins. *Trends Biochem. Sci.* **18:** 343.

Nakafuka, M., T. Satoh, and Y. Kaziro. 1992. Differentiation factors, including nerve growth factor, fibroblast growth factor, and interleukin-6, induce an accumulation of an active Ras-GTP complex in rat pheochromocytoma PC12 cells. *J. Biol. Chem.* **267:** 19488.

Noda, M., M. Ko, A. Ogura, D. Liu, T. Armano, T. Takano, and Y. Ikawa. 1985. Sarcoma viruses carrying Ras oncogenes induce differentiation-associated properties in a neuronal cell line. *Nature* **318:** 73.

Obermeier, A., R.A. Bradshaw, K. Seedorf, K. Choidas, J. Schlessinger, and A. Ullrich. 1994. Neuronal differentiation signals are controlled by nerve growth factor receptor/Trk binding sites for Shc and PLCγ. *EMBO J.* **13:** 1585.

Pawson, T. and J. Schlessinger. 1993. SH2 and SH3 domains. *Curr. Biol.* **3:** 434.

Pendergast, A.M., L.A. Quilliam, L.D. Cripe, C.H. Bassing, Z. Dai, N. Li, A. Batzer, K.M. Rabun, C.J. Der, J. Schlessinger, and M.L. Gishizky. 1993. BCR-ABL-induced oncogenesis is mediated by direct interaction with the SH2 domain of the GRB-2 adaptor protein. *Cell* **75:** 175.

Qiu, M.S. and S.H. Green. 1992. PC12 cell neuronal differentiation is associated with prolonged p21^{ras} activity and consequent prolonged ERK activity. *Neuron* **9:** 705.

Rotin, D., B. Margolis, M. Mohammadi, R.J. Daly, G. Daum, N. Li, E.H. Fischer, W.H. Burgess, A. Ullrich, and J. Schlessinger. 1992. SH2 domains prevent tyrosine phosphorylation of the EGF receptor: Identification of Tyr992 as the high affinity binding site for the SH2 domains of phospholipase Cγ. *EMBO J.* **11:** 559.

Rozakis-Adcock, M., R. Fernley, S. Wade, T. Pawson, and D. Bowtel. 1993. The SH2 and SH3 domains of mammalian Grb2 couple the EGF-receptor to the Ras activator mSos1. *Nature* **363:** 83.

Rozakis-Adcock, M., J. McGlade, G. Mbamalu, G. Pelicci, R. Daly, W. Li, A. Batzer, S. Thomas, J. Brugge, P.G. Pelicci, J. Schlessinger, and T. Pawson. 1992. Association of the Shc and Grb2/Sem5 SH2-containing proteins is implicated in activation of the Ras pathway by tyrosine kinases. *Nature* **360:** 689.

Satoh, T., M. Nakafuka, and Y. Kaziro. 1992. Function of Ras as a molecular switch in signal transduction. *J. Biol. Chem.* **267:** 24149.

Schlessinger, J. 1988. Signal transduction by allosteric receptor dimerization. *Trends Biochem. Sci.* **13:** 443.

———. 1993. How receptor tyrosine kinases activate Ras. *Trends Biochem. Sci.* **18:** 273.

———. 1994. SH2/SH3 signaling proteins. *Curr. Opin. Genet. Dev.* **4:** 25.

Schlessinger, J. and A. Ullrich. 1992. Growth factor signaling by receptor tyrosine kinases. *Cell* **9:** 383.

Silvennoinen, O., C. Schindler, J. Schlessinger, and D.E. Levy. 1993. Ras independent growth factor signaling by transcription factor tyrosine phosphorylation. *Science* **261:** 1739.

Skolnik, E.Y., A. Batzer, N. Li, C.H. Lee, E. Lowenstein, B. Margolis, and J. Schlessinger. 1991. Cloning of P13 kinase-associated p85 utilizing a novel method for expression/cloning of target proteins for receptor tyrosine kinases. *Cell* **65:** 83.

Stephens, R.M., D.M. Loeb, T.D. Copeland, T. Pawson, L. Green, and D. Kaplan. 1994. Trk receptors use redundant signal transduction pathways involving Shc and PLCγ1 to mediate NGF responses. *Neuron* **12:** 1.

Traverse, S., K. Seedorf, H. Paterson, C. Marshall, P. Cohen, and A. Ullrich. 1994. EGF triggers neuronal differentiation of PC12 cells that overexpress the EGF receptor. *Curr. Biol.* **4:** 694.

Ullrich, A. and J. Schlessinger. 1990. Signal transduction by receptors with tyrosine kinase activity. *Cell* **61:** 203.

Wood, K.W., C. Sarnecki, T.M. Roberts, and J. Blenis. 1992. Ras mediates nerve growth factor receptor modulation of three signal-transducing protein kinases: MAP kinase, Raf-1 and RSK. *Cell* **68:** 1041.

Yoon, H.S., P.S. Hajduk, A.M. Petros, E.T. Oiejniczak, R.P. Meadows, and S.W. Fesik. 1994. Solution structure of pleckstrin homology domain. *Nature* **369:** 672.

Ras Partners

L. VAN AELST, M.A. WHITE, AND M.H. WIGLER
Cold Spring Harbor Laboratory, Cold Spring Harbor, New York 11724

THE RAS PROBLEM

The *RAS* oncogenes were first discovered as the transforming elements of acutely oncogenic retroviruses. Subsequently, cellular *RAS* genes were found to be frequently activated by mutation in a wide variety of human cancers, providing the first example of a common oncogenic mechanism. As a consequence, Ras proteins have been studied extensively (for review, see Barbacid 1987).

RAS genes are found ubiquitously in eukaryotic organisms. They encode low-molecular-weight guanine nucleotide-binding proteins that hydrolyze GTP and localize to the inner surface of the plasma membrane after undergoing elaborate carboxy-terminal processing. Proteins that are involved in processing Ras, or in regulating its activity, e.g., by accelerating guanine nucleotide hydrolysis or exchange, have been largely conserved in evolution (for review, see Wigler 1993).

Understanding the function of Ras has been more difficult. Genetic and biochemical studies point to the involvement of Ras in regulating protein kinases in vertebrate cells and in the yeasts, *Schizosaccharomyces pombe* and *Saccharomyces cerevisiae*. In *S. cerevisiae*, this occurs by the direct regulation of adenylyl cyclase (Wigler et al. 1988). This particular biochemical function is not observed in other organisms. In both *S. pombe* and vertebrate cells, Ras appears to activate a conserved protein kinase cascade that we call the MAP kinase module (Neiman et al. 1993). Parallels between these two systems indicated the possibility of a conserved biochemical function. There is strong evidence for other Ras functions in both *S. pombe* and *S. cerevisiae* (Toda et al. 1987; Wigler et al. 1988; Wang et al. 1991; Chang et al. 1994).

In this paper, we address the identification of Ras binding (Rsb) proteins and experimental approaches to the analysis of their function. We have employed a powerful genetic tool, the "two-hybrid system" of Fields and Song, which enables the experimenter to determine when two proteins form complexes within yeast cells (Fields and Song 1989). The method comprises expressing the test proteins as "hybrid" proteins, fused to DNA-binding and transcription-activating proteins, such that when the test proteins interact, a reporter gene is transcribed. We describe the use of this system to discover new Ras-binding proteins and to analyze interactions between Ras and these candidate targets.

SEARCHING FOR RAS PARTNERS

S. pombe byr2 is a protein kinase homologous to the Ste11 protein kinase of *S. cerevisiae*. Both Byr2 and Ste11 function on the sexual differentiation pathway in these yeasts. Genetic analysis in *S. pombe* suggests that byr2 is a downstream target for ras1. Overexpression of *byr2* could bypass defects in $ras1^{null}$ cells; disruption of *byr2* led to defects common to $ras1^{null}$ cells; and dominant interfering mutants of *byr2* partially block the effects of activated *ras1* (Wang et al. 1991; Neiman et al. 1993; Xu et al. 1994). Similar analysis in mammalian cells suggested that Raf1 was a potential downstream target of Ras proteins. Raf1, like Byr2 and Ste11, is a protein kinase that participates in the conserved MAP kinase cascade (for review, see Blumer and Johnson 1994; Cook and McCormick 1994). Activation of Raf1 is oncogenic and bypasses the effects of blocking Ras function; moreover, interfering mutants of Raf1 block activated Ras mutants (Smith et al. 1986; Cai et al. 1990; Kolch et al. 1991). Although Raf1 and Byr2 share no significant homology outside of their kinase domains, they have a similar relationship to Ras (Van Aelst et al. 1993).

We thus sought genetic evidence, utilizing the two-hybrid system, that Ras can form complexes with either protein kinase (Table 1). As we reported previously (Van Aelst et al. 1993), Ras proteins can complex with both proteins, and this interaction appears to require an intact effector loop, that region of Ras that has been suggested, by both genetic and crystallographic studies, to be a domain required for effector interactions (for review, see Polakis and McCormick 1993). Mutants of Ras that affect guanine nucleotide binding also fail to interact with Raf1 and Byr2 (Van Aelst et al. 1993). Mutants lacking the Ras CAAX box, the carboxy-terminal residues that are required for Ras processing, still interact with Raf1 and Byr2. In contrast, whereas an intact effector loop and intact guanine nucleotide binding are also required for complex formation between Ras and its target in *S. cerevisiae*, adenylyl cyclase (Vojtek et al. 1993), this interaction does require an intact CAAX box (Table 1). This result is

Table 1. Interaction between Hras Proteins and Known Ras Targets

	Cyr1[a]	Byr2	Raf1
Hras	+	+	+
Hras(C186S)	−	+	+
Hras(G15A)	−	−	−
Hras(T35A)	−	−	−

Hras wild-type and mutant proteins were expressed as LexA-binding domain (LBD) fusions. Ras targets were expressed as GAL4 activation domain (GAD) fusions. Pairs were co-expressed in the yeast reporter strain L40 (Vojtek et al. 1993). β-Galactosidase was monitored using a filter assay (for description, see Van Aelst et al. 1993). (+) represents a positive induction of β-galactosidase activity. At least 4 independent transformants were tested for each pair. Previously we reported no interaction between Hras^{wt} expressed as a GAL4 DNA-binding (GBD) fusion with Raf1 expressed as a GAD fusion using the YBP2 strain as reporter, and we only observed interaction with Hras proteins carrying the C186S mutation. However, using the LBD fusion and the yeast reporter strain L40, we do observe this interaction.

[a] *S. cerevisiae* adenylyl cyclase.

consistent with the biochemical experiments of Kataoka and coworkers showing that processed Ras is a more effective activator of adenylyl cyclase than is unprocessed Ras (Kurodi et al. 1993).

We next tested whether the two-hybrid system could be used to screen cDNA libraries for Ras binding partners. In particular, we screened libraries in which cDNA inserts were fused to the carboxyl terminus of the *trans*-activating domain of Gal4 to find genes encoding proteins that could interact with either Ras or Raf1 fused to the DNA-binding domain of Gal4 or to the DNA-binding domain of LexA (Fields and Song 1989; Vojtek et al. 1993). Using a variety of different cDNA libraries, on the order of 1 in 50,000 clones was found able to encode a protein that interacts with these targets. Some of the cDNAs we isolated encoded members of the Ras family (interacting with Raf1) or members of the Raf family (interacting with Ras). Some encoded new proteins. We discuss here only those putative proteins that interact with Ras.

Table 2 contains the results of screening several libraries, a total of about 10^7 cDNA clones. cDNAs encoding Raf1 or RafA were found multiple times. In addition, five other genes were identified, each multiple times. Among these were RalGDS, a protein that was previously identified as a guanine nucleotide exchange factor for Ral, a member of the RAS superfamily of guanine nucleotide binding proteins (Albright et al. 1993). The smallest binding fragment comprised amino acids 770 to the carboxy-terminal end, the region just carboxy-terminal to the conserved nucleotide exchange domain. In addition, another cDNA encoded a product, which we call Rsb3, that was homologous to RalGDS (67% homology between amino acids 580 and 768). Other workers have since reported finding these two targets for Ras (Hofer et al. 1994; Kikuchi et al. 1994). On the basis of these results, it is apparent that Rsb3 (called RGL by Kikuchi et al. 1994) is a global homolog of RalGDS. Therefore, interaction with Ras appears to be a conserved feature of these proteins. These results suggest the hypothesis that Ras can regulate Ral. Previous studies have indicated that members of the RAS superfamily can regulate each other (Chang et al. 1994; Horii et al. 1994).

Rsb1 was another peptide identified by Ras binding. Its sequence is completely contained within a protein called AF6, which has been previously described in a single example as a fusion partner for ALL-1 in acute lymphoblastic leukemias (Prasad et al. 1993). ALL-1 is a homolog of *Drosophila* trithorax and is found fused to various partners in acute lymphoblastic leukemia (Gu et al. 1992; Tkachuk et al. 1992; Domer et al. 1993; Morrissey et al. 1993; Nakamura et al. 1993). Rsb1 comprises amino acids 1–180 of AF6. These sequences overlap with the sequences of AF6 that are deleted when fused to ALL-1 (1–35). We do not know the biological significance, if any, of the interaction between AF6 and Ras, nor whether this relates to its involvement in leukemias.

All these targets, Byr2, Cyr1, Raf1, RafA, RalGDS, Rsb3, and Rsb1, share the property that they do not bind to Ras mutated in the effector loop Hras(T35A) or impaired in guanine nucleotide binding Hras(G15A). We infer from this that these proteins interact with Ras in its active, GTP-bound state. In contrast, two other proteins were found, Rsb2 and Rsb4, which putatively bind Ras containing these mutations, but not Ras with mutations in the -CAAX box. One of these, Rsb4, is identical to the catalytic (β) subunit of the Rab geranyl-geranyl transferase (Armstrong et al. 1993).

Table 2. Interaction between Hras Proteins and New Ras Binding Partners

	Raf1	RafA	Ral GDS	Rsb3	Rsb1	Rsb2	Rsb4
Hras	+	+	+	+	+	+	+
Hras(C186S)	+	+	+	+	+	−	−
Hras(G15A)	−	−	−	−	−	+	+
Hras(T35A)	−	−	−	−	−	+	+

Hras proteins were expressed as LBD fusions. Ras targets were expressed as GAD fusions. Pairs were co-expressed in the yeast reporter strain L40. (+) represents a positive induction of β-galactosidase activity using filter assays. The Ras binding partners were isolated by co-transforming mouse brain, rat embryonic, and Jurkat cDNA libraries in which cDNA inserts were fused to the carboxyl terminus of the *trans*-activating domain of GAL4 with either Hras(C186S) mutant fused to GBD or Hras^{wt} fused to LBD. Raf1, RafA, Ral GDS, Rsb3, and Rsb1 were isolated several times both in the Gal4 and LexA systems. RSB2 and RSB4 were isolated only when Hras^{wt} fused to LBD was used. The tester strain used for the screens was either L40 or HF7e (Vojtek et al. 1993; Feilotter et al. 1994). Primary positives were isolated by selecting for colonies that could grow in the absence of histidine and subsequently screened for β-galactosidase activity.

Table 3. Interaction between Ras-like Proteins and Ras Binding Partners

	Cyr1	Byr2	Raf1	RafA	RalGDS	Rsb3	Rsb1	Rsb2	Rsb4
Hras	+	+	+	+	+	+	+	+	+
Sc.Ras2		+	+	+	+	+	+	+	
Sp.Ras1		+	+	+	+	+	+	+	
Rap1A	−	+	+	+	+	+	+	−	−
R-Ras	+	+	+	+	+	+	+		
Rab6	−	−	−	−	−	−	−	−	−
Rho1A	−	−	−	−	−	−	−	−	−
Cdc42	−	−	−	−	−	−	−	−	−
Rac1	−	−	−	−	−	−	−	−	−

Ras and Ras-like proteins were expressed as LBD fusions. All proteins were wild type except for R-Ras(C215R). Ras targets were expressed as GAD fusions. (+) represents a positive induction of β-galactosidase activity using filter assays. At least 4 independent transformants were tested for each pair. Positive interactions in the two-hybrid system have been observed between Rab6, Rho1A, Cdc42, and Rac1 with other proteins (data not shown).

Rab proteins are members of the RAS superfamily. Although the βRabGGtase is homologous to the β subunit of the Ras farnesyl transferase, the carboxy-terminal consensus processing signal for Rab is -CXC or -CC, and the Rab geranyl-geranyl transferase does not process Ras (Khosravi-Far et al. 1992). These results raise fundamental questions about what features the transferases recognize in their substrates, and what their role is in cellular compartmentalization. Curiously, we do not detect interaction between the βRabGGtase and Rab proteins in the two-hybrid system (Table 3). We believe that the productive enzymatic interaction between βRabGGtase and Rab may lead to a very transient interaction.

Rsb2 has been sequenced and does not show homology with known proteins. It does have homology with a protein predicted from nucleotide sequencing of the *Caenorhabditis elegans* genome, and we have found a homolog in *S. cerevisiae*. Curiously, although Rsb2 will bind mammalian H-ras, *S. cerevisiae* Ras2, and *S. pombe* ras1, it does not bind other members of the Ras superfamily that we have tested, including Rap1a (Table 3), which is otherwise very homologous to Ras proteins.

EVALUATING PARTNERS

To evaluate the physiological role of candidate Hras targets, we next sought mutant Hras proteins that could discriminate among the candidates. Libraries of vectors expressing mutant Hras proteins fused to GBD were created by PCR mutagenesis (White et al. 1995) and screened against Raf1 and byr2 fused to GAD. Two mutants were identified. One, Hras(T35S), could bind Raf1 but not Byr2, and one, Hras(E37G), could bind Byr2 but not Raf1. These mutants contain mutations in the conserved effector loop of Hras.

In keeping with the hypothesis that Raf1 is a critical target for Hras, the Hras(G12V, E37G) mutant failed to induce transformed foci of NIH-3T3 cells (Fig. 1). To further test this hypothesis, we identified mutants in Raf1 that restore interaction with Hras(E37G). These mutant Raf1s can cooperate with Hras(G12V, E37G) in the induction of transformed foci (Fig. 1). Such results provide further genetic evidence for the physiological importance of Hras/Raf1 interactions. A more detailed description of these results can be found in White et al. (1995).

More puzzling is that the Hras(G12V, T35S) mutant is greatly attenuated in focus induction. Perhaps its interaction with Raf1 is altered in some way that is not reflected by the two-hybrid assay. Alternatively, or in addition, the interaction of Hras(G12V,T35S) with another effector, perhaps one homologous to Byr2, is adversely affected. In keeping with this idea, Hras(G12V,E37G) is capable of cooperating with Hras(G12V,T35S) in focus induction (White et al. 1995). These results suggest that two different effectors of Hras can contribute to mammalian cell transformation. Unfortunately, none of the candidate mammalian targets we have identified to date, except Raf1, discriminates between our Hras mutants. Continued screening of cDNAs in the two-hybrid system, using the mutant Hras proteins, may identify candidates for this putative additional effector of Hras transformation. Alternatively, this effector may be among our candi-

Table 4. Interaction between New Ras Mutants and Ras Binding Partners

	Cyr1	Byr2	Raf1	RalGDS	Rsb3	Rsb1	Rsb2	Rsb4
Ras(G12V)	+	+	+	+	+	+	+	+
Ras(G12V,E37G)	+	+	−	+	+	+	+	+
Ras(G12V,T35S)	−	−	+	+	+	+	+	+

The Ras mutants were expressed as LBD fusions. Isolation and characterization of these mutants is described elsewhere (White et al. 1995). Ras targets were expressed as GAD fusions. (+) represents a positive induction of β-galactosidase activity using filter assays. At least 4 independent transformants were tested.

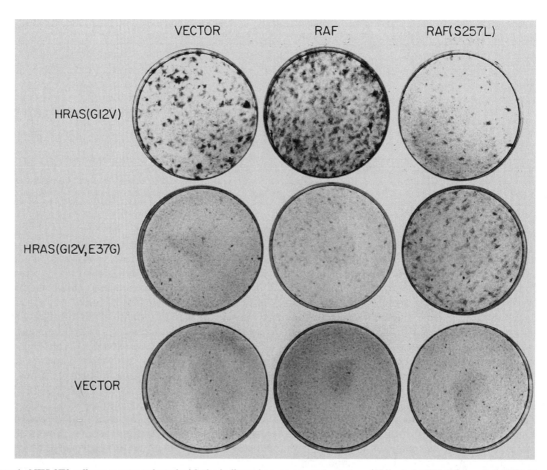

Figure 1. NIH-3T3 cells were cotransfected with the indicated genes or empty vectors (Shih et al. 1979; Wigler et al. 1979; Perucho et al. 1981). Transfected cells were grown in reduced serum for 14 days, then fixed in 10% formaldehyde and stained with Giemsa. Foci of transformed cells appear as diffuse darkly staining spots. Transfection efficiencies, as measured by growth in selective media, were similar for each transfection (White et al. 1995).

dates and interact in the two-hybrid system with Hras(T35S) but interact nonproductively with it in mammalian cells.

THE RAS PROBLEM REVISITED

From our work and that of others, we conclude that Raf1 is a direct mediator of Ras in mammalian cells (Moodie et al. 1993; Van Aelst et al. 1993; Vojtek et al. 1993; Warne et al. 1993; Zhang et al. 1993; White et al. 1995). From our genetic studies, we know that Ras has multiple functions in yeasts (Wigler et al. 1988; Chang et al. 1994) and appears to have multiple functions that can contribute to mammalian cell transformation. We do not know what these other functions may be, although we have defined approaches that may ultimately lead to their identification.

We do not know at the biochemical level how Ras activates effectors. Some data support the idea that Ras merely causes Raf1 to colocalize to the plasma membrane, where Raf1 can encounter other regulators or substrates. For example, Raf1 is activated by fusion to peptides that direct it to the plasma membrane (Leevers et al. 1994; Stokoe et al. 1994), and the function of Ras is severely impaired by mutations that block the processing events which direct it to the plasma membrane (Willumsen et al. 1984). Although Ras may function to cause protein localization, this is certainly not its only function in evolution. In *S. cerevisiae*, Ras activates the enzymatic activity of adenylyl cyclase in a cell-free system (Broek et al. 1987; Field et al. 1988), and in *S. pombe*, Ras can promote the cooperative binding of proteins involved in morphogenesis (Chang et al. 1994). None of the mechanisms of productive interaction between Ras and its effectors has been solved.

ACKNOWLEDGMENTS

We thank Anne Vojtek, Chaker N. Adra, and Isabelle Janoueix-lerosey for providing some of the plasmids and yeast strains used in this study. We thank Jacques Camonis for his help in constructing the cDNA libraries. We thank Linda Rodgers, Michael Riggs, Kim Farina, and Gigi Asouline for technical assistance; and Patricia Bird for preparing the manuscript. This work was supported by the American Cancer Society and by the National Cancer Institute of the National Institutes of Health. L.V.A. is a member of the Belgian National

Fund for Scientific Resarch. M.H.W. is an American Cancer Society Research Professor.

REFERENCES

Albright, C.F., B.W. Giddings, J. Liu, M. Vito, and R.A. Weinberg. 1993. Characterization of guanine nucleotide dissociation stimulator for a *ras*-related GTPase. *EMBO J.* **12:** 339.

Armstrong, S.A., M.C. Seabra, T.C. Sudhof, J.L. Goldstein, and M.S. Brown. 1993. cDNA cloning and expression of the α and β subunits of rat Rab geranylgeranyl transferase. *J. Biol. Chem.* **268:** 12221.

Barbacid, M. 1987. ras genes. *Annu. Rev. Biochem.* **56:** 779.

Blumer, K.J. and G.L. Johnson. 1994. Diversity in function and regulation of MAP kinase pathways. *Trends Biochem. Sci.* **19:** 236.

Broek, D., T. Toda, T. Michaeli, L. Levin, C. Birchmeier, M. Zoller, S. Powers, and M. Wigler. 1987. The *S. cerevisiae* CDC25 gene product regulates the RAS/adenylate cyclase pathway. *Cell* **48:** 789.

Cai, H., J. Szeberenyi, and G.M. Cooper. 1990. Effect of a dominant inhibitory Ha-*ras* mutation on mitogenic signal transduction in NIH 3T3 cells. *Mol. Cell. Biol.* **10:** 5314.

Chang, E.C., M. Barr, and Y. Wang. 1994. Cooperative interaction of *S. pombe* proteins required for mating and morphogenesis. *Cell* **79:** 131.

Cook, S. and F. McCormick. 1994. Ras blooms on sterile ground. *Nature* **369:** 361.

Domer, P.H., S.S. Fakharzadeh, C.-S. Chen, J. Jockel, L. Johansen, G.A. Silverman, J.H. Kersey, and S.J. Korsmeyer. 1993. Acute mixed-lineage leukemia t(4;11)-(q21;23) generates an *MLL-AF4* fusion product. *Proc. Natl. Acad. Sci.* **90:** 7884.

Feilotter, H.E., G.J. Hannon, C.J. Ruddell, and D. Beach. 1994. Construction of an improved host strain for two hybrid screening. *Nucleic Acids Res.* **22:** 1502.

Field, J., J. Nikawa, D. Broek, B. MacDonald, L. Rodgers, I.A. Wilson, R.A. Lerner, and M. Wigler. 1988. Purification of a RAS-responsive adenylyl cyclase complex from *Saccharomyces cerevisiae* by use of an epitope addition method. *Mol. Cell. Biol.* **8:** 2159.

Fields, S. and O.-K. Song. 1989. A novel genetic system to detect protein-protein interactions. *Nature* **340:** 245.

Gu, Y., T. Nakamura, H. Alder, R. Prasad, O. Canaani, G. Cimino, C.M. Croce, and E. Canaani. 1992. The t(4:11) chromosome translocation of human acute leukemias fuses the *ALL-1* gene related to *Drosophila trithorax*, to the AF-4 gene chromosome translocations in acute leukemias. *Cell* **71:** 701.

Hofer, F., S. Fields, C. Schneider, and G.S. Martin. 1994. Activated Ras interacts with the Ral guanine nucleotide dissociation stimulator. *Proc. Natl. Acad. Sci.* **91:** 11089.

Horii, Y., J.F. Beeler, K. Sakaguchi, M. Tachibana, and T. Miki. 1994. A novel oncogene, *ost*, encodes a guanine nucleotide exchange factor that potentially links Rho and Rac signaling pathways. *EMBO J.* **13:** 4776.

Khosravi-Far, R., G.J. Clark, K. Abe, A.D. Cox, T. McLain, R.J. Lutz, M. Sinensky, and C.J. Der. 1992. Ras (CXXX) and Rab (CC/CXC) prenylation signal sequences are unique and functionally distinct. *J. Biol. Chem.* **267:** 24363.

Kikuchi, A., S.D. Demo, Z.-H. Ye, Y.-W. Chen, and L.T. Williams. 1994. ralGDS family members interact with the effector loop of ras p21. *Mol. Cell. Biol.* **14:** 7483.

Kolch, W., G. Heidecker, P. Lloyd, and U. Rapp. 1991. Raf-1 protein kinases is required for growth of induced NIH/3T3 cells. *Nature* **349:** 426.

Kurodi, Y., N. Suzuki, and T. Kataoka. 1993. The effect of posttranslational modifications on the interaction of Ras2 with adenyl cyclase. *Science* **259:** 683.

Leevers, S.J., H.F. Paterson, and C.J. Marshall. 1994. Requirement for Ras in Raf activation is overcome by targeting Raf to the plasma membrane. *Nature* **369:** 411.

Moodie, S.A., B.M. Willumsen, M.J. Weber, and A. Wolfman. 1993. Complexes of Ras-GTP with Raf-1 and mitogen-activated protein kinase kinase. *Science* **260:** 1658.

Morrissey, J., D.C. Tkachuk, A. Milatovich, U. Francke, M. Link, and M.L. Cleary. 1993. A serine/proline-rich protein is fused to *HRX* in t(4;11) acute leukemias. *Blood* **81:** 1124.

Nakamura, T., H. Alder, Y. Gu, R. Prasad, O. Canaani, N. Kamada, R.P. Gale, B. Lange, W.M. Crist, P.C. Nowell, C.M. Croce, and E. Cannani. 1993. Genes on chromosome 4,9 and 19 involved in 11q23 abnormalities in acute leukemia share sequence homology and/or common motifs. *Proc. Natl. Acad. Sci.* **90:** 4631.

Neiman, A., B. Stevenson, H.-P. Xu, G.F. Sprague, Jr., I. Herskowitz, M. Wigler, and S. Marcus. 1993. Functional homology of protein kinases required for sexual differentiation in *Schizosaccharomyces pombe* and *Saccharomyces cerevisiae* suggests a conserved signal transduction module in eukaryotic organisms. *Mol. Biol. Cell* **4:** 107.

Perucho, M., M. Goldfarb, K. Shimizu, C. Lama, J. Fogh, and M. Wigler. 1981. Human-tumor-derived cell lines contain common and different transforming genes. *Cell* **27:** 467.

Polakis, P. and F. McCormick. 1993. Structural requirements for the interaction of p21^{ras} with GAP, exchange factors, and its biological effector target. *J. Biol. Chem.* **268:** 9157.

Prasad, R., Y. Gu, H. Alder, T. Nakamura, O. Canaani, H. Saito, K. Huebner, R.P. Gale, K. Nowell, K. Kuriyama, Y. Miyazaki, C.M. Croce, and E. Canaani. 1993. Cloning of the *ALL-1* fusion partner, the *AF-6* gene, involved in acute myeloid leukemias with the t(6;11) chromosome translocation. *Cancer Res.* **53:** 5624.

Shih, C., B. Shilo, M. Goldfarb, A. Dannenberg, and R. Weinberg. 1979. Passage of phenotypes of chemically transformed cells via transfection of DNA and chromatin. *Proc. Natl. Acad. Sci.* **76:** 5714.

Smith, M.R., S.J. DeGudicibus, and D.W. Stacey. 1986. Requirement for c-ras proteins during viral oncogene transformation. *Nature* **320:** 540.

Stokoe, D., S.G. MacDonald, K. Caddwallader, M. Symons, and J.F. Hancock. 1994. Activation of Raf as a result of recruitment to the plasma membrane. *Science* **264:** 1463.

Tkachuk, D., S. Kohler, and M. Cleary. 1992. Involvement of a homolog of *Drosophila trithorax* by 11q23 chromosome translocations in acute leukemias. *Cell* **71:** 691.

Toda, T., S. Cameron, P. Sass, and M. Wigler. 1987. Three different genes in the yeast *Saccharomyces cerevisiae* encode the catalytic subunits of the cAMP dependent protein kinase. *Cell* **50:** 277.

Van Aelst, L., M. Barr, S. Marcus, A. Polverino, and M. Wigler. 1993. Complex formation between RAS and RAF and other protein kinases. *Proc. Natl. Acad. Sci.* **90:** 6213.

Vojtek, A., S.M. Hollenberg, and J.A. Cooper. 1993. Mammalian Ras interacts directly with the serine/threonine kinase Raf. *Cell* **74:** 205.

Wang, Y., H.-P. Xu, M. Riggs, L. Rodgers, and M. Wigler. 1991. *byr2*, an *S. pombe* gene encoding a protein kinase capable of partial suppression of *ras1* mutant phenotype. *Mol. Cell. Biol.* **11:** 3554.

Warne, P.H., P.R. Viciana, and J. Downward. 1993. Direct interaction of Ras and the amino-terminal region of Raf-1 in vivo. *Nature* **364:** 352.

White, M.A., C. Nicolette, A. Minden, A. Polverino, L. Van Aelst, M. Karin, and M. Wigler. 1995. Multiple RAS functions can contribute to cell transformation. *Cell* (in press).

Wigler, M.H. 1993. The Ras system in yeasts. In *The ras superfamily of GTPases* (ed. J.C. Lacal and F. McCormick), p. 155. CRC Press, Boca Raton, Florida.

Wigler, M., A. Pellicer, S. Silverstein, R. Axel, G. Urlaub, and L. Chasin. 1979. DNA mediated transfer of the *APRT* locus into mammalian cells. *Proc. Natl. Acad. Sci.* **76:** 1373.

Wigler, M., J. Field, S. Powers, D. Broek, T. Toda, S. Cameron, J. Nikawa, T. Michaeli, J. Colicelli, and K. Ferguson. 1988. Studies of RAS function in the yeast *Saccharomyces cerevisiae*. *Cold Spring Harbor Symp. Quant. Biol.* **53:** 649.

Willumsen, B.M., A. Christensen, N.L. Hubbert, A.G. Papageorge, and D.R. Lowy. 1984. The p21 *ras* C-terminus is required for transformation and membrane association. *Nature* **310:** 583.

Xu, H.-P., M. White, S. Marcus, and M. Wigler. 1994. Concerted action of RAS and G proteins in the sexual response pathways of *Schizosaccharomyces pombe*. *Mol. Cell. Biol.* **14:** 50.

Zhang, X.-F., J. Settleman, J.M. Kyriakis, E. Takeuchi-Suzuki, S.J. Elledge, M.S. Marshall, J.T. Bruder, U.R. Rapp, and J. Avruch. 1993. Normal and oncogenic p21^{ras} proteins bind to the amino-terminal regulatory domain of c-raf-1. *Nature* **364:** 308.

Proteins of the 14-3-3 Family Associate with Raf and Contribute to Its Activation

E. FREED, F. MCCORMICK, AND R. RUGGIERI
Onyx Pharmaceuticals, Richmond, California 94806-5206

The Raf serine/threonine kinase is activated in a Ras-dependent manner during mitogenic cell signaling (Avruch et al. 1994). Ras proteins in their active, GTP-bound states bind to Raf (Finney et al. 1993; Moodie et al. 1993; Van Aelst et al. 1993; Vojtek et al. 1993; Warne et al. 1993; Zhang et al. 1993; Hallberg et al. 1994) and translocate it to the plasma membrane (Leevers et al. 1994; Traverse et al. 1993; Stokoe et al. 1994). Once in the membrane, Raf kinase becomes activated through mechanisms that are poorly understood. A number of phosphorylation sites on Raf have been identified that appear necessary for the activation process (Fabian et al. 1993; Morrison et al. 1993), and indeed, protein kinase C can activate Raf in cells under certain conditions (Kolch et al. 1993; Carroll and May 1994). However, the precise role of these phosphorylation events is not yet clear; nor is it clear whether phosphorylation is the critical step that is necessary for activation. To identify proteins that may interact with Raf and participate in its activation, we have performed a two-hybrid screen in yeast cells, using full-length Raf as the target. cDNAs encoding two members of the 14-3-3 family were identified in this screen, and their biological significance was evaluated using a functional assay that detects proteins capable of contributing to the activation of Raf expressed in yeast cells. In this assay, expression of the 14-3-3 proteins resulted in the activation of Raf to a similar extent as that seen with expression of Ras. In mammalian cells, 14-3-3 proteins were found to be associated both with cytosolic Raf and with Raf that was membrane-associated in the presence of activated Ras. From this work we conclude that 14-3-3 proteins are a part of the active Raf complex and may be necessary for the process of Ras-dependent Raf activation.

RESULTS

Two-hybrid Screen for Proteins Interacting with Raf

Full-length human c-Raf-1 (Raf) fused to the Gal4 DNA-binding domain was used as the target of a screen for Raf-binding proteins, using a library of cDNAs fused to the Gal4 activation domain (Fields and Song 1989; Chien et al. 1991). Ten clones were identified that interacted specifically with Raf. They did not react with a fragment of APC (Fig. 1B), nor with the Gal4 DNA-binding domain alone. Three of the positive clones encoded members of the 14-3-3 family, of which two were distinct isolates of 14-3-3 β (Leffers et al. 1993) and one was 14-3-3 ζ (Zupan et al. 1992; Fu et al. 1993). In additional control experiments, the 14-3-3 proteins showed no interaction with Ras (Ser-186) nor with p53 in the two-hybrid system (data not shown).

The region of Raf that interacts with these proteins was determined using protein hybrids consisting of each of three regions of the Raf protein. The results are summarized in Figure 1. A Ras (Ser-186) fusion interacted with full-length Raf and with the amino-terminal region including CR1, but not with the middle region containing CR2, nor with the carboxy-terminal kinase domain (CR3) (Fig. 1). Both of the 14-3-3 proteins also interacted with the amino-terminal domain of Raf, and 14-3-3 β displayed a similar interaction with the middle portion of Raf and with the carboxy-terminal kinase domain. Although interactions of 14-3-3 ζ with both the middle portion of Raf and the carboxy-terminal kinase domain were barely detectable in the β-galactosidase assay, they were evident using the more sensitive measure of growth on medium lacking histidine ($-$His) (Fig. 1B). In general, 14-3-3 β interacted more strongly with Raf than did 14-3-3 ζ. Note also that the two distinct readouts of interaction in the two-hybrid system resulted in different impressions of the relative strength of interaction between 14-3-3 and different portions of the Raf protein. Some of this difference may be due to the fact that expression of the construct encoding the amino-terminal portion of Raf containing CR1 appeared to be toxic to the cells and might selectively affect the readout of cell growth on medium lacking histidine. In summary, both of the 14-3-3 proteins displayed some interaction with all three of the Raf fragments we studied. Given that there is some overlap between these Raf fragments, we conclude that at least two distinct regions of Raf appear to be involved in the interaction with 14-3-3 proteins: some portion of the amino-terminal regulatory region, and the carboxy-terminal region containing the kinase domain. In an in vitro assay, 14-3-3 β did not compete with Ras p21 for binding to the Ras binding domain (Vojtek et al. 1993) of Raf (Fig. 2). Therefore, the region of the amino-terminal Raf fragment that interacts with 14-3-3 proteins is distinct from the Ras-binding domain. These results suggest that 14-3-3 does not compete with Ras for binding to Raf and that Ras, Raf, and 14-3-3 proteins may be part of a multimeric complex in vivo.

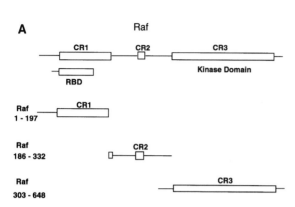
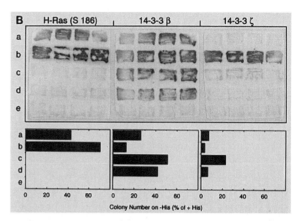

Figure 1. Interaction between portions of Raf and 14-3-3 proteins in the two-hybrid system. (*A*) Diagrammatic representation of the portions of the Raf kinase encoded by the fusion plasmids used to determine the regions of Raf which interact with 14-3-3 β and 14-3-3 ζ. Raf 1–197 includes conserved region 1 (CR1) which contains both the Ras-binding domain (amino acids 51–131) and the zinc-finger-like region. Raf 186–332 includes conserved region 2 (CR2). Raf 303–648 includes the entire kinase domain (CR3) (Freed et al. 1994). (*B*) Plasmids encoding Ha-Ras (Ser-186), or 14-3-3 β, or 14-3-3 ζ fused to the Gal4 activation domain as isolated in the two-hybrid screen (Freed et al. 1994), were transformed into yeast with DNA-binding domain fusions containing: (*a*) full-length Raf, (*b*) Raf amino acids 1–197, (*c*) Raf amino acids 186–332, (*d*) Raf amino acids 303–648, or, as a negative control, (*e*) adenomatous polyposis coli (APC) protein amino acids 1034–2130 (Rubinfeld et al. 1993). Aliquots of the cells from each transformation were plated both on medium lacking histidine (−His) and on medium containing histidine (+His). Four independent colonies were picked from the nonselected (+His) plates from each transformation, patched on +His plates, and grown for 3 days before they were tested for production of β-galactosidase as described previously (Hannon et al. 1993). The colonies on the selected (−His) plates were counted, and their number was expressed as a percentage of the number on nonselected plates. The numbers given here represent the data from one representative transformation. (Reprinted, with permission, from Freed et al. 1994 [copyright AAAS].)

Interaction between Raf and 14-3-3 Proteins in Higher Eukaryotes

To look for in vivo association between Raf and 14-3-3 proteins, Raf was immunoprecipitated from insect Sf9 cells expressing Raf alone, Raf plus v-Src, or catalytically inactive Raf (K375A). Western blotting with an antibody recognizing multiple 14-3-3 isoforms indicated that 14-3-3 proteins were present in all of these preparations of Raf (Fig. 3A). These data suggest that the catalytically inactive Raf binds less well to 14-3-3 proteins than do the other Raf samples, perhaps because its conformation is altered. Preparations of unrelated proteins, such as H-Sos and human papillomavirus (HPV) E2, were similarly isolated from Sf9 cells and found not to contain 14-3-3 proteins. Only a portion of the Raf molecules expressed with v-Src in insect cells is active. We have chromatographically separated active and inactive populations of Raf from these preparations and found 14-3-3 proteins associated with both of these fractions (S.G. Macdonald et al., unpubl.).

To determine if 14-3-3 proteins associated with Raf in mammalian cells, epitope (Glu-Glu)-tagged Raf was expressed in COS7 cells alone or together with 14-3-3 β and immunoprecipitated with antibody to the Glu-Glu tag. The immunoprecipitates from cells that expressed tagged Raf contained 14-3-3 proteins whether or not 14-3-3 β was also overexpressed (Fig. 3B). In contrast, anti-Glu-Glu immunoprecipitates from cells that expressed 14-3-3 β alone did not contain Raf or 14-3-3 proteins (Fig. 3B). These data indicate that the binding observed in the yeast two-hybrid system reflects a true interaction that occurs in vivo.

Raf that is directed to the plasma membrane by the addition of a CAAX motif becomes activated without the need for activated Ras (Leevers et al. 1994; Stokoe et al. 1994). Other proteins that participate in Raf

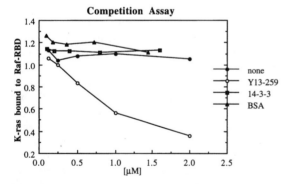

Figure 2. Competition assay for Ki-ras and 14-3-3 binding to Raf-RBD. 96-well plates were coated with the fusion protein GST-Raf 51 to 131 (the Ras-binding domain of Raf) and then incubated with Ki-Ras (0.16 μg /well) preloaded with GTPγS alone or together with one of the following competitors: the monoclonal antibody to Ras, Y13-259, the 14-3-3 β protein, or BSA at the indicated concentrations in binding buffer (50 mM Tris-HCl [pH 8.0], 50 mM NaCl, 5 mM $MgCl_2$, 1% NP-40, 0.15 BSA). After one hour at 37°C, the plates were washed, and the Ki-Ras bound to the coated plate was measured through an ELISA reader after incubation with anti-Glu-Glu antibodies conjugated with alkaline phosphatase (Tropix, Inc.). Results are from one representative experiment.

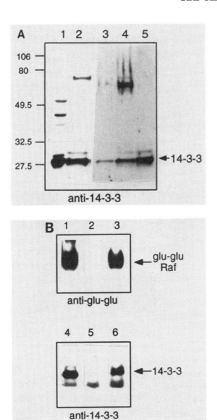

activation must therefore either be localized in the plasma membrane or be translocated there under conditions that translocate Raf. For this reason, it was of interest to determine if Raf isolated from the membrane fraction was associated with 14-3-3 protein. We addressed the localization of 14-3-3 in COS7 cells by immunoprecipitation of Raf overexpressed alone or in combination with activated Ras. When Raf was expressed alone, it was found mainly in the soluble fraction (S100) and was accompanied by 14-3-3 protein (Fig. 4). When Raf was expressed with activated Ras, a larger proportion was recovered in the membrane or pellet fraction (P100) and both membrane-associated and cytosolic Raf contained co-immunoprecipitated 14-3-3 protein (Fig. 4). Analysis of the S100 and P100 fractions prior to immunoprecipitation indicated that only a small proportion of total 14-3-3 protein was associated with Raf, and that most of the 14-3-3 protein was present in the soluble fraction (S100) whether or not Raf was translocated to the membrane (data not shown). Consistent with the fractionation data, immunofluorescence microscopy indicated that 14-3-3 β expressed alone in MDCK cells was localized in the cytosol, whereas a portion of 14-3-3 β expressed together with activated Ras could be detected at the

Figure 3. Binding of Raf to 14-3-3 proteins in insect cells and COS cells. (*A*) Raf tagged with the Glu-Glu epitope (Met Glu Tyr Met Pro Met Glu) at its carboxyl terminus was purified from insect Sf9 cells with monoclonal antibody to the Glu-Glu epitope coupled to Sepharose (anti-Glu-Glu Sepharose) as described previously (Macdonald et al. 1993) and eluted with excess Glu-Glu peptide. The resultant preparations were resolved by polyacrylamide gel electrophoresis (PAGE) on 10% gels (NOVEX) and transferred to immobilon membrane. The membrane was then incubated with polyclonal antiserum raised against bovine brain 14-3-3 proteins (Ichimura et al. 1991) followed by horseradish peroxidase (HRP)-conjugated goat antibody to rabbit IgG. Bands were visualized with enhanced chemiluminescence (ECL) reagents (NEN). Samples that contained approximately equal amounts of Raf as judged by Coomassie staining (not shown) were loaded in lanes 3–5. (Lane *1*) A whole-cell lysate from MDCK cells; (lanes *2* and *5*) two different preparations of tagged Raf from Sf9 cells that were co-infected with v-*src* baculovirus; (lane *3*) catalytically inactive (Lys-375Ala) Raf also tagged with the Glu-Glu epitope; (lane *4*) catalytically competent tagged Raf from cells that were infected with the Raf virus alone. (*B*) COS cells were transiently transfected as described previously (Huang et al. 1993) with plasmids encoding Glu-Glu-tagged Raf alone (lanes *1* and *4*) or in combination with 14-3-3 β (lanes *3* and *6*), or with 14-3-3 β alone (lanes *2* and *5*), and allowed to grow for 2 days before they were serum-deprived for ~16 hr. Cells were then lysed in MAP kinase immunoprecipitation buffer (Freed et al. 1994), and the lysates were subjected to clarification by centrifugation followed by immunoprecipitation with anti-Glu-Glu Sepharose. The resultant proteins were resolved by SDS-PAGE on 10% gels and analyzed by Western blotting with anti-Glu-Glu antibody conjugated to alkaline phosphatase (lanes *1–3*) or with rabbit polyclonal antiserum to 14-3-3 followed by HRP-conjugated goat antibody to rabbit IgG (lanes *4–6*) as described in Fig. 2A. The anti-Glu-Glu blot was visualized with the Western light protein detection kit (Tropix, Inc.). (Reprinted, with permission, from Freed et al. 1994 [copyright AAAS].)

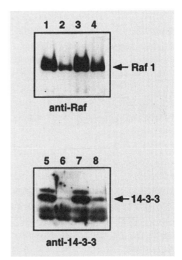

Figure 4. Association of 14-3-3 protein with Raf at the plasma membrane. COS cells transiently transfected with Glu-Glu-tagged Raf alone (lanes *1, 2, 5,* and *6*) or in combination with Ha-Ras (Val-12) (lanes *3, 4, 7,* and *8*) were lysed in hypotonic lysis buffer and subjected to fractionation by centrifugation at 100,000g and the pellet (P100) was then solubilized in a buffer containing Triton X100 as described previously (Freed et al. 1994). The supernatant (S100) (lanes *1, 3, 5,* and *7*) fraction and the detergent-soluble pellet (P100) (lanes *2, 4, 6,* and *8*) fraction were then immunoprecipitated with anti-Glu-Glu Sepharose. Immunoprecipitated proteins were resolved by SDS-PAGE on 10% gels (NOVEX), transferred to immobilon membrane, and subjected to immunoblotting with anti-C20 polyclonal antibody to c-Raf (Santa Cruz Biotech.) (lanes *1–4*) or with polyclonal anti-14-3-3 (lanes *5–8*). Blots were then incubated with HRP-conjugated goat antibody to rabbit IgG and bands were visualized with ECL reagents (NEN). (Reprinted, with permission, from Freed et al. 1994 [copyright AAAS].)

plasma membrane (M. Symons et al., unpubl.). Therefore, it is likely that 14-3-3 protein is translocated to the plasma membrane with Raf in the presence of activated Ras and is associated with Raf in the membrane where activation occurs.

Activation of Raf by 14-3-3 Proteins in Yeast

To address the functional significance of the interaction between Raf and the 14-3-3 proteins, we used a yeast system in which Raf is coupled through MEK to one of the yeast MAP kinase pathways. This is the mating pheromone signaling pathway that consists of a cascade of protein kinases homologous to that involved in the mammalian MAP kinase pathway. They include Ste11, the yeast homolog of MEK kinase (Lange-Carter et al. 1993); Ste7, which is homologous to MEK; and two MAP kinases, Fus3 and Kss1 (Errede and Levin 1993). The latter kinases phosphorylate and activate the transcription factor Ste12, which induces expression of genes required for mating (Errede and Ammerer 1989), such as *FUS1*. Fusion of the *FUS1* promoter to the coding sequences of the *HIS3* gene renders a strain that is defective at the *HIS3* locus dependent on the activation of the mating pathway for survival in the absence of histidine (−His). We have used a strain that is also defective in the *STE11* gene and, therefore, is unable to grow on −His media. An active form of Raf, such as its constitutively active kinase domain, together with its substrate MEK, can complement this growth defect (B. Yashar et al., in prep.; R. Ruggieri and F. McCormick, unpubl.). In this system, full-length Raf does not support growth on −His; however, overexpression of Ha-ras activates Raf and allows signaling through the pathway (Fig. 5A) (B. Yashar et al., in prep.).

When either 14-3-3 β or 14-3-3 ζ was expressed in this strain, this allowed growth on −His to a similar extent as did Ras (Fig. 5A). This phenotype was dependent on the presence of both Raf and MEK in the cell, as cells carrying either Raf or MEK alone did not grow on −His. This result indicated that 14-3-3 proteins affected Raf activity. We confirmed this observation by measuring the activity of Raf proteins in vitro. These were immunoprecipitated from the active cells and tested for their ability to phosphorylate recombinant MEK. Figure 5B shows that Raf preparations extracted from cells overexpressing either form of 14-3-3 protein had a kinase activity that was two- to threefold over background. This level of activation was similar to that observed for Raf preparations from cells overexpressing Ras (Fig. 5B). Therefore, in this system, Ras and 14-3-3 behave similarly in contributing to Raf activation.

Yeast express endogenous Ras proteins that could contribute to the effect of 14-3-3 proteins on Raf. In an attempt to assess the requirement for Ras in the 14-3-3 phenotype, we constructed a strain that lacked both genes encoding the yeast Ras proteins and carried, on a plasmid, a gene (TPK1) for one of the catalytic subunits

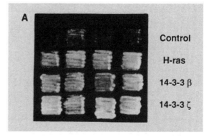

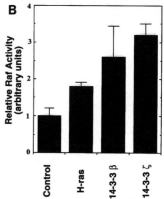

Figure 5. Activation of Raf by 14-3-3 proteins in yeast. *Saccharomyces cerevisiae* strain SY1984R-L (MAT α *ste11Δ pep4Δ his3Δ FUS1::HIS3 leu2 ura3 trp1 can1 RAF::LEU2*) was derived from strain SY1984 (B. Yashar et al., in prep.) by integrating the Raf gene preceded by the ADH promoter into the genome at the LEU2 locus (Freed et al. 1994). This strain was transformed with plasmid YEpMEK (Freed et al. 1994), and then retransformed with one of the following plasmids: the pNV11 vector (a 2 μm-based control plasmid carrying the URA3 marker for selection), pNV11-Ha-Ras (Ha-Ras), pNV11-14-3-3 β, or pNV11-14-3-3 ζ. (*A*) Four independent isolates of each derivative were plated on medium lacking histidine and growth was monitored after 3 days. (*B*) The Raf protein was affinity-purified on anti-Glu-Glu Sepharose from solubilized P100 fractions prepared from each of the above derivatives as described previously (Freed et al. 1994). Immune complexes were assayed for kinase activity and protein immunoblot analysis. A portion of each preparation was incubated with saturating amounts of recombinant MEK (8 μg) and [γ-^{32}P]ATP (5 μCi) for 20 min at 30°C. The reactions were stopped by addition of Laemmli buffer and boiled for 5 min. The phosphorylated products were separated by SDS-PAGE and then quantitated with an AMBIS radioisotope detector (AMBIS). The counts incorporated into MEK were corrected for the amount of Raf proteins as measured by protein immunoblotting. The linearity range was established with a calibration curve as described previously (Freed et al. 1994). The relative Raf activity was normalized to the basal activity of the control. Vertical bars represent standard errors calculated on means of three experiments. (Reprinted, with permission, from Freed et al. 1994 [copyright AAAS].)

of protein kinase A (pkA), which renders the cells viable in the absence of endogenous Ras (Toda et al. 1987). We expressed 14-3-3 or Ha-Ras together with Raf in this strain and tested the level of Raf activity on immunoprecipitation, as described above. Figure 6 shows that the level of activation of Raf from cells expressing TPK1 was very low, whether or not the cells also expressed either Ras or 14-3-3. In contrast, cells that contained Ha-Ras for viability instead of TPK1

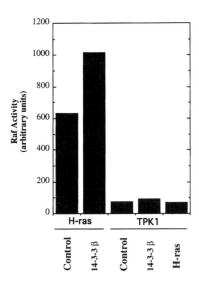

Figure 6. Effect of overexpression of Ha-Ras, 14-3-3, and TPK1 on activation of Raf in yeast. Strain RRY9-2Du⁻ (MAT α ura3 leu2 his3 trp1 ade8 ras1::HIS3 ras2::ura3Δ) was maintained viable either by plasmid AAH5-Ha-Ras (Ruggieri et al. 1992) or by plasmid YEpTPK1 (Toda et al. 1987). Plasmids pADU-RAF and YEpMEK were introduced in each of these strains and then one of the following plasmids was additionally introduced: pMR11 (*Control*) which is a high-copy-number plasmid carrying the *ADE8* gene, pMR11-14-3-3 β, or pMR11-Ha-Ras. The Raf proteins were immunoprecipitated from each strain and tested for activity as described in Fig. 5. Values are from a representative experiment.

displayed higher Raf activity that could be further increased by overexpression of 14-3-3. These results pointed to a possible inhibitory effect of pkA on Raf activity, probably similar to that recently described for mammalian pkA (Burgering et al. 1993; Cook and McCormick 1993; Graves et al. 1993; Wu et al. 1993). In addition, these data showed that the effects of Ha-ras and 14-3-3 on Raf activation in the yeast system are additive, which suggests that Ras and 14-3-3 act on Raf through different mechanisms. These results, however, do not clarify whether Ras is needed for 14-3-3 to act on Raf. We are currently addressing this point by using a different strategy.

DISCUSSION

The 14-3-3 proteins comprise a well-conserved family of proteins present in mammalian cells as well as in flies, yeast, and plants (Aitken et al. 1992). Several members of the family have been described in mammalian cells. These proteins have been implicated in biochemical processes such as inhibition of protein kinase C, activation of tyrosine and tryptophan hydroxylases, and the control of exocytosis from adrenal chromaffin cells. 14-3-3 proteins have also recently been found to associate with polyoma middle T antigen (Pallas et al. 1994) and to be involved in the control of mitosis in fission yeast (Ford et al. 1994). These results suggest a role for 14-3-3 proteins in cellular growth control mechanisms.

We have shown that Raf is present in mammalian cells in association with members of the 14-3-3 protein family. Our data suggest that one of the regions of Raf that interacts with 14-3-3 proteins may be the zinc-finger-like region just carboxy-terminal to the Ras-binding domain. PK-C contains similar zinc-finger-like regions and associates with RACKs (receptors for activated PK-C), with annexin I, and with K-CIP (*k*inase *C* *i*nhibitor *p*rotein), a member of the 14-3-3 protein family (Mochly-Rosen et al. 1991a; Aitken et al. 1992). The region of these proteins which is thought to mediate their interaction with PK-C is homologous to a region contained in all members of the 14-3-3 protein family and at the carboxyl terminus of proteins of the annexin family (Mochly-Rosen et al. 1991b; Aitken et al. 1992; Ron et al. 1994). This annexin-like domain may take part in an association with the zinc-finger-like region of Raf. In addition, our data indicate that other regions of Raf must participate in the interaction with the 14-3-3 proteins. It will be of interest to more clearly define the regions of interaction and to determine if homologous regions in other proteins mediate similar protein-protein interactions.

The mechanism by which Ras proteins lead to activation of Raf kinase has been unresolved until recently. The role of Ras in this process is now clear: The Ras protein recruits Raf to the plasma membrane, where activation takes place (Leevers et al. 1994; Stokoe et al. 1994). Here we discuss one possible component of the activation process, the 14-3-3 protein associated with Raf. Binding to 14-3-3 proteins alone is not sufficient to activate the kinase activity of Raf. 14-3-3 proteins are associated with both inactive Raf in the cytosol and active Raf in the membrane fraction and thus can be considered to be a structural component of the Raf kinase. Because 14-3-3 proteins accompany Raf to the membrane in the presence of activated Ras, it is likely that Ras recruits a complex of Raf and 14-3-3. 14-3-3 proteins are then present with Raf at the plasma membrane where activation occurs. Therefore, an activation mechanism involving 14-3-3 proteins will also depend on activated Ras for membrane localization. When BMH1 (van Heusden et al. 1992) (one of two yeast 14-3-3 homologs) is deleted, Ras expression no longer leads to increased activity of Raf in the yeast system (Irie et al. 1994). Taken together, these observations suggest that Ras and 14-3-3 work together in a common pathway for Raf activation. In the yeast system, overexpression of 14-3-3 β or 14-3-3 ζ is sufficient to activate Raf kinase threefold over background and to replace the need for exogenous Ras. We propose that when Raf is expressed in yeast, where no known homolog exists, Raf plasma membrane localization and activation are limited by the endogenous levels of Ras and 14-3-3 proteins. Introduction of Raf alone allows a basal level activity that is dependent on these endogenous activators (Ras and BMH1). Expression of Ras would then increase Raf activity by increasing the population of the Raf/14-3-3 complex at the plasma membrane. Expression of 14-3-3 proteins would, on the other hand, allow for the activation of Raf already localized to the membrane by endogenous Ras. This interpretation is supported by the finding that the effects of Ras and 14-3-3 in yeast are additive.

In mammalian cells, our analysis demonstrates that Raf is bound to 14-3-3 proteins constitutively and that 14-3-3 proteins are present in excess of Raf. Preliminary experiments indicate that overexpression of 14-3-3 β in COS7 cells does not measurably affect the stoichiometry of binding or the activity of the Raf kinase in these cells. These results may explain why 14-3-3 proteins have not yet been identified as regulators of the Ras pathway in *Drosophila* or *Caenorhabditis elegans*, in which relatively small changes in gene dosage affect Ras signaling. One hypothesis which is consistent with the data from both yeast and mammalian cells is that 14-3-3 proteins participate in Raf activation after membrane localization has been achieved. Other investigators have shown that Raf also forms a complex with hsp90 and a p50 protein and that these proteins are present with Raf at the plasma membrane (Stancato et al. 1993; Wartmann and Davis 1994). These observations support the conclusion that active Raf is present in a multiprotein complex. The extent to which different members of the complex participate in the activation process has not yet been determined.

The biochemical nature of the role 14-3-3 proteins play in the regulation of Raf function remains to be addressed. It will be of interest to determine which members of the 14-3-3 protein family interact with Raf in mammalian cells and whether different members of the 14-3-3 family have distinct functions with respect to the modulation of Raf function. Phosphorylation or other modification of the 14-3-3 proteins may also contribute to regulating Raf activation, and this will be the subject of future work.

ACKNOWLEDGMENTS

We thank K. Matsumoto for communicating results prior to publication; T. Isobe for providing antiserum to the 14-3-3 proteins; B. Rubinfeld for providing pGBT8 APC; M. Callow and T. Vuong for technical assistance; and J. Huang, D. Ramirez, and J. Fitzsimmons for secretarial assistance.

REFERENCES

Aitken, A., D.B. Colinge, B.P.H. van Heusden, T. Isobe, P.H. Roseboom, G. Rosenfeld, and J. Soll. 1992. 14-3-3 proteins: A highly conserved, widespread family of eukaryotic proteins. *Trends Biochem. Sci.* **17:** 498.

Avruch, J., X.-F. Zhang, and J.M. Kyriakis. 1994. Raf meets Ras: completing the framework of a signal transduction pathway. *Trends Biochem. Sci.* **19:** 279.

Burgering, B.M.T., G.J. Pronk, P.C. van Weeren, P. Chardin, and J.L. Bos. 1993. cAMP antagonizes p21^{ras}-directed activation of extracellular signal-regulated kinase 2 and phosphorylation of mSos nucleotide exchange factor. *EMBO J.* **12:** 4211.

Carroll, M.P. and W.S. May. 1994. Protein kinase C-mediated serine phosphorylation directly activates Raf-1 in murine hematopoietic cells. *J. Biol. Chem.* **269:** 1249.

Chien, C.-T., P.L. Bartel, R. Sternglanz, and S. Fields. 1991. The two-hybrid system: A method to identify and clone genes for proteins that interact with a protein of interest. *Proc. Natl. Acad. Sci.* **88:** 9578.

Cook, S.J. and F. McCormick. 1993. Inhibition by cAMP of Ras-dependent activation of Raf. *Science* **262:** 1069.

Errede, B. and G. Ammerer. 1989. STE12, a protein involved in cell-type-specific transcription and signal transduction in yeast, is part of protein-DNA complexes. *Genes Dev.* **3:** 1349.

Errede, B. and D.E. Levin. 1993. A conserved kinase cascade for MAP kinase activation in yeast. *Curr. Opin. Cell Biol.* **5:** 254.

Fabian, J.R., I.O. Daar, and D.K. Morrison. 1993. Critical tyrosine residues regulate the enzymatic and biological activity of Raf-1 kinase. *Mol. Cell Biol.* **13:** 7170.

Fields, S. and O.-K. Song. 1989. A novel genetic system to detect protein-protein interactions. *Nature* **340:** 245.

Finney, R.E., S.M. Robbins, and J.M. Bishop. 1993. Association of pRas and pRaf-1 in a complex correlates with activation of a single transduction pathway. *Curr. Biol.* **3:** 805.

Ford, J.C., F. Al-Khodairy, E. Fotou, K.S. Sheldrick, D.J.F. Griffiths, and A.M. Carr. 1994. 14-3-3 Protein homologs required for the DNA damage checkpoint in fission yeast. *Science* **265:** 533.

Freed, E., M. Symons, S.G. Macdonald, F. McCormick, and R. Ruggieri. 1994. Binding of 14-3-3 proteins to the protein kinase Raf and effects on its activation. *Science* **265:** 1713.

Fu, H., J. Coburn, and R.J. Collier. 1993. The eukaryotic host factor that activates exoenzyme S of *Pseudomonas aeruginosa* is a member of the 14-3-3 protein family. *Proc. Natl. Acad. Sci.* **90:** 2320.

Graves, L.M., K.E. Bornfeldt, E.W. Raines, B.C. Potts, S.G. Macdonald, R. Ross, and E.G. Krebs. 1993. Protein kinase A antagonizes platelet-derived growth factor-induced signaling by mitogen-activated protein kinase in human arterial smooth muscle cells. *Proc. Natl. Acad. Sci.* **90:** 10300.

Hallberg, B., S. Rayter, and J. Downward. 1994. Interaction of Ras and Raf in intact mammalian cells upon extracellular stimulation. *J. Biol. Chem.* **269:** 3913.

Hannon, G.H., D. Demetrick, and D. Beach. 1993. Isolation of the Rb-related p130 through its interaction with CDK2 and cyclins. *Genes Dev.* **7:** 2378.

Huang, D.C.S., C.J. Marshall, and J.F. Hancock. 1993. Plasma membrane-targeted *ras* GTPase-activating protein is a potent suppressor of p21^{ras} function. *Mol. Cell. Biol.* **13:** 2420.

Ichimura, T., H. Sugano, R. Kuwano, T. Sunaya, T. Okuyama, and T. Isobe. 1991. Widespread distribution of the 14-3-3 protein in vertebrate brains and bovine tissues: Correlation with the distributions of calcium-dependent protein kinases. *J. Neurochem.* **56:** 1449.

Irie, K., Y. Gotoh, B. Yashar, B. Errede, E. Nishida, and K. Matsumoto. 1994. Stimulatory effects of yeast and mammalian 14-3-3 proteins on the Raf protein kinase. *Science* **265:** 1716.

Kolch, W., G. Heidecker, G. Kochs, R. Hummel, H. Vahidi, H. Mischak, G. Finkenzeller, D. Marmé, and U.R. Rapp. 1993. Protein kinase Cα activates Raf-1 by direct phosphorylation. *Nature* **364:** 249.

Lange-Carter, C.A., C.M. Pleiman, A.M. Gardner, K.J. Blumer, and G.L. Johnson. 1993. A divergence in the MAP kinase regulatory network defined by MEK kinase and Raf. *Science* **260:** 315.

Leevers, S.J., H.F. Paterson, and C.J. Marshall. 1994. Requirement for Ras in Raf activation is overcome by targeting Raf to the plasma membrane. *Nature* **369:** 411.

Leffers, H., P. Madsen, H.H. Rasmussen, B. Honore, A.H. Andersen, E. Walbum, J. Vandekerckhove, and J.E. Celis. 1993. Molecular cloning and expression of the transformation sensitive epithelial marker stratifin. A member of a protein family that has been involved in the protein kinase C signalling pathway. *J. Mol. Biol.* **231:** 982.

Macdonald, S.G., C.M. Crews, L. Wu, J. Driller, R. Clark, R.L. Erikson, and F. McCormick. 1993. Reconstitution of the raf-1-MEK-ERK signal transduction pathway *in vitro*. *Mol. Cell. Biol.* **13:** 6615.

Mochly-Rosen, D., H. Khaner, and J. Lopez. 1991a. Identification of intracellular receptor proteins for activated protein kinase C. *Proc. Natl. Acad. Sci.* **88:** 3997.

Mochly-Rosen, D., H. Khaner, J. Lopez, and B.L. Smith. 1991b. Intracellular receptors for activated protein kinase C. Identification of a binding site for the enzyme. *J. Biol. Chem.* **266:** 14866.

Moodie, S.A., B.M. Willumsen, M.J. Weber, and A. Wolfman. 1993. Complexes of Ras·GTP with Raf-1 and mitogen-activated protein kinase kinase. *Science* **260:** 1658.

Morrison, D.K., G. Heidecker, U.R. Rapp, and T.D. Copeland. 1993. Identification of the major phosphorylation sites of the Raf-1 kinase. *J. Biol. Chem.* **268:** 17309.

Pallas, D.C., H. Fu, L.C. Haehnel, W. Weller, R.J. Collier, and T.M. Roberts. 1994. Association of polyomavirus middle tumor antigen with 14-3-3 proteins. *Science* **265:** 535.

Ron, D., C.H. Chen, J. Caldwell, L. Jamieson, E. Orr, and D. Mochly-Rosen. 1994. Cloning of an intracellular receptor for protein kinase C: A homolog of the β subunit of G proteins. *Proc. Natl. Acad. Sci.* **91:** 839.

Rubinfeld, B., B. Souza, I. Albert, O. Muller, S.H. Chamberlain, F.R. Masiarz, S. Munemitsu, and P. Polakis. 1993. Association of the APC gene product with β-catenin. *Science* **262:** 1069.

Ruggieri, R., A. Bender, Y. Matsui, S. Powers, Y. Takai, J.R. Pringle, and K. Matsumoto. 1992. *RSR1*, a ras-like gene homologous to Krev-1(smg21A/rap1A): Role in the development of cell polarity and interactions with the ras pathway in *Saccharomyces cerevisiae*. *Mol. Cell. Biol.* **12:** 758.

Stancato, L.F., Y.-H. Chow, K.A. Hutchison, G.H. Perdew, R. Jove, and W.B. Pratt. 1993. Raf exists in a native heterocomplex with hsp90 and p50 that can be reconstituted in a cell-free system. *J. Biol. Chem.* **268:** 21711.

Stokoe, D., S. Macdonald, K. Cadwallader, M. Symons, and J.F. Hancock. 1994. Activation of Raf as a result of recruitment to the plasma membrane. *Science* **264:** 1463.

Toda, T., S. Cameron, P. Sass, M. Zoller, and M. Wigler. 1887. Three different genes in *S. cerevisiae* encode the catalytic subunits of the cAMP-dependent protein kinase. *Cell* **50:** 277.

Traverse, S., P. Cohen, H. Paterson, C.J. Marshall, U. Rapp, and R.J. Grand. 1993. Specific association of activated MAP kinase kinase kinase (Raf) with the plasma membranes of ras-transformed retinal cells. *Oncogene* **8:** 3175.

Van Aelst, L., M. Barr, S. Marcus, A. Polverino, and M. Wigler. 1993. Complex formation between RAS and RAF and other protein kinases. *Proc. Natl. Acad. Sci.* **90:** 6213.

van Heusden, G.P.H., T.J. Wenzel, E.L. Lagendijk, H.Y. de Steensma, and J.A. van den Berg. 1992. Characterization of the yeast *BMH1* gene encoding a putative protein homologous to mammalian protein kinase II activators and protein kinase C inhibitors. *FEBS Lett.* **302:** 145.

Vojtek, A.B., S.M. Hollenberg, and J.A. Cooper. 1993. Mammalian Ras interacts directly with the serine/threonine kinase Raf. *Cell* **74:** 205.

Warne, P.H., P. Rodriguez Viciana, and J. Downward. 1993. Direct interaction of Ras and the amino-terminal region of Raf-1 in vitro. *Nature* **364:** 352.

Wartmann, M. and R.J. Davis. 1994. The native structure of the activated Raf protein kinase is a membrane-bound multi-subunit complex. *J. Biol. Chem.* **269:** 6695.

Wu, J., P. Dent, T. Jelinek, A. Wolfman, M.J. Weber, and T.W. Sturgill. 1993. Inhibition of the EGF-activated MAP kinase signaling pathway by adenosine 3′, 5′-monophosphate. *Science* **262:** 1065.

Zhang, X.-F., J. Settleman, J.M. Kyriakis, E. Takeuchi-Suzuki, S.J. Elledge, M.S. Marshall, J.T. Bruder, U.R. Rapp, and J. Avruch. 1993. Normal and oncogenic p21^{ras} proteins bind to the amino-terminal regulatory domain of c-Raf-1. *Nature* **364:** 308.

Zupan, L.A., D.L. Steffens, C.A. Berry, M. Landt, and R.W. Gross. 1992. Cloning and expression of a human 14-3-3 protein mediating phospholipolysis. *J. Biol. Chem.* **267:** 8707.

Regulation of the Cryptic Sequence-specific DNA-binding Function of p53 by Protein Kinases

T.R. Hupp AND D.P. Lane

Cancer Research Campaign Laboratories, Cell Transformation Group, Department of Biochemistry, Dundee University, Dundee DD1 4HN, Scotland

p53 protein functions as a tumor suppressor and appears to be involved in regulation of the cellular response to DNA-damaging agents or oxidative stress (Maltzman and Czyzyk 1984; Kastan et al. 1991, 1992; Kuerbitz et al. 1992; Lane 1992, 1993; Lu et al. 1992; Hall et al. 1993; Lu and Lane 1993; Tishler et al. 1993; Zhan et al. 1993). One biochemical activity of p53 required for its tumor suppressor function involves its ability to function as a sequence-specific DNA-binding protein and a transcription factor (Bargonetti et al. 1991; Kern et al. 1991, 1992; El-Deiry et al. 1992; Funk et al. 1992; Scharer and Iggo 1992). p53 also exhibits a remarkably strong affinity for single-stranded and double-stranded DNA or RNA; this activity is made manifest in the ability of carboxy-terminal tetrameric peptides derived from p53 to bind to DNA nonspecifically (Foord et al. 1991; Wang et al. 1993) and in the ability of p53 to catalyze single-stranded DNA or RNA reassociation into duplex molecules (Oberosler et al. 1993; Bakalkin et al. 1994).

p53 protein is composed of at least four separate functional domains which are involved in modulating its biochemical activity (Fig. 1). p53 represents an apparently unique class of DNA-binding protein on the basis of the following: (1) Its primary amino acid sequence (Soussi et al. 1990) lacks significant homology with the active sites of well-characterized sequence-specific DNA-binding proteins (Pabo and Sauer 1992); (2) it assembles into large oligomeric complexes (Kraiss et al. 1988; Milner and Medcalf 1991; Sturz-

A. Functional Domains of p53

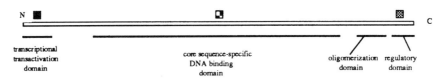

B. Regulatory site of human p53

Figure 1. (*A*) Functional domains within human p53. The amino terminus harbors the transcriptional activation domain (Fields and Jang 1990; O'Rouke et al. 1990; Raycroft et al. 1990; Funk et al. 1992; Kern et al. 1992; Scharer and Iggo 1992) and the binding site of the monoclonal antibody DO-1 (■) (Vojtesek et al. 1992). Regulation of p53 through modification of the amino-terminal domain has been shown to occur through its interaction with Mdm-2 (Oliner et al. 1993) and is suggested by multiple sites for phosphorylation (Lees-Miller et al. 1992; Meek and Street 1992; Milne et al. 1992b, 1994); the conserved central core (Soussi et al. 1990) contains the site-specific DNA-binding domain (Bargonetti et al. 1993; Pavletich et al. 1993; Wang et al. 1993) and a conformationally flexible region, recognized by monoclonal antibody PAb240, whose structure is often altered by point mutations of the type found in human cancer (◨) (Gannon et al. 1990; Stephen and Lane 1992). The oligomerization domain of human p53 resides between amino acids 334 and 356 (Sturzbecher et al. 1992) and mouse p53 between amino acids 330 and 344 (Shaulian et al. 1992); and the domain encoded by exon 11 (Lamb and Crawford 1986) with the PAb421 binding site (▩), between amino acids 368 and 393, which allosterically regulates site-specific DNA binding (Hupp et al. 1992). (*B*) Binding sites of proteins within the carboxy-terminal allosteric domain of human p53. The binding site of PAb421 (Wade-Evans and Jenkins 1985; C.W. Stephen and D.P. Lane, unpubl.), casein kinase II (Bischoff et al. 1992), predicted sites of protein kinase C phosphorylation at serine residues within the PAb421 epitope (Baudier et al. 1992; this paper), and the endpoint of the p53Δ30 protein (Hupp et al. 1992) are indicated.

becher et al. 1992), mostly tetramers or multimers of tetramers (Stenger et al. 1992; Friedman et al. 1993); and (3) it exhibits the propensity to assemble into a homo-oligomeric and biochemically latent state in mammalian and prokaryotic cells (Hupp et al. 1992). p53 is also multiply phosphorylated at amino- and carboxy-terminal sites (Meek and Street 1992), suggesting that kinases and phosphatases may be involved in the regulation of the tumor suppressor function of the protein. Downstream gene products apparently under transcriptional regulation by p53 include *gadd45* (Kastan et al. 1992), *mdm-2* (Barak and Oren 1992; Momand et al. 1992; Oliner et al. 1992; Barak et al. 1993), and p21^{WAF-1} (El-Deiry et al. 1993; Harper et al. 1993).

A relatively large number of cellular proteins are now known to interact with p53, including the stress-70 family of proteins, protein kinases, DNA-binding proteins, and viral or cellular oncogenes. A growing number of these enzymes and proteins can regulate posttranslationally the biochemical function of p53 itself. These include the oncogenes *mdm-2* and SV40 T antigen, which negatively regulate p53 function (Farmer et al. 1992; Momand et al. 1992; Oliner et al. 1993; Zauberman et al. 1993) and the WT1 tumor suppressor protein, which enhances p53-mediated transcriptional activation from the creatine kinase promoter/reporter constructs (Maheswaran et al. 1993). A carboxy-terminal regulatory site of p53 itself is implicated in negative autoregulation of sequence-specific DNA binding (Hupp et al. 1992), and native casein kinase II from rabbit muscle or recombinant human casein kinase II activates the latent DNA-binding function of p53 by modifying the carboxy-terminal regulatory site (Hupp et al. 1993). Regulation of p53 by casein kinase II may have physiological significance, as the casein kinase II site is modified by phosphate in vivo (Meek et al. 1990; Herrmann et al. 1991; Bischoff et al. 1992) and appears to be required for maximal growth-suppressive activity of mouse p53 in rodent cell lines (Milne et al. 1992a). Regulation of endogenous wild-type p53 through modification of the carboxyl terminus may also occur through alternate splicing, giving rise to a protein lacking the carboxy-terminal negative regulatory domain encoded by exon 11 (Arai et al. 1986; Kulesz-Martin et al. 1994), or through the covalent attachment of an RNA moiety (Samad and Carroll 1991).

We have previously taken a biochemical approach to determine how the sequence-specific DNA-binding activity of p53 is regulated. By studying unmodified forms of recombinant human p53, a regulatory site that controls sequence-specific DNA binding has been identified in the extreme carboxyl terminus (Hupp et al. 1992). Casein kinase II targets the negative regulatory domain within the carboxy-terminal 30 amino acids and unmasks the cryptic activity of p53, providing direct evidence that p53 site-specific DNA-binding activity is subject to positive enzymatic regulation. Presently, two highly conserved cellular proteins, whose binding sites lie within the carboxy-terminal regulatory domain, can both activate the latent function of p53. These two enzymes are casein kinase II and protein kinase C. Not only does modification of p53 by protein kinase C activate the latent function of the protein, implicating a second protein kinase in the direct control of p53 activity, but modification also coincides with a striking loss of the PAb421 epitope. PAb421 nonreactivity with p53 has previously been shown to coincide with growth arrest of glioblastoma cells upon overexpression of the wild-type p53 protein (Ullrich et al. 1992a) and in resting lymphocytes (Milner 1984; Gamble and Milner 1988). This suggests that phosphorylation within the carboxy-terminal regulatory site by an enzyme with the specificity of protein kinase C may regulate the growth suppressor function of p53 and that PAb421 serves as a noninvasive probe to monitor the extent of protein-kinase-C-like modification in the carboxy-terminal regulatory site.

EXPERIMENTAL PROCEDURES

Reagents, proteins, and enzymes. Heparin-Sepharose and Superose-12 columns were obtained from Pharmacia. Cellulose phosphate P11 was from Whatman. Casein kinase II from rabbit muscle was purified as described previously (Hupp et al. 1992). Rat brain protein kinase C and human recombinant casein kinase II were obtained from Boehringer Mannheim. Monoclonal antibodies DO-1 and PAb421 were purified on protein A, followed by dialysis into Buffer T (containing 10% glycerol, 0.1 M KCl, and 20 mM HEPES [pH 7.6]). Overproduction of recombinant human p53 in baculovirus-infected insect cells was performed as described previously (Bargonetti et al. 1991).

Purification of latent p53 from recombinant strains of Escherichia coli. p53 was purified to near homogeneity by a derivation of the published protocol (Hupp et al. 1992). A phosphocellulose step was added between heparin-Sepharose and Superose-12 gel filtration chromatography. After pooling of the active heparin-Sepharose fractions, the protein was diluted to the conductivity of Buffer B (15% glycerol, 20 mM HEPES [pH 7.6], 0.1 mM EDTA, 5 mM DTT, 1 mM benzamidine, and 0.1% Triton X-100) containing 0.1 M KCl. This was applied to a phosphocellulose column at a ratio of protein (mg) to bed volume (ml) of resin of 0.5, and bound protein was eluted with a 20-column volume linear gradient from 0.1 M to 1 M KCl in Buffer B. p53 was assayed for sequence-specific DNA binding in the absence and presence of PAb421. The latent form of p53 eluted at approximately 0.55 M KCl. A phosphocellulose step results in a substantial purification of the p53 protein and is required to purify the protein to near homogeneity using low-level overexpression systems. Fractions from phosphocellulose were concentrated using a Centricon-30 and applied to a Pharmacia Superose-12 gel filtration column in Buffer B. p53Δ30 was purified using the same three columns used for full-length p53.

Immunological assays. Enzyme-linked immunosorbent assays were performed using pure antibodies as indicated previously (Harlow and Lane 1988) using p53-coated wells (2 μg/ml in carbonate buffer) or antibody-coated wells (2 μg/ml of pure IgG in carbonate buffer). The immune complexes were formed with either p53 or polyclonal antibodies (CM-1) to human p53 in phosphate-buffered saline + 0.1% Tween-20 + 3% bovine serum albumin. Secondary antibodies conjugated to horseradish peroxidase were incubated in the same buffer.

Phosphorylation and activation of latent p53. p53 or p53Δ30 (Fraction IV, 12 ng) was incubated in 10 μl of a buffer (10% glycerol, 10 mM $MgCl_2$, 20 mM HEPES [pH 7.6], 0.1 mM ATP [containing [γ-^{32}P]ATP where indicated], 0.1 mM EDTA, 5 mM dithiothreitol, 0.1% Triton X-100) with casein kinase II from rabbit muscle or recombinant human casein kinase II obtained from Boehringer Mannheim (0.1 mU, Hupp et al. 1993) or in a buffer with rat brain protein kinase C (0.02 mU, 20 mM HEPES [pH 7.6], 10 mM $MgCl_2$, 0.5 mM $CaCl_2$, 100 μg/ml phosphatidyl serine, 20 μg/ml diacylglycerol, 10 μM ATP [and containing [γ-^{32}P]ATP where indicated]) and the indicated antibodies. Incubations were performed for the indicated times at 30°C and reactions were stopped by the addition of SDS protein sample buffer as described to measure the moles of phosphate per mole of p53, or activity of p53 in DNA-binding assays was measured by adding the reactions to 10-μl of a buffer containing radioactive target DNA as described previously (Hupp et al. 1993). The incorporation of radioactive phosphate into p53 was quantified by cutting out gel slices and counting radioactivity using a liquid scintillation counter. Activation of p53 with casein kinase II or PAb421 and DNA-binding reactions were carried out according to the method of Hupp et al. (1993).

RESULTS

Purification of the Latent Form of p53

A pure source of homogeneous, unphosphorylated p53 is essential in order to make accurate correlations between unique modifications and their subsequent effects on biochemical function. Wild-type human p53 overexpressed in *E. coli* is a good source of unmodified p53, as *E. coli* does not harbor the protein kinases which are known to modify p53, including casein kinase I isozyme (Milne et al. 1992b), casein kinase II (Meek et al. 1990; Herrmann et al. 1991; Bischoff et al. 1992), DNA-activated protein kinase (Lees-Miller et al. 1990; Wang and Eckhart 1992), protein kinase C (Baudier et al. 1992), cdc2 (Addison et al. 1990; Bischoff et al. 1990), and microtubule-associated protein kinase (Milne et al. 1994).

Human p53 was purified from an overproducing strain of *E. coli* to near homogeneity using heparin-Sepharose, phosphocellulose, and Superose-12 gel filtration chromatography (see Experimental Procedures; Hupp et al. 1992). This latent form of p53, with a Stokes radius of 65 Å and a sedimentation coefficient of 6.2 S, is homotetrameric in solution (Hupp and Lane 1994). Thus, the latent form of p53 has a subunit structure identical to the active protein produced in insect-cell expression systems and immunoaffinity-purified using a PAb421 matrix (Friedman et al. 1993).

p53 Has a Cryptic DNA-binding Function

The unphosphorylated form of p53 is inactive as a site-specific DNA-binding protein (Fig. 2, lower panel). Fractions of pure p53 eluting from gel filtration (Fig. 2, fractions 15–22) were also assayed in the presence of the carboxy-terminal-specific monoclonal antibody, PAb421. Antibody PAb421 binds near the phosphorylation sites of both casein kinase II and protein kinase C (Fig. 1) and can activate the latent function of the protein (Fig. 2, middle panel, fractions 18 and 19). A clue into the biochemical mechanism of activation of latent p53 by monoclonal antibody PAb421 came from the observation that deletion of the carboxy-terminal domain encoded by exon 11 (Lamb and Crawford 1986) constitutively activated the protein (Hupp et al. 1992). Thus, the carboxy-terminal domain harbors a negative regulatory motif, the neutralization of which unmasks the cryptic DNA-binding function of p53.

Casein Kinase II Activation of Latent p53

The biochemical evidence implicating the existence of a novel regulatory site within the carboxyl terminus

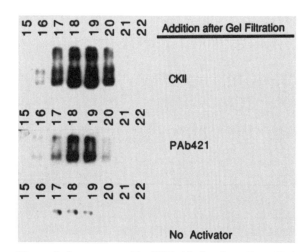

Figure 2. Sequence-specific DNA binding by p53 purified using gel filtration chromatography, with and without the activating proteins casein kinase II or PAb421. Sequence-specific DNA-binding reactions containing Mg/ATP were performed with the indicated Superose-12 fractions containing (1) p53 (1 μl of indicated fraction) with casein kinase II from rabbit muscle (125 ng), (2) p53 (0.25 μl of indicated fraction) and PAb421 (500 ng), or (3) p53 (1 μl of indicated fraction) without activator, and the products of the reaction were separated on a nondenaturing polyacrylamide gel as described in Experimental Procedures.

of p53 suggests that this domain may play a role in controlling the sequence-specific DNA-binding function of the protein. As such, the possible existence of cellular factors that could replace the antibody PAb421 in the activation of latent p53 was compelling. Given the established requirement for an intact casein kinase II motif at Ser-392 for the growth suppressor function of mouse p53 (Milne et al. 1992a), and the proximity of the casein kinase II motif to the PAb421-binding site (Fig. 1), casein kinase II was purified from rabbit muscle and tested for the ability to activate latent p53. In the presence of ATP/Mg and casein kinase II, activation of latent p53 function is observed in fractions eluting from the gel filtration column (Fig. 2, lanes 18 and 19, top panel).

The activation of p53 by casein kinase II was GTP- or ATP-dependent (Fig. 3A, lanes 6–19 versus lane 2; from Hupp et al. 1993), consistent with the ability of the kinase to utilize either GTP or ATP as a phosphate donor (Meisner and Czech 1991), whereas CTP could not be used as a phosphate donor for the casein-kinase-II-dependent activation of p53 (lanes 3–5 versus lane 2). Recombinant human casein kinase II could activate latent p53 (Fig. 3B), although the large-molecular-weight-protein/DNA complexes produced were quite distinct from those observed after activation using native casein kinase II in vitro (lanes 1-3 versus lanes 7 and 8) or using cellularly activated p53 (see Fig. 8). These results suggest that the native enzyme may be modified or may contain cofactors such that it gives rise to a more physiological activation of p53, compared to the recombinant kinase, which may be unmodified or devoid of specific cofactors.

Protein Kinase C Phosphorylates within the Carboxy-terminal Regulatory Site of p53

Casein kinase II is a nuclear, signal-independent kinase that phosphorylates a variety of DNA-binding proteins involved in regulation of gene expression and chromosomal DNA integrity (Meisner and Czech 1991). Interest in this presumably constitutively active kinase springs from the observation that its activity is stimulated by growth signals or mitogens (Sommercorn

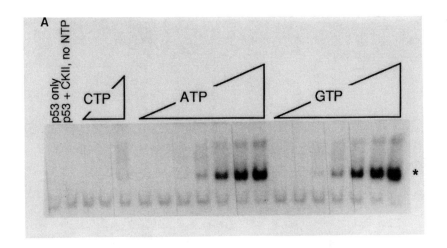

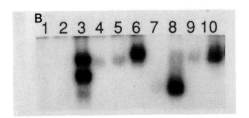

Figure 3. Casein kinase II activation of latent p53. (*A*) GTP and ATP can be used to activate p53 with casein kinase II. Complete reactions containing p53 and rabbit muscle casein kinase II (without an ATP regeneration system and DNA) were assembled without nucleotide. The indicated nucleotide was added and incubations were continued at 30°C for 40 min. After the addition of the DNA-binding mixture, reaction products were analyzed as indicated in Experimental Procedures. From left to right: lane *1* (p53 only with 1 mM ATP), lane *2* (p53 and casein kinase II, without NTP), lanes *3–5* (as lane 2 with 62, 250, and 1000 μM CTP), lanes *6–12* (as lane 2 with 0.24, 0.97, 3.4, 15, 62, 250, and 1000 μM ATP, respectively), and lanes *13–19* (as in lanes *6–12*, but using GTP in place of ATP). (*B*) Recombinant human casein kinase II activates p53. p53 was activated for DNA binding by phosphorylation with increasing amounts of rabbit muscle casein kinase II or human recombinant casein kinase II. Reaction products were separated on a native polyacrylamide gel as indicated in Experimental Procedures. Lanes *1–3* (p53 and 1.2 ng, 12.5 ng, and 125 ng of human recombinant casein kinase II); lanes *4–6* (as in lanes *1–3*, but with DO-1 added after DNA binding); lanes *7* and *8* (37 and 375 ng of rabbit muscle casein kinase II); lanes *9* and *10* (as in lanes *7* and *8*, but with DO-1 added after DNA binding).

et al. 1987; Carroll et al. 1988; Klarlund and Czech 1988). In contrast, protein kinase C is a signal-dependent kinase containing up to 12 family members with differing subcellular localizations, some of which require phospholipid, diacylglycerol, and calcium ions for maximal activation of kinase function (for review, see Dekker and Parker 1994). p53 is an effective substrate for protein kinase C (Baudier et al. 1992). As the consensus phosphorylation site of protein kinase C on p53 encompasses the epitope of the activating monoclonal antibody PAb421 in the carboxy-terminal regulatory site of p53 (Fig. 1), we investigated the effects of protein kinase C phosphorylation of p53 on biochemical activity to localize the modification site of this potential p53 regulator.

Casein kinase II phosphorylation of p53 results in up to 1.45 moles of phosphate per mole of p53 (Fig. 4B). The rate of phosphorylation is reduced by the prior inclusion of PAb421, which binds near the casein kinase II site and may reduce the rate of phosphorylation through steric hindrance. Similarly, protein kinase C phosphorylates p53 with 0.80 moles of phosphate per mole of p53 (Fig. 4A). This modification is also reduced by the prior inclusion of PAb421, which is consistent with the presumed map position of protein kinase C on p53. The major site of phosphorylation appears to reside within the carboxyl terminus, as p53Δ30 is not phosphorylated compared to full-length p53 (Fig. 4C). Phospho-amino acid analysis of full-length p53 or synthetic peptides derived from the carboxyl terminus, which had been phosphorylated by protein kinase C, indicated that a serine residue(s) in the carboxy-terminal regulatory domain is the major site of modification by protein kinase C (T.R. Hupp and D.P. Lane, unpubl.); potential residues within the carboxy-terminal 30 amino acids of human p53 that therefore harbor potential protein kinase C phosphorylation sites include serines 371, 376, and 378 (Fig. 1).

Protein Kinase C Phosphorylation of the Carboxyl Terminus of p53 Inhibits Binding of the Monoclonal Antibody PAb421

Given the presumed overlapping binding sites of the PAb421 epitope (amino acids 372–381) and the site(s) of protein kinase C modification on p53 (amino acids 371, 376, and 378), we investigated whether the prior phosphorylation of p53 increases or decreases the relative affinity of PAb421 for its epitope.

p53 was first phosphorylated by protein kinase C or casein kinase II, the products of the reaction were separated using SDS-gel electrophoresis, and denatured protein was blotted to nitrocellulose. Although PAb421 bound equally well to unphosphorylated or casein-kinase-II-modified p53, PAb421 bound with reduced affinity to protein-kinase-C-phosphorylated p53 (Fig. 5, top panel). Loss of PAb421 binding was notably coincident with stoichiometric phosphorylation of p53.

A two-site ELISA was also used to investigate the effects of protein kinase C phosphorylation of p53 on

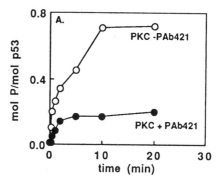

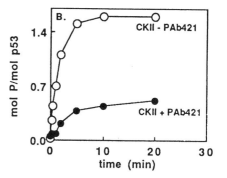

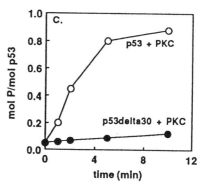

Figure 4. Protein kinase C phosphorylation of the carboxyl terminus of p53. The rate of p53 phosphorylation (A–C) or p53Δ30 phosphorylation (C) by (A) protein kinase C ± PAb421; (B) recombinant human casein kinase II ± PAb421; and (C) protein kinase C. Reactions contained radiolabeled ATP, p53 (12 ng) and, as indicated, PAb421 (50 ng), protein kinase C (0.02 mU), or casein kinase II (0.1 mU) and were incubated at 30°C for the indicated times. Reactions were processed as indicated in Experimental Procedures.

PAb421 reactivity to nondenatured p53 (Fig. 5, bottom panels). p53, was modified with either casein kinase II or protein kinase C and bound to DO-1 (epitope from amino acids 19–26) or PAb421 (epitope from amino acids 372–381) -coated ELISA wells. Following time to allow for immune complex formation, rabbit polyclonal antibody specific for p53 (Midgley et al. 1992) was used as a probe to quantify the amount of p53 captured by the monoclonal antibody. Although casein kinase II phosphorylation did not affect the binding of p53 to DO-1- or PAb421-coated wells (C and D), protein-

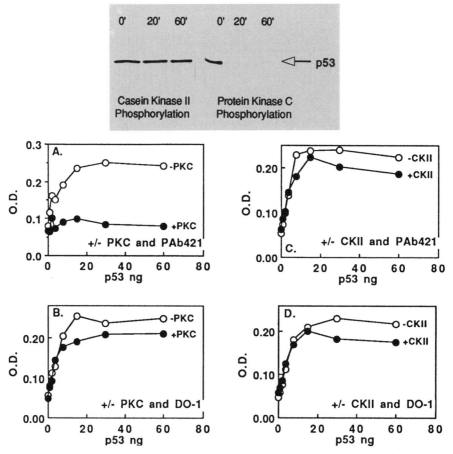

Figure 5. (*Top*) PAb421 does not bind to protein-kinase-C-modified p53 in a denaturing protein blot. p53 was phosphorylated for the indicated times with recombinant human casein kinase II (lanes *1–3*, left panel) or protein kinase C (lanes *4–6*, right panel). Reaction products were separated on a denaturing SDS-polyacrylamide gel and after blotting protein to nitrocellulose, reactivity of phospho-p53 to PAb421 was investigated as described in Experimental Procedures. (*Bottom*) PAb421 does not bind to native protein-kinase-C-modified p53. p53 was phosphorylated with either recombinant human casein kinase II (*C,D*) or protein kinase C (*A,B*) and adsorbed to ELISA wells coated with the amino-terminal-specific antibody DO-1 (*B,D*) or the carboxy-terminal-specific antibody PAb421 (*A,C*). p53 bound to the respective monoclonal antibody was detected using the p53-specific antiserum CM-1 as indicated in Experimental Procedures.

kinase-C-phosphorylated p53 failed to bind to PAb421-coated ELISA wells (panel A). As a control, the protein-kinase-C-modified p53 did bind to DO-1-coated wells (panel B). This confirms that protein-kinase-C-modified p53 fails to bind specifically to PAb421 using either denaturing methods or nondenaturing antibody-capture methods and suggests that phosphate modification may occur near or within the PAb421 epitope in full-length p53.

Protein Kinase C Activates the Latent Function of p53

Having established that protein kinase C specifically phosphorylates within the carboxy-terminal negative regulatory domain of p53 as the kinase (1) does not modify p53Δ30 and (2) inhibits PAb421 binding to full-length p53, the effects of modification on the latent biochemical character of p53 were investigated. A time course of casein kinase II phosphorylation shows that after 60 minutes, maximal DNA binding can be observed, despite the fact that phosphorylation reaches a maximum in 15–20 minutes (Fig. 6, lanes 3 and 4 versus lane 1). Thus, phosphorylation is a rapid step followed by a presumably rate-limiting conformational change that leads to activation of p53.

The rate of protein kinase C phosphorylation is increased dramatically by specific lipids and calcium ions, and these have been used in the previous assay (Figs. 3–5) to maximize the incorporation of phosphate into p53, which was important to demonstrate complete loss of PAb421 binding. Interestingly, the lipids used to increase the rate of protein kinase C activity actually inactivate the sequence-specific DNA-binding function of PAb421-activated p53 (T.R. Hupp and D.P. Lane, unpubl.); therefore, to investigate the effects of protein kinase C phosphorylation on the biochemical function of p53, reactions were performed in the absence of activating lipid for longer times.

After 60 minutes of phosphorylation of p53 by protein kinase C in the absence of lipid, up to 30% phosphorylation of p53 can be achieved, compared to the

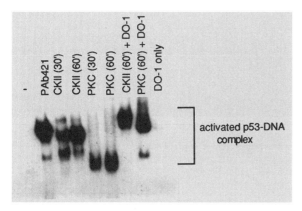

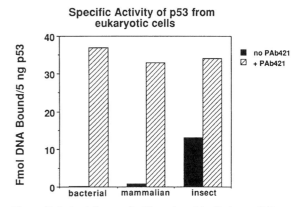

Figure 6. Protein kinase C activates the latent DNA-binding function of p53. p53 was treated with the indicated reagents at 30°C in standard activation reactions and at 0°C in subsequent DNA-binding reactions; latent p53 only (lane *1*, far left), p53 and PAb421 (lane *2*), p53 and recombinant human casein kinase II (lanes *3* and *4*, 30′ and 60′ of phosphorylation, respectively; and supershifted by DO-1 [lane *7*]), p53 and protein kinase C (lanes *5* and *6*, 30′ and 60′ of phosphorylation, respectively; and supershifted by DO-1 [lane *8*]) and latent p53 with DO-1 (lane *9*, far right). Reaction products were separated using native acrylamide gels as indicated in Experimental Procedures.

Figure 7. Latent forms of p53 produced in distinct cell lines. The specific activity of p53 with and without PAb421 purified from recombinant *E. coli*, in mammalian C6 cell lysates, or in recombinant Sf9 insect cell lysates. The activity of p53, with and without PAb421, from the indicated source was determined by quantifying the amount of radiolabeled target DNA bound using native gel electrophoresis.

complete reaction (T.R. Hupp and D.P. Lane, unpubl.). Phosphorylation of p53 by protein kinase C under these conditions was also coincident with the activation of latent p53 (Fig. 6, lanes 5 and 6 versus lane 1). Both kinase-activated forms of p53 were shifted by the amino-terminal antibody DO-1 (lane 7 versus lane 4, for casein kinase II, and lane 8 versus lane 6, for protein kinase C), compared to the control reaction with DO-1 and p53 (lane 9). The specific activity of protein-kinase-C-activated p53 was 3.5-fold lower than casein-kinase-II-activated p53, which may be due to the substoichiometric modification of p53 by protein kinase C in the absence of lipid.

The Latent Form of p53 Is Produced in a Variety of Cell Backgrounds

As observed in *E. coli*, the bulk of p53 overproduced in mammalian cells (C6 cell line) is latent (Fig. 7) (Hupp et al. 1992), and activity can only be observed after allosteric activation by the binding of PAb421. These results suggest that activation of p53 in vivo is rate-limiting. We have recently established that recombinant insect cells overproducing human p53 also produce large amounts of latent protein, yet unlike that observed in mammalian cells, a significant fraction of the p53 is active in the absence of PAb421 (Fig. 7) and can be chromatographically separated from latent p53 (Hupp and Lane 1994). These latter data indicate that there is a physical basis for the latent and activated biochemical phenotype of p53 and establish that insect cells contain a factor(s) that primes p53 in vivo for sequence-specific DNA binding. Based on in vitro studies using pure enzymes and proteins, activation of

latent p53 in eukaryotic cells could be catalyzed by two distinct protein kinases.

Cellularly Activated p53 Comprises Two Distinct Populations

The biochemical character of the cellularly activated p53 was investigated for evidence that protein kinases, and in particular a protein-kinase-C-like enzyme that modifies the PAb421 epitope, may modify p53 and activate the protein in vivo. All of the cellularly activated p53 (Fig. 8, lane 10) is shifted using the DO-1 antibody to produce one intermediate and one limit species containing two DO-1 antibodies for each p53 tetramer (lanes 1–5 versus lane 10). The molecular weight of the cellularly activated p53/DNA complex (lane 10) is identical to that observed when native casein kinase II from rabbit muscle or protein kinase C is used to activate latent p53 in vitro.

A large fraction of the cellularly activated p53/DNA complexes is unaffected in mobility by PAb421, and a second fraction is shifted by PAb421 antibody to produce a species that has one PAb421 molecule bound per p53 tetramer (lanes 6–9 versus lane 10). Thus, two populations of activated p53 exist when purified from insect cells; these data suggest that two distinct kinases may activate p53 in insect cell lines. Complete loss of the carboxy-terminal PAb421 epitope in one population of cellularly activated p53 is consistent with the effects of protein kinase C modification on p53 function in vitro, including (1) activation of the latent biochemical function of p53 and (2) loss of PAb421 reactivity. Reactivity of PAb421 to the second population of cellularly activated p53, as detected by the supershifting of p53/DNA complexes by PAb421 during native gel electrophoresis, is consistent with a casein-kinase-II-like activation of p53. However, as Hsc70/Hsp70 binds to the carboxyl terminus of p53 (Hainaut and Milner 1992) and *E. coli* Hsp70 can activate latent p53 through

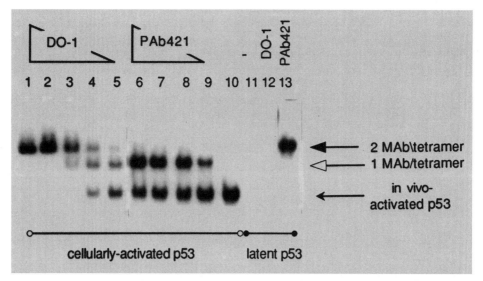

Figure 8. Modification of p53 at the PAb421 epitope in vivo. p53 protein was expressed in recombinant insect cells and the latent and activated proteins were separated by chromatography (T.R. Hupp et al., in prep.) and assayed for sequence-specific DNA binding using native gel electrophoresis. The latent p53 (25 ng/reaction in lanes *11–13*; lane *11*, p53 only) is activated by PAb421 (100 ng, lane *13*) but not by DO-1 (100 ng, lane *12*). Cellularly activated p53 (25 ng/reaction in lanes *1–10*; cellularly activated p53 only, lane *10*) is shifted by DO-1 to a limit species containing two antibody molecules/tetramer (200, 100, 50, 25, 12 ng of IgG, lanes *1–5*) and is only partially shifted by PAb421 (200, 100, 50, 25 ng of IgG, lanes *6–9*) to yield a population containing one PAb421 molecule per p53 tetramer or a second population which is not bound at all by PAb421, due to modification near or within the PAb421 epitope. The closed arrow indicates the position of PAb421-activated p53/DNA complexes bound by two monoclonal antibodies per tetramer or DO-1-shifted, cellularly activated p53/DNA complexes bound by two monoclonal antibodies per tetramer. The open arrow marks the migration of complexes of cellularly activated p53 bound by one DO-1 antibody per tetramer or cellularly activated p53 bound to one PAb421 molecule per tetramer; the small arrow marks the migration of cellularly activated p53/DNA complexes unbound by antibody or unaffected by PAb421 binding (lanes *6–10*).

binding to the carboxyl terminus (Hupp et al. 1992 and unpubl.), we cannot rule out the possibility that the PAb421-reactive, active p53 has been activated by an Hsp70-like factor in vivo.

As a control establishing that the latent p53 produced in recombinant insect cells, which has been chromatographically separated from cellularly activated p53, resembles latent p53 purified from *E. coli*, (1) the latent protein can be activated by the binding of two PAb421 molecules per p53 tetramer to produce the species with an identical mobility to the cellularly activated p53 shifted by DO-1 (lane 13 versus lane 11) and (2) the latent state is unaffected by DO-1 (lane 12 versus lane 11).

DISCUSSION

The Carboxy-terminal Allosteric Regulatory Site of p53

Understanding the regulation of wild-type p53 function and the signals that can lead to a biochemically competent form of the protein are important steps in unraveling the p53-dependent apoptotic and growth-arrest pathways. Site-specific DNA binding by p53 (Bargonetti et al. 1991; Kern et al. 1991) is essential for its tumor suppressor activity (Kern et al. 1992). By studying unmodified forms of recombinant p53, a regulatory site that controls sequence-specific DNA binding in vitro has been identified in the extreme carboxyl terminus (Hupp et al. 1992). Modification of p53 at this site by a relevant metabolic signal, i.e., phosphorylation by casein kinase II or protein kinase C, activates p53 sequence-specific DNA binding. As deletion of this carboxy-terminal domain functionally mimics the effects of phosphorylation (Hupp et al. 1992), the regulatory motif appears to function solely in a negative autoregulatory manner as it modulates p53 protein structure and subsequent function. The biochemical mechanism of activation of latent p53 tetramers has not been established; however, data from our laboratory indicate that both activated and latent forms of the protein are tetrameric in solution and when bound to target DNA (Hupp and Lane 1994).

The carboxy-terminal regulatory site encoded by exon 11 has the hallmarks of an allosteric domain defined by Monod et al. (1963), distinguishing this carboxy-terminal domain from an essential cofactor-binding site. The carboxy-terminal regulatory motif of p53 is a domain distinct from the active site which solely controls protein function, and the ability to desensitize p53 to negative control by its regulatory domain through partial proteolysis or deletion establishes the dispensability of the carboxy-terminal regulatory domain for sequence-specific DNA binding (Hupp et al. 1992) and tumor suppression (Unger et al. 1993; Pietenpol et al. 1994).

p53 has been known to be a conformationally flexible

protein (Gannon et al. 1990; Milner and Medcalf 1991); however, the existence of an allosteric regulatory site has only been uncovered after studying the recently identified biochemical activity of the protein. Evidence that p53 may be subject to allosteric regulation has also stemmed from the studies of Halazonetis and Kandil (1993). Using murine p53 synthesized in reticulocyte lysates, PAb246 binding to an amino-terminal domain of p53 was shown to inhibit the sequence-specific DNA-binding function of PAb421-bound mouse p53 in vitro, suggesting that conformational changes are reversibly transmitted between the amino- and carboxy-terminal domains (Halazonetis and Kandil 1993). However, given the proximity of the PAb246 epitope to the central conserved core domain and active site (Bargonetti et al. 1993; Pavletich et al. 1993) of p53, it remains equally possible that PAb246 inhibition of p53 may be steric. Direct evidence that this carboxy-terminal domain *allosterically* alters polypeptide conformation and subsequent site-specific DNA binding by p53 has not yet been reported.

Protein Kinases Regulate the Biochemical Activity of p53 In Vitro

Protein kinases play an important role in growth control (Hunter and Karin 1992). Not surprisingly, p53 is a substrate for a variety of protein kinases (Meek and Street 1992), suggesting that its tumor suppressor activity may be controlled by reversible phosphorylation (Ullrich et al. 1992b). Casein kinase II and protein kinase C both modify the carboxy-terminal regulatory domain of p53 (Meek et al. 1990; Herrmann et al. 1991; Bishoff et al. 1992; Baudier et al. 1992; this paper) and control sequence-specific DNA binding. Reported here are three novel observations concerning the effects of the modification of p53 by protein kinase C. First, covalent modification by protein kinase C activates the latent function of p53, consistent with the hypothesis that activating agents must modify a negative regulatory motif in the carboxyl terminus, presumably responsible for p53 latency. Second, protein kinase C phosphorylation of the carboxyl terminus of full-length p53 inhibits the binding of the activating antibody PAb421. Finally, a large proportion of cellularly activated p53 produced in insect cells fails to bind to PAb421 (Fig. 8), indicating that a kinase has modified the protein kinase C site in vivo. However, not all of the cellularly activated p53 is nonreactive with PAb421, suggesting that activation can be catalyzed by another factor(s) at a distinct site.

Although PAb421 has been used routinely as a broad-specificity immunochemical reagent to study p53, direct phosphorylation at the PAb421 epitope in vitro can inhibit its binding to p53. The use of PAb421 as an immunoaffinity purification reagent (Bargonetti et al. 1991) would preclude the identification and study of the protein-kinase-C-like, cellularly activated form of p53. Given the close proximity of the presumed phosphorylation site (Ser-371, -376, and -378) to two lysine residues crucial for PAb421 recognition (C.W. Stephen and D.P. Lane, unpubl.), it appears likely that phosphate modification of full-length p53 by protein kinase C would interfere with PAb421/p53 complex formation by neutralizing the lysine or arginine residues required for PAb421 binding. Casein kinase II phosphorylation of p53 fails to inhibit binding of PAb421, indicating that phosphorylation and subsequent activation do not induce a dramatic conformational change in the carboxyl terminus that inhibits all antibody binding to the region.

Protein Kinase Phosphorylation of p53 In Vivo

Phosphorylation of p53 at both amino- and carboxy-terminal domains can influence the activity of the protein in cells. Mutation of a phosphorylation site within the amino-terminal transcriptional *trans*-activation domain (Ser-15 to Ala-15) reduces the ability of p53 to inhibit progression through the cell cycle (Fiscella et al. 1993). Modification of the carboxy-terminal regulatory site of p53 at two different positions also appears to be associated with the proliferation state of the cell. Mutating the carboxy-terminal casein kinase II site of mouse p53 from a serine to an alanine reduces the growth-suppressor function of p53 in rodent cell lines (Milne et al. 1992a). However, it has not been established if this directly results in an inability to signal either the apoptotic or growth-arrest pathways. The regulation of casein kinase II activity in vivo has not been established, although the activity of the kinase has been reported to increase in response to mitogens (Sommercorn et al. 1987; Carroll et al. 1988; Klarlund and Czech 1988). In addition, an androgen-dependent translocation of casein kinase II from the nucleus to cytoplasm suggests that the kinase may be subject to regulation by nuclear transport (Ahmed et al. 1993). Although there is direct evidence that the casein kinase II site of p53 is bound by the kinase and modified by phosphate in vivo (Meek et al. 1990; Herrmann et al. 1991; Bischoff et al. 1992), direct evidence that the protein kinase C site in full-length human or mouse p53 is modified by a protein-kinase-C-like enzyme in vivo comes not from radiolabeling studies, but from the conclusion that PAb421 binding to p53 is inhibited in vitro by protein kinase C phosphorylation (see below).

The positive effects on p53 function through protein kinase C phosphorylation provide the first instance in which this kinase is known to *allosterically* activate the site-specific DNA-binding function of a protein and suggest a direct link between the induction of an enzyme with the specificity of protein kinase C and activation of a tumor suppressor protein. Because many isoforms of protein kinase C exist, some of which are signal-independent, it is not yet known if the phorbol-ester-dependent protein kinase C pathway would be a relevant controller of p53 function. However, in favor of this possibility, the treatment of HeLa and A549 tumor cell lines with phorbol myristate acetate (PMA) induces a G_1/S growth arrest (Skouv et al. 1994), consistent with an activation of the growth suppressor function of wild-type p53 in these cell lines.

Milner (1984) first reported that appearance of the PAb421 epitope on murine p53 occurs during serum stimulation of growth-arrested lymphocytes, suggesting that p53 undergoes conformational changes dependent on the growth state of the cell. Using glioblastoma cell lines, there is evidence that two forms of p53 exist in vivo; one form expresses the PAb421 epitope and the other form fails to bind to PAb421 in its native state. A dramatic loss of the PAb421 epitope on exogenously supplied wild-type p53 is coincident with a growth arrest of the transformed cells (Ullrich et al. 1992a). This loss of the PAb421 epitope also coincides with an increase in phosphorylation within the amino-terminal transcriptional activation domain. We also show in this paper that loss of PAb421 epitope is observed in vitro and in vivo and is coincident with activation of latent p53 sequence-specific DNA-binding function. Together, these results indicate that an enzyme with the specificity of protein kinase C may modify the carboxy-terminal regulatory site of p53 and lead to an activation of the latent biochemical function of the protein. We also suggest that loss of PAb421 epitope on p53 under different growth conditions is not due to a conformational change in p53 but occurs through a direct steric inhibition of PAb421 binding to phospho-p53. Although we have shown that protein kinase C can activate p53 and cause loss of the PAb421 epitope in vitro, we have not established that it is the enzyme responsible for these effects in vivo. Positive identification of the presumed growth-regulating kinase as a protein kinase C family member in vivo will be of utmost interest.

SUMMARY

p53 is an allosterically regulated protein with a latent DNA-binding activity. Posttranslational modification of a carboxy-terminal regulatory site in vitro, by casein kinase II and protein kinase C, can activate the sequence-specific DNA-binding function of the wild-type protein. The latent form of p53 is produced in a variety of eukaryotic and prokaryotic cell lines, including E. coli, Sf9 insect cells, and C6 cells, indicating that the activation of p53 in vivo is rate-limiting. In addition, phosphorylation of p53 at the protein kinase C site and activation in vivo correlate with the loss of reactivity of active p53 protein to the carboxy-terminal antibody, PAb421. These results suggest that two highly conserved protein kinases modify polypeptide structure through a common biochemical mechanism and that different enzymatic pathways may channel information into the carboxy-terminal regulatory site of p53, allosterically activating its function as a tumor suppressor.

ACKNOWLEDGMENTS

We thank our colleagues in Dundee for support and friendly supervision. This work was funded by the Cancer Research Campaign and Pfizer. D.P. Lane is a Gibb Fellow of the Cancer Research Campaign and a Howard Hughes International Scholar.

REFERENCES

Addison, C., J. Jenkins, and H.W. Sturzbecher. 1990. The p53 nuclear localization signal is structurally lined to a $p34^{cdc2}$ kinase motif. *Oncogene* **5:** 423.

Ahmed, K., S. Yenice, A. Davis, and S.A. Goueli. 1993. Association of casein kinase 2 with nuclear chromatin in relation to androgenic regulation of rat prostate. *Proc. Natl. Acad. Sci.* **90:** 4426.

Arai, N., D. Nomura, K. Yokota, D. Wolf, E. Brill, O. Shohat, and V. Rotter. 1986. Immunologically distinct p53 molecules generated by alternate splicing. *Mol. Cell. Biol.* **6:** 3232.

Bakalkin, G., T. Yakovela, G. Selivanova, K.P. Magnusson, L. Szekely, E. Kiseleva, G. Klein, L. Terenius, and K.G. Wiman. 1994. p53 binds single-stranded DNA ends and catalyzes DNA renaturation and strand transfer. *Proc. Natl. Acad. Sci.* **91:** 413.

Barak, Y. and M. Oren. 1992. Enhanced binding of a 95kDa protein to p53 in cells undergoing p53-mediated growth arrest. *EMBO J.* **11:** 2115.

Barak, Y., T. Juven, R. Haffner, and M. Oren. 1993. mdm2 expression is induced by wild type p53 activity. *EMBO J.* **12:** 461.

Bargonetti, J., P.N. Friedman, S.E. Kern, B. Vogelstein, and C. Prives. 1991. Wild-type but not mutant p53 proteins bind to sequences adjacent to the SV40 origin of replication. *Cell* **65:** 1083.

Bargonetti, J., J.J. Manfredi, X. Chen, D.R. Marshak, and C. Prives. 1993. A proteolytic fragment from the central region of p53 has marked sequence-specific DNA-binding activity when generated from wild-type but not from oncogenic mutant p53 protein. *Genes Dev.* **7:** 2565.

Baudier, J., C. Delphin, D. Grunwald, S. Khochbin, and J.J. Lawrence. 1992. Characterization of the tumor suppressor protein p53 as a protein kinase C substrate and a S100b-binding protein. *Proc. Natl. Acad. Sci.* **89:** 11627.

Bischoff, J.R., D. Casso, and D. Beach. 1992. Human p53 inhibits growth in *Schizosaccharomyces pombe*. *Mol. Cell. Biol.* **12:** 1405.

Bischoff, J.R., P.N. Friedman, D.R. Marshak, C. Prives, and D. Beach. 1990. Human p53 is phosphorylated by p60-cdc2 and cyclin B-cdc2. *Proc. Natl. Acad. Sci.* **87:** 4766.

Carroll, D., N. Santoro, and D.R. Marshak. 1988. Regulating cell growth: Casein-kinase-II-dependent phosphorylation of nuclear oncoproteins. *Cold Spring Harbor Symp. Quant. Biol.* **53:** 91.

Dekker, L.V. and P.J. Parker. 1994. Protein kinase C—A question of specificity. *Trends Biochem. Sci.* **19:** 73.

El-Deiry, W.S., S.E. Kern, J.A. Pietenpol, K.W. Kinzler, and B. Vogelstein. 1992. Definition of a consensus binding site for p53. *Nat. Genet.* **1:** 45.

El-Deiry, W.S., T. Tokino, V.E. Velculescu, D.B. Levy, R. Parsons, J.M. Trent, D. Lin, W.E. Mercer, K.W. Kinzler, and B. Vogelstein. 1993. WAF1, a potential mediator of p53 tumor suppression. *Cell* **75:** 817.

Farmer, G., J. Bargonetti, H. Zhu, P. Friedman, R. Prywes, and C. Prives. 1992. Wild-type p53 activates transcription in vitro. *Nature* **358:** 83.

Fields, S. and S.J. Jang. 1990. Presence of a potent transcription activating sequence in the p53 protein. *Science* **249:** 1046.

Fiscella, M., S.J. Ullrich, N. Zambrano, M.T. Shields, D. Lin, S.P. Lees-Miller, C. Anderson, W.E. Mercer, and E. Appella. 1993. Mutation of the serine 15 phosphorylation site of human p53 reduces the ability of p53 to inhibit cell cycle progression. *Oncogene* **8:** 1519.

Foord, O.S., P. Bhattacharya, Z. Reich, and V. Rotter. 1991. A

DNA binding domain is contained in the C-terminus of wild type p53. *Nucleic Acids Res.* **19:** 5191.
Friedman, P.N., X. Chen, J. Bargonetti, and C. Prives. 1993. The p53 protein is an unusually shaped tetramer that binds directly to DNA. *Proc. Natl. Acad. Sci.* **90:** 3319.
Funk, W.D., D.T. Pak, R.H. Karas, W.E. Wright, and J.W. Shay. 1992. A transcriptionally active DNA-binding site for human p53 protein complexes. *Mol. Cell. Biol.* **12:** 2866.
Gamble, J. and J. Milner. 1988. Evidence that immunological variants of p53 represent alternate protein conformations. *Virology* **162:** 452.
Gannon, J.V., R. Greaves, R. Iggo, and D.P. Lane. 1990. Activating mutations in p53 produce a common conformational effect. A monoclonal antibody specific for the mutant form. *EMBO J.* **9:** 1595.
Hainaut, P. and J. Milner. 1992. Interaction of heat-shock protein 70 with p53 translated *in vitro*; evidence for interaction with dimeric p53 and for a role in the regulation of p53 conformation. *EMBO J.* **11:** 3513.
Halazonetis, T.D. and A.N. Kandil. 1993. Conformational shifts propagate from the oligomerization domain of p53 to its tetrameric DNA binding domain and restore DNA binding to select p53 mutants. *EMBO J.* **12:** 5057.
Hall, P.A., P.H. McKee, H. Menage, R. Dover, and D.P. Lane. 1993. High levels of p53 protein in UV irradiated human skin. *Oncogene* **8:** 203.
Harlow, E.E. and D.P. Lane. 1988. *Antibodies: A laboratory manual*. Cold Spring Harbor Laboratory, Cold Spring Harbor, New York..
Harper, J.W., G.R. Adami, N. Wei, K. Keyomarsi, and S. Elledge. 1993. The p21 Cdk-interacting Cip1 is a potent inhibitor of G1 cyclin-dependent kinases. *Cell* **75:** 805.
Herrmann, C.P.E., S. Kraiss, and M. Montenarh. 1991. Association of casein kinase II with immunopurified p53. *Oncogene* **6:** 877.
Hunter, T. and M. Karin. 1992. The regulation of transcription by phosphorylation. *Cell* **70:** 375.
Hupp, T.R. and D.P. Lane. 1994. Allosteric activation of latent p53 tetramers. *Curr. Biol.* **4:** 865.
Hupp, T.R., D.W. Meek, C.A. Midgley, and D.P. Lane. 1992. Regulation of the specific DNA binding function of p53. *Cell* **71:** 875.
———. 1993. Activation of the cryptic DNA binding function of mutant forms of p53. *Nucleic Acids Res.* **21:** 3167.
Kastan, M.B., O. Onyekwere, D. Sidransky, B. Vogelstein, and R.W. Craig. 1991. Participation of p53 protein in the cellular response to DNA damage. *Cancer Res.* **51:** 6304.
Kastan, M.B., Q. Zhan, W.K. El-Deiry, F. Carrier, T. Jacks, W.V. Walsh, B.S. Plunkett, B. Vogelstein, and A.J. Fornace. 1992. A mammalian cell cycle checkpoint pathway utilising p53 and gadd45 is defective in Ataxia-Telangiectasia. *Cell* **71:** 587.
Kern, S., K. Kinzler, A. Bruskin, D. Jarosz, P. Friedman, C. Prives, and B. Vogelstein. 1991. Identification of p53 as a sequence specific DNA binding protein. *Science* **252:** 1708.
Kern, S.E., J.A. Pietenpol, S. Thiagalingam, A. Seymour, K.W. Kinzler, and B. Vogelstein. 1992. Oncogenic forms of p53 inhibit p53 regulated gene expression. *Science* **256:** 827.
Klarlund, J.K. and M.P. Czech. 1988. Insulin-like growth factor I and insulin rapidly increase casein kinase II activity in BALB/c 3T3 fibroblasts. *J. Biol. Chem.* **263:** 15872.
Kraiss, S., A. Quaiser, M. Oren, and M. Montenarh. 1988. Oligomerization of the oncoprotein p53. *J. Virol.* **62:** 4737.
Kuerbitz, S.J., B.S. Plunkett, W.V. Walsh, and M.B. Kastan. 1992. Wild type p53 is a cell cycle checkpoint determinant following irradiation. *Proc. Natl. Acad. Sci.* **89:** 7491.
Kulesz-Martin, M.F., B. Lisafield, H. Huang, N.D. Kisiel, and L. Lee. 1994. Endogenous p53 protein generated from wild-type alternatively spliced p53 RNA in mouse epidermal cells. *Mol. Cell. Biol.* **14:** 1698.
Lamb, P. and L. Crawford. 1986. Characterization of the human p53 gene. *Mol. Cell. Biol.* **6:** 1379.
Lane, D.P. 1992. p53, Guardian of the genome. *Nature* **358:** 15.
———. 1993. A death in the life of p53. *Nature* **362:** 786.
Lees-Miller, S.P., Y. Chen, and C.W. Anderson. 1990. Human cells contain a DNA-activated protein kinase that phosphorylates simian virus 40 T-antigen, mouse p53, and the human Ku autoantigen. *Mol. Cell. Biol.* **10:** 6472.
Lees-Miller, S.P., K. Sakaguchi, S.J. Ullrich, E. Appella, and C.W. Anderson. 1992. Human DNA-activated protein kinase phosphorylates serines 15 and 37 in the amino-terminal transactivation domain of human p53. *Mol. Cell. Biol.* **12:** 5041.
Lu, X. and D.P. Lane. 1993. Differential induction of transcriptionally active p53 following UV or ionizing radiation: Defects in chromosome instability syndromes? *Cell* **75:** 765.
Lu, X., S.H. Park, T.C. Thompson, and D.P. Lane. 1992. *ras*-Induced hyperplasia occurs with mutation of p53, but an activated *ras* and *myc* together can induce carcinoma without p53 mutation. *Cell* **70:** 153.
Maheswaran, S., S. Park, A. Bernard, J.F. Morris, F.J. Rauscher, D.E. Hill, and D.A. Haber. 1993. Physical and functional interaction between WT1 and p53 proteins. *Proc. Natl. Acad. Sci.* **90:** 5100.
Maltzman, W. and L. Czyzyk. 1984. UV irradiation stimulates levels of p53 cellular tumor antigen in nontransformed mouse cells. *Mol. Cell. Biol.* **4:** 1689.
Meek, D.W. and A.J. Street. 1992. Nuclear protein phosphorylation and growth control. *Biochem. J.* **287:** 1.
Meek, D.W., S. Simon, U. Kikkawa, and W. Eckhart. 1990. The p53 tumour suppressor protein is phosphorylated at serine 389 by casein kinase II. *EMBO J.* **9:** 3253.
Meisner, H. and M.P. Czech. 1991. Phosphorylation of transcriptional factors and cell-cycle-dependent proteins by casein kinase II. *Curr. Opin. Cell Biol.* **3:** 474.
Midgley, C.A., C.J. Fisher, J. Bartek, B. Vojtesek, D.P. Lane, and D.M. Barnes. 1992. Analysis of p53 expression in human tumours: An antibody raised against human p53 expressed in *E.coli*. *J. Cell Sci.* **101:** 183.
Milne, D.M., R.H. Palmer, and D.W. Meek. 1992a. Mutation of the casein kinase II phosphorylation site abolishes the anti-proliferative activity of p53. *Nucleic Acids Res.* **20:** 5565.
Milne, D.M., D.G. Campbell, B.F. Caudwell, and D.W. Meek. 1994. Phosphorylation of the tumor suppressor protein p53 by mitogen-activated protein kinases. *J. Biol. Chem.* **269:** 9253.
Milne, D.M., R.H. Palmer, D.G. Campell, and D.W. Meek. 1992b. Phosphorylation of the p53 tumor-suppressor protein at three N terminal sites by a novel casein kinase 1-like enzyme. *Oncogene* **7:** 1361.
Milner, J. 1984. Different forms of p53 detected by monoclonal antibodies in non-dividing and dividing lymphocytes. *Nature* **310:** 143.
Milner, J. and E.A. Medcalf. 1991. Cotranslation of activated mutant p53 with wild type drives the wild-type p53 protein into the mutant conformation. *Cell* **65:** 765.
Momand, J., G.P. Zambetti, D.C. Olson, D. George, and A.J. Levine. 1992. The mdm-2 oncogene product forms a complex with the p53 protein and inhibits p53-mediated transactivation. *Cell* **69:** 1237.
Monod, J., J.P. Changeux, and F. Jacob. 1963. Allosteric proteins and cellular control systems. *J. Mol. Biol.* **6:** 306.
Oberosler, P., P. Hloch, U. Ramsperger, and H. Stahl. 1993. p53-catalyzed annealing of complementary single-stranded nucleic acids. *EMBO J.* **12:** 2389.
Oliner, J.D., K.W. Kinzler, P.S. Meltzer, D.L. George, and B. Vogelstein. 1992. Amplification of a gene encoding a p53-associated protein in human sarcomas. *Nature* **358:** 80.
Oliner, J.D., J.A. Pietenpol, S. Thiagalingam, J. Gyuris, K.W. Kinzler, and B. Vogelstein. 1993. Oncoprotein MDM2 conceals the activation domain of tumour suppressor p53. *Nature* **362:** 857.
O'Rouke, R.W., C.W. Miller, G.J. Kato, K.J. Simon, D.-L. Chen, C.V. Dang, and H.P. Koeffler. 1990. A potential transcription activation element in the p53 protein. *Oncogene* **5:** 1829.
Pabo, C.O. and R.T. Sauer. 1992. Transcription factors:

Structural families and principles of DNA recognition. *Annu. Rev. Biochem.* **61:** 1053.

Pavletich, N.P., K.A. Chambers, and C. Pabo. 1993. The DNA-binding domain of p53 contains the four conserved regions and the major mutation hot spots. *Genes Dev.* **7:** 2556.

Pietenpol, J.A., T. Tokino, S. Thiagalingam, W. El-Deiry, K. Kinzler, and B. Vogelstein. 1994. Sequence-specific transcriptional activation is essential for growth suppression by p53. *Proc. Natl. Acad. Sci.* **91:** 1998.

Raycroft, L., H.Y. Wu, and G. Lozano. 1990. Transcriptional activation by wild type but not transforming mutants of the p53 anti-oncogene. *Science* **249:** 1049.

Samad, A. and R.B. Carroll. 1991. The tumour suppressor p53 is bound to RNA by a stable covalent linkage. *Mol. Cell. Biol.* **11:** 1598.

Scharer, E. and R. Iggo. 1992. Mammalian p53 can function as a transcription factor in yeast. *Nucleic Acids Res.* **20:** 1539.

Shaulian, E., A. Zauberman, D. Ginsberg, and M. Oren. 1992. Identification of a minimal transforming domain of p53: Negative dominance through abrogation of sequence-specific DNA binding. *Mol. Cell. Biol.* **12:** 5581.

Skouv, J., P. Ostrup-Jensen, J. Forchhammer, J.K. Larsen, and L.R. Lund. 1994. Tumor-promoting phorbol ester transiently down-modulates the p53 level and blocks the cell cycle. *Cell Growth Differ.* **5:** 329.

Sommercorn, J., J.A. Mulligan, F.J. Lozeman, and E.G. Krebs. 1987. Activation of casein kinase II in response to insulin and to epidermal growth factor. *Proc. Natl. Acad. Sci.* **84:** 8834.

Soussi, T., C. Caron de Fromentel, and P. May. 1990. Structural aspects of the p53 protein in relation to gene evolution. *Oncogene* **5:** 945.

Stenger, J.E., G.A. Mayr, K. Mann, and P. Tegtmeyer. 1992. Formation of stable p53 homotetramers and multiples of tetramers. *Mol. Carcinog.* **5:** 102.

Stephen, C.W. and D.P. Lane. 1992. Mutant conformation of p53: Precise epitope mapping using a filamentous phage epitope library. *J. Mol. Biol.* **225:** 577.

Sturzbecher, H.-W., R. Brain, C. Addison, K. Rudge, M. Remm, M. Grimaldi, E. Keenan, and J.R. Jenkins. 1992. A C-terminal α-helix plus basic region motif is the major structural determinant of p53 tetramerization. *Oncogene* **7:** 1515.

Tishler, R.B., S.K. Calderwood, C.N. Coleman, and B.D. Price. 1993. Increases in sequence-specific DNA binding by p53 following treatment with chemotherapeutic and DNA damaging agents. *Cancer Res.* **53:** 2212.

Ullrich, S.J., W.E. Mercer, and E. Appella. 1992a. Human wild-type p53 adopts a unique conformational and phosphorylation state in vivo during growth arrest of glioblastoma cells. *Oncogene* **7:** 1635.

―――. 1992b. The p53 tumor suppressor protein, a modulator of cell proliferation. *J. Biol. Chem.* **267:** 15259.

Unger, T., J.A. Mietz, M. Scheffner, C.L. Yee, and P.M. Howley. 1993. Functional domains of wild type and mutant p53 proteins involved in transcriptional regulation, transdominant inhibition, and transformation suppression. *Mol. Cell. Biol.* **13:** 5186.

Vojtesek, B., J. Bartek, C.A. Midgley, and D.P. Lane. 1992. An immunochemical analysis of human p53: New monoclonal antibodies and epitope mapping using recombinant p53. *J. Immunol. Methods* **151:** 237.

Wade-Evans, A. and J.R. Jenkins. 1985. Precise epitope mapping of the murine transformation-associated protein, p53. *EMBO J.* **4:** 699.

Wang, Y. and W. Eckhart. 1992. Phosphorylation sites in the amino-terminal region of mouse p53. *Proc. Natl. Acad. Sci.* **89:** 4231.

Wang, Y., M. Reed, P. Wang, J.E. Stenger, G. Mayr, M.E. Anderson, J.F. Schwedes, and P. Tegtmeyer. 1993. p53 domains: Identification and characterization of two autonomous DNA-binding regions. *Genes Dev.* **7:** 2575.

Zauberman, A., Y. Barak, N. Ragimov, N. Levy, and M. Oren. 1993. Sequence-specific binding by p53: Identification of target sites and lack of binding to p53-MDM2 complexes. *EMBO J.* **12:** 2799.

Zhan, Q., F. Carrier, and A.J.J. Fornace. 1993. Induction of cellular p53 activity by DNA-damaging agents and growth arrest. *Mol. Cell. Biol.* **13:** 4242.

DNA-binding Properties of the p53 Tumor Suppressor Protein

C. Prives,* J. Bargonetti,* G. Farmer,* E. Ferrari,*† P. Friedlander,*
Y. Wang,* L. Jayaraman,* N. Pavletich,‡ and U. Hubscher†

*Department of Biological Sciences, Columbia University, New York, New York 10027;
†Department of Veterinary Biochemistry, University of Zurich-Irchel, Zurich, Switzerland;
‡Cellular Biochemistry and Biophysics Program, Memorial Sloan-Kettering Cancer Center, New York, New York 10021

The p53 tumor suppressor gene product plays a pivotal role in cells by transmitting a signal from damaged DNA to genes that regulate the cell cycle and apoptosis (Donehower and Bradley 1993; Levine 1993; Prives and Manfredi 1993). Central to its function in this capacity is its ability to bind specifically to DNA (Vogelstein and Kinzler 1992). p53 activates transcription from promoters that contain p53-binding sites (Vogelstein and Kinzler 1992). Identified genes that contain such p53 response elements, which are activated in cells upon DNA damage and subsequent accumulation of p53, include *GADD45* (Kastan et al. 1992), *mdm2* (Wu et al. 1993), and *WAF1/p21* (El-Deiry et al. 1993). The interest in p53 that stems from its unique and critical role in a DNA damage signal transduction pathway is reinforced by, and indeed possibly subsidiary to, the fact that mutation of the p53 coding sequence occurs with extraordinarily high frequency in many of the major forms of human cancer (see, e.g., Hollstein et al. 1991). Importantly, tumor-derived mutant forms of p53 display defective or aberrant DNA binding and activation properties (Bargonetti et al. 1991; Kern et al. 1991; Vogelstein and Kinzler 1992).

p53 genes cloned from several vertebrate species reveal both divergent and highly homologous regions (Soussi et al. 1990). There are five portions of homology, one of which is within the amino terminus, and the remaining four of which are within the center of the polypeptide. The vast majority of tumor-derived p53 mutations are located within the central portion of the protein, and these tend to be clustered within its four conserved regions. Discrete functional domains whose independent properties contribute to the function of p53 as a regulatory protein have been identified by deletion, swap, and protease analysis. These include (1) an activation domain at the amino terminus (residues 1–43) (Unger et al. 1992 and reference therein), (2) a sequence-specific DNA-binding domain within the center (residues 100–300) (Bargonetti et al. 1993; Halazonetis and Kandil 1993; Pavletich et al. 1993; Wang et al. 1993), (3) an oligomerization region (residues 335–355) that dictates that the protein form tetramers (Sturzbecher et al. 1992; Pavletich et al. 1993), and (4) an autonomous nonspecific DNA-binding region within the carboxyl terminus (Wang et al. 1993).

Identification of the factors involved in the regulation of p53 levels in cells in response to DNA damage and discovery of additional target genes that transmit its message to the DNA damage-sustaining cell are worthy goals. However, it will be equally important to understand the properties of the p53 protein as it functions within this pathway. We have focused on the DNA-binding properties of p53 in the belief that this aspect of the protein is central to its tumor suppressor function.

MATERIALS AND METHODS

Proteins. Wild-type and mutant p53 proteins were immunopurified from Sf21 insect cells infected with recombinant baculoviruses as reported previously (Bargonetti et al. 1991). SV40 and polyoma large T antigens were also immunopurified from baculovirus-infected Sf21 cells according to previously published procedures (Wang et al. 1989). Bacterially expressed p53 fragments were expressed in and purified from *Escherichia coli* as described previously (Pavletich et al. 1993). To prepare cyclin-B-dependent cdc2 kinase, Sf21 insect cells were coinfected with recombinant baculoviruses expressing human cyclin B and cdc2 kinase tagged with the 12CA5 antibody epitope (Field et al. 1988) that were generously provided by D. Morgan. Extracts of infected insect cells were passed over a column containing the antibody 12CA5 and eluted in buffer containing 1 mg/ml peptide containing the sequence YPYDVPDYA according to the same protocol used for immunopurifying p53 proteins. Sp1 protein was purchased from Promega-Biotech (Madison).

DNA-binding and reannealing assays. DNase I footprinting assays were performed as described previously (Bargonetti et al. 1991). Gel mobility shift assays (EMSA) were carried out as described by Chen et al. (1993). The oligonucleotides used for EMSA reactions were GADD45, which contains the sequence identified by Kastan et al. (1992): 5'-TGG TAC AGA ACA TGT CTA AGC ATG CTG GGG TCTG-3', and SCS, which contains a very strong p53 consensus binding site 5'-TCG AGC CGG GCA TGT CCG GGC ATG TCC

GGG CAT GGTC-3' derived from Halazonetis et al. (1993). DNA reannealing assays followed the procedure of Oberosler et al. (1993), using as substrate a labeled 120-base polyoma origin fragment (nts 5267–90) that was heated for 5 minutes at 95°C immediately before use.

RESULTS AND DISCUSSION

The Nature of a p53-binding Site Determines Whether It Can Function as a p53 Response Element

We previously identified, using DNase I footprinting, a binding site for p53 that was within the SV40 regulatory region (Bargonetti et al. 1991). This site, which is adjacent to the SV40 origin of replication, is within the "21 BP repeat" region that contains six binding sites for the well-characterized transcriptional activator Sp1. The region of SV40 that was the most well protected by p53 was located within SV40 (nts 34–73), whose sequence, 5'-GTACCCCGCCTCTTACCCGCCTT GACCCGCCTCAATCCCCGCCCTACCCGCC-3', contains four Sp1-binding sites. However, although p53 binds specifically there, this region does not contain a sequence closely related to the consensus site derived for p53. The p53 cognate site consists of two copies of the symmetrical 10-mer: 5'-R R R C A/T T/A G Y Y Y-3' (El-Deiry et al. 1992). It is, however, possible to identify a potential half-site within the SV40 site that contains the sequence 5'-CG CCTTG ACC-3' (nts 51–60) which contains a good "core" CTTG and is overall a 7 out of 10 match to the consensus of a p53 half-site. Additionally, within the Sp1 sites are several repeats of the sequence **CCCG** which might also provide some specificity for p53 binding, since C(4) and G(7) are the most invariant nucleotides within the p53 consensus site (El-Deiry et al. 1992). Given that p53 does display marked specificity for binding to the site within SV40, can it activate transcription from a promoter in which this site is present?

To answer this question, a p53 expression plasmid was transfected into *Drosophila* Schneider cells along with reporter constructs containing either the RGC site or the SV40 site upstream of the *fos* basal promoter (Fig. 1). These cells were chosen because they have proven useful for examining the transcriptional activity of several mammalian and insect regulatory proteins and promoters (Colgan and Manley 1992). Furthermore, insect cells do not express p53 or Sp1, and therefore, reporter plasmids containing promoters bearing these binding sites should by themselves display only basal activity. There was a marked difference in the relative responses of the two plasmids to coexpressed p53: Whereas the RGCfos reporter was activated dramatically by p53, there was virtually no effect of p53 on the SVfos reporter. Thus, the SV40-binding site is unresponsive to p53 under conditions where p53 is capable of strong activation of a reporter with a classical p53 response element. Since p53 binding to the

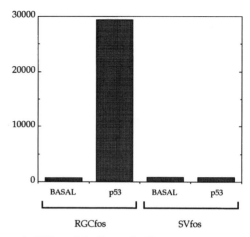

Figure 1. Differential effects of p53 on two promoters containing p53-binding sites. Plasmids (1 μg), RGCfos or SVfos, containing either the RGC or SV40 site (Bargonetti et al. 1991), respectively, upstream of the c-*fos* TATA box and CAT gene (Farmer et al. 1992), were transfected into Schneider cells either alone (basal) or with 3 μg of a plasmid containing wild-type human p53 under the control of the insect actin promoter, Act-hp53 (p53) (Chen et al. 1993). CAT assays were performed as described previously (Chen et al. 1993). The amount of the RGCfos plasmid used was previously determined to give plateau levels of activation by p53. CAT activity is expressed as counts per minute of tritium-labeled acetylated chloramphenicol.

SV40 site is clearly specific, as determined by three different DNA-binding assays (DNase I footprinting [Bargonetti et al. 1991], EMSA, and filter binding [Bargonetti et al. 1992]), and indeed, since binding to the SV40 site is only less than binding to the RGC site by a factor of 2–4, the lack of any activation whatever of SVfos by p53 is striking. Clearly, there is more to activation by p53 than the ability to display sequence-specific binding. p53, by contacting a half-site within the SV40 region, may have different or insufficient interactions with the basal transcription machinery to allow for transcriptional activation.

p53 and Sp1: Altered Interactions with SV40 DNA

The location of the p53 site within the Sp1-binding region of the SV40 regulatory zone raises another question: Do p53 and Sp1 influence each others' interactions with the SV40 origin-proximal region? To address this question, either purified wild-type or mutant human p53 or Sp1 proteins were bound either separately or together to an end-labeled fragment of SV40 DNA that contains both the Sp1- and p53-binding sites. DNA binding was determined by DNase I footprinting analysis (Fig. 2). The results showed that the interactions of Sp1 with this region, which is normally completely protected from DNase I digestion (Gidoni et al. 1984), were dramatically altered in the presence of p53 (Fig. 2, cf. lanes 3 with lanes 4 and 5). Rather than the

SV40 T Antigen Inhibits the p53 Central Region from Binding to DNA

It is widely acknowledged that many oncogenic DNA tumor viruses have evolved mechanisms to abrogate p53 function (for review, see Levine 1990). In the case of SV40, the viral large T antigen binds to p53 and blocks its ability to bind to DNA (Bargonetti et al. 1992) and to activate transcription in vitro (Farmer et al. 1992). The region of p53 that is bound by SV40 T antigen was shown previously to be within the central portion of the tumor suppressor molecule (Jenkins et al. 1988; Ruppert and Stillman 1993). Since this is the region that contains the p53 sequence-specific DNA-binding domain, we predicted that this region alone would also be blocked from binding to DNA by SV40 large T antigen. A human p53 fragment (core fragment, residues 94–312) expressed in and purified from bacteria was previously shown to bind specifically to DNA (Pavletich et al. 1993). The ability of either immunopurified SV40 T antigen or polyoma T antigen to affect binding by this core fragment to an oligonucleotide containing the GADD45-binding site p53 response element (Kastan et al. 1992) was examined by EMSA (Fig. 3). Our data indicate that this site is bound specifically and with high affinity by both full-length and p53 core (data not shown). We determined that

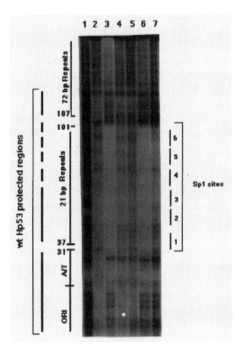

Figure 2. Wild-type and mutant p53 proteins change Sp1 interactions with SV40 DNA. DNase I footprinting reactions were carried out as described previously (Bargonetti et al. 1991). The SV40 fragment, labeled such that interactions with the late strand are detected, was also as in Bargonetti et al. (1991). Landmarks within the SV40 regulatory region and nucleotide positions are marked on the left, individual Sp1 sites on the right. Binding reaction mixtures contained no protein (lane *1*), 60 ng wild-type human p53 (lane *2*), or 10 ng Sp1 either alone (lane *3*) or with 30 ng (lane *4*) or 60 ng (lane *5*) wild-type human p53, or 30 ng (lane *6*) or 60 ng (lane *7*) mutant (His-175) human p53.

typical pattern of protection brought about by Sp1, overall binding was reduced, and yet each Sp1 site was individually protected to some extent. p53 interactions were changed as well, such that there was less protection in the primary p53-binding region, and the characteristic pattern of hypercutting and protection extending from the primary binding site was reduced.

One interpretation of this result is that p53 is simply occupying the Sp1 site and therefore Sp1 binding is occluded. However, the ability of p53 to bind to DNA is not essential for this inhibition, since a mutant p53 (Arg-175 → His) that fails to bind at all to DNA (Bargonetti et al. 1991) also changed the protection pattern by Sp1. This conclusion is further supported by the fact that Sp1 binds with higher affinity to this region than does p53, whose binding is relatively weak. Therefore, simple competition for this region by the two proteins is unlikely. These data suggest, therefore, that p53 interacts with Sp1 in such a way that its DNA binding is altered and reduced, a conclusion that is presently being investigated. It is interesting to note that complexes containing p53 and Sp1 have been detected in human erythroleukemia cells (Borellini and Glazer 1993).

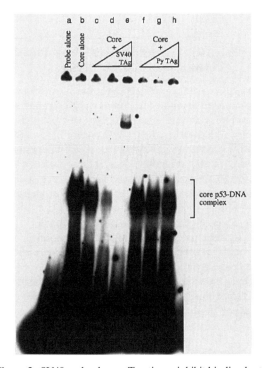

Figure 3. SV40 and polyoma T antigens inhibit binding by the p53 core DNA-binding domain. DNA-binding reaction mixtures containing 3 ng ^{32}P-labeled GADD45 oligonucleotide (all lanes) and purified p53 core domain (100 ng) (lanes *b–h*) with 12 (lane *c*), 24 (lane *d*), or 48 (lane *e*) ng SV40 T antigen, or 24 (lane *f*), 48 (lane *g*), or 96 (lane *h*) ng polyoma T antigen. DNA binding was analyzed by EMSA.

adding SV40 T antigen to p53-binding reaction mixtures dramatically inhibited the formation of the core/DNA complex. At the highest concentration of SV40 T antigen, there was another more slowly migrating complex that was less abundant, which may indicate either a DNA complex containing SV40 alone or SV40 and p53. We had used immunopurified polyoma T antigen as a control since it fails to bind detectably to p53 (Wang et al. 1989). However, we were surprised to observe that polyoma T antigen also caused a significant reduction in p53 core/DNA binding. Nevertheless, the binding by the p53 core at all concentrations of polyoma T antigen was greater than that with SV40 T antigen, and the extent to which p53 binding was inhibited by polyoma T antigen reached a plateau and was not reduced further. It is not known how polyoma T antigen inhibits p53 binding. Possibly there is an interaction of polyoma T antigen with p53 that either is undetected by coimmunoprecipitation protocols or is dependent on the presence of DNA. This result is of interest because we have identified within the polyoma replication origin a binding site for p53 that actually fits the p53 consensus sequence significantly better than the p53 site in SV40 (data not shown). Additionally, we have determined that polyoma T antigen binding to the polyoma replication origin inhibits the binding by p53 to its site within this region. However, in this case, it is not possible to rule out competition by polyoma T antigen for binding to this region. It is intriguing that within the vicinities of both SV40 and polyoma origins are binding sites for p53.

Phosphorylation by Cyclin B/cdc2 Kinase Complex Stimulates p53 DNA Binding

Several well-defined phosphorylation sites within p53 are located within either its amino or carboxyl terminus (Prives and Manfredi 1993). That no sites have as yet been identified within the central portion is consistent with physical studies indicating that the compact folded structure of this domain is less solvent-exposed and therefore less accessible to cellular enzymes such as kinases (Bargonetti et al. 1993; Pavletich et al. 1993). Nevertheless, since DNA binding is central to the tumor suppressor role of p53, it is reasonable to ask whether this activity is affected by the phosphorylation state of the protein. Indeed, it was previously shown that phosphorylation of the casein kinase II (CKII) protein kinase site at the extreme carboxyl terminus (Ser-392) stimulates specific DNA binding by bacterial p53 (Hupp et al. 1992). This stimulation is presumably related to the observation that the 30 carboxy-terminal amino acids, a region rich in basic residues, can negatively regulate p53 DNA binding (Hupp et al. 1992). Thus, there is precedent for regulation of DNA binding of p53 by phosphorylation of sites distal to the central region. Little is known about the function of the cdk kinase site that was mapped at Ser-315 in human p53 (Bischoff et al. 1990). To examine whether phosphorylation of p53 at that site regulates DNA binding, immunopurified cyclin B/cdc2 kinase was incubated with p53 protein in the presence of kinase buffer containing $[\gamma\text{-}^{32}P]ATP$. Purified p53 was an excellent substrate for the purified cyclin B/cdc2 kinase (Fig. 4). The specificity of phosphorylation of p53 by this protein kinase was confirmed by performing two-dimensional phosphopeptide analysis of p53 after treatment with kinase (data not shown). Massive phosphorylation of p53 was evidenced both by transfer of ^{32}P from $[\gamma\text{-}^{32}P]ATP$ and also by the reduced electrophoretic mobility of p53 after SDS-PAGE caused by such treatment (Fig. 4A, compare lanes a and b showing kinase-treated and untreated protein). As a control, p53 incubated with kinase that was boiled prior to its addition to reaction mixtures was not phosphorylated by the above criteria. Parallel samples of p53 were incubated with cyclin B/cdc2 kinase in kinase buffer (but lacking $[\gamma\text{-}^{32}P]ATP$) under identical conditions that led to extensive phosphorylation of p53. When these samples were added to DNA-binding reaction mixtures containing ^{32}P-labeled GADD45 oligonucleotides and binding was examined

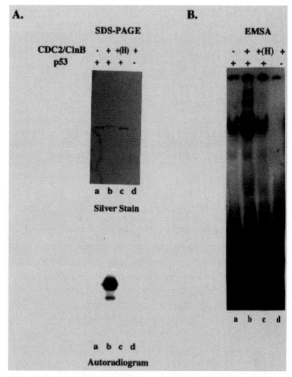

Figure 4. Cyclin B/cdc2 kinase phosphorylates p53 and stimulates its binding to a p53-binding site oligonucleotide. (*A*) Immunopurified human p53 (100 ng, lanes *a–c*) and cyclin B/cdc2 kinase (10 ng, lanes *b–d*) were incubated in kinase buffer (50 mM HEPES [pH 7.5], 10 mM MgCl$_2$, 1 mM DTT, 100 μM ATP) containing 2.5 μCi $[\gamma\text{-}^{32}P]ATP$ at 20°C for 30 min. Reaction mixtures were then separated by SDS-PAGE, and proteins were visualized by silver staining (*top* panel) or by autoradiography of the dried gel (*bottom* panel). In lanes c, the cyclin B/cdc2 kinase was boiled prior to adding to reaction mixtures. (*B*) Reaction mixtures as in *A*, but lacking [^{32}P]ATP, were incubated for 30 min at 20°C after which ^{32}P-labeled GADD45 oligonucleotide (1 ng) was added and incubation continued at 4°C for 30 min before analysis by EMSA.

by EMSA, there was a marked increase in the ability of the phosphorylated p53 to bind to DNA. Not only was the major and most rapidly migrating p53/DNA complex increased, but slower migrating complexes, perhaps containing multiples of p53 tetramers, were increased to a relatively greater extent, suggesting that p53 protein-protein interactions possibly leading to increased cooperative binding were activated by phosphorylation at the cdk kinase site. These data thus show that p53 purified from baculovirus-infected insect cells is efficiently phosphorylated by cdk kinase and that such phosphorylation stimulates DNA binding by p53. Examination of the mechanism by which cdk kinases stimulate binding is presently under way.

Temperature-sensitive DNA-binding by Mutant p53 Is Stabilized by UTP

Tumor-derived human p53 mutants are impaired in binding to DNA. However, we and other workers have found that many such mutants are conditionally defective such that DNA binding occurs at temperatures lower than within the physiological range (Bargonetti et al. 1993; Pietenpol et al. 1994; Zhang et al. 1994). One such mutant (Ala-143) was reported to be temperature-sensitive for binding, transcriptional activation, and growth suppression, such that it functions like wild-type p53 at 32°C but not at 37°C (Zhang et al. 1994). During analysis of the nature of temperature-sensitive DNA binding by mutant p53, we identified a mode by which binding by Ala-143 human mutant p53 is stabilized to some extent at 37°C (Fig. 5). Experiments were initially designed to determine whether nucleotides are capable of affecting p53 DNA binding. One nucleotide, UTP, was uniquely able to stimulate binding by the human mutant p53, Ala-143 (Fig. 5). Binding to DNA at 20°C was tested by EMSA after preincubation of p53 at 37°C for 0, 2, 4, or 7 minutes. Within 2–4 minutes after heating Ala-143 p53 at 37°C, the mutant irreversibly lost the ability to bind to an oligonucleotide containing a strong consensus p53 binding site, SCS. However, in the presence of increasing amounts of UTP, there was a marked relative increase in binding to this site, such that at the optimum concentration of UTP (8 mM), good binding to DNA was seen even after 4 minutes of preheating. In fact, even after heating this mutant p53 for 7 minutes, there was detectable binding at 8 mM UTP. The stimulation was peculiar to Ala-143 mutant p53. Several other nucleotides tested failed to stimulate this mutant. Furthermore, neither wild-type p53 nor other mutant forms of p53 were significantly affected by the presence of this nucleotide.

Irradiation of cells containing wild-type p53 leads to arrest in G_1 or apoptosis (for review, see Donehower and Bradley 1993; Prives and Manfredi 1993). In contrast, mutant p53-containing cells are radiation resistant (Clarke et al. 1993; Lee and Bernstein 1993; Lowe et al. 1993). If mutant p53 in tumor cells can be manipulated such that it is capable of binding to p53 response elements and activating transcription of the appropriate target genes, this could have important implications for possible therapeutic approaches. Although the UTP stimulatory effect is specific for Ala-143, this might provide a basis for further experiments designed to find a more general inducer of DNA binding by mutant p53 at physiological temperatures.

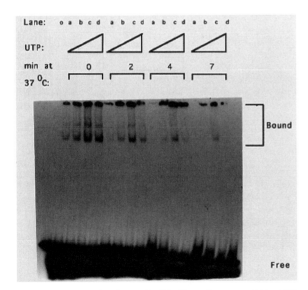

Figure 5. UTP stimulates the binding of mutant (Ala-143) human p53 to DNA. Reaction mixtures (lanes a–d) containing mutant p53 (80 ng) and 8 ng ^{32}P-labeled SCS oligonucleotide with 0 mM (lanes a), 4 mM (lanes b), 8 mM (lanes c), or 10 mM (lanes d) UTP were incubated at 37°C for the times indicated and then at 20°C for a total of 30 min in each case prior to analysis by EMSA.

The Carboxyl Terminus of p53 Reanneals Complementary DNA Single Strands

p53 contains within its carboxyl terminus several functions, including an oligomerization domain, an autonomous DNA-binding domain, and a basic region that negatively regulates p53-specific DNA binding. One property of p53 that may be related to the DNA-binding domain at the carboxyl terminus is its ability to reanneal complementary DNA single strands and to bind to DNA ends (Oberosler et al. 1993; Bakalkin et al. 1994). Since the p53 carboxyl terminus can bind nonspecifically to DNA, it was predicted that this region might be independently capable of the reannealing activity. To test this, we compared the ability of wild-type p53 and two bacterially expressed fragments of the p53 carboxyl terminus—tetra + basic (residues 310–393) or tetra (residues 310–365)—to reanneal a heat-denatured DNA fragment (Fig. 6). Both full-length p53 and the intact carboxy-terminal fragment, tetra + basic, efficiently renatured the DNA single strands. However, the carboxy-terminal fragment, tetra, which lacked the 30-amino-acid basic region, was completely incapable of this reannealing activity. Additionally, when the

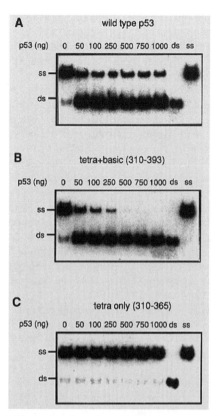

Figure 6. The carboxyl terminus of p53 renatures complementary DNA single strands. DNA reannealing reactions containing purified p53 proteins as indicated and ^{32}P-labeled heat-denatured DNA fragment were incubated for 25 min at 37°C prior to analysis on native gels according to the method of Oberosler et al. (1993). The position of the heat-denatured (ss) and renatured (ds) fragments are indicated at left and also in lanes containing only ds or ss DNA.

central core region of p53 was tested, it displayed only very inefficient reannealing activity (data not shown). Therefore, the carboxyl terminus of p53 is largely, if not entirely, responsible for its DNA reannealing activity. Although this activity requires the presence of the basic region at the extreme carboxyl terminus, it is not known whether the presence of the tetramerization domain is also a necessity.

What is the significance of this activity? We have observed that not only wild-type p53, but also several tumor-derived mutant forms of p53 can reanneal DNA (data not shown). Although the different mutant p53s tested displayed DNA renaturation with varying efficiencies that ranged from equivalent to less than that of the wild-type form, this indicates that DNA reannealing is not responsible for the tumor suppressor activity of p53. Several speculative suggestions can be made about this ability of p53 to bind to DNA ends and to reanneal complementary single strands: (1) Since these sorts of activities are found in proteins involved in DNA repair, p53 may not only function to activate genes that tell the cell to stop until DNA is repaired, it may also be able to actually participate in the repair process. (2) p53, being able to block DNA helicase activities, reanneal DNA, and bind to DNA ends (Oberosler et al. 1993; Bakalkin et al. 1994; E. Ferrari et al., in prep.), may play a direct role in inhibiting DNA replication. (3) Oberosler et al. (1993) showed that p53 can reanneal complementary RNA as well as DNA chains. Therefore, p53:RNA interactions may be significant; perhaps, p53 interacts with its own mRNA and thus regulates its translation. (4) Finally, the p53 reannealing activity may be indicative of another aspect of the protein that is directly relevant to its role as a DNA damage-sensing transcriptional activator: p53 may use the broken ends of DNA as an entry point from which it may then track and find its response elements. Until genetic and biochemical approaches are further developed, it will not be possible to fathom the significance of the p53 denaturation activity.

ACKNOWLEDGMENTS

We thank Ella Freulich for expert technical assistance. The research described in this paper was supported by National Institutes of Health grants CA-58316 and CA-26905; by grant CN-85 from the American Cancer Society; and by the Schweizerische Bankgenellschaft (SVZG-MRY).

REFERENCES

Bakalkin, G., T. Yakovleva, G. Selivanova, K.P. Magnusson, L. Szekely, E. Kiseleva, G. Klein, L. Terenius, and K.G. Wiman. 1994. p53 binds single-stranded DNA ends and catalyzes DNA renaturation and strand transfer. *Proc. Natl. Acad. Sci.* **91:** 413.

Bargonetti, J., I. Reynisdottir, P.N. Friedman, and C. Prives. 1992. Site-specific binding of wild-type p53 to cellular DNA is inhibited by SV40 T antigen and mutant p53. *Genes Dev.* **6:** 1886.

Bargonetti, J., P.N. Friedman, S.E. Kern, B. Vogelstein, and C. Prives. 1991. Wild-type but not mutant p53 immunopurified proteins bind to sequences adjacent to the SV40 origin of replication. *Cell* **65:** 1083.

Bargonetti, J., J.J. Manfredi, X. Chen, D.R. Marshak, and C. Prives. 1993. A proteolytic fragment from the central region of p53 has marked sequence-specific DNA-binding activity when generated from wild-type but not from oncogenic mutant p53 protein. *Genes Dev.* **7:** 2565.

Bischoff, J.R., P.N. Friedman, D.R. Marshak, C. Prives, and D. Beach. 1990. Human p53 is phosphorylated by p60-cdc2 and cyclin B-cdc2. *Proc. Natl. Acad. Sci.* **87:** 4766.

Borellini, F. and R.I. Glazer. 1993. Induction of Sp1-p53 DNA-binding heterocomplexes during granulocyte/ macrophage colony-stimulating factor-dependent proliferation in human erythroleukemia cell line TF-1. *J. Biol. Chem.* **268:** 7923.

Chen, X., G. Farmer, H. Zhu, R. Prywes, and C. Prives. 1993. Cooperative DNA binding of p53 with TFIID (TBP): A possible mechanism for transcriptional activation. *Genes Dev.* **7:** 1837.

Clarke, A.R., C.A. Purdie, D.J. Harrison, R.G. Morris, C.C. Bird, M.L. Hooper, and A.H. Wyllie. 1993. Thymocyte apoptosis induced by p53-dependent and independent pathways. *Nature* **362:** 849.

Colgan, J. and J.L. Manley. 1992. TFIID can be rate limiting in vivo for TATA-containing but not TATA-lacking RNA polymerase promoters. *Genes Dev.* **6:** 304.

Donehower, L.A. and A. Bradley. 1993. The tumor suppressor p53. *Biochim. Biophys. Acta* **1155:** 181.

El-Deiry, W.S., S.E. Kern, J.A. Pietenpol, K.W. Kinzler, and B. Vogelstein. 1992. Definition of a consensus binding site for p53. *Nat. Genet.* **1:** 45.

El-Deiry, W.S., T. Tokino, V.E. Velculescu, D.B. Levy, R. Parsons, J.M. Trent, D. Lin, W.E. Mercer, K.W. Kinzler, and B. Vogelstein. 1993. *WAF1*, a potential mediator of p53 tumor suppression. *Cell* **75:** 817.

Farmer, G., J. Bargonetti, H. Zhu, P. Friedman, R. Prywes, and C. Prives. 1992. Wild-type p53 activates transcription in vitro. *Nature* **358:** 83.

Field, J., J.-I. Nikawa, D. Broek, B. MacDonald, L. Rodgers, L.A. Wilson, R.A. Lerner, and M. Wigler. 1988. Purification of a *ras*-responsive adenyl cyclase complex from *Saccharomyces cerevisiae* by use of an epitope addition method. *Mol. Cell. Biol.* **8:** 2159.

Gidoni, D., W.S. Dynan, and R. Tjian. 1984. Multiple specific contacts between a mammalian transcription factor and its cognate promoters. *Nature* **312:** 409.

Halazonetis, T.D. and A.N. Kandil. 1993. Conformational shifts propagate from the oligomerization domain of p53 to its tetrameric DNA binding domain and restore DNA binding to select p53 mutants. *EMBO J.* **12:** 5057.

Halazonetis, T.D., L.J. Davis, and A.N. Kandil. 1993. Wild-type p53 adopts a "mutant"-like conformation when bound to DNA. *EMBO J.* **12:** 1021.

Hollstein, M., D. Sidransky, B. Vogelstein, and C.C. Harris. 1991. p53 mutations in human cancers. *Science* **253:** 49.

Hupp, T.R., D.W. Meek, C.A. Midgley, and D.P. Lane. 1992. Regulation of the specific DNA binding function of p53. *Cell* **71:** 875.

Jenkins, J.R., P. Chumakov, C. Addison, H.-W. Sturzbecher, and A. Wade-Evans. 1988. Two distinct regions of the murine p53 primary amino acid sequence are implicated in stable complex formation with simian virus T antigen. *J. Virol.* **62:** 3903.

Kastan, M.B., Q. Zhan, W.S. El-Deiry, F. Carrier, T. Jacks, W.V. Walsh, B.S. Plunkett, B. Vogelstein, and A.J. Fornace, Jr. 1992. A mammalian cell cycle checkpoint pathway utilizing p53 and GADD45 is defective in ataxia-telangiectasia. *Cell* **71:** 587.

Kern, S.E., K.W. Kinzler, A. Bruskin, D. Jarosz, P. Friedman, C. Prives, and B. Vogelstein. 1991. Identification of p53 as a sequence-specific DNA binding protein. *Science* **252:** 1708.

Lee, J.M. and A. Bernstein. 1993. p53 mutations increase resistance to ionizing radiation. *Proc. Natl. Acad. Sci.* **90:** 5742.

Levine, A.J. 1990. The p53 protein and its interactions with the oncogene products of the small DNA tumor viruses. *Virology* **177:** 419.

———. 1993. The tumor suppressor genes. *Annu. Rev. Biochem.* **62:** 623.

Lowe, S.W., E.M. Schmitt, S.W. Smith, B.A. Osborne, and T. Jacks. 1993. p53 is required for radiation-induced apoptosis in mouse thymocytes. *Nature* **362:** 847.

Oberosler, P., P. Hloch, U. Ramsperger, and H. Stahl. 1993. p53-catalyzed annealing of complementary single-stranded nucleic acids. *EMBO J.* **12:** 2389.

Pavletich, N.P., K.A. Chambers, and C.O. Pabo. 1993. The DNA-binding domain of p53 contains the four conserved regions and the major mutation hot spots. *Genes Dev.* **7:** 2556.

Pietenpol, J.A., T. Tokino, S. Thiagalingam, W.S. El-Deiry, K.W. Kinzler, and B. Vogelstein. 1994. Sequence-specific transcriptional activation is essential for growth suppression by p53. *Proc. Natl. Acad. Sci.* **91:** 1998.

Prives, C. and J.J. Manfredi. 1993. The p53 tumor suppressor protein: Meeting review. *Genes Dev.* **7:** 529.

Ruppert, J.M. and B. Stillman. 1993. Analysis of a protein-binding domain of p53. *Mol. Cell. Biol.* **13:** 3811.

Soussi, T., C. Caron de Fromentel, and P. May. 1990. Structural aspects of the p53 protein in relation to gene evolution. *Oncogene* **5:** 945.

Sturzbecher, H.W., R. Brain, C. Addison, K. Rudge, M. Remm, M. Grimaldi, E. Keenan, and J.R. Jenkins. 1992. A C-terminal α-helix plus basic region motif is the major structural determinant of p53 tetramerization. *Oncogene* **7:** 1513.

Unger, T., M.M. Nau, S. Segal, and J.D. Minna. 1992. p53: A transdominant regulator of transcription whose function is ablated by mutations occurring in human cancer. *EMBO J.* **11:** 1383.

Vogelstein, B. and K.W. Kinzler. 1992. p53 function and dysfunction. *Cell* **70:** 523.

Wang, E.H., P.N. Friedman, and C. Prives. 1989. The murine p53 protein blocks replication of SV40 DNA in vitro by inhibiting the initiation functions of SV40 large T antigen. *Cell* **57:** 379.

Wang, Y., M. Reed, P. Wang, J.E. Stenger, G. Mayr, M.E. Anderson, J.F. Schwedes, and P. Tegtmeyer. 1993. p53 domains: Identification and characterization of two autonomous DNA-binding regions. *Genes Dev.* **7:** 2575.

Wu, X., J.H. Bayle, D. Olson, and A.J. Levine. 1993. The p53-mdm-2 autoregulatory feedback loop. *Genes Dev.* **7:** 1126.

Zhang, W., X.-Y. Guo, G.-H. Hu, W.-S. Liu, J.W. Shay, and A.S. Deisseroth. 1994. A temperature sensitive mutant of human p53. *EMBO J.* **13:** 2535.

Functions of the p53 Protein in Growth Regulation and Tumor Suppression

J. LIN, X. WU, J. CHEN, A. CHANG, AND A.J. LEVINE
Department of Molecular Biology, Princeton University, Princeton, New Jersey 08544-1014

The wild-type p53 protein is the product of a tumor suppressor gene that has been shown to block oncogenic transformation (Eliyahu et al. 1989; Finlay et al. 1989), negatively regulate progress through the cell cycle (Baker et al. 1990; Martinez et al. 1991), and induce apoptosis in some cell types (Yonish-Rouach et al. 1991; Shaw et al. 1992). The wild-type p53 protein is a transcription factor (Fields and Jang 1990; Raycroft et al. 1990), and that property may mediate its tumor suppressor phenotype, because mutant p53 proteins from human cancers have lost both their transcriptional *trans*-activation activity (Raycroft et al. 1990) and their tumor suppressor activity (Hinds et al. 1990). The transcriptional activation domain of p53 has been mapped to amino acid residues 1–42 (Raycroft et al. 1990), and the sequence-specific DNA-binding domain of the p53 protein has been shown to reside within residues 120–290 (Pavletich et al. 1994).

Several viral and cellular oncogene products bind to the p53 protein and inhibit its activities. Two of these oncoproteins, Mdm2 (Chen et al. 1993; Oliner et al. 1993) and the adenovirus 5 E1B 55-kD protein (Kao et al. 1990), bind to the amino-terminal domain of p53. Both the Ad5 E1B 55-kD protein and Mdm2 block the ability of p53 to function as a transcription factor (Momand et al. 1992; Yew and Berk 1992), and this could then result in oncogene-mediated transformation or tumorigenesis. The amino-terminal domain of p53 is thought to promote transcriptional *trans*-activation by interacting with the transcriptional machinery of the cell, and so the oncogene products could well block this functional contact.

Mutations in the p53 gene in human cancer cells have been extensively sequenced and mapped (Levine et al. 1994). Out of 1447 mutations sequenced from 43 different human cancers, 85.6% are missense mutations and 92% of those cluster in the DNA-binding domain of the p53 protein (see Fig. 1). When selected mutant proteins are tested for several phenotypes, they fail to bind to p53 sequence-specific DNA elements as efficiently as the wild-type protein and also fail to transcriptionally *trans*-activate a test gene (Bargonetti et al. 1992; Kern et al. 1992; Zambetti et al. 1992). These are the expected "loss of function" phenotypes of a tumor suppressor gene. On the other hand, some p53 mutant proteins possess novel phenotypes not exhibited by the wild-type p53 protein; i.e., "gain of function" phenotypes (Dittmer et al. 1993). Mutant p53 proteins can transcriptionally *trans*-activate the multi-drug resistance gene-1 enhancer-promoter element (Chin et al. 1992) and can also enhance the tumorigenic potential of some cell lines in culture (Dittmer et al. 1993). These gain-of-function mutations could explain why there appears to be a selection for p53 missense mutations in 85.6% of cancerous tissues, possibly leaving an altered protein with "growth promoting" activity.

The experiments described in this paper provide support for the idea that the loss-of-function mutations select for the loss of transcription factor activity. The results reported here identify two critical amino acids in the amino-terminal domain required for transcriptional *trans*-activation, probably via physical contacts with the transcriptional machinery. These same amino acids are required for the gain-of-function phenotypes exhibited by mutant p53 proteins. In addition, these two amino acid residues are critical for the binding of the Mdm2 and Ad5 E1B 55-kD oncogene products (Lin et al. 1994). Thus, the oncogene products target the same amino acid residues that appear to contact the cellular transcriptional machinery.

METHODS

Plasmids and site-specific mutagenesis. pBKS-p53 consists of a Bluescript KS(-) vector (Stratagene) and the entire wild-type human p53 cDNA. pRC/CMVp53 consists of a PRC/CMV vector (Invitrogen) and the entire wild-type human p53 cDNA. p53-281G consists of a CMV-Bam$_3$-Neo vector (Karasuyama et al. 1989) and the genomic human p53 DNA with a mutation at amino acid 281 (Hinds et al. 1990). Site-specific mutations were generated in pBKS-p53 or PRC/CMVp53 as described previously by Kunkel (1985) and in p53-281G as described previously by Deng and Nickoloff (1992). p53 mutants generated in pBKS-p53 were recloned to the expression vector pRC/CMV.

For CAT assays, four reporters were used: pWAF1CAT, Cosx1CAT, p50-2CAT, and pMDR1CAT. pWAF1CAT contains a p53-responsive element from the WAF1 promoter (El-Deiry et al. 1993). Cosx1CAT contains a p53-responsive element from the murine *mdm2* gene (Wu et al. 1993). p50-2CAT consists of two copies of p53-responsive element from the murine muscle creatine phosphokinase promoter (Zambetti et al. 1992). pMDR1CAT contains a 1.8-kb enhancer-promoter region from the multi-drug

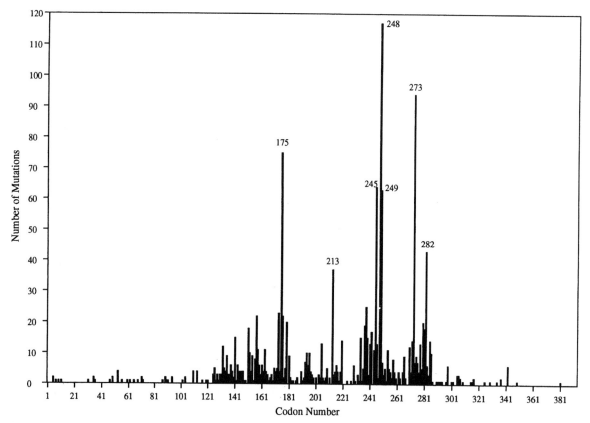

Figure 1. The distribution of p53 mutations from human tumors. The codon number containing the mutation in the p53 gene from 1447 different human cancers (53 different cell or tissue types) is compared to the frequency of these mutations.

resistance gene (Chin et al. 1992). The E2CAT plasmid was employed to measure E2F-1 levels in a cell (Wu and Levine 1994).

Cell culture and DNA transfection. The human osteogenic sarcoma Saos-2 (Masuda et al. 1987), murine embryo fibroblast (10)1, and (10)3 (Harvey and Levine 1991) cells that do not express endogenous p53 were maintained in DMEM with 15% (Saos-2) or 10% ([10]1 and [10]3) fetal bovine serum. The DNA transfections were carried out as described previously (Graham and van der Eb 1973). CAT activity was assayed as described previously (Zambetti et al. 1992). The growth suppression experiments in Saos-2 cells were as described by Chu (1991). The soft-agar growth assays in Saos-2 cells and tumorigenicity assays in Saos-2 and (10)3 cells were carried out as described previously (Dittmer et al. 1993).

In vitro complex formation assays and immunoprecipitations. Human p53 mutants and Mdm2 were generated by in vitro translation, and each mutant p53 protein was tested for its ability to bind Mdm2 as described previously (Lin et al. 1994). For the p53-Ad E1B 55-kD binding assays, the protocol of Kao et al. (1990) was modified as described previously (Lin et al. 1994). Mutant or wild-type p53 transiently expressed in Saos-2 cells was determined for their half-life or conformation as described in Lin et al. (1994). The immunoprecipitation assays were carried out as described previously (Finlay et al. 1989).

Gel electrophoretic mobility-shift assay. Human wild-type p53 and the 22-23 mutant were expressed in high-five insect cells (Invitrogen) from baculovirus vectors, and the p53 proteins were purified. The p53 proteins were incubated with the p53 consensus oligonucleotide (Funk et al. 1992), and a gel mobility-shift assay was employed to test for p53 binding to this DNA sequence (Lin et al. 1994).

Methods for measuring apoptosis. Apoptosis was measured by trypan blue dye exclusion and DNA degradation as described by Wu and Levine (1994).

RESULTS

Phenotype of p53 Mutations That Do Not Naturally Occur in Cancers

Figure 1 presents the frequency of p53 mutations (1447 mutations) as a function of the codon that is altered in the gene. 85.6% of these mutations are missense mutations and 92% of those cluster in the DNA-binding domain (codons 120–290) of this protein. These data are consistent with the selection of mutations in this gene to reduce DNA binding by the protein and eliminate transcriptional activation by the p53 protein. In the DNA-binding domain of p53, there are 24 codons which have never been reported as mutants (out of the 1447 mutants sequenced to date). They reside at codons 122, 123, 124, 150, 180, 185, 188, 189, 200, 202, 206, 210, 212, 221, 222, 225, 226, 228, 231, 243, 264, 268, 269, and 288. If indeed selection for p53 mutations is to eliminate or reduce the transcription factor activity of p53 mutant proteins, then mutations at these codons, not selected in human cancers, would be expected to have little or no effect on the ability of these mutant proteins to function as transcription factors. To test this idea, site-specific mutations were introduced into four different codons: 122, 180, 221, and 243. These codons were chosen because their amino acids are conserved over evolutionary time scales in different p53 molecules from different species: 122 is in conserved region II; 180 in conserved regions III; 243 in conserved region IV; and 221 resides in a cluster of codons (221, 222, 225, 226, 228, 231) that are never found as mutations in cancerous tissues (Soussi et al. 1990). Both transition and transversion mutations were created at these sites so as to introduce both conservative and more radical changes in the type of amino acid substitutions. Finally, four different mutant alleles were produced at codon 180 to examine the impact of a range of diverse amino acid substitutions at this codon. Table 1 presents the novel p53 mutant cDNA clones generated by site-specific mutagenesis. Each of these mutant p53 cDNA clones, as well as a p53 wild-type cDNA (WT), two p53 mutants which are commonly found in cancerous tissues (175H and 273H), and an expression vector with no p53 cDNA (vector alone) were individually transfected into Saos-2 cells, a human osteogenic sarcoma cell line with no endogenous p53 protein. In addition, a p53-responsive DNA element regulating the chloramphenicol acetyl transferase (CAT) gene was also transfected into these cells. Two different p53-responsive elements were employed in these reporter plasmids, one from the *p21-pic1* or *WAF1* gene (El-Deiry et al. 1993) and a second from the *mdm2* gene (Wu et al. 1993), both of which are genes normally regulated by p53 protein levels in cells.

Cell extracts were prepared and CAT assays were carried out to assess the ability of mutant p53 proteins to transcriptionally *trans*-activate these p53-responsive elements. The level of wild-type p53 was set at 100% (Table 1). Vector alone was 5–12%, and mutant p53 proteins that commonly occur in cancerous cells (175H, 273H) were 5–14% of the wild-type p53 levels. All of the mutant p53 proteins that were not selected for in cancerous tissues (Table 1) had good transcriptional activities with some up to 3.5-fold higher than wild type or, in the least efficient case, 71% of the wild-type activity. Clearly, the mutants that were never selected for in cancerous cells retained their transcriptional *trans*-activation activity.

As a second test or assay for growth suppression by these mutants, the mutant cDNA clones were transfected into Saos-2 cells and selected for G418-resistant colonies. When this was done with wild-type p53, it resulted in a 10-fold reduction in G418-resistant colonies due to the growth-suppressive properties of this p53 protein (Table 1, cf. wild-type versus vector alone). The mutant p53 cDNA for 122A, 180V, 221G, and 243T, which are never selected in cancer cells, reduced Saos-2 colony formation by 2.5- to 10-fold. Mutants that were selected for in cancerous cells, such as 175H or 273H, stimulated cell plating efficiencies or growth in agar (Dittmer et al. 1993).

These data are consistent with the idea that the mutants produced and tested in Table 1 are not selected for in cancerous cells because such mutations do not eliminate transcription factor activity. The loss of function selected for by p53 mutations may well then be a loss of transcriptional *trans*-activation.

Site-specific Mutagenesis in the Amino-terminal Domain of the p53 Protein

The clustering of mutations in the DNA-binding domain of p53 (Fig. 1) clearly results in a reduction of DNA binding and transcriptional activity. The amino terminus of p53 is also involved in transcriptional *trans*-activation, probably by making contacts with compo-

Table 1. Phenotypes of Novel Human p53 Mutations in the DNA-binding Domain Which Have Not Been Observed in Cancers

Codon	Amino acid		Transcription from enhancer (% of WT)		Growth suppression Saos-2 cells
	from	to	WAF1-CAT	MDM2-CAT	
122A	Val	Ala	113	192	0.11
180V	Glu	Val	290	195	0.34
180G	Glu	Gly	320	196	
180K	Glu	Lys	293	95	
180Q	Glu	Gln	268	71	
221G	Glu	Gly	226	168	0.37
243T	Met	Thr	351	91	0.41
WT p53	—		100	100	0.12
Vector alone	—		12	5	1.0
175H	Arg	His	14	5	
273H	Arg	His	—	7	

nents of RNA polymerase or the transcriptional machinery of the cell. The question then arises, If mutations in the p53 gene are selecting against a transcription factor function, why are there so few mutations in this amino-terminal domain? Which amino acids in this domain are critical for contacts with the transcriptional machinery?

To test these questions, a wide variety of site-specific mutations were made in the first 32 codons of the p53 cDNA, and both single and double mutants were constructed as shown in Table 2. These mutants were then tested for their ability to transcriptionally *trans*-activate two p53-responsive elements: p50-2 CAT from the enhancer of the muscle form of the creatine phosphokinase gene (Zambetti et al. 1992) and Cosx1-CAT from the *mdm2* gene-responsive element. The wild-type p53 cDNA activity was set at 100% (Table 2), and most of these mutations had little or no impact on the ability of p53 to transcriptionally *trans*-activate these two reporter genes in Saos-2 cells. Only the double mutation at codons 22 and 23 eliminated the ability of the p53 protein to actively promote transcription of both p50-2 CAT and *mdm2* CAT (Table 2). Single mutations at codons 22 or 23 had less of an impact on this phenotype.

The codon 22, 23 mutant protein was stable in Saos-2 cells, based on pulse-chase analysis. In addition, the codon 22, 23 mutants preserve their conformational sensitive epitopes as detected by monoclonal antibodies which recognize the wild-type p53 protein but not a mutant protein. Finally, the codon 22, 23 mutant protein binds to DNA with a p53-responsive element and has an affinity identical to the wild-type protein (Lin et al. 1994). It therefore appears that codons 22 and 23 are required for contacts with the transcriptional machinery of the cell, and this mutant protein is otherwise normal in its functions. This result may explain why so few mutations cluster in the amino-terminal domain of the p53 protein. To eliminate the transcription factor activity of p53 with mutations at the amino terminus requires a double mutation. This would occur with the square of the probability of single mutations in the DNA-binding domain which can result in the same phenotype.

Binding of Mdm2 and Ad5 E1B 55-kD Proteins to p53

The series of mutant p53 cDNAs and proteins generated in Table 2 were tested for their ability to bind to the human Mdm2 (hMdm2) protein or the adenovirus E1B 55-kD protein, both of which had been previously shown to bind to the p53 amino-terminal domain (Kao et al. 1990; Chen et al. 1993). Either codons 14, 19, or

Table 2. Phenotypes of p53 Mutations in the Amino-terminal Domain

Position	Change from	Change to	p50-2 CAT	*mdm2* CAT	hMdm-2 binding	E1B 55-kD binding
WT p53			100	100	100	100
2	Glu	Lys	79	72	66	84
3	Glu	Lys				
7	Asp	His	58	64	81	
11	Glu	Lys				
13	Pro	Thr	56	63	85	141
15	Ser	Gly	300	117	66	
14	Leu	Gln	56	41	1	61
19	Phe	Ser				
16	Gln	Leu	219	210	124	41
18	Thr	Ile				
17	Glu	Lys	40	41	80	228
21	Asp	His				
22	Leu	Gln	17	59	56	49
23	Trp	Ser	22	74	22	3
22	Leu	Gln	5	8	2	6
23	Trp	Ser				
24	Lys	Thr	127	94	181	8
25	Leu	Gln	143	34	39	18
26	Leu	His				
27	Pro	Tyr	126	73	779	4
28	Glu	Lys	56	36	162	47
31	Val	Ser	96	95	101	35
32	Leu	Arg				

22, 23 double mutants eliminated the hMdm2 binding to p53, whereas codons 23, the 22, 23 double mutant, 24, and 27 were critical residues for the E1B 55-kD protein binding to p53 (Table 2). Thus, the same amino acids, at residues 22 and 23, which are involved in transcriptional activation are also targeted by two oncoproteins that are negative regulators of p53 function.

Codon 22, 23 Residues Are Essential for the Mutant p53 Gain of Function Phenotype

Some p53 mutant proteins have at least two phenotypes not expressed by p53 wild-type proteins. First, they can transcriptionally *trans*-activate the multidrug resistance gene, *mdr1* (Chen et al. 1993; Dittmer et al. 1993). Second, these mutant p53 proteins can enhance the tumorigenic potential of cells in culture (Dittmer et al. 1993). Mutant p53 proteins can accomplish this despite the fact that they fail to bind to DNA transcriptional elements and positively regulate genes with such elements. One hypothesis to explain how mutant p53 proteins could regulate transcription of new genes and alter the growth potential of cells is to assume that a mutant p53 protein can bind to a transcription factor and the resulting complex binds to DNA-specific sequences by virtue of that transcription factor. Next, the combination (p53 and the transcription factor) activates transcription of a gene by virtue of the p53 amino-terminal domain that *trans*-activates by interacting with the transcriptional machinery of the cell. Thus, the gain of function would arise from a mutant p53-transcription factor complex that utilizes a discrete function (DNA binding or *trans*-activation) from each partner. This is a testable hypothesis. When a wild-type p53 cDNA was transfected into cells, it failed to transcriptionally activate an *MDR1* CAT test gene (Table 3). When a p53 mutant cDNA at codon 281 (in the DNA-binding domain) was transfected into cells with the *MDR1* CAT reporter gene, large levels of CAT activity were detected (Table 3). When the double mutant at codon 281 and codons 22, 23 (which eliminates the *trans*-activation function) was used to test for activity on a *MDR1* CAT reporter, this mutant had little or no CAT activity. These results (Table 3) indicate that the amino-terminal domain of p53 involved in *trans*-activation is required for this gain of function by the mutant p53 proteins.

In a different test for the gain-of-function activity, the codon 281 mutant cDNA was transfected into murine (10)3 cells (with no endogenous p53 protein;

Table 3. Transcriptional *trans*-Activation by Mutant p53 Proteins with *MDR1* CAT

Codon	Relative activity (%)
Vector alone	2
WT p53	6
281	100
281, 22, 23	11

Cell line: Human Saos-2 cells.

Table 4. Tumorigenic Potential of Mutant p53 Proteins in (10)3 Murine Cells

Codon	Tumorigenicity in nude mice number with tumors/total
Vector alone	0/1
281	5/9
281, 22, 23	0/14

Harvey and Levine 1991) which are not by themselves tumorigenic in nude mice (see Table 4) (Dittmer et al. 1993). Of 9 independent cell lines of (10)3 cells that express the codon 281 p53 mutant protein, 5 now made tumors in nude mice (Table 4). Of 14 cell lines that expressed codon 281 and codon 22, 23 double mutants, none produced tumors in nude mice (Table 4). Thus, based on this biological assay for a gain-of-function mutation in the p53 gene, the amino-terminal *trans*-activation function is required for this new phenotype.

Overexpression of E2F-1 Plus p53 Mediates Apoptosis in Cells in Culture

The DNA tumor viruses—SV40, some human adenoviruses, and the human papillomaviruses—each encode oncogene products that target and inactivate both the retinoblastoma susceptibility protein, Rb, and p53. The Rb protein regulates the E2F transcription factor by binding to it and inactivating its function, which is to stimulate the transcriptional expression of a set of genes promoting entry into S phase. Clearly, the DNA tumor viruses target the Rb protein for inactivation so as to move cells into S phase and enhance the replicative potential of the viruses to synthesize their own DNA and cellular histones which, in the case of SV40 and the papillomaviruses, are used to package the viral DNA. Why then is it necessary for these same viruses to also encode functions to inactivate p53 in virus-infected cells? p53 responds to DNA damage by placing the cells in either G_1 arrest or apoptosis. Perhaps the results of high E2F levels or viral DNA replicative intermediates are perceived by p53 as a signal for activation, similar to DNA damage, and the conflicting signals of high E2F levels and activated p53 could limit virus replication.

To test these ideas, as shown in Table 5, the (10.1)Val5 cell line (with no endogenous p53 protein; Harvey and Levine 1991) containing a temperature-sensitive mutant of p53 (a valine to alanine change at codon 135), which behaves like wild-type p53 at 32°C and mutant p53 at 37.5°C, was transfected with an E2F-

Table 5. Relative Levels of E2F Activity in the ts p53 (10.1)Val5 Cell Lines

Cell lines	Relative E2F activity
(10.1)Val5	1.0
C11	2.5
C12	3.3
C18	3.1

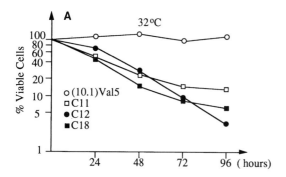

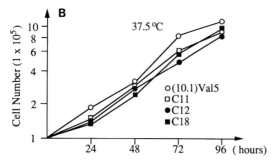

Figure 2. Growth curves of cells containing a temperature-sensitive p53 gene product and excessive E2F-1 proteins. The parental cell line (10.1)Val5 contains temperature-sensitive p53 gene product whereas the C11, C12, and C18 cells contain in addition an E2F-1 plasmid overexpressing the E2F-1 transcription factor. (A) Cells at 37.5°C (p53 is mutant). (B) Cells at 32°C (p53 is wild type).

1 cDNA clone at 37.5°C so as to overexpress this transcription factor in these cells. Three cell lines were isolated, called C11, C12, and C18, and each was compared for its level of expression of E2F-1 using an E2CAT reporter plasmid, compared with the parental (10.1)Val5 cells. Table 5 presents the relative E2F activity in these four cell lines where the (10.1)Val5 cell line is set at 1.0 (see Table 5). The C11, C12, and C18 cell lines expressed about 2.5- to 3.3-fold higher E2F activities at 37.5°C in these cell lines compared with the parental cell line.

The (10.1)Val5, C11, C12, and C18 cell lines grew in culture at similar rates at 37.5°C where p53 was in the mutant conformational form (see Fig. 2A). When these cells were shifted to 32°C (Fig. 2B), the p53 protein took on a wild-type conformation and the parental (10.1)Val5 cells ceased dividing. These cells were blocked in the G_1 phase of the cell cycle by excess levels of wild-type p53 (Wu and Levine 1994), and these cells remain viable for longer than 96 hours at 32°C. Viability can be measured by trypan blue exclusion, plating efficiency at 37.5°C, or the ability to shift to 37.5°C and resume cell division with these cultures (Martinez et al. 1991). In contrast, the C11, C12, and C18 cells with excess E2F-1 and wild-type p53 (at 32°C) rapidly lost viability by all of these criteria (Fig. 2B). The C11, C12, and C18 cells kept at 32°C rounded up and eventually failed to remain attached to the substratum (Fig. 3D). By 18 hours at 32°C, the DNA in these cells was degraded with a typical nucleosome

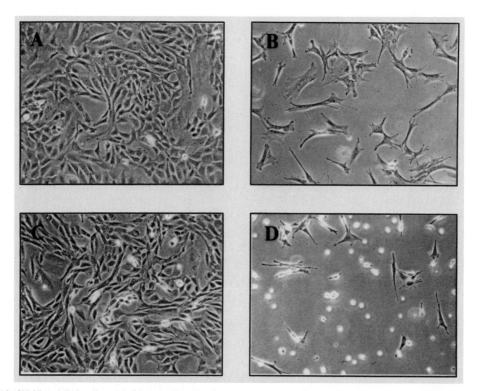

Figure 3. (10.1)Val5 and C12 cells at 37.5°C and 32°C. The (10.1)Val5 cells with a ts p53 replicate at 37.5°C (A) and enter G_1 arrest at 32°C (B). The C12 cells with a ts p53 and high levels of E2F-1 grow at 37.5°C (C) and undergo apoptosis at 32°C (D).

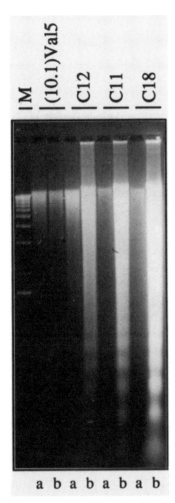

Figure 4. DNA degraded in cells undergoing apoptosis. Cells from the cell lines (10.1)Val5, C12, C11, and C18 grown at 37.5°C (lanes *a*) or 32°C (lanes *b*) were collected at 18 hr after the temperature shift. The DNA was extracted and samples were analyzed in a 1% agarose gel. Lane *M* contains a molecular-weight marker for a 1-kb ladder (GIBCO/BRL).

spacing ladder (Fig. 4) indicative of apoptosis (Wu and Levine 1994). In contrast, the (10.1)Val5 cells at 32°C or the C11, C12, or C18 cells at 37.5°C showed little or no indication of such DNA fragmentation. Thus, enhanced levels of E2F-1, signaling cells to enter S phase, and the overexpression of wild-type p53 occurring simultaneously in a cell set up a conflicting set of growth signals resulting in apoptosis. The overexpression of p53 with lower levels of E2F in a cell results in a G_1 arrest and retention of viability.

DISCUSSION

On the basis of a variety of experiments (Pietenpol et al. 1991; Ginsberg et al. 1991; Kastan et al. 1992; Martinez et al. 1992), p53 appears to act as a transcription factor whose function is to provide a checkpoint for the cell cycle. Depending on the signals from other negative regulations, such as Rb-E2F, p53 can mediate either a G_1 arrest or apoptosis (see Fig. 2). The experiments presented here suggest that p53 can monitor and regulate Rb function in two ways: (1) high p53 levels or activity can induce the expression of *p21-pic1* or *WAF1* (El-Deiry et al. 1993, 1994), which in turn inhibits the cyclin E (or D)-dependent cdk2 (cdk4 or 6) phosphorylation of Rb. Inhibition of Rb phosphorylation results in a stable Rb-E2F complex and low E2F activity and G_1 arrest. If, on the other hand, (2) E2F levels rise while p53 levels or activity is high, then the conflict between these two signaling pathways results in apoptosis (Fig. 5). In this way, p53 acts as a checkpoint backup for the Rb regulatory signal, resulting in one of two different outcomes. Mdm2 is a negative regulator of p53 (Momand et al. 1992), binding to amino acids 22 and 23 in the *trans*-activation domain of p53 (Lin et al. 1994) and blocking p53 from functioning as a transcription factor. Clearly then, the regulation of Mdm2 levels or activity (Wu et al. 1993) plays a critical role in p53 checkpoint function, just as the cyclin-dependent kinases play a central role in Rb regulation and function. This model (Fig. 5) functionally relates two central tumor suppressor gene products in the cell cycle.

p53 carries out these checkpoint functions of G_1 arrest or apoptosis by modulating the transcription of several sets of genes. All of the evidence presented here is consistent with the idea that mutations in the p53 gene in cancerous cells select for a loss of transcription factor activity, as is also the case for the viral oncogene products. Whereas SV40 large T antigen targets the p53 DNA-binding domain, the adenovirus E1B 55-kD protein and the Mdm2 protein target the *trans*-activation domain. Indeed, these oncoproteins bind to the same amino acids, at codons 22 and 23, that appear to interact with the transcriptional machinery of the cell.

Finally, p53 mutations in cancerous tissues appear to select for missense mutant proteins altered in the p53 DNA-binding domain (Fig. 1). These faulty proteins appear to gain a new set of functions in that they are now able to transcriptionally activate the MDR1 enhancer-promoter (Chin et al. 1992; Dittmer et al. 1993) and enhance the tumorigenic potential of cells in culture (Dittmer et al. 1993). The experimental evidence presented in this paper suggests a model to explain this gain of function phenotype. If mutant p53 proteins defective in their DNA-binding domains (codons 120–290) form complexes with transcription factors with functional DNA-binding recognition sites, then the p53 *trans*-activation domain will be brought to a new enhancer element and a novel set of genes will be expressed. This could explain the transcriptional activation of the MDR1 CAT element and even the enhanced tumorigenic potential of cells that express p53 mutant proteins. To test this idea, the p53 mutant in codon 281, defective in its DNA-binding domain, was compared with the double mutant codon 281 plus codons 22, 23, defective in both its DNA binding and its *trans*-activation domains. The 281-22, 23 triple mutant failed to transcriptionally *trans*-activate the MDR1 CAT element and failed to enhance the tumorigenic potential of cells in culture. These data are consistent with the gain-

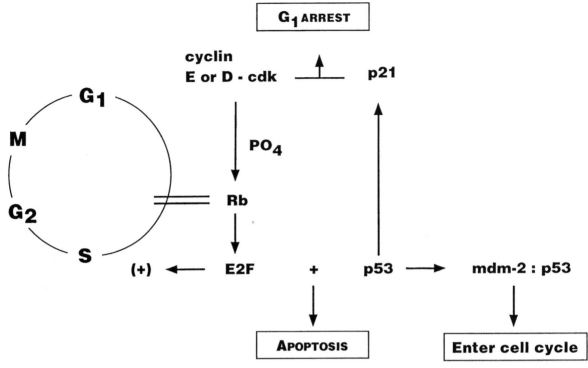

Figure 5. A model that relates the p53 checkpoint control to the Rb cell cycle regulation. The stages of the cell cycle (G_1, S, G_2, M) progress by the expression of the cyclin-dependent kinases at specific times in the cycle. Cyclin E-cdk-2 or D-cdk-4 or 6 phosphorylate Rb in G_1, releasing E2F transcription factors (along with DP-1) to enhance expression of a variety of genes required to enter S phase. In the presence of DNA damage, p53 levels rise and p53 enhances the transcription of p21 (WAF1/CIP1/pic1), which in turn blocks the cyclin-dependent kinase activity, thus keeping the cells in G_1. If, on the other hand, DNA damage occurs after E2F levels are high, p53 in its wild-type conformation initiates a pathway for apoptosis or cell death. Mdm-2, as a negative regulator of p53, can reverse both the G_1 arrest and the apoptosis pathway mediated by p53.

of-function model presented here and make it more likely that the gain-of-function hypothesis of p53 missense mutants is correct.

ACKNOWLEDGMENTS

The authors thank Drs. W.G. Kaelin, Jr. and D.M. Livingston for providing the E2F-1 clone, and Drs. Perry and Dittmer for helpful technical contributions. Drs. Perry, Finlay, and Quartin contributed with helpful discussions, and K. James helped in the preparation of this manuscript. Dr. C. Harris of the National Institutes of Health kindly shared a database of p53 mutations he and others have assembled at the National Institutes of Health. This research was supported by grants from the National Cancer Institute.

REFERENCES

Baker, S.J., A.C. Preisinger, J.M. Jessup, C. Paraskeva, S. Markowitz, J.K. Willson, S. Hamilton, and B. Vogelstein. 1990. p53 gene mutations occur in combination with 17p allelic deletions or late events in colorectal tumorigenesis. *Cancer Res.* **50:** 7717.

Bargonetti, J., I. Reynisdottir, P.N. Friedman, and C. Prives. 1992. Site-specific binding of wild-type p53 to cellular DNA is inhibited by SV40 T antigen and mutant p53. *Genes Dev.* **6:** 1886.

Chen, J., V. Marechal, and A.J. Levine. 1993. Mapping of the p53 and mdm-2 interaction domains. *Mol. Cell. Biol.* **13:** 4107.

Chin, K.V., K. Ueda, I. Pastan, and M.M. Gottesman. 1992. Modulation of activity of the promoter of the human *MDR1* gene by ras and p53. *Science* **255:** 459.

Chu, S. 1991. "Expression of the hot spot mutants of p53 in SOAS-2 cells." Senior thesis, Princeton University, Princeton, New Jersey.

Deng, W.P. and J.A. Nickoloff. 1992. Site-directed mutagenesis of virtually any plasmid by eliminating a unique site. *Anal. Biochem.* **200:** 81.

Dittmer, D., S. Pati, G. Zambetti, S. Chu, A.K. Teresky, M. Moore, C. Finlay, and A.J. Levine. 1993. p53 gain of function mutations. *Nat. Genet.* **4:** 42.

El-Deiry, W.S., T. Tokino, V.E. Velculescu, D.B. Levy, R. Parsons, J.M. Trent, D. Lin, W.E. Mercer, K.W. Kinzler, and B. Vogelstein. 1993. *WAR1*, a potential mediator of p53 tumor suppression. *Cell* **75:** 817.

El-Deiry, W.S., J.W. Harper, P.M. O'Connor, V.E. Velculescu, C.E. Canman, J. Jackman, J.A. Pietenpol, M. Burrell, D.E. Hill, Y. Want, K.G. Wiman, W.E. Mercer, M.B. Kastan, K.W. Kohn, S.J. Elledge, K.W. Kinzler, and B. Vogelstein. 1994. *WAF1/CIP1* is induced in *p53*-mediated G_1 arrest and apoptosis. *Cancer Res.* **54:** 1169.

Eliyahu, D., D. Michalovitz, S. Eliyahu, O. Pinhasi-Kimhi, and M. Oren. 1989. Wild-type p53 can inhibit oncogen-mediated focus formation. *Proc. Natl. Acad. Sci.* **86:** 8763.

Fields, S. and S.K. Jang. 1990. Presence of a potent transcription activating sequence in the p53 protein. *Science* **249:** 1046.

Finlay, C.A., P.W. Hinds, and A.J. Levine. 1989. The p53 proto-oncogene can act as a suppressor of transformation. *Cell* **57:** 1083.

Funk, W.D., D.J. Pak, R.H. Karas, W.E. Wright, and J.W. Shay. 1992. A transcriptionally active DNA binding site for human p53 protein complexes. *Mol. Cell. Biol.* **12:** 2866.

Ginsberg, D., D. Michael-Michalovitz, D. Ginsberg, and M. Oren. 1991. Induction of growth arrest by a temperature-sensitive p53 mutant is correlated with increased nuclear localization and decreased stability of the protein. *Mol. Cell. Biol.* **11:** 582.

Graham, F.L. and A.J. van der Eb. 1973. A new technique for the assay of infectivity of human adenovirus 5 DNA. *Virology* **52:** 456.

Harvey, D. and A.J. Levine. 1991. p53 alteration is a common event in the spontaneous immortalization of primary BALB/C murine embryo fibroblasts. *Genes Dev.* **5:** 2375.

Hinds, P.W., C.A. Finlay, R.S. Quartin, S.J. Baker, E.R. Fearon, B. Vogelstein, and A.J. Levine. 1990. Mutant p53 cDNAs from human colorectal carcinomas can cooperate with ras in transformation of primary rat cells: A comparison of the "hot spot" mutant phenotypes. *Cell Growth Differ.* **1:** 571.

Kao, C.C., P.R. Yew, and A.J. Berk. 1990. Domains required for *in vitro* association between the cellular p53 and the adenovirus 2 E1B 55K proteins. *Virology* **179:** 806.

Karasuyama, H., N. Tohyama, and T. Tada. 1989. Autocrine growth and tumorigenicity of interleukin 2-dependent helper T-cells transformed with IL-2 gene. *J. Exp. Med.* **169:** 13.

Kastan, M.B., Q. Zhan, W.S. El-Deiry, F. Carrier, T. Jacks, W.V. Walsh, B.S. Plunkett, B. Vogelstein, and A.J. Fornace, Jr. 1992. A mammalian cell cycle checkpoint pathway utilizing p53 and GADD45 is defective in ataxia-telangiectasia. *Cell* **71:** 587.

Kern, S., J.A. Pietenpol, S. Thiagalingam, A. Seymour, K. Kinzler, and B. Vogelstein. 1992. Oncogenic forms of p53 inhibit p53-regulated gene expression. *Science* **256:** 827.

Kunkel, T.A. 1985. Rapid and efficient site-specific mutagenesis without phenotype selection. *Proc. Natl. Acad. Sci.* **82:** 488.

Levine, A.J., M.E. Perry, A. Chang, A. Silver, D. Dittmer, M. Wu, and D. Welsh. 1994. The role of the p53 tumor suppressor gene in tumorigenesis. *Br. J. Cancer* **69:** 409.

Lin, J., J. Chen, B. Elenbaas, and A.J. Levine. 1994. Several hydrophobic amino acids in the p53 N-terminal domain are required for transcriptional activation, binding to mdm-2 and the adenovirus 5 E1B 55kd protein. *Genes Dev.* **8:** 1235.

Martinez, J., I. Georgoff, J. Martinez, and A.J. Levine. 1991. Cellular localization and cell cycle regulation by a temperature sensitive p53 protein. *Genes Dev.* **5:** 151.

Masuda, H., C. Miller, H.P. Koeffler, H. Battifora, and M.J. Cline. 1987. Rearrangement of the p53 gene in human osteogenic sarcomas. *Proc. Natl. Acad. Sci.* **84:** 7716.

Momand, J., G.P. Zambetti, D.C. Olson, D. George, and A.J. Levine. 1992. The mdm-2 oncogene product forms a complex with the p53 protein and inhibits p53 mediated transactivation. *Cell* **69:** 1237.

Oliner, J.D., J.A. Pietenpol, S. Thiagalinzam, J. Gyures, K.W. Kinzler, and B. Vogelstein. 1993. Oncoprotein MDM2 conceals the activation domain of tumour suppressor p53. *Nature* **362:** 857.

Pavletich, N.P., K.A. Chambers, and C.O. Pabo. 1994. The DNA binding domain of p53 contains the four conserved regions and the major mutation hotspots. *Genes Dev.* **7:** 2556.

Pietenpol, J.A., T. Takashi, S. Thiagalingam, W.S. El-Deiry, K.W. Kinzler, and B. Vogelstein. 1994. Sequence-specific transcriptional activation is essential for growth suppression by p53. *Proc. Natl. Acad. Sci.* **91:** 1998.

Raycroft, L., H. Wu, and G. Lozano. 1990. Transcriptional activation by wild-type but not transforming mutants of the p53 anti-oncogene. *Science* **249:** 1049.

Shaw, P., R. Bovey, S. Tardy, R. Sahli, B. Sordat, and J. Costa. 1992. Induction of apoptosis by wild-type p53 in a human colon tumor-derived cell line. *Proc. Natl. Acad. Sci.* **89:** 4495.

Soussie, T., C. Caron de Fromentel, and P. May. 1990. Structural aspects of the p53 protein in relation to gene evolution. *Oncogene* **5:** 945.

Wu, X. and A.J. Levine. 1994. p53 and E2F-1 cooperate to mediate apoptosis. *Proc. Natl. Acad. Sci.* **91:** 3602.

Wu, X., J.H. Bayle, D. Olson, and A.J. Levine. 1993. The p53-mdm-2 autoregulatory feedback loop. *Genes Dev.* **7:** 1126.

Yew, P.R. and A.J. Berk. 1992. Inhibition of p53 transactivation required for transformation by adenovirus E1B 55 Kd protein. *Nature* **357:** 82.

Yonish-Rouach, E., D. Resnitzky, J. Lotem, L. Sachs, A. Kimchi, and M. Oren. 1991. Wild-type p53 induces apoptosis of myeloid leukaemic cells that is inhibited by interleukin-6. *Nature* **352:** 345.

Zambetti, G.P., J. Bargonetti, K. Walker, C. Prives, and A.J. Levine. 1992. Wild-type p53 mediates positive regulation of gene expression through a specific DNA sequence element. *Genes Dev.* **6:** 1143.

Targets for Transcriptional Activation by Wild-type p53: Endogenous Retroviral LTR, Immunoglobulin-like Promoter, and an Internal Promoter of the *mdm2* Gene

Y. Barak,* A. Lupo,* A. Zauberman,* T. Juven,* R. Aloni-Grinstein,†
E. Gottlieb,* V. Rotter,† and M. Oren*

*Departments of *Chemical Immunology and †Cell Biology, Weizmann Institute of Science, Rehovot 76100, Israel*

The p53 protein is the product of a tumor suppressor gene which is subject to an extremely high frequency of structural alterations in human cancer (for review, see Hollstein et al. 1991; Levine et al. 1991; Oren 1992; Donehower and Bradley 1993; Harris and Hollstein 1993; Berns 1994). The main selective advantage of such alterations is probably the elimination of the tumor suppressor functions of the wild-type (wt) p53 protein. However, at least some of the p53 point mutations commonly encountered in tumor cells may also entail an oncogenic gain of function (Wolf et al. 1984; Chen et al. 1990; Michalovitz et al. 1991; Dittmer et al. 1993).

Attempts to elucidate the normal functions of wt p53, which probably underlie its ability to act as a tumor suppressor, have suggested that this protein may be involved in a variety of biological processes. Most prominent among those are the ability to arrest cell proliferation at the G_1 phase of the cell cycle, to activate apoptotic cell death (for review, see White 1993; Oren 1994), to promote the differentiation of a variety of cell types (for review, see Rotter et al. 1993), and to mediate the response to DNA damage and perhaps other types of genomic stress (for review, see Lane 1992; Berns 1994).

The molecular basis for the diverse activities of p53 is still not fully understood. It appears most likely, however, that many of the effects of p53 are due to its ability to function as a potent transcriptional regulator. In fact, p53 possesses distinct properties of a sequence-specific transcriptional activator, including the presence of an efficient acidic *trans*-activation domain and a potential for sequence-specific DNA binding (for review, see Prives and Manfredi 1993; Zambetti and Levine 1993). In addition, the wt p53 protein can also repress the expression of a large variety of genes, presumably through inhibitory interactions with a number of limiting general transcription factors. Consistent with the proposed importance of p53's transcriptional activities for its tumor suppressor function, the coding region mutations that are commonly found in human tumors are greatly impaired both in their ability to activate specific target genes and in their ability to repress transcription from other promoters.

The realization that wt p53 is a sequence-specific transcriptional regulator has prompted the search for potential target genes, whose activation by p53 may account for the various biological effects of this protein. The first gene that was found to harbor a p53-binding site and to possess a potential for being activated by p53 is the muscle-type creatine kinase gene (Weintraub et al. 1991; Zambetti et al. 1992). The biological significance of this observation remains to be established. Subsequent studies have led to the identification of several putative p53 target genes, most notably the growth arrest and DNA damage-induced *gadd45* gene (Kastan et al. 1992; Zhan et al. 1994), and the *WAF1/cip1* gene (El-Deiry et al. 1993; Harper et al. 1993). The latter is of particular interest, since it encodes an inhibitor of cyclin-dependent protein kinases. Increased production of this inhibitor, $p21^{waf1}$, could explain at least some of the antiproliferative effects of wt p53 (Hunter 1993).

The present study describes the characterization of three additional promoters that appear to be targets for regulation by wt p53 and tries to place these promoters within the broader context of the biological activities of p53.

EXPERIMENTAL PROCEDURES

Isolation of p53-binding sites from genomic DNA. The approach employed to isolate new p53-binding sites from genomic DNA is described in detail by Zauberman et al. (1993) and is schematically outlined in Figure 1. Essentially, purified mouse or rat genomic DNA was digested to completion with *Sau*3A1, resulting in a mixture of restriction fragments of relatively small molecular sizes. This mixture was then ligated into *Bam*HI-digested pBlueScript DNA, and the ligation products were electroporated into electrocompetent *Escherichia coli* cells. Following amplification of the resultant recombinant plasmid library in liquid culture, plasmid DNA was extracted and incubated with an extract of clone 6 cells maintained at 32°C. Clone 6 cells harbor a temperature-sensitive (ts) p53 mutant, p53val135. At 37.5°C, the mutant protein lacks detectable wt p53 activity; however, upon shift to

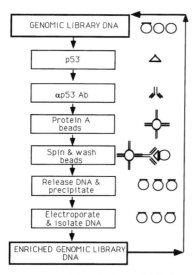

Figure 1. Strategy used for cloning p53-binding sites from genomic DNA. See Experimental Procedures for details.

32°C, it rapidly regains a variety of properties characteristic of authentic wt p53 (Michalovitz et al. 1990). Hence, the extract derived from cells that had been maintained at 32°C contains a high concentration of active wt p53. Following incubation of the plasmid DNA with the clone 6 extract, a mixture of p53-specific monoclonal antibodies (PAb248, PAb421, and RA3.2C2, mouse DNA) or antibody PAb421 only (rat DNA) was added to the reaction, and immune complexes were collected on immobilized protein A. In some experiments, the antibody was first coupled covalently to protein A–Sepharose before incubation with the DNA/protein complexes. The bound plasmid DNA was released from the immune complexes by the addition of SDS, followed by phenol extraction, ethanol precipitation, and retransformation of electrocompetent *E. coli*.

This binding site selection step was repeated twice more. Plasmid clones emerging from the third round of enrichment/transformation were then individually tested for sequence-specific DNA binding to p53.

DNA, RNA, and protein analysis. Electrophoretic mobility shift analysis (EMSA), DNA sequencing, CAT analysis, and in vitro transcription and translation were carried out as described previously (Shaulian et al. 1992, 1993; Zauberman et al. 1993). Luciferase assays were performed with the aid of a Turner Design luminometer. Luciferase activity values are presented in arbitrary units. RNase protection, using total cellular RNA and a synthetic RNA probe, was performed as described by Barak et al. (1994). The generation of plasmids serving as templates for the production of various synthetic *mdm2* RNA species, under control of the T7 promoter, is described by Barak et al. (1994).

RESULTS

The LTR of an Endogenous Murine Retrovirus-like Element Is Transcriptionally Activated by wt p53

The screening of a mouse genomic DNA library for p53-binding sites gave rise to a series of potential positive clones. Analysis of a representative group of clones by cross-hybridization revealed that the majority of the cloned DNA fragments were closely related, and most probably shared a high degree of sequence complementarity (data not shown). Yet, they had different insert sizes, indicating that they represented distinct clones rather than multiple isolates of the same single clone. Hence, it appeared most probable that all these cross-hybridizing clones originated in a relatively large family of mouse genomic DNA repetitive elements.

A representative member of this group, designated clone 15, was subjected to DNA sequence determination. The deduced nucleotide sequence is shown in Figure 2A. Comparison of this sequence with the EMBL database revealed that it was derived from the 5′ region of an endogenous murine retrovirus-like element, known as the GLN element (see Zauberman et al. 1993). The mouse genome contains about 10 intact GLN elements, plus approximately 1500 copies of the corresponding solo long terminal repeats (GLN LTRs). The position of the p53-binding motif (underlined) within the GLN LTR suggests that it may actually constitute part of the enhancer of this retrovirus-like transcriptional control region (Zauberman et al. 1993).

The p53-binding motif of the cloned GLN LTR constitutes a perfect match (20/20 nucleotides) with the consensus p53-binding site defined by El-Deiry et al. (1992) (Fig. 2C). Such a match is expected to result in high-affinity binding of wt p53 to this sequence. This prediction was indeed borne out by performing DNase I footprinting analysis, as well as by comparing the performance of this site to that of other p53-binding sites in an EMSA (Zauberman et al. 1993).

We wanted to find out whether the cloned LTR could indeed function as an independent transcriptional control unit, and whether the presence of a p53-binding site endowed it with the ability to be activated by wt p53. To that end, the intact GLN LTR was placed in front of the bacterial chloramphenicol acetyl transferase (CAT) reporter gene, to generate plasmid GLN-LTR-CAT (Fig. 3A). This plasmid was transfected transiently into p53-deficient Saos-2 cells, together with expression plasmids encoding either wt or tumor-derived mutant p53. After 36 hours, cell extracts were prepared, and the amount of CAT enzymatic activity was determined. A representative result is depicted in Figure 4A, lanes 9–13. The results of such transient transfection assays demonstrated clearly that the cloned GLN LTR is indeed functional, at least in the context of a free recombinant plasmid. Moreover, the transcriptional activity emanating from this LTR was strongly p53-dependent and was induced specifically by wt (lanes 9, 12) but not mutant (lanes 10, 13) protein.

The p53-binding region of the GLN LTR could also

A

```
5'  GATCCACACATTTGCCTGGAAATTTTGTGGAGTCATTGTTTTTAATAGC
    AGATTAGTACTCTATTGTGTAAACATATCACATTTTCTGTATCCATTCCT
    CTTTTGAGAAACATATGGGTTGTTTCCAGCTTCTGGCTATTATAAATATG
    GTTGCTCTGAACACAATGAAAGGAAATAAGACTGTAATTCATGTAATGTA
    TGTTAAATAGCCCAAAGAGTTGTTTGTGAGCTTTGAAACCTGGGGCTGAG
    AACATAGCAGAACAGACCAGGACATGCCCGGGCAAGCCCATCGCCTCCCT
    AGCTCCCACCCCTCTCGACCTAAGTTAAATGTTACAGGCTGCTGATGTTT  3'
```

B

```
                          Percent identity : 78 %

              1         p53 BE
   7R   GATCGTGTCTAGACATGTTTCCTGTGATTCCTACAGTCGTCATACACTCT          AACTATA
   MOPC       TCTAGATGTGTTTCCTGTGATTTCTAAAGTCTTATTGCTCTCTtattggagactcaCACTATA
                   .....

   7R   GGAAACCAGAGAGAATGGTGGCCTTATTTTAATCACCATGTGCAGTTACTACATAC
   MOPC GGAAGCCAGAGACCATGATGGTCTTACTTTAATAACCAAGGGCATTCATTATTTAC

                                               OCTA
   7R   CTTTCCCAATTATGAAGGCAAGGCTAACTCCCACATGCAAATGTATCTTCTAGCTC
   MOPC CTTCCCCAATTATGAAGGCTGGGCTGTCTCCCTGCATGCAAATGCTTCCTCTAACTC

             TATA                                              226
   7R   TTAGTTAAATCCCTTTGGTGGTGGTCCATATCACATACCTCTCACTGGAGGCTGATC
   MOPC TAAGTTAAATCCCTCTTGGGGT GTAAAGCTCACATCTCTCTCATTAGAGGTTGATC
```

C

```
   G G A C A T G C C C G G G C A A G C C C      GLN-LTR
   R R R C W W G Y Y Y R R R C W W G Y Y Y      CONSENSUS
       *   *
   G A T C G T G T C T A G A C A T G T T T      Clone 7R
```

Figure 2. Sequences of p53-regulated promoters. (*A*) Sequence of the cloned mouse genomic DNA fragment containing the GLN LTR. The underlined region indicates the 20-bp p53-binding element (for additional information, see Zauberman et al. 1993). (*B*) Comparison between the clone 7R rat genomic DNA fragment and a known mouse immunoglobulin heavy-chain variable region gene promoter, derived from the MOPC315 plasmacytoma. (p53 BE) p53-binding element (underlined). The octamer motif and the TATA box are also underlined. The dots indicate the position of the *Xba*I site in the MOPC315 clone. (*C*) Comparison between the p53-binding elements of the GLN LTR and clone 7R, and the p53-binding consensus site defined by El-Deiry et al. (1992). (R) Purine; (Y) pyrimidine; (W) A or T. Asterisks indicate mismatches.

confer p53 responsiveness on an unrelated promoter. Three tandem copies of a 100-bp region, centered around the p53-binding element (p53 BE) of the GLN LTR, were inserted in front of the basal SV40 promoter (100/T-CAT, Fig. 3A). As seen in Figure 4A, lanes 4–6, this was sufficient to turn this otherwise very weak promoter into a strong transcriptional control unit, in the presence of wt but not mutant p53.

The transcriptional analysis of transfected recombinant plasmids is prone to a wide variety of artifacts. It was therefore of importance to determine whether wt p53 could also activate the endogenous, nonmanipulated chromosomal GLN LTR. Such analysis was made possible through the use of LTR-6 cells (Yonish-Rouach et al. 1993), which express stably the temperature-sensitive p53val135 mutant. RNA was prepared from LTR-6 cells maintained at either 32°C (excess wt p53 activity) or 37.5°C (greatly reduced wt p53 activity) and subjected to Northern blotting and hybridization with a GLN LTR probe. In view of the fact that interleukin-6 (IL-6) exhibits pronounced effects on the response of LTR-6 cells to p53-mediated signals (Yonish-Rouach et al. 1991, 1993; Levy et al. 1993), we also wanted to test whether this cytokine affected the activation of potential p53 target genes. The results (Fig. 4B) clearly demonstrated that whereas LTR-6 cells expressed only very low levels of GLN element transcripts at 37.5°C, a marked accumulation of such transcripts could be observed following the induction of wt p53 activity through a temperature downshift to 32°C. Accumulation was rather slow and continued for many hours, presumably reflecting the

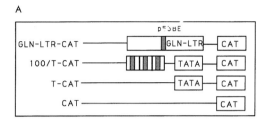

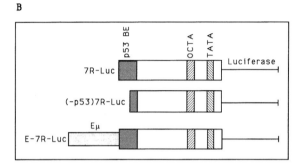

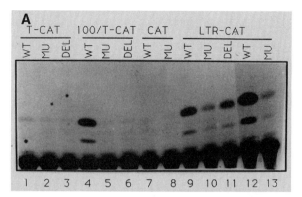

Figure 3. Expression plasmids employed for measuring p53-dependent *trans*-activation. (*A*) GLN LTR-based plasmids. Hatched bars indicate the p53-binding elements (p53BE). Plasmid GLN-LTR-CAT contains the entire GLN LTR in front of an otherwise promoterless CAT gene. Plasmid 100/T-CAT contains 3 tandem copies of a 100-bp fragment from the GLN LTR, including the p53BE, inserted in front of a core promoter (TATA). (*B*) Clone 7R (rat immunoglobulin gene)-based plasmids. (OCTA) Octamer motif; (TATA) TATA box. Plasmid E-7R-Luc contains an immunoglobulin heavy-chain enhancer in front of the putative heavy-chain variable region promoter.

relatively large size of the transcript (~ 8 kb) and the fact that this RNA, once made, is rather stable. These data argue strongly that the GLN element can be transcriptionally activated by wt p53 also within its natural chromosomal context, and thus the GLN LTR qualifies as a bona fide target of p53. It is noteworthy that IL-6, which inhibits the apoptotic effects of wt p53 in LTR-6 cells (Yonish-Rouach et al. 1991, 1993; Levy et al. 1993), does not interfere with the activation of the GLN LTR. Similar observations have also been made for another target of p53, the *mdm2* gene (data not shown). Hence, IL-6 does not appear to interfere directly with the transcriptional activity of p53, and it may rather protect cells from p53-mediated apoptosis by acting at a more downstream stage of the process.

Sequence-specific Activation of an Immunoglobulin Heavy-chain Gene-like Promoter by wt p53

Attempts to isolate a spectrum of p53-binding sites from the mouse genomic DNA library were hampered by the prevalence of multiple clones corresponding to GLN element sequences (data not shown). This was most probably due to the high p53-binding affinity of the perfect consensus p53 BE of the GLN LTR, coupled with the relatively large number (1500) of GLN LTRs within the mouse genome. Consequently, the vast majority of clones isolated through this search

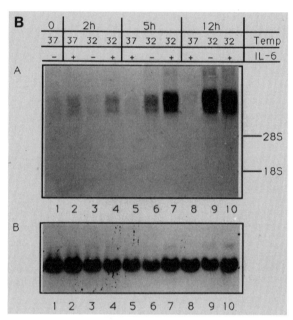

Figure 4. Induction of the GLN LTR by wt p53. (*A*) Induction of transiently transfected GLN LTR-derived plasmids by cotransfected wt p53. The various reporter plasmids are described in Fig. 3A. Transfections were done in primary rat embryo fibroblasts (REF). The following expression vectors were used as effectors: (WT) a plasmid encoding wt murine p53; (MU) a plasmid encoding a mutant p53 derived from the MethA fibrosarcoma; (Del) a plasmid expressing a p53 molecule from which the bulk of the coding sequence has been deleted (vector control). Lanes 12 and 13 are essentially similar to lanes 9 and 10, except that they are derived from an independent transfection. (*B*) Northern analysis of GLN element transcripts (panel *A*) in clone LTR-6, harboring the ts p53 mutant p53val135 (Yonish-Rouach et al. 1991), at various times after induction of wt p53 activity by a temperature downshift to 32°C. Cells were maintained either in the absence (−) or presence (+) of IL-6, which protects LTR-6 cells from the induction of p53-mediated apoptosis at 32°C (Yonish-Rouach et al. 1991). In *B*, the same blot was rehybridized with a GAPDH probe, to normalize for RNA inputs.

contained GLN LTR sequences. A single unrelated clone, designated clone 255, did not exhibit any significant identity with EMBL database entries (Zauberman et al. 1993) and was not studied further.

To overcome the masking of potentially interesting

single-copy p53-binding clones by this repetitive element, we applied a similar site-selection approach to a rat genomic DNA library. Preliminary analysis (data not shown) indicated that GLN LTRs were not present in the rat genome, even though the existence of other repetitive p53-binding elements could not be excluded.

The screening of a total of 7×10^6 rat genomic DNA plasmid clones finally resulted in the isolation of 19 distinct p53-binding clones. Sequencing of the genomic DNA insert of each revealed that only a single one was significantly homologous to a previously characterized DNA region. This clone, designated 7R, exhibited 78% identity with an immunoglobulin heavy-chain variable region gene, cloned from mouse MOPC315 plasmacytoma cells. Specifically, the homology between the cloned rat DNA and the mouse immunoglobulin gene mapped to a domain containing the promoter of the latter (Fig. 2B). A number of sequence hallmarks of this promoter, including the octamer motif (OCTA) and the proposed TATA box, were perfectly conserved in clone 7R. It thus appeared likely that we may have cloned a rat immunoglobulin heavy-chain variable region promoter, which binds specifically to p53. Further examination of the sequence (Fig. 2B) confirmed the presence of a recognizable potential p53-binding element (p53BE), exhibiting a 18/20 match with the p53 consensus DNA-binding site (Fig. 2C). The position of this potential p53BE suggested that it could serve as a promoter-proximal regulatory element, potentially conferring p53 responsiveness on this putative promoter.

To test whether the DNA insert of clone 7R indeed possessed promoter activity and whether this activity was modulated by p53, the plasmids illustrated in Figure 3B were constructed. 7R-Luc contains the entire putative promoter in front of a luciferase reporter gene; (−p53) 7R-Luc lacks most of the p53BE and is expected to neither bind nor respond to wt p53; in E-7R-Luc, an immunoglobulin heavy-chain enhancer has been inserted in front of the 7R DNA insert. As a target for transfection, we chose L12 early pre-B cells; L12 cells do not produce any detectable p53 mRNA or protein, as a consequence of a proviral insertion into the p53 gene (Wolf and Rotter 1984).

Each plasmid was cotransfected transiently, along with an expression plasmid for either wt or mutant mouse p53, into L12 cells. After 36 hours, extracts were prepared from each culture and subjected to a quantitative determination of luciferase activity. The results of two independent experiments are summarized in Table 1. It is evident that the cloned DNA fragment could indeed function as a promoter in early pre-B cells, albeit at a rather low efficiency (construct 7R-Luc). This transcriptional activity, however, was strongly dependent on the presence of wt p53, and was hardly detectable when mutant, rather than wt, p53 was overexpressed. Incapacitation of the putative p53BE (plasmid [−p53] 7R-Luc) practically ablated the ability of this promoter to be activated by wt p53. On the other hand, the much stronger RSV promoter was not induced by

Table 1. Summary of Transfection Experiments with Luciferase Reporter Genes Driven by the p53-responsive Immunoglobulin-like Promoter from Clone 7R

Plasmid	Experiment 1		Experiment 2	
	wt	mu	wt	mu
7R-Luc	0.151	0.023	0.132	0.028
(−p53)7R-Luc	0.006	0.043	0.026	0.023
E-7R-Luc	1.25	0.164	0.53	0.11
RSV-Luc	n.d.	n.d.	94.1	80.16

The plasmids used in these experiments are described in Fig. 3B. Each plasmid (5 μg per transfection) was introduced into L12 early pre-B cells together with 2 μg of a plasmid encoding either wild-type or mutant mouse p53 (pCMVp53wt and pCMVp53m, respectively; Eliyahu et al. 1989). Transfection was by the DEAE dextran-chloroquin procedure. Luciferase activity was determined with cell extracts prepared 36 hr after transfection. n.d. indicates not determined.

cotransfection with a wt p53 expression plasmid. These data argue strongly that clone 7R contains a p53-responsive transcriptional control unit. As expected, transcriptional activation with or without the p53BE was further augmented by inclusion of an immunoglobulin enhancer (E-7R-Luc). However, it is noteworthy that under our experimental conditions the effect of the enhancer alone (E-7R-Luc + mutant p53) was not much greater than that of the p53BE alone (7R-Luc + wt p53).

In conclusion, the promoter contained within clone 7R can activate transcription, in a p53-dependent manner, even in the absence of a linked immunoglobulin enhancer.

Wild-type p53 Activates an Endogenous p53-responsive Promoter within the Murine *mdm2* Gene

A search for p53-associated proteins has revealed that polypeptides encoded by the *mdm2* proto-oncogene can bind tightly to both wt and mutant p53 (Barak and Oren 1992; Momand et al. 1992; Barak et al. 1993). The major product of this gene is an Mdm2 protein exhibiting an apparent molecular mass of approximately 95 kD (p95). Cotransfection experiments demonstrated that $p95^{mdm2}$ can counteract the antiproliferative effects of wt p53 (Finlay 1993). The underlying molecular mechanism appears to rely on the ability of $p95^{mdm2}$ to block p53-mediated transcriptional activation (Momand et al. 1992), presumably through masking the *trans*-activation domain of p53 (Oliner et al. 1993), and perhaps also by interference with sequence-specific DNA binding by p53 (Zauberman et al. 1993).

The relationship between p53 and *mdm2* became more complex when it was realized that the *mdm2* gene is actually a target for positive transcriptional regulation by wt p53 (Barak et al. 1993; Juven et al. 1993; Wu et al. 1993). Induction of *mdm2* gene transcription by excess wt p53 was subsequently reported in a wide variety of cell types, including those where p53 activity was triggered by DNA damage (Otto and Deppert 1993; Perry et al. 1993; Chen et al. 1994). This led to the suggestion that *mdm2* is part of a negative feedback loop, through which the cell restricts the duration of p53-dependent signals (Barak et al. 1993; Picksley and Lane 1993; Wu et al. 1993). According to this model (illustrated in Fig. 7), activation of wt p53 induces transcription from a variety of relevant target genes. In parallel with the accumulation of other proteins, which presumably mediate the antiproliferative effects of wt p53, the *mdm2* mRNA starts giving rise to $p95^{mdm2}$. The resultant polypeptide, in turn, binds to the p53 protein and inhibits its function, so that p53-induced signal transduction is terminated rapidly. Such a mechanism is believed to act as a safeguard against excess wt p53 activity, which may be disadvantageous to the cell. Moreover, in cells that undergo a transient growth arrest in response to DNA damage or other stimuli, negative regulation of p53 function by $p95^{mdm2}$ may be required for the ability of the cell to re-enter the proliferative cycle once the damage has been repaired or the stimulus eliminated (Perry et al. 1993).

Analysis of the murine *mdm2* gene resulted in the identification of a p53 response element (p53RE, Fig. 5A), located within the first intron of the gene (Juven et al. 1993; Wu et al. 1993). This element comprises two closely spaced p53-binding motifs, each exhibiting a 17/20 match with the p53-binding site consensus. This domain can bind directly, in a sequence-specific manner, to wt but not mutant p53 (Juven et al. 1993 and unpubl.; Wu et al. 1993). Moreover, inclusion of this p53 RE in an expression vector endows upon the latter the ability to be transcriptionally induced by wt p53. These observations suggested a simple model, wherein the p53 RE functions as an intronic enhancer. Following the accumulation of active wt p53, such an enhancer is expected to act on the upstream promoter located at the 5' end of the *mdm2* gene and bring about its transcriptional activation.

A more detailed analysis of the 5' part of the murine *mdm2* gene led, however, to an observation that was inconsistent with this simple model. Specifically, plasmids in which a reporter gene was placed directly behind various internal segments of the *mdm2* gene, in the absence of any defined promoter, retained strong p53-dependent transcriptional activation (Juven et al. 1993).

This unexpected finding raised the possibility that the *mdm2* gene contains, in addition to the recognized upstream promoter, an independent internal p53-inducible promoter. However, given the very artificial nature of transient transfections with recombinant reporter plasmids, it was necessary to prove that this conclusion also held true for the actual *mdm2* gene. This was addressed by the use of an RNase protection assay. As a template for the in vitro synthesis of the radiolabeled riboprobe, we used an *mdm2* cDNA clone extending from exon 3 to the beginning of exon 1 (Fig. 5B). Exon 1 is located upstream of the intronic p53 RE (Fig. 5A) and should only be present in transcripts initiating from the upstream, "standard" promoter P1.

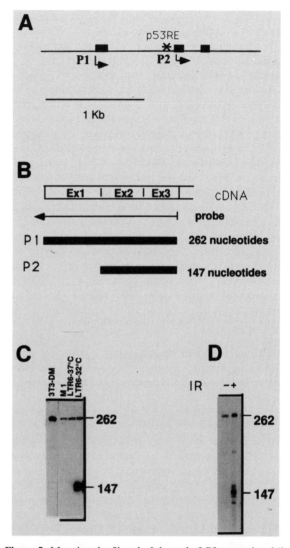

Figure 5. Mapping the 5' end of the *mdm2* P2 transcript. (*A*) Schematic representation of the 5' part of the murine *mdm2* gene. P1 and P2 are the constitutive and p53-dependent promoters, respectively (see text for details). The asterisk denotes the p53 response element (p53RE), which is composed of two adjacent degenerate p53-binding motifs (Juven et al. 1993; Wu et al. 1993). Filled boxes indicate the first 3 exons of the *mdm2* gene. (*B*) The probe employed for RNase protection, along with the protected fragments expected to result from hybridization of this probe with the P1 and P2 transcripts, respectively. Exons 1–3, as represented in the *mdm2* cDNA used for the preparation of the radiolabeled probe, are shown on top. (*C*) RNase protection analysis of *mdm2* transcripts using RNA from the following cell lines: (3T3-DM) containing multiple copies of the *mdm2* gene on double-minute chromosomes (Cahilly-Snyder et al. 1987); (M1) p53-null murine myeloid leukemia (Yonish-Rouach et al. 1991); (LTR6) a line derived from M1 by stable transfection with the ts p53 mutant p53val135. The sizes of the protected portions of the probe, deduced from comparison with co-electrophoresed markers, are indicated. (*D*) RNase protection analysis of *mdm2* transcripts using RNA from wt p53-containing DA-1 cells (Gottlieb et al. 1994), without (−) or with prior exposure to 500 Rad of ionizing γ irradiation (IR).

RNA from various sources was hybridized with the radiolabeled probe, followed by RNase treatment and gel electrophoresis of the digestion products. RNA initiated from P1 and containing exon 1 is expected to yield a protected band of 262 nucleotides (Fig. 5B). On the other hand, RNA initiated from a putative internal promoter will lack all of exon 1 and perhaps also sequences from exon 2 and/or exon 3, depending on the nature of the 5' ends of the transcripts. Such RNA should give rise to much shorter protected products.

Figure 5C depicts the results of such an RNase protection experiment, in which the following cell types were used for the preparation of RNA: 3T3-DM cells, harboring multiple copies of an amplified *mdm2* gene (Cahilly-Snyder et al. 1987; Fakharzadeh et al. 1991); M1 cells, which do not express any detectable p53 (Yonish-Rouach et al. 1991); and LTR-6 cells, derived from M1 and expressing stably the p53val135 temperature-sensitive mutant (Yonish-Rouach et al. 1993). RNA from 3T3-DM cells, in which the large excess of Mdm2 proteins is expected to render the endogenous p53 inactive, protected exclusively a fragment of 262 nucleotides. Hence, in those cells all *mdm2* transcripts include the entire exon 1, indicating that P1 (or additional promoters located upstream of it) is used for the generation of most, if not all, *mdm2* mRNA molecules. Essentially the same pattern, albeit in much reduced quantities, emerged from the analysis of the p53-negative M1 cells, as well as of LTR-6 cells maintained at 37°C (where the ts p53 does not exhibit wt activity). Hence, in the absence of functional wt p53 there was no evidence for the existence of an intronic *mdm2* promoter located downstream from exon 1.

A totally different picture emerged when the ts p53 of LTR-6 cells became activated at 32°C. The intensity of the protected 262-nucleotide band was practically unaltered, indicating that overexpression of wt p53 had no significant effect on transcription from P1. On the other hand, a much more prominent band, corresponding to a protected fragment of 147 nucleotides, now became apparent (Fig. 5C). Primer extension analysis, using RNA from LTR-6 cells maintained at 32°C, established that the 5' end of this 147-nucleotide fragment corresponded to the actual 5' end of the major *mdm2* RNA species (Barak et al. 1994). The size of these transcripts indicated that they are initiated within the "standard" exon 2, a few nucleotides downstream from its 5' end. Such an initiation site implies the existence of an intronic promoter, P2, located in the vicinity of the p53RE (Fig. 5A). As LTR-6 cells can express *mdm2* mRNA only from the endogenous, nonmanipulated *mdm2* gene, this internal p53-responsive promoter must exist also in vivo.

Although the *mdm2* gene in LTR-6 cells appears to be normal, the mode of activation of p53 in these cells is rather artificial. Thus, the extent of wt p53 overexpression attained in these cells at 32°C may well be far higher than ever encountered in nature. Moreover, the activation of p53 occurs in the absence of other physiological signals that may normally be associated with p53-mediated responses. We therefore wanted to find out whether transcription from P2 took place also when endogenous p53 was activated by an appropriate signal.

Exposure of wt p53-expressing cells to various DNA-damaging agents induces a substantial increase in p53 levels (Maltzman and Czyzyk 1984; Kastan et al. 1991; Fritsche et al. 1993). This increase is consistent with the proposed role of p53 in preventing the accumulation of cells with damaged DNA (Kuerbitz et al. 1992; Lane 1992; Oren 1992). Moreover, the p53 protein that is induced in response to DNA damage is biochemically active, as reflected by its ability to *trans*-activate various p53-responsive genes (Kastan et al. 1992; Lu and Lane 1993; Perry et al. 1993; Zhan et al. 1993; Chen et al. 1994). Cells of the lymphoma-derived line DA-1 express functional wt p53, whose inactivation renders them greatly resistant to DNA-damage-induced apoptotic cell death (Gottlieb et al. 1994). Exposure of DA-1 cells to ionizing radiation elicits a rapid and pronounced accumulation of active p53 (G. Gat-Yablonsky and M. Oren, unpubl.). RNA was therefore prepared from DA-1 cells with or without prior exposure to ionizing radiation (500 Rad), and subjected to RNase protection analysis with the same probe as above. The result is shown in Figure 5D. There was no evidence for the utilization of P2 in unirradiated DA-1 cells (−), even though those cells do express basal levels of wt p53. On the other hand, upon exposure to DNA damage, there was a clear induction of transcripts that gave rise to a cluster of protected bands centered around 147 nucleotides. Such a protection pattern is indicative of the activation of P2. Thus, at least in DA-1 cells, P2 becomes recruited as a promoter only when the endogenous p53 is activated in response to an appropriate signal.

In conclusion, the experiments described in Figure 5 demonstrate that the *mdm2* gene can be transcribed from two distinct promoters. The 5' promoter, P1, directs the production of longer transcripts. It is not significantly affected by overproduction of wt p53 (Fig. 5C), and exposure to DNA damage induces it only slightly, as suggested from Figure 5D. On the other hand, the more 3' promoter, P2, acts essentially as a "private" target of wt p53. It is practically inactive in the absence of wt p53, or even when the latter is produced in low, unstimulated amounts. On the other hand, P2 is strongly triggered following the induction of excess wt p53 activity, indicating that the p53RE (Fig. 5A) is indeed not an intronic enhancer element, but rather a component of a distinct, p53-dependent promoter.

Different *mdm2* Transcripts Have Nonidentical Translation Profiles

The *mdm2* mRNA emanating from P2 differs from that generated through the use of P1. The P2-derived transcripts lack the entire exon 1, plus a few nucleotides from the 5' end of exon 2 (Fig. 6A). However, the missing sequences do not affect the coding capacity of the resulting P2-derived mRNA, because the first in-frame AUG codon is located only at the beginning of

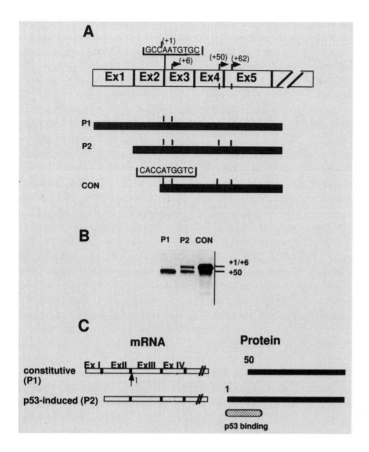

Figure 6. Translational potential of various forms of *mdm2* RNA. (*A*) Schematic representation of the analyzed transcripts. The structure of the 5' portion of the "constitutive" *mdm2* RNA is shown on top. Indicated are the positions of the first 4 in-frame AUG codons, as well as the sequence around the first AUG. P1 and P2 denote synthetic RNA molecules corresponding to *mdm2* transcripts initiated from either P1 or from P2. CON denotes a synthetic transcript where the sequences around the first AUG have been modified to yield a favorable Kozak consensus (Kozak 1986). (*B*) Polypeptides obtained by translation of the three synthetic *mdm2* transcripts depicted in *A* in a rabbit reticulocyte lysate. The translation start sites corresponding to each polypeptide are indicated on the right. Translation was performed under standard reaction conditions, using 5 picomoles RNA. (*C*) Scheme of the P1 and P2 transcripts (*left*), along with the truncated (+50) and full-length (+1) Mdm2 polypeptides generated by in vitro translation (*right*). The p53-binding domain of $p95^{mdm2}$ is indicated.

exon 3 (Fig. 6A, +1). Nevertheless, alternative 5'-untranslated regions might have a significant effect on the translatability of the corresponding mRNAs. To find out whether this is also the case for *mdm2*, several synthetic RNAs were produced by in vitro transcription from appropriate plasmids. These included a molecule closely resembling the natural P1-derived transcript (Fig. 6A, P1), and one essentially similar to the P2-promoted transcript (P2). In addition, we noted that the first in-frame AUG was situated within an unfavorable context for translation initiation (Kozak 1986). This suggested that it might be skipped by the translation machinery and not be translated efficiently. To address this possibility, we also generated another synthetic RNA in which sequences flanking the first AUG were mutated in order to generate a favorable consensus context (Fig. 6A, CON).

Each synthetic RNA was translated in a rabbit reticulocyte lysate under standard conditions (5 pmole RNA per reaction). Analysis of the translation products by SDS-PAGE (Fig. 6B) revealed that the different RNA species gave rise to distinct sets of polypeptides. The relationship between the *mdm2* transcripts and their in vitro translation products is illustrated schematically in Figure 6C. Under standard reaction conditions, the longer P1 transcript generates essentially only polypeptides lacking the amino terminus of the protein. This is due to translational initiation at codon

50, located within a relatively favorable RNA sequence context. On the other hand, the shorter P2 RNA produces, in addition to amino-terminally truncated proteins, substantial amounts of full-length Mdm2 protein (p95), initiated at codon 1. One crucial difference between these two forms of Mdm2 proteins is that the form initiated at position +50 lacks the major part of the p53-binding domain (Chen et al. 1993; Oliner et al. 1993; Haines et al. 1994), and is therefore unlikely to bind p53 and interfere with its function. In fact, the inability of this amino-terminally truncated Mdm2 to bind p53 has been confirmed experimentally in coprecipitation experiments (Barak et al. 1994).

Analysis of cellular Mdm2 proteins revealed that the situation in intact cells is quite different from that observed in the reticulocyte lysate. Irrespective of whether or not wt p53 was overexpressed, the full-length Mdm2 protein (p95) appeared more abundant than the amino-terminally truncated p85 (Barak et al. 1994). Hence, both forms of *mdm2* mRNA can be translated preferentially in vivo into Mdm2 proteins initiated at position 1, even though activation of wt p53 does sometimes shift the ratio further in favor of the full-length protein. It should be kept in mind, however, that the total amounts of Mdm2 proteins made upon induction of wt p53 activity are typically much higher than before induction (Barak et al. 1993, 1994; Perry et al. 1993; Chen et al. 1994).

The fact that the P1 mRNA, too, can encode full-length Mdm2 (p95) in vivo raised the possibility that its inability to do so in the reticulocyte lysate is related to the large mRNA excess used in standard in vitro translation assays. We therefore analyzed the translation products programmed in vitro by lower amounts of the different synthetic *mdm2* RNA species. This analysis revealed that when only 2% of the standard amount of RNA was used per reaction, both P1 and P2 RNA gave rise primarily to full-length Mdm2 proteins; yet, the total amounts of resultant proteins were consistently higher for P2 RNA than for P1 RNA (Barak et al. 1994).

In conclusion, the data indicate that the P1 RNA is less efficiently translated into intact Mdm2 protein than the P2 transcripts. At high RNA concentrations, this can lead (at least in vitro) to preferential translation of the P1 transcripts into molecules lacking the ability to bind p53. At low RNA concentrations, the implications are mainly quantitative, with P2 RNA generating more p95 molecules per RNA molecule than the P1 RNA. Under those conditions, the main outcome of the p53-induced promoter activation is thus a more efficient production of protein molecules capable of binding p53 and restricting its activity.

DISCUSSION

The present study reports the analysis of three very different p53-inducible promoters. The first drives the expression of an endogenous retrovirus-like element. The biological relevance, if any, of this aspect of p53-mediated transcription is hard to assess. There is no evidence that the proteins encoded by the GLN element affect in any way the properties of the cell. It thus appears more likely that the presence of a p53-responsive element in the GLN LTR reflects a feature that was of relevance to the predecessor of this retrovirus-like element. It is possible that, at some earlier time in the evolution of the mouse, this predecessor existed as a replication-competent retrovirus. In that case, it may have capitalized on the strong transcriptional activity of p53 by evolving a p53-responsive enhancer, which provided it with high levels of expression in p53-containing cells. It is remarkable that the p53-responsive element of the GLN LTR exhibits an extremely high affinity for p53, owing to its perfect match with the p53-binding consensus. This is unlike all cellular p53-responsive promoters described thus far, where the match is less perfect. Such an imperfect match may be required in order to enable the effective fine tuning of these promoters. On the other hand, as is often the case with viral enhancers, a high-affinity site such as the one found in the GLN LTR may ensure strong constitutive activity even in the presence of relatively low concentrations of p53 in the cell.

The second p53-responsive element resides in the 5' region of a rat immunoglobulin-like promoter. The ability of p53 to activate this promoter may indicate a direct role for p53 in the control of B-cell differentiation. In fact, such a role has already been suggested on the basis of earlier studies. First, the stable introduction of wt p53 expression into early pre-B cells, otherwise lacking p53, could activate heavy-chain gene expression (Shaulsky et al. 1991). Furthermore, it was shown that activation of pre-B cells by exposure to lipopolysaccharide elicits a marked increase in p53 levels and that wt p53 overexpression in various cell types results in increased transcription from a transfected immunoglobulin κ light-chain promoter (Aloni-Grinstein et al. 1993).

It is still unknown whether this p53-binding promoter indeed resides within the immunoglobulin heavy chain gene cluster. If it does, it is tempting to speculate that this serves to recruit p53 to the immunoglobulin gene loci, so that it is now poised to respond to any genomic damage that may occur during subsequent immunoglobulin gene rearrangement.

The third p53-responsive promoter discussed above is the one located within the murine *mdm2* gene. The "standard" *mdm2* promoter, P1, is only minimally affected by p53 overexpression, although it is conceivable that P1 may be regulated by other transcription factors, responding to different regulatory signals (Fig. 7). On the other hand, the novel promoter described here, P2, functions as a "private" target of p53. It is noteworthy that the p53-responsive domain of the P2 is actually composed of two adjacent low-affinity binding sites (Juven et al. 1993). This is unlike other p53-inducible promoters, which typically contain a single site with a relatively high affinity. In fact, both P2 sites must be present in order to confer significant activation by wt p53 (T. Juven, unpubl.). This suggests that P2 activation requires some sort of cooperative binding between two adjacent p53 tetramers positioned at the two adjacent p53-binding motifs. Such a requirement is fulfilled only when the cellular concentration of p53 becomes high enough to enable the frequent occupation of both sites simultaneously. As a result, P2 will stay practically

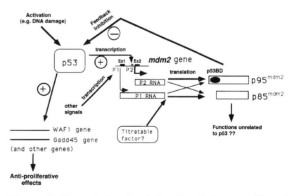

Figure 7. Transcriptional relationships between p53 and *mdm2*. See Discussion for further details. (p53BD) p53-binding domain of the full-length Mdm2 polypeptide ($p95^{mdm2}$). $p85^{mdm2}$ refers to the amino-terminally truncated forms of Mdm2, initiating at either codon 50 or codon 62 (see Fig. 6). The putative positions of the P1 and P2 promoters are indicated. Ex1 and Ex2 refer to the first 2 exons of the P2 transcript.

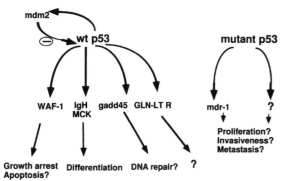

Figure 8. Identified transcriptional target genes of p53, and biological processes in which their activation by p53 may be involved.

inactive as long as p53 activity does not reach a minimal threshold, whereas other p53-regulated promoters may already become induced by milder increases in p53.

The model presented in Figure 7, which is an extension of the one proposed by Wu et al. (1993), depicts the putative regulatory network. Activation of wt p53 by signals such as DNA damage leads to increased transcription from a series of p53-responsive genes. Some of those, such as *WAF1* and *gadd45*, are likely to mediate the antiproliferative effects of p53. Once sufficiently high levels of p53 have been attained, the *mdm2* P2 promoter is activated, giving rise to a specific new transcript. This P2 transcript is translated efficiently into the p95 polypeptide, which possesses a p53-binding domain and can therefore act in a negative feedback loop to down-modulate the *trans*-activation function of p53. In addition, both p95 and the p53-nonbinding p85 may have additional functions, possibly unrelated to p53. The model in Figure 7 also proposes that whereas the P2 transcript gives rise predominantly to p95, the ability of the P1 RNA to generate p95 depends on the availability of a putative titratable factor. Such a factor could, for instance, modify the conformation of the 5′ end of the P1 RNA so that the first in-frame AUG is recognized well. In the absence of sufficient amounts of this factor, as occurs in reticulocyte lysates primed by excess P1 RNA, translation starts predominantly at the third AUG, resulting in the preferential synthesis of p85.

The array of p53-responsive promoters described here and in a growing number of other studies suggests that transcriptional activation by p53 may play a role in a wide spectrum of biological processes (Fig. 8). Some of the target genes may mediate the antiproliferative effects of wt p53, whereas others may participate in distinct, equally important processes. Moreover, mutant forms of p53 may also possess a transcriptional regulatory potential, even though the target genes are likely to be very different from those of wt p53.

ACKNOWLEDGMENTS

We thank Dr. D. George for the gift of genomic *mdm2* plasmids and the 3T3-DM cell line. This work was supported in part by grant RO1 CA-40099 from the National Cancer Institute and by grants from the Minerva Foundation (Munich); the National Council for Research and Development, Israel; the DKFZ, Heidelberg; the Laub Foundation; and the Leo and Julia Forchheimer Center for Molecular Genetics.

REFERENCES

Aloni-Grinstein, R., I. Zan-Bar, I. Alboum, N. Goldfinger, and V. Rotter. 1993. Wild type p53 functions as a control protein in the differentiation pathway of the B-cell lineage. *Oncogene* **8:** 3297.

Barak, Y. and M. Oren. 1992. Enhanced binding of a 95-kDa protein to p53 in cells undergoing p53-mediated growth arrest. *EMBO J.* **11:** 2115.

Barak, Y., E. Gottlieb, T. Juven-Gershon, and M. Oren. 1994. Regulation of *mdm2* expression by p53: Alternative promoters produce transcripts with non-identical translation potential. *Genes Dev.* **8:** 1739.

Barak, Y., T. Juven, R. Haffner, and M. Oren. 1993. *mdm2* expression is induced by wild type-p53 activity. *EMBO J.* **12:** 461.

Berns, A. 1994. Is p53 the only real tumor suppressor gene? *Curr. Biol.* **4:** 137.

Cahilly-Snyder, L., T. Yang-Feng, U. Francke, and D.L. George. 1987. Molecular analysis and chromosomal mapping of amplified genes isolated from a transformed mouse 3T3 cell line. *Somatic Cell Mol. Genet.* **13:** 235.

Chen, C.-Y., J.D. Oliner, Q. Zhan, A.J. Fornace, B. Vogelstein, and M.B. Kastan. 1994. Interactions between p53 and Mdm2 in a mammalian cell cycle checkpoint pathway. *Proc. Natl. Acad. Sci.* **91:** 2684.

Chen, J.D., V. Marechal, and A.J. Levine. 1993. Mapping of the p53 and mdm-2 interaction domains. *Mol. Cell. Biol.* **13:** 4107.

Chen, P.-L., Y. Chen, R. Bookstein, and W.-H. Lee. 1990. Genetic mechanisms of tumor suppression by the human p53 gene. *Science* **250:** 1576.

Dittmer, D., S. Pati, G. Zambetti,, S. Chu., A.K. Teresky, M. Moore, C. Finlay, and A.J. Levine. 1993. Gain of function mutations in p53. *Nat. Genet.* **4:** 42.

Donehower, L.A. and A. Bradley. 1993. The tumor suppressor p53. *Biochim. Biophys. Acta* **1155:** 181.

El-Deiry, W.S., S.E. Kern, J.A. Pietenpol, K.W. Kinzler, and B. Vogelstein. 1992. Definition of a consensus binding site for p53. *Nat. Genet.* **1:** 45.

El-Deiry, W.S., T. Tokino, V.E. Velculescu, D.B. Levy, R. Parsons, J.M. Trent, D. Lin, W.E. Mercer, K.W. Kinzler, and B. Vogelstein. 1993. *WAF1*, a potential mediator of p53 tumor suppression. *Cell* **75:** 817.

Eliyahu, D., D. Michalovitz, S. Eliyahu, O. Pinhasi-Kimhi, and M. Oren. 1989. Wild-type p53 can inhibit oncogene-mediated focus formation. *Proc. Natl. Acad. Sci.* **86:** 8763.

Fakharzadeh, S.S., S.P. Trusko, and D.L. George. 1991. Tumorigenic potential associated with enhanced expression of a gene that is amplified in a mouse tumor cell line. *EMBO J.* **10:** 1565.

Finlay, C.A. 1993. The mdm-2 oncogene can overcome wild-type p53 suppression of transformed cell growth. *Mol. Cell. Biol.* **13:** 301.

Fritsche, M., C. Haessler, and G. Brandner. 1993. Induction of nuclear accumulation of the tumor-suppressor protein p53 by DNA-damaging agents. *Oncogene* **8:** 307.

Gottlieb, E., R. Haffner, T. Von-Ruden, E.F. Wagner, and M. Oren. 1994. Down regulation of wild-type p53 activity interferes with apoptosis of IL-3-dependent hematopoietic cells following IL-3 withdrawal. *EMBO J.* **13:** 1368.

Haines, D.S., J.E. Landers, L.J. Engle, and D.L. George. 1994. Physical and functional interaction between wild-type p53 and *mdm2* proteins. *Mol. Cell. Biol.* **14:** 1171.

Harper, J.W., G.R. Adami, N. Wei, K. Keyomarsi, and S.J. Elledge. 1993. The p21 Cdk-interacting protein Cip1 is a potent inhibitor of G1 cyclin-dependent kinases. *Cell* **75:** 805.

Harris, C.C. and M. Hollstein. 1993. Clinical implications of the p53 tumor suppressor gene. *N. Engl. J. Med.* **329:** 1318.

Hollstein, M., D. Sidransky, B. Vogelstein, and C. C. Harris. 1991. p53 mutations in human cancers. *Science* **253:** 49.

Hunter, T. 1993. Braking the cycle. *Cell* **75:** 839.

Juven, T., Y. Barak, A. Zauberman, D.L. George, and M. Oren. 1993. Wild type p53 can mediate sequence-specific transactivation of an internal promoter within the *mdm2* gene. *Oncogene* **8:** 3411.

Kastan, M.B., O. Onyekwere, D. Sidransky, B. Vogelstein, and R.W. Craig. 1991. Participation of p53 protein in the cellular response to DNA damage. *Cancer Res.* **51:** 6304.

Kastan, M.B., Q. Zhan, W.S. El-Deiry, F. Carrier, T. Jacks, W.V. Walsh, B.S. Plunkett, B. Vogelstein, and A.J. Fornace. 1992. A mammalian cell cycle checkpoint pathway utilizing p53 and GADD45 is defective in ataxia-telangiectasia. *Cell* **71:** 587.

Kozak, M. 1986. Point mutations define a sequence flanking the AUG initiator codon that modulates translation by eukaryotic ribosomes. *Cell* **44:** 283.

Kuerbitz, S.J., B.S. Plunkett, W.V. Walsh, and M.B. Kastan. 1992. Wild-type p53 is a cell cycle checkpoint determinant following irradiation. *Proc. Natl. Acad. Sci.* **89:** 7491.

Lane, D.P. 1992. p53, guardian of the genome. *Nature* **358:** 15.

Levine, A.J., J. Momand, and C.A. Finlay. 1991. The p53 tumour suppressor gene. *Nature* **351:** 453.

Levy, N., E. Yonish-Rouach, M. Oren, and A. Kimchi. 1993. Complementation by wild type p53 of interleukin-6 effects on M1 cells: Induction of cell cycle exit and relationship to c-*myc* suppression. *Mol. Cell. Biol.* **13:** 7942.

Lu, X. and D.P. Lane. 1993. Differential induction of transcriptionally active p53 following UV or ionizing radiation—Defects in chromosome instability syndromes? *Cell* **75:** 765.

Maltzman, W. and L. Czyzyk. 1984. UV irradiation stimulates levels of p53 cellular tumor antigen in nontransformed mouse cells. *Mol. Cell. Biol.* **4:** 1689.

Michalovitz, D., O. Halevy, and M. Oren. 1990. Conditional inhibition of transformation and of cell proliferation by a temperature-sensitive mutant of p53. *Cell* **62:** 671.

———. 1991. p53 mutations: Gains or losses? *J. Cell. Biochem.* **45:** 22.

Momand, J., G.P. Zambetti, D.C. Olson, D. George, and A.J. Levine. 1992. The mdm-2 oncogene product forms a complex with the p53 protein and inhibits p53-mediated *trans*-activation. *Cell* **69:** 1237.

Oliner, J.D., J.A. Pietenpol, S. Thiagalingam, J. Gvuris, K.W. Kinzler, and B. Vogelstein. 1993. Oncoprotein MDM2 conceals the activation domain of tumour suppressor-p53. *Nature* **362:** 857.

Oren, M. 1992. p53—The ultimate tumor suppressor gene? *FASEB J.* **6:** 3169.

———. 1994. Relationship of p53 to the control of apoptotic cell death. *Semin. Cancer Biol.* **5:** 305.1.

Otto, A. and W. Deppert. 1993. Upregulation of mdm-2 expression in Meth a tumor cells tolerating wild-type p53. *Oncogene* **8:** 2591.

Perry, M.E., J. Piette, J.A. Zawadzki, D. Harvey, and A.J. Levine. 1993. The mdm-2 gene is induced in response to UV light in a p53-dependent manner. *Proc. Natl. Acad. Sci.* **90:** 11623.

Picksley, S.M. and D.P. Lane. 1993. The p53-mdm2 autoregulatory feedback loop—A paradigm for the regulation of growth control by p53? *BioEssays* **15:** 689.

Prives, C. and J.J. Manfredi. 1993. The p53 tumor suppressor protein: Meeting review. *Genes Dev.* **7:** 529.

Rotter, V., O. Foord, and N. Navot. 1993. In search of the functions of the normal p53 protein. *Trends Cell Biol.* **3:** 46.

Shaulian, E., A. Zauberman, D. Ginsberg, and M. Oren. 1992. Identification of a minimal transforming domain of p53—Negative dominance through abrogation of sequence-specific DNA binding. *Mol. Cell. Biol.* **12:** 5581.

Shaulian, E., A. Zauberman, J. Milner, E.A. Davies, and M. Oren. 1993. Tight DNA binding and oligomerization are dispensable for the ability of p53 to transactivate target genes and suppress transformation. *EMBO J.* **12:** 2789.

Shaulsky, G., N. Goldfinger, A. Peled, and V. Rotter. 1991. Involvement of wild-type p53 in pre-B cell differentiation in vitro. *Proc. Natl. Acad. Sci.* **88:** 8982.

Weintraub, H., S. Hauschka, and S.J. Tapscott. 1991. The MCK enhancer contains a p53 responsive element. *Proc. Natl. Acad. Sci.* **88:** 4570.

White, W. 1993. Death-defying acts: A meeting review on apoptosis. *Genes Dev.* **7:** 2277.

Wolf, D. and V. Rotter. 1984. Inactivation of p53 gene expression by an insertion of Moloney murine leukemia virus-like DNA sequences. *Mol. Cell. Biol.* **4:** 1402.

Wolf, D., N. Harris, and V. Rotter. 1984. Reconstitution of p53 expression in a nonproducer Ab-MuLV-transformed cell line by transfection of a functional p53 gene. *Cell* **38:** 119.

Wu, X.W., J.H. Bayle, D. Olson, and A.J. Levine. 1993. The p53 mdm-2 autoregulatory feedback loop. *Genes Dev.* **7:** 1126.

Yonish-Rouach, E., D. Resnitzky, J. Lotem, L. Sachs, A. Kimchi, and M. Oren. 1991. Wild-type p53 induces apoptosis of myeloid leukaemic cells that is inhibited by interleukin-6. *Nature* **352:** 345.

Yonish-Rouach, E., D. Grunwald, S. Wilder, A. Kimchi, E. May, J.J. Lawrence, P. May, and M. Oren. 1993. p53-mediated cell death—Relationship to cell cycle control. *Mol. Cell. Biol.* **13:** 1415.

Zambetti, G.P. and A.J. Levine. 1993. A comparison of the biological activities of wild-type and mutant-p53. *FASEB J.* **7:** 855.

Zambetti, G.P., J. Bargonetti, K. Walker, C. Prives, and A.J. Levine. 1992. Wild-type p53 mediates positive regulation of gene expression through a specific DNA sequence element. *Genes Dev.* **6:** 1143.

Zauberman, A., Y. Barak, N. Ragimov, N. Levy, and M. Oren. 1993. Sequence-specific DNA binding by p53—Identification of target sites and lack of binding to p53-Mdm2 complexes. *EMBO J.* **12:** 2799.

Zhan, Q.M., F. Carrier, and A.J. Fornace. 1993. Induction of cellular p53 activity by DNA-damaging agents and growth arrest. *Mol. Cell. Biol.* **13:** 4242.

Zhan, Q., K.A. Lord, I. Alamo, M.C. Hollander, F. Carrier, D. Ron, K.W. Kohn, B. Hoffman, D.A. Liebermann, and A.J. Fornace, Jr. 1994. The gadd and MyD genes define a novel set of mammalian genes encoding acidic proteins that synergistically suppress cell growth. *Mol. Cell. Biol.* **14:** 2361.

E6-AP Directs the HPV E6-dependent Inactivation of p53 and Is Representative of a Family of Structurally and Functionally Related Proteins

J.M. HUIBREGTSE,* M. SCHEFFNER,† AND P.M. HOWLEY*
*Department of Pathology, Harvard Medical School, Boston, Massachusetts 02115;
†Angewandte Tumorvirologie, Deutsches Krebsforschungszentrum, D-69120 Heidelberg, Germany

There are now several examples where experimental and/or epidemiologic data have implied a causative role for viruses in human cancer. Human papillomavirus (HPV) DNA has been found in approximately 90% of cervical cancers. Only a subset of the approximately 20 distinct HPV types that infect the anogenital tissues, however, are generally found in cervical cancers. This has led to the classification of anogenital-specific HPV types into "high risk" and "low risk" groups (DeVilliers 1989; zur Hausen 1991). This classification is reflected in cell culture in that only the cloned DNA of the high-risk HPVs (HPV-16 and -18, for example) is capable of immortalizing human primary genital keratinocytes, the normal host cell (Dürst et al. 1987; Münger et al. 1989). The E6 and E7 genes are the viral genes that are generally expressed in HPV-containing cancers and are together necessary and sufficient for immortalization of primary keratinocytes (Hawley-Nelson et al. 1989; Münger et al. 1989).

The HPV-16 E6 and E7 proteins are both small (151 and 98 amino acids, respectively), zinc-binding proteins. An initial understanding of the biochemical basis of the transforming activities of the E6 and E7 oncoproteins came by analogy with other DNA tumor viruses. The transforming proteins of several DNA tumor viruses, including the high-risk HPV E6 and E7 proteins, work in part by targeting and inactivating the functions of a common group of key proteins that negatively regulate cellular proliferation. The two best characterized of these are the retinoblastoma protein (pRB) and p53. The adenovirus E1A protein, the large-T antigen (TAg) of SV40, and the HPV E7 proteins share an amino acid motif which is involved in binding to and inactivating the cell cycle inhibitory activity of hypophosphorylated pRB (DeCaprio et al. 1988; Phelps et al. 1988; Whyte et al. 1988; Dyson et al. 1989). Similarly, p53 is functionally inactivated by interaction with SV40 TAg, the 55-kD E1B protein of adenovirus type 5 (Ad5), and the E6 protein of HPV-16 and -18 (Lane and Crawford 1979; Linzer and Levine 1979; Sarnow et al. 1982; Werness et al. 1990). These proteins abrogate the transcriptional *trans*-activation activity of wild-type p53, which is the biochemical activity of p53 most tightly correlated with growth or tumor suppression (Mietz et al. 1992; Yew and Berk 1992; Pietenpol et al. 1994).

Although SV40 TAg, E1B, and the high-risk HPV E6 proteins all functionally inactivate p53, the mechanisms by which they do so are distinct. In normal primary cells, the metabolic half-life of p53 is relatively short, between 20 and 40 minutes in most cell types. In SV40-immortalized cells, the half-life is greatly extended and the steady-state level of p53 is elevated. Ad5 E1B has a similar effect on p53 stability and overall concentration. In contrast, the level of p53 in HPV-containing cervical carcinoma cell lines and HPV-immortalized keratinocytes is generally lower than the level seen in primary cells, and the half-life is decreased. p53 in normal human keratinocytes in culture has a half-life of approximately 1–4 hours, whereas in HPV-immortalized keratinocytes, the half-life is 15–30 minutes (Hubert and Lambert 1993). A biochemical basis for this observation was suggested by in vitro experiments which showed that E6 could stimulate the degradation of p53 when the proteins were translated and mixed in a rabbit reticulocyte lysate (RRL) system (Scheffner et al. 1990). The E6-dependent degradation of p53 was shown to utilize the cellular ubiquitin proteolysis system, the components of which are abundant in RRL. Therefore, a model for high-risk HPV E6 protein function in carcinogenesis is that E6 promotes cellular proliferation through targeted degradation of the p53 tumor suppressor.

In the course of studying the requirements for association of high-risk HPV E6 proteins with p53, it was discovered that in the original coimmunoprecipitation experiments, which utilized RRL-translated proteins, a component that was essential for the interaction was contributed by the reticulocyte lysate (Huibregtse et al. 1991). It was shown that the factor was a protein of approximately 100 kD which is present in all mammalian cell extracts that have been examined, but not in plant or yeast extracts. Since the protein can stably interact with high-risk HPV E6 proteins in the absence of p53, it was designated E6-AP for E6-*A*ssociated *P*rotein. Neither E6 nor E6-AP by itself can stably associate with p53; only the ternary complex with p53 is stable. E6-AP was purified, and a cDNA encoding human E6-AP has been cloned (Huibregtse et al. 1993a). A structure/function analysis has defined three functional domains of E6-AP (Fig. 1) (Huibregtse et al. 1993b). The E6-binding site has been localized to an

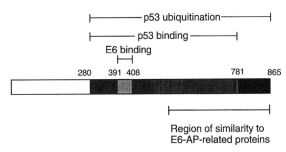

Figure 1. Functional domains of E6-AP. The minimal regions necessary for stable interaction with E6 and p53 are indicated, as well as the minimal region necessary to stimulate ubiquitination of p53 (Huibregtse et al. 1993b). The region of similarity between E6-AP and several other proteins in current databases is indicated.

18-amino-acid segment in the central portion of the molecule. The minimal region necessary for association with p53 spans the E6-binding domain and covers an approximately 500-amino-acid region of the protein. Finally, the minimal region necessary for ubiquitination of p53 includes determinants within the carboxy-terminal 84 amino acids. The primary sequence itself did not provide insight into the normal cellular function of E6-AP, since it is not similar to any proteins of defined function. It was, however, similar to several proteins of unknown function over the carboxyl terminus, as discussed below.

The ubiquitin proteolysis system is a major pathway for the intracellular degradation of proteins (Hershko and Ciechanover 1992). Ubiquitin is a 76-amino-acid protein that is highly conserved among all eukaryotic organisms. The hallmark of the pathway is the covalent linkage of the carboxyl terminus of ubiquitin to the amino group of lysine side chains, forming a stable isopeptide bond. Additional ubiquitin moieties can be linked sequentially to each other via a lysine residue at amino acid 48 of ubiquitin. Multi-ubiquitinated proteins are then specifically recognized and degraded by the proteasome, a large multisubunit complex of proteins. Although there is genetic and biochemical evidence that the ubiquitin system plays a major role in selective protein degradation, only a few cellular substrates have been identified so far, including cyclin B (Glotzer et al. 1991) and the yeast MATα2 transcription repressor (Hochstrasser et al. 1991; Rechsteiner 1991).

Three classes of proteins are involved in targeting a substrate protein for ubiquitination: the E1 ubiquitin-activating enzyme, the E2 ubiquitin-conjugating enzymes, and the E3 ubiquitin protein ligases. First, the E1 ubiquitin-activating enzyme, in a reaction that requires ATP, catalyzes the formation of a high-energy thioester between the carboxyl terminus of ubiquitin and the thiol group of a cysteine side chain on the E1 enzyme itself. The activated ubiquitin is then transferred to a thiol group of the active-site cysteine of an E2 ubiquitin-conjugating enzyme. There are several distinct E2 enzymes (at least 10 in *Saccharomyces cerevisiae*), all of which appear to be recognized by a single type of E1 enzyme. Most recently proposed models have suggested that the E2 enzymes directly catalyze formation of the stable isopeptide bond between ubiquitin and the substrate protein, the substrate specificity being provided by the E2 itself and, in some cases, by the activity of E3 ubiquitin protein ligases. The E3 proteins until now have been poorly characterized and have been simply defined as activities that are required by an E2 enzyme to ubiquitinate a specific substrate. On the basis of work with yeast UBR1, the E3 involved in N-end rule ubiquitination, E3s have been proposed to contain binding sites for both E2 enzymes and substrates, thereby facilitating the ubiquitination of the substrate by the E2 enzyme (Dohmen et al. 1991).

To fully characterize the E6/E6-AP-dependent ubiquitination of p53, which does not involve the N-end rule pathway, we have fully reconstituted the reaction in vitro with purified and/or recombinant proteins (Scheffner et al. 1993). This has revealed that E6-AP has E3 ubiquitin protein ligase activity, even in the absence of E6. Furthermore, E6-AP is shown to form a thioester with ubiquitin (M. Scheffner and J.M. Huibregtse, in prep.), suggesting that the order of transfer of ubiquitin to p53 is from E1 to E2 to E6-AP, in the form of a high-energy thioester, and finally to p53 in the form of a stable isopeptide linkage. This represents a previously undescribed mechanism of action for an E3. We have identified a cysteine near the carboxyl terminus of E6-AP that is critical for ubiquitin thioester formation. Interestingly, this cysteine and residues surrounding it are highly conserved among a group of proteins related to E6-AP, all of unknown function. This suggests that these E6-AP-related proteins are also E3 ubiquitin protein ligases.

MATERIALS AND METHODS

Plasmids. Bacterial expression plasmids encoding wild-type (wt) p53, the 75-kD form of E6-AP, and ubiquitin, respectively, were constructed using the pGEX 2TK vector. The plasmids used for expression of wt p53 and E6-AP have been described previously (Werness et al. 1990; Scheffner et al. 1992; Huibregtse et al. 1993a). Genomic DNA of *S. cerevisiae* (Promega) was used for polymerase chain reaction (PCR) of the ubiquitin cDNA. The pGEX 2T HPV-16 E6 expression plasmid has been described previously (Huibregtse et al. 1991). The cysteine residue at position 833 of the E6-AP open reading frame (Huibregtse et al. 1993a) was altered to either serine or alanine by PCR-directed mutagenesis.

Protein expression. GST fusion proteins were expressed in *Escherichia coli* DH5α and affinity-purified on glutathione-Sepharose (Pharmacia). Bound proteins were eluted with 10 mM glutathione. p53 and ubiquitin fusion proteins were radioactively labeled while bound to glutathione-Sepharose using protein kinase A (Sigma) in the presence of $[\gamma\text{-}^{32}P]ATP$ (Kaelin et al. 1992).

E1, At UBC8, and At UBC1 were expressed in *E. coli* BL21 using the pET expression system as described previously (Sullivan and Vierstra 1991; Hatfield and Vierstra 1992; Girod et al. 1993). For ubiquitination assays, E1 was either purified by ubiquitin affinity chromatography as described previously (Haas and Bright 1988) or partially purified by anion exchange chromatography as follows. Crude bacterial extracts were loaded onto a MonoQ column (Pharmacia), the column was washed with 25 mM Tris-HCl (pH 7.5), 100 mM NaCl, 1 mM dithiothreitol (DTT), and bound proteins were eluted with 300 mM NaCl. As the source of UBC8 and UBC1 in the ubiquitination assays, either crude bacterial extracts containing UBC8 or UBC1 were used, or UBC8 and UBC1 were purified by ubiquitin affinity chromatography.

The construction of the recombinant baculovirus expressing the 95-kD form of E6-AP has been described previously (Huibregtse et al. 1993b). For expression of E6-AP, a 10-cm plate of Sf9 cells at 50% confluency was infected at a high moi with the recombinant baculovirus expressing E6-AP. Protein extracts were prepared 48 hours after infection, and E6-AP was partially purified by MonoQ chromatography as described below.

Protein extracts and chromatography. The C-33A cell line was obtained from the American Type Culture Collection. Protein extracts were prepared from confluent cells in 1% Nonidet P-40, 20 mM Tris-HCl (pH 8.0), 100 mM NaCl, 1 mM DTT, 0.01% phenylmethylsulfonyl fluoride (PMSF), and 1 μg/ml aprotinin and leupeptin.

For fractionation of the components involved in E6-mediated ubiquitination, 0.8 ml RRL was chromatographed on a 1-ml MonoQ column. Columns were washed with 25 mM Tris-HCl (pH 7.6), 125 mM NaCl, 1 mM DTT, and bound proteins were eluted with a linear salt gradient (30 ml, 1-ml fractions) from 125 mM NaCl to 400 mM NaCl in 25 mM Tris-HCl (pH 7.6), 1 mM DTT. Fractions containing E6-AP were determined by the ability of these fractions to reconstitute binding of HPV-16 E6 translated in wheat germ extract (WGE) to bacterially expressed p53 (Huibregtse et al. 1991). The same procedure was used for the partial purification of E6-AP expressed in Sf9 cells.

For further purification of the mammalian homolog of At UBC8, the flowthrough of the MonoQ column was dialyzed against 50 mM HEPES (pH 6.5), 20 mM NaCl, 1 mM DTT. The dialyzed flowthrough was then chromatographed on a Bio-Rad CM column. Bound proteins were eluted in 1-ml fractions with a 20-ml linear salt gradient from 50 mM NaCl to 300 mM NaCl in 50 mM HEPES (pH 6.5), 1 mM DTT. The fraction containing the E2 activity was further purified by ubiquitin affinity chromatography as described previously (Haas and Bright 1988) in the presence of bacterially expressed E1.

For fractionation of radioactively labeled C-33A cells, extract from two 10-cm plates was chromatographed on MonoQ. Bound proteins were sequentially eluted with 0.6 ml of 200 mM NaCl, 300 mM NaCl, 400 mM NaCl, and 500 mM NaCl, respectively, in 25 mM Tris-HCl (pH 7.6), 1 mM DTT.

Ubiquitination assays. Ubiquitination assays using ^{32}P-labeled glutathione-*S*-transferase (GST)-p53 or ^{35}S-labeled cell extract as substrates (~5–10 ng) were performed in 150-μl volumes under the following conditions: 25 mM Tris-HCl (pH 7.6), 120 mM NaCl, 3 mM DTT, 2 mM ATPγS, 1 mM MgCl$_2$, and 6 μg ubiquitin (Sigma). After 3 hours incubation at 25°C, the reactions were stopped by adding 50 mM Tris-HCl (pH 6.8), 100 mM DTT, 2% SDS, and 10% glycerol (final concentrations), and boiling for 5–10 minutes at 97°C. Total reaction mixtures were then electrophoresed on SDS-polyacrylamide gels, and radioactively labeled fusion proteins were visualized by autoradiography. For identification of the MonoQ RRL fractions (Fig. 2) that are necessary to reconstitute the ubiquitination of GST-p53, 40 μl of each fraction was added to reaction mixtures containing 20 μl of the E6-AP-containing fraction. Ubiquitination reactions with bacterially expressed proteins contained approximately 20 ng of E1, 20 ng of UBC8, 50 ng of the 75-kD form of E6-AP, and 50 ng of bacterially expressed GST-HPV-16 E6.

E6-AP-induced ubiquitination of ^{35}S-labeled cellular proteins was studied using 200 ng of baculovirus-expressed E6-AP, 8 μg of GST-ubiquitin, 20 ng of E1, 20 ng of UBC8, and 20 μl of the MonoQ fractions obtained from ^{35}S-labeled C-33A cells as described above. After 3 hours, glutathione-Sepharose was added and ubiquitinated proteins were precipitated. Precipitated proteins were separated by SDS-PAGE and visualized by fluorography.

Thioester assays. The construction of the recombinant baculoviruses expressing the 95-kD form or the 76-kD form of E6-AP has been described previously (Huibregtse et al. 1993b). Protein extracts were prepared 48 hours after infection and fractionated by chromatography on a MonoQ column. Since Sf9 cells do not contain any detectable E6-AP activity, fractions containing E6-AP were determined by the ability of these fractions to reconstitute ubiquitination of GST-p53 (wt). Thioester reactions contained approximately 1 μg ^{32}P-labeled GST-ubiquitin, 5 ng of E1, and 200 ng of At UBC8 in 20 mM Tris-HCl (pH 7.6), 50 mM NaCl, 4 mM ATP, 10 mM MgCl$_2$, and 0.2 mM DTT. Reactions were terminated either by incubating the mixtures for 15 minutes at 30°C in 50 mM Tris-HCl (pH 6.8), 2% SDS, 4 M urea, 10% glycerol or by boiling the mixtures in the buffer above containing 100 mM DTT instead of urea. The whole reaction mixtures were separated on 10% SDS polyacrylamide gels at 4°C.

RESULTS

As shown previously, all of the components necessary to facilitate HPV-16 E6-dependent degradation of

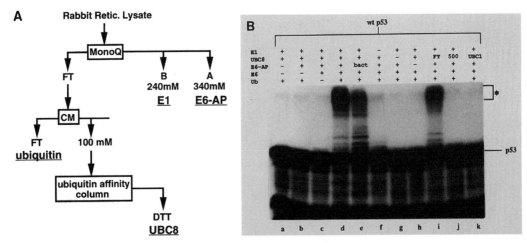

Figure 2. (*A*) Chromatographic properties of components in RRL necessary to reconstitute the E6/E6-AP-dependent ubiquitination of p53. The chromatographic procedures are described in Materials and Methods. The necessary factors and the recombinant proteins that they could be replaced with are indicated in bold and underlined. (*B*) Reconstitution of the E6-mediated ubiquitination of p53 with bacterially expressed proteins. A fusion protein was used consisting of wt p53 linked by a protein kinase A (PKA) site to the carboxyl terminus of GST. The fusion protein was expressed in *E. coli*, purified by affinity chromatography, and radioactively labeled by PKA in the presence of [γ-^{32}P]ATP. The ubiquitination reactions were carried out as described in Materials and Methods. E1, UBC8, and HPV-16 E6 were expressed and purified from *E. coli*. The presence or absence of these proteins in each reaction is indicated. In some cases UBC8 was replaced by UBC1 (lane *k*) or by the MonoQ flowthrough of RRL (lane *i*) or by the 500 mM NaCl MonoQ eluate of RRL (lane *j*). As a source of E6-AP, either the 75-kD form expressed in *E. coli* was used (lane *e*) or fraction A from RRL (where indicated, except lane *e*). (Reprinted, with permission, from Scheffner et al. 1993 [copyright Cell Press].)

p53 are present in RRL (Scheffner et al. 1990). To fully characterize the components required for this reaction, RRL was fractionated and the reaction was reconstituted with purified or recombinant factors. Figure 2A shows a schematic of the chromatographic properties of the necessary components. Initially, three fractions were identified that were separable by chromatography of RRL on a MonoQ column. The flowthrough fraction contained a necessary activity, and two salt fractions (A and B, 340 mM and 240 mM NaCl fractions, respectively) each contained an activity. From our previous purification of E6-AP, it seemed likely that fraction A contained E6-AP, and as expected, fraction A could be substituted with recombinant E6-AP, produced either in bacteria or insect cells (which lack endogenous E6-AP). The E1 ubiquitin-activating enzyme has been previously reported to elute from DEAE columns at a salt concentration close to that of fraction B. Since E1 was predicted to be required for reconstitution of p53 ubiquitination, bacterially expressed E1 was tested and shown to substitute for fraction B. Thus, two of the three MonoQ activities were identified.

The other fraction necessary to reconstitute E6-dependent p53 ubiquitination was in the MonoQ flowthrough. The flowthrough was expected to contain free ubiquitin; however, purified ubiquitin could not substitute for the flowthrough fraction, indicating that another factor was present. The MonoQ flowthrough fraction was further fractionated by CM cation exchange chromatography. The flowthrough of the CM column contained an activity that could substitute for commercially available purified ubiquitin, but another necessary factor bound to the column and eluted with 100 mM NaCl. An E2 ubiquitin conjugation enzyme was expected to be necessary for ubiquitination; however, all previously identified E2s had been shown to bind to anion exchange columns, suggesting that if the CM fraction factor was an E2, it was of a novel variety. The factor in the 100 mM NaCl fraction did bind to a ubiquitin affinity column in the presence of ATP and recombinant E1 enzyme and could be eluted with DTT, a characteristic of previously isolated E2 proteins. At the time this work was being carried out, there was a report of an E2 called At UBC8 that had been cloned and isolated from *Arabidopsis thalania* (Sullivan and Vierstra 1991), which had similar chromatographic properties to the putative E2 in the CM eluate, and indeed, bacterially expressed At UBC8 was shown to substitute for the CM fraction. Figure 2B shows the complete in vitro reconstitution of E6-dependent p53 ubiquitination using recombinant factors and bacterially expressed ^{32}P-labeled GST-p53 (Scheffner et al. 1993). E1, a specific E2 (At UBC8), E6-AP, HPV-16 E6, ubiquitin, and ATP were necessary and sufficient to stimulate the ubiquitination of p53. At UBC8 could not be replaced with UBC1, a previously characterized E2 that binds to anion exchange columns. We have subsequently identified and cloned a novel human E2, called hUbc5, which has similar chromatographic properties to At UBC8 and can function in E6-AP-dependent ubiquitination of p53 (Scheffner et al. 1994a).

Since E3 activities have been operationally defined as activities that are required for ubiquitination of a substrate by an E2, an implication of the reconstitution experiments was that the combination of E6 and E6-AP was acting as an E3 activity. Since the normal function of E6-AP was not known, experiments were carried out to determine if E6-AP has E3 activity in the absence of HPV E6 protein. Figure 3 shows one type of experiment that demonstrates that E6-AP has E3 activity. A [^{35}S]methionine-labeled extract from C-33A cells was passed over a MonoQ column and bound proteins were sequentially eluted with 200 mM, 300 mM, 400 mM, and 500 mM NaCl. Bacterially expressed E1, UBC8, and unlabeled GST-ubiquitin (which can substitute for free ubiquitin in ubiquitination reactions) were added to the C-33A fractions, and the mixtures were incubated in the presence or absence of partially purified baculovirus-expressed E6-AP. After 3 hours, glutathione-Sepharose was added and ubiquitinated proteins were selectively precipitated through the GST portion of the GST-ubiquitin protein. Bound proteins were released by boiling in SDS-containing buffer, separated by SDS-PAGE, and visualized by fluorography. Figure 3 shows the results using the 200 mM NaCl fraction as substrate. This fraction did not contain significant endogenous E3 activity. A significant increase in the amount of ubiquitinated proteins was detected on addition of MonoQ-purified E6-AP, whereas addition of the corresponding MonoQ fraction from either uninfected Sf9 cells or Sf9 cells infected with wt baculovirus had no stimulating effect. This demonstrates that E6-AP, in the absence of E6, is able to stimulate the ubiquitination of cellular proteins (so far undefined), suggesting that the normal function of E6-AP is that of an E3 protein or a component of an E3 activity (Scheffner et al. 1993).

Only one other gene encoding an E3 enzyme has been previously cloned: UBR1 of *S. cerevisiae* (Bartel et al. 1990). This protein, which has no significant sequence similarity to E6-AP, has separate binding sites for E2 proteins and substrates. We have not been able to demonstrate direct binding of UBC8 to E6-AP, nor have we observed an effect of E6-AP on the ability of UBC8 to form a thioester with ubiquitin (data not shown). In the course of these experiments, however, it was noted that perhaps E6-AP itself was forming a thioester with ubiquitin. This was demonstrated in the following manner. Baculovirus-expressed E6-AP was partially purified on a MonoQ column. The MonoQ fractions were mixed with bacterially expressed At UBC8, E1, ATP, and ^{32}P-labeled GST-ubiquitin. After incubation for 5 minutes at room temperature, SDS loading buffer was added to the samples which either did or did not contain DTT, and the samples were separated on SDS-polyacrylamide gels. A band migrating with an apparent molecular mass of approximately 130-kD was seen in fractions containing E6-AP when the sample was run without DTT (Fig. 4). 130 kD is the approximate mass predicted for E6-AP (95 kD) linked to a single GST-Ub molecule (30 kD). This band was not seen when DTT was added to the sample buffer, indicating that the linkage was covalent but unstable to

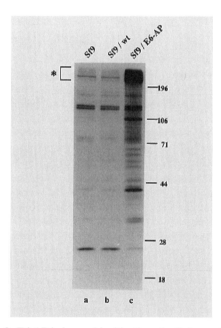

Figure 3. E6-AP induces ubiquitination of cellular proteins in the absence of HPV-16 E6. Radioactively labeled cellular proteins from metabolically ^{35}S-labeled C-33A cells were chromatographed on a MonoQ column, and bound proteins were eluted with 200 mM NaCl. Ubiquitin was expressed as a GST fusion protein in *E. coli*. Bacterially expressed E1, UBC8, and the ubiquitin fusion protein (GST-Ub) were added to the 200 mM NaCl fraction of the C-33A extract in the absence or in the presence of baculovirus-expressed E6-AP partially purified by anion exchange chromatography. As control, corresponding fractions of extracts from uninfected Sf9 cells or from wt virus-infected Sf9 cells were used. After 3 hr at 25°C, glutathione-Sepharose was added, and the mixture was rotated for an additional 30 min. The beads were collected and bound proteins were released by boiling in SDS-containing buffer. The supernatants were separated on 10% SDS-polyacrylamide gels, and proteins were visualized by fluorography. Bands containing presumably ubiquitinated proteins are indicated on the left, the running positions of molecular-mass markers are indicated on the right. (Reprinted, with permission, from Scheffner et al. 1993 [copyright Cell Press].)

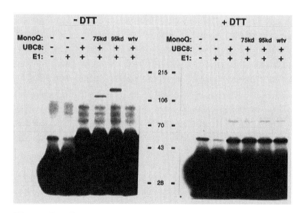

Figure 4. E6-AP forms a thioester with ubiquitin. Thioester reactions contained ^{32}P-labeled GST-ubiquitin, ATP, and, as indicated, E1, *Arabidopsis thaliana* (At) UBC8, and MonoQ fractions containing the 75-kD E6-AP, 95-kD E6-AP, or an equivalent fraction from Sf9 cells infected with wt baculovirus (wtv). After 5 min at 25°C, reactions were stopped in the absence (*left*) or presence (*right*) of DTT and subjected to SDS-PAGE followed by autoradiography. Due to the low amount of E1 used, the formation of a thioester between ubiquitin and E1 was not detectable.

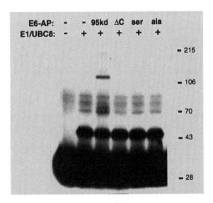

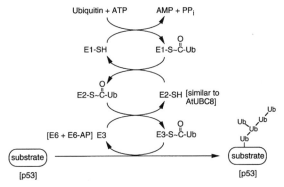

Figure 5. A cysteine residue in the carboxyl terminus of E6-AP is essential for thioester formation. The 95-kD form of E6-AP, a truncated form of the 95-kD E6-AP lacking the carboxy-terminal 84 amino acids (ΔC), or mutants of the 95-kD E6-AP in which the cysteine at position 833 was altered to either serine (ser) or alanine (ala) were expressed and partially purified from insect cells. Approximately equivalent amounts of the different E6-APs were tested in the thioester assay as described in Fig. 4. Reactions were stopped in the absence of a reducing agent.

Figure 6. A ubiquitin thioester cascade model for E6/E6-AP-dependent ubiquitination of p53. Ubiquitin is activated by the E1 enzyme in an ATP-dependent step and forms a high-energy thioester with E1. E1 transfers the ubiquitin to a UBC8-like E2 ubiquitin-conjugating enzyme (hUbc5), maintaining a thioester linkage between ubiquitin and a cysteine on the E2. E2 then transfers ubiquitin to a cysteine on E6-AP, again in the form of a thioester. E6-AP, in conjunction with HPV-16 E6, then recognizes p53 and catalyzes the formation of an isopeptide bond between the carboxyl terminus of ubiquitin and a lysine side chain of p53.

reducing agents, as are the ubiquitin thioester bonds with E1 and E2 proteins. A further indication that the 130-kD band represented a ubiquitin adduct with E6-AP was that a 76-kD amino-terminally truncated form of E6-AP, which behaves like wt E6-AP in p53 ubiquitination, gave rise to a band of 110 kD in the same assay (Fig. 4). This was also labile to reducing agents.

A carboxy-terminally truncated E6-AP, lacking the last 84 amino acids, does not function in p53 ubiquitination (Fig. 1) (Huibregtse et al. 1993b). Consistent with this, the baculovirus-expressed carboxy-truncated E6-AP did not form an adduct with GST-ubiquitin (Fig. 5). Given that the carboxy-terminal 84 amino acids of E6-AP are important for its E3 activity and its ability to form a thioester, and that there is only a single cysteine residue within this region (33 amino acids from the end of the protein), we mutated this cysteine to either a serine or an alanine and tested it for its ability to ubiquitinate p53 and its ability to form a thioester with ubiquitin. Both of these activities were completely abolished by the mutation (Fig. 5), even though the mutated proteins were expressed at comparable levels and retained the ability to stably associate with p53 (not shown). This demonstrates the requirement for this cysteine in ubiquitin thioester formation.

DISCUSSION

The study of the inactivation of the p53 tumor suppressor by the product of the HPV-16 E6 oncogene has provided insights into the mechanisms of the ubiquitin proteolysis system. The data presented here demonstrate that E6-AP is an E3 ubiquitin protein ligase and that it has the ability to form a thioester with ubiquitin. This suggests a model for ubiquitination in which ubiquitin, after first being activated by the E1 enzyme, is transferred to an E2 enzyme, then to an E3, which then catalyzes the final isopeptide bond formation between ubiquitin and the targeted substrate protein (Fig. 6). This model differs from most previously proposed models in that the E3, rather than the E2, is proposed to directly catalyze the isopeptide bond formation between ubiquitin and a lysine side chain of the substrate. Previous models have proposed that E3 proteins function in substrate recognition by binding both E2 and substrate, so that the E2 can ubiquitinate the substrate. It is quite possible that there are distinct families of E3s which follow one model or the other, or which function by yet other mechanisms. This is a strong possibility since E3s, as mentioned earlier, have been defined very broadly as any activity that stimulates the E2-dependent ubiquitination of a substrate.

A major challenge in the ubiquitin field has been to determine how substrate specificity is controlled. E3s have been invoked as key players in this control, and this is supported by studies with E6-AP. There is little E6-AP-dependent ubiquitination of p53 in vitro in the absence of E6, although there is some evidence that p53 degradation is mediated by the ubiquitin system (Ciechanover et al. 1991). It is therefore possible that E6-AP is not involved in E6-independent ubiquitination of p53 or that additional cellular proteins are needed to allow it to function in p53 ubiquitination. Such cellular E6 analogs might normally function by directing the substrate specificity of E3s. The E6-binding site of E6-AP, a small region in the central portion of the molecule (Fig. 1), might also be the site of interaction of such proteins.

There are still several open questions regarding the E6/E6-AP recognition of p53. It is not known whether E6 activates E6-AP so that it binds p53, or vice versa, or if both proteins have contact sites for p53. In addition, we do not know how the E2 protein recognizes E6-AP

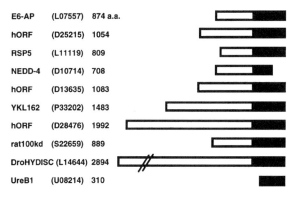

Figure 7. Proteins similar to E6-AP. Protein or gene designations are given with their GenBank accession numbers. The size of each protein is given in number of amino acids, and the schematic indicates their relative sizes and the region of each protein that is similar to E6-AP. The blackened area of E6-AP represents approximately 350 amino acids.

or the E6/E6-AP/p53 complex. UBC8 can stimulate E6-AP-dependent ubiquitination of cellular proteins in the absence of E6, so presumably the E2 protein is able to recognize E6-AP. A stable interaction with E6-AP, however, has not yet been demonstrated.

There are now nine DNA sequences in data banks that encode proteins with similarity to E6-AP (Fig. 7). Little is known of the normal biochemical function of most of these proteins. The three human open reading frames (hORFs) were identified through sequencing of random cDNA clones, and one of the two yeast sequences, YKL162, was identified in a genome sequencing project (Pascolo et al. 1992). RSP5 (also called SYGP-ORF41 in the GenBank entry) was isolated in a yeast genetic screen, but its function is unknown (F. Winston, pers. comm.). The cDNA encoding the rat 100-kD protein, which was sequenced in the course of a search for an unrelated protein (Müller et al. 1992), has no known biochemical activity but has regions of similarity to certain RNA-binding proteins, in regions outside the region of similarity to E6-AP. UreB1 is another rat protein, which is reported to be a DNA-binding protein that affects transcription of the dynorphin gene in brain (Gu et al. 1994). NEDD-4 was identified by sequencing of murine mRNAs expressed in embryonal neuronal precursor cells but down-regulated in adult brain (Kumar et al. 1992). To our knowledge, there is nothing published in a peer-reviewed journal regarding the Drosophila hyperplastic discs protein (DroHYDISC).

As shown in Figure 7, the regions of similarity correspond to the carboxyl terminus of E6-AP and the carboxyl terminus of the E6-AP-related proteins. Over this approximately 350-amino-acid region, the degree of similarity ranges from a low of 20% to a high of 40%. Figure 8 shows the amino acid sequences aligned that correspond to the carboxy-terminal 50 amino acids of E6-AP. Most interestingly, the cysteine that is required for thioester formation is conserved at a very similar position relative to the carboxyl terminus of each protein. Other amino acids in the immediate vicinity of the cysteine are also conserved. Not shown in Figure 8 is NEDD-4, which is carboxy-terminally truncated relative to the others. It is possible, since the NEDD-4 cDNA was not isolated on a functional basis, that the cDNA sequence is incorrect. This possibility is currently being investigated. Except for UreB1 and NEDD-4, the E6-AP-related proteins are large proteins, with molecular masses of at least 95 kD, up to about 300 kD for the Drosophila protein. None of the proteins shows significant similarity to the amino-terminal half of E6-AP, and none has significant similarity to E6-AP over the E6-binding domain, suggesting that E6 probably does not affect their activity. RSP5 and NEDD-4 do show some similarity over much of their length, and some of these proteins are clearly more related to each other over the carboxy-terminal domain than they are to E6-AP (the rat 100-kD protein and the Drosophila hyperplastic discs protein, for example). The UreB1 protein sequence is unique among the E6-AP-related proteins in that it lacks an amino-terminal extension beyond the region of similarity to E6-AP. Experiments are in progress to determine whether these E6-AP-related proteins have E3 activity and the ability to form thioesters with ubiquitin and, if so, whether the conserved cysteine is critical to their activity. A further challenge will be to determine the substrate specificity of the putative E3 proteins and how it is controlled.

There is a substantial amount of biochemical data which demonstrate that the high-risk HPV E6 and E7 proteins affect the activities of the p53 and pRB tumor suppressor proteins (Scheffner et al. 1994b). This, in parallel with genetic evidence demonstrating the role of E6 and E7 in cellular immortalization, has begun to give a view of how these proteins function in carcinogenesis. Just as the HPV E7 proteins have additional targets besides pRB, such as the pRB-related proteins p107 and p130, it will be interesting to determine if the E6/E6-AP complex targets proteins for ubiquitin-mediated degradation in addition to p53. Experiments directed in this way will not only continue to give us insight into how the E6 proteins function in carcinogenesis, but also into the mechanisms that govern the ubiquitin proteolysis pathway.

Figure 8. The amino acid sequences of the E6-AP-related proteins are aligned relative to the carboxy-terminal 50 amino acids of E6-AP. The last amino acid in each case represents the last amino acid encoded by the ORF. The conserved cysteine essential for E6-AP thioester formation is indicated in bold. Other invariant or highly conserved amino acids are boxed in gray.

```
E6-AP:   AK-NGPDTERLPTSHTCFNVLLLPEYSSKEKLKERLLKAIT-YAKGFGML   (865)
hORF2:   QS-TASGEEYLPVAHTCYNLLLDLPKYSSKEILSARLTQALD-NYEGFSLA   (1054)
RSP5 :   IE-KAGEVQQLPKSHTCFNRVDLPQYVDYDSMKQKLTLAVE-ETIGFGQE   (809)
hORF1:   HN-GGSDLERLPTASTCMNLLLKLPEFYDETLLRSKLLYAIE-CAAGFELS   (1083)
hORF3:   ES-TENPDDYLPSVMTCVNYLKLPDYSSIEIMREKLLIAAREGQQSFHLS   (2894)
YKL162   AEDGLTADEYLPSVMTCANYLKLPKYTSKDIMRSRLCQAIEEGAGAFLLS   (1482)
r100K:   DG-RPPDDQHLPTANTCISRLYVPLYSSKQILKQKLLLAIK--TKNFGFV   (889)
DrHDP:   ---RPADDSHLPTANTCISRLYIPLYSSKSILRSKNLMAIK--SKNFGFV   (2894)
UreB1:   HR-DDRSTDRLPSAHTCFNQLDLPAYESFEKLRHMLLAIQECSEGFGLANK (310)
```

ACKNOWLEDGMENT

J.M.H. was supported by a postdoctoral fellowship from the American Cancer Society (grant PF-3583).

REFERENCES

Bartel, B., I. Wünning, and A. Varshavsky. 1990. The recognition component of the N-end rule pathway. *EMBO J.* **9:** 3179.

Ciechanover, A., J.A. DiGiuseppe, B. Bercovich, A. Orlan, J.D. Richter, A.L. Schwartz, and G.M. Brodeur. 1991. Degradation of nuclear oncoproteins by the ubiquitin proteolysis system in vitro. *Proc. Natl. Acad. Sci.* **88:** 139.

DeCaprio, J.A., J.W. Ludlow, J. Figge, J.-Y. Figge, C.-M. Huang, W.-H. Lee, E. Marsilio, E. Paucha, and D.M. Livingston. 1988. SV40 large tumor antigen forms a specific complex with the product of retinoblastoma susceptibility gene. *Cell* **54:** 275.

DeVilliers, E.M. 1989. Heterogeneity of the human papillomavirus group. *J. Virol.* **63:** 4898.

Dohmen, R.J., K. Madura, B. Bartel, and A. Varshavsky. 1991. The N-end rule is mediated by the UBC2(RAD6) ubiquitin-conjugating enzyme. *Proc. Natl. Acad. Sci.* **88:** 7351.

Dürst, M., R.T. Dzarlieva-Petrusevska, P. Boukamp, N.E. Fusenig, and L. Gissmann. 1987. Molecular and cytogenetic analysis of immortalized human primary keratinocytes obtained after transfection with human papillomavirus type 16 DNA. *Oncogene* **1:** 251.

Dyson, N., P.M. Howley, K. Münger, and E. Harlow. 1989. The human papillomavirus-16 E7 protein is able to bind to the retinoblastoma gene product. *Science* **243:** 934.

Girod, P.-A., T.P. Carpenter, S. van Nocker, M.L. Sullivan, and R.D. Vierstra. 1993. Homologs of the essential ubiquitin conjugating enzymes UBC1, 4, and 5 in yeast are encoded by a multigene family in *Arabidopsis thaliana*. *Plant J.* **3:** 545.

Glotzer, M., A.W. Murray, and M.W. Kirschner. 1991. Cyclin is degraded by the ubiquitin pathway. *Nature* **349:** 132.

Gu, J., K. Ren, R. Dubner, and M.J. Iadarola. 1994. Cloning of a DNA binding protein that is a tyrosine kinase substrate and recognizes an upstream initiator-like sequence in the prodynorphin promoter. *Brain Res. Mol. Brain Res.* (in press).

Haas, A.L. and P.M. Bright. 1988. The resolution and characterization of putative ubiquitin carrier protein isozymes from rabbit reticulocytes. *J. Biol. Chem.* **263:** 13258.

Hatfield, P.M. and R.D. Vierstra. 1992. Multiple isoforms of ubiquitin-activating enzyme E1 from wheat. *J. Biol. Chem.* **267:** 14799.

Hawley-Nelson, P., K.H. Vousden, N.L. Hubbert, D.R. Lowy, and J.T. Schiller. 1989. HPV16 E6 and E7 proteins cooperate to immortalize human foreskin keratinocytes. *EMBO J.* **8:** 3905.

Hershko, A. and A. Ciechanover. 1992. The ubiquitin system for protein degradation. *Annu. Rev. Biochem.* **61:** 761.

Hochstrasser, M., M.J. Elison, V. Chau, and A. Varshavsky. 1991. The short-lived MATα2 transcriptional regulator is ubiquitinated in vivo. *Proc. Natl. Acad. Sci.* **88:** 4606.

Hubert, W.G. and P.F. Lambert. 1993. The 23-kilodalton E1 phosphoprotein of bovine papillomavirus type 1 is nonessential for stable plasmid replication in murine C127 cells. *J. Virol.* **67:** 2932.

Huibregtse, J.M., M. Scheffner, and P.M. Howley. 1991. A cellular protein mediates association of p53 with the E6 oncoprotein of human papillomavirus types 16 or 18. *EMBO J.* **10:** 4129.

———. 1993a. Cloning and expression of the cDNA for E6-AP, a protein that mediates the interaction of the human papillomavirus E6 oncoprotein with p53. *Mol. Cell. Biol.* **13:** 775.

———. 1993b. Localization of the E6-AP regions that direct human papillomavirus E6 binding, association with p53, and ubiquitination of associated proteins. *Mol. Cell. Biol.* **13:** 4918.

Kaelin, W.G., Jr., W. Krek, W.R. Sellers, J.A. DeCaprio, F. Ajchenbaum, C.S. Fuchs, T. Chittenden, Y. Li, P.J. Farnham, M.A. Blanar, D.M. Livingston, and E.K. Flemington. 1992. Expression cloning of a cDNA encoding a retinoblastoma-binding protein with E2F-like properties. *Cell* **70:** 351.

Kumar, S., Y. Tomooka, and M. Noda. 1992. Identification of a set of genes with developmentally down-regulated expression in the mouse brain. *Biochem. Biophys. Res. Commun.* **185:** 1155.

Lane, D.P. and L.V. Crawford. 1979. T antigen is bound to a host protein in SV40-transformed cells. *Nature* **278:** 261.

Linzer, D.I.H. and A.J. Levine. 1979. Characterization of a 54K dalton cellular SV40 tumor antigen present in SV40-transformed cells and uninfected embryonal carcinoma cells. *Cell* **17:** 43.

Mietz, J.A., T. Unger, J.M. Huibregtse, and P.M. Howley. 1992. The transcriptional transactivation function of wild-type p53 is inhibited by SV40 large T-antigen and by HPV-16 E6 oncoprotein. *EMBO J.* **11:** 5013.

Müller, D., M. Rehbein, H. Baumeister, and D. Richter. 1992. Molecular characterization of a novel rat protein structurally related to poly(A) binding proteins and the 70K protein of the U1 small nuclear ribonuclear particle (snRNP). *Nucleic Acids Res.* **20:** 1471.

Münger, K., W.C. Phelps, V. Bubb, P.M. Howley, and R. Schlegel. 1989. The E6 and E7 genes of the human papillomavirus type 16 together are necessary and sufficient for transformation of primary human keratinocytes. *J. Virol.* **63:** 4417.

Pascolo, S., M. Ghazvini, J. Boyer, L. Colleaux, A. Thierry, and B. Dujon. 1992. The sequence of a 9.3 kb segment located on the left arm of the yeast chromosome XI reveals five open reading frames including the CCE1 gene and putative products related to MYO2 and the ribosomal protein L10. *Yeast* **8:** 987.

Phelps, W.C., C.L. Yee, K. Münger, and P.M. Howley. 1988. The human papillomavirus type 16 E7 gene encodes transactivation and transformation functions similar to adenovirus E1a. *Cell* **53:** 539.

Pietenpol, J.A., T. Tokino, S. Thiagalingam, W.S. El-Deiry, K.W. Kinzler, and B. Vogelstein. 1994. Sequence-specific transcriptional activation is essential for growth suppression by p53. *Proc. Natl. Acad. Sci.* **91:** 1998.

Rechsteiner, M. 1991. Natural substrates of the ubiquitin proteolytic pathway. *Cell* **66:** 615.

Sarnow, P., Y.S. Ho, J. Williams, and A.J. Levine. 1982. Adenovirus E1b-58kD tumor antigen and SV40 large tumor antigen are physically associated with the same 54kD cellular protein in transformed cells. *Cell* **28:** 387.

Scheffner, M., J.M. Huibregtse, and P.M. Howley. 1994a. Identification of a human ubiquitin-conjugating enzyme that mediates the E6-AP-dependent ubiquitination of p53. *Proc. Natl. Acad. Sci.* (in press).

Scheffner, M., J.M. Huibregtse, R.D. Vierstra, and P.M. Howley. 1993. The HPV-16 E6 and E6-AP complex functions as a ubiquitin-protein ligase in the ubiquitination of p53. *Cell* **75:** 495.

Scheffner, M., K. Münger, J.M. Huibregtse, and P.M. Howley. 1992. Targeted degradation of the retinoblastoma protein by human papillomavirus E7-E6 fusion proteins. *EMBO J.* **11:** 2425.

Scheffner, M., H. Romanczuk, K. Münger, J.M. Huibregtse, and P.M. Howley. 1994b. Functions of human papillomavirus proteins. *Curr. Top. Microbiol. Immunol.* **186:** 83.

Scheffner, M., B.A. Werness, J.M. Huibregtse, A.J. Levine, and P.M. Howley. 1990. The E6 oncoprotein encoded by human papillomavirus types 16 and 18 promotes the degradation of p53. *Cell* **63:** 1129.

Sullivan, M.L. and R.D. Vierstra. 1991. Cloning of a 16-kDa ubiquitin carrier protein from wheat and *Arabidopsis thaliana*. *J. Biol. Chem.* **266:** 23878.

Werness, B.A., A.J. Levine, and P.M. Howley. 1990. Association of human papillomavirus types 16 and 18 E6 proteins with p53. *Science* **248:** 76.

Whyte, P., K.J. Buchkovitch, J.M. Horowitz, S.H. Friend, M. Raybuck, R.A. Weinberg, and E. Harlow. 1988. Association between an oncogene and an antioncogene: The adenovirus E1a proteins bind to the retinoblastoma gene product. *Nature* **334:** 124.

Yew, P.R. and A.J. Berk. 1992. Inhibition of p53 transactivation required for transformation by adenovirus early 1B protein. *Nature* **357:** 82.

zur Hausen, H. 1991. Human papillomaviruses in the pathogenesis of anogenital cancer. *Virology* **184:** 9.

p53-Dependent Apoptosis In Vivo: Impact of *p53* Inactivation on Tumorigenesis

H. SYMONDS,* L. KRALL,* L. REMINGTON,† M. SÁENZ ROBLES,§
T. JACKS,†‡ AND T. VAN DYKE*

*Department of Biochemistry and Biophysics, University of North Carolina, Chapel Hill, North Carolina 27599;
†Center for Cancer Research, Massachusetts Institute of Technology, Cambridge, Massachusetts 02139;
‡Howard Hughes Medical Institute and §Department of Pediatrics, University of Pittsburgh Medical School, Pittsburgh, Pennsylvania 15213*

Loss of tumor suppressor function clearly contributes to the multistep process of tumorigenesis, as evidenced by the frequent mutation of these genes in human cancer (Marshall 1991; Hinds and Weinberg 1994). Investigation into the mechanisms by which inactivation of tumor suppressors contributes to tumorigenesis is therefore central to understanding cancer. The *p53* gene is presently the most frequently mutated tumor suppressor gene in human cancers. Moreover, tumors of many different cell types harbor *p53* mutations. Exactly how the disruption of *p53* contributes to the genesis of these tumors and whether the contribution is the same or different in cancers of different cell types is not understood.

On the basis of recent in vitro studies of *p53* function, at least two models can be proposed (Fig. 1). In cultured fibroblasts, *p53* appears to have a role in mammalian cell cycle checkpoint control, inducing G_1 arrest in response to DNA damage (Kuerbitz et al. 1992). Consistent with this function, cells lacking *p53* fail to arrest in G_1 following DNA damage (Kastan et al. 1991), and this may lead to genetic instability (Livingstone et al. 1992; Yin et al. 1992). *p53* has also been shown to induce programmed cell death, or apoptosis, when overexpressed in some cultured cells (Yonish-Rouach et al. 1991, 1993; Shaw et al. 1992) and is required for DNA damage-induced apoptotic cell death of mouse thymocytes (Clarke et al. 1993; Lowe et al. 1993). Because of its role in DNA damage-induced growth arrest and apoptosis, *p53* has been described as a "guardian of the genome" functioning to selectively inhibit the growth of or eliminate damaged cells (Lane 1992). Thus, as a checkpoint regulator in tumorigenesis, *p53* would indirectly suppress tumor formation by reducing the occurrence of oncogenic mutations in other genes (Fig. 1, I). In this capacity, the loss of *p53* function would be predicted to occur in early stages of tumorigenesis.

A second more direct role for *p53* in tumorigenesis is predicted from the observation that *p53*-dependent apoptosis can be induced in cultured cells by first altering their growth properties (Fig. 1, II). For example, adenovirus E1A-induced proliferation of primary fibroblasts leads to *p53*-dependent apoptosis (Debbas and White 1993; Lowe et al. 1994). This observation suggests that *p53* could directly suppress the growth of a developing tumor via death of abnormally proliferating cells. In this case, selective pressure (cell survival) would exist for the loss of *p53* function. *p53* mutation would thus be observed subsequent to other oncogenic events and would contribute directly to tumor progression rather than to initiation. Such a model could account for the observation that *p53* mutations arise late in the development of some human cancers. For example, in human colorectal cancer, *p53* mutations are rarely observed in benign adenomas but are frequent in the more aggressive malignant carcinomas (Fearon and Vogelstein 1990).

To test these predictions for the role of *p53*, we have

I. DNA Damage Checkpoint

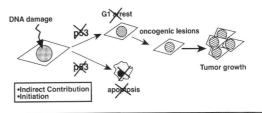

II. Cell Growth Checkmate

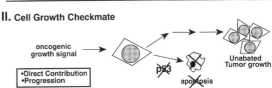

Figure 1. Predictions for *p53* inactivation in tumorigenesis. Recent cell culture observations have provided information concerning the possible roles for *p53* in tumorigenesis. As a checkpoint control (*I*), *p53* would be required in cells which have undergone DNA damage. *p53* would normally function on a pathway to arrest these cells in G_1 or to commit them to an apoptosis pathway. Loss of *p53* would indirectly lead to tumorigenesis, since cells with oncogenic mutations would be propagated. Thus, mutation or loss of *p53* would be an initiating factor in tumorigenesis. However, *p53* may play a more direct role in cells that have suffered an oncogenic growth signal (*II*). In cell-growth checkmate, *p53* would have the direct role of eliminating abnormally proliferating cells, thus slowing tumor growth. Loss of *p53* would allow unabated proliferation leading to rapid tumor development. Thus, *p53* would be a major factor in the progression or rate of tumorigenesis.

explored *p53* function in transgenic mouse models of tumorigenesis (Van Dyke 1994). We have used SV40 T antigen to induce tumorigenesis in specific cell types and have examined mutants of T antigen in combination with *p53* null mice to deduce the contribution of *p53* loss to tumorigenesis. We have focused on two cell types with distinct differences in their normal growth patterns: thymocytes and brain choroid plexus (CP) epithelial cells.

Previous experiments showed that expression of a *p53*-binding form of T antigen in thymocytes induces a *p53* null phenotype (Clarke et al. 1993; Lowe et al. 1993) in that these cells become resistant to DNA damage-induced apoptosis and are predisposed to tumorigenesis (McCarthy et al. 1994). In this cell type, the loss of *p53*-dependent apoptosis appears to be an important event in the initiation of tumorigenesis (McCarthy et al. 1994), and secondary oncogenic mutations are clearly required for the development of tumors (McCarthy et al. 1994). Thus in thymocytes, which may frequently encounter DNA damage as a result of aberrant T-cell-receptor gene rearrangement, *p53* mutation may predispose cells to tumorigenesis as predicted by model I (Fig. 1).

To test the role of *p53* in response to proceeding oncogenic changes, we have induced abnormal proliferation in the normally nondividing brain CP epithelial cells using a mutant of SV40 T antigen that disturbs pRB family function but not *p53* function. This protein is sufficient to induce abnormal proliferation of the CP cells in vivo, resulting in slow-growing tumors (Sáenz Robles et al. 1994). In contrast, tumors induced by wild-type T antigen, which is capable of binding *p53*, are much more aggressive and lead to rapid death of the mice (Chen and Van Dyke 1991). Here we show that aggressive tumor growth is directly attributable to the inactivation of *p53* and further define the mechanism by which *p53* acts to suppress the progression of tumors arising from abnormally proliferating cells in vivo. These studies provide in vivo evidence for a direct role for *p53* in tumor progression (Fig. 1, II). The studies described here also indicate that different cell-specific roles for *p53* inactivation in tumorigenesis exist in vivo.

EXPERIMENTAL PROCEDURES

Transgenic mice. The generation, screening, and preliminary characterization of transgenic mice were described previously. The TgT$_{121}$ transgenic line described here was previously referred to as LST1137-5 (Sáenz Robles et al. 1994). The TgT$_{121}$ transgenic mice harbor the dl1137 mutant T-antigen gene under the control of the lymphotropic papovavirus (LPV) transcriptional signals (Chen et al. 1989). Under LPV control, expression of T antigen is targeted uniformly to the CP epithelium (CPE) and to B and T lymphoid cells. The dl1137 mutant (here referred to as T$_{121}$) encodes a truncated protein consisting of the first 121 T-antigen amino acids followed by 11 missense residues due to a small out-of-frame deletion (Pipas et al. 1983). Transgenic mice that express a derivative of T$_{121}$ (dl1137K1, here referred to as T$_{121}$K1) with an additional amino acid change (E107 to L) were described previously (Sáenz Robles et al. 1994). The wild-type and mutant T-antigen transgenic mice are maintained on a C5BL6/DBA2 hybrid background. Mice are screened using a polymerase chain reaction (PCR) assay of tail DNA. The primers, 5'-CTA GTG ATG ATG AGG CTA CTG-3' and 5'-TTC TTG TAT AGC AGT GCA GC-3', amplify a 162-bp mutant fragment from the dl1137 mutant transgene and a 191-bp fragment from the wild-type T-antigen gene. Two types of transgenic mice that express wild-type T antigen in the CP were used in the present study. TgT founder mice (formerly referred to as LST mice) expressed T antigen under the control of the LPV transcriptional signals (Chen and Van Dyke 1991) and developed CP tumors at a young age (Fig. 2B). Mice in the SV11 family express T antigen from the SV40 transcriptional signals, and tumors develop at about 15 weeks of age (Van Dyke et al. 1987); slower tumor development is due to inefficient expression in the CP by this promoter (Chen and Van Dyke 1991). The expression of small t antigen was eliminated in the TgT, SV11, TgT$_{121}$, and TgT$_{121}$K1 mice, as described previously.

Generation of T-antigen transgenic mice defective at the p53 locus. Mice harboring a homozygous null mutation at the *p53* locus (129/sv background) have been described previously (Jacks et al. 1994). TgT$_{121}$ transgenic mice heterozygous for the *p53* mutant allele were generated by crossing TgT$_{121}$ homozygous mice with *p53*$^{-/-}$ mice, thus creating a hybrid 129/sv/BDF2 genetic background in the progeny. TgT$_{121}$ mice homozygous for the *p53* null allele were generated by further crossing TgT$_{121}$ *p53*$^{+/-}$ mice with *p53*$^{-/-}$ mice. TgT$_{121}$ mice homozygous or heterozygous for the *p53* null allele were identified by PCR analysis using three primers specific for the presence of the wild-type and/or the Neo-inserted *p53* mutant alleles (Jacks et al. 1994). TgT$_{121}$ mice were also crossed with *p53*$^{+/+}$ mice of the 129/sv background as a control for strain effects.

Histology. For routine histology, brain tissues were fixed in 10% formalin for 7–16 hours, washed three times in phosphate-buffered saline (PBS: 0.14 M NaCl, 2.7 mM KCl, 10 mM NaPO4, and 1.75 mM KPO$_4$, pH 7.3), dehydrated in graded ethanols for 15 minutes each, and embedded in paraffin. Sections (6 μ) were stained with hematoxylin and eosin as described previously (Sáenz Robles et al. 1994).

In situ detection of apoptosis. Paraffin-embedded brain sections (7 μ) were deparaffinized in xylene (Richard-Allan) twice and rehydrated through graded ethanols. Slides were incubated with proteinase K (20 μg/ml) in PBS for 15 minutes at 25°C and were then washed 2× 5 minutes in PBS and incubated in PBS containing 0.006% H$_2$O$_2$ to inhibit endogenous peroxidase activity. Slides were washed 4× 5 minutes in

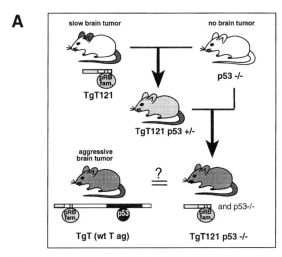

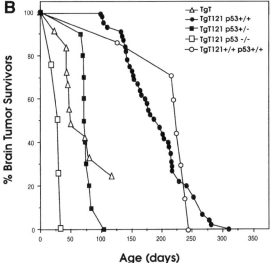

Figure 2. Survival of T-antigen transgenic mice in relation to *p53* function. The mouse crosses performed to obtained TgT_{121} $p53^{+/-}$ and TgT_{121} $p53^{-/-}$ mice as described in Experimental Procedures are shown in *A*. The transforming functions of the T-antigen fragment expressed in TgT_{121} mice are indicated. The possibility that TgT_{121} $p53^{-/-}$ mice may demonstrate a brain tumor phenotype similar to TgT mice is illustrated. Linear diagrams of the wild type and TgT_{121} T antigens are illustrated with designated regions involved in binding to pRB family proteins and *p53* (see Fig. 6). The survival (in days) of mice is shown in *B*. The TgT_{121} family of hemizygous transgenic mice was established previously (Sáenz Robles et al. 1994). The survival of TgT mice is based on independently derived founder mice which were described previously (Chen and Van Dyke 1991); early death of founder mice expressing the wild-type T-antigen transgene precluded establishment of a representative family. A line of mice homozygous for the expression of T_{121} ($+/+$) was generated as a control. Survival times reflect the time of sacrifice, which was dependent on the appearance of a bulged cranuim indicating the presence of a brain tumor.

reaction buffer (Oncor) for 15–30 minutes at 25°C. Enzymatic reactions were stopped by incubating the slides in $1 \times$ SSC and Apoptag stop/wash buffer (Oncor) at 25°C for 30 minutes. Slides were washed $4 \times$ 5 minutes with dH_2O and then incubated with anti-digoxigenin horseradish peroxidase (HRP) (Oncor) for 30 minutes at 25°C. Slides were washed with PBS ($4 \times$ 5 min) and incubated with 0.05% DAB (Sigma) for 2.5 minutes at 25°C. After counterstaining for 10 minutes with 0.5% methyl green in 1 M NaAc (pH 4.0) at 25°C, slides were washed $4 \times$ 1 minute with PBS, dehydrated twice in 100% butanol, soaked twice in 100% xylene, and mounted with Eukitt mounting solution (Vitromed) for analysis. All aqueous washes were performed with diethylpyrocarbonate-treated dH_2O. Apoptosis levels were quantitated by determining the percentage of HRP-stained cells within a given field of view at a magnification of $400 \times$. A total of 10–20 randomly chosen fields were counted per slide assayed, and the counts were averaged to obtain the apoptotic index.

Southern hybridization analysis of the p53 locus. Genomic DNA was purified from CP tumor tissue or tail tissue and was digested with *Eco*RI and *Stu*I. The digested DNA was analyzed by Southern hybridization and assayed for the presence of wild-type and mutant *p53* alleles as described previously (Jacks et al. 1994).

RESULTS

Inactivation of *p53* Promotes Rapid Progression of Brain Tumors

The T_{121} SV40 T antigen mutant (previously referred to as dl1137) encodes a fragment consisting of the T-antigen amino-terminal 121 residues (Pipas et al. 1983), which interacts with the pRB family proteins but which lacks the *p53*-binding domain (Fig. 2A). When expressed in the brain CPE of transgenic mice, T_{121} is weakly oncogenic (Sáenz Robles et al. 1994). Although it is sufficient to induce proliferation of the CP cells, tumors develop significantly more slowly than those induced by wild-type T antigen (Chen et al. 1992; Sáenz Robles et al. 1994), and see below). This difference in tumor development is related to differences in the T-antigen proteins, since both wild-type and mutant forms are expressed uniformly throughout the tissue under the same transcriptional controls (see Experimental Procedures) (Chen et al. 1989; Chen and Van Dyke 1991). Thus, while T_{121} is sufficient for the initiation of tumors in this cell type, the carboxy-terminal region of T antigen, which harbors the *p53*-binding domain, appears to be involved in tumor progression.

To test whether inactivation of *p53* alone is responsible for the increased tumor growth rate observed in mice expressing wild-type T antigen, we examined the effect of expressing the T_{121} truncated protein in *p53* null mice. Mice of the TgT_{121} lineage, which express

dH$_2$O and then incubated for 10 minutes in a humid chamber with Oncor equilibration buffer at 25°C as described in the Apoptag in situ apoptosis kit (Oncor). After blotting, slides were incubated with terminal deoxynucleotide transferase (Oncor) and Apoptag

T_{121} in the CP and develop slow-growing tumors (Sáenz Robles et al. 1994), were backcrossed to $p53$ null mice (which do not develop CPE abnormalities) for two generations to obtain TgT_{121} $p53^{-/-}$ mice (as diagrammed in Fig. 2A). As a control for background strain effects, TgT_{121} mice were also crossed to normal 129/sv mice, the strain from which the $p53$ null mice were originally derived (Jacks et al. 1994).

Survival times of the TgT_{121} mice with and without intact $p53$ loci are shown in Figure 2B. Survival time was measured as the age at which animals were sacrificed due to the appearance of a bulged cranium indicating the presence of a brain tumor. Founder transgenic mice expressing wild-type T antigen (TgT mice) developed CP tumors very rapidly, with a t_{50} of 6 weeks (the time at which 50% of the animals had died; Chen and Van Dyke 1991). In contrast, mice of the TgT_{121} lineage shown in Figure 2B developed tumors much more slowly, with a t_{50} of 26 weeks. The extended survival of mice within this family is representative of founder TgT_{121} animals as well as of another independently derived TgT_{121} family (Sáenz Robles et al. 1994). Moreover, increasing the level of T_{121} expression by generating mice homozygous for the transgene did not alter survival time.

When expressed in $p53$ null mice, however, the T_{121} protein induced tumors at a rate comparable to those induced by wild-type T antigen (Fig. 2B). No increase in tumor growth rate was observed when TgT_{121} mice were crossed with the 129/sv $p53^{+/+}$ control mice (data not shown). The apparent shorter survival of TgT_{121} $p53^{-/-}$ mice (3.5 weeks) compared to the wild-type T antigen mice (6 weeks) was most likely due to the fact that sacrifice occurred at the earliest sign of illness to minimize the risk of losing critical samples. Histological examination clearly indicated that tumors were present by this time.

Analyses of CP tumor tissue from TgT_{121}, TgT, and TgT_{121} $p53^{-/-}$ mice show that neoplastic growth of grossly similar morphology occurred in all animals. Figure 3 shows representative hematoxylin and eosin stained brain sections from $p53^{-/-}$ mice (normal CP, panels a and e), TgT_{121} K1 mice (expressing a mutant of T_{121} that does not bind pRB nor induce CP tumors, inset, panel a), TgT_{121} mice (terminal, t_{50} = 26 weeks, panels b and f), TgT mice (terminal, t_{50} = 6 weeks, panels c and g) and TgT_{121} $p53^{-/-}$ mice (terminal, t_{50} = 3.5 weeks, panels d and h). The CP tissue in all TgT_{121} mice lacking a functional $p53$ gene was uniformly neoplastic by 3 weeks of age (Fig. 3, panels d and h). For comparison, the CP of a 10-week-old TgT_{121} mouse (with the $p53$ alleles intact), although clearly abnormal, shows little increase in size (Fig. 5, panel a). These results indicate that loss of $p53$ function has no effect on tumor initiation, but leads to rapid tumor progression, and they further show that the tumor-promoting function in the carboxy-terminal region of T antigen (residues 122–708) inactivates $p53$.

$p53$-Dependent Apoptosis Suppresses Tumor Growth in Abnormally Proliferating Brain Epithelium In Vivo

We previously showed that although the CP tumors induced by T_{121} in $p53^{+/+}$ mice develop more slowly than those induced by wild-type T antigen, the per-

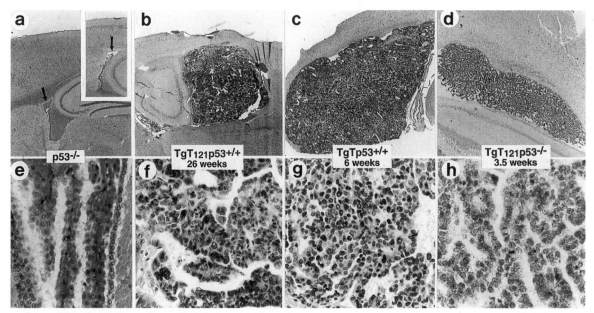

Figure 3. CP tumor morphology. Brain tissue sections were stained with hematoxylin and eosin and photographed at magnifications of 25 × (a–d) and 400 × (e–h). Brain sections showing CP tissue from a 2.5-month-old $p53^{-/-}$ mouse are shown in a and e. $p53^{-/-}$ mice assayed to 7 months of age were found to have the same normal CP morphology. Mice expressing the TgT121K1 mutant unable to complex with the pRB family also show normal CP morphology at all ages (inset in a). Representative views of lateral CP tissue showing full tumor development in the TgT_{121} (b,f), TgT (c,g), and TgT_{121} $p53^{-/-}$ (d,h) transgenic mice are given, along with the age of 50% survival for each mouse genotype.

centage of abnormally proliferating cells during both early and late stages of tumor growth is the same in each case (Sáenz Robles et al. 1994). Thus, the difference in tumor growth rate cannot be accounted for by a difference in the percentage of proliferating cells. As an alternative explanation, we explored the possibility that the abnormal proliferation induced by T_{121} was balanced by the induction of *p53*-dependent cell death causing the tumors to develop slowly. Inactivation of *p53* would then eliminate or reduce apoptosis, and tumor growth would proceed more rapidly, as observed. We examined brain epithelium from several mice for evidence of apoptosis in situ and quantified the percentage of apoptotic cells as described in Experimental Procedures (Fig. 4).

Indeed, when *p53* function was intact, expression of T_{121} induced high levels of apoptosis (Fig. 4, a–c). The percentage of apoptotic cells in TgT_{121} $p53^{+/+}$ CP tissue of two independent mice averaged 8.3% and 6.2%. These values represent a minimum estimate of the apoptotic index, since cells undergoing apoptosis are rapidly destroyed in vivo (Cicala et al. 1992). No apoptotic cells were detected in normal CP from nontransgenic, $p53^{+/-}$ or $p53^{-/-}$ mice (data not shown). The apoptosis induced by T_{121} is due, at least in part, to the interaction of this T antigen fragment with one or more of the pRB family member proteins, since expression of the same fragment with a point mutation in the pRB-binding site leaves the epithelial cells unperturbed (Sáenz Robles et al. 1994).

The high level of apoptosis observed in TgT_{121} mice was dependent on functional *p53*, since a much reduced level of apoptosis was observed in TgT_{121} $p53^{-/-}$ mice (Fig. 4e,f). Several sections from three independent TgT_{121} $p53^{-/-}$ mice were assayed. In two mice, the average percentage of apoptotic cells was less than 1% (0.8% and 0.5%). In a third mouse, the average value was higher (1.9%) but was significantly lower than that of T_{121}-expressing $p53^{+/+}$ tissue. Like the TgT_{121} $p53^{-/-}$ tumors, tumors expressing wild-type T antigen

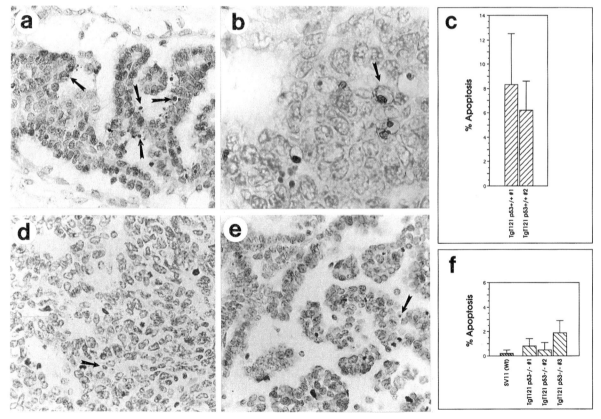

Figure 4. Level of apoptosis depends on *p53* function. Terminal deoxynucleotide transferase was used to detect fragmented DNA characteristic of apoptotic nuclei as described in Experimental Procedures. Apoptotic CP cells (*brown*) were stained with HRP and the tissue was counterstained with 0.5% methyl green. Panels *a* (400×) and *b* (1000×) show representative CP tissue from a 3.5-week-old TgT_{121} $p53^{+/+}$ transgenic mouse brain. Apoptotic cells detected by this method, some of which are indicated by arrows, show the characteristic nuclear condensation associated with physiological cell death (arrows in *b*). The percentage of cells staining positive for apoptosis in two TgT_{121} $p53^{+/+}$ mice (3.5 weeks old) is shown in *c*. Representative areas of an SV11 CP tumor (expressing wild-type T antigen, *c*) and a TgT_{121}-expressing CP tumor from a $p53^{-/-}$ mouse (3 weeks old, *d*) demonstrate the decreased level of apoptosis observed. The percentage of apoptotic cells in one SV11 and three TgT_{121} $p53^{-/-}$ mice (3–4 weeks old) is shown in *f*. Apoptotic cells were detected only in the abnormal CP tissue. The percentage of cells staining positive for apoptosis was determined for multiple fields of CP tissue from each mouse assayed (see Experimental Procedures).

contained few apoptotic cells (Fig. 4d,f). These data are consistent with the view that *p53*-dependent cell death limits tumor growth in TgT_{121} mice.

Progression to Rapid Tumor Growth upon Loss of Heterozygosity at the *p53* Locus

In familial cancers in which one mutant tumor suppressor allele is inherited, loss or mutation of the second allele is often observed in tumors, suggesting the existence of selective pressure for complete loss of function. Although the experiment described above in $p53^{-/-}$ mice indicates that the loss of *p53*-dependent apoptosis leads to rapid tumor progression, it does not address whether cell survival provides a selective pressure for the loss of *p53* function during the natural evolution of a tumor. To examine the role of *p53* loss during tumor progression, we characterized tumor development in TgT_{121} mice which carried only a single wild-type *p53* allele.

The TgT_{121} $p53^{+/-}$ mice survived to an intermediate time (t_{50} of 11 weeks) compared to the survival of TgT_{121} mice (26 weeks) and TgT_{121} $p53^{-/-}$ (3.5 weeks). Thus, inactivating a single *p53* allele significantly increased the rate at which tumors developed (Fig. 2B). The regulation of tumor growth rate by *p53* may be dose-dependent, or, alternatively, stochastic loss of the wild-type *p53* allele may occur, resulting in progression to rapid tumor development. We distinguished between these possibilities by analyzing the morphological development of the TgT_{121} $p53^{+/-}$ tumors and by examining the *p53* gene within tumors. If lower levels of *p53* result in intermediate tumor growth rates, morphological examination would show uniformly growing $p53^{+/-}$ tumor tissue at all times. Alternatively, if loss of the wild-type *p53* allele is required for the generation of rapidly growing tumors, focal areas of aggressive tumor growth would be evident. Moreover, these tumors would not contain a functional *p53* gene.

Morphological examination of 10 different brains from TgT_{121} $p53^{+/-}$ mice at several ages showed a distinct progression to aggressive tumor growth. At a young age (3.5 weeks), the CP tissue was similar in appearance to that of $p53^{+/+}$ mice expressing T_{121} (data not shown). In this hyperplastic tissue, most cells of the epithelial layer were crowded and abnormal, having reduced cytoplasm and darkly stained nuclei sometimes of variable size. Usually at this early stage, the overall tissue organization was retained. In $p53^{+/+}$ mice expressing T_{121}, this overall appearance persisted well beyond 10 weeks of age (Fig. 5a), although the tissue mass was often somewhat increased. In contrast, by 10 weeks of age, the TgT_{121} $p53^{+/-}$ mice had developed large nodular tumors consisting of densely crowded and disorganized cells (Fig. 5b,e). These distinct tumor masses were evident in all three TgT_{121} $p53^{+/-}$ mice examined between 9 and 10 weeks of age. The tissue adjacent to these focal masses had the hyperplastic appearance of TgT_{121} $p53^{+/+}$ CP at this age (Fig. 5, cf. a and b). In addition, in some mice, more than one such nodule was observed (data not shown). This consistent progression to more aggressive tumor growth suggested that secondary stochastic changes in some cells provided a selective growth advantage.

To determine if inactivation of the wild-type *p53* allele had occurred, the *p53* alleles in terminal TgT_{121} $p53^{+/-}$ tumors were examined by Southern hybridization analysis (Fig. 5c). As a control, the same analysis was performed on a CP tumor induced by wild-type T antigen in a $p53^{+/-}$ mouse (SV11 $p53^{+/-}$; see Experimental Procedures). As demonstrated above, wild-type T antigen inactivates *p53* in CP cells, and there should be no selection for loss of the wild-type *p53* locus during the development of this tumor. As predicted, both the germ-line TgT_{121} null and wild-type *p53* alleles were unaltered in the wild-type T-antigen-induced tumor (Fig. 5c, lane 2). However, tumors from the TgT_{121} $p53^{+/-}$ mice had lost the wild-type *p53* allele, whereas the mutant *p53* allele was unaffected (lanes 1 and 3). These results suggest that complete loss of the *p53* gene product provides a growth advantage to cells in these tumors, resulting in progression to rapid tumor growth and earlier death of the animals.

Reduction of Apoptosis Levels upon Progression to Aggressive Tumor Growth

To determine whether loss of the remaining wild-type *p53* allele in TgT_{121} $p53^{+/-}$ mice caused a reduction in the apoptotic index as observed in TgT_{121} $p53^{-/-}$ mice, we analyzed two TgT_{121} $p53^{+/-}$ brain samples with morphologically distinct nodular tumors for the presence of apoptotic cells. Analyses of several different sections through these regions showed a clear distinction in the percentage of apoptotic cells that correlated with the observed regional morphological differences described above. The results for one of these mice are shown in Figure 5,d-g. In hyperplastic regions outside the tumor nodule (see panel e), an average of 4.8% of the cells were apoptotic (Fig. 5d,g), consistent with the level observed in TgT_{121} $p53^{+/+}$ mice. However, in the larger focal tumors (panel e), few apoptotic cells (0.2%) were observed, characteristic of the TgT_{121} $p53^{-/-}$ tumors (Figure 5f,g). These data indicate that within the same brain tissue, the level of apoptosis is dependent on the presence or absence of functional *p53*. When *p53* is functional, the level of apoptosis is high, and T_{121}-induced tumor growth is suppressed; the loss of *p53* function results in a focal reduction in apoptosis, and tumor growth becomes more aggressive.

DISCUSSION

In an experimental transgenic mouse system, we show that *p53*-dependent apoptosis, in response to abnormal cell proliferation, acts to suppress the growth of a developing tumor. Inactivation of *p53* leads to the attenuation of apoptosis and to rapid tumor progres-

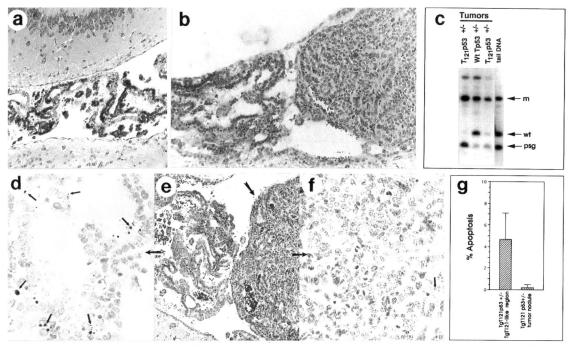

Figure 5. Tumor progression in $TgT_{121}\ p53^{+/-}$ mice. Histological abnormalities present in the CP tissue of $TgT_{121}\ p53^{+/+}$ (a) and $TgT_{121}\ p53^{+/-}$ (b) mice at 10 weeks of age were detected by staining with hematoxylin and eosin (200×). Uniformly hyperplastic tissue is present in $TgT_{121}\ p53^{+/+}$ mice, and this appearance persists throughout tumor development. In contrast, focal development of tumor nodules is evident in 10-week-old $TgT_{121}\ p53^{+/-}$ mice. Two examples are shown (b and e, 200×). In c, the p53-specific fragments found in genomic DNA purified from terminal CP tumors from two $TgT_{121}\ p53^{+/-}$ transgenic mice (11–12 weeks old, lanes 1 and 3) and from a $p53^{+/-}$ SV11 transgenic mouse expressing wild-type T antigen (lane 2) were detected by Southern analysis as described previously (Jacks et al. 1994). The positions of the wild-type (wt), mutant (m), and pseudogene (psg) p53 bands are indicated. Analysis of tail DNA from a $TgT_{121}\ p53^{+/-}$ mouse served as a control for the presence of both wild type and null p53 alleles. Cells undergoing apoptosis in $TgT_{121}\ p53^{+/-}$ CP tissue were stained, and the tissue was counterstained as described in Fig. 4. Similar sections from the same 10-week-old $TgT_{121}\ p53^{+/-}$ transgenic mouse shown in e were stained for evidence of apoptosis. Representative fields showing in situ apoptosis in the $p53^{+/-}$ hyperplastic tissue surrounding the tumor nodule (d) and within the tumor nodule (f) are shown. Some apoptotic cells are identified by arrows. The percentages of cells undergoing apoptosis within and surrounding the tumor nodule were quantitated as in Fig. 4 and are shown in g.

sion. The results described here for CP tumorigenesis further support the "multiple hit" hypothesis in the development of cancer; in this system, both the loss of proliferation control and the disruption of cell death pathways which normally destroy cells that have lost this control are required. Hence, this in vivo model of brain tumorigenesis provides specific support for a direct role for p53 in tumor progression, as predicted by model II of Figure 1. The results of this study are summarized in Figure 6. Also shown is a comparison with previous analyses of T-antigen-induced T-cell tumors in vivo. Below we discuss the key features of this system, the relevance of this mouse model to human cancer, and the broader implications of cell-specific roles for p53 function.

p53-Dependent Apoptosis Is Triggered by Abnormal Proliferation

An important aspect of this work is that the abnormal state of proliferation triggers p53-dependent apoptosis. An amino-terminal fragment consisting of the first 121 amino acids of SV40 T antigen, able to interact with the pRB family proteins but not with p53 (Fig. 6), induced abnormal proliferation in the brain CPE (Fig. 7, II). Although the CP normally proliferates rapidly early in embryonic life, it subsequently undergoes progressive lengthening of the cell cycle and then stops proliferating by 2 weeks after birth (Schultze and Korr 1981; Van Dyke 1993). p53 is clearly not required for these events, since this tissue develops normally in p53 null mice. The negative growth signals required to initiate and/or maintain the terminal nonproliferating state, however, must involve one or more of the pRB family members, since binding of these proteins by the T_{121} T-antigen fragment is required to elicit abnormal proliferation (Sáenz Robles et al. 1994). Only after this abnormal state is induced is p53-dependent cell death triggered (Fig. 7, II).

Previous in vitro studies showed that expression of E1A, another pRB-binding viral oncoprotein (Whyte et al. 1988; Moran 1993), induced p53-dependent apoptosis in cultured fibroblasts which was shown to be dependent on wild-type p53 (Debbas and White 1993; Lowe et al. 1994). Our studies provide in vivo evidence that such p53-dependent apoptosis is a natural response

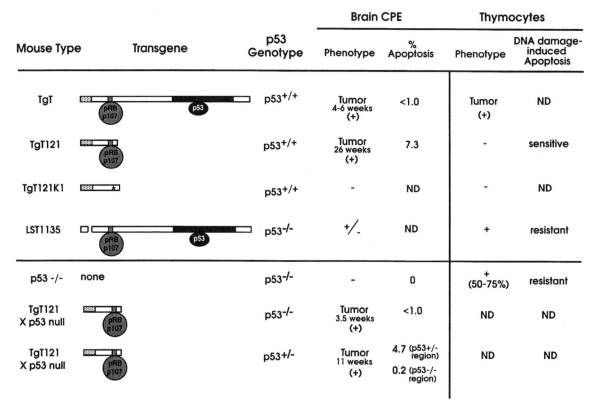

Figure 6. Summary of T-antigen-induced tumorigenesis in vivo. Diagrams of mutant proteins, genotypes, and cell-type specificities of phenotypes in various T antigen-expressing transgenic mice are shown. Established transformation-related activities of the T-antigen proteins are represented by shaded regions, from amino terminus to carboxyl terminus, as follows: the genetically defined amino-terminal transforming activity, the pRB-family-binding region, and the $p53$-binding region. The $p53$ genotype of each mouse is also indicated. T-antigen transgenic mice with a $p53^{+/+}$ background were described in the following references: Van Dyke et al. 1987; Chen et al. 1989; Chen and Van Dyke 1991; Symonds et al. 1993; Sáenz Robles et al. 1994. Mice expressing the T_{121} transgene in altered $p53$ backgrounds (this study) are listed in the bottom section. The phenotypes induced by each protein with respect to CPE tumorigenesis are shown in addition to the average percentage of cells undergoing apoptosis quantitated as in Figs. 4 and 5. The phenotypes of each protein with respect to T-cell tumorigenesis are also summarized along with previously determined susceptibility to irradiation-induced apoptosis (see text). Summary of data for $p53^{-/-}$ mice is taken from Donehower et al. (1992), Clarke et al. (1993), Lowe et al. (1993), and Jacks et al. (1994).

to aberrant proliferation. Substantial cell death occurred in T121-expressing CP cells where $p53$ function was intact. In contrast, expression of T121 in $p53$ null mice resulted in a significant reduction in the level of apoptosis (Fig. 6). Recent evidence that the transcription factor E2F-1 elicits $p53$-dependent apoptosis in cultured fibroblasts (Wu and Levine 1994) suggests that the apoptosis observed in vivo, which is dependent on disruption of pRB family functions, could be mediated through E2F regulation. Of interest, recent in vivo evidence suggests that the human papillomavirus E7 protein, which also interacts with the pRB family proteins, elicits $p53$-dependent apoptosis in the eye lens (Pan and Griep 1994) and retina (Howes et al. 1994) of transgenic mice. Whether a variety of different abnormal growth signals can serve to induce $p53$-dependent apoptosis in vivo remains to be determined.

Role of $p53$ in Tumorigenesis

Figure 1 describes two possible roles for $p53$ in tumor suppression. As a DNA damage checkpoint control in vitro, the loss of $p53$ has been proposed to indirectly contribute to tumorigenesis by permitting the propagation of cells that have suffered mutations, some of which may be oncogenic (Lane 1992). $p53$ mutation in this case would be an initiating event in tumorigenesis. Alternatively, the demonstration that $p53$-dependent apoptosis is induced as a consequence of abnormal proliferation predicts that loss of $p53$ instead may enhance the progression of tumors initiated by other events. The studies described here indicate that loss of $p53$-dependent apoptosis can play a direct role in the progression of tumors. In this model of CP tumorigenesis, $p53$ inactivation is not required for tumor initiation (for references, see Chen et al. 1992; Sáenz Robles et al. 1994; and this paper). Rather, $p53$ loss results in an increased rate of tumor development—which is largely accounted for by reduced apoptosis rather than enhanced proliferation.

A critical role for $p53$ inactivation in the progression of tumors, as demonstrated in the studies reported here, is consistent with the observation that $p53$ mutations are often observed in late stages of human cancer

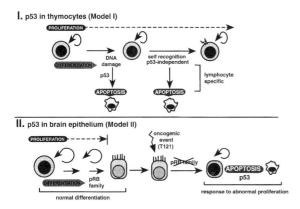

Figure 7. Model for cell-specific roles of *p53*-dependent apoptosis in tumor suppression. Deduced roles for *p53* and the pRB family proteins in cell growth regulation for thymocytes and the CPE are depicted. *p53*-dependent apoptosis is necessary to remove thymocytes suffering DNA damage during maturation (Model I). During early stages of differentiation, maturing thymocytes undergo gene rearrangement in order to express specific T-cell receptors. The model proposes that cells suffering DNA damage from this process are removed via a *p53*-dependent apoptosis pathway. Other signals leading to apoptosis, such as those involved in clonal deletion, are *p53*-independent (McCarthy et al. 1994). Loss of *p53* would eliminate the DNA damage checkpoint and would allow the accumulation of cells that have suffered oncogenic mutations as predicted by model I of Fig. 1. In contrast, *p53*-dependent apoptosis is not necessary for the normal development of the nonproliferating CPE (Model II). Proliferation and differentiation of normal CP begin during embryonic development. Cell proliferation decreases at birth and is undetectable 2 weeks after birth. One or more of the pRB family proteins are required for the maintenance of this nonproliferative state. Neither proliferation nor cellular apoptosis is detectable in the differentiated CP throughout the life of a normal mouse. When abnormal proliferation is induced, however, it is balanced by the concomitant induction of *p53*-dependent apoptosis. Cell death in conjunction with proliferation leads to the slow tumor development observed in TgT$_{121}$ *p53*$^{+/+}$ mice. As predicted in model II of Fig. 1, when *p53* function is inactivated, *p53*-dependent apoptosis is lost, leading to the aggressive development of tumors observed in TgT$_{121}$ *p53*$^{-/-}$ and wild-type T-antigen transgenic mice as well as the tumor progression observed in TgT$_{121}$ *p53*$^{+/-}$ mice.

(Fearon and Vogelstein 1990; Bartek et al. 1991; Davidoff et al. 1991; Crook and Vousden 1992; Ichikawa et al. 1992; Jaros et al. 1992; Sidransky et al. 1992; Visakorpi et al. 1992; Horio et al. 1993) and are frequently associated with aggressive tumors. Hence, the animal model described here may accurately recapitulate the cause and effect relationships occurring in the development of human cancers where *p53* mutations arise only after other oncogenic changes have accumulated. If this model indeed accurately reflects similar events in the development of human cancers, restoration of *p53*-dependent apoptosis may provide a viable approach to cancer therapy.

Although loss of *p53*-dependent apoptosis appears to be a rate-limiting step in tumor progression, aggressive tumor growth in the absence of *p53*-dependent checkpoint control may subsequently serve to rapidly fix stochastic mutations that provide additional selective advantages to tumor propagation, such as increased invasive and metastatic potential. The *p53*$^{+/-}$ mice described here which develop aggressive focal tumors as a result of stochastic *p53* loss will serve as a good model to test this hypothesis.

Cell-specific Roles for *p53*

Whether loss of *p53* plays a role in initiation, as predicted by the checkpoint hypothesis, or in progression, as predicted by the experiments described here, may depend on the cell type (Fig. 7). Although the *p53*-binding property of T antigen is dispensable for tumor induction in the CPE, it appears essential for the production of lymphoid tumors (summarized in Fig. 6). All T-antigen mutants so far tested that are defective for binding *p53* are unable to induce lymphoid abnormalities (examples are shown in Fig. 6; for a more complete list, see Van Dyke 1994), whereas the mutant which retains the ability to bind *p53* (dl1135) efficiently induces T-cell tumors (Symonds et al. 1993). Consistent with these observations, *p53* null mice are predisposed to thymocyte tumorigenesis with a high frequency (Donehower et al. 1992; Jacks et al. 1994). Indeed, like thymocytes from *p53* null mice (Clarke et al. 1993; Lowe et al. 1993), T-antigen-expressing thymocytes are resistant to irradiation-induced apoptosis (McCarthy et al. 1994). These results suggest a mechanism for thymocyte tumorigenesis which differs from that described for CPE tumorigenesis. Although *p53*-dependent apoptosis appears to play a critical role in both cases, in thymocytes, apoptosis is triggered in response to DNA damage.

Hence, in thymocytes, *p53* may be part of a surveillance pathway which, under normal conditions, would lead to the death of the T cells which have accumulated DNA mutations (Fig. 7, I). These cells may require such an apoptosis pathway due to the potential for DNA damage during T-cell receptor gene rearrangements. After inactivation of *p53*, survival of damaged cells that have accumulated oncogenic mutations in other genes may lead to tumor initiation, as predicted by model I of Figure 1. This hypothesis would predict that inactivation of *p53*, although required for the initiation of tumorigenesis, would not be sufficient to render thymocytes tumorigenic. Indeed, T-antigen-induced thymomas are clonal, indicating the requirement for secondary tumorigenic events (Symonds et al. 1993).

Together, these studies of tumorigenesis in different cell types appear to indicate that the specific roles for tumor suppressors, such as *p53*, could depend on the specific abnormal circumstances encountered by a given cell type, which in turn would depend on its normal pattern of cell growth and differentiation. T cells which proliferate as a normal aspect of their differentiation may not sense proliferation as an abnormal state, but appear to require *p53*-dependent apoptosis due to the specific potential for DNA damage (Fig. 7, I). In contrast, terminally differentiated CP cells which normally are not likely to encounter DNA damage may be equipped with regulatory pathways that sense abnormal cell proliferation and respond by the execution of regulated *p53*-dependent apoptosis (Fig. 7, II). In this circumstance, interference with such a pathway leads to the survival of aberrantly proliferating cells and may ultimately be the rate-limiting step in the development of malignancy.

ACKNOWLEDGMENTS

We are grateful to Jim Pipas for the dl1137 mutant and for critical discussion. We thank Rachele Kauffman for carrying out some initial crosses and Mike Wade, Vicky Bautch, Jude Samulski, Jim Leisy, and Kay McCloud for useful editorial or technical advice. Kinuko Suzuki's assessment of pathological abnormalities is much appreciated. We also appreciate technical assistance from Lisa Edwards-Moore, Stacy Bridge, Deepthimau Gowda, and Kimberly Burns. This work was supported by a grant to T.V.D. from the National Cancer Institute (CA-46283) and a General Cinemas Charitable Trust grant to T.J. T.V.D. is the recipient of a Research Career Development Award from the National Cancer Institute. T.J. is a Lucille P. Markey Scholar.

REFERENCES

Bartek, J., J. Bartkova, B. Vojtesek, Z. Staskova, J. Lukas, A. Rejthar, J. Kovarik, C.A. Midgley, J.V. Gannon, and D.P. Lane. 1991. Aberrant expression of the *p53* oncoprotein is a common feature of a wide spectrum of human malignancies. *Oncogene* **6**: 1699.

Chen, J. and T. Van Dyke. 1991. Uniform cell-autonomous tumorigenesis of the choroid plexus by papovavirus large T antigens. *Mol. Cell. Biol.* **11**: 5968.

Chen, J., K. Neilson, and T. Van Dyke. 1989. Lymphotropic papovavirus early region is specifically regulated in transgenic mice and efficiently induces neoplasia. *J. Virol.* **63**: 2204.

Chen, J., G. Tobin, J.M. Pipas, and T.A. Van Dyke. 1992. T antigen mutant activities in transgenic mice: Roles of *p53* and pRB-binding in tumorigenesis of the choroid plexus. *Oncogene* **7**: 1167.

Cicala, C., F. Pompetti, P. Nguyen, K. Dixon, A.S. Levine, and M. Carbone. 1992. SV40 small t deletion mutants preferentially transform mononuclear phagocytes and B lymphocytes in vivo. *Virology* **190**: 475.

Clarke, A.R., C.A. Purdie, D.J. Harrison, R.G. Morris, C.C. Bird, M.L. Hooper, and A.H. Wyllie. 1993. Thymocyte apoptosis induced by *p53*-dependent and independent pathways. *Nature* **362**: 849.

Crook, T. and K.H. Vousden. 1992. Properties of *p53* mutations detected in primary and secondary cervical cancers suggest mechanisms of metastasis and involvement of environmental carcinogens. *EMBO J.* **11**: 3935.

Davidoff, A.M., J.E. Herndon, N.S. Glover, B.-J.M. Kerns, J.C. Pence, J.D. Iglehart, and J.R. Marks. 1991. Relation between *p53* overexpression and established prognostic factors in breast cancer. *Surgery* **110**: 259.

Debbas, M. and E. White. 1993. Wild-type *p53* mediates apoptosis by E1A, which is inhibited by E1B. *Genes Dev.* **7**: 546.

Donehower, L.A., M. Harvey, B.L. Slagle, M.J. McArthur, C.A.J. Montgomery, J.S. Butel, and A. Bradley. 1992. Mice deficient for *p53* are developmentally normal but susceptible to spontaneous tumours. *Nature* **356**: 215.

Fearon, E.R. and B. Vogelstein. 1990. A genetic model for colorectal tumorigenesis. *Cell* **61**: 759.

Hinds, P.W. and R.A. Weinberg. 1994. Tumor suppressor genes. *Curr. Biol.* **4**: 135.

Horio, Y., T. Tadahaski, T. Kuroishi, K. Hibi, M. Suyama, T. Niimi, K. Shimokata, K. Yamakawa, Y. Nakamura, R. Ueda, and T. Takahashi. 1993. Prognostic significance of *p53* mutations and 3p deletions in primary resected non-small cell lung cancer. *Cancer Res.* **53**: 1.

Howes, K.A., N. Ransom, D.S. Papermaster, J.G.H. Lasudry, D.M. Albert, and J.J. Windle. 1994. Apoptosis or retinoblastoma: Alternative fates of photoreceptors expressing the HPV-16 E7 gene in the presence or absence of *p53*. *Genes Dev.* **8**: 1300.

Ichikawa, A., T. Hotta, N. Takagi, K. Tsushita, T. Kinoshita, H. Nagai, Y. Murakami, K. Hayashi, and H. Saito. 1992. Mutations of *p53* gene and their relation to disease progression in B-cell lymphoma. *Blood* **79**: 2701.

Jacks, T., L. Remington, B. Williams, E. Scmitt, S. Halachmi, R. Bronson, and R. Weinberg. 1994. Tumor spectrum analysis in *p53*-mutant mice. *Curr. Biol.* **4**: 1.

Jaros, E., R.H. Perry, L. Adam, P.J. Kelly, P.J. Crawford, R.M. Kalbag, A.D. Mendelow, R.P. Sengupta, and A.D.J. Pearson. 1992. Prognostic implications of *p53* protein, epidermal growth factor receptor, and Ki-67 labelling in brain tumours. *Br. J. Cancer* **66**: 373.

Kastan, M.B., O. Onyekwere, D. Sidransky, B. Vogelstein, and R.W. Craig. 1991. Participation of *p53* protein in the cellular response to DNA damage. *Cancer Res.* **51**: 6304.

Kuerbitz, S.J., B.S. Plunkett, W.V. Walsh, and M.B. Kastan. 1992. Wild-type *p53* is a cell cycle checkpoint determinant following irradiation. *Proc. Natl. Acad. Sci.* **89**: 7491.

Lane, D.P. 1992. *p53*, guardian of the genome. *Nature* **358**: 15.

Livingstone, L.R., A. White, J. Sprouse, E. Livanos, T. Jacks, and T.D. Tlsty. 1992. Altered cell cycle arrest and gene amplification potential accompany loss of wild-type *p53*. *Cell* **70**: 923.

Lowe, S., T. Jacks, D. Housman, and H. Ruley. 1994. Abrogation of oncogene-associated apoptosis allows transformation of *p53*-deficient cells. *Proc. Natl. Acad. Sci.* **91**: 2026.

Lowe, S.W., E.M. Schmitt, S.W. Smith, B.A. Osborne, and T. Jacks. 1993. *p53* is required for radiation-induced apoptosis in mouse thymocytes. *Nature* **362**: 847.

Marshall, C.J. 1991. Tumor suppressor genes. *Cell* **64**: 313.

McCarthy, S.A., H.S. Symonds, and T. Van Dyke. 1994. Regulation of apoptosis in transgenic mice by SV40 T antigen-mediated inactivation of *p53*. *Proc. Natl. Acad. Sci.* **91**: 3979.

Moran, E. 1993. Interaction of adenoviral proteins with pRB and *p53*. *FASEB J.* **7**: 880.

Pan, H. and A.E. Griep. 1994. Altered cell cycle regulation in the lens of HPV-16 E6 or E7 transgenic mice: Implications for tumor suppressor gene function in development. *Genes Dev.* **8**: 1285.

Pipas, J.M., K.W. Peden, and D. Nathans. 1983. Mutational analysis of simian virus 40 T antigen: Isolation and characterization of mutants with deletions in the T-antigen gene. *Mol. Cell. Biol.* **3**: 203.

Sáenz Robles, M.T., H. Symonds, J. Chen, and T. Van Dyke. 1994. Induction versus progression of brain tumor development: Differential functions for the pRB- and *p53*-targeting domains of SV40 T antigen. *Mol. Cell. Biol.* **14**: 2686.

Schultze, B. and H. Korr. 1981. Cell kinetic studies of different cell types in the developing and adult brain of the rat and the mouse: A review. *Cell Tissue Kinet.* **14**: 309.

Shaw, P., R. Bovey, S. Tardy, R. Sahli, B. Sordat, and J. Costa. 1992. Induction of apoptosis by wild-type *p53* in a human colon tumor-derived cell line. *Proc. Natl. Acad. Sci.* **89**: 4495.

Sidransky, D., T. Mikkelsen, K. Schwechheimer, M.L. Rosenblum, W. Cavenee, and B. Vogelstein. 1992. Clonal expansion of *p53* mutant cells is associated with brain tumour progression. *Nature* **355:** 846.

Symonds, H., S.A. McCarthy, J. Chen, J.M. Pipas, and T. Van Dyke. 1993. Use of transgenic mice reveals cell-specific transformation by a simian virus 40 T-antigen amino-terminal mutant. *Mol. Cell. Biol.* **13:** 3255.

Van Dyke, T.A. 1993. Tumors of the choroid plexus. In *Molecular genetics of nervous system tumors* (ed. A.J. Levine and H.H. Schmidek), p. 287. Wiley-Liss, New York.

———. 1994. Analysis of viral-host protein interactions and tumorigenesis in transgenic mice. *Semin. Cancer Biol.* **5:** 106.1.

Van Dyke, T.A., C. Finlay, D. Miller, J. Marks, G. Lozano, and A.J. Levine. 1987. Relationship between simian virus 40 large tumor antigen expression and tumor formation in transgenic mice. *J. Virol.* **61:** 2029.

Visakorpi, T., O.-P. Kallioniemi, A. Heikkinen, T. Koivula, and J. Isola. 1992. Small subgroup of aggressive, highly proliferative prostatic carcinomas defined by *p53* accumulation. *J. Natl. Cancer Inst.* **84:** 883.

Whyte, P., K. Buchkovich, J.M. Horowitz, S.H. Friend, M. Raybuck, R.A. Weinberg, and E. Harlow. 1988. Association between an oncogene and an anti-oncogene: The adenovirus E1A proteins bind to the retinoblastoma gene product. *Nature* **334:** 124.

Wu, X. and A.J. Levine. 1994. p53 and E2F-1 cooperate to mediate apoptosis. *Proc. Natl. Acad. Sci.* **91:** 3602.

Yin, Y., M.A. Tainsky, F.Z. Bischoff, L.C. Strong, and G.M. Wahl. 1992. Wild-type *p53* restores cell cycle control and inhibits gene amplification in cells with mutant *p53* alleles. *Cell* **70:** 937.

Yonish-Rouach, E., D. Resnitzky, J. Lotem, L. Sachs, A. Kimchi, and M. Oren. 1991. Wild-type *p53* induces apoptosis of myeloid leukaemic cells that is inhibited by interleukin-6. *Nature* **352:** 345.

Yonish-Rouach, E., D. Grunwald, S. Wilder, A. Kimchi, E. May, J. Lawrence, P. May, and M. Oren. 1993. *p53*-mediated cell death: Relationship to cell cycle control. *Mol. Cell. Biol.* **13:** 1415.

Cell Cycle Checkpoints, Genomic Integrity, and Cancer

L. Hartwell,* T. Weinert,† L. Kadyk,‡ and B. Garvik*
*Department of Genetics, University of Washington, Seattle, Washington 98195;
†Department of Molecular and Cellular Biology, University of Arizona, Tucson, Arizona 85721;
‡Bock Laboratories, University of Wisconsin, Madison, Wisconsin 53706

Cancer cells typically exhibit a large number of phenotypic differences from normal cells, including loss of contact inhibition, changes in adhesion, loss of gap junctions, loss of differentiation characteristics, decreased proteolysis, increased invasiveness, angiogenesis, metastasis, restoration of telomerase, decreased apoptosis, and decreased drug sensitivity. Underlying these phenotypic changes must be a large number of genomic changes. How do such massive changes in the genome occur? In 1976, Peter Nowell suggested that an early change in precancerous cells may be a mutation that creates genomic instability (Nowell 1976). This prediction is being borne out.

There are about 10^{13} cells in an adult human organism, and at least a thousand times that many arise during the normal life span in renewing tissues (Cairns 1975). Considering that the mutation rate for most genes is around 10^{-6} per gene per cell division, clearly, a cell number this large poses special problems. The main defense of multicellular organisms against the origin of cancer through the mutation of tumor suppressor genes must be diploidy. By maintaining two copies of every wild-type gene, the rate of recessive mutations causing loss of gene function would be on the order of 10^{-12} per cell per division.

Evidence for the role of genetic instability in cancer comes from a large variety of sources. For example, many of the cancer-prone syndromes, such as ataxia telangiectasia (AT), Bloom's syndrome, Fanconi's syndrome, and Werner syndrome, all show greatly elevated increases in the loss of chromosomes or regions of chromosomes (Hanawalt and Sarasin 1986). Studies on colon cancer reveal that many cancers have undergone extensive loss of chromosomal regions (Vogelstein et al. 1989). The recent identification of homologs of the bacterial *mutS* and *mutL* genes as genetic lesions that predispose to colon cancer demonstrates the strong role of mutation in the generation of cancer (Leach et al. 1993; Bronner et al. 1994; Papadopoulos et al. 1994).

CELL CYCLE CHECKPOINTS

The foregoing raises the question of how the cell controls the integrity of its genome throughout the many cell divisions required to make an adult organism. We and other investigators have been studying this question in yeast cells. Yeast cells are providing a paradigm not only for identifying the genes that control cell division, but also for understanding how the logic of cell cycle control is responsible for maintaining the integrity of the genome. During the cell division cycle, three macromolecular processes must be coordinated. The centrosomes (or in yeast, the spindle pole bodies) that provide the pole of the mitotic spindle must be replicated and segregated. The mitotic spindle itself must be formed, attached to chromosomes, and activated for elongation and sister chromatid separation at anaphase. The DNA must be replicated, condensed, and segregated. These processes are fundamentally independent of one another and require control mechanisms to coordinate their activity so that they occur in synchrony during the cell cycle. Recent findings have shown that a series of signal transduction systems, called checkpoints, coordinate these three different processes (Hartwell and Weinert 1989). Loss of at least one of these checkpoints is frequent in cancer and may be the leading cause of genomic instability in tumor cells.

RAD9 CHECKPOINT

We initially discovered one of these negative signal transduction systems while studying how the cell cycle responded to DNA damage (Weinert and Hartwell 1988). When yeast cells are treated with ionizing radiation, they arrest in the G_2 phase of the cell cycle. Mammalian cells do the same. During this G_2 arrest, chromosome breaks are being repaired so that the chromosomes will be intact when mitosis is resumed. Hittleman and Rao (1974) used premature chromosome condensation during the G_2 arrest of mammalian cells to show that many DNA breaks were repaired in G_2 before the irradiated cell entered mitosis. We looked for and discovered mutants that are no longer capable of this G_2 arrest. That is, after radiation, rather than arresting in G_2, they proceeded to divide, forming a small microcolony of dead cells. Among the X-ray-sensitive mutants of yeast, mutations in the genes *RAD9*, *RAD17*, and *RAD24* all behaved this way (Weinert et al. 1994). By irradiating *RAD9* cells and holding them in G_2 with the microtubule inhibitor, methylbenzimidazole carbamate, it was possible to show that *RAD9* mutant cells were capable of repairing X-ray-induced DNA damage (Weinert and Hartwell 1990). Thus, the defect in RAD9 mutant cells is an inability to arrest cell division so that DNA damage can be repaired before the cell enters mitosis. Other radiation-sensitive mutants are not defective in cell cycle control, but rather in the machinery that is needed to effect repair; mutants in these genes, for example, *rad18* and *rad52*, do arrest in G_2 but fail to repair DNA. By arresting wild-type cells in G_2, adding cycloheximide to inhibit protein synthesis, and then exposing them to X-irradiation, we were able to show that all of the processes elicited by the *RAD9* pathway, i.e., cell division arrest in G_2, DNA repair, and recovery from cell division arrest, can occur in the absence of protein synthesis (Weinert and Hartwell 1990).

Temperature-sensitive cell cycle mutants of yeast have shown that the events of the cell cycle are organized into a dependent pathway (Pringle and Hartwell 1981). For example, a temperature-sensitive mutation in DNA polymerase I, whose primary mode of action is to block DNA replication, also results in a failure of the cell to undergo mitosis. We wondered whether the *RAD9* pathway was also responsible for the dependent relationships observed in the cell cycle when temperature-sensitive blocks to cell cycle progression were imposed. We examined this question by making double mutants between *RAD9* and a variety of temperature-sensitive *cdc* mutants (Weinert and Hartwell 1993). Most mutants arrested synchronously in the cell cycle whether the *RAD9* gene was mutant or not. However, mutants defective in some aspects of DNA replication failed to arrest if they were defective in *RAD9*. Thus, the *RAD9* pathway appears to be important not only for arrest in response to DNA damage, but also for arrest of mitosis in response to a block in DNA replication. An interesting twist on this story was the finding that the *RAD9* gene was important if cells were blocked late in the S phase but not if they were blocked early in the S phase. For example, a DNA polymerase I mutant, *cdc17*, was dependent on an intact *RAD9* gene for arrest at 34°C, where it arrested with a nearly G_2 content of DNA. However, when the same mutant was arrested at 38°C, with a nearly G_1 DNA content, its arrest was not *RAD9*-dependent. A search for additional mutations like *RAD9* revealed a total of six genes that participate in the control of the G_2/M transition in response to DNA damage or defects in DNA replication: *RAD9*, *RAD17*, *RAD24*, *MEC1*, *MEC2*, and *MEC3* (Weinert and Hartwell 1993). At least two of these six genes also participate in controlling the G_1/S transition in response to single-strand DNA gaps generated during excision repair (Siede et al. 1993).

RELATIONSHIP BETWEEN CELL CYCLE PROGRESSION AND DNA REPAIR

These checkpoint controls act like gates that permit passage to the next phase of the cycle only if previous processes have been properly completed and if the cell has not incurred damage. We wanted to know whether passage from one stage of the cycle to another influenced the pathways of DNA repair. We constructed substrates for recombinational repair that distinguished whether a homologous chromatid or a sister chromatid was used as the template for repair. By comparing diploid cells X-irradiated in G_1 or G_2, we found that whereas G_1 cells (of necessity) employed the homolog as template, the G_2 cells employed almost exclusively the sister chromatid (Kadyk and Hartwell 1992). Moreover, when cells experienced UV irradiation in G_1 they delayed entry into S phase, presumably to ensure excision repair prior to replication (T.W. Seeley and L.H. Hartwell, unpubl.). However, if excision repair was blocked by a defect in the *RAD1* gene, then cells eventually entered S phase with unexcised UV dimers. In the latter case, a distinct type of sister chromatid exchange was induced, replication-dependent sister chromatid exchange (Kadyk and Hartwell 1993). Replication-dependent sister chromatid exchange differed from sister chromatid exchange induced in G_2 by its confinement to S phase and by the fact that essentially all replication-dependent sister chromatid exchange events were conversional, whereas a substantial proportion of sister chromatid exchange events induced during G_2 were reciprocal. These experiments suggest that cell cycle checkpoints are important not only in preventing the cell from executing cell cycle events in the presence of unrepaired damage, but also in orchestrating the most appropriate type of repair pathway.

PROPERTIES OF CHECKPOINT CONTROLS

Several checkpoint controls have been identified in the cell cycle, as enumerated below. They all appear to share a set of properties. They are signal transduction systems consisting of multiple components that must communicate between the cell cycle machinery and the cell cycle engine. Their function is to arrest the cell cycle in order to permit repair. They are nonessential in that some components of these pathways can be deleted without affecting cell cycle progress in an unperturbed cell. They increase the fidelity of the cell cycle even in unperturbed cells, probably because intrinsic stochastic errors require the occasional participation of checkpoint controls to assure accurate chromosome replication and segregation. Finally, they exhibit adaptation; that is, even in the absence of repair, after a delay, the cell eventually tries to proceed through the cell cycle. We illustrate these principles with the *RAD9* pathway.

We think the signal for the *RAD9* pathway may be single-stranded DNA. This conclusion comes from studies of a mutant, *cdc13*, that arrests in G_2 in a *RAD9*-dependent manner (B. Garvik and L.H. Hartwell, unpubl.). The *RAD9* pathway is particularly sensitive to the damage generated by the *CDC13* mutant. This fact is evidenced by the observation that *cdc13 RAD9* cells can grow at higher temperature than *CDC13 RAD9* cells, indicating that the damage which accumulates in *CDC13* is sufficient to stimulate the *RAD9* arrest even when it is sublethal. *CDC13* mutants at the restrictive temperature cause lesions that induce mitotic recombination. By mapping these mitotic recombination events, we find that the lesions are clustered near the end of the chromosome. By examining DNA extracted from *cdc13* cells by two-dimensional gel electrophoresis (Brewer and Fangman 1987), we find an unusual species of DNA located near telomeres that is single-stranded DNA. Strand-specific probes show that this DNA is specific for the ends of chromosomes and for the GT-rich telomeric strand. That is, the AT-rich telomere-associated strand is missing. Single-stranded DNA can extend about 50 kb down the chromosome.

Although we have no direct evidence in *Saccharomyces cerevisiae*, by analogy with *Schizosaccharomyces pombe*, we suspect that the *RAD9* pathway arrests the cell cycle by inhibiting the cyclin-dependent kinase necessary for the transition to mitosis. Mutants in the *S. pombe CDC2* gene are capable of overcoming the dependence of mitosis on DNA replication (Enoch and Nurse 1991).

The *RAD9* gene is nonessential for yeast cells. If we delete one copy of the gene from a diploid cell and then sporulate the diploid to produce four haploid spores, all four spores produce viable colonies, including the two that contain the deletion (Weinert and Hartwell 1990). The gene is, however, essential for the high fidelity of mitotic chromosome segregation that is normally observed in wild-type cells. Cells with the deleted gene lose chromosomes about 20-fold more frequently than normal cells (Weinert and Hartwell 1990). This result indicates that there is stochastic, endogenous damage that requires the function of the *RAD9* gene in order to elicit cell cycle arrest and repair.

Studies by Sandell and Zakian (1993) illustrate that the *RAD9* pathway undergoes adaptation. They induced a double-strand break near the end of a chromosome in a cell that was incapable of repairing double-strand breaks. They found that those cells arrested in G_2 for 8–10 hours, but then began dividing again. By growing colonies from the cells that went through division, they discovered that the broken chromosome was transmitted in an unstable state to progeny. These results indicate that although a single double-strand break is capable of arresting the yeast cell cycle for many hours, adaptation eventually does occur and allows the cells to continue dividing in the presence of a broken chromosome.

OTHER CHECKPOINTS

These properties of the *RAD9* checkpoint are shared by other checkpoints in the cell cycle. There is a checkpoint that controls the initiation of DNA synthesis in response to DNA strand breaks (Kuerbitz et al. 1992), a checkpoint that controls the transition from metaphase to anaphase in response to defects in spindle assembly (Hoyt et al. 1991; Li and Murray 1991), and indications of a checkpoint that controls metaphase to anaphase in response to defects in spindle pole body duplication (Winey and Byers 1993). The first and second of these three are nonessential, are necessary for the fidelity of chromosome transmission in nonperturbed cells, exhibit adaptation, are multicomponent systems, and probably all impinge on cyclin-dependent kinase transitions in response to different cell cycle signals.

RELEVANCE OF CHECKPOINTS TO KARYOTYPIC CHANGES IN TUMOR CELLS

We can think of the cell cycle as composed of three independent macromolecular processes, controlling the replication and segregation of the spindle pole body, the DNA, and the mitotic spindle. Tumor cells exhibit three kinds of chromosome instability: rearrangements in the structure of individual chromosomes, whole chromosome aneuploidy, and the generation of polyploid cells. It is possible that these three types of genomic changes are the result of defects in checkpoints that control, respectively, the response of the cell cycle to defects in DNA replication, to defects in chromosome attachment to the spindle, or to defects in the replication cycle of the centrosome or spindle pole body.

What evidence is there for checkpoints in mammalian cells? It is well known that mammalian cells undergo a delay in the transition from G_1 to S and the transition from G_2 to M in response to DNA damage. Recent evidence indicates that the p53 gene, defective in many types of tumors, is involved in a checkpoint control over the transition from G_1 to S phase (Kuerbitz et al. 1992). p53 mutant cells grow normally and, in fact, are able to form a completely normal mouse, indicating that the p53 gene is not essential for cell cycle progression (Donehower et al. 1992). It is, however, necessary for controlling the transition from G_1 to S in response to DNA damage (Livingstone et al. 1992; Yin et al. 1992). Moreover, the fidelity of mitotic chromosome transmission in p53 mutant cells is reduced. The rate of gene amplification in p53 mutant cells is several orders of magnitude higher than that in normal cells (Livingstone et al. 1992; Yin et al. 1992). p53 mutant mice have an abnormally high frequency of cancer (Donehower et al. 1992).

Studies on the development of esophageal cancer in patients with acid reflux, known as Barrett's syndrome, has provided strong evidence that loss of the p53 gene is an early event in the development of genetic instability leading to cancer (Neshat et al. 1994). These studies show that the first observable transition in esophageal cells from Barrett's patients is from a normal population of G_0 cells to a population of G_1 cells that are slowly or not proliferating. These cells then give rise to clones of cells that exhibit G_2 delay, probably in response to DNA damage. Subsequent clones arise that are defective for the p53 gene. These clones then give rise to aneuploid cell populations that often develop into cancer cells.

WHAT IMPLICATIONS WILL AN UNDERSTANDING OF CHECKPOINT CONTROLS HAVE FOR TUMOR THERAPY AND PREVENTION?

Loss of checkpoint controls leads to genomic instability and increases the likelihood that the unstable cell will progress to cancer. Detection of cells exhibiting genomic instability may therefore provide an early prognosis of susceptibility to cancer. Since checkpoints influence the susceptibility of cells to chemotherapeutic agents, either by affecting their ability to repair damage to cell cycle processes or by affecting their likelihood of undergoing programmed cell death (Clarke et al. 1993; Lowe et al. 1993), agents that target checkpoint controls may have therapeutic utility. Moreover, since tumor cells have often lost checkpoint controls that are present in normal cells, this difference may permit the design of therapeutic strategies with a high therapeutic index. Finally, since loss of checkpoint controls permits the rapid evolution of cells to malignancy, their restoration in cells that have impaired checkpoints may offer the possibility of delaying the onset of cancer.

NEGATIVE CONTROLS ON CELL CYCLE PROGRESSION AND GENOMIC INSTABILITY

We are becoming increasingly aware of a large number of controls that inhibit cell proliferation. In addition to those discussed above that respond to cell cycle perturbations, the cellular programs that lead to differentiation, senescence, and cell death all inhibit cell cycle progression, probably through signal transduction systems that act at the level of cyclin-dependent kinases. Moreover, just as loss of the cell cycle checkpoints can produce genomic instability, so can loss of the controls that lead cells to differentiate, senesce, or die. Differentiation of immune cells leads to programmed genomic rearrangements, senescence leads to loss of telomeres, and programmed cell death leads to endonucleolytic attack on chromosomes. If cells escape the controls on proliferation that accompany any one of these programs while retaining the genomic insults, then such cells are prime candidates for producing cancer.

ACKNOWLEDGMENTS

The work from our laboratory was supported by grants from the Institute of General Medical Sciences, National Institutes of Health (GM-17709). T.W. was supported by a fellowship from the Jane Coffin Childs Memorial Fund for Medical Research, and L.K. by a training grant from the Institute of General Medical Sciences. L.H. is a Research Professor of the American Cancer Society.

REFERENCES

Brewer, B.J. and W.L. Fangman. 1987. The localization of replication origins on ARS plasmids in *S. cerevisiae*. *Cell* **51:** 463.

Bronner, C.E., S.M. Baker, P.T. Morrison, G. Warren, L.G. Smith, M.K. Lescoe, M. Kane, C. Earabino, J. Lipford, A. Lindblom, P. TannergÅrd, R.J. Bollag, A.R. Godwin, D.C. Ward, M. Nordenskjøld, R. Fishel, R. Kolodner, and R.M. Liskay. 1994. Mutation in the DNA mismatch repair gene homologue hMLH1 is associated with hereditary non-polyposis colon cancer. *Nature* **368:** 258.

Cairns, J. 1975. Mutation selection and the natural history of cancer. *Nature* **255:** 197.

Clarke, A.R., C.A. Purdie, D.J. Harrison, R.G. Morris, C.C. Bird, M.L. Hooper, and A. H. Wyllie. 1993. Thymocyte apoptosis induced by p53-dependent and independent pathways. *Nature* **362:** 849.

Donehower, L.A., M. Harvey, B.L. Slagle, M.J. McArthur, C.A. Montgomery, J.S. Butle, and A. Bradley. 1992. Mice deficient for p53 are developmentally normal but susceptible to spontaneous tumors. *Nature* **356:** 215.

Enoch, T. and P. Nurse. 1991. Coupling M phase and S phase: Controls maintaining the dependence of mitosis on chromosome replication. *Cell* **65:** 921.

Hanawalt, P.C. and A. Sarasin. 1986. Cancer-prone hereditary diseases with DNA processing abnormalities. *Trends Genet.* **2:** 124.

Hartwell, L.H. and T.A. Weinert. 1989. Checkpoints: Controls that ensure the order of cell cycle events. *Science* **246:** 629.

Hittelman, W.N. and P.N. Rao. 1974. Premature chromosome condensation. I. Visualization of X-ray-induced chromosome damage in interphase cells. *Mutat. Res.* **23:** 251.

Hoyt, M.A., L. Totis, and B.T. Roberts. 1991. *Saccharomyces cerevisiae* genes required for cell cycle arrest in response to loss of microtubule function. *Cell* **66:** 507.

Kadyk, L.C. and L.H. Hartwell. 1992. Sister chromatids are preferred over homologs as substrates for recombinational repair in *Saccharomyces cerevisiae*. *Genetics* **132:** 387.

———. 1993. Replication-dependent sister chromatid recombination in *rad1* mutants of *Saccharomyces cerevisiae*. *Genetics* **133:** 469.

Kuerbitz, S.J., B.S. Plunkett, W.V. Walsh, and M.B. Kastan. 1992. Wild-type p53 is a cell cycle checkpoint determinant following irradiation. *Proc. Natl. Acad. Sci.* **82:** 7491.

Leach, F.S., N.C. Nicolaides, N. Papadopoulos, B. Liu, J. Jen, R. Parsons, P. Peltomäki, P. Sistonen, L.A. Aaltonen, M. Nyström-Lahti, X.-Y. Guan, J. Zhang, P.S. Meltzer, J.-W. Yu, F.-T. Kao, D.J. Chen, K.M. Cerosaletti, R.E.K. Fournier, S. Todd, T. Lewis, R.J. Leach, S.L. Naylor, J. Weissenbach, J.-P. Mecklin, H. Järvinen, G.M. Petersen, S.R. Hamilton, J. Green, J. Jass, P. Watson, H.T. Lynch, J.M. Trent, A. de la Chapelle, K.W. Kinzler, and B. Vogelstein. 1993. Mutations of a *mutS* homolog in hereditary non-polyposis colorectal cancer. *Cell* **75:** 1215.

Li, R. and A.W. Murray. 1991. Feedback control of mitosis in budding yeast. *Cell* **66:** 519.

Livingstone, L.R., A. White, J. Sprouse, E. Livanos, T. Jacks, and T.D. Tlsty. 1992. Altered cell cycle arrest and gene amplification potential accompany loss of wild-type p53. *Cell* **70:** 923.

Lowe, S.C., E.M. Schmitt, S.W. Smith, B.A. Osborne, and T. Jacks. 1993. p53 is required for radiation-induced apoptosis in mouse thymocytes. *Nature* **362:** 847.

Neshat, K., C.A. Sanchez, P.C. Galipeau, D.S. Levine, and B.J. Reid. 1994. Barrett's esophagus: The biology of neoplastic progression. *Gastroenterol. Clin. Biol.* **18:** D71.

Nowell, P.C. 1976. The clonal evolution of tumor cell populations. *Science* **194:** 23.

Papadopoulos, N., N.C. Nicolaides, Y.-F. Wei, S.M. Ruben, K.C. Carter, C.A. Rosen, W.A. Haseltine, R.D. Fleischmann, C.M. Fraser, M.D. Adams, J.C. Venter, S.R. Hamilton, G.M. Petersen, P. Watson, H.T. Lynch, P. Peltomäki, J.-P. Mecklin, A. de la Chapelle, K.W. Kinzler, and B. Vogelstein. 1994. Mutation of a *mutL* homolog in hereditary colon cancer. *Science* **263:** 1625.

Pringle, J.R. and L.H. Hartwell. 1981. The *Saccharomyces cerevisiae* cell cycle. In *Molecular biology of the yeast Saccharomyces: Life cycle and inheritance* (ed. J.N. Strath-

ern et al.), p. 97. Cold Spring Harbor Laboratory, Cold Spring Harbor, New York.
Sandell, L.L. and V.A. Zakian. 1993. Loss of a yeast telomere: Arrest, recovery, and chromosome loss. *Cell* **75:** 729.
Siede, W., A.S. Friedberg, and E.C. Friedberg. 1993. *RAD9* dependent G_1 arrest defines a second checkpoint for damaged DNA in the cell cycle of *Saccharomyces cerevisiae*. *Proc. Natl. Acad. Sci.* **90:** 7985.
Vogelstein, B., E.R. Fearon, E.K. Scott, S.R. Hamilton, A.C. Preisinger, Y. Nakamura, and R. White. 1989. Allelotype of colorectal carcinomas. *Science* **244:** 207.
Weinert, T.A. and L.H. Hartwell. 1988. The *RAD9* gene controls the cell cycle response to DNA damage in *Saccharomyces cerevisiae*. *Science* **241:** 317.
———. 1990. Characterization of *RAD9* of *Saccharomyces cerevisiae* and evidence that its function acts posttranslationally in cell cycle arrest after DNA damage. *Mol. Cell. Biol.* **10:** 6554.
———. 1993. Cell cycle arrest of *cdc* mutants and specificity of the *RAD9* checkpoint. *Genetics* **134:** 63.
Weinert, T.A., G.L. Kiser, and L.H. Hartwell. 1994. Mitotic checkpoint genes in budding yeast and the dependence of mitosis on DNA replication and repair. *Genes Dev.* **8:** 652.
Winey, M. and B. Byers. 1993. Assembly and functions of the spindle pole body in budding yeast. *Trends Genet.* **9:** 300.
Yin, Y., M.A. Tainsky, F.Z. Bischoff, L. Strong, and G.M. Wahl. 1992. Wild-type p53 restores cell cycle control and inhibits gene amplification in cells with mutant p53 alleles. *Cell* **70:** 937.

Genomic Integrity and the Genetics of Cancer

T.D. TLSTY,† A. WHITE,† E. LIVANOS,* M. SAGE,*
H. ROELOFS,* A. BRIOT,* AND B. POULOSE*

*Lineberger Comprehensive Cancer Center, Department of Pathology and †Curriculum in Genetics,
School of Medicine, University of North Carolina at Chapel Hill, North Carolina 27599-7295*

One method of maintaining genomic integrity involves the action of cell cycle checkpoint genes. These genes integrate proper sensing of environmental signals and appropriate cellular responses. Recent work suggests that *p53* performs such an integrator function in mammalian cells. In this paper, we describe how, in mortal cells, the E6 and E7 viral oncoproteins of type 16 human papillomavirus (HPV) each disrupts the integration of these signals in its own unique way, resulting in two distinct types of genomic alterations and cell death as the cells respond to negative growth signals. Furthermore, these alterations in response to negative growth signals can precede any overt phenotypes displayed by the oncoprotein-expressing cells and suggest a novel way to examine cells from individuals that are predisposed to malignancy. These studies implicate loss of cell cycle control in tumor progression.

Genomic integrity is maintained by a network of cellular activities that assesses the status of the genome at a given point in time, provides signals to proceed with or halt cell cycle progression, and provides for repair of damaged DNA. Mutations in any part of these pathways can have the ultimate effect of disturbing chromosomal architecture. This loss of genomic integrity is a hallmark of neoplastic cells and fuels the multistep process of carcinogenesis. Data to support this idea come from several sources, most importantly, studies of genes in yeast which control cell cycle progression. Hartwell (Hartwell and Weinert 1989) has demonstrated that some of these genes, called checkpoint genes, coordinate cell cycle progression with cellular signals and allow for the maintenance of chromosomal integrity. In mammalian cells, the wild-type p53 protein is part of a pathway that allows cells to arrest in their cell cycle progression when conditions for growth are unfavorable and thereby prevents the generation of drug-resistant colonies (Livingstone et al. 1992; Yin et al. 1992). Studies by Kastan and coworkers have implicated *p53* in a similar activity, the cellular response to γ radiation (Kastan et al. 1991; Kuerbitz et al. 1992). In this capacity, it is suggested that *p53* receives information that DNA damage has occurred and mediates, through the possible action of *gadd45*, the growth arrest that occurs in normal cells. Hence, *p53* exhibits several functions of a checkpoint gene in mammalian cells and is involved in maintaining genomic integrity. Further studies demonstrated that some human tumorigenic cells showed loss of genomic integrity despite the presence of only wild-type *p53*, thus suggesting that alternate pathways can bypass the role of *p53*, and that, not surprisingly, *p53* is not the sole determinant for the regulation of genomic integrity (Livingstone et al. 1992).

In addition to mediating growth arrest, *p53* has been implicated in the triggering of apoptosis, an active death process (Debbas and White 1993; Yonish-Rouach et al. 1993). This function is readily observed in the adenovirus-mediated transformation of cells. These elegant experiments by Eileen White and coworkers demonstrate that expression of E1A stimulates cell proliferation and, in the presence of wild-type *p53* (the absence of E1B), the apoptotic response is triggered. If wild-type *p53* is removed by E1B or if expression of E1A occurs in a *p53* null cell line, no apoptosis is detected and cells proliferate. Efficient transformation requires that induction of cell proliferation be coupled with suppression of apoptosis (Debbas and White 1993). Hence, the processing of cellular growth signals can have different consequences depending on the genetic composition of the cell. When aberrations occur in the mechanisms that allow a cell to sense and respond to negative growth signals, an uncoupling of cell cycle control with those cellular signals occurs. This uncoupling can result in the generation of genomic alterations in one genotype or the triggering of cell death in a different genotype.

Changes in genomic material are seen when cells are transformed spontaneously, virally, or with chemicals. In studying the loss of genomic integrity, our previous work has focused on spontaneously transformed cells (both rodent and human) and their ability to exhibit one form of genetic instability, gene amplification. Gene amplification is an increase in gene copy number of a given locus and occurs spontaneously at a high frequency (10^{-6} to 10^{-3}) in preneoplastic and neoplastic cells (Sager et al. 1985; Otto et al. 1989; Tlsty et al. 1989; Jonczyk et al. 1993). In contrast, gene amplification is undetectable in normal cells ($<10^{-9}$) as measured by the ability to become resistant to the drug *N*-(phosphonoacetyl)-L-aspartate (PALA) (Tlsty 1990; Wright et al. 1990). Genetic analysis of this property indicated that, in the cells examined, the ability to amplify is a recessive genetic trait (Tlsty et al. 1992). These data suggest that normal cells have a gene or set of genes which suppresses amplification. In our search for possible candidate genes, the *p53* gene emerged as an important modulator of gene amplification frequency, as well as other genomic alterations.

Ideally, to investigate the events that occur when cells become genomically unstable, we wanted a model system that started with a normal cell and disrupted genomic integrity at a defined point. The DNA tumor viruses provided us with such a system. We knew from previous studies (Jonczyk et al. 1993) that virally transformed cells possessed the ability to amplify. We asked when, after infection, this ability was acquired. In this study we used the human papillomavirus (HPV) type 16 and type 6. The HPVs induce tumors of the epithelial or fibroepithelial components of skin or mucosa. Much clinical, epidemiological, and molecular evidence associates specific viral types with the formation of carcinomas of the genital and oral mucosa (zur Hausen 1991). Some HPV types (16, 18, 31, 33, 39) are referred to as "high risk" because of their association with cervical carcinoma, whereas other types (6 and 11) associated with benign lesions are termed "low risk." Although ubiquitous, HPV type 16 or type 18 is not sufficient to trigger malignant progression. The virus may be present with a long latency, and only rarely does progression of genital lesions into carcinomas occur (zur Hausen 1991). Expression of two viral proteins, E6 and E7, is important for the transformation of mammalian cells, and expression of these proteins is found in the majority of cervical carcinomas and derived cell lines (Schwarz et al. 1985; Smotkin and Wettstein 1986; Baker et al. 1987; Schlegel et al. 1988; Wilczynski et al. 1988). Expression of E6 and E7 is sufficient to efficiently transform and immortalize human keratinocytes and extend the normal life span of human fibroblasts (Pirisi et al. 1987; Munger et al. 1989a; Watanabe et al. 1989; Halbert et al. 1991; Dhanwada et al. 1992). Recently, these viral oncoproteins have been shown to bind a set of host-cellular proteins that are the products of known tumor suppressor genes. The HPV-16 E6 gene product binds and aids in the degradation of *p53* (Scheffner et al. 1990; Werness et al. 1990), and the E7 gene product binds a set of proteins that includes hypophosphorylated pRb, p107, p130, p300, and cyclin A (Dyson et al. 1989, 1992; Munger et al. 1989b; Barbosa et al. 1990). The mechanistic results of these protein couplings on the neoplastic process are under intense investigation. The data presented in this review suggest a physiological consequence of these interactions and their importance in viral carcinogenesis; both results related to the maintenance of genomic integrity. Viral oncoproteins not only alter the proliferative capacity of human cells, but also alter the control of genomic integrity.

GENERATION OF HPV16-INFECTED NHF

Introduction of the viral oncoproteins from HPV-16 into normal human fibroblasts (NHF) was accomplished by use of a recombinant retroviral system (White et al. 1994). Retroviral constructs contained the gene coding for neomycin resistance alone, or in combination with the open reading frames (ORF) for HPV-16 E6/E7 together, or each of the viral oncoproteins individually. NHF at population doubling (PDL) 33 (neo, E6) and PDL 36 (E6/E7, E7) were infected at high titer with the described retroviral vectors and selected in the neomycin analog G418 sulfate. Similar numbers of G418 sulfate-resistant colonies arose from each retroviral infection. Resistant cells were pooled, resulting in the cell populations NHF neo, NHF16 E6/E7, NHF16 E6, and NHF16 E7. Radioimmunoprecipitation of HPV-16 E6 and E7 proteins in the cell populations demonstrated their expression, with the E6/E7-infected population expressing both viral oncoproteins and the E6- and E7-infected populations expressing the appropriate individual proteins (data not shown; White et al. 1994). These populations of cells demonstrated the expected alterations of their cellular proteins; p53 protein was dramatically diminished in the E6-expressing cells, and pRb protein was diminished in the E7-expressing cells (data not shown). Oncogene-expressing cells were examined for the effect of the viral proteins on genomic integrity and the cellular potential to amplify.

CHARACTERIZATION OF HPV-INFECTED NHF

Chromosome complements were examined in the uninfected and HPV-infected NHF. The karyotypic stability of the five cell populations was examined during their extended propagation in culture, using samples taken soon after the viral oncogene-expressing populations were selected, as well as prior to their senescence in culture. Normal, diploid karyotypes were observed in the parental NHF both at early passage and at late passage close to senescence. Consistent with the literature is the observation of some chromosomal abnormalities in senescent fibroblasts (Benn 1976). At early passage after infection with the indicated oncoprotein-coding genes, the NHF neo, NHF16 E6, and NHF16 E7 cells produced karyotypes that were indistinguishable from the parental NHF. All three had a basic normal karyotype with no rearrangements or telomeric association. The karyotypes of the NHF16 E6/E7 cells at early passage were found to have a small number of telomeric associations and rearrangements. When the karyotypes of the E6-expressing cells (E6 and E6/E7) were examined late in passage, they were found to have numerous genomic rearrangements. Few of the cells were diploid; most contained multiple rearrangements, telomeric associations, and gross aneuploidy (data not shown; Y. Xiong et al., in prep.). In contrast to the E6-expressing cells, the karyotypes of the E7-expressing cells at late passage were completely diploid. The differences in these late-passage karyotypes may be due to the activity of wild-type *p53*. In the E7-expressing cells, the p53 protein is still active. It is reasonable to assume that in these cells chromosome breakage is still sensed and that the response is the clearing of the population of damaged cells. This activity would result in an ostensibly "normal" karyotype.

Cell morphology, population doubling time, mortality, and behavior at confluence were examined in the uninfected and HPV-16-infected NHF. Morphologically, the HPV protein-expressing cells were indistinguishable from each other (Fig. 1). Although similar to the uninfected NHF, all three of the HPV-infected cell populations were smaller and less extended than the NHF parental cells. The population doubling time of each cell population was determined and found to be similar (White et al. 1994). Past studies have indicated that individually, the E6 or E7 proteins of HPV-16 do not influence the life span nor immortalize NHF, but together, the proteins can efficiently extend the life span of NHF compared with that of normal controls

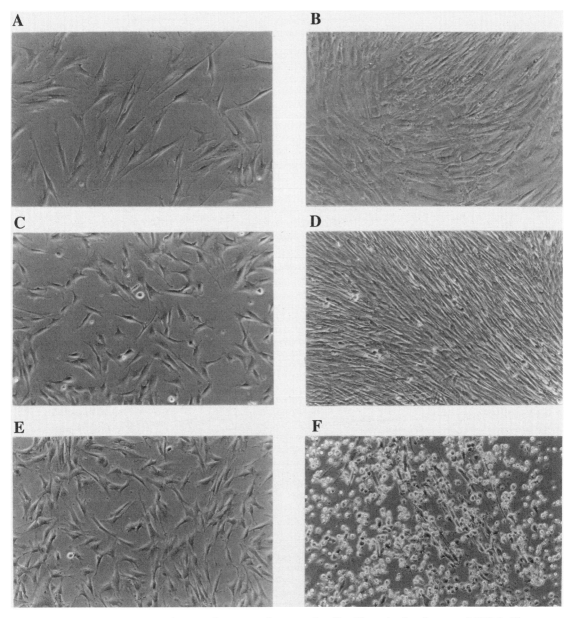

Figure 1. Morphology of normal and HPV-16 oncoprotein-expressing fibroblasts in the absence of PALA. Phase-contrast photographs of cells at subconfluence and confluence, grown in the absence of PALA, are shown. Cells were plated at 2×10^5 cells per 60-cm petri dish and grown until they attained confluence. Cells were photographed at subconfluence after 3 days in culture and after 6 days at confluence. At subconfluence, the HPV protein-expressing cells were indistinguishable from each other (NHF16 E6/E7, C; E; NHF16 E7, E), but distinct from the control cell populations (NHF, A; NHFneo, not shown). NHF expressing the HPV proteins were smaller in size and had less extended processes compared to the parental cells. At confluence, control cell populations (NHF, B; NHFneo, not shown) formed a tight monolayer of nondividing, viable cells. In contrast, the NHF expressing both the E6 and E7 proteins or only the E6 protein continued to proliferate after reaching confluence, as indicated by the presence of mitotic cells (NHF16E6/E7, D; NHF16E7, F). Massive cell death occurred in the E7-expressing NHF after 2–3 days at confluence. Death was verified by cellular incorporation of the dye trypan blue. Magnification, $63\times$.

(Pirisi et al. 1987; Watanabe et al. 1989). In agreement with these studies, we found the E6- and E7-expressing cells senesced between 78 and 88 PDL, whereas in the E6/E7-expressing cells an extension of life span of about 20 PDL compared to NHF was noted. Immortalization of oncoprotein-expressing fibroblasts is rare or negligible and was not observed in this set of experiments.

The oncoprotein-expressing cells demonstrated rather striking responses to growth conditions at confluence (Fig. 1). As normal diploid fibroblasts, such as NHF, become confluent, they become tightly packed in a monolayer and cease proliferation. This contact inhibition results in a cessation of progression through the cell cycle, with the cells arrested in a G_0 state indefinitely. The cells containing E6/E7 together or E6 alone demonstrated responses that were similar to each other but slightly different from that exhibited by parental NHF. The cells formed a monolayer and for the most part arrested in growth, but infrequent mitotic figures could be observed even several days after the cells achieved confluence (White et al. 1994). The cells could be held in this state for long periods of time (months). Interestingly, the E7-expressing cells demonstrated an entirely different response; they died (White et al. 1994). The cells formed a tight monolayer as they approached confluence, and mitotics could be detected, similar to those seen with the E6-expressing cells. However, after approximately 5 days at confluence, these cells first became thin, elongated, and eventually rounded and refractile as they lifted from the plate and died, as measured by trypan blue exclusion. This apoptotic response is similar to that observed when adenovirus E1A is expressed in cells.

CELLULAR RESPONSE TO METABOLIC INHIBITOR

We also examined the effect of a second negative growth signal. Incubation of cells in the drug PALA results in nutrient deprivation. Placing cells in this negative growth signal also allows the determination of mutation rate by measuring the emergence of drug-resistant colonies. To analyze amplification potential in these cells, we incubated them in the drug PALA under the standard conditions of the clonogenic assay (White et al. 1994). The normal human fibroblasts and those expressing the viral oncoproteins demonstrated dramatic differences in both the cellular response to the drug and the generation of PALA-resistant colonies. Uninfected NHF and neo-containing NHF exhibited a response to PALA that we have found is typical of NHF. The cells remain on the plate and fail to increase their cell number. There was no observable outgrowth of PALA-resistant colonies. Hence, uninfected NHF, as described previously (Tlsty 1990; Wright et al. 1990; Livingstone et al. 1992), and those containing the neo control lacked a detectable frequency of CAD gene amplification. When E6/E7- and E6-expressing cells are incubated in PALA, the vast majority gradually enlarged and sloughed from the plate over a period of 1–2 weeks, resulting in clearing of cells from the background. Infrequent, nondividing cells remained in the background while actively growing PALA-resistant colonies emerged during the selection process. Those cell populations containing the E6/E7 proteins together or the E6 protein alone generated PALA-resistant colonies at a frequency of 3.8×10^{-5} and 1.8×10^{-5}, respectively.

In contrast, when E7-expressing cells were incubated in the presence of PALA, there was a distinctly different response to the drug which resulted in the clearing of cells from the plate within approximately 6 days. The E7-expressing cells became thin, elongated, and refractile, and after 3 days in PALA, they began demonstrating a condensed, rounded morphology as they loosened from the substrate of the tissue culture plate. As cells lifted off the plate, they underwent death, as evidenced by the lack of exclusion of trypan blue, a vital dye. This morphological change is identical to that seen when these cells are held at confluence and is reminiscent of an apoptotic sequence of events that is initiated in certain cells under adverse conditions (Fig. 1). The characteristic nuclear degradation that occurs during apoptosis was not detected. At the termination of the PALA selection with the E7-expressing cells (~6 weeks later), there was no background of nondividing cells. The plate was cleared of cells except for the rare PALA-resistant colonies. Cells that expressed E7 alone generated PALA-resistant colonies at a frequency that was tenfold lower than that observed for the E6/E7-containing cells. The frequencies of the generation of PALA-resistant colonies obtained with the viral oncoprotein-expressing cells are within the range of those previously observed in preneoplastic and neoplastic cell lines (10^{-6} to 10^{-3}) (Sager et al. 1985; Otto et al. 1989; Tlsty et al. 1989; Jonczyk et al. 1993).

Individual PALA-resistant subclones from each population were isolated and propagated in culture. These cells were used to determine the mechanism of drug resistance and to examine the proliferative capacity of the cells. All PALA-resistant subclones grew for a period of time and then entered senescence. In summary, all PALA-resistant clones were mortal. These data indicate that these human cells acquire genomic instability prior to immortalization. The loss of genomic integrity is a very early step in viral carcinogenesis.

MECHANISM OF DRUG RESISTANCE

Several PALA-resistant subclones from each of the viral oncogene-expressing cell populations were analyzed for amplification of the *CAD* gene. Fluorescent in situ hybridization (FISH) analysis of the subclones verified intrachromosomal *CAD* gene amplification as the mechanism of resistance to PALA in the E6/E7- and E6-expressing cells. In all cases, the amplified copies were clustered with no distinct patterns observable (White et al. 1994). Karyotypic analysis of PALA-resistant subclones from each of the cell populations indicated a variety of chromosome rearrangements and telomeric associations, similar to that seen in the unselected populations after extended passage in culture. Analysis of PALA-resistant subclones from the E7-expressing population was difficult because of their limited proliferative capacity. Although complete karyotypic analysis was not possible because of the difficulty in obtaining adequate metaphase spreads, several samples were noted to have excessive numbers of chromosomes (<100). As an alternative to metaphase analysis, two-color FISH analysis was performed on interphase preparations using probes specific for human chromosome 2 centromeric sequences and for the human CAD gene. Most often, multiple copies of centromeric chromosome 2 and of the CAD gene (4–7 copies) were detected, indicating aneuploidy as the primary mechanism of resistance to PALA (White et al. 1994). Hybridization with sequences specific for chromosome 6 also demonstrated multiple signals (data not shown), indicating possible polyploidy.

PERTURBED CELL CYCLE PROGRESSION

Our previous studies had shown that NHF arrested in the G_1 and G_2 phases of the cell cycle when incubated with PALA (Livingstone et al. 1992). Figure 2A–G shows the cell cycle distribution of cells at various times after placement in drug (0, 1, 2, 3, 4, 5, 6 days). Hours after exposure to the metabolic inhibitor PALA, an exponentially growing population of normal, diploid cells will accumulate in the majority of the cells in the S phase of the cell cycle. This increased S fraction gradu-

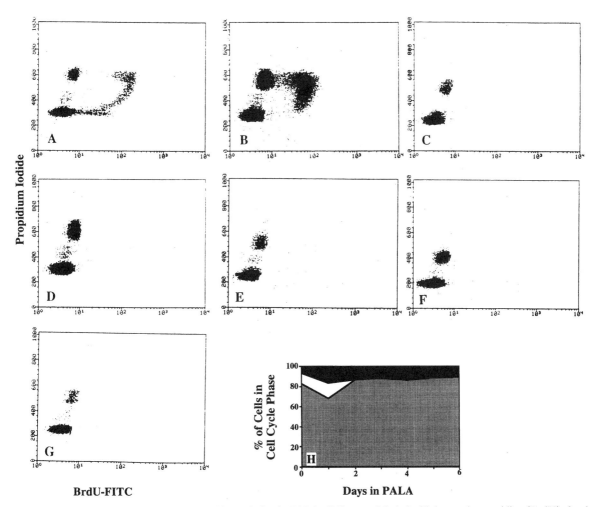

Figure 2. Flow cytometric analysis of NHF cell population in PALA. Cells were labeled with bromodeoxyuridine (BrdU), fixed, and stained with anti-BrdU-FITC and propidium iodide, then analyzed using a Becton Dickinson FACScan instrument. The x axis represents cells undergoing DNA synthesis. Cells at the right have more BrdU incorporated and are therefore in the S phase of the cell cycle. The y axis represents DNA content as measured by propidium iodide staining. The upper population has a G_2 phase DNA content, and the lower population has a G_1 phase DNA content. Panels show NHF with no PALA exposure (A), NHF days 1–6 in PALA (B–G). The cell cycle data were compiled and presented in the form of an area graph for each cell population, providing a characteristic cell cycle pattern that could be conveniently compared among the cells (H). Each graph indicates the percentage of cells in each phase of the cell cycle (relative cell phase) after 0, 1, 2, 3, 4, 5, and 6 days in PALA. The G_2, S, and G_1 phases of the cell cycle are represented by the top, middle, and bottom rows of the graph, respectively. There was no significant portion of the parental NHF in S phase after 2 days in PALA.

ally decreases until, at day 2, there is a complete absence of cells in the S phase, indicating an arrest in cell cycle progression. The population is distributed between the G_1 and G_2 phases of the cell cycle for the duration of the selection. This temporal distribution of cells within each phase of the cell cycle is presented as a scatter plot (data directly from the flow cytometer) in Figure 2A–G and as a compilation over time in Figure 2H. This visual representation of the relative fractions of the population in G_1, S, or G_2/M phase is helpful for comparing the response of normal cells in drug with that of the oncoprotein-expressing cells in drug. The white area on the graph in Figure 2H represents the percentage of cells in the S-phase fraction; the lower gray area represents those in G_1; and the upper dark area represents those in G_2. Deviations from this pattern are seen in the NHF expressing HPV-16 viral oncoproteins. Figure 3A–D portrays the cell cycle progression of control cells (A), as well as those expressing E6/E7 together (B), E6 alone (C), and E7 alone (D) while they are incubated in PALA. Analysis of cells expressing low-risk HPV-6 proteins showed a complete arrest when the cells were incubated in PALA (Fig. 3E,F).

In this study, we demonstrate that distinct alterations in the growth control circuitry can lead to the emergence of different types of genomic instability. We begin with cell populations that are isogenic and, by expression of HPV E6 and E7 oncoproteins, disrupt specific circuits that allow cells to respond to altered growth conditions. It is important to note that the differential response of the cells depends on their need to respond to outside stimuli. The two stimuli described here are the negative growth signals that cells produce as they attain confluence or when the medium is deficient in nutrients (in this case, nucleotide precursors). In the absence of either of these signals, each of the cell populations grows in an exponential fashion with no overt difficulties. Only when the cell must modulate that pattern do the differences between the cells emerge and generation of rare variants is observed.

When exposed to the metabolic inhibitor PALA, NHF arrest in the G_1 and G_2 phases of the cell cycle. The virally altered cell populations failed to arrest under these conditions and continued to cycle. Although the majority of these cells die, a few generate drug-resistant colonies. In the cells where *p53* function is compromised, cells with E6/E7 or E6 protein only, the majority of cycling cells died gradually, leaving the PALA-resistant colonies to expand. The altered karyotypes of these drug-resistant clones manifested *CAD* gene amplification in addition to other genomic changes. Rare PALA-resistant colonies arising from a background of gradually dying cells is the typical observation in most of the immortalized and tumorigenic cell lines we have assayed to date. A plausible scenario begins to emerge as to how *p53* participates in the generation of the rearrangements that fuel neoplasia. Cells suffer either endogenous or exogenous damage to their DNA, which brings about breakage of their genetic material. Broken chromosomes are usually detected in a normal cell, and cell cycle progression is arrested while the damage is repaired or the cell is removed. In cells which have mutated parts of the circuitry that senses, repairs, or responds to damage in DNA, cell cycle progression fails to arrest. Broken chromosomes can replicate and fuse, leading to the formation of dicentric chromosomes (Trask and Hamlin 1989; Toledo et al. 1992). The formation of a dicentric chromosome and the subsequent fusion-bridge-breakage cycle originally described by McClintock more than 40 years ago (McClintock 1939, 1942) generates duplications, deletions, and other abnormalities. These dicentric intermediates are seen quite often in cells with amplified genes (Stark and Wahl 1984; Smith et al. 1990; Windle et al. 1991; Toledo et al. 1992; Ma et al. 1993; White et al. 1994). These types of rearrangements activate oncogenes and inactivate tumor suppressor genes.

Whereas the elimination of checkpoint function of *p53* may account for the tolerance of broken chromosomes and the emergence of cells with amplified DNA sequences in the E6-expressing cells, the E7-expressing cells also generate PALA-resistant colonies. Here the cellular and genomic response is totally different. The cells expressing only the E7 protein exhibited massive, immediate cell death after approximately 3 days in PALA, with very rare resistant colonies arising on a cleared background. These drug-resistant cells contained increased copies of *CAD* gene sequences along with increased copies of chromosome 2 centromeric

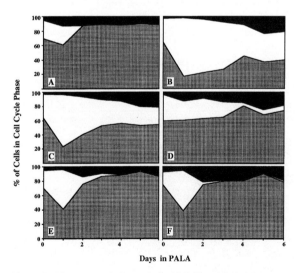

Figure 3. Flow cytometric analysis of NHF retroviral-infected cell populations in PALA. Cells were treated and data are displayed as in Fig. 2. There was no significant portion of cells in S phase after 2 days in PALA for cells expressing only the neomycin resistance gene and cells expressing the non-oncogenic HPV6E6 or HPV6E7 proteins. All of the HPV-16 protein-expressing cells continue to cycle in the presence of PALA, as indicated by a percentage of cells in the S phase throughout PALA exposure. (*A*) NHF neo, (*B*) NHF16 E6/E7, (*C*) NHF16 E6, (*D*) NHF16 E7, (*E*) NHF6E6, (*F*) NHF6E7.

sequences, indicating that aneuploid cells had been generated. These data suggest that aneuploidy is a condition that is tolerated by cells that have active p53 protein. As demonstrated in Figure 3, these cells fail to arrest in cell cycle progression when placed in PALA, even though they contain a wild-type *p53* gene. This *p53*-independent abrogation of cellular arrest in some, as yet to be determined, way leads to chromosomal consequences that are clearly distinct from those generated in cells that have eliminated *p53*. One may speculate that E7, which is known to bind cyclin A, as well as cellular kinases (Tommasino et al. 1993), disrupts cyclin homeostasis and leads to elimination of G_2 or relaxation of G_2/M checkpoints. Since aneuploidy is one of the most common genomic aberrations seen in tumor cells, this mechanism deserves further investigation. Studies with polyoma TAg indicate that the expression of this gene in NHF also produces drug-resistant colonies, but the spectrum of genomic changes differs from those seen with HPV E6/E7 (Schaefer et al. 1993). These results demonstrate that diverse perturbations of molecular pathways can have different effects on chromosomal architecture.

These experiments demonstrate that expression of viral oncoproteins leads to the immediate acquisition of the ability to amplify. These cells relinquish their control of genomic integrity prior to immortalization, clearly indicating that immortalization is not necessary. Since previous studies with Syrian hamster epithelial cells identified immortal cells that lack the ability to amplify (sup^+) (Jonczyk et al. 1993), and this study documents mortal cells that have the ability to amplify, it is clear that immortalization of mammalian cells is neither necessary nor sufficient for amplification to occur. Our observation of amplification in these mortal cells addresses our initial question as to when the virally infected cells acquire the ability to amplify or undergo genomic rearrangements. This change must occur immediately with expression of the E6/E7 genes. Strikingly, the biology of in vivo cervical carcinomas points to the expression of E6 and E7 as important in tumor progression. The high-risk viruses are episomal in benign lesions, and in this state, products of the E1/E2 ORFs generate a protein that suppresses transcription of the E6 and E7 genes. Little protein is made. Genomic integrity is intact during this latency period. In carcinomas, the high-risk viruses are usually found integrated into the host genome (Durst et al. 1985; Pater and Pater 1985). This integration event is accompanied by a change in transcription activity and pattern. Reproducibly, this integration results in the disruption of the coding region between the E1 and E2 genes and consequently allows for the initiation of E6 and E7 transcription and subsequent protein production (Romanczuk and Howley 1992). Whereas in the episomal state, the entire spectrum of viral proteins used for replication, capsid formation, etc. are expressed, in the integrated state, the E6 and E7 viral oncogenic products become the dominant transcripts. Our data demonstrate that expression of these oncoproteins and their concomitant binding of cellular proteins result in genomic instability. The generation of chromosomal rearrangements ostensibly allows progression toward malignancy.

The net result of viral oncogene expression and its relaxation of checkpoint function is the generation of rare variants that have a growth advantage under any given set of circumstances. For example, incubation in drug allows the emergence of drug-resistant colonies. Likewise, holding cells at confluence allows the emergence of cells that no longer heed these negative growth signals. They grow out and become transformed foci. The observation of the E6-expressing cells as they proliferate (in the absence of any overt challenge) is educational with regard to the generation of rare variants. Genetic instability in the HPV-16 E6-expressing cells is evident in the absence as well as the presence of PALA selection. At approximately 20 PDL after retroviral infection, karyotypes of the virally infected cells propagated in the absence of PALA were diploid, like the parental, uninfected NHF strain. Upon continued culture, the cells accumulated a variety of genomic alterations. It is probable that endogenous damage and the relaxation of the *p53* checkpoint combine to generate these abnormalities. One would postulate that exogenous damage would accelerate this process.

APOPTOSIS: PARALLELS WITH E1A EFFECTS

Our observations with HPV E6/E7 expression in human cells parallel recent work with adenovirus E1A/E1B in a striking manner, suggesting further functional analogies between the various viral oncoproteins (Vousden and Jat 1989). Cells expressing E1A alone showed a rapid loss of viability immediately upon achieving saturation density. Those cells expressing E1A and E1B did not demonstrate this rapid cytocidal response (Rao et al. 1992). White and coworkers have demonstrated that transformation of rodent cells by the adenovirus oncoproteins, E1A and E1B, requires two events, E1A-dependent induction of proliferation and E1B-dependent suppression of apoptosis. The sole expression of E1A results in efficient initiation of focus formation, but the cells cannot sustain proliferation. They degenerate and die. Cell death is mediated through an up-regulation of *p53*, which results in the production of a cytocidal (cyt) and DNA degradation (deg) phenotype. If *p53* is functionally removed from these cells, as when E1B is coexpressed, the apoptotic pathway can no longer be induced. The E1A/E1B-expressing cells can now produce transformed foci with a high frequency that have a good probability of immortalizing. In our system, the sole expression of the E7 gene (the functional equivalent of E1A) in challenged cells results in a cytocidal response and the formation of drug-resistant colonies that cannot sustain proliferation. If *p53* is functionally removed from these cells, as when E6 is coexpressed, the cytocidal response is no longer induced and the drug-resistant colonies grow in a healthy manner. In each case, a negative growth signal (confluence in the rodent cells and nutrient deprivation or confluence in the human fibroblasts) elicits a physiological response that is similar in each cell type and dependent on the genotype of the cell.

It is interesting that the cyt and deg phenotypes are not always expressed in a concomitant fashion. White has demonstrated that two parameters, host cell type and expression of E1A proteins, influence the appearance of degraded DNA (deg) (White and Stillman 1987). She reports that whereas degradation of DNA is pronounced in HeLa and KB cells infected with E1A, the deg phenotype is absent from WI38 cells under the same conditions (White et al. 1986). We find a parallel phenotype with HPV E7. Normal human cells express the cytocidal response, but no DNA degradation is detectable. The molecular basis for this difference between normal and transformed cells is unknown.

Finally, the studies presented here illuminate the biological basis for the differences between high-risk and low-risk HPV types in terms of neoplastic transformation. Both the high-risk and low-risk viruses cause hyperproliferation in cells, yet only the high-risk viruses produce variants that can escape differentiation, immortalize, and progress to malignancy. The high-risk viruses are known to be much more efficient in binding and degrading p53 than their low-risk counterparts (Werness et al. 1990). Likewise, the high-risk E7 protein binds pRB with a tenfold greater affinity than the low-risk type (Munger et al. 1989b; Gage et al. 1990). We propose that one of the basic differences between these two types of viruses is their effect on cellular maintenance of genomic integrity. Three pieces of data demonstrate that cells infected with low-risk virus maintain their genomic integrity whereas those expressing the high-risk virus do not. First, no PALA-resistant colonies are generated when type-6-infected cells are challenged with drug. The frequency is at least three orders of magnitude lower than measured with type 16. Second, examination of the type 6 karyotype shows an intact diploid profile. As shown in White et al. (1994), the type 16 oncoproteins allow the disruption of the diploid profile. Finally, perhaps the most sensitive assay, the arrest in cell cycle progression when cells are exposed to PALA or confluence, demonstrates that cells expressing the type 6, low-risk virus can sense negative growth signals and respond in an appropriate manner; they arrest. The low-risk viral oncoproteins do not allow the generation of chromosomal abnormalities that fuel tumor progression. In contrast, the biology of the lesions containing the high-risk virus points to the involvement of E6 and E7 in the loss of genomic integrity and in tumor progression.

IDENTIFICATION OF INDIVIDUALS WITH GERM-LINE MUTATIONS IN THESE PATHWAYS

We and other investigators hypothesize that the molecular networks described above exist to eliminate or repair cells that have genomic lesions. These lesions can be generated by a variety of sources, such as endogenous damage, exogenous damage, or a miscoupling of cellular signals. Given the existence of these pathways and their perceived function in cellular physiology, one would predict that rare germ-line mutations of this network of genes could exist and result in an accumulation of genomic lesions (rearrangements) in cells that are not eliminated from the soma of the organism. Since these rearrangements can have profound secondary effects, such as the activation or inactivation of random genes, one would expect that individuals with these germ-line mutations would be predisposed to neoplasia. Indeed, individuals have been identified whose somatic tissues demonstrate increased rates of various genomic alterations; these individuals acquire primary neoplasias at an increased rate compared to the population as a whole. We have examined cells from individuals who are at risk for malignancy to ask if they can be identified by an alteration in a functional assay that monitors cell cycle checkpoint status (CPS). We find that a subset of individuals who are predisposed to neoplasia can be identified prior to the manifestation of disease.

Our studies with cells from patients with Li-Fraumeni syndrome (LFS) led us into this line of inquiry. We noticed, both from our own work (T. Tlsty et al., in prep.) and from the literature (Bischoff et al. 1990), that fibroblasts from individuals with LFS begin with a normal diploid karyotype when cultured in vitro. Upon continued passage of these cells in culture, an accumulation of genomic alterations and rearrangements is observed. These individuals are heterozygous at their $p53$ loci, carrying a specific germ-line mutation in addition to one wild-type allele. Given the connection between aberrant cell cycle checkpoint function and the generation of genomic alterations described earlier, we examined the checkpoint status of these heterozygous LFS cells while they still retained a normal karyotype. Using this assay, we can distinguish cells from people with two wild-type alleles of $p53$ from those with only one. Figure 4B,C shows that early-passage LFS heterozygotes continue to incorporate BrdU (i.e., continue to cycle) under conditions when normal cells cease. Earliest passages, as well as later passages, gave similar results; the population as a whole continues to cycle at a time when normal cells have arrested (note response at day 2). Notice that the percentage of cells in the G_1 and G_2 phases of the cell cycle remains constant over a span of 6 days or longer. This means that cells are not accumulating in either the G_1 or the G_2 phase of the cell cycle. If they were, the cells incorporating BrdU (represented by the white area) would eventually accumulate in one of those phases.

This assay allows us to detect individuals in the general population who have LFS, i.e., predisposition to a subset of malignancies. This is a useful way of identifying these people for several reasons. Currently, several groups are developing methods to rapidly de-

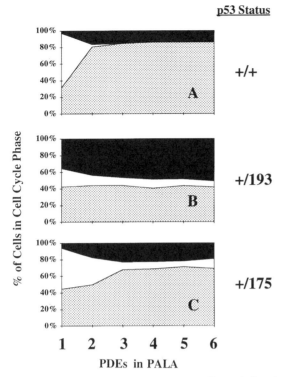

Figure 4. Flow cytometric analysis of LSF cell populations in PALA. Cells were treated and data are displayed as in Fig. 2. *A* is the profile of a normal human fibroblast; *B* and *C* represent fibroblasts from heterozygotes with LFS.

tect mutations in the *p53* gene sequence. However, there are multiple ways to lose *p53* function while retaining a wild-type genotype. Hence, alteration of the metabolism of *p53* in some cases may have similar effects to mutating the gene. Protein turnover, mutation in the gene pathways that modify (phosphorylate, etc.) *p53*, as well as alterations in genes that regulate the extent and duration of *p53* expression would affect the phenotype of the cell without altering a wild-type *p53* genotype. We are just becoming aware of cellular gene products such as mdm2 which play an important role in *p53* modulation (Finlay et al. 1989). Exogenous agents such as DNA tumor viruses can also alter *p53* function without altering its DNA sequence (Vousden and Jat 1989). The CPS assay would complement the screening techniques that use sequencing described above and, in addition, detect additional alterations because it examines the phenotype of the cell with regard to cell cycle control. The different genotypes described above can lead to the same phenotype: alteration of CPS.

The extent of application of this assay will depend on what fraction of malignancies are initiated by checkpoint aberrations. This remains to be tested. In its narrowest application, the preliminary data would suggest that a select group of neoplasias begin with an anomaly in cell cycle checkpoint control and that this flaw fuels their progression toward tumorigenicity and malignancy. The data on the checkpoint status in the LFS patients demonstrate this clearly. Likewise, the data in the case of high-risk HPV are convincing in this regard. Functional inactivation of the p53 protein by E6 and the members of the Rb family by E7 in normal cells promotes an immediate disruption of checkpoint control and allows the accumulation of genomic abnormalities such as translocations, amplifications, and deletions, which are known to activate oncogenes and inactivate tumor suppressor genes (White et al. 1994). These proteins have direct effects on the cell cycle machinery and its superimposed regulators. In a broader application, many neoplasias may have some fundamental flaw with cell cycle control either directly, as in the case of cyclin overexpression, or indirectly, such as with the sensing of DNA lesions. Both types of malfunctions would have similar endpoints: improper coordination of cell cycle progression with environmental cues. The gene products that metabolize environmental carcinogens are also integrated into this model, since it is the cellular response to these insults that maintains genomic integrity. Cellular responses to these signals unveil the underlying (sometimes subtle) flaws in the choreography of proliferative processes. Other conditions that predispose toward malignancy remain to be examined and may represent independent pathways of response.

Note Added in Proof

Much of this information was published previously by White et al. (1994) in *Genes & Development*.

ACKNOWLEDGMENTS

We thank Drs. M. Cordeiro-Stone and Denise Galloway for generously providing cells and retroviral constructs, respectively. This research was supported by National Institutes of Health grant CA-51912 to T.D.T.

REFERENCES

Baker, C.C., W.C. Phelps, V. Lindgren, M.J. Braun, M.A. Gondas, and P.M. Howley. 1987. Structural and translational analysis of human papillomavirus type 16 sequences in cervical carcinoma cell lines. *J. Virol.* **61:** 962.

Barbosa, M.S., C. Edmonds, C. Fisher, J.T. Schiller, D.R. Lowy, and K.H. Vousden. 1990. The region of the HPV E7 oncoprotein homologous to adenovirus E1a and SV40 large T antigen contains separate domains for Rb binding and casein kinase II. *EMBO J.* **9:** 153.

Benn, P.A. 1976. Specific chromosome aberrations in senescent fibroblast cell lines derived from human embryos. *Am. J. Hum. Genet.* **28:** 465.

Bischoff, F.Z., S.O. Yim, S. Pathak, G. Grant, J.J. Siciliano, B.C. Giovanella, L.C. Strong, and M.A. Tainsky. 1990. Spontaneous abnormalities in normal fibroblasts from patients with Li-Fraumeni cancer syndrome: Aneuploidy and immortalization. *Cancer Res.* **50:** 7979.

Debbas, M. and E. White. 1993. Wild-type p53 mediates apoptosis by E1A, which is inhibited by E1B. *Genes Dev.* **7:** 546.

Dhanwada, K.R., V. Veerisetty, F. Zhu, A. Razzaque, K.D. Thompson, and C. Jones. 1992. Characterization of primary human fibroblasts transformed by human papilloma virus type 16 and herpes simplex type 2 DNA sequences. *J. Gen. Virol.* **73:** 791.

Durst, M., C. Kleinheinz, M. Hotz, and L. Gismann. 1985. The physical state of human papillomavirus type 16 DNA in benign and malignant genital tumours. *J. Virol.* **66:** 1515.

Dyson, N., P. Guida, K. Munger, and E. Harlow. 1992. Homologous sequences in adenovirus E1A and human papillomavirus E7 proteins mediate interaction with the same set of cellular proteins. *J. Virol.* **66:** 6893.

Dyson, N., P.M. Howley, K. Munger, and E. Harlow. 1989. The human papillomavirus-16 E7 onco protein is able to bind to the retinoblastoma gene product. *Science* **243:** 934.

Finlay, C.A., P.W. Hinds, and A.J. Levine. 1989. The p53 proto-oncogene can act as a suppressor of transformation. *Cell* **57:** 1083.

Gage, J.R., C. Meyers, and F.O. Wettstein. 1990. The E7 proteins of the nononcogenic human papillomavirus type 6b (HPV-6b) and of the oncogenic HPV-16 differ in retinoblastoma protein binding and other properties. *J. Virol.* **46:** 723.

Halbert, C.L., G.W. Demers, and D.A. Galloway. 1991. The E6 and E7 genes of human papillomavirus type 6 have weak immmortalizing activity in human epithelial cells. *J. Virol.* **66:** 2125.

Hartwell, L.H. and T.A. Weinert. 1989. Checkpoints: Controls that ensure the order of cell cycle events. *Science* **246:** 629.

Jonczyk, P., A. White, A. Lum, J.C. Barrett, and T.D. Tlsty. 1993. Amplification potential in preneoplastic and neoplastic Syrian hamster embryo fibroblasts transformed by various carcinogens. *Cancer Res.* **53:** 3098.

Kastan, M.B., O. Onyekwere, D. Sidransky, B. Vogelstein, and R.W. Craig. 1991. Participation of p53 protein in the cellular response to DNA damage. *Cancer Res.* **51:** 6304.

Kuerbitz, S.J., B.S. Plunkett, W.V. Walsh, and M.B. Kastan. 1992. Wild-type p53 is a cell cycle checkpoint determinant following irradiation. *Proc. Natl. Acad. Sci.* **89:** 7491.

Livingstone, L.R., A. White, J. Sprouse, E. Livanos, T. Jacks, and T.D. Tlsty. 1992. Altered cell cycle arrest and gene amplification potential accompany loss of wild-type p53. *Cell* **70:** 923.

Ma, C., S. Martin, B. Trask, and J.L. Hamlin. 1993. Sister chromatid fusion initiates amplification of the dihydrofolate reductase gene in Chinese hamster cells. *Genes Dev.* **7:** 605.

McClintock, B. 1939. The behavior in successive nuclear divisions of a chromosome broken at meiosis. *Proc. Natl. Acad. Sci.* **25:** 405.

———. 1942. The fusion of broken ends of chromosomes following nuclear fusion. *Proc. Natl. Acad. Sci.* **28:** 458.

Munger, K., W.C. Phelps, V. Bubb, P.M. Howley, and R. Schlegel. 1989a. The E6 and E7 genes of the human papillomavirus type 16 together are necessary and sufficient for transformation of primary human keratinocytes. *J. Virol.* **63:** 4417.

Munger, K., B.A. Werness, N. Dyson, W.C. Phelps, E. Harlow, and P.M. Howley. 1989b. Complex formation of human papillomavirus E7 proteins with the retinoblastoma tumor suppressor gene product. *EMBO J.* **8:** 4099.

Otto, E., S. McCord, and T. Tlsty. 1989. Increased incidence of CAD gene amplification in tumorigenic rat lines as an indicator of genomic instability of neoplastic cells. *J. Biol. Chem.* **264:** 3390.

Pater, M.M. and A. Pater. 1985. Human papillomavirus types 16 and 18 sequences in carcinoma cell lines of the cervix. *Virology* **145:** 313.

Pirisi, L., S. Yasumoto, M. Feller, J. Doniger, and J.A. DiPaolo. 1987. Transformation of human fibroblasts and keratinocytes with human papillomavirus type 16 DNA. *J. Virol.* **61:** 1061.

Rao, L., M. Debbas, P. Sabbatini, D. Hockenbery, S. Korsmeyer, and E. White. 1992. The adenovirus E1A proteins induce apoptosis, which is inhibited by the E1B 19-kDa and Bcl-2 proteins. *Proc. Natl. Acad. Sci.* **89:** 7742.

Romanczuk, H. and P.M. Howley. 1992. Disruption of either the E1 or the E2 regulatory gene of human papillomavirus type 16 increases viral immortalization capacity. *Proc. Natl. Acad. Sci.* **89:** 3159.

Sager, R., I. Gadi, L. Stephens, and C.T. Grabowy. 1985. Gene amplification: An example of accelerated evolution in tumorigenic cells. *Proc. Natl. Acad. Sci.* **82:** 7015.

Schaefer, D.I., E.M. Livanos, A.E. White, and T.D. Tlsty. 1993. Multiple mechanisms of N-(phosphonoacetyl)-L-aspartate drug resistance in SV40-infected precrisis human fibroblasts. *Cancer Res.* **53:** 4946.

Scheffner, M., B.A. Werness, J.M. Huibregtse, A.J. Levine, and P.M. Howley. 1990. The E6 oncoprotein encoded by human papillomavirus types 16 and 18 promotes the degradation of p53. *Cell* **63:** 1129.

Schlegel, R., W.C. Phelps, Y.-L. Zhang, and M. Barbosa. 1988. Quantitative keratinocyte assay detects two biological activities of human papillomavirus DNA and identifies viral types associated with cervical carcinoma. *EMBO J.* **7:** 3181.

Schwarz, E., U.K. Freese, L. Gissman, W. Mayer, B. Roggenbuck, A. Stremlau, and H. zur Hausen. 1985. Structure and transcription of human papillomavirus sequences in cervical carcinoma cells. *Nature* **314:** 111.

Smith, K.A., P.A. Gorman, M.B. Stark, R.P. Groves, and G.R. Stark. 1990. Distinctive chromosomal structures are formed very early in the amplification of CAD genes in Syrian hamster cells. *Cell* **63:** 1219.

Smotkin, D. and F.O. Wettstein. 1986. Transcription of human papillomavirus type 16 early genes in a cervical cancer and a cancer-derived cell line and identification of the E7 protein. *Proc. Natl. Acad. Sci.* **83:** 4680.

Stark, G.R. and G.M. Wahl. 1984. Gene amplification. *Annu. Rev. Biochem.* **53:** 447.

Tlsty, T.D. 1990. Normal diploid human and rodent cells lack a detectable frequency of gene amplification. *Proc. Natl. Acad. Sci.* **87:** 3132.

Tlsty, T.D., B. Margolin, and K. Lum. 1989. Differences in the rates of gene amplification in nontumorigenic and tumorigenic cell lines as measured by Luria-Delbrück fluctuation analysis. *Proc. Natl. Acad. Sci.* **86:** 9441.

Tlsty, T.D., A. White, and J. Sanchez. 1992. Suppression of gene amplification in human cell hybrids. *Science* **255:** 1425.

Toledo, F., D. Le Roscouet, G. Buttin, and M. Debatisse. 1992. Co-amplified markers alternate in megabase long chromosomal inverted repeats and cluster independently in interphase nuclei at early steps of mammalian gene amplification. *EMBO J.* **11:** 2665.

Tommasino, M., J.P. Adamczewski, F. Carlotti, C.F. Barth, R. Manetti, M. Contorni, F. Cavalieri, T. Hunt, and L. Crawford. 1993. HPV16 E7 protein associates with the protein kinase p33CDK2 and cyclin A. *Oncogene* **8:** 195.

Trask, B. and J. Hamlin. 1989. Early dihydrofolate reductase gene amplification events in CHO cells usually occur on the same chromosome arm as the original locus. *Genes Dev.* **3:** 1913.

Vousden, K.H. and P.S. Jat. 1989. Functional similarity between HPV16 E7, SV40 large T and adenovirus E1a proteins. *Oncogene* **4:** 153.

Watanabe, S., T. Kanda, and K. Yoshiike. 1989. Human papillomavirus type 16 transformation of primary human embryonic fibroblasts requires expression of open reading frames E6 and E7. *J. Virol.* **63:** 965.

Werness, B.A., A.J. Levine, and P.M. Howley. 1990. Association of human papillomavirus types 16 and 18 E6 proteins with p53. *Science* **248:** 76.

White, A.E., E.M. Livanos, and T.D. Tlsty. 1994. Differential disruption of genomic integrity and cell cycle regulation in normal human fibroblasts by the HPV oncoproteins. *Genes Dev.* **8:** 666.

White, E. and B. Stillman. 1987. Expression of adenovirus E1B mutant phenotypes is dependent on the host cell and on synthesis of E1A proteins. *J. Virol.* **61:** 426.

White, E., B. Faha, and B. Stillman. 1986. Regulation of

adenovirus gene expression in human WI38 cells by an E1B-encoded tumor antigen. *Mol. Cell. Biol.* **6:** 3763.

Wilczynski, S.P., L. Pearlman, and J. Walker. 1988. Identification of HPV 16 early genes retained in cervical carcinomas. *Virology* **166:** 624.

Windle, B., B.W. Draper, Y. Yin, S. O'Gorman, and G.M. Wahl. 1991. A central role for chromosome breakage in gene amplification, deletion formation, and amplicon integration. *Genes Dev.* **5:** 160.

Wright, J.A., H.S. Smith, F.M. Watt, M.C. Hancock, D.L. Hudson, and G.R. Stark. 1990. DNA amplification is rare in normal human cells. *Proc. Natl. Acad. Sci.* **87:** 1791.

Yin, Y., M.A. Tainsky, F.Z. Bischoff, L.C. Strong, and G.M. Wahl. 1992. Wild-type p53 restores cell cycle control and inhibits gene amplification in cells with mutant p53 alleles. *Cell* **70:** 937.

Yonish-Rouach, E., D. Grinwald, S. Wilder, A. Kimchi, E. May, J. Lawrence, P. May, and M. Oren. 1993. p53-mediated cell death: Relationship to cell cycle control. *Mol. Cell. Biol.* **13:** 1415.

zur Hausen, H. 1991. Human papillomaviruses in the pathogenesis of anogenital cancer. *Virology* **184:** 9.

DNA Damage Responses: p53 Induction, Cell Cycle Perturbations, and Apoptosis

C.E. Canman, C.-Y. Chen, M.-H. Lee, and M.B. Kastan

The Johns Hopkins Oncology Center, Baltimore, Maryland 21287

Epidemiology studies have suggested that exposure to DNA-damaging agents contributes to the development of at least 80% of human cancers (Doll and Peto 1981). Therefore, understanding the various steps involved in controlling cellular responses to DNA-damaging agents is an important component to understanding mechanisms involved in human tumorigenesis. There are three potential outcomes for cells following exposure to DNA-damaging agents (Fig. 1): (1) The cell repairs the DNA damage in a timely fashion so that subsequent progression through the cell cycle results in normal daughter cells; (2) the cell dies, and therefore cannot go on to become a tumor cell; and (3) the damage results in permanent genetic alterations in the cell which are then passed on to subsequent daughter cells. When the appropriate combination of genetic changes occurs in a particular cell type, a transformed phenotype can eventually result. The gene products that determine which of these potential outcomes will occur should have a major influence on the likelihood of development of a malignant tumor.

The gene products that determine the likelihood of developing permanent genetic alterations following exposure of an individual to DNA-damaging agents can be divided into two main categories (Fig. 1): (1) those gene products that determine the amount of damage which is delivered to the DNA and (2) those gene products that dictate cellular responses to the inflicted damage. Many environmental carcinogens are not direct DNA-damaging agents and need to be metabolized to ultimate DNA-damaging agents. Efficient metabolism of procarcinogens to carcinogens would potentially

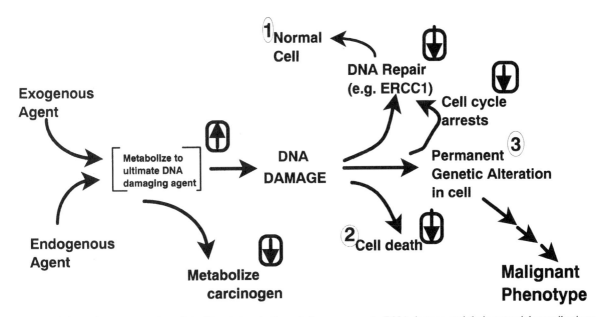

Figure 1. Schematic representation of significant steps in the cellular responses to DNA damage and their potential contributions to tumorigenesis. Either exogenous or endogenous agents can be metabolized to ultimate DNA-damaging agents or be direct-acting agents; increased activation of compounds could increase the extent of DNA damage. Some of these agents can be rapidly metabolized to inactive compounds, which would limit the extent of DNA damage. In response to the DNA damage, the cell could (1) efficiently repair the damage so that no physiologically significant permanent DNA alterations are passed on to daughter cells; (2) die (and therefore could not become malignant); or (3) survive without having adequately repaired its DNA and thus pass on potentially significant mutations to daughter cells. Cell cycle arrests theoretically permit repair of the damage prior to significant cellular events such as DNA replication or chromosome segregation. An accumulation of appropriate genetic changes in the daughter cells could result in a transformed phenotype. Variability in the efficiency of several of the steps in this scheme could account for different individual susceptibilities to various types of DNA-damaging events, ranging from metabolism to DNA repair to cell survival. Encircled arrows suggest whether the particular step would lead to increased (arrow up) or decreased (arrow down) incidence of cellular transformation.

increase the amount of DNA damage after exposure to a given level of an environmental carcinogen. Alternatively, individuals who efficiently metabolize ultimate carcinogens to inactive compounds would experience reduced amounts of DNA damage which occurs after such an exposure. It should also be recognized that much of the DNA damage which occurs to cells may not result directly from DNA damage caused by the environmental agent, but may result from damage caused by endogenous agents, such as oxidative damage (Reid et al. 1991). Although no clear examples exist yet, possibly because of the subtle physiological implications, the gene products that modulate the amount of DNA damage which occurs in such settings may eventually be categorized as "tumor susceptibility" genes.

Once cellular DNA is damaged, the cellular response to that damage will dictate the likelihood of the development of heritable genetic alterations in daughter cells (Fig. 1). The first obvious critical step is the removal of the damage and repair of the DNA with fidelity. Individuals with suboptimal DNA repair have an increased chance of developing permanent genetic alterations. In addition to the repair process itself, the timing of the repair relative to critical cellular processes, such as DNA replication and mitosis, is also critical. For example, it would not be optimal for a cell to replicate its DNA or segregate its chromosomes prior to repair of the chromosomal DNA. Cellular outcome is also influenced by gene products that determine cell survival following DNA damage. For certain cell types, such as lymphoid cells, programmed cell death (apoptosis) occurs rapidly following exposure to certain types of genotoxic agents. Therefore, survival of a cell in an inappropriate physiological situation (i.e., if the apoptosis response is missing) can also contribute to increased genetic alterations in daughter cells. As discussed below, it is intriguing that many of the gene products which dictate cell cycle progression, and thus the timing of repair relative to DNA replication and mitosis, also influence these cell death decisions.

Dysfunction of these gene products which mediate these important cellular responses to DNA damage would also seem to be a risk factor for cancer susceptibility. In fact, suboptimal responses to DNA-damaging agents appear to be responsible for the majority of known cancer susceptibility syndromes (Table 1). For example, ataxia telangiectasia, xeroderma pigmentosum, Bloom's syndrome, and Fanconi's anemia are all cancer-prone syndromes with abnormalities in DNA damage response pathways. All of these susceptibility syndromes are severe homozygous, recessive defects. However, the colon cancer susceptibility syndrome, hereditary nonpolyposis colon cancer, has recently been shown to result from mutations in mismatch DNA repair genes (Fishel et al. 1993; Leach et al. 1993), and this more subtle susceptibility syndrome results from a heterozygous germ-line abnormality. In addition, the multiple cancer susceptibility syndrome, Li-Fraumeni syndrome, appears to result from heterozygous germ-line mutations in the p53 gene (Malkin et al. 1990; Srivasta et al. 1990). Since p53 plays a critical role in DNA damage responses (see below), this provides another example of a more subtle syndrome resulting from suboptimal responses to DNA damage. Further examples of such susceptibility genes are likely to arise.

The focus of our laboratory has been to try to elucidate the signals controlling the timing of repair processes relative to critical cellular events such as DNA replication and mitosis. It has been well documented in the literature that cell cycle perturbations occur following exposure of cells to DNA-damaging agents (Little 1968; Tolmach et al. 1977; Denekamp 1986; Weinert and Hartwell 1988). Perhaps these cell cycle arrests after DNA damage occur to allow the cell to repair the damage prior to these important cellular events so there is not replication of a damaged DNA template nor segregation of damaged DNA chromo-

Table 1. Cancer Susceptibility Syndrome

Cancer susceptibility syndromes	Defect/disease/gene
Fanconi's anemia[a]	repair of cross-links/AML/FACC, chr 9[b]
Ataxia telangiectasia[a]	IR sensitivity, cell cycle/NHL/11q23[b]
Xeroderma pigmentosum[a]	excision repair/skin cancer/ERCC2,3,5
Bloom's syndrome[a]	SCEs/leukemia and other cancers/locus unknown
Li-Fraumeni syndrome[a]	multiple tumors/p53
HNPCC[a]	colon, ovary, endometrium CA/mutS, mutL
FAP	colonic polyps, colon CA/APC
Von Hippel-Lindau	renal cell CA/VHL, 3p
MEN I,II	endocrine CAs/11q, 1p, 10q
Familial Wilms' tumor	Wilms' tumor/WT-1
Familial retinoblastoma	retinoblastoma, osteosarcoma/Rb
Familial breast CA	breast CA/BRCA-1, 17q

An incomplete list of cancer susceptibility syndromes, many of which appear to arise from defective responses to DNA damage.

[a] Cancer susceptibility associated with abnormalities in responses to DNA damage.

[b] Chromosomal localization of defective genes in at least one, but maybe not all, complementation groups of the disease.

somes. The concept that such cell cycle "checkpoints" are important was first illustrated in yeast, where the mutation in the *rad9* gene resulted in loss of the G_2/M checkpoint and in increased sensitivity to ionizing radiation and increased genetic instability (Weinert and Hartwell 1988; Hartwell and Weinert 1989). Cell cycle arrest in both G_1 and G_2 has also been documented in mammalian cells. We have been studying the gene products that control cell cycle arrest in mammalian cells after DNA damage (Fig. 2). This "decision fork" for the cell (whether to arrest in the cell cycle or whether to replicate with the damage still present) is likely to impact on cellular fates, including cell survival and genetic fidelity. We have elucidated a signal transduction pathway that is critically dependent on the tumor suppressor gene, *p53*, which mediates the arrest of cells in the G_1 phase of the cycle following DNA damage. In the remainder of this paper, we discuss our current understanding of this important cell cycle checkpoint signal transduction pathway.

RESULTS AND DISCUSSION

p53 Is Required for G_1 Arrest after DNA Damage

Cell cycle perturbations following DNA damage have been documented in all eukaryotic systems from yeast to human cells. In our initial studies, we chose to use ionizing radiation (IR) as our source of DNA damage because it required the fewest manipulations of the cultured cells. Utilizing a combination of bromodeoxyuridine (BrdU) labeling to assess DNA synthesis and propidium iodide staining to assess DNA content, we evaluated cell cycle perturbations in ML-1 human myeloid leukemia cells at various times following exposures to various doses of ionizing radiation (Kastan et

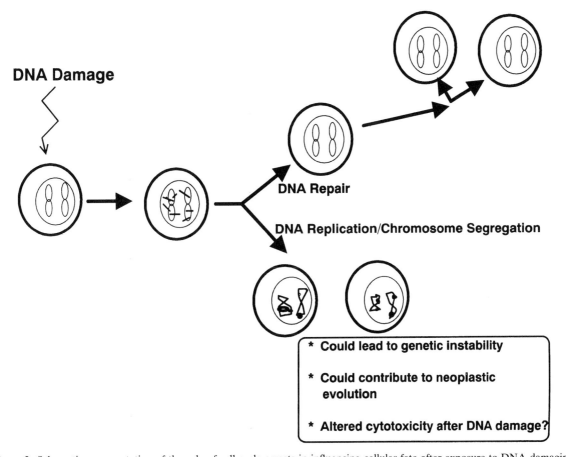

Figure 2. Schematic representation of the role of cell cycle arrests in influencing cellular fate after exposure to DNA-damaging agents. After DNA damage, cells might die or might never replicate again; these two options are not depicted here. Another possible alternative is for the cell to repair the DNA damage prior to replicative DNA synthesis and prior to chromosome segregation. Cell cycle arrests in the G_1 and G_2 phases of the cell cycle after DNA damage would increase the length of time available for the cell to repair its DNA prior to these critical cellular events of DNA replication and mitosis. The fourth alternative, a failure to adequately repair DNA prior to DNA replication and/or chromosome segregation, could result in heritable genetic alterations in daughter cells. Accumulation of such genetic alterations could contribute to the array of genetic changes that lead to a transformed phenotype in a particular cell type. This "decision fork" for the cell could also influence the likelihood of cell survival after exposure to the DNA-damaging agent; manipulation of cell survival in this situation has the potential to be useful in antineoplastic therapeutic strategies.

al. 1991). This type of cell cycle analysis provided a dynamic view of the cell cycle and allowed us to clearly delineate the various cell cycle perturbations following irradiation (Fig. 3). Almost immediately following this DNA damage, cells arrested in both the G_1 and the G_2 phases of the cell cycle. Cells that were already in S phase only transiently arrested and continued to progress through the cell cycle until they reached G_2, where they stopped. In this cell line exposed to 2 Gy of irradiation, by 24 hours there were few cells left in S phase and all the cells were arrested in G_1 and in G_2. By 48–72 hours after exposure to this dose of irradiation, the cells had recovered to a relatively normal-looking cell cycle pattern.

We reasoned that an arrest of the cell cycle following DNA damage could result from either inhibition of positive growth signals or stimulation of negative growth signals. We therefore chose to screen for changes in the levels of expression of a number of protein products that were known to mediate either positive or negative growth signals. At the time these studies were being performed, the p53 gene product had only recently been characterized as a suppressor of cell growth. One of the protein products that changed most dramatically following exposure of cells to DNA-damaging agents was the p53 protein, where we noticed a rapid increase in steady-state levels of p53 following ionizing irradiation (Kastan et al. 1991). Kinetic evaluation of the changes in p53 protein after damage revealed a very rapid increase in p53 protein levels after damage, with levels peaking 1–3 hours after the ionizing radiation. Interestingly, by 48–72 hours after irradiation, the levels of p53 protein were returning toward normal (Kastan et al. 1991). This time course of increase in p53 levels was virtually identical to the time course of decrease of DNA synthesis in these same cells at these same doses of irradiation. This was the first suggestion that these changes in p53 protein levels might be mediating the cell cycle arrest following DNA damage.

It would have been an easy task to rule out the role of p53 in cell cycle arrest after DNA damage, since many tumor cell lines have mutant *p53* genes. However, evaluation of cell cycle perturbations after irradiation in a number of human tumor cell lines revealed that cell lines with wild-type *p53* genes arrested in both G_1 and G_2 phases of the cell cycle after irradiation, whereas tumor cell lines with mutant *p53* alleles only arrested in the G_2 phase of the cell cycle after irradiation (Kastan et al. 1991, 1992; Kuerbitz et al. 1992). This observation demonstrated (1) that p53 could be mediating the G_1 arrest following irradiation and (2) that p53 was not required for the arrest of cells in G_2 following irradiation. A cause and effect relationship between expression of wild-type *p53* and the G_1 arrest that occurs after γ irradiation was then established by demonstrating (1) acquisition of the G_1 arrest after γ irradiation following transfection of wild-type *p53* genes into cells lacking endogenous *p53* genes; (2) loss of the G_1 arrest after irradiation following transfection of mutant *p53* genes into cells with wild-type endogenous *p53* genes; and (3) that primary murine fibroblasts lacking *p53* genes failed to arrest in G_1 after irradiation, whereas normal murine fibroblasts exhibited this arrest (Kastan et al. 1992; Kuerbitz et al. 1992).

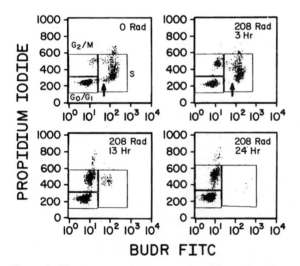

Figure 3. Time course of cell cycle changes in ML-1 cells exposed to IR. Cell cycle distributions in control cells (0 Rad) or cells at 3, 13, and 24 hr after exposure to 208 Rad. Flow cytometric dot plots display simultaneous analysis of S-phase DNA synthesis (determined after a 4-hr pulse with bromodeoxyuridine [BUDR] at the various times after IR) on the ordinate and DNA content (determined by staining with PI) on the abcissa. Cell cycle populations are characterized as G_0/G_1 (2N DNA content with no BUDR incorporation), S phase (variable DNA content with BUDR incorporation), and G_2/M (4N DNA content with no BUDR incorporation during the pulse period). Arrows denote the area containing cells which have recently entered S phase from G_1. Accumulation of cells in both G_1 and G_2 phases of the cell cycle after IR is noted. (Reprinted, with permission, from Kastan et al. 1991.)

Mechanisms Involved in p53 Induction

The previously discussed studies established the critical role of p53 induction in mediating the G_1 cell cycle arrest following irradiation. However, little was known about the mechanistic steps involved in the signaling pathways resulting in p53 induction after damage. There were no changes in the levels of p53 mRNA after irradiation, thus demonstrating that the increase in p53 protein levels after was a posttranscriptional event. It has been demonstrated by other groups (Maltzman and Czyzyk 1984; Tishler et al. 1993), and confirmed by us, that there is an increase in the half-life of p53 protein after DNA damage. However, our current data suggest that stabilization of p53 protein is not the only mechanism by which p53 levels are increasing after DNA damage. The rapidity of the increase of p53 protein suggests that there is a significant component of translational regulation as well. The observation that cycloheximide blocks both p53 induction and the G_1 arrest DNA damage (Kastan et al. 1991; Tishler et al.

1993) is consistent with, but does not prove, this concept. Further studies are under way to better characterize this translational regulation of p53 after damage.

Recent experimental results have demonstrated that the signal which initiates this signal transduction pathway following exposure to DNA-damaging agents is in fact in the DNA and appears to be a DNA strand break. First, although virtually all types of DNA-damaging agents will eventually result in p53 induction, the DNA-damaging agents that most rapidly and most efficiently induce p53 protein in all cell types are agents which directly induce DNA strand breaks, such as γ irradiation, bleomycin, and topoisomerase inhibitors. Second, DNA strand breaks were shown to be sufficient for initiating p53-dependent signal transduction by the presence of rapid p53 protein elevations following introduction of nucleases into cells by electroporation (Nelson and Kastan 1994). Although DNA strand breaks appear to be capable of triggering p53 induction, DNA lesions other than strand breaks did not. Although UV irradiation is a potent inducer of p53, we demonstrated that base damage, such as thymine dimers, is not a sufficient signal for inducing p53, but requires DNA strand breaks produced by repair of the dimers or by replication with the dimers present. This dependence of p53 induction after base damage on DNA replication and/or DNA repair probably explains some of the variability in time courses and efficiency of p53 induction in different cell lines following exposure to such base-damaging agents. Finally, the enhanced induction of p53 protein following low-dose ionizing radiation in cellular DNA substituted with halogenated pyrimidines (IdU), which only affects the number of DNA strand breaks occurring after irradiation, further demonstrates the role of DNA strand breaks in mediating this signal transduction process (Zhan et al. 1994a).

Ataxia telangiectasia (AT) is a human autosomal recessive disorder with many abnormal phenotypic characteristics, including hypersensitivity to ionizing radiation, a markedly increased incidence of cancer (particularly lymphoblastic lymphomas), progressive cerebellar ataxia with degeneration of Purkinje cells, and abnormalities in cell cycle perturbations following IR (McKinnon 1987; Gatti et al. 1991). Specifically, defects in G_1, S phase, and G_2 arrests following irradiation have all been reported in AT cells (Imray and Kidson 1983; Nagasawa et al. 1985; Rudolph and Latt 1989). We examined whether the lack of the G_1 arrest in AT cells could result from a failure of p53 induction following IR. In a number of AT lymphoblast and fibroblast cell lines, we have found little to no induction of p53 protein 1 hour following exposure of the cells to low-dose IR (Kastan et al. 1992). This suggested that the gene product which is defective in individuals with AT is required for optimal induction of p53 protein following IR. Failure to induce p53 after IR because of this defect probably accounts for the lack of the G_1 arrest in AT cells following IR. However, the AT defect is much more pleiotropic in the sense that defects in the AT gene(s) affect not only the p53-dependent G_1 checkpoint, but also S-phase arrest, G_2 arrest, and sensitivity to ionizing radiation.

Whether AT cells are actually defective in p53 induction after IR has recently been questioned by Lu and Lane (1993), since they observed increased p53 immunofluorescence of AT cells several hours after irradiation. Since Lu and Lane observed increases in p53 immunofluorescence within 1 hour after irradiation in normal cells, but did not observe increased immunostaining in AT cells for several hours, it is unclear why they concluded that AT cells induced p53 normally. DNA strand breaks following IR are repaired very rapidly; since AT cells in their experiments failed to induce p53 until several hours after the damage, this is likely to be too late to be of physiological significance. Therefore, the timing of induction of p53 is a critical feature of the optimal cellular response to DNA damage. In addition, immunofluorescence is not an accurate quantitative assessment of p53 induction, and our data and data of other laboratories suggest that some AT cells induce p53 protein after IR, but at abnormally low levels. Several other laboratories have also recently published confirmatory evidence that AT cells are defective in p53 induction (Khanna and Lavin 1993; Dulic et al. 1994; Price and Park 1994). In addition, we have extended our original studies to include numerous different AT homozygous and heterozygous cell lines, carefully evaluating time courses and dose responses of p53 induction after IR (Canman et al. 1994). In summary, the bulk of the published data strongly suggests that one manifestation of dysfunction of the gene responsible for AT is a failure to optimally induce p53 after IR. This places the AT gene products in the "proximal" arm of this pathway (Fig. 4).

Mechanisms by Which p53 Mediates Growth Arrest

Although it was clear that overexpression of p53 protein (mediated either by forced expression from a transfected expression vector [Baker et al. 1990; Diller et al. 1990; Martinez et al. 1991] or by induction of endogenous gene product following DNA damage) resulted in a G_1 cell cycle arrest, it was not clear how p53 was mediating this cell cycle arrest. Theoretically, p53 could be directly arresting the cell cycle by acting as a structural protein; for example, p53 could inhibit formation or progression of replication forks. Alternatively, p53 had recently been shown to be a DNA-binding protein and could act as a transcription factor, and the G_1 arrest could be mediated via specific transcriptional activation of certain gene products. We, and other workers, have now identified three gene products that appear to be transcriptionally activated by p53 following exposure of cells to DNA-damaging agents. The roles of these gene products in mediating cell cycle arrest after damage are still being elucidated.

The first gene product whose expression following IR was shown to be dependent on normal p53 function was the *GADD45* gene (Kastan et al. 1992). *GADD45* was

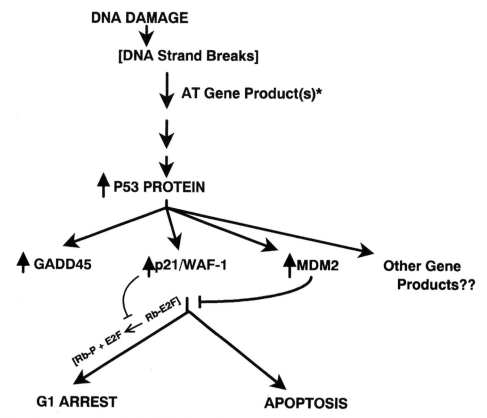

Figure 4. Schematic representation of some of the characterized steps in p53-dependent responses to DNA damage. DNA damage results in a posttranscriptional increase in levels of p53 protein, probably via induction of DNA strand breaks. Optimal induction of p53 appears to require normal function of the gene product defective in the cancer-prone disease, ataxia telangiectasia (AT), after certain types of DNA damage, particularly ionizing radiation. Accumulation of p53 protein results in transcriptional activation of several gene products, including GADD45, $p21^{WAF1/CIP1}$, and MDM2. Current studies suggest that GADD45 and $p21^{WAF1/CIP1}$ probably contribute to the G_1 cell cycle arrest which occurs following DNA damage and p53 induction, and that MDM2 may function in a "feedback" loop mechanism to limit the length or severity of the G_1 arrest. Although the mechanism by which GADD45 functions is not yet clear, it has been suggested that $p21^{WAF1/CIP1}$ inhibits cell cycle progression by preventing activation of cyclin/cdk complexes which are required for phosphorylation of Rb (and other Rb-like proteins); failure to phosphorylate Rb-like proteins presumably leads to G_1 arrest because of a failure to release the E2F family of transcription factors which are required for progression into S phase. In certain cell types and/or physiological situations, this pathway appears to result in programmed cell death or apoptosis, rather than in a G_1 arrest. Elucidation of the factors which determine whether the cell will arrest in G_1 or undergo apoptotic cell death is currently an area of intense investigation.

initially identified in subtraction libraries of cells following DNA damage compared to cells that were not exposed to DNA-damaging agents (Fornace et al. 1989). It was subsequently shown to be induced by a wide variety of DNA-damaging agents and other treatments that elicit growth arrest, such as serum reduction. Although base-damaging agents such as methyl methanesulfonate (MMS) and UV irradiation appeared to induce GADD45 in all tumor cell lines evaluated, IR seemed to induce GADD45 in only a subset of the tumor cell lines evaluated. This observation prompted us to investigate whether GADD45 induction following IR required normal p53 function. We demonstrated that significant induction of GADD45 mRNA only occurred in cells with wild-type p53 (Kastan et al. 1992). Cells with abnormal p53 function, whether it was due to *p53* gene deletion, gene mutation, or overexpression of viral gene products that interfere with p53 function (such as HPV-16E6 or SV40 large T-antigen), failed to significantly increase GADD45 levels following IR. A consensus p53-binding site was found in a conserved sequence within intron 3 of the *GADD45* gene (Kastan et al. 1992). Wild-type, but not mutant, p53 protein binds to this GADD45 intronic sequence, and cotransfection of wild-type, but not mutant, p53 expression vectors with a reporter construct containing this GADD45 sequence results in transcription of the reporter gene. Finally, electrophoretic mobility shift assays demonstrate binding activities for this GADD45 oligonucleotide in cells with wild-type p53 which have been irradiated, but not in irradiated cells that lack p53 function.

These observations demonstrated that GADD45 induction after IR is dependent on normal p53 function and is probably mediated by transcriptional enhancement resulting from direct binding of wild-type p53 protein to this GADD45 intronic sequence. It is intriguing that DNA damage caused by other types of agents, such as UV and MMS, induces GADD45 in a p53-independent manner. This induction of GADD45 by other agents appears to be mediated through GADD45 promoter sequences, rather than through the intronic sequence which we propose mediates induction following ionizing irradiation. The function of the GADD45 protein has not yet been elucidated. Overexpression of the *GADD45* gene appears to be able to decrease colony formation in transfected cells (Zhan et al. 1994b), however, its role in p53-mediated G_1 arrest has not yet been established. Experiments are currently under way to further characterize its function.

Another gene product that has recently been shown to be dependent on normal p53 function for induction after IR is MDM2 (Barak et al. 1993; Wu et al. 1993). Based on earlier observations that conversion of a temperature-sensitive p53 mutant into a wild-type conformation resulted in increased MDM2 expression (Barak et al. 1993), we investigated whether MDM2 expression would be affected in a p53-dependent manner after DNA damage. We found that MDM2 levels rise after IR in a p53-dependent fashion and that this change in MDM2 expression is regulated at the level of transcription (Chen et al. 1994). The time course of changes on MDM2 protein levels after IR lags about 2 hours behind the changes in p53 protein levels, consistent with a dependence on p53. MDM2 is an intriguing gene product because it has recently been shown to bind to the amino terminus of p53 protein and to inhibit its ability to activate transcription (Momand et al. 1992; Oliner et al. 1993). Overexpression of MDM2 has also been shown to overcome the tumor suppressor effects of p53 in transfection assays (Finlay 1993). We therefore investigated whether overexpression of MDM2 could inhibit this physiological function of p53 in the DNA damage response pathway. We found that overexpression of MDM2, whether due to natural amplification of the gene or constitutive expression of a transfected MDM2 expression vector, resulted in abrogation of this p53-dependent checkpoint (Chen et al. 1994). These two observations taken together, that p53 can induce MDM2 and that overexpression of MDM2 inhibits p53 function, suggest that MDM2 may be working in this pathway in a "feedback-loop" mechanism (Fig. 4). However, to demonstrate this more directly, experiments in which MDM2 function is abrogated, and the consequences of this manipulation on p53-mediated G_1 arrest, will have to be examined.

A third gene product dependent on normal p53 function for induction following DNA damage has recently been shown to be the $p21^{WAF1/CIP1}$ cyclin-dependent kinase inhibitor (El-Deiry et al. 1993, 1994; Dulic et al. 1994). Our investigations of $p21^{WAF1/CIP1}$ induction began with a very different question: Where in G_1 is the p53-mediated arrest occurring? To characterize where the p53-dependent cell cycle arrest was occurring, we evaluated the phosphorylation status of the retinoblastoma (Rb) protein at various times after IR. We observed accumulation of the hypophosphorylated form of Rb after IR in cells with wild-type p53, but not in cells with mutant p53 (Slebos et al. 1994). This demonstrated that the DNA-damage-induced, p53-dependent arrest was occurring at a discrete stage in G_1 itself (prior to Rb phosphorylation) and not at the G_1/S border. What was not clear at this time, however, was whether the retinoblastoma protein was a functional downstream component of this pathway. To begin to address this, we blocked Rb function by stable expression of high-risk HPV E7 protein. Although overexpression of HPV E7 did not affect p53 induction after IR, it did block the arrest of cells in G_1 (Slebos et al. 1994). This suggested that Rb or other Rb-like proteins were required for DNA-damage-induced G_1 arrest and that HPV E7 was blocking the arrest "downstream" of p53. Elimination of Rb alone (in fibroblasts from "Rb-knockout" mice) resulted in a milder effect on the G_1 checkpoint. This suggests that other Rb-like molecules are requisite downstream participants in this p53-dependent pathway.

Since Rb is phosphorylated by cyclin-dependent kinases, and since $p21^{WAF1/CIP1}$ is a p53-dependent gene product whose function is to inhibit cyclin/CDK complexes (El-Deiry et al. 1993, 1994; Harper et al. 1993; Xiong et al. 1993; Dulic et al. 1994), we asked whether $p21^{WAF1/CIP1}$ is a participant in this pathway as well. We found that DNA damage does increase $p21^{WAF1/CIP1}$ levels in a p53-dependent fashion after IR (El-Deiry et al. 1994; Slebos et al. 1994). Expression of HPV E7 did not affect induction of $p21^{WAF1/CIP1}$ after IR, thus demonstrating that the E7 abrogation of the G_1 arrest is acting downstream of $p21^{WAF1/CIP1}$ induction. These data are consistent with, but do not prove, the suggestion that $p21^{WAF1/CIP1}$ may be the critical mediator of the p53-dependent G_1 arrest after DNA damage. To prove this definitively, however, p21 function will have to be abrogated and result in failure of cells with wild-type p53 to arrest in G_1 after IR. The presumed mechanism by which $p21^{WAF1/CIP1}$ would be mediating this p53-dependent arrest is inhibition of phosphorylation of Rb and/or other Rb-like proteins, thus preventing release of E2F-like transcription factors and preventing transcription of gene products necessary for the G_1 to S transition (Fig. 4) (Slebos et al. 1994).

G_1 Arrest Versus Apoptosis

The initial description of cell cycle checkpoints was based on studies of the mutations in the *rad9* gene in yeast (Weinert and Hartwell 1988, 1990; Hartwell and Weinert 1989). The definition of a cell cycle checkpoint based on this work included that the gene products were nonessential, inhibitory, and would feed forward in a signal transduction pathway. The role of p53 in mediating the G_1 arrest in mammalian cells after IR fit these criteria. However, two manifestations of *rad9* gene dysfunction were not included in this definition. First of all, when the *rad9* gene was mutated, there was increased spontaneous and markedly increased DNA-damage-induced genetic instability in the yeast cells. In addition, loss of *rad9* gene function resulted in marked increased sensitivity to ionizing radiation. We sought to determine whether these two physiological ramifications also occurred in cells when p53 was mutated and the G_1 checkpoint was abrogated. Work from a number of different laboratories has suggested an important role for p53 in minimizing genetic instability. This includes correlations of p53 dysfunction with increased gene amplification (Livingstone et al. 1992; Yin et al. 1992), chromosomal aneuploidy (Livingstone et al. 1992; Carder et al. 1993), and increased recombination events (S. Meyn, pers. comm.).

To address the question of alterations in radiosensitivity with loss of the p53-dependent G_1 checkpoint, we evaluated clonogenic survival in cells that were isogenic except for the status of their p53 function. We observed no difference in clonogenic survival at various doses after ionizing radiation in parental RKO colorectal carcinoma cells, control transfected RKO cells, or RKO cells expressing HPV-16 E6 (which promotes p53 degradation, thereby rendering cells "p53-null" and abrogating the G_1 arrest) or expressing a dominant negative *p53* gene (Slichenmyer et al. 1993). This demonstrated that simple loss of the G_1 checkpoint did not result in increased sensitivity to IR. Treatment of the RKO cells with caffeine abrogated both the G_1 and G_2 arrests and made the cells significantly more sensitive to IR. This observation suggests, but does not prove, that the G_2 arrest may be a more important determinant of cell survival following irradiation. This would be in agreement with the significant impact that the *rad9* mutation has on survival in yeast after IR and on the sensitivity of AT cells to IR (since AT cells have lost both the G_1 and G_2 arrests).

Recent elegant work from a few different laboratories has further demonstrated that loss of p53 function does not lead to increased radiosensitivity. In certain tissues, loss of p53 function instead appears to result in marked radio*resistance*. This radioresistance was shown to result from a failure of these cells to undergo IR-induced apoptosis (Clarke et al. 1993; Lowe et al. 1993). The cell types in which these studies were performed were lymphoid in origin, and the observations probably reflect the physiological tendency of these cells to undergo apoptosis following IR. In other words, it appeared that certain cell types, such as fibroblasts and RKO colorectal carcinoma cells, tend to undergo cell cycle arrest after IR, whereas cells of hematopoietic origin tend to undergo a programmed cell death after IR. Both the G_1 arrest and the apoptosis in the different cell types appear to be dependent on normal p53 function. This cell-type specificity in cellular fate after p53 induction after IR should have been predicted on the basis of earlier work with the temperature-sensitive *p53* mutant gene in rodent cells. Moshe Oren's group initially showed that transfection of this temperature-sensitive gene into murine fibroblasts and conversion to the nonpermissive temperature resulted in a reversible cell cycle arrest (Michalovitz et al. 1990). In contrast, introduction of this gene and switching to the nonpermissive temperature in growth-factor-dependent hematopoietic cells resulted in apoptosis when growth factor was limiting (Yonisch-Rouach et al. 1991).

When the p53-dependent apoptosis observations were initially made, there were some reports hypothesizing that a little bit of DNA damage can result in cell cycle arrest, whereas a significant amount of DNA damage results in apoptosis. This hypothesis was not based on any published data. We suggested, on the basis of our data at the time and the results with the ts p53 mutant discussed above, that p53 induction by DNA damage results in cell cycle arrest in some physiological settings and in apoptosis in other physiological settings (Slichenmyer et al. 1993) (Fig. 4). (This does not mean that one cannot overwhelm the cell with a lot of DNA damage and cause cell death; it just means that at a given low amount of damage, some cells are programmed either to die or to arrest.) We have recently developed a cell culture system which demonstrates that at a given dose of DNA damage, a cell can "choose" whether to arrest in G_1 or to undergo apoptosis. In a growth-factor-dependent murine hematopoietic cell line, we have demonstrated that 4 Gy of IR results in rapidly reversible G_1 arrest if growth factor is present and results in rapid apoptosis after IR if growth factor is absent. Both the G_1 arrest and the apoptosis are dependent on normal p53 function. Therefore, the growth factor is acting as a "survival factor" for the cell, allowing it to transiently arrest and then recover following a given dose of irradiation. It is not yet clear what gene products in the cell are determining how to interpret this p53 signal to mediate either the arrest or apoptosis, but this is a system in which such a question can be investigated. At this stage, we have observed no differences in induction of p53 in the two conditions of irradiation, nor have we observed differences in the expression of Bcl-2, Bax, Bcl-x, cyclin A, or cyclin E as a function of growth factor or irradiation. We are continuing to use this system to investigate this important decision fork which determines cellular fate following DNA damage.

Implications for Human Tumorigenesis and Therapy

The studies discussed above have significant implications for our understanding of mechanisms of human tumor development. Abnormalities in these p53-dependent pathways could contribute to the development of a transformed phenotype in two ways: (1) In cell types in which p53 mediates the G_1 arrest after DNA damage, loss of p53 function and loss of the G_1 arrest would result in the replication of a damaged DNA template and likely in the development of heritable genetic changes in daughter cells; and (2) loss of p53 function and development of the signals leading to programmed cell death after DNA damage would lead to survival of cells in inappropriate physiological situations. This latter situation could be particularly detrimental, since the cell is surviving despite the DNA damage; since the p53-mutant cell would also not have the G_1 arrest signal, it is surviving the damage (because of the loss of apoptosis) and it is still replicating a damaged template. Therefore, these two concepts may in fact work in

Table 2. Association of Apoptosis Tendency and Tumor Curability

Potentially curable with chemotherapy	
childhood ALL	? *Apoptosis* +
germ-cell tumors	certain neuroblastomas
Burkitt's lymphoma (NHL)	Wilms' tumor
Hodgkin's lymphoma	~30% of myeloid leukemias
Not usually curable with chemotherapy	
colon carcinoma	? *Apoptosis* −
prostate carcinoma	breast carcinoma
esophageal CA	lung carcinomas
renal CA	ovarian CA
	pancreatic CA

There are only a subset of surgically unresectable tumors that can be cured with subsequent chemotherapy or radiation therapy. It is conceivable that one factor which contributes to the curability of these tumors is a tendency to undergo *rapid* apoptosis following exposure to these therapeutic DNA-damaging agents; in contrast, it may be difficult to induce significant, *rapid* apoptosis in the unresected incurable tumor cells. One potential mechanism of improving therapeutic outcome is the ability to induce rapid apoptosis in the latter, relatively resistant group.

concert to contribute to genetic instability in human tissues following DNA damage.

It is easy to envision how manipulation of this G_1 arrest versus apoptosis decision fork could be manipulated for therapeutic advantage in treatment of human tumors. For tumors that have wild-type *p53* genes and derive from a cell type that uses the p53 signal to undergo G_1 arrest, the tumor cells might be likely to survive following exposure to DNA-damaging chemotherapeutic agents or therapeutic radiation. In contrast, tumor cells which undergo apoptosis rapidly after DNA damage in this p53-dependent manner would be expected to be more responsive tumors. If one evaluates the tumor types that respond to radiation therapy and systemic chemotherapy, most of these tumor types derived from cells which probably use this DNA damage-induced p53 signal to undergo apoptosis (Table 2). Most of the tumor types that are historically incurable with chemotherapy and radiation therapy probably derived from cell types which do not readily undergo p53-dependent apoptosis after DNA damage. The ability to manipulate the tendency of cells to undergo apoptosis rather than G_1 arrest might be of significant therapeutic benefit for these historically resistant tumors. In other words: Can we make a breast carcinoma tumor cell "think" like a thymocyte and therefore be more sensitive to systemic chemotherapy or radiation therapy? For the approximately 50% of tumor types which have mutant p53, a more difficult task remains. In this situation, we either need to somehow restore wild-type p53 function to the cells or to induce apoptosis by initiating the signal transduction pathway "distal" to p53. Thus, further characterization of the genetically defined biochemical mediators of the signal transduction pathways will not only be very useful in improving our understanding of mechanisms of human tumorigenesis, but may also provide new options for cancer prevention and cancer therapy.

ACKNOWLEDGMENTS

This work was supported in part by grants from the National Institutes of Health (ESO-5777 and CA-61949) and the Council for Tobacco Research (3187). The excellent secretarial assistance of Debbie Stankowski is appreciated.

REFERENCES

Baker, S.J., S. Markowitz, E.R. Fearon, J.K.V. Willson, and B. Vogelstein. 1990. Suppression of human colorectal carcinoma cell growth by wild-type p53. *Science* **249:** 912.

Barak, Y., T. Juven, R. Haffner, and M. Oren. 1993. mdm2 expression is induced by wild type p53 activity. *EMBO J.* **12:** 461.

Canman, C.E., A.C. Wolff, C.-Y. Chen, A.J. Fornace, Jr., and M.B. Kastan. 1994. The p53-dependent G_1 cell cycle checkpoint pathway and ataxia-telangiectasia. *Cancer Res.* **54:** 5054.

Carder, P., A.H. Wyllie, C.A. Purdie, R.G. Morris, S. White, J. Piris, and C.C. Bird. 1993. Stabilised p53 facilitates aneuploid clonal divergence in colorectal cancer. *Oncogene* **8:** 1397.

Chen, C., J.D. Oliner, Q. Zhan, A.J. Fornace, Jr., B. Vogelstein, and M.B. Kastan. 1994. Interactions between p53 and MDM2 in a mammalian cell cycle checkpoint pathway. *Proc. Natl. Acad. Sci.* **91:** 2684.

Clarke, A.R., C.A. Purdie, D.J. Harrison, R.G. Morris, C.C. Bird, M.L. Hooper, and A.H. Wylie. 1993. Thymocyte apoptosis induced by p53-dependent and independent pathways. *Nature* **362:** 849.

Denekamp, J. 1986. Cell kinetics and radiation biology. *Int. J. Radiat. Biol.* **49:** 357.

Diller, L., J. Kassel, C.E. Nelson, M.A. Gryka, G. Litwak, M. Gebhardt, B. Bressac, M. Ozturk, S.J. Baker, B. Vogelstein, and S.H. Friend. 1990. p53 functions as a cell cycle control protein in osteosarcomas. *Mol. Cell. Biol.* **10:** 5772.

Doll, R. and R. Peto. 1981. The causes of cancer in the United States today. *J. Natl. Cancer Inst.* **66:** 1192.

Dulic, V., W.K. Kaufmann, S.J. Wilson, T.D. Tlsty, E. Lees, W. Harper, S.J. Elledge, and S.I. Reed. 1994. p53-dependent inhibition of cyclin-dependent kinase activities in human fibroblasts during radiation-induced G_1 arrest. *Cell* **76:** 1013.

El-Deiry, W.S., T. Tokino, V.E. Velculescu, D.B. Levy, R. Parsons, J.M. Trent, D. Lin, W.E. Mercer, K.W. Kinzler, and B. Vogelstein. 1993. WAF1, a potential mediator of p53 tumor suppression. *Cell* **75:** 817.

El-Deiry, W.S., J.W. Harper, P.M. O'Conner, V.E. Velculescu, C.E. Canman, J. Jackman, J.A. Pietenpol, M. Burrell, D.E. Hill, Y. Wang, K.G. Wiman, W.E. Mercer, M.B. Kastan, K.W. Kohn, S.J. Elledge, K.W. Kinzler, and B. Vogelstein. 1994. WAF1/CIP1 is induced in p53-mediated G_1 arrest and apoptosis. *Cancer Res.* **54:** 1169.

Finlay, C.A. 1993. The *mdm2* oncogene can overcome wild-type p53 suppression of transformed cell growth. *Mol. Cell. Biol.* **13:** 301.

Fishel, R., M.K. Lecoe, M.R.S. Rao, N.G. Copeland, N.A. Jenkins, J. Garber, M. Kane, and R. Kolodner. 1993. The human mutator gene homolog *MSH2* and its association with hereditary nonpolyposis colon cancer. *Cell* **75:** 1027.

Fornace, A.J., Jr., D.W. Nebert, M.C. Hollander, J.D. Luethy, M.A. Papathanasiou, J. Fargnoli, and N.J. Holbrook. 1989. Mammalian genes coordinately regulated by growth arrest signals and DNA-damaging agents. *Mol. Cell. Biol.* **9:** 4196.

Gatti, R.A., E. Boder, H.V. Vinters, R.S. Sparkes, A. Norman, and K. Lange. 1991. Ataxia-telangiectasia: An inderdisciplinary approach to pathogenesis. *Medicine* **70:** 99.

Harper, J.W., G.R. Adami, N. Wei, K. Keyormarsi, and S.J. Elledge. 1993. The p21 Cdk-interacting protein Cip1 is a potent inhibitor of G_1 cyclin-dependent kinases. *Cell* **75:** 805.

Hartwell, L.H. and T.A. Weinert. 1989. Checkpoints: Controls that ensure the order of cell cycle events. *Science* **246:** 629.

Imray, F.P. and C. Kidson. 1983. Perturbations of cell-cycle progression in gamma-irradiated ataxia telangiectasia and Huntington's disease cells detected by DNA flow cytometric analysis. *Mutat. Res.* **112:** 369.

Kastan, M.B., O. Onyekwere, D. Sidransky, B. Vogelstein, and R.W. Craig. 1991. Participation of p53 protein in the cellular response to DNA damage. *Cancer Res.* **51:** 6304.

Kastan, M.B., Q. Zhan, W.S. El-Deiry, F. Carrier, T. Jacks, W.V. Walsh, B.S. Plunkett, B. Vogelstein, and A.J. Fornace, Jr. 1992. A mammalian cell cycle checkpoint pathway utilizing p53 and GADD45 is defective in ataxia-telangiectasia. *Cell* **71:** 587.

Khanna, K.K. and M.F. Lavin. 1993. Ionizing radiation and UV induction of p53 protein by different pathways in ataxia-telangiectasia cells. *Oncogene* **8:** 3307.

Kuerbitz, S.J., B.S. Plunkett, W.V. Walsh, and M.B. Kastan. 1992. Wild-type p53 is a cell cycle checkpoint determinant following irradiation. *Proc. Natl. Acad. Sci.* **89:** 7491.

Leach, F.S., N.C. Nicholaides, N. Papadopoulos, B. Liu, J. Jen, R. Parsons, P. Peltomaki, P. Sistonen, L.A. Aaltonen, M. Nystrom-Lahti, X.-Y. Guan, J. Zhang, P.S. Meltzer, J.-W. Yu, F.-T. Kao, D.J. Chen, K.M. Cerosaletti, R.E.K. Fournier, S. Todd, T. Lewis, R.J. Leach, S.L. Naylor, J. Weissenbach, J.-P. Mecklin, H. Jarvinen, G.M. Petersen, S.T. Hamilton, J. Green, J. Jass, P. Watson, H.T. Lynch, J.M. Trent, A. de la Chapelle, K.W. Kinzler, and B. Vogelstein. 1993. Mutations of mutS homology in hereditary nonpolyposis colorectal cancer. *Cell* **75:** 1215.

Little, J.B. 1968. Delayed initiation of DNA synthesis in irradiated human diploid cells. *Nature* **218:** 1064.

Livingstone, L.R., A. White, J. Sprouse, E. Livanos, T. Jacks, and T.D. Tlsty. 1992. Altered cell cycle arrest and gene amplification potential accompany loss of wild-type p53. *Cell* **70:** 923.

Lowe, S.W., S.W. Schmitt, S.W. Smith, B.A. Osborne, and T. Jacks. 1993. p53 is required for radiation-induced apoptosis in mouse thymocytes. *Nature* **362:** 847.

Lu, X. and D.P. Lane. 1993. Differential induction of transcriptionally active p53 following UV or ionizing radiation: Defects in chromosome instability syndromes? *Cell* **75:** 765.

Malkin, D., F.P. Li, L.C. Strong, J.F. Fraumeni, Jr., C.E. Nelson, D.H. Kim, J. Kassel, M.A. Gryka, F.A. Bischoff, M.A. Tainsky, and S.H. Friend. 1990. Germ line p53 mutations in a familial syndrome of breast cancer, sarcomas, and other neoplasms. *Science* **250:** 1233.

Maltzman, W. and L. Czyzyk. 1984. UV irradiation stimulates levels of p53 cellular tumor antigen in nontransformed mouse cells. *Mol. Cell. Biol.* **4:** 1689.

Martinez, J., I. Goergoff, and A.J. Levine. 1991. Cellular localization and cell cycle regulation by a temperature-sensitive p53 protein. *Genes Dev.* **5:** 151.

McKinnon, P.J. 1987. Ataxia-telangiectasia: An inherited disorder of ionizing-radiation sensitivity in man. *Hum. Genet.* **75:** 197.

Michalovitz, D., O. Halevy, and M. Oren. 1990. Conditional inhibition of transformation and of cell proliferation by a temperature-sensitive mutant of p53. *Cell* **62:** 671.

Momand, J., G.P. Zambetti, D.C. Olson, D.L. George, and A.J. Levine. 1992. The *mdm-2* oncogene product forms a complex with the p53 protein and inhibits p53-mediated transactivation. *Cell* **69:** 1237.

Nagasawa, H., S.A. Latt, M.E. Lalande, and J.B. Little. 1985. Effects of X-irradiation on cell-cycle progression, induction of chromosomal aberrations and cell killing in ataxia-telangiectasia (AT) fibroblasts. *Mutat. Res.* **148:** 71.

Nelson, W.G. and M.B. Kastan. 1994. DNA strand breaks: The DNA template alterations that trigger p53-dependent DNA damage response pathways. *Mol. Cell. Biol.* **14:** 1815.

Oliner, J.D., J.A. Pietenpol, S. Thiagalingam, J. Gyuris, K.W. Kinzler, and B. Vogelstein. 1993. Oncoprotein mdm2 conceals the activation domain of tumor suppressor p53. *Nature* **362:** 857.

Price, B.D. and S.J. Park. 1994. DNA damage increases the levels of MDM2 messenger RNA in wtp53 human cells. *Cancer Res.* **54:** 896.

Reid, T.M., M. Fry, and L.A. Loeb. 1991. Endogenous mutations and cancer. *Proc. Int. Symp. Princess Takamatsu Cancer Res. Fund* **22:** 221.

Rudolph, N.S. and S.A. Latt. 1989. Flow cytometric analysis of X-ray sensitivity in ataxia telangiectasia. *Mutat. Res.* **211:** 31.

Slebos, R.J.C., M.H. Lee, B.S. Plunkett, T.D. Kessis, B.O. Williams, T. Jacks, L. Hedrick, M.B. Kastan, and K.R. Cho. 1994. p53-dependent G_1 arrest involves pRB-related proteins and is disrupted by the human papillomavirus 16 E7 oncoprotein. *Proc. Natl. Acad. Sci.* **91:** 5320.

Slichenmyer, W.J., W.G. Nelson, R.J. Slebos, and M.B. Kastan. 1993. Loss of a p53-associated G_1 checkpoint does not decrease cell survival following DNA damage. *Cancer Res.* **53:** 4164.

Srivasta, S., Z. Zou, K. Pirollo, W. Blattner, and E.H. Chang. 1990. Germ-line transmission of a mutated p53 gene in a cancer-prone family with Li-Fraumeni syndrome. *Nature* **348:** 747.

Tishler, R.B., S.K. Calderwood, C.W. Coleman, and B.D. Price. 1993. Increases in sequence specific DNA binding by p53 following treatment with chemotherapeutic and DNA damaging agents. *Cancer Res.* **53:** 2212.

Tolmach, L.J., R.W. Jones, and P.M. Busse. 1977. The action of caffeine on X-irradiated HeLa cells. I. Delayed inhibition of DNA synthesis. *Radiat. Res.* **71:** 653.

Weinert, T.A. and L.H. Hartwell. 1988. The *RAD9* gene controls the cell cycle response to DNA damage in *Saccharomyces cerevisiae*. *Science* **241:** 317.

———. 1990. Characterization of *RAD9* of *Saccharomyces cerevisiae* and evidence that its function acts post-translationally in cell cycle arrest after DNA damage. *Mol. Cell. Biol.* **10:** 6554.

Wu, X., H. Bayle, D. Olson, and A.J. Levine. 1993. The p53-mdm-2 autoregulatory feedback loop. *Genes Dev.* **7:** 1126.

Xiong, Y., G.J. Hannon, H. Zhang, D. Casso, R. Kobayashi, and D. Beach. 1993. p21 is a universal inhibitor of cyclin kinases. *Nature* **366:** 701.

Yin, Y., M.A. Tainsky, F.Z. Bischoff, L.C. Strong, and G.M. Wahl. 1992. Wild-type p53 restores cell cycle and inhibits gene amplification in cells with mutant p53 alleles. *Cell* **70:** 937.

Yonish-Rouach, E., D. Resnitzky, J. Lotem, L. Sachs, A. Kimchi, and M. Oren. 1991. Wild-type p53 induces apoptosis of myeloid leukemic cells that is inhibited by interleukin-6. *Nature* **352:** 345.

Zhan, Q., I. Bae, M.B. Kastan, and A.J. Fornace, Jr. 1994a. The p53-dependent gamma-ray response of *GADD45*. *Cancer Res.* **54:** 2755.

Zhan, Q., K.A. Lord, I. Alamo, Jr., M.C. Hollander, F. Carrier, D. Ron, K.W. Kohn, B. Hoffman, D.A. Liebermann, and A.J. Fornace, Jr. 1994b. The *gadd* and *MyD* genes define a novel set of mammalian genes encoding acidic proteins that synergistically suppress cell growth. *Mol. Cell. Biol.* **14:** 2361.

Lymphocyte-specific Genetic Instability and Cancer

I.R. Kirsch,* J.M. Abdallah,* V.L. Bertness,* M. Hale,†
S. Lipkowitz,* F. Lista,*‡ and D.P. Lombardi*

*National Cancer Institute-Navy Medical Oncology Branch, Bethesda, Maryland 20889-5105;
†Radiation Biophysics Department, Armed Forces Radiobiology Research Institute, Bethesda, Maryland 20889-5603;
‡Centro Studi e Ricerche della Sanità dell'Esercito 00183 Rome, Italy

DNA is a fundamentally and inherently unstable structure. It is subject to point mutation, viral insertion, deletion, amplification, translocation, and transposition. This genetic instability is a fundamental force of life, a normal process in every living organism from bacteriophage to humans. It is obviously the basis of evolution. It plays a role in normal development (e.g., bacteriophage excision, yeast mating type switching). It is also the basis of cancer, for every transformed cell that has been studied in the requisite detail has been found to carry one or more genetic alterations in genes affecting growth and/or development. These genetic alterations are tumor-specific markers that distinguish a transformed cell from the normal cells from which it arose. Thus, cancer is a disease of genetic instability. The causes of genetic instability involve factors from within and without a cell. The processes of DNA repair, DNA replication, and cell division are critical events when mutational errors and chromosome breakage can occur. A multitude of environmental factors can also exert a mutagenic effect on DNA which can overwhelm a cell's repair and restorative capabilities. These mutational events are more or less a constant for every cell type, but lymphocytes have, in addition, their own unique inherent genetic instability resulting from the presence of the VDJ and switch recombinational systems. The end product of VDJ recombination, a functional antigen receptor, is so important and compelling in terms of the necessity of immune responsiveness, that its formation via a fundamental example of genetic instability is often overlooked.

First, we present a brief primer on VDJ and switch recombination in lymphocytes. For a more extensive and detailed discussion, the reader is referred to Kirsch and Kuehl (1993).

V(D)J AND "SWITCH" RECOMBINATION IN LYMPHOCYTES

V(D)J Recombination

The DNA that will eventually comprise a mature, functional immunoglobulin (Ig) or T-cell antigen receptor (TCR) initially exists in the germ line as a number of discrete, noncontiguous segments. Although the quantity and even the quality of these segments differ from locus to locus, generically (Fig. 1) a locus comprises tens to hundreds of variable (V) segments, a few to tens of diversity (D) segments (found only in the Ig heavy [H] chain and TCR δ and β chains), one to fifty joining (J) segments, and one or two constant segments (except for the IgH locus, see below). As a prerequisite to the formation of an Ig or TCR gene, one or more site-specific recombination events occur to generate a now contiguous VJ or VDJ region from these previously disparate elements. As a consequence of this recombination, the DNA between the rearranging segments is often deleted and lost to the cell. This reaction is mediated by a group of enzymes, the V(D)J recombinase complex, some components of which are predominantly expressed in lymphoid cells. The complex recognizes specific signal sequences consisting of a heptamer with the consensus CACAGTG separated by either 12 or 23 bases from a nonamer of consensus ACAAAAACC. The enzymatic reaction involves the formation of a double-stranded break at the proximal end of the heptamer sequence. The signal sequences flanking the two rearranging elements are then ligated together and, in the process, the DNA between the rearranging segments is often looped out and deleted from the chromosome. The heptamer abuts the coding sequences of the V, D, or J elements; thus, the rearrangement event occurs such as to directly affect the amino acid sequence and potential translatability of the resultant VJ or VDJ region. In addition to the intersegment chromosomal deletion, the recombinase complex often causes insertions or deletions of a few nucleotides at the junction of the coding regions brought together at the breakpoint. Although it is site-specific, the reaction results in a shift of the translational reading frame between segments in about two-thirds of the rearrangements. Thus, about two-thirds of the rearranged loci cannot be translated into functional antigen receptor polypeptides.

This rearrangement of DNA is lymphocyte-specific and irreversible. The DNA in an Ig- or TCR-producing lymphocyte is therefore no longer the same as that in nonlymphoid cells; it has been reconfigured and, in the majority of cases, has undergone additional deletion and insertion of nucleotides.

Accessibility

In addition to the recombinase complex and appropriate signal sequences, the chromatin structure of the

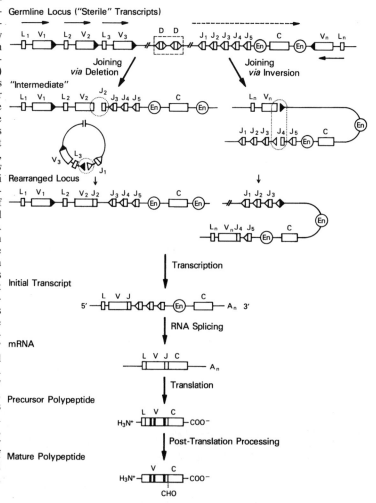

Figure 1. V(D)J recombination. A generalized schematic illustrating the structural reconfigurations of immune receptor loci that are a prerequisite to the formation of a functional polypeptide and therefore to the generation of immune diversity. The gene that will eventually constitute the immune receptor exists initially in germ-line DNA as a set of discrete, noncontiguous segments. There are multiple variable (V) segments with their upstream leader (L) exons that mediate transport of the polypeptide into or through a membrane. For certain of the immune receptor loci (IgH, TCRβ, TCRδ) there are diversity (D) segments. For all the loci, one finds joining (J) segments upstream from the constant (C) segments. Prior to gene rearrangement, "sterile" transcription (*solid* and *dashed arrows* over the first line of the schematic) of the loci about to be rearranged occurs, probably indicative of a change in the status and accessibility of the chromatin at these sites (see text). Structural and irreversible rearrangement of the DNA occurs either by deletion of the DNA between the rearranging segments or by inversion. The reaction is essentially site-specific, mediated by a recombinase enzyme complex that recognizes signal sequences (*solid* and *open triangles*) that flank the rearranging segments. Following recombination, a now contiguous V(D)J region is formed and still can be transcribed from the promoter located 5' of the rearranged V segment, with transcriptional activity "enhanced" by enhancer elements (En) that are found around the DNA flanking the C segment. A precursor RNA is transcribed and processed by the addition of a run of adenine residues (A_n). RNA intervening between coding segments is spliced out, yielding the mature messenger RNA (mRNA). The mRNA is translated into protein, processed (which includes cleavage of the leader peptide and possible glycosylation [CHO] of certain sites within the peptide), and transported to its position in the membrane. (Reprinted, with permission, from Kirsch 1993.)

loci must allow for the DNA/protein interaction that leads to rearrangement. For example, if the necessary recombination enzymes are introduced into a fibroblast, an exogenously supplied recombinogenic substrate will be rearranged, but no Ig or TCR gene rearrangement of the endogenous loci occurs, presumably because these loci are inaccessible in this cell whose function has nothing to do with the formation of immune diversity (Schatz et al. 1989; Oettinger et al. 1990). The general term given to this "recombinogenic" chromatin status is "accessibility." The loci must be accessible to the recombinase in order for recombination to occur.

Even though the nature of accessibility is not completely understood, changes in the state of the chromatin prior to V(D)J rearrangement have been observed. Segments predisposed to recombination show an increased sensitivity to DNase, decreased methylation, and the presence of "sterile" (noncoding) transcripts (Yancopoulos and Alt 1985; Blackwell et al. 1986). The transcripts do not appear to be essential for the rearrangement reaction itself (Hsieh et al. 1992).

More likely, these transcripts are markers for an increased accessibility of the locus to an RNA polymerase, and, by extension, to a V(D)J recombinase complex (Oltz et al. 1993).

Switch Recombination

The Ig heavy-chain locus manifests an additional mechanism of recombination. During the immune response, a switch often occurs between the production of an IgM (μ heavy chain) and production of an IgG, IgE, or IgA. The antibody so made has the same light-chain and the same heavy-chain V region, now linked to a different heavy-chain constant region γ, ϵ, or α. The switch of the V region occurs by a different sort of DNA breakage and rejoining event. The "switch" region located in the intervening sequence 5' of the μ constant region is broken and rejoined to one of the somewhat homologous switch regions found in the intervening sequences 5' of the γ, ϵ, or α constant regions. The switch regions comprise a few thousand nucleotides of somewhat repetitive switch signal sequence DNA.

Thus, unlike V(D)J recombination, the switch reaction occurs within an intervening sequence and is much less site-specific and more consistent with a kind of homologous recombination event. Similar to V(D)J recombination, switch recombination requires locus accessibility, and changes in chromatin status surrounding the target of the switch reaction can be observed prior to switch recombination into that locus (Blackwell and Alt 1989; Gritzmacher 1989).

INTERLOCUS REARRANGEMENTS AND HYBRID GENE FORMATION

VDJ recombination can occur, not only within a given Ig or TCR antigen receptor, but also between two distinct antigen receptors. For example, rearrangements can occur between a V segment from the Ig heavy-chain locus and a J segment from the TCRα locus, or between a V segment from the TCRγ locus and a J segment from the TCRβ locus, or between a V segment from the TCRγ locus and a J segment from the TCRδ locus or vice versa (Denny et al. 1986; Stern et al. 1989a; Tycko et al. 1989). Such rearrangements cause distinctive karyotypic abnormalities which can be observed in peripheral blood T cell metaphases from every individual (Beatty-DeSana et al. 1975; Hecht et al. 1975, 1987; Welch and Lee 1975; Zech and Haglund 1978; Mattei et al. 1979; Aurias et al. 1980; Herva 1981; Wallace et al. 1984). In this context, it is perhaps easier to appreciate VDJ recombination as a fundamental destabilizing force. The frequency of formation of one or another of these interlocus rearrangements is high enough to be observed either by labor-intensive karyotypic analyses on multiple metaphases from the same individual, or by a more sensitive, quicker, and less expensive polymerase chain reaction (PCR)-based assay that we have developed (Fig. 2). Using either of these methods, it is also clear that the frequency of these interlocus rearrangements is low enough in the general population so that differences in their occurrence between individuals or among populations can be assessed and appreciated.

Analysis of one such type of rearrangement, Vγ-Jβ, resulting in an inversion of chromosome 7, inv(7)(p13q35), has been performed in normal peripheral blood. Cloning of the PCR-amplified products followed by sequence analysis of the junctions (Lipkowitz et al. 1990) revealed that the rearrangements are "polyclonal," corresponding to multiple independent events in which different Vγ segments are rearranged to different Jβ segments. Of these rearrangements, 30–50% were "in-frame" at the DNA level, similar to that seen for the more routine rearrangements within one or another of the antigen receptor loci (see above). This suggests that these interlocus rearrangements are not selected against in the thymus simply because they occur between two loci (rather than within one antigen receptor locus). Furthermore, it suggests that such hybrid antigen receptor genes might be functional and add, in some

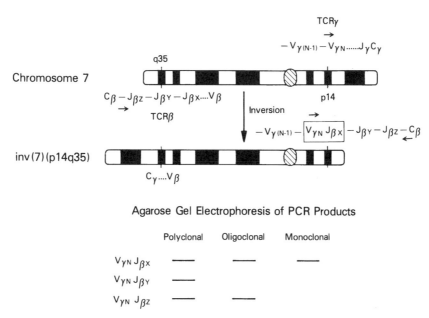

Figure 2. PCR-based assay of interlocus recombination. Consensus primers for the TCRγV segments and primers from the TCRβ locus between the Js and C segments are used to amplify hybrid TCRVγ/Jβ rearrangements. Such rearrangements in the human cause the formation of a chromosomal aberration, inv(7)(p14q35). A single such rearrangement (monoclonal) present in a lymphocyte population would yield a single amplifiable band; multiple independent rearrangements of any of the multiple Vγ segments with any of the multiple Jβ segments would yield amplified products of an array of sizes depending on which Jβ segment was utilized. The different-sized fragments can be distinguished by agarose gel electrophoresis followed by hybridization to Vγ and Jβ probes.

small way, to the diversity of the immune response. In surveys of "normal" populations it appears that 95% of the individuals tested have less than 5 Vγ-Jβ rearrangements per approximately 200,000 peripheral mononuclear cells. This is consistent with the more limited cytogenetic studies which have estimated the frequency of inversion of chromosome 7 in "normal" individuals at less than or equal to 1 in 10,000 phytohemagglutinin-stimulated peripheral T cell metaphases (Aurias 1993). Thus, assessment of the frequency of formation of interlocus rearrangements provides a measure of a kind of genetic instability that is lymphocyte-specific.

INTERLOCUS REARRANGEMENTS IN ATAXIA TELANGIECTASIA

Ataxia telangiectasia (AT) is an autosomal recessive disease of protean manifestations including progressive cerebellar degeneration, oculocutaneous telangiectasia, premature aging, immunodeficiency, radiosensitivity, chromosomal breakage, and an increased risk for the development of cancer, particularly lymphoid malignancies. Perhaps, with the cloning and characterization of the AT gene, the factor(s) that unites these various phenotypic manifestations of the disease will become instantly clear. At least until then, investigation proceeds on one or another of these consequences of the AT mutation.

When the peripheral blood mononuclear cells from individuals who are homozygous for the AT gene are assayed for the kind of lymphocyte-specific genetic instability demonstrated by interlocus rearrangement, it is found that the frequency of formation of such rearrangements is 10- to 100-fold higher than in non-AT individuals (Fig. 3). As in non-AT subjects, the rearrangements are polyclonal, site-specific, and 30–50% are in-frame. Sequence analysis of the junctions reveals no distinguishing aspects of these rearrangements compared with analogous rearrangements from the non-AT population. Although we have focused in particular on the Vγ-Jβ rearrangements, this increased level of interlocus rearrangement is found for most of the potential interlocus combinations studied (Kobayashi et al. 1991). One possible exception is the VH-Jα/δ rearrangement resulting in the inv(14)(q11.2;q32.3) or t(14;14)(q11.2;q32.3) which, although it is the most common aberration found in the peripheral blood of normal individuals (Aurias 1993), may not be significantly increased over this baseline in AT individuals. Individuals with AT also have an increased level of hybrid gene formation in their peripheral B-cell population as exemplified by an increased finding of t(2;14)(p12;q32.3) corresponding to a Igκ–IgH rearrangement (Kirsch et al. 1985; Butterworth and Taylor 1986; Fell et al. 1986). Coincidentally or not, the increased risk for lymphoid malignancy in patients with AT is of the same order of magnitude as the increased frequency of formation of interlocus rearrangement (Gatti and Good 1971; Boder 1985).

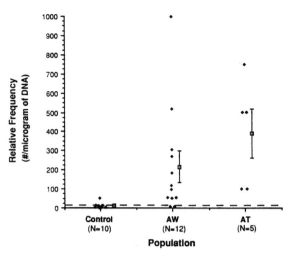

Figure 3. Frequency of TCRVγ-Jβ rearrangements in the peripheral blood mononuclear cell DNA from controls, agriculture workers (AW), and AT patients. Relative frequencies are plotted for each individual in the three populations and represent the mean of two or more independent titrations. Mean relative frequencies ± standard error for each population are also shown. The dashed line is the median value for pooled control and AW populations. Mean relative frequencies were compared by a two-tailed Student's t test. The fractions of each population greater than the pooled median value were compared by a two-tailed Fisher exact test (Lipkowitz et al. 1992).

The reason for this increased frequency of this measure of lymphocyte-specific genetic instability is not yet defined. The VDJ recombinase complex itself does not seem to be abnormal in cells from AT patients. As noted above, the molecular analysis of the end products of the interlocus rearrangement event is consistent with classic VDJ site-specific recombination (Stern et al. 1989a; Tycko et al. 1989; Lipkowitz et al. 1990; Kobayashi et al. 1991). Testing of the VDJ recombination process by introduction of test substrates into cell lines derived from AT patients does not show any basic defect in the recombinase-mediated reaction (Hsieh et al. 1993). Certain data suggest that the basis for this increased frequency of interlocus recombination in AT cells might be a function of a fundamental difference in chromatin accessibility (see above for a discussion of the role of accessibility in VDJ recombination). Homologous recombination between substrates inserted into genomic DNA has been shown to be increased in AT cells compared with non-AT cells (Meyn 1993). No difference was observed between cell lines derived from these two sources for recombination between DNA maintained episomally, suggesting that the observed difference in recombination reflected a difference in AT chromatin status. Another study suggests that there is no difference between the frequency of double strand (ds) DNA breaks in AT and non-AT DNA in response to ionizing radiation, but there is a difference in resultant chromosomal breaks (as assayed by premature chromosomal condensation under conditions designed to minimize the contribution of DNA repair) (Pandita and Hittelman 1992). One suggestion as to the basis of this observation is that AT DNA contains a larger "intermatrix" component which allows the ends of dsDNA breaks to fall away from each other as opposed to being held together by fixation to a nuclear matrix. These data are consistent with the concept of AT DNA being more structurally accessible: more free to move and recombine. It has also been demonstrated that chromosomal DNA in cell lines derived from AT patients shows increased sensitivity (compared to non-AT cells) to digestion by restriction enzymes introduced intracellularly (Liu and Bryant 1993).

Accessibility can be temporal as well as structural. Temporal accessibility implies that the DNA is not structurally different, but that given the normal accessibility of a particular developmental stage, cells that spend more time in that stage will spend more time in a particular accessible state. Either structural or temporal accessibility could be invoked to explain the observation that in TCRδ/γ interlocus rearrangements the TCR Dδ could be found rearranged to either Jγ1 or Jγ2, whereas a Vγ gene in the same normal murine thymocytes would only recombine with Jγ1 (Tycko et al. 1991). Temporal accessibility is a concept that could be invoked to explain an observation well established in tumor genotyping studies. B-cell precursor acute lymphoblastic leukemias of childhood are characterized by blasts that are remarkable for the number of antigen receptor rearrangements that they carry. As many as 80% of such leukemias of B-cell phenotype carry clonal rearrangements of the TCRδ locus; 50% TCRγ; 46% TCRα; and 29% TCRβ (Felix et al. 1990). Normal peripheral blood B cells do not carry this increased frequency of rearrangements, nor do B-cell malignancies such as sporadic Burkitt's lymphoma, follicular lymphoma, CLL, or myeloma, which are believed to be derived from cells transformed from later (and less recombinogenic) B-cell developmental stages. If in AT there was a developmental block which kept lymphocytes in a stage at which the recombinase complex was highly active, then there might be an increased frequency of all types of VDJ recombination, including interlocus rearrangements.

ACQUIRED INCREASES IN INTERLOCUS REARRANGEMENTS

Insight into the mechanism and meaning of this increased frequency of interlocus rearrangement is emerging with the realization that it is not just in one inherited disease that such an increase is seen. In a pilot study performed on a group of agriculture workers from the "leukemia-lymphoma belt" in Southern Minnesota and Northern Iowa, a significant increase in the same type of interlocus rearrangements was seen (Fig. 3) (Lipkowitz et al. 1992). The increased frequency of such rearrangements in this instance was correlated with the use by these individuals of herbicides, pesticides, and fumigants and showed a seasonal variation, increased at the time of exposure in the summer and fall, and within a more normal range in the winter. In this pilot study, a 15- to 30-fold increase in interlocus rearrangement (compared to a control population) was observed. The increased risk of developing lymphoid malignancy in this population is 3- to 7-fold (Weisenburger 1994). An increased frequency of interlocus rearrangements at a certain point in time does not mean that that individual has, or will definitely develop, a lymphoid malignancy. It rather seems to correlate with risk. In agriculture workers exposed to pesticides, the exposure is transient and intermittent. In AT patients, the "exposure" is inherent and constant.

Patients with Hodgkin's disease who are "cured" by chemotherapy face an increased risk for the development of secondary malignancies, particularly hematopoietic malignancies (Tucker et al. 1988; van Leeuwen et al. 1994). In a study in progress, it appears that Hodgkin's disease patients receiving chemotherapy show an increase in lymphocyte-specific genetic instability during the course of their therapy (Fig. 4).

It is not known what, if any, relationship exists between a finding related to environmental or iatrogenic exposure and the same finding related to part of the AT phenotype. Certainly, a variety of exogenous agents can cause changes in chromatin configuration, DNA repair, and therefore, structural accessibility (consider all the known effects of chemotherapeutic agents). It is also possible that the action of certain environmental factors could be directly targeted to the recombination process itself. For example, agents that elevate cAMP can increase the transcription of part of the recombinase complex and lead to an increase in the level of VDJ recombination (Menetski and Gellert 1990).

V(D)J REARRANGEMENT, CLONAL PROLIFERATION, AND LYMPHOID MALIGNANCY

The measure of these innocent Vγ-Jβ interlocus rearrangements may therefore be a marker of risk of development of lymphoid malignancy. If so, it is more than just a fortuitous marker, because the same mechanism responsible for interlocus rearrangement, VDJ recombination, is, at least in part, responsible for the majority of chromosomal aberrations frankly associated with the development of lymphoid malignancy. More than 50% of all lymphoid malignancies carry a rearrangement between an antigen receptor locus and a putative oncogene (Lieber 1993).

The relevance of interlocus rearrangements to malignant or "premalignant" clonal proliferations is underscored by the fact that at least 10% of AT patients will, over time, develop a clonal T-cell population in their peripheral blood that can comprise 100% of the peripheral T-cell metaphases during karyotypic analysis. Karyotypic analyses of these clonal rearrangements

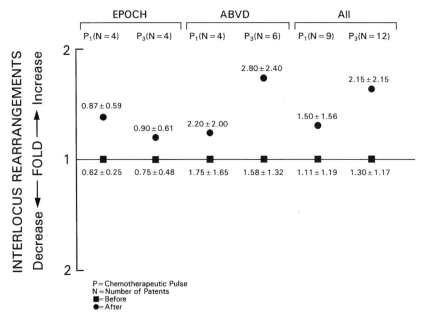

Figure 4. Average increase or decrease of Vγ-Jβ interlocus rearrangements before and after pulse 1 and pulse 3 of chemotherapy. Preliminary results are shown for all patients (All) studied to date as well as for patients stratified into two different chemotherapeutic regimens, "EPOCH" (etoposide, prednisone, vincristine, cyclophosphamide, and doxorubicin) and "ABVD" (doxorubicin, bleomycin, vinblastine, and dacarbazine). A pulse of chemotherapy is defined as all agents given sequentially during therapy and not interrupted by any "rest" days in which no therapy is received.

have revealed that they most frequently are associated with reciprocal translocations or inversions of chromosome 14 with breakpoints at 14q11.2 and 14q32.1 (Hecht et al. 1973; Kohn et al. 1982; Taylor 1982; Aurias et al. 1986). Less frequently, the clonal proliferations are associated with a t(X;14)(q28;q11.2)(McCaw et al. 1975; Canki et al. 1983; Beatty et al. 1986) or t(7;14)(Russo et al. 1988). In numerous cases, the clonal proliferations occur in the absence of frank stigmata of leukemia. The clonal proliferations appear, in general, to represent mature or "pre mature," mitogen-responsive, and TCRα/β-expressing T cells (Butterworth and Taylor 1987; Stern et al. 1989b; Stern 1993). Although many AT patients appear capable of carrying these clonal proliferations with little or no consequence, these clonal proliferations, after having been present for several years, clearly can be associated with transformation to a frankly malignant state (McCaw et al. 1975; Levitt et al. 1978; Saxon et al. 1979; Sparkes et al. 1980; Taylor and Butterworth 1986; Taylor et al. 1992). Similar types of T-cell malignancies have occurred in AT patients for whom no previous clonal expansion had been determined. In addition, morphologically, cytogenetically, and phenotypically similar T-cell malignancies have been observed in non-AT patients, with a particularly striking association of the 14;14 inversion or translocation with the occurrence of T prolymphocytic leukemia (Zech et al. 1984; Brito-Babapulle et al. 1987; Brito-Babapulle and Catovsky 1991).

Molecular cloning of the breakpoints from the AT clonal proliferations or from cytogenetically analogous T-cell malignancies from non-AT individuals clearly demonstrates involvement of the V(D)J recombinase system in the formation of the associated translocation event (Baer et al. 1987; Mengle-Gaw et al. 1987, 1988; Davey et al. 1988; Russo et al. 1988, 1989; Bertness et al. 1990; Fisch et al. 1993; Stern et al. 1993; Thick et al. 1994). The translocations involve the TCRα/δ or β loci on one side breaking site-specifically at the 5′ end of J segments. The breakpoints within 14q32.1 or Xq28 are more dispersed and more complex in terms of their sequence and structure. Some appear to break at cryptic heptamer-nonamer signal sequences, others at stretches of dinucleotide repeats consistent with "Z" DNA structures, others within clear-cut "CpG islands" possibly indicative of proximity to transcriptional regulatory sites. The location of the breakpoints within the antigen receptor loci suggests that proximity to a TCR transcriptional enhancer sequence is a likely mechanism by which the putative growth-promoting genes on Xq28 and 14q32.1 are dysregulated, although structural disruption of the growth-promoting genes has not been ruled out. A transcript unit has been defined at the Xq28 site which is likely to be involved in the process of clonal proliferation; this locus has been provisionally named MTCP1(Stern et al. 1993; Thick et al. 1994). The search for the relevant transcript unit from 14q32.1 is nearing completion (Virgilio et al. 1993; C. Croce, pers. comm.).

There are also striking examples of chromosomal aberrations in which the hallmarks of VDJ recombination are present (site specificity at signal sequences, exonucleolytic "nibbling," non-templated nucleotide addition) but in which neither partner involved in the translocation is an antigen receptor locus. One example of this is the interstitial deletion of chromosome 1p, site-specifically uniting the *sil* and *scl* (*tal-1*) genes (Aplan et al. 1990), which is found in the malignant cells of 20–30% of all patients with T-ALL (Brown et al. 1990; Aplan et al. 1992). The necessary and sufficient features for the occurrence of interlocus rearrangements are likely to be just a variation on the theme of the necessary and sufficient features for the occurrence of these oncogenic and less frequent translocations.

SUMMARY AND SPECULATION

The "innocent" hybrid antigen receptor products of V(D)J recombinase-mediated genetic instability are significantly increased in lymphocytes of patients with the autosomal recessive and cancer-prone disease ataxia telangiectasia, as well as in certain populations with an acquired increased risk for the development of lymphoid malignancy. The basis for this relationship may be that this same type of instability lies mechanistically as the root cause of the majority of chromosomal aberrations frankly associated with malignant transformation of lymphocytes. The basis for this increase may be an enhancement of expression of the V(D)J recombinase complex by environmental agents. At least as likely is the possibility that there is a structural or temporal increase in the accessibility of the genome to the action of the recombinase complex. It is also formally possible that the increase is due to the inability of the cells that we assay to "sense" their own inherent instability and invoke a programmed cell death pathway that mediates their own elimination.

Other types of assays support the suggestion presented here that patients with ataxia telangiectasia (Aurias et al. 1986; Kobayashi et al. 1991), agriculture workers (Carbonell et al. 1993), and Hodgkin's disease patients receiving chemotherapy (Mott et al. 1994) have an increased level of genetic instability. Whether the assay that we have described has any relevance to risk of nonlymphoid malignancies remains to be seen. In this regard, however, it may be relevant that a recent report from a Scandinavian cooperative group (Hagmar et al. 1994) reports that a 5- to 23-year follow-up of individuals whose peripheral blood lymphocytes had been previously studied cytogenetically demonstrated that those individuals who had earlier had the highest level of chromosomal aberrations (but not sister chromatid exchanges) had a significantly increased risk for the development of all cancers, not just lymphoid malignancy. Thus, it is possible that assays of readily available peripheral blood lymphocytes can serve as a substrate for the demonstration of genotoxic stress and cancer risk in general.

REFERENCES

Aplan, P.D., D.P. Lombardi, A.M. Ginsberg, J. Cossman, V.L. Bertness, and I.R. Kirsch. 1990. Disruption of the human SCL locus by "illegitimate" V(D)J recombinase activity. *Science* **250:** 1426.

Aplan, P.D., D.P. Lonbardi, G.H. Reaman, H. Sather, D. Hammond, and I.R. Kirsch. 1992. Involvement of the putative hematopoietic transcription factor SCL in T-cell acute lymphoblastic leukemia. *Blood* **79:** 1327.

Aurias, A. 1993. Acquired chromosomal aberrations in normal individuals. In *The causes and consequences of chromosomal aberrations* (ed. I.R. Kirsch), p. 125. CRC Press, Boca Raton, Florida.

Aurias, A., B. Dutrillaux, D. Buriot, and J. Lejeune. 1980. High frequencies of inversions and translocations of chromosome 7 and 14 in ataxia telangiectasia. *Mutat. Res.* **69:** 369.

Aurias, A., M.F. Croquette, J.P. Nuyts, C. Griscelli, and B. Dutrillaux. 1986. New data on clonal anomalies of chromosome 14 in ataxia telangiectasia: tct(14;14) and inv(14). *Hum. Genet.* **72:** 22.

Baer, R., A. Heppell, A.M.R. Taylor, P.H. Rabbitts, B. Boullier, and T.H. Rabbitts. 1987. The breakpoint of an inversion of chromosome 14 in a T-cell leukemia: Sequences downstream of the immunoglobulin heavy chain locus are implicated in tumorigenesis. *Proc. Natl. Acad. Sci.* **84:** 9069.

Beatty, D.W., L.J. Arens, and M.M. Nelson. 1986. Ataxia-telangiectasia. X, 14 translocation, progressive deterioration of lymphocyte numbers and function, and abnormal in vitro immunoglobulin production. *S. Afr. Med. J.* **69:** 115.

Beatty-DeSana, J.W., M.J. Hoggard, and J.W. Cooledge. 1975. Non-random occurrence of 7-14 translocations in human lymphocyte cultures. *Nature* **255:** 242.

Bertness, V.L., C.A. Felix, O.W. McBride, R. Morgan, S.D. Smith, A.A. Sandberg, and I.R. Kirsch. 1990. Characterization of the breakpoint of a t(14;14)(q11.2;q32) from the leukemic cells of a patient with T-cell acute lymphoblastic leukemia. *Cancer Genet. Cytogenet.* **44:** 47.

Blackwell, T.K. and F.W. Alt. 1989. Mechanism and developmental program of immunoglobulin gene rearrangement in mammals. *Annu. Rev. Genet.* **23:** 605.

Blackwell, T.K., M.W. Moore, G.D. Yancopoulos, H. Suh, S. Lutzker, E. Selsing, and F.W. Alt. 1986. Recombination between immunoglobulin variable region gene segments is enhanced by transcription. *Nature* **324:** 585.

Boder, E. 1985. Ataxia-telangiectasia: An overview. In *Ataxia-telangiectasia: Genetics, neuropathology, and immunology of a degenerative disease of childhood* (ed. R.A. Gatti and M. Swift), p.1. A.R. Liss, New York.

Brito-Babapulle, V. and D. Catovsky. 1991. Inversions and tandem translocations involving chromosome 14q11 and 14q32 in T-prolymphocytic leukemia and T-cell leukemias in patients with ataxia-telangiectasia. *Cancer Genet. Cytogenet.* **55:** 1.

Brito-Babapulle, V., M. Pomfret, E. Matutes, and D. Catovsky. 1987. Cytogenetic studies on prolymphocytic leukemia. II. T cell prolymphocytic leukemia. *Blood* **70:** 926.

Brown, L., J.-T. Cheng, Q. Chen, M.J. Siciliano, W. Crist, G. Buchanan, and R. Baer. 1990. Site-specific recombination of the *tal-1* gene is a common occurrence in human T cell leukemia. *EMBO J.* **9:** 3343.

Butterworth, S.V. and A.M.R. Taylor. 1986. A subpopulation of t(2;14)(p11;q32) cells in ataxia telangiectasia B lymphocytes. *Hum. Genet.* **73:** 346.

———. 1987. A comparison of fresh and cultured T lymphocytes from patients with ataxia telangiectasia using T-cell subset markers and chromosome translocations. *Int. J. Cancer* **39:** 678.

Canki, N., I. Tivader, N. Zupancic, and M. Debevec. 1983. Citgenetska studija sedmih bolnic z ataksijo-telangiektazijo. *Zdrav. Vestn.* **52:** 567.

Carbonell, E., N. Xamena, A. Creus, and R. Marcos. 1993. Cytogenetic biomonitoring in a Spanish group of agricultural workers exposed to pesticides. *Mutagenesis* **8:** 511.

Davey, M.P., V. Bertness, K. Nakahara, J.P. Johnson, O.W. McBride, T.A. Waldmann, and I.R. Kirsch. 1988. Juxtaposition of the T-cell receptor α-chain locus (14q11) and a region (14q32) of potential importance in leukemogenesis by a 14;14 translocation in a patient with T-cell chronic lymphocytic leukemia and ataxia-telangiectasia. *Proc. Natl. Acad. Sci.* **85:** 9287.

Denny, C.T., Y. Yoshikai, T. Mak, S. Smith, G.F. Hollis, and I.R. Kirsch. 1986. A chromosomal inversion in a T cell lymphoma is caused by site-specific recombination between immunoglobulin and T cell receptor loci. *Nature* **320:** 549.

Felix, C.A., D.G. Poplack, G.H. Reaman, S.M. Steinberg, D.E. Cole, B.J. Taylor, C.G. Begley, and I.R. Kirsch. 1990. Characterization of immunoglobulin and T-cell receptor gene patterns in B-cell precursor acute lymphoblastic leukemia of childhood. *J. Clin. Oncol.* **8:** 431.

Fell, H.P., R.G. Smith, and P.W. Tucker. 1986. Molecular analysis of the t(2;14) translocation of childhood chronic lymphocytic leukemia. *Science* **232:** 491.

Fisch, P., A. Forster, P.D. Sherrington, M.J.S. Dyer, and T.H. Rabbitts. 1993. The chromosomal translocation t(X;14)(q28;q11) in T-cell pro-lymphocytic leukaemia breaks within one gene and activates another. *Oncogene* **8:** 3271.

Gatti, R.A. and R.A. Good. 1971. Occurrence of malignancy in immunodeficiency diseases. *Cancer* **28:** 89.

Gritzmacher, C.A. 1989. Molecular aspects of heavy-chain class switching. *Crit. Rev. Immunol.* **9:** 173.

Hagmar, L., A. Brøgger, I.-L. Hansteen, S. Heim, B. Högstedt, L. Knudsen, B. Lambert, K. Linnainmaa, F. Mitelman, I. Nordenson, C. Reuterwall, S. Salomaa, S. Skerfving, and M. Sorsa. 1994. Cancer risk in humans predicted by increased levels of chromosomal aberrations in lymphocytes: Nordic Study Group on the health risk of chromosome damage. *Cancer Res.* **54:** 2919.

Hecht, F., B.K. Hecht, and I.R. Kirsch. 1987. Fragile sites limited to lymphocytes: Molecular recombination and malignancy. *Cancer Genet. Cytogenet.* **26:** 95.

Hecht, F., B. Kaiser McCaw, and R.D. Koler. 1973. Ataxiatelangiectasia—Clonal growth of translocation lymphocytes. *N. Engl. J. Med.* **289:** 286.

Hecht, F., B. Kaiser McCaw, D. Peakman, and A. Robinson. 1975. Non-random occurrence of 7-14 translocations in human lymphocyte cultures. *Nature* **255:** 243.

Herva, R. 1981. Inv(7) as a recurrent aberration in human lymphocyte cultures. *Hereditas* **95:** 163.

Hsieh, C.-L., C.F. Arlett, and M.R. Lieber. 1993. V(D)J recombination in ataxia telangiectasia, Bloom's syndrome, and a DNA ligase I-associated immunodeficiency disorder. *J. Biol. Chem.* **27:** 20105.

Hsieh, C.L., R.P. McCloskey, and M.R. Lieber. 1992. V(D)J recombination on minichromosomes is not affected by transcription. *J. Biol. Chem.* **267:** 15613.

Kirsch, I.R. 1993. Genetics of pediatric tumors: The causes and consequences of chromosomal aberrations. In *Principles and practice of pediatric oncology*, 2nd edition (ed. P.A. Pizzo and D. Poplack), p.29. J.B. Lippincott, Philadelphia.

Kirsch, I.R. and W.M. Kuehl. 1993. Gene rearrangements in lymphoid cells. In *The molecular basis of blood diseases*, 2nd edition (ed. G. Stamatoyannopoulos et al.). W.B. Saunders, Philadelphia.

Kirsch, I.R., J.A. Brown, J. Lawrence, S.J. Korsmeyer, and C.C. Morton. 1985. Translocations that highlight chromosomal regions of differentiated activity. *Cancer Genet. Cytogenet.* **18:** 159.

Kobayashi, Y., B. Tycko, A.L. Soreng, and J. Sklar. 1991. Transrearrangements between antigen receptor genes in normal human lymphoid tissues and in ataxia telangiectasia. *J. Immunol.* **147:** 3201.

Kohn, P.H., J. Whang-Peng, and W.R. Levis. 1982. Chromosomal instability in ataxia telangiectasia. *Cancer Genet. Cytogenet.* **6:** 289.

Levitt, R., R.V. Pierre, W.L. White, and R.G. Siekert. 1978. Atypical lymphoid leukemia in ataxia telangiectasia. *Blood* **52:** 1003.

Lieber, M.R. 1993. The role of site-directed recombinases in physiologic and pathologic chromosomal rearrangements. In *The causes and consequences of chromosomal aberrations* (ed. I.R. Kirsch), p. 239. CRC Press, Boca Raton, Florida.

Lipkowitz, S., V.F. Garry, and I.R. Kirsch. 1992. Interlocus V-J recombination measures genomic instability in agriculture workers at risk for lymphoid malignancies. *Proc. Natl. Acad. Sci.* **89:** 5301.

Lipkowitz, S., M.-H. Stern, and I.R. Kirsch. 1990. Hybrid T cell receptor genes formed by interlocus recombination in normal and ataxia-telangiectasia lymphocytes. *J. Exp. Med.* **172:** 409.

Liu, N. and P.E. Bryant. 1993. Response of ataxia telangiectasia cells to restriction endonuclease induced DNA double-strand breaks. 1. Cytogenetic characterization. *Mutagenesis* **8:** 503.

Mattei, M.G., S. Ayme, J.F. Mattei, Y. Aurran, and F. Giraud. 1979. Distribution of spontaneous chromosome breaks in man. *Cytogenet. Cell Genet.* **23:** 95.

McCaw, B.K., F. Hecht, D.G. Harnden, and R.L. Teplitz. 1975. Somatic rearrangement of chromosome 14 in human lymphocytes. *Proc. Natl. Acad. Sci.* **72:** 2071.

Menetski, J.P. and M. Gellert. 1990. V(D)J recombination activity in lymphoid cell lines is increased by agents that elevate cAMP. *Proc. Natl. Acad. Sci.* **87:** 9324.

Mengle-Gaw, L., D.G. Albertson, P.D. Sherrington, and T.H. Rabbits. 1988. Analysis of a T-cell tumor-specific breakpoint cluster at human chromosome 14q32. *Proc. Natl. Acad. Sci.* **85:** 9171.

Mengle-Gaw, L., H.F. Willard, C.I.E. Smith, L. Hammarstrom, P. Fischer, P. Sherrington, G. Lucas, P.W. Thompson, R. Baer, and T.H. Rabbitts. 1987. Human T-cell tumours containing chromosome 14 inversion or translocation with breakpoints proximal to immunoglobulin joining regions at 14q32. *EMBO J.* **6:** 2273.

Meyn, M.S. 1993. High spontaneous intrachromosomal recombination rates in ataxia-telangiectasia. *Science* **260:** 1327.

Mott, M.G., J. Boyse, M. Hewitt, and M. Radford. 1994. Do mutations at the glycophorin A locus in patients treated for childhood Hodgkin's disease predict secondary leukemia? *Lancet* **343:** 828.

Oettinger, M.A., D.G. Schatz, C. Gorka, and D. Baltimore. 1990. *RAG-1* and *RAG-2*, adjacent genes that synergistically activate V(D)J recombination. *Science* **248:** 1517.

Oltz, E.M., F.W. Alt, W.-C. Lin, J. Chen, G. Taccioli, S. Desiderio, and G. Rathbun. 1993. A V(D)J recombinase-inducible B-cell line: Role of transcriptional enhancer elements in directing V(D)J recombination. *Mol. Cell Biol.* **13:** 3464.

Pandita, T.K. and W.N. Hittelman. 1992. Initial chromosome damage but not DNA damage is greater in ataxia telangiectasia cells. *Radiat. Res.* **130:** 94.

Russo, G., M. Isobe, L. Pegoraro, J. Finan, P.C. Nowell, and C.M. Croce. 1988. Molecular analysis of a t(7;14)(q35;q32) chromosome translocation in a T cell leukemia of a patient with ataxia telangiectasia. *Cell* **53:** 137.

Russo, G., M. Isobe, R. Gatti, J. Finan, O. Batuman, K. Huebner, P.C. Nowell, and C.M. Croce. 1989. Molecular analysis of a t(14;14) translocation in leukemic T-cells of an ataxia telangiectasia patient. *Proc. Natl. Acad. Sci.* **86:** 602.

Saxon, A., R.H. Steven, and D.W. Golde. 1979. Helper and suppressor T-lymphocyte leukemia in ataxia telangiectasia. *N. Engl. J. Med.* **300:** 700.

Schatz, D.G., M.A. Oettinger, and D. Baltimore. 1989. The V(D)J recombination activating gene, *RAG-1*. *Cell* **59:** 1035.

Sparkes, R.S., R. Como, and D.W. Golde. 1980. Cytogenetic abnormalities in ataxia telangiectasia with T cell chronic lymphocytic leukemia. *Cancer Genet. Cytogenet.* **1:** 329.

Stern, M.-H. 1993. Clonal T-cell proliferations in ataxia telangiectasia. In *The causes and consequences of chromosomal aberrations* (ed. I.R. Kirsch), p.165. CRC Press, Boca Raton, Florida.

Stern, M.-H., S. Lipkowitz, A. Aurias, C. Griscelli, G. Thomas, and I.R. Kirsch. 1989a. Inversion of chromosome 7 in ataxia telangiectasia is generated by a rearrangement between T-cell receptor beta and T-cell receptor gamma genes. *Blood* **74:** 2076.

Stern, M.H., I. Theodorou, A. Aurias, M. Maier-Redelsperger, M. Debre, P. Debre, and C. Griscelli. 1989b. T-cell nonmalignant clonal proliferation in ataxia telangiectasia: A cytological, immunological, and molecular characterization. *Blood* **73:** 1285.

Stern, M.-H., J. Soulier, M. Rosenzwajg, K. Nakahara, N. Canki-Klain, A. Aurias, F. Sigaux, and I.R. Kirsch. 1993. *MTCP-1:* A novel gene on the human chromosome Xq28 translocated to the T cell receptor α/δ locus in mature T cell proliferations. *Oncogene* **8:** 2475.

Taylor, A.M.R. 1982. Cytogenetics of ataxia telangiectasia. In *Ataxia-telangiectasia: A cellular link between cancer, neuropathology, and immunodeficiency* (ed. B.A. Bridges and D.G. Harnden), p. 53. Wiley, New York.

Taylor, A.M. and S.V. Butterworth. 1986. Clonal evolution of T-cell chronic lymphocytic leukaemia in a patient with ataxia telangiectasia. *Int. J. Cancer* **37:** 511.

Taylor, A.M., P.A. Lowe, M. Stacey, J. Thick, L. Campbell, D. Beatty, P. Biggs, and C.J. Formstone. 1992. Development of T-cell leukaemia in an ataxia telangiectasia patient following clonal selection in t(X;14)-containing lymphocytes. *Leukemia* **6:** 961.

Thick, J., Y.F. Mak, J. Metcalfe, D. Beatty, and A.M. Taylor. 1994. A gene on chromosome Xq28 associated with T-cell prolymphocytic leukemia in two patients with ataxia telangiectasia. *Leukemia* **8:** 564.

Tucker, M.A., C.N. Coleman, R.S. Cox, A. Varghese, and S.A. Rosenberg. 1988. Risk of second cancers after treatment for Hodgkin's disease. *N. Engl. J. Med.* **318:** 76.

Tycko, B., H. Coyle, and J. Sklar. 1991. Chimeric γ-δ signal joints: Implications for the mechanism and regulation of T cell receptor gene rearrangement. *J. Immunol.* **147:** 705.

Tycko, B., J.D. Palmer, and J. Sklar. 1989. T-cell receptor gene trans-rearrangements: Chimeric γ δ genes in normal lymphoid tissues. *Science* **245:** 1242.

van Leeuwen, F.E., W.J. Klokman, A. Hagenbeek, R. Noyon, A.W. van den Belt-Dusebout, E.H.M. van Kerkhoff, P. van Heerde, and R. Somers. 1994. Second cancer risk following Hodgkin's disease: A 20-year follow-up study. *J. Clin. Oncol.* **12:** 312.

Virgilio, L., M. Isobe, M.G. Narducci, P. Carotenuto, B. Camerini, N.Kurosawa, Abbas-ar-Rushdi, C.M. Croce, and G. Russo. 1993. Chromosome walking on the *TCL1* locus in T-cell neoplasia. *Proc. Natl. Acad. Sci.* **90:** 9275.

Wallace, C., R. Bernstein, and M.R. Pinto. 1984. Non-random in vitro 7:14 translocations detected in a routine cytogenetic series: 12 examples and their possible significance. *Hum. Genet.* **66:** 157.

Weisenburger, D.D. 1994. Epidemiology of non-Hodgkin's lymphoma: Recent findings regarding an emerging epidemic. *Ann. Oncol.* (suppl.1) **5:** S19.

Welch, J.P. and C.L.Y. Lee. 1975. Non-random occurrence of 7-14 translocations in human lymphocyte cultures. *Nature* **255:** 241.

Yancopoulos, G.D. and F.W. Alt. 1985. Developmentally controlled and tissue-specific expression of unrearranged VH gene segments. *Cell* **40:** 271.

Zech, L. and U. Haglund. 1978. A recurrent structural aberration, t(7;14), in phytohemagglutinin-stimulated lymphocytes. *Hereditas* **89:** 69.

Zech, L., G. Gahrton, L. Hammarstrom, G. Juliusson, H. Mellstedt, K.H. Robert, and C.I.E. Smith. 1984. Inversion of chromosome 14 marks human T-cell chronic lymphocytic leukaemia. *Nature* **308:** 858.

Cell Cycle Checkpoint Control Is Bypassed by Human Papillomavirus Oncogenes

D.A. Galloway,*† G.W. Demers,* S.A. Foster,* C.L. Halbert,* and K. Russell‡

Fred Hutchinson Cancer Research Center, †Department of Pathology, and ‡Division of Radiation Oncology, University of Washington, Seattle, Washington 98104

Although the incidence of invasive squamous cell cancer of the cervix has declined in Western countries because of the ability of Pap smear screening to detect premalignant lesions, cervical cancer remains the most common malignancy of women in underdeveloped countries. Both epidemiological and experimental studies clearly indicate that infection by certain high-risk papillomaviruses (HPVs) is a causal step in the development of the tumor (for review, see zur Hausen 1991). Approximately 90% of cervical cancers retain HPV DNA, and the E6 and E7 viral oncogenes are almost invariably expressed. Infection by papillomaviruses precedes the development of cancer by decades, and the tumors that develop have various cytogenetic abnormalities. In culture, the E6 and E7 genes of the cancer-associated HPVs, e.g., types 16 and 18, are able in combination to efficiently immortalize primary human keratinocytes; progression to tumorigenicity requires additional alterations in cellular genes. Thus, it is likely that HPV infection plays a role in promoting the initial changes in normal epithelium that produce dysplasia and that allow subsequent genetic changes to accumulate.

The HPV oncogenes have been shown to bind to, and disrupt the function of, cellular tumor suppressor genes. The E7 proteins of the high-risk HPVs have striking homology with conserved regions 1 and 2 (CR1, CR2) of Ad E1A and with sequences in the large T antigens of SV40 and the polyomaviruses. These sequences (residues 20–29) include the phosphorylation site for casein kinase II (Firzlaff et al. 1989), as well as the binding site for the retinoblastoma protein, p105Rb, and the related p107 and p130 proteins (Dyson et al. 1989, 1992). Binding of E7 to the hypophosphorylated form of Rb releases the transcription factor E2F. The carboxy-terminal half of E7 contains a zinc finger motif. The high-risk HPV E6 proteins bind to p53 (Werness et al. 1990), a property that is in common with SV40 T antigen (but not polyomavirus T) and with adenovirus E1B 55 kD. In the case of SV40 and Ad, the viral oncogenes prolong the half-life of p53 and presumably inactivate p53 function by sequestering the protein. In the case of HPV-16 and -18, the E6 proteins target p53 for ubiquitination, which leads to the rapid degradation of the p53 protein (Scheffner et al. 1990). A cellular protein designated E6-AP is homologous to an E3 enzyme in the ubiquitin pathway and is necessary for binding of E6 to p53 and for targeting p53 for degradation (Huibregtse et al. 1991; Scheffner et al. 1993). E6 has been shown to bind to at least six other cellular proteins, although their identities are not known (Keen et al. 1994). E6 has also been shown to *trans*-activate the expression of heterologous promoters independently of p53 (Desaintes et al. 1992) and to repress transcription of other promoters (Etscheid et al. 1994).

In normal epithelium, only cells in the basal layer are capable of proliferation. Once cells leave the basal layer, they withdraw from the cell cycle and sequentially express a variety of epithelial-specific differentiation markers and are ultimately sloughed from the apical surface of the epithelium. HPV-infected cells show markers of cell division throughout the epithelium. Using an organotypic culture system in which epithelial cells differentiate on a collagen matrix (Asselineau et al. 1986; Kopan et al. 1987), we demonstrate that the loss of quiescence in suprabasal cells is attributable to HPV-16 E7 expression (Blanton et al. 1992; Halbert et al. 1992). The ability of a viral oncogene to stimulate cellular proliferation in normally quiescent cells may be a requisite step in the development of neoplasia.

The inability of a cell to maintain the integrity of its genome is likely to be involved in the development of cancer. Normal cells have checkpoints that control the progression of the cell cycle from G_1 to S and from G_2 to M, so that cells with damaged or incompletely replicated DNA can pause to correct the damage (Hartwell 1992). The molecular pathway involved in the G_1 arrest following DNA damage is beginning to be understood. Although it is not clear how DNA damage is sensed, one of the earliest responses is a posttranslational stabilization of p53 protein (Kastan et al. 1993). Cells lacking wild-type *p53* fail to arrest in G_1 in response to DNA damage (Kuerbitz et al. 1992). Wild-type *p53* has also been shown to be required to prevent gene amplification in response to PALA, another marker of genetic instability (Livingstone et al. 1992; Yin et al. 1992). *p53 trans*-activates the expression of the $p21^{cip-1,waf1,sdi1}$ gene (El-Deiry et al. 1993; Harper et al. 1993; Noda et al. 1994), which is a cyclin-dependent kinase inhibitory protein. p21 associates with cyclin/cdk complexes, inhibiting their kinase activity. Regulatory molecules such as Rb fail to become phosphorylated, and the cells pause in G_1

(Hinds et al. 1992). Recent studies have shown that either E6 (Kessis et al. 1993; Demers et al. 1994b) or E7 (Demers et al. 1994b) from HPV-16, but not HPV-6, can disrupt G_1 arrest following treatment with DNA-damaging agents, or following treatment with PALA (White et al. 1994). We present data which indicate that E6 and E7 abrogate cell cycle control; E7 prevents cells from entering G_0; E6 and E7 bypass the G_1 arrest induced by DNA damage by distinct pathways; and neither E6 nor E7 disrupts the G_2 arrest.

EXPERIMENTAL PROCEDURES

Expression of HPV oncogenes by retroviral transduction. HPV E6 or E7 genes were cloned into the retroviral vector pLXSN (Miller and Rosman 1989), and these constructs were used to produce amphotropic retroviruses (Halbert et al. 1991; Miller 1992). Briefly, the constructs were transfected by the calcium phosphate method into Psi-2 or PE501 cells (Mann et al. 1983; Miller and Rosman 1989). Virus produced from these cells was used to infect amphotropic packaging cells PA317 or PG13 (Miller and Buttimore 1986; Miller et al. 1991). Individual G418-resistant clones were isolated and confirmed to carry the HPV gene by Southern blotting. Retrovirus produced from amphotropic packaging cell lines was used to transduce HPV genes into target cells.

Organotypic culture and immunohistochemical staining. Organotypic cultures were produced by methods described previously (Asselineau et al. 1986; Kopan et al. 1987), as modified by Blanton et al. (1992). Briefly, epithelial cells were grown in a differentiation-promoting medium at the air/liquid interface on dermal equivalents consisting of a collage type I and fibroblast matrix for 12 days. Normal cells stratify and fully differentiate as they would in normal epithelial tissue. The organotypic cultures were fixed and stained with hematoxylin and eosin. Immunohistochemical staining with antibodies to p53 or PCNA was detected by the streptavidin-biotin immunoperoxidase method.

Analysis of G_1 arrest after DNA damage. Primary human foreskin keratinocytes were maintained in keratinocyte SFM medium (Gibco-BRL, Gaithersburg, Maryland). Keratinocytes were infected with pLXSN-based retroviruses (Miller and Rosman 1989; Halbert et al. 1991) containing HPV E6, E7, or E6/E7 genes and selected with 50 µg/ml G418. Subconfluent cultures of retrovirally infected keratinocytes were treated with 0.5 nM actinomycin D or left untreated. After 24 hours, the cells were treated with 10 µM bromodeoxyuridine (BrdU) for 4 hours. The cells then were fixed in ethanol and the nuclei were stained with anti-BrdU fluorescein isothiocyanate (FITC) conjugate antibody (Becton Dickinson, San Jose, California) and propidium iodide (PI) as described previously, with minor modifications (White et al. 1990; Demers et al. 1994b). The stained nuclei were analyzed on a Becton Dickinson FACScan.

Quantitation of the results was performed by using ReproMan software (TrueFacts Software, Seattle, Washington).

Analysis of p53, p21, pRb, and cyclin/cdk activity. p53 and Rb proteins were detected by Western blotting. Subconfluent cultures of retrovirally infected keratinocytes expressing various HPV genes were treated with 0.5 nM actinomycin D. After 24 hours, the cells were rinsed with PBS, lysed in 1× SDS sample buffer, and sonicated. The relative protein content of each sample was checked by SDS-PAGE followed by staining with Coomassie blue. Aliquots were analyzed by SDS-PAGE followed by immunoblotting with antibodies against p53 (p53 Ab 6, Oncogene Science, Uniondale, New York) or Rb (clone PMG3-245, PharMingen, San Diego, California). Detection was achieved by chemiluminescence (Renaissance, Du Pont NEN, Boston, Massachusetts).

$p21^{cip1,waf1,sdi1}$ RNA was detected by Northern blot from cells untreated or DNA damaged using an oligonucleotide probe (CGAAGTTCCATCGCTCACG GGCCTCCTGGATGC). Total RNA from each sample (5 µg) was separated by formaldehyde agarose gel electrophoresis and transferred to Hybond N membrane (Amersham).

Cyclin-E-associated histone H1 kinase activity was measured from cell lysates as described previously (Ohtsubo and Roberts 1993; Koff et al. 1994). Cyclin E antibodies were used to immunoprecipitate complexes in lysates from DNA-damaged and untreated cells. Immunoprecipitates were incubated with histone H1 and [γ-^{32}P]ATP, and the reaction products were analyzed by SDS-PAGE. Quantitation of the gels was performed using a Molecular Dynamics PhosphorImager.

Analysis of G_2 arrest. Asynchronously growing fibroblast cultures expressing vector or HPV oncogenes were subjected to irradiation (4 Gy). At various times after irradiation, cells were stained with DAPI and analyzed for DNA content by flow cytometry.

RESULTS

E7 Abrogates the Transition to G_0 in Suprabasal Epithelial Cells

HFE cells were infected with the amphotropic retroviruses LXSN (vector), LXSN6E6, LXSN6E7, LXSN16E6, LXSN16E7, and LXSN16E6E7. The cells were selected for neomycin resistance with G418, and pools of resistant cells were analyzed for expression of the viral genes by immunoprecipitation of radiolabeled extracts using antibodies raised against bacterially expressed trpE-HPV fusion proteins. Infected cells all expressed the expected viral protein products (data not shown). The cells were placed in organotypic culture and were analyzed for morphological alterations or for expression of proliferation markers. Cells expressing either E6 or E7 from the low-risk HPV-6 virus, or E6 from the high-risk HPV-16, formed a differentiated epithelium in organotypic culture that was indistinguishable from cells expressing the vector (Fig. 1a) or from normal epithelium (not shown). In contrast, cells expressing HPV-16 E7 alone (LXSN16E7) or HPV-16 E6 plus E7 (LXSN16E6E7) generated organotypic cultures that closely resembled premalignant squamous intraepithelial neoplasias. To examine the state of proliferation in the cells in organotypic culture, sections were stained with an antibody to the proliferating cell nuclear antigen (PCNA). Consistent with the morphological appearance observed by staining with hematoxylin and eosin (Fig. 1a), cells expressing vector, HPV-6 E6 or E7, or HPV-16 E6 showed PCNA-positive nuclei limited primarily to cells of the basal layer (Fig. 1b). In contrast, cells expressing E7, alone or in combination with E6, showed PCNA-positive nuclei throughout the suprabasal layers of the epithelium. To substantiate the presence of actively dividing cells in the epithelium, tritiated thymidine was used to supplement

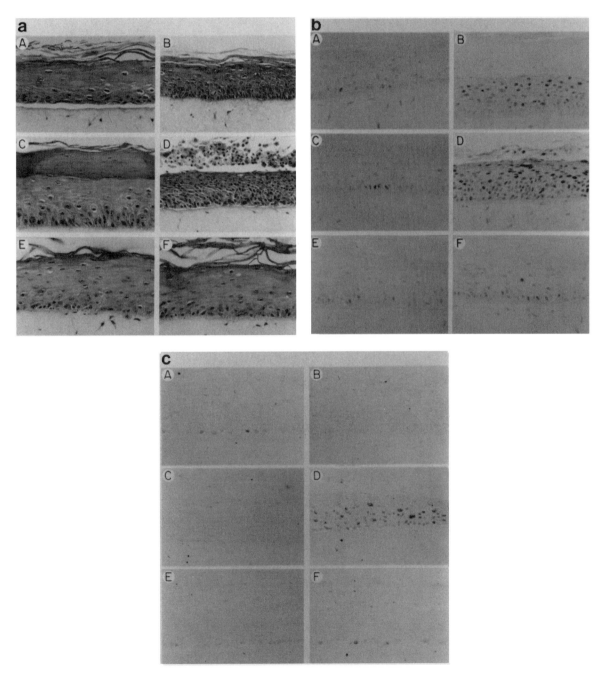

Figure 1. Proliferation of keratinocytes expressing HPV oncogenes in organotypic culture. Human foreskin epithelial cells were infected with (A) the LXSN vector; (B) LXSN16E6E7; (C) LXSN16E6; (D) LXSN16E7; (E) LXSN6E6; and (F) LXSN6E7. Panel a is stained with hematoxylin and eosin; panel b is stained with a MAb to PCNA; and panel c is stained with a MAb to p53. (Reprinted, with permission, from Halbert et al. 1992.)

the medium 24 hours prior to fixation of the cultures. Sections of the culture were exposed for autoradiography, and labeled nuclei were observed throughout the epithelium of cells expressing HPV-16 E7 (not shown). Although *p53* is a negative regulator of proliferation, its expression in cultured cells is associated with proliferating rather than quiescent cells (Reich and Levine 1984; Mercer and Baserga 1985). Sections of the organotypic cultures were examined with an antibody to p53 (Fig. 1c). In cells expressing the vector or the HPV-6 oncoproteins, p53 staining was restricted to a few cells in the basal layer, whereas cells expressing HPV-16 E7 alone showed p53 protein throughout the epithelium. No p53 was detected in the basal cells of cultures expressing HPV-16 E6 alone; more strikingly, cells expressing E6 together with E7 showed no p53 throughout the epithelium. Taken together, these data clearly indicate that stratified epithelial cells which normally become quiescent when they leave the basal layer are prohibited from entering G_0 by the HPV-16 E7 gene. In addition, p53 plays no role in establishing or suppressing suprabasal quiescence, since E7 is able to promote proliferation in the presence of high levels of p53 (LXSN16E7) or low levels of p53 (LXSN16E6E7).

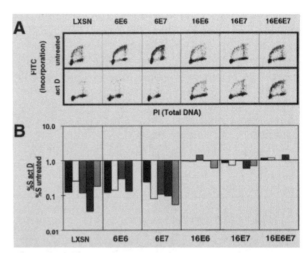

Figure 2. Flow cytometric analysis of keratinocytes after DNA damage. (*A*) Data are presented as a 2-parameter plot FITC-conjugated anti-BrdU (DNA incorporation) versus propidium iodide staining (total DNA). The top row is untreated cells and the bottom row has been treated with actinomycin D. (*B*) Data are presented as the ratio of percentage of S phase of the treated population over the percentage of S phase of the untreated population. (S phase was determined as the percentage of cells staining positive for BrdU.) Each bar represents an independent retroviral infection of different unrelated foreskin samples. (Reprinted, with permission, from Demers et al. 1994b.)

Both E6 and E7 Abrogate the G_1 Checkpoint Induced by DNA Damage

When E6 or E7 is expressed in primary epithelial cells, the cells remain diploid until senescence, or the period of crisis in culture, and then become aneuploid (Klingelhutz et al. 1994). This indicates that neither viral oncogene directly induces chromosomal alterations and suggests that perhaps the HPV oncogenes alter the checkpoints that prevent senescent cells from continuing in the cell cycle, allowing damage to the chromosomes to be replicated. To test this hypothesis, we examined the ability of cells to overcome the G_1 checkpoint in response to DNA-damaging agents. Both keratinocytes and fibroblasts expressing E6 or E7 from the high- and low-risk papillomavirus types were examined. The cells either were treated with actinomycin D, which at low doses has been shown to introduce DNA strand breaks by acting as a topoisomerase inhibitor (Trask and Muller 1988; Wassermann et al. 1990), or were irradiated with 400 rads. The cells were then pulsed for 4 hours with BrdU and stained with FITC-conjugated antibody to BrdU to label newly replicated DNA, and with propidium iodide to label total DNA. The cells were analyzed by FACS to identify the G_1, S, and G_2 populations. Figure 2 shows the results of actinomycin D treatment of keratinocytes.

A G_1 block can be seen in cells infected with the vector, HPV-6 E6 or HPV-6 E7, and treated with actinomycin D (lower panel, Fig. 2). The S-phase population was considerably reduced compared to untreated cells (upper panel). Quantitation of cells staining FITC-BrdU-positive from 3-5 independent experiments showed that cells expressing LXSN, 6E6, or 6E7 had only 3-30% of the number of cells in S phase after actinomycin D treatment compared to untreated cells. In contrast, cells infected with LXSN16E6 or LXSN16E6E7 had nearly the same number of cells in S-phase population after treatment. Cells expressing the 16E7 gene alone retained 60-90% of their S-phase population after treatment and were only slightly less effective at abrogating the G_1 block than cells containing E6. The identical patterns were seen when the breast epithelial cells were irradiated (not shown); however, when fibroblasts expressing the HPV oncogenes were either irradiated or treated with actinomycin D, only those cells expressing E6 (LXSN 16E6 or LXSN 16E6E7) were able to abrogate the G_1 block (not shown).

To determine whether the ability of HPV-16 E6 to abrogate the G_1 arrest was due to its ability to eliminate p53, mutants of E6 which had been characterized for their ability to bind and degrade p53 were analyzed for their ability to abrogate the G_1 arrest (Foster et al. 1994). Binding of E6 to p53 or to E6AP was affected by mutations scattered throughout the protein, with the exception of the carboxyl terminus of E6. All E6 proteins that bound to p53/E6AP targeted p53 for degradation. Figure 3a shows the levels of p53 protein in cells expressing the wild-type and mutated E6 proteins, and Figure 3b shows the results of actinomycin D treatment on the G_1 arrest with a selection of the mutants. Expression of HPV-16 E6 or mutated E6 proteins that bound and targeted p53 for degradation blocked ac-

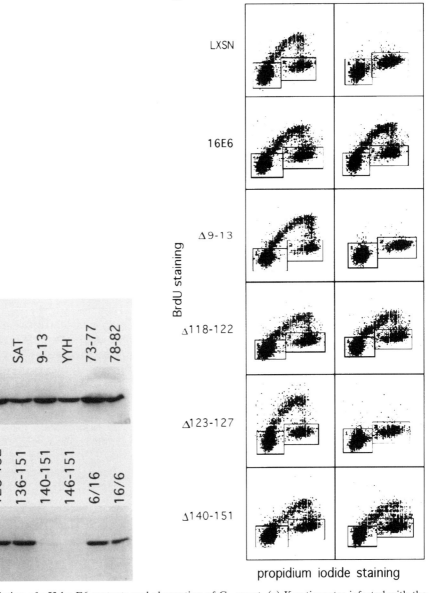

Figure 3. Correlation with degradation of p53 by E6 mutants and abrogation of G_1 arrest. (*a*) Keratinocytes infected with the retroviral constructs expressing mutated E6 proteins were treated with actinomycin D, and lysates of the cells were analyzed by Western blot for p53 protein levels. (*b*) Keratinocytes expressing the mutated E6 proteins, with or without treatment with actinomycin D, were analyzed by flow cytometry. (Reprinted, with permission, from Foster et al. 1994.)

tinomycin-D-induced cellular growth arrest, whereas expression of HPV-6 E6 or mutated E6 proteins that did not interact with p53 did not affect growth arrest. These results indicated that elimination of p53 is important for the ability of the E6 proteins to circumvent growth arrest.

The high-risk, but not the low-risk, E7 protein was also able to abrogate G_1 arrest after actinomycin D treatment or irradiation in keratinocytes. Previously, we had demonstrated that expression of HPV-16 E7 in keratinocytes elevated the level of p53 protein three- to fivefold by a posttranslational mechanism (Demers et al. 1994a). Analysis of the levels of p53 protein in cells expressing the HPV oncogenes before and after treatment with actinomycin D confirmed previous observations (Fig. 4). In cells expressing LXSN, 6E6, or 6E7, low levels of p53 were seen which increased about threefold after treatment, as had been seen by other workers after DNA damage (Kuerbitz et al. 1992). In cells expressing 16E6, alone or in combination with 16E7, barely detectable levels of p53 protein were seen, and no increase was apparent after DNA damage. Cells expressing 16E7 had elevated levels of p53 which were further increased by DNA damage. The pattern of p53 protein levels seen in fibroblasts was the same as that seen in keratinocytes (data not shown). Thus, even in the presence of high levels of p53, HPV-16 E7 was able to abrogate growth arrest.

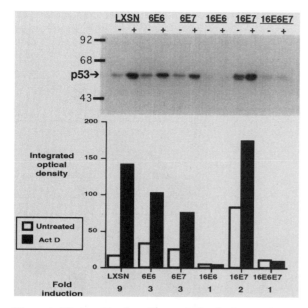

Figure 4. Western blot for p53 protein levels. Lysates from retrovirus-infected keratinocytes expressing the designated HPV genes were analyzed for p53 protein levels from untreated (−) and actinomycin-D-treated (+) cells. Below is the phosphorimager analysis of p53 protein levels. (Reprinted, with permission, from Demers et al. 1994b.)

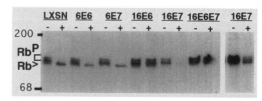

Figure 5. Western blot for Rb protein levels and phosphorylation state. Cell lysates were analyzed for Rb protein from untreated (−) and actinomycin-D-treated (+) keratinocytes after retroviral infection. The faster migrating band is the hypophosphorylated form of Rb. The slower migrating bands are the hypophosphorylated forms (Rb^P). (Reprinted, with permission, from Demers et al. 1994b.)

An inhibitor of cyclin/cdk complexes, designated p21, has been shown to be transcriptionally activated by p53. To determine whether the change in p53 protein levels seen after treatment with actinomycin D, and in response to the HPV oncogenes, was reflected in changes in the level of p21 expression, RNA was examined by Northern blot. As p53 increased, the levels of p21 RNA also increased; thus, E7-expressing cells treated with actinomycin D contained high levels of p21 (not shown). Cyclin/cdk kinase activity using histone H1 as a substrate was measured using antibodies to immunoprecipitate complexes containing cyclin E, cyclin A, cdk2, and cdc2. As expected, kinase activity was observed in untreated cycling cells and in actinomycin-D-treated keratinocytes expressing E6 or E7, and in treated fibroblasts expressing E6 (not shown). The levels of Rb protein and the state of phosphorylation were examined by Western blot. The Rb protein has multiple phosphorylated forms that run with varying mobilities on SDS-PAGE gels (Ludlow et al. 1989). The fastest migrating species is the hypophosphorylated form that appears in the cell after mitosis and becomes phosphorylated at the transit from G_1 to S. It is the underphosphorylated form that acts to inhibit proliferation and is the form to which viral oncogenes bind. Figure 5 shows that Rb protein levels dropped following actinomycin D treatment, except in cells expressing 16E6 alone. The decrease in Rb protein levels dropped most dramatically in cells expressing 16E7 alone in parallel with increased levels of p53. No change in transcription of the Rb gene was detected (not shown). In cells that underwent growth arrest, i.e., LXSN, 6E6, and 6E7, all of the Rb protein appeared to be in the hypophosphorylated state. In cells that abrogated the G_1 arrest, the Rb protein became phosphorylated. In cells expressing 16E7, the far right lanes show that the low levels of Rb that remain after actinomycin D treatment are clearly phosphorylated. In cells expressing 16E6, neither the level of Rb protein nor the phosphorylated forms changed with actinomycin D treatment. When 16E6 and 16E7 are coexpressed, the effect of E6 on Rb protein levels is dominant. These results indicate that both E6 and E7 are able to abrogate a DNA-damage-induced G_1 arrest in keratinocytes by distinct mechanisms.

HPV Oncogenes Do Not Alter the DNA-damage-induced G_2 Arrest

Following DNA damage, cells growth-arrest in both G_1 and G_2. The G_2 arrest had been shown to be unaffected in cells with mutant *p53* (Kuerbitz et al. 1992). Figures 2 and 3b showed the distribution of cells expressing in the various phases of the cell cycle after treatment with actinomycin D. After treatment, cells accumulated in G_2 as well as G_1. No alteration in the G_2 population was observed in cells expressing the oncogenes, but that presentation of data most clearly depicts changes in the S-phase populations. To examine the G_2 arrest more clearly, asynchronously growing human fibroblasts expressing the HPV-16 E6 and E7 genes were irradiated with 4 Gy and samples were analyzed for cell cycle distribution by FACS over time. Figure 6 demonstrates that the population of cells in G_2 increases from about 10% in unirradiated cells to 30–50% by 10–20 hours after irradiation. No differences were observed in cells expressing E6 or E7 compared to normal fibroblasts. These data suggest that the HPV oncogenes do not affect the DNA-damage-induced G_2 checkpoint.

DISCUSSION

The development of anogenital malignancy is a multistep process that likely involves activities encoded by papillomavirus oncogenes as well as functions encoded by cellular genes. Introduction of the E6 and E7 genes into cells by retroviral infection makes it possible to study the consequences of acute expression of the

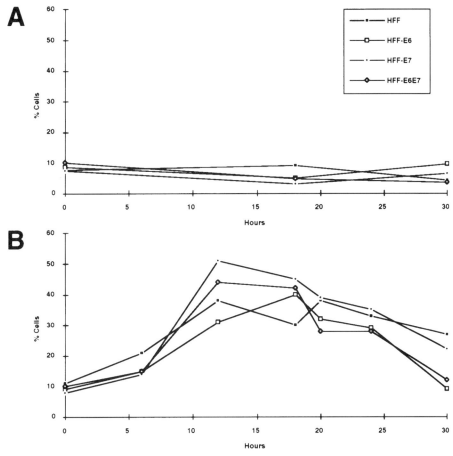

Figure 6. Analysis of DNA-damage-induced G_2 arrest. Asynchronously growing fibroblasts expressing the HPV oncogenes were untreated (*A*) or irradiated with 4 Gy (*B*). At various times after irradiation, the cells were stained with DAPI and analyzed by flow cytometry. The percentages of cells in the G_2 phase are plotted.

viral genes. Our results support the hypothesis that oncogenes from the high-risk papillomaviruses affect the control of cellular proliferation, and in doing so, disrupt the checkpoints to cell cycle progression that protect cells from incorporating damage to the genome caused by exogenous agents or from the chromosomal instability that accompanies senescence. The alterations to epithelial cell differentiation that are frequently seen in immortalized cells or in neoplastic lesions do not appear to be a direct consequence of HPV genes (Blanton et al. 1992).

HPV-16 E7 disrupts the suprabasal quiescence of normal epithelial cells, thus preventing cells from entering G_0. E7 had been shown previously to stimulate DNA synthesis in serum-starved rodent cells (Sato et al. 1989), and this activity was associated with the ability to bind Rb and to transform rodent cell lines (Banks et al. 1990). Thus, it is probable that the ability to prevent cells from entering G_0 when they leave the basal layer requires inactivation of the growth inhibitory functions of Rb. It is also likely that this function of E7 is required for immortalization of keratinocytes and for the development of malignancy. It is interesting that the E7 gene is not required for the immortalization of breast epithelial cells (Band et al. 1990; Shay et al.

1993), a type of epithelium that does not stratify. p53 expression played no role in suprabasal quiescence. E7 expression alone results in high levels of p53 and the disruption of G_0; E6E7 expression results in very low levels of p53 and the disruption of G_0. Expression of E6 alone had no effect.

Maintaining the integrity of the genome is an important feature of normal cells that is lost in cancer cells. It is becoming clear that there are several cellular proteins whose disruption allows genetic alterations to accumulate in the genome. One of these genes is the tumor suppressor *p53*, which is frequently mutated in cancers (Vogelstein 1990). Loss or mutation of *p53* results in the failure of the G_1 checkpoint to arrest in response to DNA damage. The G_1 arrest appears to be mediated by inhibitors of cyclin dependent kinases (CIPs), which prevent the cdks from phosphorylating regulators of the G_1/S transition, such as Rb. At least one CIP, p21, is responsive to *p53*, but to date loss or mutation of p21 in tumors has not been reported. Another CIP, p16, first identified as a protein found in cdk complexes, particularly cyclin D/cdk4 (Xiong et al. 1993), is not in complexes in transformed cells, but this could be due to the requirement for Rb to form cyclin D/cdk4 or cdk6 complexes (Bates et al. 1994). A third

CIP, p27, has been identified in epithelial cells arrested in G_1 in response to transforming growth factor β (TGFβ) (Koff et al. 1993; Polyak et al. 1994) and does not appear to be responsive to p53 levels (J. Roberts et al., unpubl.).

Interestingly, both HPV oncogenes have activities that lead to disruption of the DNA-damage-induced G_1 checkpoint, and neither oncogene seemed to disrupt the G_2 arrest. However, the mechanism to overcome the G_1 arrest is distinctly different for E6 compared to E7 (Fig. 7). The ability of E6 to eliminate p53, and thus to maintain very low levels of p21, appears to be responsible for abrogation of the G_1 arrest. Cyclin/cdk activity is equivalent in treated and untreated cells that have eliminated p53 by E6-mediated degradation, in both epithelial cells and fibroblasts. The mechanism of action of E7 in overcoming the DNA-damage-induced G_1 arrest is less clear and did not seem to function in fibroblasts. Despite high levels of p21 in E7-expressing cells, cdk activity in treated cells was about 80% of that seen in untreated keratinocytes, but was greatly reduced in fibroblasts. It is unclear what functions of E7 are required to overcome the inhibitory effects of p21. One possibility is that E7 displaces p21 from the cdk complex, which could arise from the association of E7 with cyclins directly or through p107 (Davies et al. 1993). Another possibility is that freeing of E2F by the association of E7 with Rb may be sufficient to drive cells into S phase and to activate cdks that are not inhibited by p21.

The basis for the cell-type specificity of E7 abrogation of the DNA-damage-induced G_1 arrest is not clear. Expression of E7, stabilization of p53 protein, and induction of p21 appeared equivalent in both cell types, yet restoration of cyclin-E-associated kinase activity was remarkably lower in actinomycin-D-treated fibroblasts than keratinocytes. Fibroblasts exposed to PALA arrest in G_1 with slower kinetics than seen after DNA damage. Expression of HPV-16 E7 eliminated the PALA-induced G_1 arrest, although the mechanism is not known (White et al. 1994).

Our data indicate that the disruption of cell cycle proliferation and the control of cell cycle checkpoints are unique to the high-risk genital tract papillomaviruses and may explain the basis for the generally benign proliferations associated with HPV-6 infection. It will be interesting to examine the effects of the E6 and E7 oncoproteins from types with an intermediate risk for progression to high-grade lesions, and to examine HPVs such as type 5 and 8 that are associated with cutaneous epithelial cancers.

ACKNOWLEDGMENTS

We thank Erik Espling for his assistance with this work; Wendy Law for examining p53 and Rb in the fibroblasts; Jim McDougall and the members of his laboratory for stimulating discussions and sharing of data; and Jim Roberts and Mark Groudine for helpful discussions. This work was supported by a grant from the American Cancer Society to D.A.G. S.A.F. was supported by a National Cancer Institute training grant.

REFERENCES

Asselineau, D., B.A. Benard, C. Bailly, and M. Darmon. 1986. Human epidermis reconstructed by culture: Is it "normal?" *J. Invest. Dermatol.* **86:** 181.

Band, V., D. Zajchowski, V. Kulesa, and R. Sager. 1990. Human papilloma virus DNAs immortalize normal human mammary epithelial cells and reduce their growth factor requirements. *Proc. Natl. Acad. Sci.* **87:** 463.

Banks, L., C. Edmonds, and K.H. Vousden. 1990. Ability of the HPV16 E7 protein to bind RB and induce DNA synthesis is not sufficient for efficient transforming activity in NIH3T3 cells. *Oncogene* **5:** 1383.

Bates, S., D. Parry, L. Bonetta, K. Vousden, C. Dickson, and G. Peters. 1994. Absence of cyclin D/cdk complexes in cells lacking functional retinoblastoma protein. *Oncogene* **9:** 1633.

Blanton, R.A., M.D. Coltrea, A.M. Gown, C.L. Halbert, and J.K. McDougall. 1992. Expression of the HPV 16 E7 gene generates proliferation in stratified squamous cell cultures which is independent of endogenous p53 levels. *Cell Growth Differ.* **3:** 791.

Davies, R., R. Hicks, T. Crook, J. Morris, and K. Vousden. 1993. Human papillomavirus type 16 E7 associates with a histone H1 kinase and with p107 through sequences necessary for transformation. *J. Virol.* **67:** 2521.

Demers, G.W., C.L. Halbert, and D.A. Galloway. 1994a. Elevated wild-type p53 protein levels in human epithelial cell lines immortalized by the human papillomavirus type 16 E7 gene. *Virology* **198:** 169.

Demers, G.W., S.A. Foster, C.L. Halbert, and D.A. Galloway. 1994b. Growth arrest by induction of p53 in DNA damaged keratinocytes is bypassed by human papillomavirus 16 E6. *Proc. Natl. Acad. Sci.* **91:** 4382.

Desaintes, C., S. Hallez, P. Van Alphen, and A. Burny. 1992. Transcriptional activation of several heterologous promoters by the E6 protein of human papillomavirus type 16. *Virology* **66:** 325.

Dyson, N., P. Guida, K. Munger, and E. Harlow. 1992. Homologous sequences in adenovirus E1A and human papillomavirus E7 proteins mediate interaction with the same set of cellular proteins. *J. Virol.* **66:** 6893.

Dyson, N., P.M. Howley, K. Munger, and E. Harlow. 1989. The

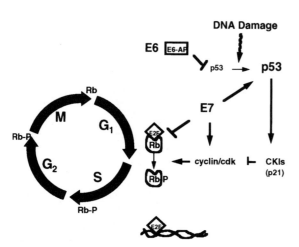

Figure 7. Mechanism of abrogation of the DNA-damage-induced G_1 checkpoint by the HPV oncoproteins.

human papilloma virus-16 E7 oncoprotein is able to bind to the retinoblastoma gene product. *Science* **243:** 934.

El-Deiry, W.S., T. Tokino, V.E. Velculescu, D.B. Levy, R. Parsons, J.M. Trent, W.E. Mercer, K.W. Kinzler, and B. Vogelstein. 1993. *WAF1*, a potential mediator of p53 tumor suppression. *Cell* **75:** 817.

Etscheid, B.G., S.A. Foster, and D.A. Galloway. 1994. The E6 protein of human papillomavirus type 16 functions as a transcriptional repressor in a mechanism independent of its capacity to bind the tumor suppressor protein p53. *Virology* (in press).

Firzlaff, J.M., D.A. Galloway, R.N. Eisenman, and B. Luscher. 1989. The E7 protein of human papillomavirus type 16 is phosphorylated by casein kinase II. *New Biol.* **1:** 44.

Foster, S.A., G.W. Demers, B.G. Etscheid, E.S. Espling, and D.A. Galloway. 1994. The ability of human papillomavirus E6 proteins to target p53 for degradation correlates with their abiltiy to abrogate actinomycin D-induced growth arrest. *J. Virol.* **68:** 5698.

Halbert, C., G. Demers, and D. Galloway. 1991. The E7 gene of human papillomavirus type 16 is sufficient for immortalization of human epithelial cells. *J. Virol.* **65:** 473.

———. 1992. The E6 and E7 genes of human papillomavirus type 6 have weak immortalizing activity in human epithelial cells. *J. Virol.* **66:** 2125.

Harper, J.W., G.R. Adami, N. Wei, K. Keyomarsi, and S.J. Elledge. 1993. The p21 CDK-interacting protein Cip1 is a potent inhibitor of G_1 cyclin-dependent kinases. *Cell* **75:** 805.

Hartwell, L. 1992. Defects in a cell cycle checkpoint may be responsible for the genomic instability of cancer cells. *Cell* **71:** 543.

Hinds, P.W., S. Mittnacht, V. Dulic, A. Arnold, S.I. Reed, and R.A. Weinberg. 1992. Regulation of retinoblastoma protein functions by ectopic expression of human cyclins. *Cell* **70:** 993.

Huibregtse, J., M. Scheffner, and P. Howley. 1991. A cellular protein mediates association of p53 with the E6 oncoprotein of human papillomavirus type 16 or 18. *EMBO J.* **10:** 4129.

Kastan, M.B., O. Onyekwere, D. Sidransky, B. Vogelstein, and R.W. Craig. 1993. Participation of p53 protein in the cellular response to DNA damage. *Cancer Res.* **51:** 6304.

Keen, N., R. Elston, and L. Crawford. 1994. Interaction of the E6 protein of human papillomavirus with cellular proteins. *Oncogene* **9:** 1493.

Kessis, T.D., R.J. Slebos, W.G. Nelson, M.B. Kastan, B.S. Plunkett, S.M. Han, A.T. Lorincz, L. Hedrick, and K.R. Cho. 1993. Human papillomavirus 16 E6 expression disrupts the p53-mediated cellular response to DNA damage. *Proc. Natl. Acad. Sci.* **90:** 3988.

Klingelhutz, A.J., S.A. Barber, P.P. Smith, K. Dyer, and J.K. McDougall. 1994. Restoration of telomeres in human papillomavirus-immortalized human anogenital cells. *Mol. Cell. Biol.* **14:** 961.

Koff, A., M. Ohtsuki, P. Kornelia, J.M. Roberts, and J. Massagué. 1993. Negative regulation of G_1 in mammalian cells: Inhibition of cyclin E-dependent kinase by TGF-β. *Science* **260:** 536.

Koff, A., A. Giordano, D. Desai, K. Yamashita, J.W. Harper, S. Elledge, T. Nishimoto, D.O. Morgan, B.R. Franza, and J.M. Roberts. 1994. Formation and activation of a cyclin E-cdk2 complex during the G_1 phase of the human cell cycle. *Science* **257:** 1689.

Kopan, R., G. Traska, and E. Fuchs. 1987. Retinoids as important regulators of terminal differentiation: Examining keratin expression in individual epidermal cells at various stages of keratinization. *J. Cell Biol.* **105:** 427.

Kuerbitz, S.J., B.S. Plunkett, W.V. Walsh, and M.B. Kastan. 1992. Wild-type p53 is a cell cycle checkpoint determinant following irradiation. *Proc. Natl. Acad. Sci.* **89:** 7491.

Livingstone, L.R., A. White, J. Sprouse, E. Livanos, T. Jacks, and T.D. Tlsty. 1992. Altered cell cycle arrest and gene amplification potential accompany loss of wild-type p53. *Cell* **70:** 923.

Ludlow, J.W., J.A. DeCaprio, C.M. Huang, W.H. Lee, E. Paucha, and D.M. Livingston. 1989. SV40 large T antigen binds preferentially to an underphosphorylated member of the retinoblastoma susceptibility gene product family. *Cell* **56:** 57.

Mann, R., R.C. Mulligan, and D. Baltimore. 1983. Construction of a retrovirus packaging mutant and its use to produce helper-free defective retrovirus. *Cell* **33:** 153.

Mercer, W.E. and R. Baserga. 1985. Expression of the p53 protein during the cell cycle of human peripheral blood lymphocytes. *Exp. Cell Res.* **160:** 31.

Miller, A.D. 1992. Retroviral vectors. *Curr. Top. Microbiol. Immunol.* **158:** 1.

Miller, A.D. and C. Buttimore. 1986. Redesign of retrovirus packaging cell lines to avoid recombination leading to helper virus production. *Mol. Cell. Biol.* **6:** 2895.

Miller, A.D. and G.J. Rosman. 1989. Improved retroviral vectors for gene transfer and expression. *BioTechniques* **7:** 980.

Miller, A.D., J.V. Garcia, N. von Suhr, C.M. Lynch, C. Wilson, and M.V. Eiden. 1991. Construction and properties of retrovirus packaging cells based on gibbon ape leukemia virus. *J. Virol.* **65:** 2220.

Noda, A., S.F. Venable, O.M. Pereira-Smith, and J.R. Smith. 1994. Cloning of senescent cell-derived inhibitors of DNA synthesis using an expression system. *Exp. Cell Res.* **211:** 90.

Ohtsubo, M. and J.M. Roberts. 1993. Cyclin-dependent regulation of G_1 in mammalian fibroblasts. *Science* **259:** 1908.

Polyak, K., J. Kato, M.J. Solomon, C.J. Sherr, J. Massagué, J.M. Roberts, and A. Koff. 1994. p27^{Kip1}, a cyclin-Cdk inhibitor, links transforming growth factor-β and contact inhibition to cell cycle arrest. *Genes Dev.* **8:** 9.

Reich, N.C. and A.J. Levine. 1984. Growth regulation of a cellular tumour antigen, p53, in nontransformed cells. *Nature* **308:** 199.

Sato, H., A. Furuno, and K. Yoshiike. 1989. Expression of human papillomavirus type 16 E7 gene induces DNA synthesis of rat 3Y1 cells. *Virology* **168:** 195.

Scheffner, M., J.M. Huibregtse, R.D. Vierstra, and P.M. Howley. 1993. The HPV-16 E6 and E6-AP complex functions as a ubiquitin-protein ligase in the ubiquitination of p53. *Cell* **75:** 495.

Scheffner, M., B.A. Werness, J.M. Huibregtse, A.J. Levine, and P.M. Howley. 1990. The E6 oncoprotein encoded by human papillomavirus types 16 and 18 promotes the degradation of p53. *Cell* **63:** 1129.

Shay, J., W.E. Wright, D. Brasiskyte, and B.A. Van Der Haegen. 1993. E6 of human papillomavirus type 16 can overcome the M1 state of immortalization in human mammary epithelial cells but not in human fibroblasts. *Oncogene* **8:** 1407.

Trask, D.K. and M.T. Muller. 1988. Stabilization of type I topoisomerase-DNA covalent complexes by actinomycin D. *Proc. Natl. Acad. Sci.* **85:** 1417.

Vogelstein, B. 1990. A deadly inheritance. *Nature* **348:** 681.

Wassermann, K., J. Markovits, C. Jaxel, G. Capranico, K.W. Kohn, and Y. Pommier. 1990. Effects of morpholinyl doxorubicins, doxorubicin, and actinomycin D on mammalian DNA topoisomerases I and II. *Mol. Pharmacol.* **38:** 38.

Werness, B.A., A.J. Levine, and P.M. Howley. 1990. Association of human papillomavirus types 16 and 18 E6 proteins with p53. *Science* **248:** 76.

White, A.E., E.M. Livanos, and T.D. Tlsty. 1994. Differential disruption of genomic integrity and cell cycle regulation in normal human fibroblasts by the HPV oncoproteins. *Genes Dev.* **8:** 666.

White, R.A., N.H.A. Terry, M.L. Meistrich, and D.P. Calkins. 1990. Improved method for computing potential doubling time from flow cytometric data. *Cytometry* **11:** 314.

Xiong, Y., G.J. Hannon, H. Zhang, D. Casso, R. Kobayashi,

and D. Beach. 1993. p21 is a universal inhibitor of cyclin kinases. *Nature* **366:** 701.

Yin, Y., M.A. Tainsky, F.Z. Bischoff, L.C. Strong, and G.M. Wahl. 1992. Wild-type p53 restores cell cycle control and inhibits gene amplification in cells with mutant p53 alleles. *Cell* **70:** 937

zur Hausen, H. 1991. Human papillomaviruses in the pathogenesis of anogenital cancer. *Virology* **184:** 9.

Telomerase, Cell Immortality, and Cancer

C.B. Harley,* N.W. Kim,* K.R. Prowse,* S.L. Weinrich,* K.S. Hirsch,*
M.D. West,* S. Bacchetti,† H.W. Hirte,† C.M. Counter,†
C.W. Greider,‡ M.A. Piatyszek,§ W.E. Wright,§ and J.W. Shay§

*Geron Corporation, Menlo Park, California 94025; †McMaster University, Hamilton,
Ontario, Canada, L8N 3Z5; ‡Cold Spring Harbor Laboratory, Cold Spring Harbor, New York 11724;
§University of Texas, Southwestern Medical Center, Dallas, Texas 75235

An important hallmark of cancer is aberrant growth control. Genetic changes that confer a growth advantage to the tumor cell are observed on numerous levels. Some of the best understood are mutations in proto-oncogenes and tumor suppressor genes that are linked to signal transduction pathways, cell cycle control, or cell-cell/cell-matrix interactions that regulate growth, movement, differentiation, survival, apoptosis, and genetic stability (Hartwell 1992; Weinberg 1992; Hunter 1993; Runger et al. 1994; Workman 1994). However, in addition to aberrant growth control, many cancer cells possess another important feature which distinguishes them from normal somatic cells: unlimited replicative capacity.

Hayflick first described the limited replicative capacity of normal human fibroblasts more than 30 years ago (for review, see Hayflick 1965; Goldstein 1990). Since then, numerous other somatic cell types, including epithelial cells, endothelial cells, myoblasts, astrocytes, and lymphocytes, have also shown evidence of a mitotic clock which limits their division capacity (Stanulis Praeger 1987; Harley 1988). The maximum number of times a young, human somatic cell can divide in culture is approximately 50–100 times, and this limit decreases as a function of donor age, presumably reflecting the replicative history of the cells in vivo. It is thought that exhaustion of this limit in certain tissues or at focal sites of high cell turnover in vivo plays a significant role in age-related pathology (Stanulis Praeger 1987; West et al. 1989; Goldstein 1990; Vaziri et al. 1993; West 1994).

Abrogation of the normal mortal phenotype of human somatic cells is very rare. Unlike rodent cells that spontaneously immortalize at a relatively high frequency, permanent lines of "normal" human cells arising without addition of viral oncoproteins or some form of mutagenesis have never been reproducibly obtained (Gonos et al. 1992; Newbold et al. 1993). Thus, it is reasonable to assume that stringent mechanisms to ensure the finite proliferative capacity of cells in long-lived organisms such as man could have evolved to dramatically reduce the probability of runaway growth of deregulated cells. Although it is still controversial (Stamps et al. 1992), there are strong arguments supporting an important role of cell immortality in advanced cancers. We believe that the key enzyme responsible for cell immortality in transformed cells is telomerase, the ribonucleoprotein enzyme that synthesizes telomeric DNA. We review data supporting a specific role for telomerase in cell immortalization and cancer, and present a model for the effects on tumor growth of reinstituting cell mortality through telomerase inhibition.

ARE CANCER CELLS IMMORTAL?

A conservative upper limit of about 50 doublings for a single somatic cell can theoretically allow a clone to generate a large biomass: 2^{50} cells/$(10^9$ cells/g$) > 1000$ kg. If the more reasonable upper limit of 100 doublings is used, a staggering 10^{18} kg of cells would be produced through exponential growth of a single clone. Thus, it has often been argued that even aggressive, metastatic tumors may not exhaust the replicative potential of somatic cells, and hence there is little need to invoke escape from cell mortality in cancer pathogenesis. However, this argument overlooks an important aspect of cancer: the independent, multi-hit nature of tumorigenesis (Weinberg 1989; Vogelstein and Kinzler 1993). If four or more independent mutations, sequentially selected, are required to generate an advanced cancer (Cairns 1975), and roughly 1 in 10^6 cells will acquire each new mutation, then exponential growth of all cells at each step would exhaust greater than 80 doublings (Fig. 1). Moreover, the generally low growth fraction, high loss fraction (necrosis or apoptosis), and asynchronous cell divisions in tumors (Norton 1991) would suggest that more than 20 doublings may be required between each mutational event. Thus, one of the steps in the pathogenesis of malignancy is likely to be acquisition of cell immortality.

The observations that some cancer cells in vivo may not be immortal, and that most tumor cells fail to establish a cell line in culture, do not weaken the argument that advanced tumors are sustained by the expansion of an immortal population of cells. The poor vascularization and homeostasis of tumors and the high degree of genomic instability in cancer cells ensure that many tumor cells will fail to thrive in vivo. Similarly, a relatively low frequency of immortal cells in tumor biopsies, genomic instability, and difficulty adapting tissue culture conditions to the growth requirements of

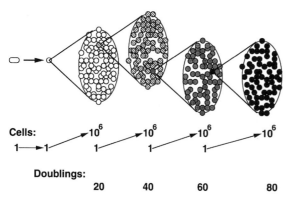

Figure 1. Sequential clonal expansions during tumorigenesis. The multi-hit process of cancer may require four or more independent mutations. The frequency of mutational events may vary significantly, but it is not unreasonable to assume, on average, at least 20 doublings between successive events. Thus, 4 events consume at least 80 doublings.

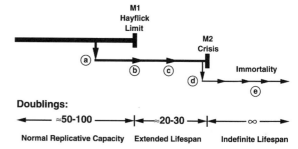

Figure 2. Immortalization of cells requires escape from two cell mortality points. Mutations (a, b) occurring before or near mortality phase 1 (M1) allow cells to escape this block in proliferation. These mutations and others (c) confer aberrant growth and an extended life span to cells, but do not themselves immortalize the population. The population as a whole undergoes crisis at mortality phase 2 (M2). Additional mutations (d), at or near crisis, allow rare clones to escape M2 yielding an immortal population. This population may also be unstable and undergo further mutational events (e).

tumor cells might account for the low frequency of immortal lines generated from most tumors.

TRANSFORMATION, IMMORTALITY, MALIGNANCY, AND METASTASIS

Immortalization does not appear to be a direct result of alterations in known oncogenes or tumor suppressor genes. The elegant work in the 1980s on transformation and immortalization of primary cells in culture through combinations of transfected oncogenes (Weinberg 1989) did not rule out the possibility that mutational events responsible for immortalization occur independently of those responsible for altered growth control, genomic instability, tumorigenicity, etc. To date, no known combination of oncogenes or mutations in tumor suppressor genes can directly immortalize a normal human somatic cell in vitro (Shay et al. 1991, 1993b; Counter et al. 1992, 1994b; Van Der Haegen and Shay 1993; Klingelhutz et al. 1994).

The timing of cell immortalization with respect to other events in tumorigenesis can be predicted and tested in experimental models of cell transformation. Since most somatic cells do not normally reach the end of their replicative life span in vivo, there is no selective advantage early in tumorigenesis for cells to mutate to the immortal phenotype. Instead, early genetic changes leading to preneoplastic lesions likely involve mutations related to loss of growth control or genomic stability. Subsequently, upon expansion of these cells, additional mutations will ultimately be required to bypass the Hayflick limit (mortality phase 1, or M1) (Fig. 2). Interestingly, these mutations may involve tumor suppressor genes such as p53 or RB that play an important role in M1 arrest (Shay et al. 1991; Counter et al. 1992; Bond et al. 1994). Eliminating these tumor suppressor genes does not immortalize cells, but instead confers an extended, but still finite, replicative capacity to the population. Thus, a second mortality phase (crisis, or M2) is seen when cells have divided beyond the normal Hayflick limit. Evidence for two mortality phases has been observed in fibroblasts, mammary epithelial cells, and embryonic kidney cells transformed with adenovirus oncogenes or SV40 T antigen (Graham et al. 1977; Shay et al. 1991; Counter et al. 1992; Van Der Haegen and Shay 1993), in fibroblasts and mammary or anogenital epithelial cells transformed with HPV E6/E7 (Shay et al. 1993b; Klingelhutz et al. 1994), and in B lymphocytes transformed with Epstein-Barr virus (EBV) (Counter et al. 1994b).

In general, crisis (M2) is associated with a significant increase in chromosomal instability, particularly telomeric associations, and cell death. At this point, there should be strong selection for a mutation(s) which confers some measure of chromosomal stability and permits continued cell growth. Thus, the immortalizing mutation(s) is predicted to arise in the population at, or near, crisis (Girardi et al. 1965; Counter et al. 1992, 1994b).

Malignancy and metastasis are often the lethal end stages of deregulated cell growth. Events which confer malignancy, the invasive character of tumors, could occur before or after events which immortalize cells. In some tissues, malignancy may require a long cascade of growth alterations and hence occur only after selection for immortality has conferred a sufficient growth capacity to the tumor. In other tissues, the cascade of growth control alterations may require fewer events, or fewer divisions between events, and hence occur within the extended, but still mortal life span of the pre-crisis cell. Metastasis, on the other hand, requires the development of malignancy in the primary tumor plus clonal expansion of cells that have migrated to the secondary site. Since clinically significant secondary growths probably require at least 20–30 additional doublings, it is likely that there would have been strong selection for cell immortality by the time metastatic lesions are detected in most cancers. An exception to this general rule might be found in certain childhood cancers, where the long, but finite, growth potential of cells may permit full transformation to the metastatic state without immortalization of cells. We would predict that unless cell immortality was achieved, such cancers would spontaneously regress when the end of the replicative capacity was reached. An example of this might be stage IV-S neuroblastomas, which are metastatic, yet have a high rate of spontaneous regression, short telomeres, and low growth fraction (Hiyama et al. 1993); it remains to be established whether these tumors express telomerase.

TELOMERES AND THE END-REPLICATION PROBLEM

The telomere hypothesis of cell aging and immortalization provides a useful framework to describe tumorigenesis (Olovnikov 1973; Harley et al. 1990; Harley 1991; Shay et al. 1993a). Telomeres are the genetic elements which stabilize the natural ends of linear eukaryotic chromosomes (Blackburn and Szostak 1984; Zakian 1989). However, the ends of linear DNA cannot be fully replicated by the conventional DNA polymerase complex, which requires a labile RNA primer to initiate DNA synthesis (Watson 1972). In the absence of a mechanism to overcome this "end-replication problem," organisms cannot pass their complete genetic complement from generation to generation, ultimately compromising further growth (Zakian 1989; Levy et al. 1992). Although different organisms have evolved various methods of avoiding the end-replication problem, most eukaryotic species utilize a specialized enzyme, telomerase, to regenerate telomeric DNA de novo, thus compensating for, rather than avoiding, terminal deletions (Greider 1990; Blackburn 1992).

TELOMERASE

Telomerase is a ribonucleoprotein enzyme first discovered in *Tetrahymena* (Greider and Blackburn 1985) and more recently in an immortal human tumor cell line (Morin 1989). The RNA component of telomerase, cloned and sequenced from several ciliates, contains a short region (the "template domain") complementary to the one or more repeats of the G-rich telomeric DNA of the species (for review, see Greider 1993). The mechanism of action of telomerase is a reiterative copying of the template domain, involving an elongation phase where deoxyribonucleotides are sequentially added to the 3' end of the telomere, followed by a slower "translocation" phase, in which the relative position of telomerase and the telomere advance one repeat, thus positioning the enzyme for another elongation phase (Fig. 3).

After its original discovery in HeLa cells (Morin 1989), human telomerase was found to be active in post-M2 (immortal) cell populations in culture (Counter et al. 1992, 1994b) and in a human tumor in vivo

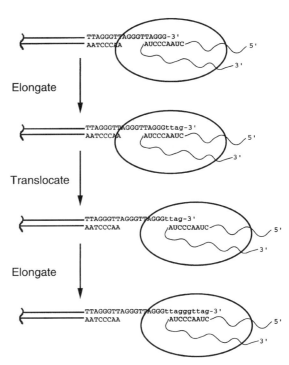

Figure 3. Model of processive telomerase action. The putative template domain of the RNA component of human telomerase is shown aligned against an arbitrary 3' end of a human telomere. This primer-template configuration allows extension of the telomere (lowercase letters) in the first round of elongation until the extended product reaches the 5' end of the template domain. Translocation then moves the extended DNA back one repeat relative to the template domain, positioning it for a another round of elongation in which a full repeat (**ggttag**) is added to the 3' end of the chromosome. (Adapted from Greider and Blackburn 1989.)

(Counter et al. 1994a), yet undetectable in normal somatic cells in vitro and in vivo, or in "extended life span" cells between M1 and M2 in vitro (Counter et al. 1992, 1994a,b). Moreover, by using a highly sensitive PCR-based assay for activity (Kim et al. 1994), telomerase has recently been detected in human testes and a large number of advanced tumors, but not in a variety of normal somatic tissues, nor in benign or early-stage tumors (Kim et al. 1994) (Table 1). Together, these data point to a striking correlation between indefinite replicative capacity and measurable telomerase on the one hand, and the mortal phenotype and lack of telomerase on the other.

Telomerase maintains long telomeres in immortal germ-line tissues but is repressed in normal somatic cells, leading to gradual telomere loss with each cell division (Fig. 4a). At M1, the Hayflick limit, permanent cell cycle exit occurs, perhaps as a consequence of a checkpoint arrest when at least one telomere becomes critically short. This form of growth control in normal somatic cells is abrogated by oncogene expression and/or tumor suppressor inactivation, conferring an extended, but still finite replicative capacity to the partially transformed cell. During this period, telomeres continue to shorten in the absence of telomerase, until cells

Table 1. Cell Immortality and Telomerase Activity in Human Cells and Tissues

Source	Phenotype	Tissue origin	Telomere dynamics	Telomerase activity	References
Cultured cells	mortal (dividing)	various (skin, lung, vascular, hematopoietic, ovary)	shorten or n.t.[d]	0 of 25 +[e]	1–5
	mortal (nondividing)	connective (skin, lung)	stable	0 of 3 +	7
	mortal (extended life span)	emb. kidney connective blood	shorten shorten shorten	0 of 5 +	5, 8
	immortal (transformed lines)[a]	various (lung, kidney, prostate, retina)	stable or n.t.	8 of 10 +	5, 8–10
	immortal (tumor lines)	various (14 different tissue origins)	stable or n.t.	0 of 70 +	5, 6, 9
Normal tissues	immortal[b]	testes	stable	2 of 2 +	1, 9
		connective	shorten		1, 2, 9
		epidermis	shorten		11, 12
		blood	shorten		3, 6, 8, 13
	mortal[c]	vascular intima	shorten	0 of >60 +	11
		brain	stable		7, 9
		others (breast, prostate, uterus, intestine, kidney, liver, lung, muscle, spleen)	n.t.		9
Tumor tissue	?	breast	n.t.		9
		prim. node neg.		1 of 4 +	
		ductal node pos.		14 of 15 +	
		ovarian carc.	stable	7 of 7 +	6
		prostate	n.t.		9
		BPH		1 of 10 +	
		PIN3		3 of 5 +	
		adenocarcinoma		2 of 2 +	
		neuroblastoma	shorten	5 of 5 +	9
		head and neck	n.t.	14 of 16 +	9
		colon			9, 13
		polyp	n.t.	0 of 1 +	
		tubular adenoma	shorten or n.t.	0 of 1 +	
		carcinoma	shorten or n.t.	8 of 8 +	
		uterine	n.t.		9
		fibroids		0 of 11 +	
		sarcoma		3 of 3 +	

References: (1) Allsopp et al. 1992; (2) Harley et al. 1990; (3) Vaziri et al. 1993; (4) Harley et al. 1992; (5) Counter et al. 1992; (6) Counter et al. 1994a; (7) Allsopp et al. 1993 and in prep.; (8) Counter et al. 1994b; (9) Kim et al. 1994; (10) Shay et al. 1993b; (11) C.B. Harley and K.R. Prowse, unpubl.; (12) Lindsey et al. 1991; (13) Hastie et al. 1990.

[a] Sublines of T-antigen-transformed and immortalized cells may become telomerase-negative, perhaps through genetic instability; it is not known whether these subclones have an indefinite replicative capacity (Kim et al. 1994)

[b] The germ-line lineage is immortal, even if specific cell types may not be.

[c] The question of whether normal tissues contain a rare immortal stem cell has not been addressed in these studies.

[d] n.t.: not tested.

[e] Some cultured hematopoietic cells may display weak telomerase activity detected with the PCR-based assay (Kim et al. 1994; N.W. Kim, unpubl.).

become nonviable at M2 (crisis), perhaps due to massive genetic instability when nearly all telomeres are critically short (Counter et al. 1992, 1994b). At or near crisis, strong selection for an additional mutation which activates telomerase may allow an immortal clone to survive with stabilized telomeres. If telomerase activation is a late event, and only minimal levels are expressed, telomeres will generally remain short in the immortal population of cells (Fig. 4a). The effects of telomerase inhibition on telomere length in these cell populations, including somatic stem cells (see below), is shown in Figure 4b.

TELOMERE LENGTH AND TELOMERASE IN NORMAL AND TUMOR CELLS

Currently, the most reliable method to assess telomere length involves Southern analysis of the terminal restriction fragment (TRF) length distribution (Allshire et al. 1989; Harley et al. 1990). However, human

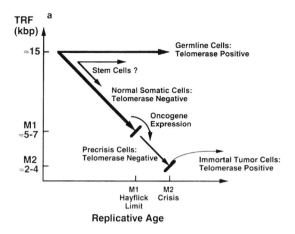

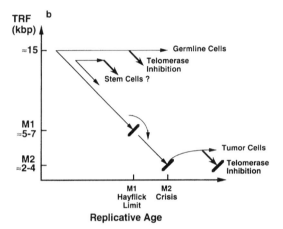

Figure 4. The telomere hypothesis of cell aging and immortalization and effect of telomerase inhibition. (*a*) Telomerase is active in the germ line, maintaining long stable telomeres (TRF ~15 kbp). Telomerase is repressed in normal somatic cells, resulting in telomere loss in dividing cells (~50–200 bp/cell doubling), until M1 (the Hayflick limit), where it is presumed that critical telomere loss on one or perhaps a few chromosomes signals irreversible cell cycle arrest (TRF ~5–7 kbp). Telomerase activity and telomere length are not known for true somatic stem cells. Oncogene expression or tumor suppressor gene inactivation allows somatic cells to bypass M1 without activating telomerase. When telomeres become critically short on nearly all chromosomes (TRF ~2–4 kbp), cells enter crisis (M2). Rare clones which activate telomerase escape M2, stabilize chromosome ends, and acquire an indefinite growth capacity. (*b*) The effect of telomerase inhibition is indicated. (Adapted from Harley 1991.)

TRFs contain DNA other than uniform telomeric TTAGGG repeat. For instance, TRFs may contain degenerate or non-TTAGGG as well as TTAGGG repeats in blocks other than at the distal end (Allshire et al. 1989; Counter et al. 1992; Levy et al. 1992). This technical limitation, together with interchromosomal variation in both non-TTAGGG and TTAGGG DNA in the TRF and variation in prior replicative histories in vitro or in vivo of cells in a population, makes it difficult to assess the relationship between cell senescence or crisis, and mean TRF length. This is especially true in cancer, where a variable and unknown number of normal cells may be present within a tumor sample. Thus, a static picture of telomere length cannot be used as an indicator of telomerase activity. Even a dynamic picture of TRF length increasing or decreasing should not be relied on as reflecting the presence or absence of telomerase, since at least three events could yield a false interpretation of telomerase activity: (1) a changing population of cells within a tissue, (2) an imbalance between telomerase activity and telomere loss, and (3) recombination.

Telomere dynamics and telomerase activity have not been extensively studied in a large number of normal and tumor cells and tissues in humans, but where studies have been conducted, the data both in vitro and in vivo are consistent with gradual telomere loss in telomerase-negative dividing cells, and generally short, stable telomeres in telomerase-positive cells (Harley et al. 1990; Hastie et al. 1990; Lindsey et al. 1991; Allsopp et al. 1992, 1993; Counter et al. 1992, 1994a,b; Shay et al. 1993a; Van Der Haegen and Shay 1993; Vaziri et al. 1993). There have been a number of studies in which telomere length has been reported for various tumors in the absence of telomerase data. Although the data are most consistent with reduced telomere length in tumors (Hastie et al. 1990; Adamson et al. 1992; Hiyama et al. 1992; Smith and Yeh 1992; Schwartz et al. 1993; Yamada et al. 1993; Mehle et al. 1994; Odagiri et al. 1994; Ohyashiki et al. 1994; Takauchi et al. 1994), there is considerable variation in TRF length, and there are some cases where tumors exhibit a trend toward longer TRFs compared to normal tissue (Nurnberg et al. 1993; Sciadini et al. 1994). However, as discussed above, caution should be exercised in interpretation of all TRF length data, especially in tumors where mixed populations of cells are likely to be present.

IS TELOMERASE NECESSARY FOR CELL IMMORTALITY?

Given the essential genetic function of telomeres, it is not surprising that telomere maintenance appears necessary for long-term cell viability (Lundblad and Szostak 1989; Yu et al. 1990; Counter et al. 1992, 1994b; Sandell and Zakian 1993). However, whether telomerase is essential for cell immortality remains unproven. Overexpression of mutant telomerase RNA in *Tetrahymena* (Yu et al. 1990), or mutation in a gene which may be a component of telomerase in yeast (Lundblad and Szostak 1989), both lead to telomere instability and a "senescent" phenotype in these otherwise immortal eukaryotic cells. However, in the latter case, an alternative pathway for telomere maintenance, possibly involving recombination, allows unstable variants to arise with extended life spans. The fact that telomerase is active in more than 100 immortal tumor cell lines and the majority of late-stage cancers (Table 1) suggests that telomerase is necessary for cell immortality. Definitive proof of this assumption awaits genetic or biochemical evidence that elimination of telomerase activity reinitiates telomere loss and irreversibly arrests cell proliferation when telomeres become critically short.

DOES TELOMERASE ACTIVITY CORRELATE WITH TUMOR PHENOTYPE?

There is a striking correlation between telomerase activity and the immortal phenotype in vitro (Table 1). Assuming this correlation extends to cells in vivo, one can determine the stage of tumorigenesis at which cells are selected for immortality by the timing of telomerase activation. In normal adult tissues, telomerase is detected only in testes, as expected, and not in somatic tissues, including those with mitotic cells, such as epidermis and intestine. However, most malignant or metastatic tumors from numerous sites, including breast, ovary, prostate, CNS, head and neck, colon, and uterus show activity (Table 1). Interestingly, telomerase activity in breast, prostate, colon, and uterus is observed to correlate with malignancy (Table 1). Similarly, we have observed that in early-stage chronic lymphocytic leukemia (CLL), telomeres were comparable to those of normal cells and telomerase was not detectable, whereas in late-stage CLL, telomeres were short and telomerase was detectable in most samples studied (C.M. Counter et al., unpubl.). Thus, the more advanced the disease, the greater the frequency of detectable telomerase in independent samples.

Selection for cell immortality (telomerase activation) and growth control mutations leading to malignancy or metastasis should be temporally, not causally, linked. Consistent with this, a small percentage of malignant or metastatic tumors test negative for telomerase (Table 1). It is unknown whether these cases reflect technical artifacts or true lack of immortal cells within the population. If it is the latter, these tumors should be mortal, and the prognosis for such patients would be better than in those for which tumors are telomerase positive. We are currently investigating this possibility. Another interesting exception may be cases of hyperplasia or benign growths that are telomerase positive. An example may be the one of ten samples from benign prostatic hypertrophy (BPH) which tested positive for telomerase (Table 1). Such growths may be more susceptible to malignant progression, since they already harbor one of the alterations associated with advanced tumors. An intriguing possibility is that a genetic defect in telomerase repression in normal somatic cells (global or tissue-specific) could account for some cases of familial predisposition to cancer. Mice possibly provide a model for this situation: They have relatively weak repression of telomerase in normal somatic tissues (K.R. Prowse and C.W. Greider; C. Chadeneau et al.; both unpubl.), a high incidence of spontaneous immortalization of cells in cultures, and an extremely high incidence of cancer in vivo (estimated at 10^4–10^5 times more frequent on a per-cell, per-year basis than in humans). Over- and underexpression of telomerase in mice by chemical or genetic approaches, including knock-out strains, will provide an interesting test of these hypotheses, including experimental validation for telomerase as an anticancer target in humans and the role of telomerase in development and differentiation of normal cells.

PROSPECTS FOR TELOMERASE INHIBITION AS A NOVEL ANTICANCER THERAPY

If cell immortality is essential for long-term growth of most cancers, and telomerase is necessary for cell immortality, it is clear that this novel enzyme is an attractive target for anticancer therapy. Moreover, since telomerase appears to play no role in growth or differentiation of normal adult somatic cells (Table 1), a telomerase inhibitor should have a high therapeutic index. Although telomerase is present in the male germ-line tissue, reproductive success is often not a concern of the cancer patient, and, if it were, sperm could be cryopreserved prior to therapy. The status of telomerase in the adult female germ line is unknown, although virtually all oocytes are produced during prenatal development. Therefore, telomerase inhibition should have minimal effect on the female germ line.

Unfortunately, there is very little known about cell kinetics, immortality, or telomerase expression in somatic stem cells. However, telomeres are lost with age in vitro and in vivo in human candidate hematopoietic stem cells (Vaziri et al. 1994), as is self-renewal capacity (Lansdorp et al. 1993). In addition, telomerase activity has not been detected with a polymerase chain reaction (PCR)-based assay in normal tissues which should contain stem cells (Table 1). Thus, it is possible that normal stem cells may lack functional telomerase activity and undergo slow replicative senescence through sequential recruitment into the proliferative pool as a function of age. If functional telomerase is present in rare, undetected stem cells, one might expect them to be deep in G_0 and to have long telomeres, and again the effects of telomerase inhibition may be minimal.

Of greater concern is the delayed efficacy of a telomerase inhibitor, since cells would enter crisis only after a number of doublings related to the telomere length at the time of treatment. The effect of telomerase inhibition on tumor growth was modeled by assuming that heterogeneity in telomere length at the time of treatment generates a distribution of replicative capacities (measured in cell doublings) within the tumor cell population (Fig. 5a). Variables within the model include not only the distribution of replicative capacities, but also the growth and cell loss fractions. These variables can only be crudely estimated, and thus the data must be interpreted with caution. For purposes of illustration, we assumed (1) a normal distribution of replicative capacities (mean 15 ± 5 doublings), (2) tumor burden at time of treatment = 10^9 cells (~1 g), (3) a continual growth fraction of 15%, and (4) no additional cell death for reasons other than exhaustion of proliferative capacity. We have also assumed that

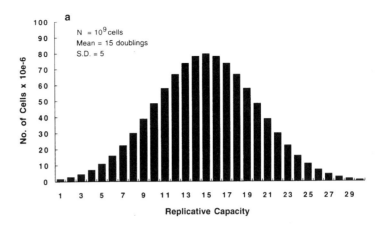

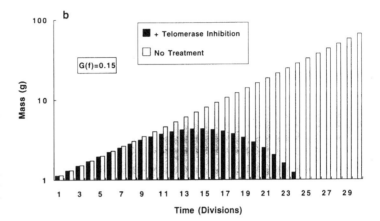

Figure 5. Tumor growth model with and without telomerase inhibition. In this simulation, the initial tumor mass is 1 g (10^9 cells), and telomere lengths at time 0 are assumed to generate a normal distribution of replicative capacities covering the range 0–30 doublings (mean = 15, S.D. = 5). (*a*) In the absence of telomerase inhibition, replicative capacities are assumed to be infinite, but the growth fraction G(f) at each cycle is 15%; all other cells are assumed to be irreversibly arrested. Tumor growth is shown as a function of time (cell division intervals) (*b*).

there are no alternative pathways for telomere maintenance. Since telomeres are lost at approximately 100 bp/doubling in vitro, 15 doublings might correspond to approximately 1.5 kbp of additional TTAGGG beyond that which initiates crisis (Counter et al. 1992, 1994b). The growth fraction of 15% was chosen by averaging that reported for a number of different tumors (ten Velde et al. 1989; Stumpp et al. 1992; Larsson et al. 1993; Porschen et al. 1993; Tinnemans et al. 1993). Under these conditions, it is clear that even with replicative capacities extending out to 20–30 doublings, the effect of telomerase inhibition on limiting tumor growth is dramatic: After 30 "doubling times," the untreated tumor is greater than 70 g, whereas the treated tumor peaks at 15 time units at about 4 g (Fig. 5b). A telomerase inhibitor would be even more effective in tumors with shorter telomeres.

In summary, telomere dynamics and telomerase expression appear to be fundamentally involved in cellular aging and cancer. Stabilization of telomeres by activation of telomerase is associated with cell immortality and appears to be a late event in many human tumors. We conclude that cell immortalization is probably required for long-term growth of malignant or metastatic tumors and that telomerase inhibition will provide effective therapy for patients diagnosed with cancer.

ACKNOWLEDGMENTS

This work was supported in part by grants from the National Institutes of Health, the Medical Research Council of Canada, and Geron Corporation.

REFERENCES

Adamson, D.J., D.J. King, and N.E. Haites. 1992. Significant telomere shortening in childhood leukemia. *Cancer Genet. Cytogenet.* **61:** 204.

Allshire, R.C., M. Dempster, and N.D. Hastie. 1989. Human telomeres contain at least three types of G-rich repeats distributed non-randomly. *Nucleic Acids Res.* **17:** 4611.

Allsopp, R.C., M. Kashefi-Aazam, K.R. Prowse, C.W. Greider, and C.B. Harley. 1993. Consistent telomere loss associated with cell division in normal human fibroblasts. *In Vitro* **29A:** 53A.

Allsopp, R.C., H. Vaziri, C. Patterson, S. Goldstein, E.V. Younglai, A.B. Futcher, C.W. Greider, and C.B. Harley. 1992. Telomere length predicts replicative capacity of human fibroblasts. *Proc. Natl. Acad. Sci.* **89:** 10114.

Blackburn, E.H. 1992. Telomerases. *Annu. Rev. Biochem.* **61:** 113.

Blackburn, E.H. and J.W. Szostak. 1984. The molecular structure of centromeres and telomeres. *Annu. Rev. Biochem.* **53:** 163.

Bond, J.A., F.S. Wyllie, and D. Wynford-Thomas. 1994. Escape from senescence in human diploid fibroblasts induced directly by mutant p53. *Oncogene* **9:** 1885.

Cairns, J. 1975. Mutation selection and the natural history of cancer. *Nature* **255:** 197.

Counter, C.M., H.W. Hirte, S. Bacchetti, and C.B. Harley. 1994a. Telomerase activity in human ovarian carcinoma. *Proc. Natl. Acad. Sci.* **91:** 2900.

Counter, C.M., F.M. Botelho, P. Wang, C.B. Harley, and S. Bacchetti. 1994b. Stabilization of short telomeres and telomerase activity accompany immortalization of Epstein-Barr virus-transformed human B lymphocytes. *J. Virol.* **68:** 3410.

Counter, C.M., A.A. Avilion, C.E. LeFeuvre, N.G. Stewart, C.W. Greider, C.B. Harley, and S. Bacchetti. 1992. Telomere shortening associated with chromosome instability is arrested in immortal cells which express telomerase activity. *EMBO J.* **11:** 1921.

Girardi, A.J., F.C. Jensen, and H. Koprowski. 1965. SV40-induced transformation of human diploid cells: Crisis and recovery. *J. Cell. Comp. Physiol.* **65:** 69.

Goldstein, S. 1990. Replicative senescence: The human fibroblast comes of age. *Science* **249:** 1129.

Gonos, E.S., A.J. Powell, and P.S. Jat. 1992. Human and rodent fibroblasts: Model systems for studying senescence and immortalisation. *Int. J. Oncol.* **1:** 209.

Graham, F.L., J.R. Smiley, W.D. Russell, and R. Nairn. 1977. Characteristics of a human cell line transformed by DNA from human adenovirus 5. *J. Gen. Virol.* **36:** 59.

Greider, C.W. 1990. Telomeres, telomerase and senescence. *BioEssays* **12:** 363.

———. 1993. Telomerase and telomere-length regulation: Lessons from small eukaryotes to mammals. *Cold Spring Harbor Symp. Quant. Biol.* **58:** 719.

Greider, C.W. and E. Blackburn. 1985. Identification of a specific telomere terminal transferase activity in *Tetrahymena* extracts. *Cell* **43:** 405.

———. 1989. A telomeric sequence in the RNA of *Tetrahymena* telomerase required for telomere repeat synthesis. *Nature* **337:** 331.

Harley, C.B. 1988. Biology and evolution of aging: Implications for basic gerontological health research. *Can. J. Aging* **7:** 100.

———. 1991. Telomere loss: Mitotic clock or genetic time bomb? *Mutat. Res.* **256:** 271.

Harley, C.B., A.B. Futcher, and C.W. Greider. 1990. Telomeres shorten during ageing of human fibroblasts. *Nature* **345:** 458.

Harley, C.B., H. Vaziri, C. Counter, and R.C. Allsopp. 1992. The telomere hypothesis of cellular aging. *Exp. Gerontol.* **27:** 375.

Hartwell, L. 1992. Defects in a cell cycle checkpoint may be responsible for the genomic instability of cancer cells. *Cell* **71:** 543.

Hastie, N.D., M. Dempster, M.G. Dunlop, A.M. Thompson, D.K. Green, and R.C. Allshire. 1990. Telomere reduction in human colorectal carcinoma and with ageing. *Nature* **346:** 866.

Hayflick, L. 1965. The limited *in vitro* lifetime of human diploid cell strains. *Exp. Cell Res.* **37:** 614.

Hiyama, E., K. Hiyama, T. Yokoyama, T. Ichikawa, and Y. Matsuura. 1992. Length of telomeric repeats in neuroblastoma: Correlation with prognosis and other biological characteristics. *Jpn. J. Cancer Res.* **83:** 159.

Hiyama, K., Y. Shirotani, S. Ishioka, E. Hiyama, T. Yokoyama, and M. Yamakido. 1993. Cancer-cell proliferation and alteration in length of telomeric repeats. *Tissue Culture Res. Commun.* **12:** 291.

Hunter, T. 1993. Oncogenes and cell proliferation, editorial review: Oncogenes and the cell cycle. *Curr. Opin. Genet. Dev.* **3:** 1.

Kim, N.W., M.A. Piatyszek, K.R. Prowse, C.B. Harley, M.D. West, P.L.C. Ho, G.M. Coviello, W.E. Wright, S.L. Weinrich, and J.W. Shay. 1994. Specific association of human telomerase activity with immortal cells and cancer. *Science* **266:** 2011.

Klingelhutz, A.J., S.A. Barber, P.P. Smith, K. Dyer, and J.K. McDougall. 1994. Restoration of telomeres in human papillomavirus-immortalized human anogenital epithelial cells. *Mol. Cell. Biol.* **14:** 961.

Lansdorp, P.M., W. Dragowska, and H. Mayani. 1993. Ontogeny-related changes in proliferative potential of human hematopoietic cells. *J. Exp. Med.* **178:** 787.

Larsson, P., G. Roos, R. Stenling, and B. Ljungberg. 1993. Tumor-cell proliferation and prognosis in renal-cell carcinoma. *Int. J. Cancer* **55:** 566.

Levy, M.Z., R.C. Allsopp, A.B. Futcher, C.W. Greider, and C.H. Harley. 1992. Telomere end-replication problem and cell aging. *J. Mol. Biol.* **225:** 951.

Lindsey, J., N. McGill, L. Lindsey, D. Green, and H. Cooke. 1991. In vivo loss of telomeric repeats with age in humans. *Mutat. Res.* **256:** 45.

Lundblad, V. and J.W. Szostak. 1989. A mutant with a defect in telomere elongation leads to senescence in yeast. *Cell* **57:** 633.

Mehle, C., B. Ljungberg, and G. Roos. 1994. Telomere shortening in renal cell carcinoma. *Cancer Res.* **54:** 236.

Morin, G.B. 1989. The human telomere terminal transferase enzyme is a ribonucleoprotein that synthesizes TTAGGG repeats. *Cell* **59:** 521.

Newbold, R.F., A.P. Cuthbert, M. Themis, D.A. Trott, A.L. Blair, and W. Li. 1993. Cell immortalization as a key, rate-limiting event in malignant transformation: Approaches toward a molecular genetic analysis. *Toxicol. Lett.* **67:** 211.

Norton, L. 1991. Clinical aspects of cell and tumor growth kinetics. In *Comprehensive textbook of oncology*, 2nd edition (ed. A.R. Moossa et al.), vol. 1, p. 409. Williams and Wilkins, Baltimore.

Nurnberg, P., G. Thiel, F. Weber, and J.T. Epplen. 1993. Changes of telomere lengths in human intracranial tumours. *Hum. Genet.* **91:** 190.

Odagiri, E., N. Kanada, K. Jibiki, R. Demura, E. Aikawa, and H. Demura. 1994. Reduction of telomeric length and c-erbB-2 gene amplification in human breast cancer, fibroadenoma, and gynecomastia. Relationship to histologic grade and clinical parameters. *Cancer* **73:** 2978.

Ohyashiki, J.H., K. Ohyashiki, T. Fujimura, K. Kawakubo, T. Shimamoto, A. Iwabuchi, and K. Toyama. 1994. Telomere shortening associated with disease evolution patterns in myelodysplastic syndromes. *Cancer Res.* **54:** 3557.

Olovnikov, A. 1973. A theory of marginotomy: The incomplete copying of template margin in enzymatic synthesis of polynucleotides and biological significance of the phenomenon. *J. Theor. Biol.* **41:** 181.

Porschen, R., S. Classen, Y. Kahle, and F. Borchard. 1993. In situ evaluation of the relationship between epidermal-growth-factor-receptor status and tumor-cell proliferation in colon carcinomas. *Int. J. Cancer* **54:** 189.

Runger, T.M., C.E. Klein, J.C. Becker, and E.B. Brocker. 1994. The role of genetic instability, adhesion, cell motility, and immune escape mechanisms in melanoma progression. *Curr. Opin. Oncol.* **6:** 188.

Sandell, L.L. and V.A. Zakian. 1993. Loss of a yeast telomere: Arrest, recovery, and chromosome loss. *Cell* **75:** 729.

Schwartz, H.S., G.A. Dahir, and M.G. Butler. 1993. Telomere reduction in giant cell tumor of bone and with aging. *Cancer Genet. Cytogenet.* **71:** 132.

Sciadini, M.F., H.S. Schwartz, L.K. Miller, and M.G. Butler. 1994. Is telomere reduction a generalized phenomenon in chromosomes of solid tissue neoplasms? *Proc. Annu. Meet. Am. Soc. Clin. Oncol.* **13:** A314.

Shay, J.W., O.M. Pereira-Smith, and W.E. Wright. 1991. A role for both RB and p53 in the regulation of human cellular senescence. *Exp. Cell Res.* **196:** 33.

Shay, J.W., W.E. Wright, and H. Werbin. 1993a. Loss of telomeric DNA during aging may predispose cells to cancer. *Int. J. Oncol.* **3:** 559.

Shay, J.W., W.E. Wright, D. Brasiskyte, and B.A. Van Der Haegen. 1993b. E6 of human papillomavirus type 16 can overcome the M1 stage of immortalization in human mammary epithelial cells but not in human fibroblasts. *Oncogene* **8:** 1407.

Smith, J.K. and G. Yeh. 1992. Telomere reduction in endometrial adenocarcinoma. *Am. J. Obstet. Gynecol.* **167:** 1883.

Stamps, A.C., B.A. Gusterson, and M.J. O'Hare. 1992. Are tumors immortal? *Eur. J. Cancer* **28A:** 1495.

Stanulis Praeger, B.M. 1987. Cellular senescence revisited: A review. *Mech. Ageing Dev.* **38:** 1.

Stumpp, J., J. Dietl, W. Simon, and M. Geppert. 1992. Growth fraction in breast carcinoma determined by Ki-67 immunostaining: Correlation with pathological and clinical variables. *Gynecol. Obstet. Invest.* **33:** 47.

Takauchi, K., S. Tashiro, M. Ohtaki, and N. Kamada. 1994. Telomere reduction of specific chromosome translocation in acute myelocytic leukemia. *Jpn. J. Cancer Res.* **85:** 127.

ten Velde, G.P., B. Schutte, M.M. Reijnders, F.T. Bosman, and G.H. Blijham. 1989. Cytokinetic analysis of lung cancer by bromodeoxyuridine labeling of cytology specimens. *Cytometry* **10:** 807.

Tinnemans, M.M., B. Schutte, M.H. Lenders, G.P. ten Velde, F.C. Ramaekers, and G.H. Blijham. 1993. Cytokinetic analysis of lung cancer by in vivo bromodeoxyuridine labelling. *Br. J. Cancer* **67:** 1217.

Van Der Haegen, B.A. and J.W. Shay. 1993. Immortalization of human mammary epithelial cells by SV40 large T-antigen involves a two step mechanism. *In Vitro Cell. Dev. Biol.* **29A:** 180.

Vaziri, H., W. Dragowska, R.C. Allsopp, T.E. Thomas, C.B. Harley, and P.M. Lansdorp. 1994. Evidence for a mitotic clock in human hematopoietic stem cells: Loss of telomeric DNA with age. *Proc. Natl. Acad. Sci.* **91:** 9857.

Vaziri, H., F. Schachter, I. Uchida, L. Wei, X. Zhu, R. Effros, D. Cohen, and C.B. Harley. 1993. Loss of telomeric DNA during aging of normal and trisomy 21 human lymphocytes. *Am. J. Hum. Genet.* **52:** 661.

Vogelstein, B. and K. Kinzler. 1993. The multistep nature of cancer. *Trends Genet.* **9:** 138.

Watson, J. 1972. Origin of concatemeric T7 DNA. *Nat. New Biol.* **239:** 197.

Weinberg, R.A. 1989. Oncogenes, antioncogenes, and the molecular bases of multistep carcinogenesis. *Cancer Res.* **49:** 3713.

———. 1992. Molecular biology of carcinogenesis: A multistep process. In *Molecular foundations of oncology* (ed. S. Broder), pp. 27. Williams and Wilkins, Baltimore.

West, M.D. 1994. The cellular and molecular biology of skin aging. *Arch. Dermatol.* **130:** 87.

West, M.D., O.M. Pereira Smith, and J.R. Smith. 1989. Replicative senescence of human skin fibroblasts correlates with a loss of regulation and overexpression of collagenase activity. *Exp. Cell Res.* **184:** 138.

Workman, P. 1994. The potential for molecular oncology to define new drug targets. In *New molecular targets for cancer therapy* (ed. D.J. Kerr and P. Workman), p. 1. CRC Press, London.

Yamada, O., K. Oshimi, and H. Mizoguchi. 1993. Telomere reduction in hematologic cells. *Int. J. Hematol.* **57:** 181.

Yu, G.-L., J.D. Bradley, L.D. Attardi, and E.H. Blackburn. 1990. *In vivo* alteration of telomere sequences and senescence caused by mutated telomerase RNAs. *Nature* **344:** 126.

Zakian, V.A. 1989. Structure and function of telomeres. *Annu. Rev. Genet.* **23:** 579.

Three Unusual Repair Deficiencies Associated with Transcription Factor BTF2(TFIIH): Evidence for the Existence of a Transcription Syndrome

W. VERMEULEN,*†† A.J. VAN VUUREN,*†† M. CHIPOULET,† L. SCHAEFFER,† E. APPELDOORN,*
G. WEEDA,* N.G.J. JASPERS,* A. PRIESTLEY,‡ C.F. ARLETT,‡
A.R. LEHMANN,‡ M. STEFANINI,§ M. MEZZINA,** A. SARASIN,**
D. BOOTSMA,* J.-M. EGLY,† AND J.H.J. HOEIJMAKERS*

*Department of Cell Biology and Genetics, Medical Genetics Centre, Erasmus University, 3000DR, Rotterdam, The Netherlands; †UPR 6520 (CNRS), Unité 184 (INSERM), Faculté de Médecine, 67085 Strasbourg Cedex, France; ‡MRC Cell Mutation Unit, University of Sussex, Falmer, Brighton BN1 9RR, United Kingdom; §Consiglio Nazionale Della Richerche, Istituto di Genetica Biochemica ed Evoluionistica, 27100 Pavia, Italy; **UPR 42 (CNRS), IFC 1101, 94801 Villejuif Cedex, France

To counteract the deleterious effects of DNA damage, a sophisticated network of DNA repair systems has evolved, which is essential for genetic stability and prevention of carcinogenesis. Nucleotide excision repair (NER), one of the main repair pathways, can remove a wide range of lesions from the DNA by a complex multistep reaction (for a recent review, see Hoeijmakers 1993). Two subpathways are recognized in NER: a rapid "transcription-coupled" repair and the less efficient global genome repair (Bohr 1991; Hanawalt and Mellon 1993). The consequences of inborn errors in NER are highlighted by the prototype repair syndrome, xeroderma pigmentosum (XP), an autosomal recessive condition displaying sun (UV) sensitivity, pigmentation abnormalities, predisposition to skin cancer, and often progressive neurodegeneration (Cleaver and Kraemer 1994). Two other excision repair disorders have been recognized, Cockayne's syndrome (CS) and trichothiodystrophy (TTD), which present different clinical features. These are, besides sun sensitivity, neurodysmyelination, impaired physical and sexual development, dental caries (in CS and TTD), ichthyosis and sulfur-deficient brittle hair and nails (in TTD)(Lehmann 1987; Nance and Berry 1992), which are difficult to rationalize on the basis of defective NER only. The NER syndromes are genetically heterogeneous and comprise at least 10 different complementation groups: 7 in XP (XP-A to XP-G), 5 in CS (CS-A, CS-B, XP-B, XP-D, and XP-G) and 2 in TTD (TTD-A and XP-D) (Hoeijmakers 1993; Stefanini et al. 1993b). Thus, considerable overlap and clinical heterogeneity are associated with a selected subset of complementation groups, of which XP-D is the most extreme, harboring patients with XP only, or combined XP and CS, or TTD (Johnson and Squires 1992).

Recently, it was discovered that the DNA repair helicase encoded by the *ERCC3* gene (Roy et al. 1994), which is mutated in XP-B (Weeda et al. 1990), is identical to the p89 subunit of the transcription factor BTF2/TFIIH (Schaeffer et al. 1993). ERCC3 in the context of TFIIH is directly involved in NER and transcription in vitro as well as in vivo (van Vuuren et al. 1994). Furthermore, another repair helicase, XPD/ERCC2, was recognized to be associated with TFIIH (Schaeffer et al. 1994), and it was shown that a partially purified TFIIH fraction is able to correct the NER deficiency of XP-D in vitro (Drapkin et al. 1994). These results reveal a link between two distinct DNA-metabolizing processes: repair and transcription.

Initiation of transcription is believed to require the formation of an elongation-competent protein complex in a highly ordered cascade of reactions. A preinitiation (DAB) complex is formed by the binding of the multisubunit TFIID factor to the TATA box element of core promoters, stabilized by TFIIA and followed by the association of TFIIB. The DAB intermediate stimulates the entry of RNA polymerase II mediated by TFIIF. Initiation is completed by the ordered association of TFIIE, TFIIH/BTF2, and TFIIJ (Roeder 1991; Gill and Tjian 1992; Drapkin et al. 1993). TFIIH/BTF2 is thought to be involved in the conversion of a closed to an open initiation complex by local melting of the transcriptional start site and phosphorylation of the carboxy-terminal repeat of the large subunit of RNA polymerase II (Lu et al. 1992; Schaeffer et al. 1993; Serizawa et al. 1993), either at the preinitiation stage, or at a step between initiation and elongation, referred to as promoter clearance (Goodrich and Tjian 1994).

Here we report that defects in the XPB/ERCC3 subunit define a new TTD complementation group extending the clinical heterogeneity associated with ERCC3. Furthermore, we demonstrate that probably the entire TFIIH complex has a dual role in transcription and repair, as it harbors now at least three NER proteins, all associated with the TTD and CS symptoms. The dual function of TFIIH, and the link between mutations in this complex and the pleiotropic features of TTD and CS, strongly support the idea that some of

††These authors have contributed equally to this work.

the clinical manifestations arise from defects in the transcription function of TFIIH. This provides for the first time a molecular explanation for the seemingly unrelated symptoms of these disorders and introduces a novel clinical entity, a heterogenous "transcription syndrome" complex, that may include many more inherited conditions.

METHODS

Purification of BTF2/TFIIH. The purification of BTF2 starting from HeLa whole-cell extract and involving sequential chromatography on Heparin-Ultrogel, DEAE-Spherodex, SP-5PW sulfopropyl, and—after ammonium sulfate precipitation—Phenyl-5PW Sepharose and hydroxyapatite, was carried out essentially as described earlier (Gerard et al. 1991). In some fractions an additional purification step using heparin chromatography was inserted in the standard protocol before the Phenyl-5PW column. The in vitro assay for following the transcriptional stimulation activity of BTF2 on an AD2MLP promoter containing template involving purified RNA polymerase II and all transcription factors except BTF2/TFIIH has been described in detail previously (Gerard et al. 1991).

Cell lines and extracts. The human cell lines used for microinjection and for complementation analysis were XP25RO and XP11PV (both XP-A), XPCS1BA and XPCS2BA (XP-B), XP21RO (XP-C), XP1BR, XP3NE, and TTD8PV (all XP-D), XP2RO (XP-E), XP126LO (XP-F), XP2BI (XP-G), and TTD1BR (TTD-A) and TTD6VI (Stefanini et al. 1993b; Vermeulen et al. 1986, 1994). The primary fibroblasts were cultured in Ham's F10 medium supplemented with antibiotics and 10–15% fetal calf serum.

For preparing cell-free extracts utilized for the in vitro repair assay, the following cell lines were used: a SV40-transformed line belonging to TTD-A (TTD1BRSV), human repair-proficient HeLa cells, mutant Chinese hamster cells 43-3B, UV5, 27.1, UV41, and UV135 assigned to complementation groups 1, 2, 3, 4, and 5, respectively (Busch et al. 1989). The cells were grown in a 1:1 mixture of F10 and DMEM medium; antibiotics and 10% fetal calf serum were added. After harvesting and washing with phosphate-buffered saline (PBS), cell-free extracts were prepared as described previously (Manley et al. 1983; Wood et al. 1988), dialyzed against a buffer containing 25 mM HEPES/KOH (pH 7.8), 0.1 M KCl, 12 mM $MgCl_2$, 1 mM EDTA, 2 mM DTT, and 17% (v/v) glycerol, and stored at $-80°C$.

In vitro DNA repair assay. Plasmid pBluescript KS^+ (3.0 kb) was damaged by treatment with 0.1 mM N-acetoxy-2-acetylaminofluorene (AAF) (a kind gift of R. Baan, TNO, Rijswijk). As a nondamaged control, plasmid pHM14 (3.7 kb) was used. Circular closed forms of both plasmids were isolated and extensively purified as described previously (Biggerstaff et al. 1991; van Vuuren et al. 1993). The reaction mixture contained 250 ng each of damaged and nondamaged control plasmids, 45 mM HEPES/KOH (pH 7.8), 70 mM KCl, 7.4 mM $MgCl_2$, 0.4 mM EDTA, 0.9 mM DTT, 2 mM ATP, 40 mM phosphocreatine, 2.5 µg creatine phosphokinase, 3.5% glycerol, 18 µg bovine serum albumin, 20 µM each of dCTP, dGTP, and TTP, 8 µM dATP, 74 kBq of $[\alpha-^{32}P]dATP$ and 200 µg of cell-free extract for the rodent extracts and 100 µg for HeLa. After 3 hours incubation at 30°C, plasmid DNAs were purified from the reaction mixture, linearized by restriction, and separated by electrophoresis on an 0.8% agarose gel. Results were quantified using an LKB Densitometer and a B&L Phospho-Imager.

Anti-p62 antibody depletion of a repair-proficient extract. Protein A-Sepharose beads in PBS were incubated with monoclonal antibodies against the p62 subunit of BTF2 (Mab3C9; Fischer et al. 1992) or against p89 for 1 hour at 0°C. The beads were washed with dialysis buffer and added to a repair-competent HeLa extract or a (partially) purified heparin or hydroxyapatite BTF2 fraction for 1 hour at 0°C (van Vuuren et al. 1993). The bound proteins were removed by centrifugation and the supernatant was tested in the in vitro repair assay or for microneedle injection. Western blot analysis following SDS-PAGE was carried out according to standard protocols (Sambrook et al. 1989).

Microinjection. Microinjection of XP homopolykaryons was performed as described earlier (Vermeulen et al. 1994). Repair activity was determined after UV irradiation (15 J/m^2), [^{3}H]thymidine incubation (10 µCi/ml; s.a.: 50 Ci/mMole), fixation, and autoradiography. Grains above the nuclei represent a quantitative measure for NER activity. RNA synthesis was detected after labeling with [^{3}H]uridine (10 µCi/ml; s.a.: 50 Ci/mMole). cDNAs were injected into one of the nuclei of the polykaryons; protein fractions and antibodies were injected into the cytoplasm of the cell. Antisera used for microinjection: polyclonal rabbit anti-ERCC1 and anti-ERCC3 (van Vuuren et al. 1994) and polyclonal rabbit anti-ERCC2 (Schaeffer et al. 1994).

Complementation analysis by cell hybridization. Fibroblasts of each fusion partner were labeled with latex beads (0.8 µm or 2.0 µm) 3 days prior to fusion by adding a suspension of beads to the culture medium. Cell fusion was performed with the aid of inactivated Sendai virus or, using polyethylene glycol, cells were seeded onto coverslips and assayed for UV-induced unscheduled DNA synthesis (UDS) as described in detail elsewhere (Stefanini et al. 1993a; Vermeulen et al. 1993).

Other procedures. Isolation of DNA, subcloning, and in vitro transcription and translation were done using established procedures (Sambrook et al. 1989).

RESULTS

Identification of Further Genetic Heterogeneity within TTD

In an effort to assess the genetic heterogeneity within the class of repair-deficient TTD patients, we have conducted a systematic complementation study of a large number of photosensitive TTD families by cell fusion and microneedle injection of cloned repair genes. This resulted recently in the identification of a second complementation group among NER-deficient TTD patients (designated TTD-A) (Stefanini et al. 1993b). Figure 1 shows the results of an exhaustive complementation analysis of fibroblasts from one of two siblings (TTD6VI and TTD4VI) with relatively mild clinical features of TTD and moderately impaired NER characteristics (detailed clinical description to be presented elsewhere). Full complementation of the repair defect was seen when fibroblasts of patient TTD6VI were fused with representatives of the known TTD complementation groups: XP group D (3 cell lines tested) and with TTD-A fibroblasts (Fig. 1) (Stefanini et al. 1993a,b). This demonstrates that this family defines a new TTD complementation group. To see whether this group is genetically identical to one of the other XP groups not previously associated with TTD, further cell-hybridization experiments were carried out. No restoration of the deficient UV-induced UDS was found when TTD6VI cells were fused with an XP-B representant (XPCS1BA), whereas normal complementation was observed with the other XP groups (Fig. 1). This indicates that the repair defect in this family resides in the *XPB/ERCC3* gene. Although TTD6VI cells appeared exceptionally sensitive to nuclear microneedle injection of DNA, still a few cells could be found corrected by *ERCC3* cDNA, in agreement with the assignment by cell fusion. Defects in the TFIIH subunit ERCC3 thus extend to the disorder TTD, increasing the already observed clinical heterogeneity among patients carrying mutations in *ERCC3* (Vermeulen et al. 1994).

Systematic Screening of TFIIH for NER Proteins

To examine whether besides XPB/ERCC3 (p89) additional NER factors are hidden in the TFIIH/BTF2 complex, we have systematically screened the existing human and rodent NER mutants for complementation by BTF2. The final fractions of TFIIH purification show at least five tightly associated proteins, including p89 (XPB/ERCC3), p62, p44, p41, and p34 (Gerard et al. 1991; Schaeffer et al. 1993). A purified BTF2-fraction (hydroxyapatite chromatography, gel pattern, and properties; see Gerard et al. 1991) was inserted by

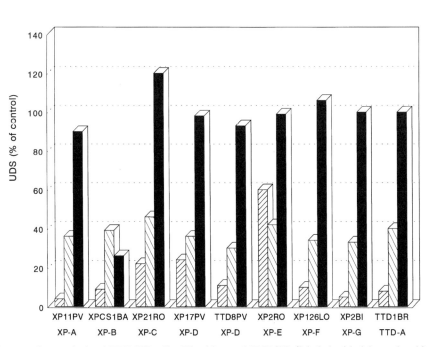

Figure 1. Complementation analysis of TTD6VI cells. Fibroblasts of TTD6VI (labeled with 0.8-μm beads) were fused with representatives of different XP and TTD complementation groups (labeled with 2.0-μm beads). After performing the UV-induced UDS assay, the average number of autoradiographic grains above the nuclei was determined. DNA repair synthesis was expressed as the percentage of control UDS (vertical axis) observed in normal fibroblasts assayed in parallel. The relative standard errors of the mean are in all cases less than 5%. The different types of binuclear cells (homodikaryons of TTD6VI and of the different fusion partners, and heterodikaryons) are recognized by their different bead content and indicated as such in the figure.

microneedle injection into fibroblasts of all known excision-deficient XP, XP/CS, and TTD complementation groups. The injected cultures were exposed to UV and incubated with [^{3}H]thymidine to permit visualization of the repair synthesis step of NER by in situ autoradiography (unscheduled DNA synthesis, UDS). The results, summarized in Table 1, show that 3 out of 8 NER-deficient human complementation groups are corrected by the purified BTF2 transcription factor. Figure 2 presents micrographs of BTF2-injected multinucleated fibroblasts of XP-D, TTD-A, and the new XP-B patient with TTD symptoms. The repair activities in injected cells reach the levels of normal fibroblasts assayed in parallel, indicating full correction of the excision defect (see also Table 2).

The restoration of NER could also be explained by correction of expression of one or more critical repair genes whose transcription is impaired by mutations in BTF2 subunits. Therefore, purified BTF2 was checked for its repair capacity in an in vitro repair assay (Wood et al. 1988; van Vuuren et al. 1993), where neither transcription nor translation can occur, because ribonucleotides and amino acids are lacking. The results in Figure 3 and Table 1 show that the BTF2 fraction not only corrects the repair defect in rodent group 3 (the equivalent of XP-B), as reported earlier (van Vuuren et al. 1994), but also the rodent group equivalent to XP-D (CHO group 2, defective in ERCC2) (Flejter et al. 1992) and an extract from a SV40-transformed TTD1BR cell line (TTD-A). In contrast, no significant NER restoration is observed in extracts of UV-sensitive mutants from groups 1, 4, and 5 (defective in ERCC1, ERCC4, and XPG/ERCC5, respectively). These findings extend the specificity of the repair-correcting activity of TFIIH and confirm the in vivo results obtained by microinjection. Furthermore, they demonstrate that besides XPB/ERCC3, XPD/ERCC2 and TTD-A must also have a direct involvement in NER. The in vitro and in vivo correction of *ERCC2* mutants confirmed the in vitro data of Drapkin et al. (1994), who used a partially purified TFIIH preparation. However, the TFIIH/BTF2 purified fractions used in this study do not contain a XPC-correcting activity, in contrast to the observations made by these authors.

Relationship of XPD/ERCC2 and TTD-A with the Core of TFIIH

To further strengthen the link between NER and the TFIIH complex, the full elution profile of hydroxyapatite chromatography, the final purification step of BTF2 (Gerard et al. 1991), was quantitatively screened for NER activity by microinjection as well as by in vitro complementation. The XP-D and TTD-A correcting activities coelute with the activities of XP-B correction and transcription initiation (Fig. 4). Identical results were obtained after cosedimentation in glycerol gradients (van Vuuren et al. 1994; W. Vermeulen et al., unpubl.).

Independent evidence for a physical association of the XPD/ERCC2, XPB/ERCC3, and TTDA proteins with each other and with other components in the TFIIH complex can be obtained from antibody depletion experiments. A crude repair-proficient HeLa extract was incubated with a monoclonal antibody against the p62 TFIIH subunit (Fischer et al. 1992), immobilized on protein A-Sepharose beads, and after centrifugation to remove bound proteins, the extract was tested for remaining repair capacity. Western blot analysis verified that the amount of p62 was strongly reduced (Fig. 5). In the in vitro correction assay, the p62^{depleted} HeLa extract had lost most of its repair activity when compared with the repair level of mutant extracts alone and a mock-treated extract (see footnote to Table 2). To examine whether other NER factors were co-depleted, the treated extract was tested for its complementing capacity using both the in vitro assay and the microneedle injection. Previously, we have demonstrated that XPB/ERCC3, but not XPG/ERCC5, is simultaneously removed with p62 (van Vuuren et al. 1994). Table 2 shows that most XPD/ERCC2 and TTDA activities are also removed, whereas XPA, ERCC1, ERCC4, and XPG/ERCC5 are not significantly eliminated (Table 2 and unpublished research). This is confirmed by the Western blot analysis revealing depletion of XPB/ERCC3(p89) at the protein level (Fig. 5, lanes 5,6). The same co-depletion patterns were observed with a monoclonal antibody against p89 (XPB/ERCC3) (Table 2). Furthermore, depletion experiments using monoclonal antibodies against two other components of TFIIH, p44 (the human homolog of yeast SSL1) and p34, revealed tight association with repair (Humbert et al. 1994; and results not shown). As with crude HeLa extracts, antibody depletion of purified BTF2 (hydroxyapatite and heparin purification fractions) using p62 monoclonal antibodies resulted in simultaneous removal of the XPB/ERCC3, XPD/ERCC2, and TTDA factors (Table 2). The depletion

Table 1. Involvement of BTF2 in Different Human and Rodent NER-deficient Complementation Groups

NER-deficient complement. groups		Correction of NER defect by BTF2
human mutants		microneedle injection
XP25RO	XP-A	−
XPCS1BA	XP-B	+
XP21RO	XP-C	−
XP1BR	XP-D	+
XP2RO	XP-E	−
XP126LO	XP-F	−
XP2BI	XP-G	−
TTD1BR	TTD-A	+
		in vitro repair assay
TTD1BRSV	TTD-A	+
rodent mutants		
43.3B	group 1	−
UV 5	group 2	+
27.1	group 3	+
UV 41	group 4	−
UV 135	group 5	−

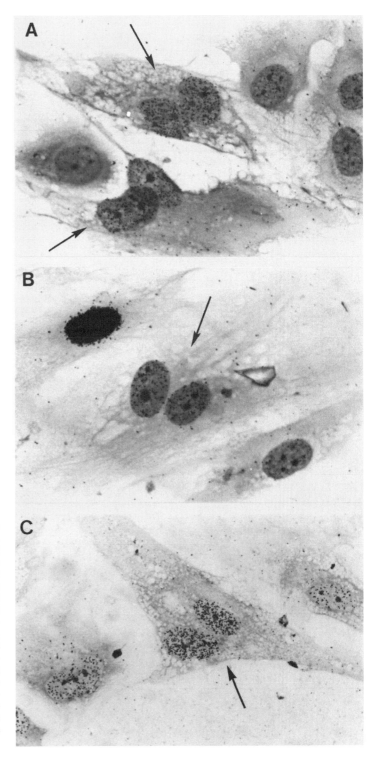

Figure 2. Microneedle injection of BTF2 in XP-D (A), TTD-A (B), and XP-B (C) fibroblasts. Micrographs showing the effect of purified BTF2 (HAP-fraction 12, containing the highest transcriptional activity; Schaeffer et al. 1993) on NER-activity of injected XP-D (cell line XP1BR), TTD-A (TTD1BR), and XP-B (TTD6VI) fibroblasts, respectively. The injected cells (binuclear fibroblasts obtained by cell fusion prior to injection) are indicated by arrows. As apparent from the number of silver grains above their nuclei, they exhibit a high (wild-type) level of UV-induced UDS when compared to the noninjected surrounding cells, which express residual UDS levels typical for these complementation groups. The heavily labeled fibroblast in panel B is a cell performing S-phase replicative DNA synthesis at the moment of the UDS assay.

and correction experiments provide strong evidence that three repair factors are physically associated with three different TFIIH subunits and, thus, likely constitute an integral part of transcription factor BTF2.

Table 2 also shows that small amounts of correcting activities persist in the anti-p62-treated extract. This is largely due to incomplete removal of p62 (as shown by the Western blot analysis in Fig. 5). However, it is not fully excluded that a small fraction of the ERCC2, ERCC3, and TTDA molecules exist dissociated from (the p62 part of) TFIIH, or reside in forms of TFIIH lacking p62. The same holds for the anti-p89 experiments. A somewhat looser association between ERCC2 and the p62-ERCC3 core of TFIIH is also derived from independent dissociation studies by Schaeffer and coworkers (1994). Further indications for exchange of

Table 2. Effect of Immunodepletion of BTF2 Components on NER Activity

Repair assay	Immunodepleted material	Tested cells/extracts (compl. group)		NER activity (% of control)			
				protA depl.	anti p62 depl.	anti p89 depl.	residual[a]
Microneedle injection	crude HeLa lysate	XP25RO	(XP-A)	99	100	94	2
		XPCS2BA	(XP-B)	103	39	33	13
		XP1BR	(XP-D)	98	41	39	21
		TTD1BR	(TTD-A)	102	29	23	8
Microneedle injection	BTF2 (heparin 5PW, fraction 12)	XPCS2BA	(XP-B)	92	37	35	12
		TTD6VI	(XP-B)	100	42	44	44
		XP1BR	(XP-D)	94	37	38	23
		TTD1BR	(TTD-A)	100	35	35	8
In vitro repair	crude HeLa lysate[b]	27.1	(group 3)	100	35	n.d.[c]	34
		UV-5	(group 2)	100	32	n.d.	24
		TTD1BR	(TTD-A)	100	32	n.d.	27

[a] Residual NER activity: amount of repair exhibited by mutated cells/extracts.
[b] p62-depleted crude extract exhibits a residual incorporation of 35% compared to a protA-treated control extract, which is close to incorporation by most mutant extracts.
[c] n.d.: Not determined.

various TFIIH subunits can be deduced from the in vitro complementation between ERCC2- and TTD-A-deficient extracts shown in Figure 3A (lanes 1–3), each of these extracts being defective in a different component of the same complex. In line with this, extracts defective in XPD/ERCC2 and XPB/ERCC3 also exhibit complementation in vitro (van Vuuren et al. 1993), and microinjection of free (recombinant) ERCC3 protein is able to induce a partial but clear correction of the repair defect of XP-B cells in vivo (van Vuuren et al. 1994). We conclude that a complex containing at least XPD/ERCC2, XPB/ERCC3, TTDA, and p62, and possibly also p44 (SSL1) and p34, is implicated in NER and that subunits of this complex can exchange in vivo and to some extent also in vitro. It is plausible that this complex represents the entire multisubunit TFIIH transcription factor.

Is TTDA Identical to Any of the Cloned TFIIH Subunits?

A prediction of the findings reported above is that one of the components of BTF2 is responsible for the repair defect in TTD-A for which no repair gene has yet been isolated. Therefore, cDNAs encoding the cloned TFIIH subunits p62, p44, and p34 (Humbert et al. 1994), and *XPD/ERCC2*(p80) and *XPB/ERCC3*(p89) were inserted into a mammalian expression vector and injected into the nucleus of TTD-A fibroblasts. Prior to injection, the cDNA-containing

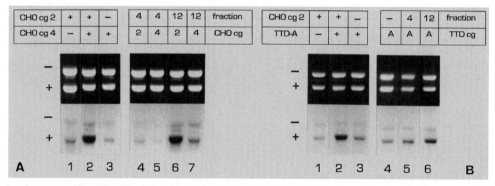

Figure 3. In vitro correction of NER defect by purified BTF2. BTF2 HAP-fraction 12 was added to an extract of rodent complementation group 2 (mutant UV5, equivalent to XP-D)(A), and to a human TTD-A lysate (TTD1BRSV)(B), and tested for its capacity to restore the NER defect using an in vitro cell-free repair assay (Wood et al. 1988). The level of correction of NER activity using purified BTF2 is of the same order (lanes 6,A and B) as the level of complementation reached when the extract of group 2 is mixed with a group 4 extract (lane 2A) or with the TTD-A extract (lane 2B). No significant correction is observed when the purified BTF2-fraction is added to rodent complementation group 4 (mutant UV47) extract (lane 7A). The upper panel shows the ethidium-bromide-stained DNA gel, the lower panel the autoradiogram of the dried gel. The presence of [^{32}P]dATP indicates repair synthesis. The positions of AAF-damaged and non-damaged DNA substrates, 250 ng each, are indicated by (+) and (−), respectively. Lanes 1–3 contain in total 200 μg of cell-free extract (in complementations, 100 μg of each extract was used); the other lanes contain 100 μg. The protein contribution by the BTF2 fraction is negligible.

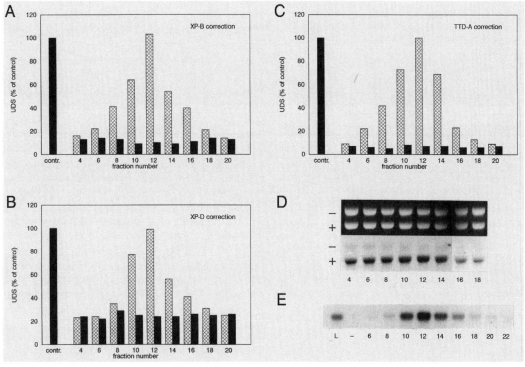

Figure 4. Correlation between transcription activation and correction of the NER defect in XP-D/ERCC2 and TTD-A mutant cells using different fractions from the HAP chromatography column. BTF2-derived in vivo NER-correcting activity was quantitatively determined, after microinjection, by counting the grains above the nuclei (UV-induced UDS) of injected cells and was expressed as the percentage of the UDS-level of repair-proficient control cells (C5RO) assayed in parallel (left black bar) (Vermeulen et al. 1986). The hatched bars indicate the average UDS level (determined by counting grains above 50 nuclei) of multikaryons injected with BTF2. The black bars represent UDS in noninjected neighboring cells. The in vivo NER activity profiles for, respectively, XP-B (XPCS1BA) (A), XP-D (XP1BR) (B), and TTD-A (TTD1BR) (C) are compared to the in vitro repair capacity for rodent complementation group 2 (ERCC2 mutant UV5) extracts (D) and to the transcription activity, as determined in a BTF2-dependent in vitro transcription runoff assay using the adenovirus late promoter (E). Note that the UV5 (group 2) extract in panel D (like the XP-D cells in panel B) has a considerable residual repair. The lower activity seen in the higher salt fractions (lanes 16–18) is attributed to an inhibition of repair incorporation by increased salt (Wood et al. 1988).

vectors were checked for their ability to specify proteins of the predicted size. None of these genes (*p62*, *p44*, *p34*), nor any of the cloned NER genes (*ERCC1*, *XPD/ERCC2*, *XPB/ERCC3*, *CSB/ERCC6*, *XPA*, *XPC*, and the genes *HHR23A* and *B*; Masutani et al. 1994), were able to exert correction of the repair defect, whereas microinjected *ERCC2* and *ERCC3* were able to restore UDS to XP-D and XP-B cells, respectively, in the same experiments. It appears that the TTDA factor is yet another non-cloned component of TFIIH.

Evidence for Direct Involvement of XPD/ERCC2 in Transcription In Vivo

The data described thus far tightly link the repair protein XPD/ERCC2 with transcription factor TFIIH/BTF2; however, they do not demonstrate that the protein is directly involved in transcription. Recently, Schaeffer and coworkers (1994) have shown that a p80 protein associated with TFIIH is identical to XPD/ERCC2 and that this protein stimulates the TFIIH-dependent in vitro transcription reaction. In addition, Drapkin and coworkers (1994) have shown that an antibody directed against ERCC2 is able to inhibit the in vitro transcription reaction. To verify whether the in vitro results can be extrapolated to the in vivo situation, we have conducted antibody microinjection experiments into living normal cells. As shown in Figure 6A and quantitatively in Table 3, introduction of antibodies against XPD/ERCC2 causes a strong inhibition of UV-induced UDS, consistent with the direct role of the protein in NER (Figs. 2 and 3) and similar to the effect observed using antibodies against another repair protein (ERCC1) that is not detectably associated with purified BTF2 (Table 3). In addition, a strong inhibition of general transcription (as measured by a 1-hour [³H]uridine labeling) is found (Fig. 6B), which is absent in the injections with ERCC1 antiserum (Table 3). Similar, but less pronounced, effects on transcription and UDS are exerted by a less powerful serum against the XPB/ERCC3 component of TFIIH (Table 3) (see also van Vuuren et al. 1994). These results indicate that XPD/ERCC2, like XPB/ERCC3, is involved in transcription in vivo.

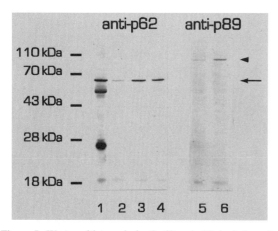

Figure 5. Western blot analysis of p62 and p89 depletion of a repair-proficient HeLa extract. Monoclonal antibody (MAb3C9) against the p62 subunit of BTF2(TFIIH) coupled to protein-A-Sepharose beads was incubated with a repair-proficient extract prepared from HeLa cells. To verify the removal of p62 and simultaneous removal of other components of BTF2, the remaining cell-free extract and the bound fraction were analyzed by immunoblotting using monoclonal antibodies against p62 itself (lanes *1–4*) and against the p89 (XPB/ERCC3) subunit of BTF2 (lanes 5 and 6). Lane *1* contains the protein fraction bound to the anti-p62 beads, released by SDS (the strong bands at 55 kD and 25 kD represent the heavy and light chains of the antibodies released from the beads). (Lanes 2 and 5) HeLa whole-cell extract treated with the anti-p62 beads. (Lanes *3* and 6) HeLa whole-cell extract treated with protein-A beads alone (control). (Lane 4) Untreated HeLa whole-cell extract. In lanes 2–6, equal amounts of sample were loaded.

DISCUSSION

The findings reported here have implications in several directions. At the molecular level, the identification of at least three repair factors in TFIIH endows this complex with a dual functionality and extends the functional overlap between two quite different processes, basal transcription and NER. At the clinical level, the identification of a link between TTD and XP/CS group B extends the association between TFIIH subunits and TTD and supports the notion that CS, TTD, and some forms of XP represent different manifestations of one large clinical continuum. The connection between TFIIH and TTD provides new clues to understand the basis of the complex clinical features displayed by these disorders and, potentially, a number of other, strikingly related syndromes.

Mechanistic Implications for Transcription and Repair

A minimum of three NER factors is found to be associated with transcription factor TFIIH. In addition to XPB/ERCC3 (Schaeffer et al. 1993), we demonstrate that XPD/ERCC2 and TTDA reside in this complex. We provide evidence for the involvement of the XPD/ERCC2 helicase in transcription in vivo. Circumstantial data support the idea that the transcription proteins p62, p44, and p34 are also implicated in NER (Humbert et al. 1994; van Vuuren et al. 1994; this paper). As previously anticipated (Bootsma and Hoeijmakers 1993), the human homolog of SSL1, a yeast repair factor which was linked to translation previously (Yoon et al. 1992), was recently identified as the p44 subunit of TFIIH (Humbert et al. 1994). Table 4 summarizes the current evidence for the involvement of different human BTF2/TFIIH and yeast factor b subunits in repair. It is possible that the TTDA protein is identical to the (non-cloned) p41 BTF2 subunit. Together, these data suggest that most, if not all, components of the transcription complex participate in excision repair, converting this complex into a functional unit participating in at least two processes.

It is still difficult to estimate how many polypeptides constitute TFIIH. Because of the absence of XPD/ERCC2 as a major protein in the most pure BTF2 preparation (Schaeffer et al. 1994; Drapkin et al. 1994), it is likely that the five predominant bands present in the final stages of BTF2 purification (Gerard et al. 1991) are derived from the most tightly associated protein fraction constituting the core of the complex. Protein profiles of the rat homolog (factor δ) suggest the presence of at least eight polypeptides (Conaway and Conaway 1989). Thus, BTF2 could well be considerably larger in vivo and form a "supercomplex" of which the more loosely bound factors have a tendency to (partly) dissociate during purification. This could also explain the absence of a XPC-correcting activity in our TFIIH purifications, whereas others found association of XPC with TFIIH, but not in the final purification fraction (Drapkin et al. 1994). Alternatively, it is possible that the XPC-correcting activity represents a spurious copurification with TFIIH. Association of XPC with TFIIH is somewhat unexpected, since XPC is thought to be selectively involved in the global genome (transcription-independent) NER pathway (Venema et al. 1990).

BTF2 possesses a bidirectional unwinding activity involving XPB/ERCC3 and XPD/ERCC2 (Schaeffer et al. 1994), which may promote transition from a closed to an open initiation complex. It is likely that TFIIH can be utilized as an independent unit in the context of transcription and NER; e.g., for loading and/or translocation of the preinitiation complex or a NER scanning/incision complex or for the repair synthesis step. The TFIIH mutants XP-B, XP-D, and TTD-A display defects in both the "transcription-coupled" and the "global genome" NER subpathways, so the complex is likely to play a role in the core of the NER reaction mechanism (Sweder and Hanawalt 1993).

Clinical Heterogeneity and Pleiotropy Associated with Mutations in TFIIH

The spectrum of diseases linked with TFIIH is heterogeneous and pleiotropic, including seemingly unrelated symptoms such as photosensitivity, brittle hair and nails, neurodysmyelination, impaired sexual

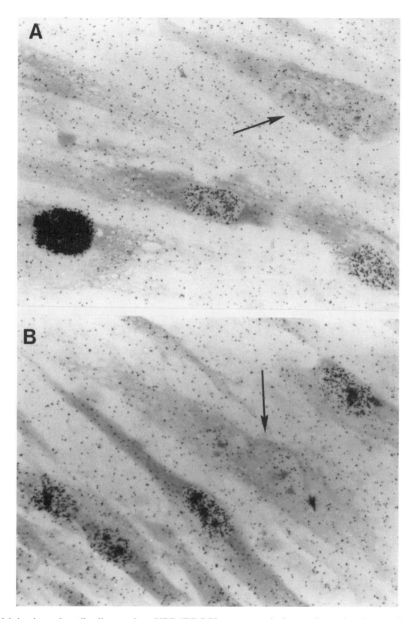

Figure 6. Effect of injection of antibodies against XPD/ERCC2 on transcription and repair of normal cells. Micrograph *A* demonstrates the effect of injection of ERCC2 antibodies on UV-induced UDS of normal fibroblasts, assayed by a 2-hr incubation in [^{3}H]thymidine immediately following UV-irradiation. Micrograph *B* shows the effect of injection of XPD/ERCC2 antiserum on RNA synthesis of control (wild-type) fibroblasts assayed by a 1-hr pulse-labeling with [^{3}H]uridine. The UDS and RNA synthesis assays were performed 20 hr after injection of the antibodies. The strong reduction of autoradiographic grains above the nuclei of injected dikaryons (*arrows*) compared to noninjected cells indicates a virtually complete inhibition of RNA and repair synthesis. The heavily labeled nucleus in *A* is from a cell in S phase at the time of incubation. Note that these preparations have a higher background labeling than the micrographs shown in Fig. 1.

development, ichthyosis, and dental caries. Both the rare XP group B and the more common group D present pronounced clinical heterogeneity: classical XP (only in group D), atypical combination of XP and CS, and TTD (Johnson and Squires 1992; Vermeulen et al. 1994; this paper). The occurrence of patients displaying TTD symptoms within XP-B extends the parallels between XP-B and XP-D and their respective gene products noted before (Weeda et al. 1990). Clinical variability in TTD is even observed within families and also apparent from the close association with at least seven disorders (shown in Fig. 7) appearing in the On-line Mendelian Inheritance in Man (OMIM) database (McKusick 1992). The occurrence of TTD in three NER-deficient complementation groups argues against a chance association between genetic loci separately involved in NER and in brittle hair. Consistent with this notion, mutations in the *XPD/ERCC2* gene have been detected recently in TTD(XP-D) patients (Broughton et al. 1994).

Table 3. Effect of Antibody Injection on Repair and Transcription

Injected antiserum	% inhibition of NER[a]	% inhibition of transcription[b]
Rabbit anti ERCC1[c]	97	0
Rabbit anti ERCC3[c]	43	48
Rabbit anti ERCC2	87	85
Preimmune rabbit serum	2	7

[a] Compared to UDS level observed in uninjected cells on the same slide.
[b] Compared to transcription level (assayed by 1-hr pulse labeling with [^{3}H]uridine) observed in uninjected cells on the same slide.
[c] van Vuuren et al. (1994).

How can we rationalize the pleiotropy and clinical heterogeneity in the above-mentioned conditions? The symptoms associated with a sole NER defect are displayed by the most common XP groups, A (totally deficient in NER) and C (defective in the "genome overall" repair subpathway). Patients in these groups present a relatively uniform clinical picture involving photosensitivity, pigmentation abnormalities, predisposition to skin cancer and, in the case of XP-A, accelerated neurodegeneration (which is not associated with neurodysmyelination), but no CS and TTD symptoms. The gene products affected in these groups are not vital and therefore do not appear to be essential for basal transcription.

Obviously, the hallmarks of a pure NER deficiency do not include the salient features of CS and TTD. It is tempting to link these with the additional transcription-related function. Indeed, it would be highly unlikely when all mutations in the three subunits of this bifunctional complex would only affect the repair function and leave the inherent transcriptional role entirely intact. This interpretation is supported by the *haywire* phenotype of the *Drosophila ERCC3* mutant, involving UV sensitivity, central nervous system abnormalities, and impaired sexual development, as found in XP-B (Mounkes et al. 1992). Spermatogenesis in *Drosophila* is very sensitive to the level of β_2-tubulin (Kemphues et al. 1982). Mutations in the *Drosophila ERCC3* gene seem to affect β-tubulin expression, causing male sterility (Mounkes et al. 1992). In mammals, β-tubulin mRNA is selectively regulated by a unique cotranslational degradation mechanism (Theodorakis and Cleveland 1992). It is possible that this renders β-tubulin expression particularly sensitive to the level of transcription and thereby to subtle mutations in BTF2, resulting in the immature sexual development found in TTD and CS. Similarly, reduced transcription of genes encoding ultrahigh sulfur proteins of the hairshaft may account for the observed reduced cysteine content in the brittle hair of TTD patients (Itin and Pittelkow 1990). Low expression of the myelin basic protein, whose transcription is known to be rate-limiting in the mouse (Readhead et al. 1987), may cause the characteristic neurodysmyelination of CS and TTD (Peserico et al. 1992; Sasaki et al. 1992). A comparable explanation is proposed for the poor enamelation of teeth in CS and TTD (Nance and Berry 1992; McCuaig et al. 1993). The skin abnormalities typical of TTD often involve ichthyosis. Various classes of ichthyoses show abnormalities in the production of filaggrin (Fleckman and Dale 1993). Thus, mutations in TFIIH which subtly disturb its transcription function may affect a specific subset of genes whose functioning critically depends on the level or fine-tuning of transcription. Recent studies indicate that the requirement for basal transcription factors may vary from promoter to promoter depending on the sequence around the initiation site, the topological state of the DNA, and the local chromatin structure (Parvin and Sharp 1993; Stanway 1993; Timmers 1994). These mechanisms can easily explain the pronounced clinical heterogeneity even within families.

On close inspection (Bootsma and Hoeijmakers 1993), many parallels can be found between CS and TTD. To a varying degree, CS patients exhibit features prominent in TTD, such as thin, dry hair and scaly skin (Nance and Berry 1992). TTD has recently been recognized to include neurodysmyelination (Peserico et al. 1992), bird-like facies, dental caries, and cataract (McCuaig et al. 1993), hallmarks normally associated with CS. This suggests that CS and TTD are manifestations of a broad clinical continuum, consistent with the notion that mutations in different subunits of the same (TFIIH) complex give rise to a similar set of

Table 4. The NER connection of BTF2/TFIIH

NER proteins associated with BTF2	NER proteins with yeast factor b	Polypeptides identified in SDS-PAGE
XPB/ERCC3	**RAD25/SSL2**[b]	**p89**
XPD/ERCC2	**RAD3**[c]	**p80**
TTD-A	?	?
p62[a]	TFB1[d]	p62
p44[a]	SSL1[d]	p44
?	?	p41[e]
?	?	p34

Factors for which the involvement in NER is unequivocally demonstrated are in boldface.
[a] NER function based on inference from (presumed) NER involvement of yeast homologs.
[b] References: Park (1992); Feaver (1993).
[c] Reference: Feaver (1993).
[d] Some alleles are UV sensitive, suggesting that they are deficient in NER.
[e] The relationship between p41 and TTD-A is not yet known.

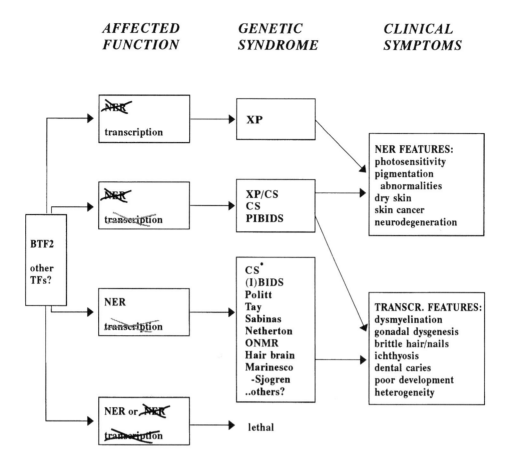

Figure 7. Model for involvement of defects in transcription and repair in human disease (see text for explanation).

phenotypic features. In this proposition, defects in the classical CS genes *CSA* and *ERCC6/CSB* (Troelstra et al. 1992), as well as *XPG/ERCC5* (Bootsma and Hoeijmakers 1993; Vermeulen et al. 1993), are expected to somehow affect basal transcription as well.

Model for Involvement of Transcription and Repair in CS and TTD: Deduction of the Existence of Transcription Syndromes

A tentative model proposed for the etiology of the defects in the conglomerate of CS, TTD, and related disorders is shown in Figure 7. In this model, mutations in BTF2 factors inactivating only the NER function result in a XP phenotype as observed in the classic XP patients of XP-D. If, in addition, the transcription function is subtly affected, the photosensitive forms of combined XP/CS and TTD are found. Theoretically, mutations causing a (still viable) transcription problem without NER impairment are predicted. The notion that the new TTD XP-B members have only a mild repair defect (see Table 2) but nevertheless display TTD features not exhibited by the more repair-deficient original XP-B cases fits perfectly into this reasoning. Indeed, a significant proportion of TTD patients, as well as clinically characteristic CS patients, is not noticeably photosensitive and has normal NER (Lehmann 1987; Nance and Berry 1992; Lehmann et al. 1993). These findings extend the implications to non-repair-defective disorders. Therefore, and in view of the pronounced heterogeneity inherent to the model, we propose that the Sjögren-Larsson (270200), RUD (308200), ICE (146720), OTD (257960), IFAP (308205), CAM(F)AK (214550), Rothmund-Thompson (268400), and KID (242150) syndromes (for references, see Baden 1991; McKusick 1992) also fall within this category. Interestingly, some of these diseases show occurrence of skin cancer.

In conclusion, our findings provide evidence for the presence of a wide class of disorders that we propose to designate collectively as "transcription syndromes." A prediction from our model is that these patients carry mutations in transcription factors that do not affect the NER process. This proposition is testable. The explanation put forward here for this class of disorders would introduce a novel concept into human genetics. It can be envisaged that similar phenomena are associated with subtle defects in translation, implying the potential existence of "translation syndromes" (as suggested earlier on completely different grounds; Fisher et al. 1990).

ACKNOWLEDGMENTS

We gratefully acknowledge S. Humbert and V. Moncollin for help in BTF2 purification, Elena Botta for help with the complementation analysis, C. Backendorf for helpful information on cornified envelopes, A. van Oudenaren for help with the densitometer scanning, and J.W. van Klaveren for help with the computer. M. Kuit is acknowledged for photography. This work was supported in part by grants from the Netherlands Foundation for Chemical Research (SON), the Dutch Cancer Society (EUR 90-20, 92-118), The Dutch Scientific Organization, Medical Scientific Research (900-501-113 and 093), the INSERM, the C.N.R.S., the Ministère de la Recherche et de l'Enseignement Supérieur, the Association pour la Recherche sur le Cancer, and the Associazione Italiana per la Ricerca sul Cancro.

REFERENCES

Baden, H. 1991. Keratinizing disorders. In *Genetic disorders of the skin* (ed. J.C. Alpen), p.170. Mosby-Year Book, St. Louis, Missouri.

Biggerstaff, M., P. Robins, D. Coverley, and R.D. Wood. 1991. Effect of exogenous DNA fragments on human cell extract-mediated DNA repair synthesis. *Mutat. Res.* **254:** 217.

Bohr, V.A. 1991. Gene specific DNA repair. *Carcinogenesis* **12:** 1983.

Bootsma, D. and J.H.J. Hoeijmakers. 1993. Engagement with transcription. *Nature* **363:** 114.

Broughton, B.C., H. Steingrimsdottin, C.A. Weber, and A.R. Lehmann. 1994. Mutations in the xeroderma pigmentosum group D DNA repair/transcription gene in patients with trichothiodystrophy. *Nat. Genet.* **7:** 189.

Busch, D., C. Greiner, K. Lewis, R. Ford, G. Adair, and L. Thompson. 1989. Summary of complementation groups of UV-sensitive *CHO* mutants isolated by large-scale screening. *Mutagenesis* **4:** 349.

Cleaver, J.E. and K.H. Kraemer. 1994. Xeroderma pigmentosum and Cockayne syndrome. In *the metabolic basis of inherited disease*, 7th edition (ed. C.R. Scriver et al.). McGraw-Hill, New York.

Conaway, R.C. and J.W. Conaway. 1989. An RNA polymerase II transcription factor has an associated DNA-dependent ATPase (dATPase) activity strongly stimulated by the TATA region of promoters. *Proc. Natl. Acad. Sci.* **86:** 7356.

Drapkin, R., A. Merino, and D. Reinberg. 1993. Regulation of RNA polymerase II transcription. *Curr. Biol.* **5:** 469.

Drapkin, R., J.T. Reardon, A. Ansari, J.C. Huang, L. Zawel, K. Ahn, A. Sancar, and D. Reinberg. 1994. Dual role of TFIIH in DNA excision repair and in transcription by RNA polymerase II. *Nature* **368:** 769.

Fischer, L., M. Gerard, C. Chalut, Y. Lutz, S. Humbert, M. Kanno, P. Chambon, and J.-M. Egly. 1992. Cloning of the 62-kilodalton component of basic transcription factor BTF2. *Science* **257:** 1392.

Fisher, E.M.C., P. Beer-Romero, L.G. Brown, A. Ridley, J.A. McNeil, J.B. Lawrence, H.F. Willard, F.R. Bieber, and D.C. Page. 1990. Homologous ribosomal protein genes on the human X and Y chromosomes: Escape from X inactivation and possible implications for Turner syndrome. *Cell* **63:** 1205.

Fleckman, P. and B.A. Dale. 1993. *Structural protein expression in the ichthyosis* (ed. B.A. Bernard and B. Shroot). Karger, Basel, Switzerland.

Flejter, W.L., L.D. McDaniel, D. Johns, E.C. Friedberg, and R.A. Schultz. 1992. Correction of xeroderma pigmentosum complementation group D mutant cell phenotypes by chromosome and gene transfer: Involvement of the human *ERCC2* DNA repair gene. *Proc. Natl. Acad. Sci.* **89:** 261.

Gerard, M., L. Fischer, V. Moncollin, J.-M. Chipoulet, P. Chambon, and J.-M. Egly. 1991. Purification and interaction properties of the human RNA polymerase B(II) general transcription factor BTF2. *J. Biol. Chem.* **266:** 20940.

Gill, G. and R. Tjian. 1992. Eukaryotic coactivators associated with the TATA box binding protein. *Curr. Opin. Genet. Dev.* **2:** 236.

Goodrich, J.A. and R. Tjian. 1994. Transcription factors IIE and IIH and ATP hydrolysis direct promoter clearance by RNA polymerase II. *Cell* **77:** 145.

Hanawalt, P. and I. Mellon. 1993. Stranded in an active gene. *Curr. Biol.* **3:** 67.

Hoeijmakers, J.H.J. 1993. Nucleotide excision repair. II. From yeast to mammals. *Trends Genet.* **9:** 211.

Humbert, S., A.J. van Vuuren, Y. Lutz, J.H.J. Hoeijmakers, J.-M. Egly, and V. Moncollin. 1994. Characterization of p44/SSL1 and p34 subunits of the BTF2/TFIIH transcription/repair factor. *EMBO J.* **13:** 2393.

Itin, P.H. and M.R. Pittelkow. 1990. Trichothiodystrophy: Review of sulfur-deficient brittle hair syndromes and association with the ectodermal dysplasias. *J. Am. Acad. Dermatol.* **22:** 705.

Johnson, R.T. and S. Squires. 1992. The XPD complementation group. Insights into xeroderma pigmentosum, Cockayne's syndrome and trichothiodystrophy. *Mutat. Res.* **273:** 97.

Kemphues, K.J., T.C. Kaufman, R.A. Raff, and E.C. Raff. 1982. The testis-specific beta-tubulin subunit in *Drosophila melanogaster* has multiple functions in spermatogenesis. *Cell* **31:** 655.

Lehmann, A.R. 1987. Cockayne's syndrome and trichothiodystrophy: Defective repair without cancer. *Cancer Rev.* **7:** 82.

Lehmann, A.R., A.F. Thompson, S.A. Harcourt, M. Stefanini, and P.G. Norris. 1993. Cockayne's syndrome: Correlation of clinical features with cellular sensitivity of RNA synthesis to UV irradiation. *J. Med. Genet.* **30:** 679.

Lu, H., L. Zawel, L. Fisher, J.-M. Egly, and D. Reinberg. 1992. Human general transcription factor IIH phosphorylates the C-terminal domain of RNA polymerase II. *Nature* **358:** 641.

Manley, J.L., A. Fire, M. Samuels, and P.A. Sharp. 1983. In vitro transcription: Whole cell extract. *Methods Enzymol.* **101:** 568.

Masutani, C., K. Sugasawa, J. Yanagisawa, T. Sonoyama, M. Ui, T. Enomoto, K. Takio, K. Tanaka, P.J. van der Spek, D. Bootsma, J.H.J. Hoeijmakers, and F. Hanaoka. 1994. Purification and cloning of a nucleotide excision repair complex involving the xeroderma pigmentosum group C protein and a human homolog of yeast RAD23. *EMBO J.* **13:** 1831.

McCuaig, C., D. Marcoux, J.E. Rasmussen, M.M. Werner, and N.E. Genter. 1993. Trichothiodystrophy associated with photosensitivity, gonadal failure, and striking osteosclerosis. *J. Am. Acad. Dermatol.* **28:** 820.

McKusick, V.A. 1992. Catalogs of autosomal dominant, autosomal recessive, and X-linked phenotypes. In *Mendelian inheritance in man*. Johns Hopkins University Press, Baltimore, Maryland.

Mounkes, L.C., R.S. Jones, B-C. Liang, W. Gelbart, and M.T. Fuller. 1992. A *Drosophila* model for xeroderma pigmentosum and Cockayne's syndrome: *haywire* encodes the fly homolog of *ERCC3*, a human excision repair gene. *Cell* **71:** 925.

Nance, M.A. and S.A. Berry. 1992. Cockayne syndrome: Review of 140 cases. *Am. J. Med. Genet.* **42:** 68.

Parvin, J.D. and P.A. Sharp. 1993. DNA topology and a minimum set of basal factors for transcription by RNA polymerase II. *Cell* **73:** 533.

Peserico, A., P.A. Battistella, and P. Bertoli. 1992. MRI of a very hereditary ectodermal dysplasia: PIBI(D)S. *Neuroradiology* **34:** 316.

Readhead, C., B. Popko, N. Takahashi, H.D. Shine, R.A. Saavedra, R.L. Sidman, and L. Hood. 1987. Expression of a myelin basic protein gene in transgenic shiverer mice: Correction of the dysmyelinating phenotype. *Cell* **48:** 703.

Roeder, R.G. 1991. The complexities of eukaryotic transcription initiation: Regulation of preinitiation complex assembly. *Trends Biochem. Sci.* **16:** 402.

Roy, R., L. Schaeffer, S. Humbert, W. Vermeulen, G. Weeda, and J.-M. Egly. 1994. The DNA-dependent ATPase activity associated with the class II basic transcription factor BTF2/TFIIH. *J. Biol. Chem.* **269:** 9826.

Sambrook, J., E.F. Fritsch, and T. Maniatis. 1989. *Molecular cloning: A laboratory manual.* Cold Spring Harbor Laboratory, Cold Spring Harbor, New York.

Sasaki, K., N. Tachi, M. Shinoda, N. Satoh, R. Minami, and A. Ohnishi. 1992. Demyelinating peripheral neuropathy in Cockayne syndrome: A histopathologic and morphometric study. *Brain Dev.* **14:** 114.

Schaeffer, L., R. Roy, S. Humbert, V. Moncollin, W. Vermeulen, J.H.J. Hoeijmakers, P. Chambon, and J.-M. Egly. 1993. DNA repair helicase: A component of BTF2 (TFIIH) basic transcription factor. *Science* **260:** 58.

Schaeffer, L., V. Moncollin, R. Roy, A. Staub, M. Mezzina, A. Sarasin, G. Weeda, J.H.J. Hoeijmakers, and J.M. Egly. 1994. The ERCC2/DNA repair protein is associated with the class II BTF2/TFIIH transcription factor. *EMBO J.* **13:** 2388.

Serizawa, H., J.W. Conaway, and R.C. Conaway. 1993. Phosphorylation of C-terminal domain of RNA polymerase II is not required in basal transcription. *Nature* **363:** 371.

Stanway, C.A. 1993. Simplicity amidst complexity in transcriptional initiation. *BioEssays* **15:** 559.

Stefanini, M., P. Lagomarisini, S. Gilliani, T. Nardo, E. Botta, A. Peserico, W.J. Kleyer, A.R. Lehmann, and A. Sarasin. 1993a. Genetic heterogeneity of the excision repair defect associated with trichothiodystrophy. *Carcinogenesis* **14:** 1101.

Stefanini, M., W. Vermeulen, G. Weeda, S. Giliani, T. Nardo, M. Mezzina, A. Sarasin, J.I. Harper, C.F. Arlett, J.H.J. Hoeijmakers, and A.R. Lehmann. 1993b. A new nucleotide-excision-repair gene associated with the disorder trichothiodystrophy. *Am. J. Hum. Genet.* **53:** 817.

Sweder, K.S. and P.C. Hanawalt. 1993. Transcription coupled DNA repair. *Science* **262:** 439.

Theodorakis, N.G. and D.W. Cleveland. 1992. Physical evidence for cotranslational regulation of beta-tubulin mRNA degradation. *Mol. Cell. Biol.* **12:** 791.

Timmers, H.T.M. 1994. Transcription initiation by RNA polymerase II does not require hydrolysis of the beta-gamma phosphoanhydride bond of ATP. *EMBO J.* **13:** 391.

Troelstra, C., A. van Gool, J. de Wit, W. Vermeulen, D. Bootsma, and J.H.J. Hoeijmakers. 1992. ERCC6, a member of a subfamily of putative helicases, is involved in Cockayne's syndrome and preferential repair of active genes. *Cell* **71:** 939.

van Vuuren, A.J., E. Appeldoorn, H. Odijk, A. Yasui, N.G.J. Jaspers, D. Bootsma, and J.H.J. Hoeijmakers. 1993. Evidence for a repair enzyme complex involving ERCC1 and complementing activities of ERCC4, ERCC11 and xeroderma pigmentosum group F. *EMBO J.* **12:** 3693.

van Vuuren, A.J., W. Vermeulen, L. Ma, G. Weeda, E. Appeldoorn, N.G.J. Jaspers, A.J. van der Eb, D. Bootsma, J.H.J. Hoeijmakers, S. Humbert, L. Schaeffer, and J.-M. Egly. 1994. Correction of xeroderma pigmentosum repair defect by basal transcription factor BTF2(TFIIH). *EMBO J.* **13:** 1645.

Venema, J., A. Van Hoffen, A.T. Natarajan, A.A. Van Zeeland, and L.H.F. Mullenders. 1990. The residual repair capacity of xeroderma pigmentosum complementation group C fibroblasts is highly specific for transcriptionally active DNA. *Nucleic Acids Res.* **18:** 443.

Vermeulen, W., P. Osseweijer, A.J. De Jonge, and J.H.J. Hoeijmakers. 1986. Transient correction of excision repair defects in fibroblasts of 9 xeroderma pigmentosum complementation groups by microinjection of crude human cell extracts. *Mutat. Res.* **165:** 199.

Vermeulen, W., J. Jaeken, N.G.J. Jaspers, D. Bootsma, and J.H.J. Hoeijmakers. 1993. Xeroderma pigmentosum complementation group G associated with Cockayne's syndrome. *Am. J. Hum. Genet.* **53:** 185.

Vermeulen, W., R.J. Scott, S. Potger, H.J. Muller, J. Cole, C.F. Arlett, W.J. Kleijer, D. Bootsma, J.H.J. Hoeijmakers, and G. Weeda. 1994. Clinical heterogeneity within xeroderma pigmentosum associated with mutations in the DNA repair and transcription gene ERCC3. *Am. J. Hum. Genet.* **54:** 191.

Weeda, G., R.C.A. Van Ham, W. Vermeulen, D. Bootsma, A.J. Van der Eb, and J.H.J. Hoeijmakers. 1990. A presumed DNA helicase encoded by ERCC-3 is involved in the human repair disorders xeroderma pigmentosum and Cockayne's syndrome. *Cell* **62:** 777.

Wood, R.D., P. Robins, and T. Lindahl. 1988. Complementation of the xeroderma pigmentosum DNA repair defect in cell-free extracts. *Cell* **53:** 97.

Yoon, H., S.P. Miller, E.K. Pabich, and T.F. Donahue. 1992. SSL1, a suppressor of a HIS4 5'-UTR stem-loop mutation, is essential for translation initiation and effects UV resistance in yeast. *Genes Dev.* **6:** 2463.

Human Mismatch Repair Genes and Their Association with Hereditary Non-Polyposis Colon Cancer

R.D. Kolodner,*‡§ N.R. Hall,¶# J. Lipford,* M.F. Kane,*
M.R.S. Rao,*‡‡‡ P. Morrison,‡ L. Wirth,‡ P.J. Finan,# J. Burn,** P. Chapman,**
C. Earabino,‡ E. Merchant,‡ D.T. Bishop,¶ J. Garber,† C.E. Bronner,††
S.M. Baker,†† G. Warren,‡‡ L.G. Smith,†† A. Lindblom,¶¶ P. Tannergard,¶¶
R.J. Bollag,‡‡ A.R. Godwin,‡‡ D.C. Ward,‡‡§§ M. Nordenskjold,¶¶
R.M. Liskay,†† N. Copeland,## N. Jenkins,## M.K. Lescoe,*** A. Ewel,***
S. Lee,††† J. Griffith,††† and R. Fishel***

*Division of Cell and Molecular Biology, †Division of Cancer Epidemiology and Control, and ‡Molecular Biology Core Facility, Dana-Farber Cancer Institute, Boston, Massachusetts 02115; §Department of Biological Chemistry and Molecular Pharmacology, Harvard Medical School, Boston, Massachusetts 02115; ¶Imperial Cancer Research Fund, Genetic Epidemiology Laboratory, St. James's University Hospital, Leeds LS9 7TF, United Kingdom; #Department of Surgery and Centre for Digestive Diseases, General Infirmary, Leeds LS1 3EX, United Kingdom; **Department of Human Genetics, University of Newcastle upon Tyne, Newcastle NE2 4AA, United Kingdom; ††Department of Molecular and Medical Genetics, Oregon Health Sciences University, Portland, Oregon 97201-3098; ‡‡Department of Genetics and §§Department of Molecular Biophysics, Yale University School of Medicine, New Haven, Connecticut 06510; ¶¶Department of Molecular Medicine, Karolinska Hospital, S-171 76 Stockholm, Sweden; ##Mammalian Genetics Laboratory, National Cancer Institute, Frederick Cancer Research and Development Center, Frederick, Maryland 21702; ***Department of Microbiology and Molecular Genetics, University of Vermont Medical School, Burlington, Vermont 05405; †††Lineberger Comprehensive Cancer Center, School of Medicine, University of North Carolina, Chapel Hill, North Carolina 27599-7295

Hereditary non-polyposis colon cancer (HNPCC) may affect up to 1 in 200 people in industrialized nations (Bishop and Thomas 1990; Lynch et al. 1991, 1993; Peltomaki et al. 1993b). Four genes have been identified in which inherited mutations appear to cause HNPCC. hMSH2 on chromosome 2p21-22 appears to account for up to 60% of HNPCC (Fishel et al. 1993; Leach et al. 1993; Sandkuijl and Bishop 1993; Nystrom-Lahti et al. 1994), hMLH1 on chromosome 3p21 appears to account for up to 30% of HNPCC (Bronner et al. 1994; Nystrom-Lahti et al. 1994; Papadopoulos et al. 1994), and hPMS1 on chromosome 2q31-33 and hPMS2 on chromosome 7p21 may account for 5% of HNPCC (Nicolaides et al. 1994).

hMSH2, hMLH1, hPMS1, and hPMS2 are members of two highly conserved families of postreplication mismatch repair genes, MutS and MutL, that are involved in increasing the fidelity of replication by specific repair of DNA polymerase misincorporation errors (Modrich 1989, 1991). Ablation of MutS, MutL, or their homologs in bacteria and yeast increases the rate of spontaneous mutation resulting in a mutator phenotype (Cox 1973, 1976; Rydberg 1978; Williamson et al. 1985; Reenan and Kolodner 1992b; Prolla et al. 1994a). In addition to the increased accumulation of point mutations, a recently appreciated mutator phenotype caused by mismatch repair defects is the accumulation of deletion and insertion mutations in simple repetitive DNA that results in microsatellite instability (Levinson and Gutman 1987; Ionov et al. 1993; Strand et al. 1993). A similar microsatellite instability mutator phenotype has been observed in tumors from HNPCC patients (Aaltonen et al. 1993; Peltomaki et al. 1993a; Risinger et al. 1993; Honchel et al. 1994), a wide variety of sporadic tumors (Gonzalez-Zulueta et al. 1993; Han et al. 1993; Ionov et al. 1993; Risinger et al. 1993; Thibodeau et al. 1993; Young et al. 1993; Burks et al. 1994; Honchel et al. 1994; Merlo et al. 1994; Rhyu et al. 1994; Schoenberg et al. 1994; Shridhar et al. 1994; Wada et al. 1994; Wooster et al. 1994; Yee et al. 1994), and several cell lines derived from tumors (Bhattacharyya et al. 1994). Consistent with an association with tumor initiation, microsatellite instability appears to occur early in the development of colorectal tumors (Aaltonen et al. 1994; Shibata et al. 1994). A proposed mechanism for microsatellite instability involves replication-induced errors that result in mismatched nucleotides within these repeat sequences (Kunkel 1993). Such replication errors would leave an integral number of repeats in an insertion/deletion loop-type mismatch in which there are more nucleotides contained in one of the DNA strands. If these mismatched nucleotides are not repaired by mismatch repair, they are subsequently fixed in the genome by a second round of replication resulting in the localized sequence expansions and/or contractions associated with microsatellite instability.

A mutator phenotype has been proposed to account for the multiple mutations required for multistage carcinogenesis (Knudson 1986; Fearon and Vogelstein 1990; Loeb 1991; Renan 1993). Within the context of the mutator hypothesis, it is proposed that mutator genes exist which increase the mutation rate of a preneoplastic cell allowing multiple mutations to accumu-

---

‡‡‡On sabbatical leave from the Indian Institutes of Science, Bangalore, India.

late. Recent work including that discussed below has provided evidence that genes like *hMSH2*, *hMLH1*, *hPMS1*, and *hPMS2* encode mismatch repair proteins and that mutant forms of these genes are in fact the mutator genes proposed to play a role in the development of cancer.

RESULTS

Cloning the Human *MutS* and *MutL* Homologs

To clone the human *MSH2*, *MLH1*, and *PMS2* genes, we utilized the degenerate PCR approach that was successful in the identification of the *Saccharomyces cerevisiae* MutS and MutL homologs *MSH1*, *MSH2*, and *MLH1* (Reenan and Kolodner 1992a,b; Prolla et al. 1994a). Degenerate oligonucleotide primers were designed to target the protein sequences TGPNM and F(ATV)TH(FY) present in the most conserved regions of the known *MutS* homologs, and KELVEN and GFRGEA present in the most conserved region of the *MutL* homologs. In the case of the *MutS* homologs, one primer was used to target the invariant TGPNM sequence, and three primers were used to individually target FATH(FY), FVTH(FY), and FTTH(FY). The FATH(FY) sequence was of particular interest because it was most specific to the bacterial and *S. cerevisiae* homologs known to be involved in the major mismatch repair pathways in these organisms. The primer sets for *MutS* and *MutL* homologs were used in a PCR reaction containing human cDNA prepared from poly(A)$^+$ RNA as template. One clone was identified using the *MutS*-specific primers which contained an open reading frame encoding a predicted amino acid sequence that had 81% identity with the *S. cerevisiae* MSH2 protein (Fishel et al. 1993). Two clones containing an open reading frame encoding a predicted amino acid sequence that was related to MutL were identified using the *MutL*-specific primers: one that showed 70% identity and another that showed 47% identity with the *S. cerevisiae* MLH1 protein (Bronner et al. 1994). These cloned segments were then used as a probe to screen a human cDNA library, and several apparently full-length cDNA clones were identified. A representative full-length clone that hybridized to the *MutS* homolog contained a 3111-bp insert that contained a 2806-bp open reading frame. This open reading frame, when translated, was found to potentially encode a 935-amino-acid protein that had 41% identity with the 966-amino-acid *S. cerevisiae* MSH2 protein and was therefore designated hMSH2 (the lowercase h signifying its human origin). The most conserved region, located between amino acids 667 and 789 of the human protein, was 85% identical with the same region of the *S. cerevisiae* protein. Two different *MutL* homologs were obtained by a similar screen. One clone contained a 2268-bp open reading frame which codes for a 756-amino-acid protein that shares 41% identity with *S. cerevisiae* MLH1 and is therefore designated hMLH1. The second clone contained a 2772-bp open reading frame which codes for a 924-amino-acid protein that is 43% identical with *S. cerevisiae* PMS1 protein and has been designated hPMS2 to be consistent with the nomenclature of Vogelstein and colleagues (Nicolaides et al. 1994; Papadopoulos et al. 1994).

Human MSH2, MLH1, and PMS2 Are Members of Mismatch Repair Protein Superfamilies

Analysis of the evolutionary relatedness of the human MSH2, MLH1, PMS1, and PMS2 amino acid sequences with the other known *MutS* and *MutL* homologs showed that they are highly related to *S. cerevisiae* proteins (Fig. 1). The evolutionary relationship between all of the known *MutS*-related proteins indicates that human MSH2 is a member of a subfamily of *MutS* homologs which are known to play a role in mismatch repair and where mutations of the respective genes have been found to cause a strong mutator phenotype (Reenan and Kolodner 1992a,b).

hMLH1 is most related to the *S. cerevisiae* *MLH1*, in which mutations have been found to cause a mutator phenotype and which is likely involved in the process of mismatch repair in yeast (Prolla et al. 1994a). *hPMS2* appears to be a relative of the *S. cerevisiae* *PMS1* gene which has been shown to play a role in yeast mismatch repair (Bishop and Kolodner 1986; Bishop et al. 1987, 1989; Kramer et al. 1989). *hPMS1* is a distant relative of both *S. cerevisiae* *MLH1* and *PMS1*, and its function in mismatch repair in human cells is uncertain.

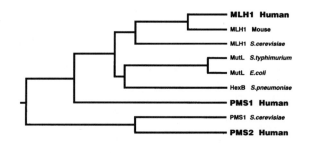

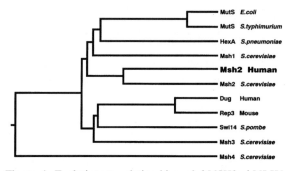

Figure 1. Evolutionary relationships of *hMSH2*, *hMLH1*, *hPMS1*, and *hPMS2*.

Chromosomal Map Location of *MSH2*, *MLH1*, and *PMS2*

Several methods have been used to determine the chromosomal location of the human *MSH2*, *MLH1*, and *PMS2* genes. These genes were first localized to chromosome 2, chromosome 3, and chromosome 7, respectively, by Southern blot analysis using NIGMS mapping panel 2 chromosome-specific cell hybrid DNAs, as well as by PCR analysis of the NIGMS mapping panel DNAs.

The map locations of *MSH2*, *MLH1*, and *PMS2* genes were further refined by mapping the location of the mouse homolog. This was possible because of the highly conserved nature of human and mouse genes, which contain large stretches of 100% amino acid identity and DNA coding sequences in several regions that contain segments as long as 100 bp that are more than 90% identical. The mouse chromosomal locations of *MSH2*, *MLH1*, and *PMS2* were determined by interspecific backcross analysis using progeny derived from matings of (C57BL/6J × *Mus spretus*)F_1 × C57BL/6J mice (Copeland and Jenkins 1991). C57BL/6J and *M. spretus* DNAs were digested with several enzymes and analyzed by Southern blot hybridization for informative restriction fragment length polymorphisms (RFLPs) using human cDNA probes to *hMSH2*, *hMLH1*, and *hPMS2*. Msh2 was located in the distal region of mouse chromosome 17 linked to *Tik*, *Msosl*, and *Lhcgr*. This distal region of mouse chromosome 17 shares a region of homology with human chromosome 2p21-p22. *Mlh1* was located to the distal region of mouse chromosome 9 linked to *Lams*, *Unp*, and *Col7a1*. This distal region of mouse chromosome 9 shares homology with human chromosome 3p21. This map location for *MLH1* was also shown in human cells by fluorescence in situ hybridization (FISH) analysis (Bronner 1994). *Pms2* was located in the distal region of mouse chromosome 5 linked to *Pdgfa* and *Gna12*. This distal region of mouse chromosome 5 shares homology with human chromosomes 7 and 13.

Hereditary Non-Polyposis Colon Cancer

A connection between *hMSH2* and *hMLH1* and carcinogenesis has been demonstrated (Fishel et al. 1993; Bronner et al. 1994; Kolodner et al. 1994). Similar results have been obtained by the Vogelstein and de la Chapelle group (Leach et al. 1993; Liu et al. 1994; Papadopoulos et al. 1994). Inherited mutations in these genes appear to be the genetic alterations responsible for 90% of HNPCC (Fishel et al. 1993; Leach et al. 1993; Sandkuijl and Bishop 1993; Bronner et al. 1994; Liu et al. 1994; Nystrom-Lahti et al. 1994; Papadopoulos et al. 1994). HNPCC has been mapped in several large kindreds to two loci: one on chromosome 2p (Peltomaki et al. 1993b; Sandkuijl and Bishop 1993; Nystrom-Lahti et al. 1994) and another on chromosome 3p (Lindblom et al. 1993; Nystrom-Lahti et al. 1994). We have shown that *hMSH2* maps to chromosome 2p21-22 (Fishel et al. 1993), and several *msh2* mutations were found to cosegregate with affected individuals of corresponding chromosome 2 kindreds (Fishel et al. 1993; Leach et al. 1993; Kolodner et al. 1994; Liu et al. 1994; Mary et al. 1994). Two such HNPCC kindreds are shown in Figure 2; HNPCC in family 1 is caused by a nonsense mutation in exon 12 of *hMSH2*, and HNPCC in family 2 is caused by a frameshift mutation in exon 12 of *hMSH2* (Kolodner et al. 1994). Both of these mutations result in the synthesis of truncated proteins predicted to be missing the highly conserved region of MutS homologs found between amino acids 640 and 762 of *hMSH2*. *hMLH1* maps to chromosome 3p21, and several mutations have been found to cosegregate with affected individuals in chromosome-3-linked HNPCC kindreds (Bronner et al. 1994; Papadopoulos et al. 1994; Kolodner et al. 1995). These genetic and physical results strongly suggest that inherited mutations in *MSH2* and *hMLH1* are the likely cause of HNPCC.

hPMS1 and *hPMS2* may also be associated with HNPCC (Nicolaides et al. 1994). A germ-line *hPMS1* mutation has been found in one person with a family history of cancer and, likewise, a germ-line *hPMS2* mutation has been found in a second person with a family history of cancer. However, there are no genetic mapping data that confirm cosegregation of HNPCC with either of these genes. Linkage studies and mutational analysis suggest that *MSH2* is responsible for up to 60%, *MLH1* is responsible for up to 30%, and *PMS1* and *PMS2* are responsible for approximately 5% of HNPCC cases (Sandkuijl and Bishop 1993; Liu et al. 1994; Nicolaides et al. 1994; Nystrom-Lahti et al. 1994); however, it should be pointed out that only a small number of families have been analyzed to date.

Purified *MSH2* Binds to Mismatched Nucleotides

We have found that introduction of the human *MSH2* gene into bacterial cells results in a dominant mutator phenotype (Fishel et al. 1993). A similar dominant mutator phenotype has been observed when the *Streptococcus pneumoniae* MutS homolog, *HexA*, gene was expressed in *Escherichia coli* (Prudhomme et al. 1991). The most likely explanation for this dominant mutator phenotype takes into account the mismatch binding function of MutS family proteins (Su and Modrich 1986) and the ability of these proteins to interact with other mismatch repair proteins (Grilley et al. 1989; Au et al. 1992; Prolla et al. 1994b). If *hMSH2* is capable of recognizing a mismatch, but the ability of a heterologous MutS homolog to interact with the *E. coli* mismatch repair machinery is impaired, then binding will effectively interfere with the repair process. In such a case, misincorporation errors will not be repaired, leading to an increase in the spontaneous mutation rate. This hypothesis suggests that *hMSH2* is capable of binding DNA containing mismatched nucleotides.

As a first step toward understanding the function of *MSH2* in human cells and correlating alterations in

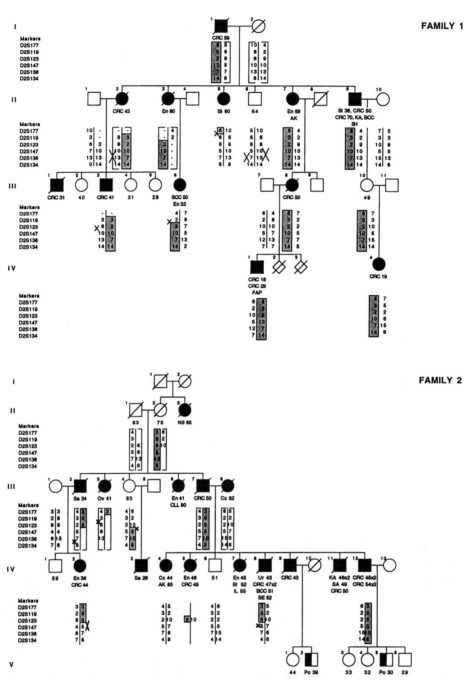

Figure 2. Pedigrees of two Muir-Torre families demonstrating 2p haplotypes. The chromosomal region bearing the disease gene in the oldest generation is shaded to show its inheritance through the family. Brackets are used to indicate inferred haplotypes, and crosses indicate recombinations. Blanks or dashes are used where the allele could not be determined. Haplotypes in some individuals are not shown to protect confidentiality of unaffected family members. Symbols: (squares) males; (circles) females; (oblique line) deceased; (full shading) confirmed cancer; (half shading) colorectal polyps. Two stillborn infants are indicated by diamonds. Under each symbol is listed the diagnosis and age at diagnosis if affected or current age/age of death if unaffected. Abbreviations: (AK) actinic keratosis; (BCC) basal cell carcinoma; (Bl) bladder carcinoma; (CRC) colorectal carcinoma; (Cx) cervical carcinoma; (En) endometrial carcinoma; (FAP) familial adenomatous polyposis (for further details, see Hall et al. 1994); (IL) ileal carcinoma; (KA) keratoacanthoma; (NS) carcinoma of the nasal septum; (Po) colorectal polyps; (Sa) sarcoma of the bone; (SA) sebaceous adenoma; (SCC) squamous cell carcinoma; (SE) sebaceous epithelioma; (St) stomach cancer; (Ur) transitional cell carcinoma of the ureter. Individuals are referred to in the text by generation (roman numeral) and number (arabic numeral). (Reprinted, with permission, from Kolodner et al. 1994.)

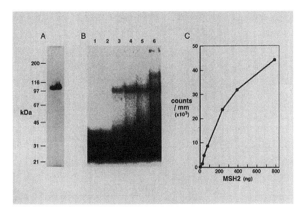

Figure 3. Purification and mismatch binding activity of hMSH2. (A) Silver-stained SDS-PAGE of purified hMSH2. (B) Mismatch binding gel shift assay using Fraction V hMSH2. (Lane 1) No protein; (lane 2) 24 ng hMSH2; (lane 3) 74 ng hMSH2; (lane 4) 239 ng hMSH2; (lane 5) 399 ng hMSH2; (lane 6) 798 ng hMSH2. (C) Quantitation of the binding activity shown in B. Counts per mm^2 of the shifted material was determined using the Molecular Analyst 2.0 software associated with the Biorad Phosphoimager (Biorad, Hercules, California).

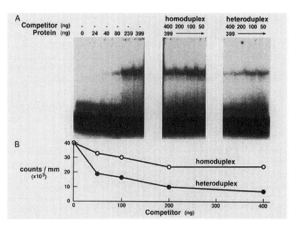

Figure 4. Competition of mismatch binding with homoduplex and heteroduplex DNA. (A) Titration of mismatch binding with purified hMSH2 followed by competition with homoduplex and heteroduplex DNA. The binding reaction was performed in 20 mM potassium phosphate (pH 7.5), 50 mM KCl, 1 mM dithiothreitol (DTT), 1 mM EDTA, and 5% glycerol at 25°C for 10 min. After binding, the shown quantity of competitor was introduced into the reaction and incubated for an additional 10 min. (B) Quantitation of bound material in counts per mm^2 was performed using the Molecular Analyst 2.0 software associated with the Biorad Phosphoimager (Biorad, Hercules, California).

function with the development of tumors in HNPCC kindreds, we have purified the normal protein from a bacterial overproduction system (Fig. 3A). A mismatch binding gel-shift assay was developed to assess MSH2 protein activities (Jiricny et al. 1988). The binding of purified MSH2 to a G-T mismatch-containing oligonucleotide is shown in Figure 3B, and quantitation of this binding activity is shown in Figure 3C. The binding appears to be cooperative and reaches 80% saturation at a protein-to-DNA ratio of approximately 4:1. These results are qualitatively similar to those found with the bacterial MutS protein (Su and Modrich 1986; Su et al. 1988) and yeast MSH1 and MSH2 proteins (Chi and Kolodner 1994; Alani et al. 1995).

To test the specificity of binding, we have carried out competition experiments in which either unlabeled homoduplex oligonucleotide DNA (containing the identical DNA sequence except that the mismatch site contained fully paired nucleotides) or unlabeled heteroduplex (G-T mismatch-containing) DNA was included in the binding reaction (Fig. 4). These data indicate that 50- to 100-fold greater homoduplex DNA competitor is required to compete MSH2 binding from the mismatch DNA substrate than unlabeled heteroduplex competitor. The purified MSH2 protein was found to efficiently bind insertion/deletion loop-type mismatched DNA containing up to 14 unpaired nucleotides on one strand of a duplex oligonucleotide. These results further suggest that the types of mismatches proposed to be intermediates in microsatellite instability are effective substrates for MSH2 protein recognition and binding.

Electron microscopy (EM) was used to directly visualize MSH2 bound to DNA containing a three-nucleotide insertion/deletion loop mismatch. Examination of a field of molecules revealed abundant DNA/ protein complexes in which a readily visible protein ball or complex was bound at the center of the mismatch-containing DNA substrate. Based on the known size of MSH2 monomer (106.4 kD), the smaller protein balls (Fig. 5A) were consistent with single monomers bound at the mismatch site, whereas the larger complexes (Fig. 5B, C) were of the size consistent with dimers or higher homopolymeric oligomers bound to the mismatch site. When the incubations were carried out in the absence of ATP, 11% of the molecules were bound with MSH2 and were located at the centrally located mismatch site ($n = 100$). The inclusion of 5 mM ATP increased this fraction to 29% ($n = 100$). We have also found that 5–10% of the DNA appeared to have MSH2 bound to the ends of the linear DNA, which is a common property of many DNA-binding proteins. These results suggest that DNA ends may be a source of nonspecific binding by MSH2. We conclude from these studies that MSH2 binds with high specificity to mismatched nucleotides, a property which is consistent with its homology to the MutS family of proteins and with the proposed role of msh2 mutations in HNPCC.

DISCUSSION

The genetic basis of human cancer has been documented over many years. During this time, a number of genes have been identified as functioning as either tumor promoter genes when mutated or as tumor suppressor genes (Knudson 1986; Stanbridge 1990). The vast majority of these oncogenes and tumor suppressors appear to be involved in cell signaling processes that affect growth control and maintenance of location (Weinberg 1989). Genetic analysis of tumor

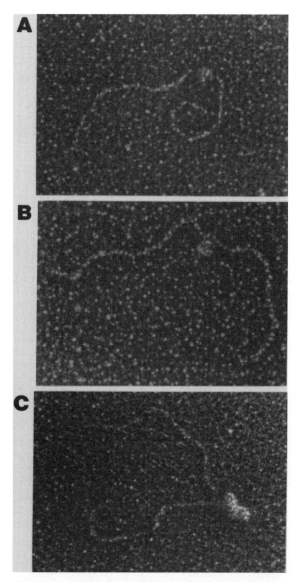

Figure 5. Visualization of *hMSH2* bound to DNA containing central insertion/deletion loop mismatches. *hMSH2* (Fishel et al. 1994) was incubated with a 1.1-kb DNA fragment containing three (3-cytosine) insertion/deletion loop mismatches at its center. DNA bound with a presumptive monomer, dimer, or homopolymeric oligomer of *hMSH2* are shown in *A*, *B*, and *C*, respectively. Samples were prepared for EM as in Griffith and Christiansen (1978).

cells has suggested that between six and twelve of these signaling genes are altered during the development of a tumor (Armitage and Doll 1954; Renan 1993). Although the detection of altered oncogenes and tumor suppressors is important from a genetic perspective, this observation represents a significant conundrum in cancer biology, since this number far exceeds the number of mutations that would be predicted by the spontaneous mutation rate in human cells (Seshadri et al. 1987; Albertini et al. 1990; Loeb 1991). Of the several theories that have been put forward to account for this discrepancy, the clonal selection model has gained the most acceptance (Nowell 1976). However, clonal selection may not completely account for the number of mutations observed in tumors during tumorigenesis (Cifone and Fidler 1981; Pitot et al. 1981; Croce 1987; Duesberg 1987).

hMSH2, *hMLH1*, *hPMS1*, and *hPMS2* are members of two highly conserved families of postreplication mismatch repair components. Alteration of these genes in all organisms studied to date results in a generalized increase in the rate of spontaneous mutation. The observation that inheritance of mutations in these genes leading to a mismatch repair defect is responsible for a significant proportion of HNPCC supports the idea that an increased mutation frequency can lead to the development of cancer. Such a mutator phenotype could explain the apparently high numbers of mutations found in tumor cells and provides a firm foundation for examining other types of mutator genes for their role in tumorigenesis.

ACKNOWLEDGMENTS

This work was supported by National Institutes of Health grants to R.D.K., R.M.L., D.C.W., J.G., and R.F.; and by National Institutes of Health Cancer Center Core grants to the Dana-Farber Cancer Institute and the Lineberger Comprehensive Cancer Center. N.R.H., P.C., and D.T.B. were supported by the Imperial Cancer Research Fund. A.L., P.T., and M.N. were supported by the Bert von Kantzow's Fund and the Stockholm Cancer Society. N.G.C. and N.A.J. were supported by a contract from the National Cancer Institute. M.R.S.R. was supported by an American Cancer Society International Cancer Research Fellowship from the International Union Against Cancer. C.E.B. was supported by an American Cancer Society fellowship, and S.M.B. was supported by the C.G. Swebillus Fund.

REFERENCES

Aaltonen, L.A., P. Peltomaki, F. Leach, P. Sistonen, S.M. Pylkkanen, J.-P. Mecklin, H. Jarvinen, S. Powell, J. Jen, S.R. Hamilton, G.M. Petersen, K.W. Kinzler, B. Vogelstein, and A. de la Chapelle. 1993. Clues to the pathogenesis of familial colorectal cancer. *Science* **260:** 812.

Aaltonen, L.A., P. Peltomaki, J.-P. Mecklin, H. Jarvinen, J.R. Jass, J.S. Green, H.T. Lynch, P. Watson, G. Tallqvist, M. Juhola, P. Sistonen, S.R. Hamilton, K.W. Kinzler, B. Vogelstein, and A. de la Chapelle. 1994. Replication errors in benign and malignant tumors from hereditary non-polyposis colorectal cancer patients. *Cancer Res.* **54:** 1645.

Alani, E., N.-W. Chi, and R. Kolodner. 1995. The *Saccharomyces cerevisiae* Msh2 protein binds to duplex oligonucleotides containing mismatched DNA base pairs and insertions. *Genes Dev.* (in press).

Albertini, R.J., J.A. Nicklas, J.P. O'Neill, and S.H. Robison. 1990. In vivo somatic mutations in humans: Measurement and analysis. *Annu. Rev. Genet.* **24:** 305.

Armitage, P. and R. Doll. 1954. The age distribution of cancer and a multi-stage theory of carcinogenesis. *Br. J. Cancer.* **8:** 1.

Au, K.G., K. Welsh, and P. Modrich. 1992. Initiation of methyl-directed mismatch repair. *J. Biol. Chem.* **267:** 12142.

Bhattacharyya, N.P., A. Skandalis, A. Ganesh, J. Groden, and

M. Meuth. 1994. Mutator phenotype in human colorectal carcinoma cell lines. *Proc. Natl. Acad. Sci.* **91:** 6319.

Bishop, D.K. and R.D. Kolodner. 1986. Repair of heteroduplex plasmid DNA after transformation into *Saccharomyces cerevisiae*. *Mol. Cell. Biol.* **6:** 3401.

Bishop, D.K., J. Andersen, and R.D. Kolodner. 1989. Specificity of mismatch repair following transformation of *Saccharomyces cerevisiae* with heteroduplex plasmid DNA. *Proc. Natl. Acad. Sci.* **86:** 3713.

Bishop, D.K., M.S. Williamson, S. Fogel, and R.D. Kolodner. 1987. The role of heteroduplex correction in gene conversion in *Saccharomyces cerevisiae*. *Nature* **328:** 362.

Bishop, D.T. and H. Thomas. 1990. The genetics of colorectal cancer. *Cancer Surv.* **9:** 585.

Bronner, C.E., S.M. Baker, P.T. Morrison, G. Warren, L.G. Smith, M.K. Lescoe, M. Kane, C. Earabino, J. Lipford, A. Lindblom, P. Tannergard, R.J. Bollag, A.R. Godwin, D.C. Ward, M. Nordenskjold, R. Fishel, R. Kolodner, and R.M. Liskay. 1994. Mutation in the DNA mismatch repair gene homologue *hMLH1* is associated with hereditary nonpolyposis colon cancer. *Nature* **368:** 258.

Burks, R.T., T.D. Kessis, K.R. Cho, and L. Hedrick. 1994. Microsatellite instability in endometrial carcinoma. *Oncogene* **9:** 1163.

Chi, N.-W. and R.D. Kolodner. 1994. The *Saccharomyces cerevisiae* Msh1 protein recognizes mispaired bases in DNA. *J. Biol. Chem.* **269:** 29984.

Cifone, M.A. and I.J. Fidler. 1981. Increased metastatic potential is associated with increasing genetic instability of clones isolated from murine neoplasms. *Proc. Natl. Acad. Sci.* **78:** 6949.

Copeland, N.G. and N.A. Jenkins. 1991. Development and applications of a molecular genetic linkage map of the mouse genome. *Trends Genet.* **7:** 113.

Cox, E.C. 1973. Mutator gene studies in *Escherichia coli*: The *mutT* gene. *Genetics* (suppl.) **73:** 67.

———. 1976. Bacterial mutator genes and the control of spontaneous mutation. *Annu. Rev. Genet.* **10:** 135.

Croce, C.M. 1987. Role of chromosome translocations in human neoplasia. *Cell* **49:** 155.

Duesberg, P.H. 1987. Cancer genes: Rare recombinants instead of activated oncogenes. *Proc. Natl. Acad. Sci.* **84:** 2117.

Fearon, E.R. and B. Vogelstein. 1990. A genetic model for colorectal tumorigenesis. *Cell* **61:** 759.

Fishel, R., A. Ewel, and M.K. Lescoe. 1994. The human MSH2 protein binds to mismatched nucleotides. *Cancer Res.* **54:** 5539.

Fishel, R.A., M.K. Lescoe, M.R.S. Rao, N. Copeland, N. Jenkins, J. Garber, M. Kane, and R. Kolodner. 1993. The human mutator gene homolog MSH2 and its association with hereditary nonpolyposis colon cancer. *Cell* **75:** 1027.

Gonzalez-Zulueta, M., J.M. Ruppert, K. Tokino, Y.C. Tsai, C.H. Spruch III, N. Miyao, P.W. Nichols, G.G. Hermann, T. Horn, K. Steven, I.C. Summerhayes, D. Sidransky, and P.A. Jones. 1993. Microsatellite instability in bladder cancer. *Cancer Res.* **53:** 5620.

Griffith, J.D. and G. Christiansen. 1978. Electron microscope visualization of chromatin and other DNA-protein complexes. *Annu. Rev. Biophys. Bioeng.* **7:** 19.

Grilley, M., K.M. Welsh, S.-S. Su, and P. Modrich. 1989. Isolation and characterization of the *Escherichia coli mutL* gene product. *J. Biol. Chem.* **264:** 1000.

Hall, N.R., V.A. Murday, P. Chapman, M.A.T. Williams, J. Burn, P.J. Finan, and D.T. Bishop. 1994. Genetic linkage in Muir-Torre syndrome to the same chromosomal site as Cancer Family syndrome. *Eur. J. Cancer* **30A:** 180.

Han, H.-J., A. Yanagisawa, Y. Kato, J.-G. Park, and Y. Nakamura. 1993. Genetic instability in pancreatic cancer and poorly differentiated type of gastric cancer. *Cancer Res.* **53:** 5087.

Honchel, R., K.C. Halling, D.J. Schaid, M. Pittelkow, and S.N. Thibodeau. 1994. Microsatellite instability in Muir-Torre syndrome. *Cancer Res.* **54:** 1159.

Ionov, Y., M.A. Peinado, S. Malkbosyan, D. Shibata, and M. Perucho. 1993. Ubiquitous somatic mutations in simple repeated sequences reveal a new mechanism for colonic carcinogenesis. *Nature* **363:** 558.

Jiricny, J., S.-S. Su, S.G. Wood, and P. Modrich. 1988. Mismatch-containing oligonucleotide duplexes bound by the *E. coli* mutS-encoded protein. *Nucleic Acids Res.* **16:** 7843.

Knudson, A.G. 1986. Genetics of human cancer. *Annu. Rev. Genet.* **20:** 231.

Kolodner, R.D., N.R. Hall, J. Lipford, M.F. Kane, P.T. Morrison, P.J. Finan, J. Burn, P. Chapman, C. Earabino, E. Merchant, and D.T. Bishop. 1995. Structure of the human MLH1 locus and analysis of a large HNPCC kindred for *mlh1* mutations. *Cancer Res.* (in press).

Kolodner, R.D., N.R. Hall, J. Lipford, M.F. Kane, M.R.S. Rao, P. Morrison, L. Wirth, P.J. Finan, J. Burn, P. Chapman, C. Earabino, E. Merchant, and D.T. Bishop. 1994. Structure of the human MSH2 locus and analysis of two Muir-Torre kindreds for *msh2* mutations. *Genomics* **24:** 516.

Kramer, W., B. Kramer, M.S. Williamson, and S. Fogel. 1989. Cloning and nucleotide sequence of DNA mismatch repair gene *PSM1* from *Saccharomyces cerevisiae*: Homology of PMS1 to procaryotic MutL and HexB. *J. Bacteriol.* **171:** 5339.

Kunkel, T. 1993. Slippery DNA and diseases. *Nature* **365:** 207.

Leach, F.S., N.C. Nicolaides, N. Papadopoulos, B. Liu, J. Jen, R. Parsons, P. Peltomaki, P. Sistonen, L.A. Aaltonen, M. Nystrom-Lahti, X.-Y. Guan, J. Zhang, P.S. Meltzer, J.-W. Yu, F.-T. Kao, D.J. Chen, K.M. Cerosaletti, R.E.K. Fournier, S. Todd, T. Lewis, R.J. Leach, S.L. Naylor, J. Weissenbach, J.-P. Mecklin, H. Jarvinen, G.M. Petersen, S.R. Hamilton, J. Green, J. Jass, P. Watson, H.T. Lynch, J.M. Trent, A. de la Chapelle, K.W. Kinsler, and B. Vogelstein. 1993. Mutations of a mutS homolog in hereditary nonpolyposis colorectal cancer. *Cell* **75:** 1215.

Levinson, G. and G.A. Gutman. 1987. High frequencies of short frameshifts in poly-CA/TG tandem repeats borne by bacteriophage M13 in *Escherichia coli* K-12. *Nucleic Acids Res.* **15:** 5313.

Lindblom, A., P. Tannergard, W. Nordenskjold, and M. Nordenskjold. 1993. Genetic mapping of a second locus predisposing to hereditary non-polyposis colon cancer. *Nat. Genet.* **5:** 279.

Liu, B., R.E. Parsons, S.R. Hamilton, G.M. Petersen, H.T. Lynch, P. Watson, S. Markowitz, J.K.V. Willson, J. Green, A. de la Chapelle, K.W. Kinzler, and B. Vogelstein. 1994. *hMSH2* mutations in hereditary nonpolyposis colorectal cancer kindreds. *Cancer Res.* **54:** 4590.

Loeb, L.A. 1991. Mutator phenotype may be required for multistage carcinogenesis. *Cancer Res.* **51:** 3075.

Lynch, H.T., T. Smyrk, P. Watson, S.J. Lanspa, B.M. Boman, P.M. Lynch, J.F. Lynch, and J. Cavalieri. 1991. Hereditary colorectal cancer. *Semin. Oncol.* **18:** 337.

Lynch, H.T., T.C. Smyrk, P. Watson, S.J. Lanspa, J.F. Lynch, P.M. Lynch, R.J. Cavalieri, and C.R. Boland. 1993. Genetics, natural history, tumor spectrum, and pathology of hereditary nonpolyposis colorectal cancer: An updated review. *Gastroenterology* **104:** 1535.

Mary, J.-L., D.T. Bishop, R. Kolodner, J.R. Lipford, M. Kane, J. Torhorst, H. Muller, M. Spycher, and R.J. Scott. 1994. Mutational analysis of the hMSH2 gene reveals a three base pair deletion in a family predisposed to colorectal cancer development. *Hum. Mol. Genet.* **3:** 2067.

Merlo, A., M. Mabry, E. Gabrielson, R. Vollmer, S.B. Baylin, and D. Sidransky. 1994. Frequent microsatellite instability in primary small cell lung cancer. *Cancer Res.* **54:** 2098.

Modrich, P. 1989. Methyl-directed DNA mismatch correction. *J. Biol. Chem.* **264:** 6597.

———. 1991. Mechanisms and biological effects of mismatch repair. *Annu. Rev. Genet.* **25:** 229.

Nicolaides, N.C., N. Papadopoulos, B. Liu, Y.-F. Wei, K.C. Carter, S.M. Ruben, C.A. Rosen, W.A. Haseltine, R.D. Fleischmann, C.M. Fraser, M.D. Adams, J.C. Venter, M.G.

Dunlop, S.R. Hamilton, G.M. Petersen, A. de la Chapelle, B. Vogelstein, and K.W. Kinzler. 1994. Mutations of two PMS homologues in hereditary nonpolyposis colon cancer. *Nature* **371:** 75.

Nowell, P.C. 1976. The clonal evolution of tumor populations. *Science* **194:** 23.

Nystrom-Lahti, M., R. Parsons, P. Sistonen, L. Pylkkanen, L.A. Aaltonen, F.S. Leach, S.R. Hamilton, P. Watson, E. Bronson, R. Fusaro, J. Cavalieri, J. Lynch, S. Lanspa, T. Smyrk, P. Lynch,T. Drouhard, K.W. Kinzler, B. Vogelstein, H.T. Lynch, A. de la Chapelle, and P. Peltomaki. 1994. Mismatch repair genes on chromosomes 2p and 3p account for a major share of hereditary nonpolyposis colorectal cancer families evaluable by linkage. *Am. J. Hum. Genet.* **55:** 659.

Papadopoulos, N., N.C. Nicolaides, Y.-F. Wei, S.M. Ruben, K.C. Carter, C.A. Rosen, W.A. Haseltine, R.D. Fleischmann, C.M. Fraser, M.D. Adams, J.C. Venter, S.R. Hamilton, G.M. Petersen, P. Watson, H.T. Lynch, P. Peltomaki, J.-P. Mecklin, A. de la Chapelle, K.W. Kinzler, and B. Vogelstein. 1994. Mutation of a *mutL* homolog in hereditary colon cancer. *Science* **263:** 1625.

Peltomaki, P., R.A. Lothe, L.A. Aaltonen, L. Pylkkanen, M. Nystrom-Lahti, R. Seruca, L. David, R. Holm, D. Ryberg, A. Haugen, A. Brogger, A.-L. Borresen, and A. de la Chappelle. 1993a. Microsatellite instability is associated with tumors that characterize the hereditary non-polyposis colorectal carcinoma syndrome. *Cancer Res.* **53:** 5853.

Peltomaki, P., L.A. Aaltonen, P. Sistonen, L. Pylkkanen, J.-P. Mecklin, H. Jarvinen, J.S. Green, J.R. Jass, J.L. Weber, F.S. Leach, G.M. Pedersen, S.R. Hamilton, A. de la Chapelle, and B. Vogelstein. 1993b. Genetic mapping of a locus predisposing to human colorectal cancer. *Science* **260:** 810.

Pitot, H.C., T. Goldsworthy, and S. Moran. 1981. The natural history of carcinogenesis: Implications of experimental carcinogenesis in the genesis of human cancer. *J. Supramol. Struct. Cell. Biochem.* **17:** 133.

Prolla, T.A., D.-M. Christie, and R.M. Liskay. 1994a. A requirement in yeast DNA mismatch repair for *MLH1* and *PMS1*, two homologs of the bacterial *mutL* gene. *Mol. Cell. Biol.* **14:** 407.

Prolla, T.A., Q. Pang, E. Alani, R.D. Kolodner, and R.M. Liskay. 1994b. MLH1, PMS1, and MSH2 interactions during the initiation of DNA mismatch repair in yeast. *Science* **265:** 1091.

Prudhomme, M., V. Mejean, B. Martin, and J.-P. Claverys. 1991. Mismatch repair genes of *Streptococcus pneumoniae*: HexA confers a mutator phenotype in *Escherichia coli* by negative complementation. *J. Bacteriol.* **173:** 7196.

Reenan, R.A.G. and R.D. Kolodner. 1992a. Isolation and characterization of two *Saccharomyces cerevisiae* genes encoding homologs of the bacterial HexA and MutS mismatch repair proteins. *Genetics* **132:** 963.

———. 1992b. Characterization of insertion mutations in the *Saccharomyces cerevisiae MSH1* and *MSH2* genes: Evidence for separate mitochondrial and nuclear functions. *Genetics* **132:** 975.

Renan, M.J. 1993. How many mutations are required for tumorigenesis? Implications from human cancer data. *Mol. Carcinog.* **7:** 139.

Rhyu, M.G., W.S. Park, and S.J. Meltzer. 1994. Microsatellite instability occurs frequently in human gastric carcinoma. *Oncogene* **9:** 29.

Risinger, J.I., A. Berchuck, M.F. Kohler, P. Watson, H.T. Lynch, and J. Boyd. 1993. Genetic instability of microsatellites in endometrial carcinoma. *Cancer Res.* **53:** 5100.

Rydberg, B. 1978. Bromouracil mutagenesis and mismatch repair in mutator strains of *Escherichia coli*. *Mutat. Res.* **52:** 11.

Sandkuijl, L.A. and T. Bishop. 1993. Results of the joint analysis of the EUROFAP linkage data: Summary. In *Genetics and Clinics of HNPCC: Proceedings of the Fourth Workshop, Copenhagen: EUROFAP* (ed. J. Mohr), p. 26.

Schoenberg, M.P., J.M. Hakimi, S. Wang, G.S. Bova, J.I. Epstein, K.H. Fischbeck, W.B. Isaacs, P.C. Walsh, and E.R. Barrack. 1994. Microsatellite mutations (CAG24→18) in the androgen receptor gene in human prostate cancer. *Biochem. Biophys. Res. Commun.* **198:** 74.

Seshadri, R., R.J. Kutlaca, K. Trainor, C. Matthews, and A.A. Morley. 1987. Mutation rate of normal and malignant human lymphocytes. *Cancer Res.* **47:** 407.

Shibata, D., M.A. Peinado, Y. Ionov, S. Malkhosyan, and M. Perucho. 1994. Genomic instability in repeated sequences is an early somatic event in colorectal tumorigenesis that persists after transformation. *Nat. Genet.* **6:** 273.

Shridhar, V., J. Siegfried, J. Hunt, M. del Mar Alonso, and D.I. Smith. 1994. Genetic instability of microsatellite sequences in many non-small cell lung carcinomas. *Cancer Res.* **54:** 2084.

Stanbridge, E. 1990. Human tumor suppressive genes. *Annu. Rev. Genet.* **24:** 615.

Strand, M., T.A. Prolla, R.M. Liskay, and T.D. Petes. 1993. Destabilization of tracts of simple repetitive DNA in yeast by mutations affecting DNA mismatch repair. *Nature* **365:** 274.

Su, S.-S. and P. Modrich. 1986. *Escherichia coli mutS*-encoded protein binds to mismatched DNA base pairs. *Proc. Natl. Acad. Sci.* **83:** 5057.

Su, S.-S., R.S. Lahue, K.G. Au, and P. Modrich. 1988. Mispair specificity of methyl-directed DNA mismatch correction *in vitro*. *J. Biol. Chem.* **263:** 6829.

Thibodeau, S.N., G. Bren, and D. Schaid. 1993. Microsatellite instability in cancer of the proximal colon. *Science* **260:** 816.

Wada, C., S. Shionoya, Y. Fujino, H. Tokyhiro, T. Akahoshi, T. Uchida, and H. Ohtani. 1994. Genomic instability of microsatellite repeats and its association with the evolution of chronic myelogenous leukemia. *Blood* **83:** 3449.

Weinberg, R.A. 1989. Oncogenes, antioncogenes, and the molecular basis of multistep carcinogenesis. *Cancer Res.* **49:** 3713.

Williamson, M.S., J.C. Game, and S. Fogel. 1985. Meiotic gene conversion mutants in *Saccharomyces cerevisiae*. I. Isolation and characterization of *pms1-1* and *pms1-2*. *Genetics* **110:** 609.

Wooster, R., A.M. Cleton-Jansen, N. Collins, J. Mangion, R.S. Cornelis, C.S. Cooper, B.A. Gusterson, B.A. Ponder, A. von Deimling, O.D. Wiester, C.J. Cornelis, P. Devilee, and M.R. Stratton. 1994. Instability of short tandem repeats (microsatellites) in human cancers. *Nat. Genet.* **6:** 152.

Yee, C.J., N. Roodi, C.S. Verrier, and F.F. Parl. 1994. Microsatellite instability and loss of heterozygosity in breast cancer. *Cancer Res.* **54:** 164.

Young, J., B. Leggett, C. Gustafson, M. Ward, J. Searl, L. Thomas, R. Buttenshaw, and G. Chenevix-Trench. 1993. Genomic instability occurs in colorectal carcinoma but not adenomas. *Hum. Mutation* **2:** 351.

Defects in Replication Fidelity of Simple Repeated Sequences Reveal a New Mutator Mechanism for Oncogenesis

M. PERUCHO, M.A. PEINADO, Y. IONOV, S. CASARES,
S. MALKHOSYAN, AND E. STANBRIDGE†

California Institute of Biological Research, La Jolla, California 92037;
†Department of Microbiology and Molecular Genetics, University of California, Irvine, California 92717

The mutational theory of cancer (Knudson 1971, 1985) has gained overwhelming support as germ-line and somatic mutations have been shown to activate the malignant potential of oncogenes and to inactivate the repressor function of tumor suppressor genes (Bishop 1991). The analysis of these tumor-specific mutations has yielded fundamental information on the mechanisms of carcinogenesis, but their etiology has remained poorly understood. There is no agreement in the estimation of the relative contribution of insults by genotoxic agents and of endogenous DNA replication errors to the genesis of these mutations in cancer (Ames and Gold 1990; Weinstein 1991).

The concept that spontaneous errors in DNA replication may be fundamental in transformation was put forward (Loeb et al. 1974; Cairns 1975) in an attempt to explain the genomic instability of cancer cells (Schimke et al. 1986; Cheng and Loeb 1993). A defective DNA replication factor could enhance the error rate in the tumor cell variants continuously selected during tumor progression (Foulds 1954; Nowell 1976). However, a critical prediction of this hypothesis, an increased mutation frequency in tumor cells, remained elusive despite intensive efforts (Harris 1991; Loeb 1991).

The importance of DNA replication and repair in transformation is demonstrated by the association of various biochemical defects in these processes with hereditary diseases that predispose to cancer (Cleaver 1967; Lindahl et al. 1991). The genes encoding factors involved in the DNA synthesis and error repair pathways are beginning to be isolated and characterized (Palombo et al. 1994; Tanaka and Wood 1994). A causal link between mutations in these factors and the origin of mutations in oncogenes and tumor suppressor genes is for the first time in sight.

Colorectal cancer is one of the best-characterized examples of the multistage nature of carcinogenesis. A dominant oncogene, c-Ki-*ras*, and at least three distinct tumor suppressor genes, *p53*, *DCC*, and *APC*, are consistently involved in colorectal tumorigenesis (Fearon and Vogelstein 1990; Fearon and Jones 1992). In addition to the mutations in these critical genes, tumors of the colon and rectum contain other apparently random genetic alterations that exhibit a remarkably heterogeneous distribution in their extent (Vogelstein et al. 1989).

DNA fingerprinting of hypervariable human minisatellite sequences is useful for the comparative analysis of multiple loci between different individuals (Jeffreys et al. 1985, 1990). DNA fingerprinting of minisatellite loci detected occasional somatic genetic alterations in gastrointestinal tumors (Thein et al. 1987; Armour et al. 1989; Vogelstein et al. 1989). The arbitrarily primed polymerase chain reaction (AP-PCR) is a DNA fingerprinting technique based in the amplification by PCR (Mullis and Faloona 1987) of multiple DNA fragments with the use of a single arbitrary primer (Welsh and McClelland 1990). The reproducible and quantitative amplification of many anonymous genomic sequences in a simple experiment permits the unbiased examination of the cell genome and provides a powerful tool for the analysis of the genetic alterations accompanying malignancy.

DNA fingerprinting by AP-PCR disclosed the presence of ubiquitous somatic mutations (USM) at simple repeated sequences (SRS) in the genome of some colorectal tumors (Peinado et al. 1992). These USM at SRS unveiled a new mutator mechanism for oncogenesis, because they accumulate after the mutational unfolding of an acquired or inherited defect in the fidelity of DNA replication, ultimately underlying tumor formation (Ionov et al. 1993). Here we describe in more detail the discovery of this mutator mechanism for cancer.

AP-PCR AS A MOLECULAR APPROACH FOR CANCER CYTOGENETICS

In PCR, the two primers that flank the targeted sequence are annealed to the template DNA at relatively high stringency to amplify only the specific sequence. In AP-PCR, DNA synthesis initiates from sites along the template with which a single oligonucleotide of arbitrary sequence matches only imperfectly. In the initial low-stringency cycles, the primer flanks many fortuitous sequences. A few rounds of primer extension result in a number of products having the original primer sequence at both ends, thereby allowing subsequent high-stringency PCR amplification. The products are then resolved by gel electrophoresis and visualized by autoradiography.

For a determined set of experimental conditions, the DNA fingerprint is reproducible and exquisitely dependent on the sequence of the arbitrary primer. In addition, the amplification is quantitative and the intensity of an amplified band is proportional to the concentration of the template sequence. Therefore, AP-PCR can identify DNA sequences that have lost their diploid state in tumor cells by either losses or gains of their corresponding chromosomal regions. Moreover, DNA fragments representing allelic losses or gains can be isolated by re-amplification with the same arbitrary primer (Peinado et al. 1992). The possibility of detecting moderate gains of genetic material constitutes a significant technical advance, because such genomic changes cannot be identified by conventional restriction fragment length polymorphism (RFLP) or microsatellite allelotyping.

DNA fingerprinting by AP-PCR thus provides a molecular approach for cancer cytogenetics (Perucho et al. 1994a). In addition to its quantitative properties, this is feasible because of two other propitious properties of the procedure: The amplified bands usually originate from single-copy sequences, with no apparent bias for their chromosomal origins. The chromosomal localization of the fingerprint bands can be simultaneously determined by AP-PCR of somatic rodent/human monochromosome cell hybrids, and fingerprints representative of the full chromosomal complement can be obtained with a few arbitrary primers (Peinado et al. 1992; Perucho et al. 1994a). The differences in the intensities of the AP-PCR bands from tumor DNA, compared to those from the normal diploid genome from the same individual, provide an estimation of the tumor cell aneuploidy. Consistent gains and losses of sequences from known chromosomal origins can be readily identified in particular types of tumors. For instance, we have detected moderate gains of sequences from chromosomes 8 and 13 in a majority of colorectal carcinomas at late stages of tumor progression, including hepatic metastases (S. Malkhosyan et al., in prep.). The relative extent of genomic damage undergone by different tumors can also be estimated by the comparative analysis of changes in their AP-PCR fingerprints. These arbitrary values, reflecting the extent of breakage of the genome integrity, may have direct application for cancer prognosis (Peinado et al. 1993a; Perucho et al. 1994a).

AP-PCR AND THE DISCOVERY OF A NEW MUTATOR MECHANISM FOR CANCER

During the course of these studies, we unexpectedly found other qualitative differences in the AP-PCR fingerprints from some tumors that were suggestive of structural somatic mutations (Peinado et al. 1992). Figure 1 shows the AP-PCR fingerprints obtained from four pairs of matched normal-tumor DNA samples

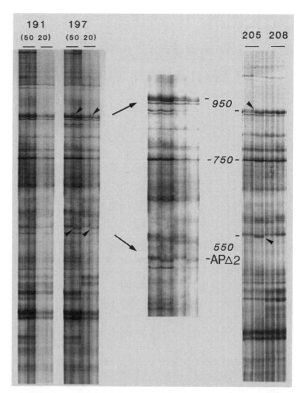

Figure 1. AP-PCR analysis of colorectal carcinomas. AP-PCR fingerprints of matched pair normal (*left*) and tumor (*right*) DNAs from colorectal carcinomas indicated at the top with arbitrary primer KpnX (Peinado et al. 1992). 50 and 20 ng of DNA were utilized. Arrowheads indicate mobility shifts in bands. The numbers in the middle indicate the molecular weights of some bands. A higher magnification detail of case 197 is shown in the center. The AP-PCR ^{32}P-labeled products were diluted 1:1 with formamide-loading buffer, denatured at 90°C for 3 min, and analyzed on a 6% polyacrylamide, 8 M urea sequencing gel for 6 hr at 50 W.

from colorectal cancer patients. Mobility shifts in bands of 950 nucleotides and 550 nucleotides (named APΔ2) were observed in tumor relative to normal tissue in cases 197 and 205, but not in cases 191 and 208. The same phenomenon was observed in other tumors analyzed with the same arbitrary primer. In all cases, the tumor tissue bands migrated faster than the normal tissue bands. These results suggested that in these tumors, small deletions had occurred in some of the sequences amplified by AP-PCR. The two bands showing mobility shifts were unrelated to each other, as they did not cross-hybridize in Southern blots of the AP-PCR gels using the cloned APΔ2 sequences as probe. This implied that these putative mutations were extremely abundant in these tumors, because of the random origin of the amplified DNA sequences due to the arbitrary nature of the primers.

These changes in the AP-PCR fingerprints were not restricted to the arbitrary primer of Figure 1, but were also observed with other unrelated primers. Figure 2 (left panels) shows the AP-PCR pattern of three cases of colorectal cancer obtained with another arbitrary primer. The difference in the migration of a 750-nu-

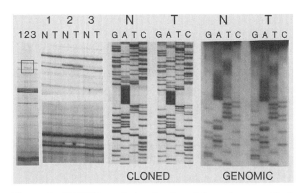

Figure 2. AP-PCR analysis of colorectal carcinomas. (*Left* panels) Autoradiograms of the AP-PCR pattern of three colorectal tumors with arbitrary primer K3US (Ionov et al. 1993). DNAs from normal (N) and tumor (T) tissues were amplified in duplicate. The ^{32}P-labeled products were analyzed as in Fig. 1. (*Center* and *right* panels) Sequencing ladders of the cloned or genomic APΔ1 sequences (see text).

cleotide polymorphic band (named APΔ1) of case 91 is highlighted by a box. The detection of additional band shifts with other arbitrary primers (6 of 10 initially analyzed) also reinforced the interpretation that these alterations were ubiquitously present in the genome of some colorectal tumor cells. The following experiments were designed to test this hypothesis.

USM IN POLY(A) TAILS OF *ALU* REPEATS

These mobility shifts were not due to mislabeling of the tumor or DNA samples, which was ruled out by the same fingerprinting properties of the procedure (see polymorphic bands in Figs. 1 and 2). Thus, the probability that the DNA from normal and tumor tissues of the relevant cases were not from the same individual was estimated to be less than 10^{-7}, assuming absence of bias for the allelic frequency for ethnic or other reasons. To investigate the nature of these shifts, we cloned and sequenced the APΔ1 and APΔ2 bands (lane 2 of Fig. 2, left panel, and Fig. 1, respectively) from normal and tumor tissues. APΔ1 contained two *Alu* repeats in a head to tail array. APΔ2 also contained an *Alu* repeat. The nucleotide sequences suggested that in both APΔ1 and APΔ2, the poly(A) tails of the *Alu* repeats were shorter in the tumor than in the normal tissue DNA (Fig. 2, center panel). However, we could not ascertain the exact number of deleted nucleotides because the allelic status of these sequences in the tumor cells was not known, and errors could have been introduced during in vitro amplification or propagation in *Escherichia coli*.

The sequence of the APΔ1 and APΔ2 bands was also determined directly from genomic DNA by cycle sequencing after PCR amplification (Peinado et al. 1993b). All tumors with deletions identified by AP-PCR contained shorter poly(A) tails in the *Alu* repeats, relative to their corresponding normal tissues. However, the exact number of deleted nucleotides was ambiguous because the bands above the poly(A) tracts were diffuse (Fig. 2, right panel). This could be due to sequence heterogeneity in the tissues or to amplification errors. The generation of "stuttered" bands after PCR amplification with Taq polymerase of repeated sequences was well documented (Weber and May 1989; Hearne et al. 1992). However, heterogeneity of minisatellite sequences also was reported in a minor fraction of cells in somatic tissues (Jeffreys et al. 1990).

To estimate more accurately the length differences in the monotonic runs of $(dA:dT)_n$ base pairs between normal and tumor tissues, we studied another DNA band (APΔ3), which did not contain an *Alu* repeat. We sequenced APΔ3 cloned in several independent plasmids after AP-PCR amplification from both normal (27 plasmids) and tumor (22 plasmids) tissues from patient 197. The heterozygosity status of APΔ3 did not represent a problem in this case because patient 197 was a male, and they were localized on chromosome X. As expected by the diffuse bands of the sequencing ladders (Fig. 2, right panel), the APΔ3 poly(A) tracts appeared heterogeneous in both normal and tumor tissues. However, the distribution of sequences in normal tissue averaged 17.3 deoxyadenines and in tumor tissue averaged only 13. This length heterogeneity was not due to instability during propagation in *E. coli*, because 10 different plasmids isolated after transfection of *E. coli* with a plasmid of 18 deoxyadenines retained this number of nucleotides.

We compared the APΔ3 PCR products amplified from genomic DNA with those from cloned DNA (Fig. 3). The band patterns generated with genomic DNA from normal and tumor tissues were indistinguishable from the patterns resulting from single plasmids containing 18 and 14 dA:dT base pairs, respectively. Because the population of plasmid molecules was at least 90% homogeneous (see above), these results implied that these sequences were equally homogeneous in the tissues. This experiment also showed that the

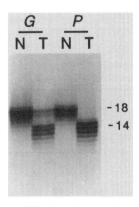

Figure 3. PCR amplification of genomic and cloned APΔ3 sequences. Autoradiogram of ^{35}S-labeled APΔ3 sequences PCR amplified from normal (N) and tumor (T) tissues from case 197. The PCR was performed for 30 cycles with 20 ng of total genomic DNA (G) or with 0.1 pg of cloned DNA (P) from plasmids containing APΔ3 sequences with 18 and 14 (dA:dT) bp from normal and tumor tissues, respectively.

APΔ3 sequences predominantly present in the normal tissue contained 18 deoxyadenines, whereas those from the tumor tissue contained only 14. Therefore, a somatic deletion of 4 base pairs had occurred in the APΔ3 sequences in the cells from colorectal carcinoma 197.

USM AT SRS IN COLORECTAL CARCINOMAS

The presence of microdeletions in several independent *Alu* repeats in the same tumors added support to the concept that these mutations were ubiquitous in the tumor cell genome. This hypothesis predicted that deletions would also occur in many other sequences containing runs of deoxyadenines of equivalent size to those present in APΔ2 and APΔ3. This prediction was confirmed by the detection of deletion mutations in two other sequences containing tracts of $(dA:dT)_n$: an *Alu* repeat with a poly(A) tail of 12 deoxyadenines in the 3' untranslated exon of the c-Ki-*ras* proto-oncogene and the 3' end region of the P450IIE1 gene. The apparent dependency of the extent of these mutations on the repeat size suggested a mechanism for these mutations, the slippage by strand misalignment (Streisinger et al. 1967). This model explained their features more satisfactorily than other recombination models. These include the proportionality between the repeat length and mutation frequency (Ripley 1990) and the bias for deletions versus insertions (Schaaper and Dunn 1987).

This hypothesis predicted that USM should also be present in other SRS, such as dinucleotide and trinucleotide microsatellites (Weber and May 1989). In agreement with this hypothesis, alterations in the mobility of microsatellite loci were subsequently found in tumors with deletions in poly(A) tracts (Ionov et al. 1993; Peinado et al. 1993b). In contrast with these last mutations, which were exclusively deletions, insertions also occurred in dinucleotide and trinucleotide repeats. The unidirectionality of the poly(A) mutations provides a molecular clock for the temporal and spatial analysis of tumor clonality and evolution (Shibata et al. 1994).

We also analyzed by *Alu* PCR (Sinnett et al. 1990) the colon carcinomas previously analyzed by PCR for mutations in the APΔ3 sequences. This was aimed at increasing the number of alterations detectable in a single experiment, because many amplified bands contain *Alu* repeats. This approach generated DNA fingerprints with apparent deletions in multiple bands (Fig. 4). In some positive cases, about one third of the amplified DNA fragments (indicated by dots) showed increased mobility. Figure 4 also shows that in many tumor DNAs, the bands of normal size were absent, or present in a minor proportion. The same result was obtained by PCR amplification of individual SRS (Ionov et al. 1993). Therefore, many of the USM at SRS were clonal because the absence of wild-type alleles implied that the mutant alleles were present in most, if not all, the tumor cells.

For all different DNA fragments analyzed, the PCR

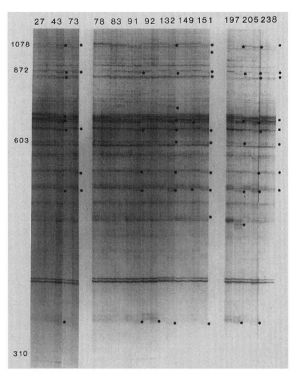

Figure 4. *Alu* PCR of colorectal carcinomas with multiple deletions. Autoradiogram of matched pairs of normal (*left*) and tumor (*right*) DNAs of colorectal tumors indicated at the top, amplified by PCR with *Alu* single primer J (Ionov et al. 1993). The numbers at left indicate the size (in nucleotides) of ΦX174 digested with *Hae*III used as molecular weight marker. The dots denote bands with deletions in tumor versus normal tissue DNA.

patterns of amplified bands from normal and tumor tissue DNA were, except for their size, very similar (see an example in Fig. 3). Therefore, the mutations in these sequences, although heterogeneous in different tumors, appeared homogeneous within individual tumors. These results also showed that mutations had often occurred in both alleles of the same amplified sequences. Altogether, these facts provided overwhelming evidence for the presence of an extraordinarily high number of clonal mutations at SRS in some tumors. If one or two AP-PCR bands contained deletions with over half of the arbitrary primers used, these results implied that deletions of a few nucleotides occurred every 1×10^4 to 3×10^4 bp, because about 10–30 kilobases of the genome are probed with each arbitrary primer. Therefore, the estimated number of mutated sequences would be $(3 \times 10^9)/(1 \times 10^4$ to $3 \times 10^4)$ = 1×10^5 to 3×10^5. The Streisinger model predicts that in SRS, changes of one repeat unit are favored over those of multiple units. We confirmed this prediction by analysis of SRS from cultured single-cell clones from mutator tumor cell lines (Shibata et al. 1994). Therefore, the deletions of several dA:dT bp observed in these tumors were most often the result of multiple, successive mutational events, each one deleting one base pair at a time. Thus, one hundred thousand mutations (Ionov et al. 1993) is probably an underestimation, because the number of deleted nucleotides in a given sequence was usually more than one and because both alleles were often targets for mutations. Altogether, these results indicated that some colorectal carcinomas had accumulated more than one million independent somatic mutations (Perucho 1994).

USM AT SRS ARE DUE TO A MUTATOR MUTATION

Although carcinogens such as proflavin and 2-(N-acetoxy-N-acetylamino) fluorene (AAF) are known to induce frameshift mutations (Ripley 1990), it appeared extremely unlikely that these USM at SRS could be due to exogenous genotoxic agents, because of their high number. It was more likely that these mutations were the result of endogenous failures in DNA replication and/or repair. Also because of their high number, it was obvious that these errors were due to a stable defect, and not to an induced failure in replication/repair (Sarasin 1985) or to a transient change in the intracellular nucleotide pool (Meuth et al. 1979).

Monotonic runs of dA:dT bp, as prototypic examples of SRS, have been shown to be genetically unstable, as judged by their increased mutation rate in *E. coli* (Benzer 1961) and by the polymorphic nature of the poly(A) tails of *Alu* repeats in the human population (Orita et al. 1990). However, because no deletions were found in the majority of the carcinomas, which represent clonal expansions of a single cell (Fearon et al. 1987), these mutations are not frequent events in the precursor cells that originate most of the colorectal tumors.

We analyzed the presence of mutations in APΔ2, APΔ3, and c-Ki-*ras Alu* sequences in a panel of colorectal tumors. In all tumors with deletions in APΔ3, there were also deletions in the other two SRS, with a few exceptions. All positive tumors for mutations in these SRS also yielded apparent deletions in the *Alu* PCR band pattern. In contrast, none of the tumors previously found negative for deletions in these three sequences exhibited noticeable deletions in the *Alu* PCR bands. A test for deviation of the Poisson distribution by these mutations indicated that they were significantly clustered in the same colorectal carcinomas, whether the calculations were made using the total number of bands with deletions, or the total number of deleted nucleotides in these three SRS (Fig. 5). Therefore, the detection by AP-PCR of USM in only some tumors could not be explained by a random distribution of mutations in all cancers, with only those with the highest number of mutations detectable with a few primers. The possibility that these mutations could also occur, albeit with lower frequency, in all (or at least some other) tumors, could not be ruled out. However, based on accumulative data of several experiments, the mutation rates of tumors with USM appeared at least two orders of magnitude higher than those without. It is noteworthy that this figure was proposed as the mini-

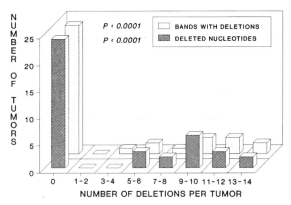

Figure 5. Frequency distribution of USM at SRS in colorectal carcinomas. Distribution of the total number of deleted nucleotides in APΔ2, APΔ3, and c-Ki-*ras Alu* sequences (Ionov et al. 1993), and the total number of bands with deletions after PCR analysis of these three DNA sequences plus those observed by *Alu* PCR with primer J (Fig. 4). Only 40 carcinomas analyzed by these four PCR amplifications are considered. Chi square for the observed frequencies of mutations compared with those expected by a random Poisson distribution (PD). The values were: PD for total number of deleted nucleotides: $m = 147/40 = 3.675$; Chi square $= 63.91$; DF $= 13$; P $= 4.38 \times 10^{-7}$. PD for total number of bands with deletions: $m = 111/40 = 2.775$; Chi square $= 67.93$; DF $= 13$; P $= 4.46 \times 10^{-7}$.

mal theoretical increase in mutation rate necessary to account for the age-dependent incidence of cancer, in the absence of selective growth advantage by the stem cells of the intestinal epithelium (Cairns 1975).

Figure 4 also shows that the number of bands with deletions varied widely among the tumors. For instance, tumor 92 showed only one deletion, whereas tumor 151 showed at least 10. The deletion size in individual sequences also varied from tumor to tumor. These differences could be related to the number of cell replications that the cell precursor of these tumors had sustained. However, there was no obvious correlation between the extent of the deletions, or the number of genomic sequences with mutations, and the time of cancer onset. These results suggested that these mutations were the result of somatic stochastic events, rather than the steady accumulation of spontaneous errors along the aging process.

On the basis of these results, we concluded first, that these USM at SRS were due to a "mutator mutation," a mutation in a gene coding for a DNA replication or repair factor resulting in decreased replication fidelity of these sequences and second, that their presence suggested the existence of a specific mutator mechanism for colorectal cancer (Ionov et al. 1993).

USM AT SRS UNDERLIE ACQUIRED AND INHERITED MUTATOR MECHANISMS FOR CANCER

This hypothesis predicted that the tumors with and the tumors without these USM should have different properties. Tumors with generalized deletions correlated negatively with mutations in the *p53* and c-Ki-*ras* genes and with tumors that had metastasized at diagnosis. On the other hand, they correlated positively with tumors exhibiting a poorly differentiated phenotype, with tumors of the right colon, and with tumors of blacks. They also appeared to be associated with an early time for cancer onset and with low recurrent rates. Indeed, only with the sex of the cancer patients did these tumors not show obvious differential association. On the basis of these distinctive characteristics in genotypic, phenotypic, biological, and clinical parameters, we deduced that USM in SRS were the genomic phenotypic manifestations of a novel molecular genetic pathway for colorectal cancer (Ionov et al. 1993; Perucho 1994), corresponding to the "cancer as a mutator phenotype" hypothesis (Loeb 1991).

Many of the characteristics that distinguished the USM tumors were strikingly coincident with tumors of hereditary non-polyposis colorectal cancer (HNPCC), or Lynch syndrome (Lynch et al. 1991). These included their preponderant localization in the proximal colon, the high frequency of carcinomas with a poorly differentiated phenotype, and their low metastatic and recurrence rates. Therefore, it appeared that USM at SRS were the common denominator of some sporadic and hereditary tumors of the colon. In the HNPCC Lynch syndrome II, tumors also develop (albeit with lower frequency) in sites other than the colon, like the endometrium, and in other sites of the gastrointestinal tract, like the stomach, esophagus, and small intestine (Lynch et al. 1991). This hypothesis predicted that USM at SRS should also be present in other tumors of the intestinal tract. We confirmed this prediction, detecting these mutations in some gastric and pancreatic carcinomas (Schaeffer et al. 1993; Perucho et al. 1993). On the other hand, no alterations in poly(A) tracts were found in lung carcinomas. This negative finding reinforced the conclusion that tumors with USM at SRS overlapped those of HNPCC patients, because a negative correlation between HNPCC patients and cancer of the lung had been described previously (Lynch et al. 1991).

The finding of a case with four synchronous colon tumors, all with USM at SRS, was also illuminating and added definitive support to the interpretation that these mutations could have a hereditary predisposition. Because these USM at SRS were present in all neoplastic areas (from the most superficial to the most invasive) of each of the four tumors from this individual, including two adenomas, we concluded that the mutator mutation played an ultimate causal role in tumorigenesis (Ionov et al. 1993).

These predictions were confirmed by the finding of a positive correlation between microsatellite instability and HNPCC tumors (Aaltonen et al. 1993). Linkage of some HNPCC families was demonstrated with loci at chromosomes 2 (Peltomaki et al. 1993) and 3 (Lindblom et al. 1993). Recently, the genes at chromosomes 2p and 3p have been cloned and characterized (Fishel et al. 1993; Leach et al. 1993; Bronner et al. 1994; Papadopoulos et al. 1994) as the human homologs to the bacterial *MutS* and *MutL*, and the yeast *MSH2* and *MLH1* genes, involved in long patch mismatch repair (Modrich 1991).

The broad but definite specificity for the organ origin of tumors in familial and sporadic cancer of the microsatellite mutator phenotype could be due to a requirement for an elevated cell replication activity of the corresponding tissues. It is likely that a minimal number of cell replications will be necessary for the mutator mutations to manifest their oncogenic potential. The high preponderance of gastrointestinal tumors with microsatellite instability (Aaltonen et al. 1993; Han et al. 1993; Thibodeau et al. 1993) could be explained because the gastrointestinal epithelium represents the tissue with highest cell turnover of the organism, undergoing more than 10^4 mitoses in normal conditions (Potten and Loeffler 1987; Armour et al. 1989). In this context, neoplasms from other tissues with intrinsically high mitotic activity, such as the endometrium, are also associated with HNPCC syndrome (Lynch et al. 1991) and with microsatellite instability (Risinger et al. 1993). However, other factors may also play a role in the tissue specificity of tumors of the microsatellite mutator phenotype, and it remains to be seen how close the tumors with microsatellite instability parallel those of HNPCC and other familial cancer syndromes.

MICROSATELLITE INSTABILITY UNFOLDS VERY EARLY AND PERSISTS IN TUMORIGENESIS

We initially favored a hypothesis to explain the presence and characteristic features of these USM at SRS, the most striking of which were (besides their high frequency) their apparent intertumor heterogeneity but intratumor homogeneity, in both the number of mutations and the number of repeat units deleted or inserted in each SRS. This hypothesis postulated that the mutator mutation occurred in a replicating stem cell of the colon epithelium, before neoplastic transformation. This cell would accumulate a high number of mutations at each round of replication and could die, frequently because mutations would irremediably occur in some essential genes. However, the mutant mutator gene could sometimes be lost, for instance, due to a mitotic error (Lasko et al. 1991). The cell would become normal again in its capacity for accurate replication, because of the remaining normal allele. However, at this point, the cell would have accumulated mutations in critical genes, like oncogenes or tumor suppressor genes, which would have led (or would eventually lead) to neoplasia. In this scenario, every SRS would have the same number of deleted or inserted nucleotides in a given tumor, but different numbers in other tumors, depending on the timing of the loss of the mutant allele of the mutator gene. A similar hypothesis was proposed more than 20 years ago, based on theoretical considerations (Nelson and Mason 1973).

By showing the continued presence of the defective

replication machinery in mutator tumor cells, however, we ruled out this "hit and run" hypothesis (Shibata et al. 1994). On the other hand, the first part of the hypothesis postulating the occurrence of USM at SRS in a stem cell of the colon epithelium before (or soon after) transformation gained support from this study. We applied a very sensitive technique, selective ultraviolet radiation fractionation (SURF), to the analysis of microsatellite instability in colorectal tumors (Shibata et al. 1994). The results showed the strong association of the activation of the underlying mutator mutation with histological transformation and showed that the manifestation of microsatellite instability occurred very early in tumorigenesis: Whereas no alterations in microsatellites were found in normal mucosa immediately adjacent to the USM$^+$ tumors, microsatellite alterations were frequently found in all areas of these carcinomas. In addition, USM at SRS were detected in all tested adenomas including a small flat adenoma and a microscopic polyp adjacent to an USM$^+$ carcinoma in a case of familial polyposis (Shibata et al. 1994). Therefore, in these patients, the underlying mutator mutation was one of the earliest events in carcinogenesis.

On the basis of these findings, we revised our initial hypothesis as follows: After the initial catastrophic mutator mutation, the cell would accumulate neutral mutations at SRS with steady and high frequency. When mutations eventually befell in critical oncogenes or suppressor genes, the clonal expansion of the corresponding cells would rapidly replace the predecessor cells with less mutated sequences and with less deleted or inserted nucleotides in each SRS. The mutation rate in these tumor cells, although still high, could be sufficiently low that the heterogeneity in the extent of these mutations would not be detectable with the sensitivity of the assay. This prediction was confirmed when we detected heterogeneity in the lengths of microsatellite loci in USM$^+$ tumors using the more sensitive SURF assay, indicating that the genomic instability at SRS persisted after transformation (Shibata et al. 1994). Therefore, the in situ degree of heterogeneity in microsatellite length can be explained by relative differences in mitotic versus mutation rates of these tumors, which in turn correlates with the presence or absence of some oncogenic mutations such as c-Ki-*ras* mutations.

THE MICROSATELLITE MUTATOR PATHWAY FOR CANCER IS RECESSIVE

Whereas in the initial hypothesis the initial catastrophic mutator mutation needed to be co-dominant, in the second hypothesis the mutation could also be recessive. Studies with somatic cell hybrids have shown that several distinct mutator genes responsible for microsatellite instability are all recessive (Perucho et al. 1994b; S. Casares et al., in prep.). A typical result of these experiments is shown in Figure 6. Although single-cell clones of a mutator cell line exhibited great length variation in a dinucleotide microsatellite, hybrids between cells with and without instability regained the fidelity of replication of these unstable sequences. The same result was also observed in hybrids between cells with high (of the MutS-L-H mismatch repair pathway) and low (of undefined mismatch repair pathway) instability at microsatellites, suggesting that the recessivity of mutator genes may be a general feature (Cheng and Loeb 1993).

These findings underscore the similarity between the mutator (Nelson and Mason 1973; Loeb et al. 1974; Cairns 1975) and tumor suppressor (Knudson 1985) pathways for cancer. In both cases, the two alleles of a recessive gene must be functionally inactivated by mutations, somatic in sporadic cases, and germ-line and somatic in hereditary cases (Knudson 1971). However, the mutator pathway exhibits a fundamental difference from the suppressor pathway: Whereas tumor suppressor mutations are by definition accompanied by an immediate capability for territorial expansion and/or a selective growth advantage, mutator mutations would not directly alter the growth properties of the cell (Fig. 7).

CONCLUSIONS

We have shown that AP-PCR is a powerful tool for the molecular genetic analysis of multistage carcinogenesis. The quantitative nature of the method permits its utilization as a molecular approach for cancer cytogenetics. This has diagnostic applications for the identification not only of genes with a negative repressor function (reflected by allelic losses), but also of those

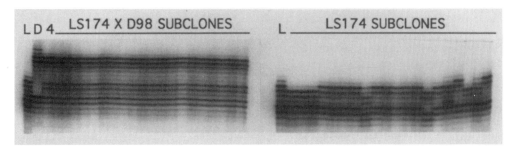

Figure 6. The microsatellite mutator phenotype is recessive. Single-cell clones were isolated from the high instability cell line LS174-T (lane *L*) and from single-cell hybrids from these cells fused with the stable cell line D98OR (lane *D*). The original cell hybrid (lane *4*) was grown for about 25 cell generations, and single-cell clones were isolated and analyzed for the presence of mutations in D1S158 microsatellite. Autoradiograms of PCR products electrophoresed in denaturing polyacrylamide gels.

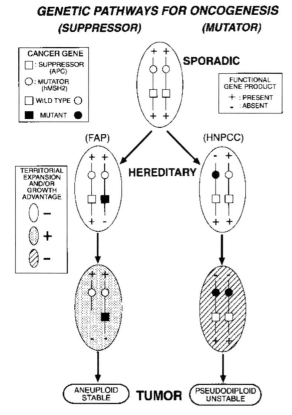

Figure 7. Mutator and suppressor pathways for oncogenesis. The hereditary syndromes and the genes in parentheses are only given as examples for the particular case of colorectal cancer. The occurrence of somatic mutations in both alleles of the mutator gene is proposed based on the instability phenotype of apparently sporadic tumors. The model is an oversimplification of the carcinogenesis process and is only intended to emphasize the similarities and differences between these two pathways for cancer.

with a positive cell growth role (reflected by allelic gains) during tumor progression. In addition, the arbitrary values for genomic damage obtained by the comparative analysis of AP-PCR fingerprints may be useful for cancer prognosis.

The large number of anonymous genomic DNA fragments that are amplified with this simple technique provides a neutral window to look at the genome of the cancer cell. Using this approach, we unexpectedly found in a specific subset of colorectal tumors consistent structural mutations in simple repeated sequences. Due to the unbiased nature of AP-PCR fingerprinting, these findings inescapably revealed the accumulation in the cancer cell genome of hundreds of thousands of somatic clonal mutations. The recognition of this genome-wide instability made possible the correct interpretation of the alterations in mobility of microsatellite loci, which were sporadically encountered during allelotyping or mapping studies. Consequently, the detection by standard PCR of alterations in a few microsatellite loci could be taken as evidence for the existence of a profound genomic instability underlying a mutator phenotype for cancer. This is illustrated by the outburst of papers describing these microsatellite alterations in a wide range of tumors.

The discovery of the microsatellite mutator phenotype for cancer was surprising both quantitatively and qualitatively. The staggering number of clonal somatic mutations present in the corresponding tumors left no doubt about the mutator nature of the underlying mechanism, and at the same time, provided evidence for its early unfolding in tumorigenesis. The current opinions on the concept of cancer as a mutator phenotype were aimed to explain mutations in cancer cells in the two-digit figures, and genomic instability as a late event in tumor progression (Nowell 1976; Fearon and Vogelstein 1990; Bishop 1991; Harris 1991; Loeb 1991). In contrast, the microsatellite mutator phenotype may lead to the accumulation of mutations in the six-digit range, and its mutagenic action occurs in tumorigenesis as early as can be possible for a mechanism underlying various forms of familial cancer.

The estimation of the relative contribution of the mutator and suppressor mechanisms for cancer, and their interrelationships depending on the relative time of occurrence of the mutator and suppressor mutations (Fig. 7), remains an open and active field of investigation. The detection of other mutator mechanisms, in addition to the long patch mismatch repair discovered by AP-PCR DNA fingerprinting, may be on the horizon, perhaps by the use of this or other unbiased genome-scanning technologies.

A causal relationship between mutator mutations and cancer development, through the induction of mutations in growth-related genes, has not been established yet. Because of the negative correlation between mutator mutations and mutations in known cancer genes, like *ras* and *p53* (Ionov et al. 1993), the "proximal" cause for neoplastic development of many tumors of the microsatellite mutator phenotype is still unknown. The existence of multiple complementation groups, not only for the manifestation of genomic instability, but also for the ability to grow in culture (S. Casares et al., in prep.), implies the absence of a common oncogenic target that could funnel the "remote" neoplastic potential of these mutator genes.

In addition, due to the recessive nature of the mutator phenotype, the etiology of the mutator mutations, both in hereditary and sporadic cases, remains unclear. Therefore, the issue of mutagenesis versus mitogenesis in the etiology of cancer still remains unsolved. Nevertheless, a reconciliation between these apparently contrary agents can be better envisioned now. A mitogenic situation such as a chronic inflammatory condition would increase the probability of occurrence not only of the mutator mutations, but also of the mutations in their oncogenic targets. Carcinogen action would not be eliminated, but augmented, in this scenario, due to the repair deficiency of carcinogen-induced DNA adducts of cells with the mutated mutator gene. In this context, mutations at microsatellites have been reported in some carcinogen-induced animal tumors (Canzian et al. 1994). It remains to be seen whether the mutator mutation in these tumor model systems is the carcinogen target.

ACKNOWLEDGMENTS

We thank Michael McClelland and John Welsh for helpful and encouraging discussions and Aaron McCarthy for his help with computer graphics programs. This work was supported by grants CA-38579, CA-63585, and CA-33021 from the National Cancer Institute.

REFERENCES

Aaltonen, L.A., P. Peltomaki, F.S. Leach, P. Sistonen, L. Pylkkanen, J.P. Mecklin, H. Jarvinen, S.M. Powell, J. Jen, S.R. Hamilton, G.M. Petersen, K.W. Kinzler, B. Vogelstein, and A. de la Chapelle. 1993. Clues to the pathogenesis of familial colorectal cancer. *Science* **260:** 812.

Ames, B.N. and L.S. Gold. 1990. Too many rodent carcinogens: Mitogenesis increase mutagenesis. *Science* **249:** 970.

Armour, J.A.L., I. Patel, S.L. Thein, M.F. Feym, and A.J. Jeffreys. 1989. Analysis of somatic mutations at human minisatellite loci in tumors and cell lines. *Genomics* **4:** 328.

Benzer, S. 1961. On the topography of the genetic fine structure. *Proc. Natl. Acad. Sci.* **47:** 403.

Bishop, J.M. 1991. Molecular themes in oncogenesis. *Cell* **64:** 235.

Bronner, C.E., S.M. Baker, P.T. Morrison, G. Warren, L.G. Smith, M.K. Lescoe, M. Kane, C. Earabino, J. Lipford, A. Lindblom, P. Tannergard, R.J. Bollag, A.R. Godwin, D.C. Ward, M. Nordenskjold, R. Fishel, R. Kolodner, and R.M. Liskay. 1994. Mutation in the DNA mismatch repair gene homologue hMLH1 is associated with hereditary non-polyposis colon cancer. *Nature* **368:** 258.

Cairns, J. 1975. Mutation selection and the natural history of cancer. *Nature* **255:** 197.

Canzian, F., T. Ushijima, T. Serikawa, K. Wakabayashi, T. Sugimura, and M. Nagao. 1994. Instability of microsatellites in rat colon tumors induced by heterocyclic amines. *Cancer Res.* **54:** (in press).

Cheng, K.C. and L. Loeb. 1993. Genomic instability and tumor progression: Mechanistic considerations. *Adv. Cancer Res.* **60:** 121.

Cleaver, J.E. 1968. Defective repair replication in xeroderma pigmentosum. *Nature* **218:** 652.

Fearon, E. and P. A. Jones. 1992. Progressing toward a molecular description of colorectal cancer development. *FASEB J.* **6:** 2783.

Fearon, E. and B. Vogelstein. 1990. A genetic model for colorectal tumorigenesis. *Cell* **61:** 759.

Fearon, E.R., S.R. Hamilton, and B. Vogelstein. 1987. Clonal analysis of human colorectal tumors. *Science* **238:** 193.

Fishel, R., M.K. Lescoe, M.R.S. Rao, N.G. Copeland, N.A. Jenkins, J. Garber, M. Kane, and R. Kolodner. 1993. The human mutator gene homolog MSH2 and its association with hereditary nonpolyposis colon cancer. *Cell* **75:** 1027.

Foulds, L. 1954. The experimental study of tumor progression: A review. *Cancer Res.* **14:** 327.

Han, H.J., A. Yanagisawa, Y. Kato, J.G. Park, and Y. Nakamura. 1993. Genetic instability in pancreatic cancer and poorly differentiated type of gastric cancer. *Cancer Res.* **53:** 5087.

Harris, C.C. 1991. Chemical and physical carcinogenesis: Advances and perspectives for the 1990s. *Cancer Res.* **51:** 5023.

Hearne, K.M., S. Ghosh, and J.A. Todd. 1992. Microsatellites for linkage analysis of genetic traits. *Trends Genet.* **8:** 288.

Ionov, Y., M.A. Peinado, S. Malkhosyan, D. Shibata, and M. Perucho. 1993. Ubiquitous somatic mutations in simple repeated sequences reveal a new mechanism for colonic carcinogenesis. *Nature* **363:** 558.

Jeffreys, A.J., R. Neumann, and V. Wilson. 1990. Repeat unit sequence variation in minisatellites: A novel source of DNA polymorphism for studying variation and mutation by single molecule analysis. *Cell* **60:** 473.

Jeffreys, A.J., V. Wilson, and S.L. Thein. 1985. Hypervariable "minisatellite" regions in human DNA. *Nature* **314:** 67.

Knudson, A.G. 1971. Mutation and cancer: Statistical study of retinoblastoma. *Proc. Natl. Acad. Sci.* **68:** 820.

———. 1985. Hereditary cancer, oncogenes and antioncogenes. *Cancer Res.* **45:** 1437.

Lasko, D., W. Cavenee, and M. Nordenskjold. 1991. Loss of constitutional heterozygosity in human cancer. *Annu. Rev. Genet.* **25:** 281.

Leach, F.S., N.C. Nicolaides, N. Papadopoulos, B. Liu, J. Jen, R. Parsons, P. Peltomaki, P. Sistonen, L.A. Aaltonen, M. Nystrom-Lahti, X.Y. Guan, J. Zhang, P.S. Meltzer, J.-W. Yu, F.-T. Kao, D.J. Chen, K.M. Cerosaletti, R.E.K. Fournier, S. Todd, T. Lewis, R.J. Leach, S. Naylor, J. Weissenbach, J.P. Mecklin, H. Jarvinen, G.M. Petersen, S.R. Hamilton, J. Green, J. Jass, P. Watson, H.T. Lynch, J.M. Trent, A. de la Chappelle, K.W. Kinzler, and B. Vogelstein. 1993. Mutations of a mutS homolog in hereditary nonpolyposis colorectal cancer. *Cell* **75:** 1215.

Lindahl, T., R.D. Wood, and P. Karran. 1991. Molecular deficiencies in human cancer-prone syndromes associated with hypersensitivity to DNA-damaging agents. In *Origins of human cancer: A comprehensive review* (ed. J. Brugge et al.), p. 163. Cold Spring Harbor Laboratory Press, Cold Spring Harbor, New York.

Lindblom, A., P. Tannergard, B. Werelius, and M. Nordenskjold. 1993. Genetic mapping of a second locus predisposing to hereditary non-polyposis colon cancer. *Nat. Genet.* **5:** 279.

Loeb, L.A. 1991. Mutator phenotype may be required for multistage carcinogenesis. *Cancer Res.* **51:** 3075.

Loeb, L.A., C.F. Springgate, and N. Battula. 1974. Errors in DNA replication as a basis of malignant changes. *Cancer Res.* **34:** 2311.

Lynch, H.T., S. Lanspa, T. Smyrk, B. Boman, P. Watson, and J. Lynch. 1991. Hereditary nonpolyposis colorectal cancer (Lynch syndrome I and II). *Cancer Genet. Cytogenet.* **143:** 160.

Meuth, M., J. L'Hereux-Huard, and M. Trudel. 1979. Characterization of a mutator gene in Chinese hamster ovary cells. *Proc. Natl. Acad. Sci.* **76:** 6505.

Modrich, P. 1991. Mechanisms and biological effects of mismatch repair. *Annu. Rev. Genet.* **25:** 229.

Mullis, K.B. and F.A. Faloona. 1987. Specific synthesis of DNA *in vitro* via a polymerase catalyzed chain reaction. *Methods Enzymol.* **155:** 335.

Nelson, R.L. and H.S. Mason. 1973. An explicit hypothesis for chemical carcinogenesis. *J. Theor. Biol.* **37:** 197.

Nowell, P. 1976. The clonal evolution of tumor cell populations. *Science* **194:** 23.

Orita, M., T. Sekiya, and K. Hayashi. 1990. DNA sequence polymorphisms in *alu* repeats. *Genomics* **8:** 271.

Palombo, F., M. Hughes, J. Jiricny, O. Truong, and J. Hsuan. 1994. Mismatch repair and cancer. *Nature* **367:** 417.

Papadopoulos, N., N.C. Nicolaides, Y.-F. Wei, S.M. Ruben, K.C. Carter, C.A. Rosen, W.A. Haseltine, R.D. Fleischmann, C.M. Fraser, M.D. Adams, J.C. Venter, S.R. Hamilton, G.M. Petersen, P. Watson, H.T. Lynch, P. Peltomaki, J.-P. Mecklin, A. de la Chapelle, K.W. Kinzler, and B. Vogelstein. 1994. Mutation of a mutL homolog in hereditary colon cancer. *Science* **263:** 1625.

Peinado, M.A., S. Malkhosyan, A. Velazquez, and M. Perucho. 1992. Isolation and characterization of allelic losses and gains in colorectal tumors by arbitrarily primed polymerase chain reaction. *Proc. Natl. Acad. Sci.* **89:** 10065.

Peinado, M.A., A. Velazquez, L. Wilson, and M. Perucho. 1993a. Determination of genomic instability in cancer cells by DNA fingerprinting. In *Genomic fingerprinting* (ed. M. McClelland and X. Estivill), vol. 20, p. 84. Instituto Juan March de Estudios e Investigaciones, Madrid, Spain.

Peinado, M.A., M. Fernandez-Renart, G. Capella, L. Wilson, and M. Perucho. 1993b. Mutations in the p53 suppressor gene do not correlate with c-K-*ras* oncogene mutations in colorectal cancer. *Int. J. Oncol.* **2:** 123.

Peltomaki, P., L.A. Aaltonen, P. Sistonen, L. Pylkkanen, J.P. Mecklin, H. Jarvinen, J.S. Green, J.R. Jass, J.L. Weber, F.S. Leach, G.M. Petersen, S.R. Hamilton, A. de la Chapelle, and B. Vogelstein. 1993. Genetic mapping of a locus predisposing to human colorectal cancer. *Science* **260:** 810.

Perucho, M. 1994. PCR and cancer diagnostics: Detection and characterization of single point mutations in oncogenes and antioncogenes. In *The polymerase chain reaction* (ed. K.B. Mullis et al.), p. 369. Birkhauser, Boston.

Perucho, M., J. Schaeffer, A. Velazquez, G. Berrozpe, and G. Capella. 1993. Genetic alterations. In *Proceedings of the International Symposium on Facing the pancreatic dilemma*, Verona (ed. P. Pederzoli et al.). Springer Verlag, New York.

Perucho, M., J. Welsh, M.A. Peinado, Y. Ionov, and M. McClelland. 1994a. Fingerprinting of DNA and RNA by arbitrarily primed PCR: Applications in cancer research. *Methods Enzymol.* (in press).

Perucho, M., S. Casares, Y. Ionov, M.A. Peinado, S. Malkhosyan, D. Shibata, and E. Stanbridge. 1994b. Somatic instability at microsatellites: A persistent and early event in cancer of the recessive mutator phenotype. In *Ras differentiation and development* (ed. J. Downward et al.), vol. 27, p. 37. Instituto Juan March de Estudios e Investigaciones, Madrid, Spain.

Potten, C.S. and M. Loeffler. 1987. A comprehensive model of the crypts of the small intestine of the mouse provides insight into the mechanisms of cell migration and the proliferation hierarchy. *J. Theor. Biol.* **127:** 381.

Ripley, L.S. 1990. Frameshift mutation: Determinants of specificity. *Annu. Rev. Genet.* **24:** 189.

Risinger, J.I., A. Berchuck, M.F. Kohler, P. Watson, H.T. Lynch, and J. Boyd. 1993. Genomic instability of microsatellites in endometrial carcinoma. *Cancer Res.* **53:** 5100.

Sarasin, A. 1985. SOS in mammalian cells. *Cancer Invest.* **3:** 163.

Schaaper, R.M. and R.L. Dunn. 1987. Spectra of spontaneous mutations in *Escherichia coli* strains defective in mismatch correction: The nature of *in vivo* DNA replication errors. *Proc. Natl. Acad. Sci.* **84:** 6220.

Schaeffer, J., G. Capella, J.M.A. Peinado, M. Fernandez-Renart, and M. Perucho. 1993. Comparative analysis of genetic alterations in pancreatic and other carcinomas. *Int. J. Pancreatol.* **14:** 75.

Schimke, R.T., S.W. Sherwood, A.B. Hill, and R.N. Johnston. 1986. Overreplication and recombination of DNA in higher eukaryotes: Potential consequences and biological implications. *Proc. Natl. Acad. Sci.* **83:** 2157.

Shibata, D., M.A. Peinado, Y. Ionov, S. Malkhosyan, and M. Perucho. 1994. Genomic instability in simple repeated sequences is an early somatic event in colorectal tumorigenesis that persists after transformation. *Nat. Genet.* **6:** 273.

Sinnett, D., J.-M. Deragon, L.R. Simard, and D. Labuda. 1990. Alumorphs-human DNA polymorphisms detected by polymerase chain reaction using *alu*-specific primers. *Genomics* **7:** 331.

Streisinger, G., Y. Okada, J. Emrich, J. Newton, A. Tsugita, E. Terzaghi, and M. Inouye. 1967. Frameshift mutations and the genetic code. *Cold Spring Harbor Symp. Quant. Biol.* **31:** 77.

Tanaka, K. and R.D. Wood. 1994. Xeroderma pigmentosum and nucleotide excision repair of DNA. *Trends Biochem. Sci.* **19:** 83.

Thein, S.L., A.J. Jeffreys, H.C. Gooi, F. Cotter, J. Flint, N.T.J. O'Connor, D.J. Weatherall, and J.S. Wainscoat. 1987. Detection of somatic changes in human cancer DNA by DNA fingerprint analysis. *Br. J. Cancer* **55:** 353.

Thibodeau, S.N., G. Bren, and D. Schaid. 1993. Microsatellite instability in cancer of proximal colon. *Science* **260:** 816.

Vogelstein, B., E.R. Fearon, S.E. Kern, S.R. Hamilton, A.C. Preisinger, Y. Nakamura, and R. White. 1989. Allelotype of colorectal carcinomas. *Science* **244:** 207.

Weber, J.L. and P.E. May. 1989. Abundant class of human DNA polymorphisms which can be typed using the polymerase chain reaction. *Am. J. Hum. Genet.* **44:** 288.

Weinstein, I.B. 1991. Mitogenesis is only one factor in carcinogenesis. *Science* **251:** 387.

Welsh, J. and M. McClelland. 1990. Fingerprinting genomes using PCR with arbitrary primers. *Nucleic Acids Res.* **18:** 7213.

Ovarian Tumors Display Persistent Microsatellite Instability Caused by Mutation in the Mismatch Repair Gene *hMSH-2*

K. ORTH,*§ J. HUNG,† A. GAZDAR,† M. MATHIS,†
A. BOWCOCK,‡ AND J. SAMBROOK‡**

*Howard Hughes Medical Institute, †Simmons Cancer Center, and ‡McDermott Center, University of Texas
Southwestern Medical School, Dallas, Texas 75235

Ovarian cancer accounts for only one-quarter of gynecological malignancies but is responsible for half of all deaths from cancer of the female genital tract. This high rate of mortality is due in large part to the lack of detectable symptoms during the early phases of the disease: Most ovarian cancers are diagnosed at a late stage by the presence either of malignant cells in the ascitic fluid, of peritoneal implants, or of metastases, which are commonly found in the liver (DiSaia and Creasman 1993).

Approximately 5–10% of ovarian cancer is hereditary, and family studies have identified three predisposing genetic syndromes: (1) site-specific ovarian cancer; (2) hereditary breast-ovarian cancer; and (3) Lynch Syndrome II (a subgroup of hereditary nonpolyposis colorectal cancer [HNPCC]) (Lynch et al. 1991c; Rubin and Sutton 1993). In recent years, Lynch and colleagues have reported that a subset of families with a high incidence of nonpolyposis colon cancer often develop malignancies in other sites, particularly the endometrium, gastric tract, and ovaries (Lynch et al. 1966; Lynch and Lynch 1993). Genetic predisposition to this constellation of tumors is inherited in an autosomally dominant fashion (Lynch et al. 1991a,b) and maps to a mismatch repair gene (*hMSH-2*) located on chromosome 2 (Fishel et al. 1993; Leach et al. 1993; Peltomäki et al. 1993b). Mutations in this gene cause widespread instability in repeated microsatellite sequences in the DNA of colon tumors in patients with HNPCC (Aaltonen et al. 1993; Ionov et al. 1993; Thibodeau et al. 1993). This instability is not limited to any one chromosomal location nor to a small subset of microsatellite markers. Instead, many microsatellite loci, scattered throughout the entire genome, display both heterogeneity and instability. On the basis of these data, it seems that, in tumors arising in HNPCC patients, the total number of mutations at microsatellite loci alone could approach 100,000 per cell. Predisposition to HNPCC is therefore associated with a susceptibility to replication errors.

The *hMSH-2* gene is homologous in both structure and function to the bacterial and yeast mismatch repair genes, *mutS* and *msh-2* (Fishel et al. 1993; Leach et al 1993). Haploid yeast or bacterial cells carrying mutant forms of these genes are defective at an early step in strand-specific mismatch repair (Strand et al. 1993), as are tumor cell lines established from HNPCC patients (Parsons et al. 1993). Genetic instability has been observed in a number of other types of cancer, including pancreatic, gastric, bladder, skin, and endometrial (Gonzalez-Zulueta et al. 1993; Han et al. 1993; Peltomäki et al. 1993a; Risinger et al. 1993; Honchel et al. 1994; Mironov et al. 1994; Rhyu et al. 1994). The incidence previously reported for ovarian tumors is very low (< 5%) (Han et al. 1993; Osborne and Leach 1994; Wooster et al. 1994). In this paper, however, we show (1) that cell lines established from ovarian cancers display a high incidence of genetic instability, (2) that genetic instability is a dynamic and ongoing process, and (3) the instability observed in one ovarian cancer cell line and its parental tumor is a consequence of a loss of heterozygosity which exposes a mutant allele of the *hMSH-2* gene.

METHODS

Cell lines. Characteristics of ovarian tumor cell lines are listed in Table 1.

Primers. Characteristics of primer sets used for microsatellite analysis are listed in Table 2 (Weber et al. 1990; Weber and May 1990; Beckman et al. 1991; Sakurai et al. 1991; van Leeuwen et al. 1991; Fountain et al. 1992; Jones and Nakamura 1992; Kwiatkowski and Diaz 1992; Weissenbach et al. 1992).

Analysis of polymorphic microsatellite markers. DNA was purified from confluent cultures of cells as described previously (Gross-Bollard et al. 1973). Microsatellites were analyzed as described previously (Wooster et al. 1994).

Clonal analysis. Single cells of the 2774 cell line

Present addresses: §Department of Internal Medicine, Division of Molecular Medicine and Genetics, University of Michigan Medical Center, Ann Arbor, Michigan 48109; **Peter MacCallum Cancer Institute, St. Andrews Place, East Melbourne, Victoria 3002, Australia.

Table 1. Characteristics of Ovarian Tumor Cell Lines

Cell line	Year estab.	Age of patient	Histology of tumor	Source	Growth media[a]
2274	1974	38	moderately differentiated serous cystadenocarcinoma	P. DiSaia U.C. Irvine	DMEM
SK-OV-3	1973	64	moderately to well differentiated cystadenocarcinoma	ATCC[b]	DME/F-12
UCI 107	1991	48	moderately differentiated serous papillary adenocarcinoma	A. Manetta U.C. Irvine	RPMI
222	1988	75	serous papillary cystadenocarcinoma	B. Bonavida U.C.L.A.	RPMI
UCI 101	1989	46	moderately to poorly differentiated papillary cystadenocarcinoma	A. Manetta U.C. Irvine	RPMI
2008[c]	1970s	—	serous cystadenocarcinoma	P. DiSaia U.C. Irvine	RPMI
OVCAR3	1974	46	poorly differentiated papillary cystadenocarcinoma	ATCC	DME/F-12
PA-1	1974	12	teratocarcinoma	ATCC	DME/F-12
Caov-3	1976	54	adenocarcinoma	ATCC	DME/F-12
Caov-4	1976	45	adenocarcinoma	ATCC	DME/F-12

[a] Media supplemented with 10% fetal bovine serum.
[b] American Type Culture Collection.
[c] Limited information on cell line 2008.

were isolated by limiting dilution as follows. 200 μl of a suspension containing about 2 viable 2774 cells/ml was dispensed into each well of a 96-well microtiter dish and incubated until the colonies were large enough to seed T-25 tissue culture flasks (~7 weeks). Confluent monolayers from the T-25 flasks were expanded into three 90-mm petri dishes. At confluency (~5×10^6 cells/plate), DNA was isolated (Gross-Bollard et al. 1973) from two 90-mm plates (zero time point). The third 100-mm plate was split into three 90-mm plates. Cells were passed for 10 weeks (80–90 cell doublings), with DNA isolated at the third, sixth, and tenth weeks. Cells from four of the original clonal lines at 10 weeks were subcloned at 10 weeks and expanded into 90-mm petri dishes by the method described above. DNA was then isolated from these expanded subclonal lines.

Table 2. Characteristics of Primer Sets Used for Microsatellite Analysis

Marker	Location	Sequence of primers	Reference
D2S123	2p	GGACTTTCCACCTATGGGAC AAACAGGATGCCTGCCTTTA	Weissenbach et al. (1992)
hTPO	2p23	CACTAGCACCCAGAACCGTC CCTTGTCAGCGTTTATTTGCC	Anker et al. (1992)
D3S1283	3p21.3	GGCAGTACCACCTGTAGAAATG GAGTAACAGAGGCATCGTGTATTC	Weissenbach et al. (1992)
C13-373	3p13	CTGCAAGGTCTGTTTAACAG ATTCCAGGGACAAGTTCCCC	Jones and Nakamura (1992)
D5S299	5q15-23	GTAAGCAGGACAAGATGACAG GCTATTCTCTCAGGATCTTG	van Leeuwen et al. (1991)
IFNA	9p13-22	TGCGCGTTAAGTTAATTGGTT GTAAGGTGGAAACCCCCACT	Kwiatkowski and Diaz (1992)
D9S126	9p13-22	ATTGAAACTCTGCTGAATTTTCTG CAACTCCTCTTGGGAACTGC	Fountain et al. (1992)
WT-1	11p13	AATGAGACTTACTGGGTGAGG TTACACAGTAATTTCAAGCAACGG	D. Haber (pers. comm.)
D15S169	15	CAGGAGAGAGCCTTGGAT GAGACATCTCTTCTGAAAGCTC	Beckmann et al. (1991)
D15S171	15	GTGACAGCAGAGAGGCCG TGATGTCTATTGAGCCCTCA	Beckmann et al. (1991)
Thra-1	17q12	AATGTCTTAAGCAGTGGGGAACC ACAGGTGACAGCGGGTGCTCCTC	Sukurai et al. (1991)
D17S261	17p12-11.1	CAGGTTCTGTCATAGGACTA TTCTGGAAACCTACTCCTGA	Weber et al. (1990)
D17S791	17q	GTTTCTCCAGTTATTCCCC GCTCGTCCTTTGGAAGAGTT	Weissenbach et al. (1992)
D18S34	18q12-21	CAGAAAATTCTCTCTGGCTA CTCATGTTCCTGGCAAGAAT	Weber and May (1990)

Sequence analysis of clonal repeats. The microsatellite located at D15S169 in the 2774 cell line was amplified with PCR primers 1591F (5'-CCG TGG CGG TCT GGT ACC CTG GCC AGT CTT CAG GA-3') and 1591R (5'-GTG CAA CGA CGG GAT CGA TCA GAT CTA GGA GAG AGC C) using the following conditions: 94°C for 4 min; 2 cycles of 94°C for 30 sec, 54°C for 30 sec, 72°C for 1 min; 30 cycles of 94°C for 30 sec, 70°C for 30 sec, 72°C for 1 min; 72°C for 4 min. The PCR product (~130 bp) was cut with *Kpn*I and *Cla*I, purified, and cloned into the pBC-SK⁻ (Stratagene). Cloned microsatellites were analyzed as described previously (Wooster et al. 1994) and sequenced with Sequenase Ver. 2.0 (USB).

Allelotyping of microdissected DNA. Minute samples of normal and tumor cells were precisely dissected from archived 22-year-old hematoxylin and eosin (H&E)-stained slides as described previously (Whetsell et al. 1992), with slight modifications. Briefly, microdissection was performed under direct observation with an inverted microscope using a microcapillary tube that had been pulled to a fine tip by a micropipette puller. The tip was introduced into the field of view at an angle of approximately 30°, and the movements in the x, y, and z axes were controlled by a joystick-operated hydraulic manipulator (Nikon-Narishige). The scraped cells (about 30 in each manipulation) were allowed to adhere to the microcapillary tip and collected in a microcentrifuge tube by breaking off the glass tip. This was followed by extraction of the DNA with proteinase K. The DNAs were stored at −20°C and allelotyped using the PCR-based assays as described previously (Wooster et al. 1994) with the following modifications. Reactions contained template DNA derived from approximately 100 microdissected cells. After initial denaturation at 94°C for 2 min, 25 cycles of PCR consisting of 10 sec at 94°C, 10 sec at the temperature appropriate for primer annealing, and 30 sec at 72°C, then 15 cycles consisting of 10 sec at 90°C, 10 sec at the annealing temperature, 30 sec at 72°C, with a final extension of 2 min at 72°C.

Sequence analysis. Direct cycle DNA sequencing was performed on PCR products amplified from cDNA copied from the *hMSH2* transcripts expressed in the 2774 cells as described previously (Leach et al. 1993). Primers for PCR analysis and sequencing of the mutated region in genomic DNA included 510F (5'-GAT TAAACTGGATTCCAGTGCAC-3') 540R (5'-CTA AAGTTTTTATTGTTACGAAGG-3') 548R (5'-GC TGTTGGTAAATTTAACACC-3').

RESULTS

Genetic Instability of Microsatellite Markers in Ovarian Cancer Cell Lines

In the course of PCR-based analysis of microsatellites in the ovarian cancer cell line, 2774, we observed instability at a number of markers containing short dinucleotide repeats. In contrast to well-behaved microsatellite markers, which typically appear in the standard PCR assay as a single distinct band with a faster running shadow (Litt 1991), microsatellite markers in DNA extracted from cultured 2774 cells are heterogeneous in size and often appear as a continuous ladder. Figure 1 shows an analysis of a typical marker (D17S79) in DNAs extracted from 2774 cells and nine other ovarian cancer cell lines (Table 1). The human histiocytic lymphoma cell line, U937, which was used as a control (lane 1), appears heterozygotic and genetically stable at the D17S791 locus, as do five of the ten ovarian cancer cell lines tested (lanes 5, 7, 8, 10, 11). In contrast, the larger of the two alleles at D17S791 in 2774 cells exhibits a ladder profile, although the smaller allele appears to be stable (lane 2). The 107 cell line exhibits a single ladder profile (lane 3), whereas the SK-OV-3, 101, and PA-1 cell lines display two diffuse ladders (lanes 2, 5, and 8, respectively) that reflect genetic instability at both alleles. Instability was not limited to D17S791 but was observed to varying degrees at the 14 additional dinucleotide markers used in this study, which are widely dispersed throughout the human genome (Fig. 2). Although some of the markers (e.g., 9p22-IFNA) appeared to retain some semblance

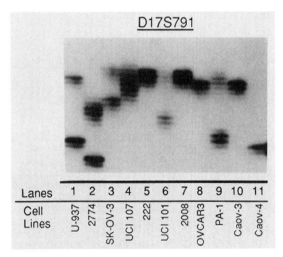

Figure 1. Genetic instability at D17S791 in ovarian tumor cell lines. DNA from 10 ovarian tumor cell lines was tested for genetic instability using the PCR-based assay. The human histiocytic lymphoma cell line U-937, used as a control, is heterozygous at this locus and exhibits two distinct alleles (lane *1*). The PCR-based assay produces a shadow band under each distinct allele. The ovarian cell lines 222 (lane *5*), 2008 (lane *7*), OVCAR3 (lane *8*), Caov-3 (lane *10*), and Caov-4 (lane *11*) appear genetically stable at D17S791. The other 5 ovarian tumor cell lines appear genetically unstable at this locus. The upper allele of 2774 is genetically unstable, whereas the lower allele exhibits only a distinct band (lane *2*); SK-OV-3 (lane *3*), UCI 101 (lane *6*), and PA-1 (lane *9*) are heterozygous at this locus, and both alleles appear genetically unstable; UCI 107 (lane *4*) appears genetically unstable at this locus, displaying one long ladder. (Reprinted, with permission, from Orth et al. 1994.)

cell lines \ marker	D2S123	D3S1283	C13-373	D5S299	9p22-IFNA	D9S126	WT-1	D15S169	D15S171	Thra-1	D17S261	D17S791	D18S34
U-937	-	-	-	-	-	-	-	-	-	-	-	-	-
2774	+	+	+	+	-	-	-	+	+	+	+	+	+
SK-OV-3	-	-	+	+	-	-	-	+	-	-	-	+	+
107	+	+	+	+	-	+	+	+	+	+	-	+	+
222	-	-	-	-	-	-	-	-	-	-	-	-	-
101	+	+	+	+	-	+	+	+	+	+	-	+	+
2008	-	-	-	-	-	-	-	-	-	-	-	-	-
OVCAR3	-	-	-	-	-	-	-	-	-	-	-	-	-
PA-1	-	+	+	+	-	+	+	+	+	+	+	+	+
Caov-3	-	-	-	-	-	-	-	-	-	-	-	-	-
Caov-4	-	-	-	-	-	-	-	-	-	-	-	-	-

Figure 2. Analysis of ovarian tumor cell lines with polymorphic microsatellites. Cell lines were assayed for genetic instability (+) or stability (-) with markers throughout the genome. (Reprinted, with permission, from Orth et al. 1994.)

of genetic respectability, the overall picture is one of genetic chaos. Instability in ovarian cancer may therefore be more prevalent than has previously been realized.

To ensure that the ladder profiles resulted from the expansion and contraction of repeated dinucleotide sequences, we cloned and analyzed the unstable D15S169 locus in 2774 DNA. As shown in Figure 3, each clonally isolated band in the ladder differed by multiples of two bases (i.e., by one dinucleotide repeat), which verified that instability was a result of expansion and/or contraction in the number of repeated sequences in microsatellites.

A Dynamic Process: Genetic Instability in Ovarian Cancer Cell Lines

We next wanted to find out whether the genetic instability observed in ovarian cancer cell lines was the result of a single cataclysmic event or was an ongoing, dynamic process. To distinguish between these alternatives, we isolated and analyzed ten clonal cell lines from the parental 2774 cell line. Whereas the uncloned parental cell line contained a complex mixture of alleles at each marker, the clonal sublines generally displayed simpler patterns consisting predominantly of two or three alleles (D5S299, Fig. 4A; D2S123, D15S168, D15S171, and D17S791, data not shown). Many of these alleles were represented in the uncloned cells. However, the subclones appeared to be enriched in rare alleles, which were present only at low frequency in the parental cell line ladder (Fig. 4A, lanes 6, 7, and 9). These results show that subcloning leads to a simplification and to an apparently biased sampling of the heterogeneous populations of alleles that are detectable in 2774 cells.

DNA isolated from 4 of the subclonal cell lines after 3, 6, and 10 weeks in culture was analyzed with markers for genetic instability. Changes were detected over time in only one clonal cell line (C1) at seven of nine microsatellite markers tested (markers changed: D2S123, C13-1136, D5299, C13S373, D15S169, D15S171, D17S791; markers unchanged: D3S1283, D3S1293) (Fig. 4B, lanes 3–6). These changes could have been due to genetic instability or, alternatively, from starting with a mixed cell line rather than a clonal cell line. To distinguish between these possibilities, another set of subclonal lines was established from the C1 clonal line at the 10-week time point and analyzed. The appearance of altered and novel alleles in the second set of subclonal lines confirmed that cells of the C1 line are genetically unstable at D17S791 (Fig. 4C) and at four other microsatellite markers (D2S123, D5S299, D15S169, D5S171; data not shown). At the heterozygous microsatellite marker D17S791 (Fig. 4C), the larger of the two alleles appears to be more unstable than the smaller allele, which was altered in only 1 of 50 different clonal and subclonal lines (Fig. 4C, lane 12). In the 3 remaining subclonal lines (C2, C3, C4) derived from 2774 cells, changes in allele size were not detected at nine different microsatellite markers during 10 weeks of growth in culture (an example for one marker is shown for C2; Fig. 4B). Although these clonal lines appear to be stable, serial subclonal analysis once again indicates otherwise. The second generation of subclonal lines contain alleles differing in size from those observed in the first generation (Fig. 4D, lanes 3–12) and, as before, alleles that are barely visible in the clonal lines are strongly represented in the subclon-

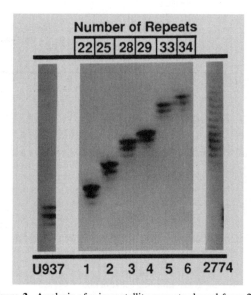

Figure 3. Analysis of microsatellite repeats cloned from 2774 cell line. The number of CA repeats for each plasmid (lanes 1–6) is listed above the microsatellite profile for each plasmid. Each band represents a difference in a multiple of one CA repeat.

Analysis of Normal, Tumor, and Cell Line 2774 DNA

The genetically unstable 2774 ovarian tumor cell line was derived from ascitic fluid from a patient who died in 1974, at age 38 (Freedman et al. 1978). Most of her medical records were destroyed in a hospital fire, but a few uninformative pages of clinical history survived along with two H&E pathology slides: one of normal epithelial tissue and one of tumor tissue. The 22-year-old pathology slides were precisely microdissected and DNA was extracted from about 1000 nuclei isolated from histologically normal tissue and from an equal number isolated from malignant tissue. The profiles of dinucleotide repeats in microsatellite markers were analyzed for genetic instability in the DNAs prepared from normal tissue, tumor tissue, and uncloned cultures of 2774 cells (Fig. 5). Figure 5A shows that both the normal (N, lane 1) and the tumor (T, lane 2) DNA loci are homozygous and stable at D17S261, but the cell line DNA is unstable, as demonstrated by the presence of a ladder profile (CL, lane 3). The normal DNA (lane 1) is homozygous and stable at D3S1283 (Fig. 5B), whereas the tumor DNA (lane 2) appears to be genetically unstable, albeit to a lesser degree than DNA isolated from 2774 cells (lane 3). In Figure 5C, the normal DNA at the D2S123 locus appears to be heterozygous (two alleles differing by single repeat) and genetically stable, whereas the profile of the tumor DNA suggests both genetic instability and a loss of heterozygosity. The DNA isolated from 2774 cells at this locus displays laddering and is therefore genetically unstable. Figure

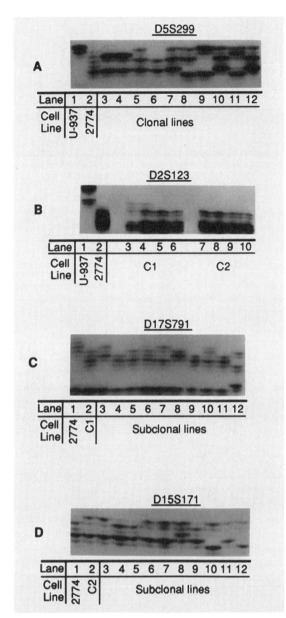

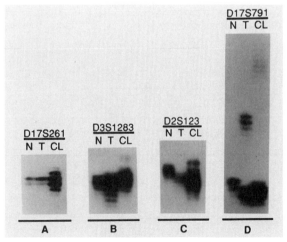

Figure 4. (*A*) Analysis at D5S299 in U-937 cell line (lane *1*), 2774 cell line (lane *2*), and 2774 clonal cell lines (lane *3–12*). (*B*) Analysis at D2S123 in 2774 clonal cell lines C1 (lanes *3–6*) and C2 (lanes *7–10*) over 10 weeks. Controls U-937 (lane *1*) and 2774 (lane *2*) cell lines are shown for this polymorphic microsatellite. (*C*) Analysis at D17S791 in C1 subclonal cell lines (lanes *3–12*) made from clonal C1 (lane *2*) at the 10-week time point. The profile for the 2774 cell lines is shown in lane *1*. (*D*) Analysis at D15S171 in C2 subclonal cell lines (lanes *3–12*) made from clonal C2 (lane *2*) at the 10-week time point. The profile for the 2774 cell lines is shown in lane *1*. (Reprinted, with permission, from Orth et al. 1994.)

al lines. These results show that genetic instability is a continuing and dynamic process and that subclonal analysis may be required to reveal genetic heterogeneity in populations of cultured tumor cells.

Figure 5. Analysis of normal, tumor, and cell line DNA at (*A*) D17S261, (*B*) D3S1283, (*C*) D2S123, and (*D*) D17S791. Normal epithelial and tumor cells were precisely microdissected from 22-year-old archival H&E slides and DNA was extracted. PCR-amplified fragments were prepared from normal (N), tumor (T), and cell line (CL) DNA, and separated by polyacrylamide gel electrophoresis, followed by autoradiography. (Reprinted, with permission, from Orth et al. 1994.)

5D shows that the normal DNA is homozygous and stable at D17S791, and that this locus is heterozygotic and genetically unstable in both the tumor and the cell line DNAs: One allele appears to be reduced in size and to be stable (compare with Fig. 4C), whereas the other allele appears to have become larger and less stable (note the dramatic difference in the normal, tumor, and cell line DNA for this allele). From these analyses, we conclude that the genetic instability observed in the DNA of the 2774 cell line is the result of a dynamic process that had originated in the primary tumor and was not an artifact of the in vitro environment of tissue culture. Furthermore, the multiple alleles detected in 2774 cells are the result of genetic instability in individual cells rather than massive polyploidy in the cell population. This conclusion is supported by karyotype analysis which demonstrates (1) that the 2774 cell line is triploid ($n = 60–66$) (Freedman et al. 1978; reconfirmed at UTSWMC, 1993) and (2) that two other genetically unstable ovarian tumor cell lines (107 and 101) are diploid with median chromosome numbers of $n = 46$ (Fuchtner et al. 1993; and data not shown). Interestingly, the karyotype of the 101 cell line shows a deletion on the short arm of chromosome 2, where the *hMSH-2* gene is located.

Mutation Identified in *hMSH2* Locus

Because mutations in the *hMSH-2* gene have been linked to genetic instability in $(CA)_n$ microsatellites (Fishel et al. 1993; Leach et al. 1993; Parsons et al. 1993), we investigated whether the *hMSH-2* gene might also be involved in the genetic instability observed in ovarian tumor cell lines. We observed a loss of heterozygosity in DNAs of both the tumor and 2774 cells (Fig. 6) at a polymorphic tetranucleotide repeat marker (human thyroid peroxidase [hTPO]; 2p23; Anker et al. 1992) and at the D2S123 locus (Fig. 5C). Both of these markers are located near the *hMSH-2* gene (2p16),

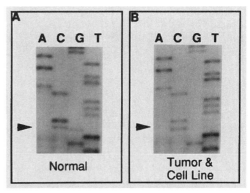

Figure 7. Sequence analysis (sense strand; 5' to 3' and top to bottom, respectively) of normal (*A*) and tumor and cell line (*B*) genomic DNA surrounding codon 524 (CGT). The normal DNA is heterozygous (CG/CT) at position two of codon 524 (marked by the arrow). The tumor and cell line DNA have lost the normal allele (CGT) and contain only the mutant allele (CTT) (marked by the arrow). (Reprinted, with permission, from Orth et al. 1994.)

which implies that loss of the 2p region is involved in tumorigenesis. Using reverse transcriptase-polymerase chain reaction (RT-PCR), cDNA fragments of the *hMSH-2* mRNA from the 2774 cell line were prepared and the *hMSH-2* cDNA was then sequenced by direct cycle sequencing (Leach et al. 1993). The *hMSH-2* sequence from this cell line was identical to the *hMSH-2* cDNA sequence of Leach et al. (1993) with the exception of codon 524, which was mutated from CGT to CCT. This transition mutation results in the change of a highly conserved arginine residue (conserved in the yeast homolog) to a proline residue (Reenan and Kolodner 1992; Leach et al. 1993).

Intron/exon borders flanking this mutation were identified in the genomic 2774 DNA using PCR and direct cycle sequencing of the amplified fragments. The exon containing the mutation was amplified from normal, tumor, and 2774 DNA and sequenced directly with a nested primer (540R). The DNA sequence of the exon amplified from microdissected tumor DNA was identical to that amplified from the 2774 cell line (Fig. 7B), whereas the sequence of the exon amplified from microdissected normal DNA exhibited heterozygosity at codon 524, with one wild-type allele (CGT) and one mutant allele (CCT) (Fig. 7A). On the basis of these results, we propose (1) that the onset or progression of the ovarian tumor in patient 2774 was the consequence of loss of heterozygosity at the *hMSH-2* locus, (2) that the mutant form of the gene codes for a protein that is unable to carry out mismatch repair, and (3) genetic instability observed in both the 2774 tumor and the cell line from which it was derived are a consequence of this lack of repair.

DISCUSSION

We have demonstrated that five of ten ovarian tumor cell lines are genetically unstable and that instability is a dynamic and ongoing process. DNA from the tumor

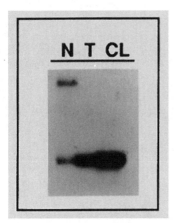

Figure 6. Analysis of the tetranucleotide marker hTPO (2p23) in normal (N), tumor (T), and cell line (CL) DNA. The normal DNA is heterozygous at this locus, whereas the tumor and cell line DNA display a loss of heterozygosity. (Reprinted, with permission, from Orth et al. 1994.)

that was the progenitor of one of these cell lines (2774) was also genetically unstable. Both the cell line and the tumor DNAs encode a mutated allele of the *hMSH-2* gene, whereas the normal DNA, precisely dissected from an archival H&E-stained slide, encodes both the wild-type and mutated alleles. The substitution of a proline residue for the conserved arginine at codon 524 is likely to introduce a turn in the polypeptide chain just upstream of the predicted nucleotide binding site in the *hMSH-2* protein. Because both the tumor and cell line DNA encode only the mutant *hMSH-2* allele, we suggest that the instability observed in 2774 is a result of replication/repair errors that stem from misfolding of the MSH-2 protein. Confirmation of the relationship between the dominant germ-line transmission of susceptibility to ovarian cancer and mutations in mismatch repair genes must await future studies on HNPCC kindreds. However, the evidence presented here strongly suggests that patient 2774 suffered and died from Lynch syndrome II ovarian cancer as result of a germ-line mutation at the *hMSH-2* locus.

It follows that tumor-susceptible tissues in HNPCC patients (colon, endometrium, gastric epithelium, and ovaries) must express target genes that are susceptible to this form of self-induced mutagenesis or mutator phenotype. Tumorigenesis is a multistep process that requires a minimum of four to five mutations in specific genes (Foulds 1958; Fearon and Vogelstein 1991). It is reasonable to believe that absence of a mismatch repair function might both accelerate this process and target specific sets of genes that are active in susceptible cells and tissues (Loeb 1991). However, because the penetrance of HNPCC is not 100%, some alleles of these target genes (i.e., tumor suppressor and/or oncogenes) may be more susceptible to genetic damage than others.

Genetic instability has previously been observed in approximately 5% of solid ovarian tumors (Han et al. 1993; Osborne and Leach 1994; Wooster et al. 1994). However, our results show that a much higher percentage (50%) of cell lines derived from ovarian tumors are genetically unstable. The disparity between these observations may be a result of selective advantage of genetically unstable cell lines in tissue culture. Alternatively, genetic instability may be more prevalent in ovarian tumors than had previously been thought, a hypothesis that is consistent with the observation that minor alleles in a heterogeneous population may be obscured in the PCR-based assay. Three of four clonal cell lines, which had been passaged 10 weeks, initially appeared to be genetically stable. However, subclonal analysis revealed that these clonal cell lines were, in fact, genetically unstable at all of the markers analyzed. Alleles in the clonal cell line that were barely detectable by the PCR-based assay were isolated in the subclonal cell lines. In addition, when we used precisely microdissected tumor DNA (~100% pure tumor cells) and the cell line DNA (~100% transformed cells), we detected genetic instability at the majority of markers analyzed. Thus, if bulk tumor tissue or crudely dissected tissue that contains a high proportion of normal tissue is used in the analysis of genetic instability, the results may be misleading due to the limited sensitivity of the PCR-based assay. This issue can be decided by further allelotyping of DNA isolated from microdissected specimens.

ACKNOWLEDGMENTS

We thank John Minna, Gail Tomlinson, Mike Lovett, and Ramon Parsons for their helpful conversations. We thank Joseph Lucci for help with obtaining the 2774 H&E slides. We thank Nancy Schneider for karyotype analysis. We thank Jill Speaker and Michele Williams for their excellent technical assistance. This work has been supported by grants from the National Institutes of Health.

REFERENCES

Aaltonen, L.A., P. Peltomäki, F.S. Leach, P. Sistonen, L. Pylkkänen, J.P. Mecklin, J. Järvinen, S.M. Powell, J. Jen, S.R. Hamilton, G.M. Petersen, K.W. Kinzler, B. Vogelstein, and A. de la Chapelle. 1993. Clues to the pathogenesis of familial colorectal cancer (see comments). *Science* **260**: 812.

Anker, R., T. Steinbrueck, and H. Donis-Keller. 1992. Tetranucleotide repeat polymorphism at the thyroid peroxidase (hTPO) locus. *Hum. Mol. Genet.* **1**: 137.

Beckman, J.S., I. Richard, D. Hillaire, O. Broux, C. Antignac, E. Bois, H. Cann, R.W. Cottingham, N. Feingold, and J. Feingold. 1991. A gene for limb-girdle muscular dystrophy maps to chromosome 15 by linkage. *C.R. Acad. Sci. Ser. III* **312**: 141.

DiSaia, P. and W. Creasman. 1993. *Clinical gynecologic oncology*. Mosby-Year Book, St. Louis, Missouri.

Fearon, E. and B. Vogelstein. 1991. A genetic model for colorectal tumorigenesis. *Cell* **61**: 759.

Fishel, R., M.K. Lescoe, M.R. Rao, N.G. Copeland, N.A. Jenkins, J. Garber, M. Kane, and R. Kolodner. 1993. The human mutator gene homolog *MSH2* and its association with hereditary nonpolyposis colon cancer. *Cell* **75**: 1027.

Foulds, L. 1958. The natural history of cancer. *J. Chronic Dis.* **8**: 2.

Fountain, J.W., M. Karayiogou, D. Taruscio, S.L. Graw, A.J. Buckler, D.C. Ward, N.C. Dracopoli, and D.E. Housman. 1992. Genetic and physical map of the interferon region on chromosome 9p. *Genomics* **14**: 105.

Freedman, R.S., E. Pihl, C. Kusyk, H.S. Gallager, and F. Rutledge. 1978. Characterization of an ovarian carcinoma cell line. *Cancer* **42**: 2352.

Fuchtner, C., D.A. Emma, A. Manetta, G. Gamboa, R. Bernstein, and S.Y. Liao. 1993. Characterization of a human ovarian carcinoma cell line: UCI 101. *Gynecol. Oncol.* **48**: 203.

Gonzalez-Zulueta, M., J.M. Ruppert, K. Tokino, Y.C. Tsai, C. Spruck, N. Miyao, P.W. Nichols, G.G. Hermann, T. Horn, and K. Steven. 1993. Microsatellite instability in bladder cancer. *Cancer Res.* **53**: 5620.

Gross-Bollard, M., P. Oudet, and P. Chambon. 1973. Isolation of high-molecular-weight DNA from mammalian cells. *Eur. J. Biochem.* **36**: 32.

Han, H.-J., A. Yanagisawa, Y. Kato, J.-G. Park, and Y. Nakamura. 1993. Genetic instability in pancreatic and poorly differentiated type of gastric cancer. *Cancer Res.* **53**: 5087.

Hauser, I.L. and C.V. Weller. 1936. A further report on the cancer family of Warthin. *Am. J. Cancer* **27**: 434.

Honchel, R., K. Halling, D. Schaid, M. Pittelkow, and S.

Thibodeau. 1994. Microsatellite instability in Muir-Torre syndrome. *Cancer Res.* **54:** 1159.

Ionov, Y., M.A. Peinado, S. Malkhosyan, D. Shibata, and M. Perucho. 1993. Ubiquitous somatic mutations in simple repeated sequences reveal a new mechanism for colonic carcinogenesis. *Nature* **363:** 558.

Jones, M.H. and Y. Nakamura. 1992. Deletion mapping of chromosome 3p in female genital tract malignancies using microsatellite polymorphisms. *Oncogene* **7:** 1631.

Kwiatkowski, D.J. and M.O. Diaz. 1992. Dinucleotide repeat polymorphism at the IFNA locus (9p22). *Hum. Mol. Genet.* **1:** 658.

Leach, F.S., N.C. Nicolaides, N. Papadopoulos, B. Liu, J. Jen, R. Parsons, P. Peltomaki, P. Sistonen, L.A. Aaltonen, M. Nystrom-Lahti, X.-Y. Guan, J. Zhang, P.S. Meltzer, J.-W. Yu, F.-T. Kao, D.J. Chen, K.M. Cerosaletti, R.E. Fournier, S. Todd, T. Lewis, R.J. Leach, S.L. Naylor, J. Weissenbach, J.-P. Mecklin, H. Jarvinen, G.M. Petersen, S.R. Hamilton, J. Green, J. Jass, P. Watson, H.T. Lynch, J.M. Trent, A. de La Chapelle, and B. Vogelstein. 1993. Mutations of a *mutS* homolog in hereditary nonpolyposis colorectal cancer. *Cell* **75:** 1215.

Litt, M. 1991. PCR of TG microsatellites. In *PCR—A practical approach* (ed. M.J. McPherson et al.), p. 85. Oxford University Press, New York.

Loeb, L.A. 1991. Mutator phenotype may be required for multistage carcinogenesis. *Cancer Res.* **51:** 3075.

Lynch, H.T. and J.F. Lynch. 1993. The Lynch syndromes. *Curr. Opin. Oncol.* **5:** 687.

Lynch, H.T., T. Conway, and J. Lynch. 1991a. Hereditary ovarian cancer (pedigree studies): Part II. *Cancer Genet. Cytogenet.* **52:** 161.

Lynch, H.T., M.W. Shaw, C.W. Magnuson, A.L. Larsen, and A.J. Krush. 1966. Hereditary factors in cancer. *Arch. Intern. Med.* **117:** 206.

Lynch, H.T., S. Lanspa, T. Smyrk, B. Boman, P. Watson, and J. Lynch. 1991b. Hereditary nonpolyposis colorectal cancer (Lynch syndrome I and II): Part I. *Cancer Genet. Cytogenet.* **53:** 143.

Lynch, H.T., P. Watson, C. Bewtra, T.A. Conway, C.R. Hippee, P. Kaur, J.F. Lynch, and B.A. Ponder. 1991c. Hereditary cancer. Heterogeneity in age at diagnosis. *Cancer* **67:** 1460.

Mironov, N.M., A.-M. Aguelon, G.I. Potapova, Y. Omori, O.V. Gorbunov, A.A. Klimenkov, and H. Yamasaki. 1994. Alterations of (CA)n DNA repeats and tumor suppressor genes in human gastric cancer. *Cancer Res.* **54:** 41.

Orth, K., J. Hung, A. Gazdar, A. Bowcock, M. Mathis, and J. Sambrook. 1994. Genetic instability in human ovarian cancer cell lines. *Proc. Natl. Acad. Sci.* (in press).

Osborne, R.J. and V. Leach. 1994. Polymerase chain reaction allelotyping of human ovarian cancer. *Br. J. Cancer* **69:** 429.

Parsons, R., G.M. Li, M.J. Longley, W.H. Fang, N. Papadopoulos, J. Jen, A. de la Chapelle, K.W. Kinzler, B. Vogelstein, and P. Modrich. 1993. Hypermutability and mismatch repair deficiency in RER$^+$ tumor cells. *Cell* **75:** 1227.

Peltomäki, P., R.A. Lothe, L.A. Aaltonen, L. Pylkkänen, M. Nyström-Lahti, R. Seruca, L. David, R. Holm, D. Ryberg, A. Haugen, A. Brøgger, A.-L. Børresen, and A. de la Chapelle. 1993b. Microsatellite instability is associated with tumors that characterize the hereditary non-polyposis colorectal carcinoma syndrome. *Cancer Res.* **53:** 5853.

Peltomäki, P., L.A. Aaltonen, P. Sistonen, L. Pylkkänen, J.P. Mecklin, H. Järvinen, J.S. Green, J.R. Jass, J.L. Weber, F.S. Leach, G.M. Petersen, S.R. Hamilton, A. de la Chapelle, and B. Vogelstein. 1993b. Genetic mapping of a locus predisposing to human colorectal cancer. *Science* **260:** 810.

Reenan, R.A.G. and R.D. Kolodner. 1992. Isolation and characterization of two *Saccharomyces cerevisiae* genes encoding homologues of the bacterial HexA and MutS mismatch repair proteins. *Genetics* **132:** 963.

Rhyu, M.G., W.S. Park, and S.J. Meltzer. 1994. Microsatellite instability occurs frequently in human gastric carcinoma. *Oncogene* **9:** 29.

Risinger, J.I., A. Berchuck, M.F. Kohler, P. Watson, H.T. Lynch, and J. Boyd. 1993. Genetic instability of microsatellites in endometrial carcinoma. *Cancer Res.* **53:** 5100.

Rubin, S.C. and G.P. Sutton. 1993. *Ovarian cancer*. McGraw-Hill, New York.

Sakurai, A., G.I. Bell, and L.J. DeGroot. 1991. Dinucleotide repeat polymorphism in the human thyroid hormone receptor a gene (*THRA1*) on chromosome 17. *Hum. Mol. Genet.* **1:** 5.

Strand, M., T.A. Prolla, R.M. Liskay, and T.D. Petes. 1993. Destabilization of tracts of simple repetitive DNA in yeast by mutations affecting DNA mismatch repair (see comments). *Nature* **365:** 274.

Thibodeau, S.N., G. Bren, and D. Schaid. 1993. Microsatellite instability in cancer of the proximal colon. *Science* **260:** 816.

van Leeuwen, C., C. Tops, C. Breukel, H. van der Klift, L. Deaven, R. Fodde, and P.M. Khan. 1991. CA repeat polymorphism within the MCC (mutated in colorectal cancer) gene. *Nucleic Acids Res.* **19:** 5805.

Weber, J.L. and P.E. May. 1990. Dinucleotide repeat polymorphism at the D18S34 locus. *Nucleic Acids Res.* **18:** 2201.

Weber, J.L., A.E. Kwitek, P.E. May, M.R. Wallace, F.S. Collins, and D.H. Ledbetter. 1990. Dinucleotide repeat polymorphisms at the D17S250 and D17S261 loci. *Nucleic Acids Res.* **18:** 4640.

Weissenbach, J., G. Gyapay, C. Dib, A. Vignal, J. Morissette, P. Millasseau, G. Vaysseix, and M. Lathrop. 1992. A second-generation linkage map of the human genome. *Nature* **359:** 794.

Whetsell, L., G. Maw, N. Nadon, D.P. Ringer, and F.V. Schaefer. 1992. Polymerase chain reaction microanalysis of tumors from stained histological slides. *Oncogene* **7:** 2355.

Wooster, R., A.-M. Cleton-Jansen, N. Collins, J. Mangion, R.S. Cornelis, C.S. Cooper, B.A. Gusterson, B.A.J. Ponder, A. vonDeimling, O.D. Wiestler, C.J. Cornelisse, P. Devilee, and M.R. Stratton. 1994. Instability of short tandem repeats (microsatellites) in human cancers. *Nat. Genet.* **6:** 152.

Loss of Imprinting in Human Cancer

A.P. FEINBERG, L.M. KALIKIN, L.A. JOHNSON, AND J.S. THOMPSON
Departments of Medicine, Molecular Biology and Genetics, and Oncology, Johns Hopkins University School of Medicine, Baltimore, Maryland 21205

Genomic imprinting in mammalian development is functionally defined as parental-origin-specific differential allele expression (Monk 1988). In contrast, the classic definition by the insect geneticist Crouse is a parental-origin-specific modification of the chromosome in the germ line (Crouse 1960). Some confusion has arisen in mouse genetics between a chromosome-centered or function-centered definition (for a thoughtful review, see Barlow 1994). Thus, imprinting in mice was first found adventitiously by both parental-origin-specific DNA methylation and monoallelic expression of transgenes (Reik et al. 1987; Swain et al. 1987). Some transgenes have shown differential methylation but not parental-origin-specific expression (Surani et al. 1988; Reik et al. 1990). Some endogenous genes have exhibited parental-origin-specific monoallelic expression in mice, as well as differential allele modification (although the developmental timing of allelic modification and monoallelic expression do not always correspond). For example, modification of the insulin-like growth factor-II receptor gene (*IGF2R*) occurs prezygotically (Stoger et al. 1993), but modification of the *H19* gene occurs at the time of implantation, after establishment of parental-origin-specific expression (Bartolomei et al. 1993; Ferguson-Smith et al. 1993). These differences between expression and modification have created controversy in the field, which we think can be resolved, at least in part, by viewing imprinting from the perspective of regulation of chromosomal domains rather than from the classic perspective of individual genes.

Our laboratory's interest in genomic imprinting derived from our studies of childhood cancer. We had mapped loss of heterozygosity (LOH) in Wilms' tumor (WT), a childhood kidney cancer, to 11p15, not 11p13 which harbors *WT1* (Reeve et al. 1989). Also mapping to 11p15 are the human homologs of two imprinted mouse genes, insulin-like growth factor-II (*IGF2*), which in man is an important autocrine growth factor in cancer (El-Badry et al. 1990; Thompson et al. 1990), and *H19*, a gene that apparently acts as an RNA and may inhibit cell growth when overexpressed in vivo (Brunkow and Tilghman 1991) and in vitro (Hao et al. 1993). A peculiar feature of 11p15 LOH is the preferential loss of maternal alleles in WT and other childhood tumors (Schroeder et al. 1987), suggesting that *WT2* might be imprinted (i.e., differentially expressed on maternal and paternal chromosomes).

We had also mapped by genetic linkage analysis a genetic syndrome predisposing to childhood cancer, to the same region of 11p15 (Ping et al. 1989). Beckwith-Wiedemann syndrome (BWS) is an autosomal dominant disorder of abnormally increased growth, characterized by an enlarged kidney, liver, and other abdominal viscera, hemihypertrophy (enlargement of one side of the body), macroglossia (large tongue), neonatal hypoglycemia (due to pancreatic hyperplasia), craniofacial and ear anomalies, genitourinary malformations, and occasionally, mental retardation (Engstrom et al. 1988). BWS patients are at increased risk of several childhood malignancies, including WT, rhabdomyosarcoma, and hepatoblastoma (Wiedemann 1983). Most cases of BWS occur sporadically, although some families show dominant inheritance with incomplete penetrance (Aleck and Hadro 1989). What is most striking about these families is that the penetrance is affected by the gender of the parent from whom the gene is inherited, with most cases being transmitted from the mother (Viljoen and Ramesar 1992). Furthermore, Junien's laboratory showed that some BWS patients have uniparental disomy, or two paternal copies, of 11p15 (Henry et al. 1991). These studies suggested that one or more genes on 11p15 were imprinted, resulting in parental-origin-specific gene expression. We hypothesized that altered imprinting of *IGF2* or *H19* could explain the non-Mendelian patterns of these forms of cancer.

METHODS

To assay for genomic imprinting, it was necessary to determine not just which patients were heterozygous for a given polymorphism, but also which alleles were transcribed into RNA. For *IGF2*, there was already a known transcribed *Apa*I polymorphism (Tadokoro et al. 1991), but the fraction of informative individuals is relatively small. To examine a relatively large number of tumors, polymerase chain reaction (PCR) primers were designed to amplify a dinucleotide repeat (DR), but initially we saw no variation in length among normal individuals. However, the DR is unusually large (800 bp), and most polymorphic DRs show length differences of only a few nucleotides, which would likely be unresolvable with this size fragment. Because of the repetitive nature of the sequence, it was not possible to design PCR primers that would amplify only a portion of the DR. However, there is a unique *Mvn*I site in the approximate middle of the DR, allowing digestion of the PCR product into two segments of approximately 400 nucleotides each and revealing a highly informative transcribed polymorphism. Both the 5' and 3' ends of the dinucleotide repeat are polymorphic, with a combined heterozygosity of 0.9 (Rainier et al. 1993). For *H19*, a published transcribed polymorphism was used (Zhang and Tycko 1992), but with primers that do not require nested amplification (Rainier et al. 1993).

To determine if either *IGF2* or *H19* was imprinted, RNA was extracted from 50 normal specimens, including 8 fetal tissues. Duplicate poly(A)$^+$ RNA samples were incubated in the presence and absence of reverse transcriptase (RT), and only those samples showing no amplification in the absence of RT were studied. As an additional control, reverse transcription was also performed with an RT-specific tailed primer, as described previously (Shuldiner et al. 1991), in order to ensure amplification of RNA. No amplification of DNA was observed in control experiments.

For Northern blots, RNA was extracted as described previously (Steenman et al. 1994) and electrophoresed on 1% agarose/5% formaldehyde gels, transferred to GeneScreen (DuPont), and hybridized with random-primed probes (Feinberg and Vogelstein 1983a). Gene expression was quantified on a PhosphorImager (Molecular Dynamics), and normalized for total mRNA by rehybridizing with the glyceraldehyde 6-phosphate dehydrogenase (GAPDH) gene. All of the blots were hybridized simultaneously with a given probe to preclude differences in probe quality (Steenman et al. 1994).

DNA methylation was assayed by digestion of 5 μg of genomic DNA with 20 units of a given methyl-cytosine-sensitive restriction endonuclease (e.g., *Hpa*II), followed by an additional 4 hours of incubation with another 20 units. The digested DNA was electrophoresed on 3% NuSieve agarose (FMC)/1% agarose gels, transferred to Hybond-N+ (Amersham), and hybridized with the appropriate random-primed probes.

RESULTS

To test for a role of genomic imprinting in cancer, we examined parental-specific allelic expression of genes on 11p15. In all 8 normal tissues informative for a transcribed polymorphism in *IGF2*, only one of the two polymorphic alleles was expressed as RNA. Blood specimens were obtained from parents of 6 individuals exhibiting monoallelic expression, and in each case, the expressed allele was paternal (Rainier et al. 1993). For example, in the experiment shown in Figure 1, the genomic DNA was polymorphic for *Apa*I, with allele **a** inherited from the patient's father and allele **b** from the patient's mother. Only allele **a** was expressed in normal tissue, indicating that the paternal allele was expressed exclusively. The patient was also heterozygous for the DR polymorphism, which verified paternal monoallelic expression.

Monoallelic expression of *H19* in human fetal tissues

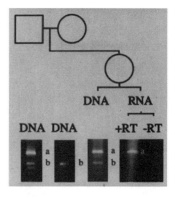

Figure 1. Imprinting of the human *IGF2* gene. Using a transcribed *Apa*I polymorphism (Tadokoro et al. 1991), two alleles are seen in DNA but only one in RNA. In 6 of 6 normal tissues examined, only the paternal *IGF2* allele was expressed (Rainier et al. 1993). (Reprinted, with permission, from Rainier et al. 1993 [copyright Macmillan].)

was reported earlier by Zhang and Tycko (1992), although in that study the parent of origin was unknown. To confirm monoallelic expression of *H19* in our tissues and to test the hypothesis of genomic imprinting of *H19*, we exploited the same polymorphism used by Zhang and Tycko at a transcribed *Rsa*I site (Zhang and Tycko 1992). However, novel primers were designed that permit a single round of PCR amplification. Consistent with the earlier study, *H19* showed monoallelic expression in all 10 normal tissues examined, including 3 fetal tissues, kidney, heart, and lung. In every case, it was always the maternal allele that was expressed (Fig. 2), indicating that the *H19* gene is imprinted (i.e., monoallelic expression is parental-origin-specific) (Rainier et al. 1993). Thus, both *H19* and *IGF2* are imprinted in man, with maternal expression of *H19* and paternal expression of *IGF2*. Two other laboratories

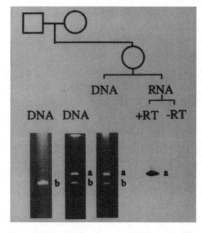

Figure 2. Imprinting of the human *H19* gene. Using a transcribed *Rsa*I polymorphism (Zhang and Tycko 1992), two alleles are seen in DNA but only one in RNA. In 10 of 10 normal tissues examined, only the maternal *H19* allele was expressed (Rainier et al. 1993). (Reprinted, with permission, from Rainier et al. 1993 [copyright Macmillan].)

also observed imprinting of *IGF2* (Ogawa et al. 1993b; Ohlsson et al. 1993).

Our initial experiments on imprinting in cancer were focused on WT, because these tumors express both *IGF2* and *H19*, and we had localized the BWS gene, which predisposes to WT, near *IGF2* and *H19*. Since approximately 30% of WTs undergo LOH of chromosome 11 (Mannens et al. 1988; Reeve et al. 1989), we examined 42 WTs to identify those retaining both parental chromosomes. We then extracted RNA from those tumors not showing LOH. Of 17 informative tumors retaining heterozygosity in the DNA, 12 expressed both maternal and paternal alleles of *IGF2* (77%), *H19* (29%), or both (Rainier et al. 1993). For example, as shown in Figure 3, both alleles of *IGF2* and *H19* were expressed. The tumors we examined included 6 from BWS patients, 5 of which showed biallelic expression of *IGF2*, and 1 of which showed biallelic expression of both *IGF2* and *H19*. Thus, biallelic expression appears to be a pathogenetic mechanism in the malignancies of BWS patients, as well. Reeve's laboratory also observed loss of imprinting (LOI) of *IGF2* in 3 samples (Ogawa et al. 1993b), bringing the total number studied (including our study) to 20. It is interesting that we observed LOI of *H19* only in cases in which *IGF2* also showed abnormal imprinting, suggesting that effects on *H19* could arise secondarily.

We have coined the term "loss of imprinting," or LOI, to refer to this novel type of mutation. By definition, LOI is the loss of parental-origin-specific gene expression, but it makes no prior assumption about the nature of the loss of this transcriptional specificity. LOI could include transcriptional activation or transcriptional repression (indeed, both may be at play, as discussed later). LOI could involve a loss of proper maintenance of imprinting, without necessarily the loss of an imprinting "mark" on the affected chromosome.

If LOI plays a causal role in neoplasia and/or in BWS, then one would predict that nontumor tissue of at least some patients should show LOI. Examination of imprinting in nontumor tissue in BWS was problematic, however (except in the case of normal tissue resected at the same time as tumor from these patients), since expressing tissues are not generally available. *IGF2* is expressed from skin fibroblasts but not significantly from lymphocytes, even as detected by PCR, and given the emotional barrier to families of skin biopsy of BWS children, we were initially limited to a small number of resected tissues from BWS patients. These included 5 adjacent nontumor kidneys and one resected tongue specimen. Two of these did show LOI. For example, as shown in Figure 4, both alleles of *IGF2* were expressed from the nontumor kidney of a BWS patient with Wilms' tumor (Steenman et al. 1994). We carefully excluded the presence of premalignant nephrogenic rests in this sample by histopathological examination. Additionally, we found LOI in a resected tongue. Both of these tissues are enlarged in BWS, so LOI could be linked to the pathogenesis of organ overgrowth. Others have also found LOI in fibroblasts of some BWS patients (Weksberg et al. 1993), and LOI has also been found in a large child with Wilms' tumor who showed none of the dysmorphic features of BWS (Ogawa et al. 1993a).

Another prediction based on a role of LOI in cancer would be measurable alterations in the expression of affected genes. To test this hypothesis, WTs with LOI were compared to WTs without LOI for quantitative expression of *IGF2* and *H19*. WTs cannot be compared to "normal kidney" of the same patient, because WT arises from persistent fetal tissue and is developmentally unlike the kidney from which it is removed. Thus, WTs with LOI were compared to WTs with normal imprinting.

WTs with LOI expressed an expected 2-fold higher level of *IGF2* than WTs without LOI, although these differences were not statistically significant, probably because of the difficulty in matching suitable control tissue noted above (Steenman et al. 1994). However, it

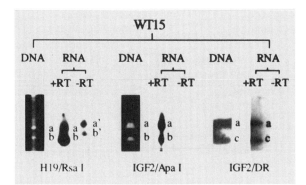

Figure 3. Loss of imprinting (LOI) in human cancer. Biallelic expression of *IGF2* was seen in 77% of Wilms' tumors (Rainier et al. 1993). WT15 is an unusual tumor showing LOI of both *IGF2* and *H19*, and it also illustrates that one can distinguish between biallelic expression and amplification of DNA (deliberately included in lane 3), because of the presence of an intron between the primers. (Reprinted, with permission, from Rainier et al. 1993 [copyright Macmillan].)

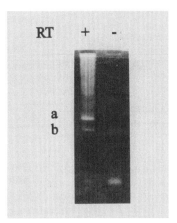

Figure 4. Loss of imprinting in Beckwith-Wiedemann syndrome. Using the *Apa*I polymorphism, biallelic expression of *IGF2* is detected in normal kidney from a BWS patient.

was surprising that in every case of LOI of *IGF2*, the level of *H19* was barely detectable, an approximate 80-fold decrease in expression compared to tumors with normal imprinting of *IGF2*. These differences in expression were highly significant ($p < 0.0001$) (Steenman et al. 1994). Thus, tumors with LOI of *IGF2*, while activating the normally silent maternal *IGF2* allele, also showed abrogation of expression of the maternal *H19* allele. It was also found that LOH involves loss of expression of *H19*, as expected, since LOH involves the loss of the maternal *H19*-expressing chromosome (Steenman et al. 1994). Tycko's laboratory has reported suppression of tumorigenicity by *H19* promoter constructs (Hao et al. 1993). Thus, the role of LOI in tumorigenesis may be the epigenetic inactivation of a tumor suppressor gene.

A clue to the mechanism of abnormal imprinting derives from recent studies in our laboratory of DNA methylation in the region of *IGF2* and *H19*. DNA methylation is a covalent modification of the nucleoside cytosine associated with gene inactivation. The role of DNA methylation in the regulation of normal gene expression has remained controversial, because alterations in DNA methylation often occur subsequent to changes in expression, and many genes show no specific alterations in methylation with changes in expression. On the other hand, many studies have shown that experimental manipulation of methylation can alter gene expression (Cedar and Razin 1990), and in vivo alteration of DNA methylation results in embryonic lethality (Li et al. 1993).

Nevertheless, DNA methylation does exhibit some interesting specific alterations in imprinted genes. For example, the laboratories of Surani and Tilghman have found parental-origin-specific, tissue-independent methylation of a CpG island in the mouse *H19* promoter (Bartolomei et al. 1993; Ferguson-Smith et al. 1993). Methylation of this island in development does not simply reflect gene expression but appears to represent an imprinting mark in the gene (Bartolomei et al. 1993; Ferguson-Smith et al. 1993). This modification occurs at the time of implantation (after establishment of parental-origin-specific gene expression) and thus cannot be the primary determinant of imprinting. A developmentally earlier modification of an imprinted gene is methylation in the first intron of the *IGF2R* gene (in mice), which occurs prezygotically (Stoger et al. 1993). That methylation at least maintains the pattern of genomic imprinting is supported by knock-out mice deficient in DNA methyltransferase, which show loss of imprinting of the *H19* and *IGF2R* genes (Li et al. 1993). More than 10 years ago, Feinberg and Vogelstein described widespread alterations in DNA methylation in human cancer (Feinberg and Vogelstein 1983b), but the mechanistic significance of this observation has been murky during that decade. Because of the correlation between methylation and gene expression, genomic imprinting represents an attractive regulator target of this ubiquitous phenomenon in human cancer.

To determine if methylation plays a role in LOI, we examined the CpG island of *H19* for changes in methylation correlating with changes in imprinting. A CpG island in the promoter of the human *H19* gene was half-methylated in all tissues (although it is hypomethylated in sperm as in mouse, consistent with modification at implantation). That hemimethylation was due to modification of the paternal allele specifically was shown by examining patients with paternal uniparental disomy (UPD). UPD was associated with complete methylation of the *H19* island, indicating that the paternal allele is methylated and the maternal allele is unmethylated, regardless of expression (Steenman et al. 1994).

When tumors with LOI were compared to tumors without LOI, the following pattern emerged. LOI of *IGF2* was uniformly associated with complete methylation of the *H19* imprint-specific island (Steenman et al. 1994). Since the same tumors also showed down-regulation of *H19*, three events had transpired over a >100-kb region of 11p15: (1) the maternal *IGF2* allele was activated as on the paternal chromosome; (2) the maternal *H19* allele was inactivated, as on the paternal chromosome; and (3) the pattern of DNA methylation on the maternal chromosome switched to a paternal-specific pattern (Steenman et al. 1994). Thus, LOI involves a switch of the maternal chromosome to a paternal epigenotype (Fig. 5).

We then examined Beckwith-Wiedemann syndrome patients in order to compare the DNA from BWS with UPD, BWS with LOI, and BWS with neither apparent abnormality. Our prediction was that as BWS with UPD involves two paternal alleles, and thus an entirely paternal methylation pattern, BWS with LOI should

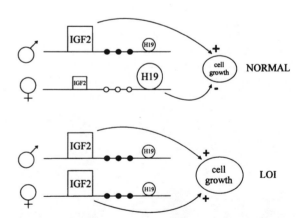

Figure 5. Switch in a domain of imprinting. LOI of *IGF2* was invariably linked to down-regulation of *H19* on the maternal chromosome (shown as a switch from large circle to small circle). Furthermore, a CpG island in the *H19* promoter showed parental-origin-specific, tissue-independent DNA methylation of the paternal allele (shown as open circles upstream of *H19* on the maternal chromosome and closed circles on the paternal chromosome). In tumors with LOI, the maternal CpG island had acquired a paternal pattern of methylation. Thus, in LOI the maternal chromosome had switched to a paternal epigenotype, which could lead to increased cell growth mediated by both *IGF2* and *H19*. (Reprinted, with permission, from Steenman et al. 1994.)

show two *apparent* paternal alleles, again with a paternal methylation pattern; and BWS with neither alteration should show a normal (hemimethylated) methylation pattern.

Our observations exactly confirmed these predictions. Four patients with UPD all demonstrated a paternal methylation pattern. Incidentally, none of these patients had developed malignancy, in contrast to Junien's observation of a 75% incidence of cancer in UPD (Steenman et al. 1994). Of six patients with BWS, two showed LOI, and both of these had undergone a switch to a paternal methylation pattern, biallelic expression of *IGF2*, and loss of expression of *H19* (Steenman et al. 1994). Differences in expression of *IGF2* and *H19*, similar to those seen in tumors with LOI, were seen in two nontumor tissues from BWS patients who showed LOI of *IGF2* in nontumor cells (Steenman et al. 1994). Thus, some BWS patients show in their nontumor cells the same LOI-associated changes seen in sporadic tumors occurring in the general pediatric population.

We and our colleagues have also isolated five balanced chromosomal translocation breakpoints from BWS patients, clustered over a region of several hundred kilobases, but all lying centromeric to *IGF2* and *H19*. Interestingly, all of the breakpoints are of maternal origin, suggesting that normal imprinting of the maternal *IGF2* and *H19* genes has been disturbed by separation from a *cis*-acting element on the same chromosome (L.M. Kalikin et al., in prep.).

DISCUSSION

In summary, our laboratory has made the first observation at the gene level of imprinting in man, and we have found that cancer involves abnormal imprinting. We have found that two genes on chromosomal band 11p15 are imprinted. The gene for insulin-like growth factor II (*IGF2*) is normally expressed only from the paternally inherited allele, whereas the gene for *H19* is normally expressed only from the maternally inherited allele. (Several imprinted genes have now been identified by others on other chromosomes [Jinno et al. 1994; Ohlsson et al. 1994]). We have also found that several types of childhood and adult tumors demonstrate LOI on the maternal chromosome of 11p15 (S. Rainier and A.P. Feinberg, in prep.). LOI is associated with both abnormal activation of *IGF2* and abnormal repression of *H19* expression from the maternal chromosome. We have identified parental-origin-specific tissue-independent methylation of the *H19* promoter in normal tissues, with imprint-specific methylation of the paternal allele in expressing and nonexpressing tissues. In LOI, the maternal chromosome becomes methylated at this site, and thus LOI is associated with a switch of the maternal chromosome to a paternal epigenotype. This "sex change" on an autosome defies conventional Mendelian thinking.

Finally, we have found that in a genetic disorder that predisposes to cancer, BWS, both LOI and UPD of 11p15 lead to a double dose of paternally expressed genes and loss of expression of maternally expressed genes, resulting in biallelic expression and increased dosage of *IGF2*, an important autocrine growth factor, and loss of expression of *H19*, a growth inhibitory gene. Thus, LOI appears to contribute directly to the organ overgrowth, related clinical stigmata, and predisposition to cancer of this syndrome.

LOI is a novel mutational mechanism in man, and its identification presents several questions: Does LOI play a causal role in malignancy? That it does is suggested by its high frequency, both in WT (5-fold more common than *WT1* mutations [Little et al. 1992]), and likely in other tumors (Zhan et al. 1994; S. Rainier and A.P. Feinberg, in prep.). It also occurs in the earliest stages of tumorigenesis, consistent with a causal role (Rainier et al. 1993). Tumorigenic consequences of LOI could include both increased expression of *IGF2* and decreased expression of *H19*, as well as effects on other genes. Hanahan and coworkers have observed that deregulated *IGF2* expression is necessary for tumor progression in transgene-induced mouse tumors (Christofori et al. 1994). Finally, BWS patients with constitutional LOI are at substantially increased risk of malignancy, just as one sees with Knudsonian tumor suppressor genes. If LOI is an important step in carcinogenesis, it will be important to determine whether this is a reversible alteration (since it is epigenetic) and thus might lead to novel forms of cancer treatment.

Is LOI limited to *IGF2* and *H19*, or could it involve other genes as well? Additional genes on 11p15 may be involved, as we have identified other transcripts in this region (L.M. Kalikin et al., in prep.). The *WT2* gene on 11p15 is distinct from *IGF2* and *H19*, as we have isolated a "subchromosomal transferable fragment" (STF) from 11p15, centromeric to *IGF2* and *H19*, that suppresses the growth of WT and other cells (Koi et al. 1993). This STF includes several balanced chromosomal translocation breakpoints from BWS patients, all of which involve the maternal chromosome, suggesting that the *WT2* gene is maternally expressed and inhibits the expression of *IGF2* in *cis* (L.M. Kalikin et al., in prep.). Other chromosomal domains on other chromosomes may also be affected, and many laboratories are searching for additional imprinted genes.

Does methylation serve as the marker for determining an imprint for parental-specific gene expression? In some cases, this may prove to be true. The parental-specific methylation patterns in the intronic *IGF2* receptor gene (in mice) are inherited and preserved from the gametes, and thus fit the strict chromosomal definition of an imprint. In contrast, the parental-specific methylation patterns observed at *IGF2* and *H19* occur later in embryonic development than does the establishment of parental-specific gene expression, suggesting that some other imprint must be present prior to methylation which determines the expression pattern. One possibility that has been popular for explaining X-chromosome inactivation has been that gene inactivation is mediated by some unidentified mechanism, with methylation serving as a means of maintaining or "locking in" the inactive state (Singer-Sam and Riggs 1993). Similarly, imprinted states may be initially established by an imprinting mark (*trans*-acting factors, chromatin structure), followed by methylation at later points in development. The role of methylation in imprinting will remain one of optimistic speculation until additional imprinting candidates can be identified.

To grasp the mechanism of imprinting and its relevance in tumorigenesis, it may be necessary to examine the problem from a chromosomal point of view, rather than from the perspective of the single gene. Imprinting may serve as a signal which determines the transcriptional state of certain large chromosomal domains, which are often described as looped, independently regulated regions of the chromosome. Like X-chromosome inactivation, imprinting may propagate outward from an organizing center, spreading over time across the domain. Imprinting may skip certain genes, perhaps due to the forms of chromosomal protein-protein interactions established over large distances, or due to particular folding patterns of the chromatin within a looped domain. Loss of imprinting in this model could be explained by the disturbance of the imprinting process at a number of different levels. LOI could result from mutations in an imprinting organization center, affecting the imprinted status of a large chromosomal region. Alternatively, mutations in *cis*-acting regions of imprinted genes could influence the local establishment of imprinting, perhaps by impairing the binding capacity of *trans*-acting imprinting factors. Furthermore, mutations in these *trans*-acting imprinting factors could alter imprinting capacity at numerous imprinted loci.

Utilizing this view of chromosomal gene regulation, we propose a unifying model (Fig. 6) that explains postzygotic allele-specific modification and transcription, LOI in tumors, and the location of a tumor suppressor locus on 11p15 within several hundred kb of imprinted genes. According to this model, an organizing center on 11p15 (and centromeric to BWS breakpoints) causes changes in chromatin structure that propagate outward from this organizing center, similar to the proposed role of the X-inactivation center in the X-inactivation (Willard et al. 1993). Modification of *H19* and *IGF2* could occur postzygotically but with parental-origin specificity, since the information is already embedded in *cis* on the same chromosome in the gamete. Loss of the imprinting center would lead to LOI, but only if it occurs on the maternal chromosome. In this manner, the imprinting center itself would represent a tumor suppressor gene, with preferential maternal LOH (since the maternal copy represses *IGF2* expression and is permissive for *H19* expression). Physical dissociation of the *IGF2/H19* locus from this inactivation center (due to disruption of the chromosomal loop) would also cause LOI. Thus, BWS translocation breakpoints could prevent propagation of the

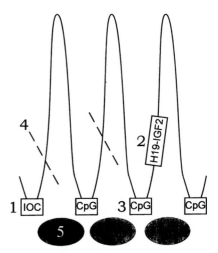

Figure 6. A unifying model of imprinting, loss of imprinting, and tumor suppression. An imprint organizing center (IOC), located several hundred kb centromeric to *H19* and *IGF2* on 11p15, exerts a *cis*-acting influence on chromatin structure, with opposite effects on maternal and paternal chromosomes (fulfilling Crouse's strict chromosomal definition of an imprint). The effect of the IOC is propagated outward during development, similar to the organizing center on the X chromosome, and specific proteins (gray) mediate formation of facultative heterochromatin. Imprinting is maintained in part by allele-specific methylation of CpG islands. According to this model, loss of imprinting could arise by any of several mechanisms (numbered in the figure): (1) Deletion or mutation in the IOC itself, which would lead to a failure of parental-origin-specific switching in the germ line; (2) local mutation of regulatory sequences controlling the target imprinted genes themselves; (3) abnormal methylation of the CpG islands; (4) separation of the IOC from the imprinted target genes (dashed lines), as seen in BWS germ-line translocation cases; (5) loss or mutation in proteins that mediate heterochromatinization, such as *su(var)* mutations in *Drosophila*. Alterations in any of these steps would behave like a tumor suppressor gene.

imprinting signal to *IGF2*. Consistent with this idea, all of the observed balanced rearrangements in BWS are of maternal origin. Imprinting could also be disrupted by mutations in proteins that help to mediate propagation of the imprinting signal, similar to "*suppressor of variegation (su[var])*" mutations in *Drosophila*.

By this model, imprinting affects a higher order of gene regulation (beyond promoter and enhancer control) that influences large chromosomal domains. The model suggests that a key region to examine in tumors includes sites that appear to regulate chromosome modification over long distances, such as has been recently described in the region of the Prader-Willi gene on chromosome 15 (Sutcliffe et al. 1994). Although there are more questions than answers with respect to imprinting at this time, these studies (and those of many other investigators) have provided the beginning of an understanding of the mechanism of several forms of cancer, as well as chromosomal domain regulation.

ACKNOWLEDGMENTS

This work was supported by the National Institutes of Health.

REFERENCES

Aleck, K.A. and T.A. Hadro. 1989. Dominant inheritance of Wiedemann-Beckwith syndrome: Further evidence for transmission of "unstable premutation" through carrier women. *Am. J. Med. Genet.* **33:** 155.

Barlow, D.P. 1994. Imprinting: A gamete's point of view. *Trends Genet.* **10:** 194.

Bartolomei, M., A.L. Webber, M.E. Brunkow, and S.M. Tilghman. 1993. Epigenetic mechanisms underlying the imprinting of the mouse H19 gene. *Genes Dev.* **7:** 1663.

Brunkow, M.E. and S.M. Tilghman. 1991. Ectopic expression of the H19 gene in mice causes prenatal lethality. *Genes Dev.* **5:** 1092.

Cedar, H. and A. Razin. 1990. DNA methylation and development. *Biochim. Biophys. Acta* **1049:** 1.

Christofori, G., P. Naik, and D. Hanahan. 1994. A second signal supplied by insulin-like growth factor II in oncogene-induced tumorigenesis. *Nature* **369:** 414.

Crouse, H.V. 1960. The controlling element in sex chromosome behavior in Sciara. *Genetics* **45:** 1429.

El-Badry, O.M., C. Minniti, E.C. Kohn, P.J. Houghton, W.H. Daughaday, and L.J. Helman. 1990. Insulin-like growth factor II acts as an autocrine growth and motility factor in human rhabdomyosarcoma tumors. *Cell Growth Differ.* **1:** 325.

Engstrom, W., S. Lindham, and P. Schofield. 1988. Wiedemann-Beckwith syndrome. *Eur. J. Pediatr.* **147:** 450.

Feinberg, A.P. and B. Vogelstein. 1983a. A technique for radiolabeling DNA restriction endonuclease fragments to high specific activity. *Anal. Biochem.* **132:** 6.

―――. 1983b. Hypomethylation distinguishes genes of some human cancers from their normal counterparts. *Nature* **301:** 89.

Ferguson-Smith, A.C., H. Sasaki, B.M. Cattanach, and M.A. Surani. 1993. Parental-origin-specific epigenetic modification of the mouse H19 gene. *Nature* **362:** 751.

Hao, Y., T. Crenshaw, T. Moulton, E. Newcomb, and B. Tycko. 1993. Tumor-suppressor activity of H19 RNA. *Nature* **365:** 764.

Henry, I., C. Bonaitijk-Pellie, V. Chehensse, C. Beldjord, C. Schwartz, G. Utermann, and C. Junien. 1991. Uniparental paternal disomy in a genetic cancer-predisposing syndrome. *Nature* **351:** 665.

Jinno, Y., K. Yun, K. Nishiwaki, T. Kubota, O. Ogawa, A.E. Reeve, and N. Niikawa. 1994. Mosaic and polymorphic imprinting of the *WT1* gene in humans. *Nat. Genet.* **6:** 305.

Koi, M., L.A. Johnson, L.M. Kalikin, P.F.R. Little, Y. Nakamura, and A.P. Feinberg. 1993. Tumor cell growth arrest caused by subchromosomal transferable DNA fragments from human chromosome 11. *Science* **260:** 361.

Li, E., C. Beard, and R. Jaenisch. 1993. Role for DNA methylation in genomic imprinting. *Nature* **366:** 362.

Little, M.H., J. Prosser, A. Condie, P.J. Smith, V. van Heyningen, and N.D. Hastie. 1992. Zinc finger point mutations within the *WT1* gene in Wilms' tumor patients. *Proc. Natl. Acad. Sci.* **89:** 4791.

Mannens, M., R.M. Slater, C. Heyting, J. Bliek, J. de Kraker, N. Coad, P. de Pagter-Holthuizen, and P.L. Pearson. 1988. Molecular nature of genetic changes resulting in loss of heterozygosity of chromosome 11 in Wilms' tumours. *Hum. Genet.* **81:** 41.

Monk, M. 1988. Genomic imprinting. *Genes Dev.* **2:** 921.

Ogawa, O., D.M. Becroft, I.M. Morison, M.R. Eccles, J.E. Skeen, D.C. Mauger, and A.E. Reeve. 1993a. Constitutional relaxation of insulin-like growth factor II gene imprinting associated with Wilms' tumour and gigantism. *Nat. Genet.* **5:** 408.

Ogawa, O., M.R. Eccles, J. Szeto, L.A. McNoe, K. Yun, M.A. Maw, P.J. Smith, and A.E. Reeve. 1993b. Relaxation of insulin-like growth factor II gene imprinting implicated in Wilms' tumour. *Nature* **362:** 749.

Ohlsson, R., K. Hall, and M. Ritzen. 1994. *Nobel conference on genomic imprinting: Causes and consequences.* Cambridge University Press, Cambridge, United Kingdom.

Ohlsson, R., A. Nystrom, S. Pfeifer-Ohlsson, V. Tohonen, F. Hedborg, P. Schofield, F. Flam, and T.J. Ekstrom. 1993. IGF2 is parentally imprinted during human embryogenesis and in the Beckwith-Wiedemann syndrome. *Nat. Genet.* **4:** 94.

Ping, A.J., A.E. Reeve, D.J. Law, M.R. Young, M. Boehnke, and A.P. Feinberg. 1989. Genetic linkage of Beckwith-Wiedemann syndrome to 11p15. *Am. J. Hum. Genet.* **44:** 720.

Rainier, S., L.A. Johnson, C.J. Dobry, A.J. Ping, P.E. Grundy, and A.P. Feinberg. 1993. Relaxation of imprinted genes in human cancer. *Nature* **362:** 747.

Reeve, A.E., S.A. Sih, A.M. Raizis, and A.P. Feinberg. 1989. Loss of allelic heterozygosity at a second locus on chromosome 11 in sporadic Wilms' tumor cells. *Mol. Cell. Biol.* **9:** 1799.

Reik, W., S.K. Howlett, and M.A. Surani. 1990. Imprinting by DNA methylation: From transgenes to endogenous gene sequences. *Development Suppl.* p. 99.

Reik, W., A. Collick, M.L. Norris, S.C. Barton, and A. Surani. 1987. Genomic imprinting determines methylation of parental alleles in transgenic mice. *Nature* **328:** 248.

Schroeder, W., L. Chao, D. Dao, L. Strong, S. Pathak, V. Riccardi, W. Lewis, and G. Saunders. 1987. Nonrandom loss of maternal chromosome 11 alleles in Wilms' tumors. *Am. J. Hum. Genet.* **40:** 413.

Shuldiner, A.R., K. Tanner, C.A. Moore, and J. Roth. 1991. RNA template-specific PCR: An improved method that dramatically reduces false positives in RT-PCR. *BioTechniques* **11:** 760.

Singer-Sam, J. and A.D. Riggs. 1993. X chromosome inactivation and DNA methylation. *EXS* **64:** 358.

Steenman, M.J.C., S. Rainier, C.J. Dobry, P. Grundy, I. Horon, and A.P. Feinberg. 1994. Loss of imprinting of IGF2 is linked to reduced expression and abnormal methylation of H19 in Wilms' tumor. *Nat. Genet.* **7:** 433.

Stoger, R., P. Kubicka, C.-G. Liu, T. Kafri, A. Razin, H. Cedar, and D.P. Barlow. 1993. Maternal-specific methylation of the imprinted mouse IGF2r locus identifies the expressed locus as carrying the imprinting signal. *Cell* **73:** 61.

Surani, M.A., W. Reik, and N.D. Allen. 1988. Transgenes as molecular probes for genomic imprinting. *Trends Genet.* **4:** 59.

Sutcliffe, J.S., M. Nakao, S. Christian, K.H. Orstavik, N. Tommerup, D.H. Ledbetter, and A.L. Beaudet. 1994. Deletions of a differentially methylated CpG island at the SNRPN gene define a putative imprinting control region. *Nat. Genet.* **8:** 52.

Swain, J.L., T.A. Stewart, and P. Leder. 1987. Parental legacy determines methylation and expression of an autosomal transgene: A molecular mechanism for parental imprinting. *Cell* **50:** 719.

Tadokoro, K., H. Fujii, T. Inoue, and M. Yamada. 1991. Polymerase chain reaction (PCR) for detection of ApaI polymorphism at the insulin like growth factor II gene (IGF2). *Nucleic Acids Res.* **19:** 6967

Thompson, M.A., A.J. Cox, R.H. Whitehead, and H.A. Jonas. 1990. Autocrine regulation of human tumor cell proliferation by insulin-like growth factor II: An in vitro model. *Endocrinology* **126:** 3033.

Viljoen, D. and R. Ramesar. 1992. Evidence for paternal imprinting in familial Beckwith-Wiedemann syndrome. *J. Med. Genet.* **29:** 221.

Weksberg, R., D.R. Shen, Y.L. Fei, Q.L. Song, and J. Squire. 1993. Disruption of insulin-like growth factor 2 imprinting in Beckwith-Weidemann syndrome. *Nat. Genet.* **5:** 143.

Wiedemann, H.R. 1983. Tumours and hemihypertrophy associated with Wiedemann-Beckwith syndrome. *Eur. J. Pediatr.* **141:** 129.

Willard, H.F., C.J. Brown, L. Carrel, B. Hendrich, and A.P. Miller. 1993. Epigenetic and chromosomal control of gene expression—Molecular and genetic analysis of chromosome inactivation. *Cold Spring Harbor Symp. Quant. Biol.* **58:** 315.

Zhan, S.L., D.N. Shapiro, and L.J. Helman. 1994. Activation of an imprinted allele of the insulin-like growth factor II gene implicated in rhabdomyosarcoma. *J. Clin. Invest.* **94:** 445.

Zhang, Y. and B. Tycko. 1992. Monoallelic expression of the human H19 gene. *Nat. Genet.* **1:** 40.

Enhanced Cell Survival and Tumorigenesis

S. Cory,* A. Strasser,* T. Jacks,† L.M. Corcoran,*
T. Metz,* A.W. Harris,* and J.M. Adams*

*Walter and Eliza Hall Institute of Medical Research, Melbourne, Victoria 3050, Australia;
†Cancer Center, Massachusetts Institute of Technology, Cambridge, Massachusetts

Until recently, molecular oncology has been preoccupied almost exclusively with the regulation of cellular proliferation. Now, however, there is a surge of interest in the control of cell survival. This shift in focus reflects a heightened appreciation that it is the balance between cell death and proliferation that governs the accumulation of cells, whether in a normal tissue or a tumor. Insights derived from developmental and cell biology have revealed that the normal demise of a cell is genetically programmed, an active rather than a passive process. Several key players in this program have recently been identified, and two are genes already implicated in cancer development: the oncogene *bcl-2* and the tumor suppressor gene *p53*. There is now good reason to believe that induced suicide is the major mechanism for eliminating cells with damaged DNA or an aberrant cell cycle, i.e., the cells likely to initiate a neoplastic clone. Increased understanding of physiological cell death is likely to reveal the basis for conventional genotoxic approaches to cancer therapy. Even more important, learning how to manipulate the cell death machinery should eventually engender new strategies for cancer treatment.

In this paper, we first sketch the background for the mounting attention to the mechanism of cell death and then review studies by ourselves and other workers exploiting genetically modified mice to investigate the in vivo consequences of deregulating cell survival. We have focused principally on the lymphoid system, in which cellular attrition plays a dominant role, but many of the conclusions are germane to other cell lineages, particularly those having high cell turnover. We discuss the impact of gain of function of Bcl-2, a negative regulator of cell death, and loss of function of p53, a positive regulator. The studies reviewed suggest that Bcl-2 can inhibit cell killing by many, but not all, cytotoxic agents and that its effectiveness varies with cell type and differentiation stage. Bcl-2 can promote the development of B-lymphoid tumors and is strongly synergistic with Myc in tumorigenesis, probably because Myc promotes cell death as well as proliferation. Bcl-2 appears to act downstream from p53, which induces cell death only in response to DNA damage. Loss or mutation of p53 is highly tumorigenic because p53 promotes apoptosis of mutant cells and perhaps also because it is needed to enforce cellular senescence. Absence of p53 is insufficient to immortalize hematopoietic cells but markedly facilitates their transformation by oncogenes. Lymphocytes that lack p53 can survive exposure to genotoxic agents but, unexpectedly, their resistance depends on cell cycle status. These and other findings prompt a model in which multiple pathways to apoptosis converge and the final common steps are regulated by Bcl-2 and related survival genes.

APOPTOSIS: ALTRUISTIC CELL SUICIDE

The seminal histological observations of Kerr, Wyllie, and their collaborators distinguished two modes of cell death: necrosis and apoptosis (Kerr et al. 1972). Necrosis represents a pathological response to severe cellular injury, whereas apoptosis is the normal cellular response to physiological signals (or lack thereof) (for reviews, see Wyllie et al. 1980; Kerr and Harmon 1991). The term apoptosis derives from the Greek for "falling off" (as of leaves from trees) and emphasizes its active, genetically controlled nature. During necrosis, chromatin clumps form in the damaged cell, its mitochondria swell, the plasma membrane eventually lyses, and the cell spills its contents, eliciting inflammation. Apoptosis follows a different course, depicted in Figure 1. Its first visible sign is condensation of the chromatin into large masses along the nuclear envelope, followed typically by shrinkage of the cytoplasm and blebbing of the plasma and nuclear membranes. Within a few minutes, the cell fragments into membrane-bound bodies containing densely packed but intact organelles. Over the next hour or so, these bodies are engulfed by neighboring cells. Importantly, the cellular demise is accomplished without disrupting tissue architecture or eliciting inflammation. The rapid clearance of the dying cells obscures their detectability in normal tissues, and this has retarded the recognition of the importance of cell death in normal physiology.

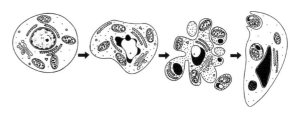

Figure 1. Morphological features of apoptosis. The first step is chromatin condensation and mild membrane convolution, followed by fragmentation into apoptotic bodies which are finally engulfed by a neighboring cell. (Adapted, with permission, from Kerr and Harmon 1991.)

Internucleosomal cleavage of DNA is a feature of apoptosis not found in necrosis (Wyllie et al. 1980), and the resulting DNA "ladder" shown by gel electrophoresis is often used as a distinguishing marker. The responsible nuclease was eagerly sought on the grounds that its activation might be the trigger for apoptosis (Wyllie et al. 1980), but recent findings suggest that more than one endonuclease participates and that DNA cleavage may not be obligatory (see Martin 1993). Indeed, enucleated cells can undergo cytoplasmic changes strongly resembling those of apoptotic cells (Jacobson et al. 1994). Thus, apoptosis may be orchestrated from the cytoplasm.

Developmentally programmed cell death, long postulated by embryologists to sculpt tissues during embryogenesis and metamorphosis, almost certainly proceeds by apoptosis, although not all of its hallmarks are always apparent (Lockshin and Zakeri 1991). Neuronal development is accompanied by extensive cell death, and perceptive studies in this system have emphasized the critical role played by cell-cell interactions in maintaining cell viability (Oppenheim 1991; Raff 1992). In the adult, apoptotic cells can be detected cytologically in many tissues, including hematopoietic organs and intestinal epithelium, as well as organs that involute on hormone withdrawal, such as the adrenal, mammary, and prostate glands. Indeed, apoptosis is probably the principal determinant of cellular homeostasis in many systems, and, as discussed further below, it can be induced experimentally in tissues and cultured cell lines by a wide variety of agents. Pertinently, tumors often contain apoptotic bodies, and cell death probably accounts for their often paradoxically slow rate of enlargement.

In keeping with its active nature, apoptosis can often be blocked by inhibitors of RNA or protein synthesis. However, this does not hold for all cell types. Furthermore, the requirement for macromolecular synthesis can vary with different apoptotic stimuli and in some instances can elicit rather than inhibit apoptosis (Martin 1993). It therefore seems likely that certain regulators of the death pathway are differentially expressed.

It has been hypothesized that apoptosis represents abortive mitosis (Ucker 1991). If so, the principal mitotic cyclin-dependent kinase, Cdc2, might participate in apoptosis. No evidence for involvement of Cdc2 could be found in one system (Lazebnik et al. 1993), but other studies have implicated its premature activation (Meikrantz et al. 1994; Shi et al. 1994), at least in the specific form of apoptosis promoted by proteases from cytotoxic T cells (see below). It would not be surprising if other members of the growing family of cyclin-dependent kinases also participated.

NEGATIVE REGULATION OF APOPTOSIS

An intrinsic suicide program is clearly advantageous to the organism for disposing of cells that are no longer needed, or cells that are effete, dysfunctional, or dangerous. Safeguards on this process are needed, however, to prevent accidental or even mass suicide of cells. The paradigm inhibitor of apoptosis is Bcl-2. The *bcl-2* gene was initially found by its frequent translocation to an immunoglobulin locus in human follicular lymphoma (for review, see Korsmeyer 1992). The presumptive oncogene proved to encode a 26-kD cytoplasmic protein associated with intracellular membranes via its hydrophobic carboxyl terminus (Chen-Levy and Cleary 1990). The Bcl-2 protein was initially thought to reside in the inner membrane of the mitochondrion (Hockenbery et al. 1990), but it is now agreed that the polypeptide lies on the cytosolic face of the nuclear membrane, the endoplasmic reticulum, and the outer mitochondrial membrane (Monaghan et al. 1992; Krajewski et al. 1993; de Jong et al. 1994; Lithgow et al. 1994).

The biological function of Bcl-2 was discovered (Vaux et al. 1988) when enforced *bcl-2* expression was found to sustain the survival of interleukin-3-dependent myeloid and lymphoid cells in the absence of the cytokine. Although factor deprivation normally promotes apoptosis of such cells, the introduced *bcl-2* gene allowed them to enter a quiescent (G_0) state and to persist there for many days (Fig. 2). Since constitutive synthesis of high levels of Bcl-2 neither promoted proliferation nor blocked re-entry of resting cells into cycle after addition of cytokine, it was apparent that cell survival and proliferation are subject to independent genetic control. Bcl-2 was subsequently shown to enhance cell survival in the face of diverse insults, including radiation, genotoxic drugs, glucocorticoids, heat shock, sodium azide, and Ca^{++} influx (for review, see Reed 1994).

Bcl-2 is found widely in the developing embryo (LeBrun et al. 1993; Novack and Korsmeyer 1994) and is readily detected in certain adult tissues such as hematopoietic and neuronal tissues, skin, intestine, and glandular epithelium of the breast, thyroid, and prostate (Pezzella et al. 1990; Hockenbery et al. 1991). Since frequent apoptosis is also a characteristic of these tissues, Bcl-2 presumably ensures the survival of cells "hard-wired" for suicide until specific physiological conditions warrant their death.

It has recently become clear that Bcl-2 has several relatives. One of these, Bax, associates with Bcl-2 (Oltvai et al. 1993), and Bax/Bcl-2 heterodimers appear to be essential for the survival-promoting function of Bcl-2 (Yin et al. 1994). High levels of Bax, however, can inhibit the survival of Bcl-2-expressing cells, perhaps because Bax homodimers form in preference to heterodimers. The gene for another homolog, *bcl-x*, produces two proteins by differential splicing (Boise et al. 1993). The larger, $Bcl-x_L$, promotes cell survival, whereas the smaller, $Bcl-x_S$, antagonizes Bcl-2 function. Thus, cell survival appears to be regulated by competition between Bcl-2-like proteins which block apoptosis and structurally related proteins which antagonize this function.

The pathway to cell death is an ancient one. Genetic studies on the nematode *Caenorhabditis elegans* have

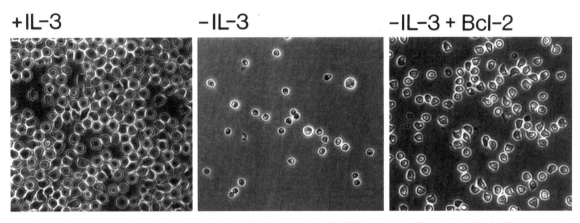

Figure 2. Bcl-2 enhances cell survival. Interleukin-3 (IL-3)-dependent FDC-P1 myeloid cells and a derivative line expressing a *bcl-2* retrovirus were washed and plated in medium with and without IL-3. The photomicrographs were taken 3 days later. In the presence of IL-3, both lines proliferated and remained viable (e.g., *left* panel). In the absence of IL-3, the uninfected cells died rapidly (*center* panel). Those expressing high levels of Bcl-2 (*right* panel) survived well in the absence of IL-3 but did not proliferate; the small size of the cells is indicative of a G_0 state.

uncovered two genes that trigger cell death (*ced-3* and *ced-4*), another that prevents it (*ced-9*), and several that mediate engulfment of the dying cell and degradation of its DNA (Ellis et al. 1991). Significantly, the Ced-9 protein has a limited sequence similarity to Bcl-2 (23% identity) (Hengartner and Horvitz 1994), and *bcl-2* can replace *ced-9* in *C. elegans* (Vaux et al. 1992b; Hengartner and Horvitz 1994); hence, they are functional homologs. As yet, almost nothing is known about the biochemical function of Ced-9/Bcl-2 or of Ced-4 but, as discussed further below, Ced-3 and its mammalian relatives are proteases (Miura et al. 1993; Yuan et al. 1993). Hence, apoptosis is most likely triggered by scission of (as yet unknown) proteins rather than by cleavage of DNA.

IMPACT OF CONSTITUTIVE Bcl-2 EXPRESSION IN LYMPHOCYTES

Cell death is essential in the lymphoid system for both the development and function of an effective immune repertoire. Lymphocytes with defective or self-reactive antigen receptors continually arise and must be removed, and even the useful clones that are expanded during an immune response must eventually be disposed of. Modulation of Bcl-2 is implicated in this process by the intriguing variation in its expression pattern during lymphopoiesis. In the thymus, for example, the level of Bcl-2 protein is high in T-cell precursors and mature medullary T cells but low in the cortical population that must run the gauntlet of selection (Pezzella et al. 1990; Hockenbery et al. 1991; Gratiot-Deans et al. 1993; Veis et al. 1993). Such observations suggest that deletion of unwanted lymphocytes involves down-regulation of Bcl-2 and, conversely, that up-regulation ensures the survival of the selected repertoire.

The impact of constitutive Bcl-2 expression on lymphoid homeostasis and selection has been investigated in mice expressing a *bcl-2* transgene specifically within the B-lymphoid compartment (McDonnell et al. 1989; Strasser et al. 1991b; Katsumata et al. 1992) or the T-lymphoid compartment (Sentman et al. 1991; Strasser et al. 1991a; Siegel et al. 1992). B and T lymphocytes from such mice proved to be remarkably robust. They were long-lived in vitro in the absence of cytokines and were impressively resistant to normally cytotoxic levels of radiation, glucocorticoids, phorbol ester, and calcium ionophore.

Homeostasis of the B-lymphoid compartment was markedly perturbed by expression of the *bcl-2* transgene. The most obvious manifestation was a severalfold excess of mature, noncycling B cells (McDonnell et al. 1989; Strasser et al. 1991b). These cells had a normal surface marker phenotype and responded normally to mitogens and cytokines. Immature pre-B cells and mature immunoglobulin-secreting cells were also elevated, and immunization of the mice provoked an amplified and prolonged humoral immune response (Strasser et al. 1991b). These findings suggest that, irrespective of differentiation stage, B-lymphoid cells expressing high levels of Bcl-2 have an increased life span.

Mice of certain *bcl-2* strains frequently developed a fatal autoimmune disease that resembled human systemic lupus erythematosus (Strasser et al. 1991b). The *bcl-2* transgene was essential for its onset, but backcrossing experiments have indicated that background genes of C57BL/6 and SJL origin also contributed to its development (Strasser et al. 1993; A.W. Harris and M.L. Bath, unpubl.). The pathological accumulation of autoreactive antibodies was attributed to enhanced longevity of self-reactive B-cell clones. Although normal mice contain some autoantibody-producing cells, their frequency may have increased in the transgenic animals due to impairment of the selection mechanisms that normally operate in the bone marrow and the germinal centers. The postulate that a high level of Bcl-2 protein hampers deletion of autoreactive B cells is supported by the analysis of mice coexpressing the *bcl-2* transgene and a self-antigen-reactive immunoglobulin transgene (Hartley et al. 1993).

In view of these B-lymphoid disturbances, it was surprising that mice expressing the transgene in T cells exhibited no alterations in T-lymphoid development (Sentman et al. 1991; Strasser et al. 1991a; Siegel et al. 1992). T-cell numbers were normal, as were the relative proportions of the major subpopulations, and thymic involution with age was unaffected. No autoimmune disorder was evident. Although their thymocytes were refractory to killing by an antibody to the T-cell receptor complex, a stimulus believed to mimic that which deletes self-reactive thymocytes, no autoreactive T cells could be detected in peripheral lymphoid organs, despite evidence of somewhat increased numbers in the thymus (Sentman et al. 1991; Strasser et al. 1991a, 1994c; Siegel et al. 1992; Linette et al. 1994). These results suggest that apoptosis of self-reactive T cells is triggered by a mechanism that either neutralizes or ignores Bcl-2.

ROLE OF *bcl-2* IN LYMPHOMA DEVELOPMENT

The consistent association of the *bcl-2* translocation with follicular lymphoma provided strong circumstantial evidence that constitutive Bcl-2 expression figures in its etiology. Direct evidence was sought by monitoring multiple lines of *bcl-2* transgenic mice for lymphoma development. Clonal B-lymphoid tumors eventually emerged at significant incidence (10–15% at 12 months of age) (McDonnell and Korsmeyer 1991; Strasser et al. 1993), validating *bcl-2* as an oncogene for that cell type, but the long latency indicated the need for additional mutation. Consistent with the failure of Bcl-2 to perturb T-lymphoid homeostasis, the incidence of T-cell lymphomas was as low as in nontransgenic mice of the same genetic background (Strasser et al. 1993).

In keeping with the evidence from transgenic mice that *bcl-2* deregulation is insufficient to precipitate tumor development, cells bearing a *bcl-2* translocation can be detected by polymerase chain reaction (PCR) in many nonlymphomatous human subjects (Limpens et al. 1991). The principal role of Bcl-2 in human follicular lymphoma presumably is to enable a B cell that has acquired a 14;18 translocation to resist the induction of apoptosis that normally deletes many B cells in the lymphoid germinal centers. Additional mutation is probably required for disease onset, but antigenic stimulation may play a role in expanding the mutant clone.

Follicular lymphoma usually follows an indolent course, with most patients surviving more than 10 years after diagnosis (Horning and Rosenberg 1984), but occasionally a more aggressive lymphoma arises which bears a *myc* t(8;14) as well as a *bcl-2* t(14;18) translocation. Intriguingly, a proportion of the tumors in *bcl-2* transgenic mice had rearranged the *myc* locus (McDonnell and Korsmeyer 1991; Strasser et al. 1993). These observations suggested that constitutive *myc* expression can cooperate with *bcl-2* to promote lymphomagenesis.

Cooperation was first tested by introducing a *bcl-2* retrovirus into bone marrow cells from healthy young *myc* transgenic mice (Vaux et al. 1988). These mice express Myc constitutively in B-lymphoid cells and, following a preleukemic phase characterized by an excess of cycling pre-B cells, succumb to clonal pre-B or B lymphoma (Adams et al. 1985; Langdon et al. 1986; Harris et al. 1988). Consistent with *bcl-2/myc* synergy, immortal pre-B cell lines reproducibly arose from the virus-infected cultures but not from mock-infected *myc* marrow (Vaux et al. 1988).

A more definitive test was provided by interbreeding the *myc* and *bcl-2* transgenic mice (Strasser et al. 1990). The blood of weanling bitransgenic progeny contained huge numbers of large cycling pre-B cells, about 25-fold more than littermates bearing only the *myc* transgene. These cells were not malignant, despite their obvious proliferative activity. However, the *bcl-2/myc* animals became terminally ill sooner and more synchronously than littermates bearing only the *myc* transgene (Fig. 3). Indeed, tumor cells could be detected by transplantation tests as early as 10 days after birth. Surprisingly, the cell-surface marker phenotype of the tumors was uniformly that of a lymphoid stem or progenitor cell rather than a pre-B cell, arguing that this primitive cell is considerably more sensitive than pre-B cells to transformation by constitutive expression of *myc* and *bcl-2*.

Recent work suggests that *myc* can promote apoptosis as well as proliferation (Askew et al. 1991; Evan et al. 1992). If induction of apoptosis were an automatic

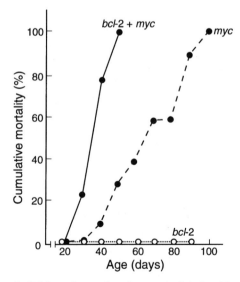

Figure 3. Bcl-2 accelerates lymphomagenesis induced by Myc. Cumulative mortality from lymphoma in transgenic mice carrying both a *bcl-2* and a *myc* transgene compared with that in littermates carrying only the single transgenes. (Redrawn, with permission, from Strasser et al. 1990 [copyright Macmillan Magazines Ltd.].)

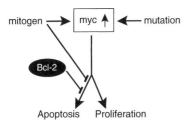

Figure 4. Model coupling proliferation and apoptosis (Harrington et al. 1994). Activation within a cell of *myc* expression, whether extrinsically by a mitogen or cytokine, or intrinsically by mutation of a proto-oncogene or a tumor suppressor gene, triggers the potential for both proliferation and apoptosis. Hence, net increase in cell number depends on inhibition of the cell death response. It seems likely that there are diverse intracellular survival signals, including elevation of Bcl-2 or Bcl-x_L protein levels and depression of Bax or Bcl-x_S levels. (Modified from Harrington et al. 1994.)

consequence of *myc* activation, increase in cell number would be dependent on active suppression of the apoptotic pathway (Harrington et al. 1994). Certain cytokines (perhaps all) are both survival and proliferation factors, implying that their receptors transduce signals inhibiting apoptosis as well as signals stimulating entry into the cell cycle. A mechanism coupling proliferation and apoptosis (Fig. 4) would provide a biological safety net; any proliferating (*myc*-expressing) cell, whether stimulated by a mitogen or an oncogenic mutation, would abort spontaneously when it exhausted the available cytokines (Askew et al. 1991; Evan et al. 1992). The danger of a mutation up-regulating the expression of a survival gene like *bcl-2* then becomes apparent: With the safety net broken, the cells would persist even in the absence of cognate growth factors.

Induction of apoptosis by Myc probably accounts for our earlier observations that the pre-leukemic expansion of pre-B cells in *myc* transgenic mice reaches a limit some weeks after birth (Langdon et al. 1986) and that, when cultured in medium containing serum but no additional cytokines, these cells died faster than normal pre-B cells (Langdon et al. 1988). Thus, the accelerated lymphoma development in *bcl-2/myc* transgenic mice is probably due to enhanced survival capacity of *myc*-expressing cells in vivo. Another oncogene with documented cytotoxic effects is *v-abl* (Ziegler et al. 1981). We have found that introducing a *bcl-2* transgene into *v-abl* transgenic mice substantially increases their incidence of lymphoid tumors (A. Strasser et al., unpubl.). It seems likely that circumvention of cell death mechanisms will prove to be a critical step in the onset or progression of many types of cancer.

DIFFERENTIAL CELLULAR RESPONSIVENESS TO Bcl-2

The survival advantage conferred by Bcl-2 varies with cell type. We expected the tumor cells derived from the *bcl-2/myc* mice to survive and proliferate when placed in culture. Surprisingly, they were instead very labile (Strasser et al. 1990). Figure 5 shows that,

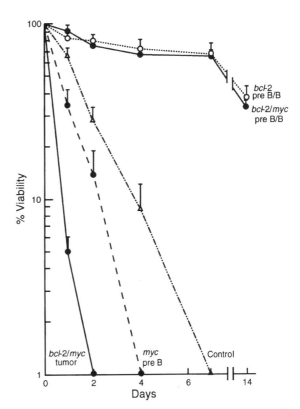

Figure 5. Efficiency of Bcl-2 in promoting survival varies with differentiation stage. Bone marrow cells from young healthy *bcl-2*, *myc*, and *bcl-2/myc* mice and nontransgenic littermates were plated in culture medium containing serum and 2-mercaptoethanol but no exogenous cytokines. Cells from a transplanted *bcl-2/myc* tumor were analyzed in parallel. Cell viability was followed by trypan blue exclusion. The tumor cells were B-lymphoid progenitor or stem cells, whereas the pre-leukemic marrow of *myc*, *bcl-2*, and *bcl-2/myc* mice was dominated by pre-B and B cells as indicated.

although the nonmalignant pre-B and B cells from *bcl-2/myc* mice survived just as well as those expressing only *bcl-2*, the *bcl-2/myc* tumor cells died even more rapidly than pre-B cells expressing only the *myc* transgene. Both cell types expressed the same level of Bcl-2 protein. Hence, under these conditions, Bcl-2 is far more effective at sustaining pre-B and B cells than the more primitive lymphoid cell type. In vivo, the tumor cells presumably receive additional survival signals not provided in culture.

Survival signals probably are conveyed to lymphocytes not only via cytokine receptors, but also via other membrane-bound molecules such as the antigen receptor and coreceptors like CD40 and CD28. Persuasive evidence that the antigen receptor conveys a survival signal comes from mutant mice unable to assemble functional antigen receptor genes, such as scid mice or those with a disrupted *rag-1* or *rag-2* gene. In these mice, lymphoid development arrests at an early stage and the nascent lymphocytes perish by apoptosis (Osmond et al. 1992). To test whether *bcl-2* expression could supplant the receptor requirement for lymphocyte survival, we introduced a *bcl-2* transgene expressed in both the B- and T-lymphoid compartments into a scid or $rag\text{-}1^{-/-}$ background. T-cell development still failed to proceed, but B-lymphoid cells increased greatly in the bone marrow and in peripheral lymphoid tissue (Strasser et al. 1994a; L.M. Corcoran and A. Strasser, unpubl.). Moreover, on culture in simple medium, the B-lymphoid cells survived far longer than those from conventional scid or $rag\text{-}1^{-/-}$ mice but the thymocytes died rapidly, despite expression of the transgene at a level sufficient to preserve more mature T cells. This intriguing difference between immature B- and T-lymphoid cells in their responsiveness to Bcl-2 suggests that a function facilitating Bcl-2 action may be expressed earlier in the B-cell than the T-cell lineage. This function may be triggered via the antigen coreceptor, since the immunoglobulin coreceptor (mb-1/B29) is prominent on the surface of pro-B cells, whereas the T-cell coreceptor (CD3) is present at very low levels on pro-T cells.

It seems that the ability of Bcl-2 to block apoptosis depends on differentiation stage; at the expression levels tested, Bcl-2 was insufficient to save either primitive B-lymphoid progenitor cells or pro-T cells but was adequate to prevent apoptosis in more mature cells of either lineage. These results may mean that lymphocyte survival normally requires two signals, only one of which up-regulates Bcl-2. The second signal may up-regulate a second survival protein such as Bcl-x_L, or reduce the levels of Bax or Bcl-x_S. Alternatively, the second signal may inhibit cell death at a step independent of the Bcl-2-like proteins. The receptors mediating these signals may vary with differentiation stage. For example, *bcl-2/myc* pre-B cells may survive in tissue-culture medium because they express a receptor for a cytokine present in fetal calf serum, whereas the more immature *bcl-2/myc* tumor cells may lack such a receptor.

p53, DNA DAMAGE, AND CANCER

The *p53* tumor suppressor gene (for review, see Donehower and Bradley 1993) encodes a transcription factor now thought to inhibit the propagation of mutant cells. The increased risk of cancer in the absence of a functional *p53* gene was first inferred from the prevalence of *p53* mutations in human tumors and the linkage of the Li-Fraumeni syndrome, an inherited predisposition to develop various malignancies, to germ-line *p53* mutations. Definitive evidence came when mice having both *p53* alleles inactivated were found to develop multiple types of malignancies, particularly thymic lymphoma (Donehower et al. 1992; Jacks et al. 1994; Purdie et al. 1994).

Important insight into *p53* function has recently been gained. The level of *p53* rises rapidly after DNA damage. In proliferating fibroblasts, this rise promotes transient arrest in G_1, presumably to facilitate repair prior to DNA replication. G_1 arrest is achieved by *p53*-induced transcription of the gene encoding p21, an inhibitor of cyclin/cdk complexes that regulate cell cycle progression (for review, see Hunter 1993). Certain other cell types, including lymphocytes, respond to *p53* induction by undergoing apoptosis (Yonish-Rouach et al. 1991; Shaw et al. 1992; Ramqvist et al. 1993; Ryan et al. 1993). The basis for *p53*-induced apoptosis is not yet known.

The higher risk of malignant transformation in the absence of *p53* probably stems from both a reduced opportunity for DNA repair and an inability to eliminate by apoptosis those cells sustaining DNA damage. In keeping with that notion, irradiated fibroblasts from $p53^{-/-}$ mice are unable to arrest in G_1 and display a heightened propensity for gene amplification (Kastan et al. 1992; Livingstone et al. 1992; Yin et al. 1992). Furthermore, their lymphoid cells are impressively resistant to ionizing radiation, unlike normal lymphocytes, which are exquisitely sensitive (see below). Such findings justify for *p53* the epithet "guardian of the genome" (Lane 1992).

RELATIONSHIP BETWEEN p53 AND Bcl-2

In the cell death process, p53 acts upstream of Bcl-2 since, by enforcing *bcl-2* expression in an erythroleukemia cell line with inducible *p53* function, we have demonstrated that Bcl-2 can block *p53*-induced apoptosis (Strasser et al. 1994b). To compare loss of p53 and gain of Bcl-2 function in countering apoptosis in highly sensitive populations of normal cells, we intercrossed *bcl-2* transgenic and $p53^{-/-}$ mice and studied progeny having both or only one of these mutations. Thymocytes from $p53^{-/-}$ mice were markedly resistant to radiation, as reported by other investigators (Clarke et al. 1993; Lotem and Sachs 1993; Lowe et al. 1993), and so were immature B-lymphoid cells. Bcl-2 also confers radioresistance (Sentman et al. 1991; Strasser et al. 1991; Siegel et al. 1992), but this was not due to interference with *p53* induction.

p53 is instrumental in the response to DNA damage but not in the apoptosis induced in lymphocytes by glucocorticoids, ionomycin, or phorbol esters (Clarke et al. 1993; Lotem and Sachs 1993; Lowe et al. 1993; Strasser et al. 1994b). Bcl-2, in contrast, inhibits apoptosis in all these circumstances (Sentman et al. 1991; Strasser et al. 1991). Lack of *p53* did not further increase the resistance of Bcl-2-expressing lymphocytes to these agents (Strasser et al. 1994b). The absence of any evidence of synergy argues that *p53* and Bcl-2 act within the same pathway.

DNA DAMAGE CAN PROMOTE APOPTOSIS BY A *p53*-INDEPENDENT ROUTE

The radioresistance of the lymphocytes of $p53^{-/-}$ mice suggested that tumors bearing mutant (inactive) *p53* alleles would be refractory to the induction of apoptosis by genotoxic agents. To test this, we established cell lines from spontaneous thymic lymphomas of $p53^{-/-}$ mice and subjected them to graded doses of γ

irradiation. Unexpectedly, the cells were remarkably sensitive to radiation (Strasser et al. 1994b). They also died rapidly when exposed to diverse cytotoxic drugs that cause DNA damage, as shown for three independent lines of lymphoma cells in Figure 6. The possibility that their sensitivity reflected other oncogenic changes was ruled out by the finding that mitogenically activated T lymphocytes from $p53^{-/-}$ mice were also radiosensitive. Thus, cycling lymphoid cells that incur DNA damage undergo apoptosis via a $p53$-independent, as well as a $p53$-dependent, mechanism. These observations have an encouraging implication for treatment of lymphoid and perhaps other tumors: $p53$ mutation need not render a tumor cell refractory to genotoxic agents.

On irradiation, asynchronously cycling normal cells arrest in both G_1 and G_2. Cell cycle analysis indicated that irradiated $p53^{-/-}$ lymphoma cells arrested in G_2 prior to undergoing apoptosis. Thus, as suggested from analysis of fibroblasts, $p53$ controls only the G_1 checkpoint (Fig. 7). G_2 checkpoint control of mammalian cells remains to be elucidated. In yeast, it is regulated by the rad-9 gene product, which presumably invokes inhibition of the cyclin/kinase complex required for progression into mitosis. An intriguing possibility is that, like $p53$, a mammalian homolog of Rad-9 promotes apoptosis as well as cell cycle arrest (Fig. 7).

When Bcl-2 expression was enforced in the $p53^{-/-}$ lymphoma cells, they acquired significant resistance to genotoxic agents. Bcl-2 did not prevent cell cycle arrest, but irradiated cells now accumulated in G_1 as well as in G_2. This observation raises the unexpected possibility

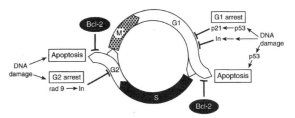

Figure 7. DNA damage triggers apoptosis of lymphocytes by $p53$-dependent and $p53$-independent mechanisms. Increase in $p53$ promotes G_1 arrest by enhanced expression of the p21 inhibitor of cyclin/cdk2 and also promotes apoptosis. Cycling lymphocytes which lack $p53$ still undergo apoptosis in response to DNA damage (see text). The prior arrest of irradiated $p53^{-/-}$ lymphoblasts in G_2 may indicate that $p53$-independent apoptosis is controlled by a mammalian Rad-9 homolog, possibly via an inhibitor (denoted here as In) of the cyclin-dependent kinase that promotes G_2/M progression. High levels of Bcl-2 block both $p53$-dependent and $p53$-independent apoptosis but do not prevent cell cycle arrest. It is unclear whether apoptosis and cell cycle arrest are coupled or are independent responses.

that a $p53$-independent checkpoint can be triggered in G_1 by DNA damage (Fig. 7). A clonogenic assay indicated that Bcl-2 did not merely delay cell death but significantly enhanced the probability of long-term survival after exposure to otherwise lethal doses of irradiation. We infer from these experiments that high expression of survival genes like bcl-2 may pose a greater impediment than loss of $p53$ to genotoxic cancer therapy.

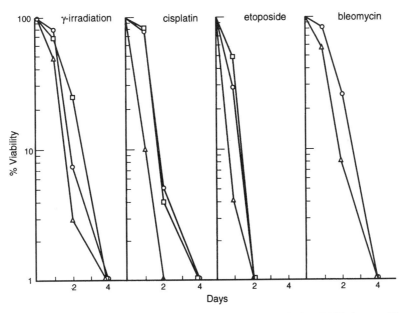

Figure 6. $p53^{-/-}$ lymphoma cells undergo apoptosis after exposure to agents that induce DNA damage. Three independent cell lines derived from thymic T-cell lymphomas arising spontaneously in $p53^{-/-}$ mice were exposed to ionizing radiation (10 Gy) or cultured in medium containing cisplatin (10 µg/ml), etoposide (10 µg/ml), or bleomycin (0.15 units/ml). Cell viability was monitored by trypan blue exclusion. Flow cytometric assay for cells with a subdiploid DNA content confirmed that death was by apoptosis.

p53 AND IMMORTALIZATION

Immortalization of fibroblasts is frequently associated with *p53* mutation (Harvey and Levine 1991). We are exploring whether lack of *p53* facilitates the immortalization of hematopoietic cells (T. Metz, unpubl.). Since induction of erythroleukemia in mice by the Friend virus is almost invariably associated with *p53* mutation (Ben-David and Bernstein 1991), we turned to erythroid cells for this study. The specific question posed was whether the absence of *p53* from primitive erythroid cells would facilitate their immortalization by oncogenes. The genes *myc* and *raf* were chosen because each stimulates cellular proliferation and their combined action is particularly potent. Fetal liver cells from normal and $p53^{-/-}$ mice were infected with a retrovirus bearing v-*myc*, v-*raf*, or both genes. Colonies of proliferating cells were readily obtained with viruses expressing *raf*, or *myc* plus *raf*, but none arose with vectors expressing *myc* alone. The absence of *p53* did not affect the number or size of the colonies, but an important difference emerged when the colony cells were subcultured. With $p53^{+/+}$ cells, no continuous lines could be established from the *raf* virus-infected colonies, and the few that arose from the *myc/raf* colonies emerged only after a "crisis" suggestive of the demise of all but a few cells that had acquired mutations. In striking contrast, nearly all *myc/raf*-expressing colonies of $p53^{-/-}$ cells yielded lines without a crisis. Significantly, these lines were diploid and remained so through many cell divisions, whereas the few lines obtained from v-*raf*-expressing colonies were frequently aneuploid.

These experiments indicate that although loss of *p53* is insufficient for immortalization, even in the context of an activated *myc* or *raf* gene, its absence facilitates the generation of stable cell lines from erythroid progenitors by the concerted activity of both oncogenes. The efficiency of their establishment argues that transformation of cells lacking *p53* function is not solely due to a failure to prevent the propagation of cells with damaged DNA. We suggest that *p53* also normally plays a role in blocking cell division once the "Hayflick limit" on clonal life span is reached. In keeping with this hypothesis, senescent human cells contain elevated levels of mRNA encoding the p21 inhibitor of cyclin-dependent kinases whose expression is stimulated by *p53* (Noda et al. 1994).

A MOLECULAR MODEL FOR APOPTOSIS

Since Bcl-2 can inhibit apoptosis induced by such a wide variety of cytotoxic insults, it seems likely that several independent routes to cell death converge and that the step regulated by Bcl-2 lies within a final common pathway (Fig. 8). Important clues to the biochemical mechanism underlying apoptosis have been gleaned from *C. elegans*. Both the *ced-3* and *ced-4* genes are required for efficient operation of the death pathway in that organism. If either is inactive, the cells

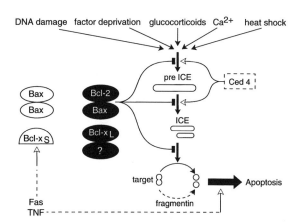

Figure 8. Model for regulation of apoptosis. The ability of Bcl-2 to inhibit cell death triggered in diverse ways argues for a final common pathway. This involves an ICE-like cysteine protease and possibly the mammalian homolog of Ced-4. Bcl-2 and its relative Bcl-x may suppress apoptosis by inhibiting activity of the protease by either preventing cleavage of the pro-enzyme, directly interfering with proteolytic activity, or sequestering a target protein. Bcl-2 apparently acts as a heterodimer with a homologous protein Bax; Bcl-x_L is postulated to have an equivalent partner. Bax/Bax homodimers and Bcl-x_S antagonize Bcl-2-promoted survival. Death provoked by the Fas or TNF receptors is not inhibited by Bcl-2 and the effector(s) may therefore act by increasing the level of Bcl-2 (Bcl-x) antagonists or by interfering with a late step in the pathway. Since Bcl-2 also fails to block killing by cytotoxic T cells, the fragmentins (or granzymes) released into the target cell may cleave critical ICE targets.

that normally would die during development are spared. The *ced-9* gene product blocks the action of *ced-3* and *ced-4*. In *ced-9* gain-of-function mutants, programmed cell death is ablated, whereas loss of its function results in attrition of cells that normally would persist. This suggests that the death mechanism is poised for action in many cells and that continuous *ced-9* expression is necessary to suppress it. Although the sequence of *ced-4* has provided little clue to its function and no mammalian homolog has yet emerged, *ced-3* is related to the mammalian gene encoding interleukin-1β converting enzyme (ICE), a cysteine protease which cleaves the cytokine precursor at specific aspartate residues. Significantly, mammalian fibroblasts overexpressing either ICE or *ced-3* undergo apoptosis which can be blocked by Bcl-2 (Miura et al. 1993). The strong inference is that apoptosis is triggered by the action of an ICE-like protease.

ICE is itself synthesized as a proenzyme which must be cleaved at aspartic acid residues to yield two subunits, suggestive of an autocatalytic cascade, and both subunits are required for apoptosis (Miura et al. 1993). There may be a family of ICE-like proteases, since the cDNA for a homologous gene isolated from neuronal precursors, denoted *NEDD2*, also promotes cell death and its action can be blocked by Bcl-2 (Kumar et al. 1994). The relevant substrates for the apoptosis-inducing proteases remain to be determined. Interleukin-1β is unlikely to be the only target because ICE is expressed in some cells that do not synthesize

IL-1. It is unclear how Bcl-2 and Bcl-x inhibit this pathway. They may, for example, prevent cleavage of the pro-enzyme, directly interfere with proteolytic activity, or sequester a relevant target protein.

High levels of Bcl-2 do not block apoptosis triggered by certain stimuli. As mentioned above, negative selection of self-reactive T lymphocytes remains largely unperturbed by *bcl-2* transgenes (Sentman et al. 1991; Strasser et al. 1991a, 1994c; Siegel et al. 1992; Linette et al. 1994). Also largely unaffected is cell death induced by engagement of Fas or the 55-kD tumor necrosis factor (TNF) receptor (Vanhaesebroeck et al. 1993; A. Strasser, unpubl.) or by the action of cytotoxic T cells (Strasser et al. 1991a; Vaux et al. 1992a). Death provoked by Fas or the TNF receptor may be resistant to Bcl-2 because a critical effector molecule acts downstream from Bcl-2 or because the receptor induces antagonists such as Bax/Bax homodimers or Bcl-x_S. Among the fragmentins (or granzymes) implicated in the killing process implemented by cytotoxic T cells is an aspartate-specific protease; although it bears no homology with ICE, this fragmentin might cleave critical ICE targets, thereby short-circuiting the pathway and escaping Bcl-2 inhibition. Thus, failure to respond to Bcl-2 does not necessarily imply an alternative effector mechanism for apoptosis. Many fundamental issues regarding the cell death circuitry clearly remain to be resolved, but the momentum of current research suggests that answers will rapidly be forthcoming—cell death is a lively field.

ACKNOWLEDGMENTS

This work was supported by the National Health and Medical Research Council of Australia, the U.S. National Cancer Institute (CA-43540), and the Howard Hughes Medical Institute (75193-531101). S.C. is an International Scholar and T.J. is an Assistant Investigator of the Howard Hughes Medical Institute. A.S. was the recipient of fellowships from the Leukemia Society of America and the Swiss National Science Foundation; L.M.C. is a Cancer Research Institute Investigator; T.M. has a fellowship from the Boehringer Ingelheim Fonds.

REFERENCES

Adams, J.M., A.W. Harris, C.A. Pinkert, L.M. Corcoran, W.S. Alexander, S. Cory, R.D. Palmiter, and R.L. Brinster. 1985. The c-myc oncogene driven by immunoglobulin enhancers induces lymphoid malignancy in transgenic mice. *Nature* **318:** 533.

Askew, D.S., R.A. Ashmun, B.C. Simmons, and J.L. Cleveland. 1991. Constitutive c-myc expression in an IL-3-dependent myeloid cell line suppresses cell cycle arrest and accelerates apoptosis. *Oncogene* **6:** 1915.

Ben-David, Y. and A. Bernstein. 1991. Friend virus-induced erythroleukemia and the multistage nature of cancer. *Cell* **66:** 831.

Boise, L.H., M. Gonzalez-Garcia, C.E. Postema, L. Ding, T. Lindsten, L.A. Turka, X. Mao, G. Nunez, and C.B. Thompson. 1993. bcl-x, a bcl-2-related gene that functions as a dominant regulator of apoptotic cell death. *Cell* **74:** 597.

Chen-Levy, Z. and M.L. Cleary. 1990. Membrane topology of the Bcl-2 proto-oncogenic protein demonstrated in vitro. *J. Biol. Chem.* **265:** 4929.

Clarke, A.R., C.A. Purdie, D.J. Harrison, R.G. Morris, C.C. Bird, M.L. Hooper, and A.H. Wyllie. 1993. Thymocyte apoptosis induced by p53-dependent and independent pathways. *Nature* **362:** 849.

de Jong, D., F.A. Prins, D.Y. Mason, J.C. Reed, G.B. van Ommen, and P.M. Kluin. 1994. Subcellular localization of the bcl-2 protein in malignant and normal lymphoid cells. *Cancer Res.* **54:** 256.

Donehower, L.A. and A. Bradley. 1993. The tumor suppressor p53. *Biochim. Biophys. Acta* **1155:** 181.

Donehower, L.A., M. Harvey, B.L. Slagle, M.J. McArthur, C.A.J. Montgomery, J.S. Butel, and A. Bradley. 1992. Mice deficient for p53 are developmentally normal but are susceptible to spontaneous tumours. *Nature* **356:** 215.

Ellis, R.E., J. Yuan, and H.R. Horvitz. 1991. Mechanisms and functions of cell death. *Annu. Rev. Cell Biol.* **7:** 663.

Evan, G.I., A.H. Wyllie, C.S. Gilbert, T.D. Littlewood, H. Land, C. Brooks, C.M. Waters, L.Z. Penn, and D.C. Hancock. 1992. Induction of apoptosis in fibroblasts by c-myc protein. *Cell* **69:** 119.

Gratiot-Deans, J., L. Ding, L.A. Turka, and G. Nunez. 1993. bcl-2 proto-oncogene expression during human T cell development. *J. Immunol.* **151:** 83.

Harrington, E.A., A. Fanidi, and G.I. Evan. 1994. Oncogenes and cell death. *Curr. Opin. Genet. Dev.* **4:** 120.

Harris, A.W., C.A. Pinkert, M. Crawford, W.Y. Langdon, R.L. Brinster, and J.M. Adams. 1988. The Eµ-myc transgenic mouse. A model for high-incidence spontaneous lymphoma and leukemia of early B cells. *J. Exp. Med.* **167:** 353.

Hartley, S.B., M.P. Cooke, D.A. Fulcher, A.W. Harris, S. Cory, A. Basten, and C.C. Goodnow. 1993. Elimination of self-reactive B lymphocytes proceeds in two stages: Arrested development and cell death. *Cell* **72:** 325.

Harvey, D.M. and A.J. Levine. 1991. p53 alteration is a common event in the spontaneous immortalization of primary BALB/c murine embryo fibroblasts. *Genes Dev.* **5:** 2375.

Hengartner, M.O. and H. R. Horvitz. 1994. *C. elegans* cell survival gene ced-9 encodes a functional homolog of the mammalian proto-oncogene bcl-2. *Cell* **76:** 665.

Hockenbery, D., G. Nunez, C. Milliman, R.D. Schreiber, and S.J. Korsmeyer. 1990. Bcl-2 is an inner mitochondrial membrane protein that blocks programmed cell death. *Nature* **348:** 334.

Hockenbery, D.M., M. Zutter, W. Hickey, M. Nahm, and S. Korsmeyer. 1991. BCL2 protein is topographically restricted in tissues characterized by apoptotic cell death. *Proc. Natl. Acad. Sci.* **88:** 6961.

Horning, S.J. and S.A. Rosenberg. 1984. The natural history of initially untreated low-grade non-Hodgkin's lymphomas. *N. Engl. J. Med.* **311:** 1471.

Hunter, T. 1993. Braking the cycle. *Cell* **75:** 839.

Jacks, T., L. Remington, B.O. Williams, E.M. Schmitt, S. Halachmi, R.T. Bronson, and R.A. Weinberg. 1994. Tumor spectrum analysis in p53-mutant mice. *Curr. Biol.* **4:** 1.

Jacobson, M.D., J.F. Burne, and M.C. Raff. 1994. Programmed cell death and Bcl-2 protection in the absence of a nucleus. *EMBO J.* **13:** 1899.

Kastan, M.B., Q. Zhan, W.S. El-Deiry, F. Carrier, T. Jacks, W.V. Walsh, B.S. Plunkett, B. Vogelstein, and A.J.J. Fornace. 1992. A mammalian cell cycle checkpoint pathway utilizing p53 and GADD45 is defective in ataxia-telangiectasia. *Cell* **71:** 587.

Katsumata, M., R.M. Siegel, D.C. Louie, T. Miyashita, Y. Tsujimoto, P.C. Nowell, M.I. Greene, and J.C. Reed. 1992. Differential effects of Bcl-2 on T and B cells in transgenic mice. *Proc. Natl. Acad. Sci.* **89:** 11376.

Kerr, J.F.R. and B.V. Harmon. 1991. Apoptosis: An historical perspective. *Curr. Commun. Cell Mol. Biol.* **3:** 5.

Kerr, J.F.R., A.H. Wyllie, and A.R. Currie. 1972. Apoptosis: A basic biological phenomenon with wide-ranging implications in tissue kinetics. *Br. J. Cancer* **26:** 239.

Korsmeyer, S.J. 1992. Bcl-2 initiates a new category of oncogenes: Regulators of cell death. *Blood* **80:** 879.

Krajewski, S., S. Tanaka, S. Takayama, M.J. Schibler, W. Fenton, and J.C. Reed. 1993. Investigation of the subcellular distribution of the *bcl*-2 oncoprotein: Residence in the nuclear envelope, endoplasmic reticulum, and outer mitochondrial membranes. *Cancer Res.* **53:** 4701.

Kumar, S., M. Kinoshita, M. Noda, N.G. Copeland, and N.A. Jenkins. 1994. Induction of apoptosis by mouse *Nedd2* gene encoding a protein similar to the product of *C. elegans* cell death gene *ced-3* and mammalian IL-1β-converting enzyme. *Genes Dev.* **8:** 1613.

Lane, D.P. 1992. p53, guardian of the genome. *Nature* **358:** 15.

Langdon, W.Y., A.W. Harris, and S. Cory. 1988. Growth of Eμ-myc transgenic B-lymphoid cells *in vitro* and their evolution toward autonomy. *Oncogene Res.* **3:** 271.

Langdon, W.Y., A.W. Harris, S. Cory, and J.M. Adams. 1986. The c-*myc* oncogene perturbs B lymphocyte development in Eμ-*myc* transgenic mice. *Cell* **47:** 11.

Lazebnik, Y.A., S. Cole, C.A. Cooke, W.G. Nelson, and W.C. Earnshaw. 1993. Nuclear events of apoptosis in vitro in cell-free mitotic extracts: A model system for analysis of the active phase of apoptosis. *J. Cell Biol.* **1234:** 7.

LeBrun, D.P., R.A. Warnke, and M.L. Cleary. 1993. Expression of bcl-2 in fetal tissues suggests a role in morphogenesis. *Am. J. Pathol.* **142:** 743.

Limpens, J., D. de Jong, J.H.J.M. van Krieken, G.A. Price, B.D. Young, G.-J.B. van Ommen, and P.M. Kluin. 1991. Bcl-2/J_H rearrangements in benign lymphoid tissues with follicular hyperplasia. *Oncogene* **6:** 2271.

Linette, G.P., M.J. Grusby, S.M. Hedrick, T.H. Hansen, L.H. Glimcher, and S.J. Korsmeyer. 1994. Bcl-2 is upregulated at the CD4$^+$CD8$^+$ stage during positive selection and promotes thymocyte differentiation at several control points. *Immunity* **1:** 195.

Lithgow, T., R. van Driel, J.F. Bertram, and A. Strasser. 1994. The protein product of the oncogene *bcl*-2 is a component of the nuclear envelope, the endoplasmic reticulum and the outer mitochondrial membrane. *Cell Growth Differ.* **5:** 411.

Livingstone, L.R., A. White, J. Sprouse, E. Livanos, T. Jacks, and T.D. Tlsty. 1992. Altered cell cycle arrest and gene amplification potential accompany loss of wild-type p53. *Cell* **70:** 923.

Lockshin, R.A. and Z. Zakeri. 1991. Programmed cell death and apoptosis. *Curr. Commun. Cell Mol. Biol.* **3:** 47.

Lotem, J. and L. Sachs. 1993. Hematopoietic cells from mice deficient in wild-type p53 are more resistant to induction of apoptosis by some agents. *Blood* **82:** 1092.

Lowe, S.W., H.E. Ruley, T. Jacks, and D.E. Housman. 1993. p53-dependent apoptosis modulates the cytotoxicity of anticancer agents. *Cell* **74:** 957.

Martin, S.J. 1993. Apoptosis: Suicide, execution or murder? *Trends Cell Biol.* **3:** 141.

McDonnell, T.J. and S.J. Korsmeyer. 1991. Progression from lymphoid hyperplasia to high-grade malignant lymphoma in mice transgenic for the t(14; 18). *Nature* **349:** 254.

McDonnell, T.J., N. Deane, F.M. Platt, G. Nunez, U. Jaeger, J.P. McKearn, and S.J. Korsmeyer. 1989. bcl-2-immunoglobulin transgenic mice demonstrate extended B cell survival and follicular lymphoproliferation. *Cell* **57:** 79.

Meikrantz, W., S. Gisselbrecht, S.W. Tam, and R. Schlegel. 1994. Activation of cyclin A-dependent protein kinases during apoptosis. *Proc. Natl. Acad. Sci.* **91:** 3754.

Miura, M., H. Zhu, R. Rotello, E.A. Hartwieg, and J. Yuan. 1993. Induction of apoptosis in fibroblasts by IL-1β-converting enzyme, a mammalian homolog of the *C. elegans* cell death gene *ced-3*. *Cell* **75:** 653.

Monaghan, P., D. Robertson, T.A.S. Amos, M.J.S. Dyer, D.Y. Mason, and M.F. Greaves. 1992. Ultrastructural localization of BCL-2 protein. *J. Histochem. Cytochem.* **40:** 1819.

Noda, A., Y. Ning, S.F. Venable, O.M. Pereira-Smith, and J.R. Smith. 1994. Cloning of senescent cell-derived inhibitors of DNA synthesis using an expression screen. *Exp. Cell Res.* **211:** 90.

Novack, D.V. and S.J. Korsmeyer. 1994. *Bcl*-2 protein expression during murine development. *Am. J. Pathol.* **145:** 1.

Oltvai, Z.N., C.L. Milliman, and S.J. Korsmeyer. 1993. Bcl-2 heterodimerizes in vivo with a conserved homolog, Bax, that accelerates programed cell death. *Cell* **74:** 609.

Oppenheim, R.W. 1991. Cell death during development of the nervous system. *Annu. Rev. Neurosci.* **14:** 453.

Osmond, D.G., N. Kim, R. Manoukian, R.A. Phillips, S.A. Rico-Vargas, and K. Jacobsen. 1992. Dynamics and localization of early B-lymphocyte precursor cells (pro-B cells) in the bone marrow of *scid* mice. *Blood* **79:** 1695.

Pezzella, F., A.G.D. Tse, J.L. Cordell, K.A.F. Pulford, K.C. Gatter, and D.Y. Mason. 1990. Expression of the *bcl*-2 oncogene protein is not specific for the 14;18 chromosomal translocation. *Am. J. Pathol.* **137:** 225.

Purdie, C.A., D.J. Harrison, A. Peter, L. Dobbie, S. White, S.E.M. Howie, D.M. Salter, C.C. Bird, A.H. Wyllie, M.L. Hooper, and A.R. Clarke. 1994. Tumour incidence, spectrum and ploidy in mice with a large deletion in the p53 gene. *Oncogene* **9:** 603.

Raff, M.C. 1992. Social controls on cell survival and cell death. *Nature* **356:** 397.

Ramqvist, T., K.P. Magnusson, Y. Wang, L. Szekely, G. Klein, and K.G. Wiman. 1993. Wild-type p53 induces apoptosis in a Burkitt lymphoma (BL) line that carries mutant p53. *Oncogene* **8:** 1495.

Reed, J.C. 1994. Bcl-2 and the regulation of programmed cell death. *J. Cell Biol.* **124:** 1.

Ryan, J.J., R. Danish, C.A. Gottlieb, and M.A. Clarke. 1993. Cell cycle analysis of p53-induced cell death in murine erythroleukemia cells. *Mol. Cell. Biol.* **13:** 711.

Sentman, C.L., J.R. Shutter, D. Hockenbery, O. Kanagawa, and S.J. Korsmeyer. 1991. *bcl*-2 inhibits multiple forms of apoptosis but not negative selection in thymocytes. *Cell* **67:** 879.

Shaw, P., R. Bovey, S. Tardy, R. Sahli, B. Sordat, and J. Costa. 1992. Induction of apoptosis by wild-type p53 in a human colon tumor-derived cell line. *Proc. Natl. Acad. Sci.* **89:** 4495.

Shi, L., W.K. Nishioka, J. Th'ng, E.M. Bradbury, D.W. Litchfield, and A.H. Greenberg. 1994. Premature p34^{cdc-2} activation is required for apoptosis. *Science* **263:** 1143.

Siegel, R.M., M. Katsumata, T. Miyashita, D.C. Louie, M.I. Greene, and J.C. Reed. 1992. Inhibition of thymocyte apoptosis and negative antigenic selection in *bcl*-2 transgenic mice. *Proc. Natl. Acad. Sci.* **89:** 7003.

Strasser, A., A.W. Harris, and S. Cory. 1991a. Bcl-2 transgene inhibits T cell death and perturbs thymic self-censorship. *Cell* **67:** 889.

———. 1993. Eμ-*bcl*-2 transgene facilitates spontaneous transformation of early pre-B and immunoglobulin-secreting cells but not T cells. *Oncogene* **8:** 1.

Strasser, A., A.W. Harris, M.L. Bath, and S. Cory. 1990. Novel primitive lymphoid tumours induced in transgenic mice by cooperation between *myc* and *bcl*-2. *Nature* **348:** 331.

Strasser, A., A.W. Harris, L.M. Corcoran, and S. Cory. 1994a. *bcl*-2 expression promotes B but not T lymphoid development in *scid* mice. *Nature* **368:** 457.

Strasser, A., A.W. Harris, T. Jacks, and S. Cory. 1994b. DNA damage can induce apoptosis in proliferating lymphoid cells via p53-independent mechanisms inhabitable by Bcl-2. *Cell* **79:** 329.

Strasser, A., A.W. Harris, H. Von Boehmer, and S. Cory. 1994c. Positive and negative selection of T cells in T cell receptor transgenic mice expressing a *bcl*-2 transgene. *Proc. Natl. Acad. Sci.* **91:** 1376.

Strasser, A., S. Whittingham, D.L. Vaux, M.L. Bath, J.M. Adams, S. Cory, and A.W. Harris. 1991b. Enforced *BCL2* expression in B-lymphoid cells prolongs antibody responses and elicits autoimmune disease. *Proc. Natl. Acad. Sci.* **88:** 8661.

Ucker, D.S. 1991. Death by suicide: One way to go in mammalian cellular development? *New Biol.* **3:** 103.

Vanhaesebroeck, B., J.C. Reed, D. de Valck, J. Grooten, T. Miyashita, S. Tanaka, R. Beyaert, F. van Roy, and W. Fiers. 1993. Effect of *bcl-2* proto-oncogene expression on cellular sensitivity to tumor necrosis factor-mediated cytotoxicity. *Oncogene* **8:** 1075.

Vaux, D.L., H.L. Aguila, and I.L. Weissman. 1992a. *Bcl-2* prevents death of factor-deprived cells but fails to prevent apoptosis in targets of cell mediated killing. *Int. Immunol.* **4:** 821.

Vaux, D.L., S. Cory, and J.M. Adams. 1988. *Bcl-2* gene promotes haemopoietic cell survival and cooperates with c-*myc* to immortalize pre-B cells. *Nature* **335:** 440.

Vaux, D.L., I.L. Weissman, and S.K. Kim. 1992b. Prevention of programmed cell death in *Caenorhabditis elegans* by human *bcl-2*. *Science* **258:** 1955.

Veis, D.J., C.L. Sentman, E.A. Bach, and S.J. Korsmeyer. 1993. Expression of the Bcl-2 protein in murine and human thymocytes and in peripheral T lymphocytes. *J. Immunol.* **151:** 2546.

Wyllie, A.H., J.F.R. Kerr, and A.R. Currie. 1980. Cell death: The significance of apoptosis. *Int. Rev. Cytol.* **68:** 251.

Yin, X.-M., Z.N. Oltvai, and S.J. Korsmeyer. 1994. BH1 and BH2 domains of Bcl-2 are required for inhibition of apoptosis and heterodimerization with Bax. *Nature* **369:** 321.

Yin, Y., M.A. Tainsky, F.Z. Bischoff, L.C. Strong, and G.M. Wahl. 1992. Wild-type p53 restores cell cycle control and inhibits gene amplification in cells with mutant p53 alleles. *Cell* **70:** 937.

Yonish-Rouach, E., D. Resnitzky, J. Lotem, L. Sachs, A. Kimchi, and M. Oren. 1991. Wild-type p53 induces apoptosis of myeloid leukaemic cells that is inhibited by interleukin-6. *Nature* **352:** 345.

Yuan, J., S. Shaham, S. Ledoux, H.M. Ellis, and H.R. Horvitz. 1993. The *C. elegans* cell death gene *ced-3* encodes a protein similar to mammalian interleukin-1β-converting enzyme. *Cell* **75:** 641.

Ziegler, S.F., C.A. Whitlock, S.P. Goff, A. Gifford, and O.N. Witte. 1981. Lethal effect of the Abelson murine leukemia virus transforming gene product. *Cell* **27:** 477.

The Genetics of Programmed Cell Death in the Nematode *Caenorhabditis elegans*

H.R. HORVITZ, S. SHAHAM, AND M.O. HENGARTNER*

Howard Hughes Medical Institute, Department of Biology, Massachusetts Institute of Technology, Cambridge, Massachusetts 02139

Cancerous growth often results from an increased rate of cell proliferation caused by the abnormal activation of a signal transduction pathway that normally stimulates cell division only in response to growth factor signals; many of the proto-oncogenes that have been characterized function in or respond to such intercellular signaling pathways (for review, see Cooper 1990). Recent findings indicate that cancerous growth also can result from a decreased rate of cell loss. The most striking example is provided by human B-cell follicular lymphomas. These cancers are often associated with t(14;18) chromosomal translocations that cause the proto-oncogene *bcl-2* (*bcl*, *B c*ell *l*ymphoma), normally located on chromosome 18, to be adjacent to and regulated by an enhancer of the immunoglobulin heavy-chain locus, normally located on chromosome 14 (Bakhshi et al. 1985; Cleary and Sklar 1985; Tsujimoto et al. 1985). The resulting overexpression of a normal Bcl-2 protein in the B-cell lineage leads to the cancerous growth of B cells (McDonnell and Korsmeyer 1991). Studies of *bcl-2* have indicated that this gene acts to protect cells from undergoing programmed cell death and that the oncogenic activity of *bcl-2* is a consequence of its allowing B cells that would normally die instead to survive and subsequently proliferate (for review, see Korsmeyer 1992).

Other findings have also associated cancerous growth with the process of programmed cell death. For example, overexpression of the tumor suppressor gene *p53* can trigger programmed cell death (Yonish-Rouach et al. 1991), and, on the basis of studies of *p53*-deficient mice, *p53* is required for at least radiation-induced programmed cell death (Lowe et al. 1993; Clarke et al. 1993). These findings suggest that a loss of *p53* function could contribute to cancerous growth by causing cells that normally die instead to live. The proto-oncogene c-*myc* also can trigger programmed cell death (Evan et al. 1992; Shi et al. 1992). This observation has led to the hypothesis that *myc* functions by preventing cellular differentiation and can cause either proliferation or death, depending on whether appropriate survival factors are or are not present (G. Evan, pers. comm.). These relationships between cancer genes and programmed cell death suggest that an understanding of the mechanisms responsible for programmed cell death could be of central importance to an understanding of human cancer.

To identify and characterize the genes and proteins involved in programmed cell death, we have been analyzing this process in the nematode *Caenorhabditis elegans*. During *C. elegans* development, the generation of the total of 959 non-germ-line nuclei present in the adult hermaphrodite is accompanied by the generation and subsequent death of 131 additional cells (Sulston and Horvitz 1977; Sulston et al. 1983). These cell deaths display certain morphological features characteristic of the apoptotic programmed cell deaths seen in mammals, including cell shrinkage, nuclear condensation, and the phagocytosis of cell corpses while cellular organelles remain intact (Wyllie et al. 1980; Robertson and Thomson 1982). Furthermore, as we discuss below, at least some of the genes that function in *C. elegans* programmed cell deaths have counterparts that function in mammalian cell deaths. These findings suggest that the pathway of programmed cell death has been conserved from nematode to human and that studies of this pathway in *C. elegans* may well facilitate an understanding of the mechanisms that can lead to cancerous growth in humans.

GENETIC PATHWAY FOR PROGRAMMED CELL DEATH

Mutations that affect programmed cell deaths in *C. elegans* have defined a four-step genetic pathway (Fig. 1). Three genes—*ces-1* (*ces*, *c*ell death *s*pecification), *ces-2*, and *egl-1* (*egl*, *egg-l*aying abnormal)—define the first of these steps (Trent et al. 1983; Ellis and Horvitz 1986; Ellis and Horvitz 1991). These genes can mutate to perturb the life-versus-death decision of only a few cells, either causing specific cells that normally die instead to live or causing specific cells that normally live instead to undergo programmed cell death. These three genes appear to regulate in a cell-specific fashion the process of programmed cell death, which involves 10 known genes and is responsible for killing cells, for removing cell corpses, and for degrading the cellular debris of these corpses.

Three genes—*ced-3* (*ced*, *c*ell *d*eath abnormal), *ced-4*, and *ced-9*—can mutate to cause the survival of all 131 cells that normally undergo programmed cell death (Ellis and Horvitz 1986; Hengartner et al. 1992). The surviving or "undead" cells generated when cell death

---

*Present address: Cold Spring Harbor Laboratory, P.O. Box 100, Cold Spring Harbor, New York 11724.

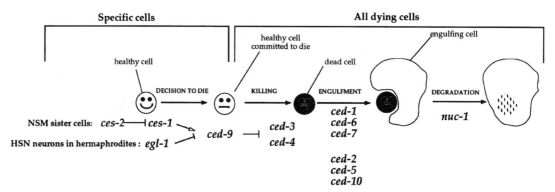

Figure 1. The genetic pathway for programmed cell death in *C. elegans*. See text for details. (Adapted from Ellis et al. 1991b.)

is blocked generally differentiate into recognizable cell types that can be, in at least some cases, capable of functioning (Ellis and Horvitz 1986; Avery and Horvitz 1987; White et al. 1991). These three genes define a killing or execution step of programmed cell death.

Six genes—*ced-1*, *ced-2*, *ced-5*, *ced-6*, *ced-7*, and *ced-10*—can mutate to disrupt the process of phagocytosis that normally acts to remove cell corpses from the body of the developing animal (Hedgecock et al. 1983; Ellis et al. 1991a). These genes define an engulfment step of programmed cell death. One gene—*nuc-1* (*nuc*, *nu*clease abnormal)—can mutate to prevent the degradation of DNA in the cell corpses formed by programmed cell death (Sulston 1976; Hedgecock et al. 1983). *nuc-1* defines a terminal step in programmed cell death in which the cellular debris of cell corpses is degraded. We discuss each step in the pathway of programmed cell death in more detail below.

SPECIFIC GENES REGULATE THE LIFE-VERSUS-DEATH DECISIONS OF SPECIFIC CELLS

The genes *ces-1* and *ces-2* control the decisions of the sister cells of the two serotonergic NSM neurons of the pharynx to undergo programmed cell death (Ellis and Horvitz 1991). Whereas these cells normally die, mutations that result in increased or altered *ces-1* function or reduced *ces-2* function cause the NSM sister cells to survive and differentiate into serotonergic cells with the morphology of the NSM neurons. Genetic analyses have suggested that *ces-1* can act to prevent NSM sister cell deaths, that *ces-2* acts as a negative regulator of *ces-1*, and that both *ces-1* and *ces-2* act genetically upstream of the cell survival gene *ced-9* and the cell killer genes *ced-3* and *ced-4* (see below). Mutations in *ces-1* that prevent the deaths of the NSM sister cells also prevent the programmed deaths of the sisters of the two pharyngeal I2 neurons. Most, and possibly all, other cell deaths appear to occur normally in *ces-1* and *ces-2* mutants, suggesting that these genes control the life-versus-death decision of only one or two cell types. However, the phenotypes that result from a complete loss of *ces-1* or *ces-2* function are unknown, so it is possible that one or both of these genes act more broadly.

Mutations in the gene *egl-1* cause the deaths in hermaphrodites of the serotonergic HSN motor neurons, which innervate the vulval muscles and stimulate egg laying (Trent et al. 1983; Ellis and Horvitz 1986). For this reason, *egl-1* mutants are defective in egg laying. Six *egl-1* alleles have been isolated, and all six have a dominant effect, causing the deaths of the HSN neurons in *egl-1/+* heterozygotes as well as in *egl-1/egl-1* homozygotes. Since in *egl-1/+* heterozygotes the HSN neurons die, but in *nDf41/+* heterozygotes, which carry a deficiency that deletes the *egl-1* gene, these cells survive, the existing *egl-1* alleles probably do not simply reduce or eliminate *egl-1* function (M. Hengartner and H.R. Horvitz, unpubl.). How these *egl-1* mutations affect *egl-1* activity and the life-versus-death decision of the HSN neurons remains to be determined.

The functions and cellular specificities of action of *ces-1*, *ces-2*, and *egl-1* are not yet known. Nonetheless, the studies of these genes to date suggest that specific genes regulate the life-versus-death decisions of specific cells and that many more genes of this class may exist, with each such gene controlling the life-versus-death decisions of only a small number of cells.

THE *ced-9* GENE PROTECTS CELLS AGAINST PROGRAMMED CELL DEATH

The gene *ced-9* was initially defined by a gain-of-function mutation, *n1950*, that causes all 131 cells that normally undergo programmed cell death instead to live (Hengartner et al. 1992). Mutations that reduce or eliminate *ced-9* function have the opposite effect, causing cells that normally live instead to undergo programmed cell death. For example, animals homozygous for a *ced-9* loss-of-function (*lf*, *l*oss of *f*unction) allele generated as the progeny of *ced-9(lf)/+* heterozygous mothers hatch and grow to normal size but lack a variety of cells normally present in wild-type animals; these cells are absent because in *ced-9(lf)/ced-9(lf)* homozygous animals they (or their progenitor cells) undergo programmed cell death (Fig. 2a). These *ced-9(lf)/ced-9(lf)* homozygous animals generate *ced-9(lf)/ced-9(lf)* embryos that arrest development during

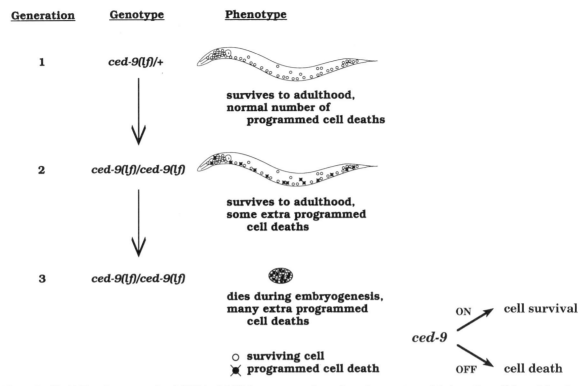

Figure 2. (*Left*) The phenotype of *ced-9(lf)/ced-9(lf)* homozygotes depends on the genotype of their mothers. *lf*, loss-of-function. (*Right*) *ced-9* acts as a binary switch to regulate programmed cell death in *C. elegans* (see text for details). (Reprinted, with permission, from Hengartner et al. 1992.)

embryogenesis; this developmental arrest occurs because large numbers of cells undergo programmed cell death. Thus, in animals homozygous for *ced-9* loss-of-function alleles, cells that normally survive instead undergo programmed cell death, and the extent to which these deaths occur depends on whether the animal is derived from a mother that was heterozygous or homozygous for the *ced-9(lf)* mutation. This maternal effect presumably reflects a contribution by a *ced-9(lf)/+* mother of wild-type *ced-9* activity (in the form of RNA or protein) to a genotypically *ced-9(lf)* oocyte, allowing a *ced-9(lf)/ced-9(lf)* embryo to survive to adulthood; in contrast, a *ced-9(lf)/ced-9(lf)* embryo derived from a *ced-9(lf)/ced-9(lf)* mother would not contain wild-type *ced-9* activity and would die.

The observation that a reduction in *ced-9* function causes cells that should live to undergo programmed cell death indicates that *ced-9* acts to protect cells from programmed cell death. Conversely, overexpression of a wild-type *ced-9* transgene under the control of a *C. elegans* heat-shock promoter can prevent programmed cell deaths (Hengartner and Horvitz 1994a). Together these findings demonstrate that *ced-9* acts as a binary switch gene to regulate programmed cell death in *C. elegans*, causing cells in the developing animal to live when active and to die when inactive (Fig. 2b). Remarkably, it seems that many, and possibly all, of the cells that survive during *C. elegans* development do so because *ced-9* protects them from undergoing programmed cell death.

ced-9 ANTAGONIZES THE ACTIONS OF *ced-3* AND *ced-4*

ced-9, which protects cells from programmed cell death, acts oppositely to *ced-3* and *ced-4*, which cause cells to undergo programmed cell death (see below). *ced-9* might function by preventing the action of either or both of the killer genes *ced-3* and *ced-4*. Conversely, *ced-3* and *ced-4* might cause programmed cell death by preventing the action of *ced-9*. To distinguish between these alternatives (which are not mutually exclusive), Hengartner et al. (1992) constructed double mutants in which both *ced-9* and *ced-3* or *ced-9* and *ced-4* activities were reduced or eliminated. If *ced-9* protected cells by antagonizing *ced-3*, then *ced-9* activity would not be important for cell survival in an animal lacking *ced-3* function, i.e., in a *ced-9; ced-3* double mutant, cells that should die would live, just as they do in the *ced-3* single mutant. If, however, *ced-3* killed by antagonizing *ced-9*, then *ced-3* activity would not affect programmed cell deaths in an animal lacking *ced-9* function; i.e., in a *ced-9; ced-3* double mutant, cells that should live would die, just as they do in the *ced-9* single mutant. These double mutant studies demonstrated that in *ced-9; ced-3* and *ced-4; ced-9* double mutants, all cells—i.e., both those that usually die and those that die when *ced-9* activity is reduced—live, strongly supporting the hypothesis that *ced-9* acts by antagonizing the actions of *ced-3* and *ced-4*.

These gene interaction studies indicate that for *ced-9*

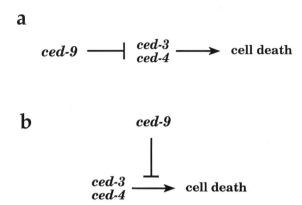

Figure 3. Two models of ways in which *ced-9* might act as a negative regulator of *ced-3* and *ced-4* function. Gene interaction studies indicate that *ced-9* inhibits the function of *ced-3* and/or *ced-4* and could do so either (*a*) by directly preventing the expression or action of *ced-3* and/or *ced-4* or (*b*) by preventing the action of any function that acts in response to *ced-3* and *ced-4* (see text for details). (Based on data from Hengartner et al. 1992.)

activity to have an effect, *ced-3* and *ced-4* must be functional, so that the effect of *ced-9* activity is to decrease the activities of *ced-3* and/or *ced-4*. Thus, *ced-9* functions in a genetic sense as a negative regulator of *ced-3* and *ced-4*. Biochemically, a CED-9 protein could function either before or after the formation of a CED-3 or CED-4 protein; e.g., a CED-9 protein might repress transcriptionally *ced-3* or *ced-4* expression, might modify and thereby inhibit posttranslationally the activity of a CED-3 or CED-4 protein, or might inhibit the activity of any function that acts in response to CED-3 and CED-4 to effect the killing of cells by programmed cell death. The first two of these alternatives are consistent with the model that these three genes act in a linear pathway for programmed cell death, whereas the third requires a branched pathway in which *ced-9* does not act directly on *ced-3* and *ced-4* but rather prevents their effect (Fig. 3).

ced-9 ENCODES A FUNCTIONAL HOMOLOG OF THE PROTO-ONCOGENE *bcl-2*

The cloning of the *ced-9* gene revealed that the inferred 280-amino-acid protein product of *ced-9* is similar in sequence to the protein product of the mammalian proto-oncogene *bcl-2* (Hengartner and Horvitz 1994a). As discussed above, *bcl-2* had been shown to delay or prevent the programmed cell deaths of a variety of cells exposed to a variety of stimuli and, for this reason, was postulated to cause B-cell follicular lymphoma by allowing the survival and subsequent proliferation of B cells in the immune system (for review, see Korsmeyer 1992). The finding that *ced-9*, the normal in vivo function of which is to protect cells from programmed cell death, and *bcl-2* were members of a gene family strongly supported the hypothesis that the normal function of *bcl-2* is to protect cells against programmed cell death.

However, the degree of overall sequence identity between the CED-9 and Bcl-2 proteins is only 23%. Is this similarity meaningful? A number of additional findings also indicate that *ced-9* and *bcl-2* are homologs (Hengartner and Horvitz 1994a). First, the sequence similarity between the protein products of these two genes is highest in those regions that are most conserved, and hence likely to be of the greatest functional significance, among all members of a growing *ced-9/bcl-2* gene family (for review, see Vaux 1994; see below). Similarly, the regions most conserved among *bcl-2* genes from different vertebrate species are also the regions most conserved between the *ced-9* genes of *C. elegans* and of another *Caenorhabditis* species, *C. briggsae*. Second, based on hydrophobicity plots, the CED-9 protein contains a hydrophobic tail like that thought to be important (Alnemri et al. 1992; Tanaka et al. 1992; Hockenbery et al. 1993) for *bcl-2* localization and/or function. Third, the last *ced-9* intron is in precisely the same position of the open reading frame as that of the last intron in many *ced-9/bcl-2* family members.

To determine if *bcl-2* is sufficiently similar to *ced-9* to function in *C. elegans* to protect worm cells against programmed cell death, we expressed a human *bcl-2* transgene in *C. elegans* under the control of a *C. elegans* heat-shock promoter (Hengartner and Horvitz 1994a). These experiments revealed that *bcl-2* can prevent the deaths of *C. elegans* cells that normally undergo programmed cell death. Similar results were obtained by Vaux et al. (1992). We further showed that expression of human *bcl-2* can prevent ectopic cell deaths that occur in *ced-9(lf)* mutants, consistent with the hypothesis that *bcl-2* is capable of substituting for *ced-9* (Hengartner and Horvitz 1994a). It should be noted that *bcl-2* is not the only gene that can act in this way: The baculovirus gene *p35*, an inhibitor of virus-induced programmed cell death in insect cells, also can act in *C. elegans* to protect against programmed cell death and prevent the ectopic cell deaths that occur in *ced-9(lf)* mutants (Sugimoto et al. 1994). *p35* is not an obvious member of the *ced-9/bcl-2* gene family, underscoring the fact that the conclusion that *ced-9* and *bcl-2* are functional homologs depends on the combined structural and functional evidence.

ced-9 is one of a growing number of *C. elegans* genes known to be transcribed as a polycistronic RNA (Hengartner and Horvitz 1994a). The *ced-9* polycistronic RNA also encodes a protein similar to cytochrome b560 of complex II of the mitochondrial respiratory chain (Yu et al. 1992), indicating that the transcription of *ced-9* and this cytochrome gene can be coordinately regulated. In this regard, it is interesting to note that Bcl-2 protein has been found localized to mitochondria (Hockenbery et al. 1990) and has been suggested to function in an antioxidant pathway to prevent cell death by acting at sites of free radical generation, such as mitochondria (Hockenbery et al. 1993; Kane et al. 1993; Veis et al. 1993).

The *bcl-2* mutations that have been associated with

follicular lymphoma are all translocations that result in the overexpression of a normal Bcl-2 protein in B cells (Bakhshi et al. 1985; Cleary and Sklar 1985; Tsujimoto et al. 1985; Cleary et al. 1986; Tsujimoto and Croce 1986). Like these mutations, the *ced-9(n1950)* mutation dominantly suppresses programmed cell death, which suggests, by analogy, that *ced-9(n1950)* may be a chromosomal rearrangement that leads to the overexpression of a normal CED-9 protein. Indeed, overexpression of a normal CED-9 protein can protect cells from programmed cell death (Hengartner and Horvitz 1994a). Sequence analysis, however, revealed that the *ced-9(n1950)* mutation is a missense mutation that causes a substitution of a glutamate for a glycine that is conserved among all *ced-9/bcl-2* gene family members (Hengartner and Horvitz 1994b). This same substitution inactivates rather than activates human *bcl-2* when assayed in both mammalian cells (Yin et al. 1994) and *C. elegans* (Hengartner and Horvitz 1994b), indicating that *ced-9* and *bcl-2* are not completely equivalent. The site of the *n1950* mutation may identify a protein domain that regulates the activities of proteins of the *ced-9/bcl-2* family, leading to *ced-9* activation and *bcl-2* inactivation when appropriately perturbed. The finding that a missense mutation can activate *ced-9* suggests that similar mutations in other members of the *ced-9/bcl-2* family might lead to oncogenic activation. In this context, it is noteworthy that a chicken *bcl-2* cDNA isolated from a B-cell lymphoma contains a valine instead of the glycine found in the genomic sequence at this site (Cazals-Hatem et al. 1992; Eguchi et al. 1992). Perhaps this chicken lymphoma resulted from an activation of *bcl-2* caused by this missense mutation.

bcl-2 is only one of a number of *ced-9/bcl-2* gene family members found in vertebrates (Boise et al. 1993; Kozopas et al. 1993; Lin et al. 1993; Oltvai et al. 1993). In contrast, *ced-9* is the only such family member as yet described for *C. elegans*. Of the vertebrate gene family members, some (e.g., *bcl-2*) can protect against programmed cell death, some (*bax*) can promote programmed cell death, and some (*bcl-x*) encode both death-inhibiting and death-promoting forms (Boise et al. 1993; Oltvai et al. 1993). Genetic analyses of *ced-9* have suggested that *ced-9* also might encode a death-promoting activity as well as its well-characterized death-inhibiting activity (Hengartner and Horvitz 1994b).

THE *ced-3* AND *ced-4* GENES CAUSE CELLS TO UNDERGO PROGRAMMED CELL DEATH

Mutations that reduce or eliminate the functions of the genes *ced-3* or *ced-4* cause cells that normally undergo programmed cell death instead to survive, showing that the activities of both of these genes are needed for the process of programmed cell death (Ellis and Horvitz 1986). Genetic mosaic experiments suggested that *ced-3* and *ced-4* are expressed within cells that will die and act within those cells to cause them to undergo programmed cell death (Yuan and Horvitz 1990). Thus, programmed cell death can be considered, at least to some extent, to be an active process on the part of dying cells. As discussed above, both *ced-3* and *ced-4* act genetically downstream from or in parallel to *ced-9* in the pathway of programmed cell death.

The cloning of *ced-4* (Yuan and Horvitz 1992) and *ced-3* (Yuan et al. 1993) indicated that each encodes a single mRNA that is expressed predominantly during embryogenesis, when 113 of the 131 programmed deaths occur. *ced-3* gene function is not required for *ced-4* gene expression, nor is *ced-4* gene function required for *ced-3* gene expression. Thus, *ced-3* and *ced-4* do not control the onset of programmed cell death by acting sequentially in a transcriptional regulatory cascade.

The sequence of the inferred 549-amino-acid CED-4 protein shows no significant similarities to the sequences of other known proteins (Yuan and Horvitz 1992). Although we previously noted that two regions of the CED-4 protein show some similarity to the EF-hand calcium-binding motif, comparison with a more recent compilation of a larger number of EF-hand sequences (A. Bairoch, pers. comm.) indicates that this similarity is unlikely to be meaningful. Furthermore, relevant residues within these two regions are not conserved in the related *Caenorhabditis* species *C. briggsae* and *C. vulgaris*, and we have mutated by site-directed mutagenesis residues that might be expected to be functionally important within these regions without disrupting *ced-4* function (S. Shaham and H.R. Horvitz, unpubl.). Thus, we do not have any evidence that *ced-4* encodes a calcium-binding protein.

ced-3 ENCODES A PROTEIN SIMILAR TO THE HUMAN CYSTEINE PROTEASE ICE

The sequence of the inferred 503-amino-acid CED-3 protein is similar to that of the human cysteine protease interleukin-1β-converting enzyme, or ICE (Yuan et al. 1993). The CED-3 and ICE proteins are 29% identical over their entire lengths and 43% identical over a 115-amino-acid region that includes a completely conserved pentapeptide (QACRG), which contains the active-site cysteine of ICE. ICE was identified and purified on the basis of its ability to cleave the inactive 31-kD precursor of interleukin-1β (IL-1β) to generate the active 17.5-kD cytokine (Cerretti et al. 1992; Thornberry et al. 1992). ICE was inferred to be a cysteine protease from inhibitor studies. Active human ICE is composed of two subunits (p20 and p10) that appear to be cleaved autoproteolytically from a single 45-kD proenzyme. Both the precursor of IL-1β and the ICE proenzyme are cleaved after aspartate residues, although the precise amino acid sequence that defines a good cleavage target site is not known.

The similarity between CED-3 and ICE strongly suggests that CED-3 functions as a cysteine protease in controlling programmed cell death in *C. elegans*. CED-3 might act either by activating another protein or set of proteins that cause cells to die or by inactivating a protein or set of proteins that protect cells from death. In the latter case, such a protective protein could not be CED-9 or a protein that acts in response to CED-9, since, as discussed above, genetic studies indicate that *ced-9* acts upstream of or in parallel with rather than downstream from *ced-3*. CED-9 could act by negatively regulating CED-3 activity, for example, either by preventing the proteolytic activation of a pro-form of CED-3 or by inhibiting the activity of a CED-3 protease once it has been generated. CED-4 could be a target of CED-3 or, alternatively, could function to activate CED-3. Targets of CED-3 might include not only proteins that function in causing or preventing programmed cell death, but also proteins that function in later aspects of the process of programmed cell death, such as the phagocytosis of cell corpses (which involves the presumptive proteins CED-1, CED-2, CED-5, CED-6, CED-7, CED-10) or in the degradation of these corpses (which involves the protein NUC-1).

The similarity between CED-3 and ICE also suggests that ICE or a cysteine protease in the CED-3/ICE family might function in controlling programmed cell death in vertebrates. This hypothesis is supported by two additional findings. First, expression of either *ced-3* or ICE in rat fibroblasts can cause these cells to undergo apoptotic cell death (Miura et al. 1993). Thus, this class of cysteine proteases can cause mammalian cells to undergo programmed cell death. Second, *crmA* (*crm*, *c*ytokine *r*esponse *m*odifier), a protein inhibitor of ICE encoded by a cowpox virus, can block the apoptotic death of chick dorsal root ganglion cells that have been deprived of nerve growth factor (Gagliardini et al. 1994). The finding that an inhibitor of CED-3/ICE-like proteases can block an endogenous pathway of vertebrate cell death indicates that a protease of this class functions in this pathway. Although these observations indicate that a CED-3/ICE-like cysteine protease functions in vertebrate programmed cell death, they do not reveal whether IL-1β is the relevant target of the cysteine protease or whether it is ICE or one or more related cysteine proteases that act in this process.

The CED-3 protein contains a region that is very rich in serines: 32 of the 99 amino acids from residue 107 to residue 205 are serines (Yuan et al. 1993). This region is not highly similar to the corresponding region of ICE, which is 68 amino acids in length, contains 10 serines, and is overall 21% identical to that of CED-3. The functional signficance of the serine-rich region in the CED-3 protein is unknown.

SIX GENES AFFECT THE PHAGOCYTOSIS OF CELL CORPSES BY NEIGHBORING CELLS

Cells that undergo programmed cell death in *C. elegans* are engulfed by neighboring cells (Robertson and Thomson 1982; Sulston et al. 1983). Six genes have been identified that have functions needed for this process of engulfment or phagocytosis to occur: *ced-1*, *ced-2*, *ced-5*, *ced-6*, *ced-7*, and *ced-10*, which we will refer to collectively as the engulfment *ced* genes (Hedgecock et al. 1983; Ellis et al. 1991a). (A seventh gene previously thought to be needed for engulfment, *ced-8*, now seems likely not to be involved in this process; G. Stanfield et al., unpubl.) Mutations in each of the engulfment *ced* genes can cause dying cells to remain unengulfed. Some or all of these genes might function in the process of engulfment per se, for example, by encoding cytoskeletal proteins that act in the extension of pseudopodia by the engulfing cell. Alternatively, these genes might function in the initiation of the process of engulfment, for example, by encoding proteins that act in an intercellular signaling system that triggers the engulfing cell to initiate phagocytosis in response to a signal from the dying cell.

Genetic studies suggest that the engulfment *ced* genes may control two parallel and partially redundant processes. These genes appear to define two sets, such that animals with mutations in genes of only one set have relatively few unengulfed cell corpses, whereas animals with mutations in genes of both sets have substantially more unengulfed corpses. One set consists of the genes *ced-1*, *ced-6*, and *ced-7*, and the other set consists of the genes *ced-2*, *ced-5*, and *ced-10*. Thus, for example, *ced-1*, *ced-2*, or *ced-6* single mutant animals have few unengulfed corpses, as do *ced-1; ced-6* double mutant animals, but *ced-1; ced-2* and *ced-6; ced-2* double mutant animals have many unengulfed corpses. One interpretation of such genetic interactions is that each of these sets of genes defines a pathway that can function to cause phagocytosis and that these two pathways are partially redundant, so that disrupting one pathway leaves the other, and hence the process of phagocytosis, mostly intact; only if both pathways are disrupted is phagocytosis blocked.

When the process of phagocytosis is blocked by appropriate double mutant combinations, those cells that normally undergo programmed cell death still die and form morphologically distinct cell corpses. This observation indicates that phagocytosis is not what kills cells during programmed cell death. Rather, phagocytosis is a downstream event that removes cell corpses that have been generated by a process that depends on the activities of *ced-3* and *ced-4*.

A NUCLEASE CONTROLLED BY THE GENE *nuc-1* DEGRADES THE DNA IN CELL CORPSES

The final step in programmed cell death is the removal of the cellular debris that constitutes the cell corpse. To date, only one gene has been identified of the presumably many genes that function in this step. The gene *nuc-1*, which controls the activity of an endodeoxyribonuclease, is required for the degradation of the DNA in dying cells (Sulston 1976; Hedgecock et al. 1983; Hevelone and Hartman 1988). In *nuc-1* mutant animals, cell death proceeds normally, except that the DNA of dying cells remains in pyknotic bodies. This observation suggests that the processes of killing and engulfment precede and are necessary for the process of DNA degradation controlled by *nuc-1*.

The observation that cell corpses form in *nuc-1* animals reveals that the process of DNA degradation controlled by this gene, like the process of phagocytosis, is not what kills cells during programmed cell death in *C. elegans*. Nuclease activity is also involved in the programmed deaths of cells in other organisms, and a basic issue concerning this activity has been whether it is causally responsible for killing cells (see, e.g., Arends et al. 1990; Oberhammer et al. 1993). If so, and if programmed cell death in *C. elegans* involves a similar mechanism, there must be an as yet unidentified nuclease in the worm that functions in the killing step of programmed cell death and acts to cleave the DNA in dying cells to a state that allows that DNA to remain visible within the pyknotic bodies present in *nuc-1* animals.

THE PATHWAY FOR PROGRAMMED CELL DEATH MAY BE CONSERVED FROM NEMATODE TO HUMAN

As discussed above, of the three genes that act in the killing step of programmed cell death in *C. elegans*, at least two—*ced-9* and *ced-3*—have counterparts that act comparably in vertebrates. Specifically, *ced-9* and the human gene *bcl-2* both act to protect cells against programmed cell death and are members of a gene family, and *ced-3* and the mammalian gene that encodes the cysteine protease ICE both can act to cause programmed cell death and are also members of a gene family. Furthermore, the human *bcl-2* gene can act in *C. elegans* to prevent worm cells from undergoing programmed cell death and can even substitute for *ced-9* in mutant worms deficient in *ced-9* function. Similarly, the *C. elegans ced-3* gene can act in mammalian cells to cause them to undergo programmed cell death. That *bcl-2* can act in worms and *ced-3* can act in mammals indicates that the pathways in which these genes function are highly similar in these different organisms. In short, these observations strongly support the hypothesis that the pathway for programmed cell death is conserved from nematode to human.

CELL DEATH GENES MAY DEFINE NEW CLASSES OF PROTO-ONCOGENES AND TUMOR SUPPRESSOR GENES

That *bcl-2* acts as a proto-oncogene by suppressing programmed cell death suggests that other genes that function in the process of programmed cell death could also be responsible for human cancers. For example, mutations in human genes that act like mutations in the *ces* genes and prevent the deaths of specific classes of cells could lead to the uncontrolled proliferation of those cell types. Similarly, mutations that eliminate the activities of cysteine proteases of the CED-3/ICE family or of proteins similar to the *C. elegans* CED-4 protein could prevent the normal process of programmed cell death and thereby promote malignancy, just as does overexpression of *bcl-2*. Thus, the CED-3/ICE family and CED-4 may define new classes of tumor suppressor genes. If indeed cell death genes define new classes of proto-oncogenes and tumor suppressor genes, one could seek as anticancer agents antagonists of genes that function to prevent programmed cell death, such as members of the *ced-9*/*bcl-2* family, as well as agonists of genes that function to promote programmed cell death, such as members of the *ced-3*/ICE family or *ced-4*. In this way, studies of the genetics of programmed cell death in *C. elegans* could lead to the identification of both drug targets and therapeutic agents that might be important for the development of novel methods for cancer therapy.

ACKNOWLEDGMENTS

We thank Michael Basson, Michael Koelle, Mark Metzstein, Gillian Stanfield, Yi-Chun Wu, and Ding Xue for comments concerning this manuscript. The studies performed in our laboratory were supported by U.S. Public Health Service research grants GM-24663 and GM-24943 to H.R.H. and by the Howard Hughes Medical Institute. H.R.H. is an Investigator of the Howard Hughes Medical Institute.

REFERENCES

Alnemri, E., N. Robertson, T. Fernandez, C. Croce, and G. Litwack. 1992. Overexpressed full-length human *BCL2* extends the survival of baculovirus-infected Sf9 insect cells. *Proc. Natl. Acad. Sci.* **89:** 7295.

Arends, M., R. Morris, and A. Wyllie. 1990. Apoptosis: The role of the endonuclease. *Am. J. Pathol.* **136:** 593.

Avery, L. and H.R. Horvitz. 1987. A cell that dies during wild-type *C. elegans* development can function as a neuron in a *ced-3* mutant. *Cell* **51:** 1071.

Bakhshi, A., J. Jensen, P. Goldman, J. Wright, O. McBride, A. Epstein, and S. Korsmeyer. 1985. Cloning the chromosomal breakpoint of t(14;18) human lymphomas; clustering around J_H on chromosome 14 and near a transcriptional unit on 18. *Cell* **41:** 899.

Boise, L., M. Gonzalez-Garcia, C. Postema, L. Ding, T. Lindsten, L. Turka, X. Mao, G. Nuñez, and C. Thompson. 1993. *bcl-x*, a *bcl-2*-related gene that functions as a dominant regulator of apoptotic cell death. *Cell* **74:** 597.

Cazals-Hatem, D., D. Louie, S. Tanaka, and J. Reed. 1992. Molecular cloning and DNA sequence analysis of cDNA encoding chicken homolog of the Bcl-2 oncoprotein. *Biochim. Biophys. Acta* **1132:** 109.

Cerretti, D.P., C.J. Kozlosky, B. Mosley, N. Nelson, K. Van Ness, T. Greenstreet, C. March, S. Kronheim, T. Druck, L. Cannizzaro, K. Huebner, and R. Black. 1992. Molecular cloning of the interleukin-1β converting enzyme. *Science* **256:** 97.

Clarke, A., C. Purdie, D. Harrison, R. Morris, C. Bird, M. Hooper, and A. Wyllie. 1993. Thymocyte apoptosis induced by p53-dependent and independent pathways. *Nature* **362:** 849.

Cleary, M. and J. Sklar. 1985. Nucleotide sequence of a t(14;18) chromosomal breakpoint in follicular lymphoma and demonstration of a breakpoint-cluster region near a transcriptionally active locus on chromosome 18. *Proc. Natl. Acad. Sci.* **82:** 7439.

Cleary, M., S. Smith, and J. Sklar. 1986. Cloning and structural analysis of cDNAs for *bcl-2* and a hybrid *bcl-2*/immunoglobulin transcript resulting from the t(14;18) translocation. *Cell* **47:** 19.

Cooper, J.A. 1990. Oncogenes and anti-oncogenes. *Curr. Opin. Cell Biol.* **2:** 285.

Eguchi, Y., D. Ewert, and Y. Tsujimoto. 1992. Isolation and characterization of the chicken *bcl-2* gene: Expression in a variety of tissues including lymphoid and neuronal organs in adult and embryo. *Nucleic Acids Res.* **20:** 4187.

Ellis, H. and H.R. Horvitz. 1986. Genetic control of programmed cell death in the nematode *C. elegans*. *Cell* **44:** 817.

Ellis, R. and H.R. Horvitz. 1991. Two *C. elegans* genes control the programmed deaths of specific cells in the pharynx. *Development* **112:** 591.

Ellis, R., D. Jacobson, and H.R. Horvitz. 1991a. Genes required for the engulfment of cell corpses during programmed cell death in *C. elegans*. *Genetics* **129:** 79.

Ellis, R., J. Yuan, and H.R. Horvitz. 1991b. Mechanisms and functions of cell death. *Annu. Rev. Cell Biol.* **7:** 663.

Evan, G., A. Wyllie, C. Gilbert, T. Littlewood, H. Land, M. Brooks, C. Waters, L. Penn, and D. Hancock. 1992. Induction of apoptosis in fibroblasts by c-myc protein. *Cell* **69:** 119.

Gagliardini, V., P.A. Fernandez, R. Lee, H. Drexler, R. Rotello, M. Fishman, and J. Yuan. 1994. Prevention of vertebrate neuronal death by the *crmA* gene. *Science* **263:** 826.

Hedgecock, E., J. Sulston, and N. Thomson. 1983. Mutations affecting programmed cell deaths in the nematode *Caenorhabditis elegans*. *Science* **220:** 1277.

Hengartner, M. and H.R. Horvitz. 1994a. *C. elegans* cell survival gene *ced-9* encodes a functional homolog of the mammalian proto-oncogene *bcl-2*. *Cell* **76:** 665.

———. 1994b. Activation of *C. elegans* cell death protein CED-9 by an amino-acid substitution in a domain conserved in Bcl-2. *Nature* **369:** 318.

Hengartner, M., R. Ellis, and H.R. Horvitz. 1992. *C. elegans* gene *ced-9* protects cells from programmed cell death. *Nature* **356:** 494.

Hevelone, J. and P.S. Hartman. 1988. An endonuclease from *Caenorhabditis elegans*: Partial purification and characterization. *Biochem. Genet.* **26:** 447.

Hockenbery, D., G. Nuñez, C. Milliman, R. Schreiber, and S. Korsmeyer. 1990. Bcl-2 is an inner mitochondrial membrane protein that blocks programmed cell death. *Nature* **348:** 334.

Hockenbery, D., Z. Otlvai, X. Yin, C. Milliman, and S. Korsmeyer. 1993. Bcl-2 functions in an antioxidant pathway to prevent apoptosis. *Cell* **75:** 241.

Kane, D., T. Sarafian, R. Anton, H. Hahn, E. Gralla, J. Valentine, T. Ord, and D. Bredesen. 1993. Bcl-2 inhibition of neural death: Decreased generation of reactive oxygen species. *Science* **262:** 1274.

Korsmeyer, S. 1992. Bcl-2: An antidote to programmed cell death. *Cancer Surv.* **15:** 105.

Kozopas, K., T. Yang, H. Buchan, P. Zhou, and R. Craig. 1993. *MCL1*, a gene expressed in programmed myeloid differentiation, has sequence similarity to *BCL2*. *Proc. Natl. Acad. Sci.* **90:** 3516.

Lin, E., A. Orlofsky, M. Berger, and M. Prystowsky. 1993. Characterization of *A1*, a novel hemopoietic-specific early response gene with sequence similarity to *bcl-2*. *J. Immunol.* **151:** 1979.

Lowe, S., E. Schmitt, S. Smith, B. Osborne, and T. Jacks. 1993. p53 is required for radiation-induced apoptosis in mouse thymocytes. *Nature* **362:** 847.

McDonnell, T. and S. Korsmeyer. 1991. Progression from lymphoid hyperplasia to high-grade malignant lymphoma in mice transgenic for the t(14;18). *Nature* **349:** 254.

Miura, M., H. Zhu, R. Rotello, E. Hartwieg, and J. Yuan. 1993. Induction of apoptosis in fibroblasts by IL-1 β-converting enzyme, a mammalian homolog of the *C. elegans* cell death gene *ced-3*. *Cell* **75:** 653.

Oberhammer, F., J. Wilson, C. Dive, I. Morris, J. Hickman, A. Wakeling, R. Walker, and M. Aikorska. 1993. Apoptotic death in epithelial cells: Cleavage of DNA to 300 and/or 50 kb fragments prior to or in the absence of internucleosomal fragmentation. *EMBO J.* **12:** 3679.

Oltvai, Z., C. Milliman, and S. Korsmeyer. 1993. Bcl-2 heterodimerizes in vivo with a conserved homolog, Bax, that accelerates programed cell death. *Cell* **74:** 609.

Robertson, A. and N. Thomson. 1982. Morphology of programmed cell death in the ventral nerve cord of *Caenorhabditis elegans* larvae. *J. Embryol. Exp. Morphol.* **67:** 89.

Shi, Y., J. Glynn, L. Guilbert, T. Cotter, R. Bissonnette, and D. Green. 1992. Role for c-*myc* in activation-induced apoptotic cell death in T-cell hybridomas. *Science* **257:** 212.

Sugimoto, A., P. Friesen, and J. Rothman. 1994. Baculovirus *p35* prevents developmentally programmed cell death and rescues a *ced-9* mutant in the nematode *Caenorhabditis elegans*. *EMBO J.* **13:** 2023.

Sulston, J. 1976. Post-embryonic development in the ventral cord of *Caenorhabditis elegans*. *Philos. Trans. R. Soc. Lond. B Biol. Sci.* **275:** 287.

Sulston, J.E. and H.R. Horvitz. 1977. Post-embryonic cell lineages of the nematode, *Caenorhabditis elegans*. *Dev. Biol.* **56:** 110.

Sulston, J.E., E. Schierenberg, J.G. White, and N. Thomson. 1983. The embryonic cell lineage of the nematode *Caenorhabditis elegans*. *Dev. Biol.* **100:** 64.

Tanaka, S., K. Saito, and J. Reed. 1992. Structure-function analysis of the Bcl-2 oncoprotein: Addition of a heterologous transmembrane domain to portions of the Bcl-2 beta protein restores function as a regulator of cell survival. *J. Biol. Chem.* **B268:** 10920.

Thornberry, N.A., H.G. Bull, J.R. Calaycay, K. Chapman, A. Howard, M. Kostura, D. Miller, S. Molineaux, J. Weidner, J. Aunins, J.K. Elliston, J. Ayala, F. Casano, J. Chin, J. Ding, L. Egger, E. Gaffney, G. Limjuco, O. Palyha, S. Raju, A. Rolando, J. Salley, T. Yamin, T. Lee, J. Shively, M. MacCross, R. Mumford, J. Schmidt, and M. Tocci. 1992. A novel heterodimeric cysteine protease is required for interleukin-1β processing in monocytes. *Nature* **356:** 768.

Trent, C., N. Tsung, and H.R. Horvitz. 1983. Egg-laying defective mutants of the nematode *Caenorhabditis elegans*. *Genetics* **104:** 619.

Tsujimoto, Y. and C. Croce. 1986. Analysis of the structure, transcripts, and protein products of Bcl-2, the gene involved in human follicular lymphoma. *Proc. Natl. Acad. Sci.* **83:** 5214.

Tsujimoto, Y., J. Gorham, J. Crossman, E. Jaffe, and C. Croce. 1985. The t(14;18) chromosome translocations involved in B-cell neoplasms result from mistakes in VDJ joining. *Science* **229:** 1390.

Vaux, D. 1994. A boom time for necrobiology. *Curr. Biol.* **3:** 877.

Vaux, D., I. Weissman, and S. Kim. 1992. Prevention of programmed cell death in *Caenorhabditis elegans* by human *bcl-2*. *Science* **258:** 1955.

Veis, D.J., C.M. Sorenson, J.R. Shutter, and S. Korsmeyer. 1993. Bcl-2-deficient mice demonstrate fulminant lymphoid apoptosis, polycystic kidneys, and hypopigmented hair. *Cell* **75:** 229.

White, J., E. Southgate, and N. Thomson. 1991. On the nature of undead cells in the nematode *Caenorhabditis elegans*. *Philos. Trans. R. Soc. Lond. B Biol. Sci.* **331:** 263.

Wyllie, A.H., J.F.R. Kerr, and A.R. Currie. 1980. Cell death: The significance of apoptosis. *Int. Rev. Cytol.* **68:** 251.

Yin, X.-M., Z. Oltvai, and S. Korsmeyer. 1994. BH1 and BH2 domains of Bcl-2 are required for inhibition of apoptosis and heterodimerization with Bax. *Nature* **369:** 321.

Yonish-Rouach, E., D. Resnitzky, J. Lotem, L. Sachs, A. Kimchi, and M. Oren. 1991. Wild-type p53 induces apoptosis of myeloid leukaemic cells that is inhibited by interleukin-6. *Nature* **352:** 345.

Yu, L., Y.-Y. Wei, S. Usui, and C.-A. Yu. 1992. Cytochrome b560 (QPs1) of mitochondrial succinate-ubiquinone reductase. *J. Biol. Chem.* **267:** 24508.

Yuan, J. and H.R. Horvitz. 1990. The *Caenorhabditis elegans* genes *ced-3* and *ced-4* act cell-autonomously to cause programmed cell death. *Dev. Biol.* **138:** 33.

———. 1992. The *C. elegans* cell death gene *ced-4* encodes a novel protein and is expressed during the period of extensive programmed cell death. *Development* **116:** 309.

Yuan, J., S. Shaham, S. Ledoux, H. Ellis, and H.R. Horvitz. 1993. The *C. elegans* cell death gene *ced-3* encodes a protein similar to mammalian interleukin-1 β-converting enzyme. *Cell* **75:** 641.

Bcl-2 Gene Family and the Regulation of Programmed Cell Death

X.-M. YIN, Z.N. OLTVAI, D.J. VEIS-NOVACK,
G.P. LINETTE, AND S.J. KORSMEYER

Division of Molecular Oncology, Departments of Medicine and Pathology, Howard Hughes Medical Institute, Washington University School of Medicine, St. Louis, Missouri 63110

The genetic control of tumorigenesis has been greatly advanced by the discovery of oncogenes. A central dogma holds that oncogenes induce an overt proliferation providing the driving force for oncogenesis. Cell death has long been recognized as a physiological event during embryonic development. Moreover, a distinct morphological form of cell destruction, apoptosis, had also been described even within cancer tissue (Kerr et al. 1972). In many respects, cancer can be viewed as a violation of normal tissue homeostasis. The maintenance of a relatively constant number of cells in normal tissues reflects a balanced equation between cell proliferation and cell death (Fig. 1). Aberrations of homeostasis that manifest as tumorigenesis would include events that promote proliferation or repress cell death. The history of cancer genetics is replete with examples of oncogenes that promote proliferation. However, *bcl-2* provided the first certain example of an oncogene that regulated cell demise. The overexpression of *bcl-2* repressed death and extended cell survival, which promoted oncogenesis. This established a new paradigm and provided the novel perspective that cancer biology would also involve the regulation of cell death.

bcl-2 INITIATES A NEW CATEGORY OF ONCOGENES: REGULATORS OF CELL DEATH

Genomic instability has been implicated in the pathogenesis of a variety of cancers. Specific interchromosomal translocations are found repeatedly within distinct types of hematologic malignancies but not in their normal cellular counterparts (Klein 1981; Rowley 1982). These sites of chromosomal breakage have proven an extremely rich source of novel proto-oncogenes integral to normal cellular development as well as neoplasia. The t(14;18)(q32;q21) is the most common translocation in human lymphoid malignancies. Approximately 85% of follicular and 20% of diffuse B-cell lymphomas possess this translocation (Yunis et al. 1987). Follicular lymphoma is initially often a low-grade disease comprising predominantly small resting IgM/IgD B cells. Over time, a histological conversion to a diffuse large-cell architecture and aggressive high-grade lymphoma occurs frequently in these patients. Molecular cloning of the t(14;18) breakpoint revealed a new putative proto-oncogene, *bcl-2*, at 18q21 (Bakhshi et al. 1985; Cleary and Sklar 1985; Tsujimoto et al. 1985). Despite the mature B-cell phenotype of t(14;18)-bearing lymphomas, the translocation appears to occur earlier in development at a pre-B-cell stage (Bakhshi et al. 1987).

To date, most of our knowledge concerning oncogenic events has concentrated on mechanisms of increased growth and proliferation. However, increased cell number as a violation of normal homeostasis in neoplastic tissue could arise from either increased proliferation or decreased death. Bcl-2 provided the first example of oncogenesis mediated by decreased cell death. Transgenic mice were created that possessed a Bcl-2-Ig (immunoglobulin) minigene that recapitulated the pathological consequence of the t(14;18). Bcl-2-Ig transgenic mice initially displayed a polyclonal follicular lymphoproliferation that selectively expanded a small resting IgM/IgD B-cell population (Fig. 2). These resting B cells accumulate because of an extended survival rather than increased proliferation (McDonnell et al. 1989, 1990). Over time, these transgenics progress from indolent follicular hyperplasia to diffuse large-cell immunoblastic lymphoma (McDonnell and Korsmeyer 1991). A long latency period and progression from polyclonal hyperplasia to monoclonal high-grade malignancy issued an indictment of secondary genetic abnormalities. Approximately half of the high-

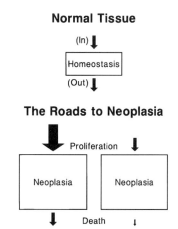

Figure 1. Schematic representation of normal tissue homeostasis with balanced input and output reactions. Alternative roads to neoplasia are depicted as either increased proliferation (In) or decreased death (Out).

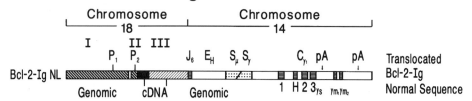

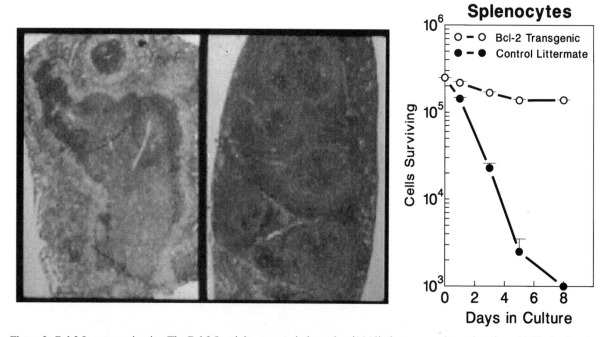

Figure 2. Bcl-2-Ig transgenic mice. The Bcl-2-Ig minigene recapitulates the t(14;18) chromosomal translocation of follicular B-cell lymphoma. Gain of function Bcl-2 leads to B-cell expansion and an enlarged white pulp in the transgenic (*left*) versus control spleen (*right*). B cells accumulate due to extended cell survival (*graph*).

grade tumors possess a c-*myc* translocation involving an immunoglobulin heavy-chain locus. These tumor cells have complemented an inherent survival advantage (*bcl-2*) with a gene that promotes proliferation (*myc*). When *bcl-2* transgenic mice were mated to *myc* transgenic mice, rapidly emerging undifferentiated hematopoietic leukemia occurred, providing further testimony for the potent synergy of this particular oncogene combination (Strasser et al. 1990). In addition to promoting cell cycle progression, *myc* has been shown to promote apoptosis (Evan et al. 1992). Thus, the overexpression of *myc* may specifically benefit from the ability of *bcl-2* to block apoptosis. The Bcl-2-Ig mice document the prospective importance of the t(14;18) in setting the stage for lymphomagenesis and tumor progression. T-cell lymphomas were also found in mice where *bcl-2* was overexpressed in thymocytes by the *lck* promoter (G.P. Linette and S.J. Korsmeyer, unpubl.). In this case, thymocytes and peripheral T cells also exhibited enhanced cell survival (Sentman et al. 1991; Strasser et al. 1991). Thus, prolonged cell survival can be a primary oncogenic event in both B and T cells. This reflects the ability of *bcl-2* to block cell death following a wide variety of apoptotic stimuli. These include glucocorticoids, γ-irradiation, and lineage-specific signals through immunoglobulin or T-cell receptors. By preventing the apoptotic demise of activated lymphocytes, Bcl-2 would enable the acquisition of additional genetic aberrations and the emergence of monoclonal neoplasms. By repressing normal apoptotic pathways, Bcl-2 could also allow cells which sustained DNA damage to avoid a suicide response.

Bcl-2 PROTEIN DISTRIBUTION AND NORMAL TISSUE HOMEOSTASIS

Bcl-2 protein displays a very restricted topographic distribution within mature tissues characterized by apoptotic cell death (Hockenbery et al. 1991). Within lymphocyte development, Bcl-2 displays an on-off-on again pattern of expression during both B- and T-cell development (Veis et al. 1993a; Merino et al. 1994). The earliest progenitors possess *bcl-2*, but it is down-regulated at the critical stage of lymphocyte development when cells with nonfunctional and autoreactive antigen receptors are eliminated by apoptosis (Fig. 3). Bcl-2 is reexpressed in mature B and T cells that have successfully traversed the selective steps. This biphasic

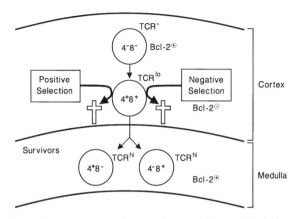

Figure 3. Developmental expression of Bcl-2 protein during T-cell development reveals an on-off-on pattern of expression during differentiation.

pattern is well defined within thymocytes in which the very immature CD44$^+$CD25$^-$ subset of double negative (CD4$^-$8$^-$) cells express Bcl-2. In contrast, the vast majority of CD4$^+$8$^+$ double positive (DP) thymocytes, the stage at which both positive and negative selection operates, lacked Bcl-2 protein. Detailed studies of T-cell receptor transgenic mice backcrossed to various major histocompatibility backgrounds demonstrated that Bcl-2 is up-regulated at the DP stage during positive selection. Bcl-2 is uniformly present in mature, single positive CD4$^+$ or CD8$^+$ T cells to help maintain homeostasis of peripheral T-cell populations (Linette et al. 1994).

In immune responses, immunohistochemical assessment of secondary germinal centers revealed that the follicular mantle, composed of long-lived recirculation IgM/IgD B cells, possessed abundant Bcl-2 (Pezzella et al. 1990; Hockenbery et al. 1991). However, Bcl-2 protein was absent from the dark zone of proliferating centroblasts and from the portion of the light zone where centrocytes are dying by apoptosis. Yet, Bcl-2 expression returned in cells within the more apical portion of the light zone where B cells are selected for survival. Bcl-2-Ig transgenic mice which overexpressed Bcl-2 in their B-cell compartment displayed prolonged secondary immune responses following antigenic challenge. This could be traced to longer-lived memory B cells. Thus, one physiological role for Bcl-2 appears to be the generation and maintenance of immune responsiveness (Nuñez et al. 1991).

Another category of tissues that express Bcl-2 is the glandular epithelium that undergoes hyperplasia or involution, usually in response to hormonal stimuli or growth factors. In organized epithelium, Bcl-2 is restricted to stem cell and proliferation zones. Bcl-2 is present in the lower crypts of the intestine and the basal layer of the epidermis (Hockenbery et al. 1991).

Within the embryo, Bcl-2 is widely expressed in tissues derived from all three germ layers (Veis-Novack and Korsmeyer 1994). The developing limb provides a clear example where Bcl-2 is restricted to zones of cell survival. Bcl-2 is present within the digital zones, but not in the interdigital zones of cell death. Bcl-2 is especially prominent in the developing nervous system, both in the proliferating neuroepithelium of the ventricular zones and in postmitotic populations of the cortical plate, cerebellum, hippocampus, and spinal cord (Merry et al. 1994). Retinal neuroepithelial cells uniformly express Bcl-2 until they begin to differentiate and then display the topographic distribution maintained into adulthood. Interestingly, the photoreceptors have lower levels of Bcl-2, perhaps explaining their susceptibility to apoptotic death in retinal degeneration (Veis-Novack and Korsmeyer 1994). Bcl-2 expression in the CNS declines with aging. In the regeneration-competent peripheral nervous system, neurons and supporting cells of sympathetic and sensory ganglia retain substantial Bcl-2 protein throughout their life. Developmental patterns of Bcl-2 expression would suggest that the susceptibility to cell death is much more widespread in the embryo than previously appreciated, or that Bcl-2 has roles beyond protecting cells from death.

GAIN OF FUNCTION: Bcl-2 PROMOTES DIFFERENTIATION

The fate of a cell can include proliferation, differentiation, or death. Genes that promote or inhibit each pathway have been described. However, the interrelationships between these decisions remain predominantly unknown. To assess the role of Bcl-2 in T-cell maturation, transgenic mice were generated utilizing the *lck* promoter, which redirected Bcl-2 to the vulnerable stage of development where more than 95% of thymocytes perish. We wanted to assess the effects of this antidote to apoptosis when it was placed proximal to positive and negative selection events of thymocytes. Bcl-2 protected the normally sensitive DP thymocytes from a wide variety of apoptotic stimuli, including glucocorticoids, radiation, and anti-CD3 treatment (Sentman et al. 1991; Strasser et al. 1991). Despite this, negative selection still occurred, eliminating self-reactive T cells. Surprisingly, Bcl-2 promoted the maturation of thymocytes at several control points during thymocyte development (Fig. 4). This even occurred in the absence of correct T-cell receptor interactions with matched major histocompatibility (MHC) loci (Linette et al. 1994). However, some selectivity was noted, in that Bcl-2 would promote maturation of CD8 but not CD4 T cells. Bcl-2 could not completely substitute for positive selection, in that the rescued CD8 thymocytes were not efficiently exported to the periphery. Most remarkably, Bcl-2 promoted differentiation to the CD4$^+$8$^+$ DP stage in RAG1$^{-/-}$ mice, which lack T-cell receptors and usually arrest maturation at the CD4$^-$8$^-$ DN stage. Similarly, enforced expression of Bcl-2 in the B-cell lineage also promotes differentiation (Strasser et al. 1994). Moreover, Bcl-2 enabled a multipotential hematopoietic cell line to survive in the absence of growth factor and differentiate toward erythroid or

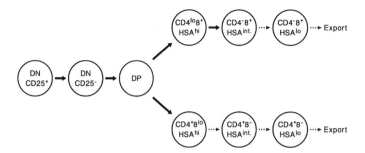

Figure 4. Control points in thymocyte differentiation. Bcl-2 can promote lineage progression at multiple stages of thymocyte maturation, indicated by a thick solid line. Bcl-2 cannot substitute for the final step of positive selection, indicated by the dotted line.

myeloid lineages (Fairbairn et al. 1993). In total, these studies suggest that if cell death is prevented, endogenous programs of differentiation can manifest even in the absence of critical signals.

LOSS OF FUNCTION: Bcl-2-DEFICIENT MICE DISPLAY FULMINANT LYMPHOID APOPTOSIS

Bcl-2 appears not to be singularly required to complete embryonic development, as deficient mice generated by gene targeting are born at the expected frequency (Veis et al. 1993b). Despite the widespread expression of Bcl-2 during embryonic development, most organs were histologically normal in $bcl-2^{-/-}$ mice. The mice appear relatively normal for the first week but then display growth retardation, small external ears, immature facial features, and early mortality postnatally. Hematopoiesis including B- and T-cell differentiation is initially normal. Over time, these mice demonstrate fulminant apoptosis of the thymus and spleen (Nakayama et al. 1993; Veis et al. 1993b) (Fig. 5). Importantly, $bcl-2^{-/-}$ thymocytes require an apoptotic stimulus to manifest their predisposition to massive cell death. Bcl-2-deficient mice also develop severe polycystic kidney disease characterized by dilated proximal and distal tubular segments and hyperproliferation of epithelium and interstitium. The kidney represents perhaps the most disordered embryonic development in this mouse. $Bcl-2^{-/-}$ kidneys are markedly hypoplastic, containing fewer nephrons and smaller nephrogenic zones. Moreover, in vitro culture of metanephroi from $Bcl-2^{-/-}$ E12 embryos confirmed that this developmental aberration was cell autonomous. These abnormalties appear to result from excessive apoptotic death of developing metanephric blastema.

Bcl-2 mice have initially normal hair pigmentation, but turn gray with the second hair follicle cycle at puberty. Melanocytes possess a unique synthetic pathway to generate the melanin responsible for hair pigmentation. A key intermediate, DOPA-quinone, is an extremely reactive compound capable of generating free-radical species. The absence of Bcl-2 could make melanocytes susceptible to an oxidative death or perhaps affect the ratio of light to dark melanin synthesis. Thus, the imbalanced cell death pathway in Bcl-2-deficient mice results in marked aberrations in homeostasis of several cell types.

REACTIVE OXYGEN SPECIES AND CELL DEATH

Bcl-2 has a carboxy-terminal hydrophobic region that functions as a signal-anchor sequence responsible for the integral membrane position of the 25-kD Bcl-2 product. Bcl-2 has been most convincingly localized to mitochondria, nuclear membrane, and endoplasmic reticulum. Subcellular fractionation studies have placed Bcl-2 predominately within the inner mitochondrial membrane, nuclear membrane, and smooth endoplasmic reticulum (Chen-Levy and Cleary 1990; Hockenbery et al. 1990; Jacobson et al. 1993; Krajewski et al.

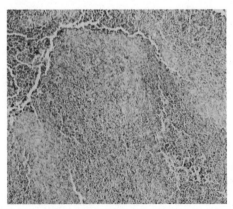 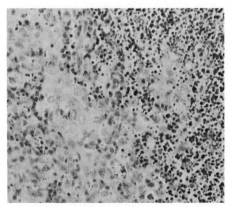

Figure 5. Fulminant lymphocyte apoptosis in Bcl-2$^{-/-}$ thymus. Low magnification (*left*) and high magnification (*right*) reveal a massive number of pyknotic cells. Only large epithelial cells remain in the medulla.

1993). Immunolocalization techniques have found Bcl-2 in the outer mitochondrial membrane, perinuclear membrane, and endoplasmic reticulum (Monaghan et al. 1992; deJong et al. 1994). A difference in subcellular distribution has been noted between cell lineages and between native versus overexpressed Bcl-2. Patterns of distribution are also somewhat antibody-dependent, suggesting a difference in epitope availability in some situations (Merry et al. 1994; Veis-Novack and Korsmeyer 1994). In-vitro-targeting studies indicate that *bcl-2* can target microsomes as well as mitochondria. However, the targeting of mitochondria appears to be selective and displays the kinetics and characteristics of an outer membrane protein (Nguyen et al. 1993). Thus, most, if not all, of Bcl-2 appears to be targeted by its carboxy-terminal signal-anchor sequence as an integral membrane protein with its amino terminus facing the cytosol.

Could reactive oxygen species (ROS) be a final mediator of cell demise in both apoptotic and necrotic cell death? Many of the effects of oxygen free radicals, including DNA strand breaks and membrane blebbing, match the hallmark features of apoptosis. Bcl-2 was noted to block γ-irradiation-induced cell death, protecting from ionizing radiation which generates the most reactive oxygen species, hydroxyl radicals (OH·). Bcl-2 also countered other oxidative cell deaths, induced by H_2O_2 or *t*-butyl-hydroperoxide (Hockenbery et al. 1993; Kane et al. 1993). Bcl-2 also interfered with cell death induced by menadione, a quinone compound that undergoes redox cycles intracellularly to generate superoxide. Importantly, Bcl-2 could not prevent the oxidative burst generated by menadione. Moreover, oximetry studies in two separate model systems detected no burst in the production of reactive oxygen species. Bcl-2 did not alter the efficiency of normal electron transport nor the endogenous rate of ROS production through peroxides. Instead, overexpressed Bcl-2 prevented the damage to vital cellular constituents including the peroxidation of lipid membranes. Consistent with these observations, overexpression of glutathione peroxidase (GSHPX), a known inhibitor of lipid peroxidation, repressed apoptosis (Hockenbery et al. 1993). Which ROS might be of importance in mediating cell death? The strongest genetic evidence implicating ROS in cell death is provided by familial amyotrophic lateral sclerosis, in which multiple mutations in the Cu/Zn SOD have been identified. Why motor neurons are selectively lost and whether these are apoptotic deaths remains uncertain. Data from the established apoptosis assays favor the importance of the more downstream peroxides as important ROS. This may reflect their diffusibility or proclivity for conversion to the most highly reactive OH· radical. Bcl-2 appears to block damage between the generation of peroxides and lipid membrane peroxidation. However, it is currently uncertain whether this effect of Bcl-2 is direct or several steps removed. The intracellular sites of oxygen free radical generation include mitochondria, endoplasmic reticulum, and perhaps nuclear membrane, also the sites where Bcl-2 has been localized. One model for further testing would be that Bcl-2 through protein interactions would focus, as well as regulate, an antioxidant pathway to these sites. Whether Bcl-2 could serve a normal physiological role regulating redox and how this might interrelate to known death effector genes like the Ced-3/ICE cysteine protease remains to be determined.

BH1 and BH2 DOMAINS ARE NOVEL DIMERIZATION MOTIFS NECESSARY FOR Bcl-2 TO REPRESS APOPTOSIS AND INTERACT WITH Bax

The mechanism by which Bcl-2 represses cell death is an important vantage point to approach the induction and repression of apoptosis. One clue was provided when Bcl-2 was found to associate in vivo with a 21-kD partner, Bax (*B*cl-2 *a*ssociated *X* protein) (Fig. 6). Bax shares homology with Bcl-2 (Oltvai et al. 1993) and, when overexpressed, counters the effect of Bcl-2. Excess Bax promotes apoptotic cell death, but only following a death signal. Of note, the Bcl-2/Bax association exists in cells prior to a death stimulus, and their ratio dictates whether a cell will live or die following an apoptotic stimulus. When Bcl-2 is in excess, Bcl-2 homodimers dominate and cells are protected. When Bax is in excess, Bax homodimers dominate and cells are susceptible to apoptosis (Fig. 7) (Oltvai et al. 1993).

Beyond Bax, additional members of a Bcl-2-related family have been found, including Bcl-x (Boise et al. 1993), A1 (Lin et al. 1993), MCL-1 (Kozopas et al. 1993), Ced-9 (Hengartner and Horvitz 1994a), BHRF-1 (Baer et al. 1984), and LWM5-HL (Neilan et al. 1993). Bcl-xL, Ced-9, and BHRF-1 (Henderson et al. 1993) have all been shown to repress cell death in various systems. The functions of other members are being explored. The most prominent regions of homology

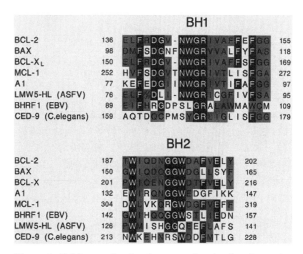

Figure 6. Bcl-2 gene family. An expanding family of proteins homologous to Bcl-2 are most highly conserved within Bcl-2 homology 1 and 2 (BH1 and BH2) domains. Identical amino acids in black background and conserved positions in gray background.

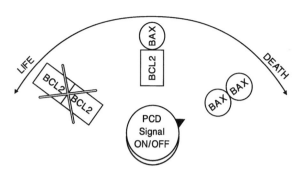

Figure 7. Cell autonomous rheostat regulates cell death. Bcl-2 and Bax form homodimers and heterodimers. Following a signal for programmed cell death (PCD), cells die if Bax is in excess, but live if Bcl-2 predominates. Mutagenesis studies suggest that Bcl-2 homodimers may not be the active moiety.

with Bcl-2 are clustered within two domains, entitled BH1 and BH2 domains (Oltvai et al. 1993; Yin et al. 1994). The fact that Bax heterodimerizes with Bcl-2 and counteracts the activity of Bcl-2 suggested that BH1 and BH2 would prove to be functional domains or regulate protein-protein interactions. Site-specific mutagenesis of Bcl-2 demonstrated that BH1 and BH2 are novel dimerization motifs (Yin et al. 1994). Single amino acid substitutions at either Gly-145 of the BH1 domain or Trp-188 of the BH2 domain completely abrogated Bcl-2's death repressor activity in either mammalian systems or in *Caenorhabditis elegans* (Hengartner and Horvitz 1994b). Such domain functions seem to be present in other Bcl-2 family members, such as Ced-9. The gain-of-function mutant of the Ced-9 protein has a mutation of Gly-169-Glu in the BH1 domain corresponding to the Gly-145 of Bcl-2 (Hengartner and Horvitz 1994b). Of note, those mutations that affected the function of Bcl-2 also disrupted its heterodimerization with Bax, yet still permitted Bcl-2 homodimerization. This implicates the crucial importance of Bcl-2/Bax heterodimerization. It favors a model in which Bcl-2 must heterodimerize with Bax to exert its death-repressing function. This suggests that the Bcl-2/Bax heterodimers may be the active pair regulating cell death. Alternatively, the principal role of Bcl-2 may be to disrupt Bax/Bax homodimers that may prove to be directly connected to a death effector pathway. A number of biological systems indicate that cells vary during differentiation in their inherent sensitivity to a given death stimulus. The ratio of Bcl-2/Bax represents one cell-autonomous rheostat that determines a cell's life or death decision (Fig. 7).

REFERENCES

Baer, B., A.T. Bankier, M.D. Biggin, P.L. Deininger, P.J. Farrell, T.J. Gibson, G. Hatfull, G.S. Hudson, S.C. Satchwell, C. Sequin, P.S. Tuffnell, and B.G. Barrell. 1984. DNA sequence and expression of the B95-8 Epstein-Barr virus genome. *Nature* **310:** 207.

Bakhshi, A., J.P. Jensen, P. Goldman, J.J. Wright, O.W. McBride, A.L. Epstein, and S.J. Korsmeyer. 1985. Cloning the chromosomal breakpoint of t(14;18) human lymphomas: clustering around J_H on chromosome 14 and near a transcriptional unit on 18. *Cell* **41:** 889.

Bakhshi, A., J.J. Wright, W. Graninger, M. Seto, J. Owens, J. Cossman, J.P. Jensen, P. Goldman, and S.J. Korsmeyer. 1987. Mechanism of the t(14;18) chromosomal translocation: Structural analysis of both derivative 14 and 18 reciprocal partners. *Proc. Natl. Acad. Sci.* **84:** 2396.

Boise, L.H., M. Gonzalez-Garcia, C.E. Postems, L. Ding, T. Lindsten, L.A. Turka, X. Mao, G. Nuñez, and C.B. Thompson. 1993. Bcl-x, a Bcl-2 related gene that functions as a dominant regulator of apoptotic cell death. *Cell* **74:** 597.

Chen-Levy, Z. and M.L. Cleary. 1990. Membrane topology of the Bcl-2 proto-oncogenic protein demonstrated *in vivo*. *J. Biol. Chem.* **265:** 4929.

Cleary, M.L. and J. Sklar. 1985. Nucleotide sequence of a t(14;18) chromosomal breakpoint in follicular lymphoma and demonstration of a breakpoint cluster region near a transcriptionally active locus on chromosome 18. *Proc. Natl. Acad. Sci.* **82:** 7439.

deJong, D., F.A. Prins, D.Y. Mason, J.C. Reed, G.B. van Ommen, and P.M. Kluin. 1994. Subcellular localization of the bcl-2 protein in malignant and normal lymphoid cells. *Cancer Res.* **54:** 256.

Evan, G.I., A.H. Wyllie, C.S. Gilbert, T.D. Littlewood, H. Land, M. Brooks, C.M. Waters, L.Z. Penn, and D.C. Hancock. 1992. Induction of apoptosis in fibroblasts by c-myc protein. *Cell* **69:** 119.

Fairbairn, L.J., G.J. Cowling, B.M. Reipert, and T.M. Dexter. 1993. Suppression of apoptosis allows differentiation and development of a multipotent hemopoietic cell line in the absence of added growth factors. *Cell* **74:** 823.

Henderson, S., D. Huen, M. Rowe, C. Dawson, G. Johnson, and A. Rickinson. 1993. Epstein-Barr virus-coded BHRF1 protein, a viral homologue of Bcl-2, protects human B cells from programmed cell death. *Proc. Natl. Acad. Sci.* **90:** 8479.

Hengartner, M.O. and H.R. Horvitz. 1994a. *C. elegans* cell survival gene *ced-9* encodes a functional homolog of the mammalian proto-oncogene bcl-2. *Cell* **76:** 665.

———. 1994b. Activation of *C. elegans* cell death protein CED-9 by an amino-acid substitution in a domain conserved in Bcl-2. *Nature* **369:** 318.

Hockenbery, D.M., G. Nuñez, C. Milliman, R.D. Schreiber, and S.J. Korsmeyer. 1990. Bcl-2 is an inner mitochondrial membrane protein that blocks programmed cell death. *Nature* **348:** 334.

Hockenbery, D.M., Z.N. Oltvai, X.-M. Yin, C.L. Milliman, and S.J. Korsmeyer. 1993. Bcl-2 functions in an anti-oxidant pathway to prevent apoptosis. *Cell* **75:** 241.

Hockenbery, D., M. Zutter, W. Hickey, M. Nahm, and S.J. Korsmeyer. 1991. Bcl-2 protein is topographically restricted in tissues characterized by apoptotic cell death. *Proc. Natl. Acad. Sci.* **88:** 6961.

Jacobson, M.D., J.F. Burne, M.P. King, T. Miyashita, J.C. Reed, and M.C. Raff. 1993. Bcl-2 blocks apoptosis in cells lacking mitochondrial DNA. *Nature* **361:** 365.

Kane, D.J., T.A. Saraflan, R. Anton, H. Hahn, E.B. Grulla, J.S. Valentine, T. Ord, and D.E. Bredesen. 1993. Bcl-2 inhibition of neural death: Decreased generation of reactive oxygen species. *Science* **262:** 1274.

Kerr, J.F.R., A.H. Wyllie, and A.R. Curie. 1972. Apoptosis: A basic biologic phenomenon with wide-ranging implications in tissue kinetics. *Br. J. Cancer* **26:** 239.

Klein, G. 1981. The role of gene dosage and genetic transpositions in carcinogenesis. *Nature* **294:** 313.

Kozopas, K.M., T. Yang, H.L. Buchan, P. Zhou, and R.W. Craig. 1993. *MCL-1*, a gene expressed in programmed myeloid cell differentiation, has sequence similarity to Bcl-2. *Proc. Natl. Acad. Sci.* **90:** 3516.

Krajewski, S., S. Tanaka, S. Takayama, M. Schibler, W. Fenton, and J.C. Reed. 1993. Investigation of the subcellular distribution of the *bcl-2* oncoprotein: Residence in the nuclear envelope, endoplasmic reticulum, and outer mitochondrial membranes. *Cancer Res.* **53:** 4701.

Lin, E.Y., A. Orlofsky, M.-S. Berger, and M.B. Prystowsky. 1993. Characterization of A1, a novel hemopoietic-specific early-response gene with sequence similarity to Bcl-2. *J. Immunol.* **151:** 1979.

Linette, G.P., M.J. Grusby, S.M. Hedrick, T.H. Hansen, L.H. Glimcher, and S.J. Korsmeyer. 1994. Bcl-2 is upregulated at the CD4$^+$CD8$^+$ stage during positive selection and promotes thymocyte differentiation at several control points. *Immunity* **1:** 197.

McDonnell, T.J. and S.J. Korsmeyer. 1991. Progression from lymphoid hyperplasia to high-grade malignant lymphoma in mice transgenic for the t(14;18). *Nature* **349:** 254.

McDonnell, T.J., N. Deane, F.M. Platt, G. Nuñez, U. Jaeger, J.P. McKearn, and S.J. Korsmeyer. 1989. Bcl-2-immunoglobulin transgenic mice demonstrate extended B cell survival and follicular lymphoproliferation. *Cell* **57:** 79.

McDonnell, T.J., G. Nuñez, F.M. Platt, D. Hockenbery, L. London, J.P. McKearn, and S.J. Korsmeyer. 1990. Deregulated Bcl-2 immunoglobulin transgene expands a resting but responsive immunoglobulin M and D-expression B-cell population. *Mol. Cell. Biol.* **10:** 1901.

Merino, R., L. Ding, D.J. Veis, S.J. Korsmeyer, and G. Nuñez. 1994. Developmental regulation of the Bcl-2 protein and susceptibility to cell death in B lymphocytes. *EMBO J.* **13:** 683.

Merry, D.E., D.J. Veis, W.F. Hickey, and S.J. Korsmeyer. 1994. Bcl-2 protein expression is widespread in the developing nervous system and retained in the adult PNS. *Development* **120:** 301.

Monaghan, P., D. Robertson, T.A.S. Amos, M.J.S. Dyer, D.Y. Mason, and M.F. Greaves. 1992. Ultrastructural localization of Bcl-2 protein. *J. Histochem. Cytochem.* **40:** 1819.

Nakayama, K.-I., K. Nakayama, I. Negishi, K. Kuida, Y. Shinkai, M.C. Louie, L.E. Fields, P.J. Lucas, V. Stewart, F.W. Alt, and D.Y. Loh. 1993. Disappearance of the lymphoid system in Bcl-2 homozygous mutant chimeric mice. *Science* **261:** 1584.

Neilan, J.G., Z. Lu, C.L. Afonso, G.F. Kutish, M.D. Sussman, and D.L. Rock. 1993. An African swine fever virus gene with similarity to the proto-oncogene *bcl-2* and the Epstein-Barr virus gene *BHRF1*. *J. Virol.* **67:** 4391.

Nguyen, M., D.G. Millar, V. Wee Yong, S.J. Korsmeyer, and G.C. Shore. 1993. Targeting of Bcl-2 to the mitochondrial outer membrane by a COOH-terminal signal-anchor sequence. *J. Biol. Chem.* **268:** 25265.

Nuñez, G., D. Hockenbery, T.M. McDonnell, C.M. Sorensen, and S.J. Korsmeyer. 1991. Bcl-2 maintains B cell memory. *Nature* **353:** 71.

Oltvai, Z.N., C.L. Milliman, and S.J. Korsmeyer. 1993. Bcl-2 heterodimerizes *in vivo* with a conserved homolog, Bax, that accelerates programmed cell death. *Cell* **74:** 609.

Pezzella, F., A.G. Tse, J.L. Cordell, K.A.F. Pulford, K.C. Guter, and R.Y. Mason. 1990. Expression of the Bcl-2 oncogene protein is not specific for the 14;18 chromosome translocation. *Am. J. Pathol.* **137:** 225.

Rowley, J.D. 1982. Identification of the constant chromosome regions involved in human hematologic malignant disease. *Science* **216:** 749.

Sentman, C.L., J.R. Shutter, D. Hockenbery, O. Kanagawa, and S.J. Korsmeyer. 1991. Bcl-2 inhibits multiple forms of apoptosis but not negative selection in thymocytes. *Cell* **67:** 879.

Strasser, A., A.W. Harris, and S. Cory. 1991. Bcl-2 transgene inhibits T cell death and perturbs thymic self-censorship. *Cell* **67:** 889.

Strasser, A., A.W. Harris, M.L. Bath, and S. Cory. 1990. Novel primitive lymphoid tumours induced in transgenic mice by cooperation between *myc* and *bcl-2*. *Nature* **348:** 331.

Strasser, A., A.W. Harris, L.M. Corcoran, and S. Cory. 1994. Bcl-2 expression promotes B- but not T-lymphoid development in scid mice. *Nature* **368:** 457.

Tsujimoto, Y., J. Gorham, J. Cossman, E. Jaffe, and C.M. Croce. 1985. The t(14;18) chromosome translocations involved in B-cell neoplasms result from mistakes in VDJ joining. *Science* **229:** 1390.

Veis, D.J., C.L. Sentman, E.A. Bach, and S.J. Korsmeyer. 1993a. Expression of the Bcl-2 protein in murine and human thymocytes and in peripheral T lymphocytes. *J. Immunol.* **151:** 2546.

Veis, D.J., C.M. Sorenson, J.R. Shutter, and S.J. Korsmeyer. 1993b. Bcl-2 deficient mice demonstrate fulminant lymphoid apoptosis, polycystic kidneys and hypopigmented hair. *Cell* **75:** 229.

Veis-Novack, D.J. and S.J. Korsmeyer. 1994. Bcl-2 protein expression during murine development. *Am. J. Pathol.* **145:** 61.

Yin, X.-M., Z.N. Oltvai, S.J. Korsmeyer. 1994. BH1 and BH2 domains of Bcl-2 are required for inhibition of apoptosis and heterodimerization with Bax. *Nature* **369:** 321.

Yunis, J.J., G. Frizzera, M.M. Oken, J. McKenna, A. Theologides, and M. Arnesen. 1987. Multiple recurrent genomic defects in follicular lymphoma. *N. Engl. J. Med.* **316:** 79.

Control of p53-dependent Apoptosis by E1B, Bcl-2, and Ha-*ras* Proteins

E. WHITE,*† S.-K. CHIOU,* L. RAO,* P. SABBATINI,* AND H.-J. LIN*
Center for Advanced Biotechnology and Medicine and †Department of Biological Sciences, Rutgers University, Piscataway, New Jersey 08854

Regulation of programmed cell death (apoptosis) has been known as a critical process in normal development and differentiation. More recently, apoptosis has been realized as a cellular response to growth deregulation that occurs during oncogenic transformation and viral infection and may indeed be a defense mechanism for the elimination of infected and transformed cells. One of the best illustrations of this activity is the cellular response to adenovirus infection and expression of the viral transforming gene products.

Expression of the adenovirus E1A and E1B oncogenes is important for sustaining a productive infection in human cells and is essential for the oncogenic transformation of primary rodent cells. The E1A proteins function to initiate cellular proliferation (Stabel et al. 1985; Kaczmarek et al. 1986), in part by functional inactivation of the retinoblastoma protein (p105 Rb) (Whyte et al. 1988), to provide a favorable cellular environment for virus replication. E1A expression also produces accumulation of the product of the *p53* tumor suppressor gene (Lowe and Ruley 1993) which is a potent inducer of apoptosis (Yonish-Rouach et al. 1991; Debbas and White 1993), which hinders virus replication (for review, see White 1994a,c). The adenovirus E1B gene, however, encodes overlapping, redundant functions to suppress p53-dependent apoptosis, the 19K and 55K proteins (for review, see White and Gooding 1994; White et al. 1994; White 1994a,c). The E1B 55K protein complexes with and inhibits p53 directly (Sarnow et al. 1982; Yew and Berk 1992), whereas the E1B 19K protein likely acts indirectly to suppress p53-dependent apoptosis and is functionally indistinguishable from Bcl-2 (Debbas and White 1993; Chiou et al. 1994a). Thus, expression of the E1B 19K, E1B 55K, or Bcl-2 proteins permits E1A expression and subsequent growth deregulation to occur unimpeded by cell death. Without inhibition of p53-dependent apoptosis by E1B or Bcl-2, transformation of rodent cells is rare and premature death of productively infected human host cells impairs virus yield.

In cells driven into apoptosis by E1A and p53, E1B 19K and Bcl-2 protein expression prevents apoptosis, but the growth arrest function of p53 remains intact (Debbas and White 1993; Chiou et al. 1994a; Sabbatini et al. 1995). Thus, the apoptotic and growth arrest functions of p53 are separable, and Bcl-2 expression may represent a cellular mechanism for controlling the activity of p53. These observations also raise the possibility that simultaneous stimulation of cell cycle progression and arrest by E1A and p53, respectively, may create conflicting growth signals that are manifested by the induction of apoptosis. The E1B 19K and Bcl-2 proteins may alleviate apoptosis by diverting cells into a growth-arrested state.

p53 is a transcription factor that can both activate and repress transcription. The E1B 19K protein does not hinder the *trans*-activation activity of p53 but does alleviate transcriptional repression (Sabbatini et al. 1995). The induction of growth arrest by p53 is, therefore, associated with activation of transcription, whereas apoptosis is associated with down-regulation of gene expression.

Activated Harvey-ras (Ha-*ras*) will cooperate with E1A in transformation assays, which indicates that, like E1B and Bcl-2, Ha-*ras* may possess the capacity to regulate p53-dependent apoptosis. Directed Ha-*ras* expression will block the induction of apoptosis by E1A and p53, but unlike E1B 19K and Bcl-2, Ha-*ras* in combination with E1A appears to overcome both growth arrest and apoptosis (H.-J. Lin et al., in prep.). Thus, E1B 19K and Bcl-2 act differently from Ha-*ras* in the regulation of p53 function. By elucidating the pathways involved in the regulation of apoptosis, we hope to enhance our understanding of not only the development of cancer, but also how these same mechanisms contribute to normal cell growth and differentiation.

METHODS

Plasmids and tissue culture. Primary Fisher baby rat kidney (BRK) cells were used as a source of primary cells for transformation assays with adenoviral oncogenes E1A and E1B. BRK cells were transfected with plasmid DNAs by electroporation as described previously (White et al. 1991). To derive cell lines that undergo controlled p53-dependent apoptosis, primary BRK cells were transfected with a cytomegalovirus (CMV) promoter construct to express the E1A (pCMVE1A, White et al. 1991) and the plasmid pLTRcGval135 (Michalovitz et al. 1990) to express murine mutant p53(val135). The p53(val135) protein is temperature sensitive and predominantly in the mutant conformation at 37.5°C–38.5°C and predominantly in the wild-type conformation at 32°C (Michalovitz et al. 1990; Gannon and Lane 1991; Martinez et al. 1991). Continuous propagation of cell lines transformed with

E1A and the temperature-sensitive p53(val135), such as p53A, and various derivatives containing the p53(val135) protein were carried out at 38.5°C. Cell lines were maintained in Dulbecco modified Eagle medium with 10% fetal bovine serum.

The E1A plus p53(val135)-transformed BRK line p53A was transfected by electroporation with a neomycin resistance marker only (pSV2neo), or with both pSV2neo and the pCMV19K expression vector (White and Cipriani 1989), to express the E1B 19K protein (Debbas and White 1993). Derivatives of p53A that express Bcl-2, 3B and 4B, and 1A (Chiou et al. 1994a) were obtained by transfecting the human Bcl-2 expression vector pSFFV*bcl-2* containing a neomycin resistance marker (Hockenbery et al. 1990). The pneoCMV/Ha-*ras* plasmid was used to express Ha-*ras* (H.-J. Lin et al., in prep.). Transformants were screened for E1B 19K, Bcl-2, or Ha-*ras* expression by Western blotting with monoclonal antibodies directed against E1B 19K, human Bcl-2, or activated Ha-*ras* (Debbas and White 1993; Chiou et al. 1994a; H.-J. Lin et al., in prep.).

Antibodies and Western blotting. Monoclonal antibodies directed against murine p53 (pAb248 and pAb2C2) were generously provided by Dr. Arnold J. Levine (Princeton University, Princeton, New Jersey). The E1A-specific monoclonal antibody M73 was generously provided by Dr. Ed Harlow (Massachusetts General Hospital, Charlestown, Massachusetts). The Bcl-2-specific monoclonal antibody was provided by Dr. S. Korsmeyer (Washington University, St. Louis, Missouri). The Pan-rasVal12 (AB-1) monoclonal antibody directed against Ha-*ras* was obtained from Oncogene Science. Cell extracts for Western analysis were prepared from subconfluent cultures, and equal quantities of protein from each cell line were analyzed by polyacrylamide gel electrophoresis and semi-dry blotting onto nitrocellulose membranes by standard procedures. Following antibody incubations, immune complexes were detected by enhanced chemiluminescence according to the manufacturers specifications (Amersham).

Viability and DNA fragmentation analysis. BRK cell lines were plated at equal density at the restrictive temperature of 38.5°C and evaluated for viability and indications of apoptotic cell death following incubation at the permissive temperature for wild-type p53 (32°C). The viable cell number was determined by trypan blue exclusion at the permissive and restrictive temperatures. DNA integrity, the loss of which is a hallmark of apoptotic cell death, was monitored by a modified Hirt procedure (Hirt 1967; White et al. 1984). Low-molecular-weight Hirt supernatant fractions from BRK lines were equalized with respect to the original viable cell number. Hirt DNA was analyzed by electrophoresis in a 1% agarose gel and visualized by ethidium bromide staining.

Flow cytometry analysis. Cells (10^6) in monolayer culture were harvested, washed, fixed, and stained with propidium iodide as described previously (White et al. 1988). As a measure of cellular DNA synthesis, cells were labeled with BrdU and stained with anti-BrdU antibodies as described previously (H.-J. Lin et al., in prep.). Fluorescence intensities were determined by quantitative flow cytometry and profiles were generated on a Coulter EPICS-PROFILE II analyzer (Chiou et al. 1994a; H.-J. Lin et al., in prep.).

RESULTS AND DISCUSSION

Adenovirus E1A Induces Both Cell Proliferation and Apoptosis

Expression of the adenovirus E1A gene is required to create a suitable environment for viral DNA replication. E1A achieves this by stimulating cellular DNA synthesis through subverting cell growth control. The interaction of E1A with a discrete set of cell-growth-negative regulatory proteins is required for stimulation of cellular DNA synthesis. One group of these E1A-associated cellular proteins has been identified as the product of the retinoblastoma susceptibility gene (p105 Rb) (for review, see Dyson and Harlow 1992; Moran 1993) and its relatives p107 (Zhu et al. 1993) and p130 (Hannon et al. 1993; Li et al. 1993; Mayol et al. 1993). Another E1A-associated protein is the CBP-related protein p300 (Stein et al. 1991; Arany et al. 1994; Eckner et al. 1994). Complex formation between Rb family members and E1A activates the transcriptional activity of the E2F family of transcription factors (for review, see Nevins 1992), which control the expression of gene products necessary for entry into S phase of the cell cycle (Johnson et al. 1993). p300 displays sequence homology with the CREB binding protein CBP, suggesting that p300 may be a component of the family of transcriptional coactivators responsible for transducing negative growth regulatory signals through cAMP (Arany et al. 1994). Thus, E1A interferes with transcriptional pathways for inhibiting cell cycle progression, and thereby acts to stimulate cellular DNA synthesis. These same activities of E1A are required for transformation (for review, see Moran and Zerler 1988; Dyson and Harlow 1992) and for the induction of cell death by apoptosis (White et al. 1991; for review, see White 1993, 1994a,b,c; White and Gooding 1994; White et al. 1994).

Although the role of E1A in stimulation of cell proliferation has been clear for some time, the induction of apoptosis was not initially apparent because adenovirus encodes two independent functions within the E1B gene to inhibit apoptosis. Thus, apoptosis is not apparent in wild-type adenovirus-infected human cells or during the transformation of rodent cells. However, infection of human cells with adenovirus mutants with loss-of-function mutations within the E1B 19K gene produced the degradation of host cell and viral DNA (*deg* phenotype), enhanced cytopathic effects (*cyt* phenotype), and premature death of the infected host cell, culminating in a reduction of virus yield (Pilder et al. 1984; Subramanian et al. 1984; Takemori et al. 1984; White et al. 1984). E1B 19K expression is apparently required to inhibit host cell destruction during viral infection. Degradation of cellular DNA into a nucleosome ladder pattern most often associated with the induction of apoptotic cell death is revealed by infection with a double mutant adenovirus carrying mutations in the E1B 19K and E2 72K DNA-binding proteins (White 1994c), the latter of which prevents replication of virus DNA. In the absence of the degraded viral DNA, which obscures the degradation pattern of cellular DNA, the degraded cellular DNA in the nucleosome pattern is observed (White 1994c). These events are all characteristic of cells undergoing apoptosis, which suggests that the E1B 19K protein may act as an inhibitor of apoptosis triggered by expression of another viral gene product (White et al. 1991).

E1A was identified as the factor responsible for triggering apoptosis in adenovirus-infected and transformed cells by mapping studies with recombinant adenoviruses and was confirmed by transient E1A expression from plasmid vectors (White and Stillman 1987; White et al. 1991, 1992; Rao et al. 1992). Thus, E1A expression is both necessary and sufficient for the induction of apoptosis. Analysis of E1A mutants suggested that the ability of E1A to stimulate cell proliferation cosegregated with the induction of apoptosis, implicating a role for the Rb and p300 proteins in the induction not only of cell proliferation, but also of cell death (White et al. 1991). Interestingly, deregulated c-*myc* has also been associated with stimulation of cell proliferation and apoptosis (Evan et al. 1992). These observations raise the possibility that untimely or deregulated DNA synthesis or cell cycle control may elicit an apoptotic response (Evan et al. 1992; Debbas and White 1993). Apoptosis may thereby serve as a default safety mechanism to eliminate cells undergoing unscheduled DNA synthesis as a consequence of sustaining transforming mutations or upon viral infection.

Induction of Apoptosis by E1A Is p53 Dependent

p53 is the most commonly mutated gene in human tumors (Vogelstein 1990; Hollstein et al. 1991) and clearly has a role in the suppression of cell growth (Finlay et al. 1989; Baker et al. 1990; Diller et al. 1990; Mercer et al. 1990; Michalovitz et al. 1990; Kastan et al. 1991; Martinez et al. 1991; Kuerbitz et al. 1992; Livingstone et al. 1992; Yin et al. 1992). Part of the ability of p53 to arrest cell cycle progression is likely due to transcriptional activation by p53 of the *p21/WAF1/cip1/sdi1* gene, which is an inhibitor of cell-cycle-dependent kinases (El-Deiry et al. 1993; Harper et al. 1993; Xiong et al. 1993; Noda et al. 1994). Another mechanism by which p53 could act as a tumor suppressor was realized with the observation that in some circumstances, p53 produced not growth arrest, but cell death by apoptosis (Yonish-Rouach et al. 1991). Thus, p53 can function as a tumor suppressor either by suppressing cell cycle progression or by inducing cell death. The induction of apoptosis is attractive as a mechanism of tumor suppression because it is irreversible, whereas growth arrest is not. The particular physiological state of the cell may determine whether growth arrest or apoptosis is the outcome of p53 function.

It is now clear that the induction of apoptosis by E1A is mediated by p53. The E1B 55K protein, which will block the induction of apoptosis and cooperate with E1A to transform primary rodent cells (White and Cipriani 1990; Rao et al. 1992), complexes with p53 and inhibits p53 function as a transcription factor (Sarnow et al. 1982; Yew and Berk 1992). E1A expression increases the half-life of the p53 protein, causing p53 accumulation to high levels followed by apoptosis (Lowe and Ruley 1993). Dominant mutant forms of p53 will block the induction of apoptosis by E1A and cooperate with E1A to transform primary rodent cells with high frequency (Debbas and White 1993; White et al. 1994). Finally, returning p53 to the wild-type conformation in E1A plus p53-transformed rodent cell lines through the use of a temperature-sensitive mutant allele of p53 produces massive cell death by apoptosis (Debbas and White 1993). Thus, p53 function is required for the induction of apoptosis by E1A. p53 may be the cellular response to the deregulation of cell growth control by E1A in the course of adenovirus infection and may trigger cell death as a means of eliminating virus-infected or transformed cells.

The Adenovirus E1B 19K Protein Inhibits p53-dependent Apoptosis

Although the function of the E1B 55K protein in subverting p53 function has been well documented, the ability of the E1B 19K protein to block the induction of apoptosis by E1A, which was p53-dependent, suggested that the E1B 19K protein may also interfere with p53 function. This was tested directly by introducing an E1B 19K expression vector into a cell line transformed by E1A and a temperature-sensitive mutant of p53, p53A, which is transformed and rapidly proliferating when p53 is in the mutant conformation, but rapidly undergoes apoptosis when p53 is returned to the wild-type conformation and the permissive temperature (Debbas and White 1993). E1B 19K protein expression completely prevented the induction of apoptosis by wild-type p53 at the permissive temperature (Debbas and White 1993). Instead of initiating apoptosis, cells were diverted into growth arrest with predominantly 4N DNA content, suggesting that progression out of the G_2 phase of the cell cycle was blocked (see below) (Sabbatini et al. 1995; H.-J. Lin et al., in prep.). Furthermore, the growth arrest and apoptotic activities of p53 are apparently separable, since the E1B 19K protein can specifically block apoptosis but not the growth arrest capacity of p53.

Functional Equivalence of E1B 19K and Bcl-2 Proteins

Establishment that the probable function of the E1B 19K protein was to inhibit apoptosis (White et al. 1991) prompted a functional comparison with the only other anti-apoptotic gene product at that time, Bcl-2 (Vaux et al. 1988; Hockenbery et al. 1990). Bcl-2 was found to cooperate with E1A in the transformation of primary BRK cells (Rao et al. 1992). Inhibition of apoptosis is therefore required for transformation by E1A, and either the E1B 19K or Bcl-2 proteins can provide that function. As the induction of apoptosis by E1A in rodent cells is mediated by the p53, this suggested that Bcl-2 may also overcome p53-dependent apoptosis. Bcl-2 expression in rodent cell lines transformed by E1A and the temperature-sensitive mutant p53, which undergo apoptosis when p53 reverts to the wild-type conformation, completely inhibited apoptosis (Chiou et al. 1994a). As with the E1B 19K-expressing cell lines, the presence of the Bcl-2 protein enabled growth arrest but not apoptosis in the presence of wild-type p53 (see below) (Chiou et al. 1994a). Thus, the E1B 19K and Bcl-2 proteins are functionally indistinguishable in transformation assays and the ability to block p53-dependent apoptosis.

The functional similarity between Bcl-2 and E1B 19K proteins suggests that they may act by a similar mechanism, and that Bcl-2 may complement the requirement for E1B 19K expression during productive infection. Infection of human HeLa cells with an E1B 19K loss-of-function mutant adenovirus produces apoptosis and results in premature host cell death, which impairs virus yield (Pilder et al. 1984; Subramanian et al. 1984; White et al. 1984). HeLa cells express extremely low levels of p53 due to expression of human papillomavirus E6 protein; however, levels of p53 were substantially increased by E1A expression during adenovirus infection (Chiou et al. 1994b). Therefore, E1A may induce p53-dependent apoptosis in HeLa cells by overriding the E6-induced degradation of p53 and promoting p53 accumulation.

HeLa cells express extremely low levels of Bcl-2. Stable Bcl-2 overexpression in HeLa cells infected with the E1B 19K$^-$ mutant adenovirus blocked the induction of apoptosis (Chiou et al. 1994b). Expression of Bcl-2 in HeLa cells also conferred resistance to tumor necrosis factor (TNF)-α and Fas antigen-mediated apoptosis (Chiou et al. 1994b), which are also established functions of the E1B 19K protein (Gooding et al. 1991; Hashimoto et al. 1991; White et al. 1992; White 1993). A comparison of the amino acid sequence of Bcl-2 family members and the E1B 19K protein indicated limited amino acid sequence homology (Chiou et al. 1994b). The highly conserved central region of the E1B 19K protein has been defined by mutational analysis to be important in transformation and regulation of apoptosis (White et al. 1992; Chiou et al. 1994b). The region that corresponds to the most highly conserved region within the Bcl-2 family members, particularly Bcl-2 homology region 1 (Oltvai et al. 1993), falls within the highly conserved central region of the E1B 19K protein (Chiou et al. 1994b). The limited sequence homology and functional relatedness of Bcl-2 and the E1B 19K proteins suggest that both may function by a similar mechanism to inhibit apoptosis (Chiou et al. 1994b). The functional significance of the limited sequence homology between the E1B 19K and Bcl-2 family members awaits the identification of the biochemical mode of action.

Other viruses have been found to encode proteins with sequence similarity to Bcl-2 (Cleary et al. 1986; Neilan et al. 1993), suggesting that avoidance of apoptosis may be advantageous to a productive viral infection, not only for adenovirus, but for viral infection in general. It is likely that the E1B 19K protein is the adenoviral functional equivalent of the cellular Bcl-2 protein (Chiou et al. 1994b).

E1B 19K and Bcl-2 Resolve the Growth Arrest from Apoptosis Functions of p53

The ability of E1B 19K and Bcl-2 proteins to suppress apoptosis by E1A, which was p53-dependent, suggested that E1B 19K and Bcl-2 may affect the function of p53. Either E1B 19K or Bcl-2 expression was sufficient to completely prevent p53-mediated apoptosis in BRK cell lines transformed by E1B and p53(val135) at the permissive temperature and caused cells to remain in a predominantly growth-arrested state (Chiou et al. 1994a; Debbas and White 1993; Sabbatini et al. 1995). There was a complete cessation of DNA synthesis with growth arrest occurring predominantly in G_2 (Chiou et al. 1994a; Sabbatini et al. 1995; H.-J. Lin et al., in prep.). Growth arrest imposed by E1B 19K or Bcl-2 expression in the presence of wild-type p53 was reversible (Chiou et al. 1994a; Sabbatini et al. 1995). Neither E1B 19K nor Bcl-2 affected the levels of p53 protein or the ability of p53 to localize to the nucleus (Debbas and White 1993; Chiou et al. 1994a). Thus, E1B 19K and Bcl-2 expression diverts the activity of p53 from induction of apoptosis to induction of growth arrest, and both may thereby act as modifiers of p53 function. The ability of E1B 19K and Bcl-2 proteins to bypass induction of apoptosis by p53 may be the means by which each contributes to oncogenesis. Furthermore, if cell cycle progression is required for the cell death process, then the E1B 19K and Bcl-2 proteins may provide escape from cell death by diverting cells into a growth-arrested state (Chiou et al. 1994a; H.-J. Lin et al., in prep.).

The E1B 19K Protein Alleviates Transcriptional Repression by p53

p53 likely functions as a tumor suppressor by binding to DNA in a site-specific fashion and activating transcription (Fields and Jung 1990; Raycroft et al. 1990; El-Deiry et al. 1992; Farmer et al. 1992; Zambetti et al. 1992). Transcriptional repression of promoters that lack a p53-specific DNA-binding element has also been reported (Ginsberg et al. 1991; Seto et al. 1992; Mack et al. 1993). The ability of the E1B 19K and Bcl-2 proteins to regulate p53-mediated transcriptional activation and repression was therefore investigated as a potential means for modulation of p53 activity. Particularly significant is the ability of E1B 19K and Bcl-2 to resolve the growth arrest from apoptotic functions of p53, which could be potentially attributable to differential regulation of p53-dependent transcriptional regulation.

In promoter-reporter assays, the E1B 19K protein did not prevent p53-mediated *trans*-activation, suggesting that the 19K protein does not inhibit apoptosis by blocking transcriptional activation by p53 (Sabbatini et al. 1995). E1B 19K expression did, however, alleviate p53-mediated transcriptional repression, suggesting that p53 may induce apoptosis by turning off survival genes, a situation that may be prevented by expression of the 19K protein (Sabbatini et al. 1995).

Analysis of E1B 19K mutants, which differ in their individual transformation potentials (White et al. 1992; Chiou et al. 1994b), demonstrated that this modulation of p53-mediated transcription cosegregated with the transformation potential of the E1B 19K protein (Sabbatini et al. 1995). The E1B 19K protein-dependent alleviation of transcriptional repression was not due to inhibition of apoptosis, since the same results were obtained in the p53-null cell line, Saos-2, which is not vulnerable to p53-mediated apoptosis (Sabbatini et al. 1995). Modulation of p53-mediated transcription is, therefore, not a secondary effect of the ability of the E1B 19K protein to overcome p53-mediated apoptosis.

The E1B 19K protein is unique among the DNA virus-transforming proteins in that it discriminates between transcriptional activation and repression by p53. The human papillomavirus E6 and the adenovirus E1B 55K proteins, for example, inhibit p53 function by totally abrogating both transcriptional activation and repression by p53 (Lechner et al. 1992; Yew and Berk 1992). The specific abrogation of transcriptional repression but not activation by the E1B 19K protein is, however, consistent with the ability of p53 to retain the growth arrest function in the presence of the E1B 19K protein. The p21/Waf1/Cip1/Sdi1 gene product has been identified as an inhibitor of cell-cycle-dependent kinases (cdk) and is transcriptionally activated by wild-type p53 (El-Deiry et al. 1993; Harper et al. 1993; Xiong et al. 1993; Noda et al. 1994). By turning on the expression of p21, p53 may achieve growth arrest through the inhibition of cdk activity. E1B 19K expression would be predicted from the promoter-recorder assays described above not to alter p21 *trans*-activation by p53, and in fact it does not prevent p53-dependent up-regulation of p21 mRNA (Sabbatini et al. 1995). Apoptosis may result, not from the activation of transcription by p53, but rather from the repression of transcription. In support of this, inhibitors of protein synthesis and transcription are potent inducers of apoptosis, supporting the possibility that global inhibition of gene expression by either drugs or wild-type p53 may cause apoptosis (Sabbatini et al. 1995). Essential survival factors may require continual transcription and translation, which if inhibited, results by default in cell death (Raff 1992).

E1A and Ha-*ras* Cooperate to Overcome Both Growth Arrest and Apoptosis by p53

The cooperation between growth stimulatory oncogenes such as E1A and c-*myc*, and those acting as inhibitors of apoptosis, such as E1B and Bcl-2, suggests that transformation requires cell proliferation to be coupled to inhibition of apoptosis. E1A and c-*myc* will also cooperate with activated Ha-*ras* oncogene (Land et al. 1983; Ruley 1983), suggesting a role for Ha-*ras* in overcoming apoptosis. To test this possibility, a Ha-*ras* expression vector pneoCMV/Ha-ras was introduced into the BRK cell line p53A that had been transformed with E1A and the temperature sensitive p53(val135), and the Ras-expressing transformants were analyzed for the induction of growth arrest and apoptosis at the permissive temperature for wild-type p53 (H.-J. Lin et al., in prep.).

Ras-expressing cell lines remained viable at the permissive temperature, suggesting that Ras expression was sufficient to overcome apoptosis by wild-type p53. Unlike the E1B 19K- and Bcl-2-expressing cell lines, however, the Ras transformants continued to synthesize DNA at the permissive temperature, indicating that Ras did not divert cells into a growth-arrested state, but rather, Ras in combination with E1A was able to overcome growth arrest inflicted by p53 (H.-J. Lin et al., in prep.). The mechanism of action of Ras is thereby distinctly different from that of E1B 19K and Bcl-2.

To evaluate the specific contribution of Ras to the transformation process, BRK cells were transformed with Ras and the temperature-sensitive p53(val135) in the absence of E1A, and the resulting cell lines were examined at the permissive temperature for growth arrest and apoptosis (H.-J. Lin et al., in prep.). The ability of Ha-*ras* to transform primary cells in cooperation with mutant p53 suggests that Ras can stimulate cell proliferation efficiently when p53 function is ablated. All Ras plus p53(val135) transformants underwent growth arrest and not apoptosis at the permissive temperature for wild-type p53, as determined by cell viability and BrdU incorporation (H.-J. Lin et al., in prep.). Ras alone is apparently not sufficient to overcome growth arrest by p53, whereas the combined action of both E1A and Ras is effective. This was confirmed by introduction of E1A, which then permitted DNA synthesis (H.-J. Lin et al., in prep.). Interestingly, although capable of driving cell proliferation as long as p53 is inactivated, Ras does not drive cells into apoptotic cell death as does E1A (H.-J. Lin et al., in prep.). All means to stimulate cell proliferation apparently do not necessarily lead to apoptotic cell death. Perhaps the inability of p53 to arrest cells expressing E1A is what leads to apoptosis, whereas p53 can effectively arrest cells expressing activated Ras.

Oncogene cooperation can be viewed as a means of evading the actions of tumor suppressor proteins, particularly p53, which may explain why p53 is the most frequently mutated gene in human tumors (Vogelstein 1990; Hollstein et al. 1991). In the examples we have

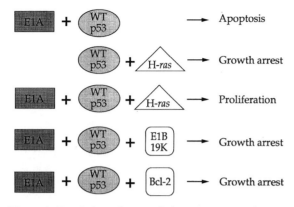

Figure 1. Regulation of apoptosis by oncogenes and tumor suppressor genes. See text for explanation.

provided, E1A alone is not sufficient for transformation because of the induction of p53-dependent apoptosis (Fig. 1). Ras alone is not sufficient because of the inability to overcome p53-dependent growth arrest (Fig. 1). The action of E1A and Ras together is sufficient to enable cell proliferation in the presence of high levels of p53 protein and, thereby, transformation (Fig. 1). E1B 19K and Bcl-2 can cooperate with E1A by suppressing the apoptosis function of p53, permitting growth arrest (Fig. 1). Presumably, E1A can overcome growth arrest imposed by p53 as long as apoptosis is inhibited. It will be of interest in the future to determine how many of the actions of other oncogene products are directed at overcoming the actions of the p53 protein.

Apoptosis directed by p53 upon loss of Rb function is likely to be a fail-safe mechanism to minimize the impact of viral infection and transformation on the host organism. Recent evidence suggests that Rb and p53 regulate apoptosis in normal growth and differentiation (Morgenbesser et al. 1994; Pan and Griep 1994; White 1994b). DNA viruses have found the means to usurp the activity of Rb and p53 for efficient propagation. By studying the mechanisms involved, we hope to gain insight into the prevention of human cancer.

ACKNOWLEDGMENTS

We thank E. Harlow, L. Laimins, A. Levine, and S. Korsmeyer for antibodies and plasmids. This work was supported by grants from the National Institutes of Health (CA-53370 and CA-60088), the American Cancer Society (MV-71975), and the New Jersey Commission on Science and Technology to E.W. S.-K.C. was supported by a National Institutes of Health biotechnology fellowship. L.R. was supported by a C.A.B.M. fellowship, and P.S. was supported by a Rutgers/UMDNJ Interdisciplinary Core Curriculum in Molecular and Cellular Biology predoctoral fellowship.

REFERENCES

Arany, Z., W. Sellers, D. Livingston, and R. Eckner. 1994. E1A-associated p300 and CREB-associated CBP belong to a conserved family of coactivators. *Cell* **77:** 799.

Baker, S.J., S. Markowitz, E.R. Fearon, J.K.V. Willson, and B. Vogelstein. 1990. Suppression of human colorectal carcinoma cell growth by wild-type p53. *Science* **249:** 912.

Chiou, S.-K., L. Rao, and E. White. 1994a. Bcl-2 blocks p53-dependent apoptosis. *Mol. Cell. Biol.* **14:** 2556.

Chiou, S.-K., C.C. Tseng, L. Rao, and E. White. 1994b. Functional complementation of the adenovirus E1B 19K protein with Bcl-2 in the inhibition of apoptosis in infected cells. *J. Virol.* **68:** 6553.

Cleary, M.L., S.D. Smith, and J. Sklar. 1986. Cloning and structural analysis of cDNAs for bcl-2 and a hybrid bcl-2/immunoglobulin transcript resulting from the t(14;18) translocation. *Cell* **47:** 19.

Debbas, M. and E. White. 1993. Wild-type p53 mediates apoptosis by E1A which is inhibited by E1B. *Genes Dev.* **7:** 546.

Diller, L., J. Kassel, C.E. Nelson, M.A. Gryka, G. Litwak, M. Geghardt, and B. Bressac. 1990. p53 functions as a cell cycle control protein in osteosarcomas. *Mol. Cell. Biol.* **10:** 5772.

Dyson, N. and E. Harlow. 1992. Adenovirus E1A targets key regulators of cell proliferation. *Cancer Surv.* **12:** 161.

Eckner, R., M.E. Ewen, D. Newsome, M. Gerdes, J.A. DeCaprio, J.B. Lawrence, and D.M. Livingston. 1994. Molecular cloning and functional analysis of the adenovirus E1A-associated 300-kD protein (p300) reveals a protein with properties of a transcription adaptor. *Genes Dev.* **8:** 869.

El-Deiry, W.S., S.E. Kern, J.A. Pietenpol, K.W. Kinzler, and B. Vogelstein. 1992. Definition of a consensus binding site for p53. *Nat. Genet.* **1:** 45.

El-Deiry, W.S., T. Tokino, V.E. Velculescu, D.B. Levy, R. Parsons, J.M. Trent, D. Lin, E. Mercer, K.W. Kinzler, and B. Vogelstein. 1993. WAF1, a potential mediator of p53 tumor suppression. *Cell* **75:** 817.

Evan, G.I., A.H. Wyllie, C.S. Gilbert, T.D. Littlewood, H. Land, M. Brooks, C.M. Waters, L.Z. Penn, and D.C. Hancock. 1992. Induction of apoptosis in fibroblasts by c-myc protein. *Cell* **69:** 119.

Farmer, G., J. Bargonetti, H. Zhu, P. Friedman, R. Prywes, and C. Prives. 1992. Wild-type p53 activates transcription *in vitro*. *Nature* **358:** 83.

Fields, S. and S.K. Jung. 1990. Presence of a potent transcription activating sequence in the p53 protein. *Science* **249:** 1046.

Finlay, C.A., P.W. Hinds, and A.J. Levine. 1989. The p53 proto-oncogene can act as a suppressor of transformation. *Cell* **57:** 1083.

Gannon, J.V. and D.P. Lane. 1991. Protein synthesis required to anchor a mutant p53 protein which is temperature-sensitive for nuclear transport. *Nature* **349:** 802.

Ginsberg, D., F. Mechta, M. Yaniv, and M. Oren. 1991. Wild-type p53 can down-modulate the activity of various promoters. *Proc. Natl. Acad. Sci.* **88:** 9979.

Gooding, L.R., L. Aquino, P.J. Duerksen-Hughes, D. Day, T.M. Horton, S. Yei, and W.S.M. Wold. 1991. The E1B-19K protein of group C adenoviruses prevents cytolysis by tumor necrosis factor of human cells but not mouse cells. *J. Virol* **65:** 3083.

Hannon, G., D. Demetrick, and D. Beach. 1993. Isolation of the Rb-related p130 through its interaction with cdk2 and cyclins. *Genes Dev.* **7:** 2378.

Harper, J.W., G.R. Adami, N. Wei, K. Keyomarsi, and S.J. Elledge. 1993. The p21 cdk-interacting protein cip1 is a potent inhibitor of G_1 cyclin-dependent kinases. *Cell* **75:** 805.

Hashimoto, S., A. Ishii, and S. Yonehara. 1991. The E1B oncogene of adenovirus confers cellular resistance to cytotoxicity of tumor necrosis factor and monoclonal anti-Fas antibody. *Int. Immunol.* **3:** 343.

Hirt, B. 1967. Selective extraction of polyoma DNA from infected mouse cultures. *J. Mol. Biol.* **26:** 365.

Hockenbery, D., G. Nuñez, C. Milliman, R.D. Schreiber, and S. Korsmeyer. 1990. Bcl-2 is an inner mitochondrial membrane protein that blocks programmed cell death. *Nature* **348:** 334.

Hollstein, M., D. Sidransky, B. Vogelstein, and C. Harris. 1991. p53 mutations in human cancers. *Science* **253:** 49.

Johnson, D.J., J.K. Schwarz, W.D. Cress, and J.R. Nevins. 1993. Expression of transcription factor E2F1 induces quiescent cells to enter S phase. *Nature* **365:** 349.

Kaczmarek, L., B. Ferguson, M. Rosenberg, and R. Baserga. 1986. Induction of cellular DNA synthesis by purified adenovirus E1A proteins. *Virology* **152:** 1.

Kastan, M.B., O. Onyekwere, D. Sidransky, B. Vogelstein, and R.W. Craig. 1991. Participation of p53 protein in the cellular response to DNA damage. *Cancer Res.* **51:** 6304.

Kuerbitz, S.J., B.S. Plunkett, W.V. Walsh, and M.B. Kastan. 1992. Wild-type p53 is a cell cycle checkpoint determinant following irradiation. *Proc. Natl. Acad. Sci.* **89:** 7491.

Land, H., L.F. Parada, and R.A. Weinberg. 1983. Tumorigenic conversion of primary embryo fibroblasts requires at least two cooperating oncogenes. *Nature* **304:** 596.

Lechner, M.S., D.H. Mack, A.B. Finicle, T. Crook, K.H. Vousden, and L.A. Laimins. 1992. Human papillomavirus E6 proteins bind p53 in vivo and abrogate p53-mediated repression of transcription. *EMBO J.* **11:** 3045.

Li, Y., C. Graham, S. Lacy, A. Duncan, and P. Whyte. 1993. The adenovirus E1A-associated 130-kd protein is encoded by a member of the retinoblastoma gene family and physically interacts with cyclins A and E. *Genes Dev.* **7:** 2366.

Livingstone, L.R., A. White, J. Sprouse, E. Livanos, T. Jacks, and T.D. Tlsty. 1992. Altered cell cycle arrest and gene amplification potential accompany loss of wild-type p53. *Cell* **70:** 923.

Lowe, S. and H.E. Ruley. 1993. Stabilization of the p53 tumor suppressor is induced by adenovirus-5 E1A and accompanies apoptosis. *Genes Dev.* **7:** 535.

Mack, D.H., J. Vartikar, J.M. Pipas, and L. Laimins. 1993. Specific repression of TATA-mediated but not initiator-mediated transcription by wild-type p53. *Nature* **363:** 281.

Martinez, J., I. Georgoff, J. Martinez, and A.J. Levine. 1991. Cellular localization and cell cycle regulation by a temperature-sensitive p53 protein. *Genes Dev.* **5:** 151.

Mayol, X., X. Grana, A. Baldi, N. Sang, Q. Hu, and A. Giordano. 1993. Cloning of a new member of the retinoblastoma gene family (pRb2) which binds to the E1A transforming domain. *Oncogene* **8:** 2561.

Mercer, W.E., M.T. Shields, M. Amin, G.J. Suave, E. Appella, J.W. Romano, and S.J. Ullrich. 1990. Negative growth regulation in a glioblastoma tumor cell line that conditionally expresses human wild-type p53. *Proc. Natl. Acad. Sci.* **87:** 6166.

Michalovitz, D., O. Halevy, and M. Oren. 1990. Conditional inhibition of transformation and of cell proliferation by a temperature-sensitive mutant of p53. *Cell* **62:** 671.

Moran, E. 1993. DNA tumor virus transforming proteins and the cell cycle. *Curr. Opin. Genet. Dev.* **3:** 63.

Moran, E. and B. Zerler. 1988. Interactions between cell growth-regulating domains in the products of the adenovirus E1A oncogene. *Mol. Cell. Biol* **8:** 1756.

Morgenbesser, S.D., B.O. Williams, T. Jacks, and R.A. DePinho. 1994. p53-dependent apoptosis produced by Rb-deficiency in the developing mouse lens. *Nature* **371:** 72.

Neilan, J.G., Z. Lu, C.L. Afonzo, G.F. Kutish, M.D. Sussman, and D.L. Rock. 1993. An African swine fever virus gene with similarity to the proto-oncogene *bcl-2* and the Epstein-Barr virus gene *BHRF1*. *J. Virol.* **67:** 4391.

Nevins, J.R. 1992. E2F: A link between the Rb tumor suppressor protein and viral oncoproteins. *Science* **258:** 424.

Noda, A., Y. Ning, S.F. Venable, O.M. Pereira-Smith, and J.R. Smith. 1994. Cloning of senescent cell-derived inhibitors of DNA synthesis using an expression screen. *Exp. Cell Res.* **211:** 90.

Oltvai, Z.N., C.L. Millman, and S.J. Korsmeyer. 1993. Bcl-2 heterodimerizes in vivo with a conserved homolog, Bax, that accelerates programmed cell death. *Cell* **74:** 609.

Pan, H. and A.E. Griep. 1994. Altered cell cycle regulation in the lens of HPV-16 E6 or E7 transgenic mice: Implications for tumor suppressor gene function in development. *Genes Dev.* **8:** 1285.

Pilder, S., J. Logan, and T. Shenk. 1984. Deletion of the gene encoding the adenovirus 5 early region 1B—21,000-molecular weight polypeptide leads to degradation of viral and cellular DNA. *J. Virol.* **52:** 664.

Raff, M.C. 1992. Social controls on cell survival and cell death. *Nature* **356:** 398.

Rao, L., M. Debbas, P. Sabbatini, D. Hockenbery, S. Korsmeyer, and E. White. 1992. The adenovirus E1A proteins induce apoptosis which is inhibited by the E1B 19K and Bcl-2 proteins. *Proc. Natl. Acad. Sci.* **89:** 7742.

Raycroft, L., H. Wu, and G. Lozano. 1990. Transcriptional activation by wild-type but not transforming mutants of the p53 anti-oncogene. *Science* **249:** 1049.

Ruley, H.E. 1983. Adenovirus early region 1A enables viral and cellular transforming genes to transform primary cells in culture. *Nature* **304:** 602.

Sabbatini, P., S.-K. Chiou, L. Rao, and E. White. 1995. Modulation of p53-mediated transcriptional repression and apoptosis by the adenovirus E1B 19K protein. *Mol. Cell. Biol.* (in press).

Sarnow, P., Y.S. Ho, J. Williams, and A.J. Levine. 1982. Adenovirus E1b-58 kd tumor antigen and SV40 large tumor antigen are physically associated with the same 54 kd cellular protein in transformed cells. *Cell* **28:** 387.

Seto, E., A. Usheva, G.P. Zambetti, J. Momand, N. Horikoshi, R. Weinmann, A.J. Levine, and T. Shenk. 1992. Wild-type p53 binds to the TATA-binding protein and represses transcription. *Proc. Natl. Acad. Sci.* **89:** 12028.

Stabel, S., P. Argos, and L. Philipson. 1985. The release of growth arrest by microinjection of adenovirus E1A DNA. *EMBO J.* **4:** 2329.

Stein, R.W., M. Corrigan, P. Yaciuk, J. Whelan, and E. Moran. 1991. Analysis of E1A-mediated growth regulation functions: Binding of the 300-kilodalton cellular product correlates with E1A repression function and DNA synthesis-inducing activity. *J. Virol.* **64:** 4421.

Subramanian, T., M. Kuppuswamy, J. Gysbers, S. Mak, and G. Chinnadurai. 1984. 19-kDa tumor antigen coded by early region E1b of adenovirus 2 is required for efficient synthesis and for protection of viral DNA. *J. Biol. Chem.* **259:** 11777.

Takemori, N., C. Cladaras, B. Bhat, A.J. Conley, and W.S.M. Wold. 1984. *cyt* gene of adenovirus 2 and 5 is an oncogene for transforming function in early region E1B and encodes the E1B 19,000-molecular-weight polypeptide. *J. Virol.* **52:** 793.

Vaux, D.L., S. Cory, and T.M. Adams. 1988. Bcl-2 promotes the survival of haemopoietic cells and cooperates with c-*myc* to immortalize pre-B cells. *Nature* **335:** 440.

Vogelstein, B. 1990. A deadly inheritance. *Nature* **348:** 681.

White, E. 1993. Regulation of apoptosis by the transforming genes of the DNA tumor virus adenovirus. *Proc. Soc. Exp. Biol. Med.* **204:** 30.

———. 1994a. Function of the adenovirus E1B oncogene in infected and transformed cells. *Semin. Virol.* (in press).

———. 1994b. p53, guardian of Rb. *Nature* **371:** 21.

———. 1994c. Regulation of p53-dependent apoptosis by E1A and E1B. *Curr. Top. Microbiol. Immunol.* (in press).

White, E. and R. Cipriani. 1989. Specific disruption of intermediate filaments and the nuclear lamina by the 19-kDa product of the adenovirus E1B oncogene. *Proc. Natl. Acad. Sci.* **86:** 9886.

———. 1990. Role of adenovirus E1B proteins in transformation: Altered organization of intermediate filaments in transformed cells that express the 19-kilodalton protein. *Mol. Cell. Biol.* **10:** 120.

White, E. and L.R. Gooding. 1994. Regulation of apoptosis by human adenoviruses. *Curr. Commun. Cell Mol. Biol.* **81:** 111.

White, E. and B. Stillman. 1987. Expression of the adenovirus E1B mutant phenotypes is dependent on the host cell and on synthesis of E1A proteins. *J. Virol.* **61:** 426.

White, E., T. Grodzicker, and B.W. Stillman. 1984. Mutations in the gene encoding the adenovirus E1B 19K tumor antigen cause degradation of chromosomal DNA. *J. Virol.* **52:** 410.

White, E., D. Spector, and W. Welch. 1988. Differential distribution of the adenovirus E1A proteins and colocalization of E1A with the 70-kilodalton cellular heat shock protein in infected cells. *J. Virol.* **62:** 4153.

White, E., R. Cipriani, P. Sabbatini, and A. Denton. 1991. The adenovirus E1B 19-kilodalton protein overcomes the cytotoxicity of E1A proteins. *J. Virol.* **65:** 2968.

White, E., P. Sabbatini, M. Debbas, W.S.M. Wold, D.I. Kusher, and L. Gooding. 1992. The 19-kilodalton adenovirus E1B transforming protein inhibits programmed cell death and prevents cytolysis by tumor necrosis factor α. *Mol. Cell. Biol.* **12:** 2570.

White, E., L. Rao, S.-K. Chiou, C.-C. Tseng, P. Sabbatini, M. Gonzalez, and P. Verwaerde. 1994. Regulation of apoptosis by the transforming gene products of adenovirus. In *Apoptosis* (ed. E. Mihich and R.T. Schimke), p. 47. Plenum Press, New York.

Whyte, P., K. Buchkovich, J.M. Horowitz, S.H. Friend, M. Raybuck, R.A. Weinberg, and E. Harlow. 1988. Association between an oncogene and an anti-oncogene: The adenovirus E1A proteins bind to the retinoblastoma gene product. *Nature* **334:** 124.

Xiong, Y., G. Hannon, H. Zhang, D. Casso, R. Kobayashi, and D. Beach. 1993. p21 is a universal inhibitor of cyclin kinases. *Nature* **366:** 701.

Yew, P.R. and A.J. Berk. 1992. Inhibition of p53 *trans*-activation required for transformation by adenovirus early 1B protein. *Nature* **357:** 82.

Yin, Y., M.A. Tainsky, F.Z. Bischoff, L.C. Strong, and G.M. Wahl. 1992. Wild-type p53 restores cell cycle control and inhibits gene amplification in cells with mutant p53 alleles. *Cell* **70:** 937.

Yonish-Rouach, E., D. Resnitzky, J. Lotem, L. Sachs, A. Kimchi, and M. Oren. 1991. Wild-type p53 induces apoptosis of myeloid leukaemic cells that is inhibited by interleukin-6. *Nature* **352:** 345.

Zambetti, G.P., J. Baronetti, K. Walker, C. Prives, and A.J. Levine. 1992. Wild-type p53 mediates positive regulation of gene expression through a specific DNA sequence element. *Genes Dev.* **6:** 1143.

Zhu, L., S. Van Den Heuvel, K. Helin, A. Fattaey, M. Ewen, D. Livingston, N. Dyson, and E. Harlow. 1993. Inhibition of cell proliferation by p107, a relative of the retinoblastoma protein. *Genes Dev.* **7:** 1111.

Apoptosis in Carcinogenesis: The Role of p53

A.H. Wyllie, P.J. Carder, A.R. Clarke, K.J. Cripps, S. Gledhill,
M.F. Greaves,* S. Griffiths,* D.J. Harrison, M.L. Hooper, R.G. Morris,
C.A. Purdie, and C.C. Bird

*CRC Laboratories, Department of Pathology, University Medical School, Edinburgh EH8 9AG, Scotland, United Kingdom; *LRF Unit, Institute for Cancer Research, London SW3 6JB, United Kingdom*

In this paper, we explore the role played by apoptosis in the genesis of cancer. The hypothesis under test is that apoptosis affords a means whereby cells that have sustained genotoxic damage can be irreversibly deleted from the generative compartment of the affected tissue. Such damaged cells might otherwise have acquired serious but survivable DNA modifications, and so could have become the precursors of malignant clones (Fig. 1). One means of testing the hypothesis is to devise systems in which a critical gene on the signaling pathway from genotoxic damage to apoptosis is selectively disabled. Following exposure to the genotoxic agent, cells bearing such disabled genes would be expected to survive, generating a population of previously "forbidden" cells. The hypothesis predicts that these cells should have undergone genomic changes similar or identical to those known to be clonally expanded in cancer, a proposition that can be addressed directly.

Here we describe two such systems. In the first, the potentially genotoxic stimulus is ionizing radiation, delivered to the tissues of animals bearing constitutively two, one, or no copies of the endogenous wild-type p53 gene. We show that this gene plays a limiting role in the response (by apoptosis) of the cell to DNA strand breakage and that cells bearing multiple genetic abnormalities appear in the irradiated tissues. These abnormalities, moreover, are compatible with continuing cell survival and proliferation. In the second system, we observe some of the characteristics of authentic human colorectal cancers relative to the status of their p53 genes. Although we do not know the nature of the genotoxic agents in human colorectal carcinogenesis, we show that absence of functional p53 appears at an early stage in the evolution of malignant behavior, and that those tumors which display loss of normal p53 expression have distinctive features, with regard to genomic stability, when compared with those that do not. These observations fall short of proof of the original hypothesis, but they are compatible with it and indicate that the previously forbidden cell, which survives because of failure in apoptosis, has characteristic features that would fit it well for a role in carcinogenesis.

METHODS

A large deletion (from exon 2 to exon 6 inclusive, and with stop codons in all three reading frames) was introduced into the endogenous p53 gene of murine ES cells by homologous recombination, transmitted through the germ line, and rendered homozygous by breeding (Clarke et al. 1993). Wild-type animals were compared with their p53-disabled heterozygous and homozygous littermates in terms of the responses to genotoxic stimuli of three tissues: cortical thymocytes, small and large intestinal enterocytes, and bone marrow pre-B cells. Apoptosis was scored in the thymocytes within 8 hours of γ-irradiation (from a ^{137}Cs source) or other genotoxic treatments in vitro by direct observation in a fluorescence microscope of the nuclear acridine-orange-staining pattern. Enterocytes were studied in situ in histological sections of the gut, using a computer-assisted microscope system (the Zeiss Highly Organized Microscope Environment) to reliably identify, summate, and audit scores of apoptosis, mitosis, and BrdU uptake in each cell position from the crypt base, using data from at least 200 crypts per mouse (Clarke et al. 1994). Pre-B cells were cultured from femoral bone marrow in McCoy's medium supplemented with recombinant human interleukin-7 (IL-7) (Griffiths et al. 1994) and exposed to X-rays from a 240-KV source. Colonies of cells in irradiated and control cultures were counted after 6 days.

For the studies on human cancer, 83 colorectal carcinomas, from unselected consecutive symptomatic patients, were studied immediately after surgical removal (Carder et al. 1993). Each tumor was sampled at multiple sites (mean of 4), and each sample was studied by routine histology to confirm the presence of carcinoma, by immunocytochemistry using the monoclonal antibody PAb1801 to detect stabilization of p53, and by flow cytometry to measure DNA ploidy. Additional

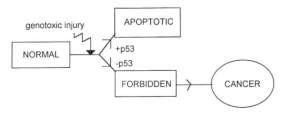

Figure 1. Hypothesis relating the process of apoptosis to carcinogenesis.

material from some samples was stored frozen for analysis of mutation or of loss of heterozygosity at various chromosomal sites (Cripps et al. 1994; P.J. Carder et al., in prep.).

RESULTS AND DISCUSSION

p53 Plays an Essential Role in the Apoptosis of Cells Exposed to Genotoxic Damage

As described by Donehower et al. (1992), live-born progeny with complete ablation of p53 function are developmentally normal, and we tested first the capacity of their thymic cortical cells to undergo apoptosis in response to a variety of stimuli. In culture, and in the absence of any added lethal stimulus, these cells enter apoptosis at the same rate and quantity as normal cells (Clarke et al. 1993). Similarly, when apoptosis was accelerated by treatment with glucocorticoids (Fig. 2) or a combination of calcium ionophore and the mitogen TPA, cells with and without functional p53 died in exactly the same numbers and with identical kinetics. In contrast with these quasi-physiological stimuli, ionizing radiation induced apoptosis in the thymocytes of wild-type animals, but not at all in cells null for p53 (Fig. 2). Interestingly, cells with only one intact copy of the gene entered apoptosis with kinetics intermediate between these two extremes, indicating that changes in expression of p53 within the physiological range are likely to influence the probability of apoptosis after injury. Etoposide, a topoisomerase II inhibitor that induces double strand breaks in DNA, also kills thymocytes in a p53-dependent fashion, at least at low doses. These experiments, and others of similar design reported independently (Lowe et al. 1993b), therefore demonstrate unequivocally that although cells without functional p53 have an intact apoptosis effector pathway, their entry to apoptosis following genotoxic damage is strictly limited by the availability of p53.

Closely similar experiments have been conducted on different cell lineages. In myeloid cells (Lotem and Sachs 1993), marrow pre-B cells (S.D. Griffiths et al., in prep.), and large and small intestinal mucosa (Fig. 3) (Clarke et al. 1994; Merritt et al. 1994), exactly the same conclusions pertain: Genotoxic damage initiates apoptosis by a p53-mediated, gene-dose-dependent mechanism.

p53 Does Not Cause Significant G_1 Arrest in Irradiated Mucosal Cells

In some cell lineages (e.g., fibroblasts), postirradiation p53 accumulation does not initiate apoptosis but rather G_1 arrest (Kastan et al. 1991; Kuerbitz et al. 1992). Recent data suggest that entry to G_1 arrest or apoptosis are mutually exclusive alternatives, the decision between them being dependent on the status of other specific genes, such as expression of the adenoviral E1A protein (Rao et al. 1992; Lowe et al. 1993a), or the *trans*-activating protein E2F (Wu and Levine 1994). The central hypothesis of this paper is that removal of potentially carcinogenic cells requires the process of apoptosis rather than merely the repair of genotoxic damage. Hence, it was important to formally evaluate the extent to which the irradiated cell populations entered G_1 arrest and, potentially, a repair pathway. Accordingly, we studied BrdU uptake simultaneously with apoptosis in the intestinal crypts within the first 4 hours after ionizing radiation in animals completely deficient in p53, or their wild-type or heterozygous littermates. Despite convincing evidence for p53-dependent apoptosis in cells in the bottom third of the crypts (in both small and large intestine overlapping the stem-cell region), differences in the fractions of BrdU-incorporating cells were never statistically significant (Fig. 3). Other investigators have also failed to find evidence of G_1 arrest in the mouse small intestine after similar doses of radiation (Chawlinski et al. 1988). It therefore appears that in the small and large intestine, the predominant p53-mediated response to genotoxic damage is apoptosis.

In contrast, using similar tissue quantitation methods, a profound reduction in the number of cells entering mitosis was evident in small and large intestinal crypts following irradiation (i.e., G_2 arrest). This effect, which is a well-known reaction to irradiation (Potten 1990; Kastan et al. 1991), was of similar amplitude in animals of all genotypes, and hence was clearly p53-independent (Table 1).

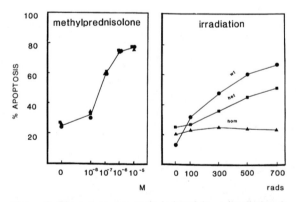

Figure 2. Thymocyte apoptosis induced by γ-irradiation is strictly p53-dependent, whereas apoptosis induced by glucocorticoid is not. Apoptosis, scored by fluorescence microscopy of cell suspensions exposed to 10 μg ml^{-1} acridine orange after fixation in 4% formaldehyde in ethanol, was studied 8 hr after radiation in cell suspensions of 10^7 ml^{-1} cultured at 37°C in MEM. Responses to 10^{-6} M methylprednisolone or to γ-irradiation (as shown) are compared, using cell suspensions from the same animals. The results are means of data gathered from littermates with the normal gene complement of wild-type p53 (●), heterozygous for disabled p53 (■), or homozygous null for p53 (▲). Each point is the mean of triplicate observations from two animals.

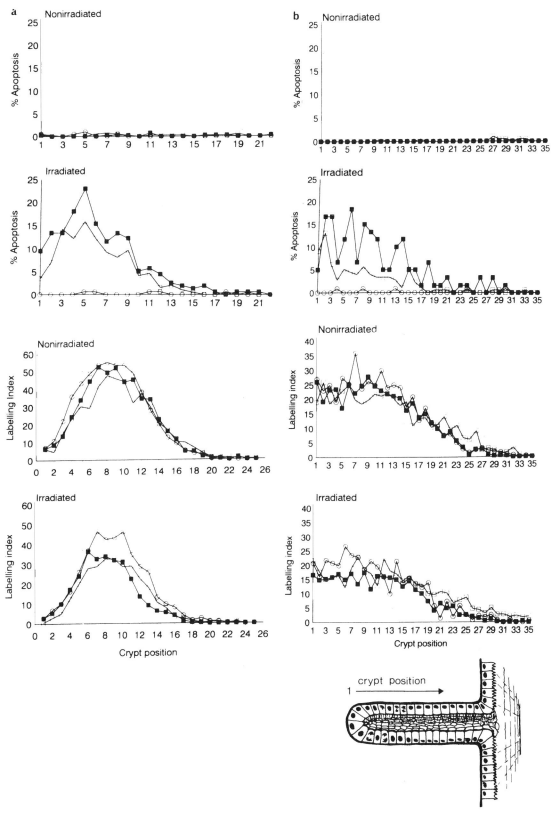

Figure 3. Responses of crypt cells from the small (*a*) and large (*b*) intestine showing apoptosis incidence (top two panels) or BrdU incorporation (bottom two panels) in each cell position in animals left unirradiated or exposed to 4 Gy whole-body radiation with ^{137}Cs γ-rays as indicated. The mice were studied 4 hr after an authentic or mock-irradiation event, and 90 min after receiving a bolus of BrdU (20 mg intraperitoneally). Each line represents pooled results from 200 half-crypt scores, from at least three wild-type (■), heterozygous (□), or homozygous null (○) mice. (Reprinted, with permission, from Clarke et al. 1994 [copyright Macmillan].)

Table 1. Response of Mitotic Index of Intestinal Crypt Cells to γ-Irradiation In Vivo

	Wild type		Heterozygote		Homozygote	
	L	S	L	S	L	S
0 Gy	1.24	2.14	1.40	2.36	1.08	2.61
5 Gy	0.15	0.18	0.04	0.22	0.15	0.35

Mean mitotic indices from groups of 3–7 mice, for large (L) and small (S) intestinal crypts. (Reprinted, with permission, from Clarke et al. 1994 [copyright Macmillan].)

Cells Lacking Functional p53 That Survive Irradiation Have Multiple Genetic Defects

The preceding data show that cells of many lineages are capable of survival following exposure to high doses of ionizing radiation. The death normally induced by radiation cannot therefore be due directly to the DNA damage inflicted by the radiation itself, but rather to the accumulation of p53 that this damage induces, and the interpretation by the cell of this accumulation of p53. The question therefore arises whether the p53 null cells that inappropriately survive genotoxic damage (the forbidden cells) have particular features that might render them precursors of cancer. One such feature might be evidence of genomic instability.

The incidence of mutations in crypt stem cells of p53 null mice has yet to be measured, but relevant data are available from analysis of bone marrow pre-B cells. Pre-B-cell stem cells, cultured in the presence of IL-7, produce scorable colonies, whose number declines with increasing radiation dose with classic two-hit kinetics. Growth of pre-B-cell colonies from wild-type animals is exquisitely sensitive to radiation (Griffiths et al. 1994), whereas that from p53-null marrow is resistant (S.D. Griffiths et al., in prep.). The frequency of mutations in these radiation-resistant colonies was assessed using hypoxanthine phosphoribosyl transferase (HPRT) as an indicator. Mutations in this X-linked gene (present in single copy in the male mice used in this experiment) were identified by the resulting resistance to the killing effect of 6-thioguanine. Thioguanine-resistant colonies were present in control cultures from normal marrow at too low an incidence to be measured accurately. Non-irradiated p53 null colonies showed a basal level of HPRT mutation of approximately 5×10^{-5}. After irradiation, this incidence rose sharply, at 4 Gy to around 10^{-3}. Preliminary analysis by Southern blotting revealed that the thioguanine resistance was associated with major rearrangements of the HPRT gene in a high proportion of the cases.

The conclusion from these experiments is therefore that p53 null cells can survive major DNA damage inflicted by ionizing radiation, but that the survivors have sustained major DNA rearrangements that are not normally permitted. In this case, the rearrangements were detected in a convenient indicator gene, but there is no reason to suppose that many other genes—including those significant in transformation—would be selectively spared. Hence the data strongly support the hypothesis that, in the absence of an appropriate apoptosis response to genotoxic damage, previously forbidden cells appear that may be the precursors of cancer. It is noteworthy, however, that the genetic defects that appear in these forbidden cells are not (or at any rate are not exclusively) the point mutations that are familiar features in oncogenes and oncosuppressor genes of both benign and malignant tumors. Rather, they tend to involve major genomic rearrangements, such as occur in the majority of human malignant tumors, but very infrequently in benign tumors. In this connection, it is significant that the tumors that arise "spontaneously" in p53 null mice (and constitute their major cause of death) are also usually aneuploid (Purdie et al. 1994).

Loss of p53 Function Appears Prior to DNA Rearrangement in Human Colorectal Carcinomas

In an attempt to relate these concepts to the development of an authentic human cancer, we studied the role of p53 in colorectal carcinogenesis. The genotoxic agents responsible for colorectal carcinogenesis in man are not well characterized; nor is it easy in man to study definitively the early evolution of cancer. Existing literature does show, however, that genomic rearrangements, suggestive of an acquired DNA instability, are a feature of malignant tumors (i.e., carcinomas), in contrast with the benign but potentially premalignant adenomas. Two types of instability are recognized. In the first, the tumors are composed of clones of aneuploid cells, often with reduplications of many chromosomes (Reichmann et al. 1981; Offerhaus et al. 1992). In the second, the tumors are often diploid or near diploid, but instability is detectable at microsatellite loci (Lothe et al. 1993; Thibodeau et al. 1993). The first type of instability has often been recorded in association with abnormalities in p53 (Monpezat et al. 1988; Delattre et al. 1989; Offerhaus et al. 1992), whereas microsatellite instability is associated preferentially with mutations in DNA mismatch repair genes (for review, see Service 1994).

Mutations in p53 are seldom recorded in adenomas (Purdie et al. 1991; Ichii et al. 1993). It is therefore conceivable that p53 abnormalities are true "late" events in the process of carcinogenesis (Baker et al. 1990), selected for in the course of what is often called tumor progression: the stepwise acquisition in time of increasingly aggressive growth features. This scenario would be incompatible with p53-induced apoptosis being a major force in the origins of this human cancer. On the other hand, the data are compatible with the view that p53 defects might prescribe a critical "window" in cancer development in which the characteristic major DNA rearrangements appear. This is exactly what one would expect if p53-induced apoptosis plays a significant role in suppressing the appearance of potential cancer cells. To distinguish between these possibilities, we studied a large series of carcinomas at multiple sites, establishing at each site the DNA ploidy status and whether p53 was in the normal or the stabilized configuration (Carder et al. 1993 and in prep.).

The data showed clearly that p53 stabilization occurred prior to the divergence of aneuploid subclones in the evolution of the overwhelming majority of these carcinomas, since subclones of different DNA ploidy within the same tumor almost always had the same p53 status (Fig. 4). In a small number, there was a villous adenoma element contiguous with infiltrative carcinoma, presumably representing a residue of the original premalignant tumor. Almost invariably, this element lacked stabilized p53, whereas the infiltrative

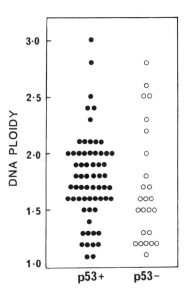

Figure 5. DNA indices, determined by propidium iodide flow cytometry, for individual nondiploid subclones of colorectal carcinomas that contained stabilized, putatively mutated p53 (p53+) or apparently wild-type p53 (p53−). (Reprinted, with permission, from Carder et al. 1993 [copyright Macmillan].)

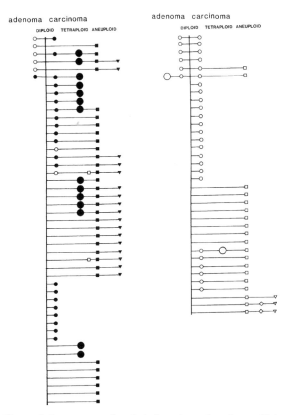

Figure 4. Immunocytochemical detection of nuclear p53 in subclones of human colorectal carcinomas, identified on the basis of their DNA ploidy. Diploid (*small circle*), tetraploid (*large circle*), or aneuploid subclones (other symbol shapes) were identified by propidium iodide flow cytometry and for each tumor are represented on the same horizontal line. Where adenoma and carcinoma elements were contiguous within the same tumor, the adenoma element is shown to the left of the vertical line, carcinoma to the right. Positive or negative staining for p53, using the monoclonal antibody PAb1801, is indicated for each subclone by closed or open symbols, respectively. (Reprinted, with permission, from Carder et al. 1993 [copyright Macmillan].)

element possessed it. Hence the data indicated that p53 stabilization occurs in a tightly defined window prior to clonal divergence in carcinomas. This type of finding was not reproduced when other common gene alterations in colorectal cancer were considered. Thus, mutations in the familial polyposis gene, *APC*, are known to occur at the same high frequency in both carcinomas and adenomas (regardless of their size) (Miyoshi et al. 1992; Powell et al. 1992; Ichii et al. 1993). Loss of heterozygosity at the *APC* locus appears in only a proportion of adenomas or carcinomas, but in carcinomas it shows a heterogeneous pattern, affecting some but not all clones within the same tumor (Carder et al. 1993).

Loss of p53 Function Is Associated with Aneuploid Clonal Divergence in Colorectal Carcinoma

Unexpectedly, however, the data showed that tumors with stabilized p53 also differed significantly from those without in possessing a larger number of divergent sublines. The p53-stabilized tumors were exactly the same in diameter and in number of sampled sites as the others; hence, it was improbable that this finding resulted from some trivial bias related to tumor size. Analysis of the DNA ploidy values for all the nondiploid subclones showed that the p53-stabilized tumors, unlike the remainder, had a large excess of subclones in the tetraploid and immediately subtetraploid range (Fig. 5). The data therefore are concordant with the view that the event that stabilizes p53 precedes and perhaps facilitates appearance of cells with grossly rearranged genomes, which are subsequently clonally expanded in the cancer. The rearrangement could well start as an inappropriate endoreduplication, producing tetraploidy, as suggested many years ago (Nowell 1976), but the additional changes required to generate the subtetraploid state might well involve chromosomal nondisjunction or fusion-bridge-breakage cycles (McClintock 1952).

These studies all depended on a classification of carcinomas based on the stability of p53, producing an immunochemically identifiable intranuclear product. With the methods used, presence of this product correlates closely, although not absolutely, with the existence of a mutation in the p53 gene (Cripps et al. 1994). In a subset of the cases, analysis by single-strand conformational polymorphism, sequencing, and a mutation-specific restriction fragment length polymorphism has provided direct evidence that p53 mutation is indeed involved in the clonally divergent carcinomas and that the same mutation is present in the divergent subclones of each carcinoma (P.J. Carder et al., in prep.).

CONCLUSIONS

The data from genetically manipulated animals strongly suggest that one stem-cell pool from which carcinomas arise is the population of previously forbidden cells that fail to undergo apoptosis after sustaining mutagen-induced DNA damage. As p53 plays a central role in coupling the apoptosis pathway to the occurrence of DNA damage, it is not surprising that p53 mutations appear with high frequency in malignant tumors. The forbidden cells that result from this p53-dependent escape from apoptosis tend to demonstrate major genomic rearrangements. It is possible that similar processes occur in the genesis of human colorectal carcinomas, since p53 inactivation events clearly precede and probably facilitate the genesis of subtetraploid subclones of cells.

REFERENCES

Baker, S.J., A.C. Preisinger, J.M. Jessup, C. Paraskeva, S. Markowitz, J.K.V. Willson, S. Hamilton, and B. Vogelstein. 1990. p53 gene mutations occur in combination with 17p allelic deletion as late events in colorectal tumorigenesis. *Cancer Res.* **50:** 7717.

Carder, P., A.H. Wyllie, C.A. Purdie, R.G. Morris, S. White, J. Piris, and C.C. Bird. 1993. Stabilised *p53* facilitates aneuploid clonal divergence in colorectal cancer. *Oncogene* **8:** 1397.

Chawlinski, S., C.S. Potten, G. Evans. 1988. Double labelling with bromodeoxyuridine and [^{3}H]-thymidine of proliferative cells in small intestinal epithelium in steady state and after irradiation. *Cell Tissue Kinet.* **21:** 317.

Clarke, A.R., S. Gledhill, M.L. Hooper, C.C. Bird, and A.H. Wyllie. 1994. *p53* dependence of early apoptotic and proliferative responses within the mouse intestinal epithelium following γ-irradiation. *Oncogene* **9:** 1767.

Clarke, A.R., C.A. Purdie, D.J. Harrison, R.G. Morris, C.C. Bird, M.L. Hooper, and A.H. Wyllie. 1993. Thymocyte apoptosis induced by *p53*-dependent and independent pathways. *Nature* **362:** 849.

Cripps, K.J., C.A. Purdie, P.J. Carder, S. White, K. Komine, C.C. Bird, and A.H. Wyllie. 1994. A study of stabilisation of p53 protein versus point mutation in colorectal carcinoma. *Oncogene* **9:** 2739.

Delattre, O., S. Olschwang, D.J. Law, T. Melot, Y. Remvikos, R.J. Salmon, X. Sastre, P. Validre, A.P. Feinberg, and G. Thomas. 1989. Multiple genetic alterations in distal and proximal colorectal cancer. *Lancet* **II:** 353.

Donehower, L.A., M. Harvey, B.L. Slagle, M.J. McArthur, C.A. Montgomery, J.S. Butel, and A. Bradley. 1992. Mice deficient for *p53* are developmentally normal but susceptible to spontaneous tumors. *Nature* **356:** 215.

Griffiths, S.D., D.T. Goodhead, S.J. Marsden, E.G. Wright, S. Krajewski, J.C. Reed, S.J. Korsmeyer, and M. Greaves. 1994. IL7-dependent B lymphocyte precursors are ultrasensitive to apoptosis. *J. Exp. Med.* **179:** 1789.

Ichii, S., S. Takeda, A. Horii, S. Nakatsuru, Y. Myoshi, M. Emi, Y. Fujiwara, K. Koyama, J. Furuyama, J. Utsunomiya, and Y. Nakamura. 1993. Detailed analysis of genetic alterations in colorectal tumors from patients with and without familial adenomatous polyposis (FAP). *Oncogene* **8:** 2399.

Kastan, M.B., O. Onyinye, D. Sidranski, B. Vogelstein, and R.W. Craig. 1991. Participation of *p53* protein in the cellular response to DNA damage. *Cancer Res.* **51:** 5304.

Kuerbitz, S.J., B.S. Plunkett, M.V. Walsh, M.B. Kastan. 1992. Wild-type *p53* is a cell cycle check-point determinant following irradiation. *Proc. Natl. Acad. Sci.* **89:** 7491.

Lotem, J. and L. Sachs. 1993. Hematopoietic cells from mice deficient in wild-type *p53* are more resistant to induction of apoptosis by some agents. *Blood* **82:** 1092.

Lothe, R.A., P. Peltomäki, G.I. Meling, L.A. Aaltonen, M. Nyström-Lahti, L. Pylkkähen, K. Heimdal, T.I. Andersen, P. Møller, T.O. Tognum, S.D. Fossæ, T. Haldorsen, F. Langmark, A. Brøgger, A. de la Chapelle, and A.-L. Børresen. 1993. Genomic instability in colorectal cancer: Relationship to clinicopathological variables and family history. *Cancer Res.* **53:** 5849.

Lowe, S.W., H.E. Ruley, T. Jacks, and D.E. Housman. 1993a. *p53*-dependent apoptosis modulates the cytotoxicity of anticancer agents. *Cell* **74:** 957.

Lowe, S.W., E.M. Schmitt, S.W. Smith, B.A. Osborne, and T. Jacks. 1993b. *p53* is required for radiation-induced apoptosis in mouse thymocytes. *Nature* **362:** 847.

McClintock, B. 1952. Chromosome organisation and genic expression. *Cold Spring Harbor Symp. Quant. Biol.* **16:** 13.

Merritt, A.J., C.S. Potten, C.J. Kemp, J.A. Hickman, A. Balmain, D.P. Lane, and P.A. Hall. 1994. The role of p53 in spontaneous and radiation-induced apoptosis in the gastrointestinal tract of normal and *p53*-deficient mice. *Cancer Res.* **54:** 614.

Miyoshi, Y., H. Nagase, H. Ando, A. Horii, S. Ichii, S. Nakatsuru, T. Aoki, Y. Miki, T. Mori, and Y. Nakamura. 1992. Somatic mutations of the APC genes in colorectal tumors: mutation cluster region in the APC gene. *Hum. Mol. Genet.* **1:** 229.

Monpezat, J.-Ph., O. Delattre, A. Bernard, D. Grunwald, Y. Remvikos, M. Muleris, R.J. Salmon, G. Frelat, B. Dutrillaux, and G. Thomas. 1988. Loss of alleles on chromosome 18 and on the short arm of chromosome 17 in polyploid colorectal carcinomas. *Int. J. Cancer* **41:** 404.

Nowell, P.C. 1976. The clonal evolution of tumor cell populations. *Science* **194:** 23.

Offerhaus, G.J.A., E.P. de Feyter, C.J. Cornelisse, K.W.F. Tersmette, J. Floyd, S.E. Kern, B. Vogelstein, and S.J. Hamilton. 1992. The relationship of DNA aneuploidy to molecular genetic alterations in colorectal carcinoma. *Gastroenterology* **102:** 1612.

Potten, C.S. 1990. A comprehensive study of the radiobiological response of the murine (BDF1) small intestine. *Int. J. Radiat. Biol.* **58:** 925.

Powell, S.M., N. Zilz, Y. Benzer-Barclay, T.M. Bryan, S.R. Hamilton, S.-N. Thibodeau, B. Vogelstein, and K.W. Kinzler. 1992. APC mutations occur early during colorectal tumorigenesis. *Nature* **359:** 235.

Purdie, C.A., J. O'Grady, J. Piris, A.H. Wyllie, and C.C. Bird.

1991. p53 expression in colorectal tumors. *Am. J. Pathol.* **138:** 807.

Purdie, C.A., D.J. Harrison, A. Peter, L. Dobbie, S. White, S.E.M. Howie, D.M. Salter, C.C. Bird, A.H. Wyllie, M.L. Hooper, and A.R. Clarke. 1994. Tumor incidence, spectrum and ploidy in mice with a large deletion in the *p53* gene. *Oncogene* **9:** 603.

Rao, L., M. Debbas, P. Sabbatini, D. Hockenbery, and S. Korsmeyer. 1992. The adenovirus E1A proteins induce apoptosis, which is inhibited by the E1B 19KDa and *bcl-2* proteins. *Proc. Natl. Acad. Sci.* **89:** 7742.

Reichmann, A., P. Martin, and B. Levin. 1981. Chromosome banding patterns in human large bowel cancer. *Int. J. Cancer* **28:** 431.

Service, R.F. 1994. Stalking the start of colon cancer. *Science* **263:** 1559.

Thibodeau, S.N., G. Brew, and D. Schaid. 1993. Microsatellite instability in cancer of the proximal colon. *Science* **260:** 816.

Wu, X. and A.J. Levine. 1994. p53 and E2F-1 cooperate to mediate apoptosis. *Proc. Natl. Acad. Sci.* **91:** 3602.

Cellular Senescence and Cancer

J.C. BARRETT, L.A. ANNAB, D. ALCORTA, G. PRESTON, P. VOJTA, AND Y. YIN

Laboratory of Molecular Carcinogenesis, Environmental Carcinogenesis Program, National Institute of Environmental Health Sciences, National Institutes of Health, Research Triangle Park, North Carolina 27709

Homeostasis in normal tissues is maintained by a balance between cell proliferation and cell death, whereas tumor growth occurs when the cellular birth rate exceeds the death rate. This can be achieved by either increasing the proliferative rate of cells, decreasing the death rate, or both (Boyd and Barrett 1990). Both positive and negative growth controls are involved in the regulation of cell proliferation and cell death. Tumor suppressor genes act to negatively regulate tumor growth either by reversibly blocking cell division or by increasing cell death or terminal arrest. Reversible controls on cell growth are important in the processes of normal development and tissue homeostasis, and in pathological conditions such as neoplasia. The majority of stem cells or basal cells in a tissue are growth arrested in a reversible state, and alterations in this growth arrest mechanism may be common in neoplastic cells. Terminal growth arrest and cell death are also important in normal and neoplastic growth. There are several forms of irreversible growth control, including terminal differentiation, cellular senescence, apoptosis, and necrosis. The signals controlling these forms of replicative cell death and their mechanisms of activation and action are only partially understood.

Terminal differentiation, cellular senescence, and apoptosis are each regulated by cellular genes and, in that sense, are forms of programmed (replicative) cell death. However, distinct features of each form of cell death exist (Table 1). Differentiation involves expression of cell-type-specific (differentiation) markers. Cellular senescence is related to the age or cell division potential of the cell. Apoptosis differs from cellular senescence and terminal differentiation in that it is a very rapid form of cell death and destruction. Apoptotic cells die within a matter of hours or days, whereas senescent cells and certain differentiated cells remain metabolically active, but nonproliferative, for up to one year.

Despite these differences, it is possible that common genetic and biochemical pathways exist in different forms of growth arrest. For example, some terminally differentiated cells and senescent cells down-regulate the same cell cycle control proteins (Richter et al. 1991). Differentiated cells in the skin and the nervous system express a protein, Bcl-2, that blocks apoptosis (Hockenbery et al. 1991). As the genes involved in the negative regulation of cell growth and death are identified, common as well as distinct mechanisms of these processes will be better understood. The topic of this review is cellular senescence, as an example of a negative growth arrest process that is controlled by specific cellular genes.

Normal cells in culture can be grown for only a limited number of cell divisions, after which they exhibit morphological changes and cease proliferation, a process termed cellular senescence or cellular aging (Hayflick 1976). Hayflick and Moorhead (1961) reported this finding with human fibroblasts more than 30 years ago, and it has been subsequently confirmed by many investigators using cells from different tissues and species. The failure of cells to grow beyond this limit is an inherent property of the cells that cannot be explained simply by inadequate medium components or growth conditions (Hayflick and Moorhead 1961; Hayflick 1976). The key determinant in the life span of cells in culture is the number of cell doublings, not the length of time in culture (Hayflick 1976). Normal cells transplanted serially in vivo also exhibit a finite life span, suggesting that cellular senescence is not a cell culture artifact (Daniel et al. 1968). Several lines of evidence suggest that the aging of cells in culture may be related to the aging of the organism (Hayflick 1976; Barrett 1993). These lines of evidence, although not conclusive, provide provocative support for the hypothesis that aging of cells is related to the aging process of the organism.

Table 1. Irreversible Growth-arrest States

Terminal differentiation	Irreversible growth-arrest state characterized by expression of differentiation-specific markers; not age-dependent
Cellular senescence	Irreversible growth-arrest state that depends on the age or cell doublings of a cell; cells are viable for over one year
Apoptosis	Irreversible growth-arrest state characterized by active process of cell death; cell death occurs within hours
Necrosis	Irreversible growth-arrest state characterized by passive process of cell death

Escape from cellular senescence is an important step in neoplastic progression of human and rodent cancers (Barrett and Fletcher 1987). Many, but not all, tumor cells can be grown indefinitely in culture and have escaped senescence and are therefore termed immortal. It is not clear whether the failure of some tumor cells to grow in culture is a technical artifact or an indication that escape from senescence is not required for these cancers. A model to explain these observations is discussed below.

Normal cells escape senescence following treatment with diverse carcinogenic agents, including chemical carcinogens, radiation, viruses, and oncogenes (Barrett and Fletcher 1987). This observation suggests that immortalization is important in carcinogenesis. Although immortality is not sufficient for neoplastic transformation, most immortal cells have an increased propensity for spontaneous, carcinogen-induced or oncogene-induced neoplastic progression (Barrett and Fletcher 1987). Therefore, escape from senescence is a preneoplastic change that predisposes a cell to neoplastic conversion. It is clear that immortal cells are further along the multistep pathway to neoplasia than normal cells. Since cellular senescence limits the growth of cells, it is reasonable that senescence might be one mechanism by which tumor suppressor genes operate (Sager 1986; Barrett and Fletcher 1987).

GENETIC BASIS FOR CELLULAR SENESCENCE: SENESCENCE GENES?

Two major theories of cellular senescence have been debated for many years (Macieira-Coelho 1988). One is the error catastrophe or damage model, which proposes that the random accumulation of damage or mutations in DNA, RNA, or proteins leads to the loss of proliferative capacity. The experimental evidence supporting the error accumulation hypothesis has been criticized (Macieira-Coelho 1988). A second hypothesis is that senescence is a genetically programmed process. Strong experimental support for a genetic basis for senescence is provided by studies of Pereira-Smith and Smith (1988) and of Sugawara et al. (1990), which are discussed below.

It is possible to fuse cells of different origins and to selectively grow the hybrid cells uniquely expressing biochemical markers for drug sensitivity or resistance, thereby eliminating the parental cells. When cells with a finite life span are fused to immortal cells with an indefinite life span, the majority of these hybrids senesce (Bunn and Tarrant 1980; Koi and Barrett 1986; Pereira-Smith and Smith 1988). Even hybridization of two different immortal human cell lines with each other can result in senescence, indicating that different complementation groups exist for the senescence function lost in these cells (Pereira-Smith and Smith 1988). By fusing different immortal human cell lines with each other, Pereira-Smith and Smith (1988) established four complementation groups, suggesting that loss or inactivation of one of multiple genes allows cells to escape from senescence. If this hypothesis is correct, it should be possible to map the genes involved in cellular senescence, and, consistent with this hypothesis, findings with hamster and human interspecies hybrids and microcell-mediated chromosome transfer experiments have mapped putative senescence genes to specific human chromosomes (Sugawara et al. 1990; Hara et al. 1991; Klein et al. 1991; Ning et al. 1991; Ogata et al. 1993).

The initial mapping by Sugawara et al. (1990) of a senescence gene to chromosome 1 was demonstrated by three independent experimental approaches: (1) interspecies hybrids of normal human cells with immortal hamster cells that showed nonrandom losses of human chromosome 1 in hybrids that escaped senescence, (2) interspecies cell hybrids with human cells carrying a t(1;X) chromosome that allowed selective pressure for the long arm of chromosome 1 and a corresponding increased frequency of senescent hybrids, and (3) microcell-mediated chromosome transfer which demonstrated that introduction of a single copy of human chromosome 1, but not other chromosomes, restored the program of senescence in certain immortal cell lines.

Using the technique of microcell-mediated chromosome transfer, Klein et al. (1991) mapped another senescence gene to the X chromosome, and Ning et al. (1991) mapped a senescence gene for HeLa cells to chromosome 4. In addition, several other studies indicate the presence of senescence genes on other chromosomes (Table 2).

These results have led us to propose the following hypothesis: Cellular senescence is controlled by genes that are activated or whose functions become manifest at the end of the life span of the cell. Defects in the function of these gene products allow cells to escape the program of senescence and become immortal. Immortalization relieves one constraint on tumor cell growth, allowing malignant progression.

According to this hypothesis, a family of senescence genes exists, and immortalization occurs due to defects in these genes. This theory is supported by the complementation studies of Pereira-Smith and Smith (1988), which show that different immortal cell lines, when fused together, can complement each other and senesce. This theory also explains our results that introduction of a specific human chromosome causes senescence in some cell lines but not others. It appears that senescence gene mutations in a specific immortal cell line are not related to tumor histology or activated oncogenes in the cell.

We propose the following criteria for defining genes that are responsible for the initiation and maintenance of cellular senescence growth arrest: (1) induction of irreversible growth arrest upon expression in a replicating cell, with preservation of cell viability; (2) down-regulation, mutation, or deletion in immortal cells or cells that have an extended proliferative capacity; (3) induction of immortalization or extended life span in normal cells by mutation or down-regulation (e.g.,

Table 2. Mapping Putative Senescence and Anti-senescence Genes

(a) *Putative senescence genes*

Chromosome	Gene	Affected cell line	References
1	unknown	Syrian hamster fibrosarcoma (10W) endometrial carcinoma (HHUA) endometrial carcinoma (Ishikawa) osteosarcoma (TE85)	Sugawara et al. (1990) Yamada et al. (1990) Sasaki et al. (1994) Hensler et al. (1994)
4	unknown	cervical carcinoma (HeLa) bladder carcinoma (J82)	Ning et al. (1991)
6	unknown	SV40 immortalized	S. Kodama et al. (in prep.); Sandhu et al. (1994)
7	unknown	immortalized human fibroblasts (KMST-6 and SUSM-1)	Ogata et al. (1993)
9	unknown	melanoma (H32941) leukemia (K562) mouse A9	M. Diaz (pers. comm.) R. Newbold (pers. comm.)
11	unknown	bladder carcinoma (H2)	Koi et al. (1993)
13	retinoblastoma	many	Hinds et al. (1992)
17	p53	many	Levine et al. (1991)
18	unknown	endometrial carcinoma (HHUA)	Sasaki et al. (1994)
X	unknown	Chinese hamster (N:2) ovarian carcinoma (Hoc8) breast carcinoma (ELC0)	Klein et al. (1991)

(b) *Putative anti-senescence gene*

Chromosome	Gene	Syndrome	Reference
8	unknown	Werner syndrome	Goto et al. (1992)

antisense); and (4) an increase in expression or activity associated with senescent cell populations compared to young cell populations. The first three criteria allow one to assign a functional activity to a candidate senescence gene and are characteristics necessary to demonstrate a regulatory role in senescence. The last criterion, by itself, will include genes that exhibit altered levels of expression as a consequence of senescence growth arrest, in addition to genes up-regulated in other types of in vitro cellular growth arrest, such as serum starvation and contact inhibition. Indeed, many genes have been identified to be deferentially expressed in proliferating states compared to growth arrest states (Fornace et al. 1989; Ciccarelli et al. 1990). The overlap between genes involved in senescence arrest and other forms of cell cycle arrest certainly includes a number of common arrest-mediator genes. However, genes that regulate these common arrest pathways are likely to be different for different growth arrest states and possibly different cell types.

We have included in Table 2 the Rb and p53 tumor suppressor genes as putative senescence genes. Operationally, these genes fit the criteria for senescence genes in certain contexts. Down-regulation of their expression by antisense methods results in extension of the life span of normal human cells (Hara et al. 1991), and reintroduction of the genes into certain immortal cells causes cessation of growth and morphological changes similar to senescent cells (Levine at al. 1991; Hinds et al. 1992). According to our hypothesis, specific genes are involved in one or more pathways leading to a program of senescence. In normal cells, Rb and p53 proteins are negative regulators of the cell cycle, which are regulated by other proteins, allowing cell cycle progression. In senescent cells, a program is activated that blocks entry into the DNA synthesis phase of the cell cycle, and the cells become irreversibly growth arrested. Rb and p53 may participate in one or more pathways that activate or affect the senescence program. Defects in the senescence program can result from inactivation or mutation of different genes. In some immortal cells, p53 and Rb can be normal and genes that control their phosphorylation or other posttranslational modifications may be defective. In other cell lines, deletions or mutations of p53 or Rb genes could result in the inability to activate or mediate the senescence program.

The gene for Werner syndrome has been mapped to chromosome 8 (Goto et al. 1992). Mutations of this gene result in premature senescence of the Werner cells and premature aging of individuals with this syndrome. The Werner gene differs from the senescence genes discussed above. In contrast to putative senescence genes, the inactivation of which causes cells to become immortal, the Werner gene should be considered an anti-senescence or anti-aging gene because its loss results in accelerated senescence, not immortalization (Table 2).

MULTIPLE PATHWAYS FOR CELLULAR SENESCENCE

Recently, we observed that introduction of different chromosomes can induce senescence in the same immortal cell line. Using a human endometrial carcinoma cell line, senescence genes for these cells were mapped to chromosomes 1 and 18 (Sasaki et al. 1994). This finding implies that multiple pathways of senescence exist and that immortal cells arise due to defects in each of these pathways. The senescence program could be activated by a single pathway and immortal cells would arise due to mutations in any of the genes that encode proteins involved in this pathway. An alternative hypothesis is that the senescence program is activated by multiple, independent pathways (Fig. 1). Immortal cells would require at least one mutation in each pathway. Mutations that affect only a single pathway would not result in cells that were immortal, but the cells might have an extended life span. For example, SV40 virus infection (which inactivates Rb and p53 proteins) results in extended life span but not immortalization of infected cells (Wright and Shay 1992). Additional genetic changes, possibly loss of chromosome 6, are required for immortalization of SV40-infected human cells (Hubbard-Smith et al. 1992). Reintroduction of a normal chromosome 6 results in senescence of SV40 immortal cells (Sandhu et al. 1994; S. Kodama and J.C. Barrett, unpubl.). Antisense down-regulation of Rb and p53 mRNAs also results in extension of the life span of human cells without immortalization (Hara et al. 1991). The multiple-pathways-to-senescence hypothesis is also consistent with the multistep nature of immortalization observed in chemically induced immortalization (Barrett and Fletcher 1987). In addition, the inability to assign certain immortal cells to a single complementation group further supports this hypothesis (Duncan et al. 1993; Berry et al. 1994).

An important aspect of the hypothesis of multiple pathways to cellular senescence is that it explains why many tumor-derived cells are not immortal. Hayflick has shown that cells from adults can be grown in culture for 14–29 population doublings (Hayflick 1976). If all the changes necessary for tumorigenic conversion were to accumulate in an adult cell without loss or gain of life-span potential (which may be unlikely), then this cell could grow to form a tumor of only 16,354 cells (14 doublings or 2^{14} cells) to 5.4×10^8 cells (29 doublings or 2^{29} cells). It is estimated that a tumor formed after 30 cell doublings would be approximately 1 cm^3 in size (Fig. 2). Interestingly, Paraskeva and coworkers have shown that colon adenomas of less than 1 cm^3 in size are rarely capable of indefinite growth in vitro whereas cells from adenomas of more than 1 cm^3 are often immortal (Paraskeva et al. 1988, 1989a,b). This observation supports the hypothesis that escape from senescence is a requirement for tumor growth beyond a certain size or cell number. A tumor of less than 1 cm^3 may be lethal in some cases, but not generally. For the tumor to expand, an extension of the life span would be necessary. Extension of the life span to 40 population doublings would yield a tumor of 1 kg, whereas an extension to 50 population doublings would yield a tumor of 1000 kg, which would very certainly be sufficient to kill the host. Thus, immortalization is not as important as extension of life span for neoplastic progression. Mutation of a senescence gene in one pathway may result in an extended life span without immortali-

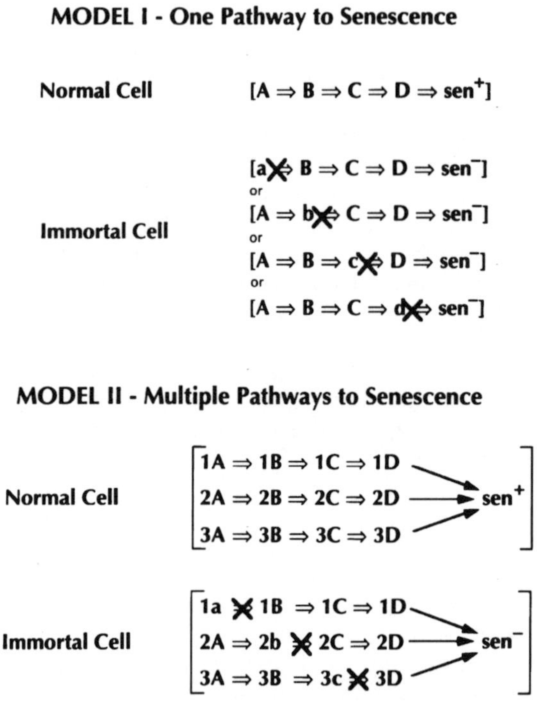

Figure 1. Alternative models for cellular senescence.

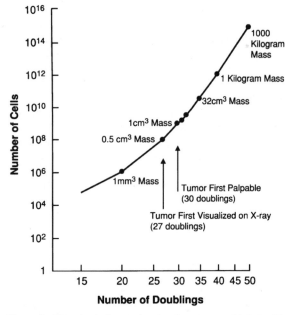

Figure 2. Hypothetical growth curve for tumors with finite life span. (Adapted from DeVita et al. 1975.)

zation, according to the multiple pathways model. This hypothesis may explain why tumor cells are not always immortal.

p21 AND p16: ROLE IN CELLULAR SENESCENCE?

The product of the retinoblastoma tumor suppressor gene may be a key regulator of cellular senescence (Stein et al. 1990; Futreal and Barrett 1991). Senescent cells express levels of Rb protein comparable to young cells; however, only the unphosphorylated form of Rb is observed. Quiescent cells, maintained in medium containing low serum, exhibit only the unphosphorylated form of the Rb protein. When the cells are stimulated with serum, phosphorylation of the Rb protein is observed between 10 and 20 hours after stimulation, which corresponds to the time course for stimulation of DNA synthesis. However, when senescent cells are stimulated with serum, the Rb protein remains unphosphorylated (Stein et al. 1990; Futreal and Barrett 1991). This result indicates that senescent cells are blocked in their ability to phosphorylate the Rb protein in response to normal growth stimuli.

Recent progress in determining the molecular basis of cell cycle control has resulted in the identification of genes that regulate and/or coordinate different phases of the cell cycle. Central to these genes are the cyclin-dependent kinases (cdk) and associated activating regulatory subunits, termed cyclins (Hunter 1993). In proliferating cells, cyclins are synthesized and degraded in systematic patterns to drive cell cycle progression in a regulated manner (Reed 1991). The number of identified cdks and cyclins has rapidly grown, with different family members harboring activity during specific cell cycle phases (Xiong and Beach 1991; Meyerson et al. 1992). With the evidence that senescent cells appear to be blocked in G_1, it is reasonable to hypothesize that modifications of G_1-regulated cyclins or cdk proteins may be pivotal to senescence-related growth arrest, and recent evidence supports this hypothesis. The failure of senescent cells to phosphorylate the Rb protein is consistent with this hypothesis.

This hypothesis implicates upstream modifiers of Rb phosphorylation as possible crucial regulatory elements in mediating cellular senescence, with the end result being a block to proliferation caused by the presence of unphosphorylated Rb protein acting on its own or through other effector molecules. Down-regulation of a Rb kinase activity in senescent cells and/or up-regulation of a Rb phosphatase are possible mechanisms for the alterations of Rb phosphorylation in senescent cells. The $p34^{cdc2}$ kinase, which is an S-phase kinase, is down-regulated in senescent human and hamster cells (Richter et al. 1991; Stein et al. 1991). However, transfection of human $p34^{cdc2}$ under the control of a constitutive promoter into hamster cells fails to allow the cells to enter into S phase even when the protein is synthesized (Richter et al. 1991). This suggests that critical events in the G_1 phase of the cell cycle are blocked in senescent cells.

The observation that a key alteration in senescent cells is the lack of a Rb kinase activity suggests that a cyclin/cdk complex with this activity may be down-regulated or inactivated in senescence, causing a subsequent inhibition of cell cycle progression via the activity of hypophosphorylated Rb protein. Transfection experiments of the Rb gene along with different cyclin genes have shown that cyclin D and E can overcome Rb-mediated cell cycle arrest via an increase in phosphorylation of exogenous and endogenous Rb protein (Hinds et al. 1992). However, exogenous overexpression of cyclin E in normal human fibroblasts is not sufficient to overcome replicative senescence (Ohtsubo and Roberts 1993). Failure of exogenously expressed cyclins to override the senescence program is consistent with the detection of G_1 cyclin mRNAs (Won et al. 1992; Afshari et al. 1993) and the accumulation of inactive cyclin E/cdk2 complexes in senescent cells (Dulic et al. 1993). Gene products that negatively regulate the kinase activity of cyclin/cdk complexes may be up-regulated in senescent cells, inhibiting cell cycle progression via posttranslational inactivation of cell cycle progression factors. Identification of expression or lack of expression of other genes that positively or negatively regulate cyclin/cdk kinase activity may explain the predominance of hypophosphorylated Rb protein and the subsequent cell cycle arrest of senescent cells.

One negative regulator of cyclin/cdk kinase activity, termed WAF1, CIP1, SDI1, or p21, has recently been identified by several different groups (El-Deiry et al. 1993; Gu et al. 1993; Harper et al. 1993; Hunter 1993; Xiong et al. 1993; Noda et al. 1994), and expression of this gene has been shown to be increased 10–20-fold in senescent fibroblasts, compared to their younger counterparts (Noda et al. 1994; D. Alcorta and J.C. Barrett, unpubl.). When overexpressed in young cells, p21 causes an inhibition of DNA synthesis (El-Deiry et al. 1993; Noda et al. 1994). In addition, increased *p21* expression is correlated with a variety of other growth-arrest states (El-Deiry et al. 1994; Johnson et al. 1994). *p21* appears to be a general inhibitor of cyclin/cdk complexes and thus may play a key role in inhibiting the activity of the cyclin/cdk complex(es) necessary for phosphorylating the Rb protein. As previously mentioned, phosphorylated Rb protein is likely to be at least partially responsible for the growth arrest of senescent cells. However, we have not observed mutation or deletion of *p21* in cell lines (L. Terry et al., unpubl.), which would be expected if *p21* were one of the senescence genes identified in the somatic cell genetics experiments described in the previous section.

$p16^{INK4}$ is another recently identified gene product with properties of a G_1 cyclin/cdk inhibitor (Serrano et al. 1993; Kamb et al. 1994; Nobori et al. 1994) that may be active during senescence induction. Mutationally inactivated $p16^{INK4}$ could cause unchecked Rb and other activated cyclin/cdk phosphorylations, leading to unregulated cell growth (Serrano et al. 1993; Kamb et al. 1994). Studies in our laboratory have revealed that the level of p16^{INK4} protein in senescent human fibroblasts is elevated compared to young fibroblasts (D. Alcorta and J.C. Barrett, unpubl.). p16 is commonly mutated or deleted in immortal cell lines (Serrano et al. 1993; Kamb et al. 1994), which is a property expected of a senescence gene. p16 maps to chromosome 9p21, which is the location of a previously mapped senescence gene (Table 2).

As discussed above, senescence activity has been mapped to more than 10 human chromosomes (Table 2). The mapping of these activities to specific chromosomes provides an important step toward the cloning of additional senescence genes. Included in this group will likely be genes that directly participate in a variety of growth-arrest pathways in addition to the regulation of senescence. As additional senescence genes are cloned and their products functionally characterized, pathways and relationships between senescence, terminal differentiation, and apoptosis programs will be amenable to analysis.

CELL SENESCENCE VERSUS APOPTOSIS: TWO FORMS OF IRREVERSIBLE GROWTH ARREST

Senescence and apoptosis represent two distinct forms of growth arrest that are irreversible (Table 3). They can be distinguished by the fact that apoptosis is rapid cell death with cell disintegration in a matter of hours. Senescent cells are irreversibly growth arrested, fail to enter S phase when stimulated by mitogens, but, in contrast to apoptotic cells, the senescent cells are stable and remain viable, intact, and metabolically active for more than one year. A second distinction between senescence and apoptosis is that apoptosis occurs in an age-independent manner and is often controlled by external signals. Senescence, in contrast, is programmed within a cell and is determined by age- or cell-division-dependent events.

The decision of a cell to enter a growth-arrest state such as terminal differentiation or an alternative program, such as programmed cell death or apoptosis, may be controlled by at least a subset of genes involved in replicative senescence. Induction of one of many optional pathways by a regulatory gene may depend on a variety of factors, including the state of differentiation of the cell and the environment in which it exists at the time of gene activation. For example, the *p53* gene has been implicated as a mediator of both apoptosis and senescence (Shay et al. 1991) and provides an example of such a regulatory gene. *p53* expression in murine myeloid leukemia cells does not affect differentiation but causes cells to die by apoptosis. This effect is inhibited by the presence of interleukin-6 (Yonish-Rouach et al. 1991). In other cell types, wild-type p53 causes reversible growth arrest without inducing cell death (Levine et al. 1991). The p21/WAF-1/CIP-1/SDI-1 gene is up-regulated when cells are induced to undergo apoptosis (El-Deiry et al. 1994), G$_1$ arrest (El-Deiry et al. 1994), or senescence (Noda et al. 1994; D. Alcorta and J.C. Barrett, unpubl.). Antisense p21 RNAs stimulate DNA synthesis insenescent cells (J. Smith, pers. comm.) and block p53-mediated apoptosis (Y. Yin and J.C. Barrett, unpubl.), which suggests a critical role for the protein in senescence and apoptosis.

Both apoptosis and senescence are genetically controlled processes, but the genes involved are only partially elucidated. Despite the differences in these two forms of irreversible growth arrest, the same genes may be involved in certain aspects of both. *p53*, c-*myc*, and c-*fos* were all identified initially as oncogenes that alter cell growth controls, often resulting in escape from cellular senescence (Weinberg 1985). Overexpression of c-*myc* (Evan et al. 1992) and c-*fos* (G. Preston and J.C. Barrett, unpubl.) proteins can induce apoptosis in certain cells. The Bcl-2 protein is effective in blocking apoptosis induced by c-*myc*, and there is a cooperative interaction in apoptotic regulation between these genes (Bissonnette et al. 1992; Fanidi et al. 1992); preliminary experiments also suggest that Bcl-2 allows cells to exhibit an extended life span (L. Annab and J.C. Barrett, unpubl.). The roles played by the genes in different processes may be different, but it is interesting, and perhaps not surprising, that the same genes are involved in distinct types of growth-arrest processes.

This analogy can be extended to other conditions of growth arrest, for example, irradiation-induced growth arrest. Wild-type p53 is required for growth arrest

Table 3. Comparison of Apoptosis and Cell Senescence

Apoptosis	Cell senescence
Rapid cell death (<24 hr)	stable, nonproliferative state (>1 year)
Not generally age-dependent	determined by age, cell doublings
Internally and externally controlled	internally controlled
Genetically controlled	genetically controlled
p53 involved	p53 involved
myc involved	myc involved
fos involved	fos involved
Bcl-2 involved	Bcl-2 involved
p21 (CIP/WAF/SDI) involved	p21 (CIP/WAF/SDI) involved
unknown genes involved	unknown genes involved

following damage of cells by irradiation (Kuerbitz et al. 1992). Some of the other genes involved in this process have recently been identified (Hartwell and Weinert 1989; Kastan et al. 1991, 1992). It is interesting to note that there are both positive and negative regulators of growth arrest in apoptosis, senescence, and response to irradiation. Based on our current knowledge, the same genes can be involved in each of these processes but, clearly, additional yet to be identified genes are implicated. This leads to the exciting prospect of new insights into each of these processes with the discovery of further genes for apoptosis, senescence, radiation, and other cellular responses.

REFERENCES

Afshari, C.A., P.J. Vojta, H.B. Bivins, L.A. Annab, T.B. Willard, A.F. Futreal, and J.C. Barrett. 1993. Investigation of the role of G_1/S cell cycle mediators in cellular senescence. *Exp. Cell Res.* **209:** 231.

Barrett, J.C. 1993. Cell senescence and apoptosis. In *Molecular genetics of nervous system tumors* (ed. A.J. Levine and H.H. Schmidek), p. 61. Wiley-Liss, New York.

Barrett, J.C. and W.F. Fletcher. 1987. Cellular and molecular mechanisms of multistep carcinogenesis in cell culture models. In *Mechanisms of environmental carcinogenesis: Multistep models of carcinogenesis* (ed. J.C. Barrett), vol. 2, p. 73. CRC Press, Boca Raton, Florida.

Berry, I.J., J.E. Burns, and E.K. Parkinson. 1994. Two human epidermal squamous cell carcinoma cell lines assign to more than one complementation group for the immortal phenotype. *Mol. Carcinog.* **9:** 134.

Bissonnette, R.P., F. Echeverri, A. Mahboubi, and D.R. Green. 1992. Apoptotic cell death induced by c-*myc* is inhibited by *bcl*-2. *Nature* **359:** 552.

Boyd, J. and J.C. Barrett. 1990. Tumor suppressor genes: Possible functions in the negative regulation of cell proliferation. *Mol. Carcinog.* **3:** 325.

Bunn, C.L. and G.M. Tarrant. 1980. Limited lifespan in somatic cell hybrids and cybrids. *Exp. Cell Res.* **127:** 385.

Ciccarelli, C., L. Philipson, and V. Sorrentino. 1990. The p53 tumor suppressor gene. *Nature* **351:** 453.

Daniel, C.W., K.B. DeOme, J.T. Young, P.B. Blair, and L.J. Faulkin, Jr. 1968. The in vivo span of normal and preneoplastic mouse mammary glands: A serial transplantation study. *Proc. Natl. Acad. Sci.* **61:** 53.

DeVita, V.T., Jr., R.C. Young, and G.P. Canellos. 1975. Combination versus single agent chemotherapy: A review of the basis for selection of drug treatment of cancer. *Cancer* **35:** 98.

Dulic, V., L.F. Drullinger, E. Lees, S.I. Reed, and G.H. Stein. 1993. Altered regulation of G_1 cyclins in senescent human diploid fibroblasts: Accumulation of inactive cyclin E-Cdk2 and cyclin D1-Cdk2 complexes. *Proc. Natl. Acad. Sci.* **90:** 11034.

Duncan, E.L., N.J. Whitaker, E.L. Moy, and R.R. Reddel. 1993. Assignment of SV40-immortalized cells to more than one complementation group for immortalization. *Exp. Cell Res.* **205:** 337.

El-Deiry, W.S., T. Tokino, V.E. Velculescu, D.B. Levy, R. Parsons, J.M. Treat, D. Lin, W.E. Mercer, K.W. Kinzler, and B. Vogelstein. 1993. *WAF1*, a potential mediator of p53 tumor suppression. *Cell* **75:** 817.

El-Deiry, W.S., J.W. Harper, P.M. O'Connor, V.E. Velculescu, C.E. Canman, J. Jackman, J.A. Pietenpol, M. Burrell, D.E. Hill, Y. Wang, K.G. Wiman, W.E. Mercer, M.B. Kastan, K.W. Kohn, S.J. Elledge, K.W. Kinzler, and B. Vogelstein. 1994. *WAF1/CIP1* is induced in *p53*-mediated G_1 arrest and apoptosis. *Cancer Res.* **54:** 1169.

Evan, G.I., A.H. Wyllie, C.S. Gilbert, T.D. Littlewood, H. Land, M. Brooke, C.M. Waters, L.Z. Penn, and D.C. Hancock. 1992. Induction of apoptosis in fibroblasts by c-*myc* protein. *Cell* **69:** 119.

Fanidi, A., E.A. Harrington, and G.I. Evan. 1992. Cooperative interaction between c-*myc* and *bcl*-2 proto-oncogenes. *Nature* **359:** 554.

Fornace, A.J., Jr., D.W. Nebert, M.C. Hollander, J.D. Luethy, M. Papathanasiou, J. Fargnoli, and N.J. Holbrook. 1989. Mammalian genes coordinately regulated by growth arrest signals and DNA-damaging agents. *Mol. Cell. Biol.* **9:** 4196.

Futreal, P.A. and J.C. Barrett. 1991. Failure of senescent cells to phosphorylate the Rb protein. *Oncogene* **6:** 1109.

Goto, M., M. Rubenstein, J. Weber, K. Woods, and D. Drayna. 1992. Genetic linkage of Werner's syndrome to five markers on chromosome 8. *Nature* **355:** 735.

Gu, Y., C.W. Turck, and D.O. Morgan. 1993. Inhibition of Cdk2 activity *in vivo* by an associated 20K regulatory subunit. *Nature* **366:** 707.

Hara, E., H. Tsurui, A. Shinozaki, S. Nakada, and K. Oda. 1991. Cooperative effect of antisense-Rb and antisense-p53 oligomers on the extension of life span in human diploid fibroblasts. *Biochem. Biophys. Res. Commun.* **179:** 528.

Harper, J.W., G.R. Adami, N. Wei, K. Keyomarsi, and S.J. Elledge. 1993. The p21 Cdk-interacting protein Cip1 is a potent inhibitor of G_1 cyclin-dependent kinases. *Cell* **75:** 805.

Hartwell, L.H. and T.D. Weinert. 1989. Checkpoints: Controls that ensure the order of cell cycle events. *Science* **355:** 735.

Hayflick, L. 1976. The cell biology of human aging. *N. Engl. J. Med.* **295:** 1302.

Hayflick, L. and P.S. Moorhead. 1961. The serial cultivation of human diploid cell strains. *Exp. Cell Res.* **25:** 585.

Hensler, P.J., L. A. Ananb, J.C. Barrett, and O.M. Pereira-Smith. 1994. A gene involved in the control of human cellular senescence localized to human chromosome 1q. *Mol. Cell. Biol.* **14:** 2291.

Hinds, P.W., S. Mittnacht, V. Dulic, A. Arnold, S.I. Reed, and R.A. Weinberg. 1992. Regulation of retinoblastoma protein functions by ectopic expression of human cyclins. *Cell* **70:** 993.

Hockenbery, D.M., M. Zutter, W. Hickey, M. Nahm, and S.J. Korsmeyer. 1991. BCL2 protein is topographically restricted in tissues characterized by apoptotic cell death. *Proc. Natl. Acad. Sci.* **88:** 6961.

Hubbard-Smith, K., P. Patsalis, J.R. Pardinas, K.K. Jha, A.S. Henderson, and H.L. Ozer. 1992. Altered chromosome 6 in immortal human fibroblasts. *Mol. Cell. Biol.* **12:** 2273.

Hunter, T. 1993. Braking the cycle. *Cell* **75:** 839.

Johnson, M., D. Dimitrov, P.J. Vojta, J.C. Barrett, A. Noda, O.M. Pereira-Smith, and J.R. Smith. 1994. Evidence for a p53 independent pathway for upregulation of SDI1/CIP1/WAF1/p21 RNA in human cells. *Mol. Carcinog.* **11:** 59.

Kamb, A., N.A. Gruis, J. Weaver-Feldhaus, Q. Liu, K. Harshman, S.V. Tavtigian, E. Stockert, R.S. Day III, B.E. Johnson, and M. H. Skonick. 1994. A cell cycle regulator potentially involved in genesis of many tumor types. *Nature* **264:** 436.

Kastan, M.D., O. Onyekwere, D. Sidransky, B. Vogelstein, and R.W. Craig. 1991. Participation of p53 protein in the cellular response to DNA damage. *Cancer Res.* **51:** 6304.

Kastan, M.B., Q. Zhan, W.S. El-Deiry, F. Carrier, T. Jacks, W.V. Walsh, B.S. Plunkett, B. Vogelstein, and A.J. Fornace, Jr. 1992. A mammalian cell cycle checkpoint pathway utilizing p53 and GADD45 is defective in ataxia-telangiectasis. *Cell* **71:** 587.

Klein, C.B., K. Conway, X.W. Wang, R.K. Bhamra, X. Lin, M.D. Cohen, L. Annab, J.C. Barrett, and M. Costa. 1991. Senescence of nickel-transformed cells by a mammalian X chromosome: Possible epigenetic control. *Science* **251:** 796.

Koi, M. and J.C. Barrett. 1986. Loss of tumor-suppressive function during chemically induced neoplastic progression of Syrian hamster embryo cells. *Proc. Natl. Acad. Sci.* **83:** 5992.

Koi, M., L.A. Johnson, L.M. Kalikin, P.F.R. Little, Y. Nakamura, and A.P. Feinberg. 1993. Tumor cell growth arrest caused by subchromosomal transferable DNA fragments from chromosome 11. *Science* **260:** 361.

Kuerbitz, S.J., B.S. Plunkett, W.V. Walsh, and M.B. Kastan. 1992. Wild-type p53 is a cell cycle checkpoint determinant following irradiation. *Proc. Natl. Acad. Sci.* **89:** 7491.

Levine A.J., J. Momand, and C.A. Finlay. 1991. The p53 tumour suppressor gene. *Nature* **351:** 453.

Macieira-Coelho, A. 1988. Biology of normal proliferating cells in vitro. Relevance for in vivo aging. *Interdiscip. Top. Gerontol.* **12:** 218 pp.

Meyerson, M., G.H. Edners, C.-L. Wu, L.-K. Su, C. Gorka, C. Nelson, E. Harlow, and L.-H. Tsai. 1992. The human cdc2 kinase family. *EMBO J.* **11:** 2909.

Nobori, T., K. Miura, D.J. Wu, A. Lois, K. Takabayashi, and D.A. Carson. 1994. Deletions of the cyclin-dependent kinase-4 inhibitor gene in multiple human cancers. *Nature* **368:** 753.

Ning, Y., J.L. Weber, A.M. Killary, D.H Ledbetter, J.R. Smith, and O.M. Pereira-Smith. 1991. Genetic analysis of indefinite division in human cells: Evidence for a cell senescence-related gene(s) on human chromosome 4. *Proc. Natl. Acad. Sci.* **88:** 5635.

Noda, A., Y. Ning, S.F. Venable, O.M. Pereira-Smith, and J.R. Smith. 1994. Cloning of senescent cell-derived inhibitors of DNA synthesis using an expression screen. *Exp. Cell Res.* **211:** 90.

Ogata, T., D. Ayusawa, M. Namba, E. Takahashi, M. Oshimura, and M. Oishi. 1993. Chromosome 7 suppresses indefinite division of nontumorigenic immortalized human fibroblast cell lines KMST-6 and SUSM-1. *Mol. Cell. Biol.* **13:** 6036.

Ohtsubo, M. and J.M. Roberts. 1993. Cyclin-dependent regulation of G_1 in mammalian fibroblasts. *Science* **259:** 1908.

Paraskeva, C., S. Finarty, and S. Powell. 1988. Immortalization of a human colorectal adenoma cell line by continuous in vitro passage: Possible involvement of chromosome 1 in tumour progression. *Int. J. Cancer* **41:** 908.

Paraskeva, C., S. Finarty, R.A. Mountford, and S.C. Powell. 1989a. Specific cytogenetic abnormalities in two new human colorectal adenoma-derived epithelial cell lines. *Cancer Res.* **49:** 1282.

Paraskeva, C., A. Harvey, S. Finarty, and S. Powell. 1989b. Possible involvement of chromosome 1 in in vitro immortalization: Evidence from progression of a human adenoma-derived cell line in vitro. *Int. J. Cancer* **43:** 743.

Pereira-Smith, O.M. and J.R. Smith. 1988. Genetic analysis of indefinite division in human cells: Identification of four complementation groups. *Proc. Natl. Acad. Sci.* **85:** 6042.

Reed, S.I. 1991. G_1-specific cyclins: In search of an S-phase-promoting factor. *Trends Genet.* **7:** 95.

Richter, K.H., C.A. Afshari, L.A. Annab, B.A. Burkhart, R.D. Owen, J. Boyd, and J.C. Barrett. 1991. Downregulation of cdc2 in senescent human and hamster cells. *Cancer Res.* **51:** 6010.

Sager, R. 1986. Genetic suppression of tumor formation: A new frontier in cancer research. *Cancer Res.* **46:** 1573.

Sandhu, A.K., K. Hubbard, G.P. Laur, K.K. Jha, H.L. Ozer, and R.S. Athwal. 1994. Senescence of immortal human fibroblasts by the introduction of normal human chromosome 6. *Proc. Natl. Acad. Sci.* **91:** 5498.

Sasaki, M., T. Honda, H. Yamada, N. Wake, J.C. Barrett, and M. Oshimura. 1994. Evidence for multiple pathways to cellular senescence. *Cancer Res.* **54:** 6090.

Serrano, M., G.J. Hannon, and D. Beach. 1993. A new regulatory motif in cell-cycle control causing specific inhibition of cyclin D/Cdk4. *Nature* **366:** 704.

Shay, J.W., O.M. Pereira-Smith, and W.E. Wright. 1991. A role for both Rb and p53 in the regulation of human cellular senescence. *Exp. Cell Res.* **196:** 33.

Stein, G.H., M. Beeson, and L. Gordon. 1990. Failure to phosphorylate the retinoblastoma gene product in senescent human fibroblasts. *Science* **249:** 666.

Stein, G.H., L.F. Drullinger, R.S. Robetorye, O.M. Pereira-Smith, and J.R. Smith. 1991. Senescent cells fail to express *cdc2*, *cdca*, and *cycB* in response to mitogen stimulation. *Proc. Natl. Acad. Sci.* **88:** 11012.

Sugawara, O., M. Oshimura, M. Koi, L. Annab, and J.C. Barrett. 1990. Induction of cellular senescence in immortalized cells by human chromosome 1. *Science* **247:** 707.

Weinberg, R.A. 1985. The action of oncogenes in the cytoplasm and nucleus. *Science* **230:** 770.

Won, K.-A., Y. Xiong, D. Beach, and M.Z. Gilman. 1992. Growth regulated expression of D-type cyclin genes in human diploid fibroblasts. *Proc. Natl. Acad. Sci.* **89:** 9910.

Wright, W.E. and J.W. Shay. 1992. The two-stage mechanism controlling cellular senescence and immortalization. *Exp. Gerontol.* **27:** 383.

Xiong, Y. and D. Beach. 1991. Population explosion in the cyclin family. *Curr. Biol.* **1:** 362.

Xiong, Y., G.J. Hannon, H. Zhang, D. Casso, R. Kobayashi, and D. Beach. 1993. p21 is a universal inhibitor of cyclin kinases. *Nature* **366:** 701.

Yamada, H., N. Wake, S. Fujimoto, J.C. Barrett, and M. Oshimura. 1990. Multiple chromosomes carrying tumor suppressor activity for a uterine endometrial carcinoma cell line identified by microcell-mediated chromosome transfer. *Oncogene* **5:** 1141.

Yonish-Rouach, E., D. Resnitzky, J. Lotem, L. Sachs, A. Kimchi, and M. Oren. 1991. Wild-type p53 induces apoptosis of myeloid leukaemic cells that is inhibited by interleukin-6. *Nature* **352:** 345.

Apoptosis and the Prognostic Significance of p53 Mutation

S.W. Lowe,* S. Bodis,†‡ N. Bardeesy,§ A. McClatchey,*
L. Remington,* H.E. Ruley,** D.E. Fisher,† T. Jacks,*††
J. Pelletier,§ AND D.E. Housman*

*Center for Cancer Research and Department of Biology, Massachusetts Institute of Technology, Cambridge,
Massachusetts 02139; †Dana Farber Cancer Institute and ‡Joint Center for Radiation Therapy, Harvard
Medical School, Boston, Massachusetts 02115; §Department of Biochemistry, McGill University, Montreal,
Canada H3G 1Y6; **Department of Microbiology and Immunology, Vanderbilt University School of Medicine,
Nashville, Tennessee 37232; ††Howard Hughes Medical Institute, Massachusetts Institute of Technology,
Cambridge, Massachusetts 02139

The use of radiation and chemotherapy in the treatment of human malignancy has had a dramatic impact on prolonging the disease-free interval but has not substantially affected long-term patient survival. One reason for this failure is that some tumors respond poorly to either form of treatment or become nonresponsive upon tumor relapse. Since the identification of therapeutic agents has been empirical, the molecular mechanisms that determine treatment effectiveness remain largely unknown.

A more complete understanding of cellular sensitivity and resistance to cancer therapy may require the elucidation of the mechanisms by which anticancer agents cause cell death. Since radiation and many chemotherapeutic agents induce DNA damage or cause disruptions in DNA metabolism, the tumor-specific cytotoxicity of these agents has been attributed to their debilitating effects on actively proliferating cells. However, in many instances, the cellular damage caused by active doses of these agents is not sufficient to explain the observed toxicity (Dive and Hickman 1991). Thus, the development of better therapeutic agents may require (1) the identification of factors that determine the tumor-specific action of anticancer agents, (2) the elucidation of the biological and biochemical processes responsible for their cytotoxic action, and (3) an understanding of the molecular basis for tumor cross-resistance.

In the absence of a detailed understanding of the therapeutic response, a major objective of clinical cancer research has been the identification of factors associated with disease stage and patient prognosis. These indicators help predict the course of a malignancy and are often used to determine the type and/or aggressiveness of cancer therapy required (see, e.g., Harris et al. 1992). Implicit in a "poor prognosis," however, is an increased probability that cancer therapy will be ineffective. Therefore, prognostic indicators also may provide insight into molecular mechanisms that determine the efficacy of anticancer agents.

An increasing amount of evidence indicates that anticancer agents induce apoptosis (Kerr et al. 1994). Apoptosis is often referred to as programmed cell death, since it is a genetically determined process that requires the active participation of the dying cell. Apoptosis occurs under physiological circumstances and is required for embryogenesis and the maintenance of tissue homeostasis (for review, see Raff 1992). Many toxic stimuli also induce apoptosis, even at doses or concentrations insufficient to cause general metabolic dysfunction (Lennon et al. 1991). These observations suggest that divergent types of cellular damage may generate signals that activate a common cell death program. In principle, defects in apoptosis could produce cross-resistance to many anticancer agents, leading ultimately to the failure of cancer therapy.

Mutations in the *p53* tumor suppressor gene are associated with aggressive cancers and poor patient prognosis. *p53* functions as an essential component of a cell cycle checkpoint that limits proliferation following genomic damage (Kastan et al. 1992). *p53* also can promote apoptosis (Yonish-Rouach et al. 1991; Shaw et al. 1992; Debbas and White 1993). Indeed, in many instances, *p53* is required for efficient apoptosis such that cells lacking functional *p53* are resistant (Clarke et al. 1993; Lotem and Sachs 1993; Lowe et al. 1993b, 1994a; Merritt et al. 1994). These observations provide a link between a prognostic indicator (*p53* mutation) and a potential determinant of anticancer agent cytotoxicity (apoptosis). In this paper, we discuss evidence obtained from model systems and from human cancer that supports the view that *p53* participates in the apoptotic program induced by many anticancer agents. Consequently, *p53* mutations may promote cellular resistance to many anticancer therapies, thereby contributing to treatment failure.

EXPERIMENTAL PROCEDURES

Cells and gene transfer. $p53^{+/+}$ and $p53^{-/-}$ mouse embryonic fibroblasts (MEFs) were obtained from 12.5-day embryos derived from crosses between mice with a disrupted *p53* allele (Jacks et al. 1994). Cells were cultured in DME containing 10% fetal bovine serum (FBS) and used between passages 3 and 5. The generation of cell lines containing the adenovirus-5 E1A gene and an activated *ras* oncogene (T24 Ha-*ras*) has been described previously (Lowe et al. 1994a).

Cell viability assays. For irradiation experiments, exponentially growing cells were detached from plates and adjusted to 10^6 cells/ml. Samples were irradiated for different times in a Gammacell 40 irradiator containing a ^{137}Cs source ($\sim$0.8 Gy/min), and 1 ml of each cell suspension was added to 100-mm dishes containing normal growth medium. Cell viability was assessed 36 hours after irradiation by pooling adherent and nonadherent cells and measuring uptake of fluorescein isothiocyanate (FITC) by flow cytometry (Shi et al. 1990). For adriamycin experiments, cells were plated at 1×10^6 cells/100-mm dish, allowed to adhere, and incubated with various concentrations of adriamycin (provided by the Dana Farber Cancer Institute). Cell viability was determined 24 hours following treatment by FITC uptake and flow cytometry. Chromatin structure was visualized by staining formalin-fixed cells with 2,4-diamidino-2-phenylindole (DAPI) 8–10 hours after treatment with anticancer agents.

Tumor growth and treatment. Minimally passaged $p53^{+/+}$ and $p53^{-/-}$ MEFs transformed by E1A and T24 Ha-*ras* were detached from tissue culture plates, washed, and resuspended in phosphate buffered saline (PBS). 2×10^6 cells were injected into each flank of athymic nude mice (aged 4–8 weeks). Upon reaching a palpable size, tumor volumes were estimated from caliper measurements of tumor length (L) and width (l) according to the following formula: $(L \times l^2)/2$. In general, tumors were allowed to expand to 0.15–0.5 cm^3 prior to treatment. Animals receiving lower doses of radiation (≤ 7 Gy) were given a single fraction to the whole body in a Gammacell 40 irradiator.

Histological analysis. Untreated or irradiated tumors were recovered from mice, fixed in formaldehyde, embedded in paraffin, and sectioned by standard methods. Sections were either stained with hematoxylin and eosin or analyzed for apoptotic cells using the terminal deoxytransferase-mediated dUTP-biotin nick end-labeling (TUNEL) assay (Gavrieli et al. 1992). Biotin-conjugated DNA ends were detected using an avidin-horseradish peroxidase-based detection method. Tissue sections were counterstained with methyl green.

Mutational analysis of p53. Tumor cells were dispersed with trypsin and cultured in medium containing 50 μg/ml hygromycin B to select for tumor-derived cells (which contain hygromycin phosphotransferase; Lowe et al. 1994a). In general, RNA was obtained from tumor cells maintained for less than 1 week in culture. Complementary DNA was prepared by standard methods using a primer corresponding to sequences within either exon 9 or 11 of murine *p53* (Bienz et al. 1984). Amplification of *p53* sequences corresponding to exons 5–8 was accomplished by polymerase chain reaction using primers directed to sequences within exons 4 and 9. Sequencing was carried out using primers directed to several sequences within exons 5–8. Mutations were verified by sequencing the opposite strand or by repeating the sequencing. The codon-131 mutation created a *Bso*F1 site, allowing independent verification of the mutation by restriction digestion of amplified products.

RESULTS AND DISCUSSION

To investigate the involvement of *p53* in the apoptotic program induced by anticancer agents, we have developed a model system utilizing cells derived from mice in which *p53* has been disrupted by gene targeting. Since cells derived from *p53*-deficient and normal animals differ only in their *p53* status, phenotypic differences can be attributed to *p53* function. Previous studies have shown that embryonic fibroblasts coexpressing the adenovirus early region 1A (E1A) and activated *ras* oncogenes are highly tumorigenic yet susceptible to apoptosis upon withdrawal of growth factors (Lowe et al. 1994a). The increased susceptibility of these cells to apoptosis correlates with the ability of E1A to promote S-phase entry (White et al. 1991) and may reflect a cellular response to aberrant proliferation (Lowe et al. 1994a). *p53*-deficient cells transformed by E1A and *ras* are resistant to apoptosis following serum depletion, indicating that apoptosis is *p53*-dependent (Lowe et al. 1994a).

Anticancer Agents Induce *p53*-dependent Apoptosis In Vitro

As illustrated in Figure 1, both γ-irradiation and adriamycin induced *p53*-dependent killing of cells transformed by E1A and *ras*. Relatively low doses of γ-irradiation or adriamycin efficiently killed $p53^{+/+}$ cells transformed by E1A and *ras* (IC$_{50}$ of $\sim$2 Gy or $\sim$0.05 μg/ml adriamycin). The dying cells had morphological and physiological features of apoptosis, including chromatin condensation and nuclear fragmentation (Fig. 2), and the cells contained DNA that was apparently degraded by internucleosomal cleavage (Lowe et al. 1993a). In contrast, considerably higher doses of γ-irradiation or adriamycin were required for killing of $p53^{-/-}$ cells transformed by E1A and *ras* (IC$_{50}$ of $\gg 10$ Gy and >1.0 μg/ml adriamycin). At doses that induced substantial apoptosis in $p53^{+/+}$ cells transformed by E1A and *ras*, $p53^{-/-}$ cells showed few signs of apoptosis (Fig. 2).

Although γ-irradiation and adriamycin induced apoptosis in a *p53*-dependent manner in oncogenically transformed cells, the untransformed fibroblasts from which they were derived were resistant to apoptosis regardless of their *p53* status (Fig. 1). Thus, physiological changes that accompanied oncogenic transformation increased the susceptibility of normal fibroblasts to *p53*-dependent apoptosis, which was triggered by γ-irradiation or adriamycin. Several other chemo-

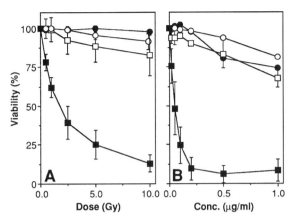

Figure 1. Effect of p53 on cell viability following treatment with anticancer agents. Viability of the untransfected MEFs and $p53^{+/+}$ and $p53^{-/-}$ clones transformed by E1A and ras was estimated 36 hr after treatment with the indicated dose of γ-irradiation or 24 hr after incubating with the indicated concentration of adriamycin. Each point represents the average and standard deviation obtained from at least 3 independent clones normalized to the viability of the corresponding untreated controls. (●) Untransfected $p53^{+/+}$ MEFs; (○) untransfected $p53^{-/-}$ MEFs; (■) $p53^{+/+}$ cells transformed by E1A and T24 Ha-ras; (□) $p53^{-/-}$ cells transformed by E1A and T24 Ha-ras.

therapeutic agents induced p53-dependent apoptosis in transformed cells but not the parental fibroblasts. These included 5-fluorouracil, etoposide, adriamycin, and cisplatin (Lowe et al. 1993a; S. Bodis, unpubl.).

Analysis of p53, Apoptosis, and Therapeutic Response In Vivo

These data provide a provocative link between defects in an apoptotic program and treatment resistance to anticancer agents. Nevertheless, although in vitro cell culture models provide simple systems for studying anticancer agent cytotoxicity and apoptosis, their relevance to tumor response in vivo remains unknown. Complicating factors such as hypoxia, nutrient supply, and pharmokinetic factors can have a substantial impact on the effectiveness of anticancer agents in vivo. To determine whether p53 status influenced *tumor* response to cancer therapy, we were interested in establishing a well-defined system in which to investigate the role of p53, apoptosis, and tumor response in an in vivo setting.

Systematic analysis of the therapeutic response in human tumors is further complicated by unknown genetic differences that exist between individual tumors or tumor-derived cell lines. For example, comparison of cells or tumors known to be different in p53 status is confounded by unknown mutations in other relevant genes. We have attempted to minimize this problem by taking advantage of the observation that the coexpression of E1A and ras is *sufficient* to transform primary cells to a malignant state irrespective of their p53 status, but cells expressing genetically normal p53 remain susceptible to apoptosis (Ruley 1983; Lowe et al. 1994a). The highly oncogenic nature of these cells implies that tumor growth could occur without strong selection for additional mutations, allowing the generation of tumors differing primarily in their p53 status.

The strategy employed in these experiments is outlined in Figure 3. Cells transformed by E1A and ras were injected subcutaneously into nude mice, and tumors were allowed to form. Although $p53^{-/-}$ tumors often appeared with a shorter latency than $p53^{+/+}$ tumors, both tumor types arose rapidly (<18 days) and with a high frequency (80–100% of injected sites) (Lowe et al. 1994a,b). Upon reaching an appropriate size, tumors were treated with γ-irradiation (7 Gy), and their volumes were monitored thereafter. A representative example of tumor growth and response is shown in Figure 4. Tumors derived from cells expressing wild-type p53 responded to γ-irradiation, typically regressing to about 10–50% of their pretreatment volume prior to regrowth (Lowe et al. 1994b). In contrast, p53-deficient tumors displayed little response and continued to grow. Moreover, increasing the dose of γ-irradiation to 12 Gy did not enhance the response in $p53^{-/-}$ tumors (Lowe et al. 1994b).

p53 status also influenced tumor response to chemotherapy. Specifically, tumors derived from $p53^{+/+}$ cells responded to adriamycin treatment whereas $p53^{-/-}$

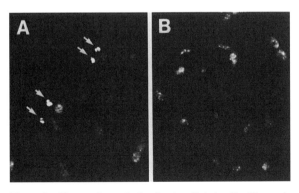

Figure 2. Chromatin analysis of γ-irradiated cells. Genomic DNA in $p53^{+/+}$ (A) and $p53^{-/-}$ (B) cells transformed by E1A and T24 Ha-ras was visualized by DAPI staining 8 hr after treatment with 5 Gy γ-irradiation. The arrows identify cells with condensed chromatin.

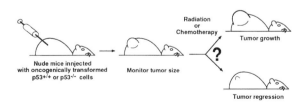

Figure 3. Generation and treatment of genetically defined tumors. Athymic nude mice were injected with $p53^{+/+}$ or $p53^{-/-}$ cells transformed by E1A and ras. Since this oncogene combination is highly tumorigenic regardless of p53 status, tumor growth can occur without strong pressure for additional mutations. Upon reaching an appropriate size, animals were treated with γ-irradiation or chemotherapy and tumors were monitored for growth or regression.

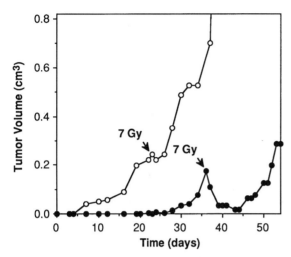

Figure 4. Effect of *p53* on tumor response to γ-irradiation. Tumor volumes were estimated at various times after injection of $p53^{+/+}$ (●) or $p53^{-/-}$ (○) fibroblasts transformed by E1A and *ras* into nude mice. Upon reaching an appropriate volume (indicated by the arrows), the mice were irradiated with 7 Gy and tumors were monitored for growth or regression.

tumors did not (Lowe et al. 1994b). Taken together, these data establish that inactivation of *p53* can promote tumor cross-resistance to anticancer therapy.

Tumor regression can occur as a result of cell death by necrosis or apoptosis. Necrosis results from cell lysis and typically occurs in distinct zones within tumors. In contrast, cells undergoing apoptosis are interspersed throughout the tumor and undergo cell shrinkage, loss of extracellular contacts, chromatin condensation, and nuclear fragmentation (Kerr et al. 1972). Apoptotic cells often activate an endonuclease that breaks genomic DNA (Wyllie 1980), generating products which are readily detected in situ using the TUNEL method (Gavrieli et al. 1992).

The model system used in these studies allowed visualization of cell death shortly after antitumor therapy. Irradiated tumors derived from $p53^{+/+}$ cells contained large numbers of pyknotic cells, and many contained fragmented nuclei (Fig. 5a). The remaining normal cells were often surrounded by large regions of extracellular space, indicative of massive cell loss. The pyknotic cells typically stained with TUNEL, indicating that they contained substantial amounts of degraded DNA (Fig. 5e). In contrast, $p53^{-/-}$ tumors contained few pyknotic cells and regions of cell loss (Fig. 5b) but displayed a small but significant increase in TUNEL staining (cf. Fig. 5d, f). Therefore, the primary mode of cell death in $p53^{+/+}$ tumors was apoptosis. *p53*-deficient tumors, which did not regress, were largely resistant to apoptosis.

Acquired *p53* Mutations Are Associated with Tumor Relapse

In human cancer, initially responsive tumors often become resistant to cancer therapy upon tumor relapse. If tumor response is determined by the ability of therapeutic agents to trigger apoptosis, anticancer therapies might enrich for tumor cells with acquired defects in apoptotic programs. We tested this hypothesis by searching for de novo *p53* mutations in tumors derived from $p53^{+/+}$ cells that had either relapsed or displayed acquired resistance to treatment.

Regions of DNA corresponding to exons 5–8 of the *p53* gene were amplified by polymerase chain reaction and sequenced. This region of *p53* contains more than 90% of the mutations responsible for *p53* inactivation in human tumors (Hollstein et al. 1991). Approximately 50% of the initially resistant or recurrent tumors tested contained acquired mutations in the *p53* gene (Lowe et al. 1994b). The growth, response, and mutational analysis of tumors derived from one $p53^{+/+}$ clone are schematically represented in Figure 6. No mutations were detected in DNA obtained from the injected cell population. It is noteworthy that the mutation at codon 131 was observed in both tumors, suggesting that it preexisted in a small percentage of the injected cells. In contrast, the mutation at codon 239 was observed in only one tumor, implying it was acquired on tumor expansion. By selecting against apoptosis, γ-irradiation may have enriched these tumors for cells harboring *p53* mutations. *p53* mutations were also detected in tumors derived from a $p53^{+/+}$ clone that underwent apoptosis in culture but were relatively resistant to γ-irradiation in vivo (Lowe et al. 1994b), indicating that tumor growth can also select for apoptosis-resistant tumors.

Apoptosis and Chemotherapy

These studies suggest a rationale for understanding the response of tumors to anticancer therapies (Fig. 7). Certain oncogenic alterations, by increasing cellular susceptibility to apoptosis, can provide the therapeutic index whereby anticancer agents specifically target tumor cells. Other lesions—one of which is functional inactivation of *p53*—can have the opposite effect, producing a general resistance to cancer therapy. Although a much more detailed analysis will be required to determine the extent to which similar mechanisms exist in human cancer, these experimental systems establish several important principles. First, tumor sensitivity and resistance to anticancer agents can be modulated by the threshold at which anticancer agents trigger apoptosis. Second, defects in apoptotic programs can cause tumor resistance to a variety of agents. Finally, genetic alterations that both accompany and promote malignant transformation also can determine the effectiveness of cancer therapy.

Does *p53* Mutation Promote Treatment Resistance in Human Tumors?

At present, clinical studies have not been designed to properly test whether *p53* status influences the therapeutic response of human tumors. The situation in

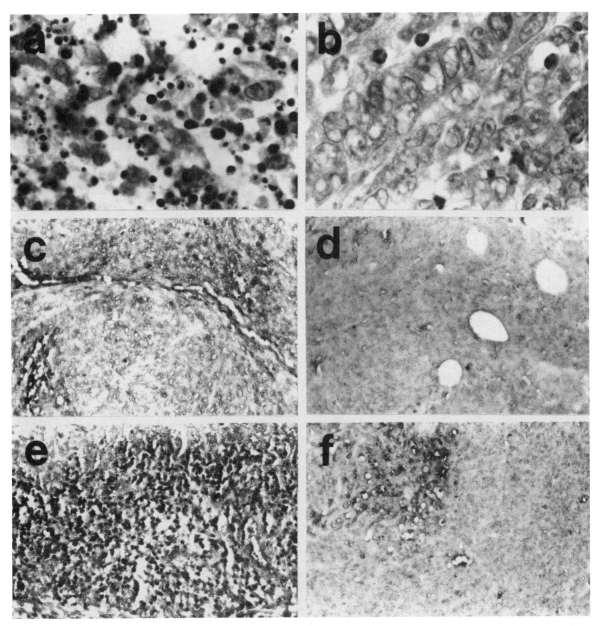

Figure 5. Histological analysis of tumors following γ-irradiation. Tumors derived from either $p53^{+/+}$ (*a, c,* and *e*) or $p53^{-/-}$ (*b, d,* and *f*) transformed fibroblasts were recovered 48 hr after irradiation with 7 Gy or from parallel untreated tumors. (*a,b*) Hematoxylin and eosin staining of tumor sections obtained from γ-irradiated tumors. (*c,d*) TUNEL staining of untreated tumors. (*e,f*) TUNEL staining of irradiated tumors. Magnifications: *a* and *b*, 1000×; *c–f*, 200×.

human cancer is probably more complex than in these model systems, and the effect of *p53* on tumor response to anticancer therapy is probably determined by additional factors, including the type of *p53* mutation, the expression of modifying genes, and the tissue of tumor origin. Nevertheless, our results predict that cancer therapy would be less effective in patients harboring tumors with *p53* mutations. Indeed, *p53* mutation is associated with poor patient prognosis in a variety of cancers, including certain lymphomas (Rodriguez et al. 1991) and leukemias (Elrouby et al. 1993; Hsiao et al. 1994; Nakai et al. 1994), carcinomas of the breast (Thompson et al. 1992; Andersen et al. 1993; Thorlacius et al. 1993) and lung (Horio et al. 1993; Mitsudomi et al. 1993), and soft-tissue sarcomas (Cordoncardo et al. 1994). Moreover, several highly curable tumor types rarely harbor *p53* mutations. These include acute lymphoblastic leukemia (Gaidano et al. 1991) and testicular cancer (Heimdal et al. 1993; Peng et al. 1993; Fleischhacker et al. 1994).

Molecular and clinical aspects of Wilms' tumor provide insight into the role of *p53*, apoptosis, and tumor response in a human cancer. Wilms' tumor is a highly curable pediatric malignancy in which chemotherapeutic regimens in combination with surgical resection have produced survival rates approaching more

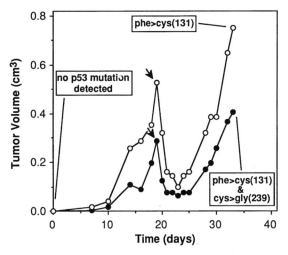

Figure 6. *p53* mutations in relapse tumors. Schematic illustration of tumor growth, response, and *p53* status of tumors derived from $p53^{+/+}$ clone C6.

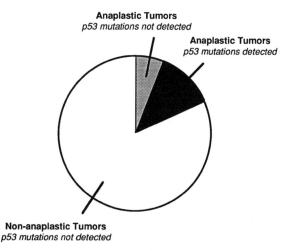

Figure 8. *p53* mutations in Wilms' tumor. *p53* genes were analyzed from 134 Wilms' tumors of known histology. *p53* mutations are strictly associated with the anaplastic form of Wilms' tumor.

than 85% for stage I and stage II (favorable histology) tumors. However, patients harboring the aggressive anaplastic subtype of Wilms' tumor (which represents ~5% of the total cases) have a considerably worse prognosis (Zuppan et al. 1988). To determine the role of *p53* inactivation in the etiology of Wilms' tumor, a large series of tumors were analyzed for the presence of *p53* mutations (Bardeesy et al. 1994). In most cases, DNA was recovered from paraffin sections and individual exons were amplified by polymerase chain reaction. The presence of point mutations was determined by single strand conformational polymorphism (SSCP) analysis and confirmed by direct sequencing. In addition, tumors were classified by their clinical grade (i.e., non-anaplastic or anaplastic).

The results obtained from the analysis of 134 tumors of known histology are summarized in Figure 8. *p53* mutations were uncommon in Wilms' tumor and occurred in approximately 12% of the tumors analyzed. However, there was a striking correlation between *p53* mutation and anaplasia—70% of anaplastic tumors contained detectable *p53* mutations, but no mutations were observed in non-anaplastic tumors. Preliminary studies indicate that anaplastic Wilms' tumors harboring *p53* mutations have a substantial decrease in spontaneous apoptosis compared to non-anaplastic tumors (N. Bardeesy et al., unpubl.). These data link *p53* mutation and attenuated apoptosis with tumor progression to anaplasia, and hence, poor prognosis.

Additional evidence supports the view that inactivation of *p53*-dependent apoptosis can have a negative impact on the therapeutic response of human tumors. In acute lymphoblastic leukemia and multiple myeloma, *p53* mutations have been identified in relapse phase tumors that were not detectable in the primary tumor (Neri et al. 1993; Hsiao et al. 1994). This suggests that *p53* mutations confer a survival advantage to cells undergoing chemotherapy. Moreover, in patients with B-cell chronic lymphocytic leukemia (B-CLL), *p53* mutations dramatically reduce the probability that patients respond to chemotherapy and enter remission (Elrouby et al. 1993). It is noteworthy that the failure of chemotherapy was not associated with increased expression of the multidrug resistance genes, implying that treatment resistance did not involve diminished drug accessibility to intracellular targets. Finally, *p53* mutation has been linked to radioresistance in human tumor lines in culture (O'Connor et al. 1993; Mcilwrath et al. 1994), and reintroduction of wild-type *p53* into a *p53*-deficient tumor line enhances apoptosis following cisplatin treatment (Fujiwara et al. 1994).

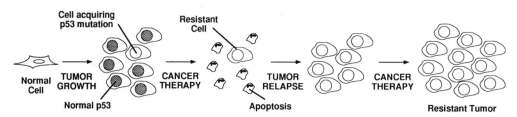

Figure 7. Example of how *p53* mutations may contribute to treatment-resistant tumors. Cancer therapy can enrich for cells harboring *p53* mutations by selectively killing tumor cells retaining an intact apoptotic program. Upon relapse, tumors demonstrate acquired resistance to therapy. This model may explain the observations in Fig. 6 and is consistent with the pattern of *p53* mutation observed in acute lymphoblastic leukemia (Hsiao et al. 1994).

FUTURE PERSPECTIVES

Although further clinical investigation will be required to verify that *p53* mutations contribute to drug resistance in human cancer, our results suggest several new directions for improving cancer therapy. First, the identification of *p53* mutations in certain tumor types may become an important factor in treatment decisions. Second, the reintroduction of normal *p53* function into tumors harboring *p53* mutations is predicted to sensitize these cells to apoptosis induced by radiation and chemotherapy. Indeed, recent studies support this view (Fujiwara et al. 1994). Finally, model systems analogous to the one used here may allow screening for agents capable of inducing apoptosis independent of *p53* function, or ultimately, their rational design.

ACKNOWLEDGMENTS

We thank M. McCurrach for assistance and helpful advice. S.W.L. is an Anna Fuller postdoctoral fellow and S.B. is supported by a fellowship from the Swiss League of Cancer. N.B. is supported by a fellowship from the Medical Research Council of Canada. This work was supported by grants R01CA-40602 (H.E.R), 5R27CA-17575 (D.E.H.), a Cancer Center Support (core) grant CA-14051 from the National Institutes of Health, a grant from the National Cancer Institute of Canada (J.P.), and a grant from the General Cinemas Charitable Foundation (T.J.).

REFERENCES

Andersen, T.I., R. Holm, J.M. Nesland, K.R. Heimdal, L. Ottestad, and A.L. Borresen. 1993. Prognostic significance of TP53 alterations in breast carcinoma. *Br. J. Cancer* **68:** 540.

Bardeesy, N., D. Falkoff, M.J. Petruzzi, N. Nowak, B. Zabel, M. Adam, M.C. Aguiar, P. Grundy, T. Shows, and J. Pelletier. 1994. Anaplastic Wilms' tumour, a subtype displaying poor prognosis, harbours p53 gene mutations. *Nat. Genet.* **7:** 91.

Bienz, B., R. Zakut-Houri, D. Givol, and M. Oren. 1984. Analysis of the gene coding for the murine cellular tumour antigen p53. *EMBO J.* **3:** 2179.

Clarke, A.R., C.A. Purdie, D.J. Harrison, R.G. Morris, C.C. Bird, M.L. Hooper, and A.H. Wyllie. 1993. Thymocyte apoptosis induced by p53-dependent and independent pathways. *Nature* **362:** 849.

Cordoncardo, C., E. Latres, M. Drobnjak, M.R. Oliva, D. Pollack, J.M. Woodruff, V. Marechal, J.D. Chen, M.F. Brennan, and A.J. Levine. 1994. Molecular abnormalities of mdm2 and p53 genes in adult soft tissue sarcomas. *Cancer Res.* **54:** 794.

Debbas, M. and E. White. 1993. Wild-type p53 mediates apoptosis by E1A, which is inhibited by E1B. *Genes Dev.* **7:** 546.

Dive, C. and J.A. Hickman. 1991. Drug-target interactions: Only the first step in the commitment to a programmed cell death? *Br. J. Cancer* **64:** 192.

Elrouby, S., A. Thomas, D. Costin, C.R. Rosenberg, M. Potmesil, R. Silber, and E.W. Newcomb. 1993. p53 Gene mutation in B-cell chronic lymphocytic leukemia is associated with drug resistance and is independent of MDR1/MDR3 gene expression. *Blood* **82:** 3452.

Fleischhacker, M., T. Strohmeyer, Y. Imai, D.J. Slamon, and H.P. Koeffler. 1994. Mutations of the p53 gene are not detectable in human testicular tumors. *Mod. Pathol.* **7:** 435.

Fujiwara, T., E.A. Grimm, T. Mukhopadhyay, W.W. Zhang, L.B. Owenschaub, and J.A. Roth. 1994. Induction of chemosensitivity in human lung cancer cells in vivo by adenovirus-mediated transfer of the wild-type p53 gene. *Cancer Res.* **54:** 2287.

Gaidano, G., P. Ballerini, J.Z. Gong, G. Inghirami, A. Neri, E.W. Newcomb, I.T. Magrath, D.M. Knowles, and R. Dalla-Favera. 1991. p53 mutations in human lymphoid malignancies: Association with Burkitt lymphoma and chronic lymphocytic leukemia. *Proc. Natl. Acad. Sci.* **88:** 5413.

Gavrieli, Y., Y. Sherman, and S.A. BenSasson. 1992. Identification of programmed cell death in situ via specific labeling of nuclear DNA fragmentation. *J. Cell Biol.* **119:** 493.

Harris, J.R., M.E. Lippman, U. Veronesi, and W. Willett. 1992. Breast cancer. *N. Engl. J. Med.* **327:** 473.

Heimdal, K., R.A. Lothe, S. Lystad, R. Holm, S.D. Fossa, and A.L. Borresen. 1993. No germline TP53 mutations detected in familial and bilateral testicular cancer. *Genes Chromosomes Cancer* **6:** 92.

Hollstein, M., D. Sidransky, B. Vogelstein, and C.C. Harris. 1991. p53 mutations in human cancers. *Science* **253:** 49.

Horio, Y., T. Takahashi, T. Kuroishi, K. Hibi, M. Suyama, T. Niimi, K. Shimokata, K. Yamakawa, Y. Nakamura, R. Ueda, and T. Takahashi. 1993. Prognostic significance of p53 mutations and 3p deletions in primary resected non-small cell lung cancer. *Cancer Res.* **53:** 1.

Hsiao, M.H., A.L. Yu, J. Yeargin, D. Ku, and M. Haas. 1994. Nonhereditary p53 mutations in T-cell acute lymphoblastic leukemia are associated with the relapse phase. *Blood* **83:** 2922.

Jacks, T., L. Remington, B.O. Williams, E.M. Schmitt, S. Halachmi, R.T. Bronson, and R.A. Weinberg. 1994. Tumor spectrum analysis in p53-mutant mice. *Curr. Biol.* **4:** 1.

Kastan, M.B., Q. Zhan, W.S. el-Deiry, F. Carrier, T. Jacks, W.V. Walsh, B.S. Plunkett, B. Vogelstein, and A. Fornace, Jr. 1992. A mammalian cell cycle checkpoint pathway utilizing p53 and GADD45 is defective in ataxia-telangiectasia. *Cell* **71:** 587.

Kerr, J.F.R., C.M. Winterford, and B.V. Harmon. 1994. Apoptosis—Its significance in cancer and cancer therapy. *Cancer* **73:** 2013.

Kerr, J.F., A.H. Wyllie, and A.R. Currie. 1972. Apoptosis: A basic biological phenomenon with wide-ranging implications in tissue kinetics. *Br. J. Cancer* **26:** 239.

Lennon, S.V., S.J. Martin, and T.G. Cotter. 1991. Dose-dependent induction of apoptosis in human tumour cell lines by widely diverging stimuli. *Cell Proliferation* **24:** 203.

Lotem, J. and L. Sachs. 1993. Hematopoietic cells from mice deficient in wild-type p53 are more resistant to induction of apoptosis by some agents. *Blood* **82:** 1092.

Lowe, S.W., T. Jacks, D.E. Housman, and H.E. Ruley. 1994a. Abrogation of oncogene-associated apoptosis allows transformation of p53-deficient cells. *Proc. Natl. Acad. Sci.* **91:** 2026.

Lowe, S.W., H.E. Ruley, T. Jacks, and D.E. Housman. 1993a. p53-dependent apoptosis modulates the cytotoxicity of anticancer agents. *Cell* **74:** 954.

Lowe, S.W., E.M. Schmitt, S.W. Smith, B.A. Osborne, and T. Jacks. 1993b. p53 is required for radiation-induced apoptosis in mouse thymocytes. *Nature* **362:** 847.

Lowe, S.W., S. Bodis, A. McClatchey, L. Remington, H.E. Ruley, D. Fisher, D.E. Housman, and T. Jacks. 1994b. p53 status and the efficacy of cancer therapy in vivo. *Science* **266:** 807.

Mcilwrath, A.J., P.A. Vasey, G.M. Ross, and R. Brown. 1994. Cell cycle arrests and radiosensitivity of human tumor cell lines: Dependence on wild-type p53 for radiosensitivity. *Cancer Res.* **54:** 3718.

Merritt, A.J., C.S. Potten, C.J. Kemp, J.A. Hickman, A. Balmain, D.P. Lane, and P.A. Hall. 1994. The role of p53 in spontaneous and radiation-induced apoptosis in the gastrointestinal tract of normal and p53-deficient mice. *Cancer Res.* **54:** 614.

Mitsudomi, T., T. Oyama, T. Kusano, T. Osaki, R. Nakanishi, and T. Shirakusa. 1993. Mutations of the p53 gene as a predictor of poor prognosis in patients with non-small-cell lung cancer. *J. Natl. Cancer Inst.* **85:** 2018.

Nakai, H., S. Misawa, M. Taniwaki, S. Horiike, T. Takashima, T. Seriu, H. Nakagawa, H. Fujii, C. Shimazaki, N. Maruo, T. Akaogi, N. Uike, T. Abe, and K. Kashima. 1994. Prognostic significance of loss of chromosome 17p and p53 gene mutations in blast crisis of chronic myelogenous leukaemia. *Br. J. Haematol.* **87:** 425.

Neri, A., L. Baldini, D. Trecca, L. Cro, E. Polli, and A.T. Maiolo. 1993. p53 gene mutations in multiple myeloma are associated with advanced forms of malignancy. *Blood* **81:** 128.

O'Connor, P.M., J. Jackman, D. Jondle, K. Bhatia, I. Magrath, and K.W. Kohn. 1993. Role of the p53 tumor suppressor gene in cell cycle arrest and radiosensitivity of Burkitt's lymphoma cell lines. *Cancer Res.* **53:** 4776.

Peng, H.Q., D. Hogg, D. Malkin, D. Bailey, B.L. Gallie, M. Bulbul, M. Jewett, J. Buchanan, and P.E. Goss. 1993. Mutations of the p53 gene do not occur in testis cancer. *Cancer Res* **53:** 3574.

Raff, M.C. 1992. Social controls on cell survival and cell death. *Nature* **356:** 397.

Rodriguez, M.A., R.J. Ford, A. Goodacre, P. Selvanayagam, F. Cabanillas, and A.B. Deisseroth. 1991. Chromosome 17- and p53 changes in lymphoma. *Br. J. Haematol.* **79:** 575.

Ruley, H.E. 1983. Adenovirus early region 1A enables viral and cellular transforming genes to transform primary cells in culture. *Nature* **304:** 602.

Shaw, P., R. Bovey, S. Tardy, R. Sahli, B. Sordat, and J. Costa. 1992. Induction of apoptosis by wild-type p53 in a human colon tumor-derived cell line. *Proc. Natl. Acad. Sci.* **89:** 4495.

Shi, Y.F., M.G. Szalay, L. Paskar, B.M. Sahai, M. Boyer, B. Singh, and D.R. Green. 1990. Activation-induced cell death in T cell hybridomas is due to apoptosis. Morphologic aspects and DNA fragmentation. *J. Immunol.* **144:** 3326.

Thompson, A.M., T.J. Anderson, A. Condie, J. Prosser, U. Chetty, D.C. Carter, H.J. Evans, and C.M. Steel. 1992. p53 allele losses, mutations and expression in breast cancer and their relationship to clinico-pathological parameters. *Int. J. Cancer* **50:** 528.

Thorlacius, S., A.L. Borresen, and J.E. Eyfjord. 1993. Somatic p53 mutations in human breast carcinomas in an Icelandic population: A prognostic factor. *Cancer Res.* **53:** 1637.

White, E., R. Cipriani, P. Sabbatini, and A. Denton. 1991. Adenovirus E1B 19-kilodalton protein overcomes the cytoxicity of E1A proteins. *J. Virol.* **65:** 2968.

Wyllie, A.H. 1980. Glucocorticoid-induced thymocyte apoptosis is associated with endogenous endonuclease activation. *Nature* **284:** 555.

Yonish-Rouach, E., D. Resnitzky, J. Lotem, L. Sachs, A. Kimchi, and M. Oren. 1991. Wild-type p53 induces apoptosis of myeloid leukaemic cells that is inhibited by interleukin-6. *Nature* **352:** 345.

Zuppan, C.W., J.B. Beckwith, and D.W. Luckey. 1988. Anaplasia in unilateral Wilms' tumor: A report from the national Wilms' tumor study pathology center. *Hum. Pathol.* **19:** 1199.

Transgenic Approaches to the Analysis of *ras* and *p53* Function in Multistage Carcinogenesis

C.J. KEMP, P.A. BURNS,* K. BROWN, H. NAGASE, AND A. BALMAIN
CRC Beatson Laboratories, Department of Medical Oncology, University of Glasgow, Bearsden, Glasgow, G61 1BD, United Kingdom

The development of cancer in humans, as in animals, is influenced by multiple genes which can act at different stages of tumor development. These genes may exert their effects as a consequence of mutation within somatic cells, subsequently conferring a selective growth advantage upon the cell in which the genetic change took place. This process may either involve a positive growth stimulus, e.g., an increase in the rate of cell proliferation, or alternatively, a loss of a normal program of cell death, e.g., a decrease in the rate of terminal differentiation or apoptosis. In either case, the consequence is that the evolving tumor cells acquire the necessary survival and proliferative capacities to enable them to sustain additional mutations and evolve to further malignancy. The genes involved in these types of somatic alterations have been broadly classified into the categories of oncogenes and tumor suppressor genes.

Some of these genes may be inherited in the germ line in mutant form and thus confer a strong predisposition to tumor development. The paradigm for this type of genetic predisposition is the familial retinoblastoma syndrome in which affected individuals carry germ-line mutations in the retinoblastoma (Rb) tumor suppressor gene (Murphree and Benedict 1984). Such individuals frequently develop bilateral retinoblastomas during childhood, all of which turn out to have arisen subsequent to additional somatic mutations leading to the functional loss of the remaining wild-type *Rb* allele, in accordance with the classic Knudson hypothesis (Knudson 1971). Many other examples are now available of genes which confer predisposition to tumor development by a similar mechanism, including neurofibromatosis (*NF1* or neurofibromin), adenomatous polyposis coli (*APC*), and Li-Fraumeni syndrome (*p53*) (Ponder 1990; Marshall 1991). Additional loci conferring susceptibility to the development of familial melanoma, familial breast cancer, and several other inherited cancer syndromes have been genetically mapped but as yet have not been cloned (Ponder 1990).

Such cases of familial predisposition are relatively rare within the general population but confer a high probability of development of disease in affected individuals. There are, however, additional categories of predisposing genes which may be present at higher frequency and probably make a substantial contribution to overall cancer risk. Such genes may encode, for example, enzymes which metabolize procarcinogens to their ultimate carcinogenic form or which contribute to DNA repair or which in a variety of ways may contribute to increased cancer risk. Because of the low penetrance of their effects on tumor development, these genes may be extremely difficult to detect and genetically map in the absence of highly extended family pedigrees and more sophisticated methods of linkage analysis.

The mouse offers an alternative system in which to study predisposition genes. The availability of a large series of inbred mouse strains which exhibit genetic susceptibility to the development of tumors of different kinds, including the skin, liver, lung, and lymphoid system, is well known. With the mapping of thousands of highly polymorphic microsatellite markers, which are distributed throughout the mouse genome (Dietrich et al. 1994), it is now feasible to carry out linkage analysis on multigenic quantitative traits such as cancer.

Our studies on mouse skin carcinogenesis have therefore been aimed at the characterization of the genetic alterations which occur in somatic cells throughout the various stages of tumor development. In parallel, we have studied the biological consequences of these genetic changes. Finally, we are now attempting to identify loci that influence predisposition to skin tumor development in different mouse strains.

RESULTS AND DISCUSSION

The model system we have chosen for these studies is the classic skin carcinogenesis model (Yuspa 1994). Studies carried out over the past decade have identified a series of genetic events that take place during the initiation, promotion, and progression phases of skin tumor development (Fig. 1). If carcinogenesis is initiated by treatment with the carcinogen dimethylbenzanthracene (DMBA), many cells suffer specific mutations at codon 61 of the Ha-*ras* gene. These initiated cells can be promoted to develop benign papillomas by subsequent treatment of the skin with tumor promoting agents such as 12-0-tetradecanoylphorbol-13-acetate (TPA). This appears to give rise to two classes of benign papilloma, one that shows only a very low risk for progression to malignancy and one

---

*Present address: Jack Birch Unit for Environmental Carcinogenesis, Department of Biology, University of York, Heslington, York, YO1 5DD, United Kingdom.

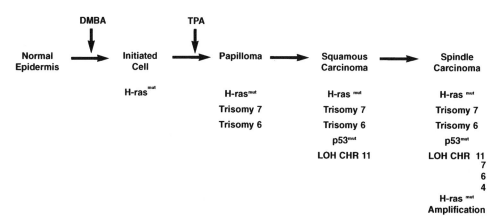

Figure 1. Summary of genetic changes observed during mouse skin carcinogenesis. Mouse skin treated with a single dose of DMBA and promoted for 15 weeks with TPA gives rise to papillomas, a small percentage of which progress to squamous cell carcinomas and spindle cell carcinomas. Genetic changes were identified by mutation analysis and allelotype analysis of tumor DNA.

that exhibits a distinctly higher risk (Hennings et al. 1985; Yuspa 1994). Tumors of the latter category can progress to squamous carcinomas, which invade the dermis and subsequently lead to death of the host organism. This stage of tumor development has been shown to involve, at least in a substantial proportion of cases, loss of heterozygosity (LOH) and mutation in the *p53* tumor suppressor gene (Burns et al. 1991; Ruggeri et al. 1993).

An additional change observed in papillomas is the development of trisomies of chromosomes 7 and 6 (Aldaz et al. 1989; Kemp et al. 1993a). Trisomy of chromosome 7 serves to duplicate the mutant Ha-*ras* gene that is located on this chromosome. In studies of a large number of benign tumors, we and other investigators have never detected duplication of the chromosome carrying the normal Ha-*ras* allele (Bianchi et al. 1990; Bremner and Balmain 1990; Bremner et al. 1994). We have also shown that tumors in which no evidence of Ha-*ras* mutation can be detected also do not exhibit trisomy of chromosome 7 (Bremner and Balmain 1990; Bremner et al. 1994). Thus, a single copy of mutant Ha-*ras* is insufficient to initiate carcinogenesis, but subsequent chromosomal alterations, possibly induced by the genetic instability caused by treatment with the tumor promoter TPA (Furstenberger et al. 1989; see discussion in Kemp et al. 1993a), result in trisomy of 7 and duplication of the mutant *ras* allele, which in turn leads to selective outgrowth from the normal population. The reasons for trisomy of 6 are more obscure and may be due to a selective advantage caused by duplication of a number of candidate genes on this chromosome, including TGF-α, Ki-*ras*, and c-*raf-1*. A particularly interesting candidate is c-*raf-1*, which has been shown to lie downstream from *ras* in the signal transduction pathway leading from growth factor stimulation of cell-surface receptors to the activation of transcription in the nucleus. Studies by Marshall and colleagues have shown that c-*raf-1* is at limiting concentrations in some cell types and that an increase in the amount of c-*raf-1* can potentiate the effects of stimulation of the *ras* signal transduction pathway (Howe et al. 1992). It is therefore tempting to speculate that duplication of c-*raf-1* on chromosome 6 in mouse skin tumors could synergize with the higher levels of Ha-*ras*, leading to an amplified growth stimulatory signal through this signal transduction pathway. It is interesting in this regard that another potent tumor promoter, okadaic acid, can also intervene in a different stage of the same pathway. Okadaic acid inhibits protein phosphatase 2A (Fujiki and Suganuma 1993) which, itself, is known to reverse the activation of MAP kinase kinase (Gomez and Cohen 1991). This would also have the capacity to synergize with a mutant Ha-*ras* gene in stimulating signal transduction.

Additional genetic changes are also found in mouse skin tumors at subsequent stages of tumor progression. These include LOH on chromosomes 6, 7, and 4 (Kemp et al. 1993a). Each of these has areas syntenic with regions of the human genome thought to harbor suppressor genes (Kemp et al. 1993a). Additional studies are in progress to identify the crucial loci which are involved in these late stages of tumor progression.

Identification of Tumor Predisposition Loci in the Mouse

Many of the above genetic alterations were identified using hybrid mouse strains that carry appropriate genetic polymorphisms which facilitate studies of LOH. During the course of these studies it was found that hybrids between two different species of mouse, *Mus spretus* and *Mus musculus*, are highly resistant to the development of tumors in a variety of tissues (R. Bremner and A. Balmain, unpubl.). Crosses between *spretus* and NIH mice (which are highly susceptible to skin carcinogenesis) showed substantial resistance to the normal initiating and promoting doses of DMBA and TPA. Sufficient numbers of tumors could, however, be induced in such F_1 hybrids by the use of higher and more frequent TPA doses, suggesting that lack of tumorigenic response was due to dominant resistance genes in *spretus* mice which render them less susceptible to the effects of tumor promoters. This suggested that it should be possible to map and characterize the genetic loci responsible by carrying out classic backcross mapping studies. This has now been carried out for over 300 backcross animals which have been treated with DMBA and TPA and had tumors quantified. Linkage analysis using multiple microsatellite markers is in progress, but initial results look extremely promising and suggest that it will be possible to map at least some of the skin tumor susceptibility loci (H. Nagase and A. Balmain, unpubl.). Moreover, initial results lead to the conclusion that the loci which confer predisposition to the development of papillomas are not necessarily the same as those which confer predisposition to carcinoma development. If confirmed, this result will have important implications for our understanding of the genetic pathways leading to tumor development.

It has been assumed in the past that the development of papillomas goes through a sequential series of events leading from low risk to high risk and, subsequently, to carcinomas. If, however, these tumors are under separate genetic control, there must, at some stage, be a bifurcation in the pathways leading to terminally benign, low-risk papillomas and high-risk papillomas that can progress to carcinomas. An additional possibility is that the low-risk and high-risk papillomas could arise from different cell populations within the skin. This would support the notion that these two categories of tumors may be distinct, both in terms of the cell of origin and in some of the genetic pathways to progression.

The Target Cell in Carcinogenesis

The use of transgenic mice offers unique opportunities to direct expression of specific genes which have putative roles in initiation or progression of carcinogenesis to particular subpopulations of cells. We have previously demonstrated that it is possible to express a mutant *ras* gene in the differentiating cells of mouse epidermis using the promoter of the keratin 10 gene (Bailleul et al. 1990). Transgenic animals expressing mutant human *ras* in this subpopulation of cells developed extensive hyperkeratosis of the skin over much of the body surface. However, the focal induction of hyperplasia at specific sites that were exposed to mild mechanical irritation induced by wounding or scratching led to the development of benign papillomas at these sites. During the course of these experiments, however, we saw no evidence for progression of these papillomas to malignancy. Similar results have subsequently been obtained by Roop and coworkers, who have shown that expression of a viral Ha-*ras* gene from a human keratin 1 promoter gives rise to benign papillomas of very low progression risk (Greenhalgh et al. 1993). Both these types of transgenic mice therefore appear to recapitulate the early stages of skin tumor development in the mouse but give rise specifically to that category of papilloma that can be classified as of low risk for malignant progression.

In subsequent studies, we have attempted to induce benign tumors of the high-risk category by targeting the same mutant Ha-*ras* gene as used in the initial studies to a different cell population within the skin. This was achieved using a keratin 5 promoter which normally directs expression to the basal cells of the epidermis but also within the hair follicles. Our initial attempts to generate such mice were unsuccessful, presumably due to lethality of the construct. However, using a short keratin 5 promoter, we were able to generate transgenic animals. These animals developed, in some cases, lesions of the skin during the first week after birth which, upon histological examination, were shown to be due to the induction of hair follicle growth (K. Brown and A. Balmain, unpubl.). Animals that survived to reach adulthood developed a high frequency of benign keratoacanthomas or papillomas, many of which subsequently progressed to squamous and, in certain cases, spindle cell carcinomas. These particular transgenic mice are therefore capable of recapitulating the complete process of skin carcinogenesis in the absence of application of carcinogens or tumor promoters. An important conclusion from these experiments is that the target cell for expression of the *ras* oncogene, even within a given tissue, can influence the phenotype of the resulting tumor. Expression of mutant *ras* in cells that are relatively far along the lineage within the skin, i.e., within the interfollicular or even suprabasal cell compartment, can induce predominantly excessive differentiation followed by the appearance of only benign or low-risk papillomas. Expression of the same gene in an area of the hair follicle that includes the putative stem cell population (Sun et al. 1991) leads to a significantly different outcome. Cell growth is induced without the necessity for treatment with tumor-promoting agents or mechanical irritation, giving rise to benign tumors of high progression probability. These results support the suggestion that the categories of high-risk and low-risk papillomas which arise during chemical carcinogenesis may be a consequence of mutation of the Ha-*ras* gene in different cell populations within the skin.

Genetic Events Associated with Tumor Progression: The Role of *p53*

Genetic analyses of many types of human tumors have shown that mutations in the *p53* tumor suppressor gene are often found more frequently in the more malignant stages of cancers compared with their benign counterparts, suggesting that loss of function of *p53* is a relatively late event and important in malignant progression (Nigro et al. 1989; Fearon and Vogelstein 1990; von Deimling et al. 1992; Tanaka et al. 1993). An animal model in which mutations of the *p53* gene occur at a similar stage of tumor development would therefore be very useful for investigations of the causal role of *p53* in tumorigenesis, and also for the development of therapeutic approaches based on restoration of *p53* function. In our initial studies on LOH in mouse skin tumors, it was demonstrated that the *IL-3* locus on mouse chromosome 11 showed LOH in a small but significant number of carcinomas but was never observed in their benign counterparts (Bremner and Balmain 1990; Kemp et al. 1993a). As the *p53* tumor suppressor gene is located on this chromosome, sequencing studies were carried out and it was demonstrated that all tumors with LOH of chromosome 11 (~25%) had also suffered point mutations in the *p53* gene, in agreement with other results (Burns et al. 1991; Ruggeri et al. 1991). Interestingly, most of the tumors that arose after initiation with DMBA and promotion with TPA exhibited mutations characteristic of loss of function; i.e., frameshift mutations, small deletions, or point mutations which generate stop codons. In all these cases, the remaining wild-type allele was lost. On the basis of these results, it was reasoned that if mutation of *p53* played a causal role in tumor progression, tumors in which progression was positively induced by treatment with mutagenic agents might show the appropriate carcinogen-specific alterations in the *p53* gene. Previously, we and other workers had shown that initiation of carcinogenesis by agents such as DMBA induces A-T mutations at codon 61 of the Ha-*ras* gene (Bizub et al. 1986; Quintanilla et al. 1986), whereas the direct-acting methylating agents MNNG or MNU induce G → A transition mutations at the second base of codon 12 (Zarbl et al. 1985; Brown et al. 1990). Figure 2 summarizes the results of a series of experiments designed to test the possibility that similar mutations can be induced in the *p53* gene when papillomas are induced to progress to carcinomas by treatment with similar specific mutagens (P.A. Burns et al., in prep.). Five mutations were detected in tumors induced with multiple DMBA: One was a deletion and four were transversion mutations. In contrast, four mutations were detected in tumors induced by multiple MNNG treatments, and all four were G → A transitions. Other groups have also shown carcinogen-specific mutations in *p53* in mouse skin tumors induced with UV-B radiation (Kress et al. 1992) or benzo[a]pyrene (Ruggeri et al. 1993). Together, these results clearly support the proposal that *p53* can be a direct target for mutagenesis when benign tumors are subjected to mutagen exposure and further corroborate the association between etiologic agent and *p53* mutation spectrum seen in human tumors of the skin (Brash et al. 1991) and liver (Bressac et al. 1991; Hsu et al. 1991), apparently induced by UV light or aflatoxin, respectively.

A Causal Role for *p53* in Tumor Progression

A major advantage of mouse model systems is that the causal role of mutations found in primary tumors in inducing a particular stage of carcinogenesis can be tested using transgenic or knockout mice. We therefore obtained *p53*-deficient mice (Donehower et al. 1992) in order to test the hypothesis that *p53* is specifically implicated in progression to malignancy. A series of wild-type, heterozygous *p53* $^{+/-}$ and homozygous *p53* null mice were initiated with DMBA and promoted with TPA for 15 weeks (Kemp et al. 1993b). It was clearly demonstrated that a constitutive 50% reduction in *p53* gene dosage did not influence either the number or the growth rate of benign papillomas (Fig. 3). The papilloma multiplicity in the *p53* null mice was surprisingly low, for reasons which are as yet not completely clear. We have repeated this experiment, and again there was a lower yield of papillomas in the null mice, although not as dramatic (C. Kemp and A. Balmain, unpubl.). We also noted a significant decrease in body weight gain in the null mice compared to wild types, and this therefore provides one possible explanation for the lower tumor yield. An alternative explanation, which we are currently exploring, is that *p53* null

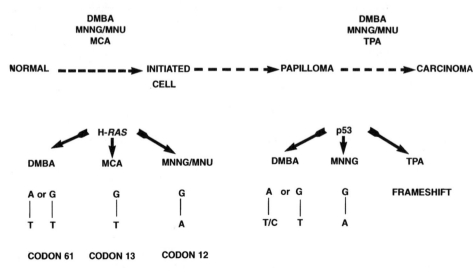

Figure 2. Carcinogen-specific mutations in *ras* and *p53* genes. Papillomas and carcinomas initiated with different carcinogens exhibit different spectra of point mutations in the Ha-*ras* (Brown et al. 1990) proto-oncogene, whereas carcinomas promoted (or progressed) with different agents show carcinogen-specific mutation in the *p53* genes (Burns et al. 1991; P.A. Burns, unpubl.).

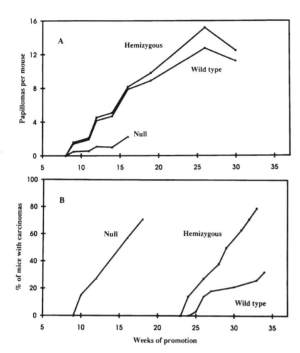

Figure 3. Rate of development of papillomas (*A*) and carcinomas (*B*) in *p53*-deficient mice. Tumors were induced with DMBA and TPA in wild-type, *p53* heterozygous, and *p53* null mice as described previously (Kemp et al. 1993b). (Reprinted, with permission, from Kemp et al. 1993b [copyright held by Cell Press].)

keratinocytes may be hypersensitive to the effects of TPA.

p53 heterozygous mice demonstrated a threefold increase in progression frequency of papillomas to carcinomas compared to wild-type mice. Moreover, loss of the remaining wild-type allele was specifically associated with the benign to malignant transition (Kemp et al. 1993b). Tumor progression in the null mice was dramatically accelerated, with some carcinomas appearing after only 10 weeks of tumor promotion, despite the relatively low papilloma yield in these mice. In addition, the carcinomas from null mice were much less well differentiated and more malignant than those from wild-type mice and showed a greater tendency to metastasize (Kemp et al. 1993b). Null mice treated with TPA alone did not develop any skin tumors, indicating that inactivation of *p53* does not contribute to initiation. These studies therefore clearly supported a functional role for wild-type *p53* specifically in the prevention of tumor progression.

Selection of Cells Harboring Mutant *p53* Alleles within Benign Tumors

Why should loss of *p53* function within benign tumor cells provide such a strong and consistent selective advantage leading to malignant progression? One possibility in line with the ascribed "guardian of the genome" role for *p53* (Lane 1992) is that the sequential accumulation of genetic changes within benign tumors,

for example, trisomies of chromosomes 7 and 6 discussed earlier, could register as "DNA damage" and consequently induce expression of wild-type *p53*. Other possibilities are that the presence of high levels of mutant Ha-*ras* expression (see Hicks et al. 1991), or suboptimal growth conditions existing within the developing tumor mass, may also induce *p53*. Since the normal function of *p53* in these cells might then be to induce growth arrest or apoptosis, one means of escape from this protection system would be for the cell to suffer somatic alterations in one or both *p53* alleles. This explanation is supported by the relatively high incidence of overexpression of *p53* in some tumor types that do not appear to have *p53* mutations, suggesting that there is constant induction of wild-type *p53* in such cells (Vojtesek and Lane 1993). Further studies are in progress to determine whether *p53* is overexpressed in such benign tumors and if the rates of apoptosis in these cells are greater than those in the corresponding malignancies, as predicted from the above model.

A consequence of loss of *p53* function in such cells might be not only an increase in genetic instability and thus rapid evolution toward the malignant phenotype, but also increased proliferation, and/or survival, in suboptimal growth conditions such as might be encountered in the foreign microenvironment of the dermis.

Radiation Oncogenesis in *p53* Null Mice

A prediction from the above discussion is that cells in which *p53* has been functionally inactivated might show enhanced rates of mutation and/or malignant transformation upon exposure to mutagenic agents. However, although documented evidence exists for no change (Slichenmyer et al. 1993), or even an increased survival of cells with abnormal *p53* function after exposure to radiation (Lee and Bernstein 1993), a clear increase in oncogenic transformation or indeed mutation frequency in vivo has not been demonstrated. Tlsty and colleagues have clearly demonstrated an increased frequency of gene amplification in *p53* null cells treated with metabolic inhibitors (Livingstone et al. 1992). However, studies carried out over the past two decades on tumor cell lines, many of which are likely to harbor *p53* mutations, have shown little or no increase in the frequency of point mutations (Barrett et al. 1990). Indeed, the above studies on carcinogenesis in the skin of *p53* null mice show that the incidence of point mutations in the Ha-*ras* gene, at least as measured by papilloma yield after treatment with TPA, is not elevated and even shows a reduction with respect to wild-type mice (Kemp et al. 1993b). Moreover, further studies carried out in this laboratory have shown that *p53* heterozygous mice treated with the liver carcinogen diethylnitrosamine, which is a strong mutagenic agent, do not exhibit an increase in the multiplicity or progression frequency of liver tumors over that seen in the corresponding wild-type animals (C.J. Kemp, in prep.). These results suggest that the effects of mutagen treatment in *p53*-deficient animals may be highly tissue-specific and/or related to the particular type of mutagenic agent chosen. We therefore decided to investigate the rate of tumor development in *p53* heterozygous and null mice after exposure to low LET ionizing radiation.

The tumorigenic response of these groups of animals to a single whole-body dose of 1 Gy or 4 Gy γ radiation given at 7–12 weeks of age is shown in Figure 4. None of the irradiated wild-type mice developed tumors. *p53* heterozygous mice spontaneously develop tumors, generally sarcomas or lymphomas, with a significantly longer latency than null mice (Harvey et al. 1993). A single exposure to 4 Gy of radiation dramatically reduced the latency for tumor development in the heterozygotes from more than 70 weeks to approximately 40 weeks of age. Molecular analysis of the resulting tumors indicated that an extremely high proportion (>95%) exhibited loss of the wild-type *p53* allele (Kemp et al. 1994). Since only 60–70% of spontaneously arising tumors exhibit loss of the wild-type allele (Harvey et al. 1993; Kemp et al. 1994; Purdie et al. 1994), this suggests that *p53* may be a direct target for radiation-induced genetic alterations.

Irradiation of *p53* null mice, however, indicated there were additional radiation-sensitive targets. Exposure of null mice to doses as low as 1 Gy given shortly after birth clearly led to a reduction in tumor latency and, in some cases, the appearance of multiple primary tumors (Fig. 4) (Kemp et al. 1994). Taken together, these results indicate an extreme sensitivity to relatively low doses of radiation in both *p53* heterozygous and null

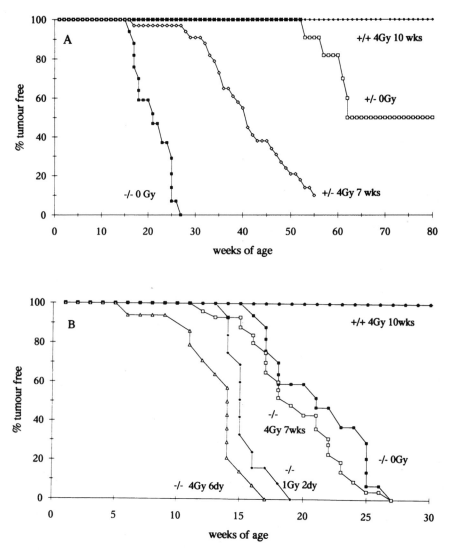

Figure 4. Radiation-induced tumorigenesis in *p53*-deficient mice. Groups of wild-type (+/−), *p53* heterozygous (+/−), and null (−/−) mice were exposed to the indicated dose of whole-body γ radiation at the indicated age (Kemp et al. 1994). The kinetics of tumor development are plotted against age. (*A*) Tumor yield in irradiated wild-type and heterozygous mice, in comparison to untreated heterozygous and null mice. (*B*) Effects of different doses and time of irradiation of null mice compared with untreated null mice. Note the difference in time scale. (Reprinted, with permission, from Kemp et al. 1994.)

mice. They further suggest that the *p53* gene itself may have been a target for radiation-induced mutational loss.

The nearly ubiquitous tumorigenic response of the heterozygotes to a single dose of radiation was somewhat surprising, given the frequency of single locus mutations expected by this dose (for discussion, see Kemp et al. 1994), and suggests an unusually high rate of transformation of cells in these mice. Previous studies from the laboratory of Little and colleagues have proposed that radiation can induce an unusually high-frequency of malignant transformation of somatic cells, which they attributed to the induction of a "global mutational response" (Kennedy et al. 1980; Yi et al. 1992). Similar proposals have recently been made by Wright and colleagues, who observed a high-frequency, apparently heritable, genetic instability phenotype, revealed by cytogenetic alterations in bone marrow stem cells, many generations after exposure to high LET radiation in vitro (Kadhim et al. 1992).

Studies on *p53*-deficient mice may therefore provide important mechanistic insights into the effects of low-level radiation exposure within the human population. Indeed, it may be expected that a proportion of human individuals carrying germ-line mutations either in the *p53* gene itself, as in Li-Fraumeni syndrome, or in other genes implicated in the *p53* response pathways could be susceptible to tumor induction by relatively low doses of radiation (Li et al. 1988). One recommendation from these studies is that careful monitoring should be carried out of the levels of radiation to which, for example, Li-Fraumeni patients are exposed in undergoing examination for the presence of possible tumors of the breast or other tissues.

CONCLUSIONS

Studies of multistage carcinogenesis in animal models such as the mouse have provided important insights into mechanisms of oncogenesis in vivo. It has been shown that tumorigenesis in the mouse, as in humans, can proceed in well-defined stages involving mutational events in both proto-oncogenes and tumor suppressor genes. Transgenic and knockout mice, moreover, demonstrated the causality of these events in initiation, promotion, or progression of tumor development. The next stage of these investigations is to use such animals constructively for the development of novel therapies based on the types of genetic events detected. The similarity in the classes of genes involved and the stages at which they are found to be mutated between mouse skin tumors and some types of human cancers, for example, colon cancer (Fearon and Vogelstein 1990), lead to a great deal of optimism that such studies can be fruitfully pursued in the near future.

ACKNOWLEDGMENTS

We especially thank the Cancer Research Campaign of Great Britain for their continued support throughout the course of these studies. C.J.K. was supported by a grant to A.B. and T. Wheldon from the UKCCCR. We also thank many colleagues for useful discussions and excellent technical assistance.

REFERENCES

Aldaz, C.M., D. Trono, F. Larcher, T.J. Slaga, and C.J. Conti. 1989. Sequential trisomization of chromosomes 6 and 7 in mouse skin premalignant lesions. *Mol. Carcinog.* **2:** 22.

Bailleul, B., M.A. Surani, S. White, S.C. Barton, K. Brown, M. Blessing, J. Jorcano, and A. Balmain. 1990. Skin hyperkeratosis and papilloma formation in transgenic mice expressing a ras oncogene from a suprabasal keratin promoter. *Cell* **62:** 697.

Barrett, J.C., T. Tsutsui, T. Tlsty, and M. Oshimura. 1990. *Genetic mechanisms in carcinogenesis and tumor progression* (ed. C.C. Harris and L.A. Liotta), p. 97. Wiley-Liss, New York.

Bianchi, A.B., C.M. Aldaz, and C.J. Conti. 1990. Non-random duplication of the chromosome bearing a mutated Ha-*ras*-1 allele in mouse skin tumours. *Proc. Natl. Acad. Sci.* **87:** 6902.

Bizub, D., A.W. Wood, and A.M. Skalka. 1986. Mutagenesis of the Ha-*ras* oncogene in mouse skin tumors induced by polycyclic aromatic hydrocarbons. *Proc. Natl. Acad. Sci.* **83:** 6048.

Brash, D.E., J.A. Rudolph, J.A. Simon, A. Lin, G.J. McKenna, H.P. Baden, A.J. Halperin, and J. Ponten. 1991. A role for sunlight in skin cancer: UV-induced p53 mutations in squamous cell carcinoma. *Proc. Natl. Acad. Sci.* **88:** 10124.

Bremner, R. and A. Balmain. 1990. Genetic changes in skin tumour progression: Correlation between presence of a mutant ras gene and loss of heterozygosity on mouse chromosome 7. *Cell* **61:** 407.

Bremner, R., C.J. Kemp, and A. Balmain. 1994. Different classes of chemical carcinogens induce different genetic changes during progression of mouse skin tumours. *Mol. Carcinog.* **11:** (in press).

Bressac, B., M. Kew, J. Wands, and M. Ozturk. 1991. Selective G to T mutations of p53 gene in hepatocellular carcinoma from southern Africa. *Nature* **350:** 429.

Brown, K., A. Buchmann, and A. Balmain. 1990. Carcinogen-induced mutations in the mouse c-Ha-*ras* gene provide evidence of multiple pathways for tumour progression. *Proc. Natl. Acad. Sci.* **87:** 538.

Burns, P.A., C.J. Kemp, J.V. Gannon, D.P. Lane, R. Bremner, and A. Balmain. 1991. Loss of heterozygosity and mutational alterations of the p53 gene in skin tumors of interspecific hybrid mice. *Oncogene* **6:** 2363.

Dietrich, W.F., J.C. Miller, R.G. Steen, M. Merchant, D. Damron, R. Nahf, A. Gross, D.C. Joyce, M. Wessel, R.D. Dredge, A. Marquis, L.D. Stein, N. Goodman, D.C. Page, and E.S. Lander. 1994. A genetic map of the mouse with 4,006 simple sequence length polymorphisms. *Nat. Genet.* **7:** 220.

Donehower, L.A., M. Harvey, B.L. Slagle, M.J. McArthur, C.A. Montgomery, Jr., J.S. Butel, and A. Bradley. 1992. Mice deficient for p53 are developmentally normal but susceptible to spontaneous tumours. *Nature* **356:** 215.

Fearon, E.R. and B. Vogelstein. 1990. A genetic model for colorectal tumorigenesis. *Cell* **61:** 759.

Fujiki, H. and M. Suganuma. 1993. Tumor promotion by inhibitors of protein phosphatases 1 and 2A: The okadaic acid class of compounds. *Adv. Cancer Res.* **61:** 143.

Furstenberger, G., B. Schurich, B. Kaina, R.T. Petrusevska, N.E. Fusenig, and F. Marks. 1989. Tumor induction in initiated mouse skin by phorbol esters and methyl methanesulfonate: Correlation between chromosomal damage and conversion ("stage I of tumor promotion") in vivo. *Carcinogenesis* **10:** 749.

Gomez, N. and P. Cohen. 1991. Dissection of the protein kinase cascade by which nerve growth factor activates MAP kinases. *Nature* **353:** 170.

Greenhalgh, D.A., J.A. Rothnagel, M.I. Quintanilla, C.C. Orengo, T.A. Gagne, D.S. Bundman, M.A. Longley, and D.R. Roop. 1993. Induction of epidermal hyperplasia, hyperkeratosis, and papillomas in transgenic mice by a targeted v-Ha-*ras* oncogene. *Mol. Carcinog.* **7:** 99.

Harvey, M., M.J. McArthur, C.A. Montgomery, Jr., J.S. Butel, A. Bradley, and L.A. Donehower. 1993. Spontaneous and carcinogen-induced tumorigenesis in p53-deficient mice (see comments). *Nat. Genet.* **5:** 225.

Hennings, H., R. Shores, P. Mitchell, E.F. Spangler, and S.H. Yuspa. 1985. Induction of papillomas with a high probability of conversion to malignancy. *Carcinogenesis* **6:** 1607.

Hicks, G.G., S.E. Egan, A.H. Greenberg, and M. Mowat. 1991. Mutant p53 tumor suppressor alleles release *ras*-induced cell cycle growth arrest. *Mol. Cell. Biol.* **11:** 1344.

Howe, L.R., S.J. Leevers, N. Gomez, S. Nakielny, P. Cohen, and C.J. Marshall. 1992. Activation of the MAP kinase pathway by the protein kinase raf. *Cell* **71:** 335.

Hsu, I.C., R.A. Metcalf, T. Sun, J.A. Welsh, N.J. Wang, and C.C. Harris. 1991. Mutational hotspot in the p53 gene in human hepatocellular carcinomas (see comments). *Nature* **350:** 427.

Kadhim, M.A., D.A. Macdonald, D.T. Goodhead, S.A. Lorimore, S.J. Marsden, and E.G. Wright. 1992. Transmission of chromosomal instability after plutonium alpha-particle irradiation. *Nature* **355:** 738.

Kemp, C.J., F. Fee, and A. Balmain. 1993a. Allelotype analysis of mouse skin tumours using polymorphic microsatellites: Sequencial genetic alterations on chromosomes 6, 7 and 11. *Cancer Res.* **53:** 6022.

Kemp, C.J., T. Wheldon, and A. Balmain. 1994. p53 deficient mice are extremely susceptible to radiation-induced tumorigenesis. *Nat. Genet.* (in press).

Kemp, C.J., L.A. Donehower, A. Bradley, and A. Balmain. 1993b. Reduction of p53 gene dosage does not increase initiation or promotion but enhances malignant progression of chemically induced skin tumors. *Cell* **74:** 813.

Kennedy, A.R., M. Fox, G. Murphy, and J.B. Little. 1980. Relationship between x-ray exposure and malignant transformation of C3H 10 T1/2 cells. *Proc. Natl. Acad. Sci.* **77:** 7262.

Knudson, A.G. 1971. Mutation and cancer: Statistical study of retinoblastoma. *Proc. Natl. Acad. Sci.* **68:** 820.

Kress, S., C. Sutter, P.T. Strickland, H. Mukhtar, J. Schweizer, and M. Schwarz. 1992. Carcinogen-specific mutational pattern in the p53 gene in ultraviolet B radiation-induced squamous cell carcinomas of mouse skin. *Cancer Res.* **52:** 6400.

Lane, D.P. 1992. p53, guardian of the genome. *Nature* **358:** 15.

Lee, J.M. and A. Bernstein. 1993. p53 mutations increase resistance to ionizing radiation. *Proc. Natl. Acad. Sci.* **90:** 5742.

Li, F.P., J.F. Fraumeni, J.J. Mulvihill, W.A. Blattner, M.G. Dreyfus, M.A. Tuacker, and R.W. Miller. 1988. A cancer family syndrome in twenty-four kindreds. *Cancer Res.* **48:** 5358.

Livingstone, L.R., A. White, J. Sprouse, E. Livanos, T. Jacks, and T.D. Tlsty. 1992. Altered cell cycle arrest and gene amplification potential accompany loss of wild-type p53. *Cell* **70:** 923.

Marshall, C.J. 1991. Tumor suppressor genes. *Cell* **64:** 313.

Murphree, A.L. and W.F. Benedict. 1984. Retinoblastoma: Clues to human oncogenesis. *Science* **223:** 1028.

Nigro, J.M., S.J. Baker, A.C. Preisinger, J.M. Jessup, R. Hostetter, K. Cleary, S.H. Bigner, N. Davidson, S. Baylin, P. Devilee, T. Glover, F.S. Collins, A. Weston, R. Modali, C.C. Harris, and B. Vogelstein. 1989. Mutations in the p53 gene occur in diverse human tumour types. *Nature* **342:** 705.

Ponder, B.A.J. 1990. Inherited predisposition to cancer. *Trends Genet.* **6:** 213.

Purdie, C.A., D.J. Harrison, A. Peter, L. Dobbie, S. White, S.E.M. Howie, D.M. Salter, C.C. Bird, A.H. Wyllie, M.L. Hooper, and A.R. Clarke. 1994. Tumour incidence, spectrum and ploidy in mice with a large deletion in the p53 gene. *Oncogene* **9:** 603.

Quintanilla, M., K. Brown, M. Ramsden, and A. Balmain. 1986. Carcinogen-specific mutation and amplification of Ha-*ras* during mouse skin carcinogenesis. *Nature* **322:** 78.

Ruggeri, B., M. DiRado, S.Y. Zhang, B. Bauer, and T. Goodrow. 1993. Benzo[a] pyrene-induced murine skin tumors exhibit frequent and characteristic G-to-T mutations in the p53 gene. *Proc. Natl. Acad. Sci.* **90:** 1013.

Ruggeri, B., J. Caamano, T. Goodrow, M. DiRado, A. Bianchi, D. Trono, C.J. Conti, and A.J.P. Klein-Szanto. 1991. Alterations of the p53 tumor suppressor gene during mouse skin tumor progression. *Cancer Res.* **51:** 6615.

Slichenmyer, W.J., W.G. Nelson, R.J. Slebos, and M.B. Kastan. 1993. Loss of p53-associated G1 checkpoint does not decrease cell survival following DNA damage. *Cancer Res.* **53:** 4164.

Sun, T., G. Cotsarelis, and R.M. Lavker. 1991. Hair follicular stem cells: The bulge-activation hypothesis. *J. Invest. Dermatol.* **96:** 77S.

Tanaka, S., Y. Toh, E. Adachi, T. Matsumata, R. Mori, and K. Sugimachi. 1993. Tumor progression in hepatocellular carcinoma may be mediated by p53 mutation. *Cancer Res.* **53:** 2884.

Vojtesek, B. and D.P. Lane. 1993. Regulation of p53 protein expression in human breast cancer cell lines. *J. Cell Sci.* **105:** 607.

von Deimling, A., R.H. Eibl, H. Ohgaki, D.N. Louis, K. von Ammon, I. Petersen, P. Kleihues, R.Y. Chung, O.D. Wiestler, and B.R. Seizinger. 1992. p53 mutations are associated with 17p allelic loss in grade II and grade III astrocytoma. *Cancer Res.* **52:** 2987.

Yi, C.H., D.W. Yandell, and J.B. Little. 1992. Evidence for coincident mutations in human lymphoblast clones selected for functional loss of a thymidine kinase gene. *Mol. Carcinog.* **5:** 270.

Yuspa, S.H. 1994. The pathogenesis of squamous cell cancer: Lessons learned from studies of skin carcinogenesis (33rd G.W.A. Clowes Memorial Award Lecture). *Cancer Res.* **54:** 1178.

Zarbl, H., S. Sukumar, A.V. Arthur, D. Martin-Zanca, and M. Barbacid. 1985. Direct mutagenesis of Ha-*ras*-1 oncogenes by *N*-nitroso-*N*-methylurea during initiation of mammary carcinogenesis in rats. *Nature* **315:** 382.

Mouse Model Systems to Study Multistep Tumorigenesis

A. Berns, N. van der Lugt, M. Alkema, M. van Lohuizen, J. Domen,
D. Acton, J. Allen, P.W. Laird, and J. Jonkers

*Division of Molecular Genetics of the Netherlands Cancer Institute, Plesmanlaan 121,
1066 CX Amsterdam, The Netherlands*

Cancer is a complex process in which a normal cell has to accumulate multiple genetic and epigenetic alterations in order to become a highly malignant metastasizing tumor cell. Over the years, efforts have concentrated on identifying the sequence of events leading to the malignant phenotype of tumor cells. Our understanding of this process is steadily growing with the identification of an increasing number of genes that are frequently modified in tumors and with the characterization of their role in specific processes of cell proliferation and transformation. During the last decade, genetically manipulated animal models, primarily transgenic mice, have become an invaluable tool to identify and characterize the in vivo function of genes involved in the different steps of tumorigenesis. Presently, transgenic mice can be designed such that they are predisposed to defined tumors, thereby allowing the investigator to focus on these various steps in the tumorigenic process: acquisition of growth factor independence; escape from the control conferred by inhibitory signals from neighboring cells; induction of angiogenesis; breaking through the confinement of the basal membrane; and acquisition of migratory capacity permitting the cell to enter and exit the circulation to form metastases while at the same time escaping or preventing destruction by the immune system. Many of these processes are difficult to study in spontaneously occurring cancers, as the earliest events are poorly accessible for analysis and it is virtually impossible to reconstruct the sequence of events and to assess the significance of each event for the transforming process at the end. Transgenic mouse systems allow a more systematic dissection of this complex series of events in tumorigenesis.

We have adopted a strategy that permits the identification of consecutive steps in lymphomagenesis in mice. We have utilized "transposon tagging" to identify new oncogenes involved in this process. The methodology is based on the observation that integration of proviruses into host genomic DNA can cause the (in)activation of genes (Berns 1988). Due to the relative randomness of the proviral integration machinery, insertions can occur in almost any site in the genome. If a selective growth or survival advantage is associated with a specific insertion, a cell carrying that insertion may grow out preferentially and the provirus will be present in all cells of the expanded cell population. If the same insertion site is found in independently induced tumors, this site of insertion is called a "common" insertion site and marks, by definition, a locus relevant for tumorigenesis or any other feature the cells were selected for. The presence of the provirus at a locus greatly facilitates the molecular cloning of genes conferring the selective advantage (Berns et al. 1989; van Lohuizen and Berns 1990). A large number of genes that can contribute to tumorigenesis have been identified by this approach (for review, see van Lohuizen and Berns 1990). By applying proviral insertional mutagenesis in genetically engineered mice, this method has even greater potential, as one can search specifically for genes that can collaborate with a resident transgene. Using transgenic mice as a starting point, we have identified a series of genetic alterations that can contribute to the initiation and progression steps in the tumorigenic process. Moreover, in mice deficient for a particular proto-oncogene (and overexpressing another), it is possible to specifically identify genes that act parallel to or downstream from the gene product that is depleted as a result of the gene disruption.

We describe here, utilizing the above-mentioned methodologies, the identification and partial characterization of genes that synergize with the *myc* oncogene in tumorigenesis: *bmi-1*, a gene that shows hallmarks of the Polycomb group of genes involved in regulation of homeotic selector genes, and *pim-1* and *pim-2*, which represent the most potent collaborators of *myc* in lymphoid transformation recorded to date. In addition, we describe approaches to identify genes, e.g., *frat-1*, involved in later stages of the tumorigenic process which are likely involved in processes such as invasion and metastasis.

MATERIALS AND METHODS

Mice and lymphoma induction. Transgenic mice and null mutant mice were generated as described previously (van Lohuizen et al. 1989, 1991b; Laird et al. 1993; van der Lugt et al. 1994). The Eμ-*bmi-1* transgene was constructed as follows: *bmi-1* sequences were derived from the *bmi-1* genomic clone 6 (van Lohuizen et al. 1991b). A *Bss*HII/*Xho*I genomic *bmi-1* fragment was cloned in a transgene cassette vector (van Lohuizen et al. 1989) linking the *bmi-1* gene to the *pim-1* promoter, the immunoglobulin heavy-chain enhancer (Eμ), and a Moloney murine leukemia virus (Mo-MLV) long terminal repeat (LTR). A purified SalI fragment was used for microinjection into FVB fertilized oocytes as described previously (van Lohuizen et al. 1989).

Breeding of $pim-1^{-/-}$ (129 OLA–BALB/c) outbred mice (Laird et al. 1993) with FVB or B/CBA mice resulted in pim-1 mutant and wild-type mice (FVB–129 OLA–BALB/c or B/CBA–129 OLA–BALB/c), which were used for the induction of lymphomas. The genotype of the mice was monitored as described previously (Laird et al. 1993). Mice of this same genetic background were used to cross in the Eμ-myc transgene (van Lohuizen et al. 1991b), resulting in pim-1 mutant and wild-type Eμ-myc transgenic mice on a mixed genetic background (FVB–B/CBA–B6–129 Ola–BALB/c). These mice were used for the induction of lymphomas: one-day-old mice were injected with 10^4–10^5 pfu of Mo-MLV clone 1A as described previously (Jaenisch et al. 1975). Mice were sacrificed when moribund and tumor tissues were frozen at $-80°C$.

Tissue culture. Colony assays were performed as described previously (Domen et al. 1993a). Briefly, 5×10^4 bone marrow cells were plated in 3.5-cm dishes in 1 ml medium containing 0.25% agar. Cells were incubated in Fishers medium with 20% horse serum (Gibco) in the presence of 20% L-cell conditioned medium (containing M-CSF) or 20% WEHI conditioned medium (containing IL-3) or in RPMI 1640 medium (Flow) with 20% fetal calf serum (Gibco), 5×10^{-5} M β-mercaptoethanol and either 10 ng/ml rIL7 (gift from Amgen Biologicals, Thousand Oaks, California), 10 ng/ml rIL7 + 35 ng/ml SF (rSCF^{164}, gift from Amgen), or 30 μg/ml LPS (Difco, Detroit, MI). Plates were incubated in a fully humidified incubator with 5% CO_2 for 8 days and colonies were counted. Every incubation was performed in triplicate.

Flow cytometric analysis. Cells (1×10^6) were incubated in 96-well plates for 30–45 minutes at 4°C in 20 μl of PBS^{++} (phosphate buffered saline with 1% BSA and 0.1% sodium azide) and saturating amounts of monoclonal antibody. Cells were washed two times with PBS^{++} and incubated with streptavidin-phycoerythrin for biotinylated antibodies or PBS^{++}. The cells were analyzed on a FACSscan (Becton Dickinson). The following antibodies were used: CD45R/B220 (6B2), CD3 (145-2C11), both from Pharmingen (San Diego, California); and sIg/goat anti mouse immunoglobulin (from Tago, Burlingame, California).

Anatomical analysis. Skeletal whole mounts of newborn mice were performed on completely eviscerated animals with the skin removed. The livers were used for DNA isolation and genotype analysis by Southern blot hybridization. Corpses were fixed in 96% ethanol for 5 days and transferred to acetone for 2 days. Staining was performed in 0.005% alizarin red S, 0.015% alcian blue 8GS in 5% acetic acid, 5% H_2O, and 90% ethanol for 3 days at 37°C. Samples were washed in H_2O and cleared for 2 days in 1% KOH followed by clearing steps in 0.8% KOH and 20% glycerol, in 0.5% KOH and 50% glycerol, and in 0.2% KOH and 80% glycerol for at least 1 week each. Cleared skeletons were stored in 100% glycerol.

In situ hybridization. 12.5 days post-coitum (dpc) embryos from $bmi-1^{+/-}$ intercrosses were used for RNA in situ hybridization according to the method of Wilkinson, with modifications as described by Kress et al. (1990). Complete series of sections were hybridized. Yolk sacs were used for DNA isolation and subsequent genotype analysis as described previously (van der Lugt et al. 1994). Hox ^{35}S-labeled antisense RNA probes were transcribed by T7 or SP6 RNA polymerase from cDNA fragments (Hoxc4, c5, and c6 probes from P. Sharpe, London, United Kingdom; Hoxa-4, b2, and b8 probes from J. Deschamps, Utrecht, The Netherlands; Hoxa-7 and c-8 from P. Gruss, Max Planck Institut, Tübingen, Germany) and partially hydrolyzed to an average length of 100 nucleotides.

RESULTS

Oncogenes Collaborating with *myc*

Transgenic mice overexpressing the c-*myc* oncogene in their B-cell compartment by placing it under the control of the immunoglobulin heavy-chain enhancer (Eμ) are predisposed to B lymphomas (Adams et al. 1985; van Lohuizen et al. 1991b). They develop monoclonal tumors after a variable latency period, indicating that tumorigenesis in these mice requires additional stochastic events. To define these additional events, lymphomagenesis was accelerated by infecting mice with the non-acutely transforming Mo-MLV. We expected that proviral tagging in transgenic Eμ-*myc* mice would uncover genes that efficiently synergize with c-*myc* in transformation. This appeared to be the case. By probing for known insertion sites and by molecular cloning of DNA flanking the proviruses in tumor chromosomal DNA, a number of insertion sites were identified that were frequently occupied by proviruses in independently induced tumors, indicating that integration in these (common) sites conferred a selective growth advantage to the tumor cells. The common insertion sites that were found in the B-cell tumors in Eμ-*myc* mice were *pim-1*, *bmi-1*, *bla-1*, and *pal-1* (Haupt et al. 1991; van Lohuizen et al. 1991b). In some of the lymphomas, more than one oncogene appeared to be activated by proviral insertion, indicating that proviral insertional mutagenesis can mediate genetic changes required for consecutive steps in tumorigenesis. This is represented by the regions along the vertical axis that overlap between the different bars in Figure 1a. Interestingly, some of these genes, such as *bmi-1* and *pal-1*, were not found to be coactivated within the same tumor clone, suggesting that they act on the same or on similar downstream targets for transformation (with only one exception, possibly as a result of oligoclonality of the tumor).

Complementary observations were made when tu-

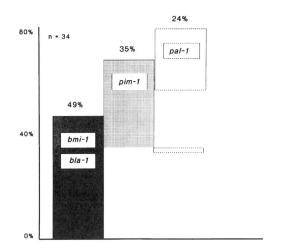

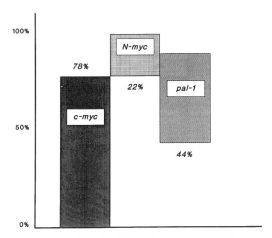

Figure 1. Identification of collaborating oncogenes in Eμ-myc and Eμ-pim-1 transgenic mice by proviral tagging. (a) Collaborating oncogenes in Eμ-myc transgenic mice. (b) Collaborating oncogenes in Eμ-pim-1 transgenic mice. Note that the regions in which the different bars overlap represent the fraction of tumors in which both proto-oncogenes have been activated by proviral insertion.

morigenesis was accelerated in Eμ-pim-1 transgenic mice that overexpressed the pim-1 transgene in both B and T cells. The spontaneous tumor incidence of Eμ-pim-1 transgenic mice was found to be very low, resulting in only 5% T-cell lymphomas at 6 months of age (van Lohuizen et al. 1989). However, the mice were highly susceptible to tumor induction by either carcinogen treatment (Breuer et al. 1989b, 1991) or Mo-MLV infection (van Lohuizen et al. 1989). When tumorigenesis was accelerated by Mo-MLV infection, insertion of proviruses was found near c-myc, N-myc, pal-1 (see Fig. 1b), and a number of other sites in the resulting tumors (see below). The absence of integrations near myc in the B lymphomas in Eμ-myc mice and near pim-1 in T lymphomas in Eμ-pim-1 transgenic mice indicates that selective advantage is no longer present when that oncogene is already overexpressed as a transgene. Loci that were identified by proviral insertional mutagenesis appear to mark strongly collaborating genes. This was illustrated by crossbreeding experiments between Eμ-myc and Eμ-pim-1 transgenic mice: Bitransgenic mice developed B lymphomas in utero (Verbeek et al. 1991). A similar strong collaboration was noted in mice bitransgenic for the myc and bmi-1 oncogenes (M. Alkema et al., in prep.) (Haupt et al. 1993).

Identification of Genes Involved in Tumor Progression

Although tumors arise very fast in bitransgenic Eμ-myc/Eμ-pim-1 mice, additional events are required for tumor progression, as is evident from the clonality of tumors arising in syngeneic mice inoculated with lymphoma cells found in 17-day-old pim-1/c-myc bitransgenic fetuses (Verbeek et al. 1991). How can we get access to the genes involved in these later events? We reasoned that prolonging the life span of Mo-MLV-induced tumors by transplanting them to syngeneic hosts might permit us to provirally "tag" these "progression genes" by molecular cloning of common proviral insertion sites that are clonal in transplanted tumors but subclonal in primary tumors (the majority of proviral insertions in the primary and transplanted tumor are shared, illustrating that the transplanted tumor is a derivative of the predominant primary tumor cell clone). In this protocol, it is unclear for what progression features we are selecting the tumor cells. They might be features relating to immortalizing functions as we select for expanded outgrowth of the cells; they might represent functions related to growth factor independence; or they might encompass alterations required for invasion and metastatic potential of the tumor cells. Obviously, this approach can also be used with other selection protocols. This has been recently illustrated by the identification of the tiam-1 gene. The tiam-1 gene was provirally tagged using an assay that permits the enrichment of lymphoma cells that have acquired invasive potential in vitro (Habets et al. 1994).

By cloning flanking DNA of proviruses that were additionally and reproducibly found in the transplanted tumors, but not in primary tumors, we identified a new locus: frat-1. We have identified the responsible gene, frat-1, which was up-regulated as a result of the proviral insertion (Fig. 2) (J. Jonkers et al., in prep.). We do not yet know what the selective advantage is that frat-1 confers to cells. The sequence of frat-1 has not provided us with any clues about its function. A series of in vivo and in vitro assays are now being carried out to gain insight into the function of this evolutionarily highly conserved gene.

The methodology described above made evident that in Eμ-pim-1 transgenic mice the activation of c-myc or N-myc, pal-1, and frat-1 can contribute to the

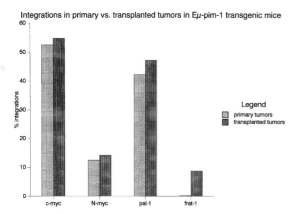

Figure 2. Selective outgrowth of primary tumor subclones with a provirally activated *frat-1* gene. The frequency of proviral insertions in the c-*myc*, N-*myc*, *pal-1*, and *frat-1* loci are indicated for primary and transplanted tumors.

tumorigenic process. It is also clear that activation of *frat-1* is preceded by activation of the other loci. This means that we have marked four different oncogenes in this process: the "initiating" transgene and the three (putative) proto-oncogenes activated by proviral insertional mutagenesis.

Functional Characterization of *pim-1* and *bmi-1*

Many questions remain with respect to the role of these genes in tumorigenesis. Biochemical analysis, as well as advanced genetic analysis, has been applied. We discuss in some detail what we have learned so far about *pim-1* and *bmi-1* using genetic approaches. We also describe a genetic search method to identify genes that act on the same downstream targets or represent the downstream targets of these known oncogenes.

Pim-1

pim-1 was originally found by proviral tagging in nontransgenic mice (Cuypers et al. 1984; Selten et al. 1986). Proviral insertions near the gene resulted in the overexpression of *pim-1* mRNA and (wild-type) protein. Both enhanced transcription and mRNA stabilization, due to the removal of AU-rich motifs from the 3'-untranslated region by proviral insertion, likely contributed to the high level of *pim-1* mRNA (Selten et al. 1985). Pim-1 is a serine/threonine protein kinase with a predominant cytoplasmic localization (Saris et al. 1991) and, when expressed from the Mo-MLV-LTR, no obvious transforming capacity in vitro in NIH-3T3 or in embryo fibroblasts (either in conjunction with *myc* or mutant *ras*; M. van Lohuizen, unpubl.). Its role in signal transduction is still unresolved. To gain insight into its biological function, we have inactivated the *pim-1* gene by homologous recombination in embryonic stem cells (te Riele et al. 1990; Laird et al. 1993) and studied the phenotype of null mutant mice. In one *pim-1*-deficient strain, most of the coding region of the gene was deleted and, therefore, lines carrying this particular disruption should represent null mutants. No obvious phenotype was found (Laird et al. 1993). Only when various hematopoietic cell types of *pim-1*$^{-/-}$ mice were examined with respect to their in vitro growth characteristics were differences observed. Mast cells from *pim-1*$^{-/-}$ mice showed reduced proliferation in response to IL-3, but grew normally in the presence of *Steel* factor (SF) (Domen et al. 1993b). Similarly, early pre-B cells showed reduced growth capacity in liquid culture and impaired colony-forming capacity in response to IL-7 + SF. Interestingly, the response to IL-7 + SF was dependent on the dose of Pim-1 protein present. The level of Pim-1 correlated with the colony-forming capacity of bone marrow cells; the colony-forming capacity of bone marrow was lowered in knock-out mice and increased in transgenics (Domen et al. 1993a). Probably, a similar dosage effect of Pim-1 can be expected to affect the T-cell lineage. This made us reinvestigate the Pim-1 dose dependence with respect to tumor formation. Therefore, we compared the tumor incidence between heterozygous and homozygous Eμ-*pim-1* transgenic mice. Indeed, a remarkable acceleration of tumorigenesis was found in mice expressing higher levels of the Pim-1 protein (see Fig. 3) (Domen et al. 1993a). When subsequently transgenic founders were generated in which the level of Pim-1 protein was further boosted by optimization of the translational initiation sequence (Kozak 1992), we were even unable to breed the mice due to the early onset of lymphomagenesis (our unpublished results). This potent oncogenic potential on the one hand, and the lack of phenotype in Pim-1-deficient mice on the other, might be indicative of parallel signaling pathways that can function independently of Pim-1.

Identification of Other Gene Products Acting in the *pim-1* Pathway

Through the recent shotgun cloning approach of mRNA in *Caenorhabditis elegans*, two relatives of Pim-1 were found, making it likely that also in vertebrates

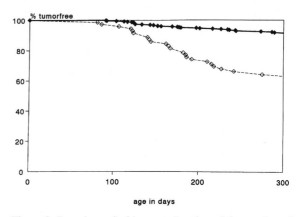

Figure 3. Lymphoma incidence as function of the number of *pim-1* transgenes. The spontaneous tumor incidence between heterozygous and homozygous Eμ-*pim-1* transgenic mice is depicted.

more than one *pim* gene exists. By taking advantage of the sequence similarity of these distant family members and by using RNA from *pim-1* null mutant mice as a starting point for degenerate primer PCR cloning, we isolated a clone with more than 50% amino acid sequence similarity to Pim-1 (J. Domen et al., in prep.). We have named this gene *pim-2* and renamed the previously identified common insertion site with that name *tic-1* (Breuer et al. 1989a). The *pim-2* gene is localized to the X chromosome (courtesy of N. Copeland and N. Jenkins). Its expression characteristics differ from those of *pim-1* in that high levels are found in brain and kidney and somewhat lower levels in thymus and spleen. When tumors induced by Mo-MLV in Eμ-*myc* mice were screened for the activation of the *pim-2* gene, similar frequencies of *pim-1* and *pim-2* activations were found. Activations of *pim-1* and *pim-2* were mutually exclusive, supporting the notion that they belong to the same complementation group in transformation.

In a parallel approach, we have followed a different strategy to clone genes that are acting on the same downstream targets as *pim-1*. We reasoned that, since *pim-1* and *myc* are such potent collaborators in tumorigenesis, proviral tagging in Eμ-*myc*/*pim-1*$^{-/-}$ mice might specifically uncover genes that act downstream from or parallel to *pim-1*. Having the *pim-2* gene cloned before this experiment was completed, we were able to test the validity of this hypothesis by screening the different Eμ-*myc*/*pim-1* genotypes for proviral activation of *pim-1* and *pim-2*. The results of this experiment are depicted in Figure 4. Whereas Eμ-*myc* transgenic mice in a wild-type *pim-1* background show a relatively low frequency of *pim-1* and *pim-2* activation, tumors in Eμ-*myc*/*pim-1*$^{-/+}$ mice showed a significant increase in the frequency of activation of both *pim-1* and *pim-2*, indicating that the reduced level of Pim-1 protein in the heterozygotes became a limiting factor. An even more pronounced effect was seen in the *pim-1*-deficient mice carrying the Eμ-*myc* transgene: Activation of *pim-2* rose to more than 80%, a dramatic increase as compared to the Eμ-*myc* transgenic mice in a wild-type *pim-1* background. This experiment validated this approach as a method to focus on genes encoding products that act on Pim-1 targets. It will be interesting to see what genes are activated or mutated in the remaining 20% of the tumors in Eμ-*myc*/*pim-1*$^{-/-}$ mice that do not carry an activated *pim-2* gene. They might act parallel to Pim-1/Pim-2 or represent downstream effectors.

Bmi-1

The *bmi-1* gene was found as a collaborator of *myc* in B lymphomagenesis through proviral tagging (Haupt et al. 1991; van Lohuizen et al. 1991b). Two integration clusters were found, one close to or in the *bmi-1* transcription unit and the other, which was originally named *bla-1*, at some distance from *bmi-1*. Proviral insertions in either cluster resulted in overexpression of *bmi-1* (B. Scheijen et al., unpubl.). As the open reading frame of the *bmi-1* transcription unit is not affected by the proviral insertions, this results in increased levels of the wild-type Bmi-1 protein. To gain insight into the role of *bmi-1*, both gain- and loss-of-function mutant mice were generated. In view of the homology of *bmi-1* with the *Drosophila Posterior sex combs* (*Psc*) Polycomb group gene (*Pc-G*) (Brunk et al. 1991; van Lohuizen et al. 1991a), we hoped to determine whether the function of these genes was conserved between mice and flies.

To inactivate the *bmi-1* gene in ES cells, two replacement-type targeting constructs were made (van der Lugt et al. 1994). The deletion of the majority of the coding region, including the AUG start codon, the Zn finger, and part of the helix-turn-helix motif, is expected to result in a genuine null allele of the gene. Heterozygous *bmi-1* mutant mice were obtained from different ES cell clones at the expected frequency and appeared healthy and fertile. Crossbreeding of heterozygous *bmi-1* mutant mice resulted in *bmi-1*$^{-/-}$ offspring. Northern blot analysis with a *bmi-1* cDNA probe (comprising exons 2–9) revealed no *bmi-1* transcripts in spleen, testis, liver, and brain of *bmi-1*$^{-/-}$ mice, and no Bmi-1 protein could be detected by Western blot analysis of nuclear extracts of brain and testis. Northern analysis did not reveal changes in the expression level of the *mel-18* and *bup* (*bmi-1* upstream; Haupt et al. 1992) genes in brain, liver, testis, and spleen, indicating that disruption of *bmi-1* does not affect the expression of these genes.

Of the offspring from heterozygote crosses observed at birth, 25% were of the *bmi-1*$^{-/-}$ genotype. This was the case in both inbred and outbred crosses of heterozygotes. However, when animals were genotyped after weaning, *bmi-1*$^{-/-}$ mice were found at a lower frequency, in both inbred and outbred crosses. This is mainly caused by death of *bmi-1*$^{-/-}$ mice shortly after birth. *bmi-1*$^{-/-}$ mice that do survive the first few days (±50%) grow more slowly than their control littermates. With increasing age, Bmi-1-deficient mice dis-

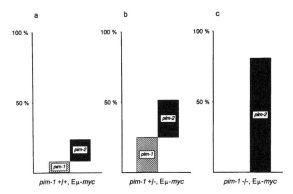

Figure 4. Selective pressure for the activation of *pim* genes in mice with a reduced endogenous level of Pim-1. Frequency of proviral insertions near *pim-1* and *pim-2* genes is indicated for the different genotypes. Note that *pim-1* and *pim-2* activations are mutually exclusive.

play waves of sickness and usually die between 3 and 20 weeks of age. Histological analysis of moribund mice often revealed pneumonia, anemia, and opportunistic infections of the intestinal tract. At 2–4 weeks, Bmi-1-deficient mice start displaying an ataxic gait. The ataxia becomes more severe with age and is accompanied by sporadic epileptic-like seizures and tremors.

Histological analysis of the cerebral cortex, the spinal cord, and the peripheral nerves of the hindlegs did not reveal obvious alterations in the $bmi\text{-}1^{-/-}$ mice. The cerebellum, however, the major motor-coordination center, showed clear abnormalities in Bmi-1-deficient mice. The cell density in the different layers of the cerebellum was significantly decreased in $bmi\text{-}1^{-/-}$ mice compared to control littermates.

Hematopoietic Defects in *bmi-1* Mutant Mice

Macroscopic examination of adult (2–3 months) $bmi\text{-}1^{-/-}$ mice showed a severely involuted thymus and a significantly smaller spleen size as compared to wild-type or heterozygous littermates. Histological analysis showed severe hypoplasia of the thymic cortex and of both the red and white pulp regions of the spleen. In bone marrow, regions of hematopoiesis were severely hypoplastic and occupied by adipocytes. This is observed both in adult and in 2-week-old $bmi\text{-}1^{-/-}$ mice. However, the erythroid lineage appeared not affected in $bmi\text{-}1^{-/-}$ mice, as normal red blood cell counts were found consistently. The absolute number of immature B cells ($B220^+/HSA^+$ and $B220^+/BP1^+$), as well as the number of mature B cells ($B220^{++}/sIg^+$), was significantly reduced in $bmi\text{-}1^{-/-}$ mice as compared to wild-type mice. Interestingly, the most immature cell population ($B220^+/S7^+$) is decreased to a lesser extent, indicating that more mature B-cell populations are affected more severely. However, it is important to stress that B cells of all maturation stages were found in *bmi-1*-deficient mice. FACS analysis of splenocytes of $bmi\text{-}1^{-/-}$ mice shows that both the B- and T-cell compartments are reduced in size, although the ratio of the different subpopulations has not changed dramatically. In contrast, when the lymphoid compartments of Eμ-*bmi-1* transgenic mice were analyzed, an expansion of an early pre-B-cell compartment was observed with the concomitant reduction in the number of more mature cells, suggesting that overexpression of *bmi-1* results in impaired or aberrant differentiation of these progenitors (M. Alkema et al., in prep.).

Bone marrow colony assays were performed to compare growth factor responses of hematopoietic cells of $bmi\text{-}1^{-/-}$ mice and control littermates. Bone marrow cells of $bmi\text{-}1^{-/-}$ mice did not respond to IL-7, which normally induces pre-B-cell colonies (Lee et al. 1989; Suda et al. 1989). The response to IL-7 + SF, which also induces more primitive cells to differentiate into pre-B cells (McNiece et al. 1991), was changed: In the absence of Bmi-1, few colonies were present, and these were composed predominantly of neutrophils. This is also seen with wild-type bone marrow in response to SF only. Furthermore, no colonies were obtained in response to LPS, indicating that functional mature B cells are virtually absent from $bmi\text{-}1^{-/-}$ bone marrow. The number of myeloid (macrophage) colonies induced by macrophage colony-stimulating factor (M-CSF) (Stanley et al. 1983), was reduced 4- to 10-fold in $bmi\text{-}1$-deficient mice as compared to wild-type, whereas the number of IL-3-responsive colonies was unaltered. An impaired response to IL-7 and M-CSF is also observed in colony assays performed with cells from fetal liver, indicating that defects in hematopoiesis in $bmi\text{-}1^{-/-}$ mice start early in development and are not a consequence of overall growth retardation of the animals. Despite the major growth defects in colony assays in vitro of cells from fetal liver, these cells appear to undergo rather normal expansion and maturation in vivo during fetal development. Only later on do severe phenotypic effects become apparent, illustrating the progressive character of the hematopoietic defect.

The impaired bone marrow hematopoiesis in $bmi\text{-}1^{-/-}$ mice leads to decreased cell numbers and substitution of large areas of hematopoiesis in bone by adipocytes. To determine whether this effect is caused by a cell-autonomous defect in $bmi\text{-}1^{-/-}$ progenitor cells or by a progressive degeneration of the microenvironment in $bmi\text{-}1^{-/-}$ mice, long-term bone marrow transplantation experiments will have to be performed.

Skeletal Abnormalities Correlate with Bmi-1 Dosage

The analysis of the skeleton of newborn $bmi\text{-}1^{-/-}$ mice revealed morphological alterations at several positions along the anterior-posterior (A-P) axis, but no obvious abnormalities of the limbs and the occipital bones of the skull were observed (see Fig. 5). The changes along the A-P axis included an extra piece of bone rostral to the atlas (C1), where in control mice no vertebra is present; the transformation of the 7th cervical vertebra (C7) to the 1st thoracic vertebra (T1), illustrated by the presence of ribs at C7, which then fuse on the ventral side with the ribs associated with T1; the transformation of T7 to T8, resulting in the presence of 6 vertebrosternal ribs instead of 7 in control mice; the transformation of T13 to the 1st lumbar vertebra (L1), as was evident from the absence or degeneration of the ribs normally associated with T13; the transformation of L6 to the 1st sacral vertebra (S1), as shown by the association of the ilial bones with L6 instead of S1; and a number of more subtle alterations. Interestingly, overexpression of *bmi-1* in transgenic mice results in the opposite phenotypic alteration: anterior transformation along the A-P axis (see Fig. 5) (M. Alkema et al., in prep.). Striking abnormalities were found in the cervical region of the transgenic mice. Newborn Eμ-*bmi-1* transgenic mice showed thickening of the neural arch of the axis. A comparable phenotype is observed in *Hoxd-3* (Condie and Capecchi 1993) and *Hoxb-4* (Ramirez-Solis et al. 1993) nullizygous mice and can be interpreted as a partial transformation of C2 into the likeness of an atlas (C1). In contrast to the

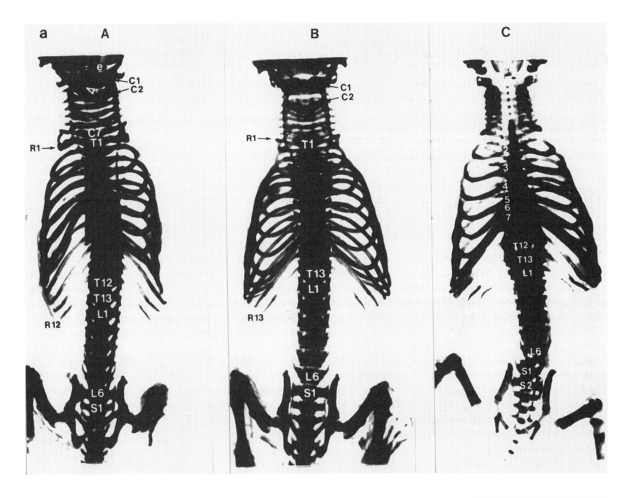

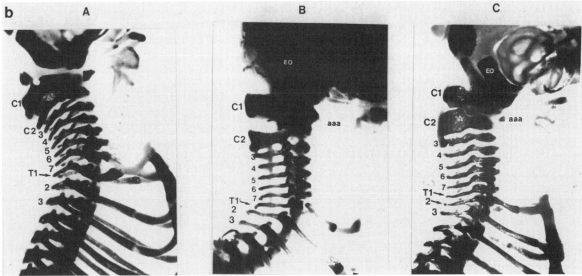

Figure 5. Posterior and anterior transformation in mice that lack or overexpress the *bmi-1* gene. (*A*) Skeleton of a *bmi-1*-deficient mouse. (*B*) Skeleton of wild-type mouse. (*C*) Skeleton of Eμ-*bmi-1* transgenic mice. Note the shifts along the complete anterior-posterior axis. (*a*) Overview of axial skeleton. (*b*) Cervical region.

Hoxb-4 mutant mice, the transformation does not involve the ventral aspect of the vertebra, since the characteristic C1 anterior arch is lacking. Whereas in wild-type mice the first thoracic vertebra (T1) has a rib that is attached to the sternum, in Eμ-*bmi-1* transgenic mice only a residual rib is present on T1 that is no longer connected to the sternum. The costal cartilage of the first rib that is connected to the sternum is now derived from T2. The third thoracic vertebra has transformed to the identity of T2, as evident from the presence of the processus spinosus on T3. However, the normal number of vertebrosternal and free ribs are present in Eμ-bmi transgenic mice. As a result, the ribs of T8 are now attached to the sternum and ribs are attached to the first lumbar vertebra (L1). The first sacral vertebra is transformed toward L6 and the ilial bones are attached to the second sacral vertebra, which is indicative of a transformation of S2 to S1. Interestingly, the anterior transformation of T8 to T7 and L1 to T13 is also observed in Hoxc-8 mutant mice (Le Mouellic et al. 1992).

Bmi-1 Acts on Homeotic Genes

As the posterior transformation in $bmi-1$-deficient mice and the anterior transformation in the Eμ-bmi-1 transgenic mice show similarities with the phenotype of several Hox gain- and loss-of-function and mutants (Kessel et al. 1990; Chisaka and Capecchi 1991; Lufkin et al. 1991, 1992; Jegalian and De Robertis 1992; Le Mouellic et al. 1992; McLain et al. 1992; Pollock et al. 1992; Ramirez-Solis et al. 1993), we wanted to determine whether Hox gene expression is affected by the absence or overexpression of Bmi-1. Therefore, the expression of several Hox genes in prevertebrae was determined by RNA in situ hybridization on sections of 11.5 or 12.5 dpc embryos that were either wild type ($bmi-1^{+/+}$) or null-mutant ($bmi-1^{-/-}$) (N. van der Lugt et al., in prep.). Analysis of four genes of the Hox-c cluster (c-4, c-5, c-6, c-8) showed that the expression boundaries of these genes had shifted one prevertebra anteriorly in the absence of $bmi-1$. These shifts are schematically summarized in Figure 6. The expression boundaries of the Hoxa-4, Hoxa-7, Hoxb-2, and Hoxb-8 genes, however, were identical in $bmi-1^{+/+}$ mice and $bmi-1^{-/-}$ mice. Although we have not yet tested the boundaries of Hox expression in Eμ-bmi-1 transgenic mice, it seems likely that in the latter mice the anterior boundary of Hox gene expression has shifted to a more posterior position.

DISCUSSION

We have utilized retroviral tagging in gain- and loss-of-function mutant mice to identify genes that collaborate in tumorigenesis. Although the mouse system is less accessible than simpler eukaryotes to resolving pathways relating to cell cycle control and cytoplasmic signaling, it is the system of choice to understand and integrate all the different parameters that are relevant for malignant transformation: Only in vertebrates can one study mechanisms underlying angiogenesis, highly specific cell-cell interactions, cell invasion, cell migration into and out of the vasculature, growth at metastatic sites, and escape of tumor cells from the immune system. Still, the question remains of how useful mouse models utilizing retrovirally induced tumors are for human cancer. It is worth stressing that although cancer genes identified in human tumors are almost invariably oncogenic in animal models, the specific mutations or translocations detected in man are often not found in tumors induced in animal models. The reverse appears to be true as well: Quite a number of genes identified as oncogenes in the experimental systems might not be a primary target for mutation in man, e.g., neither $pim-1$ nor $bmi-1$, both potent oncogenes in murine lymphoid tumors, have so far been implicated in any human malignancy. However, we can presume that the same genes, if mutated in man, would contribute to tumor-

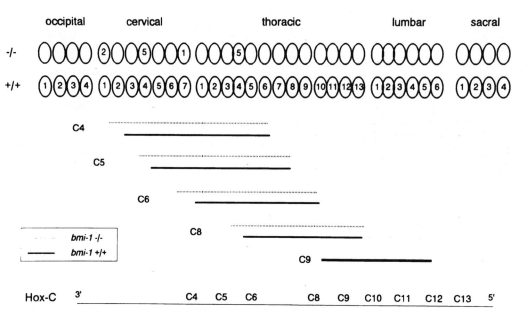

Figure 6. Schematic representation of the shifts in the anterior boundaries of expression of Hox-c cluster genes c4, c5, c6, and c8.

igenesis. In view of the conservation of signal transduction pathways, they can be expected to act in the same pathways that need to be deregulated in tumorigenesis in man. Therefore, unraveling of the mechanism of action of, e.g., *pim-1*, *bmi-1*, and *frat-1*, will not only help to elucidate these pathways and the relative importance of their constituents in mediating aberrant signaling, but will also create a basis to design strategies to selectively interfere in such a "diseased" pathway, e.g., through administration of drugs that can specifically interfere with the functioning of one of the components in the pathway. Once the upstream regulators and downstream targets of these genes have been identified, we will be closer to understanding what cellular functions are controlled by these genes. They might, therefore, serve as a target for intervention in cases where the primary defects occur upstream in the pathway.

Marking of Consecutive Events in Lymphomagenesis

Proviral tagging in Eμ-*pim-1* transgenic mice allowed us to define four consecutive events in tumorigenesis: (1) the activated transgene, which served as a starting point; (2) the activation of c-*myc* or N-*myc*, genes that are never found to be coactivated in the same tumor cell clone and therefore belong to the same complementation group; (3) the activation of *pal-1*, a locus that is identical to *Gfi-1* (Gilks et al. 1993) and *Evi-5* (N. Copeland, pers. comm.); (4) the activation of *frat-1*, a gene involved in tumor progression. We also observed activation of the *tiam-1* gene (Habets et al. 1994) at high frequency in primary tumors. Most probably, *tiam-1* belongs to a still different complementation group. Primary tumors did not contain, or were subclonal for, insertions at *frat-1*, indicating that activation of *frat-1* is a relatively late event in tumorigenesis (Fig. 2). Upon transplantation into a syngeneic host, cells carrying insertions near *frat-1* show a selective growth advantage. Since independent transplantations from a primary tumor cell population result in the outgrowth of the same tumor cell clone in independent recipients, the mutant *frat-1* cells preexisted in the primary tumor as subclones derived from the major clone constituting the primary tumor mass. Transplantation apparently provided the condition for the further selective outgrowth of these subclones. We can now explore what selective advantage the activated *frat-1* confers to the tumor cells.

When a similar Mo-MLV acceleration experiment was performed with Eμ-*myc* transgenic mice, which are predisposed to primarily pre-B-cell lymphomas, we found activation of *pim-1* or *pim-2* (see below) and *pal-1* or *bmi-1* in the pre-B-cell lymphomas that arose (Fig. 1a). It appears that genes belonging to these three complementation groups (*myc*, *pim*, and *pal*) can mediate transformation of both the T- and B-cell lineage. We have not yet determined whether *frat-1* can also contribute to progression of B lymphomas.

Oncogenes Identified by Proviral Tagging Show Strong Collaboration

To assess whether genes identified in this fashion are strong collaborators in lymphomagenesis, we have performed crossbreeding experiments between transgenic mice carrying activated oncogenes that were found by proviral tagging. Bitransgenic mice overexpressing the c-*myc* and *pim-1* oncogenes develop lymphomas in utero (Verbeek et al. 1991), the strongest collaboration between c-*myc* and another oncogene reported to date. Strong collaboration was also found between c-*myc* and *bmi-1* (M. Alkema et al., in prep.). Interestingly, Mo-MLV infection of transgenic mice carrying a particular oncogene does not always lead to tumor acceleration, suggesting that specific conditions have to be met in order to allow Mo-MLV infection to be effective. For example, Mo-MLV infection of *bcl-2* transgenic mice does not show an appreciable tumor acceleration as compared to control mice infected with Mo-MLV, although mice bitransgenic for *bcl-2* and *pim-1* show a moderate acceleration in tumorigenesis (Acton et al. 1992). Similarly, transgenic mice carrying the Rhombotin transgene show no accelerated tumorigenesis upon Mo-MLV infection (Fisch et al. 1992). One explanation for these observations is that Mo-MLV acceleration is only seen when the oncogene present as a transgene causes the expansion of a cell compartment that can serve as a target for Mo-MLV infection. However, one might expect that tumors arising in these mice after Mo-MLV infection show specific characteristics not seen in tumors induced by Mo-MLV in wild-type mice.

Function of Pim-1

The *pim-1*, as well as the recently cloned *pim-2*, oncogene (note the previously published *pim-2* locus [Breuer et al. 1989a] has been renamed *tic-1*), shows tumorigenic potential when overexpressed in hematopoietic cells. Overexpression of either *pim-1* or *pim-2*, which both carry the hallmarks of serine-threonine protein kinases (Saris et al. 1991), predisposes primarily to lymphoid malignancies (van Lohuizen et al. 1989); J. Allen et al., in prep.), although *pim-1* has been shown to be oncogenic in other lineages of the hematopoietic system as well (Dreyfus et al. 1990). Transforming activity of *pim-1* in embryo fibroblasts in conjunction with either *myc* or *ras* was not observed (M. van Lohuizen et al., unpubl.). Although the spontaneous tumor incidence in Eμ-*pim-1* transgenic mice is very low (van Lohuizen et al. 1989), the mice are highly predisposed to both carcinogen and retrovirally induced lymphomagenesis (Breuer et al. 1989b, 1991; van Lohuizen et al. 1989). Interestingly, the oncogenic potential of the *pim-1* gene is highly dose-dependent (Domen et al. 1993a), being reduced in *pim-1* knockout mice (N. van der Lugt, in prep.) and increased in mice homozygous for the Eμ-*pim-1* transgene (Fig. 3) (Domen et al. 1993a). Significantly, transgenic lines overexpressing the Pim-1 protein to very high levels by optimization of the translational initiation signals (Kozak 1992) could not be established due to the rapid tumor onset in the founder animals (our unpublished results).

Knock-out of the *pim-1* proto-oncogene showed no obvious phenotype in vivo (Laird et al. 1993). Whether this is due to redundancy in the Pim-1 pathway has yet to be assessed. The identification of a close family member of the *pim-1* proto-oncogene makes this a likely explanation. However, in vitro, where cells are only exposed to a subset of the growth stimuli that they might encounter in vivo, impaired responses were found to IL-3-mediated growth of bone-marrow-derived mast cells (Domen et al. 1993b) and to IL-7-stimulated growth of lymphocyte precursors (Domen et al. 1993a). Apparently, Pim-1 influences both the IL-3 and IL-7 growth factor pathways, seemingly by functioning as a gain-setter of these pathways. How Pim-1 performs this function is unclear.

Proviral Tagging of Genes in the *pim-1* Signal Transduction Pathway

Our experiments show that one can exploit the strong collaborative activity between gene products to search for downstream targets or genes acting in parallel pathways. Mo-MLV-induced tumor acceleration in Eμ-*myc* transgenic mice that were crossbred with *pim-1*-deficient mice resulted in B lymphomas with the genotypes Eμ-*myc*/*pim-1*$^{+/+}$, Eμ-*myc*/*pim-1*$^{+/-}$, and Eμ-*myc*/*pim-1*$^{-/-}$. We monitored the proviral activation frequency of *pim-1* and *pim-2*, a homolog of *pim-1* that was identified by degenerated primer PCR cloning using RNA from *pim-1*$^{-/-}$ mice (J. Domen et al., in prep.). Proviral activation of *pim-1* and *pim-2* occurred at a similar frequency in *pim-1*$^{+/+}$/Eμ-*myc* mice, although the frequency was lower than observed in earlier studies, probably as a result of the different genetic background. Remarkably, tumors in mice in which one allele of *pim-1* was inactivated showed a significant increase in the activation of both *pim-1* and *pim-2*, indicating that a reduction in the amount of Pim-1 increased the requirement for the activation of one of the *pim* genes. This effect was even more pronounced in *pim-1*-deficient mice: The frequency of *pim-2* activation mounted to over 80%, 6-fold higher then in *pim-1*$^{+/+}$/Eμ-*myc* mice. This indicates that wild-type levels of Pim-1 in the Eμ-*myc* background permit lymphoma development but that significantly lower levels would become prohibitive for tumor outgrowth, in support of a pivotal role for Pim-1 in lymphomagenesis. In addition, this experiment illustrates that, if *pim-2* had not been cloned by PCR, it would have been easy to clone it from the tumors in *pim-1*$^{-/-}$/Eμ-*myc* mice. Similarly, it might be expected that cloning of the proviral insertion sites in the remaining 20% of the tumors that show no activation of *pim-2* will lead to genes that again act parallel to *pim-1* and *pim-2* or, more interestingly, to genes acting downstream from these kinases.

Function of Bmi-1

The *bmi-1* gene was identified as a strong collaborator of *myc* in lymphomagenesis (Haupt et al. 1991; van Lohuizen et al. 1991b). Proviral activation of *bmi-1* leads to overexpression of the wild-type Bmi-1 protein. Bmi-1 is a nuclear phosphoprotein of 324 amino acids which harbors several motifs found in transcriptional regulators, including an unusual Zn finger motif shared by several diverse nuclear proteins (Haupt et al. 1991; van Lohuizen et al. 1991b). The endogenous *bmi-1* gene is ubiquitously expressed during development of the mouse embryo starting at least at 10.5 dpc (van der Lugt et al. 1994). In most adult tissues, low levels of the *bmi-1* mRNA are detected, whereas higher mRNA levels are found in thymus, heart, brain, and in testis. *bmi-1* is also expressed in embryonic stem cells.

A close relative is *mel-18*, a gene cloned from a melanoma cell line (Tagawa et al. 1990). The mouse Bmi-1 protein has 70% identity with the Mel-18 protein (Goebl 1991). *mel-18* mRNA is primarily expressed in brain, testes, lung, and kidney (M. Alkema, unpubl.). Bmi-1-specific motifs are conserved in the Posterior Sex Combs (Psc) protein from *Drosophila melanogaster* (Jürgens 1985; Adler et al. 1991; Brunk and Adler 1991). The conserved domains between Bmi-1 and *Psc* are located in the amino-terminal part of the much larger *Psc* protein and are also found in the *Suppressor-2 of Zeste* (*Su(z)2*) protein from *Drosophila* (Adler et al. 1989; Wu et al. 1989; Brunk et al. 1991; van Lohuizen et al. 1991a). Loss-of-function mutants of *Psc* are embryonic lethals displaying head defects and partial transformation of several segments into more posterior ones, as well as partial ventral-to-dorsal transformations (Jürgens 1985; Adler et al. 1989, 1991). *Psc* belongs to the *Polycomb* (*Pc*) group, the members of which are involved in maintaining homeotic genes in the suppressed state after their local down-regulation at a specific time during development (Lewis 1978; Paro 1990; Zink et al. 1991). These data suggest that Bmi-1 could be involved in the regulation of expression of murine homeotic genes. We have shown this to be the case.

Two major phenotypic abnormalities are found in mice either lacking or overexpressing the Bmi-1 protein: disturbances in the hematopoietic system and developmental abnormalities of the axial skeleton. The dramatic phenotypic aberrations seen in the hematopoietic system in Bmi-1-deficient mice are illustrated by the severe progressive deficiencies in the lymphoid system in vivo and the nonresponsiveness to IL-7 and strongly reduced response to M-CSF in in vitro assays of bone marrow and fetal liver cells. The lack of response to IL-7 is not due to the absence of potentially IL-7-responsive pro-B cells, as FACS analysis shows that these cells are present. Therefore, the defect already resides in early pro-B cells. This suggests a role for Bmi-1 in early B-cell development, possibly in the IL-7 signal transduction pathway, via which it supports the expansion of committed precursor B cells. This function is in line with its proposed role in lymphomagenesis. In transgenic mice overexpressing the *bmi-1* gene, this early B-cell compartment is expanded with the concomitant impairment of differentiation of more mature cells. This is accompanied by an aberrant cell-surface marker profile, such as a reduced B220 expression.

A recent study (Grabstein et al. 1993) shows the importance of IL-7 during in vivo B-cell differentiation. The treatment of mice with anti-IL-7 monoclonal antibodies completely inhibits the development of B-cell progenitors from the pro-B-cell stage onward. However, $bmi-1^{-/-}$ progenitor-B cells can fully mature to $B220^+/sIg^+$ B cells in vivo, although with a reduced efficiency. This suggests that in vivo some segment of the IL-7 pathway is still functional and that, if Bmi-1 acts in this pathway, its function can be at least partially rescued. We hypothesize that during embryonic development, parallel pathways can compensate for the absence of Bmi-1 in hematopoietic cells, but that, later on, proper hematopoietic development becomes more dependent on the pathways affected by *bmi-1*. Given the high degree of similarity between Bmi-1 and Mel-18, the latter would be a good candidate to perform such parallel function.

The posterior transformation along the axial skeleton in *bmi-1*-deficient mice and the anterior transformation in Eμ-*bmi-1* transgenic mice demonstrate that *bmi-1* plays an important role in the specification of vertebral identity along the A-P axis in a dose-dependent manner. The transformations observed in *bmi-1*-deficient and Eμ-*bmi* transgenic mice are indicative of a role for *bmi-1* in *Hox* gene regulation, as was shown for *Psc* (Adler et al. 1991; Martin and Adler 1993). However, whereas, in general, *Hox* mutant mice display their transforming phenotype at a restricted region of the vertebral column, in $bmi-1^{-/-}$ and transgenic mice, we observe transformation along the complete antero-posterior axis. In that respect, the transformations along the axial skeleton show more resemblance to the phenotypic alterations observed in mice in which retinoic acid (RA) signaling is affected either by treatment with RA (Kessel and Gruss 1991) or by disruption of the RARγ gene (Lohnes et al. 1993). Since RA treatment early in gestation can affect the expression boundaries of a number of *Hox* genes (Kessel and Gruss 1991), *bmi-1* might have a similar effect on *Hox* gene expression. We have shown this indeed to be the case: In *bmi-1*-deficient mice, a one-prevertebra anterior shift of expression was found for four *Hox-c* cluster genes, whereas the anterior expression boundary of two *Hox-a* and two *Hox-b* cluster genes was unaltered (N. van der Lugt et al., in prep.). This situation is reminiscent of *Pc-G* loss-of-function mutants in *Drosophila*, although shifts in HOM-C gene expression boundaries are more extensive in the fly (Adler et al. 1991). It is thought that in flies *Pc-G* genes exert their effect by maintaining HOM-C genes in their repressed state by changing the local chromatin structure after their expression has been down-regulated through other transcriptional regulators. A number of studies (Paro 1990; Franke et al. 1992; Rastelli et al. 1993) have led to the proposition of a model in which *Pc-G* gene products act in multimeric complexes that stably repress *Hox* gene expression by compacting the chromatin into a condensed inaccessible structure (Locke et al. 1988; Paro 1990). In agreement with this hypothesis, it was recently shown that overexpression of *Psc* results in enhanced polytene chromosome condensation (Sharp et al. 1994), whereas *Enhancer of Zeste* loss-of-function mutants give rise to a decondensed polytene chromosome appearance (Rastelli et al. 1993).

In vertebrates, mechanisms similar to those in *Drosophila* might be required to maintain a correct *Hox* gene expression throughout development. The recent characterization of M33 (Pearce et al. 1992), a mammalian *Pc*-like gene, and two mammalian genes that show homology with members of the *Drosophila trx* group (Gu et al. 1992; Tkachuk et al. 1992; Khavari et al. 1993; Muchardt and Yaniv 1993) supports this notion. In view of the homology of Bmi-1 with the Psc protein, we expect Bmi-1 to be a component of a murine *Polycomb* complex that is involved in the repression of homeotic selector genes by local compaction of chromatin of homeotic loci. In view of the potent tumorigenicity of homeotic genes in a number of systems (Perkins et al. 1990; Maulbecker and Gruss 1993), deregulation of their expression through mutation of their upstream regulators might be an effective and frequently occurring mechanism of tumor induction. The involvement of *Hox* genes and their upstream regulators in an increasing number of tumors in man supports this notion.

ACKNOWLEDGMENTS

This work was supported by the Netherlands Organization for Scientific Research (NWO) (N.v.d.L, P.W.L.), by the Netherlands Cancer Foundation (KWF) (D.A, J.J., M.A.), and by the American Leukemia Society (J.A.).

REFERENCES

Acton, D., J. Domen, H. Jacobs, M. Vlaar, S. Korsmeyer, and A. Berns. 1992. Collaboration of *pim-1* and *bcl-2* in lymphomagenesis. *Curr. Top. Microbiol. Immunol.* **182:** 293.

Adams, J.M., A.W. Harris, C.A. Pinkert, L.M. Corcoran, W.S. Alexander, S. Cory, R.D. Palmiter, and R.L. Brinster. 1985. The c-*myc* oncogene driven by immunoglobulin enhancers induces lymphoid malignancy in transgenic mice. *Nature* **318:** 533.

Adler, P.N., J. Charlton, and B. Brunk. 1989. Genetic interactions of the *Suppressor two of zeste* region genes. *Dev. Genet.* **10:** 249.

Adler, P.N., E.C. Martin, J. Charlton, and K. Jones. 1991. Phenotypic consequences and genetic interaction of a *null*

mutant in the *Drosophila Posterior Sex Combs* gene. *Dev. Genet.* **12:** 349.
Berns, A. 1988. Provirus tagging as an instrument to identify oncogenes and to establish synergism between oncogenes. *Arch. Virol.* **102:** 1.
Berns, A., M. Breuer, S. Verbeek, and M. van Lohuizen. 1989. Transgenic mice as a means to study synergism between oncogenes. *Int. J. Cancer* (suppl.) **4:** 22.
Breuer, M.L., H.T. Cuypers, and A. Berns. 1989a. Evidence for the involvement of *pim-2*, a new common proviral insertion site, in progression of lymphomas. *EMBO J.* **8:** 743.
Breuer, M., E. Wientjens, S. Verbeek, R. Slebos, and A. Berns. 1991. Carcinogen-induced lymphomagenesis in *pim-1* transgenic mice: Dose dependence and involvement of *myc* and *ras*. *Cancer Res.* **51:** 958.
Breuer, M., R. Slebos, S. Verbeek, M. van Lohuizen, E. Wientjens, and A. Berns. 1989b. Very high frequency of lymphoma induction by a chemical carcinogen in *pim-1* transgenic mice. *Nature* **340:** 61.
Brunk, B.P. and P.N. Adler. 1991. The sequence of the *Drosophila* regulatory gene *Suppressor two of zeste*. *Nucleic Acids Res.* **19:** 3149.
Brunk, B.P., E.C. Martin, and P.N. Adler. 1991. *Drosophila* genes *Posterior Sex Combs* and *Suppressor two of zeste* encode proteins with homology to the murine *bmi-1* oncogene. *Nature* **353:** 351.
Chisaka, O. and M.R. Capecchi. 1991. Regionally restricted developmental defects resulting from targeted disruption of the mouse homeobox gene *hox-1.5* [see comments]. *Nature* **350:** 473.
Condie, B.G. and M.R. Capecchi. 1993. Mice homozygous for a targeted disruption of *Hox*d-3 (*Hox-4.1*) exhibit anterior transformations of the first and second cervical vertebrae, the atlas and the axis. *Development* **119:** 579.
Cuypers, H.T., G. Selten, W. Quint, M. Zijlstra, E. Robanus-Maandag, W. Boelens, P. van Wezenbeek, and A. Berns. 1984. Murine leukemia virus-induced T-cell lymphomagenesis: Integration of proviruses in a distinct chromosomal region. *Cell* **37:** 141.
Domen, J., N.M.T. van der Lugt, D. Acton, P.W. Laird, K. Linders, and A. Berns. 1993a. Pim-1 levels determine the size of early B-lymphoid compartments in bone marrow. *J. Exp. Med.* **178:** 1665.
Domen, J., N.M. van der Lugt, P.W. Laird, C.J. Saris, A.R. Clarke, M.L. Hooper, and A. Berns. 1993b. Impaired interleukin-3 response in Pim-1-deficient bone marrow-derived mast cells. *Blood* **82:** 1445.
Dreyfus, F., B. Sola, S. Fichelson, P. Varlet, M. Charon, P. Tambourin, F. Wendling, and S. Gisselbrecht. 1990. Rearrangements of the *pim-1*, c-*myc*, and p53 genes in Friend helper virus-induced mouse erythroleukemias. *Leukemia* **4:** 590.
Fisch, P., T. Boehm, I. Lavenir, T. Larson, J. Arno, A. Forster, and T.H. Rabbitts. 1992. T-cell acute lymphoblastic lymphoma induced in transgenic mice by the RBTN1 and RBTN2 LIM-domain genes. *Oncogene* **7:** 2389.
Franke, A., M. DeCamillis, D. Zink, N. Cheng, H.W. Brock, and R. Paro. 1992. Polycomb and polyhomeotic are constituents of a multimeric protein complex in chromatin of *Drosophila melanogaster*. *EMBO J.* **11:** 2941.
Gilks, C.B., S.E. Bear, H.L. Grimes, and P.N. Tsichlis. 1993. Progression of interleukin-2 (IL-2)-dependent rat T cell lymphoma lines to IL-2-independent growth following activation of a gene (*Gfi-1*) encoding a novel zinc finger protein. *Mol. Cell Biol.* **13:** 1759.
Goebl, M.G. 1991. The *bmi-1* and *mel-18* gene products define a new family of DNA-binding proteins involved in cell proliferation and tumorigenesis. *Cell* **66:** 623.
Grabstein, K.H., T.H. Waldschmidt, F.D. Finkelman, B.W. Hess, A.R. Alpert, N.E. Boiani, A.E. Namen, and P.J. Morrissey. 1993. Inhibition of murine B and T lymphopoiesis in vivo by anti-interleukin-7 monoclonal antibody. *J. Exp. Med.* **178:** 257.

Gu, Y., T. Nakamura, H. Alder, R. Prasad, O. Canaani, G. Cimino, C.M. Croce, and E. Canaani. 1992. The t(4;11) chromosome translocation of human acute leukemias fuses the ALL-1 gene, related to *Drosophila* trithorax, to the AF-4 gene. *Cell* **71:** 701.
Habets, G.G.M., E.H.M. Scholtes, D. Zuydgeest, R.A. van der Kammen, J.C. Stam, A. Berns, and J.G. Collard. 1994. Identification of an invasion-inducing gene, *Tiam-1*, that encodes a protein with homology to GDP-DTP exchangers for rho-like proteins. *Cell* **77:** 537-549.
Haupt, Y., G. Barri, and J.M. Adams. 1992. Nucleotide sequence of *bup*, an upstream gene in the *bmi-1* proviral insertion locus. *Mol. Biol. Rep.* **17:** 17.
Haupt, Y., M.L. Bath, A.W. Harris, and J.M. Adams. 1993. *bmi-1* transgene induces lymphomas and collaborates with *myc* in tumorigenesis. *Oncogene* **8:** 3161.
Haupt, Y., W.S. Alexander, G. Barri, S.P. Klinken, and J.M. Adams. 1991. Novel zinc finger gene implicated as *myc* collaborator by retrovirally accelerated lymphomagenesis in Eμ-*myc* transgenic mice. *Cell* **65:** 753.
Jaenisch, R., H. Fan, and B. Croker. 1975. Infection of preimplantation mouse embryos and of newborn mice leukemia virus: Tissue distribution of viral DNA and RNA and leukemogenesis in the adult animal. *Proc. Natl. Acad. Sci.* **72:** 4008.
Jegalian, B.G. and E.M. De Robertis. 1992. Homeotic transformations in the mouse induced by overexpression of a human Hox3.3 transgene. *Cell* **71:** 901.
Jürgens, G. 1985. A group of genes controlling the spatial expression of the Bithorax complex in *Drosophila*. *Nature* **316:** 153.
Kessel, M. and P. Gruss. 1991. Homeotic transformations of murine vertebrae and concomitant alteration of *Hox* codes induced by retinoic acid. *Cell* **67:** 89.
Kessel, M., R. Balling, and P. Gruss. 1990. Variations of cervical vertebrae after expression of a *Hox-1.1* transgene in mice. *Cell* **61:** 301.
Khavari, P.A., C.L. Peterson, J.W. Tamkun, D.B. Mendel, and G.R. Crabtree. 1993. BRG1 contains a conserved domain of the SWI2/SNF2 family necessary for normal mitotic growth and transcription. *Nature* **366:** 170.
Kozak, M. 1992. Regulation of translation in eukaryotic systems. *Annu. Rev. Cell Biol.* **8:** 197.
Kress, C., R. Vogels, W. De Graaff, C. Bonnerot, F. Meijelink, J.F. Nicolas, and J. Deschamps. 1990. *Hox-2.3* upstream sequences mediate LacZ expression in intermediate mesoderm derivatives of transgenic mice. *Development* **109:** 775.
Laird, P.W., N.M.T. van der Lugt, A. Clarke, J. Domen, K. Linders, J. Mcwhir, A. Berns, and M. Hooper. 1993. In vivo analysis of *pim-1* deficiency. *Nucleic Acids Res.* **21:** 4750.
Le Mouellic, H., Y. Lallemand, and P. Brulet. 1992. Homeosis in the mouse induced by a null mutation in the *Hox-3.1* gene. *Cell* **69:** 251.
Lee, G., A.E. Namen, S. Gillis, L.R. Ellingsworth, and P.W. Kincade. 1989. Normal B-cell precursors responsive to recombinant murine IL-7 and inhibition of IL-7 activity by transforming growth factor-β. *J. Immunol.* **142:** 3875.
Lewis, E.B. 1978. A gene complex controlling segmentation in *Drosophila*. *Nature* **276:** 565.
Locke, J., M.A. Kotarski, and K.D. Tartof. 1988. Dosage-dependent modifiers of position effect variegation in *Drosophila* and a mass action model that explains their effect. *Genetics* **120:** 181.
Lohnes, D., P. Kastner, A. Dierich, M. Mark, M. LeMeur, and P. Chambon. 1993. Function of retinoic acid receptor gamma in the mouse. *Cell* **73:** 643.
Lufkin, T., A. Dierich, M. LeMeur, M. Mark, and P. Chambon. 1991. Disruption of the *Hox1.6* homeobox gene results in defects in a region corresponding to its rostral domain of expression. *Cell* **66:** 1105.
Lufkin, T., M. Mark, C. Hart, P. Dolle, M. LeMeur, and P. Chambon. 1992. Homeotic transformation of the occipital

bones of the skull by ectopic expression of a homeobox gene in transgenic mice. *Nature* **359:** 835.

Martin, E.C. and P.N. Adler. 1993. The Polycomb group gene *Posterior Sex Combs* encodes a chromosomal protein. *Development* **117:** 641.

Maulbecker, C.C. and P. Gruss. 1993. The oncogenic potential of deregulated homeobox genes. *Cell Growth Differ.* **4:** 431.

McLain, K., C. Schreiner, K.Y. Yager, J.L. Stock, and S.S. Potter. 1992. Ectopic expression of *Hox2.3* induces craniofacial and skeletal malformations in transgenic mice. *Mech. Dev.* **39:** 3.

McNiece, K., K.E. Langley, and K.M. Zsebo. 1991. The role of recombinant stem cell factor in early B cell development. *J. Immunol.* **146:** 3785.

Muchardt, C. and M. Yaniv. 1993. A human homologue of *Saccharomyces cerevisiae* SNF2/SWI2 and *Drosophila brm* genes potentiates transcriptional activation by the glucocorticoid receptor. *EMBO J.* **12:** 4279.

Paro, R. 1990. Imprinting a determined state into the chromatin of *Drosophila*. *Trends Genet.* **6:** 416.

Pearce, J.J., P.B. Singh, and S.J. Gaunt. 1992. The mouse has a Polycomb-like chromobox gene. *Development* **114:** 921.

Perkins, A., K. Kogsuwan, J. Visvader, J.M. Adams, and S. Cory. 1990. Homeobox gene expression plus autocrine growth factor production elicits myeloid leukemia. *Proc. Natl. Acad. Sci.* **87:** 8392.

Pollock, R.A., G. Jay, and C.J. Bieberich. 1992. Altering the boundaries of *Hox3.1* expression: Evidence for antipodal gene regulation. *Cell* **71:** 911.

Ramirez-Solis, R., H. Zheng, J. Whiting, R. Krumlauf, and A. Bradley. 1993. *Hoxb-4* (*Hox-2.6*) mutant mice show homeotic transformation of a cervical vertebra and defects in the closure of the sternal rudiments. *Cell* **73:** 279.

Rastelli, L., C.S. Chan, and V. Pirrotta. 1993. Related chromosome binding sites for *zeste*, *suppressors of zeste* and Polycomb group proteins in *Drosophila* and their dependence on *Enhancer of zeste* function. *EMBO J.* **12:** 1513.

Saris, C.J.M., J. Domen, and A. Berns. 1991. The *pim-1* oncogene encodes two related protein-serine/threonine kinases by alternative initiation at AUG and CUG. *EMBO J.* **10:** 655.

Selten, G., H.T. Cuypers, and A. Berns. 1985. Proviral activation of the putative oncogene *pim-1* in MuLV-induced T-cell lymphomas. *EMBO J.* **4:** 1793.

Selten, G., H.T. Cuypers, W. Boelens, E. Robanus-Maandag, J. Verbeek, J. Domen, C. Van Beveren, and A. Berns. 1986. The primary structure of the putative oncogene *pim-1* shows extensive homology with protein kinases. *Cell* **46:** 603.

Sharp, E.J., E.C. Martin, and P.N. Adler. 1994. Directed overexpression of *Suppressor 2 of zeste* and posterior Sex Combs results in bristle abnormalities in *Drosophila melanogaster*. *Dev. Biol.* **161:** 379.

Stanley, E.R., L.J. Guilbert, R.J. Tushinski, and S.H. Bartelmez. 1983. CSF-1 mononuclear phagocyte lineage-specific hematopoietic growth factor. *J. Cell. Biochem.* **21:** 151.

Suda, T., S. Okada, J. Suda, Y. Miura, M. Ito, T. Sudo, S.-I. Hayashy, N.-I. Nishikawa, and H. Nakauchi. 1989. A stimulatory effect of recombinant interleukin-7 on B-cell colony formation and an inhibitory effect of interleukin-1α. *Blood* **74:** 1936.

Tagawa, M., T. Sakamoto, K. Shigemoto, H. Matsubara, Y. Tamura, T. Ito, I. Nakamura, A. Okitsu, K. Imai, and M. Taniguchi. 1990. Expression of novel DNA-binding protein with zinc finger structure in various tumor cells. *J. Biol. Chem.* **265:** 20021.

te Riele, H., E. Robanus Maandag, A. Clarke, M. Hooper, and A. Berns. 1990. Consecutive inactivation of both alleles of the pim-1 proto-oncogene by homologous recombination in embryonic stem cells. *Nature* **348:** 649.

Tkachuk, D.C., S. Kohler, and M.L. Cleary. 1992. Involvement of a homolog of *Drosophila* trithorax by 11q23 chromosomal translocations in acute leukemias. *Cell* **71:** 691.

van der Lugt, N.M.T., J. Domen, K. Linders, M. van Roon, E. Robanus Maandag, H. te Riele, M. Van der Valk, J. Deschamps, M. Sofroniew, M. van Lohuizen, and A. Berns. 1994. Posterior transformation, neurological abnormalities, and severe hematopoietic defects in mice with a targeted deletion of the *bmi-1* proto-oncogene. *Genes Dev.* **8:** 757.

van Lohuizen, M. and A. Berns. 1990. Tumorigenesis by slow-transforming retroviruses—An update. *Biochim. Biophys. Acta* **1032:** 213.

van Lohuizen, M., M. Frasch, E. Wientjens, and A. Berns. 1991a. Sequence similarity between the mammalian bmi-1 proto-oncogene and the *Drosophila* regulatory genes Psc and Su(z)2. *Nature* **353:** 353.

van Lohuizen, M., S. Verbeek, B. Scheijen, E. Wientjens, H. van der Gulden, and A. Berns. 1991b. Identification of cooperating oncogenes in Eμ-*myc* transgenic mice by provirus tagging [see comments]. *Cell* **65:** 737.

van Lohuizen, M., S. Verbeek, P. Krimpenfort, J. Domen, C. Saris, T. Radaszkiewicz, and A. Berns. 1989. Predisposition to lymphomagenesis in *pim-1* transgenic mice: Cooperation with c-*myc* and N-*myc* in murine leukemia virus-induced tumors. *Cell* **56:** 673.

Verbeek, S., M. van Lohuizen, M. Van der Valk, J. Domen, G. Kraal, and A. Berns. 1991. Mice bearing the Eμ-*myc* and Eμ-*pim-1* transgenes develop pre-B-cell leukemia prenatally. *Mol. Cell Biol.* **11:** 1176.

Wu, C.T., R.S. Jones, P.F. Lasko, and W.M. Gelbart. 1989. Homeosis and the interaction of *zeste* and *white* in *Drosophila*. *Mol. Gen. Genet.* **218:** 559.

Zink, B., Y. Engström, W.J. Gehring, and R. Paro. 1991. Direct interaction of the *Polycomb* protein with *Antennapedia* regulatory sequences in polytene chromosomes of *Drosophila melanogaster*. *EMBO J.* **10:** 153.

Tumorigenic and Developmental Effects of Combined Germ-line Mutations in *Rb* and *p53*

B.O. WILLIAMS,* S.D. MORGENBESSER,† R.A. DEPINHO,† AND T. JACKS*

*Howard Hughes Medical Institute, Center for Cancer Research, Department of Biology, Massachusetts Institute of Technology, Cambridge, Massachusetts 02139;
†Department of Microbiology and Immunology, Albert Einstein College of Medicine, Bronx, New York 10461

During the past 20 years, considerable progress has been made in elucidating the genetic changes that occur in cancer. It is clear that emerging tumor cells acquire mutations in at least two classes of genes: oncogenes and tumor suppressor genes (Vogelstein and Kinzler 1993). Oncogenes are typically activated during tumorigenesis, and, therefore, they are thought to encode positive regulators of cell growth. In contrast, the commonly observed mutations in tumor suppressor genes are inactivating, leading to the suggestion that these genes encode negative regulators of cell growth or in some other way negatively affect tumorigenic progression (Knudson 1993). Due to the presence of two functional alleles of each tumor suppressor gene in most individuals, the complete loss of these functions requires two independent somatic mutational events (Knudson 1993). In addition, inheritance of a mutant allele of seven of the eight known tumor suppressor genes results in a greatly increased tumor risk (Knudson 1993), presumably because each cell in the affected individual is just one mutational event away from completely lacking the function of the given tumor suppressor gene. Although there is ample documentation of the high frequency of mutation of tumor suppressor genes in a wide variety of human tumor types (Knudson 1993), the selective advantage conferred on a tumor cell that eliminates the function of one of these genes is not well understood.

Of the different tumor suppressor genes, the retinoblastoma gene (*RB*) and *p53* have been the most intensively studied. Both genes are mutated in a wide range of sporadic human tumor types, and germ-line mutations in *RB* and *p53* lead to a greatly increased cancer risk (Levine 1992; Weinberg 1992). The *RB* gene encodes a nuclear phosphoprotein which is thought to act through protein-protein interactions as a regulator of the G_1 to S phase cell cycle transition (Weinberg 1992). Thus, elimination of *RB* function might be expected to lead to less restricted cell cycle progression. The *p53* gene, which encodes a transcriptional regulator (Levine 1992), has been linked to cell cycle arrest (Baker et al. 1990; Diller et al. 1990; Mercer et al. 1990; Michalovitz et al. 1990; Martinez et al. 1991), cellular response to DNA damage (Kastan et al. 1991, 1992), and the induction of apoptosis (Yonish-Rouach et al. 1991; Shaw et al. 1992; Clarke et al. 1993; Lowe et al. 1993a,b). The abrogation of any or all of these processes through *p53* mutation could accelerate tumorigenesis.

Several lines of evidence suggest that the coordinate elimination of *RB* and *p53* function might be selected for during the transformation of certain cell types. For example, a variety of human cancers carry mutations in both genes (Friend et al. 1986; T'Ang et al. 1988; Shew et al. 1989; Varley et al. 1989, 1991; Diller et al. 1990; Hensel et al. 1990; Prosser et al. 1990; Rygaard et al. 1990; Stratton et al. 1990; Crook et al. 1991; Davidoff et al. 1991; Hensel et al. 1991; Scheffner et al. 1991; Ruggeri et al. 1992). In addition, many DNA tumor viruses encode a protein or proteins which bind to (and presumably inactivate) pRb and p53 (Lane and Crawford 1979; Linzer and Levine 1979; DeCaprio et al. 1988; Whyte et al. 1988; Dyson et al. 1989; Werness et al. 1989; White et al. 1991; Debbas and White 1993). Both of these functions are required for efficient viral replication and cellular transformation in vitro and in vivo (Van Dyke 1994).

In an effort to study tumor suppressor gene function in the whole animal, we and other workers have constructed a series of mouse strains carrying targeted mutations in their murine homologs. Mice heterozygous for an *Rb* mutation do not develop retinoblastoma (Clarke et al. 1992; Jacks et al. 1992; Lee et al. 1992) (the childhood tumor of the eye for which the human gene was named), but they are highly susceptible to the formation of other tumor types (Jacks et al. 1992). Homozygous inactivation of *Rb* leads to mid-gestational embryonic lethality, with associated defects in erythropoiesis and neurogenesis (Clarke et al. 1992; Jacks et al. 1992; Lee et al. 1992). Like humans with Li-Fraumeni syndrome (Malkin 1993; Malkin et al. 1990), mice heterozygous for a *p53* mutation are prone to the formation of sarcomas as well as several other tumor types (Donehower et al. 1992; Harvey et al. 1993; Jacks et al. 1994; Purdie et al. 1994). The majority of *p53* homozygotes survive gestation apparently normally, but go on to develop one or more tumors within the first several weeks of postnatal life (Donehower et al. 1992; Harvey et al. 1993; Jacks et al. 1994; Purdie et al. 1994). However, we recently discovered that a small but reproducible percentage of *p53*-deficient embryos undergo abnormal neural tube closure and develop a severe neural malformation known as exencephaly (V. Tan et al., unpubl.). These animals appear to develop to term and are then cannibalized at birth.

In this paper, we describe the phenotypic effects of combined germ-line mutations in *Rb* and *p53*. These data establish that germ-line mutations in these two tumor suppressor genes can cooperate in tumorigenesis. In addition, the developmental effects of deficiency for both genes suggest models to explain the consequences of the mutation of these genes during tumor development. These experiments also have been described in other manuscripts which are currently in press (Morgenbesser et al. 1994; Williams et al. 1994a,b).

METHODS

The methods used in the generation and analysis of the mutant mice described in this paper have been described elsewhere (see Morgenbesser et al. 1994; Williams et al. 1994a,b).

RESULTS

Tumorigenic Effects of Combined Mutations in *Rb* and *p53*

Germ-line mutations in *Rb* and *p53* each predispose mice to cancer. Animals heterozygous for an *Rb* mutation develop intermediate lobe pituitary tumors within the first year and one half of age (Fig. 1) (Jacks et al. 1992). Medullary carcinoma of the thyroid is also observed in approximately three quarters of these animals (Table 1). Approximately 25% of mice heterozygous for a *p53* mutation develop tumors by 18 months of age (Fig. 1), with the most common tumor types being sarcoma and lymphoma (Table 1) (Donehower et al. 1992; Harvey et al. 1993; Jacks et al. 1994; Purdie et al. 1994). Tumor development in both of these mutant strains is usually accompanied by loss of the remaining wild-type allele of *Rb* (Jacks et al. 1992) or *p53* (Harvey et al. 1993; Jacks et al. 1994; Purdie et al. 1994). Animals homozygous for a *p53* mutation show an even more pronounced cancer susceptibility: 90% of *p53*-deficient animals develop one or more tumors by 3–6

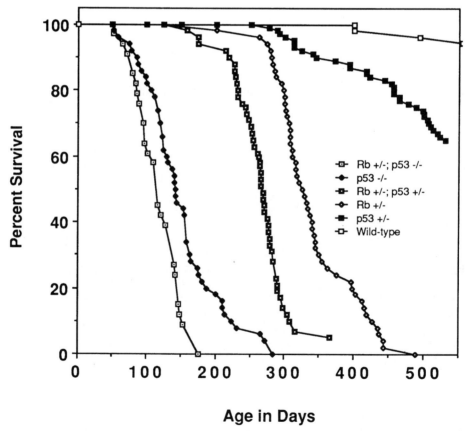

Figure 1. Life span of mice with mutations in *p53* and *Rb*. Graph summarizes the viability of 33 $Rb^{+/-}$ $p53^{-/-}$ animals; 50 $p53^{-/-}$ animals; 50 $Rb^{+/-}$ $p53^{+/-}$ animals; 100 $p53^{+/-}$ animals; and 12 wild-type animals. The mean age of survival (the point at which 50% of the animals had died) was 115 days for $Rb^{+/-}$ $p53^{-/-}$; 142 days for $p53^{-/-}$; 267 days for $Rb^{+/-}$ $p53^{+/-}$; and 325 days for $Rb^{+/-}$ mice. At 500 days, ~75% of $p53^{+/-}$ mice remained alive and ~90% of wild-type mice were still alive at 2 years of age.

Table 1. Pathological Lesions Associated with Different Combinations of *Rb* and *p53* Mutations in Mice

Genotype	Principal tumor type(s)	Approx. age of onset
$Rb^{+/-}$	pituitary and thyroid	11 months
$p53^{+/-}$	sarcoma	18 months
$p53^{-/-}$	lymphoma and sarcoma	5 months
$Rb^{+/-}\ p53^{+/-}$	pituitary and thyroid	9 months
$Rb^{+/-}\ p53^{-/-}$	lymphoma, sarcoma, pituitary, thyroid, pinealoblastoma, islet cell, bronchial epithelial dysplasia, retinal dysplasia	4 months

For each genotype, the common pathology observed is listed as well as the average age of onset (the time at which 50% of the animals remain alive). For a more complete description of the types and frequency of pathology, refer to the indicated references: Donehower et al. (1992); Harvey et al. (1993); Jacks et al. (1994); Purdie et al. (1994); and Williams et al. (1994a).

months of age (Fig. 1) (Donehower et al. 1992; Harvey et al. 1993; Jacks et al. 1994; Purdie et al. 1994). Since *Rb*-deficient embryos die at mid-gestation (Clarke et al. 1992; Jacks et al. 1992; Lee et al. 1992), it is not possible to generate *Rb*-deficient adult animals for analysis.

To investigate potential cooperative tumorigenic effects of germ-line mutations in *Rb* and *p53*, we mated animals heterozygous for the Rb^{x3t} (Jacks et al. 1992) mutation with animals homozygous for the $p53^{\Delta}$ (Jacks et al. 1994) mutation to create a series of animals heterozygous for both mutations ($Rb^{+/-}\ p53^{+/-}$). These animals were subsequently intercrossed to yield mice heterozygous for the *Rb* mutation and homozygous for the *p53* mutation ($Rb^{+/-}\ p53^{-/-}$). The health of both types of double-mutant mice was carefully monitored, and any ill animals were euthanized and subjected to complete necropsy.

As shown in Figure 1, the mean age of survival of the $Rb^{+/-}\ p53^{+/-}$ animals was slightly reduced compared to animals heterozygous for the *Rb* mutation alone (9 months versus 11 months). However, the tumor spectrum in these animals was quite similar to that observed in *Rb* heterozygous mutants. All 67 $Rb^{+/-}\ p53^{+/-}$ animals developed intermediate lobe pituitary tumors, and approximately 75% of these mice were also found to have medullary thyroid tumors (Table 1). Additional tumors were detected in six $Rb^{+/-}\ p53^{+/-}$ animals: three anaplastic sarcomas, two islet cell carcinomas, and one pinealoblastoma. The development of sarcomas was not particularly surprising, since animals heterozygous for just *p53* are quite susceptible to this tumor type, but the sarcomas observed in the double heterozygotes were more aggressive and arose at an earlier age than those occurring in the single-mutant animals. Tumors of the islet cells of the pancreas or the pineal gland have not been previously detected in animals with individual mutations in *Rb* or *p53* (Table 1). In humans, pancreatic tumors and certain types of sarcoma are known to have high frequency of mutation of both *RB* and *p53* (Friend et al. 1986; Shew et al. 1989; Diller et al. 1990; Stratton et al. 1990; Ruggeri et al. 1992). Although neither of these genes has been clearly implicated in human pinealoblastoma, it is striking that approximately 1% of familial retinoblastoma patients develop this tumor type. These cases are referred to as tri-lateral retinoblastoma (Jakobiec et al. 1977; Stannard et al. 1985; Pesin and Shields 1989), owing to the histological and physiological similarities between the retina and the pineal gland (Wurtman and Moskowitz 1977).

The state of the wild-type alleles of *Rb* and *p53* in tumors from $Rb^{+/-}\ p53^{+/-}$ animals was determined using Southern blotting. We have previously shown that the wild-type allele of *Rb* is lost during the development of nearly all pituitary and thyroid tumors in *Rb* heterozygotes (Jacks et al. 1992; Williams et al. 1994a), and this was also found in the double heterozygotes. The DNA fragment diagnostic of the wild-type *Rb* allele was absent from 16/16 pituitary tumors and 8/8 thyroid tumors isolated from these animals (data not shown). In contrast, *p53* mutation was uncommon in these tumor types, as just 1 of 16 pituitary tumors and 1 of 8 thyroid tumors showed loss of the wild-type allele of *p53* (data not shown). However, in each of the 6 distinctive tumors that arose in this population of animals (3 anaplastic sarcomas, 2 islet cell tumors, and 1 pinealoblastoma), the wild-type alleles of both *Rb* and *p53* were lost (data not shown). Thus, the elimination of function of these two tumor suppressor genes appears to accelerate tumorigenesis in certain tissues.

The cooperative tumorigenic effect of germ-line mutations in *Rb* and *p53* was more apparent from the phenotype of $Rb^{+/-}\ p53^{-/-}$ animals. These animals survived for an average of four months, and none of them lived beyond six months of age (Fig. 1). Moreover, the $Rb^{+/-}\ p53^{-/-}$ animals developed numerous distinct tumor foci, with one animal displaying 15 separate lesions affecting 8 different tissues. $Rb^{+/-}\ p53^{-/-}$ animals were susceptible to those tumor types associated with mutation of either *Rb* or *p53*, as well as the novel tumors noted in the double heterozygotes that showed inactivation of both genes (Table 1). Islet cell tumors and pinealoblastomas were observed in approximately 20% and 40% of these animals, respectively. In the one islet cell tumor tested and in 8/8 pinealoblastomas, loss of the wild-type allele of *Rb* accompanied tumorigenesis (data not shown). In addition, two anaplastic sarcomas from the $Rb^{+/-}\ p53^{-/-}$ mice had lost the wild-type Rb allele. In contrast, inactivation of Rb occurred rarely in the lymphomas which developed in these animals (1/7 tumors examined), and none of seven cases of hemangiosarcoma tested had loss of the wild-type allele of Rb (data not shown).

Beyond their overt neoplasms, $Rb^{+/-}\ p53^{-/-}$ animals also exhibited more subtle lesions that were only noticed upon histological analysis. Approximately 40% of these animals had small tufts of cells located in the bronchial epithelium of the lung. These regions of bronchial hyperplasia are histologically similar to human small cell lung cancer, in which a high frequency of mutation of RB and $p53$ has been documented (Hensel et al. 1990, 1991; Rygaard et al. 1990). In addition, nearly half of the $Rb^{+/-}\ p53^{-/-}$ mice showed retinal pathology (Table 1) consisting of areas of dysplasia (or abnormal retinal architecture) as opposed to hyperplasia. It is unclear whether these lesions would develop into retinoblastoma, since the $Rb^{+/-}\ p53^{-/-}$ animals succumb to other tumors by six months of age. However, retinal expression of the SV40 T antigen (which binds to both pRb and $p53$; [Lane and Crawford 1979; Linzer and Levine 1979; DeCaprio et al. 1988]) has been shown to cause retinoblastoma in the mouse (Windle et al. 1990; al-Ubaidi et al. 1992). Neither retinoblastoma nor retinal dysplasia has been observed in the $Rb^{+/-}$ or $p53^{-/-}$ mice (Table 1).

p53 Is Required for Apoptotic Death of Rb-deficient Lens Fiber Cells

Among the several potential explanations for the cooperative tumorigenic effects of the mutation of both Rb and $p53$ is that the loss of both genes is required to achieve unbridled proliferation without compensatory cell death. It is known that expression of the adenovirus E1A oncoprotein (which binds to and inactivates pRB) causes both increased proliferative capacity and $p53$-dependent cell death (Debbas and White 1993; Lowe et al. 1994). Abrogation of $p53$ function, either through expression of proteins that antagonize it (adenovirus E1B, dominant-negative forms of p53) or $p53$ gene deletion, eliminates the apoptotic response and allows cellular transformation (Debbas and White 1993; Lowe et al. 1994). Moreover, there is widespread apoptosis in Rb-deficient embryos (Clarke et al. 1992; Jacks et al. 1992; Lee et al. 1992), suggesting that loss of function of this tumor suppressor gene can also lead to cell death.

To address the interrelationship of inappropriate proliferation brought on by loss of Rb function and apoptotic death dependent on $p53$, we have examined the developing lens in embryos deficient for Rb alone ($Rb^{-/-}$) or deficient for both Rb and $p53$ ($Rb^{-/-}\ p53^{-/-}$). The lens was chosen as a model system because previously we had observed that chimeric mice, made in part from cells homozygous for the Rb mutation, developed cataracts (Williams et al. 1994b).

As shown in Figure 2, the lens of a newborn chimera contains considerable cellular debris and large numbers of cells with pyknotic nuclei. The appearance of these cells was indicative of apoptosis, and this result then suggested that the absence of Rb function in the cells of the developing lens led to programmed cell death. Moreover, because these effects were observed in all chimeras, it was probable that the death of the Rb-deficient cells was cell autonomous; that is, the presence of wild-type cells was not able to rescue this developmental defect.

To examine the death of lens fiber cells lacking Rb function in more detail and to determine whether this process was dependent on $p53$, we carried out an analysis of $Rb^{-/-}$ and $Rb^{-/-}\ p53^{-/-}$ embryos. Midgestation embryos of the relevant genotypes were derived from appropriate crosses of the double-mutant animals described above, and the lenses were examined for apoptotic cells using the terminal transferase-based TUNEL assay (Gavrieli et al. 1992). This protocol allows detection of cells with increased numbers of free 3'-OH DNA ends, which is an indication of apoptotic death. As shown in Figure 3, although sections of lenses from wild-type embryos at day 13.5 pc showed few if any TUNEL-positive cells, such cells were readily apparent in the Rb-deficient lenses at this stage. The apoptotic nature of these cells was further confirmed using propidium iodide staining and confocal microscopy (not shown). Therefore, even at this early stage of development, absence of Rb function results in a high level of apoptosis in the lens.

Several recent experiments performed with cultured cells have implicated $p53$ function in forms of apoptosis associated with deregulated growth (Lowe et al. 1993a, 1994). However, it is clear that $p53$-independent apoptotic pathways also exist (Clarke et al. 1993; Lowe et al. 1993b). To determine whether apoptosis resulting from the lack of Rb function in the lens was dependent on $p53$, we examined similarly staged $Rb^{-/-}\ p53^{-/-}$ embryos using the TUNEL assay. As shown in Figure 3, the percentage of TUNEL-stained cells in the $Rb^{-/-}\ p53^{-/-}$ lenses was significantly diminished compared to that observed in $Rb^{-/-}$ embryos, indicating that the increased death of Rb-deficient lens fiber cells is indeed $p53$-dependent.

We next addressed whether inappropriate proliferation of Rb-deficient lens fiber cells might trigger the observed cell death. Pregnant females were injected with BrdU on days 12.5 to 13.5 pc, and 1 hour later they were sacrificed and the embryos were processed for immunodetection of BrdU-labeled nuclei (Miller and Nowakowski 1988). In sharp contrast to the lenses of wild-type embryos in which labeled cells were confined to a narrow band comprising the anterior epithelial layer, the Rb-deficient lens contained BrdU-labeled cells throughout the epithelial layer and lens fiber cell compartments (Fig. 3). Thus, in the absence of Rb function, the tight regional compartmentalization of proliferation in the lens is lost. This result suggests that Rb might be required for cells of the lens epithelium to

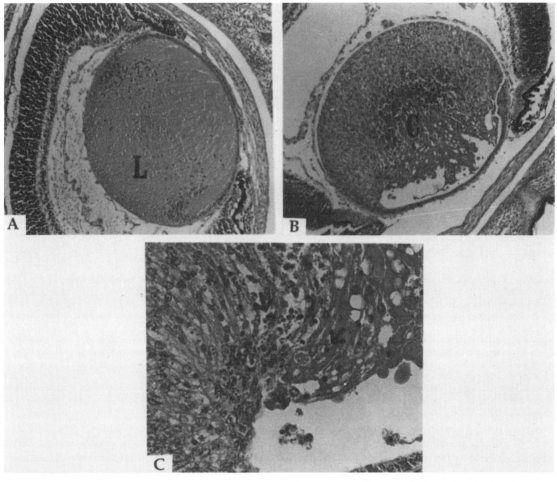

Figure 2. Lens defects in chimeras made from *Rb*-deficient ES cells. (*A*) Hematoxylin and eosin (H & E) stained section of an eye from a newborn control chimera (made from ES cells heterozygous for the *Rb* mutation). The normal structure of the lens (L) is shown. Magnification, 100×. (*B*) H & E stained eye section from a newborn chimera derived from *Rb*-deficient ES cells showing abnormal lens structure and an early cataract (C). Magnification, 100×. (*C*) Higher magnification of lens shown in *B*. Disorganized lens fiber cells are shown at the left, dead cells with pyknotic nuclei (indicative of apoptosis) are shown in the center (*arrow*), and cellular debris is shown on the right. Magnification, 400×.

withdraw from cycle prior to their initiating end-stage differentiation. Importantly, we observed the same degree of ectopic proliferation in the lenses of embryos doubly mutant for *Rb* and *p53* (Fig. 3).

We conclude from these studies that proper function of the *Rb* tumor suppressor gene is required for proper cell cycle arrest in the developing ocular lens. It is likely that loss of *Rb* function in other cells in the animal, either during development or in adult life, would also lead to inappropriate proliferation. Moreover, in the lens (and perhaps other cells as well), this abnormal state of proliferation is followed by the induction of *p53*-dependent apoptotic cell death.

DISCUSSION

In this paper, we have described some of the phenotypic effects of combined germ-line mutations in the *Rb* and *p53* tumor suppressor genes in the mouse. This study was motivated by the observation that a variety of human tumor types acquire mutations in both *Rb* and *p53* (Friend et al. 1986; T'Ang et al. 1988; Shew et al. 1989; Varley et al. 1989, 1991; Diller et al. 1990; Hensel et al. 1990, 1991; Prosser et al. 1990; Rygaard et al. 1990; Stratton et al. 1990; Crook et al. 1991; Davidoff et al. 1991; Scheffner et al. 1991; Ruggeri et al. 1992) and that certain DNA tumor viruses encode proteins that target the products of these two genes (Lane and Crawford 1979; Linzer and Levine 1979; DeCaprio et al. 1988; Whyte et al. 1988; Dyson et al. 1989; Werness et al. 1989; White et al. 1991; Debbas and White 1993). Thus, cellular transformation appears to be strongly affected by the coordinate elimination of both functions. Here, we have shown that adult animals which harbor a heterozygous mutation of *Rb* and are either heterozygous for a *p53* mutation or completely *p53*-deficient are susceptible to a wider range of tumor types than is observed for either mutation separately. That is, germ-line mutations in *Rb* and *p53* cooperate in the transformation of certain cell types in the mouse, particu-

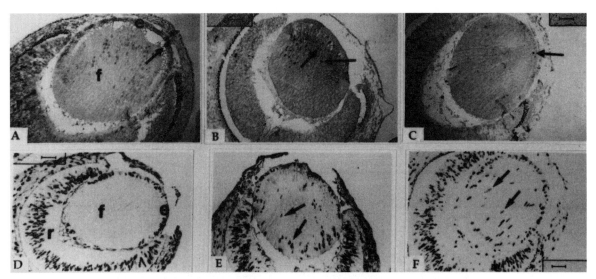

Figure 3. Apoptotic cell death and proliferation in E13.5 wild-type, $Rb^{-/-}$, and $Rb^{-/-}$ $p53^{-/-}$ lenses. Wild-type (A), $Rb^{-/-}$ (B), and $Rb^{-/-}$ $p53^{-/-}$ (C) eye sections from embryos were assayed for the presence of excessive DNA 3' OH ends (indicative of internucleosomal chromatin cleavage) using the TdT-mediated dUTP-biotin nick end-labeling (TUNEL) method. As shown in A, there was almost a complete absence of TUNEL-stained nuclei in wild-type lenses. In contrast, nuclear staining in many lens fiber cells (see arrows) of $Rb^{-/-}$ lenses confirmed the presence of apoptotic cell death (B), whereas in $Rb^{-/-}$ $p53^{-/-}$ lenses, apoptotic cell death was greatly diminished relative to $Rb^{-/-}$ lenses. To analyze lens cell proliferation, wild-type (D), $Rb^{-/-}$ (E), and $Rb^{-/-}$ $p53^{-/-}$ (F) eye sections from embryos exposed to BrdU in utero were assayed by indirect immunoperoxidase methods for the incorporation of BrdU. Note that in the case of the wild-type lens (D), the BrdU-positive nuclei are found solely in the anterior epithelial layer, whereas in both the $Rb^{-/-}$ (E) and $Rb^{-/-}$ $p53^{-/-}$ (F) lenses, a large number of BrdU-positive nuclei are found throughout the epithelial and lens fiber cell compartments (see arrows). All panels are oriented with the anterior surface of the lens facing the upper right corner. Abbreviations used are (e) anterior lens epithelial layer, (f) lens fiber cells, and (r) retina. Magnification, 56.5 ×; scale bar, 80μ.

larly cells of the pineal and pancreas and possibly also the lung and retina. Furthermore, Southern blot analysis confirms that the novel tumors arise from the functional elimination of both genes.

There are several possible explanations to account for the observed cooperative tumorigenic effects of Rb and $p53$ mutations. For example, the two genes could separately regulate a common growth arrest pathway in certain cell types, such that escape from negative growth regulation would require mutation of both genes. Alternatively, given that $p53$ has been linked to the normal cellular response to DNA damage (Kastan et al. 1991, 1992), it is possible that the absence of $p53$ function in animals that are heterozygous for the Rb mutation greatly accelerates the loss of the remaining wild-type allele of Rb. In this case, the tumor suppressive function of $p53$ can be viewed as acting indirectly through an overall effect on the cellular mutation rate (Lane 1992). Finally, the accelerated tumorigenesis observed in the double mutant animals may reflect a more direct interaction between these two mutations. In this model, for which we provide supportive evidence here, the loss of $p53$ function is required to prevent the apoptotic death of cells which inactivate Rb. According to this view, an important role of $p53$ in tumor suppression is the elimination of cells that have acquired growth-promoting mutations and have thereby become a danger to the organism.

There are two important components of this final model which we have addressed in our studies of the developing lens in single and double mutant animals. First, the model predicts that for some cell types, the loss of Rb function results in the failure to withdraw from the cell cycle and subsequent inappropriate proliferation. The BrdU-incorporation data presented here indicate that cells of the developing lens do require Rb function for proper cell cycle withdrawal. The second critical aspect of this model is that this type of inappropriate proliferation can trigger $p53$-dependent-cell death. We have shown that the Rb-deficient lenses have increased numbers of apoptotic cells and, furthermore, that this effect is greatly suppressed in the absence of $p53$ function. Thus, at least with respect to the cells of this tissue at this time in development, there is a clear link between proliferation, death, escape from death, and the function of these two tumor suppressor genes.

Figure 4 represents this model in terms of conflicting growth signals and cell cycle regulation. We suggest that cells which receive a signal to arrest but are unable to properly respond due to the presence of a growth-promoting mutation invoke a fail-safe program of apoptosis. In this schematic, we have included as growth-promoting mutations loss of Rb function, expression of adenovirus E1A (Debbas and White 1993; Lowe et al. 1994), and overexpression of E2F1 (Wu and Levine 1994) (all of which have been shown to cause $p53$-dependent apoptosis) and overexpression of c-myc (Evan et al. 1992) (which as yet has not). The signals for growth arrest can be the product of specific differentia-

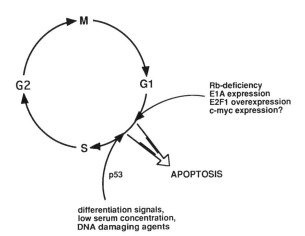

Figure 4. "Conflicting growth signals" model of apoptosis. According to this model, apoptosis can result when a cell carrying a growth-promoting mutation encounters a negative growth signal. Examples of growth-promoting mutations which have been shown to induce apoptosis are loss of *Rb* function, adenovirus E1A expression, and unregulated expression of E2F1 or c-*myc*. Negative growth signals can be produced by differentiation factors, low concentration of growth factors, or DNA damage. *p53*, which has been linked to apoptosis in response to all of these conditions except for c-*myc* expression, may act directly in the negative growth signal pathway or otherwise monitor this signal. Although the model suggests that the conflicting signals affect cells in late G_1, the resulting apoptosis could occur at any cell cycle stage.

tion factors, the absence of sufficient concentrations of growth factors, or certain forms of DNA damage. Here, we have indicated that this conflict and the resulting apoptosis occur near the G_1 to S-phase cell cycle transition. Similar conflicting signals could impinge on the cell at other phases, and the execution of the apoptotic program need not be limited to G_1-S.

The model of conflicting growth signals and *p53*-dependent apoptosis has potentially important implications for our understanding of tumorigenesis. In the course of tumor development, cells acquire growth-promoting mutations in the form of the activation of oncogenes and the loss of function of some tumor suppressor genes (Vogelstein and Kinzler 1993). These mutations are thought to increase the proliferative capacity of the cell, and, provided favorable conditions exist, produce an increase in the size of the emerging tumor clone. However, according to this model, when these mutant cells encounter negative growth signals, they will fail to respond appropriately (i.e., arrest) and instead die by apoptosis. *p53* function is required for efficient apoptosis to occur (either because the negative signaling pathway involves *p53* directly or because it is in some way monitored by *p53*), and in this way, *p53* acts as a true tumor suppressor. Moreover, the mutation of *p53* would be strongly selected for during tumor development in order to eliminate this apoptotic program. This model is consistent with the observation that *p53* mutations tend to occur at a late stage in human tumorigenesis (Fearon and Vogelstein 1990; Ruggeri et al. 1991; Harris and Hollstein 1993). Recent experiments with transgenic animals expressing different viral oncoproteins in a wild-type or *p53*-deficient background also strongly support this view (Howes et al. 1994; Symonds et al. 1994).

There are several conditions that might serve to induce apoptosis in a precancerous cell that would otherwise cause growth arrest in a normal cell; for example (as suggested by the present study), signals to withdraw from the cell cycle prior to initiating endstage differentiation, conditions of growth factor limitation, or oxygen stress. Moreover, radiation as well as various drugs used in standard chemotherapy have been shown to have this effect (Lowe et al. 1993a). This latter result might help explain why these anticancer treatments can show selectivity for tumor cells and, furthermore, why *p53* mutation has been linked to poor patient prognosis (Baker et al. 1989; Sidransky et al. 1992; Harris and Hollstein 1993; Bardeesy et al. 1994).

Many questions remain to be addressed concerning the interrelationship of growth-promoting mutations, *p53*, and cell death during tumor development. For example, the signal that actually triggers the death program, which we envision to be the product of a growth arrest signal in an unresponsive cell, has yet to be determined. The mechanism of action of *p53* in this process, whether it be through transcriptional induction (El-Deiry et al. 1994) or some other function (Caelles et al. 1994), is also not well understood. In addition, it is clear that *p53*-independent apoptotic pathways exist (Clarke et al. 1993; Lowe et al. 1993b), and cell death in tumorigenesis or in response to anticancer agents may not always involve *p53*. Still, given the high frequency of mutation of this gene in human cancer, the elucidation of these and related questions has obvious importance to the development and treatment of cancer.

Note Added in Proof

Following submission of this manuscript, Hermeking and Eick (1994) have published that c-*myc*-induced death of mouse fibroblasts is dependent on *p53*. Thus, this oncogenic signal, like E1A and E2F1 overexpression as well as *Rb* deficiency, induces *p53*-dependent apoptosis (Fig. 4).

ACKNOWLEDGMENTS

The authors thank Scott Lowe and Kay Macleod for helpful discussion, and Elizabeth Farrell for assistance with the preparation of the manuscript. B.O.W. and S.D.M. were supported by National Institutes of Health predoctoral training grants. This work was supported by grants from the National Eye Institute and the National Institute for Child Health and Human Development and National Cancer Institute cancer core grant (R.A.D.), as well as the Searle Scholars Program/ The Chicago Community Trust and the National Cancer Institute (T.J.).

REFERENCES

al-Ubaidi, M.R., R.L. Font, A.O. Quiambao, M.J. Keener, G.I. Liou, P.A. Overbeek, and W. Baehr. 1992. Bilateral retinal and brain tumors in transgenic mice expressing simian virus 40 large T antigen under control of the human interphotoreceptor retinoid-binding protein promoter. *J. Cell Biol* **119:** 1681.

Baker, S.J., S. Markowitz, E.R. Fearon, J.K.V. Willson, and B. Vogelstein. 1990. Suppression of human colorectal carcinoma cell growth by wild-type *p53. Science* **249:** 912.

Baker, S.J., E.R. Fearon, J.M. Nigro, S.R. Hamilton, A.C. Preisinger, J.M. Jessup, P. van Tuinen, D.H. Ledbetter, D.F. Barker, and Y. Nakamura. 1989. Chromosome 17 deletions and *p53* gene mutations in colorectal carcinoma. *Science* **244:** 217.

Bardeesy, N., D. Falkoff, M.J. Petruzzi, N. Nowak, B. Zabel, M. Adam, M.C. Aguiar, P. Grundy, T. Shows, and J. Pelletier. 1994. Anaplastic Wilms' tumour, a subtype displaying poor prognosis, harbors *p53* gene mutations. *Nat. Genet.* **7:** 91.

Caelles, C., A. Helmberg, and M. Karin. 1994. *p53*-dependent apoptosis in the absence of transcriptional activation of *p53*-target genes. *Nature* **370:** 220.

Clarke, A.R., C.A. Purdie, D.J. Harrison, R.G. Morris, C.C. Bird, M.L. Hooper, and A.H. Wyllie. 1993. Thymocyte apoptosis induced by *p53*-dependent and independent pathways. *Nature* **362:** 849.

Clarke, A.R., E.R. Maandag, M. van Roon, N.M.T. van der Lugt, M. van der Valk, M.L. Hooper, A. Berns, and H. te Riele. 1992. Requirement for a functional *Rb-1* gene in murine development. *Nature* **359:** 328.

Crook, T., C. Fisher, and K.H. Vousden. 1991. *p53* point mutation in HPV negative human cervical carcinoma cell lines. *Oncogene* **6:** 873.

Davidoff, A.M., P.A. Humphrey, J.K. Iglehart, and J.R. Marks. 1991. Genetic basis for *p53* overexpression in human breast cancer. *Proc. Natl. Acad. Sci.* **88:** 5006.

Debbas, M. and E. White. 1993. Wild-type *p53* mediates apoptosis by E1A, which is inhibited by E1B. *Genes Dev.* **7:** 546.

DeCaprio, J.A., J.W. Ludlow, J. Figge, J.-Y. Shew, C.-M. Huang, W.H. Lee, E. Marsilio, E. Paucha, and D.M. Livingston. 1988. SV40 large tumor antigen forms a specific complex with the product of the retinoblastoma susceptibility gene. *Cell* **54:** 275.

Diller, L., J. Kassel, C.E. Nelson, M.A. Gryka, G. Litwak, M. Gebhardt, B. Bressac, M. Ozturk, S.J. Baker, B. Vogelstein, and S.H. Friend. 1990. *p53* functions as a cell cycle control protein in osteosarcomas. *Mol. Cell. Biol.* **10:** 5772.

Donehower, L.A., M. Harvey, B.L. Slagle, M.J. McArthur, C.A. Montgomery, J.S. Butel, and A. Bradley. 1992. Mice deficient for *p53* are developmentally normal but susceptible to spontaneous tumours. *Nature* **356:** 215.

Dyson, N., P.M. Howley, K. Munger, and E. Harlow. 1989. The human papilloma virus-16 E7 oncoprotein is able to bind to the retinoblastoma gene product. *Science* **243:** 934.

El-Deiry, W.S., J.W. Harper, P.M. O'Connor, V.E. Velculescu, C.E. Canman, J. Jackman, J.A. Pietenpol, M. Burrell, D.E. Hill, Y. Wang, K.G. Wiman, W.E. Mercer, M.B. Kastan, K.W. Kohn, S.J. Elledge, K.W. Kinzler, and B. Vogelstein. 1994. WAF1/CIP1 is induced in *p53*-mediated G1 arrest and apoptosis. *Cancer Res.* **54:** 1169.

Evan, G.I., A.H. Wyllie, C.S. Gilbert, T.D. Littlewood, H. Land, M. Brooks, C. Waters, L.Z. Penn, and D.C. Hancock. 1992. Induction of apoptosis in fibroblasts by c-myc protein. *Cell* **69:** 1083.

Fearon, E.R. and B. Vogelstein. 1990. A genetic model for colorectal tumorigenesis. *Cell* **61:** 759.

Friend, S.J., R. Bernards, S. Rogelj, R.A. Weinberg, J.M. Rapaport, D. M. Albert, and T.P. Dryja. 1986. A human DNA segment with properties of the gene that predisposes to retinoblastoma and osteosarcoma. *Nature* **323:** 643.

Gavrieli, Y., Y. Sherman, and S.A. BenSasson. 1992. Identification of programmed cell death in situ via specific labelling of nuclear DNA fragmentation. *J. Cell Biol.* **121:** 899.

Harris, C.C. and M. Hollstein. 1993. Clinical implications of the *p53* tumor-suppressor gene. *N. Engl. J. Med.* **329:** 1318.

Harvey, M., M.J. McArthur, C.A. Montgomery, J.S. Butel, A. Bradley, and L.A. Donehower. 1993. Spontaneous and carcinogen-induced tumors in *p53*-deficient mice. *Nat. Genet.* **5:** 225.

Hermeking, H. and D. Eick. 1994. Mediation of c-*myc*-induced apoptosis by *p53. Science* **265:** 2091.

Hensel, C.H., R.H. Xiang, A.Y. Sakaguchi, and S.L. Naylor. 1991. Use of the single strand conformation polymorphism technique and PCR to detect *p53* gene mutations in small cell lung cancer. *Oncogene* **6:** 1067.

Hensel, C.H., C.L. Hsieh, A.F. Gazdar, B.E. Johnson, A.Y. Sakaguchi, S.L. Naylor, W.-H. Lee, and E.Y.-H.P. Lee. 1990. Altered structure and expression of the human retinoblastoma susceptibility gene in small cell lung cancer. *Cancer Res.* **50:** 3067.

Howes, K.A., N. Ransom, D.S. Papermaster, J.G.H. Lasudry, D.M. Albert, and J.J. Windle. 1994. Apoptosis or retinoblastoma: Alternative fates of photoreceptors expressing the HPV-16 E7 gene in the presence or absence of *p53. Genes Dev.* **8:** 1300.

Jacks, T., A. Fazeli, E. Schmidt, R. Bronson, M. Goodell, and R. Weinberg. 1992. Effects of an *Rb* mutation in the mouse. *Nature* **359:** 295.

Jacks, T., L. Remington, B.O. Williams, E.M. Schmitt, S. Halachmi, R.T. Bronson, and R.A. Weinberg. 1994. Tumor spectrum analysis in *p53*-mutant mice. *Curr. Biol.* **4:** 1.

Jakobiec, F.A., M.O.M. Tso, L.E. Zimmerman, and P. Danis. 1977. Retinoblastoma and intracranial malignancy. *Cancer* **39:** 2048.

Kastan, M.B., O. Onyekwere, D. Sidransky, B. Vogelstein, and R.W. Craig. 1991. Participation of *p53* protein in the cellular response to DNA damage. *Cancer Res.* **51:** 6304.

Kastan, M., Q. Zhan, W.S. El-Deiry, F. Carrier, T. Jacks, W.V. Walsh, B.S. Plunkett, B. Vogelstein, and A.J. Fornace, Jr. 1992. A mammalian cell cycle checkpoint pathway utilizing *p53* and GADD45 is defective in ataxia-telangiectasia. *Cell* **71:** 587.

Knudson, A.G. 1993. Antioncogenes and human cancer. *Proc. Natl. Acad. Sci.* **90:** 10914.

Lane, D.P. 1992. *p53*, guardian of the genome. *Nature* **358:** 15.

Lane, D.P. and L.V. Crawford. 1979. T antigen is bound to a host protein in SV40-transformed cells. *Nature* **278:** 261.

Lee, E.Y.-H.P., C.-Y. Chang, N. Hu, Y.-C.J. Wang, C.-C. Lai, K. Herrup, and W.-H. Lee. 1992. Mice deficient for *Rb* are nonviable and show defects in neurogenesis and haematopoiesis. *Nature* **359:** 288.

Levine, A.J. 1992. The *p53* tumour suppressor gene and product. *Cancer Surv.* **12:** 59.

Linzer, D.I.J. and A.J. Levine. 1979. Characterization of a 54K dalton cellular SV40 tumor antigen in SV40 transformed cells. *Cell* **17:** 43.

Lowe, S.W., T. Jacks, D.E. Housman, and H.E. Ruley. 1994. Abrogation of oncogene-associated apoptosis allows transformation of *p53*-deficient cells. *Proc. Natl. Acad. Sci.* **91:** 2026.

Lowe, S.M., H.E. Ruley, T. Jacks, and D.E. Housman. 1993a. *p53*-dependent apoptosis modulates the cytoxicity of anticancer agents. *Cell* **74:** 957.

Lowe, S.W., E.S. Schmitt, S.W. Smith, B.A. Osborne, and T. Jacks. 1993b. *p53* is required for radiation-induced apoptosis in mouse thymocytes. *Nature* **362:** 847.

Malkin, D. 1993. *p53* and the Li-Fraumeni syndrome. *Cancer Genet. Cytogenet.* **66:** 83.

Malkin, D., F.P. Li, L.C. Strong, J.F.J. Fraumeni, C.E. Nelson, and D.H. Kim. 1990. Germline *p53* mutations in a familial syndrome of breast cancer, sarcomas, and other neoplasms. *Science* **250:** 1233.

Martinez, J., I. Georgoff, J. Martinez, and A.J. Levine. 1991.

Cellular localization and cell cycle regulation by a temperature-sensitive *p53* protein. *Genes Dev.* **5:** 151.

Mercer, W.E., M.T. Shields, M. Amin, G.J. Sauve, E. Appella, J.W. Romano, and S.J. Ullrich. 1990. Negative growth regulation in a glioblastoma tumor cell line that conditionally expressed human wild-type *p53*. *Proc. Natl. Acad. Sci.* **87:** 6166.

Michalovitz, D., O. Halevy, and M. Oren. 1990. Conditional inhibition of transformation and of cell proliferation by a temperature-sensitive mutant of *p53*. *Cell* **62:** 671.

Miller, M.W. and R.S. Nowakowski. 1988. Use of bromodeoxyuridine-immunohistochemistry to examine the proliferation, migration and time of origin of cells in the central nervous system. *Brain Res.* **457:** 44.

Morgenbesser, S.D., B.O. Williams, T. Jacks, and R.A. DePinho. 1994. *p53*-dependent apoptosis produced by *Rb*-deficiency in the developing mouse lens. *Nature* **371:** 72.

Pesin, S.R. and J.A. Shields. 1989. Seven cases of trilateral retinoblastoma. *Am. J. Ophthalmol.* **107:** 121.

Prosser, J., A.M. Thompson, G. Cranston, and H.J. Evans. 1990. Evidence that *p53* behaves as a tumor suppressor gene in sporadic breast tumors. *Oncogene* **5:** 573.

Purdie, C.A., D.J. Harrison, A. Peter, L. Dobbie, S. White, S.E.M. Howie, D.M. Salter, C.C. Bird, A.H. Wyllie, M.L. Hooper, and A.R. Clarke. 1994. Tumour incidence, spectrum and ploidy in mice with a large deletion in the *p53* gene. *Oncogene* **9:** 603.

Ruggeri, B., S.-Y. Zhang, J. Caamano, M. DiRado, S.D. Flynn, and A.J.P. Klein-Szanto. 1992. Human pancreatic carcinomas and cell lines reveal frequent and multiple alterations in the *p53* and *Rb-1* tumor-suppressor genes. *Oncogene* **7:** 1503.

Ruggeri, B., J. Camano, T. Goodrow, M. DiRado, A. Bianchi, D. Trono, C.J. Conti, and A.J. Klein-Szanto. 1991. Human pancreatic carcinomas and cell lines reveal frequent and multiple alterations in the p53 and Rb-1 tumor-suppressor genes. *Cancer Res.* **51:** 6615.

Rygaard, K., G.D. Sorenson, O.S. Pettengill, C.C. Cate, and T.M. Spang. 1990. Abnormalities in structure and expression of the retinoblastoma gene in small cell lung cancer cell lines and xenografts in nude mice. *Cancer Res.* **50:** 5312.

Scheffner, M., K. Munger, J.C. Byrne, and P.M. Howley. 1991. The state of the *p53* and retinoblastoma genes in human cervical carcinoma cell lines. *Proc. Natl. Acad. Sci.* **88:** 5523.

Shaw, P., R. Bovey, S. Tardy, R. Sahli, B. Sordat, and J. Costa. 1992. Induction of apoptosis by wild-type *p53* in a human colon tumor-derived cell line. *Proc. Natl. Acad. Sci..* **89:** 4495.

Shew, H.-Y., N. Ling, X. Yang, O. Fodstad, and W.-H. Lee. 1989. Antibodies detecting abnormalities of the retinoblastoma susceptibility gene product (pp110^{Rb}) in osteosarcomas and synovial sarcomas. *Oncogene Res.* **1:** 205.

Sidransky, D., T. Mikkelsen, K. Schwechheimer, M.L. Rosenblum, W. Cavenee, and B. Vogelstein. 1992. Clonal expansion of *p53* mutant cells is associated with brain tumor progression. *Nature* **355:** 846.

Stannard, C., B.K. Knight, and R. Sealy. 1985. Pineal malignant neoplasm in association with hereditary retinoblastoma. *Br. J. Ophthalmol.* **69:** 749.

Stratton, M.R., S. Moss, W. Warren, H. Patterson, J. Clark, C. Fisher, C.D.M. Fletcher, A. Ball, M. Thomas, B.A. Gusterson, and C.S. Cooper. 1990. Mutation of the *p53* gene in human soft tissue sarcomas: Association with abnormalities of the *RB1* gene. *Oncogene* **5:** 1297.

Symonds, H., L. Krall, L. Remington, M. Saenz-Robles, S. Lowe, T. Jacks, and T. Van Dyke. 1994. *p53*-dependent apoptosis suppresses tumor growth and progression in vivo. *Cell* **78:** 703.

T'Ang, A., J.M. Varley, S. Chakraborty, A.L. Murphree, and Y.-K.T. Fung. 1988. Structural rearrangement of the retinoblastoma gene in human breast carcinoma. *Science* **241:** 1797.

Van Dyke, T.A. 1994. Analysis of viral-host protein interactions and tumorigenesis in transgenic mice. *Semin. Cancer. Biol.* **5:** 106.

Varley, J.M., W.J. Brammar, D.P. Lane, J.E. Swallow, C. Dolan, and R.A. Walker. 1991. Loss of chromosome 17p13 sequences and mutation of *p53* in human breast carcinomas. *Oncogene* **6:** 413.

Varley, J.M., J. Armour, J.E. Swallow, A.J. Jeffreys, B.A. Ponder, A. T'Ang, Y.K. Fung, W.J. Brammar, and R.A. Walker. 1989. The retinoblastoma gene is frequently altered leading to loss of expression in primary breast tumours. *Oncogene* **4:** 725.

Vogelstein, B. and K.W. Kinzler. 1993. The multistep nature of cancer. *Trends Genet.* **9:** 138.

Weinberg, R.A. 1992. The retinoblastoma gene and gene product. *Cancer Surv.* **12:** 43.

Werness, B.A., A.J. Levine, and P.M. Howley. 1989. Association of human papillomavirus types 16 and 18 E6 proteins with *p53*. *Science* **248:** 76.

White, E., R. Cipriani, P. Sabbatini, and A. Denton. 1991. Adenovirus E1B 19-kilodalton protein overcomes the cytoxicity of E1A proteins. *J. Virol.* **65:** 2968.

Whyte, P., K.J. Buchkovich, J.M. Horowitz, S.J. Friend, M. Raybuck, R.A. Weinberg, and E. Harlow. 1988. Association between an oncogene and an anti-oncogene: The adenovirus E1A proteins bind to the retinoblastoma gene product. *Nature* **334:** 124.

Williams, B.O., L. Remington, R.T. Bronson, S. Mukai, D.M. Albert, T. Dryja, and T. Jacks. 1994a. Cooperative tumorigenic effects of germline mutations in *Rb* and *p53*. *Nat. Genet.* **7:** 480.

Williams, B.O., E.M. Schmitt, L. Remington, R.T. Bronson, D.M. Albert, R.A. Weinberg, and T. Jacks. 1994b. Extensive contribution of *Rb*-deficient cells to adult chimeric mice with limited histopathological consequences. *EMBO J.* **13:** 4251.

Windle, J.J., D.A. Albert, J.M. O'Brien, D.M. Marcus, C.M. Distecheh, R. Bernards, and P.L. Mellon. 1990. Retinoblastoma in transgenic mice. *Nature* **343:** 665.

Wu, X. and A.J. Levine. 1994. *p53* and E2F-1 cooperate to mediate apoptosis. *Proc. Natl. Acad. Sci.* **91:** 3802.

Wurtman, R.J. and M.A. Moskowitz. 1977. The pineal organ. *N. Engl. J. Med.* **296:** 1329,1383.

Yonish-Rouach, E., D. Resnitzky, J. Lotem, L., Sachs, A. Kimchi, and M. Oren. 1991. Wild-type *p53* induces apoptosis of myeloid leukaemic cells that is inhibited by interleukin-6. *Nature* **352:** 345.

Insulin-like Growth Factor II Is Focally Up-regulated and Functionally Involved as a Second Signal for Oncogene-induced Tumorigenesis

P. Naik,* G. Christofori,† and D. Hanahan*

*Department of Biochemistry and Biophysics, and Hormone Research Institute, University of California, San Francisco, California 94143-0534; †Research Institute of Molecular Pathology, 1030 Vienna, Austria

The histopathology and epidemiology of human and animal cancers have clearly implicated a developmental process by which a normal cell is converted into a malignant tumor through a series of steps manifested in distinguishable histological and temporal stages (Foulds 1969; Henson and Albores-Saavedra 1993). The advanced stages have in some cases proved accessible to analysis, resulting in theories of tumorigenesis based on genetic change (Knudson 1986; Klein 1987; Fearon and Vogelstein 1990), involving activation of oncogenes, inactivation of tumor suppressor genes, and altered expression of cell surface molecules. Yet the early stages of tumorigenesis have remained largely inaccessible and the critical determinants undefined. A powerful system to investigate the process of tumorigenesis has resulted from the ability to manipulate the mouse germ line, both through the introduction of new genetic information (Hanahan 1989) and through the disruption of existing genes by homologous recombination (Capecchi 1988; Zimmer 1992). Lines of transgenic mice have been produced with a heritable predisposition to develop specific cancers, as a result of the expression of oncogenes or the abrogation of tumor suppressor genes (Hanahan 1988; Adams and Cory 1991; Clarke et al. 1992; Donehower et al. 1992; Jacks et al. 1992; Lee et al. 1992). These mice recapitulate the ontogeny of tumorigenesis, in that normal cells in their natural environment are converted into tumors, often through a series of reproducible temporal and histological stages that are accessible to molecular analysis of the events in tumor development.

In one transgenic model of tumorigenesis, the insulin gene regulatory region has been used to target expression of an oncogene, SV-40 large T-antigen (Tag), to the β cells of the pancreatic islets, of which the RIP1-Tag2 line has been the prototype (Hanahan 1985). The Tag oncoprotein possesses a number of functional activities, including the ability to bind and inactivate two tumor suppressors, pRb and p53 (Ludlow 1993). In RIP-Tag transgenic mice, expression of Tag protein begins at embryonic day 9 in the pancreatic diverticulum and persists throughout islet development and in all adult β cells (Alpert et al. 1988). Every mouse develops a few islet cell carcinomas by 14 weeks of age (Hanahan 1985). Several preneoplastic stages have been identified by temporal, histological, and statistical criteria. Remarkably, every islet expresses Tag, and yet there is initially no hyperplasia. Subsequently, focal activation of hyperproliferation ensues, as evidenced by the incidence of S-phase cells; over time, 50–70% of the islets become hyperplastic (Teitelman et al. 1988). Similarly, the capillary endothelial cells that vascularize the islets are initially quiescent. Then some are activated subsequent to the onset of hyperplasia, again in a focal pattern; approximately 10% of the islets become angiogenic (Folkman et al. 1989). Finally, 1–2% of the islets progress into highly vascularized solid tumors. Thus, three distinct steps are evident: the switch to hyperproliferation, the activation of angiogenesis, and the progression to solid tumors. A perplexing dichotomy has been the first step, since 100% of the β cells express the Tag oncoprotein, and yet hyperproliferation is induced sporadically after a period of normality. We now report that the insulin-like growth factor II (IGF-II) gene is focally activated concomitant with this first step, and that IGF-II is causally associated with the tumor phenotype.

IGF-II is a polypeptide growth factor that binds to two receptors, the transmembrane tyrosine kinase IGF type 1 receptor (IGF-1R) and the IGF type 2 receptor (IGF-2R), which also binds mannose-6-phosphate residues (Rechler and Nissley 1990; Nissley et al. 1991; Rechler 1991; Werner et al. 1991). A related growth factor, IGF-I, also binds to the IGF-1R, and both IGF-I and IGF-II are known to signal through this receptor (Gustafson and Rutter 1990; Zhang and Roth 1991; Germain-Lee et al. 1992). The IGF-2R does not contain a kinase activity and may serve as a competitive sink for IGF-II protein (Haig and Graham 1991; Filson et al. 1993), although G-protein signaling has been proposed (Murayama et al. 1990). IGF-I or IGF-II, and IGF-1R have been found in a variety of tumors and tumor cell lines, including, for example, tumors arising from mammary epithelium (Osborne et al. 1989; Yee et al. 1991), hepatocytes (Schirmacher et al. 1992), embryonic kidney (Reeve et al. 1985; Scott et al. 1985), and peripheral neurons (El-Badry et al. 1989). Interference experiments using receptor antibodies or antisense oligonucleotides have suggested that IGF/IGF-1R autocrine or paracrine signaling circuits are involved in tumor cell proliferation (Conover et al. 1986; Osborne et al. 1989; Mathieu et al. 1990; Thompson et al. 1990; Pietrzkowski et al. 1992; Porcu et al. 1992). In a screen that compared expression of about 30 growth factors, receptors, and oncogenes between normal pancreatic islets and the β-cell tumors arising in several lines of RIP-Tag transgenic mice, we found that IGF-II was selectively up-regulated in the tumors. This observation has motivated the analyses presented below; some of the results have been reported previously (Christofori et al. 1994).

EXPERIMENTAL PROCEDURES

RNA extraction. Islets of Langerhans and β-cell tumors were isolated by retrograde perfusion of the common bile duct with a collagenase/DNase solution as described previously (Lacy and Kostianovsky 1967). After collagenase digestion at 37°C, islets and tumors were washed in RPMI plus 10% calf serum and collected by micropipetting under a dissecting microscope. One hundred islets and multiple tumors from each control or transgenic mouse at the indicated age were pooled, and RNA was extracted with guanidinium thiocyanate using RNAzol (Cinna/Biotexc Labs International, Friendswood, Texas) according to the protocol supplied by the manufacturer.

Semi-quantitative RNA-PCR. cDNA synthesis and PCR reactions for IGF-I and IGF-II were carried out according to the method of Kandel et al. (1991) with the modification that the standard PCR reaction mixture contained 20 μCi [α-^{32}P]dATP, cDNA derived from 50 ng RNA, primers for β_2-microglobulin (5': GCTATCCAGAAGAAACCCCTCAAATTC; 3': CATGTCTCGATCCCAGTAGACGGTC), and primers for either IGF-II (5': CTTCTACTTCAG CAGGCCTTC; 3': GTATCTGGGGAAGTCGT CCGG) or IGF-I (5': CCGAGGGGCTTTTACTTC AACAAGC; 3': CACAGTACATCTCCAGTCTCC TCAG). Samples (8 μl) of the reaction mixture were removed after 15, 18, 21, 24, 27, and 30 cycles and electrophoresed on 8% polyacrylamide gels run in Tris/borate/EDTA buffer. The relative incorporation of radioactivity into the β_2-microglobulin and the IGF PCR products was determined by PhosphorImager analysis (Molecular Dynamics, Sunnyvale, California). The ratio (R) between the signals for IGF-II or IGF-I and β_2-microglobulin was determined at the logarithmic phase of product generation and plotted in the graph. Experiments were repeated at least twice. For cases where there was significant change in gene expression, three or four analyses were performed. Primers were designed to span regions that are known to contain introns in the genomic sequence. Control PCRs performed on cDNA samples prepared without reverse transcriptase were consistently negative. The identity of the PCR products was confirmed by cloning and sequencing. Semi-quantitative RNA-PCR for IGF-IR and IGF-IIR was performed and analyzed as described above with the exception that primers for β_2-tubulin (5': CAACGTCAAGACGGCCGTGTG; 3': GACA GAGGCAAACTGAGCACC), instead of β_2-microglobulin, were used as internal standard in combination with primers specific for either IGF-1R (5': CTCAGC CTTGTGTCCTGAGTGTCT; 3': TTCCAGGAGGT CTCCTTCTACTAC) or IGF-2R (5': ACAGATGTT GATGTAGAAGACAGG; 3': TGTACACTCTTCT TCTCCTGGCA).

Northern blot analysis. RNA from normal tissues and tumors was isolated as described above and poly(A) RNA was purified according to published protocols (Sambrook et al. 1989). Poly(A) RNA (5 μg) was loaded in each lane of a formaldehyde gel and blotted onto Hybond N membrane. The blot was hybridized with a randomly primed ^{32}P-labeled DNA fragment spanning the coding region for mouse IGF-II, from 34 bp upstream of the ATG initiation codon to 858 bp downstream from the stop codon (1500 bp total length). After exposure, the membrane was stripped and rehybridized with a randomly primed ^{32}P-labeled DNA fragment spanning the coding region for mouse IGF-I, extending from 177 bp upstream of the ATG initiation codon to 169 bp downstream from the stop codon (710 bp total length).

Radioreceptor assay. Radioreceptor assays with either ^{125}I-labeled IGF-I or ^{125}I-labeled IGF-II were performed as described previously (Osborne et al. 1989) using βTC3 cells. Briefly, βTC3 cells which had been grown to about 80% confluence in 24-well plates were incubated for 3–4 hours in ^{125}I-labeled IGF-I or ^{125}I-labeled IGF-II. In some wells, insulin, IGF-I, or IGF-II was added to compete for the binding of ^{125}I-labeled IGF-I or ^{125}I-labeled IGF-II. The amount of radioactivity bound to the cells at the end of the incubation period was quantitated on a Beckman gamma counter.

Purification of IGF-II from conditioned medium. IGF-II activity in the conditioned medium of βTC3 cells was assayed as described previously (Osborne et al. 1989), with some modifications. Cells were conditioned at 60% confluence in CG serum-free medium (Camon, Germany) for 72 hours. The conditioned medium was then collected, filtered, and processed as described previously (Osborne et al. 1989). Processed samples were resuspended in binding buffer (Hams F-12, 0.1% BSA, and 25 mM HEPES [pH 7.4]), and 15 or 250 μl was added to wells containing BRL-3A cells (Lee et al. 1986) to compete for the binding of ^{125}I-labeled IGF-II. Increasing amounts of insulin, IGF-I, or IGF-II were used as controls in the competition assay.

In situ hybridization. In situ hybridization on mouse pancreas tissues was performed as described by Smith et al. (1990) with some modifications. Briefly, after perfusion fixation with 4% paraformaldehyde, the pancreas was removed and post-fixed in 4% paraformaldehyde. Cryosections (10 μm) were prepared for in situ hybridization. ^{35}S-labeled sense and antisense RNA probes were synthesized from a linearized Bluescript-IGF-II subclone using the T3 or T7 polymerase, respectively; the transcripts spanned the entire mouse IGF-II coding region, including 34 nucleotides upstream of the start codon and 858 nucleotides downstream from the stop codon. Cryosections were treated with proteinase K (5 mg/ml, 10 min), acetylated with 0.25% acetic anhydride in 0.1 M triethanolamine buffer (10 min), and hybridized with RNA probes (600,000 cpm/section) at 55°C overnight in hybridization solution. Following hybridization, sections were washed and then treated with RNase A (20 mg/ml, 30 min), followed by a wash in $0.1 \times$ SSC/10 mM β-mercaptoethanol (55–58°C, 2 hr). Vacuum-desiccated slides were dipped in Kodak NTB2 emulsion and developed a week later, followed by counterstaining with hematoxylin and eosin.

Immunohistochemistry. For Tag staining, tissue sections were rehydrated in PBS and then incubated in 5% H_2O_2/MeOH (20 min). Retrieval of the antigen was achieved by microwaving the tissue sections 2×5 min in Antigen Retrieval *Citra* solution (Biogenex, San Ramon, California). The tissues were then permeabilized in PBS/0.25% Nonidet P40 (15 min), followed by blocking in 10% goat serum (30 min). A rabbit polyclonal antibody specific against Tag was placed on the tissue sections at a dilution of 1:1000 in PBS/1% goat serum (overnight). After three washes in PBS/1% goat serum, a peroxidase-conjugated goat anti-rabbit IgG was applied at 1:200 in PBS/1% goat serum for 45 minutes. The sections were then washed three times in PBS/1% goat serum and the immunocomplexes were visualized by treatment with diaminobenzidine, nickel sulfate, and H_2O_2. After brief washes in PBS and dehydration through graded alcohol, the sections were mounted in Entellan New mounting medium (EM Science, Gibbstown, New Jersey). For PCNA staining, rehydration of tissue sections, blocking of endogenous peroxidase, and antigen retrieval were performed as for Tag staining. Sections were blocked in PBS/10% goat serum/0.3% Triton X-100 (30 min). Following a brief wash in PBS, a mouse monoclonal antibody against PCNA (Biogenex, San Ramon, California) was applied at 1:100 dilution in PBS/2% goat serum/0.3% Triton X-100 (2 hr). Following three washes in PBS, a biotinylated goat-anti-mouse IgM was placed on the sections at a 1:200 dilution in PBS/2% goat serum/ 0.3% Triton X-100 (45 min). Again, after three washes in PBS, Vectastain *Elite* ABC reagent was applied (45 min). The sections were then washed and the avidin-biotin-immunocomplexes were visualized by treatment with diaminobenzidine, nickel sulfate, and H_2O_2. Dehydration and mounting were performed as for Tag staining. For BrdU staining, rehydration of tissue sections and blocking of endogenous peroxidase were performed as for Tag staining. Tissues were then treated with 2N HCl (1 hr), washed, and treated with proteinase K (90 sec). A directly biotinylated mouse monoclonal antibody against BrdU (CalTag, San Francisco) was then placed on the sections overnight at 4°C. After extensive washing, Vectastain *Elite* ABC reagent was applied (45 min), and the avidin-biotin immunocomplexes were visualized as for PCNA staining. Dehydration and mounting were performed also as for Tag staining.

Preparation of tissues for BrdU staining. Mice were injected intraperitoneally with 100 μg of BrdU per gram of body weight 2 hours prior to sacrifice. After perfusion fixation with 4% paraformaldehyde, the pancreas was removed and post-fixed overnight in 4% paraformaldehyde, then immersed in 30% sucrose. The tissues were embedded in OCT compound by liquid nitrogen freezing.

Determination of proliferation index. The proliferation index as revealed by expression of PCNA was determined by calculating the number of PCNA-positive nuclei as a percentage of the total number of nuclei in individual islets. The statistical analysis was done by assigning the proliferation index of a particular islet to one of the four classes of IGF-II expression, as revealed by in situ hybridization analysis of the same islet on an adjacent tissue section.

The tumor BrdU proliferation index was determined by calculating the number of BrdU-staining nuclei as a percentage of the total number of nuclei in each field. 10 fields from 4 Rip-Tag, IGF-II$^{+/+}$ tumors and 8 fields from 4 Rip-Tag, IGF-II$^{-/-}$ tumors were counted.

Tumor histopathology. Mice were sacrificed by cervical dislocation. Pancreas and tumors were removed and immersion-fixed for 16 hours in 4% paraformaldehyde/PBS at 4°C. The tissues were dehydrated through graded alcohol and xylene and embedded in paraffin. 5-μm sections were cut and stained with hematoxylin and eosin.

Determination of apoptotic index. To quantify the incidence of apoptosis, histological sections of 10 tumors from Rip-Tag, IGF-II$^{+/+}$ and 7 from Rip-Tag, IGF-II$^{-/-}$ mice were analyzed for apoptotic cells in 210 and 119 fields, respectively. Apoptotic cells were identified by their pyknotic nuclei and detachment from neighboring cells.

Analysis of tumor phenotype. Mice were sacrificed by cervical dislocation at ages as indicated. Their body weight was determined and tumors were isolated either by retrograde perfusion with collagenase solution (Lacy and Kostianovsky 1967) or by microdissection. The number of tumors and their diameters were determined and the total tumor volume was calculated as a function of an approximate spheric tumor shape. Tumor volumes were blotted as a function of age. The difference between the two groups was statistically significant ($p < 0.001$; Student's t-test). In addition, the ratio of the average tumor volume to the average body weight of 9 RIP1-Tag2; IGF-II$^{+/+}$ and 12 RIP1-Tag2; IGF-II$^{-/-}$ mice was determined (see inset, Fig. 5).

RESULTS

IGF-II Is Focally Up-regulated during β Cell Tumorigenesis

IGF-II is widely expressed in the developing mouse embryo. Beginning at birth, IGF-II expression is progressively extinguished in virtually all tissues, except for the leptomeninges and choroid plexus of the brain (Stylianopoulou et al. 1988; DeChiara et el. 1991), and the renal medulla (P. Naik, unpubl.). The pancreatic islets represent natural focal nodules of cells, which in the RIP1-Tag2 mice uniformly express the Tag oncoprotein. The islets can be isolated from the larger mass of the exocrine pancreas by collagenase digestion, density gradient centrifugation, and micropipetting under darkfield illumination using a dissecting microscope. A semi-quantitative RNA-PCR technique was used to analyze IGF-II expression in control and transgenic islets at increasing ages from 4 to 10 weeks, in comparison to the tumors that arise beginning at 12 weeks. As shown in Figure 1a, IGF-II RNA is not expressed in the islets of 4- to 10-week-old control mice. However, IGF-II is clearly expressed in islets of age-matched transgenic mice, as well as in all tumors, fractionated tumor cells, and tumor-derived cell lines (βTC; Efrat et al. 1988). In contrast, IGF-I expression is progressively decreased in both control and transgenic islets with increasing age, and the levels are even lower in primary tumors, fractionated tumor cells, and βTC (Fig. 1a). The results of the RNA-PCR analysis were substantiated by Northern blotting (Fig. 1b), which confirmed that IGF-II mRNA is abundant in primary tumors and βTC lines but is not evident in pancreas or other adult tissues of nontransgenic control mice, excepting the brain and the kidney. Again, IGF-I levels are low or undetectable in primary tumors or βTC lines. The IGF-II mRNA detected by these two assays reflects production and secretion of active IGF-II, as shown by a radioreceptor competition assay using conditioned medium from cultured tumor cell lines (Fig. 1c). Although the RIP1-Tag2 line has been the primary focus of this analysis, IGF-II up-regulation has also been detected in islets and tumors of two other independent transgenic lines that carry the SV40 T antigen under the

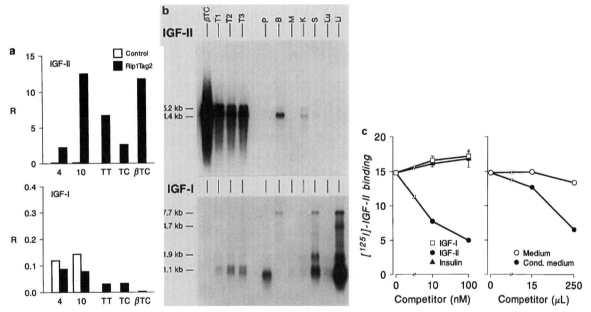

Figure 1. IGF-II is up-regulated during β-cell tumorigenesis. (*a*) Semiquantitative RNA-PCR analysis of IGF-I and IGF-II expression in isolated islets and tumors. IGF-II is not expressed in adult islets of control (C57BL/6J) mice analyzed at 4 and 10 weeks. However, IGF-II mRNA can be detected in 4- and 10-week transgenic islets, in RNA isolated from an intact tumor (TT), in fractionated tumor cells separated from stroma and endothelial cells by differential sedimentation (TC), and in cultured β tumor cell lines (exemplified by βTC3). In contrast, IGF-I mRNA is expressed in both control and transgenic islets at 4 and 10 weeks of age, and at lower levels in total tumors, purified tumor cells, and βTC lines. The relative levels (R) of IGF-II or IGF-I mRNAs are presented as a ratio between IGF-II or IGF-I and β_2-microglobulin PCR products. (see Experimental Procedures). (*b*) Northern blotting analysis of IGF-I and IGF-II mRNA levels in normal control tissues and tumors. Three primary tumors (T1–T3) and the βTC3 cell line express high levels of IGF-II mRNA (*upper panel*). Among RNAs isolated from a series of tissues in a control (nontransgenic) mouse, brain (B) and kidney (K) express IGF-II, as expected, whereas pancreas (P), muscle (M), spleen (S), lung (Lu), and liver (Li) do not. A similar analysis using a probe for IGF-I revealed low-level expression in the three primary tumors, as well as expression in total pancreas, spleen, and liver of a nontransgenic control mouse. The blot was also analyzed for the presence of β_2-tubulin mRNA which established that equal amounts of RNA were loaded into each lane (not shown). (*c*) Radioreceptor assay demonstrating that cultured β tumor cells (βTC) produce IGF-II protein. BRL3A cells express IGF-IIR on their surface (Schirmacher et al. 1992) and bind radiolabeled IGF-II in a dose-dependent fashion. The left panel illustrates the assay, whereby cold IGF-II competes for binding of radiolabeled IGF-II, whereas IGF-I and insulin do not. The right panel compares competition by culture medium (CG serum-free medium) for IGF-II binding to that of medium conditioned by βTC3 cells, demonstrating the presence of IGF-II protein in the βTC conditioned medium.

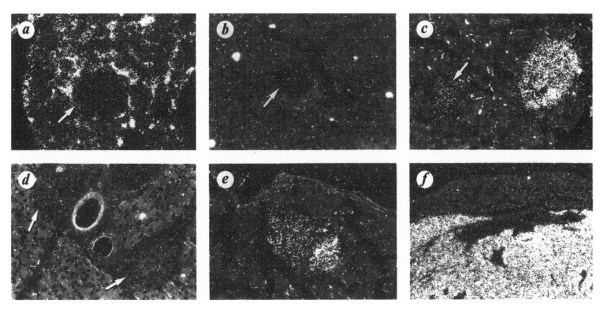

Figure 2. IGF-II expression is focally activated in hyperplastic islets and further up-regulated in tumors. In situ hybridization analysis using a ^{35}S-labeled IGF-II probe was performed on tissue sections of RIP-Tag2 transgenic and C57BL/6J control mice at varying ages. Darkfield illumination micrographs are shown. (*a*) IGF-II expression in the pancreas of a 12-day-old juvenile mouse; the islet core of β cells shows little or no signal. (*d*) An islet in a 6-week-old control mouse, which does not express detectable IGF-II. The remaining panels illustrate the major classes of IGF-II expression seen in islets from 4–14-week old RIP1-Tag2 transgenic mice: (*b*) a nonexpressing islet; (*e*) an islet with focal expression; (*c*) a pair of islets, one a hyperplastic islet uniformly expressing IGF-II, and the second expressing in a weaker, more focal pattern; and (*f*) a tumor expressing high levels of IGF-II mRNA. Arrows indicate islets with low or no detectable IGF-II expression. Magnifications: *f*, 25×; all others, 50×. (Reprinted, with permission, from Christofori et al. 1994 [copyright Macmillan Magazines Ltd.].)

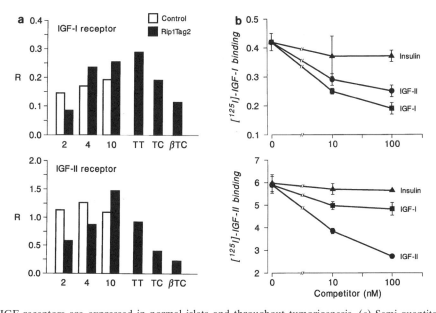

Figure 3. Both IGF receptors are expressed in normal islets and throughout tumorigenesis. (*a*) Semi-quantitative RNA-PCR analysis of IGF type 1 receptor (*top* panel) and IGF type 2 receptor (*bottom* panel) expression in control islets and the stages of tumorigenesis. Analogous to Fig. 1, expression is evaluated in 2-, 4-, and 10-week islets, primary tumors, purified tumor cells, and established tumor cell lines, documenting expression in all cases. The relative levels (R) of IGF-1R or IGF-2R mRNAs are presented as a ratio between IGF-1R or IGF-2R and $β_2$-tubulin PCR products. (see Experimental Procedures and Fig. 1). (*b*) Radioreceptor assay demonstrating that β tumor cell lines express functional type 1 and type 2 IGF receptors on the cell surface. Radiolabeled IGF-I (*top* panel) was incubated with βTC3 cells in the presence of increasing concentrations of unlabeled IGF-I, IGF-II, or insulin. Both IGF-I and IGF-II competed similarly for radioactive IGF-I binding, indicating the IGF-1R is present on βTC cells. To assess IGF-2R (*bottom* panel), radiolabeled IGF-II was used. Increasing concentrations of unlabeled IGF-II competed more effectively than unlabeled IGF-I for binding of the radiolabeled IGF-II, revealing that there is also functional IGF-2R on βTC cells.

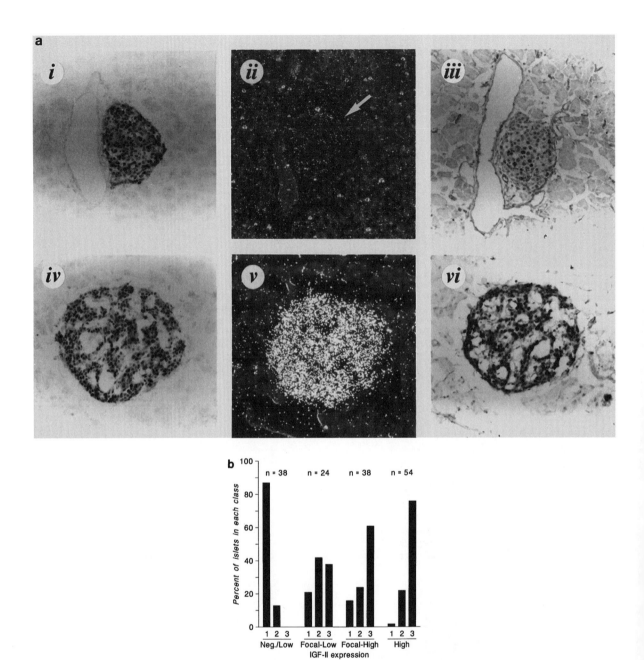

Figure 4. Activation of IGF-II correlates with the switch to proliferative hyperplasia. (*a*) IGF-II expression coincides with that of PCNA, a marker of proliferating cells. The top row shows a transgenic islet that expressed the Tag oncoprotein (i) but not IGF-II (ii; arrow); this islet is not hyperplastic, as indicated by the low number of nuclei intensely staining for PCNA (iii). The bottom row shows serial sections of a hyperproliferative islet immunostained for the Tag oncoprotein (iv), and for PCNA (vi), and analyzed for IGF-II expression by in situ hybridization (v). Only nuclei strongly reactive for the anti-PCNA antibody represent proliferating cells, as determined by comparison with nonproliferating cells in nontransgenic islets. Note that both islets contain β cells expressing the Tag oncoprotein at similar levels. Bright field (i, iii, iv, vi) and dark field (ii, v) illumination photomicrographs are shown. Magnification, $50\times$. (*b*) Statistical analysis of the correlation between IGF-II activation and the switch to the hyperproliferative stage. Four classes of islets were identified in analysis of tissue sections of RIP1-Tag2 transgenic mice: islets with no detectable IGF-II mRNA; islets with low focal IGF-II expression; islets with high focal IGF-II expression; and islets uniformly positive for IGF-II. Altogether, 154 islets were analyzed by IGF-II in situ hybridization and PCNA immunostaining of adjacent tissue sections from 6 transgenic mice at ages varying from 4 weeks to 14 weeks. Each class of islet was given a score for its proliferation index, as assessed by PCNA staining. The data are summarized as the fractions of each IGF-II expression class that show a distinctive proliferation (PCNA) index: (1) off/low, 0–15%; (2) medium, 15–25%; or (3) high, >25% PCNA-positive cells. (Reprinted, with permission, from Christofori et al. 1994 [copyright Macmillan Magazines Ltd.].)

control of the rat insulin promoter (Rip1Tag3 and Rip3Tag2), indicating this result is not a consequence of the transgene integration site.

The onset of hyperproliferation among the approximately 400 islets expressing the Tag oncoprotein occurs sporadically over time, beginning at about 3–4 weeks. Therefore, in situ RNA hybridization to pancreatic tissue sections was used to investigate the pattern of IGF-II transcription in the transgenic islets as the mice aged and tumors developed (Christofori et al. 1994). IGF-II is expressed in the pancreases of 12- to 14-day-old juvenile, nontransgenic mice (Fig. 2a). However, the core of the islets where the β cells reside shows little or no signal, indicating either that IGF-II is not normally expressed in β cells or that its expression is extinguished by this time. IGF-II mRNA cannot be detected anywhere in the pancreas of adult (>4-week-old) control mice, including the islets (Fig. 2d), consistent with the RNA-PCR results (Fig. 1a). Analysis of 12–14-day transgenic mice revealed a similar pattern of IGF-II expression in the pancreas, although an occasional islet showed a clear IGF-II signal in the islet core where the β cells are located (not shown). By 4 weeks, IGF-II was also extinguished throughout the exocrine pancreas, and many islets did not express IGF-II (Fig. 2b). However, other islets were positive (2c and 2e), as was every tumor (2f). An RNA-PCR analysis performed on single transgenic islets confirmed that only a fraction expressed IGF-II (not shown). In some islets, IGF-II expression was clearly localized (2c and 2e), whereas in larger hyperplastic islets, the pattern was more uniform (2c). The data indicate that IGF-II is focally activated in oncogene-expressing islets and further up-regulated in all tumors, motivating an analysis of the two known receptors for IGF-II in this tumorigenesis pathway.

Both IGF Receptors Are Expressed Throughout Tumorigenesis

Expression of the IGF type-1 and type-2 receptors was evaluated by semi-quantitative RNA-PCR analysis of isolated control and transgenic islets at increasing ages and of tumors, analogous to that performed with IGF-I and IGF-II. Both receptors were expressed in nontransgenic control islets at all ages tested (Fig. 3a). Similarly, both receptors were expressed at comparable levels in transgenic islets, and in primary tumors, fractionated tumor cells, and βTC lines (Fig. 3a). Functional receptors were present on the cell surface of β tumor cells, as revealed by radioreceptor competition assays (Fig. 3b). Both IGF-I and IGF-II bind to the IGF-1R, and thus both should compete for IGF-I binding, which is the case (Fig. 3b, top panel). In contrast, only IGF-II can bind to the IGF type-2 receptor. Since unlabeled IGF-II selectively competed for binding of radiolabeled IGF-II (Fig. 3b, bottom panel), we infer this receptor is also present. The presence of the two IGF receptors suggests that activation of IGF-II could result in autocrine or paracrine growth stimulation of

the β cells, which led us to analyze the correlation between β-cell proliferation and activation of IGF-II.

IGF-II Activation Correlates with the Onset of Hyperproliferation

Serial tissue sections of pancreas from RIP1-Tag2 mice at different ages and stages of tumor development were analyzed for IGF-II gene expression by in situ hybridization, and for hyperproliferation by immunostaining with antibodies to PCNA (Christofori et al. 1994), a subunit of DNA polymerase δ expressed in the late G_1 and S phases of proliferating cells (Bravo et al. 1987; Garcia et al. 1989). In some cases, adjacent sections were also immunostained to detect the Tag oncoprotein. This analysis revealed a clear correlation between hyperproliferation and IGF-II expression; islets with extensive PCNA staining had high levels of IGF-II mRNA (Fig. 4a, panels iv–vi). In contrast, the Tag oncoprotein was expressed in all islets, including those with no detectable IGF-II mRNA and a low index of PCNA expression (Fig. 4a, panels i–iii). An intermediate stage, with focal expression of IGF-II mRNA in the islet (e.g., panel 2e), showed more localized PCNA staining (not shown), supporting a model in which activation of IGF-II expression and release of IGF-II protein cause a local induction of cell proliferation. A statistical analysis (Fig. 4b) confirms the correlation between hyperproliferation and IGF-II expression. Thus, IGF-II is activated in islets wherein the β cells have begun to proliferate, consistent with its involvement in this step of tumorigenesis. To address that possibility, we have conducted tests of the functional contributions of IGF-II.

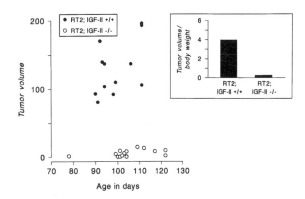

Figure 5. Transgenic complementation analysis with IGF-II null mice produces a dramatic impairment in tumor growth. The histogram shows the dramatic reduction in tumor volume at autopsy in a series of individual RIP1-Tag2, IGF-II$^{-/-}$ mice, as compared to a set of RIP1-Tag2, IGF-II$^{+/+}$ age-matched controls. The inset box compares the average tumor volumes, corrected for body weight; the body weight for the IGF-II null mice is typically 60% of age-matched controls. We infer from the monitoring of serum glucose levels that the small tumors which develop in the IGF-II null mice are sufficient to elicit lethal hypoglycemia. (Reprinted, with permission, from Christofori et al. 1994 [copyright Macmillan Magazines Ltd.].)

Absence of IGF-II Impairs Tumorigenesis

The contribution of IGF-II to the process of β-cell tumorigenesis has been examined by genetic complementation analysis, in which the RIP1-Tag2 transgenic mice were cross-bred into a line of gene-knockout mice developed by Robertson, Efstratiadis, and their colleagues. These mice carry a targeted disruption of the IGF-II gene that abrogates synthesis of IGF-II (DeChiara et al. 1990, 1991). Remarkably, the IGF-II$^{-/-}$ mice develop normally, are fertile, and have typical longevity. The one evident phenotype of the IGF-II null mice is reduced body mass, approximately 60% of an IGF-II-competent mouse (DeChiara et al. 1990, 1991). Two backcrosses to IGF-II null mice produced double transgenic mice that carried the RIP-Tag oncogene and were homozygous for the IGF-II gene disruption (RIP1-Tag2, IGF-II$^{-/-}$). These mice were monitored in comparison to age-matched RIP1-Tag2, IGF-II$^{+/+}$ mice. Although tumors developed in the RIP1-Tag2, IGF-II$^{-/-}$ mice, the tumor size was dramatically reduced (Fig. 5), and the number of tumors per mouse was lower (data not shown). Histological comparison of tumors from IGF-II wild-type and IGF-II null mice revealed significant differences (Fig. 6B) (Christofori et al. 1994). The tumor cells in the IGF-II null mice were less densely packed, with obviously smaller nuclei and more cytoplasm than those which typify the tumors in wild-type IGF-II, RIP1-Tag2 transgenic mice. By these histological criteria, the tumors represent a lower grade malignancy.

The collective tumor volume that develops in a RIP1-Tag2, IGF-II$^{-/-}$ mouse is on average about 7% of that produced by a mouse carrying functional IGF-II genes (corrected for body weight; see inset of Fig. 5). These small tumors eventually proved lethal, probably as a result of hypoglycemia elicited by the insulin-producing β cells. These data demonstrate that IGF-II makes important contributions to tumor growth in vivo (Fig. 5), much as shown for tumor cell proliferation in vitro by transfection of antisense oligonucleotides (Christofori et al. 1994).

The Apoptotic Index, but Not the Proliferation Index, Distinguishes the IGF-II Null Tumors

The significant reduction in tumor mass indicates some relative deficiency in tumor growth in IGF-II null

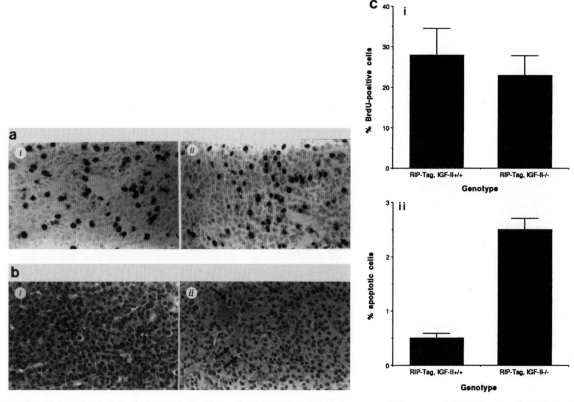

Figure 6. Comparison of the incidences of cells in S phase and apoptosis between wild type and IGF-II null mice. (*a*) BrdU incorporation into Rip-Tag, IGF-II$^{+/+}$ (*i*), and Rip-Tag, IGF-II$^{-/-}$ (*ii*) tumors, revealing a similar high proliferation index in both. (*b*) Hematoxylin and eosin stained tumors from Rip-Tag, IGF-II$^{+/+}$ (*i*), and Rip-Tag, IGF-II$^{-/-}$ (*ii*) mice, respectively. Apoptotic cells are indicated by arrows. (*c*) Statistics of *i*, proliferation index, and *ii*, apoptotic index in Rip-Tag, IGF-II$^{+/+}$ and Rip-Tag, IGF-II$^{-/-}$ tumors. The proliferation indices are 28% ± 6.5% for Rip-Tag, IGF-II$^{+/+}$, and 23% ± 4.8% for Rip-Tag, IGF-II$^{-/-}$ tumors, respectively. Although the proliferation index is similar between the two, there is a statistically significant ($p < 0.001$; Student's *t*-test) 5-fold increase in the apoptotic index of IGF-II null tumors compared to IGF-II wild-type tumors (an increase from 0.5% to 2.5% as determined by this method).

mice. Given the nature of the switch in RIP-Tag2, IGF-II$^{+/+}$ mice from quiescence to a high proliferation index, in which more than 25% of the cells are in S phase, we reasoned that the absence of IGF-II could result in a reduced proliferation index. To address that possibility, mice were injected with BrdU and sacrificed 2 hours later, and then pancreases were collected for immunohistological analysis. Surprisingly, the data indicated that the IGF-II null tumors also had a high proliferation index, similar to that of the IGF-II$^{+/+}$ tumors (Fig. 6a,c). Moreover, a temporal analysis of the proliferative switch indicated that the timing and focal activation of hyperproliferation occurs in similar numbers of islets in the IGF-II null mice (L. Daneshvar and P. Naik, unpubl.). Thus, the switch to a condition of a high proliferation index occurs normally in the absence of IGF-II, and yet the net growth in tumor mass was reduced. We then reasoned that an alteration in the rate of cell death could account for this difference. Indeed, we observed a statistically significant 5-fold increase in the incidence of apoptotic cells in histological sections of the IGF-II null tumors, as determined by counting the number of apoptotic cell bodies in multiple tumors from both IGF-II wild-type and null mice. Tissue sections of tumors from RIP1-Tag2, IGF-II$^{+/+}$ or IGF-II$^{-/-}$ mice were stained with hematoxylin and eosin. The tumors in the IGF-II null mice show a markedly higher incidence of apoptotic cells than their IGF-II$^{+/+}$ counterparts (Fig. 6b; arrows illustrate apoptotic bodies in both tumors). To quantitate the incidence of apoptosis, histological sections of tumors from RIP1-Tag2, IGF-II$^{+/+}$ and RIP1-Tag2, IGF-II$^{-/-}$ mice were analyzed for apoptotic cells. The analysis revealed an increase in apoptotic index from 0.5% in RIP1-Tag2, IGF-II$^{+/+}$ to 2.5% in RIP1-Tag2, IGF-II$^{-/-}$ tumors (Fig. 6c); there were on average 0.88 ± 0.84 and 4.50 ± 1.63 apoptotic cells per field in tumors from RIP1-Tag2, IGF-II$^{+/+}$ and RIP1-Tag2, IGF-II$^{-/-}$, respectively, indicating a 5.2-fold increase in apoptosis ($p < 0.001$; Student's t-test). This conclusion is supported by in situ assays for DNA degradation using the terminal transferase extension assay (data not shown).

DISCUSSION

IGF-II Is a Critical Tumor Growth Factor

The switch to the hyperproliferative stage in the multistep pathway leading to pancreatic β-cell tumors statistically correlates with activation of IGF-II expression, in contrast to the uniform expression of the Tag oncoprotein. Elimination of IGF-II expression by targeted disruption of the IGF-II gene or attenuation of its expression with antisense oligonucleotides impairs tumor growth in vivo and tumor cell proliferation in vitro, respectively. Together the data argue that an autocrine or paracrine signaling pathway involving IGF-II and its receptors is an important component of the circuit that controls tumor cell proliferation. This model is consistent with studies implicating IGF-I or IGF-II in proliferation of a variety of tumor cell types in vitro (Conover et al. 1986; Osborne et al. 1989; Mathieu et al. 1990; Thompson et al. 1990; Pietrzkowski et al. 1992; Porcu et al. 1992). Here for the first time we have analyzed IGF-II expression in vivo throughout a tumor development pathway. The pattern of expression implicates IGF-II in a rate-limiting step of multistage tumorigenesis, and the functional analyses reveal it to be crucial for rapid tumor growth. The small tumors that arise in the IGF-II null mice are clearly less malignant, with smaller, more normal nuclei, lower cell density, and increased incidence of apoptotic cell bodies.

IGF-II is widely expressed during embryogenesis, and then extinguished in most tissues of the adult, including the pancreas. We cannot detect IGF-II expression in the majority of β cells in 12- to 14-day-old mice, indicating that if IGF-II is expressed in developing β cells, it is extinguished by this time. It is evident that IGF-II cannot be strictly required for β-cell development, since the IGF-II null mice have normal islets. Moreover, islet development is not perturbed by expression of the Tag oncoprotein in either IGF-II wild-type or IGF-II null mice. Yet both IGF type-1 and type-2 receptors are expressed, and IGF-II can demonstrably affect β cells, since its focal activation correlates with the switch to the proliferative stage in tumorigenesis, and its absence dramatically impairs tumor cell proliferation. Thus, the β cells seem primed to receive a signal from IGF-II. However, some redundancy in the signaling system is evident, since small, less malignant tumors do arise in the IGF-II null mice. One partial substitute for IGF-II might be IGF-I, which is present at low levels (Fig. 1a,b) but binds with higher affinity than IGF-II to the type-1 receptor. Another possibility is insulin itself, which is expressed at high levels in the β cells and known to bind to the IGF-1R, although with very low affinity (Gustafson and Rutter 1990; Zhang and Roth 1991; Germain-Lee et al. 1992). However, the observation that both of these factors are expressed in the quiescent oncogene-expressing β cells prior to the switch instead argues for a distinct signal. Therefore, we postulate that a parallel circuit is supplying the modest growth stimulus in the absence of IGF-II. Some component of this supplementary signal must also be focally activated in order to explain the switch from quiescence to hyperproliferation in the RIP1-Tag2, IGF-II$^{-/-}$ mice; perhaps the same mechanism serves to activate both IGF-II and this collaborator.

The mechanism of IGF-II gene regulation in this transgenic mouse model of tumorigenesis is a clear priority for future research. A preliminary analysis by Southern blotting has not detected rearrangements or amplification of the IGF-II gene, suggesting an epigenetic mechanism. In this regard, it is notable that the IGF-II gene is imprinted; only the paternal allele is expressed in the developing embryo outside of the brain (DeChiara et al. 1991; Ferguson-Smith et al. 1991; Rappolee et al. 1992). In an ongoing study, we have discovered that both alleles of IGF-II are activated during the switch to the proliferative stage in this pathway, including the maternal allele, which is imprinted and not expressed in most tissues (G. Christofori et al., in prep.).

Autocrine Growth Factors as General Cofactors for Oncogene Action?

Dominant oncogenes are implicated in many cancers, and their ability to instruct tumorigenesis has been amply documented in transgenic mouse studies whereby oncogenes established in the mouse germ line elicit heritable development of specific cancers (Hanahan 1988, 1989; Adams and Cory 1991). However, the persistent result in these models has been that tumors arise after periods of normality through a series of distinguishable stages, and, furthermore, only a small fraction of the oncogene-expressing cells develop into tumors. Similarly, genetic disruption of a tumor suppressor (p53) allows normal development and results in sporadic tumorigenesis after a period of normality (Donehower et al. 1992). It is provocative that upregulation or activation of an oncogene (or inactivation of a tumor suppressor) does not necessarily elicit immediate, continuous cell proliferation. That dichotomy is particularly apparent in the model described here, where we have documented that the Tag oncoprotein is expressed in all β cells, beginning with the appearance of the first progenitors within the pancreatic diverticulum from the gut at embryonic day 9 and persisting in all β cells thereafter. Despite the uniform and persistent expression of Tag, the first signs of β-cell hyperproliferation are evident only 3–4 weeks later. The switch correlates with activation of IGF-II, whose absence impairs β-cell proliferation.

The data suggest a two-condition model for oncogene-induced proliferation, whereby both an oncogene and a second signal are necessary to elicit hyperproliferation. IGF-II appears to be a component of that second signal for the switch. In the absence of this second signal, or in the presence of an incomplete signal, there appear to be several alternative responses to oncogene expression: continued quiescence, programmed cell death (apoptosis), or benign proliferation (Fig. 7). Consistent with this model, the Tag oncogene is initially expressed without consequence in both IGF-II wild-type and IGF-II null mice. Then a focal switch to hyperproliferation occurs, and, in both cases, a similar high proliferation index is apparent (by PCNA staining and BrdU incorporation). Remarkably, however, the incidence of apoptotic cell bodies is 5-fold higher in the IGF-II null tumors, suggesting that the second signal is inadequate, such that a cell death program is more frequently activated. Raff and colleagues have shown that the absence of serum growth factors results in apoptosis of developing oligodendrocytes, whereas addition of IGF-II (or IGF-I or PDGF) serves to prevent apoptosis. Of the 20,000 oligodendrocytes born per day, 50% die. Yet the incidence of apoptotic cells seen in histological sections is only 0.25%, indicating a rapid time course of apoptosis and engulfment (Barres et al. 1992). By analogy, the 5-fold increase in apoptotic cells from a 0.5% index in IGF-II wild-type tumors to a 2.5% index in the IGF-II null tumors represents a significant increase in dying cells that could likely explain the reduced tumor mass. The existence of a second signal for oncogene-induced proliferation has been previously implicated in cell culture models of oncogene action. Evan and colleagues have shown that both the c-*myc* oncogene and serum growth factors are necessary for fibroblast proliferation; in the absence of serum the oncogene-expressing cells undergo apoptosis (Evan et al. 1992). Recently, IGF-II, IGF-I, and PDGF have each been shown to be capable of blocking apoptosis in this assay (Harrington et al. 1994). Although these studies implicate IGF-II as a survival factor, the markedly increased malignancy seen in the IGF-II (+/+) tumors indicates IGF-II can have additional effects on the tumor cell phenotype.

There is logic in requiring two signals to initiate unscheduled proliferation of differentiated cells, much as there is in the necessity for confirmatory signals to activate antigen-dependent immune responses (Bretscher and Cohn 1970). In the adult, cell proliferation in most tissues is infrequent and highly constrained, and it would seem disadvantageous for the

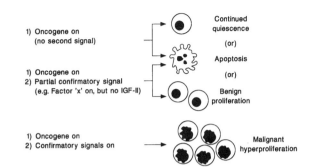

Figure 7. A model: Two signals, both from an oncogene and confirmatory growth/survival factors, are necessary to elicit hyperproliferation. Analysis of the focal switch to hyperproliferation, subsequent to activation of the Tag oncogene, and concomitant with up-regulation of IGF-II, suggests that two signals are necessary to instruct a normal β cell in its natural environment to proliferate. The Tag oncogene provides one of these, and IGF-II is a component of the second signal. In the absence of a confirmatory second signal, expression of the oncogene may result in continued quiescence (as seen here) or induction of apoptosis, as revealed by experiments with conditional oncogenes in tissue culture (Evan et al. 1992). Tumors still develop in the IGF-II null mice, suggesting there is an incomplete second signal, in which another component (factor "x") is activated. However, this putative factor x is not sufficient to induce a high-grade malignancy in the absence of IGF-II; the Tag-expressing cells exhibit a less malignant phenotype with higher incidence of apoptosis and impaired tumor growth. (Reprinted, with permission, from Christofori et al. 1994 [copyright Macmillan Magazines Ltd.].)

organism to allow mere up-regulation of an oncogene to elicit hyperproliferation, with the danger of progression to a cancerous condition. A protective mechanism could be activated in the absence of the second signal, involving active maintenance of quiescence, or induction of apoptosis if proliferation ensues. Again the analogy can be seen to the alternatives of cell paralysis (i.e., quiescence) or apoptosis induced by immune recognition without a confirmatory second signal. The data presented here suggest that IGF-II is one of the key components of the second signal for oncogene action in this model tumorigenesis pathway. It can be anticipated that other growth and survival factors will provide the necessary second signal for oncogene activation of hyperproliferation in different cell types, e.g., lymphokines in hematopoeitic malignancies. Nevertheless, given the numerous reports of their expression in tumor cells, IGF-I and IGF-II may serve as second signals in many naturally occurring cancers and thus represent targets for interventive strategies intended to restrain proliferation or drive oncogene-expressing cells into programmed cell death instead of malignant proliferation.

ACKNOWLEDGMENTS

We thank Liz Robertson and Arg Efstratiadis for the IGF-II gene knockout mice; Liz Robertson for continuing advice and encouragement; Peter Howley, Erwin Wagner, Keith Yamamoto, Ira Herskowitz, and members of the Hanahan laboratory for comments and advice in the preparation of the manuscript; Kathy Smith, Ryo Hirose, and Jeff Arbeit for technical advice; and Juliana Karrim, Frank Loeffler, and Anne Neill for technical assistance. This research was supported by a grant from the National Cancer Institute and benefited from core support for the transgenic mouse facility from the Markey Charitable Trust. G.C. was supported by an American Heart Association postdoctoral fellowship.

REFERENCES

Adams, J. and S. Cory. 1991. Transgenic models of tumor development. *Science* **254:** 1161.

Alpert, S., D. Hanahan, and G. Teitelman. 1988. Hybrid insulin genes reveal a development lineage for pancreatic endocrine cells and imply a relationship with neurons. *Cell* **53:** 295.

Barres, B.A., I.K. Hart, H.S. Coles, J.F. Burne, J.T. Voyvodic, W.D. Richardson, and M.D. Raff. 1992. Cell death and control of cell survival in the oligodendrocyte lineage. *Cell* **70:** 31.

Bravo, R., R. Frank, P.A. Blundell, and H. Macdonald-Bravo. 1987. Cyclin/PCNA is the auxiliary protein of DNA polymerase-delta. *Nature* **326:** 515.

Bretscher, P. and M. Cohn. 1970. A theory of self-nonself discrimination. *Science* **167:** 1042.

Capecchi, M.R. 1988. Altering the genome by homologous recombination. *Science* **244:** 1288.

Christofori, G., P. Naik, and D. Hanahan. 1994. A second signal supplied by insulin-like growth factor II in oncogene-induced tumorigenesis. *Nature* **369:** 414.

Clarke, A.R, E.R. Maandag, M. van Roon, N.M. van der Lugt, M. van der Valk, M.L. Hooper, A. Berns, and H. te Riele. 1992. Requirement for a functional Rb-1 gene in murine development. *Nature* **359:** 328.

Conover, C.A., P. Misra, R.L. Hintz, and R.G. Rosenfeld. 1986. Effect of an anti-insulin-like growth factor I receptor antibody on insulin-like growth factor II stimulation of DNA synthesis in human fibroblasts. *Biochem. Biophys. Res. Commun.* **139:** 501.

DeChiara, T.M., A. Efstratiadis, and E.J. Robertson. 1990. A growth-deficiency phenotype in heterozygous mice carrying an insulin-like growth factor II gene disrupted by targeting. *Nature* **345:** 75.

DeChiara, T.M., E. J. Robertson, and A. Efstratiadis. 1991. Parental imprinting of the mouse insulin-like growth factor II gene. *Cell* **64:** 849.

Donehower, L.A., M. Harvey, B.L. Slagle, M.J. McArthur, C.A. Montgomery, Jr., J.S. Butel, and A. Bradley. 1992. Mice deficient for p53 are developmentally normal but susceptible to spontaneous tumours. *Nature* **356:** 215.

Efrat, S., S. Linde, H. Kofod, D. Spector, M. Delannoy, S. Grant, D. Hanahan, and S. Baekkeskov. 1988. β-Cell lines derived from the transgenic mice expressing a hybrid insulin gene-oncogene. *Proc. Natl. Acad. Sci.* **85:** 9037.

El-Badry, O.M., J.A. Romanus, L.J. Helman, M.J. Cooper, M.M. Rechler, and M.A. Israel. 1989. Autonomous growth of a human neuroblastoma cell line is mediated by insulin-like growth factor II. *J. Clin. Invest.* **84:** 829.

Evan, G.I., A.H. Wyllie, C.S. Gilbert, T.D. Littlewood, H. Land, M. Brooks, C.M. Waters, L.Z. Penn, and D.C. Hancock. 1992. Induction of apoptosis in fibroblasts by c-myc protein. *Cell* **69:** 119.

Fearon, E.R. and B. Vogelstein. 1990. A genetic model for colorectal tumorigenesis. *Cell* **61:** 759.

Ferguson-Smith, A.C., B.M. Cattanach, S.C. Barton, C.V. Beechey, and M.A. Surani. 1991. Embryological and molecular investigations of parental imprinting on mouse chromosome 7. *Nature* **351:** 667.

Filson, A.J., A. Louvi, A. Efstratiadis, and E. Robertson. 1993. Rescue of the T-associated maternal effect in mice carrying *null* mutations in *Igf-2* and *Igf2r*, two reciprocally imprinted genes. *Development* **118:** 731.

Folkman, J., K. Watson, D. Ingber, and D. Hanahan. 1989. Induction of angiogenesis during the transition from hyperplasia to neoplasia. *Nature* **339:** 58.

Foulds, L., ed. 1969. *Neoplastic development*, vol. 1. Academic Press, London.

Garcia, R.L., M.D. Coltrera, and A.M. Gown. 1989. Analysis of proliferative grade using anti-PCNA/cyclin monoclonal antibodies in fixed, embedded tissues. Comparison with flow cytometric analysis. *Am. J. Pathol.* **134:** 733.

Germain-Lee, E.L., M. Janicot, R. Lammers, A. Ullrich, and S.J. Casella. 1992. Expression of a type I insulin-like growth factor receptor with low affinity for insulin-like growth factor II. *Biochem. J.* **281:** 413.

Gustafson, T.A. and W.J. Rutter. 1990. The cysteine-rich domains of the insulin and insulin-like growth factor I receptors are primary determinants of hormone binding specificity. *J. Biol. Chem.* **265:** 18663.

Haig, D. and C. Graham. 1991. Genomic imprinting and the strange case of the insulin-like growth factor II receptor. *Cell* **64:** 1045.

Hanahan, D. 1985. Heritable information of pancreatic beta-cell tumours in transgenic mice expressing recombinant insulin/simian virus 40 oncogenes. *Nature* **315:** 115.

———. 1988. Dissecting multistep tumorigenesis in transgenic mice. *Annu. Rev. Genet.* **22:** 479.

———. 1989. Transgenic mice as probes into complex systems. *Science* **246:** 1265.

Harrington, E.A., M.R. Bennett, A. Fanidi, and G.I. Evan. 1994. C-Myc-induced apoptosis in fibroblasts is inherited by specific cytokines. *EMBO J.* **13:** 3286.

Henson, D.E. and J. Albores-Saavedra, eds. 1993. *The pathology of incipient neoplasia*. Saunders, Philadelphia.

Jacks, T., A. Fazeli, E.M. Schmitt, R.T. Bronson, M.A. Goodell, and R.A. Weinberg. 1992. Effects of an Rb mutation in the mouse. *Nature* **359:** 295.

Kandel, J., E. Bossy-Wetzel, F. Radvanyi, M. Klagsbrun, J. Folkman, and D. Hanahan. 1991. Neovascularization is associated with a switch to the export of bFGF in the multistep development of fibrosarcoma. *Cell* **66:** 1095.

Klein, G. 1987. The approaching era of the tumor suppressor genes. *Science* **238:** 1539.

Knudson, A.G. 1986. Genetics of human cancer. *Annu. Rev. Genet.* **20:** 231.

Lacy, P.E. and M. Kostianovsky. 1967. Method for the isolation of intact islets of Langerhans from the rat pancreas. *Diabetes* **16:** 35.

Lee, E.Y., C.Y. Chang, N. Hu, Y.C. Wang, C.C. Lai, K. Herrup, W.H. Lee, and A. Bradley. 1992. Mice deficient for Rb are nonviable and show defects in neurogenesis and haematopoiesis. *Nature* **359:** 288.

Lee, P.D., D. Hodges, R.L. Hintz, J.H. Wyche, and R.G. Rosenfeld. 1986. Identification of receptors for insulin-like growth factor II in two insulin-like growth factor II producing cell lines. *Biochem. Biophys. Res. Commun.* **134:** 595.

Ludlow, J.W. 1993. Interactions between SV40 large-tumor antigen and the growth suppressor proteins pRB and p53. *FASEB J.* **7:** 866.

Mathieu, M., H. Rockefort, B. Barenton, C. Prebois, and F. Vignon. 1990. Interactions of cathepsin-D and insulin-like growth factor II (IGF-II) and possible consequences in mitogenic activity of IGF-II. *Mol. Endocrinol.* **4:** 1327.

Murayama, Y., T. Okamoto, E. Ogata, T. Asano, T. Iiri, T. Katada, M. Ui, J.H. Grubb, W.S. Sly, and I. Nishimoto. 1990. Distinctive regulation of the functional linkage between the human cation-independent mannose 6-phosphate receptor and GTP-binding proteins by insulin-like growth factor II and mannose 5-phosphate. *J. Biol. Chem.* **265:** 17456.

Nissley, P., W. Kiess, and M.M. Sklar. 1991. The insulin-like growth factor II/mannose 6-phosphate receptor. In *Insulin-like growth factors: Molecular and cellular aspects* (ed. D. LeRoith), p. 111. CRC Press, Boca Raton, Florida.

Osborne, C.K., E.B. Coronado, L.J. Kitten, C.I. Arteaga, S.A. Fuqua, and K. Ramaharma. 1989. Insulin-like growth factor-II: A potential autocrine/paracrine growth factor for human breast cancer acting via the IGF-I receptor. *Mol. Endocrinol.* **3:** 1701.

Pietrzkowski, Z., D. Wernicke, P. Porcu, B.A. Jameson, and R. Baserga. 1992. Inhibition of cellular proliferation by peptide analogues of insulin-like growth factor 1. *Cancer Res.* **52:** 6447.

Porcu, P., A. Ferber, Z. Pietrzkowski, C.T. Roberts, M. Adamo, D. LeRoith, and R. Baserga. 1992. The growth-stimulatory effect of simian virus 40 T antigen requires the interaction of insulin-like growth factor 1 with its receptor. *Mol. Cell. Biol.* **12:** 5069.

Rappolee, D.A, K.S. Sturm, O. Behrendtsen, G.A. Schultz, R.A. Pederson, and Z. Werb. 1992. Insulin-like growth factor II acts through an endogenous growth pathway regulated by imprinting in early mouse embryos. *Genes Dev.* **6:** 939.

Rechler, M.M. 1991. Insulin-like growth factor II: Gene structure and expression into messenger RNA and protein. In *Insulin-like growth factors: Molecular and cellular aspects* (ed. D. LeRoith), p. 87. CRC Press, Boca Raton, Florida.

Rechler, M.M. and S.P. Nissley. 1990. Peptide growth factors and their receptors I. In *Handbook of experimental pharmacology* (eds. M.B. Sporn and A.B. Roberts), vol. 95, p. 263. Springer-Verlag, Berlin, New York.

Reeve, A.E., M.R. Eccles, R.J. Wilkins, G.I. Bell, and L.J. Millow. 1985. Expression of insulin-like growth factor-II transcripts in Wilms' tumour. *Nature* **317:** 258.

Sambrook, J., E.F. Fritsch, and T. Maniatis. 1989. *Molecular cloning: A laboratory manual.* Cold Spring Harbor Laboratory, Cold Spring Harbor, New York.

Schirmacher, P., W.A. Held, D. Yang, F.V. Chisari, Y. Rustum, and C.E. Rogler. 1992. Reactivation of insulin-like growth factor II during hepatocarcinogenesis in transgenic mice suggests a role in malignant growth. *Cancer Res.* **52:** 2549.

Scott, J., J. Cowell, M.E. Robertson, L.M. Priestley, R. Wadey, B. Hopkins, J. Pritchard, G.I. Bell, L.B. Rall, and C.F. Graham. 1985. Insulin-like growth factor-II gene expression in Wilms' tumour and embryonic tissues. *Nature* **317:** 260.

Smith, K.M., R. M. Lawn, and J. N. Wilcox. 1990. Cellular localization of apolipoprotein D and lecithin:cholesterol acyltransferase mRNA in rhesus monkey tissues by in situ hybridization. *J. Lipid Res.* **31:** 995.

Stylianopoulou, F., J. Herbert, M.B. Soares, and A. Efstratiadis. 1988. Expression of the insulin-like growth factor II gene in the choroid plexus and the leptomeninges of the adult rat central nervous system. *Proc. Natl. Acad. Sci.* **85:** 141.

Teitelman, G., S. Alpert, and D. Hanahan. 1988. Proliferation, senescence, and neoplastic progression of β cells in hyperplasic pancreatic islets. *Cell* **52:** 97.

Thompson, M.A., A.J. Cox, R.H. Whitehead, and H.A. Jonas. 1990. Autocrine regulation of human tumor cell proliferation by insulin-like growth factor II: An in-vitro model. *Endocrinology* **126:** 3033.

Werner, H., M. Woloschak, B. Stannard, Z. Shen-Orr, C.T. Roberts, Jr., and D. LeRoith. 1991. The insulin-like growth I receptor: Molecular biology, heterogeneity and regulation. In *Insulin-like growth factors: Molecular and cellular aspects* (ed. D. LeRoith), p. 17. CRC Press, Boca Raton, Florida.

Yee, D., N. Rosen, R.E. Favoni, and K.J. Cullen. 1991. The insulin-like growth factors, their receptors, and their binding proteins in human breast cancer. *Cancer Treat. Res.* **53:** 93.

Zhang, B. and R.A. Roth. 1991. Binding properties of chimeric insulin receptors containing the cysteine-rich domain of either the insulin-like growth factor I receptor or the insulin receptor related receptor. *Biochemistry* **30:** 5113.

Zimmer, A. 1992. Manipulating the genome by homologous recombination in embryonic stem cells. *Annu. Rev. Neurosci.* **15:** 115.

Angiostatin: A Circulating Endothelial Cell Inhibitor That Suppresses Angiogenesis and Tumor Growth

M.S. O'Reilly,* L. Holmgren,* Y. Shing,* C. Chen,* R.A. Rosenthal,* Y. Cao,*
M. Moses,* W.S. Lane,† E.H. Sage,‡ and J. Folkman*

*Children's Hospital, and Harvard Medical School, Boston, Massachusetts 02115; †Harvard Microchemistry Facility, Harvard University, Cambridge, Massachusetts 02138; ‡Department of Biological Structure, University of Washington, Seattle, Washington 98195

INTRODUCTION

Prevascular and Vascular Stages of Tumor Growth

A central problem in tumor growth is how tumor cells switch to the angiogenic phenotype. Virtually all solid tumors are angiogenic and neovascularized by the time they are detectable in animals and humans. However, at the time they originate, spontaneously arising tumor cells are not usually angiogenic (Folkman et al. 1989). Tumors formed by these cells are of small volumes limited to a few cubic millimeters and restricted to an existence in this in situ stage for months or years. Further expansion of such a tumor depends on its induction of new capillary blood vessels that converge toward the tumor. This switch to angiogenesis is usually accomplished by a subset of tumor cells within the in situ lesion. The new microvessels that are recruited provide a neovascular meshwork which supports the growth, and facilitates invasion and metastasis of the rapidly expanding tumor mass.

The new microvessels not only *perfuse* the tumor with blood, which transports nutrients and oxygen to the tumor and catabolites away from it, but the endothelial cells of these vessels also produce a spectrum of growth factors which have a *paracrine* stimulatory effect on the tumor cells (Fig. 1) (Nicosia et al. 1986; Hamada et al. 1992; Rak et al. 1994). This combined perfusion and paracrine effect of neovascularization may permit a tumor to expand by as much as 1,000–16,000 times its in situ volume in weeks or months (Gimbrone et al. 1972). The neovascular network also provides an increased vascular surface for tumor cells to escape into the main circulation and to metastasize to remote organs. The evidence (direct and indirect) that tumor growth and metastasis are angiogenesis-dependent has been reviewed recently (Folkman 1994a).

How Do Tumors Switch to the Angiogenic Phenotype?

The mechanism by which tumor cells *switch* to the angiogenic phenotype has been a subject of great interest in our laboratory. However, it was inaccessible to analysis until a report in 1985 by Douglas Hanahan (1985). He developed transgenic mice in which the large "T" oncogene is hybridized to the insulin promoter. The β cells in the islets of the pancreas undergo hyperplasia and then progress to tumors by a reproducible and predictable multistep process (Hanahan 1985). One of these steps occurs at 6–7 weeks, when angiogenesis is switched on in approximately 10% of pre-neoplastic islets. Vascularized tumors arise from these islets and are fatal to the mice by 12 weeks. This animal system closely resembles the pattern of onset of angiogenic activity in human tumors. We subsequently found that the islets become chemotactic to capillary endothelial cells in vitro when they become angiogenic (Folkman et al. 1989). A similar pattern was found in another transgenic tumor, murine fibrosarcoma, driven by the bovine papilloma virus genome (Kandel et al. 1991).

Net Balance of Positive and Negative Regulators of Tumor Angiogenesis

Conditioned media from these vascularized tumors contained proteins that were mitogenic to endothelial cells. The tumors also produced proteins that inhibited endothelial proliferation (Shing et al. 1993; Y. Shing, unpubl.). This led us to view the angiogenic process as a *net balance* between positive and negative regulators of angiogenesis. Thus, the switch to the angiogenic pheno-

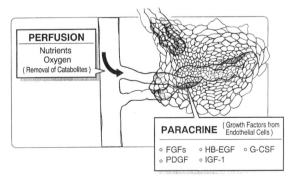

Figure 1. Tumor neovascularization. Diagram to illustrate the concept that new microvessels recruited by a tumor can stimulate tumor growth both by perfusion of nutrients (and extraction of catabolites) and by paracrine release of growth factors from invading endothelial cells.

type could not be explained simply in terms of increased expression, export, or mobilization of angiogenic factors. Rather, it appeared that positive mediators of capillary blood vessel growth (e.g., basic fibroblast growth factor [bFGF], vascular permeability factor/vascular endothelial growth factor [VPF/VEGF], IL-8, hepatocyte growth factor [HGF]), must also overcome a variety of *negative* regulators of angiogenesis. The physiological function of these negative regulators is not clear, but they may serve to defend vascular endothelium against mitogens and chemotactic agents under normal conditions.

Some negative regulators such as platelet factor 4 (Taylor and Folkman 1982) (Maione et al. 1990) and interferon-α are found in the circulation. Others operate at the tissue level; e.g., pericyte suppression of endothelial proliferation by transforming growth factor-β (TGF-β) (Antonelli-Orlidge et al. 1989; Satoh et al. 1990), growth control of endothelial cells by cell shape (Folkman and Moscona 1978), and storage of endothelial mitogens in extracellular matrix (Folkman et al. 1988). These endothelial inhibitors appear to be constitutive; i.e, they exist independently of whether a primary tumor is present. However, Shing et al. (1993) found *tumor-derived* endothelial inhibitors by analyzing the conditioned medium of β-tumor cells isolated from vascularized tumors. These inhibitors included tissue inhibitor of metalloproteinase (TIMP), TGF-β, and an endothelial cell inhibitor which has been purified but not sequenced at this writing.

However, Noel Bouck and her colleagues were the first to actually demonstrate that the angiogenic phenotype is a net balance between positive and negative regulators of angiogenesis. They reported that when hamster cells are transformed to angiogenic tumor cells, their production of thrombospondin is down-regulated to approximately 4% of the production by the normal parental cells (Rastinejad et al. 1989; Bouck 1990; Good et al. 1990). Thrombospondin inhibits endothelial proliferation and migration in vitro and angiogenesis in vivo (Good et al. 1990). Bouck has recently shown that the transcriptional regulation of thrombospondin in human fibroblasts (from cancer-prone patients with the Li-Fraumeni syndrome) is under the control of the wild-type *p53* tumor suppressor gene (Dameron et al. 1994a,b). Taken together with our studies in transgenic mice, these data may be explained by a model in which tumors in the prevascular in situ stage produce endothelial inhibitors at sufficient strength to neutralize their elaboration of angiogenic proteins such as bFGF or VEGF. In the final analysis, a tumor cell may be required to decrease production of its endothelial inhibitor(s) in order to become angiogenic.

Discovery of a Novel Circulating Angiogenesis Inhibitor That Controls Tumor Growth

This view that a tumor becomes neovascularized when it elaborates angiogenic factors in *excess* of angiogenic inhibitors was a new perspective of the start-up of angiogenesis in a primary tumor. What then was the fate of the angiogenic stimulators and inhibitors elaborated by a primary tumor when these proteins diffused into the circulation? Previous studies demonstrated that the most commonly produced angiogenic proteins have a short half-life in the blood after a bolus intravenous injection: less than 30 minutes for bFGF (Thompson et al. 1990) and approximately 3 minutes for VEGF (Richardson et al. 1994). If an angiogenic inhibitor elaborated by a primary tumor remained in the circulation longer (i.e., hours or days) than the angiogenic stimulator, then conceivably the blood could become increasingly inhibitory to endothelial proliferation as tumor mass increased.

Therefore, we began a search for other tumor phenomena in which negative regulators of angiogenesis might reveal themselves. One example is the problem of how certain tumors suppress the growth of their metastases. A return to this old problem led us to the discovery of a novel angiogenesis inhibitor that appears in the blood of mice bearing the murine Lewis lung carcinoma. This inhibitor suppresses the angiogenesis and growth of secondary metastases. A brief review of the problem of suppression of tumor growth by tumor mass is necessary here, because it forms the basis of our experimental design.

The Phenomenon of Inhibition of Tumor Growth by Tumor Mass

The inhibition of tumor growth by tumor mass has been repeatedly studied, but a satisfactory molecular mechanism has not been uncovered. In cancer patients, the removal of certain tumors, such as breast or colon cancer, is often followed by the rapid growth of distant metastases (Clark et al. 1989; Woodruff 1990). The inhibition of metastases of one tumor can be seen with a different type of primary tumor; i.e., a breast carcinoma can inhibit metastases of a melanoma. Furthermore, if one portion of a primary tumor is removed, the residual tumor increases its growth rate (Lange et al. 1980).

A variety of animal studies reveal a similar process. A primary tumor inhibits the growth, but not the number, of its metastases. Removal of the tumor results in explosive growth of the metastases. This is unrelated to the wounding effect, because sham-operated animals with an intact primary tumor do not have visible metastases (Marie and Clunet 1910; Tyzzer 1913; Schatten 1958; Greene and Harvey 1960; Milas et al. 1974; Gorelik et al. 1978, 1980; Bonfil et al. 1988). A primary tumor can also suppress the growth of a second tumor implant (Ehrlich and Apolant 1905; Bashford et al. 1907; Gershon et al. 1967; Lausch and Rapp 1969; Gunduz et al. 1979; Gorelik et al. 1981; Gorelik 1983; Ruggiero et al. 1985). The "resistance" to a second tumor challenge is directly proportional to the size of the tumor inoculum and inversely proportional to the size of the first tumor. A threshold size is necessary for the inhibitory effect to occur. As in humans, a primary tumor in a mouse can inhibit a secondary tumor of a different type (Gorelik et al. 1981; Gorelik 1983; Himmele et al. 1986).

At least three hypotheses have been proposed to explain this phenomenon (for review, see Gorelik 1983; Prehn 1991, 1993): (1) "concomitant immunity" (Bashford et al. 1907); (2) depletion of available nutrients by the primary tumor; and (3) production of antimitotic factors by primary tumor cells that directly inhibit proliferation of tumor cells in a metastasis (Mohr et al. 1968; Rounds 1970; Bichel 1971; DeWys 1972; Elleman and Eidinger 1977; Rothbarth et al. 1977; Ruggiero et al. 1985). These interesting ideas have, unfortunately, not led to a general theory of the suppression of tumor growth by tumor mass.

Hypothesis: Can a Primary Tumor Stimulate Angiogenesis in Its Own Vascular Bed, but Inhibit It in the Vascular Bed of a Remote Metastasis?

We recently proposed a new hypothesis that a primary tumor, although capable of stimulating angiogenesis in its own vascular bed, could inhibit angiogenesis in the vascular bed of a metastasis or other secondary tumor (O'Reilly et al. 1993). In its simplest terms, the hypothesis states that a primary tumor initiates its own neovascularization by generating angiogenic stimulator(s) in excess of angiogenesis inhibitor(s). However, the putative angiogenesis inhibitor, by virtue of its longer half-life in the circulation, reaches the vascular bed of a secondary tumor in excess of angiogenic stimulators escaping from the primary tumor or generated by the secondary tumor. As a result, growth of the metastasis (i.e., secondary tumor) is inhibited. This model is not limited to a single inhibitor or stimulator of angiogenesis.

The hypothesis depends on (1) the larger mass (and thus the greater production of inhibitor) achieved by a primary tumor *before* a secondary tumor begins its growth; (2) the absorption of the angiogenesis inhibitor into the circulation in excess of angiogenesis stimulator; and (3) the significantly prolonged retention of angiogenesis inhibitor in the circulation relative to angiogenic stimulator.

We have recently reported the isolation, purification, and amino acid sequence of a protein generated by a primary tumor, and we have demonstrated that it inhibits angiogenesis and growth in a secondary metastasis (O'Reilly et al. 1994). This discovery and its potential scientific and clinical applications are reviewed below.

EXPERIMENTAL METHODS

Choice of a mouse tumor that potently suppresses metastases. Several different mouse tumors were compared for the ability of a tumor implanted in the dorsal subcutaneous space to inhibit growth of its distant metastases. The primary tumor was removed under methoxyflurane anesthesia by sterile technique after it had reached 1.5 cm^3 (~14 days), and the skin was closed with sutures. After 5 days and at subsequent intervals, the liver and lungs were checked for the presence of visible metastases, and these were counted by stereomicroscopy at 10×. A Lewis lung carcinoma most potently suppressed lung metastases (designated LLC-LM, low metastatic). A variant of this tumor suppressed metastases only weakly (designated LLC-HM, high metastatic).

Detection of antimetastatic activity in serum and urine of mice bearing a primary tumor. Blood was obtained by sterile cardiac puncture (No. 22 gauge hypodermic needle) of anesthetized mice bearing 1.5 cm^3 tumors. Serum which was obtained from pooled blood that had been clotted, chilled to 4°C, and centrifuged (20 min × 3000 rpm), was filtered (0.22 μm) and stored at 4°C if used within 24 hours, or frozen. For urine collection, mice with 600-mm^3 tumors were kept in metabolic cages until the tumors were 3500 mm^3. During this period, urine was collected into a central chilled reservoir and cages were cleaned daily. To promote diuresis, 10% sucrose was added to the drinking water. Every 48 hours, urine was centrifuged at 4°C, filtered (0.45 μm), and dialyzed (molecular weight cutoff 6-8000) against distilled H$_2$O (changed every 6 hours) at 4°C for 48 hours. Dialyzed urine was pooled and frozen until used or lyophilized for animal experiments. The same method was employed to obtain urine from normal mice without tumors.

Histological studies of metastases. To highlight blood vessels, histological sections in paraffin were stained with antibody to von Willebrand's factor (Dako, Seattle, Washington), a specific marker for vascular endothelium, as described by Weidner et al. (1991). DNA synthesis in these metastases was determined by incorporation of bromodeoxyuridine (BrdU). Animals received an intraperitonal injection of BrdU (0.75 mg/mouse) 6 hours before sacrifice. Before immunohistochemical staining, sections were permeabilized with 0.2 M HCl for 10 minutes and digested with 1 μg/ml proteinase K in 0.2 M Tris-HCl, 2 mM CaCl$_2$ at 37°C for 15 minutes. Labeling index was estimated by counting the percentage of positive nuclei at 250×.

The apoptotic rate was determined by in situ terminal deoxynucleotidyl transferase (TdT) labeling with digoxigenin-labeled dUTP in lungs that had been fixed overnight in 4% formaldehyde in PBS. The stained paraffin sections were developed with diaminobenzidine (Gavrieli et al. 1992).

Our preliminary studies show that the apoptotic index of tumor cells in the nongrowing lung metastases in the presence of the primary tumor was 7.4% ± 0.3. In contrast, in the exponentially growing metastases, 5 days after removal of the primary tumor, the apoptotic index was 1.9% ± 0.3.

Angiogenesis in the cornea. A corneal pocket was created in anesthetized mice as described previously

(Muthukkaruppan and Auerbach 1979). A 0.34-mm² sustained-release pellet containing bFGF was implanted into the pocket 1–1.2 mm from the corneal limbus. (The pellet contained 80–100 ng of bFGF mixed with sucrose aluminum sulfate [Bukh Meditec, Copenhagen, Denmark] and coated with hydron polymer type NCC [Interferon Sciences, New Brunswick, New Jersey] [B. Kenyon et al., unpubl.].) Pellets were implanted into the eyes of mice with growing primary Lewis lung carcinomas of greater than 2000 mm³ or into control mice without a primary tumor. The corneas of all mice were observed daily and photographed using a slit-lamp stereomicroscope at 10× magnification.

Inhibition of capillary endothelial cells in vitro by serum or urine. Bovine capillary endothelial cells (Folkman et al. 1979) adjusted to 25,000 cells/ml were plated onto gelatinized 24-well culture plates (0.5 ml/well) and incubated (37°C, 10% CO_2) for 24 hours in DMEM with 10% heat-inactivated bovine calf serum (BCS), antibiotics, and 3 ng/ml recombinant human bFGF (Scios Nova Inc., Mountainview, California). The medium was then replaced with 0.25 ml of DMEM/5% calf serum/1% antibiotics, and the test sample of serum was added to each well, 5 μl–25 μl per well, and cells were counted 72 hours later by Coulter counter. Urine samples were tested similarly, except that urine was dialyzed against distilled water, lyophilized, and brought to a 50-fold concentration before addition to the endothelial cell cultures.

Purification of an endothelial inhibitor from serum and urine. Urine or serum from tumor-bearing animals was applied to a heparin–Sepharose column (30 × 1.5 cm) equilibrated with 50 mM NaCl/10 mM Tris [pH 7.2] and eluted with a continuous gradient of 50 mM NaCl–2 M NaCl followed by 100 ml of 2 M NaCl (described by O'Reilly et al. 1994). An aliquot of each of the collected fractions was applied to endothelial cells. Inhibitory fractions were vacuum concentrated and dialyzed. The pooled and concentrated sample from heparin–Sepharose chromatography was applied to a Bio-gel A-0.5-m column and eluted with 0.15 M NaCl at 8 ml/hour. Fractions were handled as above. The pooled and concentrated sample from Bio-gel A-0.5-m chromatography was applied to a SynChropak RP-4 (100 × 4.6 mm) (C4) column equilibrated with H_2O/0.1% TFA, and the column was eluted with a gradient of acetonitrile/0.1% TFA, and fractions were collected. An aliquot of each was evaporated by vacuum centrifugation, resuspended in H_2O, and applied to endothelial cells. Inhibitory activity was further purified by two subsequent cycles with the C4 column.

Purification of inhibitory fragments of human plasminogen. Plasminogen lysine binding site I (Sigma) was resuspended in water. Alternatively, human plasminogen obtained from Sigma, or from human plasma as described previously (Deutsch and Mertz 1970; Bok and Mangel 1985), was digested with elastase. The samples were applied to the C4 high-performance liquid chromatography (HPLC) column and eluted as above, and an aliquot of each fraction was evaporated by vacuum centrifugation, resuspended in water, and applied to capillary endothelial cells. Inhibitory activity was further purified by two subsequent cycles with the C4 column.

Extraction of inhibitory activity from SDS-PAGE. Inhibitory fractions from the serum, urine, and human plasminogen purifications were resolved by SDS-PAGE under nondenaturing conditions. Areas corresponding to silver-stained bands were cut from the gel and incubated in phosphate-buffered saline at 4°C for 12 hours. The supernatant was dialyzed twice against saline for 6 hours (MWCO = 6-8000) and twice against distilled water for 6 hours. The product was evaporated by vacuum centrifugation and applied to endothelial cells.

Trypsin digestion, HPLC separation, and protein microsequencing. The 38-kD inhibitor of capillary endothelial cell proliferation from urine of tumor-bearing animals and the 40-kD, 42.5-kD, and 45-kD inhibitory fragments of human plasminogen were purified to homogeneity. Each was resolved by SDS-PAGE, electroblotted onto PVDF (Bio-Rad, Richmond, California), detected by Ponceau S stain, and individually excised from the membrane. Amino-terminal sequence of each was determined by automated Edman degradation on an ABI Model 477A protein sequencer. In similar fashion, sequence analyses of peptides resulting from a tryptic digest of a separate sample of the 38-kD inhibitor from the urine of tumor-bearing mice and of the 40-kD inhibitory fragment of human plasminogen were isolated and submitted to in situ digestion with trypsin. The resulting peptide mixture was separated by narrow-bore HPLC (Vydac C18 reverse-phase column). Optimum fractions were chosen on the basis of symmetry, resolution, UV absorbance and spectra, and were further screened for length and homogeneity by matrix-assisted laser desorption mass spectrometry on a Finnigan Lasermat and submitted to automated Edman degradation as above. Strategies for peak selection, reverse-phase separation, and protein microsequencing have been described previously (Lane et al. 1991). Sequence library searches and alignments were performed against combined GenBank, Brookhaven Protein, SWISS-PROT, and PIR databases. Searches were performed at the National Center for Biotechnology Information using the BLAST network service. The 38-kD peptide from urine of tumor-bearing mice and the 40-kD internal fragment of human plasminogen to which it has near identity were named "angiostatin."

Antiangiogenic activity of angiostatin on the chick chorioallantoic membrane. Human angiostatin purified from elastase-treated human plasminogen was applied to the chorioallantoic membrane of 6-day-old chick embryos at concentrations of 0.1 μg to 100 μg per embryo in 10-μl pellets of air-dried methylcellulose as described previously (Crum et al. 1985). At 48 hours, the presence of avascular zones was quantitated under a stereomicroscope at $10\times$–$20\times$.

Anti-metastatic activity of angiostatin. Primary tumors were removed from mice after they had reached a volume of 1.5–2.0 cm^3. Mice received angiostatin (1.2 mgm/kg [24 μg]) by intraperitoneal injection on the day of the operation, followed by 0.6 mgm/kg/day (12 μg) for the next 13 days, after which treated and untreated mice were sacrificed, and the lungs were examined grossly and by stereomicroscopy and weighed.

Angiostatin therapy of primary tumors. Lewis lung carcinoma cells (LLC-LM) (1×10^6) were injected subcutaneously on the dorsal surface between the scapulae. When the tumors became palpable and visible ($\sim$80–160 mm^3, day 3), the mice received a daily intraperitoneal injection of angiostatin (obtained by elastase digestion of human plasminogen) at doses of 1.2 mgm/kg/day to 10 mgm/kg/day. Tumor volume was measured every other day and calculated according to a standard formula, width^2 $\times$ length $\times$ 0.52.

RESULTS

A Primary Tumor Inhibits Growth of Metastases

Within 5 days of removal of a primary Lewis lung carcinoma of 0.8–2.0 cm^3, there was a first significant increase in the number of visible lung metastases. By 15 days, the number of visible metastases had increased 10-fold (50 ± 5) compared to control mice with an intact tumor (4 ± 2) ($p<0.001$). At this time, lung weight had increased 400% in mice with the primary tumor removed compared to mice with the primary tumor intact ($p<0.001$). Similar results were observed in immunodeficient SCID mice (data not shown).

Inhibition of Metastatic Growth Correlates with Lack of Neovascularization and Increased Apoptosis

Histological sections of lungs from mice with an intact tumor showed microscopic metastases of approximately 100–150 μm diameter, mainly in the form of perivascular cuffs with a radius of 4–6 cell layers around a preexisting venule. There were also a few thin pleural implants of 2 layers of tumor cells (Fig. 2). These micrometastases were not neovascularized, as shown by staining with antibodies to von Willebrand's factor, a specific marker for vascular endothelial cells. Normal preexisting vessels in the lung were stained. In contrast, numerous new capillary sprouts were found in the metastases as early as 5 days after removal of the primary tumor.

Despite the fact that the presence of an intact primary tumor restricted lung metastases to a microscopic size, they contained 25–30% replicating tumor cells (as determined by incorporation of BrdU). Surprisingly, the rapidly growing metastases (with the primary tumor removed) also showed a 25–30% labeling index, i.e., a high rate of proliferation, equivalent to the nonenlarging, growth-suppressed metastasis (Holmgren et al. 1995). The major difference between the two types of metastases was the high death rate of individual tumor cells; i.e., an apoptosis rate of greater than 7.4% $\pm$ 0.3 in the growth-inhibited metastasis. In contrast, only 1.9% $\pm$ 0.3 of cells were apoptotic in the rapidly growing metastases, a highly significant difference (Holmgren et al. 1995). Thus, once metastatic cells have reached the lung and have begun early tumor formation, the primary tumor appears to inhibit their neovascularization and growth. Removal of a primary tumor is associated not only with increased neovascularization, but also with decreased apoptosis. This diminished apoptosis appears to be a critical event for rapid growth of metastases, because there was no significant difference in the rate of DNA synthesis (replication rate) between the inhibited metastases and the growing metastases.

Serum and Urine from Tumor-bearing Mice Inhibits Growth of Metastases

When the primary tumor was removed at day 15 and the mice were treated with daily intraperitoneal injections of serum (Fig. 3A) or dialyzed and concentrated urine from tumor-bearing mice (Fig. 3B), there were less than 5 visible surface metastases per mouse after 13 days of treatment. Lung weight was within normal limits. These results were similar to mice with an *intact* primary tumor treated only with saline injections. In contrast, when mice were treated with normal serum or urine after removal of the primary tumor, there was a 10-fold increase in visible surface metastases in the lung, and a 4-fold increase in lung weight ($p<0.001$).

The Primary Tumor Inhibits Corneal Angiogenesis

Because angiogenesis was virtually absent in metastases in the presence of a primary tumor and appeared rapidly after removal of the primary tumor, we determined whether the primary tumor could directly inhibit angiogenesis induced in the cornea by an exogenous pellet of bFGF. In mice without a primary tumor, new capillary vessels grew through the cornea and into the bFGF pellet within 6 days. In contrast, in mice with a primary tumor of at least 2000 mm^3, corneal neovascularization was completely prevented.

Serum or Urine from Tumor-bearing Animals Specifically Inhibits Proliferation of Vascular Endothelial Cells

Proliferation of bFGF-stimulated bovine capillary endothelial cells was inhibited down to 70% of untreated controls, in a dose-dependent and reversible manner, by serum of tumor-bearing mice but not by normal serum (Fig. 4). Upon removal of the primary

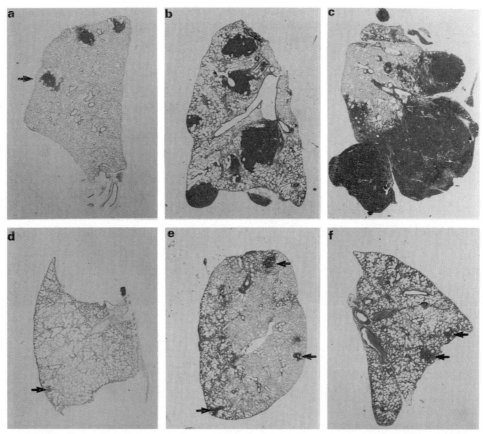

Figure 2. H & E staining of mouse lungs at 5× magnification in the presence or absence of a primary Lewis lung carcinoma. Upper panels (a–c) show growth of metastases at 5, 10, and 15 days after the primary tumor was removed from the back. Lower panels (d–f) show suppression of growth of metastases in the presence of an intact primary Lewis lung carcinoma on the back. In the presence of the primary tumor there are focal microscopic metastases that do not grow (i.e., arrow in d). (Reprinted, with permission, from O'Reilly et al. 1994 [copyright held by Cell Press].)

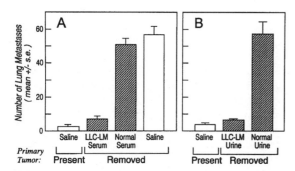

Figure 3. Inhibition of the growth of lung metastases after primary tumor removal by systemic treatments with serum or urine from tumor-bearing mice. Mice were treated with serum (A) or urine (B) from tumor-bearing or normal animals, or with saline after primary tumor removal. One group had its tumors left intact. Metastatic growth was inhibited by leaving the tumor intact. After tumor removal, metastases were inhibited only in mice treated with serum or urine from tumor-bearing mice. (LLC-LM) Lewis lung carcinoma-low metastatic. (Reprinted, with permission, from O'Reilly et al. 1994 [copyright held by Cell Press].)

tumor, serum endothelial inhibitory activity declined gradually, with half-maximal inhibitory activity present at 2.5 days and no detectable activity at 5 days (Table 1).

Dialyzed and concentrated urine from tumor-bearing mice also inhibited endothelial proliferation in vitro in a dose-dependent and reversible manner. Normal urine had no effect on endothelial cells.

Both serum and urine from tumor-bearing mice appeared to specifically inhibit only vascular endothelial cells. Thus, proliferation of bovine capillary endothelial cells, aortic endothelial cells, and endothelial cells from murine hemangioendothelioma (Obeso et al. 1990) was inhibited in a dose-dependent manner. Proliferation was not inhibited in epithelial cells (mink lung epithelium, retinal pigment epithelium), vascular smooth muscle cells (bovine aortic), fibroblasts (WI38 fetal lung, murine fetal fibroblasts), or tumor cells (Lewis lung carcinoma). Serum from tumor-bearing animals weakly and variably inhibited 3T3 fibroblasts; i.e., either no inhibition or up to 30% inhibition.

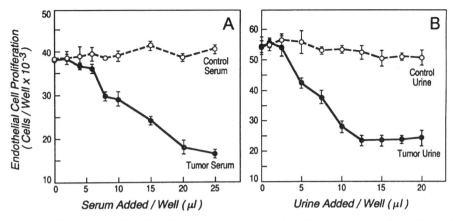

Figure 4. Inhibition of capillary endothelial cell proliferation by serum or urine of tumor-bearing animals. Serum (A) or urine (B) (dialyzed and concentrated) from tumor-bearing (●) or normal animals (○) was applied to bovine capillary endothelial cells in the presence of 1 ng/ml bFGF in a 72-hr proliferation assay. Both serum and urine from tumor-bearing animals inhibited capillary endothelial proliferation in a dose-dependent manner as compared to controls. Each point represents mean and standard error. Inhibition was specific for endothelial cells. (Reprinted, with permission, from O'Reilly et al. 1994 [copyright held by Cell Press].)

Table 1. Decrease of Serum Endothelial Inhibitory Activity after Removal of a Primary Tumor

Days after tumor removal	Endothelial proliferation (% of control)
0	−70
1	−50
3	−29
5	− 8
7	0

The Endothelial Inhibitory Activity of Serum or Urine from a Tumor-bearing Animal Appears to Be Independent of the Immune System

When primary tumors were grown in athymic (nu/nu) mice or in SCID mice, antiproliferative activity against capillary endothelial cells was similar to serum or urine from tumor-bearing immunocompetent C57BL/6 mice (data not shown). No inhibitory activity was found in the serum or urine of athymic or SCID mice that lacked a primary tumor. These results indicate that the endothelial inhibitory activity operates independently of T or B lymphocytes.

Serum and Urine from a High Metastatic Variant of Lewis Lung Carcinoma Does Not Inhibit Endothelial Cell Proliferation

Repeated transplantation of Lewis lung carcinoma yielded a subline which gave rise to primary tumors that were highly metastatic even when the primary tumor was intact. This tumor cell line, LLC-HM, could not suppress metastatic growth at all, or at the best, by less than 10% of the inhibitory activity of its parent line (LLC-LM). Serum and urine from mice carrying the LLC-HM subline did not inhibit endothelial cell proliferation in vitro (data not shown).

Purification of a 38-kD Protein from Serum and Urine of Tumor-bearing Mice Inhibits Endothelial Proliferation

Endothelial inhibitory activity was purified from pooled serum and urine as described in Experimental Methods. Inhibitory activity from both serum and urine eluted at 30–35% acetonitrile from the C4 column. After the final C4 HPLC, a single band of endothelial inhibitory activity corresponding to a protein of apparent reduced M_r of 38 kD was extracted from an SDS polyacrylamide gel (data not shown). Endothelial inhibitory activity of the purified product, as determined by a 72-hour in vitro assay of bFGF-stimulated bovine capillary endothelial cells, was heat (100°C × 10 min) and trypsin sensitive and was associated with the aqueous phase of a chloroform extraction (data not shown). When equivalent volumes of pooled serum or urine obtained from *normal* animals were fractionated by heparin-Sepharose and Biogel A-0.5m chromatography, no inhibitory activity was found (data not shown).

Microsequence Analysis of the 38-kD Protein

One sample of the purified 38-kD protein (from 20 liters of mouse urine) was submitted for amino-terminal analysis, and another sample (from 100 liters of mouse urine) was submitted for tryptic digest and sequence analysis of the resultant peptides (William S. Lane, Harvard Microchemistry, Cambridge, Massachusetts). Microsequence analysis of both samples of the endothelial inhibitor revealed more than 98% identity to an internal fragment of plasminogen with an amino terminus at amino acid 98 and a predicted approximate carboxyl terminus at amino acid 440 (Fig. 5a). The inhibitory fragment includes the first four of the five triple loop structures, or kringle regions of plasminogen (Lerch et al. 1980; Ponting et al. 1992). We named this fragment "angiostatin."

An Internal Fragment of Human Plasminogen Is Homologous to Mouse Angiostatin and Has Similar Endothelial Inhibitory Activity In Vitro

When fragments of human plasminogen were generated by degradation with elastase, purified, and electrophoresed on SDS-acrylamide, there were three distinct bands of apparent reduced M_r of 40 kD, 42.5 kD, and 45 kD, all three of which inhibited proliferation of bFGF-stimulated capillary endothelial cells. Amino-terminal microsequence analysis of each band and of peptides resulting from a tryptic digest of the 40-kD band revealed identity to an internal fragment of plasminogen with an amino terminus at amino acid 97 or 99. This region of the plasminogen molecule has been described as lysine-binding site I (Fig. 6) (Lerch et al. 1980; Ponting et al. 1992). Full-length plasminogen had no effect on the growth of capillary endothelial cells. Interestingly, fragments of plasminogen corresponding to lysine-binding site II (Sottrup-Jensen et al. 1978; Lerch et al. 1980), which includes kringle 4 alone, *stimulated* proliferation of capillary endothelial cells by at least 20% above the proliferation induced by bFGF (data not shown).

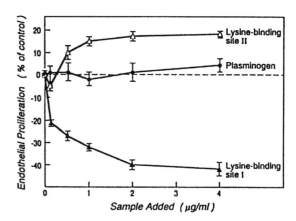

Figure 6. Inhibition of bFGF-induced capillary endothelial cell proliferation at 72 hr of incubation with the lysine-binding site I fragment of plasminogen (which includes angiostatin). Plasminogen has no effect. The lysine-binding site II fragment stimulates endothelial cell proliferation.

Human Angiostatin Inhibits Angiogenesis in the Chick Embryo

A small quantity of human angiostatin purified from elastase-digested plasminogen that inhibited endothelial cell proliferation in vitro was tested for antiangiogenic activity in the chick chorioallantoic membrane, over a concentration range of 0.1 μg to 100 μg/embryo. Human angiostatin inhibited angiogenesis, e.g., induction of avascular zones by 48 hours after implantation of the test substance on the 6-day chorioallantoic membrane, in a dose-dependent man-

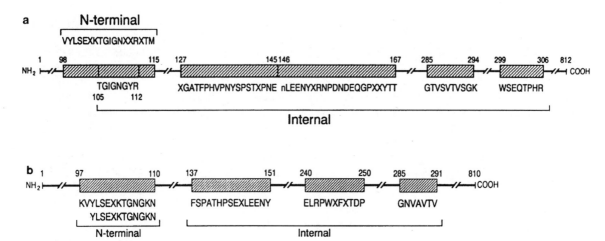

Figure 5. (*a*) Schematic of *mouse* plasminogen. Amino acid sequences of the amino terminus and of fragments resulting from a tryptic digest of the endothelial inhibitor purified from the urine of tumor-bearing mice are noted. Unidentified amino acid residues are denoted by X, and the one residue not identical to the known sequence of murine plasminogen is indicated by a lowercase letter. We have named this inhibitor "angiostatin." (*b*) Schematic of *human* plasminogen. Amino-terminal sequence and sequence of tryptic digest fragments of angiostatin purified from human plasminogen are noted. Unidentified amino acid residues are denoted by X. Angiostatin purified from human plasminogen corresponds to the angiostatin purified from the urine of tumor-bearing mice. (Reprinted, with permission, from O'Reilly et al. 1994 [copyright held by Cell Press].)

ner. The inhibition was dose dependent, beginning at 20 µg and reaching saturation at approximately 100 µg. No toxicity or inflammation was observed.

Human Angiostatin Inhibits Growth of Metastases in Mice from Which the Primary Tumor Has Been Removed

To determine whether human angiostatin inhibited the growth of metastases, primary tumors that had reached approximately 1500 mm^3 volume (on ~day 12–16 after tumor implantation), were removed from mice. Half of the mice were treated with intraperitoneal injections of purified human angiostatin at 24 µg/20 g mouse on day 1, followed by 12 µm/day for 13 days. This therapy suppressed growth of metastases with an efficacy equal to or greater than the presence of a primary tumor. Angiostatin-treated mice revealed an 18-fold reduction in the number of visible metastases ($p < 0.001$), in contrast to control mice treated with plasminogen or with saline only, in which there were more than 50 metastases per lung (Fig. 7). The angiostatin-treated mice showed no significant increase in lung weight. In contrast, there was a 3- to 5-fold increase in lung weight in control mice treated with plasminogen or saline, a direct measure of increased tumor burden in the lung.

The growing, grossly visible metastases in control mice treated only with plasminogen or saline were highly neovascularized as revealed with antibodies to von Willebrand's factor. In contrast, the majority of metastases of mice treated with angiostatin remained microscopic in size (< 100 µg diameter). They either contained no new vessels or were very sparsely vascularized.

Human Angiostatin Also Inhibits the Growth of Primary Tumors

In preliminary experiments, LLC-LM were implanted subcutaneously and allowed to grow until they were neovascularized and had reached approximately 100–150 mm^3. The mice received human angiostatin at 10 mg/kg by intraperitoneal injection once daily. Untreated controls received saline or plasminogen. At 20 days, the ratio of treated volume to control volume (T/C) was 0.23. This is more potent inhibition than AGM-1470 (TNP-470), a fumagillin angiogenesis inhibitor currently in Phase I clinical trial, which has a mean T/C of 0.35. A second tumor also was inhibited by angiostatin. Treatment of the murine fibrosarcoma T241 for 15 days resulted in a T/C of 0.26. There was no toxicity. We have not yet established an optimum dose, nor a minimum toxic dose, because of the small supply of angiostatin currently available. These data are preliminary because there are only 3 mice per group due to the scarcity of angiostatin.

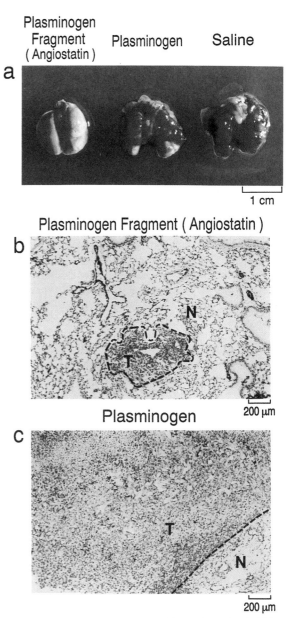

Figure 7. Systemic treatment with angiostatin inhibits neovascularization and growth of metastases after primary tumor removal. (*a*) Lungs from mice 13 days after removal of a primary tumor. Treatment with angiostatin blocked the growth of metastases. Mice treated with plasminogen or saline had almost complete replacement of the normal lung tissue by growing invasive metastases. (*b,c*) H & E staining and immunohistochemical staining with antibodies against von Willebrand's factor of lungs of mice after removal of the primary tumor. In mice treated with intact plasminogen (*c*), there was intense neovascularization and growth of the metastases. In contrast, mice treated with angiostatin (*b*) had only perivascular cuffing of metastases without neovascularization. (N) Normal lung, (T) tumor.

DISCUSSION

These experiments show that at least one mechanism by which a primary tumor suppresses its remote metastases is to release a circulating endothelial inhibitor which also inhibits angiogenesis. Angiostatin is an example of such a mediator. These studies indicate that whereas a primary tumor can *stimulate* angiogenesis in its own vascular bed, it can *inhibit* angiogenesis in remote metastases.

Angiogenesis is virtually completely abolished in the metastases. The lack of neovascularization restricts growth so that the resulting micrometastases reach a steady state in which the high replication rate (30–35% of cells undergoing S phase) is balanced by a high rate of cells dying by apoptosis. This produces a "dormant" metastasis. Removal of the primary tumor permits rapid neovascularization and a decrease in apoptosis of approximately 3-fold, so that the high replication rate now drives the rapid enlargement of the tumor. It should be noted that a difference in apoptotic index of 2- to 3-fold is highly significant. For example, an apoptotic index of only 2–3% in the rat liver results in a cell loss of 25% per day (Bursch et al. 1990). In the developing rat optic nerve, 50% of oligodendrocytes die per day; however, the maximal histologically detectable cell death (i.e., the apoptotic index) does not exceed 0.25% (Barres et al. 1992). These results suggest a new mechanism to explain tumor dormancy that is not dependent on immune cells or on tumor cells that are not cycling (G_1 or G_0). This explanation of dormancy may be applicable to other animal and human tumors; the evidence for this idea is put forward in a report by Holmgren et al. (1995). Thus, angiostatin may be the first molecular connection between the phenomena of angiogenesis, apoptosis, and tumor dormancy.

These studies provide further direct evidence that tumor growth is angiogenesis-dependent (Folkman 1971). They also emphasize the paracrine function of neovascularization; i.e., neovascularization may be considered as anti-apoptotic.

Angiostatin has more than 98% sequence homology to a 38-kD internal fragment of plasminogen with an amino terminus beginning at amino acid 98. It is surprising to find that polypeptides which regulate growth in the vascular endothelium are concealed within a parent protein, plasminogen, which itself has no direct effect on blood vessel growth. An analogous situation may be the 16-kD internal fragment of prolactin, which was found to be an angiogenesis inhibitor, whereas prolactin itself is not (Ferrara et al. 1991).

The source of angiostatin is not clear at this writing. It is associated with the presence of the primary Lewis lung tumor, but the tumor could produce an enzyme such as an elastase that would release the internal fragment of angiostatin from circulating plasminogen (Sottrup-Jensen et al. 1978; Machovich and Owen 1989; Machovich et al. 1990; Nakajima et al. 1993). Alternatively, the primary tumor cells may produce angiostatin directly. It is not yet known if Lewis lung tumor cells can produce plasminogen or its fragments. However, at least two genes for proteins which are highly homologous to other internal fragments of plasminogen have been reported (Ichinose 1992), one of which is expressed by some human carcinomas (Weissbach and Treadwell 1992).

Although angiostatin appears to account for most, if not all, of the antiangiogenic and antimetastatic properties of the primary Lewis lung carcinoma, our findings do not preclude the possibility that other antiangiogenic polypeptides may contribute to the mediation of the antimetastatic effect. Furthermore, it is likely that other angiogenesis inhibitors may be found to be generated by primary tumors as they switch to the angiogenic phenotype. Although this seems to be counterintuitive, namely, that tumors would release both positive and negative regulators of angiogenesis and that these polypeptides would be differentially cleared from the circulation, the activity of angiostatin resembles that of thrombospondin, another inhibitor of angiogenesis produced by a tumor (Iruela-Arispe et al. 1991; Tolsma et al. 1993). A different example that the onset of angiogenesis may be the outcome of a *net balance* of positive and negative regulators is the demonstration that the neovascularization in psoriasis appears to be mediated by keratinocytes that up-regulate the angiogenic stimulator interleukin-8 and down-regulate thrombospondin (Nickoloff et al. 1994).

CLINICAL IMPLICATIONS

These results provide a heuristic model for understanding certain patterns of presentation of metastases in cancer patients. On the basis of this model, it may be possible to look for endogenous inhibitors of angiogenesis in these patients. One well-recognized pattern is that metastases are not detectable at the time the primary tumor is diagnosed, but appear within a few months or a year after the tumor is removed (Folkman 1995). Thus, prolonged dormancy in human metastases could be maintained by circulating endothelial inhibitors generated by a primary tumor, or by other metastases. The loss of such an inhibitor by removal of a primary tumor could release the suppression of angiogenesis within distant metastases. Another clinical pattern is that the patient presents with multiple metastases in the presence of a primary tumor. These tumors could be low producers of angiostatin-like angiogenesis inhibitors, analogous to the high metastatic Lewis lung carcinoma (LLC-HM). A third pattern is the patient who presents with metastases but no apparent primary tumor; i.e., the "occult primary." The metastases may be suppressing the primary by a circulating angiogenesis inhibitor. A fourth pattern is the patient whose tumor is removed and/or treated with chemotherapy and radiation therapy, and in whom metastases do not appear until 5–10 years later. There are many possible mechanisms for this pattern that are unrelated to circulating angiogenesis inhibitors, but the possibility that the metastases are producing inhibitor and suppressing themselves merits consideration. Another possibility is that the metastases arose from nonangiogenic cells and remained dormant and undetectable until one or more of them switched to the angiogenic phenotype (Folkman 1994b).

Potential clinical applications of these findings suggest that a search for endothelial inhibitory activity in the blood or urine of cancer patients may be a helpful prognostic approach. Furthermore, endogenous antiangiogenic peptides such as angiostatin or related molecules may be useful as long-term therapy in suppression of growth of metastases. These agents may also prove useful in the treatment of primary carcinomas and of nonneoplastic angiogenesis, such as diabetic retinopathy.

REFERENCES

Antonelli-Orlidge A., K.B. Saunders, S.R. Smith, and P. D'Amore. 1989. An activated form of transforming growth factor-β is produced by co-cultures of endothelial cells and pericytes. *Proc. Natl. Acad. Sci.* **86:** 4544.

Barres, B.A., I.K. Hart, H.S. Coles, J.F. Burne, J.T. Voyvodic, W.D. Richardson, and M.C. Raff. 1992. Cell death and control of cell survival in the oligodendrocyte lineage. *Cell* **70:** 31.

Bashford, E.F., J.A. Murray, and W. Cramer. 1907. The natural and induced resistance of mice to the growth of cancer. *Proc. R. Soc. Lond. B. Biol. Sci.* **79:** 164.

Bichel, P. 1971. Autoregulation of ascites tumor growth by inhibition of the G-1 and the G-2 phase. *Eur. J. Cancer* **7:** 349.

Bok, R.A. and W.F. Mangel. 1985. Quantitative characterization of the binding of plasminogen to intact fibrin clots, lysine-Sepharose, and fibrin cleaved by plasmin. *Biochem.* **24:** 3279.

Bonfil, R.D., R.A. Ruggiero, O.D. Bustuoabad, R.P. Meiss, and C.D. Pasqualini. 1988. Role of concomitant resistance in the development of murine lung metastases. *Int. J. Cancer* **41:** 415.

Bouck, N.P. 1990. Tumor angiogenesis: The role of oncogenes and tumor suppressor genes. *Cancer Cells* **2:** 179.

Bursch, W., S. Paffe, B. Putz, G. Barthel, and R. Schulte-Hermann. 1990. Determination of the length of the histological stages of apoptosis in normal liver and in altered hepatic foci of rats. *Carcinogenesis* **11:** 847.

Clark, W.H., Jr., D.E. Elder, D.I.V. Guerry, L.E. Braitman, B.J. Trock, D. Schultz, M. Synnestevdt, and A.C. Halpern. 1989. Model predicting survival in stage 1 melanoma based on tumor progression. *J. Natl. Cancer Inst.* **81:** 1893.

Crum, R., S. Szabo, and J. Folkman. 1985. A new class of steroids inhibits angiogenesis in the presence of heparin or a heparin fragment. *Science* **230:** 1375.

Dameron, K.M., O.V. Volpert, and N. Bouck. 1994a. Regulation of an angiogenesis inhibitor in human fibroblasts by the p53 tumor suppressor gene. *J. Cell. Biochem.* **18C:** 187.

Dameron, K.M., O.V. Volpert, M.A. Tainsky, and N. Bouck. 1994b. Control of angiogenesis in fibroblasts by p53 regulation of thrombospondin-1. *Science* **265:** 1582.

Deutsch, D.G. and E.T. Mertz. 1970. Plasminogen: Purification from human plasma by affinity chromatography. *Science* **170:** 1095.

DeWys, W.D. 1972. Studies correlating the growth rate of a tumor and its metastases and providing evidence for tumor-related systemic growth-retarding factors. *Cancer Res.* **32:** 374.

Ehrlich, P. and H. Apolant. 1905. Beobachtungen über maligne mäusetumoren. *Berl. Klin. Wochenschr.* **42:** 871.

Elleman, C.J. and D. Eidinger. 1977. Suppressive factors in ascitic fluids and sera of mice bearing ascites tumors. *J. Natl. Cancer Inst.* **59:** 925.

Ferrara, N., C. Clapp, and R. Weiner. 1991. The 16K fragment of prolactin specifically inhibits basal or fibroblast growth factor simulated growth of capillary endothelial cells. *Endocrinology* **129:** 896.

Folkman, J. 1971. Tumor angiogenesis: Therapeutic implications. *N. Engl. J. Med.* **285:** 1182.

———. 1994a. Tumor angiogenesis. In *The molecular basis of cancer* (ed. J. Mendelsohn et al.), p. 206. W.B. Saunders, Philadelphia. (In press.)

———. 1994b. Angiogenesis and breast cancer. *J. Clin. Oncol.* **12:** 441.

———. 1995. Angiogenesis in cancer, vascular, rheumatoid, and other disease. *Nat. Med.* **1:** 27.

Folkman, J. and A. Moscona. 1978. Role of cell shape in growth control. *Nature* **273:** 345.

Folkman, J., C.C. Haudenschild, and B.R. Zetter. 1979. Long-term culture of capillary endothelial cells. *Proc. Natl. Acad. Sci.* **76:** 5217.

Folkman, J., K. Watson, D. Ingber, and D. Hanahan. 1989. Induction of angiogenesis during the transition from hyperplasia to neoplasia. *Nature* **339:** 58.

Folkman, J., M. Klagsbrun, J. Sasse, M. Wadzinski, D.E. Ingber, and I. Vlodavsky. 1988. A heparin-binding angiogenic protein–basic fibroblast growth factor–is stored within basement membrane. *Am. J. Pathol.* **130:** 393.

Gavrieli, Y., Y. Sherman, and S.A. Ben-Sasson. 1992. Identification of programmed cell death *in situ* via specific labeling of nuclear DNA fragmentation. *J. Cell Biol.* **119:** 493.

Gershon, R.K., R.L. Carter, and K. Kondo. 1967. On concomitant immunity in tumor-bearing hamsters. *Nature* **213:** 674.

Gimbrone, M.A., Jr., S. Leapman, R.S. Cotran, and J. Folkman. 1972. Tumor dormancy *in vivo* by prevention of neovascularization. *J. Exp. Med.* **136:** 261.

Good, D.J., P.J. Polverini, F. Rastinejad, M.M. Le Beau, R.S. Lemons, W.A. Frazier, and N.P. Bouck. 1990. A tumor suppressor-dependent inhibitor of angiogenesis is immunologically and functionally indistinguishable from a fragment of thrombospondin. *Proc. Natl. Acad. Sci.* **87:** 6624.

Gorelik, E. 1983. Resistance of tumor-bearing mice to a second tumor challenge. *Cancer Res.* **43:** 138.

Gorelik, E., S. Segal, and M. Feldman. 1978. Growth of a local tumor exerts a specific inhibitory effect on progression of lung metastases. *Int. J. Cancer* **21:** 617.

———. 1980. Control of lung metastasis progression in mice: Role of growth kinetics of 3LL Lewis lung carcinoma and host immune reactivity. *J. Natl. Cancer Inst.* **65:** 1257.

———. 1981. On the mechanism of tumor "concomitant immunity." *Int. J. Cancer* **27:** 847.

Greene, H.S.N. and E.K. Harvey. 1960. The inhibitory influence of a transplanted hamster lymphoma on metastasis. *Cancer Res.* **20:** 1094.

Gunduz, N., B. Fisher, and E.A. Saffer. 1979. Effect of surgical removal on the growth and kinetics of residual tumor. *Cancer Res.* **39:** 3861.

Hamada, J., P.G. Cavanaugh, and O. Lotan. 1992. Separable growth and migration factors for large-cell lymphoma cells secreted by microvascular endothelial cells derived from target organs for metastasis. *Br. J. Cancer* **66:** 349.

Hanahan, D. 1985. Heritable formation of pancreatic beta-cell tumors in transgenic mice expressing recombinant insulin/simian virus 40 oncogenes. *Nature* **315:** 115.

Himmele, J.C., B. Rabenhorst, and D. Werner. 1986. Inhibition of Lewis lung tumor growth and metastasis by Ehrlich ascites tumor growing in the same host. *J. Cancer Res. Clin. Oncol.* **111:** 160.

Holmgren, L., M.S. O'Reilly, and J. Folkman. 1995. Dormancy of micrometastases: Balanced proliferation and apoptosis in the presence of angiogenesis suppression. *Nat. Med.* **1:** 149.

Ichinose, A. 1992. Multiple members of the plasminogen-apolipoprotein(a) gene family associated with thrombosis. *Biochemistry* **31:** 3113.

Iruela-Arispe, M.L., P. Bornstein, and E.H. Sage. 1991. Thrombospondin exerts an antiangiogenic effect on cord formation by endothelial cells in vitro. *Proc. Natl. Acad. Sci.* **88:** 5026.

Kandel, J., E. Bossy-Wetzel, F. Radvanyi, M. Klagsbrun, J. Folkman, and D. Hanahan. 1991. Neovascularization is associated with a switch to the export of bFGF in the multistep development of fibrosarcoma. *Cell* **66:** 1095.

Lane, W.S., A. Galat, M.W. Harding, and S.L. Schreiber. 1991. Complete amino acid sequence of the FK506 and rapamycin binding protein, FKBP, isolated from calf thymus. *J. Prot. Chem.* **10:** 151.

Lange, P.H., K. Hekmat, G. Bosl, B.J. Kennedy, and E.E. Fraley. 1980. Accelerated growth of testicular cancer after cytoreductive surgery. *Cancer* **45:** 1498.

Lausch, R.N. and F. Rapp. 1969. Concomitant immunity in hamsters bearing DMBA- induced tumor transplants. *Int. J. Cancer* **4:** 226.

Lerch, P.G., E.E. Rickli, W. Lergier, and D. Gillessen. 1980. Localization of individual lysine-binding regions in human plasminogen and investigations on their complex-forming properties. *Eur. J. Biochem.* **107:** 7.

Machovich, R. and W.G. Owen. 1989. An elastase-dependent pathway of plasminogen activation. *Biochemistry* **28:** 4517.

Machovich, R., A. Himer, and W.G. Owen. 1990. Neutrophil proteases in plasminogen activation. *Blood Coagul. Fibrinolysis* **1:** 273.

Maione, T., G.S. Gray, J. Petro, A.J. Hunt, and A.L. Donner. 1990. Inhibition of angiogenesis by recombinant human platelet factor 4 and related peptides. *Science* **247:** 77.

Marie, P. and J. Clunet. 1910. Fréquence des métastases viscérales chez les souris cancéreuses aprés ablation chirurgicale de leur tumeur. *Bull. Assoc. Fr. Etude Cancer* **3:** 19.

Milas, L., N. Hunter, K. Mason, and H.R. Withers. 1974. Immunological resistance to pulmonary metastases in C3Hf/Bu mice bearing syngeneic fibrosarcoma of different sizes. *Cancer Res.* **34:** 61.

Mohr, U., J. Althoff, V. Kinzel, R. Suss, and M. Volm. 1968. Melanoma regression induced by "chalone": A new tumour inhibiting principle acting in vivo. *Nature* **220:** 138.

Muthukkaruppan, V. and R. Auerbach. 1979. Angiogenesis in the mouse cornea. *Science* **28:** 1416.

Nakajima, K., K. Nagata, M. Hamanoue, N. Takemoto, and S. Kohsaka. 1993. Microglia-derived elastase produces a low-molecular-weight plasminogen that enhances neurite outgrowth in rat neocortical explant cultures. *J. Neurochem.* **61:** 2155.

Nickoloff, B.J., R.S. Mitra, J. Varani, V.M. Dixit, and P.J. Polverini. 1994. Aberrant production of interleukin-8 and thrombospondin-1 by psoriatic keratinocytes mediates angiogenesis. *Am. J. Pathol.* **144:** 820.

Nicosia, R.F., R. Tchao, and J. Leighton. 1986. Interactions between newly formed endothelial channels and carcinoma cells in plasma clot culture. *Clin. Exp. Metastasis* **4:** 91.

Obeso, J., J. Weber, and R. Auerbach. 1990. A hemangioendothelioma-derived cell line: Its use as a model for the study of endothelial cell biology. *Lab. Invest.* **63:** 259.

O'Reilly, M., R. Rosenthal, E.H. Sage, S. Smith, L. Holmgren, M. Moses, Y. Shing, and J. Folkman. 1993. The suppression of tumor metastases by a primary tumor. *Surg. Forum* **44:** 474.

O'Reilly, M.S., L. Holmgren, Y. Shing, C. Chen, R.A. Rosenthal, M. Moses, W.S. Lane, E.H. Sage, and J. Folkman. 1994. Angiostatin: A novel angiogenesis inhibitor that mediates the suppression of metastases by a Lewis lung carcinoma. *Cell* **79:** 315.

Ponting, C.P., J.M. Marshall, and S.A. Cederholm-Williams. 1992. Plasminogen: A structural review. *Blood Coagul. Fibrinolysis* **3:** 605.

Prehn, R.T. 1991. The inhibition of tumor growth by tumor mass. *Cancer Res.* **51:** 2.

———. 1993. Two competing influences that may explain concomitant tumor resistance. *Cancer Res.* **53:** 3266.

Rak, J.W., E.J. Hegmann, C. Lu, and R.S. Kerbel. 1994. Progressive loss of sensitivity to endothelium-derived growth inhibitors expressed by human melanoma cells during disease progression. *J. Cell. Physiol.* **159:** 245.

Rastinejad, F., P. Polverini, and N.P. Bouck. 1989. Regulation of the activity of a new inhibitor of angiogenesis by a cancer suppressor gene. *Cell* **56:** 345.

Richardson, L., N.B. Modi, G.G. De Guzman, and T.F. Zioncheck. 1994. The in vivo fate of recombinant human vascular endothelial growth factor (rhVEGF) following intravenous administration in rats. In *The Western Regional Program of the American Assocation of Pharmaceutical Scientists* (March 1994), San Francisco. Abstr. 55, p. 30.

Rothbarth, K., G. Maier, E. Schöpf, and D. Werner. 1977. Inhibition of DNA synthesis by a factor from ascites tumor cells. *Eur. J. Cancer* **13:** 1195.

Rounds, D.E. 1970. A growth-modifying factor from cell lines of human malignant origin. *Cancer Res.* **30:** 2847.

Ruggiero, R.A., O.D. Bustuoabad, R.D. Bonfil, R.P. Meiss, and C.D. Pasqualini. 1985. "Concomitant immunity" in murine tumors of non-detectable immunogenicity. *Br. J. Cancer* **51:** 37.

Satoh, Y., R. Tsuboi, R. Lyons, H. Moses, and D.B. Rifkin. 1990. Characterization of the activation of latent TGF-β by co-cultures of endothelial cells and pericytes or smooth muscle cells: A self-regulating system. *J. Cell. Biol.* **111:** 757.

Schatten, W.E. 1958. An experimental study of postoperative tumor metastases. *Cancer* **11:** 455.

Shing, Y., G. Christofori, D. Hanahan, Y. Ono, R. Sasada, K. Igarashi, and J. Folkman. 1993. Betacellulin: A novel mitogen from pancreatic β tumour cells. *Science* **259:** 1604.

Sottrup-Jensen, L., H. Claeys, M. Zajdel, T.E. Petersen, and S. Magnusson. 1978. The primary structure of human plasminogen: Isolation of two lysine-binding fragments and one "mini"-plasminogen (MW, 38,000) by elastase-catalyzed-specific limited proteolysis. *Prog. Chem. Fibrinolysis Thrombolysis* **3:** 191.

Taylor, S. and J. Folkman. 1982. Protamine is an inhibitor of angiogenesis. *Nature* **297:** 307.

Thompson, R.W., G. Whalen, K.M. Saunders, T. Hores, and P.A. D'Amore. 1990. Heparin-mediated release of fibroblast growth factor-like activity into the circulation of rabbits. *Growth Factors* **3:** 221.

Tolsma, S.S., O.V. Volpert, D.J. Good, W.A. Frazier, P.J. Polverini, and N. Bouck. 1993. Peptides derived from two separate domains of the matrix protein thrombospondin-1 have anti-angiogenic activity. *J. Cell Biol.* **122:** 497.

Tyzzer, E.E. 1913. Factors in the production and growth of tumor metastases. *J. Med. Res.* **28:** 309.

Weidner, N., J.P. Semple, W.R. Welch, and J. Folkman. 1991. Tumor angiogenesis and metastasis—Correlation in invasive breast carcinoma. *N. Engl. J. Med.* **324:** 1.

Weissbach, L. and B.V. Treadwell. 1992. A plasminogen-related gene is expressed in cancer cells. *Biochem. Biophys. Res. Commun.* **186:** 1108.

Woodruff, M. 1990. *Cellular variation and adaption in cancer*, p. 64. Oxford University Press, New York.

The *p53* Tumor Suppressor Gene Inhibits Angiogenesis by Stimulating the Production of Thrombospondin

K.M. Dameron,* O.V. Volpert,* M.A. Tainsky,† and N. Bouck*
*Department of Microbiology-Immunology and Lurie Cancer Center, Northwestern University, Chicago, Illinois 60611;
†Department of Tumor Biology, University of Texas, M.D. Anderson Cancer Center, Houston, Texas 70030

The proteins encoded by most human tumor suppressor genes are transcription factors or components of signal transduction pathways (Levine 1993). These tumor suppressors impose their antineoplastic phenotype on cells by modulating the expression of other genes. It is the proteins encoded by these suppressor-regulated genes, acting as effector molecules for the suppressor genes, that directly alter the cell or its environment in such a way as to make progression toward malignancy less likely. For example, it is the *p53*-regulated p21 protein that actually implements the *p53*-induced cell-cycle block by binding to and inhibiting cyclin/cdk complexes (El-Deiry et al. 1993). We have sought effector proteins that implement phenotypes induced by tumor suppressor genes by studying secreted molecules that regulate new blood vessel growth (neovascularization or angiogenesis), reasoning that they may provide insight into suppressor gene action and simultaneously serve as a starting point for the development of nongenotoxic anticancer therapies.

The progressive growth of all tumors is dependent on new vessels that sprout from nearby capillaries to feed the developing tumor. Without such neovascularization, tumors remain clinically benign (<2-4 mm) and lack access to the circulation necessary for efficient metastasis (Folkman and Shing 1992). A wide variety of molecules have been identified that are able either to induce or to inhibit angiogenesis in model systems (Bouck 1990). Inhibitors predominate in normal adult tissue where there is little neovascularization, and cells cultured from normal tissues usually display an antiangiogenic phenotype, secreting high levels of inhibitors and low levels of inducers. In contrast, tumor cells are angiogenic and secrete increased levels of inducers and decreased levels of inhibitors compared to the normal cells from which they arose.

Loss of tumor suppressor genes may play a role in the switch to an angiogenic phenotype that must occur during tumor progression. In hamster cells (Rastinejad et al. 1989) and in a variety of human cells (Hsu et al. 1994; N. Bouck, unpubl.) when a tumor suppressor gene is present, the cells secrete inhibitors of angiogenesis and display an antiangiogenic phenotype. When the suppressor gene is lost, inhibitor levels fall, and as a result, the cells switch to an angiogenic phenotype (Fig. 1). In hamster fibroblasts, although the suppressor gene remains unidentified, the inhibitor whose synthesis it stimulates was identified (Good et al. 1990) as thrombospondin-1 (Frazier 1991; Bornstein 1992), a matrix glycoprotein that is a potent inhibitor of angiogenesis (Good et al. 1990; Tolsma et al. 1993). Here we show that in human fibroblasts the *p53* tumor suppressor gene (Donehower and Bradley 1993; Levine 1993) can negatively regulate the angiogenic phenotype and that it does so by stimulating the synthesis and secretion of inhibitory thrombospondin (TSP).

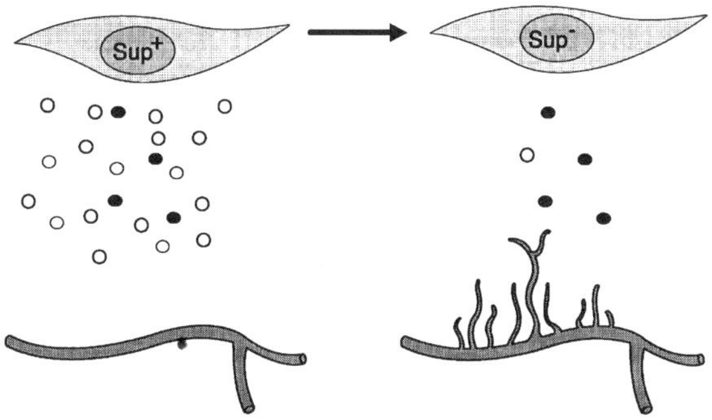

Figure 1. Model for tumor suppressor gene control of angiogenesis. Data from several systems suggest that loss of tumor suppressor genes in cells progressing toward tumorigenicity causes the cells to switch to an angiogenic phenotype, primarily by decreasing their secretion of inhibitors of nonvascularization. (○) Inhibitors of neovascularization; (●) inducers of neovascularization.

METHODS

Cells. Li-Fraumeni fibroblasts encoding mutant p53 proteins MDAH 041 (frameshift at 184) MDAH 172 (His at 175), MDAH087 (Trp at 248) and its *ras*-transfected subclones were obtained from M. Tainsky (Bischoff et al. 1990, 1991) and maintained in DME supplemented with 10% fetal bovine serum. For transfected clones, 0.2 mg/ml G418 was added. Early-passage cells were used at passage 10–21; late-passage cells were used at >207 for line 041, >62 for line 087, and >106 for line 172. Tumor lines were from the American Type Culture Collection, except the OHS osteosarcoma line which was a gift from K. G. Wiman (Stockholm). Hamster BHK cells and capillary endothelial cells have been described previously (Rastinejad et al. 1989).

Transfections. Cells were transfected using lipofectin with the temperature-sensitive (ts) plasmid pLSVp53cGVal135 (Michalovitz et al. 1990) and pSV2neo as a selectable marker, and stable neo-resistant clones were isolated. Those expressing exogenous p53 were identified by Western analysis or using Northerns probed with oligos specific for the murine p53.

Angiogenesis assays. Serum-free conditioned media were collected from cells in late log phase by incubating them after extensive washing in DME for 48–72 hours. Media were clarified and concentrated 20- to 50-fold using Amicon membranes with 10-kD molecular-weight cutoff. Protein was assayed (Biorad kit), and aliquots were stored at −80°C. To assay in vivo angiogenic activity, pellets containing test substances were implanted into the rat cornea as described previously (Polverini et al. 1991), and the induction of neovascularization was assessed 7 days later. To assay angiogenic activity in vitro, a modified Boyden chamber was used to measure the ability of test substances to induce the migration of bovine adrenal capillary endothelial cells (for detailed description, see Rastinejad et al. 1989). In these experiments, cells were suspended in DME with 0.1% bovine serum albumin, and this served as a negative control. Recombinant basic fibroblast growth factor (bFGF) was used at 10 ng/ml, media at 20 µg/ml, and A4.1 neutralizing antibodies to TSP at 20 µg/ml (Good et al. 1990). To quantitate activity, dose-response curves were performed using seven concentrations of conditioned media, and the protein concentration required to stimulate migration by 50% (EC_{50}) or to reduce migration toward bFGF by 50% (ID_{50}) was determined from the resulting graph (Tolsma et al. 1993).

Northern and Western analyses. Total RNA (10–20 µg) was isolated by the guanidine-isothiocyanate/cesium chloride method, run on a denaturing gel, blotted onto Genescreen, and hybridized with a 1.5-kb *Pst*I fragment of TSP prepared from p6SXE (Hennessy et al. 1989) and the 0.72-kb *Hin*dIII-*Bgl*I fragment of pHcGAP (Tso et al. 1985) which recognizes glyceraldehyde-3-phosphate dehydrogenase (GPD), a gene not regulated by p53 (Harvey et al. 1993).

For Western analysis, 6 µg of conditioned medium was run on a 7% SDS polyacrylamide gel, blotted to Hybond N membrane, probed with anti-TSP antibody A4.1 (Good et al. 1990), and developed with secondary antibody linked to horseradish peroxidase.

RESULTS

Fibroblasts cultured from Li-Fraumeni patients initially express both wild-type (wt) and mutant *p53*. Upon further cultivation, they spontaneously lose the wt *p53* allele and become immortal (Bischoff et al. 1990) and, if transfected with activated *ras*, some clones progress to tumorigenicity (Bischoff et al. 1991). Cells at various stages in this progression were examined for expression of TSP. Only early-passage cells that retained wt *p53* expressed high levels of TSP protein (Fig. 2A) and mRNA (Fig. 2B).

TSP is a potent inhibitor of angiogenesis (Good et al. 1990; Tolsma et al. 1993), and its expression in early-passage cells was responsible for their antiangiogenic phenotype. Media conditioned by early-passage cells cultured from three different patients were unable to induce capillary endothelial cell migration in vitro (Fig. 3) or neovascularization in vivo (example in Fig. 4C). Media conditioned by immortal cells that had lost wt *p53* and no longer expressed high levels of TSP were clearly angiogenic, since their media induced a positive response in both in vitro and in vivo assays (Figs. 3, 4A). Mixtures of media derived from early-passage cells with media conditioned by late-passage cells were not angiogenic (Figs. 3, 4B), indicating that early-pas-

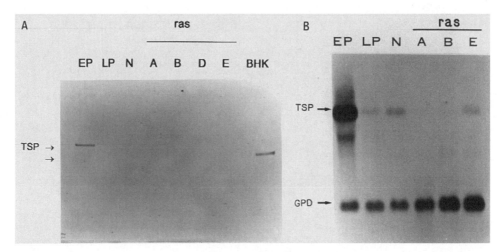

Figure 2. Synthesis of thrombospondin by Li-Fraumeni fibroblasts progressing to tumorigenicity. Fibroblasts cultured from Li-Fraumeni patient 087 were examined at early passage (EP), late passage (LP), after transfection of late-passage cells with neo (N) or with *ras* (tumorigenic lines A and B; nontumorigenic lines D and E) for ability to secrete TSP into conditioned media by Western analysis (panel *A*) and for steady-state levels of TSP RNA by Northern (panel *B*). TSP produced by BHK hamster cells, which has a lower molecular weight than human TSP, served as a positive control in *A*, and GPD served as a loading control in *B*.

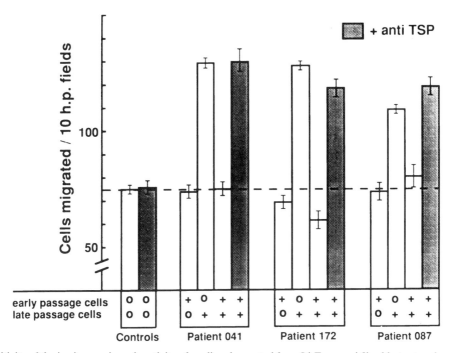

Figure 3. Sensitivity of the in vitro angiogenic activity of medium harvested from Li-Fraumeni fibroblasts at early and late passage to neutralizing anti-TSP antibodies. Serum-free medium conditioned by fibroblasts cultured from three different patients was collected and tested for ability to induce and/or inhibit migration of cultured capillary endothelial cells in the presence and absence of neutralizing antibodies against TSP.

sage cells were producing an inhibitory activity. The amount of inhibitory activity produced by Li-Fraumeni fibroblasts was similar to that produced by cultures of wt adult fibroblasts and about twice that seen in media collected from foreskin fibroblasts (Table 1). Foreskin fibroblasts also have reduced levels of inducing activity, maintaining an overall inhibitory phenotype. The inhibitory activity present in these media could be attributed to TSP, because neutralizing antibodies to TSP could block all inhibitory activity, whether measured against the classic inducer bFGF (data not shown) or against media conditioned by late-passage Li-Fraumeni fibroblasts (Fig. 3). Similar results were obtained using the in vivo cornea assay (data not shown).

When Li-Fraumeni cells lose wt p53 and immortalize, they also become aneuploid and undergo a number of other changes. The level of angiogenic inducing activity seen in their conditioned media increased severalfold (Table 2). Yet, it was the down-regulation of TSP production rather than the increase in inducing activity that was primarily responsible for the shift to an angiogenic phenotype. When TSP was inactivated with antibodies, media from early-passage cells became angiogenic (data not shown), and sufficient TSP was produced by early-passage cells to block angiogenesis induced by the more active inducers produced by late-passage cells (Fig. 2). In fact, the level of TSP produced by early-passage fibroblasts was also sufficient to block angiogenesis induced by media conditioned by several different transformed cell lines, including HBL100 from breast tissue, OSH from an osteosarcoma, and U87 from a glioblastoma, and by *ras*-transformed tumorigenic line A derived from patient 087 cells (data not shown).

Loss of the wt *p53* gene seemed to be responsible for the decrease in TSP expression that occurred as Li-Fraumeni fibroblasts were passaged and, thus, for the switch to an angiogenic phenotype, because reintroduction of wt *p53* resulted in high expression of TSP and reversion to an antiangiogenic phenotype. When late-passage cells that had lost wt *p53* were stably transfected with a plasmid encoding a p53 protein that is temperature-sensitive and assumes a wt configuration at low temperature and a mutant configuration at high temperature (Michalovitz et al. 1990), expression of the endogenous TSP gene became temperature-sensitive (Fig. 5). When the transfected cells were grown at 32°C where *p53* was wt, mRNA levels were high; when they were grown at 38.5°C where the p53 protein is mostly mutant, TSP expression was only slightly increased. This slight increase compared to the neo controls is probably due to leakiness of the ts mutant. A second clone of 087 cells, as well as pools of transfectants derived from 041 cells, behaved similarly (data not shown).

Media collected from clones and from pools of transfectants expressing ts *p53* also shifted their angiogenic phenotype with changes in temperature (Fig. 6). Media from cells growing at low temperature and expressing high levels of wt *p53* and TSP were not angiogenic and exhibited an inhibitory activity that was relieved by neutralizing anti-TSP antibodies. These in vitro assays were confirmed in vivo using the cornea assay (data not shown).

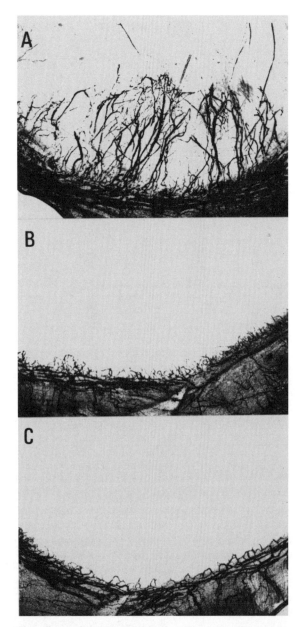

Figure 4. In vivo assay for ability to induce neovascularization in the rat cornea. A portion of a rat cornea is shown where pellets containing conditioned media had been implanted 7 days earlier. Colloidal carbon has been used to visualize vessels. Pellets contained media from (*A*) late-passage cells from patient 172, (*B*) mixture of media from early-passage and late-passage cells, or (*C*) media from early-passage cells. Vigorous neovascularization can be seen in *A* only. Similar results were obtained with media conditioned by cells from patients 087 and 041.

Table 1. Quantitation of Inhibitory Activity Present in Media Conditioned by Early-passage Li-Fraumeni Fibroblasts

Conditioned media	Inhibitory activity ID_{50} (μg/ml)
Foreskin fibroblasts	7.0
Adult fibroblasts	3.6
Li-Fraumeni	
087 EP	3.4
172 EP	3.8
041 EP	4.5

Serum-free conditioned media from early-passage fibroblasts were mixed with bFGF in an in vitro migration assay, and the amount of protein necessary to inhibit 50% of migration mediated by bFGF was determined (ID_{50}).

Table 2. Quantitation of Inducing Activity Present in Media Conditioned by Li-Fraumeni Fibroblasts

Conditioned media	p53 status	Inducing activity EC_{50} (μg/ml)
Foreskin fibroblasts	wt/wt	10
Adult fibroblasts	wt/wt	3.7, 4.0
Li-Fraumeni fibroblasts		
041		
early passage	wt/184^{fs}	3.4, 2.7
late passage	184^{fs}	0.7
172		
early passage	wt/175^{his}	4.0, 3.6
late passage	175^{his}	0.5
087		
early passage	wt/248^{trp}	3.7, 3.7
late passage	248^{trp}	0.6

Inducing activity (tested in the presence of antibodies that inactivated inhibitory thrombospondin) was calculated from a dose-response curve generated by measuring endothelial cell migration toward various concentrations of media. The concentration needed to achieve 50% maximal induction of migration was calculated and is reported as EC_{50}.

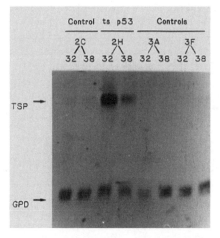

Figure 5. TSP mRNA levels in a clone of late-passage Li-Fraumeni fibroblasts expressed ts *p53*. Northern analysis of clones of late-passage fibroblasts derived from patient 087 expressing a neo plasmid alone, or ts *p53* plus neo, and grown at 32°C where *p53* is in a wild-type conformation, or at 37.5°C where *p53* is primarily in the mutant configuration. GPD served as a loading control.

In addition to increasing their expression of TSP, cells that re-expressed wt *p53* also decreased their production of inducers of angiogenesis by a factor of less than 3 (Table 3). Mixing experiments with anti-TSP antibodies (Fig. 6) confirmed that it was the variation in

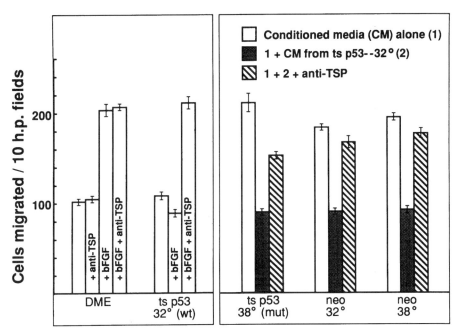

Figure 6. In vitro angiogenic activity of pools of clones expressing ts *p53*. Media conditioned by pools of clones derived from late-passage fibroblasts of patient 041 transfected with neo, or with neo and ts *p53*, and tested for ability to induce and/or inhibit capillary endothelial cell migration.

TSP and not changes in inducers that was responsible for the shift in angiogenic phenotype.

DISCUSSION

To grow in vivo to a clinically significant size, cells of a developing tumor must be able to divide without constraint and to create around themselves an environment where their growth potential can be realized. New tumor vasculature is an essential part of this permissive in vivo environment. Results presented here link the presence of a human tumor suppressor gene with the inability to induce such neovascularization. They show that in fibroblasts, wt *p53* can positively regulate the thrombospondin gene, thereby preventing the cells from expressing an angiogenic phenotype, and they define the suppression of angiogenesis as a new and unexpected function of *p53*. The *p53* gene and the TSP gene it regulates are good examples of class I and class II tumor suppressor genes, respectively, as defined by Sager (Lee et al. 1991).

The p53 protein has already been shown to inhibit growth, influence differentiation, mediate some types of apoptosis, and contribute to the maintenance of

Table 3. Inducing Activity Present in Media Conditioned by Late-passage Li-Fraumeni Cells Expressing Temperature-sensitive *p53*

Conditioned media	*p53* status	Inducing activity EC_{50} (μg/ml)
Late passage pool of 041 clones transfected with		
neo 32°	184^{fs}	0.7
neo 38.5°	184^{fs}	0.8
ts p53 32°	184^{fs} + wt	1.5
ts p53 38.5°	184^{fs} + 135^{val}	0.9
Late-passage single clones of 087 transfected with		
neo 32°	248^{trp}	0.9
neo 38.5°	248^{trp}	1.0
ts p53 32°	248^{trp} + wt	2.6
ts p53 38.5°	248^{trp} + 135^{val}	1.2

Inducing activity was calculated as described in Table 2.

genetic stability (Donehower and Bradley 1993; Zambetti and Levine 1993). The inhibition of angiogenesis by *p53* is a distinct function that is fully expressed without the overexpression of *p53*. It explains why loss of *p53* is advantageous even when it occurs late during tumor progression and offers one mechanism to explain why returning wt *p53* to tumor cells halts growth in vivo but not in vitro (Chen et al. 1990; Chen et al. 1991; Wang et al. 1993).

The majority of human fibrosarcomas have lost active wt *p53* as a result of either mutations in *p53* or amplification of *MDM2* (Leach et al. 1993; Cordon-Cardo et al. 1994), suggesting that this loss is a vital step in their development. In Li-Fraumeni cells, the loss of TSP production that accompanies loss of wt *p53* represents one necessary step in the multistep process by which these cells develop sufficient angiogenic activity to support vigorous tumor growth in vivo. In addition to decreasing secretion of inhibitory TSP, late-passage cells also increased their production of inducers of neovascularization; animal models suggest both changes are probably important in vivo. During the development of fibrosarcomas in bovine papillomavirus transgenic mice, the switch to an angiogenic phenotype in vivo has been closely correlated with the ability of the cells of the developing tumor to export bFGF (Kandel et al. 1991), an inducer of neovascularization. In this system, the down-regulation of TSP would presumably have occurred earlier in tumor progression, prior to the onset of vigorous, visible in vivo angiogenesis, possibly when the E6 protein encoded by the viral transgene began to be expressed and inactivated wt *p53* (Band et al. 1993).

The mechanism by which p53 regulates the endogenous TSP gene is under investigation. It may act directly, because when a reporter gene linked to the human TSP promoter and first intron was transfected into cells expressing ts *p53*, the reporter was expressed well only at the temperature where *p53* was in a wild-type configuration. In addition, a pair of elements that resemble the consensus sequence for *p53* response elements (Zambetti and Levine 1993) could be identified in the sequence of the first intron of TSP. These elements are not found in TSP-2, another family member that may be able to inhibit angiogenesis. Northern analyses using probes specific for TSP-1 and TSP-2 have shown that TSP-2 is not regulated by *p53* in human fibroblasts.

In cells other than fibroblasts, it is not yet clear if TSP is controlled by *p53* or acts as a barrier to the development of an angiogenic phenotype. It is up-regulated when tumorigenicity is suppressed by fusion of a human breast carcinoma line to normal breast cells (Zajchowski et al. 1990) and by introduction of a suppressor chromosome into human glioblastoma cells (Hsu et al. 1994). Both tumor types are characteristic of the Li-Fraumeni syndrome, arise from normal cells that make significant levels of TSP, and secrete an angiogenic activity that is sensitive to inhibition by TSP. The effectiveness of TSP in this context raises the possibility that it or its antiangiogenic peptides (Tolsma et al. 1993) can be developed into agents that may be of clinical utility in slowing or preventing the growth of tumors, especially in those individuals known to be at high risk due to inheritance of inactive *p53* alleles.

ACKNOWLEDGMENTS

We are grateful to M. Oren for the gift of ts p53 plasmid and to B. Thimmappaya for his early help with this project. This work was supported in part by National Institutes of Health grants CA-27306 and CA-52750 (N.B.), P01 CA-34936 (M.A.T.), and 5T32 CA-09560, which contributed to the support of K.M.D.

REFERENCES

Band, V., S. Dalal, L. Delmolino, and E.J. Androphy. 1993. Enhanced degradation of p53 protein in HPV-6 and BPV-1 E6-immortalized human mammary epithelial cells. *EMBO J.* **12:** 1847.

Bischoff, F.Z., L.C. Strong, S.O. Yim, D.R. Pratt, M.J. Siciliano, B.C. Giovanella, and M.A. Tainsky. 1991. Tumorigenic transformation of spontaneously immortalized fibroblasts from patients with a familial cancer syndrome. *Oncogene* **6:** 183.

Bischoff, F.Z., S.O. Yim, S. Pathak, G. Grant, M.J. Siciliano, B.C. Giovanella, L.C. Strong, and M.A. Tainsky. 1990. Spontaneous abnormalities in normal fibroblasts from patients with Li-Fraumeni cancer syndrome: Aneuploidy and immortalization. *Cancer Res.* **50:** 7979.

Bornstein, P. 1992. Thrombospondins: Structure and regulation of expression. *FASEB J.* **6:** 3290.

Bouck, N. 1990. Tumor angiogenesis: The role of oncogenes and tumor suppressor genes. *Cancer Cells* **2:** 179.

Chen, P.-L., Y. Chen, R. Bookstein, and W.-H. Lee. 1990. Genetic mechanisms of tumor suppression by the human p53 gene. *Science* **250:** 1576.

Chen, Y., P.-L. Chen, N. Arnaiz, D. Goodrich, and W.-H. Lee. 1991. Expression of wild-type p53 in human A673 cells suppresses tumorigenicity but not growth rate. *Oncogene* **6:** 1799.

Cordon-Cardo, C., E. Latres, M. Drobnjak, M.R. Oliva, D. Pollack, J.M. Woodruff, V. Marechal, J. Chen, M.F. Brennan, and A.J. Levine. 1994. Molecular abnormalities of mdm2 and p53 genes in adult soft tissue sarcomas. *Cancer Res.* **54:** 794.

Donehower, L.A. and A. Bradley. 1993. The tumor suppressor p53. *Biochim. Biophys. Acta* **1155:** 181.

El-Deiry, W.S., T. Tokino, V.E. Velculescu, D.B. Levy, R. Parsons, J.M. Trent, D. Lin, W.E. Mercer, K.W. Kinzler, and B. Vogelstein. 1993. WAF1, a potential mediator of p53 tumor suppression. *Cell* **75:** 817.

Folkman, J. and Y. Shing. 1992. Angiogenesis. *J. Biol. Chem.* **267:** 10931.

Frazier, W.A. 1991. Thrombospondins. *Curr. Opin. Cell Biol.* **3:** 792.

Good, D.J., P.J. Polverini, F. Rastinejad, M.M. Le Beau, R.S. Lemons, W.A. Frazier, and N.P. Bouck. 1990. A tumor suppressor-dependent inhibitor of angiogenesis is immunologically and functionally indistinguishable from a fragment of thrombospondin. *Proc. Natl. Acad. Sci.* **87:** 6624.

Harvey, M., A.T. Sands, R.S. Weiss, M.E. Hegi, R.W. Wiseman, P. Pantazis, B.C. Giovanella, M.A. Tainsky, A. Bradley, and L.A. Donehower. 1993. In vitro growth characteristics of embryo fibroblasts isolated from p53-deficient mice. *Oncogene* **8:** 2457.

Hennessy, S.W., B.A. Frazier, D.D. Kim, T.L. Deckwerth, D.M. Baumgartel, P. Rotwein, and W.A. Frazier. 1989. Complete thrombospondin mRNA sequence includes potential regulatory sites in the 3' untranslated region. *J. Cell Biol.* **108:** 729.

Hsu, S.C., O.V. Volpert, P.A. Steck, T. Mikkelsen, P.J. Polverini, W.K. Cavenee, and N.P. Bouck. 1994. Chromosome 10 regulates the angiogenic phenotype in glioblastoma cells by modulating thrombospondin. *Proc. Am. Assoc. Cancer Res.* **35:** 67.

Kandel, J., E. Bossy-Wetzel, F. Radvanyi, M. Klagsbrun, J. Folkman, and D. Hanahan. 1991. Neovascularization is associated with a switch to the export of bFGF in the multistep development of fibrosarcoma. *Cell* **66:** 1095.

Leach, F.S., T. Tokino, P. Meltzer, M. Burrell, J.D. Oliner, S. Smith, D.E. Hill, D. Sidransky, K.W. Kinzler, and B. Vogelstein. 1993. p53 mutation and MDM2 amplification in human soft tissue sarcomas. *Cancer Res.* **53:** 2231.

Lee, S.W., C. Tomasetto, and R. Sager. 1991. Positive selection of candidate tumor-suppressor genes by subtractive hybridization. *Proc. Natl. Acad. Sci.* **88:** 2825.

Levine, A.J. 1993. The tumor suppressor genes. *Annu. Rev. Biochem.* **62:** 623.

Michalovitz, D., O. Halevy, and M. Oren. 1990. Conditional inhibition of transformation and of cell proliferation by a temperature-sensitive mutant of p53. *Cell* **62:** 671.

Polverini, P.J., N.P. Bouck, and F. Rastinejad. 1991. Assay and purification of naturally occurring inhibitor of angiogenesis. *Methods Enzymol.* **198:** 440.

Rastinejad, F., P.J. Polverini, and N.P. Bouck. 1989. Regulation of the activity of a new inhibitor of angiogenesis by a cancer suppressor gene. *Cell* **56:** 345.

Tolsma, S.S., O.V. Volpert, D.J. Good, W.A. Frazier, P.J. Polverini, and N. Bouck. 1993. Peptides derived from two separate domains of the matrix protein thrombospondin-1 have anti-angiogenic activity. *J. Cell Biol.* **122:** 497.

Tso, J.Y., X.-H. Sun, T.-H. Kao, K.S. Reece, and R. Wu. 1985. Isolation and characterization of rat and human glyceraldehyde-3-phosphate dehydrogenase cDNAs: Genomic complexity and molecular evolution of the gene. *Nucleic Acids Res.* **13:** 2485.

Wang, N.P., H. To, W.-H. Lee, and E.Y.-H.P. Lee. 1993. Tumor suppressor activity of RB and p53 genes in human breast carcinoma cells. *Oncogene* **8:** 279.

Zajchowski, D.A., V. Band, D.K. Trask, D. Kling, J.L. Connolly, and R. Sager. 1990. Suppression of tumor-forming ability and related traits in MCF-7 human breast cancer cells by fusion with immortal mammary epithelial cells. *Proc. Natl. Acad. Sci.* **87:** 2314.

Zambetti, G.P. and A.J. Levine. 1993. A comparison of the biological activities of wild-type and mutant p53. *FASEB J.* **7:** 855.

Defining the Steps in a Multistep Mouse Model for Mammary Carcinogenesis

H.E. Varmus,*†‡§ L.A. Godley,†‡ S. Roy,† I.C.A. Taylor,† L. Yuschenkoff,§
Y.-P. Shi,** D. Pinkel,** J. Gray,** R. Pyle,§§
C.M. Aldaz,‡‡ A. Bradley,†† D. Medina,*** and L.A. Donehower§§

*Office of the Director, National Institutes of Health, Bethesda, Maryland 20892; †Varmus Laboratory, National Cancer Institute, Bethesda, Maryland 20892; ‡Department of Biochemistry and Biophysics, §Department of Microbiology and Immunology, and **Division of Molecular Cytometry, Department of Laboratory Medicine, University of California, San Francisco, California 94143; ††Institute for Molecular Genetics and Howard Hughes Medical Institute, Baylor College of Medicine, Houston, Texas 77030; ‡‡Department of Carcinogenesis, University of Texas M.D. Anderson Cancer Center-Science Park, Smithville, Texas 78957; §§Division of Molecular Virology, and ***Department of Cell Biology, Baylor College of Medicine, Houston, Texas 77030

Cancers arise through alterations in two general classes of genes, tumor suppressor genes and proto-oncogenes. Because cancers do not appear to develop from the derangement of a single gene, but instead require a series of changes in genes from both of these two groups, cancer development is referred to as "multistep tumorigenesis." For a few systems, such as colon cancer, a reasonable consensus exists as to which mutations occur and in what sequence (Fearon and Vogelstein 1990). Presently, many laboratories are attempting to understand which genetic changes are responsible for breast cancer, one of the most common forms of human cancers.

Mutations of tumor suppressor genes occur frequently in human breast cancer. For example, the retinoblastoma gene (Rb) undergoes somatic mutation during the development of some human breast cancers (Weinberg 1992). p53, the gene most commonly mutated in all human cancers, is mutated or deleted in at least 22% of sporadic breast tumors (Greenblatt et al. 1994). In addition, some patients with Li-Fraumeni syndrome carry germ-line mutations of p53 which predispose them to a variety of neoplasms, one of the most common of which is breast cancer (Malkin et al. 1990; Srivastava et al. 1990). When inherited in mutant form, BRCA1, a putative tumor suppressor gene on chromosome 17q, renders an individual susceptible to breast cancer. BRCA1 mutations may be present in as many as 45% of families with a high incidence of early-onset breast cancer and in 80% of families with an increased incidence of early-onset breast and ovarian cancers (Futreal et al. 1994; Miki et al. 1994). Finally, BRCA2, recently mapped to human chromosome 13q, may be another tumor suppressor gene responsible for many of the families with early-onset breast cancer that do not show mutation of BRCA1 (Wooster et al. 1994).

Two other genes when mutated confer an inherited risk for human breast cancer. First, rare mutant alleles of the androgen receptor are associated with male breast cancer (Wooster et al. 1992; Lobaccaro et al. 1993). Second, patients heterozygous for the gene mutated in ataxia telangiectasia are at an increased risk for breast cancer (Swift et al. 1991).

At least three proto-oncogenes have been implicated in human breast carcinogenesis. These include the genes c-myc, neu, and PRAD-1/cyclin D1, each of which has been found amplified and overexpressed in many human mammary tumors (Machotka et al. 1989; Lammie and Peters 1991). Genomic amplification accompanied by overexpression of neu in human breast tumors has been correlated with decreased survival (Slamon et al. 1987). The genomic amplicon that includes PRAD-1/cyclin D1 also includes the linked genes int-2/Fgf-3 and hst/Fgf-4, genes that are involved in mouse mammary tumors (see below), although they are not expressed in human breast tumors (Lammie et al. 1991).

Transgenic mouse models for breast cancer involving c-myc, neu, int-2/Fgf-3, and PRAD-1/cyclin D1 have been made using the mouse mammary tumor virus (MMTV) or the whey acidic protein promoter/enhancers which direct expression of each transgene to the mammary gland (Cardiff and Muller 1993; Wang et al. 1994). Female transgenic mice from each of these systems develop mammary tumors, although in nearly all cases it is apparent that events in addition to inheritance of the transgene are required to produce a tumor.

In addition to the usefulness of its regulatory signals to produce transgenic models of breast cancer, MMTV itself has proven to be a powerful tool for studying mammary carcinogenesis in the mouse. Infection of mice by MMTV produces mammary adenocarcinomas (Hilgers and Bentvelzen 1979), and the sites of insertion of MMTV proviruses in tumor cell DNA have been studied in an effort to understand the mode of MMTV-induced carcinogenesis. Sequences flanking MMTV integration sites provided probes that were used to isolate neighboring genes and to show that the proviral insertions transcriptionally activate those genes through the enhancer in the MMTV long terminal repeat (LTR) (Nusse and Varmus 1982). The putative proto-oncogenes activated by MMTV proviruses include two members of the *Wnt* gene family (*Wnt-1* and *Wnt-3*), three members of the fibroblast growth factor (*Fgf*) gene family (*int-2/Fgf-3*, *hst/Fgf-4*, and AIGF/*Fgf-8*), and *int-3*, a mammalian *Notch* gene.

Wnt-1 TRANSGENIC MODEL

The first MMTV integration site identified was *Wnt-1*, the mammalian homolog of the *Drosophila* segment polarity gene, *wingless* (Nusse and Varmus 1982; van Ooyen and Nusse 1984; Rijsewijk et al. 1987b). *Wnt-1* was the first cloned member of an extremely large gene family, with as many as 14 genes in the mouse, and with homologs in all metazoa tested. *Wnt-1* encodes a cysteine-rich, glycosylated secretory protein (Bradley and Brown 1990; Papkoff and Schryver 1990), which is normally expressed during mouse embryonic development and in the round spermatid of the adult testis (Shackleford and Varmus 1987; Wilkinson et al. 1987). Expression of *Wnt-1* is essential for the formation of the brain, since mice deficient in *Wnt-1* lack portions of the midbrain and cerebellum (McMahon and Bradley 1990; Thomas and Capecchi 1990; Thomas et al. 1991). Although *Wnt-1* is not expressed in the normal mammary gland, an exogenous *Wnt-1* gene can transform certain mammary epithelial cell lines in vitro (Brown et al. 1986; Rijsewijk et al. 1987a). Although two *Wnt* genes have been found as MMTV integration sites in mouse mammary tumors, *Wnt* genes have not been implicated thus far in human breast cancer.

To demonstrate that MMTV integrations into the *Wnt-1* locus were instrumental in the development of mammary tumors, transgenic mice were generated in which the transgene mimics an MMTV-activated allele, with the MMTV LTR in the opposite transcriptional orientation to the genomic *Wnt-1* sequence (Tsukamoto et al. 1988). The transgene is expressed principally in the mammary and salivary glands, and all of the mammary glands of both male and female *Wnt-1* transgenic mice (*Wnt-1 TG*) are hyperplastic. Mammary adenocarcinoma eventually develops in all female animals, although usually in only one of the hyperplastic mammary glands. 50% of virgin *Wnt-1 TG* females develop tumors by about 6 months of age (Fig. 1A), whereas only 15% of *Wnt-1 TG* males develop mammary tumors by 1 year of age (Fig. 1B). Metastases occur rarely to the lungs and lymph nodes.

The stochastic nature of tumor development suggests that *Wnt-1* expression is necessary but not sufficient for mammary tumor formation. Additional genetic or epigenetic events probably need to accompany *Wnt-1* expression to form mammary tumors in these transgenic animals. The rather slow kinetics of tumor development have allowed experiments to be designed to ask if this process can be accelerated (Table 1). These have included (1) crossing the *Wnt-1 TG* mice to other genetically altered mice to enhance susceptibility to tumor formation (Kwan et al. 1992; L.A. Donehower et al., in prep.) and (2) infecting the *Wnt-1 TG* mice with MMTV and analyzing new MMTV proviruses to identify genes that might cooperate with *Wnt-1* (Shackleford et al. 1993; C.A. MacArthur et al., in prep.). Such experiments have led to insights into the nature of the proteins that can cooperate with *Wnt-1* in tumor formation.

SYNERGY BETWEEN *Wnt-1* AND FIBROBLAST GROWTH FACTORS

Both genetic crosses between transgenic animals and MMTV infection of *Wnt-1 TG* animals have suggested that members of the *Fgf* gene family cooperate with *Wnt-1* during tumorigenesis. A genetic cross between *Wnt-1 TG* and *int-2/Fgf-3* transgenic animals was inspired by the observation that some MMTV-induced tumors from nontransgenic animals show transcriptional activation of both *Wnt-1* and *int-2/Fgf-3* (Peters et al. 1986). Although *Wnt-1 TG* females readily develop tumors (as described above), mammary tumors occur infrequently in *int-2/Fgf-3* transgenic females and not at all in males (Muller et al. 1990). Females transgenic for both *Wnt-1* and *int-2/Fgf-3* develop tumors faster than animals bearing either transgene alone (Fig. 1A) (Kwan et al. 1992). A more dramatic phenotype is seen in males, with tumors appearing much earlier and in many more animals than in those carrying only the *Wnt-1* transgene (Fig. 1B) (Kwan et al. 1992). The fact that tumors appear earlier in bitransgenic animals than in either parental line suggests that the two genes, *Wnt-1* and *int-2/Fgf-3*, both of which encode secretory proteins, cooperate to form mammary tumors.

This observation was strengthened by experiments in which *Wnt-1 TG* animals were infected with MMTV (Shackleford et al. 1993). MMTV accelerated the kinetics of tumor formation (Figs. 1A,B), and provirus tagging allowed the identification of cooperating genes. Approximately half of the mammary tumors from MMTV-infected *Wnt-1 TG* female mice showed insertional activation of *int-2/Fgf-3*, *hst/Fgf-4*, or both, as well as insertions into AIGF/*Fgf-8* (Shackleford et al. 1993; C.A. MacArthur et al., in prep.), further confirming a strong cooperative interaction between Wnt-1 and members of the Fgf family during mammary carcinogenesis in this model.

EFFECT OF *p53* DEFICIENCY ON *Wnt-1* TUMORIGENESIS

The importance of *p53* as a tumor suppressor gene has been reinforced by genetically engineered mice containing germ-line null alleles of *p53* (Donehower et al. 1992; Jacks et al. 1994; Purdie et al. 1994). Approximately half of the mice that are heterozygous for a null *p53* allele develop sarcomas and lymphomas by 18 months of age, whereas 100% of mice nullizygous for *p53* develop tumors by 10 months of age, principally T-cell lymphomas or, less frequently, sarcomas (Harvey et al. 1993; Jacks et al. 1994). Despite evidence for the

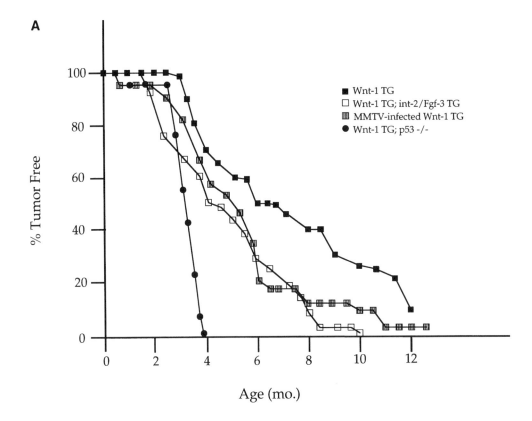

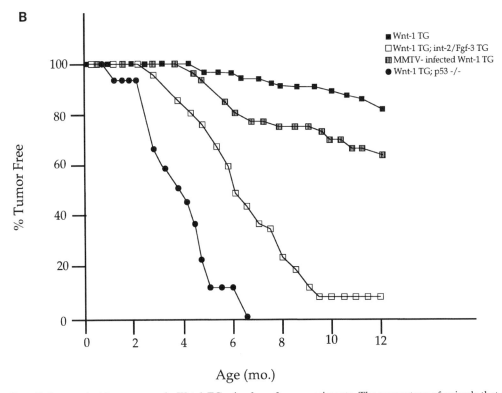

Figure 1. Compiled tumor incidence curves in *Wnt-1 TG* mice from four experiments. The percentage of animals that remained free of mammary tumors is plotted as a function of age. (*A*) *Wnt-1 TG* females; (*B*) *Wnt-1 TG* males. The references for each curve are as indicated: *Wnt-1 TG* mice (Tsukamoto et al. 1988); *Wnt-1 TG*; *int-2/Fgf-3 TG* (Kwan et al. 1992); MMTV-infected *Wnt-1 TG* (Shackleford et al. 1993); *Wnt-1 TG*; p53 −/− (L. Donehower et al., in prep).

Table 1. Experimental Approaches to Find Genes That Cooperate with a Transgene in Tumorigenesis

Experimental approaches	Experiments involving Wnt-1 TG mice
1. Cross to other transgenic and gene-targeted mouse lines to look for a change in the kinetics of tumorigenesis	1. Cross with int-2/FGF3 TG mice and p53-deficient mice; acceleration of tumorigenesis seen in progeny from each cross (cf. Fig. 1)
2. Retroviral infection to search for additional insertionally activated genes	2. MMTV infection accelerates tumorigenesis; proviral insertion mutations of int-2/Fgf-3, hst/Fgf-4, AIGF/Fgf-8
3. Cross to other mouse strains to look in tumors of F_1 animals for LOH throughout the mouse genome (Dietrich et al. 1992)	3. Cross with M. castaneus; results pending on tumors from F_1 transgenics
4. Use PCR-based methods to clone sequences related to a gene family of interest that might be implicated in mammary tumorigenesis	4. Identification of Sky, a transmembrane protein-tyrosine kinase, expressed in C57MG cells and mammary tumors
5. Differential display or subtractive hybridization to seek genes differentially expressed during stages of mammary tumorigenesis	5. Identification of gene for demilune cell-specific protein, a putative secretory protein, expressed in mammary gland hyperplasia

On the left, we present the general approaches that can be taken to identify genes which cooperate with a given transgene. On the right, we present the application of these approaches that we have used in the study of the Wnt-1 transgenic mice.

involvement of *p53* in human breast cancer, these mice rarely develop mammary tumors.

To seek an effect of *p53* loss on mammary tumorigenesis, we crossed *p53*-deficient mice to the *Wnt-1 TG* mice, which are predisposed to breast cancer (L.A. Donehower et al., in prep.). The F_1 and F_2 offspring from these crosses fell into twelve categories, depending on sex, *p53* genotype, and the presence or absence of the *Wnt-1* transgene. Mammary tumors appeared in all of the female mice with the *Wnt-1* transgene before 1 year of age, regardless of *p53* status (Table 2). *Wnt-1 TG p53 +/+* and *p53 +/−* females developed mammary tumors at similar rates, with an average age of 22.5 weeks and 23 weeks of age, respectively (Table 2, column 2). In contrast, the average age of onset of mammary carcinomas in *Wnt-1 TG* females nullizygous for *p53* was 11.5 weeks of age, and all of these mice formed tumors by 15 weeks (Fig. 1A and Table 2, columns 2 and 3).

Male *Wnt-1 TG* mice wild-type or heterozygous for *p53* showed modest and not significantly different rates of mammary tumor development by 1 year of age (32% and 43%, respectively). In contrast, mammary tumors appeared in *Wnt-1 TG p53 −/−* males at a rate almost as rapid as that observed in females of the same genotype (Fig. 1B). Tumors appeared in all of the *Wnt-1 TG p53 −/−* males by 6 months of age. Of these, 75% were mammary tumors and 25% were lymphomas and sarcomas similar to those found in the *p53 −/−* mice without the *Wnt-1* transgene. (Animals with non-mammary tumors are not included in Fig. 1B.)

Male and female mice without the *Wnt-1* transgene, but either heterozygous or homozygous for a *p53* null allele, developed a variety of non-mammary tumors with incidences comparable to those described previously (Donehower et al. 1992). Mice without the *Wnt-1* transgene and with two wild-type *p53* alleles did not form any tumors during the monitoring period.

LOSS OF HETEROZYGOSITY IN Wnt-1 TG p53 +/− MAMMARY TUMORS

DNA isolated from tumors arising in *Wnt-1 TG* mice heterozygous for a null *p53* allele was examined for retention or loss of the remaining wild-type allele. Approximately half of *Wnt-1 TG p53 +/−* tumors showed loss of the wild-type *p53* allele (data not shown), implying selection for the growth of cells lacking *p53* during tumor development. However, tumors which displayed loss of heterozygosity (LOH) of the

Table 2. Incidence, Histopathology, and Chromosomal Instability of Tumors from Female Wnt-1 Transgenic Mice in the Presence or Absence of p53

p53 Genotype	50% Tumor incidence[a] (weeks)	100% Tumor incidence[b] (weeks)	Tumors with fibrosis[c]	Aneuploid tumors[d]	Tumors with chromosome abnormalities[e]	Average number of chromosome abnormalities per tumor[f]
p53 +/+	22.5	40	8/11	0/5	2/6	0.3
p53 +/− (no LOH)	23.0[g]	41[g]	6/8	0/1	2/4	1.0
p53 +/− (LOH)			1/7	2/3	8/8	4.2
p53 −/−	11.5	15	0/6	3/6	7/7	1.7

[a] Age by which 50% of all mice of a given p53 genotype had developed mammary tumors.
[b] Age by which 100% of all mice of a given p53 genotype had developed mammary tumors.
[c] Number of tumors with extensive regions of fibrosis following histopathology/number of tumors of that p53 genotype analyzed.
[d] Number of tumors with >40% of cells displaying aneuploid karyotype by standard cytogenetics/number of tumors of that p53 genotype analyzed.
[e] Number of tumors of a group with abnormal CGH profiles/number of tumors of that p53 genotype analyzed.
[f] Average number of CGH copy number increases and decreases per tumor of a given p53 genotype.
[g] Not all Wnt-1 TG/p53 +/− tumors were assayed for p53 LOH; 23 and 41 weeks are the ages at which 50% and 100% of all p53 +/− mice developed mammary tumors, respectively.

wild-type *p53* allele did not appear earlier on average than *Wnt-1 TG p53 + / −* mammary tumors that retained the wild-type *p53* allele. *p53* cDNAs from four of the *Wnt-1 TG p53 + / −* mammary tumors without *p53* LOH were sequenced and found to have the normal nucleotide sequence.

HISTOPATHOLOGY OF MAMMARY TUMORS DEPENDS ON *p53* GENOTYPE

Dramatic histopathological differences were noted between those tumors that contained one or two wild-type *p53* genes and those without a wild-type *p53* gene. Tumors that had at least one wild-type *p53* allele tended to be more organized, fibrotic, and more differentiated, whereas tumors that lacked wild-type *p53* (*p53 − / −*, and *p53 + / −* with *p53* LOH) were significantly more anaplastic, less differentiated, and less fibrotic (Table 2, column 4).

MEASURING GENOMIC INSTABILITY IN *Wnt-1 TG* MAMMARY TUMORS

We were interested in understanding some of the mechanisms by which the lack of *p53* affected mammary tumorigenesis in the *Wnt-1 TG* mice. Previous studies have indicated that wild-type p53 may operate as a cell cycle checkpoint protein that mediates G_1 arrest or apoptosis following DNA damage or abnormal proliferation (Kastan et al. 1991, 1992; Kuerbitz et al. 1992; Clarke et al. 1993; Lowe et al. 1993, 1994). Models have been proposed in which loss of wild-type *p53* leads to a loss of cell cycle control and consequently, genomic instability (Lane 1992). Increased rates of DNA rearrangements or other mutations could contribute to oncogenesis. Support for such a model has been provided by Yin et al. (1992) and Livingstone et al. (1992), who showed that cells in culture that lose wild-type *p53* have higher levels of genomic instability as measured by more abnormal chromosome numbers and increased susceptibility to gene amplification under selective conditions. To test whether *p53* deficiency predisposes cells to increased genomic instability in the context of an animal tumor model, we examined the mammary tumors from the study described above by three methods: standard karyotype analysis, comparative genomic hybridization, and Southern blotting.

KARYOTYPES OF CULTURED CELLS FROM *Wnt-1 TG* MAMMARY TUMORS

In the karyotyping analyses, we examined chromosome numbers from cultured cells derived from fifteen *Wnt-1 TG* mammary tumors of different *p53* genotypes. Of five *p53 + / +* tumors analyzed, all contained tumor cells that displayed predominantly diploid karyotypes (Table 2, column 5). This was also true of one *p53 + / −* tumor which retained its wild-type *p53* allele and of one *p53 + / −* tumor which displayed *p53* LOH. In contrast, two of three *p53 + / −* tumors with *p53* LOH

and three of six *p53 − / −* tumors had a high percentage of cells with aneuploid karyotypes (Table 2, column 5). The remaining three *p53 − / −* tumors had karyotypes that resembled those from *p53 + / +* tumors. Thus, lack of *p53* was necessary for a nondiploid karyotype, but was not sufficient for it. This confirms that *p53* deficiency predisposes cells in this tumor model to increased genomic instability at the chromosomal level.

COMPARATIVE GENOMIC HYBRIDIZATION

Comparative genomic hybridization (CGH) detects regions of increased and decreased DNA copy number throughout the entire genome (Kallioniemi et al. 1992). Application of this technique to the *Wnt-1 TG* mammary tumors showed that loss of *p53* produced increased genomic instability at the chromosomal and subchromosomal level. Four of six *p53 + / +* and two of four *p53 + / −* mammary tumors with retention of wild-type *p53* displayed normal CGH profiles (Table 2, column 6). Each of the two *p53 + / +* tumors with abnormal CGH profiles exhibited a DNA copy number change in only a single chromosomal region (Table 2, column 6). Two of four *p53 + / −* tumors that retained wild-type *p53* had CGH abnormalities of one and three chromosomal regions each. In contrast, the *p53 + / −* tumors with *p53* LOH showed the highest levels of chromosomal abnormalities. All eight of these tumors displayed abnormal CGH profiles, with an average of 4.2 abnormalities (mostly DNA copy number decreases) per tumor (Table 2, columns 6 and 7). Finally, all seven *p53 − / −* tumors displayed CGH abnormalities, with a mean number of abnormalities per tumor of 1.7 (Table 2, columns 6 and 7).

For the 21 tumors analyzed by CGH, the frequency with which each chromosome contained a region of DNA gain or loss is displayed graphically in Figure 2. Note that chromosomes 8, 9, 11, 13, 14, and X showed frequent loss of DNA sequences. The losses on chromosome 11 were mostly observed in the *p53 + / −* tumors that lost wild-type *p53*. This result is consistent with the location of the *p53* gene on mouse chromosome 11. Interestingly, a number of the other implicated regions contain genes that have been associated with mammary cancer in humans or in other mouse models (Fig. 2, right-hand panel).

Southern blotting of normal and mammary tumor DNA from *Wnt-1 TG/p53 − / −* animals was used to test for DNA copy increases of specific loci. The loci that were chosen were three that are amplified in some human breast cancers: c-*myc*, *neu*, and *int-2/Fgf-3* (Machotka et al. 1989; Lammie and Peters 1991). Such analysis revealed one tumor (W177B) with an increased DNA copy number at the *int-2/Fgf-3* locus. This result was consistent with the CGH profile for that tumor, which showed a DNA increase in the telomeric region of mouse chromosome 7, the chromosomal localization of *int-2/Fgf-3*. Quantitation of the South-

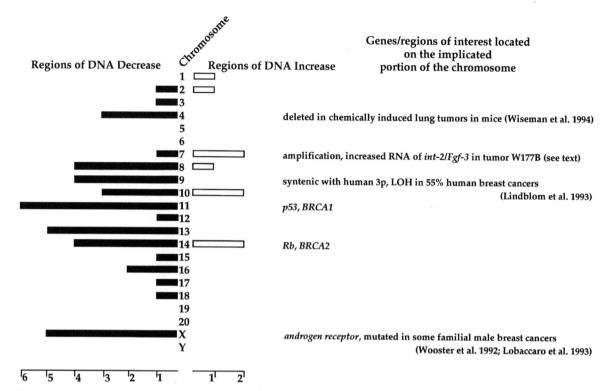

Figure 2. Summary of the chromosomal changes seen by CGH in *Wnt-1*-induced mammary tumors. The length of each solid bar indicates the number of tumors for which a decrease in DNA copy number was found on that chromosome. The length of each outlined bar indicates the number of tumors for which an increase in DNA copy number was found on that chromosome. Genes located in the implicated region which might play a role in tumor formation are listed on the right of the figure.

ern blot revealed that this locus was present in 6–8 copies, and Northern blot analysis showed abundant *int-2/Fgf-3* RNA as well. However, in human breast tumors with DNA increases at this locus, *int-2/Fgf-3* RNA is not detectable, and the DNA amplification appears to contribute to tumorigenesis by expression of the linked *PRAD-1/cyclin D1* locus (Lammie et al. 1991). *PRAD-1/cyclin D1* was not expressed at augmented levels in tumor W177B, suggesting that selective pressures may favor expression of different genes, such as *int-2/Fgf-3*, in the *Wnt-1* transgenic context.

Taken together, the CGH results confirm the increased genomic instability in tumors lacking *p53* at the chromosomal and subchromosomal level. In addition, recurring regions of chromosomal loss and gain suggest that the observed genomic instability may not be random but may reflect selection for genetic changes that promote tumor progression. The increased levels of *int-2/Fgf-3* RNA in one *Wnt-1 TG p53 –/–* tumor, for example, are consistent with the known synergy between Wnt-1 and fibroblast growth factors in mammary carcinogenesis.

OTHER APPROACHES TO FINDING Wnt-1 COOPERATING GENES

Since loss of heterozygosity of tumor suppressor genes is a frequent event during tumorigenesis, interspecies crosses can be used to search the entire genome for loci which have undergone LOH (Dietrich et al. 1992; Copeland et al. 1993). In such an interspecies cross, the F_1 animals have completely heterozygous genomes. By examining the heterozygosity profile in F_1-derived mammary tumors as compared to other somatic tissues, the entire genome can be scanned for regions that have undergone LOH. Such an analysis is currently under way using mammary tumors from F_1 animals resulting from a cross between *Wnt-1 TG* and *Mus castaneus* mice in collaboration with Karen Hong and Eric Lander at the Massachusetts Institute of Technology.

The stochastic nature of tumor development in the *Wnt-1 TG* mice suggests that other genetic or epigenetic events must accompany *Wnt-1* overexpression to lead to tumor formation. The ability to separate tumor development into stages of hyperplasia, tumor, and metastasis should allow the identification of genes whose expression is necessary for tumor progression. One approach to finding genes that might be functionally relevant in the development of a particular stage is to identify genes that are differentially expressed among these different stages. Two genes have been identified which show differential expression within the *Wnt-1* transgenic system. They are *Sky*, expressed in mammary tumors, but not in hyperplastic glands, and the gene for demilune cell-specific protein, expressed in hyperplastic glands, but not in mammary tumors.

EXPRESSION OF *SKY* IN MAMMARY TUMORS

As part of our efforts to identify the still unknown Wnt-1 receptor, we used a polymerase chain reaction (PCR)-based strategy to isolate receptor tyrosine kinases in the mouse mammary epithelial cell line C57MG. We reasoned that the Wnt-1 receptor may be a tyrosine kinase, since these types of signaling molecules are key players in the control of cell growth, differentiation, and division. C57MG cells are presumed to express the Wnt-1 receptor, since they undergo morphological transformation, divide past confluence, and form foci in the presence of Wnt-1 (Brown et al. 1986).

Our screen identified five cDNA clones encoding receptor tyrosine kinases for which the ligand is known (fibroblast growth factor receptor, platelet-derived growth factor receptor, epithelial growth factor, insulin receptor, insulin-like growth factor receptor), two putative receptor tyrosine kinases for which the ligand remains to be identified (the products of *ryk* and the mouse *klg* homolog), and a novel tyrosine kinase. We isolated cDNAs encoding both the murine and human homologs of this kinase, the sequences of which were subsequently published under the names *Sky* (Ohashi et al. 1994) and *Rse* (Mark et al. 1994).

Our analysis (I.C.A. Taylor et al., in prep.) has revealed that mouse *Sky* RNA levels are abundant in mammary tumors derived from transgenic mice that express *Wnt-1*, *int-2/Fgf-3*, or both oncogenes in their mammary glands. However, little or no expression of *Sky* is detected in mammary glands from virgin animals or in preneoplastic mammary glands from *Wnt-1 TG* mice. Moreover, the human homolog of *Sky* is expressed at elevated levels when normal human mammary epithelial cells are rendered tumorigenic by the introduction of two viral oncogenes, v-*Ha-ras* and SV40 large T antigen. Transient transfection of a *Sky* expression vector into the quail fibrosarcoma cell line QT6 demonstrated that Sky is an active tyrosine kinase that is glycosylated and forms dimers, characteristics that are hallmarks of the receptor class of tyrosine kinases. The introduction of *Sky* into NIH-3T3 and Rat B1a fibroblasts results in morphological transformation, growth in soft agar, and tumor formation when injected into nude mice, suggesting that overexpression of *Sky* may be sufficient to promote tumorigenesis. Although it remains to be determined whether Sky is the receptor for Wnt-1, its expression pattern and transforming properties in cell culture raise the intriguing possibility that it is involved in the development and/or progression of mammary tumors.

DIFFERENTIALLY EXPRESSED GENES

Genes that are differentially expressed may be functionally significant for the development of a particular stage of tumor development. One technique designed to identify such genes, named differential display (Liang and Pardee 1992), allows the resolution of reverse transcriptase (RT)-PCR products on a sequencing gel. By comparing the cDNAs generated with RNAs from nontransgenic mammary gland, *Wnt-1 TG* hyperplastic gland, *Wnt-1 TG* mammary tumor, or metastatic foci, candidate cDNAs are found which appear to be differentially expressed. These cDNAs are then used as probes on Northern blots to confirm differential expression.

One such cDNA has been identified and is expressed in *Wnt-1 TG* hyperplastic glands but not in tumors. The sequence of this cDNA matched that of an unpublished cDNA that had been cloned from a mouse salivary gland cDNA library (I. Bekhor et al., in prep.). This cDNA encodes demilune cell-specific protein, a protein with a primary sequence of 170 amino acids, which is highly expressed in the demilune cells of the mouse salivary gland. Although its sequence is not homologous to any protein of known function, the presence of a canonical signal sequence and the absence of a predicted transmembrane domain suggest that it is secreted. The significance of its expression for *Wnt-1*-induced hyperplasia for tumor formation is under investigation.

SUMMARY AND PROSPECTS

The *Wnt-1* transgenic system has provided a rich model for studying mammary tumorigenesis in the mouse. Crossing these animals to other transgenic animals and infecting them with MMTV have forcefully demonstrated the synergy between Wnt-1 and members of the Fgf family. Mammary tumors from *Wnt-1 TG*/*p53*-deficient mice present novel features: a less fibrotic histology, rapid kinetics of tumor formation, and increased genomic instability. Understanding the molecular basis for such phenotypes may provide insights into the behavior of human tumors, especially those in which *p53* has been affected.

ACKNOWLEDGMENTS

This work was supported by National Cancer Institute grants CA-54897 (L.A.D.), CA-59967 (C.M.A.), CA-25215 (D.M.), CA-39832 (H.E.V.), CA-45919 (D.P.), a grant from the U.S. Army Breast Cancer Program (L.A.D.), and a grant from the Melanie Mann Bronfman Trust (H.E.V.). L.A.D. is the recipient of a Research Career Development Award from the National Cancer Institute, and H.E.V. was an American Cancer Society Research Professor.

REFERENCES

Bradley, R.S. and A.M.C. Brown. 1990. The proto-oncogene *int-1* encodes a secreted protein associated with the extracellular matrix. *EMBO J.* **9:** 1569.

Brown, A.M.C., R.S. Wildin, T.J. Prendergast, and H.E. Varmus. 1986. A retrovirus vector expressing the putative mammary oncogene *int-1* causes partial transformation of a mammary epithelial cell line. *Cell* **46:** 1001.

Cardiff, R.D. and W.J. Muller. 1993. Transgenic models of mammary tumorigenesis. *Cancer Surv.* **16:** 97.

Clarke, A.R., C.A. Purdie, D.J. Harrison, R.G. Morris, C.C.

Bird, M.L. Hooper, and A.H. Wyllie. 1993. Thymocyte apoptosis induced by p53-dependent and independent pathways. *Nature* **352:** 849.

Copeland, N.G., N.A. Jenkins, D.J. Gilbert, J.T. Eppig, L.J. Maltais, J.C. Miller, W. Dietrich, A. Weaver, S.E. Lincoln, R.G. Steen, L.D. Stein, J.H. Nadeau, and E.S. Lander. 1993. A genetic linkage map of the mouse: Current applications and future prospects. *Nature* **262:** 57.

Dietrich, W., H. Katz, S.E. Lincoln, H.-S. Shin, J. Friedman, N.C. Dracopoli, and E.S. Lander. 1992. A genetic map of the mouse suitable for typing intraspecific crosses. *Genetics* **131:** 423.

Donehower, L.A., M. Harvey, B.L. Slagle, M.J. McArthur, C.A. Montgomery, Jr., J.S. Butel, and A. Bradley. 1992. Mice deficient for p53 are developmentally normal but susceptible to spontaneous tumours. *Nature* **356:** 215.

Fearon, E.R. and B. Vogelstein. 1990. A genetic model for colorectal tumorigenesis. *Cell* **61:** 759.

Futreal, P.A., Q. Liu, D. Shattuck-Eidens, C. Cochran, K. Harshman, S. Tavtigian, L.M. Bennett, A. Haugen-Strano, J. Swensen, Y. Miki, K. Eddington, M. McClure, C. Frye, J. Weaver-Feldhaus, W. Ding, Z. Gholami, P. Soderkvist, L. Terry, S. Jhanwar, A. Berchuck, J.D. Iglehart, J. Marks, D.G. Ballinger, J.C. Barrett, M.H. Skolnick, A. Kamb, and R. Wiseman. 1994. BRCA1 mutations in primary breast and ovarian carcinomas. *Science* **266:** 120.

Greenblatt, M.S., W.P. Bennett, M. Hollstein, and C.C. Harris. 1994. Mutations in the p53 tumor suppressor gene: Clues to cancer etiology and molecular pathogenesis. *Cancer Res.* **54:** 4855.

Harvey, M., M.J. McArthur, C.A. Montgomery, Jr., J.S. Butel, A. Bradley, and L.A. Donehower. 1993. Spontaneous and carcinogen-induced tumorigenesis in p53-deficient mice. *Nat. Genet.* **5:** 225.

Hilgers, J. and P. Bentvelzen. 1979. Interaction between viral and genetic factors in murine mammary cancer. *Adv. Cancer Res.* **26:** 143.

Jacks, T., L. Remington, B.O. Williams, E.M. Schmitt, S. Halachmi, R.T. Bronson, and R.A. Weinberg. 1994. Tumor spectrum analysis in p53-mutant mice. *Curr. Biol.* **4:** 1.

Kallioniemi, A., O.-P. Kallioniemi, D. Sudar, D. Rutovitz, J.W. Gray, F. Waldman, and D. Pinkel. 1992. Comparative genomic hybridization for molecular cytogenetic analysis of solid tumors. *Science* **258:** 818.

Kastan, M.B., O. Onyekwere, D. Sidransky, B. Vogelstein, and R.W. Craig. 1991. Participation of p53 protein in the cellular response to DNA damage. *Cancer Res.* **51:** 6304.

Kastan, M.B., Q. Zhan, W.S. El-Deiry, F. Carrier, T. Jacks, W.V. Walsh, B.S. Plunkett, B. Vogelstein, and A.J. Fornace, Jr. 1992. A mammalian cell cycle checkpoint pathway utilizing p53 and GADD45 is defective in ataxia-telangiectasia. *Cell* **71:** 587.

Kuerbitz, S.J., B.S. Plunkett, W.V. Walsh, and M.B. Kastan. 1992. Wild type p53 is a cell cycle checkpoint determinant following irradiation. *Proc. Natl. Acad. Sci.* **89:** 7491.

Kwan, H., V. Pecenka, A. Tsukamoto, T.G. Parslow, R. Guzman, T.-P. Lin, W.J. Muller, F.S. Lee, P. Leder, and H.E. Varmus. 1992. Transgenes expressing Wnt-1 and int-2 proto-oncogenes cooperate during mammary carcinogenesis in doubly transgenic mice. *Mol. Cell. Biol.* **12:** 147.

Lammie, G.A. and G. Peters. 1991. Chromosome 11q13 abnormalities in human cancer. *Cancer Cells* **3:** 413.

Lammie, G.A., V. Fantl, R. Smith, E. Schuuring, S. Brookes, R. Michalides, C. Dickson, A. Arnold, and G. Peters. 1991. D11S287, a putative oncogene on chromosome 11q13, is amplified and expressed in squamous cell and mammary carcinomas and linked to BCL-1. *Oncogene* **6:** 439.

Lane, D.P. 1992. p53, guardian of the genome. *Nature* **358:** 15.

Liang, P. and A.B. Pardee. 1992. Differential display of eukaryotic messenger RNA by means of the polymerase chain reaction. *Science* **257:** 967.

Lindblom, A., L. Skoog, S. Rotstein, B. Werelius, C. Larsson, and M. Nordenskjold. 1993. Loss of heterozygosity in familial breast carcinomas. *Cancer Res.* **53:** 4356.

Livingstone, L.R., A. White, J. Sprouse, E. Livanos, T. Jacks, and T. Tlsty. 1992. Altered cell cycle arrest and gene amplification potential accompany loss of wild type p53. *Cell* **70:** 923.

Lobaccaro, J.M., S. Lumbroso, C. Belon, F. Galtier-Dereure, J. Bringer, T. Lesimple, J.F. Heron, H. Pujol, and C. Sultan. 1993. Male breast cancer and the androgen receptor gene. *Nat. Genet.* **5:** 109.

Lowe, S.W., T. Jacks, D.E. Housman, and H.E. Ruley. 1994. Abrogation of oncogene-associated apoptosis allows transformation by p53-deficient cells. *Proc. Natl. Acad. Sci.* **91:** 2026.

Lowe, S.W., E.M. Schmitt, S.W. Smith, B.A. Osborne, and T. Jacks. 1993. p53 is required for radiation-induced apoptosis in mouse thymocytes. *Nature* **362:** 847.

Machotka, S.V., C.T. Garrett, A.M. Schwartz, and R. Callahan. 1989. Amplification of the proto-oncogenes int-2, c-erbB-2, and c-myc in human breast cancer. *Clin. Chim. Acta.* **184:** 207.

Malkin, D., F.P. Li, L.C. Strong, J.F. Fraumeni, Jr., C.E. Nelson, D.H. Kim, J. Kassel, M.A. Gryka, F.Z. Bischoff, M.A. Tainsky, and S.H. Friend. 1990. Germ line p53 mutations in a familial syndrome of breast cancer, sarcomas, and other neoplasms. *Science* **250:** 1233.

Mark, M.R., D.T. Scadden, Z. Wang, Q. Gu, A. Goddard, and P.J. Godowski. 1994. Rse, a novel receptor-type tyrosine kinase with homology to Axl/Ufo, is expressed at high levels in the brain. *J. Biol. Chem.* **269:** 100720.

McMahon, A.P. and A. Bradley. 1990. The wnt-1 (int-1) proto-oncogene is required for development of a large region of the mouse brain. *Cell* **62:** 1073.

Miki, Y., J. Swensen, D. Shattuck-Eidens, P.A. Futreal, K. Harshman, S. Tavtigian, Q. Liu, C. Cochran, L.M. Bennett, W. Ding, R. Bell, J. Rosenthal, C. Hussey, T. Tran, M. McClure, C. Frye, T. Hattier, R. Phelps, A. Haugen-Strano, H. Katcher, K. Yakumo, Z. Gholami, D. Shaffer, S. Stone, S. Bayer, C. Wray, R. Bogden, P. Dayananth, J. Ward, P. Tonin, S. Narod, P.K. Bristow, F.H. Norris, L. Helvering, P. Morrison, P. Rosteck, M. Lai, J.C. Barrett, C. Lewis, S. Neuhausen, L. Cannon-Albright, D. Goldgar, R. Wiseman, A. Kamb, and M.H. Skolnick. 1994. A strong candidate for the breast and ovarian cancer susceptibility gene BRCA1. *Science* **266:** 66.

Muller, W.J., F.S. Lee, C. Dickson, G. Peters, P. Pattengale, and P. Leder. 1990. The int-2 gene product acts as an epithelial growth factor in transgenic mice. *EMBO J.* **9:** 907.

Nusse, R. and H.E. Varmus. 1982. Many tumors induced by the mouse mammary tumor virus contain a provirus integrated in the same region of the host genome. *Cell* **31:** 99.

Ohashi, K., K. Mizuno, K. Kuma, T. Miyata, and T. Nakamura. 1994. Cloning of the cDNA for a novel receptor tyrosine kinase, Sky, predominantly expressed in brain. *Oncogene* **9:** 699.

Papkoff, J. and B. Schryver. 1990. Secreted int-1 protein is associated with the cell surface. *Mol. Cell. Biol.* **10:** 2723.

Peters, G., A.E. Lee, and C. Dickson. 1986. Concerted activation of two potential proto-oncogenes in carcinomas induced by mouse mammary tumour virus. *Nature* **320:** 628.

Purdie, C.A., D.J. Harrison, A. Peter, L. Dobbie, S. White, S.E.M. Howie, D.M. Salter, C.C. Bird, A.H. Wyllie, M.L. Hooper, and A.R. Clarke. 1994. Tumour incidence, spectrum and ploidy in mice with a large deletion in the p53 gene. *Oncogene* **9:** 603.

Rijsewijk, F., L. van Deemter, E. Wagenaar, A. Sonnenberg, and R. Nusse. 1987a. Transfection of the int-1 mammary oncogene in cuboidal RAC mammary cell line results in morphological transformation and tumorigenicity. *EMBO J.* **6:** 127.

Rijsewijk, F., M. Schuermann, E. Wagenaar, P. Parren, D. Weigel, and R. Nusse. 1987b. The *Drosophila* homologue of the mouse mammary tumor oncogene int-1 is identical to the segment polarity gene *wingless*. *Cell* **50:** 649.

Shackleford, G.M. and H.E. Varmus. 1987. Expression of the proto-oncogene *int-1* is restricted to postmeiotic male germ cells and the neural tube of mid-gestational embryos. *Cell* **50:** 89.

Shackleford, G.M., C.A. MacArthur, H.C. Kwan, and H.E. Varmus. 1993. Mouse mammary tumor virus infection accelerates mammary carcinogenesis in *Wnt-1* transgenic mice by insertional activation of *int-2/Fgf-3* and *hst/Fgf-4*. *Proc. Natl. Acad. Sci.* **90:** 740.

Slamon, D.J., G.M. Clark, S.G. Wong, W.J. Levin, A. Ullrich, and W.L. McGuire. 1987. Human breast cancer: Correlation of relapse and survival with amplification of the HER-2/neu oncogene. *Science* **235:** 177.

Srivastava, S., Z. Zou, K. Pirollo, W. Blattner, and E.H. Chang. 1990. Germ-line transmission of a mutated p53 gene in a cancer prone family with Li-Fraumeni syndrome. *Nature* **348:** 747.

Swift, M., D. Morrell, R.B. Massey, and C.L. Chase. 1991. Incidence of cancer in 161 families affected by ataxia-telangiectasia. *N. Engl. J. Med.* **325:** 1831.

Thomas, K.R. and M.R. Capecchi. 1990. Targeted disruption of the murine *int-1* proto-oncogene resulting in severe abnormalities in midbrain and cerebellar development. *Nature* **346:** 847.

Thomas, K.R., T.S. Musci, P.E. Neumann, and M.R. Capecchi. 1991. *Swaying* is a mutant allele of the proto-oncogene *Wnt-1*. *Cell* **67:** 969.

Tsukamoto, A.S., R. Grosschedl, R.C. Guzman, T. Parslow, and H.E. Varmus. 1988. Expression of the *int-1* gene in transgenic mice is associated with mammary gland hyperplasia and adenocarcinomas in male and female mice. *Cell* **55:** 619.

van Ooyen, A. and R. Nusse. 1984. Structure and nucleotide sequence of the putative mammary oncogene *int-1*; proviral insertions leave the protein-encoding domain intact. *Cell* **39:** 233.

Wang, T.C., R.D. Cardiff, L. Zukerberg, E. Lees, A. Arnold, and E.V. Schmidt. 1994. Mammary hyperplasia and carcinoma in MMTV-cyclin D1 transgenic mice. *Nature* **369:** 669.

Weinberg, R.A. 1992. The retinoblastoma gene and gene product. *Cancer Surv.* **12:** 43.

Wilkinson, D.G., J.A. Bailes, and A.P. McMahon. 1987. Expression of the proto-oncogene *int-1* is restricted to specific neural cells in the developing mouse embryo. *Cell* **50:** 79.

Wiseman, R.W., C. Cochran, W. Dietrich, E.S. Lander, and P. Soderkvist. 1994. Allelotyping of butadiene-induced lung and mammary adenocarcinomas of B6C3F1 mice: Frequent losses of heterozygosity in regions homologous to human tumor-suppressor genes. *Proc. Natl. Acad. Sci.* **91:** 3759.

Wooster, R., J. Mangion, R. Eeles, S. Smith, M. Dowsett, D. Averill, P. Barrett-Lee, D.F. Easton, B.A. Ponder, and M.R. Stratton. 1992. A germline mutation in the androgen receptor gene in two brothers with breast cancer and Reifenstein syndrome. *Nat. Genet.* **2:** 132.

Wooster, R., S.L. Neuhausen, J. Mangion, Y. Quirk, D. Ford, N. Collins, K. Nguyen, S. Seal, T. Tran, D. Averill, P. Fields, G. Marshall, S. Narod, G.M. Lenoir, H. Lynch, J. Feunteun, P. Devilee, C.J. Cornelisse, F.H. Menko, P.A. Daly, W. Ormiston, R.McManus, C. Pye, C.M. Lewis, L.A. Cannon-Albright, J. Peto, B.A.J. Ponder, M.H. Skolnick, D.F. Easton, D.E. Goldgar, and M.R. Stratton. 1994. Localization of a breast cancer susceptibility gene, BRCA2, to chromosome 13q12-13. *Science* **265:** 2088.

Yin, H., M.A. Tainsky, F.Z. Bischoff, L.C. Strong, and G.M. Wahl. 1992. Wild type p53 restores cell cycle control and inhibits gene amplification in cells with mutant p53 alleles. *Cell* **70:** 937.

The Adenomatous Polyposis Coli Gene of the Mouse in Development and Neoplasia

W.F. Dove,* C. Luongo,* C.S. Connelly,* K.A. Gould,*
A.R. Shoemaker,* A.R. Moser,* and R.L. Gardner†

McArdle Laboratory for Cancer Research and Laboratory of Genetics, The Medical School, University of Wisconsin, Madison, Wisconsin 53706; †Imperial Cancer Research Fund, Developmental Biology Unit, Oxford University, Oxford, OX1 3PS, United Kingdom

The intestinal epithelium is the most mitotically active tissue in the adult mammal (see Potten and Morris 1988). Crypts, the sites of mitotic activity, surround regions of differentiation, finger-like villi in the small intestine. As cells migrate away from the proliferative zone, they differentiate to Paneth cells at the base of the crypts or to enterocytes, goblet cells, or enteroendocrine cells in the villi. Overall, in the lifetime of the mouse, the number of mitoses in this tissue is on the order of 10^{13}; in the human it is 10^{16}. The mechanisms that largely protect the intestinal epithelium from the consequences of errors, including mutation and nondisjunction, remain a fundamental mystery (Cairns 1975).

The intestinal epithelium is one of the most common sites for human cancer. Inactivating mutations of the adenomatous polyposis coli gene, *APC*, are involved in a majority of cases, both sporadic and familial. These include the hereditary non-polyposis kindreds in which a germ-line mutator element acts somatically on a normal *APC* gene (Aaltonen et al. 1993). The Min (multiple intestinal neoplasia) mouse is heterozygous for a nonsense allele at codon 850 of *Apc*, the mouse homolog to *APC* (Su et al. 1992). This strain closely models the human familial adenomatous polyposis (FAP) cancer syndrome. The *Min* mutation was discovered at Wisconsin by phenotypic screening after random germ-line point mutagenesis with ethylnitrosourea (ENU) (Moser et al. 1990). C57BL/6-*Min*/+ (B6 Min) mice develop numerous adenomas in the colon and in the small intestine (Fig. 1). Within these tumors, a small number of cells expressing normal differentiation markers of the intestinal epithelium can be found: lysozyme characteristic of Paneth cells; serotonin characteristic of enteroendocrine cells; and the isotype of fatty acid-binding protein appropriate to the enterocytes from the corresponding region of the intestine (Moser et al. 1992). On a sensitive B6 genetic background, in excess of 50 tumors develop per animal, leading to an average life span of 120 days. Strains carrying resistant modifier alleles at loci unlinked to *Apc* develop smaller numbers of tumors and thereby live longer. Given time in this way, the *Min*-induced adenomas progress to become locally invasive adenocarcinomas that continue to contain cells expressing appropriate differentiation markers.

LOSS OF THE WILD-TYPE *Apc* ALLELE IN EARLY *Min*-INDUCED ADENOMAS

Note that only a finite number of adenomas form in an intestinal tract in which all cells carry the *Min* mutation. This pattern indicates that further events, genetic or epigenetic, beyond the germ-line *Min* lesion, are needed for adenoma formation. Is loss of the wild-type *Apc* allele involved in *Min*-induced adenoma formation?

In the corresponding human condition, it has been observed that up to 71% of tumors have incurred a detectable somatic mutation in the remaining normal allele of *APC* (Ichii et al. 1992). Thus, it is possible that tumors in individuals from FAP kindreds can form by two or more alternative pathways. Perhaps the heterogeneous genetic background within these families permits multiple pathways of tumorigenesis or perhaps the methods of detection involve a significant fraction of false negatives.

To address this issue on homogeneous genetic backgrounds in the mouse, we have developed a method to detect both the wild type and the *Min* allele of *Apc* in DNA isolated from histological sections (Luongo et al. 1994). Tumors were dissected, fixed, and serially sectioned. The region containing tumor cells was then judged by staining 1 section out of every 10. Cells were harvested from the tumor region of the intervening unstained sections, and DNA was isolated. With primers that introduce a diagnostic *Hin*dIII site into amplimers of the wild-type allele, the region flanking the mutant site was amplified by polymerase chain reaction (PCR), using radiolabeled nucleotides. The amplified DNA was cut with *Hin*dIII, and the wild-type and mutant bands were quantitated on a Phosphorimager.

Even for tumors no more than one millimeter in diameter, the normal *Apc* allele is extensively lost from tumor but not from control tissue (Fig. 2). This result has been found in 100% of tumors analyzed. Daniel Levy and his colleagues at Johns Hopkins University have shown 100% loss frequencies in *Min*-induced

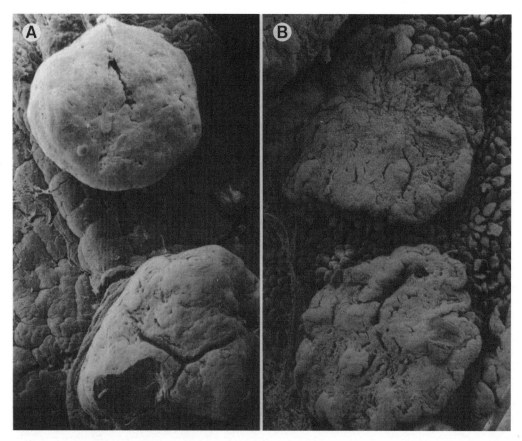

Figure 1. Scanning electron micrographs of intestinal epithelium from C57BL/6-*Min*/+ mice. (*A*) Colon; (*B*) small intestine. Multiple adenomas are found in each region of the intestinal tract.

tumors by a different method and extended the resolution to dysplastic lesions two crypts in size (Levy et al. 1994).

The *Apc* locus lies on mouse chromosome 18 (Justice et al. 1992; Luongo et al. 1993). The allelic loss in *Min*-induced adenomas extends to the entire mouse chromosome 18 (Luongo et al. 1994). Tumors from (AKR × B6)F$_1$ *Min*/+ heterozygotes show loss of AKR alleles all along chromosome 18. No such losses were observed in normal tissue. Quantitation of the intensity of the PCR products amplified from the residual chromosome 18 alleles indicated that only a single copy of this chromosome remains in the tumors: the B6 homolog carrying the *Min* allele. Examination of five additional chromosomes showed no such somatic losses in these tumors. Therefore, adenomas in human FAP kindreds differ from those in the Min mouse in two respects: (1) The loss of function of the normal allele occurs by intragenic inactivating mutations or interstitial deletions in the human, but by loss of the entire chromosome in the mouse; and (2) loss has been detected in no more than 71% of tumors in the human, but in 100% in the mouse. One interpretation of these differences is that the mouse genome permits hemizygosity, because inbred strains of mice have been bred to lack recessive cell lethal mutations. Under these conditions, the simplest way to lose the wild-type allele may be chromosome loss. For the human, in contrast, the ascertainment of intragenic somatic mutations that inactivate *APC* may currently be incomplete.

The overall conclusion from these studies is that adenoma formation in the Min mouse requires loss of normal *Apc* function. There is no evidence to date for a parallel alternative pathway to adenoma formation. Because *Apc* loss correlates completely with adenoma formation, this gene qualifies as a tumor suppressor gene controlling at least tumor establishment. Accordingly, restoration of wild-type activity should prevent tumor formation.

It remains open whether other events—genetic or epigenetic—are also necessary to establish adenomas in the Min mouse. Indeed, a second plausible interpretation of the prevalence of chromosome 18 hemizygosity in *Min*-induced adenomas is a requirement for reduction or total elimination of the function of another tumor suppressor gene encoded on this chromosome—for example, *Mcc* or *Dcc* (Justice et al. 1992; Luongo et al. 1993).

How does the frequency of *Min*-induced adenomas compare with the rate expected for a single somatic event? We make a series of arbitrary assumptions to address this issue in general terms. The small intestine of the adult mouse contains on the order of 10^6 crypts, each with one permanent stem cell dividing once per

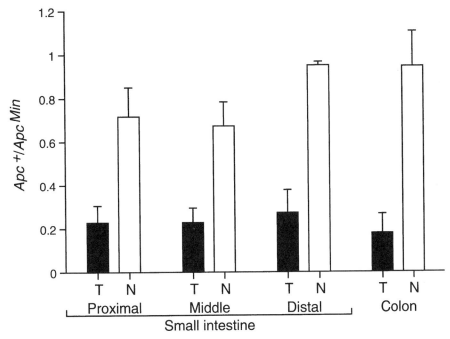

Figure 2. Quantitation of allelic frequencies for Apc^+ and Apc^{Min} from DNA isolated from unstained sections of tumor (T; filled bars) and adjacent normal (N; open bars) tissue. (Adapted from Luongo et al. 1994.)

day (Ponder et al. 1985; Potten and Morris 1988). On the single somatic hit hypothesis, a tumor load of 50 adenomas per animal is plausible for 30 days of normal crypt proliferation with a rate of somatic nondisjunction of 10^{-6} per chromosome per cell division. Thus, within this crude estimation, it is not necessary at this stage of our knowledge to invoke more than a single further somatic event for adenoma formation in the Min/+ mouse.

An interesting and open question is whether the Min/+ heterozygote has an enhanced probability of mitotic chromosome loss. The heterozygous cell might show such an aberrant phenotype either by haploinsufficiency (Bodmer et al. 1987; Fisher and Scambler 1994) or by a neomorphic dominant-negative activity of the mutant gene product (Herskowitz 1987). A heterozygous cellular phenotype would provide a driving mechanism for the evolution of Min-induced neoplasms that would contrast with the mutation-driven process summarized by Bodmer and his colleagues (1994) for hereditary nonpolyposis kindreds.

WHAT TISSUES ARE AFFECTED IN $M_{IN}/+$ HETEROZYGOTES?

The APC gene is transcribed ubiquitously (Groden et al. 1991). The neoplastic phenotype of Min/+ heterozygotes is expressed in the intestinal epithelium of sensitive animals with 100% penetrance. On resistant backgrounds, longer-lived animals also develop other types of neoplastic lesions, including mammary adenocarcinomas (Moser et al. 1993). The penetrance of these lesions is far less than 100% of Min/+ animals. However, the penetrance can be increased even on the short-lived sensitive genetic background by appropriately timed somatic ENU treatment (Moser et al. 1993). Recent experiments have shown that ENU treatment of B6 Min females aged 30–35 days can increase the incidence of mammary tumors to 82% (A.R. Moser and A.R. Shoemaker, in prep.).

A second class of neoplastic lesion in non-endodermal tissue is the desmoid fibrosarcoma, found in the peritoneal muscle wall (A.R. Moser, unpubl.). Altogether, in both the human and the mouse, a heterozygous defect in the APC or Apc gene predisposes to a broad range of hyperplastic and neoplastic lesions in an embryonically diverse range of tissues (Table 1). It remains to be seen whether each of these focal lesions requires loss of the remaining normal allele at the locus and what factors, genetic or environmental, control the penetrance of these extracolonic Min-induced lesions.

Table 1. Range of Heterozygous Phenotypes in Human Carriers of APC Mutations and Mouse Carriers of the Min Mutation

Human	Mouse
Multiple colon adenomas (100–1000s)	multiple colon adenomas
Small intestine adenomas	small intestine adenomas
Desmoid tumors	desmoid tumors
Gastric polyps	mammary adenocarcinomas
Epidermoid cysts	mammary keratoacanthomas
Osteomas of skull and mandible	
Hypertrophy of retinal pigment epithelium	
Abnormal dentition	

WHAT OTHER LOCI INTERACT WITH MIN IN INTESTINAL NEOPLASIA?

With a familial cancer determinant such as *Min*, it is feasible to scan the gene pool of the mouse to uncover genes for which allelic variants affect the formation and/or growth of neoplasms. The two rich sources of allelic variation in the mouse are targeted mutations of cloned genes and natural variants (carried in inbred strains). The targeted mutations can include recessive lethals, whereas the natural polymorphisms are viable as homozygotes. We have been studying two loci unlinked to *Apc*: *p53* on chromosome 11 and a resistant modifier locus on chromosome 4, *Mom-1*.

Min-induced adenomas involve somatic hemizygosity of chromosome 18. Defects in *p53* can cause aneuploidy in mouse embryo fibroblast cultures (Carder et al. 1993). Furthermore, as intestinal tumors progress in the human, *p53* is often mutated (see Fearon and Vogelstein 1990). However, neither Li-Fraumeni kindreds in human nor targeted *p53* mutant lines of mice—heterozygous or homozygous—show a significant elevation in the incidence of intestinal neoplasms (see Donehower and Bradley 1993; Harvey et al. 1993). Indeed, the levels of p53 polypeptide are very low in the intestinal tract and in adenomas (Purdie et al. 1991), except after ionizing radiation (Merritt et al. 1994). Does the status of the *p53* locus affect the incidence of *Min*-induced intestinal tumors? In the progeny of a cross involving *Min* and a targeted *p53* mutant allele (Donehower et al. 1992), the multiplicity of adenomas in *Min*/+ mice was independent of the *p53* genotype (Table 2). Because these crosses have been performed on a sensitive genetic background, animals were scored at 80 days, precluding long-term studies of progression. However, within the initial time window of these experiments, the absence of p53 function does not seem to affect *Min*-induced intestinal adenoma formation.

In contrast, the Modifier-of-Min locus, *Mom-1*, has been detected by its strong effect on the multiplicity of *Min*-induced intestinal adenomas. When the *Min* mutation is carried on the sensitive B6 background, the average tumor multiplicity is near 30 in the regions of intestine counted, as shown in the open bars of the histogram in Figure 3. In contrast, after outcrossing to strain AKR, the average tumor multiplicity in F_1 animals drops to 5, shown in solid bars. A backcross of these F_1 animals to the sensitive parent gave a segregating population of animals shown in striped bars on this histogram. To localize the genetic factors that modify *Min*-induced tumor multiplicity, the mouse genome has been scanned, using simple sequence length polymorphism (SSLP) markers. Backcross animals inheriting the AKR region on mouse chromosome 4 showed reduced tumor multiplicities (Dietrich et al. 1993). This region of the genome is responsible for a major part of the reduction in tumor multiplicity found in F_1 hybrids with three different strains: AKR, MA, and Castaneus.

Two of the fundamental points concerning *Mom-1* about which we are currently ignorant are (1) we do

Table 2. Multiplicity of Adenomas Versus *p53* Genotype in *Min*/+ Mice

Cross: $Apc^+/Apc^+\ p53^-/p53^+ \times Apc^{Min}/Apc^+\ p53^-/p53^+$			
	\multicolumn{3}{c}{*p53* genotype}		
	+/+	+/−	−/−
Number of *Min*/+ mice	16	31	7
Average number of tumors	24 ± 11	28 ± 19	23 ± 11

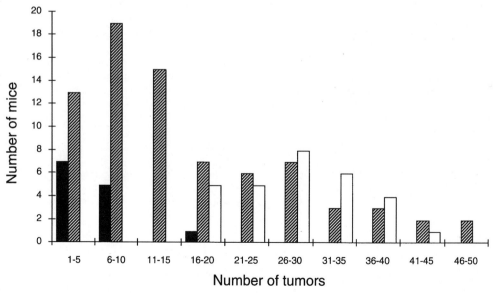

Figure 3. Histograms of tumor multiplicities of *Min*/+ mice from C57BL/6 (*open bars*), (AKR × C57BL/6)F_1 (*filled bars*), and C57BL/6 × (AKR × C57BL/6)F_1 (*striped bars*) populations. (Reprinted, with permission [Rockefeller University Press] from Moser et al. 1992.)

not know which allele of *Mom-1* is active—the resistant AKR allele blocking *Min*-induced tumor formation, or the sensitive B6 allele acting in a dosage-sensitive manner to enhance tumor formation and (2) we do not know the gene product of the *Mom-1* locus, although certain intriguing candidates have been mapped at low resolution to the *Mom-1* region (Dietrich et al. 1993; Lyon et al. 1993; Mock et al. 1993).

Our studies of genes that interact with *Min* are summarized by considering subcategories of tumor suppressor genes (Fig. 4). How does loss of function of the gene affect the tumor phenotype? With familial cancer predispositions, one is detecting genes whose loss of function affects a rate-limiting step in carcinogenesis. One could call these establishment suppressor genes. As with *Apc*, such genes often show somatic losses of heterozygosity as the familial tumors form.

In multistage cancer, one may expect to find losses of heterozygosity for loci other than the familial, rate-limiting set. For example, the *APC* locus shows frequent loss of heterozygosity in human small-cell lung cancer, which is apparently not a part of the familial FAP syndrome in humans (D'Amico et al. 1992). For these tumors, *APC* may be a "progression suppressor gene": Loss of its function may affect the severity of each tumor but not change tumor multiplicity. In this case, *APC* loss should be associated with tumor grade. Explicit evidence of this sort has indicated that *p53* controls progression but not initiation in skin carcinogenesis (Kemp et al. 1993).

It seems that *Mom-1* controls, at least, tumor multiplicity. Thus, we argue that it acts during the initial stages of tumorigenesis (see Fig. 4). The *Mom-1* locus could control either the rate of somatic variation (mitotic loss of the Apc^+ allele) or the rate of proliferation during focal tumor establishment. Finally, *Mom-1* could regulate the survival, rather than the establishment, of the initiated unit.

We believe that it will be biologically instructive to use the familial cancer syndrome of the Min mouse to explore the mammalian gene pool for unlinked modifier (*Mom*) loci. The information content will be enhanced as the analysis of the mouse genome moves effectively toward closure of the genetic and physical maps, permitting somewhat more facile identification of the molecular entity encoded by any one *Mom* locus (Dietrich et al. 1994).

A third fundamental point of which we currently are ignorant is the realm of action of the modifier gene *Mom-1*. Does it act intrinsically or extrinsically to the tumor lineage? We have begun to investigate this issue by asking whether the resistant AKR region around the *Mom-1* locus is lost when tumors form in *Mom-1* heterozygotes. Since we do not yet have molecular probes for the *Mom-1* gene itself, we must follow the region that flanks the *Mom-1* locus. In contrast to the region containing the *Apc* locus, the *Mom-1* region remains heterozygous in tumors (Table 3) (C. Luongo, in prep.). Thus, somatic loss of the $Mom-1^{AKR}$ region from the tumor lineage is not necessary for tumor formation. One interpretation of this result is that the *Mom-1* gene acts extrinsically to the tumor lineage (see Paraskeva and Williams 1990). We hope to design experiments to determine explicitly whether the *Mom-1* locus acts from outside or inside the tumor lineage.

WHAT CELL TYPES DEVELOP IN *MIN/MIN* EMBRYOS?

Min-induced intestinal adenomas are hemizygous for the *Min* allele of *Apc*. What is the developmental fate of embryos homozygous for the *Min* allele of *Apc*?

When *Min*/+ heterozygotes have been intercrossed, no liveborn *Min/Min* progeny have been detected (Moser et al. 1990; A.R. Moser et al., in prep.). In such intercrosses, abnormal embryos are found as early as the seventh day of gestation, before gastrulation. From the seventh through the eleventh day of gestation, these abnormal embryos are found at close to 25% of the intercross progeny expected for the *Min/Min* class. In control crosses that do not produce the *Min/Min* class, abnormal embryos are much more rare and do not resemble the presumptive *Min/Min* embryos. Figure 5B gives an example of a longitudinally sectioned decidual swelling containing a typically abnormal egg-cylinder-stage embryo from the seventh day of gesta-

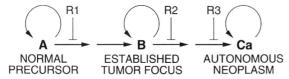

Figure 4. Formal stages in carcinogenesis at which a resistant modifier allele such as $Mom-1^{AKR}$ might act. Possible negative regulators (direct or indirect) are shown as R1, R2, and R3. The possibility that successful tumor focus formation requires cell proliferation after initiation is indicated by the positive feedback loop at stage B (Rous and Kidd 1941; for analogous considerations for normal development, see Gardner 1993; Gurdon et al. 1993). In this formalism, progression (Foulds 1964; Nowell 1976) from stage B to stage C does not affect tumor multiplicity.

Table 3. Retention of the *Mom-1* Region from AKR in *Min*-induced Adenomas

	Number of tumors showing loss of the AKR region/total scored			
	small intestine			
Locus	proximal	middle	distal	large intestine
D4Mit12	0/9	0/9	0/1	0/5
D4Mit13	0/9	0/9	0/1	0/5

tion. Note that, in contrast to the normal embryo in Figure 5A, the primitive ectoderm at the distal end of the egg cylinder is markedly deficient, whereas the endoderm layers and the extraembryonic ectoderm are less severely affected.

It is one thing to describe abnormal embryos arising in *Min/+* × *Min/+* intercrosses at 25% frequency, but quite another to demonstrate that the abnormal embryos are indeed *Min/Min* homozygotes. The deciduum comprises maternal tissue, which is heterozygous for *Min* regardless of the genotype of the embryo inside. One must dissect out the pregastrulation embryo, free it of maternal tissue, and then analyze the genotype in the *Apc* region of sectioned embryos. This rigorous analysis is being pursued (A.R. Moser et al., in prep.).

Clearly, there are cell types that survive in the complete absence of normal Apc function, such as extraembryonic ectoderm in the egg cylinder (Fig. 5B). Recall that markers for Paneth cells, enterocytes, and enteroendocrine cells are found in *Min*-induced adenomas that have lost the remaining normal *Apc* allele (Moser et al. 1992). In summary, the *Apc* gene does not directly control cell viability. Instead, it may control the interaction between cells. The interaction of the Apc molecule with catenins (Rubinfeld et al. 1993; Su et al. 1993) suggests that Apc may have an intercellular function.

The early embryonic phenotype of *Min/Min* zygotes contrasts with the developmental lesions found after organogenesis in the mouse in the cases of three other mammalian tumor suppressor genes of the developmental class: retinoblastoma (*Rb*; Jacks et al. 1992; Lee et al. 1992); neurofibromatosis (*Nf1*; Brannan et al. 1994); and α-inhibin (Matzuk et al. 1992). The *Apc* gene may be essential earlier in development than these other tumor suppressor genes. Alternatively, the *Min* allele may encode a product with neomorphic domi-

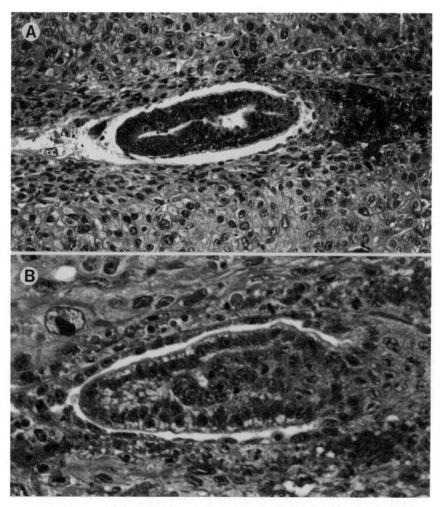

Figure 5. Egg-cylinder-stage mouse embryos longitudinally sectioned and stained with hematoxylin and eosin include ~75% normal (*A*) and 25% abnormal (*B*) embryos from *Min/+* × *Min/+* crosses. The magnification in *B* is twice that in *A*.

nant-negative activity (Herskowitz 1987), so that the mutant homozygote is more severely affected than either the heterozygote or the null homozygote.

CONCLUSIONS

Germ-line mutations in the *APC* gene of the human and the *Apc* gene of the mouse predispose to intestinal tumors. The Min mouse thus presents a good approximation to familial adenomatous polyposis in the human. The pathway to the *Min*-induced adenoma seems to be completely correlated with loss of wild-type Apc function. As in humans, tumors in tissues other than the intestine are found in *Min/+* mice, but are incompletely penetrant. Thus, on the *Min/+* background, other steps are rate-limiting for these extracolonic lesions. Although the *p53* gene seems to be mutated frequently in advanced intestinal neoplasms, germ-line lesions in this gene do not perceptibly enhance the probability of tumor formation in the Min mouse. In contrast, the *Mom-1* gene does control tumor multiplicity in these mice; the molecular identity and mechanism of action of this modifier locus are not yet known. Finally, normal Apc function is necessary before gastrulation in the development of the mouse. Despite the early developmental lesion, certain cell types outside of the embryonic ectoderm do persist in presumptive homozygous mutant embryos.

ACKNOWLEDGMENTS

Our research has been made both fruitful and personally enjoyable by several collaborative efforts. Jeffrey Gordon and Kevin Roth (Washington University, St. Louis, Missouri) have worked on the assessment of the differentiation status of *Min*-induced intestinal tumors. Li-Kuo Su, Bert Vogelstein, and Kenneth Kinzler (The Johns Hopkins University, Baltimore, Maryland) have carried out the DNA sequence analysis necessary for assigning the *Min* mutation to the *Apc* gene. William Dietrich, Jennifer Smith, and Eric Lander (The Whitehead Institute, Massachusetts Institute of Technology) have implemented the genomic SSLP scan to map the *Mom-1* locus. Michael Gould and Jill Haag (Clinical Cancer Center, University of Wisconsin) have collaborated on studies of mammary tumorigenesis in the Min mouse. Lawrence Donehower (Baylor University, Houston, Texas) has provided mice carrying the targeted mutation in *p53* and helped in genotyping mice segregating for this mutation. Our McArdle colleagues Norman Drinkwater and Henry Pitot have concretely helped by consulting on statistical considerations and on histopathology, respectively. Finally, we are deeply grateful for the fastidious work in molecular genetics by Natalie Borenstein and in mouse genetics by Melanie Woch McNeley and Darren Katzung. Linda Clipson, Kristen Adler, Mary Jo Markham, and Ilse Riegel and Alexandra Shedlovsky have made critical improvements in the manuscript. Our research has been supported by grants CA-07175, CA-23076, CA-50585, and CA-63777 from the National Cancer Institute; by a professorship from the Royal Society to R.L.G.; and by an International Fellowship from the UICC to W.F.D.

REFERENCES

Aaltonen, L.A., P. Peltomäki, F.S. Leach, P. Sistonen, L. Pylkkänen, J.-P. Mecklin, H. Järvinen, S.M. Powell, J. Jen, S.R. Hamilton, G.M. Petersen, K.W. Kinzler, B. Vogelstein, and A. de la Chapelle. 1993. Clues to the pathogenesis of familial colorectal cancer. *Science* **260:** 812.

Bodmer, W., T. Bishop, and P. Karran. 1994. Genetic steps in colorectal cancer. *Nat. Genet.* **6:** 217.

Bodmer, W.F., C.J. Bailey, J. Bodmer, H.J.R. Bussey, A. Ellis, P. Gorman, F.C. Lucibello, V.A. Murday, S.H. Rider, P. Scambler, D. Sheer, E. Solomon, and N.K. Spurr. 1987. Localization of the gene for familial adenomatous polyposis on chromosome 5. *Nature* **328:** 614.

Brannan, C.I., A.S. Perkins, K.S. Vogel, N. Ratner, M.L. Nordlund, S.W. Reid, A.M. Buchberg, N.A. Jenkins, L.F. Parada, and N.G. Copeland. 1994. Targeted disruption of the neurofibromatosis type-1 gene leads to developmental abnormalities in heart and various neural crest-derived tissues. *Genes Dev.* **8:** 1019.

Cairns, J. 1975. Mutation selection and the natural history of cancer. *Nature* **255:** 197.

Carder, P., A.H. Wyllie, C.A. Purdie, R.G. Morris, S. White, J. Piris, and C.C. Bird. 1993. Stabilised p53 facilitates aneuploid clonal divergence in colorectal cancer. *Oncogene* **8:** 1397.

D'Amico, D.D., D.P. Carbone, B.E. Johnson, S.J. Meltzer, and J.D. Minna. 1992. Polymorphic sites within the *MCC* and *APC* loci reveal very frequent loss of heterozygosity in human small cell lung cancer. *Cancer Res.* **52:** 1996.

Dietrich, W.F., E.S. Lander, J.S. Smith, A.R. Moser, K.A. Gould, C. Luongo, N. Borenstein, and W. Dove. 1993. Genetic identification of *Mom-1*, a major modifier locus affecting *Min*-induced intestinal neoplasia in the mouse. *Cell* **75:** 631.

Dietrich, W.F., J.C. Miller, R.G. Steen, M. Merchant, D. Damron, R. Nahf, A. Gross, D.C. Joyce, M. Wessel, R.D. Dredge, A. Marquis, L.D. Stein, N. Goodman, D.C. Page, and E.S. Lander. 1994. A genetic map of the mouse with 4,006 simple sequence length polymorphisms. *Nat. Genet.* **7:** 220.

Donehower, L.A. and A. Bradley. 1993. The tumor suppressor p53. *Biochim. Biophys. Acta* **1155:** 181.

Donehower, L.A., M. Harvey, B.L. Slagle, M.J. McArthur, C.A. Montgomery, Jr., J. Butel, and A. Bradley. 1992. Mice deficient for p53 are developmentally normal but susceptible to spontaneous tumours. *Nature* **356:** 215.

Fearon, E.R. and B. Vogelstein. 1990. A genetic model for colorectal tumorigenesis. *Cell* **61:** 759.

Fisher, E. and P. Scambler. 1994. Human haploinsufficiency— One for sorrow, two for joy. *Nat. Genet.* **7:** 5.

Foulds, L. 1964. Tumour progression and neoplastic development. In *Cellular control mechanisms and cancer* (ed. P. Emmelot and O. Mühlbock), p. 242. Elsevier, Amsterdam.

Gardner, R.L. 1993. Extrinsic factors in cellular differentiation. *Int. J. Dev. Biol.* **37:** 47.

Groden, J., A. Thliveris, W. Samowitz, M. Carlson, L. Gelbert, H. Albertsen, G. Joslyn, J. Stevens, L. Spirio, M. Robertson, L. Sargeant, K. Krapcho, E. Wolff, R. Burt, J.P. Hughes, J. Warrington, J. McPherson, J. Wasmuth, D. Le Paslier, H. Abderrahim, D. Cohen, M. Leppert, and R. White. 1991. Identification and characterization of the familial adenomatous polyposis coli gene. *Cell* **66:** 589.

Gurdon, J.B., P. Lemaire, and K. Kato. 1993. Community

effects and related phenomena in development. *Cell* **75:** 831.

Harvey, M., M.J. McArthur, C.A. Montgomery, Jr., A. Bradley, and L.A. Donehower. 1993. Genetic background alters the spectrum of tumors that develop in p53-deficient mice. *FASEB J.* **7:** 938.

Herskowitz, I. 1987. Functional inactivation of genes by dominant negative mutations. *Nature* **329:** 219.

Ichii, S., A. Horii, S. Nakatsuri, J. Furuyama, J. Utsunomiya, and Y. Nakamura. 1992. Inactivation of both *APC* alleles in an early stage of colon adenomas in a patient with a familial adenomatous polyposis (FAP). *Hum. Mol. Genet.* **1:** 387.

Jacks, T., A. Fazeli, E.M. Schmitt, R.T. Bronson, M.A. Goodell, and R.A. Weinberg. 1992. Effects of an *Rb* mutation in the mouse. *Nature* **359:** 295.

Justice, M.J., D.J. Gilbert, K.W. Kinzler, B. Vogelstein, A.M. Buchberg, J.D. Ceci, Y. Matsuda, V.M. Chapman, T. Muramatsu, N.A. Jenkins, and N.G. Copeland. 1992. A molecular genetic linkage map of mouse chromosome 18 reveals extensive linkage conservation with human chromosomes 5 and 18. *Genomics* **13:** 1281.

Kemp, C.J., L.A. Donehower, A. Bradley, and A. Balmain. 1993. Reduction of p53 gene dosage does not increase initiation or promotion but enhances malignant progression of chemically induced skin tumors. *Cell* **74:** 813.

Lee, E.Y.-H.P., C.-Y. Chang, N. Hu, Y.-C.J. Wang, C.-C. Lai, K. Herrup, W.-H. Lee, and A. Bradley. 1992. Mice deficient for Rb are nonviable and show defects in neurogenesis and haematopoiesis. *Nature* **359:** 288.

Levy, D.B., K.J. Smith, Y. Beazer-Barclay, S.R. Hamilton, B. Vogelstein, and K.W. Kinzler. 1994. Inactivation of both *APC* alleles in human and mouse tumors. *Cancer Res.* **54:** 5953.

Luongo, C., A.R. Moser, S. Gledhill, and W.F. Dove. 1994. Loss of Apc^+ is necessary for intestinal adenoma formation in Min mice. *Cancer Res.* **54:** 5947.

Luongo, C., K.A. Gould, L.-K. Su, K.W. Kinzler, B. Vogelstein, W. Dietrich, E.S. Lander, and A.R. Moser. 1993. Mapping of multiple intestinal neoplasia (*Min*) to proximal chromosome 18 of the mouse. *Genomics* **15:** 3.

Lyon, M.F., B.W. Ogunkolade, M.C. Brown, D.J. Atherton, and V.H. Perry. 1993. A gene affecting Wallerian nerve degeneration maps distally on mouse chromosome 4. *Proc. Natl. Acad. Sci.* **90:** 9717.

Matzuk, M.M., M.J. Finegold, J.-G.J. Su, A.J.W. Hsueh, and A. Bradley. 1992. α-Inhibin is a tumour-suppressor gene with gonadal specificity in mice. *Nature* **360:** 313.

Merritt, A.J., C.S. Potten, C.J. Kemp, J.A. Hickman, A. Balmain, D.P. Lane, and P.A. Hall. 1994. The role of *p53* in spontaneous and radiation-induced apoptosis in the gastrointestinal tract of normal and p53-deficient mice. *Cancer Res.* **54:** 614.

Mock, B.A., M.M. Krall, and J.K. Dosik. 1993. Genetic mapping of tumor susceptibility genes involved in mouse plasmacytomagenesis. *Proc. Natl. Acad. Sci.* **90:** 9499.

Moser, A.R., H.C. Pitot, and W.F. Dove. 1990. A dominant mutation that predisposes to multiple intestinal neoplasia in the mouse. *Science* **247:** 322.

Moser, A.R., W.F. Dove, K.A. Roth, and J.I. Gordon. 1992. The *Min* (multiple intestinal neoplasia) mutation: Its effect on gut epithelial cell differentiation and interaction with a modifier system. *J. Cell Biol.* **116:** 1517.

Moser, A.R., E.M. Mattes, W.F. Dove, M.J. Lindstrom, J.D. Haag, and M.N. Gould. 1993. Apc^{Min}, a mutation in the murine *Apc* gene, predisposes to mammary carcinomas and focal alveolar hyperplasias. *Proc. Natl. Acad. Sci.* **90:** 8977.

Nowell, P.C. 1976. The clonal evolution of tumor cell populations. *Science* **194:** 23.

Paraskeva, C. and A.C. Williams. 1990. Are different events involved in the development of sporadic versus hereditary tumours? The possible importance of the microenvironment in hereditary cancer. *Br. J. Cancer* **61:** 828.

Ponder, B.A.J., G.H. Schmidt, M.M. Wilkinson, M.J. Wood, M. Monk, and A. Reid. 1985. Derivation of mouse intestinal crypts from single progenitor cells. *Nature* **313:** 689.

Potten, C.S. and R.J. Morris. 1988. Epithelial stem cells *in vivo*. *J. Cell Sci. Suppl.* **10:** 45.

Purdie, C.A., J. O'Grady, J. Piris, A.H. Wyllie, and C.C. Bird. 1991. p53 expression in colorectal tumors. *Am. J. Pathol.* **138:** 807.

Rous, P. and J.G. Kidd. 1941. Conditional neoplasms and subthreshold neoplastic states. *J. Exp. Med.* **73:** 365.

Rubinfeld, B., B. Souza, I. Albert, O. Müller, S.H. Chamberlain, F.R. Masiarz, S. Munemitsu, and P. Polakis. 1993. Association of the *APC* gene product with β-catenin. *Science* **262:** 1731.

Su, L.-K., B. Vogelstein, and K.W. Kinzler. 1993. Association of the APC tumor suppressor protein with catenins. *Science* **262:** 1734.

Su, L.-K., K.W. Kinzler, B. Vogelstein, A.C. Preisinger, A.R. Moser, C. Luongo, K.A. Gould, and W.F. Dove. 1992. Multiple intestinal neoplasia caused by a mutation in the murine homolog of the APC gene. *Science* **256:** 668.

Genetic Instability and Apoptotic Cell Death during Neoplastic Progression of v-*myc*-initiated B-cell Lymphomas in the Bursa of Fabricius

P.E. Neiman, J. Summers, S.J. Thomas, S. Xuereb, and G. Loring

Fred Hutchinson Cancer Research Center and the Departments of Medicine and Pathology, University of Washington, Seattle, Washington 98104

B-cell lymphoma arising in the bursa of Fabricius of chickens is both a naturally occurring neoplasm of commercial importance and an experimental system which has contributed significantly to knowledge about the role of retroviruses and oncogenes in tumorigenesis. The salient features of this tumor system have recently been reviewed (Neiman 1994). Briefly, the lymphoid population of embryonic follicles of the bursa is composed of large cycling lymphoblasts expressing surface IgM (Thompson et al. 1987). This population includes a stem-cell compartment of cells that can clonally reconstitute ablated bursal follicles in transplantation experiments (Pink et al. 1985). Deregulated expression of a *myc* oncogene within this stem-cell compartment, either due to integration of an avian leukosis virus (ALV) near c-*myc* (Hayward et al. 1981), or due to ex vivo infection of embryonic bursal lymphoblasts with a v-*myc* transducing avian myelocytomatosis virus (Neiman et al. 1985; Thompson et al. 1987), initiates neoplastic change in bursal follicles. This change is signaled by the replacement of the lymphoid cells of the follicle with a monomorphic population of pyroninophillic lymphoblasts producing a histologically distinctive lesion called a transformed follicle (TF) (Cooper et al. 1960; Neiman et al. 1980; Baba and Humphries 1985). TF cells demonstrate the cardinal properties of bursal stem cells in that they reconstitute bursal follicles efficiently in transplantation experiments and diversify their light-chain immunoglobulin genes by gene conversion at a rate similar to that of the immunoglobulin genes of normal bursal follicles (Neiman et al. 1985; Thompson et al. 1987). Unlike normal bursal stem cells, they persist after hatching and are preneoplastic, giving rise to invasive bursal lymphomas. Therefore, deregulated expression of *myc* genes appears to produce an arrest of B-cell maturation in the bursa at the stem-cell stage and either permits or induces overgrowth of bursal stem cells within the follicular architecture. In this paper, we address the question of what mechanisms might lead these *myc*-induced bursal stem cells to further neoplastic progression.

METHODS

Bursal transplantation assay. The basic technique used for these studies was reconstitution of cyclophosphamide-ablated bursal follicle populations with donor cells transduced with a retrovirus expressing a v-*myc* oncogene. The methodology for this approach has been described previously (Neiman et al. 1985; Thompson et al. 1987) and is shown schematically in Figure 1: (1) Recipient chick embryos between 18 days of embryogenesis and day-of-hatch were treated with cyclophosphamide to selectively destroy the bursal lymphoid population (Eskola and Toivanen 1974). (2) Bursal follicles from syngeneic 18-day chick embryo donors were dispersed and co-cultivated for 8–16 hours in the presence of 100 ng/ml phorbol dibutyrate to inhibit cell death (Ewart and Weber 1987; Asakawa et al. 1993) on adherent monolayers producing the v-*myc*-transducing retrovirus and injected intravenously into the cyclophosphamide-prepared recipients. (3) Stem cells in the donor population reconstitute the ablated follicles. Those successfully transduced and expressing v-

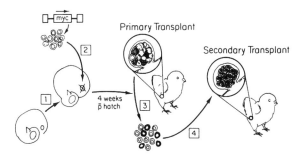

Figure 1. Schematic of the bursal transplantation technique for analysis of expression of *myc* oncogenes in bursal stem cells. The details of ablation of bursal follicle populations (*1*), infection of donor stem cells with v-*myc* transducing retroviruses or viral vectors (*2*), analysis of reconstituted bursas in primary transplants (*3*), and transplantation of *myc*-induced preneoplastic stem cell (TF cells) into secondary recipients (*4*) are given in Methods and Neiman et al. (1985) and Thompson et al. (1987).

myc show TF morphology (dark cross hatch and Fig. 2). By 4 weeks after hatching, normal stem cells have disappeared and the only transplantable stem cells in the bursa are in TF. These can be dispersed in culture and transplanted again (4) into secondary cyclophosphamide-ablated recipients. Reconstituted follicles in these secondary transplants are composed only of TF. Malignant lymphomas derived from TF, both in the bursa and metastatic to the liver, can be observed following primary and secondary transplants.

Induction and analysis of apoptotic cell death. Apoptotic bursal cell death was documented by DNA fragmentation into oligonucleosomal ladders observed by agarose gel electrophoresis and formation of characteristic apoptotic bodies on cytological examination as described previously (Neiman et al. 1991). Cytological detection of early apoptotic change in nuclei was accomplished by thymidine terminal transferase (TdT) labeling of DNA strand breaks with biotinylated nucleotides (TUNEL) as described previously (Gavrieli et al. 1992). Avidin conjugated with alkaline phosphatase (Boeringer) was used to detect biotinylated DNA ends.

RESULTS

Induction of Transformed Follicles and Bursal Lymphomas with Helper-free Retroviral Vectors

The retroviral system we have used previously to introduce v-*myc* into bursal follicles by the transplantation technique (Fig. 1) has been a "recovered" avian myelocytomatosis virus called HB1 (Enrietto et al. 1983). HB1 is defective and requires a subgroup C helper virus (tdB-77). Cocultivation of embryonic bursal lymphoblasts on monolayers producing high titers (about 10^6 focus-forming units per ml) of HB1(tdB-77) yielded 5–90% TF in primary transplants and, essentially, 100% TF in secondary transplants. Invasive lymphomas were observed in the bursa in 6% of primary transplants examined 4 weeks after hatching and about 20% of secondary transplants in which the serially transplanted populations represented bursal cells about 8 weeks after the original infection with HB1. Since HB1(tdB-77) is a spreading viral complex that causes a variety of lethal tumors in other organs (Enrietto et al. 1983), we could not easily determine the rate of metastatic disease arising from the bursal lymphomas separately from primary HB1-induced neoplasia in extrabursal sites such as the liver. Furthermore, since the replication-competent helper virus could act potentially as an insertional mutagen, we could not determine whether tumor progression in the bursa was the result of genetic instability generated by deregulated *myc* expression or reflected mutagenic activity of the spreading helper virus.

To address the relative roles of v-*myc* and helper virus on tumor progression in the bursa, we constructed a helper-free vector, LvMYCSN, in the vector system LXSN (Miller and Rosman 1989) to use for transducing the HB1 v-*myc* gene into bursal stem cells. The construction of LXSN HB1-9, which expresses HB1 v-*myc* from the constitutive promoter in the retroviral long terminal repeat (LTR) region and a neomycin resistance gene from an internal SV40 promoter, is shown in Figure 2. This vector was expressed, as helper-free virus, from the PG-13 helper cell line as described previously (Miller et al. 1991), and a clone producing virus with an infectious titer on HeLa cells of about 10^5 colonies/ml resistant to the neomycin analog G418 was selected for further studies. To demonstrate transduction of v-MYC protein expression with the vector, helper-free virus was used to infect a bursal B-cell line,

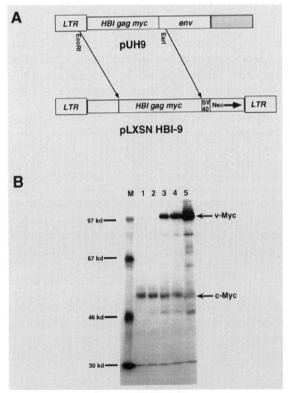

Figure 2 (*A*) Structure of a partial proviral clone, pUH9, of avian myelocytomatosis virus HB1 (Smith et al. 1985) and the construct virus LXSN HB1-9 prepared by inserting an *Eco*RI–*Ear*I fragment of pUH9, including the entire v-*myc* gene, into the cloning site of the LXSN vector virus (Miller and Rosman 1989). LTRs are viral long terminal repeats; SV is the SV40 viral promoter, and NEO is the neomycin transferase gene conferring resistance to the neomycin analog G418. The shaded box in pUH9 represents host cell sequences flanking the 3' end of the partial HB1 provirus. (*B*) Immunoprecipitation analysis of v-MYC protein expression in cells transduced by LXSN HB1-9. A *rev*-transformed bursal B-cell line called TF-32 was cocultivated with LvMYCSN producing monolayers and G418-resistant TF-32 clones were selected. (Lanes *1*–*5*) Immunoprecipitation analysis with ab5042 of MYC proteins labeled for 30 min with [^{35}S]methionine. (Lanes *1* and *2*) G418-resistant TF-32 clones lacking v-MYC expression; (lanes *3* and *4*) clones expressing high v-MYC levels. (Lane *5*) A positive control from a sister cell line (TF-26) infected with HB1 myelocytomatosis virus and expressing high levels of HB1 v-MYC.

TF-32, immortalized by infection with reticuloendotheliosis virus as described previously (Barth and Humphries 1988). Figure 2 also shows a radioimmune precipitation assay with antibody 5042 to MYC (Hann et al. 1983) of LXSN HB1-9-infected, G418-selected clones of TF-32, some of which demonstrated high levels of v-MYC protein expression. Transplantation assays were then carried out following cocultivation of embryonic bursal lymphoblasts with LXSN HB1-9-producing cells. Table 1 summarizes the results of this experiment. The proportion of TF among reconstituted follicles in primary transplants, 5–90%, was no less than we had observed previously with HB1. Importantly, we observed small lymphomas in three of four bursas and metastatic lymphomas in the liver in two birds sacrificed 8 and 10 weeks after hatching. Figure 3 shows mixed normal and transformed follicles from primary transplants and a metastatic lymphoma in the liver. As also shown in Table 1, we carried out multiple secondary transplants with cells from the bursas of two of the primary transplants sacrificed about a month after hatching. The malignant potential after only 1 month of some of the TF cells in the primary transplants was demonstrated by the high proportion of intrabursal and hepatic lymphomas observed in the secondary transplants. We confirmed the absence of infectious virus in these tissues by plating up to 10^8 TF cells on HeLa indicator cells. We obtained no G418-resistant colonies in this infectious center assay. Therefore, neoplastic progression of v-myc-induced TF cells occurred rapidly in the absence of spreading retroviral infection, presumably as an inevitable consequence of deregulated v-myc expression.

Apoptotic Cell Death and Genetic Instability

Since we have ruled out insertional mutagenesis by spreading retroviral infection as a cause for the rapid tumor progression, we have begun to seek an alternative explanation for the genetic instability that might lead to progressive malignant change in TF cells. One possible mechanism we have been investigating relates to the exceptionally active apoptotic cell death pathway present in bursal follicles. We have observed previously that DNA-damaging agents such as γ radiation, as well as disruption of cell-cell contact by dispersion of bursal follicles in short-term tissue culture, induce apoptotic cell death in virtually all of the bursal lymphoid population within a few hours (Neiman et al. 1991). In this study, we employed the TdT DNA break labeling technique (TUNEL) to examine bursal follicles for spontaneous apoptosis in situ. Figure 4 shows that normal follicles from 18-day embryos show little or no spontaneous apoptosis by TUNEL assay. In contrast, the figure shows that TF, i.e., follicles filled with preneoplastic v-myc-induced bursal stem cells, demonstrated a high rate of spontaneous apoptosis. This finding is entirely consistent with our earlier observation that TF were hypersensitive (in comparison with normal embryonic bursal populations) to induction of apoptosis induced by either dispersion or γ radiation (Neiman et al. 1991), as well as with the observation of myc enhancement of apoptotic cell death in cultured cell lines (Evan et al. 1992).

We wanted to know, therefore, whether the intensity of cell death events in preneoplastic bursal cells could somehow contribute to genetic instability and tumor progression in this system. Recently, a considerable body of data has accumulated implicating the tumor suppresser gene p53 in maintaining genetic integrity by establishing a checkpoint at the G_1/S border of the cell cycle (Kastan et al. 1991, 1992; Livingstone et al. 1992; Yin et al. 1992). These studies indicate that cells respond to DNA strand breaks, however they are introduced, by acute elevation of p53 protein levels. Downstream events, dependent on this p53 induction, include inhibition of cyclin-dependent kinases (Gu et al. 1993; Harper et al. 1993) and arrest in the G_1 stage of the cell cycle. We therefore exposed normal embryonic bursal follicles to graded doses of γ radiation. Low doses, 0.4 to 1.0 Gy, which produce no effect on bursal follicles (Neiman et al. 1991), produced a significant induction of p53 level by 1 hour after radiation (Fig. 5B, lanes 1–3). Higher doses of radiation from 1.6 to 4.8 Gy, which induced extensive apoptosis at 4 hours in normal embryonic bursal follicles (Neiman et al. 1991), failed to induce acute p53 responses. Furthermore, dispersed bursal lymphoblasts entering apoptosis also failed to show a p53 response either to apoptosis-induced DNA breakage or to radiation (Fig. 5A). Therefore, early events in the cell death pathway may inhibit p53-mediated checkpoint responses to DNA damage.

DISCUSSION

The experiments described in this paper, together with previous work, document that deregulated overexpression of the v-myc gene of HB1 in bursal stem cells results in a block to differentiation at the stem-cell stage of development and sets up a condition of instability within these cells resulting, rapidly and inevitably, in neoplastic progression. The v-myc gene of HB1 encodes a gag-myc fusion protein. Therefore, it is

Table 1. Analysis of Birds with Primary and Secondary Bursal Transplants Infected with LvMYCSN

Primary transplants

Bird No.	Age (weeks)	%TF	Metastatic lymphoma in liver
94001	4.3	45	no
94002	4.6	90	no
94006	8.0	5	yes
94009	7.7	25	yes

Secondary transplants

Donor 94001	Donor 94002
No. recipients: 5	No. recipients: 7
age (weeks): 4.6–8	age (weeks): 7.1
No. with TF: 5	No. with TF: 7
No. with liver mets.: 5	No. with liver mets: 1

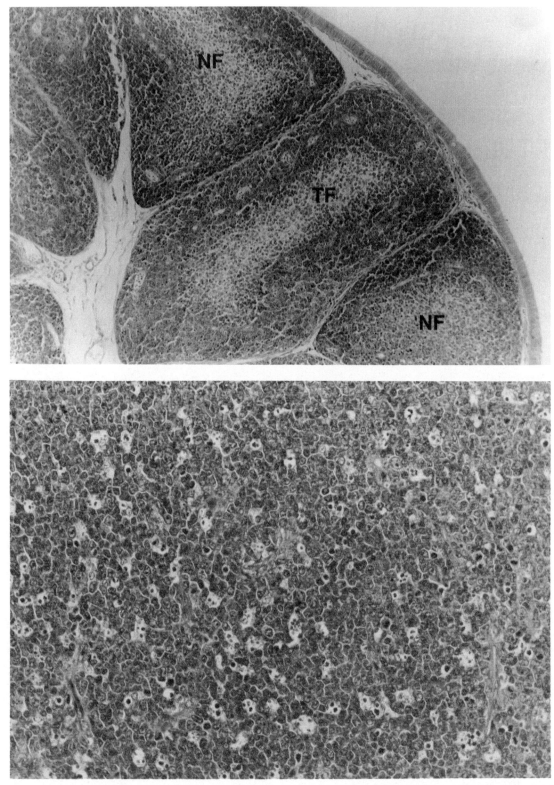

Figure 3. Transformed follicles and metastatic lymphomas from bursas transduced with LvMYCSN. (*Top*) Transformed follicles (TF) and normal follicles (NF) from a primary bursal transplant stained with methyl green pyronin. Magnification, 85×. (*Bottom*) A metastatic bursal lymphoma in the liver stained with hematoxylin and eosin demonstrating characteristic starry sky appearance. Magnification, 170×.

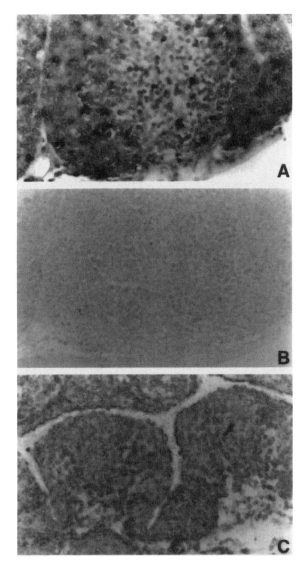

Figure 4. Spontaneous apoptosis in normal embryonic and v-*myc*-induced preneoplastic bursal follicles. (*A*) TdT DNA end labeling (TUNEL) of apoptotic cells in a transformed follicle. (*B*) Same follicle showing control in which incubation was done with buffer lacking TdT. (*C*) TUNEL analysis of normal 18-day embryonic bursal follicles showing lack of spontaneous apoptosis.

possible that either the *gag* portion of the transforming gene, and/or mutations within the *myc* portion of the gene, play some role, as has been observed in transformation of cultured fibroblasts (Tikhonenko et al. 1992). However, deregulated expression of c-*myc* produces precisely the same preneoplastic change (i.e., transformed follicles) leading to intrabursal and metastatic lymphomas histologically identical to the lesions induced with HB1 v-*myc*. At this point, there is no evidence to suggest that properties other than those inherent in the protein encoded by wild-type chicken c-*myc* are required to explain our observations in the bursa.

Mice expressing a deregulated *myc* transgene in the B-cell lineage also develop a preneoplastic expansion of

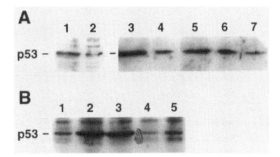

Figure 5. Immunoblot analysis of p53 protein levels in normal embryonic bursal follicles after DNA damage induced by γ radiation and/or apoptosis. (*A*) p53 levels in extracts of 5×10^6 bursal lymphoblasts from 18-day embryos at time 0 (lane *1*) and 2 hr after initiation of apoptosis by dispersion in culture (lane *2*). On a separate immunoblot, lanes *3–7* show p53 levels in dispersed embryonic bursal lymphoblasts, again at time 0 (lane *3*), and 1 hr after dispersion and 0.4, 1.0, 1.6, and 4.8 Gy of γ radiation (lanes *4–7*, respectively). (*B*) p53 levels from intact bursal follicles 1 hr after 0, 0.4, 1.0, 1.6, and 4.8 Gy of total body γ radiation of newly hatched chicks (lanes *1–5*, respectively). (Reprinted, with permission, from Neiman et al. 1994 [copyright American Society for Cell Biology].)

B-cell precursor population (Adams et al. 1985). In the absence of additional mutational events, progression to fully developed neoplasms in the large majority of these mice requires a considerable fraction of the life span of the affected animal. This process can be markedly speeded up by infection of the mice with replication-competent retroviruses, which act as insertional mutagens activating a series of oncogenes that cooperate in the development of the B-cell tumors (Haupt et al. 1991; van Lohuizen et al. 1991). In fact, this mechanism may also be operative following ALV induction of bursal lymphomas, where a second retroviral insertion near a gene called *bic* has been implicated in progression to metastatic growth (Clurman and Hayward 1989). Nevertheless, the data reported here support the notion that *myc* alone is sufficient to initiate the instability required for further rapid oncogenic change in bursal stem cells without additional retrovirus-mediated mutagenesis. Whether the targeted genes in the pathway to full tumorigenesis are different in the presence or absence of viral mutagenesis remains to be determined.

What properties of bursal stem cells might explain the extraordinarily rapid progression of neoplasia initiated by v-*myc*? The answer to this question remains to be found. However, it is intriguing that p53 responses to radiation and/or apoptosis-induced DNA damage appear to be suppressed early after induction of bursal apoptosis. This observation suggests a regulatory interaction between some component(s) of the cell death pathway and the p53-dependent G_1 checkpoint thought to play a major role in maintenance of genomic integrity. It only remains to imagine this interaction occurring in cells which are not irreversibly committed to die acutely in order reasonably to postulate a role for the cell death pathway in genetic instability and resultant tumor progression. For example, the rate of cell death in our hypothetical, genetically unstable cell population need only be less than the rate of cell division in order for net population growth to occur. This balanced rate of cell division and cell death is a common feature of the growth kinetics of many naturally occurring tumor systems.

What, then, might be the role of *myc* oncogenes in initiating this cascade? The *myc*-induced maturation arrest and overproduction of bursal stem cells might be an important element of the underlying mechanism. Physiological stem division is, characteristically, unequal, giving rise to one daughter which is still a stem cell and one differentiated daughter cell. In *myc*-induced preneoplastic bursal follicles, we observed only massive and apparently equal stem-cell division. This experimental system, in which large numbers of these cells can be recovered for analysis, might provide an opportunity to define the mechanism of unequal stem-cell division and to determine how *myc* deregulation can perturb this process. Finally, does *myc* hyperexpression play a role in the hyperactive cell death pathway we observed in preneoplastic TF? It is conceivable that a high rate of cell death is an intrinsic property of equally dividing bursal stem cells. Alternatively, *myc* overexpression has been seen to enhance apoptotic cell death in cultured cells lacking required signals mediated by survival/growth factors (Evan et al. 1992) and, therefore, could be directly inducing increased cell death through genetic pathways controlled by *myc*. One such pathway might involve driving cells through cell cycle progression. We have observed that the majority of DNA damage and cell death in bursal lymphoblasts occurs during the S phase of the cell cycle (Neiman et al. 1994). If *myc*-enhanced transit from G_1 into S phase occurs in early apoptotic cells, i.e., in cells in which G_1 checkpoint responses have been suppressed, the effect of carrying DNA strand breaks into S phase could be to enhance lethality, or if the full death pathway is not triggered, propensity for strand-break-induced genomic change.

ACKNOWLEDGMENTS

This work was supported by grants CA-20068 and CA-57215 from the National Cancer Institute.

REFERENCES

Adams, J.M., A.W. Harris, C.A. Pinkert, L.M. Corcoran, W.S. Alexander, S. Cory, R.D. Palmiter, and R.L. Brinster. 1985. The c-*myc* oncogene under control of an immunoglobulin enhancer induces lymphoid malignancy in transgenic mice. *Nature* **318:** 533.

Asakawa, J., V.K. Tsiagbe, and G.J. Thorbecke. 1993. Protection against apoptosis in chicken bursa cells by phorbol ester *in-vitro*. *Cell. Immunol.* **147:** 180.

Baba, T.W. and E.H. Humphries. 1985. Formation of a transformed follicle is necessary but not sufficient for development of an avian leukosis virus-induced lymphoma. *Proc. Natl. Acad. Sci.* **82:** 213.

Barth, C.F. and E.H. Humphries. 1988. Expression of v-*rel* induces mature B-cell lines that reflect the diversity of avian immunoglobulin heavy and light chain rearrangements. *Mol. Cell. Biol.* **8:** 5358.

Clurman, B.E. and W.S. Hayward. 1989. Multiple proto-oncogene activations in avian leukosis virus-induced lymphomas: Evidence for stage specific events. *Mol. Cell. Biol.* **9:** 2657.

Cooper, M.D., L.N. Payne, P.B. Dent, B.R. Burmester, and R.A. Good. 1960. Pathogenesis of avian lymphoid leukosis. 1. Histogenesis. *J. Natl. Cancer Inst.* **41:** 373.

Enrietto, P.J., L.N. Payne, and M.J. Hayman. 1983. A recovered avian myelocytomatosis virus that induces lymphomas in chickens: Pathologic properties and their molecular basis. *Cell* **35:** 369.

Eskola, J. and P. Toivanen. 1974. Cell transplantation in immunodeficient chickens. *Cell. Immunol.* **13:** 459.

Evan, G.I., A.H. Wyllie, C.S. Gilbert, T.D. Littlewood, H. Land, M. Brooks, C.M. Waters, L.Z. Penn, and D.C. Hancock. 1992. Apoptosis in fibroblasts by c-*myc* protein. *Cell* **69:** 119.

Ewart, D.L. and W.T. Weber. 1987. Effect of phorbol myristate acetate on *in-vitro* growth of lymphocytes from the chicken thymus and bursa of Fabricius. *Prog. Clin. Biol. Res.* **238:** 67.

Gavrieli, Y., Y. Sherman, and S.A. Ben-Sasson. 1992. Identification of programmed cell death *in-situ* via specific labeling of nuclear DNA fragmentation. *J. Cell Biol.* **119:** 493.

Gu, Y., C.W. Turck, and D.O. Morgan. 1993. Inhibition of CDK2 activity *in vivo* by an associated 20k regulatory subunit. *Nature* **399:** 707.

Hann, S.R., H.D. Abrams, L.R. Rohrschneider, and R.N. Eisenman. 1983. Proteins encoded by v-*myc* and c-*myc* oncogenes: Identification and localization in acute leukemia virus transformants and bursal lymphoma cell lines. *Cell* **34:** 789.

Harper, J.W., G.R. Adami, N. Wei, K. Keyomarsi, and S.J. Elledge. 1993. The p21 cdk-interacting protein Cip-1 is a potent inhibitor of G1 cyclin-dependent kinases. *Cell* **75:** 805.

Haupt, Y., W.S. Alexander, G. Barri, S.P. Klinken, and J.M. Adams. 1991. Novel zinc finger gene implicated as *myc* collaborator by retrovirally accelerated lymphomagenesis in Eu-*myc* transgenic mice. *Cell* **65:** 753.

Hayward, W.S., B.G. Neel, and S.M. Astrin. 1981. Activation of a cellular *onc* gene by promoter intersection in ALV-induced lymphoid leukosis. *Nature* **290:** 475.

Kastan, M.B., O. Onyekwere, D. Sidransky, B. Vogelstein, and R.W. Craig. 1991. Participation of p53 protein in the cellular response to DNA damage. *Cancer Res.* **51:** 6304.

Kastan, M.B., Z. Qimin, W.S. El-Deiry, F. Carrier, T. Jacks, W.V. Walsh, B.S. Plunkett, B. Vogelstein, and A.J. Fornace. 1992. A mammalian cell cycle checkpoint pathway utilizing p53 and *GADD45* is defective in ataxia telangiectasia. *Cell* **71:** 587.

Livingstone, L.R., A. White, J. Sprouse, E. Livanos, T. Jacks, and T.D. Tlsty. 1992. Altered cell cycle arrest and gene amplification potential accompany loss of wild-type p53. *Cell* **70:** 923.

Miller, A.D. and G.J. Rosman. 1989. Improved retroviral vectors for gene transfer and expression. *BioTechniques* **7:** 980.

Miller, A.D., J.V. Garcia, N. vonSuhr, C.M. Lynch, C. Wilson, and M. Eiden. 1991. Construction and properties of retrovirus packaging cells based on gibbon ape leukemia virus. *J. Virol.* **65:** 2220.

Neiman, P.E. 1994. Retrovirus-induced B-cell neoplasia in the bursa of Fabricius. *Adv. Immunology* **56:** 467.

Neiman, P.E., S.J. Thomas, and G. Loring. 1991. Induction of apoptosis during normal and neoplastic development in the bursa of Fabricius. *Proc. Natl. Acad. Sci.* **88:** 5857.

Neiman, P.E., L. Jordan, R.A. Weiss, and L.N. Payne. 1980. Malignant lymphoma of the bursa of Fabricius: Analysis of early transformation. *Cold Spring Harbor Conf. Cell Prolif.* **7:** 519.

Neiman, P., C. Wolf, P.J. Enrietto, and G.M. Cooper. 1985. A retroviral *myc* gene induces preneoplastic transformation of lymphocytes in a bursal transplantation assay. *Proc. Natl. Acad. Sci.* **82:** 222.

Neiman, P.E., C. Blish, C. Heydt, G. Loring, and S.J. Thomas. 1994. Loss of cell cycle controls in apoptotic lymphoblasts of the bursa of Fabricius. *Mol. Biol. Cell* **5:** 763.

Pink, J.R., O. Vainio, and A.-M. Rijnbeek. 1985. Clones of B lymphocytes in individual bursal follicles of the bursa of Fabricius. *Eur. J. Immunol.* **15:** 83.

Smith, D.R., B. Vennstrom, M.J. Hayman, and P.J. Enrietto. 1985. Nucleotide sequences of HB1, a novel recombinant MC-29 derivative with altered pathogenic properties. *J. Virol.* **56:** 969.

Tikhonenko, A.T. and M.L. Linial. 1992. *gag* as well as *myc* sequences contribute to the transforming phenotype of the avian retrovirus FH3. *J. Virol.* **66:** 946.

Thompson, C.B., E.H. Humphries, L.M. Carlson, C.-L.H. Chen, and P.E. Neiman. 1987. The effect of alterations in *myc* gene expression on B-cell development in the bursa of Fabricius. *Cell* **51:** 371.

van Lohuizen, M., S. Verbeek, B. Scheijen, E. Wientjens, H. van der Gulden, and A. Berns. 1991. Identification of cooperating oncogenes in Eu-*myc* transgenic mice by provirus tagging. *Cell* **65:** 737.

Yin, Y., M.A. Tainsky, F.Z. Bischoff, L.C. Strong, and G.M. Wahl. 1992. Wild-type p53 restores cell cycle control and inhibits gene amplification in cells with mutant p53 alleles. *Cell* **70:** 937.

Colorectal Cancer and the Intersection between Basic and Clinical Research

B. VOGELSTEIN AND K.W. KINZLER

The Johns Hopkins Oncology Center, Baltimore, Maryland 21231

Colorectal cancer provides an excellent example of a human tumor type that can be productively studied. The reasons for this are simple: Such tumors are prevalent and progress through easily recognizable stages ranging from very small benign polyps (adenomas) to large malignant cancers (carcinomas). Tumors representing each of these stages can be easily obtained, examined biochemically and genetically, and compared with appropriate control cells from normal colorectal epithelium. In the past few years, the study of colorectal cancer has also provided an example of a different sort—demonstrating how research in basic areas of cell physiology and biochemistry can have a direct and unexpected impact on understanding common human diseases. In this review, we focus on two recent developments that illustrate this intersection between basic and clinical research.

BACKGROUND

Our current model for colorectal tumorigenesis is illustrated in Figure 1. The process appears to result from waves of clonal expansions, each driven by mutation of an oncogene or tumor suppressor gene. In addition to *APC*, *p53*, *K-RAS*, and *DCC*, other genes may play a role in some tumors, in part accounting for the heterogeneity that is observed histologically and clinically. One of the most important predictions of the model shown in Figure 1 is that the order of mutations, not just their accumulation, is important. Recent experiments suggest that the nature of the colorectal lesions initiated by mutations of *K-RAS* or *APC* are markedly different (Jen et al. 1994). Mutation of *K-RAS*, in the absence of *APC* mutation, leads to hyperplastic lesions with little potential for tumor progression (aberrant crypt foci). In contrast, *APC* mutation, in the absence of *K-RAS* mutation, leads to dysplastic lesions, disorganized both at the intracellular and intercellular levels, and represents the beginning of a process that eventually culminates in cancer. The other genes shown in the figure (*hMSH2*, *hMLH1*, *hPMS1*, and *hPMS2*) are members of a class that does not intrinsically control cell growth, but rather controls the rate of mutation of growth-regulating genes.

HEREDITARY NON-POLYPOSIS COLORECTAL CANCER

HNPCC was one of the first forms of familial cancer to be described in the biomedical literature (Warthin 1913). Because no physical stigmata other than cancer are found in affected individuals, the disease was difficult to study epidemiologically. The diagnosis could be made only tentatively, on the basis of a family history of colon and other cancers, with at least one cancer occurring in a patient less than 50 years of age (Vasen et al. 1991).

Beginning in 1989, de la Chapelle and colleagues, in collaboration with our group, began a linkage study in an attempt to demonstrate that this disease was inherited in Mendelian fashion. Two large families were ascertained for this analysis, one identified by Jass in New Zealand and the other by Green in Newfoundland. After many candidate loci were excluded, and more than 400 chromosomal loci were tested, significant linkage with microsatellite markers from chromosome 2p16 was demonstrated in both families (Peltomaki et al. 1993). These studies strongly suggested that HNPCC could be caused by a single, inherited mutant gene on chromosome 2 and opened the door to physical mapping of the relevant region. This region was subsequently cloned into YAC and P1 vectors, and the cloned DNA was used to derive and orient numerous microsatellite markers for more precise mapping (Leach et al. 1993). Linkage analysis in several families, including those carefully ascertained by Lynch and colleagues, demarcated a region of 0.8 Mb that contained the culprit gene.

Concurrent with the physical mapping, a functional approach was used to help identify the responsible gene. This approach was based on a serendipitous observation made during the study of tumors from HNPCC patients (Aaltonen et al. 1993). We initially suspected that HNPCC was caused by mutations in a tumor suppressor gene. Indeed, the prototype hereditary colorectal cancer predisposition syndrome, familial adenomatous polyposis (FAP), is caused by mutations in the *APC* tumor suppressor gene (Groden et al. 1991; Nishisho et al. 1991). According to the hypothesis formulated by Knudson, one expects to find loss of the

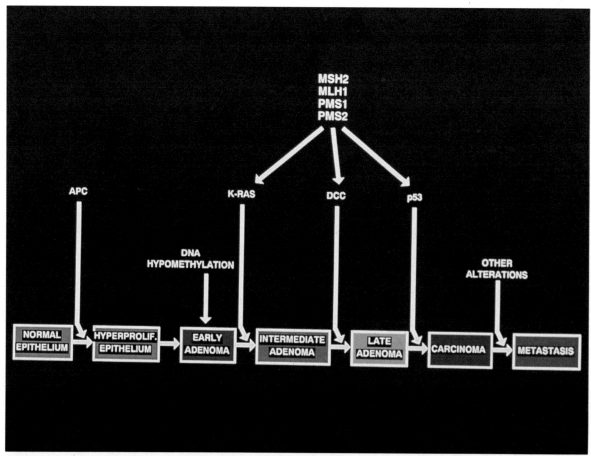

Figure 1. The process of colorectal tumorigenesis is driven by sequential mutations in tumor suppressor genes (*APC*, *p53*, and *DCC*) and oncogenes (*K-RAS*). Inactivation of genes which control the rate of mutations (*MSH2*, *MLH1*, *PMS1*, and *PMS2*) accelerates the tumorigenic process. Other genetic and epigenetic events play a role in tumor heterogeneity.

allele inherited from the nonaffected parent in cancers developing in patients with such diseases (Knudson 1985). We therefore examined the tumors of HNPCC patients for loss of chromosome 2p sequences, using as probes the microsatellite markers that initially demonstrated the linkage. Allelic losses proved difficult to document. However, this analysis revealed the presence of alleles in the tumors not found in the same patients' normal colorectal epithelium. It was soon determined that these alterations were not limited to markers on chromosome 2p and that dinucleotide and trinucleotide microsatellites throughout the genome were altered in these cancers (Aaltonen et al. 1993).

The alterations in the cancers from the HNPCC patients were similar to those originally observed in a subset of sporadic (i.e., non-familial) colorectal cancers by Perucho and colleagues (Peinado et al. 1992; Ionov et al. 1993) and independently by Thibodeau's group (Thibodeau et al. 1992). The microsatellite alterations have been dubbed "ubiquitous somatic mutations" (Ionov et al. 1993), "microsatellite instability" (Thibodeau et al. 1993), or "replication errors" (Aaltonen et al. 1993). Some of the sporadic cancers with microsatellite instability had clinical features in common with those occurring in HNPCC patients, leading to the speculation that similar molecular mechanisms were involved (Aaltonen et al. 1993; Ionov et al. 1993). Additional experiments soon showed that there was a true genetic instability of microsatellites and other sequences in these cancers, measured by fluctuation analyses (Parsons et al. 1993; Shibata et al. 1994).

Although the causes of this instability were not evident to us, several investigators studying mismatch repair (MMR) in prokaryotes and simple eukaryotes immediately recognized defective MMR as a likely contributor to the HNPCC tumor phenotype. In fact, a microsatellite instability similar to that seen in HNPCC tumors had been observed in prokaryotes deficient in MMR activity (Levinson and Gutman 1987). Modrich at Duke University pointed this out to us, and thus began a series of collaborative experiments that eventually demonstrated that HNPCC tumor extracts were virtually devoid of biochemically definable mismatch repair activity (Parsons et al. 1993), a result which was independently confirmed by T. Kunkel and colleagues (Umar et al. 1994). Kolodner, Liskay, and Fishel had a similar insight and approached the question by attempting to identify human mismatch repair genes that were mutated in the germ line of HNPCC patients; this strategy proved successful (Fishel et al. 1993; Bronner et al. 1994). Petes' and Liskay's groups provided a definitive conceptual basis for all these studies by documenting that a eukaryote (*Saccharomyces cerevisiae*) mutant in any one of three genes involved in MMR had a microsatellite instability remarkably similar to that observed in human colorectal cancers (Strand et al. 1993).

The results of all of these studies have now been published, and in aggregate have led to a detailed understanding of the genetic basis of HNPCC. Four human mismatch repair genes have been cloned that, when mutated in the germ line, lead to HNPCC (Fishel et al. 1993; Leach et al. 1993; Bronner et al. 1994; Nicolaides et al. 1994; Papadopoulos et al. 1994). In each case, the mechanism appears to be similar genetically to that proposed by Knudson for tumor suppressor genes (Knudson 1985). The cells of patients affected with HNPCC have one inactive and one active copy of the relevant mismatch repair gene in each cell. During tumorigenesis, the allele inherited from the unaffected parent is inactivated, leading to a cell with no mismatch repair activity. This in turn results in a rapid accumulation of mutations, speeding up the neoplastic process. HNPCC thereby can be thought of as a disease of accelerated tumor progression.

p53 AND CELL CYCLE REGULATION

The second area highlighted here relates to the mechanism of *p53* action. *p53* mutations play a role in diverse types of human cancer (Hollstein et al. 1991; Zambetti and Levine 1993). In the last 3 years, it has become clear that one of its key biochemical properties involves the binding of p53 to specific sequences of the form $(PuPuPuC[A/T])_n$, with the pentamers arranged head to head. In addition to binding these sequences, p53 transcriptionally activates adjacent genes. Virtually all mutants of p53 found in vivo have lost much or all of this sequence-specific *trans*-activation (Pietenpol et al. 1994). These studies led to the hypothesis that *p53* expression activates genes that in turn mediate its growth suppressor functions (Vogelstein and Kinzler 1992). But what are these putative downstream genes? One good candidate has recently emerged. Mercer's group had shown that a hormone-inducible wild-type *p53* gene could inhibit the growth of human brain cancer cells (Mercer et al. 1990). Using a subtractive cDNA approach, our group identified a gene, named *WAF1* (*w*ild-type *a*ctivated *f*ragment), that was induced to high levels in this cell line in the presence of hormone (El-Deiry et al. 1993). *WAF1* proved to be a highly conserved gene that was transcriptionally induced in both humans and other mammalian cells under conditions in which functional p53 was provoked (after DNA damage; see, e.g., El-Deiry et al. 1994). *WAF1* has p53 binding sites in its regulatory region, and these sites are highly conserved among mammals (T. Tokino et al., unpubl.). Furthermore, when transfected into a variety of cancer cell lines, *WAF1* exerted a potent growth-suppressive effect. *WAF1* thus became an excellent candidate for mediating the tumor suppressor activities of *p53*.

We initially had no idea of the function of this gene. This situation was precipitously changed by the experiments of Harper, Elledge, and colleagues (Harper et al. 1993). They identified a gene (*CIP1*) whose product interacted with cyclin-dependent kinases using a yeast two-hybrid system. The protein encoded by this gene was 21 kD, suggesting it was the polypeptide which bound to cyclin complexes previously identified by Beach's group (Xiong et al. 1992). This protein not only bound to CDK2 and related kinases, but also inhibited their enzymatic activities (Harper et al. 1993; Xiong et al. 1993). *CIP1* and *WAF1* turned out to be the same gene. This gene was also identified as one controlling senescence in human fibroblasts through a clever transfection approach (Noda et al. 1994). Altogether, these results firmly linked tumor suppressor genes, not only with regulation of the cell cycle, but perhaps also with a critical component of the aging process.

PERSPECTIVE

Although progress has been made in understanding the molecular basis of colorectal cancer, many questions remain unanswered. Is *WAF1/CIP1/p21* essential for *p53* tumor suppression in the colon, or is it redundant with other effectors? What is the relationship between growth arrest, apoptosis, and *WAF1/CIP1/p21*? How exactly do *p53* mutations mediate the transition from a benign growth to a tumor with invasive properties? What is the relationship between defects in growth control, which represent the only known biological effects of mutated oncogenes and tumor suppressor genes, to the metastatic process that is actually responsible for the death of most cancer patients? Why are HNPCC patients only at risk for tumors of the colon and a few other organs, whereas mismatch repair genes seem to play an identical role in every cell type? A significant portion (5–40%) of sporadic cancers of the colorectum, stomach, breast, lung, and bladder display microsatellite instability. Are all these tumors associated with defects in mismatch repair genes, or are other genes involved?

These questions provide fertile areas for future research. At the same time, the fact that these questions can even be asked—with specific reference to the pathogenesis of colorectal cancer—shows how basic research on seemingly unrelated biological phenomena can dramatically illuminate the pathogenesis of a major human disease. The challenge for the future is to translate these recent gains in understanding—made possible by decades of research on the processes of DNA repair and cell cycle control in simpler organisms—into similar gains in the prevention and treatment of human cancer.

ACKNOWLEDGMENTS

Work in the authors' laboratories is supported by the Preuss Foundation, the Clayton Fund, and the National Institutes of Health. The authors thank T. Gwiazda for preparation of this manuscript. B.V. is an American Cancer Society research professor.

REFERENCES

Aaltonen, L.A., P. Peltomaki, F.S. Leach, P. Sistonen, L. Pylkkanen, J.-P. Mecklin, H. Jarvinen, S.M. Powell, J. Jen, S.R. Hamilton, G.M. Peterson, K.W. Kinzler, B. Vogelstein, and A. de la Chapelle. 1993. Clues to the pathogenesis of familial colorectal cancer. *Science* **260:** 812.

Bronner, C.E., S.M. Baker, P.T. Morrison, G. Warren, L.G. Smith, M.K. Lescoe, M. Kane, C. Earabino, J. Lipford, A. Lindblom, P. Tannergard, R.J. Bollag, A.R. Godwin, D.C. Ward, M. Nordenskjold, R. Fishel, R. Kolodner, and R.M. Liskay. 1994. Mutation in the DNA mismatch repair gene homologue *hMLH1* is associated with hereditary nonpolyposis colon cancer. *Nature* **368:** 258.

El-Deiry, W.S., T. Tokino, V.E. Velculescu, D.B. Levy, R. Parsons, J.M. Trent, D. Lin, W.E. Mercer, K.W. Kinzler, and B. Vogelstein. 1993. *WAF1*, a potential mediator of p53 tumor suppression. *Cell* **75:** 817.

El-Deiry, W.S., J.W. Harper, P.M. O'Connor, V.E. Velculescu, C.E. Canman, J. Jackman, J.A. Pietenpol, M. Burrell, D.E. Hill, Y. Wang, K.G. Wiman, W.E. Mercer, M.B. Kastan, K.W. Kohn, S.J. Elledge, K.W. Kinzler, and B. Vogelstein. 1994. *WAF/CIP1* is induced in p53-mediated G_1 arrest and apoptosis. *Cancer Res.* **54:** 1169.

Fishel, R., M.K. Lescoe, M.R.S. Rao, N.G. Copeland, N.A. Jenkins, J. Garber, M. Kane, and R. Kolodner. 1993. The human mutator gene homolog *MSH2* and its association with hereditary nonpolyposis colon cancer. *Cell* **75:** 1027.

Groden, J., A. Thliveries, W. Samowitz, M. Carlson, L. Gelbert, H. Albertsen, G. Joslyn, J. Stevens, L. Spirio, M. Robertson, L. Sargeant, K. Krapsho, E. Wolff, R. Burt, J.P. Hughes, J. Warrington, J. McPherson, J. Wasmuth, D. Le Paslier, H. Abderrahim, D. Cohen, M. Leppert, and R. White. 1991. Identification and characterization of the familial adenomatous polyposis coli gene. *Cell* **66:** 589.

Harper, J.W., G.R. Adami, N. Wei, K. Keyomarsi, and S.J. Elledge. 1993. The p21 Cdk-interacting protein Cip1 is a potent inhibitor of G_1 cyclin-dependent kinases. *Cell* **75:** 805.

Hollstein, M., D. Sidransky, B. Vogelstein, and C.C. Harris. 1991. p53 mutations in human cancers. *Science* **253:** 49.

Jen, J., S.M. Powell, N. Papadopoulos, K.J. Smith, S.R. Hamilton, B. Vogelstein, and K.W. Kinzler. 1994. Molecular determinants of dysplasia in colorectal lesions. *Cancer Res.* **54:** 5523.

Ionov, Y.M., A. Peinado, S. Malkhosyan, D. Shibata, and M. Perucho. 1993. Ubiquitous somatic mutations in simple repeated sequences reveal a new mechanism for colonic carcinogenesis. *Nature* **363:** 558.

Knudson, A.G. 1985. Hereditary cancer, oncogenes, and antioncogenes. *Cancer Res.* **45:** 1437.

Leach, F.S., N.C. Nicolaides, N. Papadopoulos, B. Liu, J. Jen, R. Parsons, P. Peltomaki, P. Sistonen, L.A. Aaltonen, M. Nystrom-Lahti, X-Y. Guan, J. Zhang, P.S. Meltzer, J-W. Yu, F-T. Kao, D.J. Chen, K.M. Cerosaletti, R.E.K. Fournier, S. Todd, T. Lewis, R.J. Leach, S.L. Naylor, J. Weissenbach, J.-P. Mecklin, H. Jarvinen, G.M. Petersen, S.R. Hamilton, J. Green, J. Jass, P. Watson, H.T. Lynch, J.M. Trent, A. de la Chapelle, K.W. Kinzler, and B. Vogelstein. 1993. Mutations of a *mutS* homolog in hereditary nonpolyposis colorectal cancer. *Cell* **75:** 1215.

Levinson, G. and G.A. Gutman. 1987. High frequencies of short frameshifts in poly-A/TG tandem repeats borne by bacteriophage M13 in *Escherichia coli* K-12. *Nucleic Acids Res.* **15:** 5323.

Mercer, W.E., M.T. Shields, M. Amin, G.J. Sauve, E. Appella, J.W. Romano, and S.J. Ullrich. 1990. Negative growth regulation in a glioblastoma tumor cell line that conditionally expresses human wild-type p53. *Proc. Natl. Acad. Sci.* **87:** 6166.

Nicolaides, N.C., N. Papadopoulos, B. Liu, T.-F. Wei, K.C. Carter, S.M. Ruben, C.A. Rosen, W.A. Hazeltine, R.D. Fleischmann, C.M. Fraser, M.D. Adams, J.C. Venter, M.G. Dunlop, S.R. Hamilton, G.M. Petersen, A. de la Chapelle, B. Vogelstein, and K.W. Kinzler. 1994. Mutations of two PMS homologues in hereditary nonpolyposis colon cancer. *Nature* **371:** 75.

Nishisho, I., Y. Nakamura, Y. Miyoshi, Y. Miki, H. Ando, A. Horii, K. Koyama, J. Utsunomiya, S. Baba, P. Hedge, A. Markham, A.J. Krush, G. Petersen, S.R. Hamilton, M.C. Nilbert, D.B. Levy, T.M. Bryan, A.C. Preisinger, K.J. Smith, L.-K. Su, K.W. Kinzler, and B. Vogelstein. 1991. Mutations of chromosome 5q21 genes in FAP and colorectal cancer patients. *Science* **253:** 665.

Noda, A., Y. Ning, S.F. Venable, O.M. Pereira-Smith, and J.R. Smith. 1994. Cloning of senescent cell-derived inhibitors of DNA synthesis using an expression screen. *Exp. Cell Res.* **211:** 90.

Papadopoulos, N., N.C. Nicolaides, Y.-F. Wei, S.M. Rube, K.C. Carter, C.A. Rosen, W.A. Haseltine, R.D. Fleischmann, C.M. Fraser, M.D. Adams, J.C. Venter, S.R. Hamilton, G.M. Petersen, P. Watson, H.T. Lynch, P. Peltomaki, J.-P. Mecklin, A. de la Chapelle, K.W. Kinzler, and B. Vogelstein. 1994. Mutation of a *mutL* homolog in hereditary colon cancer. *Science* **263:** 1625.

Parsons, R., G.-M. Li, M.J. Longley, W.-H. Fang, N. Papadopoulos, J. Jen, A. de la Chapelle, K.W. Kinzler, B. Vogelstein, and P. Modrich. 1993. Hypermutability and mismatch repair deficiency in RER^+ tumor cells. *Cell* **75:** 1227.

Peinado, M.A., S. Malkhosyan, A. Velazquez, and M. Perucho. 1992. Isolation and characterization of allelic losses and gains in colorectal tumors by arbitrarily primed polymerase chain reaction. *Proc. Natl. Acad. Sci.* **89:** 10065.

Peltomaki, P., L.A. Aaltonen, P. Sistonen, L. Pylkkanen, J.-P. Mecklin, H. Jarvinen, J.S. Green, J.R. Jass, J.L. Weber, F.S. Leach, G.M. Petersen, S.R. Hamilton, A. de la Chapelle, and B. Vogelstein. 1993. Genetic mapping of a locus predisposing to human colorectal cancer. *Science* **260:** 810.

Pietenpol, J.A., T. Tokino, S. Thiagalingam, W.S. El-Deiry, K.W. Kinzler, and B. Vogelstein. 1994. Sequence-specific transcriptional activation is essential for growth suppression by p53. *Proc. Natl. Acad. Sci.* **91:** 1998.

Shibata, D., M.A. Peinado, Y. Ionov, S. Malkhosyan, and M. Perucho. 1994. Genomic instability in repeated sequences is an early somatic event in colorectal tumorigenesis that persists after transformation. *Nat. Genet.* **6:** 273.

Strand, M., T.A. Prolla, R.M. Liskay, and T.D. Petes. 1993. Destabilization of tracts of simple repetitive DNA in yeast by mutations affecting DNA mismatch repair. *Nature* **365:** 274.

Thibodeau, S.N., G. Bren, and D. Schaid. 1993. Microsatellite instability in cancer of the proximal colon. *Science* **260:** 816.

Umar, A., J.C. Boyer, D.C. Thomas, D.C. Nguyen, J.I. Risinger, J. Boyd, Y. Ionov, M. Perucho, and T.A. Kunkel. 1994. Defective mismatch repair in extracts of colorectal and endometrial cancer cell lines exhibiting microsatellite instability. *J. Biol. Chem.* **269:** 14367.

Vasen, H.F., J.P. Mecklin, P. Meera Kahn, and H.T. Lynch. 1991. The International Collaborative Group on Hereditary Non-Polyposis Colorectal Cancer (ICG-HNPCC). *Dis. Colon Rectum* **34:** 424.

Vogelstein, B. and K.W. Kinzler. 1992. p53 Function and dysfunction. *Cell* **70:** 523.

Warthin, A.S. 1913. Heredity with reference to carcinoma: As shown by the study of the cases examined in the pathological laboratory of the University of Michigan, 1895-1913. *Arch. Int. Med* **12:** 546.

Xiong, Y., H. Zhang, and D. Beach. 1992. D type cyclins associate with multiple protein kinases and the DNA replication and repair factor PCNA. *Cell* **71:** 505.

Xiong, Y., G.J. Hannon, H. Zhang, D. Casso, R. Kobayashi, and D. Beach. 1993. p21 is a universal inhibitor of cyclin kinases. *Nature* **366:** 701.

Zambetti, G.P. and A.J. Levine. 1993. A comparison of the biological activities of wild-type and mutant p53. *FASEB J.* **7:** 855.

Green Pigs, Red Herrings, and a Golden Hoe: A Retrospective on the Identification of *BRCA1* and the Beginning of Its Characterization

E.A. OSTERMEYER, L.S. FRIEDMAN, E.D. LYNCH, C.I. SZABO, P. DOWD, M.K. LEE, S.E. ROWELL, AND M.-C. KING

Department of Molecular and Cell Biology and School of Public Health, University of California, Berkeley, California 94720

Linkage of early-onset breast cancer to chromosome 17q21 was discovered four years ago (Hall et al. 1990), and in the improbability of that linkage lies the origin of this paper's title. The epidemiological complexities of breast cancer led to the nearly inescapable conclusion that linkage, or mapping inherited predisposition in high-risk families, would not be a fruitful approach for identifying breast cancer genes. The same skepticism was appropriately expressed by a student in an introductory genetics course who, when asked to describe briefly the experiments of Mendel, wrote "Gregor Mendel was an Australian bishop. It is said he worked with pigs. It is further said he worked with smooth and wrinkled pigs. It is even said he worked with yellow and green pigs. This last I do not believe." Fortunately, neither Mendel's factors nor those identified by linkage in families with breast cancer proved to be green pigs, and the latter story is described here.

The existence of a gene that increased susceptibility to breast and ovarian cancer was subsequently verified by linkage analysis with odds greater than $10^{26}:1$ (Narod et al. 1991; Easton et al. 1993) and the region of linkage refined by meiotic recombination to an interval of about 650 kb (Friedman et al. 1994b). *BRCA1* has recently been cloned (Futreal et al. 1994; Miki et al. 1994), and experiments have been carried out to verify that the announced candidate is *BRCA1* (Castilla et al. 1994; Friedman et al. 1994a; Simard et al. 1994).

BRCA1 appears to explain most of inherited breast cancer and inherited ovarian cancer. In our series, predisposition to breast and/or ovarian cancer is linked to *BRCA1* in 60% of families with five or more affected relatives (Friedman et al. 1995a). More than 90% of families with multiple relatives with breast and ovarian cancer trace susceptibility to *BRCA1* (Easton et al. 1993). Furthermore, among families with at least three cases of ovarian cancer and no early-onset breast cancer, the proportion with cancer linked to *BRCA1* is 78% if risk is assumed to be restricted to ovarian cancer, and 100% if predisposition to both breast and ovarian cancer is assumed (Steichen-Gersdorf et al. 1994).

Risks of breast cancer to women inheriting *BRCA1* are extremely high, exceeding 50% before age 50 and reaching 80% by age 65 (Newman et al. 1988; Hall et al. 1992; Easton et al. 1993; Ford et al. 1994). However, families who have participated in genetic studies were selected because multiple relatives developed breast cancer and may not be representative of all families with *BRCA1* mutations. That is, there may exist other families with breast cancer linked to less severe mutations in *BRCA1*. Now that the *BRCA1* gene is cloned, it will be possible to identify women carrying variant alleles and to begin to evaluate the risk associated with each mutant sequence.

Inherited breast cancer is not restricted to early-onset disease. Inherited predisposition influences a higher proportion of breast cancer at younger ages, accounting for an estimated 30% of breast cancer diagnosed before age 40, compared to less than 5% of breast cancer diagnosed after age 60 (Claus et al. 1991). However, it is likely that the absolute number of patients with inherited disease is greater among older patients. Re-analysis of the original *BRCA1* linkage data with a different liability class structure suggested that inherited breast cancer occurs in all age classes, and that linkage of *BRCA1* to early-onset disease was easier to identify because far fewer sporadic cases appear (Margaritte et al. 1992). This hypothesis can now be tested directly.

Other genes responsible for inherited breast cancer have been identified. *BRCA2*, on chromosome 13q12-13, is linked to breast cancer in families with both females and males affected, as well as in some families with only female breast cancer (Wooster et al. 1994). In our series, breast cancer is linked to *BRCA2* in approximately 10% of families (that is, ~30% of high-risk families with breast cancer not linked to *BRCA1*), including families with breast cancer in males and females and families with ovarian cancer (Friedman et al. 1995a). The first gene identified for inherited breast cancer was *p53*, which is responsible for the Li-Fraumeni syndrome (Malkin et al. 1990). In addition to female breast cancer, families with Li-Fraumeni syndrome have extremely high rates of brain tumors and adrenocortical cancers among children with a mutant *p53* allele. About 1% of women diagnosed with breast cancer before age 30 have germ-line mutations in *p53*. Additionally, a rare point mutation in the androgen receptor gene on the X chromosome can lead to breast cancer and androgen insufficiency among males (Wooster et al. 1992). Breast cancer in other families with multiple cases is probably due to other, as-yet-unidentified predisposing genes. However, given the very high rate of noninherited breast cancer, many remaining families with multiple cases of breast cancer may represent coincidental occurrences of breast cancer, rather than inherited predisposition due to other genes.

The great majority of breast and ovarian cancers are due solely to acquired mutations: Only 5–10% of breast cancer patients, and 8–15% of ovarian cancer patients, have inherited mutations leading to the disease. However, although inherited breast cancer is a small fraction of the breast cancer burden, it is a common genetic disease (King et al. 1993). Five percent of a disease affecting 1 in 10 women over the life span means that roughly 1 in 200 women, or more than 600,000 women in the United States, will develop breast cancer by reason of inherited susceptibility. Therefore, as an inherited trait, breast cancer is one of the most common genetic diseases in the world.

Incidence rates of breast cancer have increased 25% in the past 20 years and are now 10% by age 80 (Reis et al. 1990). Mortality rates have not increased, because survival of breast cancer patients has improved. The increase in breast cancer risk almost certainly has nothing to do with changes in frequencies of susceptibility alleles. Rather, breast cancer risk has probably increased as the result of gradual changes in the prevalence of established risk factors for breast cancer, particularly the length of the interval between menarche and first full-term pregnancy (Henderson et al. 1991).

At the beginning of this century, the average age at menarche was 14.5 years and the average maternal age at first birth was 21 years (Pike et al. 1983). Among young women in the United States now, the average age at menarche is 11 years, and childbearing frequently begins in the late 20s or 30s. Therefore, the interval of exposure of dividing breast ductal cells to hormonal stimulation is twice as long now as a century ago. Epidemiological studies of women exposed to radiation as teenagers indicate that the first mutations leading to breast cancer probably occur in this period (Boice and Monson 1977; Tokunaga et al. 1994). An association between increased breast cancer risk and an extended interval of menstruation before pregnancy is biologically plausible, because all breast ductal cells, whether normal or mutant, are dividing rapidly during this period, so abnormal cells would have every opportunity for clonal growth.

The link between inherited predisposition and increasing risk of breast cancer lies in the application of genetics to diagnosis and prevention. Creating molecular tools for earlier diagnosis and developing ways to reverse the first steps of tumorigenesis may be the most effective means of breast cancer control. In what follows, we present our approach to the search for BRCA1, the 22 genes cloned and sequenced in the BRCA1 region, the mutations we have identified so far in BRCA1, and the implications of the cloning of BRCA1 for evaluating risk.

METHODS

Methods for extracting DNA from lymphoblasts and paraffin blocks; genotyping markers; constructing a physical map and contig of the region; isolating and amplifying cDNA clones; consolidating cDNA clones into genes and mapping them to the region; sequencing cloned genes; and screening for mutations using genomic DNA and RNA are described in detail elsewhere (Hall et al. 1992; Friedman et al. 1994a, 1995b).

RESULTS

Indirect Evidence That *BRCA1* Is a Tumor Suppressor Gene

Analysis of tumors from families with *BRCA1*-linked breast and ovarian cancer suggests that *BRCA1* may act as a tumor suppressor gene. Loss of heterozygosity (LOH) for markers near *BRCA1* was evaluated for seven tumors from patients in high-risk families with cancer linked to chromosome 17q21 (Friedman et al. 1994b). For these tumors, the chromosomes carrying the presumptive *BRCA1* mutations were identifiable from linkage data. In all cases, the chromosome 17q21 carrying the normal allele at *BRCA1* was lost and the chromosome carrying the mutant *BRCA1* was retained, confirming results previously reported for other *BRCA1*-linked families (Smith et al. 1992).

BRCA1 may also act as a tumor suppressor in the far more common cases of sporadic, rather than inherited, breast or ovarian cancer. Breast and ovarian tumors from patients not selected for family history frequently lack the *BRCA1* region of 17q21. The frequency of LOH in the *BRCA1* region ranges from 40% to 80% among sporadic breast carcinomas (Futreal et al. 1992; Cropp et al. 1993; Saito et al. 1993) and from 30% to 60% among sporadic ovarian carcinomas (Russell et al. 1990; Cliby et al. 1993; Yang-Feng et al. 1993).

Patients with no evident family history of breast or ovarian cancer may nevertheless have germ-line *BRCA1* mutations. Breast and ovarian carcinomas from 25 patients with primary tumors at both sites were evaluated for LOH in the *BRCA1* region (Schildkraut et al. 1995). Of the ovarian tumors, 20 (80%) had lost part or all of chromosome 17q, always including the *BRCA1* region. Of the breast tumors from the same patients, 10 (40%) had lost the *BRCA1* region. Of the 10 patients whose breast and ovarian tumors both revealed LOH, losses were always of the same chromosome 17q. These patients may have inherited a mutant *BRCA1* allele on the retained chromosome, even in the apparent absence of a family history of breast or ovarian cancer.

An alternative to the hypothesis that *BRCA1* is a

tumor suppressor is the possibility that *BRCA1* is a dominant predisposing gene located very near a still-unknown tumor suppressor. If this were true, tumor development would require that acquired alterations in the tumor suppressor be associated with selection for retention of the mutant *BRCA1* in the tumor. However, the observation that most *BRCA1* mutations observed so far appear to lead to loss of function (Castilla et al. 1994; Friedman et al. 1994a; Simard et al. 1994) suggests that this alternative hypothesis is unlikely.

Red Herrings: Attempts to Find Shortcuts to *BRCA1*

Positional cloning typically involves two types of activities in parallel. One pathway is physical mapping, contig construction, library screening, and characterization of candidate genes. At the same time, one searches for critical patients or tumors that would provide shortcuts to the gene. Although none of these parallel efforts revealed *BRCA1*, the same approaches may reveal other cancer genes.

Identification and characterization of triplet repeats. More recent generations of women in breast cancer families develop breast cancer younger than did their ancestors. This observation probably represents a combination of ascertainment bias and a cohort effect due to changes in prevalence of noninherited risk factors, rather than biological anticipation. However, the discovery that unstable triplet repeats might play an important role in inherited disease (Richards and Sutherland 1992) motivated a search for triplet repeats in the *BRCA1* region. CAG, CCG, GAG, AAG, and AAT triplets were identified from cosmids in the region, sequenced, and screened for polymorphism. A few repeats detected by this method were polymorphic, but none underwent large expansions.

Search for chromosome 17q alterations in patients. A patient with severe developmental disabilities and early-onset breast cancer might have a chromosomal rearrangement involving both *BRCA1* and developmentally important genes. Three patients were identified with severe developmental disabilities and breast cancer at ages 28, 32, and 33, respectively. None of the patients had a family history of either breast cancer or retardation. The three patients had chromosomal aberrations on three different chromosomes, none of which involved chromosome 17. We will return to these alterations to search for other breast cancer genes.

A balanced translocation involving *BRCA1* might be revealed as both breast cancer and a history of miscarriage. Six patients were identified who were diagnosed with breast cancer before age 30, and who had, or whose mothers had, frequent miscarriages. No translocations were detected in the three patients karyotyped so far. We tried the same idea in reverse, screening records of amniocenteses for alterations in chromosome 17q, then interviewing parents of pregnancies with 17q alterations to determine family history of breast cancer. Two families were identified who had both balanced translocations involving chromosome 17q21 and family history of breast cancer. The informative relatives in both families were sampled and karyotyped, but in neither family did the balanced translocation segregate with breast cancer.

A possible contiguous gene syndrome. An extended kindred in which breast and ovarian cancer were coinherited with palmoplantar keratoderma was reported 6 years ago (Blanchet-Bardon et al. 1987), and palmoplantar keratoderma subsequently mapped to chromosome 17q12-q21 (Reis et al. 1992). Hence, the family with both conditions might carry an inversion or deletion involving *BRCA1*. Pulsed field filters and Southern blots with DNA from both affected and unaffected members of the family were probed with keratin genes, cosmids, and other genes in the *BRCA1* region. No rearrangements were detected on chromosome 17q by Southern or by further karyotyping or fluorescence in situ hybridization (FISH) analysis. Subsequently, a point mutation in keratin 9 was identified as the cause of palmoplantar keratoderma in this family (Torchard et al. 1994). This family carries independent mutations on chromosome 17q, one leading to breast and ovarian cancer and the other to palmoplantar keratoderma, certainly not impossible given the estimated frequency of 1 in 200 for carriers of *BRCA1* mutations.

FISH analysis of possible chromosomal rearrangements in tumors. Two ovarian cancer patients whose tumors had possible rearrangements of chromosome 17q were identified. The patients were diagnosed at ages 68 and 79, with no family history of breast or ovarian cancer. Both tumors were aneuploid, with three and four copies of chromosome 17, respectively. In one tumor, a translocated 17 was observed in one preparation only. In the other tumor, one of the four copies of chromosome 17 was consistently observed to carry a translocation, which by FISH analysis was proximal to the region of linkage.

Identification of 22 Genes by Positional Cloning

Candidate genes in the *BRCA1* region were isolated by screening fibroblast and ovarian cDNA libraries with eight YACs and four pools of cosmids from the partial contig (Couch et al. 1995; Friedman et al. 1995b). More than 400 independent cDNA clones were isolated from the libraries after the first round of screening. The 17q21 chromosomal locations of the cDNA clones were confirmed by hybridization to multiple chromosome 17 YACs, to chromosome 17 cosmids, and to hybrid somatic cell lines. About 60% of cDNAs identified with YACs mapped back to chromosome 17q21. cDNA clones that mapped back to 17q21 were cross-hybridized with each other and fell into clusters that defined 22 genes (Table 1).

cDNA clones representing each gene were sequenced, and the databases were searched for possible homologies. Ten genes identified by positional cloning were known human genes: Ras-related protein Rab5c (Albertsen et al. 1994); stearoyl-CoA desaturase (Li et al. 1994); 17-β estradiol dehydrogenase (EDH17B2) (Peltoketo et al. 1992); α-N-acetylglucosaminidase (Zhao et al. 1994); γ tubulin (Zheng et al. 1991); Ki nuclear autoantigen (Nikaido et al. 1990); a B-box protein IAI3B (Campbell et al. 1994); vaccinia virus VH1-related dual-specificity tyrosine/serine phosphatase (Ishibashi et al. 1992; Kamb et al. 1994); HRH1, the human homolog of yeast PRP22 (Ono et al. 1994); and adenovirus E1A enhancer binding protein (Higashino et al. 1993).

Five genes are newly cloned human homologs of genes known in other species. (1) cDNA clones B66 and B60 encode a protein homologous to proton pump cloned from rat (M58758) and *Caenorhabditis elegans* (Z11115), as well as to bovine vacuolar H$^+$ ATPase (L31770), and a mouse immune suppressor factor (M31226). (2) cDNA clones B52 and LF96 encode a protein homologous to the *Drosophila melanogaster* transcription factor enhancer of zeste (U00180) (Jones and Gelbart 1993). (3) cDNA clone B7F appears to encode a pseudogene of the high mobility group protein HMG17 (M12623). HMG17 maps to chromosome 1p and may have more than 50 pseudogenes (Srikantha et al. 1987). (4) cDNA clone LF113 encodes the human homolog of Pacific electric ray vesicle amine transport protein VAT1 (P19333) (Linial et al. 1989). (5) cDNA clone EL107 encodes a homolog of human endogenous retrovirus-related pol and protease polyproteins (sp P10266, M14123) and mouse gag polyprotein (pir S31034) (Ono et al. 1986).

Seven apparently novel genes were identified. (1) B102, LF98, and other cross-hybridizing clones encode a 6.4-kb message with patches of sequence similarity to *Xenopus laevis* Xotch, *D. melanogaster* Notch, human TAN-1, rat neurexin, and a human trithorax-homologous protein (Tkachuk et al. 1992). (2) EL76 encodes a 1.0-kb message with no significant similarities to known genes. (3) B32 and several cross-hybridizing clones encode a protein with coiled coils and similarity to tropomyosin, myosin, nuclear mitotic apparatus protein, ezrin, and merlin. (4) B169 and B213 encode a novel gene with two transcript sizes (3.0 and 3.8 kb) in lymphoblasts and no similarity to known genes. (5) LF91 encodes part of a 2.5-kb message with a motif homologous to a GATA-binding transcription factor. (6) LF79 encodes part of a 2.2-kb message with possible zinc finger and DNA-binding domains. (7) LF74 encodes part of a 4.0-kb message with similarity to collagens.

These 22 genes were analyzed for mutations (Friedman et al. 1995b). α-N-acetylglucosaminidase contained somatic mutations in three breast tumors, including a stop, a missense, and an intronic mutation that probably destroyed a splice site. The E1A-binding protein E1A-F contained missense mutations cosegregating with breast cancer in two *BRCA1*-linked families. HRH1, the human homolog of the yeast helicase PRP22, contained mutations cosegregating with breast cancer in three *BRCA1*-linked families. In retrospect, the inherited mutations are almost certainly very rare polymorphisms and the somatic mutations in tumors secondary events in tumorigenesis. Thirty-eight additional polymorphisms were identified by Southern

Table 1. Genes in the *BRCA1* Region

Gene	Genbank numbers	Primary clones	Sibling clones	Transcript (kb)
Rab5c	U11293	EL46	2	1.5
Stearoyl-CoA desaturase	S70284	LF26	31	
Homolog of rat proton pump (M58758)	U17977, U18001	B66	5	3.5
α-N-acetylglucosaminidase	M84472	LF77	14	3.2
17-β-estradiol dehydrogenase	M84472	LF78	9	2.3
Novel gene with GATA-binding domain	U18015, U18016	LF91	6	2.5
Novel gene with zinc finger domain	U18012, U18013, U18014	LF79	4	2.2
Novel gene with collagen motifs	U18010, U18011	LF74	2	4.0
γ Tubulin	M61764	B105	4	1.5
Novel gene with trithorax-like motif	U17905, U17976, U18000, U18017	B102, B47, B64, LF98	24	6.4
Homolog of *D. melanogaster* enhancer of zeste (U00180)	U18007, U18003	B52	3	5.0 and 6.2
Pseudogene of HMG17 (M12623)	U18008	B7F	1	
Novel gene	U18006	EL76	4	1.0
Novel gene with coiled coils	U17999, U17909	B32	5	2.2
Ki antigen	A60537	B55	8	2.5
Homolog of pacific electric ray VAT1 (P19333)	U18009	LF113	45	2.8
IAI3B, CA125-related B-box protein	X76952	EL86	4	4.3
Homolog of endogenous retrovirus (P10266)	U18004, U18005	EL107	1	
Serine-tyrosine VHR phosphatase	L05147	B168	9	4.1
Novel gene	U17906, U17907, U17908	B169, B213	8	3.0 and 3.8
Ets-related, adenovirus E1A enhancer binding protein	D12765, U18018	Q1	2	4.1
HRH1, homolog of *S. cerevisiae* PRP22 (X58681)	Ono et al. (1994)	B154	6	4.2 and 6.5

blot, by single-strand conformational polymorphism (SSCP) and mismatch cleavage analysis, and by direct sequencing (Friedman et al. 1995b).

Mutations in *BRCA1* (the "Golden Hoe")

The Golden Hoe refers to the inadvertent, but fitting, hybridization of our whimsical identification of *BRCA1* with the mythological Golden Bough and the more down-to-earth analysis of our low-technology approach to positional cloning as reminiscent of searching for a pot of gold in the backyard with a hoe (M. Olson, pers. comm.). In any case, in October 1994, a candidate gene for *BRCA1* was identified (Miki et al. 1994). In order to confirm that the candidate was *BRCA1*, we screened for germ-line mutations in the families that originally defined the linked phenotype, as well as in other families with breast cancer, and often ovarian cancer, linked to chromosome 17q21. For this purpose, the *BRCA1* gene was screened for single-strand conformational variants in 20 *BRCA1*-linked families using both genomic DNA and cDNA prepared from lymphoblast RNA (Friedman et al. 1994a). Criteria for distinguishing cancer-predisposing mutations from polymorphisms in *BRCA1* were (1) cosegregation of the variant with breast cancer, and with ovarian cancer, if it appeared; (2) absence of the variant in control chromosomes; and (3) amino acid substitution in, or truncation of, the *BRCA1* protein encoded by the variant sequence.

Two types of cancer-predisposing mutations in *BRCA1* occurred in the families in our series (Table 2). In families 4 and 84, a missense mutation leads to substitution of glycine for the penultimate cysteine in the C3HC4 zinc-binding motif. The families are unrelated, with the same mutation appearing in different genetic backgrounds. The inherited alterations in seven other families are chain terminations, caused by nonsense mutations (family 74), small insertions or deletions (families 3, 1, 102, 2, and 77), or a single base substitution that activated a cryptic splice acceptor (family 82). These variants did not appear in 120 control chromosomes. Ten polymorphisms were also detected in the gene, seven in the coding sequence and three in introns (Friedman et al. 1994a). Five polymorphisms altered amino acid residues and could be mistaken for disease-related mutations if they appeared in members of high-risk families.

DISCUSSION

Most of the *BRCA1* mutations identified so far result in premature termination of protein translation (Castilla et al. 1994; Friedman et al. 1994a; Futreal et al. 1994; Miki et al. 1994; Simard et al. 1994). Such truncated proteins could conceivably compete with wild-type protein, thereby acting as dominant mutations. On the other hand, loss of the wild-type allele in the tumors of patients with *BRCA1* germ-line mutations would suggest the opposite role; i.e., that the termination mutation was recessive.

The number of families with *BRCA1* mutations identified is still very small, so associations between genotype and phenotype are not at all certain. Protein truncation mutations occur throughout the *BRCA1* sequence, including termination of the *BRCA1* protein only 11 amino acids from the carboxy-terminal end in a family with very early-onset breast cancer. Bilateral breast cancer was associated with mutations throughout the sequence, although ascertainment of bilaterality is biased in very high-risk families in that some women choose bilateral mastectomy following their first diagnosis of breast cancer. In this series, there is no association between age at onset and locale or type of mutation, but these families were selected for early age at breast cancer diagnosis and probably do not represent *BRCA1* patients generally. Better evaluation of outcomes associated with *BRCA1* mutations of different types and at different locations in the sequence will be based on identifying germ-line *BRCA1* mutations in population-based series of cases not selected for family history.

Mutations were identified in only 9 of 20 families in our series with convincing linkage to *BRCA1*, although the entire gene was screened using both genomic DNA and RNA. The families for which *BRCA1* mutations have not yet been identified are as likely to carry mutations as those successfully screened. The two groups of families are similar in number and ages at diagnosis of breast and ovarian cancers linked to *BRCA1*; ovarian cancer occurred in 4 of 9 families with mutations found and in 5 of 11 families with mutations still undetected. Additional mutations may prove detectable by mismatch cleavage analysis, or by Southern or Northern hybridization. However, it is likely that a large fraction of mutations will only be detectable when genomic sequence of the entire gene is available for

Table 2. Inherited Mutations in *BRCA1* Linked to Breast and Ovarian Cancer

Family	*BRCA1* nt	Exon	Mutation	Amino acid change	Predicted effect
4	300	5	T to G substitution (TGT→GGT)	Cys61Gly	lose zinc-binding motif
84	300	5	T to G substitution (TGT→GGT)	Cys61Gly	lose zinc-binding motif
82	332	intron 5	T to G substitution → 59 bp insertion	75 Stop	protein truncation
3	2415	11	AG deletion	Ser766Stop	protein truncation
1	2800	11	AA deletion	901Stop	protein truncation
102	2863	11	TC deletion	Ser915Stop	protein truncation
74	3726	11	C to T substitution (CGA→TGA)	Arg1203Stop	protein truncation
2	4184	11	TCAA deletion	1364Stop	protein truncation
77	5677	24	A insertion	Tyr1853Stop	protein truncation

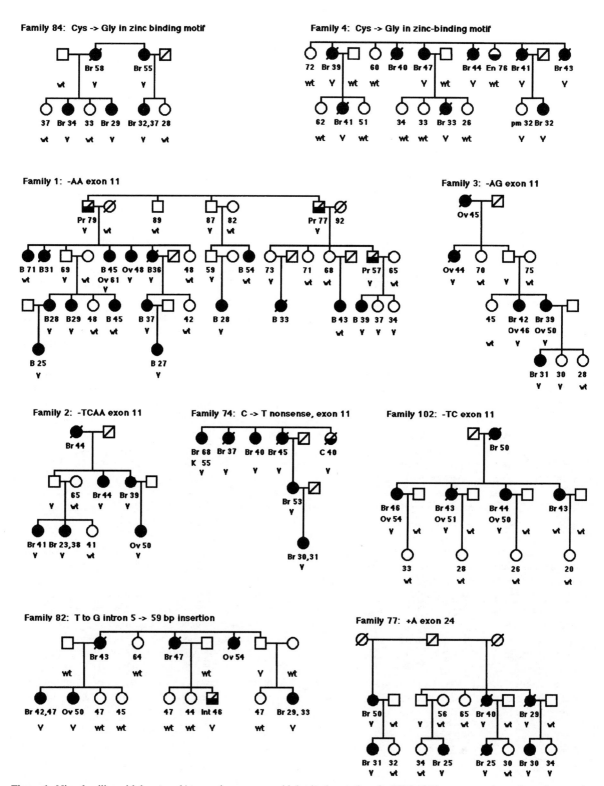

Figure 1. Nine families with breast and/or ovarian cancer and inherited mutations in *BRCA1*. V represents the variant characteristic of the family; wt represents wild-type sequence.

identifying alterations in introns and promoter regions. The alternative, screening breast or ovarian RNA for mutations using multiple techniques, will not be feasible on a wide scale.

Most *BRCA1* variants are not predisposing mutations. Most individuals, including women at high risk of breast cancer, are heterozygous at one or more sites in the *BRCA1* gene. Most of these variants are polymorphisms, some of which will be revealed by their sequence to be noncoding or silent. However, in this series, four missense mutations were revealed that altered peptide sequence. Because they occurred at similar frequencies in cases and controls, we concluded they were polymorphisms. Even by screening cases and controls, some rare polymorphisms may be mistakenly identified as predisposing mutations.

Finally, there are many predisposing mutations in *BRCA1*, and the risks associated with most are not known. The mutations identified so far have high penetrance, because they were identified in families ascertained for very high lifetime risk of breast cancer. Whether other *BRCA1* mutations are associated with more moderate risk, possibly influenced by other genetic or environmental factors, remains to be seen.

ACKNOWLEDGMENTS

This work was supported in part by National Institutes of Health grant R01-CA27632, by an American Cancer Society Professorship to M.-C.K., by a Komen Foundation Fellowship to L.S.F., by the Breast Cancer Fund, and by the Ensign and Lewis Foundation.

REFERENCES

Albertsen, H.M., S.A. Smith, S. Mazoyer, E. Fujimoto, J. Stevens, B. Williams, P. Rodriguez, C.S. Cropp, P. Slijepcevic, M. Carlson, M. Robertson, P. Bradley, E. Lawrence, T. Harrington, Z. Mei Sheng, R. Hoopes, N. Sternberg, A. Brothman, R. Callahan, B.A.J. Ponder, and R. White. 1994. A physical map and candidate genes in the *BRCA1* region on chromosome 17q12-21. *Nat. Genet.* **7:** 472.

Blanchet-Bardon, C., V. Nazzaro, J. Chevrant-Breton, M. Espie, P. Kerbrat, and B. LeMarec. 1987. Hereditary epidermolytic palmoplantar keratoderma associated with breast and ovarian cancer in a large kindred. *Br. J. Dermatol.* **117:** 363.

Boice, J.D., Jr., and R.R. Monson. 1977. Breast cancer in women after repeated fluoroscopic examinations of the chest. *J. Natl. Cancer Inst.* **59:** 823.

Campbell, I.G., H.M. Nicolai, W.D. Foulkes, G. Senger, G.W. Stamp, G. Allan, C.M. Boyer, K. Jones, R.C. Bast, E. Solomon, J. Trowsdale, and D.M. Black. 1994. A novel gene encoding a B-box protein within the *BRCA1* region at 17q21.1. *Hum. Mol. Genet.* **3:** 589.

Castilla, L.H., F.J. Couch, M.R. Erdos, K.F. Hoskins, K. Calzone, J.E. Garber, J. Boyd, M.B. Lubin, M.L. Deshano, L.C. Brody, F.S. Collins, and B.L. Weber. 1994. Mutations in the *BRCA1* gene in families with early-onset breast and ovarian cancer. *Nat. Genet.* **8:** 387.

Claus, E.B., N. Risch, and W.D. Thompson. 1991. Genetic analysis of breast cancer in the cancer and steroid hormone study. *Am. J. Hum. Genet.* **48:** 232.

Cliby, W., S. Ritland, L. Hartmann, M. Dodson, K.C. Halling, G. Keeney, K.C. Podratz, and R.B. Jenkins. 1993. Human epithelial ovarian cancer allelotype. *Cancer Res.* **53:** 2393.

Couch, F.J., L.H. Castilla, J. Xu, K.J. Abel, P. Welsch, S.E. King, L. Wong, P.P. Ho, S. Merajver, L.C. Brody, G.Y. Yin, S.T. Hayes, L.A. Geiser, W.L. Flejter, T.W. Glover, L.S. Friedman, E.D. Lynch, J.E. Meza, M.-C. King, D.J. Law, L. Deaven, A.M. Bowcock, F.S. Collins, B.L. Weber, and S.C. Chandrasekharappa. 1995. A YAC, P1, and cosmid-based physical map of the *BRCA1* region on chromosome 17q21. *Genomics* **25:** 264.

Cropp, C.S., M.H. Champeme, R. Lidereau, and R. Callahan. 1993. Identification of three regions on chromosome 17q in primary human breast carcinomas which are frequently deleted. *Cancer Res.* **53:** 5617.

Easton, D.F., D.T. Bishop, D. Ford, G.P. Crockford, and the Breast Cancer Linkage Consortium. 1993. Genetic linkage analysis in familial breast and ovarian cancer: Results from 214 families. *Am. J. Hum. Genet.* **52:** 678.

Ford, D., D.F. Easton, D.T. Bishop, S.A. Narod, and D.E. Goldgar. 1994. Risks of cancer in *BRCA1*-mutation carriers. Breast Cancer Linkage Consortium. *Lancet* **343:** 692.

Friedman, L.S., E.A. Ostermeyer, E.D. Lynch, C.I. Szabo, S.E. Rowell, and M.-C. King. 1995a. Inherited breast cancer: Patterns emerging from genetic complexity. *J. Am. Med. Assoc.* (in press).

Friedman, L.S., E.A. Ostermeyer, C.I. Szabo, P. Dowd, E.D. Lynch, S.E. Rowell, and M.-C. King. 1994a. Confirmation of *BRCA1* by analysis of germline mutations linked to breast and ovarian cancer in ten families. *Nat. Genet.* **8:** 399.

Friedman, L.S., E.A. Ostermeyer, E.D. Lynch, C.I. Szabo, L.A. Anderson, P. Dowd, M.K. Lee, S.E. Rowell, J. Boyd, and M.-C. King. 1994b. The search for *BRCA1*. *Cancer Res.* **54:** 6374.

Friedman, L.S., E.A. Ostermeyer, E.D. Lynch, P. Welsch, C.I. Szabo, J.E. Meza, L.A. Anderson, P. Dowd, M.K. Lee, S.E. Rowell, J. Ellison, J. Boyd, and M.-C. King. 1995b. 22 genes from chromosome 17q21: Cloning, sequencing, and characterization of mutations in breast cancer families and tumors. *Genomics* **25:** 256.

Futreal, P.A., P. Soderkvist, J.R. Marks, J.D. Iglehart, C. Cochran, J.C. Barrett, and R.W. Wiseman. 1992. Detection of frequent allelic loss on proximal chromosome 17q in sporadic breast carcinoma using microsatellite length polymorphisms. *Cancer Res.* **52:** 2624.

Futreal, P.A., Q. Liu, D. Shattuck-Eidens, C. Cochran, K. Harshman, S. Tavtigian, L.M. Bennett, A. Haugen-Strano, J. Swensen, Y. Miki, K. Eddington, M. McClure, C. Frye, J. Weaver-Feldhaus, W. Ding, Z. Gholami, P. Soderkvist, L. Terry, S. Jhanwar, A. Berchuck, J.D. Iglehart, J. Marks, D.G. Ballinger, J.C. Barrett, M.H. Skolnick, A. Kamb, and R. Wiseman. 1994. *BRCA1* mutations in primary breast and ovarian carcinomas. *Science* **266:** 120.

Hall, J.M., L. Friedman, C. Guenther, M.K. Lee, J.L. Weber, D.M. Black, and M.-C. King. 1992. Closing in on a breast cancer gene on chromosome 17q. *Am. J. Hum. Genet.* **50:** 1235.

Hall, J.M., M.K. Lee, B. Newman, J.E. Morrow, L.A. Anderson, B. Huey, and M.-C. King. 1990. Linkage of early-onset familial breast cancer to chromosome 17q21. *Science* **250:** 1684.

Henderson, B.E., R.K. Ross, and M.C. Pike. 1991. Toward the primary prevention of cancer. *Science* **253:** 1131.

Higashino, F., K. Yoshida, Y. Fujinaga, K. Kamio, and K. Fujinaga. 1993. Isolation of a cDNA encoding the adenovirus E1A enhancer binding protein: A new human member of the *ets* oncogene family. *Nucleic Acids Res.* **21:** 547.

Ishibashi, T., D.P. Bottaro, A. Chan, T. Miki, and S.A. Aaronson. 1992. Expression cloning of a human dual-specificity phosphatase. *Proc. Natl. Acad. Sci.* **89:** 12170.

Jones, R.S. and W.M. Gelbart. 1993. The *Drosophila* polycomb-group gene *enhancer of zeste* contains a region with sequence similarity to trithorax. *Mol. Cell. Biol.* **13:** 6357.

Kamb, A., A.P. Futreal, J. Rosenthal, C. Cochran, D. Harshmank, Q. Liu, R.S. Phelps, S.V. Tavtigian, T. Tran, C. Hussey, R. Bell, Y. Miki, J. Swenson, M.R. Hobbs, J. Marks, L.M. Bennet, C. Barrett, R.W. Wiseman, and D. Shattuck-Eidens. 1994. Localization of the VHR phosphatase gene and its analysis as a candidate for *BRCA1*. *Genomics* **23:** 163.

King, M.-C., S.E. Rowell, and S.M. Love. 1993. Inherited breast and ovarian cancer: What are the risks? What are the choices? *J. Am. Med. Assoc.* **269:** 1975.

Li, J., S.F. Ding, N.A. Habib, B.F. Fermor, C.B. Wood, and R.S. Gilmour. 1994. Partial characterization of a cDNA for human stearoyl-CoA desaturase and changes in its mRNA expression in some normal and malignant tissues. *Int. J. Cancer* **57:** 348.

Linial, M., K. Miller, and R.H. Scheller. 1989. VAT-1: An abundant membrane protein from torpedo cholinergic synaptic vesicles. *Neuron* **2:** 1265.

Malkin, D., F.P. Li, L.C. Strong, J.F. Fraumeni, Jr., C.E. Nelson, D.H. Kim, J. Kassel, M.A. Gryka, F.Z. Bischoff, M.A. Tainsky, and S.H. Friend. 1990. Germ line p53 mutations in a familial syndrome of breast cancer, sarcomas, and other neoplasms. *Science* **250:** 1233.

Margaritte, P., C. Bonaiti-Pellie, M.-C. King, and F. Clerget-Darpoux. 1992. Linkage of familial breast cancer to chromosome 17q21 may not be restricted to early-onset disease. *Am. J. Hum. Genet.* **50:** 1231.

Miki, Y., J. Swensen, D. Shattuck-Eidens, P.A. Futreal, K. Harshman, S. Tavtigian, Q. Liu, C. Cochran, L.M. Bennett, W. Ding, R. Bell, J. Rosenthal, C. Hussey, T. Tran, M. McClure, C. Frye, T. Hattier, R. Phelps, A. Haugen-Strano, H. Katcher, K. Yakumo, Z. Gholami, D. Shaffer, S. Stone, S. Bayer, C. Wray, R. Bogden, P. Dayananth, J. Ward, P. Tonin, S. Narod, P.K. Bristow, F.H. Norris, L. Helvering, P. Morrison, P. Rosteck, M. Lai, J.C. Barrett, C. Lewis, S. Neuhausen, L. Cannon-Albright, D. Goldgar, R. Wiseman, A. Kamb, and M.H. Skolnick. 1994. A strong candidate for the breast and ovarian cancer susceptibility gene *BRCA1*. *Science* **266:** 66.

Narod, S.A., J. Feunteun, H.T. Lynch, P. Watson, T. Conway, J. Lynch, and G.M. Lenoir. 1991. Familial breast-ovarian cancer locus on chromosome 17q12-q23. *Lancet* **338:** 82.

Newman, B., M.A. Austin, M. Lee, and M.-C. King. 1988. Inheritance of human breast cancer: Evidence for autosomal dominant transmission in high-risk families. *Proc. Natl. Acad. Sci.* **85:** 3044.

Nikaido, T., K. Shimada, M. Shibata, M. Hata, M. Sakamoto, Y. Takasaki, C. Sato, T. Takahashi, and Y. Nishida. 1990. Cloning and nucleotide sequence of cDNA for Ki antigen, a highly conserved nuclear protein detected with sera from patients with systemic lupus erythematosus. *Clin. Exp. Immunol.* **79:** 209.

Normand, T., S. Narod, F. Labrie, and J. Simard. 1993. Detection of eleven polymorphisms in the estradiol 17 beta-hydroxysteroid dehydrogenase II gene at the human EDH17B2 locus on 17q11-q21. *Hum. Mol. Genet.* **2:** 479.

Ono, M., T. Yasunaga, T. Miyata, and H. Ushikubo. 1986. Nucleotide sequence of human endogenous retrovirus genome related to the mouse mammary tumor virus genome. *J. Virol.* **60:** 589.

Ono, Y., M. Ohno, and Y. Shimura. 1994. Identification of a putative RNA helicase (HRH1), a human homolog of yeast PRP22. *Mol. Cell. Biol.* **14:** 7611.

Peltoketo, H.E., V. Isomma, and R. Vihko. 1992. Genomic organization and DNA sequences of human 17 beta-hydroxysteroid dehydrogenase genes and flanking regions; localization of multiple alu sequences and putative *cis*-acting elements. *Eur. J. Biochem.* **209:** 459.

Pike, M.C., M.D. Kraile, B.E. Henderson, J.T. Casagrande, and D.G. Hoel. 1983. "Hormonal" risk factors, "breast tissue age" and the age-incidence of breast cancer. *Nature* **303:** 767.

Reis, A., W. Kuster, R. Eckardt, and K. Sperling. 1992. Mapping of a gene for epidermolytic palmoplantar keratoderma to the region of the acidic keratin gene cluster at 17q12-q21. *Hum. Genet.* **90:** 113.

Reis, L.A.G., B.F. Hankey, and B.K. Edwards, eds. 1990. Cancer statistics review, 1973–1987. *NIH Publ.* 90-2789.

Richards, R.I. and G.R. Sutherland. 1992. Heritable unstable DNA sequences. *Nat. Genet.* **1:** 7.

Russell, S.E., G.I. Hickey, W.S. Lowry, P. White, and R.J. Atkinson. 1990. Allele loss from chromosome 17 in ovarian cancer. *Oncogene* **5:** 1581.

Saito, H., J. Inazawa, S. Saito, F. Kasumi, S. Koi, S. Sagae, R. Kudo, J. Saito, K. Noda, and Y. Nakamura. 1993. Detailed deletion mapping of chromosome 17q in ovarian and breast cancers: 2-cM region on 17q21.3 often and commonly deleted in tumors. *Cancer Res.* **53:** 3382.

Schildkraut, J.M., N.K. Collins, G.A. Dent, J.A. Tucker, J.C. Barrett, A. Berchuck, and J. Boyd. 1995. Loss of heterozygosity on chromosome 17q11-21 in cancers of women who have developed both breast and ovarian cancer. *Am. J. Obstet. Gynecol.* (in press).

Simard, J., P. Tonin, F. Durocher, K. Morgan, J. Rommens, S. Gingras, C. Samson, J.F. Leblanc, C. Belanger, F. Dion, Q. Liu, M. Skolnick, D. Goldgar, D. Shattuck-Eidens, F. Labrie, and S.A. Narod. 1994. Common origins of *BRCA1* mutations in Canadian breast and ovarian cancer families. *Nat. Genet.* **8:** 392.

Smith, S.A., D.F. Easton, D.G.R. Evans, and B.A.J. Ponder. 1992. Allele losses in the region 17q12-q21 in familial breast and ovarian cancer non-randomly involve the wild-type chromosome. *Nat. Genet.* **2:** 128.

Srikantha, T., D. Landsman, and M. Bustin. 1987. Retropseudogenes for human chromosomal protein HMG-17. *J. Mol. Biol.* **197:** 405.

Steichen-Gersdorf, E., H.H. Gallion, D. Ford, C. Girodet, D.F. Easton, R.A. DiCioccio, G. Evans, M.A. Ponder, C. Pye, S. Mazoyer, T. Noguchi, F. Karengueven, H. Sobol, A. Hardouin, Y.J. Bignon, M.S. Piver, S.A. Smith, and B.A.J. Ponder. 1994. Familial site-specific ovarian cancer is linked to *BRCA1* on 17q12-21. *Am. J. Hum. Genet.* **55:** 870.

Tkachuk, D.C., S. Kohler, and M.L. Cleary. 1992. Involvement of a homolog of *Drosophila trithorax* by 11q23 chromosomal translocations in acute leukemias. *Cell* **71:** 691.

Tokunaga, M., C.E. Land, S. Tokuoka, I. Nishimori, M. Soda, and S. Akiba. 1994. Incidence of female breast cancer among atomic bomb survivors, 1950–1985. *Radiat. Res.* **138:** 209.

Torchard, D., C. Blanchet-Bardon, O. Serova, L. Langbein, S. Narod, N. Janin, A.F. Goguel, A. Bernheim, W.W. Franke, G.M. Lenoir, and J. Feunteun. 1994. Epidermolytic palmoplantar keratoderma cosegregates with a keratin 9 mutation in a pedigree with breast and ovarian cancer. *Nat. Genet.* **6:** 106.

Wooster, R., J. Mangion, R. Eeles, S. Smith, M. Dowsett, D. Averill, P. Barrett-Lee, D.F. Easton, B.A. Ponder, and M.R. Stratton. 1992. A germline mutation in the androgen receptor gene in two brothers with breast cancer and Reifenstein syndrome. *Nat. Genet.* **2:** 132.

Wooster, R., S.L. Neuhausen, J. Mangion, Y. Quirk, D. Ford, N. Collins, K. Nguyen, S. Seal, T. Tran, D. Averill, P. Fields, G. Marshall, S. Narod, G. Lenoir, H. Lynch, J. Feunteun, P. Devilee, C.J. Cornelisse, F.H. Menko, P.A. Daly, W. Ormiston, R. McManus, C. Pye, C.M. Lewis, L.A. Cannon-Albright, J. Peto, B.A.J. Ponder, M.H. Skolnick, D.F. Easton, D.E. Goldgar, and M.R. Stratton. 1994. Localization of a breast cancer susceptibility gene, *BRCA2*, to chromosome 13q12-13. *Science* **265:** 2088.

Yang-Feng, T.L., H. Han, K.C. Chen, S.B. Li, E.B. Claus, M.L. Carcangiu, S.K. Chambers, J.T. Chambers, and P.E. Schwartz. 1993. Allelic loss in ovarian cancer. *Int. J. Cancer* **54:** 546.

Zhao, H.G., R. Lopez, J. Rennecker, and E.F. Neufeld. 1994. Sanfilippo syndrome type B: cDNA and gene encoding human α-N-acetylglucosaminidase. *Am. J. Hum. Genet.* **55:** A252.

Zheng, Y., M.K. Jung, and B.R. Oakley. 1991. γ-tubulin is present in *Drosophila melanogaster* and *Homo sapiens* and is associated with the centrosome. *Cell* **65:** 817.

Progress toward Isolation of a Breast Cancer Susceptibility Gene, *BRCA1*

B.L. Weber,*† K.J. Abel,‡ F.J. Couch,* S.D. Merajver,§ S.C. Chandrasekharappa,††
L. Castilla,†† D. McKinley,†† P.P. Ho,* K. Calzone,*
T.S. Frank,** J. Xu,§ L.C. Brody,†† and F.S. Collins††

*Departments of *Internal Medicine and †Genetics, University of Pennsylvania, Philadelphia, Pennsylvania 19104;
Departments of ‡Genetics, §Medicine, and **Pathology, University of Michigan, Ann Arbor, Michigan 48109;
and ††National Center for Human Genome Research, Bethesda, Maryland 20892*

Chromosome 17q21 is known to contain *BRCA1*, a gene that is critical in the determination of susceptibility to early-onset breast cancer and ovarian cancer (Hall et al. 1990; Narod et al. 1991). It is estimated that women harboring germ-line mutations in *BRCA1* (thought to be a tumor suppressor gene) carry an 85% lifetime risk of developing breast cancer (Easton et al. 1993). Although estimates of the percentage of breast cancer cases that occur as a direct result of germ-line mutations in *BRCA1* vary from 5% to 20% (Colditz et al. 1993; Slattery and Kerber 1993), what appears certain is that the estimate is a function of age. Thus, although less than 5% of all breast cancers in elderly women are likely to result from *BRCA1* germ-line mutations, this fraction may be much greater for women diagnosed with breast cancer under the age of 45 (Easton et al. 1993). *BRCA1* germ-line mutations appear to confer a risk of developing ovarian cancer in female carriers of 10–50% (Easton et al. 1993). Recent data also suggest that individuals with *BRCA1* germ-line mutations may have a fourfold increased risk of colon cancer and a threefold increased risk of prostate cancer in males (Ford et al. 1994). Although estimates of the frequency of *BRCA1* germ-line mutations in the general population are difficult to determine with accuracy, as many as 1 in 200 women in the United States may harbor *BRCA1* germ-line mutations and the associated increased risk of developing neoplastic disease.

Since the demonstration of linkage between early-onset breast cancer and genetic markers on chromosome 17q21 by King and colleagues in 1990 (Hall et al. 1990), efforts to isolate *BRCA1* have been under way in many laboratories. A positional cloning approach utilizing genetic and physical mapping strategies, tumor analysis, yeast artificial chromosome (YAC) and cosmid cloning, and now complementary transcript-searching strategies have been employed by all groups. As a result of this effort, many families now have been studied, and genetic recombination events in linked families continue to narrow the candidate interval. Genetic and physical maps of the *BRCA1* candidate region have been constructed and are becoming more detailed as new genetic markers and new sequence-tagged sites are added. Analyses of tumors from individual members of *BRCA1*-linked families have suggested that *BRCA1* is a tumor suppressor gene, and identification of interstitial deletions detected by loss of heterozygosity (LOH) analysis of tumors has contributed to efforts to narrow the *BRCA1* candidate interval. Attention is now being focused on the identification and analysis of candidate genes from within this region. In the discussion that follows, we present a summary of the data we have collected in our effort to identify *BRCA1*. The culmination of this extensive project ultimately will be the identification and isolation of *BRCA1*, providing a basis for increased understanding of the pathogenesis of breast and ovarian cancer, the development of targeted diagnostic and therapeutic approaches, and a means of screening individuals at risk of being *BRCA1* mutation carriers.

EXPERIMENTAL PROCEDURES/RESULTS

Genetic Mapping

Polymorphic markers. Generation of new genetic markers is critical to the construction of a high-resolution genetic map of the *BRCA1* region. We have identified 13 new polymorphic markers that have been useful in defining critical recombinants in families we have collected, as well as the original seven breast cancer families (M.-C. King, pers. comm.). The first marker we isolated, D17S409 (Chamberlain et al. 1993), was isolated by screening a chromosome 17 *Not*I linking library previously constructed from total chromosome 17 material (Wallace et al. 1989) with a synthetic oligonucleotide containing a $(CA)_{22}$ repeat. Evaluation of 80 CEPH parents demonstrated the presence of eight alleles, 57.7% heterozygosity, and a PIC = 0.60.

Subsequently, 3 tetranucleotide repeat polymorphisms were isolated by parallel screening of an arrayed chromosome 17 cosmid library (prepared by L. Deaven, Los Alamos National Laboratory from flow-sorted human chromosomes) and a chromosome microdissection library prepared from the *BRCA1* region (Trent et al. 1994) with a pool of oligonucleotides containing 12 unique tetranucleotide repeats. Cosmid clones which hybridized to both probe pools were isolated and prepared for fluorescence in situ hybridization (FISH) mapping. Clones that hybridized to the *BRCA1* region of chromosome 17 were then rehybridized using a two-color FISH method, which allowed determination of chromosomal position relative to probes of known location.

Examination of recently published recombination events in *BRCA1*-linked families suggests that *BRCA1* is located between D17S857 and D17S78, a region currently estimated at 1–1.5 Mb in length (Simard et al. 1993; Kelsell et al. 1993). We have recently screened approximately 400 cosmids known to map within this small region for the presence of dinucleotide repeats with a $(CA)_{22}$ oligonucleotide probe. Cloned *Sau*3AI cosmid fragments identified as containing dinucleotide repeats by this method were sequenced to allow the development of polymerase chain reaction (PCR) primers. These markers were screened for polymorphism and heterozygosity levels by typing on a panel of CEPH parent DNAs. Seven of ten markers were identified as having a heterozygosity value of more than 50%. The markers were also localized on the physical map of the *BRCA1* region by PCR amplification of cosmids from the region that had previously been assembled into small contigs. All ten newly derived markers were observed to map between D17S856 and D17S855 (Couch et al. 1994). The localization of these new polymorphisms in relation to markers from the known genetic map allows integration of the genetic and physical maps of the region, while offering the possibility of additional narrowing of the *BRCA1* region by identification of meiotic recombination events in *BRCA1*-linked families and by LOH analysis of breast tumor DNA.

Family analysis. We have identified more than 400 families with three or more cases of early-onset breast cancer and/or ovarian cancer. Only a small percentage of these families are suitable for linkage analysis based on family size, number of affected individuals available for sampling, and mean age of onset of breast or ovarian cancer. At present, 19 families have been genotyped and analyzed. Individuals were sampled by collection of peripheral blood and subsequent isolation of leukocyte DNA by standard techniques. Seven deceased affected individuals were studied using paraffin-embedded normal breast tissue as a source of DNA. In this case, 5-μm sections were obtained from paraffin blocks. Normal tissue was dissected away from tumor tissue by a surgical pathologist, and the paraffin was dissolved by 5-minute incubation in xylene followed by pelleting in 1:1 xylene:100% ethanol. The tissue pellets were dried, resuspended in 200 ng/μl Proteinase K/50 mM Tris (pH 8.3), and incubated at 37°C overnight. The samples were then boiled and used without further purification. Aliquots of the resulting aqueous solution and DNA derived from leukocyte DNA were used as substrate for PCR amplification of polymorphic markers flanking the *BRCA1* locus. Forward primers were end-labeled with [^{32}P]dCTP (New England Nuclear) to allow detection of products by autoradiography. PCR products were analyzed on 6% denaturing polyacrylamide gels using CEPH parental DNA as size markers.

Linkage analysis was performed using the method of LOD scores as described previously (Chamberlain et al. 1993). Ten families show evidence of linkage between breast cancer and genetic markers flanking *BRCA1* with maximum LOD scores ranging from 0.57 to 2.79 with Z_{max} at $\theta = 0.00$. All recombination events in these families support the localization of *BRCA1* between D17S857 (Kelsell et al. 1993) and D17S78 (Simard et al. 1993), with one recombinant further narrowing the interval to D17S1139 (F.J. Couch et al., in prep.). Four families are inconclusive ($Z_{max} < 0.5$) and five appear to be unlinked ($Z_{max} < 0$).

Tumor Analysis

A high frequency of LOH at a specific chromosomal location in tumor DNA can be interpreted as a marker for loss of one allele of a functional tumor suppressor gene normally present in the chromosomal region adjacent to the genetic marker being analyzed. Interstitial deletions, as detected by LOH, have been identified in a number of common human cancers in regions containing known tumor suppressor genes. In the case of inherited mutations in putative tumor suppressor genes, it is generally accepted that the germ-line mutation constitutes the "first hit" required for malignant transformation. In this setting, the subsequent somatic loss of the non-mutant allele represents the "second hit" (Knudson 1971). Thus, germ-line mutations in tumor suppressor genes create a highly susceptible group of individuals for whom only one somatic event is required to produce inactivation of tumor suppressor activity at a given locus. Recent evidence (Smith et al. 1992; Chamberlain et al. 1993; and data not shown) demonstrating loss of wild-type alleles of polymorphic markers flanking *BRCA1* in breast tumors from affected members of *BRCA1*-linked families strongly supports the hypothesis that *BRCA1* is a tumor suppressor gene.

Previous studies of LOH in breast tumors reported loss of large regions on chromosome 17, but the majority have been carried out on tumors derived from individuals without known linkage to *BRCA1* or evidence of family history of breast or ovarian cancer. These LOH events could, therefore, not be unequivocally ascribed to *BRCA1*. We have studied 29 breast and 6 ovarian tumors from families with strong evidence of linkage between breast cancer and genetic markers flanking *BRCA1*. These tumors were examined for LOH using genetic markers flanking and within the *BRCA1* region, including THRA1, D17S856, EDH17B1, EDH17B2, and D17S183. Tumor and normal cells from formalin-fixed, paraffin-embedded tissue sections were prepared as described above. Of the tumors, 40% (14/35) exhibit LOH that includes the *BRCA1* candidate interval and suggests that the minimal region of loss is between D17S856 and D17S183. The origin of the allele lost in 12 of 14 tumors was assessed by evaluation of haplotypes in associated family members; in 100% of these cases, the wild-type allele was lost. These data support the hypothesis that *BRCA1* is a tumor suppressor gene and provide additional support for the current genetic boundaries of the *BRCA1* candidate region.

Physical Mapping

Sequence-tagged sites. Sequence-tagged sites (STS), although not usually polymorphic and therefore not useful for genetic mapping, are critical to the generation of a physical map. STS-based PCR assays for all cloned genes and polymorphic markers in the *BRCA1* interval have been developed for screening of the three YAC libraries we are currently utilizing. As YACs are identified, they are arrayed with all other YACs from the *BRCA1* region and rescreened with PCR primers for flanking STSs (if known) or all *BRCA1* region STSs if the map location is not known. This method of STS content mapping is a rapid and reliable way to identify overlapping clones and generate contiguous segments of cloned DNA (contigs). In contrast to YAC end-cloning techniques, STS content mapping is not affected by the presence of chimeric clones and does not require subcloning or sequencing. This strategy is the cornerstone of our contig-building strategy, but it is dependent on the availability of large numbers of closely spaced STSs in the *BRCA1* region.

A direct and elegant approach to isolating numerous STSs from a chromosomal subregion involves microdissection and PCR-microcloning of DNA from banded chromosomal regions. We have utilized microdissection of the *BRCA1* chromosomal subregion, PCR amplification of the microdissected material using universal primers, and subcloning of these PCR products (Trent et al. 1994). The resultant library contains approximately 100,000 recombinant "microclones." FISH mapping has confirmed the location of the microdissected material just centromeric to D17S579 and overlapping the *BRCA1* region. Microclones containing only single-copy sequence have been sequenced for PCR primer design and analyzed for map location using a panel of somatic cell hybrids. The STSs identified by these microclones were valuable aids to the identification of additional YACs and cosmids from the *BRCA1* region and have also been useful in the generation of the cosmid contig described below.

YAC/cosmid cloning. Using the STS content-mapping strategy described above, a total of 46 YACs have been isolated, characterized, and assembled into a contig using STS content mapping. Although YAC cloning is a convenient way of rapidly isolating large segments of DNA, several problems arise when YAC clones are used as reagents for marker generation or transcript identification. Total yeast DNA preparations are complex and difficult to use as probes, requiring the use of hybridization conditions that work close to the limits of sensitivity. The larger the YAC clone, and the smaller the target sequence, the less likely a signal will be detected. Insert isolation and subcloning are tedious and time-consuming. Additionally, the use of chimeric YAC clones for transcript searching is likely to result in the identification of significant numbers of transcripts from outside the *BRCA1* region. One way of circumventing these problems is conversion of YACs into cosmids by direct screening of the chromosome 17 cosmid library (L. Deaven, Los Alamos National Laboratory) using YAC Alu-PCR products, whole YAC DNA, or STS PCR products as probes. We have used all three approaches with success. When STS PCR products are used as probes, the procedure can be simplified further by screening the cosmid library with pools of probes, regridding the clones identified during the bulk hybridization, and rescreening this sublibrary with individual probes. This dramatically reduces the number of times the entire library must be screened, preserving filters and decreasing the number of high-volume hybridizations necessary to isolate multiple cosmids for each marker in the *BRCA1* region. More than 500 cosmids have been identified by this strategy, combined with end-cloning and "walking" techniques. We have ordered these cosmids using STS content mapping in conjunction with cross-hybridization and restriction mapping and selected a minimal set in order to provide a framework physical map of greater definition than is allowed by the YAC map. These cosmids are the backbone of our transcript-searching strategies described below. In addition, we will use this cosmid contig to provide initial structure to a large-scale cosmid-sequencing effort and target high-priority areas early in the sequencing effort. We expect that sequence data will refine our current contig and eventually replace it with a more accurately ordered minimal set of cosmids.

Transcript Identification

The final step in positional cloning of any disease-related gene involves the isolation and characterization of transcribed sequences from the genomic region that defines the candidate interval. Until recently, methods of finding transcripts tended to be laborious and cumbersome when the candidate interval spanned several hundred thousand to a few million base pairs. With the advent of several new transcript identification techniques, the effort required for isolation of transcripts from large genomic regions has been considerably lessened. We have employed three complementary transcript-searching strategies in our efforts to isolate *BRCA1*.

Direct cDNA screening. Direct screening of cDNA libraries is conceptually the most straightforward of the three methods. In brief, direct screening involves radioactive labeling of large regions of cloned genomic DNA (40–200 kb) and the use of this material to probe relevant cDNA libraries as a direct means of identifying region-specific cDNA.

It is estimated that transcribed genes make up no more than 10% of the human genome. The remainder of the human genome consists of nontranscribed DNA between and within genes, and many families of repetitive elements. This noncoding DNA is not useful for screening cDNA libraries because it is either not transcribed, or transcribed but not unique to a given region of the genome (i.e., transcribed repetitive elements). Early positional cloning methods stripped the genomic DNA of these "extraneous" sequences by first isolating nonrepetitive DNA from genomic clones; however, this method is cumbersome and time-consuming for large chromosomal regions. If the candidate interval is defined by genetic recombination, as is the case for *BRCA1*, it is likely to be a megabase or larger. For genomic regions this large, it is most efficient to directly label large clones such as cosmids (~50 kb) or YACs (100–1000 kb) and eliminate the time-consuming step of subcloning single-copy sequences. The tradeoff is that complex probes are more difficult to use for cDNA library screens. YACs have the additional complication that they are difficult to purify from the contaminating yeast genomic DNA. In the case of the *BRCA1* region, we are using cosmids that have been individually labeled and pooled in groups of two to six cosmids each. To eliminate hybridization to transcribed repetitive sequences represented in a cDNA library, the probes are preannealed to a vast excess of unlabeled human genomic DNA. The preannealing conditions are chosen such that any highly repetitive probe sequences are bound to the cold competitor DNA and therefore not available to hybridize to the filter-bound cDNAs.

Using direct screening, transcribed sequences may not be identified for the following reasons: (1) Bias toward the isolation of genes with large exons. In most cases, the 3'-terminal exon is the largest single exon in a gene. We have chosen to screen oligo dT-primed cDNA libraries to maximize the number of clones containing a complete 3'-terminal exon. (2) Low signal-to-noise ratio when compared to traditional probes. In many cases, positive phage plaques are barely detectable above the background, therefore cDNAs hybridizing at, or just below, this threshold are also missed. False-positive clones are primarily repetitive transcribed sequences. These sequences might not be sufficiently blocked by the preannealing procedure and will be present in many cDNA libraries. For this reason, every candidate clone is hybridized to a somatic cell hybrid panel to confirm that it maps to the *BRCA1* region of chromosome 17 with a hybridization pattern suggestive of single-copy DNA. To date, direct screening has allowed isolation of 14 independent clones, all of which derive from the *BRCA1* candidate region.

Exon amplification. Exon amplification (Buckler et al. 1991), also called exon trapping, is a powerful technique for isolating gene sequences from genomic clones. This technique relies on the normal processing of most primary RNA transcripts, whereby noncontiguous sequences destined to make up the mature messenger RNA (exons) are accurately spliced together and the intervening sequences (introns) are excised. The exons are flanked by specific sequences which serve as signposts for the cellular splicing machinery, identifying the precise sites where cutting and joining of the exons occur. Exon amplification permits the selective recovery of sequences present in cloned genomic DNA that are flanked by functional splice sites, yielding subclones from genomic DNA that are highly enriched for gene sequences.

Briefly, cosmid DNAs are digested and ligated into a plasmid splicing vector. The fragments are inserted into an intron of the HIV *tat* gene, which is flanked by functional donor and acceptor splice sites. The resulting constructs are transfected into COS7 cells where in vivo processing pairs exons present in the genomic fragment with the *tat* donor and acceptor splice sites in the vector. The *tat* intron sequences are excised and the exon is now present in the mature mRNA. After a brief period in culture, cytoplasmic RNA is isolated from the cells and used as a template to generate cDNA using reverse transcriptase. The second strand of cDNAs is synthesized and the double-stranded cDNA is amplified by PCR using nested, vector-specific primers, then cloned into plasmids and propagated in bacteria.

In two experiments, exon amplification was performed using DNAs of approximately 170 cosmids spanning 1–2 Mb of this region. Cosmid DNAs were pooled in groups of 6–10 each. More than 2000 candidate exon clones from these experiments have been arrayed in microtiter dishes. The average size determined for nearly 400 cloned inserts was approximately 200 base pairs. Ongoing efforts to identify and eliminate clone redundancy have thus far yielded approximately 140 unique exon clones. Less than 10% of the clones were found to be repetitive or to be artifacts resulting from cryptic splicing involving sequences present in the splicing vector. Thus, the great majority of clones were found to contain human single-copy sequence derived from the correct chromosomal location. These exons, like all partial cDNAs we have isolated, have been used as hybridization probes to isolate cDNA clones derived from normal breast, ovary, and fetal brain tissue. The cloned exons and corresponding cDNAs are being localized within developing cosmid contigs in order to assemble a transcription map of this region. Combining the results of cDNA screening and database searches, exon amplification has led to the identification of as many as 25 genes in this region.

The technique of exon amplification has a number of inherent advantages. First, large numbers of candidate gene sequences can be generated from regionally localized genomic DNA clones in a relatively short period of time. Second, coding sequences can be obtained irrespective of the sites or levels of expression of the corresponding gene, theoretically permitting isolation of genes expressed at very low levels within a highly restricted set of cell types. Third, more than 90% of clones have been free of repetitive sequences and are therefore useful as hybridization probes without additional competitor DNA. Fourth, the exon clones are typically short enough (average <200 bp) to permit rapid sequence analysis of the cloned inserts, enabling high-volume sequencing of many clones simultaneously. In an effort to predict potential functions of the corresponding genes, the sequences determined for each representative clone can be used to screen databases of nucleic acid and protein sequences for homologies with known genes. Finally, the ability to pool cosmids means that large genomic regions can rapidly be surveyed for genes by this technique.

There are also a number of limitations unique to this method. First, several types of artifactual products can arise from aberrant splicing events including pairing of only the vector splice sites or cloning of noncoding sequences which are flanked by "cryptic" splice sites within the insert and/or vector. In the original system using pSPL1, these cloning artifacts represented nearly half of the clones generated. However, recent modifications to the vector (Buckler et al. 1993) have greatly reduced the frequency of vector-only and cryptic splicing artifacts, and efficient ligation of genomic fragments and vector can nearly eliminate the background of *tat* intronic clones. Second, although only a few percent of our clones were found to contain highly repetitive sequences, it appears that highly repetitive Alu sequences may continue to be cloned as a result of splicing events. We and other workers (North et al. 1993) have observed that specific sequences within Alu repeats can act as donor and acceptor splice junctions. Third, because of the short average length of the cloned exons, some of the clones may be difficult to use as hybridization probes. Finally, although the requirement for two flanking splice sites provides greater reliability for amplifying internal exons, this technique is not designed to isolate the 5'- or 3'-most exons of genes, and thus is not expected to efficiently amplify sequences from genes with less than three exons. However, a version of this technique has been reported recently that targets the 3' exons of genes (Krizman and Berget 1993).

Magnetic bead capture of cDNA. Magnetic bead capture is based on the principle of solution hybridization of cDNAs to immobilized region-specific genomic DNA fragments. Originally, genomic DNA was immobilized on nylon filters (Lovett et al. 1991); this technique has subsequently been modified to take advantage of magnetic beads as the genomic fragment support system (Tagle et al. 1993). Briefly, cosmid DNA from the *BRCA1* candidate interval was digested with a frequent-cutter restriction enzyme and ligated to linkers that contain the appropriate restriction site and unique sequence complementary to the biotinylated PCR primers being used for subsequent amplification. cDNAs from appropriate libraries (breast and ovary in this case) were also PCR-amplified using vector primers and hybridized at high stringency to the biotinylated genomic fragments. The biotinylated genomic-cDNA complexes were captured with streptavidin-coated paramagnetic beads (Dynal) and washed to remove unbound and nonspecific cDNAs. The captured cDNAs, enriched for sequences from the chromosomal region the cosmids are derived from, were eluted by boiling and PCR-amplified with vector primers. Following purification and size selection of fragments greater than 200 bp in length, cDNAs were subcloned as a region-specific sublibrary. Larger cDNAs were analyzed to determine if they derive from the original cloned genomic DNA or contain repetitive and/or vector sequences.

We have successfully utilized this technique to isolate 23 clones that derive from the *BRCA1* candidate interval. In this case, small pools of 7–10 overlapping cosmids served as the genomic DNA template. Of those tested, 72% of cDNAs have been shown to map back to the original genomic clones, 12% contained only repetitive sequences, 6% were cosmid vector fragments, and 10% did not map back, suggesting that these clones originated in other genomic regions and were captured nonspecifically. As both genomic and cDNA clones are PCR-amplified, a singular advantage of this technique is that only small quantities of starting materials are required. An additional advantage of this approach is the speed with which large DNA segments can be screened for the presence of transcribed genes. In general, the technique is highly efficient and may be applied to several sources of cDNA simultaneously or in parallel reactions. When random-primed cDNA libraries are used as starting material, cDNA fragments obtained are randomly distributed over each mRNA, aiding in rapid generation of full-length cDNA clones. Furthermore, the method may normalize transcript levels, with enrichment of rare messages.

Several limitations are associated with this method. First, the small size of the captured subclones necessitates rescreening of cDNA libraries to obtain full-length cDNA. Second, genomic clones that contain expressed repeat sequences can result in capture of a large number of unrelated cDNAs containing these repeats. Once identified, these repetitive sequences can be used to prescreen the sublibrary of captured clones or can be added as a blocking agent to any future capture experiments in the region. Third, because the technique is based on hybridization, pseudogenes and multigene family members may be enriched. These clones can be excluded at an early stage by the more stringent hybridization conditions associated with mapping the cDNA back to the original parent genomic clones.

Transcript Characterization

All cDNAs isolated by the methods described above are being subjected to intensive characterization in order to identify germ-line mutations in affected individuals from families linked to *BRCA1* that would identify the candidate gene being analyzed as *BRCA1*. Once clones have been verified as having been derived from the *BRCA1* candidate interval and representing single-copy sequences, cDNAs are arrayed in microtiter plates and spotted onto nylon membranes. Membranes are then probed with other transcribed sequence clones. This approach allows determination of overlap between clones and, therefore, identification of sister clones. In addition, all clones are sequenced. Sequence data are being used to search nucleic acid and protein databases using the programs BLASTN and BLASTX to verify and characterize overlap with other cDNAs. Although many cDNA sequences appear to be unique, these searches have also identified genes which either had not been localized or were mapped previously to proximal 17q, or which appeared to be homologs of genes in other species. The cDNAs are then used as probes to (1) screen Southern blots containing DNA from individuals from *BRCA1*-linked families; (2) screen Southern blots containing DNA from 200 sporadic breast tumors (provided by Y. Nakamura); (3) screen Northern blots containing RNA from a variety of tissues to determine transcript size and tissue expression pattern; and (4) rescreen cDNA libraries in order to isolate additional segments of partially cloned genes. As we approach the isolation of a full-length cDNA, PCR primers are designed for single-strand conformational polymorphism (SSCP) analysis (Orita et al. 1989). RNA prepared from immortalized lymphocytes of affected members of *BRCA1*-linked families is used as substrate to detect germ-line mutations in candidate genes. To date, we have not utilized direct sequencing of candidate genes from putative carriers of *BRCA1* germ-line mutations. If necessary, this approach will be used to establish the identity of *BRCA1*.

SUMMARY

The high incidence of breast cancer and/or ovarian cancer in some families appears to be due to germ-line mutations in *BRCA1*. Genetic analysis of such families suggests that the *BRCA1* candidate region lies between D17S857 and D17S78 on chromosome 17q21(Kelsell et al. 1993; Simard et al. 1993). To identify and isolate *BRCA1*, we have used linkage and meiotic recombination analysis, characterized regions displaying LOH in tumor DNA from *BRCA1*-linked families, performed YAC and cosmid clone isolation and ordering, and used three complementary transcript-searching strategies. We have identified as many as 28 genes from the *BRCA1* candidate region, and we are searching for constitutive mutations in these candidate genes by several methods in an attempt to identify *BRCA1*.

REFERENCES

Buckler, A.J., D.D. Chang, S.L. Graw, J.D. Brook, D.A. Haber, P.A. Sharp, and D.E. Housman. 1991. Exon amplification: A strategy to isolate mammalian genes based on RNA splicing. *Proc. Natl. Acad. Sci.* **88:** 4005.

Buckler, A.J., C.J. Statler, J.L. Rutter, J. Murrell, J.A. Trofatter, M.K. McCormick, L. Deaven, R.K. Mayzis, and J.F. Gusella. 1993. Chromosome specific libraries of exons and high resolution transcription maps. *Am. J. Hum. Genet.* **53:** A192.

Chamberlain, J.S., M. Boehnke, T.S. Frank, S. Kiousis, J. Xu, S-W. Guo, E.R. Hauser, R.A. Norum, E.A. Helmbold, D.S. Markel, S.M. Keshavarzi, C.E. Jackson, K. Calzone, J.E. Garber, F.S. Collins, and B.L. Weber. 1993. *BRCA1* maps proximal to D17S579 on chromosome 17q21 by genetic analysis. *Am. J. Hum. Genet.* **52:** 792.

Colditz, G.A., W.C. Willett, D.J. Hunter, M.J. Stampfer, J.E. Manson, C.H. Hennekens, and B.A. Rosner. 1993. Family history, age and risk of breast cancer. *J. Am. Med. Assoc.* **270:** 338.

Couch, F.J., S. Kousis, L.H. Castilla, J. Xu, J.S. Chamberlain, F.S. Collins, and B.L. Weber. 1994. Characterization of ten new polymorphic dinucleotide repeats and generation of a high-density microsatellite-based physical map of the *BRCA1* region of chromosome 17q21. *Genomics* **243:** 419.

Easton, D.F., D.T. Bishop, D. Ford, G.P. Crockford, and the Breast Cancer Linkage Consortium. 1993. Genetic linkage analysis in familial breast and ovarian cancer—Results from 214 families. *Am. J. Hum. Genet.* **52:** 678.

Ford, D., D.F. Easton, D.T. Bishop, S.A. Narod, and D.E. Goldgar. 1994. Risks of cancer in *BRCA1* mutation carriers. *Lancet* **343:** 692.

Hall, J.M., M.K. Lee, B. Newman, J.E. Morrow, L.A. Anderson, B. Huey, and M.-C. King. 1990. Linkage of early onset breast cancer to chromosome 17q21. *Science* **250:** 1684.

Kelsell, D.P., D.M. Black, D.T. Bishop, and N.K. Spurr. 1993. Genetic analysis of the BRCA1 region in a large breast/ovarian family: Refinement of the minimal region containing *BRCA1*. *Hum. Mol. Genet.* **2:** 1823.

Knudson, A.G. 1971. Mutation and cancer: Statistical study of retinoblastoma. *Proc. Natl. Acad. Sci.* **68:** 820.

Krizman, D.B. and S.M. Berget. 1993. 3'-Terminal exon trapping: Identification of genes from vertebrate DNA. *Focus* **15:** 106.

Lovett, M., J. Kere, and L.M. Hinton. 1991. Direct selection: A method for the isolation of cDNAs encoded by large genomic regions. *Proc. Natl. Acad. Sci.* **88:** 9628.

Narod, S.A., J. Feuteun, H.T. Lynch, P. Watson, T. Conway, J. Lynch, and G.M. Lenoir. 1991. Familial breast-ovarian cancer locus on chromosome 17q12-23. *Lancet* **338:** 82.

North, M.A., P. Sanseau, A.J. Bucker, D. Church, A. Jackson, K. Patel, J. Trowsdale, and H. Lehrach. 1993. Efficiency and specificity of gene isolation by exon amplification. *Mamm. Genome* **4:** 466.

Orita, M., Y. Suziki, T. Sekiya, and K. Hayashi. 1989. Rapid and sensitive detection of point mutations and DNA polymorphisms using the polymerase chain reaction. *Genomics* **5:** 874.

Simard, J., J. Feuteun, G. Lenoir, P. Tonin, T. Normand, V. Luu The, A. Vivier, D. Lasko, K. Morgan, G.A. Rouleau, H. Lynch, F. Labrie, and S.A. Narod. 1993. Genetic mapping of the breast-ovarian cancer syndrome to a small interval on chromosome 17q12-21: Exclusion of candidate genes EDH17B2 and RARA. *Hum. Mol. Genet.* **2:** 1193.

Slattery, M.L. and R.A. Kerber. 1993. A comprehensive evaluation of family history and breast cancer risk: The Utah Population Database. *J. Am. Med. Assoc.* **270:** 1563.

Smith, S.A., D.F. Easton, D.G.R. Evans, and B.A.J. Ponder. 1992. Allele losses in the region 17q12-21 in familial breast and ovarian cancer involve the wild-type chromosome. *Nat. Genet.* **2:** 128.

Tagle, D.A., A. Swaroop, M. Lovett, and F.S. Collins. 1993. Magnetic bead capture of expressed sequences within large genomic segments. *Nature* **161:** 751.

Trent, J.M., B.L. Weber, X.Y. Guan, J. Zhang, F.S. Collins, K.A. Abel, A. Diamond, and P. Meltzer. 1994. Microdissection and microcloning of chromosomal alterations in breast cancer. *Breast Cancer Res. Treat.* (in press).

Wallace, M.R., J.W. Fountain, A.M. Brereton, and F.S. Collins. 1989. Direct construction of a chromosome-specific NotI linking library from flow-sorted chromosomes. *Nucleic Acids Res.* **17:** 1665.

RNA Genetics of Breast Cancer: Maspin as Paradigm

R. Sager,* S. Sheng,* A. Anisowicz,* G. Sotiropoulou,*
Z. Zou,* G. Stenman,† K. Swisshelm,‡ Z. Chen,*
M.J.C. Hendrix,§ P. Pemberton,** K. Rafidi,* and K. Ryan*

*Cancer Genetics, Dana Farber Cancer Institute, Boston, Massachusetts 02115; †Laboratory of Cancer Genetics, Department of Pathology, University of Goteborg, Sahlgrenska Hospital, S-413 45 Goteborg, Sweden; ‡University of Washington, Department of Pathology, Seattle, Washington 98195; §Pediatric Research Institute, Cardinal Glennon Children's Hospital, St. Louis University School of Medicine, St. Louis, Missouri 63110; **LXR Biotechnology, Inc., Richmond, California 94804

The strong influence of fashions in scientific thought continues to play an inhibitory role in scientific progress, even in the molecular genetics of cancer, the subject of this Symposium. As an example, we discuss here genes that are altered in expression in cancer cells compared with normal, but are not themselves mutated. As we show here, such genes are plentiful in cancer. Thus, they merit a sharpened focus: to wit, RNA genetics.

From a historical perspective, the genetic basis of cancer was still being hotly debated when I (R.S.) applied for a grant entitled "Genetic Analysis of Tumorigenesis" with a start date in 1978. It was necessary to provide a detailed justification for research on this theme including the clonal origin of cancer, correlations between cancer and chromosome aberrations, the tumorigenic effects of radiation and chemical mutagens, the integration of viral nucleic acids into the genome of virally transformed cells, the high predisposition to cancer in heritable diseases associated with defective DNA repair, and the concept that the multistep mutation hypothesis underlies the clinical time course of the disease.

Even more difficult to fund was a grant application which in 1985 explicitly proposed to screen for tumor suppressor genes, using the newly available methods of subtractive hybridization and differential cloning. This approach was roundly rejected, but fortunately, my research was rescued by the award of an Outstanding Investigator Grant for 7 years. This award supported the development of a human breast cancer experimental system with which my laboratory has been investigating tumor suppressor genes in the context of breast cancer. The existence of tumor suppressor genes is now rather widely accepted, but the list of mutated genes in this category remains slim in comparison to the hundred-odd oncogenes currently recognized. The paucity of tumor suppressor genes in the literature to date results in part from the difficulty of designing functional assays for the normal phenotype, i.e., negative selection at the cellular level.

My laboratory has taken a molecular approach to the identification and cloning of tumor suppressor genes: *comparative expression cloning*. This method is based on expression at the mRNA level of tumor suppressor genes from normal progenitor cells compared with decreased or no expression of the same genes in tumor cells recovered from surgical specimens, using subtractive hybridization (Sager 1989; Lee et al. 1991, 1992a,b) and differential display (Liang and Pardee 1992; Liang et al. 1992, 1993). This methodology, although not all-encompassing, has provided us with more than 40 candidate tumor suppressor genes. In the process of investigating these genes, we found to our surprise that they consist (to date) of down-regulated genes that are not themselves mutated.

The plan of this paper is (1) to summarize our findings which focus on the identification of genes that are expressed in normal mammary epithelial cells but not in mammary carcinomas and to show how this focus has led to the identification of a large number of candidate tumor suppressor genes, approaching the number of known oncogenes; (2) to describe as an example of a down-regulated gene, discovered in our laboratory by subtractive hybridization, *maspin*, which encodes a novel serine protease inhibitor and functions as a tumor suppressor gene; and (3) to consider the implications of the fact that so many genes are simultaneously down-regulated but not mutated in cancer cells.

DOWN-REGULATED GENES IN BREAST CANCER

In this section, a summary is provided of selected genes that have been identified in our laboratory by methods of differential gene cloning: either by subtractive hybridization (Lee et al. 1991, 1992a,b) or by differential display (DD) (Liang and Pardee 1992; Liang et al. 1992; Sager et al. 1993). This list includes previously known genes, new members of known gene families, and novel genes. Although the functions of novel genes are unknown, the presumed activities of new family members can often be inferred.

Initial differential cloning has been carried out with human breast epithelial cells, both normal and tumor-derived (Band and Sager 1989; Band et al. 1989, 1990). The normal cells grown in culture are derived from reduction mammoplasty specimens and represent, in our estimation, epithelial stem cells with traits typical of both basal cells and luminal cells (Trask et al. 1990). Basal cells lie in a discontinuous layer under the basement membrane that separates the epithelium from the stroma; these cells contain myofibrils that contract to move the milk along the ducts; they rarely if ever form tumors. Luminal cells line the ducts, produce milk during lactation, and are the progenitors of tumor cells. The tumor cells used for differential cloning come from cell lines established in our laboratory or from the American Type Culture Collection. (MCF-7 is from the Michigan Cancer Foundation.)

Growth of cells to be used for differential cloning requires precautions to guard against differential expression resulting from different growth rates or conditions of growth other than those under investigation. Our normal and tumor cell populations to be compared are grown in the same medium (Band et al. 1989) at similar growth rates and harvested for RNA at close to 70% confluence.

In subtractive hybridization, subtracted cDNAs recovered in the experiments are sometimes full length, but often require further cloning from cDNA libraries to recover the full-length molecule. In DD, only short sequences of 200–600 bp are recovered, and further cloning is required (Liang and Pardee 1992; Liang et al. 1993). Before recovering the full-length cDNAs, however, we utilize the partial cDNAs as probes in Northern analysis against a bank of RNAs from other carcinoma cell lines to verify the differential expression patterns.

A composite summary of results with a subset of 26 genes is given in Figure 1. These genes were identified by differential expression in normal versus tumor cells and examined by Northern analysis using total RNAs isolated from a standard group of test cells: 3 normal strains from mammoplasty patients (70, 76, and 81) on the left, 10 estrogen receptor (ER)-negative lines in the middle of the chart (21NT, PT, MT-1, MT-2, MDA-MB-157, -231, -435, -436, -468, BT549), and 5 ER-positive mammary carcinoma cell lines (MCF-7, ZR-75-1, T47D, MDA-MB-361, and MDA-MB-474) on the right. Since the Northern blots, which were each done more than once, differed in time of exposure from gene to gene, comparisons of expression levels are valid within each gene, but only approximate between genes.

To answer the question of how well cells in culture mimic their expression in vivo, we have carried out immunostaining of tissues. To date, we have examined five of the genes listed in Figure 1 by immunostaining: maspin, connexin 26, connexin 43, α6 integrin, and

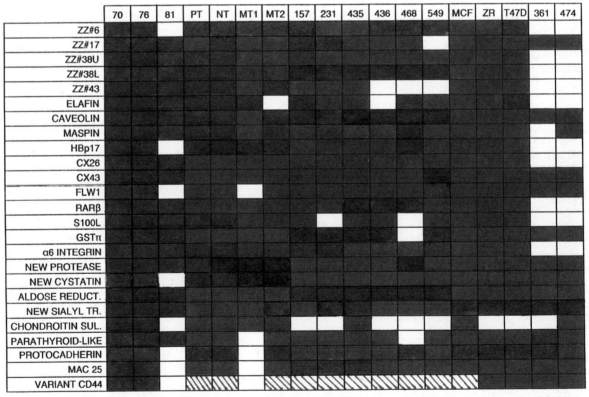

Figure 1. Composite summary of mRNA expression (Northern blots) of 26 genes in each of 18 cell populations. Total RNA was extracted from cells at ~70% confluence, prepared and hybridized (Lee et al. 1992b) using ^{32}P-labeled cDNA probes. (Green) Expression, not quantitative. (Red) Negative, no expression. (White) Not done. (Cross hatch) CD44 standard pattern (green is alternatively spliced isoform). Cx43 on chr. 6; Cx26 on chr. 13; SD100L on chr. 1q. ZZ#6 on chr. 15; ZZ#17 on chr. 10; ZZ#38 on chr. 11; ZZ#43 on chr. 7; maspin on chr. 18q; elafin on chr. 20; caveolin on chr. 7. Mapped by G. Stenman and E. Roijer (unpubl.).

tissue factor. In these examples, the presence of the protein decreases during progression in the primary tumors and is absent in metastases. The data are still preliminary, in terms of possible diagnostic or prognostic applications, but adequate to confirm the differential expression data from cell culture.

The genes listed include known genes whose sequences were identified by computer comparison using AUTOSEARCH, as well as novel genes with sequences related to known genes, and unknown genes. In addition to novel genes we have already published (Lee et al. 1992a,b; Zou et al. 1994), they include a new protease related to trypsin, a new cystatin (protease inhibitor), a new sialyl transferase, a parathyroid-hormone-related gene, and a novel CD44 variant. References are cited in the text for genes published by ourselves or others; the remainder are unpublished. Unidentified genes are listed by code (ZZ or FLW). Chromosomal locations are given for those we have determined.

The most striking features of the grid are (1) the uniformity of expression in normal cell populations from different individuals, (left-most 3 lanes); (2) the uniformity of nonexpression among the ER^+ cell lines (right-most 5 lanes); and (3) the relatively less consistent expression patterns in the ER^- cell lines (lanes 4–13). Obvious exceptions to the overall picture are genes whose expression appears to be related to ER status. The new sialyl transferase and protocadherin genes are expressed by the ER^+ cell lines but largely not by the ER^- lines, suggesting positive regulation by estrogen. Aldose reductase, GSTπ, caveolin, and CD44 are expressed in several ER^- but not in ER^+ cell lines, suggesting negative regulation by the ER. (The growth medium [Band and Sager 1989] contains estradiol.)

The genes are grouped according to their likely functions: Cx43 and Cx26 encode gap junction proteins (Lee et al. 1992a; Tomasetto et al. 1993), S100L (formerly CaN19) (Lee et al. 1992b) and NB1 (Yaswen et al. 1990) encode calcium-binding proteins, the parathyroid-hormone-like gene and HBp17 (Wu et al. 1991) probably encode proteins that regulate growth, as does Mac25 (Murphy et al. 1991). GSTπ is a well-known enzyme with detoxifying activity against many lipophilic toxic agents (Moscow et al. 1989). Genes encoding proteins involved in adhesion, cytostructure, and the extracellular matrix include the novel sialyl transferase, protocadherin (Sano et al. 1993), caveolin (Rothberg et al. 1992), $\alpha6$ integrin (Sager et al. 1993), keratin 5 (Trask et al. 1990), and CD44 (Günthert 1993). Three protease inhibitors have been identified: maspin (Zou et al. 1994); elafin, a novel elastase inhibitor (Saheki et al. 1992); a novel cysteine protease inhibitor; and a novel protease. Other genes whose expression is lost in mammary carcinoma cell lines include tissue factor (Mackman et al. 1989), aldose reductase (Graham et al. 1989), RARβ (Swisshelm et al. 1994), and as yet unidentified genes of the ZZ and FLW series.

These genes represent a subset of the 40-odd genes we have identified, but the list is open-ended. Other genes not shown in Figure 1 include fibronectin (Sager 1989), GRO (Anisowicz et al. 1988), thrombospondin (Zajchowski et al. 1990), spr-1 (DNA damage inducible), and amphiregulin. The known functions of these 40-odd genes themselves indicate that many other genes with related functions are yet to be found. Nonetheless, the list is already impressive for the diversity of activities that are represented.

As differentially expressed genes were recovered and cloned, they were checked by restriction fragment analysis in Southern blots to look for changes at the DNA level. To date, the following genes have been examined in Southerns after digestion of DNA of normal controls and several tumor cell lines by two or more restriction enzymes: maspin, Cx26, Cx43, elafin, S100L (CaN19), GRO, $\alpha6$ integrin, new protease, new cystatin, new sialyl transferase, and parathyroid hormone-like. A few polymorphisms were identified in single cell lines, but no evidence was found of gross rearrangements or deletions. None of our genes has yet been found to show evidence of changes at the genome level by this method. Further evidence that the genes described here are down-regulated rather than mutated comes from studies showing reexpression induced by azadeoxycytidine (Lee et al. 1992a; R. Sager, unpubl.) or PMA (Lee et al. 1992b; R. Sager, unpubl.).

Thus, the results reported here open a window on the magnitude of phenotypic changes in cancer cells that result from the simultaneous down-regulation of so many genes. These results raise the possibility that a large fraction of genes with tumor-suppressing activity discovered by expression cloning might not be mutated genes, but rather genes whose expression is modulated at the transcription level. These genes, which we refer to as Class II tumor suppressor genes (TSGs) (Lee et al. 1991), may play a decisive role in cancer through the phenotypic changes they determine.

MASPIN, A NOVEL TUMOR SUPPRESSOR GENE

Discovery of Maspin

Maspin was recovered by subtractive hybridization as one of a group of novel candidate TSGs expressed in normal human mammary epithelial cells at the mRNA level, but not expressed in the mammary carcinoma cells used for comparison (Zou et al. 1994). The generality of this expression pattern was shown by Northern analysis with a number of normal and breast-carcinoma-derived cell lines. A typical Northern pattern, using cDNA (open reading frame) as probe, is shown in Figure 2. Maspin is expressed in three normal strains isolated from three patients undergoing cosmetic surgery (mammoplasty) and weakly in two primary tumor lines, but not in any of the cell lines of metastatic origin. This result is in accord with in vivo studies discussed below.

Southern analyses of restriction fragment patterns (*Xba*I in Fig. 3, *Eco*RI, *Bam*HI, *Pst*I, unpubl.) show

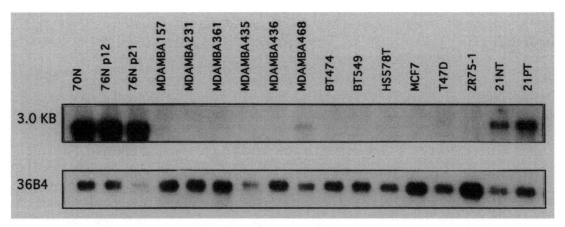

Figure 2. mRNA expression of maspin in normal and tumor-derived mammary epithelial cells. Total cellular RNA was isolated from exponentially growing cells cultured in DFCI-1 medium. Total RNA (20 mg of RNA per lane) was subjected to electrophoresis on a 1% formaldehyde-agarose gel, transferred to nylon membrane, and hybridized with a ^{32}P-labeled maspin probe. Lanes labeled 70N and 76N contain RNA from normal breast epithelial cells; all other lanes contain RNA from breast tumor cells. 21NT and 21PT are primary tumor lines; 36B4 is a loading control.

that no major changes in maspin DNA have occurred in the tumor cell lines compared with normal cells. This result is in accord with later studies showing that the gene can be reexpressed in tumor cells treated with phorbol-12-myristate-13-acetate (PMA). Thus, loss of expression in these cell lines is a consequence of regulatory changes, not of genomic changes in maspin DNA. Maspin is thus a Class II gene (Lee et al. 1991). These results do not preclude the possibility that genomic changes in maspin may also arise in some tumors. Studies of maspin regulation to determine the mechanism of differential regulation between normal and tumor cells are in progress.

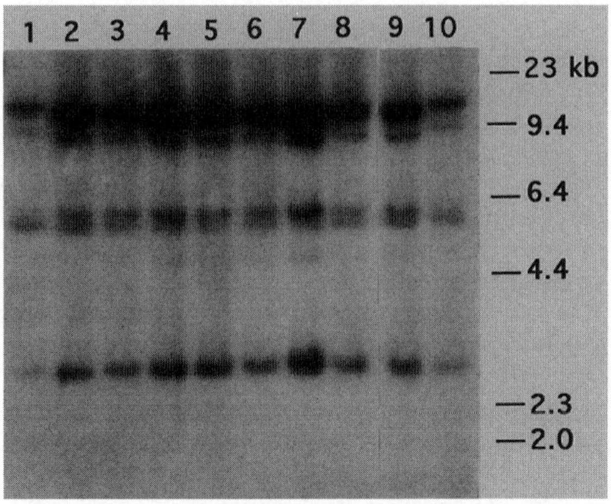

Sequence Comparison Identifies Maspin as a Serpin

Comparisons of the cDNA and predicted amino acid sequences of maspin with its closest related proteins were carried out with the AUTOSEARCH program, and plotted by the PRETTY PLOT program as shown in Figure 4. The comparison places maspin in the family of serpins, i.e., serine protease inhibitors, of which α_1 protease inhibitor (α_1 PI) is the prototype (Huber and Carrell 1989). Overall amino acid similarities (including conservative changes) compared to maspin are about 40% for PAI-2, neutrophil elastase inhibitor, and ovalbumin, and about 30% for α_1 PI and PAI-1. Most of the amino acid substitutions do not alter the overall serpin conformation. However, significant amino acid differences are located in the hinge region, a peptide stretch located 9–15 residues amino-terminal to the P1-P'1 peptide bond at the reaction center.

The reaction center is indicated in Figure 4, with the cleavage site marked by the arrow. In a typical reaction between a serine protease inhibitor and its protease target, the protease binds at the reaction center, the peptide bond between residues P1 and P' is cleaved, and the resulting 1:1 complex is relatively stable, thereby inactivating the protease. Until recently, it had been postulated (Carrell and Evans 1992) that the ability of the hinge region to fold over and insert into β-sheet A was essential for inhibitor activity. In maspin, the amino acid substitutions in this region appear to preclude this conformational shift. Now, however, the solution of the crystallographic structure of α_1-antichymotrypsin has revealed that this insertion of the hinge region is not required for full inhibitor activity (Wei et al. 1994). Thus, current understandings of serpin structure do not

Figure 3. Southern blot analysis of maspin gene. DNA (20 mg) was digested with *Xba*I, fractionated on 1% agarose gel, and transferred to nylon membrane. The blot was hybridized with ^{32}P-labeled maspin cDNA. (Lanes *1,2*) Normal breast 70N and 76N; (*3–10*) breast tumor cell lines 21MT1, 21MT2, 21NT, 21PT, MDA-MB-231, MDA-MB-435, MCF7, and ZR75-1.

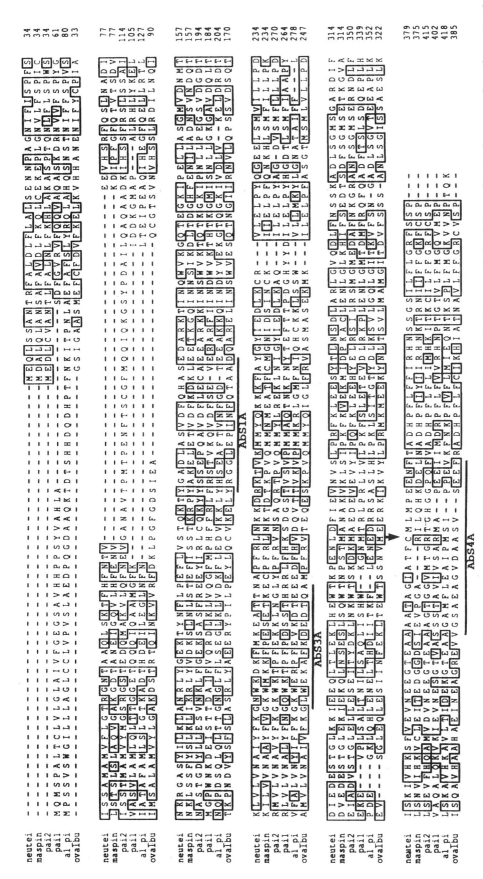

Figure 4. Comparison of maspin with other serpins. Identical residues are boxed. Three regions used for antibody production are underlined. Arrow denotes the proposed reactive center of maspin. (neutei) Human monocyte/neutrophil elastase inhibitor; (pai2) human plasminogen activator inhibitor type 2; (pai1) human plasminogen activator inhibitor type 1; (a1-pi) α₁-protease inhibitor; (ovalbu) ovalbumin.

exclude the possibility that maspin may have protease inhibitor activity.

Chromosome Localization of Maspin

A panel of 24 human-rodent somatic cell hybrids (obtained from NIGMS as mapping panel 2) was used to map maspin. All hybrids retained a single human chromosome except one line that contained both chromosome 20 and a low percentage of chromosome 4, and one that contained both chromosomes 1 and X. In human DNA, the maspin probe detected a major 5.4-kb HindIII fragment, which was clearly resolved from a weakly hybridizing Chinese hamster fragment of about 20 kb. The presence of 5.4-kb human maspin sequence in the hybrid clones correlated only with the presence of human chromosome 18 (data not shown). Only 1 of the 24 hybrids analyzed was positive for maspin; this hybrid contained chromosome 18 as the sole human DNA.

The maspin gene was regionally localized by analysis of a panel of human hamster somatic cell hybrids retaining defined regions of chromosome 18 (obtained from NIGMS; Markie et al. 1992). The chromosome 18 content of the six hybrid lines is shown in Figure 5. This panel allows assignment of genes to at least five different intervals on chromosome 18. Maspin was localized to an interval between 18q21.3 and 18q23 defined by the breakpoints in hybrids GM12086 and GM12085, respectively (Fig. 5). This localization places maspin within the same region as *PLANH2*, a closely related gene that encodes plasminogen activator inhibitor-2, the tumor suppressor gene *DCC*, and the *BCL2* gene commonly rearranged in the t(14;18)(q32;q21) in non-Hodgkin's lymphoma (LeBeau and Geurts van Kessel 1991; Markie et al. 1992; LeBeau et al. 1993). Markie et al. (1992) have mapped D18S8, a sequence known to reside within the *DCC* gene, proximal to the breakpoint in hybrid GM12084A, which places *DCC* proximal to maspin. Previous studies have also shown that *DCC* is proximal to *BCL2* (Markie et al. 1992; LeBeau et al. 1993). Since the translocation breakpoint in GM12086 is known to occur within the *BCL2* gene, and this hybrid is negative for maspin, maspin must be distal to *BCL2* (Kluin-Nelemans et al. 1991; Markie et al. 1992). The relative order of maspin and *PLANH2* is not known. The deduced gene order within this region is thus 18cen-*DCC-BCL2-(PLANH2, MASPIN)*-18qter.

Anti-maspin Antibodies

Polyclonal antibodies were produced against synthetic polypeptides representing three regions of low sequence homology, identified in Figure 4 as S2A, S3A, and S4A (Zou et al. 1994). All three antibodies reacted with proteins in fixed normal cells grown in culture, and in normal cells in tissue sections. However, S4A immunostaining was located in the cell membrane, whereas the other two were also cytoplasmic. All subsequent immunostaining studies were carried out with S4A. Since the peptide used for S4A production corresponds to the reactive center of the protein, which is the region of minimal sequence similarity with other serpins, it seems likely that S2A and S3A may recognize related serpins in the cytoplasm, and that maspin is uniquely located in the membrane of normal cells.

Immunostaining of Tissue Sections

As reported previously (Zou et al. 1994), immunostaining of normal breast and carcinomas from surgical specimens clearly demonstrated the progressive loss of expression of maspin during progression of primary tumor cells from ductal carcinoma in situ to invasive carcinoma. Maspin was rarely seen in invasive cells, and its expression was absent in lymph node and distant metastases. Thus, the presence of maspin correlates with normality and premalignant lesions. Further studies are in progress to expand the study and, in particular, to determine the value of maspin expression as a prognostic marker in clinical practice.

Properties of Tumor Transfectants Expressing Maspin

Transfectants were produced using a CMV-driven vector containing maspin and neomycin cDNAs as well as control vectors containing neomycin but not maspin

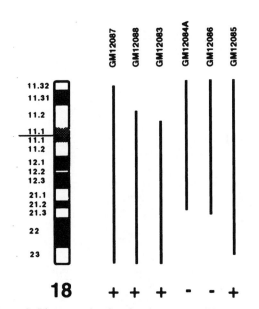

Figure 5. Ideogram showing the chromosome 18 content of the somatic cell hybrid panel (obtained from NIGMS) used for sublocalization of maspin. The breakpoints in hybrids GM12088 and GM12083 are only approximate (Markie et al. 1992). Southern blot analysis of DNAs from the six hybrids (Hagiwara et al. 1991) showed that the presence of the 5.4-kb HindIII fragment specific for maspin correlated with the presence of the region 18q21.2-q23 defined by the breakpoints in hybrids GM12086 and GM12085. The presence (+) or absence (−) of maspin sequences in the hybrid lines is indicated below each hybrid.

(Zou et al. 1994). The results of the first transfection demonstrated that maspin expression was responsible for a partial inhibition of invasive ability of tumor cells through Matrigel, a reconstituted basement membrane. This inhibition was reversed by anti-maspin antibody. In preliminary nude mouse assays, transfectants were shown to be inhibited in tumor growth and metastasis (Zou et al. 1994). These dramatic results provided the first functional evidence for the tumor suppressor role of maspin.

More recently, a new series of transfectants has been selected and characterized (S. Sheng, unpubl.). Of the 48 double transfectants initially selected, 17 expressed maspin protein as identified on Northern and Western blots. A subset of these transfectants has been examined for their invasive ability in the same Matrigel system as was previously used (Zou et al. 1994). The results shown in Figure 6 demonstrate that all 10 of the new transfectants exhibit strong inhibition of invasion, compared with the parental tumor cell line and with two neomycin transfectant controls. No new transfectants with aberrant properties have been detected. Additionally, these transfectants are stabler in the absence of G418 selection than were the first set, which is important for animal testing, which is carried out in the absence of G418.

Production and Properties of Recombinant Maspin

Recombinant maspin was produced in three organisms: in *Escherichia coli* as a GST-maspin fusion protein, in baculovirus-infected Sf9 cells (Sheng et al. 1994), and in yeast (P. Pemberton, unpubl.). The fusion protein, rGST-Maspin, accounted for 30–40% of the total protein, and after sonication, 10–20% of the protein was recovered in the lysis buffer. The protein was purified to almost 100% homogeneity on a glutathione affinity column. The identity of the fusion protein was shown by Western blot analysis using the AbS4A polyclonal antibody. rMaspin(i), produced in insect cells, was purified by heparin affinity chromatography. The protein bound to a heparin column and was eluted with 0.10 M sodium chloride. The purified protein had a molecular mass of approximately 42 kD, in good agreement with its predicted amino acid sequence.

The recombinant maspins were cleaved at their predicted cleavage sites by limiting amounts of trypsin to give two fragments: 38 kD and 4.2 kD. Although the large fragment was easily observed, the expected small fragment was not detectable on 10% SDS-PAGE gels. Further analysis was carried out by mass spectroscopy, where a peak with the expected molecular mass of 4.2 kD was found. Sequence analysis of the 11 amino-terminal amino acids matched precisely the predicted sequence of the region located carboxy-terminal to the P1 arginine. Thus, the identity of the recombinant maspin was confirmed, as was the susceptibility of the reactive site loop to proteolysis.

Recombinant Maspin in Invasion Studies

We showed previously that tumor transfectants producing maspin were effective in inhibiting invasion across reconstituted membranes, as compared with the parental tumor line and with neomycin transfectants not expressing maspin (Zou et al. 1994). These experiments were then repeated using the recombinant maspins, to examine their biological activity (Sheng et al. 1994).

As shown in Figure 7, all three recombinant maspins were active as inhibitors of invasion, assayed by treating two breast carcinoma cell lines: MDA-MB-435, the line used to produce transfectants, and MDA-MB-231. Both cell lines were inhibited when maspin was added to the cell culture medium. The fusion protein, rGST-Maspin, showed a linear dose response in the concentration range of 0.04–0.77 µM protein. The pure proteins, rMaspin(i) and rMaspin(y), however, showed maximal activity at 0.17 µM, and the effective activity decreased to about 1/3 of the peak activity at 1.11 µM. The explanation for these differences in dose response may lie in the ability of maspin to polymerize as the concentration is increased, a property that is apparently not expressed by the fusion protein.

Search for Protease Target of Maspin Activity

The postulated anti-protease activity of maspin is based on its structural similarity to other protease inhibitors in the serpin family, as well as on its functional activity in blocking invasion, motility, and metastasis in cell culture and in animal model systems. To look for the target of maspin, recombinant maspin from

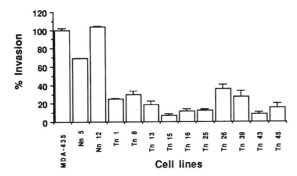

Figure 6. Effects of maspin transfection on invasion potential of MDA-MB-435 cells. The in vitro invasion assays using MICS chambers were performed as described previously (Hendrix et al. 1987). A total of 50,000 cells from each cell line was seeded on 10 mm polycarbonate membranes coated with 4 mg/ml growth-factor-reduced Matrigel in DFCI-1 medium. After incubation at 37°C with 6.5% CO_2 in a humidified incubator for 48 hr, the cells that had penetrated through the Matrigel barrier were collected and counted. The invasion data from parental MDA-MB-435 cells were normalized as 100%, and the invasion data from neo transfectants (Nn5 and Nn12) and maspin transfectants (Tn1, Tn8, n13, Tn15, Tn16, Tn25, Tn26, Tn39, Tn43, and Tn45) are expressed as a percentage of this control. The error bars represent the standard error of the mean of three parallel experiments.

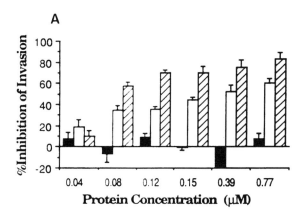

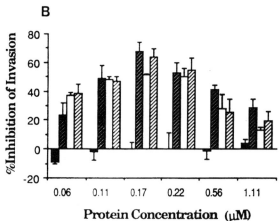

Figure 7. The effects of rGST-Maspin (*A*) and rMaspin (*B*) on invasion by two tumor cell lines, MDA-435 and MDA-231. The invasion data from untreated tumor cells were normalized as 100%, and invasion data from treated tumor cells are expressed as a percentage of this control. Each value represents the average of triplicate results and the error bars represent the standard errors of the means. (*A*) (*Black bars*) Invasion by MDA-435 in the presence of recombinant glutathione-*S*-transferase; (*open bars*) invasion by MDA-435 in the presence of rGST Maspin; (*hatched bars*) invasion by MDA-231 in the presence of rGST-Maspin. (*B*) (*Black bars*) Invasion by MDA-435 in the presence of trypsin-cleaved rMaspin (y); (*darker hatched bars*) invasion by MDA-435 in the presence of rMaspin(i); (*open bars*) invasion by MDA-435 in the presence of rMaspin (y); and (*lighter hatched bars*) invasion by MDA-231 in the presence of rMaspin(i). (Reprinted, with permission, from Sheng et al. 1994.)

yeast was tested for its ability to inhibit a series of proteases at various concentrations in PBS. In each assay, enzyme (Eo = 1 μM) was mixed with a 1- to 100-fold molar excess of maspin for periods of 5 minutes to 24 hours at room temperature prior to addition of substrate (0.1 mM final concentration). The candidate proteinases tested were trypsin, chymotrypsin, thrombin, elastase, plasmin, urokinase, and tissue plasminogen activator. The corresponding substrates were EGR-pNA (Glu-Gly-Arg-paranitroanilide) for trypsin, thrombin, urokinase, and tissue plasminogen activator; VLK-pNA for plasmin, MeOSucc-AAPF-pNA for elastase, and N-Succ-AAPF-pNA for chymotrypsin. Residual enzyme activities were measured, and the reaction mixtures were analyzed by SDS-PAGE to look for generation of maspin cleavage products (P. Pemberton, unpubl.).

In addition, the inhibitory effect of maspin on each protease was tested in the presence of the following potential ligands: chondroitin sulfate-A, -B, -C, hyaluronic acid, heparan sulfate, keratan sulfate, fibronectin, vitronectin, and Matrigel (reconstituted basement membrane). All ligands were 1 mg/ml final concentration.

The results were essentially negative. No inhibitory activity was detected under any of the specified conditions. However, proteolytic activity was shown by five of the proteinases, degrading maspin at ratios of 1:10 (E:I) for thrombin and elastase, 1:00 for trypsin and chymotrypsin, and 1:1000 for plasmin. Thus, the target of maspin has not yet been identified. Since maspin itself appears to be tissue-specific in its expression, the same may be true for its target protein. Experiments along these lines are in progress.

In summary, maspin is a new Class II tumor suppressor gene, down-regulated in breast cancer. Structurally, it is closely related to a number of serine protease inhibitors. Functionally, the recombinant protein, as well as tumor cells transfected with the maspin cDNA, are inhibitory in invasion assays in cell culture, and in preliminary nude mouse tests, transfectants show inhibition of growth and of metastatic spread. Thus, maspin has strong potential in therapeutic applications. In addition, immunostaining studies show promise for maspin as a tumor marker for diagnostic and prognostic tests.

POSING THE PROBLEM

We subdivide cancer genes, i.e., genes influencing the cancer phenotype, into those that act directly as a result of changes at the DNA level, called *Class I genes*; and the *Class II genes* which affect the phenotype by modulation at the expression level (Lee et al. 1991). A simple example is the paradigmal gene, β-galactosidase, whose expression can be changed either by gene mutation (Class I) or by altered regulation (Class II).

The importance of distinguishing between Class I and Class II genes became increasingly evident with the progress of work in our laboratory, as discussed above, in which so many genes were found to be down-regulated in human mammary carcinoma cells compared with closely related normal human mammary epithelial cells. Work of other laboratories substantially increases this number. Surely, these genes are not independently regulated in a 1:1 manner by more than 50 mutated genes. In contrast to proto-oncogenes, of which only a few are overexpressed in any one cancer cell, these tumor suppressors are all simultaneously down-regulated in the same cells.

The simplest mechanism of down-regulation is loss of expression of an upstream positive transcription factor; that factor may be encoded by a Class I tumor suppressor gene—*p53* and *WT1* are examples. Individual transcription factors may participate in many transcription complexes; the final activity depends on the presence of other members of the complex and suitable induction conditions. For some genes, the result will be coordinate regulation. A more complicated scenario would not involve a Class I gene, but rather a network of interacting transcription complexes analogous to a differentiation program. On either model, we ask, Can this coordinate down-regulation be coordinately reversed?

Our preliminary evidence of genes coordinately upregulated by treatment either with deoxyazacytidine or with PMA suggests that this approach is feasible. Deoxyazacytidine induces reexpression of S100L (formerly CaN19) (Lee et al. 1992b) and of several other mRNAs so far examined, demonstrating that as described by Achten et al. (1991) and others, changes in methylation of many genes occur in the tumorigenic process. The phorbol ester PMA induces transient expression of Cx26 in several tumor cell lines (Lee et al. 1992a) and may indicate a transcriptionally relevant pathway that operates via protein kinase C. Some of the remarkable reversals of the cancer phenotype that have been accomplished with retinoids (Sporn et al. 1994) or by transfection with integrins (Giancotti and Ruoslahti 1990) provide further paradigms.

Thus, exciting opportunities for therapy could result from a network of coordinated regulation in which reversing the down-regulation of one gene could also reexpress others in a coordinate manner, leading to a substantial normalization of the phenotype. Focusing on mechanisms of phenotypic reversion guided by reexpression of known down-regulated genes offers substantial hope for the success of nontoxic cancer chemotherapies.

ACKNOWLEDGMENTS

We thank Margarita Connolly for preparing the manuscript. This work was supported by the National Cancer Institute, the National Institute of Aging, and LXR Biotechnology.

REFERENCES

Achten, S., A. Behn-Krappa, M. Jücker, J. Sprengel, I. Holker, B. Schmitz, H. Tesch, V. Diehl, and W. Doerfler. 1991. Patterns of DNA methylation in selected human genes in different Hodgkin's lymphoma and leukemia cell lines and in normal human lymphocytes. *Cancer Res.* **51:** 3702.

Anisowicz, A., D. Zajchowski, G. Stenman, and R. Sager. 1988. Functional diversity of *gro* gene expression in human fibroblasts and mammary epithelial cells. *Proc. Natl. Acad. Sci.* **85:** 9645.

Band, V. and R. Sager. 1989. Distinctive traits of normal and tumor-derived human mammary epithelial cells expressed in a medium that supports long-term growth of both cell types. *Proc. Natl. Acad. Sci.* **86:** 1249.

Band, V., D. Zajchowski, G. Stenman, C.C. Morton, V. Kulesa, J. Connolly, and R. Sager. 1989. A newly established metastatic breast tumor cell line with integrated amplified copies of c-erbB-2 and double minute chromosomes. *Genes Chromosomes Cancer* **1:** 48.

Band, V., D. Zajchowski, K. Swisshelm, D. Trask, V. Kulesa, C. Cohen, J. Connolly, and R. Sager. 1990. Tumor progression in four mammary epithelial cell lines derived from the same patient. *Cancer Res.* **50:** 7351.

Carrell, R.W. and D.L.I. Evans. 1992. Serpins: Mobile conformations in a family of proteinase inhibitors. *Curr. Opin. Cell Biol.* **2:** 438.

Giancotti, F. and E. Ruoslahti. 1990. Elevated levels of fibronectin receptor suppress the transformed phenotype of CHO cells. *Cell* **60:** 849.

Graham, A., P.J. Hedge, S.J. Powell, J. Riley, L. Brown, A. Gammack, F. Carey, and A.F. Markham. 1989. Nucleotide sequence of cDNA for human aldose reductase. *Nucleic Acids Res.* **17:** 8368.

Günthert, U. 1993. CD44: A multitude of isoforms with diverse functions. *Curr. Top. Microbiol. Immunol.* **184:** 47.

Hagiwara, K., G. Stenman, H. Honda, P. Sahlin, A. Andersson, K. Miyazono, C.H. Heldin, F. Ishikawa, and F. Takaku. 1991. Organization and chromosomal localization of the human platelet-derived endothelial cell growth factor gene. *Mol. Cell. Biol.* **11:** 2125.

Hendrix, M.J.C., E.A. Seftor, R.E.B. Seftor, and I.J. Fidler. 1987. A simple quantitative assay for studying the invasive potential of high and low human metastatic variants. *Cancer Lett.* **38:** 137.

Huber, R. and R.W. Carrell. 1989. Implications of the three-dimensional structure of a1-antitrypsin for structure and function of serpins. *Biochemistry* **28:** 8951.

Kluin-Nelemans, H.C., A.J. Limpens, J. Meerabux, G.C. Beverstock, J.H. Jansen, D. de Jong, and P.M. Kluin. 1991. A new non-Hodgkin's B-cell line (DoHH2) with a chromosomal translocation t(14;18)(q32;q21). *Leukemia* **5:** 221.

LeBeau, M.M. and A.H.M. Geurts van Kessel. 1991. Report of the committee on the genetic constitution of chromosome 18. Human gene mapping 11: Eleventh international workshop on human mapping. *Cytogenet. Cell Genet.* **58:** 739.

LeBeau, M.M., J. Overhauser, R.E. Strab, G. Silverman, T.C. Gilliam, J. Ott, P. O'Conell, U. Francke, and A. Geurts van Kessel. 1993. Report of the first international workshop on human chromosome 18 mapping. *Cytogenet. Cell Genet.* **63:** 78.

Lee, S., C. Tomasetto, and R. Sager. 1991. Positive selection of candidate tumor suppressor genes by subtractive hybridization. *Proc. Natl. Acad. Sci.* **88:** 2825.

Lee, S.W., C. Tomasetto, D. Paul, K. Keyomarsi, and R. Sager. 1992a. Transcriptional down-regulation of gap junction proteins blocks junctional communication in human mammary tumor cell lines. *J. Cell Biol.* **118:** 1213.

Lee, S., C. Tomasetto, K. Swisshelm, K. Keyomarsi, and R. Sager. 1992b. Down regulation of a new member of the S100 gene family in mammary carcinoma cells and reexpression by azadeoxycytidine treatment. *Proc. Natl. Acad. Sci.* **89:** 2504.

Liang, P. and A.B. Pardee. 1992. Differential display of eukaryotic messenger RNA by means of the polymerase chain reaction. *Science* **257:** 967.

Liang, P., L. Averboukh, and A.B. Pardee. 1993. Distribution and cloning of eukaryotic mRNAs by means of differential display: Refinements and optimization. *Nucleic Acids Res.* **21:** 3269.

Liang, P., L. Averboukh, K. Keyomarsi, and R. Sager. 1992. Differential display and cloning of mRNAs from human breast cancer versus mammary epithelial cells. *Cancer Res.* **52:** 6966.

Mackman, N., J.H. Morrissey, B. Fowler, and T.S. Edgington. 1989. Complete sequence of the human tissue factor gene, a highly regulated cellular receptor that initiates the coagulation protease cascade. *Biochemistry* **28:** 1755.

Markie, D., T.A. Jones, D. Sheer, and W.F. Bodmer. 1992. A somatic cell hybrid panel for regional mapping of human chromosome 18. *Genomics* **14:** 431.

Moscow, J.A., C.R. Fairchild, M.J. Madden, D.T. Ransom, H.S. Wieand, E.E. O'Brien, D.G. Poplack, J. Cossman,

C.E. Myers, and K.H. Cowan. 1989. Expression of anionic glutathione-S-transferase and P-glycoprotein genes in human tissue and tumors. *Cancer Res.* **49:** 1422.

Murphy, M., M.J. Pykett, P. Harnish, K.D. Zang, and D.L. George. 1991. Identification and characterization of genes differentially expressed in meningiomas. *Cell Growth Differ.* **4:** 715

Rothberg, K.G., J.E. Heuser, W.C. Donzell, Y.-S. Ying, J.R. Glenney, and R.G.W. Anderson. 1992. Caveolin, a protein component of caveolae membrane coats. *Cell* **68:** 673.

Sager, R. 1989. Tumor suppressor genes: The puzzle and the promise. *Science* **246:** 1406.

Sager, R., A. Anisowicz, P. Liang, and G. Sotiropoulou. 1993. Identification by differential display of alpha 6 integrin as a candidate tumor suppressor gene. *FASEB J.* **7:** 964.

Saheki, T., F. Ito, H. Hagiwara, Y. Saito, J. Kuroki, S. Tachibana, and S. Hirose. 1992. Primary structure of the human elafin precursor preproelafin deduced from the nucleotide sequence of its gene and the presence of unique repetitive sequences in the prosegment. *Biochem. Biophys. Res. Commun.* **185:** 240.

Sano, K., H. Tanihara, R.L. Heimark, S. Obata, H. Davidson, T. St. John, S. Taketani, and S. Suzuki. 1993. Protocadherins: A large family of cadherin-related molecules in central nervous system. *EMBO J.* **12:** 2249.

Sheng, S., P.A. Pemberton, and R. Sager. 1994. Production, purification and characterization of recombinant maspin proteins. *J. Biol. Chem.* (in press).

Sporn, M.B., A.B. Roberts, and D.S. Goodman. 1994. *The retinoids: Biology, chemistry and medicine.* Raven Press, New York.

Swisshelm, K., K. Ryan, X. Lee, X. H.C. Tsou, M. Peacocke, and R. Sager. 1994. Down-regulation of retinoic acid receptor β in mammary carcinoma cell lines and its up-regulation in senescing normal mammary epithelial cells. *Cell Growth Differ.* **5:** 1.

Tomasetto, C., M. Neveu, J. Daley, P.K. Horan, and R. Sager. 1993. Specificity of gap-junction communication among human mammary cells and connexin transfectants in culture. *J. Cell Biol.* **122:** 157.

Trask, D., T. Suh, V. Band, D. Zajchowski, and R. Sager. 1990. Keratin proteins as markers of normal and tumorigenic breast epithelial cells. *Proc. Natl. Acad. Sci.* **87:** 2319.

Wei, A., H. Rubin, B.S. Cooperman, and D.W. Christianson. 1994. Crystal structure of an uncleaved serpin reveals the conformation of an inhibitory reactive loop. *Struct. Biol.* **1:** 251.

Wu, D., M. Kan, G.H. Sato, T. Okamoto, and J.D. Sato. 1991. Characterization and molecular cloning of a putative binding protein for heparin-binding growth factors. *J. Biol. Chem.* **266:** 16778.

Yaswen, P., A. Smoll, D.M. Peehl, D.K. Trask, R. Sager, and N.R. Stampfer. 1990. Down-regulation of a calmodulin-related gene during transformation of human mammary epithelial cells. *Proc. Natl. Acad. Sci.* **87:** 7360.

Zajchowski, D.A., V. Band, D.K. Trask, D. Kling, J.L. Connolly, and R. Sager. 1990. Suppression of tumor-forming ability and related traits in MCF-7 human breast cancer cells by fusion with immortal mammary epithelial cells. *Proc. Natl. Acad. Sci.* **87:** 2314.

Zou, Z., A. Anisowic, M.J. Hendrix, A. Thor, M. Neveu, S. Sheng, K. Rafidi, E. Seftor, and R. Sager. 1994. Identification of a novel serpin with tumor suppressing activity in human mammary epithelial cells. *Science* **263:** 526.

Acute Myeloid Leukemia with Inv(16) Produces a Chimeric Transcription Factor with a Myosin Heavy Chain Tail

P. LIU,* N. SEIDEL,* D. BODINE,* N. SPECK,†
S. TARLÉ,‡ AND F.S. COLLINS*

*Laboratory of Gene Transfer, National Center for Human Genome Research, National Institutes of Health,
Bethesda, Maryland 20892; †Department of Biochemistry, Dartmouth Medical School, Hanover,
New Hampshire 03755-3844; ‡Department of Internal Medicine, University of Michigan,
Ann Arbor, Michigan 48109

The M4Eo subtype of acute myeloid leukemia (AML) is a variant characterized by an increased number of morphologically abnormal eosinophils in bone marrow and peripheral blood, in addition to the accumulation of immature blasts common to all acute leukemias (Bennett et al. 1985). It constitutes about 8% of AML (Mitelman and Heim 1992). The leukemic cells of most M4Eo patients are characterized by a pericentric inversion of chromosome 16 (inv[16][p13;q22]) (Arthur and Bloomfield 1983; Le Beau et al. 1983), although other chromosome 16 abnormalities involving p13 and q22, e.g., t(16;16)(p13;q22), are also occasionally seen (Hogge et al. 1984; Testa et al. 1984; Holmes et al. 1985; Larson et al. 1986). Given the absence of other karyotypic abnormalities in many of these patients, a pathogenic relationship between inversion 16 and the M4Eo leukemia has been suggested.

Through positional cloning, we recently cloned both the long- and short-arm chromosome breakpoints of inv(16), and identified the genes directly involved. The gene located at the q arm breakpoint is *CBFB*, a human homolog of the mouse transcription factor PEBP2/CBFβ which, in combination with CBFα, binds to specific DNA sequences and regulates expression of a number of genes transcribed in hematopoietic cells. A smooth muscle myosin heavy-chain gene (*MYH11*) was found at the p arm breakpoint. We demonstrated that a fusion transcript is generated in the leukemic cells between the *CBFB* gene on 16q and *MYH11* on 16p (Liu et al. 1993a,b).

In this paper, we describe construction of a full-length cDNA of the *CBFB/MYH11* fusion found in the majority of leukemia patients with inv(16). We show that the fusion protein is capable of participating with CBFα in DNA binding. We also demonstrate the differential effects of the inversion 16 fusion protein on enhancer/promoters of several relevant genes. We propose three different models consistent with the available data.

MATERIALS AND METHODS

cDNA constructs. All restriction enzymes and DNA T4 ligase were purchased from Boehringer Mannheim Biochemicals (BMB) and used according to manufacturer's instructions. Reverse transcription-polymerase chain reaction (RT-PCR) was conducted essentially as described previously (Liu et al. 1993a) with Moloney murine leukemia virus (Mo-MLV) reverse transcriptase from GIBCO BRL and AmpliTaq enzyme from Perkin-Elmer Cetus. pBluescript KS$^+$ was from Stratagene. The sense primer used for both CK9 and KL1 (defined in the Results section) is 5'ATATGAATTCGGGAAGATGCCGCGCGTCGTGCCCGAC, and the antisense primer for CK9 is: 5'CCGGAAGCTTCAAAACCATGGCAGTTTGTG. The detailed steps for the cloning of the full-length cDNAs are described in the Results section.

Electrophoretic mobility shift assay (EMSA). In vitro transcription/translation using plasmid DNA as template was performed with a kit from Promega (TNT Coupled Reticulocyte Lysate System). Oligonucleotides used in the EMSA are the following: core site high-affinity oligo: HA–GGATATTTGCGGTTAGCA; HAC–TGCTAACCGCAAATAT; core site mutation oligos: MUT–GGATATCTGCCGTAAGCA; MUT-C–TGCTTACGGCAGATAT. Core site sequences are underlined. Equimolar amounts of complementary oligonucleotides were annealed by heating to 70°C for 10 minutes followed by gradual cooling to room temperature overnight in 10 mM Tris (pH 8.0), 100 mM NaCl, and 1 mM EDTA. The annealed oligos were labeled with [^{32}P]dCTP by Klenow enzyme (BMB).

EMSA was performed as described previously (Gumucio et al. 1988). Briefly, the proteins were added to EMSA buffer for 10 minutes at room temperature along with 500 ng of dIdC and 5 mg of BSA, in the presence or absence of 100 ng of unlabeled double-stranded competitor DNA. One nanogram of the labeled binding site probe (about 20,000 cpm) was then added to each reaction, the reactions were further incubated at room temperature for 20 minutes and then fractionated through a 6% nondenaturing polyacrylamide gel run in 0.5 × TBE. The dissociation rate for the protein/DNA complex was determined as described previously (Wang et al. 1993). DNA in the protein/DNA complex was quantified with a Phosphor Imager (Molecular Dynamics).

Generation of retrovirus-producing cell lines. A 550-bp fragment containing the human X-linked phosphoglycerate kinase (PGK) gene promoter (Lim et al. 1989) was fused to the insert from both KL1 and CK9 plasmids at an *Eco*RI site located in the polylinkers 5′ to both genes. The *PGK-KL1/PGK-CK9* genes were excised from the intermediate vector with *Cla*I and *Hin*dIII and cloned between the *Cla*I and *Hin*dIII sites in the polylinker of the retroviral vector pG1Na, generating pG1Na-PGK-KL1 and pG1Na-PGK-CK9. The pG1Na vector (graciously provided by P. Tolstoshev of Gene Therapy Inc.) is a derivative of the previously described pLN vector (Miller 1992), containing a polylinker.

Ecotropic retrovirus producer cell lines were generated by standard techniques (Miller et al. 1986). Briefly, pG1Na-PGK-KL1 and pG1Na-PGK-CK9 plasmid DNAs were transfected into Ψ-CRIP amphotropic packaging cells (Danos and Mulligan 1988) by calcium phosphate coprecipitation for 4 hours followed by a glycerol shock. After 48 hours of incubation, virus-containing supernatant from the transfected Ψ-CRIP cells was exposed to actively dividing GP + E86 ecotropic packaging cells (Markowitz et al. 1988). After 48 hours of incubation, the GP + E86 cells were plated at limiting dilutions in medium containing 450 mg/ml G418 (active). Individual colonies were isolated and screened for the production of the recombinant viruses by Southern blot analysis of infected NIH-3T3 cells. High-titer clones were identified and used in transient transfection studies. All cells were grown in Dulbecco's modified Eagle's medium supplemented with 10% heat-inactivated newborn calf serum (Gibco BRL) at 37°C in an incubator.

Transient transfection. Transient transfection experiments were carried out essentially as described previously (Koh et al. 1993). Plasmid DNA was prepared by a plasmid preparation kit (Wizard Maxipreps, Promega) followed by purification with $CsCl_2$-EtBr gradient centrifugation. Plasmid DNA was transfected into NIH-3T3 cells with lipofectamine (Gibco BRL) according to the manufacturer's instructions. Cells were harvested 48 hours after the transfection and lysed by freeze-thaw cycles in a buffer containing 100 mM KCl and 1 mM dithiothreitol. Soluble cellular proteins were then obtained by centrifugation. For CAT assays, a CAT-ELISA kit was used (from BMB) and the data were recorded in a Bio-Rad model 3550 microplate reader.

RESULTS

Cloning of the Inversion 16 Breakpoints and the Identification of the Affected Genes

Cloning of the inv(16) breakpoint and identification of the fusion gene were described previously (Liu et al. 1993a,b). Briefly, we first identified yeast artificial chromosomes (YACs) and cosmids containing the p-arm inversion breakpoint by studying patient samples with fluorescent in situ hybridization (FISH) and Southern blot hybridization. Junction DNA sequences, which represent the fusion between p- and q-arm DNA of chromosome 16 resulting from the inversion, were isolated from a library constructed from leukemic cells of one inversion 16 patient and were used to identify cosmids spanning the q-arm breakpoint. The specific genes affected by the chromosome 16 inversion were identified by screening cDNA libraries with cosmids from both breakpoints. The human homolog of a mouse transcription factor PEBP2/CBFβ was detected by the q-arm breakpoint cosmid, and a smooth muscle myosin heavy-chain cDNA (*MYH11*) was isolated with the p-arm breakpoint cosmid (Fig. 1a).

Fusion transcripts between these two genes were detected in 6 of 6 leukemic samples by RT-PCR. All 6 had the same breakpoint in the *CBFB* gene at the mRNA level, removing only the last 17 amino acids of the coded protein. Three breakpoints were identified in the *MYH11* gene in these 6 patients with one frequent breakpoint (4 patients) and 2 infrequent ones (1 patient each). All three breakpoints are in the tail region of the gene and all give in-frame fusions with the *CBFB* gene (Fig. 1b). More recent studies have revealed three more infrequent breakpoints in *MYH11* and one infrequent breakpoint in the *CBFB* gene, all of them generating in-frame fusion RNA transcripts (Claxton et al. 1994; S.A. Shurtleff et al., in prep.). The potential reciprocal fusion of 5′-*MYH11* to 3′-*CBFB* is probably not relevant since (1) only the last 17 amino acids of the *CBFB* gene, which is not involved in the binding to the α subunits (A. Hajra et al., in prep.), will be retained in this type of fusion and (2) the *MYH11* promoter is unlikely to be active at a significant level in the myeloid cells. In fact, primers designed to amplify the potential reciprocal fusion transcript revealed no products in RT-PCR with RNA from three inv(16) samples (Liu et al. 1993a).

The *CBFB* gene encodes a subunit of the transcription factor complex PEBP2/CBF. PEBP2 was orginally identified as a protein in NIH-3T3 cells that bound the mouse polyomavirus enhancer (Kamachi et al. 1990), and CBF was identified as a protein predominantly expressed in T cells that bound to a conserved "core" site in murine leukemia virus enhancers (Thornell et al. 1988; Boral et al. 1989; Wang and Speck 1992). Both PEBP2 and CBF recognize the same DNA sequence, PuACCPuCA. Purification and cloning of PEBP2 and CBF identified them as the same proteins, and they have recently been given the combined name of

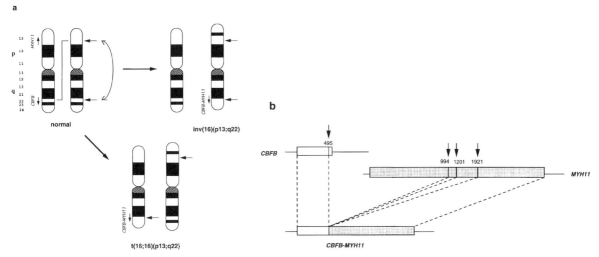

Figure 1. Diagrammatic illustration of the inv(16) event at cytogenetic and transcript levels. (*a*) Cytogenetic level. Both the chromosome 16 inversion and translocation are shown. The directions of transcription of the genes are indicated by arrows. (*b*) Transcript level. Transcription and splicing of the inversion region results in the generation of fusion transcripts between *CBFB* and *MYH11*. Depending on the breakpoint position in *MYH11* in patients, different proportions of the *MYH11* coding region are included. The numbering of the nucleotide positions is the same as in our previous publication (Liu et al. 1993a).

PEBP2/CBF (Ogawa et al. 1993a; Wang et al. 1993). The recognition sequence for PEBP2/CBF, which we refer to as the core site, has been identified in a number of viral and cellular genes that are specifically expressed in lymphoid and myeloid cells, including Rous sarcoma virus LTR, several T-cell receptor genes, interleukin genes, and the myeloperoxidase gene (MPO).

PEBP2/CBF consists of two protein subunits, an α subunit that binds the DNA target, and a β subunit that does not bind DNA directly. The β subunit forms a heterodimer with the α subunit and stablizes the protein/DNA complex (Ogawa et al. 1993a; Wang et al. 1993). The β subunit may have unidentified *trans*-activation functions as well (Fig. 2).

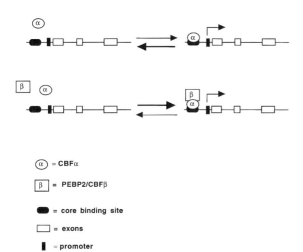

Figure 2. Interactions between the PEBP2/CBF protein complex and core site. (*Upper* drawing) CBFα by itself can bind to the core binding site, but the association is labile. (*Lower* drawing) In the presence of CBFβ, the protein/DNA complex is stabilized.

At least three different α subunits have been identified (Bae et al. 1993; Ogawa et al. 1993b; C. Wijmenga et al., in prep.), and they all share a so-called Runt domain in their sequence, which is responsible for both DNA binding and interaction with the β subunit (Kagoshima et al. 1993). One of the α subunits is encoded by the *AML1* gene, which is disrupted by t(8;21) in the AML M2 subtype (Miyoshi et al. 1991). As noted above, the β subunit is also involved in the pathogenesis of leukemia through its involvement in inversion 16. Since almost all of the coding region of *CBFB* is retained after the inversion, it seemed likely that the inv(16) fusion protein (referred to as the inv16 protein from this point on) would still be able to form heterodimers with the α subunit.

The smooth muscle myosin heavy chain (SMMHC) gene *MYH11* is a member of the myosin II family. A typical myosin II protein has an ATPase head (which is responsible for actin binding and mechanical movement), a hinge region, and a long tail with tandem copies of a coiled-coil domain that facilitates filament assembly. All the breakpoints identified in patients so far are within the tail region, fusing the carboxy-terminal portion of the tail to PEBP2/CBFβ protein.

As mentioned above, we predict that the inv16 protein should still be capable of associating with α subunits. The long coiled-coil domain on the inv16 protein may cause it to form multimers or filaments. The inv16 protein might well have a different effect on the regulation of genes containing the core site than the wild-type PEBP2/CBFβ. The inv16 protein could behave as a dominant positive, by augmenting the function of the wild-type PEBP2/CBF protein. Alternatively, it could act as a dominant negative, for example, by sequestering the α subunits in nonfunctional complexes.

Generation of Full-length *CBFB/MYH11* Fusion cDNA

To begin to address the mechanism by which the inv16 protein transforms cells, we first made plasmid constructs containing the full-length *CBFB-MYH11* fusion cDNA. RT-PCR was performed with RNA from an inv(16) patient with the most common type of fusion (Liu et al. 1993a) with sense primer from the 5' end of *CBFB* and antisense primer 3' to the breakpoint in the *MYH11* gene. The sense primer includes the first 21 bp of coding sequence of the *CBFB* gene preceded by the 6-nucleotide Kozak consensus sequence 5' of the start codon, which enhances translation in mammalian cells (Kozak 1987). The antisense primer has been described previously (M1 in Liu et al. 1993a) and is located about 150 bp 3' to the fusion point of the most common type of the p-arm breakpoints. PCR products were digested with *Eco*RI (at the 5' end of the sense primer) and *Apa*I (between the M1 primer sequence and the inversion breakpoint) and subcloned into pBluescript KS$^+$. The rest of the *MYH11* cDNA sequence was derived from a partial *MYH11* cDNA clone L11a (Liu et al. 1993a) as an *Apa*I and *Xba*I fragment (3' to the stop codon of *MYH11* on the vector), which was ligated with the aforementioned *Eco*RI-*Apa*I PCR fragment, and cloned into the *Eco*RI/*Xba*I sites of pBluescript KS$^+$ (two ligation steps were involved, since a second *Apa*I site is present just 5' to the stop codon of *MYH11*). The final construct containing the complete coding sequence of *CBFB-MYH11* fusion cDNA was given the name KL1. The full-length wild-type *CBFB* cDNA was generated by RT-PCR with RNA from inv(16) patients, using the same sense primer as used above and an antisense primer that contains the reverse complementary sequence of the last 18 bp of the coding region plus the stop codon. The PCR product was digested with *Eco*RI and *Hin*dIII (also created by the primer design) and cloned into the *Eco*RI/*Hin*dIII sites of pBluescript KS$^+$. This construct is called CK9. Both KL1 and CK9 were sequenced to confirm that no mutations or rearrangements were introduced during the cloning processes.

Association of Inv16 Protein with CBFα and Core Site

An in vitro transcription/translation system (Promega) was used to make proteins from plasmids CK9 and KL1. The production of proteins of the correct molecular weight was monitored by adding [^{35}S]methionine in the in vitro transcription/translation reactions and separating the protein products on a SDS-PAGE gel (data not shown).

EMSA was performed as described previously (Koh et al. 1993) using ^{32}P-labeled oligonucleotides containing a high-affinity (HA) PEBP2/CBF-binding site. The CBFα subunit used in the reactions was derived from affinity-purified bovine PEBP2/CBF that was first fractionated through an SDS polyacrylamide gel and subjected to a cycle of denaturation/renaturation (Wang et al. 1993). The α subunit forms a protein/DNA complex on the HA probe that could be specifically competed by an excess of unlabeled HA site but not by a mutant core site (MUT) (Fig. 3a). Addition of in-vitro-translated wild-type PEBP2/CBFβ protein to the binding reaction generated a more slowly migrating band corresponding to the α/β complex. When in-vitro-translated inv16 protein was added to the bovine PEBP2/CBFα protein, a novel band was observed migrating even more slowly than the wild-type α/β complex. Formation of both the α/β and α/inv16 protein/DNA complexes was specifically competed by excess unlabeled HA site (Fig. 3a). Radioactivity was also consistently observed at the origin of the lanes with inv16 protein present (indicated by the arrow in Fig. 3a) and ap-

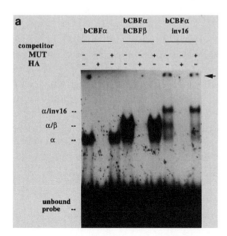 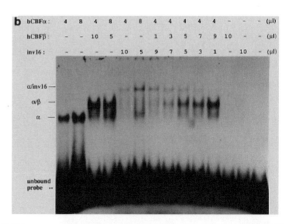

Figure 3. Association of inv16 protein with CBFα and its DNA target. (*a*) Purified bovine CBFα protein was incubated with the HA core site probe either alone or in the presence of in-vitro-transcribed/translated human PEBP2/CBFβ or inv16 protein. A 100-fold excess of either a high affinity (HA) or mutant (MUT) unlabeled binding site was used to establish specificity of the protein/DNA complexes. The migration positions for α alone, α/β, α/inv16, and unbound probe are indicated. The arrow shows the position of the gel origin. (*b*) Mixing experiment. Same EMSA condition as in *a* except that different quantities of the human CBFβ and inv16 proteins were mixed in the same reactions as indicated on top of the gel.

peared to be specific by competition analysis. Such high-molecular-weight aggregates might be produced by filament formation catalyzed by the myosin heavy-chain component at the carboxy-terminal half of the inv16 protein.

To investigate whether the fusion between PEBP2/CBFβ and myosin heavy chain affected the binding properties of PEBP2/CBFβ, we performed an additional competition experiment (Fig. 3b). Different proportions of inv16 protein and wild-type CBFβ protein were mixed together with a fixed amount of bovine CBFα and labeled HA. No extra bands were generated when all proteins were present, suggesting that no heteromultimers formed between PEBP2/CBFβ and the inv16 protein. Attaching the myosin heavy-chain tail to PEBP2/CBFβ protein probably does not interfere with binding to the α subunit, since wild-type PEBP2/CBFβ and inv16 protein compete with each other for α binding (Fig. 3b).

Addition of the inv16 protein to binding reactions reproducibly diminished the amount of HA probe retained in protein/DNA complexes relative to addition of wild-type PEBP2/CBFβ (cf., e.g., lanes 5 and 3 in Fig. 3b). This could not simply be an effect of adding less inv16 protein; were that the case there would be a commensurate increase in the intensity of the α protein/DNA complex. Phosphorimager analysis indicates that less HA probe is retained in all protein/DNA complexes, thus, the inv16 protein appears to inhibit the association of the α subunit with DNA.

We also measured the relative stability of the α/β protein/DNA complexes obtained with PEBP2/CBFβ and inv16 by measuring the rate of dissociation of the α/β protein complex from DNA. Formation of protein/DNA complexes was allowed to reach equilibrium, then excess unlabeled HA-binding site was added to binding reactions to trap protein as it dissociated from the DNA (Fig. 4a). The amount of protein/DNA complex that remained as a function of time is plotted in Figure 4b. As shown previously, addition of wild-type PEBP2/CBFβ significantly decreased the rate of dissociation of the PEBP2/CBF protein/DNA complex (Wang et al. 1993), and the inv16 protein similarly decreased the dissociation rate. This result indicates that the dissociation rate of CBFα + inv16 protein is about the same as that of CBFα + the wild-type PEBP2/CBFβ protein. Unless the association and dissociation constants are different in this case, we have to conclude that the affinity of inv16 protein is comparable with that of the wild-type PEBP2/CBFβ. Therefore, the apparent decrease of total bound core site probe observed in the previous experiment cannot be explained by a decrease of the affinity of the inv16/CBFα complex to the probe. These results suggest the possibility that the inv16/CBFα complex, with or without binding to the probe, forms large insoluble aggregates which precipitate during the incubation (and are perhaps responsible for the labeled complex consistently seen in the well).

Differential Effects of Inv16 Protein on Core Site-containing Enhancers

NIH-3T3 cell lines expressing wild-type human *CBFB* and *CBFB-MYH11* were generated by retroviral transduction (CK9-27 and KL1-10, respectively). Expression of *CBFB* and *CBFB-MYH11* in these cell lines was confirmed by RNA dot blot hybridization and Western blot analysis (data not shown).

CAT reporter constructs, driven by the Mo-MLV promoter/enhancer, were transfected into these cell lines transiently and assayed for CAT activity. One of the constructs, pMoCAT, contains a 384-bp fragment from the Mo-MLV LTR driving expression of the CAT gene (Speck et al. 1990). pMo(CORE)CAT contains a 2-bp mutation in both core sites in the Mo-MLV enhancer repeat and was used as a negative control (Fig. 5).

The Mo-MLV enhancer was transcriptionally active in the packaging cell line (GP + E86), and this activity

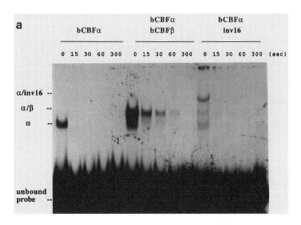
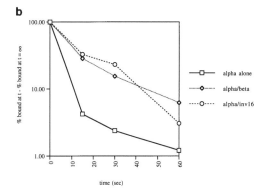

Figure 4. Measurement of dissociation rates. (*a*) Binding reactions were performed and then 50 ng of unlabeled HA was added to trap the dissociated protein. Aliquots were removed at the indicated time points and loaded onto the gel. (*b*) The radioactivity of the gel in *a* was quantitated with a Phosphorimager, and plotted as log (% bound relative to time zero - % bound at time ∞) versus time (seconds) for bovine CBFα alone, bovine CBFα plus human CBFβ, and bovine CBFα plus the inv16 protein.

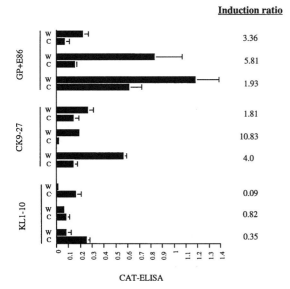

Figure 5 Graphic illustration comparing the *trans*-activation effect of inv16 versus CBFβ. Packaging cells (GP + E86), wild-type *CBFB*-expressing cells (CK9-27), and *CBFB-MYH11*-expressing cells (KL1-10) (200,000 cells/transfection) were transfected with 1 µg of plasmid DNA, pMoCAT (*W*), and pMo(CORE)CAT (*C*, containing a mutation of the CBF core site). CAT activity was measured by CAT-ELISA assay. Three independent transfections for each cell line and plasmid were performed in each experiment, and the averages of each are shown as horizontal bars (error bars are standard deviations). The induction ratio is the ratio between pMoCAT and pMo(CORE)CAT for each experiment.

was decreased approximately fourfold by mutation of the core sites (pMo[CORE]CAT). Thus, an endogenous core-binding factor, possibly endogenous PEBP2/CBF, contributes to transcription of the Mo-MLV enhancer in GP + E86. The levels of this endogenous protein appear to be saturating, since the addition of exogenous PEBP2/CBFβ (cell line CK9-27) did not further increase transcriptional activity from the Mo-MLV enhancer. In contrast, the inv16 protein (cell line KL1-10) markedly inhibited activity of the wild-type Mo-MLV enhancer, and this inhibition was partially relieved by mutation of the core site. The release of inhibition upon mutation of the core site suggests that the inv16 protein is not simply sequestering CBFα subunits in nonproductive complexes that cannot bind DNA, but is actively inhibiting transcription from this promoter in a sequence-specific manner. This inhibitory effect of the inv16 protein seen with the Mo-MLV enhancer was not found using the enhancers from either the myeloperoxidase (MPO) gene (Suzow and Friedman 1993) or the T-cell receptor β-chain gene (data not shown).

DISCUSSION

The transforming phenotype caused by the chromosome 16 inversion is dominant in vivo, since only one chromosome 16 is affected in a leukemia patient. At least three possible mechanisms could be postulated to explain this apparent dominant effect of the *CBFB/MYH11* fusion gene. One is dominant superactivation, i.e., the fusion protein is performing the same function as wild-type PEBP2/CBFβ but in an augmented fashion. For example, inv16 may more effectively stabilize the association of PEBP2/CBF with the DNA, although the binding data do not support this hypothesis. Alternatively, the myosin tail may function as a potent *trans*-activation domain. However, the transient transfection data suggest the opposite, that in fact inv16 represses transcription from one of its target genes. Taken together, these results suggest that dominant superactivation is an unlikely mechanism for inv16 function.

The second possible mechanism for transformation by the inv16 protein is through a dominant negative effect. We observed that the in-vitro-translated inv16 protein appears to form high-molecular-weight structures, evidenced by their very slow mobility in the EMSA. The formation of these structures, presumably mediated through the myosin tail, may sequester a certain amount of CBFα, making it unavailable to bind its targets in the DNA. The reduced transcriptional activity of the Mo-MLV enhancer in inv16 protein-producing NIH-3T3 cells seems to support this hypothesis.

A third possible mechanism is that the α/inv16 complex may alter the assembly of sequence-specific transcription factors on adjacent sites in the enhancers of certain target genes. Interactions between PEBP2/CBFβ and other protein targets may be abolished by the myosin tail, or alternatively, the myosin tail may participate in novel interactions with other proteins. These effects may depend on the configuration of binding sites for other transcriptional activators, which could explain why the activity of one of the enhancers we tested (Mo-MLV) was affected by inv16, whereas others showed no significant change.

In summary, we have cloned and expressed the *CBFB-MYH11* fusion gene which is found in almost all patients with the M4Eo subtype of acute myelocytic leukemia. The fusion protein produced is still capable of complexing with CBFα and binding to the PEBP2/CBF core recognition site, although indirect evidence suggests that high-molecular-weight aggregates may also be forming. The precise mechanism by which this disruption of the normal PEBP2/CBF transcription complex leads to leukemic transformation will require additional biochemical and genetic studies.

ACKNOWLEDGMENTS

The authors thank Alan Friedman for the MPO constructs, and Amit Hajra, Cisca Wijmenga, Paula Gregory, and Trevor Blake for helpful discussions.

REFERENCES

Arthur, D.C. and C.D. Bloomfield. 1983. Partial deletion of the long arm of chromosome 16 and bone marrow eosinophilia

in acute nonlymphocytic leukemia: A new association. *Blood* **61:** 994.
Bae, S.C., Y. Yamaguchi-Iwai, E. Ogawa, M. Maruyama, M. Inuzuka, H. Kagoshima, K. Shigesada, M. Satake, and Y. Ito. 1993. Isolation of PEBP2αB cDNA representing the mouse homolog of human acute myeloid leukemia gene, *AML1*. *Oncogene* **8:** 809.
Bennett, J.M., D. Catovsky, M.-T. Daniel, G. Flandrin, D.A.G. Galton, H.R. Gralnick, and C. Sultan. 1985. Proposed revised criteria for the classification of acute myeloid leukemia. A report of the French-American-British co-operative group. *Ann. Intern. Med.* **103:** 626.
Boral, A.L., S.A. Okenquist, and J. Lenz. 1989. Identification of the SL3-3 virus enhancer core as a T lymphoma cell-specific element. *J. Virol.* **63:** 76.
Claxton, D.F., P. Liu, H.B. Hsu, P. Marlton, J. Hester, F. Collins, A.B. Deisseroth, J.D. Rowley, and M.J. Siciliano. 1994. Detection of fusion transcripts generated by the inversion 16 chromosome in acute myelogenous leukemia. *Blood* **83:** 1750.
Danos, O. and R.C. Mulligan. 1988. Safe and efficient generation of recombinant retroviruses with amphotropic and ecotropic host ranges. *Proc. Natl. Acad. Sci.* **85:** 6460.
Gumucio, D.L., K.L. Rood, T.A. Gray, M.F. Riordan, C.I. Sartor, and F.S. Collins. 1988. Nuclear proteins that bind the human γ-globin promoter: Alterations in binding produced by point mutations associated with hereditary persistence of fetal hemoglobin. *Mol. Cell. Biol.* **8:** 5310.
Hogge, D.E., S. Misawa, N.Z. Parsa, A. Pollak, and J.R. Testa. 1984. Abnormalities of chromosome 16 in association with acute myelomonocytic leukemia and dysplastic bone marrow eosinophils. *J. Clin. Oncol.* **6:** 550.
Holmes, R., M.J. Keating, A. Cork, Y. Broach, J. Trujillo, W.T. Dalton, K.B. McCredie, and E.J. Freireich. 1985. A unique pattern of central nervous system leukemia in acute myelomonocytic leukemia associated with inv(16) (p13;q22). *Blood* **65:** 1071.
Kagoshima, H., K. Shigesada, M. Satake, Y. Ito, H. Miyoshi, M. Ohki, M. Pepling, and P. Gergen. 1993. The Runt domain identifies a new family of heteromeric transcriptional regulators. *Trends Genet.* **9:** 338.
Kamachi, Y., E. Ogawa, M. Asano, S. Ishida, Y. Murakami, M. Satake, Y. Ito, and K. Shigesada. 1990. Purification of a mouse nuclear factor that binds to both the A and B cores of the polyomavirus enhancer. *J. Virol.* **64:** 4808.
Koh, J., T. Sferra, and F.S. Collins. 1993. Characterization of the cystic fibrosis transmembrane conductance regulator promoter region. *J. Biol. Chem.* **268:** 15912.
Kozak, M. 1987. At least six nucleotides preceding the AUG initiator codon enhance translation in mammalian cells. *J. Mol. Biol.* **196:** 947.
Larson, R.A., S.F. Williams, M.M. Le Beau, M.A. Bitter, J.W. Vardiman, and J.D. Rowley. 1986. Acute myelomonocytic leukemia with abnormal eosinophils and inv(16) or t(16;16) has a favorable prognosis. *Blood* **68:** 1242.
Le Beau, M.M., R.A. Larson, M.A. Bitter, J.W. Vardiman, H.M. Golomb, and J.D. Rowley. 1983. Association of an inversion of chromosome 16 with abnormal marrow eosinophils in acute myelomonocytic leukemia. A unique cytogenetic-clinicopathological association. *N. Engl. J. Med.* **309:** 630.
Lim, B., J.F. Apperly, S.H. Orkin, and D.A. Williams. 1989. Long-term expression of human adenosine deaminase in mice transplanted with retrovirus infected hematopoietic stem cells. *Proc. Natl. Acad. Sci.* **86:** 8892.
Liu, P., S.A. Tarlé, A. Hajra, D.F. Claxton, P. Marlton, M. Freedman, M.J. Siciliano, and F.S. Collins. 1993a. Fusion between transcription factor CBFβ/PEBP2β and a myosin heavy chain in acute myeloid leukemia. *Science* **261:** 1041.
Liu, P., D.F. Claxton, P. Marlton, A. Hajra, J. Siciliano, M. Freedman, S.C. Chandrasekharappa, K. Yanagisawa, R.L. Stallings, F.S. Collins, and M.J. Siciliano. 1993b. Identification of yeast artificial chromosomes containing the inversion 16 p-arm breakpoint associated with acute myelomonocytic leukemia. *Blood* **82:** 716.
Markowitz, D., S. Goff, and A. Bank. 1988. A safe packaging line for gene transfer: Separating viral genes on two different plasmids. *J. Virol.* **62:** 1120.
Miller, A.D. 1992. Retroviral vectors. *Curr. Top. Microbiol. Immunol.* **158:** 1.
Miller, A.D., D.R. Trauber, and C. Buttimore. 1986. Factors involved in production of helper virus-free retroviral vectors. *Somatic Cell Mol. Genet.* **12:** 175.
Mitelman, F. and S. Heim. 1992. Quantitative acute leukemia cytogenetics. *Genes Chromosomes Cancer* **5:** 57.
Miyoshi, H., K. Shimizu, K.J. Tomoko, N. Maseki, Y. Kaneko, and M. Ohki. 1991. t(8;21) breakpoints on chromosome 21 in acute myeloid leukemia are clustered within a limited region of a single gene, *AML1*. *Proc. Natl. Acad. Sci.* **88:** 10431.
Ogawa, E., M. Inuzuka, M. Maruyama, M. Satake, M. Naito-Fujimoto, Y. Ito, and K. Shigesada. 1993a. Molecular cloning and characterization of PEBP2β, the heterodimeric partner of a novel *Drosophila runt*-related DNA binding protein PEBP2α. *Virology* **194:** 314.
Ogawa, E., M. Maruyama, H. Kagoshima, M. Inuzuka, J. Lu, M. Satake, K. Shigesada, and Y. Ito. 1993b. PEBP2/PEA2 represents a new family of transcription factor homologous to the products of the *Drosophila runt* and the human *AML1*. *Proc. Natl. Acad. Sci.* **90:** 6859.
Speck, N.A., B. Renjifo, and N. Hopkins. 1990. Point mutations in the Moloney murine leukemia virus enhancer identify a lymphoid-specific viral core motif and 1,3-phorbol myristate acetate-inducible element. *J. Virol.* **64:** 543.
Suzow, J. and A.D. Friedman. 1993. The murine myeloperoxidase promoter contains several functional elements, one of which binds a cell type-restricted transcription factor, myeloid nuclear factor 1 (MyNF1). *Mol. Cell. Biol.* **13:** 2141.
Testa, J.R., D.E. Hogge, S. Misawa, and N. Zandparsa. 1984. Chromosome 16 rearrangements in acute myelomonocytic leukemia with abnormal eosinophils. *N. Engl. J. Med.* **310:** 468.
Thornell, A., B. Hallberg, and T. Grundstrom. 1988. Differential protein binding in lymphocytes to a sequence in the enhancer of the mouse retrovirus SL3-3. *Mol. Cell. Biol.* **8:** 1625.
Wang, S. and N.A. Speck. 1992. Purification of core-binding factor, a protein that binds the conserved core site in murine leukemia virus enhancers. *Mol. Cell. Biol.* **12:** 89.
Wang, S., Q. Wang, B.E. Crute, I.N. Melnikova, S.R. Keller, and N.A. Speck. 1993. Cloning and characterization of subunits of the T-cell receptor and murine leukemia virus enhancer core-binding factor. *Mol. Cell. Biol.* **13:** 3324.

Genetic Alterations in the Chromosome 22q12 Region Associated with Development of Neuroectodermal Tumors

G. Thomas,* O. Delattre,* J. Zucman,* P. Merel,* C. Desmaze,*
T. Melot,* M. Sanson,* K. Hoang-Xuan,* B. Plougastel,*
P. Dejong,‡ G. Rouleau,† and A. Aurias*

*Laboratoire de Génétique des Tumeurs, Institut Curie, 75231-Paris, France;
†Centre de Recherche en Neuroscience, McGill University, Montreal, Quebec, Canada, H3G 1A4;
‡Human Genome Center, Lawrence Livermore National Laboratory, Livermore, California 94550

Although it is the second smallest autosome after chromosome 21, chromosome 22 is one of the most frequent sites of cytogenetic abnormality observed in human cancer. Not only was it the first chromosome shown to be recurrently involved in a balanced translocation (Nowel and Hungerford 1960), it was also the first chromosome shown to be recurrently deleted in a specific neoplasm (Zang and Singer 1967).

More recently, chromosome alterations in the 22q12 region were shown to be involved in a large spectrum of neuroectodermal tumors. A specific t(11;22)(q24;q12) translocation was demonstrated in Ewing's sarcoma (Aurias et al. 1983; Turc-Carel et al. 1983), the second most frequent primary bone tumor of childhood. Although Ewing's sarcoma cells are uniformly bland and undifferentiated, they can be induced to express markers of neuronal differentiation, suggesting a neuroectodermal origin. Furthermore, the same translocation was observed in peripheral neuroepithelioma, a tumor which is clearly derived from the neural crest (Whang-Peng et al. 1984). Ewing's sarcoma and peripheral neuroepithelioma are now considered to be the major members of a group of neuroectodermal tumors referred to as the Ewing family of tumors. In this group, the characteristic t(11;22) translocation is present in approximately 80% of the cases. Variant or complex translocations involving chromosome 22 are observed in an additional 10% of the tumors.

Independently, from the work of many laboratories and the development of allelic loss technology, it became apparent that chromosome 22 was frequently deleted in many types of solid tumors. However, the most consistent losses were observed for meningiomas (Seizinger et al. 1987; Dumanski et al. 1990) and schwannomas (Seizinger et al. 1986; Couturier et al. 1990), two tumors of neuroectodermal origin which account together for about 30% of all nervous system tumors. The development of these tumors is known to be predisposed in a dominantly inherited condition called neurofibromatosis type 2 (NF2) (Martuza and Eldridge 1988). Affected NF2 patients tend to develop multiple independent primary tumors mainly in close contact with the 8th pair of cranial nerves, thus leading to progressive deafness and eventually to compression of vital structures of the central nervous system. In gene carriers, the symptoms do not usually appear before the second decade and are almost systematically present by 40 years of age. Because NF2 is a well-characterized monogenic disease with almost complete penetrance, it was possible to perform precise cosegregation analysis of the disease phenotype with cloned polymorphic loci. The genetic alteration responsible for NF2 was shown to be on chromosome 22 and could be placed in the 22q12 band (Rouleau et al. 1990). This localization, in association with the demonstration of deletion of the chromosome 22 carrying the alteration in a tumor predisposed by NF2 (Fontaine et al. 1991), was indicative of the presence of a tumor suppressor gene in this chromosomal region.

Taken together, these observations suggested that two different genetic events associated with the tumorigenic processes of subgroups of neuroectodermal tumors could take place in the 22q12 region: (1) the activation of a dominant oncogene mediated by chromosome translocations in the Ewing's family of tumors; (2) the inactivation of a tumor suppressor gene in the subgroup of tumors predisposed by NF2.

To contribute to our understanding of the detailed genetic events underlying these alterations and to the identification of the genes that might be involved, we developed a physical map of part of the 22q12 region. As described below, further analysis of the region revealed the presence of two previously unknown genes which were each shown to be altered in one of the subgroups of tumors mentioned above.

METHODS

Isolation of overlapping cosmids. Chromosome 22-specific libraries were constructed by cloning partially digested *Mbo*I DNA isolated from flow-sorted human chromosome 22 in the Lawrist 5 cosmid vector (de Jong et al. 1989). In toto, these libraries, which were gridded in microtiter plates, contain approximately 7 genome equivalents of chromosome 22. The Lawrist 5 vector has two *Sfi*I sites that flank the cloning site.

After digestion of the recombinant cosmid DNA with *Sfi*I, the human DNA insert was isolated from the cosmid vector on low melting agarose. The isolated insert was labeled by random priming using [α-^{32}P]dCTP. Human repetitive sequences and residual vector contamination were competed by incubation of the denatured probe with a large excess of sonicated denatured total human DNA and vector DNA. The preincubated probe was then used to screen the library according to standard procedures. The overlapping regions of newly retrieved clones were determined. The least overlapping cosmid was precisely mapped using a panel of somatic cell hybrids (Delattre et al. 1991) and subsequently used for a new step (Zucman et al. 1992b). A cosmid library was also constructed from DNA extracted from a Ewing's sarcoma cell line in order to clone the region of chromosome 11 involved in the t(11;22) translocation (Zucman et al. 1992b).

Fluorescent in situ hybridization with cosmid probes. Relevant cosmid DNAs were labeled either with biotin-11-dUTP or with digoxygenin-11-dUTP by the nick translation method. The labeled products were hybridized at a final concentration of 3 µg/ml in the presence of 300–500 µg/ml of sonicated human DNA. Post-hybridization washes were performed at 45°C. The hybridized probes were revealed by specific FITC or Rhodamin antibodies to either biotin or digoxygenin, respectively (Desmaze et al. 1994).

Identification of transcribed regions. Each cosmid to be screened was digested with *Eco*RI. Aliquots were double digested with the following enzymes containing two CG dinucleotides in their target sequence (*Bss*H2, *Sac*2, *Not*I, *Nru*I, and *Mlu*I), and alteration in the migration pattern was analyzed. Individual *Eco*RI fragments were also gel purified and used as probes on Southern blots made from *Eco*RI-digested human, rat, and mouse DNAs. To reveal phylogenetic conservation, washing was performed under the two following stringency conditions (2 × SSC, 50°C and 0.1 × SSC, 65°C, in both cases for 5 min). Northern blots were performed using standard conditions with RNAs extracted from the following cell lines: HL60 (promyelocytic leukemia), HeLa (cervix carcinoma), SKNBE (neuroblastoma), 2102 EP and Tera 2 (teratocarcinomas), CCL225 (colon carcinoma), OZ (glioma), EW24 (Ewing's sarcoma), and Bewo (choriocarcinoma) (Zucman et al. 1992b; Rouleau et al. 1993).

Reverse transcription PCR. Total RNA from tumors was isolated using the RNAzol extraction kit (Bioprobe systems, France). It was reverse transcribed, and the resulting cDNA was subjected to PCR reaction with couples of primers which would enable the amplification of either EWS cDNA (positive control for the reverse transcription step) or EWS/FLI-1 cDNA or EWS/ERG cDNA (Delattre et al. 1994). Amplified products were analyzed by electrophoresis on agarose gel. The amplified fragments were also purified on Centricon 100 ultrafiltration devices (Amicon, Epernon, France), and direct sequencing was performed using a Taq polymerase Kit (PRISM, Applied Biosystem) with fluorescent primers or dideoxynucleotides. Sequencing reactions were analyzed with an Applied Biosystem model 373A automatic sequencer (Mérel et al. 1994).

Denaturing gradient gel electrophoresis. Samples were electrophoresed on polyacrylamide gel containing an appropriate denaturant gradient parallel to the direction of electrophoresis (usually 30–80%; 100% denaturant is 7 M urea and 40% formamide). The gels maintained at 60°C in a Hoeffer 2001 vertical electrophoresis unit (LKB) were run at 160 V. Gradient concentrations and running conditions for each amplified DNA fragment were determined using the Meltmap and SQHTX computer programs. After electrophoresis, the gels were stained with ethidium bromide and photographed (Hamelin et al. 1993). Direct sequencing of the amplified product was performed as described above.

RESULTS

Construction of Cosmid Contigs in 22q12

Sublocalization of chromosome 22 probes using an extended panel of somatic cell hybrids mapped 12 probes (Delattre et al. 1991; Zucman et al. 1992a) within the region bounded by D22S1 and D22S28, where the NF2 gene lies (Fig. 1A,B) (Rouleau et al. 1990). This region also contains the chromosome 22 breakpoint of two cell lines derived from tumors of the Ewing's family (Fig. 1B). These probes were used to progressively expand the corresponding loci by the recurrent isolation of overlapping cosmids using chromosome-22-specific cosmid libraries (Zucman et al. 1992b). In the course of this procedure, several contigs were connected so that a total of seven cosmid contigs was finally obtained (Fig. 1B). One cosmid, B6 (Fig. 1D), contained the chromosome 22 breakpoints of the two cell lines, since one of its ends, but not the other, yielded a positive signal on DNAs from two cell hybrids containing their der(11) chromosome (Zucman et al. 1992b). As expected, fluorescent in situ hybridization (FISH) on metaphases of several Ewing's sarcoma cell lines revealed that cosmids from the telomeric and centromeric halves of the corresponding contig yielded signals on the der(11) and der(22) chromosomes, respectively.

Southern analysis using single copy probe isolated from cosmid B6 showed that the chromosome 22 breakpoint of the Ewing's family of tumors was systematically occurring within a 7K region which was termed *EWSR1* (for *Ew*ing's *s*arcoma *r*egion *1*). A cosmid library was generated from a cell line derived from a peripheral neuroepithelioma. Probes derived from

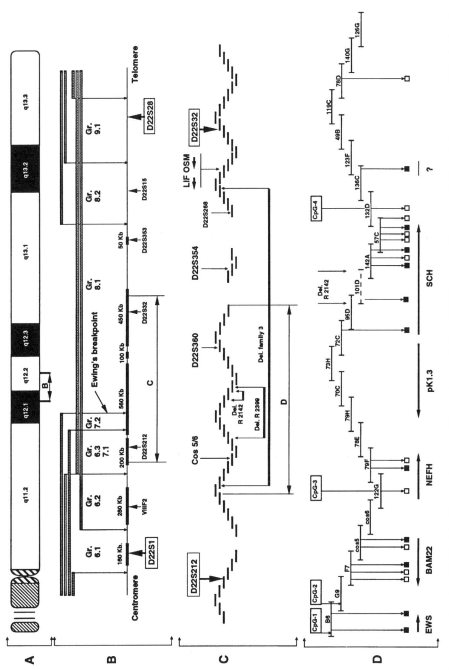

Figure 1. Map of the chromosomal region 22q12. *A* to *D* presents maps of increasing resolution. (*A*) Diagrammatic representation of chromosome 22 indicating the position of the region which is enlarged in *B*. (*B*) Gray bars represent portions of chromosome 22 that are present in different somatic cell hybrids which define the subregions Gr.6.1 to Gr.9.1. D22S1 and D22S28 are the initial flanking markers identified for the *NF2* gene. The thick black bars represent the cosmid contigs that have been constructed starting from the indicated loci. Over each contig is mentioned its size. (*C*) The tiling path (smallest number of overlapping cosmids) of the contigs constructed in the region D22S212 and D22S32 is schematized together with the position of the loci that were initially expanded. The extent and boundaries of the 3 germ-line deletions found in NF2 patients are shown. The position of the polymorphic marker D22S268, which enabled the detection of the deletion in family 3, is shown. *LIF* and *OSM* are the genes for *leukemia inhibiting factor* and *oncostatin M*. They are transcribed from telomere to centromere. (*D*) Detail of the region that has been investigated in search of genes. CpG-1 to -4 indicate the positions of putative unmethylated CG-rich islands. Open and black squares represent phylogenetically conserved regions that failed to (□) or did (■) detect transcripts on Northern blots. The position and direction of transcription for the identified genes are shown by horizontal arrows. Gene pK1.3 was detected by Xie et al. (1993)

EWSR1 enabled three groups of overlapping cosmids to be retrieved from this library (Zucman et al. 1992b). They were identified as corresponding to the normal chromosome 22 and to both derivatives of the translocation. Non-chromosome-22 fragments from these cosmids were used to identify a clone which was shown by FISH to be derived from band q24 of chromosome 11. Comparison of the restriction maps of the intact and rearranged regions of chromosome 11 and 22 indicated that the translocation was simple and reciprocal. The chromosome 11 locus was expanded to 75 kb by the recurrent isolation of cosmids. Selected EcoRI restriction fragments from the chromosome 11 locus were then used to search for rearrangement on EcoRI-digested DNA from a set of tumors of the Ewing's family. The chromosome 11 breakpoints were identified for 66 tumors (Zucman et al. 1993b). Their positions were spread, with no evident clustering, over a 40-kb region, which was termed EWSR2.

Further cosegregation analyses in NF2 families had narrowed down the position of the NF2 gene, which could now be placed between two new flanking markers, D22S212 and D22S32 (Fig. 1C) (Ruttledge et al. 1993; G. Rouleau et al., unpubl.). Further expansion of the previously established contigs between these loci was undertaken until a major part of the region was cloned. Isolation of yeast artificial chromosomes and Southern analysis of total human DNA provided an estimate of the distance between D22S212 and D22S32 at about one million base pairs and indicated that more than 90% of the region had been cloned (Fig. 1C). A highly informative microsatellite locus, D22S268, was identified in a cosmid within the region D22S212/D22S32 (Marineau et al. 1993). This marker was shown to map in close proximity of the NF2 gene, since cosegregation analyses in NF2 families failed to reveal recombination (pairwise LOD score of 14.6). Furthermore, in one NF2 family (family 3), identification of affected members that were obligate hemizygotes for this locus suggested that the NF2 disease could be associated with an interstitial germ-line deletion (Sanson et al. 1993).

Screening for rearrangements in the D22S212/D22S32 region using pulsed-field gel electrophoresis was performed for 1 patient of this family and for an additional 42 unrelated NF2 propositi. Abnormal migrating bands were revealed for 3 individuals. To document further the existence of constitutional deletions, FISH was performed on interphase nuclei or metaphase chromosomes of lymphoblastoid cell lines from these 3 patients. In contrast to controls, which systematically demonstrated labeling of the two chromosomes 22 in metaphases and the presence of two dots in interphase nuclei, cell lines from these three patients demonstrated, for some cosmids, labeling of a single chromosome 22 or of a single dot, demonstrating that these cosmids were hemizygously deleted (Rouleau et al. 1993; Sanson et al. 1993). Each deletion was thus precisely mapped, showing that they were nested within each other (Fig. 1C). The position of the two smallest deletions strongly suggested that the NF2 gene was telomeric to the Ewing's sarcoma breakpoint on the same cosmid contig.

Identification of Transcribed Regions

A systematic analysis of more than 100 EcoRI fragments isolated from the cosmids which encompassed the EWSR1 region and the immediately adjacent telomeric 350-kb region was undertaken (Zucman et al. 1992b; Rouleau et al. 1993). To identify clusters of CG dinucleotides, which frequently mark the 5' region of genes, each fragment was examined for the presence of restriction sites with CG in their recognition sequence. Four clusters of at least three different restriction sites were found (Fig. 1D). Each EcoRI fragment was also tested for its ability to hybridize to rodent DNA and, when phylogenetic conservation was found, to detect transcripts on Northern blots. When the different data generated by this analysis were put together on the map, the positions of four genes became evident (Fig. 1D). The most centromeric gene, associated with the CpG-1 cluster, overlaps EWSR1. This gene was called EWS because of its constant rearrangement in Ewing's sarcoma. The closest identified gene on the telomeric side of EWS is expressed as a 4-kb transcript. It has recently been characterized (Peyrard et al. 1994). It codes for a β-adaptin and has been called BAM22. The third gene has a neuronal pattern of expression and a transcript size identical to that of the neurofilament heavy chain gene, NEFH, a gene previously located on chromosome 22. Use of NEFH cDNAs confirmed this identification and showed that CpG-3 is located in the 5' region of this gene. The fourth gene was detected by 5 different EcoRI fragments that hybridized to transcripts of identical sizes. It was the only gene which was altered by all three deletions. This gene was called SCH because of its putative involvement in schwannomas, the most prominent tumor predisposed by NF2. It was subsequently shown that a gene, pK1.3, of unknown function, located between NEFH and SCH had escaped this gene identification procedure (Xie et al. 1993).

Characterization of the EWS Gene

Overlapping cDNA clones hybridizing with the genomic probes that had identified the EWS gene were isolated (Delattre et al. 1992). The largest clone contained a 1968-bp open reading frame. Its deduced translation product has several structural characteristics. Its first 285-amino-acid sequence, termed NTD-EWS, contains a degenerate repeated motif with the frequent occurrence of a Ser-Tyr dipeptide. It is also rich in glutamine, threonine, and proline. The region between amino acids 157 and 262 shares 40% homology with the carboxy-terminal domain of the large subunit of eukaryotic RNA polymerase II proteins (CTDpolII). However, due to the degeneracy of the repeats and the lack of proline at crucial positions, it is unlikely that NTD-EWS adopts the regular overlapping β-turn structure predicted for CDT-polII. The rest of the EWS protein contains three stretches rich in glycine, arginine, and proline that share homologies with glycine-rich proteins and, in particular, with single-stranded nucleic-acid-binding proteins. Between the first and second stretch lies an 85-amino-acid sequence which shares homologies with a consensus RNA-binding domain. Altogether, these homologies of the EWS protein suggest that it possesses two different functional domains that could participate in RNA synthesis or processing. The EWS gene was subsequently shown to be divided into 17 exons spread over about 40 kb of DNA (Plougastel et al. 1993). The first 7 exons encode the NTD-EWS, and exons 11, 12, and 13 encode the putative RNA-binding domain. The DNA sequence in the 5' region of the gene lacks canonical promoter elements such as TATA and CCAAT consensus sequences. EWS is highly conserved in the mouse and maps to chromosome 11 in a synteny group which also includes LIF and NEFH (Plougastel et al. 1994).

EWS Alterations in the Ewing Family of Tumors

A phylogenetically conserved DNA fragment from *EWSR2* was used to retrieve overlapping cDNA clones (Delattre et al. 1992). The longest clone contained a 1356-bp open reading frame. Analysis of the deduced amino acid sequence revealed that it encoded the human homolog of the mouse *Fli1* gene (97% identity). The entire human *FLI1* gene, which has now been cloned in a single yeast artificial chromosome (Selleri et al. 1994), was shown to be divided into 9 exons (J. Zucman et al., unpubl.). *EWSR2* was located between exon 3 and exon 9 (Zucman et al. 1993b).

In situ hybridization on metaphase chromosomes from an Ewing's tumor showed that both *EWS* and *FLI1* are transcribed in a centromeric to telomeric orientation. Since the translocation was shown to be reciprocal with no marked loss of genetic material, on each of the two derivative chromosomes the 5' end of one gene is juxtaposed to the 3' end of the other gene. The putative chimeric gene generated on the der(11) chromosome is not expressed to a measurable level on Northern blots and is not expected to be involved in the

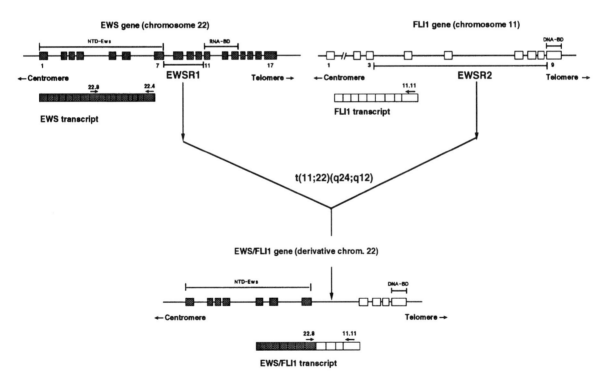

Figure 2. Formation of a hybrid gene between *EWS* and *FLI1*. Representations of the *EWS* gene on chromosome 22, of the *FLI1* gene on chromosome 11, and of the most common chimeric *EWS/FLI1* gene on the der(22) generated by the t(11;22) translocation are shown, together with their corresponding transcripts. Exons of *EWS* and *FLI1* are represented by gray and open boxes, respectively. Exons coding for the NTD-EWS, the putative RNA-binding domain (RNA-BD), and the DNA-binding domain of *FLI1* (DNA-DB) are indicated. *EWSR1* and *EWSR2* refer to the regions on chromosomes 22 and 11 in which the breakpoints resulting from the t(11;22) translocation have been mapped. The primers used for the RT-PCR amplification of the fusion transcript and of the normal *EWS* transcript are indicated by arrows. Primers 22.8 and 22.4 enable amplification of the normal *EWS* transcript. This amplification is used as an internal control. Primer 11.11 does not enable amplification unless a hybrid transcript is generated by the translocation. In this case, together with primer 22.8, it promotes the amplification of the junction region on the hybrid transcript.

tumor phenotype, since the der(11) chromosome can secondarily be lost in close to 10% of Ewing's tumors. In contrast, the fusion transcript initiated on chromosome 22 and ending on chromosome 11 is readily seen on Northern blots (Zucman et al. 1992b).

To investigate the junction of the two genes in the chimeric transcript, oligonucleotides homologous to exon 7 of *EWS* and exon 9 of *FLI1* were used to reverse-transcribe and amplify by PCR (RT-PCR) RNAs from various sources (Fig. 2) (Delattre et al. 1992, 1994; Zucman et al. 1993b). In contrast to RNAs from various tumors and normal tissues, RNAs from tumors of the Ewing family produced a variety of different products (usually a single product per tumor), which were all shown by direct sequencing to result from an in-phase junction of one exon of *EWS* with one exon of *FLI-1*. However, in about 10% of the cases, although rearrangement within *EWSR1* had been detected, no *EWSR2* rearrangement was evidenced and no amplified product was obtained by RT-PCR under standard conditions. Since the *FLI1* oligonucleotide used in the PCR reaction was taken from the most conserved region of the Ets family, it was possible to test successfully whether another gene from the same family might be involved in a fusion with *EWS*. Indeed, by lowering the annealing condition of the PCR reaction from 68°C to 62°C, amplified products were obtained in those refractory cases and direct sequencing showed that instead of *FLI1* being involved, it was *ERG*, its closest member in the Ets family. Subsequent use of oligonucleotides which perfectly matched the *ERG* sequences enabled the production of amplified products under stringent conditions. Furthermore, FISH with specific cosmids derived from the *EWS* and *ERG* loci showed that, in one tumor, the *ERG* gene and the *EWS* gene were brought together by a complex chromosome rearrangement.

ERG is located on chromosome 21. It forms a hybrid gene with *EWS* in about 10% of tumors of the Ewing family. It may appear surprising that a recurrent t(21;22) translocation had never been described in these tumors. However, a recent physical map of the *ERG*-containing region on chromosome 21 suggests that *ERG* is transcribed from telomere to centromere. The transcription of *EWS* is in the reverse orientation, precluding the generation of an *EWS/ERG* fusion gene through a simple and balanced translocation. Therefore, only complex rearrangements involving chromosome 21 and 22 can generate an *EWS/ERG* fusion transcript (Zucman et al. 1993b).

Characterization of the *SCH* Gene

A phylogenetically conserved sequence from the *SCH* locus was used to isolate a single 2.0-kb cDNA clone, N1.1, from a human fetal brain cDNA library (Rouleau et al. 1993). Sequence analysis revealed an open reading frame encoding a putative 595-amino-acid protein, which was named schwannomin. Sequence homology searches indicated it was a novel gene exhibiting significant amino acid homology with ezrin, radixin, and moesin, collectively known as the ERM family (Fig. 3) (Sato et al. 1992). Because of this homology, Trofatter et al., who independently identified the same gene, named its product merlin (moesin-, ezrin-, radixin-like protein) (Trofatter et al. 1993). The ERM family is included in the superfamily of the erythrocytic band 4.1 protein. The predicted secondary structure of schwannomin is similar to that of other members of the ERM family: a large amino-terminal domain followed by α-helical domains and a small carboxyl terminus (Fig. 3). The N1.1 cDNA was shown to be encoded by 16 exons that are spread over a region of approximately 120 kb; the sequences of the splicing sites were determined for each exon (Rouleau et al. 1993).

Search for Point Mutations in the *SCH* gene

Demonstration that, on rare occasions, the *SCH* gene could be altered by large genomic deletions is not sufficient evidence for its causal involvement in NF2 disease. To obtain further evidence of *SCH* alteration in NF2 patients, we chose to screen the coding sequence of the gene using denaturing gradient gel electrophoresis (DGGE). The efficiency of this method relies on its ability to discriminate between a perfectly matched double-stranded DNA fragment and an almost identical fragment with one or several mismatched bases. Since in a dominantly inherited disease only one of the alleles carries the causal mutation, a denaturation/renaturation cycle at the end of the PCR procedure should generate heteroduplex molecules, provided that this mutation is contained within the amplified product. Optimized experimental conditions can be determined using computer programs which predict, as a function of the run time, the expected difference in the gradient level between a homoduplex wild-type DNA molecule and a related heteroduplex with a single-base mismatch at each selected position along the sequence. These programs have previously been validated in a model system (Hamelin et al. 1993). They were exploited to search for optimal couples of primers that would enable detection of a single-base pair substitution anywhere in the coding sequence of the *NF2* gene or in the adjacent splicing sites. Fully adequate conditions were found for 90% (27 of 30) of the splicing sites and 95% (569 of 595 codons) of the coding sequence of the *NF2* gene (Mérel et al. 1994).

Using these conditions, the screening of 91 unrelated NF2 patients revealed 32 individuals for whom one of the 16 PCR-amplified products exhibited two to four bands on the DGGE profile (Mérel et al. 1994). Direct sequencing of these abnormally migrating products showed that 9 of the 32 DNA variants altered either the first or the second intronic base of an intron/exon boundary (Table 1). Such variation is predicted to interfere with the normal splicing process. Sequencing of the 23 other variants showed a mutation within the coding sequence. Of the mutations, 14 were deletions

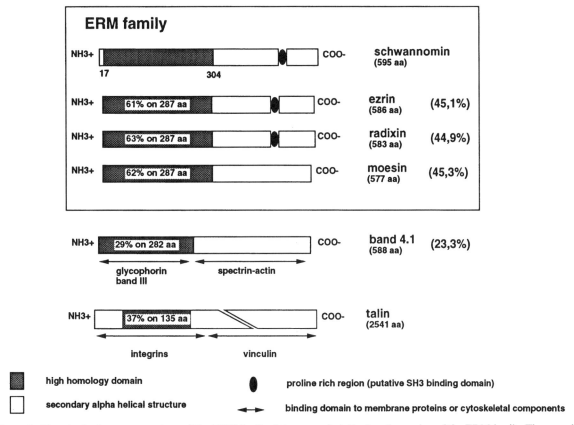

Figure 3. Homologies between members of the ERM family. Schwannomin is the fourth member of the ERM family. The overall percentage of amino acid identity with respect to the schwannomin sequence is indicated on the right. All four members have a very similar amino-terminal half (% identity indicated in the gray box) and a carboxy-terminal half that is predicted to adopt essentially an α-helical structure. In schwannomin, ezrin, and radixin, this structure is interrupted by a series of prolines which may serve as an SH3-binding domain. The ERM family is included within the band 4.1 superfamily. Two of its members, band 4.1 and talin, have been particularly well characterized. The amino-terminal domain of band 4.1 and talin have been shown to interact with glycophorin-band 3 and integrins, respectively. Their carboxy-terminal portions interact with spectrin-talin and vinculin, respectively.

Table 1. Summary of SCH Mutations Found Using DGGE

(a) *Germ-line mutations*

Mutation type	Number of mutations
Frame shift	14 (including 4 new mutants)
Splicing sites	9 (including 1 new mutant)
Nonsense mutation	8 (including 3 new mutants)
Missense	1

Number of unrelated NF2 patients screened: 91

(b) *Somatic mutations*

Tumor type	Number of cases	Number of mutations
Meningioma	56	17
Schwannoma	87	32
Ependymoma	17	0
Medulloblastoma	6	0
Neuroblastoma	15	0
Pheochromocytoma	30	0
Melanoma	23	0
Colorectal carcinoma	15	0
Breast carcinoma	15	0
Brain metastasis	6	0

or insertions, causing translational frame shifts. Eight mutations were single nucleotide substitutions which led to nonsense mutations. Finally, a single mutation was determined to be missense, resulting in a leucine to proline substitution at position 360. This amino acid substitution should disrupt a highly conserved helical region predicted in all members of the ERM family.

Importantly, 8 patients carried a new mutation, since in each case it was possible to show that both parents were clinically unaffected and, as shown by DGGE and sequencing, did not exhibit the mutation in their constitutional DNAs (Rouleau et al. 1993; Mérel et al. 1994). These observations provided the strongest evidence of a causal relationship between the occurrence of a new mutation and the development of the disease. Taken together, these data led to the conclusion that the *SCH* gene is the *NF2* gene.

The same strategy was used to screen for somatic mutations in tumors known to exhibit allelic losses on chromosome 22 (Bijlsma et al. 1994; P. Mérel et al., in prep.). A total of 270 tumors was investigated (Table 1). Mutations were only found in schwannomas and meningiomas, the two tumors that are most prominently predisposed by *NF2*. Interestingly, most mutations were also predictive of the synthesis of an altered protein with reduced molecular weight. In all, meningiomas and 50% of schwannomas with *SCH* mutation demonstrate losses of heterozygosity for chromosome 22 alleles, indicating that both copies of *SCH* are inactivated in these tumors. Therefore, functional inactivation of *SCH* by a two-hit mechanism should operate in the tumorigenic process of both schwannomas and meningiomas. In contrast, screening by DGGE of tumors different from meningioma and schwannoma failed to reveal mutation, suggesting that somatic alteration in the *SCH* gene might be mainly restricted to tumors predisposed by *NF2* (Table 1).

DISCUSSION

The physical mapping of a subregion of chromosome 22q12 has led to the identification of two genes that are recurrently altered in the tumorigenic process, enabling us to clarify the molecular events underlying the chromosome alterations observed in two subgroups of neuroectodermal tumors. Although these genes are closely associated on the physical map, they encode unrelated proteins that undergo distinct classes of oncogenic alterations, each associated with a specific subgroup of neuroectodermal tumors.

EWS is a putative RNA-binding protein that undergoes oncogenic conversion by chromosome translocations, leading to a variety of chimeric proteins. In the Ewing's family of tumors, the partners of EWS can be either FLI1 or ERG, two close members of the Ets family. These aberrant products can be described as EWS proteins in which the putative RNA-binding domain has been replaced by the DNA-binding domain of the Ets-related factor. It has recently been shown that some of these fusion proteins have a nuclear localization and may have *trans*-activating properties (May et al. 1993; Bailly et al. 1994). Identification of the genes altered in their transcriptional regulation by these aberrant factors can now be attempted. It should also be possible to investigate the consequences of the inhibition of the expression of these hybrid genes and whether this inhibition could address new therapeutic approaches of these aggressive tumors.

It has recently been shown that EWS may form hybrid proteins with different families of transcription factors in at least two sarcoma types not belonging to the Ewing's family of tumors (Fig. 4). In malignant melanoma of soft parts, a tumor which may be of neuroectodermal origin, EWS forms a hybrid protein with ATF1, a transcription factor binding to DNA through a basic domain-leucine zipper structure (Zucman et al. 1993a). In desmoplastic tumors, sarcomas with controversial nosologic origin, EWS fuses with the transcription factor WT1, which has several zinc fingers in its DNA-binding domain (Ladanyi and Gerald 1994). In both cases, the hybrid proteins may be described as aberrant transcription factors generated from the linking of NTD-EWS to the DNA-binding domains of ATF1 or WT1. Interestingly, the *TLS* gene,

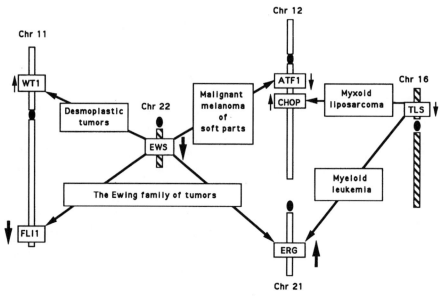

Figure 4. Tumor-associated translocations involving *EWS* and *TLS* in humans. In human tumors, *EWS* can fuse with four different genes: *FLI1* and *ERG* in the Ewing family of tumors, *WT1* in desmoplastic tumors, and *ATF1* in malignant melanoma of soft parts. *TLS*, a gene demonstrating striking homologies with *EWS*, has been shown to be fused to *CHOP* in myxoid liposarcoma and to *ERG* in a subtype of myeloid leukemia. All partners of *EWS* and *TLS* in these translocations are known transcription factors. In all cases, the product of the fusion gene is composed of the amino-terminal domain of *EWS* or *TLS* linked to the DNA-binding domain of one transcription factor. When known, the direction of transcription of the genes is indicated by a large arrow. With the exception of *EWS* and *ERG*, all other fusion genes are formed by cytogenetically well characterized balanced translocations, an observation which favors a specific orientation for the transcription of the other genes (small arrow).

the closest known human homolog of *EWS*, is also involved in chromosome translocations. In myxoid liposarcoma, TLS forms a fusion protein with the transcription factor CHOP (Crozat et al. 1993; Rabbitts et al. 1993), and in a subtype of myeloid leukemia with a t(16;21) translocation, TLS forms a hybrid with ERG (Ichikawa et al. 1994). The hybrid proteins are structurally similar to those involving EWS. Thus, the positional cloning of *EWS* and the identification of its rearrangements in the Ewing family of tumors have provided the prototypic model of an oncogenic mechanism that has been found in at least five different types of human tumors (Fig. 4).

The *SCH* gene is the *NF2* gene. It encodes a protein, schwannomin, which has high homology with the ERM family of proteins. Members of this family are known to localize at cell-cell adhesion sites, microvilli, and cleavage furrows, where actin filaments are densely associated with the plasma membrane. Inhibition of their expression by antisense oligonucleotide induces the destruction of both cell-cell interaction and cell-substrate adhesion and the disappearance of microvilli (Takeuchi et al. 1994). Although the subcellular localization of schwannomin has not been documented yet, it may be hypothesized that this protein participates in the linking of actin filaments to membrane proteins. The *NF2* gene is expressed in a wide variety of tissues. However, the observation that its functional inactivation in tumors is restricted to meningioma and schwannoma suggests that schwannomin may play a specific role in the progenitor cells of these tumors, where it may participate in the transduction of signals inhibiting cell proliferation. So far, nothing is known about the nature of these signals. Interestingly, it has recently been shown that the products of several other genes, which when altered cause a predisposition to cancer development in humans, may participate in signal transduction pathways involving G proteins (Xu et al. 1990; European Chromosome 16 Tuberous Sclerosis Consortium 1993) or may have a similar cytoplasmic localization (Rubinfeld et al. 1993; Su et al. 1993). Finally, the observation that mutations in the *NF2* gene are restricted to *NF2*-predisposed tumors and are not found in other tumor types which frequently lose chromosome 22 may indicate the presence of additional tumor suppressor genes on this chromosome.

ACKNOWLEDGMENTS

We thank J. Philippon, J.M. Sterkers, F. Resche, P. Vialat, A. Mazabraud, J.M. Zucker, J.P. Thiery, C. Petit, D. Chretien, C. Le Chalony, and G. Goubin for providing samples and biological reagents; and M. Giovannini and L. Selleri for their contribution to YAC isolation and characterization. We are particularly grateful to P. Dessen for his help in the database search. This work was supported by grants from the European Community Commissions, the Ministère de la Recherche et de la Technologie, the Ligue Nationale contre le Cancer, the Association pour la Recherche sur le Cancer, the Fondation pour la Recherche Médicale, the Association Française contre les Myopathies, the Association Neurofibromatose France. K.H.X., P.M. and C.D. are recipients of fellowships from Comité Départemental de l'Yonne de la Ligue Contre le Cancer, Ministère de la Recherche et de l'Enseignement Supérieur, and from the Association pour la recherche sur le Cancer, respectively.

REFERENCES

Aurias, A., C. Rimbaut, C. Buffe, J. Dubousset, and A. Mazabraud. 1983. Chromosomal translocations in Ewing's sarcoma. *N. Engl. J. Med.* **309:** 496.

Bailly, R.A., R. Bosselut, J. Zucman, F. Cormier, O. Delattre, M. Roussel, G. Thomas, and J. Ghysdael. 1994. DNA binding and transcriptional activation properties of the EWS/FLI-1 protein resulting from the t(11;22) translocation in Ewing's sarcoma. *Mol. Cell. Biol.* **14:** 3230.

Bijlsma, E., P. Mérel, D.A. Bosch, A. Westerveld, O. Delattre, G. Thomas, and T.J.M. Hulsebos. 1994. Analysis of mutations in the SCH gene in schwannomas. *Genes Chromosomes Cancer* **11:** 7.

Couturier, J., O. Delattre, M. Kujas, J. Philippon, M. Peter, G. Rouleau, A. Aurias, and G. Thomas. 1990. Assessment of chromosome 22 anomalies in neurinomas by combined karyotype and RFLPs analyses. *Cancer Genet. Cytogenet.* **45:** 55.

Crozat, A., P. Aman, N. Mandahl, and D. Ron. 1993. Fusion of CHOP to a novel RNA-binding protein in human myxoid liposarcoma. *Nature* **363:** 640.

de Jong, P.J., K. Vokobata, C. Chen, F. Lohman, L. Pederson, J. McNinch, and M. van Dilla. 1989. Human chromosome-specific partial digest libraries in human lambda and cosmid vectors (A2333). *Cytogenet. Cell Genet.* **51:** 985.

Delattre, O., C.J. Azambuja, A. Aurias, J. Zucman, M. Peter, F. Zhang, M.C. Hors-Cayla, G. Rouleau, and G. Thomas. 1991. Mapping of human chromosome 22 with a panel of somatic cell hybrids. *Genomics* **9:** 721.

Delattre, O., J. Zucman, T. Melot, X. Sastre, J.M. Zucker, G. Lenoir, P. Ambros, D. Sheer, C. Turc-Carel, T. Triche, A. Aurias, and G. Thomas. 1994. The Ewing family of tumours: A subgroup of small round cell tumours defined by specific chimeric transcripts. *N. Engl. J. Med.* **331:** 294.

Delattre, O., J. Zucman, B. Plougastel, C. Desmaze, T. Melot, M. Peter, H. Kovar, I. Joubert, P. De Jong, G. Rouleau, A. Aurias, and G. Thomas. 1992. Gene fusion with an ETS domain caused by chromosome translocation in human tumours. *Nature* **359:** 162.

Desmaze, C., J. Zucman, O. Delattre, T. Melot, G. Thomas, and A. Aurias. 1994. Interphase molecular cytogenetics of Ewing's sarcoma and peripheral neuroepithelioma t(11;22) translocation with flanking and overlapping cosmid probes. *Cancer Genet. Cytogenet.* **7:** 13.

Dumanski, J., G.A. Rouleau, M. Nordenskjold, and V.P. Collins. 1990. Molecular genetic analysis of chromosome 22 in 81 cases of meningioma. *Cancer Res.* **50:** 5863.

European Chromosome 16 Tuberous Sclerosis Consortium. 1993. Identification and characterization of the tuberous sclerosis gene on chromosome 16. *Cell* **75:** 1305.

Fontaine, B., M. Sanson, O. Delattre, A.G. Menon, G. Rouleau, B.R. Seizinger, A.F. Jewell, M.P. Hanson, A. Aurias, R.L. Martuza, J.F. Gusella, and G. Thomas. 1991. Parental origin of chromosome 22 loss in sporadic and NF2 neuromas. *Genomics* **10:** 280.

Hamelin, R., N. Jego, P. Laurent-Puig, M. Vidaud, and G. Thomas. 1993. Efficient screening of p53 mutations by denaturing gradient gel electrophoresis in colorectal tumors. *Oncogene* **8:** 2213.

Ichikawa, H., K. Shimizu, Y. Hayashi, and M. Ohki. 1994. An RNA-binding protein gene, TLS/FUS, is fused to ERG in human myeloid leukemia with t(16;21) chromosomal translocation. *Cancer Res.* **54:** 2865.

Ladanyi, M. and W. Gerald. 1994. Fusion of the EWS and WT1 genes in the desmoplastic small round cell tumor. *Cancer Res.* **54:** 2837.

Marineau, C., C. Baron, O. Delattre, J. Zucman, G. Thomas, and G.A. Rouleau. 1993. Dinucleotide repeat polymorphism at the D22S268 locus. *Hum. Mol. Genet.* **2:** 336.

Martuza, R.L. and R. Eldridge. 1988. Neurofibromatosis 2 (Bilateral acoustic neurofibromatosis). *N. Engl. J. Med.* **318:** 684.

May, W.A., M.L. Gishozky, S.L. Lessnick, L.B. Lunsford, B.C. Lewis, O. Delattre, J. Zucman, G. Thomas, and C.T. Denny. 1993. Ewing sarcoma 11;22 translocation produces a chimeric transcription factor that requires DNA-binding domain of FLI1 for transformation. *Proc. Natl. Acad. Sci.* **90:** 5752.

Mérel, P., K. Hoang-Xuan, M. Sanson, E. Bijlsma, P. Laurent-Puig, S. Pulst, G. Rouleau, G. Lenoir, J.M. Sterkers, J. Philippon, T. Hulsebos, A. Aurias, O. Delattre, and G. Thomas. 1994. Screening for germ-line mutations in the NF2 gene. *Genes Chromosomes Cancer* (in press).

Nowell, P.C. and D.A. Hungerford. 1960. A minute chromosome in human CML. *Science* **132:** 1497.

Peyrard M., I. Fransson, Y.G. Xie, F.Y. Han, M.H. Ruttledge, S. Swahn, J.E. Collins, I. Dunham, V.P. Collins, and J. Dumanski. 1994. Characterization of a new member of the human β-adaptin gene family from chromosome 22q12, a candidate meningioma gene. *Hum. Mol. Genet.* **3:** 1393.

Plougastel, B., MG. Mattei, G. Thomas, and O. Delattre. 1994. Cloning and chromosome localization of the mouse EWS gene. *Genomics* **23:** 278.

Plougastel, B., J. Zucman, M. Peter, G. Thomas, and O. Delattre. 1993. Genomic structure of the EWS gene and its relationship to EWSR1, a site of tumor-associated chromosome translocation. *Genomics* **18:** 609.

Rabbitts, T.H., A. Forster, R. Larson, and P. Nathan. 1993. Fusion of the dominant negative transcription regulator CHOP with a novel gene FUS by translocation t(12;16) in malignant liposarcoma. *Nat. Genet.* **4:** 175.

Rouleau, G.A., B.R. Seizinger, W. Wertelecki, J.L. Haines, D.W. Superneau, R.L. Martuza, and J.F. Gusella. 1990. Flanking markers bracket the neurofibromatosis type 2 (NF2) gene on chromosome 22. *Am. J. Hum. Genet.* **46:** 323.

Rouleau, G., P. Mérel, M. Lutchman, M. Sanson, J. Zucman, C. Marineau, K. Hoang-Xuan, S. Demczuk, C. Desmaze, B. Plougastel, S. Pulst, G. Lenoir, E. Bijlsma, R. Fashold, J. Dumanski, P. de Jong, D. Parry, R. Eldrige, A. Aurias, O. Delattre, and G. Thomas. 1993. Alteration in a new gene encoding a putative membrane-organizing protein causes neurofibromatosis type 2. *Nature* **363:** 515.

Rubinfeld, B., B. Souza, I. Albert, O. Muller, S.H. Chamberlain, F.R. Masiarz, S. Munemitsu, and P. Polakis. 1993. Association of the APC gene product with β-catenin. *Science* **262:** 1731.

Ruttledge, M.H., S.A. Narod, J.P. Dumanski, D.M. Parry, R. Eldridge, W. Wertelecki, J. Parboosingh, M.-C. Faucher, G.M. Lenoir, V.P. Collins, M. Nordenskjold, and G. Rouleau. 1993. Presymptomatic diagnosis for neurofibromatosis type 2 with chromosome 22 markers. *Neurology* **43:** 1753.

Sanson, M., C. Marineau, C. Desmaze, M. Luchtman, C. Baron, S. Narod, O. Delattre, G. Lenoir, G. Thomas, A. Aurias, and G. Rouleau. 1993. Germ-line deletion in a neurofibromatosis type 2 kindred inactivates the NF2 gene and a candidate meningioma locus. *Hum. Mol. Genet.* **2:** 1215.

Sato, N., N. Funayama, A. Nagafuchi, S. Yonemura, S. Tsukita, and S. Tsukita. 1992. A gene family consisting of ezrin, radixin and moesin. Its specific localization at actin filament/plasma membrane association sites. *J. Cell Sci.* **103:** 131.

Seizinger, B.R., R.L. Martuza, and J.F. Gusella. 1986. Loss of genes on chromosome 22 in tumorigenesis of human acoustic neuroma. *Genomics* **4:** 1.

Seizinger, B.R., S. De La Monte, L. Atkins, J.F. Gusella, and R.L. Martuza. 1987. Molecular genetic approach to human meningioma: Loss of genes on chromosome 22. *Proc. Natl. Acad. Sci.* **84:** 5419.

Selleri, L., M. Giovannini, A. Romo, J. Zucman, O. Delattre, G. Thomas, and G.A. Evans. 1994. Cloning of the entire FLI1 gene, disrupted by the Ewing's sarcoma translocation breakpoint on 11q24, in a yeast artificial chromosome. *Cytogenet. Cell Genet.* **67:** 129.

Su, L.K., B. Vogelstein, and K.W. Kinzler. 1993. Association of the APC tumor suppressor protein with catenins. *Cell* **262:** 1734.

Takeuchi, K., N. Sato, H. Kasahara, N. Funayama, A. Nagafuchi, S. Yonemura, S. Tsukita, and S. Tsukita. 1994. Perturbation of cell adhesion and microvilli formation by antisense oligonucleotides to ERM family members. *J. Cell Biol.* **125:** 1371.

Trofatter, J.A., M.M. MacCollin, J.L. Rutter, J.R. Murell, M.P. Duyao, D.M. Parry, R. Eldridge, N. Kley, A.G. Menon, K. Pulaski, V.H. Haase, C.M. Ambrose, D. Munroe, C. Bove, J.L. Haines, R.L. Martuza, M.E. MacDonald, B.R. Seizinger, M.P. Short, A.J. Buckler, and J.F.A. Gusella. 1993. A novel moesin-, ezrin-, radixin-like gene is a candidate for the neurofibromatosis 2 tumor suppressor. *Cell* **72:** 791.

Turc-Carel, C., I. Philip, M.P. Berger, T. Philip, and G.M. Lenoir. 1983. Chromosomal translocations in Ewing's sarcoma. *N. Engl. J. Med.* **309:** 497.

Turc-Carel, C., A. Aurias, F. Mugneret, I. Lizard Sidaner, C. Volk, J.P. Thiery, S. Olschwang, I. Philip, M.P. Berger, T. Philip, G.M. Lenoir, and A. Mazabraud. 1988. Chromosomes in Ewing's sarcoma. An evaluation of 85 cases and remarkable consistency of t(11;22)(q24;q12). *Cancer Genet. Cytogenet.* **32:** 229.

Whang-Peng, J., T.J. Triche, T. Knutsen, J. Miser, E.C. Douglass, and M.A. Israel 1984. Chromosomal translocations in peripheral neuroepithelioma. *N. Engl. J. Med.* **311:** 584.

Xie, Y.G., F.Y. Han, M. Peyrard, M.H. Ruttledge, I. Fransson, P. Dejong, J. Collins, I. Dunham, M. Nordenskjold, and J.P. Dumanski. 1993. Cloning of a novel, anonymous gene from a megabase-range YAC and cosmid contig in the neurofibromatosis 2/meningioma gene region on human chromosome 22q12. *Hum. Mol. Genet.* **2:** 1361.

Xu, G., B. Lin, K. Tanaka, D. Dunn, D. Wood, R. Gesteland, R. White, R. Weiss, and F. Tamanoi. 1990. The catalytic domain of the neurofibromatosis type 1 gene products stimulates ras GTPase and complements ira mutants of S. cerevisiae. *Cell* **63:** 835.

Zang, K.D. and H. Singer. 1967. Chromosomal constitution of meningiomas. *Nature* **216:** 84.

Zucman, J., O. Delattre, C. Desmaze, C. Azambuja, G. Rouleau, P. de Jong, A. Aurias, and G. Thomas. 1992a. Rapid isolation of cosmids from defined subregions by differential Alu-PCR hybridization on chromosome 22 specific library. *Genomics* **13:** 395.

Zucman, J., O. Delattre, C. Desmaze, A.L. Epstein, G. Stenman, F. Speleman, C.D.M. Fletchers, A. Aurias, and G. Thomas. 1993a. EWS and ATF-1 gene fusion induced by t(12;22) translocation in malignant melanoma of soft parts. *Nat. Genet.* **4:** 341.

Zucman, J., O. Delattre, C. Desmaze, B. Plougastel, I. Joubert, T. Melot, M. Peter, P. De Jong, G. Rouleau, A. Aurias, and G. Thomas. 1992b. Cloning and characterization of the Ewing's sarcoma and peripheral neuroepithelioma t(11;22) translocation breakpoints. *Genes Chromosomes Cancer* **5:** 271.

Zucman, J., T. Melot, C. Desmaze, J. Ghysdael, B. Plougastel, M. Peter, J.M. Zucker, T. Triche, D. Sheer, C. Turc-Carel, P. Ambros, V. Combaret, G. Lenoir, A. Aurias, G. Thomas, and O. Delattre. 1993b. Combinatorial generation of variable fusion proteins in the Ewing family of tumours. *EMBO J.* **12:** 4481.

Molecular Genetic Changes Found in Human Lung Cancer and Its Precursor Lesions

A.F. Gazdar,*† S. Bader,* J. Hung,* Y. Kishimoto,* Y. Sekido,*
K. Sugio,* A. Virmani,* J. Fleming,* D.P. Carbone,*‡ and J.D. Minna*‡§

*Simmons Cancer Center, and Departments of †Pathology, ‡Internal Medicine, and §Pharmacology,
University of Texas Southwestern Medical Center, Dallas, Texas 75235-8590*

LUNG CANCER IS CAUSED BY MUTATIONS IN DOMINANT AND RECESSIVE ONCOGENES

Lung cancer is the most common cancer killer of men and women in the United States. Current treatment options lead to cure in only 10%, a rate that has not altered in more than a decade (Minna 1992). Lung cancers frequently metastasize and are seldom diagnosed at an early curative stage. In contrast, lung cancer is largely preventable by stopping carcinogen exposure from cigarette smoke. Although it is highly desirable to stop all cigarette smoking, it is also realistic to target antismoking and chemoprevention efforts to those persons with the highest risk of developing lung cancer. Because a large fraction of lung cancers develop in former cigarette smokers, it first may be most practical to target these efforts to exsmokers, who for many years will still have a significantly increased risk of developing lung cancer.

Clinically evident lung cancer arises because of multiple changes (estimated to be between 10 and 20) in known and putative recessive oncogenes and several dominant oncogenes (Minna 1993; Gazdar and Carbone 1994). They are frequently mutated in human and experimental lung cancers, and in some cases, correction of the mutation causes reversal of the malignant phenotype in lung cancer cells (Mukhopadhyay et al. 1991; Takahashi et al. 1992; Fujiwara et al. 1993). Our approach is to determine which of these genetic events are common in primary invasive and metastatic cancer and then to search for the same changes in preneoplastic lesions often found in the lungs of persons at increased risk for the development of lung cancers. Such patients would be candidates for chemoprevention regimens, and the mutations could serve as intermediate markers to assess the efficacy of such regimens.

A list of common, recurring genetic changes in known or suspected dominant and recessive oncogenes involved with lung cancers is given in Table 1. These include changes in *p53*, *rb*, $p16^{ink4}$/MTS1 (9p21), and new candidate recessive oncogenes in chromosome regions 3p21.3, 3p25, and 3p12-14. Studies of large numbers of lung cancers have shown patterns of involvement with mutations in *rb*, *p53*, and 3p21 allele loss found in >90% of small-cell lung cancers (SCLC) but only uncommon mutation of $p16^{ink4}$, whereas *p53*, 3p21, and $p16^{ink4}$ mutations/deletions occur frequently in non-SCLC (NSCLC) cancers (Hayashi et al. 1994). This appears reasonable, given the proposed role of $p16^{ink4}$ in regulating Rb protein phosphorylation.

Table 1. Known and Postulated Dominant and Recessive Oncogenes Frequently Involved in Human Lung Cancer

Dominant oncogenes
  c-*myc*, N-*myc*, L-*myc* (deregulated expression)
  Ki-*ras*, Ha-*ras*, N-*ras* (activating mutation)
  Her-2/*neu* (overexpression)
  EGF receptor (overexpression)
  bcl-2 (overexpression)

  Additional sites of increased copy number as detected by CGH[a]
    1q24
    3q22-25, 3q27-28
    5p14-15
    5q24
    11q13-22
    19q13.1
    Xq26 (CGH)

Recessive oncogenes ("tumor suppressor" genes)
  3p12-13, 3p21.3, 3p25 (von Hippel-Lindau gene)
  5q (MCC, FAP region)
  8p
  9p21 (MTS1 or $p16^{ink4}$)
  11p15, 11p13
  13q14 (retinoblastoma gene)
  17p13 (p53)

  Additional sites of decreased copy number as detected by CGH[a]
    1p
    4p, 4q24-26
    10q26
    15q11-14
    16p11.2 (CGH), 16q11.22
    19p13.3 (CGH), 19q13.3-13.4
    22q12.1-13.1

[a] CGH, loss of genetic material documented by comparative genomic hybridization (Levin et al. 1994; Reid et al. 1994).

LUNG CANCER HAS MANY GENETIC ABNORMALITIES

As previously mentioned, multiple genetic changes are found in clinically evident lung cancer. Evidence for this finding is provided by cytogenetic and loss of heterozygosity studies (Table 1). Several important testable predictions arise from these observations: (1) bronchial epithelium is very well protected against becoming malignant; (2) early molecular detection is possible, because clinically evident lung cancer must be preceded by multiple molecular events; and (3) the potential for chemoprevention exists.

Widespread microsatellite instability has been reported in colorectal, ovarian, and other cancers that occurs because of mutations in the DNA mismatch repair genes *msh2* and *hMLH1* (Fishel et al. 1993; Orth et al. 1994). A somewhat different form of genomic instability (a change in allele size) has been reported in both SCLC and NSCLC (Merlo et al. 1994; Shridhar et al. 1994), presumably due to abnormalities in as yet unknown repair mechanisms. However, in our experience with both resected tumors and cell lines, widespread microsatellite instability is uncommon (our unpublished data). Nevertheless, the large number of mutations present in lung cancers indicates that there is some form of "genomic instability."

PROGNOSTIC SIGNIFICANCE OF ONCOGENE MUTATIONS IN LUNG CANCER

A variety of studies have asked whether the presence or absence of a particular mutation conveys altered prognosis in lung cancer patients. Some years ago, we demonstrated that the variant subtype of SCLC, frequently associated with c-*myc* amplification and overexpression, conveyed a poor prognosis (Radice et al. 1982; Johnson et al. 1992). However, the variant form of SCLC is relatively uncommon, especially in previously untreated patients. Because the prognoses for SCLC and advanced stage NSCLC are very poor, most of the other studies have correlated oncogene mutations with survival in NSCLC treated with curative intent by surgical resection.

Several studies have consistently shown an association between *ras* gene mutations and impaired prognosis in both early and advanced stage lung cancer (Mitsudomi et al. 1991; Sugio et al. 1992; Rosell et al. 1993). Her-2/*neu* (erbB-2, p185-*neu*) overexpression also has been associated with shortened survival in lung adenocarcinoma, especially when combined with *ras* mutations (Kern et al. 1990, 1994). There appears to be little or no effect on survival of loss of heterozygosity (LOH) at chromosomal regions 3p and 9p (Horio et al. 1993; Y. Kishimoto et al., unpubl.).

Studies correlating alterations in *p53* with prognosis have given complex and contradictory results (Hiyoshi et al. 1992; McLaren et al. 1992; Quinlan et al. 1992; Horio et al. 1993; Mitsudomi et al. 1993; Mrkve et al. 1993; Carbone et al. 1994). Unfortunately, many of the studies cannot be compared directly, because they have determined either overexpression of protein or the presence of mutations, but not both parameters. Protein overexpression, as demonstrated by positive immunostaining of tumor cells, is usually a reflection of missense mutations resulting in an increased p53 protein half-life with accumulation of p53 protein in tumors (Bodner et al. 1992). In one of the few studies that measured both parameters, we found that protein overexpression, but not the presence of mutations, is predictive of significantly shortened survival in resected NSCLC patients (Carbone et al. 1994). Clearly, further studies are required to determine the role of the *p53* gene as a prognostic indicator.

SEARCH FOR NEW DOMINANT AND RECESSIVE ONCOGENES IN LUNG CANCER

Because lung cancers exhibit multiple cytogenetic alterations and sites of allelic imbalance, they provide a rich source for finding new mutant genes involved in human cancers. Many of the genetic changes listed in Table 1 represent target areas to find new, predominantly recessive oncogenes. After discovering and cloning putative new dominant or recessive oncogenes in other human cancers, lung cancer cells can then be tested for mutations or abnormalities of expression. Alternatively, genetic abnormalities detected by cytogenetic or molecular techniques point to the location of genes which then have to be isolated by the laborious process of positional cloning.

Cytogenetics and the new technique of comparative genomic hybridization (CGH) (Levin et al. 1994; Ried et al. 1994) have identified many areas of chromosomal deletion and nonreciprocal translocation in lung cancers (indicating the presence of new recessive oncogenes), as well as areas of amplification (indicating positions of new dominant or recessive oncogenes) (Table 1). In addition, comparison of tumor and DNA from normal tissue from the same patients, when tested with probes detecting polymorphic markers, has also demonstrated loss of heterozygosity (allele loss) at many of these same loci (Table 1). These studies have suggested that loss of function mutations as seen in recessive ("tumor suppressor") genes are more common than those in dominant oncogenes. In some cases, such as in chromosome regions 3p13-14, 3p21.3, and 9p21, homozygous deletions have been found (Rabbitts et al. 1990; Daly et al. 1993; Olopade et al. 1993; Yamakawa et al. 1993). These homozygous deletions help limit the search to a reasonable number of cDNAs encoding genes in the region of interest. These can be systematically tested for mutations in lung cancer cells.

Currently, the most important new tumor suppressor genes to be cloned in lung cancer are those that reside on the short arm of chromosome 3. This region is also involved in the pathogenesis of several other types of cancer, including mesotheliomas, renal cell, head and neck, cervical, and breast carcinomas. The 3p region appears to be particularly important, since it is one of the first to undergo change in preneoplastic tissues (see below). We, and other investigators, have found evidence for cytogenetic abnormalities and loss of heterozygosity in this area in all forms of lung cancer (Whang-Peng et al. 1982; Kok et al. 1987; Rabbitts et al. 1989; Brauch et al. 1990). There appear to be three or four separate regions involved at 3p12-13, 3p14, 3p21.3, and 3p25 (Brauch et al. 1990; Hibi et al. 1992). The von Hippel-Lindau (VHL) tumor suppressor gene at 3p25 is frequently mutated in sporadic renal cancer and is occasionally mutated in lung cancers (Sekido et al. 1994). Thus, it is possible that an additional recessive oncogene lies in the 3p25 region near the location of the VHL gene.

PRENEOPLASIA AND THE DEVELOPMENT OF LUNG CANCER

The most effective way to prevent lung cancer is to avoid smoking cigarettes. Smoking cessation and treatment with retinoids result in decreased numbers of preneoplastic lesions (Lippman et al. 1994). However, it remains to be determined if these steps result in a reversal of molecular lesions. Many lung cancers occur in exsmokers, and although lung cancer risk starts to decrease after smoking cessation, it takes a number of years (15 or more) to return to a near normal risk status. Thus, the full beneficial effects of smoking cessation may take nearly two decades to achieve. Exsmokers represent the most important increased risk group to focus early diagnosis and chemoprevention efforts.

Patients at increased risk for the development of lung cancer (current or previous heavy smokers, and patients who have been cured of cancer of the upper aerodigestive tract) frequently develop a series of histologically identified preneoplastic and neoplastic changes in their respiratory epithelium (Saccomanno et al. 1974; Auerbach et al. 1978). These changes include (from earliest to most advanced) hyperplasia, metaplasia, dysplasia of increasing severity, carcinoma in situ (CIS), invasive cancer, and metastatic cancer. Multiple lesions are frequently present throughout the respiratory epithelium, a phenomenon referred to as "field cancerization" (Strong et al. 1984).

Although our knowledge of the molecular events in invasive cancers is relatively extensive, until recently we knew nothing about the sequence of events in preneoplastic lesions. The reasons for this ignorance include the small, focal nature of the lesions (much smaller than the polyploid preneoplastic lesions associated with colorectal cancer) and the inability to identify them accurately except by histological examination. A few studies have provided suggestions that molecular lesions can be identified at early stages of the pathogenesis of lung cancer (Sozzi et al. 1991, 1992; Sundaresan et al. 1992). To further understand these changes, we have developed a scheme to systematically search for mutations in preneoplastic lesions using archival paraffin-embedded materials (Fig. 1). Sections from lung cancer resections are examined for the presence of preneoplastic foci. Using a microdissection technique, 200–400 cells from these different areas are precisely isolated along with invasive primary and metastatic cancer, histologically normal epithelium, and stromal lymphocytes (as a source of normal constitutional DNA). Using PCR-based techniques, these different specimens are typed for *ras* and *p53* mutations, for allele loss for 3p, 5q, 9p, and *rb* gene markers, and for evidence of genomic instability. Our preliminary

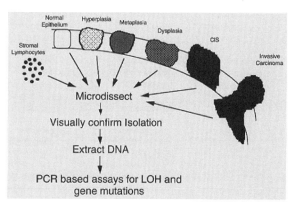

Figure 1. Schema outlining procedures for investigating molecular lesions in preinvasive lesions and invasive cancers of the respiratory tract.

results in the most common lung cancer, adenocarcinoma, indicate that 3p and 9p allele losses occur at the earliest (hyperplasia) stages, *ras* mutations occur at the time of noninvasive cancer (CIS), and *p53* mutations appear at variable times (Table 2). Progressively worse histological stages accumulate progressively larger numbers of lesions. In addition, some abnormalities, such as the degree of aneuploidy, become progressively more severe. Although not listed, increased proliferation rate and genomic instability probably begin at the stage of hyperplasia and become progressively more advanced with histological stage.

Finally, it was of great interest to find that the precise molecular changes (specific base substitution or allele loss) found in primary cancers were also found in the preneoplastic lesions even at a distance from the primary cancer. These findings suggest that multiple, widespread areas of the respiratory epithelium have undergone genetic changes predisposing to the future development of cancer. Finding loss of heterozygosity of 3p and 9p in early lesions confirms that oncogenes at these two chromosome regions play a crucial role in lung cancer pathogenesis.

MOLECULAR EARLY DETECTION OF LUNG CANCER

To date, most studies on preneoplastic lesions have been done in patients who also have clinically evident lung cancer. Although preneoplastic lesions can be found in smokers who have not yet developed lung cancer, and in other groups at increased risk, their malignant potential has not been fully defined. In collaboration with Dr. Stephen Lam, British Columbia Cancer Center, Vancouver, we are studying the molecular alterations in preneoplastic lesions in patients at increased risk. The Vancouver group has developed a fluorescence bronchoscope that greatly increases the number of preinvasive lesions that can be visualized and biopsied (Lam et al. 1993). By correlating our studies with the clinical course of the patients, we will be able to determine the prognostic impact of molecu-

Table 2. Progressive Increase in Mutations in Preneoplastic and Invasive Lung Cancers

Histological lesion	Mutation/abnormality frequently present
Hyperplasia	del 3p, (aneuploidy), (del 9p)
Metaplasia/dysplasia	del 3p, aneuploidy, del 9p, (*p53* mutations)
Carcinoma in situ	del 3p, aneuploidy, del 9p, *p53* mutations, *ras* mutations
Invasive or metastatic carcinoma	del 3p, aneuploidy, del 9p, *p53* mutations, *ras* mutations, multiple other mutations

Mutations frequently found in preneoplastic foci and invasive cancers are listed (Sugio et al. 1994; Hung et al. 1995; and our unpublished data). Lesions found occasionally are in parentheses. Progressively worse histological lesions accumulate progressively larger numbers of lesions. In addition, some abnormalities, such as the degree of aneuploidy, become progressively more severe. Although not listed, increased proliferation rate and genomic instability probably begin at the stage of hyperplasia and become progressively more advanced with stage.

lar changes in preneoplasia and follow their modulation by chemoprevention regimens. In addition, there have been anecdotal reports where a known *p53* or *ras* gene mutation in a clinically evident lung cancer has been identified and then, using this molecular information, a retrospective search has identified the same mutant sequences in sputum samples harvested several months before a clinical diagnosis was made (Mao et al. 1994). Application of molecular techniques to sputum, blood, and other clinical specimens may play an important role in early diagnosis.

EARLY DETECTION OF LUNG CANCER THROUGH IMMUNE RESPONSES

Cancer patients are capable of mounting immune responses against mutant oncogene products, and this finding can be used as a prognostic indicator as well as for novel therapeutic approaches (see below). Antibodies against p53 have been detected in the serum of about 15% of lung cancer patients (Winter et al. 1993). Although the antibodies only arise in patients whose tumors have *p53* mutations, they are not specific for mutant p53 protein. In addition to anti-p53 antibodies, about 40–50% of lung cancer patients develop antibodies, detected by immunoblotting, that react with other proteins in their autologous tumor cell extracts (Winter et al. 1993).

GENETIC PREDISPOSITION AND MOLECULAR EPIDEMIOLOGY

The majority of heavy smokers do not develop lung cancer, suggesting that genetic factors may play an important role. In families with inherited mutations in the *rb* or *p53* recessive oncogenes who survive childhood or young adult cancers, there appears to be a significantly increased risk of developing lung cancer (Sanders et al. 1989; Malkin 1993). First-degree relatives of lung cancer patients are at increased risk (Ambrosone et al. 1993), suggesting that there are as yet unidentified lung-cancer-associated genes. Certain alleles of drug-metabolizing enzyme genes such as the P450 enzymes or other polymorphic markers are associated with an increased risk of developing lung cancer after adjusting for cigarette smoking (Caporaso et al. 1992; Nakachi et al. 1993).

SPECTRUM OF ONCOGENE MUTATIONS

The *p53* and *ras* genes have a mutational spectrum consistent with carcinogens present in cigarette smoke, in particular, a predominance of G to T transversions. The mutational spectrum in nonsmokers with lung cancer is different from that found in smoking-associated cancers (Takeshima et al. 1993). In addition, there is DNA strand asymmetry in mutagenesis, with mutations occurring in the noncoding strand (the "sense" strand) consistent with damage by bulky carcinogens. As more genes are identified that undergo point mutations, these base and strand specificity changes can be confirmed. It will be of interest to study lung cancers that occur long after cigarette smoking cessation to determine the mutational pattern.

RECENTLY IDENTIFIED GENES AND REGULATORY SYSTEMS OF POTENTIAL INTEREST

Genes involved in the regulation of the cell cycle may play important roles in cancer biology. Loss of function of these genes should deprive cells of vital growth controls, and, thus, they are new candidate tumor suppressor genes. Some of the newly identified genes that may be involved in lung cancer include the cyclins, cyclin-dependent kinases, and the cyclin kinase inhibitors (such as $p21^{waf1/cip1}$, $p27^{kip}$, and $p16^{ink4}$) (El Deiry et al. 1993; Harper et al. 1993; Kamb et al. 1994; Nobori et al. 1994; Polyak et al. 1994; Toyoshima and Hunter 1994). Overexpression of cyclin D (the *prad-1* gene) may lead to loss of cell cycle control and thus represent a new type of dominant oncogene (Wu et al. 1994). Of particular interest, the cyclin kinase inhibitor $p16^{ink4}$ (also known as MTS1) is included in the 9p21 homozygous deletions frequently found in lung cancers and other human tumors and has been nominated as a tumor suppressor gene under the name *MTS1* (multiple tumor suppressor gene 1) (Kamb et al. 1994). Although

Table 3. Homozygous Deletions of p16^{INK4} Gene (*MTS1*) in Human Lung Cancer Cell Lines

Histological type	No. deleted[a]/ Total studied	Percent deleted
SCLC	4/57	7
Non-SCLC	21/51	41
Mesothelioma	5/5	100
Carcinoid	0/4	0
Extrapulmonary Small cell	1/5	20
Total	31/122	25

[a] 500 bp PCR exon 2 product.

Table 4. Correlation between *rb* Gene Mutations and p16^{ink4} Gene Deletions in 65 Lung Cancer Cell Lines

		p16^{ink4}	
		intact	deleted
rb	wild type	19	13
	mutated	33	0

homozygous deletions of MTS1 are frequent in lung cancer cell lines (Table 3), mutations are more common in fresh tumors (Hayashi et al. 1994). Because *rb* gene mutations and MTS1 deletions are part of the same pathway, they are seldom found in the same tumor cell lines (Table 4).

NEWER THERAPEUTIC APPROACHES

The past few years have resulted in a tremendous increase in our knowledge about the basic biology of cancer cells and their associated genetic lesions. Already translational research is having an impact on all aspects of the clinical management of the lung cancer patient, ranging through chemoprevention, early diagnosis, risk assessment, and novel therapeutic strategies. A partial list of some of these exciting applications is presented in Table 5. Some of these approaches are discussed below.

Mutations in dominant and recessive oncogenes represent highly specific and absolute differences between tumor cells and their normal counterparts. Thus, they are ideal targets for therapeutic interventions. Even nuclear proteins such as p53 protein can be targeted, because they are broken down, processed, and presented by MHC molecules on the cell surface. Vaccination of mice against mutant p53 peptides causes the generation of a CD8$^+$, MHC-restricted cytotoxic T-cell response (Yanuck et al. 1993). The immunization is done by using a synthetic oligopeptide that includes the mutant amino acid sequence. The peptide is adsorbed to lymphocytes that bind the peptide to MHC molecules and, following irradiation, are used for vaccination. A highly specific cytotoxic T-cell response is directed against tumor cells expressing the endogenous mutant p53 protein. Similar data also exist for mutant Ras peptides, and of interest, cytotoxic T-cell responses can be generated against overexpressed but nonmutant Her-2/*neu* proteins (Yoshino et al. 1994). Currently, we (in collaboration with Jay Berzofsky, NCI) are conducting a clinical trial of mutant p53 and Ras peptide vaccination in patients with lung cancer and other malignancies (protocol number NCI T93-0148).

Overexpression (with or without gene amplification) of the Her-2/*neu* and EGF receptor dominant oncogenes occurs in lung cancer (Arteaga 1994). Antibodies against these receptors have been developed, and these antibodies in experimental systems (alone or coupled to toxins) can exhibit antitumor effects. However, it has been found recently that antibody-receptor binding leads to an increase in tumor cell sensitivity to certain cytotoxic agents, such as etoposide and cisplatin. Thus, it may be possible to target antibodies to tumor cells because of overexpression of these dominant oncogene products and thereby "sensitize" the tumor cells to chemotherapy. In addition, overexpression of HER-2/*neu* by itself is a marker for intrinsic multidrug resistance in NSCLC cell lines (Tsai et al. 1993). If these findings are confirmed, HER-2/*neu* gene expression may be used as a predictor of therapeutic failure in NSCLC.

Mutations in *ras* and *p53* genes offer targets for gene therapy. Introduction of antisense *ras* or wild-type *p53* into lung cancers mutant for the gene in question leads to a suppression of tumorigenicity in experimental systems (Mukhopadhyay et al. 1991; Takahashi et al. 1992; Fujiwara et al. 1993, 1994). There also appears to be killing of "bystander" tumor cells not transfected with the particular gene by an unknown mechanism. These findings have resulted in a clinical trial to treat locally recurrent, endobronchial tumors by direct instillation of the genes into the tumor (Roth 1993).

Table 5. Potential Translational Research Applications of Mutations in Dominant and Recessive Oncogenes That Can Be Applied to Human Lung Cancer

- Detection of genetic changes in preneoplastic lesions
- Monitoring of genetic changes in bronchial epithelium during chemoprevention trials
- Early lung cancer diagnosis by detection of genetic changes in sputum, blood, and other clinical specimens
- Family cancer genetics, genetic epidemiology, inherited predisposition
- Immunologic early detection of cancer by detecting immune responses against oncogene products
- Prognostic information from specific mutations present
- Mutational spectrum as a clue to carcinogens
- Vaccination with mutant oncopeptide products to generate a cytotoxic T-cell response against tumor cells
- Pharmacological intervention to inhibit dominant oncogene function or replace recessive oncogene function
- Gene therapy to inhibit dominant oncogene function or replace recessive oncogene function
- Antibodies against cell-surface oncogene receptors or growth factors

The rapidly emerging knowledge about the structure and function of dominant and recessive oncogene products makes it possible to develop rational pharmacological approaches to lung cancers. For example, we now know that *p53* functions as a transcription factor and that gene mutations occurring in lung and other human tumors may abolish transcriptional ability (Chen et al. 1993). However, these lung cancer *p53* mutants are often temperature-sensitive and can regain a wild-type conformation and some wild-type transcriptional activity when shifted to a lower temperature (Medcalf et al. 1992; Chen et al. 1993). The *p53* gene functions as part of a DNA-damage-sensing pathway, including responses to cytotoxic agents, to turn on the transcription of genes that will lead to G_1/S cell cycle arrest and programmed cell death (apoptosis) (Kastan et al. 1992; Oren 1992; Lowe et al. 1993). Since the mutant p53 protein is frequently expressed at high levels in lung cancer cells, it may be possible to develop therapeutic approaches that cause the mutant *p53* to switch to a wild-type conformation and thereby induce programmed cell death.

The *ras* gene is occasionally mutated in lung cancer, especially adenocarcinomas (Slebos and Rodenhuis 1992). To be functional, the Ras protein has to be able to anchor to the inner cell membrane; this occurs in part because of farnesylation of the Ras protein. The enzyme responsible for this addition (farnesyltransferase) has been characterized, and several groups have made inhibitors of this enzyme. It is of great interest that these inhibitors reverse the transformation of rodent embryonic fibroblasts transformed by a mutant *ras* gene (James et al. 1993). The inhibitors also inhibit the in vitro growth of some lung cancer cells (our unpublished data). Clinical trials of these inhibitors are in the planning stages.

ACKNOWLEDGMENTS

We thank our many collaborators who have contributed generously to many aspects of our studies. They include Jacqueline Whang-Peng, Jay Berzofsky, Michael Yanuck, Bruce Johnson, Frederick Kaye, and Ilona Linnoila (National Cancer Institute, Bethesda), Farida Latif and Michael Lerman (NCI-Frederick Cancer Research Center, Frederick, Maryland), Olufunmilayo Olopade (Department of Medicine, University of Chicago, Illinois), and Michael Christman and Nikki Levin (University of California, San Francisco).

REFERENCES

Ambrosone, C.B., U. Rao, A.M. Michalek, K.M. Cummings, and C.J. Mettlin. 1993. Lung cancer histologic types and family history of cancer. Analysis of histologic subtypes of 872 patients with primary lung cancer. *Cancer* **72:** 1192.

Arteaga, C.L. 1994. Epidermal growth factor receptors and erbB-2 in human lung cancer. In *Lung cancer: Principles and practice* (ed. H. Pass et al.). J.B. Lippincott, Philadelphia, Pennsylvania.

Auerbach, O., G. Saccomanno, M. Kuschner, R.D. Brown, and L. Garfinkel. 1978. Histologic findings in the tracheobronchial tree of uranium miners and non-miners with lung cancer. *Cancer* **42:** 483.

Bodner, S.M., J.D. Minna, S.M. Jensen, D. D'Amico, D. Carbone, T. Mitsudomi, J. Fedorko, D.L. Buchhagen, M.M. Nau, A.F. Gazdar, and R.I. Linnoila. 1992. Expression of mutant p53 proteins in lung cancer correlates with the class of p53 gene mutation. *Oncogene* **7:** 743.

Brauch, H., K. Tory, F. Kotler, A.F. Gazdar, O.S. Pettengill, B. Johnson, S. Graziano, T. Winton, C.H. Buys, G.D. Sorenson, J. Minna, and B. Zbar. 1990. Molecular mapping of deletion sites in the short arm of chromosome 3 in human lung cancer. *Genes Chromosomes Cancer* **1:** 240.

Caporaso, N.E., P.G. Shields, M.T. Landi, G.L. Shaw, M.A. Tucker, R. Hoover, H. Sugimura, A. Weston, and C.C. Harris. 1992. The debrisoquine metabolic phenotype and DNA-based assays: Implications of misclassification for the association of lung cancer and the debrisoquine metabolic phenotype. *Environ. Health Perspect.* **98:** 101.

Carbone, D.P., T. Mitsudomi, I. Chiba, S. Paintadosi, V. Rusch, J.A. Nowak, D. McIntire, D. Slamon, A. Gazdar, and J.D. Minna. 1994. p53 immunostaining positivity is associated with reduced survival and is imperfectly correlated with gene mutations in resected non-small cell lung cancer. *Chest* (in press).

Chen, J.Y., W.D. Funk, W.E. Wright, J.W. Shay, and J.D. Minna. 1993. Heterogeneity of transcriptional activity of mutant p53 proteins and p53 DNA target sequences. *Oncogene* **8:** 2159.

Daly, M.C., R.H. Xiang, D. Buchhagen, C.H. Hensel, D.K. Garcia, A.M. Killary, J.D. Minna, and S.L. Naylor. 1993. A homozygous deletion on chromosome 3 in a small cell lung cancer cell line correlates with a region of tumor suppressor activity. *Oncogene* **8:** 1721.

El Deiry, W.S., T. Tokino, V.E. Velculescu, D.B. Levy, R. Parsons, J.M. Trent, D. Lin, W.E. Mercer, K.W. Kinzler, and B. Vogelstein. 1993. *WAF1*, a potential mediator of p53 tumor suppression. *Cell* **75:** 817.

Fishel, R., M.K. Lescoe, M.R. Rao, N.G. Copeland, N.A. Jenkins, J. Garber, M. Kane, and R. Kolodner. 1993. The human mutator gene homolog *MSH2* and its association with hereditary nonpolyposis colon cancer. *Cell* **75:** 1027.

Fujiwara, T., E.A. Grimm, T. Mukhopadhyay, D.W. Cai, L.B. Owen-Schaub, and J.A. Roth. 1993. A retroviral wild-type p53 expression vector penetrates human lung cancer spheroids and inhibits growth by inducing apoptosis. *Cancer Res.* **53:** 4129.

Fujiwara, T., E.A. Grimm, T. Mukhopadhyay, W.W. Zhang, L.B. Owen-Schaub, and J.A. Roth. 1994. Induction of chemosensitivity in human lung cancer cells in vivo by adenovirus-mediated transfer of the wild-type p53 gene. *Cancer Res.* **54:** 2287.

Gazdar, A.F. and D.P. Carbone. 1994. *The biology and molecular genetics of lung cancer*. R.G. Landes, Austin, Texas.

Harper, J., G. Adami, N. Wei, K. Keyomarsi, and S. Elledge. 1993. The p21 Cdk-interacting protein Cip1 is a potent inhibitor of G_1 cyclin-dependent kinases. *Cell* **75:** 805.

Hayashi, N., Y. Sugimoto, E. Tsuchiya, M. Ogawa, and Y. Nakamura. 1994. Somatic mutations of the MTS (multiple tumor suppressor) 1/CDK41 (cyclin-dependent kinase-4 inhibitor) gene in human primary non-small cell lung carcinomas. *Biochem. Biophys. Res. Commun.* **202:** 1426.

Hibi, K., T. Takahashi, K. Yamakawa, R. Ueda, Y. Sekido, Y. Ariyoshi, M. Suyama, H. Takagi, Y. Nakamura, and T. Takahashi. 1992. Three distinct regions involved in 3p deletion in human lung cancer. *Oncogene* **7:** 445.

Hiyoshi, H., Y. Matsuno, H. Kato, Y. Shimosato, and S. Hirohashi. 1992. Clinicopathological significance of nuclear accumulation of tumor suppressor gene p53 product in primary lung cancer. *Jpn. J. Cancer Res.* **83:** 101.

Horio, Y., T. Takahashi, T. Kuroishi, K. Hibi, M. Suyama, T. Niimi, K. Shimokata, K. Yamakawa, Y. Nakamura, and R. Ueda. 1993. Prognostic significance of p53 mutations and

3p deletions in primary resected non-small cell lung cancer. *Cancer Res.* **53:** 1.

Hung, J., Y. Kishimoto, K. Sugio, A. Virmani, D.D. McIntire, J.D. Minna, and A.F. Gazdar. 1995. Allele specific chromosome 3p deletions occur early in lung carcinoma pathogenesis. *J. Am. Med. Assoc.* (in press).

James, G.L., J.L. Goldstein, M.S. Brown, T.E. Rawson, T.C. Somers, R.S. McDowell, C.W. Crowley, B.K. Lucas, A.D. Levinson, and J.C. Marsters. 1993. Benzodiazepine peptidomimetics: Potent inhibitors of *ras* farnesylation in animal cells. *Science* **260:** 1937.

Johnson, B.E., J.F. Brennan, D.C. Ihde, and A.F. Gazdar. 1992. *myc* family DNA amplification in tumors and tumor cell lines from patients with small cell lung cancer. *J. Natl. Cancer Inst. Monogr.* **13:** 39.

Kamb, A., N.A. Gruis, J. Weaver-Feldhaus, Q. Liu, K. Harshman, S.V. Tavtigian, E. Stockert, R.S. Day, B.E. Johnson, and M.H. Skolnick. 1994. A cell cycle regulator potentially involved in genesis of many tumor types (see comments). *Science* **264:** 436.

Kastan, M.B., Q. Zhan, W.S. El-Deiry, F. Carrier, T. Jacks, W.V. Walsh, B.S. Plunkett, B. Vogelstein, and A.J. Fornace. 1992. A mammalian cell cycle checkpoint pathway utilizing p53 and GADD45 is defective in ataxia-telangiectasia. *Cell* **71:** 587.

Kern, J.A., D.A. Schwartz, J.E. Nordberg, D.B. Weiner, M.I. Greene, L. Torney, and R.A. Robinson. 1990. p185*neu* expression in human lung adenocarcinomas predicts shortened survival. *Cancer Res.* **50:** 5184.

Kern, J.A., R.J. Slebos, B. Top, S. Rodenhuis, D. Lager, R.A. Robinson, D. Weiner, and D.A. Schwartz. 1994. C-*erb*B-2 expression and codon 12 K-*ras* mutations both predict shortened survival for patients with pulmonary adenocarcinomas. *J. Clin. Invest.* **93:** 516.

Kok, K., J. Osinga, B. Carritt, M.B. Davis, A.H. van der Hout, A.Y. van der Veen, R.M. Landsvater, L.F. de Leij, H.H. Berendsen, P.E. Postmus, S. Poppema, and C.H. Buys. 1987. Deletion of a DNA sequence at the chromosomal region 3p21 in all major types of lung cancer. *Nature* **330:** 578.

Lam, S., C. MacAulay, J. Hung, and B. Palcic. 1993. Detection of dysplasia and carcinoma *in situ* using a lung imaging fluorescence endoscope (LIFE) device. *J. Thorac. Cardiovasc. Surg* **105:** 1035.

Levin, N., P. Brzoska, N. Gupta, J. Minna, J. Gray, and M. Christman. 1994. Identification of novel genetic alterations in small cell lung carcinoma. *Cancer Res.* **54:** 5086.

Lippman, S.M., S.E. Benner, and W.K. Hong. 1994. Retinoid chemoprevention studies in upper aerodigestive tract and lung carcinogenesis. *Cancer Res.* **54:** 2025.

Lowe, S.W., H.E. Ruley, T. Jacks, and D.E. Housman. 1993. p53-dependent apoptosis modulates the cytotoxicity of anticancer agents. *Cell* **74:** 957.

Malkin, D. 1993. p53 and the Li-Fraumeni syndrome. *Cancer Genet. Cytogenet.* **66:** 83.

Mao, L., R.H. Hruban, J.O. Boyle, M. Tockman, and D. Sidransky. 1994. Detection of oncogene mutations in sputum precedes diagnosis of lung cancer. *Cancer Res.* **54:** 1634.

McLaren, R., I. Kuzu, M. Dunnill, A. Harris, D. Lane, and K.C. Gatter. 1992. The relationship of p53 immunostaining to survival in carcinoma of the lung. *Br. J. Cancer* **66:** 735.

Medcalf, E.A., T. Takahashi, I. Chiba, J. Minna, and J. Milner. 1992. Temperature-sensitive mutants of p53 associated with human carcinoma of the lung. *Oncogene* **7:** 71.

Merlo, A., M. Mabry, E. Gabrielson, R. Vollmer, S.B. Baylin, and D. Sidransky. 1994. Frequent microsatellite instability in primary small cell lung cancer. *Cancer Res.* **54:** 2098.

Minna, J. 1992. Neoplasms of the lung. In *Harrison's principles of internal medicine* (ed. E. Braunwald et al.), p. 1221. McGraw-Hill, New York.

———. 1993. The molecular biology of lung cancer pathogenesis. *Chest* **103:** 449.

Mitsudomi, T., T. Oyama, T. Kusano, T. Osaki, R. Nakanishi, and T. Shirakusa. 1993. Mutations of the p53 gene as a predictor of poor prognosis in patients with non-small-cell lung cancer. *J. Natl. Cancer Inst.* **85:** 2018.

Mitsudomi, T., S.M. Steinberg, H.K. Oie, J.L. Mulshine, R. Phelps, J. Viallet, H. Pass, J.D. Minna, and A.F. Gazdar. 1991. *ras* gene mutations in non-small cell lung cancers are associated with shortened survival irrespective of treatment intent. *Cancer Res.* **51:** 4999.

Mrkve, O., O.J. Halvorsen, R. Skjaerven, L. Stangeland, A. Gulsvik, and O.D. Laerum. 1993. Prognostic significance of p53 protein expression and DNA ploidy in surgically treated non-small cell lung carcinomas. *Anticancer Res.* **13:** 571.

Mukhopadhyay, T., M. Tainsky, A.C. Cavender, and J.A. Roth. 1991. Specific inhibition of K-*ras* expression and tumorigenicity of lung cancer cells by antisense RNA. *Cancer Res.* **51:** 1744.

Nakachi, K., K. Imai, S. Hayashi, and K. Kawajiri. 1993. Polymorphisms of the CYP1A1 and glutathione S-transferase genes associated with susceptibility to lung cancer in relation to cigarette dose in a Japanese population. *Cancer Res.* **53:** 2994.

Nobori, T., K. Miura, D.J. Wu, A. Lois, K. Takabayashi, and D.A. Carson. 1994. Deletions of the cyclin-dependent kinase-4 inhibitor gene in multiple human cancers. *Nature* **368:** 753.

Olopade, O.I., D.L. Buchhagen, K. Malik, J. Sherman, T. Nobori, S. Bader, M.M. Nau, A.F. Gazdar, J.D. Minna, and M.O. Diaz. 1993. Homozygous loss of the interferon genes defines the critical region on 9p that is deleted in lung cancers. *Cancer Res.* **53:** 2410.

Oren, M. 1992. p53: The ultimate tumor suppressor gene? *J. FASEB* **6:** 3169.

Orth, K., J. Hung, A. Gazdar, A. Bowcock, J.M. Mathis, and J. Sambrook. 1994. Genetic instability in human ovarian cancer cell lines. *Proc. Natl. Acad. Sci.* **91:** 9495.

Polyak, K., M.-H. Lee, H. Erdjument-Bromage, A. Koff, J.M. Roberts, P. Tempst, and Massagué. 1994. Cloning of p27^{Kip1}, a cyclin-dependent kinase inhibitor and a potential mediator of extracellular antimitogenic signals. *Cell* **78:** 59.

Quinlan, D.C., A.G. Davidson, C.L. Summers, H.E. Warden, and H.M. Doshi. 1992. Accumulation of p53 protein correlates with a poor prognosis in human lung cancer. *Cancer Res.* **52:** 4828.

Rabbitts, P., J. Bergh, J. Douglas, F. Collins, and J. Waters. 1990. A submicroscopic homozygous deletion at the D3S3 locus in a cell line isolated from a small cell lung carcinoma. *Genes Chromosomes Cancer* **2:** 231.

Rabbitts, P., J. Douglas, M. Daly, V. Sundaresan, B. Fox, P. Haselton, F. Wells, D. Albertson, J. Waters, and J. Bergh. 1989. Frequency and extent of allelic loss in the short arm of chromosome 3 in nonsmall-cell lung cancer. *Genes Chromosomes Cancer* **1:** 95.

Radice, P., M.J. Matthews, D.C. Ihde, A.F. Gazdar, D.N. Carney, P.A. Bunn, M.H. Cohen, B.E. Fossieck, R.W. Makuch, and J.D. Minna. 1982. The clinical behavior of "mixed" small cell/large cell bronchogenic carcinoma compared to "pure" small cell subtypes. *Cancer* **50:** 2894.

Ried, T., I. Petersen, H. Holtgreve-Grez, M.R. Speicher, E. Schrock, S. du Manoir, and T. Cremer. 1994. Mapping of multiple DNA gains and losses in primary small cell lung carcinomas by comparative genomic hybridization. *Cancer Res.* **54:** 1801.

Rosell, R., S. Li, Z. Skacel, J. Mate, J. Maestre, M. Canela, E. Tolosa, P. Armengol, A. Barnadas, and A. Ariza. 1993. Prognostic impact of mutated K-*ras* gene in surgically resected non-small cell lung cancer. *Oncogene* **8:** 2407.

Roth, J.A. 1993. Targeting lung cancer genes: A novel approach to cancer prevention and therapy. *Semin. Thorac. Cardiovasc. Surg.* **5:** 178.

Saccomanno, G., V.E. Archer, O. Auerbach, R.P. Saunders, and L.M. Brennan. 1974. Development of carcinoma of the lung as reflected in exfoliated cells. *Cancer* **33:** 256.

Sanders, B.M., M. Jay, G.J. Draper, and E.M. Roberts. 1989. Non-ocular cancer in relatives of retinoblastoma patients. *Br. J. Cancer* **60:** 358.

Sekido, Y., S. Bader, F. Latif, J.R. Gnarra, A.F. Gazdar, W.M. Linehan, B. Zbar, M.I. Lerman, and J.D. Minna. 1994. Molecular analysis of the von Hippel-Lindau disease tumor suppressor gene in human lung cancer cell lines. *Oncogene* **9:** 1599.

Shridhar, V., J. Siegfried, J. Hunt, M. del Mar Alonso, and D.I. Smith. 1994. Genetic instability of microsatellite sequences in many non-small cell lung carcinomas. *Cancer Res.* **54:** 2084.

Slebos, R.J. and S. Rodenhuis. 1992. The *ras* gene family in human non-small-cell lung cancer. *Natl. Cancer Inst. Monogr.* **13:** 23.

Sozzi, G., M. Miozzo, R. Donghi, S. Pilotti, C.T. Cariani, U. Pastorino, G. Della Porta, and M.A. Pierotti. 1992. Deletions of 17p and p53 mutations in preneoplastic lesions of the lung. *Cancer Res.* **52:** 6079.

Sozzi, G., M. Miozzo, E. Tagliabue, C. Calderone, L. Lombardi, S. Pilotti, U. Pastorino, M.A. Pierotti, and G. Della Porta. 1991. Cytogenetic abnormalities and overexpression of receptors for growth factors in normal bronchial epithelium and tumor samples of lung cancer patients. *Cancer Res.* **51:** 400.

Strong, M.S., J. Incze, and C.W. Vaughan. 1984. Field cancerization in the aerodigestive tract—Its etiology, manifestation, and significance. *J Otolaryngol.* **13:** 1.

Sugio, K., Y. Kishimoto, A. Virmani, J.Y. Hung, and A.F. Gazdar. 1994. K-*ras* mutations are a relatively late event in the pathogenesis of lung carcinomas. *Cancer Res.* **54:** 5811.

Sugio, K., T. Ishida, H. Yokoyama, T. Inoue, K. Sugimachi, and T. Sasazuki. 1992. *ras* gene mutations as a prognostic marker in adenocarcinoma of the human lung without lymph node metastasis. *Cancer Res.* **52:** 2903.

Sundaresan, V., P. Ganly, P. Hasleton, R. Rudd, G. Sinha, N.M. Bleehen, and P. Rabbitts. 1992. p53 and chromosome 3 abnormalities, characteristic of malignant lung tumours, are detectable in preinvasive lesions of the bronchus. *Oncogene* **7:** 1989.

Takahashi, T., D. Carbone, T. Takahashi, M.M. Nau, T. Hida, I. Linnoila, R. Ueda, and J.D. Minna. 1992. Wild-type but not mutant p53 suppresses the growth of human lung cancer cells bearing multiple genetic lesions. *Cancer Res.* **52:** 2340.

Takeshima, Y., T. Seyama, W.P. Bennett, M. Akiyama, S. Tokuoka, K. Inai, K. Mabuchi, C.E. Land, and C.C. Harris. 1993. p53 mutations in lung cancers from non-smoking atomic-bomb survivors. *Lancet* **342:** 1520.

Toyoshima, H. and T. Hunter. 1994. p27, a novel inhibitor of G_1 cyclin-Cdk protein kinase activity, is related to p21. *Cell* **78:** 67.

Tsai, C.M., K.T. Chang, R.P. Perng, T. Mitsudomi, M.H. Chen, C. Kadoyama, and A.F. Gazdar. 1993. Correlation of intrinsic chemoresistance of non-small-cell lung cancer cell lines with HER-2/neu gene expression but not with *ras* gene mutations. *J. Natl. Cancer Inst.* **85:** 897.

Whang-Peng, J., S.C. Kao, E.C. Lee, P.A. Bunn, D.N. Carney, A.F. Gazdar, and J.D. Minna. 1982. A specific chromosome defect associated with human small cell lung cancer. *Science* **215:** 181.

Whang-Peng, J., T. Knutsen, A. Gazdar, S.M. Steinberg, H. Oie, I. Linnoila, J. Mulshine, M. Nau, and J.D. Minna. 1991. Nonrandom structural and numerical chromosome changes in non-small-cell lung cancer. *Genes Chromosomes Cancer* **3:** 168.

Winter, S.F., Y. Sekido, J.D. Minna, D. McIntire, B.E. Johnson, A.F. Gazdar, and D.P. Carbone. 1993. Antibodies against autologous tumor cell proteins in small cell lung cancer patients are associated with improved survival. *J. Natl. Cancer Inst.* **85:** 2012.

Wu, F., K.C. Bui, S. Buckley, and D. Warburton. 1994. Cell cycle-dependent expression of cyclin D1 and a 45 kD protein in human A549 lung carcinoma cells. *Am. J. Respir. Cell Mol. Biol.* **10:** 437.

Yamakawa, K., T. Takahashi, Y. Horio, Y. Murata, E. Takahashi, K. Hibi, S. Yokoyama, R. Ueda, T. Takahashi, and Y. Nakamura. 1993. Frequent homozygous deletions in lung cancer cell lines detected by a DNA marker located at 3p21.3-p22. *Oncogene* **8:** 327.

Yanuck, M., D.P. Carbone, C.D. Pendleton, T. Tsukui, S.F. Winter, J.D. Minna, and J.A. Berzofsky. 1993. A mutant p53 tumor suppressor protein is a target for peptide-induced CD8 + cytotoxic T-cells. *Cancer Res.* **53:** 3257.

Yoshino, I., P.S. Goedegebuure, G.E. Peoples, A.S. Parikh, J.M. DiMaio, H.K. Lyerly, A.F. Gazdar, and T.J. Eberlein. 1994. HER2/*neu*-derived peptides are shared antigens among human non-small cell lung cancer and ovarian cancer. *Cancer Res.* **54:** 3387.

Molecular Characterization of *QM*, a Novel Gene with Properties Consistent with Tumor Suppressor Function

E. STANBRIDGE,* A. FARMER,* A. MILLS,* T. LOFTUS,*
D. KONGKASURIYACHAI,* S. DOWDY,† AND B. WEISSMAN‡

*Department of Microbiology and Molecular Genetics, College of Medicine, University of California, Irvine, California 92717; †Department of Pathology, Washington University of Medicine, St. Louis, Missouri 63110; ‡Department of Microbiology, University of North Carolina School of Medicine, Chapel Hill, North Carolina 27599

QM is a novel gene that was isolated via a subtractive DNA hybridization procedure (Dowdy et al. 1991a) during a search for a Wilms' tumor suppressor gene.

Sporadic Wilms' tumor (WT), a childhood nephroblastoma, has been associated with constitutional deletions in the p13 region of chromosome 11. A subset of WT cases occur in association with aniridia, a defect in the development of the iris, as well as urogenital abnormalities and mental retardation, and is referred to as the WAGR syndrome. Using positional cloning, a candidate TS gene, *WT1*, has been cloned (Call et al. 1990). This gene is deleted or mutated in a fraction of Wilms' tumors and is also the germ-line mutation in a portion of Denys Drash syndrome patients (Pelletier et al. 1991). These patients occasionally present with Wilms' tumors. The WT1 gene product is a transcriptional factor that seems to have complementary and antagonistic functions with the EGR gene product (Madden et al. 1991). It clearly seems to function in urogenital development, but its role in Wilms' tumor development remains unclear.

The simple association of 11p13 deletions in WT has not survived further scrutiny. There is now evidence for loss of heterozygosity at 11p15 without involvement of 11p13. Using monochromosome transfer as a functional assay for tumor suppression, we found that a normal copy of chromosome 11 suppresses the tumorigenic phenotype of a Wilms' tumor cell line, G401 (Weissman et al. 1987). More recently, we have generated copies of normal chromosome 11 that have discrete interstitial deletions involving either 11p13 or 11p15 (Dowdy et al. 1990). When copies of a normal human chromosome 11, deleted in the region 11p13 or 11p15 (via radiation-reduction), were introduced into the WT cell line G401 via microcell fusion, the chromosome 11 deleted in region p13 (which included the *WT1* gene) retained its tumor-suppressing effect, whereas the chromosome 11 deleted in region p14-15 no longer suppressed tumor formation (Dowdy et al. 1991b). This single example of tumor-suppressing activity in the 11p14-15 region obviously does not preclude a role for *WT1* as a tumor suppressor gene but does raise the possibility of genetic heterogeneity in sporadic Wilms' tumors.

The tumorigenic G401 parental cell and nontumorigenic G401/chromosome 11 microcell hybrid were used in a subtractive cDNA hybridization strategy designed to isolate the putative tumor suppressor gene mapping to chromosome 11p15. *QM* was isolated in this procedure, and using the partial cDNA as a probe, it was found that the nontumorigenic G401/11 microcell hybrid had an elevated expression of *QM* mRNA relative to the parental G401 cells (Dowdy et al. 1991a).

QM IS A HIGHLY CONSERVED GENE

Sequencing of the full-length *QM* cDNA revealed that it was a novel gene lacking homology with any sequence present in the DNA databases. The cDNA coded for a 214-amino-acid peptide rich in charged amino acids, most notably arginine and lysine, with a predicted pI of 10.5. Northern and Southern blot analysis showed *QM* to be conserved in mammals and to be a member of a large multigene family composed of at least one expressed gene (Dowdy et al. 1991a). We have now analyzed *QM* sequences from a diverse array of eukaryotes, including mammals, birds, plants, and yeast, and have discovered a remarkable degree of conservation (Farmer et al. 1994). Alignment of these sequences indicated a high degree of conservation throughout the first 175 residues of the protein and revealed several interesting features (Fig. 1). Human and mouse *QM* are highly conserved with only one amino acid change, at codon 202. Throughout the various eukaryotic species there is considerable conservation of charged amino acids within specific regions of the protein. Secondary structure analysis suggests that two of these regions form amphipathic 2-helices, one basic and one acidic (Fig. 2). A third conserved charged domain, comprising the amino-terminal 30 amino acids, is both basic and proline rich. The carboxy-terminal region, spanning approximately 40 amino acids, contains only four residues that are 100% conserved. However, many of the changes among residues 176–200 are conservative. The longest single block of conserved residues is a 25-amino-acid stretch in the center of the primary sequence. Intriguingly, this sequence is rich in glycine and alanine residues and has similarity to the glycine-rich, nucleotide-binding P-loop of the *ras* protein superfamily. Also of interest are four conserved cysteine residues located at amino acid positions 8, 49, 71 (72 in plants), and 105. Whether the

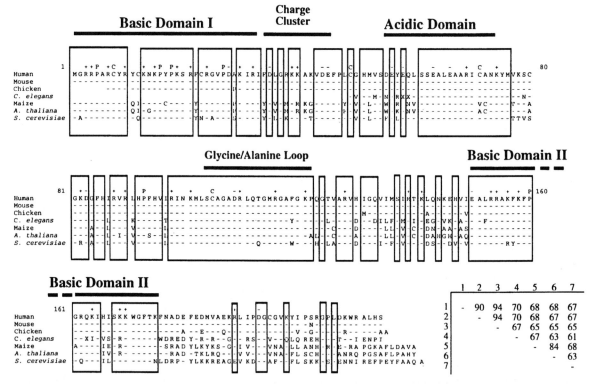

Figure 1. Alignment of QM peptide sequences. The complete sequence for human QM is shown at top, with amino acid identities between human QM and that from other species being denoted by dashes. Conserved charged residues, prolines, and cysteines are denoted +/−, P or C, respectively, above the sequence. An X denotes an ambiguous codon and a space denotes missing data. Blocks of strong conservation are boxed, and domains within the protein are detailed above the sequence. (*Insert*) Similarity matrix showing percent identity among the various sequences (1, human; 2, mouse; 3, chicken; 4, *Caenorhabditis elegans*; 5, maize; 6, *Arabidopsis thaliana*; 7, *Saccharomyces cerevisiae*).

cysteines can form disulfide bonds in the reducing intracellular environment or whether they serve other crucial functions remains to be determined. Remarkably, *QM* homologs from a diverse array of eukaryotes, ranging from human to yeast, contain no insertions or deletions within the protein itself.

Analysis of the rates of sequence divergence of the various homologs was found to be slow—on the order of 1% change every 22 million years. This is comparable to rates of divergence for enzymes of primary metabolism as well as with the highly conserved core domain of the TATA-box-binding factor TFIID. This slow rate of change is consistent with a critical functional role for *QM* in eukaryotic cells.

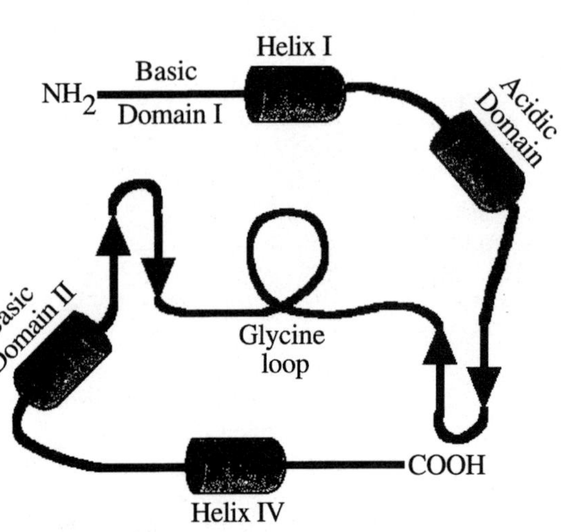

QM FUNCTION

The strategy used to isolate *QM* was based on the prediction that it would function as a tumor suppressor gene. Our preliminary data are consistent with this prediction. The first indication of *QM* function came from the studies of Monteclaro and Vogt (1993). They screened a λgt11 chicken cDNA library with avian c-Jun protein to isolate proteins that complexed with c-Jun. One of the proteins so isolated, termed Jif-1, is the chicken homolog of *QM*. This protein was found to interact with c-Jun and inhibit transcriptional activation of the collagenase promoter, which contains an AP-1 consensus site. We have confirmed this interaction and inhibition of transcriptional activation of the collage-

Figure 2. Stylized model of secondary structure domains within the QM protein, in which helices are shown as cylinders, and the β-sheet regions are represented as pairs of opposing arrows.

nase promoter with human *QM* (T. Loftus et al., in prep.). We have also noted that *QM* binds independently and specifically to DNA sequences. Thus, it is reasonable to believe that *QM* may have other functions in the cell besides the negative regulation of AP-1-activated transcription.

c-Jun is a component of the AP-1 transcription factor complex, forming heterodimers with c-Fos, and is thought to play an important role in activation of genes whose products are involved in stimulation of cell growth. Inhibition of transcription of such genes would be expected to inhibit cell proliferation. Our experience with *QM* transfection studies supports this notion. Although preliminary, our experiments indicate that overexpression of *QM* leads to cessation of growth potential in mammalian cells (T. Loftus, unpubl.). There are also preliminary indications that yeast *QM* mutants behave as cell cycle mutants (B. Trumpower, pers. comm.).

QM IS A MEMBER OF A MULTIGENE FAMILY

Southern blot analysis with the *QM* cDNA revealed at least 14 different fragments following digestion of genomic DNA with *Eco*RI restriction endonuclease (Dowdy et al. 1991a). Subsequent studies revealed that the genomic copy of *QM* consists of seven exons spanning approximately 3.0 kb (van den Ouweland et al. 1992). There are no *Eco*RI sites within the gene. These data suggest that the *QM* gene is a member of a multigene family. Possibilities could range from a single transcribed gene plus a series of pseudogenes to multiple closely related actively transcribed genes. We are currently engaged in cloning and sequencing other members of the family. To date, three homologs have been sequenced. All three are retrotransposons. Interestingly, all three retain a complete open reading frame (ORF). These sequences have been subjected to sequence divergence analysis. In Table 1 are depicted the rates of sequence divergence with respect to human *QM* in the predicted coding sequence and flanking (untranslated) regions. Mouse *QM* is also included for comparative purposes. For two of the retrotransposons (C2D and 14B), there are too few changes to determine any bias. These two retrotransposons may, therefore, have been formed relatively recently and the retention of a complete ORF is not significant. The third retrotransposon (C12D), however, shows a significant bias against changes in the coding sequence, as does mouse *QM*. Analysis of the rates of amino-acid-altering and silent nucleotide changes within the coding sequences (Table 2) again show no bias for C2D and 14B but strong evidence of a bias against mutations that change the amino acid sequence of C12D. Taken together, these data indicate that C12D may be a processed gene that is functionally active. In support of this, we have found that the 5'-flanking sequences of C12D are functionally active in luciferase reporter gene assays (A. Farmer, unpubl.). We have recently obtained evidence for other, intron-containing homologs of *QM* in both mouse and human genomes (A. Mills and D. Kongkasuriyachai, unpubl.).

IS *QM* A TUMOR SUPPRESSOR GENE?

This, of course, is an important question, although it should not detract one from the fact that *QM* is an intrinsically important gene whose function(s) is likely to be critical for the well-being of the cell. The subtractive cDNA hybridization strategy that resulted in the isolation of *QM* was designed to identify the *WT2* Wilms' tumor suppressor gene. *QM* is clearly not *WT2*, since it maps to chromosome Xq28 (van den Ouweland et al. 1992). This might tempt one to discard *QM* as a candidate tumor suppressor gene. However, the possibility remains that *QM* is a downstream effector of *WT2*, and that its expression is regulated by *WT2* in a fashion analogous to that observed between *WAF-1* and *p53* (El-Deiry et al. 1993). The functional prop-

Table 1. Rates of Sequence Divergence with Respect to Human *QM* in the Predicted Coding Sequence and Flanking (Untranslated) Regions of Three Human *QM* Family Members and Mouse *QM*

Sequence compared to *QM*	% Change in noncoding sequence	% Change in coding sequence	Probability (p) that rate of change is the same	Comment
C2D	1.0	0.6	>0.1	too few changes to determine any bias
C19-14B	2.4	1.1	>0.1	too few changes to determine any bias
C12D	36.7	10.5	<0.002	significant bias against changes in the coding sequence
Mouse	53.4	9.4	<0.002	significant bias against changes in the coding sequence

These data were generated using the program TRANS (Strong et al. 1993).

Table 2. Rates of Amino-acid-altering and Silent Nucleotide Changes within the Coding Sequences of Three Human *QM* Family Members and Mouse *QM*

Sequence compared to human *QM*	Silent divergence	Replacement divergence	Comment
C2D	0.006	0.007	too few changes to determine any bias
C19-14B	0.006	0.013	too few changes to determine any bias
C12D	0.388	0.027	strong evidence of a bias against mutations that change the amino acid sequence
Mouse	0.479	0.009	strong evidence of a bias against mutations that change the amino acid sequence

Values given are corrected divergences as calculated by the method of Prager and Wilson (Brown et al. 1982).

erties ascribed so far to *QM* are consistent with a growth suppressor function in tumors. We have begun a search for such evidence, focusing first on genes that map to Xq28 which may be involved in familial predisposition to cancer. One such syndrome is dyskeratosis congenita (Drachtman and Alter 1992), which is characterized by epithelial dysplasia and predisposition to malignancy. Preliminary sequencing of *QM* in patients with this syndrome has not revealed any mutations (T. Loftus et al., in prep.). Again, it should be noted that *WAF-1* mutations are extremely rare in human tumors. We are at a very early stage in our analysis of *QM*, but we can be fairly certain that it is a novel gene whose protein product functions as a transcription factor and may be intimately involved in control of cell proliferation. Its role in human cancer remains to be determined.

ACKNOWLEDGMENT

The studies described here were supported by National Institutes of Health grant CA-19401.

REFERENCES

Brown, W., E. Prager, A. Wang, and A. Wilson. 1982. Mitochondrial DNA sequences of primates: Tempo and mode of evolution. *J. Mol. Evol.* **18:** 225.

Call, K., T. Glaser, C. Ito, A. Buckler, J. Pelletier, D. Haber, E. Rose, A. Kral, H. Yeger, W. Lewis, C. Jones, and D. Housman. 1990. Isolation and characterization of a zinc finger polypeptide gene at the human chromosome 11 Wilms' tumor locus. *Cell* **60:** 509.

Dowdy, S., C. Fasching, D. Scanlon, G. Casey, and E. Stanbridge. 1990. Irradiation microcell-mediated chromosome transfer (XMMCT): The generation of specific chromosomal arm deletions. *Genes, Chromosomes Cancer* **2:** 318.

Dowdy, S., K. Lai, B. Weissman, Y. Matsui, B. Hogan, and E. Stanbridge. 1991a. The isolation and characterization of a novel cDNA demonstrating an altered mRNA level in nontumorigenic Wilms' microcell hybrid cells. *Nucleic Acids Res.* **19:** 5763.

Dowdy, S., C. Fasching, D. Scanlon, D. Araujo, E. Livanos, K. Lai, B. Weissman, and E. Stanbridge. 1991b. Suppression of tumorigenicity in Wilms' tumor by the p14:p15 region of chromosome 11. *Science* **254:** 293.

Drachtman, R. and B. Alter. 1992. Dyskeratosis congenita: Clinical and genetic heterogeneity. Report of a new case and review of the literature. *Am. J. Pediatr. Hematol. Oncol.* **14:** 297.

Farmer, A., T. Loftus, A. Mills, K. Sato, J. Neill, T. Tron, M. Yang, B. Trumpower, and E. Stanbridge. 1994. Extreme evolutionary conservation of the QM, a novel c-Jun associated transcription factor. *Hum. Mol. Genet.* **3:** 723.

El-Deiry, W., T. Tokino, V. Velculescu, D. Levy, R. Parsons, J. Trent, D. Lin, W. Mercer, K. Kinzler, and B. Vogelstein. 1993. *WAF1*, a potential mediator of p53 tumor suppression. *Cell* **75:** 817.

Madden, S., D. Cook, J. Morris, A. Gashler, V. Sukhatme, and F. Rauscher III. 1991. Transcriptional repression mediated by the WT1 Wilms' tumor gene product. *Science* **253:** 1550.

Monteclaro, F. and P. Vogt. 1993. A Jun-binding protein related to a putative tumor suppressor. *Proc. Natl. Acad. Sci.* **90:** 6726.

Pelletier, J., W. Bruening, F. Li, D. Haber, T. Glaser, and D. Housman. 1991. *WT1* mutations contribute to abnormal genital system development and hereditary Wilm's tumor. *Nature* **353:** 431.

Strong, M., K. Chandy, and G. Gutman. 1993. Molecular evolution of voltage-sensitive ion channel genes: On the origins of electrical excitability. *Mol. Biol. Evol.* **10:** 221.

van den Ouweland, A., P. Kioschis, M. Verdijk, F. Tamanini, D. Toniolo, A. Poustka, and B. van Oost. 1992. Identification and characterization of a new gene in the human Xq28 region. *Hum. Mol. Genet.* **1:** 269.

Weissman, B., P. Saxon, S. Pasquale, G. Jones, A. Geiser, and E. Stanbridge. 1987. Introduction of a normal human chromosome 11 into a Wilms' tumor cell line controls its tumorigenic expression. *Science* **236:** 175.

Barrett's Esophagus: A Model of Human Neoplastic Progression

K. NESHAT,* C.A. SANCHEZ,* P.C. GALIPEAU,* D.S. COWAN,*
S. RAMEL,† D.S. LEVINE,* AND B.J. REID*

Division of Gastroenterology RG-24, Department of Medicine, University of Washington Medical Center, Seattle, Washington 98195; †Department of Surgery, Ersta Hospital, Stockholm, Sweden

In 1976, Nowell hypothesized that cancer develops as a consequence of an acquired genomic instability that predisposes to the development of abnormal clones of cells with accumulated genetic errors (Nowell 1976). Some clones gain selective proliferative advantages, and eventually a subclone evolves that has acquired the capacity for invasion, becoming an early carcinoma. There is now substantial evidence that human cancers develop in association with a process of clonal evolution that leads to the accumulation of genetic errors (Vogelstein et al. 1988; Huang et al. 1992; Raskind et al. 1992; Sidransky et al. 1992). Many studies have investigated genetic abnormalities in advanced carcinomas in surgical specimens. Although the study of advanced cancers is useful for identifying genetic abnormalities that have accumulated in the cancer, this approach is limited in its ability to determine the order in which genetic and other abnormalities develop during earlier stages of neoplastic progression.

Premalignant lesions can be used to investigate intermediate events in the progression to cancer, but most premalignant lesions are removed when they are detected, making it difficult to study serial changes in the same patient over time. For example, colonic adenomas are precursors of colon cancer, and a series of elegant studies has identified genetic abnormalities in different classes of adenomas (Vogelstein et al. 1988; Baker et al. 1990; Fearon and Vogelstein 1990; Powell et al. 1992). However, adenomas are removed when they are detected at colonoscopy, and therefore, the sequence of events in individual patients during the progression to cancer cannot be determined by serial biopsies of the same patient over time.

Barrett's esophagus is a condition in which the normal squamous epithelium of the esophagus is replaced by a metaplastic columnar epithelium (Phillips and Wong 1991). It develops as a complication in approximately 10–12% of patients with chronic gastroesophageal reflux and predisposes to the development of adenocarcinomas of the esophagus and gastric cardia (Hamilton et al. 1988; Phillips and Wong 1991; Reid 1991). In the 1970s and 1980s, the incidence of these two Barrett's-associated cancers increased more rapidly than that of any other cancer in the United States (Blot et al. 1991). Unfortunately, Barrett's adenocarcinomas are rarely discovered in time for cure, and 93% of patients who develop an esophageal adenocarcinoma will eventually die of their disease (Miller et al. 1992).

Barrett's esophagus is a unique model system in which to investigate intermediate events of human epithelial neoplasia. Because patients with Barrett's esophagus typically have symptoms of gastroesophageal reflux, such as heartburn, they frequently seek medical attention before they develop cancer (Phillips and Wong 1991). The Barrett's epithelium can be visualized and biopsied safely during upper gastrointestinal endoscopy (Levine and Reid 1992). Total removal of Barrett's epithelium requires esophagectomy, a procedure with substantial morbidity and mortality (Muller et al. 1990). Therefore, the standard of care for many patients includes periodic endoscopic biopsy surveillance for the early detection of cancer (Spechler 1987; Levine and Reid 1992). Thus, intermediate events in neoplastic progression can be evaluated by serial biopsies of the same patient over time and related to progression (Reid et al. 1992). Furthermore, in addition to cancer, esophagectomy specimens frequently contain the surrounding premalignant epithelium in which the cancer arose, permitting the study of multiple stages of neoplastic progression in a single esophagectomy specimen (Rabinovitch et al. 1988; Reid et al. 1988a).

GENETIC AND CELL CYCLE ABNORMALITIES IN BARRETT'S ADENOCARCINOMAS

Barrett's adenocarcinoma, like many other human malignancies, has a high prevalence of genetic and cell cycle abnormalities (Table 1) (Reid et al. 1987, 1993; Blount et al. 1991; Boynton et al. 1992; Huang et al. 1992; Neshat et al. 1994). None of these abnormalities is present in control biopsies from columnar epithelium of

Table 1. Barrett's Adenocarcinomas

	Prevalence of abnormalities
abnormality	prevalence (%)
↑G_1 fractions	15/17 (88%)
↑S-phase fractions	24/28 (86%)
↑G_2 fractions	18/28 (64%)
Aneuploid	25/28 (89%)
17p allelic loss	34/36 (94%)
$p53$ mutations	14/16 (88%)
5q allelic loss	15/20 (75%)

the upper gastrointestinal tract; they all develop as somatic events in Barrett's metaplastic epithelium during the progression to cancer.

ANEUPLOIDY IN BARRETT'S ESOPHAGUS

We, and other workers, have used DNA content flow cytometry to investigate aneuploid cell populations in different histological stages of neoplastic progression in Barrett's esophagus (McKinley et al. 1987; Reid et al. 1987; Fennerty et al. 1989; Reid 1991; Robaszkiewicz et al. 1991). The prevalence of aneuploid cell populations increases with increasing histological risk of malignancy. Aneuploid cell populations were not detected in biopsies from 44 patients who had gastroesophageal reflux disease without Barrett's metaplastic epithelium. However, aneuploid cell populations were found in biopsies from 3 of 70 patients with metaplasia (4%), 2 of 32 patients whose biopsies were in the indefinite for dysplasia or low-grade dysplasia range (6%), 5 of 8 patients with high-grade dysplasia (63%), and 25 of 28 cancers (89%).

We investigated the distribution of aneuploid cell populations in Barrett's esophagus by taking endoscopic biopsy specimens at different levels of the metaplastic epithelium and by sampling the mucosa of esophagectomy specimens in a grid array (Reid et al. 1987; Rabinovitch et al. 1988). Many aneuploid cell populations are localized to a single region of the esophageal mucosa, but some spread to involve large areas of the esophagus. For example, one patient had the same 2.2N aneuploid cell population at each level of a 10-cm length of metaplastic epithelium. Control biopsies from gastric and squamous mucosa were diploid, indicating that this aneuploid cell population developed in the metaplastic epithelium as the result of a somatic genetic event. Cytogenetic analysis of endoscopic biopsies from this patient confirmed that the aneuploid cell population contained clonal karyotypic abnormalities that were found at multiple levels of the metaplastic epithelium (Raskind et al. 1992). Flow cytometric and cytogenetic analyses of biopsies from this patient and others indicate that abnormal clones of cells can spread by a process of cell division to involve large regions of esophageal mucosa in persons with Barrett's esophagus (Rabinovitch et al. 1988; Raskind et al. 1992; Reid et al. 1992).

Some patients have multiple aneuploid cell populations in their endoscopic biopsies or in their esophagectomy specimens (Rabinovitch et al. 1988; Reid et al. 1992). We found multiple (2–14) aneuploid cell populations in 12 of 14 patients (86%) who had high-grade dysplasia, adenocarcinoma, or both, in Barrett's esophagus. Different aneuploid cell populations occupied defined, but sometimes overlapping, spatial distributions in the Barrett's epithelium, suggesting that they represented abnormal clones of cells that had spread to involve variable regions of esophageal mucosa. In 7 patients, we investigated the relationship between multiple aneuploid cell populations that were present in premalignant epithelium and the ploidy of the invasive carcinoma that developed within the multiple aneuploidies (Blount et al. 1990). In 6 of the 7 patients, the cancer contained only one of the multiple aneuploid cell populations present in the premalignant epithelium. The seventh and largest cancer contained 5 different ploidies in the cancer itself.

Our results indicate that endoscopic biopsies from most patients with Barrett's esophagus are diploid, but some patients develop genomic instability within the metaplastic epithelium. This instability can lead to the evolution of abnormal clones of cells with large changes in DNA content or ploidy. Some of these clones can spread to involve large regions of esophageal mucosa. With continued instability, multiple aneuploid cell populations can evolve in the premalignant epithelium, and one of the aneuploid cell populations may acquire the capacity for invasion, becoming an early carcinoma. If the cancer is not resected at this stage, continued instability can lead to multiple aneuploid cell populations within the cancer itself.

THE CELL CYCLE IN BARRETT'S ESOPHAGUS

Cell cycle checkpoints cause arrest at specific stages of the cell cycle in response to genotoxic injury or a failure to complete previous events of the cell cycle (Weinert and Hartwell 1988; Hartwell and Weinert 1989; Dasso and Newport 1990; Enoch and Nurse 1990; Brown et al. 1991). After DNA damage, one checkpoint causes arrest in G_1 and another in G_2 (Weinert and Hartwell 1988; Hartwell and Weinert 1989; Kastan et al. 1991; Kuerbitz et al. 1992). Loss or adaptation of these checkpoints permits cell division in the presence of genotoxic injury, leading to aneuploidy and the propagation of genetic errors (Weinert and Hartwell 1988; Hartwell and Weinert 1989; Livingstone et al. 1992; Yin et al. 1992).

Previous studies have shown that some, but not all, patients with Barrett's esophagus have increased proliferative fractions in the metaplastic epithelium (Herbst et al. 1978; Pellish et al. 1980; Reid et al. 1987; Jankowski et al. 1991; Gray et al. 1992). However, it has been difficult to assess the transitions from G_0 to G_1 to S phase during neoplastic progression in vivo because most assays of proliferation cannot simultaneously measure all three intervals in human biopsies. Therefore, we used a multiparameter flow cytometric assay that simultaneously measures Ki67 (a proliferation-associated nuclear antigen present in cells in late G_1, S, G_2, and mitosis, but not in G_0) and DNA content to assess proliferation in biopsies from patients with Barrett's esophagus (Reid et al. 1993). Biopsies from control columnar epithelium of the upper gastrointestinal tract (fundic gland mucosa and cardiac gland mucosa) had low G_1 and S-phase fractions, suggesting that the cells were predominantly in G_0. For example, increased S-phase fractions ($>7\%$) were detected in only 4 of 420 control biopsies (1%) from fundic gland mucosa.

Increased Ki67-positive G_1 fractions were found in Barrett's metaplastic epithelium at an early stage of neoplastic progression, but S-phase fractions typically remained normal, suggesting that regulatory mechanisms at the G_1/S-phase transition prevented uncontrolled progression of G_1 cells into S phase even if G_1 fractions increased to very high levels. At later stages of neoplastic progression, increased S-phase fractions developed, usually in association with aneuploidy, high-grade dysplasia, or carcinoma. Only 19 of 73 diploid biopsies (26%) from Barrett's esophagus at all histological stages of progression had increased S-phase fractions compared with 21 of 22 aneuploid cell populations (95%) ($p < 0.001$).

Some patients also develop elevated G_2 fractions, and the prevalence of elevated G_2 fractions increases with increasing histological risk of cancer (Reid et al. 1987; Reid 1991). Increased G_2 fractions were found in none of 44 patients who had gastroesophageal reflux disease without Barrett's metaplastic epithelium and none of 70 patients who had metaplasia without dysplasia, but they were detected in 7 of 32 patients with indefinite/low-grade dysplasia (22%), 7 of 8 patients with high-grade dysplasia (88%), and 18 of 28 patients with cancer (64%).

These results are not consistent with a model of cell cycle regulation that is dependent on a single control step between G_1 and S phase, as is often presumed in neoplasia. Rather, neoplastic progression in Barrett's esophagus seems to be associated with at least three types of cell cycle abnormalities: (1) increased G_1 fractions; (2) increased S-phase fractions; and (3) accumulation of cells in G_2.

ANEUPLOIDY AND INCREASED G_2 FRACTIONS AS PREDICTORS OF PROGRESSION

We prospectively evaluated 62 patients for a mean of 34 months (Reid et al. 1992). Of these 62, 13 had either aneuploidy or increased G_2 fractions as their initial flow cytometric abnormality. Of these 13, 9 progressed to develop high-grade dysplasia or cancer that was not present at the initial endoscopy. None of 49 patients without aneuploidy or increased G_2 fractions progressed to high-grade dysplasia or cancer ($p < 0.0001$). Patients who had increased G_2 fractions were predisposed to develop aneuploid cell populations, and patients who had a single aneuploid cell population were predisposed to develop multiple aneuploid cell populations. The temporal course of progression from aneuploidy or increased G_2 fractions to cancer was variable, ranging from 18 to 84 months. Patients whose biopsies were diploid and who had normal G_2 fractions did not progress to high-grade dysplasia or cancer during the time course of this study, but 7 of the 49 patients subsequently progressed to develop aneuploid cell populations or increased G_2 fractions during prospective follow-up.

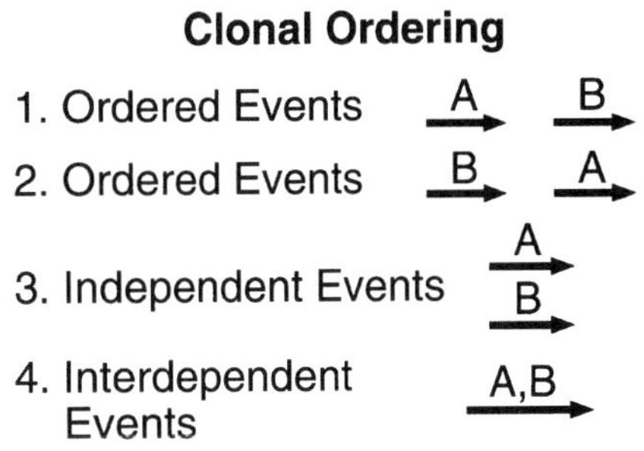

Clonal Ordering

1. Ordered Events A → B
2. Ordered Events B → A
3. Independent Events A → / B →
4. Interdependent Events A,B →

Figure 1. Four possible relationships between two events, A and B, of a multistep pathway.

ORDERING EVENTS OF NEOPLASTIC PROGRESSION IN BARRETT'S ESOPHAGUS

Any two events, A and B, of a multistep pathway can be related to each other in one of four ways (Fig. 1): (1) A can consistently precede B; for example, A could cause B, or A could create a condition that is permissive for B to occur. (2) B can consistently precede A. (3) A and B can be independent of each other; for example, A may occur before B in some cases, but after B in others. (4) A and B may be interdependent; for example, A and B may affect the same step (Hereford and Hartwell 1974). We have used this formalism to investigate the order in which two events occur relative to each other in patients with Barrett's esophagus by evaluating serial endoscopic biopsies and "mapped" biopsies from esophagectomy specimens.

p53 IN BARRETT'S ESOPHAGUS

p53 mutations and 17p allelic losses are among the most common abnormalities in human cancers (Hollstein et al. 1991; Harris and Hollstein 1993). In addition, *p53* mutations, 17p allelic losses, and p53 protein overexpression (a surrogate for mutations) have been reported in premalignant conditions of the colon, lung, and squamous esophagus (Vogelstein et al. 1988; Baker et al. 1990; Bennett et al. 1992; Sozzi et al. 1992). Studies have shown that 17p allelic losses are correlated with aneuploidy in colorectal carcinomas and that p53 protein overexpression can be detected before aneuploidy during neoplastic progression in the colon (Offerhaus et al. 1992; Carder et al. 1993). *p53* is a component of a signal transduction pathway that responds to DNA damage by causing arrest in G_1 (Kastan et al. 1991, 1992; Kuerbitz et al. 1992; Lu and Lane 1993). Inactivation of this *p53*-mediated G_1 checkpoint results in inappropriate entry into S phase, aneuploidy, and gene amplification in vitro (Livingstone et al. 1992; Yin et al. 1992).

We used *p53*/DNA content multiparameter flow cytometry to assess p53 protein overexpression at all stages of neoplastic progression in patients with Barrett's esophagus (Ramel et al. 1992). p53 protein overexpression was found in 1 of 21 patients (5%) with Barrett's metaplasia negative for dysplasia, 2 of 13 patients (15%) with indefinite/low-grade dysplasia, 5 of 11 patients (45%) with high-grade dysplasia, and 8 of 15 patients (53%) with Barrett's adenocarcinoma. These results, and those of others, suggest that *p53* abnormalities develop as relatively early events in the progression to cancer in Barrett's esophagus (Casson et al. 1991; Ramel et al. 1992; Younes et al. 1993; Blount et al. 1994; Neshat et al. 1994). Therefore, we investigated the relationships among 17p allelic losses, *p53* mutations, and other events of neoplastic progression in Barrett's esophagus.

THE ORDER OF 17p (*p53*) ALLELIC LOSSES AND CANCER IN BARRETT'S ESOPHAGUS

17p allelic losses were found in 34 of 36 Barrett's adenocarcinomas (94%). The order in which 17p allelic losses and cancer developed could not be determined in 21 cases because the cancer had overgrown its precursors. In the remaining 13 patients, 17p allelic losses were detected in premalignant epithelium in 11 cases (85%), cancer and the 17p allelic loss were detected simultaneously in 1 case (7.5%), and cancer preceded the 17p allelic loss in 1 case (7.5%).

We also investigated *p53* mutations in Barrett's adenocarcinomas and premalignant epithelium. *p53* mutations were detected in 14 of 16 cancers (88%) and in 6 of 9 patients with high-grade dysplasia (67%). In three cases, the same *p53* mutation was found in both premalignant epithelium and cancer, indicating that both *p53* mutations and 17p allelic losses develop before cancer in Barrett's esophagus (Fig. 2).

Patient 10490

Histology	Normal	→	HGD	→	Cancer
Ploidy	2.0N	→	2.0N	→	3.2N
#17p Alleles	2	→	1	→	1
p53 gene	wt	→	mutation	→	mutation

Figure 2. Development of *p53* mutation, 17p allelic loss, aneuploidy, and cancer in a patient with Barrett's esophagus. This patient's normal epithelium was diploid and contained two wild-type *p53* alleles. In diploid cell populations in premalignant epithelium, one allele of the *p53* gene was lost and the second had a CGA → TGA mutation in codon 306 that led to an Arg → Stop change in the amino acid sequence. Inactivation of the *p53* gene was followed by the development of an aneuploid cell population with a DNA content of 3.2N and cancer. (HGD) High-grade dysplasia; (wt) wild type.

THE ORDER OF 17p (*p53*) ALLELIC LOSSES AND ANEUPLOIDY IN BARRETT'S ESOPHAGUS

17p allelic losses were found in the aneuploid cell populations of 41 of 44 patients (93%) who had aneuploidy in Barrett's metaplastic epithelium. Because both aneuploidy and 17p allelic losses can be detected before cancer in Barrett's esophagus, we investigated the order in which the two events occurred relative to each other in the premalignant epithelium of 17 patients. 17p allelic losses were detected in diploid cells in the premalignant epithelium of 16 of the 17 patients (94%). In the remaining patient, the 17p allelic loss and aneuploidy were detected simultaneously. We sequenced the *p53* gene in the diploid and aneuploid cell populations of 3 patients to confirm that the 17p allelic losses were associated with *p53* mutations. In all 3 patients, diploid and aneuploid cell populations contained the same *p53* mutation.

THE ORDER OF 17p (*p53*) ALLELIC LOSSES AND INCREASED G_1 FRACTIONS IN BARRETT'S ESOPHAGUS

Increased G_1 fractions and 17p allelic losses can both be detected in diploid cells in Barrett's esophagus. Therefore, we investigated the order in which 17p allelic losses and increased G_1 fractions developed relative to each other. In 8 of 9 patients (89%), we detected increased G_1 fractions before 17p allelic losses. In the remaining patient, increased G_1 fractions and the 17p allelic loss were detected simultaneously.

THE ORDER OF 17p ALLELIC LOSSES AND 5q ALLELIC LOSSES IN BARRETT'S ESOPHAGUS

Chromosome 5q is the second most frequent known site of allelic loss in Barrett's adenocarcinoma (Table 1). We investigated the order in which 17p allelic losses and 5q allelic losses occurred in 38 aneuploid cell populations from 14 patients with Barrett's esophagus (Blount et al. 1993). 17p allelic losses developed before 5q allelic losses in 7 patients (50%). The two allelic losses were detected simultaneously in 4 patients (29%), and 17p allelic losses occurred without 5q allelic losses in 3 patients (21%). The order of genetic events in the esophagectomy specimen of a patient who developed a Barrett's adenocarcinoma is illustrated in Figure 3.

THE ORDER OF 5q ALLELIC LOSSES AND CANCER IN BARRETT'S ESOPHAGUS

The above results indicate that both 5q allelic losses and cancer typically occur after 17p allelic losses during neoplastic progression in Barrett's esophagus. There-

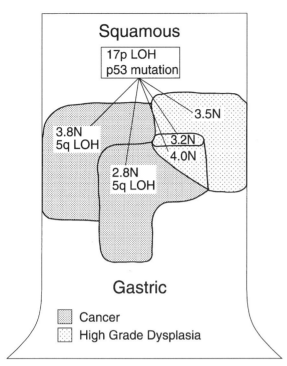

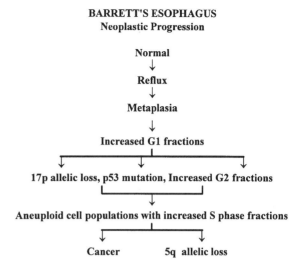

Figure 3. An esophagectomy specimen from a patient who developed an esophageal adenocarcinoma in high-grade dysplasia in Barrett's esophagus. The adenocarcinoma has two aneuploid cell populations with DNA contents of 2.8N and 3.8N. The premalignant epithelium has three aneuploid cell populations with DNA contents of 3.2N, 3.5N, and 4.0N. The same *p53* mutation, a 15-bp deletion extending from codon 209 to codon 213, was found in both high-grade dysplasia (3.2N aneuploid cell population) and cancer (2.8N and 3.8N aneuploid cell populations). All five aneuploid cell populations had 17p allelic losses (LOH) and p53 protein overexpression, indicating that inactivation of the *p53* gene by mutation and allelic loss occurred as early events. The 5q allelic loss (LOH) developed as a later event.

Figure 4. A model of neoplastic progression in Barrett's esophagus.

fore, we investigated the order in which 5q allelic losses and cancer developed in 21 patients with Barrett's esophagus. 5q allelic losses preceded cancer in 3 patients (14%), cancer preceded 5q allelic losses in 4 patients (19%), the two events were detected simultaneously in 8 patients (38%), cancer occurred without 5q allelic losses in 5 patients (24%), and 5q allelic losses occurred without cancer in 1 patient (5%). These results suggest that 5q allelic losses are independent of the events that cause invasion during neoplastic progression in Barrett's esophagus.

A MODEL OF NEOPLASTIC PROGRESSION IN BARRETT'S ESOPHAGUS

These results suggest a model of neoplastic progression in Barrett's esophagus (Fig. 4). In response to chronic reflux of gastric contents into the esophagus, approximately 10% of patients will develop Barrett's metaplastic epithelium. Increased G_1 fractions develop as an early event in the metaplastic epithelium, but S-phase fractions are typically normal at this stage of neoplastic progression. Subsequently, the *p53* gene is inactivated in diploid cells by a combination of mutation and allelic loss, and increased G_2 fractions develop in diploid cells. These events precede the appearance of aneuploid cell populations, which almost universally have increased S-phase fractions when they are detected. Cancer and 5q allelic losses develop later. Even though 5q allelic losses are detected in 75% of Barrett's adenocarcinomas, they appear to be independent of the events that lead to invasion in the sense that they can be detected before invasion, simultaneously with invasion, or after invasion.

The development of increased G_1 fractions is probably a consequence of the injury associated with chronic reflux and the need to replace the injured epithelium by cellular proliferation. At this stage of neoplastic progression, cell cycle controls still prevent the unregulated entry of cells into S phase, and the increased G_1 fractions may be reversible (data not shown). Inactivation of the *p53* gene by mutation and allelic loss presumably eliminates the G_1 checkpoint, because loss of *p53* leads to a condition that is permissive for the subsequent evolution of aneuploid cell populations. Increased diploid G_2 fractions are found frequently in biopsies of patients in whom the *p53* gene has been inactivated, suggesting that cells may accumulate at the G_2 checkpoint as a backup mechanism for monitoring DNA damage when the *p53*-mediated G_1 checkpoint is lost. If so, the G_2 checkpoint must either adapt, as it does in *Saccharomyces cerevisiae* (Sandell and Zakian 1993), or be inactivated, because inactivation of *p53* and the development of increased G_2 fractions are followed by progression to aneuploidy, other allelic losses, and cancer in vivo.

SOMATIC GENETICS AND CANCER SURVEILLANCE IN BARRETT'S ESOPHAGUS

Periodic endoscopic biopsy surveillance is recommended for the early detection of cancer in Barrett's esophagus, but the histological indicators that are commonly used for cancer surveillance lack the precision necessary to accurately identify patients who are at increased risk of progression (Reid et al. 1988b). Therefore, an understanding of the genetic and cell cycle abnormalities that lead to cancer in Barrett's esophagus has clinical implications for the management of the cancer risk in these patients. Only a small subset of Barrett's patients have increased G_2 fractions or aneuploid cell populations in their Barrett's epithelium, but this subset has an increased risk of progression to high-grade dysplasia and cancer. Once high-risk patients are identified by flow cytometric or molecular tests, they can be placed in more frequent surveillance for the detection of adenocarcinoma at an early stage in which it is curable (Levine et al. 1993). The majority of patients, who have not yet entered a pathway of genetic instability, can be followed by less frequent surveillance. Thus, genetic and cell cycle tests that can accurately distinguish patients who have an increased risk of progressing to cancer from those whose risk is low can improve the cure rate of patients who develop cancer, while simultaneously decreasing health care costs by safely permitting less frequent surveillance of low-risk patients.

ACKNOWLEDGMENTS

We thank Ellen Morrison for administrative assistance, Susan Irvine, RN, for assistance in patient care, and Rodger C. Haggitt, MD, for histological interpretation. This research was supported by American Cancer Society grant EDT-21E, National Institutes of Health grant R01 CA-55814, Ryan Hill Research Foundation, and Ersta Hospital Research Fund.

REFERENCES

Baker, S.J., A.C. Preisinger, J.M. Jessup, C. Paraskeva, S. Markowitz, J.K.V. Willson, S. Hamilton, and B. Vogelstein. 1990. p53 gene mutations occur in combination with 17p allelic deletions as late events in colorectal tumorigenesis. *Cancer Res.* **50:** 7717.

Bennett, W.P., M.C. Hollstein, R.A. Metcalf, J.A. Welsh, A. He, S. Zhu, I. Kusters, J.H. Resau, B.F. Trump, D.P. Lane, and C.C. Harris. 1992. p53 mutation and protein accumulation during multistage human esophageal carcinogenesis. *Cancer Res.* **52:** 6092.

Blount, P.L., P.S. Rabinovitch, R.C. Haggitt, and B.J. Reid. 1990. Early Barrett's adenocarcinoma arises within a single aneuploid population. *Gastroenterology* **98:** A273.

Blount, P.L., S.J. Meltzer, J. Yin, Y. Huang, M.J. Krasna, and B.J. Reid. 1993. Clonal ordering of 17p and 5q allelic losses in Barrett dysplasia and adenocarcinoma. *Proc. Natl. Acad. Sci.* **90:** 3221.

Blount, P.L., S. Ramel, W.H. Raskind, R.C. Haggitt, C.A. Sanchez, P.J. Dean, P.S. Rabinovitch, and B.J. Reid. 1991. 17p allelic deletions and p53 protein overexpression in Barrett's adenocarcinoma. *Cancer Res.* **51:** 5482.

Blount, P.L., P.C. Galipeau, C.A. Sanchez, K. Neshat, D.S. Levine, J. Yin, H. Suzuki, J.M. Abraham, S.J. Meltzer, and B.J. Reid. 1994. 17p allelic losses in diploid cells of patients with Barrett's esophagus who develop aneuploidy. *Cancer Res.* **54:** 2292.

Blot, W.J, S.S. Devesa, R.W. Kneller, and J.F. Fraumeni, Jr. 1991. Rising incidence of adenocarcinoma of the esophagus and gastric cardia. *J. Am. Med. Assoc.* **265:** 1287.

Boynton, R.F., P.L. Blount, J. Yin, V.L. Brown, Y. Huang, Y. Tong, T. McDaniel, C. Newkirk, J.H. Resau, W.H. Raskind, R.C. Haggitt, B.J. Reid, and S.J. Meltzer. 1992. Loss of heterozygosity involving the *APC* and *MCC* genetic loci occurs in the majority of human esophageal cancers. *Proc. Natl. Acad. Sci.* **89:** 3385.

Brown, M., B. Garvik, L. Hartwell, L. Kadyk, T. Seeley, and T. Weinert. 1991. Fidelity of mitotic chromosome transmission. *Cold Spring Harbor Symp. Quant. Biol.* **56:** 359.

Carder, P., A.H. Wyllie, C.A. Purdie, R.G. Morris, S. White, J. Piris, and C.C. Bird. 1993. Stabilised p53 facilitates aneuploid colonal divergence in colorectal cancer. *Oncogene* **8:** 1397.

Casson, A.G., T. Mukhopadhyay, K.R. Cleary, J.Y. Ro, B. Levin, and J.A. Roth. 1991. p53 gene mutations in Barrett's epithelium and esophageal cancer. *Cancer Res.* **51:** 4495.

Dasso, M. and J.R. Newport. 1990. Completion of DNA replication is monitored by a feedback system that controls the initiation of mitosis in vitro: Studies in *Xenopus*. *Cell* **61:** 811.

Enoch, T. and P. Nurse. 1990. Mutation of fission yeast cell cycle control genes abolishes dependence of mitosis on DNA replication. *Cell* **60:** 665.

Fearon, E.R. and B. Vogelstein. 1990. A genetic model for colorectal tumorigenesis. *Cell* **61:** 759.

Fennerty, M.B., R.E. Sampliner, D. Way, R. Riddell, K. Steinbronn, and H. Garewal. 1989. Discordance between flow cytometric abnormalities and dysplasia in Barrett's esophagus. *Gastroenterology* **97:** 815.

Gray, M.R., P.A. Hall, J. Nash, B. Ansari, D.P. Lane, and A.N. Kingsnorth. 1992. Epithelial proliferation in Barrett's esophagus by proliferating cell nuclear antigen immunolocalization. *Gastroenterology* **103:** 1769.

Hamilton, S.R., R.R.L. Smith, and J.L. Cameron. 1988. Prevalence and characteristics of Barrett esophagus in patients with adenocarcinoma of the esophagus or esophagogastric junction. *Hum. Pathol.* **19:** 942.

Harris, C.C. and M. Hollstein. 1993. Clinical implications of the p53 tumor-suppressor gene. *N. Engl. J. Med.* **329:** 1318.

Hartwell, L.H. and T.A. Weinert. 1989. Checkpoints: Controls that ensure the order of cell cycle events. *Science* **246:** 629.

Herbst, J.J., M.M. Berenson, D.W. McCloskey, and W.C. Wiser. 1978. Cell proliferation in esophageal columnar epithelium (Barrett's esophagus). *Gastroenterology* **75:** 683.

Hereford, L.M. and L.H. Hartwell. 1974. Sequential gene function in the initiation of *Saccharomyces cerevisiae* DNA synthesis. *J. Mol. Biol.* **84:** 445.

Hollstein, M., D. Sidransky, B. Vogelstein, and C.C. Harris. 1991. p53 mutations in human cancers. *Science* **253:** 49.

Huang, Y., R.F. Boynton, P.L. Blount, R.J. Silverstein, J. Yin, Y. Tong, T.K. McDaniel, C. Newkirk, J.H. Resau, R. Sridhara, B.J. Reid, and S.J. Meltzer. 1992. Loss of heterozygosity involves multiple tumor suppressor genes in human esophageal cancers. *Cancer Res.* **52:** 6525.

Jankowski, J., R. McMenemin, D. Hopwood, J. Penston, and K.G. Wormsley. 1991. Abnormal expression of growth regulatory factors in Barrett's oesophagus. *Clin. Sci.* **81:** 663.

Kastan, M.B., O. Onyekwere, D. Sidransky, B. Vogelstein, and R.W. Craig. 1991. Participation of p53 protein in the cellular responses to DNA damage. *Cancer Res.* **51:** 6304.

Kastan, M.B., Q. Zhan, W.S. El-Deiry, F. Carrier, T. Jacks,

W.V. Walsh, B.S. Plunkett, B. Vogelstein, and A.J. Fornace. 1992. A mammalian cell cycle checkpoint pathway utilizing p53 and *GADD45* is defective in ataxia-telangiectasia. *Cell* **71:** 587.

Kuerbitz, S.J., B.S. Plunkett, W.V. Walsh, and M.B. Kastan. 1992. Wild-type p53 is a cell cycle checkpoint determinant following irradiation. *Proc. Natl. Acad. Sci.* **89:** 7491.

Levine, D.S. and B.J. Reid. 1992. Endoscopic diagnosis of esophageal neoplasms. In *Gastrointestinal endoscopy clinics of North America* (ed. M. Sivak et al.), p. 395. W.B. Saunders, Philadelphia.

Levine, D.S., R.C. Haggitt, P.L. Blount, P.S. Rabinovitch, V.W. Rusch, and B.J. Reid. 1993. An endoscopic biopsy protocol can differentiate high-grade dysplasia from early adenocarcinoma in Barrett's esophagus. *Gastroenterology* **105:** 40.

Livingstone, L.R., A. White, J. Sprouse, E. Livanos, T. Jacks, and T.D. Tlsty. 1992. Altered cell cycle arrest and gene amplification potential accompany loss of wild-type p53. *Cell* **70:** 923.

Lu, X. and D.P. Lane. 1993. Differential induction of transcriptionally active p53 following UV or ionizing radiation: Defects in chromosome instability syndromes? *Cell* **75:** 765.

McKinley, M.J., D.R. Budman, D. Grueneberg, R.L. Bronzo, G.S. Weissman, and E. Kahn. 1987. DNA content in Barrett's esophagus and esophageal malignancy. *Am. J. Gastroenterol.* **82:** 1012.

Miller, B.A., L.A.G. Ries, B.F. Hankey, C.L. Kosary, and B.K. Edwards. 1992. Cancer statistics review, 1973–1989. U.S. Department of Health and Human Services. *NIH Publ.* No. 92-2789.

Muller, J.M., H. Erasmi, M. Stelzner, U. Zieren, and H. Pichlmaier. 1990. Surgical therapy of oesophageal carcinoma. *Br. J. Surg.* **77:** 845.

Neshat, K., C.A. Sanchez, P.C. Galipeau, P.L. Blount, D.S. Levine, G. Joslyn, and B.J. Reid. 1994. p53 mutations in Barrett's adenocarcinoma and high-grade dysplasia. *Gastroenterology* **106:** 1589.

Nowell, P.C. 1976. The clonal evolution of tumor cell populations. *Science* **194:** 23.

Offerhaus, G.J.A., E.P. De Geyter, C.J. Cornelisse, K.W.F. Tersmette, J. Floyd, S.E. Kern, B. Vogelstein, and S.R. Hamilton. 1992. The relationship of DNA aneuploidy to molecular genetic alterations in colorectal carcinoma. *Gastroenterology* **102:** 1612.

Pellish, L.J., J.A. Hermos, and G.L. Eastwood. 1980. Cell proliferation in three types of Barrett's epithelium. *Gut* **21:** 26.

Phillips, R.W. and R.K.H. Wong. 1991. Barrett's esophagus: Natural history, incidence, etiology and complications. *Gastroenterol. Clin. North Am.* **20:** 791.

Powell, S.M., N. Zilz, Y. Beazer-Barclay, T.M. Bryan, S.T. Hamilton, S.N. Thilbodeau, B. Vogelstein, and K.W. Kinzler. 1992. APC mutations occur early during colorectal tumorigenesis. *Nature* **359:** 235.

Rabinovitch, P.S., B.J. Reid, R.C. Haggitt, T.H. Norwood, and C.E. Rubin. 1988. Progression to cancer in Barrett's esophagus is associated with genomic instability. *Lab. Invest.* **60:** 65.

Ramel, S., B.J. Reid, C.A. Sanchez, P.L. Blount, D.S. Levine, K. Neshat, R.C. Haggitt, P.J. Dean, K. Thor, and P.S. Rabinovitch. 1992. Evaluation of p53 protein expression in Barrett's esophagus by two-parameter flow cytometry. *Gastroenterology* **102:** 1220.

Raskind, W.H., T. Norwood, D.S. Levine, R.C. Haggitt, P.S. Rabinovitch, and B.J. Reid. 1992. Persistent clonal areas and clonal expansion in Barrett's esophagus. *Cancer Res.* **52:** 2946.

Reid, B.J. 1991. Barrett's esophagus and esophageal adenocarcinoma. *Gastroenterol. Clin. North Am.* **20:** 817.

Reid, B.J., R.C. Haggitt, C.E. Rubin, and P.S. Rabinovitch. 1987. Barrett's esophagus: Correlation between flow cytometry and histology in detection of patients at risk for adenocarcinoma. *Gastroenterology* **93:** 1.

Reid, B.J., C.A. Sanchez, P.L. Blount, and D.S. Levine. 1993. Barrett's esophagus: Cell cycle abnormalities in advancing stages of neoplastic progression. *Gastroenterology* **105:** 119.

Reid, B.J., P.L. Blount, C.E. Rubin, D.S. Levine, R.C. Haggitt, and P.S. Rabinovitch. 1992. Flow cytometric and histologic progression to malignancy in Barrett's esophagus: Prospective endoscopic surveillance of a cohort. *Gastroenterology* **102:** 1212.

Reid, B.J., W.M. Weinstein, K.J. Lewin, R.C. Haggitt, G. Van Deventer, L. DenBesten, and C.E. Rubin. 1988a. Endoscopic biopsy can detect high-grade dysplasia or early adenocarcinoma in Barrett's esophagus without grossly recognizable neoplastic lesions. *Gastroenterology* **94:** 81.

Reid, B.J., R.C. Haggitt, C.E. Rubin, G. Roth, C.M. Surawicz, G. VanBelle, K. Lewin, W.M. Weinstein, D.A. Antonioli, H. Goldman, W. MacDonald, and D. Owen. 1988b. Observer variation in the diagnosis of dysplasia in Barrett's esophagus. *Hum. Pathol.* **19:** 166.

Robaszkiewicz, M., E. Hardy, A. Volant, J.B. Nousbaum, J.M. Cauvin, G. Calament, F.X. Robert, J.P. Saleun, and H. Gouerou. 1991. Analyse du contneu cellulaire en ADN par cytometrie en flux dans les endobrachyoesopheags. *Gastroenterol. Clin. Biol.* **15:** 703.

Sandell, L.L. and V.A. Zakian. 1993. Loss of a yeast telomere: Arrest, recovery, and chromosome loss. *Cell* **75:** 729.

Sidransky, D., T. Mikkelsen, K. Schwechheimer, M.L. Rosenblum, W. Cavenee, and B. Vogelstein. 1992. Clonal expansion of p53 mutant cells is associated with brain tumour progression. *Nature* **355:** 846.

Sozzi, G., M. Miozzo, R. Donghi, S. Pilotti, C.T. Cariani, U. Pastorino, G. Della Porta, and M.A. Pierotti. 1992. Deletions of 17p and p53 mutations in preneoplastic lesions of the lung. *Cancer Res.* **52:** 6079.

Spechler, S.J. 1987. Endoscopic surveillance for patients with Barrett esophagus: Does the cancer risk justify the practice? *Ann. Intern. Med.* **106:** 902.

Vogelstein, B., E.R. Fearon, S.R. Hamilton, S.E. Kern, A.C. Preisinger, M. Leppert, Y. Nakamura, R. White, A.M.M. Smits, and J.L. Bos. 1988. Genetic alterations during colorectal tumor development. *N. Engl. J. Med.* **319:** 525.

Weinert, T.A. and L.H. Hartwell. 1988. The *RAD9* gene controls the cell cycle response to DNA damage in *Saccharomyces cerevisiae*. *Science* **241:** 317.

Yin, Y., M.A. Tainsky, F.Z. Bischoff, L.C. Strong, and G.M. Wahl. 1992. Wild-type p53 restores cell cycle control and inhibits gene amplification in cells with mutant p53 alleles. *Cell* **70:** 937.

Younes, M., R.M. Lebovitz, L.V. Lechago, and J. Lechago. 1993. p53 protein accumulation in Barrett's metaplasia, dysplasia, and carcinoma: A follow-up study. *Gastroenterology* **105:** 1637.

Detection of Genetic Loss in Tumors by Representational Difference Analysis

N.A. LISITSYN,* F.S. LEACH,† B. VOGELSTEIN,† AND M.H. WIGLER*
*Cold Spring Harbor Laboratory, Cold Spring Harbor, New York 11724;
†The Johns Hopkins Oncology Center, Baltimore, Maryland 21231

A variety of genetic lesions are found in tumors, including DNA losses, point mutations, gene amplifications, and rearrangements (Lasko et al. 1991; Salomon et al. 1991). Frequent losses of both alleles at a given locus or losses of one allele with functional inactivation of the other have been detected in many tumor types. These genetic lesions, manifesting themselves as loss of heterozygosity (LOH) and hemizygous and homozygous deletions, have been found to be the hallmarks of the presence of tumor suppressor genes. Many approaches have been taken in the past to identify these genes, but recently we have developed a new method that is both general and efficient (Lisitsyn et al. 1993, 1995). The method, called representational difference analysis, or RDA, is designed for analyzing the differences between complex but highly related genomes and combines three elements: representation, subtractive enrichment, and kinetic enrichment. The first stage of the procedure comprises the preparation of representations from the genomes, during which DNAs are cut with restriction endonuclease, ligated to oligonucleotide adapters, and amplified by the polymerase chain reaction (PCR). Since only small fragments (<1 kb in length), called ARFs, are efficiently amplified by standard PCR procedures, representations have at least tenfold lower complexity than initial DNAs. This enormously increases the efficiency of the second stage, comprising the reiterative hybridization/selection steps during which ARFs present in one sample, the tester, but not in the other, the driver, are selectively enriched. We describe here the application of RDA to discover sequences that are lost in tumors.

CLONING SEQUENCES LOST IN TUMORS

We performed RDA on 16 individual pairs of tumor DNA (used to derive driver) and matched normal DNA (used to derive tester) from the same patient. We isolated 15 DNAs from tumor cell lines (including 9 renal cell and 6 colon cancer cell lines) with normal DNAs derived from unaffected blood or tissue. In one case, we used a fluorescent activated cell sorter to fractionate nuclei from an esophageal cancer biopsy into aneuploid and diploid fractions used for preparation of driver and tester DNA, respectively.

In each application of RDA, difference products were cloned and analyzed by blot hybridization. The "informative" probes hybridized to DNA from the normal representation but not the tumor representation. Some of these probes mapped to the Y chromosome. Loss of Y chromosome information is frequently observed in renal cell carcinomas (Presti et al. 1991). Other probes detected binary polymorphisms at BglII sites and were presumed to reflect loss of heterozygosity in tumor. Finally, some probes did not hybridize at all to total genomic DNA from the tumor. Probes of this type were sequenced, and oligonucleotides were derived for use in PCR screening of genomic DNA from the tester and driver sources, and from panels of normal human and tumor cell lines. Occasionally, we found probes that did not hybridize to several normal human DNAs. We presume that these probes reflect hemizygous loss in the tumor of a deletion polymorphism common in the human population (see Table 1, footnote c). Table 1 summarizes the types of probes we obtained (for further details, see Lisitsyn et al. 1995).

HOMOZYGOUS LOSS ON CHROMOSOME 3p

From 16 comparisons, 6 pairs yielded probes that appear to detect homozygous loss in the tumor used as driver. Three probes were found to be homozygously deleted in other DNAs isolated from a collection of 100 tumor cell lines established from different types of cancer but not from normals (N. Lisitsyn and R. Lucito, unpubl.). Probe 758-6, derived from a patient with Barrett's esophagus, detected frequent losses in cancers of the digestive tract. Loss detected with this probe has also been observed in breast, bladder, and lung tumors. This probe has been analyzed in greatest detail.

758-6 was mapped to chromosomal region 3p by PCR analysis of monochromosomal human/rodent cell hybrids. The probe was used for screening a chromosome-3 cosmid library, and a cosmid contig was built by chromosome walking (see Fig. 1). Single-copy probes were derived from this contig and used to screen DNAs from a collection of human colon cancer cell lines and xenografts. Of 175 tumor DNAs, 20 (11%) lacked sequences from at least one of the probes from this region. In contrast, losses were observed in 1 of 122 lung cancer cell lines (S. Bader and J. Minna, unpubl.). Figure 1 shows the different patterns of loss that were observed in colorectal cancer cell lines and xenografts.

Additional probes derived from a cosmid contig were used in hybridizations to Southern blots containing DNAs harboring deletions. Several hybridizing restric-

Table 1. Analysis of RDA Probes Derived Using Tumor DNA as Driver

	Selected for initial characterization	Found to be informative[a]
Renal cell carcinoma cell lines		
UOK 112 (male)	13[b]	13 (0/13/0)
UOK 114 (female)	12[b]	4 (3/0/1)
UOK 124 (female)	12[b]	4 (4/0/0)
UOK 132 (male)	10[b]	9 (3/6/0)
UOK 108 (female)	2	2 (2/0/0)
UOK 111 (female)	5	5 (5/0/0)
UOK 127 (male)	3	3 (2/1[c]/0)
UOK 146 (female)	3	3 (1/1[c]/1)
UOK 154 (female)	5	1 (1/0/0)
Colon cancer cell lines		
VACO 429 (male)	2	1 (0/0/1)
VACO 441 (female)	3	3 (1/0/2)
VACO 432 (male)	2	1 (1/0/0)
VACO 456 (female)	2	1 (1/0/0)
VACO 576 (female)	2	2 (2/0/0)
RBX (male)	2	1 (1/0/0)
Barrett's esophagus		
BE 758[d] (male)	5	5 (0/4/1[e])
Total:	83	58 (27/25/6)

[a] Entries are a(b, c, d), where a is the total number of probes detecting DNA loss in tumors, judged to be: b, loss-of-heterozygosity; c, hemizygous loss; d, presumably homozygous loss (see Discussion). All but two probes judged to detect hemizygous loss were derived from the Y chromosome. The difference between quantities of initially selected probes (83) and informative probes (58) was due to the presence of the repeat sequences (9 cases), nonhuman DNA contaminating tester (5 cases), and single-copy sequences present in both tester and driver DNAs (11 cases).

[b] The difference products after two rounds of hybridization/selection were cloned; in all the rest of the experiments cloning was performed after three rounds.

[c] Probes 127-1 and 146-1 were found to be deletion polymorphisms, absent on both autosomes of 7 out of 35 and 3 out of 35 of normal humans, respectively.

[d] Nuclei from a biopsy were sorted by flow cytometry into aneuploid (tumor) and diploid (normal) fractions.

[e] This result is presumed, but was not confirmed because of the small amount of sorted tumor nuclei available.

tion fragments were observed in tumors that were absent in normals, presumably as a result of rearrangements occurring at the ends of some of the deletions. This observation rules out the possibility that probe 758-6 detects a deletion polymorphism and that the loss of sequences is caused by the same types of mechanisms that underlie loss of heterozygosity at polymorphic markers. Work is in progress to identify transcribed sequences from this region.

DISCUSSION

The RDA methodology may be successfully applied to detection of DNA losses in tumors, readily providing probes that detect homozygous deletions. As we were able to demonstrate, some of these deletions are relatively small (<50 kb) and, thus, positional cloning of genes that must be inactivated in tumors becomes much more efficient, as compared to other techniques used for this purpose (e.g., allelotyping, linkage analysis of predispositions in families, and cytogenetic studies followed by microdissection). Since some of these probes are deleted in more than one DNA isolated from tumor cells, it is possible that the deleted locus contains a gene that is commonly inactivated in tumors.

It is well documented that some genes known to be disrupted by homozygous deletions in tumors regulate cellular growth, differentiation, and genomic stability. Some of these genes have strong tumor suppressor phenotypes after transfection into tumor cell lines. Our screening technique based on RDA methodology holds promise for the identification of new genes participating in these or other processes. We have taken a similar approach for the cloning of dominant oncogenes. These are frequently amplified in tumors, and probes detecting amplifications can be efficiently cloned by RDA when tumor DNA is used to derive tester (Lisitsyn et al. 1995). Although amplified regions are usually rather large, one can map candidate oncogenes more precisely by finding the minimal region common to all amplifications at a given locus. The use of RDA for the analysis of cancers thus opens up new avenues for understanding the etiology of the disease and for the development of prognostic and diagnostic markers.

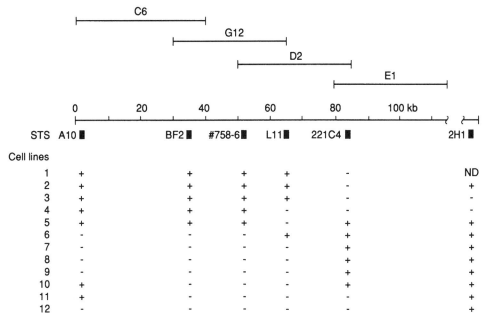

Figure 1. Physical map of a chromosome 3p region. On top are shown cosmids from a region isolated by chromosomal walking. Patterns of the homozygous loss of STSs (thick bars) detected by PCR analysis are depicted on the bottom. Pluses and minuses indicate presence or absence of the probe in DNAs from colorectal cancer cells.

ACKNOWLEDGMENTS

We thank J. Garnes for providing the chromosome-3 library constructed at the Human Genome Center, Lawrence Livermore National Laboratory, Livermore, California, under the auspices of the National Laboratory Gene Library Project sponsored by the U.S. Department of Energy; J. Gnarra and W.M. Linehan for providing DNA samples and cell lines from renal cell carcinomas; J. Willson and S. Markowitz for preparation of DNA from colorectal cancers; C. Sanchez and B. Reid for providing DNAs from FACS sorted nuclei; S. Bader and J. Minna for PCR screening of a collection of DNAs from lung cancers; Natalia Lisitsina for performing RDA; Linda Rodgers and Michael Riggs for DNA sequencing; and Patricia Bird for preparation of this manuscript.

REFERENCES

Lasko, D., W. Cavenee, and M. Nordenskjold. 1991. Loss of constitutional heterozygosity in human cancer. *Annu. Rev. Genet.* **25:** 281.

Lisitsyn, N.A., N.M. Lisitsina, and M. Wigler. 1993. Cloning the differences between two complex genomes. *Science* **259:** 946.

Lisitsyn, N.A., N.M. Lisitsina, G. Dalbagni, P. Barker, C.A. Sanches, J. Gnarra, W.M. Linehan, B.J. Reid, and M. Wigler. 1995. Comparative genomic analysis of tumors: Detection of DNA losses and amplification. *Proc. Natl. Acad. Sci.* (in press).

Presti, J.C., Jr., P.H. Rao, Q. Chen, V.E. Reuter, F.P. Li, W.R. Fair, and S.C. Jhanwar. 1991. Histopathological, cytogenetic, and molecular characterization of renal cortical tumors. *Cancer Res.* **51:** 1544.

Salomon, E., J. Bozzow, and A.D. Goddard. 1991. Chromosome aberrations and cancer. *Science* **254:** 1153.

Genetic Approaches to Defining Signaling by the CML-associated Tyrosine Kinase BCR-ABL

D.E.H. AFAR,* A. GOGA,† L. COHEN,* C.L. SAWYERS,§
J. MCLAUGHLIN,* R.N. MOHR,* AND O.N. WITTE*†‡

*Department of Microbiology and Molecular Genetics, †Molecular Biology Institute, ‡Howard Hughes Medical Institute, and §Department of Medicine, Division of Hematology-Oncology, University of California, Los Angeles, California 90024-1662

Human chronic myelogenous leukemia (CML) is a cancer that is associated with a specific cytogenetic marker known as the Philadelphia (Ph^1) chromosome (Nowell and Hungerford 1960; Rowley 1973; for review, see Kurzrock et al. 1988). Cells of multiple lineages carry Ph^1, suggesting that the disease originates in a pluripotent stem cell. Ph^1 is the result of a reciprocal translocation of chromosomes 9 and 22, which results in the fusion of two cellular genes encoding BCR and c-ABL (Shtivelman et al. 1985; Clark et al. 1988). The resulting fusion gene (see Fig. 1) encodes BCR-ABL, an activated protein tyrosine kinase. The tyrosine kinase activity of BCR-ABL is deregulated when compared to c-ABL (Konopka and Witte 1985) and is the most essential feature for its transforming activity. In nature, two fusion products are detected. P210 BCR-ABL is associated with CML and P185 BCR-ABL is associated with the more aggressive acute lymphocytic leukemia (ALL). The tyrosine kinase activity of P185 is more potent than P210 kinase activity, which also correlates with their respective transforming activities (Lugo et al. 1990).

The association of BCR-ABL with CML suggests that BCR-ABL plays a role in the initiation of the disease. However, cells expressing BCR-ABL during the chronic phase of CML, which may endure for 3–5 years, are still capable of differentiation. The terminal phase of CML is an acute leukemia, characterized by expansion of predominantly myeloid or lymphoid precursor cells (for review, see Kantarijan et al. 1993). This time course of events suggests that additional genetic alterations are necessary to trigger blast crisis.

In an attempt to recapitulate human CML, several experimental animal systems have been developed. These involved generating transgenic mice with BCR-ABL (Heisterkamp et al. 1990) or reconstituting lethally irradiated mice with bone marrow infected with BCR-ABL-containing retrovirus (Daley et al. 1990; Elefanty et al. 1990; Kelliher et al. 1990, 1991; Scott et al. 1991; Gishizky et al. 1993). In both systems, mice expressing BCR-ABL in hematopoietic cells developed various malignancies, including CML-like syndromes. In most cases, the disease developed within 20 weeks and could not be propagated when transplanted into syngeneic recipient mice. This suggested that the experimentally induced syndrome was due to stimulation of the proliferative capacity of committed myeloid progenitor cells by BCR-ABL. In cases where the disease developed with a longer latency period (>20 weeks), however, the leukemia was transplantable (Gishizky et al. 1993). Molecular and cellular analysis demonstrated that this CML syndrome was a consequence of the disease originating in a multipotent stem cell that gave rise to cells of various lineages (Gishizky et al. 1993). This situation more closely reflected the etiology and pathology of human CML.

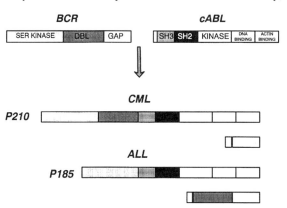

Figure 1. Schematic representation of BCR, c-ABL, and the fusion product BCR-ABL. BCR exhibits a serine kinase domain at the amino terminus (Maru and Witte 1991), a DBL homology domain that encodes a putative GTP/GDP exchange factor (Ron et al. 1991), and a carboxy-terminal GTPase activating domain (GAP) (Diekmann et al. 1991). The c-ABL tyrosine kinase contains SH3 and SH2 domains at the amino terminus, a tyrosine kinase domain, and DNA- (Kipreos and Wang 1992) and actin-binding sequences (McWhirter and Wang 1991). Both BCR-ABL fusion products, CML-associated P210 and ALL-associated P185, contain BCR sequences fused upstream of the SH3 domain of c-ABL. P210 includes the serine kinase and DBL regions from BCR. P185 only contains the serine kinase region from BCR.

While providing valuable information on the development of CML, animal experiments have contributed little insight into the genes that cooperate with BCR-ABL in the disease. Earlier work with in vitro systems showed that BCR-ABL is sufficient for transforming primary bone marrow and fibroblasts (McLaughlin et al. 1987; Lugo and Witte 1989). The in vivo setting differs from this situation in that BCR-ABL alone is not capable of inducing blast crisis, for which additional genetic lesions are required. However, the in vitro systems provide us with a valuable tool in understanding BCR-ABL signaling. Using three approaches, we developed genetic tests to identify components and pathways involved in BCR-ABL signaling. These include: (1) the use of dominant negative forms of genes believed to cooperate with BCR-ABL in transformation; (2) a complementation strategy to rescue signaling-defective mutants of BCR-ABL for transformation; (3) an approach to identifying new genes associated with BCR-ABL-induced malignancy. These strategies provide us with knowledge on genes acting downstream from BCR-ABL signaling, which are potentially important in accelerating the disease.

DOMINANT NEGATIVE FORMS OF MYC BLOCK BCR-ABL-MEDIATED TRANSFORMATION

Dominant negative genes have been extensively used to demonstrate the involvement of specific genes in certain signaling pathways (Feig and Cooper 1988; Dang et al. 1989; DeClue et al. 1991; Stacey et al. 1991a,b; Thomas et al. 1992; Wood et al. 1992). Their modes of action vary depending on the gene in question. In general, dominant negative molecules form nonfunctional (hetero- or homodimeric) complexes with binding partners, effector molecules, or regulatory molecules. Regardless of the mechanism, they function to down-regulate the activity of the wild-type protein.

The proto-oncogene c-*myc* encodes a nuclear transcription factor that synergizes with BCR-ABL in transforming rat-1 fibroblasts (Lugo and Witte 1989). The activity of MYC depends on its ability to heterodimerize with its binding partner MAX (Blackwood and Eisenman 1991; Prendergast et al. 1991). MYC competes for MAX binding with MAD, another MAX-binding protein, which opposes the action of MYC (Ayer et al. 1993). The MYC-MAX complex formation is mediated by helix-loop-helix and leucine zipper domains in both binding partners. The complex is capable of binding to specific DNA sequences via a carboxy-terminal basic region in MYC. Transcriptional activation is mediated by an activation domain at the amino terminus (Kato et al. 1990).

To analyze the importance of MYC expression during BCR-ABL-mediated transformation, we used molecules with a deletion within the activation domain (D106-143) and molecules with a linker insertion adjacent to the DNA-binding basic region (ln373) (Sawyers et al. 1992). These MYC mutants retain the capacity to interact with its binding partner MAX, but are deficient in transcriptional activation. Therefore, expression of dominant negative MYC should compete with wild-type MYC for interaction with MAX and block a MYC-dependent signal.

To test the effect of dominant negative MYC on the transforming potential of BCR-ABL, rat-1 indicator cell lines expressing the different MYC mutants were generated. These cell lines expressed about 10 times more dominant negative MYC over endogenous wild-type MYC (Sawyers et al. 1992). Retroviruses containing BCR-ABL were then used to infect the different cell lines. The transforming activity was assessed by plating infected cells into soft agar. The growth of cellular colonies in soft agar signifies a transformed phenotype. Using this assay, we determined that BCR-ABL-mediated transformation of fibroblasts was reduced by a factor of 10 in the presence of dominant negative MYC (Sawyers et al. 1992).

Similar results were obtained with primary mouse bone marrow infected with BCR-ABL retrovirus. In this case, the retroviruses contained both BCR-ABL and the different forms of MYC. The advantage of having both genes in the same retrovirus is that both proteins will be expressed in the same infected cells. Using this strategy, BCR-ABL with dominant negative MYC was unable to efficiently transform hematopoietic cells (Sawyers et al. 1992). Together with the rat-1 data, these results demonstrated that c-MYC is essential for a BCR-ABL-generated signal critical for transformation.

COMPLEMENTATION OF SIGNALING DEFECTIVE MUTANTS OF BCR-ABL WITH c-MYC

To genetically define the minimum number of signals needed for oncogenesis, we developed a genetic complementation strategy (Afar et al. 1994) (see Fig. 2). Point mutations were generated in domains of the gene of interest that might be expected to play a role in connecting to downstream effector molecules. To define its signaling pathways, these inactive point mutants are assayed in the presence of a gene known to function downstream from a signal generated by the oncogene. If the downstream gene is unable to rescue any of the mutants, the experiment becomes uninformative. If all mutants are rescued by the downstream gene, then the same signaling pathway is inactivated by the different mutations. Complementation of one point mutant, but not the other, would signify that each mutation blocks a different pathway.

Point mutations were generated in the *src*-homology 2 (SH2) region of BCR-ABL, the major tyrosine auto-

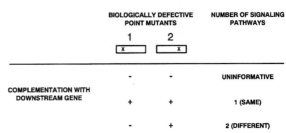

Figure 2. Complementation strategy. Signaling-defective point mutants of an oncogene are tested for restoration of transformation in the presence of a downstream gene. Information about the signaling pathways can be gained with the differential rescue of the point mutants (see text).

phosphorylation site of the kinase domain, and a tyrosine phosphorylation site in the BCR region known to interact with Grb-2, an adapter molecule involved in Ras activation (Pendergast et al. 1993a,b; Afar et al. 1994). The SH2 region is a molecular domain that mediates protein-protein interactions by binding specifically to tyrosine phosphorylated residues (for review, see Koch et al. 1991; Pawson and Gish 1992). The residue targeted for mutagenesis in the SH2 domain was the arginine in the FLVRES motif, which is the single most important amino acid for phosphotyrosine binding (Waksman et al. 1992).

All BCR-ABL point mutants retained kinase activity but were incapable of transforming rat-1 cells, as determined by soft agar growth (Pendergast et al. 1993a,b; Afar et al. 1994). Since c-MYC cooperates with BCR-ABL in transformation, and since dominant negative MYC blocks BCR-ABL signaling, we decided to test the ability of c-MYC to complement the BCR-ABL point mutants for transformation. Rat-1 cell lines were generated that stably expressed the various BCR-ABL forms. These cells were then superinfected with retrovirus containing c-MYC and plated into soft agar. The results showed that only cells expressing the SH2 mutant of BCR-ABL were rescued for transformation with c-MYC hyperexpression (summarized in Fig. 3) (Afar et al. 1994). The number and size of colonies generated by the SH2 mutant with c-MYC overexpression was comparable to those generated by wild-type BCR-ABL.

These results support the concept that BCR-ABL generates at least two signals for transformation. MYC complementation of the FLVRES mutant, but not the other mutants, segregates these pathways. Although the Grb-2-binding mutant is defective in signaling to Ras (Pendergast et al. 1993b), the defect in signaling by the autophosphorylation mutant is unknown. The possibility that the signals generated by Grb-2 binding and autophosphorylation ultimately feed into the same pathway, however, cannot be ruled out.

Rescue of transformation by MYC further supports the concept that c-MYC functions downstream from a BCR-ABL signal. Our data provide genetic evidence that the FLVRES motif in the SH2 domain is responsible for activating the BCR-ABL signal via the MYC pathway. Since Grb-2 binding was shown to link tyrosine kinases to *ras* activation (for review, see Pawson and Gish 1992; Boguski and McCormick 1993), these data also provide functional evidence to segregate *ras* activation from MYC transcription.

IDENTIFICATION OF A NOVEL Ras-RELATED GENE ASSOCIATED WITH BCR-ABL-INDUCED MALIGNANCY

An important issue in cancer biology is knowledge about specific genes that are expressed in response to an oncogenic signal. This question seems particularly critical in the case of BCR-ABL-induced leukemia, where knowledge of genetic changes that precipitate blast crisis would be very valuable in understanding and managing the disease. To address this issue, we developed an approach to identify genes that are associated with BCR-ABL-mediated transformation of hematopoietic cells.

Previous work has shown that wild-type BCR-ABL, as well as the autophosphorylation mutant of BCR-ABL, can induce pre-B-cell lines to proliferate in the absence of growth factor (Pendergast et al. 1993a). When tested in vivo, however, only cells expressing wild-type BCR-ABL formed highly metastatic and lethal leukemias (Pendergast et al. 1993a). Since cells expressing wild-type BCR-ABL and mutant BCR-ABL would be expected to express similar genes associated with growth in culture, we anticipated a difference in the expression of genes associated with malignant transformation. Therefore, both cell lines were screened using the method of differential display of mRNA by polymerase chain reaction, first described by Liang and Pardee (1992).

Using this technique, five cDNA clones were identified that displayed differential expression by Northern analysis. The clone that showed the highest quantitative difference was chosen for sequence analysis. Sequencing of the full-length cDNA revealed a 40–75% homology at the amino acid level with the *ras* superfamily of genes (Cohen et al. 1994). This novel gene was named *kir*, for *k*inase-*i*nducible *r*as. Using S1 nuclease protection of isolated RNA, *kir* expression is induced in pre-B cells transformed by wild-type BCR-ABL and v-ABL. Its expression is absent in the parental cell line and in cells expressing the autophosphorylation mutant of BCR-ABL.

kir displayed the highest homology with *rad*, another new member of the *ras* superfamily that is induced in skeletal muscle of type II diabetes (Reynet and Kahn 1993). Interestingly, *rad* is a gene that is expressed in cells which are defective in a tyrosine kinase signal, whereas *kir* is induced in cells with a hyperactive tyrosine kinase signal.

How does BCR-ABL induce the expression of *kir*? We began to address this issue by examination of the DAGM myeloid cell line which was transformed with

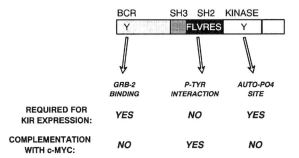

Figure 3. Summary of the signaling pathways activated by BCR-ABL. A signal dependent on the SH2 domain functions upstream of MYC but is not required for induction of *kir* expression. Signals dependent on Grb-2 binding and autophosphorylation of the kinase domain are required for *kir* expression but function independently of MYC.

the different BCR-ABL point mutants. Expression of *kir* was monitored by S1 nuclease protection (Cohen et al. 1994) and was shown to be induced in cells expressing only wild-type BCR-ABL and the SH2 mutant of BCR-ABL (summarized in Fig. 3). The Grb-2 binding mutant and the autophosphorylation mutant showed undetectable or highly reduced levels of *kir* expression. This result reinforces our model of multiple signaling pathways generated by BCR-ABL and once again segregates a signal dependent on the SH2 domain from signals dependent on Grb-2 binding and autophosphorylation. These results also suggest that signals dependent on autophosphorylation and Grb-2 binding ultimately converge at a point upstream of transcriptional activation of *kir*.

ALTERNATIVE MODES OF SIGNALING BY ABL

The disruption of c-ABL in the genesis of BCR-ABL may contribute to the abnormal cell physiology in CML. Knowledge of normal c-ABL function will add to our understanding of the transformation process. c-ABL exhibits some homology with the SRC family of tyrosine kinases. Similar to SRC, c-ABL contains a myristylation site at the amino terminus, an SH3 domain, an SH2 region, and a tyrosine kinase domain. It differs from the SRC family by exhibiting a long carboxyl terminus with a nuclear localization signal and DNA- and actin-binding capabilities (see Fig. 1). Unlike SRC, which is cytoplasmic, c-ABL is predominantly nuclear in localization (Van Etten et al. 1989) and is postulated to play a role in cell cycle regulation (Kipreos and Wang 1992; Welch and Wang 1993; Sawyers et al. 1994). A direct interaction was recently demonstrated between c-ABL and the retinoblastoma gene product (RB) (Welch and Wang 1993), a growth suppressor acting in the cell cycle. The binding of RB to c-ABL is postulated to control c-ABL tyrosine kinase activity. More recently, we have shown that c-ABL itself functions to suppress cell growth by arresting cells in G_1 (Sawyers et al. 1994). This means that in the case of BCR-ABL, a disruption of the normal *c-abl* gene may have a dual effect: (1) the generation of an activated oncogene and (2) the loss of a growth suppressor. Whether the loss of a single copy of *c-abl* contributes to the malignancy induced by BCR-ABL remains to be tested.

To further understand the biological function of c-ABL, a random mutagenesis scheme was developed to isolate transformation-active mutants of c-ABL (Goga et al. 1993). An in-frame deletion mutant was identified, which was capable of transforming fibroblasts, but not lymphoid cells in culture (Goga et al. 1993). This mutant was named ΔLX, because the deletion occurred in the last exon of c-ABL, which includes the DNA-binding region. There are several features which distinguish ΔLX from other activating ABL mutants, such as BCR-ABL and v-ABL. BCR-ABL and v-ABL are the result of mutations at the 5' end of c-ABL

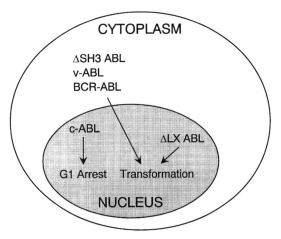

Figure 4. Positive and negative growth effects of the ABL tyrosine kinase. The nuclear c-ABL proto-oncogene is believed to control cell growth in the cell cycle. Activated mutants of ABL, including ΔSH3 ABL, v-ABL, and BCR-ABL, signal via cytoplasmic pathways to cause transformation. In contrast, the ΔLX mutant of ABL initiates a nuclear signal for transformation.

sequences. They are cytoplasmic in localization, exhibit elevated autophosphorylation activity, and are capable of transforming lymphoid cells (for review, see Rosenberg and Witte 1988). In contrast, ΔLX is both nuclear and cytoplasmic in localization, does not exhibit detectable autophosphorylation activity, and is incapable of transforming lymphoid cells (Goga et al. 1993). These data suggest that ΔLX transforms cells via a different mechanism than BCR-ABL and v-ABL (see Fig. 4). Removal of the nuclear localization signal in ΔLX blocks its transformation activity (A. Goga and O.N. Witte, unpubl.), suggesting that the mechanism of action involves a nuclear signal. The details of ΔLX signaling are not clear, but they may involve a disruption of normal c-ABL activity in growth suppression.

CONCLUSION AND FUTURE PROSPECTS

Our studies with dominant negative target genes, complementation of signaling-defective point mutants, identification of new target genes, and elucidation of c-ABL biology provide us with important functional data on the complexity of BCR-ABL signaling. With the identification of more pathways and downstream genes involved in BCR-ABL signaling, we hope to gain an understanding of oncogenic signaling that is crucial to the development of therapeutic intervention. These strategies should be useful for delineating signaling pathways generated by other oncogenes and signal transduction molecules.

ACKNOWLEDGMENTS

This work was supported by the National Cancer Institute Physician Scientist Award CA-01551 (C.L.S.), Jonsson Comprehensive Cancer Center seed fund

(C.L.S.), grant CA-53867 from the National Institutes of Health (O.N.W.), and fellowship support from the Medical Research Council of Canada (D.E.H.A.), the UCLA Medical Scientist Training Program (A.G.), Human Frontier Science Program (L.C.), and a U.S. Public Health Service NRSA award GM-07104 (R.N.M.). O.N.W. is an Investigator of the Howard Hughes Medical Institute.

REFERENCES

Afar, D.E.H., A. Goga, J. McLaughlin, O. Witte, and C.L. Sawyers. 1994. Differential rescue of Bcr-Abl point mutants with c-Myc. *Science* **264:** 424.

Ayer, D.E., L. Kretzner, and R.N. Eisenman. 1993. Mad: A heterodimeric partner for max that antagonizes Myc transcriptional activity. *Cell* **72:** 211.

Blackwood, E.M. and R.N. Eisenman. 1991. Max: A helix-loop-helix zipper protein that forms a sequence-specific DNA-binding complex with *myc*. *Science* **251:** 1211.

Boguski, M. and F. McCormick. 1993. Proteins regulating Ras and its relatives. *Nature* **366:** 643.

Clark, S.C., J. McLaughlin, M. Timmons, A.M. Pendergast, Y. Ben-Neriah, L.W. Dow, W. Crist, G. Rovera, S.D. Smith, and O.N. Witte. 1988. Expression of a distinctive BCR-ABL oncogene in Ph1-positive acute lymphocytic leukemia (ALL). *Science* **239:** 775.

Cohen, L., R. Mohr, Y.-Y. Chen, M. Huang, R. Kato, D. Dorin, F. Tamanoi, A. Goga, D. Afar, N. Rosenberg, and O. Witte. 1994. Transcriptional activation of a novel *ras*-like gene (*kir*) by oncogenic tyrosine kinases. *Proc. Natl. Acad. Sci.* **91:** 12448.

Daley, G.Q., R.A. Van Etten, and D. Baltimore. 1990. Induction of chronic myelogenous leukemia in mice by the P210$^{bcr/abl}$ gene of the Philadelphia chromosome. *Science* **247:** 824.

Dang, C.V., M. McGuire, M. Buckmire, and W.M.F. Lee. 1989. Involvement of the "leucine zipper" region in the oligomerization and transforming activity of human c-myc protein. *Nature* **337:** 664.

DeClue, J.E., K. Zhang, P. Redford, W.C. Vass, and D.R. Lowy. 1991. Suppression of *src* transformation by overexpression of full-length GTPase-activating protein (GAP) or of the GAP C terminus. *Mol. Cell. Biol.* **11:** 2819.

Diekmann, D., S. Brill, M.D. Garrett, N. Totty, J. Hsuan, C. Monfries, C. Hall, L. Lim, and A. Hall. 1991. *Bcr* encodes a GTPase activating protein for p21^{rac}. *Nature* **351:** 400.

Elefanty, A.G., I.K. Hariharan, and S. Cory. 1990. *bcr-abl*, the hallmark of chronic myeloid leukaemia in man, induces multiple haemopoietic neoplasms in mice. *EMBO J.* **9:** 1069.

Feig, L.A. and G.M. Cooper. 1988. Inhibition of NIH 3T3 cell proliferation by a mutant ras protein with preferential affinity for GDP. *Mol. Cell. Biol.* **8:** 3235.

Gishizky, M.L., J. Johnson-White, and O.N. Witte. 1993. Efficient transplantation of BCR/ABL induced CML like syndrome in mice. *Proc. Natl. Acad. Sci.* **90:** 3755.

Goga, A., J. McLaughlin, A.M. Pendergast, K. Parmar, A. Muller, N. Rosenberg, and O.N. Witte. 1993. Oncogenic activation of c-ABL by mutation within its last exon. *Mol. Cell. Biol.* **13:** 4967.

Heisterkamp, N., G. Jenster, J. ten Hoeve, D. Zovich, P.K. Pattengale, and J. Groffen. 1990. Acute leukaemia in *bcr/abl* transgenic mice. *Nature* **344:** 251.

Kantarijan, H.M., A. Deisseroth, R. Kurzrock, Z. Estrov, and M. Talpaz. 1993. Chronic myelogenous leukemia: A concise update. *Blood* **82:** 691.

Kato, G.J., J. Barrett, M. Villa-Garcia, and C.V. Dang. 1990. An amino-terminal c-Myc domain required for neoplastic transformation activates transcription. *Mol. Cell. Biol.* **10:** 5914.

Kelliher, M.A., J. McLaughlin, O.N. Witte, and N. Rosenberg. 1990. Induction of a chronic myelogenous leukemia-like syndrome in mice with v-*abl* and BCR/ABL. *Proc. Natl. Acad. Sci.* **87:** 6649.

Kelliher, M., A. Knott, J. McLaughlin, O.N. Witte, and N. Rosenberg. 1991. Differences in oncogenic potency but not target cell specificity distinguish the two forms of the BCR/ABL oncogene. *Mol. Cell. Biol.* **11:** 4710.

Kipreos, E.T. and J.Y.J. Wang. 1992. Cell cycle-regulated binding of c-Abl tyrosine kinase to DNA. *Science* **256:** 382.

Koch, C.A., D. Anderson, M.F. Moran, C. Ellis, and T. Pawson. 1991. SH2 and SH3 domains: Elements that control interactions of cytoplasmic signaling proteins. *Science* **252:** 668.

Konopka, J.B. and O.N. Witte. 1985. Detection of c-*abl* tyrosine kinase activity in vitro permits direct comparison of normal and altered *abl* gene products. *Mol. Cell. Biol.* **5:** 3116.

Kurzrock, R., J.U. Gutterman, and M. Talpaz. 1988. The molecular genetics of Philadelphia chromosome-positive leukemias. *N. Engl. J. Med.* **319:** 990.

Liang, P. and A.B. Pardee. 1992. Differential display of eukaryotic messenger RNA by means of the polymerase chain reaction. *Science* **257:** 967.

Lugo, T. and O.N. Witte. 1989. The BCR/ABL oncogene transforms Rat-1 cells and cooperates with v-myc. *Mol. Cell. Biol.* **9:** 1263.

Lugo, T.G., A. Pendergast, A.J. Muller, and O.N. Witte. 1990. Tyrosine kinase activity and transformation potency of *bcr-abl* oncogene products. *Science* **247:** 1079.

Maru, Y.M. and O.N. Witte. 1991. The BCR gene encodes a novel serine/threonine kinase activity within a single exon. *Cell* **67:** 459.

McLaughlin, J., E. Chianese, and O.N. Witte. 1987. In vitro transformation of immature hematopoietic cells by the P210 BCR/ABL oncogene product of the Philadelphia chromosome. *Proc. Natl. Acad. Sci.* **84:** 6558.

McWhirter, J.R. and J.Y.J. Wang. 1991. Activation of tyrosine kinase and microfilament-binding functions of c-*abl* by *bcr* sequences in *bcr/abl* fusion proteins. *Mol. Cell. Biol.* **11:** 1553.

Nowell, P.C. and D.A. Hungerford. 1960. A minute chromosome in human chronic granulocytic leukemia. *Science* **132:** 1497.

Pawson, T. and G.D. Gish. 1992. SH2 and SH3 domains: From structure to function. *Cell* **71:** 359.

Pendergast, A.M., M.L. Gishizky, M.H. Havlik, and O.N. Witte. 1993a. SH1 domain autophosphorylation of P210 BCR/ABL is required for transformation but not growth factor-independence. *Mol. Cell. Biol.* **13:** 1728.

Pendergast, A.M., L.A. Quilliam, L.D. Cripe, C.H. Bassing, Z. Dai, N. Li, A. Batzer, K.M. Rabun, C.J. Der, J. Schlessinger, and M.L. Gishizky. 1993b. BCR-ABL-induced oncogenesis is mediated by direct interaction with the SH2 domain of the GRB-2 adaptor protein. *Cell* **75:** 175.

Prendergast, G.C., D. Lawe, and E.B. Ziff. 1991. Association of Myn, the murine homolog of Max, with c-Myc stimulates methylation-sensitive DNA binding and Ras cotransformation. *Cell* **65:** 395.

Reynet, C. and C.R. Kahn. 1993. Rad: A member of the Ras family overexpressed in muscle of type II diabetic humans. *Science* **262:** 1441.

Ron, D., M. Zannini, M. Levis, R.B. Wickner, L.T. Hunt, G. Graziani, S.R. Tronick, S.A. Aaronson, and A. Eva. 1991. A region of proto-dbl essential for its transforming activity shows sequence similarity to a yeast cell cycle gene, CDC 24, and the human breakpoint cluster gene, *bcr*. *New Biol.* **3:** 372.

Rosenberg, N. and O.N. Witte. 1988. The viral and cellular forms of the Abelson (*abl*) oncogene. *Adv. Virus Res.* **39:** 39.

Rowley, J.D. 1973. A new consistent chromosomal abnormality in chronic myelogenous leukemia identified by quinacrine fluorescence and giemsa staining. *Nature* **243:** 290.

Sawyers, C.L., W. Callahan, and O.N. Witte. 1992. Dominant negative *myc* blocks transformation by ABL oncogenes. *Cell* **70:** 901.

Sawyers, C.L., J. McLaughlin, A. Goga, M. Havlik, and O. Witte. 1994. The nuclear tyrosine kinase c-Abl negatively regulates cell growth. *Cell* **77:** 121.

Scott, M.L., R.A. Van Etten, G.Q. Daley, and D. Baltimore. 1991. v-*abl* causes hematopoietic disease distinct from that caused by bcr-abl. *Proc. Natl. Acad. Sci.* **88:** 6506.

Shtivelman, E., B. Lifshitz, R.P. Gale, and E. Canaani. 1985. Fused transcript of *abl* and *bcr* genes in chronic myelogenous leukaemia. *Nature* **315:** 550.

Stacey, D.W., L.A. Feig, and J.B. Gibbs. 1991a. Dominant inhibitory Ras mutants selectively inhibit the activity of either cellular or oncogenic Ras. *Mol. Cell. Biol.* **11:** 4053.

Stacey, D.W., M. Roudebush, R. Day, S.D. Mosser, J.B. Gibbs, and L.A. Feig. 1991b. Dominant inhibitory Ras mutants demonstrate the requirement for Ras activity in the action of tyrosine kinase oncogenes. *Oncogene* **6:** 2297.

Thomas, S.M., M. DeMarco, G. D'Arcangelo, S. Halegoua, and J.S. Brugge. 1992. Ras is essential for nerve growth factor- and phorbol ester-induced tyrosine phosphorylation of MAP kinases. *Cell* **68:** 1031.

Van Etten, R.A., P. Jackson, and D. Baltimore. 1989. The mouse type IV c-*abl* gene product is a nuclear protein, and activation of transforming ability is associated with cytoplasmic localization. *Cell* **58:** 669.

Waksman, G., D. Kominos, S.C. Robertson, N. Pant, D. Baltimore, R.B. Birge, D. Cowburn, H. Hanafusa, B.J. Mayer, M. Overduin, M.D. Resh, C.B. Rios, L. Silverman, and J. Kuriyan. 1992. Crystal structure of the phosphotyrosine recognition domain SH2 of v-*src* complexed with tyrosine-phosphorylated peptides. *Nature* **358:** 646.

Welch, P.J. and J.Y.J. Wang. 1993. A C-terminal protein-binding domain in the retinoblastoma protein regulates nuclear c-Abl tyrosine kinase in the cell cycle. *Cell* **75:** 779.

Wood, K.W., C. Sarnecki, T.M. Roberts, and J. Blenis. 1992. Ras mediates nerve growth factor receptor modulation of three signal-transducing protein kinases: Map kinase, Raf-1 and RSK. *Cell* **68:** 1041.

The *AML1* Gene in the 8;21 and 3;21 Translocations in Chronic and Acute Myeloid Leukemia

G. Nucifora and J.D. Rowley

Section of Hematology/Oncology, Department of Medicine, The University of Chicago, Chicago, Illinois 60637

Recurring chromosomal translocations are relatively common in certain neoplasms, particularly leukemias, lymphomas, and some sarcomas (Heim and Mitelman 1987). Cloning of the translocation breakpoints has identified the genes involved in the breakpoint junctions, and this in turn has led to a determination of the manner in which the genes are altered. The genes can be altered in the level of their expression or in the properties of the encoded proteins. These alterations appear to play an integral role in the development of the malignant phenotype.

In B and T cells, recurring chromosomal translocations disrupt the normal growth regulation by placing a gene important in this regulation under the control of another gene, most often the immunoglobulin genes in B cells or the T-cell receptor genes in T cells. Classic examples of these translocations involve the *MYC* proto-oncogene and are seen in Burkitt's lymphoma or B-cell acute lymphoblastic leukemia (ALL) and in T-cell leukemia with the t(8;14). The MYC protein is not altered in these translocations. Experiments with transgenic mice have shown directly that the deregulated overexpression of MYC can indeed lead to neoplasia (Adams et al. 1985). In contrast, a few translocations in B-cell leukemia lead to fusion of the two genes, and thus to the production of a fusion protein, e.g., t(1;19), t(4;11), and t(11;19).

In myeloid leukemias, chromosomal translocations disrupt the normal genes by altering the normal properties of proteins through gene fusion (Table 1). A classic example is the *BCR/ABL* fusion in the 9;22 translocation, seen in almost all patients with chronic myelogenous leukemia (CML), and in at least 30% of adult patients with ALL. In many acute myeloid leukemias (AML), the genes involved are most often transcription factors, frequently not normally active in hematopoietic cells. In some instances, the same gene is involved in fusion with more than one chromosomal partner (Kamps et al. 1990; Inaba et al. 1992; Thirman et al. 1993). One of the genes involved with multiple chromosomal translocation partners is *AML1*, located at 21q22. *AML1* is split in the t(8;21)(q22;q22) and in the t(3;21)(q26;q22). In addition, preliminary fluorescence in situ hybridization (FISH) results indicate that *AML1* is also involved in rare translocations such as a t(5;21)(q13;q22) or t(17;21)(q11;q22) seen in patients with myelodysplastic syndrome (MDS) or AML (Roulston et al. 1993).

The (8;21)(q22;q22) translocation, reported in about 40% of karyotypically abnormal cases of AML-M2 (Nucifora et al. 1993c), is one of the most common examples of association between a nonrandom reciprocal translocation and a specific subtype of AML. The t(8;21) usually predicts a good response to chemotherapy with a high remission rate and relatively long median survival. The translocation breakpoint has been cloned, and the gene that is involved with *AML1* is *ETO* (Erickson et al. 1992; Nisson et al. 1992). Both genes are transcribed from telomere to centromere. A human *AML1* cDNA was isolated that includes an open reading frame encoding a 250-amino-acid polypeptide. We and other investigators performed a BLAST search of GenBank DNA sequences and detected a 67% identity between *AML1* (from nucleotides 915 to 1301) with nucleotides 564 to 950 from the *Drosophila melanogaster* segmentation gene, *runt* (Daga et al. 1992; Erickson et al. 1992). The homology extended over 118 residues (69% identity), and most of the differences were conservative changes.

AML1 belongs to a new family of heterodimeric transcriptional regulatory proteins with important roles in processes ranging from segmentation in *D. melanogaster* to hematopoiesis and leukemia in humans (Kagoshima et al. 1993). The founding protein of this family is the *D. melanogaster* Runt. The *runt* gene is an early acting pair-rule segmentation gene in *D. melanogaster* development and regulates the expression of other segmentation genes such as *hairy*, *even-skipped*, and *fushi tarazu* (Frasch and Levine 1987; Gergen and Butler 1988; Ingham and Gergen 1988; Edgar et al. 1989). Runt is a nuclear DNA-binding protein (Kania et al. 1990), but lacks homology with any well-characterized DNA-binding motif.

Little is known about the function of *AML1*, but the identification of the homology with *runt* suggested that the conserved region might correspond to a functional domain. *AML1* is expressed in hematopoietic cells (Erickson et al. 1992; Miyoshi et al. 1993a), suggesting that it might function in normal hematopoietic development. Bae et al. (1993) recently cloned the murine homolog of *AML1* that encodes a protein of 451 amino acids with an amino-terminal region of 241 residues identical to AML1 but with a carboxy-terminal extension of 210 amino acids that may represent an alternatively spliced form of the gene. Other human cDNA clones of *AML1* have also been obtained which are

Table 1. Functional Classification of Transforming Genes at Translocation Junctions

	Location	Translocation	Disease
SRC Family (Tyr protein kinases)			
ABL	9q34	t(9;22)	CML/ALL
LCK	1p34	t(1;7)	T-ALL
ALK	5q35	t(2;5)	NHL
Serine protein kinase			
BCR	22q11	t(9;22)	CML/ALL
Cell-surface receptor			
TAN1	9q34	t(7;9)	T-ALL
Growth factor			
IL2	4q26	t(4;16)	T-NHL
IL3	5q31	t(5;14)	PreB-ALL
Mitochondrial membrane protein			
BCL2	18q21	t(14;18)	NHL
Cell-cycle regulator			
CCND1 (BCL1-PRAD1)	11q13	t(11;14)	CLL/NHL
Myosin family			
MYH11	16p13	inv(16), t(16;16)	AML-M4Eo
Ribosomal protein			
EAP (L22)	3q26	t(3;21)	t-AML/CML BC
Unknown			
DEK	6p23	t(6;9)	AML-M2/M4
DNA-binding factors			
Homeobox			
PBX	1q23	t(1;19)	PreB-ALL
HOX11	10q24	t(10;14)/t(7;10)	T-ALL
Helix-loop-helix			
CAN	9q34	t(6;9)	AML
LYL1	19p13	t(7;19)	T-ALL
MYC#	8q24	t(8;14)	B-ALL/T-ALL
TAL1(SCL)	1p32	t(1;14)	T-ALL
TAL2	9p32	t(7;9)	T-ALL
TCF3(E2A)#	19p13	t(1;19)	PreB-ALL
Zinc finger			
ETO	8q24	t(8;21)	AML-M2
MLL	11q23.3	t(11q23)	ALL/AML
PLZF	11q23.1	t(11;17)	APL
PML	15q22	t(15;17)	APL
RARA	17q12	t(15;17)	APL
EVI1	3q26	inv(3), t(3;3)	AML
BCL6	3q27	t(3;14)	NHL
Lim			
RBTN1(TTG1)	11p15	t(11;14)	T-ALL
RBTN2	11p13	t(11;14)	T-ALL
Other			
AML1 (*runt* homology)	21q22	t(8;21), t(3;21)	AML-M2
LYT10 (*rel* homology)	10q24	t(10;14)	B-NHL
Undefined			
MDS1	3q26	t(3;21)	AML
Transcriptional modulators			
BCL3	19q13	t(14;19)	B-CLL
CBFB	16q22	inv(16), t(16;16)	AML-M4Eo

alternatively spliced following the *runt* homology (Nucifora et al. 1993d; Bae et al. 1994). Expression of this gene is detected in most tissues and cell lines tested, although not in the brain, lung, ovaries, and testis. Northern blot analysis of *AML1* has shown several bands corresponding to transcripts of 2–8 kb (Erickson et al. 1992; Miyoshi et al. 1993a), and it has been suggested that differential splicing may constitute an important regulatory mechanism of gene expression of *AML1* (Kagoshima et al. 1993). Preliminary evidence from other groups has suggested that the human AML1 lacks a strong activation domain (Meyers et al. 1993).

However, because multiple forms of AML1 exist, an activation domain is very likely to be present in one of the larger AML1 proteins, as has been determined for the murine homolog.

The latest member of the Runt family to be identified is the α subunit of the murine transcription factor Pebp2. Pebp2 interacts with the core enhancer region of the polyoma DNA tumor virus, and it is important in viral transcription and replication (Speck and Renjifo 1990). Pebp2 is probably identical or related to factors in T-cell nuclear extracts variously referred to as CBF, SEF-1, S/A-CBF, and NF-dE3A (Kagoshima et al.

1993). Pebp2/CBF binds to a conserved site within the core enhancers of murine leukemia viruses (Wang and Speck 1992). Several studies have shown that this conserved core site, TGTGGTT, is important for the activity of these enhancers in T cells. Pebp2/CBF-binding sites have also been identified in T-cell promoters, including the *LCK* proximal promoter (Allen et al. 1992), and in the enhancers of the T-cell receptor genes (Gottschalk and Leiden 1990; Redondo et al. 1990, 1992; Kappes et al. 1991). Purified Pebp2/CBF is a complex of two unrelated polypeptide subunits (Ogawa et al. 1993). Pebp2α/CBFα is the DNA-binding subunit. Sequencing of Pebp2α/CBFα subunit revealed that it encoded a region highly homologous to the shared domain of the Runt and AML1 proteins. The second subunit of Pebp2/CBF, Pebp2β/CBFβ, does not itself bind to DNA, but its association with the α subunit results in a heterodimeric complex with greater affinity to DNA (Kagoshima et al. 1993). An inversion in human chromosome 16 that is associated with the M4Eo subtype of AML has recently been found to disrupt the *Pebp2β/CBFβ* gene by fusion to the myosin heavy-chain gene *MYH11* (Liu et al. 1993). Both Pebp2α and AML1 dimerize with Pebp2β/CBFβ. The lack of homology with protein motifs known to be involved in protein-protein interaction suggests that interaction between the Runt domain and Pebp2β/CBFβ is mediated by a novel type of protein dimerization motif.

The partner gene of *AML1* on chromosome 8, *ETO*, encodes a zinc finger protein (Miyoshi et al. 1993b). Two full-length clones were isolated (Miyoshi et al. 1993b) that encode a 577- and a 604-amino-acid polypeptide, respectively, differing at their amino end, where one of the clones extends for an additional 27 amino-terminal residues. The breakpoint of *ETO* is at the beginning of the gene after either codon 2 (for the 577-residue ETO) or after codon 29 (for the 604-residue ETO), and the fusion product contains the 575 carboxy-terminal residues of ETO fused to the Runt segment of *AML1*. The predicted amino acid sequence shows no significant overall homology with any known protein sequences, but the carboxy-terminal region contains two putative zinc finger DNA-binding motifs, one of the type Cys-Cys/Cys-Cys, found in the glucocorticoid receptors, and the second of the type Cys-Cys/His-Cys, found in the retroviral nucleic-acid-binding proteins. These two potential zinc fingers are similar to those of a cell-death-associated protein RP8. In addition, the carboxy-terminal region contains a PEST region that confers rapid intracellular degradation (Miyoshi et al. 1993b). The predicted polypeptide has three proline-rich regions, similar to those noted in transcription activation domains (Miyoshi et al. 1993b). The expression of *ETO* was detected by Northern blot analysis at a high level only in the brain and at low levels in the lung, testis, and ovaries. It is of interest that no detectable level of *ETO* was seen in hematopoietic cells, in the spleen, or in the thymus. The predicted chimeric AML1/ETO protein retains the ability to recognize the AML1 consensus binding site and to dimerize with the Pebp2β/CBFβ subunit (Meyers et al. 1993). A reciprocal translocation product from the derivative 21 (der[21]) chromosome has not yet been identified.

In reciprocal translocations, the identification of the derivative chromosome containing the critical junction is essential. Consistent data from cytogenetic and FISH analyses indicated that the der(8) chromosome contains the junction always conserved in two-way or in variant translocations (Lindgren and Rowley 1977; Rowley 1982).

AML1 is also involved in a second, more rare recurring translocation, the t(3;21)(q26;q22), which occurs in therapy-related AML (t-AML) patients, previously treated with drugs including topoisomerase II inhibitors, and in patients with CML in blast crisis (CML-BC) (Rubin et al. 1987, 1990). Chromosomal band 3q26 is also the location of *EVI1*, whose transcription can be activated by translocation 13 to 330 kb upstream from the 5' end of the gene, as well as by rearrangement 150 kb downstream from the 5' end of the gene (Morishita et al. 1992b). Analysis of leukemic blasts from 116 AML patients showed expression of *EVI1* in 8 of them, 7 of whom had cytogenetically detectable translocations of chromosome 3q26, where *EVI1* resides (Morishita et al. 1992b). The *EVI1* gene encodes a 145-kD DNA-binding protein with seven zinc finger motifs at the amino terminus and three zinc finger motifs at the carboxyl terminus (Morishita et al. 1990a). The gene is essential for normal development of kidney, brain, and peripheral nervous system (Morishita et al. 1990b). Inappropriate expression of the gene in fibroblasts and macrophages has no obvious effect, but in hematopoietic cells it blocks granulocytic and erythrocytic differentiation (Morishita et al. 1992a). Similarly, inappropriate expression of *EVI1* in normal bone marrow progenitors blocks their ability to differentiate terminally in vitro. The mechanism by which inappropriate expression of *EVI1* interferes with differentiation seems to reside in the specificity of the zinc fingers at the amino terminus to bind GATA-containing motifs. It has been suggested that *EVI1* might activate or repress promoters containing this motif, and thereby interfere with a family of factors which recognize the GATA motif and which have been implicated in the regulation of stem cell differentiation. Mitani et al. (1994) recently reported on the isolation of a fusion *AML1/EVI1* cDNA from the library of a t(3;21)(q26;q22) human CML-BC cell line.

MATERIALS AND METHODS

Patient samples and cell lines. The AML-M2 patients with a t(8;21), the CML-BC and t-AML/MDS patients with a t(3;21), and the cell lines used in this study have been described previously (Nucifora et al. 1993a,b,c,d, 1994b).

DNA and RNA isolation. Standard procedures for

the isolation of high-molecular-weight DNA have been described previously (Nucifora et al. 1993c). Poly(A) RNA was prepared from 1×10^7 to 4×10^7 cells with the Fast Track mRNA isolation kit (Invitrogen, San Diego, California) according to the manufacturer's instructions. All of the poly(A) RNA isolated (0.1–1 μg) was used for the synthesis of the first-strand cDNA necessary for PCR amplification. The synthesis was performed with the cDNA Cycle First Strand Synthesis kit (Invitrogen, San Diego, California) in a total reaction volume of 20 μl according to the manufacturer's instructions.

Polymerase chain reaction amplification. The conditions and the primers used for reverse transcription polymerase chain reaction (RT-PCR) have been described by Nucifora et al. (1993c) (for detection of the *AML1/ETO* junction) and by Nucifora et al. (1994b) (for detection of the *AML1/EAP*, *AML1/MDS1*, and *AML1/EVI1* junctions).

FISH analysis. The procedure for FISH analysis, including labeling of the probes, capture of the image with a CCD camera, and image analysis, have been described previously (Nucifora et al. 1993d).

RESULTS

Analysis of the *AML1* Gene in the t(8;21)

Southern blot analysis. The cloning and molecular analyses of translocation breakpoint junctions have provided very sensitive tools to detect chromosomal abnormalities in patients either by FISH or by Southern blot analyses, and by RT-PCR amplification when the translocation leads to a fusion gene with a known sequence. These molecular techniques allow specific detection of the two genes that are involved at the breakpoint. We and other workers have cloned several probes from the intron in which the breakpoints on chromosome 21 consistently occur to use for Southern blot analysis. The probes (wc-1, wc-2 and wc-3, Fig. 1) are located on either side of a *Bam*HI site contained in the 25-kb intron, and they detect abnormal *Bam*HI fragments in DNA from t(8;21) cells. In addition, we subcloned a probe containing genomic sequences from chromosome 8 using a t(8;21) junction clone (Nucifora et al. 1993c). Generally, DNA rearrangements are observed by Southern blot analysis in only 80–100% of the cases with cytogenetically detected 8;21 translocations. Figure 2 shows the results of a representative Southern blot analysis of AML-M2 patients with the t(8;21). The results for patient 5 (samples 5-I, diagnosis; 5-II, remission; and 5-III, relapse, Fig. 2) are of particular interest as they were obtained at various stages of the disease in this patient (Nucifora et al. 1993c). Samples 5-I and 5-III have the same rearranged pattern in the Southern blot analysis, indicating that the relapse was due to the expansion of the original malignant

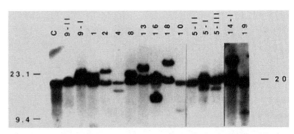

Figure 2. Representative Southern blot analysis of 15 samples from 12 AML t(8;21) patients, including two remission samples (9-II and 5-II). Human placenta (C) was used as control. The size of the germ-line fragment is indicated on the right. DNA size markers are indicated on the left (Nucifora et al. 1993c).

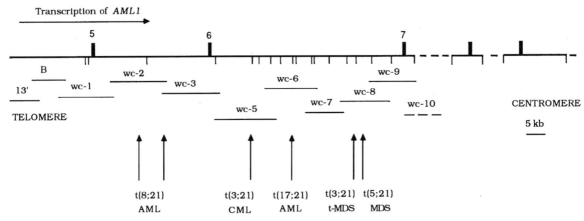

Figure 1. *Bam*HI restriction map of chromosome 21 in the region containing the breakpoints of the t(8;21) (AML), t(3;21) (t-MDS/t-AML), t(3;21) (CML), t(5;21) (MDS), and t(17;21) (AML). The breakpoints are indicated by vertical arrows. Vertical bars below the horizontal line indicate the *Bam*HI sites. Black boxes above the horizontal line indicate the numbered *AML1* exons. The direction of transcription of *AML1* is indicated by a horizontal arrow. The recombinant phage clones 13' through wc-10 obtained from the YAC-Not-42 library and used for FISH and Southern blot analyses are indicated by horizontal lines below the map. wc-10 was not precisely mapped, so is indicated by a dashed line. Genomic regions indicated by a dashed line have not been mapped (Nucifora and Rowley 1994).

clone detected by the Southern blot analysis of sample 5-I, although analysis of the remission sample (5-II) showed no detectable rearranged band.

RT-PCR analysis. In recent years, RT-PCR has been successfully used as a very sensitive tool to identify fusion genes resulting from chromosomal translocations in many leukemias, to monitor the patients' response to therapy, and to detect minimal residual disease in leukemic patients. We used RT-PCR to analyze patients who had a t(8;21) cytogenetically, but who had a normal pattern on Southern blot analysis. All of the patients' samples that were analyzed were positive for the chimeric junction (not shown).

The sensitivity of the RT-PCR technique is ideal to follow the persistence of leukemic cells expressing fusion genes in patients during remission. So far, we have only studied three t(8;21) patients who had sequential samples from diagnosis or relapse through long-term remissions of up to 8 years. Our results (Fig. 3) show that even after 6 to 8 years in remission, these patients still have circulating leukocytes with the t(8;21) producing the chimeric *AML1/ETO* transcript (Nucifora et al. 1993a). We have studied one additional patient after 1 month and 4.5 years in remission whose cells had the t(8;21) detected at diagnosis by cytogenetic analysis only. Several analyses of this patient's early remission bone marrow and of a more recent blood sample failed to detect an amplified chimeric message on RT-PCR. We did not have a sample from this patient at diagnosis to analyze for the presence of the chimeric transcript; thus, it is possible that some long-term survivors may actually be cured.

Leukemic cells with a t(8;21) have a unique morphology that has been described previously (Nucifora et al. 1994a). To investigate whether there was a correlation between the morphology of the t(8;21) cells and the expression of the chimeric *AML1/ETO* gene, we used RT-PCR to analyze samples from leukemic patients without a t(8;21) but whose neoplastic cells had the features of the t(8;21) cells. We detected the fusion transcript in two of the AML patients whose leukemic cells were virtually identical in morphology to the t(8;21) cells although the patients' cells had cytogenetically normal chromosomes 8 and 21 (Fig. 4) (Nucifora et al. 1994a). The results indicated a close correlation between the characteristic morphology of t(8;21) cells and the products of the translocation.

Analysis of the *AML1* Gene in the t(3;21)

Mapping of **AML1** *breakpoint in the t(3;21).* We used probes obtained from the lambda library of the yeast artificial chromosome (YAC) Not-42, containing the *AML1* gene, to confirm that the YAC was also split in cytogenetic samples from patients with a t(3;21) (Nucifora et al. 1993d). By analyzing a number of phage subclones for their location relative to the breakpoint, probes were identified that were centromeric and telomeric to the breakpoint, and additional overlapping genomic phage clones were isolated that covered an *AML1* region of about 60 kb downstream from the t(8;21) breakpoint (Nucifora et al. 1993d). Southern blot analysis of material from t(3;21) patients indicated that the breakpoint in *AML1* was approximately 40 kb downstream from the t(8;21) breakpoint (Nucifora et al. 1993d).

Isolation of the **AML1/EAP** *fusion cDNA.* By using a cDNA library made from leukemic cells of a t-MDS patient with the t(3;21) and an *AML1*-specific probe, chimeric cDNA clones were isolated that contained the 5' region of *AML1* fused to a previously characterized gene, *EAP* (Nucifora et al. 1993b). EAP is a highly expressed small protein of 129 residues with a putative nuclear localization signal and a long stretch

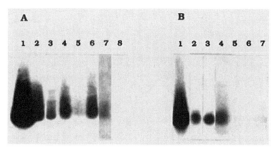

Figure 3. (*A*) Autoradiogram of PCR-amplified cDNA from the cell line Kasumi-1 (lane *1*) and from bone marrow and peripheral blood samples of t(8;21) patients early in hematological and cytogenetic remission (lanes *2–7*) and the myeloid cell line ML-1 (lane *8*). (*B*) Autoradiogram of the PCR-amplified products from the cell line Kasumi-1, carrying the t(8;21) (lane *1*), the peripheral blood of the patients in long-term remissions (lanes *2–4*), the cell line TK-6 (lane *5*), and two control individuals (lanes *6* and *7*) (Nucifora et al. 1993a).

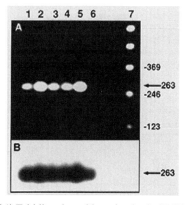

Figure 4. (*A*) Ethidium-bromide-stained gel of PCR-amplified products of an AML patient without a t(8;21) but with cells with the t(8;21) morphology. The samples were obtained at diagnosis (lane *1*, peripheral blood [PB] and lane *2*, bone marrow [BM]), during complete remission (lane *3*, BM), at the time of surgical removal of granulocytic sarcoma (lane *4*, PB and lane *5*, BM), and during the extramedullary relapse but PB and BM remission (lane *6*, BM). (Lane *7*) 123-bp ladder DNA and BM remission. (*B*) Autoradiogram after Southern blot analysis of the PCR-amplified products (Nucifora et al. 1994a).

of acidic residues at the carboxyl end. EAP was initially identified as a protein associated with Epstein-Barr virus (EBV) small RNA and was later recognized as the ribosomal protein L22 (Toczyski et al. 1994). *EAP* belongs to a family of pseudogenes and expresses three transcripts of about 1 kb, 1.5 kb, and 2 kb in all tissues and cell lines (Nucifora et al. 1993b). The 3;21 fusion clones contained the *runt* homology that is fused to *ETO* in the t(8;21) and, in addition, one other *AML1* exon (exon 6, Fig. 1). After the breakpoint junction, the 3' end of *AML1* was replaced by the last 96 codons of *EAP*, followed by about 0.9 kb of *EAP* 3' untranslated exon. However, in contrast to the fusion genes produced from other chromosomal translocations, the fusion transcript did not maintain the reading frame of the second gene, and the translation of the fusion mRNA was terminated shortly after the *AML1/EAP* junction by an out-of-frame stop codon introduced by the *EAP* sequence (Nucifora et al. 1993b). These results were confirmed in four other t(3;21) patients (Nucifora et al. 1994b). As a consequence, the predicted AML1/EAP polypeptide did not contain any amino acid homology with EAP, and it was suggested that most likely, AML1/EAP was devoid of *trans*-activation activity and could act as a constitutive repressor inhibiting the function of the normal AML1.

Isolation of the **AML1/MDS1** *fusion cDNA.* In all five patients, a second additional chimeric transcript was identified that fused *AML1* to sequences not previously reported centromeric (downstream) to *EAP*. This second fusion clone contained the same region of *AML1* found in the *AML1/EAP* cDNA fused to new sequences that were localized on chromosome band 3q26 with FISH. We named these sequences *MDS1* (for *m*yelo*d*ysplasic *s*yndrome) (Nucifora et al. 1994b). The region of *MDS1* that was cloned contained approximately 400 bp in frame with the *AML1* sequence, followed by about 1 kb of untranslated region. There was no significant homology between *MDS1* and the sequences deposited in GenBank. Southern blot analysis of higher eukaryote DNA showed that these new sequences were evolutionarily conserved, but in contrast to *EAP*, they were represented only once in the genome (Nucifora et al. 1994b). In adult human tissues, *MDS1* was expressed only in kidney, lung, and pancreas, with two major transcripts of about 5.6 kb and 6.2 kb (Nucifora et al. 1994b).

A comparison of the fusion proteins predicted from the two chimeric cDNAs obtained from the same library is shown in Figure 5. Whereas *AML1/EAP* produces a truncated form of AML1 consisting mostly of the Runt homology region of the protein, the fusion of *AML1* with *MDS1* adds 127 residues with a high percentage of proline, serine, and acidic residues, suggestive of a transcription regulation domain, to the Runt domain. Analysis of the t(3;21) patients by RT-PCR with primers specific for the *AML1/EAP* junction and for the *AML1/MDS1* junction showed that in addition to the expected *AML1/EAP* or *AML1/MDS1* chimeric junctions, smaller bands were also consistently amplified. These bands represented isoforms of the chimeric junctions that did not contain exon 6 of *AML1*. These results indicated that the production of two chimeric transcripts between *AML1* and two unrelated genes was a unique but consistent consequence of this 3;21 translocation, and that two different chimeric mRNAs coexisted in these patients. Figure 6 (top) shows a diagram with the position of the primers used for amplification and the various chimeric junctions identified in the patients.

Isolation of the **AML1/EVI1** *fusion cDNA.* Mitani et al. (1994) reported on the detection of an *AML1/EVI1* fusion transcript in t(3;21) CML-BC patients and in one t(3;21) cell line. By RT-PCR analysis, we detected a high level of *EVI1* message in the peripheral blood cells of one of the five t(3;21) patients (Nucifora et al. 1994b). To determine whether *EVI1* was also expressed as a fusion gene with *AML1* in this patient, we synthesized *EVI1* primers to use with the *AML1* primer in RT-PCR. One of the *EVI1* primers amplified four major bands that were cloned and sequenced (Nucifora et al. 1994b). The two smaller fragments contained the junction between *AML1* and either exon 2 or exon 3 of *EVI1*; the two larger fragments represented complex fusion clones, including *AML1* (plus or minus exon 6) fused to *MDS1*, which was fused to exon 2 of *EVI1* (Fig. 6, top). All of these fusion transcripts maintained the same reading frame, leading to complex fusion polypeptides. Although 5' RACE and S1 mapping have shown that in the *EVI1* locus the transcriptional start sites of *EVI1* are a few hundred bp upstream of exon 1 (Bartholomew and Ihle 1991), we cannot rule out that the *MDS1* sequences represent even more 5' previously unreported exon(s) of *EVI1*. The remaining four patients failed to express *EVI1* or any of the fusion transcripts between *AML1* and *EVI1*.

Mapping of the genes at 3q26. Each of these genes, *EAP*, *MDS1*, and *EVI1*, is oriented 5' to 3' telomere to centromere. Previous pulsed field gel electrophoresis (PFGE) and Southern blot studies had shown that a CpG island exists 5' of the *EVI1* gene which contains sites for *BSS*HII, and that the next *BSS*HII site occurs approximately 600 kb upstream (telomeric) of the gene (Morishita et al. 1992b). By PFGE and Southern blot analyses, we determined that both the *EAP* and *MDS1* probes were localized on a 300-kb *Sfi*I fragment contained within the *BSS*HII fragment that is detected by a 5' *EVI1* probe. The 300-kb *Sfi*I fragment was mapped about 170 kb upstream of *EVI1*, therefore, both *EAP* and *MDS1* probes are at least 170 kb upstream (telomeric) of *EVI1*. To determine the order of the two genes, we mapped them relative to the breakpoint found in a case of AML with t(3;3)(q21;q26) for which somatic cell hybrids existed that contained the derivative chromosomes (Nucifora et al. 1994b). Previous studies had shown that this breakpoint occurs within the region 400–170 kb 5' of *EVI1* (Morishita et al.

```
AML1/EAP    1    M N P C R D V H D A S T S R R F T P P S T A L S P G K M S E A L P
AML1        1    M R I P V * * * * * * * * * * * * * * * * * * * * * * * * * * *
AML1/MDS1   1        M R * * * * * * * * * * * * * * * * * * * * * * * * * * *

AML1/EAP    34   L G A P D A G A A L A G K L R S G D R S M V E V L A D H P G
AML1        31   * * * * * * * * * * * * * * * * * * * * * * * * * * * * * *
AML1/MDS1   28   * * * * * * * * * * * * * * * * * * * * * * * * * * * * * *

AML1/EAP    64   E L V R T D S P N F L C S V L P T H W R C N K T L P I A F K
AML1        61   * * * * * * * * * * * * * * * * * * * * * * * * * * * * * *
AML1/MDS1   58   * * * * * * * * * * * * * * * * * * * * * * * * * * * * * *

AML1/EAP    94   V V A L G D V P D G T L V T V M A G N D E N Y S A E L R N A
AML1        91   * * * * * * * * * * * * * * * * * * * * * * * * * * * * * *
AML1/MDS1   88   * * * * * * * * * * * * * * * * * * * * * * * * * * * * * *

AML1/EAP    124  T A A M K N Q V A R F N D L R F V G R S G R G K S F T L T I
AML1        121  * * * * * * * * * * * * * * * * * * * * * * * * * * * * * *
AML1/MDS1   118  * * * * * * * * * * * * * * * * * * * * * * * * * * * * * ▼

AML1/EAP    154  T V F T N P P Q V A T Y H R A I K I T V D G P R E P R R H R
AML1        151  * * * * * * * * * * * * * * * * * * * * * * * * * * * * * *
AML1/MDS1   148  * * * * * * * * * * * * * * * * * * * * * * * * * * * * * *

AML1/EAP    184  Q K L D D Q T K P G S L S F S E R L S E L E Q L R R T A M R
AML1        181  * * * * * * * * * * * * * * * * * * * * * * * * * * * * * *
AML1/MDS1   178  * * * * * * * * * * * * * * * * * * * * * * * * * * * * * *

AML1/EAP    214  V S P H H P A P T P N P R A S L N H S T A F N P Q P Q S Q M
AML1        211  * * * * * * * * * * * * * * * * * * * * * * * * * * * * * *
AML1/MDS1   208  * * * * * * * * * * * * * * * * * * * * * * * * * * * * * *
                                       ▼
                                       ▼
AML1/EAP    244  Q E S W M L P I L S S F C K K G S K -
AML1        241  * D A R Q I Q P S P -
AML1/MDS1   238  * N E C V Y G N Y P E I P L E E M P D A D G V A S T P S L N

AML1/MDS1   268  I Q E P C S P A T S S E A F T P K E G S P Y K A P I Y I P D

AML1/MDS1   298  D I P I P A E F E L R E S N M P G A G L G I W T K R K I E V

AML1/MDS1   328  G E K F G P Y V G E Q R S N L K D P S Y G W E V R M Y F E L

AML1/MDS1   358  T H L Y I A A V -
```

Figure 5. Amino acid alignment of AML1 and of the two chimeric polypeptides AML1/EAP and AML1/MDS1. Asterisks indicate identities. The segment homologous to the *D. melanogaster* segmentation protein Runt is indicated in italics. The single arrowhead indicates the consistent site of the fusion with the protein ETO in the 8;21 translocation. The double arrowhead indicates the beginning of the *AML1* intron containing the t(3;21) breakpoint. In our patients with a t(3;21), cDNA fusion junctions have been detected between *AML1* (at the sites indicated with the single arrowhead and with the double arrowhead; see text) and *EAP*, *MDS1*, or *EVI1*. The region included between the single and double arrowheads corresponds to exon 6 of *AML1*, as shown in Fig. 6 (Nucifora et al. 1994b).

1992b). By Southern blot analysis, only the *EAP*-specific probe hybridized with cells containing the 3q-derivative chromosome, demonstrating that *EAP* was telomeric of the breakpoint, whereas the *MDS1* probe was centromeric of the breakpoint (Fig. 6, bottom). Thus, the order of the probes from the telomere is *EAP-MDS1-EVI1*. We also used FISH analysis to confirm that *EVI1* was centromeric to *MDS1*.

DISCUSSION

Significance and Consequences of the *AML1/ETO* Fusion in Cells with the t(8;21)

Our extensive Southern blot, RT-PCR, and FISH analyses of patients with the t(8;21) have shown that the same exons encoding the *runt* homology telomeric to the breakpoint are consistently conserved and translocated to the der(8) chromosome (Nucifora et al. 1993c), and our work has been confirmed independently by others. Thus, the t(8;21) should juxtapose the 5' region of the *AML1* gene, containing the entire DNA-binding *runt* homology region to sequences from the *ETO* gene on chromosome 8, leading to a fusion gene product.

The *AML1/ETO* cDNAs that have been cloned and sequenced show a fusion of the 5' region of *AML1* with the 3' region of *ETO* (Erickson et al. 1992; Nisson et al. 1992; Miyoshi et al. 1993b). The *AML1/ETO* junction on the der(8) chromosome has been consistently de-

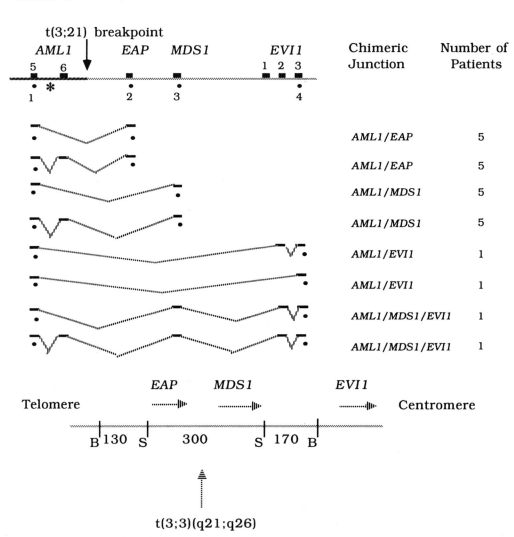

Figure 6. (*Top*) Diagram illustrating the chimeric junctions detected by RT-PCR and confirmed by nucleotide sequencing. The heavily shaded portion of the horizontal line represents the translocated region of chromosome 21 containing the promoter and the 5' region of *AML1*. The lightly shaded portion of the horizontal line represents the region of chromosome 3 containing the *EAP*, *MDS1*, and *EVI1* probes. The vertical arrow indicates the genomic junction. Black boxes represent the exons involved in the chimeric junctions and identified by nucleotide sequencing; numbers above the black boxes indicate the published numbering of the *AML1* and *EVI1* exons. The asterisk between exons 5 and 6 of *AML1* indicates the intron where the breakpoints of the t(8;21) have been consistently detected. Black dots below the horizontal line indicate the positions of the RT-PCR primers. The short black horizontal segments joined by the V-shaped dotted lines represent the exons amplified by PCR and sequenced. The two columns on the right side indicate the type of junction and the number of patients who had the chimeric junction. (*Bottom*) Illustration of the PFGE mapping results of the genes on chromosome 3. The horizontal line represents the region on 3q26 where the genes are located. Horizontal arrows indicate the direction of transcription. Short vertical bars indicate the restriction site for *BSS*HII (B) and *Sfi*I (S), and the numbers below the horizontal lines indicate the size of the restriction fragments. The dashed vertical arrow indicates the approximate position of the t(3;3) breakpoint used for mapping of the genes (Nucifora et al. 1994b).

tected in patients with the t(8;21). By analogy with other translocations leading to fusion proteins, the resultant fusion protein is likely to be involved in the pathogenesis or progression of AML. Based on the homology with the pebp2a and Runt proteins, AML1 is thought to be a transcription factor that is highly expressed in hematopoietic cells and that, therefore, it could play an important role in hematopoietic cell growth and differentiation. The ETO protein also contains functional domains characteristic of transcription factors. Thus, it might act as a transcription factor by binding to specific DNA target sequences through the zinc finger regions and regulating transcriptional activity through interaction with the proline-rich regions. As a consequence, the chimeric AML1/ETO protein could very well be a hybrid transcription factor, capable of binding to the DNA not only through the Runt domain, but also through the zinc finger motifs, and causing inappropriate activation of gene expression through the proline-rich domains of ETO. ETO is not normally expressed in hematopoietic cells; therefore, its inappropriate expression may also contribute to leukemogenesis.

The detection of a chimeric transcript in cells which are t(8;21)-negative is similar to the detection of *BCR-ABL* transcripts in Ph-negative CML patients with a normal karyotype (Bartram et al. 1985; Morris et al. 1986). As in the case of Ph-negative, *BCR-ABL*-positive CML patients, our results show that *AML1* and *ETO* can be fused in t(8;21)-negative AML without an apparent chromosomal translocation; presumably this is the result of a different mechanism which inserts DNA, in this instance *AML1*, adjacent to *ETO* on chromosome 8, leading to the expected fusion transcript. These observations confirm the essential importance of the *AML1/ETO* transcript in this form of AML.

Our RT-PCR results with the long-term remission patients raise the question of the role of the translocation product in the process of malignant transformation. The expression of the chimeric protein could represent only one of the modifications necessary for the development of cancer and leukemia, and the affected clone could require additional mutations to express the transformed phenotype. Alternatively, it is possible that these t(8;21) cells are in fact completely transformed, but that cell proliferation is repressed by an unknown mechanism. Clearly, there is an increasing need to perfect sensitive techniques such as PCR amplification that can identify a unique nucleotide sequence in 10^5 cells to allow precise detection of genetic abnormalities far exceeding the sensitivity of standard techniques. However, the results of these new techniques are sometimes unexpected, and they force us to redefine accepted clinical stages such as relapse and complete remission in cancer patients. At present, the clinical significance of detecting cells carrying specific abnormalities associated with cancer in remission patients is still unclear.

Significance of Multiple Transcripts in the t(3;21)

It is generally accepted that, in most translocations associated with leukemia, the genes that are fused across the breakpoint are those directly modified in their expression or in their molecular nature and function, and thus are critically involved in leukemogenesis. In fact, there are only a few examples of genes that are affected at a long distance by a translocation, and the effect is limited to altering the expression of the involved gene. The cases of the t(3;21) that we have investigated are unique in containing chimeric transcripts between *AML1* and two or three unrelated genes located over a region of at least 400 kb at 3q26. Although we have no direct evidence that *AML1* and the partner genes at 3q26 can be transcribed as a single unprocessed nuclear message, the identification of chimeric junctions containing the same exons of *AML1* individually spliced to exons from the other genes would suggest that such nuclear transcripts exist and undergo differential splicing. The cases which we have examined contain complex fusion transcripts, and the roles of the individual transcripts in leukemia are not known. The common occurrence of carboxy-terminal truncations in AML1 in the t(3;21) and t(8;21) strongly implicates alterations in this gene as an important etiologic component in these cases of AML. It is worth noting that the carboxyl end of the Runt domain is defined by an exon-intron splicing consensus that is utilized in the t(8;21) chimeric cDNA and in the shorter isoforms of the t(3;21) chimeric cDNA, indicating the need to maintain this domain of AML1 in the fusion protein. On the basis of insights obtained from the analysis of other fusion transcripts, the sequences that are contributed by the other fusion partner are also important in leukemogenesis. The fact that the fusions with EAP alter the reading frame and only contribute 17 amino acids is of unknown functional significance. That this complex fusion event is an important component of the leukemogenesis process is supported by the detection of multiple fusions in all of the patients studied. How these two (or three) genes contribute the leukemogenic transformation is unclear. The disruption of *AML1* in *AML1/EAP* may eliminate the ability of the locus to splice exons encoding the regulatory domain of AML1 onto the DNA-binding domain, thus creating a constitutive repressor. On the other hand, AML1/MDS1 might act as a positive or negative *trans*-activator factor, depending on whether the fused section of MDS1 has a *trans*-activation function. In either case, such an ectopic transcriptional regulation could be a possible mechanism of leukemogenesis caused by the t(3;21).

It is possible that the translocations also alter the chromosomal structure to allow the inappropriate expression of *MDS1* and *EVI1*. In the case of *MDS1*, we have found that this transcript is not normally expressed in myeloid cells, and therefore it will be important to determine whether *MDS1* is expressed in other cases of AML. In the one case in which we detected *EVI1* transcripts, levels of transcripts detected with internal probes were much higher than those detected with the probes for the fusion transcript. Therefore, in the case of the *EVI1* gene, transcriptional activation may occur at sites other than the *AML1* promoter.

The mechanisms by which genes such as *AML1*, *ETO*, *MDS1*, and *EVI1* contribute to leukemogenesis are likely to be closely related to normal pathways of proliferation and differentiation. As a result, analysis of such genes is very likely to aid in understanding the physiology of normal cells. Conversely, the growing knowledge of leukocyte function and development is likely to clarify the potential contribution of abnormalities in these genes to malignant transformation.

ACKNOWLEDGMENTS

Several people have collaborated in the work described: Drs. Michelle LeBeau, Richard Larson,

Jerome Dickstein, and James Vardiman, and Debbi Birn, Catherine Begy, and Rafael Espinosa at the University of Chicago; Drs. Harry Drabkin and Paul Erickson, from the University of Colorado, Denver; and Dr. James N. Ihle, from the St. Jude's Children Research Hospital, Memphis, Tennessee. This work was supported by The G. Harold and Leila Y. Mathers Charitable Foundation (J.D.R.), by Department of Energy grant DE-FG02-86ER60408 (J.D.R.), National Institutes of Health grant CA-42557 (J.D.R.), National Institutes of Health grant PO1-HD017449 (H.D.), and National Cancer Institute grant CA-20180 (J.N.I.). G.N. is a Special Fellow of the Leukemia Society of America.

REFERENCES

Adams, J.M., A.W. Harris, C.A. Pinkert, L.M. Corcoran, W.S. Alexander, S. Cory, R.D. Palmiter, and R.L. Brinster. 1985. The c-*MYC* oncogene driven by immunoglobulin enhancers induces lymphoid malignancy in transgenic mice. *Nature* **318:** 533.

Allen, J.M., K.A. Forbush, and R.M. Perlmutter. 1992. Functional dissection of the *lck* proximal promoter. *Mol. Cell. Biol.* **12:** 2758.

Bae, S.C., Y. Yamaguchi-Iwai, E. Ogawa, M. Maruyama, M. Inuzuka, H. Kagoshima, K. Shigesada, M. Satake, and Y. Ito. 1993. Isolation of *Pebp2αA* cDNA representing the mouse homolog of acute myeloid leukemia gene *AML1*. *Oncogene* **8:** 809.

Bae, S.C., E. Ogawa, M. Maruyama, H. Oka, K. Shigesada, M. Satake, N. Jenkins, D. Gilbert, N. Copeland, and Y. Ito. 1994. *Pebp2αB*/mouse *AML1* consists of multiple isoforms that possess differential transcription potentials. *Mol. Cell. Biol.* **14:** 3242.

Bartholomew, C. and J.N. Ihle. 1991. Retroviral insertions 90 kilobases proximal to the *EVI1* myeloid transforming gene activate transcription from the normal promoter. *Mol. Cell. Biol.* **11:** 1820.

Bartram, C.R., E. Kleihauer, A. de Klein, G. Grosveld, J.R. Teyssier, N. Heisterkamp, and J. Groffen. 1985. C-ABL and BCR are rearranged in a Ph-negative CML patient. *EMBO J.* **4:** 683.

Daga, A,. J.E. Tighe, and F. Calabi. 1992. Leukemia/*Drosophila* homology. *Nature* **356:** 484.

Edgar, B.A., G.M. Odell, and G. Schubiger. 1989. A genetic switch based on negative regulation sharpens stripes in *Drosophila* embryo. *Dev. Genet.* **10:** 124.

Erickson, P., J. Gao, K.-S. Chang, T. Look, E. Whisenant, S. Raimondi, R. Lasher, J. Trujillo, J.D. Rowley, and H. Drabkin. 1992. Identification of breakpoints in t(8;21) AML and isolation of a fusion transcript with similarity to *Drosophila* segmentation gene *runt*. *Blood* **80:** 1825.

Frasch, M. and M. Levine. 1987. Complementary patterns of *even-skipped* and *fushi tarazu* expression involve their differential regulation by a common set of segmentation genes on *Drosophila*. *Genes Dev.* **1:** 981.

Gergen, J.P. and P.A. Butler. 1988. Isolation of the *Drosophila* segmentation gene *runt* and analysis of its expression during embryogenesis. *Genes Dev.* **2:** 1179.

Gottschalk, L.R. and J.M. Leiden. 1990. Identification and functional characterization of the human T-cell receptor β gene transcriptional enhancer: Common nuclear proteins interact with the transcriptional regulatory elements of the T-cell receptor α and β genes. *Mol. Cell. Biol.* **10:** 5486.

Heim, S. and F. Mitelman. 1987. *Cancer cytogenetics*. A.R. Liss, New York.

Inaba, T., W.M. Roberts, L.H. Shapiro, K.W. Jolly, S.C. Raimondi, S.D. Smith, and T.A. Look. 1992. Fusion of the leucine zipper gene HLF to the E2A gene in human acute B-lineage leukemia. *Science* **257:** 531.

Ingham, P.W. and J.P. Gergen. 1988. Interactions between the pair-rule genes *runt*, *hairy*, *even-skipped* and *fushi tarazu* and the establishment of periodic pattern in the *Drosophila* embryo. *Development* **104:** 51060.

Kagoshima, H., K. Shigesada, M. Satake, Y. Ito, M. Miyoshi, M. Ohki, M. Pepling, and P. Gergen. 1993. The Runt domain identifies a new family of heterodimeric transcriptional activators. *Trends Genet.* **9:** 338.

Kamps, M.P., C. Murre, X. Sun, and D. Baltimore. 1990. A new homeobox gene contributes the DNA binding domain of the t(1;19) translocation protein in pre-B ALL. *Cell* **60:** 547.

Kania, M.A., A.S. Bonner, J.B. Duffy, and J.P. Gergen. 1990. The *Drosophila* segmentation gene *runt* encodes a novel nuclear regulatory protein that is also expressed in the developing nervous system. *Genes Dev.* **4:** 1701.

Kappes, D.J., C.P. Browne, and S. Tonegawa. 1991. Identification of a T-cell-specific enhancer at the locus encoding T-cell antigen receptor δ chain. *Proc. Natl. Acad. Sci.* **88:** 2204.

Lindgren, V. and J.D. Rowley. 1977. Comparable complex rearrangements involving 8;21 and 9;22 translocations in leukemia. *Nature* **266:** 744.

Liu, P., S.A. Tarle, A. Hajra, D.F. Claxton, P. Marlton, M. Freedman, M.J. Siciliano, and F.S. Collins. 1993. Fusion between transcription factor CBFβ/Pebp2β and a myosin heavy chain in acute myeloid leukemia. *Science* **261:** 1041.

Meyers, S., J.R. Downing, and S.W. Hiebert. 1993. Identification of AML1 and the 8;21 translocation protein (AML1/ETO) as sequence specific DNA-binding proteins: The *runt* homology domain is required for DNA-binding and protein-protein interaction. *Mol. Cell. Biol.* **13:** 6336.

Mitani, K., S. Ogawa, T. Tanaka, H. Miyoshi, M.K. Kurokawa, H. Mano, Y. Yazaki, M. Ohki, and H. Hirai. 1994. Generation of the *AML1/EVI1* fusion gene in the t(3;21)(q26;q22) causes blastic crisis in chronic myelocytic leukemia. *EMBO J.* **13:** 504.

Miyoshi, H., K. Shimizu, T. Kozu, N. Maseki, Y. Kaneko, and M. Ohki. 1993a. The t(8;21) breakpoints on chromosome 21 in acute myeloid leukemia clustered within a limited region of a novel gene, *AML1*. *Proc. Natl. Acad. Sci.* **88:** 10431.

Miyoshi, H., T. Kozu, K. Shimizu, K. Enomoto, N. Maseki, Y. Kaneko, N. Kamada, and M. Ohki. 1993b. The t(8;21) translocation in acute myeloid leukemia results in production of an *AML1-MTG8* transcript. *EMBO J.* **12:** 2715.

Morishita, K., E. Parganas, E.C. Douglass, and J.N. Ihle. 1990a. Unique expression of the human *EVI1* gene in an endometrial carcinoma cell line: Sequence of cDNA and structure of alternatively spliced transcripts. *Oncogene* **5:** 963.

Morishita, K., E. Parganas, T. Matsugi, and J.N. Ihle. 1992a. Expression of the *EVI1* zinc finger gene in 32Dcl3 myeloid cells blocks granulocytic differentiation in response to granulocyte colony-stimulating factor. *Mol. Cell. Biol.* **12:** 183.

Morishita, K., E. Parganas, D. Parham, T. Matsugi, and J.N. Ihle. 1990b. The *EVI1* zinc finger myeloid transforming gene is normally expressed in the kidney and in the developing oocytes. *Oncogene* **5:** 1419.

Morishita, K., E. Parganas, C.L. William, M.H. Whittaker, H. Drabkin, J. Oval, R. Taetle, M.B. Valentine, and J.N. Ihle. 1992b. Activation of *EVI1* gene expression in human acute myelogenous leukemias by translocations spanning 300-400 kilobases on chromosome 3q26. *Proc. Natl. Acad. Sci.* **89:** 3937.

Morris, C.M., A.E. Reeve, P.H. Fitzgerald, P.E. Hollings, M.E. Beard, and D.C. Heaton. 1986. Genomic diversity correlates with clinical variation in Ph-negative chronic myeloid leukemia. *Nature* **320:** 281.

Nisson, P.E., P.C. Watkins, and N. Sacchi. 1992. Transcriptionally active chimeric gene derived from the fusion of the *AML1* gene and a novel gene on chromosome 8 in the t(8;21) leukemic cells. *Cancer Genet. Cytogenet.* **63:** 81.

Nucifora, G. and J.D. Rowley. 1994. The *AML1* and *ETO* genes in acute myeloid leukemia with a t(8;21). *Leuk. Lymphoma* **14:** 353.

Nucifora, G., R.A. Larson, and J.D. Rowley. 1993a. Persistence of the t(8;21) in AML-M2 long-term remission patients. *Blood* **82:** 712.

Nucifora, G., C.R. Begy, P. Erickson, H.A. Drabkin, and J.W. Rowley. 1993b. The 3;21 translocation in myelodysplasia results in a fusion transcript between the *AML1* gene and the gene for EAP, a highly conserved protein associated with the Epstein-Barr virus small RNA EBER-1. *Proc. Natl. Acad. Sci.* **90:** 7784.

Nucifora, G., J.I. Dickstein, V. Torbenson, J.D. Rowley, and J.W. Vardiman. 1994a. Correlation between cell morphology and expression of the *AML1/ETO* chimeric transcript in patients with acute myeloid leukemia without the t(8;21). *Leukemia* **8:** 1533.

Nucifora, G., D. Birn, P. Erickson, J. Gao, M. LeBeau, H. Drabkin, and J.D. Rowley. 1993c. Detection of DNA rearrangements in the *AML1* and *ETO* loci and of an *AML1/ETO* fusion mRNA in patients with t(8;21) AML. *Blood* **81:** 888.

Nucifora, G., C.R. Begy, H. Kubayashi, D. Claxton, J. Pedersen-Bjergaard, E. Parganas, J. Ihle, and J.D. Rowley. 1994b. Consistent intergenic splicing and production of multiple transcripts between *AML1* at 21q22 and unrelated genes at 3q26 in the (3;21)(q26;q22) translocation. *Proc. Natl. Acad. Sci.* **91:** 4004.

Nucifora, G., D. Birn, R. Espinosa, M. LeBeau, P. Erickson, D. Roulston, T.W. McKeithan, H. Drabkin, and J.D. Rowley. 1993d. Involvement of the *AML1* gene in the t(3;21) in therapy-related leukemia and in chronic myeloid leukemia in blast crisis. *Blood* **81:** 2728.

Ogawa, E., M. Inuzuka, M. Maruyama, M. Satake, M. Fujimoto, J. Lu, Y. Ito, and K. Shigesada. 1993. Molecular cloning and characterization of Pebp2β, the heterodimeric partner of a novel *Drosophila runt*-related binding protein Pebp2α. *Virology* **194:** 314.

Redondo, J.M., J.L. Pfohl, and M.S. Krangel. 1992. Identification of an essential site for transcriptional activation within the human T-cell receptor delta enhancer. *Mol. Cell. Biol.* **11:** 5671.

Redondo, J.M., S. Hata, C. Brocklehurst, and M.S. Krangel. 1990. A T-cell-specific transcriptional enhancer within the human T-cell receptor δ locus. *Science* **247:** 1225.

Roulston, D., G. Nucifora, J. Dietz-Band, M.M. LeBeau, and J.D. Rowley. 1993. Detection of rare 21q22 translocation breakpoints within the *AML1* gene in myeloid neoplasms by fluorescence in situ hybridization. *Blood* (suppl. 1) **82:** 532a.

Rowley, J.D. 1982. Identification of the constant chromosome regions involved in human hematologic malignant disease. *Science* **216:** 749.

Rubin, C.M., R.A. Larson, M.A. Bitter, J.J. Carrino, M.M. LeBeau, M.O. Diaz, and J.D. Rowley. 1987. Association of a chromosomal translocation with the blast phase of chronic myelogenous leukemia. *Blood* **70:** 1338.

Rubin, C.M., R.A. Larson, J. Anastasi, J.N. Winter, M. Thangavelu, J. Vardiman, J.D. Rowley, and M.M. LeBeau. 1990. t(3;21)(q26;q22): A recurring chromosomal abnormality in therapy-related myelodysplastic syndrome and acute myeloid leukemia. *Blood* **76:** 2594.

Speck, N.A. and B. Renjifo. 1990. Mutation of the core of adjacent LVb elements of the Moloney murine leukemia virus enhancer alters disease specificity. *Genes Dev.* **4:** 233.

Thirman, M.J., H.J. Gill, R.C. Burnett, D. Mbangkollo, N.R. McCabe, H. Kobayashi, S. Ziemin van der Poel, Y. Kaneko, R. Morgan, A.A. Sandberg, R.S.K. Chaganti, R.A. Larson, M.M. LeBeau, M.O. Diaz, and J.D. Rowley. 1993. Rearrangement of the *MLL* gene in acute lymphoblastic and acute myeloid leukemias with 11q23 chromosomal translocations. *N. Eng. J. Med.* **329:** 909.

Toczyski, D., A.G. Matera, D.C. Ward, and J.A. Steitz. 1994. The Epstein-Barr virus (EBV) small RNA EBER1 binds and relocalizes ribosomal protein L22 in EBV-infected human B lymphocytes. *Proc. Natl. Acad. Sci.* **91:** 3436.

Wang, S. and N.A. Speck. 1992. Purification of core-binding factor, a protein that binds the conserved core site in murine leukemia virus enhancers. *Mol. Cell. Biol.* **12:** 89.

Expression of P-Glycoprotein in Normal and Malignant Rat Liver Cells

C.H. LEE,*† G. BRADLEY,*‡ AND V. LING*†
*The Ontario Cancer Institute, Princess Margaret Hospital; †Department of Medical Biophysics, University of Toronto;
‡Faculty of Dentistry, University of Toronto, Toronto, Ontario, Canada, M4X 1K9*

Resistance to multiple chemotherapeutic drugs is observed in many advanced human cancers. The molecular basis of such clinical resistance is not understood. One hypothesis is that subpopulations of malignant cells acquire a multidrug-resistant (MDR) phenotype during tumor progression and that such subpopulations have a survival advantage during chemotherapy leading to a nonresponsive disease. Different MDR mechanisms have been identified in cell lines and transplantable tumors (Morrow and Cowan 1990; Cole et al. 1992; Beck et al. 1993; Gottesman and Pastan 1993; Childs and Ling 1994). One of the best-characterized mechanisms involves the overexpression of the putative drug efflux pump, P-glycoprotein (Pgp) (Endicott and Ling 1989). Pgp has been shown to cause resistance to a variety of unrelated anticancer drugs such as the anthracyclines, the vinca-alkaloids, taxol, epipodophyllotoxins, dactinomycin, and some alkylating agents. Such drugs constitute a significant proportion of our anticancer drugs armamentarium. Thus, it is of interest to determine whether or not the Pgp mechanism of MDR occurs in human cancers.

There are two Pgp genes in humans, *MDR1*, associated with multidrug resistance, and *MDR2/3*, not associated with drug resistance but which may be important for phosphatidylcholine translocation in liver (Smit et al. 1993; Ruetz and Gros 1994). Detection of an increased class I Pgp (*MDR1*) associated with drug resistance can be observed in a wide range of human cancers (Fojo et al. 1987; Goldstein et al. 1989; Arceci 1993). An increased expression of the *MDR2/3* gene in human cancer is rare. There is accumulating evidence that detection of Pgp in tumors at any time during the course of a patient's disease is associated with poor prognosis. In a number of studies, correlation between Pgp expression and nonresponse to chemotherapeutic treatments has been established (Chan et al. 1991; Arceci 1993; Rischin and Ling 1993). In addition, a number of chemosensitizing compounds able to reverse the Pgp-mediated MDR phenotype have been identified (Ford and Hait 1990; Georges et al. 1990). The possibility of using such agents with conventional chemotherapy to improve therapeutic outcome appears promising (Dalton et al. 1989; Sonneveld et al. 1992; List et al. 1993).

Factors important for regulating Pgp expression in human tumors have not been delineated. An increased expression of Pgp in biopsy material obtained from patients relapsing from chemotherapy has frequently been observed. Such findings are consistent with the hypothesis that subpopulations of Pgp-positive MDR cells are selected by treatment with anticancer drugs. Several studies have also shown that Pgp may be detected in tumor cells of various histological types prior to chemotherapy (Bradley and Ling 1994). The detection of Pgp in tumors at the time of diagnosis often correlates with poor response to subsequent treatment. It has been hypothesized that expression of Pgp in such instances may reflect an association between the MDR phenotype and tumor progression (Bradley and Ling 1994). This hypothesis is supported by studies with children's sarcoma and neuroblastoma, where it was observed that at the time of diagnosis, patients with localized tumors had no detectable Pgp, whereas a significant proportion of nonlocalized tumors showed Pgp-positive cells (Chan et al. 1990, 1991). The incidence of Pgp detected was particularly high for the late-stage tumors, often in metastatic lesions (Chan et al. 1991). The basis for the apparent association of Pgp and tumor progression is not understood.

We have used a well-characterized rat liver carcinogenesis system to investigate systematically the expression of Pgp during tumor progression (Bradley et al. 1992). We examined preoplastic hepatic foci and nodules, hepatic carcinomas, and lung metastases for expression of Pgp. An increased expression of Pgp during rat liver carcinogenesis was detected initially at the protein level by immunohistochemistry and confirmed at the mRNA level. We observed no remarkable increase in the level of Pgp in early preneoplastic liver lesions; however, large hyperplastic nodules and hepatocarcinoma showed a heterogeneous increase in Pgp expression. A striking finding was that lung metastases were consistently positive for Pgp. This up-regulation of Pgp appears to involve predominantly one specific Pgp isoform, the class II Pgp. In rats, Pgp is encoded by a small family of three highly homologous genes. In normal liver, the class III Pgp gene not associated with drug resistance is expressed at the highest level. The class I and class II genes which are associated with MDR are expressed at lower levels, with the class II gene being barely detectable. The basis for the increase in class II Pgp expression in rat liver cancer is not understood; however, it may be mechanistically similar to a dramatic increase in the class II Pgp expression observed in cultured rat hepatocytes. In the latter system, the level of the class II Pgp mRNA increased more than 20-fold during a 48-hour period, predominantly as a result of an increase in mRNA stability. This increased stability appears to be specific for the class II Pgp, as the other homologous Pgp genes (classes I and III) are not similarly affected. On the other hand, genes unrelated to Pgp, such as actin and tubulin, are up-regulated in this system due also to an increased mRNA stability in this system. Studies using cytoskeleton-disruptive drugs indicate that this increased mRNA stability may be dependent on the integrity of the cytoskeleton. We are currently investigating the possibility that this novel mechanism of mRNA stabilization underlies a regulatory mechanism of broad specificity which may be exploited by tumor cells for malignant progression.

EXPERIMENTAL PROCEDURES

Rat liver carcinogenesis. Male Fischer 344 rats (Charles River Breeding Laboratories, Wilmington, Massachusetts) weighing 100–120 g were maintained on Purina No. 5001 Rodent Laboratory Chow Diet and daily cycles of alternating 12-hour periods of light and darkness for 1 week before the start of the experiment. Water and food were freely available. Liver carcinogenesis was initiated by a single intraperitoneal (i.p.) injection of dimethylhydrazine at 100 mg/kg, 18 hours after two-thirds partial hepatectomy. After 1 week, the rats were changed from the basal diet to one containing 1% orotic acid (OA) and were maintained on this diet until the time of sacrifice (Fig. 1).

Culture of primary rat hepatocytes. Hepatocytes were isolated and cultured as described previously (Lee et al. 1993). In most experiments, unless otherwise stated, hepatocytes were seeded onto type I collagen. No serum or any other additional factors were added to seed the cells. The medium was replaced with fresh medium 4 hours after cell attachment and, thereafter, was replaced daily.

RNA analysis. Total RNA from cultured cells was isolated by an acid guanidinium isothiocyanate-phenol-chloroform extraction procedure (Chomczynski and Sacchi 1987), and RNA from tissues was extracted with guanidinium thiocyanate followed by centrifugation in cesium chloride solutions (Sambrook et al. 1989). Poly(A)$^+$ RNA selection was done using an oligo(dT)-cellulose column (Sambrook et al. 1989). RNA was analyzed either by blot (Northern or slot blot) analysis according to standard procedures (Lee et al. 1993) or by RNase protection, as described by Winter et al. (1985). Blots were hybridized with Pgp gene-specific probes or other random-primed probes as described previously (Lee et al. 1993).

The plasmid constructs used for RNase protection contain a *Tha*I-*Bsa*AI (blunted) 3′ rat Pgp2 genomic fragment (positions 120–327) (Deuchars et al. 1992) cloned in *Sma*I (blunted)-digested pGEM4Z (Promega) and a *Bsp*MI-*Dra*I (blunted) 3′ rat Pgp3 genomic fragment (positions 113–292) (Deuchars et al. 1992) cloned in *Sma*I-digested pGEM4Z. To synthesize antisense RNA probes, we linearized the Pgp2 and Pgp3 plasmid templates with *Eco*RI (within the vector; transcription with T7RNA polymerase). The 3′ Pgp2 antisense probe (272 nt) contains 65 nucleotides of vector sequence in addition to a 207-nucleotide *Tha*I-*Bsa*AI fragment of Pgp2. Because a 122-nucleotide sequence within this fragment is identical to that in the *pgp1* mRNA, this probe detects the rat *pgp1* transcript in addition to the rat *pgp2* mRNA. The 3′ Pgp3 antisense probe (244 nt) also contains 65 nucleotides of vector sequence in addition to 179 nucleotides from 3′ rat genomic Pgp3 and detects a 179-nucleotide fragment of rat *pgp3* mRNA. The pT7 18S template was purchased from Ambion Inc., and when in-vitro-transcribed with T7 RNA polymerase, it produced a 109-nucleotide antisense probe, of which 80 nucleotides is complementary to 18S ribosomal RNA. Labeled RNA was hybridized with 30 μg of total RNA, and protected fragments were visualized by electrophoresis through a

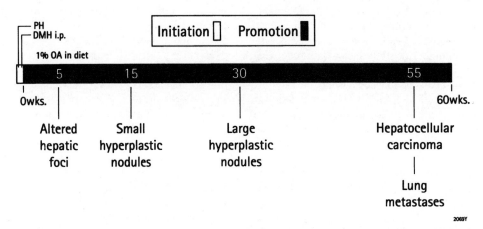

Figure 1. Protocol for rat liver carcinogenesis showing sequential development of preneoplastic liver lesions and liver cancer over a period of ~60 weeks. (PH) Two-thirds partial hepatectomy; (DMH) dimethylhydrazine dihydrochloride; (OA) orotic acid.

denaturing 6% polyacrylamide gel, followed by autoradiography.

Immunohistochemical detection of Pgp. Immunohistochemical staining for Pgp was performed on frozen sections fixed for 10 minutes in cold acetone. After rinsing with phosphate-buffered saline (PBS), endogenous peroxidase activity and endogenous biotin-like activity were blocked by appropriate pretreatments, and immunohistochemical staining was carried out as described previously (Bradley et al. 1990). Antibody C219 was used at 1.5 µg/ml in 1% bovine serum albumin (BSA) in an overnight incubation (16 hr) at 4°C.

RESULTS

Rat Liver Carcinogenesis

Rat liver carcinogenesis models are attractive for investigating cellular and molecular events during tumor development. Various models of rat liver carcinogenesis have been developed in which sequential steps of the carcinogenic process can be observed, and some of the histological and biochemical characteristics of these sequential events have been elucidated (Farber and Cameron 1980). Analysis of Pgp expression during the stepwise process of liver cancer development in such models may lead to a better understanding of the relationship between increased Pgp expression and the carcinogenic process. We have used the OA model of rat liver carcinogenesis initiated by partial hepatectomy coupled with an injection of dimethylhydrazine (DMH) followed by 50–60 weeks of promotion with a 1% OA diet (Fig. 1). Close to 100% of the animals developed hepatocarcinoma, and the sequential development of hepatic foci, hyperplastic nodules, and liver carcinoma can be observed over this period. Between 30% and 60% of the animals develop metastases to the lungs (Laurier et al. 1984; Bradley et al. 1992). Using histological staining to follow the progressive development of hepatic lesions in this system, it may be seen that after 5 weeks, hepatic foci were detectable by increased levels of γ-glutamyl transpeptidase (GGT) and glutathione S-transferase 7.7 (GST-P). These enzymes are thought to be involved in hepatic detoxification pathways. There was no detectable change in Pgp expression as stained with monoclonal antibody C219 in the hepatic foci. The hepatocytes of the foci displayed polarized staining for P-glycoprotein on the bile canalicular face of the plasma membrane similar to that observed in normal liver. In the small hyperplastic nodules (<1 cm in diameter), a heterogeneous pattern of Pgp staining was observed. The majority of the hepatocytes showed little or no Pgp staining; this was interspersed with groups of hepatocytes which stained strongly for Pgp. Large hyperplastic nodules (1 cm or more in diameter) showed stronger staining for Pgp, although there was again notable heterogeneity in Pgp expression within each nodule and among nodules. In these nodules, an association was observed between the presence of prominent vascular sinusoids separating trabeculae of hepatocytes and high levels of Pgp staining. After 50 or more weeks of promotion, the liver was typically occupied by a large, irregular tumor. Within the liver tumors, areas with moderate to strong Pgp staining were observed, and intense staining for Pgp was often found in areas of trabecular cancer with wide intervening sinusoids (Fig. 2). Pgp expression was consistently observed in lung metastases. Of more than 400 metastases examined, every metastasis contained cells that were positive for Pgp, regardless of its size. The surrounding lung tissue did not stain for Pgp (Fig. 2).

The above data suggest that an increased expression of Pgp is associated with relatively late stages in rat liver carcinogenesis. Similar findings were obtained in a study of Pgp mRNA expression by in situ hybridization of preneoplastic and neoplastic liver lesions induced by the Solt-Farber model of hepatocarcinogenesis (Nakatsukasa et al. 1992). The pattern of Pgp overexpression thus appears to be distinct from that of GST-P and GGT, which are commonly used markers of rat liver carcinogenesis and are clearly overexpressed in the earliest foci of altered hepatocytes. Increased expression of Pgp may be one of the molecular events that contribute to the growth of hyperplastic nodules and, subsequently, liver cancer. The data from the immunohistochemical study also suggest a possible role for Pgp in the vascularization of preneoplastic liver nodules and liver cancer, and in the process of hematogenous metastasis to the lungs. Direct investigations into these possibilities may further our understanding of the role of Pgp in carcinogenesis.

Expression of P-Glycoprotein Isoforms in Rat Liver Cancers

P-Glycoprotein in rat is encoded by three Pgp genes (Deuchars et al. 1992). In mammalian cells, the class I and II Pgp genes confer multidrug resistance, whereas the class III gene does not. These three classes of genes are expressed differentially in normal tissues, usually with one class predominating (Croop et al. 1989; Bradley et al. 1990). These findings suggest that Pgp isoforms have distinct physiological roles associated with specialized cell functions. In the rat liver, the class III isoform predominates, although the class I Pgp is also detectable. The class II Pgp is not readily detectable by standard techniques.

To determine what Pgp isoform might be expressed during rat liver carcinogenesis, we employed the 3' untranslated region of rat Pgp genes to differentiate between the expression of the three classes of Pgp genes. Currently, antibodies specific for each of the rat Pgp isoforms are not available. As shown in Figure 3, the class III Pgp mRNA is expressed at the highest level in normal rat liver. We estimate that the class I Pgp gene is expressed at about 10% of the class III Pgp gene, whereas class II Pgp expression is barely detectable in this assay. When a number of different hepato-

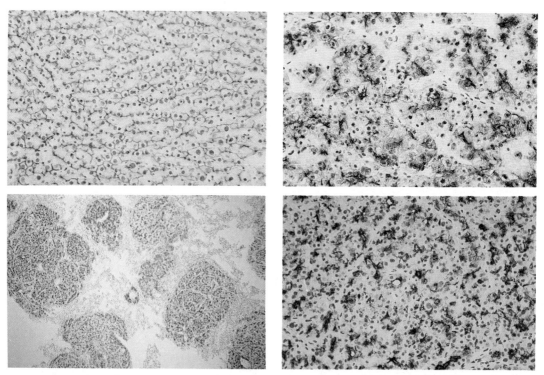

Figure 2. Immunoperoxidase staining for P-glycoprotein using C219 as primary antibody, diaminobenzidine as the final substrate, and hematoxylin as nuclear counterstain. (*Top left*) Normal rat liver, showing Pgp on the bile canalicular face of the hepatocytes. Magnification, 125×. (*Top right*) Primary hepatocarcinoma, showing overexpression of Pgp which remains predominantly in a polarized membrane distribution. Magnification, 125×. (*Bottom left*) Multiple metastases of hepatocarcinoma in the lungs, showing Pgp in the metastases but not in the surrounding lung. Magnification, 20×. (*Bottom right*) High-power view of one of the metastases in *bottom left* panel, showing strong staining for Pgp in many of the metastatic cells. Magnification, 125×.

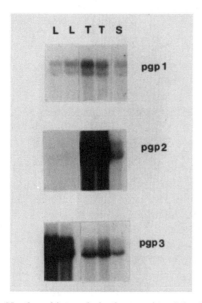

Figure 3. Northern blot analysis of expression of the three Pgp genes (*pgp 1, 2,* and *3*) in normal and malignant rat liver. Lanes marked *L* indicate two separate samples of normal male rat liver. Lanes marked *T* are two samples taken from different areas of a large liver carcinoma. Lane *S* is a sample taken from nontumorous liver in the same rat as *T*. Each lane has been loaded with 10 µg of poly A$^+$ RNA and the blots were probed with gene-specific oligonucleotide probes derived from the 3'-untranslated region of each *pgp* gene.

carcinomas were examined, there appeared to be a dramatic increase in the class II Pgp gene expression, as shown in Figure 3. There is no comparable increase in expression of class I and class III Pgp in the hepatocarcinomas. A comparison of the findings by immunohistochemical staining and those by Northern blot analysis of a large number of preneoplastic liver lesions and liver cancer suggest that overexpression of the class II Pgp isoform must account for virtually all the increase in immunohistochemical staining for Pgp during rat liver carcinogenesis. The specific and marked overexpression of one Pgp isoform (class II Pgp) during rat hepatocarcinogenesis has been reported for different models of liver carcinogenesis (Teeter et al. 1993; Fardel et al. 1994), supporting the hypothesis that this aberrant pattern of expression of the Pgp isoforms is characteristic of the hepatocarcinogenic process itself, irrespective of the agents used for carcinogenesis. At present, it is not known by what mechanism the class II P-glycoprotein is turned on during rat hepatocarcinogenesis. It is clear, however, that the up-regulation of Pgp in this process appears to be specific for the class II gene, as the class I and class III genes are not similarly affected.

In various systems, P-glycoprotein expression appears to be highly responsive to a variety of factors. Treatment of cultured cells with a wide range of chemicals, as well as heat shock or X-irradiation, can lead to an increase in Pgp expression (Kohno et al. 1989; Arceci et al. 1990; Chin et al. 1990; Hill et al. 1990). In some systems, treatment of cell lines with so-called differentiating reagents (such as sodium butyrate or retinoic acid) results in increased expression of Pgp (Bates et al. 1989; Mickley et al. 1989). The transcription of the Pgp gene may be increased through the activation of specific oncogenes. The c-Ha-*ras-1* oncogene and a mutant *p53* gene have been shown to stimulate activity of the human MDR1 promoter in NIH-3T3 cells (Chin et al. 1992). In another study, a mutant murine *p53* gene was found to stimulate the activity of the hamster Pgp1 promoter which had been transfected into Chinese hamster ovary cells, whereas the wild-type *p53* gene suppressed its activity (Zastawny et al. 1993). Although mutations in the Ha-*ras* oncogene and the *p53* gene have been identified in hepatocellular carcinomas, we do not know whether or not such mutations have occurred in our system or if they affect Pgp expression. Such studies raise several possibilities for the mechanism of Pgp overexpression during carcinogenesis. One possibility is that malignant cells activate in an aberrant manner physiological pathways which cause elevated Pgp expression that are not normally seen in a particular tissue; for example, a high level of class II Pgp in liver. Alternatively, the tumor cells may be exposed to external factors that trigger an increased expression of Pgp. Studies to investigate these possible mechanisms of Pgp expression in tumor cells will contribute significantly to our understanding of molecular alterations associated with tumor progression.

Expression of Pgp Genes in Primary Rat Hepatocytes

Primary rat hepatocytes exhibit many properties of normal liver, and they have been used as a model system for investigating liver function and gene expression (Berry et al. 1991). We and other workers have investigated Pgp expression in primary hepatocyte cultures under different conditions (Fardel et al. 1992; Lee et al. 1993; Schuetz and Schuetz 1993). It was observed that under a simple culture even in the absence of serum, there was a dramatic increase in the level of the class II Pgp (*pgp2*) expressed upon culturing of freshly isolated hepatocytes (Fig. 4). After 48 hours in culture, there was a greater than 20-fold increase in the level of *pgp2* mRNA detected. The levels of the class I Pgp mRNA (*pgp1*) also increased in hepatocyte cultures, but the change was clearly less significant than that of *pgp2* (Fig. 4). In contrast, the most abundant Pgp in the liver, the class III Pgp (*pgp3*), exhibited a decline. These results indicated a differential regulation of the different Pgp genes and that the primary hepatocytes provide an opportunity to further dissect these mechanisms. Significantly, a dramatic increase in *pgp2* expression is also observed during rat liver carcinogenesis

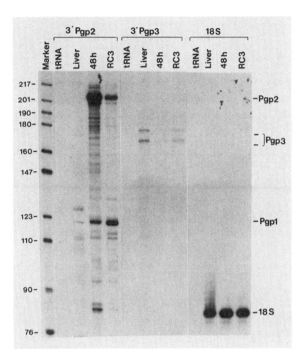

Figure 4. RNase protection analysis of total RNA from rat liver, 48-hr cultured rat hepatocytes, and rat hepatoma RC3 cells. Autoradiography showing the protected fragments of the *pgp2* (207 nt), *pgp1* (122 nt), and *pgp3* (179 nt) transcripts. The transfer RNA control hybridization in each panel is shown. As size marker, an end-labeled *Hpa*II digest of pBBR322 DNA was used.

(Fig. 3), suggesting that a similar mechanism may be operating in these two systems.

Changes in the expression of other genes are also observed in hepatocyte cultures. Although less dramatic than *pgp2*, the expression of cytoskeletal genes (α-tubulin and β-actin) was also found to increase with time in culture (Fig. 5). This increase in mRNA level is specific only for a subset of mRNAs, as mRNA for other genes such as albumin (data not shown) and connexin32 (Fig. 5) decreased as a function of time in culture. Further correlation between *pgp2* and cytoskeletal (actin and tubulin) gene expression was observed in a variety of culture conditions, suggesting that there may be a common mechanism regulating these apparently unrelated genes (Lee et al. 1993). Changes in the expression of these genes also appear to be closely associated with the cell shape of hepatocytes. For instance, hepatocytes cultured on matrigel were round, compact, and expressed significantly lower levels of *pgp2* than hepatocytes cultured on collagen-coated dishes (Fig. 6). Since cell shape is determined to a large extent by the organization of cytoskeletal network, we examined the importance of an intact cytoskeleton in the increased *pgp2* expression. We disrupted the hepatocyte microfilaments and microtubules with cytochalasin D and colchicine, respectively. As shown in Figure 7, the gradual increase in *pgp2* expression was greatly reduced in the presence of either drug, suggesting that the increase in *pgp2* expression in

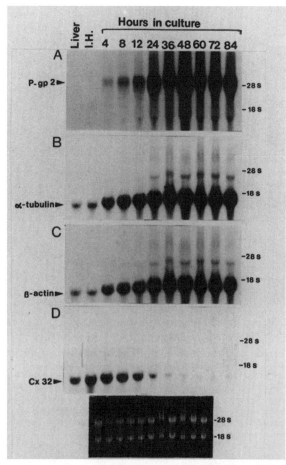

Figure 5. Northern analysis of mRNAs from rat liver, isolated rat hepatocytes, and cultured hepatocytes. Each lane contained 20 μg of total RNA from liver, isolated hepatocytes (I.H.), and hepatocytes cultured for various lengths of time as indicated. The same membrane was sequentially stripped and reprobed for α-tubulin (panel *B*), β-actin (panel *C*), and connexin 32 (panel *D*). Ethidium-bromide-stained gel at the bottom of the figure confirms that approximately equal amounts of RNA were loaded per lane. (Adapted from Lee et al. 1993.)

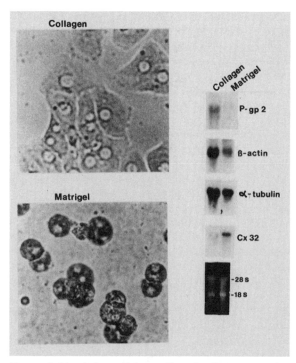

Figure 6. Morphology (*left* panel) and mRNA expression (*right* panel) of hepatocytes cultured on collagen-coated dishes and matrigel. Northern analysis of mRNAs from hepatocytes cultured for 48 hr on collagen-coated dishes or matrigel (*right* panel). Each lane contains 20 μg total RNA, and ethidium bromide stain of gel is shown at the bottom of the figure. The *left* panel shows morphology of hepatocytes cultured for 24 hr on collagen-coated dishes and matrigel. Phase contrast, magnification 175 ×. (Adapted from Lee et al. 1993.)

hepatocyte cultures is dependent on the integrity of the cytoskeleton. It is interesting to note that the parallel rise in mRNA levels for both actin and tubulin was also significantly affected.

Stability of *pgp2* mRNA in Primary Rat Hepatocytes

The underlying mechanism for the large increase in *pgp2* expression in primary rat hepatocytes was investigated. Nuclear run-on studies were conducted to directly measure the transcription rate of *pgp2* as a function of time in culture. We could not detect any significant change in *pgp2* transcription that would parallel the change in mRNA levels (C.H. Lee et al., in prep.). We then set out to establish whether *pgp2* mRNA stability is altered. Transcription was arrested in hepatocyte cultures at several time points by actinomycin D at concentrations effective in inhibiting more than 98% of new RNA synthesis. *pgp2* mRNA was then followed over a 16-hour period by slot blot analysis to determine its half-life. The half-life of the *pgp2* transcript for 8-, 12-, 36-, and 48-hour cultures is approximately 4.5, 6, 13, and 16 hours, respectively (C.H. Lee et al., in prep.). A linear relationship is obtained when the log of relative *pgp2* mRNA content is plotted against the half-life of *pgp2* mRNA (Fig. 8). This indicated that the dramatic elevation of *pgp2* mRNA in cultured hepatocytes is paralleled quantitatively by a dramatic increase in mRNA half-life. An increase in mRNA stability thus appears to be responsible for the elevation of *pgp2* mRNA in primary rat hepatocyte cultures.

Since the data in Figures 6 and 7 suggest that cytoskeletal components are involved in regulating *pgp2* expression in the hepatocyte culture, we determined if the stability of *pgp2* mRNA was different in the presence or absence (control) of cytochalasin D. In a 36-hour hepatocyte culture, the half-life of *pgp2* transcript in control cultures was approximately 14 hours, but it was 5 hours in the presence of cytochalasin D (Fig. 9). These results therefore indicated that the reduced *pgp2* transcript levels seen in the presence of cytochalasin D (Fig. 7) are most likely due to a decrease in stability of the mRNA.

DISCUSSION

Human cancers are composed of heterogeneous populations of tumor cells, of which only a portion are considered to be malignant stem cells capable of self-renewal and progression. A major challenge in cancer biology is to identify molecular alterations associated with malignant progression and to delineate the role such molecules play. Accumulating evidence suggests that detection of P-glycoprotein in tumor biopsies coincides with a therapy-resistant disease in a variety of human cancers. There is little controversy that overexpression of Pgp in malignant cells would confer resistance to many anticancer drugs and thus provide such cells with a growth advantage during treatment with chemotherapy. What additional roles Pgp may play during tumor progression are not understood. Pgp is likely to transport a broad range of substrates in addition to anticancer drugs (Gottesman and Pastan 1993; Childs and Ling 1994). These substrates may include steroid hormones, peptides, ATP, and ions. It is possible that in some cancers, the overexpression of Pgp provides growth advantages to the tumor cells through secretion of autocrine growth factors, removal of toxic metabolites, or secretion of angiogenic factors to promote vascularization. In support of such a hypothesis is the observation that an increased level of Pgp is frequently detected in advanced human cancers at diagnosis prior to chemotherapy. This observation in human cancers is paralleled by findings in the rat liver carcinogenesis system, where overexpression of Pgp occurred during late stages of carcinogenesis and where

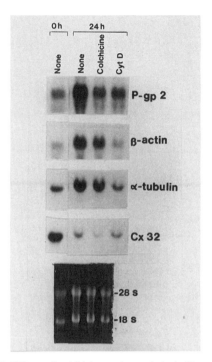

Figure 7. Effects of colchicine and cytochalasin D on class II Pgp expression. After 4 hr for cell attachment, cultures were either harvested for total RNA (None, 0 h) or changed to fresh medium containing no addition (None, 24 hr), 0.6 µg/ml colchicine, or 0.8 µg/ml cytochalasin D. After a further 24-hr incubation, cultures were harvested for total RNA. Each well had 20 µg total RNA, and ethidium bromide stain of gel is shown at the bottom of the figure to indicate equal loading of RNA. (Adapted from Lee et al. 1993.)

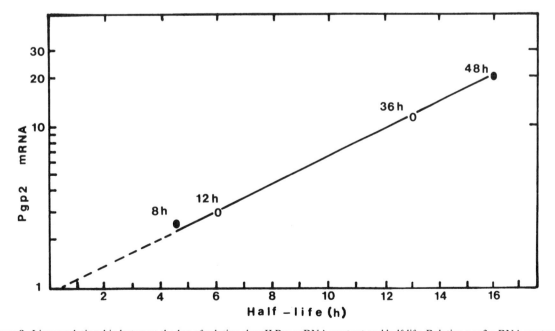

Figure 8. Linear relationship between the log of relative class II Pgp mRNA content and half-life. Relative *pgp2* mRNA content is obtained from multiple experiments, including those in Fig. 5. The half-lives of *pgp2* mRNA at different culture times were obtained by arresting transcription with 5 µg/ml actinomycin D, and following the decay of the transcripts, by slot blot analysis. Extrapolation of data is shown by broken line.

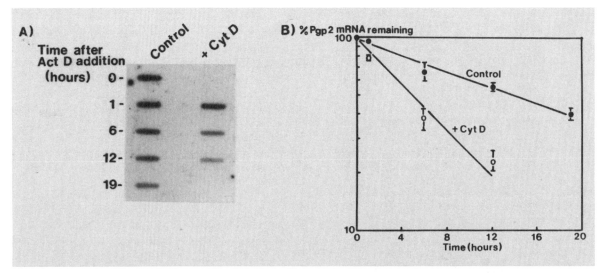

Figure 9. Rapid decay of *pgp2* mRNA in the presence of cytochalasin D. After 36 hr in culture, hepatocytes were treated with 5 μg/ml actinomycin D in the presence or absence (*control*) of 0.8 μg/ml cytochalasin D. Total RNA (10 μg) was subjected to slot blot analysis as shown (*A*). (*B*) Results of densitometry of the data in *A*. Error bars represent the range of two independent experiments using different hepatocyte preparations.

metastases to the lungs consistently expressed a high level of Pgp. Future studies involving inactivation of Pgp function or down-regulation of Pgp expression may provide further insights into the different roles Pgp may play in chemotherapy and during tumor progression.

How P-glycoprotein is regulated in human cancers is not understood; however, insights into this question may reveal factors important for mediating expression of Pgp and other malignancy-associated genes. Such factors, along with Pgp, may be useful targets for modulation to improve therapeutic efficacy. We have used a well-characterized rat liver carcinogenesis system to systematically examine the association between Pgp expression and carcinogenesis. The fact that the class II Pgp is highly overexpressed in rat liver tumors argues against a model in which tumor cells are simply expressing a differentiated phenotype of their tissue of origin, since the class II Pgp is usually not detectable in liver. Rather, the level of Pgp observed in rat liver tumor cells during carcinogenesis represents an aberrant up-regulation of an MDR-associated Pgp gene. The fact that in cultured rat hepatocytes the class II Pgp gene was also dramatically up-regulated suggests that this latter system may be a model for understanding the regulation of Pgp in rat liver tumor cells. The findings that the increased level of the class II Pgp gene of hepatocytes in culture is due to a dramatic increase in mRNA stability, that this mechanism is dependent on the integrity of the cytoskeleton, and that other non-Pgp-related genes such as actin and tubulin may be regulated in a similar manner raise the possibility that this is a novel mechanism which coordinately affects many apparently unrelated genes. We speculate that such a mechanism which is capable of regulating the expression of multiple genes, including Pgp, through mRNA stabilization may play an important role in the development of malignant disease.

ACKNOWLEDGMENTS

We thank our colleagues at the Ontario Cancer Institute for helpful discussions. This work was supported by grants from the National Cancer Institute of Canada.

REFERENCES

Arceci, R. 1993. Clinical significance of P-glycoprotein in multidrug resistance malignancies. *Blood* **81:** 2215.

Arceci, R.J., F. Baas, R. Raponi, S.B. Horwitz, D. Housman, and J.M. Croop. 1990. Multidrug resistance gene expression is controlled by steroid hormones in the secretory epithelium of the uterus. *Mol. Reprod. Dev.* **25:** 101.

Bates, S.E., L.A. Mickley, Y. Chen, N. Richert, J. Rudick, J.L. Biedler, and A.T. Fojo. 1989. Expression of a drug resistance gene in human neuroblastoma cell lines: Modulation by retinoic acid-induced differentiation. *Mol. Cell. Biol.* **9:** 4337.

Beck, W.T., M.K. Danks, J.S. Wolverton, R. Kim, and M. Chen. 1993. Drug resistance associated with altered DNA topoisomerase II. *Adv. Enzyme Regul.* **33:** 113.

Berry, M.N., A.M. Edwards, and G.J. Barritt. 1991. Monolayer culture of hepatocytes. In *Isolated hepatocytes preparation, properties, and applications* (ed. R.H. Burdon and P.H. Van Knippenberg), p. 265. Elsevier, The Netherlands.

Bradley, G. and V. Ling. 1994. P-glycoprotein, multidrug resistance and tumor progression. *Cancer Metastasis Rev.* **13:** 223.

Bradley, G., E. Georges, and V. Ling. 1990. Sex-dependent and independent expression of the P-glycoprotein isoforms in Chinese hamsters. *J. Cell. Physiol.* **145:** 398.

Bradley, G., R. Sharma, S. Rajalakshmi, and V. Ling. 1992. P-glycoprotein expression during tumor progression in the rat liver. *Cancer Res.* **52:** 51.

Chan, H.S.L., P.S. Thorner, G. Haddad, and V. Ling. 1990. Immunohistochemical detection of P-glycoprotein: Prognostic correlation in soft tissue sarcoma of childhood. *J. Clin. Oncol.* **8:** 689

Chan, H.S.L., G. Haddad, P.S. Thorner, G. DeBoer, Y.P. Lin, N. Ondrusek, H. Yeger, and V. Ling. 1991. P-glycoprotein expression as a predictor of the outcome of therapy for neuroblastoma. *N. Engl. J. Med.* **325:** 1608.

Childs, S. and V. Ling. 1994. The MDR superfamily of genes and its biological implications. In *Important advances in oncology* (ed. V.T. DeVita et al.), p. 21. J.B. Lippincott, Philadelphia, Pennsylvania.

Chin, K.V., K. Ueda, I. Pastan, and M.M. Gottesman. 1992. Modulation of activity of the promoter of the human *MDR1* gene by Ras and p53. *Science* **255:** 459.

Chin, K.V., S. Tanaka, G. Darlington, I. Pastan, and M.M. Gottesman. 1990. Heat shock and arsenite increase expression of the multidrug resistance (*MDR1*) gene in human renal carcinoma cells. *J. Biol. Chem.* **265:** 221.

Chomczynski, P. and N. Sacchi. 1987. Single-step method of RNA isolation by acid guanidinium thiocyanate-phenol-chloroform extraction. *Anal. Biochem.* **162:** 156.

Cole, S.P.C., G. Bhardwaj, J.H. Gerlach, J.E. Mackie, C.E. Grant, K.C. Almquist, A.J. Stewart, E.U. Kurz, A.M.V. Duncan, and R.G. Deeley. 1992. Identification of a novel ATP-binding cassette transporter gene overexpressed in a multidrug resistant human lung cancer cell line. *Science* **258:** 1650.

Croop, J.M., M. Raymond, D. Haber, A. Devault, R.J. Arceci, P. Gros, and D.E. Housman. 1989. The three mouse multidrug resistance (*mdr*) genes are expressed in a tissue specific manner in normal mouse tissues. *Mol. Cell. Biol.* **9:** 1346.

Dalton, W.S., T.M. Grogan, P.S. Meltzer, R.J. Scheper, B.G.M. Durie, C.W. Taylor, T.P. Miller, and S.E. Salmon. 1989. Drug-resistance in multiple myeloma and non-Hodgkin's lymphoma: Detection of P-glycoprotein and potential circumvention by addition of verapamil to chemotherapy. *J. Clin. Oncol.* **7:** 415.

Deuchars, K.L., D. Duthie, and V. Ling. 1992. Identification of distinct P-glycoprotein gene sequences in rat. *Biochim. Biophy. Acta* **1130:** 157.

Endicott, J.A. and V. Ling. 1989. The biochemistry of P-glycoprotein-mediated multidrug resistance. *Annu. Rev. Biochem.* **58:** 137.

Farber, E. and R. Cameron. 1980. The sequential analysis of cancer development. *Adv. Cancer Res.* **31:** 125.

Fardel, O., P. Loyer, V. Lecureur, D. Glaise, and A. Guillouzo. 1994. Constitutive expression of functional P-glycoprotein in rat hepatoma cells. *Eur. J. Biochem.* **219:** 521.

Fardel, O., P. Loyer, F. Morel, D. Ratanasavanh, and A. Guillouzo. 1992. Modulation of multidrug resistance gene expression in rat hepatocytes maintained under various culture conditions. *Biochem. Pharmacol.* **44:** 2259.

Fojo, A.T., K. Ueda, D.J. Salmon, D.G. Poplack, M.M. Gottesman, and I. Pastan. 1987. Expression of a multidrug-resistance gene in human tumors and tissues. *Proc. Natl. Acad. Sci.* **84:** 265.

Ford, J.M. and W.N. Hait. 1990. Pharmacology of drugs that alter multidrug resistance in cancer. *Pharmacol. Rev.* **42:** 155.

Georges, E., F.J. Sharom, and V. Ling. 1990. Multidrug resistance and chemosensitization: Therapeutic implications for cancer chemotherapy. *Adv. Pharmacol.* **21:** 185.

Goldstein, L.J., H. Galski, A. Fojo, M. Willingham, A.-L. Lai, A. Gazdar, R. Pirker, A. Green, W. Crist, G.M. Brodeur, M. Lieber, J. Cossman, M.M. Gottesman, and I. Pastan. 1989. Expression of a multidrug resistance gene in human tumors. *J. Natl. Cancer Inst.* **81:** 116.

Gottesman, M.M. and I. Pastan. 1993. Biochemistry of multidrug resistance mediated by the multidrug transporter. *Annu. Rev. Biochem.* **62:** 385.

Hill, B.T., K. Deuchars, L.K. Hosking, V. Ling, and R.D. Whelan. 1990. Overexpression of P-glycoprotein in mammalian tumor cell lines after fractionated X irradiation in vitro. *J. Natl. Cancer Inst.* **82:** 607.

Kohno, K., S. Sato, H. Takano, K. Matsuo, and M. Kuwano. 1989. The direct activation of human multidrug resistance gene (*MDR1*) by anticancer agents. *Biochem. Biophys. Res. Commun.* **165:** 1415.

Laurier, C., M. Tatematsu, P.M. Rao, S. Rajalaksmi, and D.S.R. Sarma. 1984. Promotion by orotic acid of liver carcinogenesis in rats initiated by 1,2-dimethylhydrazine. *Cancer Res.* **44:** 2186.

Lee, C.H., G. Bradley, J.-T. Zhang, and V. Ling. 1993. Differential expression of P-glycoprotein genes in primary rat hepatocyte culture. *J. Cell. Physiol.* **157:** 392.

List, A.F., C. Spier, J. Greer, S. Wolf, J. Hutter, R. Dorr, S. Salmon, B. Futcher, M. Baier, and W. Dalton. 1993. Phase I/II trial of cyclosporine as a chemotherapy-resistant modifier in acute leukemia. *J. Clin. Oncol.* **11:** 1652.

Mickley, L.A., S.E. Bates, N.D. Richert, S. Currier, S. Tanaka, F. Foss, N. Rosen, and A.T. Fojo. 1989. Modulation of the expression of a multidrug resistance gene (*mdr*-1/P-glycoprotein) by differentiating agents. *J. Biol. Chem.* **264:** 18031.

Morrow, C.S. and K.H. Cowan. 1990. Glutathione-S-transferases and drug resistance. *Cancer Cells* **2:** 15.

Nakatsukasa, H., R.P. Evarts, R.K. Burt, P. Nagy, and S.S. Thorgeirsson. 1992. Cellular pattern of multidrug-resistance gene expression during chemical hepatocarcinogenesis in the rat. *Mol. Carcinog.* **6:** 190.

Rischin, D. and V. Ling. 1993. Multidrug resistance in leukemia. In *Leukemia: Advances, research and treatment* (ed. E.J. Freireich and H. Kantarjian), 269. Kluwer Academic, The Netherlands.

Ruetz, S. and P. Gros. 1994. Phosphatidylcholine translocase: A physiologic role for the *mdr2* gene. *Cell* **77:** 1071.

Sambrook. J., E.F. Fritsch, and T. Maniatis. 1989. *Molecular cloning: A laboratory manual*, 2nd edition. Cold Spring Harbor Laboratory, Cold Spring Harbor, New York.

Schuetz, J.D. and E.G. Schuetz. 1993. Extracellular matrix regulation of multidrug resistance in primary monolayer cultures of adult rat hepatocytes. *Cell Growth Differ.* **4:** 31.

Smit, J.J.M., A.H. Schinkel, R.P.J. Oude Elferink, A.K. Groen, E. Wagenaar, L. Van Deemter, C.A.A.M. Mol, R. Ottenhofer, N.M.T. Van der Lugt, M.A. Van Roon, M.A. Van der Valk, G.J.A. Offerhaus, A.J.M. Berns, and P. Borst. 1993. Homozygous disruption of the murine *mdr2* P-glycoprotein gene leads to a complete absence of phospholipid from bile and to liver disease. *Cell* **75:** 451.

Sonneveld, P., B.G.M. Durie, H.M. Lokhorst, J. Marie, G. Solbu, S. Suciu, R. Zittoun, B. Lowenberg, and K. Nooter. 1992. Modulation of multidrug-resistant multiple myeloma by cyclosporin. *Lancet* **340:** 255.

Teeter, L.D., M. Estes, J.Y. Chan, H. Atassi, S. Sell, F.F. Becker, and M.T. Kuo. 1993. Activation of distinct multidrug-resistance (P-glycoprotein) genes during rat liver regeneration and hepatocarcinogenesis. *Mol. Carcinog.* **8:** 67.

Winter, E., F. Yamamoto, C. Almoguera, and M. Perucho. 1985. A method to detect and characterize point mutations in transcribed genes: Amplified and overexpression of the mutant c-Ki-*ras* allele in human tumor cells. *Proc. Natl. Acad. Sci.* **82:** 7575.

Zastawny, R.L., R. Salvino, J. Chen, S. Benchimol, and V. Ling. 1993. The core promoter region of the P-glycoprotein gene is sufficient to confer differential responsiveness to wild-type and mutant p53. *Oncogene* **8:** 1529.

Genes Coding for Tumor-specific Rejection Antigens

T. Boon, B. Van den Eynde, H. Hirsch,* C. Moroni,*
E. De Plaen, P. van der Bruggen, C. De Smet, C. Lurquin,
J.-P. Szikora, and O. De Backer

*Ludwig Institute for Cancer Research, Brussels Branch, B1200 Brussels, Belgium, and Cellular Genetics Unit,
Université Catholique de Louvain, B1200 Brussels, Belgium;
Institute for Medical Microbiology, University of Basel, CH-4053 Basel, Switzerland

Mouse tumors induced with chemical carcinogens or UV light are immunogenic: Tumor cells killed by irradiation or living tumor cells later removed by surgery confer to the inoculated animal a lasting protection against a challenge with the same tumor cells (Klein et al. 1960; Kripke and Fisher 1976). This specific protection was the first evidence for the existence of specific tumor rejection antigens. T lymphocytes were shown to carry the specific memory involved in these responses. Later, it became apparent that many tumors, including most, if not all, spontaneous tumors, do not elicit any immune rejection response (Hewitt et al. 1976). However, from such nonimmunogenic tumor cell lines that were treated in vitro with a mutagen, it was possible to derive variants that expressed potent new antigens (Van Pel et al. 1983). These "tum$^-$" variants were rejected by syngeneic mice, and these mice were protected against a challenge with the original tumor cells, indicating that, even though they are not immunogenic, these tumors also carry tumor rejection antigens that can constitute targets for immune rejection responses. These results suggested that most, if not all, mouse tumors bear antigens that can be targets for immune rejection mediated by syngeneic T lymphocytes. They also indicated that tumor antigens that do not elicit any immune response can be made to do so by artificial means.

Antigens recognized by T lymphocytes are small peptides inserted in a groove of major histocompatibility complex (MHC) molecules (Townsend et al. 1986; Bjorkman et al. 1987). Peptides derived from exogenous proteins captured by cells are presented by class II MHC molecules and are recognized by CD4$^+$ T lymphocytes. Peptides derived from proteins synthesized inside the cell are predominantly presented by class I MHC molecules and are recognized by CD8$^+$ T lymphocytes, but some of these endogenous peptides appear to be presented by class II MHC molecules to CD4$^+$ T lymphocytes.

In our study of mouse tumor rejection antigens, we focused on those that are recognized by cytolytic CD8$^+$ T lymphocytes (CTL). Mouse tumor mastocytoma P815 proved to be particularly favorable for the study of antitumor CTL. We obtained CTL clones directed against antigens borne by tum$^-$ antigenic variants, and we identified the genes coding for these antigens (De Plaen et al. 1988; Lurquin et al. 1989). This led to the notion that new antigens encoded by the cellular genome can arise as a result of point mutations in genes that are expressed ubiquitously. These point mutations either enable a new peptide to bind to a MHC molecule or confer a new epitope to a peptide that already binds but is not recognized because of natural tolerance (Fig. 1). Considering that mutated oncogenes, such as *ras* or mutated *p53*, are found in many tumors, it is possible that responses against specific tumor antigens encoded by the mutated region of these genes will be observed in some patients. In vitro stimulation with these mutated peptides has produced specific CTL (Jung and Schluesener 1991; Gedde-Dahl et al. 1993), but specific CTL have also been obtained in vitro by stimulation

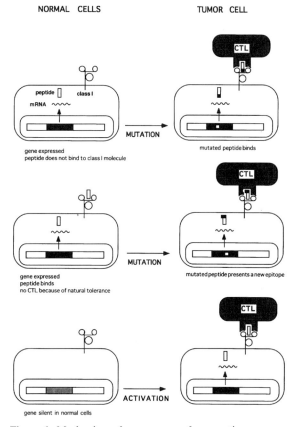

Figure 1. Mechanism of appearance of new antigens recognized by CTL.

with normal ras peptides (Skipper and Stauss 1993). Recently, a tumor rejection antigen of Lewis lung carcinoma was found to be encoded by a mutated form of protein connexin 37 (Mandelboim et al. 1994).

Another mechanism that generates new antigens recognized by CTL on tumor cells is the activation in the tumor of a gene that is silent in normal cells. The first gene coding for a mouse tumor rejection antigen that was identified, namely gene *P1A*, was specifically activated in tumor P815 (Van den Eynde et al. 1991). More recently, we have identified a gene that codes for a tumor rejection antigen of spontaneous leukemia LEC (de Bergeyck et al. 1994). The coding sequence is the *gag* gene from a defective endogenous retroviral sequence. This intracisternal A particle (IAP) sequence has undergone a transposition in the tumor cells, and it appears likely that the antigen is due to the transcriptional activation of the IAP sequence following its transposition.

For human tumors, it is possible to obtain autologous antitumor CTL by cocultivating irradiated tumor cells with blood lymphocytes of the tumor-bearing patient (Vose and Bonnard 1982; Mukherji and MacAlister 1983; Knuth et al. 1984). Many studies have been carried out with melanoma because metastatic melanoma tumors can often be adapted to culture. Antitumor CTL clones with high activity and specificity have been obtained (Hérin et al. 1987). These clones were used to select in vitro antigen-loss tumor cell variants (Van den Eynde et al. 1989; Wölfel et al. 1989). When these stable variants, which were resistant to lysis by the selective CTL, were tested for their sensitivity to other autologous CTL clones, they were found to be sensitive to most of them. This demonstrated that the latter CTL recognized different antigens on the same tumor. This analysis led to the conclusion that many human tumors carry several tumor rejection antigens that can be recognized by autologous CTL.

The search for genes coding for these human tumor antigens involved the approach that had been followed for the identification of the mouse tumor antigens (Traversari et al. 1992). Briefly, cosmid libraries were prepared from the antigen-bearing cells, and the DNA was transfected into an antigen-loss variant. The transfectants were screened for their ability to stimulate tumor necrosis factor (TNF) release by the antitumor CTL. Once antigenic transfectants were obtained, the relevant gene was selectively retrieved from the DNA of the transfectants by using the flanking cosmidic sequences to ensure packaging into bacteriophage λ heads. More recently, an alternative approach has been used. COS cells are cotransfected with cDNA libraries prepared from the antigen-bearing cells and with the *HLA* gene coding for the presenting molecule. This system provides considerable episomal multiplication of the transfected plasmid and, therefore, high expression of the antigen (Seed and Aruffo 1987). The transfectants are tested for their ability to stimulate TNF release by the CTL, and the gene can then be isolated by subcloning the bacteria of the relevant pool of the cDNA library.

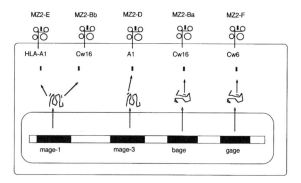

Figure 2. Antigens identified on melanoma MZ2-MEL which are encoded by genes whose expression is tumor-specific. The presenting class I molecules are indicated.

These approaches have led to the identification of four human genes that code for tumor-specific antigens recognized on a melanoma by autologous CTL (Fig. 2). These genes have been named *MAGE-1*, *MAGE-3*, *BAGE*, and *GAGE* (van der Bruggen et al. 1991; Gaugler et al. 1994). Like the *P1A* gene, they are selectively activated in the tumors, thereby causing the appearance of the antigen. Thus, the analysis of tumor rejection antigens has led in both mouse and man to the identification of genes whose expression is very selective for tumor cells. These genes are unrelated to any known oncogenes and, not unexpectedly, to known tumor-suppressor genes. Below follows a description of what we know about gene *P1A* and about *MAGE* genes.

MOUSE GENE *P1A*

Mouse mastocytoma P815 is a tumor induced in a DBA/2 mouse by methylcholanthrene. Gene *P1A* directs the expression of antigen P815A.B, which is recognized by CTL specific for the P815 tumor cells (Van den Eynde et al. 1991). This antigen constitutes a major target of the rejection response occurring in vivo, since P815 cells that escaped partial rejection were found to be resistant to lysis by these CTL (Uyttenhove et al. 1983). A peptide of nine amino acids encoded by gene *P1A* is presented by class I molecule H-2 L^d to anti-P815 CTL (Lethé et al. 1992). This peptide contains two distinct epitopes, one being recognized by the anti-P815A and the other by the anti-P815B CTL.

The sequence of gene *P1A* present in the tumor cell is identical to that found in normal DBA/2 cells. The expression of the antigen by tumor P815 results from the activation in the tumor of a gene that is nearly silent in all normal mouse tissues with the exception of testis and placenta (Table 1). Gene *P1A* is also active in several other mastocytomas, but not in all. We have not found it to be active in any other mouse tumor type.

The activation of gene *P1A* in tumor cells suggests a possible involvement of this gene in tumor generation or progression. To address this question, we used a model system of tumorigenesis of the mast cell lineage (Fig. 3) (Nair et al. 1989, 1992; Hirsch et al. 1993). By cultivating bone marrow cells of DBA/2 mice in the

Table 1. Expression of Gene *P1A*

Tumor lines (mastocytomas)	
P815	+++
15V4T3	+++
V4D6	++++
R56VT	−
15-abl	−
20-abl	−
L138.8A	+++
Mast cell lines	
MC/9	−
PB-3c-15	−
PB-3c-15V4	−
Normal tissues	
spleen	−
kidney	−
heart	−
lung	−
liver	−
stomach	−
colon	−
brain	−
bone marrow	−
thymus	−
adrenals	−
testis	+
placenta	++++

The expression of gene *P1A* was measured by RT-PCR. Samples are indicated as negative when their expression was less than 1% of P815.

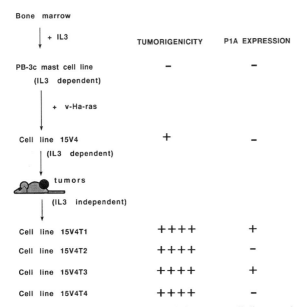

Figure 3. Expression of gene *P1A* by mast cell lines and tumors.

presence of interleukin-3 (IL-3), a mast cell line called PB-3c was obtained. It was dependent on the presence of IL-3 for in vitro growth, and it failed consistently to form tumors after subcutaneous injection into syngeneic mice. This line was infected with a retrovirus containing the *v-Ha-ras* gene, and line 15V4 expressing the *ras* p21 protein was obtained. This line was still dependent on the presence of IL-3, but contrary to PB-3c, it formed tumors in most mice but after long latencies. Four independent tumors (15V4T1-4) were obtained. Once adapted to in vitro culture, all these tumors grew in the absence of IL-3, due to autocrine secretion of this growth factor. They were highly tumorigenic, producing progressive tumors within 2–4 weeks. We tested the expression of gene *P1A* in these different cell lines, by Northern blot and by reverse transcription-PCR. Precursor line PB-3c was negative, as was *ras*-expressing line 15V4. Two of the four tumor lines, 15V4T1 and 15V4T3, expressed *P1A* at a high level, and the two others were completely negative. Among other mastocytomas obtained by the same procedure with line PB-3c, two were positive for *P1A* and two were negative. Thus, these experiments established that tumoral transformation and *P1A* activation are often linked.

The activation of the IL-3 autocrine loop appears to be necessary for the tumorigenicity of the mast cell in this model system (Nair et al. 1992). In some tumors, IL-3 activation is due to the insertion of an IAP sequence into the IL-3 promoter, whereas in others it is related to the loss of a negative regulator of the IL-3 gene. Both *P1A*-positive and *P1A*-negative tumors have been found among tumors with each type of IL-3 activation mechanism. It would be compatible with our data to assume that mastocytoma tumorigenesis requires at least two events, one being invariably the activation of the IL-3 autocrine loop. For the second event, there would be at least two alternative possibilities, one of which would involve the activation of gene *P1A*. The activation of *P1A* may not require that of specific transcription factors, since demethylation of normal cells results in *P1A* activation.

THE *MAGE* GENE FAMILY

Human gene *MAGE-1* codes for antigen MZ2-E, which is recognized on melanoma cell line MZ2-MEL by autologous CTL. The gene is unrelated to any known human gene. It comprises 3 exons spread over 4.5 kb. The last exon contains a large open reading frame coding for a protein of 309 amino acids. Its sequence is the same in the melanoma cell line and in the normal lymphocytes of patient MZ2. *MAGE-1* is silent in normal tissues, except testis (van der Bruggen et al. 1991; De Smet et al. 1994). It is expressed in a significant fraction of tumors of various histological types such as melanoma, head and neck squamous cell carcinoma, non-small-cell lung cancer, and bladder carcinoma.

Gene *MAGE-1* is a member of a family that comprises at least 12 closely related genes which are located on the long arm of chromosome X (De Plaen et al. 1994). The percentage of identity of the last exon of *MAGE-1* with that of the other *MAGE* genes varies between 64% and 85%. The exon-intron structure of 6 *MAGE* genes was established by aligning the genomic sequence with the corresponding cDNA (Fig. 4). All six genes code for proteins of approximately 300 amino acids. The lengths of the last exons are similar and the sequences are conserved. These genes, namely *MAGE-1, -2, -3, -4, -6,* and *-12*, showed significant expression in a number of tumors of various histological types. A

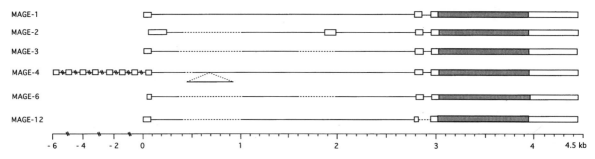

Figure 4. Structure of six *MAGE* genes found to be expressed in tumors. Exons are indicated by boxes, with the open reading frames in gray.

panel of normal adult tissues and some tissues from >20-week-old fetuses were also tested for the expression of these genes. All were negative, with the exception of testis and of placenta. Placenta expresses *MAGE-3* and *MAGE-4*.

The function of the MAGE proteins is unknown. They are devoid of a signal sequence, and the general pattern of hydrophobic and hydrophilic domains is remarkably conserved across the family, suggesting conservation of function. A computer search revealed a moderate homology (31% identity) of MAGE-10 protein with mouse protein necdin. Necdin, whose function is also unknown, has been reported to be a nuclear protein, expressed in neuronally differentiated embryonal carcinoma cells and in the brain of adult mice (Maruyama et al. 1991). We isolated from the mouse genome by hybridization with a *MAGE-1* probe two homologous genes, which we named *SMAGE-1* and *SMAGE-2*. They are more than 99% identical to each other and, like the *MAGE* genes, they are located on chromosome X. They encode proteins of 330 amino acids that show 40% identity to the MAGE-10 protein.

Although the exons containing the coding region of the *MAGE* genes are very similar, much more variation is found at the 5' ends of the genes. Compared to the initial part of *MAGE-1*, genes *MAGE-2, -3, -6*, and *-12* present deletions, and gene *MAGE-4* contains an insertion (Fig. 4). *MAGE-2* has an additional exon that is homologous to a sequence located in the first intron of *MAGE-1*. The first exon of *MAGE-2* extends 175 bases beyond that of *MAGE-1, -3, -6*, and *-12*, and this extension is homologous to the initial intron region of *MAGE-1, -3, -6*, and *-12*. In *MAGE-4*, we identified 8 sequences dispersed over 6 kb that can serve as alternative first exons in placenta, testis, and in a sarcoma cell line. These first exons are alternatively spliced to exon 2. The promoter regions of the six *MAGE* genes mentioned above display considerable differences in sequences. *SMAGE-1* and *SMAGE-2* exhibit a similar organization with several alternative first exons.

Because the overall structure of the MAGE proteins appears to be very conserved, it seems likely that they exert the same function. Accordingly, the duplication of the *MAGE* genes into a family of 12 genes may have as a main consequence to put this function under the control of a large number of different promoters (19 considering those of gene *MAGE-4*), thereby ensuring very selective expression in certain tissues at appropriate moments.

To understand the basis of the tumor-specific expression of the *MAGE* genes, we have studied the *MAGE-1* promoter. The promoter activity of various subregions was studied by insertion in front of the luciferase gene and evaluation of the transient luciferase expression after transfection into MZ2-MEL melanoma cells, which express gene *MAGE-1*. Five regions containing positive regulatory elements were identified (Fig. 5). Two important regulatory elements contain consensus binding sites for Ets-related factors. Gel retardation experiments indicated that these regions bind to factors contained in MZ2-MEL extracts. Binding was inhibited by competition with oligonucleotides containing consensus binding sequences for transcription factors of the Ets family and not with an oligonucleotide carrying a mutation that abolishes Ets binding. This confirmed that the binding factors belong to the Ets family.

The *MAGE-1* promoter/luciferase constructs were expressed in a number of melanoma cell lines that did not express *MAGE-1*. This suggested that methylation might play an important role in preventing *MAGE-1* expression. A *Hpa*II restriction site located at position −53 in the *MAGE-1* promoter was found to be methylated in normal blood cells, in tumor cell lines that did not express *MAGE-1*, but not in tumor cell lines that expressed *MAGE-1*.

The search to characterize tumor rejection antigens recognized by autologous CTL has led to the identification of several new genes that are not related to either known oncogenes or known tumor suppressor genes. Although the elucidation of the function of these genes

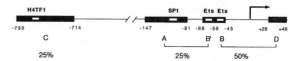

Figure 5. The promoter of gene *MAGE-1*. Five active regions of the promoter are delineated. Their role was assessed with the luciferase system in MZ2-MEL melanoma cells. The percentage of total transcription due to each region is indicated. Region A is only active in the presence of region B' and vice versa. Region D is inactive in the absence of region B. Consensus binding sites for transcription factors are indicated. The origin of transcription is indicated by an arrow.

and of their role in promoting tumoral transformation and progression may require much additional study, one can hope that it will provide interesting new insights into the cancerous phenotype.

ACKNOWLEDGMENTS

This work was partially supported by the Belgian programme on Interuniversity Poles of attraction initiated by the Belgian State, Prime Minister's Office, Office for Science, Technology and Culture (OSTC). The scientific responsibility is assumed by the authors. This research was partially subsidized by the Fonds Maisin, the Dr. Steiner prize (Switzerland), and by the Association contre le Cancer (Belgium). Charles De Smet was supported by a grant "TELEVIE" from the Fonds National de la Recherche Scientifique (Brussels, Belgium).

REFERENCES

Bjorkman, P.J., M.A. Saper, B. Samraoui, W.S. Bennett, J.L. Strominger, and D.C. Wiley. 1987. The foreign antigen binding site and T cell recognition regions of class I histocompatibility antigens. *Nature* **329:** 512.

de Bergeyck, V., E. De Plaen, P. Chomez, T. Boon, and A. Van Pel. 1994. An intracisternal A particle sequence codes for an antigen recognized by syngeneic cytolytic T lymphocytes on a mouse spontaneous leukemia. *Eur. J. Immunol.* **24:** 2203.

De Plaen, E., C. Lurquin, A. Van Pel, B. Mariamé, J.-P. Szikora, T. Wölfel, C. Sibille, P. Chomez, and T. Boon. 1988. Immunogenic (tum⁻) variants of mouse tumor P815: Cloning of the gene of tum⁻ antigen P91A and identification of the tum⁻ mutation. *Proc. Natl. Acad. Sci.* **85:** 2274.

De Plaen, E., K. Arden, C. Traversari, J.J. Gaforio, J.-P. Szikora, C. De Smet, F. Brasseur, P. van der Bruggen, B. Lethé, C. Lurquin, R. Brasseur, P. Chomez, O. De Backer, W. Cavenee, and T. Boon. 1994. Structure, chromosomal localization and expression of twelve genes of the MAGE family. *Immunogenetics* **40:** 360.

De Smet, C., C. Lurquin, P. van der Bruggen, E. De Plaen, F. Brasseur, and T. Boon. 1994. Sequence and expression pattern of the human *MAGE-2* gene. *Immunogenetics* **39:** 121.

Gaugler, B., B. Van den Eynde, P. van der Bruggen, P. Romero, J.J. Gaforio, E. De Plaen, B. Lethé, F. Brasseur, and T. Boon. 1994. Human gene *MAGE-3* codes for an antigen recognized on a melanoma by autologous cytolytic T lymphocytes. *J. Exp. Med.* **179:** 921.

Gedde-Dahl, T.I., B. Fossum, J.A. Eriksen, E. Thorsby, and G. Gaudernack. 1993. T cell clones specific for p21 *ras*-derived peptides: Characterization of their fine specificity and HLA restriction. *Eur. J. Immunol.* **23:** 754.

Hérin, M., C. Lemoine, P. Weynants, F. Vessière, A. Van Pel, A. Knuth, R. Devos, and T. Boon. 1987. Production of stable cytolytic T-cell clones directed against autologous human melanoma. *Int. J. Cancer* **39:** 390.

Hewitt, H., E. Blake, and A. Walder. 1976. A critique of the evidence for active host defense against cancer based on personal studies of 27 murine tumors of spontaneous origin. *Br. J. Cancer* **33:** 241.

Hirsch, H.H., A.P.K. Nair, and C. Moroni. 1993. Suppressible and nonsuppressible autocrine mast cell tumors are distinguished by insertion of an endogenous retroviral element (IAP) into the interleukin-3 gene. *J. Exp. Med.* **178:** 403.

Jung, S. and H.J. Schluesener. 1991. Human T lymphocytes recognize a peptide of single point-mutated, oncogenic ras proteins. *J. Exp. Med.* **173:** 273.

Klein, G., H. Sjögren, E. Klein, and K.E. Hellström. 1960. Demonstration of resistance against methylcholanthrene-induced sarcomas in the primary autochthonous host. *Cancer Res.* **20:** 1561.

Knuth, A., B. Danowski, H.F. Oettgen, and L. Old. 1984. T-cell mediated cytotoxicity against autologous malignant melanoma: Analysis with interleukin-2-dependent T-cell cultures. *Proc. Natl. Acad. Sci.* **81:** 3511.

Kripke, M.L. and M.S. Fisher. 1976. Immunologic parameters of ultraviolet carcinogenesis. *J. Natl. Cancer Inst.* **57:** 211.

Lethé, B., B. Van den Eynde, A. Van Pel, G. Corradin, and T. Boon. 1992. Mouse tumor rejection antigens P815A and P815B: Two epitopes carried by a single peptide. *Eur. J. Immunol.* **22:** 2283.

Lurquin, C., A. Van Pel, B. Mariamé, E. De Plaen, J.-P. Szikora, C. Janssens, M. Reddehase, J. Lejeune, and T. Boon. 1989. Structure of the gene coding for tum⁻ transplantation antigen P91A. A peptide encoded by the mutated exon is recognized with Ld by cytolytic T cells. *Cell* **58:** 293.

Mandelboim, O., G. Berke, M. Fridkin, M. Feldman, M. Eisenstein, and L. Eisenbach. 1994. CTL induction by a tumour-associated antigen octapeptide derived from a murine lung carcinoma. *Nature* **369:** 67.

Maruyama, K., M. Usami, T. Aizawa, and K. Yoshikawa. 1991. A novel brain-specific mRNA encoding nuclear protein (necdin) expressed in neurally differentiated embryonal carcinoma cells. *Biochem. Biophys. Res. Commun.* **178:** 291.

Mukherji, B. and T.J. MacAlister. 1983. Clonal analysis of cytotoxic T cell response against human melanoma. *J. Exp. Med.* **158:** 240.

Nair, A.P.K., H.H. Hirsch, and C. Moroni. 1992. Mast cells sensitive to v-H-*ras* transformation are hyperinducible for interleukin 3 expression and have lost tumor-suppressor activity. *Oncogene* **7:** 1963.

Nair, A.P.K., I.D. Diamantis, J.-F. Conscience, V. Kindler, P. Hofer, and C. Moroni. 1989. A v-H-*ras*-dependent hemopoietic tumor model involving progression from a clonal stage of transformation competence to autocrine interleukin 3 production. *Mol. Cell. Biol.* **9:** 1183.

Seed, B. and A. Aruffo. 1987. Molecular cloning of the CD2 antigen, the T-cell erythrocyte receptor, by a rapid immunoselection procedure. *Proc. Natl. Acad. Sci.* **84:** 3365.

Skipper, J. and H.J. Stauss. 1993. Identification of two cytotoxic T lymphocyte-recognized epitopes in the Ras protein. *J. Exp. Med.* **177:** 1493.

Townsend, A., J. Rothbard, F. Gotch, G. Bahadur, D. Wraith, and A. McMichael. 1986. The epitopes of influenza nucleoprotein recognized by cytotoxic T lymphocytes can be defined with short synthetic peptides. *Cell* **44:** 959.

Traversari, C., P. van der Bruggen, B. Van den Eynde, P. Hainaut, C. Lemoine, N. Ohta, L. Old, and T. Boon. 1992. Transfection and expression of a gene coding for a human melanoma antigen recognized by autologous cytolytic T lymphocytes. *Immunogenetics* **35:** 145.

Uyttenhove, C., J. Maryanski, and T. Boon. 1983. Escape of mouse mastocytoma P815 after nearly complete rejection is due to antigen-loss variants rather than immunosuppression. *J. Exp. Med.* **157:** 1040.

Van den Eynde, B., B. Lethé, A. Van Pel, E. De Plaen, and T. Boon. 1991. The gene coding for a major tumor rejection antigen of tumor P815 is identical to the normal gene of syngeneic DBA/2 mice. *J. Exp. Med.* **173:** 1373.

Van den Eynde, B., P. Hainaut, M. Hérin, A. Knuth, C. Lemoine, P. Weynants, P. van der Bruggen, R. Fauchet, and T. Boon. 1989. Presence on a human melanoma of multiple antigens recognized by autologous CTL. *Int. J. Cancer* **44:** 634.

van der Bruggen, P., C. Traversari, P. Chomez, C. Lurquin, E. De Plaen, B. Van den Eynde, A. Knuth, and T. Boon. 1991.

A gene encoding an antigen recognized by cytolytic T lymphocytes on a human melanoma. *Science* **254:** 1643.

Van Pel, A., F. Vessière, and T. Boon. 1983. Protection against two spontaneous mouse leukemias conferred by immunogenic variants obtained by mutagenesis. *J. Exp. Med.* **157:** 1992.

Vose, B.M. and G.D. Bonnard. 1982. Specific cytotoxicity against autologous tumour and proliferative responses of human lymphocytes grown in interleukin-2. *Int. J. Cancer* **29:** 33.

Wölfel, T., E. Klehmann, C. Müller, K.-H. Schütt, K.-H. Meyer zum Büschenfelde, and A. Knuth. 1989. Lysis of human melanoma cells by autologous cytolytic T cell clones. Identification of human histocompatibility leukocyte antigen A2 as a restriction element for three different antigens. *J. Exp. Med.* **170:** 797.

Pathogenesis of Cancer of the Cervix

H. zur Hausen and F. Rösl
Deutsches Krebsforschungszentrum, Im Neuenheimer Feld 280, 69120 Heidelberg, Germany

The concept of virus-induced carcinogenesis in humans has a long history. Data from animal systems became available in the first and second decade of this century (for review, see Gross 1983). Yet, it is only 30 years ago that the first human pathogenic virus, the Epstein-Barr virus (EBV), subsequently suspected to be linked to specific human tumors, was seen in the electron microscope (Epstein et al. 1964). It took almost these 30 years to accept EBV as a genuine tumor virus, causing at least a proportion of B-cell lymphomas in immunosuppressed individuals (for review, see zur Hausen 1991a). Its role in those malignancies in which EBV genomes had been initially identified (zur Hausen et al. 1970) is even today not fully understood.

During the past 15 years, three new virus groups have been identified which play a role in human cancers: the human T-lymphotropic retrovirus (HTLV), the hepatitis B virus (HBV), and a number of different genotypes of human papillomaviruses (HPV). Particularly the latter two groups, HBV and HPV, have been linked to major human cancers, accounting for a substantial proportion of the worldwide cancer burden (for review, see zur Hausen 1991a). Specifically, hepatocellular carcinoma and anogenital cancers (predominantly cancer of the cervix) emerge as the most important human cancers caused by virus infections.

Cancer of the cervix, although declining in incidence in affluent societies, still ranks number 2 worldwide in cancers of women (Parkin et al. 1993). With an estimated rate of 440,000 new cases per year with high mortality in the developing world and, by including the precancerous conditions, also a high morbidity everywhere, this cancer deserves special attention as a major public health problem. At the same time, cancer of the cervix can be considered as the first human cancer in which

1. the most significant risk factors have been identified
2. we know the cancer-effecting genes and start to understand their intracellular function
3. we can identify first cellular genes, functioning apparently in two independent signaling pathways whose interruption emerges as a precondition for growth-stimulating and mutagenic functions of the viral effector genes and the eventual development of invasive and metastasizing cancer (see below)

High-risk HPV E6/E7 genes have been identified as necessary but not sufficient factors for the proliferation of HPV-immortalized cells, and also for the malignant phenotype of HPV-positive cervical cancer cell lines (von Knebel Doeberitz et al. 1988, 1992, 1994). Particularly interesting questions arise from the function of the viral oncoproteins and their interaction with host cell proteins, and from observed differences in their regulation in HPV-immortalized nonmalignant cells when compared to malignant tumors and cell lines derived therefrom.

A number of studies addressed the issue of E6 and E7 functions in immortalized and malignant cells. The present data indicate that E7 activates cyclins E and A, binds Rb, and facilitates its release after phosphorylation from E2F complexes (K. Zerfass et al., in prep.). At the same time, viral oncoproteins seem to suppress cyclin D_1 (Lukas et al. 1994; Spitkovsky et al. 1994; K. Zerfass et al., in prep.), which could suggest a similar functional interaction with cell cycle regulatory mechanisms by viral oncoproteins as exerted by cyclin D_1. Modifications of the latter, or its overexpression, may result in oncogenic functions (Lovec et al. 1994; Lukas et al. 1994).

The E6 oncoprotein of high-risk HPVs effectively binds and degrades p53 (Scheffner et al. 1990; Werness et al. 1990). This may represent the prime reason for the E6-mediated prevention of a G_1 arrest after DNA damage (Kessis et al. 1993) and the increased chromosomal instability (Hashida and Yasumoto 1991; White et al. 1994) postulated at a Cold Spring Harbor meeting in 1990 (zur Hausen 1991b) as the major endogenous progression factor for high-risk HPV infections. It seems that E7 also contributes to this E6 function (White et al. 1994), although the mechanism remains to be elucidated.

HOST CELL REGULATION OF VIRAL ONCOGENES

Our group became specifically interested in the question of differential regulation of HPV expression in nonmalignant and malignant cells even more so after the demonstration that E6/E7 oncoprotein expression was necessary for the immortalized and also for the malignant phenotype of HPV genome-carrying cells (von Knebel Doeberitz et al. 1988, 1992, 1994; Crook et al. 1989). The hypothesis had been put forward that a failing host cell regulation of viral oncoprotein transcription/expression or function is the key driving force for the eventual malignant conversion of viral genome positive cells (zur Hausen 1977, 1986). This was flanked by the assumption that the mutagenic activity of E6/E7 oncoproteins of high-risk HPVs generates these modifications of host cell genes acting as endogenous progression factors (zur Hausen 1991a,b).

Support for these speculations resulted from several lines of evidence: A number of groups had noted that cells immortalized by SV40, HPV-16, or HPV-18 DNA are able to complement each other after somatic cell hybridization for senescence (Pereira-Smith and Smith 1981, 1988; Whitaker et al. 1992; Chen et al. 1993; Seagon and Dürst 1994). At least four complementation groups have been established. Previous studies had shown that viral oncoprotein expression is necessary for the immortalized state of the cells (Münger et al. 1989). In view of the absence of detectable modifications in the viral transforming genes and their promoters, it is a likely interpretation of these studies that besides the necessary but not sufficient viral oncoprotein expression, modifications of one allelic set out of at least four host cell genes contribute to unlimited in vitro growth of the infected or transfected cells. The low efficiency of immortalizing events further underlines this interpretation.

Control of the viral oncogenes by strong heterologous promoter elements did not significantly increase the immortalization frequency and, at the same time, did not result in malignant growth of the immortalized cells (Halbert et al. 1991; Blanton et al. 1992). This supported the suspicion that viral oncoprotein expression is not sufficient for immortalization and a promoter control not responsible for an immortalized but nonmalignant state of the transfected cells.

It had been noted that the senescent hybrid cells continue to synthesize SV40 T antigen or transcribe HPV E6/E7 mRNA (Whitaker et al. 1992; Chen et al. 1993). Since studies in temperature-sensitive T-antigen mutants (Randa et al. 1989) or on hormone-dependent E6/E7 expression (von Knebel Doeberitz et al. 1994) demonstrated the need for viral oncoprotein activity for the immortalized phenotype, cellular proteins complemented in these somatic hybridization studies should interfere with viral oncoprotein function rather than with their expression (zur Hausen 1994). The cellular genes involved in this process have not yet been identified. The recently characterized kinase inhibitors p21 (WAF-1, CIP-1), p16, or p27 may represent good candidates (El-Deiry et al. 1993; Harper et al. 1993; Xiong et al. 1993; Kamb et al. 1994; Polyak et al. 1994). Immortalization thus emerges as the consequence of an interrupted intracellular control of viral oncoprotein function.

In contrast to most HPV-containing cervical carcinoma cell lines, HPV-immortalized human keratinocytes fail to grow as tumors when inoculated into immunosuppressed animals (Dürst et al. 1991). Upon prolonged cultivation, however, particularly HPV-18-immortalized cells spontaneously may convert to malignant growth, permitting studies on the mode of progression from the state of immortalization to a malignant phenotype (Hurlin et al. 1991; Pecoraro et al. 1991). As previously mentioned, a number of studies had indicated that the expression of viral oncoproteins is required not only for the maintenance of the immortalized state, but also for the malignant phenotype (von Knebel Doeberitz et al. 1988, 1992, 1994; Crook et al. 1989).

Since a functional repression of viral oncoproteins appears to have been abolished already by reaching the stage of immortalization, another mode of interference obviously regulates progression toward malignant growth. This became the major subject of our research during the past few years.

Earlier studies had indicated that treatment of some HPV-positive nonmalignant and malignant cells with 5-azacytidine led to a differential response in HPV transcription. In nonmalignant cells, here specifically in HeLa-fibroblast or HeLa-keratinocyte hybrids, but also in HPV-immortalized human keratinocytes, HPV transcription was efficiently suppressed, whereas no effect was seen in the HeLa parental line or in malignant revertants obtained from such hybrids (Rösl et al. 1988). Although under regular tissue culture conditions the transcriptional rates did not differ significantly between malignant and nonmalignant lines, this early observation pointed to a differential regulation of viral transcription under selective conditions, possibly related to demethylation. This was further underlined by studies showing that fusion of a malignant HPV-16-positive cervical carcinoma cell line, SiHa, constitutively expressing a CAT gene controlled by an HPV-18 promoter, with normal human keratinocytes or an established keratinocyte line resulted in suppression of CAT gene expression under nonselective conditions. The reporter gene was not modified in the hybrid cells, and CAT message was induced by cycloheximide treatment of these cells for several hours (Rösl et al. 1991).

These studies led to the hypothesis that malignant cells had lost responsiveness to signals down-regulating HPV transcription, triggered in this case by their nonmalignant somatic fusion partners, although the latter were in part already immortalized prior to cell fusion. It was tempting to assume that these signals, although not activated under regular tissue culture conditions, would be stimulated upon heterografting HPV-positive nonmalignant cells into immunoincompetent animals by a paracrine mechanism (zur Hausen 1994).

Indeed, in situ hybridizations with antisense E6/E7 probes revealed suppression of HPV transcription selectively in nonmalignant cells after inoculation into nude mice, whereas malignant cells continued to express these genes at a high level (Bosch et al. 1990; Dürst et al. 1991). HPV-immortalized cells formed epithelial cysts, differentiating into the inner lumen, with a low degree of E6/E7 transcription within the basal cell-like layer at the periphery of the cyst. Nonmalignant HeLa-hybrid cells drastically reduced HPV transcription within 3 days following heteroimplantation.

These studies were correlated by analyses of clinical lesions with low and high grades of dysplasia (Dürst et al. 1992; Higgins et al. 1992; Stoler et al. 1992; Böhm et al. 1993). In situ hybridizations revealed a very low degree of high-risk HPV E6/E7 transcription within low-grade lesions containing HPV-16 or HPV-18 DNA, whereas high-grade lesions generally exhibited abundant transcripts derived from these genes.

The low HPV expression in heterografted immortalized cells and in low-grade dysplasias suggests that the clinical state of low-grade dysplasia corresponds to the state of immortalization under tissue culture conditions. This appears to be underscored by organotypic cultures initiated with HPV-immortalized cells which, at least in early phases of immortalization, correspond histologically to low-grade lesions (for review, see McDougall 1994). Since immortalization, as pointed out previously, requires modification of host cell genes besides the expression of viral oncoproteins, this conclusion predicts the clonal development of low-grade lesions, which at this stage has not yet been proven experimentally. The major assumption derived from these studies was, however, that immortalized cells respond to paracrine signals which down-regulate HPV transcription, whereas in malignant cells, this signal-response pathway became interrupted.

HUMAN MACROPHAGES SUPPRESS HPV TRANSCRIPTION

To develop an experimental system to further substantiate this hypothesis, nonmalignant HPV-containing cells (HPV-immortalized cells as well as HeLa-fibroblast hybrids) and malignant cells, respectively, were exposed to human macrophage preparations in tissue culture (Rösl et al. 1994). The results were remarkably uniform: One of our HPV-16-immortalized human foreskin keratinocyte lines had converted into a malignant line after more than 100 passages following exposure to X-irradiation (M. Dürst et al., in prep.). When the nonmalignant early-passage cells of this line were exposed to increasing concentrations of macrophages, HPV transcription was effectively suppressed in a dose-dependent manner (Rösl et al. 1994). The malignant convertants of the same line, however, did not respond with transcriptional suppression of HPV-16 DNA. The same was true for the nonmalignant HeLa-fibroblast line 444 which, upon addition of human macrophages, also revealed a dose-dependent suppression of HPV-18 transcription. Neither the parental HeLa cells nor the malignant revertant line originating from 444 cells responded to macrophage treatment with HPV transcriptional suppression. Thus, macrophages were able to selectively trigger in nonmalignant cells transcriptional suppression of HPV by a paracrine mechanism.

Interestingly, the nonmalignant cells upon macrophage treatment started to synthesize mRNA encoding the macrophage chemoattractant protein MCP-1 (Rösl et al. 1994). This protein appears to possess tumor-suppressive properties (Rollins and Sunday 1991; K. Kleine et al., in prep.). No MCP-1 message induction was noted in malignant cells, although the gene and its promoter remained unaltered in the latter. Thus, the induction of this protein in nonmalignant cells by activated macrophages may further enhance the suppression of HPV transcription by attracting additional macrophages.

Attempts that were initiated to analyze the potential cytokines engaged in the selective HPV suppression led to the identification of tumor necrosis factor α (TNFα) as the most effective cytokine (Rösl et al. 1994). This corresponds to observations made in other nonmalignant HPV-containing lines revealing TNFα as an effective suppressor for HPV transcription (Malejczyk et al. 1992; Kyo et al. 1994). Malignant cells had not been included in these studies. In our experiments, malignant cells did not respond to TNFα treatment with transcriptional repression. Besides TNFα, interleukin-1 was also effective, although to a lesser extent (F. Rösl and H. zur Hausen, unpubl.). It appears that both cytokines interact cooperatively in mediating HPV transcriptional suppression.

Although additional cytokines and other cell types may also be involved in the reported suppression of HPV gene activity, these studies clearly indicate the existence of an intracellular pathway triggered by paracrine signals which regulates viral transcription in HPV-immortalized nonmalignant cells. The same signaling pathway is obviously interrupted in malignant cells. Thus, malignant cells apparently acquired genetic modifications in genes regulating this pathway. This is in line with complementation studies showing (besides senescence as the consequence of cell fusion between some partners) an immortalized, but nonmalignant, phenotype after somatic cell hybridization between two malignant fusion partners (Seagon and Dürst 1994).

Host cell genes regulating this pathway emerge from two types of studies: Smits et al. (1992) demonstrated that the loss of the short arm of chromosome 11 resulted in activation of HPV transcription in human fibroblasts. Normal human fibroblasts, in contrast, suppress the HPV promoter. At the same time, they observed that fibroblasts deleted in chromosome 11p reveal an up-regulation of the regulatory subunit of protein phosphatase 2A (PP2A), which in turn down-regulated the catalytic subunit of PP2A and reduced enzyme activity. A role for PP2A in the regulation of HPV transcription was suggested by further studies showing that specific inhibitors of PP2A (okadaic acid or SV40 small t) also caused substantial up-regulation of HPV transcription in human fibroblasts (Smits et al. 1992).

Our group studied transcription factors as possible regulators of the differential activity of HPV transcription in nonmalignant HeLa hybrid cells and the parental HeLa line (Bauknecht et al. 1992). The *Ying-Yang-1* protein (YY1) was identified as one factor suppressing high-risk HPV transcription in the nonmalignant hybrids, but activating it in HeLa cells. Since a construct containing only the proximal promoter region of HPV-18 with one identified YY1-binding site is suppressed in both types of cells, the differential effect obviously resulted from an interaction of this YY1-binding site with an upstream binding factor. By deletion mapping, this site was identified, located 250 bp upstream of the YY1-binding sequence (Bauknecht et al. 1994).

MODELS OF THE HOST CELL/VIRAL ONCOGENE INTERACTION

Models of regulatory pathways controlling HPV oncogene function and transcription have been discussed previously (zur Hausen 1994; zur Hausen and de Villiers 1994). By considering the interaction of high-risk HPV E6/E7 genes with p53 (and, as a consequence, the resulting failing induction of p21) and pRB, the latter cellular proteins appear to act close to the proximal arm of two potentially independent signaling pathways negatively interfering with cell growth. DNA tumor virus infections seem to interrupt the functions of both cascades, although this is obviously not sufficient for cell immortalization. It is therefore tempting to speculate that a third signaling cascade exists which also negatively controls viral oncoprotein function (besides interfering with cell growth). p16 could be a candidate for this inhibitory function in view of the modifications of p16/cdk4 complexes in E6/E7-expressing human cells (see Tlsty et al., this volume). Therefore, at least three signaling pathways seem to be engaged in cellular growth regulation (schematically outlined in Fig. 1A). The inactivation of two of these regulatory pathways via the physical interaction of the viral oncoprotein E6 with p53 (Scheffner et al. 1990; Werness et al. 1990) and E7 with Rb (Dyson et al. 1989) may disturb the balanced equilibrium of the cell cycle, which, in combination with the postulated inhibitory effect of p16 on viral oncoprotein function, may lead to a prolonged life span of the infected cell (Fig. 1B). Consequently, genetic modifications within this remaining signaling cascade should lead to immortalization (Fig. 1C). These modifications should be facilitated by the observed clastogenic effects of high-risk HPV E6/E7 gene functions (Hashida and Yasumoto 1991; White et al. 1994). The four complementation groups identified in SV40-immortalized or high-risk HPV-immortalized cells (Pereira-Smith and Smith 1981, 1988; Whitaker et al. 1992; Chen et al. 1993; Seagon and Dürst 1994) should identify the third signaling pathway. Immortalized cells are nonmalignant upon heterotransplantation into nude mice, pointing to the existence of a fourth pathway which has recently partially been identified: It is triggered by macrophage infiltration and cytokine induction and negatively controls HPV oncogene transcription (Rösl et al. 1994). Only the eventual disruption of this remaining control finally leads to malignant growth (Fig. 1D). Thus, besides uptake of high-risk HPV DNA, modifications of at least four "master" genes in the hierarchy controlling the cell cycle appear to be required for the development of invasive growth. Therefore, viral genome persistence and gene activity emerge as the most important preconditions in the progression toward malignancy and in the maintenance of the malignant state.

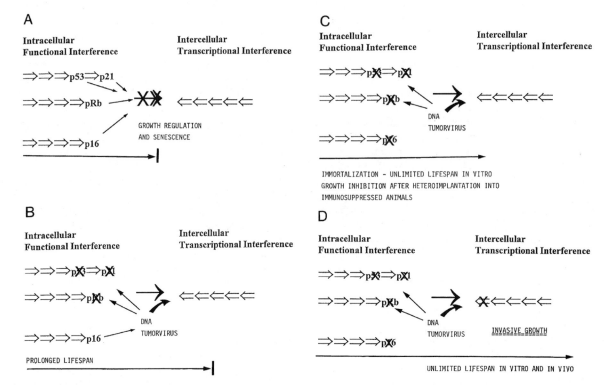

Figure 1. Schematic outline of intracellular and intercellular interactions of signaling pathways in DNA tumor virus infected cells. Further explanations are provided in the text.

REFERENCES

Bauknecht, T., P. Angel, H.-D. Royer, and H. zur Hausen. 1992. Identification of a negative regulatory domain in the human papillomavirus type 18 promoter: Intersection with the transcriptional repressor YY1. *EMBO J.* **11:** 4607.

Bauknecht, T., F. Jundt, I. Herr, T. Oehler, H. Delius, Y. Shi, P. Angel, and H. zur Hausen. 1994. A "switch region" determines cell type specific positive or negative action of YY1 on the activity of the HPV-18 promoter. *J. Virol.* (in press).

Blanton, R.A., M.D. Coltrera, A.M. Gown, C.L. Halbert, and J.K. McDougall. 1992. Expression of the HPV 16 E7 gene generates proliferation in stratified squamous cell cultures which is independent of endogenous p53 levels. *Cell Growth Differ.* **3:** 791.

Böhm, S., S.P. Wilczynski, H. Pfister, and T. Iftner. 1993. The predominant m-RNA class in HPV 16-infected genital neoplasias does not encode the E6 or the E7 protein. *Int. J. Cancer* **55:** 791.

Bosch, F., E. Schwarz, P. Boukamp, N.E. Fusenig, D. Bartsch, and H. zur Hausen. 1990. Suppression in vivo of human papillomavirus type 18 E6-E7 gene expression in nontumorigenic HeLa fibroblast hybrid cells. *J. Virol.* **64:** 4743.

Chen, T.-M., G. Pecoraro, and V. Defendi. 1993. Genetic analysis of in vitro progression of human papillomavirus-transfected human cervical cells. *Cancer Res.* **53:** 1167.

Crook, T., J.P. Morgenstern, L. Crawford, and L. Banks. 1989. Continued expression of HPV 16 E7 protein is required for maintenance of the transformed phenotype of cells cotransformed by HPV 16 plus EJ-ras. *EMBO J.* **8:** 513.

Dürst, M., D. Glitz, A. Schneider, and H. zur Hausen. 1992. Human papillomavirus type 16 (HPV 16) gene expression and DNA replication in cervical neoplasia: Analysis by in situ hybridization. *Virology* **189:** 132.

Dürst, M., F.X. Bosch, D. Glitz, A. Schneider, and H. zur Hausen. 1991. Inverse relationship between human papillomavirus (HPV) type 16 early gene expression and cell differentiation in nude mouse epithelial cysts and tumors induced by HPV positive human cell lines. *J. Virol.* **65:** 796.

Dyson, N., P.M. Howley, K. Münger, and E. Harlow. 1989. The human papillomavirus-16 E7 oncoprotein is able to bind to the retinoblastoma gene product. *Science* **243:** 934.

El-Deiry, W.S., T. Tokino, V.E. Velculescu, D.B. Levy, R. Parsons, J.M. Trent, D. Lin, E. Mercer, K.W. Kinzler, and B. Vogelstein. 1993. WAF-1 a potential mediator of p53 tumor suppression. *Cell* **75:** 817.

Epstein, M.A., B.G. Achong, and Y.M. Barr. 1964. Virus particles in cultured lymphoblasts from Burkitt's lymphoma. *Lancet* **I:** 702.

Gross, L. 1983. *Oncogenic viruses*, 3rd edition, Pergamon. Oxford.

Halbert, C.L., G.W. Demers, and D.A. Galloway. 1991. The E7 gene of human papillomavirus type 16 is sufficient for immortalization of human epithelial cells. *J. Virol.* **65:** 473.

Hashida, T. and S. Yasumoto. 1991. Induction of chromosomal abnormalities in mouse and human epidermal keratinocytes by the human papillomavirus type 16 E7 oncogene. *J. Gen. Virol.* **72:** 1569.

Harper, J.W., G.R. Adami, N. Wei, K. Keyomarsi, and S.J. Elledge. 1993. The p21 cdk interacting protein Cip-1 is a potent inhibitor of G1 cyclin-dependent kinases. *Cell* **75:** 805.

Higgins, G.D., D.M. Uzelin, G.E. Phillips, P. McEnvoy, R. Marin, and C.J. Burrell. 1992. Transcription patterns of human papillomavirus type 16 in genital intraepithelial neoplasia: Evidence for promoter usage within the E7 open reading frame during epithelial differentiation. *J. Gen. Virol.* **73:** 2047.

Hurlin, P.J., P. Kaur, P.P. Smith, N. Perez-Reyes, R.A. Blanton, and J.K. McDougall. 1991. Progression of human papillomavirus type 18-immortalized human keratinocytes to a malignant phenotype. *Proc. Natl. Acad. Sci.* **88:** 570.

Kamb, A., N.A. Gruis, J. Weaver-Feldhaus, Q. Liu, K. Harshman, S.V. Tavtigian, E. Stockert, R.S. Day III, B.E. Johnson, and M.H. Scolnick. 1994. A cell cycle regulator potentially involved in genesis of many tumor types. *Science* **264:** 436.

Kessis, T.D., R.J. Slebos, W.G. Nelson, M.B. Kastan, B.S. Plunkett, S.M. Hau, A.T. Lörincz, L. Hedrick, and K.R. Cho. 1993. Human papillomavirus 16 E6 expression disrupts the p53-mediated cellular response to DNA damage. *Proc. Natl. Acad. Sci.* **90:** 3988.

Kyo, S., M. Inoue, N. Hayasaka, T. Inoue, M. Yutsudo, O. Tanizawa, and A. Hakura. 1994. Regulation of early gene expression of human papillomavirus type 16 by inflammatory cytokines. *Virology* **200:** 130.

Lovec, H., A. Sewing, C. Lucibello, R. Müller, and T. Möröy. 1994. Oncogenic activity of cyclin D1 revealed through cooperation with Ha-*ras*: Link between cell cycle control and malignant transformation. *Oncogene* **9:** 323.

Lukas, J., M. Pagano, Z. Staskova, G. Draetta, and J. Bartek. 1994. Cyclin D1 protein oscillates and is essential for cell cycle progression in human tumour cell lines. *Oncogene* **9:** 707.

Malejczyk, J., M. Malejczyk, A. Urbanski, A. Köck, S. Jablonska, G. Orth, and T.A. Luger. 1992. Autocrine growth limitation of human papillomavirus type 16-harbouring keratinocytes by constitutively released tumor necrosis factor-α. *J. Immunol.* **149:** 2702.

McDougall, J.K. 1994. Immortalization and transformation of human cells by human papillomavirus. *Curr. Top. Microbiol. Immunol.* **86:** 101.

Münger, K., W.C. Phelps, V. Bubb, P.M. Howley, and R. Schlegel. 1989. The E6 and E7 genes of human papillomavirus type 16 are necessary and sufficient for transformation of primary human keratinocytes. *J. Virol.* **63:** 4417.

Parkin, D.M., P. Pisani, and J. Ferlay. 1993. Estimates of the worldwide incidence of eighteen major cancers in 1985. *Int. J. Cancer* **54:** 594.

Pecoraro, G., M. Lee, D. Morgan, and V. Defendi. 1991. Evolution of in vitro transformation and tumorigenesis of HPV-16 and HPV-18 immortalized primary cervical epithelial cells. *Am. J. Pathol.* **138:** 1.

Pereira-Smith, O.M. and J.R. Smith. 1981. Expression of SV40 antigen in finite lifespan hybrids of normal and SV40-transformed fibroblasts. *Somatic Cell Genet.* **7:** 411.

———. 1988. Genetic analysis of indefinite division in human cells: Identification of four complementation groups. *Proc. Natl. Acad. Sci.* **85:** 6042.

Polyak, K., J.-Y. Kato, M.J. Solomon, C.J. Sherr, J. Massagué, J.M. Roberts, and A. Koff. 1994. p27^{Kip1}, a cyclin-Cdk inhibitor, links transforming growth factor-β and contact inhibition to cell cycle arrest. *Genes Dev.* **8:** 9.

Randa, R.L., Y. Caton, K.K. Jha, P. Kaplan, P. Li, F. Traganos, and H.L. Ozer. 1989. Growth of immortal simian virus 40 ts A$^-$ transformed human fibroblasts is temperature dependent. *Mol. Cell. Biol.* **9:** 3093.

Rollins, B.J. and M.E. Sunday. 1991. Suppression of tumor formation in vivo by expression of the JE gene in malignant cells. *Mol. Cell. Biol.* **11:** 3125.

Rösl, F., M. Dürst, and H. zur Hausen. 1988. Selective suppression of human papillomavirus transcription in nontumorigenic cells by 5-azacytidine. *EMBO J.* **7:** 1321.

Rösl, F., T. Achtstetter, K.-J. Hutter, T. Bauknecht, G. Futterman, and H. zur Hausen. 1991. Extinction of the HPV 18 upstream regulatory region in cervical carcinoma cells after fusion with non-tumorigenic human keratinocytes under non-selective conditions. *EMBO J.* **10:** 1337.

Rösl, F., M. Lengert, J. Albrecht, K. Kleine, R. Zawatzky, B. Schraven, and H. zur Hausen. 1994. Differential regulation of the JE gene encoding the monocyte chemoattractant protein (MCP-1) in cervical carcinoma cells and derived hybrids. *J. Virol.* **68:** 2142.

Scheffner, M., B.A. Werness, J.M. Huibregtse, J.M. Levine, and P.M. Howley. 1990. The E6 oncoprotein encoded by human papillomavirus types 16 and 18 promotes the degradation of p53. *Cell* **63:** 1129.

Seagon, S. and M. Dürst. 1994. Genetic analysis of an in vitro model system for human papillomavirus type 16-associated tumorigenesis. *Cancer Res.* **54:** 21.

Smits, P.H.M., H.L. Smits, R. Minnaar, B.A. Hemmings, R.E. Mayer-Jaekel, R. Schuurman, J. van der Noordaa, and J. ter Schegget. 1992. The *trans*-activation of the HPV 16 long control region in human cells with a deletion in the short arm of chromosome 11 is mediated by the 55kDa regulatory subunit of protein phosphatase 2A. *EMBO J.* **11:** 4601.

Stoler, M.H., C.R. Rhodes, A. Whitbeck, S.M. Wolinsky, L.T. Chow, and T.R. Broker. 1992. Human papillomavirus type 16 and 18 gene expression in cervical neoplasias. *Hum. Pathol.* **23:** 117.

Spitkovsky, D., P. Steiner, J. Lukas, E. Lees, M. Pagano, A. Schulze, S. Joswig, D. Picard, M. Tommasino, M. Eilers, and P. Jansen-Dürr. 1994. Modulation of cyclin gene expression by adenovirus E1A in a cell line with E1A-dependent conditional proliferation. *J. Virol.* **68:** 2206.

von Knebel Doeberitz, M., T. Oltersdorf, E. Schwarz, and L. Gissmann. 1988. Correlation of modified human papillomavirus early gene expression with altered growth properties in C4-1 cervical carcinoma cells. *Cancer Res.* **48:** 3780.

von Knebel Doeberitz, M., C. Rittmüller, H. zur Hausen, and M. Dürst. 1992. Inhibition of tumorigenicity of cervical cancer cells in nude mice by HPV E6-E7 antisense RNA. *Int. J. Cancer* **51:** 831.

von Knebel Doeberitz, M., C. Rittmüller, F. Aengeneyndt, P. Jansen-Dürr, and D. Spitkovsky. 1994. Reversible repression of papillomavirus oncogene expression in cervical carcinoma cells: Consequences for the phenotype and E6-p53 and E7-pRB interactions. *J. Virol.* **68:** 2811.

Werness, B.A., A.J. Levine, and P.M. Howley. 1990. Association of human papillomavirus types 16 and 18 E6 proteins with p53. *Science* **248:** 76.

Whitaker, N.J., E.L. Kidston, and R.R. Reddel. 1992. Finite lifespan of hybrids formed by fusion of different simian virus 40-immortalized human cell lines. *J. Virol.* **66:** 1202.

White, A.E., E.M. Livanos, and T.D. Tlsty. 1994. Differential disruption of genomic integrity and cell cycle regulation in normal human fibroblasts by the HPV oncoproteins. *Genes Dev.* **8:** 666.

Xiong, Y., G.J. Hannon, H. Zhang, D. Casso, R. Kobayashi, and D. Beach. 1993. p21 is a universal inhibitor of cyclin kinases. *Nature* **366:** 701.

zur Hausen, H. 1977. Cell-virus gene balance hypothesis of carcinogenesis. *Behring Inst. Mitt.* **61:** 23.

———. 1986. Human genital cancer: Synergism between two virus infections or synergism between a virus infection and initiating events. *Lancet* **2:** 1370.

———. 1991a. Viruses in human cancers. *Science* **254:** 1167.

———. 1991b. Papillomavirus/host cell interactions in the pathogenesis of anogenital cancer. In *Origins of human cancer: A comprehensive review* (ed. J. Brugge et al.), p. 685. Cold Spring Harbor Laboratory Press, Cold Spring Harbor, New York.

———. 1994. Disrupted dichotomous intracellular control of human papillomavirus infection in cancer of the cervix. *Lancet* **343:** 955.

zur Hausen, H. and E.-M. de Villiers. 1994. Human papillomaviruses. *Annu. Rev. Microbiol.* **48:** 427.

H. zur Hausen, H. Schulte-Holthausen, G. Klein, W. Henle, G. Henle, P. Clifford, and L. Santesson. 1970. EBV DNA in biopsies of Burkitt tumours and anaplastic carcinoma of the nasopharynx. *Nature* **228:** 1056.

Autocrine Mechanism for *met* Proto-oncogene Tumorigenicity

S. Rong and G.F. Vande Woude

ABL-Basic Research Program, NCI-Frederick Cancer Research and Development Center, Frederick, Maryland 21702

The mature form of the Met receptor is generated by cleavage of p170^{Met} to yield a disulfide-linked 140-kD β subunit (p140^{Met}) and a 45-kD α subunit (p45^{Met}). The α subunit is extracellular, whereas the β subunit encompasses a major portion of the extracellular domain, the transmembrane domain, and the intracellular tyrosine kinase domain (Chan et al. 1988; Giordano et al. 1988; Gonzatti-Haces et al. 1988). Met is widely expressed in adult and embryonic tissues as well as in primary cells and cell lines (Gonzatti-Haces et al. 1988; DiRenzo et al. 1991; Rong et al. 1993a) but is predominantly expressed in epithelial tissue and cells.

The ligand for Met is hepatocyte growth factor (HGF) or scatter factor (SF) (Gherardi and Stoker 1990; Bottaro et al. 1991; Furlong et al. 1991; Konishi et al. 1991; Naldini et al. 1991; Weidner et al. 1991). HGF/SF is synthesized as a single-chain precursor that is proteolytically processed to yield a heterodimer of a 69-kD α chain and a 34-kD β chain (Nakamura et al. 1987). HGF/SF is produced predominantly by cells of mesenchymal origin (Stoker and Perryman 1985; Rosen et al. 1989), is a paracrine effector of mesenchymal-epithelial interactions, and induces a variety of biological responses in epithelial cells, including stimulation of cell motility, proliferation, and differentiation (Stoker and Perryman 1985; Nakamura 1992; Rosen et al. 1991; Montesano et al. 1991), all of which are mediated through the Met receptor.

The *met* proto-oncogene is frequently amplified and overexpressed in spontaneous NIH-3T3 cell transformants (Cooper et al. 1986; Hudziak et al. 1990), and overexpression of *met* in NIH-3T3 cells directly leads to cell transformation and tumorigenicity (Iyer et al. 1990). We have shown that NIH-3T3 cells express murine HGF/SF (HGF/SF^{mu}) endogenously and that, with ectopic murine Met expression, these cells become tumorigenic by an autocrine mechanism (Rong et al. 1992). Although HGF/SF^{mu} does not efficiently activate human Met (Met^{hu}) to transform NIH-3T3 cells, coexpression of Met^{hu} with the human ligand (HGF/SF^{hu}) efficiently transforms these cells (HMH cells), rendering them tumorigenic in nude mice (Rong et al. 1992). We have also demonstrated that these autocrine tumors are invasive and metastatic in nude mice (Rong et al. 1994). We found that, as in the mouse model systems, Met is overexpressed in most human mesenchymal tumor cell lines and perhaps in primary tumors of this type (Rong et al. 1993a). These tumors arise frequently in patients with Li-Fraumeni syndrome (Malkin et al. 1990), and we show that Met is also overexpressed in these tumors.

MATERIALS AND METHODS

Cell line and immunoprecipitation. NIH-3T3 cells transformed by *met*^{mu} (Met^{mu} cells) or by the coexpression of *met*^{hu} and HGF/SF^{hu} (HMH cells) (Rong et al. 1992) have been described previously. Met^{mu} and HMH cells are either parental cells or primary (1°) or secondary (2°) tumor explants. All cells were maintained as described previously (Rong et al. 1992, 1993b). Most of the human cell lines used in this study were obtained from the American Type Culture Collection (ATCC) and grown as recommended (Rong et al. 1993a).

The 19S Met^{hu} monoclonal antibody was generated against a bacterially expressed p50 form of Met^{hu} (Faletto et al. 1991). The Met^{hu}-specific C28 antipeptide antibody was raised by immunization of rabbits with the 28-amino acid carboxy-terminal peptide of Met^{hu} (Gonzatti-Haces et al. 1988). A3.1.2 is a monoclonal antibody against human recombinant HGF (Rong et al. 1992; a gift from Dr. T. Nakamura, Osaka, Japan). The SP260 peptide antibody is a rabbit antiserum directed against the carboxy-terminal 21 amino acids of Met^{mu} (Iyer et al. 1990). Anti-rhHGF is a rabbit polyclonal antibody against human recombinant HGF (Montesano et al. 1991; a gift from T. Nakamura, Osaka, Japan). 4G10 is a monoclonal phosphotyrosine antibody (anti-P-Tyr) (Morrison et al. 1989).

Immunoprecipitation analysis. Immunoprecipitation analysis for both Met and HGF/SF was carried out as described previously (Rong et al. 1992, 1993a,b).

Western immunoblot analysis. Western analysis was performed as described previously (Rong et al. 1993a).

Nude mouse tumor assay. The assay was performed as described previously (Blair et al. 1982; Rong et al. 1992) using weanling athymic nude mice (Harlan Sprague Dawley, Inc.).

Nude mouse metastasis assay. The assay was carried out as described previously (Rong et al. 1994).

RESULTS

Tumorigenicity of Met in NIH-3T3 Cells

NIH-3T3 cell cultures overexpressing Met^{mu} (but not Met^{hu}) are transformed and tumorigenic in nude mice (Iyer et al. 1990) (Table 1A). By Northern hybridization analyses (data not shown), HGF/SF mRNA was detected in nontransfected NIH-3T3 cells as well as in cells expressing met^{mu}, suggesting that transformation by met^{mu} is mediated by an autocrine mechanism (Rong et al. 1992). Met^{hu} was shown to be highly tumorigenic in NIH-3T3 cells only when transfected with HGF/SF^{hu} cDNAs (Table 1B). Moreover, the tumor explant of these HMH cells showed increased expression of both Met^{hu} and HGF/SF^{hu} compared to the parental lines (Fig. 1A and C, respectively; lanes 2, 4, and 5) and did not express increased levels of endogenous Met^{mu} (Fig. 1B). As further support for an autocrine mechanism of transformation, we showed that when the mouse external ligand-binding domain was linked to the human transmembrane and tyrosine kinase domains, the chimeric receptor displayed tumorigenic activity equivalent to that of Met^{mu} (Rong et al. 1992). In contrast, the reciprocal human amino-terminal/mouse carboxy-terminal chimera was poorly tumorigenic unless this chimera was cotransfected with HGF/SF^{hu} cDNA (Rong et al. 1992).

Expression of Met^{hu} and HGF/SF^{hu} in Tumor Explants of the Second Passage

Whereas tumors arose in nude mice from primary transfected HMH cells after 4–6 weeks, the same number of injected cells from tumor explants produced tumors in less than 4 weeks (data not shown). We observed a significant increase in HGF/SF^{hu} expression after tumor passage (Fig. 2B, lanes 3–6). The Met^{hu} receptor was also highly expressed in the same tumor explants (Fig. 2A, lanes 2 and 3). Furthermore, the level of Met^{hu} detected with anti-P-Tyr (Fig. 2C, upper panel, lanes 1–3) increased with the increase in the receptor (Fig. 2C, lower panel, lanes 1–3) after tumor passage. No amplification of the endogenous Met^{mu} was found in these tumor explants (Fig. 2A, lanes 5 and 6).

Metastatic Activity of HMH- and Met^{mu}-transformed NIH-3T3 cells

HGF/SF has been shown to induce invasive activity in vitro (Stoker 1989; Rosen et al. 1991; Giordano et al. 1993; Weidner et al. 1993), and histopathological examination of tumors generated by Met^{mu} and HMH cells showed that the tumors invaded adjacent tissue (data not shown). We tested the various met-transformed NIH-3T3 cells for spontaneous metastatic activity in

Table 1. Tumorigenicity of NIH-3T3 Cells Transfected with met^{hu} or met^{mu} cDNA

Transfected genes	Mice with tumors/ no. tested	Latency (weeks)
A		
neo^r	0/19	—
neo^r, met^{hu}	2[a]/41	5
neo^r, met^{mu}	17/17	3–5
B		
neo^r	0/6	—
$neo^r, met^{hu}, HGF/SF^{hu}$ [b]	17/19	4–6
$neo^r, HGF/SF^{hu}$	3/7	7

Cells (10^6) were washed twice with serum-free medium and injected subcutaneously on the back of weanling athymic nude mice. Tumor formation was monitored each week for up to 10 weeks. (Reprinted, with permission, from Rong et al. 1992.)

[a] These two tumors displayed elevated levels of endogenous Met^{mu} and reduced levels of Met^{hu} compared to those of the parental line.

[b] Cells derived from this transfection are referred to as HMH cells.

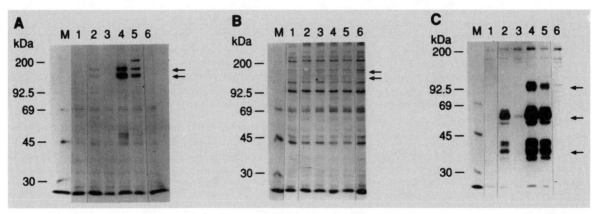

Figure 1. Characterization of Met and HGF/SF in NIH-3T3 tumor explants. Tumor cells were explanted and metabolically labeled with [^{35}S]methionine and [^{35}S]cysteine for 6 hr. Cell lysates were immunoprecipitated with either 19S monoclonal antibody (A) or SP260 peptide antibody (B). A 0.25-ml aliquot of 6-hr supernatants was concentrated threefold in the Centricon apparatus (Amicon; 10K cut-off); the volumes were adjusted to 0.3 ml with RIPA buffer, and the samples were immunoprecipitated with HGF monoclonal antibody A3.1.2 (C) and autoradiographed for 16 hr. Lanes 1 and 3 are samples from two different lines of cotransfected cells before injection. Lane 2 is a tumor explant derived from the cells analyzed in lane 1; lanes 4 and 5 are tumor explants derived from the cells analyzed in lane 3. Lane 6 is a sample prepared from control NIH-3T3 cells. Arrows indicate the positions of p170^{Met} and p140^{Met} (A and B) and the positions of the 87-kD (precursor), 69-kD, and 34-kD HGF polypeptides (C). (Reprinted, with permission, from Rong et al. 1992.)

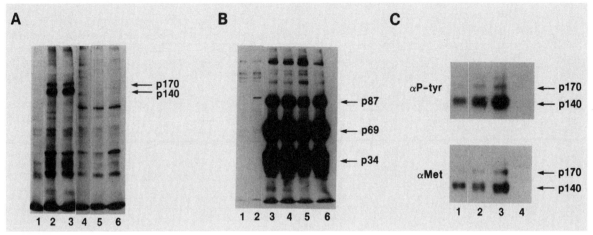

Figure 2. Characterization of Met and HGF/SF in cells of the second tumor passage. (*A*) Immunoprecipitation analysis for Met. Cells were metabolically labeled with [^{35}S]methionine and [^{35}S]cysteine for 6 hr. Cell lysates were immunoprecipitated with C28 peptide antibody for Met^{hu} (lanes *1–3*) or SP260 peptide antibody for Met^{mu} (lanes *4–6*). Lanes *1* and *4*, NIH-3T3 490 cells; lanes *2*, *3*, *5*, and *6*, cells of the second tumor passage. Arrows indicate the positions of p170^{Met} and p140^{Met}. (*B*) Immunoprecipitation analysis for HGF/SF. A 0.5-ml aliquot of medium was concentrated threefold in the Centricon apparatus (Amicon; 10K cut-off); the volumes were adjusted to 0.3 ml with RIPA buffer (lacking reducing agent) to allow efficient immunoprecipitation with HGF monoclonal antibody A3.1.2. (Lane *1*) NIH-3T3 490; (lane *2*) cells cotransfected with *met*^{hu}:HGF/SF^{hu}; (lanes *3–6*) cells of the second tumor passage. Arrows indicate the positions of the 87-kD (precursor), 69-kD, and 34-kD HGF polypeptides. (*C*) Western analysis for Met^{hu}. Near-confluent cells were lysed in lysis buffer and immunoprecipitated with C28 peptide antibody. After being dissolved in SDS-PAGE sample buffer, proteins were separated by SDS-PAGE on a 7.5% acrylamide gel, transferred to immobilon-P (Millipore), and probed with anti-P-Tyr (*upper* panel), or 19S monoclonal antibody (*lower* panel). (Reprinted, with permission, from Rong et al. 1993b.)

nude mice. We observed multifocal lung metastases with the HMH and Met^{mu} primary tumor cell explants (Table 2A). NIH-3T3 cells transformed by activated *ras* or *src* oncogenes are known to generate efficient lung metastasis (Greenberg et al. 1989) and served as positive controls (Table 2A).

We also tested HMH and Met^{mu} cells for metastatic activity after inoculating cells into the mammary fat pad of nude mice. Both cell types induced spontaneous lung metastases (Table 2B), but again, only the 1° and 3° athymic nude mouse tumor explants were active in this assay. The above experiments show that autocrine Met-HGF/SF activation is tumorigenic and metastatic in NIH-3T3 cells (Rong et al. 1994).

Met and HGF/SF Expression in Human Fibroblast Cell Cultures and Sarcoma Cell Lines

We examined human cell lines established from various human sarcomas for Met expression (Table 3) (Rong et al. 1993a). Whereas low levels of Met protein, p140^{Met}, and its precursor, p170^{Met}, were detected in several of the primary fibroblast cultures (Table 3), higher levels were present in most of the human sarcoma cell lines tested. These levels were similar to the high levels of Met protein observed in NIH-3T3 cells transformed by Met^{mu}. The high levels of Met in the sarcoma cell lines and its presence in fibroblast cells represented novel findings, since Met expression was previously thought to be preferentially present in epithelial cells, whereas the ligand was restricted primarily to mesenchymal cells (Gherardi and Stoker 1991; Birchmeier et al. 1993).

Abundant levels of HGF/SF were observed in all of the primary fibroblast cell cultures (Table 3). The level of HGF/SF was also determined by scatter assays performed on growth medium that was conditioned on confluent cells for 72 hours. Again, high levels of scatter activity were detected in the primary fibroblast cultures, but not in any of the sarcoma lines (Table 3). A similar marked reduction in the endogenous HGF/SF was observed in NIH-3T3 cells overexpressing mouse Met (Met^{mu}), which was presumably due to suppression of expression or depletion of the ligand by the receptor (Rong et al. 1993b). Evidence for the HGF/SF receptor activation was indicated by the high reactivity of Met^{mu} with anti-phosphotyrosine (anti-P-Tyr) antibody (Rong et al. 1992). Immunoblot analyses of Met in the sarcoma cell lines showed that Met was highly reactive with anti-P-Tyr antibody, similar to the model autocrine system when Met is expressed in NIH-3T3 cells (Table 3) (Rong et al. 1992). We also found that Met was weakly reactive with anti-P-Tyr antibody in the primary fibroblasts, suggesting that the receptor was activated in an autocrine fashion by endogenous HGF/SF (Rong et al. 1993a).

Autocrine Interaction of Met and HGF/SF in Primary Fibroblast Cells

We tested the possibility that an autocrine Met-HGF/SF stimulatory pathway exists in the primary fibroblast cell cultures (Rong et al. 1993a). Hems and

Table 2. Met-mediated Spontaneous Metastasis in Athymic Mice

Cells	Metastasis/total	Organs	Cells injected (10^5)	Latency (weeks)
(A) Subcutaneous injection				
Neo^r	0/4	—	10	—
Met^{mu}	1/9	lung	10	12
	1/3	lung	3	10
	0/3	—	1	—
Met^{mu} 1° explant	2/6	lung	10	5
	3/3	lung	3	6–7
	2/3	lung/salivary gland/retroperitoneum	1	6
HMH	0/8	—	10	—
	0/3	—	3	—
	0/2	—	1	—
HMH 1° explant	2/7	lung	10	6–9
	1/2	lung	3	9
	2/3	lung/diaphragm/heart	1	5–8
ras	2/3	lung	10	4–5
src	1/3	lung	10	2.5
(B) Mammary fat pad injection				
Neo^r	0/5	—	4	—
Met^{mu}	1/5	lung	4	6
Met^{mu} 1° explant	1/5	lung	4	3–4
HMH	0/5	—	4	—
HMH 1° explant	2/4	lung	4	1–3
HMH 3° explant	3/4	lung	4	3

Cells were washed twice with serum-free medium and either injected subcutaneously (A) or injected into the mammary fat pad (B) of weanling athymic nude mice. When the mice seemed distressed, major organs were analyzed by histopathology for the presence of metastasis. (Reprinted, with permission, from Rong et al. 1994.)

HEL299 cells expressed low levels of Met and high levels of HGF/SF compared to other cell lines tested (Table 3). To test whether the level of Met was being down-modulated by HGF/SF, we added anti-rhHGF, an HGF/SF-neutralizing antibody (Montesano et al. 1991), to the growth medium of HEL299 and Hems cells. After 48 hours, cell lysates were analyzed for levels of Met by immunoblot analysis (Fig. 3). This analysis showed that there was a significant increase in the amount of p140^{Met} in HEL299 cells (Fig. 3, lanes 1–4) and Hems cells (Fig. 3, lanes 5–6) in the presence of the antibody, indicating that Met was down-regulated by extracellular-autocrine activation. The requirement for HGF/SF to be activated by extracellular proteolytic cleavage (Naka et al. 1992; Miyazawa et al. 1993) is consistent with these results.

Expression of Met and HGF/SF in Human Primary Tumors

Elevated expression of Met in sarcoma cell lines compared to primary fibroblast cultures, and our previous demonstration that NIH-3T3 cells overexpressing Met were tumorigenic (Rong et al. 1992), suggested that the high expression of Met contributes to the formation of sarcomas in vivo. We therefore examined paraffin-embedded human sarcoma sections stained for Met and HGF/SF by confocal laser scan microscopy. Several were positive for both Met and HGF/SF staining: One leiomyosarcoma and one of two chondrosarcoma examined expressed both Met and HGF/SF, and three osteosarcomas showed significant Met and HGF/SF staining (Rong et al. 1993a).

Loss of p53 and Met Expression

An unusually high frequency of sarcomas occurs in p53-deficient mice as well as in patients with Li-Fraumeni syndrome (Malkin et al. 1990; Donehower et al. 1992). Since we had shown that Met is inappropriately expressed in mesenchymal tumors, we examined a number of sarcomas from p53-deficient animals (Donehower et al. 1992), as well as several sarcoma specimens from Li-Fraumeni patients. Most of the tumors from p53-deficient mice were high in both Met and HGF/SF staining (data not shown). Moreover, tumor sections from Li-Fraumeni patients had abundant levels of Met (data not shown). We therefore conclude that Met plays an important role in mesenchymal cell transformation and malignancy.

DISCUSSION

Cell lines that express growth factors simultaneously with their cognate receptors elicit tumorigenic behavior by autocrine stimulation (Sporn and Robert 1985; Ullrich and Schlessinger 1990). The transforming activity of epidermal growth factor receptor (EGFR) (Di-Fore et al. 1987; Riedel et al. 1988) and colony-stimulating factor-1 receptor (CSF-1R) (Roussel et al. 1987) in

Table 3. Met and HGF/SF Expression in Human Fibroblast and Sarcoma Cell Lines

	Met[a]	met[b]	HGF/SF[c]	Scatter activity[d]	P-Tyr-Met[e]
Human diploid fibroblast					
HEL299 (fetal lung)	+	+	++++	++++	–
Hems (fetal muscle)	++	+++	+++	+++	+
Hs68 (newborn fibroblast)	++	++	++++	+	–
Malme-3 (skin fibroblast)	+	n.d.[f]	++++	+	n.d.
Fibrosarcoma					
8387	++++	++++	++	+	+++
HT1080	++++	+++	+	+	+++
Hs913T	++++	++	++++	+	++++
SW684	++++	n.d.	++	–	+++
Osteogenic sarcoma					
HOS	+++	+	–	–	+
Saos-2	+++	++++	+	+	++
U-2 OS	+++	+	–	–	++
Chondrosarcoma					
SW1353	++	n.d.	–	n.d.	n.d
Rhabdomyosarcoma					
RD	+++	++	–	+	++
RD-1	++	+	+	+	–
A204	–	n.d.	–	n.d.	–
A673	+	n.d.	–	–	+
Hs729	+++	n.d.	+	+	+++
Leiomyosarcoma					
SK-LMS-1	++++	n.d.	+	+	++++
SK-UT-1B	+	n.d.	–	+	+
Liposarcoma					
SW872	+++	+++	–	n.d.	–
Mesodermal tumor					
SK-UT-1	++++	n.d.	–	–	+++
Synovial sarcoma					
SW982	+	n.d.	+++	–	+
Melanoma					
Malme-3M	+++	n.d.	–	–	+++
WM115	+	n.d.	+	–	+
WM266-4	+	n.d.	–	–	–

(Reprinted, with permission, from Rong et al. 1993a.)

[a] Met protein level was assessed by immunoprecipitation and Western analysis.
[b] met gene expression was detected by Northern analysis.
[c] HGF/SF protein level was assessed by immunoprecipitation analysis.
[d] Scatter activity was assayed with MDCK cells.
[e] P-Tyr-Met was analyzed by Western analysis with anti-P-Tyr.
[f] Not determined.

NIH-3T3 cells has been shown to be factor dependent. NIH-3T3 cells express endogenous mouse *HGF/SF*, which suggests that an autocrine mechanism is involved in the transforming activity of the met^{mu} proto-oncogene (Iyer et al. 1990; Rong et al. 1992). We have also shown that Met^{mu} expressed in NIH-3T3 cells is readily recognized by anti-P-Tyr, suggesting that transformation can occur by autocrine activation of the receptor. In contrast, NIH-3T3 cells expressing Met^{hu} are generally not tumorigenic in athymic nude mice, and Met^{hu} reacts poorly with anti-P-Tyr. However, Met^{hu} is activated when it is coexpressed with HGF/SF^{hu}. One explanation for these results is that Met^{hu} and HGF/SF^{mu} interact with low affinity (Bhargava et al. 1992; Rong et al. 1992; Faletto et al. 1993). This explanation is supported by the results of Bhargava et al. (1992), who found that purified SF^{mu} induces scattering response in the mouse mammary carcinoma cell line EMT6 but does not induce scattering in the human carcinoma cell lines A253, FaDu, and YaOVBix 2NMA. Similar species specificity of HGF/SF has been reported for the activation of Met and its potential substrates in signal transduction studies (Faletto et al. 1993). However, very high levels of ligand could activate the heterologous receptor (Tsarfaty et al. 1994). Furthermore, mouse CSF-1 does not bind to the human CSF-1R with high affinity (Roussel et al. 1988), and even though NIH-3T3 cells synthesize CSF-1, only coexpression of both human CSF-1 and CSF-1R can lead to transformation of NIH-3T3 cells (Roussel et al. 1987).

We have shown that Met^{hu} and HGF/SF^{hu} are both overexpressed in NIH-3T3 cells after tumorigenic selection in nude mice. Quite dramatically, independently derived explants of the second tumor passage expressed approximately the same high levels of HGF/SF^{hu} and Met^{hu}. Tumor explants of the third passage did not show further increase in either ligand or receptor expression compared to the tumor explants of the second passage. This result suggests that there is an optimal

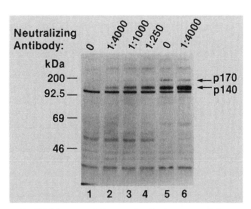

Figure 3. HGF-neutralizing antibody increases Met protein abundance. Primary fibroblast HEL299 (lanes *1–4*) and Hems (lanes *5–6*) cells were incubated for 48 hr with or without HGF-neutralizing antibody (anti-rhHGF), added at 0 hr and 24 hr at the following dilutions: 1:4000 (lanes *2* and *6*); 1:1000 (lane *3*); 1:250 (lane *4*); no antibody (lanes *1* and *5*). After cells were lysed in RIPA buffer, 100 µg of protein was resolved by SDS-PAGE on a 7.5% acrylamide gel and immunoblotted with 19S Met^{hu} monoclonal antibody. (Reprinted, with permission, from Rong et al. 1993a.)

level of expression of receptor and ligand for tumorigenicity.

One possible explanation for the high level of Met^{hu} and HGF/SF^{hu} expression is that the human receptor is inefficient in signaling in mouse cells. High levels of Met would be required for tumorigenesis and, correspondingly, high levels of HGF/SF^{hu} would be required to mediate the autocrine signal. Because HGF/SF^{hu} is also tumorigenic in this assay (Rong et al. 1992), it is possible that the secretion of the ligand contributes to tumorigenesis, e.g., by enhancing neovascularization (Rong et al. 1994).

Stoker (1989) first proposed that HGF/SF is involved in tumor invasion and metastasis, since the ligand elicits scattering and motility in epithelial cells in vitro. HGF/SF and other cell motility factors are postulated to be involved in inflammatory reactions, tissue repair, and interactions in the immune system (Gherardi and Stoker 1991). Moreover, HGF/SF induces blood vessel formation and contributes to tumor angiogenesis in vivo (Bussolino et al. 1992; Grant et al. 1993) and stimulates endothelial cell migration and capillary-like tube formations in vitro (Rosen et al. 1991). Indeed, we have observed abundant neovascularization in Met tumors (data not shown). The inappropriate or ectopic expression of Met in NIH-3T3 cells leads to tumorigenesis through an autocrine mechanism (Rong et al. 1992, 1993b). We have shown that the same autocrine interaction in NIH-3T3 cells confers the phenotypes of invasiveness in vitro and experimental and spontaneous metastases in vivo (Rong et al. 1994). Abundant Met expression was found in human sarcoma cell lines and tumors (Rong et al. 1993a), raising the possibility that similar mechanisms of tumorigenesis function in these human malignancies.

HGF/SF serves as a paracrine effector of mesenchymal-epithelial interactions (Gherardi and Stoker 1991; Birchmeier et al. 1993). When epithelial cells scatter in vitro, they resemble fibroblasts in culture (Stoker 1989). HGF/SF has been detected in various carcinoma cell lines (Yoshinaga et al. 1992; Tsao et al. 1993); furthermore, epithelial carcinoma cell lines grow as single cells when expressing both Met and HGF/SF, but as colonies when HGF/SF signaling is blocked (Adams et al. 1991). Moreover, we recently demonstrated that autocrine Met-HGF/SF expression can mediate mesenchymal to epithelial cell transition (Tsarfaty et al. 1994). These observations suggest that alternating between high and low levels of Met and/or HGF/SF expression could provide the phenotypes of migration and colonization necessary for metastasis.

Mice deficient in p53, or patients with Li-Fraumeni syndrome, frequently develop sarcomas (Malkin et al. 1990; Donehower et al. 1992). We find that elevated Met expression is associated with these tumors in both the animal model system and the human syndrome. Based on these findings, we propose that Met-HGF/SF autocrine signaling contributes to the tumorigenic process in human sarcomas.

Spontaneous transformants of NIH-3T3 fibroblasts often have the *met* proto-oncogene amplified and overexpressed (Cooper et al. 1986; Hudziak et al. 1990). Furthermore, Met overexpression in NIH-3T3 fibroblasts induces fibrosarcomas by an autocrine mechanism (Rong et al. 1992, 1993b). The same autocrine mechanism induces an invasive and metastatic phenotype of NIH-3T3 cells (Rong et al. 1994). The fact that Met and HGF/SF are abundantly expressed in various types of human sarcomas suggests that this ligand-receptor pair contributes to human malignancy the same way as in the mouse model systems.

ACKNOWLEDGMENTS

We thank Anne Arthur for editing and Lori Summers for preparation of the manuscript. Research was sponsored by the National Cancer Institute, DHHS, under contract NO1-CO-74101 with ABL. The contents of this publication do not necessarily reflect the views or policies of the Department of Health and Human Services, nor does mention of trade names, commercial products, or organizations imply endorsement by the U.S. Government.

REFERENCES

Adams, J.C., R.A. Fulong, and F. Watt. 1991. Production of scatter factor by ndk, a strain of epithelial cells, and inhibition of scatter factor activity by suramin. *J. Cell Sci.* **96:** 385.

Bhargava, M., A. Joseph, J. Knesel, R. Halaban, Y. Li, S. Pan, I. Goldberg, E. Setter, M.A. Donovan, R. Zaraegar, G.A. Michalopoulos, T. Nakamura, D. Faletto, and E.M. Rosen. 1992. Scatter factor and hepatocyte growth factor: Activities, properties and mechanisms. *Cell Growth Differ.* **3:** 11.

Birchmeier, C., E. Sonnenberg, K.M. Weidner, and W. Birchmeier. 1993. Tyrosine kinase receptors in the control of epithelial growth and morphogenesis during development. *BioEssays* **15:** 1.

Blair, D.G., C.S. Cooper, M.K. Oskarsson, L.A. Eader, and G.F. Vande Woude. 1982. New method for detecting cellular transforming genes. *Science* **218:** 1122.

Bottaro, D.P., J.S. Rubin, D.L. Faletto, A.M.-L. Chan, T.E. Kmiecik, G.F. Vande Woude, and S.A. Aaronson. 1991. The hepatocyte growth factor receptor is the c-*met* protooncogene product. *Science* **152:** 802.

Bussolino, F., M.F. DiRenzo, M. Ziche, E. Bacchietto, M. Olivero, L. Naldini, G. Gaudino, L. Tamagnone, A. Coffer, and P.M. Comoglio. 1992. Hepatocyte growth factor is a potent angiogenic factor which stimulates endothelial cell motility and growth. *J. Cell Biol.* **119:** 628.

Chan, A.M-L., H.W.S. King, E.A. Peakin, P.R. Tempest, J. Hilkens, V. Kroozen, D.R. Edwards, A.J. Wills, C.S. Cooper, and P. Brooke. 1988. Characterization of the mouse *met* protooncogene. *Oncogene* **2:** 593.

Cooper, C.S., P.R. Tempest, P.M. Beckman, C.-H. Helden, and P. Breakers. 1986. Amplification and overexpression of *met* gene in spontaneously transformed NIH/3T3 mouse fibroblasts. *EMBO J.* **5:** 2623.

DiFore, P.P., J.H. Pierce, T.P. Fleming, R. Hazan, A. Ullrich, C.R. King, J. Schlessinger, and S.A. Aaronson. 1987. Overexpression of the human EGF receptor confers an EGF-dependent transformed phenotype to NIH/3T3 cells. *Cell* **51:** 1063.

DiRenzo, M.F., R.P. Narsimhan, M. Olivero, S. Bretti, S. Giordano, E. Medico, P. Gaglia, P. Zara, and P.M. Comoglio. 1991. Expression of the Met/HGF receptor in normal and neoplastic human tissues. *Oncogene* **6:** 1997.

Donehower, L.A., M. Harvey, B.L. Slagle, M. McArthur, C.A. Montgomery, J.S. Butel, and A. Bradley. 1992. Mice deficient for p53 are developmentally normal but susceptible to spontaneous tumors. *Nature* **356:** 215.

Faletto, D.L., D.R. Kaplan, D.O. Halverson, E.R. Rosen, and G.F. Vande Woude. 1993. Signal transduction in c-*met* mediated mutagenesis. In *Hepatocyte growth factor-scatter factor (HGF/SF) and the c-Met receptor* (ed. I.D. Goldberg and E.M. Rosen), p. 107. Birkhauser Verlag, Basel, Switzerland.

Faletto, D.L., I. Tsarfaty, T.E. Kmiecik, M. Gonzatti, T. Suzuki, and G.F. Vande Woude. 1991. Evidence for noncovalent clusters of the c-*met* proto-oncogene product. *Oncogene* **7:** 1149.

Furlong, R.A., T. Takehara, W.G. Taylor, T. Nakamura, and J.S. Rubin. 1991. Comparison of biological and immunochemical properties indicates that scatter factor and hepatocyte growth factor are indistinguishable. *J. Cell Sci.* **100:** 173.

Gherardi, E. and M. Stoker. 1990. Hepatocyte and scatter factor. *Nature* **346:** 228.

———. 1991. Hepatocyte growth factor-scatter factor: Mitogen, motogen, and Met. *Cancer Cells* **3:** 227.

Giordano, S., Z. Zhen, E. Medico, and G. Gaudino. 1993. Transfer of motogenic and invasive response to scatter factor/hepatocyte growth factor by transfection of human MET protooncogene. *Proc. Natl. Acad. Sci.* **90:** 649.

Giordano, S., M.F. DiRenzo, R. Ferracini, L. Chiado-Plat, and P.M. Comoglio. 1988. p145, a protein with associated tyrosine kinase activity in a human gastric carcinoma cell line. *Mol. Cell. Biol.* **8:** 3510.

Gonzatti-Haces, M., A. Seth, M. Park, T. Copeland, S. Oroszlan, and G.F. Vande Woude. 1988. Characterization of the *tpr-met* oncogene p65 and the *met* protooncogene p140 protein-tyrosine kinase. *Proc. Natl. Acad. Sci.* **85:** 21.

Grant, D.S., H.K. Kleinman, I.D. Goldberg, M.M. Bhargava, B.J. Nickoloff, J.L. Kinsella, P. Polverini, and E.M. Rosen. 1993. Scatter factor induces blood vessel formation *in vivo*. *Proc. Natl. Acad. Sci.* **90:** 1937.

Greenberg, A.H., S.E. Egan, and J.A. Wright. 1989. Oncogenes and metastatic progression. *Invasion Metastasis* **9:** 360.

Hudziak, R., G.D. Lewis, W.E. Holmes, A. Ullrich, and H.M. Shepard. 1990. Selection for transformation and *met* protooncogene amplification in NIH/3T3 fibroblasts using tumor necrosis factor α. *Cell Growth Diff.* **1:** 129.

Iyer, A., T.E. Kmiecik, M. Park, I. Daar, P. Blair, K.J. Dunn, P. Sutrave, J.N. Ihle, M. Bodescot, and G.F. Vande Woude. 1990. Structure, tissue-specific expression, and transforming activity of the mouse *met* protooncogene. *Cell Growth Differ.* **1:** 87.

Konishi, T., T. Takehara, T. Tsuji, K. Ohsato, K. Matsumoto, and T. Nakamura. 1991. Scatter factor from human embryonic lung fibroblasts is probably identical to hepatocyte growth factor. *Biochem. Biophys. Res. Commun.* **180:** 765.

Malkin, D., F.P. Li, L.C. Strong, J.F. Fraumeni, C.E. Nelson, D.H. Kim, J. Kassel, M.A. Gryka, F.Z. Bischoff, M.A. Tainsky, and S.H. Friend. 1990. Germ line p53 mutations in a familial syndrome of breast cancer, sarcomas, and other neoplasms. *Science* **250:** 1233.

Miyazawa, K., T. Shimomura, A. Kitamura, and J. Kondo. 1993. Molecular cloning and sequence analysis of the cDNA for a human serine protease responsible for activation of hepatocyte growth factor. Structural similarity of the protease precursor to blood coagulation factor XII. *J. Biol. Chem.* **268:** 10024.

Morrison, D.K., D.R. Kaplan, J.A. Escobedo, J.R. Rapp, T.M. Roberts, and L.T. Williams. 1989. Direct activation of the serine/threonine kinase activity of *Raf*-1 through tyrosine phosphorylation by the PDGF-β receptor. *Cell* **58:** 649.

Montesano, R., K. Matsumoto, T. Nakamura, and L. Orci. 1991. Identification of a fibroblast-derived epithelial morphogen as hepatocyte growth factor. *Cell* **67:** 901.

Naka, D., T. Ishii, Y. Yoshiyama, and K. Miyazawa. 1992. Activation of hepatocyte growth factor by proteolytic conversion of a single chain form to a heterodimer. *J. Biol. Chem.* **267:** 20114.

Nakamura, T. 1992. Structure and function of hepatocyte growth factor. *Prog. Growth Factor Res.* **3:** 67.

Nakamura, T., K. Nawa, A. Ichihara, A. Kaise, and T. Nishino. 1987. Subunit structure of hepatocyte growth factor from rat platelets. *FEBS Lett.* **224:** 311.

Naldini, L., E. Vigna, R.P. Narsimhan, G. Gandino, R. Zarnegar, G.K. Michalopoulos, and P.M. Comoglio. 1991. Hepatocyte growth factor (HGF) stimulates the tyrosine kinase activity of the receptor encoded by the protooncogene c-*met*. *Oncogene* **6:** 501.

Riedel, H., S. Massozlia, J. Schlessinger, and A. Ullrich. 1988. Ligand activation of overexpressed epidermal growth factor receptors transforms NIH/3T3 mouse fibroblasts. *Proc. Natl. Acad. Sci.* **85:** 1477.

Rong, S., S. Segal, M. Anver, J. Resau, and G.F. Vande Woude. 1994. Invasiveness and metastasis of NIH/3T3 cells induced by Met-hepatocyte growth factor/scatter factor autocrine stimulation. *Proc. Natl. Acad. Sci.* **91:** 4731.

Rong, S., M. Jeffers, J. Resau, I. Tsarfaty, M. Oskarsson, and G.F. Vande Woude. 1993a. Met expression and sarcoma tumorigenicity. *Cancer Res.* **53:** 5355.

Rong, S., M. Bodescot, D. Blair, T. Nakamura, K. Mizuno, M. Park, A. Chan, S. Aaronson, and G.F. Vande Woude. 1992. Tumorigenicity of the *met* protooncogene and the gene for hepatocyte growth factor. *Mol. Cell. Biol.* **12:** 5152.

Rong, S., M. Oskarsson, D.L. Faletto, I. Tsarfaty, J. Resau, T. Nakamura, E. Rosen, R. Hopkins, and G.F. Vande Woude. 1993b. Tumorigenesis induced by coexpression of human hepatocyte growth factor and the human *met* protooncogene leads to high levels of expression of the ligand and receptor. *Cell Growth Differ.* **4:** 563.

Rosen, E.M., I.D. Goldberg, B.M. Kacinski, T. Buckholz, and D.W. Vinter. 1989. Smooth muscle releases an epithelial scatter factor which bids to heparin. *In Vitro Cell Dev. Biol.* **25:** 163.

Rosen, E.M., O. Grant, H. Kleinman, S. Jaken, M.A. Donovan, E. Setter, P.M. Luckett, and W. Carley. 1991. Scatter factor stimulates migration of vascular endothelium and capillary-like tube formation. In *Cell motility factors* (ed. I.D. Goldberg and E.M. Rosen), p. 76. Birkhauser Verlag, Basel, Switzerland.

Roussel, M.F., J.R. Downing, C.W. Rettenmier, and C.J. Sherr. 1988. A point mutation in the extracellular domain of the human CSF-1 receptor (c-*fms* protooncogene product) activates its transforming potential. *Cell* **55:** 979.

Roussel, M.F., T.J. Dull, C.W. Rettenmier, P. Ralph, A. Ullrich, and C.J. Sherr. 1987. Transforming potential of the c-*fms* protooncogene (CSF-1 receptor). *Nature* **325:** 549.

Sporn, M.B. and A.B. Robert. 1985. Autocrine growth factors and cancer. *Nature* **313:** 745.

Stoker, M. 1989. Effect of scatter factor on motility of epithelial cells and fibroblasts. *J. Cell. Physiol.* **139:** 565.

Stoker, M. and M. Perryman. 1985. An epithelial scatter factor released by embryo fibroblasts. *J. Cell Sci.* **77:** 209.

Tsao, M.-S., H. Zhu, A. Giaid, I. Viallet, T. Nakamura, and M. Park. 1993. Hepatocyte growth factor/scatter factor is an autocrine factor for human normal bronchial epithelial and lung carcinoma cells. *Cell Growth Differ.* **4:** 571.

Tsarfaty, I., S. Rong, J.H. Resau, S. Rulong, P. Pinta da Silva, and G.F. Vande Woude. 1994. Met mediated signaling in mesenchymal to epithelial cell conversion. *Science* **263:** 98.

Ullrich, A. and J. Schlessinger. 1990. Signal transduction by receptors with tyrosine kinase activity. *Cell* **61:** 203.

Weidner, K.M., G. Hartmann, L. Naldini, P.M. Comoglio, M. Sachs, C. Fonatsch, H. Rieder, and W. Birchmeier. 1993. Molecular characteristics of HGF-SF and its role in cell motility and invasion. In *Hepatocyte growth factor-scatter factor (HGF/SF) and the c-Met receptor* (ed. I.D. Goldberg and E.M. Rosen), p. 311. Birkhauser Verlag, Basel, Switzerland.

Weidner, K.M., N. Arakaki, G. Harmann, J. Vanderkerckhove, S. Weingart, H. Rieder, C. Fonatsch, H. Tsubouchi, T. Hishida, Y. Daikuhara, and W. Birchmeier. 1991. Evidence for the identity of human scatter factor and human hepatocyte growth factor. *Proc. Natl. Acad. Sci.* **88:** 7001.

Yoshinaga, Y., S. Fukita, M. Gotoh, T. Nakamura, M. Kikuchi, and S. Hiroashi. 1992. Human lung cancer cell line producing hepatocyte growth factor/scatter factor. *Jpn. J. Cancer Res.* **83:** 1257.

Studies of the Deleted in Colorectal Cancer Gene in Normal and Neoplastic Tissues

E.R. Fearon,* B.C. Ekstrand,*‡ G. Hu,*‡ W.E. Pierceall,*
M.A. Reale,§ and S.H. Bigner†

*Departments of *Pathology, ‡Pharmacology, and §Medicine, Yale University School of Medicine, New Haven, Connecticut 06536-0812;*
†Department of Pathology, Duke University School of Medicine, Durham, North Carolina 27710

A current view is that cancer results from the accumulation of multiple mutations in a single cell. These mutations include gain of function mutations in proto-oncogenes and loss of function mutations in tumor suppressor genes and DNA repair genes. A large number of oncogenic alleles in human tumors have been identified, perhaps because direct strategies for identifying them have been possible. In contrast, a rather limited number of tumor suppressor genes have been identified at the molecular level thus far (Weinberg 1991). Although the mutations in and functions of several tumor suppressor genes, including the retinoblastoma, *p53*, and adenomatous polyposis coli (*APC*) genes, are becoming increasingly well-characterized, mutational analyses and functional studies on other established and candidate tumor suppressor genes are still in rather preliminary stages. Moreover, the existence of many additional and as yet unidentified tumor suppressor genes is based on rather indirect and limited evidence, including the frequent loss of specific chromosomal regions in various human tumor types. These losses can be detected using either cytogenetic techniques or recombinant DNA-based methods. Usually the losses involve one of the two parental chromosomal sets, or alleles, present in normal cells. Thus, the loss is termed an allelic loss or loss of heterozygosity (LOH). It has been well established that some allelic losses can result in the inactivation of a tumor suppressor gene in the affected chromosomal region; the remaining tumor suppressor allele often harbors a more localized mutation inactivating its function (Weinberg 1991).

LOH affecting chromosome 18q can be detected in more than 70% of primary colorectal cancers, in about 50% of advanced adenomas, and infrequently in earlier stage adenomas (Vogelstein et al. 1988). Patients whose colorectal cancers have 18q LOH have an increased likelihood of distant metastasis and death from their disease (Kern et al. 1989; Jen et al. 1994). In addition, the prevalence of 18q LOH rises to nearly 100% in colorectal carcinoma metastases to the liver (Ookawa et al. 1993). In an effort to localize the 18q tumor suppressor gene, a common region of LOH on 18q was first identified (Fearon et al. 1990). Subsequently, markers from this region were studied in an effort to detect more localized alterations in colorectal cancers. Two cases were found to have somatic mutations involving or immediately flanking an anonymous marker (*p15-65*) from the region. One tumor had a homozygous loss that affected *p15-65*, but that did not involve any other 18q markers studied. A second tumor had a point mutation approximately 5 kb from the *p15-65* marker that appeared to generate a novel splice acceptor site. To identify a transcription unit from this region of chromosome 18q that might be targeted for inactivation by 18q LOH and affected by the somatic mutations in these two cases, approximately 380 kb of contiguous DNA sequences were obtained from a bidirectional chromosome walk from the *p15-65* marker. A large transcriptional unit spanning more than 1350 kb was then identified and cDNAs for this gene, termed *DCC* (for deleted in colorectal cancer), were obtained from human fetal brain cDNA libraries (Fearon et al. 1990; Cho et al. 1994). The *DCC* gene encodes a transmembrane protein of the immunoglobulin superfamily, and recent immunohistochemical and in situ hybridization studies have provided evidence that *DCC* may be expressed in subsets of cells within the nervous system and the colon (Hedrick et al. 1994). In this paper, we describe some of our recent studies of the structure and expression of the *DCC* gene in normal and neoplastic tissues.

EXPERIMENTAL PROCEDURES

Reverse transcription-polymerase chain reaction (RT-PCR) assay of DCC gene expression. Total RNA was prepared using the acid guanidium thiocyanate method (Chomczynski and Sacchi 1987). Random hexamers and the BRL Superscript Preamplification Kit (Gibco/BRL Life Technologies, Grand Island, New York) were used to prepare first strand cDNA. The cDNA was then amplified by PCR using primer pairs derived from the human DCC cDNA sequence (primer location is indicated relative to the initiating ATG at bp 604). Sense primer (for both extracellular pairs a and b) - DCK2834S - 5'-CCCAGACTAAC TGCATCATCATGAG-3', antisense for pair a - DCK3088A - 5'-CGAGGTGGGGAAATCATCAA GCA-3', antisense for pair b - DCK3 151A - 5'-CA CCTACTGGTGGGAGCAT-3'. The RT-PCR products were electrophoresed on 1.2% agarose gels and visualized with UV light following ethidium bromide staining.

Ribonuclease (RNase) protection assay. The RNase protection assay was carried out essentially as described previously (Pierceall et al. 1994). In brief, 20 µg of total RNA from each sample was incubated overnight at 48°C with 5.0×10^4 cpm of a 480-bp antisense *DCC* riboprobe and 5.0×10^4 cpm of a 273-bp γ-actin antisense riboprobe. The samples were then treated with RNase T2, precipitated with isopropanol, and run on a sequencing gel. Autoradiography was carried out with Hyperfilm (Amersham).

Preparation of tissue and cell line protein lysates. Cell lines and tissue homogenates were solubilized in Tris-buffered saline (TBS) (25 mM Tris [hydroxymethylaminomethane], pH 8) with detergents (1% deoxycholate 1% NP-40, 0.1% SDS) and protease inhibitors (antipain 50 µg/ml, aprotinin 5 µg/ml, leupeptin 2 µg/ml, PMSF 100 µg/ml, EDTA 1 mM). Protein concentrations were determined with the BCA protein assay (Pierce, Rockford, Illinois).

Western blot and immunoprecipitation analyses of DCC expression. Cell line/tissue lysates or lysates that had been first immunoprecipitated underwent SDS-PAGE on 8% polyacrylamide gels and Transblot semi-dry transfer (Biorad, Hercules, California) to Immobilon (Millipore, Bedford, Massachusetts) membranes. The primary DCC-specific antisera were used at 10–100 ng/ml, and the secondary goat-anti-rabbit or goat-anti-mouse antiserum coupled to horseradish peroxidase (Pierce, Rockford, Illinois) was used at a 1:20,000 dilution. The antigen-antibody complex was detected by enhanced chemiluminescence (ECL) (Amersham) and subsequent exposure to Hyperfilm (Amersham). For immunoprecipitation, lysates were precleared by incubation with purified rabbit immunoglobulin and protein A–Sepharose (Immunopure Plus, Pierce, Rockford, Illinois). The supernatant was then incubated with the purified DCC-specific antisera (10 µg/ml). DCC-specific immunoprecipitates were recovered following incubation with protein A–Sepharose. Following washing, the immunoprecipitates were resuspended in Laemmli's sample buffer and subjected to the ECL-immunoblot assay described above.

RESULTS AND DISCUSSION

Structure and Sequence of the *DCC* Gene

On the basis of Northern blot studies of adult human brain tissues, the *DCC* gene appears to encode a transcript of approximately 12 kb (Fearon et al. 1990). cDNA clones spanning more than 6 kb and predicted to encode a 1447-amino-acid protein with similarity to the neural cell adhesion molecule (N-CAM) family of cell-surface molecules have been isolated from human fetal brain libraries (Fearon et al. 1990; Hedrick et al. 1994; Reale et al. 1994). The transmembrane protein en-

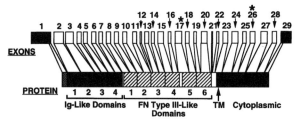

Figure 1. *DCC* gene exon map and predicted protein product. The 29 exons encoding the 4341-bp open reading frame of *DCC* are indicated, with their size proportional to the amount of cDNA sequence encoded. Analysis of the 5' and 3' untranslated regions of the *DCC* cDNA has not been completed, and thus exons 1 and 29 are indicated by filled boxes. Exons 17 and 26 are affected by alternative splicing (noted by *). The *DCC* protein is predicted to have 4 immunoglobulin-like (Ig-like) domains, 6 fibronectin (FN) type III-like domains, a transmembrane (TM) region, and ~325 amino acids of cytoplasmic sequence.

coded by the *DCC* gene is predicted to contain four immunoglobulin (Ig)-like and six fibronectin (FN) type III-like domains in the extracellular region and approximately 325 amino acids in the cytoplasmic domain (Fig. 1). There are 29 exons for the coding region, and sequences of intron-exon boundaries for all exons have been identified (Cho et al. 1994). Yeast artificial chromosome (YAC) clones spanning the *DCC* gene have been obtained, and the YAC contig demonstrated that the *DCC* gene spans more than 1350 kb (Fig. 2).

The sequence of the entire *DCC* open reading frame has been confirmed by sequencing of cDNAs obtained from the IMR32 human neuroblastoma cell line (Reale et al. 1994). In *DCC* transcripts from IMR32, one predicted amino acid substitution in exon 19 (Phe to Leu at codon 951) was observed; this substitution is likely to be a polymorphism. In addition, two alterative splice sites were identified: One is an alternative splice acceptor site located within exon 17 and the other is an alternative splice donor site at the 3' end of exon 26. Use of the alternative splice acceptor within exon 17 results in the loss of 20 amino acids between the fourth and fifth FN type III-like domains of DCC (Reale et al. 1994). Alternative splicing of N-CAM transcripts in sequences encoding FN type III-like domains has been noted previously, and, as we observed for the DCC extracellular domain sequences (see below), the alternative splicing occurs in a tissue-specific fashion (Cunningham et al. 1987; Dickson et al. 1987; Edelman and Crossin 1991; Hamshere et al. 1991). In particular, a muscle-specific domain (MSD) exon lies between N-CAM exons 12 and 13 and encodes an additional 37 amino acids that lie amid FN type III-like domains. This region is known to undergo significant O-linked glycosylation (Walsh et al. 1989), and it has been speculated to introduce a "hinge" into the molecule (Dickson et al. 1987; Edelman and Crossin 1991). The 20-codon region in the *DCC* extracellular domain sequences that is alternatively spliced is not similar at the amino acid level to the predicted product of the MSD

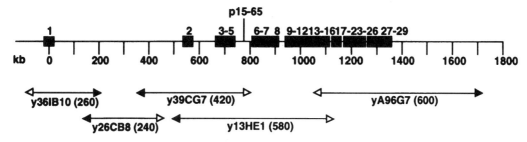

Figure 2. Genomic organization of the *DCC* gene. The relative positions of the 29 *DCC* exons, the *p15-65* marker, and five YACs spanning the *DCC* locus are indicated. The scale is shown in kb, and the size of each YAC in kb is indicated. Note that the boxes for the *DCC* exons are not drawn to scale.

exon, but the similarity in their locations might reflect some functional similarity.

The other alternative splice noted (exon 26) affects six nucleotides in the *DCC* cytoplasmic domain sequences. *DCC* transcripts lacking these sequences have a loss of two codons and an amino acid substitution at a third codon, with a resultant loss of a potential serine phosphorylation site for casein kinase II. Of interest, transcripts for the N-CAM family member L1 are affected by a 12-bp deletion in the cytoplasmic domain, with resultant loss of a potential casein kinase II phosphorylation site (Miura et al. 1991). Phosphorylation of the cytoplasmic domain of N-CAM and L1 has been demonstrated, and in all cases, it has been serine phosphorylation (Sorkin et al. 1984; Sadoul et al. 1989). Given the abundance of potential phosphorylation sites in the DCC cytoplasmic domain (e.g., 10 casein kinase II consensus sites) and evidence that alternative splicing affects one of these sites, it is possible that phosphorylation may play a role in the biological function of the DCC cytoplasmic domain.

DCC Gene and Protein Expression in Adult Tissues

We have used RT-PCR-based assays to characterize the relative levels of *DCC* transcripts in adult mouse tissues/organs, as well as to identify tissue-specific differences in splicing. As has been observed previously in studies of rat and human tissues (Fearon et al. 1990; Nigro et al. 1991), *DCC* transcripts containing extracellular domain sequences are present at low levels in many adult tissues. The alternatively spliced transcript derived from use of the internal splice acceptor in exon 17, and thus lacking 60 nucleotides (20 codons), was only detected in bladder tissue and the IMR32 neuroblastoma cell line (Fig. 3). The expression of transcripts containing DCC cytoplasmic domain sequences and the pattern of alternative splicing for exon 26 have also been studied. Transcripts containing cytoplasmic domain sequences were found to be present in all adult mouse tissues studied (data not shown). Our preliminary results suggest that although both alternatively spliced forms are present, transcripts containing the 6

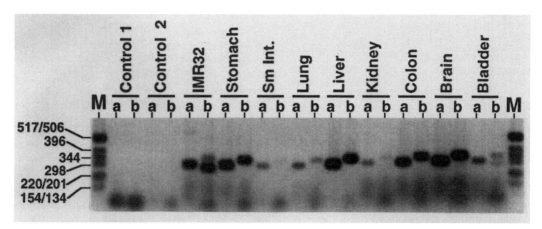

Figure 3. RT-PCR analysis of *DCC* gene expression in adult mouse tissues. First strand cDNA was prepared from total RNA using random hexamers and reverse transcriptase. PCR was carried out with oligomers derived from the human *DCC* sequence (see Experimental Procedures). The RT-PCR products were then electrophoresed on agarose gels and visualized by UV light following ethidium bromide staining. (Control 1) No template; (Control 2) nonspecific RNA control; the RNA source is indicated for the remainder of the lanes. IMR32 is a human neuroblastoma cell line. All tissues were obtained from an adult mouse. Size markers (M) in bp are indicated. Oligomer pair "a" detects a 301-bp product in IMR32 and all mouse tissues; oligomer pair "b" detects a 341-bp product in all mouse tissues, as well as a 281-bp alternatively spliced product in IMR32 (lane *3b*) and mouse bladder (lane *11b*).

additional nucleotides at the 3' end of exon 26 are more abundant than those lacking these sequences (data not shown).

Polyclonal rabbit antisera reactive with the DCC protein have been generated against recombinant bacterially produced DCC proteins as well as three DCC peptides (Reale et al. 1994). These antisera have been used to characterize DCC expression in normal mouse and rat tissues, using ECL-immunoblotting and a combined immunoprecipitation and ECL-immunoblotting strategy. Among the tissues studied were brain, lung, stomach, small intestine, colon, liver, kidney, and bladder. We were able to detect DCC expression in brain and bladder (Fig. 4). DCC expression was below the level of detection in all other rodent tissues studied, although we have detected expression in *Xenopus* stomach, lung, and kidney by the combined immunoprecipitation/immunoblotting strategy (data not shown).

DCC: A Candidate Tumor Suppressor Gene in Multiple Tumor Types

Whereas 18q LOH affects the *DCC* gene in more than 70% of colorectal cancers, the remaining *DCC* allele has been shown to be affected by localized somatic mutations in only a subset of cases. Approximately 10–15% of cases have insertions in a microsatellite sequence located downstream from one of the exons (Fearon et al. 1990). These insertions might affect *DCC* transcription or processing of transcripts; however, definitive functional support for this proposal has not yet been obtained. In addition to the one case (described above) with a point mutation detected by the *p15-65* probe, two other cases have been found to have somatic point mutations—one in *DCC* exon 28 and one in intron 13 (Cho et al. 1994). In the majority of colorectal tumor cell lines, tumor xenografts, and primary tumor specimens, localized mutations in *DCC* have not been identified; however, given the extremely large size of the gene (>1350 kb), only a small subset of its sequences have been studied. Expression of *DCC* transcripts has consistently been found to be greatly reduced or absent in 60% to 85% of these samples (Fearon et al. 1990; Kikuchi-Yanoshita et al. 1992). Although the mechanisms underlying the loss of *DCC* expression have not been elucidated in the majority of colorectal cancers, the demonstrated loss of gene expression and somatic mutations designate the *DCC* gene as a promising candidate tumor suppressor gene from 18q.

Inactivation of the *DCC* gene may also occur in several other tumor types, including cancers of the stomach, pancreas, endometrium, breast, prostate, and esophagus, as well as in some leukemias. LOH affecting chromosome 18q and the *DCC* locus has been demonstrated in 61% of gastric cancers, 35–70% of breast cancers, 45% of prostate cancers, and 24% of esophageal cancers (Cropp et al. 1990; Devilee et al. 1991; Huan et al. 1992; Uchino et al. 1992; Gao et al. 1993; Thompson et al. 1993). Using sensitive RT-PCR assays, decreased or absent *DCC* gene expression has also been shown in 63% and 86% of pancreatic and prostatic cancers, respectively (Hohne et al. 1992; Gao et al. 1993). Whereas *DCC* expression has been detected in 14 of 34 primary breast cancers using Northern blot analysis, aberrantly sized transcripts were noted in 5 of these 14 cases, and the remaining 20 cases did not express *DCC* at detectable levels (Thompson et al. 1993). Moreover, two independent reports have provided data suggesting that *DCC* may also be inactivated in some leukemias. Porfiri et al. (1993) studied bone marrow specimens from 4 patients whose leukemia

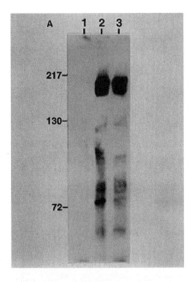
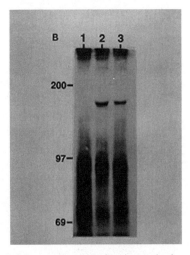

Figure 4. DCC protein expression in rat brain and mouse bladder tissues. Combined immunoprecipitation and enhanced chemiluminesence (ECL)-immunoblot analysis of DCC expression in rat brain (*A*) and mouse bladder (*B*). Lysates from the tissues were immunoprecipitated with control antisera (lane *1*) or two different DCC-specific polyclonal rabbit antisera (lanes *2* and *3*). Immunoprecipitates were then analyzed by an ECL-immunoblot assay using a third DCC-specific polyclonal rabbit antisera. The DCC protein has a molecular weight of about 175,000–190,000. The relative mobility of molecular weight markers ($\times 10^3$) is indicated.

cells were monosomic for chromosome 18. Using RT-PCR and four different primer pair sets, *DCC* expression was seen in control specimens, such as normal tonsillar tissue and several leukemic bone marrow specimens without 18q abnormalities, but no *DCC* expression was seen in the monosomy 18 leukemia specimens. Miyake et al. (1993) have also studied *DCC* expression in bone marrow specimens from 24 patients with myelodysplastic syndrome (MDS) and 7 overt leukemias arising in patients with prior MDS. Using an RT-PCR assay, they found *DCC* expression was absent or extremely reduced in 2 of the 24 MDS specimens; these 2 patients developed leukemia within 6 months. Moreover, in 5 of the 7 cases of overt leukemia arising from MDS, *DCC* expression was absent or greatly reduced. These data suggest that inactivation of *DCC* may also contribute to the development of some leukemias.

To begin to address the role of the *DCC* gene in the pathogenesis of other tumor types, we have undertaken studies of *DCC* gene and protein expression in a panel of 30 human brain tumor xenografts, the majority of which were derived from glioblastomas. Our preliminary findings, using a ribonuclease protection assay, suggest that *DCC* transcripts can be detected in nearly two-thirds of cases (examples shown in Fig. 5). Studies of DCC protein expression in these tumors using an ECL-immunoblotting approach have demonstrated that less than 50% of the samples synthesize detectable levels of DCC protein (Fig. 6 and data not shown). Moreover, several tumor samples that synthesize *DCC* transcripts fail to synthesize DCC protein (e.g., specimens 317 and 617). Sequence analysis of *DCC* cDNAs should be useful for determining whether the absence of detectable levels of DCC protein in cases with *DCC* transcripts results from mutations (e.g., missense or nonsense mutations) in the *DCC* coding sequences.

We have also analyzed DCC protein expression in 37

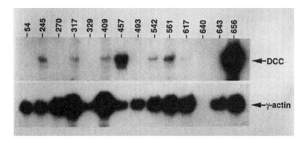

Figure 5. *DCC* gene expression in human brain tumor xenograft specimens. The results of a RNase protection assay with antisense *DCC* and γ-actin riboprobes are shown. Approximately 25 µg of total RNA from each sample was incubated with both probes simultaneously. Following treatment with RNase T2, the samples were electrophoresed on sequencing gels and autoradiography was carried out. The RNA in sample 640 was degraded, as judged by the absence of a protected fragment with the γ-actin probe. A specific *DCC*-protected fragment was seen in samples 245, 317, 409, 457, 542, 561, 617, and 656. The exposure time for the *DCC*-protected fragments was 10 days and the exposure time for the γ-actin-protected fragments was 4 hr.

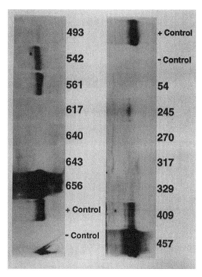

Figure 6. DCC protein expression in human brain tumor xenograft specimens. An ECL-immunoblot analysis with DCC-specific rabbit polyclonal antisera was carried out on lysates prepared from xenograft specimens and control cell lines. The xenograft specimen numbers correspond to those shown in Fig. 5. The " + control" cell line is a Chinese hamster ovary (CHO) cell line transfected with and expressing the full-length DCC cDNA. The " − control" cell line is a CHO cell line transfected with and expressing a mutant DCC cDNA encoding only the first 40 amino acids of DCC (Hedrick et al. 1994; Reale et al. 1994). DCC protein(s) with molecular weight of approximately 175,000–190,000 can be seen in the + control cell line and brain tumor xenograft samples 245, 409, 457, 542, 561, and 656.

cancer cell lines by either immunoblot or combined immunoprecipitation/immunoblot analysis. The DCC protein of 175–190 kD was found to be expressed at detectable levels in only 3 human neuroblastoma cell lines. The other 34 lines, including 8 other neuroblastoma cell lines, did not express DCC at detectable levels in our assays (Reale et al. 1994). The data are consistent with previous RT-PCR studies that have shown DCC expression is absent or greatly reduced in the majority of human cancer cell lines studied (Fearon et al. 1990).

Role of DCC in Mediating Alterations in Cell Fate

At least three different classes of cell-surface proteins—integrins, cadherins, and immunoglobulin superfamily CAMs—have been shown to have critical roles in mediating cell-cell recognition and differentiation in the nervous system in vivo and in in vitro model systems of neuronal differentiation and axonal outgrowth (Neugebauer et al. 1988; Edelman and Crossin 1991; Reichardt and Tomaselli 1991; Takeichi 1991; Hynes 1992; Walsh and Doherty 1992). Although in most studies purified preparations of N-CAM have not been found to be active in promoting neurite outgrowth, NIH-3T3 fibroblast monolayers expressing high levels of N-CAM or N-cadherin will stimulate neurite outgrowth from rat PC12 pheochromocytoma cells (Doherty et al. 1991; Walsh and Doherty 1992). Furthermore, the studies of Doherty, Walsh, and their colleagues have suggested that neurite-stimulating activities of N-CAM and N-cadherin involve activation of G-protein-dependent and calcium-channel-dependent signaling pathways.

Two independent studies have suggested that DCC may have a critical role in differentiation pathways and cell fate determination in neuronal cells. It has been shown that NGF-mediated morphological differentiation of PC12 cells was inhibited in cells transfected with and expressing high levels of an antisense *DCC* expression construct, or when PC12 cells induced to differentiate with NGF were incubated with antisense DCC oligonucleotides (Lawlor and Narayanan 1992). These data suggest that DCC expression in PC12 cells may be necessary for induction and maintenance of morphological differentiation. In addition, we have shown that transfected NIH-3T3 cells expressing the full-length DCC protein can stimulate morphological differentiation of PC12 cells through signaling pathways similar to those through which N-CAM and N-cadherin stimulate differentiation of PC12 cells, but distinct from those utilized in NGF-mediated differentiation (Pierceall et al. 1994). Thus, DCC may function in differentiation pathways and cell fate determination through cell-cell and/or cell-extracellular matrix interactions. Nevertheless, at the present time, there is no evidence that DCC functions through homotypic interactions (B.C. Ekstrand et al., unpubl.), and other DCC ligands have yet to be identified. Further studies of the mechanisms by which DCC stimulates PC12 neurite outgrowth may provide insights into its function in other tissues, including the means by which DCC may mediate differentiation and tumor suppression in epithelial cells.

SUMMARY

Chromosome 18q is among the chromosomal regions thought to harbor a tumor suppressor gene(s) that is frequently inactivated during the development of several cancer types, particularly those of the gastrointestinal tract. Moreover, preliminary data suggest that colorectal cancers with 18q LOH have a more aggressive clinical behavior than those cancers without 18q LOH. A candidate tumor suppressor gene from 18q, termed *DCC* for deleted in colorectal cancer, has been identified. The *DCC* gene is contained within the common region of LOH on 18q, its expression is markedly decreased or absent in colorectal cancers and cell lines, and a subset of colorectal cancers have been shown to have somatic mutations within the *DCC* gene. Thus, *DCC* represents the most promising candidate tumor suppressor gene from 18q.

At present, however, many questions remain regarding the mechanisms underlying the inactivation of *DCC* in the majority of colorectal cancers. In addition, although studies of 18q LOH and *DCC* gene expression in other cancer types suggest that *DCC* inactivation may contribute to the pathogenesis of other tumor types, few studies have provided definitive data to demonstrate that *DCC* inactivation is a critical genetic event in these tumors. Moreover, little is known about the function of *DCC* in the regulation of normal cell growth and tumor suppression. The predicted structural similarity of DCC to the N-CAM family of cell-surface proteins suggests that it may function through cell-cell and/or cell-extracellular matrix interactions. Many reports have detailed the phenotypic alterations in cancer cells that may result from alterations in cell-surface protein expression, including changes in cell morphology and tissue architecture, loss of differentiated phenotype, decreased cell adhesion and aggregation, increased motility, and invasive behavior. These altered biological properties are likely to account, in part, for the invasive and metastatic properties of cancer cells in the patient. Perhaps, inactivation of *DCC* is responsible for some of the altered growth properties observed in advanced cancer cells. Increased understanding of the prevalence and nature of the genetic alterations involving 18q and the *DCC* gene in cancers, as well as increased knowledge of the normal function of *DCC*, should not only prove of value for our understanding of the pathogenesis of cancer, but should also provide novel insights for improving the diagnosis and management of patients with cancer.

ACKNOWLEDGMENTS

This work was supported by a James S. McDonnell Foundation Molecular Medicine in Cancer Research Award, the Lucille P. Markey Charitable Trust, a Patrick and Catherine Weldon Donaghue postdoctoral fellowship (M.A.R.), and National Institutes of Health grant 5P30-CA16359-19 and Medical Scientist Training Program grant GM 07205.

REFERENCES

Cho, K.R., J.D. Oliner, J.W. Simons, L. Hedrick, E.R. Fearon, A.C. Preisinger, P. Hedge, G.A. Silverman, and B. Vogelstein. 1994. The *DCC* gene: Structural analysis and mutations in colorectal carcinomas. *Genomics* 19: 525.

Chomczynski, P. and N. Sacchi. 1987. Single-step method of RNA isolation by acid guanidium thiocyanate-phenol-chloroform extraction. *Anal. Biochem.* 162: 156.

Cropp, C.S., R. Lidereau, G. Campbell, M.H. Champene, and R. Callahan. 1990. Loss of heterozygosity on chromosomes 17 and 18 in breast carcinoma: Two additional regions identified. *Proc. Natl. Acad. Sci.* 87: 7737.

Cunningham, B.A., J.J. Hemperly, B.A. Murray, E.A. Prediger, R. Brackenbury, and G.M. Edelman. 1987. Neural cell adhesion molecule: Structure, immunoglobulin-like domains, cell surface modulation, and alternative RNA splicing. *Science* 236: 799.

Devilee, P., M. van Vliet, N. Kuipers-Dijkshoorn, P.L. Pearson, and C.J. Cornelisse. 1991. Somatic genetic changes on chromosome 18 in breast carcinomas: Is the *DCC* gene involved? *Oncogene* 6: 311.

Dickson, G., H.J. Gower, C.H. Barton, H.M. Prentice, V.L. Elsom, S.E. Moore, R.D. Cox, C. Quinn, W. Putt, and F.S. Walsh. 1987. Human muscle neural cell adhesion molecule

(N-CAM): Identification of a muscle-specific sequence in the extracellular domain. *Cell* **50:** 1119.

Doherty, P., S.V. Ashton, S.E. Moore, and F.S. Walsh. 1991. Morphoregulatory activities of N-CAM and N-cadherin can be accounted for by G protein-dependent activation of L- and N-type neuronal Ca^{2+} channels. *Cell* **67:** 21.

Edelman, G.M. and K.L. Crossin. 1991. Cell adhesion molecules: Implications for a molecular histology. *Annu. Rev. Biochem.* **60:** 155.

Fearon, E.R., K.R. Cho, J.M. Nigro, S.E. Kern, J.W. Simons, J.M. Ruppert, S.R. Hamilton, A.C. Preisinger, G. Thomas, K.W. Kinzler, and B. Vogelstein. 1990. Identification of a chromosome 18q gene that is altered in colorectal cancers. *Science* **247:** 49.

Gao, X., K.V. Honn, D. Grignon, W. Sakr, and Y.Q. Chen. 1993. Frequent loss of expression and loss of heterozygosity of the putative tumor suppressor gene *DCC* in prostatic carcinomas. *Cancer Res.* **53:** 2723.

Hamshere, M., G. Dickson, and I. Eperon. 1991. The muscle specific domain of mouse N-CAM: Structure and alternative splicing patterns. *Nucleic Acids Aes.* **19:** 4709.

Hedrick, L., K.R. Cho, E.R. Fearon, T.-C. Wu, K.W. Kinzler, and B. Vogelstein. 1994. The *DCC* gene product in cellular differentiation and colorectal tumorigenesis. *Genes Dev.* **8:** 1174.

Hohne, M. W., M.-E. Halatsch, G.F. Kahl, and R.J. Weinel. 1992. Frequent loss of expression of the potential tumor suppressor gene *DCC* in ductal pancreatic adenocarcinoma. *Cancer Res.* **52:** 2616.

Huan, Y., R.F. Boynton, P.L. Blount, R.J. Silverstein, J. Yin, Y. Tong, T.K. McDaniel, C. Newkirk, J.H. Resau, R. Sridhara, B.J. Reid, and S. Meltzer. 1992. Loss of heterozygosity involves multiple tumor suppressor genes in human esophageal cancers. *Cancer Res.* **52:** 6525.

Hynes, R.O. 1992. Integrins: Versatility, modulation, and signaling in cell adhesion. *Cell* **69:** 11.

Jen, J., H. Kim, S. Piantadosi, Z.-F. Liu, R.C. Levitt, P. Sistonen, K.W. Kinzler, B. Vogelstein, and S.R. Hamilton. 1994. Allelic loss of chromosome 18q and prognosis in colorectal cancer. *N. Engl. J. Med.* **331:** 213.

Kern, S.E., E.R. Fearon, K.W.F. Tersmette, J.P. Enterline, M. Leppert, Y. Nakamura, R. White, B. Vogelstein, and S.R. Hamilton. 1989. Allelic loss in colorectal carcinoma. *J. Am. Med. Assoc.* **261:** 3099.

Kikuchi-Yanoshita, R., M. Konishi, H. Fukunari, K. Tanaka, and M. Miyaki. 1992. Loss of expression of the *DCC* gene during progression of colorectal carcinomas in familial adenomatous polyposis and non-familial adenomatous polyposis patients. *Cancer Res.* **52:** 3801.

Lawlor, K.G. and R. Narayanan. 1992. Persistent expression of the tumor suppressor gene *DCC* is essential for neuronal differentiation. *Cell Growth Differ.* **3:** 609.

Miura, M., M. Kobayashi, H. Asou, and K. Uyemura. 1991. Molecular cloning of cDNA encoding the rat neural cell adhesion molecule Ll. Two Ll isoforms in the cytoplasmic region are produced by differential splicing. *FEBS Lett.* **289:** 91.

Miyake, K., K. Inokuchi, K. Dan, and T. Nomura. 1993. Expression of the *DCC* gene in myelodysplastic syndromes and overt leukemia. *Leukemia Res.* **17:** 785.

Neugebauer, K.M., K.J. Tomaselli, J. Lilien, and L.F. Reichardt. 1988. N-cadherin, N-CAM, and integrins promote retinal outgrowth on astrocytes in vitro. *J. Cell Biol.* **107:** 1177.

Nigro, J.M., K.R. Cho, E.R. Fearon, S.E. Kern, J.M. Ruppert, J.D. Oliner, K.W. Kinzler, and B. Vogelstein. 1991. Scrambled exons. *Cell* **64:** 607.

Ookawa, K., M. Sakamoto, S. Hirohashi, Y. Yoshida, T. Sugimura, M. Terada, and J. Yokota. 1993. Concordant p53 and *DCC* alterations and allelic losses on chromosome 13q and 14q associated with liver metastases of colorectal carcinoma. *Int. J. Cancer* **53:** 382.

Pierceall, W.E., K.R. Cho, R.H. Getzenberg, M.A. Reale, L. Hedrick, B. Vogelstein, and E.R. Fearon. 1994. The deleted in colorectal cancer (*DCC*) tumor suppressor gene product stimulates neurite outgrowth in rat PC12 pheochromocytoma cells. *J. Cell Biol.* **124:** 1017.

Porfiri, E., L.M. Secker-Walker, A.V. Hoffbrand, and J.F. Hancock. 1993. *DCC* tumor suppressor gene is inactivated in hematologic malignancies showing monosomy 18. *Blood* **81:** 2696.

Reale, M.A., G. Hu, A.I. Zafar, R.H. Getzenberg, S.M. Levine, and E.R. Fearon. 1994. Expression and alternative splicing of the deleted in colorectal cancer (*DCC*) gene in normal and malignant tissues. *Cancer Res.* **54:** 4493.

Reichardt, L.F. and K.J. Tomaselli. 1991. Extracellular matrix molecules and their receptors: Functions in neural development. *Annu. Rev. Neurosci.* **14:** 531.

Sadoul, R., F. Kirchhoff, and M. Schachner. 1989. A protein kinase activity is associated with and specifically phosphorylates the neural cell adhesion molecule Ll. *J. Neurochem.* **53:** 1471.

Sorkin, B.C., S. Hoffman, G.M. Edelman, and B.A. Cunningham. 1984. Sulfation and phosphorylation of the neural cell adhesion molecule, N-CAM. *Science* **225:** 1476.

Takeichi, M. 1991. Cadherin cell adhesion receptors as a morphogenetic regulator. *Science* **251:** 1451.

Thompson, A.M., R.G. Morris, M. Wallace, A.H. Wyllie, C.M. Steel, and D.C. Carter. 1993. Allele loss from 5q21 (*APC/MCC*) and 18q21 (*DCC*) and *DCC* mRNA expression in breast cancer. *Br. J. Cancer* **68:** 64.

Uchino, S., H. Tsuda, M. Noguchi, J. Yokota, M. Terada, T. Saito, M. Kobayashi, T. Sugimura, and S. Hirohashi. 1992. Frequent loss of heterozygosity at the *DCC* locus in gastric cancer. *Cancer Res.* **52:** 3099.

Vogelstein, B., E.R. Fearon, S.R. Hamilton, S.E. Kern, A.C. Preisinger, M. Leppert, Y. Nakamura, R. White, A.M.M. Smits, and J.L. Bos. 1988. Genetic alterations during colorectal tumor development. *N. Engl. J. Med.* **319:** 525.

Walsh, F.S. and P. Doherty. 1992. Second messengers underlying cell-contact-dependent axonal growth stimulated by transfected N-CAM, N-cadherin, or Ll. *Cold Spring Harbor Symp. Quant. Biol.* **57:** 431.

Walsh, F.S., R.B. Parekh, S.E. Moore, G. Dickson, C.H. Barton, H.J. Gower, R.A. Dwek, and T.W. Rademacher. 1989. Tissue specific O-linked glycosylation of the neural cell adhesion molecule (N-CAM). *Development* **105:** 803.

Weinberg, R.A. 1991. Tumor suppressor genes. *Science* **254:** 1138.

Molecular Cytogenetics of Human Breast Cancer

J.W. Gray,*‡ C. Collins,*‡ I.C. Henderson,** J. Isola,*§ A. Kallioniemi,*§
O.-P. Kallioniemi,*§ H. Nakamura,*† D. Pinkel,*‡
T. Stokke,*‡‡ M. Tanner,§ and F. Waldman*

*Division of Molecular Cytometry, Department of Laboratory Medicine, and **Department
of Medicine, University of California, San Francisco, California 94143-0808;
‡Life Sciences Division, Lawrence Berkeley Laboratory, Berkeley, California;
§Laboratory of Cancer Genetics, Tampere University Hospital, Tampere, Finland;
†Department of Surgery, Tokyo Medical College Hospital, Shinjuku-ku, Tokyo, Japan;
‡‡Department of Biophysics, Institute for Cancer Research, Montebello, 0310 Oslo, Norway

Progression of human solid tumors through accumulation of multiple advantageous genetic abnormalities is well established. Genes that may be involved include those that influence proliferation rate (Murray 1992; Weinert and Lydall 1993), apoptosis (Green et al. 1994; Kerr et al. 1994), genetic stability and/or DNA repair fidelity (Hartwell 1992; Bronner et al. 1994; Fishel et al. 1993), differentiation (Sell and Pierce 1994), adhesion (Juliano and Varner 1993; Bernstein and Liotta 1994), and angiogenesis (Harris and Horak 1993; Horak et al. 1993; Weinstat-Saslow and Steeg 1994). These genes may be either dominant or recessive, and they may be differentially regulated through amplification, structural rearrangement, or translocation, or inactivated through loss and/or mutation. In addition, there is evidence that some abnormalities, on average, occur earlier than others (Fearon and Vogelstein 1990; Bullerdiek et al. 1994). This is important since screening for early abnormalities may facilitate cancer detection, and detection of later abnormalities may provide information about the extent of disease progression. Characterization of these genes may reveal information about the biology of the progression process that will suggest new therapeutic possibilities.

The genetic progression process is best understood in colon cancer where abnormalities involving *APC* or *KRS* seem, on average, to occur early, followed by abnormalities involving *p53* and *DCC* (Fearon and Vogelstein 1990). It is less well understood for other tumors, such as breast cancer, because morphological and histological indicators of cancer progression are not as distinct. Nonetheless, cytogenetic studies, analyses of loss of heterozygosity (LOH), and gene dosage analyses by Southern analysis have revealed numerous abnormalities that may be involved in the progression of human breast cancer. Genes so far implicated include amplifications of *Her-2/neu* (Slamon et al. 1987; Lupu and Lippman 1993), c-*MYC* (Gaffey et al. 1993), cyclin-D (Gillett et al. 1994), and EMS (Karlseder et al. 1994) and inactivation of cell cycle regulatory genes such as *p53* and *RB1* (Walker and Varley 1993). However, many, if not most, of the genetic abnormalities involved in breast cancer progression remain to be discovered. Gene amplification is one common mechanism leading to up-regulation of gene expression (Stark et al. 1989). In fact, a recent cytogenetic study indicates that significant amplification occurs in more than 50% of human breast cancers and that most of the involved regions are not associated with known oncogenes (Saint-Ruf et al. 1990). Analyses of LOH show loss at 1p and q (Borg et al. 1992; Bieche et al. 1993; Leger et al. 1993), 3p13-14 (Chen et al. 1994), 11p13 (Winqvist et al. 1993), 13q (Thorlacius et al. 1991), 16q22-23 (Cleton-Jansen et al. 1994; Tsuda et al. 1994), 17p13 (Matsumura et al. 1992; Casey et al. 1993), 17q21 (Cropp et al. 1993), and 18q. Some of these regions, e.g., 17p13, 13q14, and 18q (Thompson et al. 1993), contain known cancer genes, e.g., *p53*, *RB1*, and *DCC*, respectively. In addition, linkage has been demonstrated between heritable breast cancer and genetic markers at 17q21 (*BRCA1*; Cropp et al. 1994). However, the involved genes are unknown for most regions of abnormality.

Fluorescence in situ hybridization (FISH) (Lichter et al. 1988; Pinkel et al. 1988) and comparative genomic hybridization (CGH) (A. Kallioniemi et al. 1992a; Ried et al. 1994) yield additional information about genetic changes in human breast cancer. CGH is useful because it provides a genome-wide view of unit changes in gene copy number abnormalities that involve more than about 10 Mb of DNA and high-level amplification of smaller regions, whereas FISH provides quantitative, locus-specific information about gene copy number in single cells. Thus, CGH is well suited to identification of abnormalities involving most of the cells of a tumor, for investigation of correlations between abnormalities, and for identification of corrections between specific abnormalities and biological and clinical endpoints. FISH allows precise aberration definition, reveals cell to cell heterogeneity, enables detection of small subpopulations of genetically distinct cells, and allows correlation between genotype and phenotype.

GENOME-WIDE ASSESSMENT OF DNA SEQUENCE COPY NUMBER CHANGES USING CGH

CGH is performed by hybridizing differentially labeled genomic DNA from a tumor and from a normal cell population to normal metaphase spreads (A. Kallioniemi et al. 1992a). The ratio of the intensities of the hybridized tumor and normal DNAs along the metaphase chromosomes is measured using digital imaging microscopy as an indication of the relative tumor DNA sequence concentration (Piper et al. 1994). DNA sequences that are overrepresented in the tumor produce high tumor:reference hybridization intensity ratios on the target chromosomes, and those that are underrepresented produce low tumor:reference ratios. CGH is powerful because it maps regions of DNA sequence copy number abnormality onto a normal representation of the human genome. Figure 1 shows the DNA sequence copy number increases that have been detected in 33 primary human breast cancers and 15 breast cancer cell lines using CGH (Kallioniemi et al. 1994). Remarkably, 85% of the primary tumors and all of the cell lines show increased DNA sequence copy number at one or more loci. Regions of decreased copy number were also prevalent, so that almost all breast cancers show increased or decreased DNA sequence copy number at one or more loci. The number of copy number abnormalities varies considerably among tumors, with a few having almost no abnormalities and others having more than 15. We speculate that the number of abnormalities is a rough indication of the degree of genetic instability, since unstable tumors may be expected to accumulate more abnormalities than those that maintain genomic integrity. If this is the case, the number of abnormalities may be a useful indicator of clinical behavior, since unstable tumors may evolve and therefore progress more rapidly than stable tumors. Recent studies by Isola and Waldman support this idea, showing decreased survival with increasing number of abnormalities (J. Isola et al., in prep.). This linkage suggests a genetic basis for the instability and motivates a search for the involved genes. It is unlikely that mutations in *MSH2*, *MLH1*, and related genes (Bodmer et al. 1994) play this role, since microsatellite repeat instability is not commonly found in spontaneous breast cancers (Wooster et al. 1994). Mutations in *p53* may account for some of the observed instability (Livingstone et al. 1992), but other genes may be involved as well (Hartwell 1992; Weinert and Lydall 1993).

The challenge now is to identify the specific genes that, when activated or inactivated by these copy number abnormalities, contribute to cancer progression. A first step is to distinguish between copy number abnormalities that contribute to cancer progression through activation or inactivation of oncogenes and suppressor genes and those that accumulate by chance as a result of structural genetic instability (e.g., as a result of

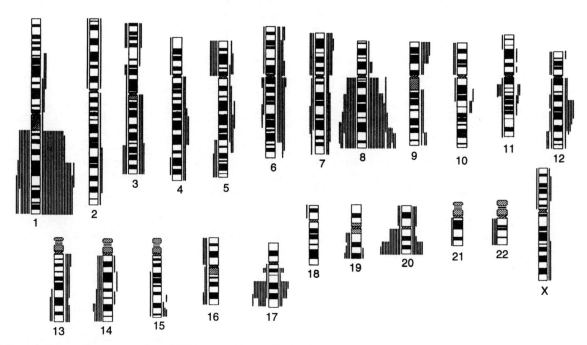

Figure 1. Schematic representation of DNA sequence copy number increases found using CGH in 15 breast cancer cell lines and 33 primary breast cancers. The extents of regions of increased copy number are shown as lines next to chromosome ideograms. Each line represents the region of increase found in one tumor. Data for the cell lines are shown to the left of each ideogram; data for the primary tumors are shown to the right (Kallioniemi et al. 1994).

chromosome breakage at fragile sites, through loss of DNA fragments lacking centromeres, or whole chromosome loss or gain during division). Our approach has been to focus on the most frequent abnormalities and to define small regions that appear aberrant in a high fraction of breast tumors. The existence of such regions increases confidence that they contain one or a few genes that are selectively activated or inactivated during tumor progression. The most common regions of increased copy number in breast cancer revealed by CGH include 1q, 8q, 17q22-q24, and 20q13 (Kallioniemi et al. 1994). Some increases were much higher than others, so that multiple mechanisms of copy number increase may be operating. These increases are found in 20–40% of late-stage human breast cancers. The region of increase at 8q spans the c-*MYC* locus. However, the region of increase at 17q22-q24 is distal to the *Her-2/neu* locus, and the other loci are not associated with known oncogenes. These data suggest the existence of several dominantly acting but still unknown oncogenes that play a role in breast cancer progression. Regions of decreased DNA sequence copy number detected by CGH that occur at frequencies exceeding 15% include 3p, 6q, 8p, 11p, 12q, 13q, and 17p (J. Isola et al., in prep.). These data suggest the existence of additional recessive, tumor suppressor genes that, when inactivated, play a role in breast cancer progression.

Interestingly, our CGH studies and published LOH studies have not revealed any abnormalities that occur in most or all breast cancers. Thus, it seems unlikely that a strict linear genetic progression exists. It appears more likely that tumors progress through accumulation of advantageous phenotypic characteristics (e.g., increased proliferation, reduced apoptosis, increased angiogenesis, reduced cell adhesion), each produced by any of several different genetic abnormalities. For example, reduced cell adhesion might be produced by inactivation of any of the genes that code for proteins involved in the cell adhesion process or of genes that code for regulatory proteins. Thus, it may be necessary for optimal diagnosis and/or prognostication to develop assays that detect any of several different genetic aberrations that result in the same phenotypic change.

The genetic heterogeneity between tumors revealed using CGH also may have an impact on cancer therapy, since it is apparent that response to therapy is determined to some extent by the genetic characteristics of the tumors. Gene amplification is well known to contribute to drug resistance, and inactivation of *p53* seems to enable the amplification process (Livingstone et al. 1992). The recent report of increased effectiveness of high-dose adriamycin-based therapies in breast cancers that are amplified for *Her-2/neu* (Muss et al. 1994) suggests that other genetic characteristics will affect therapy response as well. Thus, stratification of patients according to genotype is likely to be important in the application of existing therapies. Genetic analysis also may contribute to cancer treatment through identification of new abnormalities that result in aberrant proteins and/or metabolic pathways that can be attacked therapeutically. It now remains to develop the genetic basis for therapy optimization.

INTRATUMOR HETEROGENEITY ASSESSED USING FISH

Analysis of genetic aberrations in human tumors using molecular techniques such as CGH and Southern analysis may convey the impression that individual tumors are relatively homogeneous genetically. FISH with locus-specific probes shows that for most breast cancers, and indeed for most solid tumors, this is usually not the case. Figure 2, a and b, for example, shows the distribution of hybridization signals in two human breast cancers after FISH with probes for the centromeres of chromosomes 1, 7, 11, 15, 17, and X (F. Waldman et al., in prep.). One tumor appears nearly normal with most cells showing 2 signals for all probes except that specific for chromosome 1. The other tumor appears wildly abnormal with few cells showing 2 signals and about 20% showing 12 signals for chromosome 17. Most breast cancers fall between these extremes, and almost all are aneuploid at one or more loci. This calls into question the utility of partitioning cells according to whether they are diploid or aneuploid in efforts to improve prognostication. In fact, almost all breast cancers are aneuploid at some level. It seems more productive to consider the specific locus (or loci) that is aneuploid.

FISH with probes to specific genes known to be involved in cancer progression also shows variations in copy number. Figure 2c, for example, shows variation in the number of hybridization signals produced by FISH with a probe to the retinoblastoma gene *RB1* in human breast cancers (A. Kallioniemi et al. 1992b). The number of copies per tumor varies between 0 and 4. We found similar variation using a probe on chromosome 17p13 near the *p53* locus (Matsumura et al. 1992) and using a probe to E-cadherin on chromosome 16q22.1 (H. Nakamura et al., in prep.). Variation in copy number for genes commonly found to be amplified appears to be even larger. Our studies of *Her-2/neu* show extreme variation, with some tumors showing *Her-2/neu* copy number ranging from a few per cell to almost 90 per cell (O.-P. Kallioniemi et al. 1992). Occasionally, tumors that do not appear to be amplified by slot blot analysis show subpopulations with more than tenfold amplification.

Copy number variability revealed using FISH may be another reflection of the level of genetic instability. Thus, it will be interesting to learn whether the degree of intratumor variability measured using FISH correlates with the number of copy number abnormalities measured using CGH. If so, either measure of instability may be useful as an indicator of rapid tumor progression. In addition, this would suggest that both may be determined by the same genetic mechanism. The

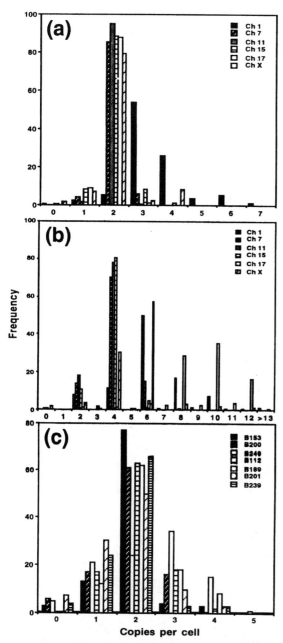

Figure 2. Frequencies of cells showing hybridization domain numbers ranging from 0 to >12/cell detected in primary human breast cancers after FISH with chromosome-specific probes. (*a*) Frequencies measured after hybridization to a near diploid tumor with probes for the centromeres of chromosomes 1, 7, 11, 15, 17, and X (F. Waldman et al., in prep.). (*b*) Frequencies measured after hybridization to an aneuploid tumor with probes for chromosomes 1, 7, 11, 15, 17, and X (F. Waldman et al., in prep.). (*c*) Frequencies of cells showing 0–4 hybridization domains in 7 primary breast cancers after hybridization with a probe to RB1 (A. Kallioniemi et al. 1992b).

existence of copy number heterogeneity also is an issue in the design of therapies that attack specific loci (e.g., overexpression of the tyrosine kinase receptor *Her-2/neu*; Bacus et al. 1992; Crews et al. 1992; Pietras et al. 1992) since subpopulations that are not amplified may not be sensitive. This may not be a problem for *Her-2/neu*, since expression seems relatively homogeneous in tumors showing amplification. However, it may be a problem at other loci. Also of concern is the simultaneous amplification of multiple regions of the genome (Lafage et al. 1992; A. Kallioniemi et al., in prep.). If coamplified genes result in similar phenotypes, an attack on one may fail as a result of selection of cells carrying another, phenotypically equivalent copy number increase. The issue will be decided only by further investigations of the clinical and biological behavior of genetically distinct tumor subpopulations during targeted therapy.

DEFINITION OF REGIONS OF COMMON GENETIC ABNORMALITY USING FISH

Precise definition of the extent of gene copy number abnormalities detected using CGH is necessary for identification of the involved gene(s). This can be accomplished using FISH with mapped probes distributed over the area of interest. The number of hybridization signals produced using FISH is a good estimate of the number of copies of the probed sequence in the target cell population. This is illustrated in Figure 3b, which shows the strong correlation between relative gene copy number of *Her-2/neu* determined using FISH and relative copy number measured by quantitative Southern analysis. FISH is advantageous for several reasons: (1) It reveals abnormalities in subpopulations of cells. (2) It can be applied to analysis of paraffin-embedded tumors where clinical outcome is known. (3) Only a few hundred cells are needed for analysis. (4) It provides information about mechanisms of aberration formation. Disadvantages of FISH include the labor intensive "spot counting" that is required and the need to analyze tumors one or a few loci at a time.

Increased Copy Number

One abnormality identified using CGH is a region of consistent increased copy number on chromosome 20q13. We defined the region of increased copy number using a collection of mapped cosmid and P1 clones (T. Stokke et al., in prep.). The collection of these probes and their use in the definition of the region of 20q amplification in the breast cancer cell line BT474 are illustrated in Figure 4. A region of highly increased copy number is clearly visible at 20q13.2. However, a much larger region of moderately increased copy number also is apparent, as is variation from locus to locus along the chromosome. We believe that the quantitative nature of copy number analysis using FISH will allow precise localization of putative oncogenes, since we expect that regions harboring these genes will be most highly increased in copy number. Analysis of five breast cancer cell lines and three primary tumors shows the region defined by the probe RMC20C001 to be consistently most highly increased in copy number (Tanner et al. 1994). Several known genes map near this site, including *PTPN1*, *PCK1*, and *MC3R*. Figure 5 shows

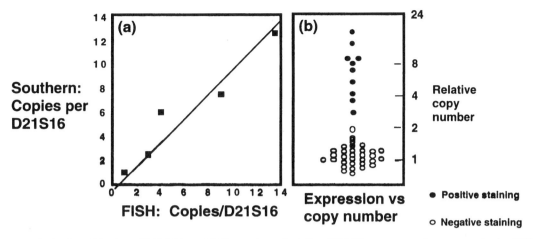

Figure 3. Analysis of amplification of *HER-2/neu*. (*a*) Comparison of relative *HER-2/neu* copy number measured using FISH and Southern analysis in 5 different cell lines. (*b*) Assessment of *HER-2/neu* overexpression using immunohistochemistry in breast cancers carrying different numbers of *HER-2/neu* assessed using FISH (O.-P. Kallioniemi et al. 1992).

the cumulative analysis of copy number for these genes. The probe RMC20C001 is consistently present at higher copy number than other known genes in this region. We believe that this indicates the presence of a novel, dominantly acting oncogene at or near RMC20C001. However, it remains to be proven that the gene(s) advantageous for breast cancer progression will be consistently most highly increased in copy number.

FISH also provides information about the process by which the copy number increases occur. Figure 5 shows that the number of copies of chromosome 20 sequences increases and decreases along chromosome 20 in a complex pattern. This suggests that several breakage and/or insertion events have contributed to the copy number increase. This is supported by analyses of the distribution of the FISH signals produced by RMC20C001 in metaphase spreads of BT474. The RMC20C001 sequence is distributed over more than 10 marker chromosomes and at some sites is closely associated with *ERBB2* (also increased in copy number in

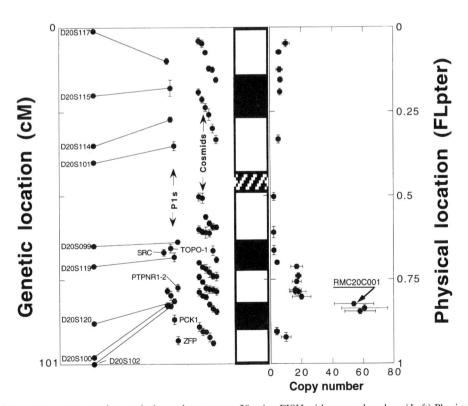

Figure 4. DNA sequence copy number analysis on chromosome 20 using FISH with mapped probes. (*Left*) Physical and genetic locations of cosmid and P1 clones along chromosome 20. (*Right*) Number of copies of these probes measured using FISH in interphase nuclei in the breast cancer cell line BT474 (Tanner et al. 1994).

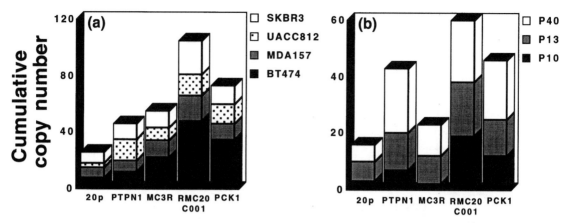

Figure 5. Cumulative copy number for *PTPN1*, *MC3R*, and *PCK1* and for anonymous DNA sequences on 20p and 20q13.2 (RMC20C001) assessed using FISH. (*a*) Cumulative copy number analysis for 4 breast cancer cell lines. (*b*) Cumulative copy numbers in 3 primary breast cancers (Tanner et al. 1994).

this line; T. Stokke et al., in prep.). The phenomenon of close genomic association between normally separate sequences (including known oncogenes) seems to be a common phenomenon (A. Kallioniemi et al., in prep.). This raises the interesting possibility that coordinate copy number increase and overexpression of several different oncogenes may be especially advantageous for tumor progression.

Decreased Copy Number

Localization of tumor suppressor genes that contribute to human tumor progression often is accomplished by analysis of LOH, since allelic loss is often one step in the inactivation of these genes (Knudson 1991). However, this may be difficult if polymorphic markers suitable for analysis of LOH are not polymorphic in important tumors. FISH can be used as an adjunct to analysis of LOH, since LOH is often caused by physical loss of one allele. Probes suitable for FISH can be produced at essentially any locus defined by a sequence-tagged site, and FISH may be informative when LOH is not. Detection of a decrease in DNA sequence copy number from two to one is straightforward using FISH. Unfortunately, the use of FISH for analysis of DNA sequence loss is complicated by the occurrence of aneuploidy and genetic instability in breast cancer and other solid tumors. For example, loss of one allele through physical deletion followed by tetraploidization will result in a cell with two copies of the remaining allele, which will appear normal by FISH. This problem may be overcome using dual color FISH with one probe to the test region and another to the centromere of the same chromosome, since we have found a high concordance between LOH and a reduction in the ratio of the number of test probes to the number of centromeric probes. This concordance is illustrated in Figure 5, which shows a concordance of about 85% between LOH and a reduced test:centromeric signal ratio at 17p13 near *p53* (Matsumura et al. 1992). This same concordance has been observed at other loci as well (e.g., on 16q in breast cancer; H. Nakamura et al., in prep.). One explanation for this correlation is that the structural events leading to loss of one allele occur before the tumor becomes tetraploid. In any case, analysis of the test:centromeric hybridization signal ratio is useful for analysis of chromosome loss leading to inactivation of tumor suppressor genes.

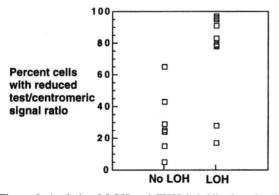

Figure 6. Analysis of LOH and FISH hybridization signal number for a probe near *p53* relative to the number of hybridization signals produced using a probe to the centromere of chromosome 17. Tumors showing LOH and no LOH by polymorphic probe analysis are separated on the X axis (Matsumura et al. 1992).

SUMMARY

Our studies of human breast cancer using CGH and FISH with precisely mapped probes have revealed a surprising amount of intratumor and intertumor heterogeneity in human breast cancers. However, some regions of abnormal copy number are found frequently. We believe that the regions of frequent abnormality harbor novel cancer genes, and that identification of these is important for diagnosis, prognostication, and the development of new therapies.

ACKNOWLEDGMENTS

Preparation of this paper was supported by U.S. Public Health Service grant CA-58207, the U.S. Department of Energy under contract DE-AC-03-76SF00098, and Imagenetics.

REFERENCES

Bacus, S.S., I. Stancovski, E. Huberman, D. Chin, E. Hurwitz, G.B.Mills, A. Ullrich, M. Sela, and Y. Yarden. 1992. Tumor-inhibitory monoclonal antibodies to the HER-2/Neu receptor induce differentiation of human breast cancer cells. *Cancer Res.* **52:** 2580.

Bernstein, L.R. and L.A. Liotta. 1994. Molecular mediators of interactions with extracellular matrix components in metastasis and angiogenesis. *Curr. Opin. Oncol.* **6:** 106.

Bieche, I., M.H. Champeme, F. Matifas, C.S. Cropp, R. Callahan, and R. Lidereau. 1993. Two distinct regions involved in 1p deletion in human primary breast cancer. *Cancer Res.* **53:** 1990.

Bodmer, W., T. Bishop, and P. Karran. 1994. Genetic steps in colorectal cancer. *Nat. Genet.* **6:** 217.

Borg, A., Q.X. Zhang, H. Olsson, and E. Wenngren. 1992. Chromosome 1 alterations in breast cancer: Allelic loss on 1p and 1q is related to lymphogenic metastases and poor prognosis. *Genes Chromosomes Cancer* **5:** 311.

Bronner, C.E., S.M. Baker, P.T. Morrison, G. Warren, L.G. Smith, M.K. Lescoe, M. Kane, C. Earabino, J. Lipford, A. Lindblom, P. Tannergard, R. Bollag, A. Godwin, D. Ward, M. Nordenskjold, R. Fishel, R. Kolodner, and R. Liskay. 1994. Mutation in the DNA mismatch repair gene homologue hMLH1 is associated with hereditary non-polyposis colon cancer.

Bullerdiek, J., U. Bonk, B.Staats, E. Leuschner, G. Gohla, T. Ebel, and S. Bartnitzke. 1994. Trisomy 18 as the first chromosome abnormality in a medullary breast cancer. *Cancer Genet. Cytogenet.* **73:** 75.

Casey, G., S. Plummer, G. Hoeltge, D. Scanlon, C. Fasching, and E.J. Stanbridge. 1993. Functional evidence for a breast cancer growth suppressor gene on chromosome 17. *Hum. Mol. Genet.* **2:** 1921.

Chen, L.C., K. Matsumura, G. Deng, W. Kurisu, B.M. Ljung, M.I. Lerman, F.M. Waldman, and H.S. Smith. 1994. Deletion of two separate regions on chromosome 3p in breast cancers. *Cancer Res.* **54:** 3021.

Cleton-Jansen, A.M., E.W. Moerland, N.J. Kuipers-Dijkshoorn, D.F. Callen, G.R. Sutherland, B. Hansen, P. Devilee, and C.J. Cornelisse. 1994. At least two different regions are involved in allelic imbalance on chromosome arm 16q in breast cancer. *Genes Chromosomes Cancer* **9:** 101.

Crews, J.R., L.A. Maier, Y.H. Yu, S. Hester, K. O'Briant, D.S. Leslie, K. DeSombre, S.L. George, C.M. Boyer, Y. Argon, and R.C. Bast. 1992. A combination of two immunotoxins exerts synergistic cytotoxic activity against human breast-cancer cell lines. *Int. J. Cancer* **51:** 772.

Cropp, C.S., M.H. Champeme, R. Lidereau, and R. Callahan. 1993. Identification of three regions on chromosome 17q in primary human breast carcinomas which are frequently deleted. *Cancer Res.* **53:** 5617.

Cropp, C.S., H.A. Nevanlinna, S. Pyrhonen, U.H. Stenman, P. Salmikangas, H. Albertsen, R. White, and R. Callahan. 1994. Evidence for involvement of BRCA1 in sporadic breast carcinomas. *Cancer Res.* **54:** 2548.

Fearon, E.R. and B. Vogelstein. 1990. A genetic model for colorectal tumorigenesis. *Cell* **61:** 759.

Fishel, R., M.K. Lescoe, M.R. Rao, N.G. Copeland, N.A. Jenkins, J. Garber, M. Kane, and R. Kolodner. 1993. The human mutator gene homolog MSH2 and its association with hereditary nonpolyposis colon cancer. *Cell* **75:** 1027.

Gaffey M.J., H.F. Frierson, Jr, and M.E. Williams. 1993. Chromosome 11q13, c-erbB-2, and c-myc amplification in invasive breast carcinoma: Clinicopathologic correlations. *Mod. Pathol.* **6:** 654.

Gillett, C., V. Fantl, R. Smith, C. Fisher, J. Bartek, C. Dickson, D. Barnes, and G. Peters. 1994. Amplification and overexpression of cyclin D1 in breast cancer detected by immunohistochemical staining. *Cancer Res.* **54:** 1812.

Green, D.R., R.P. Bissonnette, and T.G. Cotter. 1994. Apoptosis and cancer. *Important Adv. Oncol.* **37:** 52.

Harris, A.L. and E. Horak. 1993. Growth factors and angiogenesis in breast cancer. *Recent Results Cancer Res.* **127:** 35.

Hartwell, L. 1992. Defects in a cell cycle checkpoint may be responsible for the genomic instability of cancer cells. *Cell* **71:** 543.

Horak, E.R., A.L. Harris, N. Stuart, and R. Bicknell. 1993. Angiogenesis in breast cancer. Regulation prognostic aspects and implications for novel treatment strategies. *Ann. N.Y. Acad. Sci.* **698:** 71.

Juliano, R.L. and J.A. Varner. 1993. Adhesion molecules in cancer: The role of integrins. *Curr. Opin. Cell Biol.* **5:** 812.

Kallioniemi, A., O.P. Kallioniemi, D. Sudar, D. Rutovitz, J.W. Gray, F. Waldman, and D. Pinkel. 1992a. Comparative genomic hybridization for molecular cytogenetic analysis of solid tumors. *Science* **258:** 818.

Kallioniemi, A., O.-P. Kallioniemi, F.M. Waldman, L.C. Chen, L.-C. Yu, Y.K. Fung, H.S. Smith, D. Pinkel, and J.W. Gray. 1992b. Detection of retinoblastoma gene copy number in metaphase chromosomes and interphase nuclei by fluorescence in situ hybridization. *Cytogenet. Cell Genet.* **60:** 190.

Kallioniemi, A., O.-P. Kallioniemi, J. Piper, M. Tanner, T. Stokke, L. Chen, H.S. Smith, D. Pinkel, J.W. Gray, and F.M. Waldman. 1994. Detection and mapping of amplified DNA sequences in breast cancer by comparative genomic hybridization. *Proc. Natl. Acad. Sci.* **91:** 2156.

Kallioniemi, O.-P., A. Kallioniemi, W. Kurisu, A. Thor, L.C. Chen, H.S. Smith, F.M. Waldman, D. Pinkel, and J.W. Gray. 1992. ERBB2 amplification in breast cancer analyzed by fluorescence in situ hybridization. *Proc. Natl. Acad. Sci.* **89:** 5321.

Karlseder, J., R. Zeillinger, C. Schneeberger, K. Czerwenka, P. Speiser, E. Kubista, D. Birnbaum, P. Gaudray, and C. Theillet. 1994. Patterns of DNA amplification at band q13 of chromosome 11 in human breast cancer. *Genes Chromosomes Cancer* **9:** 42.

Kerr, J.F., C.M. Winterford, and B.V. Harmon. 1994. Apoptosis. Its significance in cancer and cancer therapy. *Cancer* **73:** 2013.

Knudson, A. 1991. Overview: Genes that predispose to cancer. *Mutat. Res.* **247:** 185.

Lafagc, M., F. Pedeutour, S. Marchetto, J. Simonetti, M.-T. Prosperi, P. Gaudray, and D. Birnbaum. 1992. Fusion and amplification of two originally non-syntenic chromosomal regions in a mammary breast carcinoma. *Genes Chromsomes Cancer* **5:** 40.

Leger, I., M. Thoma, X. Ronot, and G. Brugal. 1993. Detection of chromosome 1 aberrations by fluorescent in situ hybridization (FISH) in the human breast cancer cell line MCF-7. *Anal. Cell. Pathol.* **5:** 299.

Lichter, P., T. Cremer, J. Borden, L. Manuelidis, and D.C. Ward. 1988. Delineation of individual human chromosomes in metaphase and interphase cells by in situ suppression hybridization using recombinant DNA libraries. *Hum. Genet.* **80:** 224.

Livingstone, L.R., A. White, J. Sprouse, E. Livanos, T. Jacks, and T.D. Tlsty. 1992. Altered cell cycle arrest and gene amplification potential accompany loss of wild-type p53. *Cell* **70:** 923.

Lupu, R. and M.E. Lippman. 1993. William L. McGuire Memorial Symposium: The role of erbB2 signal transduction pathways in human breast cancer. *Breast Cancer Res. Treat.* **27:** 83.

Matsumura, K., A. Kallioniemi, O.-P. Kallioniemi, L. Chen,

H.S. Smith, D. Pinkel, J.W. Gray, and F.M. Waldman. 1992. Deletion of chromosome 17p loci in breast cancer cells detected by fluorescence in situ hybridization. *Cancer Res.* **52:** 3474.

Murray, A.W. 1992. Creative blocks: Cell-cycle checkpoints and feedback controls. *Nature* **359:** 599.

Muss, H.B., A.D. Thor, D.A. Berry, T. Kute, E.T. Liu, F. Koerner, C.T. Cirrincione, D.R. Budman, W.C. Wood, M. Barcos, and I.C. Henderson. 1994. c-erbB-2 expression and response to adjuvant therapy in women with node-positive early breast cancer. *N. Engl. J. Med.* **330:** 1260.

Pietras, R.J., B.M. Fendly, V.R. Chazin, M.D. Pegram, S.B. Howell, and D.J. Slamon. 1992. Antibody to HER-2/neu receptor blocks DNA repair after cisplatin in human breast and ovarian cancer cells. *Oncogene* **9:** 1829.

Pinkel, D., J. Landegent, C. Collins, J. Fuscoe, R. Seagraves, J. Lucas, and J.W. Gray. 1988. Fluorescence in situ hybridization with human chromosome-specific libraries: Detection of trisomy 21 and translocations of chromosome 4. *Proc. Natl. Acad. Sci.* **85:** 9138.

Piper, J., D. Rutovitz, D. Sudar, A. Kallioniemi, O.-P. Kallioniemi, F. Waldman, J.W. Gray, and D. Pinkel. 1994. Computer image analysis of comparative genomic hybridization. *Cytometry* (in press).

Ried, T., I. Petersen, H. Holtgreve-Grez, M.R. Speicher, E. Schrock, S. du Manoir, and T. Cremer. 1994. Mapping of multiple DNA gains and losses in primary small cell lung carcinomas by comparative genomic hybridization. *Cancer Res.* **54:** 1801.

Saint-Ruf, C., M. Gerbault-Seureau, E. Viegas-Pequignot, B. Zafrani, R. Cassingena, and B. Dutrillaux. 1990. Protooncogene amplification and homogeneously staining regions in human breast carcinomas. *Genes Chromosomes Cancer* **2:** 18.

Sell, S. and G.B. Pierce. 1994. Maturation arrest of stem cell differentiation is a common pathway for the cellular origin of teratocarcinomas and epithelial cancers. *Lab. Invest.* **70:** 6.

Slamon, D., G. Clark, S. Wong, W. Levin, A. Ullrich, and W. McGuire. 1987. Human breast cancer: Correlation of relapse and survival with amplification of the HER-2/neu oncogene. *Science* **235:** 177.

Stark, G., M. Debatisse, E. Giulotto, and G. Wahl. 1989. Recent progress in understanding mechanisms of mammalian DNA amplification. *Cell* **75:** 901.

Tanner, M., M. Tirkkonon, A. Kallioniemi, C. Collins, T. Stokke, R. Karhu, D. Kowbel, F. Shadravan, M. Hintz, W.-L. Kuo, F. Waldman, J. Isola, J.W. Gray, and O.-P. Kallioniemi. 1994. Increased copy number at 20q13 in breast cancer: Defining the critical region and exclusion of candidate genes. *Cancer Res.* **54:** 4257.

Thompson, A.M., R.G. Morris, M. Wallace, A.H. Wyllie, C.M. Steel, and D.C. Carter. 1993. Allele loss from 5q21 (APC/MCC) and 18q21 (DCC) and DCC mRNA expression in breast cancer. *Br. J. Cancer* **68:** 64.

Thorlacius, S., O. Jonasdottir, and J.E. Eyfjord. 1991. Loss of heterozygosity at selective sites on chromosomes 13 and 17 in human breast carcinoma. *Anticancer Res.* **11:** 1501.

Tsuda, H., D.F. Callen, T. Fukutomi, Y. Nakamura, and S. Hirohashi. 1994. Allele loss on chromosome 16q24.2-qter occurs frequently in breast cancers, irrespectively of differences in phenotype and extent of spread. *Cancer Res.* **54:** 513.

Walker, R.A. and J.M. Varley. 1993. The molecular pathology of human breast cancer. *Cancer Surv.* **16:** 31.

Weinert, T. and D. Lydall. 1993. Cell cycle checkpoints, genetic instability and cancer. *Semin. Cancer Biol.* **4:** 129.

Weinstat-Saslow, D. and P.S. Steeg. 1994. Angiogenesis and colonization in the tumor metastatic process: Basic and applied advances. *FASEB J.* **8:** 401.

Winqvist, R., A. Mannermaa, M. Alavaikko, G. Blanco, P.J. Taskinen, H. Kiviniemi, I. Newsham, and W. Cavenee. 1993. Refinement of regional loss of heterozygosity for chromosome 11p15.5 in human breast tumors. *Cancer Res.* **53:** 4486.

Wooster, R., A.M. Cletonjansen, N. Collins, J. Mangion, R. Cornelis, C. Cooper, B. Gusterson, B. Ponder, A. von Deimling, O. Wiestler, C. Cornelisse, P. Devilee, and M. Stratton. 1994. Instability of short tandem repeats (microsatellites) in human cancers. *Nat. Genet.* **6:** 152.

Genetic Alterations in Prostate Cancer

W.B. Isaacs, G.S. Bova, R.A. Morton, M.J.G. Bussemakers,
J.D. Brooks, and C.M. Ewing

Brady Urological Institute, Johns Hopkins Medical Institutions, Baltimore, Maryland 21287

Prostate cancer is currently the most commonly diagnosed malignancy in American men. Approximately 38,000 men die from this disease annually in this country (Boring et al. 1994). Despite this, the study of the molecular biology of this disease has lagged behind that of other common tumors.

A number of features of the prostate gland and its neoplasms are somewhat unique. The prostate is probably the only organ which continues to grow throughout life, such that the size of a prostate gland from a 65-year-old man will, on the average, be two- to threefold larger than that of a 20-year-old man (Berry et al. 1984). The factors responsible for this characteristic growth pattern are essentially unknown. Although most of this age-related increase in prostatic mass is due to frequent hyperplasia of the periurethral or transition zone of the prostate, lesions that are histologically identifiable as cancer can be found with a frequency of similar magnitude. Although, fortunately, few of these latter lesions progress to clinically manifest disease, these features of the natural history of the prostate emphasize the tremendous predilection of this organ to undergo abnormal growth. Understanding this characteristic in molecular terms is an important goal in the study of the molecular biology of the prostate.

With respect to prostate cancer, a particularly important question from a clinical point of view is, What are the molecular events responsible for the progression of prostate cancer, or, in other words, why and how does prostate cancer evolve from an indolent to a life-threatening disease? Furthermore, is this evolution inevitable, or are some prostate cancers destined never to progress to advanced disease, let alone clinically detectable disease, regardless of the time frame provided? As mentioned above, the *initiation* of prostate cancer appears to be a very frequent event, occurring in nearly one-third of men over age 45 (Dhom 1977). Geographically, this rate of histological cancer incidence is roughly the same worldwide (Breslow et al. 1977; Yatani et al. 1984). The number of clinically manifest cases, however, differs widely among various populations, suggesting important environmental factors in triggering this progression (Carter et al. 1990a). In contrast, studies of familial aggregation of this disease have suggested that between 5% and 10% of prostate cancers may be directly attributable to the inheritance of prostate cancer susceptibility alleles (Carter et al. 1992), which may act as genetic factors driving this progression. Thus, as with numerous other cancers, there is strong evidence for both genetic and environmental factors in the etiology of prostate cancer, with the majority of disease most likely being a result of the interaction of the two. We have pursued a number of approaches to identify and characterize genetic alterations in both sporadic and inherited prostate cancer, and these efforts are reviewed here. It is anticipated that these efforts will provide markers capable of identifying individuals at high risk of developing prostate cancer, and which may be of prognostic value in evaluating the aggressiveness of the disease once diagnosed.

RAS IN PROSTATE CANCER

On the basis of evidence suggesting a potentially important role for activated *ras* genes in prostate cancer (Peehl et al. 1987; Treiger and Isaacs 1988; Thompson et al. 1989), we examined prostate cancers for the presence of *ras* gene mutations using differential oligo-deoxy-nucleotide hybridization of DNA amplified by means of the polymerase chain reaction (PCR) (Carter et al. 1990b). Overall, together with our subsequent studies, we have assayed 51 samples of prostate cancer, including both acinar and ductal carcinomas, primary tumors and metastatic deposits, and prostate cancer cell lines. Only two mutations have been detected: an A to G transition causing a glutamate to arginine substitution at codon 61 of the Ha-*ras* gene in a primary prostatic duct adenocarcinoma and a G to T transversion causing a glycine to valine substitution at codon 12 of the Ha-*ras* gene in a prostate cell line (TSU-Pr1) derived from a lymph node metastasis. Two subsequent reports have found similar low frequencies of *ras* gene mutations in prostate cancer (Gumerlock et al. 1991; Moul et al. 1992). When these three independent studies are combined, a total of 3 *ras* gene mutations (all in Ha-*ras*) were found in 94 samples assayed, for an overall frequency of less that 4%. These results, all obtained in American patients, demonstrate that activation of *ras* oncogenes via point mutation is not a common event in either the initiation or progression of prostatic neoplasia. In contrast to these studies, however, two reports have appeared which suggest that *ras* gene mutations *do* occur at significant frequencies (~25%) in prostate cancer, both in latent carcinoma (Konishi et al. 1992) and in clinically manifest disease (Anwar et al. 1992). Both of these studies examined prostate tissue from Japanese men, raising the intriguing possibility that significant differences may exist in the genetic events associated with prostate cancer in American men versus Japanese men (wherein the latter have an ~5-fold lower incidence rate of clinically manifest disease). Curiously, in latent prostate carcinoma, the mutations were found in the Ki-*ras* gene, whereas in the clinically manifest cancers, mutations were found predominantly in the Ha-*ras* gene, and were associated with late stage. In fact, the same Ha-*ras* mutation, a A→T transversion in codon 61, accounted for each mutation found in this gene. These results would suggest that Ki-*ras* mutations do not appear to be associated with progression of latent lesions to clinically manifest disease, and in fact, may be associated with just the opposite, i.e., lack of progression. A clearer picture of this point may emerge upon additional, confirmatory studies.

Additionally, in this latter study (Anwar et al. 1992), human papillomavirus (HPV) DNA was detected in more than 40% of the prostate cancer samples analyzed. The presence of HPV DNA in prostate cancer from North American men is controversial (McNicol and Dodd 1991; Sarkar et al. 1993), but the most recent evidence suggests that this family of viruses is not commonly found in prostate cancer (Effert et al. 1992; Serfling et al. 1992), at least not at levels comparable to those found in cervical cancer, where a definitive etiologic link has been established.

LOH STUDIES

We have carried out loss of heterozygosity (LOH) studies to search for sites of consistent chromosomal deletion, possibly uncovering the location of tumor suppressor genes which become inactivated in prostatic carcinogenesis. Cytogenetic analyses of prostate cancer have not revealed consistent chromosomal deletions (see Brothman et al. 1990; Lundgren et al. 1992; Sandberg 1992), which might provide information regarding the location of such genes. An early study by Atkin and Baker (1985), however, showed that 4 of 4 patients with late-stage prostate carcinomas exhibited chromosome 10q deletions, and 3 of 4 exhibited chromosome 7q deletions. Since that time, alterations of the long arm of chromosome 10, although by no means ubiquitous, have been the most consistently observed karyotypic abnormality in prostate cancer.

To pursue these observations at a molecular level, we studied allelic loss in prostate cancer using polymorphic DNA probes for chromosomes 10q and 7q as well as probes for chromosomes containing documented and putative tumor suppressor genes (3p, 9q, 11p, 13q, 17p, 18q). We examined 28 primary prostate cancer specimens, most of which were localized, stage B lesions, measuring less than 2 cm in diameter. Briefly, we found that the majority (61%) of the tumors in the study exhibited allelic loss on at least one of the chromosomes examined (Carter et al. 1990c). The regions most frequently deleted are found on the long arms of chromosomes 10 and 16. These results suggested that clonal populations of cells with gross chromosomal deletions are common in prostate cancer specimens, even at a relatively early stage. This contrasts with most cytogenetic analyses which reveal primarily normal diploid karyotypes (see above references), suggesting that perhaps in culture, prostate cancer cells, or least those with chromosomal deletions, are at a growth *dis*advantage.

In addition to our study, Kunimi et al. (1991) examined ten primary tumor specimens, as well as eight metastatic lesions, for allelic loss on every non-acrocentric autosomal chromosomal arm. Interestingly, chromosomes 10 and 16, along with chromosome 8, most frequently exhibited allelic loss in this study. Taken together, these results provide evidence for the specificity of allelic loss on chromosomes 8, 10, and 16 in prostate cancer.

Critical deletion mapping studies performed by Bergerheim et al. (1991) demonstrated that chromosome 8 LOH in prostate cancer was restricted to the short arm. To further refine this observation, we carried out a deletion mapping study of this chromosome in more than 50 samples of prostate cancer. Similar to Bergerheim et al. (1991), we find high rates of loss of alleles on chromosome 8p in these samples (Bova et al. 1993b). In fact, we find a locus at chromosome 8p22 (*MSR*) which is deleted in almost 70% of the prostate cancers examined, and in one case, a homozygous deletion at this locus was observed. A chromosome 8p deletion map defines the smallest region of overlap to a 14-cM interval between D8S163 and LPL, flanking the *MSR* locus. Frequent loss of markers in this region has also been observed by Macoska et al. (1994) using a combination of PCR and in situ techniques, and by MacGrogan et al. (1994), who have used a large number of microsatellite repeat polymorphisms to map losses on this chromosome. These data strongly suggest the presence of a prostate tumor suppressor gene in this region of chromosome 8p. Concomitantly, evidence for *increased* copy number of sequences on the long arm of chromosome 8 has been observed in a subset of prostate tumors with loss of 8p sequences (Bova et al. 1993b), suggesting the presence of an activated oncogene on this chromosome as well. The finding of a higher frequency of 8p loss compared to gain of 8q sequences suggests that the former alteration may precede the latter, and that this latter event may be more important in tumor progression. Further studies of advanced prostate cancer will be required to clarify this issue. Interestingly, the short arm of chromosome 8 shows frequent LOH in a variety of common human tumors, including colorectal and lung cancer (Vogelstein et al. 1989; Emi et al. 1992). Whether the same gene is the target of these LOH events in prostate, colorectal, and lung cancer is unclear, but it is interesting to speculate that a gene which inactivated in a large fraction of human cancers may reside in this region of the genome. In any case, the identification of the chromosome 8p gene(s) involved in prostate carcinogenesis should provide important insight into this process.

To confirm and extend the results of our initial study of LOH on chromosome 16q in prostate cancer, we have examined this chromosomal arm in 52 additional samples of prostate cancer (47 primary, 5 metastases) using five different polymorphic probes for this region. Overall, of 39 tumors for which the markers were informative, 14 (36%) showed loss of alleles on chromosome 16q. Interestingly, whereas 32% of primary tumors had chromosome 16q deletions, about 60% of metastatic tumors showed deletions on this arm, suggesting a possible role for inactivation of a gene(s) in this region in tumor metastasis.

E-CADHERIN/α CATENIN IN PROSTATE CANCER

The finding of allelic loss on chromosome 16q led to an examination of putative candidate genes in this region. The E-cadherin gene, mapped to chromosome 16q22.1 (Natt et al. 1989), is of particular interest. The product of this gene, originally termed uvomorulin by Peyrieras et al. (1983), Arc-1 by Behrens et al. (1985), and E-cadherin by Takeichi (see review, 1991), had been demonstrated to play a critical role in embryo- and organogenesis by mediating epithelial cell-cell recognition and adhesion processes. Indeed, the appearance of E-cadherin protein on the surface of cells of the early mouse embryo was shown to be required for normal development. The cadherin family of proteins has now been determined to be quite large, with expression found in numerous cell types. The role of E-cadherin in maintaining various aspects of cell differentiation led a number of investigators to examine E-cadherin in adenocarcinoma cells. Such studies found that E-cadherin protein levels were frequently reduced or absent in cancer cell lines, and such lines were often more fibroblastic in morphology and invasive in experimental assays (Frixen et al. 1991; Sommers et al. 1991). Vleminckx et al. (1991) demonstrated that experimental inactivation of E-cadherin either with antibodies or antisense RNA resulted in the acquisition of invasive potential, and that transfection of invasive adenocarcinoma cells with E-cadherin cDNA rendered the expressing cells noninvasive.

The first studies to examine E-cadherin in prostate cancer were carried out by Bussemakers et al. (1992) in the Dunning rat model. These studies found a strong correlation between the lack of E-cadherin and metastatic and/or invasive potential. This correlation was strengthened by the direct observation of the progression of a noninvasive, E-cadherin-positive tumor to an E-cadherin-negative, highly metastatic tumor. To extend these observations to human prostate cancer, in collaboration with J. Schalken's laboratory in Nijmegen, we examined a series of more than 90 samples of prostate cancer for E-cadherin protein levels by immunohistochemistry (Umbas et al. 1992). Whereas all benign samples stained with uniform, strong intensity at cell-cell borders, approximately one-half of the tumors examined showed reduced or absent E-cadherin protein staining. When compared with Gleason grade, there was a strong association between high-grade and aberrant E-cadherin staining. In fact, no tumors with Gleason score 9 and 10 had normal staining, whereas all tumors with a score of less than 6 had normal staining patterns. Importantly, a subsequent study of the potential prognostic significance of E-cadherin staining shows that, indeed, aberrant E-cadherin staining is a powerful predictor of poor outcome, in terms of both disease progression and overall survival (Umbas et al. 1994).

To extend our understanding of the role of E-cadherin-mediated cell-cell adhesion in prostate cancer, we have examined the synthesis and organization of this protein and its associated proteins in prostate cancer cell lines (Morton et al. 1993b). Using immunoprecipitation experiments, we find that cultured normal prostatic epithelial cells make abundant E-cadherin, which can be observed to form complexes with a pair of cytoplasmic proteins termed catenins α and β. These proteins, in particular α catenin, are thought to link E-cadherin to the microfilament cytoskeleton, and the interaction of E-cadherin with these proteins has been shown to be required for induction of cell-cell adhesion (Ozawa et al. 1989). In cultured prostatic cancer cell lines, E-cadherin levels are frequently reduced or absent, as in surgical specimens of prostate cancer. Furthermore, this study provided the unexpected finding that the levels of the E-cadherin-associated protein, α catenin, are also frequently reduced in prostate cancer cells. Loss of this protein in PC3 prostate cancer cells is associated with their decreased ability to aggregate when compared to cells expressing normal levels of E-cadherin and α catenin. Thus, the E-cadherin-mediated pathway of cell-cell adhesion may become inactivated in a variety of ways. The lack of α catenin expression in PC3 cells is due to a homozygous deletion within the α catenin gene (Morton et al. 1993b). We have recently mapped the α catenin gene to chromosome 5q (the gene is located at 5q31, and a pseudogene is located at 5q22; McPherson et al. 1994 and unpubl.). Preliminary results suggest that LOH of this chromosomal arm occurs in approximately 25% of prostate cancers analyzed. Finally, the recent observation that the APC gene product interacts with β catenin (Rubinfeld et al. 1993; Su et al. 1993) suggests a potential novel mechanism by which a cancer susceptibility gene may act, i.e., at the level of regulation of cell-cell adhesion.

MICROCELL TRANSFER STUDIES

As a functional test for the presence of genes on various chromosomes which are capable of suppressing tumor growth, we have used the microcell transfer technique to insert a normal copy of a given chromosome into prostate cancer cell lines. To date, we have successfully transferred chromosomes 5, 8, 10, 11, and 17 into a number of prostate cancer cell lines, including

TSU-Pr1, DU145, and PC3. The presence of transferred chromosomes has been confirmed by in situ hybridization with centromeric probes and Southern and CA repeat analysis with polymorphic markers which can discriminate the endogenous from exogenous chromosomes. The most striking results are observed after reintroduction of a normal copy of chromosome 5 into PC3 cells (C.M. Ewing et al., in prep.). Not only do the cells undergo a dramatic change in morphology to form tightly packed, more epithelioid colonies in vitro, but also the tumorigenicity of the cells is essentially abolished when placed into nude mice. These alterations in growth characteristics are accompanied by dramatic alterations in the organization of the actin-based cytoskeleton. Importantly, this reversion to a more normal phenotype is correlated with the reexpression of α catenin with restoration of E-cadherin-mediated cell-cell adhesion (C.M. Ewing et al., in prep.).

In experiments with chromosome 10, in the TSU-Pr1 cells, in vivo growth suppression is seen following microcell transfer, but only in clones of cancer cells which have taken up two or more copies of chromosome 10. The growth suppression is not observed in vitro, and is not complete in vivo, but rather is manifest as an increase in lag time before tumors appear. Experiments with the PC3 have been more dramatic, in that, following microcell transfer and selection of resistant clones in neomycin, two types of colonies appear. The most numerous are small colonies in which cells undergo only a limited number of divisions and subsequently appear to senesce. Less frequently, colonies of rapidly growing cells form; these cells are fully tumorigenic when injected into nude mice. Experiments with chromosome 17 show a definite tumor suppressive activity (Bova et al. 1993a), which, curiously, appears to be independent of restoration of p53 function, suggesting the presence of an alternative prostate cancer suppressor on this chromosome (see below). To date, no consistent suppression of tumor formation has been observed upon transfer of either chromosomes 8 or 11 into prostate cancer cells.

p53 AND Rb

In addition to detection of allelic loss on chromosomes 10q and 16q, about one fifth of the prostatic tumors analyzed in our initial study were deleted at chromosomes 17p, 18q, and 13q. These regions harbor the tumor suppressor genes *p53*, *DCC*, and *Rb*, respectively. Thus, the possibility exists that inactivation of these genes plays a role in at least a subset of prostatic neoplasms. The importance of *Rb* gene inactivation in prostate cancer has been suggested by the studies of Bookstein et al. (1990a,b). To examine the *Rb* gene more closely, we examined additional tumors for allelic loss of the *Rb* gene using a PCR-based detection of a variable number of tandem repeats (VNTR) in intron 20 of the gene. Of 41 informative tumors, 11 (27%) demonstrated allelic loss of one copy of the *Rb* gene (Brooks et al. 1994), suggesting that inactivation of the *Rb* gene may be important in a significant subset of prostatic adenocarcinomas.

With regard to the *p53* gene, we have found that three of five prostate cancer cell lines (TSUPr-1, PC3, and DU145) contain mutations in the coding sequence of the *p53* gene, and that the growth of these lines can be suppressed by the introduction of a wild-type copy of this gene (Isaacs et al. 1991). Studies by others have revealed that the frequency of mutations in primary prostate cancer is low (10–20%) (Visakorpi et al. 1992; Bookstein et al. 1993; Dinjens et al. 1994; Voeller et al. 1994) in comparison to many other common cancers (Soini et al. 1992), whereas evidence from several laboratories now suggests that bone metastatic deposits of prostate cancer may frequently exhibit mutant *p53* (Navone et al. 1993; Aprikian et al. 1994) and that mutations in this gene may be associated with the transition to hormone refractory disease (Navone et al. 1993). These results suggest that *p53* mutations are truly late events in the progression of prostate cancer.

To pursue this question further, we analyzed 71 primary prostate cancers for allelic loss of chromosome 17p and the *p53* gene. Of the 56 informative tumors, 11 (20%) demonstrated allelic loss. Allelic loss was directly correlated with increasing Gleason grade and appeared correlated with disease stage. Curiously, however, sequence analysis of the remaining allele in the 11 primary tumors with LOH on chromosome 17p reveals a lack of mutations in exons 5–8 of the remaining *p53* allele, suggesting existence of a different gene which may be the target of these LOH events. Obviously, a closer examination of the roles of this gene in prostatic cancer is warranted, particularly with respect to progression of this disease.

DNA METHYLATION

In addition to deletions, mutations, and chromosomal translocations, alterations in DNA methylation can affect gene expression. Hypermethylation of CpG islands (areas of the genome rich in the sequence CpG, associated with the 5' regulatory regions of genes) has been associated with gene inactivation, and this mechanism has been proposed by Baylin and coworkers (1991) as a means to inactivate tumor suppressor genes in a nonmutational fashion. Work by Sakai et al. (1991) has demonstrated that the CpG island in the *Rb* gene is methylated in retinoblastoma, and that this results in repression of this gene (Ohtani-Fujita et al. 1993). Furthermore, methylation of a CpG island in the promoter region of the estrogen receptor has been found in ER-negative breast cancers (Ottaviano et al. 1994). We have examined the methylation status of a highly polymorphic locus, D17S5, located at chromosome 17p13.3 with the probe YNZ22.1. This locus lies within a cluster of *Not*I restriction sites (GC–GGCCGC), a finding consistent with the presence of a CpG island. Previous studies by Makos et al. (1992) have observed hypermethylation at this locus in both colon and lung cancer. Furthermore, deletions of the D17S5 locus at 17p13.3 have been identified in some breast cancers when other commonly deleted areas of 17p, including the *p53* gene at 17p13.1, are retained (Sato et al. 1990). In prostate cancer, probing *Not*I-digested DNA with the probe YNZ22.1, we found hypermethylation at this site in 14 of 15 tumors and 4 of 5 prostate cancer cell lines (Morton et al. 1993a). These findings further suggest that there may be a tumor suppressor gene on chromosome 17p, separate from *p53*, that plays a role in prostate carcinogenesis and is frequently inactivated via hypermethylation. Further studies, such as identification of the affected gene, will be necessary to clarify the role of this methylation process in the development of prostate cancer. In related studies, recent work by Lee et al. (1994) has demonstrated the promoter of the glutathione S-transferase (GST) π gene is methylated in each of more than 30 prostate cancers examined. This invariant methylation pattern was not observed in either normal or hyperplastic prostate tissue, and correlates with the loss of GST π gene expression, which is also restricted to the cancer tissue. From these types of studies, it is apparent that altered DNA methylation patterns may play a crucial role in determining the overall pattern of gene expression in prostate cancer tissue, and that this mechanism rather than a mutational mechanism may account for inactivation of a number of growth regulatory genes.

HEREDITARY PROSTATE CANCER

Finally, Carter et al. (1992, 1993) have recently demonstrated that in a study of familial aggregation of prostate cancer, this clustering can be best explained by the Mendelian inheritance of a rare autosomal dominant allele which predisposes men to develop this disease. From the study population examined, it was estimated that the gene responsible for this predisposition accounts for approximately 9% of the total cases of prostate cancer observed, and 45% of the cases observed in men under 55 years of age. In collaboration with Drs. Patrick Walsh, Terri Beaty, Roy Levitt, and Deborah Meyers at our institution, we are initiating efforts to map this gene by linkage studies. Analysis of candidate regions, including *BRCA1* and chromosomes 8p22 and 16q22, are under way. Although prostate cancer poses a number of somewhat unique obstacles to such an analysis, it is anticipated that such a study will provide critical information regarding the genetic mechanisms not only of hereditary prostate cancer, but also of the more common sporadic disease as well.

SUMMARY

A number of genetic changes have been documented in prostate cancer, ranging from allelic loss to point mutations and changes in DNA methylation patterns (summarized in Fig. 1). To date, the most consistent changes are those of allelic loss events, with the majority of tumors examined showing loss of alleles from at least one chromosomal arm. The short arm of chromosome 8, followed by the long arm of chromosome 16, appear to be the most frequent regions of loss, suggesting the presence of novel tumor suppressor genes. Deletions of one copy of the *Rb* and *p53* genes are less frequent, as are mutations of the *p53* gene, and accumulating evidence suggests the presence of an additional tumor suppressor gene on chromosome 17p,

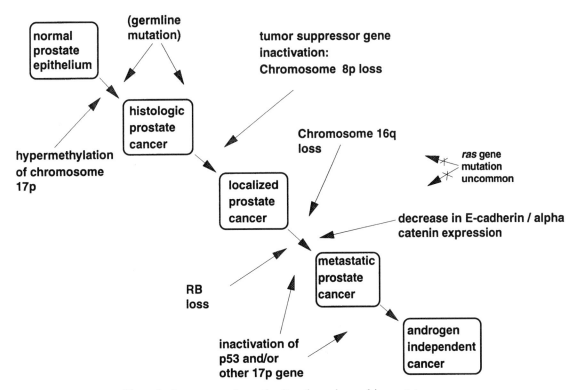

Figure 1. A summary of genetic alterations observed in prostate cancer.

which is frequently inactivated in prostate cancer. Alterations in the E-cadherin/α-catenin-mediated cell-cell adhesion mechanism appear to be present in almost half of all prostate cancers and may be critical to the acquisition of metastatic potential of aggressive prostate cancers. Finally, altered DNA methylation patterns have been found in the majority of prostate cancers examined, suggesting widespread alterations in methylation-modulated gene expression. The presence of multiple changes in these tumors is consistent with the multistep nature of the transformation process. Finally, efforts to identify prostate cancer susceptibility loci are under way and may elucidate critical early events in prostatic carcinogenesis.

ACKNOWLEDGMENTS

This work was supported by the U.S. Public Health Service (CA-55231, CA-58236).

REFERENCES

Anwar, K., K. Nakakuki, T. Shiraishi, H. Naiki, R. Yatani, and M. Inuzuka. 1992. Presence of ras oncogene mutations and human papilloma virus DNA in human prostate carcinoma. *Cancer Res.* **52:** 5991.

Aprikian, A.G., A.S. Sarkis, W.R. Fair, Z.F. Zhang, Z. Fuks, and C. Cordon-Cardo. 1994. Immunohistochemical determination of p53 protein nuclear accumulation in prostatic adenocarcinoma. *J. Urol.* **151:** 1276.

Atkin, N.B. and M.C. Baker. 1985. Chromosome study of five cancers of the prostate. *Hum. Genet.* **70:** 359.

Baylin, S.B., M. Makos, J.J. Wu, R.W. Yen, A. de Bustros, P. Vertino, and B.D. Nelkin. 1991. Abnormal patterns of DNA methylation in human neoplasia: Potential consequences for tumor progression. *Cancer Cells* **3:** 383.

Behrens, J., W. Birchmeier, S.L. Goodman, and B.A. Imhof. 1985. Dissociation of Madin-Darby canine kidney epithelial cells by the monoclonal antibody anti-Arc1: Mechanistic aspects and identification of the antigen as a component related to uvomorulin. *J. Cell Biol.* **101:** 1307.

Bergerheim, U.S., K. Kunimi, V.P. Collins, and P. Ekman. 1991. Deletion mapping of chromosomes 8, 10, and 16 in human prostate carcinoma. *Genes Chromosomes Cancer* **3:** 215.

Berry, S.J., D.S. Coffey, P.C. Walsh, and L.L. Ewing. 1984. The development of human benign prostatic hyperplasia with age. *J. Urol.* **132:** 474.

Bookstein, R., D. MacGrogan, S.G. Milsenbeck, F. Sharkey, and D. Allred. 1993. p53 is mutated in a subset of advanced-stage prostate cancer. *Cancer Res.* **53:** 3369.

Bookstein, R., J. Shew, P. Chen, P. Scully, and W.-H. Lee. 1990a. Suppression of tumorigenicity of human prostate carcinoma cells by replacing a mutated Rb gene. *Science* **247:** 712.

Bookstein, R., P. Rio, S. Madreperla, F. Hong, C. Allred, W. Grizzle, and W.-H. Lee. 1990b. Promoter deletion and loss of retinoblastoma gene expression in human prostate carcinoma. *Proc. Natl. Acad. Sci.* **87:** 7762.

Boring, C.C., T.S. Squires, T. Tong, and S. Montgomery. 1994. Cancer statistics, 1994. *CA Cancer J. Clin.* **44:** 7.

Bova, G.S., J.C. Robinson, and W.B. Isaacs. 1993a. Chromosome 17 suppresses tumorigenicity of a human prostate cancer cell line. *Proc. Am. Assoc. Cancer Res.* **34:** 540.

Bova, G.S., B.S. Carter, M.J.G. Bussemakers, M. Emi, Y. Fujiwara, N. Kyprianou, S.C. Jacobs, J.C. Robinson, J.I. Epstein, P.C. Walsh, and W.B. Isaacs. 1993b. Homozygous deletion and frequent allelic loss of chromosome 8p22 loci in human prostate cancer. *Cancer Res.* **53:** 3869.

Breslow, N., C. Chan, G. Dhom, R. Drury, L. Franks, B. Gellei, Y. Lee, S. Lundberg, B. Sparke, N. Sternby, and H. Tulinius. 1977. Latent carcinoma of the prostate at autopsy in seven years. *Int. J. Cancer* **20:** 680.

Brooks, J.D., G.S. Bova, and W.B. Isaacs. 1994. Allelic loss of the retinoblastoma gene in prostate cancer. *Prostate* (in press).

Brothman, A., D. Peehl, A. Patel, and J. McNeal. 1990. Frequency and pattern of karyotypic abnormalities in human prostate cancer. *Cancer Res.* **50:** 3795.

Bussemakers, M.J.G., R.J.A. Van Moorselaar, L.A. Giroldi, T. Ichikawa, J.T. Isaacs, M. Takeichi, F.M.J. Debruyne, and J.A. Schalken. 1992. Decreased expression of E-cadherin in the progression of rat prostatic cancer. *Cancer Res.* **52:** 2916.

Carter, B.S., H.B. Carter, and J.T. Isaacs. 1990a. Epidemiologic evidence regarding predisposing factors to prostate cancer. *Prostate* **16:** 187.

Carter, B.S., J.I. Epstein, and W.B. Isaacs. 1990b. Ras gene mutations in human prostate cancer. *Cancer Res.* **50:** 6830.

Carter, B.S., T.H. Beaty, G.D. Steinberg, B. Childs, and P.C. Walsh. 1992. Mendelian inheritance of familial prostate cancer. *Proc. Natl. Acad. Sci.* **89:** 3367.

Carter, B.S., G.S. Bova, T.H. Beaty, G.D. Steinberg, B. Childs, W.B. Isaacs, and P.C. Walsh. 1993. Hereditary prostate cancer: Epidemiology and clinical features. *J. Urol.* **150:** 797.

Carter, B.S., C.M. Ewing, S.W. Ward, B.F. Treiger, T.W. Aalders, J.A. Schalken, J.I. Epstein, and W.B. Isaacs. 1990c. Allelic loss of chromosomes 16q and 10q in human prostate cancer. *Proc. Natl. Acad. Sci.* **87:** 8751.

Dhom, G. 1977. Classification and grading of prostatic carcinoma. *Recent Results Cancer Res.* **60:** 14.

Dinjens, W.N., M.M. van der Weiden, F.H. Schroeder, F.T. Bosman, and J. Trapman. Frequency and characterization of p53 mutations in primary and metastatic human prostate cancer. *Int. J. Cancer* **56:** 630.

Effert, P.J., R.A. Frye, A. Neubauer, E.T. Liu, and P.J. Walther. 1992. Human papillomavirus types 16 and 18 are not involved in human prostate carcinogenesis: Analysis of archival human prostate cancer specimens by differential polymerase chain reaction. *J. Urol.* **147:** 192.

Emi, M., Y. Fujiwara, T. Nakajima, E. Tshuchiya, H. Tsuda, S. Hirohashi, Y. Maeda, K. Tsuruta, M. Miyaki, and Y. Nakamura. 1992. Frequent loss of heterozygosity for loci on chromosome 8p in hepatocellular carcinoma, colorectal cancer, and lung cancer. *Cancer Res.* **52:** 5368.

Frixen, U.H., J. Behrens, M. Sachs, G. Eberle, B. Voss, A. Warda, D. Lochner, and W. Birchmeier. 1991. E-cadherin-mediated cell-cell adhesion prevents invasiveness of human carcinoma cells. *J. Cell Biol.* **113:** 173.

Gumerlock, P.H., U.R. Poonamallee, F.J. Meyers, R.W. de Vere White. 1991. Activated ras alleles in human prostate cancer are rare. *Cancer Res.* **51:** 1632.

Isaacs, W.B., B.S. Carter, and C.M. Ewing. 1991. Wild-type p53 suppresses growth of human prostate cancer cells containing mutant p53 alleles. *Cancer Res.* **51:** 4716.

Konishi, N., T. Enomoto, G. Buzard, M. Oshima, J.M. Ward, and J.M. Rice. 1992. K-ras activation and ras p21 expression in latent prostatic carcinoma in Japanese men. *Cancer* **69:** 2293.

Kunimi, K., U.S.R. Bergerheim, I.-L. Larsson, P. Ekman, and V.P. Collins. 1991. Allelotyping of human prostatic adenocarcinoma. *Genomics* **11:** 530.

Lee, W.-H., R.A. Morton, J.I. Epstein, J.D. Brooks, P.A. Campbell, G.S. Bova, W.-S. Hsieh, W.B. Isaacs, and W.G. Nelson. 1994. Methylation-inactivation of the GSTP1 gene accompanies human prostatic carcinogenesis. *Proc. Natl. Acad. Sci.* (in press).

Lundgren, R., N. Mandahl, S. Heim, J. Limon, H. Henrikson, and F. Mitelman. 1992. Cytogenetic analysis of 57 primary prostatic adenocarcinomas. *Genes Chromosomes Cancer* **4:** 16.

MacGrogan, D., A. Levy, D. Bostwick, M. Wagner, D. Wells, and R. Bookstein. 1994. Loss of chromosome arm 8p loci in prostate cancer: mapping by quantitative allelic imbalance. *Genes Chromosomes Cancer* **10:** 151.

Macoska, J.A., T.M. Trybus, W.A. Sakr, M.C. Wolf, P.D. Benson, I.J. Powell, and J.E. Pontes. 1994. Fluorescence in situ hybridization analysis of 8p allelic loss and chromosome instability in human prostate cancer. *Cancer Res.* **54:** 3824.

Makos, M., B.D. Nelkin, M.I. Lerman, F. Latif, B. Zbar, and S.B. Baylin. 1992. Distinct hypermethylation patterns occur at altered chromosome loci in human lung and colon cancer. *Proc. Natl. Acad. Sci.* **89:** 1929.

McNicol, P.J. and J.G. Dodd. 1991. High prevalence of human papillomavirus in prostate tissues. *J. Urol.* **145:** 850.

McPherson, J.D., R.A. Morton, C.M. Ewing, J.J. Wasmuth, J. Overhauser, A. Nagafuchi, S. Tsukita, and W.B. Isaacs. 1994. Assignment of the human α catenin gene to chromosome 5q21-q22. *Genomics* **19:** 188.

Morton, R.A., G.S. Bova, and W.B. Isaacs. 1993a. Hypermethylation of chromosome 17p13.3 in human adenocarcinoma of the prostate. *J. Urol.* **149:** 376A. (Abstr.)

Morton, R.A., C.M. Ewing, A. Nagafuchi, S. Tsukita, and W.B. Isaacs. 1993b. Reduction of E-cadherin levels and deletion of the α catenin gene in human prostate cancer cells. *Cancer Res.* **53:** 3585.

Moul, J.W., P.A. Friedrichs, R.S. Lance, S.M. Theune, and E. Chung. 1992. Infrequent *ras* oncogene mutations in human prostate cancer. *Prostate* **20:** 327.

Natt, E., R.E. Magenis, J. Zimmer, A. Mansouri, and G. Scherer. 1989. Regional assignment of the human loci for uvomorulin and chymotrypsinogen B with the help of two overlapping deletions on the long arm of chromosome 16. *Cytogenet. Cell Genet.* **50:** 145.

Navone, N.M., P. Troncoso, L.L. Pisters, T.L. Goodrow, J.L. Palmer, W.W. Nichols, A.C. von Eschenbach, and C.J. Conti. 1993. p53 protein accumulation and gene mutation in the progression of human prostate carcinoma. *J. Natl. Cancer Inst.* **85:** 1657.

Ohtani-Fujita, N., T. Fujita, A. Aoike, N.E. Osifchin, P.D. Robbins, and T. Sakai. 1993. CpG methylation inactivates the promoter activity of the human retinoblastoma tumor-suppressor gene. *Oncogene* **8:** 1063.

Ottaviano, Y.L., J.P. Issa, F.F. Parl, H.S. Smith, S.B. Baylin, and N.E. Davidson. 1994. Methylation of the estrogen receptor gene CpG island marks loss of estrogen receptor expression in human breast cancer cells. *Cancer Res.* **54:** 2552.

Ozawa, M., H. Baribault, and R. Kemler. 1989. The cytoplasmic domain of the cell adhesion molecule uvomorulin associates with three independent proteins structurally related in different species. *EMBO J.* **8:** 1711.

Peehl, D.M., N. Wehner, and T.A. Stamey. 1987. Activated Ki-*ras* oncogene in human prostatic adenocarcinoma. *Prostate* **10:** 281.

Peyrieras, N., F. Hyafil, D. Louvard, H.L. Ploegh, and F. Jacob. 1983. Uvomorulin: A nonintegral membrane protein of early mouse embryo. *Proc. Natl. Acad. Sci.* **80:** 6274.

Rubinfeld, B., B. Souza, I. Albert, O. Muller, S.H. Chamberlain, F.R. Masiarz, S. Munemitsu, and P. Polakis. 1993. Association of the APC gene product with β-catenin. *Science* **262:** 1731.

Sakai, T., J. Toguchida, N. Ohtani, D.W. Yandell, J.M. Rapaport, and T.P. Dryja. 1991. Allele-specific hypermethylation of the retinoblastoma tumor-suppressor gene. *Am. J. Hum. Genet.* **48:** 880.

Sandberg, A.A. 1992. Chromosomal abnormalities and related events in prostate cancer. *Hum. Pathol.* **23:** 368.

Sarkar, F.H., W.A. Sakr, Y.W. Li, P. Sreepathi, and J.D. Crissman. 1993. Detection of human papillomavirus (HPV) DNA in human prostatic tissues by polymerase chain reaction (PCR). *Prostate* **22:** 171.

Sato, T., A. Tanigami, K. Yamakawa, F. Akiyama, F. Kasumi, G. Sakamoto, and Y. Nakamura. 1990. Allelotype of breast cancer: Cumulative allele losses promote tumor progression in primary breast cancer. *Cancer Res.* **50:** 7184.

Serfling, U., G. Ciancio, W.Y. Zhu, C. Leonardi, and N.S. Penneys. 1992. Human papillomavirus and herpes virus DNA are not detected in benign and malignant prostatic tissue using the polymerase chain reaction. *J. Urol.* **148:** 192.

Soini, Y., P. Paakko, K. Nuorva, D. Kamel, D.P. Lane, and K. Vahakangas. 1992. Comparative analysis of p53 protein immunoreactivity in prostatic, lung and breast carcinomas. *Virchows Arch. A Pathol. Anat. Histopathol.* **421:** 223.

Sommers, C.L., E.W. Thompson, J.A. Torri, R. Kemler, E.P. Gelmann, and S.W. Byers. 1991. Cell adhesion molecule uvomorulin expression in human breast cancer cell lines: Relationship to morphology and invasive capacities. *Cell Growth Differ.* **2:** 365.

Su, L.K., B. Vogelstein, and K.W. Kinzler. 1993. Association of the APC tumor suppressor protein with catenins. *Science* **262:** 1734.

Takeichi, M. 1991. Cadherin cell adhesion receptors as a morphogenetic regulator. *Science* **251:** 1451.

Thompson, T.C., J. Southgate, G. Kitchener, and H. Land. 1989. Multistage carcinogenesis induced by *ras* and *myc* oncogenes in a reconstituted organ. *Cell* **56:** 917.

Treiger, B.F. and J.T. Isaacs. 1988. Expression of a transfected v-Ha-*ras* oncogene in a Dunning rat prostate adenocarcinoma and the development of high metastatic ability. *J. Urol.* **140:** 1580.

Umbas, R., W.B. Isaacs, P.P. Bringuier, H.E. Schaafsma, F.M. Karthaus, G.O. Oosterhof, F.M.J. Debruyne, and J.A. Schalken. Decreased E-cadherin expression is associated with poor prognosis in patients with prostate cancer. *Cancer Res.* **54:** 3929.

Umbas, R., J.A. Schalken, T.W. Aalders, B.S. Carter, H.F.M. Karthaus, H.E. Schaafsma, F.M.J. Debruyne, and W.B. Isaacs. 1992. Expression of the cellular adhesion molecule, E-cadherin, is reduced or absent in high grade prostate cancer. *Cancer Res.* **52:** 5104.

Visakorpi, T., O.P. Kallioniemi, A. Heikkinen, T. Koivula, and J. Isola. 1992. Small subgroup of aggressive, highly proliferative prostatic carcinomas defined by p53 accumulation. *J. Natl. Cancer Inst.* **84:** 883.

Vleminckx, K., L. Vakaet, M. Mareel, W. Fiers, and F. Van Roy. 1991. Genetic manipulation of E-cadherin expression by epithelial tumor cells reveals an invasion suppressor role. *Cell* **66:** 107.

Voeller, H.J., L.Y. Sugars, T. Pretlow, and E.P. Gelmann. 1994. p53 oncogene mutations in human prostate cancer specimens. *J. Urol.* **151:** 492.

Vogelstein, B., E.R. Fearon, S.E. Kern, S.R. Hamilton, A.C. Preisinger, Y. Nakamura, and R. White. 1989. Allelotype of colorectal carcinomas. *Science* **244:** 207.

Yatani, R., I. Chigusa, K. Akazaki, G. Stemmerman, R. Welsh, and P. Correa. 1984. Geographic pathology of latent prostatic carcinoma. *Int. J. Cancer* **29:** 611.

Multicellular Resistance: A New Paradigm to Explain Aspects of Acquired Drug Resistance of Solid Tumors

R.S. Kerbel, J. Rak, H. Kobayashi, M.S. Man, B. St. Croix, and C.H. Graham

Division of Cancer Research, Sunnybrook Health Science Centre, and Department of Medical Biophysics, University of Toronto, Toronto, Ontario, Canada M4N 3M5

The study of the cellular and molecular basis of acquired drug resistance in cancer has been dominated by a single experimental approach for the past two decades. This approach consists of exposing established tumor cell lines—usually obtained from solid tumors—to increasing concentrations of a single cytotoxic drug. Conventional two-dimensional monolayer cell culture systems, in which the cells are exposed over prolonged periods of time, are generally used so that drug-resistant mutant subpopulations expressing very high levels of resistance (e.g., 10- to 1000-fold) in a stable or semistable fashion are selected. This invariably leads to the isolation of so-called "unicellular" resistant mutants, that is, tumor cells which express their resistance properties at the single-cell level, as a result of biochemical alterations having an underlying genetic basis. The assumption has always been that a similar process takes place in cancer patients exposed to chemotherapy whose tumors may initially respond (i.e., regress, or partially regress) to chemotherapy, only to eventually grow back as a result of acquiring resistance to the same and perhaps even unrelated drugs. Whereas this is undoubtedly true in many instances, there are a number of reasons to suggest that such unicellular drug-resistance mechanisms are not always the underlying factor in explaining some clinical forms of acquired drug resistance expressed by solid tumors after drug exposure in vivo.

One of the most compelling of these reasons is the time generally required to isolate drug-resistant mutants in tissue culture, in contrast to clinical drug resistance. In the case of human tumor cell lines, it is not unusual for investigators to continuously expose the cells to increasing concentrations of drug for periods of 1 or 2 years (Frei et al. 1985; Cadman 1989). However, as summarized by Cadman (1989), patients can often manifest symptoms of resistance over much shorter periods of time. This is an unexpected paradox, given the following considerations: (1) Established human tumor cell lines generally have population doubling times of 24–48 hours in tissue culture, whereas solid tumors in patients have population doubling times measured in weeks, or more often, months; (2) in contrast to cell cultures, patients cannot be exposed continuously to high doses of chemotherapeutic drug; (3) in dispersed monolayer cell cultures, every cell can be exposed to high concentrations of drug, whereas problems of drug access and penetration would prevent many of the cells in solid tumor masses from being exposed to effectively toxic concentrations of drug. For these reasons, one would reasonably anticipate that the conditions for selecting rare drug-resistant mutants in culture should be vastly more favorable than those in patients or animals with solid tumors. Consequently, it might be expected that acquired drug resistance would develop much more slowly in the clinical setting than is frequently observed. It is probable that in some cases this paradox can be resolved by assuming that in patients only very low levels of drug resistance, e.g., 2- to 5-fold, are all that is often required to result in a lack of response (Young 1989; Kuboda et al. 1991) to a given drug or set of drugs; selection and analysis in vitro of such low-level resistant variants (which is rarely done) might be achievable over much shorter periods of time than the typical high-level variants which are normally isolated and studied.

There is, however, an alternative explanation which, if correct, has wide-ranging implications for the study of many aspects of tumor biology, including drug resistance. It is that the mechanisms of acquired drug resistance which develop in multicellular (three-dimensional) solid tumors may sometimes involve types of intercellular interaction which would not necessarily be operative in two-dimensional monolayer culture systems containing a high fraction of dispersed single tumor cells. If this is so, the basic mechanisms of resistance which operate and/or develop in multicellular solid tumors may sometimes be fundamentally different from those in unicellular, dispersed tumor cell systems such as leukemias in vivo or monolayer cell culture systems in vitro. As it turns out, there is a quite sound rationale for this possibility, some of which is summarized below.

BIOLOGY OF MULTICELLULAR TUMOR SPHEROIDS

In an effort to more accurately recreate the in vivo growth conditions and behavior of solid tumors in a tissue culture context, Sutherland and coworkers developed the "multicellular tumor spheroid" model (Inch et al. 1970). The basic approach is to place tumor cell lines, which normally grow as sheets of flattened monolayers, on tissue culture plastic in an in vitro context where such a type of monolayer growth is inhibited. For example, the cells may be grown in spinner flasks where the culture medium is constantly rotated, thus preventing the cells from adhering to the culture vessel. Alternatively, the cells may be grown in petri dishes or multi-well tissue-culture plates in which the plastic surface has been precoated with a thin layer of nonadhesive material, e.g., 1% agarose (Rofstad 1986; Rofstad et al. 1986). Under these conditions, the adhesive conditions *between* the cells may become greater than between the cells and the substrate on which they are plated, and thus result in spontaneous homotypic cell aggregation. These aggregates may then form multilayered spheroidal structures up to about 1 mm in diameter. Growth beyond this size in vitro is severely limited by availability of nutrients and oxygen, since the spheroids lack a vasculature. However, growth to larger sizes, even 1–1.5 cm, can sometimes be achieved by simulating microgravity conditions and reducing shear forces, using specially designed rotating wall ("bioreactor") culture vessels (Goodwin et al. 1992).

Multicellular tumor spheroids, which are roughly the equivalent of avascular microscopic metastases, manifest several characteristics commonly associated with solid tumors in vivo (with the obvious exception that they lack a stroma). For example, they generally contain an outer crust, or shell, of viable proliferating cells and an inner, less dense, core containing many dead or dying cells (Muller-Klieser 1987; Sutherland 1988). The inner regions generally have lower pH values than the outer regions, and also tend to be hypoxic (Muller-Klieser 1987; Sutherland 1988). This latter feature is particularly crucial with respect to radiation sensitivity; i.e., tumor cells in regions of hypoxia tend to be (more) radioresistant than those found in oxygen-rich regions (Muller-Klieser 1987; Sutherland 1988). Not surprisingly, radiation biologists have been most responsible for advancing the use of multicellular tumor spheroids in basic studies of cancer research, including experimental therapeutics. Many of these studies have shown that the response of tumor cells grown in vitro as multicellular spheroids more faithfully mimics the response of the same cells grown in vivo as solid tumors than does analysis of the cells grown in vitro as conventional monolayers (Muller-Klieser 1987; Sutherland 1988).

Around 1980, a small number of investigators began to employ spheroid models to study aspects of drug sensitivity and resistance of tumor cells in tissue culture. These studies revealed several important findings (Sutherland et al. 1979). One such finding was that the sensitivity of many tumor cell lines to the toxic effects of a given chemotherapeutic drug could be reduced, sometimes dramatically, by exposing the cells in the form of 0.5-mm to 1-mm spheroids, when compared to monolayer cell cultures of the same cell lines (Muller-Klieser 1987; Sutherland 1988). This is especially true for high-molecular-weight drugs, suggesting that in many cases this observation can be explained simply by a relative lack of drug access to the inner regions of the spheroids (Muller-Klieser 1987; Sutherland 1988); i.e., there is a penetration gradient. This is an important physiological explanation to account for some manifestations of drug "resistance." However, other factors, which appear to incriminate three-dimensional intercellular contact, also appear to contribute to the relative "intrinsic" resistance of multicellular tumor spheroids. For example, the layers of cells comprising the very outermost regions of such tumor spheroids, which are accessible to drugs, nevertheless tend to express a much greater degree of resistance (e.g., 10-fold) to drugs such as etoposide than the same cells grown as conventional monolayer cell cultures (Olive et al. 1993). Eventually, such spheroid-derived cells grown in monolayer culture begin to express a drug-sensitivity phenotype characteristic of the same cells grown in long-term monolayer cell cultures. This kind of observation led Sutherland, Durand, and Olive to propose many years ago the hypothesis that a cell-cell or cell-ECM (extracellular matrix) "contact effect" may endow cells in solid tumors with an elevated ability to resist the toxicity associated with exposure to anticancer chemotherapeutic drugs (Durand and Sutherland 1972, 1973). We would emphasize that the contact effect hypothesis has been mainly studied in the context of *intrinsic* drug resistance, whereby tumor cells are exposed once to a given drug and then their survival response is assessed. However, as discussed in a subsequent section, versions of the contact effect may also help explain some aspects of *acquired* or *induced* resistance of solid tumors to certain anticancer drugs, such as alkylating agents. Assuming that many manifestations of drug resistance, especially of the acquired variety, have a mutational or epigenetic basis, it would seem reasonable to suppose that fundamental differences in the control of gene expression may be mediated by intercellular or tumor cell-ECM contact effects associated with cells growing as three-dimensional multicellular aggregates, i.e., as solid tumors. Indeed, there is some evidence to support this, as described below.

ALTERED GENE EXPRESSION IN THREE-DIMENSIONAL MULTICELLULAR SPHEROID SYSTEMS

Although the literature is by no means extensive, there are a number of reports which have shown interesting differences in gene expression, or the control of gene expression, in cultured cells grown as multicellular spheroids compared to conventional monolayers. For example, Laderoute et al. (1992) found mRNA and protein levels for transforming growth factor-α (TGF-α) were elevated (about 3-fold) in A549 human squamous carcinoma cells grown as multicellular tumor spheroids compared to the same cells grown as monolayers, even as densely packed cell cultures (Laderoute et al. 1992). Similar results were reported by the same group in relation to expression of the heme oxygenase gene (Murphy et al. 1991). Previous studies in our laboratory by Theodorescu et al. (1993) were undertaken to examine expression of mRNA levels for TGF-β_1 in two different human tumor cell lines, namely A549 squamous carcinoma cells and MDA-MB-231 human breast carcinoma cells. Both constitutive levels and TGF-β_1-inducible mRNA levels were examined in monolayer and spheroid cultures of these cell lines. Essentially opposite patterns of expression were observed when this type of comparison was undertaken. An example of this is shown in Figure 1 for MDA-MB-231 human breast carcinoma cells. Thus TGF-β_1 mRNA was readily detectable in monolayer cell cultures, and the level was not significantly changed by exposing the cells to TGF-β_1 protein. In contrast, TGF-β_1 mRNA levels, relative to GAPDH controls, were much lower in spheroid cultures of the same cells, but these levels could be inducibly up-regulated by exposure of the spheroids to TGF-β_1 (Theodorescu et al. 1993). The increase peaked at about 24 hours posttreatment and returned to control levels by 72 hours.

The basis of altered gene expression in multicellular tumor spheroids is unknown but conceivably could be traced to differential function of promoters and transcription factors in cells within spheroids. An interesting example of the latter possibility was recently provided by studying the expression of a transfected gene (Klunder and Hulser 1993). The bacterial β-galactosidase (LacZ) gene under the control of a human β-actin promoter was transfected into mouse Ltk$^-$ fibrosarcoma cells and the expression of LacZ was monitored in several transfected clones grown either as monolayers or multicellular spheroids. As expected, LacZ was strongly expressed in monolayer cell cultures, with almost 100% of the cells being positive. However, only 2% of the same cells expressed LacZ activity when grown long-term as spheroids. Recultivation of disaggregated spheroids as dispersed single cell monolayers resulted in a gradual resurrection of full expression of LacZ activity over a 10- to 14-day period (Klunder and Hulser 1993). In another study, the activity of the human metallothionein promoter in cell populations isolated from varying depths in multicellular spheroids of HT-29 human colon carcinoma cells was examined following exposure to cadmium. Cells obtained from inner regions did not manifest promoter activity, whereas cells from the outer regions did. However, the results were apparently attributable to limited diffusion of cadmium into the deeper regions of the spheroids (Untawale and Mulcahy 1992).

There is thus some preliminary evidence which indicates that the function of regulatory promoter sequences may be different in the tissue-like environment of a multicellular mass when compared to cells grown as dispersed monolayer cell cultures. This could help account for altered protein expression in cells within multicellular spheroids, e.g., enhanced or suppressed expression of certain integrin receptors for ECM proteins (Walch et al. 1994).

Altered gene expression in cells grown as multicellular spheroids is not restricted to tumor cell populations. For example, normal liver hepatocytes can be grown in vitro as spheroidal aggregates or as monolayer cell cultures (Flouriot et al. 1993). When the hepatocytes are grown three-dimensionally, the pattern of expression of liver-specific genes more faithfully mimics that of hepatocytes in vivo: Stable and long-term expression of several liver-specific or associated genes, e.g., the vitellogenin gene, was observed in spheroids, but not in monolayer cell culture (Flouriot et al. 1993). Similarly, expression of mammary-gland-specific proteins by normal mammary epithelial cells is restricted to cells cultured three-dimensionally on a reconstituted basement membrane, matrigel (Barcellos-Hoff et al. 1989).

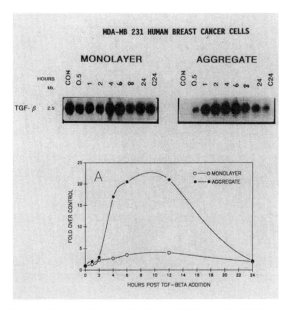

Figure 1. Contrasting constitutive or TGF-β_1-inducible levels of TGF-β_1 mRNA in MDA-MB-231 human breast cancer cells grown as monolayer cell cultures or multicellular tumor spheroids. The upper left-hand panel shows that the cells grown as monolayers expressed readily detectable TGF-β_1 mRNA, the level of which was minimally increased by exposing the cells to exogenous recombinant human TGF-β_1 (as shown by densitometry estimates in the bottom panel). In contrast, constitutive TGF-β_1 mRNA was not readily detectable in the same cells grown as multicellular tumor spheroids (as shown in CON or control lane) but was observed in the cells after exposure of the spheroids to TGF-β_1. There was up to a 20-fold induction which peaked between 8 and 12 hr. (Adapted, with permission, from Theodorescu et al. 1993 [copyright Tissue Culture Association].)

RECAPITULATION OF SOLID TUMOR PHENOTYPES EXPRESSED IN VIVO BY TUMOR CELLS GROWN THREE-DIMENSIONALLY IN CELL CULTURE

There are a number of situations where a particular phenotype expressed by solid tumors growing in vivo cannot readily be reproduced in cell culture unless the cells are grown in a three-dimensional manner, e.g., as multicellular tumor spheroids or as so-called native-state "histocultures" in floating collagen gels (Guadagni et al. 1991; Hoffman 1991b). The latter systems, unlike the spheroids, contain a stroma and will, for example, express patterns of tumor cell antigen expression characteristic of the same tumors growing in vivo, as summarized by Hoffman (Hoffman 1991a,b; Vescio et al. 1991).

One interesting example of resurrection of an in vivo phenotype by growth as multicellular spheroids is metastatic cell "clonal dominance" of primary tumors. This refers to the phenomenon in which rare, metastatically competent cellular variants found within primary tumors at an early stage of development can sometimes manifest a growth advantage at this site and thus, over time, become the dominant functional cell population in advanced primary tumors (Kerbel et al. 1988; Kerbel 1990). As such, it represents a potentially effective means by which some types of primary tumors, even small ones such as most cutaneous human melanomas, can generate and seed sufficient numbers of metastatic tumor cells into the circulation, and thus increase the probability of distant metastases (Kerbel et al. 1988; Kerbel 1990).

This phenomenon was first discovered by mixing a large number of genetically "tagged" tumor cell clones in which only a small minority of the clones are metastatically competent, and then injecting the mixture into an experimental animal host to grow as a solid tumor (Kerbel et al. 1987, 1989; Korczak et al. 1988). In some cases, clones of the metastatic subpopulation can be observed to overgrow their initially more numerous nonmetastatic counterparts at the site of inoculation (Waghorne et al. 1988). An example of this is shown in Figure 2. In this case, clones of a mouse mammary carcinoma were tagged by transfection with the plasmid pSV2neo: The random integrations of the plasmid DNA can be exploited as unique heritable genetic markers for each and every clone, and these can be detected by Southern blotting. As shown in Figure 2, a clone which was metastatically competent and present initially in only 1 in 50 to 100 of the pooled clones (obtained from the SP1 mouse mammary carcinoma) came to dominate the primary tumors within 6 weeks. This clone, as expected, was also responsible for the formation of distant lung metastases recovered in the same animals. Paradoxically, when the same mixture of genetically marked clones was grown in conventional monolayer cell cultures for an extended period of time (e.g., 3 months) there was no evidence of dominance of the metastatic clone (Waghorne et al. 1988) as shown in Figure 2. However, in later studies this apparent paradox was solved when it was found that less hospitable in vitro growth conditions—especially growth as multicellular tumor spheroids—could recapitulate the clonal dominance patterns observed within solid tumors of the same mixtures grown in mice (Rak and Kerbel 1993). The most logical interpretation is that the selective growth advantage of the metastatic subclone could not be expressed in a cell culture environment in which every clone finds itself under ideal (i.e., nonlimiting) growth conditions, e.g., in subconfluent monolayer cell cultures containing a high concentration of serum. However, the environment of a spheroid, especially the deeper regions, would present far more restrictive growth conditions, and thus recreate the required selective conditions for clonal dominance observed in vivo.

These results raise an obvious question: Could selection and overgrowth of drug-resistant subpopulations *in*

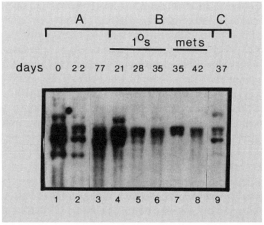

Figure 2. Clonal dominance of a metastatically competent tumor subline in vivo during solid tumor growth but not in vitro monolayer cell culture. A mouse mammary carcinoma called SP1 was transfected with pSV2neo, and between 50 and 100 G418-resistant clones were pooled. Southern blot analysis of this pooled population using pSV2neo as a hybridization probe revealed a heterogeneous pattern of multiple bands as shown in lane *1*. When these cells were grown in monolayer tissue culture for up to 77 days, there was little resolution of the bands, as shown in lanes *2* and *3*. If, however, the pooled population was injected subcutaneously into syngeneic CBA/J mice and the tumors which arose at the site of inoculation were removed on day 21 (lane *4*), day 28 (lane *5*), or day 35 (lane *6*), a resolution was noted indicating overgrowth (selection) of a single dominant clone (Kerbel et al. 1987; Waghorne et al. 1988). Unlike the majority of the other clones (SP1 is normally nonmetastatic when the cells are injected subcutaneously), the dominant clone within the primary tumor is metastatically competent. This is indicated by the presence of the same clone in solitary and independent lung metastases which arose in the same animals, and which were removed 35 days (lane *7*) or 42 days (lane *8*) after tumor cell injection. Lane *8* is the Southern pattern obtained from a number of pooled lung colonies which arose after intravenous injection of the cells. These are so-called "artificial metastases" and most are derived from clones which are distinct from the dominant clone observed giving rise to distant "spontaneous" lung metastases after subcutaneous inoculation. Although the pattern of in vivo clonal dominance was not observed in monolayer cell cultures using high serum conditions, we found that growth under more restrictive conditions, especially as multicellular tumor spheroids, resulted in clonal dominance similar to that observed in vivo (Rak and Kerbel 1993). (Reprinted, with permission, from Waghorne et al. 1988.)

solid tumors be affected in a similar way? We believe the answer is yes, based on results described below.

TUMOR CELL LINES SELECTED FOR RESISTANCE TO ALKYLATING AGENTS IN VIVO EXPRESS THEIR RESISTANT PHENOTYPE IN VITRO WHEN GROWN AS MULTICELLULAR SPHEROIDS

From the above discussion, it is apparent that solid tumors could conceivably acquire drug resistance properties in vivo based on mechanisms which might not necessarily be expressed unless the cells are grown as three-dimensional cell aggregates. If correct, this could have some important implications for understanding the clinical basis of acquired drug resistance, as attempts to select drug-resistant mutants in vivo by exposing solid-tumor-bearing animals to chemotherapeutic drugs are relatively rare. Thus, resistance mechanisms which are operative in vivo might not always be uncovered by selection procedures and analysis in vitro (Schecter et al. 1991). This was the rationale for the experiments undertaken by Teicher et al. (1990) when they exposed tumor-bearing mice to one of a number of alkylating agents in order to select for drug-resistant sublines. The tumor was the BALB/c mouse EMT-6 mammary carcinoma grown as a solid tumor after subcutaneous inoculation. The tumor was serially passaged in ten successive BALB/c recipients and the mice were exposed each time to drug by intraperitonal injection of maximum tolerated doses of cisplatin (CDDP), cyclophosphamide (CTX), thiotepa, or carboplatin. The tumors were removed 24 hours after each exposure and passaged into new animals. The process was repeated 10 times over a 6-month period (Teicher et al. 1990). This resulted in the derivation of a series of sublines which expressed resistance preferentially to the drug used for the selection as well as lower levels of resistance to the other alkylating agents. The resistance was stable for at least 3 months in vivo in the absence of drug. Surprisingly, the resistance phenotype was only expressed in vivo: When the tumors were removed and grown in tissue culture (as monolayer cell cultures), they quickly expressed a wild-type drug-sensitive EMT-6 parental phenotype. If these cell lines were returned to animals by subcutaneous injection, the resultant solid tumors which developed reexpressed their resistance properties, thus suggesting a form of drug resistance which was "operative only in vivo" (Teicher et al. 1990).

It is unclear how common this type of phenomenon is, since there are so few reports of generating drug-resistant mutants or variants by exposing animals with *solid* tumors to chemotherapeutic drugs. Thus, it is noteworthy that there is at least one other report, showing that drug-resistant sublines generated in vivo from solid tumors do not express their resistance properties in vitro (Starling et al. 1990). In this case, a human lung adenocarcinoma cell line was grown subcutaneously in nude mice, and subpopulations resistant to a vinca alkaloid-monoclonal antibody conjugate were isolated in vivo by a process involving multiple drug exposures of the tumor in a succession of nude mouse recipients. Variants were isolated which expressed resistance to the drug-antibody conjugate or to the unconjugated drug in nude mice, but their resistance properties were not expressed in monolayer tissue culture (Starling et al. 1990). However, it is also clear that there are other examples in which human or mouse solid tumors made drug resistant in vivo *will* express their resistance properties in monolayer cell culture (Jones et al. 1993).

Given the literature on multicellular tumor spheroids, including our own previous results (as discussed above), we initiated a collaboration with Teicher to determine whether growth of the EMT-6 sublines selected for alkylating agent resistance in vivo would express their resistance properties if grown in three-dimensional cell culture systems (Kobayashi et al. 1993). The results we obtained did indeed show this to be the case. Thus, in agreement with Teicher, we found that the sublines expressed resistance in vivo but not in vitro when grown as monolayer cell cultures. However, similar levels and patterns of cross-resistance were observed if the cell lines were grown as multicellular spheroids on plastic surfaces precoated with 1% agarose (Kobayashi et al. 1993), as shown in Figure 3. Resistance was monitored by exposing BALB/c tumor-bearing mice to various concentrations of CDDP or CTX and then evaluating the surviving fractions of clonogenic (viable colony-forming) cells. We have also monitored resistance in vivo by monitoring solid tumor growth delays subsequent to drug exposure of tumor-bearing animals, as shown in Figure 4.

These preliminary results suggested a new possible paradigm to explain some forms of acquired drug resistance in solid tumors, namely that drug resistance, in some cases, may be expressed at the multicellular

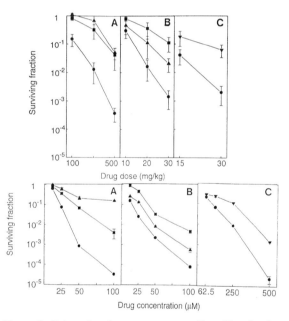

Figure 3. Colony-forming assays representing either in vivo (*top* panel) or in vitro (*bottom* panel) resistance of EMT-6 cells selected in vivo for resistance to cyclophosphamide (CTX; ▲), cisplatin (CDDP; ■) or thiotepa (THIO; ▼), or control cells (●) passaged in vivo without drug treatment. (*Top* panel) 24 hr after i.p. injections of CTX, CDDP, or THIO (*A*, *B*, and *C*, respectively) into tumor-bearing mice, tumors were removed and single cell suspensions were prepared and plated for colony-forming assays. (*Bottom* panel) Colony-forming assay after 1-hr drug exposure of each tumor line in three-dimensional culture. Following drug exposure, multicellular aggregates were dispersed into single-cell suspensions which were then plated to determine colony-forming ability. All results are expressed as surviving fractions ±S.E.M. of cells from treated groups, compared with untreated control groups. (Adapted, with permission, from Kobayashi et al. 1993.)

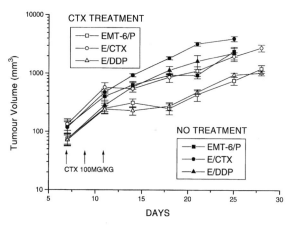

Figure 4. In vivo growth delay response of EMT-6 mammary carcinoma cells or drug-resistant sublines isolated from the EMT-6 cell line in vivo, by Teicher et al. (1990). Syngeneic BALB/c mice were used as recipients for a subcutaneous injection of 10^6 cells of each of the cell lines. The cell lines used were the drug-sensitive EMT-6 parental cells (EMT-6/P) a cisplatin (DDP)-resistant subline (E/DDP), or a cyclophosphamide (CTX)-resistant subline (E/CTX). When the tumors reached a size of about 100 mm³, the mice were injected intraperitoneally (on days 6, 8, and 12) with 100 mg/kg CTX and the response of the tumors was examined by measuring their size and rate of growth. The CTX-resistant subline exhibited no growth delay after CTX treatment, whereas the EMT-6/P- and DDP-resistant subline did so. The growth delay of the cisplatin-resistant subline took place despite the fact that it shows some cross-resistance to CTX in the colony formation assay in vivo and in vitro, as shown in Fig. 3 and described by Kobayashi et al. (1993).

(group) level, and not only at the (uni)cellular level. This possibility was strengthened by disaggregating the drug-resistant spheroids by trypsinization and subsequently testing the released cells in monolayer cell culture for their relative drug sensitivity. These experiments revealed a major reversal of drug resistance (Kobayashi et al. 1993).

During the course of these experiments we also noted, unexpectedly, that the morphology of the drug-resistant EMT-6 sublines differed quite strikingly from that of the parental cells (Kobayashi et al. 1993). Thus, the resistant cells formed dense, compact spheroids, whereas the parental drug-sensitive cells formed much looser (and larger) aggregates (Kobayashi et al. 1993). Whether the resistance properties of the drug-resistant spheroids are in some way related to this increase in cell density remains a matter of speculation. For example, although there are precedents for this "compaction" phenomenon in drug-treated multicellular spheroids (Vasanthi et al. 1986), there are also instances where drug-resistant sublines grow in a less—not more—compact fashion as multicellular spheroids (Soranzo et al. 1989). Nevertheless, the results suggested that in vivo acquired resistance to alkylating agents may involve a nonconventional mechanism in which the basic unit of resistance is not necessarily the single cell but rather a group of interacting cells. This is probably another version of the contact effect (Olive and Durand 1994), but in this case, acquired, as opposed to intrinsic, drug resistance is involved.

A TRANSIENT FORM OF "MULTICELLULAR" RESISTANCE TO ALKYLATING AGENTS CAN BE INDUCED BY SINGLE DRUG EXPOSURE

We subsequently undertook a series of experiments to evaluate whether resistance to alkylating agents can be induced rapidly by single drug exposures when the drug is applied to tumor cells grown as three-dimensional cell cultures (Graham et al. 1994). The rationale for doing so stems from the fact that the isolation of drug-resistant sublines in vitro often requires exposure to increasing concentrations of drug over very long periods of time, as summarized above. This is especially true for human cell lines when trying to isolate alkylating-agent-resistant variants (Frei et al. 1985). We therefore sought to determine whether mouse EMT-6 or human breast cancer cells exposed in three-dimensional culture to a single dose of CDDP or 4-hydroperoxycyclophosphamide (an activated form of cyclophosphamide) would express a detectable degree of drug resistance when the surviving cells are reexposed to the same drug as cell aggregates, but not when reexposed as conventional cell monolayers. This was indeed found to be the case (Graham et al. 1994). An example of this is shown in Figure 5 using human MDA-MB-231 cells. This resistance was found to be unstable and disappeared with five population doublings. The cells also did not express cross-resistance to other alkylating agents. These results suggest yet again that multicellular organization can confer a survival/protective advantage to cells that is not necessarily observed in single cell systems. Because this type of multicellular resistance can be induced rather quickly, it may be relevant to explaining the seemingly rapid development of drug resistance in some patients with solid tumors who initially are responsive to chemotherapy (Graham et al. 1994; Steeg et al. 1994).

MULTICELLULAR CONTACT AS A REGULATOR OF PROGRAMMED CELL DEATH

The notion that acquired drug resistance may sometimes be expressed preferentially or exclusively at the multicellular level, and that changes in intercellular contact may influence the degree of resistance expressed, is something that deserves serious consideration for several reasons. First, there is now considerable evidence that anticancer chemotherapeutic drugs (including alkylating agents) often kill tumor cells by induction of apoptosis or programmed cell death (Barry et al. 1990; Eastman 1990; Reed 1993). Second, there is evidence that the process of programmed cell death can be strongly influenced by cell-cell interactions. As Raff (1992) recently summarized, individual cells normally require signals from other cells in order

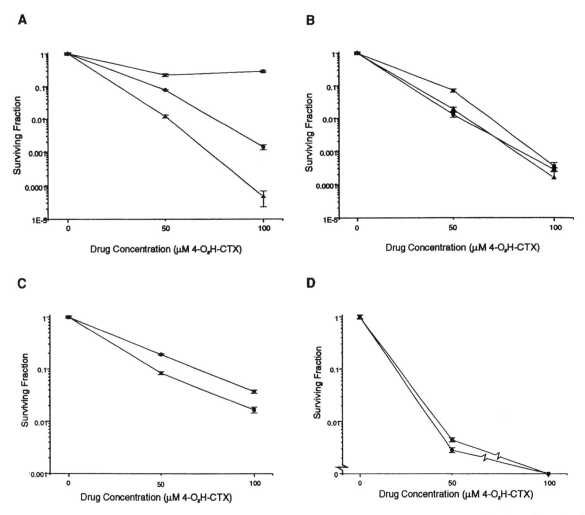

Figure 5. Rapid induction of transient resistance to cyclophosphamide in EMT-6 mammary tumor cells using three-dimensional culture systems (Graham et al. 1994). This figure summarizes the results of experiments in which the in vitro survival of EMT-6 cells was assessed after a 1 hr exposure to the activated form of CTX (4-O$_2$H-CTX) applied to either spheroid (*A* and *C*) or monolayer (*B* and *D*) cultures. The cells used in *A* and *B* had been preexposed to a single dose of 25 μM 4-O$_2$H-CTX for one (●) or three (■) days, or had not received any previous drug treatment (▲). Cells in *C* and *D* were either not previously treated (■) or had been preexposed to 10 μM 4-O$_2$H-CTX for one day (●). Symbols represent the mean surviving fraction (S.F.) ±S.E. of cells plated in triplicate wells. (Reprinted, with permission, from Graham et al. 1994.)

$$\text{S.F.} = \frac{\text{Plating efficiency (P.E.) of drug-treated cells}}{\text{P.E. of cells from tumors without drug treatment}}$$

to proliferate and, likewise, appear to need signals from other cells in order to survive; in their absence, individual cells may self-destruct by activating intrinsic suicide programs. An interesting illustration of this hypothesis, from an oncologic perspective, was provided by Bates et al. (1994), who recently reported that apoptosis can be induced in solid tumors by inhibition of intercellular contact. They studied a human colon carcinoma cell line (called LIM 1863) which grows spontaneously in vitro as a three-dimensional spheroid or "organoid" composed of goblet and columnar cells around a central lumen. However, when the organoids were transferred to calcium-depleted medium, they underwent disaggregation into a single cell suspension. Reformation of organoids took place if the single cells were placed in calcium-containing medium, and this could be prevented by antibodies to the α_v integrin subunit (Bates et al. 1994). If single tumor cells were prevented from reforming organoids, they spontaneously underwent apoptosis, suggesting that cell survival ultimately depended on establishing and maintaining close intercellular homotypic contacts.

Support for the idea that cell adhesion may function as a regulator (suppressor) of apoptosis has also been investigated in heterotypic cell contact systems. For example, the survival and growth of a mouse T-cell lymphoma cell line called CS-21 was shown to depend on contact with lymph node stromal cells (Fujita et al. 1993). This type of heterotypic cell contact could be disrupted by monoclonal antibodies to lymphoma-associated adhesion molecules, and this resulted in apoptosis of the lymphoma cells (Fujita et al. 1993). Such results suggest that promotion of tumor cell survival within solid tumors may depend on the expression of certain cell-surface molecules which help regulate and maintain multicellular tumor growth. A variation of this theme comes from recent results which show the process of apoptosis can be induced in normal cells when certain interactions involving cell-surface integrin receptors with their corresponding ECM ligands are prevented or disrupted (Meredith et al. 1993; Frisch and Francis 1994; Ruoslahti and Reed 1994).

RELATIONSHIP OF CELL SURVIVAL AND MOLECULES INVOLVED IN CELL-CELL ADHESION IN SOLID TUMORS

Little is known about the molecular basis of multicellular organization of solid tumors, including tumor spheroid formation in vitro. Some obvious possible molecular mediators include the cadherin family of calcium-dependent homophilic adhesion molecules, such as E (epithelial) cadherin (Behrens et al. 1993), members of the immunoglobulin supergene family of adhesion molecules such as caranoembryonic antigen (CEA) (Benchimol et al. 1988), and recently described, the epithelial adhesion molecule known as EpCAM as well (Litvinov et al. 1994). Various CD44 isoforms are also candidate molecules (St. John et al. 1990), as are the heterodimeric integrin receptors for ECM proteins (Walch et al. 1994), as discussed above (Bates et al. 1994). Expression of such molecules may endow cells with a survival advantage, and sometimes this may occur preferentially or exclusively in a multicellular context. Moreover, their expression may be altered by exposure to certain cytotoxic agents. A possible paradigm for this comes from studies on a protein known as "clusterin," a serum glycoprotein endowed with cell-aggregating ability (Fritz et al. 1983; Jenne and Tschopp 1992). Clusterin has also been considered to be a specific marker of cells undergoing programmed cell death (Jenne and Tschopp 1992), including tumor cells exposed to cytotoxic chemotherapeutic drugs (Buttyan et al. 1989). However, very recent evidence has implicated that clusterin may actually be a *survival* molecule. Thus, upon closer examination, human clusterin gene expression appears to be confined to surviving cells during various types of in vitro programmed cell death (French et al. 1994). Clusterin could, therefore, turn out to be an example of a molecule involved in cell adhesion, the altered expression of which in drug-treated tumor cells promotes their survival by inducing changes in intercellular contact.

CAN "ANTI-ADHESIVES" FUNCTION AS CHEMOSENSITIZERS FOR CONVENTIONAL CYTOTOXIC ANTICANCER CHEMOTHERAPEUTIC DRUGS?

The main points we have tried to make thus far include: (1) The ability of tumor cells from solid tumors to *acquire* resistance to cytotoxic drugs can be strongly influenced by multicellular growth; indeed, there are situations where induced or acquired resistance to alkylating agents is not observed unless the tumor cells are assayed as multicellular aggregates. (2) This type of "multicellular" drug resistance is sometimes accompanied by changes in the three-dimensional structure of surviving drug-exposed cells or drug-resistant sublines, such as increased adhesivity and aggregation. (3) Disruption of close intercellular homotypic or heterotypic contacts in a variety of tumor systems can promote tumor cell death.

Taken together, these observations suggest that, under certain circumstances, agents that possess "anti-adhesive" functions, even if not toxic on their own, may enhance the ability of conventional cytotoxic drugs to kill tumor cells in solid tumors by minimizing, preventing, or reversing the acquisition of drug resistance (Kerbel et al. 1994). Anti-adhesives might also be used with other types of anticancer agents such as interferon or suramin, since induced/acquired resistance against such agents may develop at the multicellular level as well (Gorlach et al. 1994; Lelievre and Larsen 1994). Agents that could be used as anti-adhesives in conjunction with chemotherapy include neutralizing antibodies to cell adhesion molecules (or other types of surface molecules involved in cell-cell communication) and synthetic peptides of ECM components, or non-peptide mimetics (Ruoslahti et al. 1994). Indeed, at least one antitumor monoclonal antibody called 17.1A, which has been used with some success (on its own) in phase I, II, or III clinical trials in patients with solid carcinomas (LoBuglio et al. 1988; Reithmüller et al. 1994), is now known to react with a calcium-independent homophilic cell adhesion molecule known as EpCAM (Litvinov et al. 1994). If such antisera have anti-adhesive effects on solid tumors in vitro or in vivo, it would be of considerable interest to determine if they could increase the effectiveness of conventional anticancer chemotherapeutic drugs when combined with them in combination therapy protocols. Could they, in short, function as chemosensitizers, or reversal agents, either to augment the activity of chemotherapeutic drugs or to reverse resistance acquired to such drugs? There is already evidence that certain monoclonal antibodies to cell-surface EGF receptors can render tumor cells more sensitive to cisplatin (Fan et al. 1993), possibly by impairing DNA repair processes in the antibody-treated cells (Pietras et al. 1994). With this precedent in mind, it would be of considerable interest to evaluate whether antibodies to adhesion molecules, integrins, etc. could bring about similar effects, and whether their capacity to do so is observed only, or preferentially, in multicellular systems such as solid tumors.

SUMMARY: IMPLICATIONS OF MULTICELLULAR RESISTANCE FOR DRUG SCREENING AND CLINICAL TRANSLATION OF BASIC RESEARCH IN CANCER BIOLOGY

The results summarized in this paper raise the specter of a new paradigm to explain aspects of acquired drug resistance in solid tumors. Among the implications of the concept of acquired "multicellular" resistance is the possibility that molecules involved in multicellular organization may contribute to this type of resistance. These include various cell adhesion molecules, integrin receptors for ECM constituents, the ECM itself, gap junctional proteins, and intracellular molecules which contribute to cell adhesion and shape, e.g., microtubules, intermediate filaments, and the catenins (Hirano et al. 1992). Consequently, agents that interfere with the expression and/or function of such molecules may be useful as chemosensitizers for some conventional anticancer chemotherapeutic drugs.

Multicellular resistance may also be relevant to intrinsic drug resistance, as well as acquired (or intrinsic) resistance to other types of anticancer agents, e.g., cytostatic cytokines such as interferon or tumor necrosis factor-α (Gorlach et al. 1994). With respect to the issue of intrinsic drug resistance, there is now considerable evidence that multicellular spheroids are often more intrinsically resistant to various cytotoxic drugs than corresponding monolayers of the same tumor cell lines (Muller-Klieser 1987; Sutherland 1988; Olive and Durand 1994). The same may be true for certain growth inhibitory cytokines (Gorlach et al. 1994). As summarized above, this may sometimes be due to reduced drug (or cytokine) penetration into the inner regions of spheroids, but other factors involving intercellular contact and increased cell survival (which in turn is probably related to decreased apoptosis) also appear to be involved. This may explain some recent results obtained in our laboratory in which tumor cell lines exposed to highly toxic concentrations of alkylating agents can be rescued from the imminent and massive cell death observed in monolayer cell cultures by placing the drug-treated cells into three-dimensional culture (C.H. Graham et al., in prep.). This is similar to the results of Miller et al. (1985), who found that tumor cells treated in monolayer with toxic doses of various drugs could be protected substantially from cell death by incorporating the drug-treated cells into three-dimensional floating collagen gels immediately after drug exposure.

Such results may be one explanation for the high degree of intrinsic resistance which most solid tumors express in vivo. As discussed in a preceding section, this raises a number of interesting questions about the interrelationships of cell survival, apoptosis, and multicellular organization, including the issue of how to screen for new anticancer drugs. The latter is traditionally done using monolayer tumor cell culture systems, the wisdom of which can be called into question with respect to drug treatment of solid tumors (Durand 1990; Johnson 1990; Steeg et al. 1994). Thus, some agents that appear to be highly toxic based on monolayer cell culture screens may be ineffective on three-dimensional aggregates of the same cells. Conversely, there may be drugs which might actually be more effective on tumor cells grown as solid tumors rather than dispersed single cell monolayers. Such agents would be missed using current cell culture screening methods. If true, this could be one important factor in helping to explain why so few new drugs for the treatment of solid tumors have been discovered during the last 20 years. The same argument can probably be made with respect to the discovery and study of drug-resistance-reversal agents or chemosensitizers. A possible illustration of this comes from studies on P-glycoprotein-mediated multidrug resistance.

It is now well known that overexpression of P-glycoprotein is a major molecular mechanism of multidrug resistance involving natural product compounds including antibiotics, anthracyclines, vinca alkaloids, and taxol (Bradley et al. 1988; Borst 1991). P-glycoprotein is thought to act as a cell-surface energy-dependent drug efflux pump to reduce the intracellular levels of drug that can be attained in individual cells (Bradley et al. 1988). This is a classic example of a "unicellular" drug resistance mechanism. Interest in P-glycoprotein-mediated multidrug resistance was further stimulated by the discovery that its function as a drug resistance molecule could be suppressed or reversed by certain pharmacological agents, e.g., calcium channel blockers such as verapamil, or cyclosporin (Bradley et al. 1988; Fan 1994). Such reversal agents work exceptionally well in monolayer cell culture systems in vitro and in "liquid" tumors in vivo. Indeed, this is almost without exception how they have been experimentally studied (Fan 1994; Houghton and Kaye 1994).

On the basis of such encouraging results, many clinical trials have been undertaken to evaluate the effects of P-glycoprotein reversal agents in the treatment of cancer (Bellamy et al. 1990; Fan 1994; Houghton and Kaye 1994). A definite trend has emerged from the results of such trials; namely, that such reversal agents seem to have little (if any) clinical benefit for the treatment of most adult *solid* tumors, whereas enhanced clinical benefit may be realized in certain hematologic cancers, e.g., multiple myeloma and myelogenous leukemias (Fan 1994; Houghton and Kaye 1994). What is the reason for this respective pattern of apparent failure and success? Durand has speculated that one major factor may be the way P-glycoprotein functions in single cells versus multicellular solid tumor masses (Durand 1990). Thus, in single cells bathed in an excess of fluid, the drug may be pumped into this surrounding fluid and the cells protected. In contrast, in solid tumors, intracellular drug might be pumped back and forth between neighboring tumor cells existing in close multicellular contact, thus cancelling the apparent benefits of the P-glycoprotein pump in this type of physiological situation (Durand 1990). This speculative but disturbing possibility merely serves to emphasize the point that chemosensitizers for reversing drug resistance of solid tumors should be analyzed in vitro using three-dimensional cell culture systems and in vivo using solid tumors—not just monolayer cell culture systems, or leukemias and ascites tumors in vivo, respectively, as has commonly been the practice. The few attempts to do so have indeed demonstrated that reversal agents which effectively reverse P-glycoprotein-mediated multidrug resistance in monolayer may be completely ineffective against the same cells grown as multicellular tumor spheroids (Sakata et al. 1994). Similarly, the phenomenon of "verapamil hypersensitivity" of multidrug-resistant cells sometimes observed in monolayer cell cultures of P-glycoprotein-positive tumor cells is not necessarily observed in spheroid cultures of the same tumor cells (Anderson and Warr 1990).

The overall message from such studies, including those from our own laboratory, is that understanding the behavior of solid tumors, including the nature of their responses to therapy, will require in vitro experimental approaches different from those commonly used. This may seem at odds with the current use of "reductionist" approaches to study the complex biology of cancer, but it is necessary if we are to make more substantial progress in the therapeutic treatment of solid tumors.

ACKNOWLEDGMENTS

The excellent secretarial assistance of Lynda Woodcock is gratefully acknowledged. Dr. Kerbel's work is supported by grants from the Medical Research Council of Canada (MRC), and the National Cancer Institute of Canada (NCIC) through the Canadian Cancer Society. R.S.K. is a Terry Fox Career Scientist of the NCIC, supported by funds from the Terry Fox Run. C.H.G. is a Research Fellow of the NCIC, and H.K. is a Fellow of the MRC.

REFERENCES

Anderson, M. and J.R. Warr. 1990. Expression of verapamil hypersensitivity in multidrug-resistant cells grown as multicellular spheroids. *Cancer Chemother. Pharmacol.* **26:** 151.

Barcellos-Hoff, M.H., J. Aggeler, T.G. Ram, and M.J. Bissell. 1989. Functional differentiation and alveolar morphogenesis of primary mammary cultures on reconstituted basement membrane. *Development* **105:** 223.

Barry, M.A., C.A. Behinke, and A. Eastman. 1990. Activation of programmed cell death (apoptosis) by cisplatin, other anticancer drugs, toxins and hyperthermia. *Biochem. Pharmacol.* **40:** 2353.

Bates, R.C., A. Buret, D.F. van Helden, M.A. Horton, and G.F. Burns. 1994. Apoptosis induced by inhibition of intercellular contact. *J. Cell Biol.* **125:** 403.

Behrens, J., L. Vakaet, R. Friis, E. Winterhager, F. Van Boy, and M.M. Mareel. 1993. Loss of epithelial differentiation and gain of invasiveness correlates with tyrosine phosphorylation of the E-Cadherin/β-catenin complex in cells transformed with a temperature-sensitive v-*src* gene. *J. Cell Biol.* **120:** 757.

Bellamy, W.T., W.S. Dalton, and R.T. Dorr. 1990. The clinical relevance of multidrug resistance. *Cancer Invest.* **8:** 547.

Benchimol, S., A. Fuks, S. Jothy, N. Beauchemin, K. Shirota, and C.P. Stanners. 1988. Carcinoembryonic antigen, a human tumor marker, functions as an intercellular adhesion molecule. *Cell* **57:** 327.

Borst, P. 1991. Genetic mechanisms of drug resistance. *Acta Oncol.* **30:** 87.

Bradley, G., P.F. Juranaka, and V. Ling. 1988. Mechanism of multidrug resistance. *Biochem. Biophys. Acta* **948:** 87.

Buttyan, R., C.A. Olsson, J. Pintar, C. Chang, M. Bandyk, P.-Y. Ng, and I.S. Sawdzuk. 1989. Induction of the *TRPM-2* gene in cells undergoing programmed death. *Mol. Biol. Cell* **9:** 3473.

Cadman, E.C. 1989. The selective transfer of drug-resistant genes in malignant cells. In: *Resistance to antineoplastic agents* (ed. D. Kessel), p. 168. CRC Press, Boca Raton, Florida.

Durand, R.E. 1990. Slow penetration of anthracyclines into spheroids and tumors: A therapeutic advantage? *Cancer Chemoth. Pharmacol.* **26:** 198.

Durand, R.E. and R.M. Sutherland. 1972. Effects of intercellular contact on repair of radiation damage. *Exp. Cell Res.* **71:** 75.

———.1973. Growth and radiation survival characteristics of V79-171b Chinese hamster cells: A possible influence of intercellular contact. *Radiat. Res.* **56:** 513.

Eastman, A. 1990. Activation of programmed cell death by anticancer agents: Cisplatin as a model system. *Cancer Cells* **2:** 275.

Fan, D. 1994. Reversal of multidrug resistance. In *Reversal of multidrug resistance in cancer* (ed. J.A. Kellen), p. 93. CRC Press, Boca Raton, Florida.

Fan, Z., J. Baselga, H. Masui, and J. Mendelsohn. 1993. Antitumor effect of anti-epidermal growth factor receptor monoclonal antibodies plus *cis*-diamminedichloroplatinum on well established A431 cell xenografts. *Cancer Res.* **53:** 4637.

Flouriot, G., C. Vaillant, G. Salbert, C. Pelissero, J.M. Guiraud, and Y. Valotaire. 1993. Monolayer and aggregate cultures of rainbow trout hepatocytes: Long-term and stable liverspecific expression in aggregates. *J. Cell Sci.* **105:** 407.

Frei, E., C.A. Cucchi, A. Rosowsky, R. Tantravahi, S. Bernal, T.J. Ervin, R.M. Ruprecht, and W.A. Haseltine. 1985. Alkylating agent resistance: *In vitro* studies with human cell lines. *Proc. Natl. Acad. Sci.* **82:** 2158.

French, L.E., A. Wohlwend, A.-P. Sappino, J. Tschopp, and J.A. Schifferi. 1994. Human clusterin gene expression is confined to surviving cells during in vitro programmed cell death. *J. Clin. Invest.* **93:** 877.

Frisch, S.M. and H. Francis. 1994. Disruption of epithelial cell-matrix interactions induces apoptosis. *J. Cell Biol.* **124:** 619.

Fritz, I.B., K. Burdzy, B. Setchell, and O. Blaschuk. 1983. Ram rete testis fluid contains a protein (clusterin) which influences cell-cell interactions in vitro. *Biol. Reprod.* **28:** 1173.

Fujita, N., S. Kataoka, M. Naito, Y. Heike, N. Boku, M. Nakajima, and T. Tsuruo. 1993. Suppression of T-lymphoma cell apoptosis by monoclonal antibodies raised against cell surface adhesion molecules. *Cancer Res.* **53:** 5022.

Goodwin, T.J., J.M. Jessup, and D.A. Wolf. 1992. Morphologic differentiation of colon carcinoma cell lines HT-29 and HT-29KM in rotating-wall vessels. *In Vitro Cell. Dev. Biol.* **28A:** 47.

Gorlach, A., P. Herter, H. Hentschel, P.J. Frosch, and H. Acker. 1994. Effects of nIFN β and rIFN γ on growth and morphology of two human melanoma cell lines: Comparison between two- and three-dimensional culture. *Int. J. Cancer* **56:** 249.

Graham, C.H., H. Kobayashi, K.S. Stankiewicz, S. Man, S.J. Kapitain, and R.S. Kerbel. 1994. Rapid acquisition of multicellular drug resistance after a single transient exposure of mammary tumor cells to alkylating agents. *J. Natl. Cancer Inst.* **86:** 975.

Guadagni, F., M. Roselli, and R.M. Hoffman. 1991. Maintenance of expression of tumor antigens in three-dimensional *in vitro* human tumor gel-supported histoculture. *Anticancer Res.* **11:** 543.

Hirano, S., N. Kimoto, Y. Shimoyama, S. Hirohashi, and M. Takeichi. 1992. Identification of a neural α-catenin as a key regulator of cadherin function and multicellular organization. *Cell* **70:** 293.

Hoffman, R.M. 1991a. Three-dimensional gel-supported

native-state histoculture for evaluation of tumor-specific pharmacological activity: Principles, practice and possibilities. *J. Cell Pharmacol.* **2:** 189.

———. 1991b. Three-dimensional histoculture: Origins and applications in cancer research. *Cancer Cells* **3:** 86.

Houghton, P.J. and S.B. Kaye. 1994. Multidrug resistance is not an important factor in therapeutic outcome in human malignancies. *J. NIH Res.* **6:** 55.

Inch, W.R., J.A. McCredie, and R.M. Sutherland. 1970. Growth of nodular carcinomas in rodents compared with multi-cell spheroids in tissue culture. *Growth* **34:** 271.

Jenne, D.E. and J. Tschopp. 1992. Clusterin: The intriguing guises of a widely expressed glycoprotein. *Trends Biochem. Sci.* **17:** 154.

Johnson, R.K. 1990. Screening methods in antineoplastic drug discovery. *J. Natl. Cancer Inst.* **82:** 1082.

Jones, M., J. Siracky, L.R. Kelland, and K.R. Harrap. 1993. Acquisition of platinum drug resistance and platinum cross resistance patterns in a panel of human ovarian carcinoma xenografts. *Br. J. Cancer* **67:** 24.

Kerbel, R.S. 1990. Growth dominance of the metastatic cancer cell: Cellular and molecular aspects. *Adv. Cancer Res.* **55:** 87.

Kerbel, R.S., I. Cornil, and B. Korczak. 1989. New insights into the evolutionary growth of tumors revealed by Southern gel analysis of tumors genetically tagged with plasmid or proviral DNA insertions. *J. Cell Sci.* **94:** 381.

Kerbel, R.S., B. St. Croix, J. Rak, and C. Graham. 1994. Is there a role for "anti-adhesives" as chemosensitizers in the treatment of solid tumors by chemotherapy? *Bull. Inst. Pasteur* (in press).

Kerbel, R.S., C. Waghorne, B. Korczak, A. Lagarde, and M. Breitman. 1988. Clonal dominance of primary tumors by metastatic cells: Genetic analysis and biological implications. *Cancer Surv.* **7:** 597.

Kerbel, R.S., C. Waghorne, S.M. Man, B. Elliott, and M.L. Breitman. 1987. Alterations of the tumorigenic and metastatic properties of neoplastic cells is associated with the process of calcium phosphate-mediated DNA transfection. *Proc. Natl. Acad. Sci.* **84:** 1263.

Klunder, I. and D.F. Hulser. 1993. β-Galactosidase activity in transfected Ltk-cells is differentially regulated in monolayer and in spheroid cultures. *Exp. Cell Res.* **207:** 155.

Kobayashi, H., S. Man, S.J. Kapitain, B.A. Teicher, and R.S. Kerbel. 1993. Acquired multicellular-mediated resistance to alkylating agents in cancer. *Proc. Natl. Acad. Sci.* **90:** 3294.

Korczak, B., I.B. Robson, C. Lamarche, A. Bernstein, and R.S. Kerbel. 1988. Genetic tagging of tumor cells with retrovirus vectors: Clonal analysis of tumor growth and metastasis in vitro. *Mol. Cell. Biol.* **8:** 3143.

Kuboda, H., T. Sugimoto, K. Ueda, S. Tsuchida, Y. Horii, J. Inazawa, K. Sato, and T. Sawada. 1991. Different drug sensitivity in two neuroblastoma cell lines established from the same patient before and after chemotherapy. *Int. J. Cancer* **47:** 732.

Laderoute, K.R., B.J. Murphy, S.M. Short, T.D. Grant, A.M. Knapp, and R.M. Sutherland. 1992. Enhancement of transforming growth factor-α synthesis in multicellular tumor spheroids of A431 squamous carcinoma cells. *Br. J. Cancer* **65:** 157.

Lelievre, S. and A.K. Larsen. 1994. Development and characterization of suramin-resistant Chinese hamster fibrosarcoma cells. Morphological changes and increased metastatic potential. *Cancer Res.* **54:** 3993.

Litvinov, S.V., M.P. Velders, H.A.M. Bakker, G.J. Fleuren, and S.O. Warnaar. 1994. Ep-CAM: A human epithelial antigen is a homophilic cell-cell adhesion molecule. *J. Cell Biol.* **125:** 437.

LoBuglio, A.F., M.N. Saleh, J. Lee, M.B. Khazaeli, R. Carrano, H. Holden, and R.H. Wheeler. 1988. Phase I trial of multiple large doses of murine monoclonal antibody CO17-1A. I. Clinical aspects. *J. Natl. Cancer Inst.* **80:** 932.

Meredith, J.E., Jr., B. Fazeli, and M.A. Schwartz. 1993. The extracellular matrix as a cell survival factor. *Mol. Biol. Cell* **4:** 953.

Miller, B.E., F.R. Miller, and G.H. Heppner. 1985. Factors affecting growth and drug sensitivity of mouse mammary tumor lines in collagen gel cultures. *Cancer Res.* **45:** 4200.

Muller-Klieser, W. 1987. Multicellular spheroids: A review on cellular aggregates in cancer research. *Eur. J. Cancer Res. Clin. Oncol.* **113:** 101.

Murphy, B.J., K.R. Laderoute, H.J. Vreman, T.D. Grant, N.S. Gill, D.K. Stevenson, and R.M. Sutherland. 1991. Enhancement of heme oxygenase expression and activity in A431 squamous carcinoma multicellular tumor spheroids. *Cancer Res.* **53:** 2700.

Olive, P.L. and R.E. Durand. 1994. Drugs and radiation resistance in spheroids: Cell contact and kinetics. *Cancer Metastasis Rev.* **13:** 121.

Olive, P.L., J.P. Banath, and H.H. Evans. 1993. Cell killing and DNA damage by etoposide in Chinese hamster V79 monolayers and spheroids: Influence of growth kinetics, growth environment and DNA packaging. *Br. J. Cancer* **67:** 522.

Pietras, R.J., B.M. Fendly, V.R. Chazin, M.D. Pegram, S.B. Howell, and D.J. Slamon. 1994. Antibody to HER-2/neu receptor blocks DNA repair after cisplatin in human breast and ovarian cancer cells. *Oncogene* **9:** 1829.

Raff, M.C. 1992. Social controls on cell survival and cell death. *Nature* **356:** 397.

Rak, J.W. and R.S. Kerbel. 1993. Growth advantage ("clonal dominance") of metastatically-competent tumor cell variants expressed under selective two or three dimensional tissue culture conditions. *In Vitro Cell. Dev. Biol.* **29A:** 742.

Reed, J. 1993. Bcl-2 and the regulation of programmed cell death. *J. Cell Biol.* **124:** 1.

Reithmüller, G., E. Schneider-Gädicke, G. Schlimok, W. Schmiegel, R. Raab, R. Höffken, R. Gruber, H. Pichlmaier, H. Hirche, R. Pichlamayr, P. Buggisch, and J. Witte. 1994. Randomized trial of monoclonal antibody for adjuvant therapy of resected Dukes' colorectal carcinoma. *Lancet* **343:** 1177.

Rofstad, E.K. 1986. Growth and radiosensitivity of malignant melanoma multicellular spheroids initiated directly from surgical specimens of tumours in man. *Br. J. Cancer* **54:** 569.

Rofstad, E.K., A. Wahl, C. Davies, and T. Brustad. 1986. Growth characteristics of human melanoma multicellular spheroids in liquid-overlay culture: Comparisons with the parent tumour xenografts. *Cell Tissue Kinet.* **19:** 205.

Ruoslahti, E. and J.C. Reed. 1994. Anchorage dependence, integrins, and apoptosis. *Cell* **77:** 477.

Ruoslahti, E., M.D. Pierschbacher, and V.L. Woods. 1994. The $\alpha 5 \beta 1$ integrin in tumor suppression. *Bull. Inst. Pasteur* (in press).

Sakata, K., T.T. Kwok, G.R. Gordon, N.S. Waleh, and R.M. Sutherland. 1994. Resistance to verapamil sensitization of multidrug resistant cells grown as multidrug spheroids. *Int. J. Cancer* **59:** 282.

Schecter, R.L., A. Woo, M. Duong, and G. Batist. 1991. In vivo and in vitro mechanisms of drug resistance in a rat mammary carcinoma model. *Cancer Res.* **51:** 1434.

Soranzo, C., A. Ingrosso, M. Buffa, G.D. Torre, R.A. Gambetta, and F. Zunino. 1989. Changes in the three-dimensional organization of LoVo cells associated with resistance to doxorubicin. *Cancer Lett.* **48:** 37.

St. John, T., J. Meyer, R. Idzerda, and W.M. Gallatin. 1990. Expression of CD44 confers a new adhesive phenotype on transfected cells. *Cell* **80:** 45.

Starling, J.J., R.S. Maciak, N.A. Hinson, J. Hoskins, B.C. Laguzza, R.A. Gadski, J. Strnad, L. Rittmann-Grauer, S.V. DeHerdt, T.F. Bumol, and R.E. Moore. 1990. In vivo selection of human tumor cells resistant to monoclonal antibody-vinca alkaloid immunoconjugates. *Cancer Res.* **50:** 7634.

Steeg, P.S., M.C. Alley, and M.R. Grever. 1994. An added

dimension: Will three-dimensional cultures improve our understanding of drug resistance? *J. Natl. Canc. Inst.* **86:** 953.

Sutherland, R.M. 1988. Cell and environment interactions in tumor microregions: The multicell spheroid model. *Science* **240:** 177.

Sutherland, R.M., H.A. Eddy, B. Bareham, K. Reich, and D. Vanantwerp. 1979. Resistance to adriamycin in multicellular spheroids. *Int. J. Radiat. Biol.* **5:** 1225.

Teicher, B.A., T.S. Herman, S.A. Holden, Y. Wang, M.R. Pfeffer, J.W. Crawford, and E. Frei III. 1990. Tumor resistance to alkylating agents conferred by mechanisms operative only in vivo. *Science* **247:** 1457.

Theodorescu, D., C. Sheehan, and R.S. Kerbel. 1993. Tumor cells grown as monolayers or multicellular spheroids indicate autoregulation of TGF-β gene expression is dependent on tissue architecture. *In Vitro Cell. Dev. Biol.* **29:** 105.

Untawale, S. and R.T. Mulcahy. 1992. Activity of the human metallothionein promoter (hMT-II$_A$) in cell populations isolated from varying depths in multicell spheroids following exposure to cadmium. *Chem.-Biol. Interact.* **83:** 171.

Vasanthi, L.G., V.J. Bakanauskas, and A.D. Conger. 1986. Early changes accompanying development of adriamycin resistance in spheroids of a mouse mammary carcinoma line. *Anticancer Res.* **6:** 567.

Vescio, R.A., K.M. Connors, T. Kubota, and R.M. Hoffman. 1991. Correlation of histology and drug response of human tumors grown in native-state three-dimensional histoculture and in nude mice. *Proc. Natl. Acad. Sci.* **88:** 5163.

Waghorne, C., M. Thomas, A.E. Lagarde, R.S. Kerbel, and M.L. Breitman. 1988. Genetic evidence for progressive selection and overgrowth of primary tumors by metastatic cell subpopulations. *Cancer Res.* **48:** 6109.

Walch, N.S., J. Gallo, T.D. Grant, B.J. Murphy, R.H. Kramer, and R.M. Sutherland. 1994. Selective down-regulation of integrin receptors in spheroids of squamous cell carcinoma. *Cancer Res.* **54:** 838.

Young, R.C. 1989. Drug resistance: The clinical problem. In *Drug resistance in cancer therapy* (ed. R.F. Ozols), p. 1. Kluwer Academic, Boston.

Overcoming Complexities in Genetic Screening for Cancer Susceptibility

S.H. Friend,*§ R. Iggo,† C. Ishioka,* M. Fitzgerald,*
I. Hoover,* E. O'Neill,* and T. Frebourg‡

*Molecular Genetics, Massachusetts General Hospital Cancer Center, Charlestown, Massachusetts 02129;
†Swiss Institute for Experimental Cancer Research, 1066 Epalinges, Switzerland;
‡Unite de Genetique Moleculaire, CHU de Rouen, France

Eight years ago, we cloned the first human cancer susceptibility gene, the retinoblastoma gene (Friend et al. 1986). During the intervening time, much has been learned about this gene, and a series of other cancer susceptibility genes have been identified. This paper does not focus on the molecular details of how these susceptibility genes function, but instead provides an overview of how knowledge regarding mutations in these genes can provide clues to identify individuals at high risk for developing specific malignancies. As shown below, although it is possible to find mutations in cancer susceptibility genes using current techniques, there are still significant hurdles that need to be overcome before patients can routinely benefit from this knowledge.

It is well recognized that germ-line mutations in cancer susceptibility genes represent the strongest risk factors for cancer. This is because individuals who are born with mutations that inactivate one of their two alleles for these growth-limiting genes need only suffer a single further inactivating event in a susceptible cell to have lost an important control for maintaining normal growth. The single example where screening for individuals who carry a genetic predisposition to develop cancer is performed frequently is in the case of patients with retinoblastoma. Germ-line mutations in the retinoblastoma gene are quite potent, shifting an individual's risk for retinoblastoma from 1:20,000 to more than a 90% probability of developing the tumor. The importance of identifying such individuals is high because approximately one half of the several hundred children in the United States who develop the embryonal eye tumor, retinoblastoma, carry a germ-line mutation in this gene. The high penetrance for tumors occurring in carriers (>90%), the defined site of malignancy, the small number of patients to be screened, and the limited time frame when these children are at risk (confined primarily to the first few years of life) have made the screening for patients with an inherited risk for this cancer a routine part of the work-up of new retinoblastoma patients. Unfortunately, as we show below, screening for individuals at high risk to develop other cancers has been much more difficult.

STUDIES ON LI-FRAUMENI SYNDROME

In an effort to better understand the nature of cancer susceptibility genes, we turned several years ago to a rare disorder in which more than five different tumor types clustered in specific families. This syndrome, called Li-Fraumeni syndrome, was identified by requiring that a proband have a sarcoma, and that two other first-degree relatives also develop cancer by age 45 (Li et al. 1988). So defined, affected members in families with Li-Fraumeni syndrome have an extraordinarily high frequency of breast cancers, brain tumors, leukemias, adrenocortical carcinomas, and of course, sarcomas (Li et al. 1988). At the time we began the search to identify the cancer susceptibility gene responsible for this syndrome, the first $p53$ mutations were being identified in diverse tumor types. In an effort to reduce the otherwise lengthy process of positional cloning, we chose to take a candidate gene approach, and tested for germ-line $p53$ mutations in the Li-Fraumeni syndrome families. We found germ-line mutations in affected members for many of the families with Li-Fraumeni syndrome (Malkin et al. 1990). This suggested that a single mutation in the $p53$ gene could produce a high susceptibility to a broad range of malignancies.

Beyond the obvious biological implications, this created a significant difficulty for those interested in clinically following patients with inborn cancer susceptibilities. When there is a single site for increased cancers, for example, the eye in patients with germ-line mutations in the retinoblastoma gene, or the colon in families with germ-line mutations in the APC gene, the surveillance techniques can be focused in a single area. In contrast, the surveillance of patients with germ-line $p53$ mutations requires tumors to be anticipated throughout most of the body. The resultant difficulties in properly using the knowledge of increased cancer susceptibility in families with Li-Fraumeni syndrome to lower cancer incidence is obvious. To help these patients, it will be important to develop better methods for early tumor detection. Until then, the benefits of ascertaining the carriers will remain indeterminate.

The second perspective that has arisen from studying these patients with germ-line $p53$ mutations pertains to the variability in their clinical presentation. Since the

---

Present address: §Department of Clinical Medicine, Fred Hutchinson Cancer Research Center, Seattle, Washington 98104.

original Li-Fraumeni families were ascertained by virtue of the intense clustering of tumors in rare families, it should not be surprising that the penetrance for developing tumors among the carriers in these families is quite high. Within these families, the penetrance approaches 90%. To test how representative this penetrance would be for other carriers of germ-line *p53* mutations, we have chosen to survey consecutive children with childhood osteosarcomas. This tumor was chosen because it is quite frequently the tumor by which the probands in families with Li-Fraumeni syndrome are ascertained. By analyzing 235 consecutive osteosarcoma patients for *p53* mutations, we found that 7 of the children carried germ-line *p53* mutations. This means that approximately 3% of children with osteosarcomas have germ-line *p53* mutations (McIntyre et al. 1994). None of these had classic Li-Fraumeni syndrome. More importantly, it can be seen that only 3 of the 7 children with germ-line *p53* mutations had any first-degree relatives with cancer under age 45. This means that more than half of the children with germ-line *p53* mutations would have been missed if only patients with a history of cancer among first-degree relatives with cancer at an early age were screened. Because such a large fraction of patients carrying inborn susceptibilities to cancer may not present with a positive history of cancer, the number of patients that might eventually benefit from genetic screening for mutations in cancer susceptibility genes may be quite large. The particular problem with the goal to screen large numbers of patients is that the current mutation detection methods, although somewhat adequate for the analysis of limited numbers of samples, are insufficient to analyze large numbers of samples. For this reason, during the past 2 years we have begun to focus on the development of better detection methods. The rest of this paper concerns two new methods which we have developed to accomplish this goal of identifying high-risk individuals.

DNA-BASED SCREENING METHODS

Because the mutations inactivating tumor suppressor genes that provide increased cancer susceptibility only disrupt one of the two alleles in the constitutional cells, the detection of these changes requires the identification of heterozygous mutations. Such heterozygous changes are difficult to find because these inactivating mutations often occur at multiple sites spaced throughout the genes. Furthermore, quite frequently the mutations are not large deletions but a result of single basepair changes. The most sensitive method for identifying these often subtle mutations is to analyze the actual sequencing reactions. Unfortunately, this process is so time-consuming that alternative screening methods have been developed. Such alternate mutation screening methods are primarily focused on scanning fragments of DNA for properties that reflect DNA alterations within the fragment. Good examples of such techniques would be scanning for single strand conformation polymorphisms (SSCP) (Orita et al. 1989) and the use of constant denaturant gel electrophoresis (CDGE) (Borresen et al. 1991). Recognizing the power of new software technologies, we chose to return to the option of looking directly at DNA sequences as a method to detect heterozygous mutations.

Through the work of the software engineers at Applied Biosystems Inc., programs have been developed that are now able to scan the DNA sequencing reactions and analyze the first and second peak heights of fluorescence emissions for the nucleotides at each basepair position. By comparison of wild-type DNA first and second peak heights for the sequencing reactions to the first and second peak heights for a reaction containing an unknown sample template, the software program is able to mark sequencing reactions that contain potential heterozygous mutations. This allows an entire sequencing gel with more than 30 samples to be scanned by this program so that only those specific sample reactions potentially containing mutations are then called up on the screen for further direct analysis. The program is then able to display the specific region of the particular reaction that potentially contains a mutation. Figure 1 shows a representative example of a portion of a sequencing reaction that is highlighted because it might contain a mutation. As shown, it is possible to display both the standard template reaction and the unknown template reaction and to directly compare the peak intensities. Not shown is the fact that it is also possible to display the reactions for both the standard and unknown samples for the opposite strand to confirm the presence of a mutation. This allows one to quite accurately identify heterozygous mutations in unknown samples. Using this technology, we have tested more than 50 cell lines for *p53*, and in that instance, performed a second analysis to confirm that the technique had greater than a 95% sensitivity in correctly identifying mutations. We believe that this technique of automated heterozygous sequence analysis can function as a significant tool in the screening for germ-line cancer susceptibility gene mutations.

FUNCTIONAL ASSAYS FOR CANCER SUSCEPTIBILITY

The method described above helps improve the efficiency of detecting mutations, and yet it does not allow the type of large-scale screening that will eventually be needed. An alternate new approach to screening that might provide such an improvement has come out of our efforts to develop direct assays for inactivations in the gene products of these cancer susceptibility genes. This work was begun as a result of the recognition that germ-line mutations often represent polymorphisms and other mutations that do not disrupt function (Frebourg et al. 1992). Since cancer risks are only

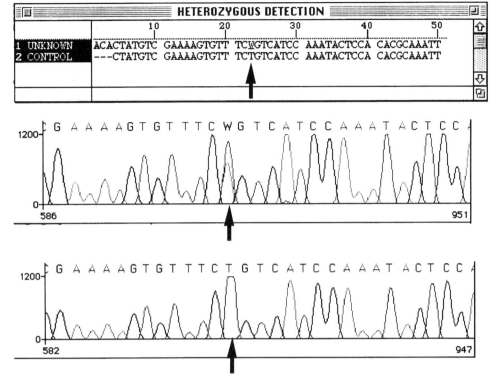

Figure 1. Software-assisted detection of heterozygous mutations. This is a representative computer screen from the Sequence Navigator software (ABI, Foster City, California) showing a heterozygous mutation A/T at the second base pair of codon 208 in the unknown sample (Lane *1*). The control sequence has a thymidine at the same position. The computer highlights both the text sequence position and the chromatogram position by arrows.

associated with alterations that disrupt the wild-type function of these susceptibility genes, it seemed that an initial screen that ascertained whether those wild-type functions were intact would offer an advantage. Because the germ-line mutations occur in only one allele, it was necessary to design assays that were capable of separating the two alleles. Initially this was done by subcloning each allele and expressing the alleles in cells that lacked function for the gene being analyzed (Frebourg et al. 1992). The success of this method was outweighed by the large effort required for it to be performed. Recognizing that much of the effort went into subcloning separated alleles, Thierry Frebourg took an alternative approach that used techniques already standard in work on yeast. Because it is possible to select for specific homologous recombination events, it is possible to perform in vivo subcloning of the patient cDNA samples directly into yeast expression vectors. As a model system we chose to develop a *p53* functional assay (Ishioka et al. 1993). As is shown in Figure 2, the first part of the assay involves the synthesis of a pool of patient cDNAs, followed by transformation along with gapped yeast expression vectors containing selectable markers so that only the yeast which can undergo homologous recombination to repair the gap is able to grow. This allows efficient in vivo subcloning of the human cDNAs into yeast expression vectors. Because these expression vectors also contain centromeric elements, this transformation results in a lawn of yeast colonies, each expressing a single *p53* allele.

Taking into account that *p53* is a strong *trans*-activator in its wild-type form, we engineered a yeast strain that would only allow growth when *p53* through its wild-type *trans*-activating properties allowed expression of a positive selectable marker (histidine). By replica plating from the initial plate where each yeast colony expresses a single allele of *p53* of indeterminate function onto a plate lacking histidine, it is possible to distinguish wild-type from mutant *p53* by the ability of the clone to grow on both plates (Ishioka et al. 1993).

We have now tested the validity of this screen for *p53* function by making a panel of 48 cell lines, and, in parallel, both sequenced them and performed this *p53* functional assay. In all but one of the cell lines, those cell lines that contained mutant *p53* by sequencing scored as mutant in the *p53* functional assay. In the one exception, a mutation found at amino acid 285 did not score as mutant at 30°C. Upon examination at 37°C, this mutation also scored as functionally inactive. As a result of this temperature-sensitive allele, we have modified the functional assays so that they are performed at 37°C. The ease of performing this assay is obvious enough that this functional assay for *p53* is now in use in several laboratories.

During the last few months, we have gone on to test

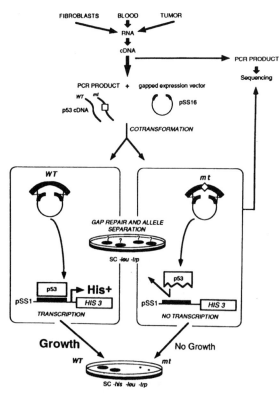

Figure 2. Schematic diagram of the *f*unctional *a*ssay on *s*eparated *a*lleles in *y*east (FASAY) method. (WT) Wild-type *p53*, (mt) mutant *p53*. pSS1 is a *p53* reporter vector containing the HIS3 open reading frame (nucleotides +19 to +642) under the control of a *p53* consensus *p53* binding site upstream of a minimal *GAL1* promoter. This vector also contains *CEN4* and *ARS1* for stable low copy number replication and *TRP1* for plasmid maintenance. (Reprinted, with permission, from Ishioka et al. 1993.)

the extent to which other functional assays might be feasible. We have found that it is possible to design a functional assay for the GTPase activating portion of the *NF1* tumor suppressor gene. We have begun to work on functional assays for the DNA mismatch repair genes. In each instance, there are two hurdles to developing such assays. The first obvious one pertains to identifying the specific function associated with the increased risk of malignancies. The second hurdle relates to modifying the yeast so that the human gene product will be able to be assayed in yeast. In the case of the DNA mismatch repair gene assays, the complexity of the multicomponent system that must be coordinated to achieve function has been such that it appears necessary to replace much of the yeast mismatch repair system with their human counterparts before functional assays for the human genes are possible.

In summary, there are a multitude of complexities that must be addressed before most patients will benefit from cancer susceptibility screening. Some of these complexities are due to the fact that cancer susceptibilities depend on a multitude of potentially mutually interactive genes. Other important factors concern the need for rapid methods to screen for these defects. Significant advances need to occur in both areas.

REFERENCES

Borresen, A.L., E. Hovig, B. Smith-Sorensen, D. Malkin, S. Lystad, T.I. Andersen, J.M. Nesland, K.J. Isselbacher, and S.H. Friend. 1991. Constant denaturant gel electrophoresis as a rapid screening technique for p53 mutations. *Proc. Natl. Acad. Sci.* **88:** 8405.

Frebourg, T., N. Barbier, J. Kassel, Y.S. Ng, P. Romero, and S.H. Friend. 1992. A functional screen for germ line p53 mutations based on transcriptional activation. *Cancer Res.* **52:** 6976.

Friend, S.H., R. Bernards, S. Rogelj, R.A. Weinberg, J.M. Rapaport, D.M. Albert, and T.P. Dryja. 1986. A human DNA segment with properties of the gene that predisposes to retinoblastoma and osteosarcoma. *Nature* **323:** 643.

Ishioka, C., T. Frebourg, Y.-X. Yan, M. Vidal, S.H. Friend, S. Schmidt, and R. Iggo. 1993. Screening patients for heterozygous p53 mutations using a functional assay in yeast. *Nat. Genet.* **5:** 124.

Li, F.P., J.F. Fraumeni, Jr., J.J. Mulvihill, W.A. Blattner, M.G. Dreyfus, M.A. Tucker, and R.W. Miller. 1988. A cancer family syndrome in twenty-four kindreds. *Cancer Res.* **48:** 5358.

Malkin, D., F.P. Li, L.C. Strong, J.F. Fraumeni, Jr., C.E. Nelson, D.H. Kim, J. Kassel, M.A. Gryka, F.Z. Bischoff, M.A. Tainsky, and S.H. Friend. 1990. Germ line p53 mutations in a familial syndrome of breast cancer, sarcomas, and other neoplasms. *Science* **250:** 1233.

McIntyre, J.F., B. Smith-Sorensen, S.H. Friend, J. Kassel, A.L. Borresen, Y.-X. Yan, C. Russo, J. Sato, N. Barbier, J. Miser, D. Malkin, and M.C. Gebhardt. 1994. Germline mutations of the p53 tumor suppressor gene in children with osteosarcoma. *J. Clin. Oncol.* **12:** 925.

Orita, M., H. Iwahana, H. Kanazawa, K. Hayashi, and T. Sekiya. 1989. Detection of polymorphisms of human DNA by gel electrophoresis as single-strand conformation polymorphisms. *Proc. Natl. Acad. Sci.* **86:** 2766.

Exploiting Multidrug Resistance to Treat Cancer

M.M. Gottesman,* S.V. Ambudkar,*‡ B. Ni,* J.M. Aran,†
Y. Sugimoto,* C.O. Cardarelli,† and I. Pastan†

*Laboratory of Cell Biology, †Laboratory of Molecular Biology, National Cancer Institute,
National Institutes of Health; ‡Department of Medicine, Division of Nephrology,
The Johns Hopkins University School of Medicine, Baltimore, Maryland 21205

Of the 1,000,000 patients who present to their physicians each year in the United States with cancer, approximately 500,000 have tumors that have spread beyond their primary origin. Currently, the most effective therapy for metastatic and disseminated cancers is the use of chemotherapy, but chemotherapy eliminates only 5–10% of these cancers, including germ-cell cancers, leukemias, lymphomas, and childhood tumors. In most solid tumors, chemotherapy is either ineffective or, after a brief remission, becomes ineffective. The molecular basis of resistance to modern multiagent chemotherapy (multidrug resistance, MDR) has been the subject of considerable investigation. One major conclusion from these studies is that most resistance is cell-based, occurring because of the expression in cancer cells of drug resistance genes; or loss of pathways which sensitize cells to chemotherapy, such as those involved in metabolic activation of cytotoxins; or programmed cell death. Thus, drug-resistance phenomena can be modeled in cultured cancer cells selected for resistance to anticancer drugs. In Figure 1, we summarize a variety of different mechanisms thought to play a role in the intrinsic expression or acquisition of drug resistance in cancer.

When cultured cancer cells are exposed to a variety of natural product anticancer drugs, including anthracyclines (doxorubicin, daunorubicin), *Vinca* alkaloids (vincristine, vinblastine), epipodophyllotoxins (etoposide, teniposide), antibiotics (actinomycin D), and others (e.g., taxol), the most frequent mechanism of drug resistance is the appearance of an ATP-dependent drug efflux pump in the plasma membrane (for review, see Gottesman and Pastan 1993). This pump is a 170,000-dalton phospho-glycoprotein, termed P-glycoprotein, or the multidrug transporter. This pump detects and ejects a wide range of natural product drugs, including the anticancer agents listed above, other cytotoxic agents such as colchicine, puromycin, and the cytotoxic peptide gramicidin D, as well as a wide range of amphipathic noncytotoxic agents commonly used in the clinic (calcium channel blockers such as verapamil, immune suppressive agents such as cyclosporine A, antihistamines, antibiotics, antidepressants, antihypertensives, anti-arrhythmics, etc.). These noncytotoxic drugs and their analogs which are substrates for P-glycoprotein can be used as competitive inhibitors to block the action of the pump and sensitize MDR cells to anticancer drugs. Such agents have been termed "reversing agents." Recently, a second ATP-dependent transporter, termed MRP for multidrug-resistance associated protein, has been detected in some human cells resistant to multiple natural product anticancer drugs (Cole et al. 1992), and its frequency of involvement in multidrug resistant human cancers is currently under investigation.

The cloning of cDNAs for the human *MDR1* and mouse *mdr* genes (Chen et al. 1986; Gros et al. 1986) revealed that the *mdr* genes were the prototype for a superfamily of ATP-dependent transporters characterized by strong sequence homology in their ATP-binding cassettes (ABCs) and the presence of multiple transmembrane domains (Higgins et al. 1992). In the case of the human *MDR1* gene product, P-glycoprotein was predicted to consist of 1280 amino acids, comprising two ABCs, each associated with six transmembrane domains (Chen et al. 1986).

Normal Physiology and Pathophysiology of the Multidrug Transporter

Why should mammalian genes encode a transporter able to efflux anticancer drugs? The first hint as to the normal function of P-glycoprotein came from analyses of the distribution of P-glycoprotein in human tissues. Three major localizations were found (Thiebaut et al.

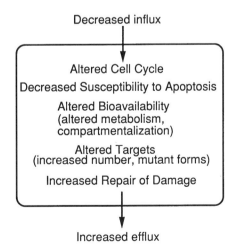

Figure 1. Schematic illustration of ways in which cells can become resistant to anticancer drugs.

1987, 1989; Cordon-Cardo et al. 1989): (1) luminal surfaces of epithelia including brush border of the kidney, mucosal surface of the large and small intestine, apical surface of pancreatic ductules, and apical surface of biliary hepatocytes; (2) luminal surfaces of capillary endothelial cells of the brain and testes; and (3) plasma membranes of steroid-secreting cells in the adrenal cortex and pregnant endometrium. These localization studies suggested that P-glycoprotein might be involved in normal excretion of xenobiotics in the diet (most anticancer drugs recognized by the multidrug transporter are natural products), handling of steroids in the adrenal cortex and pregnant uterus, and as part of the blood-brain and blood-testis barriers, which keep many hydrophobic natural products from entering these organs. The recent report of a transgenic knockout mouse lacking the mouse *mdr1b* gene (Schinkel et al. 1994), which is normally expressed in mouse brain capillary endothelial cells and epithelia, demonstrated increased blood levels of vinblastine and high toxic brain levels of the antihelminthic agent ivermectin in these mice compared to their isogenic sibs. This study strongly supports a role for the multidrug transporter for excretion of xenobiotics and as part of the blood-brain barrier.

Evidence that P-glycoprotein plays a significant role in drug resistance in human cancers came primarily from the demonstration that about 50% of human cancers express levels of *MDR1* mRNA sufficient to confer drug resistance in in vitro systems (Goldstein et al. 1989). Cancers found to express P-glycoprotein at the time of presentation include many tumors derived from epithelia that normally express P-glycoprotein, including cancers of the colon, adrenal, liver, kidney, and pancreas. In addition, tumors initially sensitive to chemotherapy which are selected in patients after multiple rounds of chemotherapy frequently show increased expression of P-glycoprotein, including leukemias, multiple myeloma, lymphomas, neuroblastoma, and childhood sarcomas. Finally, some cancers seem to show increased expression of P-glycoprotein during the development of malignancy, as if malignant transformation itself correlates to some extent with *MDR1* gene expression. For example, chronic myelogenous leukemia, when it enters a more malignant, blastic phase, usually shows increased expression of P-glycoprotein regardless of chemotherapy. There is some evidence that the activation of *MDR1* gene expression which may accompany malignant transformation is related to an effect of oncogene activation on the *MDR1* promoter; in vitro mutant p53 and Ras both stimulate the *MDR1* promoter in *MDR1*-promoter-CAT cotransfections (Chin et al. 1992).

The preponderance of evidence favors the interpretation that increased *MDR1* gene expression in tumors contributes to multidrug resistance. Cancer cells grown ex vivo which express P-glycoprotein are drug resistant, and their resistance is reversed by agents known to inhibit P-glycoprotein. A transgenic mouse in which the human *MDR1* gene is expressed in bone marrow cells at levels comparable to those found in human tumors has bone marrow cells resistant to anticancer drugs, suggesting that levels of *MDR1* gene expression found in cancers are sufficient to confer multidrug resistance in vivo (Galski et al. 1989; Mickisch et al. 1991). Most convincingly, initial clinical trials using agents that reverse drug resistance in vitro have also been found to sensitize human tumors to anticancer drugs, especially in multiple myeloma and lymphoma (for review, see Sikic 1993). Whether this strategy will be sufficient to sensitize intrinsically resistant tumors to chemotherapy, and whether prolonged response to anticancer drugs used in combination with MDR reversing agents will be achieved, remains to be shown in controlled clinical trials.

Mechanism of Action of the Multidrug Transporter

An understanding of the mechanism by which the energy of ATP is harnessed to the expulsion of drugs by the multidrug transporter would help in the development of agents to circumvent drug resistance and would also help explain the biology of the large family of ABC transporters. Several lines of evidence have suggested that the multidrug transporter detects and ejects drugs directly from the lipid bilayer as they are crossing the plasma membrane to enter the cell (Higgins and Gottesman 1992; Gottesman and Pastan 1993). First, virtually all MDR substrates are hydrophobic compounds that have substantial lipid solubility. Second, evidence exists for reduced concentrations of drugs within the lipid bilayer of cells which have an active multidrug transporter (Raviv et al. 1990). Third, kinetic analysis of the uptake and efflux of drugs by P-glycoprotein indicates that the transporter reduces the rate of drug influx (consistent with its detection of drug before it crosses the bilayer) and also increases the rate of efflux. These data also suggest that the interaction of drugs with the transporter might occur in both the external leaflet and the internal leaflet of the plasma membrane bilayer (Stein et al. 1994). Finally, acetoxymethyl ester derivatives of fluorescent compounds, such as the calcium reporter FURA2-AM, are blocked from entering MDR cells, presumably because they are extruded as they enter the plasma membrane, but once inside cells they are rapidly hydrolyzed to their carboxyl derivatives and cannot be pumped by the transporter (Homolya et al. 1993).

Several different approaches have been taken to evaluate the means by which P-glycoprotein detects a broad variety of substrates and removes them from the plasma membrane. Photoaffinity labeling studies have demonstrated at least two sites for interaction of substrates, such as azidopine, and substrate analogs, such as iodoazidoprazosin or iododeoxyforskolin, with P-glycoprotein (Greenberger 1993; Morris et al. 1994). These sites have been identified by immunoprecipitation of proteolytic fragments of P-glycoprotein labeled with radioactive azidopine, forskolin, or prazosin and reside within a region including the 6th and 12th transmembrane domain and associated extracytoplasmic and cytoplasmic regions. Analysis of mutant forms of P-glycoprotein with altered substrate specificity have identified several mutations within or close to most of the transmembrane regions, including a substitution of Gly-185 for valine, which increases resistance to colchicine and etoposide and decreases resistance to vinblastine and vincristine. The preliminary picture that emerges from these studies is of a transporter of one or several subunits in which the transmembrane regions acting concertedly form a pathway for extrusion of substrates entering from within the plasma membrane.

MATERIALS AND METHODS

Multidrug-resistant cell lines. All cells were grown in Dulbecco's MEM supplemented with 10% fetal bovine serum and drugs, as appropriate, dissolved in DMSO, including vinblastine (Sigma). KB-V1 cells are a subclone of HeLa which were selected for resistance to 1 µg/ml vinblastine and express high levels of P-glycoprotein. For expression of human P-glycoprotein in yeast, *Saccharomyces cerevisiae* strain BKY36-6B-T1, (a, his 4-15, Leu 2-3, 112, ura 3-52, erg 6 D:: LEU2) were transformed with pBN11 or pBN12 plasmids carrying either wild-type (G185) or mutant (V185) P-glycoprotein, respectively. Approximately equal levels of expression of both wild-type and mutant P-glycoprotein were obtained with these two plasmids, and these levels were about 10% of the level per plasma membrane protein seen in KB-V1 cells.

Purification and reconstitution of P-glycoprotein. P-glycoprotein was partially purified as described previously (Ambudkar et al. 1992). The column fraction enriched in P-glycoprotein was incorporated into proteoliposomes composed of lipids from *Escherichia coli*, cholesterol, phosphatidylcholine, and phosphatidylserine by detergent dialysis. Proteoliposomes containing P-glycoprotein were further purified by Sephadex G-50 chromatography. ATPase and vinblastine transport were measured as described previously (Horio et al. 1988; Ambudkar et al. 1992).

Preparation of MDR1 retroviral vectors. The Harvey sarcoma virus retroviral backbone has been used for most of our *MDR1* retroviruses (Pastan et al. 1988). To produce bicistronic vectors in which the *MDR1* cDNA is translated independently from a second cDNA on the same mRNA, the Harvey retroviral backbone was engineered to contain an encephalomyocarditis internal ribosome entry site (IRES), and *MDR1* cDNAs were cloned either upstream or downstream from this site (Aran et al. 1994; Sugimoto et al. 1994).

RESULTS

Model Systems for Studying Mechanism of Action of P-Glycoprotein

The major reason for slow progress in understanding the mechanism of action of P-glycoprotein is the difficulty in molecular analysis of mutant transporters in mammalian cells. For this reason, we have searched for a more tractable system for introducing mutant forms of P-glycoprotein in which relatively normal function is preserved and in which biochemical analysis of the products can be achieved. Despite reports of successful expression of low levels of functional mouse P-glycoprotein in bacterial cells (Bibi et al. 1993), our experience has been that expression of wild-type human P-glycoprotein in growing *E. coli* inevitably leads to cell death (G. Evans et al., unpubl.). However, expression of functional P-glycoprotein in yeast is possible (Kuchler and Thorner 1992; Raymond et al. 1992, 1994; and our unpublished data).

Recently, we have been able to show that expression of wild-type and mutant (V185) forms of P-glycoprotein in drug-sensitive variants of *S. cerevisiae* results in resistance to anticancer drugs, including actinomycin D and daunorubicin (Fig. 2). Furthermore, membrane vesicles prepared from yeast expressing human P-glycoprotein are capable of ATP-dependent drug transport, indicating that this system will be usable for more detailed biochemical analysis (B. Ni et al., unpubl.).

Crude and purified preparations of P-glycoprotein (Ambudkar et al. 1992; Sarkadi et al. 1992) either purified from multidrug-resistant mammalian cells or from insect cells infected with an MDR baculovirus (Germann et al. 1990) demonstrate drug-dependent ATPase activity which is only seen with drugs known to

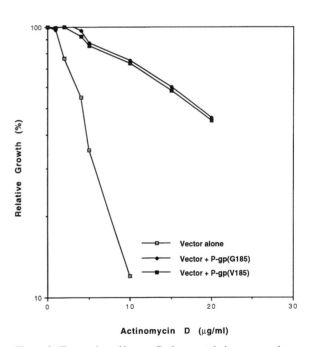

Figure 2. Expression of human P-glycoprotein in yeast confers resistance to actinomycin D. Yeast transformants were grown in modified synthetic liquid medium to midexponential phase, diluted into the same medium to an OD_{600} of about 0.1, and allowed to grow for 1 hr with shaking before addition of the drug. The cell growth at different concentrations of the drug as indicated was optically monitored during a 22-hr incubation. The relative growth percentage was expressed as cell growth in the presence of a given concentration of the drug divided by that in the absence of the drug.

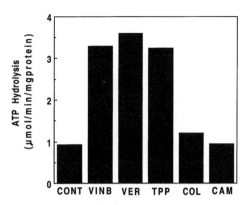

Figure 3. Effect of various drugs on ATPase activity of reconstituted P-glycoprotein. The partially purified human P-glycoprotein was reconstituted into proteoliposomes by detergent-dilution, and the ATPase activity in the presence and absence of 100 μM vanadate was measured. (CONT) Control, dimethyl sulfoxide-treated; (VINB) 20 μM vinblastine; (VER) 100 μM verapamil: (TPP) 100 μM tetraphenyl phosphonium bromide; (COL) 100 μM colchicine; (CAM) 100 μM camptothecin. Only the vanadate-sensitive activities are shown. (Adapted from Ambudkar et al. [1992] and unpublished data of S.V. Ambudkar et al.)

be substrates for the multidrug transporter (Fig. 3). The definitive proof that P-glycoprotein is a complete multidrug transporter has come from evidence that highly purified protein is capable of ATP-dependent drug transport. Protein was purified from highly multidrug-resistant KB-V1 cells in the presence of octylglucoside. Reconstitution into proteolipid vesicles results in vesicles capable of ATP-dependent transport of vinblastine which is blocked by vanadate, a known inhibitor of the drug transport and ATPase activity of P-glycoprotein (S.V. Ambudkar et al., unpubl.). Furthermore, transport of vinblastine is inhibited by known substrates of P-glycoprotein including the potent inhibitor, verapamil, and the anticancer drug, doxorubicin. Thus, both the transport and ATPase activity of the multidrug transporter can be studied in purified preparations of P-glycoprotein.

Retroviral Vectors for Gene Therapy Using the *MDR1* cDNA as a Selectable Marker

The *MDR1* transgenic mouse (Galski et al. 1989; Mickisch et al. 1991) established the principle that expression of the human *MDR1* cDNA in bone marrow could confer resistance to anticancer drugs on bone marrow cells in vivo. Retroviral vectors carrying an *MDR1* cDNA based on the Harvey sarcoma virus backbone (Pastan et al. 1988) were shown to transfer the *MDR1* cDNA to bone marrow cells, which could then be transplanted and selected for their multidrug resistance in mice (Podda et al. 1992; Sorrentino et al. 1992). The success of this strategy has made it possible to consider the initiation of clinical trials in which a human *MDR1* cDNA is transduced via a retrovirus into bone marrow cells of cancer patients undergoing au-

tologous bone marrow transplantation so that they can receive high-dose chemotherapy. The immediate goal of these studies is to demonstrate efficient marking of human bone marrow cells by the *MDR1* cDNA; ultimately, as these patients undergo further chemotherapy with MDR anticancer drugs such as taxol and vinblastine, it should be possible to determine whether the presence of the transduced *MDR1* cDNA protects their bone marrow cells against the toxic effects of anticancer drugs.

In addition to the protection of bone marrow cells from the toxic effects of high-dose chemotherapy, introduction of vectors carrying the *MDR1* cDNA into bone marrow cells may facilitate bone marrow transplantation without the need for complete marrow ablation, since the presence of cells with selective advantage may allow them to gradually replace normal marrow elements under more gentle selective conditions. Furthermore, vectors carrying other genes in addition to the selectable *MDR1* cDNA should allow for coselection of these otherwise nonselectable genes. For example, correction of inborn errors of metabolism in which bone marrow expression of genes is sufficient (as demonstrated by the success of bone marrow transplantation in curing these patients) should be possible. Three different kinds of retroviral vector systems for coexpression of the *MDR1* gene with a second gene have been constructed (Fig. 4). Retroviral vectors with two promoters tend to be inefficient for expression of the nonselectable gene. Vectors encoding translational fusions (protein chimeras) in which the unselected protein is fused to the carboxyl terminus of P-glycoprotein work well for proteins that do not mind being tethered to the undersurface of the plasma membrane, such as adenosine deaminase (Germann et al. 1989), but do not function at all for proteins with nuclear localization (such as thymidine kinase) or function at reduced efficiency for lysosomal proteins, such as glucocerebrosidase (our unpublished data).

The most efficient vector systems we have con-

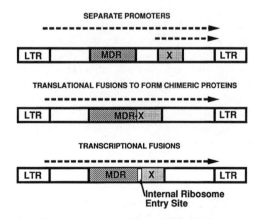

Figure 4. Schematic representation of three kinds of retroviral vectors using the *MDR1* gene as a selectable marker for gene therapy. (LTR) Long terminal repeat; (X) second cDNA whose expression is desired. Dashed lines with arrows show transcription units.

structed to date employ transcriptional fusions in which there is a single mRNA driven by the Harvey sarcoma retroviral LTR promoter, but in which translation is either cap-dependent for the first cDNA, or cap-independent for the second cDNA because of the presence of an IRES from encephalomyocarditis virus (Aran et al. 1994; Sugimoto et al. 1994). We have constructed such bicistronic vectors with *MDR1* cDNAs and cDNAs for herpes simplex virus (HSV) thymidine kinase, human glucocerebrosidase (defective in Gaucher disease, in collaboration with the laboratories of Ed Ginns and Roscoe Brady, NIH), human α-galactosidase (defective in Fabry disease, in collaboration with the laboratory of Roscoe Brady, NIH), and the 91-kD subunit of the human NADPH oxidase complex (defective in chronic granulomatous disease, in collaboration with the laboratory of Harry Malech, NIH). In all cases, selection for multidrug resistance results in substantial expression of the second cDNA of interest. Selection for higher levels of drug resistance increases expression of both the *MDR1* cDNA and the nonselectable gene, as illustrated in Figure 5 for glucocerebrosidase. The position of the *MDR1* cDNA (upstream or downstream from the IRES) determines its relative level of expression; as shown in the α-galactosidase vectors in which the *MDR1* cDNA has been placed in both locations, upstream location increases the level of expression compared to the downstream cDNA (our unpublished data).

In the case of the HSV thymidine kinase bicistronic vector, cells selected for expression of the *MDR1* cDNA are efficiently killed by ganciclovir, a toxic antiviral nucleoside whose phosphorylation (and activation) are HSV thymidine kinase dependent. Thus, the *MDR1*/HSV thymidine kinase vector system is a suicide vector which facilitates destruction of cells inadvertently transduced with *MDR1* vectors, as could happen if a cancer cell is made multidrug resistant during attempts to introduce the *MDR1* cDNA into normal bone marrow elements.

DISCUSSION

These studies on the mechanism of action of P-glycoprotein and on the construction of vectors employing the *MDR1* cDNA as an in vivo selectable marker suggest two new kinds of approaches for improved treatment of cancer: (1) ways to circumvent multidrug resistance in cancer cells whose failure to respond to chemotherapy is primarily due to expression of the *MDR1* gene and (2) gene therapy approaches to the treatment of cancer.

Circumvention of Multidrug Resistance

Understanding the mechanism of interaction of substrates and inhibitors with P-glycoprotein should facilitate the design of new transport inhibitors. The demonstration of drug transport and ATPase activity of isolated, purified P-glycoprotein should improve the possibility of finding such inhibitors (Sarkadi et al. 1994), including suicide inhibitors which bind to substrate sites on P-glycoprotein but cannot be transported, and may also allow the development of inhibitors of the ATPase activity specific to P-glycoprotein. Alternative approaches to drug delivery, such as the use of liposomes which bypass the requirement for drugs to pass through the plasma membrane, may prove useful. Chemical modification of existing drugs so they are no longer substrates for P-glycoprotein may also be of value.

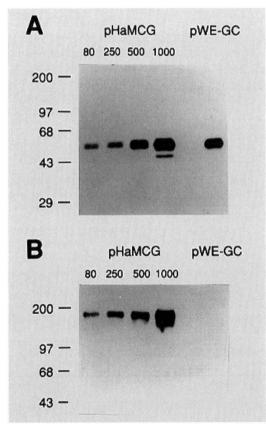

Figure 5. pHaMCG, an MDR-IRES-GC (glucocerebrosidase cDNA) bicistronic retroviral vector, was transfected into mouse fibroblasts. Cells obtained by single-step selection with the indicated concentrations of colchicine (in ng/ml) were disrupted, electrophoresed in 8% SDS-polyacrylamide gels (10 μg protein per lane), and analyzed by Western blotting. pWE-GC refers to G-418-selected transfectants derived from a retroviral vector that expresses high levels of human GC driven by the chicken β-actin promoter. (*A*) Human GC detected with anti-GC antiserum. (*B*) P-glycoprotein detected with C219 MAb. (Reprinted, with permission, from Aran et al. 1994.)

Gene Therapy Approaches

The use of vectors encoding the *MDR1* cDNA to protect human bone marrow during high-dose chemotherapy is an obvious clinical application of the ability of the *MDR1* gene to protect cells from the toxic effects of anticancer drugs. Two problems with this approach are the likelihood of encountering multidrug-resistant tumors whose resistance would need to be reversed, and the possibility of inadvertently transducing tumor cells present in bone marrow. The latter problem may be obviated with the use of "suicide vectors" such as the HSV thymidine kinase vector. The former difficulty can be dealt with by using engineered *MDR1* cDNAs in which sensitivity to "reversing agents" is reduced compared to endogenous *MDR1* gene products expressed in tumors. The feasibility of this approach is suggested by recent studies in which the sensitivity of the Val-185 mutant of P-glycoprotein to reversing agents such as cyclosporine A and verapamil has been shown to differ from that of the Gly-185 wild type (our unpublished observations).

The ability to introduce a second gene along with the selectable *MDR1* cDNA raises the possibility of using *MDR1* vectors to introduce genes into cells whose synthesis or secretion might be detrimental to the growth of cancer. For example, tumor infiltrating lymphocytes might be engineered to produce higher levels of cytokines with antitumor properties. Such an approach would allow the concurrent use of chemotherapy and immunotherapy, which is currently not possible because of the sensitivity of most lymphoid cells to anticancer agents.

Although the *MDR1* vectors described here represent first-generation approaches to the gene therapy of cancer, there are many aspects of the MDR system which make it an attractive candidate for such gene therapy. The extreme flexibility of the system with respect to cytotoxic substrates and the presence of P-glycoprotein on the cell surface as an easily identifiable marker for transduction and isolation of transduced cells are two clear advantages of this system. Preliminary studies in mice suggest no significant toxicity associated with overexpression of the *MDR1* cDNA in bone marrow cells. Finally, molecular genetic analysis points to the feasibility of designing P-glycoprotein molecules with altered substrate specificity and inhibitor sensitivity, which should facilitate gene therapy of cancer and other diseases.

ACKNOWLEDGMENTS

We thank Ms. Paula R. Morgan and Ms. Joyce Sharrar for secretarial support, and all of our former and current colleagues who contributed to the ideas presented in this manuscript.

REFERENCES

Ambudkar, S.V., I.H. Lelong, J. Zhang, C.O. Cardarelli, M.M. Gottesman, and I. Pastan. 1992. Partial purification and reconstitution of the human multidrug-resistance pump: Characterization of the drug-stimulatable ATP hydrolysis. *Proc. Natl. Acad. Sci.* **89:** 8472.

Aran, J.M., M.M. Gottesman, and I. Pastan. 1994. Drug-selected coexpression of human glucocerebrosidase and P-glycoprotein using a bicistronic vector. *Proc. Natl. Acad. Sci.* **91:** 3176.

Bibi, E., P. Gros, and H.R. Kaback. 1993. Functional expression of mouse *mdr1* in *Escherichia coli*. *Proc. Natl. Acad. Sci.* **90:** 9204.

Chen, C.-J., J.E. Chin, K. Ueda, D. Clark, I. Pastan, M.M. Gottesman, and I.B. Roninson. 1986. Internal duplication and homology with bacterial transport proteins in the *mdr1* (P-glycoprotein) gene from multidrug-resistant human cells. *Cell* **47:** 381.

Chin, K.-V., K. Ueda, I. Pastan, and M.M. Gottesman. 1992. Modulation of activity of the promoter of the human *MDR*1 gene by Ras and p53. *Science* **255:** 459.

Cole, S.P.C., G. Bhardwaj, J.H. Gerlach, J.E. Mackie, C.E. Grant, K.C. Almquist, A.J. Stewart, E.U. Kurz, A.M.V. Duncan, and R.G. Deeley. 1992. Overexpression of a transporter gene in a multidrug-resistant human lung cancer cell line. *Science* **258:** 1650.

Cordon-Cardo, C., J.P. O'Brien, D. Casals, L. Rittman-Grauer, J.L. Biedler, M.R. Melamed, and J.R. Bertino. 1989. Multidrug-resistance gene (P-glycoprotein) is expressed by endothelial cells at blood-brain barrier sites. *Proc. Natl. Acad. Sci.* **86:** 695.

Galski, H., M. Sullivan, M.C. Willingham, K.-V. Chin, M.M. Gottesman, I. Pastan, and G.T. Merlino. 1989. Expression of a human multidrug-resistance cDNA (*MDR*1) in the bone marrow of transgenic mice: Resistance to daunomycin-induced leukopenia. *Mol. Cell. Biol.* **9:** 4357.

Germann, U.A., M.M. Gottesman, and I. Pastan. 1989. Expression of a multidrug resistance-adenosine deaminase fusion gene. *J. Biol. Chem.* **264:** 7418.

Germann, U.A., M.C. Willingham, I. Pastan, and M.M. Gottesman. 1990. Expression of the human multidrug transporter in insect cells by a recombinant baculovirus. *Biochemistry* **29:** 2295.

Goldstein L.J., H. Galski, A. Fojo, M.C. Willingham, S.-L. Lai, A. Gazdar, R. Pirker, A. Green, W. Crist, G. Brodeur, C. Grant, M. Lieber, J. Cossman, M.M. Gottesman, and I. Pastan. 1989. Expression of a multidrug resistance gene in human cancers. *J. Natl. Cancer Inst.* **81:** 116.

Gottesman, M.M. and I. Pastan. 1993. Biochemistry of multidrug resistance mediated by the multidrug transporter. *Annu. Rev. Biochem.* **62:** 385.

Greenberger, L.M. 1993. Major photoaffinity drug labeling sites for iodoarylazidoprazosin in P-glycoprotein are within or immediately C-terminal to, transmembrane domain-6 and domain-12. *J. Biol. Chem.* **268:** 11417.

Gros, P., J. Croop, and D.E. Housman. 1986. Mammalian multidrug resistance gene: Complete cDNA sequence indicates strong homology to bacterial transport proteins. *Cell* **47:** 371.

Higgins, C.F. and M.M. Gottesman. 1992. Is the multidrug transporter a flippase? *Trends Biochem. Sci.* **17:** 18.

Higgins, C.F., I.D. Hiles, G.P.C. Salmonel, D.R. Gill, J.A. Downie, I.J. Evans, I.B. Holland, L. Gray, S.D. Buckel, A.W. Bell, and M.A. Hermodson. 1992. ABC transporters—From microorganisms to man. *Annu. Rev. Cell Biol.* **8:** 67.

Homolya, L., A.Z. Holló, U.A. Germann, I. Pastan, M.M. Gottesman, and B. Sarkadi. 1993. Fluorescent cellular indicators are extruded by the multidrug resistance protein. *J. Biol. Chem.* **268:** 21493.

Horio, M., M.M. Gottesman, and I. Pastan. 1988. ATP-dependent transport of vinblastine in vesicles from human multidrug-resistant cells. *Proc. Natl. Acad. Sci.* **85:** 3580.

Kuchler, K. and J. Thorner. 1992. Functional expression of human *mdr*1 in the yeast *Saccharomyces cerevisiae*. *Proc. Natl. Acad. Sci.* **89:** 2302.

Mickisch, G., G.T. Merlino, H. Galski, M.M. Gottesman, and I. Pastan. 1991. Transgenic mice that express the human multidrug resistance gene in bone marrow enable a rapid identification of agents that reverse drug resistance. *Proc. Natl. Acad. Sci.* **88:** 547.

Morris, D.I., L.M. Greenberger, E.P. Bruggemann, C.O. Cardarelli, M.M. Gottesman, I. Pastan, and K.B. Seamon. 1994. Localization of the labeling sites of forskolin to both halves of P-glycoprotein: Similarity of the sites labeled by forskolin and prazosin. *Mol. Pharmacol.* **46:** 329.

Pastan, I., M.M. Gottesman, K. Ueda, E. Lovelace, A.V. Rutherford, and M.C. Willingham. 1988. A retrovirus carrying an *MDR*1 cDNA confers multidrug resistance and polarized expression of P-glycoprotein in MDCK cells. *Proc. Natl. Acad. Sci.* **85:** 4486.

Podda, S., M. Ward, A. Himelstein, C. Richardson, E. de la Flor-Weiss, L. Smith, M.M. Gottesman, I. Pastan, and A. Bank. 1992. Transfer and expression of the human multiple drug resistance gene into live mice. *Proc. Natl. Acad. Sci.* **89:** 9676.

Raviv, Y., H.B. Pollard, E.P. Bruggemann, I. Pastan, and M.M. Gottesman. 1990. Photosensitized labeling of a functional multidrug transporter in living drug-resistant tumor cells. *J. Biol. Chem.* **265:** 3975.

Raymond, M., P. Gros, M. Whiteway, and D.Y. Thomas. 1992. Functional complementation of yeast ste6 by a mammalian multidrug resistance *mdr* gene. *Science* **256:** 232.

Raymond, M., S. Ruetz, D.Y. Thomas, and P. Gros. 1994. Functional expression of P-glycoprotein in *Saccharomyces cerevisiae* confers cellular resistance to the immunosuppressive and antifungal agent FK520. *Mol. Cell Biol.* **14:** 277.

Sarkadi, B., E.M. Price, R.C. Boucher, U.A. Germann, and G.A. Scarborough. 1992. Expression of the human multidrug resistance cDNA in insect cells generates a high activity drug-stimulated ATPase. *J. Biol. Chem.* **267:** 4854.

Sarkadi, B., M. Müller, L. Homolya, Z. Holló, U.A. Germann, M.M. Gottesman, E.M. Price, and R.C. Boucher. 1994. Interaction of bioactive hydrophobic peptides with the human multidrug transporter. *FASEB J.* **8:** 766.

Schinkel, A.J., J.J.M. Smit, O. van Tellingen, J.H. Beijnen, E. Wagenaar, L. van Deemter, C.A.A.M. Mol, M.A. van der Valk, E.C. Robanus-Maandag, H.P.J. te Riele, A.J.M. Berns, and P. Borst. 1994. Disruption of the mouse mdr1a P-glycoprotein gene leads to a deficiency in the blood-brain barrier and to increased sensitivity to drugs. *Cell* **77:** 491.

Sikic, B. 1993. Modulation of multidrug resistance: At the threshold. *J. Clin. Oncol.* **11:** 1629.

Sorrentino, B.P., S.J. Brandt, D. Bodine, M.M. Gottesman, I. Pastan, A. Cline, and A.W. Nienhuis. 1992. Retroviral transfer of the human *MDR*1 gene permits selection of drug resistant bone marrow cells *in vivo*. *Science* **257:** 99.

Stein, W.D., C.O. Cardarelli, I. Pastan, and M.M. Gottesman. 1994. Kinetic evidence suggesting that the multidrug transporter differentially handles influx of its substrates. *Mol. Pharmacol.* **45:** 763.

Sugimoto, Y., I. Aksentijevich, M.M. Gottesman, and I. Pastan. 1994. Efficient expression of drug-selectable genes in retroviral vectors under control of an internal ribosome entry site. *Bio/Technology* **12:** 694.

Thiebaut, F., T. Tsuruo, H. Hamada, M.M. Gottesman, I. Pastan, and M.C. Willingham. 1987. Cellular localization of the multidrug resistance gene product P-glycoprotein in normal human tissues. *Proc. Natl. Acad. Sci.* **84:** 7735.

———. 1989. Immunohistochemical localization in normal tissues of different epitopes in the multidrug transport protein, P170: Evidence for localization in brain capillaries and cross-reactivity of one antibody with a muscle protein. *J. Histochem. Cytochem.* **37:** 159.

Toxicity and Immunologic Effects of In Vivo Retrovirus-mediated Gene Transfer of the Herpes Simplex-Thymidine Kinase Gene into Solid Tumors

K.W. Culver, D.W. Moorman, R.R. Muldoon, R.M. Paulsen, Jr.,
J.L. Lamsam, H.W. Walling, and C.J. Link, Jr.

Human Gene Therapy Research Institute, Iowa Methodist Medical Center, Des Moines, Iowa 50309

Our advancing understanding of the molecular basis of cancer and gene transfer offers new possibilities for the treatment of cancer. The liver is a common site of metastasis from a number of primary sites including the colon, prostate, skin, and pancreas. As a result, tens of thousands of patients die of metastatic liver disease. In addition, more than 400,000 people will succumb to primary liver cancer annually worldwide. Therefore, we are in the process of developing in situ gene transfer for the treatment of hepatic malignancies.

The widespread application of gene therapy to cancer is limited primarily by a lack of efficient in situ gene transfer methodologies, not toxicities. The ability to selectively deliver genes into all malignant cells in the body could theoretically effect a cure without damage to normal tissues. Likewise, the delivery of a gene into premalignant tissues could theoretically prevent malignancy. However, the problems associated with the efficient and selective delivery of genes into malignant cells is a significant problem, and no currently available system has a 100% efficiency in vivo. As a result, a number of scientists have been removing tissues for ex vivo manipulation in order to improve gene transfer efficiency. In this process, tumor cells, skin fibroblasts, or T lymphocytes are removed from the patient and transduced with immunostimulatory genes (Culver and Blaese 1994). The gene-modified cells are then reimplanted into the patient (mixed with irradiated non-gene-modified autologous tumor cells in the case of skin fibroblasts) in an effort to stimulate a vigorous antitumor immune response. Early human clinical trials using cytokine genes have not been reported to generate a potent systemic immune response against the autologous tumor cells at distant sites. This is in part due to the debilitated status of patients with disseminated cancer who have received prior immunosuppressive chemotherapy and whose tumor cells produce immunosuppressive compounds (e.g., TGF-β and IGF-1) (Roszman et al. 1991; Trojan et al. 1993). However, one encouraging study utilizing the transfer of a foreign HLA gene (B7) with liposomes has induced a clinically significant immune response in 1 of 5 patients with metastatic melanoma (Nabel et al. 1993).

Another approach is the transfer of genes that will be directly cytotoxic to tumor cells and do not necessarily require the immune system for tumor destruction. One example is the herpes simplex-thymidine kinase (*HS-tk*) gene. *HS-tk* is a "sensitivity" or "suicide" gene that confers a sensitivity to the prodrugs acyclovir (Zovirax; ACV) and ganciclovir (Cytovene; GCV). When ACV or GCV enters an *HS-tk*-containing cell, it is converted to GCV-monophosphate (MP). Endogenous cellular phosphorylases convert the GCV-MP to GCV-triphosphate (GCV-TP). Studies in herpes simplex virus-infected cells suggest that ACV-TP and GCV-TP inhibit DNA polymerase and incorporate into the elongating DNA strand, resulting in chain termination and cell death (Elion 1980). Likewise, the transfer of the *HS-tk* gene into tumor cells results in tumor cell destruction with either of the anti-herpes drugs (Moolten 1986; Moolten and Wells 1990).

Since ACV and GCV are generally nontoxic, a number of investigators have pursued the development of this prodrug approach for the treatment of cancer (Moolten et al. 1990; Ezzeddine et al. 1991; Huber et al. 1991; Plautz et al. 1991; Barba et al. 1993; Freeman et al. 1993; Takamiya et al. 1993; Vile and Hart 1993; Chen et al. 1994). Our initial studies confirmed previous observations that the *HS-tk*/GCV system was very effective at destroying cancer cells. Surprisingly, our in vivo investigations demonstrated that only ≥10% of the tumor cells need to contain *HS-tk* to result in complete tumor ablation in 9 out of 15 mice (Culver et al. 1992). This phenomenon of tumor cell killing has been termed the "bystander" tumor killing effect and is thought to result from the passage of GCV or its phosphorylated derivatives via gap junctions into neighboring tumor cells (Bi et al. 1993). The next step was the development of an in vivo gene transfer method that could achieve at least a 10% gene transfer efficiency in vivo.

We have used murine retroviral vectors (MRVs) as our *HS-tk* gene delivery system because these vectors stably insert and express their genes exclusively in dividing tissues (Miller et al. 1990). This feature allows the targeting of vector genes selectively to tumor cells, due to their greater mitotic rate compared to surrounding normal tissues. The brain was chosen as the first site because of the low basal rate of proliferation in normal cells. The intra-tumoral implantation of retroviral vector producer cells (VPC) should allow continuous production of vector within the tumor compared to the direct injection of MRVs, which only transduces cells that are actively synthesizing DNA at that time of injection.

Experiments in a rat model have demonstrated that the vector genes can be selectively delivered into tumors of the CNS following the direct intra-tumoral injection of MRV-producing cells (Short et al. 1990; Culver et al. 1992; Ram et al. 1993a,b). The studies have also shown that the injected VPC can survive for up to 2 weeks in the brain of xenogeneic animals with gene transfer efficiencies of 10–55% (Ram et al. 1993a). Early results from human brain tumor clinical trials using this gene transfer approach have confirmed the animal studies, demonstrating no discernible toxicities with evidence of selective destruction of tumor cells in some patients.

These early, encouraging results in the human brain tumor studies have led us to begin to examine this form of in vivo gene delivery in non-CNS solid tumors. Through a series of in vivo studies, we have demonstrated the following: (1) The direct injection of VPC into the liver or peritoneal cavity does not transfer genes into hepatocytes, peritoneum (Moorman et al. 1994), or ovarian tissues; (2) the injection of *HS-tk* VPC into the skin, liver, and abdomen results in no evidence of toxicity with GCV treatment; (3) human tumors can be transduced with VPC in vivo in athymic nude mice; (4) human hepatoma and colorectal cell lines manifest the bystander effect in vitro; and (5) preexisting immunity to the VPC does not preclude tumor regression. Together, these results suggest that the in vivo transfer of the *HS-tk* gene into non-CNS tissues can be accomplished safely and with potent antitumor efficacy even in the presence of preexisting systemic immunity against the VPC.

METHODS

Animal models. Female 6- to 8-week-old C57BL/6 mice and athymic nude (nu/nu) mice were obtained from either the National Cancer Institute (Frederick, Maryland) or Harlan Sprague Dawley (Indianapolis, Indiana). The nude mice were housed in laminar flow animal isolation units with microisolator tops. Food, water, and bedding were autoclaved. Mice received metofane (Pitman-Moore; Mundelein, Illinois) by inhalation for anesthesia. Once the mice were anesthetized, an upper midline laparotomy was performed, the left anterior hepatic lobe of the liver was isolated, and the VPC were injected through a 27g needle into the parenchyma of the liver. The laparotomy and skin were closed with interrupted 4-0 silk suture ligatures followed by skin clips. All animals were allowed to recover under a warming light and were returned to standard chow and water ad libitum. GCV was injected twice daily, 8–12 hours apart, for 14 days. For each injection, the GCV was reconstituted in saline or Hank's balance salt solution and injected intraperitoneally through a 25 gauge needle at a dose of 150 mg/kg/dose.

Tumor cell lines. The human hepatocellular carcinoma cell lines Hep3B (ATCC; HB 8064) and HepG2 (HB 8065), the human colorectal carcinoma Hct 15 (ATCC; CCL 225), and the human melanoma A375 (ATCC; CRL 1619) were utilized in the nude mouse experiments. The syngeneic MCA 205 fibrosarcoma (NCI; Bethesda, Maryland) was used in the immunization experiments. The tumor cells were maintained in complete medium consisting of RPMI-1640 with 10% fetal calf serum, 2 mM L-glutamine, 100 units/ml Penicillin G sodium, 100 μg streptomycin sulfate, and 2.5 μg/ml Fungizone, all products of Quality Biological Inc. (Gaithersburg, Maryland). The cells were grown in T-175 flasks (COSTAR, Cambridge, Massachusetts). The cells were removed with trypsin (2 ml/flask at 2 μg/ml; Quality Biological Inc.), washed, and counted. The cells were passaged at least once per week depending on the cell growth rate.

Tumor analysis. Tumor tissues were processed with three different methods. For histological analysis, the tumor tissue was fixed in 4% formalin, and sections were stained with hematoxylin and eosin. Replicate frozen sections were incubated with 5-bromo-4-chloro-3-indolyl β-D-galactoside (X-GAL) according to previously described methods to detect β-galactosidase produced from the vector reporter gene, *LacZ* (Matsuoka et al. 1991). X-GAL sections were counterstained with neutral red. Multiple sites were sectioned from injected areas, and single sites were sectioned from adjacent PBS-injected (control) areas and other visceral organs. Third, freshly resected tumor tissue was enzymatically digested, placed in culture using complete medium, and stained with X-GAL on day 7 in culture (Culver et al. 1992; Ram et al. 1993a).

Retroviral vectors and tumor cell transduction. The STK/PA317 vector producer cell line was generously provided by Dr. Fred Moolten (Boston, Massachusetts). STK expresses the herpes simplex-thymidine kinase type 1 cDNA, which is transcribed from an internal SV40 promoter, and expresses the *NeoR* gene from the 5' viral long terminal repeat (LTR) (LTR–*NeoR*–SV40–*HS-tk*–LTR). The *NeoR* gene encodes for neomycin phosphotransferase II (NPT II), an enzyme that protects STK-expressing cells from the toxic effects of G418, a neomycin analog. The G1TkSvNa/PA317 and G1BgSvNa/PA317 vectors were provided by Genetic Therapy Inc. (Gaithersburg, Maryland). G1TkSvNa is the reverse construction of STK with the *HS-tk* gene under the control of the 5' LTR and *NeoR* under the direction of an internal SV40 promoter. The G1BgSvNa vector is quite similar to G1TkSvNa, with the *Escherichia coli LacZ* gene under the control of the 5' LTR and the *NeoR* under control of an internal SV40 promoter. Transduction of these ecotropic vectors into the amphotropic packaging cell line PA317 resulted in the production of amphotropic, replication-incompetent MRVs. The titers of these VPC range from 1×10^3 to 1×10^4.

The VPC were maintained in Dulbecco's modified

Eagle's medium (DMEM) with 4.5 gm/l glucose with the same additives as were used for tumor cell growth above (Quality Biological, Inc.). The monolayer cultures are grown in T-175 flasks at 37°C with 5% CO_2. Supernate from the VPC was harvested from the flask when the cells were 80–100% confluent. Protamine sulfate (Elkins-Sinn; Cherry Hill, New Jersey) was added at 5 µg/ml to the supernate and filtered through a 0.45-µm cellulose acetate filter (Nalgene; Rochester, New York). The filtered supernate was then transferred into flasks containing tumor cells at a moi of ≤5 twice daily for 2 or 3 days. After the final transduction, the tumor cells were placed into complete RPMI medium. S^+L^- assays for replication-competent virus in the supernate from the producer cell line and transduced tumor cells were negative for each cell line. G418 (GIBCO) was added to the cultures at 1 mg/ml for 7–14 days. This selection condition resulted in the destruction of an estimated 75–90% of the tumor cells. Once the remaining cells had recovered from the selection process and were growing well, they were used for in vitro and in vivo assays.

In vitro HS-tk sensitivity and bystander assays. Tumor cells were trypsinized, washed, and counted. *HS-tk*-transduced and non-transduced tumor cells were cultured alone or in mixtures in triplicate at 10,000 cells per well in 96-well flat-bottom plates (Costar). The following day, GCV was added at increasing concentrations to determine their GCV sensitivity or at a final concentration of 5.1 µg/ml for the bystander assay. After 44 hours, [^{3}H]thymidine was added to the wells and, 4 hours later, the plates were harvested. The quantity of incorporated [^{3}H]thymidine (counts per minute) was determined with a Betaplate reader (LKB Wallac). The results are expressed as a mean percent of control (no GCV) of triplicate wells.

Induction of immunity to the VPC. Ten million G1TkSvNa/PA317 VPC were injected subcutaneously into the shaved right flank of C57BL/6 mice. Growth of the VPC was monitored twice weekly for 6 weeks by measuring the site of injection in 3 dimensions. 8 weeks after the VPC had been injected, the animals were rechallenged with either tumor cells alone (1×10^6), VPC alone (10^7), or a 1:1 mixture of tumor cells and VPC (1×10^6) each on the contralateral side of the abdomen from the first injection of cells. The site of tumor injection was followed weekly for 8 weeks after the challenge.

RESULTS

GCV Sensitivity and Bystander Effect of Human Liver/Colon Tumors In Vitro

To determine the effects of *HS-tk* gene transfer on tumor sensitivity to GCV, the STK retroviral vector was transferred into two human hepatocellular carcinomas (Hep3B and HepG2) and one human colorectal carcinoma (Hct 15). In each case, transfer of the *HS-tk*

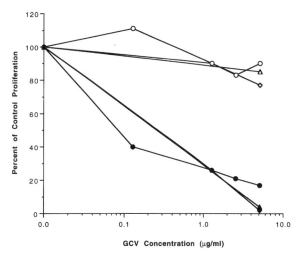

Figure 1. In vitro GCV sensitivity of human hepatocellular and colorectal carcinomas. Non-transduced Hep3B (○), HepG2 (△), and Hct 15 (◇) and *HS-tk*-transduced Hep3B (●), HepG2 (▲), and Hct 15 (◆) cell lines were cultured in parallel in increasing concentrations of GCV. The amount of [^{3}H]thymidine incorporation was determined and is expressed as a percent of control proliferation (no GCV). Each of the *HS-tk*-transduced cell lines demonstrated a substantial sensitivity to GCV, whereas the nontransduced cells were minimally inhibited by the drug.

gene conferred a significant increase in sensitivity to GCV in vitro at 5.1 µg/ml as measured by [^{3}H]thymidine incorporation into DNA (Fig. 1). The proliferative rate of both HepG2 and Hct 15 decreased to baseline. The Hep3B cell line appears to be somewhat less sensitive than the others tested. GCV was used at this dose because it is an easily attainable serum level in humans with FDA approved dosing.

Mixtures of STK-transduced tumor cell lines and non-transduced tumor cells were used to evaluate the presence and degree of the bystander effect in vitro (Fig. 2). If GCV produced no bystander effect, the degree of proliferation would be expected to be directly proportional to the number of *HS-tk*-containing cells in the mixture. Each of the tumor cell lines manifested a greater than expected decrease in proliferation. These findings suggest that these three human tumor cell lines manifest an in vitro bystander effect with a >50% decrease in proliferation with only 10% of the cells containing *HS-tk*-transduced tumor cells.

In Vivo Gene Transfer Studies into Normal and Malignant Tissues

Elucidation of the toxicities associated with this form of gene transfer requires an assessment of gene transfer efficiencies into both normal tissues and tumor cells. In an effort to determine the relative in vivo gene transfer efficiency of this system in normal tissues, we have evaluated the gene transfer efficiency following the direct implantation of *LacZ* VPC. The injection of G1BgSvNa/PA317 VPC directly into the skin, liver

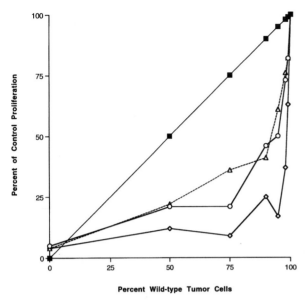

Figure 2. The in vitro bystander effect of the Hep3B (○) and HepG2 (△) hepatocellular carcinomas and the Hct 15 (◇) colorectal carcinoma. Mixtures of *HS-tk*-transduced and wild-type tumor cells were cultured with and without GCV, and proliferation was measured by [³H]thymidine uptake. The degree of proliferation is expressed as a percent of control proliferation (no GCV). If no bystander effect is present, the percent of *HS-tk*-positive cells would be expected to correlate with the percent decrease in proliferation, since only the *HS-tk*-positive cells should be affected (■). In contrast, all three human tumor cell lines manifested a much greater decrease in proliferation than expected, confirming inhibition of proliferation of *HS-tk*-negative cells in this assay.

parenchyma, or peritoneal cavity resulted in no evidence of *LacZ* gene transfer by X-GAL staining in surrounding normal tissues (data not shown).

Toxicity of G1TkSvNa/PA317 VPC injection was evaluated in athymic nude mice. This animal model system was chosen for safety studies because the nude mouse is unable to reject the VPC, providing an opportunity for sustained vector production in vivo. This should represent a worst-case scenario for VPC-associated toxicity. Growth of the VPC in nude animals was allowed to continue for up to 30 days (Table 1). Subsequent GCV therapy did not result in detectable damage to the organ injected or other visceral organs in any of the mice. There were occasional local bacterial infections in these immunodeficient mice related to the twice daily intraperitoneal injections of GCV.

Table 1. In Vivo Toxicity Experiments with the Direct Implantation of *HS-tk* VPC

Injection route	Time prior to GCV	Result[a]
Subcutaneous	30 days	no acute toxicity
Intraperitoneal	14 days	no acute toxicity
Intrahepatic	21 days	no acute toxicity

[a] In each case, complete resolution of the VPC tumor was noted.

In Vivo Gene Transfer Efficiency in Human Tumors

To assess the gene transfer efficiency of direct injection of VPC into human tumors growing in nude mice, 1.5×10^7 A375 tumor cells were injected subcutaneously into athymic nude mice ($N = 5$) and allowed to grow to 1.0 cm in diameter. 2×10^7 *LacZ* VPC were then injected into each of the tumors. After 5 days, the animals were treated with a 7-day course of GCV to eliminate the G1BgSvNa/PA317 VPC, since an *HS-tk* gene was transfected into PA317 to facilitate production of the packaging cell line. At the end of GCV treatment, the tumors were excised, minced, and enzymatically digested into a single cell suspension. The tumor cell suspension was cultured for 7 days and stained with X-GAL. The number of X-GAL-positive cells ranged from 22% to 30% (24.8 ±3.1). These results confirm that human tumors can be transduced in vivo via this method.

Effect of Preexisting Immunity to the VPC on Tumor Growth

In an attempt to develop this gene transfer method for non-CNS tumors, we have evaluated the survival and antitumor effect of *HS-tk* VPC in mice with or without preexisting immunity to the VPC. Table 2 depicts the results of an experiment in which preimmunized and naive animals were injected with a mixture of syngeneic MCA 205 fibrosarcoma tumor cells and G1TkSvNa/PA317 VPC. Tumor cells injected alone grew regardless of GCV administration. The previously immunized mice demonstrated immunity to the VPC since no evidence of growth of VPC was noted compared to transient growth in naive animals. Comparison of the antitumor effect in mice receiving the tumor/VPC mixtures demonstrated that prior immunization of the animals did not adversely affect the antitumor effect in this animal model.

Table 2. Antitumor Effect of *HS-tk* VPC in Naive and VPC-immune Mice

Group	N	Preimmunized	GCV	Antitumor response
Tumor alone	5	no	yes	none
	3	yes[a]	yes[b]	none
VPC alone	5	no	no	transient growth
	4	yes	no	no growth
Tumor + VPC	5	yes	no	none
	4	yes	yes	100% response
Tumor + VPC	5	no	no	none
	5	no	yes	100% response

[a] Animals were injected subcutaneously before tumor challenge with 1×10^6 *HS-tk* VPC.
[b] Animals were treated with GCV twice daily for 14 days.

DISCUSSION

The development of efficient, selective in vivo gene transfer methods is crucial for the widespread application of gene therapy to cancer. Ex vivo gene transfer systems will only be applicable on a limited basis. As a result, we have investigated the toxicities associated with the use of a MRV in vivo gene transfer system. Due to the fact that MRVs require a proliferating cell target, the first applications have been in the CNS where the resident normal tissues have a very low mitotic index. This feature allows the selective targeting of the vector genes into tumor cells in animal brain tumor models (Short et al. 1990; Culver et al. 1992; Ram et al. 1993a). However, tumors of the brain are not as easily accessible to repeated intratumoral injections, and they are much less common than non-CNS tumors. Therefore, we have been developing this in vivo gene delivery system for solid tumors in the liver, skin, and peritoneal cavity.

Since the mitotic index of non-CNS tissues is generally greater than CNS tissues, we first examined the extent of gene transfer of the *LacZ* gene in normal tissues. Second, the toxicity of *HS-tk* VPC was evaluated following the direct injection of VPC into the liver, the peritoneal cavity, and the skin. These studies demonstrate no evidence of gene transfer and no evidence of toxicity to the surrounding normal tissues with GCV treatment in either location. These results are consistent with the findings of other investigators who demonstrated a lack of efficient gene transfer of MRVs into the liver without stimulation of hepatic cell division (Ferry et al. 1991; Kaleko et al. 1991; Caruso et al. 1993; Moorman et al. 1994). Even athymic nude mice bearing a *HS-tk* VPC tumor in the liver or peritoneal cavity did not develop evidence of clinically significant toxicity with GCV treatment. These findings suggest that the intratumoral injection of VPC may safely avoid gene transfer into normal cells in visceral organs.

We next examined the gene transfer efficiency of the *LacZ* gene into subcutaneous human tumor xenografts growing in athymic nude mice. The direct injection of VPC into growing tumors resulted in a 22–30% level of gene transfer. Further enhancement of the level of gene transfer might be enhanced by increasing the number of VPC injected, using VPC with a higher titer (the *LacZ* producer has a low titer), and/or by repeated injections of VPC. Another possibility would be to attempt to improve the duration of VPC survival by constructing them from autologous tumor cells.

The early animal experiments using in vivo gene transfer with *HS-tk* VPC were conducted in the CNS, where the immunologic response to foreign tissues is diminished due to the immunoprivileged nature of the CNS. However, the immune response outside the CNS may be sufficient to block gene transfer and prevent repeated treatments without using either autologous VPC or immunosuppression. Therefore, we have evaluated the effect of prior immunization with *HS-tk* VPC in immunocompetent mice on the antitumor response to GCV. The results are consistent with the induction of preexisting anti-VPC immunity, since the VPC did not grow in any of the previously immunized mice. However, the immunized mice mounted the same degree of tumor rejection as nonimmunized mice. This suggests that foreign VPC survived long enough to transduce surrounding tumor cells and/or provide enough bystander effect prior to rejection to mediate a comparable antitumor response in vivo.

The *HS-tk*/GCV system has potent, selective tumor destruction capabilities. Its utility as a cancer therapy is limited by gene transfer efficiency. Our gene delivery approach appears to be limited to the local area of injection. Therefore, our investigations have been expanded to include the evaluation of *HS-tk* in combination with cytokines (e.g., IL-2) (Link et al. 1994). It is hoped that combinations of cytokine genes and cytotoxic genes will generate significant local tumor destruction and a potent immune response against metastatic disease. This approach may convert a local therapy into a regional response that might effectively treat metastases limited to certain tissues such as the liver or to the peritoneal cavity.

REFERENCES

Barba, D., J. Hardin, J. Ray, and F.H. Gage. 1993. Thymidine kinase-mediated killing of rat brain tumors. *J. Neurosurg.* **79:** 729.

Bi, W.L., L.M. Parysek, R. Warnick, and P.J. Stambrook. 1993. In vitro evidence that metabolic cooperation is responsible for the bystander effect observed with HSV tk retroviral gene therapy. *Hum. Gene Ther.* **4:** 725.

Caruso, M., Y. Panis, S. Gagandeep, D. Houssin, J.-L. Salzmann, and D. Klatzmann. 1993. Regression of established macroscopic liver metastases after in situ transduction of a suicide gene. *Proc. Natl. Acad. Sci.* **90:** 7024.

Chen, S.-H., H.D. Shine, J.C. Goodman, R.G. Grossman, and S.L.C. Woo. 1994. Gene therapy for brain tumors: Regression of experimental gliomas by adenovirus-mediated gene transfer in vivo. *Proc. Natl. Acad. Sci.* **91:** 3054.

Culver, K.W. and R.M. Blaese. 1994. Gene therapy for cancer. *Trends Genet.* **10:** 174.

Culver, K.W., Z. Ram, S. Walbridge, H. Ishii, E.H. Oldfield, and R.M. Blaese. 1992. In vivo gene transfer with retroviral vector producer cells for treatment of experimental brain tumors. *Science* **256:** 1550.

Elion, G.B. 1980. The chemotherapeutic exploitation of virus-specified enzymes. *Adv. Enzyme Regul.* **18:** 53.

Ezzeddine, Z.D., R.L. Martuza, D. Platika, M.P. Short, A. Malick, B. Choi, and X.O. Breakfield. 1991. Selective killing of glioma cells in culture and in vivo by retrovirus transfer of the herpes simplex virus thymidine kinase gene. *New Biol.* **3:** 608.

Ferry, N., O. Duplessis, D. Houssin, O. Danos, and J.M. Heard. 1991. Retroviral-mediated gene transfer into hepatocytes in vivo. *Proc. Natl. Acad. Sci.* **88:** 8377.

Freeman, S.M., C.N. Abboud, K.A. Whartenby, C.H. Packman, D.S. Koeplin, F.L. Moolten, and G.N. Abraham. 1993. The "bystander effect": Tumor regression when a fraction of the tumor mass is genetically modified. *Cancer Res.* **53:** 5274.

Huber, B.E., C.A. Richards, and T.A. Krenitsky. 1991. Retroviral-mediated gene therapy for the treatment of hepatocellular carcinoma: An innovative approach for cancer therapy. *Proc. Natl. Acad. Sci.* **88:** 8039.

Kaleko, M., J.V. Garcia, and A.D. Miller. 1991. Persistent gene expression after retroviral gene transfer into liver cells in vivo. *Human Gene Ther.* **2:** 27.

Link, C.J., E. Beecham, L. McCann, R. Muldoon, R. Paulsen, and K.W. Culver. 1994. In vivo combination gene therapy with the herpes thymidine kinase and interleukin-2 genes for the treatment of ovarian cancer. *Proc. Am. Soc. Clin. Oncol. Annu. Meet.* **13:** 810.

Matsuoka, M., F. Nagawa, K. Okazaki, L. Kingsbury, K. Yoshida, U. Muller, D.T. Larue, J.A. Winer, and H. Sakano. 1991. Detection of somatic DNA recombination in the transgenic mouse brain. *Science* **254:** 81.

Miller, D.G., M.A. Adam, and A.D. Miller. 1990. Gene transfer by retrovirus vectors occurs only in cells that are actively replicating at time of infection. *Mol. Cell Biol.* **10:** 4239.

Moolten, F.L. 1986. Tumor chemosensitivity conferred by inserted herpes thymidine kinase genes: Paradigm for a prospective cancer control strategy. *Cancer Res.* **46:** 5276.

Moolten, F.L. and J.M. Wells. 1990. Curability of tumors bearing herpes thymidine kinase genes transferred by retroviral vectors. *J. Natl. Cancer Inst.* **82:** 297.

Moolten, F.L., J.M. Wells, R.A. Heyman, and R.M. Evans. 1990. Lymphoma regression induced by ganciclovir in mice bearing a herpes thymidine kinase transgene. *Human Gene Ther.* **1:** 125.

Moorman, D.W., D.A. Butler, J.D. Stanley, J.L. Lamsam, M.R. Ackermann, C.D. Jacobson, and K.W. Culver. 1994. Survival and toxicity of xenogeneic murine retroviral vector producer cells in liver. *J. Surg. Oncol.* **57:** 152.

Nabel, G.J., E.G. Nabel, Z. Yang, Z-Y. Yang, B.A. Fox, G.E. Plautz, X. Gao, L. Huang, S. Shu, D. Gordon, and A.E. Chang. 1993. Direct gene transfer with DNA-liposome complexes in melanoma: Expression, biologic activity, and lack of toxicity in humans. *Proc. Natl. Acad. Sci.* **90:** 11307.

Plautz, G., E.G. Nabel, and G.J. Nabel. 1991. Selective elimination of recombinant gene in vivo with a suicide vector. *New Biol.* **3:** 709.

Ram, Z., K.W. Culver, S. Walbridge, R.M. Blaese, and E.H. Oldfield. 1993a. In situ retroviral-mediated gene transfer for the treatment of brain tumors in rats. *Cancer Res.* **53:** 83.

Ram, Z., K.W. Culver, S. Walbridge, J.A. Frank, R.M. Blaese, and E.H. Oldfield. 1993b. Toxicity studies of retroviral-mediated gene transfer for the treatment of brain tumors. *J. Neurosurg.* **79:** 400.

Roszman T, L. Elliott, and W. Brooks. 1991. Modulation of T-cell function by gliomas. *Immunol. Today* **12:** 370.

Short, M.P., J.K. Choi, A. Lee, A. Malick, X.O. Breakfield, and R.L. Martuza. 1990. Gene delivery to glioma cells in rat brain by grafting of a retrovirus packaging cell line. *J. Neurosci. Res.* **27:** 427.

Takamiya, Y., M.P. Short, F.L. Moolten, C. Fleet, T. Mineta, X.O. Breakfield, and R.L. Martuza. 1993. An experimental model of retrovirus gene therapy for malignant brain tumors. *J. Neurosurg.* **79:** 104.

Trojan, J., T.R. Johnson, S.D. Rudin, J. Ilan, M.L. Tykochinski, and J. Ilan. 1993. Treatment and prevention of rat glioblastoma by immunogenic C6 cells expressing antisense insulin-like growth factor 1 RNA. *Science* **259:** 94.

Vile, R.G. and I.R. Hart. 1993. Use of tissue-specific expression of the herpes simplex virus thymidine kinase gene to inhibit growth of established murine melanomas following direct intratumoral injection of DNA. *Cancer Res.* **53:** 3860.

Gene Transfer and Bone Marrow Transplantation

M.K. Brenner,*‡§ H.E. Heslop,*‡ D. Rill,* C. Li,* T. Nilson,* M. Roberts,*‡
C. Smith,* R. Krance,*‡ and C. Rooney†

*Department of Hematology/Oncology, Division of Bone Marrow Transplantation, and †Department of Virology
and Molecular Biology, St. Jude Children's Research Hospital, Memphis, Tennessee 38105;
Departments of ‡Pediatrics and §Medicine, University of Tennessee, Memphis, Tennessee 38163*

Hematopoietic progenitor cells are appealing targets for gene transfer (Anderson 1992; Miller 1992), since their genetic modification could be effective in a variety of malignant and nonmalignant disorders. In addition to their suitability as targets for a wide range of clinical applications, hematopoietic progenitor cells are attractive for gene transfer because of their logistical characteristics. Both marrow and peripheral blood progenitor cells are readily obtained, manipulated, and returned to the host. Moreover, in principle, successful gene transfer into a single pluripotent stem cell could be sufficient to repopulate each individual with modified cells for his entire life span. In almost every other type of somatic gene therapy, transfer is required to a vast number of cells which may be spread over a wide anatomical area.

To take full advantage of these characteristics, it is necessary to use vectors that permanently integrate in the DNA of the hematopoietic stem cell and are therefore present in all the progeny of this cell (Hughes et al. 1992). At present, retroviruses are the only integrating vectors approved for clinical use (Anderson 1992; Miller 1992). One characteristic of these vectors is that integration occurs almost exclusively in cells which are progressing through the cell cycle. Since most mammalian hematopoietic stem cells are in G_0, the efficiency of gene transfer into these cells is generally far below the levels that would be expected to produce benefit in the great majority of gene therapy settings (Schuening et al. 1991; Anderson 1992; Miller 1992; Smith 1992; Van Beusechem et al. 1992). In animal models, a number of techniques have been shown to induce hematopoietic stem cells to cycle and thereby increase the efficiency of retrovirus-mediated gene transfer (Hughes et al. 1992; Rill et al. 1992a). These include prior treatment of the animal with cycle-specific cytotoxic drugs (Wieder et al. 1991) and in vivo/ex vivo exposure to hematopoietic growth factors (Dick et al. 1991; Otsuka et al. 1991a,b). At present, however, the human stem cell has not been precisely identified, and no assays can directly analyze the number or activity of these cells. It has therefore been impossible to devise preclinical experiments capable of showing that retroviruses can transfer genes to human stem cells or that manipulation of those stem cells enhances their susceptibility to retrovirus-mediated gene transfer. This has been a major constraint to the development of therapeutic protocols. No clinical gene transfer study can be entirely devoid of risk, and when the probability of benefit is unknown and likely to be small, then the risk/benefit ratio becomes unacceptably high (Cornetta et al. 1991; Anderson 1992; Miller 1992; Smith 1992). On the other hand, until human studies are performed, it is impossible to make an accurate assessment of the true risks and the true probability of success.

One way of resolving this impasse is to revise our concept of the respective roles of gene transfer and bone marrow transplantation. Instead of gene transfer representing the therapy and bone marrow transplantation the means of delivery, we can consider the transplant itself to be the therapeutic procedure, the safety and efficacy of which can be improved by gene transfer. One example of this application is to use gene transfer simply as a means of marking hematopoietic progenitor cells prior to transplantation, to track their subsequent fate in vivo (Rill et al. 1992a,b; Brenner et al. 1993a).

GENE MARKING OF HEMATOPOIETIC PROGENITOR CELLS

Gene marking can be used to determine whether harvested progenitor cells in marrow or blood are contaminated with tumorigenic cells and whether attempts to remove such cells are successful (Rill et al. 1992a,b; Brenner et al. 1993a). It may also be used to learn about the contribution of harvested stem cells to long-term engraftment and the impact of ex vivo growth factor treatment on the subsequent behavior of these cells in vivo (Brenner et al. 1993b).

Are Harvested Hematopoietic Progenitor Cells Contaminated by Tumorigenic Cells?

Although the dose intensification permitted by autologous hematopoietic stem cell (HSC) rescue has shown promise as effective treatment for malignant diseases, including leukemias, lymphomas, and breast cancer (Burnett et al. 1984; Appelbaum and Buckner 1986; Goldstone et al. 1986; Brugger et al. 1994), disease recurrence remains the major cause of treatment failure. When the malignancy originates from or involves the marrow, relapse could originate from malignant cells persisting in the patient, in the rescuing HSC, or in both (Appelbaum and Buckner 1986; Rill et al. 1992a,b; Brugger et al. 1994). The origin of relapse is an important issue to resolve. Concern that the HSC may contain residual malignant cells has led to extensive evaluation of techniques for purging these cells, prior to storage and subsequent reinfusion (Gorin et al. 1986; De Fabritiis et al. 1989; Gambacorti-Passerini et al. 1991). Animal and preclinical human studies have shown that these methods do reduce contamination with malignant cells that have been deliberately added to marrow, but no method has been unequivocally shown to reduce the risk of relapse in naturally occurring disease (Yeager et al. 1986; Santos et al. 1989; Gribben et al. 1991).

HSC for autologous transplants are usually harvested at a time when, by definition, no malignant cells are detectable in the marrow (Campana et al. 1991; Gribben et al. 1991; Uckun et al. 1993). It is therefore impossible to undertake any form of quality control after purging to determine whether the putative residual malignant cells have genuinely been eradicated. Unfortunately, these unproven purging techniques almost invariably damage normal progenitor cells, so that engraftment of purged HSC is typically far slower than that of untreated marrow (Kaizer et al. 1985; Gorin et al. 1986; Yeager et al. 1986). Morbidity and mortality from the complications of hematopoietic and immune system failure are correspondingly increased.

If marker genes could be transferred into residual malignant cells in the HSC prior to reinfusion, and gene-marked malignant cells subsequently became detectable in the marrow or peripheral blood in patients who relapsed following autologous transplant, this would be powerful evidence that the harvested HSC contributes to disease recurrence. Moreover, the finding of marked cells at relapse would permit subsequent evaluation of ex vivo purging techniques for their ability to eradicate clonogenic cells. Thus, for the first time it would be possible to directly compare the efficacy of the numerous available techniques for HSC purging.

Method of marking. Marking studies began in September 1991, using two closely related retroviral vectors, GINa and LNL6, provided by Genetic Therapy, Inc. Both contain the neomycin phosphotransferase gene (Neo^R) and therefore confer resistance to neomycin and its analogs such as G418. Two thirds of harvested marrow was frozen directly, and one third was exposed to one or other of the vectors for 6 hours at a multiplicity of infection of 10:1. The cells were then recovered, washed, and frozen. At the time of transplant, both modified and unmodified cells were thawed and infused.

Marker genes to determine source of relapse in AML and neuroblastoma. Twelve patients with acute myeloblastic lymphoma (AML) have entered the study and, so far, three have relapsed. In two of these patients, the marker gene was present in the malignant population. For example, Patient 2 had $CD34^+$ $CD56^+$ dual positive blast cells in the peripheral blood at 80 days after transplantation. This is a phenotypic combination not present on normal progenitor cells and allows separation of the leukemic clone by flow cytometry (Brenner et al. 1993a). The clone also contained a genotypic marker, the 1:8:21 complex translocation, the 8:21 portion of which produces a fusion transcript and protein, AML:ETO (Miyoshi et al. 1991; Erickson et al. 1992; Nucifora et al. 1992). The leukemic clone contained products of both the malignant gene marker (AML:ETO) and the transferred gene marker (Neo^R).

Nine patients with neuroblastoma were also studied, three of whom relapsed (Rill et al. 1994). In two, recurrence was detected in routine marrow biopsies taken at 6 and 9 months posttransplantation. These samples contained 15% and 45% neuroblastoma cells, respectively. In each case, the marker gene was detected by polymerase chain reaction (PCR) analysis in a DNA sample from 10^4 $GD2^+$ $CD45^-$ neuroblasts and from clonogenic cells grown from the neuroblast population (Fig. 1).

The third patient had evidence of metastatic disease in the liver on computed tomography scans performed 6 months after marrow transplantation. The hepatic lesion was removed surgically. Immunohistology confirmed the presence of a neuroblastoma deposit and $GD2^+$ $CD45^-$ cells were separated from a specimen of the excised material. PCR analysis of DNA from 10^4 or more of the isolated neuroblasts demonstrated the presence of the marker gene. Histological examination of a concurrent marrow biopsy failed to reveal tumor cells; however, the patient subsequently developed a marrow relapse and $GD2^+$ $CD45^-$ neuroblastoma colonies containing the neomycin-resistance gene were grown in methylcellulose (Fig. 1).

We used a limiting cycle PCR technique to estimate the proportion of resurgent neuroblasts that were marker-gene-positive in each of these three patients. The neomycin resistance gene was present in 0.05–1% of the $GD2^+$ $CD45^-$ cell population at sites of relapse. Since only 1% of clonogenic neuroblasts present in the autologous marrow transplant were genetically marked (Rill et al. 1992a), the proportion of marked malignant

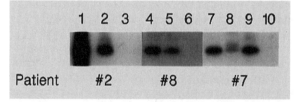

Figure 1. PCR analysis of $GD2^+$ $CD45^-$ neuroblastoma cells and colonies at the time of relapse. PCR amplification was performed on DNA from 10^4 sorted $GD2^+$ $CD45^-$ cells from the site of relapse and from individual, G418-resistant colonies grown in methylcellulose (see text). Patient 2: (lane *1*) marrow tumor, (lane *2*) G418-resistant colony, (lane *3*) negative control. Patient 8: (lane *4*) marrow tumor, (lane *5*) G418-resistant colony, (lane *6*) negative control. Patient 7: (lane *7*) liver tumor, (lanes *8,9*) G418-resistant colonies (liver, marrow), (lane *10*) negative extraction control nontransduced neuroblastoma cells. The band shown is the 720-bp neomycin-resistance gene amplification product.

cells in vivo at the time of relapse was similar to the proportion marked ex vivo and infused with the autologous marrow transplant (Rill et al. 1992a; Brenner et al. 1993a). These data show that residual tumor cells in marrow have considerable clonogenic potential in vivo and suggest that they make a substantial contribution to resurgent disease.

Marker genes to determine source of relapse in other malignancies. Related studies are under way to analyze the source of relapse in chronic myelogenous leukemia and breast cancer (Deisseroth et al. 1991; O'Shaughnessy et al. 1993).

Conclusions about source of relapse. These data directly demonstrate that autologous marrow harvested in apparent clinical remission from patients with hematologic and nonhematologic malignancies may still harbor tumorigenic cells capable of contributing to relapse. Moreover, these cells may contribute to relapse both within the marrow and at extramedullary sites. Although we could not formally assess the contribution of residual, unharvested tumor cells to disease recurrence, our results clearly suggest that effective methods of purging will be needed to improve the outcome of intensified treatment followed by autologous bone marrow transplantation, at least in the two malignancies we have studied.

Clonality of relapse. The provirus may integrate in any of a vast number of possible sites in the host cell genome. It is therefore possible to estimate the clonality of relapse by enumerating these sites, using restriction enzymes which cut once within the proviral genome and once in adjacent DNA. Vector integration events can then be distinguished by amplifying the variably sized products (Ochman et al. 1988; Rill et al. 1994). We have detected two integrants among 10^6 purified neuroblasts from each of 2 patients who relapsed with neuroblastoma. If we assume that neuroblasts are marked with an overall efficiency of 1%, then a *minimum* of 200 "clonogenic" tumor cells were reinfused with the autologous marrow transplant. Although the precision of this estimate is dependent on the accuracy of the assumption that we mark neuroblasts which are tumorigenic in vivo with the same frequency that we mark neuroblasts which form colonies ex vivo, the data establish that a multiplicity of cells in the marrow transplant contribute to disease recurrence.

Use of marking to analyze purging. The overall implication of these results is that *effective* marrow purging will need to be incorporated into protocols of autologous HSC transplantation. It should be possible to use the same sensitive marking technique described here to assess the efficacy of conventional and novel methods of marrow purging. Indeed, as new vectors become available for clinical application, the scope of marker gene investigation should increase greatly. By distinctively marking different aliquots of marrow from the same patient, one could compare several purging technologies simultaneously. Since the end point of such purging studies would be the absence of marked relapse, rather than a prolongation of survival or a reduction in relapse risk, even small-scale clinical trials could be informative. To illustrate this application, our current AML marker/purging protocol is shown in Figure 2. Marrow is divided into three aliquots. One is frozen as a safety backup, one is marked with G1N, one is marked with LNL6. These vectors can be distinguished because each produces a PCR fragment of different size. The marked portions are then randomized to purging with a pharmaceutical agent, 4HC, or an immunological agent, interleukin 2 (IL-2). The purged cells are frozen, and the marrow is subsequently reinfused in the standard manner. If patients relapse with marked cells, it is possible to determine which purging technique failed.

Can Genes Be Transferred to Normal Human Stem Cells and Expressed Long Term in Their Progeny?

During these relapse studies (Brenner et al. 1993b), we also had the opportunity to assess in vivo the ability of retrovirus vectors to effect gene transfer to normal hematopoietic stem cells following ex vivo culture. We were also able to determine whether the transferred gene was expressed in the mature progeny of these stem cells.

Studies at one month showed that the infused marrow had contributed to early hematopoietic reconstitution. Semiquantitative PCR at this time detected the Neo^R gene in the marrow mononuclear cells of 14/18 patients, with gene copy numbers that ranged from approximately 0.005 to 0.05 per cell. The presence of the gene in hematopoietic progenitor cells was confirmed by clonogenic assays which showed gene-marked progenitor cells in 15/18 patients. We obtained evidence for the contribution of infused marrow to long-term hematopoietic recovery from the nine patients who were evaluable 6 months or more after transplantation. We used a clonogenic assay to quantitate the presence of the Neo^R gene in progenitor cells at increasing intervals after transplantation. G418-resistant progenitor cells persisted for 6 months in 9/9 patients

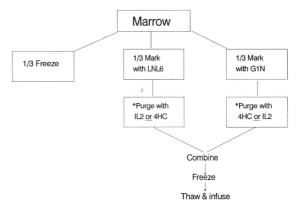

Figure 2. Outline of AML purging protocol. *IL-2 or 4HC assignment will be randomized to LNL6 or G1N marking.

and for 1 year in 5/5; marked progenitor cells remained detectable in marrow from the 2 patients who are more than 2 years posttransplant. The gene was also present and expressed long term in the mature progeny of marrow precursor cells, T and B lymphocytes, monocytes, and neutrophils.

These results indicate a major contribution to marrow reconstitution from autologous transplants and strongly suggest that these stem cells can be successfully transduced by retrovirus vectors; gene-marked marrow progenitor cells were detected for as long as 30 months after autologous bone marrow transplantation; multilineage granulocyte-erythroid-monocyte/macrophage (GEMM) colonies become positive at 6 months, implying transfer into a progenitor at least one stage earlier in differentiation; detection of the marker gene in both T and B lymphocytes for at least 30 months is consistent with the transfection of a primitive lympho-hematopoietic stem cell. The efficiency of transfer to long-lived progenitor cells was higher than would have been predicted from studies in large animals using similar transduction protocols. This increased transfer rate may be a consequence of harvesting marrow during the hematopoietic recovery phase that follows multiple cycles of intensive (marrow-ablative) chemotherapy. During this recovery period, there is substantial proliferation of early marrow progenitor cells that may favor integration of the proviral genome (Otsuka et al. 1991a,b).

These marker gene studies (Brenner et al. 1993a,b) suggest that modest levels of gene transfer may be effected into normal progenitor cells. They also confirm the need to persist with efforts to increase gene transfer efficiency, if most gene *therapy* applications are to be effective. A number of animal models have suggested that ex vivo treatment with growth factor combinations, with or without the addition of marrow stromal cells or extracellular matrix, can induce stem cells to enter the cell cycle and thereby increase both their number and the efficiency with which they can be transduced (Dick et al. 1991; Otsuka et al. 1991a,b). In humans, it certainly appears possible to use growth factors such as IL-1, IL-3, and stem-cell factor to increase the numbers of hematopoietic progenitor cells by 10- to 50-fold and to increase the efficiency of gene transfer to levels that may exceed 50% (Hughes et al. 1992). Unfortunately, it is by no means certain that such ex vivo data will be reflected by results in vivo. In primate studies, transplantation of marrow treated ex vivo with the growth factor combinations shown to greatly augment both progenitor numbers and gene transfer rates have been followed by disappointingly low levels of long-term gene expression in vivo (D. Bodine and C. Dunbar [NIH/NHLBI], unpubl.). Similarly, in human studies in which stem-cell factor, IL-3, and IL-6 were used to stimulate marrow progenitor cells before reinfusion, expansion of colony-forming units (CFUs) and a high gene-transfer rate ex vivo were associated with in vivo expression that was both scanty and of short duration (A. Nienhuis, pers. comm.). The likeliest explanation for this apparent paradox is that many of the growth factors intended only to induce cycling in marrow stem cells also induce their differentiation and the loss of their self-renewal capacity.

In the absence of any proven ex vivo surrogate method for studying the effects of growth factors on stem cell expansion and transducibility, it is possible to use the marker-gene technique to evaluate whether any increase in progenitor cell numbers and transducibility produced by growth factor combinations and cell culture devices ex vivo has an effect in vivo. By using two distinguishable vectors to mark each patient's marrow, we are able to compare treatment regimens *within* a patient. Because each patient acts as his own control, it should be possible to discern the effects of any given growth factor regimen on stem cells, even in a cohort as small as 10 patients (Fig. 3). By comparing the proportion of clonogenic cells marked long term with each vector, we will have a measure of the impact of ex vivo growth factor treatment on stem-cell *transducibility*.

Is Gene Transfer to Hematopoietic Progenitor Cells Safe?

One of the major concerns about any use of long-lived marrow progenitor cells for retrovirus-mediated gene transfer or therapy is that insertional mutagenesis will occur (Cornetta et al. 1991). This event could occur at the time of initial exposure to the vector, or subsequently—if wild-type retrovirus contaminates the vector or is formed by recombination. This concern has been given extra weight following the discovery of thymomas in monkeys injected with a vector contaminated with wild-type virus (Donahue et al. 1992). In the clinical studies reported here, helper-free vector was used, and no evidence has been obtained for any recombinational events with endogenous retroviral sequences that have generated infectious virus. Although no events to date have occurred that are attributable to mutagenesis, all patients will have prolonged follow-up and genetic analysis of any tumors that do appear within the next 15 years.

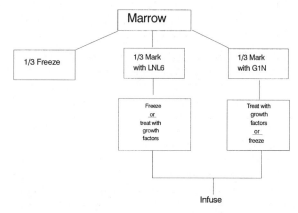

Figure 3. Outline of growth factor treatment.

Gene Transfer in Allogeneic Bone Marrow Transplantation

Marker-gene transfer may also be used to improve the outcome of allogeneic bone marrow transplantation, for example, by facilitating the introduction of adoptive immunotherapy programs to treat viral and malignant disease.

Following allogeneic bone marrow transplantation, graft versus host disease is a major cause of morbidity and mortality. This complication is particularly frequent and severe if the recipient is given marrow from an HLA-mismatched family member or from an unrelated donor, but it can be largely prevented if T lymphocytes are removed from the donor marrow before transplantation. However, there is little change in overall survival, because of the profound immunocompromise experienced by recipients of marrow so treated. Epstein-Barr virus (EBV) infection in particular has proved problematic in these individuals. EBV is a herpesvirus that infects the majority of people and persists in an asymptomatic state by a combination of chronic replication in the mucosa and latency in peripheral blood B cells (Straus et al. 1992). These EBV-infected B cells are highly immunogenic and are normally susceptible to killing by specific cytotoxic T lymphocytes (CTLs). However, immunocompromised patients may develop EBV-driven lymphoproliferation, and the incidence may be as high as 30% in the recipients of T-cell-depleted allografts from HLA-mismatched or unrelated donors (Papadopoulos et al. 1994). Although it has recently been shown that unmanipulated T cells are effective therapy for this complication (Papadopoulos et al. 1994), these cells may also induce graft versus host disease (Heslop et al. 1994). We are evaluating whether adoptive transfer of EBV-specific CTLs is effective as prophylaxis and therapy for this complication (Heslop et al. 1994). To learn more about the survival, distribution, and activity of these cells after administration, they are first marked with the Neo^R gene.

To date, eight patients have received gene-marked, EBV-specific CTL lines. Semiquantitative PCR analysis showed the marker to be present in between 0.5 and 5% of the lymphocytes, levels which previous experience suggested would be adequate to enable the survival and behavior of these cells to be followed. The fate of the eight gene-marked donor CTL lines was followed in their respective marrow recipients. The Neo^R gene was detected in peripheral blood lymphocytes for 10 weeks or more after injection of gene-marked cells.

Three lines of evidence suggest that these transferred cells were functional in vivo. First, cytotoxic activity against EBV-infected target cells was evident in a population of neomycin-resistant and marker-gene-positive peripheral blood lymphocytes obtained from the patients 1–4 weeks after adoptive transfer. Second, in two patients, it has been possible to show antiviral efficacy in vivo. Finally, in an additional patient, both antiviral and antilymphoproliferative activity were demonstrable. In this individual, cervical, hilar, and retroperitoneal lymphadenopathy developed before the CTL were given. Immunohistology was typical of post-transplant lymphoproliferative lymphoma. Administration of CTL was associated with resolution of the lymphadenopathy over a 6-week period; to date there has been no resurgence of his disease (Rooney et al. 1994).

These gene-marking studies demonstrate that adoptively transferred antigen-specific T cells can survive well in vivo in the absence of exogenous cytokines, retain their cytotoxic activity, and are able to eradicate a lymphoproliferative disorder. The information gained from these studies will allow future gene therapy interventions to be planned. For example, incorporation of cytokine genes may increase their cytotoxic activity, whereas incorporation of genes encoding antigen-specific receptor molecules may alter or enhance their antitumor specificity (Rosenberg 1992).

CONCLUSION

Gene marking is providing information that is of value in improving the outcome of autologous bone marrow transplantation by identifying the source of relapse and the efficiency of purging. Marking is also valuable for assessing the safety and efficacy of adoptive immunotherapy after allogeneic bone marrow transplantation. Of more general importance, the technique is providing insights into ways to optimize gene transfer into hematopoietic progenitor cells and to maximize expression in their progeny.

ACKNOWLEDGMENTS

We thank all our clinical and laboratory coworkers who made these studies possible. We also thank Nancy Parnell for word processing. These studies were supported by National Institutes of Health grants CA-20180, CA-9228, CA-61386, and by ALSAC (American Lebanese Syrian Associated Charities).

REFERENCES

Anderson W.F. 1992. Human gene therapy. *Science* **256:** 808.
Appelbaum, F.R. and C.D. Buckner. 1986. Overview of the clinical relevance of autologous bone marrow transplantation. *Clin. Haematol.* **15:** 1.
Brenner M.K., D.R. Rill, R.C. Moen, R.A. Krance, J. Mirro, Jr., W.F. Anderson, and J.N. Ihle. 1993a. Gene-marking to trace origin of relapse after autologous bone marrow transplantation. *Lancet* **341:** 85.
Brenner, M.K., D.R. Rill, M.S. Holladay, H.E. Heslop, R.C. Moen, M. Buschle, R.A. Krance, V.M. Santana, W.F. Anderson, and J.N. Ihle. 1993b. Gene marking to determine whether autologous marrow infusion restores long-term haemopoiesis in cancer patients. *Lancet* **342:** 1134.
Brugger, W., K.J. Bross, M. Glatt, F. Weber, R. Mertelsmann, and L. Kanz. 1994. Mobilization of tumor cells and hematopoietic progenitor cells into peripheral blood of patients with solid tumors. *Blood* **83:** 636.
Burnett, A.K., P. Tansey, R. Watkins, M. Alcorn, D. Maharaj,

C.R. Singer, S. McKinnon, G.A. McDonald, and A.G. Robertson. 1984. Transplantation of unpurged autologous bone-marrow in acute myeloid leukaemia in first remission. *Lancet* **II:** 1068.

Campana, D., E. Coustan-Smith, and F.G. Behm. 1991. The definition of remission in acute leukemia with immunologic techniques. *Bone Marrow Transplant.* **8:** 429.

Cornetta, K., R.A. Morgan, and W.F. Anderson. 1991. Safety issues related to retrovirus-mediated gene transfer in humans. *Hum. Gene Ther.* **2:** 5.

De Fabritiis, P., D. Ferrero, A. Sandrelli, C. Tarella, G. Meloni, A. Pulsoni, P. Pregno, R. Badoni, L. DeFelice, E. Gallo, S. Amadori, F. Mandelli, and A. Pileri. 1989. Monoclonal antibody purging and autologous bone marrow transplantation in acute myelogenous leukemia in complete remission. *Bone Marrow Transplant.* **4:** 669.

Deisseroth, A.B., H. Kantarjian, M. Talpaz, R. Champlin, C. Reading, B. Andersson, and D. Claxton. 1991. Autologous bone marrow transplantation for CML in which retroviral markers are used to discriminate between relapse which arises from systemic disease remaining after preparative therapy versus relapse due to residual leukemia cells in autologous marrow: A pilot trial. *Hum. Gene Ther.* **2:** 359.

Dick, J.E., S. Kamel-Reid, B. Murdoch, and M. Doedens. 1991. Gene transfer into normal human hematopoietic cells using in vitro and in vivo assays. *Blood* **78:** 624.

Donahue, R.E., S.W. Kessler, D. Bodine, K. McDonagh, C. Dunbar, S. Goodman, B. Agricola, E. Byrne, M. Raffeld, R. Moen, J. Bacher, K.M. Zsebo, and A.W. Nienhuis. 1992. Helper virus induced T cell lymphoma in nonhuman primates after retroviral mediated gene transfer. *J. Exp. Med.* **176:** 1125.

Erickson, P., J. Gao, K.-S. Chang, T. Look, E. Whisenant, S. Raimondi, R. Lasher, J. Trujillo, J. Rowley, and H. Drakbin. 1992. Identification of breakpoints in t(8;21) AML and isolation of a fusion transcript with similarity to *Drosophila* segmentation gene *runt*. *Blood* **80:** 1825.

Gambacorti-Passerini, C., L. Rivoltini, M. Fizzotti, M. Rodolfo, M.L. Sensi, C. Castelli, A. Orazi, N. Polli, M. Bregni, S. Siena, L. Borin, E. Pogliani, G.M. Gorneo, and G. Parmiani. 1991. Selective purging by human interleukin-2 activated lymphocytes of bone marrows contaminated with a lymphoma line or autologous leukaemic cells. *Br. J. Haematol.* **78:** 197.

Goldstone, A.H., C.C. Anderson, D.C. Linch, I.M. Franklin, B.J. Boughton, J.C. Cawley, and J.D. Richards. 1986. Autologous bone marrow transplantation following high dose chemotherapy for the treatment of adult patients with acute myeloid leukaemia. *Br. J. Haematol.* **64:** 529.

Gorin, N.C., L. Douay, J.P. Laporte, M. Lopez, J.Y. Mary, A. Najman, C. Salmon, P. Aegerter, J. Stachowiak, R. David, F. Pene, G. Kantor, J. Deloux, E. Duhamel, J. van den Akker, J. Gerota, Y. Parlier, and G. Duhamel. 1986. Autologous bone marrow transplantation using marrow incubated with Asta Z 7557 in adult acute leukemia. *Blood* **67:** 1367.

Gribben, J.G., A.S. Freedman, D. Neuberg, D.C. Roy, K.W. Blake, S.D. Woo, M.L. Grossbard, S.N. Rubinoew, F. Coral, G.J. Freeman, J. Ritz, and L.M. Nadler. 1991. Immunologic purging of marrow assessed by PCR before autologous bone marrow transplantation for B-cell lymphoma. *N. Engl. J. Med.* **325:** 1525.

Heslop, H.E., M.K. Brenner, C.M. Rooney, R.A. Krance, W.M. Roberts, R. Rochester, C.A. Smith, V. Turner, J. Sixbey, R. Moen, and J.M. Boyett. 1994. Administration of neomycin resistance gene marked EBV specific cytotoxic T lymphocytes to recipients of mismatched-related or phenotypically similar unrelated donor marrow grafts. *Hum. Gene Ther.* **5:** 381.

Hughes, P.F.D., J.D. Thacker, D. Hogge, H.J. Sutherland, T.E. Thomas, P.M. Lansdorp, C.J. Eaves, and R.K. Humphries. 1992. Retroviral gene transfer to primitive normal and leukemic hematopoietic cells using clinically applicable procedures. *J. Clin. Invest.* **89:** 1817.

Kaizer, H., R.K. Stuart, R. Brookmeyer, W.E. Beschorner, H.G. Braine, W.H. Burns, D.J. Fuller, M. Korbling, K. Mangan, R. Saral, L. Sensenbrenner, R.K. Shadduck, A.C. Shende, P.J. Tutschka, A.M. Yeager, W.H. Zinkham, O.M. Colvin, and G.W. Santos. 1985. Autologous bone marrow transplantation in acute leukemia: A phase I study of in vitro treatment of marrow with 4-hydroperoxycyclophosphamide to purge tumor cells. *Blood* **65:** 1504.

Miller, A.D. 1992. Human gene therapy comes of age. *Nature* **357:** 455.

Miyoshi, H., K. Shimizu, T. Kozu, N. Maseki, Y. Kaneko, and M. Ohki. 1991. t(8;21) breakpoints on chromosome 21 in acute myeloid leukemia are clustered within a limited region of a single gene, *AML1*. *Proc. Natl. Acad. Sci.* **88:** 10431.

Nucifora, G., D.J. Birn, P. Erickson, J. Gao, M.M. LeBeau, H.A. Drabkin, and J.D. Rowley. 1992. Detection of DNA rearrangements in the *AML1* loci and of an AML1/ETO fusion mRNA in patients with t(8;21) AML. *Blood* **81:** 883.

O'Shaughnessy, J.A., K.H. Cowan, W. Wilson, G. Bryant, B. Goldspiel, R. Gress, A.W. Nienhuis, C. Dunbar, B. Sorrentino, F.M. Stewart, R. Moen, M. Fox, and S. Leitman. 1993. Pilot study of high dose ICE (ifosfamide, carboplatin, etoposide) chemotherapy and autologous bone marrow transplant (ABMT) with neo-R-transduced bone marrow and peripheral blood stem cells in patients with metastatic breast cancer. *Hum. Gene Ther.* **4:** 331.

Ochman, H.A., S. Gerber, and D.L. Hart. 1988. Genetic application of an inverse polymerase chain reaction. *Genetics* **120:** 621.

Otsuka, T., J.D. Thacker, and D.E. Hogge. 1991a. The effects of interleukin 6 and interleukin 3 on early hematopoietic events in long-term cultures of human marrow. *Exp. Hematol.* **19:** 1042.

Otsuka, T., J.D. Thacker, C.J. Eaves, and D.E. Hogge. 1991b. Differential effects of microenvironmentally presented interleukin 3 versus soluble growth factor on primitive human hematopoietic cells. *J. Clin. Invest.* **88:** 417.

Papadopoulos, E.B., M. Ladanyi, D. Emanuel, S. MacKinnon, F. Boulad, M.H. Carabasi, H. Castro-Malaspina, B.H. Childs, A.P. Gillio, T.N. Small, J.W. Young, N.A. Kernan, and R.J. O'Reilly. 1994. Infusions of donor leukocytes to treat Epstein-Barr virus-associated lymphoproliferative disorders after allogeneic bone marrow transplantation. *N. Engl. J. Med.* **330:** 1185.

Rill, D.R., M. Buschle, N.K. Foreman, C. Bartholomew, R.C. Moen, V.M. Santana, J.N. Ihle, and M.K Brenner. 1992a. Retrovirus mediated gene transfer as an approach to analyze neuroblastoma relapse after autologous bone marrow transplantation. *Hum. Gene Ther.* **3:** 129.

Rill, D.R., R.C. Moen, M. Buschle, C. Bartholomew, N.K. Foreman, J. Mirro, Jr., R.A. Krance, J.N. Ihle, and M.K. Brenner. 1992b. An approach for the analysis of relapse and marrow reconstitution after autologous marrow transplantation using retrovirus-mediated gene transfer. *Blood* **79:** 2694.

Rill, D.R., V.M. Santana, W.M. Roberts, T. Nilson, L.C. Bowman, R.A. Krance, H.E. Heslop, R.C. Moen, J.N. Ihle, and M.K. Brenner. 1994. Direct demonstration that autologous bone marrow transplantation for solid tumors can retain a multiplicity of tumorigenic cells. *Blood* **84:** 380.

Rooney, C.M., C.A. Smith, C.Y. Mg, S. Loftin, C.F. Li, R.A. Krance, M.K. Brenner, and H.E. Heslop. 1994. Use of gene-modified virus-specific T lymphocytes to control Epstein-Barr virus-related lymphoproliferation. *Lancet* (in press).

Rosenberg, S.A. 1992. The immunotherapy and gene therapy of cancer. *J. Clin. Oncol.* **10:** 180.

Santos, G.W., A.M. Yeager, and R.J. Jones. 1989. Autologous bone marrow transplantation. *Annu. Rev. Med.* **40:** 99.

Schuening, F.G., K. Kawahara, A.D. Miller, R. To, S. Goehle, S. Stweard, K. Mullally, L. Fisher, T.C. Graham, F.R. Appelbaum, R. Hackman, W.R.A. Osborne, and R. Storb.

1991. Retrovirus-mediated gene transduction into long-term repopulating marrow cells of dogs. *Blood* **78:** 2568.

Smith, C. 1992. Retroviral vector-mediated gene transfer into hematopoietic cells: Prospects and issues. *J. Hematother.* **1:** 155.

Straus, S.E., J.I. Cohen, G. Tosato, and J. Meier. 1992. Epstein-Barr virus infections: Biology, pathogenesis and management. *Ann. Intern. Med.* **118:** 45.

Uckun, F.M., J.H. Kersey, R. Haake, D. Weisdorf, M.E. Nesbit, and N.K.C. Ramsay. 1993. Pretransplantation burden of leukemic progenitor cells as a predictor of relapse after bone marrow transplantation for acute lymphoblastic leukemia. *N. Engl. J. Med.* **329:** 1296.

Van Beusechem, V.W., A. Kukler, P.J. Heidt, and D. Valerio. 1992. Long-term expression of human adenosine deaminase in rhesus monkeys transplanted with retrovirus-infected bone-marrow cells. *Proc. Natl. Acad. Sci.* **89:** 7640.

Wieder, R., K. Cornetta, S.W. Kessler, and W.F. Anderson. 1991. Increased efficiency of retrovirus-mediated gene transfer and expression in primate bone marrow progenitors after 5-fluorouracil-induced hematopoietic suppression and recovery. *Blood* **77:** 448.

Yeager, A.M., H. Kaizer, G.W. Santos, R. Saral, O.M. Colvin, R.K. Stuart, H.G. Braine, P.J. Burke, R.F. Ambinder, W.H. Burns, D.J. Fuller, J.M. Davis, J.E. Karp, W.S. May, S.D. Rowley, L.L. Sensenbrenner, G.B. Vogelsang, and J.R. Wingard. 1986. Autologous bone marrow transplantation in patients with acute nonlymphocytic leukemia, using ex vivo marrow treatment with 4-hydroperoxycyclophosphamide. *N. Engl. J. Med.* **315:** 141.

Molecular Genetic Interventions for Cancer

G.J. Nabel,* E.G. Nabel,† Z. Yang,* B.A. Fox,†
G.E. Plautz,† X. Gao,‡ L. Huang,‡
S. Shu,† D. Gordon,† and A.E. Chang†

*Howard Hughes Medical Institute and †University of Michigan Medical Center, Ann Arbor, Michigan 48109;
‡University of Pittsburgh, Pittsburgh, Pennsylvania 15261

Recent progress in understanding the genetic basis for human cancer has led to a variety of new perspectives regarding the pathogenesis of these diseases. From many studies, it is now clear that there is considerable genetic instability in malignant cells, and progressive genetic changes occur as carcinogenesis proceeds (for review, see chapter 18 in Watson et al. 1992). To utilize molecular genetic interventions for the treatment of malignancy, it is important to understand the relevant underlying genetic abnormalities and to develop strategies appropriate for the genetic lesions. Although a variety of cellular genes regulate growth control and contribute to oncogenesis, it would be difficult presently to complement deficiencies of specific genes, e.g., tumor suppressor or cell cycle regulatory genes, in such cells. To complement such defects would require the transduction of a large percentage of cells within malignancies, and current methods of gene transfer are relatively inefficient. It is, therefore, preferable to seek strategies that can serve to catalyze antitumor responses without the need to complement missing or defective genes in every cell of the tumor population.

The discovery of mutations in DNA mismatch repair enzymes confirms that genetic instability contributes to the malignant phenotype (Parsons et al. 1993; Fishel et al. 1994; Kant et al. 1994; Papadopoulos et al. 1994). Such defects in repair enzymes could lead to the generation of mutations in many gene products within these cells, giving rise to novel polypeptides potentially recognizable by the immune system. Another implication of mutations in DNA repair enzymes is that the types of mutations would vary among malignancies and among patients with the same malignancy. Therefore, it would be desirable to use genetic interventions to facilitate the development of immunity against mutant gene products arising within the tumor. One method to accomplish this goal is to design strategies to improve immunologic recognition within specific tumor nodules.

Several approaches can be taken to enhance immune recognition of tumors by genetic interventions. First, genes encoding immunogenic cell-surface proteins could be introduced into malignant cells to elicit immune responses that might catalyze further release of immunologically active molecules. This process would facilitate antigen presentation and lymphocyte proliferation directed to tumor-associated antigens. Second, the secretion of immunologic effector molecules, specifically cytokines, could also initiate and sustain this cascade of events. Finally, since it has become increasingly clear that the immune system can recognize antigens that may not be recognized by the immune system, it is essential to neutralize any local immunosuppression which might interfere with activation and the immune response to the malignancy in vivo.

In this paper, we describe our efforts to introduce a cell-surface antigen to achieve this goal. The introduction of foreign histocompatibility antigens can provide a potent means to elicit a local immune response. We have used direct gene transfer to introduce such genes into malignancies in vivo, and we describe our efforts to develop this system to enhance the immune response to malignancy in humans.

EXPERIMENTAL BACKGROUND

Increasing knowledge of molecular genetics and of the protective role played by the immune system is facilitating a more sophisticated form of immunology. For example, it is now well known that the cellular immune system, through specific antitumor responses, is capable of mediating direct cytolysis of tumor cells. Animal models have shown that both cytolytic T cells and lymphokines can destroy tumor cells. However, advanced age can be associated with decreased ability to respond to antigens, as demonstrated in studies showing that old mice fail to respond to immunotherapy that is protective in young mice (Dunn and North 1991a,b). Furthermore, the observation that most malignancies arise in apparently immunocompetent hosts indicates that stratagems for evading host defenses probably have been developed by the tumor cells.

Studies in which transfection of transforming growth factor β_1 (TGF-β_1) into tumor cells converted a highly immunogenic regressor tumor into a progressively growing tumor and inhibited the tumor-specific cytotoxic T lymphocyte (CTL) response have shown that TGF-β_1, secreted by several tumor types, has immunosuppressive properties (Torre-Amione et al. 1990; Inge et al. 1992) Inadequate presentation of antigens by major histocompatibility complex (MHC) class I on tumor cells is another mechanism for evading an effective immune response; indeed, malignant cells are frequently observed to have low or absent expression of MHC class I molecules (Elliott et al. 1989). In addition, expression of MHC class I in metastatic foci is often decreased, compared to primary tumor lesions (Gopas et al. 1989; Cordon-Cardo et al. 1991; Pantel et al. 1991), either because of altered binding of transcription factors to enhancer regions of MHC genes (Blanchet et al. 1992), abnormal expression of β_2-microglobulin (Gattoni-Celli et al. 1992), or proteins involved in processing and transport of peptide fragments (Restifo et al. 1992). Tumorigenicity in some tumor lines can be eliminated or reduced by transfection of transporter genes (Franksson et al. 1993) or by β_2-microglobulin (Glas et al. 1992). Conversely, tumors lacking or having low-level expression of MHC class I can acquire an immunogenic phenotype, which induces rejection, by transfection of the appropriate MHC class I gene (Hui et al. 1984; Tanaka et al. 1985; Wallich et al. 1985). In some cases, these immunogenic MHC class I-expressing cells can protect against rechallenge with the MHC-deficient parental tumor cells. These experiments, as well as others, suggest that despite poor immunogenicity or immunosuppressive properties of tumor cells, provision of appropriate immune stimulation may enhance the host cellular defenses against invading and proliferating cancer cells.

Conventional approaches to immunotherapy have involved the administration of nonspecific immunomodulating agents such as Bacillus Calmette-Guerin (BCG), cytokines, and/or adoptive T-cell transfer. These agents have shown promise in many animal models (Zbar et al. 1971; Rosenberg et al. 1985; Shu et al. 1985; Spiess et al. 1987; Chou et al. 1988; Yoshizawa et al. 1991a) and in man (Morton et al. 1974; Rosenberg et al. 1988, 1989; Kradin et al. 1989). Other approaches include the stimulation of cytotoxic T cells that demonstrate MHC class I-restricted lysis (Kern et al. 1986); introduction of tumor-infiltrating lymphocytes to selectively lyse the tumor from which they were derived (Rosenberg et al. 1986; Muul et al. 1987; Topalian et al. 1989); stimulation of natural killer cells (Oldham 1983; Herberman 1985); and enhancement of macrophages.

In both animal studies and human trials, augmentation of the immune response has been attempted by tumor immunization (Shu et al. 1985), administration of cytokines (Lafraniere and Rosenberg 1985), or infusion of lymphoid cells sensitized to tumor-associated antigens (Lafraniere and Rosenberg 1985; Rosenberg et al. 1988; Yoshizawa et al. 1991b). Introduction of new antigens into tumor cells by drug treatment or virus infection has also been attempted (Itaya et al. 1987; Pucetti et al. 1987; Sugiura et al. 1988).

Gene therapy approaches to stimulating the immune system to recognize viruses, parasite-infected cells, and cancerous cells have followed the rapid development of gene transfer protocols. Ex vivo transfer of recombinant genes was employed in early attempts to augment the naturally occurring immune response to malignancies. Expression of cytokine genes by genetically modified tumor cells has been shown to induce an antitumor effect mediated by lymphocytes or other inflammatory cells (Tepper et al. 1989; Fearon et al. 1990; Gansbacher et al. 1990a,b; Esumi et al. 1991).

These experiments demonstrated that modification of tumor cells by gene transfer can increase their immunogenicity by causing tumor-specific, immune-mediated rejection and protecting against rechallenge. In some instances, the genetically modified tumor cells have produced an immune response that eliminates previously established microscopic tumors (Golumbek et al. 1991; Chen et al. 1992), suggesting that patients with preexisting tumors might be treated. However, in many cases, the clinical applicability of cell-mediated gene transfer is limited by the length of time required and high cost of the method, as well as by the inability to establish cell lines from human tumors. These difficulties might be circumvented for certain tumors if there were common tumor antigens presented by frequent HLA alleles. Indeed, active immunotherapy using interleukin-2 (IL-2) or interferon-γ (IFN-γ) expression by HLA-A1 or HLA-A2 melanoma cells, for which HLA-restricted common tumor antigens are known, is under development (Gansbacher et al. 1992). However, this approach may not be applicable for many tumors for which the frequency of tumor antigen expression and distribution on normal tissues and HLA restriction are not characterized. It is clear that direct, in vivo modification of tumor cells would greatly expand the provision of immunotherapy by recombinant gene expression.

Direct Gene Transfer of Allogeneic MHC Class I Genes

MHC class I proteins bind proteolytically derived peptide fragments from endogenously synthesized proteins (Jardetzky et al. 1991). This complex, displayed on the surface of nearly all somatic cells in association with β_2-microglobulin, is recognized by T cells through steric interaction with unique T-cell receptors and is restricted to CD8$^+$ T cells by conserved interaction with MHC class I (Townsend and Bodmer 1989). A combination of residues from the MHC class I in the peptide-binding groove and the bound peptide fragment constitute the structure recognized by each individual T-cell receptor. MHC alleles are highly polymorphic, with most of the variation between alleles concentrated in residues that compose the peptide-binding portion of the molecule (Parham et al. 1988). Consequently, transplanted tissue expressing allogeneic MHC molecules is highly immunogenic, due to the altered shape of the MHC in the peptide-binding region. Because of the high precursor frequency of CD8$^+$ lymphocytes capable of reacting against allogeneic cells (Skinner and Marbrook 1976; Lindahl and Wilson 1977), introduction of an allogeneic MHC class I gene into tumors should stimulate a strong immune response.

Transplantable murine tumors inoculated subcutaneously often exist as discrete masses having a pseudocapsule. This configuration facilitates the precise delivery of recombinant genes to the tumor cells in vivo by direct intratumoral injection. Initial studies of direct gene transfer into tumors, from which our gene therapy protocols against melanoma have been derived, employed direct transfer of an allogeneic MHC class I gene into the poorly immunogenic, transplantable CT26, colon adenocarcinoma ($H\text{-}2K^d$), or MCA 106 tumors of mice to determine whether protective or therapeutic immune responses could be generated (Fig. 1) (Plautz et al. 1993). Generation of a protective response in mice without tumors was studied by subcutaneous injection of CT26 cells expressing $H\text{-}2K^s$ protein into BALB/c ($H\text{-}2K^d$) mice sensitized to this antigen; no tumors appeared during the 8-week follow-up period. The protective effects of this procedure against the further growth of preestablished CT26 tumors was then studied using direct injection into tumors of a recombinant $H\text{-}2K^s$ gene or β-galactosidase control gene complexed with liposome or retroviral vectors. Approximately 14 days later, when the tumors were 5–8 mm in diameter, they were directly transduced with the $H\text{-}2K^s$ gene. Splenic lymphocytes obtained from these animals 14 days after gene transfer demonstrated lytic activity against CT26 targets expressing $H\text{-}2K^s$, whereas injection of β-galactosidase did not induce cytolytic T cells. More importantly, in the experimental animals, cytotoxic lymphocytes were also generated against the parental nontransduced CT26 cells. Cytolytic activity against both transduced and parental tumor cells could be blocked by antibody against $CD8^+$ lymphocytes, but not against $CD4^+$ lymphocytes. The immune response was not protective in

mice having no prior sensitization to $H\text{-}2K^s$, since the tumors, which were relatively large at the time of gene transfer, continued to grow during the generation of the antitumor response (Plautz et al. 1993).

Significant attenuation of tumor growth and, in some cases, complete regression of existing tumors were achieved following $H\text{-}2K^s$ gene transfer when mice were presensitized by irradiated tumor cells expressing $H\text{-}2K^s$ or by splenic lymphocytes from an $H\text{-}2^s$ mouse strain. Since two histological types of tumors, the colon adenocarcinoma CT26 and the fibrosarcoma MCA 106, from two different mouse strains (BALB/c and C57/Bl6, respectively), responded to tumor transduction with the $H\text{-}2K^s$ gene, the immunologic response to $H\text{-}2K^s$ was not a tumor- or strain-specific response (Plautz et al. 1993).

In nude mice treated with the $H\text{-}2K^s$ gene, the growth of CT26 or MCA 106 tumors was identical to the growth in untreated mice or in mice receiving β-galactosidase. This indicates that the antitumor response is mediated by T lymphocytes. Indeed, preimmunized mice depleted of $CD8^+$ T cells prior to $H\text{-}2K^s$ gene transfer did not respond to $H\text{-}2K^s$ gene transfer, whereas $CD4^+$ depletion had no effect on the antitumor response. Direct gene transfer of the $H\text{-}2K^s$ gene into the poorly immunogenic murine melanoma, B16BL6, of naive mice generated pre-effector cells in tumor-draining lymph nodes. When stimulated ex vivo with anti-CD3 monoclonal antibody for 2 days, followed by culture in low-dose IL-2 (10 units/ml) for 3 days, these effector cells had no direct cytolytic activity against B16BL6 targets. However, adoptive transfer of these cells to mice bearing 3-day-old lung metastases resulted in a marked decrease in lung metastases (Wahl et al. 1992).

These studies demonstrated to us that actively growing, macroscopic tumors could be transduced with recombinant genes in vivo by intratumoral gene transfer. Direct transfection of the appropriate DNA liposome complex could generate an immunologic response against an allogeneic MHC protein expressed within actively growing tumors, and this strong immune response can stimulate a host immune response to weak tumor antigens. Augmentation of this antitumor immune response can mediate regression of preexisting tumors.

Vector Safety in Direct Gene Transfer

The DNA liposome complex is the preferred transfection vehicle, since the effectiveness of retroviral vectors partially depends on active replication of the target cells. Other potential concerns for in vivo gene therapy have been addressed by studies demonstrating that direct gene transfer and expression of recombinant genes using DNA/liposome complexes do not induce generalized autoimmunity, toxicity, or transmission to germ cells (E.G. Nabel et al. 1992b; Stewart et al. 1992). In studies involving delivery of retroviral vectors and DNA liposome complexes to anatomically localized

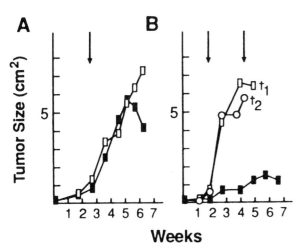

Figure 1. Introduction of $H\text{-}2K^s$ into CT26 or MCA 106 tumors in vivo inhibits tumor growth and prolongs survival. (A) Naive BALB/c mice bearing CT26 tumors were injected at the indicated times (arrows) with the β-galactosidase reporter plasmid (□) or pLJ-H-$2K^s$ (■) retroviral vector. (B) DNA/liposome complexes with pLJ-H-$2K^s$ (■) or the β-galactosidase plasmid (□) were injected into CT26 tumors preimmunized to H-$2K^s$ in BALB/c mice, or these mice were uninjected (○). Mice treated with β-galactosidase DNA/liposome injection or untreated expired on day 35(t_1) and day 30 (t_2), respectively. (Reprinted, with permission, from Plautz et al. 1993.)

sites in the vasculature encompassed by the protected space of a double-balloon catheter (E.G. Nabel et al. 1990, 1992a), we found expression of the recombinant gene to be restricted to the site of introduction in both the intratumoral and intravascular gene transfer, with occasionally very low levels noted at other sites (E.G. Nabel et al. 1992a; Plautz et al. 1993). In additional studies on rabbits and pigs receiving intravenous or intra-arterial injections of DNA liposome complexes, transfected DNA was never detected by polymerase chain reaction (PCR) in gonadal tissue despite its presence in the transduced artery segments and occasionally in other tissues (E.G. Nabel et al. 1992b). It is possible that expression of a foreign MHC class I gene on normal tissues could induce autoimmunity; however, mice receiving the allogeneic MHC gene all generated a specific CTL response to the foreign gene product with no histopathological or serological evidence of organ toxicity (E.G. Nabel et al. 1992b).

CLINICAL TRIALS OF GENE THERAPY FOR MELANOMA

Having established the cellular effectors mediating the immune response and the signals that provide sufficient stimulus for their activation, as well as identifying immune reactivity to melanoma cells, we needed to address the third requirement for effective immunotherapy, that of developing a reliable, clinically applicable system to deliver an adequate activation signal to the immune system without excessive toxicity. Because of the lower cost and greater ease of in vivo gene delivery, as well as the availability of the safe and relatively efficient cationic liposome vector, we decided to focus on direct gene therapy as our approach to immunotherapy against melanoma. Accordingly, our first human gene therapy trial employing direct gene transfer to treat melanoma was approved by the Recombinant Advisory Committee of the National Institutes of Health in February, 1992 (G.J. Nabel et al. 1992). The overall objectives of this and a subsequent trial have been to (1) establish a safe and effective dose of the recombinant gene, HLA-B7, to be introduced into tumors by DNA liposome transfection and (2) confirm expression of this MHC class I gene by tumor cells following direct gene transfer in vivo. Secondarily, we have analyzed the immune response against HLA-B7 and tumor antigens.

Five patients with metastatic melanoma and subcutaneous tumor nodules refractory to all previous treatment received intratumoral injections of the human MHC class I gene, HLA-B7, complexed with the DC-cholesterol liposome vector. All the patients suffered from Stage IV melanoma refractory to all conventional forms of therapy, which included surgery, radiation, chemotherapy, tumor vaccine plus IL-2, IFN-γ, and T cells plus IL-2. There were three females, aged 61, 62, and 65, and two males, aged 52 and 68. Three patients received the lowest dose at each of three treatments, one 0.2-ml injection (0.29 μg of DNA) into the same tumor nodule, for a cumulative dose of 0.86 μg DNA. Three patients received the next highest dose at each of three treatments, 0.2 ml injected into the same tumor nodule, for a cumulative dose of 2.58 μg DNA. One patient with pulmonary metastases received escalating treatments to the right lung delivered by catheter (E.G. Nabel et al. 1993; G.J. Nabel et al. 1993). The latter series comprised two treatments of ~5 ml of the DNA-liposome solution (1.43 μg DNA each) delivered to the pulmonary artery of the posterior segment of the right lower lobe via percutaneous catheter.

Core needle biopsy samples taken 3–7 days after each treatment were analyzed by PCR to confirm recombinant HLA-B7 gene expression. Limiting dilu-

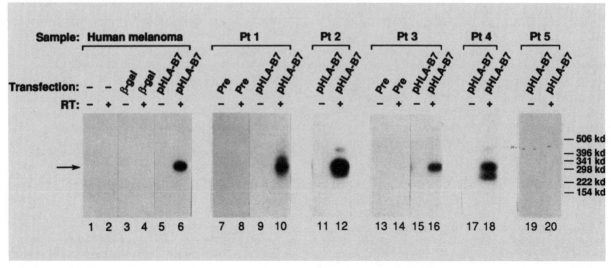

Figure 2. Confirmation of gene expression in tumor nodules transduced by direct HLA-B7 gene transfer. Recombinant HLA-B7 imRNA was detected using a reverse transcriptase (RT)-PCR technique on nucleic acids from biopsy samples. Total RNA was incubated in the presence (+) or absence (−) of RT and analyzed. (Reprinted, with permission, from G.J. Nabel et al. 1993.)

tion analysis of the HLA-B7 cytolytic T-lymphocyte response was performed in some cases comparing autologous Epstein-Barr-virus-transformed B cells to a subline transduced with an HLA-B7 retroviral vector.

All five patients participating in this phase of the protocol tolerated the treatment well, with no pattern of systemic abnormalities detected after treatment (G.J. Nabel et al. 1993). The single patient with pulmonary metastases experienced no toxicities or complications following catheter-based gene delivery. In four patients, DNA was detected within the injected nodule 3–7 days following injection (Fig. 2). Although neither recombinant DNA nor RNA could be detected in one patient, this probably was related to inhibition of the PCR analysis, since HLA-B7 was immunohistochemically detected in tumor cells of this patient (Fig. 3). The patient who received both treatment protocols plus catheter-based gene transfer showed complete regression of the treated nodule and, at the same time, complete regression of metastatic lesions at distant sites (Fig. 4). Detection of tumor-specific CTLs in the blood of three patients suggests that expression of a foreign MHC gene after direct gene transfer can alter the reactivity of the immune system. In all patients, plasmid DNA was not detected in the blood at any time following gene transfer, and there were no increases in anti-DNA antibodies or clinical evidence of autoimmune phenomena. Analysis of the mechanism of the antitumor response is now proceeding.

With this protocol indicating that expression of a foreign MHC gene after direct gene transfer could safely alter the reactivity of the immune system, we decided to investigate this mode of immunotherapy in an enlarged protocol employing a newly developed, more efficient liposome vector. Accordingly, a subsequent clinical protocol (G.J. Nabel et al. 1994) was

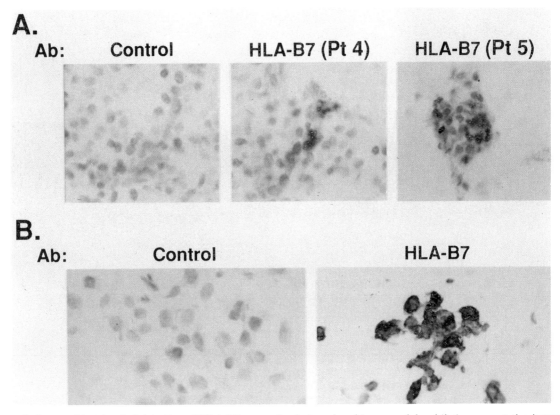

Figure 3. Immunohistochemical detection of HLA-B7 expression in transduced tumor nodules. (*A*) A representative immunohistochemical stain on biopsies from patients (Pt) 4 and 5 is shown, with a control (IgG1 monoclonal) or anti-HLA-B7 (ME-1/HB119) antibody (IgG1 isotype) as indicated. Because this antibody can be cross-reactive with HLA-B27, expression was also confirmed with another antibody nonreactive with HLA-B27 (MB40.2/HB59; patient 5). The frequency of positive cells was estimated at ~1% in both patients, but could be influenced by sampling. (*B*) Expression of HLA-B7 protein was evaluated in an HLA-B7-positive Epstein-Barr-virus-transformed line, JY. Additional monoclonal antibodies used to human HLA-B7 were ME-1 and MB 4.0.2. Immunostaining was performed using a horseradish peroxidase-coupled rabbit anti-mouse second antibody (Vector Cloning Systems). Samples were analyzed in some cases by immunofluorescence with the BB7.1 monoclonal antibody. Indirect immunofluorescent staining of mechanically dispersed unfixed cells was evaluated in patients 1,2,3A, and 3B (see G.J. Nabel et al. 1993) and revealed a frequency of positive cells of ~1% for patients 1 and 2 and 1–10% for patient 3, treatments A and B, in the biopsy sample. (Reprinted, with permission, from G.J. Nabel et al. 1993.)

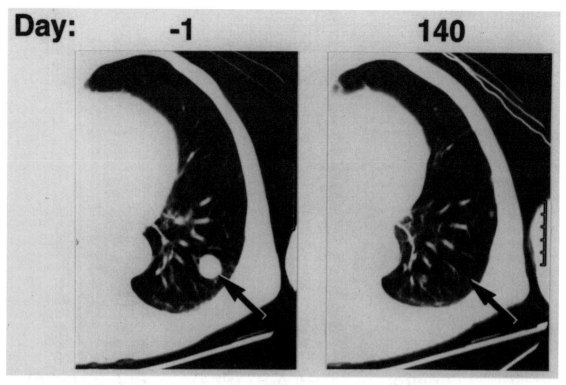

Figure 4. Computerized tomographic evidence of tumor regression at a distant pulmonary site is indicated, showing a responsive pulmonary nodule at the indicated times after treatment 1 (Patient 3A) began. (Reprinted, with permission, from G.J. Nabel et al. 1993.)

approved by the National Institutes of Health in June, 1993. This study, now under way, incorporates four advances in gene transfer technology:

- improved transfection efficiency through use of DMRIE/DOPE, a more effective liposome vector that has a gene transfer ability in vitro 10-fold that of the DC-cholesterol vector used in the initial clinical protocol (San et al. 1993)
- improved expression of the HLA-B7 vector by the addition of a consensus translation initiation sequence and removal of an intron, and by the inclusion of the β_2-microglobulin gene, with which MHC class I genes normally associate, to allow synthesis of the complete histocompatibility molecule (these two gene products are normally cotransported to the cell surface; since some human melanoma cells do not express endogenous β_2-microglobulin, their ability to stably express class I on the cell surface is limited)
- delivery of recombinant genes by catheter, which enables transduction of a larger percentage of cells within the tumor microcirculation, thus improving the extent of gene expression; catheter delivery was well tolerated by the one patient with pulmonary metastasis in the initial protocol (Fig. 5) (G.J. Nabel et al. 1993; E.G. Nabel et al. 1994)
- inclusion of other human cancers, such as colon, renal cell, and breast carcinoma, for treatment, primarily catheter-mediated

The most effective and desirable immunotherapy

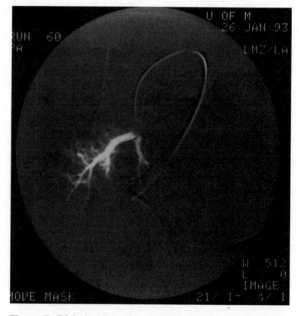

Figure 5. Digital subtraction angiography of right posterior basal segment artery. After selective catheterization of the right posterior basal segment artery, a digital subtraction angiogram was performed. The intravenous contrast agent demonstrates displacement of the posterior basal segmental arteries by the tumor. The contrast agent was removed by instillation of 30 ml sterile saline solution followed by the introduction of 1.43 μg pHLA-B7 plasmid DNA complexed to DC-cholesterol liposome. (Reprinted, with permission, from E.G. Nabel et al. 1994.)

requires not only effectiveness, but also safety and efficiency. The first protocol showed the safety and feasibility of a gene therapy approach to melanoma. The efficiency of this method is likely to be increased by the use of the DMRIE/DOPE vector. This cationic liposome, described by Dr. Philip Felgner (Felgner et al. 1994), has improved transfection efficiency in vitro. Of greater note is the failure of this cationic liposome to aggregate at high concentrations, in contrast to the DC-cholesterol liposome. Because of this property, greater concentrations of DNA and liposomes can be introduced without toxicity. Use of DMRIE/DOPE should significantly improve not only the safety, but also the efficacy of this immunotherapy protocol. The level of gene transfer and expression required for successful treatment of human melanoma is expected to be 1–5%. Previous studies using the DC-cholesterol vector have shown the percentage of cells containing recombinant DNA following gene transfer to be only ~1%. The greater concentrations of DNA allowed by use of DMRIE/DOPE should allow a greater percentage of DNA transfer to be achieved.

SUMMARY AND FUTURE DEVELOPMENTS

Genetic instability within malignant cells gives rise to mutant proteins which can be recognized by the immune system, and this recognition by T lymphocytes could contribute to the elimination of the neoplastic cells. We have developed a molecular genetic intervention for the treatment of malignancies based on the knowledge that foreign MHC proteins expressed on the cell surface are efficient at stimulating an immune response. Initially, in a murine model, foreign MHC genes were introduced directly into malignant tumors in vivo in an effort to stimulate an immune response and subsequent rejection of the tumor cells. Expression of this foreign MHC gene within the tumors did induce a cytotoxic T-cell response to the introduced gene. More importantly, the immune system recognized tumor-specific antigens on unmodified tumor cells as foreign. Growth of the tumors diminished, and in many cases there was complete regression. This evidence that direct gene transfer in vivo can induce cell-mediated immunity against specific gene products offers the potential for effective immunotherapy for the treatment of human cancer and infectious diseases.

We have successfully conducted a human clinical study to determine the safety and efficacy of this treatment and are now conducting additional studies with improved vectors and liposomes. Since several studies have shown that the introduction of cytokines improves antitumor effect (Fearon et al. 1990; Gansbacher et al. 1990a,b; Dranoff et al. 1993), future modifications of gene therapy protocols against cancer could include the added expression of a cytokine gene, such as IL-2 or GM-CSF. Although combined vectors are not yet available, inclusion of a cytokine in the treatment protocol could further stimulate T-cell immunity against tumors locally and improve recognition of tumor-associated antigens.

Immunotherapy has shown promise as a primary treatment for some forms of cancer, particularly melanoma and renal cell carcinoma, which are relatively more responsive to modulation of immune function. The development of more efficient viral and nonviral vectors has also enhanced the attractiveness of gene therapy for cancer and facilitates the application of molecular biology to clinical settings. Finally, progress in developing safe and effective methods of direct, in vivo gene transfer brings this approach to cancer therapy well within the constraints of clinical medicine. It is possible that within a few years gene therapy will provide alternative treatments for cancer.

ACKNOWLEDGMENTS

We thank Fran Frueh and Donna Gschwend for their editorial and secretarial assistance. This work is supported in part by a grant from the National Institutes of Health (P01 CA-59327 to A.E.C.).

REFERENCES

Blanchet, O., J.-F. Bourge, H. Zinszner, A. Israel, P. Kouilsky, J. Dausset, L. Degos, and P. Paul. 1992. Altered binding of regulatory factors to HLA class I enhancer sequence in human tumor cell lines lacking class I antigen expression. *Proc. Natl. Acad. Sci.* **89:** 3488.

Chen, W., D.J. Peace, D.K. Rovira, S.-G. You, and M.A. Cheever. 1992. T-cell immunity to the joining region of p210BCR-ABL protein. *Proc. Natl. Acad. Sci.* **89:** 1468.

Chou, T., A.E. Chang, and S. Shu. 1988. Generation of therapeutic T lymphocytes from tumor-bearing mice by in vitro sensitization: Culture requirements and characterization of immunologic specificity. *J. Immunol.* **140:** 2453.

Cordon-Cardo, C., Z. Fuks, M. Drobnjak, C. Moreno, L. Eisenbach, and M. Feldman. 1991. Expression of HLA-A,B,C antigens on primary and metastatic tumor cell populations of human carcinomas. *Cancer Res.* **51:** 6372.

Dranoff, G., E. Jaffee, A. Lazenby, P. Golumbek, H. Levitsky, K. Brose, V. Jackson, H. Hamada, D. Pardoll, and R. C. Mulligan. 1993. Vaccination with irradiated tumor cells engineered to secrete murine granulocyte-macrophage colony-stimulating factor stimulates potent, specific, and long-lasting anti-tumor immunity. *Proc. Natl. Acad. Sci.* **90:** 3539.

Dunn, P.L. and R.J. North. 1991a. Effect of advanced aging on the ability of mice to cause tumor regression in response to immunotherapy. *Immunology* **74:** 355.

———. 1991b. Effect of advanced aging on ability of mice to cause regression of an immunogenic lymphoma in response to immunotherapy based on depletion of suppressor T cells. *Cancer Immunol. Immmunother.* **33:** 421.

Elliott, B.E., D.A. Carlow, A.-M. Rodricks, and A. Wade. 1989. Perspectives on the role of MHC antigens in normal and malignant cell development. *Adv. Cancer Res.* **52:** 181.

Esumi, N., B. Hunt, T. Itaya, and P. Frost. 1991. Reduced tumorigenicity of murine tumor cells secreting gamma-interferon is due to nonspecific host responses and is unrelated to class I major histocompatibility complex expression. *Cancer Res.* **51:** 1185.

Fearon, E.R., D.M. Pardoll, T. Itaya, P. Golumbek, H.I. Levitsky, J.W. Simons, H. Karasuyama, B. Vogelstein, and P. Frost. 1990. Interleukin-2 production by tumor cells

bypasses T helper function in the generation of an antitumor response. *Cell* **60:** 397.
Felgner, J.H., R. Kumar, C.N. Sridhar, C.J. Wheeler, Y.-J. Tsai, R. Border, P. Ramsey, M. Martin, and P.L. Felgner. 1994. Enhanced gene delivery and mechanism studies with a novel series of cationic lipid formulation. *J. Biol. Chem.* **269:** 2550.
Fishel, R., M.K. Lescoe, M.R. Rao, N.G. Copeland, N.A. Jenkins, J. Garber, M. Kane, and R. Kolodner. 1994. The human mutator gene homolog MSH2 and its association with hereditary nonpolyposis colon cancer. *Cell* **77:** 167.
Franksson, L., E. George, S. Powis, G. Butcher, J. Howard, and K. Kärre. 1993. Tumorigenicity conferred to lymphoma mutant by major histocompatibility complex-encoded transporter gene. *J. Exp. Med.* **177:** 201.
Gansbacher, B., R. Bannerji, B. Daniels, K. Zier, K. Cronin, and E. Gilboa. 1990a. Retroviral vector-mediated gamma-interferon gene transfer into tumor cells generates potent and long lasting anti tumor immunity. *Cancer Res.* **50:** 7820.
Gansbacher, B., K. Zier, B. Daniels, K. Cronin, R. Bannerji, and E. Gilboa. 1990b. Interleukin 2 gene transfer into tumor cells abrogates tumorigenicity and induces protective immunity. *J. Exp. Med.* **172:** 1217.
Gansbacher, B., K. Zier, K. Cronin, P. A. Hantzopoulos, B. Bouchard, A. Houghton, E. Gilboa, and D. Golde. 1992. Retroviral gene transfer induced constitutive expression of interleukin-2 or interferon-γ in irradiated human melanoma cells. *Blood* **80:** 2817.
Gattoni-Celli, S., K. Kirsch, R. Timpane, and K.J. Isselbacher. 1992. β_2 Microglobulin gene is mutated in a human colon cancer cell line (HCT) deficient in the expression of HLA class I antigens on the cell surface. *Cancer Res.* **52:** 1201.
Glas, R., K. Sturmhöfel, G. J. Hämmerling, K. Kärre, and H.-G Ljunggren. 1992. Restoration of a tumorigenic phenotype by β_2-microglobulin transfection to EL-4 mutant cells. *J. Exp. Med.* **175:** 843.
Golumbek, P.T., A.J. Lazenby, H.I. Levitsky, L.M. Jaffee, H. Karasuyama, M. Baker, and D.M. Pardoll. 1991. Treatment of established renal cancer by tumor cells engineered to secrete interleukin-4. *Science* **254:** 713.
Gopas, J., B. Rager-Zisman, M. Bar-Eli, G. Hämmerling, and S. Segal. 1989. The relationship between MHC antigen expression and metastasis. *Adv. Cancer Res.* **53:** 89.
Herberman, R.B. 1985. Multiple functions of natural killer cells, including immunoregulation as well as resistance to tumor growth. *Concepts Immunopathol.* **1:** 96.
Hui, K., F. Grosveld, and H. Festenstein. 1984. Rejection of transplantable leukemia cells following MHC DNA-mediated cell transformation. *Nature* **311:** 750.
Inge, T.H., S.K. Hoover, B.M. Susskind, S.K. Barrett, and H.D. Bear. 1992. Inhibition of tumor-specific cytotoxic T-lymphocyte responses by transforming growth factor β_1. *Cancer Res.* **52:** 1386.
Itaya, T., S. Yamagiwa, R. Okada, T. Oikawa, N. Kuzumaki, N. Takeichi, M. Hosokawa, and H. Kobayashi. 1987. Xenogenization of a mouse lung carcinoma (3LL) by transfection with an allogeneic class I major histocompatibility complex gene (H-2L^d). *Cancer Res.* **47:** 3136.
Jardetzky, T.S., W.S. Lane, R.A. Robinson, D.R. Madden, and D.C. Wiley. 1991. Identification of self peptides bound to purified HLA-B27. *Nature* **353:** 326.
Kant, A., N.A. Gruis, J. Weaver-Feldhaus, Q. Liu, K. Harshman, S.L. Tavtigian, E. Stockert, R.S. Day III, B.E. Johnson, and M.H. Skolnik. 1994. A cell cycle regulator potentially involved in genesis of many tumor types. *Science* **264:** 436.
Kern, O., J. Klarnet, M. Jensen, and P.J. Greenberg. 1986. Requirement for recognition of class II MHC molecules and processed antigen for optimal generation of syngeneic tumor specific class I MHC restricted cytotoxic T lymphocytes. *J. Immunol.* **136:** 4303.
Kradin, R.L., D.S. Lazarus, and S.M. Dubinett. 1989. Tumor-infiltrating lymphocytes and interleukin-2 in treatment of advanced cancer. *Lancet* **I:** 577.
Lafraniere, R. and S.A. Rosenberg. 1985. Successful immunotherapy of murine experimental hepatic metastases with lymphokine-activated killer cells and recombinant interleukin 2. *Cancer Res.* **45:** 3735.
Lindahl, K.F. and D.B. Wilson. 1977. Histocompatibility antigen-activated cytotoxic T lymphocytes. I. Estimates of the absolute frequency of killer cells generated in vitro. *J. Exp. Med.* **145:** 500.
Morton, D.L., F.R. Eilber, and E.C. Holmes. 1974. BCG immunotherapy of malignant melanoma: Summary of a seven-year experience. *Ann. Surg.* **180:** 635.
Muul, L.M., P.J. Spiess, E.P. Director, and S.A. Rosenberg. 1987. Identification of specific cytolytic immune response against autologous tumor in humans bearing malignant melanoma. *J. Immunol.* **138:** 989.
Nabel, E.G., G. Plautz, and G.J. Nabel. 1990. Site-specific gene expression in vivo by direct gene transfer into the arterial wall. *Science* **249:** 1285.
———. 1992a. Transduction of a foreign histocompatibility gene into the arterial wall induces vasculitis. *Proc. Natl. Acad. Sci.* **89:** 5157.
Nabel, E.G., A. Yang, S. Liptay, H. San, D. Gordon, C.C. Haudenschild, and G.J. Nabel. 1993. Recombinant platelet-derived growth factor B gene expression in porcine arteries induces intimal hyperplasia in vivo. *J. Clin. Invest.* **91:** 1822.
Nabel, E.G., Z.-Y Yang, D. Muller, A.E. Chang, X. Gao, L. Huang, K. U. Cho, and G.J. Nabel. 1994. Safety and toxicity of catheter gene delivery to the pulmonary vasculature in a patient with metastatic melanoma. *Hum. Gene Ther.* **5:** 1089.
Nabel, E.G., D. Gordon, Z.-Y Yang, L. Xu, H. San, G.E. Plautz, B.-Y Wu, X. Gao, L. Huang, and G.J. Nabel. 1992b. Gene transfer in vivo with DNA-liposome complexes: Lack of autoimmunity and gonadal localization. *Hum. Gene Ther.* **3:** 649.
Nabel, G.J., A. Chang, E.G. Nabel, G. Plautz, B.A. Fox, L. Huang, and S. Suyu. 1992. Clinical protocol: Immunotherapy of malignancy by in vivo gene transfer into tumors. *Hum. Gene Ther.* **3:** 399, 705.
Nabel, G.J., A.E. Chang, E.G. Nabel, G.E. Plautz, W. Ensminger, B.A. Fox, P. Felgner, S. Shu, and K. Cho. 1994. Clinical protocol: Immunotherapy for cancer by direct gene transfer into tumors. *Hum. Gene Ther.* **5:** 57.
Nabel, G.J., E.G. Nabel, Z.-Y Yang, B.A. Fox, G.E. Plautz, X. Gao, L. Huang, S. Shu, D. Gordon, and A.E. Chang. 1993. Direct gene transfer with DNA liposome complexes in melanoma: Expression, biologic activity, and lack of toxicity in humans. *Proc. Natl. Acad. Sci.* **90:** 11307.
Oldham, R.K. 1983. Natural killer cells: Artifact to reality: An odyssey in biology. *Cancer Metastasis Rev.* **2:** 323.
Pantel, K., G. Schlimok, D. Kutter, G. Schaller, T. Genz, B. Wiebecke, R. Backmann, I. Funke, and G. Reithmüller. 1991. Frequent down-regulation of major histocompatibility class I antigen expression of individual micrometastatic carcinoma cells. *Cancer Res.* **51:** 4005.
Papadopoulos, N., N.C. Nicolaides, Y.-F Wei, S.M. Ruben, K.C. Carder, C.A. Rosen, W.A. Haseltine, R.D. Fleischmann, C.M. Fraser, M.D. Adams, J.C. Venter, S.R. Hamilton, G.M. Petersen, P. Watson, H.T. Lynch, P. Peltomaki, J.-P. Mecklin, A. de la Chapelle, K.W. Kinzler, and B. Vogelstein. 1994. Mutation of a *mutL* homolog in hereditary colon cancer. *Science* **263:** 1525.
Parham, P., C.E. Lomen, D.A. Lawlor, J.P. Ways, N. Holmes, H.L. Coppin, R.D. Salter, A.M. Wan, and P.D. Ennis. 1988. Nature of polymorphism in HLA-A, -B, -C molecules. *Proc. Natl. Acad. Sci.* **85:** 4005.
Parsons, R., G.M. Li, M.J. Longley, W.H. Fang, N. Papadopoulos, J. Jen, A. de la Chapelle, K.W. Kinzler, B. Vogelstein, and P. Modrich. 1993. Hypermutability and mismatch repair deficiency in RER$^+$ tumor cells. *Cell* **75:** 1227.
Plautz, G.E., Z.Y. Yang, X. Gao, L. Huang, and G.J. Nabel.

1993. Immunotherapy of malignancy by in vivo gene transfer into tumors. *Proc. Natl. Acad. Sci.* **90:** 4645.

Pucetti, P., L. Romani, and M.C. Fioretti. 1987. Chemical xenogenization of experimental tumors. *Cancer Metastasis Rev.* **6:** 93.

Restifo, N.P., P.J. Spiess, S.E. Karp, J.J. Mule, and S.A. Rosenberg. 1992. A nonimmunogenic sarcoma transduced with the cDNA for interferon γ elicits CD8$^+$ T cells against the wild-type tumor: Correlation with antigen presentation capability. *J. Exp. Med.* **175:** 1423.

Rosenberg, S.A., J.J. Mule, and P.J. Speiss. 1985. Regression of established pulmonary metastases and subcutaneous tumor mediated by the systemic administration of high-dose recombinant interleukin-2. *J. Exp. Med.* **161:** 1169.

Rosenberg, S.A., M.T. Lotze, J.C. Yang, P.M. Aebersold, W.M. Linehan, C.A. Seipp, and D.E. White. 1989. Experience with the use of high-dose interleukin-2 in the treatment of 652 cancer patients. *Ann. Surg.* **210:** 474.

Rosenberg, S.A., P. Spiess, and R. Lafreniere. 1986. A new approach to the adoptive immunotherapy of cancer with tumor-infiltrating lymphocytes. *Science* **223:** 1318.

Rosenberg, S.A., B.S. Packard, P.M. Aebersold, D. Solomon, S.L. Topalian, S.T. Toy, P. Simon, M.T. Lotze, J.C. Yang, and C.A. Seipp. 1988. Use of tumor-infiltrating lymphocytes and interleukin-2 in the immunotherapy of patients with metastatic melanoma. *N. Engl. J. Med.* **319:** 1676.

San, H., Z.-Y Yang, V.J. Pompili, M.L. Jaffe, G.E. Plautz, L. Xu, J.H. Felgner, C.J. Wheeler, P.L. Felgner, X. Gao, L. Huang, D. Gordon, G.J. Nabel, and E.G. Nabel. 1993. Safety and toxicity of a novel cationic lipid formulation for human gene therapy. *Hum. Gene Ther.* **4:** 781.

Shu, W., T. Chou, and S.A. Rosenberg. 1985. Adoptive immunotherapy of newly induced murine sarcomas. *Cancer Res.* **45:** 1657.

Skinner, M.A. and J. Marbrook. 1976. An estimation of the frequency of precursor cells which generate cytotoxic lymphocytes. *J. Exp. Med.* **143:** 1562.

Spiess, P.J., J.C. Yang, and S.A. Rosenberg. 1987. In vivo antitumor activity of tumor-infiltrating lymphocytes expanded in recombinant interleukin-2. *J. Natl. Canc. Inst.* **79:** 1067.

Stewart, M.J., G.E. Plautz, L. Del Buono, Y. Yang, L. Xu, X. Ga, L. Huang, E.G. Nabel, and G.J. Nabel. 1992. Gene transfer in vivo with DNA-liposome complexes: Safety and acute toxicity in mice. *Hum. Gene Ther.* **3:** 267.

Sugiura, C., T. Itaya, N. Kondoh, T. Oileawa, N. Kuzamaki, N. Takeichi, M. Hosokawa, and H. Kobayashi. 1988. Xenogenization of tumor cells by transfection with plasmid containing env gene of Friend leukemia virus. *Jpn. J. Cancer Res.* **79:** 1259.

Tanaka, K., K.J. Isselbacher, G. Khoury, and G. Jay. 1985. Reversal of oncogenesis by the expression of H-2K antigen following H-2 gene transfection. *Science* **228:** 26.

Tepper, R.I., P.K. Pattengale, and P. Leder. 1989. Murine interleukin-4 displays potent anti-tumor activity *in vivo*. *Cell* **57:** 503.

Topalian, S.L., D. Solomon, and S.A. Rosenberg. 1989. Tumor-specific cytolysis by lymphocytes infiltrating human melanomas. *J. Immunol.* **142:** 3714.

Torre-Amione, G., R.D. Beauchamp, H. Koeppen, B.H. Park, H. Schreiber, H.L. Moses, and D.A. Rowley. 1990. A highly immunogenic tumor transfected with a murine transforming growth factor type β_1 cDNA escapes immune surveillance. *Proc. Natl. Acad. Sci.* **87:** 1486.

Townsend, A. and H. Bodmer. 1989. Antigen recognition by class I-restricted T lymphocytes. *Annu. Rev. Immunol.* **7:** 601.

Wahl, W.L., G.E. Plautz, B.A. Fox, G.J. Nabel, S. Shu, and A.E. Chang. 1992. Generation of therapeutic lymphocytes after in vivo transfection of tumor with a gene encoding allogeneic class I MHC antigen. *Surg. Forum* **63:** 476.

Wallich, R., N. Bulbuc, G. Hämmerling, S. Karzav, S. Segal, and M. Feldman. 1985. Abrogation of metastatic properties of tumor cells by *de novo* expression of H-2K antigen following H-2 gene transfection. *Nature* **315:** 301.

Watson, J.D., M. Gilman, J. Witkowski, and M. Zoller. 1992. *Recombinant DNA*, 2nd edition. Scientific American, New York.

Yoshizawa, H., A.E. Chang, and S. Shu. 1991a. Specific adoptive immunotherapy mediated by tumor-draining lymph node cells sequentially activated with anti-CD3 and IL-2. *J. Immunol.* **147:** 729.

Yoshizawa, H., K. Sakai, A.E. Chang, and S.Y. Shu. 1991b. Activation by anti-CD3 of tumor-draining lymph node cells for specific adoptive immunotherapy. *Cell. Immunol.* **134:** 473.

Zbar, B., I.D. Bernstein, and H.J. Rapp. 1971. Suppression of tumor growth at the site of infection with living bacillus Calmette-Guerin. *J. Natl. Cancer Inst.* **46:** 831.

An Introduction to the Puzzle

E. Harlow

Massachusetts General Hospital Cancer Center, Charlestown, Massachusetts 02129

5FU 600 mg/m^2 ivb q week x 6 weeks
Interrupting after 1 hr
LCV 500 mg/m^2 ivci over 2 hr q week x 6 weeks
Colon cancer with liver metastasis.

In the United States, approximately 150,000 new cases of colorectal cancer were reported in 1994. About one-quarter of these patients presented with distant metastases, most often in the liver. Current treatment for these patients begins with removal of the primary tumor by the surgeon, and then the chemotherapy listed above is begun. In this regimen, 5-fluorouracil (5-FU) is the key antitumor agent. It is metabolized to 5-FdUMP, which binds covalently to thymidylate synthetase in the presence of the reduced folate 5,10-methylenetetrahydrofolate. Including LCV (leucovorin, 5-formyltetrahydrofolate) in the therapy drives the formation of this ternary complex and consequently leads to increased toxicity of 5-FU. 5-FU is also converted into various fluoro-nucleotide analogs that are incorporated into RNA and DNA. The toxicity of RNA incorporation is unknown, but incorporation into DNA inhibits DNA elongation and leads to single strand breaks. These toxic effects are seen in all dividing cells and establish the limits to the highest dose that can be delivered safely.

Effective treatment is often measured by a decrease in the level of carcinoembryonic antigen in the blood and confirmed by a decrease in the size of the metastases. This decrease occurs at best in about one-half of the cases. If no change is seen, no further treatment is likely to have much effect. If shrinkage of the metastases occurs, the treatment is repeated with 2-week rest periods between each treatment cycle. In 6 months, 1 year, or, in unusual circumstances perhaps as long as 2 years, the metastases will stop shrinking and begin to enlarge. With treatment or without, in all but the rarest cases the patient will eventually die.

Far more is known about the molecular causes for colorectal cancer than for any other adult tumor. We know the names and some of the biochemical roles for the cancer genes and their products that contribute to this tumor development. These changes have been painstakingly mapped out, and for colon cancers we may know the majority of the necessary changes that drive tumorigenesis. We can put names to HNPCC, APC, ras, p53, DCC, and others. Our knowledge of these molecular defects has come from a dedicated 20 years of the best molecular biological research. Yet this information does not translate to any effective treatment nor to an understanding of why many of the current treatments succeed or why others fail.

It is a frustrating time.

I introduce this summary of the 1994 Cold Spring Harbor Symposium on the Molecular Genetics of Cancer with an example of a commonly used cancer therapy not to invoke the emotional response that it will bring, but to point out how great the gap is between the worlds of patient and researcher, between the detailed understanding of the pharmacology of current treatments and the molecular biology of the underlying genetic causes, and between the cultures of the oncologist and the researcher. There is a great distance between our knowledge and our need. Filling these gaps provides a daunting task for all of us in this research community.

FRAMING THE QUESTION

Put in its simplest terms, tumor development is a multistep, multilocus genetic selection. Over time and under selective pressure for clonal survival, tumor cells acquire mutations that promote their proliferation and persistence—a type of cellular "natural selection" occurring on a tissue landscape. During this selection process, tumor cells alter a large number of characteristics, many of which we now recognize, and gradually acquire phenotypes that threaten their host. Our ultimate task in studying oncogenesis must be to understand the selection pressures that reward the surviving cells, to discover the mutations that promote the incremental changes in cell phenotype, and if possible, to intervene. Of these, only in the discovery of mutations can we claim much progress.

Selection. Although we have learned a great deal about which mutations are found in human tumors, even more about how known genes can promote tumorigenesis in animal models, and remarkable volumes about how mutations promote transformation of cells in culture, we know painfully little about the actual selection pressures that drive tumor development. It is clear that each tissue poses a different set of hurdles for genetic selection. These differences and the difficulties of reproducing the selective settings in a model setting

---

All authors cited here without dates refer to papers in this volume.

have made our learning process slow and cumbersome. Only in the development of colon carcinoma (Vogelstein and Kinzler) and Barrett's esophageal carcinoma (Neshat et al.) have we begun to string together an ordered picture of the molecular events of human tumorigenesis, and even here the story is still myopic. Our view is one of a series of events and not about the selection pressures themselves. Are there common problems that all tumors must solve? What particular problems does tumor development in an individual tissue pose? When is the order of mutations important? What are the selective pressures that these cells solve? These are massively complex biological questions, but without this level of sophistication, it will be difficult to understand how tumor development is driven.

Since it is now clear that the tissue origin of a tumor is one of the key considerations in which mutations participate in the tumor development, we need to expand on the limited base provided by the colon carcinoma and Barrett's esophageal models. In addition, the mouse model systems that have been started (see the work described in Naik et al; Berns et al; Kemp et al.; Varmus et al.; Cory et al; Van Dyke et al.; Williams et al.; and Dove et al.) need to be exploited to include more known and tissue-specific changes in order to take our knowledge of the selective pressures to a more complete picture. This has been an extremely difficult area to develop, but it promises to be one of the most important, as our knowledge of cancer genetics expands to meet the world of tumor biology.

Cancer genes. As for any genetic selection, the mutations that promote tumor development can be broadly classified as either gain-of-function or loss-of-function changes. Although the origins of the terms *oncogene* or *tumor suppressor gene* were not based on the genetic distinction between a gain-of-function or loss-of-function change, over time these terms have become synonymous with these mutational classifications. We often succumb to the simple analogies of jammed accelerators or faulty brakes to convey the concepts of oncogenes or tumor suppressor genes to the uninitiated. For this review, I follow this simple-minded distinction and use oncogene to mean any allele that is selected by a gain-of-function mutation and tumor suppressor gene to be its genetic antithesis, a loss-of-function mutation.

Implicit in the decision to let oncogenes and tumor suppressor genes be defined as gain-of-function and loss-of-function mutations is the realization that these changes can occur in any step of tumor development. This would include a broad spectrum of mutations as diverse as those that release a well-differentiated cell from its cell cycle arrest to a late-stage mutation that allows invasion of a metastatic cell. All of these plus innumerable other mutations that provide a selective advantage to any stage of tumorigenesis can be promoted either by a gain-of-function or loss-of-function mutation. Oncogenes can jam the accelerator or tumor suppressors can lose the brakes on any of the steps of tumor development.

At present there are more than 100 known oncogenes that score in one or another of the available tests. The number of tumor suppressor genes is about 15 or so, depending on how rigorous the proof of loss of function. Whatever this number, it seems likely that the group of tumor suppressor genes will continue to grow for some time, albeit probably never with the speed at which new oncogenes were once identified. Even from the earliest studies, it has been clear that the actions of oncogenes and tumor suppressors can occur on the same biochemical pathways. NF1 as a biochemical counter to ras is an excellent example, but the list can be extended to include others as well. Therefore, oncoproteins and tumor suppressor proteins do not appear to control radically different functions, only that their normal roles provide different types of regulation.

Intervention. Our attempts to intervene have been built on an impressive empirical test of possible therapies. Although they have led to useful treatment of some childhood tumors, our success in adult tumors has been marked largely by temporary remissions and quality-of-life improvements rather than by large advances in cures. Perhaps the most important aspect of modern cancer research is the consensus that we are on the threshold of major advances in what has been called rational therapy. This optimism is based strongly on the progress that we have made in our understanding of the genetic basis of tumorigenesis. The 1994 Cold Spring Harbor Symposium on Quantitative Biology has been an opportunity to review these advances and assess our ability to make progress on the important aspects of human tumor development and possible therapeutic intervention.

Overview

The summary below has been divided into three broad sections based on my current idiosyncratic view of the field. The three divisions are Fidelity, Decisions, and Tumor Biology. The first two are just a convenient method to subdivide the field into categories that reflect the consequence of the various mutations. At present, all of the mutations that have been cataloged in cancer genetics can be divided into mutations that lead to loss of fidelity or deregulation of external signaling. The third category of Tumor Biology has allowed me to summarize some of the more complicated issues of tumor development and model systems.

I have not made a conscious effort to cover all the talks or papers in this volume. Let me apologize to my colleagues and the readers for these omissions, but during the preparation of this summary it became clear that I was too slow in my writing to cover all of the work satisfactorily and still meet even a much extended publication deadline.

FIDELITY

When the first stereo record players were being manufactured, the desired label on their covers was

"Hi-Fi." *High fi*delity meant the ability to reproduce the recorded sound in an accurate manner. High fidelity is a remarkable feature of all biological reproductions. Our cells have intricate mechanisms to ensure their faithful duplication. Perhaps the most exciting aspect of recent cancer research has been the maturation of our understanding of how cell division maintains its fidelity.

Loss of fidelity is an aspect of tumor development that has been recognized for some time. Nowell (1976), Cairns (1975), and Loeb (Loeb et al. 1974) all are acknowledged for their insights in how important the loss of fidelity would be in generating cancer. In the last several years, their realizations and predications have taken on a biochemical face with the appreciation of the way in which some of these mammalian Hi-Fi systems maintain fidelity.

Mistake Generators

Any disabling of the fidelity apparatus continually promotes new mutations. Thus, the mutation of key elements in a fidelity maintenance system acts as a mistake generator that will continue to contribute to cellular diversity. This mutator phenotype may help explain why tumor cells have such a dramatically altered genome. Predictions from epidemiologists suggested that the minimum number of rate-limiting steps needed to generate an adult tumor would be approximately 4–6. Yet even the simplest examination of a lung tumor, for example, reveals 10 mutations in previously identified tumor genes, and this number will probably continue to grow as more genes are studied. Some estimates suggest that 20 or more mutations contribute to the phenotype of these tumor cells. A simple calculation of normal mutation rates (1 in 10^6 per cell division per gene) illustrates that the chance of randomly generating those 10 mutations in a single cell lineage is ridiculously small (1 in 10^{60} divisions). An adult will undergo approximately 10^{16} divisions in a normal life span. That would mean that 1 in 10^{44} adults would develop cancer if 10 mutations were needed. If the lowest estimate is more accurate, only 1 in 10^8 adults would get cancers. The real number is about 1 in 3. Comparing numbers in this manner is somewhat artificial (early events in tumors increase the number of divisions, most tumors do not occur in tissues that rapidly turn over, etc.), but the calculations begin to point out how important this area is. There must be mutator phenotypes driving the observed genetic diversity seen in tumor cells.

Any homeostasis system that helps ensure genetic integrity is likely to be linked to carcinogenic progression. To date, loss of fidelity has been reported in four different contexts. If the changes occur in DNA repair mechanisms, DNA mutations may remain uncorrected or DNA may be misrepaired, both yielding significantly higher mutation rates. Poor fidelity also can be generated by inappropriate manipulation of chromosomes, through problems in condensation, spindle alignment, or chromosome migration. If the loss of fidelity results in changes in cell cycle timing or inability to arrest following DNA damage, mutations that would normally be corrected before DNA replication or mitosis can be fixed in the genetic lineage. Finally, any major perturbation that would normally lead to cell death can be ignored if apoptotic pathways fail. All of these defects lead to higher mutation rates, and higher mutation rates increase the chances of a tumor cell developing a selective advantage.

Errors in DNA Repair Systems Increase Mutation Rates

Two clear links between the enzymology of DNA repair and tumorigenesis have now been established. These defects lead to loss of efficient repair and subsequent propagation of mutations in other genes. The two examples are found in excision repair, in which an abnormal nucleotide(s) is detected and replaced, and mismatch repair, where a local stretch of mismatched DNA is corrected. These processes are biochemically unrelated, but mutations in both enzymatic processes have been shown to contribute to tumorigenesis.

Excision repair. Modified nucleotides can be recognized and corrected by three types of excision repair. In one, damaged bases are recognized directly, and the sugar-base linkage is cleaved by a glycosylase. The remaining phosphate-sugar backbone is then hydrolyzed by an endonuclease, and the nucleotide is replaced using the other strand as template. This process is known as base excision repair (BER) and is limited to settings in which the glycosylase itself can physically recognize and attack the damaged nucleotide. Linkage of BER with tumor development has not been reported, but it seems likely that in the proper setting the loss of BER would lead to the inability to recognize certain nucleotide adducts and promote higher rates of mutagenesis.

The other two types of excision repair are variants of nucleotide excision repair (NER) and are considerably more complicated, each involving a large number of steps and proteins. Although a recognition process is needed to initiate both these types of repair, the recognition events are quite different and depend on different protein products. One type of NER is initiated by protein/DNA interactions occurring anywhere in the genome, the so-called global NER, and the second relies on an unknown sensing mechanism that is linked to transcriptional progression, the transcription-coupled NER. After the recognition step, these two processes proceed by related biochemical steps. The enzymology begins with two nuclease events, one cutting 5 base pairs to the 5′ side of the modification and the other 24 nucleotides to the 3′ side. This oligonucleotide product is then removed by a helicase, and the gap is filled.

The work of Hoeijmakers and colleagues (Vermeulen et al.) highlights how mutations of many of the different proteins of NER can lead to inefficient DNA repair and, in some cases, are linked to increased risk of skin carcinogenesis. Attention to NER was prompted originally through the demonstration that a clinical syndrome, xeroderma pigmentosum (XP), was linked to loss of NER. XP patients inherit a defect in one of seven complementation groups, many of which are now cloned. These lesions lead to patients with defects in NER, and, depending on the protein in NER repair that is affected, these patients may also show effects in one of two other syndromes, Cockayne's syndrome (CS) and tricothiodystrophy (TTD). All three syndromes show extreme sensitivity to UV irradiation, and each has its own set of characteristic disorders. These include pigmentation abnormalities and progressive neurodegeneration for XP, neurodysmyelination and impaired physical development for CS, and brittle hair and nails for TTD. The partially overlapping phenotypes of these syndromes suggest that they may have related causes, and this has been borne out by identification of the genetic defects.

The most surprising finding to come from this work has been the realization that several of the genes linked to the NER mutations encode proteins of the basal transcription machinery. These converging fields have centered their interest on the transcription factor TFIIH. TFIIH is a multifunctional protein complex that participates in the conversion of the closed transcriptional initiation assembly to an open complex. This conversion promotes transcriptional initiation and relies on the helicase activities found in TFIIH. Two helicases in TFIIH have now been identified; they are the products of the XP-B and XP-D genes. The product of a third complementation group, TTD-A, also has been found in the TFIIH complex. All three genes provide a link between the activities of NER and transcriptional initiation. These findings argue that the components of the TFIIH complex may be used for both transcription and repair, and that a common biochemical step links these two processes.

NER's link to carcinogenesis appears to be most clearly evident in XP patients. Here affected individuals are particularly sensitive to UV damage and have a higher rate of skin carcinogenesis. Presumably this link is due to the loss of the repair defect. What seems confusing is that with the other syndromes the link to carcinogenesis is not as clear. The loss of NER is present, but higher frequencies of skin tumors are not seen. This suggests that the various complementation groups will have different risks of carcinogenesis. In some cases, different alleles of XP may even have different risks. These relationships are still hard to establish because examples of XP, CS, or TTD mutations in the many of the various complementation groups are still very rare. Detailed biochemical studies of all these proteins needed for NER will be needed to understand the reasons for the links to carcinogenesis in some but not all cases.

Mismatch repair. Mismatch repair is a universal mechanism found in both prokaryotic and eukaryotic organisms that corrects local unpaired stretches of DNA. In mammalian cells, these mismatches appear to result primarily from unfaithful DNA replication of dinucleotide or trinucleotide repeats, apparently by some slippage of the replication machinery as it passes through these repeats. During this process, the number of repeats is occasionally expanded in the synthesis of the nascent strand. This expansion leaves local mispairing that is recognized by interactions with the products of the MSH2 gene. Here, a clever trick allows the repair system to distinguish the nascent from the template strands. The MLH and PSM proteins form complexes between the MSH2/DNA complex and a nearby methylation site, but one in which the methyl group has yet to be added to the nascent strand. This hemimethylation allows strand specificity, as an endonuclease cleaves the unmethylated nascent strand. By exonuclease and polymerase action the mutated strand is replaced, and the mismatch is repaired.

Defects in any of the proteins in this system lead to expansion of dinucleotide and trinucleotide repeats. This clue was one of the key steps in uncovering the link to tumorigenesis. Perucho (Perucho et al.) developed a rapid method that could detect such lengthening in the repeat sequences. Colon carcinomas showed unexpected high levels of these abnormally long repeats. Several groups (Perucho et al; Vogelstein and Kinzler) were able to show that the loss of mismatch repair mapped to chromosome 2p, and the gene was cloned by Kolodner and colleagues and then almost immediately confirmed by Vogelstein and his colleagues (Vogelstein and Kinzler). These workers showed that the mismatch repair gene was MSH2, a human homolog of the yeast protein that binds to the mismatched stretch. Loss of this gene makes it impossible to begin the repair process.

Previous attention to this locus had been high. The 2p gene was originally identified as a tumor suppressor locus that was linked to hereditary predisposition to nonpolyposis colon cancer (HNPCC). HNPCC mutations account for approximately 10% of all colon tumors and also predispose to endometrial, ovarian, and gastrointestinal tumors (Orth et al.; Kolodner et al.; Vogelstein and Kinzler).

Given the enzymology of the mismatch repair process, it is not surprising that mutations in other proteins in this process show similar links to HNPCC. The genes for MLH1, PSM1, and PSM2 have now been cloned and shown to be mutated in human cancers.

Puzzles. Many obvious cutting-edge problems are highlighted by these studies of repair. Perhaps the most important will be the identification of the gene targets that are affected by these types of repair defects. For excision repair, there does not appear to be any target gene that could not be affected by this lack of repair. For mismatch repair this will not be true. Frequently mutated genes will include those with dinucleotide or trinucleotide repeats, and these then are likely to be the key targets in colon carcinogenesis.

For me, the most curious question raised by these findings is, Why are the repair events so tightly tied to tissue-specific tumor development? The mutator phenotype should arise from the loss of efficient repair, and in both NER and mismatch repair the defects are inherited. All tissues must have acquired the mutated gene and have relatively equal chances of losing repair function. Yet for both NER and HNPCC, the tissue distribution is quite limited. Key tissue-specific target genes could account for the localized effects, but it seems more likely that secondary properties of affected tissues will determine the local effects. This argues for some cooperating event in the target tissues. For NER it seems likely that the cooperating element will be the source of mutagenic insult, UV. This also indicates that mutagens in other tissues are either not as active as UV or that the damage is not repaired by NER but through other mechanisms. Are we missing other repair mechanisms, or are our views of environmental mutagens misaimed?

For HNPCC, one cooperating event must be replication. Without DNA replication, changes in dinucleotide or trinucleotide repeats shouldn't arise. Why then should HNPCC-linked tumors be limited to colon, endothelium, gastric epithelium, and the ovaries? Why not hematopoeitic cells? Obviously, we just don't know enough, but the concepts laid down in these reports provide the outline of how we should begin to think about this key area of tumor research.

"No Time to Stop for Repairs"

Cell cycle checkpoints are intracellular signaling pathways which ensure that progression to the next stage in the cycle is appropriate. Although it seems likely that other checkpoints will be discovered, at present there are two general types. One group of signaling pathways respond to DNA damage and arrest cells prior to either DNA replication or mitosis. The second ensure that one stage of the cell cycle has been completed prior to the initiation of the next stage. The consequences of these two classes of checkpoint control are different, and to make them easier to discuss, they are separated into these two groups below.

Following DNA damage, one important cellular response is stopping the division cycle long enough to allow DNA repair to be completed. This is particularly true in G_1. If this arrest does not occur, the mutations can be set in the genome through duplicating the mistake during DNA replication. We still know very little about the biochemistry of cell cycle arrest, but one key protein in these processes has been identified. This is the p53 tumor suppressor protein. The biochemical function of p53 is still somewhat mysterious. Its main role appears to be as a transcriptional activator (Lin et al.), but many different biochemical properties have been ascribed to p53. For brevity in this summary, I pretend that *trans*-activation is the only p53 function. Those who are not already familiar with the p53 literature are advised to take a deep breath, relax, and consult some of the reviews cited in the text.

p53 is discussed in several contexts in this review, as it has been linked to several different cell decision processes. My suspicion is that in each case, p53 is likely to be performing exactly the same biochemical role, but that the context of this action is different. However, because we don't understand the action of p53 in sufficient detail yet, I will discuss p53 in multiple settings.

Wild-type p53 is necessary to allow cells to arrest following DNA damage that involves single- or double-strand DNA breaks. This could result from treatment with ionizing radiation, high-dose UV, topoisomerase II inhibitors, or DNA-damaging adducts. The earlier notion that p53 might act in response to many different types of mutagenic insults seems not to be correct (for an example of p53-independent responses, see Wyllie et al.). However, all events that cause DNA strand breaks appear to feed into the p53 arrest pathway.

In cells that lack p53, cell cycle arrest does not occur following DNA damage (Canman et al.; Galloway et al.). When tested in p53 nullizygous mice, mutations induced by DNA-damaging agents are found with higher frequency than in normal mice compared as controls (Wyllie et al.). The enzymology of DNA repair appears to be unaffected in these tissues, but still the mutations appear in the targeted tissues. Since many different human tumors lack p53, it seems likely that analogous events occur there. This would link the loss of cell cycle arrest with increased mutagenesis rates in human tumors. It will be almost impossible to verify this link in human tumors until we understand more about how p53 establishes the correct conditions for repair.

As expected, other proteins that function in the response to DNA strand breaks are also implicated in tumor development (Canman et al.). Ataxia telangiectasia (AT) is an autosomal recessive disorder that appears to be linked to the signaling mechanism induced by ionizing radiation. The speed with which a cell responds to ionizing radiation and the duration of the response are adversely affected in AT patients, and individuals with AT show a higher frequency of certain tumors such as non-Hodgkin's lymphoma. There appear to be multiple AT complementation groups, and we can expect that those proteins that play a rate-limiting role in responding to DNA damage will be linked to carcinogenesis.

There appear to be several types of regulation that affect the ability of p53 to respond to DNA damage. The best characterized is the increase in p53 levels following DNA damage (Canman et al.; Lin et al.). Here it appears that at least two biochemical mechanisms are involved. The half-life of p53 is increased following DNA damage, and the rate of p53 translation is also stimulated. Other steps also appear to enhance p53's action. These include a regulatory p53 phosphorylation event that is required for DNA binding (Hupp and Lane; Prives et al.). This phosphorylation stimulates p53 to bind DNA and may be essential to initiate the downstream events of cell cycle arrest.

Three potential gene targets have now been identified for the role of p53 as a transcription factor. These include GADD45 (Canman et al.), mdm2 (Lin et al.; Barak et al.), and p21 (Vogelstein and Kinzler). p21 has been suggested to be a key downstream event for p53's action in cell cycle arrest. p21 is an inhibitor of the cyclin-dependent kinases (cdk) (Zhang et al.; Vogelstein and Kinzler; Roberts et al.). In vitro, p21 binds to many of the cell-cycle-regulating cdks and inhibits their kinase activity, and in vivo, ectopic expression of p21 causes cells to arrest. Since p53 activates the p21 promoter, and high levels of p21 have been reported following treatment with ionizing radiation, a logical hypothesis would be that p53 induces a cell cycle arrest through the induction of p21 expression.

Puzzles. The cell cycle arrest in response to DNA damage provides an appealing rationale for how mutations in this pathway are linked to increased mutation rates. However, I am stymied by the finding that p21 is not a tumor suppressor gene. Many of the other steps in this pathway have shown up as mutations in tumors, yet not p21. Our view of p21's function must be too limited. Is p21 overexpression the biochemical switch for p53-mediated cell cycle arrest? If p21's only role were to respond to the p53 signal for cell cycle arrest, then it would be mutated in tumors. It isn't. We must have several surprises yet to come in the p21 story.

The biochemical mechanism of inhibition by p21 also is unusual. The work described in Zhang et al. shows that p21 appears to bind to the cyclin/cdk complexes as a monomer, and these kinase complexes are active. To inhibit kinase activity, another molecule of p21 is added to the complex. What is the role of p21 binding to the active complex? Does this p21/cyclin/cdk association contribute to the subsequent inhibition when the next molecule of p21 is added, or does it have a separate role? A first step may be to learn how the second p21 molecule is added. Where does it bind and how is the enzymology of the cdk changed? So much of the inhibition story seems to revolve around the role of p21, yet the dynamics of its interaction with the cdks are not yet understood.

I haven't spent any time discussing the aspects of G_2 arrest that occur following DNA damage. However, cells are just as likely to arrest in G_2 following DNA damage. These settings are best understood in yeast, but it is clear that similar events occur in mammalian cells, as p53 minus cells arrest exclusively in G_2. What is the consequence of loosing the ability to arrest in G_2? More directly for these discussions, does this loss contribute to cancer development?

Chromosome Mismanagement

How is chromosome integrity maintained? Normal cells never show any evidence of aneuploidy, chromosome duplication, or large deletions. Lymphocytes can rearrange their antibody or T-cell-receptor genes efficiently and without undue damage, but tumor cells get it wrong. They show little control of chromosome number. Translocations are commonplace. Large deletions occur frequently. The underlying defects that allow these changes are poorly understood. Hartwell et al. discuss the issues of genomic instability here better than I can, but a few general issues deserve repeating.

Again studies in yeast are the most advanced in defining this problem. At least three signaling pathways exist in *Saccharomyces* that help control this problem. They are checkpoint controls which ensure that cell cycle progression does not occur prior to the completion of a previous event. These "feed-forward" controls appear to be essential to keep the genome stable. The events that are measured by checkpoint controls include chromosome alignment on the metaphase plate. This appears to ensure that each daughter cell receives one set of the chromosome pairs. Assembly of the spindle is another point at which the cell can be delayed until accurate completion is seen. The improper assembly of the spindle could lead to complete duplication of the genome in the most extreme case or to any combination of chromosome pair misappropriation. Finally, the duplication of the spindle pole body (centrosome for the mammalian counterparts) is also measured by a checkpoint control. Here too, genome duplication is possible if cytokinesis occurs without proper duplication of the spindle pole body. All of these regulatory pathways help keep chromosome number constant. The gross changes seen in tumor cells resemble some of the potential outcomes anticipated by mutation of these checkpoint controls in yeast. Unfortunately, mammalian counterparts for the genes that control these events are still largely unknown.

Even though we don't know the genes that regulate these processes, we can begin to anticipate their characteristics by examining the changes one sees in tumor cells. Here the best examples come from the timing of the development of aneuploidy and the loss of control of gene duplication.

Tlsty and her colleagues report that the loss of p53 through various mechanisms increases the frequency with which cells amplify genes. These frequencies are normally astonishingly low but increase dramatically when p53 is absent. The suggestion of a link between p53 loss and the induction of aneuploidy has also been made by several workers who have correlated the appearance of p53 mutations in tumors with detection of aneuploidy. This correlation in timing must be kept well short of an actual biochemical consequence, because Reid (Neshat et al.) points out that p53 loss occurs at least one year prior to the detection of aneuploid cells in Barrett's esophageal tumors. Nonetheless, the development of genomic instability appears to be a recognizable stage in tumor development that continues the generation of diversity.

Translocations. One consequence of genomic instability is the generation of specific rearrangements that give a strong selective advantage to the affected cell. In most settings, the mechanisms that allow these translocations are not understood. In lymphocytes they may be a rampant VDJ recombination event (Kirsch et al.), but in other cell types the genesis of these translocations is not known. Whatever the driving force, the outcome is a change in the normal regulation of a target gene. Probably the most common rearrangements target transcription factors where the translocations re-assort protein domains to create novel factors that retain their DNA-binding properties but now have lost their regulatory domains (Dalla-Favera et al.; Aster et al.; Rauscher et al.; Liu et al.; Thomas et al.; Nucifora and Rowley). In the majority of these cases, the novel fusion proteins are constitutively active and presumably now drive transcription from key target genes that promote a selective advantage. Other translocations do not create novel proteins but promote deregulated expression of signaling (Afar et al.), transcription (Hurlin et al.), or anti-apoptotic proteins (Yin et al.; Cory et al.). The outcome of these translocations is inappropriate levels, timing, or regulation of key proteins. The function of the various target gene products will provide the selective advantage to developing tumor cells. However, with the exception of aberrant VDJ recombination, we still know very little about the mechanisms that govern the generation of translocations.

New techniques. At this meeting, two relatively new techniques that take advantage of genomic instability were discussed. Here the changes in the genomic makeup of a tumor cell can be used to display differences between the tumor cell and the normal cell. These techniques both demand technically advanced labs, but both promise quicker routes to identifying cancer genes.

Representational difference analysis (RDA) uses PCR technology combined with the hybridization variables to compare two highly related genomes (Lisitsyn et al.). This might be a comparison between a tumor cell and a normal cell from the same person. It requires a fairly skilled practitioner to squeeze the technique just so, but the outcome is a rapid identification of sequence differences between the two genomes. Several labs have complained privately that they have tried and failed to master this technique. However, it is clear that some groups have managed to exploit it usefully, and the results promise to lead us to many new probes to compare these genomes.

The second technique, comparative genomic hybridization (CGH), also relies on hybridization to define differences between a normal genome and one seen in a tumor cell (Gray et al.). Here the readout is determined by fluorescence intensity on a chromosome spread. Regions of 10 Mb or more that have suffered deletion or amplification can be detected. It's a computer-hungry method, but again, the comparisons allow a relatively quick look at how two genomes compare.

Can't Die, Won't Die

During normal development, many cells undergo programmed cell death or apoptosis (Horvitz et al.). Signals for apoptosis apparently occur in tumor development as well. Several oncogenes can induce apoptosis under the appropriate conditions. This has been seen following ectopic expression of nuclear oncogenes such as c-myc, E2F, adenovirus E1A, SV40 large T, or HPV E7 (for examples, see the work of White et al.; Lowe et al.; Symonds et al.; Galloway et al.). Likewise, high doses of ionizing radiation can cause apoptosis (see Neiman et al.; Canman et al.). Many tumor cells escape this selective pressure and no longer are susceptible to apoptosis. This outcome is an example of selection for survival, and its importance in tumor development has only recently been recognized. Here the pressure is not for an advantage for proliferation per se, but rather an avoidance of imposed cell death.

Two mutations have been identified in human tumors that lead to loss of programmed cell death. These are loss of p53 and deregulation of bcl-2.

p53-mediated death. In certain cell types, activation of the p53 pathway leads to apoptosis. This is seen with the activation of many different agents that drive the cell cycle. All of the examples listed above are such cases. Ectopic expression of any of these oncogenes stimulates p53-dependent apoptosis at least in some cell types. Similarly, the apoptosis mediated by high-dose irradiation is also p53 dependent. If p53 is mutated or inactivated in some manner, the apoptotic response is not activated and cell proliferation is allowed. There is still a friendly debate about the type of signaling that underlies these outcomes. Is cell death initiated because of a conflict of signals, one calling for cell cycle progression (by c-myc, for example) and one for cell cycle arrest (perhaps in the p53 setting described above)? Or do the cell cycle progression signals also set up a fail-safe system of cell death that is utilized if other cell survival signals are not present? Both views fit all of the available data.

p53-mediated death has been examined in a vast number of systems, but perhaps the most elegant is in nullizygous mice where the absence of p53 can be studied with as few complicating secondary mutations as possible. For example, examine the consequences of loss of p53 as described in the work of Symonds et al., Williams et al., Cory et al., Wyllie et al., or Lowe et al. In these studies and in many others, the clear link between p53 function and apoptosis can be seen. These studies use different agents to induce apoptosis, with nuclear oncogenes driving cell cycle progression in some instances and irradiation driving it in others. In cases where the apoptosis normally would involve p53, however, the nullizygous cells are spared. If the same events occur in human tumors, affected cells would be spared, rewarding the action of mutations that block apoptosis. One can also imagine that this selection pressure might be one of the major reasons that p53 mutations are so prevalent in human tumors.

bcl-2-mediated death avoidance. A second mutation which occurs in human tumors that helps cells avoid cell death is the activation of the bcl-2 oncogene (Yin et al.; Cory et al.). Translocations that activate this oncogene have been seen in a number of B-cell lymphomas. In this case, it is the expression of the wild-type protein that protects cells from apoptosis. Unlike p53, which is linked to only a portion of cases of apoptosis, bcl-2 expression appears to block all types of programmed cell death. bcl-2 clearly plays a role in development, as loss of this gene in knockout mice changes the ratios of maturing lymphocytes. The biochemical role of bcl-2 still remains elusive, although extensive work over the last several years has identified a large number of related and interacting proteins. It appears that the key signal in deciding whether apoptosis will be blocked or not is the ratio of bcl-2 homodimers to bcl-2 heterodimers, in this case formed with an interacting protein known as bax. Higher levels of bcl-2 homodimers protect cells from programmed cell death. How this apoptosis signal is interrupted or how protection is achieved is still unknown.

Avoiding therapy. Perhaps the most telling and best example of how tumor cells may avoid cell death is seen following chemotherapeutic treatment. Many of the empirically identified antitumor agents appear to elicit their response by activating programmed cell death. These include the DNA-damaging agents and the topoisomerase II inhibitors. When these agents are effective in therapy, they appear to kill tumor cells by apoptosis. When responses are short-lived, the surviving cells are characteristically p53 minus, even if the original tumors were p53 plus. It does not appear that the chemotherapeutic agents are inducing mutations; rather, they are selecting preexisting clones that have already lost their p53 expression. For an example of this type of selection in a mouse model, see the work of Lowe and colleagues.

Puzzles. It is curious that both a G_1 arrest and an apoptosis pathway should rely on p53. The different outcomes appear to depend on which cells are used for the experiments. Here context is the key element, and this raises a key question. What events throw the switch between these two different outcomes? If you took a G_1-arrestable cell and a killable cell and fused them, which phenotype would prevail? Can you convert a dying cell to an arrestable one? Are there ever any survivors?

Another puzzle that has caught my attention came first from a discussion at a recent meeting of the Forbeck Foundation. Several people there asked, "Why aren't there more anti-death mutations?" The selection pressures to avoid cell death must be severe, yet only two types of mutations have been identified to date, and one of these is selected only in the relatively rare B-cell tumors. One would have thought that these pathways would have been dotted with tumor-specific mutations.

Finally, is the cell death that occurs during development functionally equivalent to the cell death that provides the selective pressures in tumor development?

DECISIONS

The three Ds: divide, differentiate, or die. These are the three main decisions that a cell faces in any developmental program. Although the 59th Symposium wasn't specifically devoted to signal transduction, it is impossible to avoid the issue of signaling in the discussion of cancer genetics. Because of the subject of the Symposium, I will try to avoid the classic discussions of how different signaling pathways work and concentrate on how they are mutated in cancer.

There appear to be two main pathways that are frequently mutated in cancer, the ras pathway and the retinoblastoma (RB) pathway. Although molecular epidemiology studies still haven't answered the question, it seems possible that most adult tumors require a mutation in each of these pathways. Certainly, they are commonly targeted in many tumorigenic settings. The pathways are quite different, but the ultimate aims of both are similar. They initiate a change in transcriptional program. As cues for these changes, they use the extracellular environment to initiate their programs. They can be short-circuited in a number of ways. In the ras pathway, the most common mutations are oncogenic, and they act to drive the pathway at inappropriate levels or at inappropriate times. In the RB pathway, we now have examples of both oncogene and tumor suppressor gene mutations, either of which decouple the signaling events from cell cycle timing.

The ras pathway is clearly linked to extracellular signals. In some settings, these signals relay environmental cues about the need to differentiate, and in other settings, these signals induce mitogenesis. The distinction between these two is one of the key questions in the next phase of research on the ras pathway. In contrast, the RB pathway is an example of how the signaling of the cell cycle clock is connected both upstream to extracellular growth factors and downstream to transcriptional changes.

I haven't tried to rehash the individual steps of both pathways here. Below are the views on these two pathways that were prominent in this Symposium.

Economy of Use

Parsimony was the word that stopped me. Not surprisingly, Mike Bishop, or "J. Michael" behind his back, was the orator (Bishop et al.). Parsimony was a word I knew, at least until Bishop used it in his presentation. It stuck in my mind then, and I have turned over the implications it gives to discussions of signaling a number of times since the Symposium. The electronic version of the American Heritage Dictionary now loaded on my computer says "unusual or excessive frugality; extreme economy or stinginess." The O.E.D. claims "carefulness in the employment of money or material resources." Bishop used parsimony in conjunction with his descriptions of how a single signaling pathway could be *plastic*, how it could be used for different purposes in different settings. The economy of use in Bishop's example was a ras pathway that can be used to signal either mitogenesis or differentiation in the PC12 rat pheochromocytoma cell line.

One of the exciting developments of the last several years has been the identification of all of the steps in a pathway that links extracellular environment with transcriptional activation in the nucleus (for elegant updates on how this pathway is put together, see Freed et al.; Schlessinger and Bar-Sagi). A quick look at the ras pathway leaves one with the image of a linear decision tree. Ligand in at one end, transcriptional activation out at the other. Bishop's arguments highlight how this pathway could be used for different purposes. This is an intriguing realization that raises two possible, but nonexclusive, options. Is the biochemistry of ras different in these settings, or is the pathway wired to downstream effectors in different ways? When I heard Bishop's talk, I assumed that the biochemistry of ras was not significantly different in either setting. I suspect that is still probably true; the GTPase signaling through the effector loop of ras probably works similarly in all cases. What I had assumed when listening to the Bishop talk was that the differences lay in the transcriptional programs of the cell. The ras signal just fell on different ears, or more correctly, on different transcription factors, and this changed the outcome of the signaling events. This fit well with the observations reported by Sternberg et al. and Chang et al., where the ras pathway has been implicated in well-established differentiation decisions. Here the cells were wired to respond to a ras signal with poised differentiation genes. In other settings I thought that the signals would fall on proliferation ears. Bishop's plasticity could be explained by different secondary events that changed the transcription program.

The major question will be how those secondary events might be controlled. In the PC12 example, the ras pathway is essential for either outcome, division or differentiation. Since the ras function was common to both outcomes, I assumed that the differences lay outside of the ras pathway. Wigler (Van Aelst et al.) startled me with his talk and helped me expand on the thought process that Bishop had started. Wigler provides genetic evidence for more than one ras target, different effectors that stimulate different downstream events. The same upstream signal splits at the ras step to give different downstream effects. By having one or both of the downstream signaling systems open, the effects of the activation could be radically different. Since the Symposium, this observation has been repeated by a number of labs, and there now may be as many as three ras effectors. Forget linear signaling; context is important not only for the transcription program, but also for many of the individual steps. The ideas of plasticity that Bishop introduced are evident here.

Puzzles. The major puzzle that presents itself following this discussion is, How will we be able to learn about these branches? Here my major hope is that we can learn to incorporate the power of the genetic selections so strongly represented by the reports from Chang et al. and Sternberg et al.

Will all steps in the ras pathway be subject to multiple effectors? Or is ras more important than the others and an unusual player?

What about the other signaling pathways that link the environment with transcriptional activation events? Why haven't they been picked up in cancer epidemiological studies? Are they not as important in carcinogenesis? And will the insulin/IL-4, JAK/STAT, and dorsal pathways be equally complicated? If the idea of a signaling pathway is correct, one would expect that any deregulation should suffice to activate the pathway inappropriately and there should be only one mutation/pathway/tumor.

Cycling

There appear to be two mammalian rate-limiting steps that link extracellular signals to the cell cycle clock and transcriptional programs. One runs through the cyclin D/cdk4 kinase complex and the other through cyclin E/cdk2 (Sherr et al.; Hatakeyama et al.; Roberts et al.). Both are rate-limiting steps for the length of G_1, as judged by experiments that overexpress either the cyclin D or cyclin E and shorten the G_1 interval by 3 hours or more. Of these pathways, only the cyclin D pathway is commonly mutated in human tumors. This pathway now links four different cancer gene products including pRB, p16, cyclin D, and cdk4.

The RB pathway. The retinoblastoma gene has been the prototype for tumor suppressors. Its normal role is to serve as a transcriptional inhibitor by binding and inactivating *trans*-activators such as E2F. When the RB gene is lost in human tumors, these negative steps are lost and key transcriptional events are deregulated. In wild-type settings, pRB is regulated by phosphorylation by cdk such as the cyclin D/cdk4 complex. Although formal proof that the cyclin D/cdk4 complex is the authentic in vivo kinase for pRB is still lacking, the circumstantial evidence is compelling. The details of these arguments can be found in the papers from Hatakeyama et al. and Sherr et al. Upstream regulation of cyclin D/cdk4 activation can be controlled from multiple sources. One source is extracellular signals. Cyclin D expression is activated by growth factor treatment of cells. Here cyclin D formally should be called a delayed immediate early gene, as its expression is blocked by cycloheximide treatment.

One of the most exciting recent developments in cancer genetics has been the identification and cloning of the gene for an inhibitor of the cyclin D/cdk4 kinase.

This protein, known as p16, was originally identified as a polypeptide that interacted with cdk4 in SV40-transformed fibroblasts (Zhang et al.). It was cloned and shown to be a small-molecular-weight inhibitor, specific for cyclin D/cdk4. Recently, several groups reported that the gene for p16 mapped to the region of chromosome 9p21 that was responsible for the melanoma predisposition gene (see Kamb; Okamoto et al.). Although at the time of the Symposium there was still some question, it now appears likely that p16 is a tumor suppressor gene that is commonly mutated in a number of different human tumors. At present, the exact percentage and types of tumors that are affected are not completely established. It also appears that at least in some cases, mutations in p16 can account for the predisposition to melanoma.

Several interesting findings stem from this work. First, the RB pathway is now littered with cancer genes. p16 and pRB are tumor suppressors; cyclin D and cdk4 are commonly amplified in tumors and score as oncogenes in various assays. At the time of this writing, the rumor mill is rife with stories that E2F has been shown to be mutated in human tumors, and it has already been demonstrated that E2F scores well in standard cooperating oncogene assays.

In addition, in at least the cases that have been reported to date, only one of these genes is mutated in any one cancer, arguing for a linked pathway that is commonly lost in human tumors. If all of the percentages for mutations of these genes are correct, it appears that this pathway may be mutated in the majority of human tumors.

Puzzles. What does the RB pathway do? We now have collected an expansive biochemical picture of pRB function and the pathway in which it acts, but we still know very little about what this pathway signals. Here the best data come from the knockout mice (Williams et al.). Mice without a functional copy of the RB gene develop normally until day 13.5 of gestation. By this time, the embryos have undergone extensive cell division and differentiation, so the RB pathway can't be an integral part of the cell cycle engine. At about E13.5, RB-minus embryos appear to die from lack of oxygen caused by failures of erythrocyte differentiation. These animals also fail to undergo proper differentiation of neural cells. In addition, muscle myotubes from mutant animals appear to lack some of the negative controls that keep them from entering the cell cycle. What events do the proteins of the RB pathway control? Until we know the transcription targets of this pathway we cannot know for sure, but the clues suggest that the pathway has some role in terminal differentiation.

What is the molecular switch that represents "restriction point?" Restriction point remains one of the most interesting enigmas of mammalian biology. Defined in the mid-1970s by Pardee, restriction point is the time at which a mammalian cell becomes committed to proceed through the cell cycle (Pardee 1974). It marks the end of the dependence on extracellular signals. Cells that make it past this time are locked into progression to S phase and the remainder of the cycle. Before this time, cells may pause or reset their cell cycle clock when changes in extracellular signals occur. The identification of the molecular switch for restriction point has eluded us. The time when cells become committed to S phase seems to correspond best with the activation of the cyclin E/cdk2 kinase, but there does not seem to be a functional method to link the activation of cyclin E/cdk2 with restriction point. (See Roberts et al. and Hatakeyama et al. for excellent discussions of this still confusing event.)

Tumor Suppressors with Unknown Signaling Roles

One the most exciting sessions at this meeting featured talks on the cloning of BRCA-1 (Ostermeyer et al.; Weber et al.). At the time of the Symposium, the gene was not cloned, and BRCA-1 was probably the most sought-after human gene. These talks were a wonderful chance to watch the drama unfold. This paragraph serves just as a postscript to the cloning. Mutations in BRCA-1 appear to be responsible for approximately 5% of all breast cancer, and about 50% of all inherited cases. At the moment, it appears that it is seldom if ever mutated in sporadic cases. The protein product is large, the RNA is subject to a large number of differential splicing events, and the sequence gives no clues to its possible role. In the time since the Symposium, it has moved from being the hottest cloning project going to a tumor suppressor protein in search of a function.

After all the hard work that goes into cloning a tumor suppressor gene, it seems a little unfair that one then must learn how its loss contributes to carcinogenesis. p53, pRB, WT1, NF1, and p16 have been simple. The harder ones are DCC, APC, and now BRCA-1. Here we start from scratch. Where in the cell are they? Who are their partners? Are there any clues from the primary sequence?

Some of these genes have begun to fall together in such a way that they may give the first clues of a signaling event. The APC, E-cadherin, and α-catenin proteins all are beginning to appear in the same aspects of cell-to-cell communications. In this Symposium, some of the problems were discussed by Issacs (Isaacs et al.) as he considered the loss of E-cadherin and α-catenin in prostate tumors. APC has been shown to bind to another member of the catenin complex, β-catenin. The catenins have also been identified in flies as one of the downstream targets of the wnt oncogenes. Although this work is at its earliest stages, there now appears to be a growing suggestion that these proteins may be part of the first example of a cell-to-cell signaling pathway that is disrupted in human cancer.

Puzzles. Given the strong link between the loss of some tumor suppressor genes, particularly p53, and cancer development, why don't we have more copies of these genes? Obviously, as a species there hasn't been evolutionary pressure to duplicate these genes, but for the individual there might be tremendous advantages. Could we test this? Would an extra p53 locus either save mice from some tumor development or allow a broader range or lower doses of chemotherapeutics to be used?

A little personal note. Several speakers at the Symposium put forth the suggestion that tumor suppressor gene products should be thought of as gatekeepers, that they played a special role as regulators. From the available data, I would say that this inflates their roles. Any negative-acting protein that is functioning in a rate-limiting step will appear as a gatekeeper when it's lost. Loosing a gate, dam, brake, or even a small governor will appear catastrophic if you are standing just downstream. Tumor suppressors are just mutations, and along with the oncogenes they are like us: Some are a little pushy and some are in the way. It's tempting to get carried away just because tumor suppressors are hard to clone.

TUMOR BIOLOGY

Over the past 20 years, we have amassed an impressive body of knowledge about the physiological changes that follow individual cancer gene mutations. This research has also forced us to explore the normal roles of proto-oncoproteins and tumor suppressor proteins. In many ways, this combination of work has laid the groundwork for all of modern mammalian biology. At some point, however, the interplay between these changes needs to become our major focus. Here, we need to establish reproducible tumor systems that will allow examination. For the molecular geneticists, what is the range of the mutations that contribute to tumorigenesis, and have we missed whole classes of mutations because of our limited experimental systems? For the molecular epidemiologists, how do the mutations that we know affect the patterns of tumor development? For the tumor biologists, are there rules to these tumorigenic changes?

The ultimate test for all of these ideas will be human tumorigenesis. Two clear settings of tumor development have been discussed at this Symposium. They are colon cancer and Barrett's esophageal carcinoma (Vogelstein and Kinzler; Neshat et al.). They have similar technical advantages. The developing tumors can be seen without surgery. The developing changes can be recognized without elaborate histochemical tests, and when needed, the tumors and surrounding tissues can be sampled. The colon carcinoma studies have produced our first good view of the molecular genetics of tumor development. Although this view of carcinogenesis may not be radically different from that predicted by the epidemiologists, the stages now have molecular events that can be tied to them.

There have been some surprises. At least for several of the mutations, the order is not mandatory. For example, mutations in p53 normally occur late, but this is not required. In contrast, other orders seem to be critical. When small microscopic hyperplastic foci in the colon are examined, they can show mutations in ras or loss of APC. When a slightly later stage is examined, looking at dysplastic foci, all show loss of APC and none have ras mutations. This suggests that the loss of APC early in tumor development is a strong prognostic indicator of later developments, and it suggests that in some cases the molecular order will be critical to understand.

These examples of human cancers will continue to be essential to our understanding of tumor development. However, the numbers of tumors that will be accessible to these types of studies will be limited. In addition, the work will always be correlative. Models where these questions can be asked in experimental settings are essential to broadening our knowledge in this area.

Mouse Models for Tumor Development

Perhaps one of the most important areas of research for the next decade of cancer research will be mouse models for tumor development. In their early incarnations, these systems were proving grounds that were used to establish the validity of a particular oncogene. In more recent times, they have become one of the only methods to examine and dissect the intricate issue of tumor development in a controlled setting. At the Symposium we heard from several masters of the art. The chief problem is how to combine the multiple steps found in tumorigenesis either to learn of new mutations or to study the requirements of tumor development.

There seem to be several useful strategies, but all of them start with one common theme. Mice are engineered to be predisposed to cancer. This could be done by using a mouse carrying a deficiency in a tumor suppressor gene (Dove et al.; Williams et al.) or by expressing a known oncogene through transgenics (Naik et al.; Cory et al.; Berns et al.; Kemp et al.; Varmus et al.; Symonds et al.). Tumor progression can be allowed to develop naturally (Williams et al.; Naik et al.; Symonds et al.), helped along by mutagenesis (Kemp et al.; Cory et al.; Berns et al.) or by mating to mice carrying other mutations (Dove et al.; Varmus et al.; Williams et al.; Symonds et al.). In all cases, the tumorigenic setting is reproducible.

I highlight just a few of the ways in which these systems have been exploited. This isn't an exhaustive list, but the individual papers can be explored for more detailed information. One outcome of this work is the identification of new oncogenes. These seem easiest to find when the mutagen allows the easy identification of the affected gene. This is clearest in the case when the mutation is driven by insertion of a retrovirus. In these cases, the genes are easily cloned, identified, and studied further. Key in these studies is the demonstration that these genes are true oncogenes and, particularly, that they appear as oncogenes in human tumors that mimic the systems used in mice.

Another important finding from these systems has been the validation of aspects of normal tumor development: the expression of survival factors such as IGF-2 that rescue cells undergoing apoptosis; in an analogous finding, the rescue of apoptotic cells by the loss of p53; the obligatory loss of the wild-type allele of tumor suppressor genes during tumor development; and the increased probability of tumor progression in mice carrying alleles that change the penetrance of the original mutation.

All of these aspects of tumor development need model systems in which they can be studied. These types of systems appear to be the only methods to explore in detail the complicated events of tumorigenesis.

Angiogenesis and Metastasis

Similar to the selection for enhanced proliferation, angiogenesis and metastasis follow a multistep selection process. Of these mutations, we know only the rudiments, but they promise to be every bit as complicated as the selection process for proliferation.

A nonvascularized tumor will never grow beyond the size of about 1 mm^3. Angiogenesis then is an essential step in the development of all solid tumors, and without this step, little damage will be done by the primary tumor or by the metastases. In the mouse tumor model described by Hanahan (Naik et al.), angiogenesis is an essential step to allow large tumor growth, and in most other studies angiogenesis has been shown to be a necessary precursor to metastasis. Neovascularization is controlled at least in part by the secretion of both positive and negative angiogenic or anti-angiogenic factors. Positive factors include members of the FGF family, and negative factors include proteins such as thrombospondin (O'Reilly et al.; Dameron et al.). The expression of angiogenic factors is under both positive and negative transcriptional control. This includes the expression of the angiogenic factors themselves as well as several negative factors that inhibit neovascularization.

Most life-threatening problems that develop from solid tumors only are seen following metastasis. The development of the ability to metastasize clearly involves a large set of genetic changes, but the mutations that drive this process have been difficult to identify and to clone. There have been sporadic reports of the identification of genes that are important in these changes, but little has come from these studies. The major hurdle must be the ability to set up a model system that faithfully mimics the genetic selection that leads to metastasis. Two reports from this Symposium heralded some advances in this difficult quest. Rong and Vande Woude reported the ability to make NIH-3T3 cells that now could invade local tissue and metastasize. This was achieved by expressing both the met oncogene, a tyrosine kinase transmembrane receptor, and the met ligand, hepatocyte growth factor. This autocrine setting enables the noninvasive 3T3 cells to acquire properties of fully metastatic clones. When human mesenchymal tumor cell lines were tested, they were found to overexpress the met receptor. Here may be a setting in which a reproducible model for developing metastasis can be established.

Another approach to establishing a model of late stages of tumor development was suggested by Berns et al. They have used transgenic mice to establish reproducible tumor development models, but as has been the problem in other cases, the mouse life span is not sufficiently long to allow the selection of late-stage changes. Berns suggests the use of retroviral insertion to speed this process and has coupled this with syngeneic transfer of the tumor to younger animals. This has already allowed some later stages to be examined and may provide an approach to study complicated developmental problems such as metastasis.

Therapy

"SUPER DOC PREDICTS CANCER JAB." It's 1982. Imagine this banner headline in a London tabloid. The heralded event was the recent discovery that the v-sis oncogene was a mutated version of the PDGF gene. It marked my first encounter with the press, and we all had a nice laugh about it. But under our laughter and the furor that the headline caused was the real and honest need for progress in cancer therapy. It's a daunting, and at times a frustrating, realization that progress in cancer therapy will follow on the successes of cancer genetics, that the successes discussed at this Symposium will be the foundations for meaningful therapy.

At this Symposium we heard examples of potential new therapies at three different stages of development. Folkman (O'Reilly et al.) described the isolation and characterization of an angiogenesis inhibitor that may be used to block neovascularization in metastases. It has been called angiostatin and is a fragment of TPA. They found the inhibitor as a naturally occurring molecule secreted from a primary tumor that kept distant metastases from growing. When the primary tumor is removed, the metastases grow quickly, now that angiogenesis is no longer inhibited by angiostatin. The potential uses seem realistic. It's a naturally occurring protein, so there shouldn't be immune response problems. The action was originally detected in a setting that validates its effectiveness. Apparently surgeons see this response occasionally when they remove the primary tumor. Here, unrecognized distant metastases that then proliferate quickly after the surgery. Potential problems may be in the number of tumors that respond, the side effects of inhibiting neovascularization systemically, and the new selective pressures placed on the metastases. Presumably, these tumors will develop new mutations to avoid the effects of angiostatin.

A second area of therapeutic development relies on small-molecular-weight compounds to block the action of tumor gene products. One of these approaches targets the isoprenylation of ras. An essential modification that occurs on ras is a fatty acid acylation that helps direct ras to the plasma membrane. Several research groups have reported the isolation of compounds that block this step. The drugs are now in animal model testing but have shown surprisingly good results in tissue culture. The major worry must be specificity. Why should ras modifications in a tumor cell be different than in a normal cell? The answer may be that it isn't, but that the tumor cell relies on the mutated ras so heavily that appropriate doses of drugs can be established that distinguish tumor from normal cells.

The third group of therapeutic approaches relies on gene transfer (Boon et al.; Culver et al.; Brenner et al.; Nabel et al.). Several variations on this approach were discussed. Make the tumor cells more immunogenic by introducing vectors that express cell-surface genes. Make the tumor cells sensitive to a drug by expressing a selectable marker. Alter normal nontumorigenic cells that might be damaged by conventional therapy and then use higher doses of chemotherapy. Two technical hurdles have hampered these approaches. At present, the percentage of target cells that can be forced to express the desired gene is low, and it has been difficult to keep the expression of genes activated for long-term studies. If either or both of these problems can be overcome, the promise of gene therapy may be realized.

CHANGES

It has been about 15 years since the Cold Spring Harbor Symposium covered a subject directly related to cancer. This span of years has seen a number of major changes. Given the importance of these Symposia, it may be worthwhile to highlight just a few of the major changes.

The last 10 years have seen the emergence of tumor suppressor genes as important players in cancer genetics. The swell of interest in tumor suppressor genes began with the early insights of Harris et al. (1969) and Knudson (1971) in the early 1970s, the waves grew with the first tumor suppressor gene clonings in late 1986 and 1987, the first waves broke in 1988, 1989, and 1990, and we have been riding a long curl since then. These studies have been monstrous undertakings, requiring extensive collection of patients and often massive family histories, spectacularly intensive cloning efforts, and lucky breaks to identify the biochemical functions. However, these studies have brought us about 15 new cancer genes that, as judged by the frequency of their mutations in human tumors, are very important in tumor development. These studies have also established the need to view cancer genetics as the combination of gain-of-function and loss-of-function mutations. This importance of the tumor suppressor gene is not a radical change. It might even have been predicated by the genetic nature of the disease, but these developments now round out the picture of possible genetic changes.

Over the last 5 years, our views on signaling have changed radically. In addition to the large number of new proteins that have been added to the pathways, our concepts of signaling have changed. Where signal transduction was once seen as one enzyme, often a kinase, talking to the next enzyme in a rather loose, amplifying, step-by-step procedure, we now see solid-state signaling as commonplace. Physical interactions gate the amplitude of signals and insulate them from undesired cross-talk. Both in the recognition of large signaling complexes and in stepwise complex assembly, we have found that signaling is not always an untethered process. This change has come easily and with little fanfare, but nevertheless, it marks a new way of thinking about how signals originate and are processed.

The last several years have seen the excitement generated in the signal transduction field by the identification of all of the steps necessary to connect an external signal with transcriptional activation in the nucleus. This pathway proceeds through numerous steps and utilizes a number of different classes of signaling molecules. The pathway runs from peptide ligand to transmembrane tyrosine kinase receptor, SH2-docking protein, GTP exchange protein, small G-protein, a series of serine/threonine kinases, and finally to phosphorylation and activation of transcription factors. This pathway has been the proving ground for many of the ideas on signaling and now provides our first overview of how signaling occurs in a mammalian system. Many proteins in this pathway are known to cancer researchers as well-studied products of proto-oncogenes. One of the most exciting aspects of the work on this pathway has been the realization that this pathway, and probably many others, will be accessible to study in genetically manipulatable organisms such as the fly and worm. The combination of the biochemical studies in mammalian cells and the genetics in these organisms promises to provide an exciting pairing for future work. The signaling pathway described above will be the paradigm for these studies.

One of the most obvious changes in our views of cancer genetics has come over the last several years. This was the major change seen at this Symposium, the importance of fidelity systems in tumor development. Although many investigators have pointed out how important the loss of these systems must be in carcinogenesis, this message had been largely ignored. The message has finally sunk in. Most tumors derive a good deal of their diversity from mutator phenotypes. These studies are still in their early stages, but I think it is clear that they will continue to make a major impact on the way we view tumorigenesis.

CHALLENGES

During the writing of this summary, I have tried to list several of the surprises and puzzles that have occurred to me. I suspect the reader has made his or her own list of issues that I have missed or misinterpreted. However, there have been a few general items that I believe will be important challenges for further studies.

The first is the surprising number of times I have had to note that mutations have selective advantages only in certain tissues. These genes are likely to be mutated frequently in all tissues, but they appear in tumors only from certain sites. All cancer genes show this property, and in some cases the tissue specificity is remarkable. RB mutations are never found in colon carcinomas. Other adult tumors frequently lose RB, but never here. Inherited predisposition to loss of mismatch repair, although thought to affect DNA repair in all tissues, is a selective advantage for tumor development only in a small number of tumors of the colon, GI tract, ovary, and endometrial mucosa. Ras and p53 are apparently present and active in all cells, yet mutations are not seen in tumors from all tissues. Although some aspects of tissue specificity can be expected, I am surprised to find that tissue specificity is the rule rather than the exception.

A second problem concerns feedback loops. In every case where we understand the biology of a signaling pathway in any detail, there are important feedback loops that control the effective strength of the pathway. If these feedback or feed-forward loops are universal, it may mean that rational therapies based on specific mutations are doomed to failure or at least to very difficult dosing problems. I see no quick way to assess this issue short of empirical tests.

Perhaps one of the most difficult problems, but one that must be a key to understanding human tumorigenesis, is the identification of alleles that predispose to cancers but only with a low penetrance. The field has had remarkable success in identifying and cloning genes with high penetrance. Family studies and large-scale cloning efforts have made this possible. However, it seems clear from the literature and just from a casual understanding of the genetics that one of the biggest sources of variation will come from genes that predispose to tumor formation only weakly and will not drive tumorigenesis in every generation. This promises to be a large number of genes, but genes that will be particularly difficult to identify. It seems likely that identification of these genes will more likely come from detailed biochemical studies rather than human genetics. An important step will be a method to test for mutations of such low penetrance.

I also noted several technical challenges that need to be addressed. The first two are signaling problems that need solutions. We need methodologies that allow researchers to identify the in vivo target of a kinase or a transcription factor. As a field, we are relatively competent to move upstream in a pathway, but we are poorly equipped to look for targets. We can ask "Can a kinase phosphorylate this sequence?", but we are hard pressed to determine what are legitimate in vivo substrates. Similar problems dog our studies of transcription factors. We ask, "Can this factor bind this sequence?", not "What does it bind in the cell?" If these questions can't be answered genetically, we need sensitive ways to freeze a cell and find the authentic in vivo partners.

Last, we have an immense information problem. We need a way to store our knowledge in a reliable and useful format. We need to build an electronic model of the cell, perhaps of many different cells. Storage of this amount of relevant information in dusty journals will not satisfy the needs of the next 10 years. The storage methods used for genome-based information are just not a useful format to study a complex, multifactoral human disease like cancer. This information must be made available in a form that reflects function. It's an old lesson, but it still applies.

An Ending

I began this summary with a vignette of a patient with metastatic colon carcinoma. This served as a method to contrast the different worlds of clinical care and basic cancer research. This comparison raised two questions for me. One was whether during the course of the Symposium or in writing this review I could find signs of how this gap could be bridged. The goal is to understand the genetic changes in tumor cells and design therapies that exploit these differences. This is a difficult problem, but there are early signs of progress. As highlighted throughout this meeting, our knowledge of cancer genes continues to grow. There are several pharmaceutical and biotech companies that have taken up the charge of making drugs that target these changes. There are excellent connections between the clinic and the bench in the fields of molecular epidemiology and diagnostics where the information flows in both directions. Also, there is a trend to open the lines of conversation between the two worlds. Informal discussions and small meetings have begun to spring up to help these communications. Most encouragingly, I know of several prominent members of both fields who have personally stepped over the gap. Senior basic scientists have donned white coats and spent time in the clinic, and prominent oncologists are taking sabbaticals in basic research labs. All of these signs are encouraging, not signs of sure success, but at least a narrowing of the gap.

The second question raised by the introduction was strictly a personal one. By the time the summary was finished, would I be depressed by all we didn't know? The answer is no, I'm not. In fact I feel quite enthusiastic. The boundaries of the field of cancer genetics seem to be coming into view. I am convinced we can clone all of the cancer genes. Most of the consequences of these mutations are either known or can be learned. Several excellent tumor development models are either available or can be established. And smart people have begun to work on rational therapies. I'm not sure when the next opportunity will come for someone to write a piece like this, but I am certain that they will be amazed, as I have been, in how much progress has been made. We continue to make impressive strides, and with each one a little more of the puzzle is solved.

ACKNOWLEDGMENTS

I thank all of my colleagues at the MGH Cancer Center and elsewhere for their many challenging and enlightening conversations. In particular, I am indebted to Lee Hartwell, Doug Hanahan, Anton Berns, Steve Friend, Eric Fearon, and Frank McCormick. I have stolen unashamedly from our discussions; thank you for your time and likely conversions. Nick Dyson and Josh LaBaer were kind enough to read and comment on a late version of this text on very short notice.

I thank Cold Spring Harbor Laboratory for continuing to support meetings such as the Symposium. For me, this is the granddaddy of them all (gta-1). It holds a special place in all of biology. I was pleased and excited to be part of this Symposium. Finally, my thanks go to everyone at the Cold Spring Harbor Laboratory Press, and particularly to Joan Ebert, for their personal support during the writing of this summary. Their help, understanding, and friendship has been most encouraging. During the course of this work I have been supported by research grants from the National Institutes of Health and by an American Cancer Society Research Professorship.

REFERENCES

Cairns, J. 1975. Mutation selection and the natural history of cancer. *Nature* **255:** 197.

Harris, H., O.J. Miller, G. Klein, P. Worst, and T. Tachibana. 1969. Suppression of malignancy by cell fusion. *Nature* **223:** 363.

Knudson, A., Jr. 1971. Mutation and cancer: Statistical study of retinoblastoma. *Proc. Natl. Acad. Sci.* **68:** 820.

Loeb, L.A., C.F. Springgate, and N. Battula. 1974. Errors in DNA replication as a basis of malignant changes. *Cancer Res.* **34:** 2311.

Nowell, P.C. 1976. The clonal evolution of tumor cell populations. *Science* **194:** 23.

Pardee, A.B. 1974. A restriction point for control of normal animal cell proliferation. *Proc. Natl. Acad. Sci.* **71:** 1286.

Author Index

A

Abdallah, J.M., 287
Abel, K.J., 531
Acton, D., 435
Adams, J.M., 365
Afar, D.E.H., 589
Alcorta, D., 411
Aldaz, C.M., 491
Alkema, M., 435
Allen, J., 435
Aloni-Grinstein, R., 225
Ambudkar, S.V., 677
Anisowicz, A., 537
Annab, L.A., 411
Appeldoorn, E., 317
Aran, J.M., 677
Arany, Z., 85
Arlett, C.F., 317
Aster, J., 125
Aurias, A., 555
Ayer, D.E., 109

B

Bacchetti, S., 307
Bader, S., 565
Baker, S.M., 331
Balmain, A., 427
Baltimore, D., 125
Bar-Sagi, D., 173
Barak, Y., 225
Bardeesy, N., 419
Bargonetti, J., 207
Barrett, J.C., 411
Beach, D., 21
Beach, D.H., 49
Benjamin, L.E., 137
Bennett, W.P., 49
Berns, A., 435
Bertness, V.L., 287
Bigner, S.H., 637
Bird, C.C., 403
Bishop, D.T., 331
Bishop, J.M., 165
Bodine, D., 547
Bodis, S., 419
Bollag, R.J., 331
Boon, T., 617
Bootsma, D., 317
Bouck, N., 483
Bova, G.S., 653
Bowcock, A., 349
Bradley, A., 491
Bradley, G., 607
Brenner, M.K., 691
Briot, A., 265
Brody, L.C., 531
Bronner, C.E., 331
Brooks, J.D., 653
Brown, K., 427
Burn, J., 331
Burns, P.A., 427
Bussemakers, M.J.G., 653

C

Calzone, K., 531
Campisi, J., 67
Canman, C.E., 277
Cao, Y., 471
Capobianco, A.J., 165
Carbone, D.P., 565
Cardarelli, C.O., 677
Carder, P.J., 403
Casares, S., 339
Casso, D., 21
Castilla, L., 531
Cechova, K., 117
Chaganti, R.S.K., 117
Chaganti, S., 117
Chamberlin, H.M., 155
Chandrasekharappa, S.C., 531
Chang, A., 215
Chang, A.E., 699
Chang, C.-C., 117
Chang, H.C., 147
Chapman, P., 331
Chen, C., 471
Chen, C.-Y., 277
Chen, J., 215
Chen, P.-L., 97
Chen, W., 117
Chen, Z., 537
Chiou, S.-K., 395
Chipoulet, M., 317
Christofori, G., 459
Clarke, A.R., 403
Cohen, L., 589
Collins, C., 645
Collins, F.S., 531, 547
Collins, S., 31
Connelly, C.S., 501
Copeland, N., 331
Corcoran, L.M., 365
Cory, S., 365
Couch, F.J., 531
Counter, C.M., 307
Cowan, D.S., 577
Cripps, K.J., 403
Culver, K.W., 685

D

Dalla-Favera, R., 117
Dameron, K.M., 483
Davi, F., 125
De Backer, O., 617
De Plaen, E., 617
De Smet, C., 617
Dejong, P., 555
Delattre, O., 555
Demers, G.W., 297
Demetrick, D.J., 49
DePinho, R.A., 449
Desmaze, C., 555
Dimri, G.P., 67
Domen, J., 435
Donehower, L.A., 491
Dove, W.F., 501
Dowd, P., 523
Dowdy, S., 573

Dowdy, S.F., 1
Doyle, H.J., 165
Durfee, T., 97
Dyson, N., 75

E

Earabino, C., 331
Eckner, R., 85
Egly, J.-M., 317
Eisenman, R.N., 109
Ekstrand, B.C., 637
Enders, G.H., 75
Erba, H., 125
Ewel, A., 331
Ewen, M., 85
Ewing, C.M., 653

F

Farmer, A., 573
Farmer, G., 207
Fearon, E.R., 637
Feinberg, A.P., 357
Ferrari, E., 207
Finan, P.J., 331
Finney, R.E., 165
Firpo, E., 31
Fishel, R., 331
Fisher, D.E., 419
Fitzgerald, M., 673
Fleming, J., 565
Folkman, J., 471
Forrester, K., 49
Foster, S.A., 297
Fox, B.A., 699
Frank, T.S., 531
Frebourg, T., 673
Fredericks, W.J., 137
Freed, E., 187
Friedlander, P., 207
Friedman, L.S., 523
Friend, S.H., 673

G

Galipeau, P.C., 577
Galloway, D.A., 297
Gao, X., 699
Garber, J., 331
Gardner, R.L., 501
Garvik, B., 259
Gazdar, A.F., 349, 565
Gerwin, B., 49
Gledhill, S., 403
Godley, L.A., 491
Godwin, A.R., 331
Goga, A., 589
Golden, A., 155
Goletz, T.J., 59
Gordon, D., 699
Gottesman, M.M., 677
Gottlieb, E., 225
Gould, K.A., 501
Graham, C.H., 661

Grandori, C., 109
Gray, J.W., 491, 645
Greaves, M.F., 403
Greenblatt, M.S., 49
Greider, C.W., 307
Griffith, J., 331
Griffiths, S., 403

H

Hagiwara, K., 49
Halbert, C.L., 297
Hale, M., 287
Hall, N.R., 331
Hanahan, D., 459
Hannon, G.J., 21
Harley, C.B., 307
Harlow, E., 75, 709
Harris, A.W., 365
Harris, C.C., 49
Harrison, D.J., 403
Hartwell, L., 259
Hasserjian, R., 125
Hatakeyama, M., 1
Henderson, I.C., 645
Hendrix, M.J.C., 537
Hengartner, M.O., 377
Herrera, R.A., 1
Heslop, H.E., 691
Hirsch, H., 617
Hirsch, K.S., 307
Hirte, H.W., 307
Ho, P.P., 531
Hoang-Xuan, K., 555
Hoeijmakers, J.H.J., 317
Holmgren, L., 471
Hong, F., 97
Hooper, M.L., 403
Hoover, I., 673
Horvitz, H.R., 377
Housman, D.E., 419
Howley, P.M., 237
Hu, G., 637
Huang, L., 699
Huang, L.S., 155
Hubscher, U., 207
Huibregtse, J.M., 237
Hung, J., 349, 565
Hupp, T.R., 195
Hurlin, P.J., 109
Hussain, S.P., 49

I

Iggo, R., 673
Ionov, Y., 339
Isaacs, W.B., 653
Ishioka, C., 673
Isola, J., 645

J

Jacks, T., 1, 247, 365, 419, 449
Jaspers, N.G.J., 317
Jayaraman, L., 207

Jenkins, N., 331
Johnson, L.A., 357
Jongeward, G.D., 155
Jonkers, J., 435
Juven, T., 225

K

Kadyk, L., 259
Kalikin, L.M., 357
Kallioniemi, A., 645
Kallioniemi, O.-P., 645
Kamb, A., 39
Kane, M.F., 331
Karim, F.D., 147
Kastan, M.B., 277
Kato, J., 11
Katz, W.S., 155
Kayne, P.S., 155
Kemp, C.J., 427
Kerbel, R.S., 661
Kim, N.W., 307
King, M.-C., 523
Kinzler, K.W., 517
Kirsch, I.R., 287
Kishimoto, Y., 565
Kobayashi, H., 661
Koff, A., 31
Kolodner, R.D., 331
Kongkasuriyachai, D., 573
Korsmeyer, S.J., 387
Krall, L., 247
Krance, R., 691

L

Laird, P.W., 435
Lamsam, J.L., 685
Lane, D.P., 195
Lane, W.S., 471
Leach, F.S., 585
Lee, C.H., 607
Lee, J., 155
Lee, M.-H., 277
Lee, M.K., 523
Lee, S., 331
Lee, W.-H., 97
Lees, J.A., 75
Lehmann, A.R., 317
Lesa, G., 155
Lescoe, M.K., 331
Levine, A.J., 215
Levine, D.S., 577
Li, C., 691
Lin, H.-J., 395
Lin, J., 215
Lindblom, A., 331
Linette, G.P., 387
Ling, V., 607
Link, C.J., Jr., 685
Lipford, J., 331
Lipkowitz, S., 287
Lisitsyn, N.A., 585
Liskay, R.M., 331
Lista, F., 287
Liu, J., 155
Liu, P., 547
Livanos, E., 265
Livingston, D.M., 85
Lo Coco, F., 117
Loftus, T., 573

Lombardi, D.P., 287
Loring, G., 509
Louie, D.C., 117
Lowe, S.W., 419
Luo, B., 125
Luongo, C., 501
Lupo, A., 225
Lurquin, C., 617
Lynch, E.D., 523

M

Makela, T., 1
Malkhosyan, S., 339
Man, M.S., 661
Mancini, M.A., 97
Massagué, J., 31
Mathis, M., 349
Matsuoka, M., 11
McClatchey, A., 419
McCormick, F., 187
McKinley, D., 531
McLaughlin, J., 589
McMahon, M., 165
Medina, D., 491
Mellado, W., 117
Melot, T., 555
Merajver, S.D., 531
Merchant, E., 331
Merel, P., 555
Metz, T., 365
Mezzina, M., 317
Migliazza, A., 117
Mills, A., 573
Minna, J.D., 565
Moberg, K.H., 75
Mohr, R.N., 589
Moorman, D.W., 685
Morgenbesser, S.D., 449
Moroni, C., 617
Morris, J.F., 137
Morris, R.G., 403
Morrison, P., 331
Morton, R.A., 653
Moser, A.R., 501
Moses, M., 471
Muldoon, R.R., 685

N

Nabel, E.G., 699
Nabel, G.J., 699
Nagase, H., 427
Naik, P., 459
Nakamura, H., 645
Neiman, P.E., 509
Neshat, K., 577
Ni, B., 677
Nilson, T., 691
Niu, H., 117
Nordenskjold, M., 331
Nucifora, G., 595

O

O'Neill, E., 673
O'Neill, E.M., 147
O'Reilly, M.S., 471
Offit, K., 117
Ohtsubo, M., 31

Okamoto, A., 49
Oltvai, Z.N., 387
Oren, M., 225
Orth, K., 349
Ostermeyer, E.A., 523

P

Parsa, N.Z., 117
Pastan, I., 677
Paulsen, R.M., Jr., 685
Pavletich, N., 207
Pear, W., 125
Peinado, M.A., 339
Pelletier, J., 419
Pemberton, P., 537
Pereira-Smith, O.M., 59
Perucho, M., 339
Piatyszek, M.A., 307
Pierceall, W.E., 637
Pinkel, D., 491, 645
Plautz, G.E., 699
Plougastel, B., 555
Polyak, K., 31
Poulose, B., 265
Preston, G., 411
Priestley, A., 317
Prives, C., 207
Prowse, K.R., 307
Purdie, C.A., 403
Pyle, R., 491

Q

Quelle, D.E., 11

R

Rafidi, K., 537
Rak, J., 661
Ramel, S., 577
Rao, L., 395
Rao, M.R.S., 331
Rao, P.H., 117
Rauscher, F.J., III, 137
Reale, M.A., 637
Rebay, I., 147
Reid, B.J., 577
Remington, L., 247, 419
Riley, D., 97
Rill, D., 691
Robbins, S.M., 165
Roberts, J.M., 31
Roberts, M., 691
Roelofs, H., 265
Rong, S., 629
Rooney, C., 691
Rosenthal, R.A., 471
Rösl, F., 623
Rotter, V., 225
Rouleau, G., 555
Roussel, M.F., 11
Rowell, S.E., 523
Rowley, J.D., 595
Roy, S., 491
Rubin, G.M., 147
Ruggieri, R., 187
Ruley, H.E., 419
Russell, K., 297
Ryan, K., 537

S

Sabbatini, P., 395
Sáenz-Robles, M., 247
Sage, E.H., 471
Sage, M., 265
Sager, R., 537
Sambrook, J., 349
Samuels, M.L., 165
Sanchez, C.A., 577
Sanson, M., 555
Sarasin, A., 317
Sawyers, C.L., 589
Schaeffer, L., 317
Scheffner, M., 237
Schlessinger, J., 173
Scott, M., 125
Seidel, N., 547
Sekido, Y., 565
Sellers, W., 85
Serrano, M., 49
Shaham, S., 377
Shay, J.W., 307
Sheng, S., 537
Sherr, C.J., 11
Shi, Y.-P., 491
Shing, Y., 471
Shiseki, M., 49
Shoemaker, A.R., 501
Shu, S., 699
Sklar, J., 125
Smith, C., 691
Smith, J.R., 59
Smith, L.G., 331
Solomon, N.M., 147
Sotiropoulou, G., 537
Speck, N., 547
Spillare, E.A., 49
St. Croix, B., 661
Stanbridge, E., 339, 573
Starz, M.A., 75
Stefanini, M., 317
Stenman, G., 537
Sternberg, P.W., 155
Stokke, T., 645
Strasser, A., 365
Sugimoto, Y., 677
Sugio, K., 565
Summers, J., 509
Swisshelm, K., 537
Symonds, H., 247
Szabo, C.I., 523
Szikora, J.-P., 617

T

Tainsky, M.A., 483
Tanner, M., 645
Tannergard, P., 331
Tarlé, S., 547
Taylor, I.C.A., 491
Therrien, M., 147
Thomas, G., 555
Thomas, S.J., 509
Thompson, J.S., 357
Tlsty, T.D., 265

U

Ueng, Y.-C., 97

AUTHOR INDEX

V

Van Aelst, L., 181
Van den Eynde, B., 617
van der Bruggen, P., 617
van der Lugt, N., 435
Van Dyke, T., 247
van Lohuizen, M., 435
van Vuuren, A.J., 317
Vande Woude, G.F., 629
Varmus, H.E., 491
Veis-Novack, D.J., 387
Vermeulen, W., 317
Vetter, M., 165
Virmani, A., 565
Vogelstein, B., 517, 585
Vojta, P., 411
Volpert, O.V., 483

W

Waldman, F., 645
Walling, H.W., 685
Wang, Y., 207
Ward, D.C., 331
Warren, G., 331
Wassarman, D.A., 147
Weber, B.L., 531
Weeda, G., 317
Weinberg, R.A., 1
Weinert, T., 259
Weinrich, S.L., 307
Weissman, B., 573
West, M.D., 307
White, A., 265
White, E., 395
White, M.A., 181
Wigler, M.H., 181, 585
Williams, B.O., 449
Wirth, L., 331
Witte, O.N., 589
Wolff, T., 147
Wright, W.E., 307
Wu, C.-L., 75
Wu, X., 215
Wyllie, A.H., 403

X

Xu, J., 531
Xu, Y., 97
Xuereb, S., 509

Y

Yang, Z., 699
Ye, B.H., 117
Yin, X.-M., 387
Yin, Y., 411
Yokota, J., 49
Yoon, C.H., 155
Yuschenkoff, L., 491

Z

Zauberman, A., 225
Zhang, H., 21
Zhang, J., 117
Zhu, L., 75
Zou, Z., 537
Zucman, J., 555
zur Hausen, H., 623

Subject Index

A

ABL
  role in CML, 592
  tyrosine kinase activity, 592
α-N-Acetylglucosaminidase, mutations in breast cancer, 526
Acute myeloid leukemia (AML)
  chromosomal abnormalities, 547, 552, 595, 603
  inv16 role, 550–552
  M4Eo subtype, 547, 552
  marker genes in hematopoietic progenitor cells, 692
Acyclovir, gene therapy application, 685, 689
Adaptin complex
  role in clathrin coat assembly, 158
  structure, 158
Adenomatous polyposis coli gene (APC). *See also* Multiple intestinal neoplasia mouse
  allele loss in cancer, 501–504
  locus in mice, 502, 504
  loss of heterozygosity, 505
  Min mutation in mice, 501, 507
  mutations in HNPCC, 517–518, 719
  product interaction with catenins, 506, 718
  suppression gene activity, 505
  tissue specificity in tumorigenesis, 503
Adenovirus
  mechanism of cell transformation, 75, 85
  oncoprotein
    apoptosis regulation, 395–400, 420
    Bcl-2 functional similarity, 397–398
    binding to pRB, 75, 395
    cell proliferation induction, 396–397
    Ha-*ras* cooperativity, 399–400
    plasmid transfection, 395–396
    p300 protein binding to E1A, 85–87, 89–90
    tendency correlation with tumor curability, 285
Adriamycin, induction of apoptosis, 420–421
Alveolar rhabdomyosarcoma, PAX3 mutations, 137
AML. *See* Acute myeloid leukemia
AML1
  fusion with *ETO* in cancer, 597, 599, 601–603
  homology with *runt*, 595–596
  isolation of fusion cDNA
    EAP, 599–600
    EVI11, 600
    MDS1, 600
  mapping of breakpoints, 599
  Southern blotting, 598–599
  tissue distribution, 596
  translocations in cancer, 595
Anaplastic sarcoma, *RB/p53* mutations, 451
Angiogenesis. *See* Tumor angiogenesis
Angiostatin
  antiangiogenic activity, 474–475, 478–479
  antimetastatic activity, 475, 479
  extraction from SDS-PAGE, 474
  homology with plasminogen, 478, 480
  purification
    serum, 474, 477
    urine, 474, 477
  sequencing, 474, 477–478
  tumor therapy, 475, 479–481, 720
APC. *See* Adenomatous polyposis coli gene
Apoptosis. *See also Caenorhabditis elegans*; p53
  assay, 216, 248–249, 452, 461, 510
  cell-cell interaction role, 666–668
  characterization, 411, 416
  chemotherapy induction, 666, 716
  conservation between species, 383
  DNA cleavage, 366
  E2F transcription factor mediation, 219–221
  inhibition by Bcl-2, 366–367, 370, 373, 377, 383, 387, 716
  morphological features, 365
  tumor growth suppression, 250–256
ARMS. *See* Alveolar rhabdomyosarcoma
AT. *See* Ataxia telangiectasia
Ataxia telangiectasia (AT)
  chromosomal abnormalities, 292
  DNA breakage, 290–291, 713
  p53 defects, 281
  phenotype, 281, 290
  VDJ rearrangements, 290–292

B

Barrett's esophagus adenocarcinoma
  allelic loss progression, 580–581
  cell cycle abnormalities, 577–579
  p53 role, 579–580
  risk, 577
  aneuploid cells, 578–579
  cancer screening, 577, 582
  neoplastic progression models, 581
  tumor selection pressure, 710
Barrett's syndrome, p53 mutations, 261
Basic fibroblast growth factor (bFGF)
  angiogenesis stimulation, 472, 475
  half-life, 472
Bax
  apoptosis role, 392
  Bcl-2 association, 366, 391–392
B-cell lymphoma, bursa of Fabricius
  apoptosis, 510–511, 513
  bursal transplantation assay, 509–510
  genetic instability, 511
  myc role, 509–511, 513–514
  properties in chicken, 509
Bcl-2
  apoptosis inhibition, 366–367, 370, 373, 377, 383, 387, 391, 395–400, 411, 416, 716
  Bax association, 366, 391–392
  cell type effect on response, 369–370
  conserved domains, 391–392
  E1B functional similarity, 397–398
  HeLa cell expression, 398
  homology with CED-9, 380–381, 392
  oxidative damage protection, 391
  role in
    autoimmune disease, 367
    cell survival, 365
    chemotherapy resistance, 371
    lymphoma, 368–369, 377, 381, 390
    lymphopoiesis, 367, 388–390
    thymocyte differentiation, 389–390
  subcellular localization, 366, 390–391

Bcl-2 (continued)
  tissue distribution, 366, 388–389
  transgenic mice, 367–368
Bcl-3, tan-1 protein complex, 132, 135
BCL-6
  expression in mature B cells, 119
  product
    biological activity, 118
    structure, 118
    tissue distribution, 119
  rearrangements in DLCL, 119–121
  screening of mutations, 122
BCR-ABL
  induction of *kir*, 591–592
  MYC role in signaling, 590–591
  role in CML, 589
  tyrosine kinase fusion protein, 589
Beckwith-Wiedmann syndrome (BWS)
  cancer risk of patients, 357
  clinical presentation, 357
  DNA methylation, 360–361
  genomic imprinting, 359
  heredity, 357
bFGF. *See* Basic fibroblast growth factor
bmi-1
  effect on Hox gene expression, 442, 445
  homology with *mel-18*, 444
  role in
    hematopoiesis, 440
    skeletal abnormalities, 440–442, 445
    tumorigenesis, 439–442, 444–445
  transgenic mice, 436–437, 439, 443
Bone marrow transplantation. *See* Hematopoietic progenitor cells
BRCA1
  loss of heterozygosity, 524, 531–532
  mutations
    breast cancer, 491, 523, 525, 527–529, 531, 718
    incidence, 531
    ovarian cancer, 525, 527, 531
  polymorphic markers, 531–532
  positional gene cloning, 523, 525–527, 531, 533, 718
  sequence-tagged sites, 533
  transcript identification
    direct cDNA screening, 533–534
    exon amplification, 534–535
    magnetic bead capture of cDNA, 535
    sequence analysis, 535–536
    triplet repeat identification, 525, 531

  tumor suppression, 524–525, 531
BRCA2, breast cancer linkage, 523
Breast cancer
  chromosomal rearrangement, 525, 648–649
  comparative genomic hybridization, 645–647
  down-regulated genes, 537–539, 544–545
  effect of menstruation interval to pregnancy, 524
  FISH, 645, 647–650
  heredity, 39, 523–524, 528
  incidence, 524, 531
  linkage analysis, 532
  loss of heterozygosity, 645, 650
  mortality, 524
  proto-oncogenes, 491–492, 497
  suppressor gene mutations, 491, 523, 527–529
Bromodeoxyuridine, immunostaining, 461
BTF2
  complex with TFIIH, 317
  DNA repair, 317, 319–320
  purification, 318–319
BWS. *See* Beckwith-Wiedmann syndrome
Byr2, Ras complex, 181

C

Caenorhabditis elegans
  apoptosis
    ced-3
      apoptosis induction, 381
      homology with ICE, 381–382
    ced-9
      antagonism of other genes, 379–380
      homology with *bcl-2*, 380–381
      polycistronic RNA, 380
      protection against apoptosis, 378–379
    nuc-1 control of DNA degradation, 382–383
    pathway, 377–378, 383
    phagocytosis control genes, 382
    triggering genes, 367, 372, 377–378, 381
  hermaphrodism, 155–156
  molecular genetics characterization, 155, 161
  vulval differentiation, 155–161
CAKs. *See* Cdk-activating kinases
Carcinogens, metabolism, 277–278
Casein kinase II, regulation of p53, 196–198, 203
CBFB/MYH11 fusion
  AML association, 547, 552
  cDNA constructs, 547–548, 550

  transient transfection, 548
CDC13, cell cycle checkpoints, 260–261
Cdk-activating kinases (CAKs)
  cyclin complexes, 8
  phosphorylation sites of substrates, 32–34
  regulation of expression, 8
  R point transition regulation, 7
CDKN2
  cancer screening, 44–45
  gene therapy, 45
  mapping, 40–41
  mutations in cancer, 41–43
  product. *See* p16
  regulation, 44
  tumor suppressor gene, 39
Cdks. *See* Cyclin-dependent kinases
Cell cycle
  checkpoints in yeast, 259–262
  DNA repair, 259–260, 278, 713–714
  negative controls, 262
  phases, 3
  RB regulation, 2–3
  tumor cell defects, 37
Cell hybridization
  complementation analysis, 318
  effect on population doublings, 61
  senescence evaluation
    immortal-immortal cell hybrids, 62
    normal-immortal cell hybrids, 61–62
    normal-normal cell hybrids, 61
Cell viability, flow cytometry assay, 420
Cellular senescence
  aging model systems, 59, 67–68, 411
  cell cycle arrest, 59, 67
  cell hybrid studies, 61–62, 412
  characterization, 411, 416
  chromosome transfer studies, 62–63
  DNA synthesis, 59–60
  gene mapping, 412–414, 416
  genetic control, 412–413
  growth regulatory gene repression, 70–71
  markers. *See* Senescence-associated β-galactosidase
  pathways, 414
  role
    p16, 415–416
    p21, 415
    p53, 69–70, 413, 416
    pRB, 69–70, 72, 413–414
  senescent cell-derived inhibitors of DNA synthesis, 60–61, 64
  tumor suppression, 68–69, 414
Cervical cancer

HPV association, 237, 297, 623
incidence, 623
c-fos, repression in senescence, 70–71
Chemotherapy. See also Multidrug resistance
  induction of apoptosis, 419–422, 666
  sensitization by anti-adhesives, 668
Chromosome, 22
  alterations in cancer, 555
  FISH with cosmid probes, 556
  mapping of, 22q12, 556–558, 562
  overlapping cosmid isolation, 555–556
  RT-PCR, 556
  transcribed region identification, 556, 558
Chronic myelogenous leukemia (CML)
  BCR-ABL role, 589, 595
  Philadelphia chromosome marker, 589
CIP1. See p21
Cisplatin, apoptosis induction, 421
Clusterin, apoptosis marker, 668
CML. See Chronic myelogenous leukemia
c-myc
  CSF-1 sensitivity role, 13
  mutations in cancer, 117
Cockayne's syndrome (CS), DNA repair deficiency, 317, 326–327, 712
Colony-stimulating factor-1 (CSF-1)
  cell cycle regulation, 12
  cofactors in cell transformation, 632–633
  receptor mutants and cell proliferation, 13
Colorectal cancer. See also Adenomatous polyposis coli gene; DCC; Hereditary non-polyposis colon cancer
  DNA fingerprinting of tumors, 340–343
  DNA ploidy status, 408
  heredity, 39
  incidence, 709
  loss of heterozygosity, 637, 642
  mitotic activity of intestinal epithelium, 501
  mutator mutations, 343
  oncogenes, 339, 346, 517
  p21 role, 519
  p53 status, 403–404, 406–408
  staging, 517
  suppressor genes, 339, 346, 517, 642
  treatment, 709
Comparative genomic hybridization
  breast cancer genes, 645–647

lung cancer oncogenes, 566
sensitivity, 715
Wnt-1 and p53, 495–496
CREB-binding protein
  p300 similarity, 85, 87–88, 93
  transcriptional coactivation, 93
CS. See Cockayne's syndrome
CSF-1. See Colony-stimulating factor-1
CTL. See Cytotoxic T lymphocytes
Cyclic AMP, transcription activation, 93
Cyclin D
  binding by pRB, 4–5, 16–17
  cell cycle regulation, 11, 13, 33
  expression control, 6
  half-life, 12
  kinases, 12
  regulation of pRB phosporylation, 3–4, 13, 16
Cyclin-dependent kinases (cdks)
  assay, 22, 298
  cdk4
    cell cycle role, 43
    regulation of pRB phosphorylation, 3–4, 13
    substrate specificity, 12
  cdk6, substrate specificity, 12
  cell cycle regulation, 11–12, 31–32
  cellular senescence role, 415
  cyclin binding, 12, 16–17, 21, 32–34
  expression
    cell cycle, 32
    control, 6
  kinases, 12, 33
  Kip1 inhibition, 34
  p21
    association in active complex, 21, 23
    inhibition cooperativity, 22
    stoichiometry in binding, 24–28
  p53 as substrate, 210–211
  regulation of pRB phosphorylation, 3–4
  sequence homology, 12
Cyclin H, CAK binding, 8
Cdk-activating kinases
Cytotoxic T lymphocytes (CTL)
  oncogene stimulation, 617–618
  tumor cell antigens, 617–618

D

DCC
  cell adhesion molecule similarity of product, 642
  expression, 639–640
  loss of heterozygosity, 640, 642
  PCR assay, 637
  ribonuclease protection assay, 638

  role in cell differentiation, 641–642
  sequence, 638–639
  structure, 638–639
  tumor types and functionality, 640–641
  Western blotting of product, 638
Deoxyazacytidine, S100L induction, 545
Desmoplastic small round cell tumor
  clinical presentation, 139–140
  EWS1-WT1 fusion, 139–140, 143–144
  histology, 139–140
Diffuse large cell lymphoma (DLCL)
  BCL-6 rearrangements, 119–121
  chromosomal alterations, 118
  oncogene mutations, 117
  survival, 122
DLCL. See Diffuse large cell lymphoma
DMRIE/DOPE, liposome vector in gene therapy, 704–705
DNA damage
  cellular outcomes, 277–279
  p53-induced cell cycle arrest, 279–280, 298, 300–304, 713–714
  repair
    assay, 318
    cancer-associated defects, 278, 711
    cell cycle, 259–260, 278
    mismatch repair, 518–519, 712
    nucleotide excision repair, 317, 711–712
DNA fingerprinting. See Polymerase chain reaction
DNA methylation
  assay, 358
  genomic imprinting role, 360–362
  prostate cancer, 656–657
DNA ploidy, status in colorectal cancer, 407–408
Drosophila, Ras signal transduction in eye development, 147–152
DSRCT. See Desmoplastic small round cell tumor

E

E1B, p53 binding, 215, 218–219
E2F transcription factor
  apoptosis mediation, 219–221
  binding
    p107, 75–76, 80–81
    pRB, 5, 15, 43, 49, 75, 80–81, 99, 101
  cell immortalization role, 75–76
  repression in senescence, 71–72

E6-associated protein
  domains, 237–238
  E6 binding region, 237–238
  homology with other proteins, 243
  purification, 237–239
  ubiquitin thioester formation, 238–239, 241–243
EAP
  isolation of *AML1* fusion cDNA, 599–600
  mapping, 600–601
E-cadherin, role in prostate cancer, 655
End-replication problem, 309
Epidermal growth factor (EGF) receptor
  autophosphorylation, 173, 176
  Ras stimulation, 173
  signal transduction, 169–170, 173–174
ERCC3
  DNA repair, 317
  Drosophila mutant, 326
ETO
  chromosomal translocations and cancer, 597
  cloning, 597
  fusion with *AML1* in cancer, 597, 599
  tissue distribution, 597
Etoposide, apoptosis induction, 404, 421
EVI1
  isolation of *AML1* fusion cDNA, 600
  mapping, 600–601
Ewing's sarcoma (EWS)
  alterations in Ewing family of tumors, 559–560
  BAM22 product, 558
  chromosomal translocation, 555
  EWS1 mutations, 138–139
  fusion with other proteins in cancer, 562–563
  RNA-binding activity, 559, 562
EWS1
  characterization, 558–559
  fusion with *WT1* in cancer, 137, 139–140, 142–143
  mutations in cancer, 138–139
  protein
    structure, 138–139
    transcription activity, 138–139
EWS-WT1 protein
  DNA-binding activity, 140–141
  gene fusion in cancer, 137, 139–140, 142–143
  transcription activation, 141–142
  transformation mechanism, 144

F

Familial adenomatous polyposis.
  See Multiple intestinal neoplasia mouse
FISH. *See* Fluorescence in situ hybridization
Fluorescence in situ hybridization (FISH)
  breast cancer genes, 645, 647–648, 650
  chromosome, 22, 556
  IGF2, 460–461
5-Fluorouracil
  apoptosis induction, 421
  mechanism of action, 709
Follicular lymphoma
  Bcl-2 role, 368–369, 377, 381, 387, 390
  chromosomal translocations, 387
  course, 368, 387
5-Formyltetrahydrofolate. *See* Leucovorin
fos, product accumulation in cancer cell nuclei, 1
frat-1, tumorigenesis role, 437–438, 443

G

GADD45
  expression following ionizing radiation, 281–282
  p53 binding, 282–283
Ganciclovir, gene therapy application, 685, 687, 689
G_1 cyclins. *See also* Cyclin D
  binding by pRB, 4–5
  expression control, 6
  gene deletion effect on yeast cells, 4–5
  regulation of pRB phosphorylation, 3–4
Gene amplification
  breast cancer, 645
  frequency in tumors, 265, 585
  modulation by p53, 265
Gene therapy. *See also* Hematopoietic progenitor cells
  CDKN2, 45
  efficiency of delivery systems, 685, 688–689, 704, 721
  H-$2K^s$, 701
  HLA-B7, 702–705
  liposome vectors, 704
  MDR1 and high-dose chemotherapy, 681–682
  melanoma, 702–705
  MHC class I molecules, 700–701
  vector safety in direct gene transfer, 701–702
Genomic imprinting
  assay, 357–358
  DNA methylation, 360–362
  functional definitions, 357
  role in cancer, 358–362
Genomic instability
  cancer association, 118, 259, 265, 270, 287, 293, 339, 349, 406, 577, 699
  causes, 265, 287
  cell cycle perturbation, 270
  dynamics
    breast cancer, 495
    colorectal cancer, 406–408
    ovarian cancer, 352–355
  lymphocytes. *See* VDJ
  maintenance of integrity, 265, 714
  screening, 272–273
  types, 406
Glucocorticoids, apoptosis acceleration, 404
Grb2, role in Ras signaling, 173–175

H

H-$2K^s$, gene transfer and tumor response, 701
H19
  DNA methylation, 360
  genomic imprinting, 357–359, 361
Ha-*ras*
  apoptosis regulation, 395, 399–400
  mutations in cancer, 428–429
Hematopoietic progenitor cells
  gene marking, 691–692
  gene transfer
    allogeneic bone marrow transplantation, 695
    safety, 694
    targeting, 691
    purging, 693
    relapse clonality, 693
    retrovirus vectors, 691, 693–694
    tumor cell contamination, 691–693
Hepatocyte
  carcinogenesis model in rat, 608–609
  culture in rats, 608
  RNA analysis, 608
Her-2/neu, breast cancer role, 647–648
Hereditary non-polyposis colon cancer (HNPCC)
  gene mutations, 331, 335–336, 344, 517, 712
  incidence, 331
  microsatellite instability, 331, 344–346, 349, 518
  mismatch repair defects, 518–519
  mutator genes, 333, 344–346
  ovarian cancer relationship, 349
  pedigree, 334, 517
Herpes simplex virus (HSV)
  thymidine kinase
    animal models, 686

delivery systems, 685-686, 688-689
sensitivity to prodrugs, 685, 687
suicide vector, 681-682, 685-686
tumor cell transduction, 686-688
viral immunity effect on tumor growth, 688-689
HGF/SF
angiogenesis role, 634
expression in tumors, 632-634
met ligand, 629, 631-632
Histone H1, phosphorylation, 104
HLA-B7, gene therapy, 702-705
HNPCC. See Hereditary non-polyposis colon cancer
HPV. See Human papillomavirus
hsp70, activation of p53, 201-202
hsp90, Raf complex, 192
HSV. See Herpes simplex virus
Human papillomavirus (HPV). See also E6-associated protein
cervical cancer association, 237, 297, 623
classification, 237, 266
G_1 arrest disruption after DNA damage, 298, 300-304
host cell regulation of oncogenes, 623-626
human fibroblasts
apoptosis, 271-272
behavior at confluence, 268
doubling time, 267
drug resistance, 268-269
infection, 266
karyotypic stability, 266-267
morphology, 267
mortality, 267-268
keratinocyte immortalization, 297
mechanism of cell transformation, 75
oncogene expression by retroviral transduction, 298
oncoprotein binding to pRB, 75, 237, 243, 266, 272, 297, 623
p53 inactivation
binding, 237, 297, 623
half-life, 237
transcription effects
5-azacytidine, 624
human macrophages, 625

I

ICE. See Interleukin-1β converting enzyme
Id genes, repression in senescence, 71
IGF2. See Insulin-like growth factor-II
IL-2. See Interleukin-2

IL-3. See Interleukin-3
IL-7. See Interleukin-7
Immortalization
mechanisms, 61
p53 role, 372
spontaneous frequency, 67
telomerase role, 307, 309-312
tumor cells, 68-69, 307-308, 414
viral oncogene role, 69
Immunoglobulin, switch recombination, 288-289
Immunotherapy
approaches, 700
requirements, 702
Insulin-like growth factor-II (IGF2)
absence impairment of tumorigenesis, 466-467
apoptosis inhibition, 467-468
DNA methylation, 360-361
gene regulation in transgenic mice, 467-468
genomic imprinting, 357-359, 361
hyperproliferation association, 465, 467-468
in situ hybridization, 460-461
purification from conditioned medium, 460
quantitative RNA-PCR, 460, 462, 465
receptors
expression in tumorigenesis, 465
radioreceptor assay, 460
types, 459
up-regulation in tumors, 459-462, 465
Interleukin-1β converting enzyme (ICE)
apoptosis role, 372-373, 381-382
CED-3 homology, 381-382
cysteine protease activity, 381-382
expression, 372-373
proenzyme processing, 381
Interleukin-2 (IL-2)
gene therapy, 705
T cell mitogenesis, 34-35
Interleukin-3 (IL-3), activation in tumors, 619
Interleukin-7 (IL-7), B cell differentiation, 445
Inv16 AML role, 550-552
Ionizing radiation
apoptosis induction, 370-371, 403-404
B-stem cell sensitivity, 406
G_2 arrest after DNA damage, 404
induction of p53, 280-281, 406
p53 effect on resistance, 284
Islet cell tumor
induction by T antigen, 459
RB/p53 mutations, 451, 454
tumorigenesis, 459

J

Jak-Stat signaling pathway, 177

K

Kip1
cdk inhibition, 34
cell cycle arrest, 14, 33, 49
expression during cell cycle, 15
gene cloning, 15

L

LCV. See Leucovorin
LET-23
mutation in multivulva phenotypes, 156-157
negative regulators, 157-161
role in vulval differentiation, 156
signaling pathway, 155, 157, 160-161
structure, 156
tissue-specific action, 161
Leucovorin (LCV), combination therapy with 5-fluorouracil, 709
LFS. See Li-Fraumeni syndrome
Li-Fraumeni syndrome (LFS)
p53 defects, 272-273, 278, 370, 427, 449, 491, 523, 632, 673-674
penetrance, 674
screening
DNA methods, 674
functional assays, 674-676
genomic instability, 272-273
LIN-3
extragenic suppressors, 157
structure, 156
LIN-15, role in nematode vulval differentiation, 158, 160
Liver cancer. See also Hepatocyte
metastasis site, 685
mortality, 685
LOH. See Loss of heterozygosity
Loss of heterozygosity (LOH), cancer association, 357, 359, 428, 430, 494-495, 585, 645, 654-655
Lung cancer. See also Small-cell lung cancer
chromosomal alterations, 566-567
cure rate, 565
early detection
immune response, 568
molecular screening, 567-568
heredity, 568
oncogenes, 565, 567
comparative genomic hybridization, 566
mutation and prognosis, 566

Lung cancer (*continued*)
  mutation types, 567–568, 711
  p16 role, 568
  p53 role, 566, 568, 570
  preneoplasia, 567
  smoking relationship, 565, 567
  stages, 567
  therapeutic targets, 569–570
Lymphoma. *See* B-cell lymphoma, bursa of Fabricius; Diffuse large cell lymphoma; Follicular lymphoma; Non-Hodgkin's lymphoma

M

Mad
  expression
    keratinocyte differentiation, 111–112
    myeloid differentiation, 110–111
  Max binding, 110
  repression of transcription, 110, 113
MAGE-1
  promoter, 620
  protein function, 620
  structure of related genes, 619–620
  tissue distribution of product, 619
  tumor antigen product, 618, 620
Magnetic beads, cDNA capture, 535
Major histocompatibility complex (MHC)
  antigen presentation, 617, 699–700
  class I gene transfer into tumors, 700–701
Malignancy, mechanism, 308
Maltose-binding protein (MBP), p21 fusion protein, 21
MAP kinase. *See* Mitogen-activated protein (MAP) kinase
Maspin
  expression in breast tumor, 539–540
  gene locus, 542
  GST fusion protein, 543
  homology with serpins, 540–541
  immunostaining, 538, 542
  polyclonal antibodies, 542
  proteolytic processing, 543
  serine protease inhibition, 537, 540, 543–544
  tumor
    suppression, 537, 543
    transfectants, 542–543
Max
  binding
    E-box, 109
    heterocomplex switching and differentiation, 112–113
  Mad, 110
  Mxi1, 110
  Myc proteins, 109–110, 590
  dimerization, 109
  expression
    keratinocyte differentiation, 111–112
    myeloid differentiation, 110–111
  target genes of Myc complex, 113–115
MBP. *See* Maltose-binding protein
Mdm2
  gene
    promoters, 231, 233
    trans-activation by p53, 228–234, 283
    translation profiles, 231–234
  p53 binding, 215, 218–219, 221, 229, 283
MDR1
  gene therapy in high-dose chemotherapy, 681–682
  retroviral vectors, 679–681
  trans-activation by p53, 215, 219, 221, 678
  transgenic mice, 678, 680
MDR2/3
  expression in tumorigenesis, 607
  mRNA stability, 608, 612–613
MDS1
  isolation of *AML1* fusion cDNA, 600
  mapping, 600–601
MEK, Ras signal transduction, 190
Mek-1, role in LET-23 signaling pathway, 157
Melanoma
  CDKN2 mutations, 43
  gene therapy, 702–705
  heredity, 39
Met
  cellular distribution, 631
  expression in tumors, 632–634
  HGF/SF ligand, 629, 631–632
  immunoprecipitation, 629
  proto-oncogene, 629, 634
  receptor processing, 629
  tumorigenicity in 3T3 cells, 630–631
Metastasis
  apoptosis in tumors, 473, 475
  DNA synthesis in tumors, 473
  incidence, 677
  inhibition by tumor growth, 472–473, 475
  mechanism, 308, 720
MHC. *See* Major histocompatibility complex
Microsatellite
  clonal analysis, 349–350
  instability
    colon cancer, 331, 344–346, 349, 518–519
    ovarian cancer, 349, 351–352
    primers used in analysis, 349–350
Mitogen-activated protein (MAP) kinase
  activation by kinase, 167–168, 174
  Ras1 signal transduction, 147–148, 150, 165–167, 174, 181, 190
  translocation, 177–178
Mitotic clock, limiting of cell division capacity, 307
MLH1
  cloning, 332
  evolution, 332
  locus, 333
MLM. *See also CDKN2*
  cloning, 40
  locus, 39
  mapping, 40
  melanoma susceptibility gene, 39–40
  mutations, 40
MMTV. *See* Mouse mammary tumor virus
Mo-MLV. *See* Moloney murine leukemia virus
MO15 kinase, substrates, 8, 13
Moloney murine leukemia virus (Mo-MLV)
  enhancer, 551–552
  lymphoma induction, 435–436
  mechanism of action, 436, 438–439, 443
Mouse mammary tumor virus (MMTV), murine breast cancer model, 491–492
Mouse skin carcinogenesis
  carcinogens, 427
  genetic alterations, 427–428
  p53 role, 429–433
  radiation oncogenesis, 431–433
  tumor predisposition loci identification, 428–429
M phase, regulation, 11
MSH2
  cloning, 332
  evolution, 332
  HNPCC role, 349, 566
  locus, 333
  ovarian cancer mutations, 354, 566
  product binding to mismatched nucleotides, 333, 335, 349, 712
MTS1
  cloning, 50
  functional gene assay, 52–53
  locus, 53
  PCR, 49–50
  senescence activity, 54–55
  SSCP analysis, 50–51, 53
MTS2
  mapping, 40–41

SUBJECT INDEX

mutations in cancer, 41
Multicellular tumor spheroid
  culture, 661–662
  drug resistance, 661, 664–666
  P-glycoprotein activity, 669–670
  solid tumor phenotypes, 663–664
  structure, 662
  transforming growth factor expression, 662–663
Multidrug resistance
  mechanism, 607, 613, 669, 677
  multicellular resistance, 661, 666, 668–669
  onset time, 661
Multiple intestinal neoplasia mouse
  desmoid fibrosarcoma, 503
  genotyping, 501
  homozygous allele effect on embryogenesis, 505–507
  loss of *Apc* allele, 501–504
  mammary tumors, 503
  Modifier-of-Min locus, 504–505
  p53 role, 504
MutL
  ablation effect on DNA mutation, 331
  cloning of human homologs, 332
MutS
  ablation effect on DNA mutation, 331
  cloning of human homologs, 332
Mxi1, Max binding, 110
myc
  collaborating oncogenes, 436–437
  product accumulation in cancer cell nuclei, 1
  regulation of expression, 2
  transgenic mice, 368, 436–437
Myc proteins
  apoptosis role, 368–369, 377, 388
  cellular functions, 109
  expression
    keratinocyte differentiation, 111–112
    myeloid differentiation, 110–111
  half-life, 109
  Max association, 109–110, 113–114, 590
  subcellular localization, 113
  target genes of Max complex, 113–115

N

Necrosis
  characterization, 411
  mechanism, 365
Nematode. *See Caenorhabditis elegans*
Neovascularization. *See* Tumor angiogenesis
Nerve growth factor (NGF), signal transduction of receptor, 168–169, 177
Neuroblastoma, marker genes in hematopoietic progenitor cells, 692
Neurofibromatosis type 2 (NF2)
  course, 555
  gene mutations, 560–563
  heredity, 55
NF1, functional assay, 676
NF2. *See* Neurofibromatosis type 2
NGF. *See* Nerve growth factor
NHL. *See* Non-Hodgkin's lymphoma
Non-Hodgkin's lymphoma (NHL). *See also* Diffuse large cell lymphoma
  B-cell lineage, 117
  classification, 117
  molecular lesion distribution, 117
Notch, role in *Drosophila*, 125, 135

O

Okadaic acid, mechanism of carcinogenesis, 428
Ovarian cancer
  DNA analysis, 353–354
  dynamics of genetic instability, 352–355
  heredity, 349, 523–524
  microsatellite instability, 349, 351–352
  mortality, 349
  MSH2 mutations, 354–355

P

P1A
  activation in mastocytoma, 618–619
  tumor antigen product, 618
p16
  blot analysis, 49
  cdk inhibition, 43
  cell cycle arrest, 15, 39, 43
  cellular senescence role, 415–416
  gene. *See CDKN2; MTS1*
  immunoprecipitation, 49
  levels in tumor cells, 50–51
  lung cancer role, 568–569
  mutations in tumor cells, 51–55
  R point transition regulation, 7
p21. *See also* SDI1
  assay, 298
  cdk
    association in active complex, 21, 23
    inhibition cooperativity, 22
    stoichiometry in binding, 24–28
  cell cycle arrest, 14, 21–22, 36, 49, 714
  cellular senescence role, 415
  cyclin complex binding, 34
  maltose-binding fusion protein, 21
  phosphorylation sites, 26
  regulation by p53, 14, 22, 28, 36, 283, 297, 303, 370, 397, 483, 519, 714
  R point transition regulation, 7
p27. *See also* Kip1
  cell cycle arrest, 15, 304
  cyclin complex binding, 34
  levels during cell cycle, 35–36
  R point transition regulation, 7
p53
  angiogenesis regulation, 483–488
  apoptosis
    anticancer agents induction, 420–424
    brain epithelium, 250–253
    cell specificity, 255–256
    genotoxic-damaged cells, 403–408
    induction, 215, 219–222, 225, 247, 254–255, 265, 284, 370, 419, 425, 455, 715
    RB-deficient lens fiber cells, 452–453
    regulation by oncogenes, 395–400
    triggering by proliferation, 253–254
  assay, 196–197, 298
  carboxy-terminal monoclonal antibody, 197, 199–201, 203
  cell cycle control, 2, 12, 247, 261, 265, 270–271, 279–280
  cell survival role, 365
  DNA binding, 195, 197, 202, 204
  domains, 195–196, 207, 215
  embryogenesis role, 449–450
  G_1 arrest after DNA damage, 279–284, 297, 370, 404–405, 416–417, 713–714
  gene. *See TP53*
  half-life, 237
  immortalization role, 372
  immunohistochemical staining, 298
  inactivation and brain tumors, 249–250
  induction mechanisms, 280–281
  ligands
    E1B, 215, 218–219
    E7, 237, 297
    Mdm2, 215, 218–219, 221, 229
    Sp1 interactions, 208–209
    T antigen, 209–210, 221, 237, 248–250, 255
  mutations in cancer, 54, 207, 211, 215, 247, 255, 261, 272–273, 303, 419, 422–423, 450–451, 491, 722
  mutations with *RB* in cancer, 449–454

SUBJECT INDEX

p53 (continued)
  nucleic acid reannealing by carboxyl terminus, 211–212
  phosphorylation
    casein kinase II, 196–198, 203
    cyclin B/cdc2 kinase, 210–211
    PKC, 196, 198–201, 203
    sites, 203, 210
    stoichiometry, 199
  purification
    active and latent forms, 201–204
    recombinant protein, 196–197, 207, 238–239
  regulation
    allosterism, 202–203
    conformational change in activation, 200, 202–203
    hsp70, 201–202
    p21, 14, 22, 28, 36, 283
    regulatory site, 195–196, 202–203
  screening for mutations, 674–676
  senescence role, 69–70
  site-specific mutagenesis, 215–218
  status and lung cancer prognosis, 566, 568, 570
  thrombospondin regulation, 484–488
  transcriptional activation
    immunoglobin heavy chain, 228–229, 233
    mdm2, 228–234
    mdr1, 215, 219, 221
    murine retrovirus GLN LTR, 226–228, 233
  transgenic mice, 248
  tumor suppression, 195, 202, 207, 215, 225, 247, 254–255
  ubiquitination, 237–238, 240–243
  UTP stabilization in mutants, 211
  assays, 207–208
  binding sequence, 208
  binding site isolation from genomic DNA, 225–226
  phosphorylation effects, 210–211
  SV40, 208–209
  target genes, 207, 225
  Western blotting, 396
p84, pRB binding, 104
p107
  binding
    cyclins, 79
    E2F transcription factor, 75–76, 80–81
    viral oncoproteins, 75
  growth suppression
    assay, 77
    cell cycle arrest, 76–77, 81
  cervical cell lines, 79
  domains, 77–78, 81–82
  fibroblasts, 80
  mechanism, 79–80, 82
p300
  aggregation in overexpression systems, 93
  cDNA isolation, 85–86
  CREB-binding protein similarity, 85, 87–88, 93
  E1A binding, 85–87, 89–90
  immunoprecipitation, 86
  luciferase assay, 86
  peptide mapping, 86, 88–89
  plasmid construction, 86
  sequence motifs, 87–89
  transcriptional coactivation, 90, 92–93
PALA. See N-(phosphonoacetyl)-L-aspartate
Papillomavirus. See Human papillomavirus
Paternal uniparental disomy (UPD), DNA methylation, 360–361
PAX3, mutations in cancer, 137
PCNA. See Proliferating cell nuclear antigen
PCR. See Polymerase chain reaction
PDGF. See Platelet-derived growth factor
PEBP2/CBF
  DNA-binding site, 549
  electrophoretic mobility shift assay, 547–548, 550
  inv16 protein association, 550–551
  subunit structure, 549
P-glycoprotein
  ATP dependence, 677
  cell toxicity, 679–680, 682
  expression in tumorigenesis, 607, 609–611, 613–614
  genes. See MDR1, MDR2/3
  immunostaining, 609
  inhibitors, 681
  isoform expression in rat hepatocytes, 611–612
  liver, 609–611
  mechanism of action, 678–679
  mRNA stability, 608, 612–613
  multicell system behavior, 669–670
  multidrug resistance role, 607, 613, 669, 677–678
  photoaffinity labeling, 678
  recombinant protein
    expression in yeast, 679
    purification, 679
  substrates, 613, 677
  tissue distribution, 677–678
  xenobiotic excretion, 678
  yeast model systems, 679–680
Phospholipase C, growth factor signal transduction, 176–177
N-(Phosphonoacetyl)-L-aspartate
  cell cycle perturbation, 269–270
  HPV infection effect on resistance, 268
  mechanism of resistance, 268–269
  metabolic inhibition, 268
Phy1
  Ras1 signal transduction, 151–152
  structure, 151
pim-1
  kinase product, 438
  signal transduction pathway, 444
  transgenic mice, 436–437
  tumorigenesis role, 438–439, 443–444
pim-2, tumorigenesis role, 439
Pinealblastoma, RB/p53 mutations, 451, 454
PKC. See Protein kinase C
Plasminogen, angiogenesis inhibition, 478
Platelet-derived growth factor (PDGF)
  apoptosis inhibition, 468
  receptor
    activation, 168
    autophosphorylation, 176
    cellular distribution, 168
    signaling pathways, 168–169
PLC. See Phospholipase C
Pleckstrin homology domain
  distribution in signaling molecules, 175
  structure, 175
PMS2
  cloning, 332
  evolution, 332
  locus, 333
Polymerase chain reaction (PCR)
  arbitrarily primed PCR
    colorectal cancer, 340–343
    DNA fingerprinting, 339–340
    primers, 339
  assay of VDJ recombination, 289, 293
  p16, 49–50
PP2A. See Protein phosphatase 2A
pRB. See Retinoblastoma protein

Programmed cell death. See Apoptosis
Proliferating cell nuclear antigen (PCNA), component of cyclin kinases, 21–22
Prostate, age-related growth, 653
Prostate cancer
  heredity, 657
  initiation, 653
  loss of heterozygosity, 654–655
  microcell transfer studies, 655–656
  mortality, 653

SUBJECT INDEX

role
  catenin, 655
  DNA methylation, 656–657
  E-cadherin, 655
  p53, 656–657
  pRB, 656–657
  ras 653–654
Protein kinase C (PKC)
  p53 regulation, 196, 198–201
  Raf activation, 187
Protein phosphatase 2A (PP2A), negative regulator of Ras1 signal transduction, 148, 150
14-3-3 Proteins
  biological function, 191
  conservation, 191
  PKC association, 191
  Raf binding
    activation, 190–192
    analysis, 187
    binding regions, 187
    immunoprecipitation of complex, 188
    localization of complex, 189–190
Proto-oncogenes, plasticity in cell signaling, 165, 170–171

Q

QM
  conservation of sequence, 573–574
  functional definition, 710
  isolation, 573
  locus, 575
  multigene family, 575
  protein product
    structure, 573–574
    transcription factor, 576
  tumor suppression, 574–576

R

Rab, geranylgeranyl transferase, 182–183
Rad9
  cell cycle checkpoint in yeast, 259–261, 283
  essentiality, 261
Radiation. *See* Ionizing radiation; Radiation therapy
Radiation therapy, induction of apoptosis, 419–422
Raf
  activation by 14-3-3 proteins, 190–192
  genetic screening for modifiers, 148
  geranylgeranylation, 148, 721
  GTP binding, 167, 173, 181
  ligands
    binding regions, 187
    Byr2, 181
    Cyr1, 182

hsp90, 192
14-3-3 proteins, 187–192
Raf, 166, 181–183, 187
Ras, 166, 181–183, 187
Rsb proteins, 182–183
yeast two-hybrid system analysis, 181–182, 187–188
MAP kinase activation, 167
mutations in lung cancer, 570
phosphorylation, 167, 184, 187
protein localization function, 184, 187, 191
Ras. *See also* Ha-*ras*
  signal transduction
    conservation, 175, 181
    Drosophila eye development, 147–152
    Ets transcription factors, 150–151
    MAP kinase, 147–148, 150
    pathways, 165–168
    Phy1, 151–152
    plasticity, 165, 170
    Pnt expression, 151
    protein phosphatase 2A as negative regulator, 148, 150
    receptor tyrosine kinase pathway, 147, 155, 173–174
    redundancy, 170
    Yan activity, 151
  tissue specificity, 168
RB. *See* Retinoblastoma gene
RDA. *See* Representational difference analysis
Reactive oxygen species (ROS), apoptosis role, 391
Representational difference analysis (RDA)
  cancer gene identification, 585–586, 715
  cloning of sequences, 585
  probes, 585–586
Restriction (R) point
  feedback loops, 6–7, 31
  operational definitions, 3
Retinoblastoma gene (*RB*)
  germ-line mutations, 673
  inactivation and cancer, 97, 673
  mutations with *p53* in cancer, 449–454
Retinoblastoma protein (pRB)
  apoptosis role, 400
  assay, 298
  binding
    affinity for proteins, 100–101
    cyclins, 4–5, 16–17
    E2F transcription factors, 5, 15, 75, 80–81, 99, 101
    oligomeric forms of pRB, 101–102
    p84, 104
    viral oncoproteins, 15, 75, 100, 283
    yeast two-hybrid system analysis, 99–100

cell cycle control, 2–3, 12, 15, 43–44, 75, 98, 717–718
domains, 97
heredity of mutations, 427
mechanism of growth suppression, 79–80, 82, 98–99, 104–105
mutations in tumors, 54, 450–451, 491
phosphorylation
  effect on oligomerization, 102–103, 105
  ionizing radiation effects, 283
  pattern, 4, 15, 97–98
  regulation, 3–4, 13, 221
recombinant protein
  electron microscopy, 98–99
  immunoblotting, 98
  immunofluorescence imaging, 99, 103
  oligomerization, 98, 101
  purification, 98
  senescence role, 69–70, 72
ROK-1, role in nematode vulval differentiation, 159–160
ROS. *See* Reactive oxygen species
R point. *See* Restriction point

S

SCH
  characterization, 560
  denaturing gradient gel electrophoresis, 560
  identification, 558
  mutations in cancer, 560–562
  NF2 role, 563
  schwannomin product, 563
SCLC. *See* Small-cell lung cancer
SDI1. *See also* p21
  DNA synthesis inhibition in senescent cells, 60–61
  regulation, 61
Senescence. *See* Cellular senescence
Senescence-associated β-galactosidase
  age-dependent expression, 68
  marker of cellular senescence, 68
  pH optimum, 68
Serine protease, mechanism of protein inhibitors, 540, 542
Sevenless
  Ras1 signal transduction, 147
  role in *Drosophila* eye development, 147
Sf9 cell, culture, 21
SH2 domain
  binding sites on receptors, 173–174
  distribution in signaling molecules, 175–176, 178

SH3 domain, distribution in signaling molecules, 175–176, 178
Signal transduction, advances in research, 721
Single-strand conformation polymorphism (SSCP), analysis of p16, 50–51, 53–54
Sky, expression in mammary tumors, 497
SLI-1
  homology with CBL protein, 159
  role in nematode vulval differentiation, 159–160
  structure, 159
Small-cell lung cancer (SCLC)
  prognostic factors, 566
  suppressor gene mutations, 565
Sos, role in Ras signaling, 173–175
Sp1, effect on p53 binding to SV40, 208–209
S phase, regulation, 11
SSCP. *See* Single-strand conformation polymorphism
Start, yeast cells, 31
Ste11
  activation, 190
  Ras signal transduction, 190
Subtractive hybridization, breast cancer genes, 537–538
SV40
  immunostaining, 461
  p53 DNA binding, 208
  T antigen
    immortalization of cells, 69, 75
    islet cell cancer induction, 468
    oncoprotein binding to pRB, 75, 248, 459
    p53 interactions, 209–210, 221, 237, 248, 255, 459
    transgenic mice, 248

T

TAN-1
  bone marrow transplantation, 126, 130–132
  cDNA synthesis, 126
  mutations in leukemia, 125–126, 133–134
  protein
    bcl-3 complex, 132, 135
    GST fusion protein, 126, 132
    immunolocalization, 127, 129–130, 134
    immunoprecipitation, 127
    p120 similarity, 134
    peptide mapping, 127
    SREBP-1 similarity, 135
    structure, 125, 127–129
    transcriptional activation, 132–133
    Western blotting, 126–127
  tissue distribution, 125

T antigen. *See* SV40
T cell, mitogenic signals, 34–35
Telomerase
  activity correlation with tumor phenotype, 312
  effect on telomere length, 310–311
  inhibition as anticancer therapy, 312–313
  mechanism, 309
  PCR assay, 312
  role in cell immortalization, 307, 309–312
  tissue distribution, 309–310, 312
Telomere, length assessment, 310–311
Terminal differentiation, characterization, 411
TFIIH
  complex with BTF2, 317
  DNA repair, 317, 319–320, 712
  mutation effect on clinical heterogeneity, 324–327
  protein ligands, 320–324
  purification, 318–319
  transcription inhibitors, 323
TGF-β. *See* Transforming growth factor β
Thrombospondin
  angiogenesis inhibition, 483–485
  regulation by p53, 484–488
  Western blotting, 484
Thymidine kinase. *See* Herpes simplex virus
Thymocyte, p53 role in apoptosis, 255
TNF. *See* Tumor necrosis factor
Torso
  mutations, 165
  signaling pathway, 165–166
TP53
  mutations in cancer, 44
  transcriptional activator, 44
Transforming growth factor β (TGF-β)
  CAK inhibition, 13–14
  cell cycle arrest, 2, 14, 33
  DNA synthesis inhibition, 60
  expression in multicellular spheroids, 662–663
  receptor
    gene isolation, 2
    tumor loss, 2
    transfection into tumor cells, 699
Transgenic mice
  bcl-2, 367–368
  bmi-1, 436–437, 439
  breast cancer model, 491–492
  lymphoma
    induction, 435–436
    tumorigenesis events, 443
  myc, 368, 436–437
  oncogene identification, 435, 437–438, 442–443, 719
  p53, 248

pim-1, 436–437
Trichothiodystrophy (TTD)
  DNA repair deficiency, 317, 327, 712
  genetic heterogeneity, 319, 325
TTD. *See* Trichothiodystrophy
Tumor angiogenesis
  assay, 484
  cornea, 473–475
  frequency of phenotype switch, 471
  inhibition by metastasis, 473, 475, 480
  microvessel formation, 471
  negative regulators, 472
  p53 regulation, 483–488
  perfusion of tumors, 471
  positive regulators, 472
Tumor mass
  inhibition of metastatic tumor growth, 472–473
  limitations, 720
Tumor necrosis factor (TNF), apoptosis role, 373
Tumor suppressor genes
  functional definition, 710
  multiple mutations in cancer, 449
  plasticity in cell signaling, 165
  screening by subtractive hybridization, 537
TUNEL assay, apoptotic cells, 452, 510
Two-hybrid system
  pRB ligand analysis, 99–100
  Raf ligand analysis, 187
  Ras ligand analysis, 181–182
Tyrosine kinase
  Jak-Stat signaling pathway, 177
  Ras signal transduction, 147, 155, 173
  tissue specificity, 178

U

Ubiquitin
  assay, 239
  p53 proteolysis, 237–238, 240
  purification of recombinant protein, 238–239
  reconstitution in vitro, 240–241
  structure, 238
  substrate specificity, 242
  substrate targeting proteins, 238, 240–242
  thioester formation with E6-associated protein, 238–239, 241–243
UNC-101
  homology to adaptin complex, 158
  role in nematode vulval differentiation, 158–160
UPD. *See* Paternal uniparental disomy

V

Vascular endothelial cells, proliferation inhibition by tumors, 475–477
Vascular endothelial growth factor (VEGF)
  angiogenesis stimulation, 472
  half-life, 472
VDJ
  accessibility of loci, 287–288, 291
  interlocus recombination
    acquired increases, 290–291
    ataxia telangiectasia, 290–291
    frequency, 290
    lymphoid malignancy marker, 291–293
  PCR assay, 289, 293
  recombinase complex, 287, 293
  structure, 287–288
  switch recombination, 287
VEGF. See Vascular endothelial growth factor
Vinculin, phosphorylation, 104

W

WAF1. See p21
Werner syndrome, gene locus, 413
Wilms' tumor
  chromosomal mutations, 573
  incidence, 138
  p53 mutations, 424
  suppressor gene. See WT1
  survival rate, 423–424
Wnt-1
  breast cancer proto-oncogene, 492, 497
  loss of heterozygosity, 494–496
  p53 deficiency and tumorigenesis, 492, 494–495
  synergy
    fibroblast growth factors, 492
    Sky, 496–497
WT1
  cloning, 573
  fusion with EWS1, 137, 139–140, 142–144
  genomic imprinting, 359–360
  locus, 138
  loss of heterozygosity, 357, 359
  protein
    target genes, 138
    tissue distribution, 138

X

Xeroderma pigmentosum (XP)
  DNA repair deficiency, 317, 712
  genetic heterogeneity, 325
XP. See Xeroderma pigmentosum

Z

Zinc-finger genes
  alterations in cancer, 118
  truncation and DNA-binding affinity, 143